Some Physical Consta

Gravitational constant	G	$6.674 \times 10^{-11}\ \mathrm{N \cdot m^2/kg^2}$
Speed of light	c	$2.998 \times 10^{8}\ \mathrm{m/s}$
Fundamental charge	e	$1.602 \times 10^{-19}\ \mathrm{C}$
Avogadro's number	N_A	$6.022 \times 10^{23}\ \mathrm{particles/mol}$
Gas constant	R	$8.315\ \mathrm{J/mol \cdot K}$
		$1.987\ \mathrm{cal/mol \cdot K}$
Boltzmann's constant	$\kappa = R/N_A$	$1.381 \times 10^{-23}\ \mathrm{J/K}$
		$8.617 \times 10^{-5}\ \mathrm{eV/K}$
Unified mass unit	u	$1.661 \times 10^{-27}\ \mathrm{kg}$
Coulomb's constant	$k = \dfrac{1}{4\pi\varepsilon_0}$	$8.988 \times 10^{9}\ \mathrm{N \cdot m^2/C^2}$
Permittivity of free space	ε_0	$8.854 \times 10^{-12}\ \mathrm{C^2/N \cdot m^2}$
Permeability of free space	μ_0	$4\pi \times 10^{-7}\ \mathrm{N/A^2}$
		$1.257 \times 10^{-6}\ \mathrm{N/A^2}$
Planck's constant	h	$6.626 \times 10^{-34}\ \mathrm{J \cdot s}$
		$4.136 \times 10^{-15}\ \mathrm{eV \cdot s}$
	$\hbar = h/2\pi$	$1.055 \times 10^{-34}\ \mathrm{J \cdot s}$
		$6.582 \times 10^{-16}\ \mathrm{eV \cdot s}$
Mass of electron	m_e	$9.109 \times 10^{-31}\ \mathrm{kg}$
		$511.0\ \mathrm{keV/c^2}$
Mass of proton	m_p	$1.673 \times 10^{-27}\ \mathrm{kg}$
		$938.3\ \mathrm{MeV/c^2}$
Mass of neutron	m_n	$1.675 \times 10^{-27}\ \mathrm{kg}$
		$939.6\ \mathrm{MeV/c^2}$
Bohr magneton	$m_B = e\hbar/2m_e$	$9.274 \times 10^{-24}\ \mathrm{J/T}$
		$5.788 \times 10^{-5}\ \mathrm{eV/T}$
Nuclear magneton	$m_n = e\hbar/2m_p$	$5.051 \times 10^{-27}\ \mathrm{J/T}$
		$3.152 \times 10^{-8}\ \mathrm{eV/T}$
Magnetic flux quantum	$\phi_0 = h/2e$	$2.068 \times 10^{-15}\ \mathrm{T \cdot m^2}$
Quantized Hall resistance	$R_K = h/e^2$	$2.581 \times 10^{4}\ \Omega$
Rydberg constant	R_H	$1.097 \times 10^{7}\ \mathrm{m^{-1}}$
Josephson frequency–voltage quotient	$2e/h$	$4.836 \times 10^{14}\ \mathrm{Hz/V}$
Compton wavelength	$\lambda_C = h/m_e C$	$2.426 \times 10^{-12}\ \mathrm{m}$

Physical Data Often Used

Mass of Earth, M_E	5.97×10^{24} kg
Radius of Earth, R_E, mean	6.37×10^6 m
Mass of Moon, M_M	7.35×10^{22} kg
Radius of Moon, R_M	1.74×10^6 m
Mass of Sun, M_S	1.99×10^{30} kg
Radius of Sun, R_S	6.96×10^8 m
Earth–Sun distance, $D_{E\text{-}S}$, mean	1.496×10^{11} m
Earth–Moon distance, $D_{E\text{-}M}$, mean	3.844×10^8 m
One light year (the distance light travels in one year)	9.46×10^{15} m
Acceleration of gravity on Earth, g (standard value)	9.81 m/s^2
At sea level, at equator	9.78 m/s^2
At sea level, at poles	9.83 m/s^2
Acceleration of gravity on Moon, g_M	1.62 m/s^2
Average orbital speed:	
Earth	2.98×10^4 m/s
Moon	1.02×10^3 m/s
Period:	
Earth (sidereal year)	365.26 days
Moon (sidereal month)	27.3 days
Escape velocity:	
Earth, $v_{E\text{-esc}}$	1.12×10^4 m/s
Moon, $v_{M\text{-esc}}$	2.4×10^3 m/s
Solar constant	1.35 kW/m^2
Time it takes light to travel the distance:	
from Earth to Sun	498 s $= 8.3$ min
from Earth to Moon	1.28 s
around Earth	0.13 s
equal to one light year	1 year $= 3.16 \times 10^7$ s
Milky Way Galaxy data:	
Diameter	100 000 to 120 000 light years ($\approx 10^{21}$ m)
Disk thickness	1 000 light years ($\approx 10^{19}$ m)
Number of stars	100 to 400 billion
Estimated mass	$5.8 \times 10^{11} M_S$
Useful ratios	$\dfrac{M_S}{M_E} = 3.33 \times 10^5;\ \dfrac{M_E}{M_M} = 81.3$
	$\dfrac{R_S}{R_E} = 109;\ \dfrac{R_E}{R_M} = 3.66$
	$\dfrac{D_{E\text{-}S}}{D_{E\text{-}M}} = 389;\ \dfrac{D_{E\text{-}M}}{R_E} = 60.4;\ \dfrac{D_{E\text{-}S}}{R_E} = 2.35 \times 10^4$
	$\dfrac{g_E}{g_M} = 6.1;\ \dfrac{v_{E\text{-esc}}}{v_{M\text{-esc}}} = 4.67$

physics

For Scientists and Engineers

An Interactive Approach

Second Edition

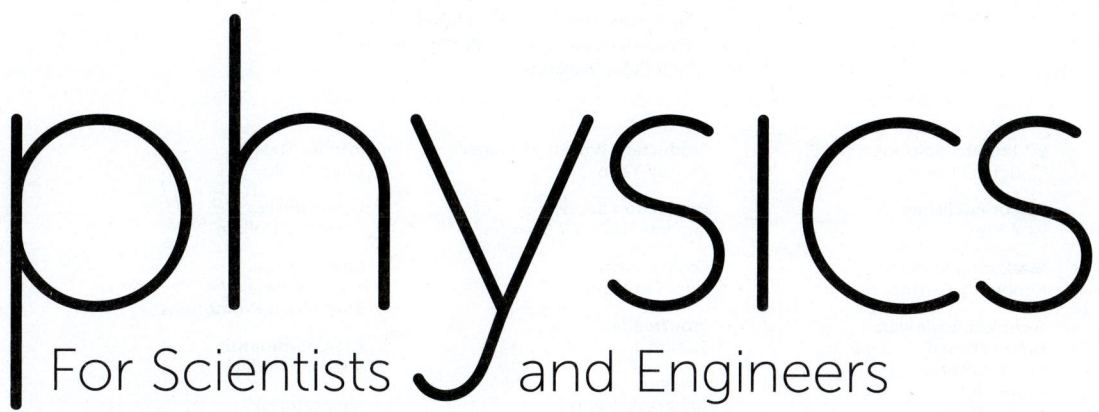

physics
For Scientists and Engineers
An Interactive Approach
Second Edition

ROBERT HAWKES
Mount Allison University

JAVED IQBAL
University of British Columbia

FIRAS MANSOUR
University of Waterloo

MARINA MILNER-BOLOTIN
University of British Columbia

PETER WILLIAMS
Acadia University

NELSON

NELSON

**Physics for Scientists and Engineers:
An Interactive Approach, Second Edition**

by Robert Hawkes, Javed Iqbal,
Firas Mansour, Marina Milner-Bolotin,
and Peter Williams

VP, Product Solutions, K–20:
Claudine O'Donnell

Senior Publisher:
Paul Fam

Marketing Manager:
Kimberley Carruthers

Technical Reviewer:
Simon Friesen
Karim Jaffer
Anna Kiefte
Kamal Mroue

Content Manager:
Suzanne Simpson Millar

Photo and Permissions Researcher:
Kristiina Paul

Production Project Manager:
Wendy Yano

Production Service:
Cenveo Publisher Services

Copy Editor:
Julia Cochrane

Proofreader:
Subash.J

Indexer:
Robert A. Saigh

Design Director:
Ken Phipps

Higher Education Design PM:
Pamela Johnston

Interior Design:
Brian Malloy

Cover Design:
Courtney Hellam

Cover Image:
Yuichi Takasaka/
Blue Moon Promotions

Art Coordinator:
Suzanne Peden

Illustrator(s):
Crowle Art Group,
Cenveo Publisher Services

Compositor:
Cenveo Publisher Services

**Library and Archives Canada
Cataloguing in Publication Data**

Hawkes, Robert Lewis, 1951–,
author
    Physics for scientists and
engineers : an interactive
approach / Robert Hawkes, Mount
Allison University, Javed Iqbal,
University of British Columbia,
Firas Mansour, University of
Waterloo, Marina Milner-Bolotin,
University of British Columbia,
Peter Williams, Acadia University.
— Second edition.

Includes index.
Issued in print and electronic
formats.
ISBN 978-0-17-658719-2
(hardcover).—ISBN 978-0-17-
680985-0 (PDF)

    1. Physics—Textbooks.
2. Textbooks. I. Iqbal, Javed,
1953–, author II. Mansour, Firas,
author III. Milner-Bolotin, Marina,
author IV. Williams, Peter
(Peter J.), 1959–, author V. Title.

QC23.2.H38 2018   530
C2017-906981-0
C2017-906982-9

ISBN-13: 978-0-17-658719-2
ISBN-10: 0-17-658719-5

Brief Table of Contents

Table of Contents

Preface

DEMYSTIFYING PHYSICS, A SCIENCE FOR LIFE

Physics is an exciting field that has changed our understanding of the world we live in and has immense implications for our everyday lives. We believe physics should be seen as the creative process that it is, and we aim to help the reader feel their own thrill of discovery.

To that end, *Physics for Scientist and Engineers: An Interactive Approach,* Second Edition, has taken a unique **student-first** development model. **Fundamental topics are developed gradually**, with great attention to the logical transition from the simple to the complex, and from the **intuitive to the mathematical**, all while highlighting the **interdisciplinary nature of physics**. This inquisitive and inspirational science is further supported with current events in Canada and beyond, and innovative pedagogy **based on Physics Education Research (PER)** such as **Interactive Activities**, **Checkpoints**, **unique problem-solving strategies** via open-ended problems, and ending **Examples** with "Making sense of the results."

HOW WE DO IT

Student-First Development Model

- The vision for this text was to develop it from the student perspective, providing the background, logical development of concepts, and sufficient rigour and challenge necessary to help students excel. It provides a significant array of engaging examples and original problems with varying levels of complexity.

- Students who are the primary users of educational textbooks have not traditionally been involved in their development. In *Physics for Scientists and Engineers: An Interactive Approach* we engaged Student Advisory Boards to evaluate the material **from a student perspective** and to develop the Peer to Peer boxes, which provide useful tips for navigating difficult concepts.

- One idea that spans a number of the PER-informed instructional strategies is **the value of student collaboration**. It is clear that learning is deeper when students develop ideas in collaboration with peers and work together both in brainstorming approaches and in developing solutions. This text has been written to encourage collaborative learning. For example, the open-ended problems and Interactive Activities are ideally suited

to a group approach. The conceptual problems in each chapter are well suited for use in studio-style classrooms or in approaches that involve peer instruction strategies or interactive lectures. In some places, we have moved derivations from chapters to problems to encourage student discovery of key relationships. The simulations and experiment suggestions will encourage students to engage with the material in a meaningful way. For example, in Chapter 3 students are asked to answer their own questions by using motion detectors on their own smartphones. And with so many PhET simulations now accessible by mobile devices, students can extend their own investigations from the Interactive Activities.

- One goal of any book is to inspire students to appreciate the beauty of the subject and even go on to contribute and become leaders in the field. For this to be achieved, students must see the relevance of the subject. The strong interdisciplinary focus throughout the book will help students achieve this goal. At the same time, it is also important that **students can see themselves as future physicists**. This is a broad-market calculus-based introductory physics text written by a Canadian author team, and we have used Canadian and international examples highlighting physics discoveries, applications, notable scientists past and present, as well as contributions from young Canadians.

- Students place high value on **learning that will help them contribute to society**. For example, service learning is more popular than ever before, and a high number of students set goals of medical or social development careers. Also, there is strong public interest in such fundamental areas as particle physics, quantum mechanics, relativity, string theory, and cosmology. Revised and additional Making Connections boxes support the view of physics as a highly relevant, modern, and socially important field.

Gradual Development of Fundamental Topics

The following are some examples of how fundamental topics are developed in a way that mirrors how a student's own learning progresses, without overwhelming them up front.

- **Motion**: Chapters 3 and 4 have been reworked with an improved flow, logical structure, more diagrams, and consistent notation. Free body diagrams are now introduced in one dimension first (Chapter 5). Chapter 9 now develops angular momentum with an easy-to-grasp approach that includes student participation. The concept of rolling motion is covered from different angles in Chapter 9. Dedicating a chapter to rolling motion has allowed us to focus on and develop the subject gradually, starting with intuitive definitions related to everyday life. Problems that are commonly used at this level are offered in multiple versions with increasing difficulty, and novel open problems walk the student through powerful concepts such as spin and momentum.

- **Forces**: In the mechanics chapters, students are urged to consider how situations would feel. For example, prior to formally stating Newton's Third Law, the idea is qualitatively treated from the perspective of what happens when two friends on ice push each other.

- **Torque**: In Chapter 8, the often problematic concept of torque is introduced in a simple representation of the product of force and distance for the case where these are perpendicular. This is done with examples from everyday life. The discussion then evolves to treating the case where the force is not perpendicular to the displacement. The factors contributing to the torque exerted by a force are developed intuitively and presented using different perspectives, leading to the concept of the moment arm and the full vector representation of torque as the cross product between two vectors.

- **Inertia**: In Chapter 8, moment of inertia is introduced using the simple case of a rotating point mass. This leads intuitively to the moment of inertia of a collection of point masses. The point mass model is used to calculate the moment of inertia of a ring which is contrasted to the moment of inertia of a disk to aid with the intuitive appreciation of the radial distribution of mass on moments of inertia for simple cases. The moment of inertia of a ring is then calculated using integration, which is also applied to the calculation of the moment of inertia of a disk, and employed in the development of the parallel axis theorem.

- Treatment of **exoplanets** in Chapter 11 begins with a qualitative discussion before moving on to quantitative treatment and end-of-chapter problem material. Unique to introductory physics textbooks on the market, coverage of this concept also includes Canadian connections in the development of the field.

- **Gauss's Law**: Chapter 20 is now devoted to Gauss's law, and provides broader range of coverage including concepts that students may not have encountered in math courses (such as vector fields and surface integrals). We invoke an approach in introducing Gauss's law that is unique among introductory physics texts in Canada: We introduce the idea of flux through closed surfaces by first considering how many electric field lines are "caught" in different situations. This semi-quantitative treatment precedes the traditional mathematical treatment developed later in the chapter.

- **Capacitance** comes to life in Chapter 22 with qualitative treatment in two Interactive Activities, which reflects the approach of PhET simulations in general, and provides opportunity for both group and individual work—and further supported responses in the solutions manual.

- **Electromagnetism**: In Chapter 24, cross products relate more strongly to their use in earlier chapters; magnetic field calculations and interactions between fields and charges have been more thoroughly developed.

- While most texts cover the idea of historical **interferometers**, our treatment through the new Making Connection boxes in Section 29-1 (LIGO) and Section 30-10 (Detecting Gravitational Waves) is highly current and combines the basic idea of interferometers with the amazing technology allowing the precision of LIGO. We then provide quantitative treatment in the details of the first black hole coalescence detected by LIGO (and this is extended with a new problem at the end of the chapter).

Physics Education through an Interdisciplinary Lens

As the Canadian Association of Physicists Division of Physics Education (CAP DPE) and others have pointed out, the work of physicists—and the use of physics by other scientists, engineers, and professionals from related fields—is increasingly interdisciplinary. We aimed to **promote the interdisciplinary nature of physics** beyond simply having problem applications from various fields. Chapter content is presented with a rich interdisciplinary feel and stresses the need to use ideas from other sciences and related professions.

The diverse backgrounds of the author team help create this rich interdisciplinary environment, and we have employed many examples related to such fields as medicine, sports, sustainability, engineering, and even music. The text is also richer than most in coverage of areas such as relativity, particle physics, quantum physics, and cosmology.

Informed by the Latest in Physics Education Research

The text is written with Physics Education Research findings in mind, encouraging and supporting PER-informed instructional strategies. The author team brings considerable expertise to the project, including direct experience with a variety of PER-informed instructional strategies, such as peer response systems, computer simulations, interactive lecture demonstrations, online tutorial systems, collaborative learning, project-based approaches, and personalized system of instruction (PSI-based) approaches.

- While the text encourages PER-informed approaches, **it does not support only a single instructional strategy.** Instructors who use traditional lecture and laboratory approaches, those who use peer-response systems, those who favour interactive lecture demonstrations, and indeed those who use other approaches, will find the text well suited for their needs.

- The **visual program** throughout the text has been improved for clarity, consistency, use of colour as an instructional tool, and symbol handling. The Pedagogical Chart on the inside front cover of the text provides a summative quick-stop for student review when confronted with a complex figure, and supports more integration between chapters.

Unique Problem-Solving Approaches

While a professional physicist can view physics as a unified, small set of concepts that can be applied to a very diverse set of problems, the novice sees an immense number of loosely related facts. To guide students through this maze, this text is **concise in wording and emphasizes unifying principles and problem-solving approaches**.

- We have made most chapters self-contained so that each instructor can select which content is addressed in a course. A carefully selected set of problems, both conceptual and quantitative, helps to reinforce mastery of key concepts.

- While all physics texts strive to provide **"real-world problems,"** we believe that we have achieved this to a higher degree. This edition provides more consistent application of data-rich and open-ended problems, as well as improvements in quality, quantity, and richness of all questions and problems.

- Our **Open Problems** are modelled on how the world really is: a key part of applying physics is deciding what is relevant and making reasonable approximations as needed. Closed-form problems, which in most textbooks are the only type used, portray an artificial situation in which what is relevant— and only that—is given to the student.

- Our **Data-Rich Problems** and encouragement of the use of graphing, statistical, and numerical solution software help reinforce realistic situations.

- Our **Making Connections** boxes help students see and identify with real-life applications of the physics.

Interactive Learning

Modern computational tools play a key role in the lives of physicists and have been shown to be effective in promoting the learning of physics concepts. Data-Rich Problems teach students how to do computations, which students use to learn concepts and principles while exploring through PhET simulations and similar animation tools. These **allow students to develop their own conceptual understanding by manipulating variables in the simulated environments**. In the second edition, we use a wider range of PhET simulations and provide more complete guidance on each activity. We also number the simulations, which makes it easier for instructors to assign them to students.

Ultimately, students must take ownership of their learning; that is essentially the goal of all education. The strong links between objectives, sections, Checkpoint questions, and Examples provide an efficient environment for students to achieve this. We view our role in terms of maximizing student interest and engagement and eliminating obstacles on the road to active engagement with physics.

Robert Hawkes
Javed Iqbal
Firas Mansour
Marina Milner-Bolotin
Peter Williams
January 2018

KEY CHANGES TO THE SECOND EDITION

Throughout the Text

Reviewer feedback over the past four years has been valuable in identifying key trends used in classrooms today. That, along with additional PER resources and our own experiences in classrooms across Canada, has culminated in this new and vastly improved second edition. For example:

- We have expanded the array of examples and added significantly more challenging, high-calibre end-of-chapter problems that engage, inspire, and challenge students to attain a high level of proficiency, mastery, and excellence. The material on electromagnetism has been overhauled in this regard.

- **Examples** have been refined to be more consistent in structure, and with a more detailed approach to "Making sense of the result." This change was made to connect different problem-solving strategies to physics examples.

- **Significant digits** are implemented more consistently across chapters.

- We have made the use of **units** and **vector notation** consistent across all chapters. The use of vectors has been significantly revised in the first part of the text.

- We have enhanced cross-references between chapters and between topics and, when needed, between examples and problems within the chapters.

- The **art** throughout the text has been improved through clearer fonts, consistent terminology and symbols, and consistent use of colour and symbol handling. (See the Pedagogical Colour chart on the inside front cover of the book.)

- **Summaries** have been improved to better align with the Learning Objectives.

- **Data-Rich Problems** and **Open Problems** have been incorporated into almost all chapters.

- **Interactive Activities** have been overhauled to make better use of online materials. In the text, they are now presented with a title and description of what is available online and what students will learn from it. If the Interactive Activity uses a PhET simulation, it is identified in the text. Once online, students will receive the interactive activity description and instructions in a detailed and segmented manner to help them work through it. Questions are asked at the end, and the solutions are provided to instructors only.

- New notations have been added to the **problems** at the end of chapters to identify when a problem involves $\frac{d}{dx}$ differentiation, $\int$ integration, $\square$ numerical approximation, and/or $\sim$ graphical analysis. This helps instructors select appropriate problems to assign.

- Heading structure has been improved, with top-level headings aligning with **Learning Objectives** in all chapters.

- More **Checkpoints** have been incorporated into the chapters.

- Each chapter now has at least one **Making Connections** box, and throughout the book this feature has been refined to reflect the latest developments in physics.

- All end-of-chapter problems have been carefully checked and improved with more detailed explanations in the Solutions Manual.

KEY CHAPTER CHANGES

Chapter 1 Introduction to Physics

- A new Making Connections on the 2015 Nobel Prize Winner in Physics, Art McDonald, has been added, as well as a Meet Some Physicists feature to show the diversity in physics-related careers.

- Sophistication of treatment of dimensional analysis and unit conversion has been improved, including additional examples and problems.

- On the suggestion of a reviewer, Approximations in Physics now has its own section and related problems.

- The number of problems and questions has approximately doubled in this chapter compared to the first edition, with a wide variety of types of problems.

Chapter 2 Scalars and Vectors

- The chapter has been improved through re-checking its examples, removing inconsistencies in notation and figures, and ensuring that all the subsections are aligned carefully with the learning objectives.

- The drawings of the free body diagrams and of corresponding life-like situations have been improved.

- One Making Connections, Longitude and Latitude on Earth, has been added.

- On the suggestion of the reviewers, the notation for vectors and their components has been changed, so vectors are always bold and italic with a vector sign over them (e.g., $\vec{a}$), while their components are just italic (e.g., x_a).

- On the suggestion of the reviewers, the difficulty level in some of the problems has been adjusted.

- All problems and solutions have been checked, and careful attention has been paid to mathematically appropriate problems. Two new problems have been added, while three problems have been significantly changed to eliminate ambiguity. The chapter now has almost 70 problems.

Chapter 3 Motion in One Dimension

- This chapter has been extensively reworked, with enhanced attention to its logical structure, conceptual understanding, accuracy of the examples, and consistency of significant figures in the examples.

- The learning objectives have been clarified and aligned carefully with the flow of the chapter.

- A few Checkpoints connecting algebraic and graphical representations of motion have been added.

- A new vignette, additional examples, six Peer to Peer boxes, and two art- and nature-related Making Connections boxes have been added to connect one-dimensional motion to real life.

- Motion diagrams have been introduced and are used consistently throughout the chapter.

- A table illustrating the connection between the relative directions of an object's velocity and acceleration and their impact on the object's motion has been introduced (Table 3-3).

- A table summarizing the relationships between kinematics quantities has been added (Table 3-5).

- Examples of using modern technologies to evaluate the scale of the universe (Interactive Activity 3-1) and analyze and visualize one-dimensional motion (e.g., Example 3-8, Section 3-6, Interactive Activities 3-3, 3-4) have been introduced.

- A video analysis technique to analyze motion is described, including a connection to the works of Eadweard Muybridge (Making Connections in Section 3-6).

- Some repetitive examples have been moved to the end-of-chapter problems or eliminated.

- Additional care has been taken regarding treatment of vector terms, and topics like the selection of the positive direction of motion have been clarified.

- The sections that require calculus (e.g., the analysis of motion with changing acceleration) have been isolated, so students will not find them distracting.

- All problems and solutions have been checked and attention paid to mathematically appropriate problems. A number of new problems have been added, and a number of others changed or eliminated. The difficulty level of all problems has been checked and adjusted where needed. The chapter has almost 140 problems.

Chapter 4 Motion in Two and Three Dimensions

- The graphical vector method for finding the trajectory of a projectile has been introduced.

- Video analysis of motion, including a data-rich problem, is utilized.

- New examples based on sports have been introduced.

- The relative motion discussion has been expanded.

Chapter 5 Forces and Motion

- This chapter has been extensively reworked, with enhanced attention to logical structure and care in explaining terms.

- The new Section 5-1 Dynamics and Forces introduces free body diagrams and net forces in one dimension, before going on to two and three dimensions.

- Additional care has been taken with vector terms and treatment, and topics like the selection of the positive direction have been clarified.

- For those who like to think of force as the derivative of linear momentum, a section (5-11 Momentum and Newton's Second Law) has been added. (Instructors who wish can delay this treatment until after momentum is covered in detail in Chapter 7.)

- A new section on component-free approaches has been added (5-5 Component-Free Solutions) that illustrates that vectors have meaning deeper than

their component representations (it can be considered optional by those who do not want to cover this in first year).

- Several new Making Connections (e.g., Higgs boson) link this classical chapter to modern physics concepts.

- Fundamental and non-fundamental forces now have their own section (at the suggestion of a reviewer).

- The section on non-inertial reference frames has been reworked.

- A total of 28 examples richly illustrate all concepts and techniques for this important material.

- All problems and solutions have been checked and attention paid to mathematically appropriate problems. Over a dozen new problems have been added, and a number of others changed or eliminated. The chapter has more than 100 problems.

Chapter 6 Work and Energy

- This chapter now has an intuitive approach to work and energy, developing the idea of work, starting with the simple 1D situation and evolving into more complex situations.

- The discussion of the work-energy theorem, while sufficiently rigorous, is also intuitive and builds on what students have seen in earlier chapters.

- Vector formalism is employed in a way that encourages students to present their discussions using mathematical formulation.

- The chapter opener poses stimulating and intriguing questions regarding energy in general in a discussion that expands students' horizons while grounding the discussion in the discourse of the field.

Chapter 7 Linear Momentum, Collisions, and Systems of Particles

- The common form of the elastic collision equations is used.

- More examples, including two-dimensional inelastic collision, have been added.

- The centre of mass discussion has been expanded to include an example with a continuous mass distribution.

Chapter 8 Rotational Kinematics and Dynamics

- An intuitive development of torque has been added, examining the representations of torque in great detail using a variety of engaging illustrations and formualtions.

- The key concept of the moment arm is now fully developed in the chapter.

- A detailed exposure of the right-hand rule is now included in the chapter and connects well with the discussion on magnetic fields.

Chapter 9 Rolling Motion

- The chapter now develops the concepts of spin and orbital angular momentum using an intuitive and easy-to-grasp approach that allows active participation by students, but still with sufficient mathematical rigour. Sufficient emphasis is given to the power of the approach.

- Problems and examples now tie better into one another when it comes to considering more realistic approaches to a given scenario. Higher levels of complexity and rigour are included as needed.

Chapter 10 Equilibrium and Elasticity

- The topic of equilibrium is now introduced from an intuitive point of view, using real-life examples, and is exposed in a more complete fashion.

- The connection to the fully developed approach to torque in Chapter 8 is brought out more clearly. This is also summarized in the chapter for easy reference. The chapter now makes it easier to teach static equilibrium before rotational dynamics, as needed.

Chapter 11 Gravitation

- More quantitative treatment of elliptical orbits, and new derivations of Kepler's laws, have now been included.

- Classical treatment of black holes is now included in this chapter (this was in Chapter 29 only in the first edition).

- An expanded exoplanet section includes calculation of their masses.

- About 20 new problems and 4 new examples have been added in this chapter, with improvements in a number of others.

Chapter 12 Fluids

- The subsection "Solids and Fluids under Stress" has been added in Section 12-1.

- The subsection "A Simple Barometer" has been added in Section 12-3.

- Example 12-8 Weighing an Object Immersed in a Fluid has been added in Section 12-5.

- Example 12-9 Blood Flow through a Blocked Artery has been added in Section 12-8.

- Example 12-10 Water Pressure in a Home (Example 12-8 in the first edition) has been rewritten.

- We have replaced Example 12-11 (first edition) with a new example (Example 12-13 Pumping Blood to an Ostrich's Head) in the second edition.

- The subsection "Derivation of Poiseuille's Equation" has been added.

- Eleven new end-of-chapter problems have been added.

Chapter 13 Oscillations

- Example 13-1 (first edition) has been deleted.

- Section 13-6 The Simple Pendulum has been rewritten and expanded.

- Making Connections "Walking Motion and the Physical Pendulum" has been rewritten.

- The subsection "The Quality Factor or the Q-value" has been added in Section 13-9.

- Optional Section 13-11 Simple Harmonic Motion and Differential Equations has been added.

- Fourteen new end-of-chapter problems have been added.

Chapter 14 Waves

- A summary of the main results is provided at the start of Section 14-8.

- Optional Section 14-14 String Musical Instruments has been added.

- Optional Section 14-15 The Wave Equation in One-Dimension has been added.

- Eight new end-of-chapter problems have been added.

Chapter 15 Sound and Interference

- The art for many topics, including resonating columns, has been improved.

- A new section on the role of standing waves in musical instruments has been included.

- The discussion of determining sound levels due to multiple sources has been improved and expanded.

Chapter 16 Temperature and the Zeroth Law of Thermodynamics

- Consistency of wording has improved by use of the word "heat" for the energy that is transferred from one object to another.

Chapter 17 Heat, Work, and the First Law of Thermodynamics

- The sign of work and the convention adopted in the text have been clarified.

Chapter 18 Heat Engines and the Second Law of Thermodynamics

- Figure 18-3 is a detailed illustration showing a steam turbine in a CANDU nuclear power plant.

- Consistent colouring of heat flows in diagrams has been achieved.
- The discussion of the operation of a refrigerator expansion valve has been improved.

Chapter 19 Electric Fields and Forces

- Some topics have been reorganized and a new section added on charging objects by induction.
- Superposition has been added to the titles of Sections 19-4 and 19-7 as part of the enhanced treatment of vector superposition for electric forces and fields.
- A different symbol is used for linear charge density to agree with most other books.
- The electric field vector and field line diagrams are now in a section devoted just to that topic, with significantly enhanced treatment of electric field lines compared to the first edition.
- The number of example problems has more than doubled, as has the number of end-of-chapter problems and questions.
- A new short final section uses a new Checkpoint to clarify electric field misconceptions.

Chapter 20 (part of Chapter 18 in first edition) Gauss's Law

- A full chapter is now devoted to just this topic.
- A strong semi-quantitative base for electric flux is developed prior to the formal introduction of Gauss's law.
- Necessary math concepts such as vector fields, open and closed surfaces, symmetry types, and surface integrals are developed within the chapter for those who have not yet encountered them in their math courses.
- Common Gauss's law misconceptions are addressed through many additional Checkpoints.
- Symbols now differentiate calculation of surface integrals for open and closed surfaces.
- Section 20-9 introduces Gauss's law for gravity to illustrate application of the ideas in another area of physics.
- The chapter structure gives flexibility to instructors in how much of the subject is treated and how.
- There is now a good variety in types and difficulty level in questions and problems.
- In our opinion, we have one of the most complete and innovative treatments of Gauss's law in any introductory text.

Chapter 21 (part of Chapter 20 in first edition) Electric Potential Energy and Electric Potential

- The opening image relates the material of this chapter to the Large Hadron Collider.
- Rather than start right off with electrical potential energy, this chapter now opens with detailed calculations of work to move charges in electric fields. Both the work done by an external agent and the work done by an electric field are introduced, and the relationship between the two views is stated.
- A number of new Peer to Peer boxes and Checkpoints help eliminate misconceptions.
- The material has been enhanced and extended almost everywhere.
- We now include the method of images in the final section (21-9 Electric Potential: Powerful Ideas), but those who prefer not to cover this topic in first year can readily omit it without loss of continuity.
- The number of problems has been significantly expanded, with more than 100 in this chapter.

Chapter 22 (Chapter 21 in first edition) Capacitance

- While the overall structure of this chapter is only slightly changed from the first edition, there have been a large number of small improvements at the suggestion of reviewers and readers.
- We now use two different approaches to derive the electric field between the plates of an ideal parallel plate capacitor in Section 22-2 (one uses superposition and one does not). In this way, we establish where the electric charge must be on the plates as one of the important points summarized in bullet form at the end of the chapter.
- The notation for combining capacitors has been made consistent with that used later for combining resistors in Chapter 23.
- The Applications section has been altered, with a few topics that require resistance ideas eliminated.
- Almost 30 new problems have been added (and a few others changed).

Chapter 23 (Chapter 22 in the first edition) Electric Current and Fundamentals of DC Circuits

- The chapter has been improved through revising its examples by removing inconsistencies in notation and figures.
- The applications of Kirchhoff's laws have been clarified by using additional examples and improving the table clarifying the sign convention for the directions

of currents and the signs of potential differences across the circuit elements (Table 23-4).

■ Nine new end-of-chapter problems have been added. The chapter now has more than 70 problems.

Chapter 24 (Chapter 23 in the first edition) Magnetic Fields and Magnetic Forces

■ This chapter has undergone major revisions in terms of its content, examples, end-of-chapter problems, and solutions in the Solutions Manuals.

■ The topic of cross products is developed intuitively as it relates to the chapter material and is closely linked to the development and use of cross products in earlier chapters.

■ The presentation of magnetic field calculations and interactions between magnetic fields and moving charges is now done in much greater detail, evolving from the simple to the complex, and more comprehensively highlights the utility of the right-hand rule.

■ The learning objectives have been edited and the sections are now better aligned with them.

■ One new Checkpoint, two expanded examples, and two Making Connections boxes have been added, including a discussion of Canadian astronomer T. Victoria Kaspi and applications of magnetism to the animal kingdom.

■ More than 30 figures in the chapter have either been added or edited and significantly improved.

■ The discussion of the Hall effect has been significantly improved.

■ More than 20 end-of-chapter problems of various complexity have been added, including a number of problems requiring differentiation and integration. The chapter now has more than 100 end-of chapter problems.

Chapter 25 (Chapter 24 in the first edition) Electromagnetic Induction

■ While this chapter has not undergone major revisions, it has been edited for clarity and accuracy.

■ The learning objectives have been edited, and the sections are now better aligned with these objectives.

■ One new example in the chapter has been added, while all other examples have been edited for clarity, accuracy, and meaningful connections to everyday life and students' experiences.

■ The figures and tables in the chapter have been clarified and improved.

■ The chapter has more than 80 end-of-chapter problems of a wide range of difficulty, including a number of problems requiring differentiation and integration.

Chapter 26 Alternating Current Circuits

■ Voltage is used in place of emf in this chapter, and this is explicitly discussed.

■ Energy usage statistics have been updated.

■ A new Checkpoint testing understanding of phase shifts has been added.

Chapter 27 Electromagnetic Waves and Maxwell's Equations

■ In Section 27-8, we have added the Making Connections box "Polarization and 3D Movies."

■ We have added five new end-of-chapter problems.

Chapter 28 (Chapter 27 in the first edition) Geometric Optics

■ We have made relatively minor changes from a well-received first edition chapter.

■ One extra Checkpoint question was added, and three examples have been improved.

■ One Making Connections about image formation in plane mirrors has been edited and improved.

■ All the tables summarizing sign conventions of geometric optics and properties of images created by mirrors and thin lenses have been improved.

Chapter 29 (Chapter 28 in first edition) Physical Optics

■ We have made relatively minor changes from a well-received first edition chapter.

■ The strategy for thin film interference problems is made explicit.

■ Links with modern physics have been extended (e.g., a new Making Connections on the LIGO detector).

■ More than 45 new end-of-chapter questions and problems have been added that are well distributed over all topics.

Chapter 30 (Chapter 29 in first edition) Relativity

■ We retained consideration of both special relativity and some aspects of general relativity in this chapter, ending with the well-received quantitative example on the two relativistic corrections in the GPS system.

- As suggested by reviewers, we have provided more on the experimental evidence for relativity, including the new Making Connections box on the Hafele–Keating experiment ("Testing Time Dilation with Atomic Clocks").

- Lorentz transformations are now covered in depth with their own section. Those who prefer not to teach Lorentz transformations can skip Section 30-5 and the derivation in Section 30-8 and still cover the rest of the chapter.

- Matrix formulations are used for Lorentz transformations, which are also expressed without this notation for instructors who prefer not to use matrices in first year.

- At the suggestion of one reviewer, the relativistic velocity addition relationship is now fully derived in the text.

- Through a new qualitative problem we urge students to express arguments for and against the concept of relativistic mass.

- The spacetime diagram and interval coverage have been expanded.

- The relativistic Doppler shift is rigorously derived and has its own section.

- The term *four vector* is explained in the chapter.

- The relationship between total energy, relativistic momentum, and rest mass energy is now a key equation (30-38) and not simply part of a problem derivation, as it was in the first edition.

- An extensive new Making Connections quantitatively explains the evidence from the recent LIGO detection of black hole coalescence.

- There are about 30 new problems, along with changes and a few deletions from the first edition. There are four new examples.

- The Solutions Manuals have been totally reworked to make both the approach and the notation consistent between the Solutions Manuals and the chapter.

- We feel that we have one of the most comprehensive relativity treatments of any first-year textbook.

Chapter 31 Fundamental Discoveries of Modern Physics

- We have expanded the Fundamental Concepts and Relationships section to synthesize the results from the chapter, making it clearer why some new physics was needed.

Chapter 32 Introduction to Quantum Mechanics

- In Section 32-3, we have expanded the subsection "The Physical Meaning of the Wave Function."

- We have added Section 32-6 The Finite Square Well Potential, which includes the concept of the parity operation in quantum mechanics.

- We have added three new end-of-chapter problems.

Chapter 33 Introduction to Solid-State Physics

- We have replaced the formal derivation of the density of states at the Fermi surface with a more physical argument.

Chapter 34 Introduction to Nuclear Physics

- In Section 34-6, the subsection "Gamma Decay" from the first edition has been rewritten and is now called "Nuclear Levels and Gamma (γ) Decay."

- The new Section 34-7 Nuclear Stability has been added. The effect of Coulomb repulsion on the nuclear levels is discussed in this section.

- The new Section 34-10 Nuclear Medicine and Some Other Applications has been added.

Chapter 35 Introduction to Particle Physics

- Section 35-12 Beyond the Standard Model has been greatly expanded and contains the subsections "Dark Matter" and "Dark Energy."

About the Authors

ROBERT HAWKES Dr. Robert Hawkes is a Professor Emeritus of Physics at Mount Allison University. In addition to having extensive experience in teaching introductory physics, he has taught upper-level courses in mechanics, relativity, electricity and magnetism, electronics, signal processing, and astrophysics, as well as education courses in science methods and technology-enhanced learning. His astrophysics research program is in the area of solar system astrophysics, using advanced electro-optical devices to study atmospheric meteor ablation, as well as complementary lab-based techniques such as laser ablation. He is the author of more than 80 research papers. Dr. Hawkes received his B.Sc. (1972) and B.Ed. (1978) at Mount Allison University, and his M.Sc. (1974) and Ph.D. (1979) in physics from the University of Western Ontario. He has won a number of teaching awards, including a 3M STLHE National Teaching Fellowship, the Canadian Association of Physicists Medal for Excellence in Undergraduate Teaching, and the Science Atlantic University Teaching Award, as well as the Atlantic Award for Science Communication. He was an early adopter of several interactive physics teaching techniques, in particular collaborative learning in both introductory and advanced courses. The transition from student to professional physicist, authentic student research experiences, and informal science learning are recent research interests. He was a co-editor of the 2005 *Physics in Canada* special issue on physics education, and a member of the Canadian physics education revitalization task force. Minor planet 12014 is named Bobhawkes in his honour.

Outside physics and education, he combines walking and hiking with photography, and volunteers at a community non-profit newspaper. He treasures exploring the joy and fun of learning with his grandchildren.

JAVED IQBAL Dr. Javed Iqbal is the director of the Science Co-op Program and an Adjunct Professor of Physics at the University of British Columbia (UBC). At UBC he has taught first-year physics for 20 years and has been instrumental in promoting the use of clickers at UBC and other Canadian universities. In 2004, he was awarded the Faculty of Science Excellence in Teaching Award. In 2012, he was awarded the Killam Teaching Prize. His research areas include theoretical nuclear physics, computational modelling of light scattering from nanostructures, and computational physics. Dr. Iqbal received his Doctoral Degree in Theoretical Nuclear Physics from Indiana University.

FIRAS MANSOUR As a lecturer in the Department of Physics and Astronomy at the University of Waterloo since 2000, Firas Mansour has gained respect and praise from his students for his exceptional teaching style. He currently teaches first-year physics classes to engineering, life science, and physical science students, as well as upper-year elective physics courses in the past. He is highly regarded for his quality of teaching, his enthusiasm in teaching, and his understanding of students' needs. His dedication to teaching is exemplary, as is his interest in outreach activities in taking scientific knowledge beyond the university boundary. He is a 2012 Distinguished Teaching Award recipient at the University of Waterloo. He has overseen the creation of high-quality material for online learning and face-to-face instruction and has implemented various PER–established practices ranging from flipped and blended classroom instruction to peer instruction and assessment and group work.

MARINA MILNER-BOLOTIN
Dr. Marina Milner-Bolotin is an Associate Professor in Science (Physics) Education at the Department of Curriculum and Pedagogy at the University of British Columbia. She holds an M.Sc. in theoretical physics from Kharkiv National University in Ukraine (1991), a teaching certification in physics and mathematics from Bar-Ilan University in Israel (1994), and a Ph.D. in mathematics and science education from the University of Texas at Austin (2001). She educates future physics and mathematics teachers and studies how modern technologies can be used to support physics learning and teaching, increasing student' interest in physics and their understanding of physics concepts and principles.

For the last 25 years, she has been teaching physics in Israel, the United States (Texas and New Jersey), and Canada (UBC and Ryerson University). She has taught physics and mathematics to a wide range of students, from gifted elementary students to university undergraduates and future physics teachers. She has also led a number of professional development activities for physics, science, and mathematics teachers in Ontario, British Columbia, and abroad. She is often invited to conduct professional development activities with science and mathematics teachers in China, the Republic of Korea, the United States, Iceland, Germany, Denmark, Israel, and other countries. In addition, Dr. Milner-Bolotin has led many science outreach events engaging the general public in physics. She founded the UBC Faculty of Science Faraday Christmas Lecture in 2004 and the UBC Faculty of Education Family Mathematics and Science Day in 2010.

She has published more than 50 peer-reviewed papers and 9 book chapters, and she led the development of online resources for mathematics and science teaching used by thousands of teachers and students: scienceres-edcp-educ.sites.olt.ubc.ca/.

She has served as the President of the British Columbia Association of Physics Teachers and as a member of the Executive Board of the American Association of Physics Teachers. She has received many teaching, research, and service awards, including the National Science Teaching Association Educational Technology Award (2006), the UBC Department of Physics and Astronomy Teaching Award (2007), the Ryerson University Teaching Excellence Award (2009), the Canadian Association of Physicists Undergraduate Teaching Medal (2010), the UBC Killam Teaching Award (2014), and the American Association of Physics Teachers Distinguished Service Citation (2014) and Fellowship (2016).

PETER WILLIAMS Dr. Peter Williams is Professor of Physics at Acadia University, where he also served as Dean of the Faculty of Pure and Applied Science between 2010 and 2016. He has received numerous awards for his teaching, including the 2006 Canadian Association of Physicists (CAP) Medal for Excellence in Teaching.

He played a critical role in the introduction of studio physics modes of instruction at Acadia University and has developed several innovative courses, including most recently a Physics of Sound course. He is very interested in effectively combining the best of technology-enhanced educational techniques while maintaining a strong personal approach to teaching. He is also a strong proponent of applying research methodology to the evaluation of the effectiveness of different modes of physics instruction and has published several articles in teaching journals.

When he is not busy with physics, he loves to play his upright bass, go sailing with his family, and cook.

TEXT WALKTHROUGH

Physics for Scientists and Engineers: An Interactive Approach, Second Edition, is carefully organized so you can stay focused on the most important concepts and explore with strong pedagogy.

Learning Objectives are brief numbered and directive goals or outcomes that students should take away from the chapter. Listed at the beginning of each chapter, these also correspond to major sections within that chapter.

Opening Vignettes These narratives at the beginning of each chapter introduce topics through an interesting and engaging real-life example that pertains to the chapter topics. An engaging entry into the chapter, these vignettes also provide students with the opportunity to read about historical and very recent current events in physics.

Examples Each example is numbered and corresponds to each major concept introduced in the section. Examples are now more consistently structured across all chapters, with a title, a statement of the problem, a solution, and a paragraph titled "Making sense of the result." Within the example, the authors have modelled desired traits, such as care with units and consideration of appropriate significant figures. "Making sense of the result" is one of the most important features, in which authors model the idea of always considering what has been calculated to determine whether it is reasonable.

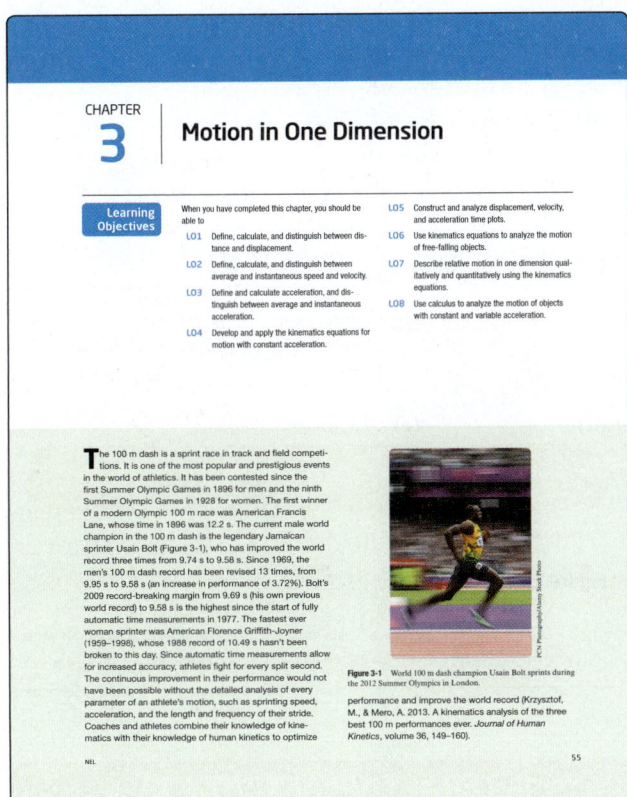

CHAPTER 3 | Motion in One Dimension

EXAMPLE 15-2

Wave Amplitude

Calculate the displacement amplitude of a 1000 Hz sound wave whose pressure amplitude is 100.0 μPa.

Solution

This is a simple application of the relationship expressed in Equation 15-11. Using the bulk modulus for air found in Table 15-1, we rearrange Equation 15-11 to find

$$A = \frac{\Delta p}{Bk} = \frac{\Delta p_m}{B\dfrac{2\pi f}{v}} = \frac{0.000\,100\,0 \text{ Pa}}{1.41 \times 10^5 \text{ Pa} \dfrac{2\pi (1000 \text{ s}^{-1})}{343 \text{ m}\cdot\text{s}^{-1}}}$$

$$= 3.87 \times 10^{-11} \text{ m}$$

Making sense of the result

We have a pressure amplitude that is about four times that shown in Figure 15-4. Since the pressure amplitude is proportional to the displacement amplitude, we should find a displacement amplitude that is four times that shown in Figure 15-4.

Peer to Peer Written by students for students, Peer to Peer boxes provide useful tips for navigating difficult concepts.

PEER TO PEER

In doing relativistic trip type problems, I find the most important thing is to keep in mind the definitions of proper length and proper time. The person on the trip measures the proper time (if the time interval is the trip), but a different observer not moving with respect to the end points of the trip measures the proper length.

Making Connections   Making Connections boxes are provided in a narrative format and contain concise examples from international contexts, the history of physics, daily life, and other sciences.

Checkpoints   Each learning objective has a Checkpoint box to test students' understanding of the material they have just read. Checkpoint boxes include questions in different formats, followed immediately by the answer placed upside down at the end of the box. While different formats are used, these Checkpoints are meant to be self-administered, so they all have a single clear answer so that students know whether they have mastered the concept before moving on to dependent material. The close linking of sections, learning objectives, and Checkpoints is a major feature of the text.

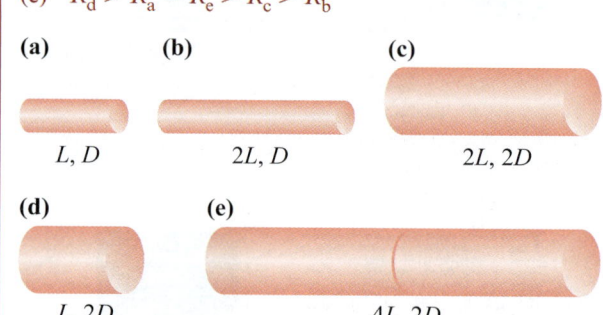
Interactive Activities   provide activities, such as computer simulations, that help with concept development. Many of these are matched to the research-validated PhET simulations. Students are introduced to an Interactive Activity in the text, and then when online, they will see a full description and set of instructions embedded with the activity, so they can adjust variables or diagrams provided. Questions are provided at the end. Answers are available to instructors only.

Key Equations It is important for students to differentiate fundamental relationships from equations that are used in steps of derivations and examples. Key equations are clearly indicated.

KEY EQUATION

$$T^2 = \frac{4\pi^2}{G(m_1 + m_2)} a^3 \qquad (11\text{-}37)$$

Key Concepts and Relationships provide a summary at the end of each chapter. This section provides students with an opportunity to review the key concepts discussed in the chapter. Care has been taken to make these concise and yet at the same time cover all core ideas and correspond to major sections in the chapter. Applications and Key Terms introduced in that chapter are also listed here for student reference.

KEY CONCEPTS AND RELATIONSHIPS

Kinematics is the study of motion. In kinematics, we study the relationships between an object's position, displacement, velocity, and acceleration and their dependence on time. We also examine relative motion.

End-of-Chapter Questions and Problems Questions, Problems by Section, Comprehensive Problems, Data-Rich Problems, and Open Problems are provided at the end of each chapter to test students' understanding of the material. The volume of exercises and problems has been significantly expanded in this edition.

For the problems, star ratings are used (*, **, or ***), with more stars indicating more-challenging problems. New to this edition, problems now include notation to identify if they involve $\frac{d}{dx}$ differentiation, $\int$ integration, ▢ numerical approximation, and/or 〰 graphical analysis.

QUESTIONS

1. A sound wave is a longitudinal wave. True or false?
2. The displacement and pressure amplitudes are
   (a) in phase
   (b) out of phase by 90°
   (c) out of phase by 180°
3. When we double the frequency of a sound wave, by what factor does the wavelength change?

PROBLEMS BY SECTION

For problems, star ratings will be used (★, ★★, or ★★★), with more stars meaning more challenging problems. The following codes will indicate if $\frac{d}{dx}$ differentiation, $\int$ integration, ▢ numerical approximation, or 〰 graphical analysis will be required to solve the problem.

Section 15-1   Sound Waves

15. ★ A wave is observed to have a frequency of 1000 Hz in air. What is the wavelength?

COMPREHENSIVE PROBLEMS

42. ★ At large concerts, it is sometimes disconcerting to observe the musicians moving apparently out of sync with the music. This results from the time it takes the sound to travel from the stage to you. When the musicians are playing at 100 beats/min, at what distance from the stage will they appear to be one full beat behind?

DATA-RICH PROBLEM

68. ★★★ You have been hired by an environmental consulting firm to do a noise analysis for a quarry operation. The operation uses two trucks, a drill, and a crusher. The manufacturers of the equipment provided the specifications in Table 15-5 for the sound level of the various pieces of equipment. Local bylaws specify that the maximum sound level at the perimeter of the quarry property not exceed 85 dB. How close to the perimeter can the quarry operate all these devices simultaneously?

Table 15-5   Data for Problem 68

Equipment	Sound Level at 10 m
Truck	85 dB
Drill	110 dB
Crusher	110 dB

OPEN PROBLEM

69. ★★★ Many of us have heard the effect that can be produced by inhaling helium and speaking. The speaker's voice is shifted to higher frequencies. Discuss the physics behind this effect.

ACKNOWLEDGMENTS

The Nelson Education team has been amazing—this book would never have been completed without their expertise, attention to detail, flexibility, and above all emphasis on producing a high-quality and innovative text. Particular credit goes to Paul Fam, Senior Publisher—Higher Education, who has so enthusiastically guided and supported the project from the earliest days. Content Manager Suzanne Simpson Millar's extensive experience and professionalism were critical in moving us from rough drafts to finished manuscript.

A text written by a team of physicists poses a challenge in making the final book have a common voice and a consistent approach. The success we have achieved in that regard is due in large part to our copyeditor, Julia Cochrane. Words cannot adequately express the debt we owe. The production stage was complex, and we thank the many people who helped us through this process—often under tight deadlines—especially Production Project Managers Wendy Yano and Natalia Denesiuk Harris, who had primary responsibility for overall production issues. Kristiina Paul, our photo researcher, worked hard to get permissions for our first choices for images and, when they were not available, to find suitable alternatives. The publisher and the author team would also like to convey their thanks to Simon Friesen, University of Waterloo; Karim Jaffer, John Abbott College; Anna Kiefte, Acadia University; and Kamal Mroue, University of Waterloo, for their technical edits, which ensured consistency in key areas, such as the use of significant digits, accuracy in the figures, and making sure all of the steps were accounted for in the examples presented and solutions prepared.

Thanks go to those who reviewed the text. Collectively, these professionals offered ideas, and occasional corrections, that helped make the book more accurate and clear. The diversity of their views of physics and how it should be taught—while occasionally resulting in not all suggestions being able to be incorporated in this printing—provided us with a broader view than that of the five authors alone. We give thanks to the following individuals:

Daria Ahrensmeier, Simon Fraser University

Jake Bobowski, University of British Columbia

Jonathan Bradley, Wilfrid Laurier University

David Crandles, Brock University

Jason Donev, University of Calgary

Richard Goulding, Memorial University of Newfoundland

Stanley Greenspoon, Capilano University

Jason Harlow, University of Toronto

Stanislaw Jerzak, York University

Mark Laidlaw, University of Victoria

Robert Mann, University of Waterloo

Ryan D. Martin, Queen's University

Ben Newling, University of New Brunswick

Ralph Shiell, Trent University

Zbigniew M. Stadnik, University of Ottawa

Salam Tawfiq, University of Toronto

Members of the UBC Student Advisory Board (SAB) helped us remain grounded in what sort of text students wanted and would use, and most of them also contributed to the Peer to Peer boxes, ensuring that the material in this text is presented from a truly "student" perspective.

We thank Kimberley Carruthers, Marketing Manager at Nelson Education, for her skilled promotion of the book, along with the team of publisher's sales representatives across the country.

To be honest, this book has taken more of our time and energy than any of the authors ever anticipated. All of the authors combined the writing of this text with other career demands, and as a result many weekends and evenings were devoted to this text. We thank our family members for their understanding and encouragement. We thank our colleagues for their support in various ways during the course of this project. The authors would also like to acknowledge Rohan Jayasundera and Simarjeet Saini for enlightening discussions. The authors would like to thank Olga Myhaylovska for her valuable input and advice. The authors would also like to thank Reema Deol and Renee Chu for conducting background research on some of the material used in the text.

The authors would like to thank professor David F. Measday (UBC) for providing valuable feedback for Chapter 33 Introduction to Nuclear Physics.

Daily in our classrooms we learn from our students and are rejuvenated by their enthusiasm, creativity, and energy. Our view of the teaching and learning of physics owes a great deal to them, probably more than they realize. Similarly, our interactions with colleagues, both at our own institutions and beyond, have shown us new and more effective ways to approach difficult concepts and helped in our own education as physicists. We would like to acknowledge the valuable work done by organizations such as the Canadian Association of Physicists Division of Physics Education (CAP DPE) and the American Association of Physics Teachers (AAPT).

INSTRUCTOR RESOURCES

 The **Nelson Education Teaching Advantage** (NETA) program delivers research-based instructor resources that promote student engagement and higher-order thinking to enable the success of Canadian students and educators. Visit Nelson Education's **Inspired Instruction** website at nelson.com/inspired/ to find out more about NETA.

The following instructor resources have been created for *Physics for Scientists and Engineers: An Interactive Approach*, Second Edition. Access these ultimate tools for customizing lectures and presentations at nelson.com/instructor.

NETA Test Bank

This resource was written by Karim Jaffer, John Abbot College. It includes more than 1,000 multiple-choice questions written according to NETA guidelines for effective construction and development of higher-order questions. Also included are 500 true/false questions.

 The NETA Test Bank is available in a new, cloud-based platform. **Nelson Testing Powered by Cognero®** is a secure online testing system that allows instructors to author, edit, and manage test bank content from anywhere Internet access is available. No special installations or downloads are needed, and the desktop-inspired interface, with its drop-down menus and familiar, intuitive tools, allows instructors to create and manage tests with ease. Multiple test versions can be created in an instant, and content can be imported or exported into other systems. Tests can be delivered from a learning management system, the classroom, or wherever an instructor chooses. Nelson Testing Powered by Cognero for *Physics for Scientists and Engineers: An Interactive Approach*, Second Edition, can be accessed through nelson.com/instructor.

NETA PowerPoint

Microsoft® PowerPoint® lecture slides for every chapter have been developed by Sean Stotyn, University of Calgary. There is an average of 55 slides per chapter, many featuring key figures, tables, and photographs from *Physics for Scientists and Engineers: An Interactive Approach*, Second Edition. Notes are used extensively to provide additional information or references to corresponding material elsewhere. NETA principles of clear design and engaging content have been incorporated throughout, making it simple for instructors to customize the deck for their courses.

Image Library

This resource consists of digital copies of figures, short tables, and photographs used in the book. Instructors may use these jpegs to customize the NETA PowerPoint or create their own PowerPoint presentations. An Image Library Key describes the images and lists the codes under which the jpegs are saved.

TurningPoint® Slides

TurningPoint® classroom response software has been customized for *Physics for Scientists and Engineers: An Interactive Approach*, Second Edition. Instructors can author, deliver, show, access, and grade, all in PowerPoint, with no toggling back and forth between screens. With JoinIn, instructors are no longer tied to their computers. Instead, instructors can walk about the classroom and lecture at the same time, showing slides and collecting and displaying responses with ease. Anyone who can use PowerPoint can also use JoinIn on TurningPoint.

Instructor's Solutions Manual

This manual, prepared by the textbook authors, has been independently checked for accuracy by Simon Friesen, University of Waterloo; Karim Jaffer, John Abbott College; Anna Kiefte, Acadia University; and Kamal Mroue, University of Waterloo. It contains complete solutions to questions, exercises, problems, Interactive Activities, and Data-Rich Problems.

Möbius

 Möbius allows you to integrate powerful, dynamic learning and assessment tools throughout your online course materials, so your students receive constant feedback that keeps them engaged and on track.

- Integrate meaningful, automatically graded assessment questions into lessons and narrated lectures, in addition to formal assignments, so students can test their understanding as they go.

- Provide interactive applications for exploring concepts in ways not available in a traditional classroom.

- Leverage powerful algorithmic questions to provide practice for students as they master concepts, as well as individual summative assessments.

- Incorporate engaging, enlightening visualizations of concepts, problems, and solutions, through a wide variety of 2D and 3D plots and animations that students can modify and explore.

- Bring your online vision to life, including online courses, open-access courses, formative testing, placement and remediation programs, independent learning, outreach programs, and flipped or blended classrooms.

- Provide exactly the content you want, from individual lessons and textbook supplements, to full courses, remedial materials, enrichment content, and more.

- Choose the learning experience by allowing students open access to your course material or guiding them along a specific learning path.

- Stay in control of your content, creating and customizing materials as you wish to suit your needs.

STUDENT ANCILLARIES

Student Solutions Manual (ISBN 978-0-17-677046-4)

The Student Solutions Manual contains solutions to selected odd-numbered exercises and problems.

Möbius

 Möbius is an HTML5-native online courseware environment that takes a "learn-by-doing" philosophy to STEM education, utilizing the highly interactive Maple visualization engine that drives online applications for immediate learning outcome development and assessment. It also harnesses the power of the Maple TA™ platform, enabling over 15 different types of algorithmic assessments that can be posed to a student at any time within the courseware environment. The power of the assessment is immediate confirmed understanding of difficult STEM-based topics in real time. This type of power is necessary to ensure the high level of learning outcomes that is possible within the environment. Instructors can easily create and share their own assessments and modify any lesson, assessment, or interactive activity and share with their students or the wider Möbius user community. In addition, unlike traditional learning technologies, textbook exposition, interactives (i.e., PhET simulations), and assessment are all "in line" so that students are presented with a unified learning environment, keeping them firmly focused on the topic at hand.

CHAPTER

1 | Introduction to Physics

When you have completed this chapter, you should be able to

LO1 Define what we mean by physics in your own words.

LO2 List the types of possible errors, and differentiate between precision and accuracy.

LO3 Calculate the mean, standard deviation, and standard deviation of the mean (SDOM) for data sets, and correctly use $\pm$ notation and error bars.

LO4 Correctly apply significant digits rules to calculated quantities.

LO5 Convert quantities to and from scientific notation.

LO6 State SI units, and write the units and their abbreviations correctly.

LO7 Apply dimensional analysis to determine if a proposed relationship is possible.

LO8 Perform unit conversions.

LO9 List and explain reasons why we make approximations in physics.

LO10 Make reasonable order-of-magnitude estimates and solve open problems.

Why should you study physics? One reason is that physics helps us answer amazing questions. For example, physics has provided a remarkably detailed picture of what the universe is like and how it has developed over time. The Hubble Space Telescope produced the image in Figure 1-1. The image shows an area of the sky equivalent to what you would cover if you held a 1 mm square at arm's length. Yet this image shows about 10 000 galaxies, and each galaxy typically contains 100 billion stars. In the hundred years since the first evidence of the existence of galaxies, observations and theoretical calculations by numerous physicists have provided strong evidence that the evolution of the universe started in a "big bang" about 13.8 billion years ago, when the universe was infinitesimally small, almost infinitely dense, and incredibly hot.

Like cosmology, aspects of research in particle physics, quantum mechanics, and relativity can be fascinating because they challenge our common-sense ideas. However, discoveries in these fields have also led to numerous extremely useful applications. For example, atomic physics underlies medical imaging technologies from simple X-rays to the latest MRIs and CAT scans. In fact, physics concepts are the basis for almost all technologies, including energy production and telecommunications. As you explore this book, each chapter will bring a new answer to the question, "Why should physics matter to me?"

Figure 1-1 A tiny part of the night sky captured with incredible detail by the Hubble Space Telescope.

Robert Williams and the Hubble Deep Field Team (STScI) and NASA

1-1 What Is Physics?

We will start this chapter by considering the nature of physics, and what differentiates physics from other areas of study.

The domain of physics is the physical universe The domain of physics extends from the smallest subatomic particles to the universe as a whole. Physics does not seek to answer questions of religion, literature, or social organization. While physics is creative, and we may refer to the art of physics and recognize artistic beauty in conceptual frameworks, there is a fundamental difference between art and physics. Art can be created in any form envisioned by the artist, but physics must comply with the nature of the physical universe. Nor is mathematics or philosophy the same as physics. Most argue that for something to be considered physics, it must, at least potentially, be validated through observations and measurements. Not everyone in physics agrees about this last point. There has been recent debate about whether aspects of string theory and multiverses (parallel universes) are properly considered physics.

Physics is a quantitative discipline Although there are a few topics in physics where our understanding is currently mainly qualitative, overall, measurement and calculation play critical roles in developing and testing physics ideas. Most physicists spend more time performing computations than they spend on any other single aspect of physics. Although all sciences and engineering are increasingly mathematically sophisticated, most would agree that physics is the most mathematical of the sciences.

Associated with quantitative reasoning must be the recognition that there is an inherent uncertainty in any measured quantity. Later in this chapter we will explore how to determine the uncertainty in common situations and to express that uncertainty in how you write a number.

Physics uses equations extensively to express ideas You should view physics equations as a shorthand notation for the theories and relationships they represent. While we can express physics concepts in words, it is more efficient, particularly in situations simultaneously involving a number of different physics ideas, to use equation notation. You will need to develop proficiency in manipulating equations and deriving relationships from basic principles. Applying physics, though, is not simply selecting from a large pool of established equations. You should always ask yourself whether a relationship is applicable to the situation, and what assumptions are inherent in using any particular equation. It is a good idea to start every problem by considering the physics concepts that

---

MAKING CONNECTIONS

Meet Fabiola Gianotti

Italian physicist Fabiola Gianotti (Figure 1-2) was until recently the scientific spokesperson for the ATLAS experiment at the LHC at CERN (Conseil Européen pour la Recherche Nucléaire) and she is now Director General at CERN, arguably the world's most important scientific undertaking. Even during her university studies, Fabiola Gianotti was undecided between a career in the creative arts, in philosophy, in other sciences, or in physics. She is a skilled pianist and studied piano at the Milan Conservatory. She is quoted as saying that her interest in philosophy helped her see that asking the right questions was critical, a view that has shaped her success in physics. She feels that it is sometimes misunderstood how close physics is to the arts: "... art and physics are much closer than you would think. Art is based on very clear, mathematical principles like proportion and harmony. At the same time, physicists need to be inventive, to have ideas, to have some fantasy." She is excited about the progress that physics has made in understanding our universe but realizes that much remains to be done: "... what we know is really very, very little compared to what we still have to know."

Courtesy Mike Struik

Figure 1-2 Director General at CERN Fabiola Gianotti.

MAKING CONNECTIONS

The CCD: Applied Physics and a Nobel Prize

The 2009 Nobel Prize in Physics was awarded to three scientists: the late Canadian Willard S. Boyle and the American George E. Smith for the invention of the charge coupled device (CCD) (see Figure 1-3), and Charles Kuen Kao from China for work leading to fibre-optic communication. Born in Amherst, Nova Scotia, Willard Boyle studied at McGill University before working at Bell Laboratories in New Jersey, where he and Smith made the first CCD.

The CCD is a semiconductor device with many rows, each consisting of a large number of tiny cells that accumulate an electric charge proportional to the light intensity at each cell (see Chapter 22). The CCD is the heart of digital cameras. A fundamental obstacle to digital imaging was that it was not practical to connect one wire to each of the millions of pixels that make up the digital image. This problem was overcome through the CCD invented by Dr. Boyle and colleagues.

The key idea is that the electric charge, representing the brightness of the image, is passed from one cell to the next. It is as though you have a line of people, each with a number written on a piece of paper, and you want to read out the codes from all of the papers. One approach is to have each person hand their paper to the person beside them in sequence, all down a line, and collect all the papers at a single point. The cells in a CCD do this with electric charge—a sequence of voltage pulses applied to the CCD cells causes the charge in each cell to transfer to the next cell in the row. The charge sequence leaving the last cell produces a signal that corresponds to the light that was focused on all the different cells in the line (signals can be moved from line to line in a similar manner). This signal is amplified to make an electronic record of the image that was stored as charge on the CCD.

CCD imaging has many advantages over film, including substantially greater sensitivity, linearity (meaning twice as much light produces twice as much signal), and the ability to be remotely operated, essential for applications such as space cameras. CCDs are the heart of all space telescopes and many medical instruments, as well as consumer devices containing digital cameras.

(a)

(b)

OLIVIER MORIN/Staff/Getty Images

Figure 1-3   (a) An image of Comet 67P/Churyumov–Gerasimenko taken with a charge coupled device (CCD) digital camera on the ESA Rosetta mission. (b) Willard Boyle (left) and George Smith in 1970, shortly after their invention of the CCD.

Source: (a) European Space Agency – ESA. This work is licenced under the Creative Commons Attribution-ShareAlike 3.0 IGO (CC BY-SA 3.0 IGO) licence: http://creativecommons.org/licenses/by-sa/3.0/igo/

may be helpful, rather than starting with equations. Although equations form the language of physics, the heart of physics is made up of the physical concepts the equations represent. An analogy might be you and your name; it is efficient for others to refer to you by your name, but the important thing is who you are, not your name.

Models, predictions, and validation Physicists develop hypotheses and models based on patterns recognized in observations and experiments. From these

hypotheses and models they develop predictions that can be tested with further measurements. If the additional measurements are not consistent with the predictions, our model must be wrong, or at least inadequate. It is important to realize that "proof" in physics is never absolute. We can prove that a model or hypothesis is wrong through predictions and experiments, but we cannot prove it is absolutely right. We do develop confidence in models that have been used for many predictions, all of which have been found consistent with experiment, but that is not the same as saying we are sure the model will hold up in all possible future experiments and situations. For example, in Chapter 30 you will learn about general relativity and see that it has been used to predict a number of results that are contrary to common sense. Most of these have now been tested, and passed those tests, so we do have confidence in general relativity, but that is not the same as saying we are sure the theory is necessarily complete.

Physics seeks explanations with the greatest simplicity and widest realm of application Those from outside physics often view physics, incorrectly, as a collection of a large number of laws. Rather, physics seeks to explain the physical universe and all that it contains using a limited number of relationships. For example, we only need to invoke four types of interactions to explain all forces in physics: gravitation, electromagnetism, weak nuclear forces, and strong nuclear forces. Many physicists believe that ultimately these can be brought together as different aspects of a single unified theory. In your study of physics it is critical to keep in mind this goal of applying core theoretical ideas in a wide variety of situations. We suggest that at the end of each chapter you try to express the key concepts as concisely as possible, repeating this exercise for the entire book near the end of your course.

MAKING CONNECTIONS

The Neutrino and the 2015 Nobel Prize

The 2015 Nobel Prize in Physics went to another Canadian scientist with deep Nova Scotia roots. Art McDonald (Figure 1-4) was born in Sydney, Nova Scotia, and, following B.Sc. and M.Sc. degrees in physics at Dalhousie University, he completed a Ph.D. at the California Institute of Technology. After positions at Chalk River, Princeton, and Queen's University, he became the director of the SNO (Sudbury Neutrino Observatory).

The Sun is powered by nuclear fusion processes deep in its core. These nuclear reactions are predicted to produce tiny, electrically uncharged particles called neutrinos (the word comes from the Italian for "little neutral one"). Neutrinos are very difficult to detect, since they pass through most objects without interaction. For example, many billions of neutrinos from the Sun pass through your fingernail every second! Later in this book (Chapters 30, 34, and 35), you will learn much more about neutrinos and nuclear reactions. The early neutrino detection measurements consistently revealed a lower number of neutrinos than predicted by nuclear models, and this was called the solar neutrino problem.

Located deep underground within a former nickel mine in Sudbury, the SNO collaboration built a sensitive detector for neutrinos (Figure 1-4). Ultimately, researchers there were able to show that the resolution of the solar neutrino problem was that neutrinos could change from one variety to another during passage from the Sun to Earth (there are three types of neutrinos, and detectors are usually sensitive only to one type).

The 2015 Nobel Prize in Physics was awarded jointly and equally to Art McDonald of SNO and to Takaaki Kajita of Japan, who had studied neutrinos produced from cosmic rays using the Super-Kamiokande neutrino detector. Together the two groups clearly showed that neutrino oscillations took place, with one type of neutrino transforming into another. This in turn implied that even though the neutrino mass is very tiny, it must not be zero.

Figure 1-4 On left is a portion of the neutrino detector at SNO, and at right is Art McDonald, co-recipient of the 2015 Nobel Prize in Physics.

Physicists need to be creative Physicists design experiments, find applications for physical principles, and develop new models and theories. Some philosophers of science have asked whether the electron was invented or discovered. Such questioning stresses that, while there is a part of physics that is independent of the observer, the specific models we develop to help understand nature critically depend on the creativity and imagination of physicists. It is not surprising that many physicists are also interested in other creative pursuits, such as music and art.

Physics is a highly collaborative discipline Most physicists routinely work with colleagues from other countries, often using international research facilities. Pick up a physics research journal and you will see that the majority of papers are written by collaborations of scientists from different institutions and countries. For example, the ATLAS (A Toroidal LHC Apparatus) LHC (Large Hadron Collider) experiment is a collaboration of more than 3000 researchers from more than 40 different countries. The LIGO (Laser Interferometer Gravitational-Wave Observatory) scientific collaboration includes more than 1000 scientists. Because of its collaborative nature, interpersonal and leadership skills are critical for success in physics.

Physics is both deeply theoretical and highly applied Some physicists work exclusively in fundamental areas that have no immediate application, whereas others concentrate on solving applied problems. Very often work initially deemed to have little practical use turns out to have important applications. For example, in 1915, Albert Einstein published a new theory of gravitation called general relativity. When general relativity was developed, it had no foreseeable practical applications. Today, however, the Global Positioning System (GPS) would be hopelessly inaccurate without corrections for the gravitational effects predicted by general relativity theory (see Chapter 30).

Physics involves many skills Through the study of physics you can learn critical thinking, computational, and analytical skills that can be applied beyond the sciences and engineering. You will use leading-edge technologies such as 3-D models and printing, digital signal analysis, automated control systems, digital image analysis, visualization, symbolic algebra, and powerful computational software in physics, learning techniques that have broad application. For example, a number of physicists find employment developing economic and investment models for financial institutions, while others find positions in computing and technological fields, including game development, media special effects, quality control, and advanced manufacturing support.

Physics interfaces strongly with other sciences The study of physics is increasingly interdisciplinary, with many physicists working in areas that span physics and other disciplines, for example, medical physics, biophysics, chemical physics, materials science, geophysics, or physical environmental science. Also, many scientists in other fields use physics as part of their everyday work.

Successful physicists are good communicators Whether writing experiment reports, scientific papers, or grant applications, or communicating with classes or the general public, scientists must have effective and flexible communication skills. You will probably be surprised at how much of your time as a physicist is spent in some form of communication, and also at the breadth of audiences you will serve. For example, in a typical month you may find yourself presenting to a policy institute, speaking at a local school, presenting at a scientific conference, and providing comment to reporters on a scientific development. Indeed, a number of physicists are well-known communicators of science, people such as Brian Cox, Brian Greene, Michio Kaku, Lawrence Krauss, Lisa Randall, and Neil deGrasse Tyson.

In this textbook, we provide a way for you to check your understanding of key concepts, relationships, and techniques. Each section of every chapter will have at least one checkpoint, the first of which (on the nature of physics) follows. You should test your understanding before reading further and then check your answer with the upside-down response at the bottom of the checkpoint. Physics is a highly sequential subject, and mastery of one concept is often needed to understand the next concept.

CHECKPOINT 1-1

What Is Physics?

Which of the following activities most accurately describes the realm of physics?

(a) Making conjectures about the universe that can be neither proved nor disproved

(b) Proposing and testing physical models by collecting experimental data

(c) Suggesting a theory that is only applicable to interacting galaxies

(d) Defining virtual environment properties in a game environment

ANSWER: (b) is the best answer, incorporating aspects of data collection, models, prediction, and testing. The realm of physics is concerned with matters that can, at least potentially, be proved or disproved. Therefore, the first answer is eliminated. While physicists certainly study galaxy interactions, a theory that only applies to one part of the universe would not be physics, where we seek relationships with broad application. A physics background would be useful for those developing virtual environments for games, but the work itself is not physics.

Meet Some Physicists

You already know several physicists at your university and perhaps beyond. Here we introduce you to several other physicists with varied career paths to demonstrate the many opportunities offered by a degree in physics. We hope you will follow this up by investigating additional physics career stories at www.cap.ca/careers/careers.html.

Courtesy of Leslie Anne Rogers

Leslie Rogers followed an undergraduate physics degree at the University of Ottawa with exoplanet graduate research at the Massachusetts Institute of Technology (MIT). She is now a faculty member in the Department of Astronomy and Astrophysics at the University of Chicago. Her research group develops models to validate with observations of super-Earth and sub-Neptune sized exoplanets. Ultimately her research will constrain the interior structure, formation, evolution, and habitability these planets.

Sara Desjardins Photography

Dan Falk draws on his physics background for work as an award-winning author, speaker, journalist, and broadcaster. His undergraduate degree in physics was from Dalhousie University, followed by a graduate degree in journalism from Ryerson. His book, *The Science of Shakespeare*, considers how the scientific environment of the time influenced Shakespeare. Earlier books include *In Search of Time: Journeys along a Curious Dimension*. Of that work one reviewer wrote "Falk's book is what (Stephen Hawking's) *Brief History of Time* should have been."

THE CANADIAN PRESS/Guelph Mercury-Nicki Corrigall

Diane Nalini de Kerckhove is both a Rhodes Scholar physicist and a highly acclaimed jazz musician. Her physics has covered many areas, including proton-induced X-ray emission, semiconductors, ion optics, and trace element analysis of human hair samples. Her current work is in climate change policy with Environment Canada. She has performed as a jazz singer on national radio and has released several CDs, including *Kiss Me Like That*, which is based on music created for the International Year of Astronomy.

Courtesy of Ingrid Stairs

Ingrid Stairs draws upon a number of branches of physics for her astrophysics research at the University of British Columbia. She uses radio telescopes to detect and do precise timing of pulsars, rapidly rotating dense stellar remnants. Recently she has studied a binary system of dense stellar remnants that are so close together that the orbits just take a few hours. Changes in orbits due to the curvature of space-time allow her to help constrain models of gravity such as general relativity.

Courtesy of Dianna Cowern

You probably know **Dianna Cowern** as "The Physics Girl" and have watched some of her highly popular physics videos online (www.youtube.com/c/physicsgirl). She has worked in science outreach at the University of California, San Diego (UCSD) and has won numerous awards for her work. Originally from Hawaii, she conducted research on dark matter while an undergraduate at MIT, later moved on to stellar astrophysics, and then worked as an engineer at General Electric (GE). We hope some readers of this book will take a passion for sharing physics with children and the public as The Physics Girl does so effectively!

Figure 1-5 P.Phys. and its associated logo are registered trademarks of the Canadian Association of Physicists (CAP). The logo is reproduced here with the express permission of the CAP.

We hope that as you study physics in this text you will consider becoming a physicist, or using physics in some other science or engineering career. The Canadian Association of Physicists (CAP) has a useful website (www.cap.ca) that includes a section on careers in physics. The corresponding organization in the United States is the American Physical Society (APS) (www.aps.org). An international organization with particular strengths in Europe is the Institute of Physics (IOP) (www.iop.org).

The CAP has a professional certification program (P.Phys.) (see Figure 1-5). To qualify for P.Phys. certification, you must complete minimum education and experience requirements, agree to a code of ethical conduct, and pass a professional examination that tests your ability to interpret physics in applied situations, as well as your sensitivity to ethical aspects of the profession. See the CAP website (www.cap.ca) for additional information.

It is important to realize that many physicists work as part of interdisciplinary teams in areas such as medical physics, astrophysics, biophysics, geophysics, materials science, environmental science, and engineering. Others who study physics will go on to careers not directly applying physics, but the skills you will learn in physics, such as critical quantitative problem solving and powerful computational techniques, will be very valuable.

The typical route to becoming a research physicist is to complete an undergraduate degree (usually an honours degree) in physics or a closely related area, such as applied mathematics, geophysics, or astrophysics, followed by graduate degrees (masters and doctorate). Many physicists then complete several years of postdoctoral work before taking a permanent position in research, development, teaching, or administration. Yes, that is a long road. However, graduate students in physics (and other sciences) are almost always fully supported at a subsistence level by graduate scholarships and assistantships, so following your undergraduate degree you will probably not go into further student debt. Of course you should only go into physics if you are passionate about the subject!

While most research positions require a doctorate, a B.Sc. is sufficient for many careers, including technical positions in research or development, application of physics in professional programs such as patent law,

teaching at the high school level, marketing of physics-related materials, and science outreach or journalism. A career you might not have considered is politics or more broadly public policy. Many of the economic, environmental, and social problems we face as a society benefit from the expertise and quantitative analysis skills physicists possess. Angela Merkel, the Chancellor of Germany, brings to her position skills learned in her former positions as a research scientist in quantum physics and chemistry.

Of course the majority of students who study first-year university physics will quite rightly go on to careers in other disciplines. Even if you do not complete a degree in physics, the skills you learn in physics courses, such as quantitative analysis, problem solving, model formation, and experimental design, will be valuable in your future career.

1-2 Experiments, Measurement, and Uncertainties

Now we want to develop tools that you will use as a physics student. Let us first consider why we conduct experiments and measure. Often we make measurements to test hypotheses. For example, if someone proposes that a new material expands less in response to temperature changes than some other material, we can use physics measurements to test this claim.

Sometimes, we make measurements to learn about relationships between quantities and to help develop models and theories. For example, in the 1920s, astronomers Edwin Hubble and Milton Humason measured the light from 46 galaxies and found that the light from the more distant galaxies has a greater redshift. (A redshift is a shift to longer wavelengths caused by motion away from the observer.) This relationship, in turn, led to the startling conclusion that the universe is expanding.

Sometimes we have a proposed mechanism or hypothesis in mind that we are using an experiment to test, while at other times we are using the observations to help guide us to develop that hypothesis. Modern science provides enormous amounts of data, so powerful data visualization techniques play key roles.

At other times, we make measurements to obtain precise values for quantities such as the mass of a celestial object or fundamental physical constants such as the gravitational constant (see Chapter 11), which establishes the strength of the gravitational force between two masses.

Most physicists spend a significant amount of their professional lives designing, conducting, and interpreting their own experiments, as well as evaluating the claims from experiments conducted by others. This is part of an evidence-based approach to science. Any claim must be supported by a careful and unbiased interpretation of all relevant data. Every measurement that you will ever make will have some associated uncertainty, and scientific evidence should always include an estimate of that uncertainty. In the rest of this section and the following one we will develop tools describing these uncertainties.

Suppose that measurements of the expansion of two materials show that for a given increase in temperature, the length of one material increases by 0.0025% and the length of the other material increases by 0.0030%. Do these measurements mean that the first material has a lower expansion coefficient? The answer depends not only on the measurements, but also on the uncertainties of those measurements.

There are three different classes of potential errors in measurements of physical quantities: true errors, systematic errors, and inherent uncertainties.

True errors are mistakes made in the measurement. These errors can in general be identified by conducting more measurements, ideally by experimenters using different techniques. Examples of true errors are an experimenter reading a scale on a measurement incorrectly, or transcribing a reading into a notebook incorrectly, or connecting electrical equipment incorrectly so that the measurement made is not the one intended.

Systematic errors result from some bias in the way that the quantity is being measured. For example, if you are measuring an electrical current with a meter that is not accurately calibrated, all the measured values might be somewhat higher (or lower) than the true values.

Inherent uncertainties (also called **random errors**) result from limitations in the precision of the measurements. No matter which technique is used, a measurement cannot be made with absolute precision, so there is always some uncertainty. In Chapter 32, you will learn how the uncertainty principle determines the minimum uncertainty inherent in all measurements.

We need to distinguish between the terms *precision* and *accuracy*. **Accuracy** is how close the measurement of some quantity is to the true value. **Precision** is a measure of how close to one another repeated measurements are. For example, if you measure the length of a box and get 100 values all between 1.259 m and 1.260 m, then the precision is quite good. However, if the actual length of the box is 1.204 m, then the accuracy is rather poor (and you need to look for the cause of the systematic error).

The best way to state the uncertainty in a measured quantity is to provide an estimate of the uncertainty. That is, we could write a length measurement as 1.15 ± 0.08 m, which means that the true value is likely between 1.07 m and 1.23 m. Often the $\pm$ value is selected so that the true value is within the stated range about two-thirds of the time. In some fields, the criterion is 95% certainty, meaning, in this case, that 95 times out of 100 the true value will be between 1.07 m and 1.23 m. We will consider these criteria in more detail in the next section.

Sometimes the uncertainty is not symmetric about the best estimate value. For example, in 2016 LIGO announced the detection of gravitational waves from two merging black holes. They reported the mass of the larger black hole as 36^{+5}_{-4} times the mass of the Sun. This means that it is believed that the black hole mass is between 32 and 41 times the mass of the Sun.

We will see in the following section how to estimate the uncertainty in a mean, or average, value. The mean is obtained by simply adding the values and dividing by the number of values. Whether a value has been obtained as a mean, or estimated in some other manner, the uncertainty in that value should be stated explicitly or implied in how we write the number. This information can be used to estimate whether the difference between two values is **statistically significant**. We say the difference between two quantities is statistically significant if random errors are unlikely to account for that difference.

Before we conclude this section on the significance of differences, we must consider one additional

---

CHECKPOINT 1-2

What Type of Error?

Classify each of the following as a random error, a systematic error, or a true error.

(a) You notice in a table of values of outside air temperature over a few hours that one value is much different than the others: 4.5 °C, 5.2 °C, 5.5 °C, 8.8 °C, 6.2 °C, 6.9 °C.

(b) You use four different ammeters to measure the electrical currents flowing in the same circuit, noticing that one of the meters always reads a higher value.

(c) You carefully measure the weight of some fruit using a weigh scale at the supermarket, noticing that the values cluster over a small range of values.

(d) You and your friend measure the length of a table with the same metre stick, and you obtain a value of 2.255 m while she measures 2.260 m.

ANSWER: (a) is a true error, since the 8.8 °C value would not seem consistent with the situation and other values, probably a result of incorrectly transcribing 5.8 °C. (b) This is an example of a systematic error. (c) Any reading has some uncertainty, reflected by repeated measurements clustering over a small range, so this is an example of a random error (also called inherent uncertainty). (d) Depending on how different the values are, this might be an example of a true error or a random error. Given that the difference is just a few millimetres over a table length, it is most likely that this is a random error.

CHECKPOINT 1-3

Is the Difference Significant?

The mean (average) of a set of length measurements for object A is 1.25 ± 0.05 m, and the mean of similar measurements for the length of object B is 1.32 ± 0.20 m. Can we conclude that B is longer than A? Explain your reasoning.

ANSWER: No. When the range of uncertainties is taken into account, A could be between 1.20 m and 1.30 m long, whereas B is between 1.12 m and 1.52 m long. Therefore, we cannot conclude that B is longer than A.

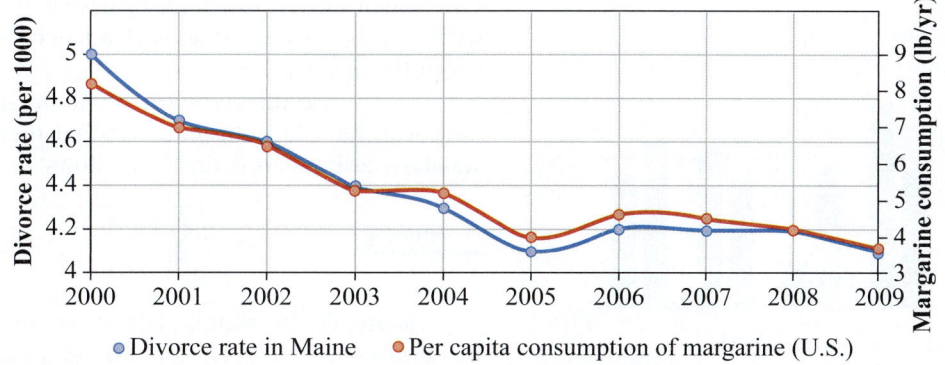

Figure 1-6 Plot of divorce rate in Maine and the U.S. consumption of margarine from Tyler Vigen, at www.tylervigen.com. Correlation does not imply causation!

Source: Divorce rate in Maine correlates with Per capita consumption of margarine (US), http://www.tylervigen.com/view_correlation?id=1703

point. If we plot two parameters, they may be highly correlated (i.e., they may both change in a similar way with some other variable, such as time). That does not mean that we can assume that one causes the other. For example, if you plot the divorce rate in the state of Maine versus use of margarine over the past decades, the correlation is very high (Figure 1-6). That does not mean that eating margarine causes divorce! **Correlation** means that two quantities vary in a similar way when plotted together (or against a third parameter). **Causation** means that some result is the result of a change in another quantity. Best practice in science requires that we always question whether correlation implies causation, or simply that some other factor is independently affecting both.

1-3 Mean, Standard Deviation, and SDOM

The **mean**, also called the arithmetic average, is calculated by adding individual measurements and then dividing the sum by the number of measurements. The most common notation for mean is to place an overbar over the expression, $\overline{x}$. We can write this as Equation 1-1, where the symbol Σ is shorthand for taking the sum of the series of values that follow. Here x_i represents the ith data value and the summation is taken of the N data points:

KEY EQUATION

$$\overline{x} = \frac{\sum_{i=1}^{N} x_i}{N} \qquad (1\text{-}1)$$

It is important to realize that different equations in physics have different roles: some are definitions, while others represent a fundamental physical relationship between physical variables. Equation 1-1 is shorthand for the definition of the mean. When the equation is a definition, we sometimes use the symbol $\equiv$ for equals. All math symbols used in the book are summarized in Appendix E.

Sometimes in physics when we estimate the mean for a set of values we weight different values according to the uncertainty in each value. Values that are more precisely known are given additional weight in computing the mean. There are other ways to obtain a best measure for a quantity in physics, although the mean is the most widely used.

Consider the two sets of measurements shown in Table 1-1 and plotted in Figure 1-7. They both have the

CHECKPOINT 1-4

What is $\overline{x}$?

A set of length measurements for an object is $x_i = 4.25$ m, 4.35 m, 4.20 m, 4.30 m, 4.45 m, 4.35 m, 4.30 m. What is the $\overline{x}$ value?

$$\overline{x} = \frac{\sum_{i=1}^{N} x_i}{N}$$

$$= \frac{4.25 \text{ m} + 4.35 \text{ m} + 4.20 \text{ m} + 4.30 \text{ m} + 4.45 \text{ m} + 4.35 \text{ m} + 4.30 \text{ m}}{7}$$

$$\overline{x} = \frac{30.2 \text{ m}}{7} = 4.31 \text{ m}$$

lowing result.

We have implicitly assumed that the values are to be equally weighted.

we obtain by adding the values and dividing by the number of data points, 7 in this case.

ANSWER: $\overline{x} = 4.31$ m. The symbol $\overline{x}$ represents the mean of the set of values, which

Table 1-1 Two sets of data with the same mean, but very different distributions.

Set A	Set B
5.5	3.0
5.0	8.5
4.7	7.0
4.8	1.5
4.4	5.5
5.6	4.5
Mean = 5.0	**Mean = 5.0**

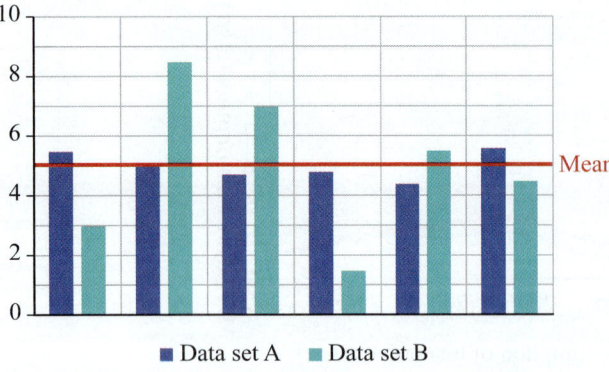

Figure 1-7 The two distributions have the same mean value (5.0), but the data shown with green bars have much more scatter around this mean value.

same mean value (5.0), but they have very different distributions: set A has values with a much smaller spread about the mean value.

We can use the standard deviation, σ, to specify how near the values cluster around the mean value. The **standard deviation** is defined as follows:

KEY EQUATION
$$\sigma = \sqrt{\sum_{i=1}^{N} \frac{(x_i - \overline{x})^2}{N - 1}} \qquad (1\text{-}2)$$

The smaller the standard deviation, the more closely the measured values cluster around their mean. For example, data set A in Table 1-1 has a standard deviation of 0.47, and data set B has a standard deviation of 2.57.

EXAMPLE 1-1

Mean and Standard Deviation

Measurements for a certain distance are 2.6 km, 2.8 km, 3.1 km, and 2.7 km. Find the mean and standard deviation.

Solution

We have four values, so $N = 4$. To find the mean we simply add the values and divide by N:

$$\overline{x} = \frac{\sum_{i=1}^{N} x_i}{N} = \frac{2.6 \text{ km} + 2.8 \text{ km} + 3.1 \text{ km} + 2.7 \text{ km}}{4}$$

$$\overline{x} = \frac{11.2 \text{ km}}{4} = 2.8 \text{ km}$$

To find the standard deviation we now use this mean value in the relationship.

$$\sigma = \sqrt{\sum_{i=1}^{N} \frac{(x_i - \overline{x})^2}{N - 1}}$$

$$\sigma = \sqrt{\frac{(2.6 \text{ km} - 2.8 \text{ km})^2 + (2.8 \text{ km} - 2.8 \text{ km})^2 + (3.1 \text{ km} - 2.8 \text{ km})^2 + (2.7 \text{ km} - 2.8 \text{ km})^2}{4 - 1}}$$

$$\sigma = \sqrt{\frac{(-0.2 \text{ km})^2 + (0.0 \text{ km})^2 + (0.3 \text{ km})^2 + (-0.1 \text{ km})^2}{3}}$$

$$\sigma = \sqrt{\frac{0.04 \text{ km}^2 + 0.00 \text{ km}^2 + 0.09 \text{ km}^2 + 0.01 \text{ km}^2}{3}}$$

$$\sigma = \sqrt{\frac{0.14 \text{ km}^2}{3}} = 0.22 \text{ km}$$

In the next section we will see that this should be rounded to $\sigma = 0.2$ km, so we should write the final result as 2.8 ± 0.2 km, where 2.8 km is the mean and 0.2 km is the standard deviation.

Making sense of the result

This is a reasonable value for the mean, since we have two data values slightly less than the mean, and one value significantly more. Note that the standard deviation and the mean have the same units. We see that in this case three of the four data values lie within the range suggested by the final answer, and the other only slightly outside. We will see why this is reasonable later in the section.

For many types of measurements, intrinsic uncertainties (random errors) follow a **normal distribution** (also called a Gaussian distribution). Most values we obtain will fall very close to the mean value, with progressively fewer as we get farther away from the mean. The normal distribution is plotted in Figure 1-8. Approximately 68% of the measurements from a normally distributed set lie within ± 1 standard deviation of the mean value, about 95% of the values lie within ± 2 standard deviations of the mean, and about 99.7% lie within ± 3 standard deviations of the mean.

The more measurements we take, the closer the mean of the sampled values in our set of measurements will come to the true mean of the distribution. Now,

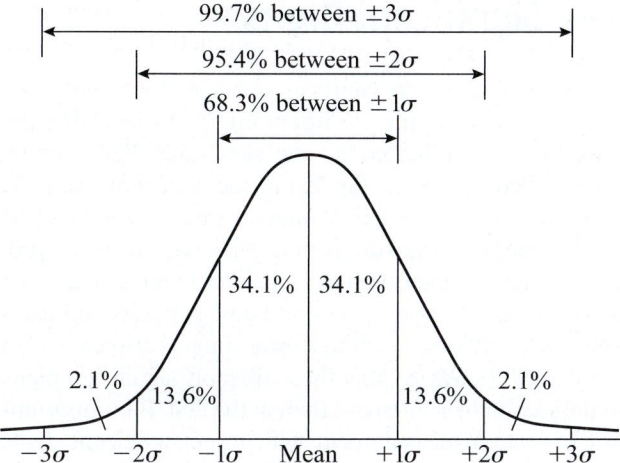

Figure 1-8 The normal distribution (also called the Gaussian distribution). The *y*-value represents how frequently a certain *x*-value occurs, with *x*-values plotted according to how they vary from the mean value.

CHECKPOINT 1-5

Standard Deviation

The results of a public opinion poll (assumed a normal distribution) are reported as being accurate to within 4%, 19 times out of 20. Expressed as a percentage, what is the implied standard deviation?

ANSWER: Nineteen times out of twenty is about 95%, so it must correspond to two standard deviations according to Figure 1-8. Therefore, if two standard deviations represent 4% uncertainty, one standard deviation is 2% uncertainty.

---

we consider how to determine whether the difference between two mean values is significant. You might think we could use the standard deviations of the two sets of data, but these standard deviations are actually quite misleading for comparing different sets of data. If the mean is based on a larger number of observations, its value is more certain, even if the standard deviation of the set is the same as the standard deviation of a smaller set.

Therefore, we use a quantity called the **standard deviation of the mean (SDOM)**, which is defined in terms of the number of observations (N) used for the mean and the standard deviation (σ). The SDOM is also called the standard error of the mean;

KEY EQUATION

$$\text{SDOM} = \frac{\sigma}{\sqrt{N}} \qquad (1\text{-}3)$$

For example, in the data in Table 1-1, set B has a SDOM of $\frac{2.57}{\sqrt{6}}$, or about 1.0.

It is easy to make mistakes when interpreting the standard deviation and the SDOM. Remember that the standard deviation is a measure of the uncertainty of a single value, and the SDOM is a measure of the uncertainty of a mean. For example, about 68% of the time, a single measurement will be within ±1 standard

deviation of the mean value, and about 68% of the time, the mean of a set of measurements will be within ±1 SDOM of the true value. Therefore, the SDOM is used when you have taken the average value from multiple samples of two different populations, and you want to determine whether the sets are statistically different. Most literature in physics uses the notation with the form mean ± SDOM when expressing a quantity, but you will sometimes see a notation with the SDOM in parentheses after the mean value. Note also that sometimes in scientific literature the ± refers to two, rather than one, standard deviations (when this is done

EXAMPLE 1-2

Significant Difference?

In an experiment, the decay time of a subatomic particle (the time the particle takes to break down into other particles) was measured repeatedly, producing the following values: 0.45 μs, 0.37 μs, 0.67 μs, 0.84 μs, 0.38 μs, 0.55 μs, and 0.48 μs. A different experiment measured decay times of 0.68 μs, 0.48 μs, 0.88 μs, 0.87 μs, and 0.46 μs. Can we conclude that the decay times in the two experiments are statistically different? Note: In this chapter we will use a criterion of ±1 standard deviation unless otherwise stated.

Solution

First, we calculate the mean of each set of measurements using Equation 1-1. For set A (plotted in dark blue in Figure 1-9), we have a mean decay time of 0.53 μs, and for set B we have a mean decay time of 0.67 μs.

Next, we calculate the standard deviation of each set, using Equation 1-2, and we obtain the values 0.17 μs and 0.20 μs.

Now, we calculate the two SDOM values. Set A has $N = 7$, and set B has $N = 5$. Using Equation 1-3, we obtain SDOM values of 0.06 μs for set A and 0.09 μs for set B.

The mean of set A can be written as 0.53 ± 0.06 μs, and the mean of set B can be written as 0.67 ± 0.09 μs. So, we are reasonably confident that the true value of the mean of A lies between 0.47 μs and 0.59 μs, and also reasonably confident that the true value of the mean of B lies between 0.58 μs and 0.76 μs. Since these two ranges overlap, we cannot conclude that the difference between the sets of data is statistically significant.

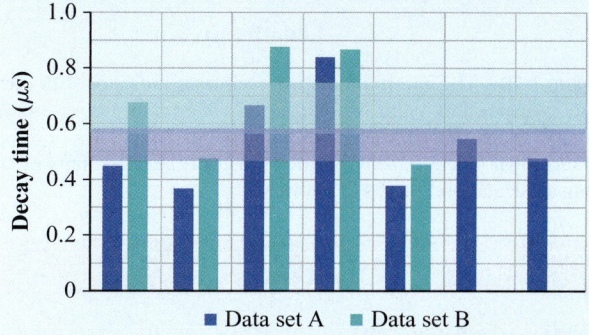

Figure 1-9 Comparison of the two sets of data for particle decay times. The blue bars represent the values from set A, and the green bars represent the values from set B. The shaded rectangles show the areas defined by the mean ± SDOM for each distribution.

(continued)

Making sense of the result

Note that, although we cannot conclude that the difference between the means is significant, we have not proven that the means are the same. If we took more measurements, or more precise measurements, we might find a significant difference in those data.

it is usually called a 95% confidence interval). In this book, unless otherwise noted, we will use ± values corresponding to one standard deviation.

It is important to always graph your data and not depend simply on the mean and the standard deviation to characterize the dataset. See Problem 37 for an example of how very different datasets can have the same mean and standard deviation.

When graphing data, we can indicate the uncertainty in the data with **error bars**. In Figure 1-10, the error bars are symmetric about the data points, and the uncertainty is the same for all the data values. However, uncertainties are often not symmetric, and they can vary with different data points. In some instances, there are significant uncertainties in the *x*-values as well, in which case the graph would have both vertical and horizontal error bars.

Error bars are important enough to be included in most experimental results published in physics. These bars give a clear graphical indication of uncertainty in the individual values and how much confidence we can have in any apparent trend in the data.

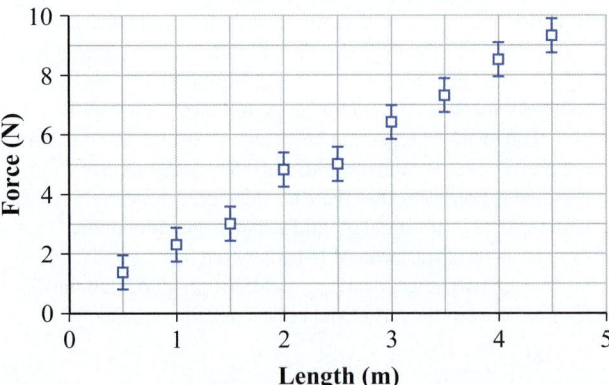

Figure 1-10 The uncertainty in force values is indicated with error bars. When a data point is obtained by taking a mean, we use the SDOM to determine the length of the error bar. Sometimes, however, the error bar is determined from an estimate of the uncertainty rather than a calculation.

INTERACTIVE ACTIVITY 1-1

Curve Fitting and Error Bars

Perform the interactive activity, which uses the PhET simulation "Curve Fitting," to reinforce and extend your skills in understanding error bars and lines of best fit for a collection of data points.

1-4 Significant Digits

When uncertainty in values is not stated explicitly (for example, by using the ± notation), we assume that the uncertainty is indicated by the **significant digits** (sometimes called *significant figures*) in the data. For example, two measurements of a time interval are stated as 3.6 s and 3.598 s; the latter measurement implies a much greater precision. The first measurement indicates that the true value is somewhere between 3.55 s and 3.65 s; the second measurement implies a much narrower range, between 3.5975 s and 3.5985 s. We say that there are more significant digits in the second measurement than in the first. It is important not to state a measurement with more significant digits than justified by the true uncertainty in the value.

Unfortunately, conventions for determining the number of significant digits vary somewhat. In this book, we use the following rules:

1. All non-zero digits are always significant. For example, 345.6 has 4 significant digits.

2. Zeros between non-zero digits are always significant. For example, 1204 has 4 significant digits.

3. Zeros at the end of a number are significant only if they are to the right of the decimal point *or* followed by a decimal point. For example, both 450 and 45 000 have 2 significant digits; 450. has 3 significant digits; and both 450.0 and 1.560 have 4 significant digits.

4. Zeros at the beginning of a number are not significant. This rule makes the number of significant digits consistent when unit prefixes are changed. For example, the measurements 0.067 m and 6.7 cm express the same length, and both have 2 significant digits. Example 1-3 demonstrates the application of these rules.

EXAMPLE 1-3

Significant Digits

How many significant digits are in each of the following?
(a) 5.46 m
(b) 0.024 575 m
(c) 320 m
(d) 24.260 m
(e) 204 m

Solution

(a) The value 5.46 m has 3 significant digits according to rule 1.
(b) The value 0.024 575 m has 5 significant digits (neither 0 is significant according to rule 4).
(c) The value 320 m has 2 significant digits according to rule 3.
(d) The value 24.260 has 5 significant digits according to rules 1 and 3.
(e) The value 204 m has 3 significant digits according to rule 2.

When performing mathematical operations on numbers, it is important to reflect the appropriate precision of the result by rounding the final answer to the correct number of significant digits. Normally we keep additional precision in the intermediate steps and only round when the final answer is stated. There are two basic rules for dealing with significant digits in calculations.

Multiplication and division When multiplying and dividing quantities, the number of significant digits in the final answer should be the same as the least precise quantity used in the calculation, as demonstrated in Example 1-4. Note that integers and mathematical constants are assumed to have infinite significant digits. For example, in the expression for the surface area of a sphere, $4\pi r^2$, the factors 4 and π do not affect the number of significant digits in the calculated area, which will be determined by the precision of r. For example, if the r-value is expressed to 3 significant digits, the area will have 3 significant digits.

EXAMPLE 1-4

Significant Digits in Multiplication and Division

Express the answer to the following calculation with the correct number of significant digits:

$$\frac{32.55 \times 0.021}{465}$$

Solution

When we enter these numbers into a calculator, the answer will likely be displayed as 0.00147. However, the number 32.55 has 4 significant digits, the number 0.021 has 2 significant digits, and the number 465 has 3 significant digits. Therefore, the final answer should have only 2 significant digits, and the calculated number should be rounded to 0.0015.

Making sense of the result

A final answer of 0.0015 implies that we have confidence (assuming that the original values were expressed appropriately for their precision) that the real value is somewhere between 0.001 45 and 0.001 55.

Addition and subtraction When adding and subtracting quantities, we keep the same number of significant digits after (or in front of) the decimal point as in the least precise of the quantities used in the calculation. For example, the number 2.45 is assumed to have an uncertainty of the order of 0.01. If we subtract the number 2.30 from 2.45, we obtain 0.15—the

uncertainty is still of the order of 0.01. Note that with this example, although the input quantities had 3 significant digits, after the subtraction the result is only accurate to 2 significant digits. Example 1-5 demonstrates the application of this rule.

EXAMPLE 1-5

Significant Digits in Addition and Subtraction

You measure the total length of two objects together as 2.45 m. You know that one of the objects has a length of 2.248 m. What is the length of the other object?

Solution

A calculator will give an answer of 0.202 m. However, the measurement 2.45 m has 2 significant digits to the right of the decimal point. Therefore, the final answer should be written as 0.20 m.

Making sense of the result

The measured lengths have 3 and 4 significant digits, but the calculated length has only 2. When we subtract two quantities that are approximately equal, the result will have fewer significant digits than the original quantities.

PEER TO PEER

When I started university physics I had to correct some bad habits I had developed. If I measured a distance in an experiment that just happened to be, for example, four metres, I would wrongly write my lab result as 4 m. However, really I had measured it to a precision of about a centimetre, so it should have been written as 4.00 m.

It is important to use significant digits appropriate to the $\pm$ value included with the statement of the value. Test your reasoning with the following checkpoint. Note that the uncertainty given with a $\pm$ value specifies the uncertainty more completely than just using an appropriate number of significant digits.

CHECKPOINT 1-7

What Is Wrong?

What is wrong with writing a result as follows? 3.459 ± 0.4 m How should the number be expressed?

ANSWER: The ± 0.4 m suggests that the value is uncertain in the first place after the decimal point, and therefore it should be consistently written as 3.5 ± 0.4 m.

As we demonstrate in Example 1-6, it is best practice to keep additional precision in the intermediate steps and only round to the correct number of significant digits when the final answer is stated (or multiple final answers for problems with more than one part). This is the approach that we will take in this book, and it is naturally what will happen if you use a calculator or numerical program for your computations. When there are multiple parts to a question (i.e., parts (a), (b), etc.), you should round to the correct number of significant digits when you state the answer for each part, but if you use the answer from that part in subsequent computations, use additional precision. The purpose of significant digits is to not imply more precision than is present, and it is the final answers that must reflect that. It is also good practice in physics to use symbols, rather than the numerical values, as far into the computation as feasible.

EXAMPLE 1-6

Writing Your Speed

You run a trail and want to correctly report your distance and speed. Several friends have each measured the length of a portion of the trail (you need to add the three to get the total distance). Jill measures the first segment, and tells you it is 236 m. James measures the second segment, reporting a result of 75.5 m. Mingkai measures the last segment and reports that it is 234.2 m long. You run the trail in a total time of two minutes and twenty-five seconds, with an assumed precision of one second. What are the total distance, and your average speed in metres per second, both expressed to the correct number of significant digits? Average speed is total distance divided by time (Chapter 3).

Solution

First find the total distance by adding the three values:

$$\text{distance} = 236 \text{ m} + 75.5 \text{ m} + 234.2 \text{ m} = 545.7 \text{ m} = 546 \text{ m}$$

Since Jill's estimate is written without any digit after the decimal point, the distance is 546 m, and it is known to 3 significant digits.

We will use the extra precision of 545.7 m for the speed calculation but must remember in later rounding that we only knew it to a precision of 3 significant digits.

The time for the run expressed in seconds is $(2 \times 60 \text{ s}) + 25 \text{ s} = 145 \text{ s}$. Note that the 2 does not mean a one-significant-digit answer because it is a pure number. We are told that the time is precise to ± 1 s, which supports the idea that the 145 s time is known to 3 significant digits.

Now we can calculate the average speed.

$$\text{speed} = \frac{\text{distance}}{\text{time}} = \frac{545.7 \text{ m}}{145 \text{ s}} = 3.763 \text{ m/s} = 3.76 \text{ m/s}$$

Since both numerator and denominator are known to 3 significant digits (even though we wrote the numerator to 4 significant digits in the intermediate step), the final speed should be expressed to 3 significant digits.

Making sense of the result

This is a reasonable speed, since it implies you could run a kilometre in a bit under four and a half minutes. Note that normally we use a more precise value in intermediate calculations, only rounding the stated answers (here the distance and the speed) to the correct number of significant digits.

1-5 Scientific Notation

In physics we encounter very large and very small numbers. To get a sense of the range of dimensions encountered in physics, explore The Scale of the Universe (htwins.net/scale/).

It would be awkward to use standard fixed notation in all cases. For example, the North Star, Polaris, is at a distance of roughly 4 100 000 000 000 000 000 m. **Scientific notation** is widely used and generally preferable. In scientific notation we write the number 46 500 as 4.65×10^4. To convert a number to scientific notation, we determine the power of 10 factor that will let us express the number in a form with just one non-zero digit to the left of the decimal point. The corresponding operation is used for small numbers. For example, 0.000 567 is 5.67×10^{-4}. We can write the distance to Polaris far more conveniently in scientific notation as 4.1×10^{18} m.

Scientific notation assists in calculating and checking results without a calculator as well. For example, 2.00×10^5 multiplied by 3.00×10^4 yields 6.00×10^9, obtained simply by multiplying the first parts (called the significands) and adding the exponents.

Another important reason to use scientific notation is that it eliminates any confusion over whether zeros at the end of a number are significant digits. For example, 5.0×10^4 has two significant digits, while 5.00×10^4 has three significant digits.

EXAMPLE 1-7

Scientific Notation

(a) The approximate distance from Earth to the Sun is 150 000 000 000 m. Express this distance in scientific notation.
(b) What is 0.0230 in scientific notation?
(c) Write the number 2.45×10^{-3} in standard notation.

Solution

(a) We move the decimal point 11 places to the left to get 1.5×10^{11} m.
(b) We move the decimal point 2 places to the right to get 2.30×10^{-2}.
(c) We move the decimal point 3 places to the left to eliminate the multiplier, giving 0.002 45.

Making sense of the result

It is not obvious from part (a) of the question how many significant digits are represented in the number. This illustrates the advantage of scientific notation that the number of significant digits is always clear. In (b), make sure you respect the number of significant digits and write it in the form shown, and not as 2.3×10^{-2}, which is incorrect.

1-6 SI Units

In this text, we almost exclusively use the International System of Units, or **SI**. The abbreviation comes from the French title, *le Système internationale d'unités*. This system is used worldwide now; in fact, only three countries have not adopted SI for commerce as well as for science: Burma (also known as Myanmar), Liberia, and the United States.

Base SI Units

SI is an integrated metric system designed so that it can readily be used throughout the sciences as well as engineering and commerce. SI has seven base units, chosen such that all quantities can be expressed in some combination of the base units. SI also has a set of derived units, defined in terms of the base units, to provide convenient measures of particular quantities. Table 1-2 lists SI base units and the quantities they measure.

Table 1-2 The Base SI Units

Unit Name	Unit Symbol	Measures
metre	m	length
second	s	time
kilogram	kg	mass
kelvin	K	temperature
mole	mol	amount of substance
ampere	A	electric current
candela	cd	luminous intensity

The *metre*, *kilogram*, and *second* are units that you likely know well from everyday use. For historical reasons, the kilogram—not the gram—is the SI base unit of mass. The kilogram is the only base unit that has a prefix.

A *kelvin* is equal to a Celsius degree, but the zero point of the Kelvin temperature scale, denoted 0 K, is absolute zero, the lowest temperature possible (approximately $-273\ ^\circ$C).

A *mole* is the amount of a substance that contains the same number of elementary entities (e.g., atoms or molecules) as exactly 12 g of the isotope carbon-12, about 6.022×10^{23}. In other words, 1 mol contains Avogadro's number of atoms or molecules.

The *ampere* is a measure of electric current, the flow of electric charge past a given point. When you study electromagnetism later in this book, you will see why the designated base unit for electrical quantities is based on current rather than charge.

The *candela* is the unit of luminous intensity, a measure of the light output of a source. As you might guess from the name, the candela is based on the light output of a standard candle with a specified composition and rate of combustion. In fact, a typical candle has brightness in the order of 1 cd. The candela is a precisely defined unit that replaces older units.

Other Units

As well as the base units, there are a number of fundamental derived SI units. We list the most important ones used in this book in Table 1-3.

Some frequently used units are not shown in Table 1-3. Some of these are simply combinations of the units listed. For example, speed and velocity have units of m/s, and acceleration (the rate of change of velocity) has units of m/s^2.

There are also commonly used non-SI units, including the minute (min); the solar day (d); the year (a or y or yr); the astronomical unit (AU), which is the mean distance between the centres of Earth and the Sun; and the light year (ly), which is the distance that light travels in one year.

Table 1-3 Derived SI Units Commonly Used in Introductory Physics Courses

Unit Name	Unit Symbol	Quantity	Base Unit Equivalent
newton	N	force	$m \cdot kg \cdot s^{-2}$
joule	J	energy, work	$m^2 \cdot kg \cdot s^{-2}$
watt	W	power	$m^2 \cdot kg \cdot s^{-3}$
hertz	Hz	frequency	s^{-1}
radian	rad	angle	$m \cdot m^{-1}$ (dimensionless)
steradian	sr	solid angle	$m^2 \cdot m^{-2}$
pascal	Pa	pressure	$m^{-1} \cdot kg \cdot s^{-2}$
coulomb	C	electric charge	$s \cdot A$
volt	V	electric potential difference	$m^2 \cdot kg \cdot s^{-3} \cdot A^{-1}$
ohm	Ω	electric resistance	$m^2 \cdot kg \cdot s^{-3} \cdot A^{-2}$
farad	F	electric capacitance	$m^{-2} \cdot kg^{-1} \cdot s^4 \cdot A^2$
henry	H	electromagnetic inductance	$m^2 \cdot kg \cdot s^{-2} \cdot A^{-2}$
tesla	T	magnetic field strength	$kg \cdot s^{-2} \cdot A^{-1}$
weber	Wb	magnetic flux	$m^2 \cdot kg \cdot s^{-2} \cdot A^{-1}$
becquerel	Bq	radioactivity	s^{-1}
lumen	lm	luminous flux	cd
lux	lx	luminous flux per unit area	$m^{-2} \cdot cd$

SI Prefixes

SI has a set of prefixes that indicate different powers of 10; for example, a kilometre is 1000 m, and a millimetre is 0.001 m. Table 1-4 lists the names and abbreviations for the most commonly used SI prefixes. Note that the prefix symbols greater than "kilo" are capitalized.

CHECKPOINT 1-9

How Many mm Is That?

How many mm are in 1 km?

(a) 1000
(b) 10 000
(c) 1 000 000
(d) 1 000 000 000

ANSWER: (c) There are 1000 m in 1 km, and there are 1000 mm in 1 m, so there must be 1 000 000 mm in 1 km.

Writing SI

SI specifies rules for writing measurements. These rules combine aspects of traditional styles used in various countries into a single, unambiguous convention that is recognized worldwide.

EXAMPLE 1-8

Magnetic Units

As you will see in Chapter 24, the tesla (T) is a unit of magnetic field strength, while another unit, the weber (Wb), is the unit of magnetic flux. Using data from Table 1-3, suggest how these two units must be related.

Solution

We see from the table that the tesla, expressed in terms of the base SI units, is $kg \cdot s^{-2} \cdot A^{-1}$.

The weber, again in terms of the base SI units, is $m^2 \cdot kg \cdot s^{-2} \cdot A^{-1}$.

These only differ by the m^2 factor. Therefore it must be true that

$$1 \, T = 1 \, \frac{Wb}{m^2}$$

Making sense of the result

Note that we have been able to determine this result without knowing anything about magnetism. This is an example of dimensional analysis, which will be more formally covered in the next section.

Table 1-4 The Most Common SI Prefixes

Name	Symbol	Multiplication
zetta	Z	10^{21}
exa	E	10^{18}
peta	P	10^{15}
tera	T	10^{12}
giga	G	10^{9}
mega	M	10^{6}
kilo	k	10^{3}
hecto	h	10^{2}
deca	da	10^{1}
deci	d	10^{-1}
centi	c	10^{-2}
milli	m	10^{-3}
micro	μ	10^{-6}
nano	n	10^{-9}
pico	p	10^{-12}
femto	f	10^{-15}
atto	a	10^{-18}
zepto	z	10^{-21}

- We always use a space between the number and the unit symbol (e.g., 10 m, not 10m).

- Abbreviations for units are capitalized when they are named after a person but not otherwise (e.g., N for newton but kg for kilogram).

- Units are not capitalized, even when named after a person, when written out in full (e.g., newton not Newton). This rule prevents confusion over whether one is referring to the person or the unit.

- SI unit symbols do not end with a period (e.g., W and cd, not W. and cd.).

- Unit symbols stand for both single units and multiples of the unit, so we never add an "s" to make a symbol plural (e.g., 10 m means ten metres, but 10 ms means ten milliseconds).

- The unit symbols are written in normal (roman or upright) type to distinguish them from variables, which are written in italic (e.g., "m" always represents the unit metre, while m usually represents the variable mass).

- When two (or more) units are multiplied together, we use a small dot to separate them (e.g., N·m).

- Normally we write simple units in the denominator using the slash (e.g., kg·m/s^2). When the situation is complex, it is sometimes more clear to use negative exponents (e.g., C^2·N^{-1}·m^{-2}). Either format may be used.

- When writing numbers with many digits, sets of 3 digits on either side of the decimal point are separated by spaces instead of commas. In a number with 4 digits, the space is optional (e.g., 52 469 450, 7.630 941 23, and 4123 or 4 123).

- Normally we use the symbol for the unit when it is along with a numerical value (e.g., 12.5 N), but we write the unit in full otherwise (e.g., measure the force in newtons). The full form is also used when we write the number out (e.g., a force of five newtons).

When we use SI units consistently, the results will be in the corresponding SI units. Even though we frequently carry units throughout a computation as a check, and show that the units reduce to the expected units for the final answer, if we use SI consistently we can predict the units of the final answer whether we do this or not. The website of the U.S. National Institute of Standards and Technology is a helpful reference for learning more about the definitions, derivations, and uses of SI units (physics.nist.gov/cuu/Units/), as is the similar site by Public Works and Government Services Canada (www.btb.termiumplus.gc.ca/tcdnstyl-chap?lang =eng&lettr=chapsect1&info0=1.23).

1-7 Dimensional Analysis

In this section we will define what **dimensional analysis** is, and then we will show how to use it for three different purposes. In dimensional analysis we work in terms of the units, e.g., metres (or their corresponding physical meaning such as length), to develop possible relationships, or to test for the dimensional correctness of proposed relationships. When checking dimensional correctness, note that expressions on the two sides of the equation must always have the same units, as must any quantities added (or subtracted). The first person to use what we now consider dimensional analysis was the French physicist and mathematician Jean-Baptiste Joseph Fourier (1768–1830). Interestingly, he was also the first to discover the greenhouse effect of Earth's atmosphere. Let us now consider the three ways we can use dimensional analysis.

If we know a relationship, we can use dimensional analysis to determine the units for a calculated quantity. For example, if we are told that pressure is defined as force per unit area, then we can conclude that the SI units of pressure must be N/m^2 (this is also called a pascal, Pa). In Example 1-9 we use dimensional analysis to find the units of a fluid mechanics term.

EXAMPLE 1-9

Units of Dynamic Viscosity

An important parameter in fluid mechanics that comes into everything from automobile design to blood flow is the dimensionless Reynold's number, Re. It is defined by the following relationship, where ρ is the density of the fluid (mass per unit volume), v is the speed of the object in the fluid, L is the characteristic dimension of the object moving in the fluid, and μ is the dynamic viscosity:

$$Re = \frac{\rho v L}{\mu}$$

Use the information provided to determine the SI units for dynamic viscosity.

(continued)

CHECKPOINT 1-10

Writing SI

Identify all the errors in this statement: "The applied force was 23 Ns. and the displacement was 4.5 *cm.*"

ANSWER:

- We never make abbreviations of units plural, so it should be N, not Ns.
- There should be a space between the number and the unit, i.e., 23 N.
- There should not be a period (except for end of sentence) after the symbol.
- The unit cm should not be italicized (only variables are italicized).
- There should be a space between 4.5 and cm.
- The corrected format of the statement is: "The applied force was 23 N and the displacement was 4.5 cm."

Solution

We are told that *Re* is dimensionless, so dynamic viscosity must have the same units as the combination ρvL, so the units of dynamic viscosity must be the following:

$$\frac{kg}{m^3}\frac{m}{s}m = \frac{kg}{m \cdot s}$$

Making sense of the result

Note that we found the units for a physical quantity simply on dimensional grounds without actually having knowledge of that area of physics.

We can also use dimensional analysis to decide if a proposed relationship is *possible*. Usually we use this when some other insight suggests a relationship. Note, however, that dimensional analysis determines only whether a relationship is *possible*; further data are needed to determine constants and to prove that the relationship is actually valid. We demonstrate this technique in Checkpoint 1-11 and in Example 1-10. Dimensional analysis can be used in this way as a check to detect mistakes in results that you have derived.

CHECKPOINT 1-11

Dimensional Analysis in Kinematics

Later in the text (Chapter 3), you will use kinematic relationships between displacement (in m), velocity, acceleration, and time. Use dimensional analysis to determine whether the following relationship could be correct (*x* is in m, v_0 is in m/s, and *a* is in m/s²). If the relationship is not correct, suggest how it could be altered:

(a) $x \overset{?}{=} x_0 + v_0 t^2 + at$

(b) $x \overset{?}{=} x_0 + v_0 t + at^2$

ANSWER: Relationship (a) is not possible. The term $v_0 t^2$ has units of m · s, but the term at has units of m/s, and x and x_0 have units of m. Relationship (b) is possible, since all three terms can be reduced to units of m. That does not mean that relationship (b) is correct. As you will see in later chapters, the actual expression (for constant-acceleration situations) is $x = x + v_0 t + \frac{1}{2}at^2$. Remember that dimensional analysis cannot give you numerical factors, such as $\frac{1}{2}$.

EXAMPLE 1-10

Dimensional Analysis for a Proposed Kinematic Relationship

When acceleration (*a*) is constant, one can derive a relationship for straight-line motion between speed, *v*, initial (time $t = 0$) speed v_0, acceleration (*a*), and the displacement $(x - x_0)$. Is the following proposed relationship potentially correct in terms of dimensional analysis?

$$v^2 = v_0^2 + 2a(x - x_0)$$

Solution

The SI unit for length is the metre (m). As you will see in Chapter 3, the unit of velocity (speed) is distance per time, so the two speeds squared must have SI units of $(m/s)^2$. Acceleration (as we will see in Chapter 3) has units of m/s². Therefore, when we consider the proposed relationship

$$v^2 = v_0^2 + 2a(x - x_0)$$

the units of the last term are

$$\frac{m}{s^2} \cdot m = \frac{m^2}{s^2} = \left(\frac{m}{s}\right)^2$$

Therefore, we have proven that the units are consistent in all three terms, and the relationship is possible.

Making sense of the result

Note that we have not established that the relationship is correct, only that it is dimensionally plausible. Nor does the technique provide the factor of 2 in the relationship. While dimensional analysis can check or suggest relationships, deductive logic and/or experiments are needed to establish relationships in physics.

We can also use dimensional analysis and physical insight to propose relationships with consistent units. Of course experimentation is required to prove whether the relationship really holds. Just because the units are correct does not necessarily make the proposed relationship valid.

Sometimes dimensional analysis can be used by inspection if the proposed relationships are relatively simple. When the relationship involves many different units, and in particular when they are raised to different powers, this simple inspection technique usually breaks down. In these cases one approach is to assume that different quantities appear to some power, with those unknown powers appearing as unknown variables. One then solves simultaneous equations to make sure that each unit matches on the two sides of the equal sign. This is best illustrated with an example (see Example 1-11).

In applying dimensional analysis, it is most effective to work in symbolic terms, rather than with numbers (i.e., keep *v* representing a velocity or speed in m/s, rather than the numerical value of that speed). This is good practice in general in physics, and most of the time we will take that approach, only inserting numerical values in the final step of a problem solution.

EXAMPLE 1-11

Dimensional Analysis for the Period of a Pendulum

Simple observations show that the period, T, of a simple pendulum depends on the length of the pendulum and on the acceleration due to gravity, g (about 9.81 m/s^2 on Earth's surface). Use dimensional analysis to determine a possible equation for period.

Solution

The period, T, has units of time, s. The SI unit for length, L, is the metre, and acceleration due to gravity, g, has units of m/s^2. Let's assume that period depends on length, L, and acceleration due to gravity, g, with unknown powers a and b, as shown below:

$$T \sim L^a g^b$$

Inserting the SI units, we have

$$s \sim m^a \left(\frac{m}{s^2} \right)^b$$

This can be written as

$$s^1 \sim m^{a+b} s^{-2b}$$

To make the dimensions match on the two sides of the equation, we require that the following two relationships hold:

$$a + b = 0$$
$$1 = -2b$$

This gives the solution

$$b = -\frac{1}{2}, \quad a = \frac{1}{2}$$

We can see that the only way to get units of time by combining the units of length and acceleration is as follows:

$$T \sim \sqrt{\frac{L}{g}}$$

In this case, the real expression has a numerical factor (see below).

Making sense of the result

Observations confirm that a longer pendulum will oscillate more slowly and hence have a longer period, and a faster acceleration due to gravity will make the period shorter. In fact, the actual relationship for the period does have the form we determined above, but the relationship also contains a dimensionless numerical factor:

$$T = 2\pi \sqrt{\frac{L}{g}}$$

We have the interesting result that the period only depends on the length of the pendulum (assuming g is constant), so all pendulums of grandfather clocks with a 1 s period have the same length.

1-8 Unit Conversion

You can easily do some **unit conversions** in your head; for example, 25 mm is the same as 2.5 cm, and 12 km is the same as 12 000 m. However, for conversions that involve more than a simple power of 10, we recommend that you use the formal unit conversion method outlined in Example 1-12. This method is based on multiplying by factors in which the quantities in the numerator and denominator are equivalent. Since these factors equal 1, the multiplication does not change the value of the expression. We can cancel units to confirm that the conversion is correct. Note that this method can also be used for converting from or to units that are not SI.

EXAMPLE 1-12

Unit Conversion

Express the speed 72.0 km/h in SI base units.

Solution

Speed is a distance (length unit) divided by a time, so the SI base units are metres per second. We do the conversion by writing factors so that our units cancel to the desired final units of m/s, with each factor having quantities in the numerator and the denominator that are equivalent:

$$\frac{72.0 \text{ km}}{1 \text{ h}} \times \frac{1000 \text{ m}}{1 \text{ km}} \times \frac{1 \text{ h}}{60 \text{ min}} \times \frac{1 \text{ min}}{60 \text{ s}} = 20.0 \text{ m/s}$$

Making sense of the result

If we travelled at 1.0 m/s, we would go 3600 m, or 3.6 km, in an hour. Thus, it makes sense that a speed of 20.0 m/s corresponds to 72.0 km/h.

MAKING CONNECTIONS

Costly Mistakes in Unit Conversion

In 1999, the Mars Climate Orbiter (Figure 1-11) provided a spectacular and very expensive demonstration of the critical importance of unit conversions. This spacecraft was intended to orbit Mars and collect data on the planet's climate. However, on September 23, 1999, the orbiter angled far too low into the Martian atmosphere and disintegrated. The reason for the trajectory error was a misunderstanding between NASA staff and contractors on the units used in thrust calculations.

On July 3, 1983, an Air Canada jetliner ran out of fuel while on a flight from Montreal to Edmonton. The fuel gauge system on the aircraft was malfunctioning, so the ground crew and pilots had to calculate the fuel for the flight manually. These calculations involved both the volume and the mass of the fuel. The aircraft was a new model calibrated with metric units, but the paperwork for refuelling was based on gallons and pounds, the units used by most aircraft at the time. Although the pilots double-checked the calculations, one of the conversion factors was inverted. Fortunately, the captain was an experienced glider pilot and managed to land, with only minor damage to the plane, on a drag strip at a former military base at Gimli, Manitoba. That aircraft has gone down in aviation history as the Gimli Glider.

Figure 1-11 The ill-fated Mars Climate Orbiter (artist's rendition).

1-9 Approximations in Physics

In physics you will often need to make approximations both in values and in relationships. As we have seen in this chapter, whenever you make an experimental measurement, there is uncertainty in the value that you obtain. We represent the true value for a physical quantity with our best estimate, based on experimental measurements. We have developed tools in this chapter to estimate and express the uncertainty in that estimate. For example, when we write 4.56 ± 0.03 s, we are expressing the view (for a normal distribution) that the true value lies between 4.53 s and 4.59 s about two-thirds of the time, and between 4.50 s and 4.62 s about 95% of the time. We can use the SDOM to help estimate uncertainties in the mean of a set of measurements. Significant digits provide an alternative way to represent the confidence that we have in our estimate, with, for example, 4.5 s meaning that we expect the true value to be no better than between 4.45 s and 4.55 s.

Sometimes in physics you need to make an estimate without having any specific experimental measurements. The problem may involve a typical, rather than a specific, object. For example, you need to estimate the dimensions or mass of a typical human, small car, or asteroid. At other times you are doing a problem involving a specific object, such as a certain star, but you do not have mass information for that star. In this case you can use the mass range of stars that has been computed, along with information about whether the star you are studying is probably larger or smaller than an average star.

As you develop an expert view in physics, one important trait is to always think about the assumptions inherent in your calculation. Even when doing a textbook problem, there are often assumptions inherent in an approach. For example, are you to assume when using the speed of sound that the room is at room temperature of about 20 °C if no value is given? If the problem mentions only electrical quantities, does it mean for you to ignore gravitational forces when finding the force between two charges? Is friction important, or is it to be ignored in a problem? Is air resistance to be ignored in a problem involving a projectile? When you make estimates in the absence of measurements, it is particularly important to be clear on your assumptions.

At other times you need to make approximations because the actual situation is too complicated to solve exactly. Earth is really an oblate spheroid, but we often assume that it is a sphere. In later physics courses, you will see how to solve situations involving air resistance, but it is much easier to deal with problems mathematically when it can be neglected. If we have a sufficiently long line of electric charges, we can approximate it as an infinite line of charge (more on this in Chapters 19 and 20). See Example 1-13 for how we can use approximation to estimate the surface area of a human body.

EXAMPLE 1-13

Surface Area of a Human

Sometimes, for example, when we are estimating evaporation rate from the skin or thermal radiation from a naked person, we need an estimate of the surface area of a human. Use reasonable estimates to approximate the surface area of a typical human.

Solution

This is a case where the actual form of a human body is extremely difficult to precisely measure or calculate, even if all dimensions are known. So we must approximate the human with shapes that we can calculate. Also, we are asked for a typical human, so we do not have specific values for the various dimensions.

We will assume that the head is a sphere, that the two arms are cylinders, and that the two legs are cylinders of different dimension, and that the torso can be approximated as another cylinder. Next, we need to estimate the dimensions. We will assume that the head has a radius of 10. cm. We will assume that each arm is 55 cm long, and when approximated as a cylinder the radius is 4.0 cm. For the legs we assume a length of 70. cm and a radius of 8.0 cm. For the torso we assume a length of 45 cm and a radius of 14 cm. We demonstrate the actual calculations below. The surface area of a cylinder is the circumference of the circle times the length of the cylinder:

$$A_{head} = 4\pi r_{head}^2 = 4\pi (0.10 \text{ m})^2 = 0.126 \text{ m}^2$$

$$A_{arm} = 2\pi r_{arm} L_{arm} = 2\pi (0.040 \text{ m})(0.55 \text{ m}) = 0.138 \text{ m}^2$$

$$A_{leg} = 2\pi r_{leg} L_{leg} = 2\pi (0.080 \text{ m})(0.70 \text{ m}) = 0.352 \text{ m}^2$$

$$A_{torso} = 2\pi r_{torso} L_{torso} = 2\pi (0.14 \text{ m})(0.45 \text{ m}) = 0.396 \text{ m}^2$$

$$A_{total} = A_{head} + A_{torso} + 2A_{leg} + 2A_{arm} = 1.5 \text{ m}^2$$

We kept additional digits in each part, but when adding to the final answer we reduced the number of significant digits, since our final answer is at most two significant digits.

Making sense of the result

This demonstrates that in physics we often obtain a better estimate by breaking down a situation. In this case we can approximate the surface area of each body part, and combine the surface areas to get the total surface area. Neither the legs nor the arms are actual cylinders: both taper, are oblate, and are irregular. If you needed an estimate of the uncertainty for the final value you could redo the calculation with minimum and maximum dimensions inserted. It is reasonable that the torso and legs contribute more than the head and arms.

At times we need to make approximations to permit an analytical solution to a problem. For example, when angles are small, the sine and tangent of an angle are approximately equal to the angle expressed in radians. We will use this approximation in Chapter 13. In more advanced courses in physics you will frequently use approximations such as this, as well as series approximations for functions.

Modern physics makes extensive use of numerical techniques to solve situations where analytic solutions are impossible. Equations, including those involving derivatives, can be solved using computer programs that break the situation up into many steps, using the numerical result from one tiny step in the next.

The final section of this chapter deals with Fermi problems, in which making realistic approximations plays a key role in determining a realistic value for a situation in which even a rigorous numerical solution is not possible.

CHECKPOINT 1-13

Why Approximate?

Which of the following is a valid reason for using approximations in physics?

(a) They may make an analytic solution possible.
(b) They can simplify calculations.
(c) We do not have precise values for some quantities, so we must approximate them.
(d) All of the above are valid reasons, depending on the situation.

ANSWER: (d) There are many reasons that approximations are used in physics, including these.

1-10 Fermi Problems

What Is a Fermi Problem?

In physics (and in engineering and other sciences), we often encounter situations in which we do not have enough information to calculate precise values. Even in these situations we may still be able to make useful approximations, as we saw in the previous section. Such estimates often involve reasonable guesses for the values of unknown quantities, probabilities, and dimensional analysis if a relevant physical law is not known.

The famous physicist Enrico Fermi (1901–1954) skillfully used such techniques, and problems involving such estimates are now called **Fermi problems** in his honour. When the Trinity nuclear bomb test was carried out in 1945, Fermi used the deflection of some small pieces of paper at a location well away from the blast to approximate the energy of the explosion. He obtained a value of 10 kilotons of TNT, within a factor of 2 of the accepted value of 20 kilotons of TNT. If you search on the Internet you can see various collections of Fermi problems. In this text, we include a few such problems at the end of each chapter, under the heading Open Problems.

When doing Fermi problems, scientific notation is very useful because it separates the magnitude of the number (in the exponent) from the precise value. Often, we are just seeking an order-of-magnitude estimate and can ignore everything other than the powers of 10 in coming up with an estimate.

Physicists regularly work in situations in which they cannot necessarily establish even a best estimate

EXAMPLE 1-14

The Wheels on the Bus Go Round and Round

How many revolutions do the wheels on a bus make during a trip from Toronto to Montreal?

Solution

Since this problem calls for estimates, we will only use one or two significant digits throughout. We first estimate the driving distance from Toronto to Montreal to be about 550 km.

Next, we need to estimate the radius, r, of a typical bus wheel. Tire sizes for buses vary somewhat, but a reasonable value is 50 cm.

With each revolution, the bus tire will move a distance of $2\pi r$ along the road. Using N as the number of revolutions and D as the total distance travelled, then

$$N(2\pi r) = D$$

We rearrange this equation to solve for N. Before we substitute the values for our estimates, we convert the distance and radius estimates to SI units (m) using the techniques of the earlier section. Then we solve for N:

$$N = \frac{D}{2\pi r} = \frac{550\,000 \text{ m}}{2\pi \times 0.50 \text{ m}} = 1.8 \times 10^5$$

Therefore, the wheels on the bus turn about 180 000 times during the trip. However, since our estimate for the radius is probably only good to one significant digit, we should perhaps state our answer as about 200 000.

Making sense of the result

For each kilometre travelled, the wheel makes about 350 revolutions, which seems reasonable. If the wheels are rated 60 000 km between times they need to lubricated, this would imply that the wheels turn about 20 million times between services!

for a number, but they can determine a maximum possible value (or minimum possible value). These worst-case calculations are helpful, for example, to estimate the highest possible force on an object.

Probability

One of the most famous Fermi problems, purportedly stated and solved by Enrico Fermi himself, was, "How many piano tuners are there in the city of Chicago?" This was posed in an era when household pianos were more common than at present. If you are interested, he came up with a value of about 225, and apparently the actual number of piano tuners in Chicago at that time was 290. To propose a solution to this problem, Fermi used a series of probabilities, and many Fermi problems use probability ideas.

We usually represent a **probability** using a numerical value between 0 and 1, with 0 representing no chance of something happening and 1 indicating that it is certain to happen. For example, if you flip a coin, the probability of heads is 0.5, and so is the probability of tails. Similarly, if you roll a six-sided unbiased die, the probability of any particular number, for example, a 4, is 1/6, or about 0.17.

If we want to determine the overall probability of some event that depends on multiple independent events, then we simply multiply the probabilities of each of the events. For example, the probability of rolling two 4s with a pair of unbiased dice is (1/6) (1/6), or about 0.028. Note that this technique is valid *only* if the individual probabilities are *independent*.

Many Fermi problems involve setting probabilities for events. Even though we have now confirmed several thousand exoplanets (planets around stars other than the Sun), we are still not certain how likely planets with intelligent life are, or even if there are any, other than us. We demonstrate the use of probabilities in a Fermi problem to estimate the number of habitable exoplanets in Example 1-15.

CHECKPOINT 1-15

Probability of Two Queens

In a standard deck of 52 playing cards, what is the probability of randomly drawing two consecutive queen cards? Assume that the drawn card is not returned to the deck.

(a) $\dfrac{1}{13} \times \dfrac{3}{51}$

(c) $\dfrac{1}{26}$

(b) $\dfrac{1}{13} \times \dfrac{1}{13}$

(d) $\dfrac{2}{13}$

ANSWER: (a) There are four queen cards in the 52-card deck, so the probability of the first card being a queen is $\dfrac{4}{52} = \dfrac{1}{13}$. If the first card drawn is a queen, there are now 3 queen cards left from 51 from the second draw, giving a probability of $\dfrac{3}{51}$. Therefore, the overall probability is obtained by multiplying these two independent probabilities.

EXAMPLE 1-15

Habitable Planets

Estimate the number of habitable planets in the universe.

Solution

Start with an estimate of the number of stars in our galaxy, the Milky Way. Informed estimates range from about 100 billion to 400 billion. We will use 150 billion for our calculation.

Now, consider the image of galaxies shown at the beginning of this chapter. If we count the approximate number of galaxies in this small region, and then multiply by the area of the entire sky divided by the area of just this region, we obtain an estimate for the total number of galaxies in the universe. An answer of about 100 billion is obtained. It is possible that the image does not show some of the fainter, more distant galaxies, so this estimate may be low.

Therefore, the total number of stars in the visible galaxies is roughly 150 billion times 100 billion, or about 1.5×10^{22}.

Next, we need to estimate what fraction of stars will have planetary systems. Astrophysicists believe that the same process that produces stars from clouds of dust and gas also creates planets. With detections of exoplanets, we are narrowing the uncertainty of how frequently planets are found, although this is still uncertain due to detection limits and observational biases. With considerable uncertainty, we estimate that one-fifth of the stars have planetary systems, so there could be about 3.0×10^{21} planetary systems.

Now, we must consider the probable number of habitable planets per star. We know that our solar system has eight major planets. However, only Mars and Earth have conditions that could support life-forms that depend on water. We might consider other forms of life, but many experts regard them as unlikely. In any case, we are looking only for an order-of-magnitude estimate. So, we have two habitable planets out of eight. Many stars will have energy outputs much different from that of the Sun, which could substantially reduce the number of planets with habitable

temperatures. So, we estimate that there is an average of perhaps 0.2 habitable planets per planetary system.

Then the number of habitable planets in the universe is in the order of

$$0.2 \times 3.0 \times 10^{21} = 6.0 \times 10^{20}$$

Making sense of the result

This is an almost unimaginably large number. For example, if a human could count one planet per second, and if the entire world population of about 7 billion people counts simultaneously (for 10 h/day, 6 days/week), about

$$365 \text{ days} \times (6/7) \times 10 \text{ h/day} \times 60 \text{ min/h} \times 60 \text{ s/min} \times 7 \times 10^9$$
$$= 7.9 \times 10^{16} \text{ planets}$$

could be counted per year. In other words, it would take the entire population about 7600 years to count this number of planets, assuming a count rate of one per second, 6 days a week, 10 h/day. This estimate is commonly covered in astronomy, and is called the Drake equation, in honour of Frank Drake, one of the pioneers in the search for extraterrestrial intelligence. The Drake equation usually calculates the number for our galaxy, not the entire universe as we have done here. Now exoplanet research (see Chapter 11) is providing observational data to test estimates such as this.

Advice for Learning Physics

Before finishing this chapter, we want to leave you with some suggestions for doing well in your study of physics. You only really learn physics by mentally interacting with the ideas. Do not skip suggested interactive activities—they are key to your learning. While the chapter summaries are valuable, you should read the entire chapter at least once. Pay particular attention to the learning objectives that specify what you need to learn in each section. Also, make sure that you can answer each checkpoint correctly before going on to the next section.

Remember that your goal in physics is not to remember a large number of relationships or algorithms for solving problems of a certain type. Rather, it is to get a sound grasp of a surprisingly small number of key concepts. These, along with expertise in the appropriate mathematical techniques, will allow you to tackle novel problems. Learning anything new, including physics, can, for all of us, feel frustrating at times, but the more problems you try on your own, the more your expertise will grow. It is important that you try the whole range of questions and problems, including conceptual questions, problems by section, the broader comprehensive problems, problems that involve data analysis, and the open problems that will require you to make reasonable assumptions and decide what is relevant to the situation. Interacting with others in a study group will help you master the material, as multiple viewpoints help us to see things in a more complete way.

KEY CONCEPTS AND RELATIONSHIPS

Nature of Physics

Physics deals with physical phenomena on all scales, from subatomic particles to the universe as a whole. Physics has a number of distinguishing characteristics, including its quantitative nature; the strong role of models; both theoretical and applied aspects; the need for creativity and a collaborative work environment; and, perhaps most important, the goal of explaining physical phenomena in the simplest way possible.

Precision and Accuracy

Precision specifies how narrowly the values range around a central value. Accuracy refers to how close that central value is to the "correct" value. There are three types of errors: true errors (mistakes), random errors (spread about the central value), and systematic errors (e.g., if the technique overestimated all values).

Mean and Standard Deviation

The mean, $\bar{x}$, and standard deviation, σ, of N measurements are defined by the following:

$$\bar{x} = \frac{\sum_{i=1}^{N} x_i}{N} \qquad (1\text{-}1)$$

$$\sigma = \sqrt{\sum_{i=1}^{N} \frac{(x_i - \bar{x})^2}{N - 1}} \qquad (1\text{-}2)$$

For data that are normally distributed, about two-thirds of the data lie within ± 1 standard deviation of the mean, and about 95% of the data lie within ± 2 standard deviations of the mean.

Standard Deviation of the Mean

When determining whether the mean values from two different data sets are statistically different, we use the standard deviation of the mean (SDOM), also called the standard error of the mean:

$$SDOM = \frac{\sigma}{\sqrt{N}} \qquad (1\text{-}3)$$

Significant Digits

Significant digits indicate the approximate precision of a value. When multiplying and dividing, you should have the same number of significant digits in the result as in the least significant entry number. When adding and subtracting, keep the same number of significant digits after the decimal point as in the least precise of the quantities used in the calculation.

SI Units

SI is a unified system of units recognized worldwide and applicable for all fields of science and engineering. SI has seven base units: metre, kilogram, second, ampere, candela, kelvin, and mole. Many other SI units are derived from these base units. If you use only SI units when solving problems, the answer will always be another SI unit.

Unit Conversion

To convert from one unit to another, you multiply by a series of factors, each with a numerator and a denominator that are equivalent. The factors are arranged so that the original units cancel out, leaving the desired final units.

Dimensional Analysis

Dimensional analysis can be used to determine possible relationships among physical quantities. Dimensional analysis can rule out an incorrect equation, but it is not sufficient to prove whether a proposed equation is correct.

Fermi Problems

We can combine realistic estimates, dimensional analysis, and other techniques to obtain order-of-magnitude estimates for situations that cannot be calculated precisely.

APPLICATIONS

evidence based approaches, problem solving, statistical significance, computation and interpretation of statistical measures, unit conversion, dimensional analysis, approximation, Fermi problems

KEY TERMS

accuracy, causation, correlation, dimensional analysis, error bars, Fermi problems, mean, normal distribution, precision, probability, random errors, scientific notation, SI, significant digits, standard deviation of the mean (SDOM), standard deviation, statistically significant, systematic errors, true errors, unit conversions

QUESTIONS

1. Which of the following is *not* a characteristic of physics?
   (a) Physics seeks to explain phenomena in the simplest way possible and in a way that has the widest application.
   (b) While some phenomena may be described qualitatively, in general, physics seeks quantitative descriptions.
   (c) Physics is primarily involved with developing a large number of equations so that there is one equation relevant for any given situation.
   (d) Physics requires creativity.
2. Interview or shadow a professional physicist, asking them what different tasks they do in a typical week. How would you modify the "what is physics" in the first section based on these results?

3. In what ways are physics and mathematics similar? How are they different?

4. There has been some debate recently about whether topics such as parallel universes and aspects of string theory should really be considered physics. Write a concise argument, either in favour or opposed.

5. (a) List as many occupations as possible that use the ideas and techniques of physics. Be comprehensive, open-minded, and creative in developing your list.

   (b) After your list is complete, exchange it with a classmate, and then add to your list based upon their ideas. Clearly indicate the parts of their list that are different from yours by using a different font or colour.

6. Assume that measurements of a time for some process are normally distributed, with no systematic errors. About two-thirds of the time the value is between 3.4 s and 4.8 s. What are the mean and standard deviation of the time measurement?

7. Describe an example where two quantities have a strong positive correlation, but almost certainly one does not cause the other.

8. Assume that measurements of a length are normally distributed, with no systematic errors. About 19 times out of 20, the value is between 2.2 m and 5.8 m. What are the mean and standard deviation of the length measurement?

9. How many significant digits are in the number 230 100?
   (a) 3
   (b) 4
   (c) 5
   (d) 6

10. How many significant digits are in the number 5100.?
    (a) 2
    (b) 3
    (c) 4
    (d) 5

11. Complete the calculation $\dfrac{12.0 \times 5.0}{4.000}$. According to the rules for significant digits, how should the answer be written?
    (a) 15
    (b) 15.0
    (c) 15.00
    (d) 15.000

12. Which of the following is *not* a base SI unit?
    (a) kg
    (b) s
    (c) N
    (d) cd

13. Which of the following is *not* a base SI unit?
    (a) kg
    (b) m
    (c) J
    (d) K

14. Work (or energy) is equal to force times x. Use Tables 1-2 and 1-3 to identify the units of x.

15. (a) Use a reference source to look up how the length of the metre is currently defined.

    (b) From this definition, what must be the speed of light (to the number of significant digits inherent in the definition)?

    (c) Provide a brief overview of some of the previous ways that have been used to define the length of the metre.

16. Use a reliable reference to determine how the second is currently defined.

17. (a) Use a reliable reference to determine the method currently used to define the kilogram.

    (b) Provide a brief overview of some other ways that have been used to define the kilogram.

18. Describe the current method used to define the ampere. Include a reference to the source of your information.

19. Use a reliable reference source to identify the method currently used to define the candela, as well as the method that was historically used.

20. How many cubic centimetres are in one cubic metre?
    (a) 100
    (b) 1000
    (c) 10 000
    (d) 1 000 000

21. Two data sets, A and B, have the same mean value, but data set B has a larger standard deviation by a factor of 2. There are 16 data points in data set A. How many data points are needed in set B for the SDOM to be identical for the two sets?

22. If we multiply a force (N) by a velocity (m/s), which of the following will be the units of the resulting quantity?
    (a) kg/s
    (b) J
    (c) Pa
    (d) W

PROBLEMS BY SECTION

For problems, star ratings will be used (★, ★★, or ★★★), with more stars meaning more challenging problems. The following codes will indicate if $^d/_{dx}$ differentiation, $\int$ integration, ▢ numerical approximation, or ⌁ graphical analysis will be required to solve the problem.

Section 1-1 What Is Physics?

23. ★★ Go to a journal that publishes original research in physics or a physics-related discipline (your instructor, teaching assistant, or library staff may be able to direct you). Find an article where you can understand in broad measures what the article is about (don't worry about the details). Look at the article and the distinguishing characteristics of physics in this section, and evaluate which are reflected in the research found in the article.

24. ★ Go to a news-oriented science periodical such as *New Scientist*, *Nature*, *Science News*, or *Science* and find an article that describes some new physics research. What characteristics of the nature of physics are reflected in the research?

25. ★★ (a) From popular articles and your own prior reading, establish a list of "unanswered questions in physics." Try to come up with about five important items. Now order the list in terms of most important or interesting. Finally, comment on the ones you feel are likely to be answered in the next decade.

(b) After you have completed part (a), share your list with two classmates (and have them share their lists with you). Discuss as a group the reasons for your choices.

(c) Now modify your list from (a) based on the feedback you received from your classmates. Do you feel that the peer review resulted in an improvement to your list?

26. ★★ Interview (either in person or electronically) a physicist, and try your hand at science journalism by writing an article on some aspect of their recent research. You may want to consider making this an ongoing activity by starting a blog or submitting your work to the media or to the communications team at your university.

27. ★★ Develop a short video describing the contributions of a physicist who you think should be more well-known for his or her contribution to the field.

Section 1-2 Experiments, Measurement, and Uncertainties

28. ★★ Independent measurements have established that the actual length of a room is 4.50 m. Some friends use the same metre stick and measure values from 4.45 m to 4.56 m. Another friend uses an ultrasound rangefinder electronic device, reporting values from 4.405 m to 5.014 m. Comment on the precision and accuracy of each type of measurement, and also on the possibility of systematic errors.

29. ★★ Consider some instrument from your kitchen, workshop, exercise equipment, or other everyday use. State the precision of the instrument, and also estimate the likely accuracy of the instrument.

Section 1-3 Mean, Standard Deviation, and SDOM

30. ★ A set of 16 measurements has a mean of 4.525 and a standard deviation of 1.2. How should the mean be written with an uncertainty given by the SDOM?

31. ★ The SDOM is 1.2, the mean is 4.8, and the standard deviation is 2.4. Approximately how many data points must be in the set?

32. ★★ Consider the following set of measurements. Compute the mean, standard deviation and SDOM.

3.45, 3.02, 2.05, 2.65, 4.04, 5.24, 3.33, 3.14, 3.19

33. ★ According to a journal article, experiments suggest that 19 times out of 20, the measurement obtained is within ±0.5 of 10.5. What must be the standard deviation and mean values, assuming that the data are normally distributed?

34. ★★ A friend is measuring the length of an object. Describe precisely the steps that you would take to quantitatively establish the following: true errors, random errors, and systematic errors. Note that this problem relates to this section but also uses some learning from the next section.

35. ★★ A class obtains the following measurements for the acceleration due to gravity, g. Comment on probable true errors, random errors, and systematic errors.

9.25 m/s^2, 8.95 m/s^2, 9.03 m/s^2, 9.33 m/s^2, 8.99 m/s^2, 9.07 m/s^2, 4.56 m/s^2, 8.87 m/s^2, 9.15 m/s^2

36. ★★ (a) Plot the following data points, with error bars, on a graph. Assume that there is no uncertainty in the x-value and that each y-value has an uncertainty of ±1 m: (0, 0), (1, 2), (2, 5), (3, 7).

(b) Is there a line of best fit that passes through the data points, within the uncertainty of each data point? If so, plot it approximately on your graph.

(c) Add a new data point: (4, 6). What must be the error bar for that point for the line of best fit to include that point as well?

37. ★★ (a) Look up (and record) the meaning of the term *Anscombe's quartet*.

(b) Plot four datasets that illustrate this idea.

(c) What is the important message for data analysis from this topic?

Section 1-4 Significant Digits

38. ★ How many significant digits are in each of the following?
(a) 324.2
(b) 0.032
(c) 9.004
(d) 12 000
(e) 85.0
(f) 6000.

39. ★ Suppose that the uncertainty in the length 17.386 22 m is ±0.04 m. Round the length to the correct number of significant digits.

40. ★ Perform the following calculations. Express the answer with the correct number of significant digits.
(a) 12.456 m − 11.9 m
(b) 6.542 m + 0.0092 m

41. ★ Perform the following calculation. Express the answer with the correct number of significant digits.

$$\frac{12\,400 \times 0.025\,63}{12.34}$$

42. ★ Add the values 4.2 m, 12 m, 0.75 m, 9 m. What is the sum, expressed to the correct number of significant digits?

43. ★★ In an experiment, an object is at 27.5 m (relative to some starting reference point) at $t = 0.00$ s and moves to 28.3 m at $t = 1.00$ s. What is the average speed, expressed to the correct number of significant digits? Average speed is distance travelled divided by the time interval.

44. ★★ Express the result of the following computation to the correct number of significant digits: $\dfrac{34.56 - 29.35}{8.75 - 3.23}$

45. ★★ Express the result of the following computation to the correct number of significant digits: $\dfrac{5.16 + 9.12}{2.750 + 1.232}$

46. ★★ (a) Sometimes the uncertainty in values is expressed as a percentage of the value. If a distance is expressed as 5.00 m ± 10%, is it written to the correct number of significant digits? Justify your answer.

(b) Write the number in the conventional form we use in this chapter for writing values using the $\pm$ notation.

47. ★★ (a) Sometimes the uncertainty in values is expressed as a percentage of the value. If a distance is expressed as 8.0 m $\pm$ 1%, is it written to the correct number of significant digits? Justify your answer.

(b) Write the number in the conventional form we use in this chapter for writing values using the $\pm$ notation.

Section 1-5   Scientific Notation

48. ★ Write each of the following in scientific notation.
(a) 2452
(b) 0.592
(c) 12 000
(d) 0.000 045

49. ★ (a) Write 8900. in scientific notation, making sure to express it with the correct number of significant digits.

(b) Would the answer be different if we had asked you to express 8900 in scientific notation (i.e., without the decimal point)?

50. ★ Perform the following calculation, and express the final result in scientific notation with the correct number of significant digits.

$$\frac{(3.20 \times 10^5)(1.5 \times 10^4)}{6.400 \times 10^3}$$

51. ★★ Convert each of the following from scientific to fixed notation.
(a) 3.67×10^3
(b) 2.25×10^{-12}
(c) 2.4×10^5
(d) 1.200×10^3

Section 1-6   SI Units

52. ★ Mass density is mass per unit volume. What are the units of mass density in terms of the base SI units?

53. ★★ (a) Pressure is given as a force per unit area. What are the SI units for pressure that use the newton?

(b) What is pressure in terms of the base SI units?

54. ★ Work, or energy (see Chapter 6), can be expressed as a force times a distance. How is this expressed in terms of units including the newton, and also in terms of base SI units only?

55. ★★ What are the units of RC, where R is electrical resistance and C is electrical capacitance? (Use Tables 1-2 and 1-3.) You will see in Chapter 23 the importance and meaning of this quantity.

56. ★★ You will see in Chapter 21 that a quantity called the electric field strength can be expressed in units of N/C or V/m. Use the data in the tables to express each of these in terms of base SI units to show that they are equivalent.

Section 1-7   Dimensional Analysis

57. ★★ Acceleration due to gravity, g, has dimensions of acceleration (m/s^2). Assuming that the gravitational potential energy depends only on mass, m; g; and height, h, above Earth in some way, what must be the form of the equation? Use units to verify this.

58. ★★ In thermodynamics, we have the equation $PV = nRT$, where P is pressure, V is volume, n is the number of moles, T is the temperature (in kelvin), and R is a constant. Use the data in Tables 1-2 and 1-3 to determine the units for the constant R.

59. ★★ Use dimensional analysis to determine a relationship between electric potential difference (voltage), current, and electrical resistance. Consider the units given in Tables 1-2 and 1-3.

60. ★★ The aerodynamic drag force, F, for fast relative motion (e.g., a car at highway speeds) has the form $F = -cv^2$, where v is speed (m/s). What must be the units (in terms of SI base units) of the variable c?

61. ★★ Aerodynamic drag for slow relative motion can be expressed in the form $F = -bv$, where F is a force in newtons and v is a speed (m/s). Use dimensional analysis to determine the units (in terms of SI base units) of the variable b.

Section 1-8   Unit Conversion

62. ★ You are travelling at 75.0 km/h. What is this in m/s?

63. ★ Scientists believe that continental drift in the Atlantic Ocean happens at the rate of about 1 to 2 cm/year. Convert this rate into nm/s.

64. ★★ An area is 1.00 km^2. What is this in mm^2?

65. ★★ A cell of baker's yeast is typically 4.0 μm in diameter. (Assume that the cells are spherical, although in reality they have a variety of shapes.) If you had 18 mL of yeast (i.e., about one tablespoon), about how many cells would it contain?

66. ★★ An electron microscope indicates that a rectangular object has dimensions of 2.50 μm by 4.75 μm. What is the area of this object in m^2, expressed in scientific notation?

Section 1-9   Approximations in Physics

67. ★ We mention in the section that when the angles are small, $\tan\theta \approx \theta$, as long as the angle is expressed in radians. If the angle is 4.00°, what is the percentage difference between $\tan\theta$ and θ?

68. ★★ It is a very complex job to precisely calculate the surface area of an egg, since it is not even an exact ellipsoid. Using a simple model, estimate the surface area of a typical large size egg. Express your answer in cm^2.

69. ★★ If you lived in a city with population 500 000, what would you estimate as the total hourly water use, on average, for the city? Express your answer in terms of m^3. Do this by estimating a typical person's water use, rather than just looking up a value on the Internet.

Section 1-10   Fermi Problems

70. ★★ Estimate the value of money if the classroom where you study physics is totally filled as compactly as possible with 25 cent coins. (Note: If your country does not use coins of this type, answer for a different type of coin.)

71. ★★ One of the famous Fermi problems is a calculation of how many piano tuners there were in Chicago. (If you are interested in the problem, search the Internet. There are many solutions available.) Using similar methods, estimate the number of goalies there are in Canada at any one time (from all types of hockey teams).

72. ★★ Using reasonable estimates (don't try to look up precise values) for number of universities, fraction of students who take physics, etc., estimate the number of new first-year university-level physics textbooks purchased in your country each year.

73. ★★★ Using reasonable estimates, estimate how many molecules there are in 1 L of milk.

74. ★★ While you have not yet studied thermal physics in detail, use common-sense ideas of heat loss to estimate what percentage of energy for heating could be saved if you turned down the thermostat by 3 °C during winter. Note that if you live in a region where summer cooling costs dominate energy needs, modify the question to increasing the thermostat by 3 °C during summer.

COMPREHENSIVE PROBLEMS

75. ★★ Find the names of Nobel Prize–winning physicists from your country (note that some physicists may have won the prize in another area, such as chemistry). For each, briefly outline the work cited for the prize, and then comment on what aspects of the nature of physics are reflected in this work.

76. ★ Find an example in social media where correlation is assumed to imply causation but that assumption is questionable.

77. ★★ From your physics lab or your home life, select a measuring instrument that you will use to repeatedly measure an object.
   (a) Before making your measurements, estimate the precision and the accuracy of the instrument.
   (b) Now perform 20 measurements, recording results. This is even better if you can have a variety of individuals use the instrument for the measurements.
   (c) Calculate the mean, standard deviation, and SDOM.
   (d) Based on your results in (c), do you feel that the stated accuracy from (a) should be changed?

78. ★★ Expressed to the correct number of significant digits, a speed is 560.5 cm/s. What is the speed in km/h, using the correct number of significant digits?

79. ★★ You will later learn that the quantity of heat, in energy units of joules, required to heat a mass, m, through a temperature difference, ΔT, is given by $mc\Delta T$, where c is a constant that depends on the substance. If the other units are expressed in base SI units, what must be the units for c, both in units including joules for energy and in base SI units?

80. ★★ HIV (Figure 1-12) is approximately spherical, and it has a diameter of about 130 nm. If we reasonably assume that the mass density is about the same as water, 1000. kg/m^3, what is the mass of one HIV? Express your answer in pg.

81. ★★ The active area of the A8 processor (Figure 1-13) found in some smartphones has the equivalent of about 2.0 billion transistors (the transistor is a semiconductor device that controls electronic processes).

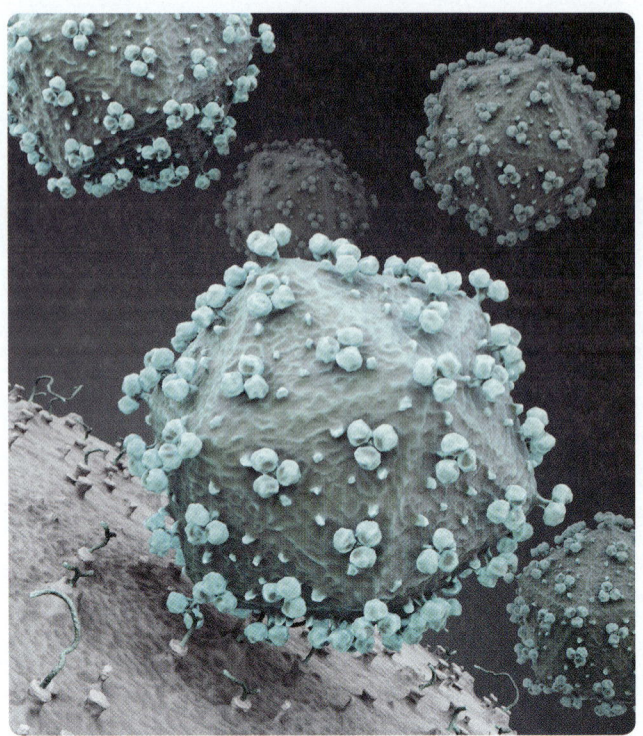

Figure 1-12   Problem 80: HIV.

The dimensions of the integrated circuit chip are approximately 9.3 mm × 9.6 mm. What are the dimensions of one transistor on the integrated circuit, assuming all of the area is transistors and each is a square?

Figure 1-13   Problem 81: A8 processor.

82. ★★★ The expansion of the universe mentioned at the beginning of this chapter is often represented by the equation $v = H_0 d$, where v is the speed of recession of the galaxy, d is the distance to the galaxy, and H_0 is the Hubble constant. In this equation, astronomers usually express v in km/s and d in mega parsecs (Mpc).
    (a) Look up the value of H_0 in terms of these units.
    (b) The parsec is equal to 3.26 ly (light years). A light year is the distance travelled by light in one year. Write the number of metres in one parsec in scientific notation.
    (c) Use unit conversion to write the value of H_0 in SI base units.

83. ★★ Scientists estimate that the flux of meteoritic material on the entire Earth is about 20 000 tonnes per year. One tonne is 1000 kg.
    (a) Convert this flux to mass per square metre of Earth's surface.
    (b) Micrometeorites are meteorites that are small enough to float to Earth without being ablated significantly. They come in a variety of sizes, but for this problem we assume that a typical micrometeorite has a diameter of 25 μm. If 25% of the total meteoritic annual mass was in the form of micrometeorites of this size, how many micrometeorites (on average) would be deposited on each square metre of Earth annually? Assume a density of 3400 m^3.

84. ★★★ Look up the values and units for the following three fundamental physical constants: G (the universal gravitational constant), c (the speed of light), and $\hbar$ (Planck's constant divided by 2π).
    (a) Use dimensional analysis to determine an expression using only G, c, and $\hbar$ that has units of length. Hint: The expression will have a square root, and one of the constants will be to a power, such as squared or cubed. The other two will be just the constant.
    (b) The expression that you obtained is called the Planck length. Calculate the numerical value of the Planck length.
    (c) Use the Internet or other resources to find objects that might have dimensions of the order of the Planck length.

85. ★★ (a) Find a combination of the same three physical constants listed in question 84 (G, $\hbar$, and c) that has units of time. This is called the Planck time. Some physicists consider the Planck time to be the time after the Big Bang, which we can study using physics.
    (b) Calculate a numerical value for the Planck time.

86. ★★ Volunteers are searching a wooded area for a piece of forensic evidence. It is estimated that searchers should use a grid with a separation no larger than 4.0 m between searchers and that the searchers average a walking speed of 0.25 m/s. How long would it take a search party of 10 people to search an area of 1 km^2?

87. ★★ When you are in a merry-go-round it feels as though you are being pushed outward. This "centrifugal force" is because we are not in an inertial reference frame (more on this in Chapter 5). If you are told that the force depends in some way on the distance you are from the centre, r; the speed you are travelling, v; and your mass, m, use dimensional analysis to find a possible formula. Note in this case that the formula you obtain will in fact be correct.

88. ★★ A typical oil molecule has an effective diameter of 0.25 nm. If 25 L (1 L is the same as 1000 cm^3) of oil is spilled on a lake, and if the oil is spread so that it is only a single molecule thick, what will be the approximate area of the oil slick? Assume that the oil molecules are just touching. Hint: Determine the cross-sectional area of an oil molecule.

89. ★★ A car is supported through the contact of the car tires with the ground. Recall that pressure is force per unit area. Use this information and reasonable estimates for the mass of the car and the contact area to predict the air pressure in the tires.

90. ★★ You have a 2.0 kg bag of medium length rice. Estimate how many grains of rice are in the bag. First make the estimate assuming that there are no spaces between grains, and then revise the estimate with a reasonable packing density.

91. ★★ What is the total length of all of the hair on your head? If you have super short hair, do this for someone with longer hair to provide a more interesting case.

DATA-RICH PROBLEM

92. ★★ Based on the data in Table 1-5, can we conclude that the teams of the years 1990–2002 were significantly better or worse than the teams of the years 2003–2014? Clearly explain your reasoning.

Table 1-5 Win–Loss Statistics for the Toronto Blue Jays Baseball Team

Year	Win–Loss Percentage	Year	Win–Loss Percentage
1990	0.531	2003	0.531
1991	0.562	2004	0.416
1992	0.593	2005	0.494
1993	0.586	2006	0.537
1994	0.478	2007	0.512
1995	0.389	2008	0.531
1996	0.457	2009	0.463
1997	0.469	2010	0.525
1998	0.543	2011	0.500
1999	0.519	2012	0.451
2000	0.512	2013	0.457
2001	0.494	2014	0.512
2002	0.481		

OPEN PROBLEMS

93. ★★★ At the top of the atmosphere the total power (energy per unit time) per unit area in all wavelengths for solar energy is about 1400 W/m². Estimate the total area that would be needed in solar photovoltaic cells to meet the total global electrical energy needs. State assumptions you have made about efficiency, energy use, etc.

94. ★★ How many professional physicists would you estimate there are in your country? Do this by making reasonable assumptions, not by looking up a number online!

95. ★★★ Associated with dimensional analysis is the idea that we can evaluate how important aerodynamic drag is in a problem by comparing the mass of the object to the mass of air intercepted by the object.

(a) First use this idea to figure out how important air drag would be if you dropped a 5.0 cm radius steel ball from a height of 2.0 m.

(b) Now estimate how important air drag is when you throw a football (note that the type of flow affects the intercepted mass, but for an order of magnitude assumption assume that the air in the region met by the cross-section of the football is the relevant amount). Use reasonable estimates for a long football pass.

96. ★★★ Assume that the atmosphere is fully mixed, and that there has not been loss of molecules from the atmosphere to space, the oceans, or the ground over the past few thousand years. Is it likely that when you take a deep breath you are breathing any molecules that would have been exhaled by William Shakespeare during his life?

2 | Scalars and Vectors

Learning Objectives

When you have completed this chapter, you should be able to

LO1 Determine whether a physical quantity is a vector or a scalar.

LO2 Describe vector quantities in terms of components, and in Cartesian and polar notation.

LO3 Add and subtract vector quantities, and find the product of a vector and a scalar.

LO4 Define a dot (scalar) product of two vectors, and use dot products to determine angles between vectors and the projection of a vector on an arbitrary axis.

LO5 Define and calculate a cross (vector) product of two vectors using either Cartesian notation or the algebraic definition and the right-hand rule.

Philosophy is written in this grand book, the universe which stands continually open to our gaze. But the book cannot be understood unless one first learns to comprehend the language and read the letters in which it is composed. It is written in the language of mathematics, and its characters are triangles, circles and other geometric figures without which it is humanly impossible to understand a single word of it; without these, one wanders about in a dark labyrinth.

—Galileo Galilei in *Assayer*

In this chapter, we introduce the concepts of scalars and vectors and develop the mathematical language that will be used throughout the book to describe these concepts.

Have you ever wondered why two different words—*speed* and *velocity*—are used in English to describe how fast you are moving? The word *speed* has Old English roots, and the word *velocity* has Latin roots. Speed refers to how fast you are moving. If you are asked about your driving speed, your answer might be 50 km/h. However, if you are asked about your velocity, the 50 km/h value does not suffice: velocity requires additional information about the direction of motion, for example, 50 km/h north on Main Street. For many physical quantities, not knowing the direction makes the information incomplete. Many scientific terms have Latin origins. Thus, we use the word *velocity* to describe both the speed and the direction of motion. For example, if you have been flying from Ottawa for 6 h with an average speed of 500 km/h, your final destination could be anywhere within the circle shown in Figure 2-1. To be able to pinpoint the exact destination, you also need to know the direction of your flight. If your plane headed directly northwest, your destination could be Yellowknife, the capital of Northwest Territories.

Figure 2-1 If you were to fly from Ottawa for 6 h with an average speed of 500 km/h, what would your final destination be?

2-1 Definitions of Scalars and Vectors

As we saw from the opening vignette to this chapter, to provide an accurate description of everyday life phenomena, sometimes we require only numbers and units (e.g., 50 km/h), while other times, we need numbers, units, and directions (e.g., 50 km/h northwest). Thus, the physical quantities we use to describe the world around us can be broadly divided into two categories. For the quantities in the first category, the direction is irrelevant—these quantities are called scalars or scalar quantities. For the quantities in the second category, the direction is very important. These quantities are called vectors or vector quantities. A mathematical generalization of scalars and vectors is called a tensor, so that a scalar is a 0th-order tensor, while a vector is a 1st-order tensor. In this textbook, we will limit our study to vectors and scalars, but in more advanced physics courses you will deal with physics objects that will require higher-rank tensors for their description.

Scalar quantities, or **scalars**, require only a number (either positive or negative) and a unit for their description. For example, temperature, mass, length, area, volume, time, distance, speed, work, and energy are all scalars. Scalar physical quantities in this book are denoted using *italics*. For example, the quantities listed above are usually represented by the symbols T, m, L, A, V, t, d, v, W, and E.

Vector quantities, or **vectors**, require a positive number, called a **vector magnitude**; a unit; and a direction for their description. For example, displacement, velocity, acceleration, force, momentum, angular momentum, impulse, and magnetic field are all vectors. By convention, the symbols for vector quantities are denoted using *italic* letters with arrows above them, bold letters, or bold letters with arrows above them. Throughout this textbook vector quantities will be denoted using bold italic letters with arrows above them, for example, $\vec{v}$ for velocity, $\vec{p}$ for momentum, $\vec{F}$ for force, and $\vec{a}$ for acceleration. The magnitude of a vector is always a positive scalar quantity and can be denoted using the absolute value sign or by using the same letter as the vector without the arrow above it and in light face italics. For example, the magnitude of the velocity vector $\vec{v}$, also called speed, can be denoted as either $|\vec{v}|$ or v.

To describe vectors we need a **frame of reference** (a convention about what is "at rest") and a coordinate system. For example, a two-dimensional (2-D) coordinate system can consist of an x-axis and a y-axis, as shown in Figure 2-2. A three-dimensional (3-D) coordinate system has an additional axis, the z-axis.

In Figure 2-2, the velocity vector $\vec{v}_1$ has a magnitude of 30 m/s (v_1 = 30 m/s) and a direction of 25° above the x-axis (θ_1 = 25°); the second velocity, $\vec{v}_2$, has a magnitude of 15 m/s (v_2 = 15 m/s) and a direction of θ_2 = −10° = 350°. By convention, a positive angle

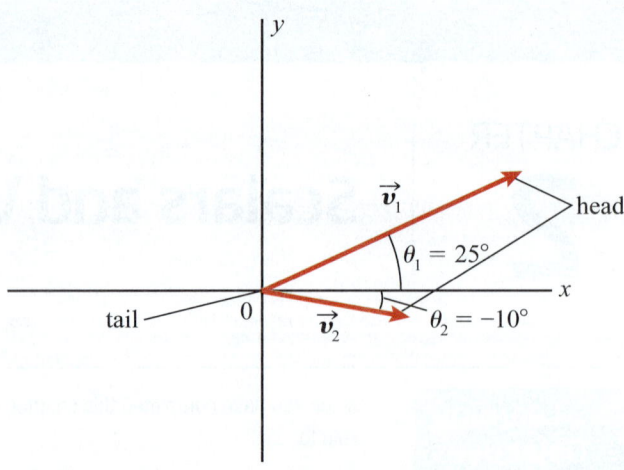

Figure 2-2 Graphical representation of two 2-D velocity vectors. Given: v_1 = 30 m/s, θ_1 = 25°, v_2 = 15 m/s, and θ_2 = −10° = 350°.

is measured in the counterclockwise sense from the positive x-axis, and a negative angle is measured in the clockwise sense from the positive x-axis. Vector magnitude is represented by the length of the arrow.

Any vector can be represented graphically using arrows. The length of the arrow represents the vector magnitude; the direction of the vector coincides with the arrow's direction. Note that a vector is not attached to any particular location in space: moving a vector parallel to itself, which is called translating a vector, does not affect it. The end point of a vector is called the **tail of the vector**, and the arrowhead end of a vector is called the **head of the vector** (Figure 2-2). The direction of the vector is from its tail to its head. To describe a vector, we have to know its magnitude (including its units) and direction. A 2-D vector can be described in terms of its magnitude and the angle it makes with the positive x-axis measured in the counterclockwise direction. Such a description is referred to as a **polar notation** or **polar coordinates**. For example, the polar notation for vector $\vec{v}_1$ in Figure 2-2 is 30 m/s with a direction of 25° or 30 m/s [25°].

CHECKPOINT 2-1

Describing Vectors

Which of the following is the most complete and precise description of a vector quantity?
(a) Vector magnitude, direction, and the location of its tail relative to the origin of the coordinate system
(b) Vector magnitude (including its units), and direction
(c) Vector magnitude (including its units), and location of its head relative to the origin
(d) Vector magnitude (including its units), direction, and location of its head relative to the origin
(e) None of the above

ANSWER: (b)

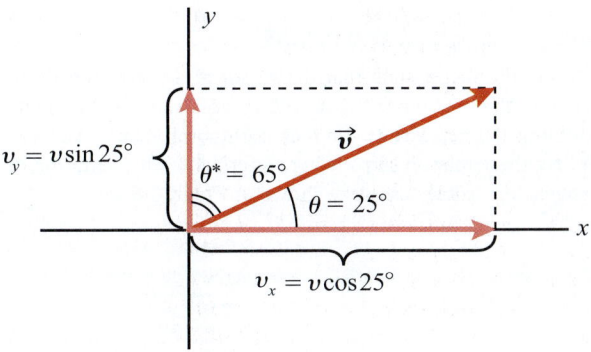

$v_y = v\sin 25°$

$\theta^* = 65°$

$\vec{v}$

$\theta = 25°$

$v_x = v\cos 25°$

Figure 2-3 Vector components of velocity vector $\vec{v}$ (30 m/s [25°]).

In addition to **polar coordinate systems**, we will also use **orthogonal coordinate systems**, where the coordinate axes are perpendicular to each other, such as the xyz-coordinate system you are most familiar with from your previous studies. The word *orthogonal* means directed at right angles. You might encounter the term *rectangular coordinate systems* as well. Using orthogonal coordinate systems, vectors can also be described in terms of their **components**, which are scalar quantities. A 2-D vector has two components, v_x and v_y, as shown in Figure 2-3. These components are **projections** of the vector onto the x- and y-axes, respectively, so vector components can be positive, negative or zero:

$$\begin{cases} v_x = v\cos\theta \\ v_y = v\sin\theta \end{cases} \Rightarrow \begin{cases} v = \sqrt{v_x^2 + v_y^2} \\ \tan\theta = \dfrac{v_y}{v_x} \end{cases} \quad (2\text{-}1)$$

where $|\vec{v}| \equiv v$ and the appropriate value of $\theta = \tan^{-1}\left(\dfrac{v_y}{v_x}\right)$ can be determined from knowing the quadrant in which the vector is located.

Using the Pythagorean theorem, we can see that the magnitude of a vector is equal to the square root of the sum of the squares of its components. By convention, all angles are measured in a counterclockwise direction from the positive x-axis. However, it is sometimes convenient to use the complementary angle θ^*, which measures the angle clockwise from the vertical y-axis, so that $\theta^* = 90° - \theta$. Hence,

$$\begin{cases} v_x = v\sin\theta^* \\ v_y = v\cos\theta^* \end{cases} \Rightarrow \begin{cases} v = \sqrt{v_x^2 + v_y^2} \\ \tan\theta^* = \dfrac{v_x}{v_y} \end{cases} \quad (2\text{-}2)$$

where $|\vec{v}| \equiv v$.

Representation of a vector using its components is often called **scalar notation**. The vector component in the x-direction is positive when the vector is fully or partially oriented in the positive x-direction. Similarly, the x-component is negative when the vector is fully or partially oriented in the negative x-direction. The same principle applies to any other axis. Therefore, the notation $v_x = 20$ m/s means that a projection of the object's velocity vector onto the x-axis is $+20$ m/s. However, an endless number of vectors have this x-component, for example, 20 m/s [0°] and 40 m/s [60°]. We need to provide both components to fully define a 2-D vector. By the same logic, it takes three components to represent a 3-D vector. In general, the description of a vector using components is called **vector resolution into scalar components**. Example 2-1 demonstrates how to resolve a vector into scalar components.

EXAMPLE 2-1

Vector Resolution

Force $\vec{F_1}$ has a magnitude of 20.0 N and is directed at $\theta_1 = 45.0°$ to the positive x-axis. What happens to the components of the force if

(a) the magnitude of the force remains the same, but the angle decreases to $\theta_2 = 30.0°$?

(b) the magnitude of the force doubles, but the direction of the force remains the same?

Solution

Let us express the components of force $\vec{F_1}$ in scalar notation, Equation 2-1. The magnitudes of forces will be denoted as $F_1, F_2,$ and F_3, respectively:

$$\begin{cases} F_{1x} = F_1\cos\theta_1 \\ F_{1y} = F_1\sin\theta_1 \end{cases} \Rightarrow \begin{cases} F_{1x} = (20.0\text{ N})\cos 45.0° = 14.1\text{ N} \\ F_{1y} = (20.0\text{ N})\sin 45.0° = 14.1\text{ N} \end{cases}$$

(a) Let $\vec{F_2}$ denote the force when the angle decreases to $\theta_2 = 30.0°$:

$$\begin{cases} F_{2x} = F_2\cos\theta_2 \\ F_{2y} = F_2\sin\theta_2 \end{cases} \Rightarrow \begin{cases} F_{2x} = (20.0\text{ N})\cos 30.0° = 17.3\text{ N} \\ F_{1y} = (20.0\text{ N})\sin 30.0° = 10.0\text{ N} \end{cases}$$

We can check that force $\vec{F_2}$ has the same magnitude as the original force:

$$F_2 = \sqrt{(F_{2x})^2 + (F_{2y})^2} = \sqrt{(17.32\text{ N})^2 + (10.0\text{ N})^2}$$
$$F_2 = 20.0\text{ N}$$

(continued)

(b) Let $\vec{F}_3$ denote the force when the magnitude of the force doubles while its direction remains unchanged. Then,

$$\begin{cases} F_{3x} = F_3 \cos\theta_1 \\ F_{3y} = F_3 \sin\theta_1 \end{cases} \Rightarrow \begin{cases} F_{3x} = (40.0 \text{ N})\cos 45.0° = 28.3 \text{ N} \\ F_{3y} = (40.0 \text{ N})\sin 45.0° = 28.3 \text{ N} \end{cases}$$

We can verify that these components describe a vector with twice the magnitude of $\vec{F}_1$:

$$F_3 = \sqrt{(F_{3x})^2 + (F_{3y})^2} = 40.0 \text{ N}$$

Making sense of the result

Rotating the vector such that the vector is closer to the x-axis increases the x-component and decreases the y-component. Doubling the magnitude doubles both components. You can confirm these relationships using a graphical representation of these vectors, such as the one shown in Figure 2-2.

2-2 Vector Addition: Geometric and Algebraic Approaches

From everyday experience, we intuitively know that force, which we experience as a pull or a push, must be a vector quantity, as it has magnitude, units, and direction. It is also reasonable that the sum of two forces is a force as well. The sum of two equal forces pulling in opposite directions is zero, and the sum of two equal forces pulling in the same direction is a force with double the magnitude and the same direction. Therefore, it is not surprising that the sum of two or more vectors depends on both their magnitudes and directions. While a force is only one example of a vector quantity, the techniques we develop in this section apply to any vectors, such as velocity, acceleration, momentum, etc.

The Geometric Addition of Vectors

There are two commonly used rules for the **geometric method for vector addition**: the **parallelogram rule** and the **triangle construction rule**. Figure 2-4 shows how these rules can be used to find the resultant vector when two students pull on the arms of a friend. Both rules use the fact that a vector can be translated (moved parallel to itself) as long as its magnitude and direction remain unchanged.

Parallelogram rule Translate the vectors so that they are joined at their tails. Then, build a parallelogram with the two vectors forming adjacent sides, as shown in Figure 2-4(b). The vector sum of $\vec{F}_1$ and $\vec{F}_2$ is represented by the diagonal of this parallelogram, $\vec{F}_R$ (the subscript R stands for resultant). The tail of $\vec{F}_R$ coincides with the tails of $\vec{F}_1$ and $\vec{F}_2$, and the head of $\vec{F}_R$ is at the opposite vertex of the parallelogram.

Triangle construction rule Translate the vectors so that the head of the first vector is joined to the tail of the second vector. Then, connect the tail of the first vector to the head of the second vector, as shown in Figure 2-4(c) (this rule is also known as the **head-to-tail rule**). The vector sum, $\vec{F}_R$, of forces $\vec{F}_1$ and $\vec{F}_2$ is represented by the third side of the triangle, where the tail of $\vec{F}_R$ coincides with the tail of $\vec{F}_1$, and its head coincides with the head of $\vec{F}_2$. The same method can be applied to add three or more vectors.

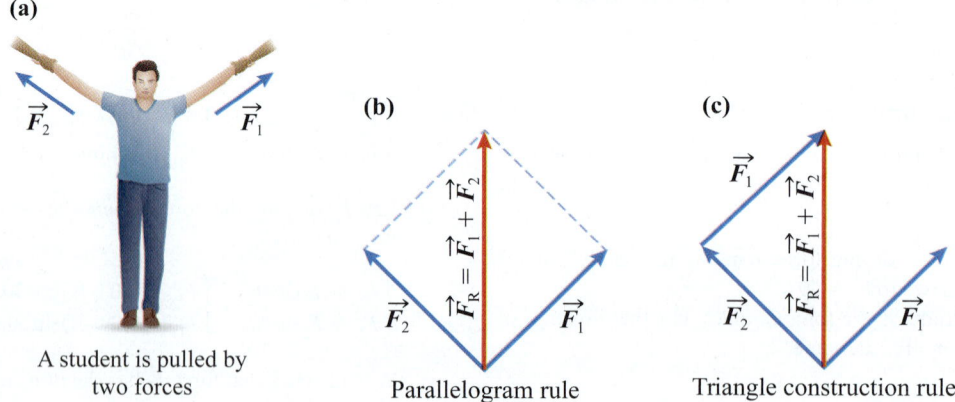

(a)

A student is pulled by two forces

(b) Parallelogram rule

(c) Triangle construction rule

Figure 2-4 Vector addition. (a) Two forces are exerted on a person. The resultant of these two forces, $\vec{F}_R$, is found using (b) the parallelogram rule and (c) the triangle construction rule.

As an exercise, prove that the parallelogram and triangle construction rules produce the same result.

In your previous physics courses you might have heard the term **net force** or $\vec{F}_{net}$. Net force is the sum of all (this is important!) the forces acting on an object, while the **resultant force** is the sum of a subset of forces acting on the object. So while every net force must be a resultant force of all the forces, not every resultant force is a net force. As you can see in Figure 2-4, the resultant force $\vec{F}_R$ includes only two forces, while ignoring other forces, such as the force of gravity or the normal force exerted on the person. The concept of net force is extremely important, and we will use it extensively in the following chapters.

INTERACTIVE ACTIVITY 2-1

Graphical and Algebraic Representations of Vector Addition

This Interactive Activity uses the PhET simulation "Vector Addition" to explore how vectors can be added or subtracted. It will help you understand how to represent vectors using polar notation (vector magnitude and angle), as well as vector components. This activity will also demonstrate different but equivalent ways to represent vector addition or subtraction and compare algebraic and graphical vector representations. Work through the simulation and accompanying questions to deepen your understanding.

Algebraic Addition of Vectors

The geometric approach to vector addition is useful for visualizing the relationships between vectors, but it is difficult to draw and measure the diagrams with high precision. Example 2-2 illustrates an algebraic approach to vector addition.

The method for finding the resultant vector for the two perpendicular 2-D vectors in Example 2-2 can be generalized to apply to any number of vectors with arbitrary orientations (Equation 2-3). The first and second equations in the equation set 2-3 below mean that the x- and y-components of the resultant vector equal the sum of the respective components of the vectors being added. The third equation shows that we can use the Pythagorean theorem to find the magnitude of the resultant vector. The fourth equation helps to find the angle of the resultant vector relative to the x-axis, as we did earlier in Equation 2-1.

CHECKPOINT 2-2

Geometric Addition of Vectors

Which diagram in Figure 2-5 correctly represents the vector sum of the three vectors $\vec{F}_1$, $\vec{F}_2$, and $\vec{F}_3$? Explain.

(a) diagram (a) only
(b) diagram (b) only
(c) diagrams (c) and (d)
(d) diagram (c) only
(e) diagram (d) only
(f) diagrams (b) and (c)
(g) diagrams (a) and (b)

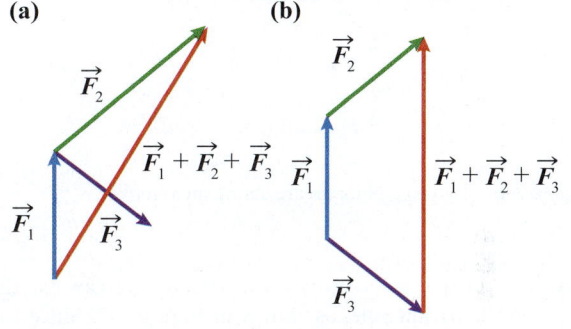

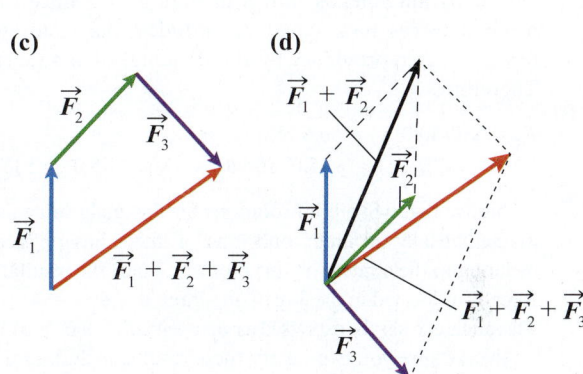

Figure 2-5 Checkpoint 2-2: Geometric addition of vectors. (Note: In subsequent chapters, vectors representing forces will be denoted in blue.)

ANSWER: (c) Only diagrams (c) and (d) properly apply the triangle construction rule and the parallelogram rule to vector addition.

$$\begin{cases} F_{R,x} = F_{1x} + F_{2x} + \cdots + F_{N,x} \\ F_{R,y} = F_{1y} + F_{2y} + \cdots + F_{N,y} \\ F_R = \sqrt{(F_{R,x})^2 + (F_{R,y})^2} \\ \tan\theta = \dfrac{F_{R,y}}{F_{R,x}} \end{cases} \quad (2\text{-}3)$$

KEY EQUATION

EXAMPLE 2-2

Finding the Resultant of Perpendicular Forces

Two 40.0 N forces, $\vec{F_1}$ and $\vec{F_2}$, directed at 90.0° to each other, act on a hook, as shown in Figure 2-6.

(a) Use the geometric approach to vector addition to find the resultant force acting on the hook.

(b) Find the components of each force along the x- and y-axes.

(c) Find the components of the net force (resultant vector).

(d) Compare your results from parts (b) and (c). What conclusions can you draw?

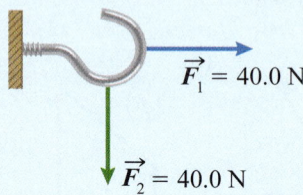

Figure 2-6   Two 40.0 N forces are acting on a hook.

Solution

(a) To add the two vectors geometrically, we can use the parallelogram rule, as shown in Figure 2-7. Since the original forces have equal magnitudes, the resultant force, $\vec{F_R}$, is represented by the diagonal of a square. Therefore,

$$F_R = \sqrt{2}(40.0\ \text{N}) = 56.6\ \text{N}$$
$$\theta_1 = \tan^{-1}(-1) = -45.0° \text{ and } \theta_2 = 180° - 45.0° = 135°$$

Notice that while both solutions for the angle value are mathematically accurate, only one of them has physical meaning in the context of this problem. Since the resultant force $\vec{F_R}$ is located in the fourth quadrant, $\theta \equiv \theta_1 = -45.0°$. This angle can also be expressed as: $\theta_1 = -45.0° + 360° = 315°$.

(Make sure you can justify these results, including the negative sign for the angle.)

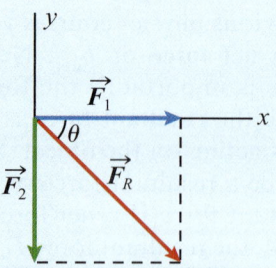

Figure 2-7   Vector addition for Example 2-2.

(b) To find the components of forces $\vec{F_1}$ and $\vec{F_2}$, we first choose a coordinate system, and then use Equation 2-1:

$$\begin{cases} F_{1x} = 40.0\ \text{N} \\ F_{1y} = 0 \end{cases} \text{ and } \begin{cases} F_{2x} = 0 \\ F_{2y} = -40.0\ \text{N} \end{cases}$$

(c) To find the components of the resultant vector $\vec{F_R}$, we use Equation 2-3:

$$\begin{cases} F_{R,x} = 40.0\ \text{N} \\ F_{R,y} = -40.0\ \text{N} \end{cases} \text{ and } \begin{aligned} F_R &= \sqrt{(40.0\ \text{N})^2 + (40.0\ \text{N})^2} \\ &= \sqrt{2}(40.0\ \text{N}) = 56.6\ \text{N} \end{aligned}$$

(d) Comparing the results of (b) and (c), we see that the x-components of the two original vectors sum to 40.0 N and their y-components sum up to −40.0 N. These sums match the components of the resultant vector $\vec{F_R}$.

Making sense of the result

Since we used a special case where the original two vectors were aligned along the x- and y-axes, we should have expected that the components of the resultant vector are equal to the sums of the corresponding components of the individual vectors. You can check this example using the PhET computer simulation "Vector Addition" mentioned in Interactive Activity 2-1. You can also check how the result would be different if the original two vectors did not have equal magnitudes or were not orthogonal to each other.

The following three steps give us an **algebraic method for vector addition** (Figure 2-8):

1. Resolve the vectors into components.

2. Add components pointing along the same axis to obtain the corresponding components of the resultant vector.

3. Use the components of the resultant vector to determine its magnitude and direction.

By resolving vectors into components, we can use scalar calculations to perform operations with vectors.

The algebraic method can be extended to 3-D vectors:

KEY EQUATION
$$\begin{cases} F_{R,x} = F_{1x} + F_{2x} + \cdots + F_{N,x} \\ F_{R,y} = F_{1y} + F_{2y} + \cdots + F_{N,y} \\ F_{R,z} = F_{1z} + F_{2z} + \cdots + F_{N,z} \end{cases} \quad (2\text{-}4)$$
$$F_R = \sqrt{(F_{R,x})^2 + (F_{R,y})^2 + (F_{R,z})^2}$$

While determining the magnitude of the 3-D vector is a relatively straightforward procedure, we will discuss later how to describe the orientation of 3-D vectors (Figure 2-13).

Example 2-3 demonstrates algebraic addition of two vectors. It also shows why physicists and engineers often prefer to draw a pictorial sketch of forces acting on an object, often called a force diagram or a **free body diagram (FBD)**. An FBD helps visualize the forces acting on an object, which will be especially useful when we want to predict and describe the motion of an object caused by the forces acting on it. Let us consider a few consequences of the vector addition rules.

Multiplying a vector by a scalar Adding a vector to itself preserves its direction while doubling its magnitude (Figure 2-10(a)). In general, adding N identical

(a)

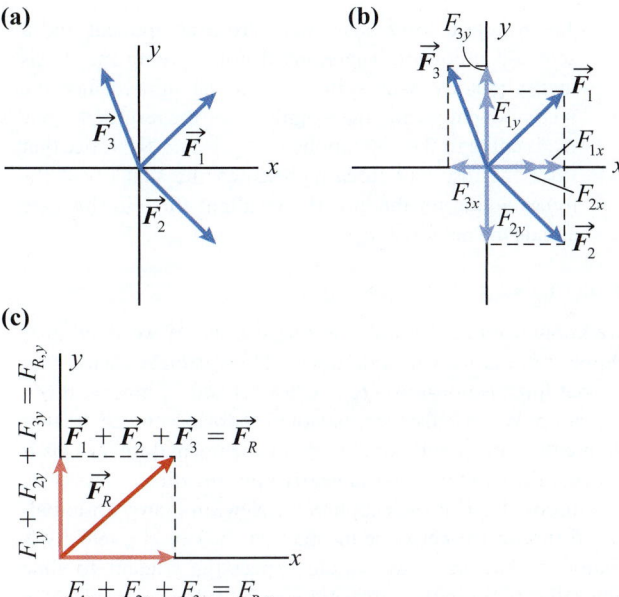

(b)

(c)

Figure 2-8 (a) Visual representation of the addition of three vectors. (b) Vector resolution of the three vectors. (c) The resultant vector $\vec{F}_R$ and its components.

vectors means preserving the original direction and multiplying the original magnitude of the vector by N, as shown in Figure 2-10(b). We can generalize this process to the multiplication of a vector by any scalar, not just integers. The product of a scalar and a vector is a vector with a magnitude equal to the product of the magnitude of the original vector and the absolute value of the scalar. The direction of the resulting vector coincides with the original direction of the vector if the scalar is positive, and is opposite to the original vector direction if the scalar is negative. If we denote a scalar as a and a vector as $\vec{F}$, we can express it as

$$|a\vec{F}| = |a||\vec{F}|$$

$$\begin{cases} \text{if } a > 0,\ a\vec{F} \text{ has the same direction as } \vec{F} \\ \text{if } a < 0,\ a\vec{F} \text{ has the opposite direction to } \vec{F} \\ \text{if } a = 0,\ a\vec{F} \text{ has a magnitude of zero and no direction} \end{cases}$$

(2-5)

EXAMPLE 2-3

Forces Acting on a Student on a Slippery Ramp

A 51.0 kg student steps onto a slippery 10.0° ramp (the force of friction between the ramp and the girl can be neglected). Two forces are acting on her: the force of gravity exerted by Earth, which is $F_g = 500$ N directed down (toward the centre of Earth), and the normal (perpendicular to the surface) force exerted by the ramp, which is $N = 493$ N, directed perpendicular to the ramp, as shown in Figure 2-9.

(a) Find the components of the force of gravity along the x- and y-axes directed along the incline and perpendicular to the incline, respectively.

(b) Find the components of the normal force along these axes.

(c) Find the resultant force acting on the student.

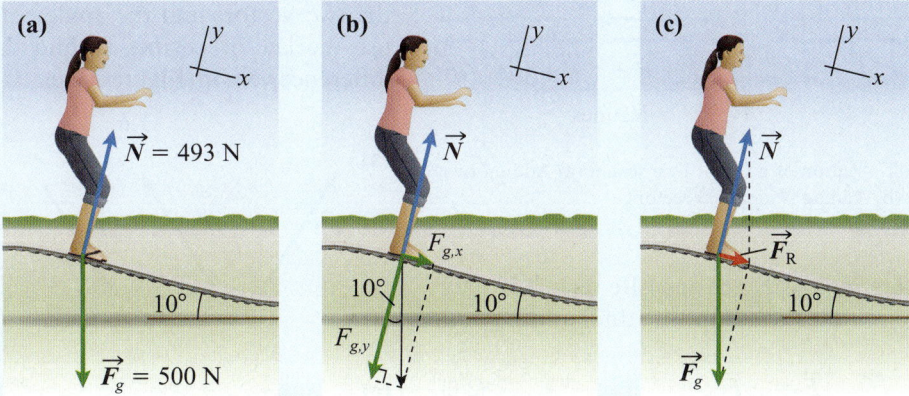

Figure 2-9 (a) A life-like problem representation. (b) The free body diagram (FBD) shows the forces acting on the student. (c) Geometric representation of the resultant force (in this case it is also the net force) acting on the student.

Solution

When dealing with two or more forces, it is often helpful to draw a detailed FBD, such as the one shown in Figure 2-9(b).

(a) To find the components of each vector, we have to find the angles that the vector forms with the coordinate axes.

The angle between the force of gravity $\vec{F}_g$ and the y-axis equals the angle of the ramp. (Make sure you can prove it.) The force of gravity vector, marked in green, forms the hypotenuse of the right-angled triangle in Figure 2-9(b). In this triangle, the side adjacent to the 10.0° angle represents the y-component of $\vec{F}_g$, and the side opposite to

(*continued*)

the 10.0° angle represents the x-component. Therefore, the components of the force of gravity are as follows, by Equation 2-1:

$$\begin{cases} F_{g,x} = F_g \sin 10° \\ F_{g,y} = -F_g \cos 10° \end{cases} \Rightarrow \begin{cases} F_{g,x} = (500 \text{ N})(0.1736) = 86.8 \text{ N} \\ F_{g,y} = -(500 \text{ N})(0.9848) = -493 \text{ N} \end{cases}$$

The negative sign indicates that the y-component of the force of gravity is directed in the negative y-direction. In this case, it is convenient to measure the angle from the negative y-axis; however, we always compare the direction of the vector with the direction of the positive x- and y-axes to determine the signs of the x- and y-components, respectively.

(b) The normal force, $\vec{N}$, is directed in the positive y-direction. Consequently, the angle between the normal force and the x-axis is 90.0°, and

$$\begin{cases} N_x = N \cos 90° = 0 \\ N_y = N \sin 90° = 493 \text{ N} \end{cases}$$

(c) From Figure 2-9(c) we can see that if the vectors are not drawn exactly to scale, or if their directions are slightly inaccurate, we will not be able to obtain a precise value of the resultant force from measuring it on the diagram. Therefore, we use the algebraic approach. Adding the x- and y-components of the gravitational and normal forces, we get

$$\begin{cases} F_{R,x} = 86.8 \text{ N} + 0 = 86.8 \text{ N} \\ F_{R,y} = -493 \text{ N} + 493 \text{ N} = 0 \end{cases}$$

The resultant force has a non-zero x-component and a zero y-component. Therefore, it points along the x-axis in the positive x-direction—down the ramp. Since the y-component is zero, the magnitude of the resultant force, $\vec{F}_R$, is equal to the x-component, $F_R = 86.8$ N. Notice that since the resultant force represented the sum of all the forces acting on the girl, the resultant force in this case equals the net force: $\vec{F}_R \equiv \vec{F}_{net}$.

Making sense of the result

We know from our everyday experience that if we stand on a slippery ramp, we will slide down. Thus, it makes sense that the net force is non-zero ($F_{net} = 86.8$ N) and is directed down the ramp. We can also see that the net force is much smaller than either the gravitational force or the normal force. This is because the two forces act in nearly opposite directions. As we will discuss later in the chapters on Newton's laws, an unbalanced (non-zero) net force means that the object's velocity is changing. Therefore, we should expect the student to slide down the ramp with increasing speed. In this example of a slippery ramp, we ignored the force of friction between the student and the ramp. If present, a friction force would have been directed up the ramp—in the direction opposite to the girl's motion. Notice that choosing an x-axis along the ramp simplified the calculations because the net force is directed down the ramp. It is often convenient to have one of the axes directed along the net force.

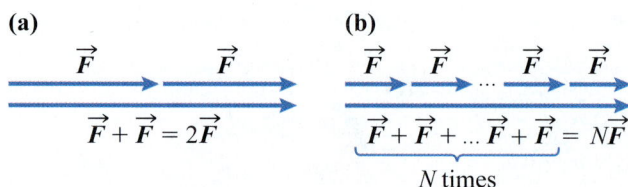

Figure 2-10 Multiplication of a vector by a scalar. (a) Adding two identical vectors. (b) Adding N identical vectors.

Opposite vectors $\vec{F}_1$ and $\vec{F}_2$ are **opposite vectors** if they have equal magnitudes and opposite directions:

$$\vec{F}_1 = -\vec{F}_2 \tag{2-6}$$

Subtracting vectors We can perform vector subtraction by adding an opposite vector, as shown in Figure 2-11:

$$\vec{F}_1 - \vec{F}_2 = \vec{F}_1 + (-\vec{F}_2) \tag{2-7}$$

We can also find the difference between two vectors, $\vec{F}_1 - \vec{F}_2$, by translating the two vectors so that their tails

coincide, and then drawing a vector from the head of $\vec{F}_2$ to the head of $\vec{F}_1$. If you know the magnitudes of the two vectors and the angle between them, you can use the law of cosines to find the magnitude of the difference vector (Figure 2-11).

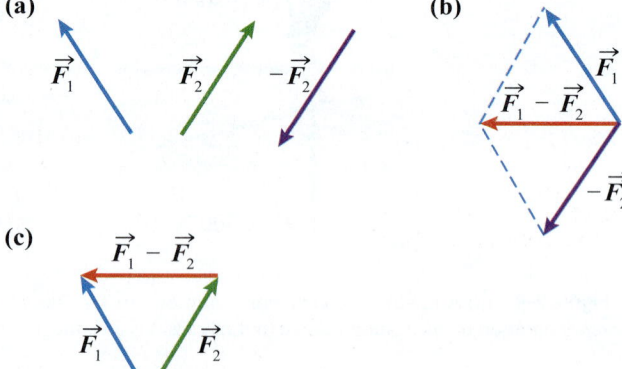

Figure 2-11 Vector subtraction. To find $\vec{F}_1 - \vec{F}_2$, first construct a vector opposite to $\vec{F}_2$ and then add $\vec{F}_1$ and $(-\vec{F}_2)$. Alternatively, translate the vectors so their tales coincide and then draw a vector from the tail of $\vec{F}_2$ to the head of $\vec{F}_1$.

2-3 Cartesian Vector Notation

Cartesian vector notation is another "language" for describing and handling vector components. This notation is based on the concept of **unit vectors**. A unit vector has a dimensionless magnitude of 1. Unit vectors in the positive x-, y-, and z-directions are denoted as $\hat{i}$, $\hat{j}$, and $\hat{k}$, respectively (Figure 2-12(a)). Some books use $\hat{x}$, $\hat{y}$, and $\hat{z}$ as an alternative notation for unit vectors. A unit vector in the direction of an arbitrary vector $\vec{F}$ is denoted as $\hat{u}_F$ (Figure 2-12(b)). A "hat" above an italic bold letter indicates a unit vector.

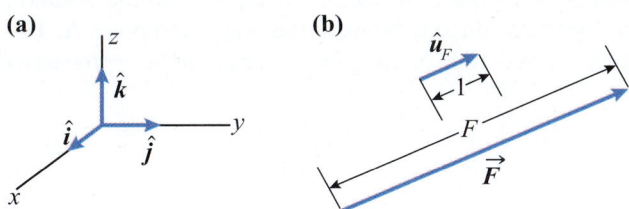

Figure 2-12 (a) Cartesian unit vectors: $\hat{i}$, $\hat{j}$, and $\hat{k}$. (b) A unit vector in the direction of the force vector $\vec{F}$.

In Cartesian notation, any 3-D vector can be represented as follows:

$$\vec{F} = F_x\hat{i} + F_y\hat{j} + F_z\hat{k}, \tag{2-8}$$

where F_x, F_y, and F_z are the usual scalar components of a vector: projections of the vector $\vec{F}$ onto the x-, y-, and z-axes, respectively.

Cartesian vector notation allows us to simplify vector addition significantly:

KEY EQUATION

$$\vec{F}_R = \vec{F}_1 + \vec{F}_2 + \vec{F}_3 + \cdots + \vec{F}_N$$

$$\vec{F}_R = (F_{1x}\hat{i} + F_{1y}\hat{j} + F_{1z}\hat{k}) + (F_{2x}\hat{i} + F_{2y}\hat{j} + F_{2z}\hat{k}) + F_{3x}\hat{i} + F_{3y}\hat{j} + F_{3z}\hat{k} + \cdots + F_{N,x}\hat{i} + F_{N,y}\hat{j} + F_{N,z}\hat{k}$$

$$\vec{F}_R = (F_{1x} + F_{2x} + F_{3x} + \cdots + F_{N,x})\hat{i} + (F_{1y} + F_{2y} + F_{3y} + \cdots + F_{N,y})\hat{j} + (F_{1z} + F_{2z} + F_{3z} + \cdots + F_{N,z})\hat{k}$$

$$\vec{F}_R = \left(\sum_{l=1}^{N} F_{l,x}\right)\hat{i} + \left(\sum_{l=1}^{N} F_{l,y}\right)\hat{j} + \left(\sum_{l=1}^{N} F_{l,z}\right)\hat{k} \tag{2-9}$$

Equation 2-9 states that the sum of all the x-components of the original vectors equals the x-component of the resultant vector, the sum of all the y-components of the original vectors equals the y-component of the resultant vector, and the sum of all the z-components of the original vectors equals the z-component of the resultant vector. If we compare Equation 2-9 with Equation 2-4, which we obtained earlier for algebraic vector addition, we can see that they are equivalent: both state that the components of the resultant vector are the sums of the relevant components of the original vectors. Example 2-4 illustrates the use of Cartesian notation.

When dealing with equations containing unit vectors, I always remember that after I simplify the equation, its left and right hand sides must have the same coefficients in front of the corresponding unit vectors. For example, in the equation $3\hat{i} + 4\hat{j} + 5\hat{k} = a\hat{i} + b\hat{j} + c\hat{k}$, the following *must* be true: $a = 3, b = 4, c = 5$.

CHECKPOINT 2-5

Identifying Unit Vectors

Which of the following vectors is a unit vector? Explain.

(a) $\dfrac{3}{2\sqrt{2}}\hat{i} + \dfrac{3}{2\sqrt{2}}\hat{j} - \dfrac{2}{2\sqrt{2}}\hat{k}$

(b) $3\hat{i} + 3\hat{j} - 5\hat{k}$

(c) $\dfrac{3}{\sqrt{14}}\hat{i} + \dfrac{3}{\sqrt{14}}\hat{j} - \dfrac{2}{\sqrt{14}}\hat{k}$

(d) $\dfrac{3}{\sqrt{22}}\hat{i} + \dfrac{3}{\sqrt{22}}\hat{j} - \dfrac{2}{\sqrt{22}}\hat{k}$

(e) $\dfrac{4}{\sqrt{12}}\hat{i} - \dfrac{4}{\sqrt{12}}\hat{j} - \dfrac{4}{\sqrt{12}}\hat{k}$

ANSWER: (d) The unit vector must have a magnitude of 1. Use Equation 2-4 to check that the only vector that has a magnitude of 1 is (d).

EXAMPLE 2-4

Finding a Unit Vector in the Direction of a Given Force

A force vector is described as $\vec{F} = (3\hat{i} + 4\hat{j} - 5\hat{k})$ N. Express the unit vector in the $\vec{F}$ direction using Cartesian notation $(\hat{i}, \hat{j}, \hat{k})$.

Solution

By definition, a unit vector must have a magnitude of 1. Therefore, if we divide the force vector $\vec{F}$ by its magnitude (which is a scalar), the resultant vector will have a magnitude of 1 and the same direction as force $\vec{F}$. To find the magnitude of the force vector, we use the Pythagorean theorem:

$$F = \sqrt{F_x^2 + F_y^2 + F_z^2} = \sqrt{(3\text{ N})^2 + (4\text{ N})^2 + (-5\text{ N})^2}$$
$$= \sqrt{50}\text{ N} = 5\sqrt{2}\text{ N}$$

Thus, the unit vector in the direction of $\vec{F}$ is

$$\hat{u}_F = \frac{\vec{F}}{F} = \frac{(3\hat{i} + 4\hat{j} - 5\hat{k})\text{ N}}{5\sqrt{2}\text{ N}} = \frac{3}{5\sqrt{2}}\hat{i} + \frac{4}{5\sqrt{2}}\hat{j} - \frac{1}{\sqrt{2}}\hat{k}$$

Making sense of the result

Since the units cancel out, $\hat{u}_F$ is dimensionless, as required for a unit vector, and its magnitude equals 1.

The method used in Example 2-4 can be applied to any vector:

KEY EQUATION $\quad \hat{u}_F = \dfrac{\vec{F}}{F} = \dfrac{F_x}{F}\hat{i} + \dfrac{F_y}{F}\hat{j} + \dfrac{F_z}{F}\hat{k}$

which is equivalent to

$$\vec{F} = F\hat{u}_F \tag{2-10}$$

where $\hat{u}_F$ is a dimensionless unit vector in the direction of vector $\vec{F}$ and F is the magnitude of vector $\vec{F}$.

CHECKPOINT 2-6

Adding Vectors Using Cartesian Notation

Which of the following vectors represents the sum of the two vectors $\vec{F}_1 = 3\hat{i} + 3\hat{j} - 5\hat{k}$ and $\vec{F}_1 = -3\hat{i} + 6\hat{j} - 5\hat{k}$? Explain.

(a) $\vec{F}_R = 9\hat{j} - 10\hat{k}$ (d) $\vec{F}_R = -10\hat{k}$

(b) $\vec{F}_R = 6\hat{i} + 6\hat{j} - 10\hat{k}$ (e) $\vec{F}_R = -3\hat{i} - 10\hat{k}$

(c) $\vec{F}_R = 6\hat{i} - 3\hat{j}$

ANSWER: (a) This follows from the vector addition rule, Equation 2-9.

We can specify the direction of a 3-D vector by giving the angles the vector makes with each of the coordinate axes. These angles are called **coordinate direction angles** and are denoted α, β, and γ, as shown in Figure 2-13. In Example 2-5, we will see that these three angles have a special relationship.

Position vector A special vector widely used in physics is the **position vector**, $\vec{r}$ (Figure 2-14). A position vector describes the location of a point, such as point A, in space relative to some fixed point (often, the origin). The magnitude of the position vector corresponds to the distance, as measured directly between the origin and point A. The position vector of point A (x_A, y_A, z_A) can be expressed as

$$\vec{r}_A = (x_A - 0)\hat{i} + (y_A - 0)\hat{j} + (z_A - 0)\hat{k}$$
$$= x_A\hat{i} + y_A\hat{j} + z_A\hat{k} \tag{2-11}$$

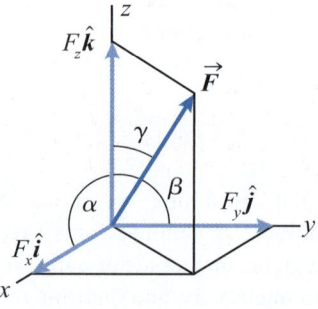

Figure 2-13 Coordinate direction angles, α, β, γ, describing the orientation of vector $\vec{F}$ relative to coordinate axes x, y, and z.

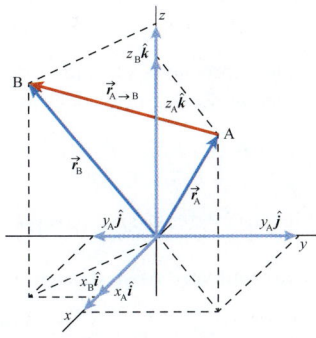

Figure 2-14 Position vectors $\vec{r}_A$ and $\vec{r}_B$ represent the locations of points A and B relative to the origin. Their magnitudes correspond to the distances, as measured directly between the origin and points A and B, respectively. Displacement vector $\vec{r}_{A \to B}$ represents the length and direction of the straight path from point A to point B.

Sometimes the position of an object is described relative to an arbitrary point, not necessarily the origin. For example, if we want to describe the position of point $B(x_B, y_B, z_B)$ as compared to point $A(x_A, y_A, z_A)$, we call this a **displacement vector** because it refers to how much an object was displaced while moving from point A to point B. Imagine a straight line connecting points A and B. Then the magnitude of the displacement vector $\vec{r}_{A \to B}$ represents the length of this path, and the direction of the displacement vector is aligned along this straight path, pointing from point A toward point B:

$$\vec{r}_{A \to B} = (x_B - x_A)\hat{i} + (y_B - y_A)\hat{j} + (z_B - z_A)\hat{k} \quad \text{(2-12)}$$

The magnitude of the displacement vector $\vec{r}_{A \to B}$ corresponds to the distance as measured directly between points A and B. However, since an object can travel from point A to point B along a curved path, we also define a concept called **distance** (or distanced travelled), denoted as d. Distance denotes how much ground an object has covered during its motion. Distance, unlike displacement, is a scalar physical quantity. For example, when you travel on a month-long cross-Canada road trip and return home, the odometer of your car will show the distance you travelled (for example, 15000 km), but since you have returned to the same location where you started, your displacement will be zero.

MAKING CONNECTIONS

Longitude and Latitude on Earth

While we are discussing coordinate systems, you might realize that you are already very familiar with a number of them. For example, a common orthogonal street grid we find in many North American cities helps us agree easily on a meeting location, such as at the intersection of Main and Broadway. The global coordinate system we use on Earth to pinpoint our location is **latitude** and **longitude**. While Earth is a curved surface, if we consider it as a 2-D object (let us forget for the moment about the altitude above its surface), we need two numbers to describe where we are located on it. These two numbers are our coordinates on Earth—latitude and longitude (Figure 2-15). Since Earth is a spherical surface, the gridlines on its surface are not straight in a sense of plain geometry. The horizontal latitude lines run parallel to the equator, so they are called **parallels**. (Most Canadians are familiar with the phrase "north of the 49th parallel," but few of us would know that Paris is located only a few kilometres south of the 49th parallel). The vertical lines—the **meridians**—can be visualized as big imaginary circles going through the poles.

It is interesting that one can find the latitude of a place from astronomical observations, such as the altitude of the Sun at noon or the position of the known stars. On the other hand, to find the longitude, one must measure time very accurately. Finding both latitude and longitude is an extremely important task for safe ocean navigation and for map making (cartography). However, until the 1730s, people didn't know how to measure time accurately, so finding an accurate value for longitude was practically impossible. Thus, it is not surprising that Columbus accidentally landed on the shores of America (notice that he never set foot in North America though), thinking he was in India! Finding an accurate and reliable way to measure longitude took humans centuries and involved great minds such as Galileo Galilei and Isaac Newton. However, it was the British watchmaker John Harrison (1693–1776) who developed the first accurate chronometer to allow explorers to measure time and determine their longitude and thus their exact location.

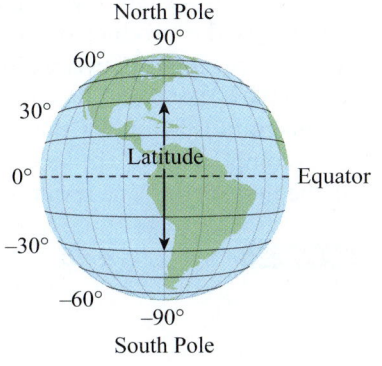
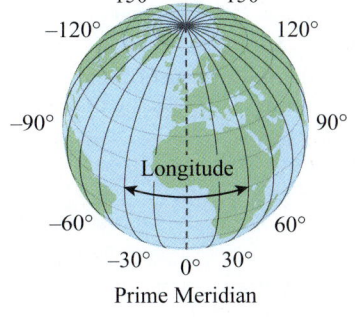

Figure 2-15 Earth's longitude and latitude.

EXAMPLE 2-5

Relationships between the Coordinate Direction Angles

Prove that the coordinate direction angles for any 3-D vector $\vec{F}$ satisfy the equation

$$\cos^2\alpha + \cos^2\beta + \cos^2\gamma = 1 \qquad (2\text{-}13)$$

Solution

From Figure 2-16 we see that

$$\cos\alpha = \frac{F_x}{F}; \cos\beta = \frac{F_y}{F}; \cos\gamma = \frac{F_z}{F} \qquad (2\text{-}14)$$

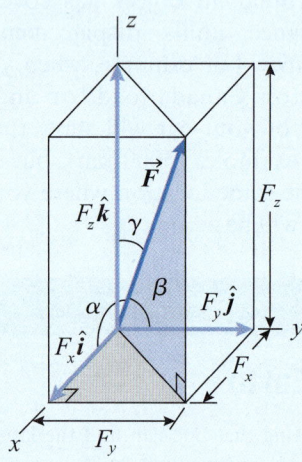

Figure 2-16 Vector $\vec{F}$ expressed using Cartesian vector notation.

Substituting the expressions in Equation 2-14 into the equation for a unit vector, Equation 2-10, gives

$$\hat{u}_F = \frac{\vec{F}}{F} = \frac{F_x}{F}\hat{i} + \frac{F_y}{F}\hat{j} + \frac{F_z}{F}\hat{k}$$
$$= (\cos\alpha)\hat{i} + (\cos\beta)\hat{j} + (\cos\gamma)\hat{k} \qquad (2\text{-}15)$$

The magnitude of the unit vector is 1. Therefore,

$$(\cos\alpha)^2 + (\cos\beta)^2 + (\cos\gamma)^2 = 1 \Rightarrow \cos^2\alpha + \cos^2\beta + \cos^2\gamma$$

Making sense of the result

This result means that the three coordinate direction angles are related. In fact, given two of the coordinate direction angles, we can find the absolute value of the cosine of the third angle, which limits the value of the unknown angle to two possible values (as coordinate direction angles must be between 0° and 180°). This result also applies to 2-D vectors. In the case of 2-D vectors, $\gamma = 90°$ and $\cos\gamma = 0$, so $\cos^2\alpha + \cos^2\beta = 1$. For 2-D vectors, the angles α and β are complementary angles; therefore, $\alpha = 90° - \beta$. Consequently,

$$(\cos\alpha)^2 + (\cos\beta)^2 + (\cos\gamma)^2 = 1 \Rightarrow \cos^2\alpha + \cos^2\beta + \cos^2\gamma$$

$$\cos^2\alpha + \cos^2\beta = \cos^2\alpha + \cos^2(90° - \alpha)$$
$$= \cos^2\alpha + \sin^2\alpha = 1 \qquad (2\text{-}16)$$

This is a well-known result. Equation 2-13 applies to both 2-D and 3-D vectors.

2-4 The Dot Product of Two Vectors

Many physics and engineering applications involve the projection of a vector onto a given axis. Although this projection can be done geometrically, it is often more practical to use an algebraic (component) approach. The **dot, or scalar, product** can be used to find a projection of a vector onto any given axis. Since the dot product is a scalar quantity, it is often called a scalar product.

The dot (scalar) product of vectors $\vec{A}$ and $\vec{B}$ is defined as the product of their magnitudes and the cosine of the angle between the vectors. In other words, to find the dot or scalar product of two vectors, multiply the magnitude of one of the two vectors by the projection (component) of the second vector in the direction of the first vector. The result is a scalar quantity:

KEY EQUATION $$\vec{A} \cdot \vec{B} = AB\cos\theta \qquad (2\text{-}17)$$

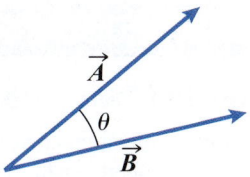

Figure 2-17 The dot (scalar) product of vectors $\vec{A}$ and $\vec{B}$ is the product of their magnitudes and the cosine of the angle between them.

and θ is the angle between the vectors, as shown in Figure 2-17. The product $\vec{A} \cdot \vec{B}$ is commonly read as "A dot B."

From the definition above, we can see that a dot product has the following properties:

- The unit of a dot product is the product of the units of the vectors being multiplied.

- A dot product of two perpendicular vectors equals zero because $\cos 90° = 0$.

- A dot product of two vectors of given magnitudes is greatest when the vectors are parallel to each other because $\cos 0° = 1$. The dot product is smallest when the vectors are antiparallel to each other because $\cos 180° = -1$.

- A dot product of an arbitrary vector $\vec{A}$ and $\hat{u}_B$, a unit vector in the direction of vector $\vec{B}$, represents the projection of vector $\vec{A}$ onto the direction of vector $\vec{B}$.

- A dot product is commutative, that is,

$$\vec{A} \cdot \vec{B} = \vec{B} \cdot \vec{A} = AB\cos\theta \qquad (2\text{-}18)$$

- Multiplication of a dot product by a scalar can be expressed as

$$a(\vec{A} \cdot \vec{B}) = (a\vec{A}) \cdot \vec{B} = \vec{A} \cdot (a\vec{B}) = (\vec{A} \cdot \vec{B})a \\ = (AB\cos\theta)a = aAB\cos\theta \qquad (2\text{-}19)$$

- A dot product is distributive:

$$\vec{A} \cdot (\vec{B} + \vec{C}) = \vec{A} \cdot \vec{B} + \vec{A} \cdot \vec{C} \\ = AB\cos\theta_{A\text{-}B} + AC\cos\theta_{A\text{-}C} \qquad (2\text{-}20)$$

where $\theta_{A\text{-}B}$ and $\theta_{A\text{-}C}$ are the angles between vectors $\vec{A}$ and $\vec{B}$, and $\vec{A}$ and $\vec{C}$, respectively.

A dot product has two important applications: it can be used to find a vector component or a projection of a vector onto a given direction, and it can be used to find the angle between any two given vectors.

The Dot Product and Unit Vectors

The dot products of the various combinations of unit vectors are as follows:

KEY EQUATION

$$\hat{i} \cdot \hat{i} = 1 \cdot 1\cos 0° = 1; \qquad \hat{i} \cdot \hat{j} = 1 \cdot 1\cos 90° = 0;$$
$$\hat{i} \cdot \hat{k} = 1 \cdot 1\cos 90° = 0;$$
$$\hat{j} \cdot \hat{j} = 1 \cdot 1\cos 0° = 1; \qquad \hat{j} \cdot \hat{i} = 1 \cdot 1\cos 90° = 0;$$
$$\hat{j} \cdot \hat{k} = 1 \cdot 1\cos 90° = 0; \qquad\qquad\qquad (2\text{-}21)$$
$$\hat{k} \cdot \hat{k} = 1 \cdot 1\cos 0° = 1; \qquad \hat{k} \cdot \hat{i} = 1 \cdot 1\cos 90° = 0;$$
$$\hat{k} \cdot \hat{j} = 1 \cdot 1\cos 90° = 0$$

These dot products can be written in a more compact form, as a dot (scalar) multiplication table:

KEY EQUATION

$\cdot$	$\hat{i}$	$\hat{j}$	$\hat{k}$
$\hat{i}$	1	0	0
$\hat{j}$	0	1	0
$\hat{k}$	0	0	1

$$(2\text{-}22)$$

Equations 2-21 and 2-22 show that the dot product of a unit vector with itself equals 1, while the dot product of two perpendicular unit vectors equals 0.

The dot product of two unit vectors can be found at the intersection of the row for the first unit vector and the column for the second unit vector. For example, the dot product of $\hat{i}$ and $\hat{i}$ is 1. The dot products of $\hat{i}$ and $\hat{j}$ and of $\hat{i}$ and $\hat{k}$ are 0. Since the dot product is commutative, the table is symmetric about its main diagonal.

Now we are ready to define the dot product for any two arbitrary vectors using Cartesian notation:

$$\vec{A} \cdot \vec{B} = (A_x\hat{i} + A_y\hat{j} + A_z\hat{k}) \cdot (B_x\hat{i} + B_y\hat{j} + B_z\hat{k}) \\ = A_xB_x(\hat{i} \cdot \hat{i}) + A_xB_y(\hat{i} \cdot \hat{j}) + A_xB_z(\hat{i} \cdot \hat{k}) \\ + A_yB_x(\hat{j} \cdot \hat{i}) + A_yB_y(\hat{j} \cdot \hat{j}) + A_yB_z(\hat{j} \cdot \hat{k}) \quad (2\text{-}23) \\ + A_zB_x(\hat{k} \cdot \hat{i}) + A_zB_y(\hat{k} \cdot \hat{j}) + A_zB_z(\hat{k} \cdot \hat{k}) \\ = A_xB_x + A_yB_y + A_zB_z$$

PEER TO PEER

I have to remind myself that the dot product of two vectors is just a number (a scalar); therefore, it does *not* have components. However, the dot product has a unit, and the unit is the product of the units of the two vectors. I find it helpful to call the dot product "the scalar product."

Therefore, the dot product of two vectors can be defined in terms of their components:

KEY EQUATION

$$\vec{A} \cdot \vec{B} = A_x B_x + A_y B_y + A_z B_z \qquad (2\text{-}24)$$

Since the magnitude of a unit vector is 1, the dot product of vector $\vec{A}$ and a unit vector gives the component of $\vec{A}$ along an axis in the direction of the unit vector:

KEY EQUATION

$$\vec{A} \cdot \hat{u}_l = A \cdot 1 \cdot \cos\theta = \underbrace{A\cos\theta}_{A_l} \cdot 1 = A_l \qquad (2\text{-}25)$$

where $\hat{u}_l$ is a unit vector in an arbitrary direction l, and A_l is the projection of vector $\vec{A}$ in the direction l.

EXAMPLE 2-6

The Dot Product and Simple Coordinate Transformations

Two forces, $\vec{F}_1$ and $\vec{F}_2$, are described as $\vec{F}_1 = (3.00\hat{i} - 2.00\hat{j})$ N and $\vec{F}_2 = (-1.00\hat{i} + 3.00\hat{j})$ N.

(a) Find the angle between the forces.
(b) Verify that the x- and y-components of the forces, F_{1x}, F_{1y}, F_{2x}, and F_{2y}, are their projections onto the x- and y-axes, respectively.
(c) Find the components of forces $\vec{F}_1$ and $\vec{F}_2$ in the orthogonal $x'y'$-coordinate system that is rotated by 30° counterclockwise about the origin, as shown in Figure 2-18(a).
(d) Calculate the magnitudes of the forces using their components in both coordinate systems, and compare the results.
(e) Suppose that we move the xy-coordinate system in a plane such that the new coordinate axes remain parallel to the original axes, but the origin O moves to a new location (Figure 2-18(b)). Find the components and the magnitudes of forces $\vec{F}_1$ and $\vec{F}_2$ in the $x'y'$-coordinate system.

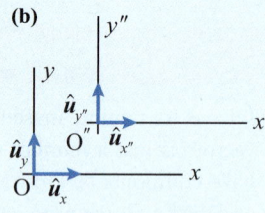

Figure 2-18 (a) The xy-coordinate system undergoes rotation in a plane (by 30°) around its origin, transforming into an $x'y'$-coordinate system. (b) The xy-coordinate system undergoes translation in a plane such that the new coordinate axes (x'', y'') remain parallel to the original coordinate axes (x, y).

Solution

(a) To find the angle between the forces, we use the dot product, described by Equation 2-17:

$$\cos\theta = \frac{\vec{F}_1 \cdot \vec{F}_2}{F_1 F_2} = \frac{F_{1x}F_{2x} + F_{1y}F_{2y}}{\sqrt{F_{1x}^2 + F_{1y}^2}\sqrt{F_{2x}^2 + F_{2y}^2}}$$

$$= \frac{(-3.00 - 6.00) \text{ N}^2}{\sqrt{13.0 \text{ N}^2}\sqrt{10.0 \text{ N}^2}} = \frac{-9.00 \text{ N}^2}{\sqrt{130 \text{ N}^2}} = -0.789$$

$$\theta = \cos^{-1}\left(\frac{-9.00}{\sqrt{130}}\right) = \cos^{-1}(-0.789) = 142°$$

(b) We can use dot products with the $\hat{i}$ and $\hat{j}$ unit vectors to find the projections of the forces onto the x- and y-axes:

$$\begin{cases} F_{1x} = \vec{F}_1 \cdot \hat{i} = [(3.00\hat{i} - 2.00\hat{j}) \cdot \hat{i}] \text{ N} = 3.00 \text{ N} \\ F_{1y} = \vec{F}_1 \cdot \hat{j} = [(3.00\hat{i} - 2.00\hat{j}) \cdot \hat{j}] \text{ N} = -2.00 \text{ N} \end{cases}$$

Dot products for $\vec{F}_2$ indicate that its x- and y-projections are -1.00 N and 3.00 N, respectively:

$$\begin{cases} F_{2x} = \vec{F}_2 \cdot \hat{i} = [(-1.00\hat{i} + 3.00\hat{j}) \cdot \hat{i}] \text{ N} = -1.00 \text{ N} \\ F_{2y} = \vec{F}_2 \cdot \hat{j} = [(-1.00\hat{i} + 3.00\hat{j}) \cdot \hat{j}] \text{ N} = 3.00 \text{ N} \end{cases}$$

(c) To find the projections of forces $\vec{F}_1$ and $\vec{F}_2$ onto the orthogonal x'- and y'-axes, we define unit vectors $\hat{u}_{x'}$ and $\hat{u}_{y'}$ directed along the x'- and y'-axes, as shown in Figure 2-18(a):

$$\hat{u}_{x'} = \cos 30° \, \hat{i} + \sin 30° \, \hat{j} = \frac{\sqrt{3}}{2}\hat{i} + \frac{1}{2}\hat{j}$$

$$\hat{u}_{y'} = -\sin 30° \, \hat{i} + \cos 30° \, \hat{j} = -\frac{1}{2}\hat{i} + \frac{\sqrt{3}}{2}\hat{j}$$

Then, we use Equation 2-25 to find the projections of the forces:

$$F_{1x'} = \vec{F}_1 \cdot \hat{u}_{x'} = (3.00\hat{i} - 2.00\hat{j}) \cdot \left(\frac{\sqrt{3}}{2}\hat{i} + \frac{1}{2}\hat{j}\right) \text{N}$$

$$= \left(\frac{3\sqrt{3}}{2} - 1\right) \text{N} = 1.60 \text{ N}$$

$$F_{1y'} = \vec{F}_1 \cdot \hat{u}_{y'} = (3.00\hat{i} - 2.00\hat{j}) \cdot \left(-\frac{1}{2}\hat{i} + \frac{\sqrt{3}}{2}\hat{j}\right) \text{N}$$

$$= \left(-\frac{3}{2} - \sqrt{3}\right) \text{N} = -3.23 \text{ N}$$

$$F_{2x'} = \vec{F}_2 \cdot \hat{u}_{x'} = (-1.00\hat{i} + 3.00\hat{j}) \cdot \left(\frac{\sqrt{3}}{2}\hat{i} + \frac{1}{2}\hat{j}\right) \text{N}$$

$$= \left(-\frac{\sqrt{3}}{2} + \frac{3}{2}\right) \text{N} = 0.634 \text{ N}$$

$$F_{2y'} = \vec{F}_1 \cdot \hat{u}_{y'} = (-1.00\hat{i} + 3.00\hat{j}) \cdot \left(-\frac{1}{2}\hat{i} + \frac{\sqrt{3}}{2}\hat{j}\right) \text{N}$$

$$= \left(\frac{1}{2} + \frac{3\sqrt{3}}{2}\right) \text{N} = 3.10 \text{ N}$$

(d) $F_1 = \sqrt{(F_{1x})^2 + (F_{1y})^2} = \sqrt{13.0} \text{ N} = 3.61 \text{ N}$ } the xy-
$F_2 = \sqrt{(F_{2x})^2 + (F_{2y})^2} = \sqrt{10.0} \text{ N} = 3.16 \text{ N}$ coordinate system

$F_1 = \sqrt{(F_{1x'})^2 + (F_{1y'})^2} = \sqrt{1.60^2 + 3.23^2} \text{ N} = 3.61 \text{ N}$ }
$F_2 = \sqrt{(F_{2x'})^2 + (F_{2y'})^2} = \sqrt{0.634^2 + 3.10^2} \text{ N} = 3.16 \text{ N}$

the $x'y'$-coordinate system

(e) When a coordinate system is moved parallel to itself—that is, the coordinate system undergoes pure translation—such that its origin is moved to a different location but the directions of its axes remain unchanged, the components of a given vector in the new coordinate system ($x''y''$) will be the same as its components in the old coordinate system (xy). This is because the magnitude of a vector, as well as the angles between the vectors and the coordinate axes, will not change under this coordinate system transformation (Figure 2-18(b)). Therefore, applying Equation 2-1 we find

$$\begin{cases} F_{1x} = F_{1x''} = 3.00 \text{ N} \\ F_{1y} = F_{1y''} = -2.00 \text{ N} \end{cases} \text{ and } \begin{cases} F_{2x} = F_{2x''} = -1.00 \text{ N} \\ F_{2y} = F_{2y''} = 3.00 \text{ N} \end{cases}$$

Notice that the position vector (Equation 2-11) does depend on the position of the origin, so when the coordinate system undergoes a translation, the components of the position vector do change. However, as we mentioned earlier, the position vector $\vec{r}$ does not appear in any physical equations. It always appears in the form of the displacement vector, $\Delta\vec{r}$, which does not change under a translation of the coordinate system.

Making sense of the result

Coordinate systems can undergo transformations, such as translation or rotation of the coordinate axes. When a coordinate system undergoes pure translation, the components of the vectors (except for the position vector) remain unchanged. When the coordinate system is rotated about the origin (the coordinate system undergoes a rotational transformation), the components of the vectors change, but the magnitudes of the vectors do not.

2-5 The Cross Product of Vectors

A number of physics and engineering applications involve vectors that are perpendicular to two given vectors. As you will see in later chapters, the angular momentum vector is perpendicular to both the position vector and the velocity vector of an object, and the vector for the magnetic force acting on a charged particle is perpendicular to both the magnetic field vector and the velocity vector of the particle. The **cross, or vector, product** of two vectors defines a vector that is perpendicular to the two given vectors.

The magnitude of the cross product, $\vec{C}$, of vectors $\vec{A}$ and $\vec{B}$ is defined as the product of the magnitudes of vectors $\vec{A}$ and $\vec{B}$ and the sine of the angle θ between them (Figure 2-19(a)). The direction of the cross product is determined using the **right-hand rule**: when you curl the fingers of your right hand from vector $\vec{A}$ to vector $\vec{B}$, your thumb will point in the same direction as the cross product, $\vec{C}$ (Figure 2-19(b)).

KEY EQUATION

$$\vec{A} \times \vec{B} = \vec{C}$$
$$C = AB\sin\theta$$

(2-26)

A cross product has the following properties:

■ The cross product is a vector.

■ The unit of a cross product is the product of the units of the vectors being multiplied.

MECHANICS

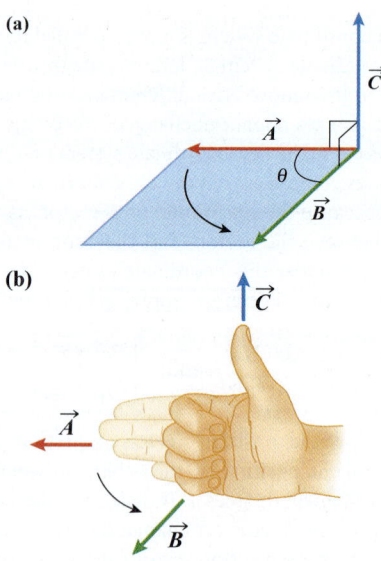

(a)

(b)

Figure 2-19 (a) Visual representation of the cross product of two vectors, $\vec{A}$ and $\vec{B}$; the area of the parallelogram represents the magnitude of the cross product. (b) The right-hand rule for determining the direction of the cross product.

■ The magnitude of the cross product corresponds to the area of a parallelogram that has vectors $\vec{A}$ and $\vec{B}$ as its sides.

■ The cross product of two parallel or two antiparallel vectors equals zero because $\sin 0° = \sin 180° = 0$. The cross product of two vectors of a given magnitude is greatest when the vectors are perpendicular to each other because $\sin 90° = 1$.

■ The cross product is anticommutative:

$$\vec{A} \times \vec{B} = -\vec{B} \times \vec{A} \qquad (2\text{-}27)$$

You can verify this property by applying the right-hand rule to the vector products in the equation above.

■ The cross product can be multiplied by a scalar. From the definition of the cross product, we can see that

$$a(\vec{A} \times \vec{B}) = (a\vec{A}) \times \vec{B} = \vec{A} \times (a\vec{B}) = (\vec{A} \times \vec{B})a \Rightarrow$$
$$|a(\vec{A} \times \vec{B})| = aAB\sin\theta \qquad (2\text{-}28)$$

■ The cross product operation is distributive:

$$\vec{A} \times (\vec{B} + \vec{C}) = (\vec{A} \times B) + (\vec{A} \times \vec{C})$$
$$= \vec{A} \times B + \vec{A} \times \vec{C} \qquad (2\text{-}29)$$

The Cross Product and Unit Vectors

Consider the vector products of the unit vectors $\hat{i}$, $\hat{j}$, and $\hat{k}$. Since the unit vectors are orthogonal and have a magnitude of 1,

$$\begin{aligned}
\hat{i} \times \hat{i} &= 0; & \hat{i} \times \hat{j} &= \hat{k}; & \hat{i} \times \hat{k} &= -\hat{j}; \\
\hat{j} \times \hat{j} &= 0; & \hat{j} \times \hat{i} &= -\hat{k}; & \hat{j} \times \hat{k} &= \hat{i}; \\
\hat{k} \times \hat{k} &= 0; & \hat{k} \times \hat{i} &= \hat{j}; & \hat{k} \times \hat{j} &= -\hat{i}
\end{aligned} \qquad (2\text{-}30)$$

As with the dot products in Section 2-4, we can represent cross products as a cross product multiplication table:

KEY EQUATION

$\times$	$\hat{i}$	$\hat{j}$	$\hat{k}$
$\hat{i}$	0	$\hat{k}$	$-\hat{j}$
$\hat{j}$	$-\hat{k}$	0	$\hat{i}$
$\hat{k}$	$\hat{j}$	$-\hat{i}$	0

$$(2\text{-}31)$$

The cross product of two unit vectors is located at the intersection of the row for the first vector and the column for the second vector. Since the cross product is anticommutative, the order in which vectors are multiplied changes the result. Consequently, unlike a dot product multiplication table (see Equation 2-22), this table is *not* symmetrical about its main diagonal.

You can use a **cross-multiplication circle** to remember the signs of the cross products of the unit vectors. Arrange the unit vectors in a counterclockwise order around a circle, as shown in Figure 2-20. To find the cross product of two unit vectors, locate the first vector and then move along the circle to the second vector of the product. If you move counterclockwise (in the direction of the arrows), the result is the third vector along the circle. If you move clockwise, the result is the negative (or opposite) of the third vector. For example, to find the cross product $\hat{i} \times \hat{j}$, you move counterclockwise along the circle, so the result is $\hat{k}$. However, to find $\hat{j} \times \hat{i}$, you move clockwise, so the result is $-\hat{k}$.

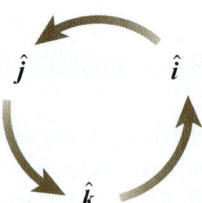

Figure 2-20 The unit cross-multiplication circle.

Similarly to what we did earlier with a dot product of two vectors, we can define a cross product of two vectors in terms of their components and unit vectors:

$$\vec{A} \times \vec{B} = (A_x\hat{i} + A_y\hat{j} + A_z\hat{k}) \times (B_x\hat{i} + B_y\hat{j} + B_z\hat{k})$$
$$= A_xB_x(\hat{i} \times \hat{i}) + A_xB_y(\hat{i} \times \hat{j}) + A_xB_z(\hat{i} \times \hat{k})$$
$$+ A_yB_x(\hat{j} \times \hat{i}) + A_yB_y(\hat{j} \times \hat{j}) + A_yB_z(\hat{j} \times \hat{k})$$
$$+ A_zB_x(\hat{k} \times \hat{i}) + A_zB_y(\hat{k} \times \hat{j}) + A_zB_z(\hat{k} \times \hat{k})$$
$$= A_xB_y(\hat{k}) + A_xB_z(-\hat{j}) + A_yB_x(-\hat{k}) + A_yB_z(\hat{i})$$
$$+ A_zB_x(\hat{j}) + A_zB_y(-\hat{i})$$
$$= (A_yB_z - A_zB_y)\hat{i} + (A_zB_x - A_xB_z)\hat{j}$$
$$+ (A_xB_y - A_yB_x)\hat{k}$$

(2-32)

Therefore, using Cartesian notation, the cross product of two vectors can be expressed as follows:

KEY EQUATION
$$\vec{A} \times \vec{B} = (A_yB_z - A_zB_y)\hat{i} + (A_zB_x - A_xB_z)\hat{j}$$
$$+ (A_xB_y - A_yB_x)\hat{k}$$
(2-33)

EXAMPLE 2-7

Finding a Unit Vector Perpendicular to Two Given Vectors

Two forces are described as $\vec{F}_1 = (\hat{i} + 2\hat{j} - 3\hat{k})$ N and $\vec{F}_2 = (-2\hat{i} + 3\hat{j})$ N. Find an expression for a unit vector directed perpendicular to both of them.

Solution

By definition, the cross product of any two vectors is a vector that is perpendicular to both vectors. Therefore, the unit vector we are looking for is directed along the line of the cross product of the two given vectors. Let us first find this cross product:

$$\vec{C} = \vec{F}_1 \times \vec{F}_2 = [(\hat{i} + 2\hat{j} - 3\hat{k}) \text{ N}] \times [(-2\hat{i} + 3\hat{j}) \text{ N}]$$
$$= [3\hat{k} + 4\hat{k} + 6\hat{j} + 9\hat{i}] \text{ N}^2 = [9\hat{i} + 6\hat{j} + 7\hat{k}] \text{ N}^2$$

Now, we find a unit vector along the direction of vector $\vec{C}$:

In linear algebra, the determinant of a matrix is a useful mathematical object that can be computed from the elements of a square matrix (a matrix that has the same number of columns as rows). The expression described by Equation 2-33 can also be represented as a determinant of the 3-3 matrix (Equation 2-34). The **determinant form of a vector product** is an elegant way of representing Equation 2-33:

KEY EQUATION
$$\vec{A} \times \vec{B} = \begin{vmatrix} \hat{i} & \hat{j} & \hat{k} \\ A_x & A_y & A_z \\ B_x & B_y & B_z \end{vmatrix}$$
(2-34)

The first row of the matrix lists the unit vectors, the second row lists the components of the first vector to be multiplied, and the bottom row lists the components of the second vector. Expand the determinant in Equation 2-34 about its first row and you will reproduce Equation 2-33.

$$\hat{u}_C = \frac{\vec{C}}{C} = \frac{[9\hat{i} + 6\hat{j} + 7\hat{k}] \text{ N}^2}{\sqrt{81 + 36 + 49} \text{ N}^2} = \frac{1}{\sqrt{166}}(9\hat{i} + 6\hat{j} + 7\hat{k})$$

Making sense of the result

We can check geometrically that our answer is correct. We can also use scalar products to confirm that the unit vector is perpendicular to the given force vectors:

$$\hat{u}_C \cdot \vec{F}_1 = \frac{9\hat{i} + 6\hat{j} + 7\hat{k}}{\sqrt{166}} \cdot (\hat{i} + 2\hat{j} - 3\hat{k}) \text{ N} = \frac{9 + 12 - 21}{\sqrt{166}} \text{ N} = 0$$

$$\hat{u}_C \cdot \vec{F}_2 = \frac{9\hat{i} + 6\hat{j} + 7\hat{k}}{\sqrt{166}} \cdot (-2\hat{i} + 3\hat{j}) \text{ N} = \frac{-18 + 18}{\sqrt{166}} \text{ N} = 0$$

Since the dot products are zero, the unit vector is perpendicular to both force vectors. Note that the vector opposite to $\hat{u}$ is also perpendicular to both force vectors.

KEY CONCEPTS AND RELATIONSHIPS

Physical quantities used in this book can be categorized as scalars and vectors. Scalars require only a number (either positive or negative) and a unit for their description. Vectors require a number, a unit, and a direction. Vector operations include addition and subtraction, multiplication of a vector by a scalar, and dot (scalar) and cross (vector) products of two or more vectors.

Vector Components

Vector components are scalar quantities that represent vector projections onto the coordinate axes. In the case of 2-D vectors:

$$\begin{cases} v_x = v\cos\theta \\ v_y = v\sin\theta \end{cases} \Rightarrow \begin{cases} v = \sqrt{v_x^2 + v_y^2} \\ \tan\theta = \dfrac{v_y}{v_x} \end{cases} \tag{2-1}$$

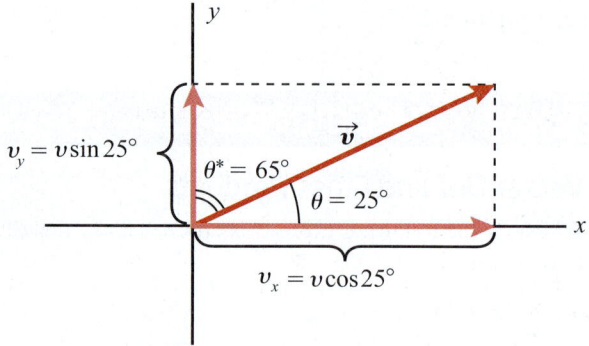

Cartesian Notation

$\vec{F} = F_x\hat{i} + F_y\hat{j} + F_z\hat{k}$, where $\hat{i}$, $\hat{j}$, and $\hat{k}$ are unit vectors (2-8)

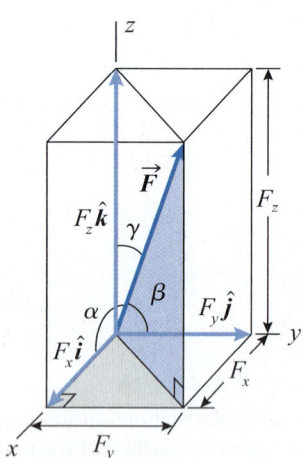

Vector Magnitude and Simple Coordinate System Transformations

When a coordinate system undergoes either pure translation or rotation, the magnitude of a vector in the transformed coordinate system remains unchanged. The vector components in the translated coordinate system do not change either. However, the vector components in a rotated coordinate system will be different from the vector components in the original coordinate system.

Vector Addition and Subtraction

Vector Addition: Vectors can be added using either a geometric or an algebraic approach.

Geometric Approach: Use the triangle construction or parallelogram rule.

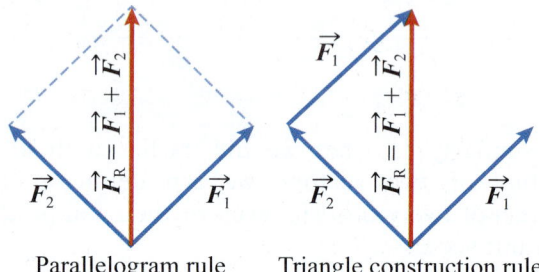

Parallelogram rule   Triangle construction rule

Vector Subtraction:

$$\vec{F}_1 - \vec{F}_2 = \vec{F}_1 + (-\vec{F}_2) \tag{2-7}$$

Algebraic Approach:

$$\begin{cases} F_{R,x} = F_{1x} + F_{2x} + \cdots + F_{N,x} \\ F_{R,y} = F_{1y} + F_{2y} + \cdots + F_{N,y} \\ F_R = \sqrt{(F_{R,x})^2 + (F_{R,y})^2} \\ \tan\theta = \dfrac{F_{R,y}}{F_{R,x}} \end{cases} \tag{2-3}$$

Vector Multiplication by a Scalar:

$$|a\vec{F}| = |a||\vec{F}|$$

$$\begin{cases} \text{if } a > 0, a\vec{F} \text{ has the same direction as } \vec{F} \\ \text{if } a < 0, a\vec{F} \text{ has the opposite direction to } \vec{F} \\ \text{if } a = 0, a\vec{F} \text{ has a magnitude of zero and no direction} \end{cases} \tag{2-5}$$

Products of Two Vectors

Scalar (Dot) Product:

$$\vec{A} \cdot \vec{B} = AB\cos\theta \tag{2-17}$$

and

$$\vec{A} \cdot \vec{B} = A_xB_x + A_yB_y + A_zB_z \tag{2-24}$$

Vector (Cross) Product:

$$\vec{A} \times \vec{B} = \vec{C}$$

$C = AB\sin\theta$, and $\vec{C}$ is perpendicular to both $\vec{A}$ and $\vec{B}$ (2-26)

$$\vec{A} \times \vec{B} = (A_yB_z - A_zB_y)\hat{i} + (A_zB_x - A_xB_z)\hat{j} + (A_xB_y - A_yB_x)\hat{k} \tag{2-33}$$

APPLICATIONS

Scalars are used to describe physical quantities that require magnitude and a unit for their description, such as mass, temperature, speed, and pressure. Vectors are used to describe physical quantities that require magnitude, unit, and direction for their description, such as velocity, force, acceleration, and momentum. The dot product is used to determine the projections of vectors onto various directions. The cross product of vectors can be used to define or calculate physical quantities, such as angular momentum, torque, and magnetic force.

KEY TERMS

algebraic method for vector addition; Cartesian vector notation; components; coordinate direction angles; cross-multiplication circle; cross, or vector, product; determinant form of a vector product; displacement vector; distance; dot, or scalar, product; frame of reference; free body diagram (FBD); geometric method for vector addition; head of the vector; head-to-tail rule; latitude; longitude; magnitude; meridians; net force; opposite vectors; orthogonal coordinate system; parallelogram rule; parallels; polar coordinates; polar coordinate system; polar notation; position vector; projections; resultant force; right-hand rule; scalar notation; scalars; tail of the vector; translating a vector; triangle construction rule; unit vectors; vector magnitude; vector resolution into scalar components; vectors

QUESTIONS

1. A given force $\vec{F}$ is located in the horizontal plane. It has a magnitude of 30 N and is directed at a 45.0° angle clockwise from the positive x-axis. Describe $\vec{F}$ using
   (a) polar notation
   (b) its x- and y-components
   (c) Cartesian notation
2. A given force $\vec{F}$ is located in the yz-plane. It has a magnitude of 15 N and is directed at a 30.0° angle clockwise from the positive y-axis. Describe $\vec{F}$ using
   (a) polar notation
   (b) its y- and z-components
   (c) Cartesian notation
3. A force acting on a 5.00 kg object is located in the xy-plane. It has a magnitude of 8.00 N and is directed parallel to the line $y = -3.00x + 4.00$. Express this force in Cartesian notation. How many solutions do you have? Explain why.
4. An object is located at the point (3.00 m, −4.00 m, 0.00 m). A position vector connects this object to the origin. Find the magnitude of the position vector and the angle it makes with the positive x-axis.
5. A displacement vector connects two points, A(−2.00 m, 2.00 m, 0.00 m) to B(3.00 m, 0.00 m, −2.00 m). Find the magnitude and direction angles of the displacement vector.
6. An object travelling in space passes five different points: A(2.00 m, 3.00 m, −1.00 m), B(2.00 m, 6.00 m, −1.00 m), C(2.00 m, 6.00 m, −8.00 m), D(−4.00 m, 6.00 m, −8.00 m), and E(−11.00 m, 6.00 m, −1.00 m). The object returns to its initial location, point A.
   (a) Rank the displacement vectors $(\vec{r}_{AB}, \vec{r}_{BC}, \vec{r}_{CD}, \vec{r}_{DE}, \vec{r}_{EA})$ from the vector that has the largest magnitude to the vector that has the smallest magnitude.
      Largest 1___2___3___4___5___Smallest
   (b) Are the displacements the same in all cases?
   (c) Is the displacement zero in all cases?
   (d) If the displacement is the same for two or more cases, clearly indicate it on the ranking scheme. Explain your reasoning.
7. Three vectors are described as $\vec{A}(1,-3,4)$, $\vec{B}(5,-2,1)$, and $\vec{C}(-2,3,6)$. Find the following vectors:
   (a) $\vec{D} = \vec{A} + \vec{B} + \vec{C}$
   (b) $\vec{E} = -\vec{A} - \vec{B} - \vec{C}$
   (c) $\vec{F} = 2\vec{A} - 4\vec{B} + 3\vec{C}$
   (d) $\vec{G} = 2(\vec{A} - 2\vec{B}) + \vec{C}$

8. Two horizontal forces of magnitude 10 N each are acting on a horizontal force board ring (Figure 2-21). A third force acting on the ring perfectly balances these two forces. What *must be* true about this force?
   (a) The third force must be 10 N, and it must be at an angle of 120° to each one of the forces.
   (b) The magnitude of the third force must be more than 10 N.
   (c) The magnitude of the third force must be less than 10 N.
   (d) The magnitude of the third force must be between −20 N and 0 N.
   (e) The third force must be directed at a 45° angle to the two given forces.

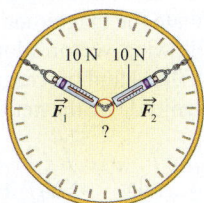

Figure 2-21 Question 8: A bird's-eye view of a force board.

9. Which of the following statement(s) do you agree with? Explain why.
   (a) Vector components must always be smaller than the vector magnitudes.
   (b) To find the vector magnitude, one must add the vector components.
   (c) Vector magnitude does not depend on the choice of the coordinate system, even though vector components do.
   (d) Vector components do not depend on the choice of the coordinate system, but vector magnitude does.
   (e) If two vectors have equal magnitudes, they must either be opposite vectors or have the same direction.

10. Which of the following vector pairs are orthogonal (perpendicular to each other)? Explain.
   (a) (2, 3, −6) and (−2, 3, −6)
   (b) (2, 3, −1) and (−2, −1, −7)
   (c) (−2, −3, −1) and (−2, −1, 7)
   (d) (1, 0, −6) and (−2, 3, 0)
   (e) (2, 3, 0) and (0, −3, 6)

11. Two students are having an argument. Student 1 claims that any two vectors located in orthogonal planes must also be orthogonal, and student 2 disagrees. Who do you agree with and why? (Hint: It is sufficient to have one counterexample to disprove a statement.)

12. Two objects are moving at constant velocities: $\vec{v}_A = (2.000\hat{i} - 5.000\hat{j})$ m/s and $\vec{v}_B = (-2.000\hat{i} + 4.000\hat{j})$ m/s.
    (a) Which object is moving faster?
    (b) Find the angle between objects' trajectories.

13. Two vectors are described as $\vec{A}(2, 2, -2)$ and $\vec{B}(-1, 3, -2)$.
    (a) Find the projections of these vectors on the x-, y-, and z-coordinate axes.
    (b) Find the scalar product of these two vectors.
    (c) Find the vector product of these two vectors.
    (d) What do the scalar and the vector products of these vectors represent?

14. Find the coordinate direction angles α, β, and γ for the position vector $\vec{r} = (2.00\text{ m})\hat{i} - (5.00\text{ m})\hat{j} - (3.00\text{ m})\hat{k}$. Verify that Equation 2-13 holds true.

15. You overhear your classmate uttering the following statement: "Since the magnitude of a unit vector equals 1, the components of a unit vector must also equal 1." Do you agree or disagree with this statement? Explain why or why not.

16. Three forces are acting on an object: $\vec{F}_1$, $\vec{F}_2$, and $\vec{F}_3$:

$$\vec{F}_1 = (3.00\text{ N})\hat{i} + (4.00\text{ N})\hat{j} - (5.00\text{ N})\hat{k}$$
$$\vec{F}_2 = (1.00\text{ N})\hat{i} - (4.00\text{ N})\hat{j} + (2.00\text{ N})\hat{k}$$
$$\vec{F}_3 = (-2.00\text{ N})\hat{i} + (3.00\text{ N})\hat{j} - (3.00\text{ N})\hat{k}$$

Find the expression for the net (resultant) force acting on it, and then calculate the magnitude of the net force.

17. Complete the following vector addition (the forces are all measured in N), and find the magnitude of each force, as well as the magnitude of the net force:

+	$\hat{i}$	$\hat{j}$	$\hat{k}$
$\vec{F}_1$	1.00	−3.00	1.00
$\vec{F}_2$	−2.00	?	3.00
$\vec{F}_3$	?	2.00	0.00
$\vec{F}_R$	0.00	−5.00	?

18. Prove that the area of a parallelogram that has vectors $\vec{A}$ and $\vec{B}$ for two of its sides can be expressed as $|\vec{A} \times \vec{B}|$.

19. Your classmate claims that, to add two vectors using Cartesian notation, you need to add corresponding vector components; therefore, the same can be done using polar notation for vectors. Do you agree with your classmate? Explain why or why not.

20. List the advantages and disadvantages of using polar versus Cartesian vector notation. Illustrate your argument with relevant examples.

21. The head-to-tail rule for adding vectors is as follows: "Start with the first vector in the sum, and then arrange the rest of the vectors so that every vector's tail touches the previous vector's head. The sum can be represented as a vector connecting the tail of the first vector in the sum to the head of the last one." Prove that when you add more than two vectors you can still use the head-to-tail rule.

22. Find the relationship between the coordinate direction angles of any two opposite vectors.

23. Explain why the scalar product is commutative and the vector product is not.

24. If the scalar product of two non-zero vectors equals zero, then the vectors are orthogonal. What does it mean when we say that the vector product of two vectors equals zero? Explain how your result corresponds to the right-hand rule for finding the vector product of two vectors.

25. Force $\vec{F}$ has a magnitude of 5.00 N; its coordinate direction angles are $\alpha = 30.0°$, $\beta = -40.0°$, and $\gamma = -60.0°$. Describe $\vec{F}$ using Cartesian notation.

PROBLEMS BY SECTION

For problems, star ratings will be used (★, ★★, or ★★★), with more stars meaning more challenging problems. The following codes will indicate if $\frac{d}{dx}$ differentiation, $\int$ integration, ▢ numerical approximation, or ∿ graphical analysis will be required to solve the problem.

Section 2-2   Vector Addition: Geometric and Algebraic Approaches

26. ★ The components of three given vectors are as follows:

$$\vec{A}(3.00, 5.00, -2.00); \vec{B}(-1.00, 2.00, 4.00); \text{ and}$$
$$\vec{C}(4.00, -2.00, -1.00)$$

Suppose that $\vec{A} + 2\vec{B} - 3\vec{C} = \vec{D}$.
    (a) Find the coordinates of vector $\vec{D}$.
    (b) Describe vector $\vec{D}$ using Cartesian notation.
    (c) Describe vector $\vec{D}$ using a polar notation: vector magnitude and its coordinate direction angles.
    (d) Find a vector anti-parallel to vector $\vec{D}$ and describe it using Cartesian notation.
    (e) Find a vector anti-parallel to vector $\vec{D}$ and describe it using polar notation.

27. ★ Use the PhET computer simulation "Vector Addition" (http://phet.colorado.edu/simulations/sims.php?sim= Vector_Addition) to illustrate the following addition operations for the three 2-D vectors $\vec{A}(5, 10)$, $\vec{B}(-5, 10)$, and $\vec{C}(10, -5)$. Then, for each case, use algebraic addition, and compare the two methods.
    (a) $\vec{A} + \vec{B} + \vec{C}$
    (b) $\vec{A} + 2\vec{B} + \vec{C}$
    (c) $2\vec{A} - \vec{B} + 3\vec{C}$
    (d) $-3\vec{A} + \vec{B} - 4\vec{C}$

28. ★ Consider that all the forces in Figure 2-22 are located in the xy-, xz-, and yz-planes.
    (a) Find the sum of the four given forces in Figure 2-22: $\vec{F}_R = \vec{F}_A + \vec{F}_B + \vec{F}_C + \vec{F}_D$.
    (b) Express the net force $\vec{F}_R$ using Cartesian notation.
    (c) Express the net force $\vec{F}_R$ using polar notation.
    (d) Find a unit vector in the direction of the net force. Express it using Cartesian notation.

Explain why it might be impractical to add these forces by hand using the geometric method.

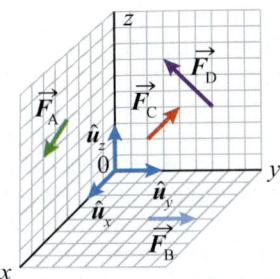

Figure 2-22 Problem 28: All the forces are located in the xy-, xz- and yz-planes. Each one of the units in the figure represents 1 N.

29. ★ Four 2-D displacement vectors describing consecutive displacements of an object moving on a flat surface are shown in Figure 2-23.
    (a) Use the geometric approach to vector addition to find the magnitude and direction of the total displacement of the object.
    (b) Estimate the components of the four displacement vectors as accurately as you can. Then use the algebraic approach to find the total displacement. Compare your results to your results in part (a).

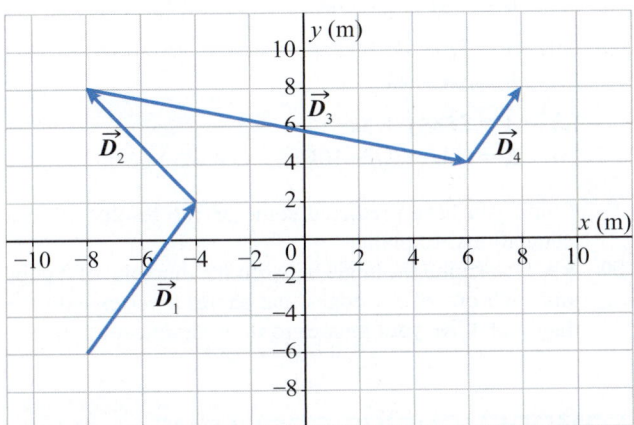

Figure 2-23 Problem 29. Consider all the values to be given to two significant figures.

30. ★ A cat walks 10.0 m in the direction of 20.0° [W of N]. Then it walks 7.00 m [W] and runs 20.0 m in the direction of 30.0° [W of S]. Find the magnitude of the total displacement of the cat using a geometric method, and then check your result using the algebraic approach.

31. ★★ A child is sliding down a steep 60.0° water slide. The magnitude of the gravitational force exerted on the child by Earth equals 300. N, and the normal force exerted on the child by the slide is 150. N.
    (a) Draw a free body diagram (FBD) representing the problem.
    (b) Find the components of the gravitational force along the x- and y-axes directed along the slide (x) and perpendicular to it (y). (Hint: Use Figure 2-9 as a guide.)

(c) Find the components of the normal force along the x- and y-axes.
(d) Find the net force acting on the child. What does your result imply? (Hint: To answer this part you might need to review Newton's second law.)

32. ★★ The magnitudes of the vectors shown in Figure 2-24 are $A = 20.0$ units, $B = 15.0$ units, and $C = 25.0$ units, and the corresponding angles are $\theta_1 = -15.0°$, $\theta_2 = 35.0°$, and $\theta_3 = 125°$.
    (a) Determine the components of each vector.
    (b) Find the sum of the three vectors, and express the sum in both Cartesian and polar notation (magnitude and direction).
    (c) Find the magnitude and direction of $\vec{D} = 2\vec{A} - 3\vec{B} + \vec{C}$.

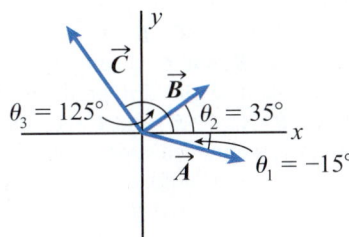

Figure 2-24 Problem 32.

Section 2-3 Cartesian Vector Notation

33. ★★ Find the magnitude and the coordinate direction angles for the following two vectors:
    (a) $\vec{F_1} = (3.00 \text{ N})\hat{i} - (5.00 \text{ N})\hat{j} + (6.00 \text{ N})\hat{k}$
    (b) $\vec{F_2} = (-3.00 \text{ N})\hat{i} + (5.00 \text{ N})\hat{j} - (6.00 \text{ N})\hat{k}$
    (c) Compare your results from (a) and (b). What does your comparison mean?
    (d) Find the sum of the squares of the cosines of the corresponding coordinate direction angles found above. How do you know if your result makes sense?

34. ★★ Find the Cartesian expression of a unit vector directed
    (a) parallel to the force $\vec{F} = (3.00 \text{ N})\hat{i} - (5.00 \text{ N})\hat{j} + (6.00 \text{ N})\hat{k}$
    (b) perpendicular to the force $\vec{F_1} = (3.00 \text{ N})\hat{i} - (5.00 \text{ N})\hat{j} + (6.00 \text{ N})\hat{k}$
    (c) parallel to the plane defined by $3x + 2y - 4z = 0$
    (d) perpendicular to the plane defined by $3x + 2y - 4z = 0$

35. ★★ The average velocity of an object as it is moving from position A to position B is defined as $\vec{v}_{avg} = \dfrac{\vec{r}_B - \vec{r}_A}{\Delta t}$,
    where Δt is the time it takes for the object to move from A to B. The object's positions are described as $\vec{r}_A = (3.00 \text{ m})\hat{i} + (2.00 \text{ m})\hat{j}$ and $\vec{r}_B = (-3.00 \text{ m})\hat{i} + (2.00 \text{ m})\hat{j}$, and it took the object 5.00 s to move from A to B along a straight line.
    (a) Find the object's average velocity.
    (b) Find the object's average speed.
    (c) What is the difference between the object's average speed and its average velocity in this case?

36. ★★ Solve the following vector equations:
    (a) $F_x\hat{i} + 3\hat{j} + \sqrt{2}\hat{i} - F_y\hat{j} + F_z\hat{k} - 5\hat{k} = 0$
    (b) $3\hat{i} - 5\hat{j} + F_x\hat{i} - 2F_y\hat{j} + F_z\hat{k} - 3\hat{k} = 0$

Section 2-4   The Dot Product of Two Vectors

37. ★ Calculate the dot product of vectors $\vec{A}(3,4,-5)$ and $\hat{B}(2,-2,4)$. What does this product represent?

38. ★ Solve the following vector equation to find F_z:
$$(3\hat{i} - 5\hat{j} + F_z\hat{k}) \cdot (4\hat{i} + 4\hat{j} + F_z\hat{k}) = 0$$

39. ★ Two vectors, $\vec{A}$ and $\vec{B}$, have magnitudes 10 units and 5 units, respectively, and are located in the xy-plane. The angle between them is 20°. Find the dot product of these two vectors. What does this dot product represent?

40. ★ (a) Without doing any calculations, rank the values of the following dot products in Figure 2-25: $\vec{A} \cdot \vec{B}$, $\vec{B} \cdot \vec{C}$, $\vec{A} \cdot \vec{C}$, $\vec{B} \cdot \vec{A}$, $\vec{C} \cdot \vec{B}$, and $\vec{C} \cdot \vec{A}$ for vectors $\vec{A}$, $\vec{B}$, and $\vec{C}$. Justify your ranking.
    (b) Now calculate the dot products $\vec{A} \cdot \vec{B}$, $\vec{B} \cdot \vec{C}$, $\vec{A} \cdot \vec{C}$, $\vec{B} \cdot \vec{A}$, $\vec{C} \cdot \vec{B}$, and $\vec{C} \cdot \vec{A}$, and check if the values you obtained support your ranking in (a).

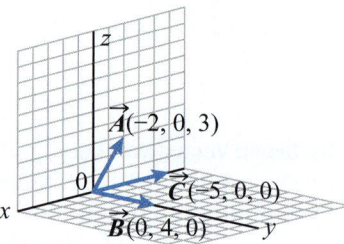

Figure 2-25   Problem 40.

41. ★★ Find a unit vector located in the xy-plane and perpendicular to the vector $\vec{A} = 3\hat{i} - 2\hat{j} + 5\hat{k}$. How many answers do you have? Explain why.

42. ★★ Find the projection of the 2-D force vector $\vec{F}(3.00, 4.00)$ N onto the line $y = -2x + 5$. (Hint: Consider how the dot product can be used to calculate the projection of a vector onto a given direction.)

43. ★★★ Suggest at least two different methods to find the angle between two vectors, $\vec{A}$ and $\vec{B}$, where $\vec{A} = -2.00\hat{i} + 3.00\hat{j} - 5.00\hat{k}$ and $\vec{B} = -2.00\hat{i} + 2.00\hat{j} - 5.00\hat{k}$. Explain each method and compare the angles you found using each one.

Section 2-5   The Cross Product of Vectors

44. ★★ Use two different methods to calculate the cross (vector) product of the vectors $\vec{A}(3,4,-5)$ and $\vec{B}(2,-2,4)$. Note that to find the cross product one must find both the magnitude and the direction of the cross product vector. Based on your calculations, which method do you prefer and why?

45. ★ Two vectors, $\vec{A}$ and $\vec{B}$, have magnitudes of 10 units and 5 units, respectively, and are located in the xz-plane. The angle between them is 20°. Find the magnitude and direction of the vector product of these two vectors. Consider all the values are given to three significant figures.

46. ★★ Three vectors, $\vec{A}(4,3,1)$, $\vec{B}(-2,4,-2)$, and $\vec{C}(-1,-3,5)$, form the sides of a parallelogram prism.
    (a) Find the volume of the prism. Assume the values are given to three significant figures.
    (b) Find the surface area of the prism.

47. ★★ For the vectors $\vec{A}$, $\vec{B}$, and $\vec{C}$ in Figure 2-25, calculate the cross products $\vec{A} \times \vec{B}$, $\vec{B} \times \vec{C}$, $\vec{A} \times \vec{C}$, $\vec{B} \times \vec{A}$, $\vec{C} \times \vec{B}$, and $\vec{C} \times \vec{A}$, using two different methods:
    (a) Use Cartesian vector notation to calculate the cross product using its determinant form.
    (b) Find the angles between the corresponding vectors and calculate the cross product using the original definition of the cross product—Equation 2-26.
    (c) Compare the results in (a) and (b). What does your comparison tell you?

48. ★★ Solve the following vector equations:
    (a) $2\hat{k} \times \left(\dfrac{1}{4}F_x\hat{i} + 7\hat{j}\right) + F_y\hat{i} - \sqrt{2}\hat{j} = 0$
    (b) $4\hat{j} \times (F_x\hat{i} + 5\hat{k}) + F_z\hat{k} + \sqrt{3}F_x\hat{i} = 0$

49. ★★ Use the determinant form for calculating cross product to find the cross product of vectors $\vec{A}$ and $\vec{B}$ $(\vec{A} \times \vec{B})$ in each of the following:
    (a) $\vec{A} = 3\hat{i} - \hat{j} + 5\hat{k}$
        $\vec{B} = -4\hat{i} + 2\hat{j} - 4\hat{k}$
    (b) $\vec{A} = 3\hat{j} - 5\hat{k}$
        $\vec{B} = -\hat{i} - 4\hat{k}$
    (c) $\vec{A} = 2\hat{i} - \hat{j} + 5\hat{k}$
        $\vec{B} = -4\hat{i} + 2\hat{j} - 10\hat{k}$

Could you have predicted some of the results without calculations? Explain.

50. ★★ Determine the angle between the diagonal of a cube and each one of the edges that shares a vertex with the diagonal. Give your answer to three significant figures.

COMPREHENSIVE PROBLEMS

51. ★★★ An orthogonal coordinate system, xy, undergoes the following transformation: it moves parallel to itself such that the new location of its origin is $O'(4, 5)$. It is then rotated 60° counterclockwise. The new coordinate system is $x'y'$.
    (a) Derive an expression for the coordinates of an arbitrary point under this transformation. Express how the new coordinates depend on the old coordinates.
    (b) Derive an expression that describes how vector components change under this transformation. Express how the new vector components depend on the old vector components.

52. ★★ For the three vectors defined as $\vec{A} = (A_x, A_y, A_z)$, $\vec{B} = (B_x, B_y, B_z)$, and $\vec{C} = (C_x, C_y, C_z)$, prove that
$$\vec{A} \cdot (\vec{B} \times \vec{C}) = \begin{vmatrix} A_x & A_y & A_z \\ B_x & B_y & B_z \\ C_x & C_y & C_z \end{vmatrix}$$

53. ★★ Prove that for any two non-zero vectors $\vec{A}$ and $\vec{B}$, both $\vec{A}$ and $\vec{B}$ are perpendicular to their cross product $\vec{A} \times \vec{B}$.

54. ★★ Prove the following relationships:
   (a) $(\vec{A} \times \vec{B}) \times \vec{C} = -(\vec{B} \times \vec{A}) \times \vec{C}$
   (b) $(\vec{A} \times \vec{B}) \times \vec{C} = -\vec{C} \times (\vec{A} \times \vec{B}) = \vec{C} \times (\vec{B} \times \vec{A})$
   (c) $(\vec{A} \times \vec{B}) \times \vec{C} \neq \vec{A} \times (\vec{B} \times \vec{C})$
   (d) $(\vec{A} \times \vec{B}) \times \vec{C} \neq \vec{A} \times (\vec{C} \times \vec{B})$

55. ★★ Prove that the algebraic expression for the cross product of two vectors using the determinant form (Equation 2-34) is equivalent to the cross product calculation using the cross product definition (Equation 2-26). (Hint: Find the magnitude of each vector, and then find the angle between them using the dot product.)

56. ★★ Prove that the algebraic expression for the cross product of two vectors using the determinant form (Equation 2-34) is equivalent to the cross product calculation using Cartesian notation and the expressions for the cross product for the unit vectors (Equations 2-30 and 2-32).

57. ★ (a) Find the cross product of the two vectors $\vec{A} = 3\hat{i} + 2\hat{j} - 5\hat{k}$ and $\vec{B} = -6\hat{i} - 4\hat{j} + 10\hat{k}$ using the determinant form, Equation 2-34.
   (b) Explain how you could have predicted this result without doing the calculation.

58. ★★ Two straight lines, a and b, are oriented in the xy-plane and described as $a: y = m_1x + b_1$ and $b: y = m_2x + b_2$, where neither m_1 nor m_2 is 0. Prove that, to be perpendicular to each other, the following must be true: $m_1m_2 = -1$. (Hint: Prove that any two arbitrary vectors directed along these lines are perpendicular to each other.)

59. ★★ The work (W_F) by a constant force $\vec{F}$ on an object can be described as a scalar product of the force and the object's displacement, $W_F = \vec{F} \cdot \Delta\vec{r}$, and is measured in joules when the force is measured in newtons and the displacement in metres. A constant force $\vec{F} = (3.00\ \text{N})\hat{i} + (4.00\ \text{N})\hat{j} - (5.00\ \text{N})\hat{k}$ acts on an object moving along the line $\vec{r} = \vec{a} + t\vec{b}$, with $\vec{a} = 2.00\hat{i} - 3.00\hat{j} - 4.00\hat{k}$ and $\vec{b} = 3.00\hat{i} - 2.00\hat{j} + 4.00\hat{k}$ (t is a parameter represented by a real number: $t \in R$).
   (a) Prove that points A(5.00 m, −5.00 m, 0.00 m) and B(8.00 m, −7.00 m, 4.00 m) are located along this line.
   (b) Find the work done on the object by force $\vec{F}$ while the object moved from point A(5.00 m, −5.00 m, 0.00 m) to point A(5.00 m, −5.00 m, 0.00 m) along the line $\vec{r} = \vec{a} + t\vec{b}$.

60. ★★★ A triangle is built on the diameter of a circle such that all three of its vertices lie on the circle's circumference. Prove that the triangle is a right triangle. (Hint: Express the sides of the triangle as vectors, and use a scalar product to prove that these vectors are orthogonal.)

61. ★★ Show that $|\vec{A} \times \vec{B}|^2 = |\vec{A}|^2|\vec{B}|^2 - (\vec{A} \cdot \vec{B})^2$.

62. ★★★ The law of cosines states that for any triangle, the following holds true: $C^2 = A^2 + B^2 - 2AB\cos\theta$, where θ is the angle between sides A and B of the triangle. Prove the law of cosines using the triangle construction method of vector addition and the fact that the square of the magnitude of a vector is equal to the dot product of the vector with itself.

63. ★★★ Prove the formula for the differentiation of the scalar product: $d(\vec{A} \cdot \vec{B}) = (d\vec{A}) \cdot \vec{B} + \vec{A} \cdot (d\vec{B})$. Use your proof to find an expression for $d(\vec{v} \cdot \vec{v})$, where $\vec{v}$ is the velocity of a particle.

64. ★★★ The speed of a particle moving in the xy-plane can be described as follows: $\vec{v} = v_{0x}\hat{i} + (v_{0y} - gt)\hat{j}$, where g is a positive constant called the acceleration due to gravity, t is time, and v_{0x} and v_{0y} are the components of the velocity on the x- and y-axes when $t = 0$. For Earth, $g \approx 9.81\ \text{m/s}^2$.
   (a) Find the expression for the speed of the particle as a function of time.
   (b) Draw a diagram that illustrates how the speed of the particle changes with time.
   (c) Determine the time when the speed of the particle is the smallest.
   (d) This mathematical model is appropriate to describe a common, everyday phenomenon. Suggest what that phenomenon might be.

65. ★★★ Newton's second law of motion states that the acceleration of a moving object is proportional to the net force acting on it and inversely proportional to the object's mass: $\vec{a} = \dfrac{\sum \vec{F}}{m} = \dfrac{\vec{F}_{\text{net}}}{m}$. There are three forces acting on a particle:

$$\vec{F}_1 = (3.00\ \text{N})\hat{i} + (2.00\ \text{N})\hat{j} - (5.00\ \text{N})\hat{k};$$
$$\vec{F}_2 = (-1.00\ \text{N})\hat{i} + (4.00\ \text{N})\hat{j} - (2.00\ \text{N})\hat{k};$$
$$\vec{F}_3 = (-2.00\ \text{N})\hat{i} + (5.00\ \text{N})\hat{j} - (7.00\ \text{N})\hat{k}$$

The particle's mass is 0.100 kg.
   (a) Find the magnitude of the particle's acceleration.
   (b) Find the coordinate direction angles of its acceleration vector.
   (c) Write the expression for the particle's acceleration in Cartesian notation.
   (d) Determine the projections of the particle's acceleration onto each of the coordinate planes xy, xz, and yz.

66. ★★★ (a) Three vectors are described as $\vec{A}(2,3,-5)$, $\vec{B}(-1,4,-2)$, and $\vec{C}(1,2,1)$. Find the following products, if possible:
   (i) $\vec{A} \times \vec{B}$
   (ii) $\vec{A} \times \vec{B} + \vec{C}$
   (iii) $2\vec{A} \times \vec{B} + \vec{C}$
   (iv) $(\vec{A} \times \vec{B}) \cdot \vec{C}$
   (v) $(\vec{A} \times \vec{B}) \times \vec{C}$
   (vi) $(\vec{A} \cdot \vec{B}) \times \vec{C}$
   (b) Explain why some products are vectors, while others are scalars. If you think it is impossible to calculate a product, explain why.

67. ★★★ A trigonometric identity involving the cosine function can be proven using the dot product operation. Imagine a unit circle, which is a circle whose radius is 1 unit. Unit vectors $\hat{u}$ and $\hat{v}$ form angles α and β, respectively, with the horizontal axis. Use the unit circle to prove that $\cos(\beta - \alpha) = \cos\alpha \cos\beta + \sin\alpha \sin\beta$.

68. ★★★ Three vectors are described as $\vec{A}(2, -3, -6)$, $\vec{B}(1, -4, 2)$, and $\vec{C}(1, -2, 3)$. Find all the vectors that are perpendicular to vectors $\vec{A}$ and $\vec{B}$ and have the same magnitude as vector $\vec{C}$.

69. ★★ The carbon tetrachloride molecule, CCl_4 (Figure 2-26(a)), has the shape of a tetrahedron. Chlorine atoms are located at the vertices of the tetrahedron, and the carbon atom is located in the centre. The bond angle (θ) is an angle formed by three connected atoms: $Cl-C-Cl$. A tetrahedron is formed by connecting alternating vertices of a cube, as shown in Figure 2-26(b). Find the bond angle (θ) of the carbon tetrachloride molecule. Express this angle to four significant figures.

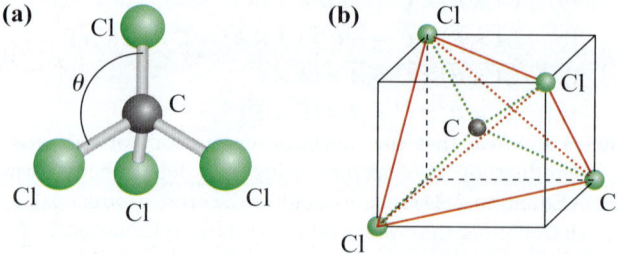

Figure 2-26 Problem 69. (a) A carbon tetrachloride molecule. (b) Atom locations in the carbon tetrachloride molecule.

Motion in One Dimension

When you have completed this chapter, you should be able to

LO1 Define, calculate, and distinguish between distance and displacement.

LO2 Define, calculate, and distinguish between average and instantaneous speed and velocity.

LO3 Define and calculate acceleration, and distinguish between average and instantaneous acceleration.

LO4 Develop and apply the kinematics equations for motion with constant acceleration.

LO5 Construct and analyze displacement, velocity, and acceleration time plots.

LO6 Use kinematics equations to analyze the motion of free-falling objects.

LO7 Describe relative motion in one dimension qualitatively and quantitatively using the kinematics equations.

LO8 Use calculus to analyze the motion of objects with constant and variable acceleration.

The 100 m dash is a sprint race in track and field competitions. It is one of the most popular and prestigious events in the world of athletics. It has been contested since the first Summer Olympic Games in 1896 for men and the ninth Summer Olympic Games in 1928 for women. The first winner of a modern Olympic 100 m race was American Francis Lane, whose time in 1896 was 12.2 s. The current male world champion in the 100 m dash is the legendary Jamaican sprinter Usain Bolt (Figure 3-1), who has improved the world record three times from 9.74 s to 9.58 s. Since 1969, the men's 100 m dash record has been revised 13 times, from 9.95 s to 9.58 s (an increase in performance of 3.72%). Bolt's 2009 record-breaking margin from 9.69 s (his own previous world record) to 9.58 s is the highest since the start of fully automatic time measurements in 1977. The fastest ever woman sprinter was American Florence Griffith-Joyner (1959–1998), whose 1988 record of 10.49 s hasn't been broken to this day. Since automatic time measurements allow for increased accuracy, athletes fight for every split second. The continuous improvement in their performance would not have been possible without the detailed analysis of every parameter of an athlete's motion, such as sprinting speed, acceleration, and the length and frequency of their stride. Coaches and athletes combine their knowledge of kinematics with their knowledge of human kinetics to optimize

PCN Photography/Alamy Stock Photo

Figure 3-1 World 100 m dash champion Usain Bolt sprints during the 2012 Summer Olympics in London.

performance and improve the world record (Krzysztof, M., & Mero, A. 2013. A kinematics analysis of the three best 100 m performances ever. *Journal of Human Kinetics*, volume 36, 149–160).

3-1 Distance and Displacement

Kinematics is the study of motion, which allows us to predict *how* an object will move, *where* it will be at a certain time, *when* it will arrive at a certain location, or *how long* it will take to cover a certain distance. In other words, in kinematics, we analyze *how* an object's position, velocity, and acceleration relate to one another, and *how* they change with time. However, in kinematics, we purposely ignore the *why* questions that will require an understanding of forces. In Chapter 5, when we consider the related topic of **dynamics**, we will focus on *why* objects move in a certain way. We restrict the discussion in this chapter to motion in one dimension (along a straight line), extending it to two and three dimensions in Chapter 4. Here, we derive the general relationships for kinematics and the equations that describe the motion of objects moving with constant (uniform) acceleration.

To describe motion, we have to define the concepts of position, displacement, and distance, originally introduced in Chapter 2 (Section 2-3). Let us expand on these concepts and describe the relationships between them. Consider the girl in Figure 3-2, who is racing a dogsled across the finish line. The figure shows her at locations corresponding to two different times, one before and one after she has crossed the finish line. We call these the initial and final positions: they describe where the object was located at the initial and final times of its motion. The SI unit for describing an object's position is the metre (m). **Displacement** is a vector physical quantity that refers to "how far out of place" (thus *dis*placement) an object is. The displacement can be represented by a vector connecting the object's initial and final positions. As a vector, displacement has both magnitude and direction. The SI unit for displacement is also the metre.

Since we are considering motion in one dimension, the direction of all the vectors describing an object's motion in this chapter will be aligned with the chosen coordinate axis. We will describe motion as directed in the positive or in the negative x-direction by using respectively positive (which can be omitted for simplicity) and negative signs before the vectors. For example, a velocity of $\vec{v} = +3$ m/s (in this chapter we will represent it as $v_x = +3$ m/s $\equiv 3$ m/s) is directed in the positive x-direction, while a velocity of $\vec{v} = -4$ m/s (we will represent it as $v_x = -4$ m/s) is directed in the negative x-direction. Index x means that we are talking about velocity, not speed. This means we won't have to use formal vector notation in this chapter, but it is still important to remember that displacements and velocities are always vectors.

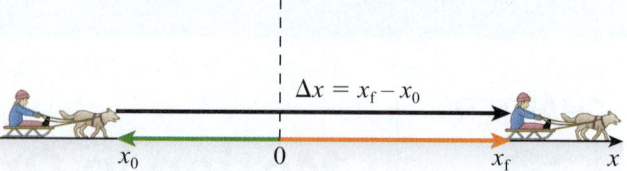

Figure 3-2 The displacement of an object (a dogsled racer in this case) represents the change in its position in terms of both its magnitude and its direction. The x-axis is chosen along the direction of the girl's motion.

In this chapter, for convenience, we label the one-dimensional axis as the x-axis.

Returning to the dogsled racer example shown in Figure 3-2, we define the x-axis as being oriented horizontally, with right representing positive, left representing negative, and the origin chosen to be the finish line. Her initial and final positions, x_0 and x_f, are measured from the origin, and her displacement is shown by the vector quantity Δx connecting x_0 and x_f (Figure 3-2). The displacement can be expressed as the difference between the final and initial positions, $\Delta x = x_f - x_0$ and its SI unit is the metre. Since the final position of an object is often denoted as x, the expression for displacement becomes

KEY EQUATION
$$\Delta x = x - x_0 \quad [\Delta x] = \text{m} \tag{3-1}$$

For this example and our choice of reference frame (the x-axis, the origin, and the unit of measurement), the initial position is negative and the final position is positive; hence, according to Equation 3-1, the displacement is positive. In the one-dimensional case, positive displacement means that the object moved in the positive x-direction, negative displacement means that it moved in the negative x-direction, and zero displacement means that the object's final position is the same as its initial position.

Let's consider an example to help differentiate the terms *distance* and *displacement*. You are sitting at your desk at home, your coffee machine indicates a fresh brew is ready, you walk 10 steps to fill your coffee mug, and then you bring it back to your desk. Your displacement in this case is zero, since your final position is the same as your initial position. However, the distance you have covered, there and back, is a total of 20 steps.

As we discussed in the dogsled racer example above, displacement is a vector that represents the straight-line path between the end points of the **trajectory** (the path in space followed by a moving object). For one-dimensional motion, that displacement can take on positive or negative values. The **distance**, on the other hand, is a positive scalar representing the actual distance travelled. The SI unit for measuring distance is also the metre.

One way to define distance is to consider the sum of the magnitudes of the small displacements that make up the journey. For example, if each step you take along a certain path is δx, we can say that the distance you cover, d, is given by

KEY EQUATION

$$d = \sum_i |\delta x_i| \qquad (3\text{-}2)$$

EXAMPLE 3-1

Get That Fly

A tarantula is on a wall 21 cm below a nail. It moves down to a point 64 cm below the nail to catch a fly, and then takes the fly straight up to a position 32 cm above the nail, as shown in Figure 3-3.

(a) Find the displacement of the tarantula.
(b) Find the distance covered by the tarantula.

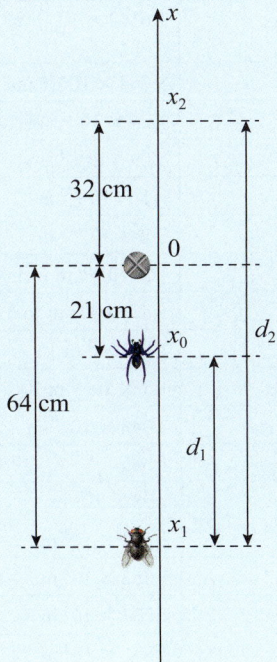

Figure 3-3 Example 3-1.

Solution

(a) We define a vertical x-axis with the origin located at the position of the nail, and up as the positive direction.

Therefore, the initial position of the tarantula as measured from the nail is negative ($x_0 = -21$ cm), and its final position is positive ($x_2 = 32$ cm). The displacement shown in Figure 3-3 is given by

$$\Delta x = x_2 - x_0 = 32 \text{ cm} - (-21 \text{ cm}) = 53 \text{ cm}$$

(b) On the way down, the tarantula moves from its initial position, 21 cm below the nail ($x_0 = -21$ cm), to the location of the fly, 64 cm below the nail ($x_1 = -64$ cm). The distance covered by the tarantula on the way down is

$$d_1 = |x_1 - x_0| = |-64 \text{ cm} - (-21 \text{ cm})|$$
$$= |-64 \text{ cm} + 21 \text{ cm}| = 43 \text{ cm}$$

On the way up, the tarantula moves from position x_1, 64 cm below the nail, to position x_2, 32 cm above the nail, for a total upward distance of

$$d_2 = |x_2 - x_1| = |32 \text{ cm} - (-64 \text{ cm})|$$
$$= |32 \text{ cm} + 64 \text{ cm}| = 96 \text{ cm}$$

The total distance covered by the tarantula is the sum of the two distances:

$$d = d_1 + d_2 = 43 \text{ cm} + 96 \text{ cm} = 139 \text{ cm} \approx 140 \text{ cm} = 1.4 \text{ m}$$

Making sense of the result

For displacement, only the initial and final positions matter. Positive displacement in this case means that the tarantula ended up above where it started. The distance travelled, however, is a positive scalar quantity that takes into account the details of the path taken, such as turns of the tarantula along its path. Lastly, notice that the final answer has only two significant figures, as the quantities given in the problem had only two significant figures.

CHECKPOINT 3-1

Race Car: Distance or Displacement?

When a Formula One race car drives a full lap around a circular track that has a radius of 210 m,
(a) the displacement of the car is the circumference of the track
(b) the distance travelled by the car is the circumference of the track
(c) the distance travelled by the car is zero
(d) the displacement and distance travelled are equal

ANSWER: (b) Displacement is the difference between the final and initial positions and is, therefore, zero. The distance covered is the circumference of the track.

INTERACTIVE ACTIVITY 3-1

The Scale of the Universe

In this activity, you will use the computer simulation called "The Scale of the Universe," designed by Cary Huang (htwins.net/scale2/) to help you gain some useful intuition about the vast range of sizes in our universe. The simulation will allow you to change the scale (zoom in and out) and experience very small and very big objects on a grand scale. Work through the simulation and accompanying questions to deepen your understanding of the scale of the universe you live in.

MAKING CONNECTIONS

The Scale of the Universe

Table 3-1 lists some common ranges of distances in nature and everyday life.

Table 3-1 Common Ranges of Distances

Description	Distance	Equivalent Distance in Metres
Planck length, the shortest distance that can theoretically be measured	1.6162×10^{-35} m	1.6162×10^{-35} m
Radius of a nucleon	0.8768 fm	8.768×10^{-16} m
Upper limit for the wavelength of gamma rays	12.4 pm	12.4×10^{-12} m
Radius of a hydrogen atom	52.9 pm	52.9×10^{-12} m
Covalent bond length	~0.1 nm	~1×10^{-10} m
Hydrogen bond length	~0.2 nm	~2×10^{-10} m
Wavelength of X-rays	0.1–10 nm	1×10^{-10} m to 1×10^{-8} m
Wavelength of yellow light	580–590 nm	5.8×10^{-7} m to 5.9×10^{-7} m
Diameter of strand of human hair	100 μm	1×10^{-4} m
Thickness of human skin on palm	1.5 mm	1.5×10^{-3} m
1 ft	30.48 cm	3.048×10^{-1} m
Distance travelled by light in 1/299 752 458 s	1 m	1 m
Wingspan of Airbus 380-800	79.8 m	7.98×10^2 m
1 mile	1.609 km	1.609×10^3 m
1 nautical mile	1.852 km	1.852×10^3 m
Mean radius of the Moon	1737 km	1.737×10^3 m
Mean radius of Earth	6371 km	6.371×10^3 m
Mean distance between Earth and the Moon	384 400 km	3.844×10^5 m
Radius of the Sun	6.955×10^8 m	6.955×10^8 m
Radius of R136a1, the most massive known star	35.4 solar radii	2.46×10^{10} m
1 astronomical unit, the mean Earth–Sun distance	149 597 871 km	$1.495\,978\,71 \times 10^{11}$ m
Radius of VY Canis Majoris, possibly the largest known star	1800–2100 solar radii	1.25×10^{12} m to 1.46×10^{12} m
1 light year (ly), the distance that light travels in one year	$9.460\,528\,4 \times 10^{15}$ m	$9.460\,528\,4 \times 10^{15}$ m
Approximate diameter of the Milky Way galaxy	100 000 light years (ly)	$9.460\,528\,4 \times 10^{20}$ m
Approximate diameter of the observable universe	93 billion light years (ly)	8.8×10^{26} m

3-2 Speed and Velocity

Motion Diagrams

As we discussed earlier, one of the main goals of kinematics is to describe how the object's position changes with time. We can do this by plotting an $x(t)$ graph. To do this we have to collect detailed information about the object's position at different times. One way of doing this is to use a video camera. A video analysis of motion allows us to collect position versus time data of the object's motion and then plot the corresponding graphs, such as $x(t)$, $v(t)$, etc. The video recording is nothing else but time-lapse photography of the object's motion. We will discuss the advantages of computerized video analysis later in the chapter. For now let us imagine a person walking along a straight line oriented in the positive x-direction. Let us record his position by making a series of snapshots of his locations along the coordinate axis at equal time intervals (for example, every 2 s) and recording the clock reading when these snapshots were taken. Then we can visualize his motion using a **motion diagram**, as shown in Figure 3-4. If he walks at a constant pace (Figure 3-4(a)), the distances between his adjacent positions on the diagram will be equal. However, when he is speeding up, the distances between his adjacent positions will continuously increase (Figure 3-4(b)). Figure 3-4(c) shows the case when the person is slowing down; in this case the distances will decrease. The motion diagram illustrates the pace of the person's motion. In physics, to describe the pace, or the rate of change of the object's position, we use the concepts of velocity and speed, discussed below.

Average Speed and Average Velocity

Two cars that start at the same location and travel for an hour at the same speed but heading in different directions will end up at different final destinations. Consequently, their displacements will also be different. To be able to predict *where* an object will end up, it is important to know not only where it started (**initial position**, x_0) and how fast it is moving, but also where it is headed. Unfortunately, in everyday life, the terms *speed* and *velocity* are often used interchangeably to describe the pace of motion. However, as discussed in Chapter 2, the meanings of these terms in physics are distinctly different (see the Chapter 2 introduction). **Speed** is a scalar quantity describing how fast an object is moving, while **velocity** is a vector indicating not only how fast an object is moving, but also where it is headed (the direction of its motion).

Average speed (a scalar) is the *distance* covered by the object divided by the time it took the object to cover it (**elapsed time**)—since distance is a positive scalar (if the object moved), so is average speed. Notice the difference between the concepts of elapsed time, Δt, and **clock reading**, t. If an object starts moving when the clock reading (initial time) was $t_0 = 5$ s and finishes moving when $t_f = 8$ s, the elapsed time for the object's motion can be calculated as $\Delta t = t_f - t_0 = 8\text{ s} - 5\text{ s} = 3\text{ s}$. As in the case of displacement (Equation 3-1), the index f in the final clock reading is often omitted, so the expression for elapsed time can be written as $\Delta t = t - t_0$. The SI unit for time measurement and for elapsed time is the second (s).

Average velocity (a vector) is given by the *displacement* divided by the elapsed time. For motion along the x-axis, with initial position x_0 and final position x, the average speed and the average velocity are given by the following expressions, where t_0 and t are the initial and final clock readings, respectively:

KEY EQUATION

$$\text{Average speed} = v_{\text{avg}} = \frac{d}{\Delta t} = \frac{d}{t - t_0}$$

$$[v_{\text{avg}}] = \frac{\text{m}}{\text{s}} \qquad (3\text{-}3)$$

$$\text{Average velocity} = v_{x,\text{avg}} = \frac{\Delta x}{\Delta t} = \frac{x - x_0}{t - t_0}$$

$$[v_{x,\text{avg}}] = \frac{\text{m}}{\text{s}} \qquad (3\text{-}4)$$

Here d is the distance, a scalar quantity that is always positive; Δx is the displacement, a vector quantity that can take on positive or negative values in one-dimensional motion; and Δt is the elapsed time. When the two cars are moving in opposite directions with the same speed, their velocities will be equal in magnitude but opposite in sign. The SI units for both speed and

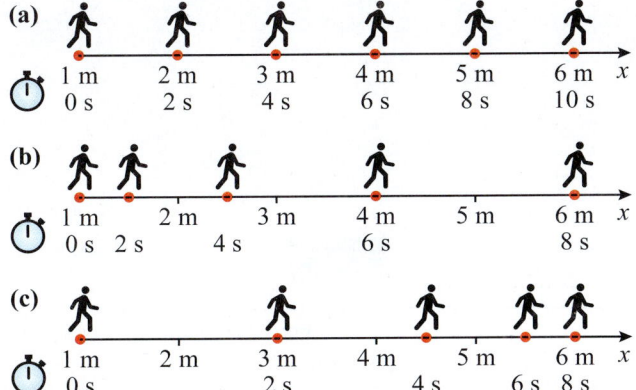

Figure 3-4 Motion diagrams representing one-dimensional motion of a person walking in the positive x-direction. (a) Walking at a constant speed (constant pace). (b) Speeding up. (c) Slowing down. The motion diagram represents both the person's position (top scale) and the corresponding clock reading (bottom scale).

velocity are metres per second (m/s), but in everyday life it is often convenient to use kilometres per hour (km/h) or miles per hour (mph or mi/h). For example, the speed limit in school zones in Canada is 30 km/h to 50 km/h, which translates to approximately 20 mi/h to 30 mi/h (1.0 mi ≈ 1.6 km). For historical reasons, the speed of aircraft and boats is measured in knots (kn): 1 knot = 1 nautical mile/h = 6072 ft/h = 1852 m/h ≈ 0.514 m/s.

Pay attention to the differences in the notations for average speed (v_{avg}) and for average velocity ($v_{x,avg}$). We use the index x for average velocity to emphasize that velocity is a vector quantity and we are considering the one-dimensional case along the x-axis. Sometimes the notation $\bar{v}$ is used to indicate *average* speed and velocity, but we will generally use v_{avg} and $v_{x,avg}$, respectively. In Chapter 4, we consider velocity in more complex two- and three-dimensional cases. There, we will express velocity in its full two- and three-dimensional vector form, $\vec{v}$.

Let us examine the difference between speed and velocity in the following scenario. You instruct the courier truck for your company to leave the garage of your office building, pick up a package from the post office 4.80 km down the road, and bring it to a lab facility that is 1.20 km away from your office (Figure 3-5). From your office, the lab is in the opposite direction down the road as the post office from which the package was picked up. It takes the driver 15.0 min to finish the task. Let us choose the positive direction of the x-axis to the right (in the direction of the post office) and the origin to be where your office is located.

As illustrated in Figure 3-5, the total displacement of the courier, the change in her position, in this case is −1.20 km. Dividing this by the total elapsed time gives us the average velocity for the trip (Equation 3-4). Since the total trip time is 15.0 min (900. s), we calculate the average velocity to be

$$v_{x,avg} = \frac{-1.20 \times 10^3 \text{ m}}{900. \text{ s}} = -1.33 \text{ m/s}$$

Since the courier's displacement was directed to the left, in the negative x-direction, the average velocity is also negative.

The average speed, however, is given by the total distance (not the displacement!) divided by the elapsed time (Equation 3-3). The total distance covered by the truck is 4.80 km + 4.80 km + 1.20 km = 10.80 km = 1.080×10^3 m, so the average speed is given by

$$v_{avg} = \frac{1.080 \times 10^3 \text{ m}}{900 \text{ s}} = 12.0 \text{ m/s}$$

Notice that since the truck turned around, the values of the average speed and the magnitude of the average velocity differ.

PEER TO PEER

I know that the speedometer in a car shows how fast I am moving but not where I am heading. This helps me remember that speed is a scalar quantity, while velocity is a vector. I also like to use the fact that a speed of 10 m/s is equivalent to 36 km/h. It allows me to do fast and easy mental calculations of speed conversions from m/s into km/h and vice versa. For example, a speed limit of 70 km/h is equivalent to about 20 m/s.

The average velocity of an object during a specific time interval, $v_{x,avg}$, can be represented on a motion diagram as an arrow pointing in the direction of motion (we use red to denote velocity vectors). The length of the arrow should be proportional to the distance between the adjacent locations of an object (Figure 3-6).

In addition to using motion diagrams, it is convenient to describe the motion of an object using a position versus time plot, $x(t)$, as shown in Figure 3-7. In an $x(t)$ plot, position is displayed along the vertical

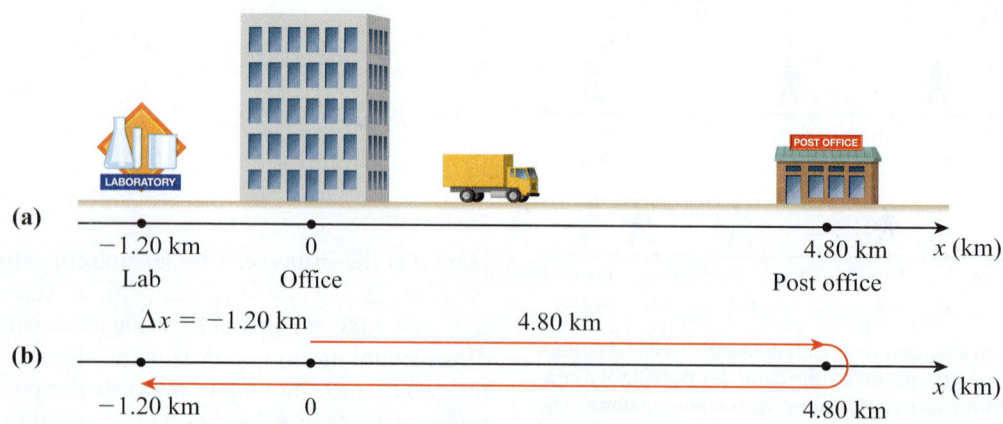

Figure 3-5 (a) Schematic representation of the problem including the coordinate system. (b) The trajectory of a truck moving from the office building, to the post office, and back to the lab. The distance covered by the truck is different from the magnitude of its displacement.

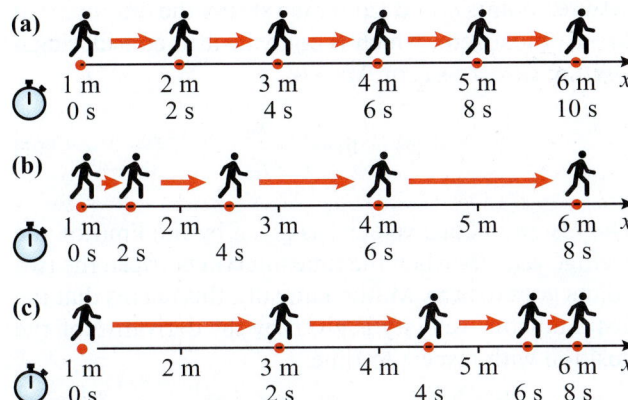

(a)

1 m 2 m 3 m 4 m 5 m 6 m x
0 s 2 s 4 s 6 s 8 s 10 s

(b)

1 m 2 m 3 m 4 m 5 m 6 m x
0 s 2 s 4 s 6 s 8 s

(c)

1 m 2 m 3 m 4 m 5 m 6 m x
0 s 2 s 4 s 6 s 8 s

Figure 3-6 Motion diagrams including average velocity vectors representing the one-dimensional motion of a person walking in the positive x-direction: (a) Walking at a constant pace. (b) Speeding up. (c) Slowing down.

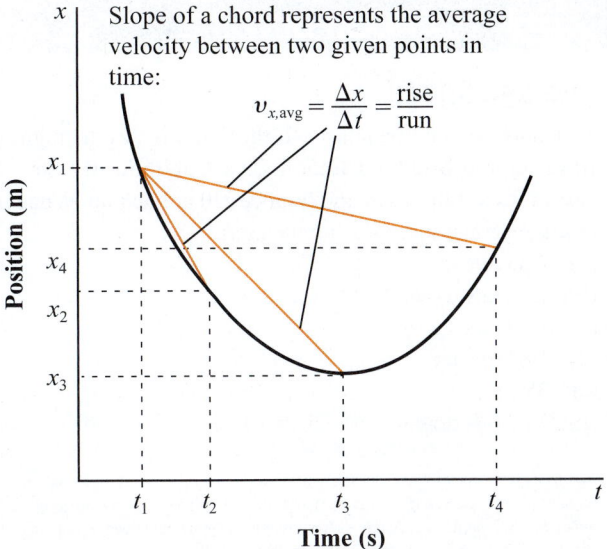

Slope of a chord represents the average velocity between two given points in time:

$$v_{x,\text{avg}} = \frac{\Delta x}{\Delta t} = \frac{\text{rise}}{\text{run}}$$

Figure 3-7 The average velocity of an object during a certain time interval is represented by the slope (rise over run) of the chord connecting the two points on the $x(t)$ plot that correspond to the endpoints of the time interval. The three chords represent the average velocities of an object during the time intervals (t_1, t_2), (t_1, t_3), and (t_1, t_4), respectively.

axis and time is displayed along the horizontal axis. Let us use the $x(t)$ plot to visualize the average velocity of the moving object between two given points in time. According to Equation 3-4, $v_{x,\text{avg}} = \dfrac{\Delta x}{\Delta t}$. Since Δx represents the rise and Δt represents the run of the $x(t)$ function, the average velocity between any two points in time is represented by the slope (rise over run) of the chord (a segment between two points on a curve) connecting corresponding points on the $x(t)$ plot. For example, in Figure 3-7 the slopes of the three chords represent the average velocities of an object between the times t_1 and t_2, t_1 and t_3, and t_1 and t_4.

EXAMPLE 3-2

Distinguishing Speed and Velocity

A meteorological observation plane flies 1240 km straight southeast in 1.70 h from Stockholm, Sweden, to Moscow, Russia. The plane then flies to Oslo, Norway, 425 km to the northwest of Stockholm. The trip to Oslo takes 2.10 h. The three cities lie along a straight line, as shown in Figure 3-8.

(a) Find the average velocity of the plane in km/h and in knots.
(b) Find the average speed of the plane for the entire trip in km/h and in knots (1 knot = 1.852 km/h).

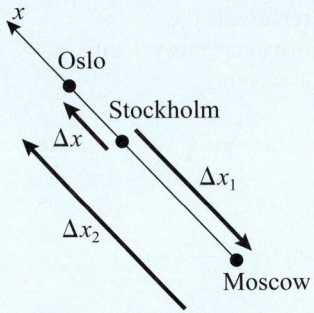

Figure 3-8 Flight path of the plane in Example 3-2.

Solution

(a) Let us pick the direction from Moscow to Oslo to be along the positive x-axis. The average velocity depends on the net displacement of 425 km and the total elapsed time:

$$v_{x,\text{avg}} = \frac{\Delta x}{\Delta t} = \frac{425\ \text{km}}{1.70\ \text{h} + 2.10\ \text{h}} = 112\ \text{km/h} \approx 60.4\ \text{knots}$$

We used the following relationship to convert km/h into knots: 1 knot = 1.852 km/h.

(b) The average speed is the total distance, d, travelled divided by the total time:

$$v_{\text{avg}} = \frac{d}{\Delta t} = \frac{1240\ \text{km} + 1240\ \text{km} + 425\ \text{km}}{1.70\ \text{h} + 2.10\ \text{h}}$$

$$= \frac{2905\ \text{km}}{3.80\ \text{h}} = 764\ \text{km/h} \approx 413\ \text{knots}$$

where we have chosen the positive x-axis to point from Moscow to Oslo.

Making sense of the result

Since the overall displacement of the plane is relatively small, the average velocity is expected to be small. The average speed, however, represents how fast the plane was going in the air, which is a typical speed for a plane. Next time you are flying, pay attention to the interactive inflight map available on many modern commercial aircraft. It will indicate a lot of interesting information about the flight, including the distance covered by the aircraft and the time it took to cover it (flight time). This will allow you to estimate your average speed during the flight.

CHECKPOINT 3-2

Average Speed

You make a bungee jump off the famous 321 m high Royal Gorge Bridge in Cañon City, Colorado. It takes you 14.0 s to fall 270 m and bounce 230 m back up. What is your average speed during the jump?

(a) 2.86 m/s
(b) 2.86 m/s down
(c) 17.9 m/s
(d) 19.3 m/s up
(e) 35.7 m/s
(f) 35.7 m/s down

ANSWER: (f) The distance covered is 270 m + 230 m = 500 m. Covering 500 m in 14.0 s gives an average speed of 35.7 m/s, which is about 130 km/h. Notice that this is a scalar quantity, so it has no direction. The average velocity is obtained by dividing the displacement by the time, in this case 40.0 m (downward) divided by 14.0 s, and gives the magnitude of the average velocity as 2.86 m/s and its direction as downward (you end up below the point where you started). Since the question asks for average speed, only (f) is correct.

Instantaneous Velocity

If we want to know the details of how an object's position is changing with time at a given point along its trajectory, we consider the concept of **instantaneous velocity**, the velocity, $v_x(t)$, of an object at a given moment in time. For example, to calculate the instantaneous velocity of an object at point t in the case described in Figure 3-9, we can find the average velocity between points t_0 and t and then shrink the time interval Δt to almost zero, which is equivalent to considering a limiting case, $t \rightarrow t_0$ or $\Delta t \rightarrow 0$:

$$v_x(t) = \lim_{t \rightarrow t_0} \frac{x - x_0}{t - t_0} \tag{3-5}$$

The instantaneous velocity is given by the limit of the average velocity when the time interval between the two points goes to zero. Mathematically, this means that the instantaneous velocity is given by the derivative of the position with respect to time:

KEY EQUATION

$$v_x(t) = \lim_{\Delta t \rightarrow 0} v_{x,\text{avg}} = \lim_{\Delta t \rightarrow 0} \frac{\Delta x}{\Delta t} = \frac{dx(t)}{dt} = x'(t) \tag{3-6}$$

Figure 3-9 shows the position versus time plot for a moving object, $x(t)$. The instantaneous velocity at a given point in time is represented by the slope of the tangent to the position versus time graph at that point. Conceptually, you can see that when the time interval between any two given points becomes very small, the average velocity becomes almost identical to the instantaneous velocity in the middle of this time interval. In Figure 3-9, when both t_1 and t_4 approach t such that $\Delta t = t_4 - t_1 \rightarrow 0$, the average velocity between t_1 and t_4 approaches the value of the instantaneous velocity at time t: $v_{x,\text{avg}(t_1, t_4)} \rightarrow v(t)$ as $\Delta t \rightarrow 0$.

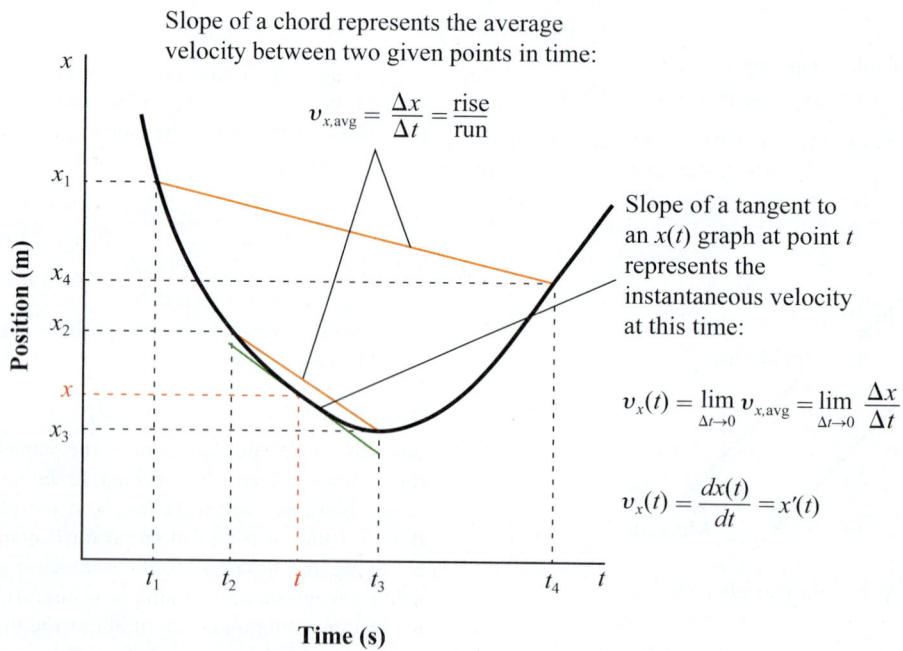

Slope of a chord represents the average velocity between two given points in time:

$$v_{x,\text{avg}} = \frac{\Delta x}{\Delta t} = \frac{\text{rise}}{\text{run}}$$

Slope of a tangent to an $x(t)$ graph at point t represents the instantaneous velocity at this time:

$$v_x(t) = \lim_{\Delta t \rightarrow 0} v_{x,\text{avg}} = \lim_{\Delta t \rightarrow 0} \frac{\Delta x}{\Delta t}$$

$$v_x(t) = \frac{dx(t)}{dt} = x'(t)$$

Figure 3-9 The average velocity of an object is represented by the slope of the chord connecting the two ends of the time interval on the $x(t)$ plot, while the instantaneous velocity at time t is represented by the slope of the tangent line to the $x(t)$ plot at time t. The instantaneous velocity is the time derivative of the $x(t)$ plot.

The **instantaneous speed** is the magnitude of the instantaneous velocity. It corresponds to the absolute value of the slope of the tangent to the $x(t)$ plot at a given point in time. The instantaneous velocity can be found as the derivative of the position versus time graph. Instantaneous velocity is a vector quantity (in the case of one-dimensional motion, this means it can be negative, positive, or zero), while instantaneous speed is always positive or zero. For example, in Figure 3-9, the instantaneous velocity at time t_3 is zero (the slope of the tangent to the curve at this point in time is zero), which means that at t_3 the object will momentarily stop.

MAKING CONNECTIONS

Range of Speeds in Nature and Everyday Life

Table 3-2 lists some common ranges of speeds in nature and everyday life.

Table 3-2   Range of Speeds in Nature and Everyday Life

Description	Speed	Speed in m/s
Theoretical rate of flow of glass at room temperature	~1 m in 10^{10} y	~3×10^{-18} m/s
Speed of continental drift	~1 cm/y	~3×10^{-10} m/s
Speed of a small earthworm	~0.2 cm/s	~2×10^{-3} m/s
Speed of a molecule in the Bose–Einstein condensate (occurs at very low temperatures)	1 cm/s	1×10^{-2} m/s
1 knot, the customary unit for the speed of ships and aircraft	1.85 km/h	0.514 m/s
Average walking speed	~5 km/h	~1.4 m/s
Sprint speed in a 100 m race	~10 m/s	~10 m/s
Top running speed of a bear	~52 km/h	~14 m/s
Top speed of cheetahs and sailfish	~112 km/h	~31.1 m/s
Top swooping speed of a peregrine falcon	~325 km/h	~90.3 m/s
Top speed of a Bugatti Veyron race car	431 km/h	120 m/s
Speed record for the MLX01 magnetic levitation train (Japan, 2003)	581 km/h	161 m/s
Maximum operating speed of an Airbus 380 at cruising altitude	945 km/h	262 m/s
Speed of sound at sea level at 15 °C	340 m/s	340 m/s
Equatorial rotation speed of Earth	1674 km/h	465 m/s
Top speed of the Concorde supersonic jet at cruising altitude	2172 km/h	603 m/s
Orbital speed of the Moon around Earth	3600 km/h	1000 m/s
Root mean square speed of a room-temperature helium atom	1352 m/s	1352 m/s
Equatorial rotation speed of the Sun	~2.0 km/s	~2000 m/s
Top speed of the X-15, the fastest piloted rocket plane	7275 km/h	2,020 m/s
Speed of the space shuttle on re-entry	~28 000 km/h	~7,800 m/s
Mean orbital speed of Earth around the Sun	29.8 km/s	2.98×10^4 m/s
Range of meteor speeds in Earth's atmosphere	11–72 km/s	1.1×10^4 m/s to 7.2×10^4 m/s
Speed of the Sun around the centre of the Milky Way galaxy	~230 km/s	~2.3×10^5 m/s
Speed of the Milky Way galaxy (relative to the cosmic microwave background)	~600 km/s	~6×10^5 m/s
Equatorial rotation speed of the Crab Pulsar neutron star	~2100 km/s	2.1×10^6 m/s
Speed of light	3.00×10^8 m/s	3.00×10^8 m/s

Relating Motion Diagrams to Position versus Time Plots

Which position versus time graph represents the motion diagram shown in Figure 3-10?

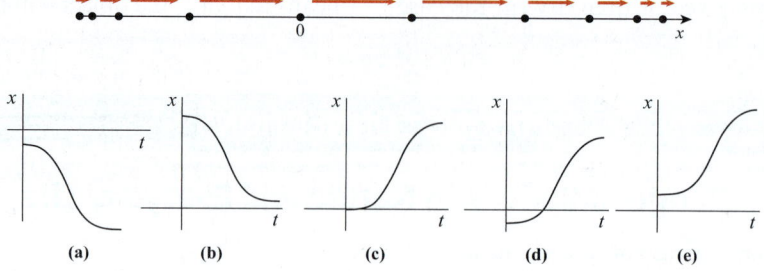

Figure 3-10 Checkpoint 3-3.

EXAMPLE 3-3

Deriving Velocity from a Position versus Time Plot

Figure 3-11 shows a position versus time plot for a car. Represent this graph algebraically and then use it to derive the velocity versus time and speed versus time graphs for the interval shown. Assume all the information is known to two significant figures.

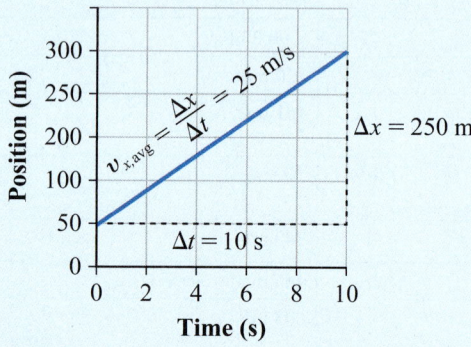

Figure 3-11 Position versus time plot for a car moving at a constant velocity.

Solution

The $x(t)$ plot is represented by a straight line. Therefore, it can be expressed algebraically using the expression for a straight line, $y = mx + b$. The y-intercept in this case represents the initial position of the car, which we will represent as x_0. According to the plot, the initial position of the car is $x_0 = 50$ m. Since the slope of the line doesn't change, the velocity of the car and consequently its speed are constant. Thus, the average velocity of the car is also equal to the instantaneous velocity of the car during this 10 s time interval: $v_{x,\text{avg}} = v_x$. Therefore, the motion of this car can be represented algebraically as $x(t) = x_0 + v_x t$. During this time interval, the displacement of the car is 250 m in the positive x-direction. Therefore, we can calculate the average velocity of the car for this time interval:

$$v_{x,\text{avg}} = \frac{\Delta x}{\Delta t} = \frac{250 \text{ m}}{10 \text{ s}} = 25 \text{ m/s}$$

This gives us

$$x(t) = x_0 + v_{x,\text{avg}}t = 50 \text{ m} + 25t \text{ m} = (50 + 25t) \text{ m}$$

The units of this expression are metres, which is the proper unit for position.

The graphs of both the instantaneous velocity and the instantaneous speed versus time will be represented by a horizontal line with a y-intercept of 25 m/s, as shown in Figure 3-12. Both the instantaneous and the average velocities for the car during this time interval can be expressed as $v_x = v_{x,\text{avg}} = 25$ m/s.

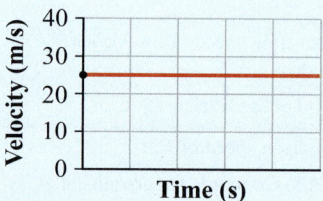

Figure 3-12 Velocity versus time graph for a car moving at a constant velocity of 25 m/s.

Making sense of the result

Notice that constant velocity does not mean that the values of speed and velocity are the same. An object can move with a constant but negative velocity. Its speed is constant but not equal to the value of its velocity (as the speed is always positive).

We also discussed earlier that the instantaneous velocity is the time derivative of $x(t)$:

$$v_x(t) = x'(t) = \frac{d}{dt}(50 + 25t) \text{ m} = 25\frac{\text{m}}{\text{s}}$$

Since the $x(t)$ plot is linear, its derivative is a constant, which means the speed and the velocity are also constant. Notice that the initial position of an object (50 m in this case) has no impact on its velocity.

EXAMPLE 3-4

Instantaneous and Average Velocities

In this example, we highlight how the average velocity approaches the instantaneous velocity as the time interval separating two events gets smaller. A toy remote-controlled car moves in a straight line along the x-axis such that its position (in metres) is given by $x(t) = (0.250t^3)$ m, where t is measured in seconds and the coefficient 0.250 has units of m/s³.

(a) Plot the car's position versus time graph for the interval $t_1 = 1.00$ s to $t_2 = 2.00$ s.

(b) Find the average velocity of the car during the interval $t_1 = 1.00$ s to $t_2 = 2.00$ s, and include the corresponding line on your graph.

(c) Find the car's average velocity during the interval $t_3 = 1.25$ s to $t_4 = 1.75$ s, and include the corresponding line on your graph.

(d) Find the car's instantaneous velocity at $t_5 = 1.50$ s, and include the corresponding line on your graph. Comment on the difference between average and instantaneous velocities as Δt gets smaller.

Solution

(a) We will take the positive x-axis to point along the direction of the car's motion. To create the plot, we can pick convenient values along the horizontal t-axis, calculate the corresponding x-coordinates, and indicate them on the vertical axis and sketch a smooth curve from the resulting points, as shown in Figure 3-13. You can also use graphing software or a spreadsheet to visualize this plot.

(b) We can apply Equation 3-4 to find the average velocity of the car between $t_1 = 1.00$ s and $t_2 = 2.00$ s:

$$v_{x,\text{avg}(t_1,t_2)} = \frac{\Delta x}{\Delta t} = \frac{x_2 - x_1}{t_2 - t_1}$$

$$= \left(\frac{0.250(2.00)^3 - 0.250(1.00)^3}{2.00 - 1.00}\right)\frac{\text{m}}{\text{s}} = 1.75 \text{ m/s}$$

As we discussed earlier, the average velocity between $t_1 = 1.00$ s and $t_2 = 2.00$ s can be represented by the slope of the chord shown in orange connecting the points on the x(t) plot corresponding to $t_1 = 1.00$ s and $t_2 = 2.00$ s. Notice, since the car is moving in the positive x-direction, the average velocity is positive. The slope of the chord is also positive.

(c) We can use the same logic to find the average velocity of the car during the time interval between $t_3 = 1.25$ s and $t_4 = 1.75$ s:

$$v_{x,\text{avg}(t_3,t_4)} = \frac{\Delta x}{\Delta t} = \frac{x_4 - x_3}{t_4 - t_3}$$

$$= \left(\frac{0.250(1.75)^3 - 0.25(1.25)^3}{1.75 - 1.25}\right)\frac{\text{m}}{\text{s}} = 1.70 \text{ m/s}$$

Once again, the average velocity corresponds to the slope of the chord connecting the points on the x(t) plot

corresponding to $t_3 = 1.25$ s and $t_4 = 1.75$ s. This chord is also shown in orange.

(d) To get the instantaneous velocity at $t_5 = 1.50$ s, we use Equation 3-6. Let us first find the expression for instantaneous velocity:

$$v_x(t) = \frac{dx(t)}{dt} = \frac{d(0.250t^3)}{dt} = 3\left(0.250\frac{\text{m}}{\text{s}^3}\right)t^2$$

Now we can find the value for the instantaneous velocity at time t_5:

$$v_x(t_5) = 3\left(0.250\frac{\text{m}}{\text{s}^3}\right)(t_5)^2 = \left(0.750\frac{\text{m}}{\text{s}^3}\right)(1.50 \text{ s})^2$$

$$= 1.69 \text{ m/s}$$

As expected, the instantaneous velocity of the car at $t_5 = 1.50$ s corresponds to the slope of the tangent to the x(t) graph at $t_5 = 1.50$ s, shown in green.

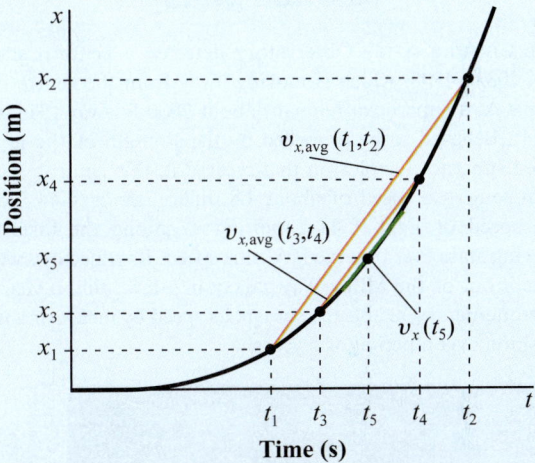

Figure 3-13 The closer the two points t_1 and t_2 are on a position versus time plot, the closer the average velocity approaches the instantaneous velocity. As the chord gets shorter, the slope of the chord approaches the slope of the tangent.

Making sense of the result

We see that as the time, Δt, separating the two points gets smaller, the chord representing the average velocity during this time interval becomes parallel to the tangent at the point in the middle of this time interval. Therefore, the average velocity during a very short time interval is for practical purposes equivalent to the instantaneous velocity in the middle of this interval. Mathematically, this means that the instantaneous velocity is represented by the derivative of the position versus time function: $v_x = x'(t) \equiv \frac{dx}{dt}$. This means that the limit of the average velocity when the time interval approaches zero approaches the instantaneous velocity at that point in time.

MAKING CONNECTIONS

Measuring the Speed of Neutron Stars

The Chandra X-ray Observatory detected a neutron star, RX J0822-4300, which is moving away from the centre of Pupis A, a supernova remnant about 7000 ly away (Figure 3-14). Believed to be propelled by the strength of the lopsided supernova explosion that created it, this neutron star is moving at a speed of about 4.8 million km/h (0.44% of the speed of light, 0.44c), putting it among the fastest-moving stars ever observed. At this speed, its trajectory will take it out of the Milky Way galaxy in a few million years. Astronomers were able to estimate its speed by measuring its position over a period of 5 years.

Chandra: NASA/CXC/Middlebury College/ F.Winkler; ROSAT: NASA/GSFC/S.Snowden et al.; Optical: NOAO/CTIO/Middlebury College/F.Winkler et al.

Figure 3-14 Supernova remnant RX J0822-4300.

3-3 Acceleration

Just as velocity represents the rate of change of position in time (Equation 3-6), acceleration represents the rate of change of velocity in time. Since velocity is a vector quantity (it has both magnitude and direction), the rate of change of velocity is also a vector. In the case of one-dimensional motion, we indicate the direction of acceleration using either positive or negative signs, as we did with velocity. The letter subscript in the notation of acceleration, such as a_x or a_y, indicates that we are talking about the vector of acceleration in one dimension. For example, for one-dimensional motion along the x-axis, a_x denotes the acceleration vector along the x-axis. However, if we want to talk about the absolute value of acceleration, we will denote it as a. We will use full three-dimensional vector acceleration notation in the following chapters.

The **average acceleration** of an object is the change in its velocity divided by the elapsed time:

KEY EQUATION

$$a_{x,\text{avg}} = \frac{\Delta v_x}{\Delta t} = \frac{v_x - v_{x,0}}{t - t_0} \quad [a] = \frac{\text{m}}{\text{s}^2} \qquad (3\text{-}7)$$

The average acceleration of an object between two points in time depends only on its final and initial velocities and the elapsed time. Since acceleration indicates by how much the velocity is changing every second, its units are metres per second per second, or m/s^2. Sometimes the notation $\overline{a}$ is used to denote average acceleration, but we will use $a_{x,\text{avg}}$, analogous to average velocity.

Figure 3-15 shows the velocity versus time plot for an object. The average acceleration, given by Equation 3-7, is the change in velocity (rise) divided by the elapsed time (run) between the two points in time. The average acceleration is represented by the slope of the chord connecting these two points on the $v_x(t)$ plot.

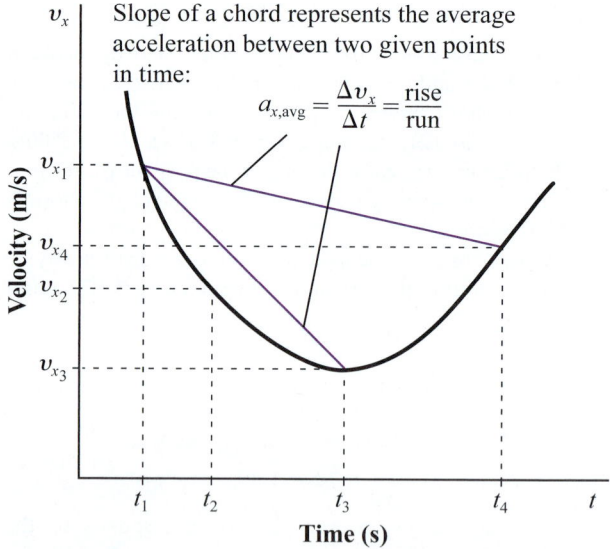

Figure 3-15 The average acceleration between any two points in time is given by the slope (rise over run) of the chord connecting the two points on the $v_x(t)$ graph.

EXAMPLE 3-5

Can You Have a Non-zero Average Acceleration while Moving at a Constant Speed?

A car races down the highway at a speed of 110. km/h, the highway loops around the base of a mountain, and 6.00 min later the car is moving at the same speed but now in the opposite direction. Calculate the average acceleration of the car.

Solution

This is a one-dimensional motion, but the car is moving along a curve, not along a straight line as before. The acceleration of the car represents the change in its velocity, which includes both its magnitude and its direction. Here the speed has not changed, but the direction of the car's motion did change. Therefore, the car must have had a non-zero average acceleration. Using Equation 3-7, and choosing the original direction of the car to be aligned with the positive x-axis, the initial velocity of the car is 110 km/h and the final velocity is −110 km/h. To calculate the acceleration of the car, we need to convert these values into SI units:

$$v_{x_1} = (110 \text{ km/h}) \times \frac{1000 \text{ m}}{1.00 \text{ km}} \times \frac{1.00 \text{ h}}{3600 \text{ s}} = 30.6 \text{ m/s}$$

$$v_{x_2} = (-110 \text{ km/h}) \times \frac{1000 \text{ m}}{1.00 \text{ km}} \times \frac{1.00 \text{ h}}{3600 \text{ s}} = -30.6 \text{ m/s}$$

Therefore, the average acceleration of the car is

$$a_{x,\text{avg}} = \frac{\Delta v_x}{\Delta t} = \frac{v_{x_2} - v_{x_1}}{t_2 - t_1} = \frac{(-30.6 \text{ m/s}) - (30.6 \text{ m/s})}{360. \text{ s} - 0 \text{ s}}$$

$$= -0.170 \text{ m/s}^2$$

Making sense of the result

The initial and final values of the speed of the car are the same. However, the car's direction of motion has changed, so the car had a non-zero average acceleration. When considering acceleration, changes in both the speed and the direction of motion matter. In this case, with our choice of positive being in the original direction of the car's motion, the velocity has gone from being positive to being negative while the speed stayed constant. As a result, the acceleration of the car is also negative. Notice that in this case we assumed the car turned around instantaneously.

Instantaneous Acceleration

Let us consider Figure 3-16 and allow a time interval, Δt, between any two points on the $v_x(t)$ graph to decrease gradually. For example, let us make times t_1 and t_3 approach each other so they come infinitely close to time t_2. When this happens, the average acceleration between times t_1 and t_3 will approach the value of the instantaneous acceleration at time t_2. Graphically, this means that the slope of the resulting chord (shown by the orange line connecting points on the graph at

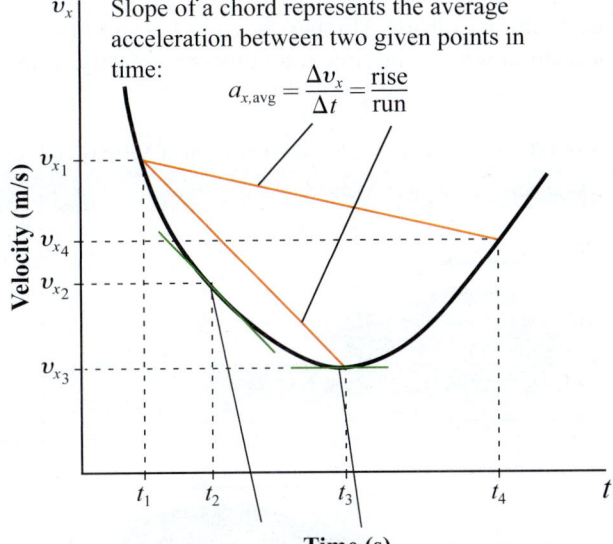

Slope of a chord represents the average acceleration between two given points in time:

$$a_{x,\text{avg}} = \frac{\Delta v_x}{\Delta t} = \frac{\text{rise}}{\text{run}}$$

Slope of a tangent to a $v_x(t)$ graph at point t represents the instantaneous acceleration at this time:

$$a_x(t) = \lim_{\Delta t \to 0} a_{x,\text{avg}} = \lim_{\Delta t \to 0} \frac{\Delta v_x}{\Delta t} = \frac{dv_x}{dt} = \frac{d^2x}{dt^2}$$

Figure 3-16 The average acceleration of an object is represented by the slope of the chord connecting the two ends of the time interval on the $v_x(t)$ plot, while the instantaneous acceleration at time t is represented by the slope (rise over run) of a tangent line to the $v_x(t)$ plot at time t. The instantaneous acceleration is the time derivative of $v_x(t)$. Therefore, $a_x(t) = \dfrac{dv_x(t)}{dt} = v'_x(t)$.

times t_1 and t_3) represents the average acceleration over smaller and smaller time intervals, eventually representing the tangent of the $v_x(t)$ graph at time t_2 shown in green. Thus, the instantaneous acceleration is the derivative of the velocity as a function of time:

KEY EQUATION
$$a_x = \lim_{\Delta t \to 0} \frac{\Delta v_x(t)}{\Delta t} = \frac{dv_x(t)}{dt} = v'_x(t) \qquad (3\text{-}8)$$

On a velocity versus time plot, the instantaneous acceleration at any point is the slope of the tangent to the plot at that point (Figure 3-16).

Since the instantaneous velocity can be represented as the derivative of position with respect to time, and the instantaneous acceleration is the derivative of velocity with respect to time, the instantaneous acceleration is equal to the *second* derivative of position with respect to time. The second derivative is simply a way of writing the derivative of a derivative:

$$a_x = \lim_{\Delta t \to 0} \frac{\Delta v_x(t)}{\Delta t} = \frac{dv_x(t)}{dt} = \frac{d}{dt}\left(\frac{dx(t)}{dt}\right) = \frac{d^2x(t)}{dt^2} = x''(t)$$

$$(3\text{-}9)$$

The relative directions of an object's velocity and acceleration indicate whether the object is speeding up, slowing down, or moving at a constant velocity. These relationships for one-dimensional motion are summarized in Table 3-3. The right column shows a $v_x(t)$ graph for the case of constant acceleration.

Table 3-3 The Effect of the Relative Direction of Velocity and Acceleration on the Object's Motion in the Case of Constant Acceleration

Visual Representation of Motion	Description of Motion	v_x	a_x	$v_x(t)$, when a_x = const
a_x v_x	Speeding up in the positive x-direction	+	+	
a_x v_x	Slowing down in the negative x-direction	−	+	
a_x v_x	Slowing down in the positive x-direction	+	−	
a_x v_x	Speeding up in the negative x-direction	−	−	

PEER TO PEER

I realize that the words I use to describe motion are very important. For example, I try to avoid using the term *deceleration* when describing motion. This word is misleading as it might mean different things depending on the direction of the object's motion. Instead, I say that the object is speeding up or slowing down. In each case, the acceleration can be positive or negative, and depending on the direction of motion, it will mean the object will be either slowing down or speeding up (Table 3-3).

I am also careful to distinguish between the words *drop* and *throw*. To drop means to let go of something (zero initial velocity relative to the person who dropped it, $v_0 = 0$); to throw means to impart an initial velocity (speed and direction) to an object (non-zero initial velocity, $v_0 \neq 0$). For example, if I drop a rock while riding my bike, the initial velocity of the rock relative to me or to my bike will be zero, but the initial velocity of the rock relative to the ground will be equal to the velocity of the bike relative to the ground.

CHECKPOINT 3-5

Acceleration

If the acceleration of a moving train points in the same direction as its motion, but the acceleration is decreasing in magnitude, the speed of the train is
(a) increasing at a constant rate
(b) increasing at a decreasing rate
(c) increasing at an increasing rate
(d) decreasing at an increasing rate
(e) decreasing at a decreasing rate

ANSWER: (b) There is a non-zero acceleration in the direction of the train's velocity, so its speed is increasing (Table 3-3). Decreasing magnitude of the acceleration indicates that the rate of change of the train's speed is going down (the train is speeding up at a decreasing rate).

INTERACTIVE ACTIVITY 3-2

The Moving Man

In this activity, you will use the PhET simulation "The Moving Man" to investigate the relationship between position, velocity, and acceleration. The simulation will allow you to experiment with changing the acceleration and the velocity of an object and examine the impact on its motion. You will also analyze acceleration, velocity, and position versus time plots.

Acceleration Due to Gravity

If an object is moving near the surface of Earth under the influence of the force of gravity only (neglecting air resistance), it accelerates in a direction toward the centre of Earth. If gravity is the only force acting upon the object, we say the object is in **free fall**. We call the acceleration that the object experiences during a free fall the **acceleration due to gravity**, also called the **gravitational acceleration**, $\vec{g}$. It is directed toward the centre of Earth. The magnitude of this acceleration depends on the location of the object on Earth's surface based on longitude, latitude, and altitude. In terms of latitude, it can vary from $g = 9.780$ m/s^2 at the equator to $g = 9.832$ m/s^2 at the poles. Unless noted otherwise, in this book we will use an average value of $g = 9.81$ m/s^2. We will discuss free fall in more detail in Section 3-6. We will use the acceleration due to gravity many times in this book, and it is worth remembering its average value. Note that the mass and shape of the object do not matter if there is no air resistance; the average value of g is 9.81 m/s^2 regardless of the object's mass and shape.

CHECKPOINT 3-6

Acceleration in Free Fall

Three objects are experiencing free fall near the surface of Earth. Object A was dropped from the third floor of a building, object B was thrown upward, and object C was thrown downward. During the motion of these objects, what is true about the magnitudes of their accelerations? (Consider air resistance to be insignificant.)
(a) $a_A > a_B = a_C$
(b) $a_A < a_B = a_C$
(c) $a_A < a_B < a_C$
(d) $a_A > a_B > a_C$
(e) $a_A = a_B = a_C$

ANSWER: (e) These three objects are experiencing free fall—they are only influenced by the gravitational attraction of Earth. Their initial velocities are different, but these velocities have no impact on the accelerations of the objects during their motion.

MAKING CONNECTIONS

Noteworthy Accelerations

Table 3-4 lists some common accelerations in nature and everyday life.

Table 3-4 Noteworthy Accelerations

Description	Acceleration	Acceleration in SI Units
Radial acceleration at Earth's equator due to Earth's rotation	0.033 73 m/s^2	0.033 73 m/s^2
Starting acceleration of Usain Bolt in his winning 100 m dash in 2009 in Berlin (0.977g)	9.58 m/s^2	9.58 m/s^2
Typical acceleration due to gravity at Earth's surface at sea level (g)	9.81 m/s^2	9.81 m/s^2
Acceleration of fastest unmodified production cars (calculated via measuring the time it takes the car to reach 100 km/h from rest)	~1.3g	~13 m/s^2
Typical transverse acceleration of a Formula One car on a curve	1.7g	17 m/s^2
Acceleration experienced by astronauts during takeoff and re-entry for both the space shuttle and the *Soyuz*	~4g	40 m/s^2

(continued)

Table 3-4 Noteworthy Accelerations (*continued*)

Description	Acceleration	Acceleration in SI Units
Typical acceleration of a luge sled at Whistler, British Columbia, during the 2012 Winter Olympics	$5.2g$	51 m/s^2
Typical acceleration of a Top Fuel drag racer	$5.3g$	52 m/s^2
Lateral acceleration, Formula One, Suzuka circuit; Maximum acceleration (while slowing down), Formula One	$6g$	60 m/s^2
Tolerable sustained acceleration during inside loop for pilots	$8g$–$9g$	80 m/s^2 to 90 m/s^2
Threshold acceleration for sustained inside loop (vertical), with specially designed G-suit	$12g$	120 m/s^2
Commercial aircraft seat standard as of 1988	$16g$	160 m/s^2
Injury threshold for upward acceleration for a duration of 100 ms	$18g$	180 m/s^2
Injury threshold for lateral acceleration	$20g$	200 m/s^2
Spinal injury threshold for upward acceleration for a duration of 100 ms	$20g$–$25g$	200 m/s^2 to 250 m/s^2
Tolerance standard for U.S. military aircraft seats	$32g$	310 m/s^2
Highest acceleration survived during a crash	$\sim214g$	~2100 m/s^2
Approximate maximum acceleration of a golf ball in competition (Jason Zuback, 2009)	$18\,000g$	$180\,000$ m/s^2
Gravitational acceleration on the surface of a neutron star	$18\,000g$	$180\,000$ m/s^2

Since instantaneous velocity and acceleration represent the rate of change of position and velocity with time, respectively, we can use the concepts of calculus to calculate these values. As we pointed out earlier, in calculus, the rate of change of a function is represented by its derivative: $\dfrac{df(t)}{dt} = f'(t)$. If you haven't studied calculus yet, you might decide to omit this example.

EXAMPLE 3-6

Using Derivatives to Calculate Velocity and Acceleration

The position in metres (as a function of time, in seconds) for a particle moving along the x-axis is given by $x(t) = -0.500t^4 + 2.50t^3 - 7.00t + 3.00$. Find

(a) the instantaneous velocity of the particle at $t_1 = 2.00$ s
(b) the instantaneous acceleration of the particle at $t_1 = 2.00$ s
(c) the average acceleration of the particle between $t_1 = 2.00$ s and $t_2 = 3.00$ s
(d) the instantaneous acceleration of the particle at $t_3 = 2.50$ s
(e) the maximum speed the particle reaches in the first 5.00 s of its motion

Solution

(a) To find the velocity from the position versus time plot, we use Equation 3-6. We first find an expression for velocity as a function of time, and then input the time. Notice, in determining significant figures, we treat all the exponents as exact numbers.

$$v_x(t) = \frac{dx(t)}{dt} = (-0.500(4t^3) + 2.50(3t^2) - 7.00)\ \text{m/s}$$

$$= (-2.00t^3 + 7.50t^2 - 7.00)\ \text{m/s}$$

At $t_1 = 2.00$ s,

$$v_{x_1} = v_x(t_1) = v_x(2.00\ \text{s})$$

$$= (-2.00(2.00)^3 + 7.50(2.00)^2 - 7.00)\ \text{m/s} = 7.00\ \text{m/s}$$

(b) To find the acceleration at t_1, we first take the derivative of the velocity with respect to time:

$$a_x(t) = \frac{dv_x(t)}{dt} = (-2.00(3t^2) + 7.50(2t))\ \text{m/s}^2$$

$$= (-6.00t^2 + 15.0t)\ \text{m/s}^2$$

At $t_1 = 2.00$ s,

$$a_{x_1} = a_x(2.00\ \text{s}) = (-6.00(2.00)^2 + 15.00(2.00))\ \text{m/s}^2$$

$$= 6.00\ \text{m/s}^2$$

(c) For the average acceleration, we use Equation 3-7:

$$a_{x,\text{avg}} = \frac{\Delta v_x}{\Delta t} = \frac{v_{x_2} - v_{x_1}}{t_2 - t_1}$$

Here, v_{x_1} and v_{x_2} are the velocities at 2.00 s at 3.00 s, respectively, so

$$v_{x_1} = v_x(t_1) = v_x(2.00\ \text{s}) = 7.00\ \text{m/s} \ (\text{found earlier})$$

$$v_{x_2} = v_x(t_2) = v_x(3.00\ \text{s})$$

$$= (-2.00(3.00)^3 + 7.50(3.00)^2 - 7.00)\ \text{m/s}$$

$$= 6.50\ \text{m/s}$$

$$a_{x,\text{avg}} = \frac{v_{x_2} - v_{x_1}}{t_2 - t_1} = \frac{6.50\ \text{m/s} - 7.00\ \text{m/s}}{3.00\ \text{s} - 2.00\ \text{s}} = -0.500\ \text{m/s}^2$$

(d) The instantaneous acceleration of the particle at $t_3 = 2.50$ s is

$$a_{x_3} = a_x(t_3) = a_x(2.50 \text{ s})$$

$$= (-6.00(2.50)^2 + 15.00(2.50)) \text{ m/s}^2 = 0 \text{ m/s}^2$$

The instantaneous value of the acceleration is not exactly the same as the average value found in (c), but if we were to shrink the time interval between the two points in time, the average acceleration value would approach its instantaneous value.

(e) The maximum (or minimum) speed occurs when the time derivative of the velocity with respect to time is zero, which occurs when the acceleration is zero:

$$a_x(t) = 0 \Rightarrow (-6.00t^2 + 15.0t) \text{ m/s}^2 = 0 \Rightarrow$$

$$t_1 = 0 \text{ s}; t_2 = 2.50 \text{ s}$$

At these times, the acceleration of the particle will be zero and the velocities of the particle will be v_{x_3} and v_{x_4}, respectively:

$$v_{x_3}(0 \text{ s}) = -7.00 \text{ m/s}$$

$$v_{x_4}(2.50 \text{ s}) = 8.62 \text{ m/s}$$

You can check that the first value is the local minimum of the velocity and the second value is the local maximum. These are the minimum and maximum values in the first 5.00 s, but not for the entire motion, as seen in Figure 3-17.

Making sense of the result

Using the visual representations of the $x(t)$, $v_x(t)$, and $a_x(t)$ graphs helps us see that the slopes of the $x(t)$ and $v_x(t)$ graphs at a certain point represent instantaneous velocity and acceleration values, respectively, at that point. For example, at the points where the $x(t)$ graph reaches its maximum or minimum values, the velocity is zero. At the points where the velocity reaches its maximum or minimum values, the acceleration is zero. Plotting the graphs by hand or using graphing software is very helpful for acquiring an intuitive understanding of kinematics.

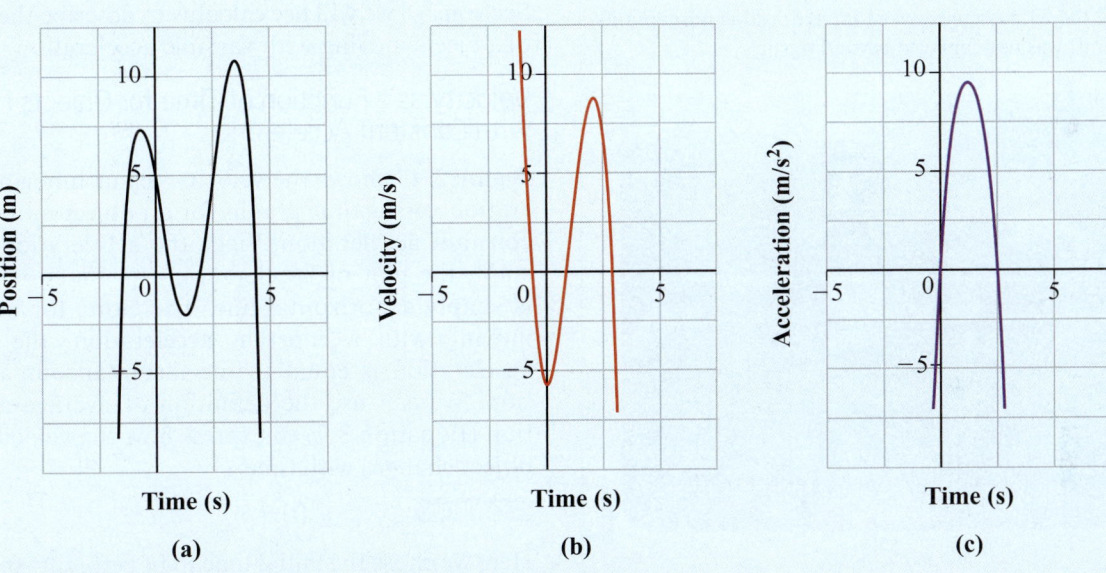

Figure 3-17 $x(t)$, $v_x(t)$, and $a_x(t)$ graphs for the particle motion in Example 3-6.

CHECKPOINT 3-7

Average Acceleration

You throw a ball straight down with an initial speed of 5.00 m/s. The ball then bounces off the ground and moves back up. You catch it 2.00 s later when it reaches the initial height having a speed of 5.00 m/s. If the positive x-axis is directed upward, what is the average acceleration of the ball during its motion?

(a) 9.81 m/s^2

(b) −9.81 m/s^2

(c) 5.00 m/s^2

(d) −5.00 m/s^2

(e) zero

ANSWER: (c) We can use Equation 3-7 to find the average acceleration. The change in velocity is $\Delta v_x = 5.00$ m/s $-$ $(-5.00$ m/s$) = 10.00$ m/s. The time elapsed is 2.00 s, so the average acceleration is 5.00 m/s^2. Notice that, during its entire motion (except for the collision with the floor), the ball has an acceleration due to gravity directed downward, -9.81 m/s^2. However, the average acceleration for the entire trip includes the collision with the ground, which generated a significant upward acceleration. As a result, the average acceleration for the entire motion is different from g. The entire motion of the ball cannot be considered as free-fall motion, as during a portion of its motion, the ball interacted with the ground. Only on the way down and on the way up was the ball in free fall. Notice that as long as the ball is influenced by the force of gravity only, we say that it is in free fall, even when it is moving up. We will discuss this in more detail in Section 3-6 of this chapter.

EMAS Runway Extensions

Modern jet aircraft require longer runways for takeoff and landing than older propeller aircraft. At most airports, runways have been upgraded and lengthened to safely handle both jets and propeller planes under different weather and equipment failure conditions. At some airports, it is not possible to fully extend the runways for geographic or infrastructure reasons. One solution used at some airports to address this is to use "foamcrete" at the ends of their runways. Foamcrete is an example of an Engineered Material Arresting System (EMAS). These types of materials are designed to crumble under the weight of an aircraft, which helps to stop the aircraft more quickly and over a shorter distance, significantly lowering the likelihood of significant damage to the aircraft or harm to the passengers inside. Figure 3-18 shows a Bombardier CRJ 200 that plowed through an EMAS after it overran a runway due to an aborted takeoff. Fortunately, none of the 34 passengers were injured in this mishap and the aircraft was not damaged beyond repair.

NTSB

Figure 3-18 A PSA Bombardier CRJ200 jet stopped in the EMAS installation at Charleston, South Carolina, on January 19, 2010.

3-4 Mathematical Description of One-Dimensional Motion with Constant Acceleration

We started this chapter by stating that the main goal of one-dimensional kinematics is to describe the motion of objects moving along a straight line. Specifically, in kinematics, provided that the initial information about an object's motion is given, we want to predict its location, speed, velocity, and acceleration at different points along its trajectory. In this section, we focus on developing a mathematical description of the motion

of objects moving with **constant (uniform) acceleration** (acceleration that is not changing with time). While in most everyday life situations, acceleration is not strictly constant, scenarios involving approximately constant acceleration are quite common. For example, gravitational acceleration is very nearly constant as long as we stay close to Earth's surface. The mathematical treatment required for constant-acceleration problems is relatively straightforward, and the resulting equations are elegant and have a beautiful graphical interpretation. In what follows we examine how the position and velocity of an object moving with constant acceleration depend on time. The equations we derive are referred to as the **kinematics equations for constant acceleration**. These equations are widely used when discussing mechanics, electricity and magnetism, and many everyday life applications. Since we consider constant acceleration, we can use simple algebraic and geometric arguments to develop these equations. In Section 3-8 we will use calculus to describe the motion of objects moving with variable acceleration.

Velocity as a Function of Time for Objects Moving with Constant Acceleration

Figure 3-19 shows the velocity versus time and acceleration versus time graphs for an object moving with constant acceleration. Since the acceleration is constant, the plot of acceleration as a function of time is simply a horizontal line. Therefore, for an object moving with a constant acceleration, the average acceleration is equal to the instantaneous acceleration. We can use the definition of average acceleration (Equation 3-7) to express how the velocity of an object changes with time:

KEY EQUATION
$$v_x(t) = v_{x_0} + a_x t \qquad (3\text{-}10)$$

Here we chose the initial time to be zero, $t_0 = 0$, and the velocity at this time to be v_{x_0}, while the final time and velocity are denoted as t and v_x, respectively.

Equation 3-10 indicates that the final velocity of an object consists of two contributions: its initial velocity, v_{x_0}, and the gain in velocity caused by the acceleration over time t, $a_x t$. The acceleration, a_x, is represented by the slope of the $v_x(t)$ graph, while the velocity axis intercept on the $v_x(t)$ graph represents the initial velocity, v_{x_0}.

Equation 3-10 allows us to predict the final velocity of an object at any given time provided we know its initial velocity and its uniform acceleration during that time period.

It is important to notice that since the acceleration of an object is constant, the average velocity equals the mean value of the initial and final velocities of the object:

$$v_{x,\text{avg}} = \frac{v_{x_0} + v_x(t)}{2} \qquad (3\text{-}11)$$

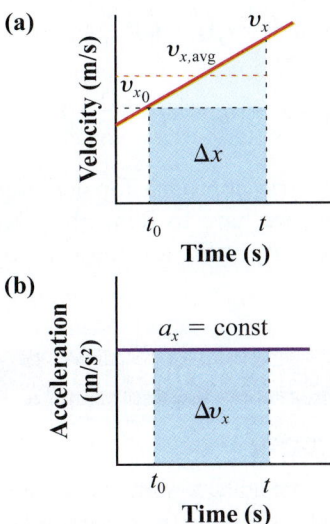

(a)

(b)

Figure 3-19 Kinematics graphs describing the motion of an object moving with constant acceleration. (a) In a $v_x(t)$ graph, the area under the graph represents the object's displacement. The average velocity is represented by the horizontal dashed red line; the dashed black line represents how the object would move if its acceleration were zero. (b) In an $a_x(t)$ graph, the area under the graph represents the change in the object's velocity.

Remember though that Equations 3-10 and 3-11 are only valid when the acceleration is constant. We can visualize these relationships using the graphs in Figure 3-19. The area under the $a_x(t)$ graph has units of velocity and it represents the change in the object's velocity between the times t_0 and t. We can check it using dimensional analysis:

$$\left[\text{Area}_{\text{under } a_x(t) \text{ graph}} \right] = \frac{m}{s^2} \times s = \frac{m}{s} \qquad (3\text{-}12)$$

This interpretation of the area under the $a_x(t)$ graph is valid for motion with constant as well as varying acceleration. It is often referred to as the **principle of graphical integration**. In this case it is applied to an $a_x(t)$ graph, but it can be applied to other graphs as well. When the acceleration is constant, it is easy to find this area:

$$\text{Area}_{\text{under } a_x(t) \text{ graph}} = a_x \Delta t \qquad (3\text{-}13)$$

The average velocity can be visualized by the dashed horizontal red line shown in Figure 3-19(a). However, when the acceleration is changing, we will need to turn to calculus and integration to replace Equations 3-10 and 3-11 with the modified expressions. We will do this in Section 3-8.

Position as a Function of Time for Objects Moving with Constant Acceleration

As we saw above, the velocity of an object moving with constant acceleration depends linearly on time (Equation 3-10, Figure 3-19). We will now derive an expression to describe how the position of an object moving with a constant acceleration depends on time,

$x(t)$. As we did before, let us take the initial time to be zero. We will denote the initial position of the object as x_0, and the later time and position will be denoted as t and x, respectively. Using the definition of average velocity (Equation 3-4), we obtain

$$v_{x,\text{avg}} = \frac{\Delta x}{\Delta t} = \frac{x(t) - x_0}{t - 0} \Rightarrow x(t) = x_0 + v_{x,\text{avg}} t \qquad (3\text{-}14)$$

Substituting the value for average velocity from Equation 3-11 into Equation 3-14, we obtain the following result:

$$x(t) = x_0 + \left(\frac{v_{x_0} + v_x(t)}{2} \right) t \qquad (3\text{-}15)$$

Now using Equation 3-10 to substitute for the final velocity, we find

$$x(t) = x_0 + \left(\frac{v_{x_0} + v_x(t)}{2} \right) t = x_0 + \left(\frac{v_{x_0} + v_{x_0} + a_x t}{2} \right) t \qquad (3\text{-}16)$$

Equation 3-16 can be further simplified to provide an expression for the position of an object as a function of time:

KEY EQUATION $\quad x(t) = x_0 + v_{x_0} t + \dfrac{1}{2} a_x t^2 \quad$ when $t_0 = 0$ (3-17)

The position of the object at time t depends on its initial position, x_0; initial velocity, v_{x_0}; and acceleration, a_x. This is a quadratic equation, which can be represented graphically as a parabola. The leading coefficient, the acceleration, represents how wide or narrow the parabola is. Positive acceleration means that the parabola opens upward, while negative acceleration means that the parabola opens downward.

Equation 3-17 could have been derived using the geometrical interpretation of the area under the $v_x(t)$ graph (Figure 3-19(a)). This area represents the displacement of an object as it moves from t_0 to t. It is often referred to as the principle of graphical integration with regard to a $v_x(t)$ graph. We can check the validity of this claim by using dimensional analysis, as we did earlier with the $a_x(t)$ graph:

$$\left[\text{Area}_{\text{under } v_x(t) \text{ graph}} \right] = \frac{m}{s} \times s = m \qquad (3\text{-}18)$$

This area, in the case of motion with constant acceleration, is the area of a trapezoid, found using the well-known geometric formula $A = \dfrac{a + b}{2} h$, where a and b are the bases of the trapezoid and h is its height. Applying this geometric formula to find the area of the trapezoid sketched out in Figure 3-19(a), we confirm the validity of Equation 3-17. It is left as an exercise for you to confirm this.

Both Equations 3-10 and 3-17 involve time explicitly. In using these two relationships, we need to know how much time has elapsed to see how the position and the velocity of the moving object have evolved. There are, however, times when the information of the time is not readily available, or needed. For these situations we develop a time-independent relationship between an object's position, its initial and final velocities, and its acceleration.

To eliminate time from Equation 3-15, we express it through initial and final velocities and acceleration, using Equation 3-11 for average velocity:

$$t = \frac{v_x(t) - v_{x_0}}{a} \tag{3-19}$$

Now let us substitute Equation 3-19 into Equation 3-15:

$$x(t) = x_0 + \left(\frac{v_{x_0} + v_x(t)}{2} \right) t$$

$$= x_0 + \frac{v_{x_0} + v_x(t)}{2} \cdot \frac{v_x(t) - v_{x_0}}{a_x} \tag{3-20}$$

$$= x_0 + \frac{v_x(t)^2 - v_{x_0}^2}{2a_x}$$

To obtain this expression, we used the well-known algebraic difference of squares identity, $(a - b)(a + b) = a^2 - b^2$.

Equation 3-20 can also be expressed in the following form, describing the motion using displacement rather than position:

$$x(t) = x_0 + \frac{v_x(t)^2 - v_{x_0}^2}{2a_x} \text{ or } \Delta x(t) = \frac{v_x(t)^2 - v_{x_0}^2}{2a_x} \tag{3-21}$$

This expression can be rearranged as:

KEY EQUATION $$v_x(t)^2 - v_{x_0}^2 = 2a_x \Delta x(t) \tag{3-22}$$

We can also consider the area under the $v_x(t)$ graph (Figure 3-19(a)) as consisting of the area of a rectangle, $A_{\text{rectangle}} = v_{x_0}(t - t_0)$, and a triangle, $A_{\text{triangle}} = \frac{1}{2}(t - t_0)(v_x - v_{x_0})$. The area of the rectangle represents the displacement of an object that is moving at constant velocity v_{x_0} for $(t - t_0)$ seconds. The area of the triangle represents the displacement of the object due to its acceleration. The area of the triangle can be expressed using Equation 3-10: $A_{\text{triangle}} = \frac{1}{2}a_x(t - t_0)^2 = \frac{1}{2}a_x \Delta t^2$. The total displacement consists of two of these contributions:

$$\Delta x = v_{x_0}(t - t_0) + \frac{1}{2}a_x(t - t_0)^2 = v_{x_0}\Delta t + \frac{1}{2}a_x\Delta t^2$$

$$t_0 = 0 \Rightarrow \Delta x = x - x_0 = v_{x_0}t + \frac{1}{2}a_xt^2$$

This was a geometric approach for deriving Equation 3-17. Once again, we have to remember that all these relationships are only valid for the case of constant acceleration.

PEER TO PEER

When I need to calculate the average speed of an object that has various speeds along its trajectory, I think of calculating my GPA. The average of my grades is not equal to the mean of my grades, as different courses have different weights (the number of credits associated with the course). The concept of course weight in terms of calculating my GPA is equivalent to the concept of time when calculating the average speed. The longer I drive at a certain speed, the more impact it has on the average speed of my entire trip, just like courses that have more credits impact my GPA more significantly.

EXAMPLE 3-7

Braking Time

A car is moving at 92 km/h when the driver realizes that there is a traffic jam 80. m ahead, and the highway is blocked. Assuming that the average reaction time for a driver is 0.75 s, determine the following.

(a) What is the minimum acceleration (in absolute value) the car must undergo to come to a stop without colliding with the cars ahead?

(b) How much time does it take for the car to stop?

Solution

(a) We know the speed of the car, the time it takes the driver to apply the brakes, and the distance between the car and the traffic jam. We also know the velocity of the car since the car didn't change its direction. The car's displacement is also known as we know the car's initial and final positions. We choose the positive x-direction to be in the direction of the car's motion.

Let us consider two parts of the car's motion. During the first part, the car keeps moving at a constant velocity. During the second part, it begins to slow down. The total distance travelled by the car in both parts must not exceed 80 m.

Part 1: Since we will be using SI units to solve the problem, let us first convert the car's initial velocity to m/s. We will also remember that this velocity is positive because it is directed in the positive x-direction (we will keep three significant figures in the intermediate calculations):

$$v_{x_0} = 92 \frac{km}{h} \times \frac{1000 \text{ m}}{1.00 \text{ km}} \times \frac{1.00 \text{ h}}{3600 \text{ s}} = 25.6 \text{ m/s}$$

Since it takes the driver 0.75 s to apply the brakes (Δt_1), and during this time the car moves at a constant velocity of 25.6 m/s, using Equation 3-3 we can find the car's displacement:

$$\Delta x_1 = v_{x_0} \Delta t_1 = \left(25.6 \frac{m}{s}\right)(0.75 \text{ s}) = 19.2 \text{ m}$$

Note that we are keeping an extra significant figure in the intermediate steps, but we will express the final answer using the correct number of significant figures. This leaves the car a distance of 80.0 m − 19.2 m = 60.8 m from the traffic jam at the beginning of the slowing-down phase. Therefore, the driver needs to apply the brakes to bring the car to a full stop over this distance.

Part 2: We know the speed of the car at the beginning of the slowing-down phase (v_{x_0} = 25.6 m/s), the distance the car has to come to a complete stop (Δx_2 = 60.8 m), and its final velocity (since it has to stop, v_{x_1} = 0 m/s). Therefore,

we can use Equation 3-22 to find the car's acceleration. By using this equation, we are implicitly assuming that the acceleration is constant:

$$v_{x_1}^2 - v_{x_0}^2 = 2a_x \Delta x_2 \Rightarrow a_x = \frac{v_{x_1}^2 - v_{x_0}^2}{2\Delta x_2}$$

Substituting the known values into this equation, we find the car's acceleration:

$$a_x = \frac{0 - (25.6 \text{ m/s})^2}{2(60.8 \text{ m})} = -5.39 \text{ m/s}^2 \approx -5.4 \text{ m/s}^2$$

The negative sign indicates that the acceleration points in the negative x-direction, which is opposite to the original direction of motion, so the car will be slowing down while moving in the positive x-direction (Table 3-3). Notice that, in its absolute value, this is the minimum acceleration needed for the car to stop. Any other negative acceleration that has a larger absolute value (for example, $a_{x_1} = -6.0 \text{ m/s}^2$) will stop the car even faster.

(b) To find the time it takes the car to stop with the acceleration we found in part (a), $a_x = -5.39 \text{ m/s}^2$, we can use Equation 3-11 since we know the car's initial and final velocities and its acceleration:

$$v_{x_1} = a_x \Delta t_2 + v_{x_0} \Rightarrow \Delta t_2 = \frac{v_{x_1} - v_{x_0}}{a_x}$$

$$\Delta t_2 = \frac{0 - 25.6 \text{ m/s}}{-5.39 \text{ m/s}^2} = 4.75 \text{ s}$$

During the calculation for total stopping time of 4.75 s + 0.75 s = 5.50 s, we kept extra digits in the intermediate steps. This is done to reduce the calculation rounding error, but the final result is precise to two significant figures only, so the final answer should be expressed as 5.5 s.

Making sense of the result

Note the significant time and distance for a car to come to a full stop. Since the stopping distance is proportional to a square of the initial velocity (Equation 3-20), the initial velocity has a very significant impact on the stopping distance. Doubling the initial velocity quadruples the stopping distance during the slowing-down phase, as shown by examining the ratio of the displacement term to the initial velocity term in Equation 3-22. This relationship between displacement and initial velocity is the reason school zone speed limits in Canada and around the world are so low. It is important to remember this relationship when driving while feeling tired. When you are tired, your reaction time increases significantly, making it harder to drive safely (see Data-Rich Problem 135 to find out how you can measure your own reaction time).

Table 3-5 Conceptual Summary of Kinematics Equations for One-Dimensional Motion with Constant Acceleration

Quantity	Meaning	Formulas	Cases When the Formula Is Useful
Displacement, Δx	Shows how much and in what direction the object was displaced by comparing its original position, x_0, to its final position, x: $\Delta x = x - x_0$ Units: $[\Delta x] = $ m	$\Delta x = v_{x_0}\Delta t + \dfrac{a_x \Delta t^2}{2}$ $t_0 = 0 \Rightarrow \Delta x = v_{x_0}t + \dfrac{a_x t^2}{2}$	When you are given the initial velocity, v_{x_0}; the elapsed time, Δt; and the acceleration, a_x; when $t_0 = 0$, Δt is equal to t.
		$\Delta x = \dfrac{v_x^2 - v_{x_0}^2}{2a_x}$	When you are given the initial and final velocities (v_0 and v) and the acceleration (a), but you do not have information about the time
		$\Delta x = v_{x,\text{avg}}\Delta t$	When you know the average velocity, $v_{x,\text{avg}}$, and the time, t
Velocity, v_x	Shows how fast the displacement is changing; the rate of change of displacement Units: $[v_x] = $ m/s	$v_x = \dfrac{\Delta x}{\Delta t}$	When the elapsed time, Δt, is large, this ratio represents an average velocity. When Δt approaches zero, the ratio represents an instantaneous velocity.
Acceleration, a_x	Shows how fast the velocity is changing; the rate of change of velocity Units: $[a_x] = $ m/s²	$a_x = \dfrac{\Delta v_x}{\Delta t}$	When the elapsed time, Δt, is large, this ratio represents an average acceleration. When Δt approaches zero, the ratio represents an **instantaneous acceleration**.

Table 3-5 summarizes the relationships discussed in this section.

3-5 Analyzing the Relationships between $x(t)$, $v(t)$, and $a(t)$ Plots

In the previous section, we discussed the graphical interpretation of the kinematics equations and the principle of graphical integration. We have shown that the areas under the $v_x(t)$ and $a_x(t)$ graphs represent the object's displacement and the change in its velocity, respectively. We have also shown that the slopes of the $x(t)$ and $v_x(t)$ graphs at a certain time represent the object's instantaneous velocity and acceleration at this time, respectively. This means that position, velocity, acceleration, and time are intimately related. As long as we can obtain the position and time values for the object's motion, for example, via videotaping the motion or using a motion sensor to collect motion data, we can find the values of the object's velocity and acceleration. Example 3-8 shows how a motion sensor coupled with its supporting software can be used to collect and analyze real-time motion data.

EXAMPLE 3-8

A Graph-Matching Game: Using a Motion Detector to Analyze Motion

A motion detector (or sonic ranger) is a device that uses ultrasound to measure the position of objects (Figure 3-20(a)). If an object is moving along a straight line, then a single motion detector is sufficient to determine its location. A motion detector is a "digital bat," as it uses **echolocation** to determine where objects are in space and then deduce how they are moving. Echolocation uses the time it takes a sound pulse emitted from a detector to reflect off an object and return to the detector to determine the distance from the object to the detector. Since a motion detector can detect sound pulses many times a second, it can provide ample data about an object's motion. Accompanying software, such as Logger *Pro*, can visualize these data points by plotting real-time $x(t)$, $v_x(t)$, and $a_x(t)$ graphs. A student is asked to walk in front of the motion detector (Figure 3-20(a)) to match the given $x(t)$ graph shown in Figure 3-20(b). Provided that the positive

x-axis is directed away from the motion detector toward the student, and the origin of the x-axis by default is located inside the motion detector, answer the following questions about each one of the four segments of the given $x(t)$ plot.

(a) Describe how the student has to move to match the graph, including her initial position, velocity, and acceleration.
(b) Plot corresponding $v_x(t)$ and $a_x(t)$ graphs for each one of the segments.
(c) What are the student's displacement and distance during her trip?
(d) What are the student's average velocity and average speed during her trip?

Solution

Since the position versus time plot consists of five linear segments, the student should walk with constant but different velocities in each one of these segments. Since the velocities are constant, the acceleration of the student is always zero (except at the brief

(a)

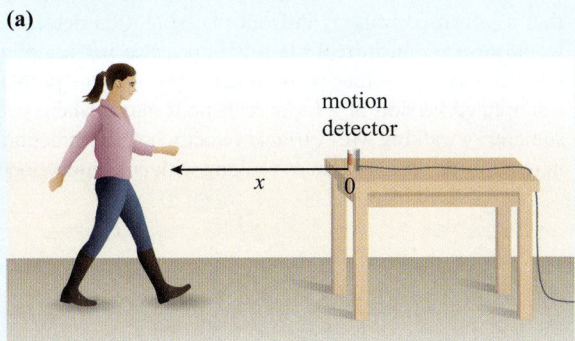

(b)

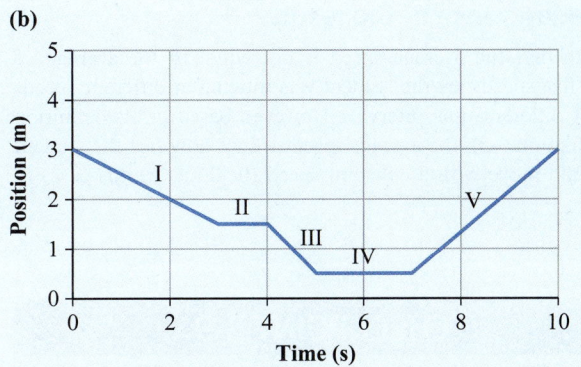

Figure 3-20 A graph-matching game. (a) A student is walking back and forth in front of a motion detector. (b) The $x(t)$ graph she is asked to match (reproduce).

instants when her motion changes, but we will not include those instants in this discussion and instead focus on the segments themselves). Since the slope of the $x(t)$ graph represents the student's velocity, we can find the slope within each one of the segments. Positive slope indicates that the student should be moving away from the detector, while negative slope means that she needs to move toward the detector. A zero slope means that the student should remain motionless (zero velocity). Notice, since the position versus time graph of her motion is represented by straight line segments, the velocities during each one of the five time intervals must be constant. Therefore, the accelerations of the girl during each one of the segments must be zero.

(a) Description of student's motion:

Segment I: The student starts from the point located 3.0 m from the front of the motion detector ($x_0 = 3.0$ m) and walks in the negative direction (toward the motion detector) with a constant velocity of

$$v_I = \frac{1.5\text{ m} - 3.0\text{ m}}{3.0\text{ s}} = -0.50\text{ m/s}.$$

Segment II: The student is not moving; she is standing still for 1.5 s at the location $x_{II} = 1.5$ m from 3.0 s to 4.0 s. Therefore, her velocity is zero: $v_{II} = 0$ m/s.

Segment III: The student is walking with constant velocity toward the motion detector (in the negative x-direction) from the point located 1.5 m in front of the motion detector to the point located 0.5 m in front of the detector.

Her velocity is $v_{III} = \dfrac{0.5\text{ m} - 1.5\text{ m}}{1.0\text{ s}} = -1.0\text{ m/s}.$

Segment IV: The student is once again not moving; she is standing still for 2.0 s at the location $x_{IV} = 0.05$ m from 5.0 s to 7.0 s. Therefore, her velocity is zero: $v_{IV} = 0$ m/s.

Segment V: The student starts walking with constant velocity in the positive x-direction (away from the motion detector) from the point located 0.50 m from the motion detector to the point located 3.0 m from the motion detector. She walks for 3.0 s (from 7.0 s to 10.0 s). Her velocity is

$$v_V = \frac{3.0\text{ m} - 0.5\text{ m}}{3.0\text{ s}} = 0.83\text{ m/s}.$$

(b) We will plot the $x(t)$ and $v_x(t)$ graphs in a stacked format (Figure 3-21), so they have aligned time axes indicated by the vertical dashed lines. The $v_x(t)$ graph of the student's motion is plotted based on the answers from part (a).

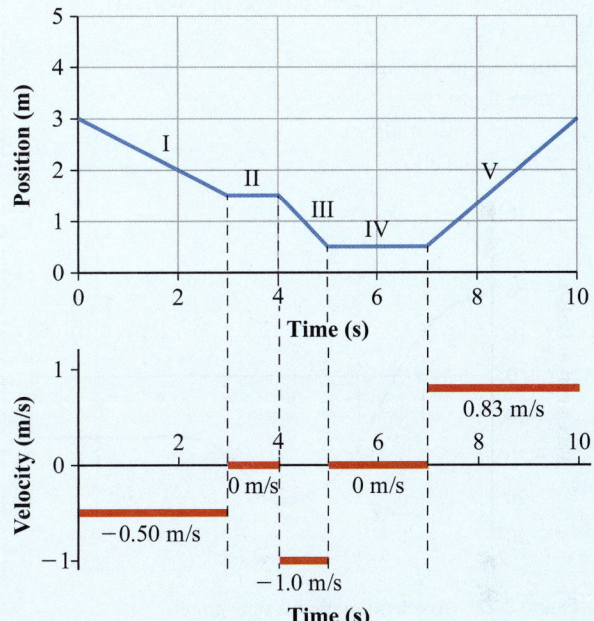

Figure 3-21 Stacked $x(t)$ and $v_x(t)$ plots for Example 3-8. Since the acceleration is zero during these segments, there is no point in plotting it.

(c) To find the student's displacement during the trip, we note where she started and where she ended her walk. Since she started at $x_0 = 3.0$ m and finished at the same position, her displacement is zero: $\Delta x = 0$ m. However, the distance she travelled is not zero. She walked 2.5 m toward the detector during segments I and III and then 2.5 m away from the detector during segment V, so her total distance travelled is 5.0 m.

(d) Since she completed her walk in 10.0 s, her average velocity is zero and her average speed is 0.50 m/s, as shown below:

$$v_{x,avg} = \frac{\Delta x}{\Delta t} = \frac{0\text{ m}}{10.0\text{ s}} = 0\text{ m/s}$$

$$v_{avg} = \frac{d}{\Delta t} = \frac{5.0\text{ m}}{10.0\text{ s}} = 0.50\text{ m/s}$$

(continued)

Making sense of the result

Note that the average speed is not equal to the average of the five speeds, as the student was moving at different speeds over different time intervals. However, based on its definition (Equation 3-3), the average speed defines how fast the student should move with a constant speed (0.50 m/s in this case) to cover the overall distance in the overall time (10.0 s). Also, note that if you tried walking in front of the motion detector, you would notice that in real life it is impossible to "jump" from one velocity to another without accelerating. This problem is a simplified version of a more realistic scenario where you are sometimes walking with variable velocity both in direction and in magnitude and sometimes walking with constant velocity.

CHECKPOINT 3-9

Acceleration versus Time Graph

Figure 3-22 shows an acceleration versus time graph for an object. Positive is taken to be to the right. At $t = 3$ s, the object is
(a) moving to the right
(b) moving to the left
(c) not moving at all
(d) any of the above

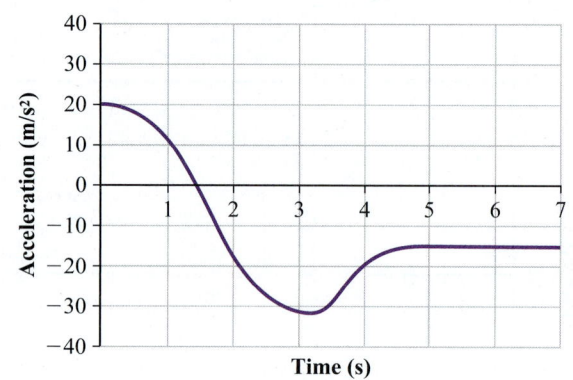

Figure 3-22   Acceleration versus time graph.

ANSWER: (d) The answer could be any of the above, as the graph is showing acceleration, not velocity. Since we do not know anything about the original velocity, we cannot know the direction of the object's motion.

Let us consider an additional example of how we can use kinematics relationships to describe an object's motion.

PEER TO PEER

It took me a while to realize that an $x(t)$ plot of motion does not represent the trajectory of that motion. For example, an $x(t)$ plot of an object accelerating uniformly (such as a ball dropped from a certain height) is a parabola. But the object is moving along a straight line. Now I always pay close attention to the axes of the plots to make sure I interpret them properly.

MAKING CONNECTIONS

Motion Detectors and Biomimicry

Scientists and engineers often use nature to come up with new devices. This approach to technology and innovation, which mimics nature to provide solutions, is called **biomimicry**. For example, a motion detector mimics how bats can "see" in complete darkness using echolocation. To echolocate, bats send out sound waves from their mouth or nose. When these waves encounter an object, they reflect off the object and produce an echo. This echo travels back to the bat's ears. The bat's brain can perceive how far away an object is (based on the speed of sound), thus allowing the bat to catch its prey in complete darkness. Check out this website to learn more about echolocation: askabiologist.asu.edu/echolocation. There are many other examples of biomimicry, such as swimsuits, Velcro, turbines, and jets.

EXAMPLE 3-9

Rocket Car Motion in Graphs

A rocket car moves with a constant speed of 32 m/s for the first 6.0 s of its motion. Following that, the rocket car accelerates forward with a constant acceleration of 18 m/s² for 5.0 s. The driver then applies reverse thrusters to produce an acceleration of −24 m/s² for the following 9.0 s. Use the positive x-direction to represent the initial direction of the rocket car's motion.

(a) Plot the acceleration versus time graph for the rocket car.

(b) Plot the velocity versus time graph for the rocket car.
(c) Find the total displacement of the rocket car from start to finish.

Solution

(a) The acceleration of the rocket car is zero for the first 6.0 s due to its constant velocity; then the acceleration jumps to a value of 18 m/s² for 5.0 s, and then it jumps to −24 m/s² for the next 9.0 s. A plot of acceleration versus time is shown in Figure 3-23.

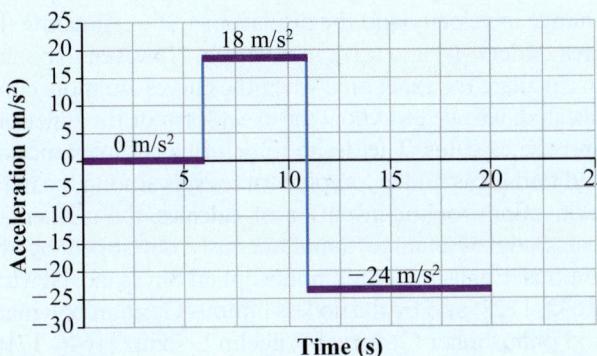

Figure 3-23 The acceleration versus time graph for the rocket car of Example 3-9.

(b) The velocity is constant for the first 6.0 s, which is shown by the zero acceleration value during that period in Figure 3-23. Between 6.0 s and 11.0 s, the change in velocity is given by the area under the acceleration versus time plot. This area is the area of the rectangle of height 18 m/s^2 and width 5.0 s, the length of this time interval, yielding a product (the area) of 90 m/s. This product represents the amount by which the velocity will change over those 5.0 s, so the velocity changes from 32 m/s to 122 m/s over that interval. For the last 9.0 s, the change in the velocity is again given by the area under the acceleration versus time plot. The acceleration is negative, so the total area under the curve is negative, $\Delta v_x = (-24 \text{ m/s}^2)(9.0 \text{ s}) = -216 \text{ m/s}$, and the change in velocity is also negative. At the beginning of this negative acceleration stage, the velocity was 122 m/s, so at the end of 20 s, the velocity will be 122 m/s − 216 m/s = −94 m/s. The entire velocity versus time plot is shown in Figure 3-24.

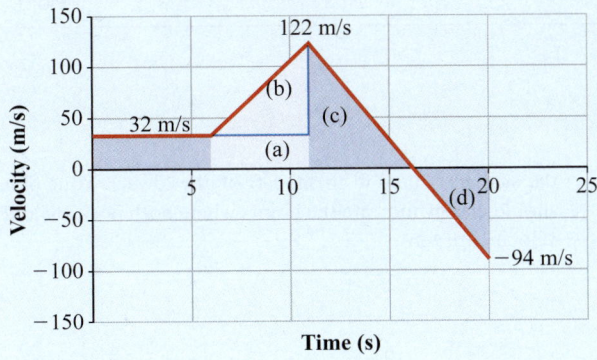

Figure 3-24 Velocity versus time plot for the rocket car of Example 3-9. The area under the graph represents the displacement of the rocket car.

(c) According to the principle of graphical integration, the displacement is given by the area under the velocity versus time curve. For the first 6.0 s, the area of the rectangle is equal to the height, which is the constant velocity of 32 m/s, multiplied by the width, which is 6.0 s. Therefore, the distance covered during the first 6.0 s (which also equal to the displacement in this case) is $\Delta x_1 = 192$ m. For the next 5.0 s, the area under the

velocity curve has two parts, (a) and (b): the shaded rectangle, (a), which has the constant-velocity contribution given by $\Delta x_{2a} = \left(32 \dfrac{\text{m}}{\text{s}}\right)(5.0 \text{ s}) = 160$ m, and the shaded triangle, (b), whose area is half its base multiplied by the height. The base of the triangle is the length of time of 5.0 s, and the height of the triangle is the change in velocity for the car, given by 90 m/s. Hence, the car covers a displacement of $\Delta x_{2b} = \dfrac{1}{2}\left(90 \dfrac{\text{m}}{\text{s}}\right)(5 \text{ s}) = 225$ m in part (b). The third segment of the trip is broken into two parts, (c) and (d). In part (c), the $v_x(t)$ graph is located above the t-axis, so the area under the graph is positive. The time when the velocity becomes zero, t_{stop} (when the car stops momentarily before changing the direction of its motion), can be found using Equation 3-10:

$$t_{\text{stop}} = 11.0 \text{ s} + \frac{0 - 122 \text{ m/s}}{-24 \text{ m/s}^2} = 11.0 \text{ s} + 5.08 \text{ s} = 16.08 \text{ s}$$

The time interval between the maximum value of the velocity and the time the car stops momentarily is 5.08 s. The displacement of the car during the time from 11 s to 16.08 s is positive, given by half the base times the height:

$\Delta x_{3c} = \dfrac{1}{2}\left(122 \dfrac{\text{m}}{\text{s}}\right)(5.08 \text{ s}) = 310$ m. Therefore, the total displacement for the motion of the rocket car while it is moving in the positive x-direction equals the sum of these four positive displacements: $\Delta x_+ = 192$ m + 160 m + 225 m + 310 m = 887 m.

For the segment where the velocity is negative, (d), the car is moving in the opposite direction and the displacement is negative. This displacement can be calculated as

$$\Delta x_{3d} = \Delta x_- = \frac{1}{2}\left(-94 \dfrac{\text{m}}{\text{s}}\right)(20.0 \text{ s} - 16.08 \text{ s})$$

$$= -184 \text{ m} \approx -180 \text{ m}$$

Therefore, the total displacement of the rocket car is $\Delta x_{\text{total}} = \Delta x_+ + \Delta x_- = 887$ m − 184 m = 703 m. Strictly speaking, each segment produces results precise to only 2 significant figures, but we have kept a third digit in these intermediate results. The final answer should thus be written as 7.0×10^2 m.

Making sense of the result

We see that graphical techniques can be readily employed to find velocity and displacement values for one-dimensional motion. We have been aided in our calculations by the linear nature of the velocity versus time graphs owing to the constant acceleration during each segment. We were thus able to extract exact velocity and position information for the object as a function of time.

Even though in this problem the acceleration is not constant overall, it is constant within each of the three segments of the motion. In Example 3-10 below we further develop and plot the dependence of the position on time for an object moving with constant acceleration. Notice, we saw here that in addition to using kinematics equation in algebraic approach to problem solving, we can also use a graphical approach.

Applicability of the Principle of Graphical Integration

Before moving on to the final problem in this section, we would like to make a note about the problem-solving approach we use in this section and its applicability. The principle of graphical integration, which allows us to link the areas under the $v_x(t)$ and $a_x(t)$ graphs and the respective displacement and change in velocity, holds for motion with constant acceleration, as well as with variable acceleration. One can use the principle of graphical integration anywhere. Despite the fact that when acceleration is not constant it is more difficult to precisely analyze the graphs, one can usually find good approximations of both the change in velocity and the displacement by estimating the area under $a_x(t)$ and $v_x(t)$, respectively. However, to be able to calculate the exact area when the curves are more complicated, we need to know the exact form of the functions and use calculus. The desire to describe complex motion and find areas under complex curves was among the main motivations for the invention of **calculus**, the mathematical study of change, simultaneously developed by the famous English natural philosopher Sir Isaac Newton (1642–1727) and by the no less famous German polymath and philosopher Gottfried Wilhelm Leibnitz (1646–1716) in the early 18th century.

CHECKPOINT 3-10

Velocity versus Time Graph

Figure 3-25 shows a velocity versus time graph for a speedboat. Over the time period when the velocity curve is below the horizontal axis,

(a) the average speed is zero
(b) the average acceleration is zero
(c) the instantaneous acceleration is always negative
(d) none of the above

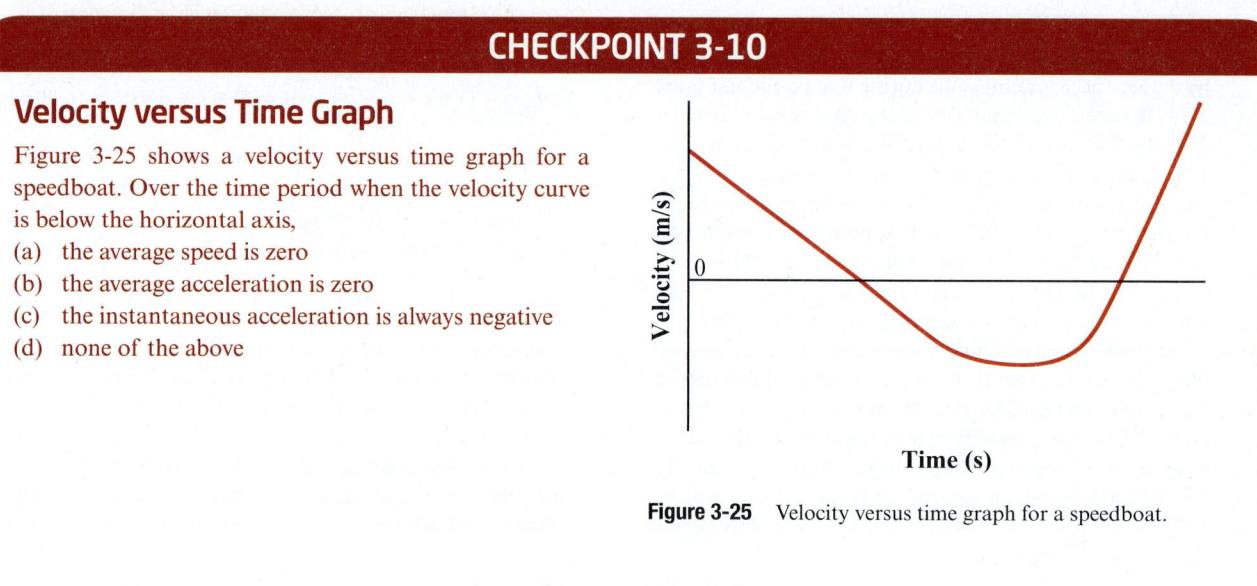

Figure 3-25 Velocity versus time graph for a speedboat.

ANSWER: (b) Since the change in velocity for that interval is zero, the average acceleration is also zero (Equation 3-7).

EXAMPLE 3-10

Rocket Sled

Figure 3-26 shows the acceleration versus time graph for a rocket sled equipped with thrusters on both ends. The sled is slowing down and starts at $t = 0$ with a speed of 250.0 m/s moving to the right. If we define our coordinate system so that the positive x-direction coincides with the direction of the initial motion of the sled (to the right), the initial velocity is positive and the acceleration is negative because it is slowing down. Also, we define the origin of our coordinate system as an observation tower that is 700.0 m to the right of the initial position of the sled. Therefore, the initial position of the sled is $x_0 = -700$ m.

(a) Plot the velocity versus time graph until the sled comes to a stop for the first time.
(b) For what value of time will the rocket momentarily come to a stop ($v_x(t_1) = 0$)?
(c) Plot the position versus time graph for the rocket sled as measured from the observation tower (recall that at $t_0 = 0$ the sled was 700.0 m to the left of the tower). Your plot should clearly indicate the points where both position and velocity are zero.

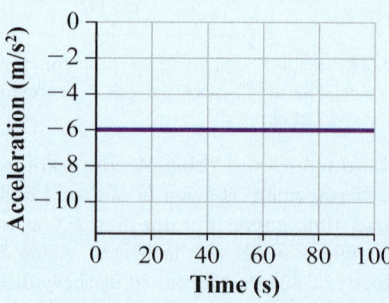

Figure 3-26 Acceleration versus time graph for a rocket sled.

Solution

(a) The $a_x(t)$ graph shows that the acceleration is constant: $a_x = -6.000$ m/s². Since the acceleration is constant, we can use Equation 3-10 to find an expression for the velocity as a function of time. We choose the positive x-axis to be along the direction of motion of the sled. Then, considering that $t_0 = 0$ s:

$$v_x(t) = v_{x_0} + a_x t = 250.0 \text{ m/s} + (-6.000 \text{ m/s}^2)t \quad (1)$$

The velocity versus time plot is linear, as shown in Figure 3-27. The constant (negative) slope of the graph represents constant and negative acceleration.

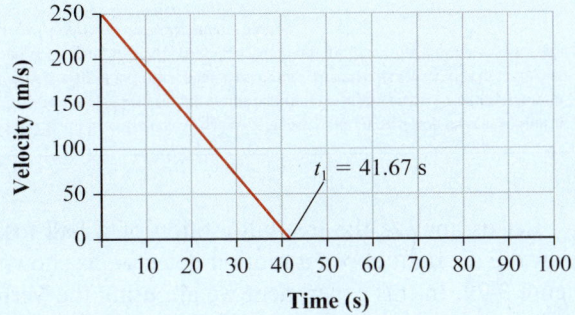

Figure 3-27 Velocity versus time plot for Example 3-10.

(b) To find out when the velocity of the sled becomes zero (when the sled stops momentarily), we set $v_x(t_1) = 0$ in (1) above:

$$v_x(t_1) = 0 \Rightarrow 250.0 \text{ m/s} + (-6.000 \text{ m/s}^2)t_1 = 0$$

$$t_1 = \frac{250.0 \text{ m/s}}{6.000 \text{ m/s}^2} = 41.67 \text{ s}$$

(c) The dependence of the position on time is given by Equation 3-17. Recall that we selected positive as to the right, and that we defined the origin of the coordinate system as an observation tower located 700.0 m to the right of the initial position of the sled, so the initial ($t = 0$) position of the sled is $x_0 = -700.0$ m (or 700.0 m to the left of the observation tower):

$$x(t) = \frac{1}{2}a_x t^2 + v_{x_0} t + x_0 \quad (2)$$

or

$$x(t) = \frac{1}{2}(-6.000 \text{ m/s}^2)t^2 + (250.0 \text{ m/s})t - 700.0 \text{ m}$$
$$= (-3.000 \text{ m/s}^2)t^2 + (250.0 \text{ m/s})t - 700.0 \text{ m} \quad (3)$$

As we expected, the graph of this $x(t)$ equation is a parabola, such that $x(0) = -700.0$ m. Let us pick a few more points to plot the position function. When this parabola intersects the time axis, the position, x, will be zero, so the expression in (3) is 0. These times will be represented by the roots of this quadratic equation. They are given by the formula for the roots of a quadratic equation:

$$ax^2 + bx + c = 0 \Rightarrow x_{1,2} = \frac{-b \pm \sqrt{b^2 - 4ac}}{2a} \quad (4)$$

Plugging the corresponding quantities from (3) into this formula, we find

$$t_{2,3} = \frac{-250 \pm \sqrt{(250)^2 - 4(-3.00)(-700)}}{-6.00} \text{ s} \quad (5)$$

So, $t_2 = 2.901$ s and $t_3 = 80.43$ s.

Since the velocity is the derivative of $x(t)$, a maximum or minimum for x will occur at the point of zero velocity. From part (b), the velocity is zero when $t_1 = 41.67$ s. At this time,

$$x(t_1) = (-3.00 \text{ m/s}^2)(41.67 \text{s})^2 + (250 \text{ m/s})(41.67 \text{s})$$
$$- 700.0 \text{ m} = 4508.33 \text{ m} \quad (6)$$

This is 4508 m to four significant figures. A plot of $x(t)$ is shown in Figure 3-28.

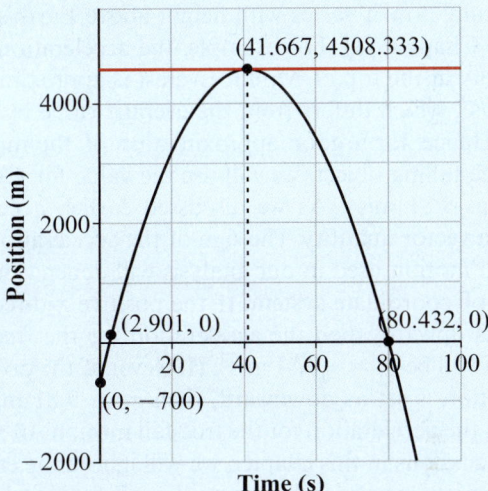

Figure 3-28 Position versus time graph for the rocket sled. The maximum value of x is 4508 m. It happens when the rocket stops momentarily ($t_1 = 41.67$ s). A momentary stop means that the slop of the tangent to the $x(t)$ graph at this point is horizontal.

Making sense of the result

While solving kinematics problems requiring you to plot graphs, we strongly recommend you to use online calculators or plotting software to visualize the graphs. For example, the Desmos online graphing calculator (www.desmos.com) is a convenient choice.

The rocket sled will be moving to the right while slowing down. It will go past the control tower for the first time at $t_2 = 2.901$ s. The speed of the sled will continue to drop until it stops momentarily at $t_1 = 41.67$ s. When the sled stops, it will be 4508 m to the right of the tower. The sled will then reverse direction and will start speeding up in the negative direction. It will pass the tower for the second time at $t_3 = 80.43$ s.

3-6 Free Fall

If you toss a ball straight up in the air, it slows down on the way up, stops momentarily when it reaches the top of its trajectory, and then speeds up on the way down. As we mentioned earlier, throughout the journey the ball experiences constant and downward acceleration due to gravity, denoted as g. For consistency, we will use g_x (provided the x-axis is directed vertically) when we talk about the acceleration vector and we will use g when we talk about the absolute value of the acceleration due to gravity. Gravity continuously accelerates the ball downward, or toward the centre of Earth. We say that an object that is acted upon by nothing other than the force of gravity is in a state of free fall. As we discussed earlier, the magnitude of the acceleration due to gravity close to Earth's surface is $g = 9.81$ m/s². This value does not change appreciably between the highest and lowest altitude points on Earth (you will see more specifically how it varies with height above Earth's surface in Chapter 11). For example, the acceleration due to gravity at the top of Mount Everest is approximately 9.76 m/s², which differs from the average value by only 0.5%. Hence, for a good approximation of the motion of a free-falling object, we will use the value for g given above as 9.81 m/s². As we discussed earlier, acceleration is a vector quantity. The sign of the acceleration for free-fall motion used in our analysis will depend on the choice of coordinate system. If the positive x-direction is set as upward, then the acceleration for the free-fall motion will be $g_x = -9.81$ m/s². However, if the positive x-direction is set as downward, then $g_x = 9.81$ m/s² is used as the acceleration for the free-fall motion. In all of our discussions in this chapter, we will ignore the effects of air resistance and will assume that all falling objects are in a state of free fall, unless otherwise indicated.

It is worth remembering that the relative directions of an object's velocity and acceleration vectors determine the direction of its motion and define whether the object is speeding up, slowing down, or moving with constant velocity at a given moment in time (Table 3-3). Moreover, when an object changes its direction of motion, as we saw in Example 3-10, it has to stop momentarily in between; thus at that moment its instantaneous velocity is zero. It is interesting to note that "free fall" does not indicate the direction of motion—it only indicates that the object is under the influence of only the force of gravity and has an acceleration downward equal to 9.81 m/s².

From a physics perspective, while moving up under the influence of gravity, we are also in free fall!

CHECKPOINT 3-11

A ball is thrown straight up in the air. Consider the positive x-direction to be downward. At the instant when the ball reaches its highest point, its velocity and acceleration are

(a) $v_x = 0$ m/s $a_x = -9.81$ m/s²
(b) $v_x = 0$ m/s $a_x = 9.81$ m/s²
(c) $v_x = 0$ m/s $a_x = 0$ m/s²
(d) $v_x = 9.81$ m/s $a_x = 9.81$ m/s²
(e) $v_x = 9.81$ m/s $a_x = -9.81$ m/s²
(f) $v_x = 9.81$ m/s $a_x = 0$ m/s²
(g) $v_x = -9.81$ m/s $a_x = 0$ m/s²

ANSWER: (b) Since the object reaches its highest point, it must stop there momentarily. Otherwise, it would keep moving up and reach an even higher point. Therefore, its instantaneous velocity at its highest point must be zero. However, its acceleration in the highest point is not zero, it is 9.81 m/s² directed downward. Since the positive x-axis is also directed downward, the acceleration is positive.

Let us analyze the free-fall motion of a ball tossed up in the air right above a motion detector, as shown in Figure 3-29. In this experiment we are using the Vernier motion detector and Logger *Pro* software to analyze the free-fall motion of the tossed ball. Notice that the motion detector is placed on the table to record the position of the ball during its motion. Based on these data, the software calculates the values of the velocity and acceleration for the ball during its motion. The graphs are stacked so it is easier to see corresponding values of position, velocity, and acceleration. The video of the ball's motion is synchronized with the recording of the motion detector, so we can see the graphs being produced in real time. While a similar analysis can be done using other technologies, the ability of Logger *Pro* to synchronize video recording with the collected data is very powerful. By analyzing these data within the software, you can see that the slope of the $x(t)$ graph gives the instantaneous velocity and the slope of the $v_x(t)$ graph gives the instantaneous acceleration. For example, for the graphs shown in Figure 3-29, the slope of the position versus time graph at $t = 0.8325$ s equals the velocity of the ball, which is 1.698 m/s. The slope of the velocity versus time graph gives the instantaneous acceleration, which is -10.107 m/s². Notice that the negative value of the acceleration means that the positive direction of the x-axis is up (away from the motion detector). The value of the acceleration of free fall we obtained is slightly different from 9.81 m/s² due to experimental error, but the difference is only 3%. Moreover, if we averaged the acceleration value over the entire free-fall motion, the two values would be even closer.

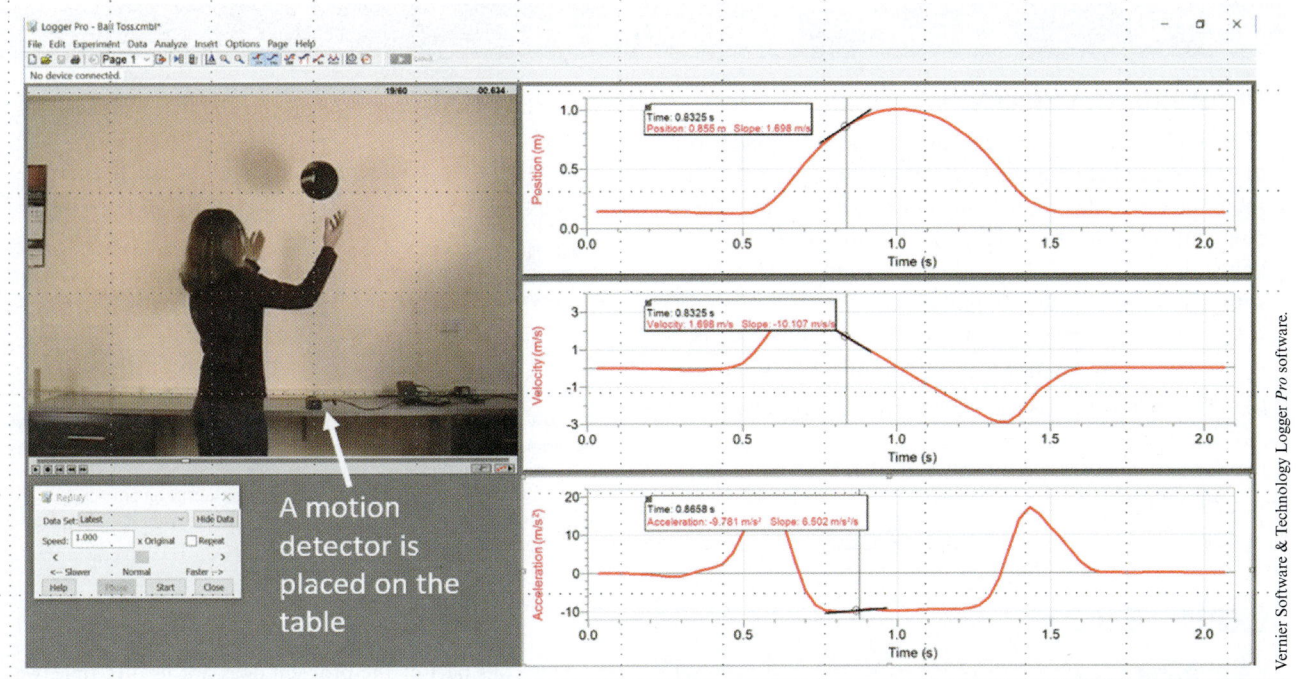

Figure 3-29 A ball toss experiment performed using Logger *Pro* motion detector and a video of the experiment. The stacked graphs show the relationship between the slopes of $x(t)$ and $v_x(t)$ graphs and the respective values of instantaneous velocity and acceleration.

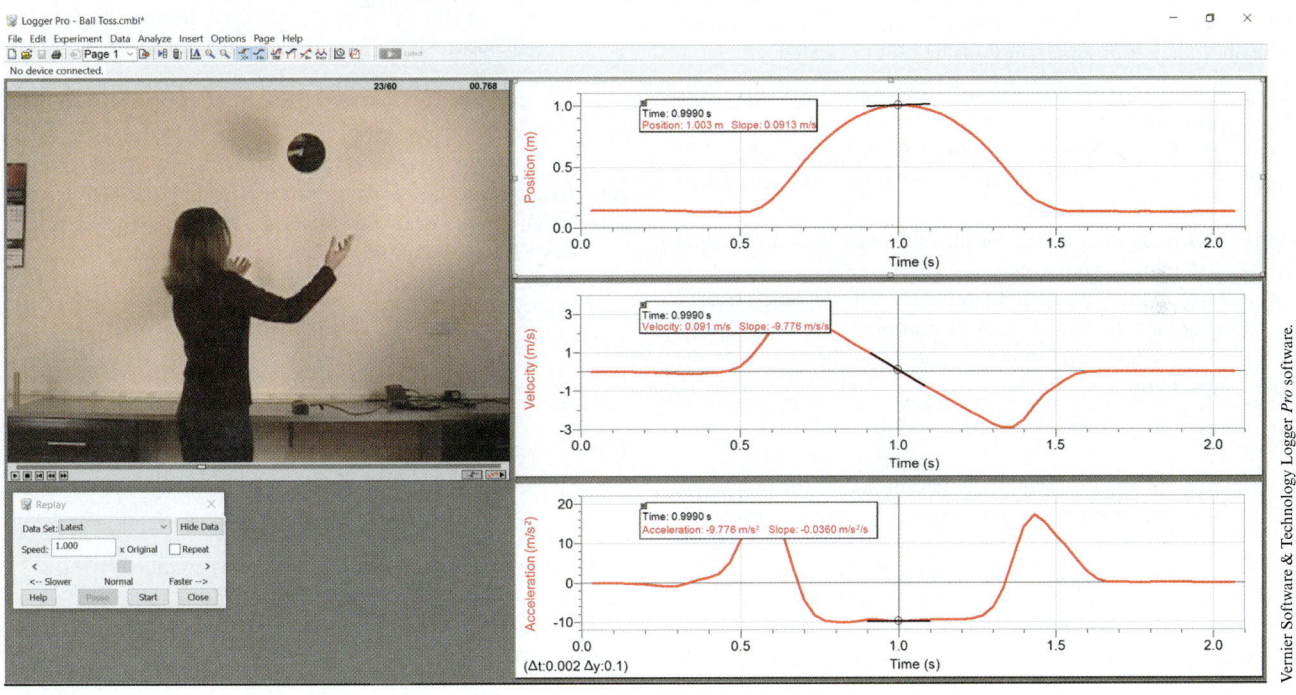

Figure 3-30 A ball toss experiment. The values for the instantaneous positon, velocity, and acceleration of the ball at the moment when it reaches its highest point are shown.

Notice that at the highest point of the ball trajectory (Figure 3-30), when $t = 0.99$ s, the velocity of the ball is zero (it stops momentarily), but its acceleration is still -9.81 m/s^2. The slope of the velocity versus time graph is constant and negative, as we expect based on our discussion of free-fall motion and the choice of the x-axis.

Difference in Velocity

The relative velocity between two objects is defined as the difference between their velocities. Imagine that you simultaneously throw two otherwise identical red and blue balls from the top of a building: you throw the red ball with a speed of 7.0 m/s straight up and the blue one with a speed of 14 m/s straight down. After 2.0 s, what is true about the relative velocity between the two balls, $v_{x,\text{red}} - v_{x,\text{blue}}$? (Assume the positive x-direction is up.)

(a) The relative velocity between the two balls has increased.

(b) The relative velocity between the two balls has decreased.

(c) The relative velocity between the two balls has not changed.

(d) The answer will depend on the choice of the coordinate axis.

ANSWER: (c) The two balls have the same acceleration and hence zero relative acceleration, which means the relative velocity at any later time must be the same as the initial relative velocity. This means that since the balls are falling for the same amount of time ($\Delta t_{\text{red}} = \Delta t_{\text{blue}}$), they each change their velocity by exactly the same amount. Mathematically, this can be expressed using Equation 3-10:

$$v_{x,\text{red}} = v_{x_0,\text{red}} + g_x \Delta t$$
$$v_{x,\text{blue}} = v_{x_0,\text{blue}} + g_x \Delta t$$
$$\Rightarrow v_{x,\text{red}} - v_{x,\text{blue}} = v_{x_0,\text{red}} - v_{x_0,\text{blue}} = \text{const}$$

The direction of the positive x-axis doesn't matter, because the difference in velocity will stay the same regardless of the initial sign choices. Finally, think how your answer would have been different if the question asked about the difference in speeds of the balls.

EXAMPLE 3-11

Launching Lunch

A construction worker standing on a beam asks his very strong friend on the ground 9.00 m below to throw his lunch box up to him. The friend wants to throw the box straight up such that it has a velocity of 0.500 m/s up when it reaches the worker on the beam, to give him some leeway as he tries to catch it.

(a) What assumption should you make to solve the problem?

(b) At what speed should the friend throw the lunch box?

(c) How long will the lunch box be in the air before it is caught by the worker?

(d) How long does it take for the lunch box to cover the first 4.50 m of its trajectory?

(e) How high above its initial position will the lunch box be after half the total time of its motion?

(f) Plot the $x(t)$, $v_x(t)$, and $a_x(t)$ graphs for the motion of the lunch box.

Solution

(a) To solve the problem we will assume that the lunch box is going straight up and the effect of air resistance on its motion can be neglected. Let us choose a positive x-axis upward (in the direction of the initial box toss) and the origin as the point from which the box was tossed.

(b) In this problem, we are given the initial and final positions of the lunch box (and thus we can calculate its displacement), its acceleration, and its final velocity. We need to find the initial speed of the ball (we know it is tossed upward, but we do not know its speed). We can record all the givens for the problem as follows, where the subscript f stands for the final values when the box is being caught (notice, the final velocity is positive, while the acceleration is negative):

$$x_0 = 0 \text{ m}; \, x_f = 9.00 \text{ m}, \, v_{x,f} = 0.500 \text{ m/s},$$
$$g_x = -9.81 \text{ m/s}^2, \text{ and } v_{x0} = ?$$

Since this part of the problem does not involve time, we can use Equation 3-22. Note that the acceleration and the initial velocity have opposite directions:

$$v_{x,f}^2 - v_{x_0}^2 = 2a_x \Delta x$$
$$(0.500 \text{ m/s})^2 - v_{x_0}^2 = 2(-9.81 \text{ m/s}^2)(9.00 \text{ m})$$
$$v_{x_0} = \sqrt{(0.500 \text{ m/s})^2 + 2(9.81 \text{ m/s}^2)(9.00 \text{ m})} \quad (1)$$
$$= 13.298 \text{ m/s} \approx 13.3 \text{ m/s}$$

(c) To determine how long the box will be in the air before it is caught (Δt_1), we can use Equation 3-10, $v_x(t) = v_{x_0} + g_x \Delta t_1$, as we know the initial and final velocities of the box and its acceleration:

$$\Delta t_1 = \frac{v_{x,f} - v_{x_0}}{g_x} = \frac{0.500 \text{ m/s} - 13.298 \text{ m/s}}{-9.81 \text{ m/s}^2} \quad (2)$$
$$= 1.305 \text{ s} \approx 1.31 \text{ s}$$

(d) To determine how long it will takes for the box to rise halfway up, or 4.5 m (Δt_2), we can use Equation 3-17, $x_{\text{height}/2} = x_0 + v_{x_0}\Delta t_2 + \frac{1}{2}a_x\Delta t_2^2$. For clarity, we will omit the units while solving this quadratic equation:

$$4.50 = 0 + 13.3\Delta t_2 + \frac{1}{2}(-9.81)\Delta t_2^2$$
$$4.905\Delta t_2^2 - 13.3\Delta t_2 + 4.50 = 0 \quad (3)$$
$$(\Delta t_2)_1 = 0.396 \text{ s}$$
$$(\Delta t_2)_2 = 2.32 \text{ s}$$

This quadratic equation has two roots. This makes sense, as the box passes the 4.50 m point on the way up at the first time, $(\Delta t_2)_1$, and then would normally pass the same point on the way back down at the second time, $(\Delta t_2)_2$, if it were not caught before returning back down. So for this example, we are only interested in the first solution, $(\Delta t_2)_1 = 0.396$ s, as the box is caught on the way up. Notice that the time it takes the box to go halfway up is less

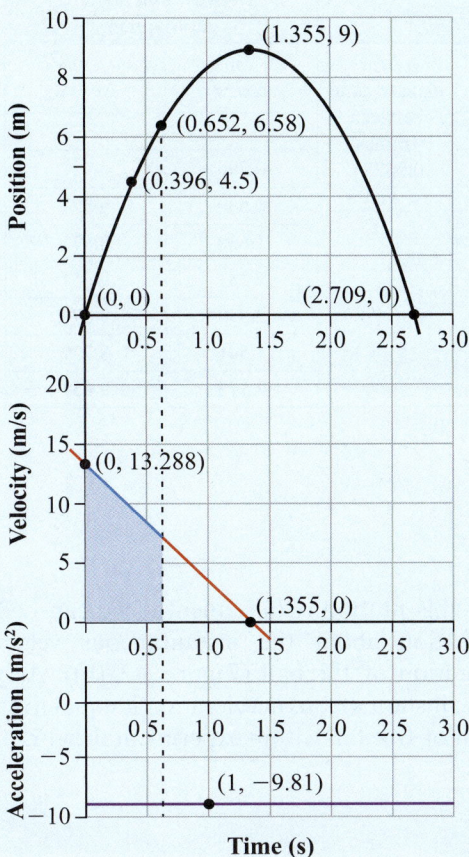

Figure 3-31 $x(t)$, $v_x(t)$, and $a_x(t)$ graphs for the motion of the lunch box in Example 3-11. The points on the $x(t)$ graph represent the locations of the box after half the time of the throw has elapsed and at the halfway point (by distance) of its journey. The shaded area under the $v_x(t)$ graph represents the distance covered by the box during the first half of the time of its journey.

than half the time of the box's total trip up: $(\Delta t_2)_1 < \frac{\Delta t_1}{2}$. This happens because the box moves faster during the first half of its trajectory on the way up than during the second half due to its slowing down. We will discuss how to visualize this result in part (e) of the solution.

(e) To find how high above the ground the box will be after half of the total time has elapsed ($\Delta t_1/2 = 0.652$ s), we use Equation 3-17 once again:

$$x_{1/2\,\text{time}} = x_0 + v_{x_0}\frac{\Delta t_1}{2} + \frac{1}{2}a_x\left(\frac{\Delta t_1}{2}\right)^2$$

$$= 0 + \left(13.3\,\frac{\text{m}}{\text{s}}\right)(0.652\,\text{s}) + \frac{1}{2}\left(-9.81\,\frac{\text{m}}{\text{s}^2}\right)(0.652\,\text{s})^2$$

$$= 6.59\,\text{m} > \frac{1}{2}(9.00\,\text{m}) = 4.500\,\text{m} \qquad (4)$$

This result also makes sense because during the first half of the time of the journey, the box moves faster than during the second half of the time. Therefore, it will cover more than half of the total height during the first half of the time of its upward motion.

(f) We can use the principle of graphical integration discussed in the previous section to visualize the results found in the previous parts by plotting the $x(t)$, $v_x(t)$, and $a_x(t)$ graphs for the motion of the lunch box (Figure 3-31).

Making sense of the result

The principle of graphical integration discussed earlier and applied to interpret part (f) is very useful in making sense of kinematics problems. We recommend that you use the Desmos graphing calculator (www.desmos.com/calculator) or another way of plotting data to model the results of kinematics problems and make sense of them. See Interactive Activity 3-3 for guidance on how to do this.

CHECKPOINT 3-13

Exploring Free Fall

Ignoring air resistance, if you drop an object, it accelerates downward at a rate of 9.81 m/s². If instead you throw it down, what is the magnitude of the object's acceleration after you release it?

(a) less than 9.81 m/s² upward
(b) exactly 9.81 m/s² upward
(c) more than 9.81 m/s² upward
(d) less than 9.81 m/s² downward
(e) exactly 9.81 m/s² downward
(f) more than 9.81 m/s² downward

ANSWER: (e) After you release the ball, you have no effect on its motion. It is in free fall, with an acceleration of exactly 9.81 m/s² directed toward the centre of Earth. Your tossing or dropping the ball affects its initial velocity, not its acceleration.

INTERACTIVE ACTIVITY 3-3

Visualizing Kinematics Problems with Graphing Software

In this activity, you will use the graphing software called "Desmos Graphing Calculator" (www.desmos.com/calculator) to help you gain an intuitive understanding of kinematics problems. The simulation will allow you to plot, manipulate, and explore graphs describing one-dimensional kinematics problems.

Video analysis is a very useful tool for analyzing free fall. According to Equation 3-17, if you drop a ball, it will cover about 5 m in a second. This is too fast for the human eye to examine the ball's motion. However, if you videotape it, you can slow it down and collect important information about the fall. For example,

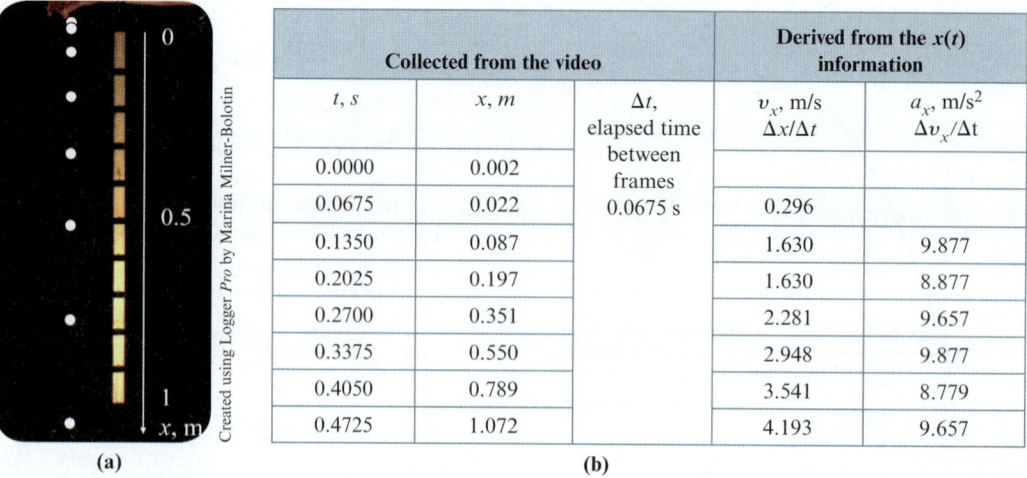

Collected from the video			Derived from the $x(t)$ information	
t, s	x, m	Δt, elapsed time between frames	v_x, m/s $\Delta x/\Delta t$	a_x, m/s^2 $\Delta v_x/\Delta t$
0.0000	0.002	0.0675 s		
0.0675	0.022		0.296	
0.1350	0.087		1.630	9.877
0.2025	0.197		1.630	8.877
0.2700	0.351		2.281	9.657
0.3375	0.550		2.948	9.877
0.4050	0.789		3.541	8.779
0.4725	1.072		4.193	9.657

(a) (b)

Created using Logger *Pro* by Marina Milner-Bolotin

Figure 3-32 Example of a video analysis of the free-fall motion of a ball.

Figure 3-32(a) shows a frame-by-frame recording of the free-fall motion of a ball. Considering that the metre stick is located in the same plane as the falling ball and that the time interval between two consecutive frames is 0.0675 s, we can obtain the $x(t)$ information about this motion. Consequently, we can deduce the information about the instantaneous velocity and acceleration of the ball (Figure 3-32(b)). As you can see, the instantaneous acceleration is equal to the acceleration of free fall within experimental error.

MAKING CONNECTIONS

Describing Motion Using Video Analysis: Eadweard Muybridge

To describe motion, we need to have information about the object's positions at different times. As we discussed earlier, we can find this information using motion detectors. However, it is not always practical, especially when we are dealing with a wide range of fast motion. We can obtain very useful information by recording the motion and then analyzing the video using a tool called video analysis. Video processing has found many applications in sports and medicine. Slow-motion replay, pattern analysis, statistical analysis, and video archiving are only a few examples of these types of applications. Video analysis is one of the major tools allowing athletes to improve their performance. For example, sprinters like Usain Bolt use video analysis to analyze different segments of their motion and improve it.

Interestingly, video analysis of motion is not a recent invention. The process of making consecutive photographs of a moving object to analyze its motion was pioneered by English photographer Eadweard Muybridge (1830–1904). He invented a method of using multiple cameras to photograph and consequently analyze the motion of moving objects. This helped him to answer a popular question of his day—whether all four feet of a galloping horse were off the ground at the same time (Figure 3-33). Muybridge found that despite the popular belief, all four hooves of a horse *are* off the ground at some point during its motion. Later Muybridge applied his method to analyze the motion of recuperating patients,

as well as for analyzing the motion of athletes. The work of Muybridge is a beautiful example of how art and science complement each other to ask and answer new questions and move our experience and understanding of the universe forward.

To learn more about the use of video analysis to analyze the motion of objects, see Antimirova, T., & Milner-Bolotin, M. 2009. A brief introduction to video analysis. *Physics in Canada*, volume 65 (April-May), 74.

Eadweard James Muybridge. Library of Congress Prints and Photographs Division Washington, D.C. 20540 USA

Figure 3-33 *The Horse in Motion*, an 1878 photo of a galloping horse by Eadweard Muybridge.

3-7 Relative Motion in One Dimension

Reference Frames

Imagine that you are at the airport waiting for your flight to board and that you are standing by a magazine shop. You notice people moving past you, walking on the ground at the hurried pace common at airports. Distinctly faster will be the people moving in the same direction on the moving sidewalk, also called a sliding sidewalk. Relative to a person walking on the moving sidewalk, a person walking on the ground in the same direction at a "normal" speed seems to be moving backward. Relative to a person walking on the ground, a person walking on the moving sidewalk seems to be walking at what one would call a normal speed. Perspectives are important when it comes to the topic of relative motion, which we examine below. In physics, to indicate our perspective precisely, we use reference frames. A **frame of reference**, or a **reference frame**, consists of a coordinate system that includes the position of the origin; the direction of the axes; and the units of measurement, such as metres, seconds, etc. Notice that a frame of reference must include both the spatial coordinates, such as x, y, and z, and a temporal coordinate, such as time, t. Therefore, in the case of one-dimensional motion, to describe the location of an object at a certain moment of time, we need two coordinates: its position and time (x, t). In general, the number of coordinates needed to define the reference frame precisely for the n-dimensional case is $n + 1$. This explains the four-dimensional space-time (three spatial dimensions and one temporal dimension) described when we discuss Einstein's theory of relativity. We will discuss this theory in Chapter 30, when we talk about objects moving with speeds comparable to the speed of light. For now, we will discuss the principles of relativity as applied to slow-moving (as compared to the speed of light) objects, such as many of the objects we encounter in everyday life.

Relative Velocity

Imagine that you are standing on a train platform waiting for a friend. You see a train going slowly past you at a velocity of 4 m/s. You notice a child on that train who is moving in one of the cars, in the same direction as the train. The velocity of the child relative to the train (as seen by someone sitting in their seat on the train) is 3 m/s. However, relative to you, the velocity of the child is 7 m/s.

One can say that the velocity of the child relative to the ground, $v_{x,\text{child–ground}}$, is equal to the velocity of the child relative to the train, $v_{x,\text{child–train}}$, plus the velocity of the train relative to the ground, $v_{x,\text{train–ground}}$:

$$v_{x,\text{child–ground}} = v_{x,\text{child–train}} + v_{x,\text{train–ground}} \quad (3\text{-}23)$$

(We call the notation above the *double-subscript notation*.) Therefore, the velocity of the child relative to the train is given by

$$v_{x,\text{child–train}} = v_{x,\text{child–ground}} - v_{x,\text{train–ground}} \quad (3\text{-}24)$$

This is the velocity of the child as seen by someone sitting in their seat on the train.

In general, for any two objects moving relative to one another, we can say that the velocity of object 1 relative to object 2, or as seen by object 2, $v_{x,12}$, is given by

KEY EQUATION $\quad v_{x,12} = v_{x,1g} - v_{x,2g} \quad (3\text{-}25)$

where v_{1g} is the velocity of object 1 relative to the reference frame (such as the ground) and v_{2g} is the velocity of object 2 relative to that reference frame. Also,

$$v_{x,1g} = v_{x,12} + v_{x,2g} \quad (3\text{-}26)$$

This relationship is called the **Galilean principle of velocity addition**, which states that the velocity of object 1 relative to the ground (or any other reference frame) equals the sum of the velocity of object 1 relative to object 2 and the velocity of object 2 relative to the ground.

Using the definitions of relative velocity provided above, and the relationships between position, velocity, and acceleration we derived earlier (Equations 3-4, 3-7), we can derive the values of the relative displacements and the relative accelerations for the two objects.

For the relative positions and displacements, we obtain

$$x_{12} = x_{1g} - x_{2g} \Rightarrow$$
$$\Delta x_{12} = v_{x,12}\Delta t = (v_{x,1g} - v_{x,2g})\Delta t = \Delta x_{1g} - \Delta x_{2g} \quad (3\text{-}27)$$

And for the relative acceleration of object 1 relative to object 2, we have

KEY EQUATION $\quad a_{x,12} = \dfrac{\Delta v_{x,12}}{\Delta t} = \dfrac{\Delta(v_{x,1g} - v_{x,2g})}{\Delta t} \quad (3\text{-}28)$

$$= \dfrac{\Delta v_{x,1g}}{\Delta t} - \dfrac{\Delta v_{x,2g}}{\Delta t} = a_{x,1g} - a_{x,2g}$$

It is important to note that the velocity of object 1 relative to object 2 is equal and opposite to the velocity of object 2 relative to object 1. Therefore, we obtain

$$x_{12} = -\Delta x_{21}$$
$$v_{x,12} = -v_{x,21} \quad (3\text{-}29)$$
$$a_{x,12} = -a_{x,21}$$

The discussion above provides an intuitive approach for describing relative motion. It implies, however, that space and time are absolute. As much as it feels obvious to us, this idea was challenged by Einstein when objects move at speeds comparable to the speed of light, as we will describe in Chapter 30. Nevertheless, for now this description and understanding will suffice. We offer a more formal development and treatment of relative quantities for slow-moving objects in Chapter 4, where the treatment is also extended to more dimensions.

EXAMPLE 3-12

Relative Velocities

You are driving east on the Trans-Siberia highway at 100 km/h. Take east to be the positive x-direction.

(a) Express the velocity of a tree on the side of the highway with respect to you, using double-subscript notation.

(b) Find the speed of a tree on the side of the highway with respect to you.

(c) Find your velocity with respect to the tree.

(d) Find the velocity, with respect to you, of a bus driving west at 115 km/h.

(e) Find the velocity of your car with respect to the bus.

Solution

We choose east as the positive x-direction, so your velocity relatively to the ground is positive, that of the trees is zero, and that of the bus is negative. We will use the subscripts c, t, and b for your car, the tree, and the bus, respectively. Since the tree is not moving relative to the ground, the tree frame of reference in this problem is equivalent to the ground frame of reference.

(a) For the velocity of the tree relative to your car, we use the relative velocity Equation 3-25:

$$v_{x,\text{tc}} = -v_{x,\text{ct}} = -100 \text{ km/h} \tag{1}$$

(b) The speed of the tree relative to you (to your car) equals the magnitude of the relative velocity found in part (a): $v_{\text{tc}} = 100$ km/h.

(c) The velocity of your car with respect to the tree is the same as the velocity of your car relative to the ground (the tree is not moving relative to the ground):

$$v_{x,\text{ct}} = 100 \text{ km/h} \tag{2}$$

(d) The velocity of the bus moving in the opposite direction with respect to your car is

$$v_{x,\text{bc}} = v_{x,\text{bt}} - v_{x,\text{ct}} = -115 \text{ km/h} - (100 \text{ km/h})$$
$$= -215 \text{ km/h}$$

(e) Similarly, your car's velocity with respect to the bus is

$$v_{x,\text{cb}} = -v_{x,\text{bc}} = v_{x,\text{ct}} - v_{x,\text{bt}}$$
$$= 100 \text{ km/h} - (-115 \text{ km/h}) = 215 \text{ km/h}$$

Making sense of the result

Vehicles coming in the direction opposite to yours often seem to be moving very fast, because relative to you, they do move much faster than they move relative to the ground.

EXAMPLE 3-13

Relative Position

At the annual auto show in Frankfurt, Germany, an Audi is 7.0 m to the left of a post and a Maserati is 12 m to the left of the same post.

(a) Find the position of the Audi relative to the Maserati and the position of the Maserati relative to the Audi. Consider the positive x-direction to be to the right.

(b) How would your answer to (a) change if the positive x-direction were to the left?

(c) How would your answer to (a) change if you changed the position of the origin of your reference frame?

Solution

(a) Let us first choose the positive x-direction to be directed to the right and the post to be the origin of the reference frame. Therefore, left of the post is the negative direction. Let the subscripts A and M represent the Audi and the Maserati, respectively. Both x_A and x_M are negative because they are both located left of the origin (Figure 3-34). Applying Equation 3-27, we find

$$x_{\text{AM}} = x_{\text{Ag}} - x_{\text{Mg}} = x_A - x_M \tag{1}$$

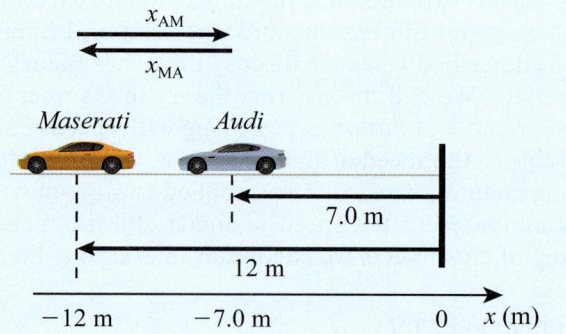

Figure 3-34 The positions of the Maserati and the Audi relative to the post for Example 3-13.

Here we drop the subscript g (ground) for simplicity as both positions are taken with respect to the ground by default.

$$x_{\text{AM}} = -7.0 \text{ m} - (-12 \text{ m}) = 5.0 \text{ m}$$
$$x_{\text{MA}} = -x_{\text{AM}} = -5.0 \text{ m} \tag{2}$$

The positive relative position of the Audi relative to the Maserati means that the Audi is located to the right of the

Maserati. The negative relative position of the Maserati relative to the Audi means that the Audi is located to the right of the Maserati.

(b) If the coordinate system were reversed—positive to the left and negative to the right—then $x_{AM} = 7.0$ m $-$ $(12.0$ m$) = -5.0$ m. The Audi is still to the right of the Maserati, but now the direction to the right is negative, so the answer is also negative. Thus, $x_{MA} = -x_{AM} = 5.0$ m.

(c) If we change the position of the origin of the coordinate system, the answers to (a) will not change, as both values will change by the same number and their difference will remain the same. For example, if the origin is chosen at the location of the Maserati, while the positive x-direction is chosen to be to the right, then

$$x_M = 0 \text{ m}$$
$$x_A = 5.0 \text{ m}$$
$$x_{AM} = x_A - x_M = 5.0 \text{ m} - 0 = 5.0 \text{ m}$$

Making sense of the result

If the positive x-direction is to the right and the Audi is to the right of the Maserati, the Audi's position relative to the Maserati should be positive. If the positive x-direction is to the left, since the Audi is to the right of the Maserati, its position relative to the Maserati should be negative. As expected, the location of the origin doesn't influence the relative positions of the cars.

CHECKPOINT 3-14

Relative Velocity

You are in the free-fall stage of a sky-dive when the sky-diver below you opens her parachute. A few seconds later, she passes you at a relative velocity of 50 km/h upward. She has a velocity with respect to the air of 80 km/h downward. What is your velocity with respect to the air?

(a) 30 km/h downward
(b) 30 km/h upward
(c) 130 km/h downward
(d) 130 km/h upward
(e) none of the above

ANSWER: (c) Your velocity is 50 km/h downward with respect to her, and her velocity is 80 km/h relative to the air.

Derivation of the General Kinematics Equations for Relative Motion

We now derive the general kinematics equations for one object with respect to another when both objects have constant accelerations. Here again we drop the subscript g (for ground), as all quantities are taken relative to ground.

Relative position The position of object 1 with respect to object 2 is given by Equation 3-27:

$$x_{12} = x_1 - x_2$$

The position of each object as a function of time is given by Equation 3-17, where it is assumed that $t_0 = 0$ s:

$$x(t) = \frac{1}{2}a_x t^2 + v_{x_0} t + x_0$$

Combining these two equations, we get

$$x_{12}(t) = x_{0,1} + v_{x_{0,1}} t + \frac{1}{2}a_{x_1} t^2 - \left(x_{0,2} + v_{x_{0,2}} t + \frac{1}{2}a_{x_2} t^2 \right)$$

$$= (x_{0,1} - x_{0,2}) + (v_{x_{0,1}} - v_{x_{0,2}})t + \left(\frac{1}{2}a_{x_1} - \frac{1}{2}a_{x_2} \right)t^2$$

$$x_{12}(t) = x_{0,12} + v_{x_{0,12}} t + \frac{1}{2}a_{x_{12}} t^2 \qquad (3\text{-}30)$$

Equation 3-30 makes sense. If the second object is not moving relative to the ground, then

$$x_{12} \equiv x_1, \quad v_{x_{0,12}} \equiv v_{x_{0,1}}, \quad \text{and } a_{x_{12}} \equiv a_{x_1}$$

This shows that Equation 3-30 is a general form of Equation 3-17 when the second reference frame is the ground.

Relative velocity Let us derive the expressions for relative velocity. We begin with Equation 3-25:

$$v_{x_{12}} = v_{x_1} - v_{x_2}$$

Then using Equation 3-10 describing the velocity of an object moving with constant acceleration, we obtain

$$v_{12}(t) = v_{0,1} + a_1 t - (v_{0,2} + a_2 t)$$
$$= v_{0,1} - v_{0,2} + (a_1 - a_2)t \qquad (3\text{-}31)$$
$$v_{x_{12}}(t) = v_{x_{0,12}} + a_{x_{12}} t$$

We leave it as an exercise in problem 97 to apply relative motion notation to the time-independent kinematics Equation 3-21 to show that

$$v_{x_{12}}^2 - v_{x_{0,12}}^2 = 2a_{x_{12}} \Delta x_{12} \qquad (3\text{-}32)$$

EXAMPLE 3-14

Relative Acceleration

You are driving a pickup truck when the driver of a motorcycle in front of you suddenly brakes to slow down for a bump on the road. Being a smaller vehicle with good brakes, the motorcycle can slow down at an acceleration of 19 m/s². Your heavy pickup truck slows at a rate of 12 m/s².

(a) What is the acceleration of the motorcycle with respect to the pickup truck?
(b) What is the acceleration of the pickup truck with respect to the motorcycle?

Solution

For this example, we use Equation 3-28 and choose the direction of motion as the positive x-direction. Therefore, both accelerations are negative. We use the subscripts p for the pickup truck, m for the motorcycle, and g for the ground.

(a) The acceleration of the motorcycle with respect to the pickup truck is

$$a_{x,\text{mp}} = a_{x,\text{mg}} - a_{x,\text{pg}} = -19 \text{ m/s}^2 - (-12 \text{ m/s}^2)$$
$$= -7.0 \text{ m/s}^2$$

(b) The acceleration of the pickup truck with respect to the motorcycle is

$$a_{x,\text{pm}} = a_{x,\text{pg}} - a_{x,\text{mg}} = 19 \text{ m/s}^2 - (12 \text{ m/s}^2) = 7.0 \text{ m/s}^2$$

Making sense of the result

Since the motorcycle is slowing down faster than the pickup truck, from the point of view of the pickup truck the motorcycle is accelerating toward it. The acceleration of the motorcycle relative to the truck will point toward the truck, which is the negative direction as per our choice of coordinate system. This result confirms our findings in Equation 3-29.

EXAMPLE 3-15

Relative Position, Velocity, and Acceleration

At the same time, you throw a red ball straight up with a speed of 21.0 m/s and a blue ball straight down with a speed of 17.0 m/s. Find the acceleration, velocity, and position of the blue ball with respect to the red ball 3.00 s later. Assume that air resistance is negligible.

Solution

Let us choose a vertical x-axis for this problem and positive x to be directed downward. Once in the air, both balls have the same constant acceleration of 9.81 m/s² directed toward the centre of Earth.

We calculate the relative velocity using Equation 3-10. We use the subscript r for the red ball and the subscript b for the blue ball:

$$v_{x,\text{br}}(t) = v_{x_0,\text{br}} + a_{x,\text{br}}t \tag{1}$$

Using Equations 3-25 and 3-27, we can rewrite (1) as

$$v_{x,\text{br}} = (v_{x_0,\text{b}} - v_{x_0,\text{r}}) + (a_{x,\text{b}}t - a_{x,\text{r}}t)$$
$$= (v_{x_0,\text{b}} - v_{x_0,\text{r}}) + (g_x t - g_x t) = v_{x_0,\text{b}} - v_{x_0,\text{r}} \tag{2}$$

The terms in the second parentheses cancel because the two balls have the same constant acceleration, g_x. Consequently, the relative velocity of the two balls is constant because it depends only on their initial velocities. This reasoning is akin to the reasoning we used in Checkpoint 3-12 above. Since positive x is directed downward, we can write

$$v_{x,\text{br}} = v_{x_0,\text{b}} - (-v_{x_0,\text{r}}) = v_{x_0,\text{b}} + v_{x_0,\text{r}} \tag{3}$$
$$= 17.0 \text{ m/s} + 21.0 \text{ m/s} = 38.0 \text{ m/s},$$

where the variables in (3) denote the magnitudes of the quantities in (2).

To express relative position, we use the final line of Equation 3-30:

$$x_{12}(t) = x_{0,12} + v_{x_0,12}t + \frac{1}{2}a_{x_{12}}t^2 \tag{4}$$

We use line 2 of Equation 3-30 to rewrite (4) as

$$x_{\text{br}} = (x_{0,\text{b}} - x_{0,\text{r}}) + (v_{x0,\text{b}} - v_{x0,\text{r}})t + \left(\frac{1}{2}a_{x,\text{b}} - \frac{1}{2}a_{x,\text{r}}\right)t^2$$

$$= (x_{0,\text{b}} - x_{0,\text{r}}) + (v_{x0,\text{b}} - v_{x0,\text{r}})t + \left(\frac{1}{2}g_x - \frac{1}{2}g_x\right)t^2 \tag{5}$$

$$= (x_{0,\text{b}} - x_{0,\text{r}}) + (v_{x0,\text{b}} - v_{x0,\text{r}})t$$

$$= (v_{x_0,\text{b}} - v_{x_0,\text{r}})t$$

Again, the acceleration terms cancel out. The first two terms also cancel because the initial position of the two balls is the same. Since we have chosen the downward direction to be positive, we have

$$x_{\text{br}} = (v_{0,\text{b}} - (-v_{0,\text{r}}))t = (17.0 \text{ m/s} + 21.0 \text{ m/s})(3.00 \text{ s})$$
$$= 114 \text{ m} \tag{6}$$

Making sense of the result

Since the acceleration due to gravity contributes to the change in velocity and position of both objects identically, the acceleration due to gravity makes no contribution to the relative quantities.

3-8 Calculus of Kinematics

General Framework for Kinematics Equations

[Optional section, requires basic integration]
In the previous sections, we developed mathematical equations to describe the motion of objects moving with constant (uniform in time) acceleration. In what follows, we use calculus, more specifically integrals and derivatives, to develop the kinematics equations to describe the motion of an object moving with variable acceleration, $a_x(t)$. So far, we have learned how to derive instantaneous velocity and acceleration from a position versus time description, $x(t)$, for a given object

(Equations 3-6, 3-8). We have done this by calculating the slopes of $x(t)$ and $v_x(t)$ curves, which can be represented mathematically by finding the derivatives of $x(t)$ and $v_x(t)$. We also discovered that the areas under the $v_x(t)$ and $a_x(t)$ graphs represent the changes of the object's position and velocity, respectively (Equations 3-12, 3-18). Finding the area under a curve mathematically can be represented by integration. When the acceleration of an object varies with time, $a_x(t)$, it is much more practical to use calculus, instead of findings slopes and areas using the previously introduced principles of graphical integration (Figure 3-35).

If the acceleration is given by a continuous smooth function $a_x(t)$, then we can rearrange Equation 3-8 and integrate to find the velocity of the object:

$$dv_x(t) = a_x(t)dt \tag{3-33}$$

$$\int_{v_0}^{v} dv_x(t) = \int_{0}^{t} a_x(t)dt$$
$$v_x(t) - v_{x_0} = \int_{0}^{t} a_x(t)dt \tag{3-34}$$

or

KEY EQUATION $$v_x(t) = v_{x_0} + \int_{0}^{t} a_x(t)dt, \tag{3-35}$$

where v_0 is the velocity of the object at time $t = 0$, the initial velocity of the object.

According to Equation 3-35, $v_x(t)$, the velocity of the object as a function of time is given by the sum of the initial velocity of the object and the change in the velocity due to the acceleration term.

Similarly, rearranging the derivative definition of velocity and integrating gives $x(t)$, the position of the object as a function of time:

$$dx(t) = v_x(t)dt \tag{3-36}$$

$$\Delta x = x(t) - x_0 = \int_{0}^{t} v_x(t)dt \tag{3-37}$$

or

KEY EQUATION $$x(t) = x_0 + \int_{0}^{t} v_x(t)dt, \tag{3-38}$$

where x_0 is the initial position of the object.

As we should have expected, in the case when the acceleration is constant and we start at $t_0 = 0$, Equation 3-35 becomes Equation 3-10:

$$v_x(t) = v_{x_0} + \int_{t_0}^{t} a_x dt = v_{x_0} + a_x(t - t_0)$$
$$= v_{x_0} + a_x \Delta t = v_{x_0} + a_x t \tag{3-10}$$

Combining Equations 3-35 and 3-38 gives us the expression for the position as a function of time:

KEY EQUATION

$$x(t) = x_0 + \int_{t_0}^{t} v_x(t)dt = x_0 + \int_{t_0}^{t} (a_x(t)t + v_{x_0})dt \tag{3-39}$$

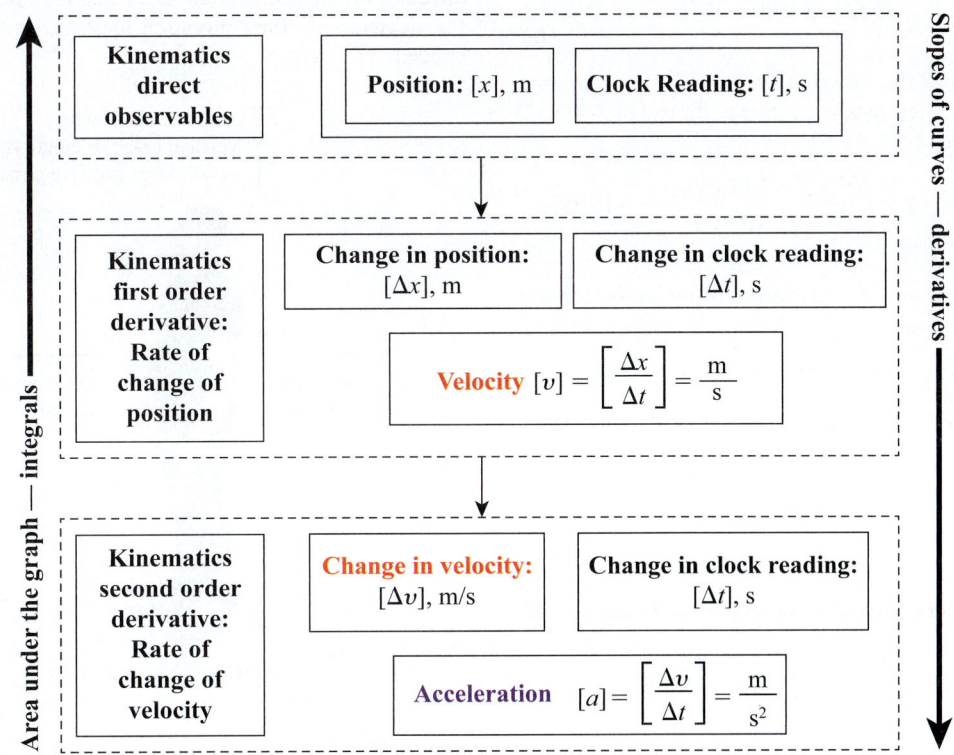

Figure 3-35   A conceptual representation of the calculus used for describing one-dimensional kinematics.

In the case when acceleration is constant and we start at $t_0 = 0$, Equation 3-39 becomes Equation 3-17:

$$x(t) = x_0 + v_{x_0}t + \frac{1}{2}a_x t^2 \text{ or } \Delta x(t) = v_{x_0}t + \frac{1}{2}a_x t^2$$

when $t_0 = 0$ \hfill (3-17)

PEER TO PEER

When thinking of mathematical relationships describing one-dimensional kinematics, I always try to remember to check the units. Time derivatives correspond to finding slopes, while integrals correspond to finding areas. Checking the units helps me make sense of the equations.

EXAMPLE 3-16

Variable Acceleration

A Formula One car is moving at a speed of 98.0 m/s along a straight track when the driver brings the car to a stop, gradually applying the brakes such that the magnitude of the acceleration increases linearly with time. The car takes 3.40 s to stop.

(a) Express the car's acceleration as a function of time.
(b) Calculate the speed of the car 3.00 s after the driver starts braking.
(c) Find the total distance covered by the car while slowing.
(d) Find the highest (in magnitude) acceleration of the car.

Solution

(a) Since the acceleration changes linearly with time and has an initial value of zero, we can express the acceleration of the car as

$$a_x(t) = ct \tag{1}$$

where c is the *rate of change* of the acceleration (often called **jerk** or **jolt**).

Substituting this acceleration function into Equation 3-35, we obtain

$$v_x(t) = v_{x_0} + \int_0^t a_x(t)dt = v_{x_0} + \frac{1}{2}ct^2 - \frac{1}{2}c(0)^2 = v_{x_0} + \frac{1}{2}ct^2 \tag{2}$$

Let us choose the direction of motion as the positive x-direction. Since the speed is zero at the end of 3.40 s, (2) becomes

$$0 = v_{x_0} + \frac{1}{2}ct^2 \Rightarrow c = \frac{-2v_{x_0}}{t^2} = \frac{-2(98.0 \text{ m/s})}{(3.40 \text{ s})^2} = -16.96 \text{ m/s}^3 \tag{3}$$

Combining (1) and (3) and expressing our answer to three significant digits,

$$a_x(t) = (-17.0 \text{ m/s}^3)t$$

(b) For the speed of the car 3.00 s into its braking phase, we use (2):

$$v_x(t) = v_{x_0} + \frac{1}{2}ct^2 = 98.0 \text{ m/s} + \frac{1}{2}(-16.96 \text{ m/s}^3)(3.00 \text{ s})^2$$

$$= 21.7 \text{ m/s}$$

(c) For the distance covered by the car, we use Equation 3-37:

$$\Delta x = x(t) - x_0 = \int_0^t v_x dt = \int_0^t \left(v_{x_0} + \frac{1}{2}ct^2\right)dt$$

$$= (v_{x_0}t - v_{x_0}(0)) + \left(\frac{1}{6}ct^3 - \frac{1}{6}c(0)\right) = v_{x_0}t + \frac{1}{6}ct^3$$

Since the direction of motion was chosen to be positive, and for clarity suppressing units until the final answer,

$$\Delta x = v_{x_0}t + \frac{1}{6}ct^3$$

$$= \left(98.0(3.40) + \frac{1}{6}(-16.96)(3.40)^3\right) \text{ m} = 222 \text{ m}$$

(d) From (1), the highest (in magnitude) acceleration is

$$a_x = (-16.96 \text{ m/s}^3)(3.40 \text{ s}) = -57.7 \text{ m/s}^2$$

If desired, we can express this in terms of the acceleration due to gravity: $a_x = -5.88g$.

Making sense of the result

The constant c has units of m/s^3, and the acceleration as a function of time is the product of that constant with time, leaving us with m/s^2 as the units for the acceleration, as expected. The initial speed of the car is about 353 km/h, so the fairly long stopping distance seems reasonable. The maximum acceleration is equal to $5.88g$, not unheard of in Formula One racing.

The body's tolerance to acceleration depends on the direction of the acceleration as well as the rate of change of the acceleration—the convention illustrated in Figure 3-36 is applied.

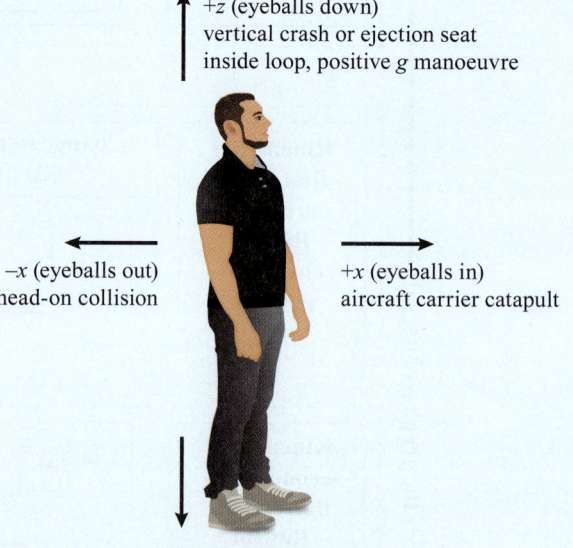

Figure 3-36 Conventions used for discussing the tolerance of the human body to acceleration.

In this chapter, we looked at motion in one dimension. We started with a discussion of how athletes, like Usain Bolt, can use the concepts of kinematics such as speed, velocity, and acceleration to analyze and improve their performance. We considered motion with constant (uniform) acceleration as well as with variable acceleration. In Chapter 4, we extend the discussion to the motion in two and three dimensions.

KEY CONCEPTS AND RELATIONSHIPS

Kinematics is the study of motion. In kinematics, we study the relationships between an object's position, displacement, velocity, and acceleration and their dependence on time. We also examine relative motion.

Position, Velocity, Acceleration, and Time

The displacement of an object is the change in its position at two different times:

$$\Delta x = x - x_0 \quad [\Delta x] = \text{m} \tag{3-1}$$

Distance is the path length covered by an object:

$$d = \sum_i |\delta x_i| \tag{3-2}$$

The average speed is the distance covered divided by the time elapsed:

$$v_{\text{avg}} = \frac{d}{\Delta t} = \frac{d}{t - t_0} \quad [v_{\text{avg}}] = \frac{\text{m}}{\text{s}} \tag{3-3}$$

The average velocity is the displacement divided by the time elapsed:

$$v_{x,\text{avg}} = \frac{\Delta x}{\Delta t} = \frac{x - x_0}{t - t_0} \quad [v_{x,\text{avg}}] = \frac{\text{m}}{\text{s}} \tag{3-4}$$

The instantaneous velocity is the derivative of the position with respect to time:

$$v_x(t) = \lim_{\Delta t \to 0} v_{x,\text{avg}} = \lim_{\Delta t \to 0} \frac{\Delta x}{\Delta t} = \frac{dx(t)}{dt} = x'(t) \tag{3-6}$$

The instantaneous speed is the magnitude of the instantaneous velocity.

The average acceleration is the change in the velocity divided by the time elapsed:

$$a_{x,\text{avg}} = \frac{\Delta v_x}{\Delta t} = \frac{v_x - v_{x_0}}{t - t_0} \quad [a_x] = \frac{\text{m}}{\text{s}^2} \tag{3-7}$$

The instantaneous acceleration is the derivative of velocity with respect to time:

$$a_x = \lim_{\Delta t \to 0} \frac{\Delta v_x(t)}{\Delta t} = \frac{dv_x(t)}{dt} = v'_x(t) \tag{3-8}$$

Kinematics Equations for Constant Acceleration

For constant acceleration, the velocity of an object as a function of time is given by

$$v_x(t) = v_{x_0} + a_x t \tag{3-10}$$

The position of an object moving with constant acceleration is given by

$$x(t) = x_0 + v_{x_0} t + \frac{1}{2} a_x t^2 \quad \text{when } t_0 = 0 \tag{3-17}$$

The time-independent kinematics equation is given by

$$v_x(t)^2 - v_{x_0}^2 = 2 a_x \Delta x \tag{3-22}$$

Relative Motion

The relative velocity of object 1 with respect to object 2 is (g stands for ground)

$$v_{x_{12}} = v_{x_{1g}} - v_{x_{2g}} \tag{3-25}$$

The relative displacement and position of object 1 with respect to object 2 are given by

$$x_{12} = x_{1g} - x_{2g} \Rightarrow$$
$$\Delta x_{12} = v_{x_{12}} \Delta t = (v_{x_{1g}} - v_{x_{2g}}) \Delta t = \Delta x_{1g} - \Delta x_{2g} \tag{3-27}$$

The relative acceleration is given by

$$a_{x_{12}} = a_{x_{1g}} - a_{x_{2g}} \tag{3-28}$$

Motion with Variable Acceleration (General Case)

$$v_x(t) = v_0 + \int_0^t a_x(t)\, dt \tag{3-35}$$

$$x(t) = x_0 + \int_0^t v_x(t)\, dt \tag{3-38}$$

$$x(t) = x_0 + \int_{t_0}^t v_x(t)\, dt = x_0 + \int_{t_0}^t (a_x(t)t + v_{x_0})\, dt \tag{3-39}$$

APPLICATIONS

analysis of the motion of objects in everyday life, traffic safety, airline safety, airline runway construction, motor vehicle safety, projectile motion, relative motion, sports and medicine

KEY TERMS

acceleration due to gravity, average acceleration, average speed, average velocity, biomimicry, calculus, clock reading, constant (uniform) acceleration, displacement, distance, dynamics, echolocation, elapsed time, frame of reference, free fall, Galilean principle of velocity addition, gravitational acceleration, initial position, instantaneous speed, instantaneous velocity, jerk, jolt, kinematics, kinematics equations for constant (uniform) motion, motion diagram, principle of graphical integration, reference frame, speed, trajectory, velocity

QUESTIONS

1. Figure 3-37 shows a plot of the position versus time for two objects.
   (a) Which object has the higher average velocity for the interval shown?
   (b) Which object has the higher speed at $t = 30$ s? Which has the higher speed at $t = 90$ s?
   (c) Do the objects ever have the same speed? If so, at what time do you estimate this will happen?

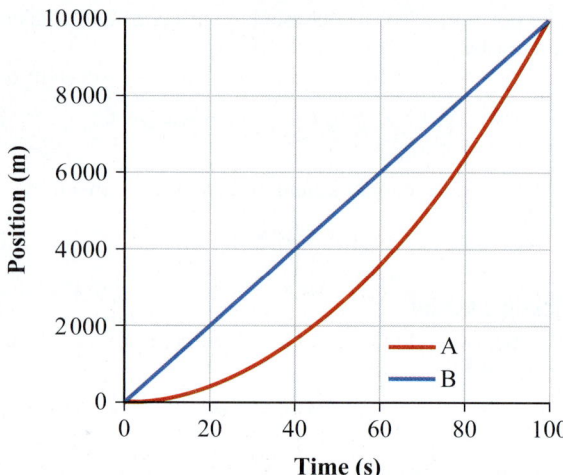

Figure 3-37   Question 1.

2. Figure 3-38 shows the plot of the position versus time for two objects, A and B.
   (a) Which object has the higher average speed for the interval shown (during the first 100 s of its motion)?
   (b) Estimate the average speeds of object A and object B for the interval shown.
   (c) At the points indicated by the arrows, do the objects have the same speed? The same acceleration?
   (d) Which object has the higher speed at the position indicated by arrow 1?
   (e) Which object has the higher speed at the position indicated by arrow 2?
   (f) Is the speed of object A ever higher than the speed of object B?

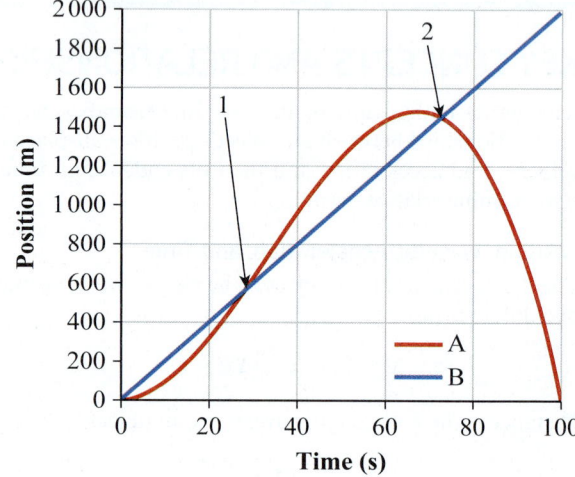

Figure 3-38   Question 2.

3. The plot of velocity versus time for two objects is shown in Figure 3-39.
   (a) Starting at the same point at the same time, are the two objects ever at the same position at the same time again?
   (b) Do the two objects have the same speed? If so, when does that occur?
   (c) Which object has the higher magnitude of average acceleration for the interval shown?
   (d) Estimate the magnitude of the acceleration for each object during the first 10 s of their motion.

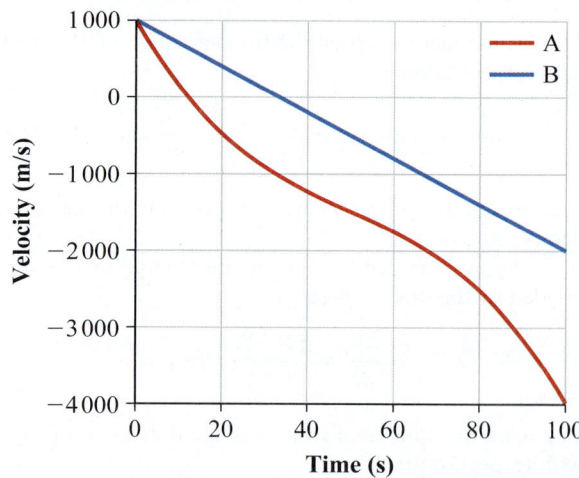

Figure 3-39   Question 3.

4. Figure 3-40 shows the plot of the velocity versus time for two boats.
   (a) For the interval shown, which boat covers more distance? Explain your reasoning.
   (b) Do the boats ever have the same acceleration?

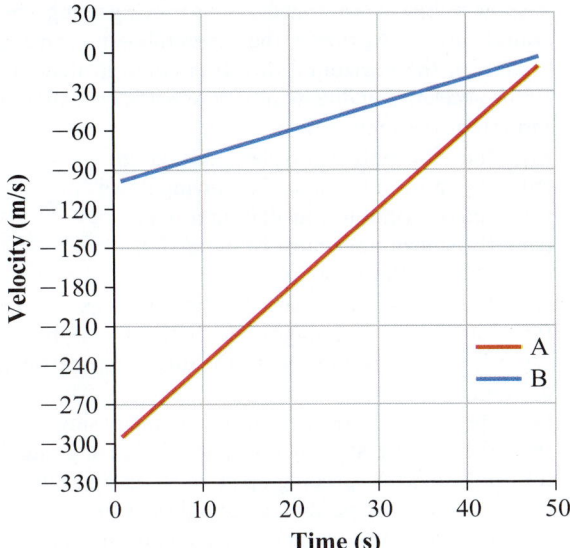

Figure 3-40   Question 4.

5. Figure 3-41 shows the velocity versus time plot for two objects; they are at the same position at $t = 0$. Explain your reasoning for each one of the questions below.
   (a) For the interval shown, which object has the higher average acceleration?
   (b) Which object covers more distance?
   (c) Which object has the higher magnitude of acceleration at the point of intersection for the two curves?
   (d) Which object is ahead of the other at the point of intersection of the two curves?
   (e) Do the objects ever have the same displacement and/or acceleration? If so, estimate where it happens.

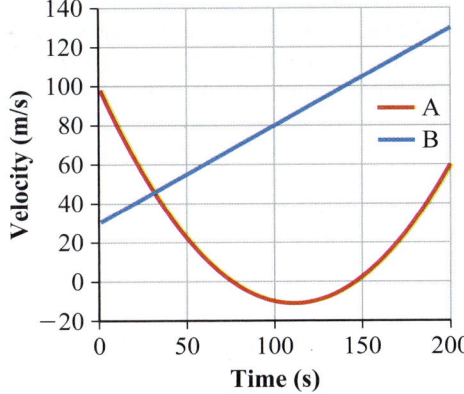

Figure 3-41   Question 5.

6. Figure 3-42 shows the plot of an object's position versus time.
   (a) Where does this object have its highest speed?
   (b) What is the object's lowest speed?
   (c) Infer where the object's acceleration is zero.
   (d) Is the object ever at rest?
   (e) Does the object cover more distance between $t = 0$ s and $t = 140$ s than it does between $t = 140$ s and $t = 200$ s? Explain your reasoning.

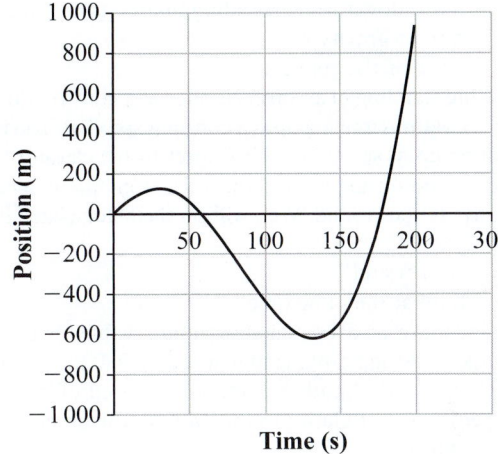

Figure 3-42   Question 6.

7. The velocity versus time graph for an object is shown in Figure 3-43.
   (a) Is the acceleration of this object ever zero?
   (b) Where is this object's magnitude of acceleration the highest? The lowest?

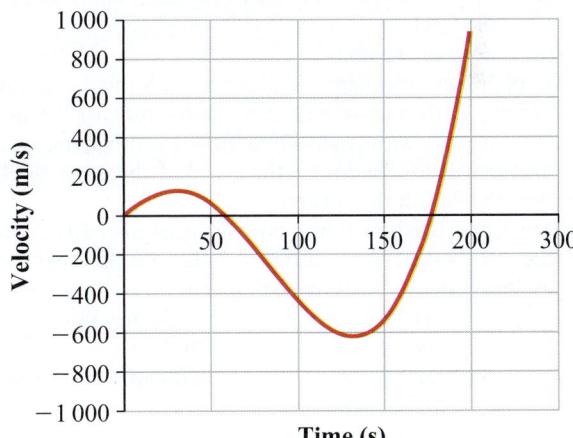

Figure 3-43   Question 7.

8. Two balls are thrown simultaneously, one up and one down, both with an initial speed of 27 m/s. They pass each other 3 s after they are thrown. What is the speed of one ball with respect to the other as they pass each other?
   (a) 27 m/s
   (b) 54 m/s
   (c) It depends on the distance the balls travel.
   (d) None of the above are correct.

CHAPTER 3 | **MOTION IN ONE DIMENSION**

MECHANICS

9. You drop a tennis ball from a hot-air balloon that is going upward at a speed of 4 m/s. Relative to the ground, the ball
   (a) falls straight down, while speeding up with the acceleration due to gravity
   (b) goes up while slowing down, stops momentarily, and then goes down while speeding up
   (c) hangs in the air for a short time, then moves up while speeding up, and then begins to move down while slowing down due to the acceleration of gravity
   (d) hangs in the air for a short time and then begins to move down while speeding up due to the acceleration of gravity
   (e) none of the above

10. You are standing in an open elevator at a construction site. The elevator is moving up at a constant speed, v. You throw a stone up at speed $2v$ with respect to the elevator, while your friend on the ground throws a stone up at speed $2v$ relative to the ground. Who will get their stone back first?
    (a) you
    (b) your friend
    (c) both at the same time
    (d) cannot be determined without a value for v

11. You are piloting a plane at a speed of 980 km/h relative to the ground. Another plane has a speed of zero with respect to you. Relative to the ground, the other plane is
    (a) stationary
    (b) moving at 980 km/h in the opposite direction
    (c) moving at 980 km/h in the same direction
    (d) moving at 980 km/h in some arbitrary direction
    (e) none of the above

12. When a Formula One race car completes two laps around a track,
    (a) the distance covered by the car is zero
    (b) the car's displacement is zero
    (c) the car's displacement is twice the circumference of the track
    (d) none of the above

13. Explain why you agree or disagree with this statement: When you throw two balls downward from a 10 m high building, one with twice the speed of the other, the speed of one will still be double the speed of the other by the time they reach the ground.

14. You throw two balls down, one with a speed of 10 m/s and the other with a speed of 6 m/s. As long as both balls are in the air, the speed of one ball relative to the other one
    (a) depends on the height from which they fall
    (b) stays the same as it was initially
    (c) changes because the faster ball gains more speed than the slower ball
    (d) none of the above

15. You throw two balls from the same height, one straight down and the other straight up at the same speed, v_0. Which of the following is correct?
    (a) As long as both balls are in the air, the ball you throw straight down will have twice the speed of the ball you throw straight up.

(b) As long as both balls are in the air, the relative speed of the two balls is $2v_0$.
(c) As long as both balls are in the air, the relative velocity of the two balls is $2v_0$.
(d) The balls will hit the floor with the same speed.
(e) Both (c) and (d) are correct.

16. A snowmobile (a motorized sled) is moving along a horizontal surface. At time t, the snowmobile has an acceleration in the horizontal direction equal in magnitude to the acceleration due to gravity. What is true about the motion of the snowmobile?
    (a) the snowmobile must be speeding up
    (b) the snowmobile must be slowing down
    (c) the snowmobile must be stationary
    (d) the snowmobile must be in free fall
    (e) none of the above

17. An object moves in a straight line with an acceleration that increases in magnitude. If the direction of the object's velocity is opposite to the direction of its acceleration, then
    (a) the object is slowing down at an increasing rate
    (b) the object is slowing down at a decreasing rate
    (c) the object is speeding up at an increasing rate
    (d) the object is speeding up at a decreasing rate
    (e) the object is slowing down at a constant rate
    (f) all the above are possible

18. Can an object's instantaneous velocity be equal to its average velocity? Explain.

19. Can an object's average acceleration be equal to its instantaneous acceleration? Explain.

20. When an object moves to the right and its acceleration points to the right, then
    (a) the object's speed will decrease
    (b) the object's speed will increase
    (c) both (a) and (b) are possible
    (d) none of the above

21. An object moves to the right with an increasing speed. Define left to be the positive x-direction. The acceleration of the object
    (a) is negative
    (b) is positive
    (c) could be either negative or positive
    (d) none of the above

22. You throw a ball straight down with a speed of 20 m/s. You wait for 1.0 s, and throw another ball down with a speed of 20 m/s. Setting down as the positive x-direction, and considering the acceleration of gravity to have a magnitude of 9.8 m/s^2, the velocity of the second ball relative to the first ball, $v_{x_{21}} = v_{x_2} - v_{x_1}$, will
    (a) always be -9.8 m/s
    (b) always be $+9.8$ m/s
    (c) increase with time
    (d) decrease with time
    (e) none of the above

23. An object moves to the right, but its acceleration is directed to the left. Define left to be the positive direction. Which of the following is true?
    (a) The object's velocity increases.
    (b) The object's velocity decreases.
    (c) The object's speed increases.
    (d) The object's speed decreases.
    (e) More one than of the above is possible.

24. Can your speed be zero when your velocity is non-zero? Explain and give an example.

25. Can your speed be zero when your acceleration is non-zero? Explain and give an example.

26. Can your acceleration be zero when your speed is non-zero? Explain and give an example.

27. When an object's instantaneous velocity is zero, its acceleration can be
    (a) positive
    (b) negative
    (c) zero
    (d) any of the above

28. A person's displacement over a given time interval is zero. From this information you can conclude that
    (a) the distance the person travelled is zero
    (b) the person's average velocity is zero
    (c) the person's average speed is zero
    (d) the person's average acceleration is zero
    (e) the person never moved, so all of the above are zero

29. You walk along a straight line. You start from rest at point A, then walk in the positive x-direction for five minutes and reach point B, where you stop for two minutes and then walk for five minutes to point C located between points A and B. Which of the following must be true?
    (a) Your velocity when you move from point A to point B is positive.
    (b) Your average speed during the entire trip is higher than the magnitude of your average velocity for the entire trip.

    (c) Your displacement during the entire trip is non-zero.
    (d) Your speed when you were walking from A to B is higher than your speed when you were walking from B to C.
    (e) All of the above must be true.

30. If your average speed over a given interval is zero, then
    (a) you did not move
    (b) you returned to the same place
    (c) your displacement can be non-zero
    (d) all of the above are possible

31. You drop a ball from the rooftop of a building. Then, 3 s later you drop another ball from the same location. While the two balls are in the air, the distance between them
    (a) increases as the square of t (the time both balls are in the air)
    (b) increases linearly with t (the time both balls are in the air)
    (c) decreases linearly with t (the time both balls are in the air)
    (d) stays constant
    (e) none of the above

32. You throw a ball straight down with a speed of 20.0 m/s. It bounces off the floor and comes up to the same height at the same speed with which you launched it. Which of the following is true regarding the ball's entire motion?
    (a) The average acceleration of the ball is 9.80 m/s^2.
    (b) The average speed of the ball is 20.0 m/s.
    (c) The average velocity of the ball is zero.
    (d) None of the above statements are true.

33. Two friends start from points that are 30 m apart and change positions. It takes them 6 s to do so. Which of the following is true?
    (a) Their relative displacement is zero.
    (b) Their average relative velocity is zero.
    (c) Their average relative velocity is 10 m/s.
    (d) Their relative displacement is 30 m.

34. In Figure 3-44, which velocity versus time graph matches the position versus time graph on the left?

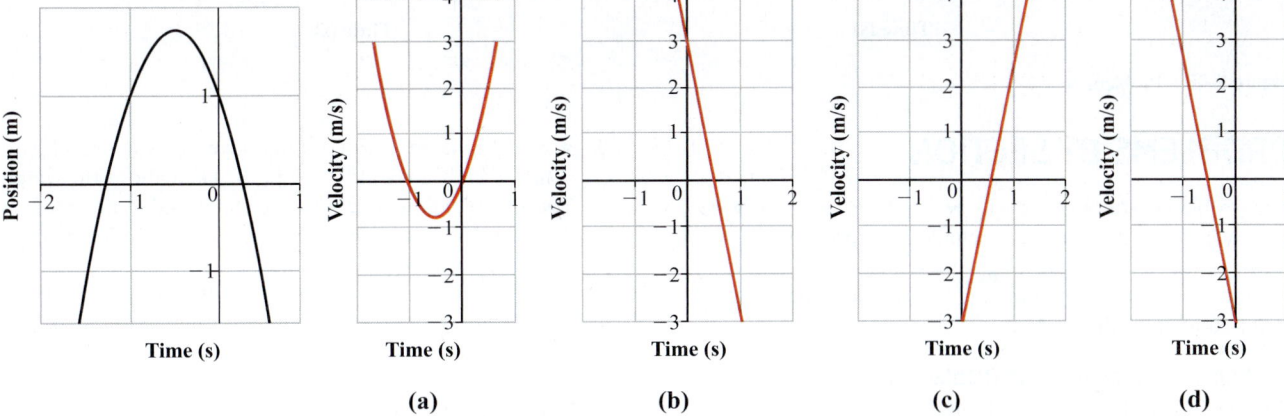

(a)   (b)   (c)   (d)

Figure 3-44 Problem 34.

35. In Figure 3-45, which position versus time graph matches the velocity versus time graph on the left? Explain.

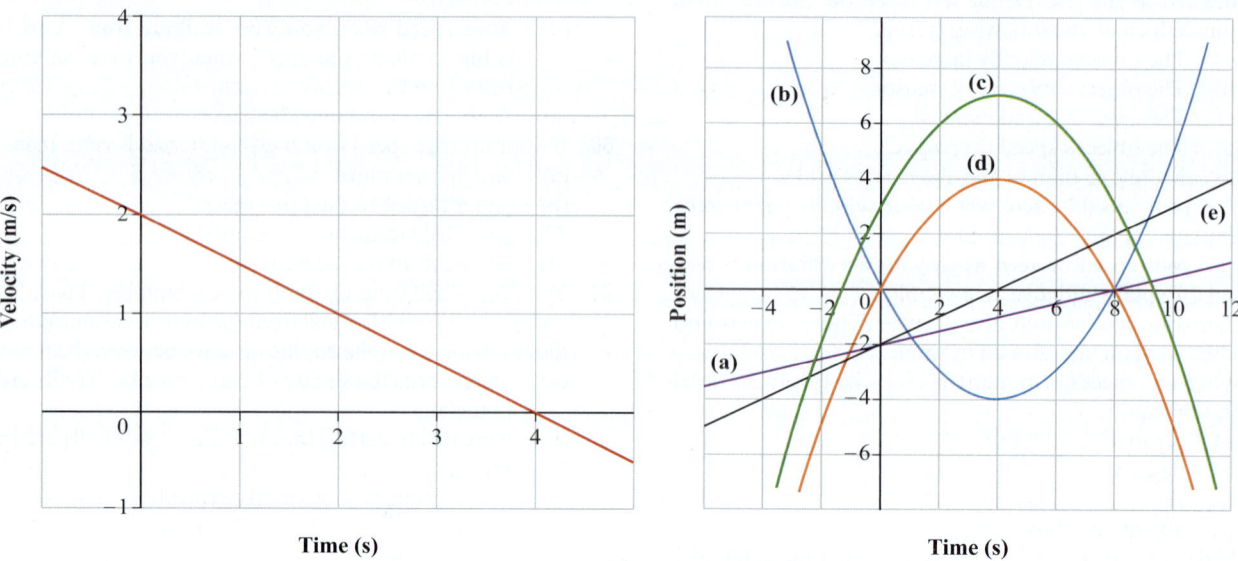

Figure 3-45   Problem 35.

36. In Figure 3-46, which position versus time graph matches the acceleration versus time graph on the left? Explain.

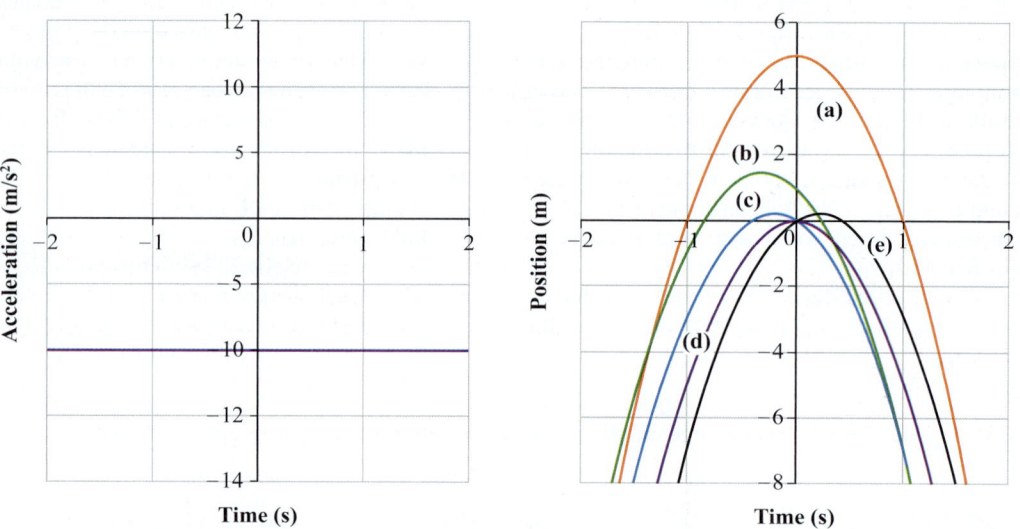

Figure 3-46   Problem 36.

PROBLEMS BY SECTION

For problems, star ratings will be used (★, ★★, or ★★★), with more stars meaning more challenging problems. The following codes will indicate if $\frac{d}{dx}$ differentiation, $\int$ integration, ▭ numerical approximation, or ∿ graphical analysis will be required to solve the problem.

Section 3-1   Distance and Displacement

37. ★ You walk from your house 1100 m down the street, and then you reverse direction and walk 1300 m up the street.
(a)   What is the distance you travelled?
(b)   What is your total displacement? Choose down the street as the positive direction.

38. ★ Starting from a trail 3.0 m above the water level, you climb straight up a cliff to a point 11 m above the water. You then shallow dive into the water and climb back to the trail. Find your displacement as well as the distance you covered (not including the swim).

39. ★ A grasshopper jumps straight up from a leaf that is 80. cm above the ground, reaches a maximum height of 75 cm above the leaf, and lands on the ground. Find the distance covered by the grasshopper as well as its displacement.

40. ★ A water taxi starts from a dock on the west bank of a channel, delivers passengers to a dock on the east bank, and then returns to its starting point on the west bank. The two docks are 230 m apart. For simplicity, we assume there is no current in the channel.
(a)   Determine the displacement of the water taxi.
(b)   Calculate the distance travelled by the taxi.

Section 3-2 Speed and Velocity

41. ★ A barge takes a load of processed fish from a fishing boat to the dock 3.2 km away, and then returns to the fishing boat along the same path. During this time, the fishing boat has moved 2.0 km away from the dock. The whole journey takes 20.0 min. Find
    (a) the displacement of the barge
    (b) the average velocity of the barge
    (c) the average velocity of the fishing boat
    (d) the average speed of the barge
    (e) the average speed of the fishing boat.

42. ★ On a distant planet, a ball is dropped from rest from a height of 21.0 m. It hits the ground and bounces back up until it is 1.00 m below its original position. The trip down and back up takes 10.0 s. Find the ball's average speed, displacement, and average velocity.

43. ★★ Sailfish have a maximum speed of 112 km/h. A sailfish moves in a straight line at its maximum speed for 3.00 s, slows down, and then comes to a stop 2.00 s later, having covered a total distance of 125 m.
    (a) Find the acceleration during the last 2.00 s, assuming constant acceleration.
    (b) Find the average acceleration during the entire 5.00 s.
    (c) Find the average velocity during the entire 5.00 s.
    (d) The sailfish then takes 8.00 s more to return to its starting point, accelerating to its maximum speed as it does so. Find its average speed and average velocity for the round trip.

44. ★★ A stone is released at the top of a 1.2 m tall barrel filled with liquid. The stone reaches a speed of 1.7 m/s just before it hits the bottom of the barrel 0.90 s later. Assuming constant acceleration, find the stone's average speed and average velocity.

45. ★ The maximum speed of an Airbus 380 at cruising altitude is 945 km/h. Estimate the minimum amount of time it would take this plane take to go from Paris to New York, a flight distance of 5833 km, if it could fly all the time with this speed? Why do you think it always takes longer than this to fly direct from Paris to New York?

46. ★★ The distance between the Sun and Earth is one astronomical unit, 149 600 000 km. Calculate the average speed of Earth around the Sun in km/h, assuming that Earth has a nearly circular orbit and that the Sun is at the centre of Earth's orbit.

47. ★★ The speed of the Moon around Earth is approximately 1 km/s. Estimate the distance between the Moon and Earth. Express it in km.

48. ★ Lightning flashes in your neighbourhood, and 3.00 s later you hear thunder. How far away was the lightning flash? Assume the speed of sound to be 345 m/s.

49. ★ The speed of the space shuttle in low Earth orbit was about 28 000 km/h. When the altitude of the shuttle was 300 km, how long did it take to orbit Earth once?

50. ★ A rocket car covers a distance of 10.0 km forward in 1.0 min, turns around, which takes 5.0 min, and goes back halfway to its starting position in 1.5 min.
    (a) Find the displacement of the rocket car.
    (b) Find the distance covered by the rocket car.
    (c) Find (i) the average velocity and (ii) the average speed of the rocket car.

51. ★★ Concorde aircraft are famous for breaking the sound barrier. The maximum speed of the original Concorde 1 aircraft at cruising altitude was 2172 km/h or Mach 1.759 (a supersonic speed). The Mach number is a dimensionless quantity representing the ratio of the aircraft speed (in this case) to the speed of sound (340 m/s). However, in 2003 Concorde 1 was retired due to technical issues. In the summer of 2015, a new Airbus design, a successor to the original Concorde 1 aircraft, was suggested. Concorde 2 will fly in 2021 at supersonic speeds reaching Mach 4.500. Engineers describe the new aircraft ("the son of Concorde 1") as "the highest rollercoaster in the world." It would rely on three different types of engine, each fuelled by different forms of hydrogen. Considering $v_{sound} = 340$ m/s, answer the following questions.
    (a) What is the Mach number of a Boeing 737 if its max speed is 583 mph? Compare this with the maximum Mach speed of Concorde 2.
    (b) What is the maximum speed of Concorde 2 in km/h?
    (c) How long would it take Concorde 2 to travel from London, England, to Montréal, Québec, a flight distance of 5238 km, assuming it can fly at its maximum speed for the whole trip?

52. ★★ 〰 Figure 3-47 shows a position versus time plot for a particle moving along the horizontal x-axis.
    (a) Find the distance covered by the particle over the first 5 s of its motion.
    (b) Find the displacement of the particle during the first 5 s of its motion.
    (c) Find the average speed and the average velocity of the particle.
    (d) Find the speed and velocity of the particle at $t = 3$ s.

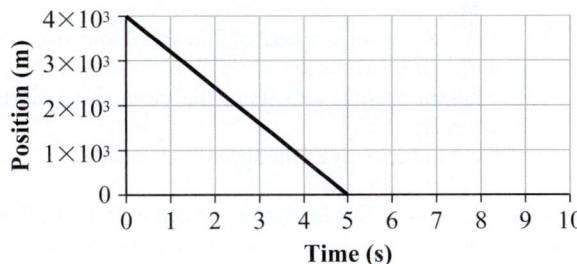

Figure 3-47 Problem 52.

53. ★ 〰 Figure 3-48 shows the velocity versus time plot for a motorcycle. In doing this problem, assume that the times are exact numbers, and express your answers to two significant digits.
    (a) Find the displacement of the motorcycle between $t = 0$ s and $t = 3$ s, between $t = 3$ s and $t = 7$ s, and between $t = 7$ s and $t = 9$ s.
    (b) What is the acceleration of the motorcycle between $t = 2$ s and $t = 3$ s?

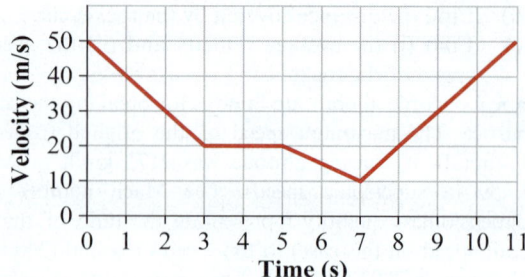

Figure 3-48   Question 53.

54. ★★ 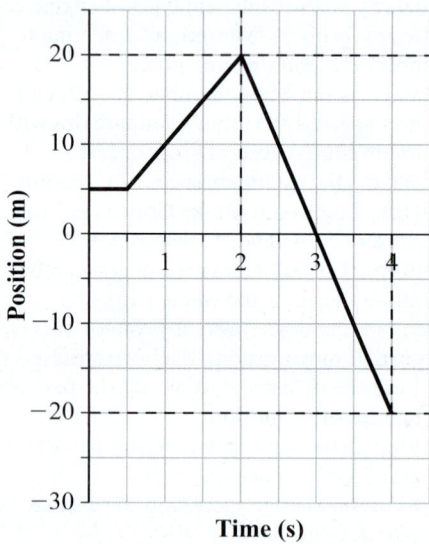 Figure 3-49 shows a position versus time plot for a speedboat.

Figure 3-49   Problem 54.

(a)   Find the average speed of the speedboat over the entire interval shown.
(b)   Find the average velocity of the speedboat over the entire interval.
(c)   Find the velocity of the speedboat between $t = 1.0$ s and $t = 2.0$ s.

55. ★ (a)   How long does light take to reach us from the Sun?
(b)   Roughly how long does it take light to reach us from the star Sirius?
(c)   How long would it take light to cross the diameter of the Milky Way galaxy?
(d)   What does the unit "light year" represent, and how can you use it to describe our galaxy—the Milky Way?

Section 3-3   Acceleration

56. ★ The space shuttle had a takeoff acceleration approximately 3.5 times the acceleration due to gravity. What was the speed of the space shuttle 100.0 s after takeoff if it maintained that acceleration?

57. ★★ A sunspot is on the Sun's equator. Find the speed of this spot relative to the Sun's centre. Look up in a reference source values for the rotation and the radius of the Sun.

58. ★★ Starting with a speed of 4.0 m/s, a stone falls toward a container of molasses, reaching a speed of 21 m/s by the time it hits the surface of the molasses. Over 2.0 s, the molasses slows the stone down to a speed of 3.0 m/s, its terminal velocity. Find the average speed of the stone during this fall from the start of the fall until the time it reaches its terminal velocity.

59. ★★ 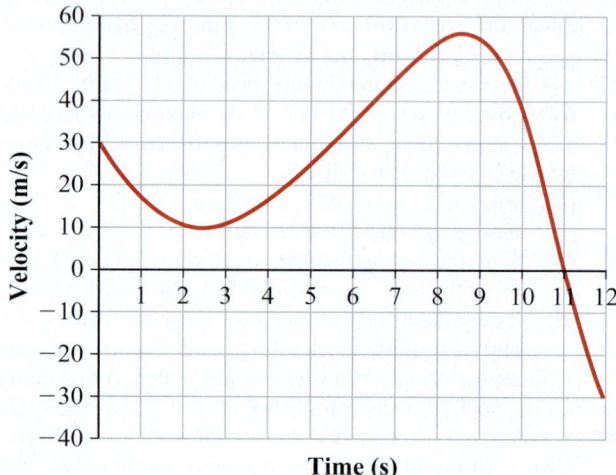 Figure 3-50 shows the velocity versus time plot for a car.
(a)   Find the average acceleration of the car over the first 4.0 s.
(b)   Find the instantaneous acceleration of the car at $t = 11$ s.

Figure 3-50   Problem 59.

(c)   Find the average acceleration of the car over the entire interval shown.
(d)   Estimate the distance covered by the car between $t = 0$ s and $t = 12$ s.

60. ★★★ 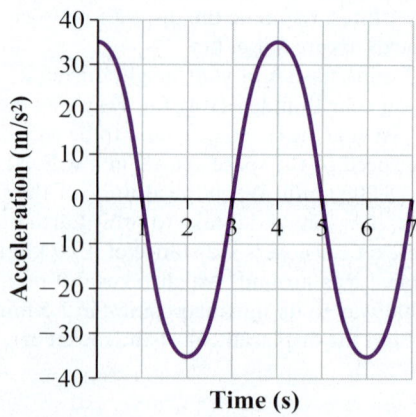 Figure 3-51 shows the acceleration versus time plot for a particle undergoing a form of cyclic motion called simple harmonic motion. The velocity is zero when $t = 0$. Answer to two significant digits.
(a)   Find the average speed of the particle between $t = 2.0$ s and $t = 3.0$ s.
(b)   Find the speed of the particle at $t = 3.0$ s.

Figure 3-51   Problem 60.

Section 3-4 Mathematical Description of One-Dimensional Motion with Constant Acceleration

For this section, some data were obtained from NASA. Convert all units to SI units and report your answers in SI units. Notation: mph = miles per hour; ft/s = feet/s.

61. ★★ During the second-stage burn of an *Apollo* launch, the rocket's speed increased from 7882.9 ft/s to 21 377.0 ft/s. This burn lasted 384.22 s. Assuming constant acceleration and a straight path, find
    (a) the average acceleration
    (b) the distance travelled during this second-stage burn

62. ★★ 〰 Derive Equation 3-17 using both an algebraic approach and a graphical approach.

63. ★★ Derive Equation 3-32 and check its validity using dimensional analysis.

64. ★ Typically, a golf club is in contact with a golf ball for about 0.50 ms. Long-drive legend Jason Zuback is reported to be able to hit a golf ball so that it is launched into the air at a speed of 357 km/h. Assuming constant acceleration while the golf club is in contact with the ball, estimate the acceleration of Zuback's golf ball and the distance his club travelled while in contact with the ball.

65. ★ A Formula One car has an acceleration 1.45 times the acceleration due to gravity. Find the time the car takes to go from 0 to 100 km/h, assuming constant acceleration.

66. ★★ The Bugatti Veyron can go from 0 to 100 km/h in 2.46 s. It takes another 7.34 s to reach a speed of 240 km/h. Assume different constant accelerations during these two different stages of the car's motion, as the driver lets up on the gas a little after a speed of 100 km/h is reached.
    (a) Calculate the acceleration of the car over the first 2.46 s.
    (b) How much total distance does the Veyron have to cover to reach a speed of 240 km/h?
    (c) The speed of the car is limited to 415 km/h to protect the tires. If the car maintained the same acceleration that it had at 240 km/h, how long would it take the car to reach 415 km/h?

67. ★★ The greatest acceleration experienced by Col. Stapp on the Gee Whiz rocket sled was $46.2g$ (i.e., an acceleration 46.2 times that of gravity) as the sled braked from a speed of 632 mph.
    (a) Over what distance would the sled come to a stop if the acceleration was constant?
    (b) How long would the sled take to come to a stop?

68. ★★ Engineered material arresting systems (EMASs) are sometimes used to stop airplanes when they overshoot a runway. Consider a worst-case scenario in which an Airbus 380 moving at its full landing speed of 250 km/h hits the EMAS. What is the minimum stopping distance that will keep the acceleration below the $16g$ (i.e., 16 times the acceleration of gravity) rating for the seats on the plane?

69. ★★ Crumple zones in cars increase the time and the distance over which a car stops during a collision. The length of one crumple zone is 1.6 m.
    (a) Find the maximum speed that a car could have when hitting a wall head-on without subjecting the passenger to an acceleration of more than $46.2g$ (i.e., 46.2 times the acceleration due to gravity).
    (b) How long would that collision last?

70. ★★ Kenny Brack survived a $214g$ car crash during an Indy car race in Fort Worth, Texas, in 2003. His speed is reported to have been 354 km/h when he crashed into a wall. If his final speed was zero, how long did the crash last and over what distance did he stop? Assume constant acceleration during the crash.

71. ★ Some ejection seats have an acceleration of $14g$ upward. How fast would the pilot be going relative to the plane at the end of a 1.3 s ejection burst?

72. ★★ The Peugeot EX1 electric sports car set a few speed records in 2010. Starting from rest, it covered the first 201.2 m (1/8 mile) in 8.89 s.
    (a) Assuming constant acceleration, find the time the EX1 takes to go from 0 to 100 km/h.
    (b) The car took a total of 16.81 s to travel 500 m starting from rest. Find the average acceleration for the period after it covered the first 1/8 mile.

73. ★★ A rocket is launched such that for the first 40.0 s its acceleration is upward at $5.00g$ (i.e., five times the acceleration due to gravity). For the following 25.0 s, its acceleration is upward at $3.50g$. The final stage lasts 90.0 s, with an acceleration of $3.20g$.
    (a) What is the speed of the rocket at the end of the second stage ($t = 65.0$ s)?
    (b) What is the total distance covered by the rocket in the third stage?
    (c) What is the total distance covered by the rocket during the three stages?

74. ★ A rocket is launched vertically such that its acceleration upward is $4.5g$. If it can maintain that acceleration, how much time (in minutes) would it take to reach the orbit of the International Space Station 300 km above Earth's surface?

75. ★★ A lunar lander 200 m above the surface of the Moon has a downward speed of 50 m/s. Assume the acceleration of the lander is constant during the landing stage of its motion and the positive x-direction coincides with the direction of the lander's motion.
    (a) Find the magnitude of the constant acceleration required for the lander to arrive at the surface with a "safe speed" of 1.1 m/s.
    (b) Find the displacement of the lander during the slowing-down stage.
    (c) If the acceleration of the lander during the slowing-down stage were doubled, how far would it have moved while slowing down from a speed of 50 m/s to the safe speed of 1.1 m/s?
    (d) If the initial and the final speeds of the lander were doubled, while the acceleration of the lander remained the same, how far would it have moved while slowing down?

Section 3-5 Analyzing the Relationships between $x(t)$, $v(t)$, and $a(t)$ Plots

76. ★★ 〰 Figure 3-52 shows the velocity versus time plot for a motorcycle.
    (a) Find the acceleration at $t = 3.0$ s.
    (b) What is the displacement between $t = 2.0$ and $t = 4.0$ s?
    (c) What is the position of the motorcycle at $t = 4.0$ s if at the initial time of $t = 0$ s, the motorcycle was at $x_0 = 30.00$ m?

(d) What is the displacement of the motorcycle between $t = 4.0$ s and $t = 6.0$ s?

(e) Sketch the position versus time plot from $t = 0$ s to $t = 10.0$ s. (Hint: Find the position at each of the "transition times" where the velocity graph changes shape, and think about what shape the position would have in between those times.)

Give all the answers to four significant digits.

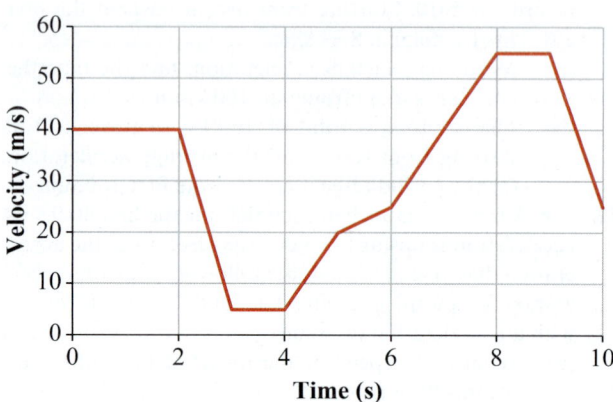

Figure 3-52   Plot of velocity versus time for problem 76.

77. ★ 〰 Figure 3-53 shows the velocity versus time plot for a toy. Consider all the information to be given to three significant digits.

(a) Calculate the area under the curve between $t = 0$ and $t = 2.0$ s.

(b) Is that area positive or negative?

(c) What does it mean to have a negative area in this context?

(d) Calculate the area under the velocity versus time plot between $t = 6.0$ s and $t = 9.0$ s.

(e) What is the position of the car at $t = 11.0$ s given that it was 3.0 m left of the origin at $t = 0$ s? Take the positive direction to be to the right.

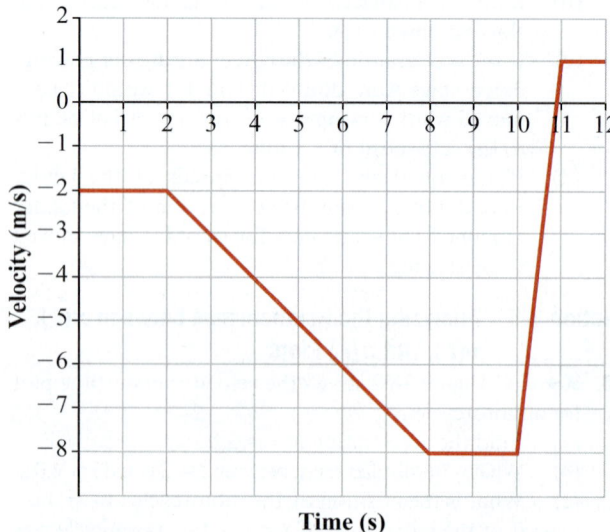

Figure 3-53   Plot of velocity versus time for the situation of problem 77.

78. ★★ 〰 Figure 3-54 shows the acceleration versus time plot for a race car. Give all the answers to two significant digits.

(a) Calculate the area under the curve between $t = 0$ s and $t = 2.0$ s. What does that area represent?

(b) If the velocity of the car was 67 m/s at $t = 0$ s, what is the velocity of the car at the end of 2.0 s?

(c) Examine the graph between $t = 0$ s and $t = 7.0$ s. What does it mean for the acceleration to be negative?

(d) Is the area between $t = 7.0$ s and $t = 10.0$ s positive or negative? Was the velocity decreasing or increasing in that time?

(e) What is the change in velocity between $t = 4.0$ s and $t = 7.0$ s?

(f) What is the change in velocity between $t = 9.0$ s and $t = 10.0$ s?

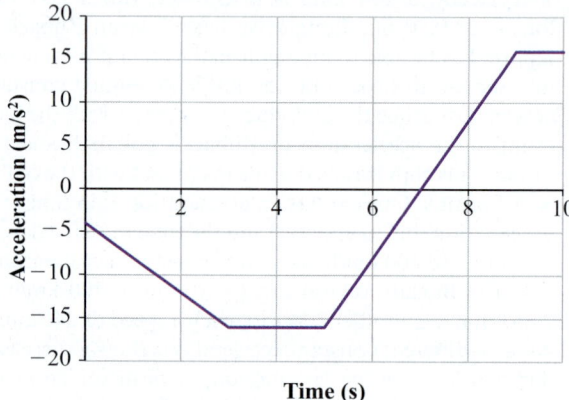

Figure 3-54   Plot of acceleration versus time for situation of problem 78.

79. ★★ 〰 Figure 3-55 shows the acceleration as function of time for a speed boat. Give all the answers to two significant digits.

(a) Given that the initial velocity of the speed boat was 32 m/s, find the velocity of the boat at the end of 5.0 s.

(b) How much does the boat's velocity change between $t = 7.0$ s and $t = 9.0$ s?

(c) What is the velocity of the boat at $t = 10$. s?

(d) Reconstruct the velocity versus time plot for the boat for the period of time shown.

(e) Find the total displacement over the period of time shown.

(f) Plot the position as a function of time for the interval shown, if the object starts at an initial position of 3.4 m.

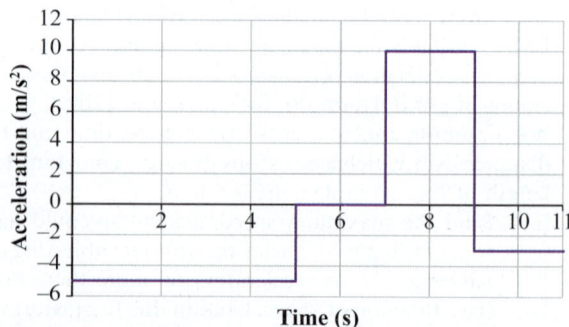

Figure 3-55   Plot of acceleration versus time for the situation of problems 79 and 80.

80. ★ ~ Figure 3-55 shows the acceleration versus time graph for a speed boat. Given that the boat had an initial velocity of 14 m/s, find the displacement of the boat between $t = 2.0$ s and $t = 4.0$ s.

81. ★★ ~ Figure 3-56 shows the acceleration versus time plot for a rocket sled equipped with rear and front thrusters. The sled had a velocity of 290 m/s at $t = 0$. Give the answers to three significant digits.
    (a) Find the displacement of the rocket sled during the first 5.0 s.
    (b) What is the velocity of the sled at $t = 5.0$ s?
    (c) Find the position of the sled at $t = 8.0$ s, if it was at $x = -300$ m at $t = 0$.
    (d) What is the displacement of the sled between $t = 8.0$ s and $t = 11.0$ s?
    (e) How fast is the sled going at $t = 10.0$ s?

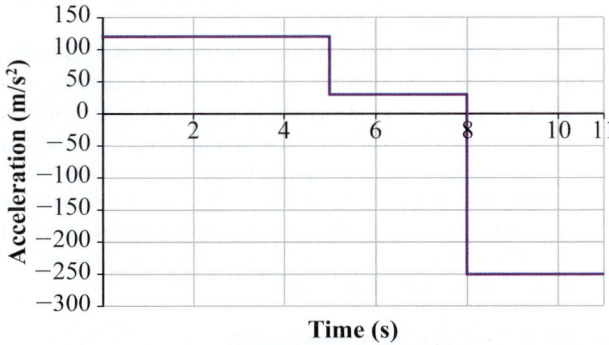

Figure 3-56 Plot of acceleration versus time for the situation of problem 81.

Section 3-6 Free Fall

82. ★ You toss a stone straight down at a speed of 8.0 m/s from a height of 70. m. What is the average acceleration of the stone while it is in the air?

83. ★★ ~ Ball A is dropped from a 30 m tall building. A second ball, B, is dropped from a 60 m tall building. How do the times the balls stay in the air compare? What assumptions did you make to solve the problem?

84. ★★ Ball A is dropped from a 10-storey building. Calculate the ratio of the time it takes the ball to fall the first five stories to the time it takes to fall the second five stories. Explain your solution using (a) an algebraic approach (kinematics equations for one-dimensional motion) and (b) a graphical approach. What assumptions did you make to solve the problem?

85. ★★ A marble is thrown downward at a speed of 10.0 m/s from the top of a 30.0 m tall building. 0.50 s later, a second marble is thrown downward. At what speed must the second marble be thrown so that the two marbles arrive at the same time?

86. ★★ You throw two balls downward from a height of 90.0 m, 1.0 s apart. You throw the first ball at a speed of 8.0 m/s. At what speed must you throw the second ball if it is to reach the ground
    (a) with the same speed as the first ball?
    (b) at the same time as the first ball?

87. ★★ Janice throws her wallet straight up in the air with an initial speed of 5 m/s, and catches it on the way down at the moment it has the same speed at which she threw it.
    (a) Find the wallet's average acceleration, average speed, and average velocity while it is in the air.
    (b) How would the answers to part (a) change if the initial speed of the wallet were doubled?

88. ★★★ Shirley (a very strong and athletic girl) throws a gift straight up in the air toward her friend Jasmine, who is 14.0 m above Shirley. Jasmine catches the gift when its speed is 3.0 m/s, on the way back down.
    (a) How long does it take for the gift to pass Jasmine on the way up?
    (b) What is the average speed of the gift between the time Shirley throws it and the time Jasmine catches it?
    (c) What is the average speed of the gift from launch to maximum height, and what is that maximum height?
    (d) Explain why the average speed on the way up (from the launching point to the apex) is different from the average speed you found in part (b).

89. ★★ A stone is tossed straight down from the top of a building. During the third second of its flight, the stone covered a distance of 45.0 m.
    (a) What was the stone's initial speed?
    (b) How much distance did the stone cover during the first 2.00 s?

90. ★★ You build a contraption that allows you to launch a marble straight up in the air. The contraption can also catch the marble when it falls back down. You measured that the marble spent exactly 7.00 s in the air during its entire motion.
    (a) Find the total distance covered by the marble and its displacement from the moment it left the contraption to the moment it came back.
    (b) Find the average speed, the average velocity, and the average acceleration of the marble during its upward motion.
    (c) Find the instantaneous speed, the instantaneous velocity, and the instantaneous acceleration of the marble when it is at its highest point.
    (d) Find the average speed, the average velocity, and the average acceleration of the marble during its entire motion.

91. ★★★ You are standing on top of a building and are challenged by a friend to throw a ball such that it covers the 1.70 m distance between the top and the bottom of his window in 0.30 s. The top of the window is 13.0 m below the point where you release the ball.
    (a) Is this throw possible? Explain.
    (b) What is the maximum amount of time during which the ball can cover the distance?

92. ★★ Two young men on a balcony are trying to attract the attention of a young lady on the ground 11 m below. One young man drops a flower. The other young man wants his flower to land first. After 0.70 s, how fast should he throw it down? Ignore air resistance.

93. ★★ A ball is dropped from the top of a 56.0 m tall building. After 2.00 s, another ball is thrown upward from the ground. When the two balls pass the same point, they have the same speed. How fast was the upward-bound ball thrown?

94. ★★★ (a) Derive an expression for the distance covered by a free-falling object during the *n*th second of its fall.
   (b) How would the expression derived in part (a) look if the initial velocity of the object was zero (the object was dropped)?
   (c) How would the expression derived in part (a) look if the object was tossed downward?
   (d) How would the expression derived in part (a) look if the object was tossed upward?
   (e) Perform dimensional analysis and limiting case analysis to make sure the expressions you found are reasonable.

Section 3-7   Relative Motion in One Dimension

95. ★ A Mini Cooper moving at 135 km/h passes a Mustang moving in the same direction at 90 km/h. Calculate the velocity of the Mustang relative to the Mini, and the speed of the Mustang relative to the Mini.

96. ★ While a car is travelling on a road at a speed of 60 km/h, a child throws a toy from the car, and 2.0 s later the car is 8.0 m ahead of the toy. How much has the toy moved relative to the ground in that time?

97. ★★ Derive the expressions for relative velocity, acceleration, displacement, and position for two objects moving with constant acceleration.

98. ★★ A canoeist can paddle at a speed of 3.0 km/h relative to the water. They paddle downstream, with the river, for a given amount of time, and then paddle upstream back to the starting point. If the upstream trip takes six times as long as the trip downstream, how fast is the river flowing? Assume the ratio is exactly six, so you can answer to two significant digits.

99. ★★ A train moves to the right at a speed of 100 km/h. On the train is a dog that ran away from its owner and moves toward the back end of the train at a speed of 3.0 m/s. What is the velocity of the dog relative to the ground, in km/h?

100. ★★★ A hot-air balloon is ascending (moving upward) at a speed of 4.0 m/s, when one of the riders drops an apple they were hoping to eat. Trying to catch the apple, all they manage to do is impart a downward velocity of 2.0 m/s to it relative to the balloon. If the apple at that moment is 45 m above the ground, how long does it take for the apple to reach the ground? How far will the balloon be above the ground when the apple reaches the ground? Answer these questions
   (a) in the frame of reference of the balloon, and
   (b) in the frame of reference of an observer standing on the ground
   Compare your results in parts (a) and (b). How do they differ and how are they the same? Explain.

101. ★★★ A boy standing on a balcony drops a marble from a height of 16 m above the ground. At exactly the same time, a girl tosses a tennis ball up from a height of 1.0 m above the ground.
   (a) What must be the initial speed of the tennis ball for the two to meet at the highest point in the ball's trajectory?
   (b) How long were the ball and the marble in the air before they met?

   (c) How far did the ball and the marble travel before they met?
   (d) What is the speed of the ball relative to the marble just before the marble hits the ground?

102. ★★★ A motorcycle is moving at a speed of 120 km/h when the cyclist drops her spare helmet. The helmet rolls and slides on the road. Ignoring air resistance, the helmet is 2.0 m behind the motorcycle 3.0 s later. Assume the acceleration of the helmet on the ground is constant and that the helmet rolls along a straight line (we consider one-dimensional motion).
   (a) How far would the helmet have moved on the ground in these 3.0 s?
   (b) How far would the helmet be from the motorcycle 6.0 s after the helmet is dropped?
   (c) How fast would the helmet be moving relative to the motorcycle at that time?
   (d) How fast would the helmet be moving relative to the ground at that time?

103. ★★★ A golf cart (cart 1) moves at a speed of 31.0 km/h, and is chased by another golf cart (cart 2) moving at a speed of 42.0 km/h. A person in cart 1 releases a pigeon to confuse the chasing driver in cart 2 at the moment when the carts are 40. m apart. The pigeon lands in cart 2 2.00 s later. Assuming that the golf carts are moving at constant velocities and the pigeon is moving at constant velocity while in flight, at what speed does the pigeon land in the chasing cart (cart 2)?

Section 3-8   Calculus of Kinematics

104. ★★★ ∫ A car slows down with an acceleration whose magnitude increases linearly with time, starting with an acceleration of zero. The car's speed drops by one third of its value during the first 2.0 s of the acceleration phase. What percentage of its original speed will it still have at the end of 3.0 s into the acceleration phase?

105. ★★★ ∫ The magnitude of a rocket's acceleration grows exponentially with time when the thrusters are fired as $a(t) = a_0 e^{3t}$. If the rocket is initially moving at 120 m/s, and the speed 3.0 s into the thrust is 320 m/s, find the initial acceleration of the rocket, a_0. Ignore any effects of gravity.

106. ★★★ ∫ A motorcycle is moving to the right (we will call this the positive direction). The motorcycle has a negative acceleration (i.e., it is slowing down) given by $a = -10t$ m/s².
   (a) If the initial speed of the motorcycle is 57 m/s, how much time will it take for the motorcycle to reach half this speed?
   (b) What is the displacement of the motorcycle during this time?

107. ★★★ ∫ A rocket is fired at an initial speed of 300 m/s. Its speed increases with an acceleration that varies in magnitude as $a(t) = \dfrac{b}{(t+1)^2}$ m/s², where b is a constant.
   (a) If the acceleration is equal to 46 m/s² at $t = 2.0$ s, find b, and report it with its units.
   (b) What is the speed at the end of 3.0 s?
   (c) What is the displacement of the rocket over the first 4.0 s?

108. ★★ $\frac{d}{dx}$ Using SI base units, a particle's position as a function of time is given by $x(t) = (7t^3 - 4t^2 + 6)$ m.
  (a) At what time will the particle's acceleration be zero?
  (b) Find the speed of the particle at that time.

109. ★★★ $\frac{d}{dx}$ A particle's position is given by $x(t) = x_0 \cos(\omega t + \phi)$, where x_0 and ϕ are constants.
  (a) Find the possible values of ϕ, including its units, assuming that $x(t = 0) = 0$.
  (b) Given that $x_0 = 20$ cm, plot the position of the particle over the interval $0 \leqslant t \leqslant 10.0$ s with (i) $\omega = \phi = \pi$, (ii) $\omega = \pi/2$ and $\phi = 0$, and (iii) $\omega = \phi = \pi/2$.
  (c) Differentiate the position function to find an expression for the particle's velocity as a function of time.
  (d) Differentiate your result from part (c) to find an expression for the particle's acceleration as a function of time.
  (e) What is the relationship (i.e., when one is positive is the other also positive) between the acceleration and the position of a particle executing this motion (called simple harmonic motion)?
  (f) Find the maximum speed and the magnitude of the maximum acceleration of the particle.

COMPREHENSIVE PROBLEMS

110. ★ (a) How much distance does a Formula One race car initially travelling at 360 km/h cover while it brakes at 6.00g for 1.20 s?
  (b) What is the car's final velocity?
  (c) What are the car's average velocity and average speed while it brakes?

111. ★★ A stone is dropped from a very high cliff such that its speed just before impact with the ground below is 190.0 m/s.
  (a) How long did the stone travel before hitting the ground?
  (b) How high was the cliff? (What was the entire distance travelled by the stone?)
  (c) How much distance did the stone cover during the interval $5.00 \text{ s} \leq t \leq 15.0$ s?
  (d) How far did the stone travel in its last second of flight?

112. ★★★ You are riding in an elevator with an open roof. You are moving down at a speed of 4.00 m/s when you toss an apple up into the air at a speed of 6.00 m/s relative to the elevator.
  (a) How long does it take for you to catch the apple? Assume that you release and catch it at the same height relative to the elevator floor.
  (b) What is the maximum height the apple reaches above the elevator floor, assuming that it was released from a height of 1.00 m above the floor? How does this height change if the elevator is moving up?
  (c) What are the speeds and the velocities of the apple when you toss it up and when you catch it on the way back down (i) in the reference frame of the elevator, and (ii) in the reference frame of a person standing on the ground?

  (d) A friend standing on the ground throws an apple up at a speed of 6.00 m/s. How long does it take for the apple to return to your friend's hand?

113. ★★ (a) A Formula One race car can go from 0 km/h to 100 km/h in 1.700 s, from 0 km/h to 200.0 km/h in 3.800 s, and from 0 km/h to 300.0 km/h in 8.600 s. Assuming constant acceleration during the intervals 0 s to 1.700 s, 1.700 s to 3.800 s, and 3.800 to 8.600 s, find the distance that the car covers in each one of these time intervals.
  (b) A Formula One race car starts from rest, reaches a speed of 100 km/h in 1.70 s, reaches a speed of 325 km/h over the next 7.30 s, and then slows with an acceleration of 4.00g for the next 100. m. Find the average speed, average acceleration, and average velocity of the car for the entire time interval.

114. ★★ You are sitting well back from the windows in the library at the Bamfield Marine station on the West Coast of Vancouver Island. A bald eagle hovers overhead with a fish in its grip. The fish wriggles loose and falls straight down. You happen to be looking out the window and notice that the fish is in sight for 0.14 s. The windows are 1.2 m high. How high was the eagle when it dropped the fish? What is the speed of the fish as it passes the window sill?

115. ★ An ocean liner approaches a pier at 20.0 km/h. When the ocean liner is 10.0 km away, the captain sends a speedboat to the pier to pick up the harbour pilot, who will help dock the ocean liner. The speedboat takes 12.0 min to get to the pier.
  (a) How long would the speedboat take to return to the ocean liner if the speedboat maintained the same speed?
  (b) Find the displacement of the speedboat during the round trip and the distance covered by the speedboat during the round trip.

116. ★★ A speedboat accelerates from rest at the start line of a race to reach a speed of 30.0 km/h over a distance of 70.0 m and then continues with the same constant acceleration for another 11.0 s before slowing down to a speed of 25.0 km/h over a distance of 100. m. Another speedboat begins to accelerate from rest at the start line and catches up to the first boat within 30.0 s.
  (a) Find the acceleration of the second speedboat.
  (b) Find the time it takes the first speedboat to cover the first 70 m.
  (c) Find the speed of the second speedboat when it catches up to the first speedboat.

117. ★★★ A little girl, at the end of a fishing trip with her family, is examining the Coho salmon she caught when the salmon (which is unfortunately dead by now) slips and falls directly into the water. The boat is moving at 16.0 km/h relative to the water, and when the salmon hits the water it accelerates at 5.00 m/s^2 due to water resistance.
  (a) Assuming the acceleration to be constant, how far is the salmon behind the boat 0.900 s later?
  (b) How much has the boat moved in that time?
  (c) How much has the salmon moved relative to the water in that time?
  (d) If that acceleration is maintained constant, how much time would pass between the salmon hitting the water and it reaching one third the speed of the boat?

118. ★★ A very athletic swimmer is swimming downstream in a river. She swims for 10.0 min and then realizes that she has lost a very expensive waterproof sports watch. She then swims back to get the watch and finds it 55.0 min later at the bottom of the river, where she started swimming. Find the speed of the swimmer relative to the water if the current speed is 1.20 m/s. Assume that once she gets back to where she started, it takes her 2.50 min of searching to find her watch.

119. ★★ James Bond is running backward on top of a long train that is moving at 130. km/h. Mr. Bond is running at a speed of 28.0 km/h relative to the train and is trying to catch Francisco Scaramanga, who is running away from Bond (also on top of the train) at a constant speed of 35.0 km/h relative to the train. At a given moment, Mr. Scaramanga is 14.25 m away from Mr. Bond, and Mr. Bond begins to accelerate at a rate of 1.20 m/s².

    (a) How far has the train moved in the time it takes James Bond to catch Mr. Scaramanga?

    (b) How far has James Bond moved during this process, relative to the ground?

120. ★★ A squirrel climbs into a box on the roof of a tall building, which causes the box to free fall off the roof. The squirrel then climbs on the inside wall of the box, such that when the box hits the ground, 3.00 s later, the squirrel has managed to climb 80.0 cm above the floor of the box. When solving the problem, assume (1) the box stays upright during the fall, and that the acceleration of the squirrel is constant; (2) the squirrel was on the floor of the box and started climbing at a constant acceleration at the same time the box started to fall; (3) the box is too heavy to be influenced by the motion of the squirrel and by air drag. Find the velocity of the squirrel relative to the ground when the box hits the ground.

121. ★ You are in a hot-air balloon that is accelerating upward at 4.00 m/s². The balloon is at a height of 47.0 m when its speed upward is 9.00 m/s. You drop an apple from the side. Find the maximum height the apple will reach above the ground, and the speed of the apple when it hits the ground.

122. ★★ A child has a tennis ball and a baseball. He lets go of the tennis ball from the top of a high building, waits, and then throws the baseball straight down with a speed such that the two balls have a relative speed of 10.0 m/s when they meet 40.0 m below (the baseball is moving 10.0 m/s faster than the tennis ball). How long did the child wait before throwing the baseball?

123. ★★ You throw a red ball up in the air with a speed such that it just reaches your friend on a balcony 30.0 m above (9 storeys high). At the same time, your friend throws another green ball down to you such that it reaches you with twice the speed with which you threw your ball. What is the velocity of the red ball relative to that of the green ball when they pass each other? Consider the positive x-direction to be down.

124. ★ Two children are playing a game in which one of them throws a baseball down from the top of his 6.0 m high tree house and the other is supposed to hit the baseball with a tennis ball. If the child in the tree house throws the baseball with a downward speed of 1.2 m/s, what is the speed the child on the ground has to throw his baseball with from a height of 1.1 m so the balls meet halfway?

125. ★★ Two children are playing a game in which one of them throws a volleyball up in the air and the other, standing very close to her friend, throws a tennis ball up to hit the volleyball as it reaches its highest point. The first child throws the volleyball at a speed of 12.0 m/s, and the second child throws the tennis ball at a speed of 25.0 m/s. How long after the volleyball is thrown should the second child wait to throw the tennis ball?

126. ★ Two friends live beside a 25.0 m high cliff, one at the top of the cliff and the other directly below, at the bottom. While they are making dinner, the friend at the top of the cliff throws a loaf of bread down with a speed of 7.0 m/s at the same time that the other friend throws a lemon up with a speed of 30.0 m/s. Find the relative speed of the bread and the lemon when they pass one another.

127. ★ Madrid, Spain, and Beijing, China, both lie on the 40th parallel. Beijing is at 116°23′ E, and Madrid is at 3°40′ W. Find the average speed of a plane that takes 7.0 h to travel from Beijing to Madrid. [Hint: You can find the shortest distance between any two points on the sphere if you know their coordinates using the haversine formula].

128. ★★ 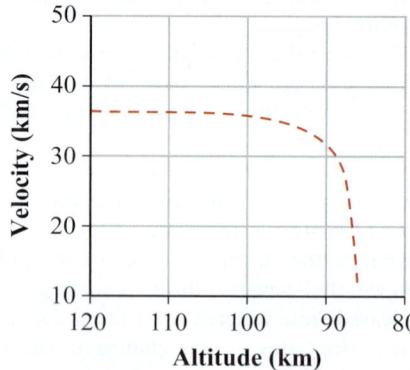 Figure 3-57 shows data for the velocity of a meteor entering the atmosphere as a function of altitude (height above Earth's surface).

    (a) For heights of 100 km and 90 km, estimate v from the graph, and then find the approximate value of the ratio $\dfrac{\Delta v_x}{\Delta h}$ for these altitudes.

    (b) We could estimate the average acceleration over an interval using the relationship

$$a_x \approx \frac{\Delta v_x}{\Delta t} \approx \frac{\Delta v_x}{\Delta h}\frac{\Delta h}{\Delta t} \approx \frac{\Delta v_x}{\Delta h}v_x$$

Using the technique developed above, estimate the *instantaneous* accelerations at heights of 100 km and 90 km.

    (c) Assuming that gravity and aerodynamic drag friction with the atmosphere are the principal forces acting on the meteor, what does the graph reveal about the relative strength of the two forces?

    (d) What does the shape of the velocity graph suggest about the density of the upper atmosphere?

Figure 3-57   Problem 128.

129. ★★  Figure 3-58 shows a graph of altitude versus time for the *Soyuz* launch vehicle.
  (a) Estimate the radial speed (i.e., the magnitude of the vertical component of velocity) of the vehicle at (i) 100 s, (ii) 200 s, (iii) 300 s, and (iv) 400 s.
  (b) Find the mean radial velocity between (i) 0 s and 100 s and (ii) between 300 s and 400 s.

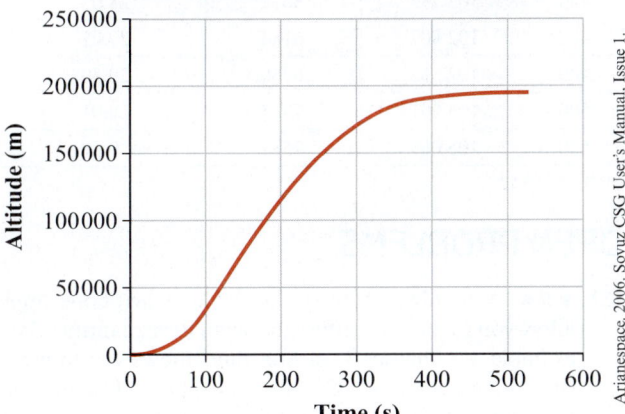

Figure 3-58 Problem 129.

130. ★★★ Figure 3-59 shows a velocity versus time plot for the *Soyuz* launch vehicle.
  (a) Estimate the peak acceleration at $t = 120$ s, $t = 280$ s, and $t = 540$ s.
  (b) What is the average acceleration of the vehicle for the entire flight time shown?

  (c) Find the average acceleration for each of the three stages, that is, from 0 to 120 s, from 120 to 280 s, and from 280 to 540 s.

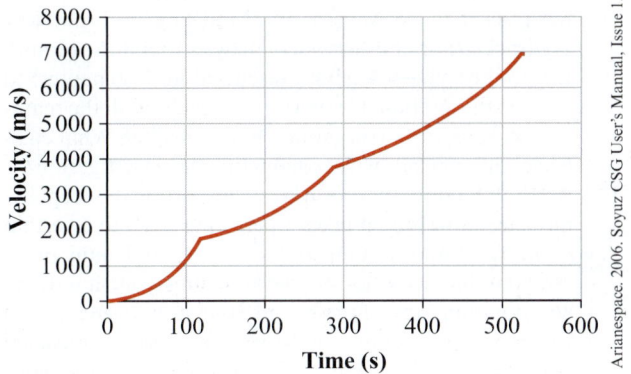

Figure 3-59 Problems 130, 131, and 132.

131. ★★★ Use Figure 3-59 to estimate the distance covered by the *Soyuz* vehicle in the first 120 s. How much distance did the vehicle cover by the end of 280 s? Compare these results to the graph of Figure 3-58, and explain the difference.

132. ★★★ 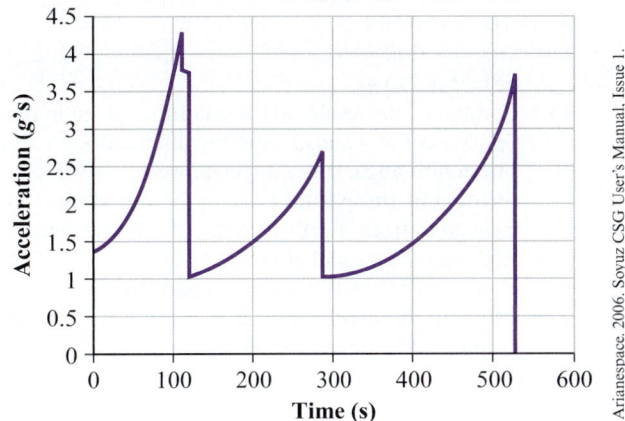 Use the velocity versus time graph for the *Soyuz* launch vehicle (Figure 3-59) to construct an acceleration versus time graph. Use two scales to show your results in m/s² and in multiples of the acceleration due to gravity, *g*. Compare your graph to the graph in Figure 3-60, below.

133. ★★★ Figure 3-60 shows an acceleration versus time plot for the *Soyuz* launch vehicle. Using the graph, calculate the speed of the vehicle at each of the three peak accelerations shown, and construct a velocity versus time graph for the vehicle. Compare your results with the graph in Figure 3-59, above.

Figure 3-60 Problem 133.

DATA-RICH PROBLEMS

134. ★★★ Video-record the motion of a moving object, such as a falling ball, a walking person, or a moving car. Use the Tracker Video Analysis and Modeling Tool (physlets.org/tracker/) or another video analysis tool (for example, Logger Pro Video Analysis, discussed in this chapter) to analyze this motion. Compare your results with the mathematical description of motion in this chapter. What challenges did you face while conducting this investigation?

135. ★★★ Reaction time is a very important concept. It measures how long it takes a person to react to a certain event. It is important for sports, driving, and many other everyday life activities. It also measures your hand–eye coordination, your fitness, and your attentiveness. You can perform a number of very simple tests to measure your own reaction time, as well as investigate how your reaction time is affected by your fatigue. Conduct a series of experiments where you measure your own reaction time using two different methods, for example, a ruler drop test (www.brianmac.co.uk/rulerdrop.htm) or an online reaction time test (faculty.washington.edu/chudler/java/redgreen.html). Explain the physics behind the ruler drop test. Analyze the data you have collected from both experiments and draw relevant conclusions.

136. ★★★ 〜 STS-121 was a 2006 NASA Space Shuttle mission to the International Space Station (ISS) flown by the space shuttle *Discovery*.
   (a) Use the data from Table 3-5 to plot the following quantities for STS-121 as a function of time:
      (i) altitude
      (ii) velocity
      (iii) acceleration
   (b) The radial velocity of the space shuttle is the component of the velocity that points in the direction of the radius connecting the shuttle to Earth's centre. From your altitude versus time graph, estimate the radial speed at
      (i) $t = 60.0$ s
      (ii) $t = 140$ s
      (iii) $t = 220$ s
   (c) Compare your results to the quantities given in the velocity column and explain the differences.
   (d) The zenith angle is the angle between the velocity vector and the position vector of the shuttle as measured from Earth's centre. Find the zenith angle at $t = 60$ s and at $t = 260$ s.
   (e) Comment on the two results in part (d).
   (f) Estimate the acceleration at $t = 140$ s from your graph, and compare it to the value in the table.

Table 3-5 Data for Mission STS-121 Showing the Altitude, Velocity, and Acceleration of the Space Shuttle Every 20 s from Liftoff to Main Engine Cut-Off (MECO)

Time (s)	Altitude (m)	Velocity (m/s)	Acceleration (m/s²)
0	−8	0	2.45
20	1 244	139	18.62
40	5 377	298	16.37
60	11 617	433	19.40
80	19 872	685	24.50
100	31 412	1026	24.01
120	44 726	1279	8.72
140	57 396	1373	9.70
160	67 893	1490	10.19
180	77 485	1634	10.68
200	85 662	1800	11.17
220	92 481	1986	11.86
240	98 004	2191	12.45
260	102 301	2417	13.23
280	105 321	2651	13.92
300	107 449	2915	14.90
320	108 619	3203	15.97
340	108 942	3516	17.15
360	108 543	3860	18.62
380	107 690	4216	20.29
400	106 539	4630	22.34
420	105 142	5092	24.89
440	103 775	5612	28.03
460	102 807	6184	29.01
480	102 552	6760	29.30
500	103 297	7327	29.01
520	105 069	7581	0.10

Source: NASA.gov

OPEN PROBLEMS

137. ★★★ A number of freely available smartphone apps allow you to measure different kinematics quantities. For example, the Physics Tool Box suite allows you to measure the acceleration of your smartphone. Use an app that you prefer to collect real-life data for an experiment on one-dimensional motion. Analyze the data you have collected and draw conclusions regarding this experiment in light of what you have learned in this chapter.

138. ★★ In 1959, Eiband compiled available information about the tolerance of humans to abrupt accelerations to create what is known today as an Eiband curve. Two main factors affect human tolerance – the magnitude of acceleration (often compared to the acceleration of gravity, *g*) and the duration of acceleration. These represent the axes of the Eiband curve. Use the Eiband curve (Figure 3-61) and the crash test data (Figure 3-62) to assess the risk level to a proposed occupant during this crash.

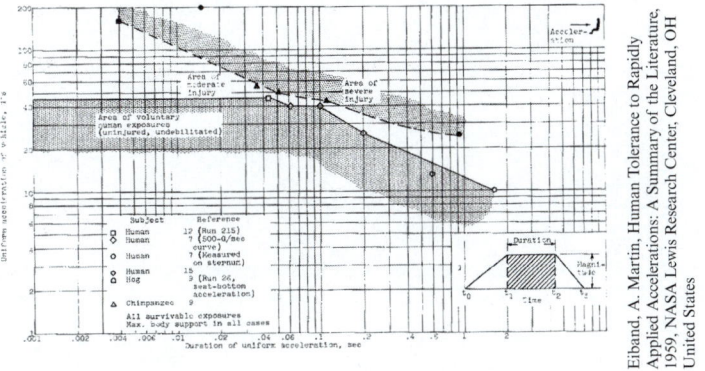

Eiband, A. Martin, Human Tolerance to Rapidly Applied Accelerations: A Summary of the Literature, 1959. NASA Lewis Research Center, Cleveland, OH United States

Figure 3-61   The Eiband curve for $-G_x$ accelerations (frontal crash).

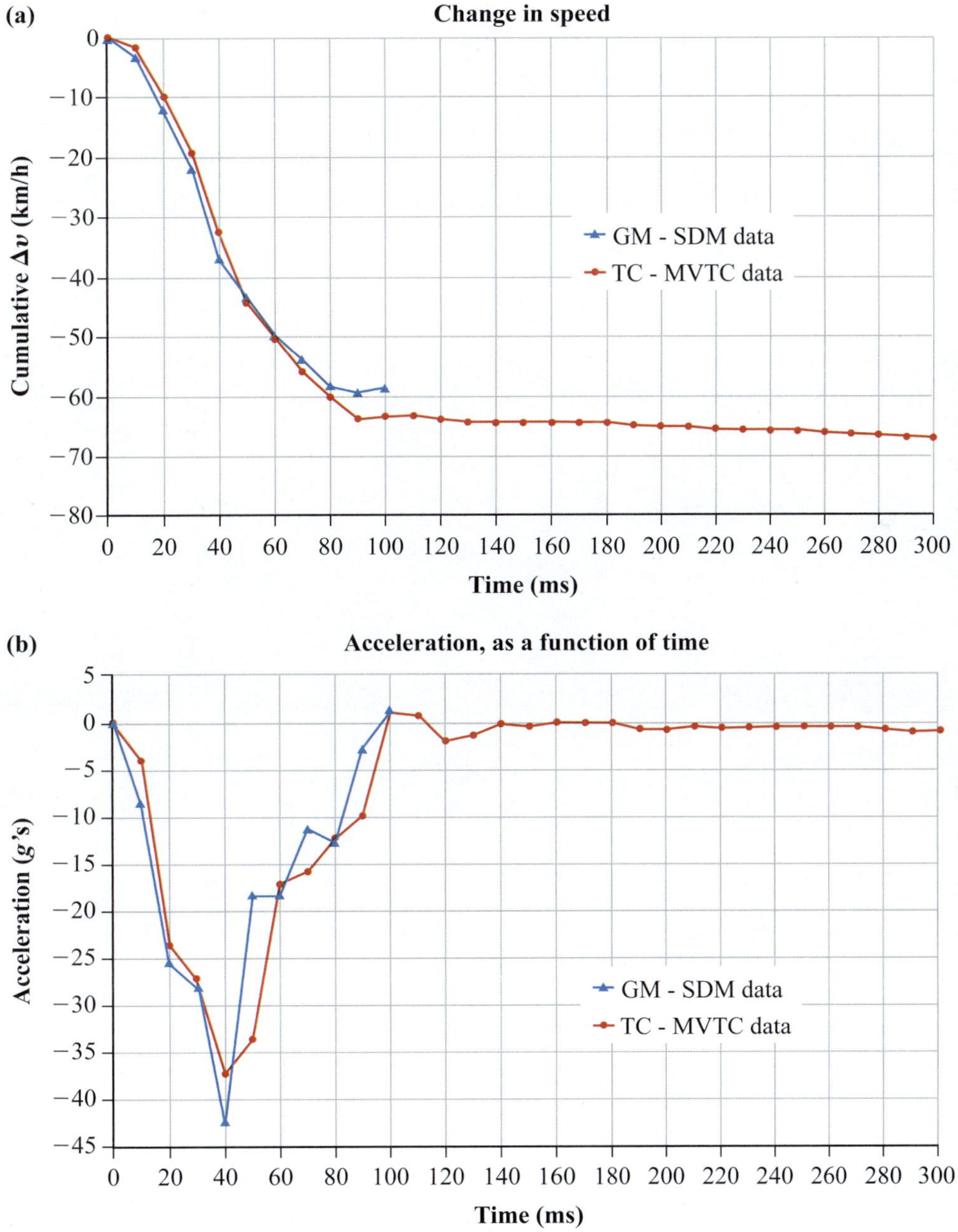

(a)   **Change in speed**

Cumulative Δv (km/h) vs Time (ms)

— GM - SDM data
— TC - MVTC data

Jean-Louis Comeau, Alan German and Donald Floyd, "Comparison of Crash Pulse Data from Motor Vehicle Event Data Recorders and Laboratory Instrumentation," Proceedings of the Canadian Multidisciplinary Road Safety Conference XIV, June 27–30, 2004, Ottawa, Ontario

(b)   **Acceleration, as a function of time**

Acceleration (g's) vs Time (ms)

— GM - SDM data
— TC - MVTC data

Comeau, German, Floyd, "Comparison of Crash Pulse Data from Motor Vehicle Event Data Recorders and Laboratory Instrumentation," "Proceedings of the Canadian Multidisciplinary Road Safety Conference XIV, June 27–30, 2004, Ottawa, Ontario

Figure 3-62   Crash test data obtained by General Motors and Transport Canada's Motor vehicle test centre for (a) change in speed and (b) acceleration, as a function of time.

When you have completed this chapter, you should be able to

LO1 Define, explain, and calculate displacement, velocity, and acceleration in two dimensions.

LO2 Analyze and represent the motion of a projectile using vectors and the two-dimensional kinematics equations.

LO3 Explain and calculate the acceleration for an object moving in a circle.

LO4 Apply relative motion principles in two and three dimensions.

Water fountains adorn many public spaces in cities around the world. As you can see in Figure 4-1, the jets of water in this fountain form beautiful parabolic arcs. The water in these jets is moving under the influence of a constant gravitational force, and as a result it also experiences a constant acceleration. Although the overall pattern is three-dimensional, each individual jet lies in a two-dimensional plane. Each individual jet is an example of **projectile motion**, which is one of the main topics of this chapter.

Figure 4-1 The Fountain of River Commerce and Navigation was built in 1840 in Paris, France.

Petr Kovalenkov/Shutterstock.com

4-1 Position, Velocity, and Acceleration

In this chapter, we will expand our study of **kinematics** to include motion in more than one dimension. You should keep in mind as we do this that the basic ideas of position, velocity, and acceleration, and the mathematical relationships between them, remain unchanged. The velocity of an object is still the derivative with respect to time of the position of an object, and the acceleration is the derivative with respect to time of the velocity. Our description becomes more complicated, however, because the motion is not constrained to occur along a line.

In one dimension, a single number is sufficient to specify the position of an object. In three dimensions, we need three numbers. In this chapter we will use a Cartesian coordinate system with three mutually perpendicular axes, labelled x, y, and z. To specify the position of an object, we then need to specify the x-, y-, and z-coordinates of the object. We can use vector notation to write the **position** vector $\vec{r}$ as

KEY EQUATION
$$\vec{r} = x\hat{i} + y\hat{j} + z\hat{k} \qquad (4\text{-}1)$$

where x, y, and z are real numbers that can have both positive and negative values. As an object moves, its position vector changes. Suppose that at time t_1 the object is at position $\vec{r}_1$, and at some later time t_2 it is at $\vec{r}_2$. We define the **displacement**, $\Delta\vec{r}$, of the object during this time interval as the difference between the two position vectors (Figure 4-2):

KEY EQUATION
$$\Delta\vec{r} = \vec{r}_2 - \vec{r}_1 \qquad (4\text{-}2)$$

The displacement can be written in unit vector notation as

$$\Delta\vec{r} = (x_2 - x_1)\hat{i} + (y_2 - y_1)\hat{j} + (z_2 - z_1)\hat{k} \qquad (4\text{-}3)$$

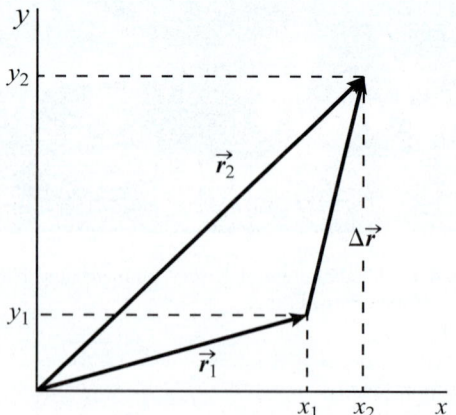

Figure 4-2 Two position vectors, $\vec{r}_1$ and $\vec{r}_2$, and the resulting displacement vector, $\Delta\vec{r} = \vec{r}_2 - \vec{r}_1$.

EXAMPLE 4-1

Distance and Displacement

You are walking briskly at a constant pace from your summer job downtown to the coffee shop. The trip takes you 400 m down one street and another 300 m along another street after turning left. What **distance** have you covered? What is your displacement?

Solution

Let us examine the quantities above in detail for this situation (Figure 4-3).

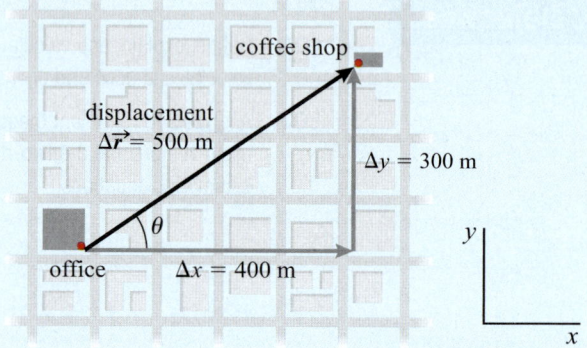

Figure 4-3 Example 4-1.

The total distance covered is the sum of the distances along the two legs of the journey:

$$d = 400 \text{ m} + 300 \text{ m} = 700 \text{ m}$$

The displacement, however, is the vector connecting the start and finish points. We choose a coordinate system in which the positive x-axis points to the right and the positive y-axis points up. We choose the origin to be at your job location. Your initial and final position vectors are then

$$\vec{r}_1 = 0, \vec{r}_2 = (400 \text{ m})\hat{i} + (300 \text{ m})\hat{j}$$

We can use Equation 4-2 to calculate the displacement:

$$\Delta\vec{r} = \vec{r}_2 - \vec{r}_1 = (400 \text{ m})\hat{i} + (300 \text{ m})\hat{j}$$

The magnitude of the displacement vector is

$$|\Delta\vec{r}| = \sqrt{\Delta r_x^2 + \Delta r_y^2} = \sqrt{(400 \text{ m})^2 + (300 \text{ m})^2} = 500 \text{ m},$$

and the angle that vector makes with the positive x-axis is given by

$$\tan\theta = \frac{\Delta r_y}{\Delta r_x} = \frac{300 \text{ m}}{400 \text{ m}} = 0.75$$

$$\theta = \tan^{-1} 0.75 = 36.9°$$

Making sense of the result

We can see that the total distance travelled is greater than the displacement. This is because we did not travel from our starting point to our destination in a straight line.

As we did in Chapter 3, we can also define an **average velocity** as the ratio of the displacement, $\Delta \vec{r}$, to the time interval, Δt, required for the displacement to occur:

KEY EQUATION
$$\vec{v}_{avg} = \frac{\Delta \vec{r}}{\Delta t} \qquad (4\text{-}4)$$

Since the average velocity is defined in terms of a vector divided by a scalar, the average velocity is also a vector. The average velocity vector is parallel to the displacement vector.

EXAMPLE 4-2

Average Speed and Average Velocity

An airplane flying at a constant speed takes 1.20 h to go from Amsterdam, Netherlands, to Berne, Switzerland, 630. km to the southwest. The airplane is then diverted before landing to Paris, France, 432. km northwest of Berne. It takes 0.900 h to get there, travelling at the same constant speed.

(a) Find the average speed of the airplane during the trip.
(b) Find the average velocity of the airplane, given that Amsterdam and Paris are 440. km apart.
(c) In a diagram, show the difference between the displacement of the airplane and the distance covered by the airplane throughout the journey.

Solution

(a) The average speed is the total *distance* covered by the airplane divided by the total time elapsed. Since distance is a scalar quantity, the total distance is the sum of the distances of the two parts of the journey:

$$d = d_1 + d_2 = 630. \text{ km} + 432. \text{ km} = 1.06 \times 10^3 \text{ km}$$

The total time taken is 1.20 h + 0.90 h = 2.10 h. Therefore, the average speed is

$$v_{avg} = \frac{1.06 \times 10^3 \text{ km}}{2.10 \text{ h}} = 506. \text{ km/h}$$

(b) Since the airplane started over Amsterdam and ended over Paris 440. km away, the displacement is 440. km pointing

directly from Amsterdam to Paris. We will choose the direction from Amsterdam to Paris as the positive x-axis. The average velocity is given by the displacement divided by the elapsed time:

$$\vec{v}_{avg} = \frac{\Delta x \hat{i}}{\Delta t} = \frac{(440. \text{ km})\hat{i}}{2.10 \text{ h}} = (210. \text{ km/h})\hat{i}$$

(c) The distance covered by the plane is shown in Figure 4-4 with the dashed line that follows the airplane's path. The displacement is the vector connecting the start and finish points.

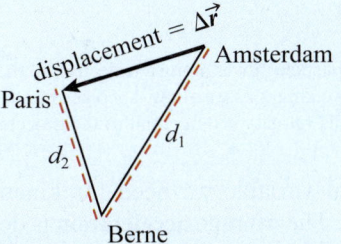

Figure 4-4 The difference between distance covered (dashed line) and the displacement of the plane.

Making sense of the result

Unlike the average speed, which tells us how fast the airplane is moving, the average velocity depends only on the direct distance between the start and finish points and the time elapsed on the way there.

If we allow the time interval $\Delta t \to 0$, we get the **instantaneous velocity**:

KEY EQUATION
$$\vec{v} = \lim_{\Delta t \to 0} \frac{\Delta \vec{r}}{\Delta t} = \frac{d\vec{r}}{dt} \qquad (4\text{-}5)$$

Note that the definition of the instantaneous velocity is simply the rate of change of the position of the particle with respect to time, as it was in Chapter 3. The only difference is that since the position is a vector in three dimensions, the velocity is also a vector in three dimensions. We can also write the velocity vector in terms of its components:

$$\vec{v} = v_x\hat{i} + v_y\hat{j} + v_z\hat{k} = \frac{dx}{dt}\hat{i} + \frac{dy}{dt}\hat{j} + \frac{dz}{dt}\hat{k} \qquad (4\text{-}6)$$

To determine the direction of the instantaneous velocity, it is useful to examine how the direction of the average velocity changes as we allow Δt to get smaller. This is illustrated in Figure 4-5. The curved path in the figure is the **trajectory** of the particle. We know that the average velocity is parallel to the displacement, and in Figure 4-5(a)–(c) we have drawn the average velocity vector pointing along the line that connects the tips of $\vec{r}_1$ and $\vec{r}_2$. We can see that as we allow the time interval to get smaller, the tip of $\vec{r}_2$ gets closer and closer to the tip of $\vec{r}_1$, and the average velocity vector approaches the tangent to the trajectory at $\vec{r}_1$. We infer from this, as shown in Figure 4-5(d), that in the limit $\Delta t \to 0$, the instantaneous velocity is indeed tangential to the trajectory of the particle.

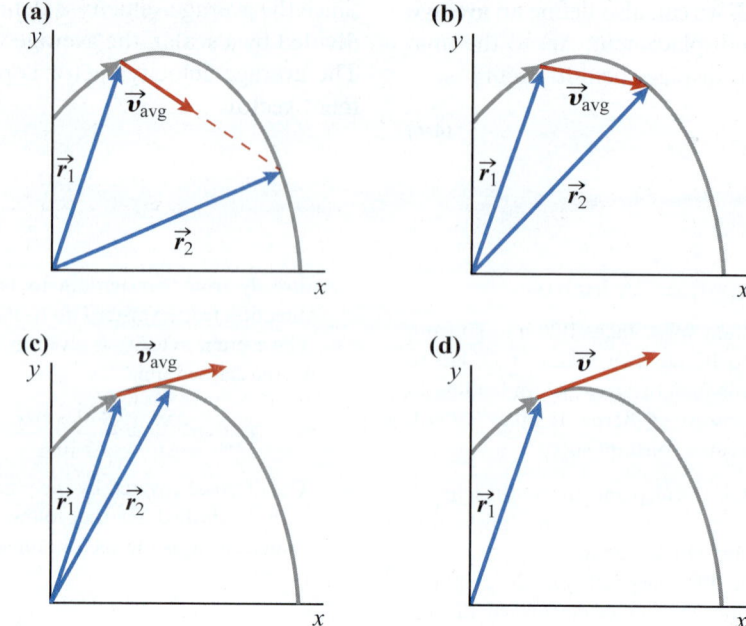

Figure 4-5 A particle moves along a trajectory in the xy-plane shown by the light grey curve. In (a) through (c), we see how the direction of the average velocity changes as we allow Δt to get smaller. We can see that it tends to get more tangential to the trajectory, and, in (d), we see that the instantaneous velocity is tangential to the trajectory.

The final variable we need for kinematics is the acceleration. The **average acceleration** is defined as

$$\vec{a}_{avg} = \frac{\vec{v}_2 - \vec{v}_1}{t_2 - t_1} = \frac{\Delta \vec{v}}{\Delta t} \qquad (4\text{-}7)$$

The **instantaneous acceleration** is

$$\vec{a} = \lim_{\Delta t \to 0} \frac{\Delta \vec{v}}{\Delta t} = \frac{d\vec{v}}{dt} \qquad (4\text{-}8)$$

The direction of the acceleration is determined by the physical environment that the particle is in—specifically the forces that act on the particle. This will be explored fully in Chapter 5. However, it is useful to examine how the velocity of a particle changes depending on the direction of the acceleration vector with respect to the velocity vector. In Chapter 3, where we dealt with one-dimensional motion, the acceleration was always directed along the direction of motion. If the acceleration was positive, the velocity of the particle increased, and if the acceleration was negative, the velocity of the particle decreased. However, the resulting motion was always along a straight line.

We now have a situation where the acceleration vector may make some angle with respect to the velocity vector. To see the results of this, it is useful to consider the average acceleration. We can solve Equation 4-7 for the change in the velocity vector to get

$$\Delta \vec{v} = \vec{v}_2 - \vec{v}_1 = \vec{a}_{avg} \Delta t$$
$$\vec{v}_2 = \vec{v}_1 + \vec{a}_{avg} \Delta t \qquad (4\text{-}9)$$

For simplicity, we will restrict the following discussion to two dimensions and assume that the initial velocity

is in the x-direction: $\vec{v}_1 = v_1 \hat{i}$. Let us first consider the case where the acceleration is also along the x-axis: $\vec{a}_{avg} = a_{avg} \hat{i}$. We then find that

$$\vec{v}_2 = v_1 \hat{i} + a_{avg} \Delta t \hat{i} = (v_1 + a_{avg} \Delta t) \hat{i} \qquad (4\text{-}10)$$

We can see that the direction of the velocity has not changed, but its magnitude has.

We now suppose that the average acceleration vector makes some angle with the initial velocity vector, which it would do if it had both x- and y-components. If $\vec{a}_{avg} = a_{avg,x} \hat{i} + a_{avg,y} \hat{j}$ then

$$\vec{v}_2 = v_1 \hat{i} + (a_{avg,x} \hat{i} + (a_{avg,y} \hat{j}) \Delta t$$
$$= (v_1 + a_{avg,x} \Delta t) \hat{i} + (a_{avg,y} \Delta t) \hat{j} \qquad (4\text{-}11)$$

We can see that in this case $\vec{v}_2$ has both an x- and a y-component, indicating that its direction has changed. It has also changed in magnitude.

Let us now consider the special case where the acceleration vector is perpendicular to the initial velocity vector. Strictly speaking, if we consider a finite time interval, then the acceleration vector is only perpendicular to the velocity vector for a very brief instant at the beginning of the time interval. The velocity vector will begin to change direction and the perpendicularity will be lost. However, for that brief initial instant, the direction of the velocity vector will change but its length will not change. As we shall see in more detail in Section 4-3, if we can arrange for the acceleration to always be perpendicular to the velocity, the resulting motion is circular at a constant speed.

MAKING CONNECTIONS

Going to the Moon

In the late 1960s and early 1970s, the Apollo space program sent numerous missions to land on the Moon. More recently, the European Space Agency landed a probe on a comet. To accomplish this, the mission planners need to select a path for the space vehicles that considers the motion of the target and the gravitational influence of the various bodies the spacecraft gets close to on its flight. Sometimes these encounters are used to advantage to "sling-shot" the spacecraft. All of this requires dealing with non-constant forces and hence accelerations—a challenging mathematical task.

Katherine Johnson, an African-American physicist and mathematician, did pioneering work at NASA using computers to calculate trajectories for spacecraft. In 2015, she was awarded the Presidential Medal of Freedom in recognition of her work.

NASA

Figure 4-6 Katherine Johnson at work at NASA in 1966.

4-2 Projectile Motion

In general, it is difficult to solve the equations for a particle that experiences an acceleration that is not constant. However, as we saw in Chapter 3, it is possible to develop kinematics equations for a particle that experiences a constant acceleration. We now consider the case of a particle that is free to move in three dimensions and experiences a constant acceleration. This motion is called **projectile motion**. For the acceleration to be constant, both its magnitude and direction must remain constant.

To simplify the analysis, we select a coordinate system so that the acceleration vector is parallel to one of the axes. For the purposes of this chapter, we will assume that the acceleration points in the y-direction. We can write the acceleration as

$$\vec{a} = a\hat{j} \tag{4-12}$$

Since the acceleration is constant, the only component of the velocity that will change is the y-component. If the initial velocity of the particle is parallel to the y-axis, then the whole problem reduces to a one-dimensional problem, as described in Chapter 3.

We are now interested in the case where the initial velocity is not parallel to the y-axis. In general, the initial velocity vector could have x-, y-, and z-components. However, we can rotate our coordinate system about the y-axis so the projection of the initial velocity vector onto the xz-plane lies along the x-axis. This results in the z-component of the velocity being zero. Since there is no acceleration in the z-direction, this component of the velocity is always zero and we can neglect it in our analysis. Thus, for an object that experiences constant acceleration, the motion all occurs in a plane, and we will select the xy-plane.

Now, suppose that the particle has some initial position at time $t = 0$ specified by

$$\vec{r}_0 = x_0\hat{i} + y_0\hat{j} \tag{4-13}$$

and an initial velocity at time $t = 0$ given by

$$\vec{v}_0 = v_{0x}\hat{i} + v_{0y}\hat{j} \tag{4-14}$$

Following Chapter 3, we can write the vector version of Equation 3-10 for the velocity:

$$\vec{v} = \vec{v}_0 + \vec{a}t \tag{4-15}$$

The vector version of Equation 3-17 for the position is

$$\vec{r}(t) = \vec{r}_0 + \vec{v}_0 t + \frac{1}{2}\vec{a}t^2 \tag{4-16}$$

These equations can be written in their component forms, but before we do that, let us examine how they behave using graphical vector addition.

A Graphical Vector Perspective

We will consider explicitly the situation of a particle that is moving under the influence of a constant force directed in the negative *y*-direction. This is the situation we encounter when we toss a ball at a moderate speed. Because the speed of the ball is low, it is reasonable to neglect any effects due to air resistance, and the ball simply experiences a constant acceleration in the negative *y*-direction:

$$\vec{a} = -g\hat{j} \qquad (4\text{-}17)$$

In this equation, $g = 9.81 \text{ m} \cdot \text{s}^{-2}$.

We will also assume that at $t = 0$, the particle is at the origin of our coordinate system, i.e., $\vec{r}_0 = 0$. We can then write for the position vector

$$\vec{r}(t) = \vec{v}_0 t + \frac{1}{2}\vec{a}t^2 \qquad (4\text{-}18)$$

The first term on the right-hand side in Equation 4-18 ($\vec{v}_0 t$) gives us the trajectory of the projectile in the absence of acceleration. This is a vector whose length grows linearly with time and remains parallel to the initial velocity, as shown in Figure 4-7. This means that the projectile would continue along a straight line in the absence of acceleration.

We note that the second term, $\frac{1}{2}\vec{a}t^2$, is also a vector, as it is the product of the acceleration vector and a scalar ($t^2/2$). In our case, this vector points in the negative *y*-direction because that is the direction of the acceleration.

Adding these two terms together, we obtain the position vector. This is illustrated graphically in Figure 4-7. The position vector of the projectile is equal to the sum of the $\vec{v}_0 t$-term, shown in red, and the $\frac{1}{2}\vec{a}t^2$-term, shown in purple.

We can perform a similar analysis for the velocity vector. As we can see from Equation 4-15, it is also

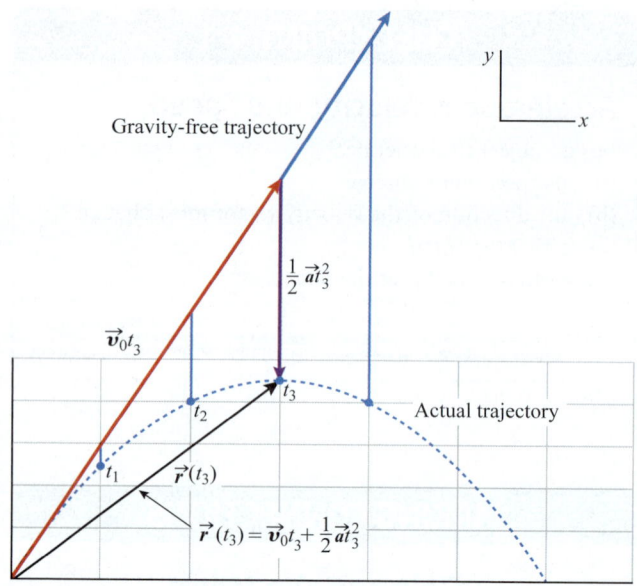

Figure 4-7 The position vector of the projectile, as measured from the origin, at any given point is the sum of the $\vec{v}_0 t$-term (shown in red) and the $\frac{1}{2}\vec{a}t^2$-term. This is demonstrated here for the position vector at t_3. Repeating the construction at the other times indicated in the figure would yield the heavy blue dots. The result is a parabolic trajectory.

the result of two terms. The first term, $\vec{v}_0$, is simply the initial velocity, which is a constant that does not change with time. The second term, $\vec{a}t$, contains the time dependence. Note that this is a vector that is parallel to the acceleration vector (for positive values of *t*). We can again perform a graphical addition of these two vectors, and this is illustrated in Figure 4-8 for various times during the motion.

We can see in Figure 4-8 that as the particle approaches the top of the trajectory, the velocity vector becomes more and more horizontal.

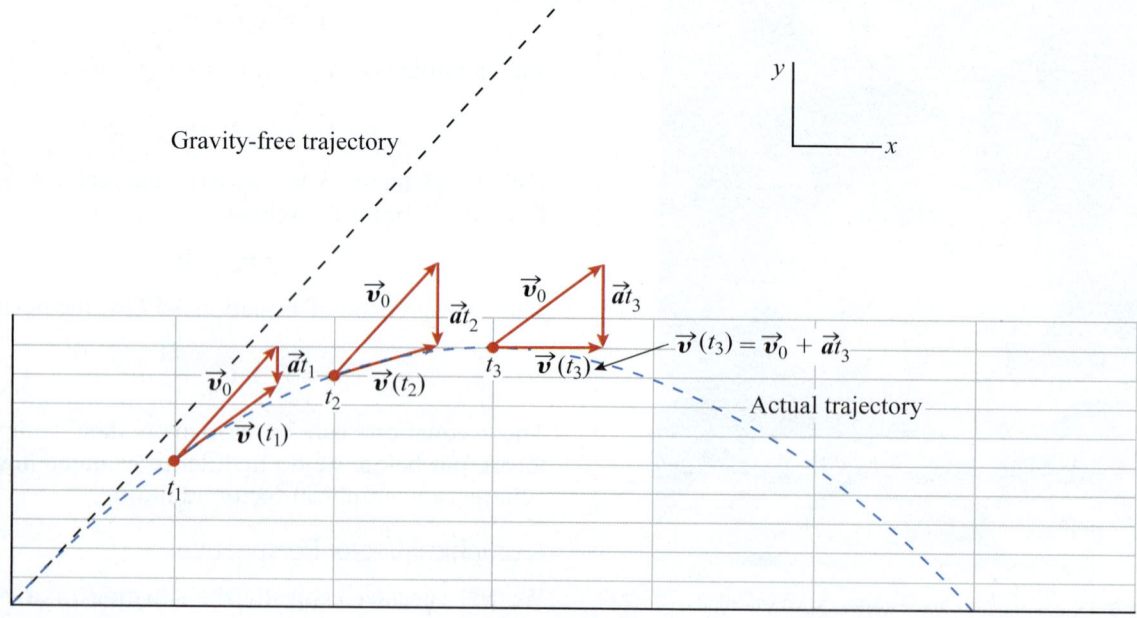

Figure 4-8 Adding the vectors $\vec{v}_0$ and $\vec{a}t$ graphically for various times during the motion of a projectile.

EXAMPLE 4-3

When Does the Velocity Become Horizontal?

Suppose a projectile is fired at a speed of $5.00 \text{ m} \cdot \text{s}^{-1}$ at an angle of $60.0°$ above the horizontal. At what time will the velocity of the projectile be horizontal?

Solution

We will solve the problem using the graphical method of adding vectors. We know the magnitude and direction of the initial velocity vector. We also know that the direction of the final velocity vector is horizontal. This allows us to make the vector diagram in Figure 4-9. We can draw in the direction and magnitude of the initial velocity. We only know the directions of the other two vectors, but that is sufficient to complete the triangle.

We can see from the triangle created in Figure 4-9 that

$$\sin 60.0° = \frac{gt}{5.00 \text{ m} \cdot \text{s}^{-1}}$$

$$t = \frac{5.00 \text{ m} \cdot \text{s}^{-1}}{9.81 \text{ m} \cdot \text{s}^{-2}} \sin 60.0° = 0.441 \text{ s}$$

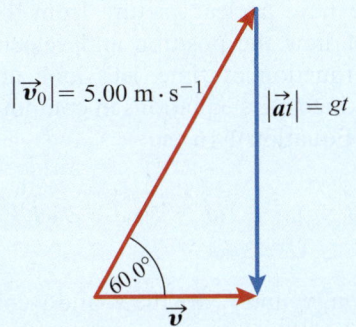

Figure 4-9 Example 4-3.

Making sense of the result

The initial velocity in the y-direction is $v_{0y} = (5.00 \text{ m} \cdot \text{s}^{-1})$ $\sin 60.0° = 4.33 \text{ m} \cdot \text{s}^{-1}$. The acceleration in the y-direction is $-9.81 \text{ m} \cdot \text{s}^{-2}$, which means that every second, the velocity in the y-direction decreases by $9.81 \text{ m} \cdot \text{s}^{-1}$. To get to zero velocity in the y-direction, we need to decrease by a little less than half this amount, so we would expect it would take a little less than half a second.

MAKING CONNECTIONS

Is It Reasonable to Neglect Air Resistance?

In this chapter, we will neglect the effects of air resistance on a projectile. This is only valid for projectiles that have relatively low speeds. In a paper titled "Maximum projectile range with drag and lift, with particular application to golf,"[1] Herman Erlichson showed that the effects of air resistance on the trajectory of a golf ball are quite significant. The fact that the ball spins is very important for extending the range. Spin is also very important in baseball, and by varying the axis of spin of the ball a pitcher can throw a fastball, a curveball, a slider, or a screwball. We will look at some of the effects of air resistance in Chapter 5.

For objects thrown at low speeds and rates of spin, a typical basketball shot, for example (Figure 4-10), it is reasonable to neglect air resistance.

Figure 4-10 Team Canada's Nirra Fields attempts a three-point shot during the 2015 Pan-Am Games.

1. *American Journal of Physics* 51(4) (1983), p. 357.

Projectile Motion in Component Form

Now that we have a clear picture from a graphical perspective of how the position and velocity vectors change as a function of time, let's look at the two-dimensional kinematics equations in component form. We can write Equation 4-16 as

$$x(t)\hat{\boldsymbol{i}} + y(t)\hat{\boldsymbol{j}} = x_0\hat{\boldsymbol{i}} + y_0\hat{\boldsymbol{j}} + (v_{0x}\hat{\boldsymbol{i}} + v_{0y}\hat{\boldsymbol{j}})t - \frac{1}{2}gt^2\hat{\boldsymbol{j}}$$

(4-19)

In this equation, x_0 and y_0 are the x- and y-components of the position vector at $t = 0$ and v_{0x} and v_{0y} are the x- and y-components of the velocity vector at $t = 0$. We can break this down even further by noting that the $\hat{\boldsymbol{i}}$-terms on the left side must equal the $\hat{\boldsymbol{i}}$-terms on the right side, and similarly for the $\hat{\boldsymbol{j}}$-terms. Thus,

KEY EQUATION

$$x(t) = x_0 + v_{0x}t$$

(4-20)

KEY EQUATION

$$y(t) = y_0 + v_{0y}t - \frac{1}{2}gt^2$$

(4-21)

We can perform a similar decomposition for the velocity, Equation 4-15, to find

KEY EQUATION

$$v_x = v_{0x}$$

(4-22)

KEY EQUATION

$$v_y(t) = v_{0y} - gt$$

(4-23)

We plot these equations in Figure 4-11.

In Figure 4-11(a), we see that the x-position increases linearly with time, which is consistent with the constant positive x-velocity shown in Figure 4-11(b). The velocity in the x-direction is constant because there is no acceleration in the x-direction.

In Figure 4-11(c), we see that the y-position is parabolic when plotted versus time, with the parabola opening downward. The parabola opens downward bvecaue the coefficient of the quadratic term (half the acceleration) is negative. The constant negative y-acceleration is reflected in the linear decrease with time of the y-velocity shown in Figure 4-11(d).

Equations 4-20 to 4-23 can be used to determine various things about the motion of a projectile, as Example 4-4 illustrates.

INTERACTIVE ACTIVITY 4-1

Projectile Motion

In this activity, you will use the PhET simulation "Projectile Motion" to explore the relationship between motion in the horizontal and vertical directions for projectile motion.

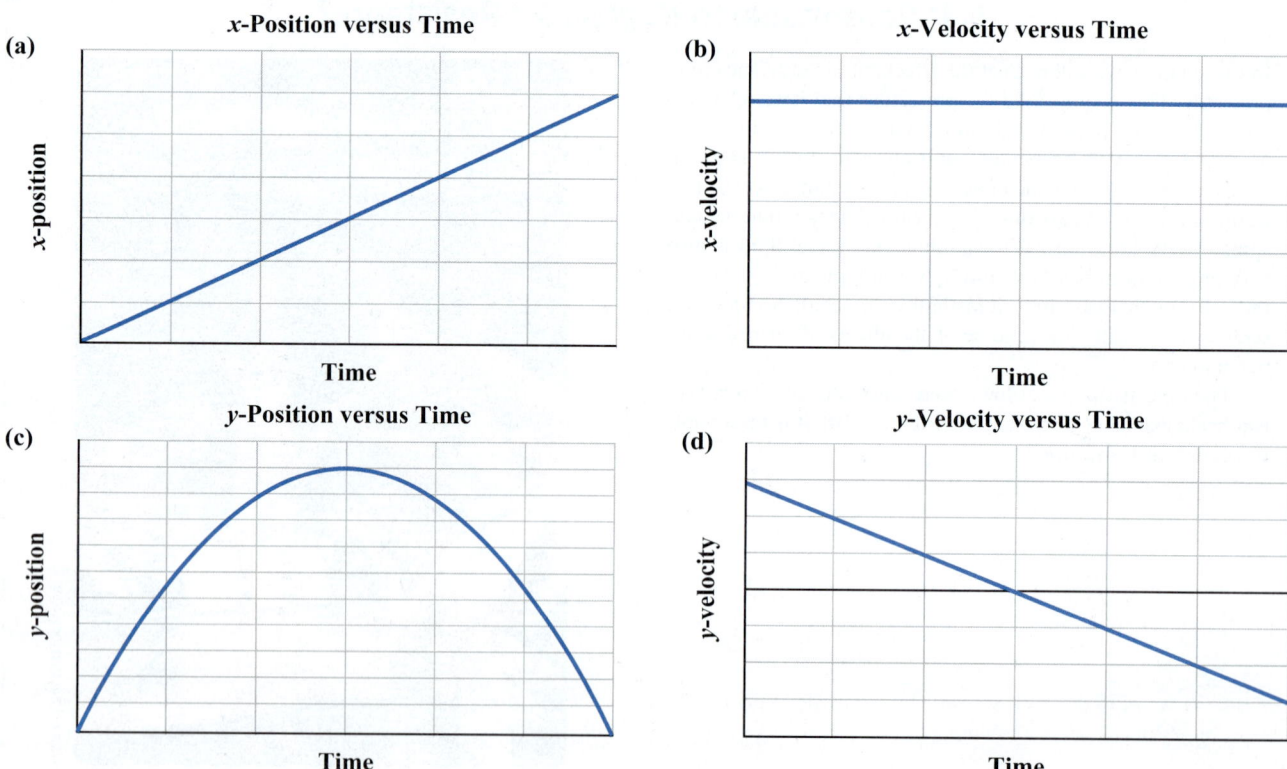

Figure 4-11 Position and velocity versus time for a projectile that experiences a constant acceleration in the negative y-direction. We plot the x-position in (a), the x-velocity in (b), the y-positon in (c), and the y-velocity in (d). We have assumed that $x_0 = 0$, $y_0 = 0$, and both v_{0x} and v_{0y} are positive.

Video Analysis of Projectile Motion

It is possible to take a video of the motion of an object and measure the position of the object versus time. We have to include an object of known length in the video to calibrate our measurements, and we use the known frame rate of the video camera to extract the time information. This can be seen in Figure 4-12, where we have collected a series of xy data points at successive frames of the video.

The data collected in Figure 4-12 are plotted below in Figure 4-13.

We can see in Figure 4-13 that, as we expect, the x-position increases linearly with time and the y-position is parabolic. Look at Data Rich Problem 78 for an opportunity to analyze these data.

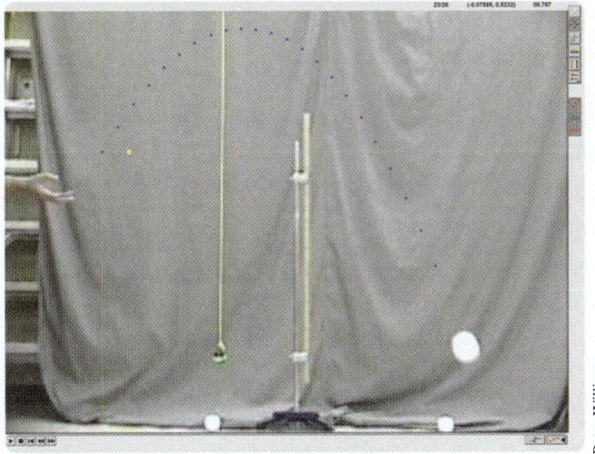

Peter Williams

Figure 4-12 Video analysis of a tossed ball. The blue dots are the points at which we collected xy data for the ball. The yellow vertical bar slightly to the right of centre in the image is a metre stick that was used to calibrate the video frame.

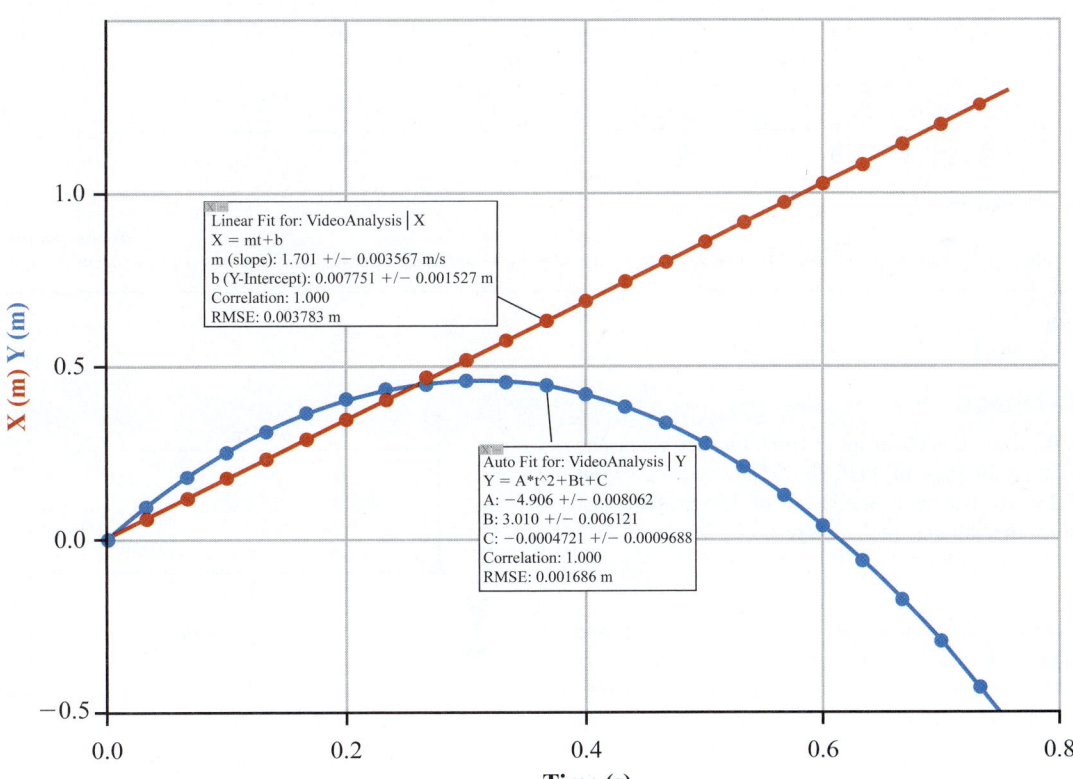

Linear Fit for: VideoAnalysis | X
X = mt+b
m (slope): 1.701 +/− 0.003567 m/s
b (Y-Intercept): 0.007751 +/− 0.001527 m
Correlation: 1.000
RMSE: 0.003783 m

Auto Fit for: VideoAnalysis | Y
Y = A*t^2+Bt+C
A: −4.906 +/− 0.008062
B: 3.010 +/− 0.006121
C: −0.0004721 +/− 0.0009688
Correlation: 1.000
RMSE: 0.001686 m

Figure 4-13 Plots of x- and y-position versus time for the motion depicted in Figure 4-12. Fits have been made to the data. The y-position is plotted in blue and the x-position in red.

Position versus Time with Constant Acceleration

Figure 4-14 shows position and velocity plots for an object that experiences constant acceleration.

(a) In which direction is the acceleration directed?
(b) Is the initial velocity of the particle directed above or below the horizontal?

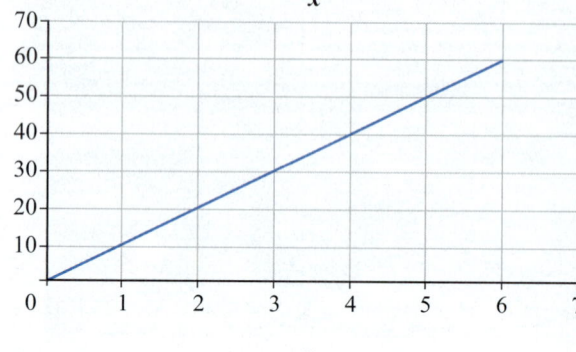

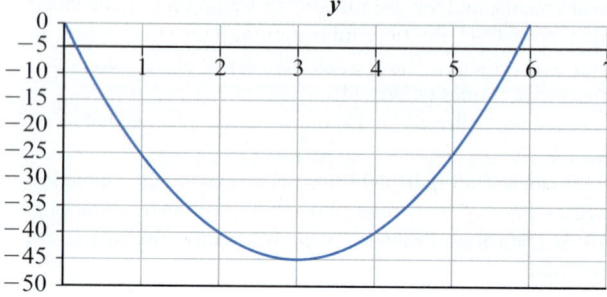

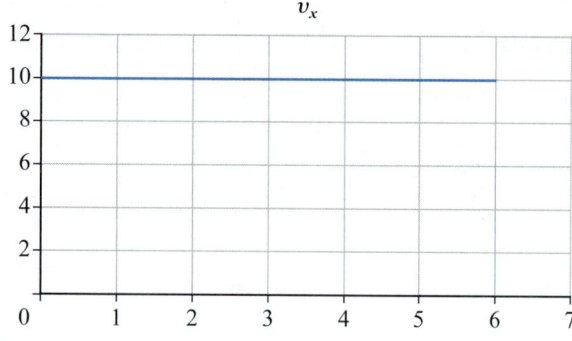

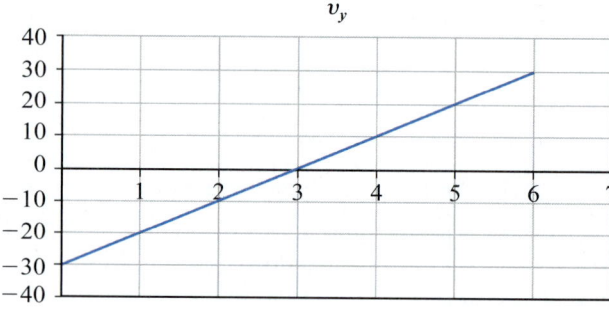

Figure 4-14 Checkpoint 4-2.

ANSWER: (a) The acceleration is in the positive y-direction. This can be seen from the positive slope of the v_y versus time plot. (b) v_{0x} is positive, but v_{0y} is negative. Therefore, the initial velocity is directed below the horizontal.

EXAMPLE 4-4

Projectile Motion

A basketball player is 6.00 m away from the net when she shoots the ball at an angle of 30.0° above the horizontal from a height of 2.00 m. The net is at a height of 3.05 m. How fast must she shoot the ball to sink the basket?

Solution

It is very useful to develop a strategy for solving projectile motion problems. If you examine Equations 4-20 to 4-23, you can see that the variables that appear are x_0, y_0, v_{0x}, v_{0y}, $x(t)$, $y(t)$, $v_x(t)$, $v_y(t)$, and t. There are in general nine variables and four equations. However, one of the equations is simply $v_x = v_{0x}$, which reduces the problem in practice to one with eight variables and three equations. This means that to solve for all of the variables, we must know at least five of them.

Creating a table can be a useful way of determining what you know and what you are trying to find out.

Let's see what we can extract from the statement of the problem in terms of these unknown variables, and then we will create our table. The situation is depicted in Figure 4-15.

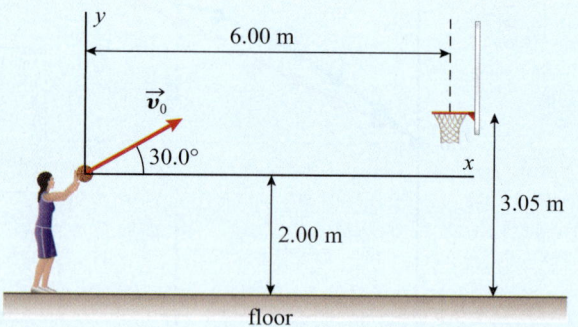

Figure 4-15 A player shoots the ball at the basket, which is 6.00 m away in the horizontal direction and 1.05 m above the launch point. She shoots the ball at a 30.0° angle.

We are free to choose the origin of the coordinate system. We will choose the initial position of the basketball to be the origin, so $x_0 = y_0 = 0$. In this coordinate system, the position of the ball as it goes through the hoop is $x = 6.00$ m, since the player is standing that far from the basket. The height of the basket is 1.05 m above the launch height of the ball, so $y = 1.05$ m when the ball goes through the hoop.

In this problem, we are told the launch angle but not the launch speed. However, if we recall that $v_{0x} = v_0 \cos\theta$ and $v_{0y} = v_0 \sin\theta$, we can recognize that there is really only one thing we do not know about the initial velocity—its magnitude.

We do not know the time at which the ball will reach the basket from the statement of the problem. With this information, we create Table 4-1.

Table 4-1 Known Values Extracted from the Statement of the Problem

	x-direction		*y*-direction
x_0	0 m	y_0	0 m
$x(t)$	6.00 m	$y(t)$	1.05 m
v_{0x}	$v_0 \cos 30.0° = ?$	v_{0y}	$v_0 \sin 30.0° = ?$
		$v_y(t)$	?
t	?	t	?

With this information, we return to examine Equations 4-20 and 4-21 to see if there is one equation that contains a single unknown. If we can find such an equation, we can determine that unknown directly by substituting the known values. $x = x_0 + v_{0x}t = x_0 + (v_0 \cos 30.0°)t$ contains two unknowns—the magnitude of the initial velocity and the time at which the ball goes through the hoop. Similarly,

$$y = y_0 + v_{0y}t - \frac{1}{2}gt^2 = y_0 + (v_0 \sin 30.0°)t - \frac{1}{2}gt^2$$ contains the

same two unknowns. We could examine the remaining equations (you should do this), but instead we recognize that we have identified two equations that contain the same two unknowns and therefore we can solve them.

We will first solve $x = x_0 + (v_0 \cos 30.0°)t$ for t:

$$t = \frac{x - x_0}{v_0 \cos 30.0°}$$

We substitute into Equation 4-21 to get

$$y = y_0 + v_0 \sin 30.0° \frac{x - x_0}{v_0 \cos 30.0°} - \frac{1}{2}g\left(\frac{x - x_0}{v_0 \cos 30.0°}\right)^2$$

$$= y_0 + (x - x_0)\tan 30° - \frac{1}{2}g\left(\frac{x - x_0}{v_0 \cos 30.0°}\right)^2 \quad (1)$$

Equation (1) contains one unknown, the magnitude of the initial velocity vector, which we solve for to find

$$v_0^2 = -\frac{1}{2}g\frac{(x - x_0)^2}{\cos^2 30.0°}\frac{1}{y - y_0 - (x - x_0)\tan 30.0°}$$

$$= -\frac{1}{2}(9.81 \text{ m·s}^{-2})\frac{(6.00 \text{ m} - 0 \text{ m})^2}{\cos^2 30.0°}$$

$$\times \frac{1}{1.05 \text{ m} - 0 \text{ m} - (6.00 \text{ m} - 0 \text{ m})\tan 30.0°}$$

$$= 97.5 \text{ m}^2\cdot\text{s}^{-2}$$

$$v_0 = 9.88 \text{ m·s}^{-1}$$

Using this result, we can calculate $v_{0x} = v_0 \cos 30.0° = 8.56$ m·s^{-1} and $v_{0y} = v_0 \sin 30.0° = 4.94$ m·s^{-1}.

Making sense of the result

It is useful to check the results we obtain to see if they make sense. We can find the time required for the shot from Equation 4-20:

$$t = \frac{x - x_0}{v_0 \cos 30.0°} = \frac{6.00 \text{ m} - 0 \text{ m}}{(9.88 \text{ m·s}^{-1})\cos 30.0°} = 0.701 \text{ s}$$

As a check on our calculations, let's see if this time gives us the correct y-position for the ball as it enters the basket:

$$y = y_0 + v_{0y}t - \frac{1}{2}gt^2$$

$$= 0 + (9.88 \text{ m·s}^{-1})\sin 30.0°(0.701 \text{ s})$$

$$- \frac{1}{2}(9.81 \text{ m·s}^{-2})(0.701 \text{ s})^2$$

$$= 1.05 \text{ m}$$

Thus, the initial velocity and time we have found place the ball in the basket.

As we saw in Example 4-4, time can be eliminated from the kinematics equations. We will now derive an additional kinematics equation that does not contain time. We start by solving Equation 4-23 for time:

$$t = \frac{v_{0y} - v_y(t)}{g} \quad (4\text{-}24)$$

which we substitute into the equation for the y-position to find

$$y = y_0 + v_{0y}\left(\frac{v_{0y} - v_y}{g}\right) - \frac{1}{2}g\left(\frac{v_{0y} - v_y}{g}\right)^2$$

$$= y_0 + \frac{v_{0y}^2}{g} - \frac{v_{0y}v_y}{g} - \frac{1}{2}g\left(\frac{v_{0y}^2 - 2v_{0y}v_y + v_y^2}{g^2}\right)$$

$$= y_0 + \frac{v_{0y}^2}{g} - \frac{v_{0y}v_y}{g} - \frac{v_{0y}^2}{2g} + \frac{v_{0y}v_y}{g} - \frac{v_y^2}{2g}$$

$$= y_0 - \frac{1}{2g}(v_y^2 - v_{0y}^2) \quad (4\text{-}25)$$

Rearranging, we get

KEY EQUATION $$v_y^2 - v_{0y}^2 = -2g(y - y_0) \quad (4\text{-}26)$$

If we compare this to Equation 3-22, $v_x(t)^2 - v_{x0}^2 = 2a_x\Delta x(t)$, we can see that they are very similar. Our result describes motion in the y-direction versus the x-direction, and we have explicitly put the acceleration in the y-direction to be $-g$.

EXAMPLE 4-5

Using the Time-Independent Kinematics Equation

A projectile is launched at an angle of 60.0° above the horizontal at a speed of 15.0 m·s⁻¹. What is the maximum height reached above the launch point?

Solution

We will approach solving the problem in two ways, using time explicitly and using the time-independent kinematics equation. We will first extract as much information as we can from the statement of the problem. A diagram always helps to visualize the situation.

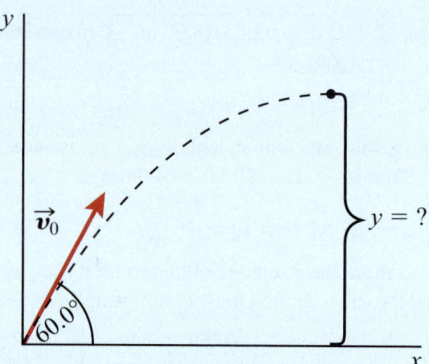

Figure 4-16 Example 4-5.

We can determine the initial velocities in the x- and y-directions:

$$v_{0x} = v_0\cos\theta = (15.0\ \text{m·s}^{-1})\cos 60.0° = 7.50\ \text{m·s}^{-1}$$

$$v_{0y} = v_0\sin\theta = (15.0\ \text{m·s}^{-1})\sin 60.0° = 13.0\ \text{m·s}^{-1}$$

We are always free to choose the origin of the coordinate system, so we will choose the launch point. Thus,

$$x_0 = y_0 = 0$$

We are asked to determine the maximum height reached by the projectile. Referring to Figure 4-16, we can see that at the top of the trajectory, the velocity vector is directed horizontally. Therefore, at this point, $v_y = 0\ \text{m·s}^{-1}$.

We place all of this information in a table as we did before (Table 4-2).

As before, we could seek an equation that has a single unknown, and since we are primarily considering motion in

Table 4-2 Known and Unknown Variables for the Projectile in Example 4-5

x-direction		y-direction	
x_0	0 m	y_0	0 m
$x(t)$	?	$y(t)$	?
v_{0x}	7.50 m·s⁻¹	v_{0y}	13.0 m·s⁻¹
		$v_y(t)$	0 m·s⁻¹
t	?	t	?

the y-direction, we will focus on those equations. The equation for the y-position has two unknowns, the maximum height and the time at which we reach the maximum height. The equation for the y-velocity has a single unknown, the time at which the projectile reaches its maximum height. We solve that equation for time to find

$$v_y = v_{0y} - gt$$

$$t = \frac{v_y - v_{0y}}{-g} = \frac{0\ \text{m·s}^{-1} - 13\ \text{m·s}^{-1}}{-9.81\ \text{m·s}^{-2}} = 1.33\ \text{s}$$

We can then substitute this time into the y-position equation to find the maximum height reached by the projectile:

$$y(t = 1.33\ \text{s}) = (13.0\ \text{m·s}^{-1})(1.33\ \text{s})$$

$$- \frac{1}{2}(9.81\ \text{m·s}^{-2})(1.33\ \text{s})^2 = 8.61\ \text{m}$$

An alternative approach is to directly use the time-independent kinematics equation, Equation 4-26:

$$v_y^2 - v_{0y}^2 = -2g(y - y_0)$$

$$0^2 - (13.0\ \text{m·s}^{-1})^2 = -2(9.81\ \text{m·s}^{-2})y$$

$$y = \frac{-(13.0\ \text{m·s}^{-1})^2}{-2(9.81\ \text{m·s}^{-2})} = 8.61\ \text{m}$$

Making sense of the result

We get the same result using either approach. The projectile is launched with a vertical velocity of 13.0 m·s⁻¹ and experiences a vertical acceleration of −9.81 m·s⁻², so the time of 1.33 s to reach the maximum height seems quite reasonable. The average vertical velocity is half the initial velocity, so this distance also makes sense.

EXAMPLE 4-6

What Launch Angle Produces the Maximum Range for a Projectile?

Derive an expression for the horizontal distance a projectile travels, and determine what launch angle produces the maximum range. Assume the projectile lands at the same height it was launched from.

Solution

To solve this problem, we need to derive an equation to express the horizontal distance travelled by a projectile as a function of the launch angle and then maximize that range with respect to the launch angle. We will assume that the projectile is launched from ground level and lands at ground level. We can introduce the launch angle into our equations by expressing the initial components of the velocity in terms of the launch speed and the launch angle. Our strategy will be to find the time taken for the projectile to go up to its maximum height and come back down again and then use that time to determine how far it has travelled horizontally.

If we choose our origin to be at the launch point, then $y_0 = 0$. When the projectile returns to the ground, $y_0 = 0$. Putting these values into the y-position equation, we find

$$0 = 0 + v_0(\sin\theta)t - \frac{1}{2}gt^2$$

$$t = \frac{2v_0\sin\theta}{g} \tag{1}$$

when the projectile returns to the ground. We can insert this time into the x-position equation to find

$$x_{max} = v_0\cos\theta\frac{2v_0\sin\theta}{g} = \frac{2v_0^2}{g}\sin\theta\cos\theta$$

We wish to maximize this with respect to the launch angle, so we differentiate with respect to θ and set the derivative to zero.

$$\frac{dx_{max}}{d\theta} = \frac{2v_0^2}{g}(\cos^2\theta - \sin^2\theta) = 0$$

$$\cos^2\theta - \sin^2\theta = \cos^2\theta - (1 - \cos^2\theta) = 2\cos^2\theta - 1 = 0$$

$$\cos\theta = \frac{1}{\sqrt{2}}$$

$$\theta = 45°$$

Making sense of the result

One might think that launching horizontally would lead to the greatest range as that produces the largest x-component of the velocity. However, doing so reduces the flight time of the projectile to zero, as you can see in Equation (1). Launching more vertically leads to an increased flight time but a decrease in the x-component of the velocity. The 45° launch angle is the compromise that achieves the largest range.

EXAMPLE 4-7

Bear-Proofing a Campsite

When you are back-country camping, it is recommended that you suspend your foodstuffs in the trees to prevent wild animals from getting into them. This is typically accomplished by tying a line to a rock and tossing the rock over a tree limb. One then ties the line to a sealed pack that contains the food and hoists it into the tree. Suppose the tree limb is 7.00 m above ground. You plan to launch your rock at 60.0° above the horizontal. What is the minimum launch speed required and how far back from the limb should you stand? Assume you launch the rock from a height of 1.50 m above the ground.

Solution

We know that to go over the branch, the rock has to reach a minimum height of 7.00 m above the ground. We can use that information to determine what the minimum y-component of the velocity must be. Since we know the launch angle, we can then find the x-component of the velocity upon launch. If we find out how long it takes to rock to reach the height of 7.00 m, we can use that time to see how far it will travel horizontally and that will tell us how far back from the limb we need to stand.

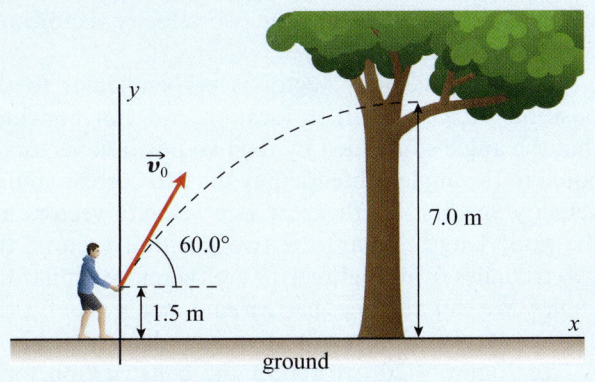

Figure 4-17 Throwing a rock over a branch.

As always, we are free to choose our origin and we will set it to be the launch point. The minimum vertical launch velocity must be large enough so that when the rock is 7.00 m above the ground, it has a velocity in the y-direction of zero.

We will again use a table to help us keep track of the various quantities (Table 4-3).

Table 4-3 The x and y Kinematic Variables for Example 4-7

	x-direction		y-direction
x_0	0 m	y_0	1.50 m
$x(t)$	?	$y(t)$	7.00 m
v_{0x}	?	v_{0y}	?
		$v_y(t)$	0 m·s⁻¹
t	?	t	?

We note that we know the initial and final y-positions and the velocity for the ascent of the projectile. This suggests we can use the time-independent kinematics equation, $v_y^2 - v_{0y}^2 = -2g(y - y_0)$, to find the initial y-velocity:

$$v_{0y} = \sqrt{2g(y - y_0) - v_y^2}$$
$$= \sqrt{2(9.81\ \text{m}\cdot\text{s}^{-2})(7.00\ \text{m} - 1.50\ \text{m}) - (0\ \text{m}\cdot\text{s}^{-1})^2}$$
$$= 10.4\ \text{m}\cdot\text{s}^{-1}$$

Since we know the launch angle, we can find the launch speed and the x-component of the velocity from

$$v_{0y} = v_0\sin\theta$$
$$v_0 = \frac{v_{0y}}{\sin\theta} = \frac{10.4\ \text{m}\cdot\text{s}^{-1}}{\sin 60°} = 12.0\ \text{m}\cdot\text{s}^{-1}$$
$$v_{0x} = v_0\cos 60° = 6.00\ \text{m}\cdot\text{s}^{-1}$$

To determine how far back we must stand, we will find the time required for the projectile to rise to the top of its trajectory

(continued)

and use that in the *x*-position equation. To find the time, it is simplest to use the *y*-velocity equation:

$$v_y = v_{0y} - gt$$

$$t = \frac{v_y - v_{0y}}{-g} = \frac{0 \text{ m·s}^{-1} - 10.4 \text{ m·s}^{-1}}{-9.81 \text{ m·s}^{-2}} = 1.06 \text{ s}$$

We can use this time in the *x*-position equation to find that the rock travels a horizontal distance of

$$x = x_0 + v_{0x}t$$
$$= 0 \text{ m} + (6.00 \text{ m·s}^{-1})(1.06 \text{ s})$$
$$= 6.36 \text{ m}$$

Making sense of the result

These are the launch speed and stand-back distance for the rock to just clear the branch. In practice, it is advisable to increase the launch speed and distance above these values. Regardless, it usually takes a few tries!

4-3 Circular Motion

Uniform Circular Motion

An object moving with **uniform circular motion** moves along a circular path of fixed radius at a constant speed. Since the direction of motion of the object is constantly changing, its velocity is also constantly changing. Therefore, the object must be continuously accelerating.

Figure 4-18 shows an object moving around a circle at a constant speed. The velocity at point A is $\vec{v}_A$, and the velocity at point B is $\vec{v}_B$. The velocities have exactly the same magnitude, v. The position vector at point A is $\vec{r}_A$, and the position vector at point B is $\vec{r}_B$. Both position vectors have magnitude r, as measured from the centre of the circle. At every point on the circle, the velocity of the object is tangential to the circle and perpendicular to the position vector. The average acceleration is defined as

$$\vec{a}_{\text{avg}} = \frac{\Delta\vec{v}}{\Delta t}$$

Therefore, the direction of average acceleration is parallel to $\Delta\vec{v}$.

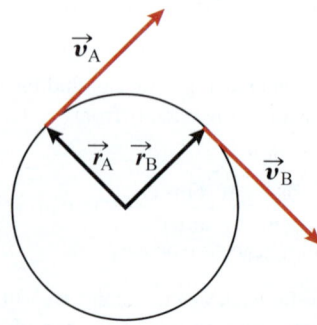

Figure 4-18 An object in uniform circular motion.

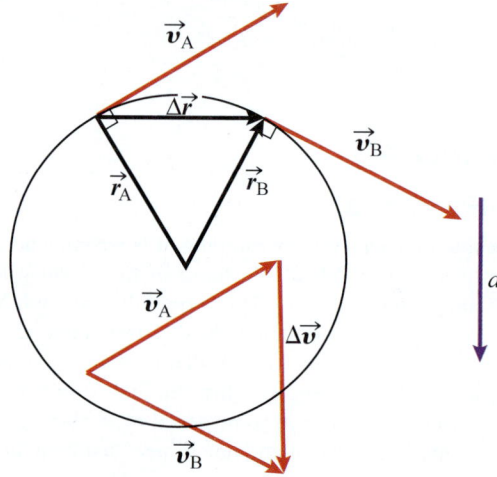

Figure 4-19 The change in the velocity vector and, therefore, the acceleration vector, are perpendicular to the change in the position vector.

To determine the direction of $\Delta\vec{v}$, we redraw the position and velocity vectors using the principle that a vector can be translated as long as its magnitude and direction do not change. First, we construct a triangle formed by the two position vectors and the change in position, $\Delta\vec{r}$, as shown in Figure 4-19. We then construct a triangle formed by the two velocity vectors and the change in velocity.

Since the velocity vector is perpendicular to the position vector at either location, we can conclude that the angle subtended by the two position vectors is equal to the angle subtended by the two corresponding velocity vectors. Further, the two velocity vectors are the same length, as are the two position vectors; the two triangles from Figure 4-19 are therefore similar triangles. We can also see that $\Delta\vec{v}$ and, consequently, the average acceleration vector are perpendicular to $\Delta\vec{r}$.

In Figure 4-20, we repeat the construction for a shorter interval of time between the two points. We can

MECHANICS

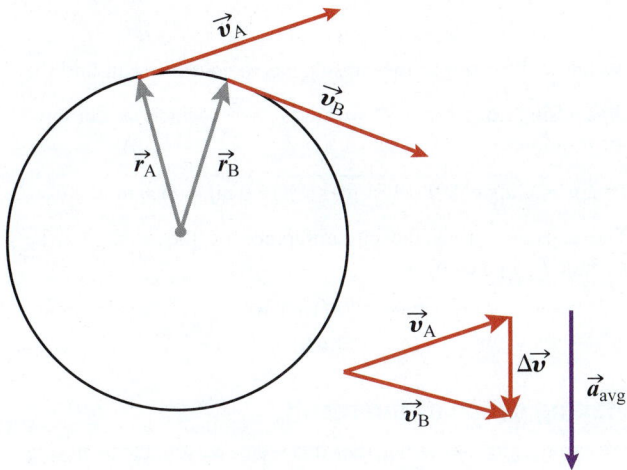

Figure 4-20 As the time elapsed gets smaller, so does the displacement between the two points; the acceleration vector, being parallel to $\Delta\vec{v}$, remains perpendicular to $\Delta\vec{r}$.

see that the conclusions we drew from our construction in Figure 4-19 still hold.

Next, we take the limit of the average acceleration as Δt goes to zero to get the instantaneous acceleration:

$$\vec{a} = \lim_{\Delta t \to 0} \frac{\Delta\vec{v}}{\Delta t} = \frac{d\vec{v}}{dt} \qquad (4\text{-}27)$$

As the time interval Δt shrinks, so does the displacement between A and B. In the limit, the two points merge. As the position triangle collapses, so does the velocity triangle. However, the acceleration, which is parallel to $\Delta\vec{v}$, remains perpendicular to $\Delta\vec{r}$, which is a chord of the circle. The acceleration vector, being perpendicular to that chord, points along the radius to the centre of the circle. In the limit as Δt goes to zero, the instantaneous acceleration points along the radius. Therefore, an object moving in a circle must continuously accelerate toward the centre of the circle. Acceleration toward the centre of the circle is called **radial acceleration**.

To find the magnitude of the acceleration, consider the ratios of the two similar triangles formed by the position and velocity vectors:

$$\frac{\Delta v}{v} = \frac{\Delta r}{r} \qquad (4\text{-}28)$$

We solve for Δv:

$$\Delta v = \frac{v\Delta r}{r} \qquad (4\text{-}29)$$

Dividing both sides by Δt, the change in time, gives

$$\frac{\Delta v}{\Delta t} = \frac{v\Delta r}{r\Delta t} \qquad (4\text{-}30)$$

In the limit as Δt goes to zero,

$$\frac{dv}{dt} = \frac{v}{r}\lim_{\Delta t \to 0}\frac{\Delta r}{\Delta t} = \frac{v}{r}v \qquad (4\text{-}31)$$

The left hand side of Equation 4-31 is the acceleration, so

KEY EQUATION

$$a_r = \frac{v^2}{r} \qquad (4\text{-}32)$$

The subscript r on the acceleration denotes that this is the radial acceleration.

We have thus established both the direction and the magnitude for the radial acceleration of an object moving around a circle at a constant speed. Figure 4-21 shows the position, velocity, and acceleration vectors for an object executing uniform circular motion. The velocity vector is always tangential to the circle. The acceleration vector is perpendicular to the velocity vector and points radially inward toward the centre of the circle. The magnitude of the acceleration vector at any point is the same: $a_r = \dfrac{v^2}{r}$.

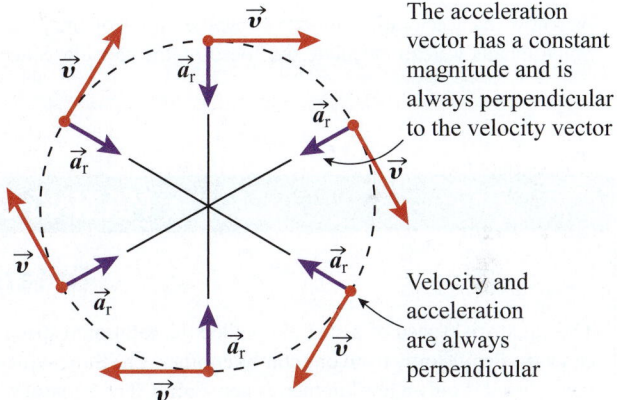

The acceleration vector has a constant magnitude and is always perpendicular to the velocity vector

Velocity and acceleration are always perpendicular

Figure 4-21 The acceleration vector is always perpendicular to the velocity vector.

CHECKPOINT 4-3

Uniform Circular Motion

For an object in uniform circular motion,
(a) the velocity of the object is constant
(b) the speed of the object is constant
(c) both (a) and (b) are true
(d) none of the above are true

ANSWER: (b) The speed must be constant for uniform circular motion, but the velocity is not constant.

EXAMPLE 4-8

Astronaut Training

Astronauts undergo large accelerations while operating the spacecraft during take-off. They train for this in centrifuges such as that shown in Figure 4-22. The astronauts sit in the training capsules located at the ends of the centrifuge arms, and the entire device is rotated about a vertical axis.

Figure 4-22 The Ames Research Center centrifuge.

Suppose a training capsule is designed to achieve radial accelerations of 11.0g. The capsule is on the end of an arm that rotates at a constant angular speed. The radius of the horizontal circular path the capsule follows is 16.0 m. Find the period of the capsule, that is, how long it takes the capsule to complete one full trip around the circle.

Solution

We know the acceleration and we know the radius of the path, which means we can calculate the speed of the capsule. Since we know the distance the capsule has to cover, we can find the time. Using Equation 4-32, we have $a_r = \dfrac{v^2}{r}$, which we can solve for v to find

$$v = \sqrt{a_r r} = \sqrt{11(9.81 \text{ m} \cdot \text{s}^{-2})(16.0 \text{ m})} = 41.6 \text{ m} \cdot \text{s}^{-1}$$

The capsule travels the circumference of the track, so the period, T, is given by

$$T = \frac{2\pi r}{v} = \frac{2\pi(16.0 \text{ m})}{41.6 \text{ m} \cdot \text{s}^{-1}} = 2.42 \text{ s}$$

Making sense of the result

We can roughly assess whether this seems reasonable by noting that in the time $T/2$, the velocity completely reverses its direction. This gives us an average acceleration of

$$a_{\text{avg}} = \frac{\Delta v}{\Delta t} = \frac{2(41.6 \text{ m} \cdot \text{s}^{-1})}{\dfrac{2.42 \text{ s}}{2}} = 68.8 \text{ m} \cdot \text{s}^{-2}$$

which is approximately seven times the acceleration due to gravity. We would expect the average acceleration to be somewhat less than the instantaneous acceleration because the average assumes a straight path and hence underestimates the distance travelled.

MAKING CONNECTIONS

Rotating Stars

The equatorial speed of a star, also called the rotational speed, can vary significantly from one star to another. The Sun's equatorial speed is only a few kilometres per second. The equatorial speed for some stars can exceed 200 km/s. Some neutron stars, which are vastly smaller than stars like our Sun, have rotational speeds as high as 38 000 km/s. With an equatorial speed of approximately 600 km/s, the fastest-spinning full-sized star observed to date is VFTS 102. VFTS 102 was discovered by astronomers using the Very Large Telescope at the European Southern Observatory in Paranal, Chile (Figure 4-23). The star is located in the Tarantula nebula within the Large Magellanic Cloud, about 160 000 light years from Earth.

Earth bulges slightly at the equator due to its rotation. The radial acceleration at the equator on Earth is about 0.03 m·s^{-2}, which is about 0.3% of the acceleration due to gravity at the surface of Earth. For VFTS 102, a rough estimate of the radial acceleration places it at about 40% of the acceleration due to gravity on the surface of the star, which suggests that the star is highly distorted about its equator.

Figure 4-23 VFTS 102, the fastest-spinning star found to date, is at the centre of the Tarantula nebula in the Large Magellanic Cloud.

Non-uniform Circular Motion

An object moving in a circle does not necessarily maintain a constant speed. For example, if you let a marble roll down the inside of a hemispherical bowl, the marble will speed up while travelling down the circular surface. When the speed of an object changes while travelling along a circle, then the velocity of the object changes along the direction of motion, which is tangential to the circle. Such an object is said to have **tangential acceleration**, $\vec{a}_t$. Therefore, the total acceleration of this object is the *vector sum* of its tangential and radial accelerations (Figure 4-24) and is given as

> **KEY EQUATION**

$$\vec{a} = \vec{a}_r + \vec{a}_t \tag{4-33}$$

The tangential acceleration is always perpendicular to the radial acceleration. The total acceleration then has a magnitude given by $a = \sqrt{a_t^2 + a_r^2}$.

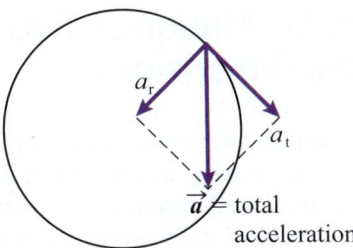

Figure 4-24 The total acceleration for an object moving in a circle at a changing speed is the vector sum of the radial and tangential acceleration vectors.

EXAMPLE 4-9

Non-uniform Circular Motion

Starting from rest, an experimental mini-drone (remote-controlled aircraft) flies around a horizontal circle of radius 2.00 m and completes three revolutions in 3.00 s. The magnitude of the tangential acceleration is constant. Find the magnitude of the mini-drone's acceleration at the end of 0.500 s.

Solution

To find the total acceleration, we need to know both the radial and the tangential accelerations. We can determine the tangential acceleration by treating the motion around the circumference as one-dimensional motion. We know it takes 3.00 s to go three times around the circle, starting from rest. We can also use this information to find the magnitude of the velocity at 0.500 s, which will then allow us to calculate the radial acceleration.

We pick the direction of motion as positive and choose s to specify the position around the circumference of the circle. To find the tangential acceleration, we apply Equation 3-17:

$$\Delta s = v_0 t + \frac{1}{2} a_t t^2$$

We solve this for a_t and substitute into that result the fact that Δs is three circumferences of the circle at $t = 3.00$ s:

$$a_t = \frac{2(\Delta s - v_0 t)}{t^2}$$

$$= \frac{2(3(2\pi)(2.00 \text{ m}) - (0 \text{ m} \cdot \text{s}^{-1})(3.00 \text{ s}))}{(3.00 \text{ s})^2}$$

$$= \frac{24.0\pi \text{ m}}{9.00 \text{ s}^2} = 8.38 \text{ m} \cdot \text{s}^{-2}$$

We will now determine the radial acceleration. To do this we need to know the speed of the drone at 0.500 s. The speed at $t = 0.500$ s is given by

$$v = v_0 + at = (0 \text{ m} \cdot \text{s}^{-1}) + (8.38 \text{ m} \cdot \text{s}^{-2})(0.500 \text{ s}) = 4.19 \text{ m} \cdot \text{s}^{-1}$$

Substituting this speed into Equation 4-32 gives

$$a_t = \frac{v^2}{r} = \frac{(4.19 \text{ m} \cdot \text{s}^{-1})^2}{2.00 \text{ m}} = 8.78 \text{ m} \cdot \text{s}^{-2}$$

The total acceleration is the vector sum of the radial and the tangential accelerations. Using Equation 4-33, we obtain

$$\vec{a} = \vec{a}_r + \vec{a}_t$$

Since these two accelerations are always perpendicular to each other,

$$a = \sqrt{a_t^2 + a_r^2} = \sqrt{(8.78 \text{ m} \cdot \text{s}^{-2})^2 + (8.38 \text{ m} \cdot \text{s}^{-2})^2}$$

$$= 12.1 \text{ m} \cdot \text{s}^{-2}$$

Making sense of the result

The drone is moving fairly rapidly around a rather tight circle, so it is not surprising that the acceleration is this large. As time goes on, the speed of the drone will continue to increase and the radial acceleration will get larger and larger, tending to dominate the total acceleration.

4-4 Relative Motion in Two and Three Dimensions

Relative motion deals with comparing the results of two observers who are moving at a constant velocity with respect to one another. An example of such a situation is encountered when ships are navigating in waters where there are currents. The ship moves relative to the surface of the water, but the water is moving relative to the ground, so the motion of the ship relative to the ground is described by the motion of the ship relative to the water plus the motion of the water relative to the ground. This can make navigation complicated.

Consider a slow ferry that needs to cross the English Channel from Calais, France, to Hastings, England, 110 km west-southwest (Figure 4-25, red arrow). However, there is a strong flood tide flowing northeast. If the captain simply sets a course straight for Hastings, the ferry, being pushed by the current, will likely not arrive at its desired destination. It will likely be far north when it reaches the English coast and could end up in Folkestone, for example, as shown in Figure 4-25 (orange arrow). To arrive at Hastings, the captain must adjust her course and follow a more southerly heading that allows the ferry to compensate for the tidal current, which can flow with speeds of up to 4 nautical miles per hour (7.4 km/h), to arrive in Hastings.

Formal Development of the Relative Motion Equations in Two Dimensions

Let's develop the equations by considering an object observed by two observers who are moving at a constant speed relative to each other. For example, one observer could be standing on the side of a road and the other observer could be travelling in a car moving down the road at constant velocity, which implies that the road is straight and they are moving at constant speed. Suppose they are observing the motion of a tractor in a field beside the road.

We will need some notation to keep track of the various quantities. Let $\vec{r}_{TG}$ denote the position vector of the tractor as measured by the observer standing on the ground. Following the same convention, the position vector of the tractor measured by the observer in the car will be $\vec{r}_{TC}$. Note that the first subscript denotes the object being observed, while the second denotes the **reference frame** in which it is being observed.

We will choose the axes of our two coordinate systems to be parallel to one another, and we will set $t = 0$ to be the time at which the origins of our coordinate system overlap, i.e., when the car passes the observer on the side of the road. Thus, the initial position of the car will be zero in both reference frames.

We will also need to describe the relative velocity of our two observers. Let $\vec{V}_{CG}$ be the velocity of the car as measured by the observer on the ground. Note that we have retained the convention of the first subscript denoting the object being observed while the second denotes the frame in which it is observed. However, we use uppercase letters to denote the relative coordinates of the reference frames. Then, $\vec{V}_{GC}$ denotes the velocity of the ground as observed by the car, and these two are equal but opposite to each other:

$$\vec{V}_{CG} = -\vec{V}_{GC} \qquad (4\text{-}34)$$

At any instant in time, the origin of the car reference frame is displaced from the origin of the ground reference frame by the vector

$$\vec{R}_{CG} = \vec{V}_{CG}t \qquad (4\text{-}35)$$

This is shown in Figure 4-26.

We can see from Figure 4-26 that the position of the tractor relative to the observer on the ground is given by

$$\vec{r}_{TG} = \vec{r}_{TC} + \vec{V}_{CG}t \qquad (4\text{-}36)$$

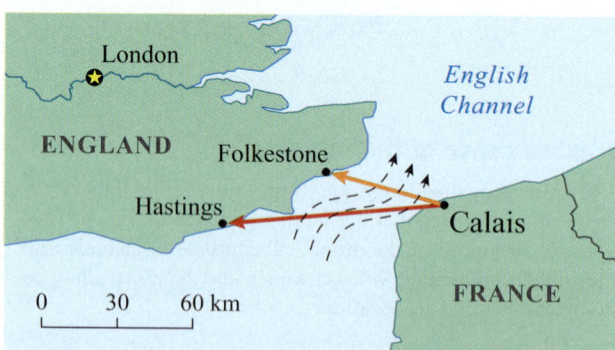

Figure 4-25 If the captain heads to Hastings (red arrow), the ferry could end up in Folkestone (orange arrow) due to the current in the English Channel. The velocity of the ferry relative to the land (the stationary frame of reference in this case) is the sum of the velocity of the ferry relative to the water and the velocity of the water relative to the land.

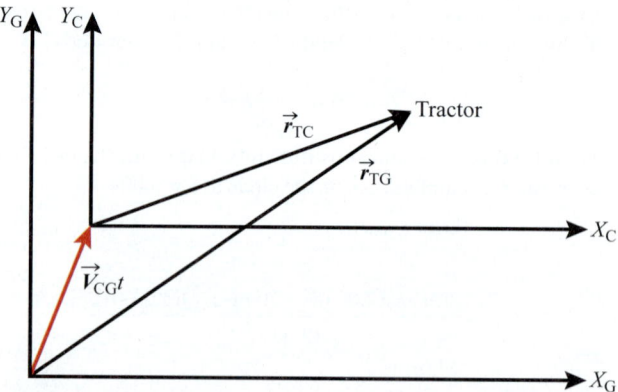

Figure 4-26 The position of the tractor relative to the ground is the sum of the tractor's position vector relative to the car, $\vec{r}_{TC}$, and that of the car relative to the ground, $\vec{R}_{CG} = \vec{V}_{CG}t$.

We can differentiate Equation 4-37 with respect to time to find the relationship between the velocities in the two reference frames:

$$\vec{v}_{TG} = \vec{v}_{TC} + \vec{V}_{CG} \qquad (4\text{-}37)$$

where $\vec{v}_{TG}$ is the velocity of the tractor in the ground reference frame, $\vec{v}_{TC}$ is the velocity of the tractor in the car reference frame, and $\vec{V}_{CG}$ is the velocity of the car with respect to the ground.

We can generalize this equation. Suppose we have a particle P that is being observed by two observers. One is in reference frame A, and the other is in reference frame B. Frame B has a velocity of $\vec{V}_{BA}$ with respect to frame A. We then write

KEY EQUATION
$$\vec{v}_{PA} = \vec{v}_{PB} + \vec{V}_{BA} \qquad (4\text{-}38)$$

PEER TO PEER

All of the subscripts can get a bit confusing in these problems. I use the following trick. Notice that in Equation 4-37 the subscripts on the left-hand side are TG while those on the right-hand side are TC and CG. The outer subscripts on the right-hand side correspond to the subscripts on the left-hand side. Does this always work? Let's rearrange Equation 4-37 as

$$\vec{v}_{TC} = \vec{v}_{TG} - \vec{V}_{CG}$$

and recall that $\vec{V}_{CG} = -\vec{V}_{GC}$. Putting this into our rearranged equation, we get

$$\vec{v}_{TC} = \vec{v}_{TG} + \vec{V}_{GC}$$

We can see that once again the outer subscripts on the right match those on the left. I do have to remember that the first subscript refers to the object being observed and the second to the frame in which it is being observed.

EXAMPLE 4-10

Crossing a River on a Kayak

A river flows from north to south. The water speed is 2.00 m/s. You are in your kayak trying to cross the river, and you keep the kayak pointing in an easterly heading that is perpendicular to the river. Your speed in still water is 4.00 m/s. Sketch the direction in which the kayak goes. What is the velocity of the kayak relative to land?

Solution

Solving these problems requires that we carefully define the various velocities in the problem. Let's choose coordinate systems where the positive y-axis points north and the positive x-axis points east. The two coordinate systems are the ground and the surface of the water. The object that is being observed is the kayak.

We are asked to find the speed of the kayak relative to the ground, and we will denote that as $\vec{v}_{KG}$. We are told what the velocity of the kayak is relative to the water, and we will denote that as $\vec{v}_{KW}$. We are also told the velocity of the water with respect to the ground, and we denote that as $\vec{V}_{WG}$. In our coordinate system,

$$\vec{v}_{KW} = (4.00 \text{ m} \cdot \text{s}^{-1})\hat{i}$$

and

$$\vec{V}_{WG} = -(2.00 \text{ m} \cdot \text{s}^{-1})\hat{j}$$

Now

$$\vec{v}_{KG} = \vec{v}_{KW} + \vec{V}_{WG} = (4.00 \text{ m} \cdot \text{s}^{-1})\hat{i} - (2.00 \text{ m} \cdot \text{s}^{-1})\hat{j}$$

The situation is illustrated in Figure 4-27.

The magnitude of the velocity of the kayak relative to the ground is given by

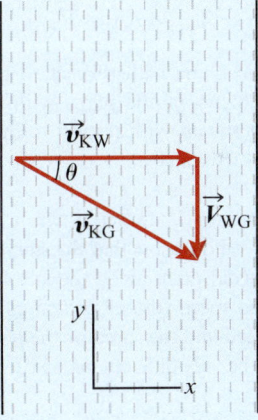

Figure 4-27 Example 4-10.

$$v_{KG} = \sqrt{v_{KW}^2 + V_{WG}^2} = \sqrt{(4.00 \text{ m} \cdot \text{s}^{-1})^2 + (2.00 \text{ m} \cdot \text{s}^{-1})^2}$$
$$= 4.47 \text{ m} \cdot \text{s}^{-1}$$

The angle that this vector makes with the positive x-axis is

$$\theta = \tan^{-1}\left(\frac{-2.00 \text{ m} \cdot \text{s}^{-1}}{4.00 \text{ m} \cdot \text{s}^{-1}}\right) = -26.6°$$

This corresponds to a compass heading of 116.6° or 26.6° South of East.

Making sense of the result

The velocity of the kayak relative to the ground is expected to be slightly larger than the velocity of the kayak relative to the water due to the contribution from the velocity of the river.

As we can see from the preceding example, a current can have a significant impact on the ultimate heading of a vessel. Similar difficulties are encountered with aircraft, as the winds in the upper atmosphere can be as high as several hundred kilometres per hour. The challenge faced by navigators is to select a course that takes these effects into account so that they arrive at their destination. This is illustrated in the following example.

EXAMPLE 4-11

Flight Plan

A plane is scheduled to fly from Casablanca, Morocco, to Lisbon, Portugal, 581 km [13° W of N]. A wind is blowing at 34.0 km/h [15° N of W]. When the plane's airspeed is 560 km/h, and the pilot flies on a direct bearing to Lisbon, how long does the flight take?

Solution

If the pilot were to point the plane along the straight line from Casablanca to Lisbon, the wind would push the plane off course to the west. To compensate, the pilot needs to point the plane somewhat to the east of the direct heading to Lisbon.

The coordinate systems we will employ are one attached to the ground and one that moves with the air at the velocity of the wind.

Let us denote the velocity of the plane with respect to the ground as $\vec{v}_{PG}$, the velocity of the plane with respect to the air as $\vec{v}_{PA}$, and the velocity of the air with respect to the ground as $\vec{V}_{AG}$. We can then write

$$\vec{v}_{PG} = \vec{v}_{PA} + \vec{V}_{AG} \qquad (1)$$

The velocity of the plane with respect to the ground equals the sum of the velocity of the plane with respect to the air and the velocity of the air with respect to the ground (i.e., the wind velocity). Let us construct a diagram of these velocities (Figure 4-28).

Although we do not know yet in which direction the plane should be oriented, we do know that the sum of the velocities of the plane and the wind point along the line connecting the two cities. Since the wind speed is small compared to the airspeed of the plane, the airspeed and the speed of the plane with respect to the ground will not be very different, as we can see from our scale diagram.

We label the angles between the velocities α, β, and γ as shown. In Figure 4-28, $\vec{v}_{PG}$ makes an angle of 13° with the vertical, and $\vec{V}_{AG}$ makes an angle of 15° with the horizontal.

Then,

$$\alpha = 90° - (13° + 15°) = 62°$$

Applying the sine law to the velocity triangle, we have

$$\frac{\sin \alpha}{v_{PA}} = \frac{\sin \beta}{V_{AG}}$$

$$\sin \beta = \frac{V_{AG}\sin \alpha}{v_{PA}} = \frac{34.0\,\text{km/h}(\sin 62.0°)}{560\,\text{km/h}} = 0.0536$$

$$\beta = 3.07°$$

Therefore,

$$\gamma = 180° - (\alpha + \beta) = 180° - (3.07° + 62°)$$
$$= 114.93°$$

To three significant digits, the value is 115°.

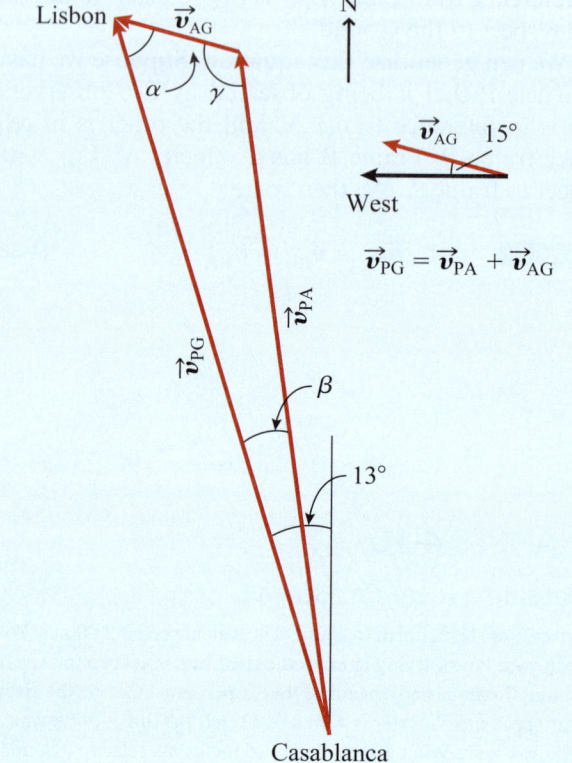

Figure 4-28 Example 4-11.

Using the sine law once again, we have

$$\frac{\sin \alpha}{v_{PA}} = \frac{\sin \gamma}{v_{PG}}$$

$$v_{PG} = \frac{v_{PA}\sin \gamma}{\sin \alpha} = \frac{560.\,\text{km/h}(\sin 115°)}{\sin 62.0°}$$

$$= 575\,\text{km/h}$$

The time it takes the plane to arrive at its destination is

$$t = \frac{581\,\text{km}}{575\,\text{km}} = 1.01\,\text{h}$$

Another way to tackle this problem is using vector components. If we choose our x-axis to point in the direction of $\vec{v}_{PG}$, we can write that vector as $\vec{v}_{PG} = v_{PG}\hat{i}$. The other two vectors will have both x- and y-components and can be written as $\vec{v}_{PA} = v_{PA,x}\hat{i} + v_{PA,y}\hat{j}$ and $\vec{V}_{AG} = v_{AG,x}\hat{i} + V_{AG,y}\hat{j}$. We substitute these into Equation (1) to get

$$v_{PG,x}\hat{i} = v_{PA,x}\hat{i} + v_{PA,y}\hat{j} + V_{AG,x}\hat{i} + V_{AG,y}\hat{j} \qquad (2)$$

Notice that the left-hand side of this equation only has x-components. This means that the y-components on the right-hand side must add up to zero:

MECHANICS

$$0 = v_{PA,y} + V_{AG,y}$$

$$v_{PA,y} = -V_{AG,y}$$

This result makes physical sense. The plane must fly slightly into the wind to counteract the effect of the wind velocity that is perpendicular to the desired direction of flight.

The desired direction of flight is [13° W of N] and the wind is blowing [15° N of W], which is [75° W of N]. Therefore, the wind makes an angle with the desired direction of 62°. The perpendicular component of the wind velocity is thus $v_{AG,y} = (34.0 \text{ km} \cdot \text{h}^{-1}) \sin 62.0° = 30.0 \text{ km} \cdot \text{h}^{-1}$ and the parallel component is $v_{AG,x} = (34.0 \text{ km} \cdot \text{h}^{-1}) \cos 62.0° = 16.0 \text{ km} \cdot \text{h}^{-1}$.

To find the time of flight, we need to know the parallel component of the velocity of the plane over the ground, which is given by the x-components of Equation (1):

$$v_{PG,x} = v_{PA,x} + V_{AG,x} \qquad (3)$$

We already know what $V_{AG,x}$ is, so we need to find $v_{PA,x}$. We know the magnitude of $\vec{v}_{PA}$ and we know that the y-component has a magnitude of $v_{PA,y} = v_{AG,y} = 30.0 \text{ km} \cdot \text{h}^{-1}$. Therefore,

$$v_{PA,x} = \sqrt{v_{PA}^2 - v_{PA,y}^2} = \sqrt{(560. \text{ km} \cdot \text{h}^{-1})^2 - (30.0 \text{ km} \cdot \text{h}^{-1})^2}$$

$$= 559 \text{ km} \cdot \text{h}^{-1}$$

We can substitute this into Equation (3) to find $v_{PG,x} = 559 \text{ km} \cdot \text{h}^{-1} + 16.0 \text{ km} \cdot \text{h}^{-1} = 575 \text{ km} \cdot \text{h}^{-1}$. We need to travel a distance of 581 km at this speed, so it takes

$$t = \frac{\text{distance}}{\text{speed}} = \frac{581 \text{ km}}{575 \text{ km} \cdot \text{h}^{-1}} = 1.01 \text{ h}$$

Making sense of the result

The plane's speed is very large compared to the speed of the wind, so the small heading correction of 3.07° is reasonable.

CHECKPOINT 4-5

Direction and Heading

If a ship needs to head west in an ocean where the current flows northwest, the captain must maintain a heading that is

(a) south of west
(b) south of east
(c) north of west
(d) north of east

ANSWER: (a) Both (a) and (b) could work, but (a) is the more practical choice.

Relative Acceleration

We can differentiate our relative velocity equation with respect to time to find

$$\frac{d\vec{v}_{PA}}{dt} = \frac{d\vec{v}_{PB}}{dt} + \frac{d\vec{V}_{BA}}{dt}$$

$$\vec{a}_{PA} = \vec{a}_{PB} \text{ since } \frac{d\vec{V}_{BA}}{dt} = 0 \qquad (4\text{-}39)$$

Thus, the accelerations are observed to be the same in both reference frames. This result only applies for inertial reference frames. As we shall see in Chapter 5, objects in inertial reference frames obey Newton's laws.

MAKING CONNECTIONS

Projectile and Relative Motion

Imagine you are sitting on a moving train and tossing a ball up into the air. If you launch the ball vertically, it will go straight up, come to rest, and then fall straight back down again. If we choose up to be the positive y-direction, the vertical position and velocity of the ball are given by

$$y_{BT} = y_{0,BT} + v_{0y,BT} - \frac{1}{2}gt^2$$

$$v_{BT,y} = v_{BT,0y} - gt$$

The x-position of the ball with respect to the train is a constant, x_0, and its x-velocity is zero.

Now let us consider what an observer standing on the platform would observe. In their reference frame, the train has a velocity relative to the platform of $\vec{V}_{TP} = V_{TP}\hat{i}$. If we then write out the equations for the position and velocity of the ball as observed from the platform, we find

$$\vec{r}_{BP} = \vec{r}_{BT} + \vec{V}_{TP}t$$

and

$$\vec{r}_{BP} = \vec{r}_{BT} + \vec{V}_{TP}$$

Let us first examine the velocity equation in component form. Since there is no x-component to the velocity in the reference frame of the train,

$$v_{BP,x} = V_{TP}$$

and

$$v_{BT,y} = v_{BT,0y} - gt$$

Both observers report exactly the same velocity in the y-direction. We obtain a similar result for the position equations. Both observers report the same y-position equation, but the platform observer reports that the ball moves at a constant speed in the x-direction

$$y_{BP} = y_{0,BT} + v_{0y,BT}t - \frac{1}{2}gt^2$$

$$x_{BP} = x_{0,BT} + V_{TP}t$$

Thus, the observer on the platform sees the motion as projectile motion. This perspective clearly illustrates the independence of the x- and y- directions in projectile motion.

KEY CONCEPTS AND RELATIONSHIPS

Position, Velocity, and Acceleration

To expand our discussion of kinematics beyond one dimension, we require the use of vectors to describe the position, velocity, and acceleration of a particle.

In a Cartesian coordinate system, the position of a particle can be written as

$$\vec{r} = x\hat{i} + y\hat{j} + z\hat{k} \qquad (4\text{-}1)$$

If the particle moves from position $\vec{r}_1$ to $\vec{r}_2$ in some time interval Δt, the displacement of the particle is given by

$$\Delta\vec{r} = \vec{r}_2 - \vec{r}_1 \qquad (4\text{-}2)$$

During that time interval, the average velocity is defined as

$$\vec{v}_{avg} = \frac{\Delta\vec{r}}{\Delta t} \qquad (4\text{-}4)$$

If we allow the time interval $\Delta t \to 0$, we get the instantaneous velocity

$$\vec{v} = \lim_{\Delta t \to 0} \frac{\Delta\vec{r}}{\Delta t} = \frac{d\vec{r}}{dt} \qquad (4\text{-}5)$$

We define average acceleration as $\vec{a}_{avg} = \dfrac{\vec{v}_2 - \vec{v}_1}{t_2 - t_1} = \dfrac{\Delta\vec{v}}{\Delta t}$, and in the limit of $\Delta t \to 0$ we get the instantaneous acceleration

$$\vec{a} = \lim_{\Delta t \to 0} \frac{\Delta\vec{v}}{\Delta t} = \frac{d\vec{v}}{dt} \qquad (4\text{-}8)$$

The relationships between these variables are exactly the same as we saw in Chapter 3. Also as in Chapter 3, if we know the acceleration that a particle experiences as a function of time, and an initial position and velocity, it is possible to completely determine the position and velocity of that particle as functions of time. The general case of a non-constant acceleration involves mathematics that are beyond the scope of this book.

Projectile Motion

For the special case of constant acceleration, it is possible to choose a coordinate system such that one of the axes is parallel to the acceleration. The other axis can then be chosen so that all of the motion occurs in the plane defined by these two axes—it is two-dimensional motion.

One special case of this motion is the motion observed for objects launched with moderate velocity near the surface of Earth, such as a basketball. In this case, we choose the y-axis to be vertical, so the acceleration is given by $\vec{a} = -g\hat{j}$, where $g = 9.81$ m·s^{-2}. We then find that the position and velocity of such an object, assuming it has an initial position (x_0, y_0) and an initial velocity (v_{0x}, v_{0y}), are given by

$$x(t) = x_0 + v_{0x}t \qquad (4\text{-}20)$$

$$y(t) = y_0 + v_{0y}t - \frac{1}{2}gt^2 \qquad (4\text{-}21)$$

$$v_x = v_{0x} \qquad (4\text{-}22)$$

$$v_y(t) = v_{0y} - gt \qquad (4\text{-}23)$$

We can solve Equation 4-23 for time and substitute it into Equation 4-21 to obtain the time-independent kinematics equation

$$v_y^2 - v_{0y}^2 = -2g(y - y_0) \qquad (4\text{-}26)$$

Developing a consistent strategy is useful when solving projectile motion problems, and a diagram of the motion and a table of the various quantities can be useful.

Circular Motion

An object that moves in a circle at a constant speed is said to be undergoing uniform circular motion. Such an object experiences a radial acceleration that is always directed toward the centre of the circle and has a magnitude, a_r, of

$$a_r = \frac{v^2}{r} \qquad (4\text{-}32)$$

If the speed of the object is not constant as it moves around the circle, the object must have, in addition to the instantaneous radial acceleration, a tangential acceleration, $\vec{a}_t$. The total acceleration is given by the vector sum of the radial and tangential accelerations:

$$\vec{a} = \vec{a}_r + \vec{a}_t \qquad (4\text{-}33)$$

Since these two are always perpendicular to one another, the magnitude of the acceleration is given by $a = \sqrt{a_t^2 + a_r^2}$.

Relative Motion

If we have two observers in reference frames that move with some constant velocity with respect to each other, it is possible to develop equations that allow us to reconcile the observations they make about the motions of objects. Suppose we have observer A and observer B, both making measurements of the motion of a particle P in their respective reference frames. The respective axes in the two frames are parallel. Further suppose that reference frame B moves with some constant velocity, $\vec{V}_{BA}$, with respect to reference frame A. The convention we adopt with the subscripts is that the first subscript denotes what is being observed, and the second denotes the observer.

We then find that the positions of particle P in the two reference frames are related via $\vec{r}_{PA} = \vec{r}_{PB} + \vec{V}_{BA}t$ and the velocities via

$$\vec{v}_{PA} = \vec{v}_{PB} + \vec{V}_{BA} \qquad (4\text{-}38)$$

If the reference frames are inertial (Newton's laws are obeyed), we find that the accelerations observed in the two frames are the same.

APPLICATIONS

projectile plotting, astronaut training, navigation

KEY TERMS

average acceleration, average velocity, displacement, distance, instantaneous acceleration, instantaneous velocity, kinematics, position, projectile motion, radial acceleration, reference frame, relative motion, tangential acceleration, trajectory, uniform circular motion

QUESTIONS

1. A particle moves with constant acceleration in the negative y-direction. Which graph in Figure 4-29 represents
   (a) x-position versus time
   (b) y-position versus time
   (c) x-velocity versus time
   (d) y-velocity versus time

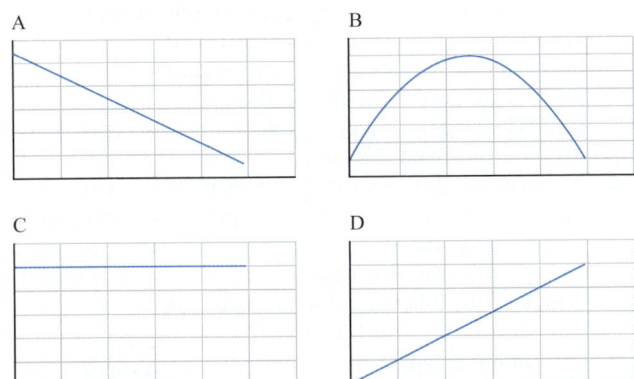

Figure 4-29 Question 1.

2. Figure 4-30 plots y-position versus time for two projectiles that experience constant acceleration in the negative y-direction. The horizontal and vertical scales are the same in each plot. The two projectiles are launched with the same initial velocity. Which projectile experiences the more negative acceleration?

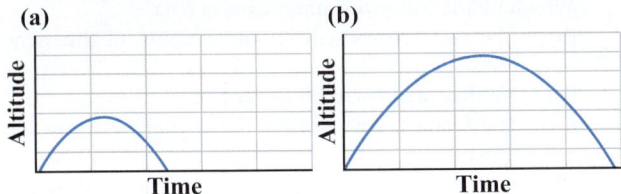

Figure 4-30 Question 2.

3. You release a marble at the top edge of a bowl. The marble rolls down the inside of the bowl and continues up the opposite side. What is the direction of the marble's acceleration when the marble comes to a momentary stop at its highest point on the side of the bowl?
   (a) down
   (b) normal to the bowl
   (c) tangent to the bowl
   (d) The acceleration is zero, so it has no direction.

4. A rocket is launched horizontally off a cliff. In addition to the negative y-acceleration due to gravity, the rocket engine provides a constant horizontal acceleration in the positive direction with a magnitude greater than the magnitude of the acceleration due to gravity. Which path in Figure 4-31 best corresponds to the trajectory of the rocket?

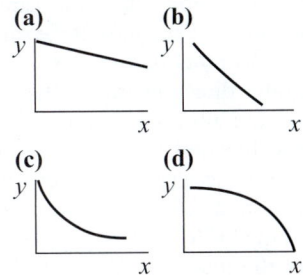

Figure 4-31 Question 4.

5. Figure 4-32 plots y versus t for two particles experiencing constant acceleration in the y-direction. Which plot corresponds to a positive y-acceleration?

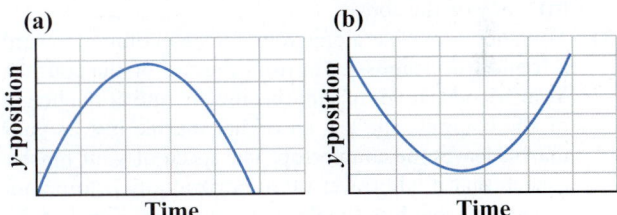

Figure 4-32 Question 5.

6. You intend to swim across a river that flows due south. Your speed with respect to the water is twice the speed of the river current. In which direction relative to the current should you swim to cross in the shortest time?
   (a) at 30° to the river
   (b) at 60° to the river
   (c) at 45° to the river
   (d) perpendicular to the river

7. You have a battery-powered model boat that has a fixed rudder so that it always moves straight ahead. You launch the boat in a pond, aiming straight at a point 60 m away on the opposite side of the pond. The boat reaches the shore 12 m north of the target point because of a current in the pond. On the next run, you aim the boat 12 m south of your target point. The boat will
   (a) arrive at the target
   (b) arrive north of the target
   (c) arrive south of the target
   (d) cannot answer without more information

8. Two objects, A and B, travelling along the same line have speeds v_A and v_B, respectively. The speed of object A with respect to object B is
   (a) $v_B - v_A$
   (b) $v_A - v_B$
   (c) $v_A + v_B$
   (d) any of the above

9. Two objects move along the same circular trajectory with the same speed. They start at the same time, one on the x-axis and the other on the y-axis. When the objects move in the same direction around the circle,
   (a) their relative position vector will always point in the same direction
   (b) their relative acceleration vector will not change
   (c) the magnitude of their relative acceleration will always equal v^2/r
   (d) none of the above

10. An object falls straight down while another object moves in a vertical circular trajectory at a speed that is not necessarily constant. The acceleration vector of the free-falling object can be equal to
    (a) the radial acceleration vector of the object in the circle
    (b) the linear acceleration vector of the object in the circle
    (c) the total acceleration vector of the object in the circle
    (d) any of the above

11. You slide a marble along your kitchen counter toward a friend who stands 1.0 m from the end of the counter. Your friend is holding a marble in her hand at the height of the counter. She lets go of her marble just as your marble leaves the countertop. The speed of your marble is such that it will travel 1.0 m horizontally before your friend's marble hits the floor, as shown in Figure 4-33. Which of the following statements is true?
    (a) The two marbles will never collide.
    (b) Your marble will pass over your friend's marble.
    (c) Your marble will pass below your friend's marble.
    (d) The marbles will collide before they hit the floor.

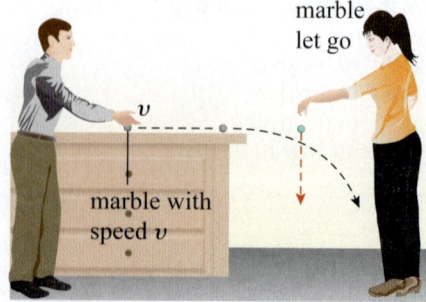

Figure 4-33 Question 11.

12. A projectile is fired at an angle of 30.0° with an initial speed of 3.40 m/s. It lands at the same elevation from which it was fired. While in the air,
    (a) the projectile's average acceleration is zero
    (b) the projectile's average acceleration is 9.81 m/s^2
    (c) the projectile's average velocity in the y-direction is zero
    (d) both (b) and (c) are true

13. An object completes 2.25 revolutions around a circle of radius $4/\pi$ m. Which of the following statements is true?
    (a) The distance covered by the object is 4 m.
    (b) The displacement of the object is $d = \dfrac{2\sqrt{2}}{\pi}$.
    (c) Both (a) and (b) are correct.
    (d) None of the above are true.

14. An object travelling at a constant speed takes 3 s to complete three full cycles around a circle of radius 1 m. Which of the following statements is true?
    (a) The object's average velocity is zero.
    (b) The object's average speed is 2π m/s.
    (c) The object's average acceleration is $4\pi^2$ m·s^{-2}.
    (d) Both (a) and (b) are true.

15. A sharpshooter has a rifle that fires bullets at a speed of 1100 m/s. A trainer 100 m away throws an apple 30 m straight up in the air. The sharpshooter tracks the apple and fires directly at it when it is at its maximum height. Which of the following statements is true?
    (a) The bullet will hit the apple at the apple's maximum height.
    (b) The bullet will pass under the apple on its way down.
    (c) The bullet will pass over the apple on its way down.
    (d) The bullet will hit the apple on its way down.

16. Two archers aim for the same coconut on a tree 14.0 m above the ground. One archer stands 50.0 m away and shoots an arrow at 60.0 m/s. The other archer stands 70.0 m away and shoots an arrow at 87.0 m/s. They both aim directly at the coconut and release their arrows at the moment the coconut begins to fall from the tree. Which of the following statements is true?
    (a) The two arrows will hit the coconut at the same time.
    (b) Neither arrow will hit the coconut.
    (c) The faster arrow will hit the coconut but not the slower one.
    (d) Both arrows will hit the coconut but at different times.

17. Two cannons, side by side on a cliff, fire projectiles at the same speed, one horizontal and one at an angle of 30° above the horizontal. Which of the following statements is true?
    (a) The vertical distance between the two projectiles is $(v \sin 30°)t$.
    (b) The vertical distance between the two projectiles is $(v \sin 30°)t - 4.9t^2$.
    (c) The horizontal distance between the two projectiles is constant.
    (d) Both (a) and (c) are correct.

18. NASA's 20g centrifuge is made to spin so the radial acceleration is 17g. Which of the following statements is true?
    (a) The acceleration is constant.
    (b) The magnitude of the acceleration is constant.
    (c) The velocity is constant.
    (d) None of the above statements are true.

19. If you throw a tennis ball straight up (relative to yourself) while running at a constant speed, the ball will land
    (a) in your hand
    (b) slightly ahead of you
    (c) slightly behind you
    (d) I cannot tell.

20. While a projectile is in the air, can its velocity and acceleration vectors ever be parallel or perpendicular? Explain.

21. Two objects are launched simultaneously. One is simply dropped while the other is launched horizontally. Ignoring air resistance, which object will hit the ground first? Explain.

22. A projectile is launched at speed v, making an angle θ with the horizontal. What is the projectile's acceleration
    (a) right after the launch?
    (b) at the top of its trajectory?

23. A bead sits on top of an overturned hemispherical bowl and begins to slide down, as shown in Figure 4-34. What are the directions of the tangential acceleration and radial acceleration of the bead as it slides down the bowl?

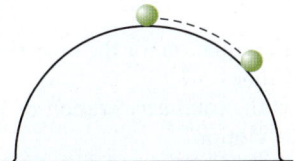

Figure 4-34 Question 23.

PROBLEMS BY SECTION

For problems, star ratings will be used (★, ★★, or ★★★), with more stars meaning more challenging problems. The following codes will indicate if d/dx differentiation, $\int$ integration, ☐ numerical approximation, or ⌐ graphical analysis will be required to solve the problem.

Section 4-1 Position, Velocity, and Acceleration

24. ★ A firefighter hears the station alarm, runs 3.0 m straight to the fire station pole in 1.7 s, and descends 4.0 m down the pole in 2.0 s.
    (a) Find the magnitude of the firefighter's overall average velocity.
    (b) Find the firefighter's overall average speed.
    (c) Find the firefighter's average speed during the run.
    (d) Find the firefighter's average speed during the descent down the pole.

(e) Can the quantities from (c) and (d) be considered components of the firefighter's overall average velocity? Explain why or why not.

25. ★★ A cheetah chases a gazelle at a speed of 112 km/h for 3.00 s, then turns 42.0° from its original direction, runs at 78.0 m/s for 2.00 s, and finally tackles the gazelle, coming to a stop in 11.0 m.
    (a) Find the cheetah's average speed.
    (b) Find the cheetah's average velocity.
    (c) Find the cheetah's average acceleration.
    (d) How long does it take the cheetah to stop in the last 11.0 m? Assume constant acceleration.

26. ★★ d/dx A particle has coordinates given by $x = (11.0 \text{ m}) \cos[(4.00 \text{ rad} \cdot \text{s}^{-1})t]$, $y = (11.0 \text{ m}) \sin[(4.00 \text{ rad} \cdot \text{s}^{-1})t]$, and $z = (0.700 \text{ m} \cdot \text{s}^{-1})t$.
    (a) Describe the trajectory of the particle.
    (b) Find the speed of the particle at $t = 4.00$ s.
    (c) Express the particle's position, velocity, and acceleration as a function of time, using Cartesian notation.
    (d) At what time will the acceleration vector first make an angle of 30° above the x-axis? Is that time affected by the z-motion? Explain.

27. ★★ d/dx, ⌐ A particle has coordinates given by $x = (8.00 \text{ m} \cdot \text{s}^{-1})t$ and $y = \left(\frac{1}{16.0} \text{ m} \cdot \text{s}\right)\frac{1}{t}$ for $t > 0$.

    Describe the trajectory of the particle, and find the particle's speed and acceleration at $t = 3.00$ s.

28. ★★ Starting from rest, a race car moves straight up an incline with a 37.0° angle at a constant acceleration for 5.00 s such that its speed reaches 23.0 m/s. The road becomes horizontal, and the car slows down with constant acceleration as it follows a circular curve that changes its direction by 90°. The radius of the curve is 121 m, and the car is moving at a speed of 7.00 m/s at the end of the curve.
    (a) Find the total displacement of the race car.
    (b) Find the average velocity of the race car from the start until the end of the circular curve.
    (c) Find the average acceleration of the race car during this trip.

Section 4-2 Projectile Motion

29. ★★ A projectile is fired from and lands at ground level 4.30 m away 2.50 s later. Find the projectile's launch speed, minimum speed, angle of launch, average acceleration while in the air, and average velocity.

30. ★ A child flings a marble so it leaves the horizontal surface of a table that is 70. cm high. The marble lands with a speed of 7.0 m/s. How fast did the child fling the marble?

31. ★ A particle is launched at an initial speed v making an angle θ above the horizontal. Prove that the trajectory of the particle is a parabola.

32. ★★ A stunt car driver drives off a 7.00 m high cliff into a lake. The car needs to clear a ledge of length d that is 5.00 m below the edge of the cliff, as shown in Figure 4-35. The car just misses the ledge on the way down. With what speed does the car hit the water if $d = 8.00$ m?

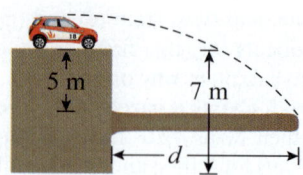

Figure 4-35 Problem 32.

Section 4-3 Circular Motion

33. ★★ The distance between São Tomé off the coast of Africa and the city of Quito, Ecuador, is approximately 9500 km. Both cities are almost on the equator.
    (a) How long does it take Earth to rotate to cover the distance between the two cities?
    (b) You board a plane in São Tomé and fly west to Quito at a speed of 970 km/h, landing 12 h later. What is your radial acceleration when the plane's altitude is 1.0×10^4 m?

34. ★ Pilots can sustain accelerations as high as $13g$ during an inside loop manoeuvre. When a plane is moving at a speed of 830 km/h, what is the minimum radius for an inside loop turn the plane can fly without its acceleration exceeding $13g$?

35. ★ Calculate the radial acceleration of a point on the Sun's equator.

36. ★ At the Istanbul Grand Prix, drivers sustain lateral (sideways) accelerations about five times the acceleration due to gravity while driving around curves at speeds of 280 km/h. Assuming a circular curve, find the radius of that section of the track.

37. ★ An ultracentrifuge produces accelerations that are 2 million times the acceleration due to gravity. Find the speed of rotation for the ultracentrifuge, which has a radius of 4.50 cm.

38. ★ The P2 centrifuge, used to enrich uranium, has a diameter of approximately 15 cm, a length of 1.0 m, and a point on the circumference has a speed of approximately 5.0×10^2 m/s.
    (a) What is the acceleration of a point on the circumference of a P2 centrifuge?
    (b) What is the acceleration of a point midway between the axis and the circumference of a P2 centrifuge?
    (c) How many turns does the P2 centrifuge complete in 1.0 s?

39. ★★ An object moves counter-clockwise around a horizontal circle of radius $r = 4.00$ m. The speed of the object is 12.0 m/s when the object is 30.0° along the circle, and 8.00 m/s when the object is 180.° along the circle. The tangential acceleration is uniform. What is the total acceleration of the object when its speed is 8.00 m/s? Hint: Recall that the distance an object travels around a circle, Δs, is related to the angular displacement, $\Delta\theta$, by $\Delta s = r\Delta\theta$.

40. ★★ A particle moves along a circular path of radius r at a constant speed v. The particle is at the angular position ϕ above the positive x-axis, as shown in Figure 4-36. Without using calculus, write expressions for
    (a) the projection of the particle's position onto the x-axis as a function of time

    (b) the projection of the velocity onto the x-axis as a function of time
    (c) the projection of the acceleration onto the x-axis as a function of time

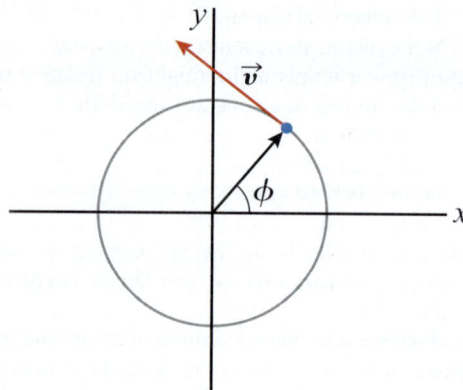

Figure 4-36 Problem 40.

41. ★★ A steel ball that is accelerated on a circular track of radius 72 cm reaches a speed of 4.0 m/s over three circuits, starting from rest. How long after the ball starts to move does its radial acceleration equal its tangential acceleration? How many turns does the ball take for that to happen? Assume constant tangential acceleration.

42. ★★★ ∫ An object moves on a circular track of radius 29.0 m such that the object's radial acceleration has a magnitude of $a_r = (0.400 \text{ m} \cdot \text{s}^{-4})t^2$. The object starts from rest.
    (a) Write the equation for the object's speed as a function of time.
    (b) What is the total acceleration of the object 6.00 s into the motion?
    (c) What is the distance covered as a function of time?

Section 4-4 Relative Motion in Two and Three Dimensions

43. ★ A parade car moving at 10. km/h shoots an acrobat out of a cannon that is mounted at 90.° to Earth's surface. The cannon fires the acrobat with a speed of 14 m/s. Where will the acrobat land relative to the parade car? Find the magnitude of the displacement of the acrobat relative to the ground. Ignore air resistance.

44. ★★ A child has a cannon mounted on a battery-powered car. While the car is moving at a speed of 2.00 m/s, the cannon fires a marble with a speed of 7.00 m/s with respect to the car, at an angle of 27.0° above the horizontal. The car's speed does not change noticeably.
    (a) What angle does the marble make with the horizontal when it lands on the ground?
    (b) How far will the marble land from the position from which it was launched?
    (c) How far is the marble from the car when it lands?

45. ★★ A swimmer swims in a river against the current for 5.0 min, and then swims with the current for another 5.0 min, and ends up 500. m away from where he started. Find the speed of the current in the river if the swimmer's speed relative to the water is 2.6 m/s.

46. ★★ A pilot wishes to fly to a destination that is 400. km directly north. Her plane has an airspeed of 300. km/h. A wind is blowing from the northwest at a speed of 75.0 km/h. In what direction must she fly to arrive at her destination, and how long will it take?

47. ★★ A director wants to produce a scene in which an archer shoots a watermelon that is dropped from a 30. m high bridge. Thinking that they need to compensate for gravity, the director asks the archer to aim 5.0 m above the watermelon, which is released at the same time that the archer releases his arrow. The initial speed of the arrow is 160 m/s, and the distance to the watermelon is 90. m. How close will the arrow get to the watermelon?

48. ★★ A sailor is piloting her boat at a speed of 5.0 knots on a heading of 45° toward her destination, which is 15 nautical miles away. After 2.0 h of sailing, she passes close by a navigational buoy, which is 0.60 nautical miles directly west of where she expected to be. What current is she experiencing, and what course should she steer to arrive directly at her destination?

49. ★★ A parade car has a cannon that shoots teddy bears into the crowd. As the car moves forward at 2.7 m/s, the cannon fires horizontally at an angle of 37° from the direction of motion of the car. The speed of the teddy bears with respect to the cannon is 6.0 m/s, the cannon is 3.0 m above the ground, and the recoil of the cannon is negligible. Find the landing speed of the teddy bear with respect to the ground.

50. ★★★ During a clay pigeon shooting match, a clay pigeon is fired with an initial speed of 82 km/h at an angle of 39° above the horizontal. Then, 1.7 s later, the shooter fires a shot that has a speed of 404 m/s. When the shooter is standing straight behind the clay pigeon launcher, at what angle should he aim?

51. ★★★ You notice an eagle soaring at an altitude of 25 m, eyeing a rabbit. You want to scare the eagle away, so 1.0 s after the eagle flies directly overhead at 34 km/h, you throw a stone at an angle of 38° above the horizontal. You want the stone to pass 50. cm in front of the eagle. With what speed should you throw the stone? Is this doable at all? Assume your hand to be 2.0 m above the ground when you throw the stone.

COMPREHENSIVE PROBLEMS

52. ★ A spinning neutron star with a radius of 16 km completes 712 rev/s.
    (a) Find the average speed of a point on the star's equator over one third of a revolution.
    (b) Find the average acceleration of a point on the star's circumference over three quarters of a revolution.
    (c) Find the distance covered by a point on the star's equator in 1.0 s.
    (d) Find the displacement of the point in part (c).

53. ★★ A monkey is 14.0 m above the ground in a tree and drops a papaya. At the same time, two other monkeys, one 11.0 m from the base of the tree and the other 16.0 m from the base of the tree, throw small stones at the papaya such that all three objects collide 0.900 s before the papaya hits the ground. Find the speed of each stone as it hits the papaya. Assume both stones were thrown from ground level.

54. ★★★ Starting from rest, a particle moves under constant acceleration in a horizontal line such that its speed 3.00 s later is 1.90×10^3 m/s. The particle is deflected by 37.0° in the horizontal plane without changing its speed. In this new direction, the particle accelerates such that it reduces its speed by half in 1.00 s. The particle is then deflected straight up, and immediately accelerates uniformly such that its velocity after 3.00 s is 9.00×10^3 m/s straight up. For the whole 7.00 s interval, find the particle's total displacement, average velocity, average speed, and average acceleration. Express these quantities using magnitude and angles measured with respect to the positive coordinate axes.

55. ★★ $\frac{d}{dx}$, 〰 A rocket is launched with an initial speed of v, making an angle θ above the horizontal. Radar tracking shows that the altitude of the rocket is given by $(60.0 \text{ m} \cdot \text{s}^{-2})t^2 - (7.00 \text{ m} \cdot \text{s}^{-3})t^3 + (4.00 \text{ m} \cdot \text{s}^{-1})t + 120.$ m, and the horizontal distance from the point of observation is given by $(1.4 \times 10^3 \text{ m}) + (3.00 \text{ m} \cdot \text{s}^{-1})t$.
    (a) Determine the maximum height reached by the rocket.
    (b) Find the speed and horizontal displacement of the rocket just before it hits the ground 120. m below the point of launch. (Hint: Solve the cubic function by graphing it.)
    (c) At what altitude and horizontal distance does the maximum acceleration of the rocket occur?

56. ★★ A sounding rocket has an engine that allows it to maintain a constant horizontal speed of 190 m/s while providing it with a constant vertical acceleration of 32 m/s^2 until it reaches a height of 60. km. The engine then shuts off, and the rocket continues under the influence of just the gravitational force. Assume that air resistance and the change in gravitational acceleration with altitude are negligible.
    (a) What is the maximum height reached by the rocket, and how long does it take to reach that height?
    (b) What angle does its velocity make with the horizontal just before the landing?

57. ★★ A projectile is fired at speed 14.0 m/s such that the projection of its velocity onto the xy-plane makes an angle of 40.0° with the positive x-axis. The initial velocity vector makes an angle of 29.0° with the positive z-axis. Express the x-, y-, and z-coordinates of the projectile as a function of time, taking the launch point as the origin (0, 0, 0).

58. ★★ A fireworks rocket is fired horizontally from a height of 17 m above the ground. The rocket has a horizontal acceleration that is twice as large as the acceleration due to gravity.
  (a) What trajectory will the rocket follow?
  (b) What will its speed be just before it hits the ground?
  (c) Find the horizontal distance between the launch and landing points.

59. ★ A child fires a tomato from an improvised catapult from a tree house 4.00 m above the ground. The tomato leaves the catapult at a speed of 35.0 m/s, making an angle of 37.0° with the horizontal.
  (a) Find the maximum height reached by the tomato.
  (b) How far from the tree house will the tomato land?
  (c) How fast is the tomato moving just before it lands?

60. ★★ A plane is approaching a remote community in northern Manitoba to drop a crate of supplies when clear-air turbulence buffets the plane, pushing it suddenly upward. The crate tears loose at an altitude of 160 m with a velocity of 290 km/h directed at an angle of 23° above the horizontal. What is the speed of the crate just before it lands? (Neglect air resistance.)

61. ★★ $\frac{d}{dx}$ Express y as a function of x for a projectile that is launched with speed v_0 at an angle θ and lands at the same level from which it was launched. Derive expressions for the angle for which the maximum distance is covered, the distance at which the maximum height occurs for that angle, and the maximum height.

62. ★★ $\frac{d}{dx}$ A particle's position is given as

$$[(2.0\,\text{m}) + (0.5\,\text{m})\sin[(0.4\,\text{rad}\cdot\text{s}^{-1})t]]\hat{i}$$
$$+ [(3.0\,\text{m}) + (0.5\,\text{m})\sin[(0.4\,\text{rad}\cdot\text{s}^{-1})t]]\hat{j}$$
$$+ (2.0\,\text{m}\cdot\text{s}^{-2})t^2\hat{k}$$

  (a) What is the position of the particle at $t = 0.00$?
  (b) Find the speed of the particle at $t = 0.20$ s.
  (c) Find the average velocity of the particle between $t = 0.20$ s and $t = 0.40$ s.
  (d) Find the average acceleration of the particle between $t = 0.20$ s and $t = 0.40$ s.
  (e) Find the displacement of the particle between $t = 0.00$ s and $t = 1.1$ s.
  (f) What is the trajectory of the particle?

63. ★★ $\frac{d}{dx}$ Two disks of radius 32 cm are placed side by side so they are touching, and the two closest points on the disks are painted red. The disks are then made to spin about fixed axes in opposite directions at 10. rpm.
  (a) Write an expression that describes the relative position of the two points.
  (b) Write expressions for the relative velocity and relative speed of the two points.
  (c) Write an expression for the relative acceleration of one red point with respect to the other.

64. ★★ $\frac{d}{dx}$ A ring lies in a horizontal plane and spins about its symmetry axis at a rate of 3 rev/s. Find the position, acceleration, and velocity of a point A on the ring with respect to a point B that is in the same plane and is a distance $3r$ from A at time t, given that the distance between A and B is $2r$ at $t = 0$.

65. ★★★ $\frac{d}{dx}$ A Boeing 747 is flying at 790 km/h at an altitude of 20. km in a direction [11° S of E]. A smaller jet at the same altitude is 12 km away [17° N of E] from the Boeing. The velocity of the smaller jet is 430 km/h [23° W of S]. Find the distance of closest approach between the planes and the time they take to reach that distance.

66. ★★ $\frac{d}{dx}$ An inexperienced drone pilot is learning how to fly and his drone makes the following rather erratic altitude profile:

$$\text{altitude: } h(t) = (2.99\,\text{m}) - (5.00 \times 10^{-5}\,\text{m}\cdot\text{s}^{-3})t^3$$
$$+ (4.00 \times 10^{-3}\,\text{m}\cdot\text{s}^{-2})t^2 + (0.100\,\text{m}\cdot\text{s}^{-1})t$$

Find the vertical position, vertical speed, and vertical acceleration of the drone 1.00 min after takeoff.

67. ★★ A test pilot is placed in a high-g training pod that follows a horizontal path of radius 20.0 m. The pod accelerates from rest to achieve an acceleration six times the acceleration due to gravity in the radial direction over five turns. The pod has constant tangential acceleration during these five turns. The pod then continues at a constant speed for 2.00 s, and then undergoes constant deceleration, coming to a stop in the next 6.00 s.
  (a) How long does the pod take to finish the first five turns?
  (b) What is the magnitude of the acceleration of the pod 3.00 s after it begins to decelerate?
  (c) What is the total number of turns the pod makes?
  (d) Find the magnitude of the average acceleration between $t = 3.00$ s and $t = 6.00$ s.
  (e) Find the magnitude of the average velocity between $t = 4.00$ s and $t = 6.00$ s.
  (f) Find the magnitude of the displacement of the pod between $t = 2.00$ s and $t = 5.00$ s.

68. ★★ You are in the passenger seat of a car driving on a country road at 43.0 km/h. You throw an acorn out the window at an angle of 26.0° above the horizontal with a speed of 4.00 m/s relative to the car, and from a height of 1.00 m from the ground. The projection of the acorn's velocity onto the horizontal makes an angle of 34.0° with the velocity of the car. Ignore wind resistance.
  (a) How far ahead of the car is the acorn 1.00 s later?
  (b) What is the speed of the acorn in flight with respect to the ground?
  (c) What angle with respect to the horizontal does the acorn's velocity make just before the acorn lands? What angle does the projection of the acorn's velocity make with the road?

69. ★★ Two monkeys standing 20. m apart are both trying to use stones to knock down a coconut hanging 7.0 m up in a tree that is directly between them. One monkey is 7.0 m from the tree, and the other is 13 m from the tree. They aim straight for the coconut, not accounting for the acceleration due to gravity. The close monkey throws a stone at speed v, and the far monkey throws a stone at $2v$. The stone thrown by the close monkey strikes the ground 3.0 s after it leaves its hand. By what vertical distance do the two stones miss each other?

70. ★★ A particle's position along the x-axis is given by $x = (20.0 \text{ m} \cdot \text{s}^{-1})\cos[(0.400 \text{ rad} \cdot \text{s}^{-1})t]$, and its position along the y-axis is given by $y = (20.0 \text{ m} \cdot \text{s}^{-1})\sin[(0.400 \text{ rad} \cdot \text{s}^{-1})t]$.
  (a) Describe the trajectory of the particle.
  (b) Find the time it takes the particle to finish a complete cycle. (This time is the period of motion.)
  (c) Find the tangential speed of the particle.
  (d) Find the magnitude of the acceleration of the particle at a given time t.

71. ★★ $\frac{d}{dx}$ Given that the particle in problem 70 is at $x = 20.0$ m at $t = 0.00$ s, find
  (a) the velocity of the particle at $t = 12.0$ s
  (b) the average velocity and the average acceleration of the particle between $t = 12.0$ s and $t = 20.0$ s

72. ★★★ ⌁ A pilot points her plane in the direction of an airfield that is 300. km east and 210 km north of her current location. Since the winds are light and varying, the pilot does not attempt to correct her bearing for wind speed. Near the end of her flight, she finds that her course will take her 34 km north of the airfield. The plane's airspeed is 270 km/h. The pilot estimates that the flight would have taken 98 min if she had corrected for wind speed. Find the bearing that would have taken her straight to the airfield. Assume the winds, on average, travel from south to north.

73. ★★★ ⌁ In a marine-rescue exercise, a small boat moves at velocity 13 km/h [37° N of E]. The rescue boat has a speed of 30. km/h and is positioned 60. m [42° S of E] from the small boat. In what direction must the rescue boat head to intercept the small boat?

74. ★★★ ⌁ Figure 4-37 is a velocity versus time graph for a meteoroid fragment entering the atmosphere. Use the graph to construct approximate distance versus time and acceleration versus time graphs for the interval $0.3 \text{ s} \leq t \leq 0.6 \text{ s}$.

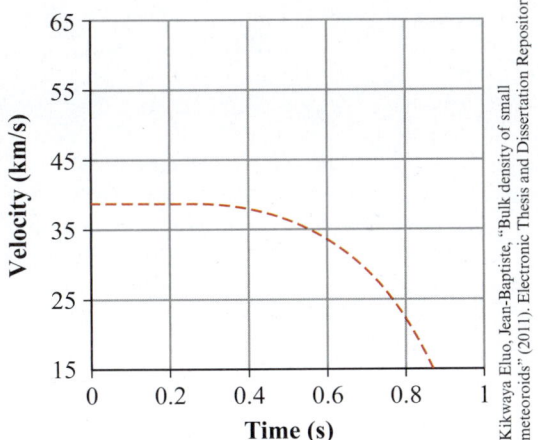

Figure 4-37 Problem 74.

Kikwaya Eluo, Jean-Baptiste, "Bulk density of small meteoroids" (2011). Electronic Thesis and Dissertation Repository.

75. ★★ ⌁ Figure 4-38 shows an altitude versus time plot for the Ariane 5's ascent to space.
  (a) Estimate the average radial velocity (i.e., vertical component of velocity) of the rocket between launch time and the separation of the second rocket stage (point H2 on the graph).
  (b) Estimate the radial speed at $t = 200$ s, $t = 400$ s, and $t = 1500$ s.

  (c) Describe the radial acceleration of the Ariane 5 between 0 s and 200 s.
  (d) Why does the slope of the graph in Figure 4-38 not correspond to the total speed of the rocket?

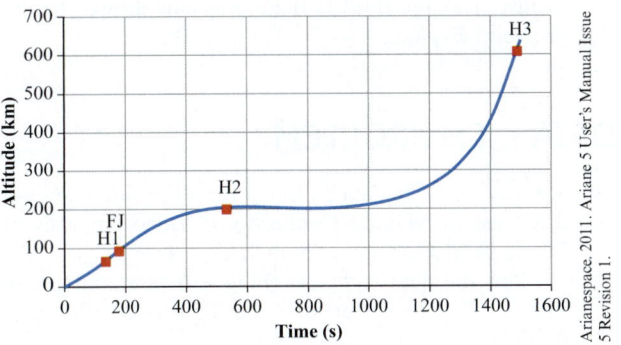

Arianespace. 2011. Ariane 5 User's Manual Issue 5 Revision 1.

Figure 4-38 Problem 75.

76. ★★ ⌁ Figure 4-39 shows the speed for a Ferrari Spider.
  (a) What is the maximum acceleration on the curve?
  (b) Calculate the acceleration at $t = 7$ s.
  (c) What is the average acceleration between $t = 0$ s and $t = 14$ s?
  (d) How far does the car go before reaching maximum speed?
  (e) How far does the car go between $t = 14$ s and $t = 16$ s?
  (f) What is the total distance covered by the car during the whole interval shown?

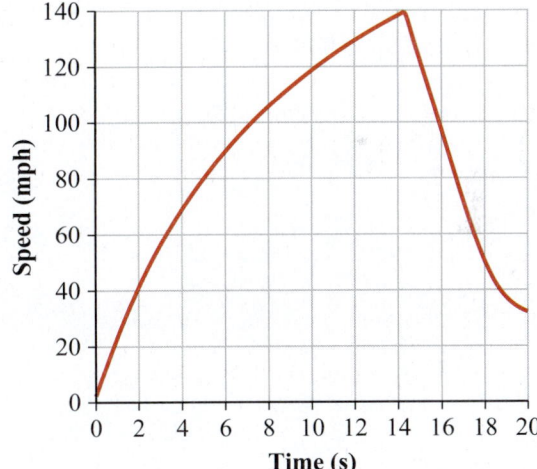

© Luca Petruzzi | Dreamstime.com

Figure 4-39 Problem 76.

CHAPTER 4 | **MOTION IN TWO AND THREE DIMENSIONS**

77. ★ You want a golf ball on its way to the hole to clear a 12.0 m tall tree 40.0 m away. You estimate that the golf ball will leave the ground at an angle of 47.0°. Neglecting air resistance, what initial speed must the golf ball have to just clear the tree? Is there only one answer to this problem? Explain.

DATA-RICH PROBLEM

78. ★★ ▢, ◡ The data collected from the video analysis discussed in Making Connections: Video Analysis of Projectile Motion are given in Table 4-4.

Plot the data using graphing software, and fit the x-data with a linear function and the y-data with a quadratic. Comment on your results.

Table 4-4 x- and y-Position versus Time Data from a Video Analysis of a Ball Toss

Time (s)	x (m)	y (m)
0	0	0
0.033	0.059072	0.092827
0.067	0.118143	0.179324
0.1	0.177215	0.248945
0.133	0.234177	0.314345
0.167	0.291139	0.364978
0.2	0.3481	0.407172
0.233	0.409282	0.436708
0.267	0.466244	0.453585
0.3	0.523206	0.459915
0.333	0.578058	0.455695
0.367	0.63502	0.443037
0.4	0.691982	0.41983
0.433	0.748943	0.383965
0.467	0.803796	0.337552
0.5	0.860758	0.27848
0.533	0.91561	0.21097
0.567	0.972572	0.128692
0.6	1.025314	0.035865
0.633	1.082276	−0.06329
0.667	1.139238	−0.17511
0.7	1.1962	−0.29536
0.733	1.248942	−0.42827

5 | Forces and Motion

Learning Objectives

When you have completed this chapter, you should be able to

LO1 Draw a free body diagram and calculate the net force in one-dimensional situations.

LO2 Calculate the force of gravity, and differentiate the terms *mass* and *weight*.

LO3 State, explain, and apply Newton's three laws of motion.

LO4 Expand the use of free body diagrams to three dimensions, and use vector components to solve Newton's second law problems in two and three dimensions.

LO5 Apply component-free methods to solve dynamics problems.

LO6 Differentiate between static and kinetic friction, and solve friction problems.

LO7 State and apply Hooke's law for ideal springs.

LO8 Differentiate between fundamental and non-fundamental forces, giving examples of each.

LO9 Solve uniform circular motion problems.

LO10 Define inertial and non-inertial reference frames, and explain the nature of fictitious forces.

LO11 Define linear momentum and impulse, and apply Newton's second law in terms of the change of momentum.

Consider the situation shown in Figure 5-1: a strong man slowly pulls a very heavy truck. This chapter is concerned with forces, and there are many forces, direct and implied, in the photo. The rope exerts a force on the truck, while the man exerts a force on the rope. Would he be able to exert as much force if the ground were slippery? The man exerts a force on the ground, and in turn the ground pushes back on the man. There are also invisible forces present, such as the force of gravity due to the mass of Earth acting on the man and on the truck. Why is it much easier to accelerate an object with a small mass? Indeed, what do we mean by mass? All of these issues are considered in this chapter. Isaac Newton developed a deep understanding of why objects move how they do, formulating his ideas into what we now call Newton's three laws of motion. While many aspects of dynamics were developed centuries ago, questions about the nature of mass are still being vigorously researched today using particle accelerators.

Figure 5-1 Can you identify the many forces, seen and implied, in this photo of a man in Europe pulling a very heavy truck?

5-1 Dynamics and Forces

In Chapters 3 and 4, you studied in detail *how* objects move, using accelerations to predict velocities and positions at various times. The study of *how* objects move is called **kinematics**. But physicists want to be able to explain *why* the motion occurs. **Dynamics** is the study of why objects move; it is concerned with forces, the central idea of this chapter.

What do we mean by a force? In everyday language we can think of a **force** as a push or a pull. For example, if you use your hand to push against your friend, then you exert a force. If you pull a wagon with a rope, that pull is a force. Some forces are not directly visible, but are still very real, such as the force of gravitational attraction of Earth on the Moon or a satellite or indeed on your body. Draw up a list of all the different types of forces that you can think of (Figure 5-1 may give you ideas to get started). There are many types of forces—you will encounter some in this chapter, and others throughout the rest of the book. The idea of force is one of the most important concepts in physics.

We can also define a force in the following way. A force, if not cancelled by another force acting in the opposite direction, results in acceleration in the direction of the applied force. An unopposed force will therefore change the velocity of the object. For example, a force applied to an object that was initially not moving will cause a velocity in the direction of the force. A force applied in the same direction as the original velocity of an object will increase that velocity, while a force in the opposite direction will decrease it.

In the preceding paragraph, we talked about the direction of the force, and indeed *forces are vector quantities*. When we describe a force, we must give both the magnitude and the direction (or use another system, such as vector components). The vector relationships you learned in Chapter 2 will be essential in this chapter.

The SI unit of force is the newton (N). An old British unit for force that you may encounter in everyday life is the pound (lb), but in this textbook we work almost exclusively with SI units. We can express the newton in terms of the fundamental SI units that you encountered in Chapter 1:

$$1\,\text{N} = 1\frac{\text{kg}\cdot\text{m}}{\text{s}^2} \qquad (5\text{-}1)$$

You will see in Section 5-3 why the relationship has this form.

In the definition of force, we mentioned the possibility of a force being opposed by another force. What is important is the **net force**, the resultant when all of the forces on an object are added as vectors. Interactive Activity 5-1 will build your expertise in finding net forces for one-dimensional situations.

CHECKPOINT 5-1

Net Force of "Squeeze"

The object in Figure 5-2 is acted upon by the two forces shown, each of magnitude 210 N.

Figure 5-2   What is the net force on the object?

The net force on the object is

(a)   zero

(b)   210 N

(c)   420 N

(d)   none of the above

ANSWER: (a) The two forces add up to zero, since one is to the left and the other to the right. So no matter what your definition of the positive direction is, one force will be positive and one will be negative, and the magnitudes are equal.

In Figure 5-2 we showed the forces pushing inward on the object. In physics we usually draw the forces starting from the centre of the object and simplify the object to a point. We call this a **free body diagram (FBD)**. We draw a simple one-dimensional FBD in Example 5-1.

You are free to select the orientation of the coordinate axes and the definitions of the positive directions for those axes, but you must be consistent throughout the question. It is good practice to state these at the beginning of a problem. It is also important to interpret the meaning of your mathematical answer, as we show in Example 5-1.

EXAMPLE 5-1

Calculating the Net Force

As shown in Figure 5-3, two forces act on a body. One force acts to the right with magnitude 1100 N, while the other force is to the left with a magnitude of 700 N. Find the net force acting on the object.

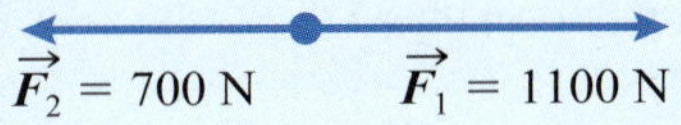

$$\overset{\leftarrow}{\vec{F_2}} = 700 \text{ N} \qquad \overset{\rightarrow}{\vec{F_1}} = 1100 \text{ N}$$

Figure 5-3   Example 5-1.

Solution

Figure 5-3 is already in the form of an FBD, with the vectors drawn relative to a point representing the object. We will choose the x-axis in the conventional horizontal sense, with

positive being to the right. We label the force to the right as $\vec{F_1}$ and the one to the left as $\vec{F_2}$. We can therefore write the sum of the forces in the following way using the unit vector $\hat{i}$ notation from Chapter 2:

$$\vec{F}_{\text{net}} = \vec{F_1} + \vec{F_2} = (1100 \text{ N})\hat{i} - (700 \text{ N})\hat{i} = (400 \text{ N})\hat{i}$$

Our net force is 400 N and is in the positive x-direction, which means toward the right.

Making sense of the result

The forces point in opposite directions, so we expect the net force to be the difference in the magnitudes of the two forces. Since the larger force points to the right, we expect the net force to be in that direction as well.

Example 5-1 dealt with forces in a one-dimensional setting, but most forces you will encounter will be in two or three dimensions. We will extend the FBD ideas to three-dimensional motion in Section 5-4.

While we kept the math simple by using one-dimensional situations in this section, the key ideas you have learned here will apply in the more complex three-dimensional situations to come. These ideas can be summarized with the following points:

- Forces are vectors and must be specified and added as vectors.

- The FBD is a core tool. In the FBD the object is represented by a point, and all forces acting on that object are drawn as vectors starting on that point.

- To find the net force on the object, add all the individual forces as vectors. If using vector components, work separately on the different components, as we saw in Chapter 2.

- You are free to define the positive direction for each axis, but you must be consistent throughout the solution.

The FBD shows only the object and all the forces acting on that object. It is a visual tool that is often a necessary and almost always a recommended step and a valuable aid in understanding and solving problems.

PEER TO PEER

When my friends make mistakes with force problems, it is usually because they forget one of the following two concepts: You find the net force by adding *as vectors* all forces acting on the object. Use only forces acting on *that object*—be careful not to add forces that act on different objects.

5-2 Mass and the Force of Gravity

Among Isaac Newton's greatest contributions was the formulation of the law of universal gravitational attraction. According to this law, every mass in the universe attracts every other mass with a gravitational force. The force of gravitational attraction between two objects is discussed in Chapter 11 in a more universal way, but in this section we will confine our consideration to the gravitational force for objects near the surface of Earth.

The attractive gravitational force, $\vec{F}_g$, that Earth exerts on an object of mass m is given by

KEY EQUATION $$\vec{F}_g = m\vec{g} \qquad (5\text{-}2)$$

Here m is the mass of the object and $\vec{g}$ is the acceleration due to gravity. As we saw in Chapter 4, acceleration is a vector quantity. In physics, both sides of an equation must be consistent: for example, if one side of the equation is a vector quantity, then the other side must also be a vector quantity.

When we write g without a vector sign, it signifies the magnitude of the acceleration due to gravity. Near the surface of Earth, $g \approx 9.81$ m/s². The value of g is simply the magnitude of the acceleration for a mass in free fall near the surface of Earth in the absence of air. The precise value of g varies slightly with position on the surface of Earth. It decreases as you go above the surface of Earth (and has far different values on different planets or on the Moon). You will learn how to calculate the acceleration due to gravity in various locations in Chapter 11.

According to universal gravitation, a mass near the surface of Earth will feel a gravitational attraction due to all of the mass elements making up the entire Earth. Each element of Earth's mass exerts a gravitational force on the mass, the direction being along the line joining the mass near the surface to that element. However, when we combine these for a symmetrical distribution of mass on Earth, the net gravitational force points toward the centre of Earth. The direction of $\vec{g}$ therefore points downward toward the centre of Earth.

We can think of **mass** as the amount of material in an object. As we saw in Chapter 1, mass is one of the fundamental SI units, with the kilogram (kg) being the unit of mass. A more quantitative way to think about mass, making reference to Equation 5-2, is that the amount of mass in an object is proportional to the gravitational force on that object. Mass defined in this manner, since it depends on gravity, is called **gravitational mass**.

MAKING CONNECTIONS

The International Standard for Mass

Currently, SI defines the base unit for mass—the kilogram—as the mass of the International Prototype Kilogram (IPK), a cylinder machined out of a platinum iridium alloy, first produced in 1879 (Figure 5-4). Although precise copies were given to other nations for reference, the original IPK is kept in Sèvre, France. The IPK is currently the only SI unit based on an actual artifact.

Courtesy of Bureau International des Poids et Mesures.

Figure 5-4 The IPK, held at the Bureau International des Poids et Mesures in France.

EXAMPLE 5-2

Gravitational Force on a Distant Planet

On a distant planet (without an atmosphere), it is observed that when a 0.250 kg object is dropped from rest, its speed after 1.0 s is 6.0 m/s, and after 3.0 s it is 18.0 m/s. What must be the gravitational force on the mass?

Solution

We will first find the acceleration due to gravity on this planet.

When the acceleration has a constant value, we know that the magnitude of the velocity as a function of time is given by $v = v_0 + at$. Since we are told the object starts from rest, we know that $v_0 = 0$. We can substitute either $t = 1.0$ s and $v = 6.0$ m/s, or $t = 3.0$ s and $v = 18.0$ m/s, to solve for the acceleration. We will use the value at 1.0 s, and we will define *downward* as the positive direction.

$$6.0 \text{ m/s} = 0 + a(1.0 \text{ s})$$
$$a = 6.0 \text{ m/s}^2$$

So the acceleration on the mass is 6.0 m/s² and its direction is downward. This acceleration is the acceleration due to gravity for this planet, and therefore we can find the magnitude of the gravitational force:

$$F_g = mg = (0.250 \text{ kg})(6.0 \text{ m/s}^2) = 1.5 \text{ N}$$

The direction of the gravitational force is the same as that of the acceleration, vertically downward.

Making sense of the result

On Earth, a dropped mass would have been travelling 9.81 m/s after 1.0 s of fall from rest, so we expect the acceleration due to gravity to have a lower value on this planet. Since we were given a second value that we did not use in the solution, it's a good idea to use it to check our solution:

$$v = v_0 + at$$

$$\text{Left side} = 18.0 \text{ m/s}$$

$$\text{Right side} = 0 + (6.0 \text{ m/s}^2)(3.0 \text{ s})$$

$$= 18.0 \text{ m/s}$$

$$\text{Left side} = \text{Right side}$$

While aspects of mass were considered long ago, a full understanding continues to be a focus of modern physics research. As we will see in Chapter 30, general relativity views gravitational mass interactions in terms of curvature of spacetime, with aspects recently verified with LIGO-observed gravitational wave detections of coalescing black holes. The question of whether fundamental particles called neutrinos have mass was key to the research cited in the 2015 Nobel Prize in Physics (see Making Connections: Neutrino Mass). Experiments at the Large Hadron Collider (LHC) have found evidence for the Higgs boson, recognized with the 2013 Nobel Prize in Physics for work contributing to our understanding of the origin of mass in subatomic particles (see Making Connections: Higgs Boson and Mass).

A **point mass** assumes an object with mass but whose physical dimensions are infinitesimally small. In most applications in this chapter, we can picture the object as though it were a point mass.

We note that the gravitational force exerted by Earth on a mass near the surface of Earth is sometimes referred to as the **weight** of the object. However, in this text we will normally avoid using the term *weight*

for this; rather we prefer the term **force of gravity** or **gravitational force**. We explain the reasons for this in the next paragraph.

There is no agreement within the physics community on how weight should be defined. If we defined weight as equal to the force of gravity, it would be a vector quantity with force units. Others define it as the magnitude of this force, so it has force units, but no direction. Still others define weight in an operational sense as the value that a weigh scale would read. If one adopts this operational definition, the weight would take on a different value if measured high on a mountain, or in an elevator that was accelerating. Some prefer to call the reading in these circumstances the **apparent weight**.

You may have noticed that in everyday commerce there is a mixture of mass and weight units. For example, items at a supermarket are often shown with both kilogram units (i.e., a unit of mass) and pound units (a unit of gravitational force or weight). There is an inherent assumption that, to permit this comparison, the weight measurement is made at the surface of Earth at a standard *g*-value.

MAKING CONNECTIONS

Neutrino Mass

Elementary particles called neutrinos are not electrically charged and therefore do not interact through the electromagnetic forces you will study in Chapter 19. They are produced in nuclear reactions, including those taking place at in the core of the Sun and other stars, and are almost massless. There was a long-standing question of whether neutrinos had mass at all, especially since the Standard Model for particle physics (see Chapter 35) predicted that they did not. Work conducted by the

team led by Takaaki Kajita at Super Kamiokande (Figure 5-5(a)) in Japan, and in Canada by Arthur B. McDonald at SNOLAB in Sudbury (Figure 5-5(b)), confirmed that neutrinos, including those from the Sun, changed "flavour" during what are known as *neutrino oscillations*. The two physicists shared the 2015 Nobel Prize in physics "for the discovery of neutrino oscillations, which shows that neutrinos have mass."

(a)

(b)

Kamioka Observatory, ICRR (Institute for Cosmic Ray Research), The University of Tokyo

Roy Kaltschmidt, Lawrence Berkeley Laboratory

Figure 5-5 Both Super Kamiokande (a) and SNOLAB (b) neutrino observatories are located deep underground. Neutrinos are not affected by the interactions that affect most other particles. Having the detectors deep underground minimizes the chances of other particles reaching the detectors, thus increasing the possibility of neutrino detection.

Higgs Boson and Mass

Researchers in various facilities and research institutes around the world engage in research in the fields of gravitation, cosmology, and particle physics to try to determine the meaning of mass, and its interactions, at a deeper level. It is believed that an experiment carried out at the LHC in 2011–2012 unveiled the existence of a key piece of the mass puzzle, namely the Higgs boson. The particle is named after Peter Higgs, who shared the 2013 Nobel Prize in Physics with François Englert "for the theoretical discovery of a mechanism that contributes to our understanding of the origin of mass of subatomic particles, and which recently was confirmed through the discovery of the predicted fundamental particle, by the ATLAS and CMS experiments at CERN's Large Hadron Collider" (see Figure 5-6). On the theoretical end, work of researchers in institutes like The Perimeter Institute for Theoretical Physics in Waterloo, Canada, aims to advance our understanding of the universe at the most fundamental level.

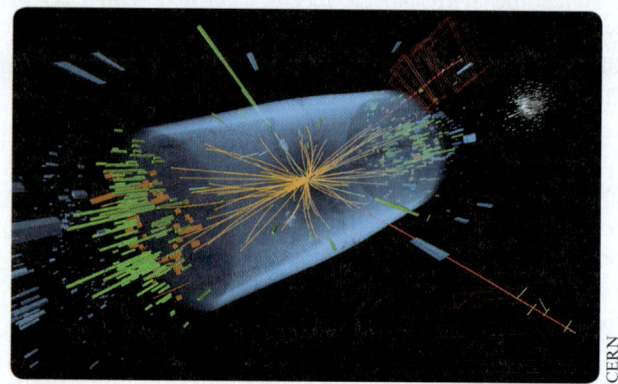

CERN

Figure 5-6 Event recorded in 2012 from a proton–proton collision observed with the Compact Muon Solenoid (CMS) experiment at LHC. The event shows characteristics expected from the decay of the Higgs boson.

5-3 Newton's Laws of Motion

English physicist Isaac Newton (1642–1726) developed three laws that provide the basis for the dynamics of mechanical situations. These three laws of motion now carry his name.

Newton's First Law

Imagine that you are a hockey player on a rink with smooth flat ice with almost no resistance. When your stick strikes a puck, the puck moves in a straight line on the ice, and its speed does not change significantly as it travels across the rink. In fact, if there were no friction between the puck and the ice, the velocity of the puck would not change at all. If you want to change the speed or the direction of the moving puck, you must exert a force on it. If you set the puck gently on the horizontal ice surface, the puck will remain where you placed it. The puck will remain at rest unless a force is exerted on it.

These examples demonstrate **Newton's first law**: *If no net force is exerted on an object, the object's velocity will not change.* Remember from Section 5-1 that the net force refers to the vector sum of all the forces acting on that object. Newton's second law considers how much force must be applied to produce a given acceleration.

Newton's Second Law

If you push a hockey puck with a horizontal force, the puck will move in the direction of the force, and the speed of the puck will continue to increase as long as you apply the force. If a larger force is applied, then the resultant acceleration will be greater. If the puck is already moving when you apply a force to it, it will accelerate in the direction of the applied force and will continue to do so as long as the force is applied. The response of the puck, or any other object already moving or stationary, to the net force applied to it is given by Newton's second law. According to **Newton's second law**, *the acceleration of a mass is proportional to the net applied force and inversely proportional to the mass.* Newton's second law is normally presented in equation form as

KEY EQUATION
$$\vec{F}_{net} = m\vec{a} \tag{5-3}$$

Sometimes we write this as follows to explicitly show that the net force is the vector sum of the various applied forces:

$$\sum_{all\ forces} \vec{F} = m\vec{a} \tag{5-4}$$

If the mass is very large, as in the case of the truck in the opening photograph of this chapter, even a large force will only produce a small acceleration.

INTERACTIVE ACTIVITY 5-2

Forces and Motion/Acceleration

In this activity, you will use the PhET simulation "Forces and Motion: Basics" to investigate the relationship between force and acceleration as you apply one-dimensional forces to objects.

PEER TO PEER

Sometimes I hear Newton's second law described as Force = Mass × Acceleration. Stated in this way, the law may be misunderstood. It is not any particular force, but rather the *net* force, obtained by taking the vector addition of *all* the forces that act on that mass.

EXAMPLE 5-3

Increasing or Decreasing Speed?

An object moving to the right is acted upon by a net force that points to the right but whose magnitude decreases linearly with time until it gets to zero. What can we say about the speed of the object: is it decreasing or increasing while the force acts on it?

Solution

Despite the fact that the magnitude of the force is decreasing with time, at any given point the force is non-zero and pointing to the right. According to Newton's second law in Equation 5-3,

the object's acceleration should also be directed to the right; therefore, the speed is increasing. Of course, this applies as long as the force does not reach a value of zero, at which point the acceleration will be zero and the puck's speed will remain constant.

Making sense of the result

The fact that the magnitude of the net force decreases with time means that the acceleration decreases with time, but as long as the acceleration is not zero, the puck will continue to accelerate and its speed will increase.

EXAMPLE 5-4

Braking Distance

A 125 tonne subway train is brought to a stop from a speed of 23.0 m/s using a constant braking force of 425 kN. Over what distance does the train come to a stop? Assume that the only force stopping the train is the constant braking force.

Solution

The forces in the vertical direction cancel, so the only net force acting on the train is the braking force. We choose the positive x-axis to be the direction of motion of the train, so the braking force is in the negative direction, as is the acceleration:

$$\vec{F}_{net} = m\vec{a}$$

$$\vec{a} = \frac{\vec{F}_{net}}{m} = \frac{(425\,000\ \text{N})(-\hat{i})}{125\,000\ \text{kg}} = (-3.40\ \text{m/s}^2)\hat{i} \qquad (1)$$

We can use the constant-acceleration kinematics equation to find the distance the train travels while braking:

$$v_f^2 - v_i^2 = 2\vec{a} \cdot \Delta\vec{x} \qquad (2)$$

The final speed is zero. Since the train is slowing down, the acceleration and the displacement have opposite directions. Therefore,

$$\vec{a} \cdot \Delta\vec{x} = a\Delta x \cos 180° \qquad (3)$$

If we substitute (3) into (2), as well as the acceleration from (1), we obtain

$$0 - (23.0\ \text{m/s})^2 = 2(3.40\ \text{m/s}^2)\Delta x(-1) \Rightarrow \Delta x = 77.8\ \text{m}$$

where Δx is the magnitude of the displacement.

Making sense of the result

425 kN is a large force, about 40 times the average weight of a car, but the train is also very massive, so the resultant acceleration is small. Note that a rather large stopping distance is required, which is reasonable.

Net Force and Direction of Motion

The direction of the net force on an object does not necessarily dictate the direction the object is moving. For example, when you throw a ball up in the air, it is under the influence of only the force of gravity as it moves up (ignoring air resistance). The net force on the ball is directed downward as the ball moves up. The force of gravity will eventually cause the ball to slow down and move downward, but at any given moment the ball could be moving up (right after you throw it), moving down, or have a speed of zero.

CHECKPOINT 5-3

Dropped Ball

A ball is dropped from rest and bounces off the floor. Which of the following statement(s) applies to the ball on its way up? Ignore air resistance. Zero, one, or more than one statement may be true.

(a) The ball's acceleration is upward because it is moving upward.

(b) There must be a force causing the ball to keep moving upward.

(c) The ball's acceleration is downward because of the force of gravity.

(d) The force of gravity is the only force acting on the ball on its way up.

ANSWER: (c) and (d) are correct.

Newton's first law is sometimes called the law of inertia, although inertia ideas are also critical to the second law. **Inertia** refers to an object's resistance to being accelerated. The quantitative measure of inertia is mass—it requires a large force to produce the same acceleration if the mass is larger. As we saw earlier, a net force of 1 N will cause a 1 kg mass to accelerate at 1 m/s^2. The more massive an object is, the more inertia it has. The very massive truck in Figure 5-1 had a large mass and therefore even a rather large force can only produce a small acceleration.

MAKING CONNECTIONS

Inertia

The word *inertia* comes from the Greek word *iners*, which means, among other things, irresponsive, idle, lazy, or sluggish. When you try to accelerate an object on a frictionless surface, you will still encounter resistance. The resistance you encounter from the object depends on its mass. Newton defined inertia as "an innate force of matter," "a power of resisting by which every body, as much as in it lies, endeavors to preserve its present state, whether it be at rest or moving uniformly forward in a straight line." Mass is a measure of an object's inertia: the more massive the object, the more inertia it has and the more difficult it is to accelerate it.

CHECKPOINT 5-4

Inertia

The term *inertia*

(a) refers to the ability of an object to resist being accelerated

(b) is related to Newton's first and second laws

(c) is represented by an object's mass

(d) all of the above

ANSWER: (d)

Now that we have considered Newton's second law and inertia, we can define mass in another way. We can define the amount of mass according to the force required to produce a certain acceleration: if twice as much force is required to produce the same acceleration, then the mass must be double. Mass defined in this way is called **inertial mass**.

The mass we defined in terms of the gravitational force in Section 5-2 is, strictly speaking, the gravitational mass. Both inertial mass and gravitational mass are measured in SI units of kilograms. A fundamental question in physics is whether the gravitational mass and the inertial mass are precisely equal. We will return to this question when we consider gravitation at a deeper level in Chapter 11, but our present understanding is that the two are exactly equal.

Newton's Third Law

You and your friend take a break while skating at the local rink; you stand opposite one another on the ice, with your skates on, and decide to see how fast you will go when one of you pushes the other person away. You start by pushing your friend. As a result of the interaction, you feel a force from your friend on your hand, and your friend feels the force of your push. Your push causes your friend to slide away from you, but the force you feel causes you to move away from your friend as well. If you and your friend both have approximately the *same mass*, someone observing you will notice, at the end of the push, that your friend and you will be moving at approximately the same speed no matter who did the pushing.

This interaction demonstrates **Newton's third law**: *If object 1 exerts a force $\vec{F}_{1 \to 2}$ on object 2, object 2 will exert a force $\vec{F}_{2 \to 1}$ on object 1 that is equal in magnitude and opposite in direction to $\vec{F}_{1 \to 2}$:*

KEY EQUATION

$$\vec{F}_{2 \to 1} = -\vec{F}_{1 \to 2} \qquad (5\text{-}5)$$

We use the arrow notation to indicate the direction of the force, e.g., $\vec{F}_{A \to B}$ means the force that object A exerts on object B. Other texts usually simplify the notation to $\vec{F}_{AB}$.

Newton's third law is true whether or not motion is involved. When you push with your finger against a wall, you feel the force of the wall on your finger. This holds for all forces. A person exerts a gravitational force on Earth, equal in magnitude and opposite in direction to the gravitational force Earth exerts on the person.

EXAMPLE 5-5

Cub Pushing Mom

A polar bear cub and his mother are standing still on a frictionless icy patch in the Gulf of Boothia, Nunavut. The 50.0 kg cub pushes the 400. kg mother bear with a constant (while in contact) force. What happens to the cub?

Solution

According to Newton's third law, the force exerted by the cub on the mother, $\vec{F}_{c \to m}$, must be equal in magnitude and opposite in direction to the force exerted by the mother on the cub, $\vec{F}_{m \to c}$ (see Figure 5-7). The acceleration of each is given by Newton's second law:

$$\vec{a} = \frac{\vec{F}_{net}}{m}$$

Since they both experience forces of equal magnitude but have different masses, their accelerations will not be the same. When the cub pushes his mother on the ice, he will end up moving backward at a higher speed than she will end up moving in the direction he pushed her.

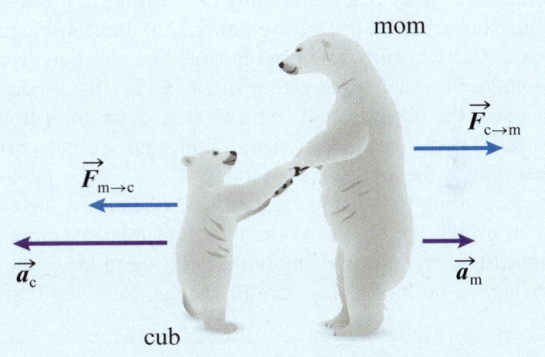

Figure 5-7 The mom pushes the cub with the same magnitude of force as the cub pushes the mom, but the accelerations are not the same.

Making sense of the result

In everyday life, when a lighter person pushes a heavier person, the lighter person does not necessarily expect to move backward. Enough friction is often present to prevent the lighter person from sliding backward. In this example, we assumed a frictionless surface.

EXAMPLE 5-6

Average Force on Hand from a Serve

A volleyball player serves a 249 g volleyball at a speed of 90.0 km/h. Assuming that the player's hand contacts the ball for 42.0 ms, find the average force exerted by the ball on the player's hand during that time.

Solution

We have no knowledge of the mass of the player's hand or its change in speed during the collision. However, Newton's third law tells us that the magnitude of the force on the player's hand equals the magnitude of the force on the volleyball. A change of speed of 90.0 km/h is

$$\Delta v = 90.0 \text{ km/h} \left(\frac{1000 \text{ m/km}}{3600 \text{ s/h}} \right) = 25.0 \text{ m/s}$$

The average acceleration of the volleyball is given by

$$a_{\text{avg}} = \frac{\Delta v}{\Delta t} = \frac{25.0 \text{ m/s}}{0.0420 \text{ s}} = 595 \text{ m/s}^2$$

We can now find the magnitude of the average force exerted by the player's hand on the ball during the serve.

$$F_{\text{avg}} = ma_{\text{avg}} = (0.249 \text{ kg})(595 \text{ m/s}^2) = 148 \text{ N}$$

By Newton's third law, the average force exerted by the ball on the player's hand will have the same magnitude, but the opposite direction.

Making sense of the result

The force on the player's hand is roughly 60 times the force of gravity on the volleyball, which is reasonable, given the sensation one feels when serving a volleyball. This question demonstrates how Newton's third law can be used to calculate a force from information on the corresponding paired force.

MAKING CONNECTIONS

Newton's Laws before Newton?

Before Isaac Newton, Western civilization's perception of motion was dominated by the Aristotelean view, dating back to the 4th century BCE. According to Aristotle, in the absence of an external force, or "motive power," an object would come to rest. Consequently, an object moving at a constant velocity would need a net force acting on it. Once that force was removed, the object would begin to slow down until it came to rest. These Aristotelean views of motion are not correct, according to the models we use today.

According to Newton's second law, when a net force acts on an object, the object's velocity must change, meaning that either the direction of its motion or the speed at which it is moving, or both of them, have to change. Newton's first law states that an object will move at a constant velocity only if no net force acts on it.

Statements summarizing the principle contained within Newton's first law date as far back as the 5th century BCE in China and the 11th century CE in parts of the Islamic world. Newton first published his laws of motion in 1687 as part of his three-volume work *Philosophiæ Naturalis Principia Mathematica*, or *Principia*. *Principia* also includes his theory of universal gravitation and a mathematical derivation of Kepler's laws of planetary motion, which are discussed in more detail in Chapter 11. Newton's laws of motion laid the foundation for classical mechanics and dominated physicists' view of the universe for the next three centuries.

PEER TO PEER

Although in Newton's third law, pairs of forces are of equal magnitude and opposite direction, they never cancel each other since they operate on *different* objects!

There are a number of examples of Newton's third law force pairs in the opening photo of this chapter (Figure 5-1). For example, the man pushes against the pavement with his feet, and the pavement exerts an equal-magnitude but opposite-direction force against his feet. List as many pairs of forces as possible from the situation in Figure 5-1.

Let us consider a soccer ball being held against a wall. When you push to exert a force on one side of the soccer ball, the wall behind the soccer ball constrains it from accelerating, and instead the ball will be deformed. The lack of acceleration is not contrary to Newton's second law, since the net force on the ball is zero in this case. The force exerted by the wall on the ball, and the force you exert on the ball, have equal magnitude, but in opposite directions on the ball. These are not a Newton's third law pair, just two forces of equal magnitude that operate on the same object. Sometimes it is said that a force causes either acceleration or deformation, but really a net force on an object always causes acceleration. It is important be clear exactly what is the object under consideration (the soccer ball in this case), and include all forces acting on that object (both your force and that exerted by the wall in this example).

5-4 Applying Newton's Laws

In Section 5-1 we learned how to calculate net forces and FBDs in one dimension, and in Section 5-3 we developed Newton's laws primarily in one dimension. Try Checkpoint 5-6 and Example 5-7 to confirm your understanding of that material. In this section, we apply Newton's second law in two- and three-dimensional situations. If necessary, review the material on adding vectors and on vector components in Chapter 2 before continuing with this section.

It is critical to remember that Newton's second law refers to the *net* force, and that we must add forces as vectors, as illustrated in Example 5-8.

CHECKPOINT 5-6

Net Force

A cat of mass 2.40 kg enters an elevator, which then moves upward with an acceleration of 4.00 m/s^2. The magnitude of the net force on the cat while the elevator accelerates is

(a) 33.1 N
(b) 23.5 N
(c) 13.9 N
(d) 9.60 N

ANSWER: (d) According to Newton's second law, we determine the magnitude of the net force on an object by multiplying its mass and acceleration.

EXAMPLE 5-7

Force and Acceleration

A Formula One race car is initially at rest at the start line. When the race starts, the car attains a speed of 135 km/h in 4.50 s. Find the net force on the driver, who has a mass of 67.0 kg. Assume constant acceleration.

Solution

Newton's second law relates the net force on the driver and her acceleration:

$$\vec{F}_{net} = m\vec{a}$$

We choose the direction of motion of the car as the positive *x*-direction.

The driver's acceleration is the same as the car's acceleration, and since this acceleration is constant, we have

$$v(t) = at + v_0$$

The initial speed is zero, and the final speed is

$$v(t) = 135 \text{ km/h} = 135 \text{ km/h}\left(\frac{1000 \text{ m/km}}{3600 \text{ s/h}}\right) = 37.5 \text{ m/s}$$

Hence the acceleration, assumed constant, is

$$a = \frac{v(t) - v_0}{t} = \frac{37.5 \text{ m/s} - 0}{4.50 \text{ s}} = 8.33 \text{ m/s}^2$$

Therefore, the magnitude of the net force is

$$F_{net} = (67.0 \text{ kg})(8.33 \text{ m/s}^2)$$
$$= 558 \text{ N}$$

The net force points in the same direction as the acceleration.

Making sense of the result

All we need to know is an object's mass and its acceleration to obtain the net force.

EXAMPLE 5-8

Adding Forces

(a) Calculate the net force on the object of Figure 5-8. Give the magnitude of the net force and the angle it makes with the direction of $\vec{F}_1$.
(b) If the mass of the block is 20.0 kg, what is the acceleration of the mass? (Assume no other forces, such as gravitation, are present.)

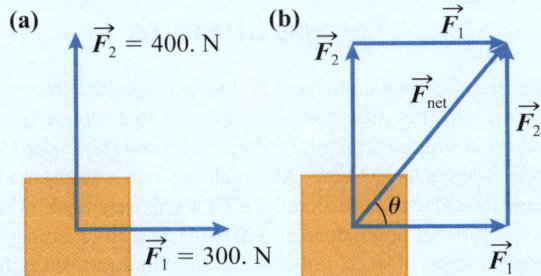

Figure 5-8   Example 5-8.

Solution

(a) The net force is obtained by the vector sum of the two forces. We use the method of vector addition from Chapter 2 to lay out the vectors as shown in Figure 5-8(a):

$$\vec{F}_{net} = \vec{F}_1 + \vec{F}_2$$

Since the forces act at 90° to each other, we can apply the Pythagorean theorem to find the magnitude of the net force:

$$F_{net}^2 = F_1^2 + F_2^2 = (300. \text{ N})^2 + (400. \text{ N})^2 = 250\,000 \text{ N}^2$$
$$F_{net} = 500. \text{ N}$$

We have written the forces with a trailing decimal point to make clear that they are known to three significant digits. From Figure 5-8(b), we can see that the angle that the net force makes with the horizontal axis is given by

$$\tan\theta = \frac{400. \text{ N}}{300. \text{ N}} \Rightarrow \theta = 53.1°$$

(continued)

Therefore, the magnitude of the net force is 500. N, and it is directed as shown in the diagram, making an angle of 53.1° relative to the horizontal axis.

(b) We can use Newton's second law to find the magnitude of the acceleration on the block:

$$a = \frac{F}{m} = \frac{500.\ \text{N}}{20.0\ \text{kg}} = 25.0\ \text{m/s}^2$$

The acceleration of the block will be in the same direction as the net force; i.e., relative to the horizontal axis, it makes an angle of an angle of 53.1°, as shown in Figure 5-8(b).

Making sense of the result

The forces make a 3:4:5 triangle, so the value of 500 N is expected for the magnitude of $\vec{F}_{\text{net}}$. Remember that the direction of the acceleration and the net force will always be the same.

In this text, we refer to the vector sum of all the forces acting on an object as the *net force*. However, it is not uncommon to refer to the sum of all the forces acting on an object as the *resultant force*. Both terms are correct.

Many types of forces may apply in Newton's second law problems. We define tension and normal forces here, and we will define frictional forces a bit later in the chapter. A **normal force** is perpendicular, or normal, to a surface. We usually use the symbol $\vec{F}_{\text{N}}$ or $\vec{N}$ for the normal force (either notation is perfectly acceptable). When you stand on a level floor, the floor exerts a normal force vertically upward on you through your feet. If no friction is present, a surface can only exert forces normal to the surface. When friction is present, it is still helpful to consider the forces at the surface as composed of a normal and a frictional part. It turns out that forces exerted at surfaces, such as these normal forces, ultimately arise from electromagnetic interactions, but for most problems it is not necessary to consider this fundamental nature of the force.

PEER TO PEER

At first I confused notation, not realizing that $\vec{F}_{\text{N}}$ means the normal (perpendicular to the surface) force, and not the net force. When we mean net force, we write it out as $\vec{F}_{\text{net}}$. I guess that's why some people prefer using $\vec{N}$ for normal forces.

When a rope or wire is being used to pull an object, we call the force exerted by the rope a **tension** force. Normally we use the notation $\vec{F}_{\text{T}}$ or $\vec{T}$ for tension forces. These tension forces are electromagnetic in nature at a fundamental level. We often assume that ropes exerting tension forces are ideal and massless, and under these conditions the tension is the same at all points in the rope. We discuss the approximation in more detail in the next paragraph.

Let us say that you use a rope to drag and accelerate a sled horizontally. If the rope's mass could not be neglected, and if the rope itself were accelerating, the tension at either end would not be the same. As a matter of fact, the net force on the string, whose magnitude would be the *difference of the magnitudes* of the

tensions at both ends, would be equal to the mass of the string multiplied by its acceleration, as per Newton's second law. Therefore, for the tension or compression forces at both ends to be equal, we require that the string have either zero acceleration or zero mass. In this chapter, and in this section, we will deal almost exclusively with strings that are inextensible (do not stretch) and whose mass can be neglected. The result is that the tensions at both ends of the string, unless otherwise specified, are the same.

MAKING CONNECTIONS

Tensile Strength

The ultimate tensile strength (UTS) indicates the maximum tension force per unit cross-sectional area that a material can withstand without structural failure. Structural steel has a UTS value of about 400 MN/m². Although concrete is widely used in elements under compression, its UTS is relatively weak, with a typical value of approximately 3 MN/m². Silk has 2.5 times the tensile strength of structural steel, and some multi-walled carbon nanotubes have been made with UTS exceeding 60 GN/m². For a more detailed discussion of this topic, see Chapter 10.

It is helpful to consider the following steps in solving Newton's second law problems. In this chapter, we limit consideration to situations in which we are only interested in the motion of small objects, and not how an extended object might rotate when forces are applied at different points on it (see Chapter 10 for that type of problem).

1. First *identify the object* under consideration, and *all of the forces* that act on that object. While the object is usually obvious, including this step will help prevent errors of assuming forces that really act on a different object.

2. Next *draw an FBD* to simplify the object as a point mass and show all the forces acting on that mass.

3. *Decide on the orientation of the axes* for splitting forces into components. Normally the direction of acceleration of the object should be along one of these axes. While we sometimes put the acceleration direction on the FBD, it should always be made clear that it is not a force. In this book,

acceleration vectors are purple and force vectors are blue.

4. *Define the direction considered positive* on these axes.

5. *Divide the forces into components* along the designated axes.

6. *Write Newton's second law for each axis direction.*

7. *Simultaneously solve* these equations. Clearly *state and interpret directions* for all quantities using your sign conventions. As for all problems, ask yourself whether your answer seems reasonable, for example, by considering limiting cases.

EXAMPLE 5-9

Drawing FBDs

In an FBD, show all the forces acting on a crate that is sliding down a frictionless incline.

Solution

The forces acting on the crate are the normal force exerted by the inclined plane on the crate, $\vec{N}$, and the force of gravity, $\vec{F}_g = m\vec{g}$. The force of gravity near the surface of Earth acts vertically downward. A normal force must be perpendicular to the surface. Therefore, the FBD for the situation is as shown in Figure 5-9.

Making sense of the result

The acceleration is along the incline in the downward direction, but it is not part of the FBD. Note that although there

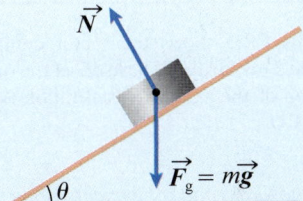

Figure 5-9   Example 5-9. The FBD of a crate sliding down a frictionless incline.

is no individual force downward along the incline, that is the direction of the net force when the forces $\vec{N}$ and $\vec{F}_g$ are added as vectors.

EXAMPLE 5-10

Motion on a Frictionless Incline

A box is free to slide on a frictionless surface. The surface is inclined at an angle θ above the horizontal. Derive an expression for the acceleration of the mass, and calculate the magnitude of the normal force exerted by the inclined surface on the mass.

Solution

We apply the analytical strategy outlined earlier in this section.

Step 1. *Identify the object and the forces on that object.*

The object is the box, and the forces acting on the box are the force of gravity vertically downward, and the normal force perpendicular to the incline (as we saw in Example 5-9).

Step 2. *Draw the FBD.*

We did this in the preceding example, with the result shown in Figure 5-9. Notice that we have shown the surface of the incline in the FBD, as well as the box, but the FBD is the box represented by the small circle, and the vectors for the two forces acting on it. Drawing the incline makes it easier to resolve the forces into components and determine the angles and the components.

(continued)

Step 3. *Establish the directions for the coordinate axes.*

We use the direction of the assumed acceleration to help guide our optimum choice. From common sense we know that the box will accelerate and slide down along the incline. Therefore, one of our axes (we will call it the *x*-axis) should be along the incline, and the other (the *y*-axis) will be perpendicular to the incline.

Step 4. *Define the positive direction for the coordinate axes.*

Which direction is positive is entirely a matter of choice, and as long as you stay consistent with your choice there is no right or wrong answer. Here we will define down along the incline as the positive *x*-direction, and upward perpendicular to the incline as the positive *y*-direction. These are demonstrated in Figure 5-10.

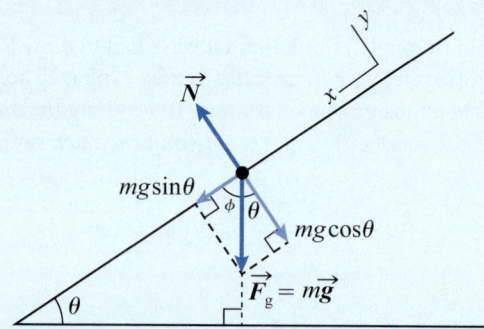

Figure 5-10 This FBD shows the vector components for the gravitational force. For clarity, we have offset the orientation of the axes. The direction of the assumed acceleration is shown, but it is not part of the FBD.

Step 5. *Resolve the forces into components along the chosen axes.*

The normal force, $\vec{N}$, from the incline on the block points perpendicular to the plane along the *y*-axis and does not need to be resolved. To resolve the force of gravity on the block into components, we draw two lines from the tip of the $\vec{F}_g$ vector, one parallel to the incline (along the *x*-axis) and one perpendicular to the incline (along the *y*-axis). These lines intersect the axes at the tips of the components of $\vec{F}_g$ (see Figure 5-10).

From the angles of the triangle, we know that $\theta + \phi + 90° = 180°$, which gives $\theta + \phi = 90°$. Therefore, we can see that

the angle between the $\vec{F}_g$ vector and the perpendicular to the incline must be θ, as shown, since this angle plus ϕ equals $90°$.

The magnitude of the force of gravity on the box of mass *m* is mg, so the component of $\vec{F}_g$ perpendicular to the incline must be $mg \cos \theta$, as shown in Figure 5-10, while the component along the incline is $mg \sin \theta$.

Step 6. *Write the equations from Newton's second law for each axis.*

Now we will use Newton's second law to write equations for the *x*-axis and the *y*-axis directions, keeping in mind that we have chosen down along the incline as the positive *x*-direction, and upward perpendicular to the incline as the positive *y*-direction:

$$x\text{-axis: } mg \sin \theta = ma \qquad (1)$$
$$y\text{-axis: } N - mg \cos \theta = 0 \qquad (2)$$

Here N without the vector sign means the magnitude of the normal force.

Step 7. *Simultaneously solve these equations of motion.*

We can obtain the magnitude of the normal force directly from (2):

$$N = mg \cos \theta \qquad (3)$$

We can divide the mass, *m*, from both sides of (1) to obtain the desired expression for the acceleration:

$$a = g \sin \theta \qquad (4)$$

Since we obtained a positive value for the acceleration, and we defined the positive direction of the *x*-axis as down along the incline, as expected the direction of the acceleration is down along the incline.

Making sense of the result

We can use limiting cases to check our result. When the mass is placed on a horizontal surface, we have $\theta = 0$, so $\sin \theta = 0$ and $\cos \theta = 1$. In this case, we obtain the expected results that the acceleration is zero and the magnitude of the normal force is simply mg. An incline angle of $90°$ means that the mass is effectively placed against a vertical wall, in which case it will simply free-fall with an acceleration equal to g. This is what we obtain from relationships (3) and (4), since now $\sin \theta = 1$ and $\cos \theta = 0$. Note that the value of the mass does not enter into the results obtained.

In Example 5-10 we independently defined the positive directions for the *x*-axis and the *y*-axis. In a conventional (right-handed) coordinate system, the relative orientation of the axes must follow this rule: the fingers of your right hand when curled through the $90°$ angle from the positive *x*-axis to the positive *y*-axis will result in your thumb pointing in the direction of the positive *z*-axis. The reason we are allowed to make

independent and arbitrary choices for the positive directions in Step 4 is that when we write the equations of motion in Step 6 we work separately in the coordinate directions. A different choice of positive would simply multiply both sides of the associated equation in Step 6 by negative one. Alternatively, we could justify the choice by imagining we viewed the *xy*-axes from the negative rather than the positive *z*-axis direction.

We want to stress that when in later chapters you deal with relationships involving vector cross products, you *must* use positive directions consistent with a right-handed coordinate system, and you are not free to independently select the positive direction for each axis.

EXAMPLE 5-11

Horizontal Push on Box on an Inclined Plane

Figure 5-11 shows a worker, wearing special boots for added traction, pushing a crate of mass m up an incline with a slope of angle θ using a horizontal force $\vec{F}$. Find the normal force from the incline on the crate, and determine the acceleration of the crate. Assume that the friction between the crate and the incline is negligible. Express your answers in terms of the variables m, g, θ, and F.

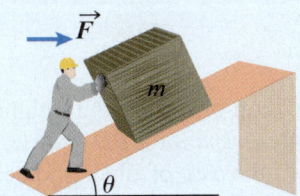

Figure 5-11 Example 5-11. The person pushes the box with an entirely horizontal force $\vec{F}$.

Solution

Here again, we invoke our strategy for solving Newton's second law problems. It is important to keep in mind that the applied force is entirely horizontal (rather than directed along the incline).

Step 1. *Identify the object under consideration and the forces acting on that object.*

We are concerned with the motion of the crate, so that is the object. There are three forces acting on the crate: the force of gravity, the normal force exerted by the incline on the crate, and the horizontal force applied by the worker.

Step 2. *Draw an FBD for the crate.*

The FBD is drawn in Figure 5-12, with the three forces acting on the crate being the normal force, $\vec{N}$; the force of gravity, $\vec{F}_g$; and the force applied by the person, $\vec{F}$.

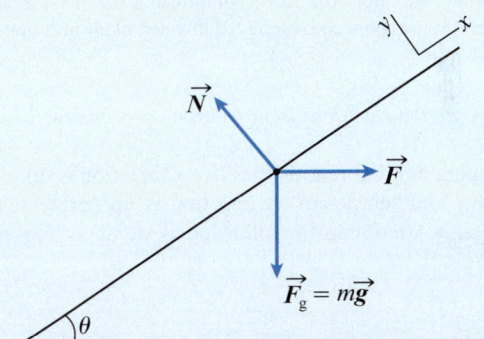

Figure 5-12 This FBD shows the three forces (normal force, force of gravity, and horizontal force applied by the person) acting on the crate.

Step 3. *Establish the directions for the coordinate axes.*

We use the direction of the assumed acceleration to help guide our choice of axes. Depending on the strength of the applied force, the crate can either accelerate up along the incline, or if the applied force is too weak, it might slide back down the incline. Therefore, one of our axes (we will call it the *x*-axis) should be along the incline, and the other (the *y*-axis) is perpendicular to the incline.

Step 4. *Define the positive direction for the coordinate axes.*

Which direction is positive is a matter of choice. We will define up along the incline as the positive *x* direction, and upward perpendicular to the incline surface as the positive *y*-direction. These are demonstrated in Figure 5-12 through the offset pair of axes.

Step 5. *Resolve the forces into components along the chosen axes.*

The normal force is already in the positive *y*-axis direction. The gravitational force and the applied force need to be divided into components along the two axes. The technique is similar to that used in the previous example.

(continued)

The applied force, $\vec{F}$, is divided into components $F \cos \theta$ along the incline and $F \sin \theta$ perpendicular to the incline. The gravitational force has a component $mg \sin \theta$ down along the incline, and a component $mg \cos \theta$ perpendicular to the incline. We show the components in Figure 5-13, along with the angles used to obtain them.

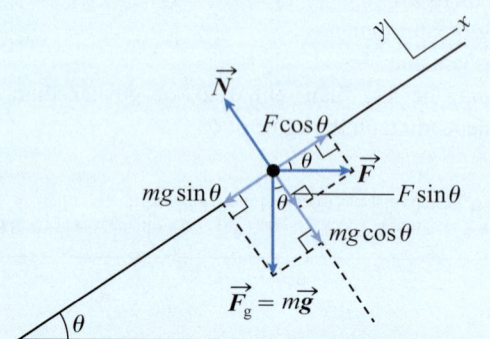

Figure 5-13 We show the force components for the forces of Example 5-11, using one axis along the inclined plane and one perpendicular to it.

Step 6. *Write the equations from Newton's second law for each axis.*

Keeping in mind that the positive x-direction is up along the incline, and the positive y-direction is up perpendicular to the incline, we obtain the following as we apply Newton's second law for each of the two axis directions. For the x-axis motion, we have

$$F\cos\theta - mg\sin\theta = ma \tag{1}$$

For the y-axis, we have

$$-F\sin\theta - mg\cos\theta + N = 0 \tag{2}$$

Step 7. *Simultaneously solve these equations of motion.*
We can solve (2) for the normal force:

$$N = F\sin\theta + mg\cos\theta \tag{3}$$

Note here that the normal force is *not equal* to $mg \cos \theta$.
We can find the magnitude of the acceleration of the crate from (1):

$$a = \frac{F\cos\theta - mg\sin\theta}{m} \tag{4}$$

Provided that the force was sufficient to make expression (4) positive, the direction of the acceleration is up along the incline, as per our sign convention.

Making sense of the result

Since the worker is pushing the crate partially against the face of the incline, the y-component of the applied force contributes to the normal force. Consequently, the magnitude of the normal force is greater than $mg \cos \theta$. If we consider the limiting cases applied to the acceleration result (4), we see that if there were no incline (i.e., $\theta = 0$), then we would simply have $a = F/m$, as expected.

With more than one mass, the technique is the same, but now we write a *separate FBD and system of equations for each mass*. In the case of objects joined by taut, ideal, connecting ropes, the accelerations must have the same magnitude. We often introduce a variable to represent the magnitude of this acceleration. If the connecting rope (and any pulleys) are ideal and massless, the tension will have the same magnitude in different parts, and again it is helpful to introduce a variable for this. We demonstrate these techniques in Example 5-12.

EXAMPLE 5-12

Simple Rope and Pulley System

An 8.00 kg mass (m_1) is resting on a horizontal frictionless surface. A massless rope that runs over a frictionless, massless pulley connects m_1 to a vertically hanging 6.00 kg mass (m_2), as shown in Figure 5-14. The two-mass system is initially at rest. Find the acceleration of m_1 and the tension in the rope.

Solution

We begin by drawing an FBD for each mass. For mass m_1, we have a force of gravity, $\vec{F}_{g1}$, vertically downward; a normal force, $\vec{N}$, exerted by the surface that is vertically upward; and a tension force, $\vec{T}_1$, from the connecting rope that pulls it to the right.

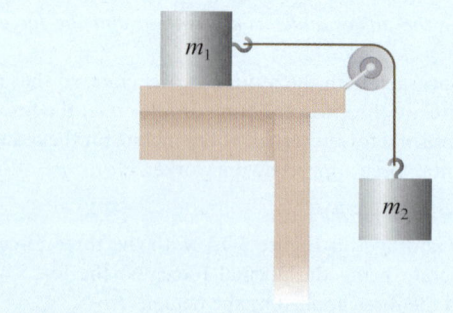

Figure 5-14 Example 5-12.

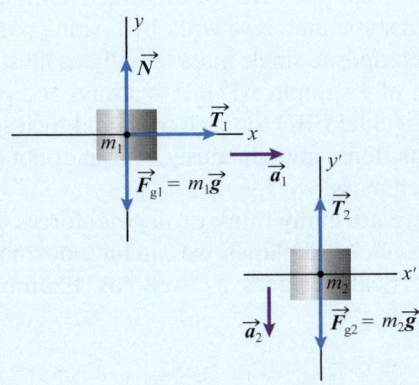

Figure 5-15 We draw a separate FBD for each of the two masses of Example 5-12.

For mass m_2, there are only two forces, a tension, $\vec{T}_2$, from the connecting rope that pulls it vertically upward, and the force of gravity, $\vec{F}_{g2} = m_2\vec{g}$, that acts vertically downward. Next we define coordinate axes and positive directions for the two masses. For mass m_1, we will define the x-axis horizontally with positive to the right, and the y-axis vertically with positive upward. For mass m_2, we only need a vertical y-axis, and we will define the positive direction as upward.

Since we are assuming a massless ideal connecting rope and a massless pulley, the tension is the same throughout the connecting rope, and we will define T as the magnitude of this tension. The downward acceleration of mass m_2 must have the same magnitude as the rightward acceleration of mass m_1, and we will let a represent the common magnitude of the acceleration.

Now we are ready to use Newton's second law. For mass m_1, the only force acting in the x-direction is the tension in the rope:

$$\sum F_x = m_1 a_x \Rightarrow T = m_1 a \qquad (1)$$

From the FBD of m_2, we have the following (remembering the upward positive direction):

$$\sum F_{y'} = m_2 a_{y'}$$
$$T - m_2 g = -m_2 a \qquad (2)$$

If mass m_1 accelerates to the right (positive), then mass m_2 must accelerate downward (negative), which is the reason for the negative sign in the acceleration term in (2).

Substituting (1) into (2), we get

$$m_1 a - m_2 g = -m_2 a$$

$$\Rightarrow a = \frac{m_2 g}{m_1 + m_2} = \frac{(6.00\ \text{kg})(9.81\ \text{m/s}^2)}{(6.00 + 8.00)\ \text{kg}} = 4.20\ \text{m/s}^2$$

To find the tension in the rope, we substitute this value for the acceleration into either (1) or (2). Using (1), we get

$$T = m_1 a = (8.00\ \text{kg})(4.20\ \text{m/s}^2) = 33.6\ \text{N}$$

Making sense of the result

The acceleration of the system is less than g because the force of gravity on m_2 is pulling both m_1 and m_2. Note that the tension is not equal to $m_2 g$, as it would have been were m_2 not accelerating. The tension in the horizontal and the vertical segments of the rope is the same (but would not be if the pulley had mass, as we will see in Chapter 8). We also note that the limiting case $m_1 = 0$ gives an acceleration of m_2 equal to g, as expected.

When pulleys are connected in a more complex fashion with multiple rope connections, you cannot assume that all parts have the same acceleration magnitude. To illustrate this, consider the situation shown in Figure 5-17. The pulley on the right is fixed, while that on the left is free to move horizontally. The rope is attached to a post on the horizontal surface, so that the movable pulley has rope above it and below it. This doubling of the rope means that when the vertical mass, m_1, moves by a distance $2d$, the horizontal mass, m_2, will only move by half as much, or a distance of d. If we differentiate the displacements to find the velocities, and then differentiate those velocities to find the accelerations, the acceleration

---

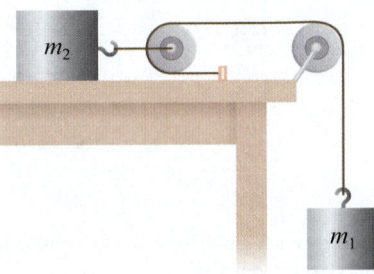

Figure 5-17 The pulley on the left can move horizontally, while that on the right is fixed.

of m_1 is twice the acceleration of m_2. Construction and mechanical systems often use a mix of fixed and movable pulleys with multiple rope segments.

Multiple Connected Objects

If multiple objects are connected, you can sometimes simplify the situation by drawing a single FBD for the composite total object. For example, if you are only interested in the acceleration of the composite object, it can be treated this way. If, however, you need to find forces or tensions between different parts of the object, you normally need to draw a separate FBD for each mass (although even then you can sometimes save work by treating part of the object as a composite single mass initially, as illustrated in the solution of Example 5-13). It is always acceptable to draw an individual FBD for each mass and then solve the resulting equations simultaneously, so if in doubt draw an FBD for each mass.

While we normally think of normal forces in situations such as inclined planes, we can have normal forces between vertical surfaces as well, as Example 5-14 illustrates.

EXAMPLE 5-13

Towing Tension

Three masses ($m_1 = 20.0$ kg, $m_2 = 50.0$ kg, and $m_3 = 30.0$ kg), resting on a horizontal frictionless surface, are connected by ideal massless taut ropes (see Figure 5-18). A horizontal force, $\vec{F}$, of 225 N is exerted on m_3 to the right.

(a) Find the tension in the rope connecting m_1 and m_2.
(b) Find the tension in the rope connecting m_2 and m_3.

Figure 5-18 Example 5-13.

Solution

(a) We will call T_{12} the magnitude of the tension in the rope connecting masses m_1 and m_2. Similarly, we will call T_{23} the magnitude of the tension in the rope that connects masses m_2 and m_3. Since they are connected with a taut ideal rope, the acceleration of all three masses must be the same, and we will use the variable a to represent the magnitude of this common acceleration.

We draw an FBD for m_1 in Figure 5-19, using horizontal and vertical for the x- and y-axes. Positive is defined to the right for the x-axis and upward for the y-axis. The normal force, $\vec{N}_1$, will be equal in magnitude and opposite in direction to the gravitational force, $\vec{F}_{1g} = m_1\vec{g}$, on that mass, but this will not enter into our solution for

motion in the x-direction. The only force acting on m_1 in the x-direction is the tension, $\vec{T}_{12}$, in the rope connecting it to m_2.

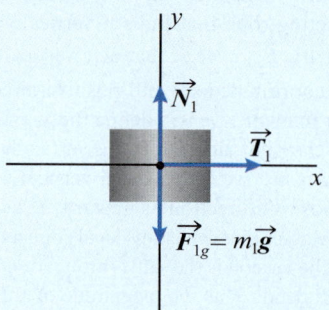

Figure 5-19 The FBD of m_1 in Example 5-13.

Therefore, applying Newton's second law in the x-direction for mass m_1 and using the tension notation and definition of positive directions introduced earlier, we have

$$\sum F_x = m_1 a_x \Rightarrow T_{12} = m_1 a \tag{1}$$

Since m_1 is known, we can solve for T_{12} if we first obtain a value for a.

The acceleration of all three masses must be the same, so while we could draw a separate FBD for each mass, as shown in Figure 5-20(a), we also have the option to draw the system as one composite mass in the FBD of Figure 5-20(b).

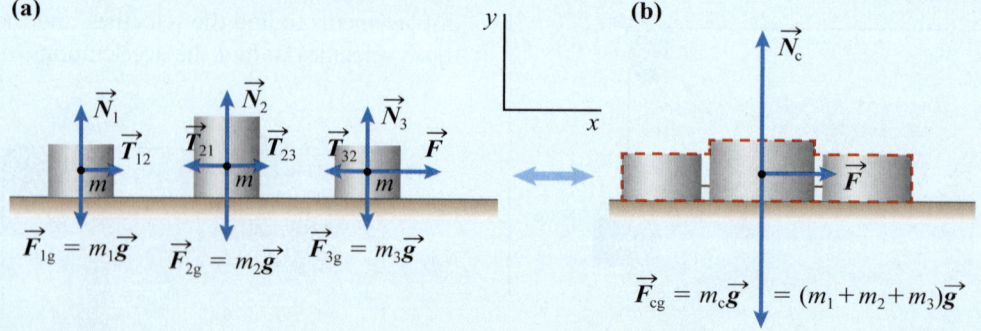

Figure 5-20 On the left we show individual FBDs of the three masses in Example 5-13, while on the right we show the FBD in terms of one composite connected mass.

Although the tension in the rope connecting m_1 and m_2 has the same magnitude at both ends, the directions of the forces are opposite, so we write the tension force on m_1 as $\vec{T}_{12}$ and that on m_2 as $\vec{T}_{21}$ (note that these are not Newton's third law pairs of forces, so we do not use the arrow notation in the subscript). These forces are of equal magnitude but opposite direction, so $\vec{T}_{21} = -\vec{T}_{12}$. Corresponding relationships hold for the tension forces between masses m_2 and m_3.

The easiest way to solve the problem is to use the FBD of the three-mass composite shown in Figure 5-20(b). We treat the three masses as one unit, with a force of gravity $\vec{F}_{cg} = m_c\vec{g} = (m_1 + m_2 + m_3)\vec{g}$, and label the sum of the normal forces as $\vec{N}_c$. Since the only force acting on the three-mass system in the x-direction is the 225 N applied force,

$$\sum F_x = m_T a_x \Rightarrow F = m_c a$$

and

$$a = \frac{F}{m_c} = \frac{225 \text{ N}}{100.0 \text{ kg}} = 2.25 \text{ m/s}^2 \qquad (2)$$

Therefore, if we substitute (2) into (1), we obtain for the magnitude of the tension in the line joining masses m_1 and m_2

$$T_{12} = m_1 a = (20.0 \text{ kg})(2.25 \text{ m/s}^2) = 45.0 \text{ N}$$

(b) The rope between m_2 and m_3 is pulling both m_1 and m_2. Since they are moving together, we consider their combined FBD as shown in Figure 5-21. Applying Newton's second law to the system of two masses, we

solve for the magnitude of the tension force between masses m_2 and m_3:

$$\sum F_x = m_{1,2} a_x \Rightarrow T_{23} = (m_1 + m_2)a$$
$$= (20.0 \text{ kg} + 50.0 \text{ kg})(2.25 \text{ m/s}^2) = 158 \text{ N}$$

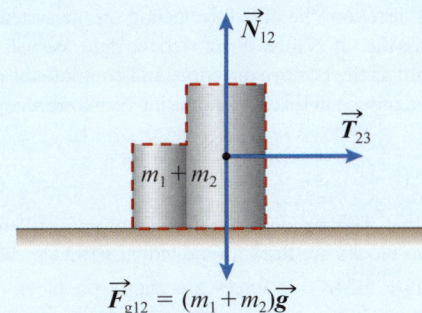

$$\vec{F}_{g12} = (m_1 + m_2)\vec{g}$$

Figure 5-21 An FBD treating m_1 and m_2 as one combined mass.

Making sense of the result

We are able to treat the three masses as one mass because they are connected by taut ideal ropes. Although the applied force acts directly on m_3 only, in effect the force is pulling all three masses. We can perform a check using our results. For the middle mass, m_2, the tension pulling it to the right has a value of 158 N while the tension pulling it to the left is 45 N. Therefore, the net force on it is 113 N. If we divide this by the mass of 50.0 kg, we obtain an acceleration of 2.26 m/s², within rounding error of the expected 2.25 m/s².

EXAMPLE 5-14

Normal Forces between Blocks

Block 1 in Figure 5-22 has a mass of 2.00 kg, and block 2 has a mass of 8.00 kg. The two blocks are sitting on a horizontal frictionless surface. You push block 1 with a horizontal force, $\vec{F}$, of magnitude 135 N directed to the right.

(a) Find the normal forces the blocks exert on one another, and verify that they are equal in magnitude, as predicted by Newton's third law.

(b) You now change the direction of the force and push block 2 to the left with a 135 N force. Find the normal force now present between the blocks.

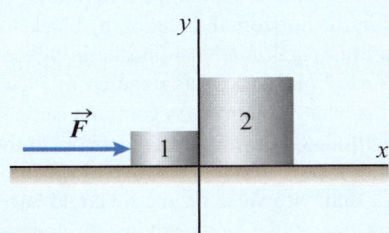

Figure 5-22 Example 5-14.

Solution

(a) We begin by drawing an FBD of the two blocks *combined*, shown in Figure 5-23. While it is always possible to draw individual FBDs for each mass, it is not always necessary to do so. You may use a combination of the masses when they move together, which is the situation here. In this problem, the solution will be simpler if we use an FBD for the combination of the two masses. In the current situation, since the two objects move as one, we can consider that force $\vec{F}$ is applied to a combined mass of $m_T = (m_1 + m_2)$. In addition

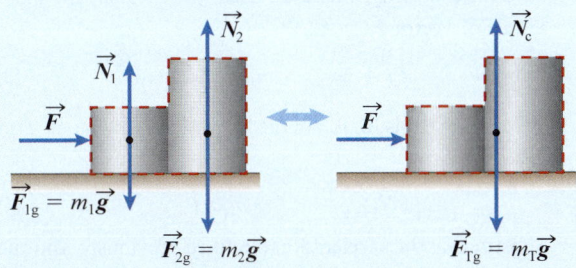

Figure 5-23 The FBD of block 1 and block 2 of Example 5-14 considered together as one unit.

(continued)

to $\vec{F}$, there is a combined upward normal force $\vec{N}_T$ (equal to the sum of the two individual normal forces) and a force of gravity corresponding to the combined mass.

In the vertical direction, the normal force from the surface must equal the force of gravity of the combined mass since there is no acceleration in this direction. In the horizontal direction, the only force acting on the system of two blocks is the 135 N force directed to the right. We will define to the right as the positive direction, and compute the acceleration magnitude in this direction using Newton's second law:

$$a = \frac{F}{m_T} = \frac{135 \text{ N}}{10.00 \text{ kg}} = 13.5 \text{ m/s}^2 \qquad (1)$$

Now that we have computed the common acceleration of the two blocks, we draw an individual FBD for each block in Figure 5-24. Two forces are acting on block 1 in the horizontal direction: the applied force $\vec{F}$ and a normal force that block 2 exerts on block 1, which we will designate $\vec{N}_{2\to1}$. Our arrow notation in the subscript reminds us that this is a force exerted on block 1 by block 2. Note that in this question we have both vertical normal forces (exerted by the surface on the blocks) and horizontal normal forces (exerted by one block on the other). For block 2, we have only a single horizontal force, that exerted by block 1, $\vec{N}_{1\to2}$. By Newton's third law, we know that the magnitudes of the horizontal forces between the blocks must be equal, but they point in opposite directions; that is, $\vec{N}_{2\to1} = -\vec{N}_{1\to2}$.

(a)   **(b)**

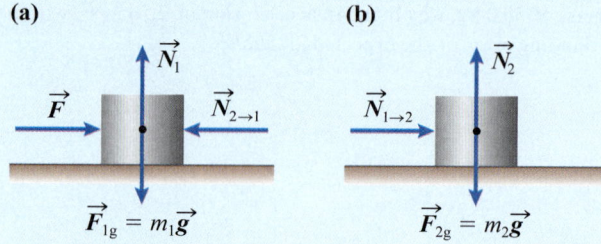

Figure 5-24 Individual FBDs for the two blocks of Example 5-14.

Since the blocks move together, they must have a common acceleration. In this part of the problem it is obvious that the acceleration will be to the right and a will have a positive value. Newton's second law for block 1 can be written as the following vector equation:

$$\vec{F} + \vec{N}_{2\to1} = m_1\vec{a} \qquad (2)$$

If we put in the known directions for the applied force and the resultant acceleration (in the positive x-direction) we have the following, where we use F to represent the magnitude of vector $\vec{F}$:

$$F\hat{i} + \vec{N}_{2\to1} = m_1 a\hat{i} \qquad (3)$$

We can solve this to obtain the force exerted by block 2 on block 1:

$$\vec{N}_{2\to1} = -(F - m_1a)\hat{i} \qquad (4)$$

Substituting the acceleration obtained previously, and the numerical value for the applied force, we obtain

$$\vec{N}_{2\to1} = -(135 \text{ N} - (2.00 \text{ kg})(13.5 \text{ m/s}^2))\hat{i}$$

$$= -(108 \text{ N})\hat{i} \qquad (5)$$

As expected, our result indicates that the horizontal force exerted by block 2 on block 1 is to the left.

The only force acting on block 2 in the horizontal direction is the normal force from block 1. Therefore, for block 2 we can write Newton's second law as

$$\vec{N}_{1\to2} = m_2\vec{a} \qquad (6)$$

Putting in numerical values, we obtain

$$\vec{N}_{1\to2} = (8.00 \text{ kg})(13.5 \text{ m/s}^2)\hat{i} = (108 \text{ N})\hat{i} \qquad (7)$$

Comparing results (5) and (7), we see that block 1 exerts a force of 108 N directed to the right on block 2, while block 2 exerts a force of 108 N directed to the left on block 1. We could have predicted this result from Newton's third law.

(b) Now we consider the situation in which we are pushing m_2 to the left with a 135 N force. Block 2 in turn pushes block 1, and they must move together with a common acceleration. Since the total mass is unchanged and the magnitude of the applied force is the same, the acceleration of the two-mass system is still 13.5 m/s^2, but this time the direction of the acceleration is to the left. The normal force exerted by block 1 on block 2 can be obtained by writing Newton's second law for mass m_1. The FBD for this mass is shown in Figure 5-25:

$$\vec{N}_{2\to1} = m_1\vec{a} = (2.00 \text{ kg})(-13.5 \text{ m/s}^2)\hat{i} = -(27.0 \text{ N})\hat{i}$$

The value of the force is negative, meaning that it points to the left. Note that the value is less than the normal force obtained in part (a).

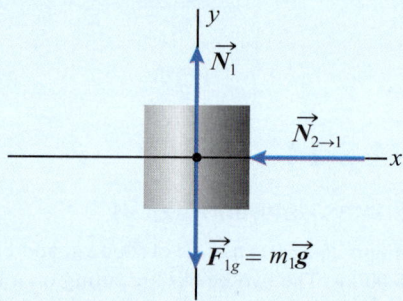

Figure 5-25 The FBD of block 1, now being pushed to the left, in Example 5-14.

Making sense of the result

Since the blocks move together, they have the same acceleration in both cases. It is important to realize that we can only employ the combined mass approach when the different parts are constrained to have a common acceleration. The normal force exerted in the horizontal direction by block 1 on block 2 must be equal in magnitude but opposite in direction to that exerted by block 2 on block 1, as required by Newton's third law, and that is what we found. As common sense would suggest, the magnitude of the normal force between the blocks is greater when you push the less massive block into the more massive block than vice versa. While we could have done this question using magnitudes, and invoking our understanding of the situation to assign directions, we have chosen to do this using a rigorous vector approach. When this is done, the directions come out of the solution.

5-5 Component-Free Solutions

Normally we use vector components when solving Newton's law problems. Newton's second law, however, does not necessarily need a coordinate system for us to appreciate and apply it. The power of vectors transcends the coordinate systems we use to represent them. The Cartesian coordinate system we have used in the examples above, like any other coordinate system, is only a tool we can choose to use to represent vectors. Are we then able to solve a vector problem independent of a coordinate system? The answer is yes, as we demonstrate in Example 5-15. As you go on to further courses in mathematics and physics, you will see that working with vector quantities without making reference to their components in a coordinate system can be both elegant and powerful.

EXAMPLE 5-15

Component-Free Solution for Motion on a Frictionless Incline

A crate of mass m is placed on a frictionless surface that is inclined at an angle θ to the horizontal (see Figure 5-26). Use a component-free technique to find the acceleration of the mass, as well as the normal force between the crate and the incline.

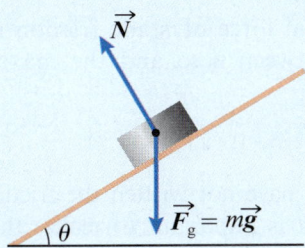

Figure 5-26 Example 5-15.

Solution

This is the same problem previously solved in Example 5-10, but this time we will not use vector components in the solution. The net force, $\vec{F}_{net}$, must be the vector sum of the two forces shown in the FBD of Figure 5-26, that is, the sum of the normal force and the gravitational force:

$$\vec{F}_{net} = \vec{F}_g + \vec{N} = m\vec{g} + \vec{N} \tag{1}$$

We use the tip to origin representation to draw this vector relationship in Figure 5-27. We earlier showed (Example 5-10) that the angle between the $\vec{N}$ and $\vec{F}_g$ vectors is the same angle, θ, of the inclined plane. Also, by definition the normal vector is perpendicular to the surface, and the net force must lie along the surface, so we know that the angle between $\vec{F}_{net}$ and $\vec{N}$ is a right angle, as shown.

We can now simply use trigonometry to obtain relationships between the magnitudes of the various vectors:

$$\sin\theta = \frac{F_{net}}{mg} \tag{2}$$

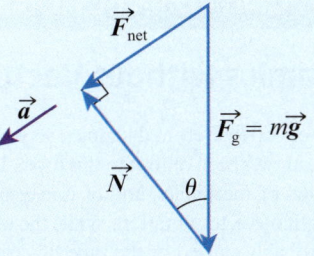

Figure 5-27 Example 5-15. A diagram representing the vector addition of the normal and gravitational forces to obtain the net force.

$$\cos\theta = \frac{N}{mg} \tag{3}$$

From Newton's second law, we know that

$$F_{net} = ma \tag{4}$$

If we combine (2) with (4), we immediately obtain the relationship for the magnitude of the acceleration of the block along the inclined plane:

$$a = g\sin\theta \tag{5}$$

From (3) we can solve for the magnitude of the normal force exerted by the inclined plane on the crate:

$$N = mg\cos\theta \tag{6}$$

Making sense of the result

As expected, we obtained the same results as for the same problem using vector components in Example 5-10. This alternative approach highlights the elegance of Newton's second law in vector form. Coordinate systems are only a vehicle, and not the only way to approach physics relationships expressed in vector form.

CHECKPOINT 5-10

Why Are the Normal and the Net Force Perpendicular?

In Example 5-15, we said that the net force vector had to be perpendicular to the normal force. What is the basis for this assumption?

ANSWER: The acceleration and the net force must be in the same direction, and we know that the object accelerates along the inclined plane. By definition, the normal force is perpendicular to this plane.

MAKING CONNECTIONS

Dynamics without Vectors?

We now use vectors extensively in dynamics, so you might think that vectors have always been central to this topic. Isaac Newton developed the laws of mechanics, and of universal gravitation, without the benefit of vector notation. While the idea of adding line segments probably started in the time of Aristotle, vector mathematics was only developed in modern form during the nineteenth century, long after the time of Newton. American scientist (he worked in physics as well as chemistry and mathematics) Josiah Willard Gibbs (1839–1903) published the first full treatment of vectors only in 1881, although aspects of modern vectors had been developed by a number of other mathematicians and physicists in the preceding 50 years. The term *vector* was apparently first used by Irish physicist William Rowan Hamilton (1805–1865).

5-6 Friction

If you push gently on the side of a heavy textbook lying on a table, the book does not move even though you are exerting a horizontal force on it. According to Newton's second law, the net force on the book must be zero because the acceleration is zero. Therefore, another force must be acting on the book to exactly balance the horizontal force you are applying. If you decrease the force you are applying, increase it slightly, or change which side of the book you push on, the book still does not move. The force that is preventing the book from moving is changing to remain exactly opposite to the force you apply. The force that prevents the book from moving is **friction**. Friction that acts between objects not moving relative to each other is called **static friction**.

If you now gradually increase the horizontal force you apply to the book, you will find that the static friction force reaches a maximum just before the book starts to move relative to the table. The maximum force of static friction between two objects in contact with each other always occurs when the objects are on the verge of moving relative to each other. After the book starts moving, you notice it takes less force than that maximum force to keep the book moving at a constant speed.

If you place another identical book on top of the first book and gradually increase the horizontal force you apply to the bottom book, you will notice that the force you have to exert on the bottom book to bring it to the verge of moving has increased. In fact, if you pull the book with a spring scale, you will find that the force required to bring the book to the verge of moving is doubled when you place the second book on top of it. The *maximum* force of static friction between two surfaces, f_s^{max}, is directly proportional to the normal force between the two surfaces:

KEY EQUATION
$$f_s^{max} = \mu_s N \qquad (5\text{-}6)$$

where μ_s is the **coefficient of static friction**, a unitless parameter normally within the range from zero (for frictionless situations) to slightly more than one (for high-friction surfaces). The value depends on the nature of both surfaces. There is no uniformity in the symbol used for the force of friction in physics, with some texts using a capital F and others a small f, as we have done here.

The actual force of static friction therefore takes on values between zero and the maximum force of static friction:

$$0 \le f_s \le f_s^{max} \qquad (5\text{-}7)$$

While we have not written the frictional force with vector signs, it is important to realize that the force of friction is a force with both direction and magnitude. The force of static friction is in a direction to oppose the tendency to move, so that if you try to push a heavy box to the right, but it does not move, the force of static friction will point to the left. However, if you try to push the box to the left, the force of static friction will then point to the right.

PEER TO PEER

I need to remember that the force of static friction is not the maximum value given by Equation 5-6, but rather the value needed to balance other forces to keep the net force at zero.

CHECKPOINT 5-11

Force of Static Friction

You place a book on the horizontal desk in front of you. The book's weight is 20.0 N, and the coefficient of static friction between the desk and the book is 0.25. You push the book with a horizontal force of 2.0 N, but the book does not move. What is the force of static friction on the book?

(a) zero
(b) 2.0 N
(c) 5.0 N
(d) none of the above

ANSWER: (b) The book is not moving, so the force of friction is equal to the force you apply.

EXAMPLE 5-16

Experimental Determination of Coefficient of Static Friction

A cube of mass 12.0 kg sits on a horizontal surface. To estimate the coefficient of static friction, you apply a steadily increasing horizontal force and note its value just before the box begins to slide on the surface. When the horizontal force is 68.0 N, the cube begins to slide. What is the coefficient of static friction between the surfaces?

Solution

We start by drawing an FBD showing the forces on the cube (Figure 5-28). There are four forces: the applied horizontal force, $\vec{F}$; the force of gravity, $\vec{F}_g$; the normal force, exerted on the crate by the surface, $\vec{N}$; and the force of static friction, $\vec{f}_s$. Since the applied force, $\vec{F}$, is attempting to move the box to the right, the force of static friction, $\vec{f}_s$, opposes this and therefore must point to the left, as shown in Figure 5-28.

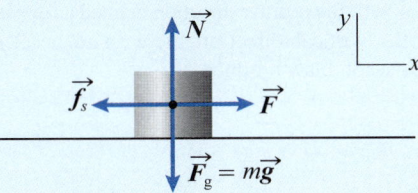

Figure 5-28 FBD for the situation of Example 5-16.

We define the x-axis horizontally, with positive to the right, and the y-axis vertically, with the positive direction upward. Newton's second law is applied in each axis direction. For the y-direction there is no acceleration, so we have the following, where N is the magnitude of the normal force:

$$\sum \vec{F}_y = m\vec{a}_y = 0$$
$$N\hat{j} - mg\hat{j} = 0 \tag{1}$$
$$N = mg$$

Therefore, in this case, the normal force is simply the following (we keep an extra significant digit at this stage but will round to the correct number in the final answer):

$$N = (12.0 \text{ kg})(9.81 \text{ m/s}^2) = 117.7 \text{ N} \tag{2}$$

Until the point just *prior* to motion, there is also no acceleration in the x-direction. If we let F and f_s^{max} represent the magnitudes of the applied horizontal force and the maximum static frictional force, we have

$$\sum \vec{F}_x = m\vec{a}_x = 0$$
$$F\hat{i} - f_s^{\text{max}}\hat{i} = 0 \tag{3}$$
$$f_s^{\text{max}} = F$$

Therefore, the maximum static friction force will equal the applied horizontal force just prior to motion, 68.0 N.

Finally, we can use the definition of the coefficient of static friction, Equation 5-6, to find the value of the coefficient:

$$f_s^{\text{max}} = \mu_s N$$
$$\mu_s = \frac{f_s^{\text{max}}}{N} = \frac{68.0 \text{ N}}{117.7 \text{ N}} = 0.578$$

Making sense of the result

The value we obtain for the coefficient of static friction is reasonable (see Table 5-1, later in the chapter, for some typical values of the coefficient of static friction). Since μ_s is the ratio of two forces, it does not have dimensional units. This method, perhaps with a spring scale, can be used to experimentally estimate the coefficient of static friction between two surfaces.

Consider a heavy object such as a thick book sitting on a horizontal surface. At first you push on the book and it does not move. When the applied force is increased, though, at some point you will reach the maximum possible force of static friction, and then the book will begin to slide on the surface. As soon as the book begins to move, you will notice a sudden drop in the force that opposes motion. To keep the book moving at a constant speed, you will have to decrease the applied force. The force of friction decreases somewhat when an object starts moving.

Friction that acts on a moving object is called **kinetic friction**. The magnitude of the kinetic friction force, f_k, is directly proportional to the magnitude of the normal force, N, between the object and the surface on which it moves:

KEY EQUATION $f_k = \mu_k N$ (5-8)

where μ_k is the **coefficient of kinetic friction**. For a given object and surface, the coefficient of kinetic friction is normally less than the coefficient of static friction.

Figure 5-29 shows a Vernier Logger *Pro* plot of the force of friction as a function of time for an object being pushed on a horizontal surface. The main peak at 1.4 s shows the maximum force of static friction just before the

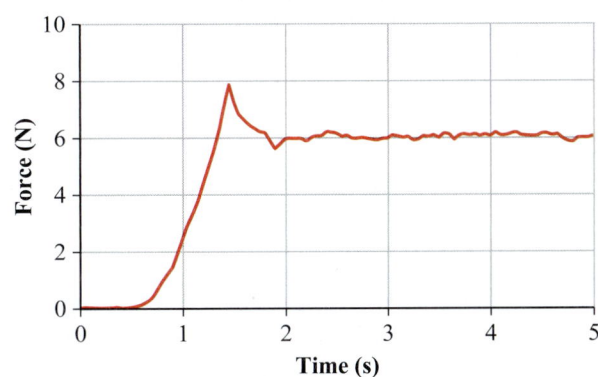

Figure 5-29 The magnitude of the force of friction reaches a maximum just before the object begins to move. When the object moves, the force of friction drops to an approximately constant value.

object starts to move. The magnitude of the force of friction drops to a lower, approximately constant, kinetic friction value as the object begins to move.

INTERACTIVE ACTIVITY 5-3

Forces and Motion: Friction

In this activity, you will use the PhET simulation "Forces and Motion: Basics" to investigate the force of friction. The simulation will allow you to experiment with applying forces to various objects in one dimension and see how the force of friction changes and how it affects the motion of objects.

MAKING CONNECTIONS

Friction, a Historical Perspective

Attempts to explain the origins of the force of friction date back as far as Leonardo da Vinci (1452–1519), who postulated that the force of friction on an object was proportional to the "load" on that object. As we now see, the "load" is represented by the normal force of contact between the object and the surface. For an object resting on a horizontal tabletop, it just equals the magnitude of the force of gravity on the object. While to date a complete understanding of the force of friction remain elusive, we know that the force is electromagnetic in nature. When two surfaces are in "contact" with one another, due to imperfections in the surfaces, only a small fraction of the area of the two surfaces can be in direct contact. It is those areas of contact that cause the friction between the surfaces. For a discussion of van der Waal's and interatomic surface forces, refer to more specialized books on the topic.

EXAMPLE 5-17

Coefficient of Kinetic Friction

Your friend is sitting on a sled on a level snow-covered surface. The sled has steel runners. You give the sled one push, which causes it to move away from you at an initial speed of 2.10 m/s. The sled stops 7.00 s later. Find the coefficient of kinetic friction, assumed constant, between the sled and the snow-covered surface.

Solution

Since we know the initial speed and elapsed time, we can calculate the acceleration. This in turn will give us information about the forces acting on the sled, which we can use in finding the coefficient of kinetic friction.

The FBD of the sled for the period after the push has ended is shown in Figure 5-30. There is one horizontal force, that of kinetic friction, $\vec{f}_k$, and two vertical forces, the force of gravity, $\vec{F}_g = m\vec{g}$, downward and a normal force, $\vec{N}$, upward. While not part of the FBD, we also show in Figure 5-30 the direction of

the velocity and the acceleration. The frictional force opposes motion so is in the opposite direction to the velocity. The acceleration is in the same direction as the frictional force.

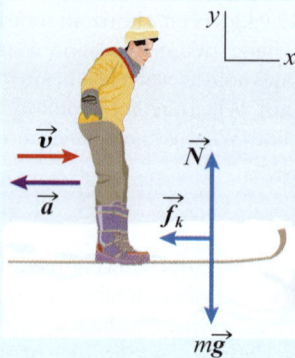

Figure 5-30   The FBD for the sled of Example 5-17.

We will adopt the horizontal direction as the x-axis, with positive being to the right. The vertical direction will be used for the y-axis, with the positive direction defined as upward.

In the vertical direction there is no acceleration, so Newton's second law produces

$$\sum F_y = ma_y = 0$$
$$N\hat{j} - mg\hat{j} = 0 \qquad (1)$$
$$N = mg$$

In the horizontal direction, the only force is that of kinetic friction. Using the definition of the coefficient of kinetic friction, Equation 5-8, we have the following, where Newton's second law is applied in the horizontal direction:

$$\sum F_x = ma_x$$
$$-f_k\hat{i} = ma_x$$
$$-\mu_k N\hat{i} = ma_x \qquad (2)$$
$$-\mu_k mg\hat{i} = ma_x$$
$$a_x = -\mu_k g$$

This relationship will allow us to find the coefficient of kinetic friction if we know the acceleration value.

We know that the initial speed is 2.10 m/s and the final speed (after 7.00 s) is 0.00 m/s. Since we assume a constant acceleration, we can use this information to find the acceleration:

$$v = v_0 + at$$
$$a = \frac{v - v_0}{t} = \frac{0.00 \text{ m/s} - 2.10 \text{ m/s}}{7.00 \text{ s}} = -0.300 \text{ m/s}^2 \qquad (3)$$

We can now combine (2) and (3) to obtain the coefficient of kinetic friction:

$$-0.300 \text{ m/s}^2 = -\mu_k(9.81 \text{ m/s}^2)$$
$$\mu_k = \frac{0.300 \text{ m/s}^2}{9.81 \text{ m/s}^2} = 0.0306$$

Making sense of the result

A sled with steel runners slides easily on snow, so a low value for μ_k is reasonable. We also note that μ_k has no units, as expected.

The coefficients of static and kinetic friction depend on both surfaces. Table 5-1 provides typical values for a number of different surfaces. The values in some cases can vary significantly with the amount of moisture present and the temperature, and with whether the surfaces are smooth and polished. Note that the coefficient of kinetic friction is normally significantly less than that of static friction for the same surfaces.

Table 5-1 Coefficients of Static and Kinetic Friction for Some Pairs of Surfaces

Material 1	Material 2	μ_s	μ_k
aluminum	steel	0.62	0.46
glass	glass	0.90	0.40
ice	ice	0.10	0.03
rubber	asphalt	0.90	0.65
rubber	concrete	1.00	0.75
steel	ice	0.03	0.02
steel	steel	0.74	0.56
Teflon	Teflon	0.05	0.04
waxed wood	dry snow	0.05	0.03
wood	wood	0.40	0.20

EXAMPLE 5-18

Will the Crate Move?

You are pushing a 26.0 kg crate on a horizontal floor, where the coefficient of static friction is 0.170 and the coefficient of kinetic friction is 0.130. You apply a 145 N force at an angle of 27.0° below horizontal.

(a) Will the crate move?
(b) If the crate does move, find its acceleration.

Solution

First, we draw the FBD of the crate in Figure 5-31. There are four forces, the applied force, $\vec{F}$; the normal force exerted upward by the surface, $\vec{N}$; the force of gravity acting vertically downward, $\vec{F}_g = m\vec{g}$; and the force of friction, $\vec{f}$. We do not yet know if the crate will move or not, so we have not designated whether the frictional force is kinetic or static. The applied force will tend to move the crate to the right (Figure 5-31), so the frictional force will act to the left. We will adopt the x-axis in the horizontal direction, with positive defined as to the right. The y-axis is vertical, with positive being defined upward.

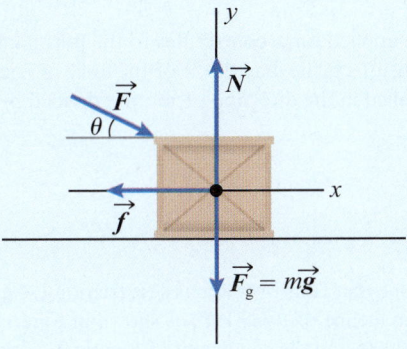

Figure 5-31 The FBD for the crate in Example 5-18.

(a) The only force that opposes the motion of the crate is the force of friction. If the horizontal component of the applied force is greater than the maximum force of static friction, then the crate will move. The maximum force of static friction depends on the normal force between the surface and the crate, which depends, in part, on the vertical component of the applied force.
   The applied force, $\vec{F}$, is the only force that requires calculation to determine its x- and y-components. Summing the forces in the x-direction gives

$$\sum F_x = ma_x \qquad (1)$$

Recalling that to the right is the positive direction,

$$F\cos\theta - f = ma \qquad (2)$$

Next we sum forces in the y-direction:

$$\sum F_y = 0 \qquad (3)$$

(continued)

Choosing up as our positive direction, we get

$$-F\sin\theta - mg + N = 0 \qquad (4)$$

$$N = mg + F\sin\theta \qquad (5)$$

The maximum force of static friction is, by Equation 5-6,

$$f_s^{max} = \mu_s N = \mu_s(F\sin\theta + mg)$$

Substituting the values, we get

$$f_s^{max} = (0.170)\left[(145\ \text{N})\sin 27.0° + (26.0\ \text{kg})(9.81\ \text{m/s}^2)\right] = 54.6\ \text{N} \qquad (6)$$

The x-component of the applied force is

$$F_x = F\cos 27° = (145\ \text{N})\cos 27.0° = 129\ \text{N}$$

Since the x-component of the applied force is greater than the maximum force of static friction, the crate will move.

(b) We use the force of kinetic friction in the calculation of the acceleration. Summing the forces in the x-direction and using (2), we can write

$$F\cos\theta - \mu_k N = ma$$

$$F\cos\theta - \mu_k(F\sin\theta + mg) = ma \qquad (7)$$

Therefore,

$$a = \frac{F\cos\theta - \mu_k(F\sin\theta + mg)}{m}$$

$$= \frac{145\ \text{N}\cos 27.0° - 0.130\left[(145\ \text{N})\sin 27.0° + (26.0\ \text{kg})(9.81\ \text{m/s}^2)\right]}{26.0\ \text{kg}}$$

$$= 3.36\ \text{m/s}^2$$

Making sense of the result

Since the vertical component of the applied force contributes to the normal force, the magnitude of the normal force is greater than mg. This, in turn, affects the magnitude of the force of friction. In the limit of no friction, the acceleration reduces to the force applied in the direction of motion divided by the mass, as expected.

EXAMPLE 5-19

Which Way Will It Move?

A 4.20 kg mass, m_1, hangs from a rope that runs over a massless frictionless pulley and is connected to a 6.70 kg mass, m_2, resting on the surface at an incline of $\theta = 31.0°$, as shown in Figure 5-32. The coefficient of static friction between the incline and m_2 is 0.0900, and the coefficient of kinetic friction is 0.0700. Initially, the system is at rest. Find the acceleration of the masses and the tension in the string.

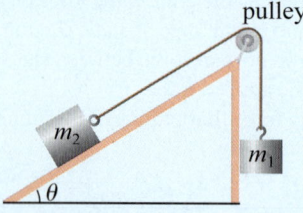

Figure 5-32   Example 5-19.

Solution

First, we determine which way the masses will move. Tending to make mass m_1 move downward, we have the gravitational force of m_1 at (4.20 kg)(9.81 m/s²) = 41.2 N. Tending to move the system in the other direction, we have the component of the force of gravity on m_2 along the incline at (6.70 kg)(9.81 m/s²)(sin 31.0°) = 33.9 N.

So the system will tend to move in a direction with mass m_1 lowering, but we also need to consider the frictional force to make sure that the system will actually move. The gravitational force on m_1 is opposed by the component of the gravitational force on m_2 along with the force of friction. The force of friction is given by $\mu_k N$. Mass m_2 is on the incline, so we can say that the normal force between m_2 and the incline is $N = m_2 g \cos\theta$. Therefore, $f_s = \mu_s m_2 g \cos\theta = 5.07$ N. The sum of the component of the weight of m_2 along the incline and the force of friction is less than the force of gravity on m_1, which means that m_1 will move down and m_2 will move up along the incline.

In problems involving friction, it is best to do the above check because the direction of motion will affect the form of the equations of motion, since we need to know the direction of the frictional force. This is not necessary in the absence of friction.

For the FBD for m_1, we have the force of gravity and the force of tension acting in the y-direction (Figure 5-33). We define upward as the positive direction:

$$\sum F_y = m_1 a_y \tag{1}$$

$$\Rightarrow T - m_1 g = -m_1 a \tag{2}$$

The acceleration of m_1 is negative because up is our choice for a positive direction and we have already shown that the mass moves and accelerates downward. Therefore the a in (2) is the magnitude of the acceleration.

The forces acting on m_2 (see Figure 5-34) are the force of friction, the normal force from the incline, the tension in the string, and the force of gravity. Since we have assumed that the rope and pulley are both massless, the tension will have the same magnitude at both ends of the rope. We have for simplicity used the same symbol, $\vec{T}$, for the tension in both Figures 5-33 and 5-34, recognizing that the magnitude of the tension is the same at both sides of the rope, but of course the directions of these two vectors are not the same. Figure 5-34 shows the force of gravity resolved into components along the two axes.

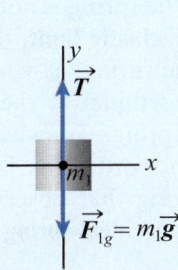

Figure 5-33   The FBD for m_1.

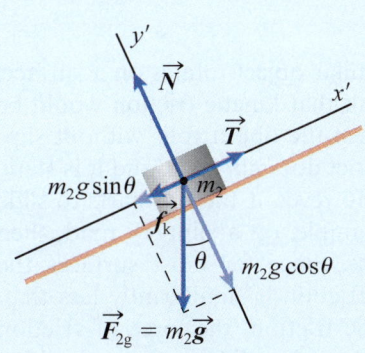

Figure 5-34   The FBD for m_2.

Summing the forces in the y-direction, we get

$$\sum F_{y'} = 0 \tag{3}$$

$$\Rightarrow N - m_2 g \cos\theta = 0$$

Hence

$$N = m_2 g \cos\theta \tag{4}$$

Summing the forces in the x-direction yields

$$\sum F_{x'} = m a_{x'} \tag{5}$$

$$\Rightarrow T - f_k - m_2 g \sin\theta = m_2 a \tag{6}$$

We multiply (2) by -1 and add it to (6) to get

$$m_1 g - f_k - m_2 g \sin\theta = m_1 a + m_2 a \tag{7}$$

Therefore,

$$a = \frac{m_1 g - \mu_k m_2 g \cos\theta - m_2 g \sin\theta}{m_1 + m_2}$$

$$= \frac{(9.81 \text{ m/s}^2)\left[(4.20 \text{ kg}) - (0.0700)(6.70 \text{ kg})\cos 31.0° - (6.70 \text{ kg})\sin 31.0°\right]}{4.20 \text{ kg} + 6.70 \text{ kg}}$$

$$= 0.313 \text{ m/s}^2 \tag{8}$$

(continued)

Thus, from (2),

$$T = m_1 g - m_1 a = m_1(g - a) = 4.20\ \text{kg}(9.81 - 0.313)\ \text{m/s}^2$$

$$= 39.9\ \text{N} \tag{9}$$

The tension in the rope is 39.9 N, in a direction up along the incline for mass m_2 and vertically upward on mass m_1. The acceleration is 0.313 m/s² directed up along the inclined plane for mass m_2 and vertically downward for mass m_1.

Making sense of the result

Since the acceleration is rather small (compared to g), we expect the value of the tension in the rope to be close to $m_1 g$. In the limiting case where the system is not moving (or is moving at a constant speed), the tension is exactly equal to $m_1 g$. It is worthwhile to determine whether the system will move when let go initially. For this, we have to look at the maximum force of static friction and see whether it is large enough to prevent the system from moving. In this case it is not. You can verify that yourself by using the value of μ_s provided.

When a tire or similar object rotates on a surface, you might at first think that kinetic friction would be relevant. However, when the object rolls without slipping, the point in contact does not slide, and it is static friction that applies. However, if the tire starts to slide on the surface, for example, on a slippery road, then kinetic friction applies. Since for most surfaces the coefficient of kinetic friction is significantly less than the coefficient of static friction, the force of friction is reduced when tires begin to slide instead of rolling without slipping. That is why you should try not to spin your wheels when going up a hill in slippery conditions. More on this topic in Chapter 9.

Look again at Figure 5-1. There is a maximum force set by the friction between the road and the boots of the man pulling the truck. By Newton's third law, the horizontal component of the force the ground exerts on the man must be equal in magnitude to the horizontal force the man exerts on the ground.

5-7 Spring Forces and Hooke's Law

In this section we will consider forces exerted by springs. While the results are directly applicable to ideal mechanical springs, they can also be applied to many other situations. The key relationship is named in honour of Robert Hooke (1635–1703), an English physicist who was a contemporary of Isaac Newton. Hooke played a key role in the publication of Newton's *Principia*.

Consider the spring–mass system of Figure 5-35. In position (a), the mass is in the **equilibrium position**, where the spring is unstretched and exerts no force on the mass. If you move the mass to the right to stretch the spring by a displacement $\vec{x}$ from the equilibrium position, the spring will pull the mass with a force opposite in direction to the displacement, as shown in Figure 5-35(b). When the mass is moved to the left, as in Figure 5-35(c), the compressed spring will push the mass to the right, again in the opposite direction to the extension of the spring.

As long the spring is not extended beyond what we refer to as its **elastic limit**, the magnitude of the force exerted by the spring, as you stretch or compress it, is directly proportional to the displacement of the free end of the spring. Translated from Latin in Robert Hooke's own words: "as is the extension, so is the force." A spring that observes this proportionality is referred to as an **ideal spring**.

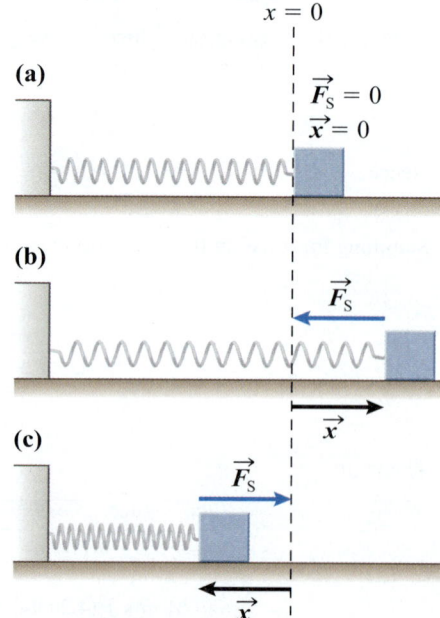

Figure 5-35 A mass on a spring that is in its equilibrium state (a), stretched (b), and compressed (c). For both extension and compression, the force on the mass is in the opposite direction to the displacement.

This relationship, applicable to ideal springs and solid deformations that have a similar response, is called **Hooke's law**, usually expressed as

KEY EQUATION $$\vec{F} = -k\vec{x} \qquad (5\text{-}9)$$

where k is the spring constant, $\vec{F}$ is the force exerted by the spring on the mass, and $\vec{x}$ is the extension or compression of the spring from the unstretched equilibrium state. The minus sign indicates that the force that the spring exerts on the mass acts in a direction that is opposite to the direction of the displacement of the mass from the equilibrium position. Hooke's law holds as an accurate approximation for deformation of many solids as long as the deformation is within their elastic limit, as we shall see in more detail in Chapter 10 on equilibrium and elasticity.

The greater the value of k, the more force is required to stretch the spring by a given amount, and, therefore, the *stiffer* the spring will be. Since Hooke's law is a good approximation for the behaviour of solids undergoing small enough deformations that they stay within their elastic limit, it is used in many applications, ranging from lattice vibrations in solid-state physics to the study of mechanical vibrations in engineering and the measurement of seismic activity.

Beyond the elastic limit, the response of the spring to elongation is no longer linear, as shown in Figure 5-36.

CHECKPOINT 5-13

Mass on a Vertical Spring

A spring with spring constant k hangs vertically from a ceiling. You attach a 2 kg mass to the free end of the spring, pull the mass down, and then release it. When will the acceleration of the mass be zero?

(a) never, because the mass is always under the effect of gravity
(b) when the mass stops momentarily at the maximum extension of the spring
(c) when the mass returns to the top and stops momentarily
(d) when the elongation of the spring from the unstretched position is mg/k

ANSWER: (d) This is the point where the net force is zero, since the force of the spring just balances the gravitational force on the mass.

CHECKPOINT 5-12

Units for Spring Constant

What are the dimensions, expressed in base SI units, for the spring constant k?

(a) N/m
(b) N·m
(c) kg/s^2
(d) it has no units

ANSWER: (c) From Equation 5-9, we can see that k must have units of N/m. However, the newton is not a base SI unit (see Chapter 1). If we write (based on Newton's second law) $1 \text{ N} = 1 \text{ kg} \cdot \text{m/s}^2$, we can reduce the units for k to N/m = (kg · m)/(m·s^2) = kg/s^2.

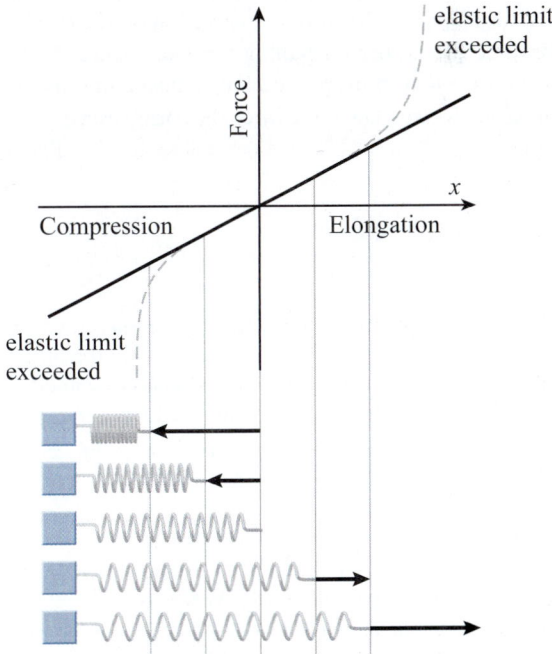

Figure 5-36 When a spring is stretched or compressed beyond its elastic limit, the force is no longer proportional to the displacement.

Source: Svjo. This file is licensed under the Creative Commons Attribution-Share Alike 3.0 Unported license, https://creativecommons.org/licenses/by-sa/3.0/deed.en

EXAMPLE 5-20

Finding the Spring Constant

A horizontal spring has one end attached to a rigid support. You stretch the free end by 10.5 cm and measure the force needed to do so as 25.0 N.

(a) What is the spring constant?
(b) What is the acceleration of a 10.0 kg mass attached to one end of the spring (with the other end attached to the rigid support) when the spring is stretched by 15.0 cm?

Solution

(a) We know the force required for stretching the spring by a given amount; this is a direct application of Hooke's law. Referring to the spring in Figure 5-37, we choose a positive x-axis that points to the right. By Hooke's law, we have

$$\vec{F} = -k\vec{x}$$

$$k = -\frac{\vec{F}}{\vec{x}} = -\frac{(-25.0 \text{ N})\hat{i}}{(0.105 \text{ m})\hat{i}} = 238 \text{ N/m}$$

Note that the $\vec{F}$ in Hooke's law is the force exerted by the spring on the mass. This will have the same magnitude as the force you apply to stretch it, but in the opposite direction (in this example you apply a force to the right, while the force exerted by the spring is to the left).

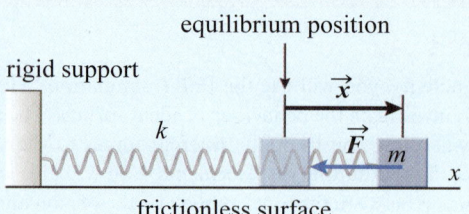

Figure 5-37 Example 5-20.

(b) We have the mass, the extension, and now the spring constant, k. The acceleration is given by the net force, in this case the force of the spring, divided by the mass. Since the spring is stretched to the right, the spring force, which is the net force on the mass, will act to the left, and hence the acceleration will also point to the left. Its magnitude is given by

$$a = \frac{F_{\text{net}}}{m} = \frac{-kx}{m} = -\frac{(238 \text{ N/m})(0.150 \text{ m})}{10.0 \text{ kg}} = -3.57 \text{ m/s}^2$$

Making sense of the result

The units for the acceleration are N/kg, which reduces to m/s^2 when we replace N with kg·m/s^2. The vector form of Newton's second law yields the right sign for the acceleration (negative), which, according to our choice of positive direction to the right, means that the acceleration of the mass is to the left.

PEER TO PEER

When we first studied springs, my friend and I were sloppy about specifying directions of forces. It is important to be clear whether it is the force of the spring on the mass, or the force of the person pulling the mass, since the two forces are equal in magnitude but opposite in direction. The $\vec{F}$ in Hooke's law is the force the spring exerts.

When the spring–mass system is on a frictionless horizontal surface, although the surface exerts a normal force that balances the force of gravity on the mass, these forces do not enter into the solution. In Example 5-21, we deal with a vertical spring, where we have to account for the effects of the force of gravity.

EXAMPLE 5-21

Vertical Spring

A massless spring hangs vertically with one end attached rigidly to the ceiling. You attach a 6.00 kg mass to the other end and let it go. Using video tracking and analysis you establish that the spring reaches its maximum speed when it is 11.0 cm below the unstretched length. Find the spring constant. Assume that the spring itself has no mass.

Solution

The acceleration is zero at the point of maximum speed since there is no change in velocity at this point. By Newton's second law, the point of zero acceleration is the point where the net force on the mass is zero. The net force acting on the mass is the vector sum of the force of gravity and the force of the spring. The net force is zero when the magnitudes of these two opposing forces are equal. When the mass is originally attached to the spring, the force on

the mass is only the force of gravity. As the spring stretches, the force that the spring exerts on the mass increases until it is equal in magnitude to the force of gravity, and that is where we have the net force and hence the acceleration equal to zero. This happens when (in terms of magnitude)

$$kx = mg$$

Therefore,

$$k = \frac{mg}{x} = \frac{(6.00 \text{ kg})(9.81 \text{ m/s}^2)}{0.110 \text{ m}} = 535 \text{ N/m}$$

Making sense of the result

These are important and powerful ideas that you will encounter again in Chapter 13 when we study simple harmonic motion. The spring constant, k, always has a positive value.

5-8 Fundamental and Non-fundamental Forces

In this chapter, we have encountered many different forces so far: normal force, kinetic and static frictional force, gravitational force, spring force, electromagnetic force, tension forces, and mechanically applied forces. This is by no means a comprehensive list, and you will encounter other forces as you continue your study of physics. We have already commented that some of these forces are really due to a different force. For example, the normal force exerted by a surface on an object is at a fundamental level electromagnetic in nature. It is helpful to divide forces into two categories: fundamental forces and non-fundamental forces.

Fundamental forces are forces that result from the fundamental interactions in nature. Interacting objects exert forces on one another and respond to these forces. For example, a pair of charges interact electromagnetically, and as a result exert an electrical force on one another (we study this in much more detail in Chapter 19). The force exerted by charges on one another results in the charges being attracted toward or repelled from one another.

Another fundamental force, and related interaction, is that of gravity. For example, Earth experiences a gravitational force due to the Sun (and vice versa) as a result of the gravitational interaction between them. These forces shape the resulting orbits—a topic you will consider in much more detail in Chapter 11.

The fundamental forces are also referred to as **field forces**. For example, one can think of an object in Earth's vicinity as being within Earth's gravitational field, so the object experiences a force of gravitational attraction. We discuss fields in more detail in Chapter 11 on gravitation and in Chapter 19 on electric fields and forces. The fundamental forces are also referred to as forces that act *at a distance* and do not require contact.

The four fundamental interactions are

I. **gravitational interaction** (affects everything that has mass)

II. **electromagnetic interaction** (affects electrically charged objects)

III. **weak interaction**, which as we will see in Chapter 35 affects subatomic particles called quarks and leptons (electrons, muons, tau particles, and their neutrinos)

IV. **strong interaction**, with its two sub-branches:

   A. the fundamental interaction between quarks and gluons

   B. the residual interaction between hadrons, such as protons and neutrons

While the summary of interactions given above represents the current state of our understanding, there remains active research that may well alter this picture. As you will see in more detail in Chapter 30 when we consider general relativity, some question whether gravity should be considered one of the fundamental interactions, or if there is a different role for the gravitational interaction. The electromagnetic force and the weak force are sometimes listed under one force, the electroweak force. In physics we are interested in common principles with the widest possible areas of application. For example, quantum gravity aims to unify general relativity and quantum theory. Work on the **Grand Unified Theory** (GUT) aims to unify the electromagnetic, weak, and strong forces, and string theory seeks to unify all interactions.

The **non-fundamental forces** depend on one or more of the fundamental forces. For example, the normal force exerted by a surface is a result of electromagnetic repulsive forces at the atomic level. Friction, elastic, tension, and normal forces (along with others) are all non-fundamental.

In this chapter, we will mostly focus on forces that require physical contact between objects (non-fundamental forces), with the exception of the force of gravity, which is a fundamental force. Such contact forces include the normal force, the tension force, the elastic force mainly exerted by a spring, and the forces of friction. The fundamental forces do not require contact (they act at a distance), while most non-fundamental forces do involve contact or near contact. What we sometimes consider as

MECHANICS

non-fundamental forces are, in fact, macroscopic manifestations of the fundamental interactions. For example, the normal force between two objects in contact results from electromagnetic repulsion and the Pauli exclusion principle, and the electromagnetic interaction is the basis for the force of friction. The force of universal gravitation (see Chapter 11) between every two point masses is a fundamental force, but the gravitational force near the surface of Earth (or another planet) given by $m\vec{g}$ is the result of many of these fundamental interactions (due to all the point masses that make up Earth). Most would still regard it as fundamental, but the situation is not as clear as for gravitational interactions between point masses.

In Section 5-10 you will be introduced to fictitious forces. These are not real forces, but rather are force-like effects due to the choice of an accelerated reference frame.

According to Newton's second law, an object moving in a circular path must be experiencing a net force that points toward the centre of the circle as well. This net force is the vector sum of all the forces acting on the object and it must point radially inward. It is sometimes called a centripetal force, although most physicists prefer not to use the term. The so-called centripetal force is *not* an independent force that just exists for an object moving in a circle; it is the sum of all the forces pointing in the radial direction, which in fact causes the object to move in its circular trajectory.

CHECKPOINT 5-14

Fundamental Forces

The fundamental forces
(a) can be referred to as field forces
(b) result from the fundamental interactions
(c) are the basis of all the non-fundamental forces
(d) all of the above

ANSWER: (d)

PEER TO PEER

At first I thought that there was a separate mysterious force called the "centripetal force" that I added (wrongly!) into my FBD with the real forces. Once I realized it was just the net force from the other forces, things were much clearer (and more correct)!

5-9 Uniform Circular Motion

An object moving with uniform speed v along a circular trajectory of radius r has a radially inward **centripetal acceleration** with a magnitude of

$$a_r = \frac{v^2}{r} \tag{5-10}$$

The word *centripetal* means "centre seeking," and it is important to remember that the centripetal acceleration is *inward* toward the centre of the circle.

CHECKPOINT 5-15

Mass on String

You swing a mass at the end of a string in a vertical circle such that the string remains taut at all times. The net force that provides the centripetal acceleration is
(a) the force of gravity
(b) the tension in the string
(c) the vector sum of the force of gravity and the tension
(d) a force other than the force of gravity and the tension

ANSWER: (c) While the tension force on the mass is downward when at the top of the loop, and upward at the bottom, in all cases you add as vectors the tension force and the gravitational force to get the net force. That net force must provide the centripetal acceleration if the mass is to move in the circular path.

EXAMPLE 5-22

Traffic Circle Speed Limit

A traffic circle has a radius of 42.0 m. The traffic circle is not banked, and it is to be designed for a minimum coefficient of static friction between the tires and the road of 0.270. Find the maximum speed that a car can travel on the traffic circle without sliding.

Solution

The centre of the traffic circle is to the left in the FBD in Figure 5-38. The force of friction, as shown, points to the centre of the circle; the normal force of the pavement on the car is vertically upward; and the gravitational force is vertically downward.

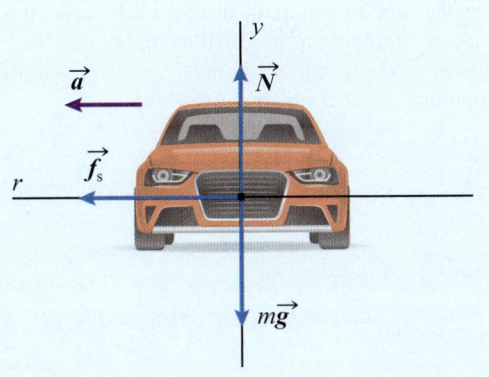

Figure 5-38 The FBD of the car in Example 5-22 shows a frictional force, a normal force, and the force of gravity.

Since the direction of the acceleration is toward the centre of the traffic circle, we choose a horizontal x-axis (or r-axis). We choose the direction pointing to the centre of the circle as the positive r-axis (i.e., to the left in the diagram). Note that we are labelling the horizontal axis as the "r-axis." One can also use "x-axis"; however, in the context of circular motion and radial accelerations, using the label "r" for the axis is more appropriate.

Summing the forces in the horizontal (or radial) direction, we get

$$\sum F_r = ma_r \tag{1}$$

The only force acting on the car is the force of friction, so static friction must provide the required centripetal force. Since the car is on the verge of slipping, the force of static friction is at its maximum value:

$$f_s^{max} = \mu_s N = \frac{mv^2}{r} \tag{2}$$

To solve for v, we need to calculate the value for N. Summing the forces in the y-direction, we get

$$\sum F_y = ma_y = 0 \tag{3}$$

Choosing up as our positive direction, we get

$$N - mg = 0 \Rightarrow N = mg \tag{4}$$

Combining (2) and (3), we get

$$\mu_s mg = \frac{mv^2}{r} \Rightarrow v^2 = rg\mu_s$$

$$\Rightarrow v = \sqrt{(42.0 \text{ m})(9.81 \text{ m/s}^2) \times 0.270} = 10.5 \text{ m/s} \tag{5}$$

Making sense of the result

Considering limiting cases, when the coefficient of friction is zero, the speed of the car, for it not to slip, must be zero! The car would not be able to turn, no matter how low its speed. The coefficient of friction for a road design must be based on the worst tires in wet conditions, that is, the lowest coefficient of friction. Traffic circles intended for higher speed use are almost always banked. You might be surprised that we used the coefficient of static friction, rather than kinetic, but the car was not sliding, and a tire that rolls without slipping dictates that it is static friction.

EXAMPLE 5-23

Staying in the Loop

You swing a 315 g metal ball at the end of a 1.25 m long string in a vertical circle. The mass has a speed of 17.0 m/s when at the lowest point in the circle.

(a) Find the tension in the string when the mass is at its lowest point.
(b) What is the minimum speed for the mass when it is at the top of the loop for the rope to stay taut?

Solution

(a) The FBD of the ball in Figure 5-39 shows that only the tension in the string and the force of gravity act on the mass. Using Newton's second law, we have

$$\sum \vec{F} = m\vec{a} \tag{1}$$

Taking the direction toward the centre of the circle to be positive, we get

$$T - mg = ma_r \Rightarrow T = mg + \frac{mv^2}{r}$$

$$T = 0.315 \text{ kg}\left(9.81 \text{ m/s}^2 + \frac{(17.0 \text{ m/s})^2}{1.25 \text{ m}}\right) = 75.9 \text{ N} \tag{2}$$

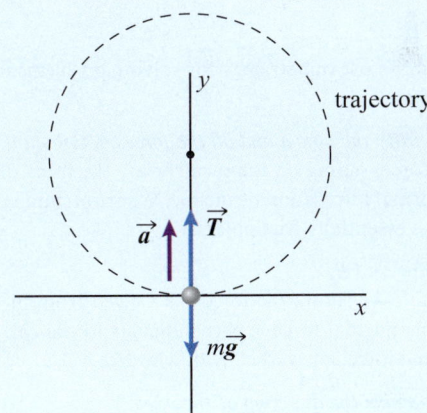

Figure 5-39 The FBD for the mass in part (a) of Example 5-23. The tension force points to the centre of the circle, as does the acceleration.

(b) When the ball is at the top, there are two forces, both acting downward, the tension from the string and the force of gravity (see Figure 5-40). Together these forces must provide the net force to drive the centripetal acceleration.

(continued)

MECHANICS

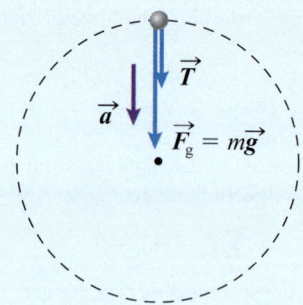

Figure 5-40 The FBD for the mass in part (b) of Example 5-23. The tension force and gravitational force both point downward, and together they provide the net force needed for the centripetal acceleration.

That is, Newton's second law applied in the vertical direction, and taking the direction toward the centre of the circle as positive (i.e., in this case downward) yields

$$\sum \vec{F}_r = m\vec{a}_r$$

$$T + mg = \frac{mv^2}{r} \tag{3}$$

For the rope to stay taut, there must be at least a slight tension remaining in it. For the limiting case we will set $T \to 0$ in (3) to obtain the following expression for the minimum speed at the top of the loop:

$$0 + mg = \frac{mv_{min}^2}{r} \tag{4}$$

$$v_{min} = \sqrt{gr}$$

Substituting in values gives

$$v_{min} = \sqrt{(9.81 \text{ m/s}^2)(1.25 \text{ m})} = 3.50 \text{ m/s}$$

Making sense of the result

The tension in the string in part (a) is much greater than the weight of the ball, as it supports the weight of the ball and provides its acceleration toward the centre to keep it moving in a circle. If the speed of the ball is zero, the tension is equal to mg, as expected. In part (b), we find the minimum speed at the top of the loop to just maintain the smallest amount of tension in the string. If the speed is reduced below this value, the mass will not complete a circle.

EXAMPLE 5-24

Frictionless Banked Racetrack

A race car is on a very slippery circular track of radius 195 m. The racetrack is banked at an angle of 27.0° with respect to the horizontal. Find the speed at which the race car will not slide up or down the track, assuming that friction with the track surface is negligible.

Solution

Here again, we use our strategy for solving problems involving Newton's laws.

Step 1. *Identify the object and all the forces acting on it.*

The forces acting on the race car are the force of gravity and the normal force from the incline. We are assuming that the racetrack is essentially frictionless.

Step 2. *Draw an FBD.*

Figure 5-41 shows the forces as viewed from in front of the car. The normal force is perpendicular to the car, and the gravitational force acts vertically downward.

Step 3. *Establish the direction of the axes.*

Since it is assumed not to slip up or down the track, the car will move in a horizontal circle. The centripetal acceleration is therefore directed horizontally toward the centre of the circle. Therefore, we choose a horizontal axis and a vertical y-axis. Either x-axis or r-axis is an appropriate label for the horizontal axis. However, since centripetal or radial acceleration is involved, we will refer to this as the r-axis, for radial (see Figure 5-41).

Step 4. *Define the positive direction for the axes.*

For the y-axis we select up as the positive direction, while for the r-axis we select to the left (toward the centre of the circle) as positive.

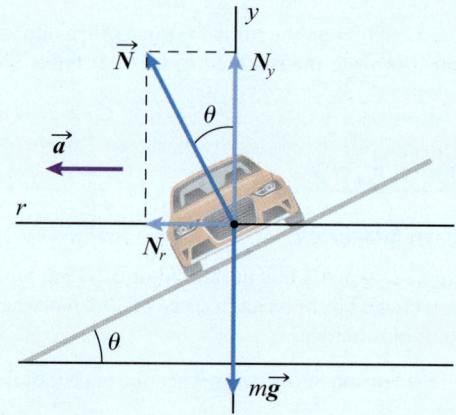

Figure 5-41 The FBD for the car on the banked frictionless track of Example 5-24.

Step 5. *Resolve the forces into components along the chosen axes.*

The force of gravity points along the vertical axis in the negative direction. We use trigonometry to resolve the normal force exerted by the incline as a radial component ($N_r = N \sin \theta$) and a vertical component ($N_y = N \cos \theta$), as shown in Figure 5-41.

Step 6. *Use Newton's second law to write equations for the motion along each axis.*

The sum of the forces along the vertical direction is

$$\sum F_y = ma_y = 0 \tag{1}$$

Remembering that up is the positive direction, we have

$$N\cos\theta - mg = 0 \tag{2}$$

Next we sum the forces along the radial direction and use the fact that the acceleration is the centripetal acceleration:

$$\sum F_r = ma_r$$

$$\Rightarrow N\sin\theta = ma_r = m\frac{v^2}{r} \qquad (3)$$

Step 7. *Solve the equations of motion.*
Equation (2) can be rewritten as

$$N\cos\theta = mg \qquad (4)$$

We have three unknowns: the mass of the car (m), the speed of the car (v), and the normal force (N). Dividing (3) by (4), the mass is eliminated, and we obtain

$$\tan\theta = \frac{v^2}{rg}$$

$$\Rightarrow v = \sqrt{rg\tan\theta} = \sqrt{(195\,\text{m})(9.81\,\text{m/s}^2)\tan 27.0°}$$

$$= 31.2\,\text{m/s}$$

Making sense of the result

Notice that the direction of the acceleration is horizontally toward the centre of the circle in this example. For a banking angle of zero, the car, without friction, will not be able to move along the track without sliding. You are asked to solve this same problem using the component-free vector approach in problem 92. Of course, the idealized case of a frictionless track is not realistic, and the next example considers the case for a banked track with friction.

EXAMPLE 5-25

Friction on a Banked Curve

A car is travelling on a curve of radius r banked at an angle θ. The coefficient of static friction between the car and the road is μ_s.

(a) Derive an expression for the maximum speed that the car can travel without slipping.
(b) Derive an expression for the normal force.

Solution

(a) **Steps 1 and 2.** *Identify all the forces acting on the object, and draw the FBD.*

The forces acting on the car are the force of gravity, the normal force from the incline, and the force of friction, with the FBD shown in Figure 5-42. Since the situation involves the maximum speed beyond which the car will slip, the force of friction will point down along the incline to oppose that.

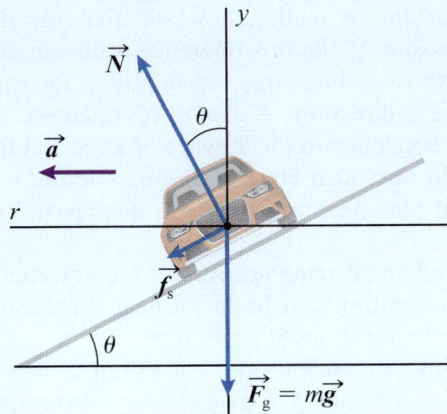

Figure 5-42 The FBD for the car as it goes around a banked curve for Example 5-25. Notice the direction of the force of friction.

Steps 3 and 4. *Establish the directions for the coordinate axes, and define the positive directions.*

The car moves in a horizontal circle, and the acceleration points toward the centre of the circular path.

Therefore, we again choose a horizontal (radial) r-axis and a vertical y-axis. We will define the positive directions as upward for the y-axis and to the left for the r-axis.

Step 5. *Resolve the forces into components along the coordinate axes.*

In Figure 5-43 we show the components of the normal force and the force of friction. For the normal force, the components are $N_r = N\sin\theta$ and $N_y = N\cos\theta$. The static friction force components are $f_r = f_s\cos\theta$ and $f_y = -f_s\sin\theta$.

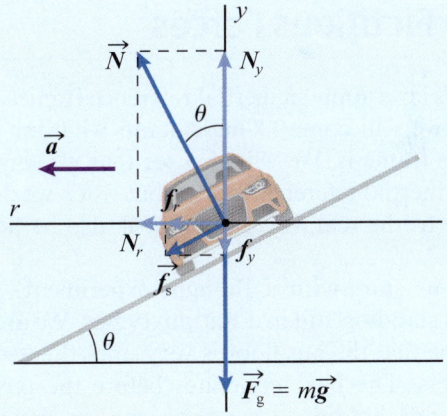

Figure 5-43 The FBD for the car of Example 5-25, now with the normal and frictional forces divided into force components.

Step 6. *Write the equations of motion along each axis.*

The sum for the forces along the vertical direction gives us

$$\sum F_y = ma_y$$

Remembering that we have defined up as the positive direction,

$$N\cos\theta - mg - f_s\sin\theta = 0 \qquad (1)$$

(continued)

If the car is travelling at the maximum speed possible without slipping, then it is on the verge of slipping, and the force of static friction is at its maximum value. Therefore, we have

$$N\cos\theta - mg - N\mu_s\sin\theta = 0 \qquad (2)$$

Next, we sum the forces along the radial direction:

$$\sum F_r = ma_r$$

We have defined the radial direction toward the centre of the circle as positive, and that is also the direction of the centripetal acceleration, so we have

$$N\sin\theta + N\mu_s\cos\theta = \frac{mv^2}{r} \qquad (3)$$

Step 7. *Solve the equations of motion.*
Rearranging (2) gives

$$N\cos\theta - N\mu_s\sin\theta = mg \qquad (4)$$

We divide (3) by (4) to eliminate N and m:

$$\frac{\sin\theta + \mu_s\cos\theta}{\cos\theta - \mu_s\sin\theta} = \frac{v^2}{rg}$$

$$v^2 = \frac{rg(\sin\theta + \mu_s\cos\theta)}{\cos\theta - \mu_s\sin\theta}$$

$$v = \sqrt{\frac{rg(\sin\theta + \mu_s\cos\theta)}{\cos\theta - \mu_s\sin\theta}}$$

(b) We can use (4) to write an expression for the normal force:

$$N = \frac{mg}{\cos\theta - \mu_s\sin\theta} \qquad (5)$$

Making sense of the result

When the coefficient of friction is zero, the value of the normal force reduces to $mg/\cos\theta$, and v^2 reduces to $rg\tan\theta$, as we obtained in Example 5-24.

Each of the examples involving inclined planes (Examples 5-10, 5-11, 5-19, 5-24, and 5-25) gives a different value for the normal force, depending on the situation involved. It is a common error to assume that the normal force is always equal to mg or $mg\cos\theta$.

5-10 Reference Frames and Fictitious Forces

Newton's laws apply in inertial reference frames. In this section, we will come to understand what an inertial reference frame is. We will also see that we sometimes use non-inertial reference frames, but when we do so, in addition to the real forces, there will also be fictitious forces.

Let us start with a thought experiment. A passenger is standing still in a stationary bus. We are going to assume that the bus floor is very smooth, essentially frictionless. The bus accelerates before the passenger (who is not holding a support) can sit down. You happen to be standing outside the bus watching what happens through the window, while a friend of yours is already sitting inside the bus. You and your friend are in different reference frames.

From your perspective, as the bus accelerates, the passenger simply stays still with respect to you (obviously this would be limited to the short time interval until the person collides with the back of the bus). Since the passenger is not holding onto anything, and since there is no friction between his shoes and the floor of the bus, the net horizontal force on the person must be zero. Your frame of reference is called an **inertial reference frame**. In an inertial reference frame an object

will only accelerate if there is a net force. There is no such horizontal force in this case, so from your inertial reference frame the person does not accelerate.

How do things look from the point of view of your friend who is sitting in her seat on the bus? What she will observe is the standing passenger accelerating backward with respect to her. It would only be natural for her to assume that the standing person is under the influence of a force pulling him backward. We call this force a **fictitious force**, since it is only present when observed in the non-inertial accelerating reference frame. The magnitude of the force will be the person's mass times the acceleration of the reference frame. It is important to realize, however, that the directions are opposite. If the bus reference frame accelerates in the positive x-direction, the fictitious force is in the negative x-direction. A frame of reference in which objects accelerate in the absence of a real net force, and in which one must employ fictitious forces to explain physical phenomena, is called a **non-inertial reference frame**.

Accelerated reference frames are non-inertial, and that acceleration can be in a linear direction, as was the case in our thought experiment, or because we have a rotating reference frame. For example, if you sit on a rapidly rotating playground merry-go-round, from your non-inertial rotating reference frame it will seem that there is a fictitious force throwing you radially outward. This fictitious force is called the **centrifugal force**. Note as before that the direction of the fictitious force (radially outward) is opposite to the direction of the acceleration of the non-inertial reference frame (radially inward). You can always eliminate the centrifugal force by choosing an inertial reference frame.

Some physicists prefer to avoid the topic of the centrifugal force since it is a fictitious force related to a non-inertial reference frame choice. However, with a non-inertial reference frame choice, the motion of objects must include the fictitious force.

We implicitly assumed throughout the earlier parts of this chapter when we applied Newton's second law that we were observing from an inertial reference frame. It is important to be clear that we have fictitious forces, such as the centrifugal force, when (and only when) we have non-inertial reference frames.

EXAMPLE 5-26

How Big Is the Fictitious Force?

A person with a mass of 64.0 kg is in a subway car that at that instant is accelerating at 1.95 m/s² in the positive x-direction. Is there a fictitious force from his reference frame, and if so what is its value?

Solution

Since the reference frame is accelerating, it is non-inertial and there will be a fictitious force. The magnitude of the fictitious force is simply given by

$$F_{\text{fict}} = ma = (64.0 \text{ kg})(1.95 \text{ m/s}^2) = 125 \text{ N}$$

The direction of the fictitious force is opposite to the direction of the acceleration. Therefore, it must, in this case, be in the negative x-direction:

$$\vec{F}_{\text{fict}} = -(125 \text{ N})\hat{i}$$

Making sense of the result

Remember that the direction of the fictitious force is always opposite to that of the acceleration of the reference frame. If you are sitting in a subway car, the horizontal force exerted by the seat on you will equal this fictitious force in magnitude, allowing you to not accelerate relative to the subway car.

EXAMPLE 5-27

Fictitious Force on a Merry-Go-Round

A child of mass 19.0 kg is sitting on the edge of a merry-go-round that has a diameter of 5.20 m. The merry-go-round makes one revolution every 3.20 s. What is the centrifugal force on the child?

Solution

Since the reference frame is accelerating, it is non-inertial and there is a fictitious centrifugal force. The magnitude of the fictitious force is simply given by the product of the mass and the acceleration, with in this case the acceleration being the centripetal acceleration.

We first need to find the speed of a point on the edge of the merry-go-round:

$$v = \frac{2\pi r}{T} = \frac{2\pi(2.60 \text{ m})}{3.20 \text{ s}} = 5.105 \text{ m/s}$$

Next we calculate the centripetal acceleration:

$$a_r = \frac{v^2}{r} = \frac{(5.105 \text{ m/s})^2}{2.60 \text{ m}} = 10.0 \text{ m/s}^2$$

(continued)

The magnitude of the centrifugal force is simply the product of this acceleration with the mass:

$$F_c = ma_r = (19.0 \text{ kg})(10.0 \text{ m/s}^2) = 190. \text{ N}$$

In terms of direction, the fictitious force is opposite to the direction of the acceleration. Therefore, it must point radially outward. We have included a decimal point at the end of the force value to indicate explicitly that it is precise to three significant digits.

Making sense of the result

Remember that the centrifugal force is not a real force, but simply a fictitious force because we chose to use an accelerated non-inertial reference frame. The centripetal acceleration points radially inward, but the centrifugal force is directed radially outward.

5-11 Momentum and Newton's Second Law

In Chapter 7, you will encounter the important physics concept of linear momentum. We briefly introduce linear momentum here, since it provides an alternative way to view Newton's second law. **Linear momentum**, $\vec{p}$, is a vector quantity defined as the product of the mass, m, and the velocity, $\vec{v}$, of an object:

$$\vec{p} = m\vec{v} \qquad (5\text{-}11)$$

If we have two objects travelling at similar speeds, the more massive object has more linear momentum. If we have two objects of similar mass, the one at the higher speed will have more linear momentum.

MAKING CONNECTIONS

The g-Force

The term **g-force** is widely used but unfortunate wording since it is not truly a force, but rather a ratio of accelerations. When we say there is a g-force of 2, we mean that the magnitude of the acceleration is double the magnitude of the acceleration due to gravity at Earth's surface.

Aerobatic pilots flying tight loops can experience considerable g-forces relating to their non-inertial reference frame. For an inside loop, the plane flies in a vertical circular path loop with the canopy (and the pilot's head) facing toward the centre of the loop (Figure 5-44). The g-forces experienced during an inside loop are called positive g-forces. Positive g-forces drive blood away from the brain. During a high positive g-manoeuvre, pilots can experience grey-outs or blackouts as a result of reduced oxygen in the brain. With special training and a G-suit, which helps maintain blood flow to the brain, a pilot can withstand a positive g-force of 9g.

During an outside loop, the top of the plane faces away from the centre of the vertical loop. Depending on the acceleration, an outside loop can cause negative g-forces, which force more blood into the brain and are often more dangerous than positive g-forces. Generally, the maximum tolerable negative g-force is about −3g.

In contrast, astronauts experience g-forces mostly during liftoff, at the end of a launch, and during re-entry, due to linear rather than rotational accelerations. The blood is "driven" to either the front or the back of the body. As the body accelerates, the blood's inertia will cause it to lag behind the body parts, owing to blood's fluid nature, which causes the blood to be pushed toward the front or the back relative to the accelerating body. The g-forces experienced in this way are called horizontal g-forces and can be as high as 12g or 17g for short periods of time, depending on the direction of the force causing

Photography Perspectives—Jeff Smith/Shutterstock.com.

Figure 5-44 A plane in a positive g loop.

the acceleration. In 1954, Colonel John Stapp is reported to have sustained a g-force of 46.2g in a rocket sled as it was slowing down.

Newton's second law can be written in terms of the time derivative of the linear momentum:

KEY EQUATION
$$\vec{F}_{\text{net}} = \frac{d\vec{p}}{dt} \qquad (5\text{-}12)$$

If we substitute Equation 5-11 into Equation 5-12, and assume that mass is constant, we readily obtain the familiar form $\vec{F}_{\text{net}} = m\vec{a}$.

This may just seem a matter of notation, but momentum is a more fundamental concept than acceleration in physics, and the form shown in Equation 5-12 is preferred as one goes on to more advanced physics courses. We can also see from Equation 5-12 that if the net force is zero, then $\frac{d\vec{p}}{dt} = 0$, and as we will see in Chapter 7, this means that linear momentum is a conserved quantity.

We could rearrange and integrate Equation 5-12 to obtain a vector quantity called **impulse**, $\vec{J}$:

$$\vec{J} = \int d\vec{p} = \int \vec{F}_{\text{net}}\, dt \qquad (5\text{-}13)$$

The units of impulse are the same as those of momentum. Impulse represents the vector change in linear momentum. Normally we apply impulse concepts when a single force dominates, and often we use the average value of that force along with the time of application of the force, Δt, as given in Equation 5-14:

KEY EQUATION
$$\vec{J} = \vec{F}_{\text{average}}\Delta t \qquad (5\text{-}14)$$

We demonstrate the use of this relationship in Example 5-28.

EXAMPLE 5-28

Average Force of a Bat on a Baseball

A baseball bat is in contact with a ball for 0.800 ms. If the baseball was travelling at 40.0 m/s before striking the bat and leaves the bat at a speed of 48.0 m/s, what must be the average force exerted by the bat on the ball? Assume that the ball leaves the bat in the exact opposite direction to the arrival direction. The mass of a baseball is 145 g.

Solution

Let us start by considering the change in linear momentum of the baseball during the collision. Initially, the ball is moving in one direction (we will define this as the negative x-direction) at 40.0 m/s. The linear momentum of the baseball at this time is

$$\vec{p}_{\text{i}} = -(0.145\ \text{kg})(40.0\ \text{m/s})\hat{i} = -(5.80\ \text{kg}\cdot\text{m/s})\hat{i}$$

The linear momentum of the baseball just after the collision is in the positive x-direction:

$$\vec{p}_{\text{f}} = +(0.145\ \text{kg})(48.0\ \text{m/s})\hat{i} = +(6.96\ \text{kg}\cdot\text{m/s})\hat{i}$$

The impulse, or change in linear momentum, has the following value:

$$\Delta\vec{p} = \vec{p}_{\text{f}} - \vec{p}_{\text{i}} = +(6.96\ \text{kg}\cdot\text{m/s})\hat{i} - (-5.80\ \text{kg}\cdot\text{m/s})\hat{i}$$
$$= (12.76\ \text{kg}\cdot\text{m/s})\hat{i}$$

Now we can calculate the average force using the change in momentum and the contact time:

$$\vec{F}_{\text{avg}} = \frac{\Delta\vec{p}}{\Delta t} = \frac{(12.76\ \text{kg}\cdot\text{m/s})\hat{i}}{0.000\,800\ \text{s}} = (15\,950\ \text{N})\hat{i} = 16.0\ \text{kN}\ \hat{i}$$

We have rounded to the correct number of significant digits in the final answer.

Making sense of the result

Remember that this is the force averaged over the entire contact time. The maximum force would be significantly higher and is sufficient to cause considerable deformation, as shown in Figure 5-45.

Figure 5-45 Example 5-28.

KEY CONCEPTS AND RELATIONSHIPS

Force is a vector quantity representing a push or pull, with the SI unit of force being the newton (N). The net force is obtained by adding as vectors all the forces acting on an object.

Newton's Laws

Newton's first law: If no net force is exerted on an object, the object's velocity will not change.

Newton's second law: The acceleration of an object depends inversely on its mass and directly on the net force:

$$\vec{F}_{net} = m\vec{a} \tag{5-3}$$

Newton's third law: Whenever one object exerts a force on another, it will experience a force from the other object that is equal in magnitude but opposite in direction to the force that it exerts. That is, if $\vec{F}_{1\to2}$ is the force that object 1 exerts on object 2, then $\vec{F}_{2\to1}$, the force that object 2 exerts on object 1, is

$$\vec{F}_{2\to1} = -\vec{F}_{1\to2} \tag{5-5}$$

A free body diagram (FBD) shows the forces acting on a body. In solving Newton's second law problems, we usually choose a set of coordinate axes and divide the forces into components along these axes. The solution is more direct when one of the axes is chosen to lie along the direction of acceleration of the object. Newton's second law is written for each axis direction, and then the resulting equations are simultaneously solved.

Types of Forces

Tension forces are present in ropes, wires, and similar objects. The tension has a constant magnitude in all parts of the ideal rope if it is assumed massless, but the direction is not the same at the two ends. In the absence of friction, a surface can only exert a force perpendicular to the surface, and this is called a normal force.

When friction is present, the frictional force is opposite to the direction of motion (or the direction the other forces would have caused motion if there were no frictional force). Static friction, when there is no motion, can take a value up to a maximum of f_s^{max} given by

$$f_s^{max} = \mu_s N \tag{5-6}$$

where N is the magnitude of the normal force between the surfaces and μ_s is the coefficient of static friction that depends on the surfaces.

When an object is sliding over a surface, there is a kinetic frictional force, f_k, according to the following relationship, where μ_k is the coefficient of kinetic friction:

$$f_k = \mu_k N \tag{5-8}$$

The force exerted by a spring on an attached mass is approximated by Hooke's law:

$$\vec{F} = -k\vec{x} \tag{5-9}$$

where k is the spring constant and $\vec{x}$ is the extension or compression of the spring from the unstretched equilibrium state. The minus sign indicates that the force the spring exerts on the mass acts in a direction opposite to the displacement of the mass from the equilibrium position.

Fundamental forces are based on the gravitational, electromagnetic, weak, or strong nuclear interaction and can be considered acting-at-a-distance and field forces. Non-fundamental forces, such as tension, normal, elastic, and frictional forces, involve direct physical contact but are ultimately traced to the fundamental forces.

Near the surface of Earth, the gravitational force, $\vec{F}_g$, is directed vertically downward and is given by

$$\vec{F}_g = m\vec{g} \tag{5-2}$$

where m is the mass and $\vec{g}$ is the acceleration due to gravity.

Mass

Inertial mass views mass as a measure of the inertia of an object to resist being accelerated. The larger the mass, the lower the acceleration for a given force. Gravitational mass views mass in terms of the gravitational force, with the strength of the gravitational force in a location being proportional to the gravitational mass. Research suggests that the gravitational and inertial mass have the same value for an object.

Reference Frames

A reference frame in which Newton's second law holds without the need to invoke fictitious forces is called an inertial reference frame. In an accelerated non-inertial reference frame, there will be fictitious forces, such as the centrifugal force for motion in a circle. These fictitious forces are in a direction opposite to that of the acceleration of the reference frame.

Impulse and Momentum

Newton's second law can be written in terms of the derivative of the linear momentum, $\vec{p}$ (defined by $\vec{p} = m\vec{v}$):

$$\vec{F}_{net} = \frac{d\vec{p}}{dt} \tag{5-12}$$

Impulse, $\vec{J}$, is a vector quantity defined by the change in momentum during an impact and can be used to calculate the average force, $\vec{F}_{average}$, during the contact time, Δt:

$$\vec{J} = \vec{F}_{average}\Delta t \tag{5-14}$$

APPLICATIONS

auto racing, body mechanics, driving safety, effects of friction, g-forces and medical safety, inertial navigation, reference frame fictitious forces, impulse collision analysis, sports physics, tension safety

KEY TERMS

apparent weight, centrifugal force, centripetal acceleration, coefficient of kinetic friction, coefficient of static friction, dynamics, elastic limit, electromagnetic interaction, equilibrium position, fictitious force, field forces, force, force of gravity, free body diagram (FBD), friction, fundamental forces, g-force, Grand Unified Theory, gravitational force, gravitational interaction, gravitational mass, Hooke's law, ideal spring, impulse, inertia, inertial mass, inertial reference frame, kinematics, kinetic friction, linear momentum, mass, net force, Newton's first law, Newton's second law, Newton's third law, non-fundamental force, non-inertial reference frame, normal force, point mass, static friction, strong interaction, tension, weak interaction, weight

QUESTIONS

1. Answer each the following as true or false.
   (a) The tension in a string is in essence electromagnetic in nature.
   (b) The force of friction is ultimately electromagnetic in nature.
   (c) All non-fundamental forces are manifestations of fundamental interactions.
   (d) The kinetic frictional force is a fundamental force.
   (e) The gravitational force is a fundamental force.
   (f) When drawing an FBD you should include the centripetal force.
   (g) The centrifugal force is always present if there is rotation.
2. The inertia of an object is determined by its
   (a) mass
   (b) net force
   (c) impulse
   (d) acceleration
3. The inertial mass of an object is used in reference to
   (a) the force of gravitational attraction between objects
   (b) the ability of an object to resist having its speed changed
   (c) the frictional force on the object
   (d) the object's weight
4. The gravitational mass depends on
   (a) the force of gravitational attraction between objects
   (b) the ability of an object to resist having its speed changed
   (c) the frictional force on the object
   (d) the centrifugal force
5. Figure 5-46 shows the FBD of an object. What is the direction of motion of the object?
   (a) The object is moving to the left.
   (b) The object is moving to the right.
   (c) The object is not moving.
   (d) There is not enough information to answer the question.

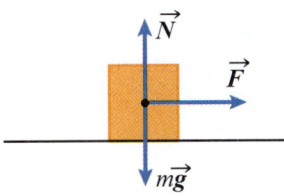

Figure 5-46   Question 5.

6. An object moves to the right while subject to a force whose magnitude increases with time but points to the left. Which of the following statements is true?
   (a) The particle will reverse direction as soon as the force is applied.
   (b) The particle will slow down and eventually reverse direction.
   (c) The particle will not slow down because the magnitude of the force decreases with time.
   (d) The particle will not reverse direction.
7. A puck is moving freely on frictionless ice along the positive x-axis when a force is applied to the puck. The direction of the force is maintained along the positive x-axis, but its magnitude drops exponentially with time. Which of the following statements is true?
   (a) The puck's speed increases with time.
   (b) The puck's speed decreases with time.
   (c) The puck's acceleration increases with time.
   (d) None of the above are true.
8. A particle moving to the right is acted upon by a constant net force pointing to the left. The particle moves along the x-axis, and the force is applied at $t = 0$, when the particle's position is to the right of the origin (positive). We can say that the magnitude of the particle's position with respect to time
   (a) changes like a parabola that opens upward
   (b) changes like a parabola that opens downward
   (c) decreases linearly
   (d) increases linearly

9. You (assume $m = 76$ kg) are riding in a car moving at a constant speed of 12 m/s on a curved road of radius 10. m. What is the net force acting on you?
   (a) The net force cannot be determined without knowing the force of friction between the road and the car.
   (b) The net force cannot be determined without knowing the force of friction between you and the car.
   (c) The net force cannot be determined without knowing the normal force from the seat cushion on you.
   (d) None of the above are true.

10. You pull your friend with a horizontal rope such that the force you apply to the rope is 160 N. Your friend stands his ground. Neither of you moves. Which of the following is true?
    (a) The net force on your friend is 160 N.
    (b) The force of friction on your friend is 160 N.
    (c) The tension in the rope is 320 N.
    (d) None of the above are true.

11. In Figure 5-47, the combined mass of the monkey, bananas, and platform is m. To keep the platform still, the monkey has to pull the rope with which force?
    (a) mg
    (b) $\dfrac{mg}{2}$
    (c) $\dfrac{mg}{4}$
    (d) $2mg$

Figure 5-47 Question 11.

12. What can be said about the relative initial accelerations of the two masses in Figure 5-48?

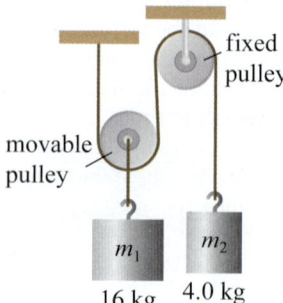

Figure 5-48 Question 12.

13. What can be said about the magnitudes of the initial accelerations of the three masses in Figure 5-49? Note that two of the pulleys are fixed, and two are movable.

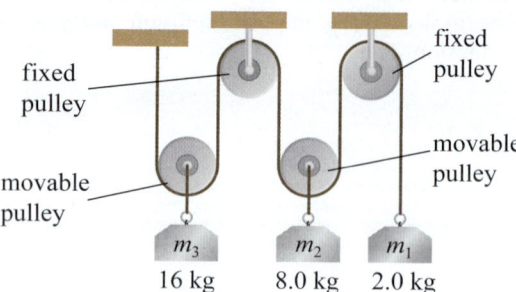

Figure 5-49 Question 13.

14. You are on an assumed frictionless skating rink. Your mass is 90. kg. You give a rope to a friend whose mass is 60. kg. You pull the rope, and your friend hangs on (without making the rope shorter by pulling on it) as she accelerates toward you at 3.0 m/s². Which of the following statements is true?
    (a) You do not move.
    (b) The force acting on you is 270 N.
    (c) Your acceleration is 3.0 m/s².
    (d) The force acting on you is 180 N.

15. Two friends ($m_1 = 52$ kg and $m_2 = 87$ kg) pull one another using a rope. They each apply a force of 230 N to the horizontal rope, but neither moves. Which of the following statements is true?
    (a) The tension in the rope is 230 N.
    (b) The tension in the rope is 460 N.
    (c) The force of friction on each friend is 460 N.
    (d) The heavier friend experiences a larger force of friction as they both pull.

16. Two friends are standing on an ice rink, one of them pulling the other with a rope with a force of 145 N. The friend being pulled ($m_1 = 50.$ kg) moves toward the friend doing the pulling ($m_2 = 75$ kg) at a constant speed of 6.0 m/s. Which of the following statements is true?
    (a) The speed of the friend doing the pulling must be 4.0 m/s.
    (b) The force of kinetic friction on the friend being pulled is 145 N.
    (c) The net force on the friend being pulled is 145 N.
    (d) The force of static or kinetic friction on the friend doing the pulling is 145 N.

17. Two boxes are in the back of a truck. The truck comes to an emergency stop, and both boxes slide with respect to the truck. Assuming the same coefficient of friction between each box and the back of the truck, which box will slide first—the lighter or the heavier box, or both at the same time? Explain your reasoning.

18. Figure 5-50 shows the FBD of an object. The object must
    (a) be moving up
    (b) be accelerating up
    (c) be momentarily stationary
    (d) be accelerating down

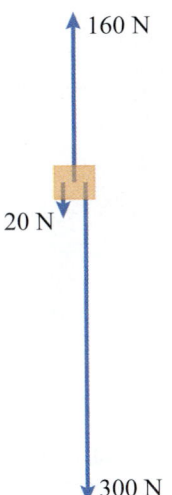

160 N

20 N

300 N

Figure 5-50 Question 18.

19. Two boxes, one twice the mass of the other, slide down the same incline, starting from rest at the same height. The coefficient of friction between each of the boxes and the incline is the same. Compare the times the boxes take to go down the incline.

20. You are standing in an elevator that is accelerating upward. The magnitude of the force that you exert on the elevator
    (a) is larger than the magnitude of the force that the elevator exerts on you.
    (b) is smaller than the magnitude of the force that the elevator exerts on you.
    (c) is equal to the force magnitude of the that the elevator exerts on you.
    (d) none of the above

21. You are sitting in a 1000. kg truck, which is slowing down at 10.0 m/s². Your mass is 50.0 kg. Which of the following is true?
    (a) The force that you exert on the truck is 500. N.
    (b) The net force on the truck is 9.5 kN.
    (c) The force that the truck exerts on you is 500. N.
    (d) The vertical force that the truck exerts on you is 490. N.

22. A crate is sitting on a horizontal surface where the coefficient of static friction is 0.60. A worker pushes horizontally on the crate, which has a mass of 45 kg. The crate does not move, and there are no other forces acting on it. Which of the following statements is true?
    (a) The force of friction on the crate is $mg\mu_s$.
    (b) The force of friction on the crate is greater than the force that the worker exerts on the crate.
    (c) The force exerted on the worker by the crate is equal in magnitude to the force of friction exerted by the surface on the crate.

23. Two friends are pushing a book in opposite directions, each pushing with a force of 320 N. One friend says the net force on the book is 640 N. Is this correct? Explain.

24. Two friends push a sled on snow in opposite directions. One pushes with a force of 190 N, and the other pushes

with a force of 130 N. The sled does not move. What is the net force on the sled?
    (a) 320 N
    (b) 60 N
    (c) zero
    (d) cannot be determined

25. You (mass m) are sitting in a car (mass M) that is rounding a banked curve of radius r at constant speed v. Which of the following statements is true?
    (a) The magnitude of the horizontal force exerted on you by the car is larger than that of the horizontal force exerted on the car by you.
    (b) The magnitude of the net force on the car is $\dfrac{Mv^2}{r} - \dfrac{mv}{r}$.
    (c) The magnitude of the net force on you is $\dfrac{mv^2}{r}$.
    (d) The magnitude of the net force on the car is $\dfrac{mv^2}{r}$.

26. A lion trainer is in a train car looking after his 125 kg lion named Pouncer. Pouncer is not in the mood to play and is sitting still. The trainer prods Pouncer gently with a force of 200. N in a direction toward the front of the train, but it is not enough to get Pouncer to move. The train is slowing down (acceleration of -5.00 m/s²). The magnitude of the net force on Pouncer is
    (a) 825 N
    (b) 625 N
    (c) 425 N
    (d) 125 N

27. An ideal massless rope connects mass M_1 hanging vertically to mass M_2 resting on top of a horizontal frictionless surface via a fixed massless pulley at the edge of the surface. The system is originally held at rest but is then released.
    (a) Is the tension in the rope larger than, smaller than, or equal to M_1g?
    (b) If the surface that M_2 rests on provides enough friction to prevent M_2 from moving, what is the tension in the rope?

28. Two buckets, each of mass m, are suspended from a massless rope that goes around the fixed massless frictionless pulley in Figure 5-51. The system is at rest. What is the tension in the rope?
    (a) mg
    (b) $2mg$
    (c) zero
    (d) mg^2

fixed pulley

Figure 5-51 Questions 28 and 29.

29. Two buckets of equal mass are suspended vertically from a rope that goes around a massless frictionless pulley with fixed centre, as shown in Figure 5-51. The pulley rotates clockwise so that the masses are moving at a constant speed. Which of the following statements is true?
    (a) The tension in the rope is equal to mg.
    (b) The tension in the rope is equal to $2mg$.
    (c) The tension in the rope is zero.
    (d) The tension cannot be determined without knowing the masses.

30. A centipede of mass m is climbing a thread that is hanging vertically from a table. When the centipede begins its climb, its acceleration is a upward. What is the tension in the thread?
    (a) mg
    (b) ma
    (c) $mg + ma$
    (d) $mg - ma$

31. A monkey of mass m uses a fixed rope to climb down from a coconut tree. The monkey races with an acceleration of $0.5g$ downward to avoid a coconut dropped by another monkey. Taking up as the positive direction, what is the tension in the fixed rope?
    (a) $0.5mg$
    (b) $1.5mg$
    (c) $2.5mg$
    (d) $-0.5mg$

32. Mass m_1 hangs from the massless movable pulley shown in Figure 5-52. One end of the rope supporting the pulley is attached to the ceiling; the other goes over the fixed pulley and is attached to mass m_2 on the frictionless surface. When the system is let go, the magnitude of the acceleration of m_1 is
    (a) half the magnitude of the acceleration of m_2
    (b) double the magnitude of the acceleration of m_2
    (c) the same as the magnitude of the acceleration of m_2
    (d) The relationship between the two accelerations depends on the masses.

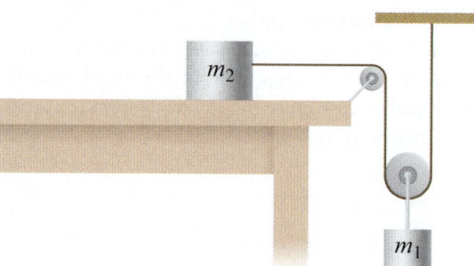

Figure 5-52   Question 32.

33. Mass m_1 hangs vertically from one end of a rope, as shown in Figure 5-53. The rope goes over the fixed pulley and around the movable pulley that is attached to m_2. The other end of the rope is fixed to the protrusion that is attached to the horizontal surface. As the system moves, the magnitude of the acceleration of m_1 is
    (a) twice the magnitude of the acceleration of m_2
    (b) half the magnitude of the acceleration of m_2
    (c) the same as the magnitude of the acceleration of m_2
    (d) More information is needed to answer the question.

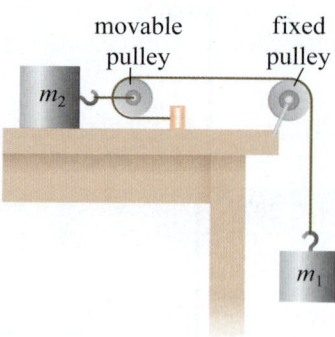

Figure 5-53   Question 33.

34. Four blocks rest on top of each other such that $m_1 = 2m_2 = 4m_3 = 8m_4$. The bottom block is m_1. The coefficients of static friction between the blocks are μ between blocks m_1 and m_2, 2μ between blocks m_2 and m_3, and 4μ between blocks m_3 and m_4. Block m_1 is pushed with a horizontal force strong enough that at least one pair of the blocks slips. Which block will be the first to slide with respect to the surface beneath it?
    (a) m_4
    (b) m_2
    (c) m_3
    (d) They will all slide simultaneously.

35. A bug is standing stationary relative to a spinning turntable. Some oil is poured on the turntable to render the surface frictionless. As soon as the surface becomes frictionless, the bug will
    (a) continue to move in a circle, but the radius of the circle will get larger and larger
    (b) move in a straight line tangent to the circular trajectory it was originally on
    (c) move radially inward toward the centre of the turntable
    (d) move directly radially outward toward the edge of the turntable

36. A child is swinging a string with a ball attached to its end in a circle that is nearly horizontal. Ignoring gravity, in which direction will the ball go if the string is cut?

PROBLEMS BY SECTION

For problems, star ratings will be used (★, ★★, or ★★★), with more stars meaning more challenging problems. The following codes will indicate if $\frac{d}{dx}$ differentiation, $\int$ integration, ▢ numerical approximation, or ∿ graphical analysis will be required to solve the problem.

Section 5-1   Dynamics and Forces

37. ★ What is the net force if there is an upward force of 125 N and a downward force of 85 N? Draw an FBD for the situation.

38. ★ An x-axis is directed in a horizontal direction with positive defined to the right. What is the net force when forces $\vec{F}_1 = -(18.5\ \text{N})\hat{i}$ and $\vec{F}_2 = -(12.5\ \text{N})\hat{i}$ are added? Interpret your answer in terms of right or left.

39. ★ The x-axis is directed horizontally with positive defined to the right. What is the net force when forces $\vec{F}_1 = +(74.5 \text{ N})\hat{i}$ and $\vec{F}_2 = -(6.5 \text{ N})\hat{i}$ are added? Interpret your answer in terms of right or left.

40. ★ The x-axis is directed horizontally with positive defined to the right. The net force when two vectors are added is given by $\vec{F}_{net} = -(25.0 \text{ N})\hat{i}$. If one of the forces is $\vec{F}_1 = -(15.0 \text{ N})\hat{i}$, what must be the other force? Draw an FBD for the situation.

Section 5-2   Mass and the Force of Gravity

41. ★ You hold a 10.0 kg object near the surface of Earth so that it is not moving. What force must you apply? Draw an FBD of the situation.

42. ★ You need to apply a 28.0 N upward force to keep an object from moving in a gravitational field near the surface of Earth. What must be the mass of the object? Draw an FBD as part of your solution.

43. ★ (a) If your weight, interpreted as the force of gravity on you, is 675 N, what is your mass?

   (b) How much would you weigh on the Moon, where the acceleration due to the Moon's gravity is approximately 1/6 the acceleration due to gravity at Earth's surface?

Section 5-3   Newton's Laws of Motion

44. ★ A constant force of 130.0 N is applied to a block of mass 9.00 kg, which is sitting on a frictionless horizontal surface. Find the speed of the mass after it has moved 14.0 m, starting from rest.

45. ★ The last part of a roller coaster's track is horizontal. The roller coaster is brought to rest using a constant braking force of 4300. N. It takes 9.00 s to bring the roller coaster to a stop from a speed of 23.0 m/s. Find the mass of the roller coaster.

46. ★ Assume that a particular crash test expert driver can withstand 7.6 times the acceleration due to gravity. Find the shortest distance over which her car can be brought to a stop from a speed of 70.0 km/h without the acceleration exceeding 7.6g.

47. ★ A car of mass 2100. kg is brought to an emergency stop from a speed of 22.0 m/s over a distance of 25.0 m. What is the average net force experienced by the car during this stop?

48. ★★ You ($m_1 = 81.0$ kg) are standing in an elevator ($m_2 = 2300.$ kg) that is moving downward with a speed of 4.50 m/s. The elevator comes to a stop with a constant acceleration over a distance of 4.00 m. Find

   (a) the net force on you
   (b) the net force on the elevator
   (c) the force that you exert on the elevator
   (d) the gravitational force on the elevator

Section 5-4   Applying Newton's Laws

49. ★★ Consider the three forces shown in Figure 5-54.

   (a) Resolve each force into its Cartesian components along horizontal and vertical axes.
   (b) Calculate the net force.

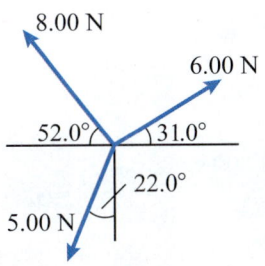

Figure 5-54   Problem 49.

50. ★ Find the magnitude and direction of the forces listed below. Report the direction as an angle measured counter-clockwise with respect to the x-axis.

   (a) $(6.00 \text{ N})\hat{i} + (2.00 \text{ N})\hat{j}$
   (b) $(-21.0 \text{ N})\hat{i} + (16.0 \text{ N})\hat{j}$
   (c) $(7.00 \text{ N})\hat{i} + (-4.00 \text{ N})\hat{j}$

51. ★ Add the following forces. Express the resultant forces in Cartesian form with unit vectors. When angles are expressed, they are measured from the x-axis in a counter-clockwise direction.

   (a) $(-4.00 \text{ N})\hat{i} + (-2.00 \text{ N})\hat{j} + (0.50 \text{ N})\hat{k}$,
       $(3.00 \text{ N})\hat{j} + (-4.00 \text{ N})\hat{k}, (3.00 \text{ N})\hat{i} + (2.00 \text{ N})\hat{j}$
   (b) 14.00 N [57.0°], $(3.00 \text{ N})\hat{i} + (-4.00 \text{ N})\hat{j}$,
       8.00 N [236°]
   (c) $(4.00 \text{ N}\,\hat{i}, 2.00 \text{ N}\,\hat{j}, -5.00 \text{ N}\hat{k})$,
       $(-9.00 \text{ N}\,\hat{i}, 14.0 \text{ N}\,\hat{j}, 11.0 \text{ N}\hat{k})$

52. ★ A child slides down a frictionless incline with a slope of angle θ. Draw an FBD, and make a good selection for coordinate axes.

53. ★ A man pushes his daughter of mass m up the frictionless surface of a slide at an angle of θ using a horizontal force of F. Draw the FBD of the daughter. Assume she is moving up the incline as a result of being pushed by her father.

54. ★★ Starting from rest, a car slides 15 m down an ice-covered road, where the friction is negligible. The speed of the car at the end of the 15 m is 3.2 m/s. Find the slope angle of the road.

55. ★★ Children are enjoying a constant-slope water slide at an amusement park. They run toward the top of the 6.0 m long slide and lie flat on the slide surface. A child who starts at the top of the slide with a speed of 1.4 m/s reaches a speed of 3.6 m/s at the bottom of the slide. Find the slope angle of the slide, assuming that its friction is negligible.

56. ★★ You are pushing four blocks of masses 4.00 kg, 7.00 kg, 11.00 kg, and 17.00 kg positioned next to each other on a horizontal frictionless surface, as shown in Figure 5-55. Find the force of contact between the 7.00 kg and 11.00 kg blocks when

   (a) you push on the lightest block with a 195 N force to the left
   (b) you push on the heaviest block with a 195 N force to the right

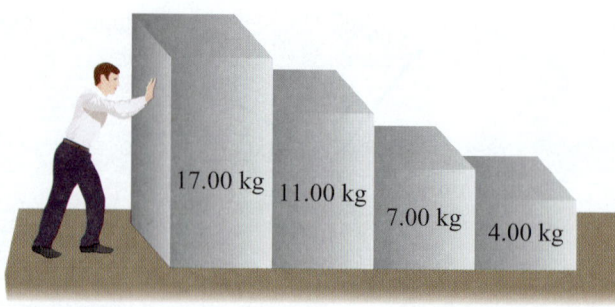

Figure 5-55   Problem 56.

57. ★★ Four food packages ($m_1 = 7.00$ kg, $m_2 = 5.00$ kg, $m_3 = 8.00$ kg, and $m_4 = 3.00$ kg) are connected to each other using ropes as shown in Figure 5-56. The four masses rest on a frozen lake surface (assume frictionless) in Labrador. A husky pulls on a rope that is attached to m_4 and pulls the four masses to the right.
   (a) The tension in the rope between m_1 and m_2 is 40.0 N. Calculate the tension in the rope connecting m_2 and m_3.
   (b) What is the magnitude of the force with which the husky is pulling on m_4?

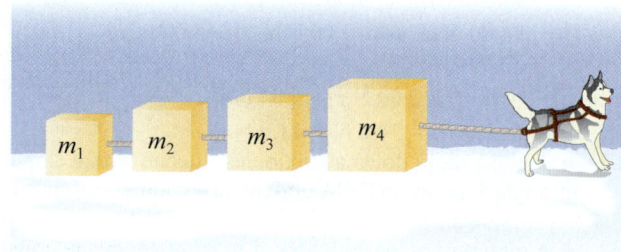

Figure 5-56   Problem 57.

58. ★★ Mass M_1, sitting on a horizontal frictionless surface, is connected by a rope that runs over a massless pulley to mass M_2, which is hanging vertically. The system is originally at rest and is then let go.
   (a) Find the tension in the rope in terms of M_1 and M_2 when $M_2 > M_1$.
   (b) How would your answer to part (a) change if the masses were interchanged?
   (c) Obtain numerical values for parts (a) and (b) using $M_1 = 2.00$ kg and $M_2 = 12.0$ kg.
59. ★★ You have just had the apples in your orchard harvested into a large number of boxes with a mass of 130.0 kg each. You wish to lift these boxes up to the fourth floor (12.0 m above ground level) of your packaging facility. Since you are the only person available, you construct a slippery surface that is inclined 30.0° above the horizontal, and then fix a pulley at the top corner, as shown in Figure 5-57. You hang from the rope at the top of the incline and let the force of gravity pull the crates up the ramp. The friction between the crates and the inclined surface is negligible.
   (a) What must your minimum mass be for this system to work?

(b) If you do not wish to hit the ground with a speed greater than 2.00 m/s, what must your maximum mass be?
(c) How long will it take to lift one crate to the fourth floor if you have the mass calculated in part (b)?

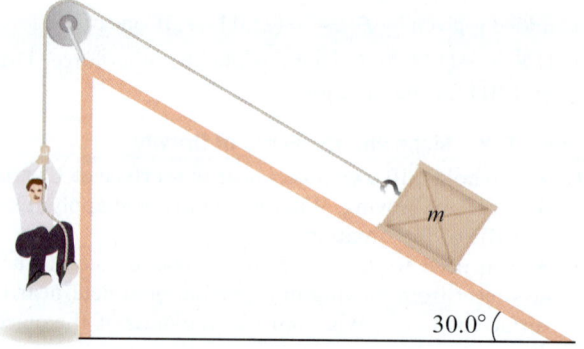

Figure 5-57   Problem 59.

Section 5-5   Component-Free Solutions

60. ★★ As illustrated in Figure 5-58, a mass, m_1, is on a frictionless incline with angle θ. A massless cord connects this to a second mass, m_2, that hangs vertically. Assume that the pulley is frictionless and massless. The system is in equilibrium and not moving. What must be the relationship between the two masses? Use the component-free vector method to solve this problem.

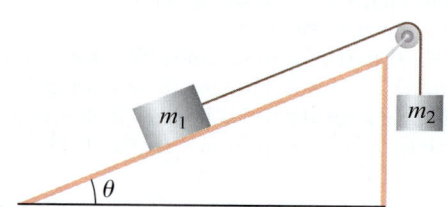

Figure 5-58   Problem 60.

Section 5-6   Friction

61. ★★ A wheel of radius 25.0 cm and mass 3.60 kg is rolling without slipping on a flat surface. The coefficient of static friction between the wheel and the surface is 0.48, while the coefficient of kinetic friction is 0.32. What will be the frictional force exerted by the surface on the wheel?
62. ★★ The coefficient of static friction between your athletic shoes and a horizontal track surface is 0.78. What is your maximum possible acceleration?
63. ★★ A polar bear cub slides (starting from rest) 31 m toward his mother down a snow-covered 37° slope starting from rest. At the end of the 31 m, the cub's speed is 4.7 m/s. Find the coefficient of friction between the cub and the snow-covered surface.
64. ★★ A hockey puck is shot across an ice rink at a speed of 3.0 m/s and comes to a stop over a distance of 11.0 m. Find the coefficient of friction between the puck and the ice.

Section 5-7 Spring Forces and Hooke's Law

65. ★ To stretch a spring by 10. cm, you have to apply a force of 120. N. How much would the spring stretch if a 10. kg mass hung from it vertically?

66. ★★ A spring is suspended from an elevator, and a 12.0 kg mass is then attached to the spring. When the elevator moves down at an acceleration of 3.00 m/s², the net elongation of the spring is 30.0 cm. Find the spring constant.

Section 5-8 Fundamental and Non-fundamental Forces

67. ★ An object moving through a fluid is subject to an aerodynamic drag force. Would you classify this force as fundamental or non-fundamental?

Section 5-9 Uniform Circular Motion

68. ★ A mass is moving at a constant speed at the end of a string rotated in a vertical circle. Draw FBDs for the situations when the mass is
   (a) at the top
   (b) at the bottom

69. ★ Draw an FBD for the situation when a car is banking a frictionless curve with radius r and banking angle θ. Derive an expression for the centripetal acceleration in terms of the banking angle.

70. ★★ You are sitting in the passenger seat of a car driving on the highway at a speed of 110.0 km/h when the driver ($m = 79.0$ kg) exits on a circular ramp of radius 63.0 m. The mass of the car is 1300. kg, and your mass is 71.0 kg. Find
   (a) the horizontal force that the road exerts on the car
   (b) the horizontal force that you exert on the car
   (c) the net horizontal force on the car

71. ★★ You swing a marble attached to the end of a string in a horizontal circle, as shown in Figure 5-59. The angle that the string makes with the vertical is 37.0°.
   (a) Find the speed of the marble if the string is 44.0 cm long.
   (b) Write an expression for the tension in the string in terms of the marble mass.

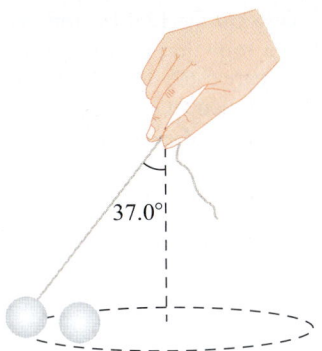

Figure 5-59 Problem 71.

72. ★★ A box of mass 20.0 kg sits on a merry-go-round 3.0 m from the centre. You push the merry-go-round faster and faster until the box is about to slip on the merry-go-round surface. The coefficient of static friction between the box and the surface is 0.45.
   (a) What is the net force acting on the box at that instant?
   (b) Find the box's speed at the moment it is about to slip.

73. ★★ You are testing a remote-controlled toy truck in an empty parking lot. The coefficient of static friction between the surface and the truck is 0.340. Find the radius of the smallest circle the truck can move in without slipping at its maximum speed of 33.0 km/h.

74. ★★ You are driving on a circular ramp of radius 72.0 m on an icy January day. The ice-covered ramp is banked at a 23.0° angle, and you anticipate that the ramp is not going to provide any significant friction to help keep your car on the road.
   (a) What must your speed be so that your car does not slide off the road?
   (b) What are the direction and magnitude of the net force acting on you at that point? Your mass is 72.0 kg.

75. ★★ Your niece finds her great grandfather's pocket watch. The light watch chain has a length of 32.0 cm, and the mass of the watch is 210.0 g. Your niece swings the watch in a vertical circle. Find the tension in the chain when it makes an angle of 43.0° with respect to the vertical (as measured from the bottom of the circle) if the speed of the watch is 2.80 m/s.

Section 5-10 Reference Frames and Fictitious Forces

76. ★★ You are sitting comfortably in a train. A passenger standing beside you is loading her luggage into the overhead compartment when the train starts moving forward with an acceleration of 5.0 m/s². Find the coefficient of static friction that is necessary to prevent the person from sliding with respect to the train. Solve the problem by considering the scenario from the train's frame of reference, and then again using the train station's frame of reference.

77. ★★ A 30.0 kg bag of mortar mix is sitting in the back of a truck rounding an unbanked traffic circle of radius 23.0 m at a speed of 67.0 km/h. Using the truck's frame of reference, find the force of contact between the bag and the side of the truck. Assume that the floor at the back of the truck is frictionless.

78. ★★ An astronaut sits in a spacecraft orbiting Earth. Explain the feeling of weightlessness that the astronaut experiences. Is it true that an astronaut living on the International Space Station (ISS) is too far from Earth to experience the force of gravity? What is the altitude of the ISS's orbit? Research the value of the force of gravity at that altitude.

79. ★★ You place a book on the seat next to you, and you then drive your car at a speed of 110. km/h on the highway. You suddenly have to slow down. The book is just beginning to move on the seat. You stop over a distance of 120. m. What is the coefficient of friction between the seat and the book?

Section 5-11 Momentum and Newton's Second Law

80. ★ $\frac{d}{dx}$ The momentum as a function of time for an object of a mass m is given by the following expression:

$$\vec{p} = \left[(35.0 - t^2) \text{ kg} \cdot \text{m/s} \right] \hat{i}$$

What must be the net applied force at the following times?
(a) 0.00 s
(b) 1.00 s

81. ★★ A bouncy ball with a mass of 150. g hits a concrete horizontal surface while the ball is travelling vertically downward at a speed of 5.50 m/s. The ball bounces back with an upward speed of 4.50 m/s. The ball is in contact with the surface for 1.00 ms.
(a) What is the impulse?
(b) What is the average force exerted by the concrete surface on the ball?

82. ★★ In 2012, Zdeno Chára of the Boston Bruins set the record for fastest hockey slapshot at 175.1 km/h. The puck has a mass of 160. g. If the stick was in contact with the puck for 38.0 ms, what was the average force on the puck while in contact with the hockey stick? Assume that the puck was stationary before being struck.

83. ★★ ∫ Initially ($t = 0.00$ s), a 500. g mass is travelling at a velocity of $(1.00 \text{ m/s}) \hat{i}$. It is then subjected to a constant net force $\vec{F} = (2.25 \text{ N}) \hat{i}$. Determine an expression for the momentum as a function of time.

COMPREHENSIVE PROBLEMS

84. ★★ Four blocks of masses 20.0 kg, 30.0 kg, 40.0 kg, and 50.0 kg are stacked on top of each other in an elevator in order of decreasing mass, with the lightest mass on top. The elevator moves down with an acceleration of 3.20 m/s². Find the contact force between the 30.0 kg block and the 40.0 kg block.

85. ★★ A woman is standing in an elevator carrying a 2.70 kg kitten. The elevator begins to accelerate upward at 3.20 m/s². Find the force that the kitten exerts on the woman.

86. ★★ A curious child finds a rope hanging vertically from the ceiling of a large storage hangar. The child grabs the rope and starts running in a circle. The length of the rope is 17.0 m. When the child runs in a circle of radius 6.0 m, the child is about to lose contact with the floor. How fast is the child running at that time?

87. ★★ A cruise ship moving at a constant speed of 30.0 km/h is equipped with a gymnasium. A gymnast is climbing a rope suspended from the ceiling of the gym. The captain of the ship puts the ship into a circular turn of radius 150. m to get closer to a pod of whales. Find the angle that the rope makes with the vertical.

88. ★★ A child sits in the back of a car with a yo-yo freely suspended from his fingers. The car is leaving the highway at a speed of 78.0 km/h on a ramp of radius 210.0 m banked at a 23.0° angle.
(a) Find the angle that the yo-yo makes with the vertical.
(b) Find the angle that the yo-yo makes with the normal to the roof of the car.

89. ★★★ The truck in Figure 5-60 is going around a circular track of radius 72.0 m, banked at a 60.0° angle. A spider rests on the inside wall of the truck, as shown. The coefficient of static friction between the truck wall and the spider is 0.910. Find the maximum speed that the truck can have before the spider begins to slip down the wall.

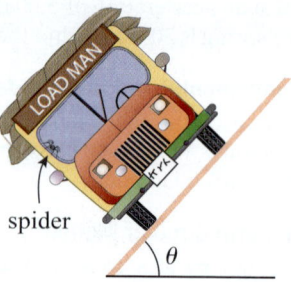

spider

Figure 5-60 Problem 89.

90. ★★ A toy car is made to run the bowl-shaped track shown in Figure 5-61. As the car's speed increases, the car moves farther and farther up the track walls. In the end, the car is able to round the vertical walls of the track, such that its height from the bottom of the track does not change. The radius of the bowl is 1.9 m. Find the minimum speed that the car needs to go to not slip when rounding the vertical walls. The coefficient of static friction between the car and the wall is 0.39.

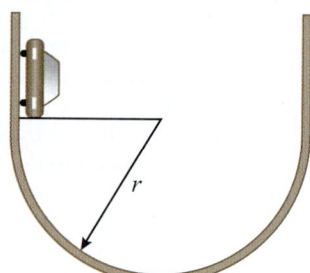

r

Figure 5-61 Problem 90.

91. ★★ Two masses ($m_1 = 12.0$ kg and $m_2 = 14.0$ kg) are hanging from a rope that passes over a massless pulley, as shown in Figure 5-62. The two masses are originally held in place and then let go.
(a) Find the speed of each mass after one has moved 4.00 m.
(b) What is the tension in the rope?

m_1

m_2

Figure 5-62 Problem 91.

92. ★★ Obtain a solution for the situation of Example 5-24 (Frictionless Banked Racetrack), but this time using the vector method that does not involve components.

93. ★★A roller coaster car is upside down at the top of a vertical circular loop with a radius of 21.0 m. The speed of the car is 24.0 m/s, and the coefficient of kinetic friction between the car and the track is 0.16. Find the radial acceleration of the car.

94. ★★ Blocks A (20.0 kg), B (10.0 kg), C (40.0 kg), and D (30.0 kg) are in contact beside each other on a horizontal frictionless surface, as shown in Figure 5-63. Block A is being pushed to the right with a horizontal force, $\vec{F}$, such that the force of contact between blocks C and D is 60.0 N. Find the force of contact between blocks B and C. Also, find the force $\vec{F}$ and the acceleration of each block.

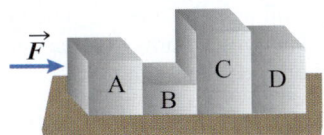

Figure 5-63   Problem 94.

95. ★★ A marble slides down a hemispherical bowl starting from rest, as shown in Figure 5-64. When the marble has dropped down from its starting point by half the radius vertically, its speed is 2.30 m/s. Write an expression for the magnitude of the total acceleration of the marble in terms of R.

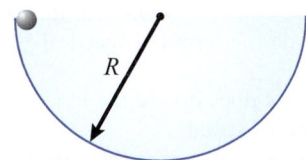

Figure 5-64   Problem 95.

96. ★★ Mass m_1 hangs from a massless movable pulley, as shown in Figure 5-65. The rope rounding the movable pulley passes over a fixed pulley and is attached to mass m_1. There is no friction between m_2 and the surface supporting it. The system is originally held at rest and then let go. Derive expressions for the acceleration of m_1 and the tension in the rope in terms of m_1 and m_2.

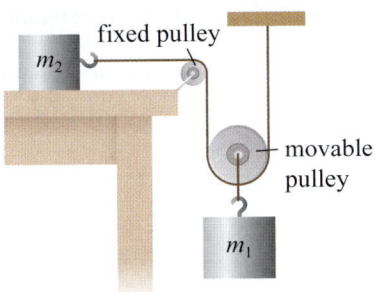

Figure 5-65   Problem 96.

97. ★★ You want to push your 53.0 kg friend up the 33.0° ramp of a slide in a nearby park. You do so by pushing your friend with a 510.0 N horizontal force. The coefficient of kinetic friction between your friend and the slide is 0.170. Find the acceleration of your friend up the slide. What is the normal force from the ramp on your friend?

98. ★★ You swing a 1.20 m long string with a metal ball attached to its end in a vertical circle, such that the speed of the ball does not change but the rope is always taut. The tension in the string when the ball is at the bottom of the circle is 65.0 N more than the tension when the ball is at the top. Find the mass of the ball.

99. ★★ You are riding in a car moving at a uniform speed of 20.0 m/s around an unbanked curve of radius 40.0 m. The mass of the car is 1300. kg, and your mass is 60.0 kg. Find the magnitude of
(a)   the force that you exert on the car
(b)   the net force on the car
(c)   the net force on you
(d)   the horizontal force that the car exerts on you
(e)   the force that the car exerts on the road

100. ★★ A 1300 kg car takes a 44 m radius roundabout at a speed of 7.0 m/s. The road is not banked. The coefficient of static friction between the car and the road is 0.41. Find
(a)   the force of friction on the car
(b)   the net force on the car
(c)   the maximum force of static friction on the car

101. ★★ A truck on level ground is taking a tight circular path of radius 54.0 m at a constant speed of 11.0 m/s. A box with a mass of 69.0 kg sits on the level cargo area of the truck. The coefficient of static friction between the box and the truck is 0.340. Find
(a)   the force of friction on the box
(b)   the force that the box exerts on the truck
(c)   the net force on the box

102. ★★★ Four blocks stacked vertically have masses as follows: $M_A = 11.0$ kg, $M_B = 7.0$ kg, $M_C = 5.0$ kg, and $M_D = 3.0$ kg. The blocks are stacked in descending order of weight, with block A on the bottom and block D on top. The coefficient of static friction between blocks A and B is 0.120, the coefficient of static friction between blocks B and C is 0.160, and the coefficient of static friction between blocks C and D is 0.180. There is no friction between block A and the surface on which it sits. A horizontal force, $\vec{F}$, is applied to block A. Find the maximum value of F for which none of the blocks slide with respect to one another.

103. ★★ A block of mass m is held in place on a ramp of mass M with a slope of angle θ, which in turn is held in place, as shown in Figure 5-66. The ramp is free to move on the surface beneath it. There is no friction anywhere. The ramp is then pushed with a horizontal force so that the block does not slide on the ramp. Derive an expression for the magnitude of this force.

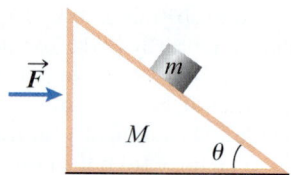

Figure 5-66 Problems 103 and 104.

104. ★★★ The coefficient of static friction between the inclined face of a wedge (mass M) and the mass m sitting on its inclined edge is μ_s. The coefficient of kinetic friction is μ_k. Derive an expression for the maximum horizontal force that you can apply to the vertical face of the wedge (Figure 5-66) so that mass m does not move with respect to the wedge.

105. ★★★ Block A of mass m_A = 25.0 kg rests on a frictionless surface, as shown in Figure 5-67. Block C of mass m_C = 6.0 kg rests on top of block A, and the coefficient of static friction between blocks A and C is 0.23. Block B is suspended from a rope that runs over the massless pulley and is then attached to block A, and the system is subsequently released from rest. Find the maximum value for the mass of block B so that when the system is let go, block C does not slide on block A.

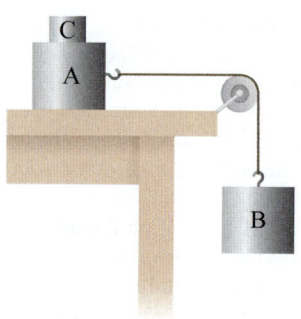

Figure 5-67 Problem 105.

106. ★★★ Block B (m_B = 1.00 kg) is in contact with block A (m_A = 3.00 kg), as shown in Figure 5-68. When block A is pushed to the right with force $\vec{F}$ of magnitude 160.0 N, block B does not slide downward, staying in place relative to block A. The coefficient of static friction between the two blocks is 0.500.
(a) Find the force of friction on block B.
(b) Find the normal force exerted by block A on block B.
(c) Find the vertical force exerted by block B on block A.

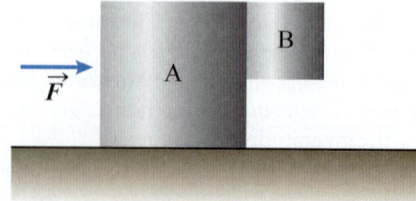

Figure 5-68 Problem 106.

DATA-RICH PROBLEM

107. ★★★ ∫, ∿, ▭ For relatively high speed motion, the magnitude of the drag force on an object moving in a fluid is given by

$$F_D = \frac{1}{2}\rho A C v^2$$

Here, F_D is the magnitude of the drag force, ρ is the density of the fluid, A is the cross section of the area intercepted by the object, C is the drag coefficient (a number normally between zero and about one that depends on the shape of the object), and v is the speed of the object relative to the fluid.

(a) Assuming that this holds for high speeds in air, develop an expression for the speed of an object in free fall as a function of time (assuming that it starts from rest). Show that the dependence of the speed on time is given by

$$v(t) = \sqrt{\frac{2gm}{\rho A C}}\tanh\left(\sqrt{\frac{\rho A C g}{2m}}t\right)$$

Hint: Start by writing the force equation from the FBD for the object.
(b) Derive an expression for the terminal speed (speed after a long time) of the object in part (a).
(c) Using a spreadsheet, plot the speed versus time for the object in part (a) for t = 0 to 200 s. Do this for a 100.0 kg long vertical cylinder 50.0 cm in radius. Assume that the drag coefficient of such a cylinder is 0.820. Use 1.225 kg/m³ for the density of air.
(d) Calculate the terminal speed of the cylinder from part (c).
(e) How long does it take the object to reach 90% of its terminal speed?
(f) How long does it take the object to reach 99% of its terminal speed?

OPEN PROBLEM

108. ★★ Using reasonable estimates for parameters and the relationships obtained in Data-Rich Problem 107, estimate the terminal speed of a sky-diver in free fall under the following circumstances.
(a) The sky-diver is "standing up" straight as she falls.
(b) She is falling in a "balled-up configuration" (curled into a ball).
(c) She has her arms and legs spread out to maximize air resistance.

MECHANICS

CHAPTER 6

Work and Energy

Recently, NASA (National Aeronautics and Space Agency) tested a new rocket, called the Space Launch System (SLS), which the agency hopes it will launch to Mars by 2030 (Figure 6-1). By mixing liquid hydrogen and liquid oxygen and burning the resulting fuel, this rocket converts chemical energy into an enormous amount of kinetic energy, generating a thrust that is equivalent to that of 14 Boeing 747 airliners taking off simultaneously. This large amount of energy is needed to accelerate the massive 85 tonne rocket from rest to a speed of over 11 km/s within a few seconds and to push the rocket away from Earth against Earth's gravitational pull. What is energy? What is kinetic energy? What is work and how are work and energy related? How much energy is needed to pull an object away from Earth's gravity? In this chapter, we take the first peek at the most fundamental and mysterious concept in the universe, energy.

Figure 6-1 A Space Launch System (SLS) rocket being tested at NASA's test facility in Nevada.

6-1 What Is Energy?

All of us are aware of the term *energy* in our daily lives. When you are tired, you do not have a lot of energy. If you are a runner, you need to eat carbohydrates and protein to store up energy for a long-distance run. A 100 W light bulb produces more heat and light than a 25 W bulb. It takes more energy to lift a 100 kg object than to lift a 10 kg object to the same height. But if you are asked to define what energy is, you will not be able to come up with a precise definition.

Like some other fundamental quantities in physics, such as the mass, the charge, and the intrinsic spin of an elementary particle (see Chapter 35), it is difficult to define energy precisely. The original word comes from the Greek words *en* (which means in) and *ergon* (which means work). So an approximate definition of energy is "the ability or the capacity to do work." There is a quantity called energy: the more we have of it, the more work we can do. But the situation is more complicated than this simple definition. As Einstein pointed out, matter can have energy by its sheer existence through the relationship $E = mc^2$, where E is the energy associated with an object of mass m and c is the speed of light in vacuum (see Chapter 30). You have undoubtedly learned in your high school physics classes that there are many different forms of energy. The energy due to the motion of an object is called kinetic energy. An object can also possess energy because of the forces exerted on it by other objects around it. This type of energy is called potential energy. Examples of potential energy are the gravitational potential energy of a ball due to Earth's gravitational pull, the elastic potential energy stored in a compressed (or stretched) spring, and the electrostatic potential energy of interaction between charges. There still are other forms of energy, including thermal energy (Chapter 17), the electromagnetic energy carried by electromagnetic waves (Chapter 27), the chemical energy released in a chemical process, and the nuclear energy released when a nuclear reaction occurs (Chapter 34).

Newton published his laws of motion in 1687, thus laying the foundation for the study of the motion of objects, large and small. The basic ingredients of these laws are the concepts of force and momentum. If all forces being exerted on an object are known, a determination of its motion simply becomes a question of following a well-defined mathematical procedure. But what does it take to generate a net force to cause an object to move? Consider the example of a space vehicle hoisted on the Space Launch System that is ready to take off. The rocket is full of liquid oxygen and hydrogen fuel and is at rest. The total momentum of the rocket and the payload is zero. When liquid hydrogen and liquid oxygen are combined and the resulting fuel is ignited, hot gases at a very high speed emerge from the back of the rocket. According to Newton's laws of motion, the total momentum of the rocket plus the emerging gases must remain zero, so the rocket must propel forward with equal and opposite momentum. The rocket is propelled upward with enormous acceleration at the expense of the burning fuel. So what is happening in the fuel-burning process to generate such an enormous force? Newton's laws of motion cannot answer this question. Once all the fuel has been used, and in the absence of gravitational pulls from heavenly objects, the space vehicle does not accelerate any more. There is no force being exerted on it (if the vehicle is far enough away that Earth's gravitational pull on it is negligible). From this point onward, the space vehicle moves with a constant velocity, a consequence of Newton's laws. We, therefore, conclude that it takes ignited fuel to generate a force and hence the motion. As the fuel burns, something is being lost in the process and at the same time a force, and hence the resulting motion, is being gained. What is being lost that results in motion? Questions similar to this (there was no space program in the nineteenth century, but steam engines did exist) led to the development of the concept of **energy**. As the fuel burns, it generates energy, which is then used to propel the rocket forward.

It is amazing that such an important property of nature was not fully realized and understood until the beginning of the nineteenth century, about 150 years after the publication of *Principia Mathematica* by Newton. The development of steam engines played an important role in understanding the relationship between the ability to perform a certain task (for example, to move a locomotive) and the amount of energy required to do so. During this time, many eminent scientists, engineers, and mathematicians contributed to the development of the concepts of heat, work, and energy, and these efforts culminated in the establishment of the laws of thermodynamics. These laws defined quantities like energy, heat, and entropy (see Chapters 17 and 18) and established mathematical relationships between them.

So we can define energy as a property of matter and waves that can be transferred among objects and can change its form. But why is this concept of energy so important? Because there is a law of nature that governs every aspect of every change in the state of the universe (it might be the emission of electromagnetic radiation from a star, sending a rocket into space, or simply lifting a book from a table to put it on a shelf) that states that *as a system changes, the total energy of the system must remain the same*. It is called the **law of conservation of energy**. This law says that if you carefully define a system and somehow measure a property called the energy of this system, then if the system changes in response to certain internal or external forces, the change must be such that the total energy of the system remains the same. Energy can change its form and move from one part of the system to another, but the amount of energy the system initially contained must remain the same—it cannot increase or decrease. The law of energy conservation for an isolated system can be written as (here energy refers to the energy of the isolated system under consideration)

$$\begin{bmatrix} \text{change in} \\ \text{kinetic energy} \end{bmatrix} + \begin{bmatrix} \text{change in} \\ \text{potential energy} \end{bmatrix} + \begin{bmatrix} \text{change in} \\ \text{thermal energy} \end{bmatrix}$$

$$+ \begin{bmatrix} \text{change in} \\ \text{electromagnetic energy} \end{bmatrix} + \begin{bmatrix} \text{change in} \\ \text{binding energy} \end{bmatrix}$$

$$+ \begin{bmatrix} \text{change in} \\ \text{nuclear energy} \end{bmatrix} + \ldots = \text{zero}$$

There is a different formula for the type of energy described above. Obviously, we have not yet defined the various forms of energy and will discuss these in this and later chapters (though you are certainly familiar with terms like *kinetic energy*, *thermal energy*, and *gravitational potential energy*). The important point is that energy comes in many forms, and one form of energy can transform into another; however, if you find that in going through a change the total energy of a system has increased or decreased, then either you were not careful enough in measuring the system's initial energy or you did not completely specify the system (i.e., you left out some parts of the system). A compressed spring has a certain amount of energy that is different from when the spring is relaxed. So the spring's energy has changed. However, an external agent (e.g., a person) compressed the spring. If we define our system to be the spring and the person, then the total energy of this system does not change when the spring is compressed. Whatever energy the external agent loses appears as the elastic potential energy of the spring plus any thermal energy that might be generated in compressing the spring.

Richard Feynman (1918–1988), the famous American physicist, described the concept of energy in his *Feynman Lectures* as follows:[1]

> There is a fact, or if you wish, a *law*, governing all natural phenomena that are known to date. There is no known exception to this law—it is exact so far as we know. The law is called the *conservation of energy*. It states that there is a certain quantity, which we call energy, that does not change in the manifold changes which nature undergoes. That is a most abstract idea, because it is a mathematical principle; it says that there is a numerical quantity which does not change when something happens. It is not a description of a mechanism, or anything concrete; it is just a strange fact that we can calculate some number and when we finish watching nature go through her tricks and calculate the number again, it is the same.

1. The Feynman Lectures on Physics, Volume I, Chapter 4. http://www.feynmanlectures.caltech.edu/I_04.html

6-2 Work Done by a Constant Force in One Dimension

To begin with, we explore the concept of work and its relationship with energy. The product of the force exerted on an object and the distance through which the point of action of the force moves is called **work**. In response to a force, an object can either move, such as when you move a ball or lift a weight, or it can deform, such as when you compress a spring or squeeze a tennis ball or both. The important point is that without motion or a change in the shape of an object, there is no work done by the force on the object. If you push against a table, but the table does not move (and does not deform), you have done no work on the table, even though you are sweating and are tired.

Let us start with the simplest situation of a rigid body on a frictionless surface that moves a distance d in a straight line under the influence of a force of constant magnitude, F, that is acting parallel to the direction of motion of the object (see Figure 6-2). Since the object is rigid, each point of the object moves by the same distance. In this case, using the definition of work as

$$\begin{matrix} \text{Work done by a force} \\ \text{on an object} \end{matrix} = \begin{matrix} \text{Force} \\ \text{applied} \end{matrix} \times \begin{matrix} \text{Distance the} \\ \text{object moves} \end{matrix}$$

a preliminary definition of work, denoted by the letter W, can be written as

KEY EQUATION
$$W_F = \pm Fd \qquad (6\text{-}1)$$

As several forces can be acting on an object, we have introduced the subscript F to clarify that the work is done by a force of magnitude F. Also, by definition, the work done by a force is positive if the force is parallel to the direction of displacement and negative if the force and the displacement are in opposite directions. The above definition is valid only if the force is acting parallel (or antiparallel) to the direction of displacement. Note that work is a scalar quantity even though both the force and the displacement are vectors. In Equation 6-1 we have multiplied the magnitudes of the force and the displacement vectors.

Units for Work

The SI unit for work is the joule (J), which equals the product of a newton (units of force) and a metre (unit of displacement). One joule is the amount of work done by a force of one newton, when the object acted upon by that force moves a distance of one metre:

$$1 \text{ J} = 1 \text{ N} \cdot \text{m}$$

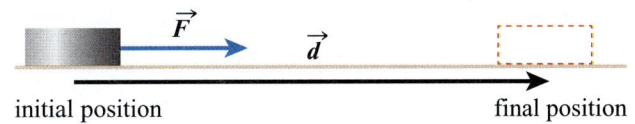

initial position   final position

Figure 6-2   Force $\vec{F}$ pushing an object over a distance $\vec{d}$.

To give you an idea, if you lift a 10 kg mass vertically up by 1 m, you are doing approximately 100 J of work on the mass. The joule is also the unit of measure for energy, as we discuss later in the chapter. It is named after English scientist James Prescott Joule (1818–1889), who did pioneering work on the relationship between heat and mechanical energy.

EXAMPLE 6-1

Pushing a Crate

You push a crate of mass 22.5 kg on a horizontal frictionless surface over a distance of 8.20 m using a constant horizontal force of 105 N starting from rest. Find the work done as you push the crate.

Solution

Since the applied force is in the direction of motion of the crate, we use Equation 6-1 with the positive sign. The work done by the force on the crate is

$$W_F = +Fd = +(105 \text{ N})(8.20 \text{ m}) = +8.61 \times 10^2 \text{ J}$$

The force does 8.61×10^2 J of positive work on the crate.

Making sense of the result

The work done by the force is given by the product of the force and the displacement, and is positive, because the force and the displacement are pointing in the same direction. In this case, the mass does not enter into the solution.

EXAMPLE 6-2

Work Done against Gravity

You lift your 2.20 kg laptop a vertical distance of 70.0 cm at a constant speed above the table on which it rests. How much work is done by you on the laptop?

Solution

To lift the laptop, you must exert a force on it that is opposite in direction to its weight, $m\vec{g}$. Let the applied force be $\vec{F}$. Since the laptop is at rest, to give it a non-zero speed, initially the magnitude of $\vec{F}$ must exceed the magnitude of $m\vec{g}$. But after that, as the laptop moves at a constant speed, the net force on it is zero and therefore the magnitudes of the two forces must be the same, that is, $F = mg$.

Figure 6-3 shows the laptop lifted upward with the two forces acting on it. To calculate the work done on the laptop by the applied force, we use Equation 6-1 with the positive sign since the applied force and displacement are in the same direction. We know the magnitude of the applied force because the weight of the laptop is known:

$$W_F = Fd = (mg)d$$
$$= (2.20 \text{ kg})(9.81 \text{ m/s}^2)(0.700 \text{ m}) = 15.1 \text{ J}$$

Figure 6-3 Force $\vec{F}$ lifts the object by a distance $\vec{d}$. The force of gravity acts vertically downward.

Making sense of the result

The work done is positive because the force is applied along the direction of displacement.

6-3 Work Done by a Constant Force in Two and Three Dimensions

In the previous section, we restricted our discussion to the case where the force is either parallel or antiparallel to the direction of displacement. We now expand our discussion to include forces that are inclined at an angle with respect to the displacement. Imagine a puck sitting on a horizontal frictionless surface of ice. If you push the puck with a force pointing vertically down, the puck will

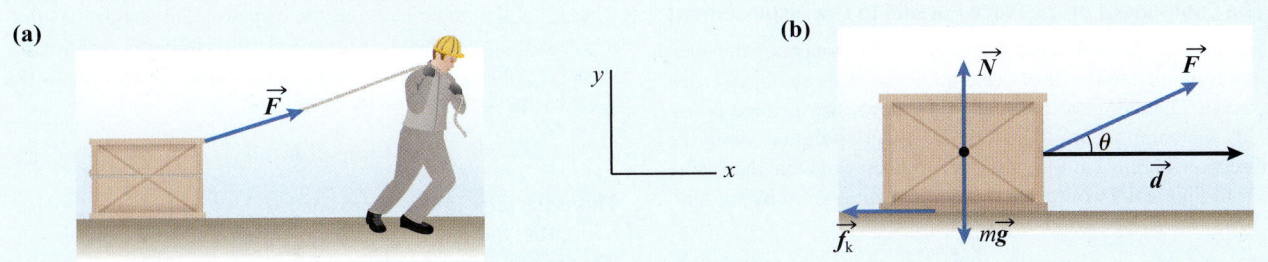

(a)

(b)

Figure 6-7 (a) A box is being pulled on a rough surface. (b) The forces being exerted on the box.

Solution

The forces acting on the box and the coordinate system are shown in Figure 6-7(b). The weight of the box ($m\vec{g}$) acts vertically downward, the normal force of the floor on the box ($\vec{N}$) acts vertically upward, the applied force ($\vec{F}$) makes an angle of 21° with respect to the positive x-axis, and the force of kinetic friction ($\vec{f}_k$) points along the negative x-axis. The net force acting on the box is zero since it moves at a constant speed.

(a) Because the box is not accelerating, the sum of all forces in the horizontal direction must be zero. Since we know the magnitude of applied force along the x-axis, we can determine the force of friction:

$$\sum F_x = F\cos\theta - f_k = 0 \qquad (1)$$

Therefore, the magnitude of the frictional force is

$$f_k = F\cos\theta = (615\text{ N})(\cos 21°) = 5.74 \times 10^2\text{ N} \qquad (2)$$

The next step is to find the normal force, which we can determine by summing the forces in the vertical direction. Since there is no motion along the vertical direction, the sum of all forces along this direction must be zero and we have

$$\sum F_y = N + F\sin\theta - mg = 0 \qquad (3)$$

Therefore,

$$N = mg - F\sin\theta = (90.0\text{ kg})(9.81\text{ m/s}^2)$$
$$- (615\text{ N})(\sin 21°) = 6.63 \times 10^2\text{ N}$$

Using Equation 5-8, we can now determine the coefficient of kinetic friction:

$$f_k = \mu_k N$$
$$\mu_k = \frac{f_k}{N} = \frac{5.74 \times 10^2\text{ N}}{6.63 \times 10^2\text{ N}} = 0.87$$

(b) To determine the work done by the person on the crate, we use Equation 6-2:

$$W_F = Fd\cos\theta = (615\text{ N})(11\text{ m})(\cos 21°) = 6.3 \times 10^3\text{ J}$$

(c) As the box is a rigid object, every point of the box is displaced by the same distance of 11 m. Therefore, we can treat the box as a point mass. The work done on the box by the force of kinetic friction is then given by

$$W_f = f_k d\cos\theta = (574\text{ N})(11\text{ m})\cos 180° = -6.3 \times 10^3\text{ J}$$

Notice that the work done by friction on the box is negative because it acts in a direction opposite to the direction of displacement of the box.

(d) The total work done on the box is equal to the sum of the work done on it by the applied force and the work done by the force of friction. Using the two previous results, we see that the total work done is equal to zero.

Making sense of the result

The magnitude of the force of friction is equal to the horizontal component of the applied force, and both act in opposite directions. Hence, the work done by each force is equal in magnitude and opposite in sign and consistent with the statement that the crate moves with a constant velocity and hence the net force on the crate must be zero.

EXAMPLE 6-4

Work Done by the Force of Gravity

Starting from rest, a block slides down a frictionless ramp of slope angle θ, as shown in Figure 6-8. Find the work done by the force of gravity on the block as it slides down a distance $\vec{d}$ along the ramp, such that its vertical height changes by an amount $\Delta\vec{h}$.

Solution

We solve this example using two perspectives.

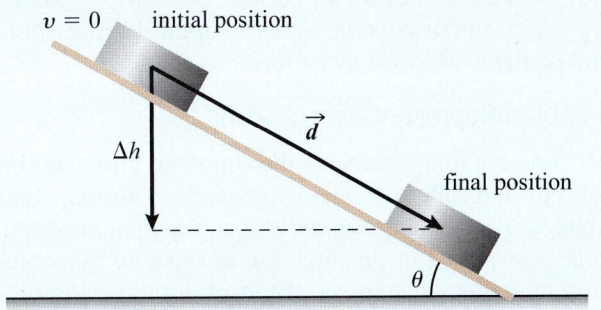

Figure 6-8 Mass sliding down an incline.

(continued)

The Component of the Force Parallel to the Displacement

We saw from Equation 6-2 that only the component of the force that is parallel to the displacement does work as an object is displaced. Therefore, as the block slides down the inclined plane, only the component of the force of gravity pointing along the direction of motion along the ramp does work on the block. From Figure 6-9 we see that this component is given by $mg\sin\theta$.

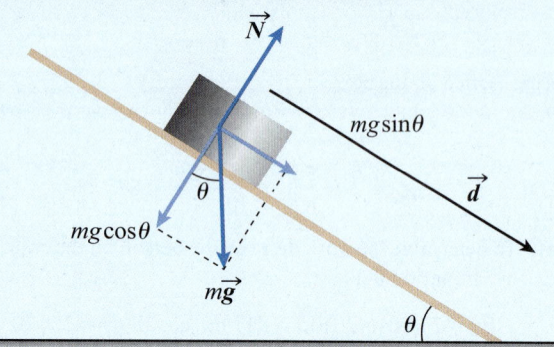

Figure 6-9 A block sliding down a frictionless inclined plane. Only the component of the force of gravity ($m\vec{g}$) along the ramp does work on the block.

Therefore, the work done by the force of gravity on the block as it slides down the incline is given by

$$W_g = (mg\sin\theta)d \tag{1}$$

Component of the Displacement Parallel to the Force

The work done on the block by the force of gravity is given by the scalar product of the force of gravity and the displacement vectors:

$$W_g = m\vec{g} \cdot \vec{d} = (mg)(d)\cos\beta \tag{2}$$

where β is the angle between the displacement vector and the force vector, as shown in Figure 6-9. Notice that $\beta = 90° - \theta$, so $\cos\beta = \cos(90° - \theta) = \sin\theta$ and the expression for the work done by the force of gravity becomes

$$W_g = (mg)(d)\sin\theta \tag{3}$$

This is the same as the expression in (1). From Figure 6-10, also notice that the vertical displacement, Δh, and the displacement d along the incline are related by $\Delta h = d\sin\theta$. Therefore, the work done by the force of gravity becomes

$$W_g = mg\Delta h \tag{4}$$

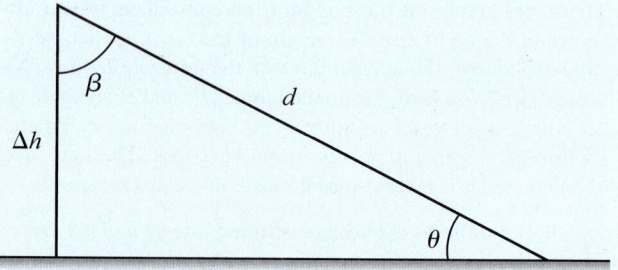

Figure 6-10 The relationship between vertical displacement and displacement along the inclined plane.

Making sense of the result

This exercise highlights the fact that the work done by a constant force on an object can be expressed as the product of the magnitude of the force with the component of the displacement parallel to the force or as the product of the component of the force in the direction of the displacement and the magnitude of the displacement.

6-4 Work Done by Variable Forces

So far, we have dealt with the work done by constant forces. However, most of the forces that we encounter in daily life are not constant. Standing up, sitting down, starting and stopping a car, skiing, and canoeing all involve forces that vary in magnitude and direction. How do we calculate the work done by a variable force? We begin our discussion with a graphical representation of the work done by a force.

Graphical Representation of Work

We choose a one-dimensional example to illustrate the concept. Consider an object moving in a straight line while being pushed over a distance Δx by a constant force that points in the direction of motion. Since the force remains constant, a graph of force versus displacement is a horizontal line, as shown in Figure 6-11.

In this graph, the work done by the force ($W_F = F\Delta x$) is therefore represented by the area of the rectangle that has height F and width Δx.

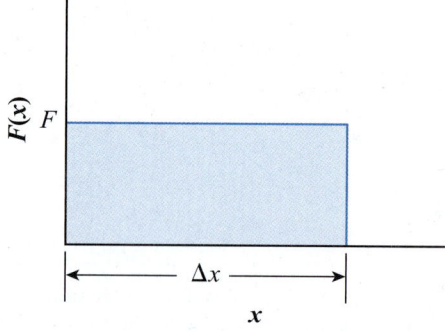

Figure 6-11 A graph of constant force versus displacement.

Figure 6-12 shows a force versus displacement graph for a force that moves an object from point $x = a$ to point $x = b$ along the x-axis. We have chosen a force of variable magnitude that is acting in the direction of displacement.

Generalizing the case for a constant force, the work done by a variable force is also given by the area under the curve between the starting point and the end point

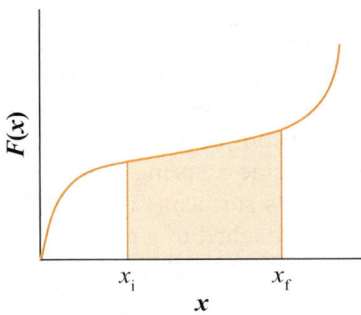

Figure 6-12 The work done by a force in displacing an object from x_i to x_f is the area under the graph of $F(x)$ between x_i and x_f.

of the force versus displacement graph. How do we calculate the area under a curve for a graph like that in Figure 6-12? We can break up the area into N very thin rectangles, find the area of each rectangle, and then sum the areas to obtain an approximate value for the area under the curve. Consider the jth rectangle. Let F_j be the magnitude of the force at $x = x_j$ and let $\Delta x_j = x_{j+1} - x_j$ be the width of the rectangle between $x = x_j$ and $x = x_{j+1}$ (Figure 6-13). The work done by the force in displacing the object from $x = x_j$ to $x = x_{j+1}$ is given by $W_j = F_j \Delta x_j$ and is equal to the area of this jth rectangle. The total work done in moving the object from $x = a$ to $x = b$ is therefore approximately equal to the sum of work done over each small displacement Δx_j:

$$W \approx F_1 \Delta x_1 + F_2 \Delta x_2 + \cdots + F_N \Delta x_N = \sum_{j=1}^{N} F_j \Delta x_j \quad (6\text{-}13)$$

If we divide the area into N rectangles of equal width ($\Delta x_j = \Delta x$ for all j; see Figure 6-13), then $N \times \Delta x = (b - a)$. Notice that for each rectangle, the magnitude of the force can be different. Therefore, this procedure allows us to calculate the work done by a variable force. In Equation 6-13 the $\approx$ sign indicates an approximate equality. As you can see from Figure 6-13, unless the force remains constant between the start and end points of displacement, the product $F_j \Delta x_j$ only approximately represents the area under the curve between these points.

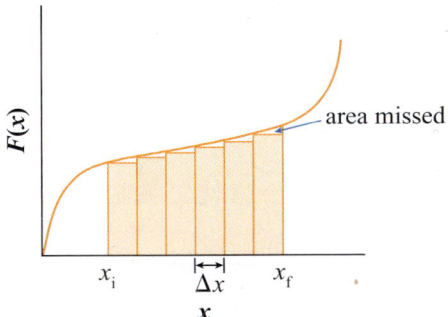

Figure 6-13 The summation of the areas of the rectangles is an approximation of the actual area under the curve. The area between the top of the rectangle and the curve is missed.

How can we improve the accuracy of this calculation? Obviously, by making N larger we can reduce the width of the rectangles. This reduces the error between the actual area under the curve and the area represented by the product $F_j \Delta x$ (see Figure 6-14).

In the limit $\Delta x \to 0$ we replace Δx by the differential dx, and the sum $\sum_{j=1}^{N} F_j \Delta x_j$ (where now $N \to \infty$) becomes the integral of the force over the displacement. The area obtained using integration is exactly equal to the work done by the variable force over the displacement from a to b:

$$W_F = \int_a^b F(x)\,dx \quad (6\text{-}14)$$

In the case of two and three dimensions, where the force may be pointing in an arbitrary direction relative to the direction of motion, we can generalize the above procedure by writing work as the dot product of force and displacement vectors. For an infinitesimal displacement $d\vec{r}$ along the trajectory of the object between points $r_i = (x_i, y_i, z_i)$ and $r_f = (x_f, y_f, z_f)$, only the component of force that is parallel to the displacement contributes to work. Using the notation of Equation 6-14, we can write the work done by a variable force in three dimensions as

KEY EQUATION

$$W_F = \int_{r_i}^{r_f} \vec{F}(\vec{r}) \cdot d\vec{r}$$
$$= \int_{r_i}^{r_f} F_x(\vec{r})\,dx + \int_{r_i}^{r_f} F_y(\vec{r})\,dy + \int_{r_i}^{r_f} F_z(\vec{r})\,dz \quad (6\text{-}15)$$

Here we have used

$$\vec{F}(\vec{r}) \cdot d\vec{r} = F_x(\vec{r})\,dx + F_y(\vec{r})\,dy + F_z(\vec{r})\,dz$$

If the motion is in two dimensions, the above integral is a two-dimensional integral.

Consider the descent of a glider under the influence of gravity along a path from an initial position $r_i = (x_i, y_i)$ to a final position $r_f = (x_f, y_f)$, as shown in Figure 6-15. Let $d\vec{r}$ be an infinitesimal displacement of the glider along this path. With the coordinate system

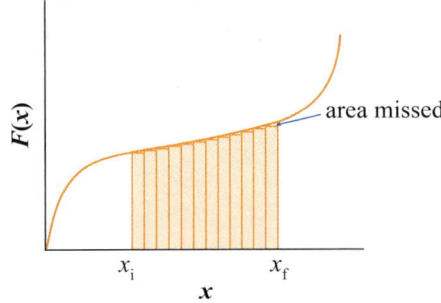

Figure 6-14 As the width of the rectangles becomes smaller, the difference between the calculated area under the curve and the actual area becomes smaller.

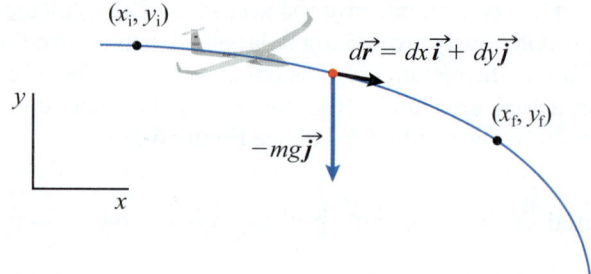

Figure 6-15 A glider descending along the shown path under the force of gravity.

shown and writing the displacement and force vectors in x- and y-coordinates, we have

$$d\vec{r} = dx\hat{i} + dy\hat{j}$$

$$\vec{F}_g = -mg\hat{j}$$

Therefore,

$$\vec{F}_g \cdot d\vec{r} = (-mg\hat{j}) \cdot (dx\hat{i} + dy\hat{j})$$

$$= -mgdx(\hat{j} \cdot \hat{i}) - mgdy(\hat{i} \cdot \hat{i}) = (-mg)dy \qquad (6\text{-}16)$$

The work done by the force of gravity on the glider is therefore

$$W_g = \int_{y_i}^{y_f} (-mg)dy$$

$$= -mgy\Big|_{y_i}^{y_f} = -mg(y_f - y_i) \qquad (6\text{-}17)$$

Since $y_f < y_i$, the work done by the force of gravity on the glider is positive, as it should be.

CHECKPOINT 6-3

Work Done by Variable Forces

Consider the force versus displacement graphs in Figure 6-16. Rank the graphs in order of work done, from most to least, by the force on the object.

(a)
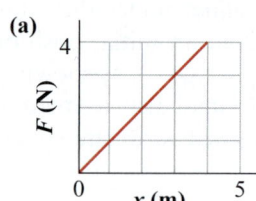

(b)
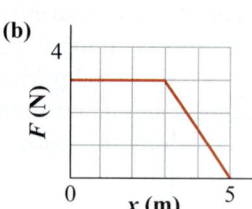

(c)
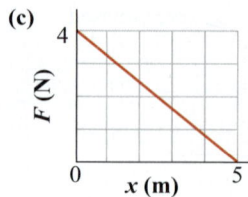

(d)
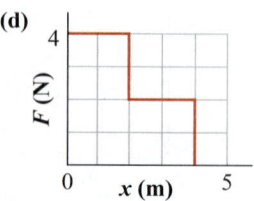

Figure 6-16 Various force and displacement graphs.

Work Done by a Spring

As an example of work done by a variable force, consider the work done by a spring on an external agent (e.g., your hand) that compresses or stretches the spring in this example. Imagine a spring that is resting on a horizontal, frictionless surface (Figure 6-17). The left end of the spring is attached to a rigid support. Let us choose a coordinate system with its origin ($x = 0$) at the free end of the unstretched spring. If this end is moved to the right by a distance x, the spring will stretch by a length x and exert a force in the opposite direction, toward the equilibrium position.

As discussed in Chapter 5, the force exerted by a spring when its length increases (stretch) or decreases (compression) by an amount x is given by Hooke's law:

$$\vec{F}_s = -k\vec{x} \qquad (6\text{-}18)$$

The negative sign indicates that the direction of this force is opposite to the direction of displacement of its free end, always tending to pull the spring back to its equilibrium length. Such a force is called a **restoring force**. The spring constant k is a measure of the stiffness of the string. The larger the value of k, the greater is the force required to stretch the spring by a given amount. Obviously, the force described by Equation 6-18 is a variable force. For a given compression or elongation, it linearly increases with the change in the length of the spring.

Consider now the situation shown in Figure 6-17. Let us calculate the work done by the spring on the external agent when its free end is moved from an initial position x_i to a position x_f. As the force changes with displacement, we use Equation 6-15 to calculate the work done by the spring:

$$W_{\text{by-spring}} = \int_{x_i}^{x_f} \vec{F}_s \cdot d\vec{x} \qquad (1)$$

Substituting Equation 6-18 in the above equation, the work done by the spring is given by

$$W_{\text{by-spring}} = \int_{x_i}^{x_f} (-k\vec{x}) \cdot d\vec{x} \qquad (2)$$

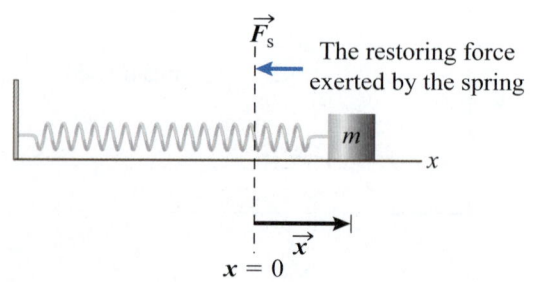

Figure 6-17 The direction of the restoring force exerted by the spring on the block is opposite to the direction of displacement of the block from the equilibrium position.

As we are dealing with one-dimensional situation,

$$\vec{x} \cdot d\vec{x} = (x\hat{i}) \cdot (dx\,\hat{i}) = xdx(\hat{i} \cdot \hat{i}) = xdx$$

Therefore, the work done by the spring is

$$W_{\text{by-spring}} = -k\int_{x_i}^{x_f} x\,dx$$

$$= -k\left(\frac{x^2}{2}\Big|_{x_i}^{x_f}\right)$$

$$= \left(-\frac{kx_f^2}{2}\right) - \left(-\frac{kx_i^2}{2}\right)$$

Therefore,

KEY EQUATION
$$W_{\text{by-spring}} = \frac{1}{2}kx_i^2 - \frac{1}{2}kx_f^2 \qquad (6\text{-}19)$$

Note that the work done by the spring in this example depends only on the initial and final positions of its free end. If the starting position is the unstretched position, $x_i = 0$, and the final position is taken to be $x_f = x$, then

$$W_{\text{by-spring}} = -\frac{1}{2}kx^2 \qquad (6\text{-}20)$$

When you stretch or compress a spring from the unstretched position, the work done by the spring is negative. Does this make sense? The force exerted by the spring with one end fixed is always directed opposite to the direction of displacement of the free end. Therefore, the work done by the spring on the external agent (attempting to stretch or compress the spring) is negative.

Work Done by the External Agent

To change the length of a spring, an external force must be exerted on its free end (for a spring with one end attached to a rigid support). Consider a spring that is stretched by a length x from its equilibrium position. In this stretched configuration, the spring is exerting a force $-k\vec{x}$ on the external agent. As the free end is not accelerating, the external force being exerted on the spring must be $+k\vec{x}$. The external force points in the direction of the displacement of the free end. Therefore, the work done by the external force on the spring in changing its length by an amount x from the equilibrium length must be positive. By following the same arguments as above, we see that

$$W_{\text{on-spring}} = +\frac{1}{2}kx^2 \qquad (6\text{-}21)$$

CHECKPOINT 6-4

Positive and Negative Work Done by a Spring

When is the work done by a spring on an external agent positive?

(a) as the spring is being compressed
(b) as the spring is being stretched
(c) as the spring returns to its equilibrium (unstretched) position
(d) when the spring is in its equilibrium (unstretched) position

ANSWER: (c) This is the only case when the force exerted by the spring points in the direction of displacement, and not opposite to it.

EXAMPLE 6-5

Stretching a Spring

A spring of spring constant $k = 500.0$ N/m sits on a horizontal surface with one end attached to a wall.

(a) How much work is done by the spring as you stretch its free end by 10.0 cm?
(b) From the point where the spring is already stretched by 10.0 cm, how much work would be done by the spring if you stretched it by another 10.0 cm?
(c) Is the work done in part (b) equal to the work done in part (a)?
(d) How much work does the spring do as you compress it from the equilibrium positon by 10.0 cm?
(e) How does this compare to the answer in part (a)?

Solution

(a) This example is an application of the expression for the work done by a spring given in Equation 6-19. The spring

is initially unstretched, and we stretch it by 10. cm, so $x_i = 0$ and $x_f = 10.0$ cm $= 0.10$ m. Substituting these values, we get

$$W_{\text{by-spring}} = \frac{1}{2}kx_i^2 - \frac{1}{2}kx_f^2$$

$$= \frac{1}{2}(500.0 \text{ N/m})(0)^2 - \frac{1}{2}(500.0 \text{ N/m})(0.10 \text{ m})^2$$

$$= -2.50 \text{ J} \qquad (1)$$

(b) We begin with a spring that is stretched by 10.0 cm and stretch it by another 10.0 cm. The final positon of the free end is 20.0 cm from the equilibrium position. In this case, $x_i = 10.0$ cm $= 0.10$ m and $x_f = 20.0$ cm $= 0.20$ m. Therefore,

$$W_{\text{by-spring}} = \frac{1}{2}kx_i^2 - \frac{1}{2}kx_f^2$$

$$= \frac{1}{2}(500.0 \text{ N/m})(0.10 \text{ m})^2 - \frac{1}{2}(500.0 \text{ N/m})(0.20 \text{ m})^2$$

$$= -7.50 \text{ J} \qquad (2)$$

(continued)

(c) The work done in part (b) is not equal to the work done in part (a), even though in each case the spring is stretched by 10.0 cm from its initial position. The force exerted by the spring changes as its length changes. It is more difficult to further stretch an already stretched spring.

(d) Here we compress the spring from the equilibrium positon. Therefore, $x_i = 0$ and $x_f = -10.$ cm $= -0.10$ m:

$$W_{\text{by-spring}} = \frac{1}{2}kx_i^2 - \frac{1}{2}kx_f^2$$

$$= \frac{1}{2}(500.0 \text{ N/m})(0)^2 - \frac{1}{2}(500.0 \text{ N/m})(-0.10 \text{ m})^2$$

$$= -2.5 \text{ J} \tag{3}$$

(e) The answer to part (e) is the same as the answer to part (a).

Making sense of the result

Whether a spring is stretched or compressed, it does negative work. The work done in compressing or stretching an unstretched spring is proportional to the square of the change in the length of the spring. Therefore, the work done by the spring (on the external agent) is the same when it is compressed or stretched by the same length from its unstretched length.

6-5 Kinetic Energy— The Work-Energy Theorem

We now consider the effect of work (positive or negative) done on an object and its state of motion, specifically its speed. Consider an object of mass m that has initial velocity $\vec{v}_i$. Consider a constant force (both the magnitude and the direction of force remain constant in this discussion) that acts on the object for a time t and causes a displacement $\vec{d}$. Let the velocity of the object at the end of this displacement be $\vec{v}_f$. We ask the following question. What is the relationship between the work done on the object and its initial and final speeds? Since the force $\vec{F}$ acts on the object over a displacement $\vec{d}$, the work done by the force on the object is $W_F = \vec{F} \cdot \vec{d}$. As the force is constant and we assume that the mass of the object does not change, the object experiences a constant acceleration, $\vec{a}$, as a result of the applied force. We can therefore use the kinematics equations for constant acceleration to relate the object's initial and final speeds to the work done on it. From Chapter 4, we know that the initial and final velocities are related to the acceleration by

$$\vec{v}_f = \vec{v}_i + \vec{a}t \tag{6-22}$$

and the displacement $\vec{d}$ is related to the initial velocity and acceleration by

$$\vec{d} = \vec{v}_i t + \frac{1}{2}\vec{a}t^2 \tag{6-23}$$

Here we have used the vector notation for the kinematics equations to keep the discussion general. Taking the dot product of Equation 6-22 with $\vec{v}_f$, we get

$$\vec{v}_f \cdot \vec{v}_f = (\vec{v}_i + \vec{a}t) \cdot (\vec{v}_i + \vec{a}t)$$

$$= \vec{v}_i \cdot \vec{v}_i + \vec{a} \cdot \vec{a}t^2 + t(\vec{a} \cdot \vec{v}_i + \vec{v}_i \cdot \vec{a})$$

Using the fact that for two given vectors $\vec{A}$ and $\vec{B}$, $\vec{A} \cdot \vec{A} = A^2$ and $\vec{B} \cdot \vec{B} = B^2$, where A and B are the

magnitudes of the two vectors, and $\vec{A} \cdot \vec{B} = \vec{B} \cdot \vec{A}$, we can write the above equation as

$$v_f^2 = v_i^2 + a^2t^2 + 2t\vec{a} \cdot \vec{v}_i$$

or

$$v_f^2 - v_i^2 = a^2t^2 + 2t\vec{a} \cdot \vec{v}_i \tag{6-24}$$

The above equation relates the initial and final speeds of the object but it involves the time for which the force acts on the object. To eliminate the time, t, from this equation we use Equation 6-23. Take the dot product of this equation with acceleration $\vec{a}$ and multiply the result by 2. We get

$$2\vec{a} \cdot \vec{d} = 2\vec{a} \cdot \vec{v}_i t + a^2t^2 \tag{6-25}$$

Here we have used the fact that $\vec{a} \cdot \vec{a} = a^2$. As the right-hand sides of Equations 6-24 and 6-25 are equal, we can write Equation 6-24 as

$$v_f^2 - v_i^2 = 2\vec{a} \cdot \vec{d} \tag{6-26}$$

You are no doubt familiar with the above equation when written in terms of the magnitudes of the acceleration and displacement vectors. However, writing this equation in terms of the vector quantities makes the sign convention for work done by a force obvious. Multiplying the above equation by m and dividing by 2, we get

$$m\vec{a} \cdot \vec{d} = \frac{m}{2}(v_f^2 - v_i^2) \tag{6-27}$$

The term $m\vec{a}$ is just the force $\vec{F}$ acting on the object. Therefore, the left side of the above equation is $\vec{F} \cdot \vec{d}$, the work done on the object by the force. We now have the desired relation that connects the work done by a force to the object's initial and final speeds:

KEY EQUATION

$$W_F = \frac{1}{2}mv_f^2 - \frac{1}{2}mv_i^2 \tag{6-28}$$

The quantity $\frac{1}{2}mv^2$ is called the **kinetic energy** and is usually denoted by the letter K:

KEY EQUATION
$$K = \frac{1}{2}mv^2 \text{ (kinetic energy)} \qquad (6\text{-}29)$$

Note that the kinetic energy depends on the square of the particle's speed. So it is independent of the direction of motion. The above equation for kinetic energy is only valid when relativistic effects can be ignored. Equation 6-28 then tells that when work is done on an object, its kinetic energy changes and that the work done is equal to the change in the object's kinetic energy. We can write Equation 6-28 as

KEY EQUATION
$$W_F = K_f - K_i = \Delta K \qquad (6\text{-}30)$$

Here K_f represents the final kinetic energy, K_i the initial kinetic energy, and ΔK the change in kinetic energy, final minus initial. Equation 6-30 is called the **work–energy theorem**: *the work done on an object is equal to the change in the object's kinetic energy.*

Let us consider two cases. In the first case, the force is parallel to the direction of displacement. Therefore,

$$W_F = \vec{F} \cdot \vec{d} = Fd\cos 0 = Fd$$

Since F and d are the magnitudes of two vectors, the product Fd is positive, so $W_F > 0$. This implies that $K_f > K_i$, and since the mass does not change (we ignore relativistic effects) with speed, we must have $v_f > v_i$. *When positive work is done on an object, its speed increases.* This is the situation when you push an object over a certain distance. The speed at the end of the push is greater than the starting speed.

In the second case, the force is applied opposite (antiparallel) to the direction of displacement. Therefore,

$$W_F = \vec{F} \cdot \vec{d} = Fd\cos 180° = -Fd$$

Since Fd is positive, in this case $W_F < 0$ and we must have $K_f < K_i$. Since the mass does not change with speed, we must have $v_f < v_i$. *When negative work is done on an object, its speed decreases.* An example of this situation is when you are standing on an ice rink and you extend your hands to try to slow down a fast-moving person who pushes on you.

The sign of the work done by a force depends on the sign of the scalar product $\vec{F} \cdot \vec{d}$. If this scalar product is positive (negative), the work done on the object is positive (negative). If the $\vec{F}$ and $\vec{d}$ vectors are perpendicular to each other, the scalar product vanishes and in this case, as we have already discussed, no work is done on the object by the force.

Although we have derived the work–energy theorem for the simple case of a constant force, the result is general. The work done by a constant or a variable force on an object is equal to the change in the object's

kinetic energy. The theorem is also true for the case when an object is subjected to a number of forces. In this case, it is the net or total work done by the net force that equals the change in the object's kinetic energy.

Total or Net Work

In any realistic case, several forces may be acting on an object. How is the work–energy theorem applied in this case? Let us consider the simple case of a force $\vec{F}$ being exerted on an object that is moving on a horizontal surface. Let us introduce a constant frictional force, $\vec{f}$, between the object and the surface. The frictional force points opposite to the direction of $\vec{F}$. So in this case the object is under the influence of two forces (Figure 6-18). Let us assume that the magnitudes of these two forces are not the same so that the object accelerates. The acceleration produced in the object is the vector sum of the two forces divided by its mass,

$$\vec{a} = \frac{\vec{F} + \vec{f}}{m} \qquad (6\text{-}31)$$

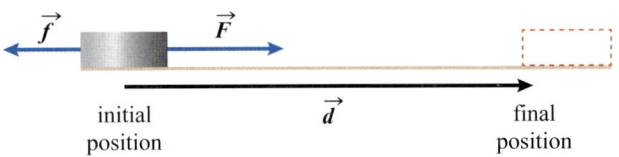

Figure 6-18 Forces $\vec{F}$ and $\vec{f}$ acting on an object that is displaced by distance $\vec{d}$.

Table 6-1   Appreciating the Joule: Energy Orders of Magnitude

Object	Energy (J)
Average kinetic energy of a molecule of an ideal gas at room temperature	6.068×10^{-21}
1 eV (electron volt);[1] the kinetic energy of an electron that has been accelerated through a potential difference of 1 V	1.6×10^{-19}
Energy range of a photon of visible light	2.614×10^{-19} to 5.28×10^{-19}
0.511 MeV (mega electron volt); the rest mass energy of an electron[2]	8.187×10^{-14}
Approximate kinetic energy gained by a 1.0 kg weight falling a distance of 10. cm	1.0
Thermochemical calorie; the energy required to raise the temperature of 1 g of water by 1 °C	4.184
1 BTU (British thermal unit);[3] the amount of energy needed to raise the temperature of 1 lb of water by 1 °F	1.0551×10^3
1 W·h (watt hour);[4] see Section 6-9	3.6×10^3
1 kcal$_\text{th}$; the energy required to raise the temperature of 1 kg of water by 1 °C	4.184×10^3
Hydrogen bond energy in water	~23 kJ/mol
Kinetic energy of a 1000 kg vehicle moving at 100 km/h	390 kJ
Nutritional energy from a can of cola	590 kJ
1 kW·h (kilowatt hour)	3.6×10^6
Cruising kinetic energy of Boeing 747 at maximum weight	1.21×10^{10}
Energy released by burning 1 T of crude oil (1 tonne of oil equivalent, or 1 toe)	4.1868×10^{10}
21 kilotonnes of TNT, "Fat Man" yield, the nuclear bomb detonated over Nagasaki	87.8×10^{12}
Energy released from the fission of 1 kg of U-235	8.8×10^{13}
Rest mass energy of a 1 lb grapefruit	5.1×10^{16}
Magnitude 7 earthquake seismic moment energy, equal to 617 times the energy released during the Hiroshima nuclear detonation	3.89×10^{19}
13 371 Mtoe (million tons of oil equivalent); the world's primary energy supply for 2012	5.59×10^{20}
Energy released during the Shoemaker Levy meteor impact	2.5×10^{23}
Energy of the meteor impact that formed the Chicxulub crater	4×10^{23}
Energy released during the Crab Supernova	~1×10^{44}

[1] The electron volt (eV) is commonly used as an energy unit in quantum physics. See Chapter 31 for details.
[2] The rest mass energy of a particle is the energy equivalent of the mass of the particle, in effect, the energy required to create the particle, using Einstein's famous equation, $E = mc^2$. See Chapter 30.
[3] The value presented here is the international table BTU.
[4] Unit commonly used in power production and consumption.

and is constant. Using this value of the acceleration in Equation 6-27, we get

$$m\left(\frac{\vec{F}+\vec{f}}{m}\right) \cdot \vec{d} = \frac{m}{2}(v_\text{f}^2 - v_\text{i}^2) \qquad (6\text{-}32)$$

which leads to

$$(\vec{F}+\vec{f}) \cdot \vec{d} = \frac{m}{2}(v_\text{f}^2 - v_\text{i}^2) \qquad (6\text{-}33)$$

The first term on the left side of the above equation is the work, W_F, done by the force $\vec{F}$ in displacing the object by $\vec{d}$. The second term, $\vec{f} \cdot \vec{d}$, represents the work, W_f, done on the object by the force of friction. The left side of the above equation is therefore

$$(\vec{F}+\vec{f}) \cdot \vec{d} = \vec{F} \cdot \vec{d} + \vec{f} \cdot \vec{d} = W_F + W_f \qquad (6\text{-}34)$$

The work done by all external forces on an object is called the net or the total work and is usually denoted by the symbol W_net. In this case,

$$W_\text{net} = W_F + W_f$$

In this case, since the two forces are pointing in opposite directions, W_F is positive ($\vec{F}$ points in the direction of $\vec{d}$) and W_f is negative.

In general, then, if N forces are being exerted on an object and as a result of these forces the object displaces by $\vec{d}$, then the net work done by all forces on the object is the sum of the work done by each force:

$$W_\text{net} = W_1 + W_2 + \cdots + W_N$$
$$= \vec{F}_1 \cdot \vec{d} + \vec{F}_2 \cdot \vec{d} + \cdots + \vec{F}_N \cdot \vec{d} \qquad (6\text{-}35)$$

The work–energy theorem then states that the *net work done by all external forces on an object is equal to the change in the kinetic energy of the object.* In mathematical form this is written as

$$W_{\text{net}} = K_f - K_i \tag{6-36}$$

The above discussion uses vector equations. For the case of motion in one dimension, we can write these equations in terms of scalar quantities. For the above example, let us assume that the force $\vec{F}$ is exerted along the positive x-axis, which is also the direction of motion of the object. In this case,

$$\vec{F} = F\hat{i}, f = f(-\hat{i}) = -f\,\hat{i}, \vec{a} = a\hat{i}, \vec{d} = d\hat{i}$$

Therefore,

$$\begin{aligned}\vec{F} \cdot \vec{d} &= (F\hat{i}) \cdot (d\hat{i}) = Fd(\hat{i} \cdot \hat{i}) = Fd \\ \vec{f} \cdot \vec{d} &= (-f\hat{i}) \cdot (d\hat{i}) = fd(-\hat{i} \cdot \hat{i}) = -fd\end{aligned} \tag{6-37}$$

Equation 6-33 then takes the form

$$(F - f)d = \frac{m}{2}(v_f^2 - v_i^2) \tag{6-38}$$

or

$$W_F + W_f = \frac{m}{2}(v_f^2 - v_i^2)$$

with $W_F = Fd$ and $W_f = -fd$.

EXAMPLE 6-6

Negative Work

In a strong-person contest, a car with a mass of 1.20×10^3 kg rolls toward a contestant at a speed of 0.800 m/s. The contestant pushes against the car with a constant force (Figure 6-19) and brings it to a stop over a distance of 0.500 m.

(a) Determine the work done by the contestant on the car.
(b) Find the magnitude of the force exerted by the contestant.

Solution

(a) The contestant is the only agent doing work on the car. The normal force ($\vec{N}$) and the force of gravity ($m\vec{g}$) act perpendicular to the direction of motion, and therefore the work done by each of these forces is zero. The initial

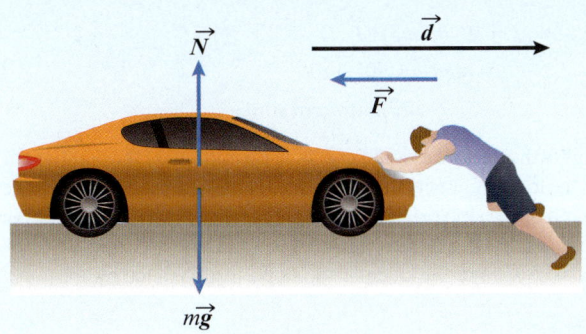

Figure 6-19 Forces acting on a car. The contestant exerts a force $\vec{F}$ on the car, and the car pushes the contestant a distance d backward before stopping. The force of gravity on the car and the normal force are perpendicular to the direction of motion of the car and the contestant.

speed of the car is 0.800 m/s and its final speed is zero. Applying the work–energy theorem, we can calculate the work done by the contestant on the car:

$$\begin{aligned}W_{\text{on-car}} &= K_f - K_i = 0 - \frac{1}{2}mv_i^2 \\ &= 0 - \frac{1}{2}(1.20 \times 10^3 \text{ kg})(0.800 \text{ m/s})^2 = -384 \text{ J}\end{aligned} \tag{1}$$

(b) Let $\vec{F}$ be the force exerted on the car by the contestant. The car moves the contestant backward by a distance $d = 0.500$ m, so the work done by the contestant on the car (be careful here, we are determining the work done by the contestant on the car and not the other way around) is given by

$$W_{\text{on-car}} = \vec{F} \cdot \vec{d} = Fd \cos 180° = -Fd \tag{2}$$

Combining (1) and (2), we can calculate the magnitude of the force exerted by the contestant on the car:

$$\begin{aligned}-Fd &= -384 \text{ J} \\ F(0.500\,\text{m}) &= 384 \text{ J} \\ F &= \frac{384 \text{ J}}{0.500 \text{ m}} \approx 768 \text{ N}\end{aligned}$$

The direction of $\vec{F}$ is opposite to the direction of motion of the car.

Making sense of the result

The final kinetic energy of the car is less than its initial kinetic energy. Therefore, $K_f - K_i$ is negative, and the work done on the car by the contestant is negative.

CHECKPOINT 6-5

Work and Kinetic Energy

You are standing on a horizontal frictionless sheet of ice when a child comes sliding toward you. Wearing shoes that prevent you from sliding, you place your hands gently against the child to slow her down and then push her back in the opposite direction with the same speed she was coming toward you, 0.5 m/s. The child's mass is 40. kg. Assuming that you exert a force of constant magnitude, what is the total work done by you on the child during the entire process?

(a) 5.0 J
(b) 10. J
(c) zero
(d) −5.0 J

ANSWER: Since the initial and the final speeds of the child are the same, the change in her kinetic energy is zero, so the total work done on the child by you is zero.

MAKING CONNECTIONS

Walking It Off

An average fast-food burger without cheese has 590 Cal (kcal) and provides 52% of the daily limit of fat for an average recommended daily caloric intake of 2000 Cal. A medium-sized cola drink (473 mL) has 210 Cal, all from sugar, and a medium order of fries has 453 Cal and provides 33% of the daily fat limit. On average, a 70 kg person burns 390 Cal/h during a very brisk 7 km/h walk. At this rate, it would take close to 21 km of walking to burn off the Calories in the food described above.

Buried in expression (2) in Example 6-7 is a very important piece of physics. The net work done on the skier is exactly the same as if the skier were dropped from height h. If the slope angle changed so the

EXAMPLE 6-7

Work Done by the Force of Gravity

A skier slides down a ski slope that makes an angle of θ with respect to the horizontal direction, as shown in Figure 6-20. The initial position of the skier is at a vertical height h above the base of the slope, and the skier starts from rest. Ignoring frictional forces, determine the work done on the skier by all the forces and calculate the skier's speed at the bottom of the slope. The total mass of the skier and the skis is m, and for this example the skier can be considered as a point particle.

Solution

There are two forces acting on the skier, the force of gravity, $m\vec{g}$, and the normal force, $\vec{N}$, between the snow and the skier. Figure 6-20(a) shows these two forces and Figure 6-20(b) shows the components of these forces along the direction of the slope and perpendicular to it. The normal force is directed perpendicular to the slope and therefore does not do any work on the skier. The force of gravity is directed vertically downward, and its component (Figure 6-20(b)) along the direction of the slope is $mg \sin\theta$ and perpendicular to the slope is $mg \cos\theta$. The slope angle, θ, does not change. Note from the figure that the angle

between the component of $m\vec{g}$ and the displacement vector, $\vec{d}$, is $90° - \theta$. Therefore, for a total displacement $\vec{d}$ from top to bottom of the slope, the net work, W_{net}, done by the force of gravity on the skier is

$$\begin{aligned} W_{net} = W_g = (m\vec{g}) \cdot \vec{d} \\ = mgd\cos(90° - \theta) \qquad (1) \\ = (mg\sin\theta)d = mg(d\sin\theta) \end{aligned}$$

This is the result we would expect if we multiplied the component of the force of gravity along the direction of slope, $mg \sin\theta$, by the length of the slope, d. From the figure, notice that the length of the slope, d; the height difference between the top and the bottom of the slope, h; and the angle of the slope, θ, are related by

$$\sin\theta = \frac{h}{d}$$

Therefore, $d \sin\theta = h$ and the net work done on the skier can be written as

$$W_{net} = mgh \qquad (2)$$

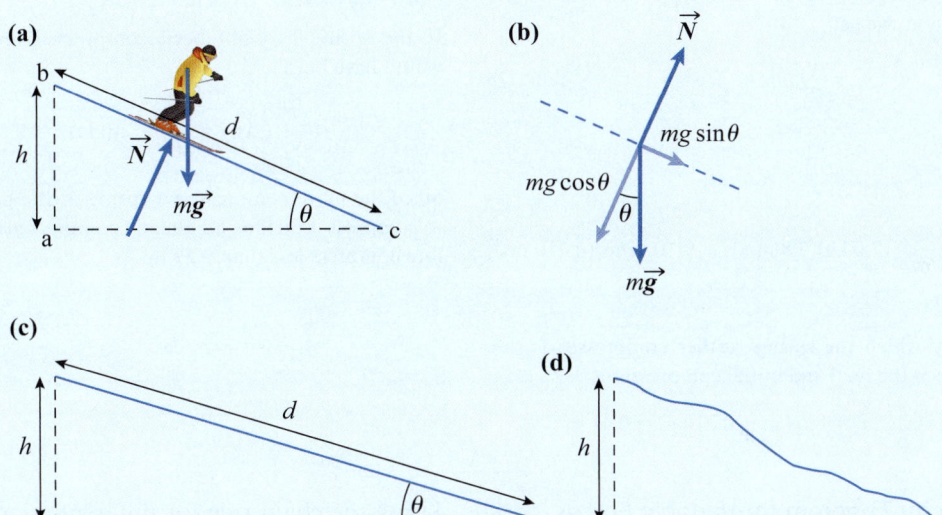

Figure 6-20 Example 6-7. (a) A person skiing down a hill of constant slope. (b) FBD of forces acting on the skier. The force of friction is ignored in this example. (c) A ski slope of the same height but less steep. (d) A ski slope with vertical bends.

Now we can use the work–energy theorem to determine the skier's final velocity. The initial velocity of the skier is zero, and let $v_f = v$ be the final velocity. Then $W_{total} = K_f - K_i$ implies

$$mgh = \frac{1}{2}mv^2 \tag{3}$$
$$v = \sqrt{2gh}$$

Making sense of the result

The speed of the skier is the same as that of an object that falls from rest, under the influence of the force of gravity, a distance h. This result is expected from the law of conservation of energy.

height h remained the same but the distance travelled from top to bottom of the slope changed (see Figure 6-20(c)), the work done by the force of gravity would remain the same. In fact, as long as h remains the same, the length or even the shape of the slope does not matter. The work done by the force of gravity is mgh. Figure 6-20(d) shows a slope that has a couple of vertical bends. If we use Newton's laws of motion to determine the skier's speed at the bottom of the hill, we need an equation for the slope, that is, an equation for height as a function of horizontal position.

We then need to use integral calculus to determine the speed at the bottom. It is a laborious procedure. However, using the work–energy theorem, we know that the final speed only depends on the vertical height by which the skier descends and is $\sqrt{2gh}$. This is a powerful technique that can greatly simplify the solution to an otherwise difficult problem. This is true as long as the force of friction is ignored. In the presence of friction, the above argument breaks down, since the work done by the force of friction depends on the distance covered by the skier.

EXAMPLE 6-8

Collision with a Spring

A 17.0 kg block is sliding on a frictionless horizontal surface at a speed of 3.8 m/s when it strikes a spring that is already compressed by 21.0 cm. Given that the spring constant is 710 N/m, find the distance by which the spring is further compressed during the collision.

Solution

Since the force exerted by the spring on the block is not constant, the deceleration of the box is also not constant. Therefore, the kinematics equations for constant acceleration

developed in Chapter 4 are not applicable in this situation. However, we can apply the work–energy theorem because the work done by the spring on the block depends only on the initial and final lengths.

The only force doing work on the block is the force exerted by the spring; therefore by work-energy theorem,

$$W_{net} = W_{by\text{-}spring} = \Delta K$$

$$W_{by\text{-}spring} = \frac{1}{2}kx_i^2 - \frac{1}{2}kx_f^2 \tag{1}$$

$$= \frac{1}{2}mv_f^2 - \frac{1}{2}mv_i^2$$

(continued)

Since the final speed of the block is zero ($v_f = 0$) at maximum spring compression, we get

$$\frac{1}{2}kx_f^2 = \frac{1}{2}mv_i^2 + \frac{1}{2}kx_i^2 \qquad (2)$$

and

$$x_f^2 = \frac{1}{k}(mv_i^2 + kx_i^2)$$

$$x_f^2 = \frac{1}{710 \text{ N/m}}[(17.0 \text{ kg})(3.8 \text{ m/s})^2 + (710 \text{ N/m})(0.210 \text{ m})^2]$$

$$x_f = 0.62 \text{ m}$$

The distance by which the spring further compresses is the difference between the final and initial compressions:

$$\Delta x = x_f - x_i = 0.62 \text{ m} - 0.210 \text{ m} = 0.41 \text{ m}$$

Making sense of the result

If the spring had not been compressed initially, the result would have been

$$\frac{1}{2}kx_f^2 = \frac{1}{2}mv_i^2 \text{ and } \Delta x = 0.59 \text{ m}$$

Since the more compressed a spring is, the harder it becomes to compress it further, an already compressed spring will therefore compress less than 0.59 m.

The Work–Energy Theorem for Variable Forces

We now present a proof that the work–energy theorem holds for variable forces. In three dimensions, the definition of the total work done on an object by a variable force in displacing the object from position a to position b is defined as

$$W_{a \to b} = \int_a^b \vec{F}(\vec{r}) \cdot d\vec{r} \qquad (6\text{-}39)$$

where $\vec{F}$ is the net (or the resultant) force acting on the object. The integrand in the above equation can be written as

$$\vec{F} \cdot d\vec{r} = m\frac{d\vec{v}}{dt} \cdot d\vec{r} \qquad (6\text{-}40)$$

Using the chain rule for differentiation and the vector identity $\vec{A} \cdot \vec{A} = A^2$, we can write the above equation as

$$\vec{F} \cdot d\vec{r} = m\frac{d\vec{v}}{dt} \cdot \frac{d\vec{r}}{dt}dt$$
$$= m\frac{d\vec{v}}{dt} \cdot \vec{v}\,dt$$
$$= \frac{m}{2}\frac{d}{dt}(\vec{v} \cdot \vec{v})dt \qquad (6\text{-}41)$$
$$= \frac{m}{2}\frac{d}{dt}(v^2)dt = \frac{d}{dt}\left(\frac{1}{2}mv^2\right)dt$$
$$= d\left(\frac{1}{2}mv^2\right)$$

MAKING CONNECTIONS

Wind Energy and Turbines on Kites

(a) Christopher Furlong/Staff/Getty Images News/Getty Images

(b) Altaeros Energies

Figure 6-21   (a) An offshore wind farm. Wind turbines convert kinetic energy to electric energy. (b) A high-altitude wind turbine.

Wind energy reportedly has the potential to provide the world's total need for energy many times over. Although renewable, the amount of energy generated from terrestrial and offshore wind turbines (Figure 6-21(a)) remains relatively modest due to the limited wind speed at low altitudes. Research is being conducted into developing technologies to harvest wind energy at high altitudes. Due to the tremendous wind speeds at higher altitudes, the energy available is much higher than what is available closer to Earth's surface. The work done by the wind on the turbine goes into the turbine's kinetic energy, which generates electricity. Figure 6-20(b) shows an airborne high-altitude wind turbine. While the European Union seems to be leading the way in terms of harvesting energy from wind, Canada and Russia have the largest potential in this regard.

Therefore, the integrand in Equation 6-39 is an exact differential and we can write that equation as

$$W_{a \to b} = \int_a^b d\left(\frac{1}{2}mv^2\right) \tag{6-42}$$

Therefore,

$$W_{a \to b} = \left(\frac{1}{2}mv^2\right)\Big|_a^b = \frac{1}{2}mv_b^2 - \frac{1}{2}mv_a^2 \tag{6-43}$$
$$= K_b - K_a$$

If $K_b > K_a$, then work is done on the object and its kinetic energy increases. If $K_b < K_a$, then work is done by the object and its kinetic energy decreases. We therefore see that the work–energy theorem applies for both constant and variable forces.

6-6 Conservative Forces and Potential Energy

In Example 6-7 we showed that the work done by the force of gravity on a skier depends only on the initial and final heights of the skier and is independent of the path taken by the skier between the two heights. A consequence of this property is that if the skier moves in a **closed path** (a path that starts and ends at the same point), the net work done on the skier by the force of gravity is zero. Suppose the skier climbs a height h straight up along path ab to the top of the hill (Figure 6-20(a)). The gravitational force and displacement vectors are pointing in opposite directions along this path and the work done by the gravitational force is $-mgh$. When the skier slides down the slope along the path bc, the work done by the gravitational force, as discussed in Example 6-7, is $+mgh$. The skier then walks back to the starting point along the path ca, and the work done by the gravitational force along this path is zero because the gravitational force is directed perpendicular to the direction of displacement. Notice in this case that as the skier climbs up, the work done by the gravitational force on the skier is given back in the form of kinetic energy when the skier descends.

Now compare this situation to that of pushing a heavy box across a horizontal surface for a distance d along a straight line from point a to b and then back. This again is a closed path. How much work is done by the force of friction along this closed path? If the coefficient of kinetic friction between the box and the surface is μ_k and the mass of the box is m, then the magnitude of the force of friction on the box is $\mu_k mg$. The force of friction is always directed opposite to the direction of motion, so the work done on the box by the force of friction is $-\mu_k mgd$. When the box is moved back from b to a, the work done by the force of friction

is again $-\mu_k mgd$. Therefore, the total work done on the box along this closed path in this case is $-2\mu_k mgd$. In this case, the work done in moving the box against the force of friction cannot be recovered in the form of the kinetic energy of the box.

The difference between these two situations suggests that forces can be divided into two classes, called conservative forces and non-conservative forces: *A force $\vec{F}$ is conservative when the net work done by the force in moving an object over any closed path is zero.* Mathematically,

KEY EQUATION
$$\oint \vec{F}(\vec{r}) \cdot d\vec{r} = 0, \tag{6-44}$$

(conservative force)

The circle symbol around the integral sign indicates that the integral is to be taken over a closed path. For motion in one dimension, the above equation takes the form

$$\oint F(x)dx = 0 \tag{6-45}$$

Suppose we move an object from point a to point b along a path along which a conservative force acts. Let $W_{a \to b}$ be the work done by the conservative force between these two points. When the object is moved back to point a, along any path, since the total work around a closed path must be zero, we must have $W_{b \to a} = -W_{a \to b}$. Therefore, the work done by a conservative force in moving an object between two points must depend on the location of the start and end points and not on the path taken between the points. Therefore, another definition of a conservative force is as follows: *The work done by a conservative force in moving an object between two points is independent of the path.*

As we discussed earlier, the work done in changing the height of an object is independent of the path followed. The gravitational force is therefore a conservative force.

Potential Energy

As we discussed in Example 6-7, the work done in moving an object against a conservative force is stored in the form of kinetic energy that can then be retrieved when the object returns to its original position. *The capacity (or the ability) of an object to do work is called the **potential energy** of the object.* Potential energy is usually represented by the symbol U and is defined so that the work done by a conservative force in moving the object from position a to position b (with no change in its kinetic energy) is equal to the change in the object's potential energy. Mathematically,

KEY EQUATION
$$\Delta U = -\int_a^b \vec{F}(\vec{r}) \cdot d\vec{r} \tag{6-46}$$

where a and b are the start and end points of the object's displacement. Here ΔU represents the change in the potential energy as a result of the object's displacement. For motion in one dimension, the above equation takes the form

$$\Delta U = -\int_a^b F(x)dx \qquad (6\text{-}47)$$

The minus sign in Equation 6-46 is introduced for the following reason. When the force is directed opposite to the displacement, the dot product of the force and the displacement vectors is negative and therefore ΔU is positive. In this case, the potential energy of the system increases. This is the case when a skier moves up a hill against the force of gravity. When the force and displacement vectors point in the same direction, the dot product of the two vectors is positive and therefore ΔU is negative, that is, the potential energy of the system decreases. This is the case when a skier slides down a hill.

If the conservative force is independent of position, that is, $\vec{F}(\vec{r}) = \vec{F}$, then $\vec{F}(\vec{r})$ can be taken out of the integral in Equation 6-46 and the equation takes the simple form

$$\Delta U = -\vec{F} \cdot \Delta \vec{r} \qquad (6\text{-}48)$$

Here $\Delta \vec{r} = \vec{r}_a - \vec{r}_b$. Note that ΔU represents a change in potential energy and is not an infinitesimal change in potential energy due to a very small displacement.

The potential energy of an object (or a system of objects) is related to the nature of the conservative force under consideration. As there are several types of conservative forces, there are several types of potential energies. If $\vec{F}$ (or F if we are considering one-dimensional motion) represents the force of gravitational attraction between objects, then the corresponding potential energy is called gravitational potential energy. If the force is the Hooke's law force of a spring, then the corresponding potential energy is called the elastic potential energy. If $\vec{F}$ represents the electrostatic Coulomb forces between charges, then the corresponding potential energy is the electrostatic potential energy. Similarly, the potential energy corresponding to nuclear forces between protons and neutrons (see Chapter 34) is called the nuclear potential energy.

In the following sections we discuss two common forms of potential energy, gravitational potential energy and elastic potential energy.

Gravitational Potential Energy near Earth's Surface

Suppose an object of mass m is lifted from an initial height h_1 to a height h_2. In this case, $\Delta h = h_2 - h_1 > 0$. The Earth exerts a force, $m\vec{g}$, that points vertically downward, on the object and the displacement is vertically upward. Therefore, the change in the **gravitational potential energy** of the object is

$$\Delta U_g = -\vec{F} \cdot \Delta \vec{r}$$
$$= -(m\vec{g}) \cdot \Delta \vec{r} = -(-mg\hat{j}) \cdot (\Delta h \hat{j}) = mg\Delta h \qquad (6\text{-}49)$$

In this case, the object is moved upward and the change in its potential energy is positive. If the object is moved

downward (so that positive work is done on the object by the force of gravity), the change in its potential energy is negative. The above formula is valid only as long as the acceleration due to gravity is constant, that is, the motion is close to Earth's surface. If an object is moved far away from Earth's surface (e.g., a rocket travelling to the Moon), we still can calculate the change in its potential energy by using Newton's law of gravitation (see Chapter 11).

We have calculated the change in an object's gravitational potential energy when it is moved between two points, but what is its potential energy at a given height? The answer depends on where we choose to take the zero of potential energy to be. If the zero of the gravitational potential energy is taken to be at the floor, then the potential energy at height h_1 from the floor is mgh_1 and at height h_2 is mgh_2, and the difference in the potential energy between the two heights is $mgh_2 - mgh_1 = mg(h_2 - h_1)$. On the other hand, if the gravitational potential energy at the floor level is taken to be U_0, then at height h_1 from the floor the potential energy is $U_0 + mgh_1$ and at height h_2 it is $U_0 + mgh_2$. The difference in the potential energy between the two heights is still the same since $(U_0 + mgh_2) - (U_0 + mgh_1) = mg(h_2 - h_1)$. If we take the zero of the gravitational potential energy with respect to a certain location, called the ground level, then we can identify the quantity mgh as the gravitational potential energy of a mass m that is at a height h with respect to the ground level.

KEY EQUATION
$$U_g = mgh \qquad (6\text{-}50)$$
(gravitational potential energy)

The units of gravitational potential energy are joules and the above relationship is valid only if it is assumed that g constant

Elastic Potential Energy

Consider an unstretched spring. If the spring is compressed or stretched from its equilibrium length, it exerts a restoring force on the external agent, that is causing it to change its length. Therefore, work must be done on a spring to change its length. This work is stored as potential energy of the spring and is referred to as the **elastic potential energy**. Since the restoring force changes with x, we must use Equation 6-47 to calculate the elastic potential energy. Let us assume that the length of a spring is changed from its initial value, x_i, to a final value, x_f (note that $x = 0$ corresponds to the unstretched length of the spring). The change in the elastic potential energy of the spring is

$$\Delta U_s = -\int_{x_i}^{x_f} F(x)dx$$
$$= -\int_{x_i}^{x_f} (-kx)dx = k\left(\frac{x^2}{2}\right)\Big|_{x_i}^{x_f}$$
$$= \frac{1}{2}kx_f^2 - \frac{1}{2}kx_i^2 \qquad (6\text{-}51)$$

CHECKPOINT 6-6

Potential Energy

Starting from the base, Jill climbs up a 300 m high hill along a gentle slope, and Jack, starting from the same point, takes a lift to the top of the same hill and meets Jill. Which of the following statements is true about the change in the gravitational potential energies of the two? Jill has a mass of 65 kg and Jack has a mass of 75 kg.

(a) Jill gains gravitational potential energy but Jack does not because he takes a lift and therefore does not do any work against gravity.

(b) They both move up by the same height and therefore gain the same amount of potential energy.

(c) Jack gains more potential energy because he has more mass.

(d) We cannot answer this question because we do not know the amount of work done by Jill against the frictional forces as she climbs up.

(e) The gravitational potential energy of the two does not change in moving from the base to the top of the hill.

ANSWER: (c) The gain in the gravitational potential energy is mgh. They both move up by the same height, but Jack has more mass and hence gains more potential energy.

EXAMPLE 6-9

Work Done by a Variable Force

This example requires calculus.

You push a block of mass m up a quarter-circle track as shown in Figure 6-22 by exerting a variable force. Starting from rest, the mass arrives at the top of the track with a speed of zero. Calculate the work done by the force of gravity and the change in the block's potential energy.

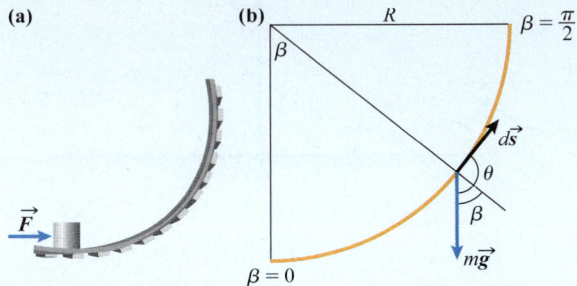

Figure 6-22 (a) A block is pushed along a semicircular track. (b) Diagram showing the force of gravity and the infinitesimal displacement along the track.

Solution

From Equation 6-46, the change in potential energy of the block is the negative of the work done by the force of gravity. The force of gravity is a constant and points

downward. As the block moves along the track therefore, the angle θ between the force of gravity vector and the displacement changes continuously. Since the angle is not constant, we have to use integration to calculate the work done by the force of gravity. Let ds be an infinitesimal displacement of the block along the track. The infinitesimal amount of work done by the force of gravity over the displacement ds is given by

$$dW_g = m\vec{g} \cdot d\vec{s} = (mg)(ds)\cos\theta \tag{1}$$

When the block is at the bottom of the track, θ (the angle between the force of gravity and the displacement) is 90°; at the top of the track, θ is 180°. We will use angle β for the angular position of the block with respect to the vertical, as measured from the centre of the circular track. The infinitesimal arc length ds is then given by

$$ds = R\,d\beta \tag{2}$$

From Figure 6-22(b), we can see that

$$\theta = \beta + \frac{\pi}{2} \tag{3}$$

Thus,

$$d\beta = d\theta$$

Therefore,

$$ds = R\,d\theta \tag{4}$$

and (1) can be written as

$$dW_g = mgR\cos\theta\,d\theta$$

The work done by the force of gravity as the mass is pushed up the track is given by

$$W_g = \int_a^b m\vec{g} \cdot d\vec{s} = \int_a^b (mg)(ds)\cos\theta \tag{5}$$

where a and b are the initial and final points, at the beginning and end of the track, respectively. At a, $\theta = \frac{\pi}{2}$, and at b, $\theta = \pi$. Substituting for ds in the integral gives

$$W_g = \int_{\pi/2}^{\pi} (mg)(R\,d\theta)\cos\theta = (mgR)\int_{\pi/2}^{\pi} \cos\theta\,d\theta$$

$$= (mgR)\sin\theta\,|_{\pi/2}^{\pi}$$

$$= mgR[\sin\pi - \sin(\pi/2)] \tag{6}$$

$$= -mgR$$

The change in potential energy of the block–Earth is equal to the negative of the work done by the force of gravity on the block. Therefore,

$$\Delta U = mgR \tag{7}$$

Making sense of the result

The change in the height of the mass is equal to R, the radius of the track. So, this example illustrates the fact that the change in gravitational potential energy is path independent and depends only on the endpoints of the path.

If we start with an unstretched spring ($x_i = 0$) and spring's length is then changed by $x(x_f = x)$, then the elastic potential energy stored in the spring is

KEY EQUATION
$$U_s = \frac{1}{2}kx^2 \qquad (6\text{-}52)$$

(elastic potential energy of a spring)

U_s is measured in joules and the above formula is valid as long as the elastic limit of the spring is not exceeded.

6-7 Conservation of Mechanical Energy

According to the work–energy theorem (Equation 6-36), the net work done on an object is equal to the change in the kinetic energy of the object. Since forces can be classified as conservative or non-conservative, the total work done on an object can also be classified into two categories: the work done by conservative forces, W_c, and the work done by non-conservative forces, W_{nc}. Therefore, we can express the statement of the work–energy theorem as follows:

$$\Delta K = W_{net} = W_{nc} + W_c \qquad (6\text{-}53)$$

As we discussed in the previous section, the work done by a conservative force is equal to the negative of the change in potential energy of the object. Substituting $W_c = -\Delta U$ (here U without a subscript represents any form of potential energy) in Equation 6-53, we get

$$\Delta K = -\Delta U + W_{nc}$$

or

KEY EQUATION
$$\Delta K + \Delta U = W_{nc} \qquad (6\text{-}54)$$

The quantity on the left side is the sum of the changes in the kinetic and potential energies of the system under consideration. It is called the **mechanical energy**. Mechanical energy is is represented by the symbol E (or E_m, where the subscript m stands for mechanical). Equation 6-54 states that the action of a non-conservative force changes the total mechanical energy of an object and the work done by non-conservative forces is equal to the change in an object's mechanical energy. When positive (negative) work is done on an object non-conservative forces, its mechanical energy increases (decreases). In the absence of non-conservative forces, Equation 6-54 reduces to

KEY EQUATION
$$\Delta K + \Delta U = 0 \qquad (6\text{-}55)$$
(in the absence of non-conservative forces)

Therefore, in the absence of non-conservative forces, the change in the sum of the kinetic and potential energies of an object (or a system) is zero. In other words, in the absence of non-conservative forces, the sum of the kinetic and potential energies of an object (or system)

remains constant. This is the statement of the **law of conservation of mechanical energy**. Let K_i and U_i be the initial kinetic and potential energies of a system and K and U be the corresponding quantities at a later time. In the absence of non-conservative forces, the sum $K_i + U_i$ must be equal to the sum $K + U$. Another form of Equation 6-55 is

KEY EQUATION
$$K_i + U_i = K + U = \text{constant} \qquad (6\text{-}56)$$

The potential energy, U, may be the sum of several types of potential energy. The kinetic or potential energy may change during a process, but the sum of of the two remains constant in the absence of non-conservative forces.

Consider the system shown in Figure 6-23. A mass M, attached to a spring, is resting on an inclined ramp. The other end of the spring is fixed to the top end of the slope. A ball of mass m is at rest at the bottom of the slope. We choose a coordinate system with the positive y-axis pointing vertically up, and for this discussion we assume the ball and the block to be point objects. Let the unstretched length of the spring be L_0 and the change in length when the mass M is attached as shown be Δl_0. Also, let the initial height of mass M be y_0. Now suppose that the ball is given an initial velocity v_0 that is sufficient for the ball to reach the top of the ramp in the absence of the mass–spring system. For this system, the initial kinetic energy is the kinetic energy of the ball and the potential energy is the sum of the gravitational potential energy of the block and the elastic potential energy of the stretched spring.

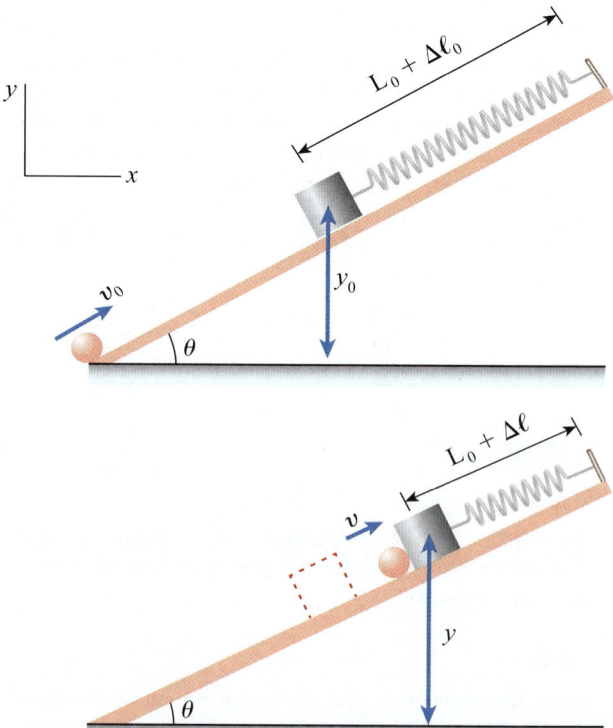

Figure 6-23 A ball rolls up a ramp at velocity v_0 and hits a spring.

$$K_i = \frac{1}{2}mv_0^2$$

$$U_i = Mgy_0 + \frac{1}{2}k(\Delta l_0)^2 \qquad (6\text{-}57)$$

Suppose at a later time the ball strikes the block and pushes the block up the ramp. Let the height of the block at some instant be y, the change in the spring's length be Δl, and the speed of the ball be v. Since the ball and the block are in contact, the speed of the block is also v. At this instant, the kinetic energy is the sum of the kinetic energies of the ball and the block and the total potential energy is the sum of the gravitational potential energies of the ball and the block plus the elastic potential energy of the spring. Therefore,

$$K = \frac{1}{2}mv^2 + \frac{1}{2}Mv^2$$

$$U = mgy + Mgy + \frac{1}{2}k(\Delta l)^2 \qquad (6\text{-}58)$$

As non-conservative forces are absent, the law of conservation of energy states that

$$\frac{1}{2}mv_0^2 + Mgy_0 + \frac{1}{2}k(\Delta l)^2$$

$$= \frac{1}{2}(m + M)v^2 + (m + M)gy + \frac{1}{2}k(\Delta l)^2 \qquad (6\text{-}59)$$

Due to conservation of energy, not all of the variables in the above equation are independent. Note that in addition to the conservation of mechanical energy, total momentum and angular momentum are also conserved in this process that introduce additional constraints on the variables in Equation 6-59. We will discuss conservation of momentum and angular momentum in later chapters.

Non-conservative forces include the force exerted by a person on an object, the force of friction between surfaces in relative motion, and the drag force exerted by a fluid. When considering the work done by non-conservative forces, we must be careful in defining the system under consideration. A non-conservative force, such as the force of friction, acts between two surfaces. Consider, for example, a puck of mass m that is given an initial speed v_i on a smooth and horizontal surface of ice. The coefficient of kinetic friction between the puck and the ice is taken to be non-zero. Therefore, the puck travels a distance d before coming to rest. The change in mechanical energy of the puck is

$$\Delta K + \Delta U = (K_f - K_i) + 0 = \left(0 - \frac{1}{2}mv_i^2\right) + 0 = -\frac{1}{2}mv_i^2$$

Let f be the magnitude of the frictional force between the puck and the ice. Then $-fd$, the work done by the force of friction, must be equal to the change in mechanical energy of the puck. If we assume the puck to be a point object of mass m that moves a distance d, then

$$-fd = -(ma)d = -\frac{1}{2}m(2ad)$$

$$= -\frac{1}{2}m(v_i^2 - v_f^2) \qquad (6\text{-}60)$$

$$= -\frac{1}{2}mv_i^2$$

However, the force of friction exists between two surfaces and therefore it does work on the puck plus the ice surface. If we carefully measure the temperature of the segment of ice on which the puck moves, we will determine that as the puck passes through, the temperature of that segment slightly increases. The loss in mechanical energy of the puck appears as thermal energy that causes the temperature of the ice and the surface of the puck in contact with the ice to rise. Therefore, we can consider $-fd$ as the work done by the force of friction as long as we keep in mind that the work is done on the object as well as the surface on which the object is moving. In this case, the system under consideration is the puck as well as the ice on which the puck moves.

EXAMPLE 6-10

Slapshot

In a slapshot competition, a player strikes a hockey puck with a stick, causing the puck to travel in a straight line on a smooth ice surface. The mass of the puck is 160.0 g and the coefficient of friction between the puck and the ice is 0.150. If 50.0 m from the starting point, the speed of the puck is measured to be 16.5 m/s, what was the initial speed of the puck?

Solution

We will use the work–energy theorem to determine the initial speed of the puck. By the work–energy theorem,

$$\Delta K + \Delta U = W_{nc} \qquad (1)$$

In this case,

$$\Delta K = \frac{1}{2}mv_f^2 - \frac{1}{2}mv_i^2 \qquad (2)$$

where $m = 0.160$ kg is the mass of the puck, $v_f = 16.5$ m/s is its speed 50.0 m from the initial position, and v_i is the puck's initial speed. Because the puck moves horizontally, there is no change in its gravitational potential energy and therefore $\Delta U = 0$. After the initial slapshot the only force acting along the direction of motion is the force of kinetic friction. From Equation 5-8, the magnitude of the normal force and kinetic friction are related by $f_k = \mu_k N = \mu_k(mg)$. The force of kinetic friction ($\vec{f}$) acts opposite to the direction of

(continued)

motion, so the angle between the force and the displacement vectors is 180° and the total work done by friction on the puck and on the ice is

$$W_{nc} = \vec{f} \cdot \vec{d} = f_k d \cos 180° = -f_k d$$

$$= -(\mu_k mg)d \qquad (3)$$

$$= -(0.150)(0.160 \text{ kg})(9.81 \text{ m/s}^2)(50.0 \text{ m}) = -11.77 \text{ J}$$

Inserting (2) and (3) into (1) yields

$$\frac{1}{2}mv_f^2 - \frac{1}{2}mv_i^2 = -f_k d \qquad (4)$$

Therefore,

$$\frac{1}{2}(0.160 \text{ kg})(16.5 \text{ m/s})^2 - \frac{1}{2}(0.160 \text{ kg})v_i^2 = -11.8 \text{ J} \qquad (5)$$

Solving for v_i, we get

$$v_i = 20.0 \text{ m/s}$$

Making sense of the result

The initial speed is greater than the final speed 50.0 m away. The puck loses kinetic energy in doing work against friction, so the answer makes sense.

EXAMPLE 6-11

Coefficient of Kinetic Friction

A box of mass 123 kg, initially at rest on a horizontal surface, is pushed in a straight line with a constant force ($\vec{F}$) of magnitude 560 N over a distance of 7.0 m. Over this distance, the speed of the box changes from zero to 4.0 m/s. The frictional force between the box and the surface cannot be ignored. Determine the coefficient of kinetic friction between the box and the surface.

Solution

Figure 6-24 shows the FBD for the box. Since there is no motion along the vertical direction, the magnitude of the normal force on the box must equal the magnitude of its weight. We also know (Equation 5-8) that the magnitude of the normal force and the force of kinetic friction are related by $f_k = \mu_k N = \mu_k (mg)$. We can then use the work–energy theorem to determine the coefficient of kinetic friction.

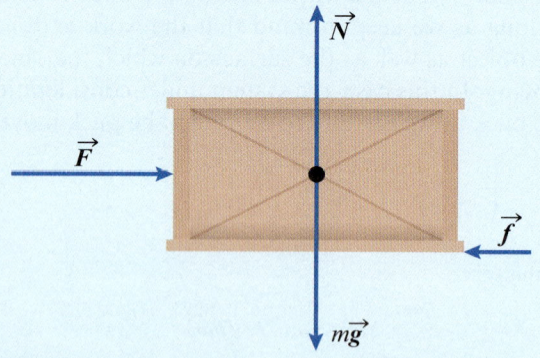

Figure 6-24 Example 6-11.

Neither the force of gravity nor the normal force does any work because they both act normal to the direction of motion. The net work done on the box is the sum of the work done by the applied force, $\vec{F}$, and the work done by the force

of friction, $\vec{f}$. The box starts from rest, so the initial kinetic energy of the box is zero. By the work–energy theorem,

$$W_F + W_f = K_f - 0 \qquad (1)$$

The applied force is in the direction of motion of the box. Therefore, the angle between the applied force and the displacement vector is zero:

$$W_F = \vec{F} \cdot \vec{d} = Fd \cos 0 = Fd \qquad (2)$$

The force of friction ($\vec{f}$) acts opposite to the direction of motion, so the angle between the force and the displacement vectors is 180°:

$$W_f = \vec{f} \cdot \vec{d} = f_k d \cos 180° = -f_k d \qquad (3)$$

Inserting (2) and (3) into (1) yields

$$Fd - f_k d = \frac{1}{2}mv_f^2 \qquad (4)$$

Therefore,

$$f_k d = Fd - \frac{1}{2}mv_f^2 \qquad (5)$$

or

$$(\mu_k mg)d = Fd - \frac{1}{2}mv_f^2$$

Solving for μ_k, we get

$$\mu_k = \frac{Fd - \frac{1}{2}mv_f^2}{mgd} = \frac{(560 \text{ N})(7.0 \text{ m}) - \frac{1}{2}(123 \text{ kg})(4.0 \text{ m/s})^2}{(123 \text{ kg})(9.81 \text{ m/s}^2)(7.0 \text{ m})} = 0.35$$

Making sense of the result

The coefficient of kinetic friction is dimensionless, and the calculated value is between zero and one.

EXAMPLE 6-12

Block on an Incline

A 1.20 kg block slides down a frictionless incline with a slope angle of 42.0°, starting from a height $h = 2.30$ m above the bottom of the incline, as shown in Figure 6-25. The incline meets a frictionless horizontal surface, at the end of which is a spring in its equilibrium position ($k = 460.$ N/m) used to stop the block. Find the maximum compression of the spring.

Solution

The only forces doing work on the block are the force of gravity and the elastic restoring force of the spring, both of which are conservative forces. Since there are no non-conservative forces, applying the principle of conservation of mechanical energy, we get

$$\Delta K = -\Delta U$$

The block starts at rest, and when the spring is at maximum compression, the block is also momentarily at rest. Therefore, the change in kinetic energy of the block and spring system between the initial and final positions is zero. Hence, the change in the potential energy of the system must also be zero.

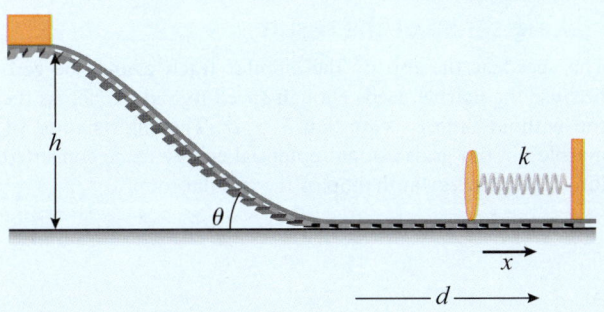

Figure 6-25   Example 6-12.

The change in potential energy has two components: the change in the elastic potential energy of the spring and the change in the gravitational potential energy of the block–Earth system:

$$\Delta U_s + \Delta U_g = 0$$

When the block reaches the bottom of the incline, the change in gravitational potential energy is

$$\Delta U_g = U_{g,\text{final}} - U_{g,\text{initial}} = 0 - mgh = -mgh$$

The change in the spring's potential energy is

$$\Delta U_s = U_{s,\text{final}} - U_{s,\text{initial}} = \frac{1}{2}kx^2 - 0$$

where we have used the fact that the spring is originally in its equilibrium position, so $x = 0$, where we have used x to denote the maximum compression of the spring. Then

$$\frac{1}{2}kx^2 - mgh = 0$$

Solving for x and substituting, we get

$$x = \sqrt{\frac{2mgh}{k}}$$

$$= \sqrt{\frac{2(1.2 \text{ kg})(9.81 \text{ m/s}^2)(2.3 \text{ m})}{460 \text{ N/m}}} = 0.34 \text{ m}$$

Making sense of the result

The gravitational potential energy is converted into kinetic energy, which is converted into the spring's elastic potential energy. However, there was no net change in kinetic energy, so we were able to bypass the kinetic energy step altogether.

EXAMPLE 6-13

Marble on a Track

The marble in Figure 6-26 starts from rest and slides down a frictionless incline from a height h above the ground, and then goes up the circular track of radius r as shown. Derive an expression for the minimum value of the height, h, so that the marble stays in contact with the circular track at the highest point on the track.

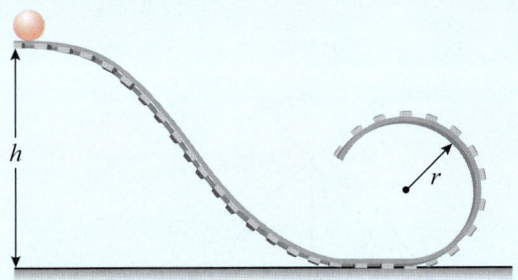

Figure 6-26   Example 6-13.

Solution

We require that the marble barely be in contact with the track at the top of the circle. The normal force between the marble and the top of the track must therefore be equal to zero. Consider the FBD of the marble at the top of the track (Figure 6-27). Applying Newton's second law to the marble, we obtain

$$\sum \vec{F} = \vec{N} + m\vec{g} = m\vec{a} \tag{1}$$

Choosing a coordinate system, as shown in Figure 6-27, and resolving forces into x- and y-components, we get

$$N(-\hat{j}) + mg(-\hat{j}) = a(-\hat{j}) \tag{2}$$

where N, mg, and a represent the magnitudes of the normal force, the force of gravity, and the acceleration, respectively. Here a is the centripetal acceleration, which is given by $a = \dfrac{v^2}{r}$. Therefore,

$$N + mg = \frac{mv^2}{r} \tag{3}$$

(continued)

MECHANICS

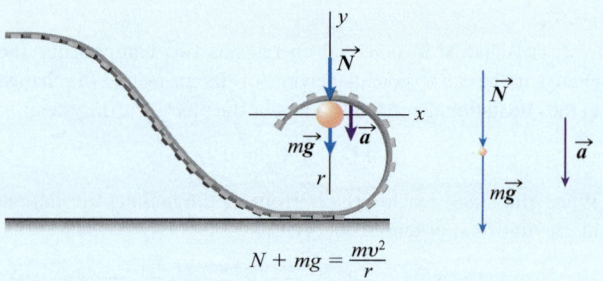

$$N + mg = \frac{mv^2}{r}$$

Figure 6-27   The marble on the track with the forces acting on it.

Since N is zero at the top of the track (the marble is about to leave the track), therefore,

$$v^2 = rg \qquad (4)$$

For the marble to have this speed at the top of the track, some of the gravitational potential energy has been converted to kinetic energy. Since there are no frictional forces, we can apply the conservation of mechanical energy:

$$\Delta K = -\Delta U_g$$

The change in height of the marble, from its initial to the final position, is $h - 2r$:

$$\Delta U = U_{g,\,final} - U_{g,\,initial}$$
$$= mg(2r) - mgh = -mg(h - 2r) \qquad (5)$$

Notice that since the marble ends up at a lower height, the gravitational potential energy decreases, and the change in potential energy is negative. The change in its kinetic energy is

$$\Delta K = \frac{1}{2}mv^2 - 0$$

where v is the speed of the marble at the top of the track. Therefore,

$$\frac{1}{2}mv^2 = mg(h - 2r) \qquad (6)$$

Using the result from (3), we get

$$\frac{1}{2}mrg = mg(h - 2r) \qquad (7)$$

Therefore,

$$\frac{1}{2}r = h - 2r \qquad (8)$$

and

$$h = 2.5r \qquad (9)$$

Hence, for the marble to be barely in contact with the track at the top of the circle, the ratio of the height to the radius must be 5/2.

Making sense of the result

The speed at the top of the circular track cannot be zero because the marble needs enough speed to make it across the top without falling. Note that $h > 2r$. This allows some of marble's initial gravitational potential energy to be converted into kinetic energy at the top of the circular loop.

EXAMPLE 6-14

Sliding on a Spherical Surface

A polar bear cub sits on top of an igloo (Figure 6-28). The surface of the igloo has a spherical curvature and is smooth enough to have negligible friction. The cub begins to slide down the igloo. Determine the angle at which the cub will leave the surface of the igloo.

Solution

As the bear cub slides down, the Earth–cub's gravitational potential energy is converted to kinetic energy and the cub gains speed. Consider the FBD of the forces acting on the cub. We measure the

angular position of the cub from the vertical. The forces acting on the bear are the force of gravity and the normal force. Since the cub's motion is circular, it will have a radial acceleration. Thus, we choose our axes so that one axis points along the radial direction and the other axis is tangent to the igloo, and we label these as the r- and t-axes, with the positive directions as shown in the FBD of Figure 6-28(b). We then resolve the forces acting on the cub into components along these axes and write the equation of motion along each axis.

In the radial direction, we have

$$\sum F_r = N - mg\cos\theta = -ma_r \qquad (1)$$

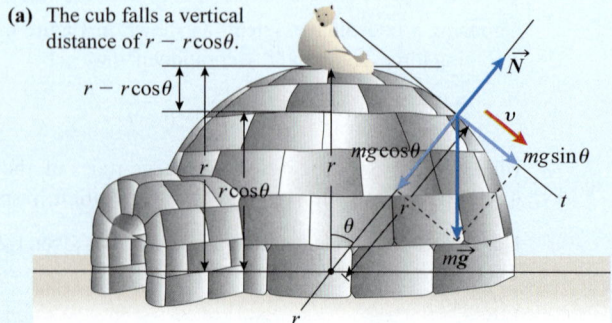

(a) The cub falls a vertical distance of $r - r\cos\theta$.

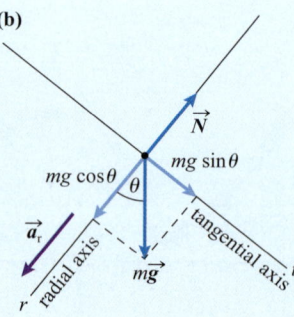

(b)

Figure 6-28   (a) Example 6-14. (b) The FBD at the point where the cub leaves the surface of the igloo.

where N denotes the magnitude of the normal force. Here we have chosen the positive radial direction to be pointing up, toward the normal force. The negative sign on the right indicates that the radial acceleration is pointing toward the centre. Since $a_r = v^2/r$,

$$mg\cos\theta - N = \frac{mv^2}{r} \tag{2}$$

When the cub is on the verge of leaving the surface, the normal force on it is zero, and at this point,

$$mg\cos\theta = \frac{mv^2}{r} \tag{3}$$

We have two unknowns: the speed, v, and the angle, θ, at which the cub leaves the surface. Let us consider the constraint imposed on these variables by the law of conservation of energy (we are ignoring the force of friction). As the cub slides down the surface, some of its potential energy is converted to kinetic energy. By conservation of energy,

$$\Delta K = -\Delta U_g \tag{4}$$

As we can see from the figure, the change in height is $r - r\cos\theta$. The change in the cub's potential energy is

$$\begin{aligned} \Delta U_g &= U_{g,\,final} - U_{g,\,initial} \\ &= mg(r\cos\theta) - mgr = -mg(r - r\cos\theta) \end{aligned} \tag{5}$$

The cub starts from rest, so the change in kinetic energy is

$$\Delta K = \frac{1}{2}mv^2 - 0 \tag{6}$$

Substituting (5) and (6) into (4), we have

$$\frac{1}{2}mv^2 = mg(r - r\cos\theta) \tag{7}$$

We see that because of the constraint imposed by conservation of energy, the two variables v and θ are not independent. Using the result from (3), we get

$$\frac{1}{2}mgr\cos\theta = mg(r - r\cos\theta) \tag{8}$$

Solving for $\cos\theta$, we get

$$\cos\theta = \frac{2}{3} \tag{9}$$

Therefore,

$$\theta = \cos^{-1}(2/3) \approx 48°$$

Thus, the cub will leave the surface at an angle of about 48° from the vertical.

Making sense of the result

As the bear slides down the hemispherical igloo it gains speed and hence the normal force on the bear gets smaller. The point where the normal force on the bear vanishes is the point where the bear leaves the surface of the igloo. This occurs when the bears falls by a height $R/3$ which corresponds to angle of $\cos^{-1}(2/3)$. The angle indicates the vertical distance the cub has fallen, this is turn is related to the change in the cub's speed. The higher the speed the lower the normal force in this case.

6-8 Force from Potential Energy

We have derived expressions for the potential energy of interaction due to the force of gravity near Earth's surface (for constant acceleration due to gravity) and that due to the restoring force of a spring. In many applications, it is convenient to actually start with the potential energy and then derive the expression for the conservative force associated with that potential energy. In this section, we develop a formula for finding the force from a given expression for potential energy.

The potential energy related to a constant conservative force, $\vec{F} = F_x\hat{i} + F_y\hat{j} + F_z\hat{k}$, that displaces an object by a distance $\Delta\vec{r} = \Delta x\hat{i} + \Delta y\hat{j} + \Delta z\hat{k}$ is given by Equation 6-48,

$$\Delta U = -\vec{F} \cdot \Delta\vec{r}$$

For the one-dimensional case, where $\vec{F} = F_x\hat{i}$, the above equation reduces to

$$\Delta U \approx -F_x\Delta x \tag{6-61}$$

The symbol $\approx$ implies "approximately equal to" since the above formula is valid if the force F_x does not change appreciably during a small displacement Δx. In the limit when $\Delta x \to 0$, the variation of F_x is negligible. Therefore, for an infinitesimal change in displacement (dx), an infinitesimal change in potential energy (dU) is given by

$$dU = -F_x dx \tag{6-62}$$

Dividing both sides of the above equation by dx, we get

KEY EQUATION

$$F_x = -\frac{dU}{dx} \qquad (6\text{-}63)$$

Therefore, the conservative force is the negative of the derivative of the potential energy with respect to the displacement. In regions where the potential energy varies rapidly with position, the force is large. In regions where the potential energy does not change, the force is zero.

In Section 6-5, we found that when the length of an unstretched spring is changed by an amount x, the elastic potential energy stored in a mass–spring system is $U = \frac{1}{2}kx^2$. Since this expression depends on a single variable x, using Equation 6-63, the force that gives rise to the elastic potential energy is

$$F_x = -\frac{dU}{dx} = -\frac{d}{dx}\left(\frac{1}{2}kx^2\right) = -kx \qquad (6\text{-}64)$$

which is the elastic restoring force (Hooke's law force) that we have already discussed. Figure 6-29 shows a plot of the elastic potential energy and the restoring force. Remember that the force is the negative of the slope of the potential energy versus displacement graph. For negative x (compressed spring), the slope of the potential energy graph is negative and therefore the restoring force is in the direction of increasing x, forcing the spring back to its equilibrium position. For positive x (stretched spring), the slope of the potential energy graph is positive and the restoring force is in the direction of decreasing x, forcing the spring back to its equilibrium position. At $x = 0$, the slope of the potential energy graph is zero, implying that the restoring force is zero. In the absence of other forces, if the mass attached to a spring is already

at rest the mass will remain at rest. So we see that when the mass is displaced to either side of $x = 0$, the resulting force pushes it back to $x = 0$, where the potential energy is minimum. We say the point $x = 0$ is the point of **stable equilibrium**. At a point of stable equilibrium, a small displacement results in a restoring force that pushes the object back to its equilibrium position.

We can generalize Equation 6-63 to three dimensions. Consider a particle moving in three dimensions from a point $\vec{r} = (x, y, z)$ to a nearby point $\vec{r} + \Delta\vec{r} = (x + \Delta x, y + \Delta y, z + \Delta z)$ under the influence of a conservative force $\vec{F} = (F_x, F_y, F_z)$, where each component of the force may depend upon all three coordinates, x, y, and z. In this case, the potential energy associated with the force will also be a function of three coordinates, that is, $U = U(x, y, z)$. Now consider the displacement Δx in the x-direction. Since F_y and F_z are perpendicular to the x direction, the work done by the force is $F_x \Delta x$ and the change in potential energy is $-F_x \Delta x$. Therefore, the relation between F_x and potential energy will be given by Equation 6-63. However, there is an important difference in this case. Now the potential energy is a function of all three coordinates. *Therefore, when taking the derivative of potential energy with respect to x we must keep y and z constant and only vary x.* Such a derivative is called a *partial derivative* and is denoted by the symbol $\frac{\partial}{\partial x}$. Therefore,

$$F_x = -\frac{\partial U}{\partial x} \qquad (6\text{-}65)$$

Similar relationships exist for displacements Δy in the y-direction and Δz in the z-direction. For motion in three dimensions, the relationships between the potential energy and the force are therefore

$$U(x) = \frac{1}{2}10x^2 \text{ (J)}$$

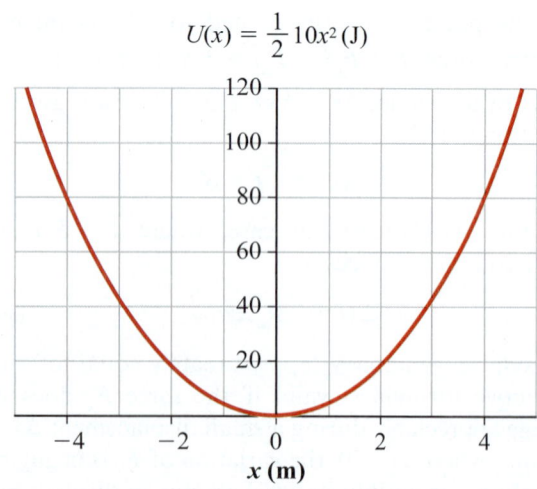

$$F(x) = -10x \text{ (N)}$$

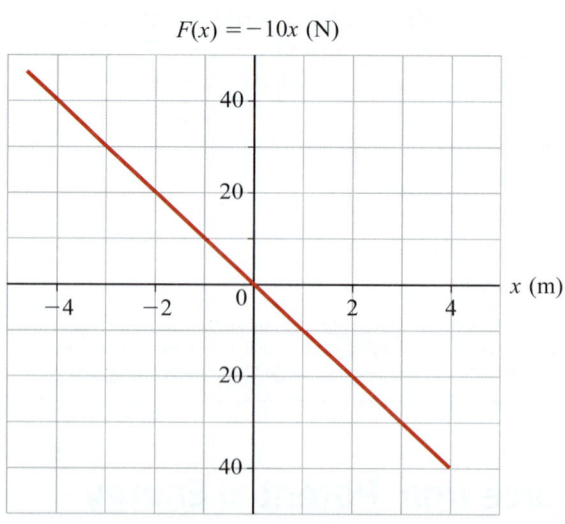

Figure 6-29 A plot of elastic potential energy $U(x) = \frac{1}{2}kx^2$ and $F(x) = -kx$ for $k = 10$ N/m and for -5 m $\leq x \leq 5$ m. For negative x, the slope of the potential energy curve is negative and the force is positive (directed toward increasing x). For positive x, the slope of the potential energy curve is positive and the force is negative (directed toward decreasing x).

KEY EQUATION $\quad F_x = -\dfrac{\partial U}{\partial x}, \; F_y = -\dfrac{\partial U}{\partial y}, \; F_z = -\dfrac{\partial U}{\partial z}$ $\quad$ (6-66)

We can write Equation 6-66 in compact notation by defining a vector differential operator, denoted by symbol $\vec{\nabla}$, called "gradient" or "del," as follows:

$$\vec{\nabla} = \frac{\partial}{\partial x}\hat{i} + \frac{\partial}{\partial y}\hat{j} + \frac{\partial}{\partial z}\hat{k} \qquad (6\text{-}67)$$

It is called an operator as it takes the derivative of a function and represent or the rate at which a function changes along coordinate axes directions. In terms of the gradient the Equation 6-68 can be written in the vector form as,

KEY EQUATION $\qquad \vec{F}(x,y,z) = -\vec{\nabla} U(x,y,z)$ $\qquad$ (6-68)

Here we have explicitly shown the x-, y- and z-dependence of both the potential energy and the corresponding conservative force. By writing the force and the gradient in terms of $\hat{i}$, $\hat{j}$, and $\hat{k}$ unit vectors you can convince yourself that Equation 6-68 is equivalent to Equation 6-66.

EXAMPLE 6-15

Potential Energy and Force in Three Dimensions

A particle is in a region in which its potential energy is given by $U = (3x^2 - 2xy + zx)$, where the coefficients 3, 2 and 1 have units J/m^2. Find an expression for the force acting on the particle, and determine the magnitude of the force at point A(0.5 m, -1.0 m, 0.5 m).

Solution

The components of the force are given by Equation 6-66. Differentiating the given formula for potential energy with respect to x, y, and z, we get

$$F_x = -\frac{\partial U}{\partial x} = -\frac{\partial}{\partial x}(3x^2 - 2xy + zx) = -(6x - 2y + z)$$

$$F_y = -\frac{\partial U}{\partial y} = -\frac{\partial}{\partial y}(3x^2 - 2xy + zx) = -(-2x) \qquad (1)$$

$$F_z = -\frac{\partial U}{\partial z} = -\frac{\partial}{\partial z}(3x^2 - 2xy + zx) = -(x)$$

Substituting the coordinates of point A gives

$$\vec{F} = -5.5\hat{i} - 1.0\hat{j} - 0.5\hat{k} \qquad (2)$$

The magnitude of the force at point A is

$$F = \sqrt{(-5.5)^2 + (-1.0)^2 + (-0.5)^2} = 5.6 \text{ N} \qquad (3)$$

Energy Diagrams

Earlier in this section we discussed the relationship between the potential energy and the force for a mass–spring system. A study of the graph of potential energy as a function of position (Figure 6-29) provides important insight into the nature of the conservative force that gives

rise to the potential energy and the resulting motion. Such an analysis can be extended to more complicated forms of potential energy. Figure 6-30 (a) shows a one-dimensional graph of potential energy. Such form of potential energy is studied in modern physics. Its mathematical form is

$$U(x) = ax^4 - bx^2 + c \qquad (6\text{-}69)$$

where a, b, and c are constants. If x is in metres, then the units of a, b and c are J/m^4, J/m^2 and J, respectively. Figure 6-30(a) shows a plot of this potential with $a = 3$, $b = 10$, and $c = 10$. The corresponding force, $F(x) = -dU(x)/dx$, is plotted in Figure 6-30(b). In the graph of potential energy, points x_1 and x_2 are the points of stable equilibrium. The derivative of $U(x)$ is zero at these points, and if a particle initially at these points is slightly displaced to either side, the direction of the resulting force is such that it brings the particle back to the equilibrium position. The derivative of $U(x)$ is also zero at x_3. However, if a particle located at x_3 is slightly displaced to the right, since the slope of $U(x)$ becomes negative to the right of x_3, the resulting force is directed toward the right (increasing x) and tends to move the particle away from x_3. If the particle is slightly displaced to the left, since the slope of $U(x)$ becomes positive, the resulting force is directed toward the left (decreasing x) and tends to move the particle away from x_3. Point x_3 is called a point of **unstable equilibrium**. It is like a skier being on top of a frictionless hill, where any slight motion will result in the skier being pulled toward a valley.

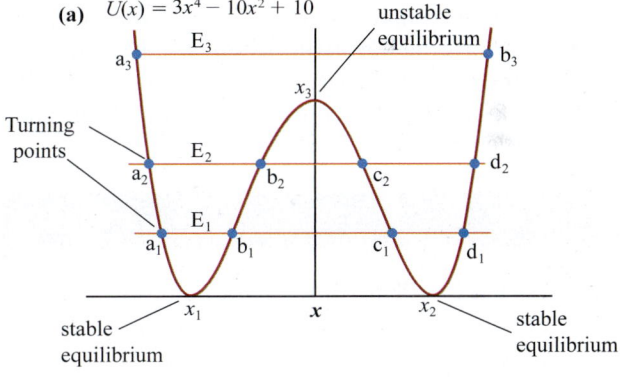

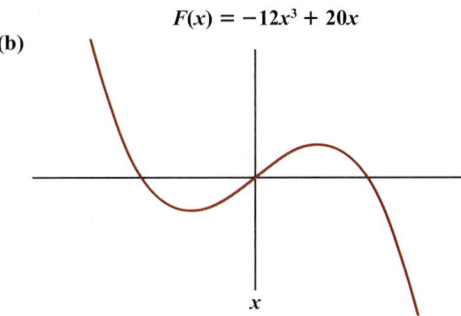

Figure 6-30 (a) A plot of the elastic potential energy $U(x) = 3x^4 - 10x^2 + 10$. Shown are the points of stable and unstable equilibrium and the turning points. (b) A plot of the corresponding force, $F(x) = -\dfrac{d}{dx}(3x^4 - 10x^2 + 10) = -12x^3 + 20x$.

If the only force acting on a particle is a conservative force (i.e., there is a potential energy associated with the force), then its mechanical energy, $E = K + U$, is conserved, that is, it is independent of x. A plot of E as a function of x on the potential energy graph is then a horizontal line. In Figure 6-30(a), E_1, E_2, and E_3 show three possible values of the total energy of a particle. Consider a particle with total energy E_1. For a given value of potential energy, U, its kinetic energy for a given position is given by $K = E_1 - U(x)$. Since kinetic energy must be positive, the maximum value of the potential energy is equal to E_1. Therefore, for the graph under consideration, the particle can only be in a region such that $a_1 \le x \le b_1$ or $c_1 \le x \le d_1$. Every other region on the graph is a *forbidden* region because in these regions the potential energy of the particle is greater than E_1, which will require the kinetic energy of the particle to be negative, which is not allowed. If a particle is originally in the region $a_1 \le x \le b_1$, it will stay in that region. Points $x = a_1$ and $x = b_1$ are called *turning points*. At the turning points, the kinetic energy of the particle is zero. Similarly, if the particle is initially in the region $c_1 \le x \le d_1$, it will stay in that region. As the mechanical energy increases, the region of space available to the particle increases. For a value of mechanical energy equal to E_3, the particle can be anywhere between $x = a_3$ and $x = b_3$.

6-9 Power

The Jamaican runner Usain Bolt ran 100 m in 9.81 s, whereas an average athletic person of the same age and weight may take twice that amount of time to run the same distance. Ignoring the air drag, they

both do the same amount of work. However, Usain Bolt does this work in half the time. We say that Usain Bolt generates more power than an average athlete.

Power is defined as the rate at which work is done. If ΔW is the amount of work done in a short interval of time, Δt, then the average power generated in this time is defined as

$$P_{avg} = \frac{\Delta W}{\Delta t} \tag{6-70}$$

In the limit when the time interval Δt goes to zero, we define power as a function of time, called the instantaneous power, as

KEY EQUATION
$$P = \frac{dW}{dt} \tag{6-71}$$

where dW is the infinitesimal amount of work done during the infinitesimal amount of time dt. If a constant force displaces an object by an infinitesimal distance dx in time dt, then the power delivered by the force is

$$P = \frac{d}{dt}(F dx) = F\frac{dx}{dt} = Fv \tag{6-72}$$

For forces acting in more than one dimension, the above relation becomes

$$P = \frac{d}{dt}(\vec{F} \cdot d\vec{r}) = \vec{F} \cdot \frac{d\vec{r}}{dt} = \vec{F} \cdot \vec{v} \tag{6-73}$$

CHECKPOINT 6-8

An Energy Diagram

For the potential energy versus position graph of a particle shown in Figure 6-31, which of the following statements are correct?

(a) Points x_1, x_2, and x_3 are points of stable equilibrium.
(b) Points x_1 and x_3 are points of stable equilibrium and point x_2 is a point of unstable equilibrium.
(c) Points x_1 and x_3 are points of unstable equilibrium and point x_2 is a point of stable equilibrium.
(d) The kinetic energy of the particle with total energy E is maximum at x_1 and x_3.
(e) The kinetic energy of the particle with total energy E is maximum at x_2.
(f) The kinetic energy of the particle with total energy E is maximum at x_3.

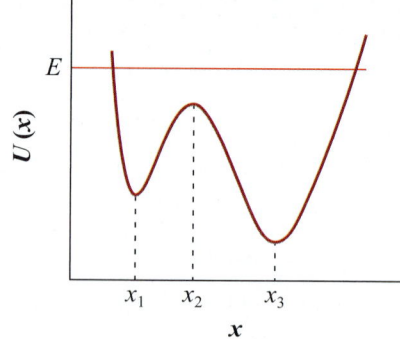

Figure 6-31 A potential energy versus position graph for a particle. E is the total energy of the particle.

ANSWER: (b), (f) At x_1 and x_3, for small displacements, the direction of the force is such that it pushes the particle back toward the points. As the potential energy is lowest at x_3, for a given total energy of a particle, the kinetic energy is therefore maximum at this point.

The SI unit for power is the watt (W), which is defined as

$$1 \text{ W} = 1 \text{ J/s}$$

Another commonly used unit to measure power is horsepower (hp). Unfortunately, the horsepower (like the calorie) has a number of different definitions. One commonly used standard for horsepower is

$$1 \text{ hp} = 746 \text{ W}$$

EXAMPLE 6-16

Hydroelectric Power

Hydroelectric dams use the potential energy of water stored in a dam to drive a turbine that converts this potential energy into electric energy. Located on the Upper Paraná River at the border between Paraguay and Brazil, the Itaipu dam is 196 m high and has an average annual water flow rate of $1.2 \times 10^4 \text{ m}^3/\text{s}$. Calculate the theoretical maximum possible power generation for the water flowing through the dam.

Solution

The work done by the force of gravity when a mass m falls through a height h is

$$W_g = mgh$$

The average power is given by

$$P_{avg} = \frac{W_g}{\Delta t}$$

The average volume of water flowing every second through the dam is $1.2 \times 10^4 \text{ m}^3$. Using 1000.0 kg/m^3 for the density of water, this amounts to $1.2 \times 10^7 \text{ kg}$ of water flowing every second. The work done by the force of gravity when this volume of water falls through a height of 196 m is

$$W_g = mgh = (1.2 \times 10^7 \text{ kg})(9.81 \text{ m/s}^2)(196 \text{ m}) = 2.3 \times 10^{10} \text{ J}$$

This amount of work is done by the force of gravity in 1 s. Therefore, the average power generated (if all of this gravitational energy is converted into electrical energy) is

$$P_{avg} = \frac{W_g}{\Delta t} = \frac{2.3 \times 10^{10} \text{ J}}{1.0 \text{ s}} = 23 \text{ GW}$$

The efficiency of a hydroelectric dam depends on the energy conversion efficiency and the turbine efficiency. The average power generated by the Itaipu dam is about 14 GW.

CHECKPOINT 6-9

Power Generated

Rank (from most to least) in the order of power generated by each of the following.
(a) a light bulb that emits 200 000 J of energy in 1 h
(b) a laser that emits 100 J of energy in 10^{-6} s
(c) a lift that pulls 2000 kg of mass to a distance of 20 m in 30 s
(d) a runner of mass 90 kg who, starting from rest, attains a speed of 10 m/s in 5 s

ANSWER: (b), (c), (d), (a)

KEY CONCEPTS AND RELATIONSHIPS

Work Done by a Constant Force in One Dimension

The work done by a constant force pointing along the direction of displacement is given by

$$W_F = \pm Fd \tag{6-1}$$

Work Done by a Constant Force in Two and Three Dimensions

The work done by a constant force making an angle theta with the direction of displacement is given by

$$W_F = (F\cos\theta)d \tag{6-2}$$

In general, if a constant force $\vec{F}$ acting on an object causes the object to be displaced by a distance $\vec{d}$, then the work, W, done by the force on the object is equal to the dot product of the force and displacement vectors:

The net work done on an object by a number of forces that displace the object by a distance d is given by

$$W_{net} = \vec{F}_{net} \cdot \vec{d} \tag{6-12}$$

where $\vec{F}_{net}$ is the sum of all forces acting on the object.

Work Done by Variable Forces

The total work done by a variable force in moving an object from position r_i to r_f is

$$W_F = \int_{r_i}^{r_f} \vec{F}(\vec{r}) \cdot d\vec{r}$$

$$= \int_{r_i}^{r_f} F_x(\vec{r}) \, dx + \int_{r_i}^{r_f} F_y(\vec{r}) \, dy + \int_{r_i}^{r_f} F_z(\vec{r}) \, dz \qquad (6\text{-}15)$$

The work done by a spring is given by

$$W_{\text{by-spring}} = \frac{1}{2}kx_i^2 - \frac{1}{2}kx_f^2 \qquad (6\text{-}19)$$

Kinetic Energy—The Work–Energy Theorem

The kinetic energy K of a mass m that has a speed v is given by

$$K = \left(\frac{1}{2}\right)mv^2 \qquad (6\text{-}29)$$

The work done by a force on an object is equal to the change in the kinetic energy of the object:

$$W = \frac{1}{2}mv_f^2 - \frac{1}{2}mv_i^2 \qquad (6\text{-}28)$$

Conservative Forces and Potential Energy

A force $\vec{F}$ is conservative when the net work done by the force in moving an object over any closed path is zero. Mathematically,

$$\oint \vec{F}(\vec{r}) \cdot d\vec{r} = 0 \ \text{(conservative force)} \qquad (6\text{-}44)$$

The change in the potential energy of an object is equal to the negative of the work done by the conservative force on the object.

$$\Delta U = -\int_a^b \vec{F}(\vec{r}) \cdot d\vec{r} \qquad (6\text{-}46)$$

Conservation of Mechanical Energy

The work done by a non-conservative force on a system changes its mechanical energy:

$$\Delta K + \Delta U = W_{\text{nc}} \qquad (6\text{-}54)$$

In the absence of non-conservative forces, mechanical energy is conserved:

$$\Delta K + \Delta U = 0 \qquad (6\text{-}55)$$

which can be stated as

$$K_i + U_i = K + U = \text{constant} \qquad (6\text{-}56)$$

From Potential Energy to Force

The conservative force is given in terms of the gradient of the potential energy:

$$\vec{F}(x,y,z) = -\vec{\nabla}U(x,y,z) \qquad (6\text{-}68)$$

In one dimension,

$$Fx(x) = \frac{dU}{dx} \qquad (6\text{-}63)$$

Power

Power is the rate at which work is done:

$$P = \frac{dW}{dt} \qquad (6\text{-}71)$$

APPLICATIONS

conservation of energy, alternative fuel sources, efficient energy use, renewable energy, global energy crisis, power generation, energy conversion, energy storage, earthquake engineering, shock absorbers, auto safety, engines, space flight, metabolism, electrostatic interactions, gravitation, friction losses, air resistance, nuclear interactions, thermodynamics, quantum mechanics.

KEY TERMS

energy, gravitational potential energy, kinetic energy, closed path, law of conservation of energy, law of conservation of mechanical energy, mechanical energy, potential energy, power, stable equilibrium, unstable equilibrium, work, work–energy theorem.

QUESTIONS

1. Does the work–energy theorem hold for forces that are not constant? Explain.
2. True or false: A spring is already stretched 10 cm when you decide to stretch it another 10 cm. The spring constant k is 1000 N/m. The work you do is positive.
3. The work done by you as you stretch a massless spring is
   (a) the change in the potential energy of the spring
   (b) the negative of the work done by the spring
   (c) parts (a) and (b)
   (d) none of the above
4. You stretch a spring that is already stretched. Is the work done by the spring positive or negative?

5. You compress a spring that is already compressed. Is the work done by the spring positive or negative?
6. Can the work done by a force ever be negative? Explain.
7. Can the total work done on an object ever be negative? Explain.
8. When you are the only external agent doing positive work on a moving car along a horizontal surface, does the car's speed increase or decrease? Explain.
9. You ride on a Ferris wheel. Your mass is 73.0 kg, and the radius of the Ferris wheel is 45.0 m. What is the total work done on you during your journey from the lowest to the highest points? What is the work done by the force of gravity?

10. What is the total work done on a 22 000 kg satellite as it moves around Earth by 180°, at a constant altitude? What is the work done by the force of gravity during that time?

11. You cause the bob (mass) at the end of a pendulum of length L to rotate in a vertical circle. What is the work done by the tension in the string as the bob is brought from the lowest to the highest position?
    (a) $-2mgL$
    (b) $2mgL$
    (c) zero
    (d) either (a) or (b), depending on the choice of positive direction

12. A hockey player intercepts a 170 g puck moving toward her on ice at 23 m/s and sends it at the same speed toward the opponent's goalie. Find the work done by the hockey player.

13. Your friend is holding onto a rope that is suspended from a tree branch such that his feet are 60.0 cm off the ground. You push him with a horizontal force of variable magnitude such that he moves at a constant speed as his feet end up being lifted by an extra 40.0 cm off the ground. Your friend's mass is 69.0 kg. Find the work done by your variable force.

14. You drop a 2.0 kg weight from a height of 43 cm above a scale. The weight comes to rest after falling 48 cm. Find the total work done on the weight over the entire journey.

15. You compress a horizontal spring from equilibrium by an amount Δx. We can say that the change in potential energy of the spring is equal to
    (a) $\frac{1}{2}k(\Delta x)^2$
    (b) $-\frac{1}{2}k(\Delta x)^2$
    (c) part (a) or (b), depending on our choice of a positive direction
    (d) We need to know either the initial or final length of the spring to answer this question.

16. You stretch a spring from an already stretched position. Which of the following statements is true?
    (a) The work you do is negative.
    (b) The work done by the spring is positive.
    (c) The change in the spring's potential energy is negative.
    (d) None of the above are true.

17. You intercept a hockey puck shot at you at a speed of 15 m/s by kicking it such that it deflects at the same speed at an angle of 65° to the incoming direction. Assume the average force that you exert on the puck is 90.0 N and that it acts over a total distance of 12 cm in one direction. What is the total work done on the puck by you? Ignore the force of gravity.
    (a) 11 J
    (b) −11 J
    (c) 5.0 N
    (d) −5.0 N
    (e) none of the above

18. A spring cannon fires a projectile at an angle of 47° above the horizontal. The cannon is fired at $t = 0$, and the projectile reaches its maximum height a time Δt later. Which of the following statements is true?
    (a) During the time Δt, the work done by the force of gravity is larger than the work done by the spring.
    (b) During the time Δt, the work done by the spring is larger than the work done by the force of gravity.
    (c) During the time Δt, the work done by the spring is equal in magnitude and opposite in sign to the work done by the force of gravity.
    (d) None of the above are true.

19. You lift your backpack of mass m a vertical distance h. What is the definition of the change in potential energy of the backpack-Earth system?
    (a) the work done by you on the backpack
    (b) the negative of the work done by you on the backpack
    (c) the negative of the work done by the force of gravity on the backpack
    (d) the work done by the force of gravity on the backpack

20. The work done by a horizontal spring as it is being compressed from its unstretched position is
    (a) positive
    (b) negative
    (c) either positive or negative, depending on where you start
    (d) depends on how fast you are moving
    (e) none of the above because work is a scalar that cannot be referred to as positive or negative

21. A horizontal spring is already stretched 20 cm from its equilibrium position (Figure 6-32). You pull it so that it is now stretched 40 cm. While the spring is being stretched,
    (a) you do negative work
    (b) the spring does positive work
    (c) the sign of the work done by the spring on you depends on whether we choose right or left to be positive
    (d) the work done by the spring is always positive because of the term $(\delta x)^2$
    (e) none of the above

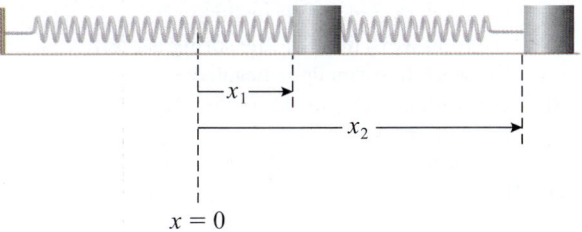

Figure 6-32   Question 21.

22. A child moves down a slide at a park at a constant speed. Find the relationship between the slope angle of the slide and the coefficient of kinetic friction between the child and the slide's surface. What can be said about the relationship between the work done by the force of gravity and the work done by friction on the child?

23. You slide down a snowy hill on a toboggan from a height of 7 m. The hill has a constant slope such that your speed remains constant at 5 m/s throughout the entire journey. Air resistance is negligible. Which of the following statements is true?
    (a) The magnitude of the work done by gravity is larger than the magnitude of the work done by friction.
    (b) The work done by friction is equal to the change in the gravitational potential energy.
    (c) The work done by friction cannot be determined.
    (d) Friction is not significant because you actually move.
    (e) None of the above are true.

24. A compressed spring is used to launch a marble onto a surface that starts out horizontal and then slopes upward at a constant angle until the marble stops. Over the entire journey,
    (a) the total work done on the marble is zero
    (b) the work done by the spring is greater than the work done by the force of gravity
    (c) the work done by friction is equal to the work done by the force of gravity
    (d) the work done by the spring is greater than the work done by friction
    (e) All the above are true except (c).

25. You move a spring such that the work done by the spring is negative. This means that you
    (a) stretched the spring from equilibrium
    (b) compressed the spring from equilibrium
    (c) stretched an already stretched spring
    (d) any of the above
    (e) none of the above

26. You move a spring such that the work done by the spring is negative. This means that
    (a) you allow the spring to move toward equilibrium
    (b) the potential energy of the spring decreases
    (c) you do negative work
    (d) There is not enough information to answer this question.
    (e) none of the above

27. A spring is compressed by 10 cm from its equilibrium position. When you move the spring so that it is now stretched by 10 cm,
    (a) the work that you do is positive
    (b) the total work done on the spring is positive
    (c) the work that you do is negative
    (d) the work done by the spring is negative
    (e) none of the above

28. You are sitting at your desk. You decide to set your desktop as the point of zero gravitational potential energy, and then you lift a book from a distance h below the desk to a distance h above the desk. Which of the following statements is true about the potential energy of the book-Earth system?
    (a) The change in potential energy is zero.
    (b) The final potential energy is $2mgh$.
    (c) The final potential energy is mgh.
    (d) The change in potential energy is $2mgh$.
    (e) Both (c) and (d) are true.

29. A spring is already stretched 10 cm when you decide to stretch it another 10 cm. The spring constant is 200 N/m. As you stretch the spring,
    (a) the work done by the spring is -1 J
    (b) the work done by the spring is -2 J
    (c) the work done by the spring is -3 J
    (d) the work done by the spring is -4 J
    (e) none of the above

30. You drop a 2.0 kg mass from a height of 7.0 m above an unstretched spring whose spring constant is 100 N/m. During the object's downward journey,
    (a) the magnitude of the work done by the force of gravity is less than $14g$
    (b) the magnitude of the work done by the spring is less than $14g$
    (c) the total work done on the mass is zero
    (d) I really don't think this makes sense.
    (e) none of the above

31. Which of the statements below is true when you catch a baseball that comes at you horizontally?
    (a) The work that you do is positive.
    (b) The work that you do is negative.
    (c) Work is not a vector and cannot be referred to as positive or negative.
    (d) Work is a scalar product expressed in terms of magnitudes, so it is never negative.
    (e) Work is expressed in terms of kinetic energy, which must be positive.

32. A spring is used to launch a marble up a frictionless incline such that the marble reaches the top of the incline with a speed of zero. Which of the following statements is true?
    (a) The total work done on the marble is positive.
    (b) The total work done on the marble is negative.
    (c) The work done on the marble by the spring is equal in magnitude and opposite in sign to the change in the gravitational potential energy.
    (d) The change in the gravitational potential energy is smaller than the change in the elastic potential energy.
    (e) None of the above are true.

33. When you drop a stone from the top of a building, does the instantaneous power delivered by the force of gravity increase, decrease, or stay the same as the stone falls?

PROBLEMS BY SECTION

For problems, star ratings will be used (★, ★★, or ★★★), with more stars meaning more challenging problems. The following codes will indicate if $\frac{d}{dx}$ differentiation, $\int$ integration, ▭ numerical approximation, or ⟿ graphical analysis will be required to solve the problem.

Section 6-2 Work Done by a Constant Force in One Dimension

34. ★ You throw a stone with a mass of 215 g straight up in the air. It reaches a maximum height of 12.0 m, and then you catch it at the same level from which you threw it.
    (a) Find the work done by the force of gravity on the stone on its way up.

(b) Find the work done by the force of gravity on the stone on its way down.

(c) Find the total work done by the force of gravity over the entire trip.

35. ★ You lift your 515 g cup of coffee from the table a total distance of 35.0 cm and hold it there.

(a) Find the work done by the force of gravity.

(b) What is the work done by you on the cup?

(c) What is the total work done on the cup?

36. ★ You push a 10.0 kg wagon, initially at rest, with a horizontal force of 145 N over a total distance of 7.00 m on a flat surface.

(a) How much work is done by you?

(b) Now you move so that you are opposing the motion of the wagon. You stop the wagon by applying a force that stops it in 2.00 m. How much work do you do to stop the wagon?

37. ★ While ice fishing, your friend slides a 5.0 kg sausage bucket over the ice surface to you at speed v. You stop the bucket over a distance of 75 cm by applying a force of 24 N on the bucket. How much work did you do?

38. ★★ An athlete picks up a 20.0 kg sandbag from the ground and throws it straight up in the air. It leaves her hands 1.50 m above the ground and reaches a height of 7.00 m.

(a) How much work did the athlete do?

(b) How much work was done by the force of gravity while the bag was in contact with the athlete?

39. ★ You pull a crate of mass 7.00 kg a distance of 3.00 m up a frictionless incline with a slope angle of 32.0° by applying a force of magnitude 325 N along the incline.

(a) Find the amount of work you do.

(b) You now pull the same crate on a frictionless horizontal surface for the same distance relative to the motion of the crate as in part (a) using a force of the same magnitude and direction as in (a). Determine the work done by you.

40. ★ A child starts from rest and slides down a snow-covered hill with a slope angle of 47° from a height of 1.7 m above the bottom of the hill. The speed of the child at the bottom of the hill is 3.2 m/s. Find the coefficient of kinetic friction between the hill and the child.

41. ★ An object of mass m is pushed along a horizontal surface by a constant horizontal force of magnitude F, over a distance d, such that the speed is constant.

(a) Find an expression for the coefficient of kinetic friction between the object and the surface in terms of F and m.

(b) Find an expression for the work done by the constant force of magnitude F.

(c) Friction is removed, and the constant force of magnitude F is now applied along an incline to cause the object to move up the incline at a constant speed. Does the force F do more or less work over the same distance?

(d) How does your answer to part (c) change if the object accelerates up the incline? (Hint: We make the incline angle smaller.)

(e) How does your answer to part (c) change if friction is brought back such that the object still moves up the incline?

Section 6-3 Work Done by a Constant Force in Two and Three Dimensions

42. ★ A block of mass 26 kg rests on a frictionless inclined surface with a slope angle of 37°. A horizontal force of 510 N is used to push the block a distance of 4.6 m along the inclined surface, as shown in Figure 6-33. Find the work done by the force $\vec{F}$.

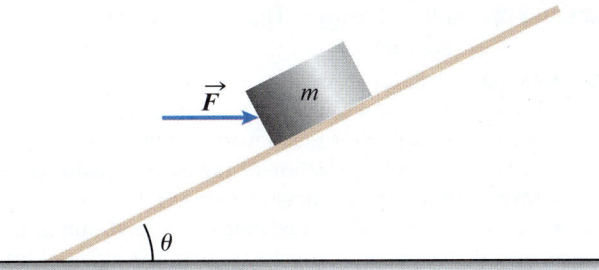

Figure 6-33   Problem 42.

43. ★ A worker leans on a crate and pushes it on a horizontal surface with a force of 230 N in a direction that makes an angle of 67° below the horizontal over a distance of 12 m. How much work is done by the worker?

44. ★★ A child uses a rope to pull her brother on a toboggan on a horizontal, snow-covered surface. The combined mass of her brother and the toboggan is 34 kg. The rope makes an angle of 41° with the surface. The horizontal projection of the rope makes an angle of 27° with the direction of motion. The child is pulling the toboggan with a force of 65 N, and the toboggan moves at a constant speed. Find the coefficient of kinetic friction between the toboggan and the snow.

45. ★★ A person applies a force $\vec{F} = (4.0\ \text{N})\hat{i} - (2.0\ \text{N})\hat{j} + (7.0\ \text{N})\hat{k}$ to move a crate on a track in the xz-plane that angles 32° above the positive x-axis. Find the work done by the person over the first 3.0 m of the motion.

Section 6-4 Work Done by Variable Forces

46. ★★ Two masses, M and m, are attached to each other with a rope that winds around a massless hanging pulley, as shown in Figure 6-34. Mass m is attached to one end of an unstretched spring whose other end is attached to the ground. The system is initially held at rest and then released.

(a) Derive an expression for the speed of mass m after M has fallen a distance x.

(b) Find the speed of m if $M = 16.8$ kg, $m = 4.8$ kg, $x = 45$ cm, and the spring constant $k = 96$ N/m.

(c) Find the maximum speed of m.

Figure 6-34   Problem 46.

47. ★★ A marble of mass m sits at the bottom of a bowl of radius R. The surface of the bowl is frictionless, and the bowl is not allowed to move. You push the marble in the horizontal direction using a force of variable magnitude until the marble is a height h above the bottom of the bowl, such that $h < R$. You bring the marble to a stop. Find the work done by the variable force.

Section 6-5 Kinetic Energy—The Work-Energy Theorem

48. ★ A tennis player strikes an incoming tennis ball of mass m and speed v such that the ball leaves the racket at speed v as well. How much work is done by the tennis player to reverse the direction of the ball? Explain. Does the tennis player do any negative work?

49. ★ You throw an 8.00 kg medicine ball straight up in the air and then catch it on the way down. Just before it hits your hands, it is moving at a speed of 3.00 m/s. You stop it over a vertical distance of 50.0 cm.
    (a) How much work did you do to stop the ball?
    (b) How much work was done by the force of gravity while you stopped the ball?

50. ★★ A 22.0 kg sled, moving at a speed of 4.00 m/s, comes loose during a sled race on ice and is stopped by an observer over a horizontal distance of 1.30 m. The observer is pushing down on the sled with a force that makes an angle of 42.0° with the horizontal. How much work does the observer do? What is the magnitude of the force used by the observer? Ignore the effects of friction.

51. ★ A child pulls a 3.00 kg wooden bus initially at rest on a frictionless horizontal surface over a distance of 3.00 m using a rope that makes a 47.0° angle above the horizontal starting from rest. The speed of the bus at the end of the pull is 5.00 m/s. Determine the work done by the child and the force exerted by the child on the bus. How much work is done by the force of gravity on the bus during the pull?

52. ★★ A particle of mass 1.1 kg moves on a straight track from point A(3.0, 2.0, −1.0) to point B(5.0, −4.0, 3.0) as a result of being pushed by a force of magnitude 170 N that rises 32° above the xy-plane such that its projection makes an angle of 195° with the positive x-axis. The given Cartesian coordinates of points A and B are in metres.

    Ignoring the effects of the force of gravity and friction, find the work done by the applied force and find the particle's speed at the end of its trajectory, knowing that the particle starts from rest.

53. ★★ ∫ Two identical springs, each of spring constant 725 N/m, are set up in parallel horizontally such that they are compressed by 32.0 cm each. The two springs are used to eject a 3.00 kg mass onto a frictionless surface. The mass then moves 1.90 m up a frictionless incline with a slope angle of 32.0°, where it encounters a combination of two springs identical to the springs used to eject it, lying along the incline. How high above the horizontal surface is the mass when it is finally stopped?

54. ★★ A 10.0 kg block slides 3.00 m down a frictionless surface inclined 30.0° above the horizontal, before being stopped by an originally unstretched spring of spring constant $k = 345$ N/m secured to the inclined surface. Find the maximum compression of the spring.

Section 6-6 Conservative Forces and Potential Energy

55. ★★ You push a crate of mass 27 kg 15 m up a ramp that is sloped 19° above the horizontal. The coefficient of kinetic friction between the ramp and the crate is 0.15, and the speed of the crate, which starts from rest, is 1.7 m/s at the end of the 15 m. To accomplish the task, you apply a variable force directed along the ramp.
    (a) Calculate the change in the crate's gravitational potential energy (crate–Earth system).
    (b) Calculate the work done by the applied force.

Section 6-7 Conservation of Mechanical Energy

56. ★ A worker tosses a 1.50 kg lunch box to a friend who is on a platform 12.0 m above. The lunch box is in contact with the worker's hands for a total distance of 35.0 cm. How much work did the worker do to accelerate the lunch box?

57. ★★★ A bead of mass 37 g moves on a frictionless vertical circular track of radius $R = 0.65$ m. The track runs through the bead (Figure 6-35). The bead is initially held at rest at the top of the track where it is tied to a spring ($k = 230$ N/m), the other end of which is fixed at a pivot at $R/2$ from the bottom of the track, as shown. The unstretched length of the spring is $R/3$. The bead is given a gentle nudge.
    (a) Find the speed of the bead when it has moved by 180° from its initial position.
    (b) Find the speed of the bead when it has moved by 90° from its initial position.

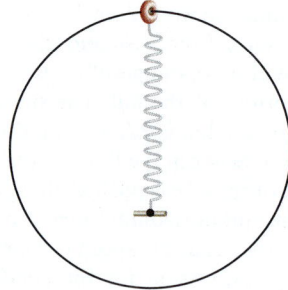

Figure 6-35 Problem 57.

58. ★★ A mass ($m = 1.20$ kg), originally at rest, sits on a frictionless surface. It is attached to one end of an unstretched spring ($k = 795$ N/m), the other end of which is fixed to a wall (Figure 6-36). The mass is then pulled with a constant force to stretch the spring. As a result, the system comes to a momentary stop after the mass moves 14.0 cm. Find the
    (a) work done by the constant force
    (b) change in potential energy of the spring

(c) change in mechanical energy of the system
(d) total work done by non-conservative forces in the problem
(e) speed of the mass 4.00 cm into the motion
(f) equilibrium position of the system
(g) maximum speed of the mass and where that occurs

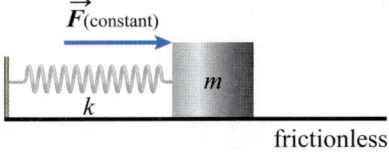

Figure 6-36   Problem 58.

59. ★★ A mass (m = 1.00 kg) rests on a 63.0° incline, where it is attached to one end of a spring, also resting on the incline, as shown in Figure 6-37. The other end of the spring is fixed. The mass is held in place so that the spring is in its unstretched position. The mass is then released. The coefficient of kinetic friction between the surface and the mass is 0.080, and the spring constant is 115 N/m. Find the maximum compression of the spring.

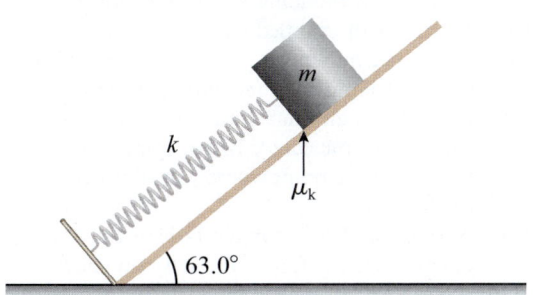

Figure 6-37   Problem 59.

60. ∫ ★★★ Newton's law of gravitation states that the force of attraction between two bodies is proportional to the product of their masses and is inversely proportional to the square of the distance between them. The force is given by $\dfrac{Gm_1m_2}{r^2}$, where G is called the gravitational constant and has a value of $G = 6.674 \times 10^{-11}$ N·m²/kg². (See Chapter 11.)
  (a) Find an expression for the gravitational potential energy between two objects. Use zero potential energy of interaction at infinite separation.
  (b) Is the potential energy expression positive or negative?
  (c) What would your escape velocity be if you weighed 800.0 N on Earth's surface? The escape velocity is the minimum velocity with which you have to be launched so you do not come back to Earth. Ignore losses due to air resistance, and assume that Earth is a sphere of radius 6371 km and a mass of 5.972×10^{24} kg.
  (d) Find the speed of a 1.00 tonne satellite orbiting Earth at an altitude of 1.00×10^3 km. (1 tonne is 1000 kg.)

(e) Find the mechanical energy for the satellite in part (d).
(f) What mechanical energy does the satellite need to just break free from Earth's gravitational field?
(g) The binding energy of a satellite is the additional energy needed for it to escape Earth's gravity. Find the binding energy for the satellite in part (d).

Section 6-8   Force from Potential Energy

61. ★★ ∫, ⤳ The force of gravitational attraction on an object beneath Earth's surface is proportional to r, the distance from the centre. Beyond Earth's surface, the force of attraction becomes inversely proportional to r^2. Plot the gravitational potential energy for an object as it moves from the centre of Earth to a point that is three times the radius of Earth as measured from Earth's centre.

62. ★★ $\frac{d}{dx}$ A particle enters a region where the potential energy function is given by $U(x, y, z) = \alpha x^{-2} + \beta xy - \gamma yz^2x$. Find an expression for the force in Cartesian unit vector notation.

63. ★★★ The electrostatic potential energy between two charges q and Q is given by KqQ/r. Charge q is brought to within a distance R of charge Q and then released from rest. Charge Q is fixed in space.
  (a) Ignoring the effects of the force of gravity, derive an expression for the speed of charge q after it has moved a distance of Δr from Q.
  (b) Using m_q = 51.0 g, q = 16.0 μC, Q = 47.0 μC, K = 8.99×10^9 N·m²/C², Δr = 1.00 cm, and R = 1.10 cm, find a value for the speed in part (a).

64. ★★ $\frac{d}{dx}$, ⤳ The potential energy of an interaction between two objects, separated by a distance r, is given by $U(r) = -A(r^3 + Br)$, where A and B are positive constants.
  (a) Is the force between these two objects repulsive or attractive? How can you tell?
  (b) Plot a potential energy versus r graph for A = 120.0 N/m² and B = 1.00 m², over the range r = 0 to r = 10.0 m.
  (c) Derive an expression for the force between the two objects as a function of r.
  (d) One of the objects is held at rest, while the other is placed a distance of 20.0 m from the fixed object, and then released from rest. The mass of the moving object is 11.0 kg. Ignoring the effects of the force of gravity, find its speed after it has moved 3.00 m.

Section 6-9   Power

65. ★ The Tupolev Tu-144 supersonic passenger jet had four Kolesov RD-36-51 afterburning turbojets that each provided a maximum thrust of 200 kN. The maximum flying speed of the Tu-144 was Mach 2.2 at an altitude of 120 000 ft. Find the force of air resistance on the airplane at that speed.

66. ★★ A tractor of mass 5.40 tonne pulls a 1.40 tonne crate up a hill with a slope angle of 12.0° above the horizontal at a constant speed of 11.0 m/s. (1 tonne is 1000 kg).
  (a) What acceleration would the tractor have on a level road at that speed? Ignore any frictional losses.

(b) Repeat part (a), but now consider that the coefficient of kinetic friction between the crate and the level road is 0.300. Assume that all frictional losses come from the friction between the crate being pulled and the road, and not from the tractor.

67. ★ A sports car has a maximum speed of 320 km/h and a maximum operating engine power output of 490 hp (1 hp = 746 W). Find the force of air resistance on the car at that speed. Ignore all other sources of friction and resistance to the motion.

68. ★★ You kick a 225 g rock from the edge of a cliff while hiking along the edge of the Grand Canyon.
    (a) Ignoring air resistance, find the average power delivered by the force of gravity during the first 4.00 s of the fall.
    (b) Find the instantaneous power delivered by the force of gravity at the end of the fifth second of the fall.

COMPREHENSIVE PROBLEMS

69. ★★★ A bead slides down a frictionless vertical helical track as shown in Figure 6-38. The helical track has 1.0 turn/cm of length of the helical axis, and the loops have a radius of 17.0 cm each. There are seven turns in the helix.

Figure 6-38   Problem 69.

(a) Estimate the slope angle experienced by the bead along the helix.
(b) Find the speed of the bead at the bottom of the helix.

70. ★★ You swing a massless string of length L with a mass m attached to its end in a vertical circle, such that the mass has enough speed to keep the string taut all the time. Derive an expression for
    (a) the difference between the maximum and the minimum tensions in the string as the mass moves in its circular path
    (b) the work done by the tension during a full revolution

71. ★★ A marble slides on the frictionless ramp starting at rest from height h as shown in Figure 6-39. The marble then moves to the circular track of radius r. Determine the minimum value of h (in terms of r) so that the marble reaches the same height as the centre of the circular track, where the magnitude of the normal force on the marble is three times that of the force of gravity on the marble.

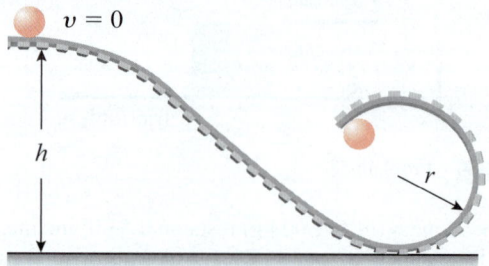

Figure 6-39   Problem 71.

72. ★★★ You happen upon what looks like an abandoned hemispherical igloo of radius R. You climb the igloo and sit at the top (Figure 6-40). A few moments later you feel a broom very gently prodding you off the roof. The push you receive from the disgruntled igloo dweller is just enough to cause you to begin to move. Assume that the surface of the igloo is smooth enough to have negligible friction.
    (a) Find the angle that your position vector as measured from the centre of the igloo makes with the vertical at the point where you slide off the surface of the igloo.
    (b) At what angle is the magnitude of the normal force equal to 1/3 the force of gravity on you?

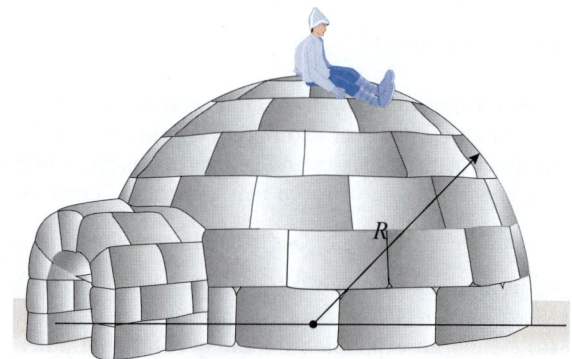

Figure 6-40   Problem 72.

73. ★★★ You happen upon a perfectly shaped ceramic hemispherical bowl at home, with a smooth, frictionless surface and of radius r. You place a small bead at the top of the inside edge of the bowl and let it slide on the inside of the bowl under the effect of the force of gravity (Figure 6-41). Locate the point along the bowl where the magnitude of the normal force from the bowl on the bead equals that of the bead's weight. What is the force of contact between the bead and the bowl when the bead is passing across the bottom of the bowl?

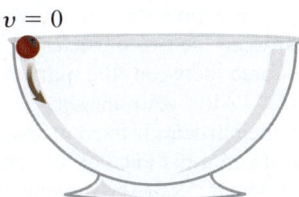

Figure 6-41 Problem 73.

74. ★★ The simple pendulum in Figure 6-42 is made to spin in a vertical circle such that it has sufficient kinetic energy to just stay taut at the top of its trajectory. When the string is in the vertical position shown, it is intercepted by a peg at a distance of 2L/3 from the centre of the vertical circle, and the mass (m) continues along the new trajectory as shown.
   (a) Find an expression for the speed of the mass (m) at its lowest point.
   (b) Find an expression for the speed of the mass at the top of the new trajectory.
   (c) Find an expression for the tension in the string when it makes an angle of 30° below the horizontal before hitting the peg.
   (d) Find an expression for the tension in the string at the top of the new trajectory.

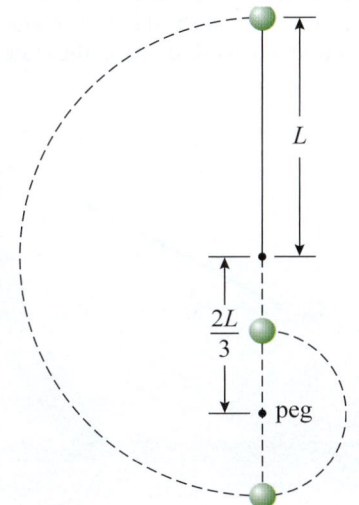

Figure 6-42 Problem 74.

75. ★★ A 225 g cube slides down a ramp starting from rest, as shown in Figure 6-43. The ramp has a 46° slope. After falling a distance of 92 cm, the cube strikes an unstretched spring of spring constant $k = 25$ N/m. Find the maximum compression of the spring when the coefficient of kinetic friction between the cube and the ramp is 0.17.

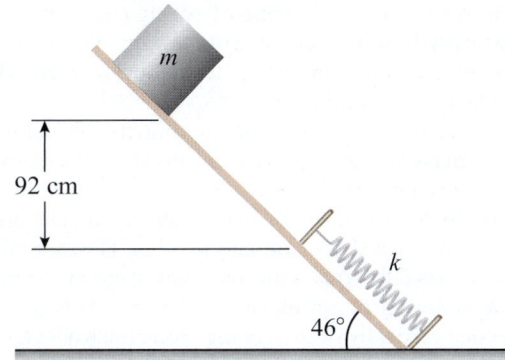

Figure 6-43 Problem 75.

76. ★★ You are sitting in a roller coaster car that begins its descent, from rest, down the main track from a height of 62.0 m. The roller coaster then climbs up the inside of a vertical circle of radius 20.0 m to a point two thirds of the circle's radius above the bottom of the circle.
   Ignoring frictional losses, find the normal force exerted on you by the seat if your mass is 57.0 kg.

77. ★★ The string in Figure 6-44 is wound around the nail as shown. The end of the string is attached to a mass, m, that is released from the horizontal position, starting from rest. Derive an expression for
   (a) the tension in the string when it makes an angle of 23° with the vertical
   (b) the total work done to bring the string to the position in part (a)
   (c) the work done by the tension

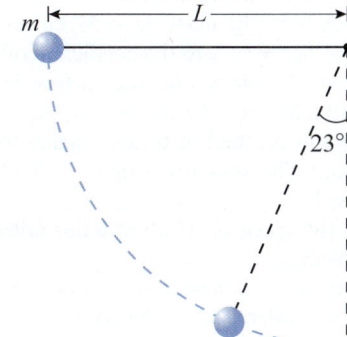

Figure 6-44 Problem 77.

78. ★★★ A block of mass M sits on top of a horizontal frictionless bench. A string attached to M runs over a massless, frictionless pulley and is attached to a vertically hanging mass m. Mass M is attached to a horizontal spring of spring constant k whose other end is fixed to a wall. The system is originally held at rest when the spring is unstretched and then released. Find an expression for
   (a) the maximum distance that m will fall
   (b) the speed of mass m after it has fallen a distance h

79. ★★ An unstretched spring of spring constant k hangs vertically from the ceiling. You now attach a mass m to the free end of the spring and let the spring extend slowly while holding the mass.
    (a) By how much does the spring stretch before the mass stops? Express your answer in terms of the given variables.
    (b) Next, you take the mass back up to the original unstretched position and let it go. How far will the mass fall? What is the maximum speed of the mass?

80. ★★ A spring of spring constant $k = 710$ N/m is suspended vertically from a ceiling. You attach a 32 kg mass to the spring and let it drop. Find the speed of the mass when it is three quarters of the way down to the position where it momentarily stops.

81. ★★ A block of mass $m = 3.2$ kg rests on top of another block of mass $M = 16.4$ kg, which rests on top of a horizontal frictionless surface. The coefficient of static friction between the two blocks is 0.23, and the coefficient of kinetic friction between them is 0.17. Block M is pushed with a horizontal force of 37 N. Find the energy lost to friction during the time it takes M to move 4.0 m, starting from rest.

82. ★★ You throw a 245 g orange straight up in the air from a height of 90.0 cm above the ground starting from rest. You let go of the orange when it is 135 cm above the ground and it reaches a maximum height of 14.0 m.
    (a) How much work is done by you?
    (b) How much work is done by the force of gravity while your hand is in contact with the orange?
    (c) Find the average power delivered by the force of gravity during the last second before the orange reaches its maximum height.

83. ★★★ $^d/_{dx}$ A block of mass $M = 37$ kg rests on top of a horizontal bench, where the coefficient of kinetic friction between the block and the surface is $\mu_k = 0.20$. A string attached to M runs over a massless, frictionless pulley and is attached to a vertically hanging mass ($m = 11$ kg). The system is originally held at rest and then released.
    (a) Find the speed of M after m has fallen a distance of 65 cm.
    (b) Using energy considerations, derive an expression for the acceleration of the system.

84. ★★ $^d/_{dx}$ A block of mass $M = 17$ kg rests on top of a frictionless horizontal bench. A string attached to M runs over a massless frictionless pulley and is attached to a vertically hanging mass ($m = 4.0$ kg). The system is originally held at rest and then released. Find the speed of M after m has fallen a distance of 35 cm. Using energy considerations, find the acceleration of the system.

85. ★★★ $\int$ The force between two particles is measured to be $F(r) = -Ar^{-3}$, where r is the distance separating the two particles and A is a positive constant.
    (a) Develop an expression for the change in the potential energy of interaction between the two particles as the distance between them changes from r_1 to r_2.
    (b) You are doing work to change the distance between the two particles. Develop an expression for the work done by you when the distance changes at a constant speed.

(c) What is the expression for the work done by the force between the two particles?
(d) Is the force between the particles attractive or repulsive? Justify your answer.
(e) One of the particles is fixed in space, and the other is let go from a distance of 115 cm from the fixed particle. The mass of the moving particle is 12.0 g. Find its speed in terms of A when it has moved 75.0 cm from the position indicated above. The experiment is conducted on a horizontal frictionless surface.

86. ★★★ $^d/_{dx}$ The worker in Figure 6-45 has to pull a crate over a concrete surface over a distance of 9.0 m, as shown. The coefficient of kinetic friction between the crate and the surface is μ_k.
    (a) Find the angle θ (in terms of μ_k) that will maximize the effectiveness of the worker in pulling the crate. (Hint: Minimize the time for the task.)
    (b) The crate's mass is 340 kg, and μ_k is 0.20. For a given angle θ, what is the minimum force needed to cause the crate to start moving when the coefficient of static friction between the crate and the surface is 0.50?
    (c) Once the crate starts moving, the worker applies just enough force to keep it moving at a constant speed. Find the work done by the worker.
    (d) Find the work done by the force of friction in part (c).
    (e) Find the work done by the force of gravity in part (c).
    (f) Find the total work done on the crate in part (c).

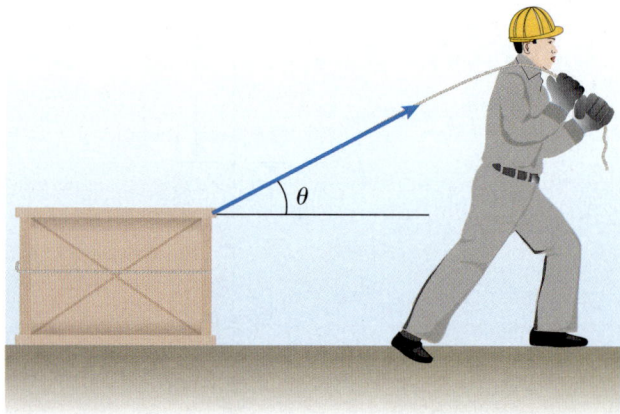

Figure 6-45 Problem 86.

87. ★★★ A 67 kg skydiver free-falls for 3.0 min, reaching 99% of her terminal speed of 210 km/h in 15 s.
    (a) Find the work done by the force of gravity during the last 2.0 min of the fall.
    (b) Find the work done by air resistance during the last 2.0 min of the fall.
    (c) Find the power delivered by the force of gravity during that time.
    (d) Find the power delivered by air resistance during that time.
    (e) Find the average force of air resistance on the skydiver during the first 15 s of the fall.

88. ★★★ $\int$ A particle with a mass of 310 g, moving in one dimension, enters a progressively damped region where the force it experiences is given by $F(x) = -kx^3$, where x is the distance travelled by the particle in the region and k is a constant. Ignoring frictional losses, find how far into the region the particle penetrates when it enters with a speed of 120 m/s. The force constant k is 320 N/m^3.

89. ★★ d/dx The position of a 25.0 g particle as a function of time is given by $x(t) = -\alpha t^3 + \beta t^2 + \gamma t^{-4}$ as a result of being under the influence of a variable force of magnitude F.
Given: $\alpha = 0.700$ m/s^3, $\beta = 1.10$ m/s^2, and $\gamma = 1.50$ m·s^4.
(a) Find the position of the particle at $t = 1.00$ s.
(b) Find the position of the particle at $t = 3.50$ s.
(c) Find the value of F at $t = 1.00$ s.
(d) Find the value of F at $t = 3.50$ s.
(e) Evidently, the acceleration is not constant. Can we still use the work–energy theorem for this particle? Explain.

90. ★★★ $\int$ Consider problem 71 in the presence of friction. The coefficient of kinetic friction between the marble and both sections of the track is $\mu_k = 0.250$, and the mass of the marble is $m = 125$ g.
(a) Knowing that the normal force on the marble is variable in the circular track and is a function of the position of the marble on the circle, use energy considerations to show that the magnitude of the normal force on the marble during its motion is given by

$$N(\theta) = 2mg\left(\frac{h}{r} - 1\right) - \frac{2\mu_k mgh\cot\varphi}{r} + 3mg\cos\theta$$

$$- 2\mu_k \int_0^\theta N(\theta)d\theta$$

where φ is the incline angle of the ramp and θ is the angle that the position vector of the marble on the circular track, as measured from the centre, makes with the vertical.
(b) Assume that the normal force, $N(\theta)$ takes the form $N(\theta) = A\cos\theta + B\sin\theta + Ce^{-2\mu_k\theta}$
Derive an expression for each of parameters A, B, and C.
(c) Find the minimum height of the ball, h (in terms of r), if the slope angle, φ, of the ramp is 26.5° and the marble reaches the same height as the centre of the circular track, where the magnitude of the normal force on the marble equals $3mg$.

91. ★★★ $\int$ Consider problem 72 when the coefficient of kinetic friction between you and the igloo is $\mu_k = 0.20$, and your mass, m, is 72 kg.
(a) Show that the magnitude of the normal force on you is given by

$$N(\theta) = 3mg\cos\theta - 2mg + 2\mu_k \int_0^\theta N(\theta)d\theta$$

where θ is the angle that your position vector, as measured from the centre of the igloo, makes with

the vertical at the point where you slide off the surface of the igloo.
(b) If the normal force takes the form

$$N(\theta) = A\cos\theta + B\sin\theta + Ce^{-D\theta}$$

find an expression for each of the parameters A, B, C, and D.
(c) Find the angle where you slide off the side of the igloo.
(d) Find the angle at which the magnitude of the normal force equals one third that of the force of gravity on you.

92. ★★★ $\int$ The resistive force of air on a free-falling stone of mass 175 g is proportional to the square of the speed of the stone. The stone falls from rest and reaches 99% of its terminal speed in the first 15 s of the fall. The terminal speed is 195 km/h.
(a) Find the coefficient of proportionality between the resistive force and the square of the speed.
(b) Find the speed of the stone after 2.00 s of falling.
Given:

$$\int \frac{dx}{a^2 - x^2} = \frac{1}{2a}\ln\left(\frac{a + x}{a - x}\right) + c$$

for $a^2 - x^2 > 0$; c is a constant
(c) Find the work done by air resistance during the first 2.00 s of the fall.
Given:

$$\int \left(\frac{e^{ax} - 1}{e^{ax} + 1}\right)dx = -x + \frac{2}{a}\ln(e^{ax} + 1) + c$$

where c is a constant.

93. ★ The Avro Arrow jet fighter had a maximum takeoff weight of 62 430 lb and an operating ceiling of 58 500 ft. (1 kg = 2.20 lb; 1 ft = 30.48 cm)
(a) Estimate the change in gravitational potential energy when a fully loaded Avro Arrow climbs to its operating ceiling.
(b) Find the combined power output of its two Pratt & Whitney turbojet engines, the thrust of which was 23 450 lb each when cruising at maximum speed of 1312 mph (miles per hour).
(c) Using the fact that the force of air resistance on an object is proportional to the square of its speed, find the maximum acceleration of the plane when at a cruising speed of 701 mph.

94. ★★ A typical banana nut muffin or a sesame seed bagel with cream cheese has around 425 Cal. Determine the height of a mountain you would have to climb to do that much work against gravity. Use a mass of 72.0 kg. How many kilometres of walking does that equal if you are climbing a 20.0° hill? This initial estimate does not take into account the fact that you actually burn calories walking on a perfectly horizontal surface. A 72.0 kg person walking briskly at 6.00 km/h will burn approximately 285 Cal on level ground in 1.00 h. How many kilometres do you have to walk up the hill if you factor that fact in?

DATA-RICH PROBLEM

95. ★★★ ▢, ⟿ According to Newton's law of universal gravitation, the attractive force between two bodies is proportional to the mass of each body and inversely proportional to the distance separating the two bodies. A meteor of mass $m = 1.000 \times 10^5$ kg starts at rest from a very distant location and strikes Earth's surface. Assume there are no other astronomical bodies in the universe, and ignore air resistance and the interaction between the meteor and the atmosphere.

Using a spreadsheet or computer program, do the following.

(a) Plot the force of gravity on the meteor as a function of r, the distance of the meteor from Earth's centre, for the range 6371 km $< r < 1.260 \times 10^5$ km.

(b) How far does the meteor have to be from Earth for the force to be 5% of its value at Earth's surface? The radius of Earth is 6371 km.

(c) Use the integration function in your spreadsheet to calculate the change in potential energy as the meteor moves from a distance of 1.260×10^5 km to Earth's surface. Alternatively, you can visually estimate the area under the curve. Do you expect the change in potential energy to be negative or positive?

(d) Plot the potential energy of interaction between Earth and the meteor as a function of r for the same range as in part (a).

(e) Compare your answer in part (c) to the value you get from the plot in part (d).

(f) What is the potential energy at Earth's surface?

(g) Calculate the speed of the meteor at Earth's surface if it started at rest from infinity.

(h) What is the acceleration of the meteor when it is 1.000×10^5 km from Earth?

(i) Calculate the acceleration of the meteor at Earth's surface.

(j) What is the minimum value of the potential energy between Earth and the meteor if the meteor started at infinity?

(k) What is the maximum value of the potential energy of interaction between Earth and the meteor if the meteor started at infinity?

OPEN PROBLEM

96. ★ The energy content (also referred to as the calorific value[2] or heat value) of gasoline is reported as 125 000 BTU per US gallon. 1 US gallon is 3.785 L.

(a) If the density of gasoline is 0.730 g/mL, find the energy content of gasoline in kJ/g.

(b) If your car does 14.0 km of highway driving per litre of gasoline, how much energy does your car burn on a 100 km trip?

(c) Assume that all the heat value is converted into kinetic energy (which is not a valid assumption, of course: typical automotive engines operate at around a 25% efficiency level). If gasoline produces 67 g of carbon dioxide per MJ, how many litres of carbon dioxide do you end up emitting during your trip? Assume the carbon dioxide volume is measured at standard temperature and pressure (STP) conditions.

(d) If your car ran on a diesel engine of the same 100% efficiency as the gas engine above, how many kilometres of highway driving would you be able to get from one litre of petro diesel? The energy content of petro diesel is 138 700 BTU per US gallon, and its density is 0.840 g/mL.

(e) Estimate your car's gas mileage on an uphill road with a slope of angle 23.0°. Assume a mass of 1250 kg for your car.

---

2. The gross calorific value is the total amount of heat released by a unit quantity of fuel, when it is burned completely with oxygen, and when the products of combustion are returned to ambient temperature. This quantity includes the heat of condensation of any water vapour contained in the fuel and of the water vapour formed by the combustion of any hydrogen contained in the fuel.

CHAPTER 7

Linear Momentum, Collisions, and Systems of Particles

Learning Objectives

When you have completed this chapter, you should be able to

LO1 Define linear momentum, and calculate and compare momenta of various objects.

LO2 Express Newton's laws in terms of rates of change of linear momentum.

LO3 Define and calculate impulse.

LO4 Find the centre of mass of a system of particles.

LO5 Identify a system and apply conservation of momentum.

LO6 Distinguish between elastic and inelastic collisions in terms of energy and momentum conservation, and apply the laws of conservation of momentum and energy appropriately.

LO7 Use the ideal rocket equation, and explain the basics of rocket propulsion.

In cars, the combination of crumple zones and strengthened passenger compartments is a direct application of Newton's laws that has proved to be a major contribution to passenger safety (Figure 7-1). The crumple zone is a part of a car that is controllably weakened so it deforms during a collision, which maximizes the dissipation of the kinetic energy from the collision. The crumpling also increases the time over which the car loses speed and, hence, decreases the acceleration and the force experienced by the passengers during an impact. First introduced by Mercedes Benz in the 1950s, crumple zones are generally installed at both the front and rear of modern cars. Cars with crumple zones generally have a hardened passenger cabin, which is designed to maintain its integrity during a collision and protect passengers against intrusions from the crumpling exterior.

Older cars have rigid bodies that do not deform much during a collision, leading to very short collision times and high accelerations. High accelerations cause more serious passenger injuries than the reduced collision accelerations experienced in cars with crumple zones.

Figure 7-1   A car's crumple zone.

cla78/Shutterstock.com

7-1 Linear Momentum

In Chapter 6, we stated that in the absence of non-conservative forces, the mechanical energy of an object is conserved.

In this chapter, we introduce another conserved quantity: **linear momentum**, $\vec{p}$. The linear momentum of an object is defined as the product of its mass and its velocity:

KEY EQUATION
$$\vec{p} = m\vec{v} \qquad (7\text{-}1)$$

Momentum is a vector quantity and has units of kilogram metres per second. We will discuss conservation of momentum in Section 7-5.

Table 7-1 lists momenta for different objects. The table indicates the wide range of possible values for momentum.

Table 7-1 Momentum of Various Objects

Object	Momentum (kg · m/s)
Argon atom at room temperature	2.123×10^{-23}
An electron moving at 1/10 the speed of light	2.733×10^{-23}
A proton moving at 1/10 the speed of light	5.01×10^{-20}
A 7.0 g marble moving at 2.0 m/s	0.014
A fully loaded passenger airliner at cruising speed	1.7×10^{8}

EXAMPLE 7-1

Mass, Velocity, and Momentum

Compare the magnitude of the momentum of a volleyball that leaves the server's hands at 80.0 km/h with the momentum of a compact car travelling at 50.0 km/h. The mass of the volleyball is about 0.270 kg and the compact car has a mass of 7.40×10^{2} kg.

Solution

We need to multiply the mass of each object by its speed to get the magnitude of the momentum. We will have to convert the speeds into m/s:

$$p_{\text{volleyball}} = (0.270 \text{ kg})\left(80.0 \text{ km} \cdot \text{h}^{-1} \frac{1000 \text{ m} \cdot \text{km}^{-1}}{3600 \text{ s} \cdot \text{h}^{-1}}\right)$$

$$= 6.00 \text{ kg} \cdot \text{m} \cdot \text{s}^{-1}$$

$$p_{\text{compact car}} = (7.40 \times 10^{2} \text{ kg})\left(50 \text{ km} \cdot \text{h}^{-1} \frac{1000 \text{ m} \cdot \text{km}^{-1}}{3600 \text{ s} \cdot \text{h}^{-1}}\right)$$

$$= 1.03 \times 10^{4} \text{ kg} \cdot \text{m} \cdot \text{s}^{-1}$$

Making sense of the result

Although the car is travelling quite a bit slower than the volleyball, the mass of the car is much greater than that of the volleyball, so it has a much larger momentum.

EXAMPLE 7-2

Momentum Is a Vector

Find the horizontal and the vertical components of the momentum of a rocket at the point following its takeoff, where the mass of the rocket (including fuel) is 1.90×10^{6} kg and its velocity is 4.20×10^{2} km/h at an angle of $11.0°$ from the vertical.

Solution

We must remember that momentum is a vector. We can use Equation 7-1 in magnitude form to get the magnitude of the momentum. We then need to resolve the vector into its components.

The magnitude of the rocket's momentum at that point is given by

$$p = mv = 1.90 \times 10^{6} \text{ kg} \times \left(4.20 \times 10^{2} \text{ km} \cdot \text{h}^{-1} \frac{1000 \text{ m} \cdot \text{km}^{-1}}{3600 \text{ s} \cdot \text{h}^{-1}}\right)$$

$$= 2.22 \times 10^{9} \text{ kg m/s}$$

The expression for the momentum vector is given by

$$\vec{p} = \vec{p}_x + \vec{p}_y$$

Using Figure 7-2, we can express the momentum as

$$\vec{p} = p\sin 11.0° \hat{i} + p\cos 11.0° \hat{j}$$

$$= (4.24 \times 10^{8} \text{ kg m/s } \hat{i} + 2.18 \times 10^{9} \text{ kg m/s } \hat{j})$$

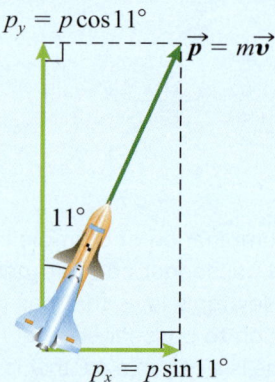

$$p_y = p\cos 11°$$

$$\vec{p} = m\vec{v}$$

$$11°$$

$$p_x = p\sin 11°$$

Figure 7-2 The horizontal and vertical components of the linear momentum of a rocket.

Making sense of the result

The vertical component is significantly greater than the horizontal component, as we would expect, because the angle from the vertical is relatively small.

EXAMPLE 7-3

Change of Momentum

A marble with a mass of 10.0 g is sliding at a constant speed of $v = 2.00$ m/s along a circular, horizontal, frictionless track, as shown in Figure 7-3(a). Find the change in the marble's momentum as it goes from point A to point B. Express the change in vector form, and find the magnitude of the momentum change. Illustrate your solution with a diagram.

Solution

The change in momentum is the difference between the final and the initial momenta. When the marble is in position A, all of its momentum is in the y-direction. Therefore,

$$\vec{p}_A = 0\hat{i} + (mv)\hat{j} = 0\hat{i} + (0.0100 \text{ kg})(2.00 \text{ m·s}^{-1})\hat{j}$$
$$= (0.0200 \text{ kg·m·s}^{-1})\hat{j} \tag{1}$$

At position B, the magnitude of the momentum has not changed, but it is now pointing in the x-direction:

$$\vec{p}_B = (mv)\hat{i} + 0\hat{j} = (0.0200 \text{ kg·m/s})\hat{i}$$

The change in momentum is

$$\Delta\vec{p} = \vec{p}_B - \vec{p}_A = 0.0200 \text{ kg·m·s}^{-1}\,\hat{i} - 0.0200 \text{ kg·m·s}^{-1}\,\hat{j}$$

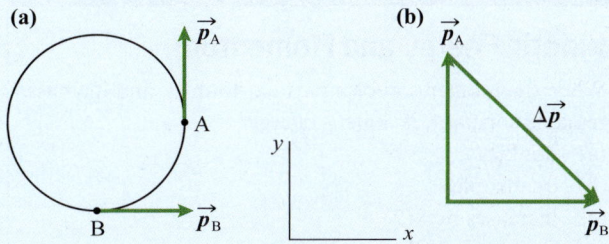

Figure 7-3 (a) The linear momentum has the same magnitude but different directions at A and B. (b) The change in momentum is shown as a difference of vectors.

The magnitude of this change in momentum is

$$\Delta p = \sqrt{(\Delta p_x)^2 + (\Delta p_y)^2}$$
$$= \sqrt{(0.0200 \text{ kg·m·s}^{-1})^2 + (0.0200 \text{ kg·m·s}^{-1})^2}$$
$$= 0.0283 \text{ kg·m·s}^{-1}$$

The diagram for the solution is shown in Figure 7-3(b).

Making sense of the result

The change in momentum must point down and to the right to give us the final momentum, which points straight to the right. It is also important to note that although the magnitude of the momentum did not change, its direction changed.

Momentum and Kinetic Energy

Massive moving objects have momentum, and we also know from Chapter 6 that they have kinetic energy. It is possible to write the kinetic energy in terms of the magnitude of the momentum. The kinetic energy of a particle is given by

$$K = \frac{1}{2}mv^2 \tag{6-31}$$

From the definition of momentum (Equation 7-1), we get

$$\vec{v} = \frac{\vec{p}}{m}$$
$$v^2 = \frac{p^2}{m^2}$$

If we substitute this into Equation 6-4, we find that

$$K = \frac{p^2}{2m} \quad \text{or}$$
$$p = \sqrt{2mK} \tag{7-2}$$

We must be careful to distinguish between energy and momentum. While they both depend upon the mass and the fact that the object is moving, momentum is a vector quantity that depends on the velocity, while kinetic energy is a scalar that depends on the speed. In addition, we have seen that there are different types of energy, and that one type can be converted into another type—kinetic into potential, for example. There is only one type of linear momentum, and it does not get converted as energy does.

CHECKPOINT 7-2

CHECKPOINT 7-2

Kinetic Energy and Momentum

When the momentum of a particle doubles, and its mass remains constant, its kinetic energy

(a) doubles
(b) quadruples
(c) increases by $\sqrt{2}$
(d) Both (a) and (b) are possible.

ANSWER: (b) It quadruples because the kinetic energy depends upon the square of the magnitude of the momentum: $K = \dfrac{p^2}{2m}$.

7-2 Rate of Change of Linear Momentum and Newton's Laws

In Chapter 5, we discussed Newton's laws in terms of forces and accelerations:

$$\vec{F}_{net} = m\vec{a} \tag{5-3}$$

However, it is also possible to write the second law in terms of momentum. Taking the derivative of Equation 7-1 with respect to time gives

$$\frac{d\vec{p}}{dt} = \frac{d}{dt}(m\vec{v}) = \frac{dm}{dt}\vec{v} + m\frac{d\vec{v}}{dt} \tag{7-3a}$$

If we consider an object whose mass does not change, then $\dfrac{dm}{dt} = 0$ and Equation 7-3a becomes

$$\frac{d\vec{p}}{dt} = m\frac{d\vec{v}}{dt} = m\vec{a} \tag{7-3b}$$

We can combine Equations 5-3 and 7-3b to write Newton's second law as

KEY EQUATION
$$\vec{F}_{net} = \frac{d\vec{p}}{dt} \tag{7-4}$$

In Chapter 5, we said that the net force that acts on an object is equal to the product of the mass and the acceleration of the object. Here we see that the net force is equal to the rate of change of linear momentum of the object with respect to time. Although we derived Equation 7-4 by assuming that the mass did not change with time, this form of the second law is also valid for objects where the mass varies. This makes it much more generally applicable. An example in which the mass of an object can vary is the squid: a squid ejects water as a means of propulsion, so the total mass of the squid decreases as it releases the water. Similarly, rockets lose mass in the form of burnt fuel as they propel themselves.

Equation 7-4 tells us that the instantaneous rate of change of momentum with respect to time is equal to the instantaneous value of the force. If we instead consider a finite change in momentum, $\Delta\vec{p}$, that occurs over a finite time interval, Δt, we can rewrite Equation 7-4 in terms of the average net force:

KEY EQUATION
$$\vec{F}_{net}^{avg} = \frac{\Delta\vec{p}}{\Delta t} \tag{7-5}$$

This form of the second law is very useful as we often cannot continuously monitor the rate of change of momentum of an object. However, we can usually measure the momentum before and after some event. If we also know the time required for the change in momentum, we can extract information about the average force involved.

CHECKPOINT 7-3

The Force on a Satellite

A satellite of mass m orbits Earth at speed v in an orbital period of T. What is the magnitude of the average force on the satellite during half an orbit?

(a) zero
(b) mv/T
(c) $2mv/T$
(d) $4mv/T$

ANSWER: (d) The change in momentum during half an orbit is $2mv$, and this occurs during a time interval of $T/2$.

EXAMPLE 7-4

Information about Forces from Collisions

Suppose we drop a basketball ($m = 0.63$ kg) from a height of 1.0 m above the floor. It falls and rebounds off the floor to a height of 0.90 m. We estimate that the ball is in contact with the floor for 0.10 s based on the length of the sound produced by the bounce.

(a) What is the change in momentum of the ball during the bounce?

(b) What average force is exerted on the ball by the floor?

Solution

(a) To determine the change in momentum of the basketball, we need to know its velocity just before the bounce and just after the bounce. We can use conservation of energy to find the speed of the ball just before the bounce:

$$mgh_i = \frac{1}{2}mv_i^2$$

$$v_i = \sqrt{2gh_i} = \sqrt{2(9.81 \text{ m}\cdot\text{s}^{-2})(1.0 \text{ m})} = 4.43 \text{ m}\cdot\text{s}^{-1}$$

The ball only rebounds to a height of 0.90 m, and the kinetic energy it has just as it leaves the floor gets converted back into gravitational potential energy. Thus,

$$mgh_f = \frac{1}{2}mv_f^2$$

$$v_f = \sqrt{2gh_f} = \sqrt{2(9.81\ \text{m}\cdot\text{s}^{-2})(0.90\ \text{m})} = 4.20\ \text{m}\cdot\text{s}^{-1}$$

The velocity before the collision is directed downward, and after the collision it is directed upward. Choosing the y-direction to be up, the change in momentum is thus

$$\Delta\vec{p} = \vec{p}_f - \vec{p}_i$$
$$= m\vec{v}_f - m\vec{v}_i$$
$$= (0.63\ \text{kg})[4.20\ \text{m}\cdot\text{s}^{-1}\hat{j} - (-4.43\ \text{m}\cdot\text{s}^{-1}\hat{j})]$$
$$= 2.65\ \text{kg}\cdot\text{m}\cdot\text{s}^{-1}\hat{j} + 2.79\ \text{kg}\cdot\text{m}\cdot\text{s}^{-1}\hat{j}$$
$$= 5.44\ \text{kg}\cdot\text{m}\cdot\text{s}^{-1}\hat{j} \approx 5.4\ \text{kg}\cdot\text{m}\cdot\text{s}^{-1}\hat{j}$$

(b) It is a straightforward use of Equation 7-5 to find the average force exerted on the ball:

$$\vec{F}_{net}^{avg} = \frac{\Delta\vec{p}}{\Delta t} = \frac{5.44\ \text{kg}\cdot\text{m}\cdot\text{s}^{-1}\hat{j}}{0.10\ \text{s}}$$

$$= \frac{5.44\ \text{kg}\cdot\text{m}\cdot\text{s}^{-1}\hat{j}}{0.10\ \text{s}} = 54.4\ \text{kg}\cdot\text{m}\cdot\text{s}^{-2}\hat{j} = 54\ \text{N}\ \hat{j}$$

Making sense of the result

The floor has to push upward on the ball to make it bounce, and we can see that both the change in momentum and the force point upward, as we would expect. The force of gravity on the ball is $mg = (0.63\ \text{kg})(9.81\ \text{m}\cdot\text{s}^2) = 6.18\ \text{N}$, and it takes $t_{fall} = \sqrt{2h/g} = \sqrt{2\cdot 1.0\ \text{m}/9.81\ \text{m}\cdot\text{s}^{-2}} = 0.45\ \text{s}$ for gravity to change the magnitude of the momentum of the ball by $2.79\ \text{kg}\cdot\text{m}\cdot\text{s}^{-1}$ as the ball falls toward the floor. The floor has to change the magnitude of the momentum of the ball by a little less than twice this amount and it has to do it in a much shorter period of time. The average force we obtained seems quite reasonable in this context.

CHECKPOINT 7-4

Bouncing Ball

When a ball hits a wall horizontally and bounces back with the same speed,
(a) the average net force on the ball is zero
(b) the momentum of the ball does not change
(c) the force on the ball is parallel to the ball's incident momentum
(d) none of the above

ANSWER: (d) Momentum does change. The force is not zero, and it is parallel to the final momentum.

MAKING CONNECTIONS

Particle Accelerators

Much of what we have learned about subatomic physics has been gleaned from studying collisions between particles in accelerators. The particles are accelerated to known energies and momenta and then collided. Detectors are used to measure the energy and momentum of the particles that emerge from the collision (see Figure 7-4).

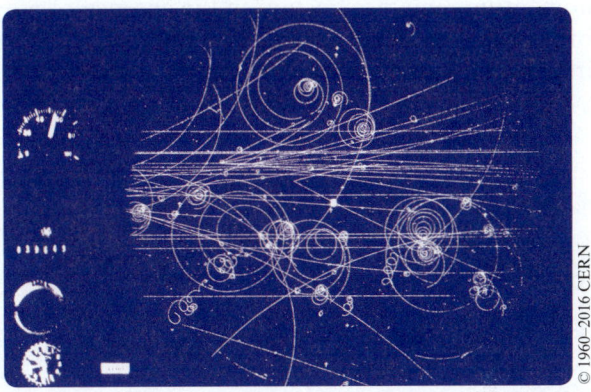

© 1960–2016 CERN

Figure 7-4 Particle tracks in a bubble chamber. The chamber is placed in a magnetic field and, as you will learn in Chapter 24, a charged particle moving in a magnetic field travels along a circular path, the radius of which depends on the momentum and charge of the particle, as well as the strength of the magnetic field. By studying such collisions, we can learn about the forces that act between fundamental particles.

7-3 Impulse

In comic books and movies, superheroes sometimes catch people in mid-fall to save them from peril. It is not uncommon to see a victim falling from the top of a skyscraper, only to have the superhero swoop up from below to reverse their direction of motion almost instantaneously. In real life, changing the person's momentum at this rate might well be fatal.

In this section, we examine in more detail the force on an object during an impact or collision. We begin by rewriting Equation 7-4 as

$$d\vec{p} = \vec{F}_{net}\ dt \qquad (7\text{-}6)$$

According to Equation 7-6, the infinitesimal change in the momentum of an object equals the product of the net force acting on that object and the infinitesimal amount of time over which the force acts on the object. Integrating Equation 7-6 gives the change in momentum, which is called the **impulse**, I:

KEY EQUATION

$$\vec{I} = \Delta\vec{p} = \int_{t_1}^{t_2} \vec{F}_{net}\ dt \qquad (7\text{-}7)$$

We can also write the impulse in terms of the average force and the collision time:

KEY EQUATION

$$\vec{I} = \Delta\vec{p} = \vec{F}_{net}^{avg}\Delta t \qquad (7-8)$$

where $\vec{F}_{net}^{avg}$ is the average net force acting on the object during the time Δt.

When the net force on an object is zero, the impulse is also zero and hence the object's momentum does not change. The fact that an object's linear momentum remains constant in the absence of a net force acting on the object is in keeping with Newton's first law.

EXAMPLE 7-5

Finding the Impulse

A hockey puck is deflected by 32.0° from its original direction of motion so its speed does not change.

(a) Create a sketch that shows the incident and scattered momenta as well as the change in momentum of the puck.

(b) Use your sketch to calculate the magnitude of the impulse experienced by the hockey puck in terms of the magnitude of the incident momentum.

Solution

(a) The situation is depicted in Figure 7-5. The incident momentum vector is shown as $\vec{p}_i$ and the "scattered" or final momentum is shown as $\vec{p}_f$. The two vectors are drawn using the origin to origin representation. Note that the magnitudes of the initial and final momenta are equal: $p_i = p_f$.

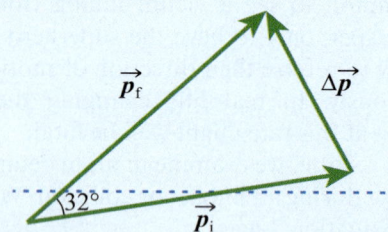

Figure 7-5 The momentum vector diagram showing the initial and final momenta as well as the change in momentum.

The impulse is the change in momentum, given by

$$\vec{I} = \Delta\vec{p} = \vec{p}_f - \vec{p}_i$$

(b) Here we need to calculate the magnitude of the impulse (the change in momentum). Rather than resolving the vectors into components, we will determine the change in momentum by analyzing the triangle. We can do this in two ways.

(i) Using the cosine law, we can see that the magnitude of the change in momentum is given by

$$\Delta p^2 = p_i^2 + p_f^2 - 2\Delta p_i \Delta p_f \cos 32°$$
$$= p_f^2 + p_f^2 - 2p_f^2 \cos 32°$$
$$= 0.304 p_f^2$$
$$\therefore \Delta p = 0.551 p_f$$

(ii) We drop down a perpendicular from the apex of the triangle to the base $\Delta\vec{p}$ (Figure 7-6). We can see that each half of the base is the opposite side in a right triangle to an angle given by $32°/2 = 16°$. The hypotenuse of the triangle is p_f (or p_i). Thus, the change in momentum is given by

$$\Delta p = 2p_f \sin 16° = 0.551 p_f$$

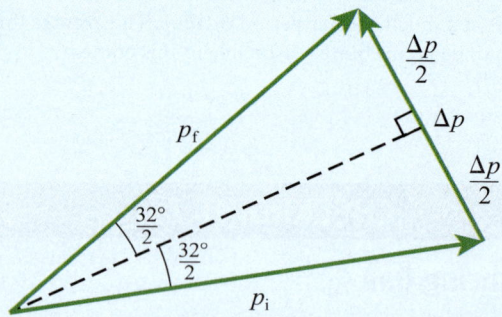

Figure 7-6 The change in momentum is made up of two equal parts, each given by $p_f \sin 16°$.

Making sense of the result

The largest possible change in momentum is $2p_f$, which would occur if the player deflected the puck by 180°. In this case, she deflected the puck by a much smaller angle and we expect a smaller change in momentum.

The Force of Impact

What force does a tennis ball experience as it bounces off a wall? For simplicity, we will consider a ball travelling horizontally when it hits a vertical wall. We also assume that the collision is elastic (neither the wall nor the tennis ball is irrecoverably deformed). Before the impact, the wall does not exert a force on the ball. As the ball strikes the wall, the force of the wall on the ball increases sharply to a peak value. During this first phase of the impact, the ball deforms somewhat. We assume that the wall is hard enough to not deform at all. The second phase of the impact starts after the force reaches its peak. The ball begins to reverse direction as it recovers from the deformation sustained during the collision. If the collision is completely elastic, the recovery of the ball will be complete. A plot of force versus time for such a collision is shown in Figure 7-7. This profile for the behaviour of the magnitude of the force with time is typical of elastic collisions.

The area under the force versus time curve corresponds to the integral of the force over the time during the collision. According to Equation 7-7, the integral is the total change in the momentum of the ball.

Linear Approximation for the Force of Impact

Since the variation of force with time during a collision is complex, integration of real-time data like those shown in Figure 7-7 is not easy, so the graph in Figure 7-7 is of limited practical use. However, we can often use a relatively simple linear model to calculate reasonable approximations.

Criteria The two parameters for our linear model are the duration of the impact and the total change in momentum. Our simplified model should produce a reasonable estimate of the average force of the impact. This model is based on a triangular graph, as shown in Figure 7-8(b). (The actual data are shown in Figure 7-8(a).) Using the properties of similar triangles, we can show that in this model, the maximum force is twice the average force. Example 7-6 will give us some experience using this approximation.

The degree to which an object can deform during a perfectly elastic collision can be quite spectacular. The sequence of images in Figure 7-10 shows a golf ball during various stages of impact with a metal plate.

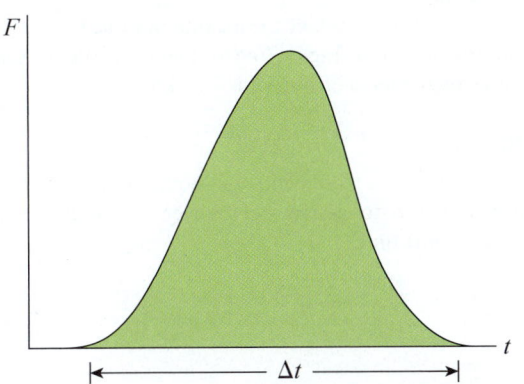

Figure 7-7 Force versus time for a ball bouncing off a wall. The green shaded area is the impulse received by the ball. When the force is at its maximum value, the ball is stationary and fully compressed against the wall.

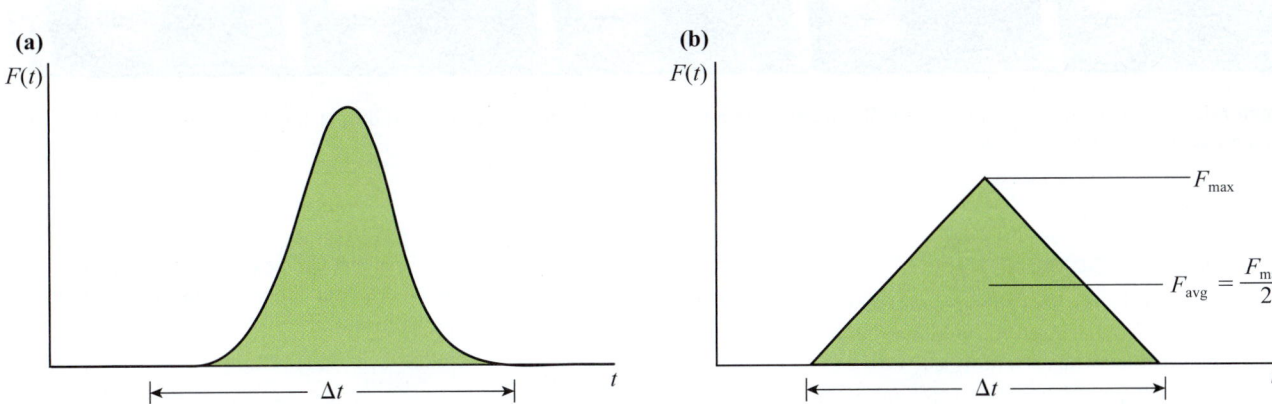

Figure 7-8 (a) Actual force versus time graph for a collision. (b) Linear model for force versus time. The linear model is constructed so that the impulse, given by the area under the force versus time curve, is the same for both. We can see that the maximum force in the linear model underestimates the actual maximum force observed in (a).

EXAMPLE 7-6

Linear Model of Impact Force

A golf ball with a mass of 46.0 g strikes a wall horizontally at a speed of 130. km/h and bounces straight back from the wall with the same speed. The collision takes 0.900 ms.

(a) Find the impulse delivered to the ball by the wall.
(b) Find the average force experienced by the ball during the collision.
(c) Assuming that the force changes linearly with time, plot a force versus time graph.
(d) Estimate the maximum value of the force of impact.

Solution

The situation is depicted in Figure 7-9(a). The initial momentum of the golf ball points to the right; after the collision, the golf ball moves to the left.

(a)   **(b)**

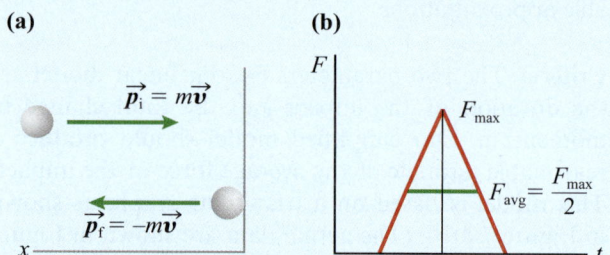

Figure 7-9 (a) The golf ball before and after the collision. (b) The linear model for the behaviour of force versus time.

(a) For convenience, we take left as the positive x-direction and convert the speed to metres per second. Then,

$$\vec{p}_i = (0.0460 \text{ kg})(-130. \text{ km} \cdot \text{h}^{-1})\left(\frac{1000 \text{ m} \cdot \text{km}^{-1}}{3600 \text{ s} \cdot \text{h}^{-1}}\right)\hat{i}$$

$$= -1.66 \text{ kg} \cdot \text{m} \cdot \text{s}^{-1}\hat{i}$$

$$\vec{p}_f = -\vec{p}_i = 1.66 \text{ kg} \cdot \text{m} \cdot \text{s}^{-1}\hat{i} \tag{1}$$

$$\vec{I} = \Delta\vec{p} = \vec{p}_f - \vec{p}_i = 3.32 \text{ kg} \cdot \text{m} \cdot \text{s}^{-1}\hat{i} \tag{2}$$

(b) From Equation 7-8, we have

$$\Delta\vec{p} = \vec{F}_{net}^{avg}\Delta t$$

$$\vec{F}_{net}^{avg} = \frac{\Delta\vec{p}}{\Delta t} = \frac{3.32 \text{ kg} \cdot \text{m} \cdot \text{s}^{-1}\hat{i}}{0.900 \times 10^{-3} \text{ s}} = 3.69 \text{ kN}\hat{i} \tag{3}$$

(c) Figure 7-9(b) shows the force as a function of time, taken to be linear.

(d) The average force is half the maximum value for the linear approximation in Figure 7-9(b); therefore, the maximum value of the force of impact is 7.38 kN.

Making sense of the result

The force calculated is tremendous—equivalent to the weight of a three quarter tonne truck. However, this force exists for only a very short time.

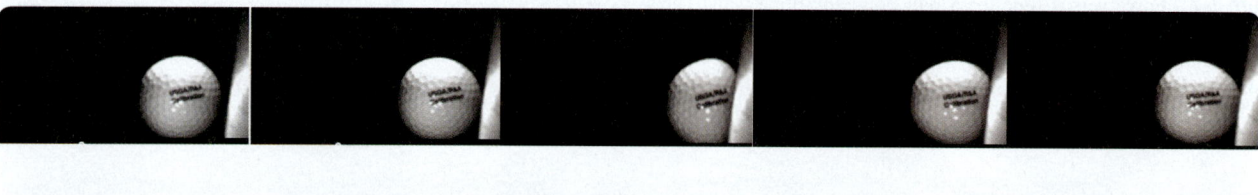

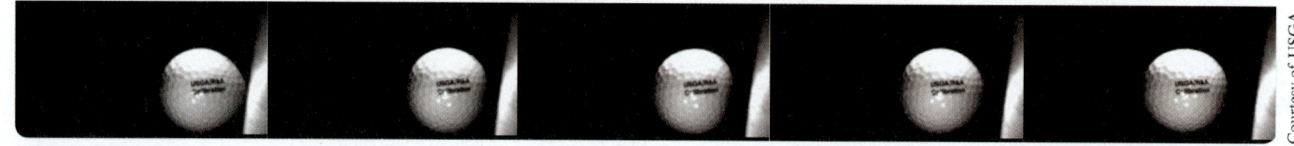

Figure 7-10 High-speed footage of a golf ball colliding with a wall. The frame-by-frame images (taken at small intervals) reveal details that human vision cannot otherwise detect.

EXAMPLE 7-7

Collision Duration

A very popular high school physics competition is the egg drop. Participants are challenged to build a structure to protect an egg from breaking when it is dropped.

(a) Calculate the impulse required to stop a 50.0 g egg that has been dropped from a height of 3.00 m.

(b) If the maximum force that can be exerted on an egg before it breaks is 100. N, determine the minimum collision time that the egg can withstand.

Solution

The idea behind a successful egg drop is to suspend/cushion the egg inside its container so that when the container hits the ground, the egg does not suddenly stop but decelerates more slowly than the container. The impulse received by the protected egg and an unprotected egg is the same—it is only the collision time that differs. We will determine the impulse by calculating the change in momentum. We can then use our simple linear impulse model, setting the maximum force to 100.0 N to find a collision time.

(a) The egg has fallen 3.00 m and thus will be travelling at a speed of

$$v_i = \sqrt{2 \cdot g \cdot 3 \text{ m}} = 7.67 \text{ m·s}^{-1}$$

Therefore, the change in momentum involved in coming to rest on the ground will be

$$\vec{p}_f - \vec{p}_i = 0 - (0.050 \text{ kg})(-7.67 \text{ m·s}^{-1} \hat{j})$$
$$= 0.384 \text{ kg·m·s}^{-1} \hat{j}$$

We have chosen upward to be the positive y-direction. Note that the impulse does not depend upon the details of the impact.

(b) We can now use Equation 7-8 to find the collision time. The average force in the linear impulse model is half the maximum force. Rearranging Equation 7-8 for the collision time, we get

$$\Delta t = \frac{|\vec{I}|}{|\vec{F}_{net}^{avg}|} = \frac{I}{\frac{1}{2} F_{max}} = \frac{0.384 \text{ kg·m·s}^{-1}}{50.0 \text{ kg·m·s}^{-2}}$$
$$= 7.68 \times 10^{-3} \text{ s}$$

Making sense of the result

We end up with a collision time of about 10 ms. Assuming the egg has a uniform acceleration during this time, it would travel a distance of

$$d = v_{avg}\Delta t = (v_i/2)\Delta t = (7.67 \text{ m·s}^{-1}/2)7.68 \times 10^{-3} \text{ s}$$
$$= 2.95 \times 10^{-2} \text{ m} \approx 3 \text{ cm}$$

Based on experience with egg drop competitions, this seems quite reasonable. 3 cm of padding is very easy to implement.

Modern safety features in cars, such as crumple zones and air bags, are designed to accomplish exactly the same thing. They increase the duration of the collision and thus decrease the maximum force experienced by the occupants of the car.

7-4 Systems of Particles and Centre of Mass

We now examine the principles of mechanics as they apply to systems of particles, or objects. So far, we have treated all objects as point masses, or particles. A **point mass** is an object with a non-zero mass but zero size. Although this approach was not realistic, it provided a simple platform for the application of physical principles. Now that we have mastered the basics of Newton's mechanics and the work–mechanical energy approach, we are in a good position to move on to a more sophisticated description of physical objects. Here, we will consider a system of point particles. Systems of particles are in effect a step in the transition from dealing with point masses to dealing with rigid objects, which we treat as continuous mass distributions in Chapter 8.

Centre of mass The first tool in our study of systems of particles is the concept of the **centre of mass**, the point corresponding to the average position of the mass in a system. For many calculations, the mass of a system or an object can be treated as a point mass, located at the centre of mass.

To find the centre of mass of the system, we add up the positions of the individual particles, with each position multiplied by the mass of that particle. We then divide by the total mass. If we denote the mass and position of the ith particle by m_i and x_i, respectively,

then in one dimension, the position of the centre of mass of a system of n particles is given by

KEY EQUATION

$$x_{cm} = \frac{m_1 x_1 + m_2 x_2 + \cdots + m_n x_n}{m_1 + m_2 + \cdots + m_n}$$

$$= \frac{\sum_{i=1}^{n} m_i x_i}{M_T} \tag{7-9}$$

where M_T is the total mass of all the particles in the system: $M_T = m_1 + m_2 + \cdots + m_n$. Thus, the centre of mass of a system of particles is the weighted average of the positions of these particles, where the weights are the masses of the particles. For example, the centre of mass of two equal masses a given distance apart is at the midpoint of the line separating them. If the mass of one of the particles is significantly larger than that of the other, the centre of mass is closer to the more massive object.

INTERACTIVE ACTIVITY 7-1

Centre of Mass

In this activity, you will use the PhET simulation "Collision Lab" to explore how the centre of mass of a two-object system depends on the relative masses of the objects.

EXAMPLE 7-8

Centre of Mass of Two Particles

Find the centre of mass of the system of two particles of masses $m_1 = 5m$ and $m_2 = m$ located a distance L apart, as shown in Figure 7-11(a).

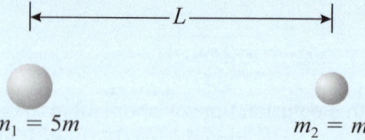

Figure 7-11(a) Two masses a distance L apart.

Solution

Our first step is to choose a coordinate system. For convenience, we choose the line connecting the two masses as the x-axis. We use the heavier of the two masses, m_1, as our origin (see Figure 7-11(b)) and choose right as the positive direction.

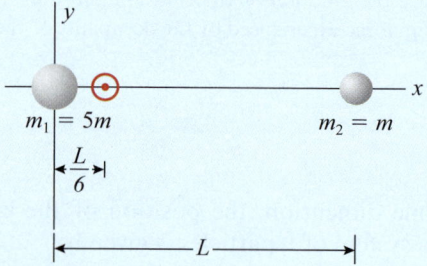

Figure 7-11(b) Mass m_1 is the origin of the coordinate system. The centre of mass is $L/6$ to the right.

Using Equation 7-9, we have

$$x_{cm} = \frac{m_1 x_1 + m_2 x_2}{m_1 + m_2} = \frac{5m(0) + m(L)}{5m + m} = \frac{L}{6}$$

We will now solve this same problem using the lighter of the two masses, m_2, as our origin (Figure 7-11(c)), but still taking right as the positive direction. Then,

$$x_{cm} = \frac{5m(-L) + m(0)}{5m + m} = -\frac{5L}{6}$$

Here, the negative value for x_{cm} tells us that the centre of mass is to the left of the origin.

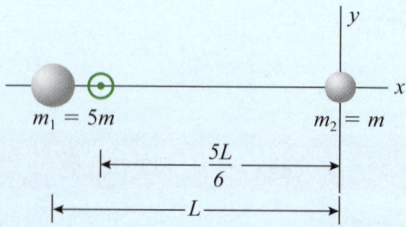

Figure 7-11(c) Mass m_2 is the origin of the coordinate system.

Making sense of the result

As we would expect, the centre of mass is between the two objects and closer to the more massive object. Whether we place our origin at the heavier or the lighter mass, the centre of mass will be a distance of $L/6$ from the heavier mass and $5L/6$ from the lighter mass.

For systems in more than one dimension, we have to use the position vector, $\vec{r}_i$, for each particle, and we get

KEY EQUATION

$$\vec{r}_{cm} = \frac{\sum_{i=1}^{n} m_i \vec{r}_i}{\sum_{i=1}^{n} m_i}$$

$$= \frac{1}{M_T}(m_1 \vec{r}_1 + m_2 \vec{r}_2 + \cdots + m_n \vec{r}_n) \qquad (7\text{-}10A)$$

We can also write Equation 7-10a in component form:

$$x_{cm} = \frac{\sum_{i=1}^{n} m_i x_i}{M_T}$$

$$y_{cm} = \frac{\sum_{i=1}^{n} m_i y_i}{M_T} \qquad (7\text{-}10b)$$

$$z_{cm} = \frac{\sum_{i=1}^{n} m_i z_i}{M_T}$$

The expression for the centre of mass of a *continuous* distribution of mass is

KEY EQUATION

$$\vec{r}_{cm} = \frac{\int_V \vec{r}\, dm}{\int_V dm} \qquad (7\text{-}11)$$

where the integration limit, V, indicates that the integral is taken over the volume of an object or a continuous distribution of particles.

It is important to choose a coordinate system to determine the centre of mass.

EXAMPLE 7-9

Three Equally Spaced Masses

Find the centre of mass of three particles located at the vertices of an equilateral triangle as shown in Figure 7-12(a). The length of each side of the triangle is L and all three particles have the same mass, m.

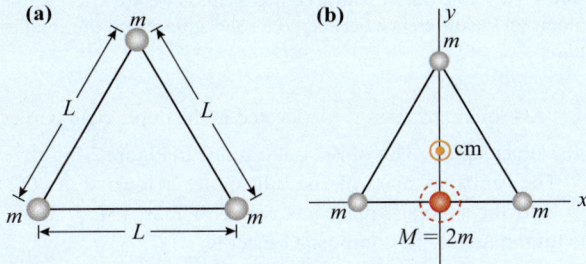

Figure 7-12 (a) Three equal masses at the corners of an equilateral triangle. (b) The centre of mass of the system of three equal masses.

Solution

First, we find the centre of mass of the two masses at the base of the triangle. From the symmetry of the triangle, we can see that this centre of mass is located in the middle of the base. To find the centre of mass of the whole system, we can replace masses m_1 and m_2 with a combined mass, $M = 2m$, at the middle of the base of the triangle. We pick this location as the origin of the coordinate system (Figure 7-12(b)).

Now, we have only two objects to deal with: mass $M = 2m$, located at the origin, and mass m_3, located at the top of the triangle. From Figure 7-12(b) we can see that the distance between m_3 and the origin is

$$y = L\cos 30° = L\frac{\sqrt{3}}{2}$$

Applying Equation 7-10b in the y-direction, we get

$$y_{cm} = \frac{2m(0) + mL\dfrac{\sqrt{3}}{2}}{3m} = \frac{\sqrt{3}}{6}L$$

Thus, the coordinates of the centre of mass of the system of three masses are $\left(0, \dfrac{\sqrt{3}}{6}L\right)$.

Making sense of the result

It makes sense for the centre of mass to be closer to the base than the apex.

EXAMPLE 7-10

A Long Thin Rod

The thin aluminum rod of a retort stand is 0.80 m in length and has a mass of 170 g. Find the centre of mass of the rod.

Solution

Because the mass is uniformly distributed in the rod, the centre of mass will lie on the axis of the rod, 40. cm from one end. However, it is useful to perform the calculation to get some experience with integration. In Equation 7-11, we are required to perform integrals over the volume of the rod. Since the rod is long and thin, we will treat it as a one-dimensional object and have it extend along the x-axis, with one end of the rod being placed at the origin, as illustrated in Figure 7-13.

We will first evaluate the integral in the denominator:

$$\int_V dm = \int_0^L \lambda dx = \lambda x \Big|_0^L = \lambda L = \frac{m}{L}L = m$$

This makes sense, because if we integrate dm over an entire object we should just get m, the mass of the object.

We will now tackle the numerator. In this one-dimensional case, the vector $\vec{r}$ reduces to the coordinate of the mass element, x:

$$\int_V \vec{r}\,dm = \int_0^L x\lambda dx = \lambda\frac{1}{2}x^2 \Big|_0^L = \frac{\lambda L^2}{2} = \frac{(\lambda L)L}{2} = \frac{mL}{2}$$

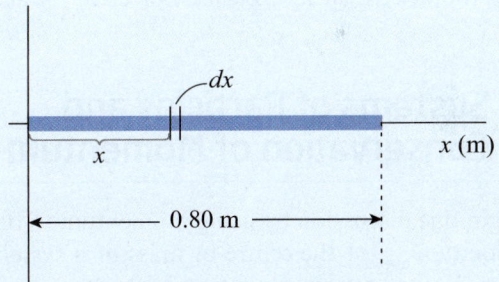

Figure 7-13 The thin aluminum rod extends along the x-axis. We consider a thin piece of it of length dx located a distance x from the origin. That element will have mass $dm = \lambda dx$, where λ is the linear mass density, defined as $\lambda = m/L$, m is the total mass of the rod, and L is the length of the rod. λ has units of $\text{kg} \cdot \text{m}^{-1}$.

Putting these two results together, we find that the position of the centre of mass for the rod is

$$x_{cm} = \frac{mL/2}{m} = \frac{L}{2} = 40.\ \text{cm}$$

Making sense of the result

As we expected, the centre of mass of this long thin rod is located at the centre of the rod.

CHAPTER 7 | **LINEAR MOMENTUM, COLLISIONS, AND SYSTEMS OF PARTICLES**

EXAMPLE 7-11

An Irregular Object

Find the centre of mass of the slab shown in Figure 7-14. The slab is homogeneous with a total mass of m and lower and left edges of length L.

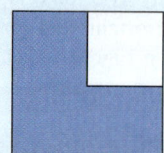

Figure 7-14   Example 7-11.

Solution

We can do this problem in various ways. We could treat the object as being made up of three identical squares. The centre of mass of each square would be in the centre of each square. To find the centre of mass of the entire object, we could treat each square as a point mass.

Choosing the lower left corner as the origin, the centres of mass of the three squares are located at ($L/4$, $L/4$), ($L/4$, $3L/4$), and ($3L/4$, $L/4$), respectively. The centre of mass of the entire object is then given by

$$x_{cm} = \frac{1}{m}\left(\frac{m}{3}\frac{L}{4} + \frac{m}{3}\frac{L}{4} + \frac{m}{3}\frac{3L}{4}\right) = \frac{5}{12}L$$

$$y_{cm} = \frac{1}{m}\left(\frac{m}{3}\frac{L}{4} + \frac{m}{3}\frac{3L}{4} + \frac{m}{3}\frac{L}{4}\right) = \frac{5}{12}L$$

Another way of doing this is to treat the object as comprising two objects: an $L \times L$ square of mass $\frac{4}{3}m$ and an

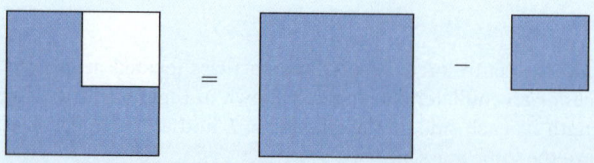

Figure 7-15   The object whose centre of mass we are after can be thought of the difference between the solid square and the smaller square.

$L/4 \times L/4$ square of mass $-\frac{1}{3}m$ located in the upper right corner of the larger square. This object is illustrated in Figure 7-15.

The centre of mass of the full square is located at ($L/2$, $L/2$), and the smaller square has centre of mass ($3L/4$, $3L/4$). The total mass of the composite object is

$$\frac{4}{3}m - \frac{1}{3}m = m$$

The coordinates of the centre of mass of this composite object are then

$$x_{cm} = \frac{1}{m}\left(\frac{4}{3}m\frac{L}{2} - \frac{1}{3}m\frac{3L}{4}\right) = \frac{5}{12}L$$

$$y_{cm} = \frac{1}{m}\left(\frac{4}{3}m\frac{L}{2} - \frac{1}{3}m\frac{3L}{4}\right) = \frac{5}{12}L$$

Making sense of the result

The centre of mass of the full square is ($L/2$, $L/2$) = ($6L/12$, $6L/12$). Removing the small square shifts the centre of mass a bit to the left and down.

7-5 Systems of Particles and Conservation of Momentum

We begin this discussion by using momentum to define the velocity, $\vec{v}_{cm}$, of the centre of mass of a system of particles. We can rewrite Equation 7-10a as

$$M_T\vec{r}_{cm} = m_1\vec{r}_1 + m_2\vec{r}_2 + \cdots + m_n\vec{r}_n$$

Taking the derivative of this with respect to time gives

$$M_T\vec{v}_{cm} = m_1\vec{v}_1 + m_2\vec{v}_2 + \cdots + m_n\vec{v}_n \quad (7\text{-}12a)$$

or

$$\vec{p}_{cm} = \vec{p}_1 + \vec{p}_2 + \cdots + \vec{p}_n \quad (7\text{-}12b)$$

From Equation 7-12a, we see that the velocity of the centre of mass of a system is really the weighted average of the velocities of the particles making up the system. Taking the derivative of Equation 7-12b with respect to time, we get

$$\frac{d\vec{p}_{cm}}{dt} = \frac{d\vec{p}_1}{dt} + \frac{d\vec{p}_2}{dt} + \cdots + \frac{d\vec{p}_n}{dt} \quad (7\text{-}13a)$$

For a system with constant mass, this is equivalent to taking the derivative of Equation (7-12a):

$$M_T\frac{d^2\vec{r}_{cm}}{dt^2} = m_1\frac{d^2\vec{r}_1}{dt^2} + m_2\frac{d^2\vec{r}_2}{dt^2} + \cdots + m_n\frac{d^2\vec{r}_n}{dt^2} \quad (7\text{-}13b)$$

In both Equations 7-13a and 7-13b, the right-hand side is the sum of all the individual net forces acting on the individual particles of the system. This sum is the net force acting on the system:

$$\frac{d\vec{p}_{cm}}{dt} = \vec{F}_{net} \quad (7\text{-}14)$$

The significance of Equation 7-14 is that, in the absence of a net force on a system of particles, the acceleration of the centre of mass of the system is zero.

Internal Forces and Systems of Particles

For a system of particles, the internal forces acting between particles do not contribute to the net force acting on the system as presented in Equation 7-14.

Internal forces between any pair of particles are action–reaction pairs. Action–reaction forces are, by definition, equal in magnitude and opposite in direction, so when we add all the forces acting between the individual particles in a system, the total is zero. Therefore, internal forces do not affect the acceleration of the centre of mass of a system, and Equation 7-14 can be written unambiguously as

KEY EQUATION
$$\frac{d\vec{p}_{cm}}{dt} = \vec{F}_{ext} \tag{7-15}$$

The implication of Equation 7-15 is that where the **external force** is zero, the momentum of the system of particles does not change:

KEY EQUATION if $\vec{F}_{ext} = 0$, then $\dfrac{d\vec{p}_{cm}}{dt} = 0 \Rightarrow \vec{p}_i = \vec{p}_f$ (7-16)

This result—**conservation of momentum**—is a very useful and important one. If we can identify a mechanical system that experiences no net external force, then the momentum of that system is conserved.

Defining the System

It is important to clearly define the system we are working with and to identify which of the forces acting on the system are external and which are internal. For the linear momentum of a system to be conserved, there can be no net external force acting on it. For problems involving conservation of momentum, we will therefore choose the system such that the forces involved are internal to the system. The total linear momentum of such a system is conserved.

During a collision between two objects, the force of the collision changes the momentum of each object. Therefore, the momentum of each object is not conserved during the collision, although the *total* momentum of the two objects is conserved. The forces exerted by the objects on each other are forces that are internal to the two-object system. To apply conservation of momentum in such a situation, we need to consider the system to comprise both objects and conserve the total momentum, which is the sum of the momenta of the two objects.

For example, when a dog walks in a canoe floating in a lake, the forces between the dog and the canoe are internal to the dog–canoe system and do not affect the system's total momentum. If we neglect the force of friction between the canoe and the water, we may use conservation of momentum. This may seem a bit odd, as clearly there is also a force of gravity and the force exerted by the dog, both of which act downward on the canoe. However, there is an equal but opposite buoyant force acting upward on the canoe, so there is no net force external to the system.

CHECKPOINT 7-9

Jumping off a Sled

You are standing on a dogsled, which sits on an icy surface just outside Oslo, Norway, in early December. You jump off the back of the sled to tend to your dogs. The sled, not tethered to the dogs, moves forward. Ignoring the friction between the sled and the ice, the centre of mass of the system consisting of you and the sled

(a) remains still
(b) moves forward
(c) moves backward
(d) may move either forward or backward depending on whether you or the sled is more massive

ANSWER: (a) There is no net external force acting on the system of you and the dogs, so conservation of momentum dictates that the centre of mass does not move.

EXAMPLE 7-12

Stranded Astronaut

An astronaut's manoeuvring unit fails while she is inspecting the nose of an orbiter in orbit. Another astronaut uses the orbiter's robotic arm to move the stranded astronaut from the nose section to the airlock. Derive an expression for how far the orbiter moves as a result of this action.

Solution

The force that the robotic arm exerts on the astronaut (and, therefore, the force that the astronaut exerts on the arm and the orbiter) is internal to the astronaut–orbiter system. According to Equation 7-15, the acceleration of the centre of mass of the astronaut–orbiter system is zero, because the external, or net, force acting on the system is zero. Since the centre of mass of the two objects is originally stationary (in their orbiting coordinate system), the centre of mass of the orbiter–astronaut system will remain stationary.

Let us draw a diagram of the orbiter and the astronaut before and after the astronaut moves (Figure 7-16).

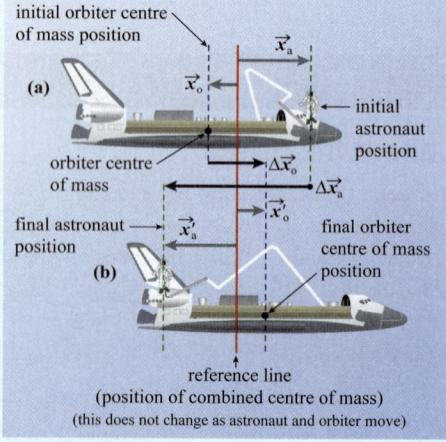

initial orbiter centre of mass position

$\vec{x}_a$

(a)

$\vec{x}_o$

orbiter centre of mass

initial astronaut position

$\Delta\vec{x}_o$

$\Delta\vec{x}_a$

final astronaut position

$\vec{x}'_a$

$\vec{x}'_o$

final orbiter centre of mass position

(b)

reference line
(position of combined centre of mass)
(this does not change as astronaut and orbiter move)

Figure 7-16 The orbiter and the astronaut (a) before and (b) after the astronaut moves.

We can use Equation 7-10 to locate the centre of mass of the astronaut and the orbiter before and after the robotic arm moves the astronaut:

$$M_T\vec{x}_{cm} = m_a\vec{x}_a + m_o\vec{x}_o \quad (1)$$

$$M_T\vec{x}'_{cm} = m_a\vec{x}'_a + m_o\vec{x}'_o \quad (2)$$

Subtract (1) from (2):

$$M_T\vec{x}'_{cm} - M_T\vec{x}_{cm} = m_a\vec{x}'_a - m_a\vec{x}_a + m_o\vec{x}'_o - m_o\vec{x}_o$$

$$M_T(\vec{x}'_{cm} - \vec{x}_{cm}) = m_a(\vec{x}'_a - \vec{x}_a) + m_o(\vec{x}'_o - \vec{x}_o) \quad (3)$$

$$M_T\Delta\vec{x}_{cm} = m_a\Delta\vec{x}_a + m_o\Delta\vec{x}_o$$

Because the total momentum is conserved, the centre of mass of the system does not move; therefore, $\Delta\vec{x}_{cm} = 0$. It appears we have one equation with two unknowns at this point. However, we have yet to incorporate the fact that we know how far the astronaut has moved on the orbiter. If we say she has moved the length of the orbiter, L, and we choose the positive direction to be to the right, then the displacement of the astronaut can be written as

$$\Delta\vec{x}_a = -L\hat{i} + \Delta\vec{x}_o \quad (4)$$

We substitute (4) into (3) to find

$$M_T\Delta\vec{x}_{cm} = 0 = m_o\Delta\vec{x}_o + m_a\Delta\vec{x}_o - m_aL\hat{i}$$

$$\Delta\vec{x}_o = \frac{m_a}{m_o + m_a}L\hat{i} \quad (5)$$

We can substitute this result into (4) to find the displacement of the astronaut:

$$\Delta\vec{x}_a = -L\hat{i} + \frac{m_a}{m_o + m_a}L\hat{i}$$

$$= \frac{-m_o - m_a + m_a}{m_o + m_a}L\hat{i}$$

$$= \frac{-m_o}{m_o + m_a}L\hat{i}$$

Making sense of the result

Since the mass of the orbiter is much greater than the mass of the astronaut, $\Delta x_o \ll \Delta x_a \approx L$.

CHECKPOINT 7-10

Sun Orbits Earth

If we assume that the external forces acting on the Sun–Earth system are negligible, we can say that, as Earth orbits the Sun,

(a) the Sun remains stationary with respect to the centre of mass of the Sun–Earth system
(b) the centre of mass of the Sun–Earth system moves back and forth
(c) the Sun orbits the centre of mass of the Earth–Sun system
(d) the motion of Earth does not affect the Sun's motion

ANSWER: (c) Both Earth and Sun orbit the centre of mass.

EXAMPLE 7-13

Applying Conservation of Momentum

A bear cub ($m = 21.0$ kg) is sitting on a wagon ($M = 31.0$ kg), which is moving at a speed of $v = 2.00$ m/s. The cub then jumps backward off the wagon with a speed of 3.00 m/s with respect to the wagon. Ignoring the effects of friction, find the speed of the cub and the wagon with respect to the ground after the cub jumps.

Solution

We will consider the system consisting of the bear cub and the wagon, as shown in Figure 7-17.

When the bear cub jumps off the wagon, the force that the cub exerts on the wagon and the force that the wagon exerts on the cub are internal to the cub–wagon system. Since we are ignoring the effects of friction and there are no other net external forces acting on the system at this time, the linear momentum of the cub–wagon system is conserved as the bear cub jumps off the wagon:

$$\vec{p}_i = \vec{p}_f \tag{1}$$

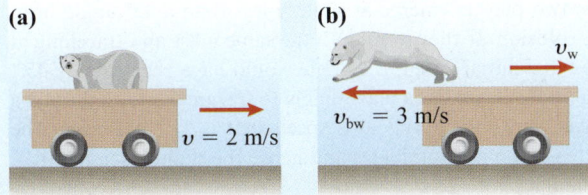

Figure 7-17 The wagon and the bear cub (a) just before the cub jumps and (b) just after the cub jumps.

Let m be the mass of the cub and M be the mass of the wagon. Initially, the bear cub and the wagon are both moving with the same velocity, $\vec{v}$, so we can write $\vec{p}_i = (m + M)\vec{v}$. If we let $\vec{v}_b$ and $\vec{v}_w$ be the velocities of the bear cub and the wagon, respectively, after the cub jumps off, then we can write $\vec{p}_f = m\vec{v}_b + M\vec{v}_w$. Using these results, (1) becomes

$$(m + M)\vec{v} = m\vec{v}_b + M\vec{v}_w \tag{2}$$

We will select our x-axis as pointing horizontally to the right. We are told that the velocity of the cub with respect to the wagon after the cub jumps is $\vec{v}_{bw} = -3 \text{ m·s}^{-1}\hat{i}$. The velocity of the bear cub with respect to the ground after the bear cub jumps is then

$$\vec{v}_b = \vec{v}_w - \vec{v}_{bw} = \vec{v}_w - 3 \text{ m·s}^{-1}\hat{i} \tag{3}$$

We substitute (3) into (2) and solve for $\vec{v}_w$:

$$(m + M)\vec{v} = m\vec{v}_b + M\vec{v}_w$$
$$= m(\vec{v}_w - 3 \text{ m·s}^{-1}\hat{i}) + M\vec{v}_w$$
$$= m\vec{v}_w - m \cdot 3 \text{ m·s}^{-1}\hat{i} + M\vec{v}_w$$
$$\vec{v}_w = \frac{(m + M)\vec{v} + m \cdot 3 \text{ m·s}^{-1}\hat{i}}{m + M}$$
$$= 3.21 \text{ m·s}^{-1}\hat{i}$$

Here we have used $\vec{v} = 2 \text{ m·s}^{-1}\hat{i}$. We can substitute our result for the velocity of the wagon into either (2) or (3) to find the final velocity of the bear cub. We will use (3):

$$\vec{v}_b = \vec{v}_w - 3 \text{ m·s}^{-1}\hat{i}$$
$$= 3.21 \text{ m·s}^{-1}\hat{i} - 3 \text{ m·s}^{-1}\hat{i}$$
$$= 0.21 \text{ m·s}^{-1}\hat{i}$$

Making sense of the result

We can check that momentum is conserved. The momentum before the jump is $(m + M)\vec{v} = (21 \text{ kg} + 31 \text{ kg})(2 \text{ m·s}^{-1}\hat{i}) = 104 \text{ kg·m·s}^{-1}\hat{i}$. The momentum after the jump is $m\vec{v}_b + M\vec{v}_w = (21 \text{ kg})(0.21 \text{ m·s}^{-1}\hat{i}) + (31 \text{ kg})(3.21 \text{ m·s}^{-1}\hat{i}) = 104 \text{ kg·m·s}^{-1}\hat{i}$, where we have rounded appropriately. Thus, momentum is conserved.

CHECKPOINT 7-11

Leaping Bear

In Example 7-13,
(a) the system had more kinetic energy after the bear cub jumped
(b) the system had more kinetic energy before the bear cub jumped
(c) the system's kinetic energy does not change when the frictional forces are negligible
(d) the bear cub exerts more force on the wagon than the wagon exerts on the bear cub

ANSWER: (a) The bear does positive work.

7-6 Collisions

Inelastic Collisions

An **inelastic collision** is a collision during which the colliding objects lose kinetic energy. The energy is lost in the form of thermal energy, sound, and irreversible deformation of the objects. In a completely inelastic collision, the two colliding objects stick together. Although kinetic energy is lost during the collision, the momentum of the system is conserved. The less elastic a collision is, the greater the proportion of the kinetic energy lost.

EXAMPLE 7-14

Rugby Tackle

Figure 7-18   Example 7-14.

A 60. kg rugby player is running at a speed of 9.0 m·s⁻¹ when she is tackled by a 70. kg player who is running at 8.0 m·s⁻¹ in a direction perpendicular to that of the first player. The players emerge from the tackle together at the same velocity. Find the magnitude and direction of the velocity with which they emerge from the tackle.

Solution

We will assume that the only forces that act in this tackle are those between the players themselves and that we can conserve momentum in the tackle. This collision occurs in two dimensions, as the player's initial velocities are perpendicular to each other. Thus, we will have to take care to select a coordinate system and keep track of the vector nature of the momenta. Finally, we identify this as a completely inelastic collision because the players stick together after the tackle.

Figure 7-19 shows the two players before and after the tackle. We have selected a coordinate system in which, before the collision, the player with the ball is running in the y-direction and the tackle is running in the x-direction.

Let m be the mass of the ball carrier and M be the mass of the tackle. The momentum of the ball carrier is $\vec{p}_{\text{ball}} = mv_{\text{ball}}\hat{j}$, while that of the tackle is $\vec{p}_{\text{tackle}} = Mv_{\text{tackle}}\hat{i}$. After the collision, the ball carrier and the tackle travel together at the same velocity, $\vec{v}$, so the final momentum can be written as $\vec{p}_{\text{ball+tackle}} = (m + M)\vec{v}_{\text{ball+tackle}}$. Applying conservation of momentum, we can write

$$\vec{p}_{\text{ball}} + \vec{p}_{\text{tackle}} = \vec{p}_{\text{ball+tackle}}$$

$$mv_{\text{ball}}\hat{j} + Mv_{\text{tackle}}\hat{i} = (m + M)\vec{v}_{\text{ball+tackle}}$$

Solving for $\vec{v}_{\text{ball+tackle}}$, we get

$$\vec{v}_{\text{ball+tackle}} = \frac{M}{m + M}v_{\text{tackle}}\hat{i} + \frac{m}{m + M}v_{\text{ball}}\hat{j}$$

$$= \frac{70.\text{ kg}}{60.\text{ kg} + 70.\text{ kg}}(8.0\text{ m·s}^{-1}\hat{i})$$

$$+ \frac{60.\text{ kg}}{60.\text{ kg} + 70.\text{ kg}}(9.0\text{ m·s}^{-1}\hat{j})$$

$$= 4.31\text{ m·s}^{-1}\hat{i} + 4.15\text{ m·s}^{-1}\hat{j}$$

The magnitude of this velocity is

$$v_{\text{ball+tackle}} = \sqrt{(4.31\text{ m·s}^{-1})^2 + (4.15\text{ m·s}^{-1})^2}$$

$$= 6.0\text{ m·s}^{-1}$$

and it makes an angle with respect to the positive x-axis of

$$\theta = \tan^{-1}\left(\frac{4.15\text{ m·s}^{-1}}{4.31\text{ m·s}^{-1}}\right) = 44°$$

Making sense of the result

The two players emerge at slightly less than a 45° angle from the collision. If they were both the same mass and travelling at the same initial speed, the angle would have been exactly 45°. The 70 kg player has a mass that is 16.6% greater than the 60 kg player, but only travels at a speed that is 11.1% slower. The increase in mass dominates and hence the slightly smaller than 45° angle.

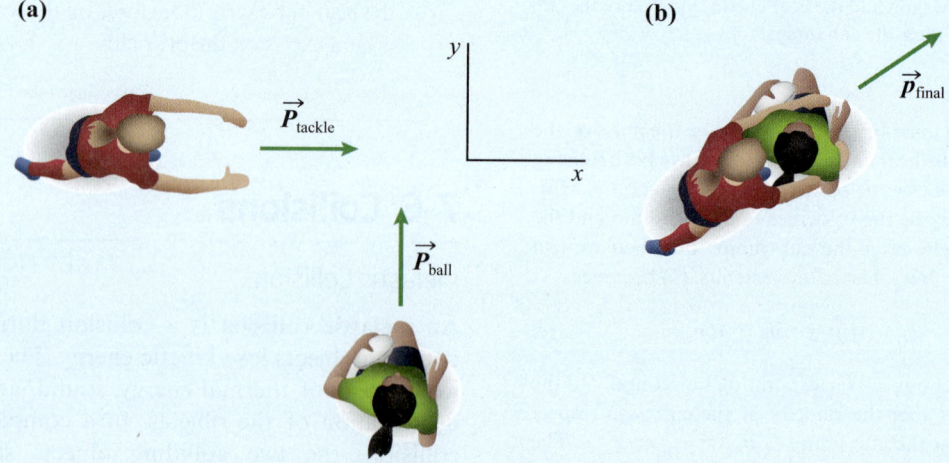

(a)   **(b)**

Figure 7-19   Two rugby players before (a) and after (b) a tackle. After the tackle, they move off together with the same velocity.

Cosmic Collisions

Stars readily form in the arms of a spiral galaxy because these regions have relatively high densities and strong gravitational forces (Chapter 11). Spiral galaxies are subclassified by their structure. As shown in Figure 7-20, the Whirlpool galaxy (catalogued as NGC 5194 and as M51) has only two prominent spiral arms, which makes it a Grand Design spiral galaxy, in contrast to spiral galaxies that have many arms and often a less-defined spiral structure. The clear spiral structure of the Whirlpool galaxy was likely enhanced by an interaction with the smaller nearby galaxy, NGC 5195. The two galaxies may have undergone two collisions, one a few hundred million years ago and the other about 100 million years ago, as NGC 5195 crossed the disk of the Whirlpool galaxy. The interaction between the two galaxies created regions of even higher gravitational attraction, promoting spectacular star formation near the lagging edge of the arms of the spirals of the Whirlpool galaxy. The Whirlpool galaxy is about 31 million light years away, in the constellation Canis Venatici (the Hunting Dogs). Its orientation with respect to Earth and its relative closeness have given astronomers a great opportunity to study the birth of stars within spiral galaxies.

Figure 7-20 An image of NGC 5194 and 5195 taken in January 2005 with the Advanced Camera for Surveys (ACS) aboard the Hubble Space Telescope. The pink colour coding represents intense star-forming regions.

EXAMPLE 7-15

Inelastic Collision

Children are playing with a small ball of steel of mass m_b suspended on a string such that the ball hangs just above a slippery plastic mat. The children slide a magnet of mass m_M along the mat to strike the steel ball. When the magnet strikes the steel ball at speed v_M, the magnet sticks to the ball, causing it to swing in an arc from its rest position. Assume that all friction in the system is negligible.

(a) Derive an expression for the height, h, that the ball and the magnet reach after the collision.

(b) Calculate the height using the following values: $m_M = 92.0$ g, $m_b = 190.$ g, and $v_M = 7.00$ m/s.

Solution

It is very useful to draw a diagram showing the situation before the magnet collides with the ball and after the ball–magnet combination has swung up to its maximum height.

We tackle the problem by analyzing the inelastic collision to find the speed, and hence the kinetic energy, of the magnet–ball combination just as it emerges from the collision. We can then find the height it rises to by converting all of that kinetic energy into potential energy. Momentum is not conserved in the magnet–ball system as it rises, because the string and gravity exert forces that are external to the system.

During the collision, the magnet and the steel ball effectively become one object of mass $m_M + m_b$ moving with velocity $\vec{v}_{M+b}$. Therefore,

$$\vec{p}_i = \vec{p}_f \tag{1}$$

$$m_M\vec{v}_M = (m_M + m_b)\vec{v}_{M+b} \tag{2}$$

Choosing the direction of motion as the positive x-axis, we have

$$m_M v_M \hat{i} = (m_M + m_b)v_{M+b}\hat{i} \tag{3}$$

After the collision, the total mechanical energy is conserved:

$$\Delta E_m = \Delta U + \Delta K = 0 \tag{4}$$

The kinetic energy of the system is converted into gravitational potential energy as the ball with the magnet attached swings upward:

$$\Delta U = -\Delta K \tag{5}$$

$$(m_M + m_b)gh = -(K_f - K_i) = -\left(0 - \frac{1}{2}(m_M + m_b)v_{M+b}^2\right) \tag{6}$$

$$= \frac{1}{2}(m_M + m_b)v_{M+b}^2$$

$$h = \frac{v_{M+b}^2}{2g} \tag{7}$$

(continued)

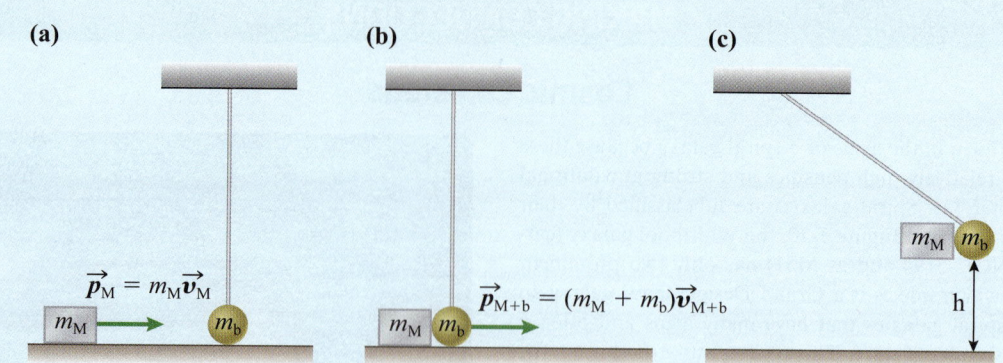

(a) $\vec{p}_M = m_M\vec{v}_M$

(b) $\vec{p}_{M+b} = (m_M + m_b)\vec{v}_{M+b}$

(c) m_M m_b h

Figure 7-21 The three stages of the process. In (a) the magnet is sliding along the frictionless surface with momentum $\vec{p}_M = m_M\vec{v}_M$. In (b) the magnet has undergone a completely inelastic collision with the ball. The magnet and ball stick together and emerge from the collision with the same velocity. We can write the momentum as $\vec{p}_{M+b} = (m_M + m_b)\vec{v}_{M+b}$. In (c), the ball–magnet combination has swung up and stopped. All of the kinetic energy in the system has been converted to gravitational potential energy, $U = (m_M + m_b)gh$.

(b) Substituting the given values into (3) gives

$$v_{M+b} = \frac{m_M v_M}{m_M + m_b} = \frac{(0.0920 \text{ kg})(7.00 \text{ m/s})}{(0.0920 + 0.190) \text{ kg}} = 2.284 \text{ m/s} \quad (8)$$

and

$$h = \frac{v_{M+b}^2}{2g} = \frac{(2.284 \text{ m/s})^2}{2(9.81 \text{ m/s}^2)} = 0.266 \text{ m} \quad (9)$$

Making sense of the result

The total initial mechanical energy is the kinetic energy of the magnet, $K_{initial} = \frac{1}{2}m_M v_M^2 = 2.25 \text{ J}$. The total final mechanical

energy is the potential energy, $U = (m_M + m_b)gh = 0.736 \text{ J}$. This seems okay, as the collision is inelastic and does not conserve energy. The total initial momentum has magnitude

$$p_M = m_M v_M = (0.0920 \text{ kg})(7.00 \text{ m·s}^{-1}) = 0.644 \text{ kg·m·s}^{-1}$$

After the collision, the momentum of the magnet-ball is

$$\begin{aligned}p_{M+b} &= (m_M + m_b)v_{M+b} \\ &= (0.0920 \text{ kg} + 0.190 \text{ kg})(2.284 \text{ m·s}^{-1}) \\ &= 0.644 \text{ kg·m·s}^{-1}\end{aligned}$$

so we have conserved momentum properly in the collision.

EXAMPLE 7-16

Dynamics Carts and Springs

A 1.45 kg dynamics cart travelling at 4.80 m·s^{-1} collides with a stationary dynamics cart of mass 1.78 kg. The second cart has a spring with spring constant 4000. N·m^{-1} attached to it to cushion the collision. Assuming that friction with the ground can be ignored, find

(a) the maximum compression in the spring
(b) the maximum force on either cart during the collision
(c) the maximum acceleration experienced by the lighter cart during the collision

Solution

(a) As the incoming cart strikes the spring attached to the stationary cart, the spring begins to compress, thus exerting a force on each cart. As a result, the incoming cart slows down, and the stationary cart speeds up until both carts reach the same speed, v. At the instant the two carts have the same speed, the distance between the two

carts does not change and the spring reaches its maximum compression.

The system comprises the two carts and the spring. Since the forces during the collision are internal to this system, the momentum is conserved, and

$$\vec{p}_i = \vec{p}_f$$

where $\vec{p}_i$ is the momentum of the system before the collision and $\vec{p}_f$ is the momentum at the instant when the spring is at maximum compression.

Then

$$m_1\vec{v}_1 = (m_1 + m_2)\vec{v}$$

$$v = \frac{m_1 v_1}{m_1 + m_2} = \frac{(1.45 \text{ kg})(4.80 \text{ m·s}^{-1})}{1.45 \text{ kg} + 1.78 \text{ kg}} = 2.15 \text{ m·s}^{-1}$$

Since the spring cushions the impact and the spring force is conservative, the total mechanical energy of the system is conserved. Some kinetic energy changes into elastic potential energy in the spring, but this energy is not lost from the system. Thus, when the carts are moving at the same speed, we can say

$$\frac{1}{2}m_1v_1^2 = \frac{1}{2}kx^2 + \frac{1}{2}(m_1 + m_2)v^2$$

$$\therefore x = \sqrt{\frac{m_1v_1^2 - (m_1 + m_2)v^2}{k}}$$

$$= \sqrt{\frac{(1.45\ \text{kg})(4.80\ \text{m}\cdot\text{s}^{-1})^2 - (1.45\ \text{kg} + 1.78\ \text{kg})(2.15\ \text{m}\cdot\text{s}^{-1})^2}{4000.\ \text{kg}\cdot\text{s}^{-2}}}$$

$$= 0.0680\ \text{m}$$

(b) The maximum force exerted by one cart on the other equals the force of the spring at maximum compression.

The magnitude of the force can be obtained using Hooke's law:

$$F = kx = (4000.\ \text{N}\cdot\text{m}^{-1})(0.0680\ \text{m}) = 272\ \text{N}$$

(c) The maximum acceleration experienced by the lighter cart is

$$a = \frac{kx}{m} = \frac{272\ \text{N}}{1.45\ \text{kg}} = 188\ \text{m/s}^2$$

Making sense of the result

The spring is quite stiff and thus the collision time is short, leading to a large force and subsequent acceleration.

Elastic Collisions

During an **elastic collision**, both the total linear momentum and the total kinetic energy of the system are conserved. We will analyze an elastic collision in one dimension.

Suppose a marble of mass m moving with velocity $\vec{v}$ collides elastically with a marble of mass M moving with velocity $\vec{u}$, as depicted in Figure 7-22(a). Depending on the masses and the initial speeds, after the collision, the two marbles will bounce away from one another (Figure 7-22(b)), both move to the left, or both move to the right.

While we have drawn Figure 7-22(a) with the two marbles initially having velocities in opposite directions, in general this is not necessary for a collision to occur. They will collide if the distance between them is decreasing, which can happen even if they are moving in the same direction. In our analysis, we will not need to make any assumptions about the directions of the initial velocities.

Since linear momentum and velocity are vectors, we need to establish a coordinate system. We take right as positive.

Conservation of Momentum

The total linear momentum of the system of two marbles is the vector sum of their individual momenta:

$$\vec{p}_i = m\vec{v} + M\vec{u}$$

After the collision, the expression for the final momentum of the system is

$$\vec{p}_f = m\vec{v}' + M\vec{u}'$$

where the primed variables indicate the final values. Because the collision takes place along a line in one

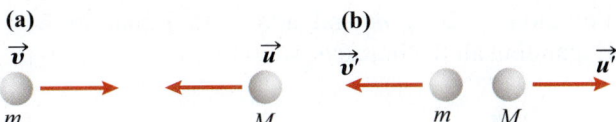

(a) **(b)**

Figure 7-22 (a) The two marbles head toward each other. (b) The two marbles bounce away from each other.

dimension, we can rewrite the equations in terms of the vector components of the momenta and velocities:

$$p_i = mv + Mu$$
$$p_f = mv' + Mu'$$

Since we have written the velocities in terms of their vector components, if a particle is moving to the right, the component of its velocity is positive; if it is moving to the left, it is negative. We have left off subscripts specifying a particular direction to keep the notation clean.

The forces involved during the collision between the two marbles are internal to the two-marble system. Since there are no external forces involved, the total linear momentum of the system is conserved during the collision, and

$$p_i = p_f$$
or   (7-17)
$$mv + Mu = mv' + Mu'$$

Conservation of energy Since the collision is elastic, kinetic energy is conserved. Therefore, the kinetic energy of the system before the collision is equal to the kinetic energy after the collision:

$$\frac{1}{2}mv^2 + \frac{1}{2}Mu^2 = \frac{1}{2}mv'^2 + \frac{1}{2}Mu'^2 \qquad (7-18)$$

This equation involves the squares of the initial and final speeds. Gathering the terms in Equation 7-18 by mass, we have

$$m(v^2 - v'^2) = M(u'^2 - u^2) \qquad (7\text{-}19)$$

Factoring the differences of squares, we get

$$m(v - v')(v + v') = M(u' - u)(u' + u) \qquad (7\text{-}20)$$

Now we regroup the terms in Equation 7-17 in the same way:

$$m(v - v') = M(u' - u) \qquad (7\text{-}21)$$

If we divide Equation 7-20 by Equation 7-21, we have our second linear equation:

$$v + v' = u' + u \qquad (7\text{-}22)$$

Between Equations 7-21 and 7-22, we now have two equations with our two unknowns. We multiply Equation 7-22 by m and add it to Equation 7-21. Expanding all the brackets, we get

$$mv + mv' + mv - mv' = Mu' - Mu + mu' + mu \qquad (7\text{-}23)$$

Solving for u', we obtain

$$u' = \frac{2mv + (M - m)u}{M + m} \qquad (7\text{-}24)$$

Similarly, if we multiply Equation 7-22 by M and subtract from Equation 7-21, we can solve for v':

$$v' = \frac{2Mu + (m - M)v}{M + m} \qquad (7\text{-}25)$$

These can be written in a more symmetric form as

KEY EQUATION

$$u' = \frac{2m}{m + M}v + \frac{M - m}{m + M}u$$
$$v' = \frac{m - M}{m + M}v + \frac{2M}{m + M}u \qquad (7\text{-}26)$$

MAKING CONNECTIONS

Laser Cooling of Atoms

Interestingly, although photons have no mass, they do carry momentum, as we show in Section 31-10. A direct application of this is the laser cooling of atoms. The higher the temperature of a gas, the more kinetic energy its atoms will have and, hence, the higher their speed. Similarly, gas molecules move more slowly at lower temperatures. Since photons have momentum, a laser can be used to cool a gas by slowing down its atoms. The laser is fired at atoms in the gas. Incoming photons at the correct wavelength are absorbed by atomic electrons that are moving toward the laser. The momentum of the absorbed photon transfers to the atom, thus slowing the atom. The target atom cools further every time it absorbs a photon. The electron that absorbs the photon moves to a higher state of energy. If the photon is subsequently re-emitted by the excited atom, the momentum of the photon has a random direction. Energy is conserved in this process, and the collision is elastic. Carefully calibrated lasers (such as those in Figure 7-23) can cool atoms to a few tenths of a microkelvin and slow them down to the point where they are effectively "trapped." Researchers began cooling atoms with lasers in 1978 and reached temperatures below 40 K. They achieved temperatures about one millionth of this just 10 years later. This technology eventually led to better atomic clocks and the observation of an ultracold state of matter. The 1997 Nobel Prize in Physics was awarded to Steven Chu, Claude Cohen-Tannoudji, and William D. Phillips for their work in developing technologies to use laser light to cool and trap atoms, and the 2001 Nobel prize in physics was awarded to Eric A. Cornell, Wolfgang Ketterie, and Carl E. Wieman for the achievement of Bose–Einstein condensation.

H. Mark Helfer, NIST

Figure 7-23 A cloud of cold sodium atoms (bright spot at centre) floats in a trap at the National Institute of Standards and Technology.

The lowest temperature ever recorded to date is 0.5 nK, reached by an MIT research team lead by Ketterie, Aaron E. Leanhardt, and David E. Pritchard.

EXAMPLE 7-17

An Elastic Collision

A 100. g marble is moving to the right at a speed of $3.00 \text{ m} \cdot \text{s}^{-1}$. It undergoes an elastic collision with a 200. g marble moving to the left at a speed of $4.00 \text{ m} \cdot \text{s}^{-1}$. Find the velocities of the marbles after the collision.

Solution

Although this is a straightforward application of Equation 7-26, we will check to ensure that both momentum and kinetic energy are conserved to verify our calculation.

The total initial momentum is

$$p_i = mv + Mu$$
$$= (0.100 \text{ kg})(3.00 \text{ m} \cdot \text{s}^{-1}) + (0.200 \text{ kg})(-4.00 \text{ m} \cdot \text{s}^{-1})$$
$$= 0.300 \text{ kg} \cdot \text{m} \cdot \text{s}^{-1} - 0.800 \text{ kg} \cdot \text{m} \cdot \text{s}^{-1}$$
$$= -0.500 \text{ kg} \cdot \text{m} \cdot \text{s}^{-1}$$

Note that the left-moving marble has a negative initial velocity. The total initial kinetic energy is

$$K_i = \frac{1}{2}mv^2 + \frac{1}{2}Mu^2$$
$$= \frac{1}{2}(0.100 \text{ kg})(3.00 \text{ m} \cdot \text{s}^{-1})^2 + \frac{1}{2}(0.200 \text{ kg})(-4.00 \text{ m} \cdot \text{s}^{-1})^2$$
$$= 0.450 \text{ J} + 1.60 \text{ J} = 2.05 \text{ J}$$

We use Equation 7-26 to find the final velocities:

$$u' = \frac{2(0.100 \text{ kg})}{0.100 \text{ kg} + 0.200 \text{ kg}}(3.00 \text{ m} \cdot \text{s}^{-1})$$
$$+ \frac{0.200 \text{ kg} - 0.100 \text{ kg}}{0.100 \text{ kg} + 0.200 \text{ kg}}(-4.00 \text{ m} \cdot \text{s}^{-1}) = 0.667 \text{ m} \cdot \text{s}^{-1}$$

$$v' = \frac{0.100 \text{ kg} - 0.200 \text{ kg}}{0.100 \text{ kg} + 0.200 \text{ kg}}(3.00 \text{ m} \cdot \text{s}^{-1})$$
$$+ \frac{2(0.200 \text{ kg})}{0.100 \text{ kg} + 0.200 \text{ kg}}(-4.00 \text{ m} \cdot \text{s}^{-1}) = -6.33 \text{ m} \cdot \text{s}^{-1}$$

We will check to see if the total momentum is the same after the collision.

$$p_f = mv' + Mu'$$
$$= (0.100 \text{ kg})(-6.33 \text{ m} \cdot \text{s}^{-1}) + (0.200 \text{ kg})(0.667 \text{ m} \cdot \text{s}^{-1})$$
$$= -0.500 \text{ kg} \cdot \text{m} \cdot \text{s}^{-1}$$

We can see that momentum is conserved. We will now check to see if the kinetic energy is conserved:

$$K_f = \frac{1}{2}mv'^2 + \frac{1}{2}Mu'^2$$
$$= \frac{1}{2}(0.100 \text{ kg})(-6.33 \text{ m} \cdot \text{s}^{-1})^2 + \frac{1}{2}(0.200 \text{ kg})(0.667 \text{ m} \cdot \text{s}^{-1})^2$$
$$= 2.003445 \text{ J} + 0.044889 \text{ J}$$
$$= 2.05 \text{ J}$$

We have kept a lot of digits at the intermediate step to avoid rounding errors.

Making sense of the result

The more massive particle is moving left at a higher speed, so it makes sense that the total momentum is negative. However, we note that there has been a significant transfer of energy and momentum from the 200. g particle to the 100. g particle, which is an important characteristic of collisions—energy gets transferred from more energetic particles to less energetic particles.

7-7 Variable Mass and Rocket Propulsion

Rocket propulsion provides an example of a system with variable mass. Here, the *system* consists of the rocket, its fuel, and the exhaust gas. In the rocket engine, the burning fuel turns into hot exhaust gases, primarily carbon dioxide and water vapour. The forces generated by the combustion of the fuel are internal to the system, so the linear momentum of the system is conserved as the fuel burns.

When this exhaust is directed out the back of the rocket, it carries with it some momentum. For momentum to be conserved, the rocket must acquire an equal but opposite momentum.

Consider a rocket that fires its engine in deep space. Assume that the engine burns fuel at a constant rate, R, and that the exit speed, v_{er}, of the exhaust gas relative to the rocket is also constant. Since the rocket is in deep space, we do not have to consider external forces, such as air drag, which would be significant in a launch

from Earth. We also assume that gravitational forces are negligible.

Since there are no external forces acting on the rocket–fuel–exhaust system and all the forces generated by burning the fuel are internal to the system, the total momentum of the system is conserved. Therefore, the momentum at time $t + dt$ is

$$\vec{p}(t + dt) = \vec{p}(t) = \text{constant} \qquad (7\text{-}27)$$

Since the rocket moves along a line, we will consider all of our velocities and momenta in one dimension and chose the direction of travel of the rocket to be positive. At a given instant, the total mass of the rocket including the remaining fuel is M, as shown in Figure 7-24(a). Figure 7-24(b) shows the rocket at a time dt later. The speed of the rocket has increased to $v + dv$, the mass of the exhaust released during the interval dt is dm, and the velocity of the exhaust gas is $\vec{v}_e$. The mass of the rocket plus fuel has decreased by dm to $M - dm$.

At time t, the momentum of the rocket and unburned fuel is Mv. After the interval dt, the

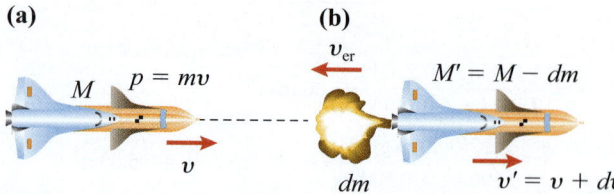

(a)

(b)

Figure 7-24 (a) A rocket moving at time t. (b) The rocket moving at time $t + dt$.

momentum of the rocket and the remaining fuel is $M'v' = (M - dm)(v + dv)$. Using these expressions to find the momenta in Equation 7-27 gives

$$p(t) = p(t + dt)$$
$$Mv = (M - dm)(v + dv) + dm(v_e) \qquad (7\text{-}28)$$

The left-hand side of this equation is the total momentum at the instant t. The right-hand side is the total momentum at time $t + dt$. Thus, Equation 7-28 expresses conservation of linear momentum for the rocket–fuel system.

The quantity v_e is the x-component of the velocity of the exhaust gas measured with respect to a stationary coordinate system. We can eliminate this from our equation by expressing it in terms of the x-component of the velocity of the rocket and the constant speed of the exhaust gas relative to the rocket:

$$v_e = v' - v_{er} = (v + dv) - v_{er} \qquad (7\text{-}29)$$

The velocity of the exhaust with respect to the rocket, $\vec{v}_{er}$, is undoubtedly negative because it points to the left. However, the direction of the velocity of the exhaust, with respect to a stationary observer, depends on how fast the rocket is moving. When the speed of the rocket is low, a stationary observer would expect to see the rocket move to the right and the exhaust gas move to the left. However, when the speed of the rocket is greater than the exit speed of the exhaust, both the rocket and the exhaust move to the right relative to a stationary observer. We assume that the exhaust is indeed moving to the right with respect to a stationary observer, hence the plus sign for the exhaust momentum in Equation 7-28.

Substituting Equation 7-29 into Equation 7-28, we get

$$Mv = (M - dm)(v + dv) + dm(v + dv - v_{er})$$
$$= Mv + Mdv - vdm - dmdv + vdm + dmdv - v_{er}dm$$
$$= Mv + Mdv - v_{er}dm$$
$$Mdv = v_{er}dm \qquad (7\text{-}30)$$

The decrease in the mass of the rocket and the remaining fuel, M, is the same as the increase in the total mass of exhaust, m. Thus, $dM = -dm$, and substituting this into Equation 7-30, we get

$$Mdv = -v_{er}dM \qquad (7\text{-}31)$$

Rearranging this gives

$$dv = \frac{-dM}{M}v_{er} \qquad (7\text{-}32)$$

To find the change in speed of the rocket after a given amount of fuel has been burned, we integrate Equation 7-32, remembering that v_{er} is a constant, with the initial and final mass as our limits of integration:

$$\Delta v = \int_{v_i}^{v_f} dv = -v_{er}\int_{M_i}^{M_f}\frac{dM}{M}$$
$$= -v_{er}\ln M \Big|_{M_i}^{M_f}$$
$$= -v_{er}(\ln M_f - \ln M_i) \qquad (7\text{-}33)$$
$$= v_{er}\ln\left(\frac{M_i}{M_f}\right)$$

Since $\Delta v = v_f - v_i$,

KEY EQUATION
$$v_f - v_i = v_{er}\ln\left(\frac{M_i}{M_f}\right) \qquad (7\text{-}34)$$

Equation 7-34 is called the **ideal rocket equation**. The change in speed is proportional to the logarithm of the ratio of the masses. Since the initial mass of the rocket with fuel is greater than the final mass, the final speed is faster than the initial speed.

We can use Equation 7-31 and Newton's second law to find the force, or thrust, that acts on the rocket:

$$M\frac{dv}{dt} = \frac{dm}{dt}v_{er} \qquad (7\text{-}35a)$$

$$Ma = Rv_{er} \qquad (7\text{-}35b)$$

The term Rv_{er} is called the thrust of the rocket engine. We can see that to get large thrust, we need a large fuel burn rate R and a large exhaust gas speed. To achieve high final velocities, Equation 7-35 tells us that in addition to a large exhaust gas speed, we also need a large mass ratio, which suggests mostly fuel and little payload.

CHECKPOINT 7-12

What Drives a Rocket?

When a rocket burns fuel for propulsion,

(a) the ejected exhaust is external to the rocket–fuel system, and the total momentum of the rocket and fuel decreases

(b) the driving force is external to the system

(c) the rocket is propelled by the speed of the exhaust, not the reduction in mass

(d) the rocket acquires a momentum that is equal but opposite to the momentum of the ejected exhaust

ANSWER: (d) by conservation of momentum

MAKING CONNECTIONS

Rockets and the Space Age

Getting into space has relied on rocket technology because all other propulsion systems employed on aircraft require an atmosphere to operate in. Despite a somewhat difficult start in the late 1950s, rocket technology is now quite reliable and is used for many purposes besides space exploration. Rockets are used to put satellites, which are used for communication and navigation purposes, into orbit. In addition, rockets are used to deploy scientific instrumentation into orbit that allows scientists to pursue various studies (Figure 7-25).

Figure 7-25 NASA launches between 20 and 30 rocket flights per year as part of its Sounding Rocket Program. This allows scientists to conduct research in various areas, including plasma physics, solar physics, planetary atmospheres, galactic astronomy, high-energy astrophysics, and microgravity research.

EXAMPLE 7-18

Applying the Ideal Rocket Equation

A space probe has a total mass of 1.30×10^3 kg, 5.20×10^2 kg of which is fuel. The probe is at rest near the edge of the solar system. The probe fires its rocket thrusters to make a final course correction, burning the fuel at a constant rate until it is gone. The burn lasts 1.00×10^2 s and produces a constant thrust of 11.0 kN. Estimate

(a) the initial acceleration of the probe
(b) the speed of the probe 80.0 s into the burn
(c) the maximum speed of the probe
(d) the maximum acceleration of the probe

Solution

If we assume that all external forces acting on the probe are negligible, we can apply the ideal rocket equation. To do so, we need to find the rate of fuel consumption, R, and the effective exhaust speed, y_{er}. Since the rate of fuel consumption is constant and the rocket burns 5.20×10^2 kg of fuel in 1.00×10^2 s,

$$R = \frac{dm}{dt} = \frac{5.20 \times 10^2 \text{ kg}}{1.00 \times 10^2 \text{ s}} = 5.20 \text{ kg} \cdot \text{s}^{-1}$$

We can determine the effective exhaust speed from the thrust, $T = Ma$ (Equation 7-35b):

$$T = Rv_{er}$$

$$\therefore v_{er} = \frac{T}{R} = \frac{1.1 \times 10^4 \text{ N}}{5.20 \text{ kg} \cdot \text{s}^{-1}} = 2.12 \times 10^3 \text{ m} \cdot \text{s}^{-1}$$

(a) Assume the probe is so far away from any celestial bodies that the only significant force acting on the probe is the thrust from the rocket. Therefore,

$$T = Ma$$

$$\therefore a = \frac{T}{M} = \frac{1.10 \times 10^4 \text{ N}}{1.30 \times 10^3 \text{ kg}} = 8.46 \text{ m} \cdot \text{s}^{-2}$$

Note that to find the initial acceleration of the probe, we have used the total initial mass of the probe.

(b) The mass of the probe and the remaining fuel after 80.0 s is

$$M(80 \text{ s}) = M(0 \text{ s}) - Rt$$

$$= 1.30 \times 10^3 \text{ kg} - (5.20 \text{ kg} \cdot \text{s}^{-1})(80.0 \text{ s})$$

$$= 8.84 \times 10^2 \text{ kg}$$

We can use this mass in Equation 7-34, recalling that the initial speed is zero, to find

$$v_f(80.0 \text{ s}) = v_{er} \ln\left(\frac{M_i}{M_f}\right)$$

$$= (2.12 \times 10^3 \text{ m} \cdot \text{s}^{-1}) \ln\left(\frac{1.30 \times 10^3 \text{ kg}}{8.84 \times 10^2 \text{ kg}}\right)$$

$$= 8.18 \times 10^2 \text{ m} \cdot \text{s}^{-1}$$

(c) The maximum speed of the rocket occurs when the rocket has burned all its fuel, at which point the mass of the probe is 1.30×10^3 kg $- 5.20 \times 10^2$ kg $= 7.80 \times 10^2$ kg. We can use Equation 7-34 again to find

$$v_f(1.00 \times 10^2 \text{ s}) = v_{er} \ln\left(\frac{M_i}{M_f}\right)$$

$$= (2.12 \times 10^3 \text{ m} \cdot \text{s}^{-1}) \ln\left(\frac{1.30 \times 10^3 \text{ kg}}{7.80 \times 10^2 \text{ kg}}\right)$$

$$= 1.08 \times 10^3 \text{ m} \cdot \text{s}^{-1}$$

(continued)

(d) Since the thrust is constant, the maximum acceleration will occur when the probe is least massive—just as the rocket runs out of fuel. At this instant, the mass of the probe is 7.80×10^2 kg and we can use our acceleration equation from part (a) to find

$$a_{max} = \frac{T}{M_{min}} = \frac{1.10 \times 10^4 \text{ N}}{7.80 \times 10 \text{ kg}} = 14.1 \text{ m} \cdot \text{s}^{-2}$$

Making sense of the result

We can calculate the average acceleration by dividing the change in speed by the time for the burn to find

$$a_{avg} = \frac{\Delta v}{\Delta t} = \frac{1.08 \times 10^3 \text{ m} \cdot \text{s}^{-1}}{1.00 \times 10^2 \text{ s}} = 10.8 \text{ m} \cdot \text{s}^{-2}.$$ This value falls somewhere between our calculated minimum and maximum values and thus they seem reasonable.

KEY CONCEPTS AND RELATIONSHIPS

Linear Momentum
Momentum is a vector:

$$\vec{p} = m\vec{v} \tag{7-1}$$

The net force on an object results in a change in the object's momentum:

$$\vec{F}_{net} = \frac{d\vec{p}}{dt} \tag{7-4}$$

or

$$\vec{F}_{net}^{avg} = \frac{\Delta \vec{p}}{\Delta t} \tag{7-5}$$

Impulse is equal to the change in an object's momentum:

$$\vec{I} = \Delta \vec{p} = \int_{t_1}^{t_2} \vec{F}_{net} dt \tag{7-7}$$

or

$$\vec{I} = \Delta \vec{p} = \vec{F}_{net}^{avg} \Delta t \tag{7-8}$$

Systems of Particles and Centre of Mass
The centre of mass in one dimension is given by

$$x_{cm} = \frac{m_1 x_1 + m_2 x_2 + \cdots + m_n x_n}{m_1 + m_2 + \cdots + m_n}$$

$$= \frac{\sum_{i=1}^{n} m_i x_i}{M_T} \tag{7-9}$$

The centre of mass in more than one dimension is given by

$$\vec{r}_{cm} = \frac{\sum_{i=1}^{n} m_i \vec{r}_i}{\sum_{i=1}^{n} m_i} \tag{7-10a}$$

For an object with a continuous mass distribution, the centre of mass is given by

$$\vec{r}_{cm} = \frac{\int_V \vec{r} \, dm}{\int_V dm} \tag{7-11}$$

Systems of Particles and Conservation of Momentum

$$\frac{d\vec{p}_{cm}}{dt} = \vec{F}_{ext} \tag{7-15}$$

Internal forces do not affect the momentum of a system of particles:

$$\text{if } \vec{F}_{ext} = 0, \text{ then } \frac{d\vec{p}_{cm}}{dt} = 0 \Rightarrow \vec{p}_i = \vec{p}_f \tag{7-16}$$

Inelastic Collisions
The momentum of a system comprising colliding particles is conserved:

$$\vec{p}_i = \vec{p}_f$$

Kinetic energy is not conserved in an inelastic collision.

Elastic Collisions
The momentum of a system comprising colliding particles is conserved:

$$\vec{p}_i = \vec{p}_f$$

Kinetic energy is conserved in an elastic collision:

$$K_i = K_f$$

The final velocities in a one-dimensional two-particle elastic collisions are given by

$$u' = \frac{2m}{m + M}v + \frac{M - m}{m + M}u$$
$$v' = \frac{m - M}{m + M}v + \frac{2M}{m + M}u \tag{7-26}$$

Rocket Propulsion
The ideal rocket equation is

$$v_f - v_i = v_{er} \ln\left(\frac{M_i}{M_f}\right) \tag{7-34}$$

Rocket engine thrust is given by

$$Ma = R v_{er} \tag{7-35b}$$

APPLICATIONS

rocket propulsion, collisions

KEY TERMS

centre of mass, conservation of momentum, elastic collision, external forces, ideal rocket equation, inelastic collision, impulse, internal forces, linear momentum, point mass, rocket propulsion

QUESTIONS

For all questions and problems involving springs, assume that the masses of the springs are negligible unless otherwise stated.

1. A bullet is stopped by a wall. Which of the following statements is true?
   (a) The change in the bullet's momentum is opposite in direction to the incoming momentum.
   (b) The change in the bullet's momentum is equal to the incoming momentum.
   (c) Both (a) and (b) are correct.
   (d) None of the above are correct.

2. A car makes a three quarter turn around a traffic circle at a constant speed. The car is originally moving in the positive x-direction. Which of the following indicates the car's change in momentum?
   (a) $mv\hat{i} + mv\hat{j}$
   (b) $mv\hat{i} - mv\hat{j}$
   (c) $-mv\hat{i} + mv\hat{j}$
   (d) $-mv\hat{i} - mv\hat{j}$

3. A hockey puck of mass m moves at speed v on ice. A hockey player deflects the puck such that its speed stays the same, but the puck is now moving at 60° with respect to its incident direction. What is the magnitude of the impulse?
   (a) zero
   (b) $mv \sin 30°$
   (c) mv
   (d) $mv \cos 30°$

4. If the puck in question 3 is moving along the positive x-axis before the shot, then the final momentum can be written as
   (a) $\dfrac{mv}{2}\hat{i} - \dfrac{\sqrt{3}mv}{2}\hat{j}$
   (b) $\dfrac{\sqrt{3}mv}{2}\hat{i} + \dfrac{mv}{2}\hat{j}$
   (c) $\dfrac{\sqrt{3}mv}{2}\hat{i} - \dfrac{mv}{2}\hat{j}$
   (d) none of the above

5. For the puck in question 4, the impulse is
   (a) $\dfrac{mv}{2}\hat{i} + \dfrac{\sqrt{3}mv}{2}\hat{j}$
   (b) $-\dfrac{mv}{2}\hat{i} + \dfrac{mv}{2}\hat{j}$
   (c) $-\dfrac{mv}{2}\hat{i} + \dfrac{\sqrt{3}mv}{2}\hat{j}$
   (d) $\dfrac{mv}{2}\hat{i} - \dfrac{\sqrt{3}mv}{2}\hat{j}$

6. A bouncy ball of mass $m = 100$ g and speed $v = 10$ m/s is incident on a wall at an angle of 30° with the normal to the wall. It then bounces off the wall at the same speed, still making an angle of 30° with the normal to the wall after the collision. The collision takes 1 ms. The average force experienced by the ball during the collision is
   (a) 1000 N into the wall
   (b) 1000 N out of the wall
   (c) zero because the speed does not change
   (d) $\sqrt{3}$ (1000) N parallel to the wall

7. A bouncy ball of mass m and speed v impacts on a wall at an angle of 60°. The component of its velocity perpendicular to the wall does not change, and the angle it makes with the wall after the collision is 30°. Is this possible? Explain.

8. A slightly deflated ball hits a wall at an angle of 60° and leaves at an angle of 30°. The ratio of the outgoing momentum to the incoming momentum is
   (a) $\dfrac{\sqrt{3}}{2}$
   (b) $\dfrac{1}{\sqrt{3}}$
   (c) $\dfrac{2\sqrt{13}}{13}$
   (d) none of the above

9. A sponge ball strikes a wall and leaves with a speed that is $\dfrac{\sqrt{13}}{4}$ of its incoming speed. If the incident ball makes an angle of 30° with the wall, then the component of its momentum perpendicular to the wall changes by factor of
   (a) $\dfrac{\sqrt{3}}{2}$
   (b) $\dfrac{\sqrt{13}}{4}$
   (c) $\dfrac{1}{2}$
   (d) none of the above

10. What happens to the centre of mass of a firecracker when it explodes?

11. When you drop a tennis ball onto Earth's surface, the collision is completely elastic and lasts a few milliseconds. Will the ball bounce back to the same height? Explain.

12. You drop a 30. g bouncy ball from a height of 11 m onto a flat, horizontal surface. The ball is in contact with the surface for 20. ms. Find the average force Earth exerts on the ball if the ball rebounds with the same speed.

13. Suppose Earth were the only planet in the solar system. As Earth orbited the Sun,
    (a) the Sun would remain stationary
    (b) the Sun would always move in the same direction as Earth
    (c) the Earth–Sun centre of mass would move back and forth
    (d) both Earth and the Sun would orbit the Earth–Sun centre of mass

14. If one of the stars in a binary system explodes without directly damaging the second star, immediately after the explosion
    (a) the centre of mass of the two stars will move closer to the star that remained intact
    (b) the centre of mass of the two stars will be at the centre of the intact star
    (c) the centre of mass will shift, depending on the explosion
    (d) the orbit of the intact star will not be affected by the explosion if the intact star does not absorb any material from the other star

15. If one of the stars in a binary star system explodes, and the intact star absorbs a good portion of the material from the first star, then
    (a) the centre of mass of the exploding star is not affected by the explosion
    (b) the centre of mass of the two-star system is not affected by the explosion
    (c) the centre of mass of the two-star system will move closer to the intact star as the material from the exploding star gets absorbed by the intact star

16. When you throw a ball straight down so that it bounces perfectly elastically from the surface of Earth, the duration of the impact will dictate how high the ball goes. True or false? Explain.

17. A flamingo lands vertically on the back of a floating hippopotamus and then takes a few steps along the hippopotamus's back toward its head. Which of the following statements is true?
    (a) The hippopotamus will move in a direction opposite to the flamingo's motion.
    (b) The hippopotamus will not move.
    (c) The centre of mass of the hippopotamus–flamingo system will move in the same direction that the hippopotamus moves.
    (d) The centre of mass of the hippopotamus–flamingo system will move in the same direction as the flamingo's motion.

18. You go for a sprint across a soccer field, heading east. Assuming that you are the only mobile object on Earth's surface, Earth's rotation would be
    (a) slightly slower than usual
    (b) slightly faster than usual
    (c) unaffected
    (d) faster or slower, depending on the coordinate system

19. A cannon fires a probe into the Moon such that the probe leaves Earth's surface at 90° to the surface. Which of the following statements is true?
    (a) Earth will not be affected at all.
    (b) Earth will recoil slightly.
    (c) Since the force of gravity is involved, momentum is not conserved.
    (d) None of the above are true.

20. A child jumps onboard a stationary wagon. The losses due to friction between the wagon and the floor and in the wagon's wheels are negligible. Neither the child's body nor the wagon sustains an irrecoverable deformation, and the child and the wagon move as one. Which of the following statements is true?
    (a) The collision is elastic because there is no irrecoverable deformation.
    (b) The two objects stick to one another, so the collision must be inelastic.
    (c) The kinetic energy of the child is reduced by an amount that is smaller than the kinetic energy gained by the wagon.
    (d) The child's momentum is reduced by an amount that is larger than the amount by which the wagon's momentum increases.

21. For question 20, is the following statement true or false? The only mechanism for the kinetic energy to be lost is friction between the child and the wagon. (Ignore losses to sound.)

22. A man standing on a wagon that is originally at rest begins to run. If there are no losses in the wagon motion to friction,
    (a) the momentum of the man relative to the ground is the same in magnitude as the momentum of the wagon with respect to the ground
    (b) the momentum of the man with respect to the wagon is the same in magnitude as the momentum of the wagon with respect to the ground
    (c) the speed of the man with respect to the ground is greater than the speed of the man with respect to the wagon
    (d) it depends on the friction between the man and the wagon

23. A well-behaved cat goes on a canoeing trip with her owners. She is left in the canoe as they set up camp. She sees a fish jump in front of the canoe and leaps forward to try to catch it. The canoe is originally at rest, and we take the resulting motion of the canoe in the horizontal direction to be negative. Which of the following statements is true?
    (a) The velocity of the cat in the horizontal direction is positive.
    (b) The velocity of the cat in the horizontal direction is negative.
    (c) Whether (a) or (b) is correct depends on how fast the canoe moves.
    (d) Whether (a) or (b) is correct depends on the ratio of the cat's mass to the canoe's mass.

24. A sled is moving on horizontal frictionless ice when a dog onboard leaps backward onto the ice. The direction of motion of the sled is taken to be positive. Whether the velocity of the dog with respect to the ground is positive or negative depends on
    (a) the masses of the sled and the dog
    (b) the original speed of the sled and the relative speed of the dog with respect to the sled
    (c) both (a) and (b)
    (d) none of the above

25. A rocket is fired vertically upward from Earth's surface. If you chose a system in which momentum is conserved during the process, the most correct system to pick would be
(a) rocket + fuel/exhaust
(b) rocket + fuel/exhaust + atmosphere
(c) rocket + fuel/exhaust + atmosphere + Earth
(d) rocket + fuel/exhaust + atmosphere + Milky Way galaxy

26. A canoeist notices her canoe drifting toward a waterfall. She has a battery-powered water pump fitted with a hose that she can use to direct the water from the pump. The pump draws water vertically up from the river. Which of the following statements is true?
(a) If the pump shoots out enough water at a high enough speed, the canoeist can angle the hose so that she can move away from the waterfall.
(b) The pump cannot propel the canoe because the momentum of the system must be conserved.
(c) Conservation of momentum does not apply here because the pump has a power source.
(d) None of the above are true.

PROBLEMS BY SECTION

For problems, star ratings will be used (★, ★★, or ★★★), with more stars meaning more challenging problems. The following codes will indicate if $\frac{d}{dx}$ differentiation, ∫ integration, ▢ numerical approximation, or ∿ graphical analysis will be required to solve the problem.

Section 7-1 Linear Momentum
27. ★ Calculate the linear momentum of a space craft of mass 2040 tonnes moving at a speed of 27 000 km/h.
28. ★ Find the momentum of a 60. g stone that starts from rest and falls to the ground from a height of 20. m.

Section 7-2 Rate of Change of Linear Momentum and Newton's Laws
29. ★★ The engine of a fire rescue boat gets damaged during a rescue operation. The captain decides to use the firefighting water-pumping system to push the boat back to the harbour. The pump is mounted on the deck and draws the water vertically out of the ocean. The crew points the nozzle of the hose toward the rear of the boat. The mass of the boat is 180 000 kg, and the pump sprays water at a rate of 330 kg/min with a muzzle speed of 20. m/s. Ignoring water resistance, find the acceleration of the rescue boat.
30. ★★ An athlete of mass 74 kg is running inside a train car of mass 2100 kg, at a speed of 9.0 m/s with respect to the ground. While the athlete is running, the car is stationary. The athlete then comes to a stop over 0.20 s.
During this time, the force of kinetic friction is the only force between her feet and the train car. Find the average force exerted on the train car by the athlete.

Section 7-3 Impulse
31. ★ A soccer player kicks a 425 g soccer ball straight on with an average force of 78.0 N. The speed of the soccer ball right after the collision is 41.0 m/s. Find the duration of the collision. Ignore the effects of gravity.

32. ★ Tennis balls, whose individual masses are 58 g, are fired at a wall by a tennis-ball launcher such that they impact the wall horizontally in perfectly elastic collisions at a speed of 117 km/h.
(a) The launcher fires the balls at a rate of 55 balls/min. Find the average force exerted on the wall as a result.
(b) The average collision time is 25 ms. Find the average force per collision.

33. ★★ A somewhat deflated volleyball ($m = 220$ g) falls on the floor from a height of 2.0 m, starting from rest. The collision between the ball and the floor takes 0.30 s and is perfectly elastic.
(a) Find the speed of the ball on the way up after the impact.
(b) Write an expression for the net force acting on the ball using the linear behaviour model.
(c) Sketch the net force acting on the ball as a function of time, using the linear behaviour model.
(d) Find the maximum value for the net force.

34. ★★ Starting from rest, a 25 g bouncy ball falls from a height of 10. m onto a flat, horizontal surface and rebounds with 99% of its incident speed. Find the amount of time that the ball was in contact with the ground if the average net force on the ball during the collision is 45 N.

35. ★ Figure 7-26 shows the net force versus time profile for an object colliding with a horizontal surface. The object's mass is 300. g, and its incident speed is 13 m/s. Find the recoil speed.

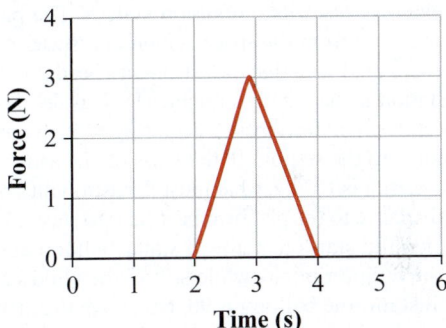

Figure 7-26 Problem 35.

Section 7-4 Systems of Particles and Centre of Mass
36. ★ Sixteen identical point masses are placed at equal intervals such that they outline the circumference of a circle of radius R. By how much does the centre of mass shift when one of the masses is removed?
37. ★ Find the centre of mass of the particles shown in Figure 7-27.

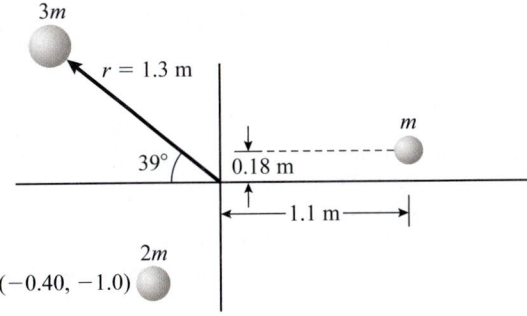

Figure 7-27 Problem 37.

38. ★ Masses m, $2m$, $3m$, and $4m$ are located at the corners of a square of side L, as shown in Figure 7-28. Locate the centre of mass of this system.

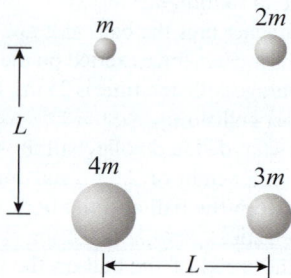

Figure 7-28 Problem 38.

39. ★ A pendulum is fashioned out of a thin bar of length 0.55 m and mass 1.9 kg, welded to a sphere of radius 0.11 m and mass 0.86 kg. Find the centre of mass of the composite object.

Section 7-5 Systems of Particles and Conservation of Momentum

40. ★★ A 4.0 kg dog runs at constant speed from one end of a 21 kg canoe to the other, a distance of 4.9 m, in 3.1 s. Assuming negligible water resistance, how far does the canoe move?

41. ★★ An astronaut finds himself stranded on a 43 kg platform designed for short hops around a space station, when the platform loses its propulsion system. The platform is moving away from the space station at a speed of 2.0 m/s and is oriented such that the astronaut's head points toward the station as he stands straight. He then decides to kick his feet down in the hopes of acquiring enough momentum to return to the station. If the mass of the astronaut with his spacesuit is 105 kg, what must the astronaut's minimum speed relative to the platform be if he is to succeed?

42. ★★ A child starts to throw a water balloon at a friend, but the balloon bursts while still in the child's hand. At that instant, the balloon is 90. cm above the ground and moving horizontally at a speed of 2.3 m/s. Where is the centre of mass of the water when all the water has hit the ground? (The balloon is very light.)

43. ★★ A 71.0 kg astronaut inside a 2.00 tonne spacecraft at rest kicks off from one end at a speed of 2.50 m/s, and stops when she reaches the other end having travelled the full length of the 9 m long spacecraft. How far does the spacecraft move?

Section 7-6 Collisions

44. ★ Two huskies collide on frictionless ice. Anyu, with a mass of 27 kg, is heading due east at a speed of 6.5 m/s. Anut, with a mass of 21 kg, is heading due north at a speed of 4.0 m/s. Find the speed of the two dogs after they collide. Assume the collision to be completely inelastic.

45. ★ Two cars in a completely inelastic head-on collision are at rest immediately after the collision. The mass of one car is 1.7 times the mass of other car. Find the ratio of their initial speeds.

46. ★★ An 11 g bullet is fired with speed v into a 68 kg ballistic pendulum. The wires holding the pendulum deviate from the vertical by 27°. The vertical distance from the point of impact to the suspension points for the wires is 92 cm. Find the speed with which the bullet hits the pendulum.

47. ★★ A tennis ball hits a wall at an angle of 48° with respect to the wall and loses 25% of its energy in the collision. Find the angle the ball makes with the wall on the way out.

48. ★★★ Derive an expression for the fractional loss of energy for an object that impacts a wall at an angle of θ and leaves it an angle of θ'.

49. ★ Two hockey players race for the puck moving approximately in the same direction. Player 1 has a mass of 69 kg and a velocity of 5.6 m/s at an angle of 32° with the rink board. Player 2 has a mass of 78 kg and a velocity of 7.1 m/s at an angle of 16° with the same board. The players become locked together in a completely inelastic collision. Find the energy lost in the collision.

50. ★★ A medicine ball of mass m and speed v is sliding on frictionless ice when it strikes a stationary medicine ball of mass $5m/4$. After the collision, the kinetic energy of the first ball is three quarters of its initial energy, and this ball moves at an angle of 21° with the incident direction. The second medicine ball moves in a direction that makes an angle of 43° with the final momentum of the first ball. Express the speed of the second ball in terms of v.

51. ★★ Two balls collide head-on; one ball (m_1) has twice the speed of the other ball (m_2). The collision is completely inelastic. Derive an expression for the speed of the two balls after the collision.

52. ★★ A 3.0 kg mass is attached to a spring of spring constant $k = 185$ N/m resting on a horizontal frictionless surface. The other end of the spring is fixed to a wall. A 1.6 kg mass is thrown toward the 3.0 kg mass and collides with it in a perfectly elastic collision. The maximum compression of the spring is 80. cm. Find the speed of the incoming mass.

53. ★★ A billiard ball of mass m is shot at speed v to strike two adjacent stationary billiard balls with the same mass as the incident ball. The two stationary balls are in contact, and the incoming ball strikes them along the line connecting their centre of mass. The collisions are perfectly elastic. Derive an expression for the final speed of each ball.

54. ★★★ Two balls, one of which is stationary, collide elastically. The speed of one ball after the collision is half the speed of the other. Find the ratio of the masses.

55. ★★★ A moving ball (speed v) collides with another at rest. After the collision, the first ball is at rest. Find the ratio of their masses if the collision is perfectly elastic. What is the speed of the second ball?

56. ★★★ Two balls collide elastically head-on; one ball has mass m and is moving at speed v, the other ball is moving at speed $3v/2$. After the collision, the lighter ball moves at speed $2v$. Find the ratio of the masses, and express the final speed of the heavier ball in terms of v.

57. ★★ A billiard ball of mass m and speed v strikes another ball of mass M and speed u, as shown in Figure 7-29. The collision is perfectly elastic. Derive an expression for the speed of each ball after the collision.

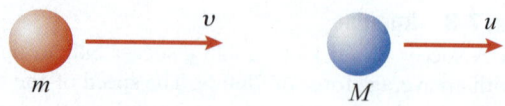

Figure 7-29 Problem 57.

Section 7-7 Variable Mass and Rocket Propulsion

58. ★★ A rocket is launched vertically. The fuel constitutes half the mass of the rocket. After 21.0 s of flight, half the fuel is burned and the speed of the rocket is 190. m/s. The rocket burns fuel at a rate of 72.0 kg/min. Find the thrust of the rocket and the relative speed of the exhaust with respect to the rocket. Ignore air resistance.

59. ★★ A light rocket's engine burns fuel at a rate of 65 kg/s. The rocket is launched in deep space with an initial speed of zero. After 20. s, only half the mass of the fuel is left, and the rocket has reached a speed of 9300 km/h.
    (a) Find the original mass of the fuel.
    (b) Find the average acceleration of the rocket over the first 10. s.
    (c) Find the rocket's thrust.

60. ★★ A giant squid of mass 230 kg propels itself horizontally by swallowing 56 kg of water and then expelling it in 6.0 s. The squid is initially at rest with zero buoyancy. Assuming that the water resistance is negligible and that the squid expels the water at a constant rate through a circular opening 7.0 cm in diameter, calculate estimates for
    (a) the acceleration of the squid when it begins expelling the water
    (b) the speed of the squid 4.0 s after it begins expelling the water
    (c) the squid's average acceleration as it expels the last sixth of the water

61. ★★ A 72 kg man is standing on a frictionless surface and exhales deeply trying to blow a wasp away. The air leaves his mouth at rate of 0.70 L/min at a speed of 9.0 m/s. What force must his neck muscles exert on his head to prevent it from moving?

COMPREHENSIVE PROBLEMS

62. ★★★ A 2.0 m long bar is held vertical on a frictionless surface. The top is tipped very gently, and the bar begins to swing. What is the speed of the bar when it is fully horizontal? Assume all potential energy becomes kinetic energy of the centre of mass, and the bar moves on the surface after falling down.

63. ★ A bar of length 1.16 m is tilted 43.0° from the vertical. Find the horizontal displacement of the bar's centre of mass.

64. ★★ A child of mass 34 kg is standing on a parade wagon (empty mass $M = 120$ kg), which is loaded with 600 gift baskets, each with a mass of 500. g. She throws the baskets off the side one at a time, such that they leave in a horizontal direction, making an angle of 57° with the length of the wagon. There are no frictional losses between the wagon and the road, and the wagon can move only forward and backward. The wagon is originally at rest. By the time the 70th basket is thrown, the speed of the wagon is 2.1 m/s. Find the magnitude of the relative *velocity* of the baskets with respect to the wagon. Assume this velocity is constant.

65. ★★ A 5 g bullet is fired at a speed of 265 m/s into a 9.2 kg watermelon sitting on a thin vertical rod. The bullet shatters the watermelon and leaves with a speed of 235 m/s. Find the speed of the centre of mass of the watermelon immediately after the collision. What is the acceleration of the centre of mass of the watermelon after the collision?

66. ★★★ A child mounts a toy spring cannon on a sled and takes this toy out to play on the icy surface of a nearby lake. The mass of the cannon plus sled is M. The projectile that the cannon shoots has mass m and leaves with muzzle speed v. The cannon faces forward at an angle of θ with respect to the horizontal. The sled-mounted cannon is moving at speed v_c when the projectile is launched.
    (a) Derive an expression for the speed of the cannon immediately after it launches the projectile.
    (b) Find the speed of the cannon, given that $M = 915$ g, $m = 32.0$ g, $v = 17.0$ m/s, $v_c = 5.00$ m/s, and $\theta = 41°$.

67. ★★★ Starting from rest, a 2.4 tonne elephant in a 7.4 tonne boxcar runs at a speed of 3.0 m/s with respect to the boxcar and then throws itself against a spring with $k = 17.5$ kN/m mounted on the other end of the boxcar, thus compressing the spring. The boxcar is mounted on smooth rails, so losses due to friction as the boxcar moves are minimal. Initially the spring is neither stretched nor compressed and the boxcar is at rest, and the spring is unstretched.
    (a) Find the maximum compression of the spring.
    (b) What is the speed of the boxcar when the spring is at maximum compression?

68. ★★★ An engineer working with an expedition in the Arctic needs to move a heavy rectangular package across the frozen ground. Initially, the package is standing on its edge. She puts two pegs in the ground at one edge of the package and lets the package drop by tilting it gently until it begins to fall. The package has a mass of 120 kg and travels a total distance of 25 m on the ground, where the coefficient of friction is 0.05. How tall is the package?

69. ★ A child of mass 19 kg lands on a 34 kg wagon. The wagon then moves with a speed of 2.11 m/s. Find the horizontal speed with which the child lands.

70. ★★ Two balls, one twice the mass of the other, are thrown straight up in the air at a speed of 20. m/s, 0.70 s apart. The heavier ball is thrown second.
    (a) Where is the centre of mass of the two-ball system 0.20 s after the second ball is thrown?
    (b) What is the acceleration of the centre of mass at the following times: (i) before the second ball is thrown, (ii) when the two balls are moving up, (iii) when the lighter ball reaches its maximum height, and (iv) 0.10 s after the lighter ball lands?

71. ★★ You are sitting on one end of a canoe ($m = 19$ kg) in calm water. You then throw your 15 kg backpack to your friend at the other end of the canoe, a distance of 3.0 m. Your friend's mass is 65 kg, and your mass is 72 kg. How much would the canoe move if the water resistance were negligible?

72. ★★★ A research team in Antarctica is launching a 32 kg weather probe into the air above the Ross Ice Shelf. They use a special cannon of mass 1600 kg, which launches the probe with a muzzle speed of 710 m/s. The cannon points at an angle of 57° above the horizontal. The team forgets to secure the cannon, which slides backward on the ice after the probe is fired. Find the angle that the probe makes with respect to the ground when it lands. Ignore air resistance, and assume that the probe lands at the same altitude from which it was launched.

73. ★★ A low Earth orbit observation station, made of 11 modules each of mass 13 000 kg, moves at a speed of 4.0 km/s. It is equipped with an emergency collision avoidance system whereby an explosive charge can be detonated to release one of the modules. The collision alarm sounds, and the commander detonates a charge, releasing one of the modules at a speed of 1400 km/h with respect to the station, directed opposite to the original direction of travel of the station. Find the resulting change in the speed of the station.

74. ★★ During the Iditarod race in Alaska, your sled gets detached from your sled dogs. The sled ($M = 200$ kg) moves at 4.1 m/s, headed for thin ice. You ($m = 70$ kg) jump backward off the sled with a speed of 2.7 m/s with respect to the sled. What is your horizontal speed with respect to the ground while you are in the air?

75. ★ A 110 kg astronaut floating in space holds one end of a 20. m long bar with one 40. kg package on each end. If she rotates the bar by 180°, by how much does she move?

76. ★ A fireworks expert launches a rocket at an angle θ from the vertical. When the rocket is at its highest point, its two parts separate by detonating a charge such that the lighter part falls straight down. The speed of the rocket before the separation is v, and the lighter part is one third the mass of the heavier part. Derive an expression for the speed of the heavier part after the separation.

77. ★★ A hockey puck on frictionless ice strikes the rink board at an angle of 65° with respect to the normal to the board. The collision is inelastic, and the puck loses 17% of its kinetic energy. Find the angle that the puck makes with the normal to the board after the collision.

78. ★★ A polar bear cub is sliding toward his sister on frictionless ice in the Arctic. The sister is stationary before the collision. After the collision, the brother is stationary. Find the ratio of their masses.

79. ★★ An archer shoots an arrow of mass 110 g through a vertical plank of cork that is firmly fixed in place on top of a stationary cart. The plank is parallel to the length of the cart, and the cart is free to move on wheels in the forward direction with frictionless bearings. The incident arrow is horizontal when it strikes the vertical plank and makes an angle of 40.° with the plane of the plank. When the arrow emerges, it makes an angle of 43° with respect to the now moving plank. The combined mass of the plank and the cart is 3.78 kg, and the incident speed of the arrow is 90. m/s.
  (a) Find the energy lost during the collision.
  (b) Find the speed of the cart immediately after the collision.

80. ★★ A pellet gun is used to shoot 0.58 g pellets at a speed of 280 m/s into a piece of wood of mass M, resting on a horizontal frictionless surface. The surface curves upward into the shape of a hemispherical bowl of radius $R = 70.$ cm. The piece of wood then moves up the bowl-shaped surface until its angular position measured from the vertical is 37°. The pellet gets completely lodged in the piece of wood. Find M.

81. ★★★ A solid disk of radius R and mass M has a hollow part of radius r, centred at distance x from the centre of the otherwise solid disk, as shown in Figure 7-30. Find the centre of mass of the disk.

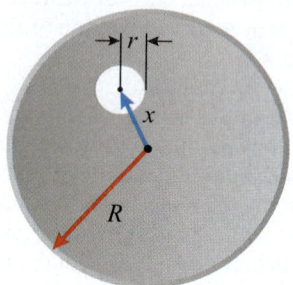

Figure 7-30 Problem 81.

82. ★★★ ∫ A bar has a linearly increasing density, such that the density at one end is five times the density at the other end. Find the location of the centre of mass as measured from the lighter end.

83. ★★★ A figure skater of mass $3m/2$ is holding onto the middle of a bar that connects two other figure skaters of masses m and M. The figure skater in the middle then spins the bar at a constant angular speed. Derive an expression for the radius of the circular trajectory that the centre figure skater will undergo.

84. ★★ A spring ($k = 430$ N/m) is compressed 9.0 cm between two masses, $m = 910$ g and $M = 1.4$ kg. While the system moves along a frictionless surface at a constant speed v, the spring is released. What is the speed of the lighter mass when the spring is stretched 7.0 cm?

85. ★★★ You drop a small ball with a mass of 32 g from the roof of a building 40 m above the ground. One second later, you throw a volleyball with a mass of 226 g down with an initial speed such that it hits the small ball 5.0 m above the ground. All the collisions are elastic.
  (a) Find the speed of the small ball immediately after the collision.
  (b) Find the speed of the small ball after it collides with the ground.

86. ★★★ Mass m_1 moves with speed v_1 toward stationary mass m_2. One end of a spring of spring constant k and unstretched length L is attached to the front of mass m_1. Mass m_2 sticks to the other end of the spring when they make contact.
  (a) Derive an expression for the maximum distance between the two masses after the collision.
  (b) Calculate this distance using the values $m_1 = 100.$ g, $m_2 = 300.$ g, $v_1 = 6.00$ m/s, $L = 28.5$ cm, and $k = 50.0$ N/m.

87. ★★★ Where is the centre of mass of the Earth–Moon system relative to the surface of Earth?

88. ★★★ A block of mass m is fitted with a horizontal spring (spring constant k) at one end. The mass is made to move at speed v, and it strikes another block of mass M that was moving toward it at speed u. The collision is cushioned by the spring. Derive expressions for
    (a) the maximum compression of the spring
    (b) the speed of m when the spring is at its maximum compression
    (c) the final speed of each mass
    (d) the speed of each mass at maximum spring compression

89. ★★★ Estimate how far the crew of a submerged submarine could move the vessel by all gathering at one end of it.

90. ★★ A child places a spring between two of his toy cars. He holds the two cars together on a frictionless surface so they compress the spring. He causes the two cars with the spring in between to move at speed v, keeping the spring compressed between the cars; then he lets the two cars go. Find the final speed of the car in the front. The spring constant is k, and the mass of each car is m.

91. ★★ A 7.45 g bullet moving at a speed of 1100 m/s strikes a 27.8 kg block of wood and emerges from the other end to become lodged in 2.0 kg plate. The plate rests on a horizontal frictionless surface and is attached to a horizontal spring ($k = 1340$ N/m) that is compressed 11 cm as a result of the collision. Find the speed of the block immediately after the bullet emerges from it.

92. ★★ Your cat, Boots ($m = 3.15$ kg), is standing in the middle of a cart with frictionless wheels. The cart is at rest on a frictionless surface. Boots then jumps to one end of the cart such that the cart is moving at 0.30 m/s while the cat is in the air. The cat is in the air for 0.80 s and the cart is 1.0 m long.
    (a) Find the mass of the cart.
    (b) What is the speed of the cart after the cat lands?

93. ★★★ Two blocks with masses m_1 and m_2 are pushed toward each other on a horizontal frictionless surface with respective speeds v_1 and v_2, respectively. The collision between them is cushioned by a spring of spring constant k that is attached to one of the masses.
    (a) Derive an expression for the maximum compression of the spring during the collision.
    (b) Calculate the maximum compression in part (a) for $m_1 = 3.20$ kg, $m_2 = 5.24$ kg, $v_1 = 2.09$ m/s, $v_2 = 3.76$ m/s, and $k = 615$ N/m.

94. ★★★ A 32 kg child runs at a speed of 3.00 m/s and jumps onto a wagon of mass $m = 167$ kg sitting on frictionless wheels. The child slides for a distance d on the wagon before coming to a stop relative to the wagon. If the coefficient of kinetic friction between the child and the wagon is 0.7, find (a) the final speed of the cart, and (b) the distance the child slides on the wagon (relative to the wagon).

DATA-RICH PROBLEM

95. ★★★ $\frac{d}{dx}$, 〜 The development of the ESA'a Vega Launch vehicle began in 1998 for the purpose of delivering small payloads to near orbits. Its targeted lift capacity is the delivery of a 1500. kg satellite to polar orbit (700. km). The first stage is powered by a Pf-80 solid fuel rocket motor that delivers a thrust of 2261 kN for 108.6 s. The speed of the rocket at the end of the first stage is $\sim$1877 m/s, at which time the Pf-80 rocket would have burned 88 365 kg of fuel and reached an altitude of 40.00 km. The liftoff mass is 137.0 tonnes.

    You will need to use a computer program or a spreadsheet for this problem.
    (a) Find the rate at which fuel is burned in the first stage, assuming a constant burn rate.
    (b) Find v_{er}, the speed of the exhaust relative to the rocket, assuming a constant burn rate.
    (c) Plot the mass of the rocket versus time for the first stage (go up to 109 s). What is the mass at the end of the first stage? Make your time steps 1 s or less on your spreadsheet.
    (d) Plot the acceleration of the rocket versus its mass if the rocket is launched far enough from any planets (no air resistance).
    (e) Plot the speed of the rocket versus time if the rocket is launched far enough from any planets. The velocity is given by

$$v(t) = v_{er}\ln\left(\frac{M_0}{M(t)}\right)$$

    where $M(t)$ is the mass of the rocket at time t and M_0 is the takeoff mass.
    (f) Find an expression for the power generated by the rocket. You may use the result that the energy output of the rocket engine is given by

$$\frac{1}{2}(M_0 - M)v_{er}^2$$

    where M is the mass of the rocket at time t and M_0 is the takeoff mass.
    (g) Estimate the work done against the force of gravity to get the rocket up to the altitude of 40.00 km. Comment on how to obtain an accurate value for the work done against gravity, and provide an approximate expression.

OPEN PROBLEM

96. Different types of nuclear reactors use different moderators to extract the energy from the high-energy neutrons that are generated as part of the fission process. The role of the moderator is to slow the neutrons down so they spend more time in the reactor, which increases the probability that they will initiate another fusion reaction. Two common moderator choices are light and heavy water. Heavy water is made from one oxygen and two deuterium atoms. Examine a collision between a neutron and (a) a hydrogen nucleus and (b) a deuterium nucleus, and discuss the relative effects on the energy of the neutron.

Rotational Kinematics and Dynamics

Learning Objectives

When you have completed this chapter, you should be able to

LO1 Define angular quantities, and relate angular variables to linear variables.

LO2 Solve kinematics equations for rotation with a constant acceleration.

LO3 Define torque, calculate it using different representations, and use the moment arm.

LO4 Explain the relationship between torque, moment of inertia, and angular acceleration.

LO5 Calculate moments of inertia for rigid objects, and solve equations of motion for rotation about a fixed axis.

LO6 Use the work–mechanical energy approach to solve rotational dynamics problems, and compare to the force–torque approach.

LO7 Define angular momentum, and apply the conservation of angular momentum.

The Crab pulsar, a pulsating neutron star within the Crab Nebula, is the result of a type II supernova that was recorded by both Chinese and Arab astronomers in 1054 CE. The light from the supernova explosion was visible to the naked eye for close to 22 months in the night sky before fading from view. The bulk of the exploding star's substance was ejected during the supernova, creating the Crab Nebula (Figure 8-1), while the rest of the stellar mass was crushed together to form the neutron star. The original star was spinning when it exploded, and the mass that became the neutron star continues to spin.

The radius of this neutron star is only 10 km, a tiny fraction (likely less than 1/100 000) of the radius of the progenitor star. As a result, the Crab pulsar spins very fast: the rotational period is only 33.1 ms. The conservation of angular momentum, the principle behind the incredibly high rotational speed of the pulsar, is discussed in this chapter. The same principle also applies to fast-spinning figure skaters and many other phenomena. A spinning neutron star (the late-stage, very dense core of a massive star) emits electromagnetic radiation along the axis of its magnetic field. If the rotation and the magnetic axes do not coincide, the radiation beam may be observed by a distant observer once during each revolution, much like the beam of light in a lighthouse. The pulsed nature of the beam and the radio-frequency component of the radiation result in the name *pulsating radio star*, or *pulsar*.

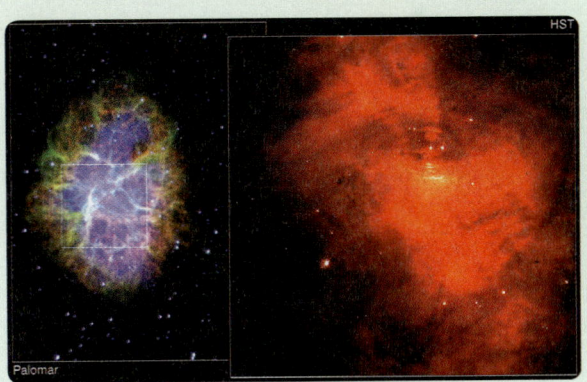

Figure 8-1 The Crab Nebula (left) and the nearby shock waves as a result of the radiation from the pulsar (right).

Jeff Hester and Paul Scowen (Arizona State University), and NASA

8-1 Angular Variables

From Translation to Rotation

You are part of an expedition studying the Macaroni penguins between the southern tip of Argentina and the Larsen Ice Shelf in Antarctica. Your research vessel travels in a "straight line" on the water surface ahead of you. The truth, however, is that your trajectory is circular as it follows the curvature of the Earth's surface. You could track your position using linear distances with respect to landmarks over short distances. However, a ship's position and displacements are given in terms of longitudes and latitudes, which are angular coordinates.

Consider a particle moving along the circumference of a circle of radius r. When the particle moves from point A to point B (Figure 8-2), the length of the arc travelled by the particle is

KEY EQUATION
$$\Delta s = r\Delta\theta \qquad (8\text{-}1)$$

where $\Delta\theta$ is the **angular displacement** expressed in radians.

In Equation 8-1, r has units of length; the SI unit for length is the metre. The SI unit for angular displacement ($\Delta\theta$) is the radian, which is dimensionless. Hence, Δs has units of length as well. However, r is a straight line, and Δs is an arc length (a segment of a circle).

The differential or infinitesimal displacement along the circumference can be written in terms of the infinitesimal angular displacement:

$$ds = rd\theta \qquad (8\text{-}2)$$

PEER TO PEER

I have to remember when relating arc lengths and angles in Equations 8-1 and 8-2 that I *must* express the angles in radians.

It follows that the instantaneous rate of change with respect to time of the position of the particle along the circumference is related to the instantaneous rate of change of the angular position:

$$\frac{ds}{dt} = r\frac{d\theta}{dt} \qquad (8\text{-}3)$$

The derivative, $\dfrac{ds}{dt}$, is the instantaneous linear speed, v, of the particle along the circumference, or the particle's instantaneous **tangential speed** (Figure 8-3).

The derivative $\dfrac{d\theta}{dt}$ is the instantaneous rate of change of the angular position with respect to time, or, simply, the **angular velocity** (its magnitude is called the **angular speed**). The conventional symbol for angular speed is ω:

KEY EQUATION
$$\omega = \frac{d\theta}{dt} \qquad (8\text{-}4)$$

The SI units for ω are radians per second. Like linear speed, angular speed can be a function of time. Equation 8-3, therefore, can be rewritten as

KEY EQUATION
$$v(t) = \omega(t)r \qquad (8\text{-}5)$$

Taking the derivative of Equation 8-3 with respect to time, we obtain

$$\frac{d^2s}{dt^2} = r\frac{d^2\theta}{dt^2} \qquad (8\text{-}6)$$

The derivative on the left-hand side of Equation 8-6 is the **tangential acceleration** of the object, not to be confused with the **radial acceleration** (see Figure 8-4). The derivative on the right-hand side is the **angular acceleration**. The standard symbol for angular acceleration is α:

KEY EQUATION
$$\alpha = \frac{d\omega}{dt} = \frac{d^2\theta}{dt^2} \qquad (8\text{-}7)$$

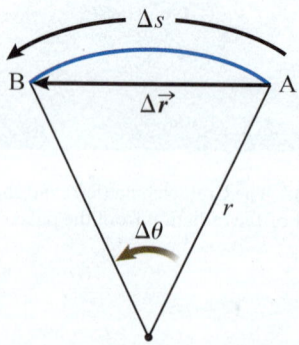

Figure 8-2 Linear displacement: the displacement vector $\Delta\vec{r}$ is a straight line connecting A and B. Tangential displacement: Δs is the full arc length covered by the particle.

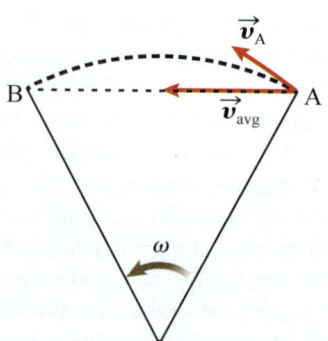

Figure 8-3 The instantaneous linear velocity, v_A, is tangential to the circle, and the average velocity, v_{avg}, points directly along the straight line from A to B.

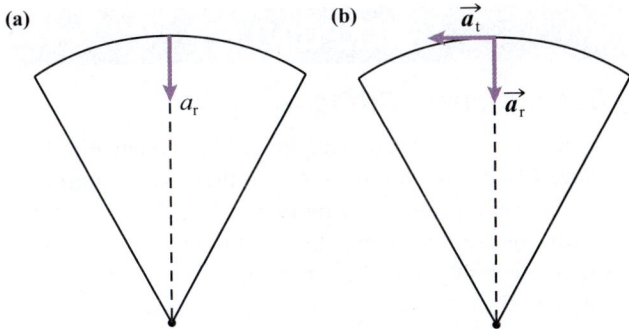

(a)

a_r

(b)

$\vec{a}_t$

$\vec{a}_r$

Figure 8-4 In (a), an object travelling around the circumference of a circle at a constant speed still has a non-zero radial acceleration, $\vec{a}_r$. In (b), the tangential speed also changes so the object will also have both a tangential acceleration, $\vec{a}_t$, and a radial acceleration, $\vec{a}_r$.

Since a and α are not necessarily constant, they are written as functions of time, t. Thus, Equation 8-6 can be rewritten as

> **KEY EQUATION** 　　$a_t(t) = \alpha(t)r$ 　　(8-8)

where a_t is the tangential acceleration, not to be confused with the radial acceleration experienced by an object moving along a circular trajectory. An object travelling at a constant tangential speed along the circumference of a circle has no tangential acceleration but still has radial acceleration pointing inward, as explained in Chapter 4. This is illustrated in Figure 8-4(a). When the object's tangential speed also changes, it will have a tangential acceleration in addition to its radial acceleration (see Figure 8-4(b)).

We have seen in Chapter 4 that any object moving along a circular path of radius r will have an instantaneous radial acceleration of $a_r = \dfrac{v^2}{r}$. The direction will be radially inward (and therefore it is sometimes called the centripetal acceleration, since the word centripetal means inward or centre pointing). We have this radial acceleration whether the speed, v, of the object is changing or not. When the speed is changing, then in addition we also have a tangential acceleration, a_t.

Using Equation 4-32, the magnitude of the radial acceleration is given in terms of the angular speed as

Table 8-1 Magnitudes of the Tangential Acceleration, a_t, and the Radial Acceleration, a_r, for Motion at Tangential Speed, v, around a Circle of Radius r

	$\omega = $ constant	$\omega \neq $ constant
a_t	0	$\alpha r = \dfrac{d\omega}{dt}r$
a_r	$\dfrac{v^2}{r} = \omega^2 r$	$\dfrac{v^2}{r} = \omega^2 r$

$$a_r = \frac{v^2}{r} = \frac{(\omega r)^2}{r} = \omega^2 r \qquad (8\text{-}9)$$

where we have used Equation 8-5 to express the tangential speed in terms of the angular speed and the radius of the circular path.

We summarize these results, for motion in a circle of radius r, in Table 8-1.

Angular velocity and angular acceleration are vector quantities. The convention for the direction of angular velocity is that you curl the fingers of your right hand in the direction of motion of the rotating object, and your thumb will point in the direction of the angular velocity. That is, the angular velocity vector is perpendicular to the plane of rotation. This is demonstrated in Figure 8-5. Since we often deal with rotation about a fixed axis, where direction just takes on a positive or negative value in a single direction, we sometimes do not write vector signs on rotational quantities (similar to the one-dimensional linear motion in Chapter 3).

The angular acceleration direction is defined in a corresponding fashion. Recall from Equation 8-7 that angular acceleration is the rate of change of angular velocity. Therefore, for example, if the angular speed is increasing, the angular acceleration will be in the same direction as the angular speed, but if the angular speed is decreasing, then the angular acceleration and angular speed will be in opposite directions.

PEER TO PEER

An object moving with a constant angular speed has an angular acceleration of zero; therefore, the tangential acceleration is also zero. The radial acceleration is not zero, however, and points radially inward with a magnitude given by $a_r = \dfrac{v^2}{r} = \omega^2 r$. I need to remember that an object moving around a circle at a constant speed still has an acceleration toward the centre of the circle.

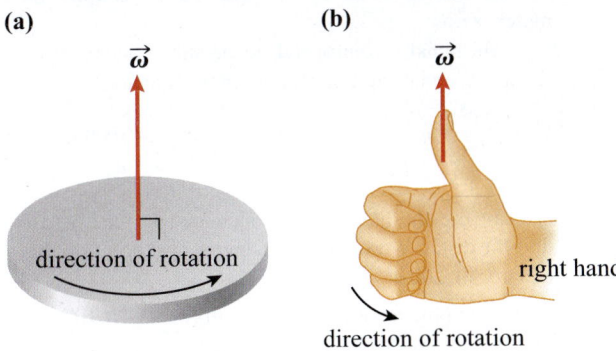

(a)

$\vec{\omega}$

direction of rotation

(b)

$\vec{\omega}$

right hand

direction of rotation

Figure 8-5 The angular velocity vector direction is perpendicular to the plane of rotation, and in the direction the thumb of your right hand points if your fingers curl in the direction of motion.

CHECKPOINT 8-1

Angular Velocity Direction

The angular velocity vector for a rotating object is in the $-\hat{k}$-direction, and the angular acceleration is also in the $-\hat{k}$-direction. The object lies in the xy-plane. If you view the object from above (i.e., from the $+\hat{k}$-direction), will the object rotate in a clockwise or a counter-clockwise fashion, and is it increasing or decreasing in angular speed?

(a) clockwise, and slowing down
(b) clockwise, and speeding up
(c) counter-clockwise, and slowing down
(d) counter-clockwise, and speeding up

ANSWER: (b) If you point your thumb in the $-\hat{k}$ (i.e., downward) direction, the fingers of your right hand curl in a sense that is clockwise as viewed from above. Since the angular acceleration and angular velocity are in the same direction, the angular velocity will increase in magnitude and be speeding up.

EXAMPLE 8-1

Rotation of Earth

(a) Determine the values for the linear speed, angular speed, tangential acceleration, and radial acceleration of a point at the equator. Assume that Earth is a sphere of radius 6340 km.

(b) Calculate how far the rotation of Earth moves a person sitting on a beach at the equator during a 2.00 h sunbath.

Solution

(a) Earth completes a full revolution (2π rad) in about 24.0 h, so the angular speed is

$$\omega = \frac{2\pi \text{ rad}}{(24 \text{ h})(3600 \text{ s/h})} = 7.27 \times 10^{-5} \text{ rad/s}$$

The linear speed for a point on the equator is given by Equation 8-5:

$$v = \omega r = (7.27 \times 10^{-5} \text{ rad/s})(6340 \text{ km})$$
$$= 0.461 \text{ km/s} = 461 \text{ m/s}$$

Earth's angular speed is nearly constant, so the tangential acceleration as given by Equation 8-7 is approximately zero.

The radial, or centripetal, acceleration points toward the centre, and its magnitude is given by Equation 8-9:

$$a_r = \frac{v^2}{r} = \omega^2 r = (7.27 \times 10^{-5} \text{ rad/s})^2 (6\,340\,000 \text{ m})$$
$$= 3.35 \times 10^{-2} \text{ m/s}^2$$

(b) We can use the linear speed calculated in part (a) to find the distance travelled (arc length) by a point on the equator in 2.00 h:

$$s = vt = (461 \text{ m/s})(2.00 \text{ h})(3600 \text{ s/h}) = 3.32 \times 10^6 \text{ m}$$

Making sense of the result

The distance covered by a point on Earth's equator during one day (24 h) is Earth's circumference, given by $2\pi r = 2\pi(6340 \text{ km})$, which is approximately 39 800 km. If we divide that by 12 we get 3320 km, the distance covered in 2.00 h, which matches our answer.

CHECKPOINT 8-2

Geostationary Orbits

Satellites in a geostationary orbit are located directly above Earth's equator and move so they appear motionless in the sky. Since the motion is synchronized with Earth, we call these orbits geosynchronous. Such orbits are particularly useful for communication and television satellites since the Earth receiver, such as a satellite television dish, can point to a single direction in the sky. Which of the following statements is true for a satellite in a geostationary orbit?

(a) The linear velocity of the satellite, with respect to Earth's centre, is the same as the linear velocity of a stationary object on Earth's surface.

(b) The angular velocity of the satellite, with respect to Earth's centre, is the same as the angular velocity of a stationary object on Earth's surface at the equator.

(c) The tangential acceleration of the satellite is non-zero.

(d) The radial acceleration of the satellite is zero.

ANSWER: (b) We require both Earth and the satellite to move at the same angular velocity so that the direction of the satellite from a point on Earth does not change. This is most easily seen for a satellite directly above you if you are at the equator, but it is true in general.

8-2 Kinematic Equations for Rotation

Constant Acceleration

In this section we will follow a treatment similar to the one we used in Chapter 3, where we derived the kinematic equations for linear motion with constant acceleration, but this time for rotational motion. The results we derive are for rotational motion with a constant angular acceleration.

We define angular acceleration as

$$\alpha = \frac{d\omega}{dt} = \frac{d^2\theta}{dt^2} \tag{8-10}$$

Integrating Equation 8-10 under the assumption that α is a constant gives the following expression for angular velocity as a function of time:

KEY EQUATION
$$\omega(t) = \alpha t + \omega_0 \tag{8-11}$$

Equation 8-11 shows that the angular speed at any time t is determined by the initial value at $t = 0$ plus the change incurred as a result of the acceleration. This is exactly analogous to the result $v(t) = at + v_0$ that we obtained in Chapter 3 for constant acceleration.

EXAMPLE 8-2

Bicycle Wheel Spin

A bicycle wheel spins about its axis at an angular speed of 6.50 rad/s and slows down because of friction with an acceleration of 8.20 rad/s². Find the time it takes the wheel to reach a speed of 3.20 rad/s.

Solution

Let us consider that wheel is spinning in the clockwise direction. We know the initial speed and final speed of the wheel, and we have the acceleration. The change in angular velocity depends on the angular acceleration and the time as given by Equation 8-11:

$$\omega(t) = \alpha t + \omega_0 \tag{1}$$

We choose the initial direction of rotation of the wheel (clockwise) as positive, so

$$\omega_0 = 6.50 \text{ rad/s and } \omega(t) = 3.20 \text{ rad/s}$$

$$\alpha = -8.20 \text{ rad/s}^2 \tag{2}$$

where the α-value is negative, since the wheel is slowing down and the acceleration is counter-clockwise.

Then

$$t = \frac{\omega(t) - \omega_0}{\alpha} = \frac{3.20 \text{ rad/s} - 6.50 \text{ rad/s}}{-8.20 \text{ rad/s}^2} = 0.402 \text{ s}$$

Making sense of the result

The time has the correct units, given by (rad/s)/(rad/s²).

When we integrate Equation 8-11, again assuming a constant value for the angular acceleration, we can show that the angular position as a function of time is given by

KEY EQUATION
$$\theta(t) = \frac{1}{2}\alpha t^2 + \omega_0 t + \theta_0 \tag{8-12}$$

We mention here that since the angular acceleration is constant, we can obtain the kinematics equations without using calculus as we did in Chapter 3.

This result is analogous to the linear motion result for constant acceleration:

$$x(t) = \frac{1}{2}at^2 + v_0 t + x_0$$

Combining Equations 8-11 and 8-12 gives us the rotational version of the time-independent kinematics equation, identical in form to Equation 3-22, $v^2 = v_0^2 + 2a\Delta x$:

KEY EQUATION
$$\omega^2 = \omega_0^2 + 2\alpha\Delta\theta \tag{8-13}$$

Note that the product of α and $\Delta\theta$ is positive when the two quantities are parallel (i.e., the object speeds up) and negative when the quantities are antiparallel (i.e., the object slows down). This is analogous to the linear kinematics situation.

EXAMPLE 8-3

Spinning Flywheel

A flywheel of radius 0.720 m is rotating clockwise at 3.60 rad/s.

(a) Find the acceleration needed to slow the wheel to an angular speed of 2.10 rad/s over 1.80 s. Assume the acceleration is constant.

(b) Find the number of revolutions the wheel will go through as it slows down.

Solution

(a) Equation 8-11 relates angular velocity, acceleration, and time:

$$\omega = \alpha t + \omega_0$$

We choose the direction of rotation (clockwise) as positive, so ω and ω_0 will be positive. Since the wheel is slowing down, α is directed counter-clockwise and is negative according to our sign convention:

$$\alpha = \frac{\omega - \omega_0}{t} = \frac{2.10 \text{ rad/s} - 3.60 \text{ rad/s}}{1.80 \text{ s}}$$

$$= \frac{-1.50 \text{ rad/s}}{1.80 \text{ s}} = -0.833 \text{ rad/s}^2$$

(b) Now we can use either Equation 8-12 or Equation 8-13 to determine the angular displacement of the wheel as it slows down. Here, we use Equation 8-13:

$$\omega^2 = \omega_0^2 + 2\alpha\Delta\theta$$

$$\Delta\theta = \frac{\omega^2 - \omega_0^2}{2\alpha} = \frac{(2.10 \text{ rad/s})^2 - (3.60 \text{ rad/s})^2}{2(-0.833 \text{ rad/s}^2)} = 5.13 \text{ rad}$$

$$= 5.13 \text{ rad} \times \frac{1.00 \text{ rev}}{2\pi \text{ rad}} = 0.817 \text{ rev}$$

Making sense of the result

We see that the units reduce to the appropriate final units as required. For part (a), the answer is of the right order of magnitude, as we are reducing the angular speed by 1.50 rad in 1.80 s, so it is reasonable that we get a magnitude for the angular acceleration of a bit less than 1 rad/s².

It is not surprising that Equations 8-11 and 8-12 are almost identical in form to Equations 3-10 and 3-17 because they are derived in the same fashion and hold for the same condition of constant acceleration. Essentially, the rotational kinematics equations developed above for *constant* angular acceleration are indeed identical to the kinematics equations that were developed in Chapter 3 for motion in one dimension at a *constant* linear acceleration. For convenience and to stress the similarity, we present in Table 8-2 a summary of the angular variables and the kinematic equations for constant-acceleration rotation about a fixed axis, along with their linear counterparts. Vector signs are suppressed for simplicity of notation, since normally we deal with rotation about a fixed axis.

Table 8-2 Comparison of Angular and Linear Quantities

Linear Quantity	Symbol/Expression	Angular Quantity	Symbol/Expression
Linear position	x	Angular position	θ
Linear velocity	v	Angular velocity	ω
Linear acceleration	a	Angular acceleration	α
Kinematic equations for constant a	$x(t) = \dfrac{1}{2}at^2 + v_0 t + x_0$ $v(t) = at + v_0$ $v^2 = v_0^2 + 2a\Delta x$	Kinematic equations for constant α	$\theta(t) = \dfrac{1}{2}\alpha t^2 + \omega_0 t + \theta_0$ $\omega(t) = \alpha t + \omega_0$ $\omega^2 = \omega_0^2 + 2\alpha\Delta\theta$

INTERACTIVE ACTIVITY 8-1

Ladybug Revolution

In this activity, you will use the PhET simulation "Ladybug Revolution" to investigate rotational kinematics. It will allow you to investigate the relation between angular velocity, angular acceleration, and angular displacement of a rotating object.

CHECKPOINT 8-3

Cranking Your Bike

On a bicycle, a large sprocket is driven by a pedal on crank arms (see Figure 8-6) and is connected to the small sprocket on the rear wheel by a chain. Assume that the radius of the large sprocket is R, the radius of the small sprocket is $R/2$, the radius of the wheel is $3R$, and the length of each pedal crank arm is $2R$.

Figure 8-6 Checkpoint 8-3.

Which of the following statements is true?
(a) The rear wheel has the same angular acceleration as the crank arms.
(b) The small sprocket has the same angular speed as the large sprocket.
(c) The radial acceleration of a point on the circumference of the small sprocket is half the acceleration of a point on the circumference of the large sprocket.
(d) The linear speed of a point on the circumference of the small sprocket is the same as the linear speed of a point on the circumference of the large sprocket.

ANSWER: (d) The speed of the chain is the same as that of the tangential speed of each sprocket.

8-3 Torque

What Is Torque?

When you open a water bottle with a twist cap, you are exerting a torque on the cap. When you use a wrench to tighten a bolt, you are exerting a torque on the bolt. When you spin a top, you are exerting a torque on the top to give it an angular speed. In the same way that a net force will cause movement, a net torque will cause rotation. Indeed, a net force causes acceleration and a net torque causes angular acceleration.

What Does Torque Depend Upon?

If you wanted to tighten a bolt using a wrench as in Figure 8-7, where would you intuitively grab the wrench to maximize your effectiveness? Would you grab the wrench at position A or position B? If you have a wrench at home, give it a try. Alternatively, you can attempt to open a spring-operated door (where the spring serves to close the door automatically, and acts to resist opening the door): once by applying a force where the handle is, and once by applying the same amount of force closer to where the hinge is. Which is easier?

If you attempt to apply force close to the door hinge, you will find it is harder for you to push the door open. If you apply the force farther away from the hinge, closer to the opposite edge of the door, where most door handles usually are, it will be much easier. We can similarly perceive that it is more efficient to apply the force at location B to turn the bolt. The distance between the point of application of the force and the bolt (or the hinge in general) is sometimes known the lever arm; we expand on this shortly. The larger the lever arm, the more effective the

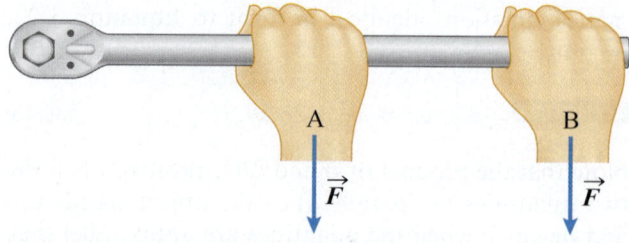

Figure 8-7 Is it more efficient to apply the same force on a wrench at A or at B to exert a torque to turn the bolt?

(a)

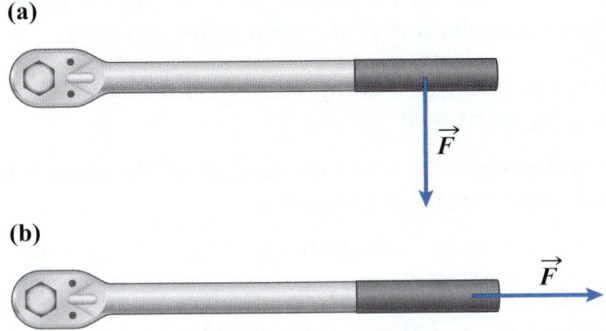

(b)

Figure 8-8 (a) Force acts to cause the wrench to turn. (b) Force cannot cause the wrench to turn.

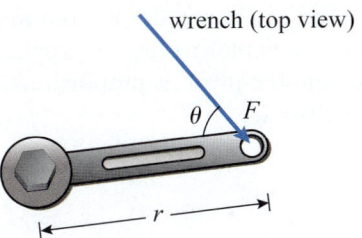

Figure 8-9 Torque depends on three parameters, the magnitude of the force (F), the length of the arm to where the force is applied (r), and the angle between the force and the arm (θ).

force in producing a torque. You have likely experienced this in many situations in life. Reflect on this the next time you try to rotate something.

What do you think would happen if, instead of pulling on the wrench to turn it, as shown in Figure 8-8(a), you pulled or pushed along the length of the wrench directly toward (or away from) the bolt, as shown in Figure 8-8(b)?

We know from common sense (give it a try if you don't agree) that this would not help the bolt to turn at all. You may be able to apply enough force to break the bolt, but not to cause it to turn. Indeed, the most efficient way to cause the bolt to turn is to apply a force perpendicular to the wrench. Therefore, the angle the force makes with the wrench length makes a difference.

Finally, of course, the torque depends on the magnitude of the applied force. The stronger the force, the larger the torque, and the more able the force will be to cause the bolt to turn.

Pivot and Axis of Rotation

We can envisage an *axis of rotation* that runs through the pivot. In the case of the wrench of Figure 8-8, the axis of rotation runs through the centre of the nut and is perpendicular to the page.

When you open a door, the axis of rotation is a vertical one that runs through the hinges. When you unscrew the lid of a water bottle, the axis of rotation runs along the length of the bottle, through its centre.

We will be using both terms, pivot and axis, in this and other chapters, depending on the context.

Torque is a measure of the effectiveness of a force in causing rotations. From the brief discussion earlier, we infer that the torque depends on

- the strength of the applied force
- the angle, θ, between the applied force and the arm to the point of application of the force
- the distance, r, between the bolt and the point where you apply the force

These quantities are shown in Figure 8-9. We examine these three quantities in more detail in what follows as we quantify our discussion.

The Force

Consider the massless bar of length r in Figure 8-10. The left end of the bar can rotate about the pivot at O, or about an axis perpendicular to the page running through the pivot at O. If the bar rotates, we say it rotates about the pivot or, more precisely, about a rotation axis that is perpendicular to the page and runs through the pivot.

Force $\vec{F}$ acts at the other end of the bar. The force is perpendicular to the bar. We designate the position vector of the force as measured from the pivot, between the force and the pivot, as $\vec{r}$.

If we double the force, we can expect the ability of the force to generate rotations, or the torque, to double. We can thus conclude that the torque is proportional to the force.

The Distance

Keeping the force perpendicular to the beam, the closer the force is to the pivot, the less effective it will be in generating rotations.

If we apply the force $\vec{F}$ at the middle of the bar, as shown in Figure 8-11, and measure the force needed to stop the bar from rotating when applied at the end, which can, for example, be done by using a spring, we will see that the force needed, $\vec{F}'$, is only half the force $\vec{F}$ in magnitude.

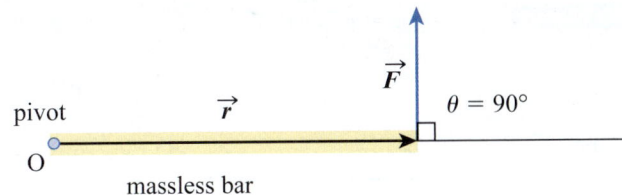

Figure 8-10 Force acting perpendicular to bar.

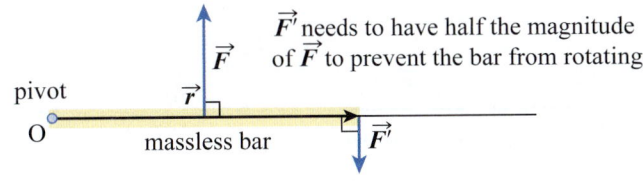

Figure 8-11 The torque doubles when we double the distance.

MECHANICS

Thus far we can conclude that the torque exerted by a force perpendicular to the bar, applied to the bar a distance r from the pivot, is proportional to both the force and the distance:

$$\tau = rF \qquad (8\text{-}14)$$

The Angle

How would things change if the force were not perpendicular to the bar? Let's examine the bar in Figure 8-12.

If the angle, θ, between the bar and the force is zero, that force will not cause the bar to rotate and thus will not exert a torque about the pivot. We can also see that the force will be most effective in causing the bar to rotate when it is perpendicular to the bar. The torque will be a maximum when the angle is 90°.

The Perpendicular Component of the Force

If we break up the force into two components, one perpendicular to the bar and one parallel to the bar, as in Figure 8-13, we conclude that only the component of the force that is perpendicular to the bar, $F_\perp$, exerts a torque about the pivot, while the component of the force parallel to the displacement exerts no torque.

Once again we remind the reader that by "torque about the pivot at O" we are referring to the ability of the force to cause the bar to rotate about the pivot at O.

Hence, the torque of the force in Figure 8-13 is given by

$$\tau = rF_\perp \qquad (8\text{-}15)$$

From Figure 8-13 we can see that the component of the force perpendicular to the bar is given by

$$F_\perp = F\sin\theta \qquad (8\text{-}16)$$

Therefore, the torque exerted by $\vec{F}$ about O becomes

$$\tau = rF_\perp = r(F\sin\theta) \qquad (8\text{-}17)$$

We can therefore say that the torque of the force $\vec{F}$ about the pivot at O is given by

KEY EQUATION $$\tau = rF\sin\theta \qquad (8\text{-}18)$$

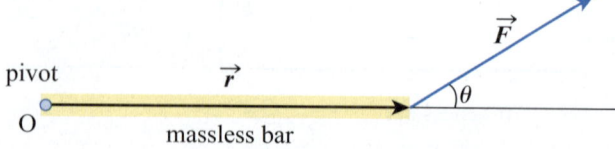

Figure 8-12 Force exerting a torque about the pivot at O.

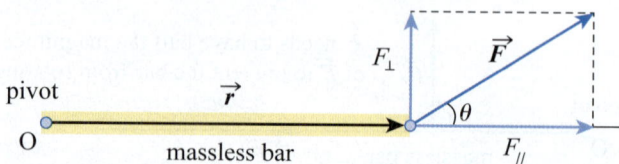

Figure 8-13 Only the perpendicular component of the force exerts a torque about the pivot.

In Equation 8-23 we expand more on this by writing the force as the sum of its components as vectors.

The Perpendicular Component of the Distance: The Moment Arm

Another way to perceive the product of Equation 8-18 is to rearrange the expression as

$$\tau = (r\sin\theta)F \qquad (8\text{-}19)$$

As shown in Figure 8-14, the product $r\sin\theta$ gives the component of $\vec{r}$ that is perpendicular to the force, $\vec{F}$. The component of $\vec{r}$ that is parallel to the force does not contribute to the torque:

$$\tau = (r\sin\theta)F = r_\perp F \qquad (8\text{-}20)$$

$r_\perp$ is referred to as the *moment arm*, the *torque arm*, or the *lever arm*. We discuss this next. We will generally be using the term *moment arm*.

The moment arm The *moment arm*, the *torque arm*, or the *lever arm* is an important notion in many applications in science and engineering. The moment arm is defined as the perpendicular distance between the point of application of the force and the pivot. For the situation in Figure 8-14, the moment arm for the force $\vec{F}$ is $r_\perp$. When we have the moment arm of a force, the calculation of the torque becomes relatively simple, as it is given by the product of the force and its moment arm.

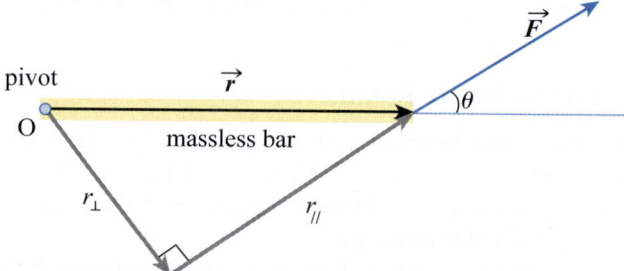

Figure 8-14 Torque calculation using the component of the displacement perpendicular to the force (the moment arm).

We expand more on the representations of torque, including the moment arm, in Chapter 10 on statics.

EXAMPLE 8-4

Supporting a Beam

The beam in Figure 8-15 is supported by a free hinge at the pivot. If left under the effect of the force of gravity, it would swing down freely. You wish to prevent it from doing so. The beam at the instant shown makes an angle of 20.0° with the vertical and you are applying your force of 98.0 N horizontally.

(a) If the top of the beam is 1.60 m above the pivot vertically, what is the torque applied by your force?
(b) Calculate the moment arm for the applied force, and use the moment arm to calculate the torque as well.

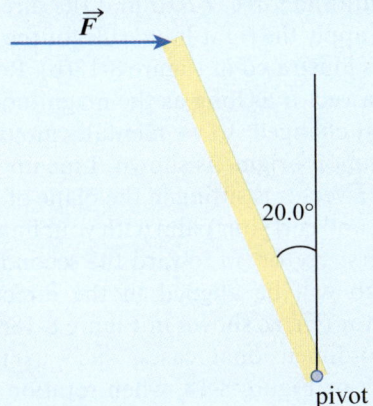

Figure 8-15   Example 8-4.

Solution

(a) To find the torque, we need the magnitude of the force, 98.0 N. We also need the distance between the point of application of the force and the pivot; this is the length of the beam, 1.60 m. The last quantity we need is the angle between the direction of the force and the vector from the hinges to where that force is applied. Since the force is horizontal and the screen makes an angle of 20.0° with the vertical, the angle between the force and the edge of the screen is 70.0°. We therefore have the following for the magnitude of the torque:

$$\tau = rF\sin\theta$$
$$= (1.60\text{ m})(98.0\text{ N})\sin 70.0°$$
$$= 147\text{ N·m (clockwise)}$$

(b) To calculate the moment arm, we *extend the line of action* of the force, as shown in Figure 8-16, and draw a perpendicular from the pivot or axis of rotation to the line of action of the force. The moment arm is identical to $r_\perp$, as we saw in the discussion following Equation 8-20.

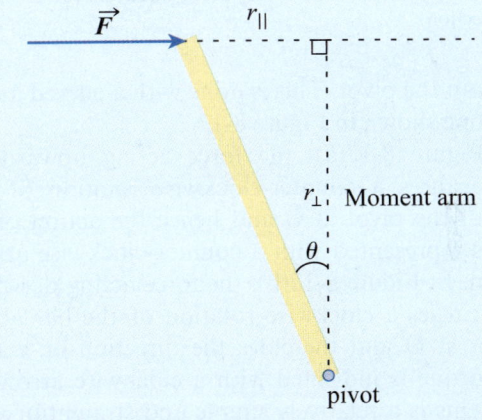

Figure 8-16   The moment arm of the force.

From the geometry of Figure 8-16, this is given by

$$r_\perp = r\cos\theta = (1.60\text{ m})\cos 20.0° = 1.50\text{ m}$$

The torque is given by the product of the force and the moment arm

$$\tau = r_\perp F = (1.50\text{ m})(98.0\text{ N}) = 147\text{ N·m}$$

This result is identical to that obtained in part (a), as expected.

Making sense of the result

Since the force is being applied almost perpendicular to the beam, the torque is almost the product of the force and the distance to application of the force. Note that there is no derived single unit for torque, so it is N·m. The units of torque, and of work, are actually the same, although the meaning is very different of course. The torque acts in the clockwise direction.

Torque Has Direction

We mentioned that torques generate rotations. In two dimensions, one can cause the bar of Figure 8-17 to rotate clockwise or counter-clockwise, as shown.

Equation 8-18 gives us the magnitude of the torque due to the force about the pivot at O but does not tell us about the direction of the torque vector. One way to indicate the direction action of the torque is to indicate the direction of the twisting or rotating action of the

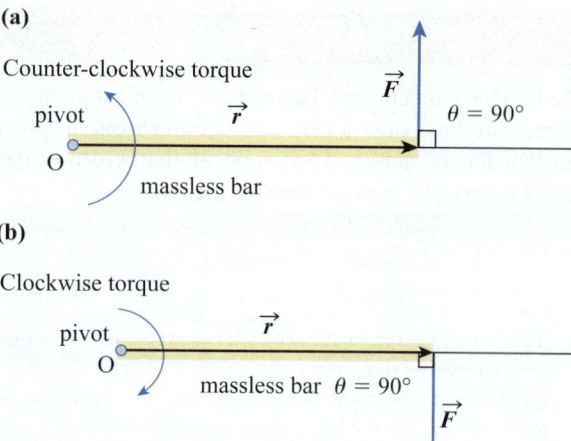

(a)

Counter-clockwise torque

pivot $\vec{r}$ $\vec{F}$

O $\theta = 90°$

massless bar

(b)

Clockwise torque

pivot $\vec{r}$

O

massless bar $\theta = 90°$

$\vec{F}$

Figure 8-17 (a) Counter-clockwise torque direction, (b) Clockwise torque direction

force about the pivot. This is done with a curved arrow like the one shown in Figure 8-17.

In Figure 8-17(a), the force acting upward on the bar causes a counter-clockwise rotation of the bar about the pivot at O and hence the action of its torque is represented with a counter-clockwise arrow, as shown. In Figure 8-17(b), the force acting down on the bar causes a clockwise rotation of the bar about the pivot at O and therefore the direction of action of the torque is indicated with a clockwise arrow, as shown. This is a relatively simple and straightforward way to show the direction of action of the torque and is used heavily in engineering applications. It is especially useful when dealing with rotations in a plane, or in two dimensions, as in Figure 8-17.

Torque Is a Vector Quantity

Torque has a direction and a magnitude and is therefore a vector quantity. Representing it as a vector becomes essential in three dimensions. We explore two ways to establish the direction of the torque vector.

Equation 8-18 gives us the magnitude of the torque as

$$(\text{Magnitude of } \vec{r})(\text{Magnitude of } \vec{F})(\sin \theta)$$

This notation is something you have encountered earlier; we have seen it Chapter 2 when evaluating the magnitude of the cross product of two vectors. As a matter of fact, torque, as a vector, is defined as the cross product of the two vectors $\vec{r}$ and $\vec{F}$:

KEY EQUATION
$$\vec{\tau}_O = \vec{r} \times \vec{F} \qquad (8\text{-}21)$$

Formally, the magnitude of the torque is then written as

KEY EQUATION
$$\tau_O = rF \sin \theta \qquad (8\text{-}22)$$

We have introduced the subscript O in Key Equations 8-21 and 8-22. In the case of Figure 8-17, which symbolizes rotation in a plane, the subscript is a label that refers to the pivot about which the force exerts a torque, or the

pivot about which the torque is calculated. More precisely, this subscript refers to an axis running through the pivot perpendicular to the plane defined by the vectors $\vec{r}$ and $\vec{F}$ (or the page). This plane is also sometimes referred to as the plane of rotation. We will sometimes drop this subscript, but it is mentioned here and in other places to highlight the central importance of the axis or pivot in torque calculations.

"Curl" Right-Hand Rule for Torque Direction

The torque vector is a cross product between two vectors. Its magnitude and direction can be determined using the cross product tools we introduced in Chapter 2. We go over this point here again briefly and discuss the "curl" approach to finding torque direction. Consider the flat object of Figure 8-18(a), which is free to rotate about a pivot point at its centre, O. A force $\vec{F}$ is applied at point P on the object. We first draw a vector $\vec{r}$ from O to the point where the force is applied, P. According to Equation 8-21, the torque is the cross product of vector $\vec{r}$ with the force, $\vec{F}$. To find the direction of the torque, we apply the right-hand rule for the vector cross product, as illustrated in Figure 8-18(b). Recall that we can move a vector as long as the magnitude and direction are not changed, so we mentally move $\vec{r}$ and $\vec{F}$ to have a common origin, as shown. Line up your thumb so that the $\vec{r}$ vector is sitting in the plane of the palm of your hand, and curl your palm with your fingers. Curling from the first vector ($\vec{r}$) toward the second vector ($\vec{F}$), your thumb will be aligned in the direction of the torque vector ($\vec{\tau}$), as shown in Figure 8-18(b).

In two-dimensional cases, such as the case of Figure 8-17 or Figure 8-18, when rotation is involved, you line up your thumb with the axis of rotation, as you *curl* your right hand in the direction of the rotation (think of the object in Figure 8-18(a) rotating as a result of the torque). The direction of the torque vector can also be determined using the Cartesian cross product rules outlined in Chapter 2. We highlight this in Example 8-5 and the discussion that follows.

(a)　　　　　　　　　　　　　　**(b)**

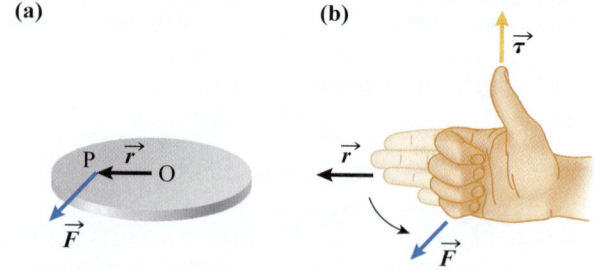

Figure 8-18 A force $\vec{F}$ is applied at point P for an object free to rotate about pivot point O. We draw a displacement vector $\vec{r}$ from O to P. To find the torque we use the curl right-hand rule to calculate the vector cross product $\vec{\tau} = \vec{r} \times \vec{F}$—curling the fingers of the right hand from $\vec{r}$ to $\vec{F}$, our thumb will point in the direction of the vector product, the torque $\vec{\tau}$.

EXAMPLE 8-5

Laptop Glare

Your laptop screen sits in the yz-plane (vertical). You wish to push it back a little by exerting a force in the horizontal direction at the top of the screen so you can read better. (The force you exert is in the negative x-direction). Establish the direction of the torque about the rotation axis (or pivot) at the bottom of the screen, and express the torque in vector form. The screen's height is l and its width is w.

Solution

A schematic of the laptop is shown in Figure 8-19 along with the axis system. By the right-hand rule, you line up your thumb with the axis of rotation at the bottom of the screen as you curl your hand in the direction of rotation of the screen. Your thumb will be pointing to the left, which is the positive y-direction.

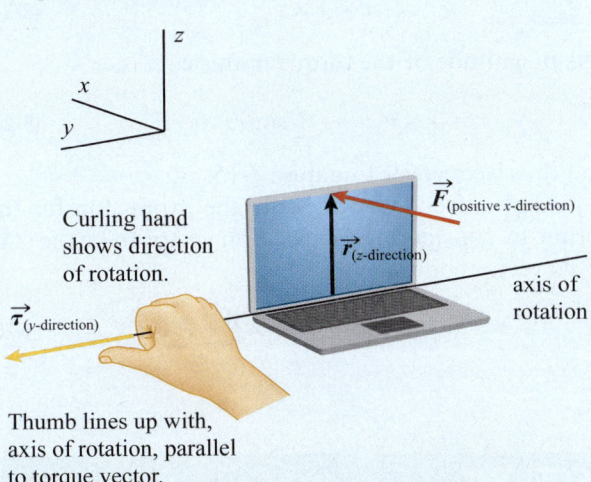

Figure 8-19 Right-hand rule applies to the rotating screen.

Next we establish the direction of the torque vector using the cross product.
The torque is given by Equation 8-21:

$$\vec{\tau} = \vec{r} \times \vec{F}$$

The $\vec{r}$-vector points from the pivot (or axis) to the point of application of the force; this is the height, l, of the screen when it is vertical and points in the $+z$-direction. Therefore,

$$\vec{\tau} = l(\hat{k}) \times F(\hat{i}) = lF(\hat{j})$$

as expected. We have used the fact that the cross product of the unit vector $\hat{k}$ with the unit vector $\hat{i}$ is the unit vector $-\hat{j}$.

Making sense of the result

The direction of rotation gives the torque vector by the right-hand rule. The magnitude of the torque is the force times the distance, as the force is perpendicular to the plane of the screen and thus is perpendicular to every straight line within the plane of the screen, including the vector $\vec{r}$.

MAKING CONNECTIONS

Supersonic Blades

The propellers on turboprop planes turn at relatively high rotational speeds. The linear speed of the tip of the propeller blades depends on the rotational speed and the distance from the tip to the propeller centre. The outer part of the blades can reach supersonic speeds. However, aircraft designers usually keep blade speeds below the speed of sound ($\sim$340 m/s) to avoid supersonic shockwaves and the increased drag they cause. The Tupolev Tu-114 (Figure 8-20) holds the record for the fastest turboprop passenger plane, with a maximum recorded airspeed of 871.4 km/h, and remains one of the fastest passenger planes in use. It has special propellers designed to sustain supersonic blade-tip speeds and cope with the resulting shockwave drag. The Tu-114 is powered by four Kuznetsoz NK-12 turboprops, driving contra-rotating propellers at 750 rpm, with a power output of 11 000 kW and a torque rating of $\sim$140 000 N·m.

Richard Seaman

Figure 8-20 The Tupolev Tu-114.

"Three-Finger" Right-Hand Rule for Torque Direction

Our earlier discussion on the direction of torque utilized what is sometimes referred to as the "curl" right-hand rule, as it involves curling one's hand from the direction of the position vector to that of the force vector. This is a convenient way to establish the direction of the torque vector or any vector given by the cross product of two vectors.

Another way to establish the direction of the torque vector is to use the right-hand "three-finger" rule. This is especially useful when one is taking the cross product of two orthogonal vectors, but we will see that the two vectors do not have to be orthogonal for us to use this rule. When the two vectors are orthogonal, you line up your thumb with the first vector in the product (in that order), and you line up your index finger with the second vector in the product. Your middle finger, if you keep all three fingers orthogonal to one another, as in an xyz orthogonal axis system, will point along the direction of the resultant, which is the cross product. This is demonstrated in Figure 8-21 for torque.

When $\vec{F}$ is not orthogonal to $\vec{r}$, we saw from Equations 8-15 and 8-20 that only orthogonal components contribute to the cross product; the direction of the torque vector is then determined by *considering perpendicular components*, as we explain next.

Torque: Vector Components as Vectors

In this part we explore the beauty and power of the full vector approach to torque. Referring to Figure 8-13, and recognizing that the force can be written as the sum of its two components expressed as vectors, the expression for the torque becomes

$$\vec{\tau} = \vec{r} \times (\vec{F}_{\parallel} + \vec{F}_{\perp}) = \vec{r} \times \vec{F}_{\parallel} + \vec{r} \times \vec{F}_{\perp} \tag{8-23}$$

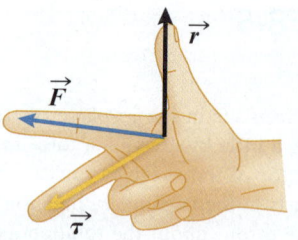

Figure 8-21 "Three-finger" right-hand rule.

The first term in the expression is identically zero because the angle between $\vec{F}_{\parallel}$ and $\vec{r}$ is zero, and only the second term survives, and the expression for the torque reduces to the cross product between $\vec{r}$ and the component of the force that is perpendicular to $\vec{r}$:

$$\vec{\tau} = \vec{r} \times \vec{F}_{\perp} \tag{8-24}$$

The magnitude of the torque in this case is

$$\tau = \left| \vec{r} \times \vec{F}_{\perp} \right| = r F_{\perp} \sin 90° = r F_{\perp} \tag{8-25}$$

and thus we recover Equation 8-15.

Alternatively, we can write the expression for the torque by considering the position vector to be the sum of its vector components:

$$\vec{\tau} = (\vec{r}_{\parallel} + \vec{r}_{\perp}) \times \vec{F} = \vec{r}_{\parallel} \times \vec{F} + \vec{r}_{\perp} \times \vec{F} \tag{8-26}$$

MAKING CONNECTIONS

Gears and Transmission Systems

A simple transmission consists of two gears coupled together (Figure 8-22). Typically, one gear (the drive gear) is connected to a power source, such as the engine of a car, and couples to the second gear (the driven gear). If the two gears are not the same size, they will rotate at different speeds. An automotive variable transmission has a mechanism that couples the gear to different-sized gears driven by the engine. When the smallest engine gear is coupled to the wheel gear, a relatively high engine rpm results in a relatively low wheel rotation speed. This gear ratio is useful when the car encounters resistance, such as when starting from rest or travelling uphill. Putting the car in low gear effectively assigns a smaller moment arm to the external resisting force, thus reducing the resistive torque (similar principles apply to multiple-gear bicycle systems). The engine can then turn the wheels with less torque. Once the car is moving, it is more efficient to couple to a larger engine gear. Coupling gears by a chain, as shown, is only one type of transmission mechanism; in cars this is seen for example in the timing chain (or belt). Other common mechanisms involve direct coupling between gears of different size.

hidesy/iStock/Thinkstock

Figure 8-22 A simple transmission with two gears coupled by a chain.

Here, again, the first term vanishes; therefore

$$\vec{\tau} = \vec{r}_{\perp} \times \vec{F} \qquad (8\text{-}27)$$

and the magnitude of the torque becomes

$$\tau = \left| \vec{r}_{\perp} \times \vec{F} \right| = r_{\perp} F \sin 90° = r_{\perp} F \qquad (8\text{-}28)$$

thus giving us back Equation 8-20.

We keep the numbers for Equations 8-25 and 8-28, despite the fact they are identical to Equations 8-15 and 8-20, for emphasis and to facilitate a vector-only approach where it is preferred.

Connection to the Right-Hand Rule

The previous discussion demonstrates that when $\vec{r}$ and $\vec{F}$ are not perpendicular, we can use the "three-finger" right-hand rule by taking the cross product of $\vec{r}$ with $\vec{F}_{\perp}$, or the cross product of the component of $\vec{r}_{\perp}$ with the force.

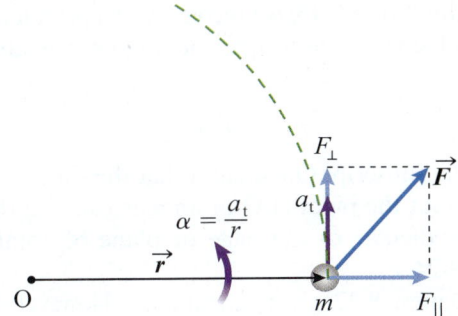

Figure 8-24 A force is applied to a point mass at the end of a rigid massless rod that rotates about point O. We can consider the force to be divided into two components, radially outward ($F_{\parallel}$) and perpendicular to the rod ($F_{\perp}$). This second component will be kept tangent to the circle. There would also be a radially inward acceleration, but it is not shown.

by considering the rotating point mass in Figure 8-24. We are seeking to find a relationship between the angular acceleration and the applied torque.

The mass in Figure 8-24 is rigidly attached to a massless rod that allows it to rotate at a constant distance from point O. The point mass and the rod are originally at rest. We now apply a force $\vec{F}$ to the mass, as shown. The mass will move upward initially, but because of the rigid rod, it will move in a circular trajectory about the axis through O perpendicular to the page. We will keep the magnitude of the force and its direction relative to the massless rod constant. Note that although the circular motion will require a radially inward acceleration, we do not focus on radial acceleration here, since that is directed inward along the rod. Here our focus is on how the angular acceleration, α, changes in response to the application of the force, and α depends on the tangential acceleration.

According to Newton's second law, $\vec{F} = m\vec{a}$. In our point mass example of Figure 8-24, the parallel component of the force does not contribute to the tangential acceleration of the mass, and the tangential acceleration, a_t, of the point mass m depends only on the force component perpendicular to the rod, marked $F_{\perp}$. By Newton's second law, we have the following relationship between the magnitudes of the force and the tangential acceleration:

$$F_{\perp} = ma_t \qquad (8\text{-}29)$$

From Equation 8-8, the magnitude of the angular acceleration is related to the magnitude of the tangential acceleration by $\alpha = a_t/r$, where r is the length of the rod. Therefore, we can say

$$F_{\perp} = m\alpha r \qquad (8\text{-}30)$$

If we multiply both sides of Equation 8-30 by r, we have

$$rF_{\perp} = rm\alpha r = m\alpha r^2 \qquad (8\text{-}31)$$

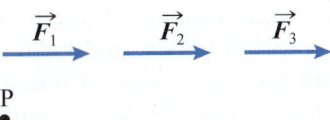
8-4 Moment of Inertia of a Point Mass

Now that we have completed our introduction to rotational kinematics and have introduced torques, we are in a good position to relate torque to angular motion through rotational dynamics. We will begin

From Equitation 8-15, we recognize the product on the left-hand side as the torque the force exerts about O; then

$$\tau_O = mr^2\alpha \qquad (8\text{-}32)$$

where the subscript O indicates that the torque is measured about the pivot at O or an axis running through O perpendicular to the page or plane of rotation in Figure 8-24.

Equation 8-32 gives magnitudes. However, torque and angular acceleration are both vector quantities. Recognizing that this is the only torque in this example, we can write the corresponding relationship in vector form:

KEY EQUATION
$$\vec{\tau}_{net} = mr^2\vec{\alpha} \qquad (8\text{-}33)$$

By writing $\vec{\tau}_{net}$, we also allow for the possibility of more than one torque contributing to the acceleration, and we take their net vector sum.

Moment of Inertia of a Point Mass

Equation 8-33 is the rotational equivalent for a point mass of Newton's second law, $\vec{F} = m\vec{a}$. The torque takes the place of the force, and the angular acceleration replaces the linear acceleration. The mass, m, in the case of a point mass is replaced by the quantity mr^2.

An object's mass is a measure of its inertia, or resistance to acceleration in a straight line. Newton's second law tells us that the net force required for a given acceleration of an object, as shown in Figure 8-25(a), depends solely on the mass of the object. However, if we hold the object at a constant distance from an axis and rotate it about the axis with a given angular acceleration, as shown in Figure 8-25(b), Equation 8-33 tells us that the torque needed to provide that angular acceleration depends not only on the mass of the object, but also on the square of the distance of the object from

the rotation axis. In Figure 8-25(b), the axis is a virtual one that runs through the spine of the person.

For a point mass, the product mr^2 is called the **rotational inertia**, or the **moment of inertia**. We use the symbol I for moment of inertia:

KEY EQUATION
$$I_O = mr^2 \text{ (point mass)} \qquad (8\text{-}34)$$

where the subscript O is used to indicate that the moment of inertia of the mass is measured about the pivot at O, or an axis running through the pivot at O perpendicular to the plane of rotation. We will sometimes drop this subscript, but it is mentioned here and in other places to highlight the central importance of the axis or pivot in moment of inertia calculations. The SI unit for moment of inertia is $kg \cdot m^2$.

The moment of inertia can be thought of as a measure of an object's resistance to rotational acceleration. In this section, we only consider point masses, but we will later see how to calculate the moment of inertia for various shapes. Now that we have defined the moment of inertia for a point mass, Equation 8-33 can be rewritten as

KEY EQUATION
$$\vec{\tau}_{net} = I\vec{\alpha} \qquad (8\text{-}35)$$

Try this simple exercise to convince yourself of the validity of what we have just considered. Hold a relatively heavy and small object, such as a stone or a dumbbell, and bring it close to your chest so it is touching your body. Stand up straight, and rotate about a virtual axis running through the length of your spine. Do this in a back and forth sense, so that you move from rest to rotation in one direction, and then reverse. Next repeat the same exercise with your arms stretched fully in front of you, still holding the object and with approximately the same angular acceleration as you did before. Now observe the amount of

(a)

(b)

Figure 8-25 (a) In linear motion, resistance to acceleration depends only on the mass of an object. (b) In rotational motion, resistance to acceleration also depends on distance: the farther the object is from the rotation axis, the greater the resistance to rotational acceleration.

resistance provided by the heavy object in your arms, and the extra pressure in the small of your back in both cases. Although the angular acceleration is pretty much the same, the amount of "resistance" you experience is very different. Please do this exercise gently so that it does not jeopardize your safety; be extra cautious not to cause injury to your back. The resistance to motion an object presents when being accelerated in a straight line on a frictionless surface depends only on its mass. In rotational motion, the resistance to angular acceleration depends on the moment of inertia. The difference between the two scenarios is illustrated in Figure 8-25.

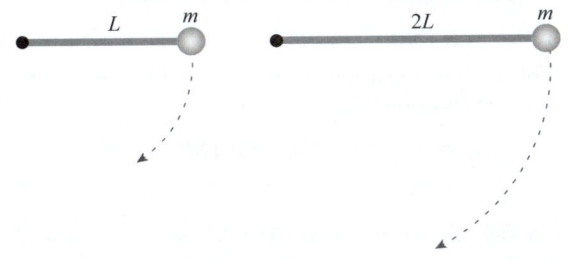
8-5 Systems of Particles and Rigid Bodies

A System of Point Masses

We saw in Chapter 5 that if we push a number of blocks stacked one behind the other to move them together,

as in Example 5-14, the net force needed is given by $\vec{F}_{net} = m_T\vec{a}$, where m_T is the total mass or sum of the masses of the individual blocks. Therefore, while we could, and often do, treat the individual blocks as point masses as long as they move together, we can treat the system of blocks as one mass being pushed with the net applied force. Does the same principle apply when it comes to rotation? Let us explore this.

Consider the system of three particles shown in Figure 8-27. The particles are attached to a massless rod that pivots around an axis perpendicular to the rotation and passing through point O.

Let us now apply a torque to the rod to cause an angular acceleration, α. The net torque on the system of three masses is the sum of the net torques on the individual masses using $\vec{\tau}_{net} = mr^2\vec{\alpha}$ from Equation 8-33 for each of the masses:

$$\vec{\tau}_{net} = \vec{\tau}_1 + \vec{\tau}_2 + \vec{\tau}_3 = m_1r_1^2\vec{\alpha} + m_2r_2^2\vec{\alpha} + m_3r_3^2\vec{\alpha} \quad (8\text{-}36)$$

We note that the angular accelerations of the three masses are the same, $\alpha_1 = \alpha_2 = \alpha_3 = \alpha$, because they are connected with a rigid rod. Therefore, Equation 8-36 can be written as

$$\vec{\tau}_O = m_1r_1^2\vec{\alpha} + m_2r_2^2\vec{\alpha} + m_3r_3^2\vec{\alpha} \quad (8\text{-}37)$$

$$\vec{\tau}_O = (m_1r_1^2 + m_2r_2^2 + m_3r_3^2)\vec{\alpha} \quad (8\text{-}38)$$

This equation can now be written in the following compact form:

$$\vec{\tau}_{net} = I_{tot}\vec{\alpha} \quad (8\text{-}39)$$

where $I_{tot} = m_1r_1^2 + m_2r_2^2 + m_3r_3^2$ is the total moment of inertia of the system of rigidly connected point masses.

In general, we can say that the moment of inertia of a two-dimensional system of particles (or point masses) about an axis a perpendicular to the plane of the particles is equal to the sum of the individual moments of inertia of each of the particles about that axis:

KEY EQUATION
$$I_a = \sum_{i=1}^{N} m_i r_i^2 \quad (8\text{-}40)$$

where N is the number of particles. The subscript a indicates that the moment of inertia is measured about the

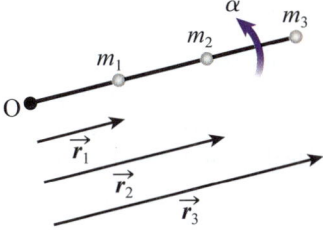

Figure 8-27 Three point masses connected to a rigid massless rod. Rotation is in the plane of the page, about a perpendicular axis through point O.

axis a. The axis about which the moment of inertia is calculated is sometimes also called the axis of rotation.

We can use the summation notation of Equation 8-40 as long as we are dealing with a system of discrete point masses held together as a **rigid body**. This is demonstrated in Example 8-6. A body is considered rigid if it does not deform when a force or torque, no matter how large its magnitude, is applied to it. A solid metal object is approximately rigid, while an object made of play dough certainly is not. Another name for a rigid body in this context is **non-deformable**. Of course, no actual bodies are exactly rigid, but many are rigid to a good approximation. You will study the deformation of objects in Chapter 10. We state here that a more complete statement on the response of an object to torque is that it either accelerates rotationally, deforms, or both.

EXAMPLE 8-6

Moment of Inertia of a Set of Point Masses

Four point masses are held in position with a set of massless, rigid connecting rods, as indicated in Figure 8-28. The masses are all located in the xy-plane. Obtain an expression for the moment of inertia, in terms of m and d, if the object rotates

(a) about the z-axis through the origin of the coordinate system
(b) about the x-axis through the origin of the coordinate system
(c) about the y-axis through the origin of the coordinate system
(d) about an axis running through the mass $(3m)$ perpendicular to the xy-plane

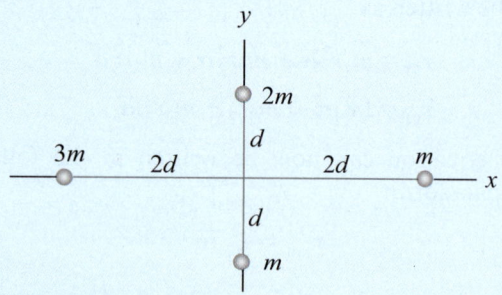

Figure 8-28   Example 8-6.

Solution

When dealing with a set of point masses, we can find the moment of inertia of the entire object by summing the moments of inertia of the individual point masses. We employ the following relationship, where r_i represents the distance from the rotation axis to the point mass.

$$I = \sum_{i=1}^{N} m_i r_i^2$$

(a) For rotation about the z-axis, we have

$$I_z = m(2d)^2 + 2md^2 + 3m(2d)^2 + md^2$$
$$= 19md^2$$

The subscript z indicates the axis of rotation about which the moment of inertia is calculated. The same goes for x, y, and a in parts (b) to (d).

(b) For rotation about the x-axis, we note that for the masses that are on the axis, the distance between the mass and the axis is zero, so

$$I_x = m(0)^2 + 2m(d)^2 + 3m(0)^2 + m(d)^2$$
$$= 3md^2$$

(c) For rotation about the y-axis, we note that the distances for the masses on the axis are zero:

$$I_y = m(2d)^2 + 2m(0)^2 + 3m(2d)^2 + m(0)^2$$
$$= 16md^2$$

(d) We call this axis the a-axis. The distance of the mass $3m$ to this axis is zero. The distance between this axis and the mass m on the x-axis is $4d$. For the other two masses, m and $2m$ on the y-axis, we use Pythagoras's theorem:

$$I_a = m(4d)^2 + 2m[(2d)^2 + d^2] + 3m(0)^2 + m[(2d)^2 + d^2]$$
$$= 16md^2 + 10md^2 + 0 + 5md^2$$
$$= 31md^2$$

Making sense of the result

The moment of inertia depends on both the mass and the square of distance from the rotation axis. Since the distance is squared, the contributions from the more distant point masses dominate. You must be clear on the rotation axis before calculating a moment of inertia.

PEER TO PEER

I must always calculate the moment of inertia about an axis. For a system of masses, I must measure the distances for the moment of inertia calculation from the axis about which I calculated the moment of inertia.

Moment of Inertia for Continuous Objects

What if we do not have point masses, but rather an object with a continuous mass distribution? In this case, we view the object as broken up into infinitesimally small mass elements dm and replace the summation of Equation 8-40 with integration over the mass of the object:

KEY EQUATION

$$I = \int_{\text{object}} r^2 \, dm \qquad (8\text{-}41)$$

Here r represents the distance from the rotation axis to that mass element. Objects with mass concentrated at larger distances from the rotation axis will have larger relative moments of inertia.

Before we use Equation 8-41, we need to define the term **uniform**, as it will come up frequently in this and other chapters. By uniform, we mean that the properties of the object, such as density, do not change over its volume. We can equivalently say that the object is **homogeneous**. For example, a ball made of brass or steel is considered to be uniform, or homogeneous, while a ball made by combining two hemispheres, one of steel and one of aluminum, is not uniform. We can write the differential mass element in terms of a differential volume element, dV, as follows, where ρ is the mass density (i.e., the mass per unit volume):

$$dm = \rho dV \qquad (8\text{-}42)$$

When the object is homogeneous, ρ does not vary with position inside the object. We will show examples of how integration is used to calculate moments of inertia shortly.

The moment of inertia of objects is a very important topic, since it helps us understand how these objects rotate in response to torques. For example, we all know that a cylinder will roll down an incline (assuming, as we will see, there is sufficient friction between the surfaces). How would two cylinders with different distributions of mass roll down an incline? As we will see in Chapter 9, to answer that question we must be able to calculate the moment of inertia for objects of different shapes. Before we get into mathematical details of calculating the moment of inertia, test your intuition with the following checkpoint.

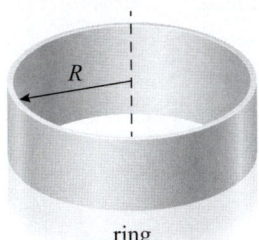 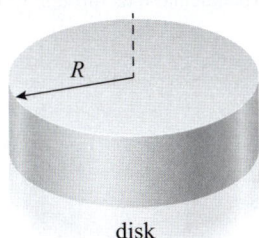
A Thin Ring

Now let us calculate the moment of inertia of a rigid uniform ring with mass M, radius R, and negligible width. Although we consider all the mass in the ring, we imagine some massless structure permits it to rotate about a perpendicular axis through its centre of mass, as shown in Figure 8-30. A bicycle wheel with thin and light spokes is a good approximation for an ideal ring.

Before we use calculus and Equation 8-41 to compute the moment of inertia for this object, we will approximate the ring with many tiny identical mass elements distributed uniformly along the rim (Figure 8-31). Let us assign a mass m_i to each of these point masses. The moment of inertia of the entire ring about an axis is simply the sum of all the moments of inertia of the individual point masses about that axis. For an axis that passes through the centre of mass of the ring and is perpendicular to the plane of the ring, the moment of inertia, I_{cm}, can be found using Equation 8-40:

$$I_{cm} = \sum_{i=1}^{n} m_i r_i^2 = m_1 r_1^2 + m_2 r_2^2 + m_3 r_3^2 + \cdots + m_n r_n^2$$

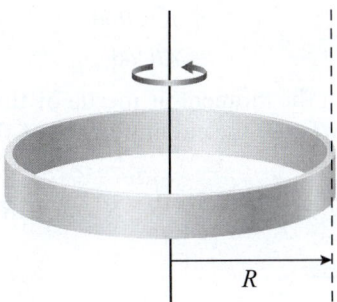

Figure 8-30 A rigid uniform ring of mass M and radius R, with a rotation axis perpendicular to it.

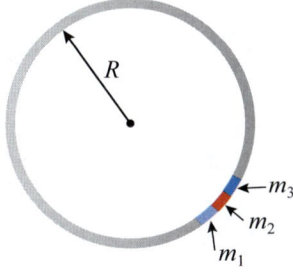

Figure 8-31 Viewing a uniform ring as a series of tiny discrete masses.

Since all the point masses are located on the circumference at a distance R from the centre of the ring, we have

$$I_{cm} = m_1R^2 + m_2R^2 + m_3R^2 + \cdots + m_nR^2$$
$$= (m_1 + m_2 + m_3 + \cdots + m_n)R^2$$

The sum of all the point masses making up the ring adds to the total mass of the ring, so we have the following result:

$$I_{cm} = MR^2 \qquad (8\text{-}43)$$

Since all of the mass is at the same distance from the rotation axis, this result is what we would expect.

The distribution of mass in the ring is continuous, so the moment of inertia of the ring is more appropriately calculated using integration. As shown in Figure 8-32, the moment of inertia of an infinitesimal small deferential element of mass dm about an axis perpendicular to the ring through its centre is

$$dI_{cm} = R^2dm \qquad (8\text{-}44)$$

Since the ring is homogeneous, it has uniform linear mass density, λ (mass per unit length). So, $dm = \lambda dl$, where dl is the infinitesimally small length of mass dm along the ring, and

$$dI_{cm} = R^2dm = R^2(\lambda dl) \qquad (8\text{-}45)$$

Now, dl can be written as an arc length as $Rd\theta$, where $d\theta$ is the corresponding differential angle. Substituting in Equation 8-44 gives

$$dI_{cm} = R^2dm$$
$$= R^2(\lambda dl)$$
$$= R^2\lambda Rd\theta$$
$$= \lambda R^3 d\theta \qquad (8\text{-}46)$$

To obtain the moment of inertia of the entire ring, we integrate dI over the circumference of the ring:

$$I_{cm} = \int dI = \int R^2dm = \int_0^{2\pi} R^3\lambda d\theta = R^3\lambda \int_0^{2\pi} d\theta$$
$$= R^3\lambda 2\pi \qquad (8\text{-}47)$$

Since the total mass of the ring is the circumference times the mass per unit length, $M = \lambda(2\pi R)$, we have

$$I_{cm} = MR^2$$

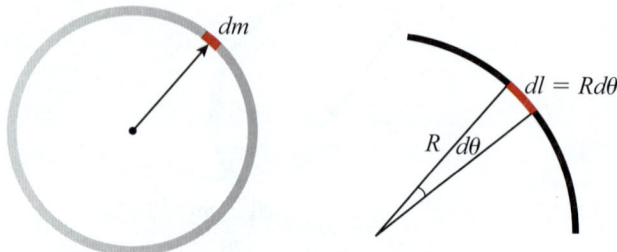

Figure 8-32 Differential elements for the calculation of the moment of inertia of a ring about a perpendicular axis through its centre of mass.

This expression is the same as Equation 8-43.

A Solid Disk

Let us now consider the moment of inertia for a thin uniform disk (Figure 8-33) about an axis perpendicular to its plane, through its centre. The mass per unit area of this disk is a constant, σ, with units of kilogram per square metre.

We will consider the disk as made up of many rings and use the previous result for the moment of inertia of a ring. The differential element we begin with is a ring of radius r and infinitesimal width dr, as shown in Figure 8-34. The area of this ring is $dA = 2\pi rdr$, and the mass of the ring is $dm = \sigma dA = \sigma 2\pi rdr$.

As shown in Equation 8-43, the moment of inertia of an infinitesimally thin ring of mass m and radius r is mr^2. Therefore, for a ring of width dr, we have

$$dI = r^2dm = r^2(2\pi r\sigma dr) = 2\pi r^3\sigma dr \qquad (8\text{-}48)$$

We integrate this differential element over the area of the disk to obtain the moment of inertia of the entire disk for rotation about a perpendicular axis through its centre of mass:

$$I_{cm} = \int dI = \int_0^R 2\pi r^3\sigma dr = 2\pi\sigma \int_0^R r^3dr = 2\pi\sigma \frac{R^4}{4} \qquad (8\text{-}49)$$

Since the total mass of the disk is $\pi R^2\sigma$, the expression for the total moment of inertia simplifies to

$$I_{cm} = \frac{MR^2}{2} \qquad (8\text{-}50)$$

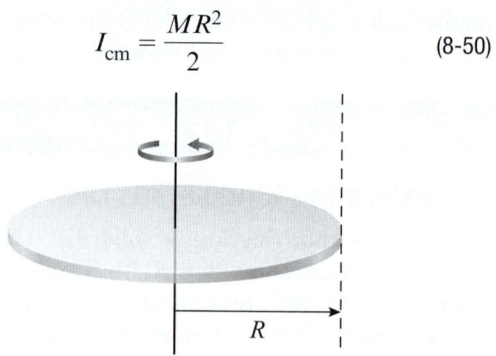

Figure 8-33 Uniform disk of mass M and radius R rotating about a perpendicular axis through its centre of mass.

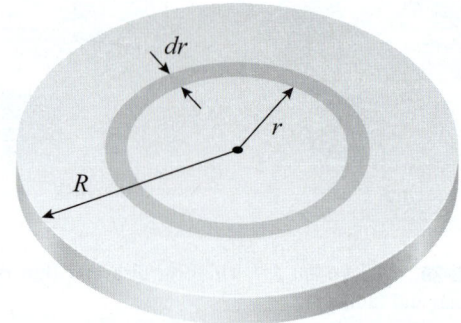

Figure 8-34 The differential element for the calculation of the moment of inertia of a uniform disk.

If we think of a disk as being made up of point masses, the point masses located at the edge of the disk will each have a moment of inertia of mR^2, the same as the point masses making up a ring. However, the rest of the point masses making up the disk are closer to the centre of the disk and will have correspondingly smaller moments of inertia. For example, the point mass located at the centre of the disk has zero moment of inertia. Thus, the moment of inertia of a disk about its centre of mass is less than the moment of inertia of a ring of the same mass and radius.

MAKING CONNECTIONS

If Earth Did Not Rotate

Our survival on Earth's surface depends, in part, on the effect that Earth's rotation has on its molten core. The outer core is a fluid mass of iron, nickel, and sulfur. A combination of Earth's rotation, convection currents, buoyancy, and compositional currents causes hotter fluid close to the inner core to rise in helical columns aligned more or less with Earth's axis (Figure 8-35).

The resulting electrical currents produce the bulk of Earth's magnetic field, which shields Earth's surface from the potentially devastating effects of the stream of charged particles ejected from the Sun (the solar wind). The interaction of the charged particles with Earth's magnetic field also causes the aurora borealis and the aurora Australis.

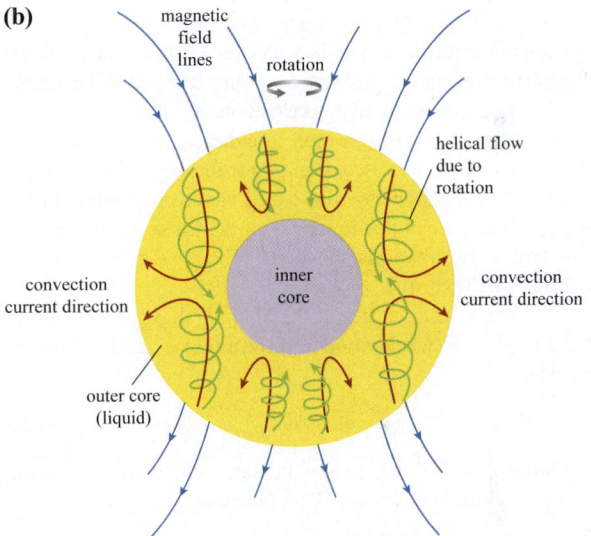

Figure 8-35   (a) The aurora borealis. (b) Earth's rotation ultimately produces currents whose magnetic fields shield Earth from the solar wind.

EXAMPLE 8-7

Clever Monkey

The solid drum in Figure 8-36(a) is free to rotate about an axis through its centre. A 23.0 kg monkey decides to ride the rope on the drum down to the ground 12.0 m below. The rope is at rest when the monkey grabs onto it. The drum is a uniform cylindrical disk of radius 0.180 m and mass 190.0 kg. Find the acceleration of the monkey and the tension in the rope. Assume that the mass of the rope is negligible.

Solution

Let us begin by drawing the FBD for the monkey (Figure 8-36(b)). The forces acting on the monkey are the force of gravity and the tension ($\vec{T}$) in the rope. The motion we are considering is in the y-direction, vertically, and we will define upward as the positive direction. We will use m for the mass of the monkey, M for the mass of the disk, and r for the radius of the disk.

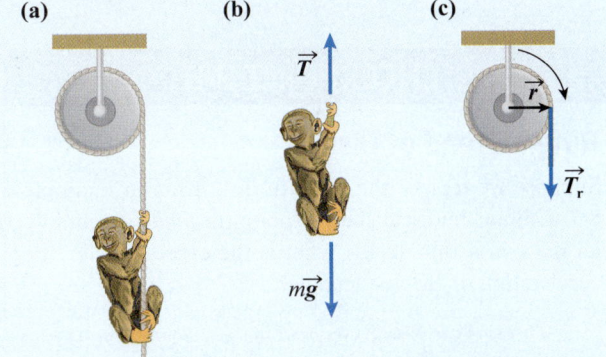

Figure 8-36   (a) Example 8-7. (b) The FBD of the monkey. (c) The force and moment arm to find the torque on the disk.

(continued)

Summing the forces in the *y*-direction for the monkey, and choosing down as the positive direction we have, by Newton's second law,

$$mg - T = ma \tag{1}$$

Since both *T* and *a* are unknown, we cannot directly solve (1).

To get a second relationship so we can solve for the two unknowns, consider the torque exerted on the disk by the rope. We call the tension force of the rope on the disk $\vec{T}_r$. The tension everywhere in the ideal, massless rope has the same magnitude, but the direction it exerts on the disk must be opposite to the direction of the force exerted by the rope on the monkey:

$$\vec{T}_r = -\vec{T} \tag{2}$$

As shown in Figure 8-36(c), we draw a vector $\vec{r}$ from the rotation axis to the point where the force $\vec{T}_r$ is applied. The torque exerted on the disk is determined from the relationship

$$\vec{\tau} = \vec{r} \times \vec{T}_r \tag{3}$$

When we apply the right-hand rule for the vector cross product (see Figure 8-36(c)), our thumb will point into the paper, so that is the direction of the torque vector. The direction of the angular acceleration and the torque are the same, according to $\vec{\tau} = I\vec{\alpha}$. Therefore, the disk will have an angular acceleration, resulting in it starting to rotate in the clockwise direction, as indicated in Figure 8-36(c).

From our earlier discussion on torque, we can also simply say that the tension in the rope exerts clockwise torque on the pulley, causing the pulley to have a clockwise angular acceleration. Both considerations above are valid ways to indicate the direction of the torque.

We know from earlier in the chapter that the magnitude of the tangential acceleration and the angular acceleration are related by

$$a_t = r\alpha \tag{4}$$

Since $\vec{r}$ and $\vec{T}_r$ are perpendicular, we have the following for the magnitude of the torque on the disk:

$$\tau = rT_r \sin 90° = rT_r = rT \tag{5}$$

Since $\tau = I\alpha$, we can use (5) to find the magnitude of the angular acceleration of the disk, where we have used the expression for moment of inertia of a solid disk (remember that *M* is the mass of the disk, while *m* is the mass of the monkey):

$$\alpha = \frac{\tau}{I} = \frac{rT}{Mr^2/2} = 2\frac{T}{Mr} \tag{6}$$

We saw from the earlier application of the right-hand rule that the torque caused an angular acceleration in a direction that results in a downward acceleration on the rope. Therefore, since the rope acceleration is the tangential acceleration, combining (4) and (6) we have

$$a = 2\frac{T}{M}$$

$$T_r = \frac{aM}{2} \text{ or } T = -\frac{aM}{2} \tag{7}$$

where we have used (2). Substituting (7) into (1), we can solve for the unknown acceleration:

$$-\frac{aM}{2} = ma$$

$$a = \frac{mg}{(m + M/2)} \tag{8}$$

We can substitute this back into (1) to solve for the acceleration:

$$a = \frac{(23.0 \text{ kg})(9.81 \text{ m/s}^2)}{\left(23.0 \text{ kg} + \dfrac{190.0 \text{ kg}}{2}\right)} = 1.912 \text{ m/s}^2 = 1.91 \text{ m/s}^2$$

We can substitute the acceleration into (7) to find the tension in the rope:

$$T = \frac{(1.912 \text{ m/s}^2)(190.0 \text{ kg})}{2} = 181.6 \text{ N} = 182 \text{ N}$$

In both cases we have kept an extra digit in the intermediate results but rounded final answers to the correct number of significant digits.

Making sense of the result

We see from result (8) that if the disk was of negligible mass, then the acceleration would just be that of *g*, as expected. When the moment of inertia is significant, the acceleration must be less, and we find a reasonable acceleration result of about one quarter of *g*. The moment of inertia of the drum slows the acceleration of the monkey. Note that the tension is not equal to *mg*, since there is acceleration and we are not in a static situation.

CHECKPOINT 8-8

Ring Instead of Disk

Suppose we replace the uniform disk drum in Example 8-7 with a cylindrical shell or hoop (i.e., with essentially all mass in a thin shell). What is the expression for the acceleration of the system now?

ANSWER: The expression for *a* is now $a = \dfrac{Mg}{m + M}$ because the moment of inertia will be $I = Mr^2$ (instead of $I = \dfrac{1}{2}Mr^2$), since all of the mass is now at distance *r*.

Moment of Inertia for Composite Objects

If you have an object that is a combination of several objects, the moment of inertia of the combined object can be found by superposition. You simply add the individual moments of inertia of all the individual parts about the axis of rotation (assuming that all rotate about the same rotation axis). We demonstrate this principle in Example 8-8.

EXAMPLE 8-8

Moment of Inertia for a Combination of Objects

As shown in Figure 8-37, a uniform disk is of radius R and mass m_1. Attached to the disk is a thin ring of mass m_2 and radius R. There is a small mass m_3 (assume a point mass) attached to the surface of the disk at a distance of $R/2$ from the rotation axis through the centre of the disk perpendicular to the plane of the disk. The direction of rotation is shown in the figure for added clarity. What is the moment of inertia of the composite object?

Solution

We simply add the moments of inertia (about the indicated rotation axis) of the three parts that make up the composite object:

$$I = \frac{1}{2}m_1R^2 + m_2R^2 + m_3\left(\frac{R}{2}\right)^2$$

$$= R^2\left(\frac{m_1}{2} + m_2 + \frac{m_3}{4}\right)$$

Making sense of the result

This technique is very useful in many situations where it is much easier to calculate the moments of inertia of the parts separately. *Remember that the objects must be attached so that they rotate together about a common rotation axis.* The overall moment of inertia will always be greater than that of any individual part.

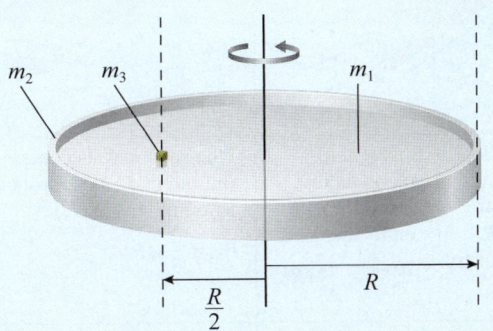

Figure 8-37 Example 8-8.

The Parallel-Axis Theorem

So far we have considered moments of inertia for objects rotating about an axis through the centre of mass of the object. In many applications, the rotation axis does not run through the centre of mass. The calculation of the moment of inertia depends on the axis of rotation, and one needs to find the correct moment of inertia for each situation.

Hold the middle of a horizontal metre stick between your thumb and index finger, and rotate the metre stick back and forth horizontally about an axis perpendicular to its length. Keep the angle by which you rotate the metre stick relatively small. You may do also try this with your pen. Gauge the amount of resistance the metre stick presents as you accelerate it. Next, hold the metre stick horizontal with one end between your thumb and index finger. Swing the metre stick back and forth through approximately the same angular displacement as when you rotated it horizontally. You will notice that you encounter more resistance and need to use more force when you hold the stick at its end. This difference indicates that the moment of inertia of the metre stick is greater about an axis running through its end than about an axis through its centre of mass. If you don't have a metre stick handy, and decide to use your pen, you might have to rotate it back and forth a bit faster to feel the effect.

The moments of inertia of commonly encountered geometrical objects about symmetry axes running through their centres of mass are given in Table 8-3. What if we wanted to rotate these objects about axes other than those through their centre of mass like the metre stick or the pen as we just discussed? This is accomplished using what is known as the *parallel-axis theorem*, which we discuss next.

To understand how the choice of axis affects the moment of inertia, we begin by calculating the moment of inertia of a thin uniform bar about an axis running through its centre of mass perpendicular to its length. The moment of inertia of the differential element of mass dm (Figure 8-38) about the centre-of-mass axis running perpendicular to the length of the bar is $r^2 dm$. Using λ as the linear density of the rod, we can express dm as λdr. The moment of inertia of the differential element is then $dI = r^2\lambda dr$. We obtain the moment of inertia of the bar by integrating over its length:

$$I_{cm}^{bar} = \int_{length} r^2 dm = \int_{-L/2}^{L/2} r^2\lambda dr = \frac{\lambda L^3}{12} \tag{8-51}$$

The total mass of the rod is given by $M = \lambda L$, so

$$I_{cm}^{bar} = \frac{1}{12}ML^2 \tag{8-52}$$

Next we consider the moment of inertia of the bar about an axis, a, perpendicular to the bar and running through one of its ends. The rotation axis is parallel to the centre-of-mass axis. The moment of inertia of the differential element of mass dm about the axis a is given by $r'^2 dm$, where r' is the distance from dm to axis a.

The distance r' can be written as $r' = r + d$, where d is the distance between the centre-of-mass axis and the axis a.

Table 8-3 Moments of Inertia for Common Objects about Different Axes

Thin rod, mass M and length L; centre-of-mass axis perpendicular to bar; $$I^{\text{bar}}_{\text{cm}} = \frac{1}{12}ML^2$$	Thin rod, mass M and length L; axis running through edge perpendicular to bar; $$I^{\text{bar}}_{a} = \frac{1}{3}ML^2$$	Rectangular block, mass M, sides lengths a and b; centre-of-mass axis perpendicular to plane of rectangle: $$I = \frac{1}{12}M(a^2 + b^2)$$	Rectangular block, mass M, side lengths a and b; axis along one of the sides of length b: $$I = \frac{1}{3}Ma^2$$

Cylinder, mass M, inner radius R_1, outer radius R_2; axis of symmetry: $$I^{\text{cyl}}_{\text{cm}} = \frac{1}{2}MR_2^2 - \frac{1}{2}MR_1^2$$	Solid cylinder, mass M, radius R, length L; axis of symmetry: $$I^{\text{cyl}}_{\text{cm}} = \frac{1}{2}MR^2$$	Ring, or hollow cylinder, thin shell, mass M and radius R; axis of symmetry: $$I^{\text{ring}}_{\text{cm}} = MR^2$$	Sphere, mass M and radius R; axis through centre of sphere: $$I^{\text{sphere}}_{\text{cm}} = \frac{2}{5}MR^2$$	Thin spherical shell, mass M and radius R; axis through centre of shell: $$I^{\text{ss}}_{\text{cm}} = \frac{2}{3}MR^2$$

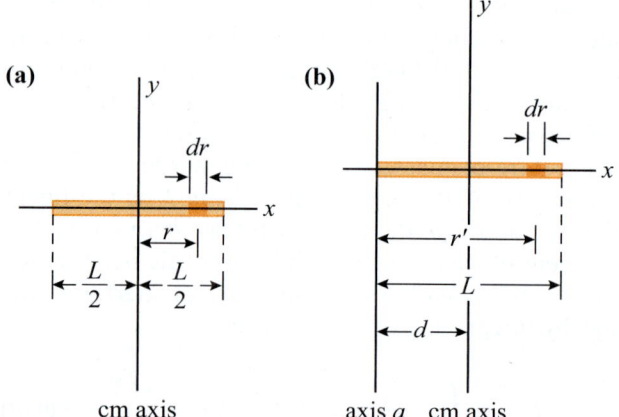

(a) cm axis

(b) axis a cm axis

Figure 8-38 The moment of inertia of a differential element of a bar about an axis (a) through the centre of mass and (b) through the end of the bar and parallel to the axis through the centre of mass.

Then, the moment of inertia of the differential element of mass dm is $(r + d)^2 dm$. We integrate this differential over the length of the bar to obtain the moment of inertia of the bar about axis a:

$$I^{\text{bar}}_{a} = \int_{\text{length}} r'^2 dm = \int_{\text{bar}} (r + d)^2 dm$$
$$= \int_{\text{bar}} (r^2 + d^2 + 2rd)\,dm \tag{8-53}$$

$$I^{\text{bar}}_{a} = \int_{\text{bar}} r^2 dm + \int_{\text{bar}} d^2 dm + \int_{\text{bar}} 2(d)(r)\,dm \tag{8-54}$$

The first part of the expression is the moment of inertia of the bar about the centre-of-mass axis, as in Equation 8-52. The integral of dm over the length of the bar gives the total mass of the bar, M; therefore, the second term in Equation 8-54 equals Md^2. The third term can be written as $2d\int_{\text{bar}} (r)\,dm$. The integral $\int_{\text{bar}} r\,dm$ is Mr_{cm}, as we saw in Chapter 7. Since we have chosen our coordinate system so that the centre of mass of the bar is at the origin, the third term in Equation 8-54 reduces to zero. So, the expression of the moment of inertia of the bar about axis a can be written as

$$I^{\text{bar}}_{a} = I^{\text{bar}}_{\text{cm}} + Md^2 \tag{8-55}$$

Equation 8-55 is an expression of the **parallel-axis theorem**. This theorem applies for any object and for any parallel axis at any arbitrary distance. The theorem can be stated as follows:

KEY EQUATION $$I^{\text{object}}_{a} = I^{\text{object}}_{\text{cm}} + Md^2 \tag{8-56}$$

Equation 8-56 states that the moment of inertia of an object about an arbitrary axis is equal to the moment of inertia of that object about a parallel axis running through its centre of mass, plus the product of the total mass of the object and the square of the distance between the centre-of-mass axis and the arbitrary axis.

For the above bar, $d = L/2$, and

$$I_a^{\text{bar}} = \frac{1}{12}ML^2 + M\left(\frac{L}{2}\right)^2 = \frac{1}{12}ML^2 + M\left(\frac{L^2}{4}\right) = \frac{1}{3}ML^2 \tag{8-57}$$

Note that this moment of inertia is four times the moment of inertia of the bar about the parallel axis passing through the centre of mass.

CHECKPOINT 8-9

Reference Axis

Given that the moment of inertia of an object about an arbitrary axis a (not the centre-of-mass axis) is I_a, can we say that the moment of inertia about another axis, axis b, parallel to axis a, is $I_b = I_a + md^2$, where d is the distance between the two parallel axes?

ANSWER: No. $I_b = I_{\text{cm}} + md^2$, where d is the distance between the centre-of-mass axis and axis b.

The Perpendicular-Axis Theorem

The **perpendicular-axis theorem** relates the moment of inertia of a planar object about perpendicular axes lying in the object's plane to an axis that is perpendicular to the plane of the object. See problem 103 for details.

8-6 Rotational Kinetic Energy and Work

Next we develop an expression for **rotational kinetic energy**, which is the kinetic energy of a rotating object. We start with a point mass connected to a massless rod, rotating freely about the pivot at point O (Figure 8-39). The linear speed of the point mass at the position shown is v. The kinetic energy of the mass is given by $K = \frac{1}{2}mv^2$. Since the particle is restricted to move in a circle, it has a rotational translational speed of $v = \omega r$.

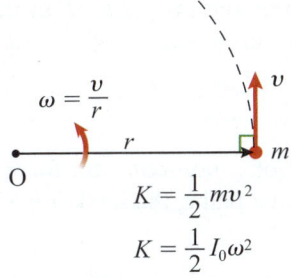

Figure 8-39 The point mass has a linear speed of v and an angular speed of $\omega = v/r$.

Substituting for v in the expression for kinetic energy gives an expression for the rotational kinetic energy of the point mass:

KEY EQUATION
$$K_{\text{rot}} = \frac{1}{2}m\omega^2 r^2 = \frac{1}{2}I_O\omega^2 \tag{8-58}$$

where I_O is the moment of inertia of the point mass about the pivot at O. We can follow the same reasoning as we did in Section 8-5 to show that Equation 8-58 applies to the kinetic energy of any rigid rotating object.

CHECKPOINT 8-10

Wheel and Disk

When a wheel and a disk with the same radius spin such that they have the same kinetic energy,
(a) they have the same mass
(b) they have the same angular speed
(c) the mass of the disk must be twice as large as the mass of the ring
(d) all of the above are possible

ANSWER: (d) All of the above are possible. We are asking for the same kinetic energy, which depends on both the moment of inertia and the angular speed.

Now let us subject the mass to a force $\vec{F}$ that remains perpendicular to the rod, as shown in Figure 8-40. For simplicity, we will keep the force magnitude constant. The work done by the force $\vec{F}$ in Figure 8-40 can be expressed in terms of angular variables. Since the magnitude of the force is constant in this example, the work done is

$$W_F = \vec{F} \cdot \Delta\vec{s} = (F)(\Delta s)\cos\theta \tag{8-59}$$

where θ is the angle between the force and the displacement.

Also for simplicity, we have considered a force that remains perpendicular to the massless rod (Figure 8-39). The angle θ in Equation 8-59 is zero and $\cos\theta = 1$. Since $\Delta s = r\Delta\theta$, the work done by force $\vec{F}$ over distance Δs is

$$W_F = F\Delta s = Fr\Delta\theta \tag{8-60}$$

The magnitude of the torque is rF in this case, so the work done by F can be expressed as (Figure 8-40)

$$W_F = F\Delta s = Fr\Delta\theta = \tau\Delta\theta \qquad (8\text{-}61)$$
$$W\tau = \tau\Delta\theta \qquad (8\text{-}62)$$

In the case of a non-constant force of magnitude F, for rotation in a plane, the work done by the varying torque is

$$W_F = \int_{\Delta s} F ds = \int_{\theta_1}^{\theta_2} \tau d\theta \qquad (8\text{-}63)$$

If the force were not perpendicular to the rod, only the perpendicular component of the force would contribute to the torque, and the parallel component would not contribute to the work done. We can consider the force $\vec{F}$ or the torque τ as doing work on the rotating point mass.

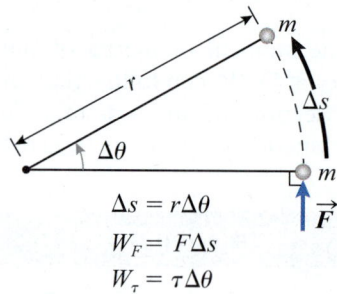

$$\Delta s = r\Delta\theta$$
$$W_F = F\Delta s$$
$$W_\tau = \tau\Delta\theta$$

Figure 8-40 Work done on a rotating point mass.

The universal relation (the work–energy theorem—see Chapter 6) between the total work done on an object and the change in its kinetic energy is

$$W_T = \Delta K \qquad (8\text{-}64)$$

Applying this relationship to rotational motion with constant torque gives the following. For a constant torque:

KEY EQUATION
$$W_T = \tau_{net}\Delta\theta = \frac{1}{2}I\omega_f^2 - \frac{1}{2}I\omega_i^2 \qquad (8\text{-}65)$$

For a varying torque:

KEY EQUATION
$$W_T = \int \tau_{net} d\theta = \frac{1}{2}I\omega_f^2 - \frac{1}{2}I\omega_i^2 \qquad (8\text{-}66)$$

For the work–energy theorem, we consider the work done by *all* the torques acting on an object, or the work done by the net torque acting on an object, as indicated in the two previous equations. It is important to realize that there can be both rotational kinetic energy (for the rotational motion about an axis) and translational kinetic energy (for the motion of the centre of mass). If the object is simply rotating about a fixed axis, it will have only rotational kinetic energy, but, for example, if you throw a hammer through the air and it rotates about some axis as well as moves as a whole, there will be both terms. We will deal with scenarios like these in more detail in Chapter 9.

EXAMPLE 8-9

Energy and Angular Acceleration

Two masses ($m_1 = 31.0$ kg and $m_2 = 23.0$ kg) are connected with a massless rope across a pulley, as shown in Figure 8-41(a). The pulley is free to rotate about an axis through its centre. There is no friction between mass m_1 and the incline. The angle of the incline is $\theta = 32.0°$. The pulley has a radius of 0.300 m and a moment of inertia of 5.00 kg·m². Determine the acceleration of the blocks.

Solution

Method A: Newton's Second Law Applied to Rotation
A quick check shows that $m_2 g$ is greater than $m_1 g\sin\theta$. Therefore, the pulley rotates clockwise, and m_1 accelerates to the right along the incline and m_2 accelerates downward (since the two are connected, the magnitudes of their accelerations must be the same). While not necessary, it is easier in this coupled question to use a positive variable a to represent the

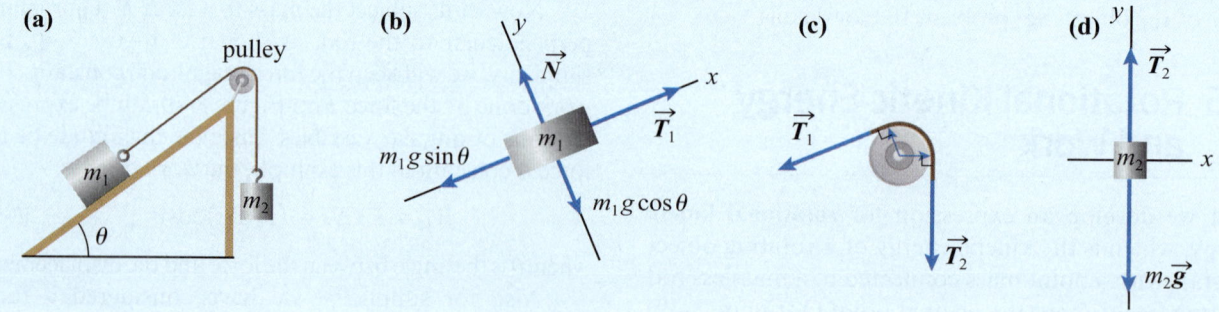

Figure 8-41   (a) Example 8-9. (b) FBD of m_1. (c) FBD of pulley. (d) FBD of m_2.

acceleration of m_1 in the assumed direction. We will use the xy-coordinates shown in Figure 8-41(b) for m_1 and choose to the right along the incline as positive x, and perpendicular to the plane and up as the positive y-direction. As shown in Figure 8-41(d), for m_2 we only need the y-coordinate system, and we define upward as positive.

We first write Newton's second law for m_2. We use $-a$ for the acceleration since we are assuming acceleration in the negative direction:

$$T_2 - m_2 g = -m_2 a \qquad (1)$$

Equation (1) has two unknowns, T_2 and a.

Next we write Newton's second law for m_1 along the incline, remembering that positive is to the right:

$$T_1 - m_1 g \sin\theta = m_1 a \qquad (2)$$

Note that we considered the tension in the rope to be different on each side of the pulley. Can you think of why this must be so?

In (1) and (2), we have three unknowns. We therefore need at least one more equation. For the pulley, we use the rotational equivalent of Newton's second law. Summing the torques about the centre of mass of the pulley, which we designate as O, we get

$$\sum \vec{\tau}_O = I_O \vec{\alpha}$$
$$\vec{\tau}_{O_1} + \vec{\tau}_{O_2} = I_O \vec{\alpha}$$

Choosing counter-clockwise as our positive direction, we get

$$-\tau_{O_1} + \tau_{O_2} = I_O \alpha$$

Since the rope applies force tangentially to the circumference of the pulley, the angle between the tension forces and the distance to the centre of rotation (the pivot point) is 90° for both forces. Therefore,

$$-rT_1 + rT_2 = I_O \alpha \qquad (3)$$

We now have a total of four unknowns: T_1, T_2, α, and a. From the diagram, we can see another relationship: the acceleration of the rope is the same as the tangential acceleration along the circumference of the pulley:

$$a = \alpha r \qquad (4)$$

Here, again, you have a choice of several methods for solving the system of equations. We will substitute the expressions for T_1 from (2), T_2 from (1), and α from (4) into (3):

$$r(m_2 g - m_2 a) - r(m_1 g \sin\theta + m_1 a) = I_O(a/r)$$

Solving for a gives

$$a = \frac{m_2 g - m_1 g \sin\theta}{m_1 + m_2 + I_O/r^2}$$

$$= \frac{(23.0 \text{ kg})(9.81 \text{ m/s}^2) - (31.0 \text{ kg})(9.81 \text{ m/s}^2)\sin 32.0°}{23.0 \text{ kg} + 31.0 \text{ kg} + [(5.00 \text{ kg}\cdot\text{m}^2)/(0.300 \text{ m})^2]}$$

$$= 0.589 \text{ m/s}^2$$

This is the magnitude of the acceleration. Block m_1 is accelerating to the right along the incline with this value, while m_2 accelerates downward with this magnitude.

Let us examine in detail why the two tension forces need to be different. According to (3), the two tensions can be equal in one of two cases: either the angular acceleration of the pulley is zero (i.e., the system is in equilibrium) or the moment of inertia of the pulley is zero (i.e., the mass of the pulley is zero). You may recall that in all the problems involving pulleys in previous chapters we assumed that the pulley was frictionless and massless. This assumption guarantees that $T_1 = T_2$. Note that friction at the axis of the pulley would add an extra torque to the equation, increasing the difference between the tension forces on either side of the pulley.

Method B: The Work–Mechanical Energy Approach
Now we will solve the same problem using an energy approach. Just as we did in Chapter 6, to use the energy approach we will allow m_2 to fall by a vertical distance h and write the energy conservation law for the system. When m_2 descends a distance h, the gravitational potential energy of interaction between m_2 and Earth's potential energy decreases by $m_2 gh$. This loss in gravitational potential energy goes into increasing the potential energy of the system comprised of Earth and m_1 by $m_1 g \sin\theta(h)$ and increasing the kinetic energies of the pulley and the two masses as shown in Figure 8-42. If the two masses are initially at rest,

$$m_2 gh = m_1 g \sin\theta(h) + \frac{1}{2}m_1 v^2 + \frac{1}{2}m_2 v^2 + \frac{1}{2}I\omega^2 \qquad (5)$$

Next we take the derivative of (5) with respect to time, noting that h is a function of time, and obtain

$$m_2 g \frac{dh}{dt} = m_1 g \sin\theta \frac{dh}{dt} + \frac{1}{2}m_1 \frac{d}{dt}(v^2) + \frac{1}{2}m_2 v^2 + \frac{1}{2}I\frac{d}{dt}(\omega^2) \qquad (6)$$

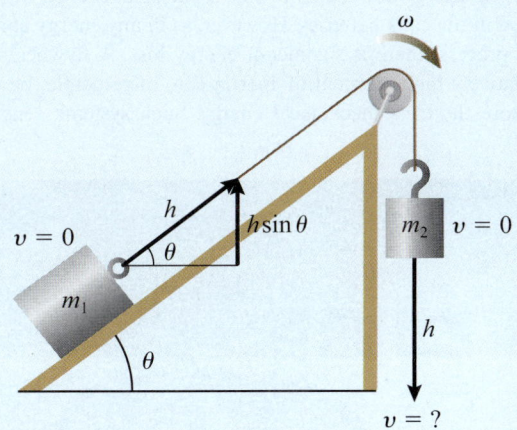

Figure 8-42 As m_2 descends and m_1 goes up the incline, some potential energy of the Earth–m_2 system is converted into potential energy for the Earth–m_2 system and into kinetic energy of the blocks and pulley.

(continued)

We can find the derivatives of v^2 and ω^2 by using the chain rule as shown below:

$$\frac{d}{dt}(v^2) = 2v\frac{dv}{dt} = 2va \qquad (7)$$

$$\frac{d}{dt}(\omega^2) = 2\omega\frac{d\omega}{dt} = 2\omega\alpha \qquad (8)$$

Also, we know that the rate of change of the height is just the speed of block m_2:

$$\frac{dh}{dt} = v \qquad (9)$$

Substituting (7), (8), and (9) into (6), we obtain

$$m_2 g v = m_1 g \sin\theta(v) + m_1 v a + m_2 v a + I\omega\alpha \qquad (10)$$

Substituting a/r for α and v/r for ω into (10) yields

$$m_2 g v = m_1 g \sin\theta(v) + m_1 v a + m_2 v a + \frac{I v a}{r^2} \qquad (11)$$

Dividing (11) by v and solving for a, we obtain the same result derived using Method A:

$$a = \frac{m_2 g - m_1 \sin\theta}{m_1 + m_2 + I_O/r^2}$$

Making sense of the result

Both approaches produced the same result. The energy approach might be a little more straightforward, as it does not require the use of vectors. As a quick check, we can verify the units for the acceleration, a. Every term in the numerator has units of $kg \cdot m/s^2$, and the units of I/r^2 are $kg \cdot m^2/m^2 = kg$, so the units for the denominator are kg. Therefore, the units for the acceleration a are m/s^2, as expected. We were able to use the principle of conservation of energy here because there are no non-conservative forces, such as friction or air resistance, involved. For a pulley with negligible mass, I is zero. If we were to substitute this zero value into the expression for a above, we would obtain the same expression that we had for the same problem in Chapter 5, where the pulley was assumed to be massless and frictionless.

MAKING CONNECTIONS

Mechanical Hybrids and Continuously Variable Transmissions

In 1900, Ferdinand Porsche developed the first electric hybrid vehicle, the Lohner-Porsche Semper Vivus (Figure 8-43). Most automotive brakes use friction to slow a car, converting most of the kinetic energy into thermal energy. Some hybrid cars use magnetic braking, where kinetic energy is converted first into electrical energy and then into electrochemical energy, which is stored in the car's batteries. However, as in any energy conversion process, there is significant energy loss. A flywheel with a relatively high moment of inertia can, in principle, be used to store the car's mechanical energy. Such systems generally require a continuously variable transmission for the precise control of energy between the car and the flywheel. Although no fully mechanical hybrids are in production at the time of writing, continuously variable transmissions are already available on many cars. Fittingly, perhaps, the Porsche 911 GT3 R is the first hybrid vehicle to use mechanical flywheel energy storage instead of the conventional electrochemical battery storage. However, this car still uses magnetic braking to transfer energy to the flywheel, which acts as a generator when called upon.

Figure 8-43 (a) The Lohner-Porsche Semper Vivus, the first electrical hybrid, on display. (b) The Porsche 911 GT3 R, the first semi-mechanical hybrid, on a racetrack.

EXAMPLE 8-10

Falling Bar

The vertical bar shown in Figure 8-44 pivots about its lower end. Starting from rest, the bar swings downward. Find the speed of the free end when the bar has rotated through angle θ.

Solution

Assuming that friction in the system is negligible, all the gravitational potential energy lost as the bar falls will be converted into kinetic energy. From Chapter 6, we have

$$\Delta K = -\Delta U \qquad (1)$$

Since the bar starts from rest, the change in kinetic energy is

$$\Delta K = K_f - K_i = \frac{1}{2}I\omega^2 - 0 = \frac{1}{2}I\omega^2 \qquad (2)$$

To calculate the change in potential energy, we need to find the distance through which the centre of mass of the bar falls, Δh_{cm}. We can see from Figure 8-44 that this distance is

$$\Delta h_{cm} = \frac{L}{2} - \frac{L}{2}\cos\theta \qquad (3)$$

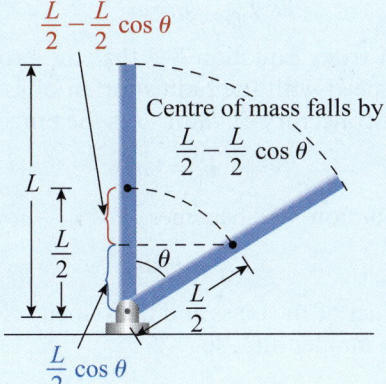

Figure 8-44 The potential energy from the fall of the bar is converted into kinetic energy.

Therefore,

$$\frac{1}{2}I\omega^2 = Mg\left(\frac{L}{2} - \frac{L}{2}\cos\theta\right) = Mg\frac{L}{2}(1 - \cos\theta) \qquad (4)$$

The moment of inertia of the bar about the pivot at its end is given by Equation 8-57:

$$I_O = \frac{1}{3}ML^2$$

Substituting into Equation (4) gives

$$\frac{1}{2}\left(\frac{1}{3}ML^2\right)\omega^2 = Mg\frac{L}{2}(1 - \cos\theta)$$

$$L^2\omega^2 = 3gL(1 - \cos\theta)$$

Since the linear speed of the free end of the bar is $v = \omega L$,

$$v^2 = 3gL(1 - \cos\theta)$$
$$v = \sqrt{3gL(1 - \cos\theta)}$$

Making sense of the result

For $\theta = 90°$, the top of the bar would have fallen by a full bar length, giving us $v^2 = 3gL$. If, instead of a swinging bar, we had a swinging pendulum with a massless rod and a point mass at its end, such that the mass falls by a height L, the result, as we saw in Chapter 6, would be $v^2 = 2gL$. Since the bar has a moment of inertia (for rotation about its centre of mass) that is less than the moment of inertia of a point mass of the same mass, we expect the tip of the bar to be easier to accelerate.

8-7 Angular Momentum

Imagine that you are a kid once again and that you approach a small merry-go-round in the park. If you push the edge a few times so that the merry-go-round rotates at a given angular speed, we say that the merry-go-round has angular momentum. If there was no friction from the bearings or resistance from the air, or interference from you and your buddies, then the merry-go-round would, in principle, continue to rotate at the same angular speed forever.

If you happened to be standing on the edge of the rotating platform and decided to walk toward the centre, making sure your feet were planted firmly on the surface and maintaining your balance, you would notice that the merry-go-round would speed up as you made your way toward the centre.

You have undoubtedly seen a figure skater spinning on his toes as he dazzles the audience, changing the rate at which he spins tremendously by simply bringing his arms in close to his torso and speeding up, or spreading them back out to slow the twirl.

The above observations all highlight the principle of conservation of angular momentum, the key idea of this section. Not only is the principle of interest on the scale of such everyday life occurrences as those above, but it also extends to quantum phenomena. The conservation of angular momentum is also evident in applications such as molecular spectroscopy, magnetic resonance imaging, and the study of particle physics. It also applies to objects on a cosmic scale, such as rapidly rotating neutron stars and the motion of galaxies themselves.

Angular momentum can be loosely described as the rotational equivalent of linear momentum. Our interest in the concept of linear momentum in Chapter 7 stemmed to a large degree from the fact that in the absence of external forces, linear momentum is a conserved quantity, another way of viewing Newton's first law. An object moving at a certain velocity will continue to do so unless it is subjected to an external force.

Linear Momentum and Angular Momentum of a Point Mass

Let us again consider a point mass attached with a massless rod to a pivot at point O and rotating with speed v, as shown in Figure 8-45. In the absence of an external force, for example, when this mass moves in a horizontal plane, the mass will continue to move at a constant speed. The tension in the rod acts perpendicular to the direction of motion of the mass and does not change its speed.

The linear momentum of the mass is given by the product of its mass and its velocity, as we saw in Chapter 7. For the case of Figure 8-45, we can say, in magnitude form,

$$p = mv$$

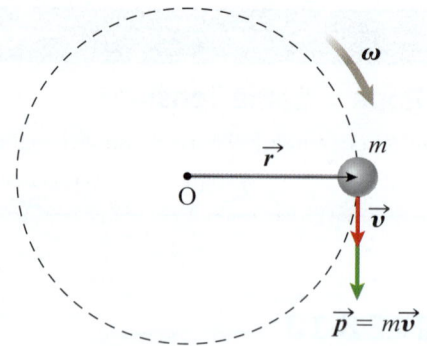

Figure 8-45 A point mass rotating about a pivot.

The rotational equivalent of linear momentum is angular momentum. Since the rotational equivalent of speed is angular speed and the rotational equivalent of mass is the moment of inertia, it is plausible to define the angular momentum of the rotating mass about the pivot at O as the product of the moment of inertia about the pivot with the angular speed:

$$L_O = I_O \omega \qquad (8\text{-}67)$$

Note that the subscript O in the symbols for angular momentum and the moment of inertia is in reference to the pivot at O.

The moment of inertia of a point mass about the pivot at O is given by Equation 8-34:

$$I_O = mr^2$$

Thus, we have

$$L_O = mr^2 \omega \qquad (8\text{-}68)$$

We recall from Equation 8-5 that the product of the angular speed with the radius for an object moving in a circular trajectory of radius r is the tangential speed:

$$v_t = \omega r$$

Thus, Equation 8-68 becomes

$$L_O = mrv \qquad (8\text{-}69)$$

The product of the mass and the velocity is nothing but the linear momentum, so

$$L_O = rp \qquad (8\text{-}70)$$

The previous discussion is an informal introduction to the topic of angular momentum. We have so far started with the notion that angular momentum is the rotational equivalent to linear momentum and arrived at Equation 8-70.

We note here the analogy between the expression for the angular momentum in Equation 8-70 and the expression for torque in Equation 8-14, which is the product of the force with the moment arm. We elaborate on this next.

Figure 8-46(a) shows a mass going around the pivot at O following a circular trajectory. From the preceding discussion, we say that the angular momentum of the mass about the pivot at O is $L_O = rp$. If instead of going around the circle, the mass travels in a straight line directly away from the pivot along the radial direction, as in Figure 8-46(b), then one can intuitively say that the mass has no angular momentum about the pivot as there is no rotation involved.

In this case, the angle between $\vec{r}$ and $\vec{p}$ is zero. While we will discuss this further shortly, right now we can deduce that the angular momentum of the mass about the pivot at O depends on the angle between the two vectors, $\vec{r}$ and $\vec{p}$. When that angle is zero, the angular momentum is zero, and when that angle is 90°, the angular momentum is $L_O = rp$. We saw a similar dependence on angle with torque. In a similar fashion to our torque discussion, it is plausible then for us to deduce that the expression for angular momentum is given by

$$L_O = rp \sin\theta \qquad (8\text{-}71)$$

This has the form of the magnitude for a cross product. As a matter of fact, the **angular momentum** of the point mass about an axis through point O, perpendicular to the page, is defined in vector form as the cross product between the position vector of the object with respect to the pivot (or axis) and the linear momentum of the object:

KEY EQUATION

$$\vec{L}_O = \vec{r} \times \vec{p} \qquad (8\text{-}72)$$

Much like torque, angular momentum is taken about a point, a pivot, or an axis.

Written formally, the magnitude of the angular momentum is given by

$$L_O = rp \sin\theta \qquad (8\text{-}73)$$

which is identical to Equation 8-71.

(a)

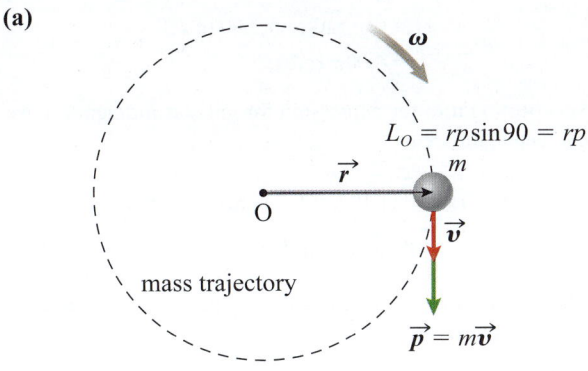

(b)

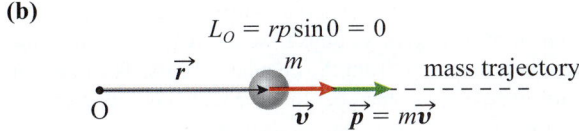

Figure 8-46 (a) Angular momentum for a mass rotating about a pivot. (b) The angular momentum of a mass translating away from the pivot is zero.

Any particle that has linear momentum also has angular momentum about a specified axis or pivot. The angular momentum is zero when $\vec{r}$ and $\vec{p}$ are collinear (or if either is equal to zero, of course). Note that a particle (or an object) can have angular momentum with respect to an axis even when the particle does not rotate about that axis.

The definition of angular momentum as a cross product in Equation 8-72, much like the definition of the torque as a cross product, implies that only perpendicular components of linear momentum and the position of the particle relative to the axis or pivot matter.

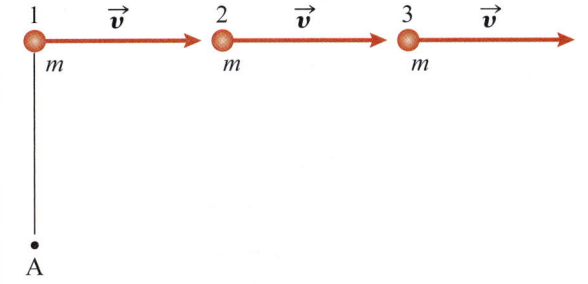

Direction of Angular Momentum

Since angular momentum is given as the cross product of two vectors, its direction can be determined by either the right-hand curl rule or the three-finger right-hand rule, as explained in the discussion on torque. The angular momentum vector is perpendicular to the plane containing $\vec{r}$ and $\vec{p}$, as illustrated in Figure 8-48.

When the point mass is following a circular trajectory, the velocity and the linear momentum will point in a direction tangential to the trajectory, and hence perpendicular to the radial position of the particle as measured from the centre of the circle. The angle θ between $\vec{r}$ and $\vec{p}$ is 90°. From Equation 8-73,

$$L_O = |\vec{r} \times \vec{p}| = rp \sin 90° = rp \tag{8-74}$$

$$L_O = rm\omega r = mr^2\omega = I_O\omega \tag{8-75}$$

where have recovered Equation 8-67.

In vector form, we can say, for rotations in two dimensions as in Figure 8-48, that

KEY EQUATION
$$\vec{L}_O = I_O\vec{\omega} \tag{8-76}$$

Angular momentum and angular velocity are in the same direction (at least for the types of cases considered in this chapter, where I is a scalar).

The subscript O on the symbol for angular momentum indicates that the angular momentum is taken about the pivot O. Recall from Section 8-3 that the direction of the angular velocity vector is determined by curling the fingers of your right hand in the direction of motion—your thumb points in the direction of the angular velocity vector. Note that Equation 8-76 is completely analogous to $\vec{p} = m\vec{v}$ for linear momentum, with moment of inertia taking the place of mass and angular velocity taking the place of velocity.

Angular Momentum of a Rotating Rigid Body

Equation 8-76 applies for *any* rigid body rotating about an axis. It is frequently written without the subscript O:

KEY EQUATION
$$\vec{L} = I\vec{\omega} \tag{8-77}$$

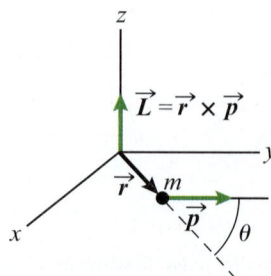

Figure 8-48 The angular momentum for a point mass.

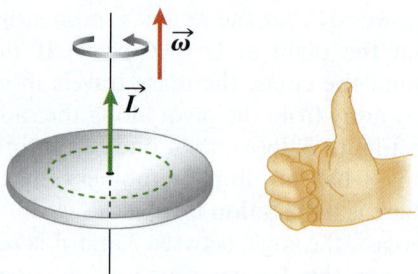

Figure 8-49 The angular momentum of a rigid object.

Figure 8-49 shows a disk spinning about its axis of symmetry at angular velocity $\vec{\omega}$. The right-hand rule gives the direction of the angular momentum vector, which, according to Equation 8-77, is parallel to the angular velocity vector.

EXAMPLE 8-11

Angular Momentum of a Spherical Shell

A child spins the brass globe on her parents' desk at 3.00 rev/s. The globe is a thin spherical shell with a mass of 1.10 kg and a radius of 23.0 cm. Find the angular momentum of the spinning globe.

Solution

The angular momentum of a rigid body about an axis of rotation is given by Equation 8-77:

$$L = I\omega$$

From Table 8-3, the moment of inertia of a spherical shell about an axis through its centre of mass is $\frac{2}{3}MR^2$. There are 2π radians in one revolution, so the angular speed is

$$\omega = (2\pi \text{ rad/rev})(3.00 \text{ rev/s})$$
$$= 6.00\pi \text{ rad/s}$$

Substituting into the expression for angular momentum, we get its magnitude as

$$L = I\omega = \frac{2}{3}(1.10 \text{ kg})(0.230 \text{ m})^2(6.00\pi \text{ rad/s})$$
$$= 0.731 \text{ kg}\cdot\text{m}^2/\text{s}$$

The direction of the angular momentum is as specified by the right-hand rule.

Making sense of the result

This question demonstrates that when we can compute the moment of inertia for an object, we can readily calculate the angular momentum from the angular velocity. The units for L are as expected (remember that the radian is a ratio of two lengths and has no real dimensional units).

The Rate of Change of Angular Momentum

A net force exerted on an object causes it to accelerate, in other words, change its linear momentum. Similarly, one can see the net torque exerted on an object in terms of the change to its angular momentum. In Chapter 7, we derived the relationship between the net force on a point mass and the change in its linear momentum:

$$\vec{F}_{net} = \frac{d\vec{p}}{dt} \text{ and } \vec{F}_{net}^{avg} = \frac{\Delta\vec{p}}{\Delta t} \qquad (8\text{-}78)$$

When we take the cross product of $\vec{r}$ with both sides of Equation 8-78, we get

KEY EQUATION

$$\vec{\tau}_{net} = \vec{r} \times \vec{F}_{net} = \vec{r} \times \frac{d\vec{p}}{dt} = \frac{d}{dt}(\vec{r} \times \vec{p}) = \frac{d\vec{L}}{dt} \qquad (8\text{-}79)$$

and

$$\vec{\tau}_{net}^{avg} = \frac{\Delta\vec{L}_{net}}{\Delta t} \qquad (8\text{-}80)$$

In Equation 8-79, using the product rule, the first term for the derivative $\frac{d}{dt}(\vec{r} \times \vec{p})$ is the cross product between the velocity and the linear momentum. Since these are parallel, this term is identically zero; only the second term survives.

Thus, a net torque acting on an object changes the angular momentum of the object. Rearranging Equation 8-80 and integrating gives

$$d\vec{L} = \vec{\tau}_{net} dt \qquad (8\text{-}81)$$

and

$$\Delta\vec{L} = \int_{interval} \vec{\tau}_{net} dt = \vec{\tau}_{net}\Delta t \qquad (8\text{-}82)$$

The change in angular momentum, $\Delta\vec{L}$, is called the **angular impulse**. The symbol J is often used for this quantity.

Conservation of Angular Momentum

If you gave a puck on completely frictionless ice a push in a given direction, it would go forever in that direction. The puck would never stop, since there would be no force to slow it down. In the absence of a net force on the puck, once in motion it would continue at the same speed in the original direction forever. This is the principle of conservation of linear momentum, really a direct manifestation of Newton's first law, as explained in Chapter 5.

The equivalent of this for a spinning or rotating object is the principle of conservation of angular momentum. If you spin a globe with very little friction impeding the rotation it will go on spinning for a very long time. If there were no external torques at all, it would spin forever at the same angular velocity. This is one example of the conservation of angular momentum, but the applications are many, and this is one of the more powerful ideas in physics. At the beginning of this chapter, we considered what happens in supernova as a result of the conservation of angular momentum, and below, we touch upon other examples.

The principle of **conservation of angular momentum** states that in the absence of a net external torque, the angular momentum will not change. For a single spinning object, this means that the product of the moment of inertia about a given axis and the rotation rate about that axis remains constant.

This principle is demonstrated by the increase in the rate at which a figure skater spins when she reduces her moment of inertia by bringing her arms in close to her torso. By doing so she reduces her moment of inertia about her axis of rotation; however, since the product $I_O\omega$ remains constant, as she brings her arms in closer, her angular speed, and thus the rate at which she spins, increases.

The principle of conservation of angular momentum can be extended to any system of objects, keeping in mind that it is the net external torque that is important. We need to clearly identify the object or the system of objects on which the external or net torque is zero to apply the principle of conservation of angular momentum.

For example, if a child starts walking along the circumference of a frictionless merry-go-round that was initially stationary, the merry-go-round will spin in the opposite direction to keep the angular momentum of the child–merry-go-round system constant, in this case, zero. The angular momentum of the merry-go-round changes as a result of the torque exerted on it by the child. The angular momentum of the child changes as a result of the torque exerted on the child by the merry-go-round. However, since these two torques are internal to the child–merry-go-round system, they do not change the angular momentum of that system. These two torques can be thought of as an action–reaction pair within the context of Newton's third law and hence can be shown to be equal in magnitude, yet opposite in direction, and they add up to zero.

If a child standing at the edge of an already freely spinning merry-go-round starts walking toward the centre, the moment of inertia of the child–merry-go-round system decreases. As a result, the spinning speed of the merry-go-round will increase to keep the angular momentum of the child–merry-go-round system constant. We expand on the previous points in the following examples.

EXAMPLE 8-12

Shuttle Payload

A European Space Agency team is designing a space shuttle orbiter so that it can carry a 3000.0 kg satellite, which has a cylindrical shape with a radius of 1.90 m (Figure 8-50). The satellite needs to be spinning about its axis before it is released into space, so just before the satellite is released, a motor in the cargo bay is to spin the satellite until the satellite has an angular speed of 15.0 rad/s. Assume a uniform cylindrical disk to calculate the moment of inertia of the satellite. The moment of inertia of the orbiter about the axis of rotation of the satellite is 4.20×10^6 kg·m².

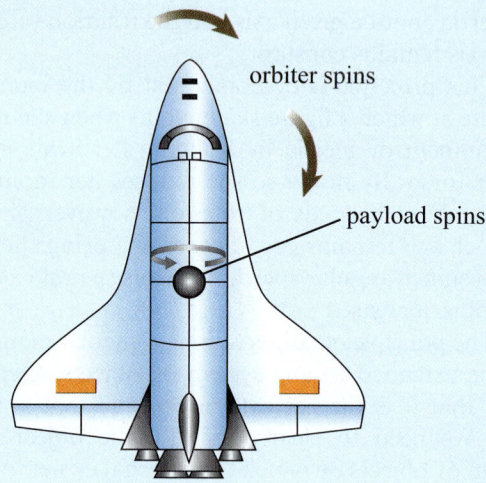

orbiter spins

payload spins

Figure 8-50 Example 8-12.

(a) Find the angular speed this orbiter would have just before the spinning satellite is released.
(b) What effect will releasing the satellite have on the angular speed of the orbiter?
(c) Would the rate at which the satellite is accelerated have an effect on the final angular speed of the orbiter?

Solution

(a) We apply the concept of conservation of angular momentum to the system consisting of the orbiter and the satellite. The force (or torque) exerted by the orbiter on the satellite is internal to the satellite–orbiter system. When the motor in the orbiter exerts a force (or torque) on the satellite, the satellite exerts an equal but opposite force (or torque) on the orbiter, in accordance with Newton's third law. Since the action–reaction pair is internal to the system, the total angular momentum is conserved. Therefore, $L_i = L_f$ for the system.

No part of the system was spinning before the motor was turned on, so the angular momentum of the system about the axis of rotation of the satellite is zero, and

$$\vec{L}_i = \vec{L}_f = 0$$

The final angular momentum is the sum of the angular momentum of the satellite and the angular momentum of the orbiter:

$$\vec{L}_f = \vec{L}_{sat} + \vec{L}_{orb} = 0$$

Therefore, the angular momenta of the satellite and the orbiter have equal magnitudes but opposite directions in magnitude, and we can say

$$L_{orb} = I_{orb}\omega_{orb} = L_{sat} = I\omega_{sat} = \frac{1}{2}M_{sat}R_{sat}^2\omega_{sat}$$

Solving for ω_{orb} and then substituting the given values, we get

$$\omega_{orb} = \frac{\frac{1}{2}M_{sat}R_{sat}^2\omega_{sat}}{I_{orb}}$$

$$= \frac{\frac{1}{2}(3000.0 \text{ kg})(1.90 \text{ m})^2(15.0 \text{ rad/s})}{4.20 \times 10^6 \text{ kg·m}^2} = 0.0193 \text{ rad/s}$$

$$= 1.11°/s$$

(b) Releasing the satellite has no effect on the angular speed of the orbiter because the satellite was already spinning when it was released.

(c) The rate at which the satellite is accelerated has no effect on the final angular speed of the orbiter. All that matters are the initial and final states.

Making sense of the result

The heavier the orbiter, the slower it will spin, according to the answer from (a). An infinitely heavy orbiter will not spin at all, and a relatively light payload will not significantly affect the orbiter's rotation.

MAKING CONNECTIONS

Tides and Precise Time

Due to tidal friction, Earth's rotation loses about 2 ms every 188 years compared to atomic clocks. In 2010, NIST (National Institute of Standards and Technology) researchers introduced a quantum logic clock (Figure 8-51), which uses the vibrations of laser-cooled aluminum and magnesium ions. The quantum clock is expected to be accurate to within 1 s over 37 billion years. Pulsars can also be used to keep time. Even though pulsar periods are about 1/100th the accuracy of the quantum clock, pulsars are useful for long-term time measurements because they will likely last much longer than the service life of a clock on Earth. Sudden changes in pulsar periods may indicate gravitational waves originating from black holes.

J. Koelemeij/NIST

Figure 8-51 The ion trap is a key component of the NIST aluminum ion clock.

INTERACTIVE ACTIVITY 8-4

Angular Momentum

In this activity, you will use the PhET simulation "Torque" to investigate angular momentum and the conservation of angular momentum. You will be able to change an object's mass and dimensions, and apply forces to the object and see the impact of these changes on the object's angular momentum and angular velocity.

CHECKPOINT 8-14

Conservation of Angular Momentum

A child steps onto a stationary merry-go-round and starts walking along its edge. Which of the following statements is true? Assume that the friction in the mechanism of the merry-go-round is negligible.

(a) The total angular momentum of the merry-go-round–child system increases as a result of the child walking.

(b) The child has no angular momentum because she walks in a tangential direction around the merry-go-round.

(c) The kinetic energy of the merry-go-round–child system does not change as a result of the child walking.

(d) The angular momentum of the merry-go-round alone does not change.

(e) The child's motion on the merry-go-round has no effect on the total angular momentum.

ANSWER: (e) The force or torque exerted by the child on the merry-go-round, and that exerted by the merry-go-round on the child, are internal to the merry-go-round–child system. Therefore, the total angular momentum is conserved.

CHECKPOINT 8-15

Angular Momentum Along a Given Direction

You are standing on a platform that is free to rotate while holding a bicycle wheel that is spinning about its axle, which is horizontal. You then turn the axle vertical (Figure 8-52).

(a) Why do you need to exert a torque to tilt the axle?

(b) When the wheel slows down, what happens to the torque you have to exert to tilt the spinning axle?

(c) Why does the platform you are standing on begin to spin as you tilt the wheel axle?

The student stands on a stationary platform holding a bicycle wheel that spins counter-clockwise; the bicycle wheel's angular momentum points horizontally.

She flips the moving wheel, reversing its angular momentum. The total angular momentum is conserved, so the wheel and the student must rotate the other way.

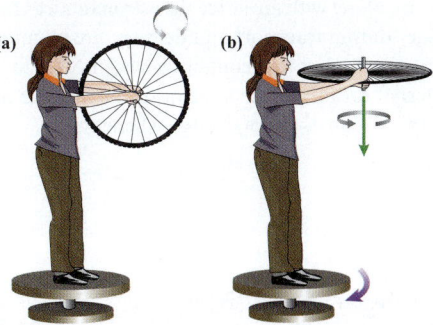

(a)   **(b)**

Figure 8-52 (a) Initially, the total angular momentum points horizontally. (b) When you turn the axis of rotation of the wheel to the vertical, the direction of the total angular momentum changes.

ANSWER: (a) Any change to the angular momentum requires a torque. (b) The torque you apply will be less if you wish to tilt the axle at the same rate. (c) Angular momentum is conserved; because the vertical component of angular momentum was zero initially, it must be zero in the final situation.

EXAMPLE 8-13

Magnetic Boots

Two astronauts are installing equipment on an orbiting platform that is rotating freely at a rate of 1.10 rad/s about its centre of mass, as shown in Figure 8-53. Astronaut Valentina is standing at the centre of the platform, and astronaut Alexander is at the edge. The platform has a radius of 11.0 m. The moment of inertia of the platform without the astronauts is I_p = 2400. kg·m^2. Astronaut Valentina walks to the edge and hands a container of mass m_c = 24.0 kg to astronaut Alexander and then returns to the centre. Both astronauts are wearing magnetic boots that keep them on the surface of the platform. Compared to the dimensions of the platform, you can consider the astronauts to be thin cylinders or point masses of mass m_a = 81.0 kg each. The container can also be considered a point mass.

(a) Find the resulting angular speed of the platform after astronaut Valentina has returned to the centre.
(b) Find the total work done by the astronaut(s) in the process.
(c) If astronaut Valentina keeps walking back and forth between the edge of the platform and its centre, will she eventually do enough work on the platform to lose all its kinetic energy and come to a stop? If so, how many times does astronaut Valentina have to walk back and forth?

Figure 8-53 Example 8-13.

Solution

(a) Any forces that astronaut Valentina exerts on the platform to walk to the edge, deliver the container, and then return, along with any force the astronauts exert on the container during transport and delivery, are all internal to the astronauts–platform–container system. There is no net external torque on the system, so the angular momentum of the system does not change. Therefore,

$$\vec{L}_i = \vec{L}_f \text{ and } I_i\vec{\omega}_i = I_f\vec{\omega}_f$$

The initial moment of inertia of the system is the sum of the moments of inertia of the platform, astronaut Valentina and the container at the centre, and astronaut Alexander at the edge. Since we can treat the astronauts and the container as point masses, their moments of inertia about the axis are simply mr^2, and r is zero for the astronaut and the container when at the centre. Therefore,

$$I_i = m_a(0)^2 + m_c(0)^2 + I_p + m_a r^2$$
$$= 0 + 0 + 2400.\ \text{kg·m}^2 + (81.0\ \text{kg})(11.0\ \text{m})^2$$
$$= 12\,201\ \text{kg·m}^2$$

The final moment of inertia is equal to the initial moment of inertia plus the moment of inertia of the container when at the edge of the platform:

$$I_f = I_p + m_a r^2 + m_c r^2$$
$$= 2400.\ \text{kg·m}^2 + (81.0\ \text{kg})(11.0\ \text{m})^2 + (24.0\ \text{kg})(11.0\ \text{m})^2$$
$$= 15\,105\ \text{kg·m}^2$$

So, our conservation of angular momentum equation becomes

$$(12\,201\ \text{kg·m}^2)(1.10\ \text{rad/s}) = (15\,105\ \text{kg·m}^2)\omega_f$$

which gives ω_f = 0.889 rad/s.

(b) The work done by astronaut Valentina is the only work done on the system. Therefore,

$$W_T = \Delta K = \frac{1}{2}I_f\omega_f^2 - \frac{1}{2}I_i\omega_i^2$$
$$= \frac{1}{2}(15\,105\ \text{kg·m}^2)(0.889\ \text{rad/s})^2$$
$$\quad - \frac{1}{2}(12\,201\ \text{kg·m}^2)(1.10\ \text{rad/s})^2$$
$$= -1410\ \text{J} = -1.41\ \text{kJ}$$

(c) Since there are no mechanisms for losing energy, such as an external force of friction, the energy of the whole system is conserved. The work done by astronaut Valentina as she walks back and forth on the platform alternates between being positive and negative, such that on each return trip the total work done is zero.

Making sense of the result

Since moving the container to the edge of the platform results in an increase in the total moment of inertia of the isolated system, we would expect the angular speed of the system to decrease to keep the total angular momentum constant.

CHECKPOINT 8-16

Negative Work?

In Example 8-13, how would the result have been different if Valentina, the astronaut on the inside, chose to slide the container on the surface of the platform to Alexander instead of delivering it herself? How would the result have changed if Valentina threw the package to Alexander, who caught it before it hit the platform?

ANSWER: The final result would be the same in each case, because the initial and final states are the same.

KEY CONCEPTS AND RELATIONSHIPS

Angular Quantities

Angular position (θ), angular velocity (ω), and angular acceleration (α) are defined in an analogous fashion to linear quantities:

$$\Delta s = r\Delta\theta \tag{8-1}$$

$$\omega = \frac{d\theta}{dt} \tag{8-4}$$

$$v(t) = \omega(t)r \tag{8-5}$$

$$\alpha = \frac{d^2\theta}{dt^2} = \frac{d\omega}{dt} \tag{8-7}$$

$$a_t(t) = \alpha(t)r \tag{8-8}$$

Angular position is measured in radians. The direction of angular velocity is perpendicular to the plane of rotation, in the direction of the thumb of the right hand when the fingers curl in the direction of motion. However, since we often deal with rotation about a fixed axis, where they just take on positive or negative values, sometimes they are written without the vector signs (as we did in Chapter 3). We can integrate to obtain kinematic solutions for angular motion, or develop equations for the special case of α = constant. See Table 8-4 for details.

Torque

The rotational quantity that plays the role of force in linear motion is torque ($\vec{\tau}$), defined by the following, where $\vec{F}$ is the applied force and $\vec{r}$ is a displacement vector from the rotation axis to the point of application of the force:

$$\vec{\tau} = \vec{r} \times \vec{F} \tag{8-21}$$

Its magnitude is

$$\tau_O = rF\sin\theta \tag{8-22}$$

where the subscript indicates that the torque is taken about a specific rotation axis.

Moment of Inertia

The rotational quantity analogous to mass in linear motion is called the moment of inertia, I. It depends on both the mass and the distribution of that mass relative to the rotation axis, a.

The moment of inertia of a point mass about a given rotation axis is

$$I_O = mr^2 \text{ (point mass)}, \tag{8-34}$$

The net torque acting on a point mass about a given axis is

$$\vec{\tau}_{net} = mr^2\vec{\alpha} \tag{8-33}$$

For a point mass or a rigid body, this can be expressed as

$$\vec{\tau}_{net} = I\vec{\alpha} \tag{8-35}$$

The moment of inertia for a set of N point masses held together in a rigid body can be calculated from the following summation, where r_i is the distance from the rotation axis for each mass m_i:

$$I_a = \sum_{i=1}^{N} m_i r_i^2 \tag{8-40}$$

For an object with a continuous distribution of mass, the moment of inertia can be calculated using the following relationship:

$$I = \int_{\text{object}} r^2 dm \tag{8-41}$$

For composite objects, we can find the moment of inertia of individual parts and then add to get the total moment of inertia.

The Parallel-Axis Theorem

The moment of inertia of an object about an arbitrary axis is equal to the moment of inertia of that object about a parallel axis running through its centre of mass, plus the product of the total mass of the object and the square of the distance between the centre-of-mass axis and the arbitrary axis:

$$I_a^{\text{object}} = I_{cm}^{\text{object}} + Md^2 \tag{8-56}$$

Work and Rotational Kinetic Energy

An object with moment of inertia I rotating with an angular speed of ω has a rotational kinetic energy given by

$$K_{rot} = \frac{1}{2}I\omega^2 \tag{8-58}$$

Used in conjunction with the work–energy theorem, this provides an alternative way to solve some rotational problems.

The work done by a constant torque is given by

$$W_T = \tau\Delta\theta = \frac{1}{2}I\omega_f^2 - \frac{1}{2}I\omega_i^2 \tag{8-65}$$

For the work done by a variable torque,

$$W_T = \int \tau d\theta = \frac{1}{2}I\omega_f^2 - \frac{1}{2}I\omega_i^2 \tag{8-66}$$

Angular Momentum

The angular momentum for a point mass is defined according to the following vector cross product, where $\vec{r}$ is a displacement vector from the rotation axis and $\vec{p}$ is the linear momentum of the point mass:

$$\vec{L}_O = \vec{r} \times \vec{p} \tag{8-72}$$

An object with moment of inertia I and an angular velocity $\vec{\omega}$ has an angular momentum given by

$$\vec{L} = I\vec{\omega} \tag{8-76}$$

Conservation of Angular Momentum

In an analogous fashion to writing Newton's second law in terms of the time derivative of the linear momentum, the net torque equals the derivative of the angular momentum:

$$\vec{\tau}_{net} = \frac{d\vec{L}}{dt} \tag{8-79}$$

If there is a zero net torque, this tells us that the angular momentum of the system must be constant or conserved.

It is very helpful to see the analogies between the quantities and concepts of rotational motion and those of linear motion studied earlier. Table 8-4 summarizes these relationships. Note that we have suppressed vector notation in the kinematic relationships.

Table 8-4 Comparison of Angular and Linear Quantities and Relationships

Linear quantity	Symbol/Expression	Angular quantity	Symbol/Expression
Linear position	$\vec{x}$	Angular position	θ
Linear velocity	$\vec{v}$	Angular velocity	$\vec{\omega}$
Linear acceleration	$\vec{a}$	Angular acceleration	$\vec{\alpha}$
Kinematics equations for constant a	$x(t) = \frac{1}{2}at + v_0 t + x_0$ $v(t) = at + v_0$ $v^2 = v_0^2 + 2a\Delta x$	Kinematics equations for constant α	$\theta(t) = \frac{1}{2}\alpha t + \omega_0 t + \theta_0 \quad (8\text{-}12)$ $\omega(t) = \alpha t + \omega_0 \quad (8\text{-}11)$ $\omega^2 = \omega_0^2 + 2\alpha\Delta\theta \quad (8\text{-}13)$
Mass	m	Moment of inertia	I
Force, average force, and net force	$\vec{F}, \vec{F}_{avg}, \vec{F}_{net}$	Torque, average torque, and net torque	$\vec{\tau}, \vec{\tau}_{avg}, \vec{\tau}_{net}$
Work done by a force	constant force $W = \vec{F}\cdot\Delta\vec{s}$ variable force: $W = \int \vec{F}\cdot d\vec{s}$	Work done by a torque	constant torque $W = \tau\Delta\theta$ variable torque: $W = \int \tau d\theta$
Work–kinetic energy theorem	$W = \Delta K$	Work–kinetic energy theorem	$W = \Delta K$
Linear momentum	$\vec{p} = m\vec{v}$	Angular momentum	$\vec{L} = \vec{r}\times\vec{p}, \vec{L} = I\vec{\omega}$
Linear impulse	$\vec{I} = \Delta\vec{p}$	Angular impulse	$\vec{J} = \Delta\vec{L}$
Conservation of linear momentum	if $\vec{F}_{ext} = 0, \vec{p}_i = \vec{p}_f, \Delta\vec{p} = 0$ (isolated system)	Conservation of angular momentum	If $\vec{\tau}_{net} = 0, \vec{L}_i = \vec{L}_f, \Delta\vec{L} = 0$ (isolated system)
Net force: change in linear momentum	$\vec{F}_{avg} = \dfrac{\Delta\vec{p}}{\Delta t}$ instantaneous $\vec{F}_{ext} = \dfrac{d\vec{p}}{dt}$	Net torque: change in angular momentum	$\vec{\tau}_{avg} = \dfrac{\Delta\vec{L}}{\Delta t}$ instantaneous $\vec{\tau}_{ext} = \dfrac{d\vec{L}}{dt}$

APPLICATIONS

navigation, GPS, flywheels, mechanical hybrids, variable transmissions, engines and propellers, tidal friction, Earth's rotation, microgravity, biophysical systems, nuclear rotation, atomic angular momentum, spin angular momentum, atomic and molecular spectra, laser cooling of atoms, nuclear magnetic resonance and MRI, neutron stars, rotating black holes

KEY TERMS

angular acceleration, angular displacement, angular impulse, angular momentum, angular speed, angular velocity, conservation of angular momentum, homogeneous, moment arm, moment of inertia, non-deformable, parallel-axis theorem, perpendicular-axis theorem, radial acceleration, rigid body, rotational inertia, rotational kinetic energy, tangential acceleration, tangential speed, torque, uniform.

QUESTIONS

1. Two spheres of equal mass and radius are spinning at the same angular speed about fixed axes through their centres. One of the spheres is solid, and the other is a spherical shell. Which of the following statements is true?
   (a) The solid sphere has the higher kinetic energy.
   (b) The solid sphere has the lower kinetic energy.
   (c) Both spheres have the same kinetic energy.
   (d) Both spheres have the same angular momentum but not the same kinetic energy.

2. Can we talk about an object's moment of inertia without specifying an axis? Explain.

3. A bicycle wheel mounted on a vertical axis is being accelerated from its initial angular speed at a constant acceleration. Compare the tangential acceleration of a point on the circumference to a point midway between the circumference and the centre. Also compare the speed and the radial acceleration of the two points.

4. Two globes of identical mass and radius are spinning about axes through their centres at the same speed. One of the globes is a solid sphere, and the other has a hollow space at its centre. Which globe is easier to stop? Explain.

5. Two disks of different radii are connected by a belt running along their circumference and driven by a motor that spins the larger disk at a constant angular speed. Which of the following statements is true?
   (a) The angular speed of both disks is the same.
   (b) The larger disk has the larger angular speed.
   (c) The smaller disk has the larger angular speed.
   (d) The tangential speed of a point on the circumference of the larger disk is higher than that of a point on the circumference of the smaller disk.

6. Consider the heavy pulley setup in Figure 8-54. Can we say that the tension in the belt is the same everywhere? Which of the two masses would you expect to have the higher acceleration?

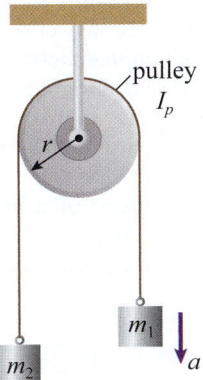

Figure 8-54 Question 6.

7. You are standing on a disk that is spinning freely while holding onto a medicine ball, with your arms extended over the edge of the disk. You then throw the medicine ball so its speed has only a vertical component when it lands on the ground.
As a result of your throwing the ball,
   (a) the angular speed of the disk increases
   (b) the angular speed of the disk decreases
   (c) the angular speed of the disk remains the same
   (d) your own angular momentum is conserved

8. Two thin rods of the same length are free to rotate about a universal hinge at their bottom end while being held vertically. One rod is heavier than the other. The two rods are let go simultaneously, and they begin to fall. Which rod will be rotating faster at the bottom of its trajectory?
   (a) The heavier rod.
   (b) The lighter rod.
   (c) The rods will have the same speed.

9. Two thin rods of the same mass are pivoted about a free hinge at their bottom edge while being held vertically. One rod is longer than the other. The two rods are let go simultaneously, and they begin to fall. Which rod will be rotating faster at the bottom of its trajectory?

10. Two pendulums are held in a horizontal position and then let go. Both pendulums are bars made of the same material, but one is thinner and has a sphere at the end. The masses and overall lengths of the pendulums are the same. Which pendulum will have a higher angular speed when it reaches a vertical position?
   (a) The thinner one.
   (b) The thicker one.
   (c) They will both have the same speed.
   (d) The thickness does not affect the moment of inertia.

11. For the rotational kinematic equations, do you have to use radians, or can you use degrees or other units of measuring angles? Explain.

12. The propeller blades of a single-engine plane spin at a constant speed. Compare the angular speed in rad/s of a point at the tip of one of the blades to the angular speed of a point at the middle of the blade. Does either of the two points accelerate? If so, compare their accelerations.

13. Your bike is upside down on the floor, and you are fixing the chain. The gear setting is such that the chain winds around the small gear at the rear wheel and around the large gear attached to the pedals (Figure 8-55).
   (a) When you spin the pedals at a constant speed, will the wheel necessarily be turning at the same rotational speed as the pedals?
   (b) You mark a point with a white marker on the rear tire, and you watch that point as you spin the wheel at a constant speed. As the wheel spins at a constant speed,
      (i) the white point has a constant velocity
      (ii) the white point does not accelerate
      (iii) the white point accelerates
      (iv) none of the above

Figure 8-55 Question 13.

14. How could you tell a raw egg from a boiled egg without cracking them open?

15. A figure skater is executing a high-speed spin routine with her arms tucked close to her body (Figure 8-56). When she spreads her arms out, is she doing positive or negative work? When she brings her arms in closer to her body again, is she doing positive or negative work? Assuming zero friction, how would her initial angular speed compare to her angular speed after extending her arms out and then bringing then back in?

Figure 8-56 Question 15.

16. Would the expression for the moment of inertia of a thin rod about an axis through its centre hold if the diameter of the rod were no longer negligible?

17. Use the moment of inertia for a ring to derive the moment of inertia for a thin cylindrical shell.

18. You are sitting on a freely spinning pedestal holding a long bar upright close to your body. You then tilt the bar until it becomes horizontal.
    (a) What happens to your angular speed?
    (b) Is the work you do as you rotate the bar positive or negative?

19. An insect ($m = 11$ g) is standing on the rim of a horizontal ring of mass $M = 0.90$ kg and radius $R = 9.0$ cm. The stationary ring is free to spin about its centre-of-mass axis. The insect then crawls around the ring and reaches a maximum speed of 15 cm/s with respect to the ring. The insect completes two full cycles around the ring and comes to a stop with respect to the ring. Find the final speed of the insect.

20. Why is the moment of inertia of a solid sphere only an approximate model for calculating the moment of inertia of Earth?

21. Using the vector model for angular momentum, explain why torque is required to tilt the axis of a spinning bicycle wheel.

22. Two cars are driving toward each other at the same speed on the highway. The lanes they are travelling in are each a distance d away from the median that divides the highway. The angular momentum of the two-car system about a point midway between them, as the cars move,
    (a) reaches a minimum when the two cars pass one another
    (b) is maximum when the two cars are the farthest they can be from one another
    (c) changes as the cars pass one another
    (d) is always the same

23. A rope over a pulley with moment of inertia I has a heavy load, m_1, attached to one end and a counterweight, m_2, tied to the other end, as in Figure 8-57. When will the tension in the segment of the rope attached to the heavier mass be the same as the tension in the segment of the rope attached to the lighter mass?

24. Mass m_2 is greater than mass m_1. Mass m_1 is given an initial speed downward, and the system slows due to gravity.
    (a) When the system momentarily stops, which part of the rope (on either side of the pulley) has the higher tension?
        (i) the rope attached to the lighter mass
        (ii) the rope attached to the heavier mass
        (iii) the tension is the same
        (iv) the answer will change when the pulley stops momentarily as it reverses direction
    (b) Does the pulley have an acceleration at that moment?

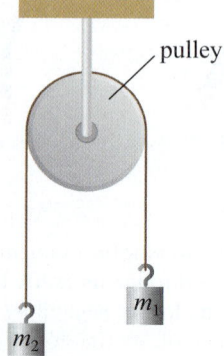

Figure 8-57   Questions 23, 24.

25. True or false? A wheel can have an angular acceleration if it is not moving. Explain.

26. Can a rotating object have an angular acceleration in one direction while rotating in the other? Explain.

27. Three people are pushing a stalled car toward a gas station. By doing so, do they create angular momentum about a tree a few hundred metres away? Explain.

28. A particle is flying in a straight line at a constant speed. Does the particle have angular momentum? Can we make conclusions regarding the direction of its angular momentum, if it exists? Explain.

29. Can we define the torque of a force without referring to a pivot point or rotation axis? Can we speak of force without referring to a coordinate system? Explain.

PROBLEMS BY SECTION

For problems, star ratings will be used (★, ★★, or ★★★), with more stars meaning more challenging problems. The following codes will indicate if $\frac{d}{dx}$ differentiation, $\int$ integration, ▢ numerical approximation, or 〰 graphical analysis will be required to solve the problem.

Section 8-1   Angular Variables

30. ★ What is the angular speed, in rad/s, of (a) a point on the equator and (b) a point in Antarctica close to the South Pole?

31. ★ Taking Earth to be a perfect sphere, find the linear speed of a point located on the 32nd parallel, as a result of Earth's rotation.

32. A car is undergoing an emissions test. The car is stationary on rollers while the gas pedal is depressed until the speedometer reads 60 km/h. The wheels are 54 cm in diameter. Find their angular speed.

33. ★ The blades of a plane propeller need to be designed such that the top speed of the blade tip is just under the speed of sound. When the maximum engine speed is 7000 rev/min, determine what length the blade should be. Take the speed of sound to be 340 m/s.

34. ★ A Ferris wheel has a radius of 37.0 m. When a point on its circumference has moved 21.0 m, by how many degrees has the Ferris wheel rotated?

35. ★ Find the angular speed of the hour hand of an analog clock.

36. ★ Astronauts often have to train in high-g (high gravity) environments before they embark on space missions. One way to simulate a high-g environment is to place astronauts in rotating capsules, such as the one in Figure 8-58, used by NASA. What angular speed is required when the radius of the track is 72 m and the acceleration target is 7.0g (i.e., seven times the normal acceleration due to gravity)?

Figure 8-58   Problem 36.

Section 8-2   Kinematic Equations for Rotation

37. ★ You have your bicycle upside down while you adjust the chain. The chain is wound around the rear wheel's small gear, which has a radius of 4.0 cm. At the pedals, the chain is wound around the large gear, which has a radius of 11.0 cm. The radius of the rear wheel is 35.0 cm. You spin the pedals initially so they finish two revolutions in 1.0 s, starting from rest. Assume constant acceleration.
 (a) Find the angular acceleration of the rear wheel in rad/s^2.
 (b) Find the magnitude of the total acceleration of a point at the edge of the small gear at the end of 1.5 s (the sum of the radial and tangential accelerations).
 (c) What is the tangential acceleration for that point during the first second?
 (d) Calculate the linear acceleration of the chain.

38. ★ A space-training module consists of a capsule that rotates inside a horizontal circular track of radius 110. m. The rotational speed increases at a constant rate from rest such that the capsule finishes its first cycle in 9.00 s.
 (a) How long does it take from the start of the motion for the radial acceleration to be equal to six times the acceleration due to gravity?
 (b) How long does it take for the radial acceleration to be equal in magnitude to the linear acceleration?

39. ★★ The propeller blades of an airplane are 2.10 m long. The plane is getting ready for takeoff, and the propeller starts turning from rest at a constant angular acceleration. The propeller blades go through two revolutions between the fifth and the seventh seconds of the rotation. Find the angular speed at the end of 8.00 s.

40. ★★ A wheel is spinning about a fixed axle at an angular frequency of 29 rad/s. A braking mechanism reduces the speed by half over the first five revolutions. Assume that the braking force is constant.
 (a) Find the angular speed 1.6 s into the braking phase.
 (b) Calculate the angular displacement during that time.
 (c) Find the angular speed at the end of 4.0 s.

41. ★★ A spinning globe is slowed at a constant acceleration of 0.750 rad/s^2 until it stops. One of the points on the equator moves 23.0° in the first 0.700 s of the slowing phase.
 (a) Find the total angular displacement of the globe during the acceleration phase.
 (b) Find the initial angular speed of the globe.

42. ★ A bicycle wheel initially spinning at 7.00 rad/s is slowed at a constant acceleration so it reaches a speed of 2.50 rad/s over an angular displacement of 11.2 rad. Find the angular acceleration.

43. ★ A horizontally mounted bicycle wheel 42.0 cm in radius is accelerated from rest at 3.50 rad/s^2 for 7.00 s, after which time it rotates at a constant speed.
 (a) Find the magnitude of the radial acceleration of a point on its circumference 3.00 s into the motion.
 (b) Find the radial acceleration of the same point 9.00 s into the motion.

Section 8-3   Torque

For problems 44 to 50, the position vector is given in metres (m) and the force in newtons (N) if not specified.

44. ★ A carpenter is driving a nail straight into the northern wall of house, with an average force of 1200 N. Find the magnitude of the torque of this force about a point on the wall 3.0 m to the right of the nail.

45. ★ Find the cross product of $\vec{r} = (2.00\hat{i}, 2.00\hat{j}, 0)$ and $\vec{F} = (4.00, 133°, 90.0°)$.

46. ★ A force $\vec{F} = (12.0\hat{i} + 4.0\hat{j} - 16.0\hat{k})$ is applied at the point A(−3.0 m, 5.0 m, 2.0 m). Find the torque of $\vec{F}$ about point B(−3.0 m, 2.0 m, −6.0 m).

47. ★ Calculate the torque about the origin of force $\vec{F} = (12.0\hat{i} + 4.00\hat{j} - 16.0\hat{k})$ applied at point A(2.00 m, −5.00 m, 7.00 m).

48. ★ Determine the force, F, needed to prevent a beam ($m = 1110$ kg) from swinging down under the effect of its own weight (Figure 8-59).

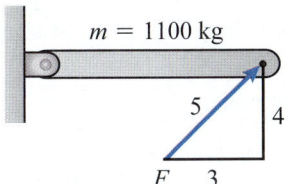

Figure 8-59   Problem 48.

49. ★ Calculate the magnitude of the torque about the origin of force $\vec{F} = (670$ N, 37°$)$ applied in the xy-plane at a point on the x-axis 2.00 m to the right of the origin.

50. ★ Find the cross product $\vec{r} \times \vec{F}$ where $\vec{r} = (17.0$ m, 90°, 33°$)$ and $\vec{F} = (430$ N, 90°, 81.0°$)$. The vectors are given in spherical coordinate notation. Report the result in both spherical and Cartesian notation.

51. ★ Calculate the vector product of $\vec{r} = (2.000\hat{i} - 3.000\hat{j} - 5.000\hat{k})$ and $\vec{F} = (-3.000\hat{i} - 5.000\hat{j} + 1.000\hat{k})$. Give the answer in both Cartesian and magnitude angle notation.

Section 8-4   Moment of Inertia of a Point Mass

52. ★ Three masses, each $m = 12$ kg, are located at the corners of an equilateral triangle. Calculate the moment of inertia of the system about an axis running through one of the point masses perpendicular to the plane of the triangle. The length of one side of the triangle is 0.94 m.

53. ★ Four 400 g masses are located at the corners of a square with sides 1.10 m long. Find the moment of inertia about one of the diagonals.

Section 8-5   Systems of Particles and Rigid Bodies

54. ★★★ ∫ Using integration, calculate the moment of inertia of a solid disk of mass M and radius R about an axis through its centre of mass when the axis lies in the plane of the disk.

55. ★ A ceiling fan is made from a cylindrical plate with a mass of 900.0 g and a radius of 11.0 cm. Three rectangular blades, 1.10 m in length and 17.0 cm in width, are attached to the circumference of the cylindrical plate. The total mass of the fan is 1.32 kg. Find the moment of inertia of the fan about an axis through its centre.

56. ★★★ ∫ A solid, uniform disk of radius R and mass M has a smaller disk of radius $R/8$ removed from it (Figure 8-60). Find the moment of inertia of the resulting partially hollow disk about an axis running perpendicular to the plane of the disk through (a) the geometrical centre of the large disk and (b) the centre of the void section, in terms of the moment of inertia of the large solid disk. The distance between the centre of the large disk and the centre of the hollow part is $R/4$.

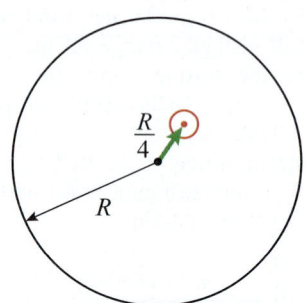

Figure 8-60   Problem 56.

57. ★★★ ∫ Use integration to calculate the moment of inertia of a sphere about an axis through its centre of mass.

Section 8-6   Rotational Kinetic Energy and Work

58. ★★ A cylindrical bar with a sphere firmly attached to one end is originally at rest in a horizontal position. The bar is pivoted about a free hinge at one end (Figure 8-61). The bar is 4.00 m long and has a mass of 231 kg. The sphere has a radius of 60.0 cm and a mass of 53.0 kg. A variable torque is applied to lift the bar by rotating it 37.0° about the pivot at a constant angular speed. Calculate the work done by the variable torque.

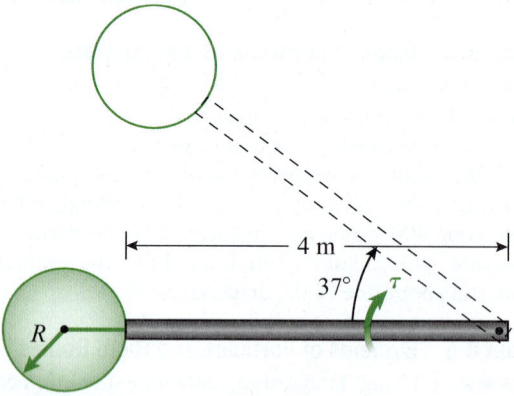

Figure 8-61   Problem 58.

59. ★★ Use energy considerations to derive an expression for the angular acceleration of the pulley of moment of inertia I in Figure 8-62. Assume the friction between the pulley and its axle are negligible. Also assume that the friction between the masses and the surfaces is negligible.

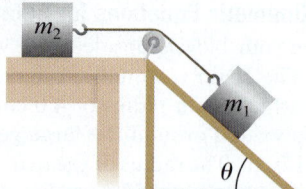

Figure 8-62   Problem 59.

60. ★★ You are trying to get a better feel for the effect of geometry and mass distribution on the moment of inertia. You have a solid disk and a thin ring, each of radius $r = 1.30$ m and mass $m = 73.0$ kg. You mount both on fixed, horizontal frictionless axes about which they can spin freely. Then you spin them both.
(a) How much work do you need to do to get each object to spin at 3.00 rad/s?
(b) Let us assume that you have been causing them to spin by using a constant force applied tangentially to their circumferences. If the above speed is to be reached within 0.700 s, what is the magnitude of the force you need to apply to each object?
(c) You next attempt to stop each object by pressing one finger on each side of each object, right at the outer edge. The coefficient of kinetic friction between each finger and the surface of each object is 0.300. Find the minimum force you have to apply to stop each object within 1.00 min.

Section 8-7   Angular Momentum

61. ★ A car is travelling in the middle lane of a highway at 133 km/h. A police officer has a speed trap set up 210 m down the road. The police officer is standing on the side of the road, 11.0 m from the centre of the middle lane. The car weighs 1250 kg. Calculate the angular momentum of the car about the officer. What would the angular momentum of the car about the officer be at the moment the car passes the officer?

62. ★★ A pellet gun is fired at a solid disk that is free to rotate about a fixed horizontal axis, as shown in Figure 8-63. The disk is 21 cm in diameter and has a mass of 370 g. The pellet ($m = 30.0$ g) gets lodged in the disk and causes it to spin at a rate of 11 rad/s. Find the speed of the pellet just before it hits the wheel.

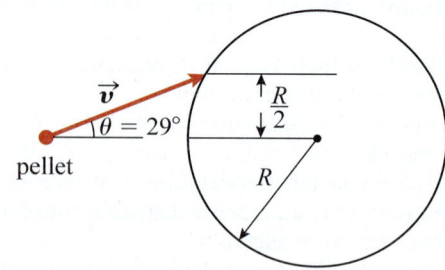

Figure 8-63   Problem 62.

63. ★★★ A paintball gun fires a ball of putty at the pendulum shown in Figure 8-64. The putty has a mass of 53.0 g and strikes the pendulum at a speed of 14.0 m/s. The pendulum is made of a thin bar that is 51.0 cm in length and has a mass of 310.0 g. The sphere fixed to the end of the pendulum is 17.0 cm in radius and has a mass of 190.0 g. The pendulum is originally at rest in the vertical position and pivots about a free hinge at its top. The putty sticks to the pendulum at the point of impact shown. Find the maximum angle that the pendulum makes with the vertical after the collision. Consider the putty to be a point mass.

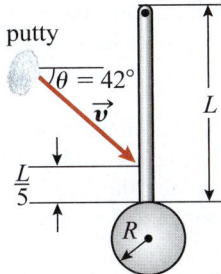

Figure 8-64 Problem 63.

COMPREHENSIVE PROBLEMS

64. ★★★ For a particle moving with uniform circular motion, show that $\vec{a}_r = \vec{\omega} \times \vec{v}$. Demonstrate this relationship for a bead moving in a circular path of radius 0.420 m and speed 3.70 m/s.

65. ★★★ Consider a bicycle wheel that consists of a thin ring of mass 200.0 g and radius 45.0 cm along with 12 thin spokes of mass 21.0 g each. The wheel is made to spin freely about its axis at an angular speed of 0.300 rad/s. A Goliath beetle with a mass of 31.0 g sits at the centre of the ring. The beetle then walks radially outward (with respect to the ring) and stops at the rim of the wheel. The beetle does not slip.
   (a) Find the final angular speed of the wheel when the beetle reaches the rim.
   (b) How much work did the beetle do to reach the edge? The coefficient of friction between the beetle and the wheel is 0.745.

66. ★★★ A solid disk of radius $R = 89$ cm is spinning about a vertical axis through its centre of mass. The disk is driven by a motor that maintains its angular speed at 4.2 rad/s. The moment of inertia of the disk about the axis is 0.30 kg·m².
   (a) How much work does a 12 g cockroach have to do to walk from the centre of the disk to its edge?
   (b) Is the work in part (a) positive or negative?
   (c) Determine the minimum coefficient of friction needed for the cockroach to make the trip in part (a). Consider the cockroach to be a point mass.

67. ★★ Calculate the cross product, $\vec{V}$, of the following two vectors: $\vec{V}_1 = 2\hat{i} + 4\hat{j} + 5\hat{k}$ and $\vec{V}_2 = 7\hat{i} + 4\hat{j} - 2\hat{k}$. Verify that $\vec{V} = \vec{V}_1 \times \vec{V}_2$ is perpendicular to the plane containing $\vec{V}_1$ and $\vec{V}_2$. Show that $\vec{V}' = \vec{V}_2 \times \vec{V}_1 = -\vec{V}$.

68. ★★ Show that the magnitude of the cross product between two vectors is equal to twice the area of the triangle they define. Assume that the two vectors have the same origin.

69. ★★ A worker is felling a 21 m high tree by cutting it at a point 1.1 m above the ground, as shown in Figure 8-65. The mass of the tree is 4200 kg. You can approximate the tree trunk as a thin rod. To consider the effect of the diameter of the tree, we will need the result of problem 103.
   (a) Find the speed of the tree's centre of mass when the tree trunk is parallel to the ground on its way down. Assume that the tree rotates about the point where it is cut.
   (b) How much kinetic energy does the tree have at that point? How fast would a 700.0 kg car have to be moving to have that kinetic energy?

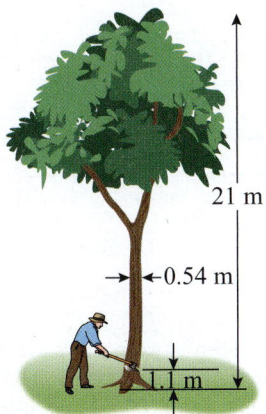

Figure 8-65 Problem 69.

70. ★★ ∫ Find the moment of inertia of a ring about an axis tangent to the ring.

71. ★ In reduced-gravity environments, astronauts suffer from loss of bone mass and muscle mass. One solution is to employ rotating sections in space stations and deep-space vehicles to simulate Earth's gravity. If a cylindrical segment has a radius of 104 m to the inside surface of its outer wall, what rotational speed is required to create a $1.1g$ (i.e., 1.1 times the gravitational acceleration on Earth) environment at this surface?

72. ★★ A movie stunt requires a tiny car with a mass of 380 kg moving at a high speed to collide with a powerful electromagnet. The electromagnet is fixed to the end of a vertical steel bar, which is free to rotate about a fixed axis through its top edge. The steel bar is 7.00 m in length and has a mass of 2.13 tonnes. The magnet is cylindrical in shape and has a mass of 210 kg. When the car collides with the magnet, it completely sticks to it. Treat the car and the electromagnet as point masses. Find approximate values for
   (a) the maximum height that the car reaches when its speed before the collision is 69.0 km/h
   (b) the maximum angle the bar makes with the vertical

73. ★★ ∫ A disk is rotating at an angular speed of 63.0 rad/s. It is then slowed according to $\alpha = -12t - 2.5t^3$. How long will it take for the disk to reach half its speed? What is the angular displacement during that time?

74. ★★ $\frac{d}{dx}$ The angular position of a point on a gear in a machine is given by $3t^3 + 5t - 32$. Determine the angular speed and angular acceleration of this point at $t = 5.00$ s.

75. ★★★ ∫ Two disks are spinning freely about axes that run through their respective centres (Figure 8-66). The larger disk ($R_1 = 1.42$ m) has a moment of inertia of 1120 kg·m² and an angular speed of 5.0 rad/s. The smaller disk ($R_2 = 0.60$ m) has a moment of inertia of 910 kg·m² and an angular speed of 8.0 rad/s. The smaller disk is rotating in a direction that is opposite to that of the larger disk. The edges of the two disks are brought into contact with each other while keeping their axes parallel. They initially slip against each other until the friction between the two disks eventually stops the slipping. How much energy is lost to friction?

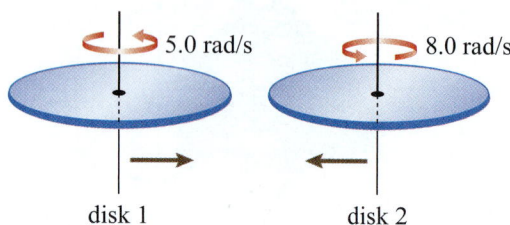

disk 1   disk 2

Figure 8-66   Problem 75.

76. ★★★ In Figure 8-67, two pulleys are mounted on fixed axles that have negligible friction. The small pulley has a moment of inertia of 9.0 kg·m²; it is made up of two cylinders welded together, one of radius 7.0 cm and the other of radius 15.0 cm. The large pulley has a radius of 41.0 cm and a moment of inertia of 84 kg·m²; the pulleys are coupled using a light belt. A 7.00 kg mass hangs from the smaller pulley by a rope that is wound around the smaller cylinder. The system is initially at rest. The mass is let go and starts to fall.
(a) Find the acceleration of the mass.
(b) Find the tension in the rope.
(c) Is the tension in the belt is the same everywhere? Explain.

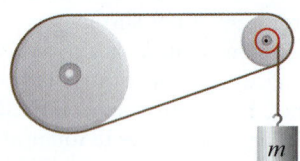

Figure 8-67   Problem 76.

77. ★★ Calculate the moment of inertia of the object shown in Figure 8-68 about the axis a through one of the five thin bars of mass m and length L each, as shown.
(a) Derive the expression for the moment of inertia of each of the bars parallel to the a axis about that axis.
(b) Can you use the parallel-axis theorem to obtain the expression in part (a)?
(c) Why is the expression in part (a) identical in form to the expression for a point mass?
(d) Find the moment of inertia of the whole object about axis a.

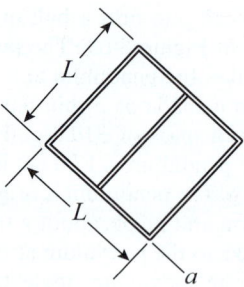

Figure 8-68   Problem 77.

78. ★★ ∫ Using calculus, find the moment of inertia of a solid cylinder about its axis of symmetry. Repeat the same calculation by considering the cylinder to be a stack of disks.

79. ★★★ ∫ Calculate the moment of inertia of a thin rectangular plate of length L and width W, about an axis parallel to W, passing through the centre of the plate. How will this change if the plate thickness is T and no longer negligible?

80. ★★ Two performers, of masses $m_1 = 62$ kg and $m_2 = 84$ kg, are standing on two fixed platforms holding onto the ends of a rope wound around the pulley shown in Figure 8-69. This pulley has a moment of inertia of 23 kg·m² and a radius of 37 cm. The platforms are then simultaneously removed, and the pulley starts to rotate.
(a) Find the angular speed of the pulley 3.0 s after the platforms are removed.
(b) After 3.0 s, the lighter performer begins to climb up the rope. What must this performer's acceleration be so that the pulley stops momentarily 7.0 s later?
(c) Find the tension in the rope for parts (a) and (b). Assume no slippage between the rope and pulley.

performer 1   performer 2

Figure 8-69   Problem 80.

81. ★★ Two gears mounted on fixed axes are connected by a belt as shown in Figure 8-70. The larger gear has a radius of 52 cm. The smaller gear has a radius of 17 cm. The smaller gear is driven by a motor that causes it to spin at 7.0 rad/s.
(a) Find the linear speed of a point on the circumference of the smaller gear.
(b) Find the linear speed of a point on the circumference of the larger gear.
(c) Find the linear speed of the belt.

(d) Find the angular speed of the larger gear.
(e) Find the ratio of the acceleration of a point on the circumference of the larger gear to the acceleration of a point on the circumference of the smaller gear.

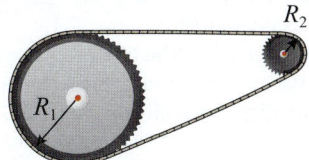

Figure 8-70   Problem 81.

82. ★★ The pulley in Figure 8-71 has a moment of inertia of 6.4 kg·m² about the axis of rotation and a radius of 0.26 m. The masses are as follows: $m_1 = 31$ kg and $m_2 = 13$ kg. Assume that the tabletop and the pulley axle are frictionless.
(a) Determine the angular speed of the pulley when the system is allowed to move from rest under the effect of gravity 3 s into the motion.
(b) Find the tension force(s) in the rope.
(c) Is the tension force the same on each side of the pulley?

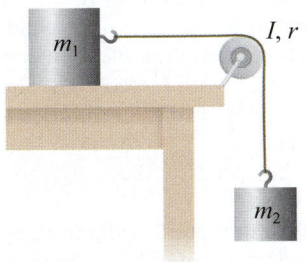

Figure 8-71   Problem 82.

83. ★★★ A solid, uniform disk of mass 12 kg and radius 0.91 m is pivoted about an axis, through its edge parallel to its axis of symmetry. The axis is perpendicular to the plane of the disk. Initially, the disk is held such that its centre of mass is directly above the axis. The highest point on the disk is marked with a dot. The disk is then let go, and it swings from rest under the effect of gravity.
(a) Find the speed of the centre of mass when the centre of mass is (i) at the same horizontal level as the axis and (ii) directly below the axis.
(b) Find the speed of the dot when the centre of mass is 40° below the horizontal.

84. ★★★ A child ($m = 34$ kg) is standing on a merry-go-round midway between the centre and the edge. Both the child and the merry-go-round are initially at rest. The child starts running around the merry-go-round without changing his distance from the centre until he reaches a speed of 3.0 m/s with respect to the merry-go-round surface underneath his feet. The moment of inertia of the merry-go-round is 1600 kg·m², and its radius is 4.0 m. Assume that friction between the merry-go-round and its mounting is negligible.
(a) Find the final angular speed of the merry-go-round when the child is running.

(b) What will the speed of the merry-go-round be when the child stops running?
(c) How much work did the child do to get the merry-go-round to rotate at the angular speed in part (a)?
(d) Calculate the total work done by the child.

85. ★★ The object shown in Figure 8-72 is made from two solid disks: $m_1 = 23.0$ kg and $m_2 = 39.0$ kg, and $r_1 = 17.0$ cm and $r_2 = 33.0$ cm. The disks are connected by a thin rod of mass $M = 11.0$ kg and length $L = 67.0$ cm. The object is held in a horizontal position and then let go. The object is free to rotate about a pivot through the centre of mass of the larger disk. Find the speed of the centre of mass of the smaller disk when it has swung 90° from its original position.

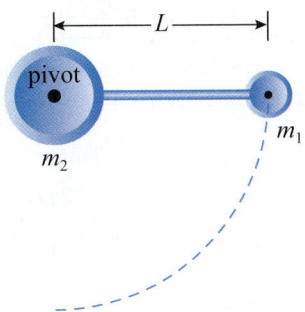

Figure 8-72   Problems 85 and 86.

86. ★★★ ⬜ For problem 85, plot the net torque about the pivot as a function of (a) the angle θ that the rod makes with the horizontal. (b) How long does it take the bar to reach the vertical position? Hint: Use the following power series expansion for the period for large-angle oscillations of a pendulum:

$$T = 2\pi\sqrt{\frac{I}{mgr_{cm}}}\left(1 + \frac{1}{16}\varphi_0^2 + \frac{11}{3072}\varphi_0^4 + \cdots\right)$$

where I is the moment of inertia of the pendulum about the pivot, m is the mass of the pendulum, and r_{cm} is the distance between the pivot and the centre of mass.

87. ★ You are trying to pry two pieces of wood apart by jamming a 37.0 cm long crowbar between them. You drive one end of the crowbar in 0.20 cm between the pieces of wood, which are held together by six nails. The frictional force from each piece of wood on a nail is 110.0 N. With the crowbar perpendicular to the edge of the pieces of wood, what is the force needed if you push at the other end of the crowbar? A clamp on the lower piece of wood keeps it in place. Assume the six nails are along the same line as the point of contact between the crowbar and the wood pieces.

88. ★★ In Figure 8-73, a block of mass with $m_1 = 16$ kg sits on the surface of a bench so the coefficient of static friction between the block and the surface is 0.20. Find the maximum weight that can hang down from the outer cylinder ($R = 34$ cm) of the pulley without causing m_1 to move. The radius, r, of the inner cylinder is 12 cm.

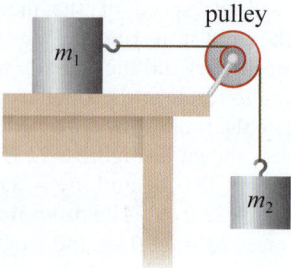

pulley

Figure 8-73 Problem 88.

89. ★ For the system shown in Figure 8-74, the coefficient of kinetic friction between the mass and the incline is μ_k, and the pulley has moment of inertia I and radius r. Derive an expression for the speed of m_1 after m_2 has fallen a distance d from rest along the incline. Assume m_2 is sufficiently larger than m_1 to cause the system to move from rest.

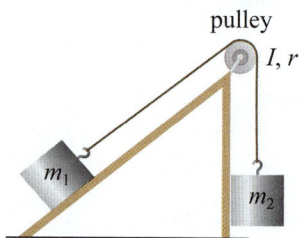

pulley
I, r

Figure 8-74 Problem 89.

90. ★★★ Two mice jump onto the edge of a record turntable, causing it to rotate at the angular speed $\omega = 2.7$ rad/s. After a few seconds, the two mice jump off the turntable along a direction that is tangential to the turntable and opposite to its direction of motion. They jump one after the other at a speed of 2.7 m/s with respect to the turntable. Consider the turntable to be a disk of mass 510 g and radius 18 cm. The mice are 37 g each. Find the final speed of the turntable after the second mouse has jumped off. Consider the mice to be point masses.

91. ★★★ A solid disk with a mass of 32 kg and a radius of 27 cm is spinning counter-clockwise at 17 rad/s in the horizontal plane about its axis of symmetry. Another disk with half the mass and half the radius is spinning at 34 rad/s in the opposite direction when it drops onto the first disk, as shown in Figure 8-75.
 (a) Find the final angular speed of the two disks when they eventually spin together.
 (b) How much energy is lost to friction?

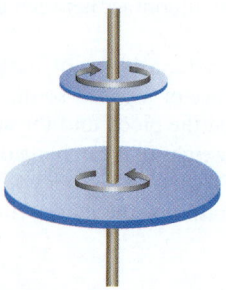

Figure 8-75 Problem 91.

92. ★★★ The beam in Figure 8-76 pivots about one end, swings down from its horizontal position, and strikes a small box, as shown. The collision is perfectly elastic. The beam has a mass of $M = 160$ kg and a length of 1.3 m. Treat the box as a point mass ($m = 32$ kg). The friction at the pivot is negligible.
 (a) Find the speed of the box after the collision.
 (b) Find the maximum angle the bar makes with the vertical after the collision.

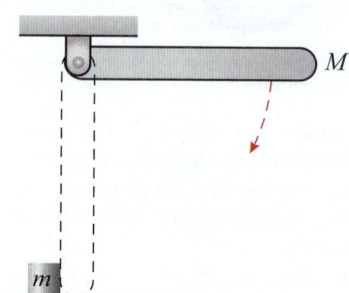

M

m

Figure 8-76 Problem 92.

93. ★★★ A child of mass $m = 28$ kg is standing at the edge of a playground merry-go-round (moment of inertia $I_{cm} = 2700$ kg·m^2 and radius $R = 3.0$ m), which is initially turning at 6.0 rev/min. The child walks opposite to the direction of motion of the merry-go-round with a speed of 2.0 m/s with respect to it. She walks along its edge and completes two full revolutions, stopping at her starting point.
 (a) Find the speed of the merry-go-round after the child stops.
 (b) The child then starts walking along the edge of the merry-go-round until the merry-go-round is spinning with an angular speed of 0.80 rad/s. She then jumps off the merry-go-round by leaping outward in a radial direction with respect to the merry-go-round. Find the angular speed of the merry-go-round just after the child leaps off. The coefficient of static friction between the child and the merry-go-round is 0.50.

94. ★★★ Simon, a young textbook technical checker, is attempting to design a catapult to use as an interesting physics problem. The catapult arm—a bar of length 3.00 m and mass 170. kg—is pulled from its vertical position against specially mounted springs fixed to the wall, as shown in Figure 8-77. At the vertical position, the springs are unstretched. The catapult arm is then pulled back until it is in the horizontal position. A stone of mass $m = 92.0$ kg is loaded into a light basket at the end of the arm, and the arm is released. At 51.0° above the horizontal, the catapult arm is stopped by a mechanism.
 (a) Consider the stone to be a point mass. What should the effective spring constant be to launch the stone at a speed of 190. m/s?
 (b) Consider the stone to be a sphere of radius 25.0 cm, and recalculate the required spring constant. Is it reasonable to approximate the stone as a point mass? What is the percent difference between the two results?

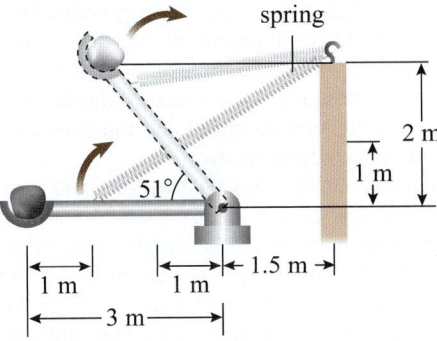

Figure 8-77 Problem 94.

95. ★★★ $\int$ A flywheel is turning at an angular speed of 35 rad/s. It is then coupled with a load that causes it to slow down with a time-dependent acceleration given by $\alpha = -604e^{-2t}$. How much time passes and how many revolutions does the wheel go through before it reaches half the initial speed? Does the flywheel ever stop? How long does it take to stop from the initial speed of 35 rad/s?

96. ★★★ A cylindrical observation station of mass 5200 kg and radius 11 m is spinning freely in space about its central axis. The orbital platform is attended by one astronaut ($m = 72$ kg). He is initially at the centre of the cylinder, but then starts slowly walking, using his magnetic boots, toward the edge to look out from one of the windows at the Sun rising behind Earth. The platform is rotating freely at 10 rev/min when the astronaut is at the centre. The cylinder's moment of inertia about the axis is 15 300 kg·m². Find the angular speed of the platform when the astronaut reaches the edge. How much work did the astronaut have to do to reach the edge? What was the net change in the rotational kinetic energy?

97. ★★ $^{d}/_{dx}$ In Figure 8-78, a rope is wound around a horizontal disk ($M_d = 2.90$ kg and $R_d = 20.0$ cm), passed across a pulley ($I_{cm} = 48.0$ kg·cm² and $R_p = 8.00$ cm), and connected to a hanging 4.00 kg mass. Initially, the mass is held in place, and the system is at rest. The mass is then released and descends in response to the force of gravity. The rope does not slip on the pulley or the horizontal disk. Both the pulley and the disk rotate on bearings that are effectively frictionless.
   (a) Find the angular speed of the pulley after the mass has descended 1.20 m.
   (b) Using energy considerations, calculate the acceleration of the mass without writing the equations of motion.

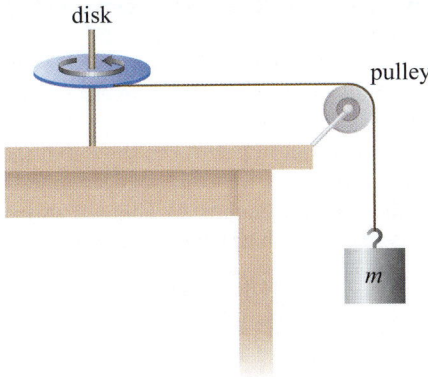

Figure 8-78 Problem 97.

98. ★★ $^{d}/_{dx}$ You are sitting on a pedestal that is free to rotate about its axis of symmetry; your arms are stretched out fully in front of you. The combined moment of inertia of you and the pedestal is 21 kg·m². You are then handed a wheel of mass $m = 15$ kg and radius $R = 27$ cm, spinning at 21 rad/s in a vertical plane. You tilt the wheel axis from the horizontal to the vertical at an angular speed of 0.15 rad/s.
   (a) Find the torque you have to apply as you tilt the wheel.
   (b) The wheel axis eventually becomes fully vertical. Find the angular speed of the pedestal at that point. When the axis of the wheel is vertical, the distance between the axis of the wheel and the axis of the pedestal is 1.1 m.

99. ★★★ Two astronauts are standing on a spinning, disk-shaped platform in space. The platform has radius $R = 17.00$ m and mass $M = 1.500 \times 10^4$ kg and is spinning at an angular speed of 1.3 rad/s. One astronaut ($m_1 = 76.00$ kg) is standing at the edge of the platform, and the other astronaut ($m_2 = 82.00$ kg) is standing at the centre. The astronaut at the edge walks toward the centre using her magnetic boots for traction.
   (a) Does the astronaut do positive or negative work while walking toward the centre?
   (b) As the astronaut walks in, would you expect the angular speed of the platform to increase or decrease?
   (c) Calculate the spinning speed of the platform when the astronaut at the edge reaches the centre.
   (d) Next, the two astronauts walk in opposite directions to the edge of the platform. Do you expect the kinetic energy of the system (platform + two astronauts) to increase or decrease?
   (e) Find the work done by the two astronauts to reach the edge.

100. ★★★ By how much does Earth's angular momentum change as a result of tidal friction over one year? What is the average torque involved?

DATA-RICH PROBLEM

101. ★★★ ⌁, ▭ Figure 8-79 shows the power delivered to a marine propeller shaft versus its rpm (revolutions per minute).

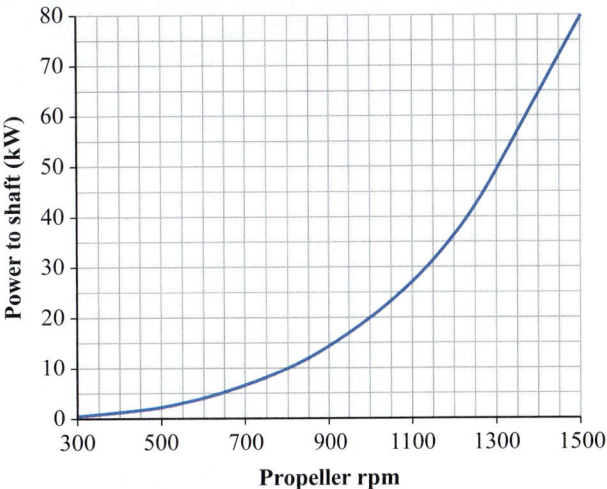

Figure 8-79 Problem 101.

(a) Express the power of a propeller in terms of the angular speed and the torque. What units must the angular speed have for the power to be in watts?

(b) Write an expression for the net torque on a propeller in terms of the driving torque and the resistive torque.

(c) If the torque is running at a constant angular speed, what is the relationship between the resistive torque and the driving torque?

(d) Either by using an online visual digitizer package or by picking points on the graph of Figure 8-79, construct a data table in xy-form from the graph for the horizontal range 500 rpm to 1500 rpm.

(e) From the data table in part (d), create a plot of power versus angular speed in radians/s.

(f) Use the fitting function in your spreadsheet to predict the power law dependence of the torque on the angular speed.

(g) Suggest another way to estimate the power law dependence of the torque on the angular speed, and carry out your estimate.

OPEN PROBLEMS

102. ★★ A red giant burns helium to produce energy. When it runs out of helium, its core begins to collapse as the outer layers expand until the star eventually becomes a red supergiant. Depending on the mass of the supergiant, enough pressure can be generated inside the star to cause heavier elements to fuse. As the star ages, its interior will comprise what is referred to as onion layers of heavier elements with a growing iron core, as the fusion of iron and heavier elements is not energetically favourable. When the iron is approximately 1.4 times the mass of the Sun, it collapses, forcing protons and electrons together to form neutrons and neutrinos. Eventually, the collapse of the core is halted when enough pressure builds as a result of the high density of neutrinos and neutrons inside. The outer layers of the star crash into the core as a result and rebound, causing shockwaves. The outer layers of the star then blow apart in a Type II supernova, which can result in the formation of a neutron star or a black hole, depending on the mass of the exploding star.

In what follows we will consider the case of the star Antares, visible to the naked eye, undergoing a Type II supernova. Assume the resulting neutron star to have a mass 1.91 times that of the Sun and a radius of 28.00 km. Assume that the nebula created as a result of the supernova has no angular momentum. You will be asked to comment on this at the end of the problem. Look up the information you need for this problem.

(a) Estimate the angular speed and period of rotation in days before the star explodes.

(b) Are we justified in saying the star's angular momentum is conserved during the explosion?

(c) Estimate the moment of inertia of the star before the explosion (assume a solid sphere).

(d) Estimate the star's angular momentum.

(e) Estimate the star's kinetic energy before the explosion.

(f) Estimate the moment of inertia of the resulting neutron star. The interior of a neutron star is a heavy liquid, but assume a solid sphere.

(g) Find the angular speed and the period of rotation of the resulting neutron star.

(h) Comment on your calculations and how realistic your answers are.

For the next part we will be roughly assuming that only the material within the core becomes the neutron star.

(i) Look up the density of the core just before the collapse and that of the neutron star after the collapse.

(j) Look up the radius of the core just before the collapse.

(k) Assuming only the material within or near the core remains of the same density as the core becomes the neutron star (i.e., the mass does not change), find the moment of inertia of the core before the collapse.

(l) Recalculate the resulting angular speed and period of the neutron star.

(m) What is the ratio of the period of the progenitor star to that of the neutron star?

(n) Look up the period of the fastest-spinning neutron star.

103. ★★ ∫ **The perpendicular-axis theorem**. Table 8-3 gives the moment of inertia of a cylinder about its principal axis of symmetry. This axis is parallel to the cylinder. In this problem, we develop the tools that will allow us to calculate the moment of inertia of the cylinder (and other objects) about an axis transverse to its length through its centre of mass. This will give us what we need to carry out a more realistic calculation for objects like the tree in problem 69.

We begin by developing what is referred to as the perpendicular axis theorem, which relates the moment of inertia of a planar object about perpendicular axes lying in the object's plane to an axis that is perpendicular to the plane of the object, using Cartesian coordinates.

(a) Consider a planar object lying in the xy-plane. Write down the moment of inertia about the z-axis of a differential element of mass dm a distance r away from the z-axis.

(b) Express that moment of inertia in terms of the x- and y-coordinates.

(c) Set up, but do not calculate, the integral that would give you the moment of inertia of the object about the z-axis.

(d) What do the terms $\int_{object} x^2 dm$ and $\int_{object} y^2 dm$ signify?

(e) Express the moment of inertia of the object about the z-axis in terms of its moment of inertia about the other two axes.

(f) Use the result in part (e) to show that the moment of inertia of a uniform solid cylinder (mass M, radius R, and length L) about a transverse axis passing through its centre of mass is given by

$$\frac{MR^2}{4} + \frac{ML^2}{12}$$

Learning Objectives

When you have completed this chapter, you should be able to

LO1 Distinguish between rotation about a fixed axis, rolling while spinning or skidding, and rolling without slipping.

LO2 Relate the translational and rotational motions of rolling objects.

LO3 Represent rolling as a rotation about a moving axis and as a rotation about a stationary point.

LO4 Apply Newton's second law to rolling motion, and identify forces, torques, and pivot points for a rolling object.

LO5 Apply work and energy considerations to rolling motion, and compare the use of Newton's laws and work–energy approaches for solving rolling motion problems.

LO6 Examine the relationship between friction and rolling.

LO7 Explain and discuss rolling friction.

During the winter in cold-climate countries, roads are often covered with snow, ice, or slush. Ongoing research into skid control, antilock braking systems (ABSs), and tire technology aims to improve traction in extreme weather conditions. Winter tires often have more "aggressive" treads, with channels that are designed to allow the snow, slush, mud, and water to leave the tires more effectively. However, even winter tires may not have an adequate grip on snowy or icy roads in mountainous terrains. In some parts of the world, it is not uncommon to see car tires with studs that bite into the ice. The term *bear claw* is sometimes used to describe sharp-edged, claw-like projections on tread blocks that enhance winter tire performance. In addition, tires with chains for added traction are required by law on mountain roads at certain times of the year in certain areas, such as the Rockies of British Columbia. Studded tires are also used in some places for the same purpose. (Figure 9-1).

Figure 9-1 Examples of enhancements used to improve winter tire performance.

Sources: Blaz Kure/Shutterstock.com, vesilvio/Shutterstock.com, Valentin Mosichev/Shutterstock.com, bofotolux/iStockphoto/Thinkstock

9-1 Rolling and Slipping

In Chapter 8, we discussed the rotation of an object about a fixed axis. However, many applications involve rotation about a moving axis, such as a ball rolling downhill. Before we begin our discussion, however, we need to define a few terms.

Rolling is rotation about an axis that is translating. For example, when you use a baker's rolling pin to flatten dough, you push the axis forward, causing the centre of mass of the rolling pin to translate forward. At the same time, the rolling pin rotates about the axis through its centre of mass.

Spinning occurs when the rotational speed of the surface of an object is too high in comparison to the translational speed of the axis about which the object rotates. For example, when a truck is trying to emerge from a muddy hole, you often see the wheels spinning at relatively high speeds while the truck moves slowly on the muddy surface or does not move at all. Similarly, when a race car driver makes a fast start, the high torque of the powerful engine causes the wheels to spin, even on dry asphalt, and the friction of the asphalt on the spinning tire can burn the surface of the rubber tire (Figure 9-2).

Skidding occurs when the translational speed of a rolling object is too high compared to the rotational speed of its surface. When you are driving a car, it is possible for the wheels to lock when you depress the brake pedal, in which case the wheels stop rotating completely while the car is still moving. The car will

Figure 9-3   Car skidding.

also skid without the wheels locking if the tires rotate at a speed slower than is necessary to keep up with the translational speed of the car (Figure 9-3).

Slipping, or the presence of kinetic friction between the tire and the road, refers to either spinning or skidding. A car moving on asphalt such that its wheels are skidding or spinning often leaves skid marks on the road. When a wheel slips, the part of the tire touching the asphalt rubs against the asphalt. The asphalt actually strips part of the black rubber from the tire in the same way that paper strips away some of a rubber eraser when you erase pencil marks on a page. Slipping often indicates trouble unless, of course, it is done deliberately, such as during a mountain rally or other types of car races (Figure 9-4).

If an object neither spins nor skids as it rolls, its motion is called **rolling without slipping** (also called **ideal rolling** or **perfect rolling**). The relative speed between the part of the wheel in contact with the road and the pavement surface is zero. Figure 9-5(a) shows the motion of a dot painted on the edge of a tire—the path traced by that dot is referred to as a **cycloid**. Notice that the dot momentarily stops with respect to the ground just as the two come in contact. In Figure 9-5(b), the car leaves clear tire marks because the part of the tire that touches the sand or dirt is actually stationary with respect to the ground as the car moves. Otherwise, the tire marks would be smeared.

Figure 9-2   (a) A Porsche Carrera 2 starting rapidly with tire burnout. (b) A Formula 1 car wheel spin.

Figure 9-4   French driver Marcel Tarres puts his Citroen Xsara into a controlled skid during the final of the famous car race on ice, *Trophée Andros*, at the Stade de France.

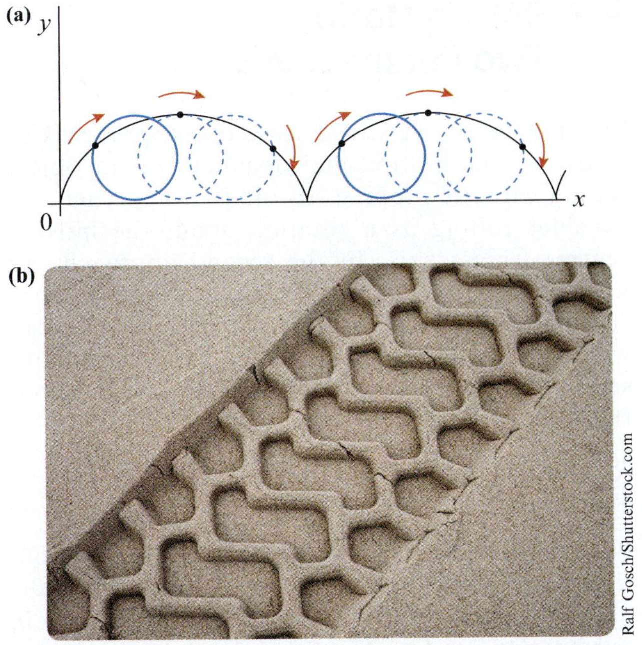
(a)
y
x
0
(b)
Ralf Gosch/Shutterstock.com

Figure 9-5 (a) The trajectory of a point on the edge of a tire. (b) Rolling without slipping produces a clear tire track in the desert.

CHECKPOINT 9-1

Moving or Not?

A car is travelling at a speed of 98 km/h, and its wheels roll without slipping. Its tires have a radius of 31 cm. The speed of a point on the tire that is momentarily touching the ground is

(a) 98 km/h
(b) −98 km/h
(c) 98 km/h + ωr
(d) zero

ANSWER: (d) The speed of the point on the tire in contact with the road is zero for rolling without slipping.

CHECKPOINT 9-2

Why Do Cars Move?

What force moves a car forward when it accelerates from rest without spinning the tires?

(a) static friction
(b) kinetic friction
(c) the torque delivered by the engine
(d) another force

ANSWER: (a) No spinning means static friction.

9-2 Relationships between Rotation and Translation for a Rolling Object

Consider a wheel of radius r rotating in a clockwise direction about a *fixed* axis through its centre of symmetry (Figure 9-6). Since the axis is fixed, the velocity of the centre of mass, v_{cm}, is zero. Choosing right as the positive direction, the velocity of a point at the top of the wheel, given by Equation 8-5, is $v_{top} = +\omega r$. The velocity of a point at the bottom of the wheel is then $v_{bottom} = -\omega r$. Since the velocity of the centre of mass is zero, the velocity of a point at the top is ωr greater than the velocity of the centre of mass, and the velocity of a point at the bottom is ωr less than the velocity of the centre of mass:

$$v_{top} = v_{cm} + \omega r$$
$$v_{bottom} = v_{cm} - \omega r$$

Rearranging and combining the above equations gives the following relationships:

$$v_{cm} = v_{bottom} + \omega r \quad (9\text{-}1)$$
$$v_{cm} = v_{top} - \omega r \quad (9\text{-}2)$$
$$v_{top} = v_{bottom} + 2\omega r \quad (9\text{-}3)$$

The same reasoning applies to the relationship between the acceleration of the centre of mass and the *tangential component* of the acceleration at the top and the bottom of any rotating object. Taking the derivatives of Equations 9-1, 9-2, and 9-3 with respect to time gives

$$a_{cm} = a_{bottom} + \alpha r \quad (9\text{-}4)$$
$$a_{cm} = a_{top} - \alpha r \quad (9\text{-}5)$$
$$a_{top} = a_{bottom} + 2\alpha r \quad (9\text{-}6)$$

These relationships apply to any rotating object, regardless of whether the rotation axis is stationary or moving.

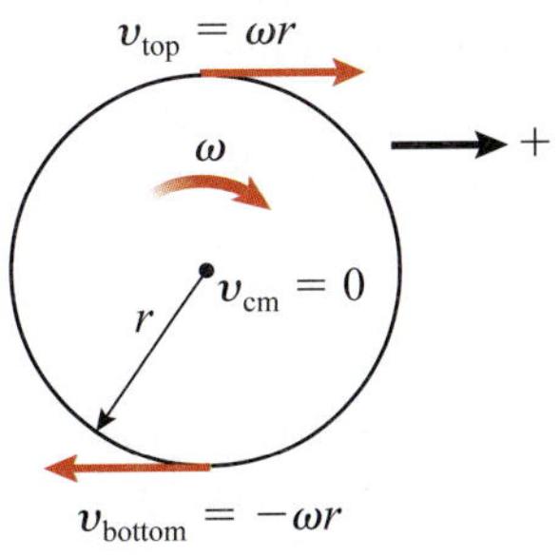
$v_{top} = \omega r$
ω
+
$v_{cm} = 0$
r
$v_{bottom} = -\omega r$

Figure 9-6 Rotation about a fixed axis.

EXAMPLE 9-1

Stuck Truck

While trying to free a dump truck stuck in the mud, you notice that all four wheels are spinning at approximately 10.0 rev/s, and that the truck is moving forward at 5.00 m/s. The diameter of the wheels is 1.50 m.

(a) Calculate the velocity, with respect to the ground, of the point on the tire that is in contact with the ground.
(b) Calculate the velocity of the centre of mass of the tire with respect to the ground.
(c) Determine the velocity of a point at the top of a wheel.

Solution

(a) From Equation 9-1,

$$v_{bottom} = v_{cm} - \omega r$$
$$= 5.00 \text{ m/s} - (10.0 \text{ rev/s})(2\pi \text{ rad/rev})(0.750 \text{ m})$$
$$= -42.1 \text{ m/s}$$

(b) The speed of centre of mass of tire is the same as that of the truck: 5.00 m/s.
(c) For the top of the wheel, we can use Equation 9-2:

$$v_{top} = v_{cm} + \omega r$$
$$= 52.1 \text{ m/s}$$

Making sense of the result

The difference between the velocity at the top and the velocity at bottom is $2\omega r = 30\pi$ m/s. The speed at the bottom of the tire is negative, which is consistent with the tires spinning.

9-3 Rolling Motion: Two Perspectives

In this section, we will discuss two approaches to analyzing the motion and dynamics of an object rolling without slipping. In the first approach, we consider rolling as a rotation about the moving centre-of-mass axis. In the second approach, we use the fact that the point on the rolling body in contact with the surface is momentarily at rest, which allows us to consider rolling as a continuous series of successive rotations of the body about that point.

Rolling as a Rotation about the Moving Centre of Mass

For an object that rolls on a horizontal surface without slipping (Figure 9-7), the velocity of the part of the object in contact with the surface is zero: $v_{bottom} = 0$. Substituting into Equation 9-1 gives

$$v_{cm} = v_{bottom} + \omega r$$
$$= 0 + \omega r$$

Hence, for an object *rolling without slipping*, we have

KEY EQUATION
$$v_{cm} = \omega r \tag{9-7}$$

Similarly, substituting into Equation 9-3 gives the speed at the top of the object or, more generally, the speed of the point opposite to the point of contact with the surface on which the object is rolling:

$$v_{top} = v_{bottom} + 2\omega r$$
$$= 0 + 2\omega r \tag{9-8}$$
$$v_{top} = 2\omega r$$

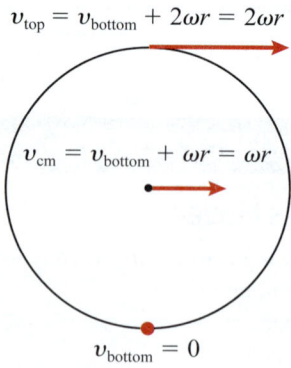

$$v_{top} = v_{bottom} + 2\omega r = 2\omega r$$

$$v_{cm} = v_{bottom} + \omega r = \omega r$$

$$v_{bottom} = 0$$

Figure 9-7 Rolling without slipping.

To determine the acceleration of the centre of mass, we can take the derivative of Equation 9-7:

KEY EQUATION
$$a_{cm} = \alpha r \qquad (9\text{-}9)$$

At the top of the object, from Equation 9-6, the tangential component of the acceleration is

$$a_{top} = 2\alpha r \qquad (9\text{-}10)$$

For the point on the object that is touching the ground, the tangential acceleration using Equation 9-4 is

$$a_{bottom} = 0 \qquad (9\text{-}11)$$

In this approach, rolling is regarded as a translation of the centre of mass combined with rotation of the object about an axis through the centre of mass. The equations of motion will have to address both the translational and the rotational components of the motion.

CHECKPOINT 9-4

Rolling without Slipping

Which of the following statements are *not* true for a wheel rolling freely without slipping on a flat horizontal surface? Pick all that apply.
(a) The total acceleration of the point touching the ground is zero.
(b) The radial acceleration of the point touching the ground is zero.
(c) The vertical velocity of the point touching the ground is zero.
(d) The total speed of the point touching the ground is zero.

ANSWER: (a) and (b) are false statements.

For the translational dynamics of the centre of mass, we can apply Newton's second law:

KEY EQUATION
$$\sum \vec{F} = m\vec{a}_{cm} \qquad (9\text{-}12)$$

For rotational dynamics, we apply the equivalent of Newton's second law for rotation about the centre of mass (Equation 8-35):

KEY EQUATION
$$\sum \vec{\tau}_{cm} = I_{cm}\vec{\alpha} \qquad (9\text{-}13)$$

As you will see in the examples that follow, Equations 9-12 and 9-13 often have to be used in conjunction with Equation 9-9.

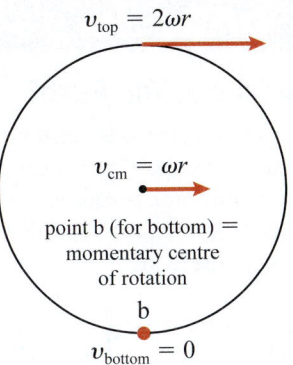

$$v_{top} = 2\omega r$$
$$v_{cm} = \omega r$$
point b (for bottom) = momentary centre of rotation
b
$$v_{bottom} = 0$$

Figure 9-8 Rolling as successive rotations about the contact point b.

Rolling as a Rotation about the Point of Contact between the Object and the Surface

As an object rolls, successive points on its circumference touch the surface on which the object is rolling. If the object does not slip, each of the successive points will be stationary with respect to the surface at the moment that the point touches the surface. Consequently, rolling can be considered as a continuous series of rotations about a contact point, *which we label point b (for bottom)*, as shown in Figure 9-8. This contact point is sometimes called a **momentary pivot**.

Since point b is at rest, we can sum the torques about this pivot point and apply the equivalent of Newton's second law for rotation. Although less intuitive than the approach using Equations 9-12 and 9-13, this approach is an effective method of analysis; it is also often the more convenient approach.

Applying the rotational equivalent of Newton's second law by summing torques about point b gives

KEY EQUATION
$$\sum \vec{\tau}_b = I_b\vec{\alpha} \qquad (9\text{-}14)$$

The object's rotational speed is ω, so the centre of mass located a distance r from the momentary pivot has a speed of $v_{cm} = \omega r$. The top of the object is located a distance $2r$ from the momentary pivot and has speed $v_{top} = 2\omega r$. These speeds are identical to the speeds obtained in Equations 9-7 and 9-8. The expressions for the tangential acceleration of the centre of mass and the top of the object in Figure 9-8 and given in Equations 9-9 and 9-10 can be obtained using the same reasoning.

Notice how we have labelled the torques and the moment of inertia in Equation 9-13. *Since we are summing the torques about point b, we must use I_b, the object's moment of inertia about point b*, rather than I_{cm}, the moment of inertia about the object's centre of mass.

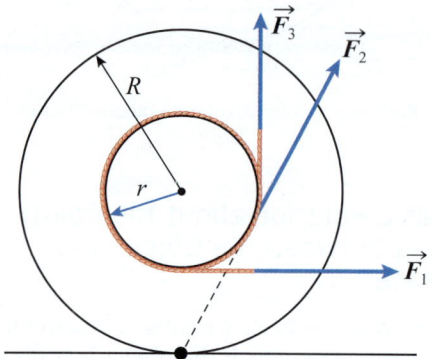
9-4 Newton's Second Law and Rolling

Next, we study the forces involved in rolling. When an object rolls downhill, in which direction does the force of friction act? Does this change when the object rolls uphill? When you drive your car down hill, does the force of friction on the front tires point in the same direction as it does on the rear tires? Is this the same for front vs. rear wheel drive cars. You will be able to answer these questions as we move deeper into the subject. Let us first analyze the motion of an object rolling freely without slipping down an inclined plane. In the following example, we examine the relationship between the slope angle, the force of friction between the object and the surface and the angular and linear accelerations of the object. We analyze the motion using the two approaches mentioned earlier. As you go through the example, pay extra attention to the value we obtain for the force of friction.

EXAMPLE 9-2

Rolling down an Incline

Consider a solid uniform disk that starts from rest and rolls without slipping down a ramp sloping at an angle θ, as shown in Figure 9-10(a). Derive expressions for the acceleration of the centre of mass of the disk and for the force of friction between the surface of the ramp and the disk.

Solution

Let us begin by drawing the FBD of the disk. The forces acting on it are the force of gravity, the normal force from the ramp,

and the force of friction from the surface of the ramp, as shown in Figure 9-10(b).

We can analyze this situation using both approaches described in Equations 9-13 and 9-14; we label these approaches as A and B.

Approach A: Rolling as a translation of the centre of mass combined with rotation of the object about the centre of mass

In the FBD of Figure 9-10(b), we choose our axes to reflect the direction of the acceleration, with the positive x-axis pointing down the incline. Applying Newton's second law to translation along the ramp gives

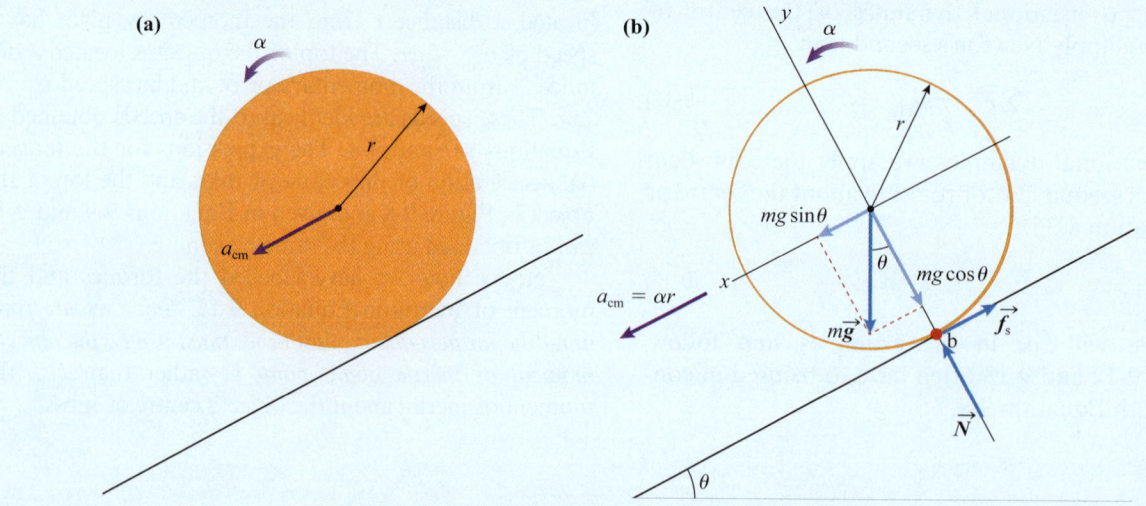

Figure 9-10   (a) A solid uniform disk rolling without slipping down an incline. (b) FBD of the disk in (a).

$$\sum F_x = ma_{\text{cm},x} \qquad (1)$$
$$mg\sin\theta - f_s = ma_{\text{cm},x}$$

Next, we apply the rotational equivalent of Newton's second law by summing the torques about the centre of mass, noting that only the force of friction exerts a torque about the centre of mass; the normal force and the force of gravity exert no torques about the centre of mass, as their respective lines of action pass through the centre of mass. Their lever arm, or moment arm, relative to the centre of mass is therefore zero:

$$\sum\vec{\tau}_{\text{cm}} = I_{\text{cm}}\vec{\alpha} \qquad (2)$$
$$rf_s = I_{\text{cm}}\alpha$$

So far, we have two equations and three unknowns. We do not know the force of friction, the linear acceleration, or the angular acceleration. The third equation is the relationship between the angular and linear accelerations for an object rolling without slipping, Equation 9-9:

$$a_{\text{cm}} = \alpha r \qquad (3)$$

As with most rotation problems, the simplest way to solve a system of three equations is substitution: we substitute the expressions for f_s and a_{cm} into (2) and then solve for a_{cm}, as we did several times in Chapter 8. The key here is to substitute into the torque equation, using the expressions obtained from the other equations. From (1),

$$f_s = mg\sin\theta - ma_{\text{cm}}$$

From result (3),

$$\alpha = \frac{a_{\text{cm}}}{r}$$

Substituting these expressions into result (2) gives

$$r(mg\sin\theta - ma_{\text{cm}}) = I_{\text{cm}}\frac{a_{\text{cm}}}{r}$$

$$a_{\text{cm}} = \frac{mg\sin\theta}{m + \dfrac{I_{\text{cm}}}{r^2}} \qquad (4)$$

The moment of inertia of a disk about its centre-of-mass axis is $I_{\text{cm}} = \dfrac{mr^2}{2}$. Substituting into the expression for a_{cm} gives

$$a_{\text{cm}} = \frac{2}{3}g\sin\theta \qquad (5)$$

To calculate the force of friction, we can substitute this expression for a_{cm} in (1) to obtain

$$f_s = \frac{1}{3}mg\sin\theta$$

Approach B: Rolling as a series of successive rotations about the momentary pivot (the point on the cylinder in contact with the surface of the incline)

Since the point of contact between the cylinder and the surface is momentarily at rest, we can treat it like a pivot about which the equivalent of Newton's second law for rotation can be applied. We call that point b:

$$\sum\vec{\tau}_b = I_b\vec{\alpha} \qquad (6)$$

From Figure 9-10(b), the only force that exerts a torque about the pivot at b is the force of gravity, and specifically the component of the force of gravity that is parallel to the incline.

The remaining forces, namely the normal, the force of friction, and the component of the force of gravity perpendicular to the incline, exert no torques about b, as their corresponding lines of action pass through b. Their lever arm, or moment arm, relative to point b is zero. The expression for torque in (6) becomes

$$rmg\sin\theta = I_b\alpha \qquad (7)$$

As we can see in (7), the advantage of this approach is that it eliminates the force of friction as a variable in the torque equation; the force of friction points straight away from b; its moment arm about b is zero and hence generates no torque about it.

Rearranging, we get

$$\alpha = \frac{rmg\sin\theta}{I_b}$$

Since $a_{\text{cm}} = \alpha r$, we get

$$a_{\text{cm}} = \alpha r = \frac{r^2 mg\sin\theta}{I_b} \qquad (8)$$

At first glance, the expression for a_{cm} obtained above does not look exactly like the one we obtained from the first approach, approach A. For quick reference, we quote that result here:

$$a_{\text{cm}} = \frac{mg\sin\theta}{m + \dfrac{I_{\text{cm}}}{r^2}} \qquad (9)$$

Why are the two expressions for a_{cm} different even though they describe the same variable? (See Checkpoint 9-6 after the example.)

To show that the two results for a_{cm} obtained from approaches A and B are identical, we use the parallel-axis theorem in result (8):

$$I_b = I_{\text{cm}} + md^2 = I_{\text{cm}} + mr^2$$

which gives us

$$a_{\text{cm}} = \frac{r^2 mg\sin\theta}{I_b} = \frac{r^2 mg\sin\theta}{I_{\text{cm}} + mr^2} = \frac{mg\sin\theta}{\dfrac{I_{\text{cm}}}{r^2} + m}$$

This is result (4), thus showing that the two approaches yield identical results.

Making sense of the result

The expression for the acceleration of the centre of mass in (4) has the form of $a = \dfrac{F_{\text{net}}}{m}$, hence has the units N/kg, as it should. The driving force is the component of the force of gravity along the ramp. The disk's ability to resist being accelerated comes from its mass, which indicates its ability to resist being accelerated linearly, and from its moment of inertia, which is a measure of its ability to resist being accelerated rotationally. The force of friction was not the maximum force of static friction because we are not told that the object is on the verge of slipping. When the object slides without rotation, we recover $a = g\sin\theta$, which we obtained in Chapter 5, Example 5-10, for a point mass sliding down an incline. This would happen here if the force of friction were zero. If the force of friction is zero, there is no torque to cause the object to rotate and hence the object does not roll and simply slides, much like the point mass we encountered in Chapter 5.

CHECKPOINT 9-6

Two Equivalent Expressions for a_{cm}?

Why are the two expressions for the acceleration in steps (8) and (9) of Example 9-2 equivalent?

ANSWER: By expanding I_b in (8) using the parallel-axis theorem, we get (9).

CHECKPOINT 9-7

Ring or Disk?

When a ring and a disk with the same mass and radius start from rest and roll down a ramp without slipping,

(a) the ring will have the greater angular acceleration
(b) the ring will have the greater linear acceleration
(c) the two objects will have the same linear acceleration because they have the same mass
(d) none of the above

ANSWER: (d) The ring has the greater moment of inertia, so its angular acceleration is less, which also means that its linear acceleration is less.

In the next example, we deal with the problem of a yo-yo.

EXAMPLE 9-3

The Yo-Yo

A yo-yo consists of a disk with a string wound around it, as shown in Figure 9-11(a). Derive an expression for the acceleration of the yo-yo and the tension in the string as the string unwinds and the yo-yo rolls down vertically downward.

Solution

In this example, we consider rolling as a rotation about the point where the string is unwinding from the disk.

Let us begin by drawing the FBD of the yo-yo, as shown in Figure 9-11(b).

The forces acting in the vertical direction are the force of gravity of the yo-yo and the tension in the string. Since this yo-yo does not slip on the string, the point at which the string leaves the yo-yo is momentarily at rest and can be treated as a temporary, or momentary, pivot, where

$$\sum \vec{\tau}_b = I_b \vec{\alpha}$$

We label this point b, but note that this pivot is not at the bottom of the yo-yo.

The only force that exerts a torque about b is the force of gravity, which acts at the centre of mass of the yo-yo, a distance r from point b. Here again we emphasize that in this approach both the torque and the moment of inertia are taken about point b and *not* about the centre of mass. Noting that mg acts at 90° to r, we have

$$rmg = I_b \alpha$$

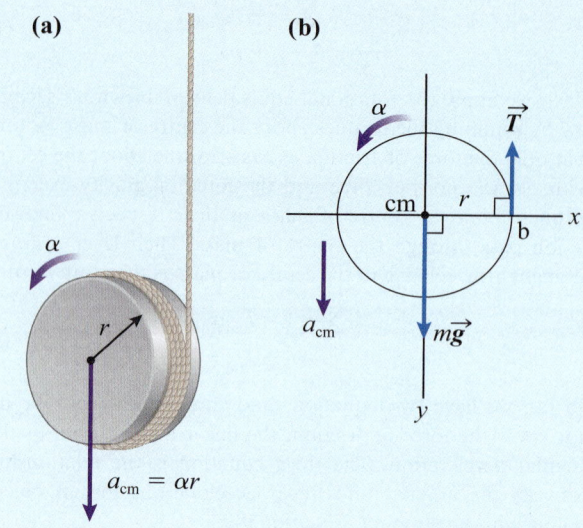

Figure 9-11 (a) A yo-yo. (b) The FBD of the forces acting on the yo-yo.

and

$$\alpha = \frac{rmg}{I_b}$$

Since $a_{cm} = \alpha r$ in magnitude, choosing the positive y-axis to point down, as shown in Figure 9-11, we have

$$a_{cm} = \alpha r = \frac{r^2 mg}{I_b}$$

Using the parallel-axis theorem gives

$$I_b = I_{cm} + md^2 = I_{cm} + mr^2$$

and

$$a_{cm} = \frac{r^2 mg}{I_b} = \frac{r^2 mg}{I_{cm} + mr^2} = \frac{mg}{\dfrac{I_{cm}}{r^2} + m} = \frac{mg}{\dfrac{m}{2} + m} = \frac{2}{3}g$$

We have used the fact that the moment of inertia of a disk about its centre of mass is $mr^2/2$. Note that by summing torques about point b, we are able to calculate the angular acceleration in a single step instead of solving a system of three equations, as we did in the first part of Example 9-2. You are encouraged to solve this problem by considering rolling as a rotation about the moving centre-of-mass axis. With this approach, you can also show that the tension in the rope is $T = \dfrac{mg}{3}$.

Making sense of the result

Again, the expression for the acceleration has the correct units of m/s². The terms in the denominator reflect the object's resistance to acceleration: mass (m) for resistance to translational acceleration, and the contribution of the moment of inertia $\left(\dfrac{I_{cm}}{r^2}\right)$ for resistance to rotational acceleration. When the moment of inertia is zero, we get a downward acceleration of g.

MAKING CONNECTIONS

Rolling, Friction, and ABSs

When applying the brakes in a car, unless the wheels slip, the external force that slows the car is static friction. The direction of the force of static friction exerted by the road on the wheels is opposite to the direction of motion of the car. However, when you accelerate, the force of static friction on the drive wheels *is in the direction* of motion of the car. In the ideal scenario of no air resistance, if the car is moving at a constant speed without slipping, the friction exerted on the tires by the road is zero. In reality, a car moves at a constant speed on the highway because the force of friction between the tires and the surface, pushing the car forward, is just enough to counter the force of air resistance opposing the motion of the car.

When a car goes into a skid while trying to stop on an icy surface (Figure 9-12), the braking distance is greatly increased because the coefficient of kinetic friction is smaller than the coefficient of static friction. ABSs are designed to maximize the force of friction between the road and the tires by keeping the tires on the verge of skidding. Ideally, the ABS will control the car's speed to keep the force of static friction close to its maximum possible value and avoid skidding, which would cause the weaker force of kinetic friction to act between the road and the tires. As described in Section 5-6, the force of static friction can take on values between zero and a maximum

value determined by the normal force and the coefficient of friction:

$$0 \leq f_s \leq f_s^{max} \tag{5-7}$$

When slipping does occur, the force of friction is given by

$$f_k = \mu_k N \tag{5-8}$$

When the tires are on the verge of skidding but not actually skidding, the force of friction between the tires and road reaches the maximum value:

$$f_s^{max} = \mu_s N \tag{5-6}$$

The ABS pumps the brakes automatically, repeatedly reducing the braking force just before the car goes into a skid, and then increasing the braking again. A feedback mechanism regulates the braking to maximize the force of static friction acting on the tires by keeping the wheels on the verge of skidding.

For rubber on ice, the coefficient of kinetic friction can be considerably less than the coefficient of static friction, which makes having an ABS quite desirable. Gabriel Voisin is credited with the introduction of the first ABS, used on airplanes in 1929. Figure 9-13 shows a sketch of the Avions Voisin AJ-T.

Figure 9-12 Extreme driving conditions: a Volvo on an ice-covered road.

Figure 9-13 Artist's rendition of the Voisin AJ-T, by Avions Voisin, a Dutch automobile manufacturing firm named in honour of Gabriel Voisin, reputed to be the inventor of the ABS system.

CHECKPOINT 9-9

Rolling and Friction

Which of the following statements is *not* correct?

(a) Rolling cannot occur without friction.

(b) Friction is often necessary to cause objects to start rolling initially, but once they do, friction is not always necessary to maintain rolling.

(c) If you throw a marble onto a surface with horizontal speed v and give it a forward rotational speed of $\dfrac{v}{r}$, then the marble will roll with or without friction.

(d) Under ideal conditions, the force of friction on a rigid object, rolling freely on a solid smooth (not frictionless) horizontal surface, will be zero.

ANSWER: (a) You can cause an object to roll without friction; see statement (c).

Kinetic Energy of a Rolling Object

The object in Figure 9-14 is rolling without slipping. Since the speed of the centre of mass is v_{cm}, the kinetic energy of translation is

KEY EQUATION
$$K_{trans} = \frac{1}{2}mv_{cm}^2 \tag{9-15}$$

The object is also rotating about the moving centre-of-mass axis at the angular frequency ω. The kinetic energy for this rotation is given by

KEY EQUATION
$$K_{rot} = \frac{1}{2}I_{cm}\omega^2 \tag{9-16}$$

The total kinetic energy is therefore the sum of the translational and the rotational kinetic energies:

KEY EQUATION
$$K = K_{trans} + K_{rot} = \frac{1}{2}mv_{cm}^2 + \frac{1}{2}I_{cm}\omega^2 \tag{9-17}$$

9-5 Mechanical Energy and Rolling

How does the kinetic energy of an object rolling without slipping on a smooth horizontal surface differ from the kinetic energy of the same object sliding without rotation on a frictionless surface, or that of the same object spinning about a fixed axis? We will first examine this question using the rotation/translation approach, and then we will apply the momentary-pivot approach.

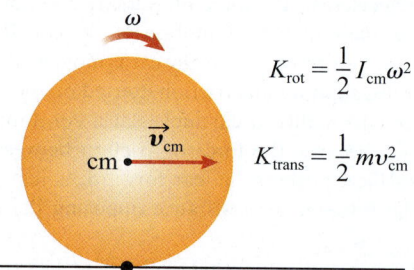

Figure 9-14 Rotation/translation analysis of the kinetic energy of a rolling object.

EXAMPLE 9-4

Kinetic Energy of a Rolling Wheel

A spoked wheel of mass m and radius r rolls without slipping. Most of the mass of the wheel is distributed along its rim, so the wheel can be treated as a hoop or a ring.

(a) Derive an expression for the total kinetic energy of the wheel when the speed of the centre of mass is v_{cm}.

(b) Starting from rest, the wheel rolls without slipping down a ramp from a height h. Derive an expression for the speed of the centre of mass of the wheel at the bottom of the ramp.

Solution

(a) Since the wheel rolls without slipping, Equation 9-7 gives us $v_{cm} = \omega r$, or $\omega = \dfrac{v_{cm}}{r}$. We can write the expression for the total kinetic energy using Equation 9-17:

$$K = \frac{1}{2}mv_{cm}^2 + \frac{1}{2}I_{cm}\omega^2 \tag{1}$$

The moment of inertia of a ring or a hoop about its centre of mass is $I_{cm} = mr^2$. Substituting this into (1) and using Equation 9-7, we get

$$K = \frac{1}{2}mv_{cm}^2 + \frac{1}{2}\frac{mr^2}{r^2}v_{cm}^2 = \frac{1}{2}mv_{cm}^2 + \frac{1}{2}mv_{cm}^2 = mv_{cm}^2 \tag{2}$$

(b) The conservation of mechanical energy requires that the gravitational potential energy lost as the wheel rolls down the ramp equal the increase in kinetic energy, which was formulated by Equation 6-55:

$$\Delta K = -\Delta U \tag{3}$$

Since gravitational potential energy is lost as the wheel rolls down the ramp,

$$\Delta U = -mgh \tag{4}$$

Substituting (2) and (4) into (3), we get

$$mv_{cm}^2 = mgh$$

Therefore,

$$v_{cm} = \sqrt{gh}$$

Making sense of the result

For a ring that rolls without slipping, the rotational kinetic energy is equal to the translational kinetic energy, which explains the value of the velocity at the bottom of the ramp.

For an object that slides without rolling, we saw in Chapter 6 that the speed at the bottom of the ramp was $v = \sqrt{2gh}$, which is $\sqrt{2}$ times the speed we get in this example. The reason the speed is lower in the case of rolling is that part of the potential energy lost as the wheel rolls down the incline goes into the rotational kinetic energy of the object, in addition to the translational kinetic energy. When the object slides down without rolling, we recover the result from Example 6-7.

Kinetic Energy Using the Momentary-Pivot Approach

We now apply the momentary-pivot approach to analyze the kinetic energy of rolling objects. As before, the point on the rolling object that contacts the ground or surface is momentarily stationary with respect to the ground and can be treated as a pivot (Figure 9-15). From that perspective, rolling motion can be regarded as a set of successive rotations about the point on the object in momentary contact with the ground, or the surface on which the object rolls. We discussed this notion in more detail toward the end of Section 9-3.

Using this method, the kinetic energy of a rolling sphere can be written as a single term:

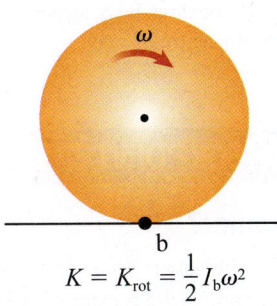

$$K = K_{rot} = \frac{1}{2}I_b\omega^2$$

Figure 9-15 Momentary-pivot analysis of the kinetic energy of a rolling object.

KEY EQUATION

$$K = \frac{1}{2}I_b\omega^2 \qquad (9\text{-}18)$$

All the rolling motion of an object is accounted for by rotations about the momentary pivot. With this approach, there is no need for a separate term for the translational motion.

Notice that Equations 9-17 and 9-18 give two different expressions for the *same* quantity: the total kinetic energy of a rolling object. Therefore, these two expressions must be equivalent (see problem 9-75).

EXAMPLE 9-5

The Speed of a Sphere Rolling Downhill

Starting from rest, a solid uniform sphere with mass m and radius r rolls without slipping down a ramp with height h (Figure 9-16). Derive an expression for the speed of the sphere at the bottom of the ramp.

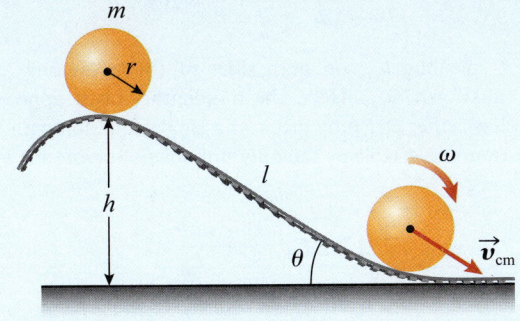

Figure 9-16 Gravitational potential energy is converted to kinetic energy as the sphere rolls down the ramp.

Solution

Here, we use the momentary-pivot approach. From Equation 9-18 we have

$$K = \frac{1}{2}I_b\omega^2 \qquad (1)$$

Conservation of mechanical energy gives

$$mgh = \frac{1}{2}I_b\omega^2 \qquad (2)$$

From the parallel-axis theorem,

$$I_b = I_{cm} + md^2 = I_{cm} + mr^2 \qquad (3)$$

$$mgh = \frac{1}{2}(I_{cm} + mr^2)\omega^2 = \frac{1}{2}I_{cm}\omega^2 + \frac{1}{2}mr^2\omega^2 \qquad (4)$$

Substituting $v_{cm} = \omega r$ gives

$$mgh = \frac{1}{2}I_{cm}\omega^2 + \frac{1}{2}mv_{cm}^2 \qquad (5)$$

(continued)

For a sphere, $I_{cm} = \frac{2}{5}mr^2$. Substituting this expression into (5), and using $v_{cm} = \omega r$, yields

$$mgh = \left(\frac{1}{2}\right)\frac{2}{5}mv_{cm}^2 + \frac{1}{2}mv_{cm}^2 = \frac{7}{10}mv_{cm}^2 \qquad (6)$$

and

$$v_{cm} = \sqrt{\frac{10}{7}gh}$$

This example can also be done by considering rolling as a rotation about the moving centre of mass.

Making sense of the result

While (1) does not explicitly contain an expression for the translational kinetic energy of the sphere, this energy does appear in (5) as a result of applying the parallel-axis theorem in (3). Note: Result (5) applies to any object of mass m and moment of inertia I starting from rest and rolling downhill without slipping. We will apply this result later in the chapter. When the object does not rotate, the rotational term in result (6) vanishes, and we recover the scenario in Example 6-7.

CHECKPOINT 9-10

Larger or Smaller?

The fraction under the square root for the velocity of the centre of mass in Example 9-5 is expected to depend on the moment of inertia of the rolling object. Would you expect this fraction to be larger or smaller if the rolling object was a ring?

ANSWER: The ring has a higher moment of inertia and will thus have a lower speed at the bottom of the hill; the fraction will be smaller.

EXAMPLE 9-6

The Acceleration of a Sphere Rolling down a Ramp Using Work and Energy

(a) Using energy considerations, derive an expression for the linear acceleration of an object of mass m, moment of inertia I, and radius r as it rolls without slipping down a ramp with a slope of angle θ.

(b) Apply your results to a solid uniform sphere and to a solid uniform disk.

(c) Compare your results for the disk to the results in Example 9-2, which was done using Newton's second law.

Solution

(a) In result (5) in Example 9-5, we applied the conservation of energy and the parallel-axis theorem to show that the increase in kinetic energy for an object rolling down a ramp without slipping is

$$mgh = \frac{1}{2}I_{cm}\omega^2 + \frac{1}{2}mv_{cm}^2 \qquad (1)$$

If the object rolls a distance l down the ramp, the change in height, from Figure 9-16, is $h = l\sin\theta$. So,

$$\Delta K = -\Delta U \qquad (2)$$

and

$$mgh = mgl\sin\theta = \frac{1}{2}mv^2 + \frac{1}{2}I_{cm}\omega^2 \qquad (3)$$

As we saw in Chapters 6 and 8, taking the derivative of the energy conservation equation with respect to

time gives an equation with acceleration as one of the terms. Because we are solving for linear acceleration, it is convenient to replace the angular quantities before taking the derivative. Note that if we were solving for angular acceleration, it would make more sense to replace the linear quantities. Since the object is rolling without slipping, the speed of the centre of mass is

$$v_{cm} = \omega r \qquad (4)$$

Substituting this expression into (3) gives

$$mgh = mgl\sin\theta = \frac{1}{2}mv_{cm}^2 + \frac{1}{2}\frac{I_{cm}}{r^2}v_{cm}^2 \qquad (5)$$

Taking the derivative of both sides with respect to time yields

$$mgv_{cm}\sin\theta = mv_{cm}a_{cm} + \frac{I_{cm}}{r^2}v_{cm}a_{cm} \qquad (6)$$

We have used the fact that the derivative with respect to time of the distance travelled by the object's centre of mass along the ramp is the velocity of the centre of mass, v_{cm}. We have also used the chain rule to obtain the derivative of the square of the velocity with respect to time:

$$\frac{dv^2}{dt} = \frac{dv^2}{dv}\frac{dv}{dt} = 2va \qquad (7)$$

Cancelling v_{cm} on both sides of (6) leaves only one unknown, a_{cm}. Here, the momentary-pivot approach has rather elegantly given us a single equation with the desired variable as the only unknown. Solving for a_{cm}, we get

$$a_{cm} = \frac{mg\sin\theta}{m + \dfrac{I_{cm}}{r^2}} \qquad (8)$$

(b) For a uniform solid sphere, $I_{cm} = \frac{2}{5}mr^2$. Substituting this expression into (8), we get

$$a_{cm} = \frac{5}{7}g\sin\theta$$

We can consider a solid uniform disk to be a solid cylinder. For a uniform solid cylinder, $I_{cm} = \frac{1}{2}mr^2$, and the acceleration becomes

$$a_{cm} = \frac{2}{3}g\sin\theta$$

(c) The expression for the acceleration of the uniform solid disk is identical to the expression obtained in Example 9-2 for a cylinder rolling downhill.

Making sense of the result

From (8), we see that if the object is a point mass and has no moment of inertia, a reduces to $g\sin\theta$, as we saw in Chapter 5, Example 5-15.

EXAMPLE 9-7

Acceleration in a Pulley System

Figure 9-17 shows a rope that is wound around a solid cylinder of mass m_c and radius R. The free end of the rope passes over a pulley of radius r and moment of inertia I^p and is attached to a hanging mass M. The system is initially at rest. Mass M is then allowed to fall, pulling the rope and causing the pulley to rotate and the cylinder to roll as the rope unwinds from the cylinder. The cylinder rolls without slipping. Derive an expression for the acceleration of the hanging mass.

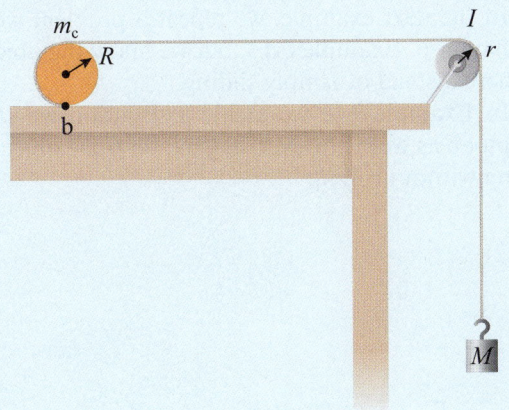

Figure 9-17 Example 9-7.

Solution

We begin by drawing FBDs for the three objects involved (Figure 9-18).

The forces acting on the hanging mass are the force of gravity and the tension in the rope, T_1. Applying Newton's second law to the hanging mass, in the vertical direction we have

$$\sum F_{y'} = Ma_{y'}$$

Choosing down as the positive y-axis,

$$Mg - T_1 = Ma \tag{1}$$

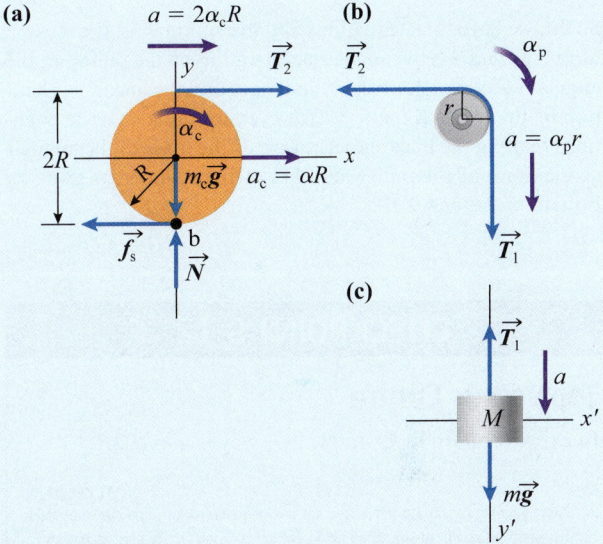

Figure 9-18 FBDs of the (a) cylinder, (b) pulley, and (c) hanging mass.

Next, we sum the torques about the pulley's centre of mass. The torques acting on the pulley are τ_1 exerted by T_1 and τ_2 exerted by T_2:

$$\sum \vec{\tau}_{cm} = I^p_{cm}\vec{\alpha}$$
$$\vec{\tau}_1 + \vec{\tau}_2 = I^p_{cm}\vec{\alpha}$$

where the superscript p refers to the pulley. Choosing clockwise as positive,

$$\tau_1 - \tau_2 = rT_1 - rT_2 = I^p_{cm}\alpha \tag{2}$$

For the rolling cylinder, the simplest approach is to consider the torques about the point of contact between the cylinder and the surface, point b, the momentary pivot:

$$\sum \tau_b = I_b\alpha_c$$

(continued)

Since the only force that exerts a torque about point b is the tension T_2,

$$2RT_2 = I_b\alpha_c \qquad (3)$$

We can use the parallel-axis theorem to determine the moment of inertia of a cylinder about point b:

$$I_b = I_{cm} + md^2$$

The moment of inertia of a cylinder about its symmetry axis is $I_{cm} = \frac{1}{2}mR^2$, and the distance between the centre of mass of the cylinder and the point b is R. Hence,

$$I_b = \frac{m_c R^2}{2} + m_c R^2 = \frac{3}{2}m_c R^2$$

Substituting for I_b in (3) gives

$$2RT_2 = \frac{3}{2}m_c R^2 \alpha_c$$

$$2T_2 = \frac{3}{2}m_c R \alpha_c \qquad (4)$$

So far, we have three equations and five unknowns: the tension forces, T_1 and T_2; the angular acceleration of the pulley, α; the angular acceleration of the cylinder, α_c; and the linear acceleration of the hanging mass, a. However, we can also write equations relating the linear acceleration of the mass to the angular acceleration of the pulley and that of the cylinder, as given by Equations 9-9 and 9-10:

For the pulley: $\qquad a = \alpha r \qquad (5)$

For the cylinder: $\qquad a = \alpha_c 2R \qquad (6)$

Note that the top of the cylinder, the rope, and the hanging mass have the same acceleration, a. The centre of mass of the cylinder has acceleration a_c, which according to Equations 9-9 and 9-10 equals $a/2$.

Substituting (6) into (3) and then substituting (1), (3), and (5) into (2) and solving for a gives

$$a = \frac{Mg}{M + \frac{3}{8}m_c + \frac{I_{cm}^p}{r^2}}$$

For the force of friction on the cylinder, we can sum the forces acting on the cylinder in the x-direction. We leave it to you to find the tension forces and the force of friction acting on the cylinder.

You can also solve this problem by considering rolling as a rotation about the moving centre of mass, or by using the work–mechanical energy approach.

Making sense of the result

The numerator in the expression for a has units of force, and the denominator has units of mass, as expected. When the pulley is light, and the cylinder does not roll, we recover the expression we obtained for the similar problem in Chapter 5, Example 5-12.

CHECKPOINT 9-12

Top versus Centre

In expression (6) in Example 9-7, why is $a = \alpha_c 2R$?

ANSWER: For an object of radius R rolling without slipping with angular acceleration α, the acceleration of the point at the top of an object (or of the point diametrically opposite to the point of contact between the object and the surface) is $\alpha 2R$. Refer to Equation 9-10.

In the next example, we repeat a problem we saw in Chapter 6, Example 6-13, but we allow the object to roll here instead of simply sliding.

In Example 9-9, we tackle a situation we saw in Chapter 6 as well; however, in Chapter 6 the object was sliding without rolling.

EXAMPLE 9-8

Ball on a Vertical Circular Track

The solid metal ball in Figure 9-19 rolls down the ramp, as shown, and then up the vertical circular track. Find the minimum height, h, for which the ball will go all the way to the highest point of the circular track. Assume perfect rolling throughout. The radius of the track is R, and the radius of the ball is r.

Solution

Let us begin by drawing the FBD of the ball at the top of the vertical circular track (Figure 9-20).

We apply Newton's second law. In the vertical (r) direction, we have

$$\sum F_r = ma_r$$

$$N + mg = ma_r$$

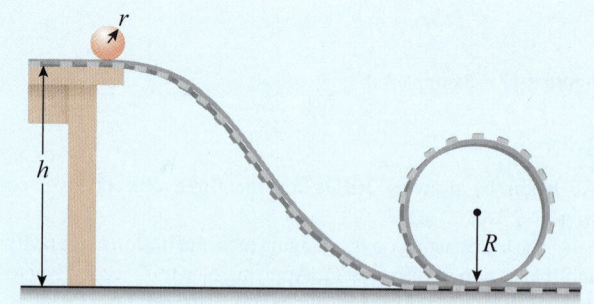

Figure 9-19 Example 9-8.

When travelling around the circular track, the centre of mass of the ball has a trajectory with radius $R - r$. Using the

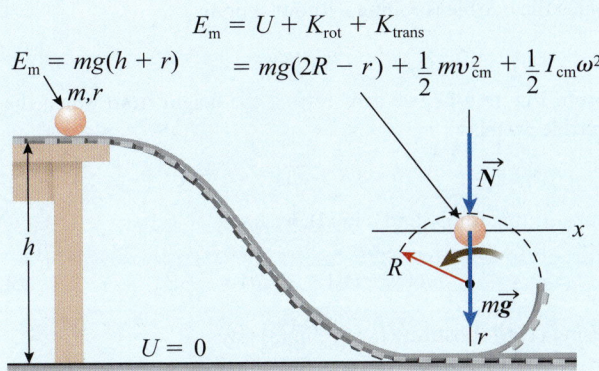

$$E_m = mg(h + r)$$
$$\overset{\bullet}{m,r}$$

$$E_m = U + K_{rot} + K_{trans}$$
$$= mg(2R - r) + \frac{1}{2}mv_{cm}^2 + \frac{1}{2}I_{cm}\omega^2$$

$$\vec{N}$$
$$R \qquad x$$
$$m\vec{g}$$
$$r$$

$$U = 0$$

Figure 9-20   FBD of the metal ball at the top of the circular track.

expression for centripetal acceleration for circular motion and choosing down as the positive r-direction, we get

$$N + mg = m\frac{v^2}{R - r}$$

The problem requires that the ball just barely touch the top of the circular track. So, at this point, the track exerts no force on the ball (i.e., the normal force is zero). Hence,

$$mg = m\frac{v_{cm}^2}{R - r} \quad \text{or}$$

$$g(R - r) = v_{cm}^2 \qquad (1)$$

Conservation of energy requires an increase in the ball's kinetic energy to match the decrease in the Earth–ball system's gravitational potential energy. The ball's centre of mass starts at a height of $(h + r)$ and ends up at the top of the track at a height of $(2R - r)$. The net change in the ball's height

is $(2R - 2r - h)$; hence, the net change in the ball's potential energy is $-mg\,[h - 2(R - r)]$. Therefore, using Equation 6-55, $\Delta K = -\Delta U$:

$$mg(h - 2(R - r)) = \frac{1}{2}mv_{cm}^2 + \frac{1}{2}I_{cm}\omega^2$$

$$= \frac{1}{2}mv_{cm}^2 + \frac{1}{2}\left(\frac{2}{5}\right)mr^2\omega^2 \qquad (2)$$

Since the ball is rolling without slipping, $v_{cm} = \omega r$. Substituting for ωr in (2) gives

$$mg(h - 2(R - r)) = \frac{7}{10}mv_{cm}^2 \qquad (3)$$

Using (1) to substitute for v_{cm}^2 in (3) gives

$$mg(h - 2(R - r)) = \frac{7}{10}mg(R - r)$$

$$h = 2.7(R - r)$$

When $r \ll R$, the expression becomes approximately $h = 2.7R$.

Making sense of the result

The result we obtained for the related example in Chapter 6, Example 6-13, showed that the minimum height required for an object *sliding* on a frictionless track was $h = 2.5R$. Thus, we can see that it takes more energy for the object to roll (rotational in addition to translational kinetic energy) along the track than to slide without rotating. For the object to reach the necessary speed, it must start at a larger height when it is rolling as opposed to just sliding. When the object does not roll or when it is a point mass ($r = 0$), we get the same result as in Example 6-13.

EXAMPLE 9-9

Marble on an Inverted Bowl

A bowl of radius R is sitting upside down on a horizontal surface (Figure 9-21). A child holds a marble of radius r at the top of the outside surface of the bowl. The child lets go of the marble, and it rolls without slipping. Derive an expression for the location at which the marble will fly off the surface of the bowl. Assume that the marble *will continue rolling and not skid* until it loses contact with the surface, which is tantamount to saying the coefficient of static friction between the marble and the bowl is very large. This assumption is put in place so you can comment on the difference between this example and Example 6-14, "Sliding on a Spherical Surface," in Chapter 6.

In this example, we assume that the marble rolls without slipping until it leaves the surface. Is this realistic? Comment on how you expect the marble to behave realistically before it leaves the surface. Compare this example to problems 9-82 and 9-91.

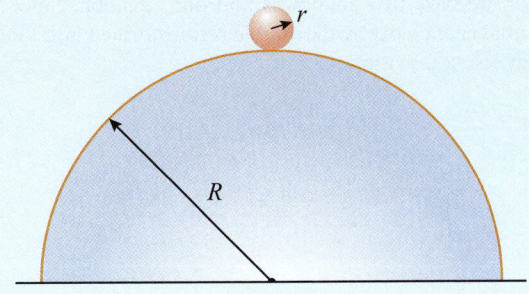

Figure 9-21   Example 9-9.

Solution

As the Earth–marble system loses gravitational potential energy, it will gain speed and kinetic energy, which might cause it to lose contact with the surface. We begin by drawing the FBD of the marble at the location where it is about to leave the surface of the bowl (Figure 9-22).

(continued)

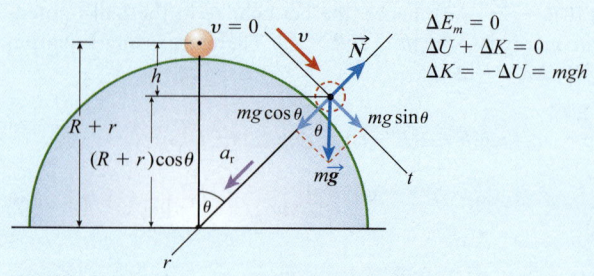

Figure 9-22 FBD of the marble about to leave the surface of the bowl.

Since the marble moves on a hemispherical bowl, the acceleration of interest is the radial acceleration. Hence, we need to determine the radial and tangential components of the forces. We apply Newton's second law. In the tangential direction we have

$$\sum F_t = ma_t$$
$$mg_t - f = ma_t$$
$$mg\sin\theta - f = ma_t$$

In the radial direction, we have

$$\sum F_r = ma_r$$
$$N_r + mg_r = ma_r$$

Choosing the radial direction toward the centre of the bowl as the positive direction, we get

$$-N + mg\cos\theta = m\frac{v_{cm}^2}{R+r}$$

Again, here N is a variable that indicates the magnitude of the normal force.

When the marble is on the verge of leaving the surface, the normal force is zero; hence,

$$mg\cos\theta = m\frac{v_{cm}^2}{R+r} \quad\Rightarrow\quad g\cos\theta = \frac{v_{cm}^2}{R+r} \quad (1)$$

So far, we have two unknowns and one equation. Since the potential energy lost by the marble is transformed into kinetic energy, using Equation 6-55,

$$mgh = \frac{1}{2}mv_{cm}^2 + \frac{1}{2}I_{cm}\omega^2$$

$$\Rightarrow mgh = \frac{1}{2}mv_{cm}^2 + \frac{1}{2}\left(\frac{2}{5}\right)mr^2\omega^2 \quad (2)$$

Since the marble is rolling without slipping,

$$v_{cm} = \omega r$$

From Figure 9-22, we can see that the height from which the marble drops is

$$h = (R+r)(1 - \cos\theta)$$

Substituting for ωr and h in (2), we have

$$mg(R+r)(1 - \cos\theta) = \frac{7}{10}v_{cm}^2 \quad (3)$$

Using (1) to substitute for v_{cm}^2 in (3) gives

$$mg(R+r)(1 - \cos\theta) = \frac{7}{10}mg\cos\theta(R+r)$$

$$\cos\theta = \frac{10}{17}$$

$$\theta = 54.0°$$

In Chapter 6, Example 6-14, we obtained a value of 48° for θ for an object sliding down a frictionless spherical surface.

Making sense of the result

If no rotational motion were involved, the lost potential energy would all be converted to translational kinetic energy, resulting in the object achieving a higher speed as it falls, and hence leaving the surface earlier, as we saw in Example 6-14.

In this example, we have assumed that the ball rolls without slipping until it reaches the point where it leaves the surface. This is, of course, not realistic; as the ball rolls down, the normal force will get smaller, and hence the maximum force of static friction will also become smaller. At some point, the force of friction will become too small to provide the angular acceleration required for the marble to roll without slipping, and the ball will begin to skid.

In problem 85, we allow the ball to skid when the normal force becomes small enough; however, we assume that the surface becomes frictionless at that point to keep the solution relatively simple yet make the scenario more realistic.

In Data-Rich Problem 91, we offer a full solution of a realistic version of the problem, where the surface does not change, and we track the details of the motion of the object until it flies off the surface, without making any assumptions.

The Angular Momentum of a Rolling Object

Consider Earth as it orbits around the Sun: as it moves in its orbit, its centre of mass has angular momentum about the centre of mass (of the solar system). This type of angular momentum is referred to as **orbital angular momentum**. Additionally, Earth possesses **spin angular momentum**, which refers to the angular momentum of Earth as it spins about its axis. The total angular momentum of an object is the sum of the two angular momenta. This concept is developed and discussed in detail in Open Problem 92 and can be used in conjunction with the principle of conservation of angular momentum to solve some of the problems in the text. See problem 92 for details.

9-6 Rolling without Friction

If you place a cylinder at rest on a frictionless incline and let it go, it will simply slide down the incline without rolling. Since there is no force of friction between the cylinder and the surface, there will be no torque due to friction about the centre of mass of the cylinder, and thus there will be no angular acceleration. The angular speed, originally at zero, will remain at zero. When you start your car in the morning and proceed to accelerate from rest, it will not move without friction between the tires and the road. The force of friction between the tires and the road is critical to allow the car to accelerate.

While friction is often needed to get an object to start rolling in the first place, there are a number of circumstances where the force of friction between a rolling object and the surface it rolls on can be zero. It is even possible to have perfect rolling motion on a completely frictionless surface, as we shall see shortly.

Rolling on an Incline with a Zero Force of Friction

Consider a rear-wheel-drive car moving freely downhill, with a speed low enough to ignore air resistance, on a road that has sufficient friction to prevent slipping. Is it possible for the engine to deliver just enough torque to cause the force of friction between the rear tires and the surface to become zero as the car moves downhill? Can the force of friction ever be zero as the car moves uphill?

On the way down, the driver must depress the gas pedal to accelerate the car; on the way up, the driver must use the brakes.

Let us examine the FBD of one of the wheels driven by the power train. Figure 9-23(a) shows the wheel rolling freely (i.e., when the engine delivers no power), and the force exerted by the car on the wheel, assumed vertical, is shown for completeness. This indicates that the wheel is supporting part of the weight of the car.

We consider the sum of the torques about the centre of mass. Since the wheel is rolling freely in Figure 9-23(a), the force of static friction provides the necessary torque to create the needed angular acceleration. If the engine delivers enough torque to the wheel τ_e, it can also cause it to accelerate at the required acceleration, as shown in Figure 9-23(b). Here there is no need for the torque due to friction, and the force of friction will be reduced to zero.

Free Rolling on a Smooth Horizontal Surface

When you drive, you speed up by depressing the gas pedal and slow down by applying the brakes, engaging the force of static friction with the surface accordingly. What if you neither depressed the gas pedal nor applied the brakes? What would the force of friction between

(a) The wheel needs a torque about the centre of mass to accelerate rotationally. Torque about the centre of mass is provided by the force of friction τ_f.

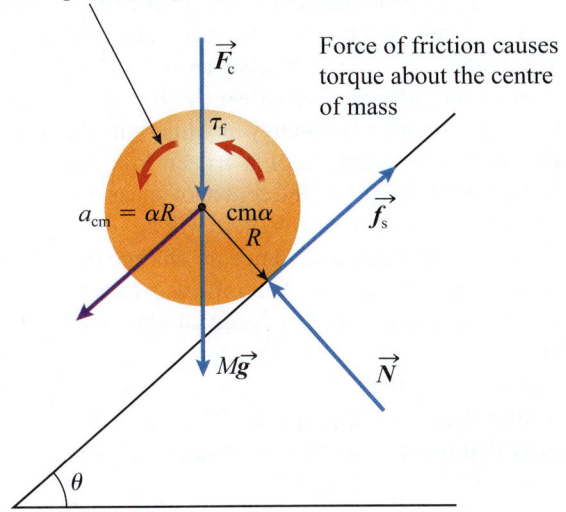

(b) The wheel needs a torque about the centre of mass to accelerate rotationally. Torque about the centre of mass is provided by the engine. No friction is necessarily needed.

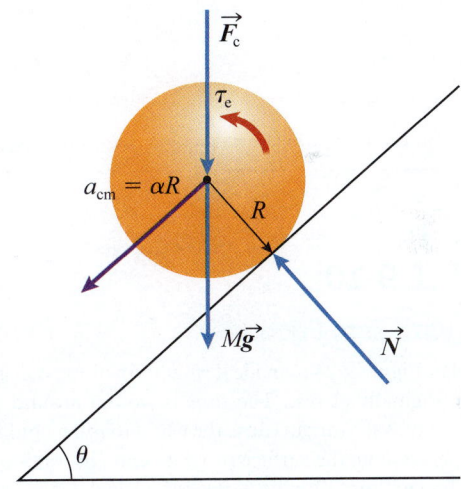

Figure 9-23 (a) FBD of a wheel rolling freely downhill. The angular acceleration is provided by the torque due to the force of friction about the centre of mass. (b) No friction, so the torque needed to accelerate the wheel comes from the engine.

your tires and the surface be? In principle, if there were no losses in the axles, bearings, etc., and if the car were in neutral as you coasted on the highway, and if neither the tire nor the surface were deformable with no adhesion between the two, the car (or a bowling ball rolling on a flat, non-deformable horizontal surface) should in principle roll indefinitely, as the force of friction would be zero, again assuming we can ignore air resistance. This is true ideally of an object rolling freely on a flat horizontal surface.

Rolling on a Frictionless Surface

On to our third example of rolling "without" friction: Can you, in principle, cause an object to roll freely on a completely frictionless surface? Assume you have a bowling ball that you wish to roll on a frictionless horizontal surface. You cause the ball to spin in your hands at angular frequency ω while you throw it onto the frictionless surface of a frozen lake such that the horizontal velocity of the centre of mass is $v = \omega r$. The absence of friction means there will be no torque about the centre of mass, and hence no change in the angular speed. The absence of friction also means that there will be no change in the speed of the centre of mass. Hence, the initial condition of $v_{cm} = \omega r$ will be maintained indefinitely. Since $v_{cm} = \omega r$ is the condition for rolling without slipping, the ball will roll indefinitely, provided there is no air resistance, of course.

CHECKPOINT 9-13

Is There Friction?

A bowling ball rolls freely without slipping on a horizontal surface. The force of friction between the ball and the surface

(a) must be zero
(b) can be zero
(c) cannot be zero
(d) none of the above

ANSWER: (a) The force of friction must be zero. See previous discussion.

EXAMPLE 9-10

Rolling with Zero Friction

The wheel in Figure 9-24 is an ideal ring/hoop of mass m and radius R, originally at rest. The rope is wound around the wheel a few times so you can cause the wheel to roll by pulling on the rope. Assume the surface provides enough friction to ensure rolling without slipping as needed. Calculate the force of friction on the hoop as you pull the rope horizontally with force $\vec{F}$; you may calculate the force of friction in terms of $\vec{F}$.

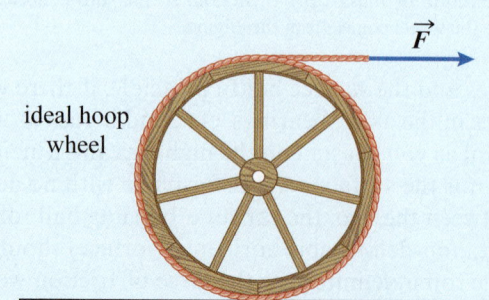

ideal hoop wheel

Figure 9-24 In which direction does the force of friction on this wheel point?

Solution

Let us begin by drawing an FBD of the wheel (Figure 9-25).

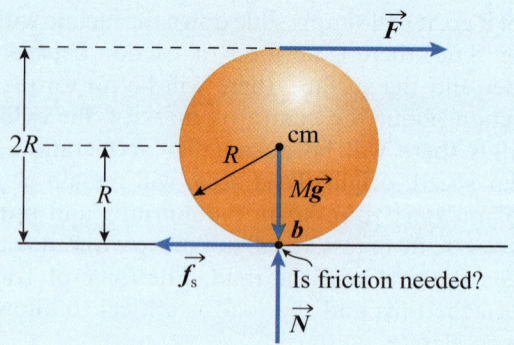

Figure 9-25 FBD of wheel.

Taking the sum of the torques about the bottom of the wheel at point b gives us

$$\sum \vec{\tau}_b = I_b \vec{\alpha} \tag{1}$$

The only force that exerts a torque about b is the force $\vec{F}$ acting at a distance of $2R$. Hence,

$$2RF = I_b \alpha \tag{2}$$

To obtain the moment of inertia of the wheel about point b, we use the parallel-axis theorem from Chapter 8:

$$I_b = I_{cm} + md^2$$

The moment of inertia of a ring about its symmetry axis is $I_{cm} = mR^2$, and the distance between the centre of mass of the ring and the point b is $d = R$. Therefore,

$$I_b = mR^2 + mR^2 = 2mR^2 \tag{3}$$

Substituting this value into (2) gives us

$$2RF = 2mR^2\alpha$$
$$F = mR\alpha$$

For an object rolling without slipping, $a_{cm} = \alpha R$. The expression for F thus becomes

$$F = ma_{cm} \tag{4}$$

Newton's second law states that $\sum \vec{F} = m\vec{a}_{cm}$, that is, the net force on an object is equal to its mass multiplied by its acceleration. This means that the applied force $\vec{F}$ is actually also the net force acting on the wheel in the x-direction, which means that the force of friction on the wheel must be zero. If you are still in doubt, let us apply Newton's second law to the wheel. In this case, we have

$$\sum \vec{F} = m\vec{a}_{cm} \tag{5}$$

Let us, for now, say that we do not know that the force of friction is zero, in which case it might make sense to assume the force of friction acts opposite to $\vec{F}$. Then (5) becomes

$$F + f_s = ma_{cm}$$

However, (4) tells us that $F = ma_{cm}$. Hence

$$ma_{cm} + f_s = ma_{cm}$$

which gives us that *the force of friction on the wheel in this particular case must be zero.*

$$\vec{f}_s = 0$$

Making sense of the result

Since all the mass is at the rim, the problem can be seen as a force accelerating the mass, and hence no other forces are needed in this case.

CHECKPOINT 9-14

Does Friction Oppose Motion?

What if the moment of inertia, I_{cm}, of the object in Figure 9-25 were $\frac{1}{3}mR^2$? In which direction would the force of friction point? Would it point with or opposite to the applied force?

ANSWER: According to the discussion in Example 9-10, we would end up with a force of friction that would actually have to help the applied force, which means it would point in the direction of the applied force!

Reflecting on Checkpoint 9-15, is it possible to have an object with a moment of inertia greater than mR^2? Can you then think of a real scenario where the moment of inertia of a rolling object of radius R and mass m is greater than mR^2? The answer is normally no; however, there are circumstances in which this can happen. For example, think of wheel rolling on a special elevated track where the spokes of the wheel go past the wheel's rim.

9-7 Rolling Friction

Why is it that an object rolling freely on a smooth, flat horizontal surface eventually slows down (ignoring air resistance)?

The processes that contribute to loss of mechanical energy during ideal rolling without slipping can be grouped in a category referred to as rolling friction (sometimes called *rolling resistance*): one cause of rolling friction is the adhesion between the surface and the object, depending on the materials they comprise. Another cause is deformation.

Figure 9-26(a) shows the profile of the normal force from a non-deformable surface acting on a non-deformable object. This normal force exerts no torque about the object's centre of mass. A non-deformable object rolling on a *deformable* surface will cause the surface to "pile up" in front of the object, as in Figure 9-26(b). An extreme example is rolling an object on a

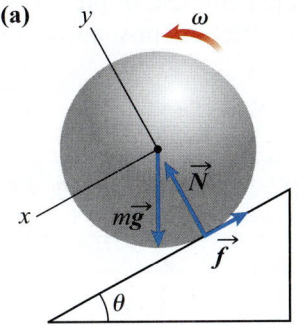

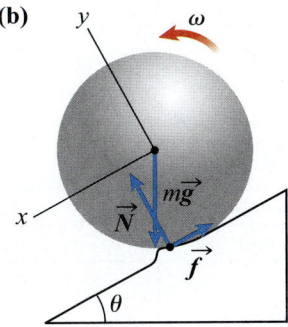

Figure 9-26 (a) For an object rolling on a non-deformable surface, the normal force points into the centre of mass and exerts no net torque about it. (b) When an object rolls on a deformable surface, part of the surface in front of the object "piles up," resulting in a normal force from the surface exerting a non-zero torque about the centre of mass axis, and hence causing the object to slow down.

thick layer of dough. Another is a child pushing a toy truck over sand. The force from the surface onto the object is no longer localized at the point of contact between the object and the surface. As shown in Figure 9-26(b), parts of the surface will exert forces on the object such that the forces have components pointing in a direction opposite to the direction of motion of the wheel. Even when the force from the surface is normal to the surface and does not directly point in a direction opposite to the motion of the object, the deformation of the surface will result in normal forces from the surface exerting a non-zero retarding torque about the centre of mass, which will result in slowing down the rotational and hence the translational speed of the object.

Rolling friction can be quantitatively approximated using a coefficient of rolling friction. This acts very similar to the coefficient of static friction that acts on a rolling wheel. The rolling friction opposes the direction of motion, acting to slow down a rolling wheel. The force of rolling friction is given by the product of the normal force and the coefficient of rolling friction:

$$f_r = \mu_r N \qquad (9\text{-}19)$$

The coefficient of rolling friction depends on the materials involved and is typically low for rigid, generally non-deformable materials. Table 9-1 shows the coefficients of rolling friction for some common objects and surfaces.

Table 9-1 Typical Values for Various Coefficients of Rolling Friction

Tire and Surface Conditions	Coefficient of Rolling Friction (μ_r)
Steel train wheel on steel rail	0.000 4
Iron on iron	0.000 51
Wood on steel	0.001 2
Wood on wood	0.001 5
Hardened steel ball bearings on steel	0.001 5
Railroad steel wheel on iron rail	0.002
Polymer on steel	0.002
Production bicycle tire on asphalt	0.005
Cast iron wheels on steel rail	0.006 5
Hard rubber on steel	0.007 7
Car tires on asphalt	0.012
Car tires on concrete	0.015
Car tire on gravel road	0.02
Hard rubber on concrete	0.02
Phenolic on steel	0.026
Cast nylon on steel	0.027
Polyurethane on steel	0.057
Tires on sand	0.3
Hard rubber on steel	0.303

In the case of a deformable object rolling on a non-deformable surface, one of the causes of energy loss, and hence rolling friction, is that the energy lost during that deformation is not all recovered in time and some of it is lost as heat. This process is called hysteresis, which is covered in more detail in more specialized texts.

For a rigid non-deformable object rolling freely on a rigid, flat, non-deformable surface, there should in principle be no loss of energy provided the surface of the object and the surface on which the object is rolling are both ideally smooth. However, in reality, even if it were possible to obtain such an ideally non-deformable surface, it is often the case that the surface is not ideally smooth. This inevitably causes "collisions" between the imperfections in the surface and those on the rolling object, resulting in a continuous loss of momentum and mechanical energy of the rolling object.

KEY CONCEPTS AND RELATIONSHIPS

Rolling

Rolling is a combination of translational and rotational motion. For rolling without slipping,

$$v_{cm} = \omega r \tag{9-7}$$

$$a_{cm} = \alpha r \tag{9-9}$$

The part of the object in touch with the surface is momentarily stationary with respect to the surface:

$$v_{bottom} = 0$$

Rolling can also be treated as a series of successive rotations about the point of contact between the object and the surface.

The Dynamics of Rolling

For the translation/rotation approach,

$$\sum \vec{F} = m\vec{a}_{cm} \tag{9-12}$$

$$\sum \vec{\tau}_{cm} = I_{cm}\vec{\alpha} \tag{9-13}$$

For the momentary-pivot approach, where b refers to the point of contact between the object and the surface,

$$\sum \vec{\tau}_{b} = I_{b}\vec{\alpha} \tag{9-14}$$

The force of static friction on an object rolling without slipping is usually an unknown:

$$0 \leq f_s \leq f_s^{max}$$

When an object is on the verge of slipping, the force of friction becomes

$$f_s^{max} = \mu_s N$$

If an object slips as it rolls, the force of friction becomes the force of kinetic friction:

$$f_k = \mu_k N$$

Work and Energy

For the translation/rotation approach,

$$K_{trans} = \frac{1}{2}mv_{cm}^2 \tag{9-15}$$

$$K_{rot} = \frac{1}{2}I_{cm}\omega^2 \tag{9-16}$$

The total kinetic energy is given by

$$K = K_{trans} + K_{rot} = \frac{1}{2}mv_{cm}^2 + \frac{1}{2}I_{cm}\omega^2 \tag{9-17}$$

For the momentary-pivot approach,

$$K = \frac{1}{2}I_{b}\omega^2 \tag{9-18}$$

APPLICATIONS

ABS, traction control, tire design, drive trains, rolling element bearings

KEY TERMS

cycloid, ideal rolling, momentary pivot, orbital angular momentum, spin angular momentum perfect rolling, rolling, rolling without slipping, skidding, slipping, spinning

QUESTIONS

In the following questions and problems, ignore the effects of rolling friction unless otherwise specified.

1. A solid uniform ring and a solid uniform disk of the same radius and mass roll without slipping at the same speed on a horizontal surface, which turns into a ramp sloping upward from the horizontal. Which object will go higher up the ramp?
   (a) the ring
   (b) the disk
   (c) both will reach the same height
   (d) it depends on the force of friction

2. A uniform solid sphere and a uniform hollow cylinder roll without slipping at the same speed on a horizontal surface. The surface then becomes frictionless and curves upward into a semi-cylindrical track of radius r. Which of these objects will go higher up the track? Assume that both objects are in contact with the track at all times and that neither of the objects has enough kinetic energy to make it to a height r above the bottom of the track.
   (a) the cylinder
   (b) the sphere
   (c) both will reach the same height
   (d) it depends on the force of friction while the objects roll

3. A uniform solid cylinder and a uniform solid sphere of the same mass and radius roll without slipping downhill from the same height.
   (a) Which object will have the higher rotational kinetic energy at the bottom of the hill?
   (b) Which object will have the higher angular speed at the bottom of the hill?

4. Consider a uniform solid sphere rolling without slipping down a ramp of angle θ. How will the force of friction on this object change if you double its mass without changing its radius?

5. What is the main function of an antilock braking system (ABS)? In the absence of the ABS, why is a driver advised to "pump" the brakes when stopping on a slippery surface?

6. As a cyclist rides her bike uphill, can you think of something she could do to end up with zero friction between her rear wheels and the road, assuming you can ignore air resistance and rolling friction?
   (a) Hit the brakes.
   (b) Accelerate faster.
   (c) Simply coast up the hill.
   (d) It is impossible; she must have friction for the wheels to roll.

7. A rear-wheel-drive car coasts on the highway at a constant speed of 130 km/h. The force of friction between the rear tires and the road is (assume you can ignore losses due to friction within the car)
   (a) kinetic and larger than the force of air resistance
   (b) kinetic and equal to the force of air resistance
   (c) static and larger than the force of air resistance
   (d) static and equal to the force of air resistance

8. The wheel in Figure 9-27 rolls freely on a flat horizontal surface without slipping. Establish the direction of the total acceleration of points A, B, and C on the wheel at the instant shown.

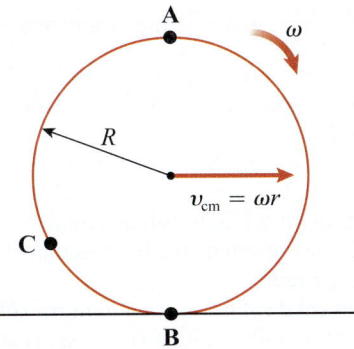

Figure 9-27 Question 8 and problem 25.

9. A solid uniform ring, disk, and sphere with equal masses and equal radii roll without slipping down a smooth ramp from the same height. Which of the three objects will reach the bottom of the ramp first?
   (a) the ring
   (b) the disk
   (c) the sphere
   (d) none of the above

10. A uniform ring rolls down a ramp without slipping. At any point, calculate the ratio of the rotational energy to the translational kinetic energy. Repeat the exercise for a uniform solid disk and for a uniform solid sphere.

11. An object rolls without slipping down a hill such that 2/3 of its kinetic energy is rotational. Determine an expression for the object's moment of inertia about an axis through its centre of mass in terms of its mass and radius.

12. Can the rotational and translational kinetic energies ever be equal for a rigid object rolling without slipping? Explain, and find the moment of inertia for such an object.

13. Two uniform solid disks, disk 1 and disk 2, of the same radius and thickness, start from rest and roll without slipping down a ramp from the same height. The mass density of disk 2 is twice the mass density of disk 1. Will one disk reach the bottom of the hill before the other? If so, which one?

14. Two uniform solid spheres, sphere 1 and sphere 2, are at rest at the top of a hill. The spheres have the same mass but different radii: the radius of sphere 1 is greater than the radius of sphere 2. They roll without slipping downhill from the same height. Will one sphere reach the bottom of the hill before the other? If so, which one?

15. Two thin cylindrical shells, shell A and shell B, are at rest at the top of a hill. The mass of shell A is 16 times the mass of shell B, and the radius of shell A is one half the radius of shell B. Both shells roll without slipping downhill from the same height. Will one shell reach the bottom of the hill before the other? If so, which one?

16. Two uniform solid spheres, sphere A and sphere B, are at rest at the same height at the top of a hill. The mass of sphere A is five times the mass of sphere B, and the radius of sphere A is $\frac{1}{\sqrt{2}}$ the radius of sphere B. Both spheres roll without slipping downhill. Will one sphere reach the bottom of the hill before the other? If so, which one?

17. Is it possible for a uniform spherical shell and a solid uniform cylinder, originally at rest and at the same height on top of a hill, to roll downhill without slipping and reach the bottom of the hill together? If so, find the ratio of their radii.

18. A uniform solid cylinder and a uniform solid sphere of the same mass roll freely down the same slope without slipping. Which of the two objects has a greater force of friction acting on it?

19. A uniform solid sphere and a uniform ring roll without slipping down a hill. The force of friction acting on the ring is 10 times the force of friction acting on the sphere. Find the mass ratio of the two objects.

20. A uniform solid cylinder and another uniform rigid object roll without slipping down a slope. The mass of the second object is twice the mass of the cylinder. The force of friction acting on the cylinder is twice the force of friction acting on the other object. Find an expression for the moment of inertia of the second object about an axis through its centre in terms of its mass and radius.

21. Consider a typical wheel rolling freely on a flat horizontal surface. What could cause the wheel to slow down in the absence of air resistance? Pick all that apply.
    (a) rolling friction
    (b) little bumps in the road
    (c) kinetic friction
    (d) static friction

22. A marble rolls on a horizontal, non-deformable, smooth glass surface. Do you expect the marble to slow down because of friction? Explain why or why not.

23. A ball rolls down a hill without slipping. Assume ideal conditions, that is, the object and surface are non-deformable and the surface is smooth. In which direction does the force of friction act on the ball?
    (a) in the same direction as the direction of motion
    (b) opposite to the direction of motion
    (c) could be (a) or (b)
    (d) none of the above

PROBLEMS BY SECTION

For problems, star ratings will be used (★, ★★, or ★★★), with more stars meaning more challenging problems. The following codes will indicate if $\frac{d}{dx}$ differentiation, $\int$ integration, ▢ numerical approximation, or ⌒ graphical analysis will be required to solve the problem.

Section 9-1 Rolling and Slipping

24. ★ A rear-wheel-drive car accelerates from rest on a flat horizontal surface, and the wheels turn (roll) without slipping. Determine the direction of the force of friction on the car's front and rear wheels.

Section 9-2 Relationships between Rotation and Translation for a Rolling Object

25. ★ The wheel in Figure 9-27 rolls freely on a horizontal surface without slipping at a constant speed of 15.0 m/s. The wheel is 74.0 cm in diameter.
    (a) Determine the speed and acceleration of point A at the top of the wheel.
    (b) Find the magnitudes of the total speed and acceleration of the point one quarter of a period later and one half of a period later.

26. ★★ A tundra buggy, which is a bus fitted with oversized wheels, is stuck in Churchill, Manitoba, on slippery ice. The wheel radius is 0.87 m. The speedometer indicates that the wheel accelerates from 0 to 27 km/h while the buggy moves a total distance of 5.0 m in 7.0 s. Find the tangential acceleration and the total acceleration of a point at the bottom of the wheel at the end of 3.0 s.

27. ★ While driving on the frozen surface of a lake, you notice that your truck is moving at 10. km/h, but the speedometer says 56 km/h. The wheel radius is 46 cm. (Hint: The speedometer reading is obtained from the rotational speed of the tires.)
    (a) Find the speed of a point at the top of a tire.
    (b) Is there a point on the tire where the velocity is zero? If so, where is it located?

28. ★★ A cylinder of radius 1.2 m rolls at a constant speed. You apply a brake mechanism that slows the cylinder. At a certain instant, the speed of the cylinder's centre of mass is 11 m/s and the speed of the top of the cylinder is 14 m/s. Calculate the angular speed of the cylinder at that instant.

29. ★★ A wheel undergoes acceleration from rest, such that it spins while rolling. The acceleration of the centre of mass is measured to be 2.30 m/s², and the angular acceleration is measured to be 7.10 rad/s². The wheel's radius is 80.0 cm.
  (a) Find the magnitude of the tangential acceleration of a point at the top of the wheel.
  (b) Find the magnitude and direction of the total acceleration of a point at the bottom of the wheel at 1.70 s into the motion.

Section 9-3   Rolling Motion: Two Perspectives

For the problems in this section, approach rolling as a rotation about point b, which is the point at which the rolling object is in contact with the surface on which it sits.

30. ★ Show that it is valid to consider rolling as a rotation about point b.

31. ★ If we consider rolling to be a rotation about point b, that leaves us with an apparent paradox: If the object were moving at a constant speed, then the centre of mass would be accelerating with respect to b. How do you resolve this apparent paradox?

32. ★★ Find the acceleration of the point on a rolling object in contact with the surface on which it rolls without slipping.

33. ★★ Calculate the maximum angle that the force, $\vec{F}$, can make with the horizontal to cause the object in Figure 9-28 to roll to the right. Assume there is enough friction to ensure rolling without slipping.

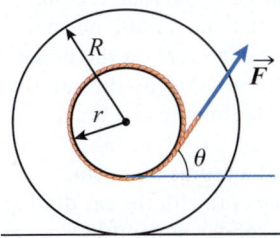

Figure 9-28   Problem 33.

34. ★★ The homogeneous object in Figure 9-29 has a mass of 1.2 kg and a moment of inertia about its axis of symmetry of 17 kg·m². The object is pulled by a force of 160 N in the direction shown. There is enough friction to prevent slipping. Calculate the angular acceleration of the object, given that $R = 0.91$ m and $r = 0.32$ m.

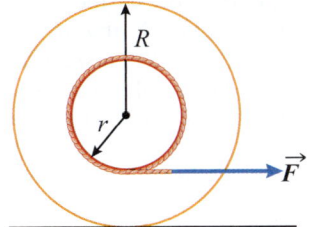

Figure 9-29   Problem 34.

35. ★★ A form of yo-yo consists of a thin cylindrical shell with a string wound around it. Find the acceleration of this yo-yo when a child lets it go while holding the end of the string.

36. ★★★ A wheel (thin hoop) of radius 72.0 cm rolls without slipping down a 32.0° slope starting from rest (Figure 9-30). Point A is shown at $t = 0$. Find the speed and acceleration of point A at 3.00 s into the motion.

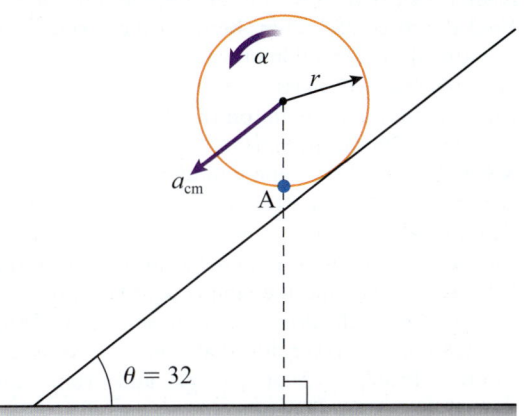

Figure 9-30   Problem 36.

37. ★ Indicate the direction in which the object in Figure 9-31 will roll for each of the given forces.

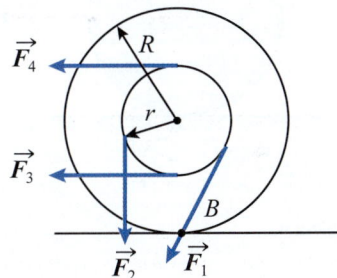

Figure 9-31   Problem 37.

38. ★★ A car is moving at a constant speed of 29 m/s on a flat horizontal road. The car wheels, 41 cm in radius, roll without slipping. You paint a dot on the edge of one of the tires and monitor the dot's motion. Find the magnitude and direction of the acceleration of the dot when
  (a) its speed is zero
  (b) its horizontal speed is zero
  (c) its vertical speed is zero
  (d) the point is at the top of the wheel

Section 9-4   Newton's Second Law and Rolling

39. ★★ A uniform thin spherical shell of mass 1.2 kg rolls from rest without slipping down a hill with a 32° slope. Find the force of friction acting on the spherical shell. What would happen to the force of friction if you tripled the radius of the shell but kept the mass the same?

40. ★★ Consider a yo-yo as a uniform solid disk of mass 210 g and radius 6.0 cm. You release the yo-yo while holding the end of the string wound around it.
(a) Find the tension in the string.
(b) Find the acceleration of the centre of mass of the yo-yo.
(c) Find the angular acceleration.

41. ★★ Consider a thin cylindrical shell of radius 21 cm and mass 1.3 kg that rolls without slipping down a 37° hill. Find the force of friction between the surface and the cylindrical shell by taking torques about
(a) the centre of mass
(b) the stationary point on the shell (the point in contact with the ground)

42. ★★★ A uniform bowling ball of mass m and radius R is given some backward spin with angular speed ω_0 and then thrown down a lane with initial linear speed v_0, as shown in Figure 9-32. The coefficient of kinetic friction between the ball and the lane is μ_k. (This problem can be solved using the principle of conservation of angular momentum. However, for that you will need to refer to Open Problem 92, on the spin and orbital angular momentum of a rolling object.)

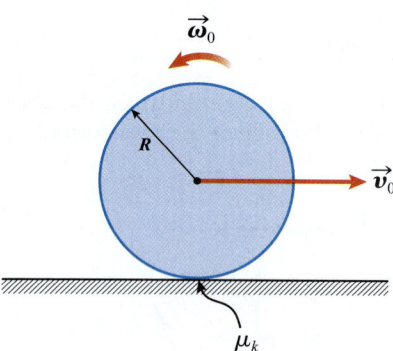

Figure 9-32 Problem 42.

(a) Determine the linear velocity of the ball's centre of mass in terms of the initial speeds when it starts to roll without slipping on the lane.
(b) How should the two initial speeds, ω_0 and v_0, be related in each of the following situations?
 (i) The ball eventually rolls forward without slipping.
 (ii) The ball eventually rolls backward without slipping.
 (iii) The ball eventually comes to rest.
(c) What is the displacement of the centre of mass before the ball comes to a perfect roll on the lane if $m = 4.50$ kg, $R = 11.0$ cm, $\mu_k = 0.150$, $v_0 = 7.50$ m/s, and $\omega_0 = 15.0$ rad/s.
(d) Does $a_{cm} = \alpha r$ before the ball starts rolling without slipping? Explain.
(e) How long does the ball spin before coming to a perfect roll in part (c)?
(f) What is the final speed of the ball's centre of mass in part (c)?
(g) How much mechanical energy is lost due to friction before the ball comes to a perfect roll in part (c)?

This problem can be solved using the principle of conservation of angular momentum. For details, refer to Open Problem 92 on the spin and orbital angular momentum of a rolling object.

43. ★★★ The ancient Egyptians may have used the method shown in Figure 9-33 to transfer large stones. A stone of mass M (the rectangular shape in Figure 9-33) is resting on two logs such that the weight is equally distributed between the logs. The stone is then pushed forward with a constant horizontal force, $\vec{F}$. Find an expression for the acceleration of the stone. The logs can be considered as ideal cylinders of mass m and radius R. Assume there is enough friction everywhere to prevent slipping.

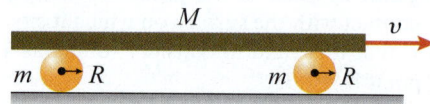

Figure 9-33 Problems 43 and 82.

44. ★★★ A sports car is cruising on a flat horizontal road when the driver hits the brakes. Two brake pads press against the brake disk of each wheel at a point 17 cm above the centre of the wheel, slowing the car down. The car has a mass of 1630 kg, and its weight is equally distributed over the four wheels. The coefficient of kinetic friction between each brake pad and brake disk on the wheels is 0.25, and the coefficient of static friction between the tires and the road is 0.56.
(a) What is the maximum normal force that each brake pad can exert on the brake disk so that the wheels do not skid as the car slows down? Each wheel (including tire and brake disk) has a moment of inertia of 34 kg·m² about an axis through its centre and a radius of 29 cm.
(b) Repeat part (a) with the car driving up a 23° slope.

45. ★★ A uniform solid cylinder rolls from rest down a 4.0 m high hill with a 39° slope. The coefficient of kinetic friction is 0.12, and the coefficient of static friction is 0.17. The mass of the cylinder is 340 g, and its radius is 3.0 cm.
(a) Will the cylinder roll without slipping? Explain.
(b) Find the angular acceleration of the cylinder.
(c) The cylinder travels from the top to the bottom of the hill. How much mechanical energy is lost due to friction?

46. ★★★ On an icy road, the coefficient of static friction between the tires of a bicycle and the road is 0.050. The wheels can be considered to be disks of radius 32 cm and mass 3.0 kg each. The combined weight of the bicycle and the cyclist is 870 N and is distributed equally over the two wheels. What is the maximum torque that can be delivered to the rear wheel without causing it to spin as the bicycle accelerates?

47. ★★ You place a uniform solid sphere on an inclined plane. The coefficient of static friction between the sphere and the plane is 0.30. Determine the maximum angle that the inclined plane can make with the horizontal so that the sphere does not skid. The sphere starts from rest.

48. ★★ A car rolls freely down a hill without slipping.
    (a) In which direction must the force of friction act on the tires?
    (b) The driver puts the car in gear and depresses the gas pedal. What must the acceleration of the car be if the force of friction on the wheels is zero? Consider the car wheels to be disks.
49. ★★★ A 52 kg cyclist rides his 7.0 kg bicycle up a 17° slope hill. The normal force from the road on the bicycle is equally distributed between the two wheels. Each wheel is 27 cm in radius and has a mass of 900. g. Two brake pads act at the rim of each of the wheels, slowing the bicycle down. The coefficient of kinetic friction between each brake pad and the rim is 0.20. The wheels can be considered as ideal rings. What is the normal force required between the brake pads and the wheels so that the force of friction between the tires and the road is zero as the bicycle slows down?
50. ★★★ The driver of an all-wheel-drive car depresses the gas pedal as soon as the light turns green at a traffic light. Find the maximum torque that can be delivered to each of the wheels without causing them to spin as the car accelerates from rest. Assume that the weight of the 2200 kg car is evenly distributed over the four wheels, and that the coefficient of static friction between the asphalt and the tires is 0.70. Each wheel has a diameter of 65 cm and a moment of inertia of 25 kg·m² about an axis through its centre.

Section 9-5   Mechanical Energy and Rolling

51. ★ A uniform solid cylinder of mass 1.20 kg rolls without slipping down a hill from a height of 3.40 m.
    (a) Calculate the cylinder's rotational and translational kinetic energies at the bottom of the hill.
    (b) Repeat the problem for a uniform ring of the same mass.
52. ★ A solid sphere and a solid cylinder, both of radius 17 cm and mass 34 kg, roll without slipping down a 4.0 m high hill. Both objects start from rest.
    (a) Calculate the total kinetic energy of each object at the bottom of the hill.
    (b) Calculate the angular speed of each object at the bottom of the hill.
    (c) Calculate the linear speed of the centre of mass of each object at the bottom of the hill.
    (d) Determine the ratio of the rotational kinetic energy to the translational kinetic energy for each object. Explain your results.
53. ★★ A solid, uniform bowling ball, 12.7 cm in diameter, starts from rest and rolls down a 7.00 m high ramp with a 37.0° slope. There is enough friction between the surface of the ramp and the bowling ball to prevent the ball from skidding. At the bottom of the ramp, the ball goes up a frictionless ramp with the same slope as the first ramp.
    (a) Determine the maximum height that the ball can reach on the frictionless ramp.
    (b) Is mechanical energy conserved? Explain.
    (c) What is the angular speed of the bowling ball when it reaches its maximum height on the frictionless ramp?
54. ★★ A 7.0 kg steel ball of radius 4.0 cm rolls without slipping from rest down a hill of height 3.0 m. The ball continues to roll without slipping on a horizontal surface for a certain distance before striking a spring whose

spring constant, k, is 72 N/m. The spring is originally unstretched. The steel ball compresses the spring until it comes to a momentary stop. The ball does not slip on the surface while it compresses the spring. The surface of contact between the ball and the spring is frictionless. Find the maximum compression of the spring.

55. ★★ A solid uniform ball of mass m and radius r rolls down a hemispherical bowl, starting from a height h above the bottom of the bowl (Figure 9-34). The surface of the left half of the bowl provides enough friction to prevent the ball from slipping on the way down. The surface of the right half of the bowl is frictionless.
    (a) Derive an expression for the maximum height the ball will reach on the frictionless side. Is energy not conserved? Explain.
    (b) What is the translational kinetic energy of the ball at its highest point on the frictionless side?
    (c) What is the rotational kinetic energy at that point?
    (d) Explain how your answer to part (a) would change if we replaced the ball with an ideal ring.

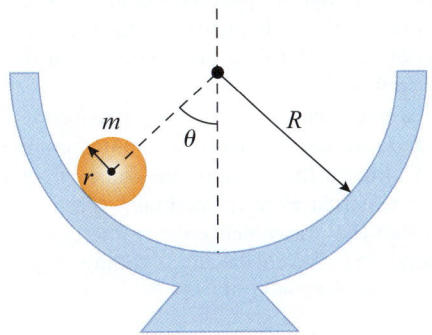

Figure 9-34   Problems 55 and 70.

56. ★★ A solid sphere of mass m starts from rest and rolls without slipping down a ramp from a height h. At the bottom of the ramp, the sphere continues to move on a flat horizontal frictionless surface and then collides with a horizontal spring whose spring constant is k. The spring is originally in its equilibrium position and is fitted with a light, frictionless plate at its end, which the sphere collides with (Figure 9-35).
    (a) Derive an expression for the maximum compression of the spring.
    (b) Derive an expression for the angular speed of the sphere when it strikes the spring.
    (c) What would the result be in part (a) if there were no friction anywhere?

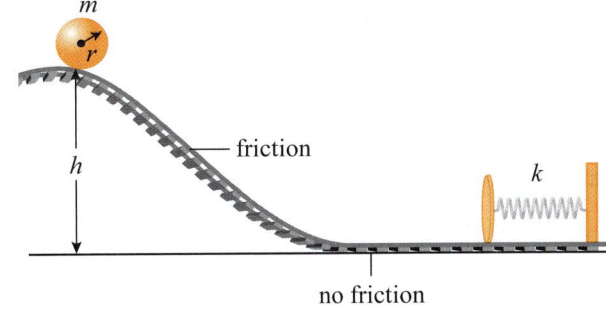

Figure 9-35   Problem 56.

57. ★★ $\frac{d}{dx}$ Use the work–mechanical energy approach to solve problem 43.

Section 9-6 Rolling without Friction

58. ★ A non-deformable bowling ball rolls down a bowling alley with a smooth non-deformable surface. Would you expect the ball to slow down?

Section 9-7 Rolling Friction

59. ★★ A steel ball and a rubber ball of the same mass and radius roll down a ramp from the same height starting at rest. Which ball will reach the bottom of the ramp faster?

60. ★★ A steel ball rolls without slipping down a steel ramp starting from rest and then up a wooden ramp with the same slope as the steel ramp. Will the ball reach the same height on the wooden ramp as the height it left from on the steel ramp?

61. Assuming the only source of rolling friction to be the deformation in the rubber tires, how far would you expect an average sedan travelling at 108 km/h to go on a concrete highway if the driver put the car in neutral? The coefficient of rolling friction between the tires and the road is 0.015.

62. ★★ Estimate the force of rolling friction on a 2250 kg car whose weight is equally distributed among the four tires. Compare this force to the force of air resistance the car experiences at it maximum speed of 216 km/h given that the power delivered by the engine is 284 hp (horsepower). The coefficient of rolling friction between the tires and the road is 0.012.

COMPREHENSIVE PROBLEMS

63. ★★ $\frac{d}{dx}$ A uniform solid sphere of mass m and radius R rolls without slipping down an ideally flat ramp with a slope of angle θ with respect to the horizontal, as shown in Figure 9-36. Use energy considerations to derive an expression for the acceleration of the sphere. Do you have to account for friction? Explain. Compare the energy approach to the Newtonian equation of motion approach.

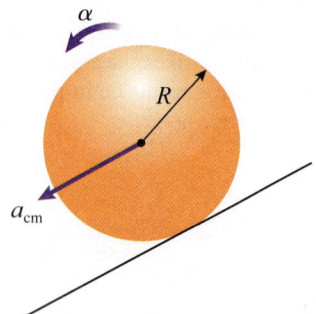

Figure 9-36 Problem 63.

64. ★★★ $\frac{d}{dx}$ In Figure 9-37, a massless string is wound around a uniform solid sphere of mass M_s and radius R_s. The string passes over a pulley with moment of inertia I_1 and radius r_1; the string is attached to the mass m on top of the frictionless table as shown. The mass m, in turn, is attached to another massless string, which passes over a second pulley with moment of inertia I_2 and radius r_2, and is attached to the centre of a solid uniform cylinder of mass M_c and radius R_c. The system starts from rest, and then the cylinder accelerates downward. Find an expression for the acceleration of the sphere's centre of mass by applying (a) Newton's second law and (b) work–mechanical energy considerations. Assume perfect rolling without slipping anywhere.

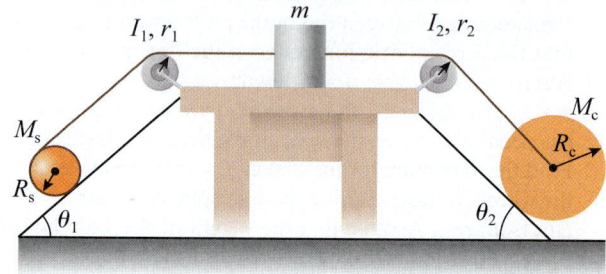

Figure 9-37 Problem 64.

65. ★★ The marble of mass m and radius R in Figure 9-38 starts from rest with its lowest point at height h above the bottom of the track and rolls down without slipping. The marble then rolls up the semicircular part of the track, which has a radius r equal to $h/4$. At the position shown, the point on the marble in contact with the track is at $r/2$ from the top of the semicircular track, and the normal force from the track on the marble is equal to 2.5 times the marble's weight.

(a) Use energy considerations to find an expression for the moment of inertia of the marble about an axis through its centre in terms of R, m, and h.

(b) Derive an expression for the force of friction on the marble at the position shown.

(c) If $h = 132R$, what is the moment of inertia?

(d) Find the limit of the moment of inertia if $h \gg R$.

(e) If the marble were a uniform ring, what would h be in terms of R?

(f) Can this marble ever be a disk? Justify your answer.

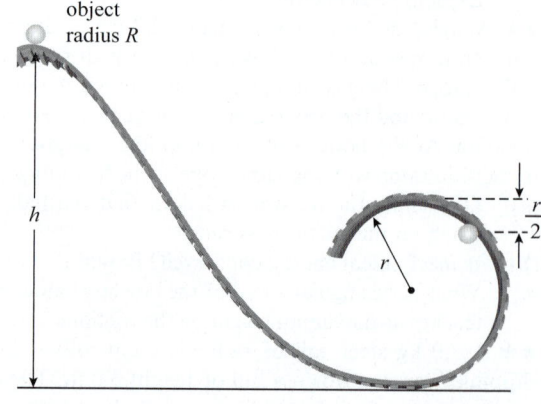

Figure 9-38 Problem 65.

66. ★★ The object in Figure 9-39 is machined out of a single piece of metal such that its moment of inertia is $\frac{mR^2}{3}$ about an axis passing through its centre. You pull horizontally on the rope as shown.
(a) In which direction will the object roll, assuming there is enough friction between the object and the surface to prevent slipping?
(b) For what value of r in terms of R will you be able to pull the rope and cause the object to roll without slipping such that the force of friction is zero?

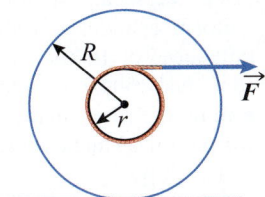

Figure 9-39 Problem 66.

67. ★★ A pilot applies the wheel brakes upon landing so that her plane skids to a stop on a runway from its landing speed of 130 km/h. The 1.9 m diameter wheels were rolling without slipping before the pilot applied the brakes. Assume constant acceleration of the plane's centre of mass and constant angular acceleration of its wheels. The plane comes to a stop over a distance of 270 m. The plane takes twice as much time to stop as it takes the wheels to lock.
(a) Find the speed of a point at the bottom of a wheel 5.4 s into the landing.
(b) Find the speed of a point at the top of a wheel 110 m after the pilot started braking.
(c) Find the total acceleration of a point on the front edge of a wheel at the same height as the centre of the wheel after 3.0 s of braking.

68. ★★★ A uniform solid ball of mass $m = 1.60$ kg and radius $r = 0.110$ m rolls down a 4.30 m hill without slipping. At the bottom of the hill, the surface is horizontal and the ball continues to roll without slipping until it strikes a stationary identical ball (Figure 9-40). Assume that the collision is perfectly elastic, and ignore any frictional losses during the collision.
(a) Find the translational and angular speeds of each ball immediately after the collision.
(b) Find the translational and angular speeds of each ball a sufficiently long time after the collision when the balls are rolling without slipping again.
(c) Find the total amount of mechanical energy lost due to friction from the instant the first ball is sent down the hill until the two balls start pure rolling again after the collision?
This problem can also be solved using the conservation of angular momentum. Refer to Open Problem 92 on the spin and orbital angular momentum of a rolling object for details.

Figure 9-40 Problem 68.

69. ★★★ You kick a soccer ball of mass 0.90 kg and radius 14 cm, as shown in Figure 9-41. The kick has a duration of 91 ms and an average force of 161 N. Your foot strikes the ball horizontally 14 cm above the floor. The coefficient of kinetic friction between the ball and the ground is 0.090. The ball can be approximated by a thin spherical shell.
(a) Find the linear speed of the ball immediately after the collision.
(b) What is the angular speed of the ball immediately after the collision?
(c) Determine the linear speed of the ball when it starts rolling without slipping.
(d) Find the total amount of mechanical energy lost due to friction.

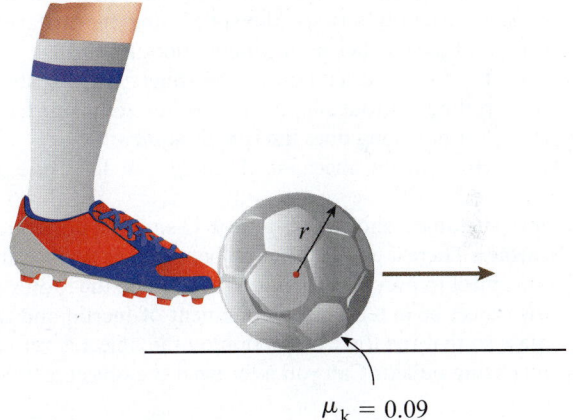

Figure 9-41 Problem 69.

70. ★★ A ball, which can be considered a uniform thin spherical shell, of mass m and radius r is placed on the surface of a hemispherical bowl of radius R and rolls down the bowl without slipping (Figure 9-34).
(a) The ball starts from rest at an angle of $\theta = 72°$ with the vertical. What is the normal force on the ball when it is at the bottom of the bowl?
(b) Repeat part (a) if the ball slides without rotating.
(c) Repeat parts (a) and (b) with $m = 0.67$ kg, $r = 0.070$ m, and $R = 9.0$ m.

71. ★★★ Starting from rest, a ball of mass 0.63 kg and radius 0.17 m slides from the top down a circular track of radius 1.1 m (Figure 9-42). The surface of the track is frictionless. The ball then moves up a ramp that is at an incline of 33° from the horizontal. The coefficient of kinetic friction between the ball and the ramp is 0.12. The ball is a uniform solid sphere.
   (a) Find the acceleration of the ball as it skids up the ramp.
   (b) What is the angular speed of the ball when it starts rolling perfectly?
   (c) What is the acceleration of the ball as it rolls perfectly?
   (d) Use energy considerations to find the maximum height the ball will reach above the bottom of the ramp.

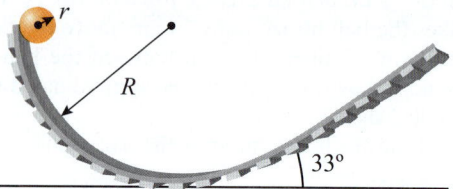

Figure 9-42 Problem 71.

72. ★★★ You throw a solid uniform sphere of radius 11 cm and mass 2.1 kg onto a smooth horizontal surface with a horizontal speed of 9.0 m/s. The sphere skids initially and then starts rolling without slipping. The coefficient of static friction between the sphere and the surface is 0.18, and the coefficient of kinetic friction is 0.14.
   (a) Find the acceleration of the sphere when it starts rolling without slipping on the horizontal surface.
   (b) For how long does the sphere skid?
   (c) How much mechanical energy is lost due to friction?

73. ★★ The object shown in Figure 9-43 sits on a horizontal surface. There is enough friction between the surface and the object to prevent slipping. What should the radius of the object be in terms of the moment of inertia and the mass so that the force of friction on the object is zero as it is being pulled? Can you infer what the object is?

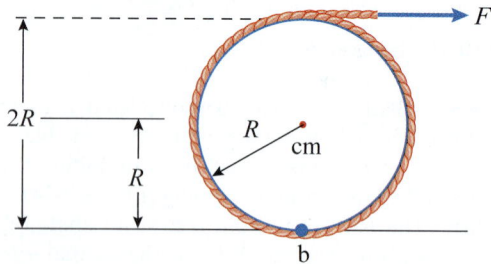

Figure 9-43 Problem 73.

74. ★★ Redo Example 9-7, this time summing the torques about the point on the cylinder's centre of mass.

75. ★★ A 2.0 kg uniform solid bowling ball of diameter 0.32 m rolls without slipping down a hill of 43° slope. At the bottom of the hill, the surface is horizontal.
   (a) Find the acceleration of the ball on its way down the hill and when it is rolling on the horizontal surface.
   (b) How would the acceleration of the ball on the horizontal surface change if the horizontal surface were covered with glossy ice such that it is almost frictionless?

76. ★★★ Is it possible for an object to roll without slipping on completely frictionless ice? Explain.

77. ★★ A solid uniform cylinder is rolling without slipping at a constant speed of 10. m/s on a horizontal surface. The cylinder then rolls 7.0 m down a frictionless 17° slope. Determine the ratio of the rotational kinetic energy to the translational kinetic energy of the cylinder at the bottom of the slope.

78. ★★ A marble starts from rest and rolls without slipping down a smooth incline from height h. At the bottom of the incline, the surface is horizontal and frictionless. Will the marble eventually stop rolling on the frictionless surface? If so, what force would be responsible for changing the marble's angular speed?

79. ★★★ You spin a uniform bowling ball of mass m and radius r in your hands until it reaches an angular frequency of ω. You then place it gently on a surface where the coefficient of kinetic friction between the surface and the ball is μ_k. The ball initially spins on the surface as it picks up linear speed because of kinetic friction (Figure 9-44).
   (a) How much time passes before the ball starts rolling without spinning? How many revolutions does the ball go through during that time?
   (b) Calculate the distance, d, that the centre of mass of the ball travels during the time you calculated in part (a).
   (c) Does $d = r(\Delta\theta)$ during the time you calculated in part (a)?
   (d) Find the linear speed of the centre of mass of the ball when it starts to roll without slipping.
   (e) What is the force of friction on the ball as it rolls on the horizontal surface without slipping?
   In a similar fashion to problem 42, this problem can be solved using the principle of conservation of angular momentum. For details, refer to Open Problem 92 on the spin and orbital angular momentum of a rolling object.

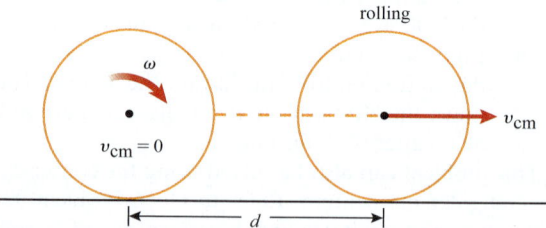

Figure 9-44 Problem 79.

80. ★ Equations 9-17 and 9-18 give two different expressions for the total kinetic energy of a rolling object. Show that these two expressions are always equivalent to each other for a rigid object rolling without slipping.

81. ★★★ Develop parametric equations to describe the cycloidal motion of a point at the edge of a rolling object.

82. ★★ A stone of mass M rests on two solid uniform cylinders, as shown in Figure 9-33. The coefficient of static friction between the stone and the cylinders is 0.35, and the coefficient of static friction between the cylinders and the horizontal surface they roll on is 0.55. Assume that the weight of the stone is equally distributed on the two cylinders.

   (a) Find an expression for the maximum horizontal force F you can push the stone with so that no slipping occurs anywhere.

   (b) What is the acceleration of the stone in part (a)?

83. ★★★ A ball of diameter 24 cm is released from rest on a slope that is inclined at 42° (Figure 9-45). The surface of the slope is frictionless for the first 7.0 m, and then it becomes uniformly rough with some friction. The ball is a uniform solid sphere.

   (a) Will the ball eventually start rolling without slipping if the rough surface has a coefficient of kinetic friction of 0.13? Explain.

   (b) What is the minimum coefficient of kinetic friction required for the ball to roll without slipping 14 m after it meets the rough surface?

   (c) How many turns will the ball in part (b) undergo before rolling without slipping?

   (d) How long will it take the ball in part (b) to start rolling without slipping?

   (e) Find the angular speed of the ball in part (b) when it starts rolling without slipping.

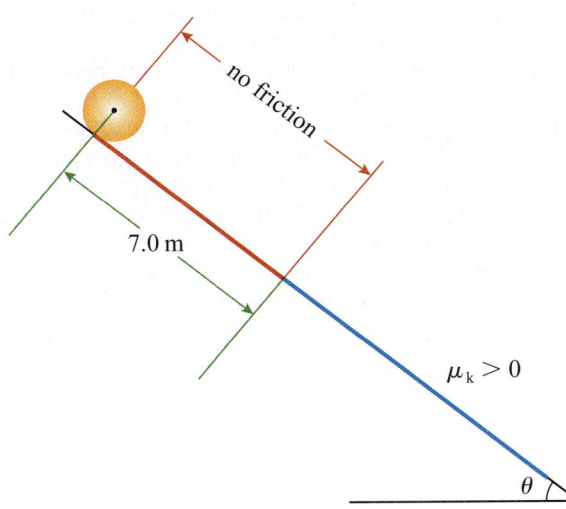

Figure 9-45 Problem 83.

84. ★★★ Suppose that the object in Figure 9-46 has moment of inertia less than mr^2 about an axis through its centre. Show that the force of friction will actually push the object forward as you pull on it horizontally with a force $\vec{F}$. Assume that there is enough friction between the surface and the object to prevent slipping.

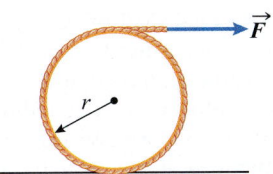

Figure 9-46 Problem 84.

85. ★★★ A ball of mass 940 g and radius 5.00 cm rolls down a hemispherical surface of radius 0.700 m, as shown in Figure 9-47. The ball begins to roll without slipping down the hemisphere, and then starts slipping along with rolling until it eventually flies off the surface. The coefficient of static friction between the surface and the ball is 0.150. The ball is a uniform solid sphere.

   (a) Determine the location at which the ball begins to slip on the way down if it started from rest at the top of the hemisphere.

   (b) What is the linear speed of the ball's centre of mass at that instant?

   (c) Assume that the surface becomes frictionless once the ball starts slipping. Find the angular position where the ball leaves the surface.

   (d) Had the surface been completely frictionless, would the ball have left it at the same position as in part (c)? Explain.

      Look at Open Problem 91 for the case in which kinetic friction is taken into account while the ball is slipping before it flies off the surface.

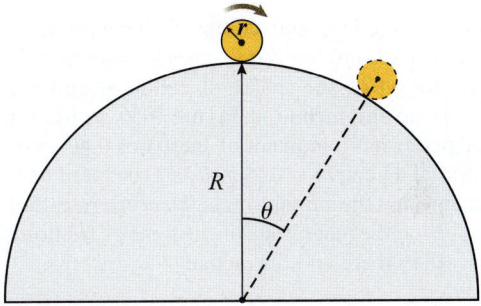

Figure 9-47 Problems 85 and 91.

86. ★★ A rolling marble of radius r enters a hemispherical bowl of radius R through a hole, as shown in Figure 9-48. The marble is moving with translational speed v and angular speed $\dfrac{v}{r}$.

   (a) What speed, v, will cause the marble to maintain a height h above the bottom of the bowl?

   (b) What would happen if the marble entered the bowl at a point slightly above or below the desired height at the same speed as in part (a)?

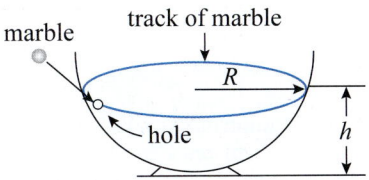

Figure 9-48 Problem 86.

87. ★★ A hemispherical shell of radius R and mass m is made to spin about a fixed vertical axis, as shown in Figure 9-49. As the spinning speed increases very slowly from zero, a small metal ball of radius r at the bottom of the shell starts to move upward. The shell's spinning speed is maintained at a constant value when the ball reaches a height h above the bottom of the bowl. The ball is considered a uniform solid sphere.

(a) Find the spinning speed of the shell at height h in terms of the other variables in the problem. Assume that the ball rolls without slipping.

(b) Find the work done on the metal ball as the shell accelerates.

(c) Why does the speed need to be increased slowly? Had the speed been changed quickly, what would you expect the ball to do, assuming it never goes above the rim?

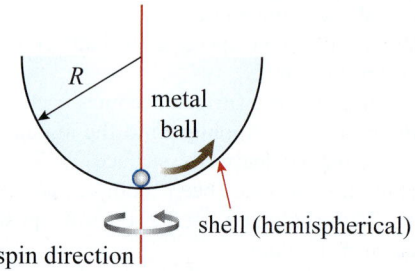

Figure 9-49 Problem 87.

88. ★★ $\frac{d}{dx}$ (a) Use and compare the work–energy and Newton's second law approaches for finding the angular acceleration of the sphere of radius R and mass m as it rolls up the incline in Figure 9-50 without slipping. The pulley has a moment of inertia of 0.40 kg·m² and a radius of 11 cm; $m = 1.1$ kg, $R = 23$ cm, and $M = 5.6$ kg. Assume that the hanging mass, M, accelerates downward.

(b) Find the magnitude of the force of friction between the sphere and the incline.

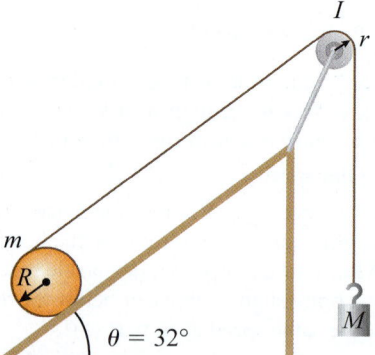

Figure 9-50 Problem 88.

89. ★★★ A solid uniform cylinder of mass 0.65 kg and radius 11 cm is held on a horizontal surface where it compresses a spring ($k = 1210$ N/m) by a distance of 41 cm from the equilibrium position. The cylinder is then let go, and it gains kinetic energy from the spring as the spring returns to equilibrium. Initially, the surface is frictionless, and then it suddenly becomes rough, with a coefficient of kinetic friction between the surface and the cylinder of 0.12.

(a) What is the speed of the cylinder when it hits the rough part of the surface? Assume the cylinder does not encounter the rough surface until the spring is fully uncompressed.

(b) How far does the cylinder go on the rough surface before it stops skidding?

(c) What is the angular displacement of the cylinder before it starts rolling without slipping?

(d) What fraction of the initial mechanical energy is lost due to friction?

(e) Find the final angular speed of the cylinder.

DATA-RICH PROBLEMS

90. ★★★ 〜 A rear-wheel-drive car's engine delivers 125 hp (horsepower) to the rear wheels. The car has a total mass of 2150 kg. Ignore the force of friction on the front wheels, and assume that the force of air resistance is proportional to the square of the speed. The car runs at a maximum speed of 145 km/h. Assume the gas pedal is depressed all the way, and that the wheels roll without slipping. Assume the wheels to be thin rings of mass $M = 25.0$ kg each.

(a) Write the equation of motion for the car.

(b) Comment on the expression of the force of friction between the rear wheels and the road at maximum speed.

(c) Write the equation of motion for the rear wheels.

(d) Express the torque of the engine in terms of the power, and substitute for the force of friction in part (c) its value obtained in part (a) in terms of the other variables in the equation.

(e) Find the constant of proportionality between the force of air resistance and the speed of the car.

(f) Using a spreadsheet, create a list of values for the velocity, with small increments of no more than 0.1 m/s. Use your answer in part (c) to obtain corresponding values for the time, and plot the speed versus time for the car.

(g) Plot the acceleration versus time for the car.

(h) What is the minimum value of the acceleration?

(i) Plot the force of friction versus time for the car.

(j) What is the minimum value you get for the force of friction?

(k) Does your answer in part (j) make sense?

91. ★★★ $\frac{d}{dx}$, $\int$, 〜 A ball of mass m and radius r is released from rest at the top of a rough hemispherical surface of radius R (Figure 9-47). The ball begins to roll without slipping down the hemisphere, and then it starts slipping *and* rolling, until it eventually flies off the surface. The coefficients of static and kinetic friction between the surface and the ball are $\mu_s = 0.150$ and $\mu_k = 0.120$, respectively. The ball is a uniform solid sphere.

(a) Determine the angle at which the ball starts to slip on the way down.

(b) Write the equations of motion in the radial and tangential directions for the ball's centre of mass once the ball starts to slip while rolling down the hemisphere. Derive an expression for the tangential acceleration in terms of the angular position, θ; the linear speed, v_{cm}; and μ_k.

(c) Using the expression for the tangential acceleration, show that the centre-of-mass speed, v_{cm}, can be expressed by the equation

$$\frac{dz}{d\theta} - 2\mu_k z = 2(\sin\theta - \mu_k\cos\theta)$$

where

$$z = \frac{v_{cm}^2}{g(R + r)}$$

The above equation is a first-order linear differential equation that can be directly solved for z by introducing an integrating factor. Alternatively, it can be solved using a simple analytical approach, as derived below.

Note that even if you do not solve this part, you can still use this result to attempt the following parts of the problem.

(d) Knowing that the form $z_1 = A\sin\theta + B\cos\theta$ is one solution of the differential equation obtained in part (c), find the constants A and B in terms of μ_k.

(e) Another simple solution can be obtained by setting the right-hand side of the differential equation in part (c) to zero and solving the left-hand side for z. Show that $z_2 = Ce^{2\mu_k\theta}$ is such a solution, where C is an integration constant.

(f) The sum $z = z_1 + z_2$ is in fact the general solution of the differential equation obtained in part (c). Verify that this is true.

(g) Determine the numerical value of the integration constant C by fitting the general solution obtained in part (f) to the initial conditions that correspond to the ball's position and speed at the instant it began slipping while rolling.

(h) Determine the angle at which the ball will fly off the surface. This part will require a graphical solution (e.g., a graphing calculator) for the final answer.

(i) Compare your result in part (h) with those obtained in problem 85 of this chapter, Example 9-9, and Example 6-14. Comment on the differences among these results.

OPEN PROBLEMS

92. ★★ As Earth orbits the Sun–Earth centre of mass, we can infer that there are two parts to its total angular momentum: one is that of Earth's centre of mass as it moves in its orbit, and the other comes from Earth spinning about its axis. In this problem, we develop a general expression to let us represent the angular momentum of any object about a given point in space as the sum of the angular momentum of the centre of mass about that point and the angular momentum of the object about an axis through is centre of mass. This general result will give us the tools to solve problems 42, 68, 69, and 79, and others like them, using the conservation of angular

momentum. It will also lay the groundwork for us to represent and analyze the motion of more interesting objects and study relatively more complex situations.

At the end of this problem, we will be able to state that the angular momentum of a rolling object about some arbitrary point P is given by the sum

$$\vec{L}_P = \vec{r}_{cm} \times \vec{p}_{cm} + I_{cm}\vec{\omega}$$

where $\vec{r}_{cm}$ is the position of the centre of mass as seen from point P, $\vec{p}_{cm} = m\vec{v}_{cm}$ is the linear momentum of the centre of mass, I_{cm} is the moment of inertia of the object about an axis through its centre of mass, and $\vec{\omega}$ is the angular speed of the object, as illustrated in Figure 9-51.

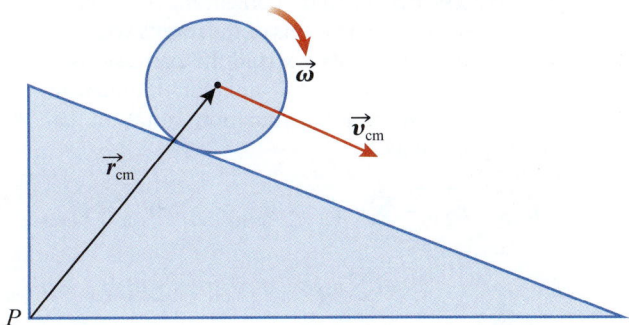

Figure 9-51   Problem 92.

The first term gives us the angular momentum of the centre of mass about the point P; this term is referred to as the orbital angular momentum of the object. The second term gives us the angular momentum of the object as it "spins" about an axis through the centre of mass and is referred to as the spin angular momentum of the object.

We walk through the derivation by considering the general case of a simple system comprising two masses, m_1 and m_2, connected via a massless bar, as shown in Figure 9-52.

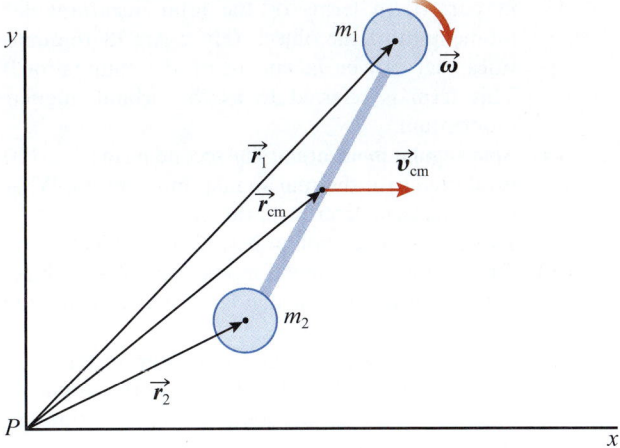

Figure 9-52   Problem 92.

(a) Write a general expression for the angular momentum of m_1 about point P in terms of the velocity of the mass and its position vector as measured from point P.

(b) Write a general expression for the angular momentum of m_2 about point P in terms of the velocity of the mass and its position vector as measured from point P.

(c) Express the velocity of each of the masses in terms of the velocity of the centre of mass of the object and the velocity of the mass relative to the centre of mass.

(d) Express the position vector of each of the masses as measured from P in terms of the position vector of the centre of mass as measured from P and the position vector of the mass as measured from the centre of mass.

(e) Express the angular momentum of each of the masses about P in terms of the results you obtained in parts (c) and (d). Expand the expressions.

(f) Show that the total angular momentum of the system using summation notation can be expressed as

$$\vec{L}_{1,P} + \vec{L}_{2,P} = \sum_{i=1}^{2} m_i \vec{r}_{cm} \times \vec{v}_{cm} + \sum_{i=1}^{2} m_i \vec{r}_{cm} \times \vec{v}_{i,cm}$$

$$+ \sum_{i=1}^{2} m_i \vec{r}_{i,cm} \times \vec{v}_{cm} + \sum_{i=1}^{2} m_i \vec{r}_{i,cm} \times \vec{v}_{i,cm}$$

Next, you will show that the second and the third terms are zero.

(g) In the second and third terms, are there any quantities that can be taken out of the summation? If so, explain why, and express the terms accordingly.

(h) Show that the second and the third terms are both identically zero.

(i) Show that the expression for the total angular momentum of the object about P is given by

$$\vec{L}_P = \vec{L}_{1,P} + \vec{L}_{2,P} = \sum_{i=1}^{2} m_i \vec{r}_{cm} \times \vec{v}_{cm} + \sum_{i=1}^{2} m_i \vec{r}_{i,cm} \times \vec{v}_{i,cm}$$

(j) **Orbital angular momentum:** Express the first term in part (i) in terms of the total linear/angular momentum of the object. (Hint: Are there quantities that can be taken out of the summation?) This term is referred to as the orbital angular momentum.

(k) **Spin angular momentum:** The second term in part (i) is referred to as the spin angular momentum. What does this term describe in this case?

(l) Extend the expression in part (i) to N objects.

(m) Express the total angular momentum of any object in terms of the orbital angular momentum and the spin angular momentum.

(n) Express the spin angular momentum in terms of the angular speed of the object and the moments of inertia of the masses about the spin axis through the centre of mass for a rolling object comprising N discrete objects.

(o) Express the total angular momentum you got in part (m) using your result from part (n).

(p) Express your result in part (o) using the total moment of inertia of the object about the spin axis through the centre of mass.

(q) For the object we have considered in this problem, the orbital angular momentum and the spin angular momentum are parallel to one another. (Refer to the right-hand rule discussed in Chapter 8 for these vectors.) This is the case for most objects rolling without slipping. Can you think of scenarios where the two vectors are not parallel?

(r) Is it possible for an object to be in orbit yet have an angular momentum of zero?

(s) How are the directions of the angular momenta related?

(t) Can you relate this to Checkpoint 8-15?

(u) Can you think of other scenarios whereby an object/system ends up having a total angular momentum of zero?

(v) Can an object gain orbital angular momentum by starting to spin?

(w) Consider the spinning bowling ball of problem 79. Find an expression for its initial total angular momentum about a point P on the surface it rolls on.

(x) What is the angular momentum about the point in part (r) when the ball starts rolling without slipping?

(y) Are the two expressions in parts (w) and (x) equal? Why?

(z) Express the equality between the two momenta using the radius of the ball, R, and the final speed of the ball's centre of mass, and comment on the solution using this approach.

93. ★★ In this chapter, when dealing with objects rolling without slipping, for example, downhill, we used the conservation of mechanical energy. Seemingly, this could be done by naively ignoring the fact that the force of friction does work on the object as it rolls downhill. A common argument is that since the force of friction for an object rolling without slipping is static, it does no work; we explore this further in this problem.

(a) Consider the disk in Figure 9-10 rolling downhill without slipping. If it starts from rest and descends a vertical distance h, can we say that

$$mgh = \frac{1}{2}Mv_{cm}^2 + \frac{1}{2}I_{cm}\omega^2?$$

(b) Is there a force of friction acting on the object? If so, what is its direction?

(c) If your answer to part (b) is yes, is this force of friction a conservative force?

(d) The answer to part (a) suggests that mechanical energy is conserved. When is mechanical energy conserved in the presence of non-conservative forces? Elaborate.

(e) Are there any other non-conservative forces acting on the object? If mechanical energy is indeed conserved, what does this say about the work done by the force of friction in this case? What must it be equal to?

(f) As the object rolls downhill, is the force of friction zero?

(g) Is the displacement of the object zero as it rolls downhill?

(h) Your answer to part (g) should have been no. In that case, how can the work done by the force of friction be zero? What is the discussion missing so far?

(i) Write an expression for the work done by the force of friction as the object rolls a distance, d, downhill. (Do not consider the work done by the frictional torque.)

(j) Write an expression for the work done by the torque caused by the force of friction about the centre of mass in terms of the angular displacement, $\Delta\theta$.

(k) Express $\Delta\theta$ in terms of the displacement of the centre of mass.

(l) Find the total work done by both the force of friction and the torque due to friction. Show that it adds up to zero.

(m) Work, as we have seen so far, results in the conversion of energy from one form to another, such as the work done by gravity, as gravitational potential energy is converted to kinetic energy for a falling object. What energy conversion can one think of in this case?

Learning Objectives

When you have completed this chapter, you should be able to

LO1 Define equilibrium; distinguish between static and dynamic equilibrium; and provide examples of stable, unstable, and neutral equilibrium.

LO2 Distinguish between centre of mass and centre of gravity, and calculate the centre of gravity for various objects.

LO3 Apply the conditions for static equilibrium to solve for unknown forces.

LO4 Identify and calculate unknown support forces acting on the parts of a structure.

LO5 Define compressive and tensile stress and strain, and describe the response of materials to stress within and beyond their elastic limit.

LO6 Distinguish between the responses of ductile and brittle materials to stress.

It is desirable to use structural materials that can sag or stretch substantially before fracturing under extreme loading, especially for buildings that may be subject to earthquakes or hurricane-force winds. Generally, the more rigid (or hard) steel alloys are, the more brittle they become. Most construction-grade steel is designed to sag a significant amount before fracturing. During a severe earthquake, a building supported by this type of steel collapses more slowly than a building with harder, more rigid steel, thus giving the occupants more time to evacuate.

Researchers at Stanford University have taken earthquake protection a step further by designing special building braces that can be incorporated into new buildings, or added to existing buildings, to make them more earthquake resistant. The blue braces in Figure 10-1 are intended to be flexible enough to allow the steel frame to rock. Strong vertical steel cables, or tendons, hold the brace frames together and help restore them to their original configuration when the frame rocks. As the frame rocks, the bulk of the energy absorbed by the structure from the earthquake goes into deforming arrays of leaf-like steel "fuses" shown in yellow. The deformed fuses can be replaced after an earthquake.

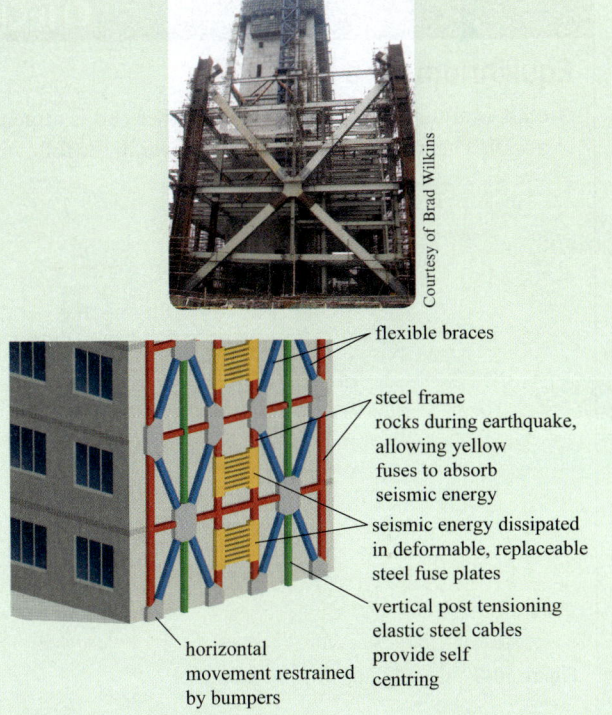

flexible braces

steel frame rocks during earthquake, allowing yellow fuses to absorb seismic energy

seismic energy dissipated in deformable, replaceable steel fuse plates

vertical post tensioning elastic steel cables provide self centring

horizontal movement restrained by bumpers

Figure 10-1 During an earthquake, the structure absorbs energy from the earthquake, and the energy deforms the replaceable fuses

10-1 The Conditions for Equilibrium

In this chapter, we will examine the conditions required for an extended object (not a point mass) to be in **equilibrium**. Our principle focus will be on static equilibrium and its application in analyzing structures.

Equilibrium for a Point Mass

Consider a glass of water sitting on your desk. We say it is in equilibrium. The net force acting on the glass, given by the sum of the force of gravity and the normal force from the desk, is zero, as shown in Figure 10-2.

In Chapter 5, we considered objects like this glass of water as a point mass, an object with no dimensions. For a point mass to be in equilibrium, the net force acting on it must be zero. If the net force acting on the point mass is zero, then its acceleration is also zero, according to Newton's second law. If the point mass is stationary, as with the glass of water resting on your desk, it will remain stationary and we say it is in **static equilibrium**. If the point mass is moving, like a hockey puck sliding with a constant speed on frictionless ice, it is in **dynamic equilibrium**. These two forms of equilibrium apply to the glass and the puck insofar as we can treat them like point masses.

Equilibrium for an Extended Object

We will now extend this equilibrium concept to an extended object that in addition to being able to translate,

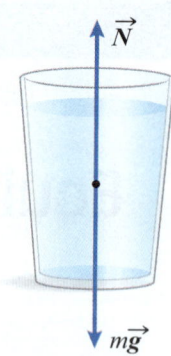

Figure 10-2 The sum of the forces on the water glass is zero.

like a point mass, is also able to rotate. Looking once again at the glass of water on your desk, we see that it has non-zero dimensions, such as height and base, and we quickly see that treating it like a point mass is too simplistic. Pushing the glass close to its bottom is not the same as pushing the glass close to its top. Can you think of a time where you knocked off the glass of water on your desk or dinner table by accidentally hitting it close to the top? The table likely provided enough friction to prevent the glass from sliding as you knocked it, but it still tipped. If the glass does not accelerate on the table, then the force of friction is equal to the "knocking" force provided by you. According to the point mass discussion, the net force on the glass is zero, and the glass is in equilibrium. Yet the fact that you caused it to tip and spill suggests that there is more to "equilibrium" than the sum of the forces on this glass being zero.

CHECKPOINT 10-1

Equilibrium for a Point Mass

A mass is attached to a spring whose other end is attached to a wall. The mass–spring system is shown in the following figures in various configurations, including snapshots of the mass as it is sliding back and forth. In which of these figures is the mass in static equilibrium?

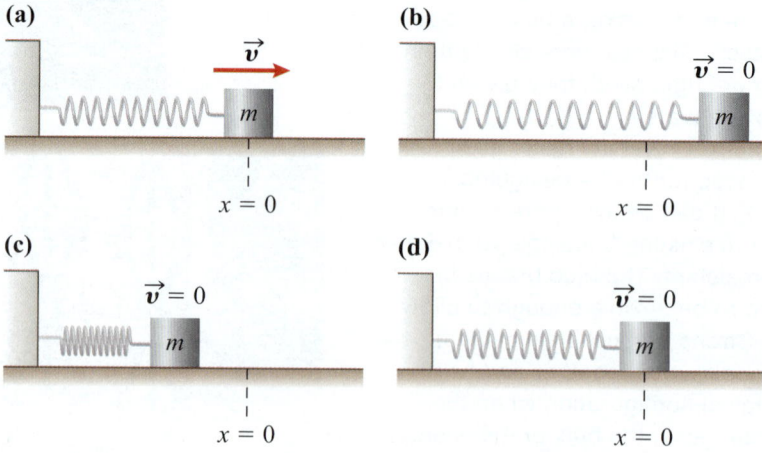

Figure 10-3 Checkpoint 10-1.

ANSWER: (d) The mass is in static equilibrium at this point because the net force acting on it is zero and it is not moving. (a) is a point of dynamic equilibrium: although the net force there is zero, the mass is moving. The mass is not moving in either (b) or (c), but the spring is stretched/compressed and thus the mass is experiencing a non-zero net force.

The force of the knock, assuming it is horizontal as shown in Figure 10-4, provides a non-zero torque about the tipping point at the opposite bottom corner. The torque created by this force about the tipping point (or pivot) has to be larger than the torque provided by the force of gravity about the tipping point to cause the glass to tip. (Notice that when the object is on the verge of tipping, the normal force acts at the pivot point, as shown in Figure 10-4. This would be the only point of contact between the surface and the glass.) The net torque on the glass results in the glass being accelerated rotationally during the "knock."

It is therefore not enough for the sum of the forces on an extended object to be zero for the object to be in equilibrium.

For an extended object to be in equilibrium, in addition to demanding that the net force acting on the object be zero, we also require that the net torque acting on the object be zero.

In summary, for an extended object to be in equilibrium, two conditions must be satisfied:

1. The vector sum of the external forces acting on the object must be zero:

KEY EQUATION
$$\sum \vec{F} = 0 \qquad (10\text{-}1)$$

2. The vector sum of all the external torques acting on the object about any pivot or axis must be zero:

KEY EQUATION
$$\sum \vec{\tau} = 0 \qquad (10\text{-}2)$$

Condition 1 is referred to as translational equilibrium, while condition 2 describes rotational equilibrium.

The two conditions for equilibrium can be restated in terms of the linear and angular momentum of the object. For an object in equilibrium:

1. The linear momentum of the object is constant:

KEY EQUATION
$$\frac{d\vec{p}}{dt} = 0 \qquad (10\text{-}3)$$

2. The angular momentum of the object about any axis is constant:

KEY EQUATION
$$\frac{d\vec{L}}{dt} = 0 \qquad (10\text{-}4)$$

Static and Dynamic Equilibrium for an Extended Object

We saw that for an extend object to be in equilibrium, the net force and the net torque acting on it must be zero. For the object to be in static equilibrium, in addition to requiring that the speed of its centre of mass be zero, we also require that its rotational speed be zero. If either of these two requirements is not met, the object is in a state of dynamic equillibrium.

Stable, Unstable, and Neutral Equilibrium

We classify states of equilibrium for an object by considering what will happen to the object if it undergoes small displacements from its equilibrium position. If the object tends to return to its equilibrium position, we refer to the state of equilibrium at that position as **stable equilibrium**. This is illustrated in Figure 10-6(a),

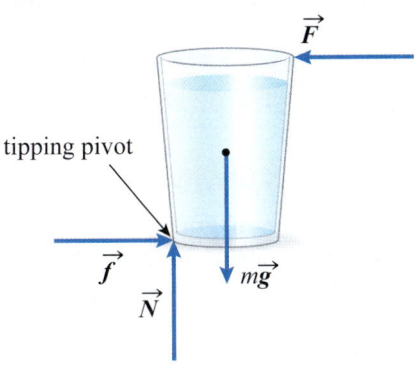

Figure 10-4 The sum of the forces acting on the mug is zero, yet it is not in equilibrium.

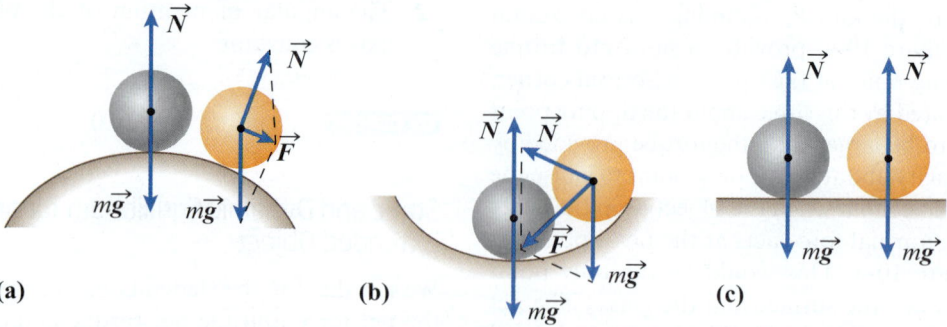

Figure 10-6 (a) Stable equilibrium. (b) Unstable equilibrium. (c) Neutral equilibrium. In (a) and (b) the grey ball is in an equilibrium position. In (c), both the grey and the orange balls are in equilibrium positions.

where we see an object resting at the bottom of a bowl. Clearly if we displace it by some small amount, it will tend to return to the bottom of the bowl.

The exact opposite will occur for the object depicted in Figure 10-6(b), where a small displacement will cause the object to roll off the bowl. This is an example of an **unstable equilibrium** position.

In contrast to both of these states of equilibrium, consider an object simply resting on a horizontal surface, as depicted in Figure 10-6(c). If we displace this object, it will simply remain in the new position, and we say that this is a position of **neutral equilibrium**.

10-2 Centre of Gravity

The force of gravity on an object, often referred to as the object's weight, is the force with which Earth and the object pull on one another. For most applications, we can treat the force of gravity as acting at a single point on the object; that point is the **centre of gravity** of the object. As we shall see below, in a uniform gravitational field, the centre of gravity coincides with the centre of mass, but this might not be the case in non-uniform gravitational fields.

Stable and Unstable Equilibrium and the Centre of Gravity

The location of the centre of gravity is important when analyzing the equilibrium conditions for a given object, as well as when determining the response of a rigid object to external forces. To see this, consider the worker holding the beam in Figure 10-8(a). Assuming that the mass is uniformly distributed throughout the beam, the centre of gravity is in the middle of the beam. If the worker lets the beam swing down gently until it is hanging vertically from the pivot (Figure 10-8(b)), the worker has moved the beam to a state of stable equilibrium. The centre of gravity of the beam is now below the pivot. A sideways nudge on the beam will move the beam slightly from its stable equilibrium position, but the force of gravity will pull the beam back to this stable position. If the worker tilts the beam up so that its centre of gravity is directly above the pivot (Figure 10-8(c)), the beam is in a state of unstable equilibrium. If the worker can get the beam to balance in this position, a sideways nudge will cause the beam to swing down.

Finding the Centre of Gravity Experimentally

We can find the centre of gravity of an object by suspending it and allowing it to come to static, stable equilibrium. As we showed in the discussion above, the centre of gravity must then be directly below the point of suspension for the object to be in stable equilibrium. Consider the triangular object shown in Figure 10-9. In Figure 10-10(a), we have suspended the object from

(a)

(b)   **(c)**

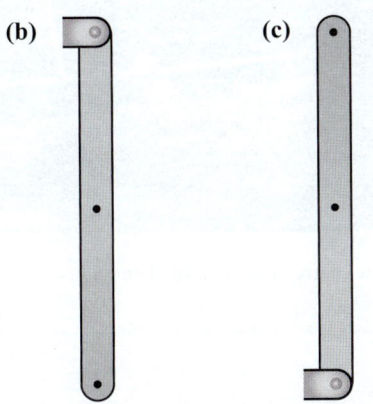

Figure 10-8 (a) A worker holding a beam. (b) The worker leaves the beam in stable equilibrium. (c) Unstable equilibrium. The centre of gravity of the beam is shown by the black dot in the middle of the beam.

the point labelled A and drawn a straight line vertically down from that point. The centre of gravity must lie somewhere along that line.

If we then suspend the object from a different point, B, as shown in Figure 10-10(b), we get a another vertical line going down from point B, along which the

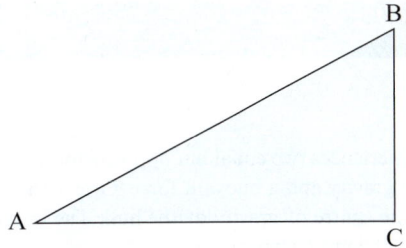

Figure 10-9 A triangular object.

centre of gravity must also lie. The point of intersection of the two lines marks the centre of gravity of the object, labelled G in Figure 10-10(c).

Centre of Gravity and Centre of Mass

In Chapter 7, we showed that the centre of mass for a collection of point masses is given by

$$\vec{r}_{cm} = \frac{\sum_{i=1}^{N} m_i \vec{r}_i}{\sum_{i=1}^{N} m_i} = \frac{1}{M_T} \sum_{i=1}^{N} m_i \vec{r}_i \qquad (7\text{-}10a)$$

where m_i and $\vec{r}_i$ are the mass and position of the ith particle and M_T is the total mass of the N-particle system.

The above equation can be expressed in component form as

$$x_{cm} = \frac{\sum_{i=1}^{N} m_i x_i}{M_T}$$

$$y_{cm} = \frac{\sum_{i=1}^{N} m_i y_i}{M_T} \qquad (7\text{-}10b)$$

$$z_{cm} = \frac{\sum_{i=1}^{N} m_i z_i}{M_T}$$

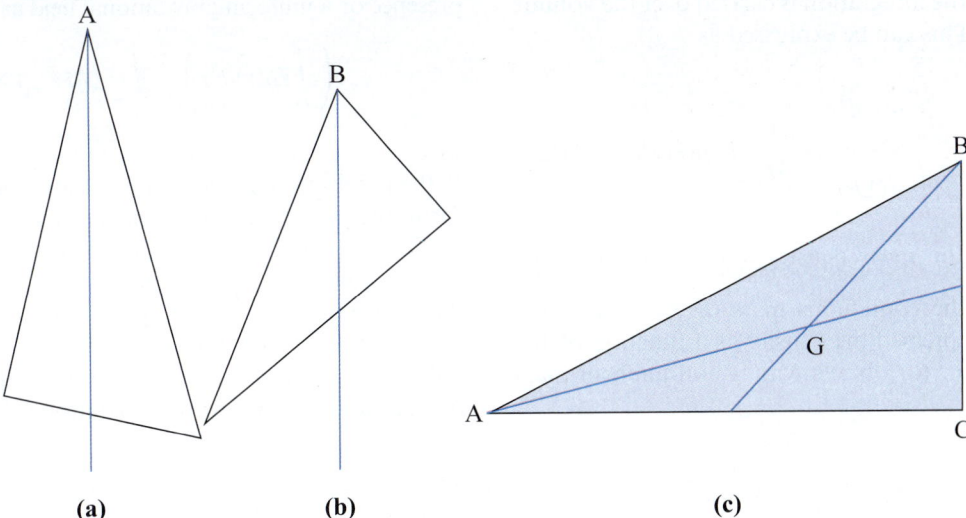

(a)   **(b)**   **(c)**

Figure 10-10 (a) Object suspended from point A: the vertical line starting at A passes through the centre of gravity. (b) Object suspended from point B: the vertical line from B also passes through the centre of gravity. (c) The centre of gravity, labelled G, is the point of intersection of the two lines.

Boat Stability

A boat experiences two equal but opposite forces in the vertical direction, gravity and a buoyant force. Gravity acts downward through the centre of gravity of the hull. The lower the centre of gravity, the more stable the boat.

Large sailboats achieve their stability by placing heavy ballast in their keels, as shown in Figure 10-11(a). Smaller racing dinghies use crew weight to maintain their balance, as shown Figure 10-11(b).

(a)

(b)

Figure 10-11 (a) The bulb at the bottom of the keel on the large boat produces a low centre of gravity. (b) The sailors in the small racing dinghy use body weight to achieve stability.

If the object can be treated as a continuous distribution of mass, we find the centre of mass using

$$\vec{r}_{cm} = \frac{\int_v \vec{r}\,dm}{\int_v dm} \qquad (7\text{-}11)$$

where the symbol v, at the lower end of the integral sign, indicates the integration is carried over the volume of the object. This can be expressed as

$$\vec{r}_{cm} = \frac{\int_v \vec{r}\rho(\vec{r})d^3r}{\int_v \rho(\vec{r})d^3r} = \frac{1}{M_T}\int_v \vec{r}\rho(\vec{r})d^3r \qquad (10\text{-}5)$$

where $\rho(\vec{r})$ is the mass density and $\int_v \rho(\vec{r})d^3r$ is the total mass of the continuous mass distribution. Note that we have expressed the density as a function of the position to allow for objects with a non-uniform mass distribution.

For objects in a uniform gravitational field, the centre of gravity coincides with the centre of mass. This is because a uniform gravitational force acts on an extended object, so it can be treated as if the total mass of the object is located at its centre of mass. Further, the total torque on an object due to a uniform gravitational field is equal to the torque exerted on the total mass of the object located at the centre of mass. Using Equation 10-5, we can write the torque exerted on continuous mass distribution in the presence of a uniform gravitational field as

$$\left(\int_v \vec{r}\rho(\vec{r})d^3r\right) \times \vec{g} = (M\vec{r}_{cm}) \times \vec{g}$$
$$= r_{cm} \times (M\vec{g}) \qquad (10\text{-}6)$$

If the gravitational field acting on an object is non-uniform, then the centre of mass and the centre of gravity of the object may not coincide. This situation is rather complicated and there is no unique way of defining the centre of gravity for objects in non-uniform gravitational fields. Since in most cases the effect of non-uniform gravitational fields is very small, we will limit our discussion to uniform gravitational fields.

Gravitational Gradients and Spaghettification

An object with dimensions that is interacting gravitationally with another will experience a gravitational gradient, whereby the gravitational field is not uniform over the dimensions of the object. As you stand on Earth's surface, the gravitational field at your feet is in principle stronger than the gravitational field at your head, although the difference is very small. In the vicinity of certain celestial bodies, such as neutron stars and black holes, the presence of significant gravitational gradients has led scientists to introduce the concept of *spaghettification*, whereby an object is stretched as a result of the gradient. Here on Earth, the interplay of gravitational gradients due to the Moon and the Sun results in the tides. See Data-Rich Problem 97 and Open Problem 98.

10-3 Applying the Conditions for Equilibrium

We can use the conditions of equilibrium to determine the forces acting on a mechanical system in equilibrium. We could apply either condition of equilibrium first; however, as we will see throughout the chapter, we can almost always simplify the solution to a given situation by carefully choosing which condition of equilibrium to consider first. We illustrate in the following example.

EXAMPLE 10-1

A Horizontal Beam

The beam in Figure 10-13 has a mass of 125 kg. Assume that the hinge at the left end of the beam is frictionless.

(a) How much force, assumed vertical, must the worker exert on the beam for it to be in equilibrium?
(b) How much force does the beam then exert on the hinge?

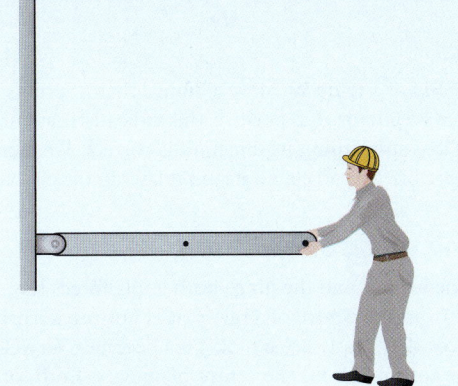

Figure 10-13   Example 10-1.

Balance and Centre of Mass

At first glance, you might think that the fork and spoon in Figure 10-12 should immediately drop off the side of the glass. However, the system consisting of the fork, spoon, and toothpick wedged together is actually in equilibrium. If you consider this system carefully, you will realize that its centre of gravity is just below the rim of the glass, so the system can remain balanced there indefinitely.

Peter Williams

Figure 10-12   How is it possible that a fork and spoon can be suspended on a toothpick totally on the outside of the edge of a glass?

Solution

(a) We start by drawing the FBD of the beam. The FBD in Figure 10-14 shows three forces acting on the beam: the force of gravity at the centre of gravity of the beam, the applied vertical force exerted by the worker at the right end, and the support force exerted by the hinge at the left end. We will use A_x and A_y to represent the horizontal and vertical components of the support force exerted by the hinge on the beam at A and we will assume that they point along the directions shown in the FBD. We note that since we are treating these as components of a force, rather than forces, you will see that, here and elsewhere, the symbols for these components will not be shown with an arrow on top, unlike other forces in the FBD. (In some texts, and in engineering applications, you might see components treated as two separate forces rather than as components of the same force. We note here that it is also common to refer to the support force at the hinge as the reaction force at the hinge. Occasionally it is also referred to as the hinge force.)

For the beam to be in equilibrium, the sum of the forces and the torques acting on the beam must be zero. Summing the forces in the x-direction gives

(continued)

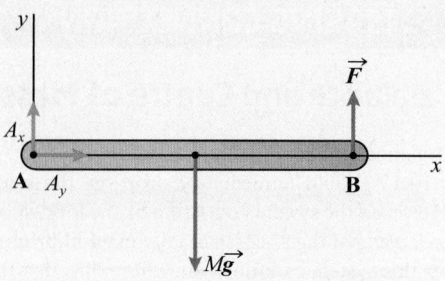

Figure 10-14 The FBD of the hinged horizontal beam.

$$\sum F_x = A_x = 0$$

and we immediately find that $A_x = 0$. We still have two unknowns: the vertical component of the support force exerted by the hinge on the beam, A_y, and the force $\vec{F}$ exerted by the worker.

We will first try applying the condition for equilibrium for forces in the y-direction:

$$\sum F_y = 0 \qquad (1)$$

Choosing up as our positive y-direction, we have

$$F - Mg + A_y = 0 \qquad (2)$$

We have two unknowns in this equation. We note that although this equation must be satisfied for the beam to be in equilibrium, it does not give us either of the unknown forces. We have deliberately chosen to begin with the sum of the forces in the y-direction to illustrate that although the equation is valid and needed, this is not the most efficient choice for a first step. We would like, whenever possible, to begin with an equation that yields one of the unknowns immediately.

We have not yet applied the equilibrium condition for torques acting on the beam. If the worker let go of the beam, it would rotate about the hinge. The force exerted by the worker prevents the beam from swinging. Summing the torques about the hinge at A gives

$$\sum \vec{\tau} = \vec{\tau}_{Mg} + \vec{\tau}_F = 0 \qquad (3)$$

From Figure 10-14, the components of the support force of the hinge on the beam exert no torque about the hinge because their moment arms (the perpendicular distance from the pivot to the force) are zero. So we only consider the torques due to the force of gravity and the force exerted by the worker. Taking clockwise to be positive, we have

$$+\!\!\Big)\sum \tau_A = \tau_{Mg} - \tau_F = 0 \qquad (4)$$

From Equation 8-20, we know that the torque due to a force about a pivot equals the product of the force and the perpendicular distance (the moment arm) between the force and the pivot. Therefore,

$$Mg\!\left(\frac{L}{2}\right) - F(L) = 0$$

$$\Rightarrow F = \frac{Mg}{2} = \frac{(125 \text{ kg})(9.81 \text{ m/s}^2)}{2} = 613 \text{ N} \qquad (5)$$

(b) Substituting the expression for F from (5) into (2) gives

$$\frac{Mg}{2} - Mg + A_y = 0$$

$$\Rightarrow A_y = \frac{Mg}{2} = F = 613 \text{ N} \qquad (6)$$

It is interesting to note that the torque equation immediately yields one of the unknown quantities; the force exerted by the worker on the beam. This is because we were able to eliminate the components of the support force of the hinge on the beam as unknowns in the torque equation by taking the sum of the torques about the hinge. The support force of the hinge on the beam exerts no torque about the hinge, therefore its components do not appear in the torque equation. As we shall see in the alternative solution below, we can use this to our advantage.

Alternative solution

The second condition for equilibrium, Equation 10-2, states that the sum of the torques about *any* point must be zero. Summing the torques about a pivot or hinge is often the simplest approach. However, we can sum the torques about any other convenient point.

Let us consider the torques about the end of the beam that the worker is holding. The worker exerts no torque at this end of the beam because the moment arm of the force exerted by the worker on the beam is zero. The forces exerting torques about this point are the force of gravity on the beam and the component A_y. The horizontal component of the support force at the hinge exerts no torque about the worker end of the beam, point B, as its moment arm is zero (its line of action goes through the point). Therefore,

$$\sum \vec{\tau}_B = \vec{\tau}_{Mg} + \vec{\tau}_{A_y} = 0$$

Taking counter-clockwise to be positive, we have

$$\Big(\!\!+\sum \tau_B = \tau_{Mg} - \tau_{A_y} = 0$$

Taking the torque to be the product of the force and the moment arm, we have

$$Mg\!\left(\frac{L}{2}\right) - A_y(L) = 0 \qquad (7)$$

and

$$A_y = \frac{Mg}{2} = 613 \text{ N} \qquad (8)$$

In the FBD of Figure 10-14 we assumed that A_y points up. Since the value we got for A_y is positive, this indicates that our assumption or guess regarding its direction is correct. We mention this point here briefly and elaborate on it later in the chapter.

Making sense of the result

Since the worker and the hinge both apply forces that are equidistant from the centre of gravity, it is not too surprising that the forces they apply are equal. This becomes very clear if we sum the torques about the centre of gravity. Both forces exert equal torques in opposite directions with equal moment arms, so the forces must be equal.

In the preceding example, we eliminated the torque due to the force applied by the worker from the torque equation by summing torques about the worker's end of the beam. This enabled us to find the force exerted by the hinge on the beam immediately.

It is important to remember that any careful application of the equilibrium conditions will yield the answers, but some choices involve a little less work than others.

CHECKPOINT 10-4

Which Equation to Start With?

You are asked to select a cable that is strong enough to support the beam in Figure 10-15.

You know the mass of the beam and the crate. Which equilibrium equation will allow you to immediately solve for the tension in the cable?

(a) sum of torques about the cable attachment point at the beam
(b) sum of torques about the centre of gravity of the beam
(c) sum of torques about the hinge at A
(d) sum of forces in the vertical direction

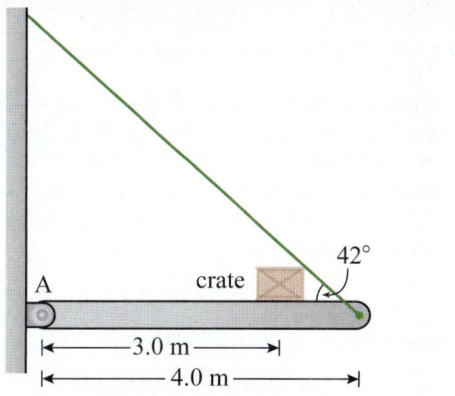

Figure 10-15 Checkpoint 10-4.

ANSWER: (c) Summing the torques about the hinge eliminates the unknown hinge forces from the equation and gives us the tension immediately.

EXAMPLE 10-2

A Sloping Beam

The beam in Figure 10-16 has a mass of 2.20×10^3 kg. It is supported at one end by a frictionless hinge, and the other end leans against a wall. Find the force that the wall exerts on the beam and the components of the support force the hinge exerts on the beam for the beam to be in equilibrium. Assume there is no friction between the wall and the beam.

Solution

We begin by drawing the FBD for the beam, which is shown in Figure 10-17. Since the wall is frictionless, there is no vertical force from the wall on the beam. Since the force exerted by the wall on the beam is normal to the wall, we label it $\vec{N}$. A_x and A_y are the horizontal and vertical components of the support force at the hinge. We will assume that these point along the directions shown in FBD. We also note that since they are treated as components they will not be shown in the FBD with arrows on top. The force of gravity, $M\vec{g}$, acts at the centre of gravity, which is located at the middle of the beam.

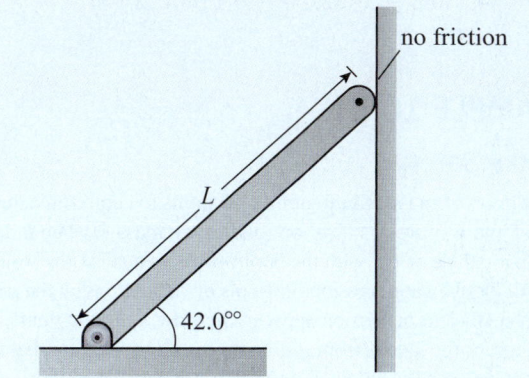

Figure 10-16 Example 10-2.

In this case, we are asked to find three unknowns—the normal force exerted by the wall on the beam and the two components of the hinge force on the beam. If we start by summing the torques about the hinge, we will eliminate the components

(*continued*)

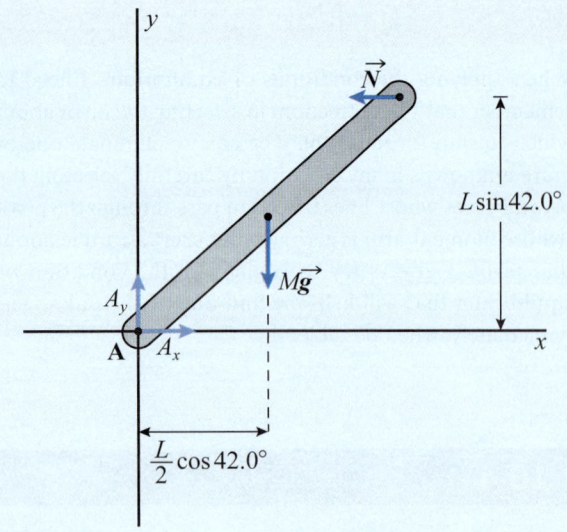

Figure 10-17 The FBD of the forces on the beam.

of the support force at the hinge from our torque equation and can immediately determine the normal force. The forces that exert torques about the hinge are the weight of the beam and the normal force from the wall (the components of the support force at the hinge exert no torque about the hinge because their moment arms are zero):

$$\sum \vec{\tau}_A = \vec{\tau}_{Mg} + \vec{\tau}_N = 0 \quad (1)$$

Taking clockwise to be positive, we have

$$+\!\!\curvearrowright \sum \tau_A = \tau_{Mg} - \tau_N = 0 \quad (2)$$

From Figure 10-17, we see that the perpendicular distance between the gravitational force and the pivot point of the hinge is $\frac{L}{2}\cos 42.0°$, and the perpendicular distance between the normal force and the pivot is $L \sin 42.0°$. We can express the sum of the torques at the pivot in terms of these moment arms and their associated forces:

$$Mg\!\left(\frac{L}{2}\cos 42.0°\right) - N(L\sin 42.0°) = 0 \quad (3)$$

Solving for N gives

$$N = \frac{Mg}{2\tan 42.0°} = 12.0 \text{ kN} \quad (4)$$

We now turn our attention to the components of the support force at the hinge. Now that we know the normal force, we can see that the only remaining unknown in the horizontal direction is A_x. Summing the forces acting on the beam in the x-direction, we obtain

$$\sum F_x = 0 \quad (5)$$

Choosing right as the positive x-direction and substituting from result (4),

$$-N + A_x = 0 \Rightarrow A_x = N = \frac{Mg}{2\tan 42.0°} = 12.0 \text{ kN} \quad (6)$$

We mention briefly here that a positive value for A_x indicates that our guess for its direction in the FBD was correct. To find the vertical component of the support force at the hinge, we sum forces in the y-direction, as there are only two forces with vertical components.

$$\sum F_y = 0 \quad (7)$$

Choosing up as our positive y-direction, we have

$$-Mg + A_y = 0 \Rightarrow A_y = Mg = 21.6 \text{ kN} \quad (8)$$

The positive value for A_y indicates that our guess for its direction in the FBD is correct.

Making sense of the result

Only A_y and the force of gravity are vertical, so they must be equal in magnitude but opposite in direction. Similarly, only A_x and $\vec{N}$ act in the horizontal direction and they too must be equal in magnitude but opposite in direction. It is interesting to note that $\vec{N}$ will get very large as the beam becomes more horizontal. The smaller the angle at the base of the beam, the smaller the moment arm of the normal force from the wall, and the greater the normal force has to be to counterbalance the torque from the force of gravity.

EXAMPLE 10-3

Ladder Safety

The worker of mass M in Figure 10-18 wants to climb three quarters of the way up a 4.0 m long ladder of mass m. The ladder makes an angle of 65° with the floor and leans against the wall as shown. Derive an expression, in terms of M and m, for the minimum coefficient of friction between the ladder and the floor that will prevent the ladder from sliding on the floor. Assume that the contact between the ladder and the wall is frictionless.

Solution

We start by drawing the FBD of the ladder (Figure 10-19). Five forces act on the ladder: the weight of the ladder ($m\vec{g}$), the weight of the worker ($M\vec{g}$), a vertical normal force ($\vec{N}_g$) exerted by the floor, the force of static friction between the ladder and the floor ($\vec{f}_s$), and a horizontal normal force ($\vec{N}_w$)

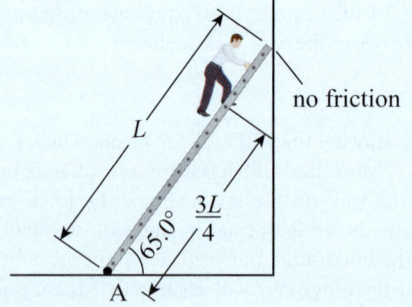

Figure 10-18 Example 10-3.

exerted by the wall. Since there is no friction between the ladder and the wall, the wall does not exert a vertical (friction) force on the ladder.

The question asks us to find the minimum coefficient of friction between the floor and the ladder so that it does not slip. Since there are only two forces with horizontal components, they must be equal in magnitude but opposite in direction, and if we can find one, we will know the other. However, recall that there is a relationship between the maximum force of static friction, the normal force, and the coefficient of static friction, μ_s:

$$f_s^{max} = \mu_s N_g \qquad (1)$$

Therefore, we will also need to determine the normal force exerted by the floor on the ladder.

Our strategy will be as follows. First, sum the torques about the point of contact between the ladder and the floor to find $\vec{N}_w$. Summing forces horizontally will then yield the magnitude of the friction force, $\vec{f}_s$, required to keep the ladder from slipping. Summing forces vertically will give us the normal force of the floor, $\vec{N}_g$, on the ladder.

We first consider the torques about point A, the point of contact between the ladder and the floor. The force of friction and the normal force from the floor do not exert a torque about point A because their moment arms are zero (their line of action goes through point A). Since the ladder is in equilibrium, as long as it does not slide, the sum of the torques of the other three forces must be zero:

$$\sum \vec{\tau}_A = \vec{\tau}_{mg} + \vec{\tau}_{Mg} + \vec{\tau}_{N_w} = 0 \qquad (2)$$

Taking clockwise to be positive,

$$+\!\bigcirc \sum \tau_A = \tau_{mg} + \tau_{Mg} - \tau_{N_w} = 0 \qquad (3)$$

The torque exerted by the worker's weight can be written as the product of Mg and its moment arm (the perpendicular distance between the force and the foot of the ladder). From Figure 10-19, we see that the distance is $\dfrac{3L}{4}\cos 65.0°$. Similarly, the moment arm of the weight of the ladder is $\dfrac{L}{2}\cos 65.0°$, and the moment arm of the normal force from the wall is $L\sin 65.0°$. Substituting these values into (3) gives

$$mg\left(\frac{L}{2}\cos 65.0°\right) + Mg\left(\frac{3L}{4}\cos 65.0°\right) - N_w(L\sin 65.0°) = 0$$

Solving for N_w, we obtain

$$N_w = \frac{mg\left(\dfrac{L}{2}\cos 65.0°\right) + Mg\left(\dfrac{3L}{4}\cos 65.0°\right)}{L\sin 65.0°} \qquad (4)$$

$$\Rightarrow N_w = \frac{\dfrac{mg}{2} + \dfrac{3}{4}Mg}{\tan 65.0°}$$

The factor of $\dfrac{3}{4}$ in this equation comes from the location of the worker on the ladder. Note that (4) indicates that the value of the normal force exerted by the wall depends on how high up the ladder the worker is. The only other force acting in the x-direction is the force of static friction between the floor and the ladder. Therefore,

$$\sum F_x = 0 \qquad (5)$$

Choosing right to be the positive direction, we have

$$-N_w + f_s = 0 \Rightarrow f_s = N_w \qquad (6)$$

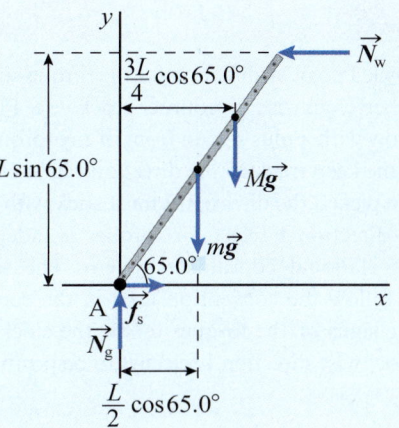

Figure 10-19 The FBD of the ladder.

To determine the coefficient of static friction between the floor and the ground, using (1), we need to find the normal force exerted by the floor on the ladder. Applying the condition of equilibrium for forces in the y-direction, we get

$$\sum F_y = 0 \qquad (7)$$

Taking up to be the positive y-direction, we have

$$-Mg - mg + N_g = 0 \Rightarrow N_g = Mg + mg \qquad (8)$$

Now that we have what we need, substituting the value of f_s from (6) and N_g from (8) into (1), we have

$$\mu_s = \frac{f_s}{N_g} = \frac{\dfrac{mg}{2} + \dfrac{3}{4}Mg}{(Mg + mg)\tan 65°} = \frac{2m + 3M}{4(M + m)\tan 65°}$$

Making sense of the result

The larger the angle that the ladder makes with the horizontal, the less friction there needs to be between the floor and the ladder to keep the ladder from falling. When the ladder is vertical, it does not exert a horizontal force on the wall or the ground, and no friction is needed. However, this position would be an unstable equilibrium because any outward force would cause the ladder to fall away from the wall. For safety, a ladder is normally positioned so the distance from the foot of the ladder to a vertical wall is one quarter of the height of the point of contact of the ladder with the wall. Step ladders have two sides with a horizontal beam connecting the two sides to eliminate the need for significant friction between the ladder and the floor, as shown in Figure 10-20.

Figure 10-20 A step ladder.

EXAMPLE 10-4

A Beam Supported by a Cable

A 500.0 kg crate sits 3.00 m from a hinge on a 4.00 m long horizontal beam, as shown in Figure 10-21. The beam is supported at one end with a cable that is attached to a wall; the cable makes an angle of 42.0° with the horizontal. The beam has a mass of 270.0 kg. Determine the tension in the cable and the force exerted by the hinge on the beam for the system to be in equilibrium.

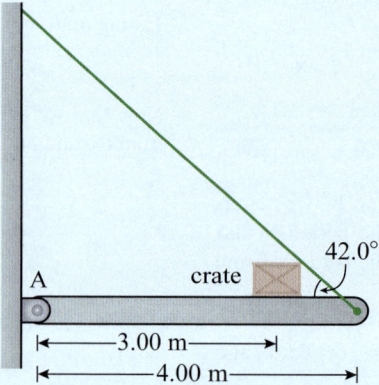

Figure 10-21   Example 10-4.

Solution

The weight of (or the force of gravity on) the beam, the weight of the crate, and the tension in the cable, as well as the horizontal and vertical components of the support force at the hinge, are shown in the FBD of Figure 10-22. Gravity exerts a downward force of $m\vec{g}$ on the crate and the beam must exert an equal but opposite force on the crate to keep the crate in equilibrium. By the third law, the crate must then exert an equal but opposite force on the beam. We call that force $\vec{N}$.

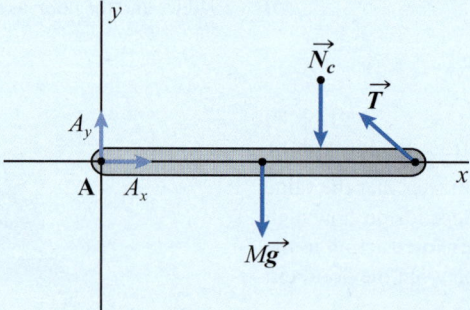

Figure 10-22   The FBD of the beam.

The x- and y-components of the force of the hinge on the beam are unknown quantities. We can find the tension force by summing torques about the hinge at point A. We can then break up the tension into components and sum forces in the horizontal and vertical directions to find A_x and A_y. An alternative approach would be to sum torques about the point where the cable is attached to find A_y; summing forces in the y-direction will give us T_y. Since we know the direction of the tension force, we can find T_x, and by summing forces in the x-direction, we can find A_x.

We apply the condition for equilibrium for the torques about the hinge at A. The forces that exert torques about A are the weight of the beam ($M\vec{g}$), the normal force of interaction between the crate and the beam ($\vec{N}$), and the tension in the cable ($\vec{T}$). The forces at the pivot, A_x and A_y, do not exert a torque about the pivot because their moment arm is zero. Therefore,

$$\sum \vec{\tau}_A = \vec{\tau}_{Mg} + \vec{\tau}_{N_c} + \vec{\tau}_T = 0$$

Taking clockwise to be positive,

$$\overset{+}{\circlearrowright}\sum \tau_A = \tau_{Mg} + \tau_{N_c} - \tau_T = 0$$

From Figure 10-22, using Equation 8-22 for the torque exerted by the tension force, given by the product of the force, the distance to the pivot and the sine of the angle, and moment arm concept for the torque exerted by $M\vec{g}$ and $\vec{N}_c$, noting that the magnitude of $\vec{N}_c$ is equal to the magnitude of the weight of the crate ($m\vec{g}$), and taking the weight of the beam to be at its centre (2.00 m from the pivot), we get

$$Mg(2.00 \text{ m}) + mg(3.00 \text{ m}) - T(4.00 \text{ m})(\sin 42.0°) = 0$$

Solving for T gives

$$T = \frac{(270.0 \text{ kg})(9.81 \text{ m/s}^2)(2.00 \text{ m}) + (500.0 \text{ kg})(9.81 \text{ m/s}^2)(3.00 \text{ m})}{(4.00 \text{ m})(\sin 42.0°)} = 7477 \text{ N}$$

$$\Rightarrow T = 7.48 \text{ kN}$$

Next, we sum the forces in the y-direction. Here, we have the weight of the beam, the normal force from the crate, the vertical component of the tension, and the vertical component (A_y) of the support force at the hinge:

$$\sum F_y = 0$$

From Figure 10-22, the y-component of the tension force is $T\sin 42.0°$. Choosing up as our positive y-direction, we have

$$-Mg - N_c + T\sin 42.0° + A_y = 0$$

We rearrange to find

$$A_y = Mg + mg - T\sin 42.0°$$
$$= (270.0 \text{ kg})(9.81 \text{ m/s}^2) + (500.0 \text{ kg})(9.81 \text{ m/s}^2) - (7477 \text{ N})\sin 42.0°$$
$$= 2551 \text{ N} = 2.55 \text{ kN}$$

Now, the only unknown is the horizontal component, A_x, of the support force at the hinge. We take the sum of the forces acting on the beam in the x-direction; these are A_x and the horizontal component of the tension, T:

$$\sum F_x = 0$$

From Figure 10-22, the x-component of the tension is $T\cos 42.0°$. Choosing right as our positive x-direction,

$$-T\cos 42.0° + A_x = 0 \Rightarrow A_x = T\cos 42.0°$$

Therefore,

$$A_x = (7477 \text{ N})\cos 42.0° = 5556 \text{ N} = 5.56 \text{ kN}$$

We note briefly that the positive values we obtained for A_y and A_x indicate that our guesses for their directions in the FBD were correct.

Making sense of the result

The cable supports part of the weight of the beam, but since the cable is on an angle, part of the force it exerts is horizontal (directed toward the wall), which results in the horizontal component of the force at the hinge. A quick check is to sum the torques about the other end of the beam, which gives $A_y = 2.55$ kN as well.

Guidelines for Approaching Equilibrium Problems

As we worked through the examples above, we discussed our approach to describing which equilibrium condition we were going to apply and why. We can summarize the approach with a few simple guidelines.

The rotational equilibrium condition provides us with a great deal of flexibility in writing our equations, because it demands that the torque be zero about any point on the object. If you can find a point on the object about which a single unknown force exerts a torque, applying the rotational equilibrium condition about that point will yield the value for that single unknown force.

The translational equilibrium condition, on the other hand, requires us to sum all of the forces to zero. Only when we have a single unknown force/force component in a particular direction will the force equilibrium equation directly yield a value for a force.

We can summarize these observations into some guidelines to keep in mind when approaching equilibrium problems:

1. Look for points on the object where a single unknown force produces a torque, and sum the torques about that point to find the unknown force.

2. Remember that the torque equilibrium condition is valid for any point on the object and can be used multiple times in any solution.

3. As you progress through the solution, monitor to see if you have a single unknown force/force component in any direction. Summing the forces in that direction will immediately tell you the value for that unknown.

10-4 Applying the Conditions for Equilibrium: Working with Unknown Forces

In Chapter 5, we rarely encountered a situation where we did not know the directions of the forces we were trying to find. Forces such as the force of gravity, the normal force, and the force of friction all have well-defined directions and we can readily represent them by vectors in FBDs. However, in this chapter we have encountered forces whose magnitude and direction are unknown, such as the force that the hinge exerts on a beam. In these examples, we assumed directions for these forces in our FBDs and obtained results that were consistent with our assumptions. Specifically, in Example 10-4 we chose the horizontal and vertical components of the force that the hinge exerts on the beam to point in the positive directions of their corresponding coordinate axes. When we solved the problem, we obtained positive results for A_x and A_y, indicating that our guess for the directions of these unknowns was correct. However, had our analysis yielded negative values for any of these unknowns, we would have concluded that our choice for the direction of that force was incorrect.

An incorrect choice of direction for an unknown force will not affect the magnitude of that force. We also emphasize that if we obtain a negative value for a given unknown, there is no need to redraw the FBD and redo the problem, as long as the negative value for that variable is used in subsequent equations in which the variable appears.

In the approach we follow here, we will always assume that an unknown force will point along the positive axis for clarity and to eliminate potential ambiguity. For action-reaction pairs, we will draw one of them so it points along the direction of the positive axis, and the other in the opposite direction. However, we point out here, and remind you below, that choosing unknown forces to point along the positive axis is not always necessary or followed elsewhere.

To illustrate the previous discussion, we select the following examples. While it is relatively easy to find the direction of the unknown forces in what follows, we will deliberately make incorrect "guesses" for the directions of some of the unknown forces and explore how this approach can be applied when needed.

EXAMPLE 10-5

Working with Unknown Forces

Assume that the weight of the V-shaped structural element shown in Figure 10-23 is negligible compared to the other forces acting on it. Find the forces exerted by the hinge and the roller on this element. Assume that the roller is frictionless and that $F_1 = 1200.0$ N, $F_2 = 400.0$ N, and $F_3 = 500.0$ N. The roller is a type of support; it is free to roll, and in this example provides a vertical force only.

Solution

We begin with the FBD of the structural element (Figure 10-24). At the hinged end, we have the support force at the hinge with horizontal and vertical components (A_x and A_y). At the roller end, we have only a normal force ($\vec{N}$). We do not know the direction of A_x or A_y, so we will assume that A_y points up and that A_x points to the right.

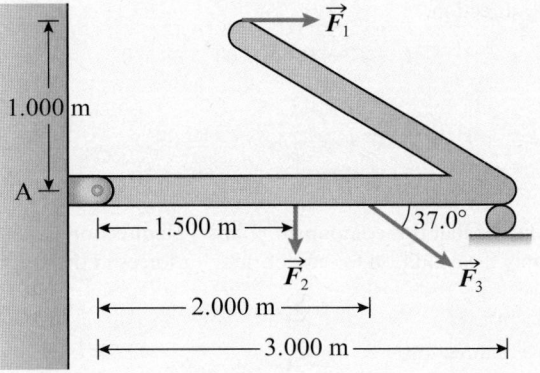

Figure 10-23 Example 10-5.

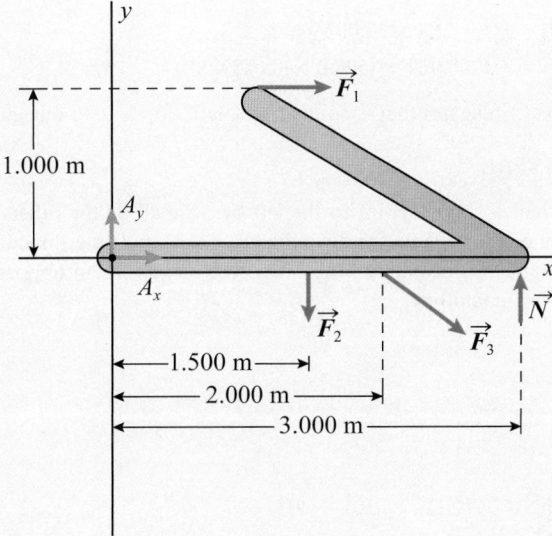

Figure 10-24 The FBD of the structural element.

To determine the unknown normal force exerted on the element by the roller, we take the sum of the torques about the hinge. The forces that exert torques about A are the forces $\vec{F}_1$, $\vec{F}_2$, $\vec{F}_3$ and the normal force at the roller end, $\vec{N}$. The components of the support force at the hinge exert no torques about A because their moment arms are zero. Thus,

$$\sum \vec{\tau}_A = \vec{\tau}_{F_1} + \vec{\tau}_{F_2} + \vec{\tau}_{F_3} + \vec{\tau}_N = 0 \tag{1}$$

Taking clockwise to be positive,

$$\curvearrowright + \sum \tau_A = \tau_{F_1} + \tau_{F_2} + \tau_{F_3} - \tau_N = 0 \tag{2}$$

$$F_1(1.000 \text{ m}) + F_2(1.500 \text{ m}) + F_3(2.000 \text{ m}) \sin 37.00° - N(3.000 \text{ m}) = 0$$

Note that for the torques exerted by $\vec{F}_1$, $\vec{F}_2$, and $\vec{N}$, we have used the moment arm concept (perpendicular distance from pivot to force) as given in Equation 8-20. For the torque exerted by $\vec{F}_3$, we have used the definition of torque as given by Equation 8-22. Solving for N, we find

$$N = \frac{(1200.0 \text{ N})(1.000 \text{ m}) + (400.0 \text{ N})(1.500 \text{ m}) + (500.0 \text{ N})(2.000 \text{ m})(\sin 37.00°)}{3.000 \text{ m}} \tag{3}$$

$$= 800.6 \text{ N}$$

N is positive, so this normal force indeed points upward, as we drew it. To determine A_y, we sum the forces in the y-direction:

$$\sum F_y = 0$$

(continued)

Choosing up as the positive y-direction,

$$A_y - F_2 - F_3 \sin 37.00° + N = 0 \tag{4}$$

Therefore,

$$A_y = F_2 + F_3 \sin 37.00° - N$$
$$= 400.0 \text{ N} + (500.0 \text{ N})(\sin 37.00°) - 800.6 \text{ N} = -99.7 \text{ N} \tag{5}$$

The negative value for A_y indicates that this component points in a direction opposite to our guess; downward. To determine A_x, we apply the condition for equilibrium for forces in the x-direction:

$$\sum F_x = 0$$

Choosing right as our positive x-direction,

$$A_x + F_1 + F_3 \cos 37.00° = 0 \tag{6}$$

Hence,

$$A_x = -F_1 - F_3 \cos 37.00°$$
$$= -1200.0 \text{ N} - (500.0 \text{ N})(\cos 37.00°) = -1599 \text{ N} \tag{7}$$

Again, the negative value for A_x indicates that it points to the left, opposite to our guess.

Making sense of the result

We could have anticipated that A_x would point to the left because all of the other x-components point to the right. The direction for A_y is not so obvious, and we illustrate how guessing directions works in this case. The normal force has a longer moment arm than all the other forces exerting an opposing torque about the hinge, which explains its relatively low magnitude.

EXAMPLE 10-6

Analyzing a Multipart Structure

The shorter beam in Figure 10-25 has mass $m_1 = 1100.0$ kg and is supported by a free hinge at point A. The longer beam has mass $m_2 = 2100.0$ kg and is supported by a free hinge at point B. The two beams are fixed to each other by a free hinge at C. Find the support forces at the three hinges, assuming that the structure is in equilibrium.

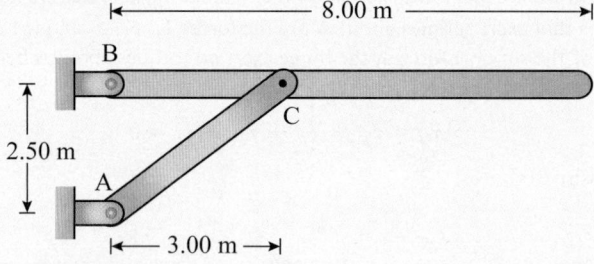

Figure 10-25   Example 10-6.

Solution

Let us first look at the FBD of the two-beam structure as a whole, which is shown in Figure 10-26. The external forces acting on this structure are the support forces at the hinge at points A and B and the weights of the two beams. We do not consider the force acting on the hinge at C because this force is internal to the system. The FBD for an object or a system shows only the external forces acting on it.

Examine Figure 10-26. Do you spot any issues with the FBD? We have obviously made an incorrect assumption about the direction of one of the horizontal components of the support forces at the hinges. They cannot both point in the same direction! As we mentioned at the beginning of this section, we do that here intentionally for illustration purposes to explore how we deal with an incorrect assumption about the direction of an unknown force and the negative value we will obtain in that case.

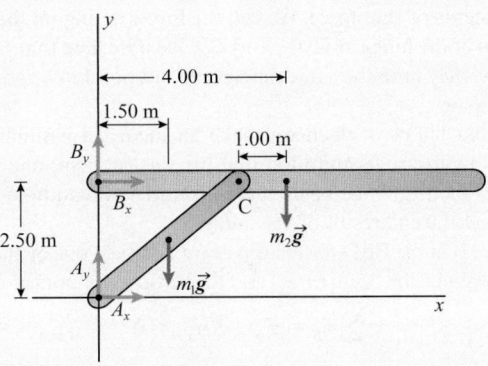

Figure 10-26 The FBD of both beams as one structure.

With some careful examination, we can perhaps already guess that the horizontal component of the support force at A should point to the right, which it does in our diagram. This can be seen by taking the torques about point B. It is perhaps also relatively easy to see that the vertical component of the support force at A must point up, which, again, it does in our diagram. This is not as evident or clear as in the case for the vertical component of the support force at B. Although with some reasoning one can arrive at the correct guess for the direction, it is advisable in many applications not to spend the time on figuring out the direction of the unknown force but simply to go with a guess and finish the solution efficiently and quickly.

The forces that exert torques about point B in Figure 10-26 are the weights of the beams, $m_1\vec{g}$ and $m_2\vec{g}$, as well as A_x, the horizontal component of the support force at the hinge at A. The components of the force at the hinge at B, B_x and B_y, exert no torque about B because the moment arm of these components about B is zero, since their lines of action pass through the pivot at B. The vertical component of the support for the hinge at A also exerts no torque about point B because the line of action of this component passes through the hinge at B; therefore, the moment arm for A_y about B is also zero. So, the sum of the torques about point B is

$$\sum \vec{\tau}_B = \vec{\tau}_{m_1g} + \vec{\tau}_{m_2g} + \vec{\tau}_{A_x} = 0$$

Taking clockwise to be positive, the torques of $m_1\vec{g}$ and $m_2\vec{g}$ are both positive, and the torque of A_x is negative. Thus,

$$+\!\!\bigcirc\!\!\sum \tau_B = \tau_{m_1g} + \tau_{m_2g} - \tau_{A_x} = 0$$

From Figure 10-26, the perpendicular distances from the pivot at B to $m_1\vec{g}$ and $m_2\vec{g}$ are 1.50 m and 4.00 m, respectively. Therefore,

$$m_1g(1.50 \text{ m}) + m_2g(4.00 \text{ m}) - A_x(2.50 \text{ m}) = 0$$

$$A_x = \frac{m_1g(1.50 \text{ m}) + m_2g(4.00 \text{ m})}{2.50 \text{ m}}$$

$$= \frac{1100.0 \text{ kg}(9.81 \text{ m/s}^2)(1.50 \text{ m}) + 2100.0 \text{ kg}(9.81 \text{ m/s}^2)(4.00 \text{ m})}{2.50 \text{ m}} \tag{1}$$

$$\Rightarrow A_x = 39\,436 \text{ N} = 39.4 \text{ kN}$$

We can now immediately determine another unknown by summing the forces in the x-direction. The only forces acting in the x-direction are the horizontal components of the hinge forces at A and B, A_x and B_x. Therefore,

$$\sum F_x = 0 \tag{2}$$

Choosing right as our positive direction,

$$B_x = -A_x = -39.4 \text{ kN} \tag{3}$$

The negative value for B_x indicates that it points in the direction opposite to what we drew in the FBD. However, in subsequent calculations we simply use the negative value we obtained for B_x.

No other unknown can be found by looking at the entire two-beam structure as one body, so we now apply the conditions for equilibrium to the components of the structure. We still need to find the vertical components of the forces at hinges A and B, as well as the forces acting on the hinge at C.

We now draw the FBD for each beam (Figure 10-27). The forces acting on the long beam are its weight, the horizontal and vertical components of the hinge force at B, B_x and B_y, and the force acting on the hinge at C. Since we do not know the direction of the force at C, we make the assumptions shown in Figure 10-27 for the

(continued)

horizontal and vertical components of that force. We call the forces acting on the top beam at C, C_x and C_y, and the forces acting on lower beam at the hinge at C, C'_x and C'_y. We note here that C_x and C'_x are an action-reaction pair, and by Newton's third law, they have the same magnitude but point in opposite directions. Similarly, C_y and C'_y are also an action-reaction pair.

Note that to illustrate what happens when we make an incorrect assumption about the direction of an unknown force, we have made an absurd assumption that the vertical component of the support force at C acts downward on the top beam. In fact, C_y must be directed upward so it can hold the top beam up. Choosing the incorrect direction will not affect the end result of the analysis.

Summing the torques about B in the FBD for the top beam will yield one of the unknowns. Since the forces that exert torques about B are the weight of the beam ($m\vec{g}_2$) and the vertical component of the support force at point C,

$$\sum \vec{\tau}_B = \vec{\tau}_{m_2 g} + \vec{\tau}_{C_y} = 0 \tag{4}$$

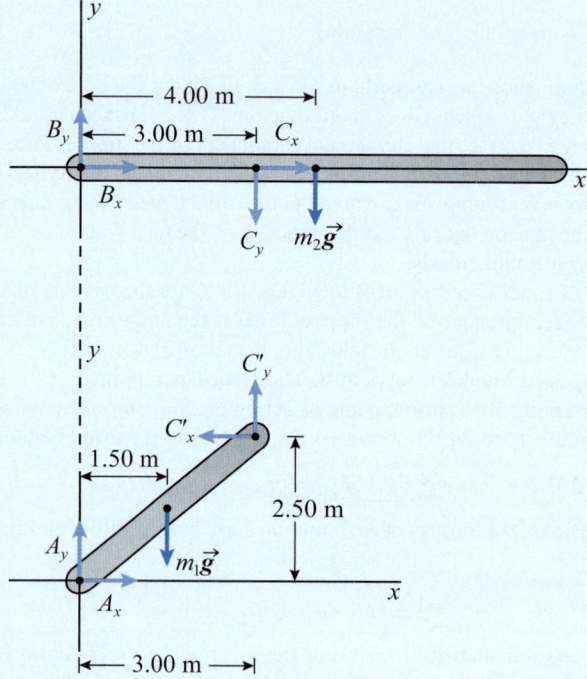

Figure 10-27 The FBD of the top and the bottom beams.

If we take clockwise to be positive, we have

$$+\!\!\sum \tau_B = \tau_{m_2 g} - \tau_{C_y} = 0 \tag{5}$$

Replacing the torques with the corresponding forces and moment arms, we obtain

$$m_2 g(4.00 \text{ m}) - C_y(3.00 \text{ m}) = 0 \tag{6}$$

This gives us

$$C_y = \frac{m_2 g(4.00 \text{ m})}{3.00 \text{ m}} = \frac{2100.0 \text{ kg}(9.81 \text{ m/s}^2)(4.00 \text{ m})}{3.00 \text{ m}} = 27468 \text{ N} = 27.5 \text{ kN} \tag{7}$$

The positive value for the variable C_y means that that our guess for its direction is correct, and it points in the direction drawn on the FBD.

To find the vertical component of the force at the hinge at B, we apply the equilibrium conditions for the vertical forces acting on the top beam. Here, we have the weight of the beam, the vertical component of the force at C, and the vertical component of the force at B. Therefore,

$$\sum F_y = 0$$

Choosing up as our positive y-direction, we have

$$-m_2 g + C_y + B_y = 0$$
$$B_y = m_2 g - C_y$$
$$B_y = 2100.0 \text{ kg}(9.81 \text{ m/s}^2) - 27468 \text{ N} \tag{8}$$
$$= -6867 \text{ N} = -6.87 \text{ kN}$$

The negative value for the variable B_y means that our assumption regarding the direction for B_y in Figure 10-27 was not correct. We must be careful to use this negative value in subsequent calculations.

Next, we find the horizontal component of the force acting on the hinge at C, C_x, by summing the forces acting on the top beam in the x-direction. Since the only forces here are the forces at B and C, we have

$$\sum F_x = 0$$

Choosing right as our positive x-direction,

$$C_x + B_x = 0$$
$$C_x = -B_x = -(-39.4 \text{ kN}) = 39.4 \text{ kN}$$

Note that we have used the value of B_x as -39.4 kN as obtained in result (3) above, as we have noted earlier. The positive value for C_x indicates that we guessed the correct direction for C_x in the FBD in Figure 10-27.

Now we can go back to the FBD for the entire system. The only unknown we have left is the vertical component of the support force at A. We take the sum of the vertical forces on the entire structure: the weight of the top beam ($m_2 \vec{g}$), the weight of the bottom beam ($m_1 \vec{g}$), and the vertical components of the support forces at A and B, A_y and B_y. Thus,

$$\sum F_y = 0$$

Choosing up as our positive y-direction,

$$-m_1 g - m_2 g + A_y + B_y = 0 \qquad (9)$$
$$A_y = m_1 g + m_2 g - B_y$$
$$= 1100.0 \text{ kg}(9.81 \text{ m/s}^2) + 2100.0 \text{ kg}(9.81 \text{ m/s}^2) - (-6867 \text{ N})$$
$$= 38.3 \text{ kN}$$

The positive value we get for A_y indicates that we have chosen the correct direction for that force. Note again that we have substituted for the variable B_y the negative value we obtained for it in result (8). We point out that wherever this variable comes up, be it in a force or a torque equation, we use substitute the negative value of the variable in the required equation.

Making sense of the result

As we anticipated, one of A_x or B_x had to point opposite to our guess for its direction. We might have guessed it would be B_x since the torques are all trying to rotate the structure in a clockwise direction. However, it is important to once again note that spending any time beforehand trying to guess these directions is unnecessary. You simply need to use a consistent approach and the results will tell you the actual directions of the forces.

10-5 Deformation and Elasticity

So far, we have treated structural components such as beams and supports as rigid objects that do not deform. However, deformation always occurs when external forces act on objects, although sometimes the deformation is not noticeable or appreciable. If sufficient force is applied, steel and concrete structural elements will deform, break, or buckle. The three types of forces of primary interest to us are tension, compression, and shear. Other types of deformation, such as torsion, are not covered here.

When you pull each end of a bar away from the centre of the bar, the bar is under **tension**. This is shown in Figure 10-28(a). When you push each end of a bar toward the centre, you are exerting a **compression force**, as in Figure 10-28(b). When you support one end of a bar, as shown in Figure 10-28(c), and push on the unsupported end as shown, you are exerting a **shear force**. An example

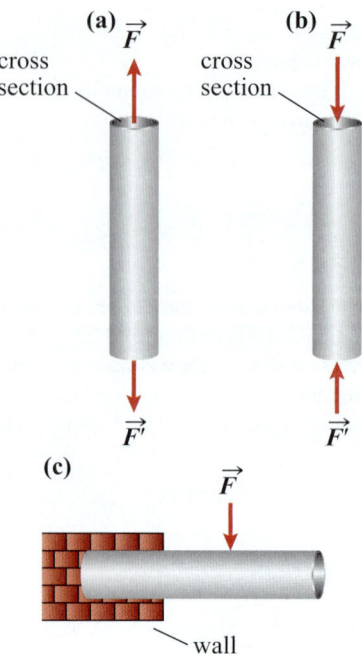

(a) cross section

(b) cross section

(c)

wall

Figure 10-28 (a) A beam under tension. (b) A beam under compression. (c) A beam experiencing a shear force.

of a shear force is that exerted by a diver on the end of a diving board or that exerted by scissors on piece of paper.

Stress

Consider the cylinder of length L and cross section A shown in Figure 10-29(a). Two forces, $\vec{F}$ and $\vec{F}'$, equal in magnitude but opposite in direction, pull the ends of the cylinder away from the centre. This tension on the cylinder causes it to elongate. Similarly, if the forces push the ends of the cylinder toward the centre, the compression causes the cylinder to shorten, as in Figure 10-29(b).

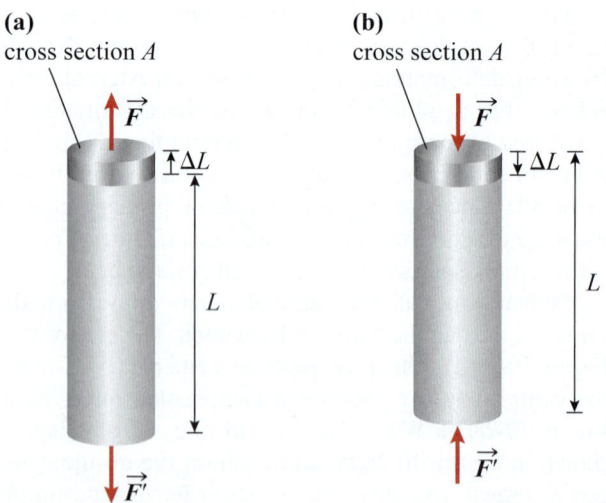

(a) cross section A

(b) cross section A

Figure 10-29 (a) Tension causes the cylinder to elongate. (b) Compression causes the cylinder to shorten.

The effect of the force exerted on a beam depends on the cross section of the beam. We are interested in the force per cross-sectional area of the beam, called the **stress**, σ:

KEY EQUATION
$$\sigma = \frac{F}{A} \qquad (10\text{-}7)$$

The SI unit for stress is the Pascal, (Pa).

Strain, Elastic Deformation, and the Proportional Limit

The deformation of an object in response to stress is characterized by a quantity called **strain**, which is the fractional change in the length of the object. A symbol commonly used for strain is the Greek letter epsilon, ε:

KEY EQUATION
$$\varepsilon = \frac{\Delta L}{L} \qquad (10\text{-}8)$$

An **elastic deformation** is a deformation that is completely reversible: when the stress is removed, the object returns to its original length. It is common for objects to deform elastically in response to applied stress as long as the stress is not too large. For many materials of interest, such as steel alloys, the elastic response region has two parts. The first part is characterized by a linear relation between stress and strain:

KEY EQUATION
$$\sigma = Y\varepsilon \qquad (10\text{-}9)$$

Combining Equations 10-7, 10-8, and 10-9, we get

KEY EQUATION
$$\frac{F}{A} = Y\frac{\Delta L}{L} \qquad (10\text{-}10)$$

The parameter Y is called **Young's modulus**, which is a measure of a material's elasticity. For many materials, especially steel alloys, the value of Young's modulus is the same for compressive and tensile stress. However, the response of a composite material, such as concrete, to tension can differ drastically from its response to compression.

A material that behaves in the fashion described by Equation 10-10 follows Hooke's law as presented in Chapter 5.

The maximum stress for which the response of the material is linear is called the **proportional limit**. This point is labelled P in the typical stress versus strain curve of Figure 10-31. Past the proportional limit, the slope of the stress versus strain curve changes: the response of the material is no longer linear, and Hooke's law does not apply. However, the deformation remains elastic past the proportional point until the stress reaches a value called the **yield point**, **yield strength**, or **elastic limit**, labelled EL in Figure 10-31.

MECHANICS

MAKING CONNECTIONS

The Bridge That Would Last Forever

Stone, like concrete, is a lot weaker in tension than it is in compression. Therefore, stone buildings are designed to have the structural elements under compression rather than tension. This is true of the stones in a classic arch. This design has been used since the times of the Etruscans and the ancient Greeks. Figure 10-30 shows Alcántara Bridge, built by Caius Julius Lacer between 104 and 106 CE over the Tagus River in Spain. The triumphal arch at the centre bears two inscriptions on marble plates. One is a dedication to the Roman emperor Trajan, who commissioned the bridge. The other says "Pontem perpetui mansurum in saecula" (I have built a bridge that will last forever). The bridge has suffered damage due to wars and has been repaired/restored more than once.

Figure 10-30 Alcántara Bridge in Spain.

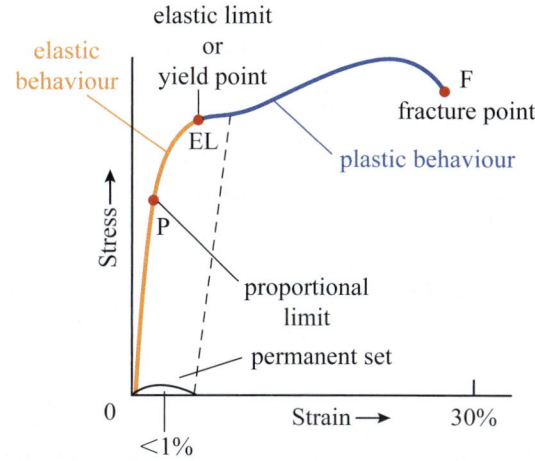

Figure 10-31 Typical stress versus strain curve for a ductile material. The elastic behaviour extends beyond the proportional limit. Between the proportional limit and the elastic limit, the response is no longer linear, but the deformation is still elastic.

CHECKPOINT 10-6

The Yield Point

The yield point of a given material is
(a) its ultimate tensile stress
(b) the maximum stress that the material will take so that its strain is directly proportional to the applied stress
(c) slightly lower than the proportional point
(d) the maximum stress for which the deformation of the material is completely, or almost completely, reversible

ANSWER: (d) This is the definition of the yield point.

EXAMPLE 10-7

Animal versus Mineral

The yield point of both A36 steel (a grade of structural steel) and some types of spider silk is 250.0 MPa, and the yield point for human tendons is 28.0 MPa. Compare the maximum mass that can be carried in tension by a steel rod or spider silk, without permanent deformation, with the mass that can be supported by a human tendon of the same size and dimensions. The rod (and spider silk) are 2.00 m long and have a radius of 2.00 cm.

Solution

Let us begin by drawing the FBD of the steel rod and the load. The forces acting on the beam are the tension forces at either end, as shown in Figure 10-32. The load is under the action of the force of gravity and the tension force from the rod.

The tension force is equal in magnitude to the weight of the load in equilibrium, and it causes the rod to stretch. According to Equation 10-7, $T = F = \sigma A = Mg$. For steel and silk,

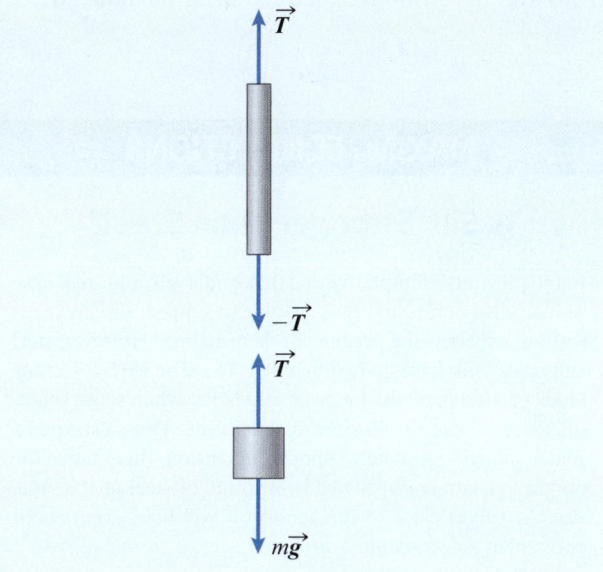

Figure 10-32 FBD of the rod and the load.

(continued)

$$M = \frac{\sigma A}{g} = \frac{\left(2.50 \times 10^8 \, \frac{\text{N}}{\text{m}^2}\right)\pi(0.0200 \text{ m})^2}{9.81 \text{ m/s}^2} = 3.20 \times 10^4 \text{ kg}$$

For the tendon,

$$M = \frac{\sigma A}{g} = \frac{\left(2.80 \times 10^7 \, \frac{\text{N}}{\text{m}^2}\right)\pi(0.0200 \text{ m})^2}{9.81 \text{ m/s}^2} = 3.59 \times 10^3 \text{ kg}$$

This is an order of magnitude less than the mass that can be carried by an equivalent piece of silk or steel.

You might find it surprising that A36 steel and spider silk have the same yield point!

Making sense of the result

The yield point for a human tendon is about 1/10 that for steel, so this result makes perfect sense. In fact, most human tendons are smaller than the one used in this example. The mean diameter of an Achilles tendon is approximately 5 mm and on average supports a load of approximately 60 kg. The piece of tendon in this example has a diameter that is 8 times as much. The load it can carry is ~64 times 60 kg, as expected.

If the material is strained beyond the yield point, the material does not return to its original length when the stress is removed. Past the yield point, the material undergoes **plastic deformation**, which is not completely reversible. The material has a **permanent set**, meaning that permanent deformation has occurred.

The greatest stress value on the graph of stress versus tensile strain is called the **ultimate tensile strength** (UTS). Increasing the strain beyond the UTS point fractures the object.

Tension and compression forces act in a direction parallel to the length of the object (perpendicular to the cross section), as shown in Figure 10-28(a) and (b), while shear forces act perpendicular to the length of the object (parallel to the cross section), as shown in Figure 10-28(c). The deformation of the object along the direction of the shear force is given by

KEY EQUATION

$$\frac{F}{A} = G\frac{\Delta L}{L} \tag{10-11}$$

The constant of proportionality between the shear stress and the strain is called the **shear modulus**, G.

10-6 Ductile and Brittle Materials

If a significant amount of plastic deformation occurs before the material fractures, the material is called a **ductile** material. If fracturing occurs shortly beyond the yield point, the material is called **brittle** (Figure 10-33). For example, the harder steel alloys tend to be more brittle, and the softer alloys tend to be more ductile.

CHECKPOINT 10-7

Ductile Metals

A ductile metal
(a) is in general harder than a brittle metal
(b) is not elastic
(c) displays more plastic behavior before fracture than a brittle material
(d) none of the above

ANSWER: (c) A ductile metal undergoes a more significant amount of plastic deformation before it fractures than a brittle metal.

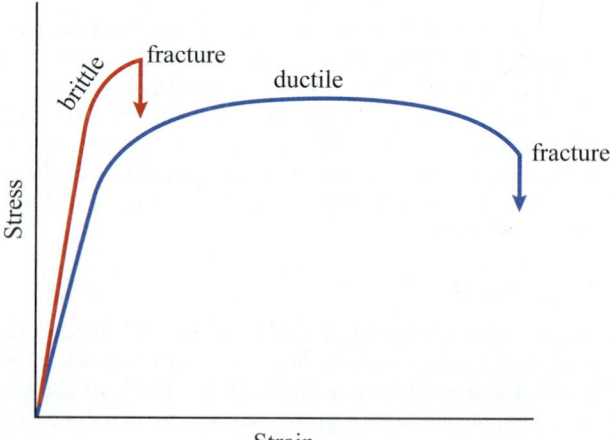

Figure 10-33 Ductile versus brittle behaviour. Stress builds more rapidly in the brittle material and it fractures at a lower strain than that for a ductile material.

MAKING CONNECTIONS

Is Silk Stronger Than Steel?

In the previous example, we saw that spider silk and steel have similar yield points and thus are able to support similar loads without experiencing permanent deformation. However, steel and spider silk have quite different UTSs. The UTS for many kinds of steel is of the order of 500 MPa, while some spider silk threads have a UTS upward of 1.5 GPa. Thus, a strand of spider silk can ultimately support, in tension, three times the weight that can be supported by a strand of steel of the same size. At values close to this tension it will have experienced permanent deformation.

MAKING CONNECTIONS

Titanic Steel

There is evidence that the steel used in the hull of the *Titanic*, while ductile at room temperature, might have become brittle at lower temperatures. This loss of ductility may have substantially increased the size of the fracture in the hull when the ship hit the iceberg in the icy Atlantic waters.

Failure Modes in Compression and Tension

Whether under compression or tension, concrete and stone structures will fail by fracturing, as shown in Figure 10-34(a). Under sufficient tension, a steel beam will eventually snap in two, as shown in Figure 10-34(b). Steel under compression has several possible failure modes. Brittle steel will likely crack, as shown in Figure 10-34(c). More ductile steel will bulge out, as shown in Figure 10-34(d), and plastic deformation

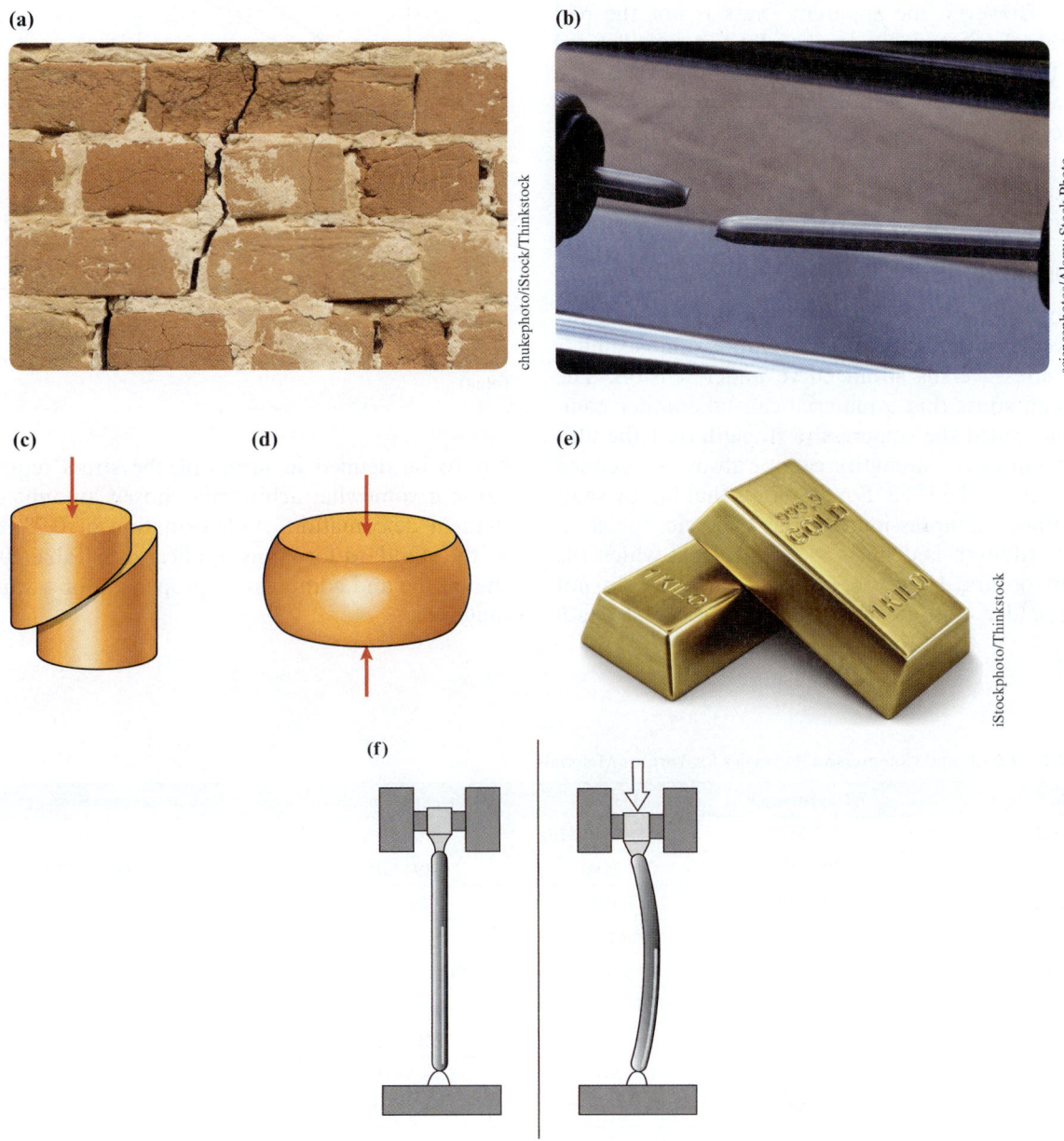

Figure 10-34 (a) Concrete failure by fracture. (b) Failure of steel under tension. (c) Failure of brittle steel by cracking. (d) Ductile failure. (e) Ductile failure used in stamping. (f) Buckling of steel.

will take place, a process called **ductile failure**. Ductile failure is applied in fabricating processes such as extruding and stamping metals, as shown in Figure 10-34(e). However, a steel beam is more likely to buckle under compression, as shown in Figure 10-34(f), before reaching its compressive strength limit and undergoing ductile failure.

Beyond the UTS, a steel bar, for example, will experience what is called necking before it eventually snaps (Figure 10-35). During **necking**, the cross section of the material shrinks as it elongates due to plastic deformation past the UTS.

Apparent stress, also called **engineering stress**, is defined as the force per unit area of the original cross section. However, the apparent stress is not the real stress that the material undergoes. **True stress** is defined as the force divided by the actual cross section as the material undergoes deformation. True stress continues to increase until fracture, and apparent stress may drop somewhat, as shown in Figure 10-36. Depending on the application, stress versus strain curves may show either the true stress or the apparent stress.

Maximum Tensile and Compressive Strength

For most metals (for example, steel and copper), the stress versus strain curve under compression is similar to the stress versus strain curve under tension. The maximum stress that a material can take under compression, called the **compressive strength** (not the ultimate compressive strength), cannot always be defined as precisely as the UTS. For materials that fail by shattering under compression, such as concrete, the compressive strength is defined as the strain at which the fracture occurs. However, for materials that do not shatter under compression, the compressive strength

Figure 10-35 A metal strip experiences necking before snapping.

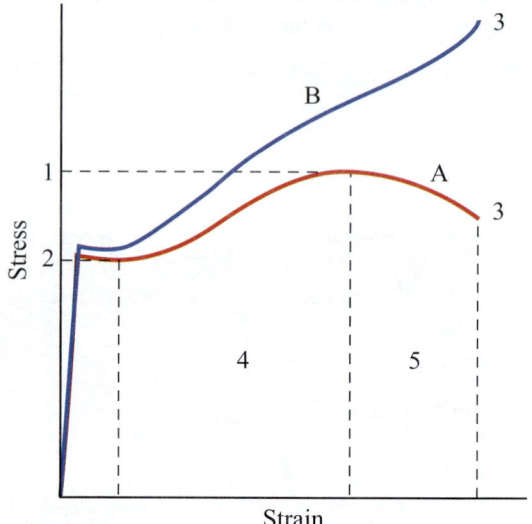

Figure 10-36 Apparent (engineering) stress (red) versus true stress (blue).

has to be defined in terms of the stress required to cause a somewhat arbitrarily chosen amount of permanent deformation. A deformation of 0.2% is used as the standard for many applications. Table 10-1 lists the tensile and compressive properties for a variety of materials.

Table 10-1 Tensile and Compressive Properties for Various Materials

Material	Yield Strength	UTS	Young's Modulus	Compressive Strength
A36 steel	250 MPa	400 MPa–550 MPa	200 GPa	152 MPa
Stainless steel	290 MPa	579 MPa	193 GPa	170 MPa A316L
Aluminum	95 MPa	110 MPa	69 GPa	
Rubber		15 MPa		
Concrete		5 MPa	30 GPa (compression)	14 MPa–42 MPa
Wood		~88 MPa–150 MPa	6 GPa–15 GPa	~14 MPa–50 MPa (parallel to grain)
Tendon			0.6 GPa–1.8 GPa	
Hair	~200 MPa	380 MPa	~5 GPa	
Spider silk	225 MPa	Up to 1.9 GPa	1.5 GPa	
Sandstone		4 MPa–25 MPa	1 GPa–20 GPa	20 MPa–170 MPa
Granite		7 MPa–25 MPa	10 GPa–70 GPa	100 MPa–250 MPa
Femur bone (along axis)	~100 MPa	135 MPa	17.4 GPa	190 MPa–205 MPa

EXAMPLE 10-8

The Strength of Human Bone

Young's modulus for the human femur bone for longitudinal stress is approximately 17.0 GPa, and the compressive strength is approximately 190.0 MPa. The average length of a female femur is 48.0 cm, and the average diameter is 25.4 mm.

(a) Compare the maximum mass that can be carried by the femur to the maximum mass that can be carried by a column of concrete that has the same dimensions and a compressive strength of 42.0 MPa.

(b) Assuming a linear relation between stress and strain until the point of maximum compression, estimate the compressive strain at which the femur will fracture.

Solution

The femur and the column of concrete are both under tension. The tension force at their bottom is equal to the weight of the load. The maximum stress each object can support will give us the maximum tension force on each object and hence the maximum weight each object can support. The FBD for this scenario is very much like that depicted in Figure 10-32.

(a) For a given stress, the force is given by Equation 10-7:

$$F = \sigma A$$

The force in this context is the weight carried by the material. So,

$$M = \frac{\sigma A}{g}$$

Since we are looking at the maximum mass, we use the maximum value of the stress, which is the compressive strength.

For the femur:

$$M = \frac{\sigma A}{g} = \frac{(1.90 \times 10^8 \text{ Pa})\pi(0.0127 \text{ m})^2}{9.81 \text{ m/s}^2} = 9.81 \times 10^3 \text{ kg}$$

For concrete:

$$M = \frac{\sigma A}{g} = \frac{(4.20 \times 10^7 \text{ Pa})\pi(0.0127 \text{ m})^2}{9.81 \text{ m/s}^2} = 2.17 \times 10^3 \text{ kg}$$

(b) Using Equation 10-9 or Equation 10-10,

$$\sigma = Y\varepsilon \quad \text{or} \quad \frac{F}{A} = \frac{Y\Delta L}{L}$$

$$\Rightarrow \varepsilon = \frac{\sigma}{Y} = \frac{190. \text{ MPa}}{17.0 \text{ GPa}} = \frac{1.90 \times 10^8 \text{ Pa}}{1.70 \times 10^{10} \text{ Pa}} = 1.12\%$$

Making sense of the result

The bone will compress by 1.12% of the length at fracture. Since the tensile strain of bone at UTS is slightly over 1%, the value makes sense.

Steel Grades

Steel is an alloy composed primarily of iron. Depending on the desired properties, steel contains various amounts of other elements, such as carbon, manganese, chromium, nickel, copper, aluminum, boron, and molybdenum. For example, 18-8 stainless steel, often used for quality cookware, contains 18% chromium and 8% nickel by mass. All types of stainless steel contain at least 10.5% chromium by mass.

The higher the carbon content of a steel alloy, the greater the strength, hardness, rigidity, and brittleness of the alloy. The carbon content of steel alloys is generally between 0.2% and 2.1%. Alloys with

Figure 10-37 A36 steel beams.

carbon content greater than 2.1% are called cast iron. Such alloys are easy to cast but rather brittle; a hard impact can crack or shatter cast iron. The alloy used for wrought iron contains minimal carbon and is quite malleable, making it easy to work but less strong.

A36 is an ASTM (American Society for Testing Materials) designation for a steel alloy widely used for construction and structural supports in the United States and Europe (Figure 10-37). The number 36 comes from the alloy's yield point of 36 000 psi, which is equivalent to 250 MPa. A36 was the preferred grade for construction steel for a long time, but it has now given way to better-performing alloys. Canada's version of A36 steel

is an alloy designated 44W. This alloy has slightly more desirable properties for construction, including greater yield strength and more elongation before snapping.

Steel will eventually fracture once the UTS is reached. The amount by which an element stretches before snapping is called **elongation at fracture**. A 5 cm long piece of A36 steel will elongate by a minimum of 23% of its original length before fracturing, and a 20 cm long piece will stretch by 20% before fracture.

We observe similar effects under compression. The stress at which a steel beam buckles also depends on its length and cross section. The buckling stress for a cylindrical A36 steel beam with a length 10 times its radius is approximately 145 MPa, but a cylindrical A36 steel beam with a length 100 times its radius buckles at as little as 52 MPa under stress.

CHECKPOINT 10-10

A36 Steel

A36 steel has
(a) a UTS of 36 000 psi
(b) a yield point of 36 000 psi
(c) a UTS of 36 MPa
(d) a yield point of 36 MPa

ANSWER: (b) The yield point of A36 steel is 36 000 psi.

KEY CONCEPTS AND RELATIONSHIPS

The Conditions of Equilibrium

When an object is in equilibrium, the vector sum of all the external forces acting on the object must be zero:

$$\sum \vec{F} = 0 \tag{10-1}$$

When an object is in equilibrium, the vector sum of all the external torques acting on the object about its centre of mass or any other given point must be zero:

$$\sum \vec{\tau} = 0 \tag{10-2}$$

The linear momentum of an object in equilibrium is constant:

$$\frac{d\vec{p}}{dt} = 0 \tag{10-3}$$

The angular momentum of an object in equilibrium about a given point is constant:

$$\frac{d\vec{L}}{dt} = 0 \tag{10-4}$$

Centre of Gravity

The centre of gravity of an object coincides with the centre of mass in a uniform gravitational field.

Applying the Conditions for Equilibrium

Keep the following guidelines in mind while tackling equilibrium problems.

1. Look for points on the object where a single unknown force produces a torque, and sum the torques about that point to find the unknown force.
2. Remember that the torque equilibrium condition is valid for any point on the object and can be used multiple times in any solution.
3. As you progress through the solution, monitor to see if you have a single unknown force/force component in any direction. Summing the forces in that direction will tell you the value for that unknown immediately.

Working with Unknown Forces

We often encounter forces in equilibrium problems whose magnitude and direction are unknown. Spending a lot of time guessing the likely direction of such unknown forces is not necessary. Introduce unknown support forces into your FBD and your solutions through the use of their vector components, which we treat as unknown variables that may be either positive or negative. For example, if you decide to call the unknown support force $\vec{A}$, introduce it into your solution in terms of its components A_x and A_y. Solve the equations you get from applying the equilibrium conditions, and the solutions will tell you what A_x and A_y are, and thus you will know both the magnitude and the direction of the unknown force. It is important to remember that when we use A_x and A_y in subsequent calculations, we use the value we obtain from our equations. Specifically, it we find one of them to be negative, we use that negative value in subsequent calculations that require it. A negative value for an unknown force initiates we have guessed the wrong direction for that force.

Elasticity

Stress, σ, is the force per unit area:

$$\sigma = \frac{F}{A} \tag{10-7}$$

Strain is the fractional change in the length of an object in response to stress:

$$\varepsilon = \frac{\Delta L}{L} \tag{10-8}$$

The linear part of an elastic deformation is characterized by a linear relation between stress and strain:

$$\sigma = Y\varepsilon \tag{10-9}$$

or

$$\frac{F}{A} = Y\frac{\Delta L}{L} \tag{10-10}$$

Young's modulus is the same for compressive and tensile stress for most applications involving elastic materials.

The response of certain materials to tension can differ drastically from the materials' response to compression.

The deformation of an object along the direction of a shear force is given by

$$\frac{F}{A} = G\frac{\Delta L}{L} \tag{10-11}$$

Plastic Deformation and Failure Modes

The yield point is the stress beyond which deformation is irreversible. Plastic deformation occurs beyond the yield point.

Ductile and Brittle Materials

Ductile materials endure significant plastic deformation before fracture. Brittle materials fracture shortly past the yield point.

Materials can fail by fracture, buckling, or ductile failure.

APPLICATIONS

earthquake-resistant structures, construction materials, cast iron and stainless steel

KEY TERMS

apparent stress, brittle, centre of gravity, compression force, compressive strength, ductile, ductile failure, dynamic equilibrium, elastic deformation, elastic limit, elongation at fracture, engineering stress, equilibrium, necking, neutral equilibrium, permanent set, plastic deformation, proportional limit, shear force, shear modulus, stable equilibrium, static equilibrium, strain, stress, tension, true stress, ultimate tensile strength, unstable equilibrium, yield point, yield strength, Young's modulus

QUESTIONS

1. Will a ball thrown straight up be in equilibrium at any point while in the air? In particular, will it be in equilibrium when it reaches its maximum height?
2. Is a car moving on level ground at a constant speed at equilibrium?
3. A soccer ball rolls downhill on a sticky surface, which causes it to have a constant speed. Which of the following is true?
   (a) The ball cannot be in equilibrium because it is going downhill.
   (b) The sum of the forces on the ball is never zero.
   (c) The ball is in equilibrium.
   (d) None of the above.
4. A particle's position as a function of time is given by $x = x_0 \cos \omega t$. Which of the following statements is true?
   (a) The particle is in equilibrium when $x = x_0$.
   (b) The particle is in equilibrium when $x = 0$.
   (c) The particle is never in equilibrium.
   (d) None of the above are true.
5. When you are standing in an elevator moving straight up at a constant speed, you are
   (a) not in equilibrium because the elevator is moving
   (b) not in equilibrium because you are moving

   (c) in equilibrium because your velocity is constant
   (d) in equilibrium because your acceleration is constant
6. You step onto a platform resting on top of a spring with spring constant k. The spring is fixed to the ground and is then compressed until it comes to a momentary stop. Which of the following statements is true?
   (a) You are in equilibrium throughout.
   (b) You are only in equilibrium when you stop momentarily.
   (c) You are in equilibrium when you are halfway to the point where you stop momentarily.
   (d) You are never in equilibrium during this process.
7. With one hand, you hold one end of a string fixed. With the other hand, you hold a small mass, which is attached to the other end of the string. You are holding the string so it is initially horizontal. Then you let the mass go. The mass
   (a) is in equilibrium
   (b) is in equilibrium at the bottom of its trajectory
   (c) is in equilibrium when it stops momentarily at either end of its trajectory
   (d) none of the above

8. A rocket is launched and accelerates vertically, lifted by the constant thrust of its engine. When the engine runs out of fuel, the rocket begins to slow down under the effect of the force of gravity. Which of the following statements is true?
   (a) After ignition, the rocket is in equilibrium at only one point in its journey upward.
   (b) The rocket is in equilibrium at two points after ignition.
   (c) The rocket is only in equilibrium when it reaches its maximum height and stops momentarily.
   (d) The rocket is in equilibrium as long as it is on the ground, and again at another point between liftoff and maximum height.
   (e) None of the above are true.

9. Is an object that is undergoing uniform circular motion in equilibrium?

10. A disk is spinning about its centre-of-mass axis at a constant angular speed. Which of the following statements is true?
   (a) The disk is in equilibrium.
   (b) Any given point on the disk is not in equilibrium.
   (c) Both (a) and (b) are true.
   (d) None of the above are true.

11. Can an object's centre of gravity lie outside the physical dimensions of the object? Explain.

12. A ball rolls downhill on a rough grassy surface at a constant speed. Aside from the centre of mass of the ball, is any other point on the ball in equilibrium?

13. Two Earth-sized planets orbiting the same star are r and $2r$ from the centre of the star. Find the centre of mass of the planets when the three objects are collinear, with both planets on the same side of the star.

14. Why do you have to lean forward when walking against the current in a river?

15. Does doubling the cross section of a steel rod increase its Young's modulus?

16. Is bone stronger than concrete? Justify your answer.

17. Which is stronger, steel or human hair? Justify your answer.

18. Which of the following statements is true?
   (a) Steel is mined in steel mines.
   (b) Steel is an alloy that can contain any amount of carbon.
   (c) The amount of carbon in steel alloys ranges between 0.2% and 2.1%; the higher the carbon content, the stronger but more brittle the steel.
   (d) High-carbon steel is more ductile than medium- and low-carbon steel.

19. Does concrete behave better in compression or in tension?

20. The compressive strength for a material that fails by shattering is the strain in which
   (a) you will get the original length back if you remove the load
   (b) the material fractures
   (c) a permanent set of 0.2% occurs
   (d) none of the above

21. The yield point for a ductile material is
   (a) the highest stress for which the material will return to its original length after the stress has been removed
   (b) the stress for which the material will return to within 0.2% of its original length after the stress has been removed
   (c) both (a) and (b)
   (d) neither (a) nor (b)

22. Ductile failure under compression is
   (a) plastic deformation in which a material shrinks in the direction of an applied force and expands along the transverse direction
   (b) deformation similar to the deformation seen when a train runs over a coin
   (c) actually "ducktile" failure, because the material makes a distinct "quack" sound a bit like that of a duck
   (d) both (a) and (b)

23. Which of the following statements is true? (Pick all that apply.)
   (a) Steel goes into plastic behaviour before reaching its UTS.
   (b) Steel goes into plastic behaviour after reaching its UTS.
   (c) Steel will begin to fail after reaching its UTS.
   (d) None of the above are true.

24. Under compression (pick all that apply),
   (a) ductile steel fails by cracking
   (b) ductile steel fails by buckling
   (c) ductile steel fails by undergoing ductile failure
   (d) Any of the above could be true.

25. A brittle steel beam can fail by
   (a) ductile failure
   (b) fracturing
   (c) buckling
   (d) none of the above

26. Which of the following statements is true?
   (a) Steel behaves in a similar way under both compression and tension.
   (b) The compressive yield point of steel is often slightly higher than its tensile yield point.
   (c) Steel will most likely buckle before reaching its compression strength.
   (d) Only (a) and (c) are true.

27. The proportional limit is
   (a) the highest point at which the stress versus strain curve is linear
   (b) the point at which a material goes into plastic deformation
   (c) outside the elastic limit
   (d) none of the above

PROBLEMS BY SECTION

For problems, star ratings will be used (★, ★★, or ★★★), with more stars meaning more challenging problems. The following codes will indicate if $\frac{d}{dx}$ differentiation, $\int$ integration, ⬚ numerical approximation, or 〰 graphical analysis will be required to solve the problem.

Section 10-1 The Conditions for Equilibrium

28. ★ Consider the simplified model of a human arm shown in Figure 10-38. What must be the force, $\vec{T}$, applied by the muscle when the arm is in static equilibrium? Treat the elbow as the pivot, and assume that all forces are perpendicular to the arm as shown. Use the following data: $L = 5.00$ cm, $c = 15.0$ cm, $d = 30.0$ cm, the weight of the forearm alone is 20.0 N, and the weight of the moving mass is 10.0 N.

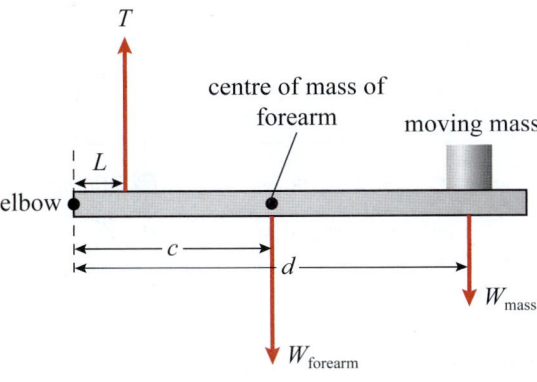

Figure 10-38 Problem 28.

29. ★ A 5.0 kg mass hangs vertically from a rope wound around a pulley, as shown in Figure 10-39. The other end of the rope is tied to a 12 kg mass sitting on the horizontal surface. Find the coefficient of static friction between the second mass and the surface it sits on, given that the mass is on the verge of slipping. Assume that the mass of the pulley is negligible.

Figure 10-39 Problem 29.

30. ★ A 10.0 kg mass hangs from a rope, as shown in Figure 10-40. The rope passes over a fixed pulley, then under a movable pulley, and is attached to the ceiling as shown. The movable pulley has a mass m attached to it and moves at a constant speed. Find the tension in the rope. Assume that the masses of the pulleys are negligible.

Figure 10-40 Problem 30.

31. ★ A 65 kg climber in a slippery crevice wedges a light, adjustable bar between the two icy, vertical walls (Figure 10-41). What must the compression force in the bar be to support the climber's weight? The coefficient of static friction between the bar and the walls is 0.12.

Figure 10-41 Problem 31.

32. ★ While climbing a chute in a cave, you can press against the opposite walls with a perpendicular force of 150.0 N for each hand. Your mass is 65 kg. The coefficient of static friction between your boots and the wall is 0.56, and that between your gloves and the wall is 0.39. What perpendicular force must you exert against the wall with each leg to avoid falling into the abyss? Assume your torso is always parallel to the vertical walls while climbing.

33. ★ The bridge in Figure 10-42 is supported at its end by forces $\vec{F}_1$ and $\vec{F}_2$. Solve for these forces, given that the mass of the bridge is 14.0 tonnes, its length is 5.00 m, and its mass is distributed uniformly. (1 tonne is 1000 kg.)

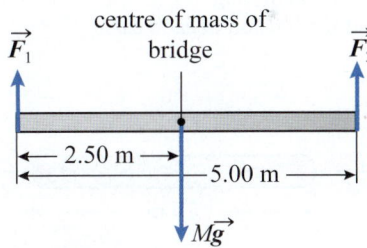

Figure 10-42 Problem 33.

34. ★ The 62.0 kg athlete in Figure 10-43 hangs from a 3.00 m long horizontal bar. The athlete is one fifth of the length from one end. Find the support force at each end of the bar, assuming that the mass of the bar is negligible.

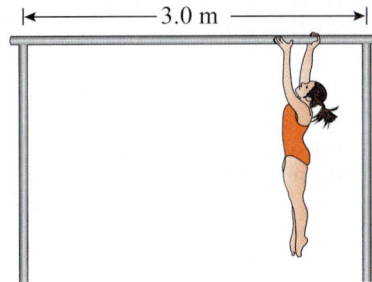

Figure 10-43 Problem 34.

35. ★★ A child rests two branches against each other, as shown in Figure 10-44. There is enough friction between the branches and the ground to prevent them from slipping. The longer of the two branches is 1.6 m in length and has a mass of 4.0 kg; the shorter branch has a mass of 2.7 kg. The shorter branch makes an angle of 67° with the horizontal ground and an angle of 105° with the longer branch. Assume that the two branches are uniform and have circular ends. What must the coefficient of static friction between the two branches be so that the structure does not collapse?

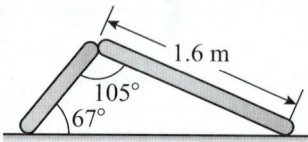

Figure 10-44　Problem 35.

36. ★ The 110.0 kg sign in Figure 10-45 is suspended by two cables as shown. Find the tension in each cable.

Figure 10-45　Problem 36.

37. ★ The 300.0 kg horizontal bar in Figure 10-46 is supported by two cables as shown.
  (a) Find the centre of mass of the bar, given that the bar is 7.00 m long
  (b) Find the tension in each cable.

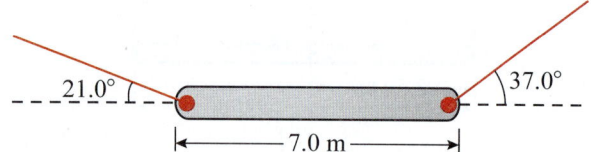

Figure 10-46　Problem 37.

38. ★ A person's leg is suspended to lie horizontally, as shown in Figure 10-47. All lengths shown are measured from the left edge of the leg, which is supported by the rest of the body. The mass of the leg alone is 6.50 kg, and its centre of mass is a distance of 0.450 m from the left end. A 2.00 kg mass is attached to the leg 0.750 m from the left end, as shown. The leg is supported by the person's body through a vertical force $\vec{F}_p$ and a tension $\vec{T}$ in a vertical cord attached at a distance of 0.650 m. Treat the left edge of the leg as the pivot.
  (a) Find the torque due to the weight of the leg alone.
  (b) Find the torque due to the 2.00 kg mass alone.
  (c) Find the magnitude of the tension in the cord, T.
  (d) Find the magnitude of the vertical force, F_p, exerted at the left end of the leg.
  (e) If the 2.00 kg mass were moved farther to the right, would the required tension increase or decrease?

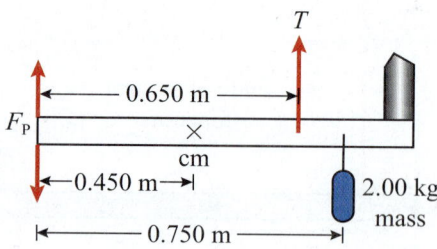

Figure 10-47　Problem 38.

Section 10-2　Centre of Gravity

39. ★★ A 75.0 kg passenger on a cruise ship leans forward into a strong horizontal headwind. When her centre of gravity is 24.0 cm in front of her ankles, she does not have to exert any effort to keep her balance. Her centre of gravity is 110.0 cm above the deck when she stands upright. What force does the wind exert on this passenger?

40. ★★ Show that the centre of mass coincides with the centre of gravity for an object if the acceleration due to gravity is constant over the dimensions of the object. (Hint: Add the torques due to the point masses that make up the object about a point outside the object.)

Section 10-3　Applying the Conditions for Equilibrium

41. ★★ For the 20.0 kg sphere shown in Figure 10-48, find the normal forces from each surface. Assume that all the surfaces are frictionless.

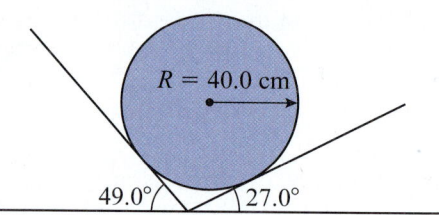

Figure 10-48　Problem 41.

42. ★★ Determine the tension in the rope in Figure 10-49. The beam is 4.0 m long, has a uniformly distributed mass of 210 kg, and is supported by a free hinge at one end and a rope at the other end, making an angle of 46° with the horizontal.

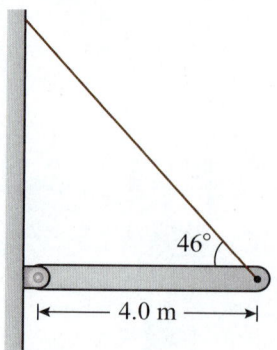

Figure 10-49　Problem 42.

43. ★★ Determine the tension in the horizontal rope supporting the beam in Figure 10-50. The 320 kg beam is uniform and is 2.4 m long.

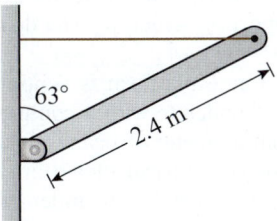

Figure 10-50 Problem 43.

44. ★ If the system in Figure 10-51 is in equilibrium, how are the masses m_1 and m_2 related? Assume that the rope and pulleys have negligible mass.

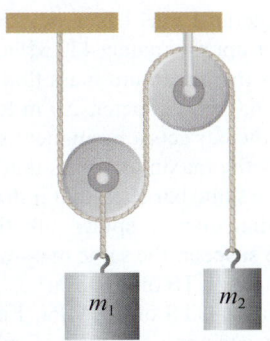

Figure 10-51 Problem 44.

45. ★ The 4.00 kg monkey in Figure 10-52 hangs from a rope that is wound around a fixed pulley and around a pulley attached to a banana crate, as shown. The crate is on the verge of moving. Find the tension in the rope. Assume that the masses of the pulleys are negligible.

Figure 10-52 Problem 45.

46. ★★ The worker in Figure 10-53 is trying to push the 120.0 kg wheel over a curb that is 20.0 cm high. What force must he apply horizontally if he pushes on the wheel at the point shown in the figure, 10.0 cm below the top of the wheel? The wheel has a radius of 40.0 cm.

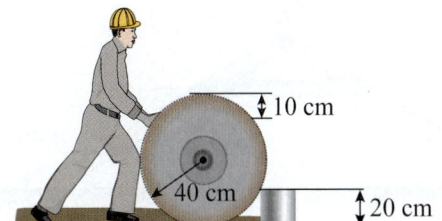

Figure 10-53 Problem 46.

Section 10-4 Applying the Conditions for Equilibrium: Working with Unknown Forces

47. ★★ A winch used to lift a car (mass 1350 kg) is supported by a cable, as shown in Figure 10-54. The 450 kg mass of the winch arm is uniformly distributed. Find the tension in the supporting cable and the forces at the hinge.

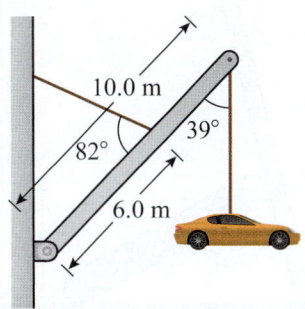

Figure 10-54 Problem 47.

48. ★★★ The mass of the upper beam in Figure 10-55 is 2100 kg and that of the lower beam is 700.0 kg, with both masses uniformly distributed. Find the horizontal and vertical action-reaction forces at the hinge at C and at the free hinges at A and B.

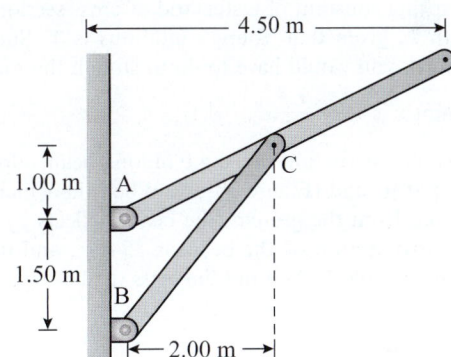

Figure 10-55 Problem 48.

49. ★★★ You are designing the crosspiece for the A-frame structure in Figure 10-56. Beams AB and AC are 4.00 m long and have a uniformly distributed mass of 300.0 kg each. How much tension must the crosspiece EF withstand? Assume that the mass of the crosspiece and the friction at points B and C are negligible.

CHAPTER 10 | **EQUILIBRIUM AND ELASTICITY**

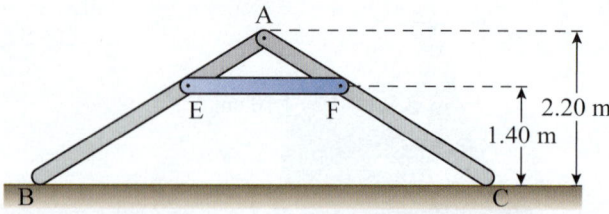

Figure 10-56 Problem 49.

50. ★★★ Each uniform beam in Figure 10-57 has a mass of 1100 kg. The length of the shorter beam is 4.0 m. The assembly is in equilibrium. Find the force of friction on each beam, and the horizontal and vertical forces exerted by one beam on the other.

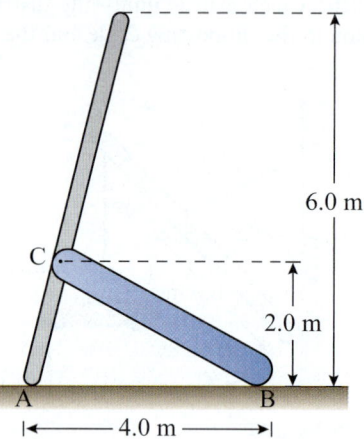

Figure 10-57 Problem 50.

Section 10-5   Deformation and Elasticity

51. ★ Below the proportional point, within the elastic limit, a steel rod can be considered like a spring. Find the effective spring constant of a steel rod of cross section A and length L, given that Young's modulus is Y. Show that the work you would have to do to stretch the rod by an amount x is $W = \dfrac{YA}{2L}x^2$.

52. ★ At a construction site, a 4.0 m long beam supports a winch at its end (Figure 10-58). When the winch lifts a crate up from the ground, the beam deflects by 13 cm. The cross section of the beam is 25 cm², and its shear modulus is 190 MPa. Find the mass of the crate.

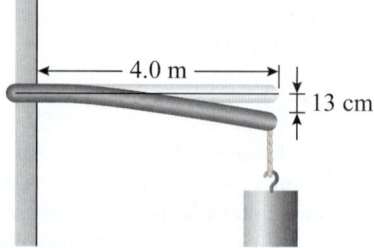

Figure 10-58 Problem 52.

53. ★ Vertebrate muscle typically produces approximately 25 N of tensile force per square centimetre of cross-sectional area. Calculate the stress.

54. ★★ The shear strength of limestone is approximately 25 MPa. Find the maximum load that can be supported at the end of a limestone beam that has a square cross section of 100.0 cm² and extends horizontally 2.3 m from its supporting structure. The density of limestone is 2500 kg/m³.

55. ★★ A 22.0 tonne platform is supported by four vertical beams of concrete with a compressive strength of 42 MPa. What diameter of beams must be used?

56. ★ A 3.0 m long cylindrical alloy beam is 21 cm in radius. It is found to stretch by 1.5 cm under a tension of 2.0 kN. Find its Young's modulus.

57. ★★ Derive an expression for the elastic potential energy stored in a metal rod in terms of its length (L), cross-sectional area (A), Young's modulus (Y), and elongation (ΔL) within the elastic limit.

Section 10-6   Ductile and Brittle Materials

58. ★★ A36 structural steel has a yield point of 250 MPa and a UTS of approximately 450 MPa.
   (a) What is the maximum mass that you can suspend from a 6.0 cm diameter, 2.0 m long A36 steel bar before the bar gets a permanent set?
   (b) What is the maximum mass that you can suspend from the same bar such that it does not break?
   (c) What diameter of spider silk thread would you need to suspend the same mass without the thread reaching its UTS of 1.0 GPa?

59. ★ Rubber has a UTS of 15 MPa. Find the maximum force that you can apply to a strip of rubber 10.0 cm long with a cross section of 5.0 mm². Assume that the cross section of the rubber strip does not change appreciably as you pull on it.

60. ★★ What proportion of the yield strength of bone is the stress on the femur of an 82 kg person standing on one leg? The yield strength of the human femur bone for longitudinal stress is approximately 100.0 MPa, and its average diameter is 25.4 mm.

61. ★★ If the concrete in a concrete-only building were replaced with enough dinosaur bone to have the same strength, how much lighter would the building be? What would the change be if the concrete were replaced with steel instead? The compressive strengths for concrete, bone, and steel are approximately 40.0 MPa, 200.0 MPa, and 150.0 MPa, respectively. The density of concrete is approximately 2400 kg/m³, and that of steel is 8050 kg/m³. Assume the average bone in the structure has a length of 48 cm, an average diameter of 25 mm, and a mass of 260 g.

62. ★★ A shipment of limestone blocks has a density of 2700 kg/m³ and a compressive strength of 250 MPa. The blocks are cubical with a side length of 30.0 cm. How many of these blocks can you stack up before the bottom one cracks?

63. ★★ Some types of granite have a Young's modulus of 70 GPa and a compressive strength of 250 MPa. Assume linear behaviour until the compressive strength point.
   (a) Find the amount by which granite compresses before shattering.
   (b) How much weight can a 1.0 m long, 20 cm radius granite beam take in compression?

64. ★★★ A sample of basalt has a compressive strength of 300.0 MPa and a Young's modulus of 73 GPa.
 (a) Assuming linear stress versus strain behaviour, estimate the amount of work it would take to break a column of basalt 2.0 m in length and 49 cm in diameter.
 (b) What is the minimum mass of the column that you can drop from a height of 20.0 m to cause it to shatter?

65. ★ The UTS of concrete is 3.0 MPa, and its compressive strength is 42 MPa. Find the maximum weight that can be supported by a concrete pillar 50.0 cm in diameter and 4.0 m long under both tension and compression.

COMPREHENSIVE PROBLEMS

66. ★★ The 900.0 kg, 3.70 m long uniform beam in Figure 10-59 is pivoted at one end, and the other end rests on a sloped, frictionless incline. A 200.0 kg mass hangs from the beam 3.00 m from the hinge. Determine the normal force from the surface on the beam and the support forces at the hinge.

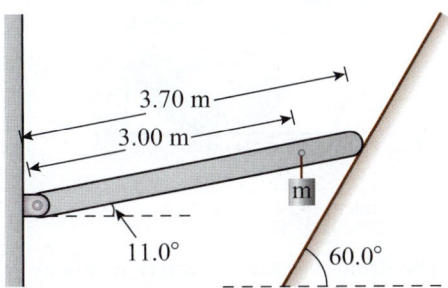

Figure 10-59   Problem 66.

67. ★★★ A disk rests on two other disks, as shown in Figure 10-60. All three disks are the same size, and each has a mass of 600.0 kg. There is enough friction among the disks and between the lower disks and the floor to prevent them from rolling apart.
 (a) Find the magnitudes of the normal force and the force of friction between the top disk and one of the lower disks when the lower disks are on the verge of moving apart.
 (b) Find the magnitudes of the force of friction between each lower disk and the floor, and the normal force from the floor on each disk.

Figure 10-60   Problem 67.

68. ★ In Figure 10-61, a sphere of radius R is placed on a frictionless slope. A rope of length $6R$ is attached to the top of the sphere. The other end of the rope is attached to the slope above the sphere. What angle does the rope make with the slope when the sphere reaches a position of stable equilibrium?

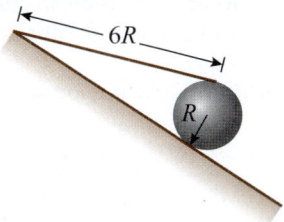

Figure 10-61   Problem 68.

69. ★★ A 125 kg wheel with a radius of 2.50 m is being pulled up an incline by a 550 kg horse (Figure 10-62). There is a 45.0 cm high protrusion in the incline, as shown. What is the tension in the rope when the wheel is just beginning to lift off the incline to go over the protrusion? What is the minimum possible coefficient of static friction between the horse and the incline for the horse to be able to pull the wheel over the protrusion?

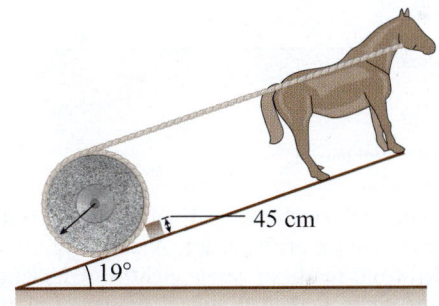

Figure 10-62   Problem 69.

70. ★ The 3.0 kg mass in Figure 10-63 rests on a horizontal surface. The mass is attached to one end of a rope, most of which is wound around a drum of radius R on the side of a pulley, as shown. Another rope is wound around the pulley, which has a radius of $2R$. The coefficient of static friction between the mass and the surface is 0.35. What is the largest mass that can hang from the vertical rope without causing the pulley to rotate?

Figure 10-63   Problem 70.

CHAPTER 10 | **EQUILIBRIUM AND ELASTICITY**

71. ★★★ The 80.0 kg painter in Figure 10-64 is using a large stepladder. The base of the ladder is 3.00 m wide, and the ladder is 7.00 m high. The painter is on a step that is 2.0 m above the floor. The horizontal bar of the ladder has a mass of 10.0 kg, and the entire ladder has a mass of 60.0 kg. Assume that the floor is frictionless.

(a) Find the force of tension in the horizontal bar and the normal forces that the floor exerts on the ladder.

(b) Find the action-reaction forces at the hinges A, B, and C.

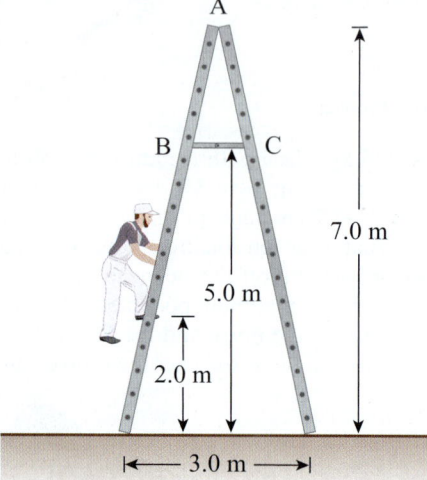

Figure 10-64 Problem 71.

72. ★★ The uniform beam in Figure 10-65 is 3.4 m long and has a mass of 760.0 kg. A 220.0 kg crate is suspended from the beam seven eighths of its length from the free hinge at the left end of the beam. A rope supports the beam as shown. Find the tension in the rope and the components of the support force at the hinge.

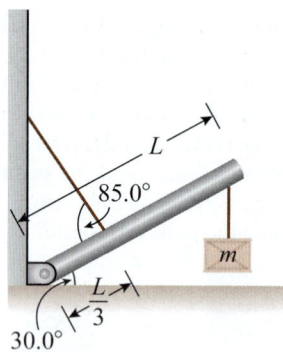

Figure 10-65 Problem 72.

73. ★★ A uniform stick of mass m and length L is supported at one end by a string, and the other end leans against a wall, as shown in Figure 10-66. The string makes an angle of θ with the wall, and the stick is horizontal. Derive expressions for the force of friction on the stick, and the minimum coefficient of static friction that will stop the stick from sliding down the wall.

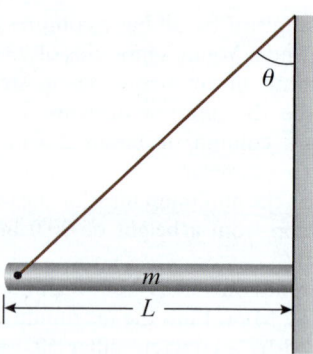

Figure 10-66 Problem 73.

74. ★★★ The sphere in Figure 10-67 rests on a frictionless floor. At the point of contact between the sphere and the wall, the coefficient of static friction is μ. Derive an expression for the maximum height d at which a horizontal force $\vec{F}$ on the sphere will not cause it to spin.

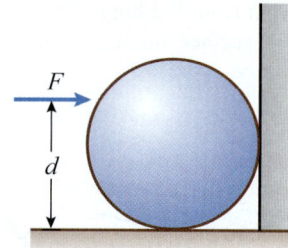

Figure 10-67 Problem 74.

75. ★★ The drum-and-axle assembly in Figure 10-68 has mass M. The radius of the drum is R, and the radius of the axle is r. You pull on a rope wound around the axle with a force $\vec{F}$ parallel to the surface of the incline. The drum does not move, but it is about to slide up the incline. Derive an expression for the coefficient of static friction between the incline and the drum.

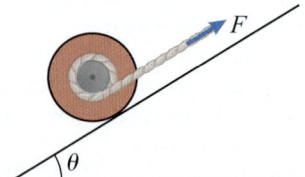

Figure 10-68 Problem 75.

76. ★★★ You pull the rope wound around the axle of the drum-and-axle assembly in Figure 10-69 with a force of 120.0 N directed horizontally as shown. The drum is trapped in a wedge-shaped opening. The coefficient of static friction between the drum and the inclined surface is 0.2000. The mass of the drum is 4.000 kg. The axle radius is 20.00 cm, and the drum radius is 50.00 cm. The drum is on the verge of spinning in place. Find the coefficient of static friction between the drum and the horizontal surface.

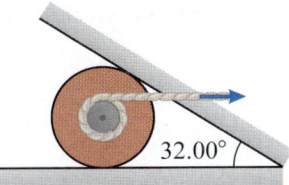

Figure 10-69 Problem 76.

77. ★★★ The coefficient of static friction between the vertical pole and the incline in Figure 10-70 is μ. A guy wire runs from the top of the pole to the top of the incline. You pull the 45.0 kg pole at its centre with a force of magnitude F parallel to the incline and directed down the incline.
(a) Derive an expression for F in terms of μ.
(b) What is the minimum value of μ so that the pole is still in equilibrium for $F = 0$?
(c) What is the value of F that will prevent the pole from sliding on the incline if $\mu = 0.650$?

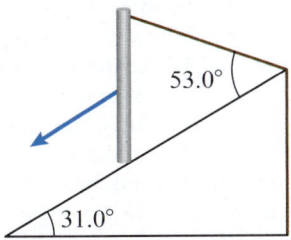

Figure 10-70 Problem 77.

78. ★★ A 4.0 m long uniform bar is supported by a string, as shown in Figure 10-71. What must be the coefficient of static friction between the bar and the floor if the bar is on the verge of slipping?

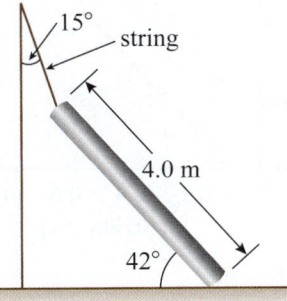

Figure 10-71 Problem 78.

79. ★★★ The vertical bar of length L in Figure 10-72 is supported at the top by a string that makes an angle of θ with the vertical. The bottom of the bar rests on a surface where the coefficient of static friction is μ_s. You have another string that you can move up and down the bar, and you use it to pull horizontally on the bar with force $\vec{F}$.
(a) Derive an expression for the maximum distance y above the floor at which you can position the string and still cause the bottom of the bar to slide on the floor.

(b) You now bring the string to the middle of the bar. Derive an expression for the maximum horizontal force exerted on the string without the bar sliding on the floor.

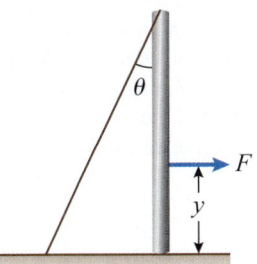

Figure 10-72 Problem 79.

80. ★★★ You are pushing a 92.0 kg refrigerator on a horizontal floor (Figure 10-73). The coefficient of static friction between the refrigerator feet and the floor is 0.34, and the refrigerator's centre of gravity is 80.0 cm above the floor.
(a) Find the horizontal force that causes the refrigerator to just move when applied 30.0 cm above the centre of gravity.
(b) Will the refrigerator tip or slide first?
(c) What is the maximum height at which you can apply the horizontal force without tipping the refrigerator?

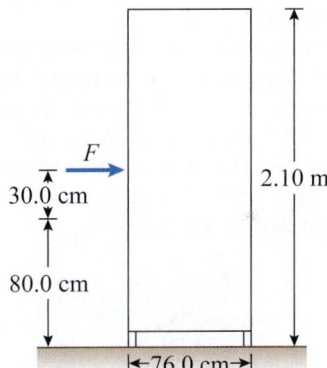

Figure 10-73 Problem 80.

81. ★★★ A wrecker is attempting to tip over a pickup truck with her forklift. She pushes against the truck with a horizontal force of $\vec{F}$ (Figure 10-74). The wheels of the truck are 1.9 m apart, and its mass is 4800 kg. The friction between the truck's wheels and the pavement prevents the truck from slipping. The point of contact between the truck and the forklift is 1.6 m above the pavement. The truck's centre of gravity is 0.9 m above the pavement. What is the minimum magnitude of the force $\vec{F}$ that will tip the truck?

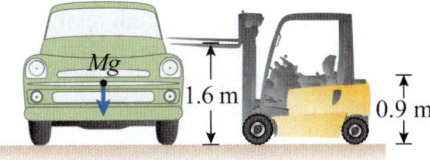

Figure 10-74 Problem 81.

CHAPTER 10 | **EQUILIBRIUM AND ELASTICITY**

82. ★★★ A child pulls on an empty 15.0 kg bookcase, as shown in Figure 10-75. She pulls on an attached shelf 1.40 m above the floor and exerts a force at angle of 30.0° with the vertical. What is the maximum force the child can exert without causing the bookcase to tip? The bookcase has a safety bracket at the back so it won't actually fall onto the child, but it might lean over a bit. Assume that the mass of the bookcase is distributed symmetrically around its geometric centre and that there is enough friction to prevent the bookcase from sliding on the floor.

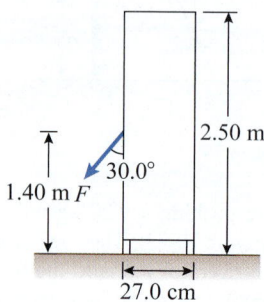

Figure 10-75 Problem 82.

83. ★★★ The uniform bar of length L in Figure 10-76 extends by a distance x over the edge of a table.
(a) What is the maximum distance x for which the bar will not fall?
(b) If we were to stack two identical bars over the edge of the table, what is the maximum distance that the right edge of a second bar could extend over the edge of the table so that the system is stable?
(c) Repeat part (b) for three identical bars. What about four bars?
(d) Repeat part (b) for N bars.

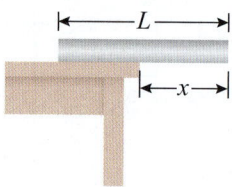

Figure 10-76 Problem 83.

84. ★★ 〰 Figure 10-77 shows a typical stress versus strain curve for spider silk.
(a) Find Young's modulus for spider silk by looking at the stress between 0 MPa and 200 MPa.
(b) Draw the force versus elongation curve for a thread diameter of 1.5 μm and an original length of 11 cm.
(c) How much energy does it take to stretch the thread by 1.0% of its original length?
(d) Identify the proportional limit on the curve.
(e) Estimate the minimum speed that a 50.0 mg fly needs to break a strand of spider silk 11 cm in length. Is this speed practical?

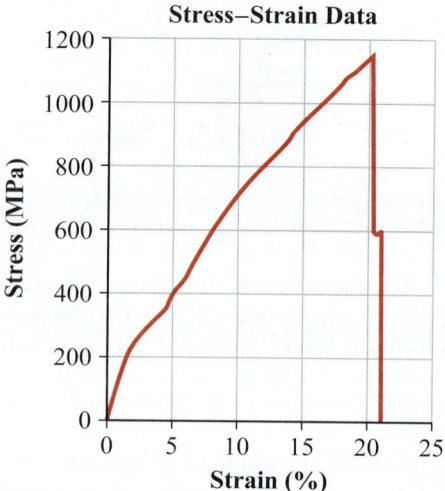

Figure 10-77 Problems 84, 85, and 87.

85. ★ For the curve in Figure 10-77, the silk snaps at a stress of approximately 1150 MPa. Find the maximum mass that can be supported by a 3.0 cm diameter silk thread. What mass can an A36 steel beam of the same diameter support? A36 has a UTS of approximately 440 MPa.

86. ★★ Figure 10-78 shows the linear region of the stress versus strain curve for a spider silk thread 15 mm long and 1.0 μm in diameter. Find the spring constant for this material. (Within the elastic limit, the material behaves like a spring.)

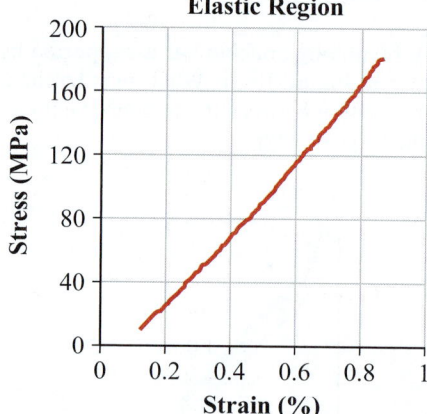

Figure 10-78 Problem 86.

87. ★★★ 〰 Use Figure 10-77 to estimate the kinetic energy needed to break a spider silk thread that is 15 mm long and 1.0 μm in diameter. (You may have to reconstruct the force versus elongation graph for the thread.)

88. ★★ Four vertical steel bars, each 2.00 m in length and 10.0 cm in diameter, support a 120 kg platform with a 4500 kg load of marble slabs on top of it. Find the stress and strain in each bar.

89. ★★ Kevlar is used in bulletproof vests. The yield point of Kevlar is approximately 3600 MPa, and its Young's modulus is approximately 100.0 GPa.
 (a) What diameter of a 30.0 cm long Kevlar thread would one need to stop a 7.5 g bullet moving at a speed of 420 m/s if the Kevlar is to stay within its elastic limit? Assume that the only stopping power comes from the stretching of the thread and that the bullet strikes the taut thread at its centre. (You may refer to Problem 51.)
 (b) What is the maximum distance that the bullet moves after striking the Kevlar?
 (c) If one did not want the bullet to move more than 7.00 mm after striking the Kevlar thread, what length of thread must be used?

90. ★★ The shear yield point of steel is approximately 0.58 of its tensile yield point, and the shear modulus of steel is approximately 0.39 of its Young's modulus. If you wanted to construct a bulletproof vest using a square A36 steel plate 4.0 cm on a side, how thick should the plate be to remain within its shear elastic limit while stopping a 7.5 g bullet moving at a speed of 390 m/s? (You may refer to Problem 51.)

91. ★★ An A36 steel cable 10.0 cm in diameter secures a ship to a bollard. Wind and wave action combine to push the ship and stretch the 12.0 m long cable by 1.00 cm. Find the elastic potential energy stored in the cable.

92. ★★★ Human leg bone typically has a Young's modulus of 17 GPa and a compressive strength of 190 MPa.
 (a) From what height can a 70.0 kg athlete jump without damaging a femur? Assume an impact time of 220 ms during which the femur remains within its elastic limit, and that the femur has a length of 52 cm and a radius of 1.7 cm. Also, assume that all the weight of the athlete is supported by both legs.
 (b) How much can the femur compress before undergoing permanent damage? Use the average force from the impact time.

93. ★★ Some types of bone have a compressive strength of 170 MPa and a tensile yield strength of 130 MPa within the elastic limit. Estimate the amount by which the femur of a 90.0 kg person compresses when the person jumps from a height of 2.00 m and lands on one foot. Use the data from problem 92, and assume that the person takes 40.0 ms to come to a stop.

94. ★★ A human femur typically has a longitudinal compressive strength of approximately 200.0 MPa, a strain at fracture of 0.019, a UTS of 135 MPa, a compressive strength in the transverse direction of 131 MPa, and a shear strength of 65 MPa to 71 MPa. Estimate the maximum longitudinal and transverse forces that a femur can support. Assume a length of 53 cm and a radius of 1.5 cm.

95. ★★★ 〜 Figure 10-79 shows the true compressive stress versus true strain curve for AISI 304L stainless steel for different stress configurations.
 (a) Find Young's modulus for this steel.
 (b) What is the maximum elongation shown on the curve?
 (c) Find the spring constant for the elastic part of this curve for a 1.7 m long cylindrical beam that is 3.0 cm in radius.

(d) Plot the force versus elongation graph for this steel.
(e) Find the energy that it takes to compress this beam by 6.0% of its original length.

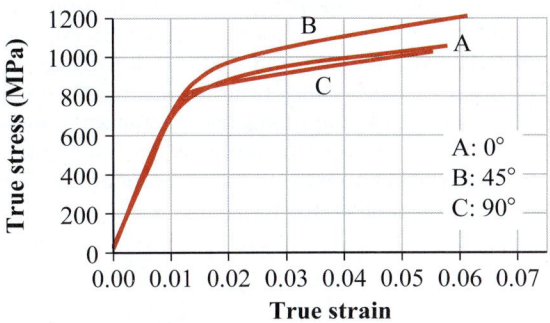

Figure 10-79 Problem 95.

Source: Reprinted from Materials Science and Engineering: A, Volume 475, Issues 1–2, 25 February 2008, S. Qua, C.X. Huanga, Y.L. Gaob, G. Yangb, S.D. Wu a, Q.S. Zanga, Z.F. Zhang, "Tensile and compressive properties of AISI 304L stainless steel subjected to equal channel angular pressing," Pages 207–216, Copyright 2008, with permission from Elsevier.

96. ★★ Figure 10-80(a) shows the stress versus strain curve of stainless steel that has undergone severe plastic deformation in a process called equal-channel angular pressing. The steel is pushed down a vertical channel and undergoes severe deformation while passing through the junction with a horizontal channel (Figure 10-80(b)).
 (a) Describe how this process changes the UTS of the steel.
 (b) Why would this change be desirable for construction grade steel?

(a)

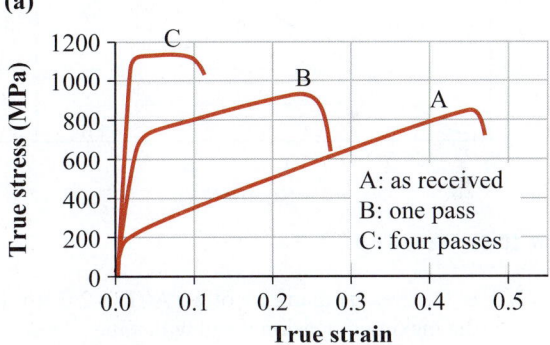

(b)

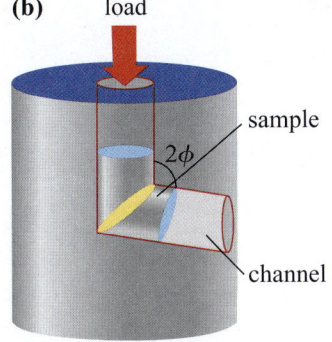

Figure 10-80 Problem 96.

Source: (a) Reprinted from Materials Science and Engineering: A, Volume 475, Issues 1–2, 25 February 2008, S. Qua, C.X. Huanga, Y.L. Gaob, G. Yangb, S.D. Wu a, Q.S. Zanga, Z.F. Zhang, "Tensile and compressive properties of AISI 304L stainless steel subjected to equal channel angular pressing," Pages 207–216, Copyright 2008, with permission from Elsevier.

DATA-RICH PROBLEM

97. ★★ 📈 Refer to the Making Connections "Gravitational Gradients and Spaghettification" in this chapter. You will need a spreadsheet for this exercise.

(a) Plot the tidal force on a 2.0 m tall, 70.0 kg person versus the distance from the centre of a neutron star, 20.0 km in radius and 1.3 times the mass of the Sun, between the surface of the star and a point 2000.0 km from the star. Use the following formula for the tidal "force":

$$g_{tidal} = \frac{2GML}{r^3}$$

Here, g_{tidal} is the tidal force per kilogram on an object of length L at a distance r from the centre of the star of mass M. G is the gravitational constant. Note that in the literature what is referred to as "tidal force" is given in units of gravitational field, or N/kg.

(b) The anterior cruciate ligament (ACL) connects the tibia to the femur. A stress versus strain curve from an ACL study is shown in Figure 10-81. Estimate the UTS from this curve.

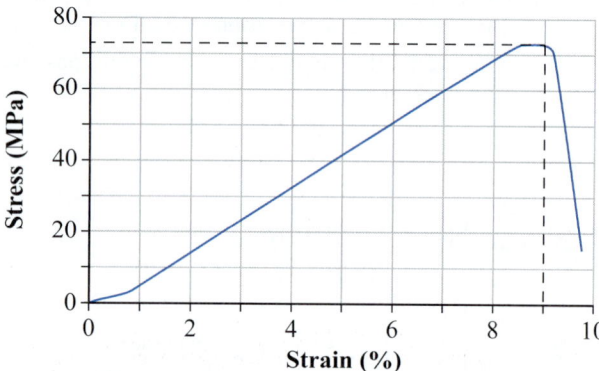

Figure 10-81 Problem 97.

(c) If the average diameter of an ACL is 2.0 cm, find the maximum tension it can withstand.

(d) How realistic is this value for the maximum stress the ACL can take?

(e) Assuming the actual value for the maximum tension force the ACL can take is 10.0 times the value you obtained in (c), how close can you get to a neutron star before you reach the maximum tension force the ACL can withstand? Refer to your spreadsheet.

OPEN PROBLEM

98. ★★ Refer to the Making Connections "Gravitational Gradients and Spaghettification."

According to Newton's law of gravitational attraction, the gravitational field due to a celestial body a distance r away from the centre of the body is given by

$$g = \frac{GM}{r^2}$$

where G is the gravitational interaction constant and M is the mass of the celestial body. Consider an object of radius a engaged in gravitational interaction with a celestial body of mass M, a distance $r > a$ from the centre of the object.

(a) Write an expression for the gravitational field at the point on the object closest to the celestial body.

(b) Write an expression for the gravitational field at the point on the object farthest away from the celestial body.

(c) Write an expression for the tidal force expressed as the difference between the gravitational fields in parts (a) and (b).

(d) Simplify the expression from (c).

(e) Find the limit of the expression in (d) for $r \gg a$.

(f) Adapt the expression in (e) to a steel bar of length L, such that its length lies along the radial direction relative to the celestial body.

(g) Estimate the tidal force on a steel bar whose mass is 1.0 kg and whose length is 1.0 m at the surface of Earth.

(h) Find the tidal force on the bar at the surface of a neutron star, 1.3 times the mass of the Sun, with a radius of 20.0 km.

(i) What is the closest this bar can get to the neutron star before snapping?

CHAPTER
11 | Gravitation

There are more than a million asteroids with diameters larger than a kilometre (Figure 11-1). Most asteroids are located in the region between Mars and Jupiter; however, some have orbits that bring them close to Earth. These near-Earth asteroids pose a small, but not insignificant, hazard to life on Earth. It is believed that, 65 million years ago, an asteroid or a comet impact caused the demise of the dinosaurs—and about three-quarters of all other species. Asteroids also have potential benefits. Within your lifetime we may well begin to harness resources from asteroids.

Understanding the physics of gravitation is essential for assessing the risks from asteroids and for exploiting their resources. Gravitational theory enables you to answer questions such as, What is the gravitational acceleration at the surface of an asteroid? How fast would you need to throw an object for it to escape from the gravitational attraction of the asteroid? How long does an asteroid take to make one orbit around the Sun? When will an asteroid be closest to Earth?

As you will learn in this chapter, gravitational theory already has many practical applications here on Earth. Beyond these applications, gravity is one of the fundamental forces in physics, and its study has been a primary focus of physics for hundreds of years. As we will see in later chapters, aspects of the nature of gravity continue to be a focus of modern physics.

Figure 11-1 Asteroid Ida is about 56 km by 24 km. Would you be able to jump from the surface fast enough to escape from Ida's gravitational pull?

11-1 Universal Gravitation

One of Isaac Newton's many contributions to physics is the **law of universal gravitation**. Every two point masses (m_1, m_2) attract each other with a gravitational force $\vec{F}_G$ that is proportional to the product of the masses and inversely proportional to the square of the distance r between the masses:

$$F_G = \frac{Gm_1 m_2}{r^2} \tag{11-1}$$

The proportionality constant G is called the **gravitational constant** and has a value of $G = 6.674 \times 10^{-11}$ N kg^{-2} m^2.

Gravity is a relatively weak force, and it is challenging to determine the value of the gravitational constant. British scientist Henry Cavendish (1731–1810) was the first person to succeed in measuring the forces between masses in key experiments conducted in 1797–1798. Go to problem 65 and follow in his footsteps as you analyze the experiment for yourself. Even today we only know the gravitational constant to a precision of about five significant digits, much less precisely than many other constants.

The gravitational force is always attractive; that is, the gravitational force exerted by mass m_1 on mass m_2 will pull m_2 directly toward m_1. We can write the law of universal gravitation in vector notation as Equation 11-2, where $\hat{r}_{1\to2}$ is a unit vector pointing in the direction from m_1 toward m_2 and $\vec{F}_{G1\to2}$ is the gravitational force that m_1 exerts on m_2. The directions of the vectors are shown in Figure 11-2:

$$\vec{F}_{G1\to2} = -\frac{Gm_1 m_2}{r_{12}^2}\hat{r}_{1\to2} \tag{11-2}$$

The law of universal gravitation applies for point masses. We can use the **principle of superposition**, however, to find the net gravitational force on a mass due to multiple other point masses by calculating the gravitational force due to each mass and then adding the individual forces as vectors. Similarly, for continuous distributions of mass, we can divide the object into many small differential mass elements and then integrate to find the net gravitational force.

For a spherically symmetric mass, such as a uniform sphere, it can be proven that the gravitational force the sphere exerts on an exterior point mass is the same as though the mass of the sphere were all located at a single point at the centre of the sphere. This helps considerably in solving many gravitational problems where direct computation by the superposition principle is mathematically challenging.

Every pair of objects in the universe exerts gravitational forces on each other (Figure 11-3), and the

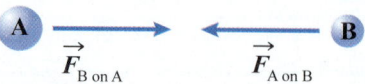

Figure 11-3 The gravitational force that mass A exerts on mass B, $\vec{F}_{A \text{ on } B}$, is equal in magnitude but opposite in direction to the force that mass B exerts on mass A, $\vec{F}_{B \text{ on } A}$.

MAKING CONNECTIONS

Isaac Newton

Sir Isaac Newton (1643–1727; Figure 11-4) developed the key ideas for the law of universal gravitation during a two-year period, shortly after graduating with his first university degree in 1665. However, his work was not published until much later. Newton had planned to go to graduate school immediately, but the University of Cambridge was closed to help prevent the spread of the plague. So, Newton worked largely alone at his home (Figure 11-5). In addition to his work on universal gravitation, he made major contributions to the development of calculus, optics, and mechanics during this time.

Godfrey Kneller, 1689

Figure 11-4 Sir Isaac Newton at age 46.

Awe Inspiring Images/Shutterstock.com

Figure 11-5 Woolsthorpe Manor, where Sir Isaac Newton developed his most important contributions to physics.

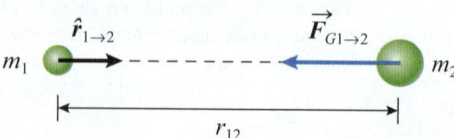

Figure 11-2 Direction of the vectors in Equation 11-2. Note that $\vec{F}_{G1\to2}$, the force mass m_1 exerts on mass m_2, is in the opposite direction from the unit vector $\hat{r}_{1\to2}$ pointing from mass m_1 to m_2 because of the negative sign in the equation.

magnitudes of these forces can be determined using Equation 11-1. For example, you can use Equation 11-1 to calculate the force that Earth exerts on your body. Since Newton's third law states that for every force exerted by one object on another, there is an equal and opposite reaction force, you can similarly calculate the force that you exert on Earth (or on any other object in the universe).

EXAMPLE 11-1

The Sun's Gravitational Force

(a) Find the gravitational force exerted by the Sun on a 50.0 kg person on Earth.
(b) Find the gravitational force exerted by Earth on the same person.

Solution

(a) We can use the law of universal gravitation, substituting the person's mass (50.0 kg) for m, the mass of the Sun (1.989×10^{30} kg) for M, and the distance from Earth to the Sun (1.496×10^{11} m) for r. Since r is approximately 10^4 times the diameter of Earth, the particular location of the person on Earth has negligible effect on the magnitude of the force in this calculation:

$$F_G = \frac{GMm}{r^2}$$

$$= \frac{(6.674 \times 10^{-11} \text{ N·m}^2/\text{kg}^2)(1.989 \times 10^{30} \text{ kg})(50.0 \text{ kg})}{(1.496 \times 10^{11} \text{m})^2}$$

$$= 0.297 \text{ N}$$

The direction of the force is toward the Sun.

(b) We use the same method as in part (a) and substitute Earth's mass and radius (6380 km) for M and r, respectively:

$$F_G = \frac{GMm}{r^2}$$

$$= \frac{(6.674 \times 10^{-11} \text{ N·m}^2/\text{kg}^2)(5.97 \times 10^{24} \text{ kg})(50.0 \text{ kg})}{(6.37 \times 10^6 \text{m})^2}$$

$$= 491 \text{ N}$$

The direction is toward the centre of the Earth.

Making sense of the result

The gravitational force of the Sun on a person is much less than the gravitational force of Earth on the person (about 0.06%). However, for high-precision calculations, the gravitational force exerted by the Sun on the person is not negligible compared to the gravitational force exerted by Earth.

CHECKPOINT 11-1

Forces on a Satellite

A satellite is orbiting Earth. Which of the following statements is true?
(a) The magnitude of the gravitational force exerted by Earth on the satellite is larger than the gravitational force exerted by the satellite on Earth.
(b) The magnitude of the gravitational force exerted by the satellite on Earth is larger than the gravitational force exerted by Earth on the satellite.
(c) Both forces have exactly the same magnitude.
(d) The relationship between the magnitudes of the two forces depends on the distance between Earth and the satellite.

ANSWER: (c) Gravitational forces exist in pairs, with the magnitudes equal but the directions of the forces on the different objects opposite.

Equivalence Principle

You encountered mass in Chapter 5 when considering Newton's second law. In that context, we had **inertial mass**, which is defined as the amount of force required to accelerate the object divided by the acceleration.

In this chapter, we have **gravitational mass**, which is a measure of the gravitational influence on other objects through the law of universal gravitation. While inertial mass and gravitational mass might be different, experiments support the view that they are equal.

The **equivalence principle** states that the inertial and gravitational masses are exactly equal. The equivalence principle plays an important role in general relativity (see Chapter 30). If this principle is correct, one could not differentiate between an apparent force in an accelerated reference frame and a gravitational force in a non-accelerated reference frame.

Celestial Terminology

In this and later chapters we will apply physics in astrophysical situations. Indeed, the gravitational ideas of this chapter were developed and tested almost entirely with astrophysical observations. The terms we will use most often are defined here.

A **star** is a celestial body that is massive enough to be held together by its own gravity and to be luminous (usually through nuclear reactions in the core). Stars range in mass from less than 10% of the mass of the Sun to about 150 times the mass of the Sun.

A **planet** is a significantly smaller object in orbit about a star, but one that is large enough to be approximately spherical and to have gravitationally "cleared" the space around it from similarly sized objects. By current convention the term *planet* is reserved for objects in orbit about our Sun. An **exoplanet** is a planet in orbit about a star other than our Sun. By the definition of planet, there are eight planets in orbit about our Sun. A **dwarf planet** is an object that would be considered a planet, but is not large enough to gravitationally clear the area of other similar sized objects. Pluto was designated a dwarf planet in 2006, along with similar sized objects in its region, as was Ceres, which orbits in the region between Mars and Jupiter.

Celestial objects called **satellites** are in gravitational orbits not directly about the Sun but rather around a planet (and indirectly move about the Sun through

MAKING CONNECTIONS

Tides

To understand ocean tides, we can reasonably approximate Earth as a rigid body that responds to gravitational forces as if its mass were concentrated at its centre. Water on the side of Earth facing the Moon is somewhat closer to the Moon than Earth's centre of mass and therefore experiences a correspondingly greater gravitational acceleration. Conversely, water on the side of Earth away from the Moon experiences a lesser gravitational acceleration and therefore gets "left behind." As shown in Figure 11-6, high tides occur at opposite points on Earth. A given point has two high tides each day as Earth rotates.

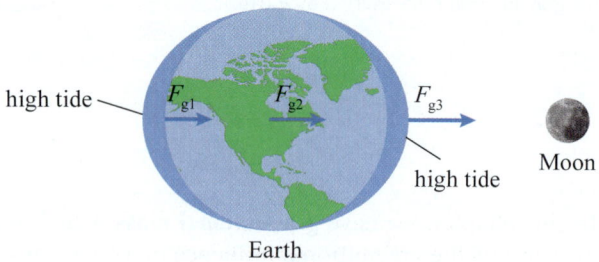

Figure 11-6 The pattern of two tides each day results from the relative magnitudes of gravitational acceleration due to the gravitational force of the Moon on water on the side of Earth nearer the Moon, the solid Earth, and water on the far side of Earth.

For a more complete understanding of tides, we have to take into account ocean currents and the gravitational force exerted by the Sun. Tides are higher at times of new and full moon because the gravitational forces of the Sun and the Moon approximately align (these are called spring tides—see Figure 11-7).

The mean tidal range of the oceans is about 1 m, but the shape and orientation of coastlines affect the local differences between the water level at low and high tide. For example, in the Bay of Fundy a combination of resonance and funnel effects produces a tidal range of up to 16 m (Figure 11-8).

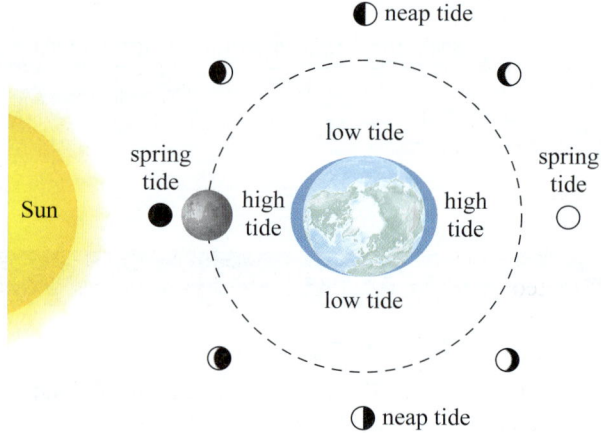

Figure 11-7 The position of the Sun, Moon, and Earth at times of high tidal range (spring tides) and low tidal range (neap tides). We have higher tidal ranges at times of full moon and new moon.

Figure 11-8 High and low tide at Alma, New Brunswick, in the Bay of Fundy.

the planet's orbit). You may think of *satellite* as a term for artificial objects in orbit about the Earth, such as weather and communications satellites. *Satellite* is a general term for objects in orbit about planets, and while including these artificial objects, it also includes our Moon and the many similar objects in orbit around other planets. For example, the planet Jupiter has at least 67 satellites, 4 of which were large enough to have been discovered by Galileo.

There is a much larger number of smaller solid objects in orbit, like planets, directly around the Sun. But these objects are not large enough to have achieved an approximately spherical shape and are called **asteroids**. It is estimated that there are about 2 million asteroids larger than 1 km in diameter, and almost 10 times that number larger than 100 m in diameter. Asteroids are generally of rocky composition. Many asteroids reside in the **main asteroid belt**, located between Mars and Jupiter. A **comet** is of a size comparable to many asteroids, but the composition is a mix of ices and rocky materials. Comets were formed farther out in the early Solar System than asteroids. As comets approach the Sun, the ices vaporize and dust and gas are released.

Smaller solid objects in orbit about the Sun are called **meteoroids**. There is not yet formal agreement on the boundary between small asteroids and large meteoroids, but usually we call objects larger than 10 m asteroids and those smaller than 1 m meteoroids, although most meteoroids are much smaller than that, extending down to dust-sized objects. The vast majority of meteoroids have origins in either comets or asteroids, although a few come from the surfaces of other planets or satellites. The term **meteor** refers to the light and ionization phenomena when a meteoroid ablates in a planetary atmosphere. A **meteorite** refers to those meteoroids that reach the surface of Earth or another planet or satellite. Most meteors vaporize entirely and never become meteorites.

We refer to the combination of the Sun and the objects directly, or indirectly, in orbit about it, as the **Solar System**. Stars, along with the planets, clouds of gas and dust, and smaller bodies in orbit around them, are part of much larger structures called **galaxies**. A typical galaxy contains millions to hundreds of billions of stars and is held together with the attractive gravitational force.

Tidal Forces

The term *tidal force* refers to the difference in gravitational forces across any body. For example, if a spacecraft were near an extremely dense star, the difference between the gravitational forces on the side facing the star and the side facing away would produce a tidal force strong enough to tear the spacecraft apart.

This difference in the gravitational force on water on the different sides of Earth causes the tides (see Making Connections: Tides). However, tides are by no means restricted to Earth. On Io, a natural satellite of Jupiter, the tidal forces actually change the solid surface of the satellite.

General Relativity

In this chapter, we consider gravity to be a force between masses. This view is adequate for most situations and can be used to calculate the orbits of planets or satellites, the motion of rockets, acceleration due to gravity, and what is required for an object to gravitationally escape from a large mass. However, Einstein realized that this force view of gravity was not consistent with relativity, and in **general relativity** (see Chapter 30) gravity is not viewed as a force but rather as a curvature of space-time near masses.

11-2 Acceleration Due to Gravity

We designate the **acceleration due to gravity** near the surface of a planet (or a satellite of a planet) as the vector quantity $\vec{g}$. This vector is directed toward the centre of the planet, for example, toward the centre of Earth if we are dealing with acceleration due to gravity near Earth's surface. Without a vector sign, g represents the magnitude of the acceleration due to gravity. On Earth, g is approximately 9.81 m/s² at sea level. We can also write the acceleration due to gravity near Earth's surface as $-g\hat{k}$, where $+\hat{k}$ is a unit vector in the upward direction.

Cartesian notation can be used over a limited region where Earth can be considered approximately flat. More generally, we have radial symmetry, and it is better to write $\vec{g} = -g\hat{r}$, where g is the magnitude of the acceleration due to gravity and $\hat{r}$ is a unit vector pointing radially outward from the centre of Earth (or other planet or satellite being considered).

In a region where the acceleration due to gravity is $\vec{g}$, we saw in Chapter 5 that the gravitational force on a mass m is given by the following relationship:

$$\vec{F}_G = m\vec{g} \qquad (11\text{-}3)$$

We now have two ways to calculate $\vec{F}_G$: When we know the local acceleration due to gravity, $\vec{g}$, we can use Equation 11-3, and when we know the mass and radius of the large object causing the gravitational force, we can use Equation 11-1. In Example 11-2, we demonstrate how these two methods are related.

EXAMPLE 11-2

Finding the Acceleration Due to Gravity

Show that the observed value of approximately 9.81 m/s² for the magnitude of the gravitational acceleration at Earth's surface is consistent with Newton's law of universal gravitation.

Solution

From Newton's law of universal gravitation, we have the following equation for the magnitude of the force due to gravity:

$$F_G = \frac{GMm}{r^2}$$

Equating this expression with $F_G = mg$ and then solving for g gives

$$g = \frac{GM}{r^2}$$

Next, we substitute values for Earth's radius and mass:

$$g = \frac{(6.674 \times 10^{-11}\,\text{N·m}^2/\text{kg}^2)(5.97 \times 10^{24}\,\text{kg})}{(6.37 \times 10^6\,\text{m})^2} = 9.82\,\text{m/s}^2$$

In fact, this value for g applies only at the equator because Earth is slightly oblate—its polar radius is approximately 6357 km, and its equatorial radius is approximately 6378 km. As a result, the value of g at Earth's surface varies slightly with latitude.

Making sense of the result

As expected, the value for g calculated using Newton's law of universal gravitation agrees with observed values. The magnitude of the force of gravity can be written as F_G or F_g. We usually use the latter only when using the acceleration due to gravity.

When we measure the gravitational force on a small test mass and then divide by that mass, we have the vector $\vec{g}$. This vector represents two equivalent quantities: the gravitational field strength (in newtons per kilogram) and the gravitational acceleration (in metres per second squared). A plot of this vector at different points around a mass shows the gravitational field created by the mass. In Chapters 19 and 20, we will apply the same concept to describe electrical fields around charged objects.

We can use the method in Example 11-2 to see how g varies at different altitudes above Earth's surface. In fact, the equation below applies for locations on or near any planet, satellite, or other body:

KEY EQUATION
$$g = \frac{GM}{r^2} \qquad (11\text{-}4)$$

In many areas of physics, such as planetary science, upper atmospheric physics, space science, and geophysics, g is used to refer to the acceleration of gravity at any point and not just at the surface, while some introductory texts use g to mean only the value at the surface. We will use the more general definition of g as the acceleration at any point in this chapter. We must make sure to use the appropriate r-value, the distance from the centre of the planet to the point where g is to be evaluated.

It is important to realize that G is a fundamental constant of physics and, as far as we know, does not depend on where or when you measure it. Some physicists are currently working on theories in which G is not strictly constant at different times, but the preponderance of evidence suggests that G is constant. On the other hand g—the magnitude of the acceleration due to gravity—is definitely not a universal constant. It is quite different on the Moon and Mars, for example, and even varies somewhat for different locations on Earth's surface.

CHECKPOINT 11-2

Acceleration Due to Gravity on Different Planets

Two spherical, homogeneous planets have the same radius, but planet B has only half the density of planet A. If g_A and g_B are the accelerations due to gravity on the surfaces of the two planets, how are they related?

(a) $g_B = \dfrac{g_A}{4}$

(b) $g_B = \dfrac{g_A}{2}$

(c) $g_B = g_A$

(d) $g_B = 2g_A$

ANSWER: (b) Since the radius is the same, the volumes will also be equal. Therefore, planet A has double the mass of planet B. Since r is the same, this means that from Equation 11-4 the g-value must be twice as large on planet A.

EXAMPLE 11-3

Finding g on an Asteroid

(a) Find the value of g on the surface of an asteroid that has a density of 3200 kg/m³ and a radius of 5.00 km. Assume that the asteroid is homogeneous and spherical.

(b) How long would it take an object starting from rest to fall through a height of 1.0 m above the surface of the asteroid? Assume that g is constant during the fall.

(c) Show that the assumption in part (b) is justified.

Solution

(a) First, find the mass of the asteroid:

$$M = \rho V = \rho \frac{4\pi r^3}{3} = 3200 \text{ kg/m}^3 \frac{4\pi (5.00 \times 10^3 \text{ m})^3}{3}$$

$$= 1.68 \times 10^{15} \text{ kg}$$

Now, calculate the acceleration due to gravity at the asteroid's surface:

$$g = \frac{GM}{r^2} = \frac{(6.674 \times 10^{-11} \text{ N} \cdot \text{m}^2/\text{kg}^2)(1.68 \times 10^{15} \text{ kg})}{(5.00 \times 10^3 \text{ m})^2}$$

$$= 4.48 \times 10^{-3} \text{ m/s}^2$$

(b) From Section 3-4 we have

$$y - y_0 = v_0 t + \frac{1}{2} a t^2$$

We choose a convention with positive representing the upward direction, so $y - y_0 = -1.0$ m. Since the object starts from rest, $v_0 = 0$. The acceleration is -4.48×10^{-3} m/s^2. The acceleration is negative because it is in the downward direction. Therefore,

$$-1.0 \text{ m} = 0 + \frac{1}{2}(-4.48 \times 10^{-3} \text{ m/s}^2)t^2$$

$$t^2 = \frac{2(-1.0 \text{ m})}{-4.48 \times 10^{-3} \text{ m/s}^2}$$

$$t = 21 \text{ s}$$

It would take 21 s for an object to fall 1.0 m above the surface of the asteroid.

(c) If we repeat part (a) with $r = 5.001 \times 10^3$ m, we obtain a value for g of 4.47×10^{-3} m/s^2. Therefore, the assumption that g is approximately constant is justified.

Making sense of the result

As expected, we found that the gravitational acceleration is much less than near the surface of Earth. Since 1.0 m is small compared to the radius of the asteroid, it is not surprising that the change in g would not affect calculations that have a precision of three significant figures.

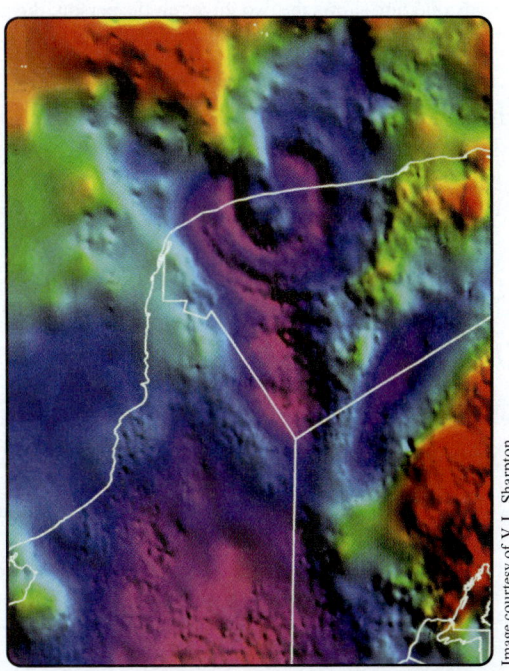
11-3 Orbits and Weightlessness

As we saw in Section 5-2, weight can be defined in an operational sense as the reading on a weigh scale. For example, if your mass is m, then on the surface of Earth a weigh scale you stand on must provide an upward force of magnitude mg to balance the gravitational force of the same magnitude exerted by Earth on your body. Therefore, your weight in that situation is mg. With this operational definition of weight it is easy to see that if you were high on a mountain, where g is a little less, then your weight would be less. On the surface of another planet or satellite, your weight would depend on the value of g there. If you were far out in deep space at a great distance from any mass, then your operational weight would be approximately zero, and we would say you are **weightless**. The term *weightless* means that a weigh scale would read zero.

Now consider the situation shown in Figure 11-10. A person of mass m is standing on a scale, but this time they are inside an elevator. If the elevator accelerates upward, then the scale will read more than mg, and if it accelerates downward, the scale will read less than mg. We can see why this is so by considering the forces on the person. There is a gravitational force downward, which we will call F_g, and the weigh scale exerts an upward force on the person,

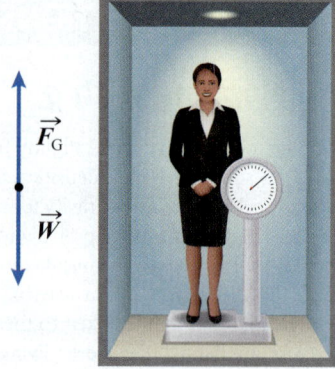

Figure 11-10 A person stands on a weigh scale inside an elevator that can accelerate in either the upward or the downward direction. A free body diagram shows the two forces on the person: the gravitational force F_g and the force exerted by the weight scale, W.

which we will call W. Now we apply Newton's second law, defining the positive direction as upward:

$$W - F_g = ma$$
$$W - mg = ma \qquad (11\text{-}5)$$
$$W = mg + ma$$

EXAMPLE 11-4

Gravity at the ISS

(a) Calculate the gravitational force exerted by Earth on a 1.00 kg object at the ISS. Assume that the ISS is in a circular orbit at an altitude of 353 km above Earth's surface.
(b) What is the value of g at the ISS?
(c) Calculate the centripetal acceleration required to keep an object in the circular orbit of the ISS, given that the orbital speed of the ISS is 7.70 km/s.

Solution

(a) We can apply the law of universal gravitation with the mass of the object (1.00 kg) for m and the mass of Earth (5.974×10^{24} kg) for M. The distance r is the sum of the radius of Earth (we will use a mean value of 6371 km) and the altitude of the ISS:

$$r = 6371 \text{ km} + 353 \text{ km} = 6724 \text{ km} = 6.724 \times 10^6 \text{ m}$$

$$F_G = \frac{GMm}{r^2}$$

We see that if the acceleration a is positive, then the weight will be more than mg. If the elevator has a negative acceleration value, the scale will read less than mg.

What if there is an accident and the elevator cable breaks, so the elevator acceleration in this case is $a = -g$? Now according to Equation 11-5 the weight is zero, and we have a state of weightlessness. The best way to think about this is that both the elevator and we are accelerating together at $a = -g$, so the scale reads zero. Note that although the weight is zero for the weightless case, the force of gravity does *not* change with the acceleration of the elevator.

You have probably heard that inside the International Space Station (ISS) (Figure 11-11) there is a state of weightlessness. As you can see in Example 11-4, there is a significant gravitational force at the altitude of the ISS. The circular orbit of the ISS requires a radially inward centripetal acceleration (see Section 4-3). As shown in Example 11-4, the gravitational force is just enough to provide this acceleration. Therefore, no additional gravitational force is left to contribute to a weight, and we have weightlessness.

$$= \frac{(6.674 \times 10^{-11} \text{N} \cdot \text{m}^2/\text{kg}^2)(5.974 \times 10^{24} \text{ kg})(1.00 \text{ kg})}{(6.724 \times 10^6 \text{ m})^2}$$

$$= 8.82 \text{ N}$$

(b) This is the force on 1.00 kg, so the g-value is 8.82 m/s^2.
(c) We apply the formula for centripetal acceleration:

$$a_c = \frac{v^2}{r} = \frac{(7.70 \times 10^3 \text{ m/s})^2}{6.724 \times 10^6 \text{ m}}$$

$$= 8.82 \; m/s^2$$

Making sense of the result

The gravitational force on the 1.00 kg object is approximately 10% less than the 9.81 N force that the object would experience on Earth's surface. The gravitational force at the ISS is certainly not zero. The centripetal acceleration is equal to the gravitational acceleration. Thus, the gravitational force is just enough to keep the ISS in a circular orbit. As a result, objects appear weightless in the ISS frame of reference.

As shown in Figure 11-12, if you start at a point high above Earth and fire projectiles at different speeds, they will, under the influence of Earth's gravity, follow the paths shown. As the speed is increased, the projectile travels farther before reaching Earth's surface. At just the right speed, the projectile will go into a circular orbit. You can think of the motion as being in perpetual free-fall, with the gravitational force exactly the amount needed to provide the centripetal acceleration of the circular motion.

Now consider throwing an object with a lesser speed near the surface of Earth. As you saw in Chapter 4, the object will follow a parabolic trajectory. What if you threw a person and a weigh scale with exactly the same initial velocity? They would both follow the same parabolic path, and one relative to the other would be in a state of weightlessness. This is used with specialized aircraft (Figure 11-13 in Making Connections: Achieving Weightlessness) to produce a few tens of seconds of weightlessness.

Figure 11-11 The ISS is an orbiting laboratory built by a consortium of many countries, including Canada.

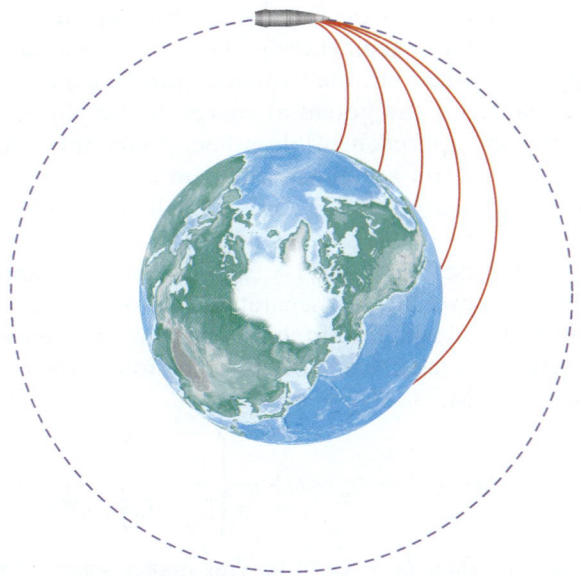

Figure 11-12 Projectiles with higher tangential speeds travel a greater distance before reaching Earth's surface. At just the right speed, the path of the projectile will curve such that the projectile remains at a constant distance from Earth.

CHECKPOINT 11-3

Weightless in Orbit

Which of the following, if any, have a zero value for an object in a circular orbit?

(a) the gravitational force on the object
(b) the acceleration of the object
(c) the net force on the object
(d) the difference between the gravitational force on the object and its mass times its acceleration

ANSWER: (d)

MAKING CONNECTIONS

Achieving Weightlessness

Aircraft like the one in Figure 11-13 produce weightless conditions more readily and at a lower cost than in an orbiting spacecraft, but only for brief periods, typically 15 s to 20 s. The aircraft accelerates steeply upward, as shown in Figure 11-14, and then the engines are idled and the aircraft follows a parabolic path like a projectile. During the parabolic portion of the flight path, passengers in the aircraft experience weightlessness, although usually not quite complete weightlessness. These flights are called parabolic flights, or sometimes suborbital zero gravity flights. The European Space Agency (ESA), NASA, the Canadian Space Agency (CSA), and others use these flights to achieve weightless conditions for short periods.

Figure 11-13 The NASA parabolic flight aircraft during ascent.

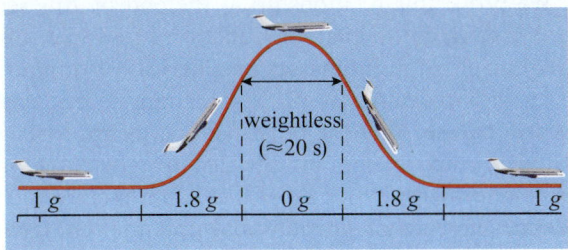

Figure 11-14 This flight path produces a brief period of weightlessness. The acceleration is expressed in units of g (equal to 9.81 m/s^2).

11-4 Gravitational Potential Energy

Gravity is a **conservative force field**, meaning that the work required to move a mass between two points is independent of the path taken. Because the field is conservative, it is possible to define a potential energy function. Potential energies are scalar quantities, and for this reason it is often easier to solve problems using energy methods instead of directly with forces.

In Section 6-5 we showed that in a region where g is approximately constant, the change in gravitational potential energy ΔU when a mass m is moved a vertical distance Δh is given by

KEY EQUATION
$$\Delta U = mg\Delta h \qquad (11\text{-}6)$$

Recall that potential energies are always relative to some defined zero point. If we define $U = 0$ at $h = 0$ (sea level on Earth), then we can write the following for potential energy relative to this zero definition:

$$U = mgh \qquad (11\text{-}7)$$

We should stress that it is perfectly acceptable to define a different zero reference; for example, one could call a point 10 m above sea level the zero point, in which case ground level would have negative potential energy.

Equation 11-7 can only be directly applied in cases where g is approximately constant. As long as we are not too far above the surface of the planet, this will be valid. However, if high in the atmosphere, or in space, we can certainly not assume that g is constant.

A general expression for the gravitational potential energy $U(r)$ when two masses, m and M, are a distance r apart is given by Equation 11-8:

KEY EQUATION
$$U(r) = -\frac{GMm}{r} \qquad (11\text{-}8)$$

We prove this relationship at the end of this section. We should note that for Equation 11-8 to be valid we require either point masses or spherically symmetric masses with m exterior to M. If you examine Equation 11-8 you can see that the potential energy approaches zero when r approaches infinite distance. In planetary science and astrophysics cases it is often inconvenient to define the zero of gravitational potential energy at the surface of the planet. In these situations the definition of zero point corresponding to infinite separation is the most logical choice.

With this definition of zero point for gravitational potential energy, the value is zero at infinite distance and is negative everywhere else. As you first encountered in Chapter 6, there are two ways of interpreting the meaning of potential energy. One is that the potential energy

difference is the work required to move the objects to that configuration. For example, to lift a mass m up a height h in a region with gravitational field g would require a constant force mg and a work mgh. This is consistent with Equation 11-6. Now if we start with $U = 0$ at infinite distance, the work required to move mass m toward the central mass M will be negative since the gravitational force is attractive. This is consistent with U being negative for all non-infinite r-values.

Equivalently, we can view the potential energy as the *negative* of the work done by the *field* (in this case the gravitational field) in moving from the zero point to that position, that is, if instead of the work done by an external agent we take the work done by the gravitational field, and then take the negative of that work (see Section 6-5). Since the two forces (gravitational force versus the force an external agent would require to move against the gravitational field) are in opposite directions, the two views produce a consistent value for the potential energy. Some physicists prefer one approach, while others prefer the other. As long as you are consistent, you can take either view of potential energy, and you will always get the same results.

If we use Equation 11-8 to calculate the potential energy for two different separations, we obtain the following useful relationship showing the potential energy difference as mass m is moved from r_a to r_b away from fixed mass M:

$$\Delta U = U_b - U_a = GMm\left\{\frac{1}{r_a} - \frac{1}{r_b}\right\} \qquad (11\text{-}9)$$

If $r_b > r_a$, then $U_b - U_a > 0$. This makes sense: if we increase the separation between two masses, the potential energy of the configuration will increase.

PEER TO PEER

It is easy to make a sign error in gravitational potential energies. I always ask myself, when dealing with potential energies: Does my final answer make sense? Would I need to do work to move it to the new situation? If so, then it should be a higher potential energy.

Equation 11-8 gives us an expression for the potential energy when we have two masses. What if we have more than two masses? In this case we can use the **superposition principle**, according to which we find the gravitational potential energy for each pair of masses and then add the gravitational potential energies to get the total potential energy.

Because energies are scalars, the addition is simply an addition. For example, if we had three masses, m_1, m_2, and m_3, the total gravitational potential energy would be the following, with the notation r_{12} meaning the distance between m_1 and m_2:

$$U = -G \left\{ \frac{m_1 m_2}{r_{12}} + \frac{m_1 m_3}{r_{13}} + \frac{m_2 m_3}{r_{23}} \right\} \qquad (11\text{-}10)$$

CHECKPOINT 11-4

Gravitational Potential Energy for Three Masses

Three point masses lie along the x-axis, each of mass m. One of these masses is at the origin of a coordinate system, one is at $x = 1$ m, and the third is at $x = 3$ m. What is the gravitational potential energy (in SI units) of the configuration, assuming the usual convention of zero potential energy for infinite separation?

(a) $-6Gm^2$

(b) $-\dfrac{3}{2}Gm^2$

(c) $-\dfrac{11}{6}Gm^2$

(d) $+\dfrac{3}{2}Gm^2$

ANSWER: (c) Add the gravitational potential energy terms for the three pairs of masses. The separations of the pairs are 1 m, 3 m, and 2 m.

The ideas can readily be applied to a larger number of masses. The superposition principle is widely applied in different areas of physics (you will encounter it again in Chapters 19, 20, and 21 for electric fields and electric potentials).

We summarize key ideas to keep in mind when dealing with gravitational potential energies.

1. Gravitational (and all other) potential energies are always defined relative to some defined zero point. It is good practice to explicitly define your zero point assumption when solving problems.

2. If you are in a region where g is constant or nearly constant, Equations 11-6 and 11-7 can be used.

3. If this is not the case, use Equation 11-8, recognizing that it implies a zero gravitational potential energy for infinite separation.

4. If there are more than two masses, find the gravitational potential energy for each pair, and then apply the superposition principle to add the results and find the total gravitational potential energy.

Optional Calculus Proof of Gravitational Potential Energy

In this optional section we will derive Equation 11-8. We start with Equation 11-6, but we will make the height difference very small; that is, we replace Δh with a differential radial distance dr. Also, we replace ΔU with dU:

$$dU = mg\,dr \qquad (11\text{-}11)$$

Next we use Equation 11-4 for g:

$$g = \frac{GM}{r^2} \qquad (11\text{-}4)$$

$$dU = m\frac{GM}{r^2}\,dr \qquad (11\text{-}12)$$

We will now integrate both sides. We define the zero point for potential energy as $U = 0$ at $r = \infty$, and therefore we have the following:

$$\int_0^{U(r)} dU = GMm \int_\infty^r \frac{dr}{r^2}$$

$$U(r) - 0 = GMm \left[-\frac{1}{r} \right]_\infty^r$$

$$U(r) = GMm \left[-\frac{1}{r} - 0 \right]$$

As expected, this gives the result from Equation 11-8:

$$U(r) = -\frac{GMm}{r} \qquad (11\text{-}8)$$

CHECKPOINT 11-5

Gravitational Potential Energy

(a) When you increase the distance between two masses, does the gravitational potential energy increase or decrease?

(b) Do you need to do work to increase the distance between two masses?

ANSWER: (a) When you increase the distance between two masses, the gravitational potential energy increases. The increased distance results in a negative energy with a smaller magnitude, so the energy increases. Since it takes work to move the objects farther apart, the potential energy must increase. (b) Yes, you need to do work to increase the distance between two masses. The force required to separate the masses has the same direction as the increase in displacement, so the work is positive.

EXAMPLE 11-5

Calculating Changes in Gravitational Potential Energy

Calculate the work required to lift a 1.00 kg object from sea level to the top of Mount Everest, which has an elevation of 8848 m. Assume that Earth is spherical with a radius of 6371 km. Perform the calculation in the following two ways:

(a) Use the *mgh* expression for potential energy, and assume that *g* remains constant at its value at sea level, 9.81 m/s².
(b) Use Equation (11-8).

Solution

(a) Since $U = 0$ at sea level,

$$\Delta U = mgh - 0 = 1.00 \text{ kg} \times 9.81 \text{ m/s}^2 \times 8848 \text{ m} = 86.8 \text{ MJ}$$

The gravitational potential energy of the object is greater at the top of the mountain. Therefore, it takes 86.8 MJ of work to lift the object to the top of Mount Everest.

(b) We find the difference between the gravitational potential energy at the top of the mountain and at sea level. The distance from the centre of Earth to the top of Mount Everest is

$$r_m = 6.371 \times 10^6 \text{ m} + 8848 \text{ m}$$

$$r_m = 6.3798 \times 10^6 \text{ m}$$

Using r_s to represent the distance from the centre of Earth to the surface at sea level, we can express the difference in gravitational potential energy as

$$\Delta U = -\frac{GMm}{r_m} - \left(-\frac{GMm}{r_s}\right)$$

$$= GMm\left(\frac{1}{r_s} - \frac{1}{r_m}\right)$$

Substituting in the known values gives

$$\Delta U = (6.674 \times 10^{-11} \text{ N} \cdot \text{m}^2/\text{kg}^2)(5.97 \times 10^{24} \text{ kg} \times 1.00 \text{ kg})$$

$$\times \left(\frac{1}{6.3710 \times 10^6 \text{ m}} - \frac{1}{6.3798 \times 10^6 \text{ m}}\right)$$

$$= 86.3 \text{ MJ}$$

Therefore, the work required to lift the 1.00 kg mass to the mountaintop is 86.3 MJ.

Making sense of the result

As expected, the two ways of calculating the work required give similar results. Since *g* decreases with height, the assumption that *g* remains equal to its value at sea level gives a result that slightly overestimates the work required but is not significant to the precision of these results.

EXAMPLE 11-6

Meteor Speeds

Estimate the speed that a meteoroid attains when it ablates as a meteor in Earth's upper atmosphere (see Figure 11-15). Use the following simplifying assumptions in obtaining the estimate:

■ The meteoroid started from rest at such a great distance from Earth that the two can be considered to have been infinitely far apart.
■ The influence of all other objects (e.g., the Sun and the Moon) is negligible.
■ Earth's upper atmosphere does not significantly slow the meteoroid up to the time of ablation, about 90 km above the surface.

While the last assumption in particular may seem unreasonable, in fact all three are valid within a few percent (see the Solution for more details).

Solution

When the meteoroid is infinitely far away, its gravitational potential energy is 0.

At meteor heights, the meteoroid's gravitational potential energy is $-\dfrac{GMm}{r}$, where *r* is the sum of the radius of Earth and 90 km, *M* is the mass of Earth, and *m* is the unknown

Figure 11-15 A meteor is the bright trail or streak caused by a meteorite entering Earth's atmosphere at high speed.

Courtesy R.L. Hawkes

mass of the meteoroid. Since we are assuming that there is no loss to aerodynamic drag, the amount of gravitational potential energy converted to kinetic energy as the meteorite falls to Earth is $\dfrac{GMm}{r}$. Using v_s to represent the speed of the meteorite at Earth's surface, we have

$$\frac{1}{2}mv_s^2 = \frac{GMm}{r}$$

$$v_s = \sqrt{\frac{2GM}{r}} = \sqrt{\frac{2(6.674 \times 10^{-11} \text{ N} \cdot \text{m}^2/\text{kg})^2(5.97 \times 10^{24} \text{ kg})}{6.461 \times 10^6 \text{ m}}}$$

$$= 1.11 \times 10^4 \text{ m/s}$$

Making sense of the result

A meteoroid that fell without air resistance to Earth's upper atmosphere would accelerate due to Earth's gravity to a speed of approximately 11.1 km/s. In fact, this approximation is within a few percent of the observed minimum speed of meteors in the upper atmosphere. Most meteors have higher speeds because they do not start from rest when at large distances from Earth. Astronomers who measure properties of meteors do need to provide a small correction for the aerodynamic drag in the upper atmosphere.

11-5 Force from Potential Energy

When a potential energy function exists, we can determine force from potential energy using derivatives, as we will show in this section (see also Section 6-8). Alternatively, we can determine a potential energy function using integration when we know the force everywhere. The concept that derivatives of potential energy functions represent the associated forces is applied in many branches of physics, for example, electromagnetism (see Chapter 21).

When a potential energy function, U, exists, we can obtain the force components through the following relationships:

$$F_x = -\frac{\partial U}{\partial x} \tag{11-13}$$

$$F_y = -\frac{\partial U}{\partial y} \tag{11-14}$$

$$F_z = -\frac{\partial U}{\partial z} \tag{11-15}$$

Note that these are partial derivatives, which means that the differentiation is only done with respect to an *explicit* appearance of the variable. This means that if, for example, U does not explicitly depend on y, then the partial derivative with respect to y will be zero and there will be no force in the y-direction. If you have only one-dimensional situations, you can think of the partial derivatives just like ordinary derivatives. We show in Example 11-7 how to obtain the gravitational force near Earth's surface from these relationships.

We can extend the relationship between force components and derivatives to situations in which a potential energy function, $U(r)$, depends only on radial distance. In the following equation, $\hat{r}$ is a unit vector in the direction of increasing r:

$$\vec{F} = -\frac{\partial U(r)}{\partial r}\hat{r} \tag{11-16}$$

EXAMPLE 11-7

Gravitational Force Near Earth's Surface

Find the gravitational force near Earth's surface where the potential energy function is given by $U = mgy$, where y is the height above the $U = 0$ point.

Solution

The y-component of the force is

$$F_y = -\frac{\partial U}{\partial y} = -mg$$

Since U does not depend on x or z, the partial derivatives yield a 0 result for these force components:

$$F_x = 0 \qquad F_z = 0$$

Making sense of the result

As expected, we found that the direction of the gravitational force is downward (negative y-direction) and the magnitude of the force is simply the weight of the object at the surface, mg.

We demonstrate in Example 11-8 how the gravitational force can be obtained from the expression for gravitational potential energy.

EXAMPLE 11-8

Deriving Gravitational Force from Gravitational Potential Energy

Determine the gravitational force, given that the function for gravitational potential energy can be written as

$$U(r) = -\frac{GMm}{r}$$

(continued)

Solution

Substituting $U(r)$ into Equation (11-16), we get

$$\vec{F} = -\frac{\partial U(r)}{\partial r}\hat{r} = -\frac{\partial\left(-\frac{GMm}{r}\right)}{\partial r}\hat{r}$$

$$= GMm\frac{\partial\left(\frac{1}{r}\right)}{\partial r}\hat{r} = -\frac{GMm}{r^2}\hat{r}$$

Making sense of the result

This approach gives us the same result as obtained using the law of universal gravitation. The force on mass m points in the $-\hat{r}$ direction, back toward the mass M because $\hat{r}$ points from M to m.

11-6 Escape Speed

In this section, we will consider the **escape speed**, which is the speed needed for an object to escape the gravitational pull of another object. What do we mean by "escape"? If you throw an object straight up, it will rise for a time, slowing as its kinetic energy decreases and its gravitational potential energy increases. The object will stop momentarily and then reverse direction. If we give the object a somewhat greater initial speed, it will go higher but will still eventually slow to a stop and then return. However, gravitational escape can occur when the object has sufficient initial speed that it does not ever stop completely and instead travels infinitely far away.

For an object to escape, the total energy must be positive: the object must have enough positive kinetic energy to overcome its negative gravitational potential energy. Then the object will still have some kinetic energy when it reaches a point infinitely far away, where its gravitational potential energy is zero.

We call the orbit or trajectory "bound" when the total orbital energy is negative and "unbound" when the total is positive. To determine the *minimum* speed for escape, we set the total orbital energy equal to zero:

$$E_{\text{total}} = T + U = \frac{1}{2}mv_{\text{esc}}^2 - \frac{GMm}{r} = 0 \qquad (11\text{-}17)$$

Solving for the escape speed, we derive the relationship given in Equation (11-18) below, where r is the initial distance of the escaping object from the centre of the larger object, and M is the mass of the larger object. Note that the mass of the escaping object, m, cancels out and does not affect the escape speed:

KEY EQUATION
$$v_{\text{esc}} = \sqrt{\frac{2GM}{r}} \qquad (11\text{-}18)$$

CHECKPOINT 11-7

Total Orbital Energy

Is the sum of the kinetic energy and the gravitational potential energy of Earth in its orbit about the Sun positive, zero, or negative?

ANSWER: The total energy is negative. We can show this starting with $F = ma$ and substituting the centripetal acceleration for a and the gravitational force for F; then compare the gravitational potential energy and the kinetic energy. We see in the next section that a circular orbit (or an elliptical orbit) has a negative total orbital energy.

EXAMPLE 11-9

Escape from Earth

Find the speed required to escape from Earth's surface. Neglect any work needed to overcome aerodynamic drag in the atmosphere.

Solution

We use Equation (11-18), substituting the mass of Earth for M and the radius of Earth for r:

$$v_{\text{esc}} = \sqrt{\frac{2GM}{r}}$$

$$= \sqrt{\frac{2(6.674 \times 10^{-11}\ \text{N}\cdot\text{m}^2/\text{kg}^2)(5.97 \times 10^{24}\ \text{kg})}{6.37 \times 10^6\ \text{m}}}$$

$$= 1.12 \times 10^4\ \text{m/s}$$

Making sense of the result

The speed needed to escape from Earth's surface is about 11.2 km/s. Note that this escape speed is very close to the speed for the falling meteoroid in Example 11-6. This makes sense because the potential energy that an object loses while falling from an infinite distance is equal to the energy needed to take the object to an infinite distance from the surface. Since the aerodynamic drag on an object passing through Earth's atmosphere is not negligible, the speed required to escape from Earth's surface is actually somewhat greater than 11.2 km/s.

Looking at Equation (11-18), we can see that the escape speed for a planet varies with $\sqrt{\dfrac{M}{r}}$, the mass of the planet divided by its radius. The volume of a sphere increases with the cube of its radius, so a large planet will have a greater escape speed than a smaller planet with the same density. In Example 11-10, we calculate the escape speed of a mid-sized asteroid.

While the calculations in this section have dealt with escape from the surface of an object, the same techniques can be applied for cases of escape starting from an altitude above the surface of an object.

EXAMPLE 11-10

Escape from an Asteroid

What speed is needed to escape from the surface of a spherical, homogeneous asteroid with a radius of 5.00 km and a density of 3200 kg/m^3? The asteroid has no atmosphere.

Solution

First, find the mass of the asteroid:

$$M = \rho V = \rho \frac{4\pi r^3}{3}$$

$$= 3200 \text{ kg/m}^3 \frac{4\pi(5.00 \times 10^3 \text{ m})^3}{3} = 1.68 \times 10^{15} \text{ kg}$$

Now, substitute values for M and r in Equation (11-18):

$$v_{esc} = \sqrt{\frac{2GM}{r}}$$

$$= \sqrt{\frac{2(6.674 \times 10^{-11} \text{ N} \cdot \text{m}^2/\text{kg}^2)(1.68 \times 10^{15} \text{ kg})}{5.00 \times 10^3 \text{ m}}}$$

$$= 6.70 \text{ m/s}$$

Making sense of the result

As expected, the escape speed from the asteroid is very small compared to the escape speed from Earth.

A **black hole** is an object that is sufficiently dense for its mass that not even light can "escape" from it. Since light and all other types of waves could not escape, if you were trapped inside a black hole, communication with the rest of the universe would be impossible. We will see in Chapter 30 that a proper view of black holes requires consideration of curved spacetime in general relativity (a black hole is essentially a region where the curvature becomes complete and the region is invisible to the rest of the universe). However, we can take a classical view of a black hole and set the escape speed to the speed of light in Equation 11-18 and solve for r, the value representing the effective size of the black hole. We call this "radius" of the black hole the Schwarzschild radius R_s in honour of the scientist who first obtained the precise general relativity solution (see more on this in Section 30-8):

$$R_s = \frac{2GM}{c^2} \tag{11-19}$$

The total mass inside the black hole is M, while c is the speed of light. In Example 11-11, we calculate the value for a large stellar black hole. While R_s represents the radius of the **event horizon** (the region invisible to the rest of the universe), it does *not* mean that the mass of the black hole is spread uniformly throughout this region, and most models have the mass concentrated into an infinitesimal point.

EXAMPLE 11-11

Schwarzschild Radius of a Stellar Black Hole

One of the stellar black holes detected by LIGO in 2015 (see Chapters 29 and 30) had a mass of about 36 times the mass of the Sun. What is the corresponding Schwarzschild radius?

Solution

We first need to express the black hole mass in SI units:

$$M = 36 \times 1.989 \times 10^{30} \text{ kg} = 7.16 \times 10^{31} \text{ kg}$$

Now we can directly apply Equation 11-19 to find the Schwarzschild radius:

$$R_s = \frac{2GM}{c^2}$$

$$= \frac{2 \times 6.674 \times 10^{-11} \text{ N} \cdot \text{m}^2 \cdot \text{kg}^{-2} \times 7.16 \times 10^{31} \text{ kg}}{(2.998 \times 10^8 \text{ m/s})^2}$$

$$= 1.06 \times 10^5 \text{ m}$$

The input mass is only given to two significant digits, so we should express the final answer as 1.1×10^5 m or 110 km.

Making sense of the result

Stellar black hole sizes are substantial, although this is still much smaller than the radius of our Sun, for example, which is about 7.0×10^5 km. The Schwarzschild radius is directly proportional to the black hole mass. Although a proper treatment of the size of black holes requires general relativity, it turns out that for a non-rotating spherically symmetric black hole in an otherwise empty universe the result obtained is precisely that given classically as Equation 11-19.

11-7 Kepler's Laws

Historically, the orbits of celestial objects were first understood empirically and only later explained in terms of Newtonian mechanics and the law of universal gravitation. The German mathematician and astronomer Johannes Kepler (1572–1630) deduced three laws of motion that describe the orbital motion of objects with negative total orbital energy. Although Kepler devised his laws to describe the orbits of planets in the Solar System, these laws apply to any object in a bound orbit.

INTERACTIVE ACTIVITY 11-1

Orbits and Speeds

In this activity, you will use the PhET simulation "My Solar System" to simulate the gravitational orbit of a mass around a much larger mass. You will see how different values of the initial speed changes the orbit. Also, you will qualitatively investigate how the speed varies in different portions of the orbit.

Kepler's Laws

1. Each planet moves in an elliptical orbit with the Sun at one of the two foci of the ellipse (Figure 11-16).

2. A line drawn from the Sun to a planet will sweep out the area inside the elliptical orbit at a constant rate (Figure 11-17).

3. The square of the period of the orbit is proportional to the cube of the semi-major axis of the orbit. The semi-major axis (a) is one-half of the longest dimension of the elliptical orbit (see Figure 11-16).

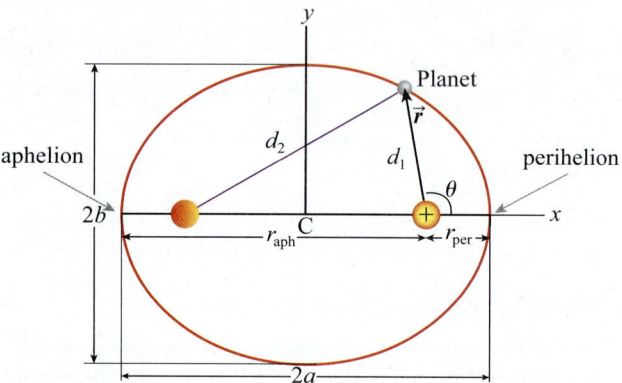

Figure 11-16 According to Kepler's first law, planets move in elliptical orbits with the Sun at one focus. The longest dimension of the ellipse is 2a (the major axis), while the smallest dimension is 2b (the minor axis). It is normal to use a coordinate system centred on the Sun, rather than the centre of the ellipse. The distance from the Sun to the nearest point in the orbit (r_{per}) is the perihelion distance, while the distance to the most distant point is the aphelion distance (r_{aph}).

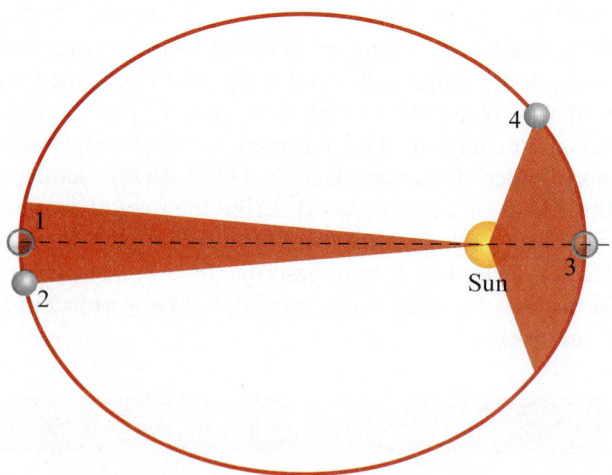

Figure 11-17 Kepler's second law: A line from the Sun to a planet will sweep equal areas in equal time periods. The areas shown in red are equal; therefore, the planet moves from point 1 to point 2 and from point 3 to point 4 in the same length of time.

Before we consider a proof of Kepler's laws, we must develop the terminology of ellipses. One-half of the longest dimension of the ellipse is called the **semi-major axis** (a), while one-half of the shortest dimension is called the **semi-minor axis** (b). If we select the centre of the ellipse (C in Figure 11-16) as the origin for a Cartesian coordinate system, the equation of the **ellipse** can be written as

$$\frac{x^2}{a^2} + \frac{y^2}{b^2} = 1 \tag{11-20}$$

Another, equivalent, way to define an ellipse is as containing all the points that satisfy the following relationship, where d_1 and d_2 are the distances from the two foci of the ellipse, as shown in Figure 11-16:

$$d_1 + d_2 = \text{constant} \tag{11-21}$$

A **circle** is a special case of an ellipse, one with $a = b$. The shape of an ellipse is determined by a parameter called the **eccentricity**, e, defined in terms of the semi-major and semi-minor axes in Equation 11-22:

$$e = \sqrt{\frac{a^2 - b^2}{a^2}} \tag{11-22}$$

The eccentricity is a dimensionless ratio bound between values of 0 and 1 for an ellipse. The nearer it is to 1, the more elongated is the ellipse.

The location where the planet is closest to the Sun is called the **perihelion** point (from two words meaning near and Sun), while the most distant point from the Sun is called the **aphelion** point. The corresponding distances are called the perihelion and aphelion distances. We see from Figure 11-16 that the sum of these distances equals the major axis of the ellipse:

$$r_{per} + r_{aph} = 2a \tag{11-23}$$

From Figure 11-16, if we take the special case of the perihelion (or aphelion) point distances, the sum must also equal $2a$, and therefore Equation 11-21 can be written as

$$d_1 + d_2 = 2a \tag{11-24}$$

The orbit depends ultimately on the angular momentum and the total energy, both of which are more readily measured in a coordinate system centred on the Sun. It also turns out to be easier to write the equation of the ellipse in polar form, with an angle θ (we define $\theta = 0$ at perihelion) and a distance from the sun r (see Figure 11-16). In this reference frame, an ellipse is the set of points that satisfy the following relationship, where C is a constant of the orbit:

$$r = \frac{C}{1 + e\cos\theta} \tag{11-25}$$

We can readily express C in terms of familiar quantities by using Equation 11-23 and recognizing that $\cos\theta = 1$ at perihelion and $\cos\theta = -1$ at aphelion:

$$\frac{C}{1 + e} + \frac{C}{1 - e} = 2a$$

$$\frac{C - Ce + C + Ce}{(1 + e)(1 - e)} = 2a$$

$$C = a(1 + e)(1 - e)$$

Therefore, for the general case of an elliptical orbit, we have the following:

KEY EQUATION
$$r = \frac{a(1 + e)(1 - e)}{1 + e\cos\theta} \tag{11-26}$$

We can readily obtain two very useful relationships by considering Equation 11-26 for perihelion ($\cos\theta = \cos 0 = 1$) and for aphelion ($\cos\theta = \cos 180° = -1$):

$$r_{per} = a(1 - e) \tag{11-27}$$

$$r_{aph} = a(1 + e) \tag{11-28}$$

Now we are ready to prove Kepler's first law. An implied part of the law is that the orbit is confined to a plane. This must be so because no external torque is exerted due to the central nature of the gravitational force (directed along the line joining the two masses). Since there is no torque, by the rotational analogue to Newton's second law, the angular momentum vector must stay constant and the motion is confined to a plane.

The elliptical nature of the orbit was historically established by carefully measuring planetary positions and showing that they fell on an ellipse. Kepler's laws were established before the time of Newton. It is possible, however, to derive Kepler's first law from Newtonian mechanics. We urge you to do problem 50, where we lead you step by step through the proof of Kepler's first law. You will see that the proof depends on only two physical principles, that angular momentum L and total energy E are conserved for the orbit. The proof in problem 50 also yields the meanings of the semi-major and semi-minor axis parameters, as stated in Equations 11-29 and 11-30:

KEY EQUATION
$$a = -\frac{GMm}{2E} \tag{11-29}$$

$$b = \sqrt{-\frac{(L/m)^2}{2(E/m)}} \tag{11-30}$$

For an elliptical orbit, the total energy, E, is negative (more on this in the next section), with the magnitude of the negative gravitational potential energy being larger than that of the positive kinetic energy. Therefore, the quantity under the square root of Equation 11-30 is positive, as is the semi-major axis value a in Equation 11-29. It is interesting that the semi-major axis depends only on the central large mass M and the orbital energy per unit mass E/m of the planet, and not the eccentricity of the orbit.

Let us now consider a proof of Kepler's second law: equal areas are swept out in equal times. In Figure 11-18 we show a small portion of an elliptical orbit, corresponding to a small time interval Δt.

As long as we consider a small interval, the area swept out will be approximately that of the triangle. Also, if the angle $\Delta\theta$ is very small, the two sides of the triangle, and the perpendicular height of the triangle,

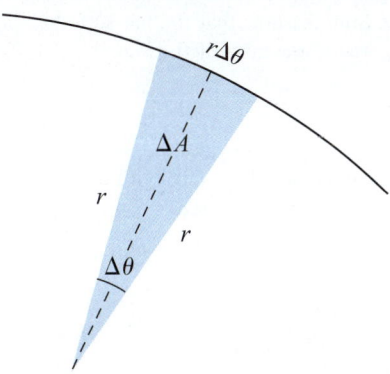

Figure 11-18 In a small portion of an elliptical orbit, the distance, r, will not change significantly for a small angular change $\Delta\theta$. The area swept out is approximately the triangle area ΔA.

will all be approximately equal to r. If $\Delta\theta$ is expressed in radians, the length of the arc will be $r\Delta\theta$, and for small angles this will also be approximately the base of the triangle. Therefore, the area of the triangle is

$$\Delta A = \frac{1}{2}r\Delta\theta r = \frac{1}{2}r^2\Delta\theta$$

Dividing both sides by the time interval, we obtain an expression for the rate at which area is being swept:

$$\frac{\Delta A}{\Delta t} = \frac{1}{2}r^2\frac{\Delta\theta}{\Delta t} \qquad (11\text{-}31)$$

If we replace this with the corresponding differentials, we have

$$\frac{dA}{dt} = \frac{1}{2}r^2\frac{d\theta}{dt} \qquad (11\text{-}32)$$

Now let us consider the angular momentum, L, for the planet. We can write this in terms of the component of the velocity perpendicular to a vector $\vec{r}$ from the Sun to the planet. That perpendicular component of the velocity can in turn be expressed in terms of the angular velocity, so we have

$$L = mv_{\perp}r$$

$$L = m\left(r\frac{d\theta}{dt}\right)r$$

$$L = mr^2\frac{d\theta}{dt}$$

If we write this in terms of the angular momentum per unit mass, we have

$$\frac{L}{m} = r^2\frac{d\theta}{dt} \qquad (11\text{-}33)$$

EXAMPLE 11-12

Kepler's Third Law

Derive Kepler's third law for a planet moving in a circular orbit around the Sun. Assume that the mass of the planet is negligible compared to the mass of the Sun.

Solution

The gravitational force provides the centripetal acceleration to maintain the circular orbit. Since both the force and the acceleration are in the same direction,

$$F_{\text{G}} = \frac{GMm}{r^2} = \frac{mv^2}{r}$$

Simplifying gives

$$\frac{GM}{r} = v^2$$

The period is the time for one orbit, and the circumference of a circle is $2\pi r$, so

Comparing Equations 11-32 and 11-33 gives us

$$\frac{dA}{dt} = \frac{1}{2}\frac{L}{m} \qquad (11\text{-}34)$$

Since, as we argued earlier, the absence of external torques requires that the angular momentum per unit mass be a conserved quantity, the area swept out per unit time is constant, and we have proven Kepler's second law.

Kepler's third law is easily proven for the special case of a circular orbit (see Example 11-12). We will outline the derivation for the general case below. We start by integrating both sides of Equation 11-34 over one orbital period:

$$\int_{\text{ellipse}} dA = \frac{1}{2}\frac{L}{m}\int_0^T dt$$

$$A_{\text{ellipse}} = \frac{1}{2}\frac{L}{m}T$$

The area of any ellipse is πab, so we have

$$\pi ab = \frac{1}{2}\left(\frac{L}{m}\right)T \qquad (11\text{-}35)$$

If we square both sides of Equation 11-35 and then substitute from Equation 11-30 for b, we obtain

$$\pi^2 a^2 b^2 = \frac{1}{4}\left(\frac{L}{m}\right)^2 T^2$$

$$4\pi^2 a^2\left(-\frac{(L/m)^2}{2(E/m)}\right) = \left(\frac{L}{m}\right)^2 T^2$$

$$4\pi^2 a^2\left(-\frac{m}{2E}\right) = T^2$$

We will multiply both sides of this by GM:

$$4\pi^2 a^2\left(-\frac{GMm}{2E}\right) = GMT^2$$

$$v = \frac{2\pi r}{T}$$

Substituting for v in the above simplified equation gives

$$\frac{GM}{r} = \frac{4\pi^2 r^2}{T^2}$$

$$T^2 = \frac{4\pi^2 r^3}{GM}$$

For a circular orbit, the radius r is the semi-major axis a, and

$$T^2 = \frac{4\pi^2}{GM}a^3$$

Making sense of the result

We have proved that the square of the period of a circular orbit is proportional to the cube of the semi-major axis, as stated in Kepler's third law.

By Equation 11-29, the expression in brackets is simply the semi-major axis a, and when we substitute that and rearrange, we obtain Kepler's third law. In fact, we have now, as well as showing that the square of the period is proportional to the cube of the semi-major axis, also found the proportionality constant:

$$T^2 = \frac{4\pi^2}{GM} a^3 \qquad (11\text{-}36)$$

Kepler's third law is universal: we can apply it to the motion of planets, asteroids, comets, or meteoroids in their orbit about the Sun, with M the mass of the Sun. We can also apply the law to satellites in their orbits about planets, but now with M being the mass of the planet. We can also apply Kepler's third law in situations involving other stars, and the exoplanets in orbit around them. It is frequently used in astrophysics to deduce an unknown mass from the observed period and semi-major axis.

To apply Kepler's law in situations with comparable mass, such as two stars in a binary system, we need to use both masses in the denominator (see Equation 11-37), and the orbit is with respect to the centre of mass of the system. You can always use this general form of Kepler's third law, with Equation 11-36 being an approximation valid when one mass is much larger than the other:

KEY EQUATION $\qquad T^2 = \frac{4\pi^2}{G(m_1 + m_2)} a^3 \qquad (11\text{-}37)$

We need one more result to complete this treatment for elliptical orbits. Equation 11-26 enables us to perfectly trace out any orbit if we know the eccentricity and semi-major axis (which can be found from the total energy and the angular momentum). But we have not yet seen how the speed varies in different parts of the orbit (except indirectly through Kepler's third law). Since total energy is conserved, the kinetic energy must be less in the outer part of the orbit where the potential energy is more. It would be convenient to have a relationship that gave us the

PEER TO PEER

My friends sometimes make mistakes and use the wrong mass in Kepler's third law. We are approximating the sum of masses from the general form of the law with just the large mass, since the smaller one is negligible in the sum. Therefore, we should always use the larger of the two masses. If it is an asteroid or planet around the Sun, use the Sun's mass. If it is a moon or satellite around a planet, use the planet's mass.

orbital speed, v, at any particular distance from the Sun, r. We will first state the relationship and then justify it:

KEY EQUATION $\qquad v = \sqrt{GM\left(\frac{2}{r} - \frac{1}{a}\right)} \qquad (11\text{-}38)$

As the first justification for Equation 11-38, we show in Example 11-13 that it is consistent with Newtonian gravitational forces for the special case of a circle.

If we square both sides of Equation 11-38 and then multiply each term by $\frac{1}{2}m$, we obtain

$$\frac{1}{2}mv^2 = \frac{1}{2}m\frac{2GM}{r} - \frac{1}{2}m\frac{GM}{a}$$
$$\frac{1}{2}mv^2 - \frac{GMm}{r} = -\frac{GMm}{2a} \qquad (11\text{-}39)$$

The left-hand side is clearly the sum of the kinetic energy and the gravitational potential energy, or in other words the total energy E. If we rearrange Equation 11-29 and solve for E we obtain

$$E = -\frac{GMm}{2a} \qquad (11\text{-}29a)$$

Therefore, both sides are the total energy, and we have shown that Equation 11-38 is consistent with conservation of total energy for any type of elliptical orbit.

EXAMPLE 11-13

Speed in Circular Orbit

Show that the speed given by Equation 11-38 is consistent with Newtonian mechanics and the law of universal gravitation for a circular orbit.

Solution

In the case of a circle, r has a constant value and $r = a$. If we substitute this into Equation 11-38 and square both sides, we have

$$v^2 = GM\left(\frac{2}{r} - \frac{1}{a}\right) = GM\left(\frac{2}{r} - \frac{1}{r}\right) = \frac{GM}{r}$$

We don't know the mass of the object in orbit, but let us call it m and multiply both sides of the above relationship by this mass:

$$mv^2 = \frac{GMm}{r}$$

Now divide each side by r:

$$\frac{GMm}{r^2} = \frac{mv^2}{r}$$

This is simply $F = ma$, with the left-hand side being the force of universal gravitation, and using centripetal acceleration on the right.

Making sense of the result

While this does not prove that Equation 11-38 is true in the general case, it does show that it is consistent with Newtonian gravity for the case of a circular orbit.

In planetary situations, distances are often expressed in astronomical units (AU). The **astronomical unit** is the mean Earth–Sun distance, so 1 AU = 1.50×10^{11} m. If using relationships such as Equations 11-26 through 11-28, you can work in AU if desired. However, when using Kepler's third law, Equation 11-38, or any relationship involving energy or angular momentum, you must work in conventional SI units.

EXAMPLE 11-14

Asteroid Orbit

An asteroid has a perihelion distance of 0.50 AU and an aphelion distance of 2.50 AU.

(a) What is the semi-major axis (in AU)?
(b) What is the period of the orbit, in years?
(c) What is the eccentricity of the orbit?
(d) What is the speed of the asteroid when it is at aphelion?

Solution

(a) The semi-major axis is defined as one-half the largest dimension of the elliptical orbit and is therefore 1.50 AU (one-half of 0.50 AU + 2.50 AU = 3.00 AU). Note by our significant digit rules for addition that the answer has 3 significant digits even though the perihelion distance has only 2.

(b) We will use Kepler's third law to calculate the orbital period from the semi-major axis. Since the mass of the Sun is huge compared to any asteroid, we can use the version applicable to the case $M \gg m$ (Equation 11-36) and do not need to know the mass of the asteroid. M is the mass of the Sun. Rearranging the equation, we have

$$T = \sqrt{\frac{4\pi^2 a^3}{GM}}$$

In SI units the semi-major axis is

$$a = 1.50 \text{ AU} \times \frac{1.496 \times 10^{11} \text{m}}{1 \text{ AU}} = 2.244 \times 10^{11} \text{m}$$

Substituting the mass of the Sun, $M = 1.989 \times 10^{30}$ kg, we obtain for the period

$$T = 5.80 \times 10^7 \text{s}$$

If we convert this to years, we have $T = 1.83$ y.

(c) We have the following relationships between perihelion distance r_{per}, aphelion distance r_{aph}, semi-major axis (a), and eccentricity (e):

$$r_{per} = a(1 - e) \qquad (11\text{-}27)$$

$$r_{aph} = a(1 + e) \qquad (11\text{-}28)$$

If we divide Equation 11-28 by Equation 11-27 we obtain

$$\frac{1 + e}{1 - e} = \frac{r_{aph}}{r_{per}} = \frac{2.50}{0.50} = 5.0$$

Note that since we are doing a ratio, it does not matter whether we leave the distances in AU or m. Solving, we obtain the following value for the eccentricity, which has no units:

$$e = 0.67$$

(d) We can use Equation 11-38 to find the speed at any particular r-value:

$$v = \sqrt{GM\left(\frac{2}{r} - \frac{1}{a}\right)}$$

In this case we have (note that quantities must be expressed in SI units to use this relationship), where M is the mass of the Sun,

$$a = 1.50 \text{ AU} = 2.24 \times 10^{11} \text{ m}$$

$$r = 2.50 \text{ AU} = 3.74 \times 10^{11} \text{ m}$$

Substituting, we determine that

$$v = 1.09 \times 10^4 \text{ m/s}$$

Making sense of the result

Since the aphelion distance is substantially more than that of Earth (2.50 AU vs 1.00 AU), we expect that the speed will be substantially less than that of Earth in its orbit (about 30 km/s). Therefore, the value of 10.9 km/s is reasonable. The orbital semi-major axis is more (1.50 AU) than that of Earth (1.00 AU), and therefore a period longer than 1 yr is expected, which we obtained (1.83 y). We expect an eccentricity value between 0 and 1 for this elliptical orbit, which we do obtain.

EXAMPLE 11-15

Mass of Saturn

Titan, the largest natural satellite of Saturn, takes 15.945 Earth days for one complete orbit. The semi-major axis of Titan's orbit around Saturn is 1.22×10^6 km. Find the mass of Saturn. Assume that the mass of Titan is much less than the mass of Saturn.

Solution

First, convert the period into SI units of seconds:

$$T = 15.945 \text{ days} \times \frac{24 \text{ h}}{1 \text{ day}} \times \frac{60 \text{ m}}{1 \text{ h}} \times \frac{60 \text{ s}}{1 \text{ m}}$$

$$= 1.3776 \times 10^6 \text{ s}$$

From Equation 11-36, when one mass is considerably smaller than the other, Kepler's third law becomes

$$T^2 = \frac{4\pi^2}{GM} a^3$$

Rearranging to solve for M and substituting the given data, we have

$$M = \frac{4\pi^2}{GT^2} a^3 = \frac{4\pi^2 (1.22 \times 10^9 \text{ m})^3}{(6.674 \times 10^{-11} \text{ N} \cdot \text{m}^2/\text{kg}^2)(1.3776 \times 10^6 \text{ s})^2}$$

$$= 5.66 \times 10^{26} \text{ kg}$$

Making sense of the result

This result agrees with published values for the mass of Saturn. In fact, the published values were obtained using Kepler's third law with measurements of various natural and artificial satellites.

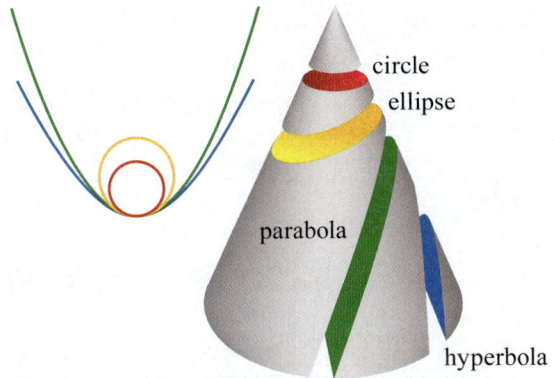

CHECKPOINT 11-9

Orbital Periods

An exoplanet around some star is at radius R and has a circular orbit of period T. The same star has a second planet in a circular orbit of radius $4R$. What is the period of the planet?

(a) $\dfrac{T}{8}$

(b) $\dfrac{T}{4}$

(c) $4T$

(d) $8T$

$$\frac{T_i^2}{R_i^3} = \frac{T_2^2}{(4R)^3}$$

ANSWER: (d) From Kepler's third law, the ratio of the square of the period to the cube of the semi-major axis (equal to radius for a circular orbit) is constant, so we have the following relationship, which yields $8T$ for the unknown period:

11-8 Types of Orbits

The total orbital energy (kinetic plus gravitational potential) is a conserved quantity and can be used to classify the type of orbit. Gravitational orbits are classified as one of the types of conic sections: ellipse, parabola, or hyperbola, with the circle being a special case of the ellipse. As the name implies, a **conic section** is a planar geometric shape corresponding to one of the ways that a cone can be cut (Figure 11-19). Cutting parallel to the base of a cone produces a circle. Cutting at an angle less than the angle of a cone's edge produces an ellipse. Cutting parallel to a cone's edge produces a parabola, and cutting at an even greater angle produces a hyperbola. The shape of each conic section can be defined by a relatively simple equation.

The type of gravitational orbit is determined by the total orbital energy:

■ When the total orbital energy is negative, the orbit is an ellipse.

■ When the total orbital energy is zero, the orbit is a parabola.

■ When the total orbital energy is positive, the orbit is a hyperbola.

A **hyperbolic orbit** is an unbound orbit, and an **elliptical orbit** is a bound orbit; a **parabolic orbit** is on

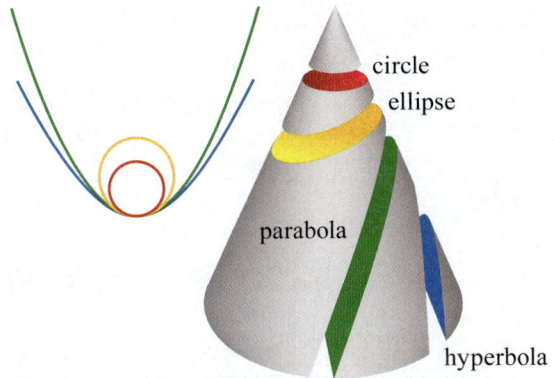

Figure 11-19 A cone can be cut at different angles to produce four types of conic sections. These sections correspond to the possible types of orbits.

INTERACTIVE ACTIVITY 11-3

Types of Orbits

In this activity, you will use the PhET simulation "My Solar System" to qualitatively explore elliptical and hyperbolic orbits.

the borderline between bound and unbound orbits. When an object is in a hyperbolic or parabolic orbit with respect to the Sun, the object will sweep past the Sun once and never return. Later chapters will describe how electromagnetic and nuclear forces can create similar types of motion for subatomic particles. In Example 11-16 we demonstrate how to determine the type of orbit.

CHECKPOINT 11-10

Total Orbital Energy

Classify the orbit of each of the following:

(a) A meteoroid approaches Earth with a geocentric (Earth-relative) speed of 25 km/s.

(b) The altitude of a satellite orbiting Earth varies from 1000 km to 3000 km.

ANSWER: (a) The meteoroid is approaching Earth with a speed higher than the escape velocity for Earth, so it is clear that the orbit must be hyperbolic (relative to Earth). (b) The satellite orbit is elliptical from the description given.

EXAMPLE 11-16

Interstellar Visitor

A spacecraft is observed to be at a distance of 2.00 AU from the Sun and is moving at a speed of 38.0 km/s.

(a) What type of orbit (relative to the gravitational field of the Sun) is the spacecraft following?

(b) Could the spacecraft have come from interstellar space?

Solution

(a) To determine the type of orbit, we need to know if the total orbital energy of the spacecraft is negative, zero, or positive. Although the mass of the spacecraft is unknown, we can calculate the total orbital energy per unit mass:

(continued)

$$\frac{E_{\text{total}}}{m} = \frac{T}{m} + \frac{U}{m} = \frac{1}{2}v^2 - \frac{GM}{r}$$

where M is the mass of the Sun, $v = 3.80 \times 10^4$ m/s, and $r = 2.00 \times 1.496 \times 10^{11}$ m $= 2.99 \times 10^{11}$ m. Substituting into the above equation gives

$$\begin{aligned}
\frac{E_{\text{total}}}{m} &= \frac{1}{2}(3.8 \times 10^4 \text{ m/s})^2 \\
&\quad - \frac{(6.674 \times 10^{-11} \text{ N} \cdot \text{m}^2/\text{kg}^2)(1.989 \times 10^{30} \text{ kg})}{2.99 \times 10^{11} \text{ m}} \\
&= +2.78 \times 10^8 \text{ J/kg}
\end{aligned}$$

The total orbital energy is positive, so the object must be in a hyperbolic orbit.

(b) Since the orbit is hyperbolic, the spacecraft would have come from interstellar space unless its engines (or booster rocket) provided enough kinetic energy to move it into its current orbit.

Making sense of the result

Earth moves at a speed of approximately 30 km/s in its near-circular orbit. Since the spacecraft is moving faster at twice the distance from the Sun, it is not unreasonable that the orbit is hyperbolic. Note that we did not need to know the direction that the spacecraft was moving to determine the type of orbit.

MAKING CONNECTIONS

Space Debris and Meteoroids

Both meteoroids and space debris pose small, but not insignificant, threats to space operations. The impact of even a tiny object can be disastrous for a satellite. Space debris impacts other objects in orbit with a lower relative speed than meteoroids, but the probability of impact is greater for space debris. The term *space debris* refers to artificial material that has ended up in Earth orbit as a result of our use of space. Such debris varies from specks of paint to much larger pieces of spacecraft. In 2008, an astronaut on a space walk accidentally released a

tool kit that ended up in Earth orbit. NASA and other space agencies track thousands of pieces of space debris larger than a centimetre across, but the majority of space debris is much smaller. Space debris eventually spirals down toward Earth. Even at orbital altitudes, there is a tenuous atmosphere that produces a tiny drag force that ultimately results in the debris falling from orbit. Relative to Earth, space debris has negative total orbital energy, and meteoroids have positive total orbital energy.

It is important to realize that both total orbital energy (kinetic plus gravitational potential) and angular momentum (see Chapter 9) are conserved

quantities. Example 11-17 illustrates the power of using conservation approaches in solving orbital problems.

EXAMPLE 11-17

Interplanetary Spacecraft Orbit

An interplanetary spacecraft has a speed of 32.0 km/s when it is at a distance of 1.20 AU from the Sun. It is at this time moving in a direction perpendicular to the line between the spacecraft and the Sun, as shown in Figure 11-20.

(a) What type of orbit is the spacecraft in? Justify your answer.
(b) When it is at a distance of $r = 2.00$ AU from the Sun, what is the speed of the spacecraft?
(c) What is the aphelion distance of the orbit (expressed in AU)?
(d) What is the angle, ϕ, between a line to the Sun and the velocity of the spacecraft (see Figure 11-20) when it is at a distance of 2.0 AU from the Sun?

Solution

(a) To find the type of orbit, we need to determine if the total orbital energy is negative, zero, or positive, resulting in an elliptical, a parabolic, or a hyperbolic orbit, respectively. We do not know the mass of the spacecraft, so we will calculate the energy per unit mass, since we only need the sign. Since the velocity vector is perpendicular to the line toward the centre of the Sun, the spacecraft must be at either the perihelion or the aphelion point, and since it goes to a larger distance later, it must be at perihelion when it is at 1.20 AU. Since some calculated quantities are used in

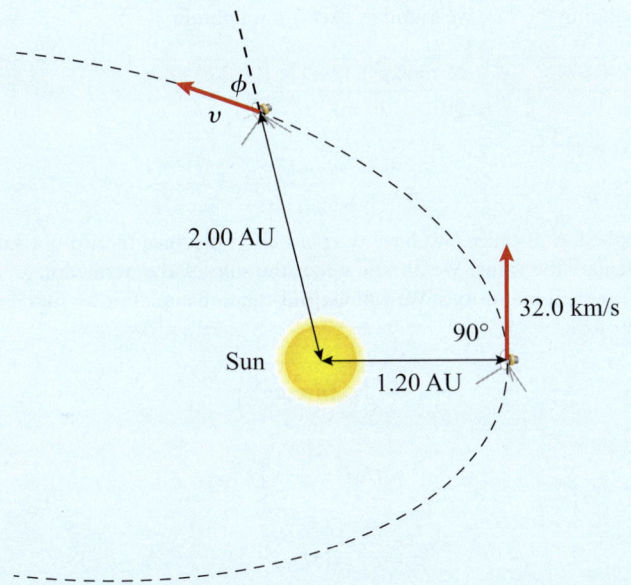

Figure 11-20 Example 11-17: A spacecraft is moving at 32.0 km/s when it is at 1.20 AU from the Sun and travelling perpendicular to a line to the Sun. You are asked to find the speed v and the angle ϕ when the spacecraft is 2.00 AU from the Sun.

later parts of the question, we will express them to extra precision, but will round all stated answers to the correct number of significant digits. Therefore, the perihelion distance is given by the following, and we will call the speed at this time v_{per}:

$$r_{per} = 1.20 \text{ AU} \times \frac{1.496 \times 10^{11} \text{ m}}{1 \text{ AU}} = 1.795 \times 10^{11} \text{ m}$$

$$v_{per} = 32.0 \text{ km/s} \times \frac{1000 \text{ m}}{1 \text{ km}} = 3.20 \times 10^4 \text{ m/s}$$

If we let E represent the total orbital energy of the spacecraft, m the mass of the spacecraft, and M the mass of the Sun, we have

$$\frac{E}{m} = \frac{1}{2}v_{per}^2 - \frac{GM}{r_{per}}$$

$$\frac{E}{m} = \frac{1}{2}(3.20 \times 10^4 \text{ m/s})^2 - \frac{(6.674 \times 10^{-11} \text{ N} \cdot \text{m}^2/\text{kg}^2)(1.989 \times 10^{30} \text{ kg})}{(1.795 \times 10^{11} \text{ m})}$$

$$\frac{E}{m} = -2.27 \times 10^8 \text{ J/kg}$$

Since the energy per unit mass is negative, the orbit must be elliptical.

(b) We can use conservation of energy to calculate what the speed must be at the new position. If we call r and v the distance (from Sun) and speed at this new position, respectively, we have

$$\frac{1}{2}v^2 - \frac{GM}{r} = \frac{E}{m}$$

where $r = 2.00 \text{ AU} \times \dfrac{1.496 \times 10^{11} \text{ m}}{1 \text{ AU}} = 2.992 \times 10^{11} \text{ m}$

We can rearrange the expression to solve for the unknown speed v:

$$v = \sqrt{\frac{2GM}{r} + 2\frac{E}{m}}$$

(continued)

Substituting values, including the E/m we found in part (a), we obtain

$$v = \sqrt{\frac{2(6.674 \times 10^{-11}\,\text{N}\cdot\text{m}^2/\text{kg}^2)(1.989 \times 10^{30}\,\text{kg})}{(2.992 \times 10^{11}\,\text{m})} + 2(-2.27 \times 10^{8}\,\text{J/kg})}$$

$$v = 2.08 \times 10^{4}\,\text{m/s}$$

$$v = 20.8\,\text{km/s}$$

(c) We are asked for the aphelion distance. We have the perihelion distance but do not know the eccentricity, so we can't directly calculate the value. We do know that the sum of the perihelion and aphelion distances must equal $2a$, double the semi-major axis. We will use this relationship after we first use Equation 11-38 to calculate the semi-major axis.

$$v = \sqrt{GM\left(\frac{2}{r} - \frac{1}{a}\right)}$$

$$v^2 = GM\left(\frac{2}{r} - \frac{1}{a}\right)$$

$$\frac{1}{a} = \frac{2}{r} - \frac{v^2}{GM}$$

Substituting the r- and v-values from part (b) (using the values in SI units), we obtain the following value for the inverse of the semi-major axis:

$$\frac{1}{a} = \frac{2}{(2.99 \times 10^{11}\,\text{m})}$$

$$- \frac{(2.080 \times 10^{4}\,\text{m/s})^2}{(6.674 \times 10^{-11}\,\text{N}\cdot\text{m}^2/\text{kg})(1.989 \times 10^{30}\,\text{kg})}$$

$$\frac{1}{a} = 3.427 \times 10^{-12}\,\text{m}^{-1}$$

Therefore,

$$a = 2.918 \times 10^{11}\,\text{m} = 2.918 \times 10^{11}\,\text{m} \times \frac{1\,\text{AU}}{1.496 \times 10^{11}\,\text{m}} = 1.951\,\text{AU}$$

From the definition of semi-major axis, we have

$$r_{\text{per}} + r_{\text{aph}} = 2a$$

$$1.795 \times 10^{11}\,\text{m} + r_{\text{aph}} = 2(2.918 \times 10^{11})\,\text{m}$$

$$r_{\text{aph}} = 4.04 \times 10^{11}\,\text{m}$$

$$r_{\text{aph}} = 4.04 \times 10^{11}\,\text{m} \times \frac{1\,\text{AU}}{1.496 \times 10^{11}\,\text{m}} = 2.70\,\text{AU}$$

(d) To find the angle, we need to use the fact that as with total energy, the angular momentum is also a conserved quantity. If we define the angle as shown in the diagram and equate the angular momentum at perihelion with the angular momentum at that point, we have

$$mv_{\text{per}}r_{\text{per}}\sin 90° = mvr\sin\phi$$

$$\sin\phi = \frac{v_{\text{per}}r_{\text{per}}}{vr}$$

$$= \frac{(3.20 \times 10^{4}\,\text{m/s})(1.795 \times 10^{11}\,\text{m})}{(2.080 \times 10^{4}\,\text{m/s})(2.992 \times 10^{11}\,\text{m})}$$

$$= 0.923$$

This gives a value of 67.4° for the angle.

Making sense of the result

It is not surprising that we find the orbit is elliptical, since the original speed at 1.20 AU is just a little more than that of Earth in its orbit. We expect the speed to be less at a greater distance (from conservation of orbital energy) since the gravitational potential energy is more, meaning the kinetic energy is less—this is what we find in part (b). Given that we are at 2.00 AU, not that much less than the aphelion distance of 2.72 AU (part (c)), it is not surprising that the angle in (d) is fairly large (it will become 90° by the time the spacecraft reaches aphelion).

Finding Debris from Other Planetary Systems

Planets form along with their parent stars from a cloud of gas and dust called a nebula. The formation process likely also produces large numbers of small meteoroids, some of which escape the gravitational pull of the star (through collisions, near collisions, and radiation processes). We can detect escaped meteoroids that reach our Solar System by finding meteoroids not gravitationally bound to the Sun. Canada is a world leader in the study of meteoroids, including the search for debris from other planetary systems. Researchers Peter Brown, Margaret Campbell-Brown, and Rob Weryk of the University of Western Ontario use the Canadian Meteor Orbit Radar (CMOR) and an automated super-sensitive digital camera system to compute meteor orbits and to isolate those on hyperbolic orbits. Figure 11-21 shows a map of meteor origins from a single day of observations with this facility. Each year, close to one million meteoroids are detected by this facility. A tiny fraction of

these are not in orbits within our Solar System but are material ejected from another planetary system.

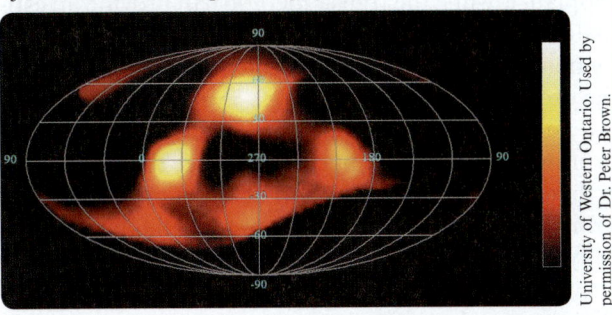

Figure 11-21 A plot of meteor origin directions from the CMOR meteor radar: the brighter colour indicates more meteors from that direction.

Sun, Planet, and Comet

In this activity, you will use the PhET simulation "My Solar System" to animate a three-body system with the Sun, a planet, and a much smaller mass comet. You will see how the orbit of the comet changes depending on how similar the planet and comet masses are. Through this activity you are introduced to the concept of gravitational **precession**, an idea that played a crucial role in the establishment of general relativity (Chapter 30).

Star and Planet Orbital Motion

In this activity, you will use the PhET simulation "My Solar System" to study how a large planet and a star move about a common centre of mass, and how the motion changes when the ratio of the planet mass to the star mass is larger. This forms the heart of the radial velocity method of exoplanet detection.

11-9 Detection of Exoplanets

The distinguished Russian American astronomer Otto Struve (1897–1963) made numerous advances, one of which is his technique for detecting exoplanets. Twenty-five years after his death, a Canadian research team used his technique to provide the first evidence that exoplanets do indeed exist. As of 2016 there were more than 3000 confirmed exoplanets, and more are being discovered every year, many of them by NASA's Kepler space telescope.

While at least seven techniques have been used to detect exoplanets, one of the most successful has been Doppler spectroscopy (also called **radial velocity**). When a planet and a star are in orbit, we often consider the position of the star to be fixed because the star is so much more massive than the planet. However, it is the centre of mass of the system that is fixed, and both the star and the planet move in elliptical orbits about this centre. The star's orbit is very small compared to the planet's orbit.

The Discovery of Exoplanets

In 1988, Canadian scientists published a paper in the prestigious *Astrophysical Journal* that presented clear evidence that seven of the stars they studied had objects in orbit with masses that could be planets. Led by Bruce Campbell, then of the University of Victoria and the Herzberg Institute of Astrophysics, the discovery team included Gordon Walker of the University of British Columbia and Stephenson Yang, who was then working at the Canada-France-Hawaii Telescope (CFHT) but is now at the University of Victoria. They found very small but regular shifts in the wavelengths of light from the stars. These shifts showed that the stars were in orbital motion, most likely with unseen planets.

In 1992, Dale Frail, a graduate of Acadia University and the University of Toronto working at the National Radio Astronomy Observatory (NRAO) in New Mexico, conclusively detected two planets in orbit around a pulsar (a rotating neutron star). Some astronomers consider 1992 to be the date of the first detection of an exoplanet because Frail made a more definitive claim of planet detection.

Another important and successful method of exoplanet detection, called the **transit method**, infers the presence of an exoplanet by the small temporary decrease in light from the star while an exoplanet passes between us and the star. As we will see, the radial velocity technique allows us to estimate the mass of the exoplanet, while the transit method can provide an estimate of the radius of the planet. If we have detection by both methods, we can also infer the mass density of the exoplanet.

In most cases we cannot directly see a planet orbiting a distant star, but we can infer that the planet is there from the slight motion of the star—as the star orbits the centre of mass of the system, the star periodically moves toward us and away from us (Figure 11-22).

The light from a star moving away from us shifts slightly to a lower frequency (and a correspondingly longer wavelength); we call this effect redshift. (This is similar to a sound having a lower frequency when the source is moving away from you.) When the star moves toward us, its light shifts to shorter wavelengths, a blueshift (Figure 11-23). Note that the term *redshift* does not mean that the light becomes red; it means the light will have a longer wavelength (more toward the red end of the spectrum). We can use redshift and blueshift to determine star motions, and from that infer the presence of orbiting exoplanets.

The Doppler shift for electromagnetic waves including light is a consequence of special relativity, and we derive the precise relationships in Chapter 30. If the speed of the star is much less than the speed of

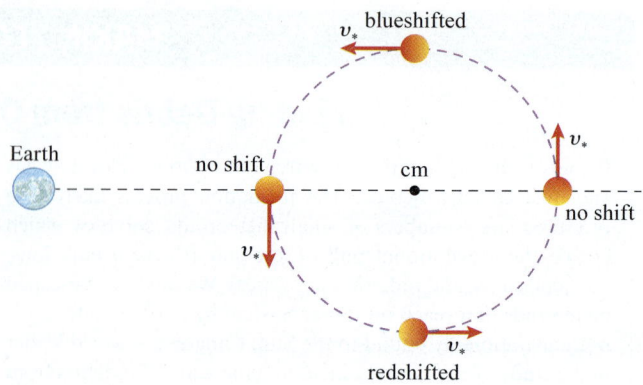

Figure 11-23 A star–planet system observed from Earth. The light from the star is blueshifted when the star moves toward us and redshifted when the star moves away from us.

light, it can be shown that the shift in wavelength, $\Delta\lambda$, compared to the wavelength emitted by the source, λ_s, is given by Equation 11-40, where v_s is the speed of the source:

$$\frac{\Delta\lambda}{\lambda_s} \approx \frac{v_s}{c} \qquad (11\text{-}40)$$

Therefore, by precisely measuring the shift in wavelength as the star moves toward and away from us, we can estimate the speed of the star.

We show in Figure 11-23 the situation to detect an exoplanet by this method. As the exoplanet orbits the centre of mass (cm), the star performs a similar, but much smaller, orbit. When the star is moving toward us, we will see the light blueshifted, and when it is moving away it will be redshifted.

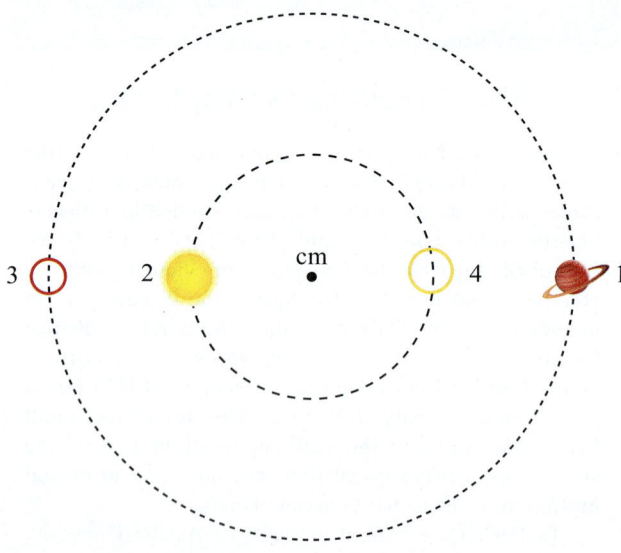

Figure 11-22 The motion of an exoplanet and a star about their centre of mass. When the planet (orange) is at point 1 on the right of the centre of mass, the star (yellow) will be on the left of the centre of mass at point 2. Similarly, when the planet is on the left at point 3, the star will be at point 4 on the right of the centre of mass. Note that the diagram is not to scale: the relative size of the orbit of the star will be much smaller than shown here.

CHECKPOINT 11-11

Exoplanet and Star

Which statement correctly describes the motion of a star and an exoplanet in orbit about a common centre of mass?
(a) They have the same period, but the semi-major axis of the star orbit is larger.
(b) They have the same period, but the semi-major axis of the star orbit is smaller.
(c) They have the same semi-major axis value, but the period of the star orbit is larger.
(d) They have the same semi-major axis value, but the period of the star orbit is smaller.

ANSWER: (b) By conservation of momentum, the larger mass must move through a smaller distance range, so the semi-major axis is less. The masses move to keep the centre of mass constant, so the period must be the same (e.g., when the star is most distant the exoplanet will be closest, and vice versa).

To calculate the mass of the exoplanet, we do the following. Note that this analysis assumes that we are observing from the plane of the exoplanet–star system, that there is a single exoplanet, that the mass of the star is much larger than the mass of the exoplanet, and that the orbit of the planet is circular.

1. Perform careful spectral measurements on the star, and plot these precise measurements of wavelength. A star with a single exoplanet will show a regular increase and decrease in wavelength as it moves in its orbit.

2. From the plot, we estimate the peak redshift and peak blueshift (they should have similar magnitudes), which become $\Delta\lambda$ in Equation 11-40. We also know (from what atomic spectral line we are studying) λ_s and the speed of light. Therefore, we can calculate the maximum orbital speed of the star, v_s, from Equation 11-40.

3. We use some means to estimate the mass of the visible star (M_s). Astronomers have various means to estimate the mass of stars from their absolute brightness and their spectral properties.

4. By conservation of linear momentum (see Chapter 7), the magnitude of the momentum of the star moving toward us equals that of the planet moving away from us. Therefore, if we call m_p the mass of the planet and v_p the speed of the planet, we have

$$M_s v_s = m_p v_p \tag{11-41}$$

5. From the plot of step 1 we can estimate the period, T, of the orbital motion. We assume that the mass of the star is much larger than the mass of the exoplanet, so we can use the simplified form of Kepler's third law. Solving for the semi-major axis a, we have

$$a = \sqrt[3]{\frac{GM_s T^2}{4\pi^2}} \tag{11-42}$$

This is the semi-major axis of the planet's orbit about the much larger star.

6. If the planet is in a circular orbit, its speed must be given by the following, since it moves the circumference of the orbit in one orbital period:

$$v_p = \frac{2\pi a}{T} \tag{11-43}$$

7. We can substitute the value for v_p from step 6 into Equation 11-41 to solve for the only remaining unknown, the mass, m_p, of the planet.

We demonstrate this for an actual exoplanet calculation in Example 11-18. The assumptions we have made are not always justified, and in those cases the analysis becomes more complicated and uncertain. In a number of cases, multiple exoplanets have been detected around a single star. An artist's conception of one of those cases is shown in Figure 11-24.

Figure 11-24 Artist's conception of the HD 7924 (the star) exoplanet system looking back toward our Sun with three of its exoplanets visible.

ESO/L. Calçada. This image is licensed under a Creative Commons Attribution 4.0 International License, https://creativecommons.org/licenses/by/4.0/

EXAMPLE 11-18

Exoplanet Mass Calculation

A Doppler spectroscopy plot for a star (already interpreted in terms of speed of the star) is shown in Figure 11-25. Assume that the mass of the star is 1.06 times the mass of our Sun, as well as the assumptions we made in this chapter.

(a) What is the wavelength shift ratio compared to the original wavelength $\left(\text{i.e., } \dfrac{\Delta\lambda}{\lambda_s}\right)$ suggested by the plotted velocity?

(b) If the original source wavelength emitted is 486.13320 nm, what is the observed range of wavelengths from the star?

(c) Calculate the mass of the exoplanet.

(d) Express your answer from (c) in terms of the mass of the planet Jupiter.

(continued)

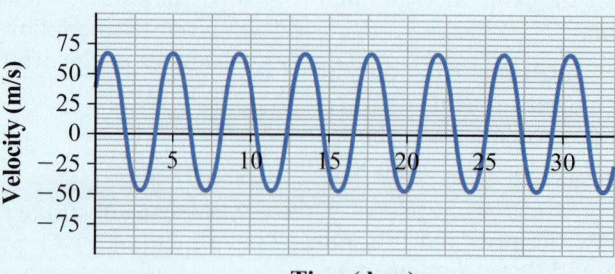

Figure 11-25 Data from an exoplanet detection. The velocity of the star is plotted versus the date (in units equivalent to days). The curve is a best sinusoidal fit to the data.

Solution

(a) There appears to be a slight systematic shift, with the lower velocities peaking at a slightly lower value (about 49 m/s) than the upper ones (about 67 m/s). Only 2 significant digits are justified in this question, and given the graphical inputs your answers may vary in the second digit. However, we will carry an extra digit in some intermediate steps. We will take an average of 58 m/s for our calculations. From Equation 11-40 we have

$$\frac{\Delta\lambda}{\lambda_s} \approx \frac{v_s}{c} \approx \frac{58 \text{ m/s}}{3.00 \times 10^8 \text{ m/s}} \approx 1.93 \times 10^{-7} \approx 1.9 \times 10^{-7}$$

(b) To find the range of received wavelengths (keeping in mind that there is both a blueshift and a redshift at different parts of the orbit), we multiply the ratio in (a) by the emitted wavelength and add and subtract that from the original wavelength:

$$\Delta\lambda \approx 1.93 \times 10^{-7} \times 486.133\,20 \text{ nm} = 9.4 \times 10^{-5} \text{ nm}$$

We can add and subtract this from the original wavelength to get the predicted range in received values of 486.133 11 nm to 486.133 29 nm.

(c) Note that in the following we are keeping one additional significant digit in the early calculations but will round to the correct two significant digits at the last step. We can estimate the orbital period from about 31 days required for 7 cycles, or 4.43 days per cycle, or $T = 3.83 \times 10^5$ s. We are told that the mass of the star is 1.06 times the Sun's mass, or $M_s = 2.11 \times 10^{30}$ kg.

We can use Kepler's third law to find the semi-major axis for the planetary orbit:

$$a = \sqrt[3]{\frac{GM_sT^2}{4\pi^2}} = \sqrt[3]{\frac{(6.674 \times 10^{-11} \text{ N·m}^2/\text{kg}^2)(2.11 \times 10^{30} \text{ kg})(3.83 \times 10^5 \text{ s})^2}{4\pi^2}} = 8.05 \times 10^9 \text{ m}$$

Now we can calculate the estimated orbital speed of the planet:

$$v_p = \frac{2\pi a}{T} = \frac{2\pi(8.05 \times 10^9 \text{ m})}{(3.83 \times 10^5 \text{ s})} = 1.32 \times 10^5 \text{ m/s}$$

Now we can use Equation 11-41, which was based on conservation of linear momentum:

$$m_p = \frac{M_s v_s}{v_p}$$

$$= \frac{(2.11 \times 10^{30} \text{ kg})(58 \text{ m/s})}{1.32 \times 10^5 \text{ m/s}}$$

$$= 9.3 \times 10^{26} \text{ kg}$$

(d) The mass of Jupiter is 1.9×10^{27} kg, so this planet is about 0.49 times the mass of Jupiter.

Making sense of the result

Part (a) shows how precise measurements for exoplanet detection need to be, since the ratio obtained is less than one part in a million. In part (b), this shows that you need to have a wavelength known to many decimal places to have any chance to detect an exoplanet by Doppler spectroscopy. While the orbital velocity of the planet calculated in (c) might at first seem large, remember that Earth has an orbital speed of about 30 km/s, and with such a short period we would expect this planet's speed to be considerably more. While exoplanets with masses comparable to Earth have been detected, the vast majority have masses comparable to Jupiter, since such planets are much easier to detect.

KEY CONCEPTS AND RELATIONSHIPS

According to the law of universal gravitation, every point object in the universe attracts every other object with a force that depends directly on the two masses and inversely on the square of the distance between the objects. The equivalence principle states that the inertial mass in Newton's second law, and the gravitational mass in the law of universal gravitation, are precisely equal. This principle also forms the basis for general relativity, a view in which gravity is a distortion of spacetime geometry rather than a force.

Universal Gravitational Force

The magnitude of the gravitational force between any pair of point masses in the universe is $F_G = \dfrac{GMm}{r^2}$, where M and m are the masses, r is the distance between them, and G is the universal gravitational constant. The force is directed toward the centre of the attracting mass.

We can use the superposition principle to add the gravitational forces of individual point masses as vectors to obtain a net total gravitational force.

Acceleration Due to Gravity

The acceleration due to gravity at a point a distance r from the centre of a spherically symmetric mass can be readily shown from Newton's second law and the universal gravitational force to have the following magnitude: $g = \dfrac{GM}{r^2}$.

Gravitational Potential Energy

The general expression for gravitational potential energy between spherically symmetric masses M and m, with centres separated by a distance r, is given by $U = -\dfrac{GMm}{r}$.

The gravitational potential energy is zero when the masses are infinitely far apart and negative everywhere else.

Force from Potential Energy

The derivatives of a potential energy function U give the components of the associated force, for example,

$$F_x = -\frac{\partial U}{\partial x}, F_y = -\frac{\partial U}{\partial y}, F_z = -\frac{\partial U}{\partial z}, \text{ and } F_r = -\frac{\partial U}{\partial r}.$$

Total Orbital Energy

By calculating the sum of an object's kinetic energy and gravitational potential energy, we can determine the type of orbit. If the total is negative, the orbit is elliptical; if the total is zero, the orbit is parabolic; if the total is positive, the orbit is hyperbolic. A positive total also corresponds to gravitational escape. The total energy and angular momentum are both conserved quantities.

Kepler's Laws

1. Planetary orbits are ellipses with the Sun at one focus.
2. A line joining a planet to the Sun sweeps out equal areas in equal times.
3. The square of the orbital period (T) is proportional to the cube of the semi-major axis (a) of the orbit. For a pair of objects with masses m_1 and m_2 orbiting around their centre of mass, this law is expressed as

$$T^2 = \frac{4\pi^2}{G(m_1 + m_2)} a^3 \qquad \text{(11-37)}$$

Orbital Speed

The speed of an object in an elliptical orbit with semi-major axis a about a mass M varies with distance r from M according to the following relationship:

$$v = \sqrt{GM\left(\frac{2}{r} - \frac{1}{a}\right)} \qquad \text{(11-38)}$$

APPLICATIONS

gravitational field mapping in geophysics, satellite orbits, interplanetary travel, predicting tides, detecting exoplanets, predicting the paths of comets and asteroids, producing weightlessness

KEY TERMS

acceleration due to gravity, aphelion, asteroids, astronomical unit, black hole, centre of mass, circle, comet, conic section, conservative force field, dwarf planet, eccentricity, ellipse, elliptical orbit, escape speed, equivalence principle, event horizon, exoplanet, galaxies, general relativity, gravitational constant, gravitational mass, hyperbolic orbit, inertial mass, Kepler's laws, law of universal gravitation, main asteroid belt, meteor, meteorite, meteoroids, parabolic orbit, perihelion, planet, precession, principle of superposition, radial velocity, satellites, semi-major axis, semi-minor axis, Solar System, star, superposition principle, transit method, weight, weightless

QUESTIONS

1. When Earth exerts a force of 550 N on a person, what is the magnitude of the force that the person exerts on Earth?
   (a) exactly 0
   (b) 550 N
   (c) 550 N times the ratio of the mass of the person to Earth's mass
   (d) 550 N times the ratio of the mass of Earth to the mass of the person

2. Consider a planet that has the same mass as Earth, but eight times the average mass density. What would your weight on this planet be if your weight on Earth is W?
   (a) W
   (b) $2W$
   (c) $4W$
   (d) $8W$

3. Weigh scales are often marked in units of kilograms. Why is this marking incorrect?

4. What is your weight at the centre of Earth?

5. Consider two spherical homogeneous planets, both with exactly the same density. The radius of one planet is twice the radius of the other planet. Which of the following statements is true?
   (a) The two planets have exactly the same gravitational acceleration at their surfaces.
   (b) The larger planet has a greater gravitational acceleration at its surface.
   (c) The smaller planet has a greater gravitational acceleration at its surface.
   (d) More information is needed to compare the gravitational accelerations.

6. On a planet that has the same mass as Earth but a smaller radius, the escape speed from the surface is
   (a) less than the escape speed on Earth
   (b) more than the escape speed on Earth
   (c) the same as on Earth
   (d) impossible to compare without more information

7. At the surface of the planet in question 6, the gravitational acceleration is
   (a) < 9.81 m/s^2
   (b) ≈ 9.81 m/s^2
   (c) > 9.81 m/s^2
   (d) impossible to compare without more information

8. Describe a simple experiment to test the validity of the equivalence principle.

9. Which of the following statements is true?
   (a) G is constant, but g varies with location.
   (b) Both G and g vary with location.
   (c) Both G and g are constant everywhere in the universe.
   (d) G varies with location, and g is constant.

10. Three objects fall to Earth from a location some distance away. Object A starts from rest, object B has an initial velocity directed toward Earth, and object C has the same initial speed but directed away from Earth. Which of the following statements correctly describes the speeds of the objects when they reach Earth's surface?
    (a) All three objects will have exactly the same speed.
    (b) Object A will have the greatest speed.

   (c) Objects B and C will have exactly the same speed, which will be greater than the speed of object A.
   (d) Object C will have the greatest speed, and object B will have the least speed.

11. Does the speed of a satellite increase or decrease when atmospheric drag causes it to lose altitude? Explain your reasoning.

12. (a) If you want to produce a straight-line graph by plotting a set of data for orbital periods and semi-major axes, what function of the periods would you plot on the y-axis? What function of the semi-major axes would you plot on the x-axis?
    (b) If you want the slope of the graph from part (a) to be 1, what units should be used for periods and semi-major axes?

13. A planet is in a circular orbit around the Sun. Its distance from the Sun is four times the average distance between Earth and the Sun. The period of this planet, in Earth years, is
    (a) 4
    (b) 8
    (c) 16
    (d) 64

14. When we are observing a planetary system (assume a single planet and star) at a certain instant in time, the star is observed to be in motion toward us. What is the direction of motion of the exoplanet at this time?
    (a) toward us
    (b) away from us
    (c) stationary relative to us
    (d) cannot be determined from the information given

15. Consider a star and an exoplanet that orbit in a plane perpendicular to the direction from the star to Earth. What shifts would we observe in the light from the star?
    (a) The light would be redshifted at all positions in the star's orbit.
    (b) The light would be blueshifted at all positions of the orbit.
    (c) There would be no observable shift from the orbital motion.
    (d) The light would be redshifted at one side of the orbit and blueshifted at the opposite side.

16. In 2005, a satellite was discovered in orbit around the distant dwarf planet Eris. What key information about Eris can we learn by observing this satellite?

17. Consider two masses in doing this question and the next. Mass 1 is located at the (x, y, z) Cartesian coordinate position $(0, 2, -2)$. Mass 2 is located at the origin of the coordinate system. If we consider the vector form of the law of universal gravitation, what is the unit displacement vector pointing from mass 1 to mass 2? Assume that the coordinates are expressed in units of metres.
    (a) $0\hat{\imath} + \sqrt{2}\hat{\jmath} - \sqrt{2}\hat{k}$
    (b) $1\hat{\imath} + \dfrac{1}{2}\hat{\jmath} - \dfrac{1}{2}\hat{k}$
    (c) $0\hat{\imath} + \dfrac{1}{\sqrt{2}}\hat{\jmath} - \dfrac{1}{\sqrt{2}}\hat{k}$
    (d) $0\hat{\imath} - \dfrac{1}{\sqrt{2}}\hat{\jmath} + \dfrac{1}{\sqrt{2}}\hat{k}$

18. Making reference to the situation of question 17, what is the vector form of the gravitational force of mass 1 on mass 2? Assume that both masses are 1 kg each, and that the (x, y, z) coordinates are in metres. G represents the universal gravitational constant.

(a) $\dfrac{G}{4}\left(0\hat{i} + \dfrac{1}{\sqrt{2}}\hat{j} - \dfrac{1}{\sqrt{2}}\hat{k}\right)$ N

(b) $\dfrac{G}{4}\left(0\hat{i} - \dfrac{1}{\sqrt{2}}\hat{j} + \dfrac{1}{\sqrt{2}}\hat{k}\right)$ N

(c) $\dfrac{G}{8}\left(0\hat{i} + \dfrac{1}{\sqrt{2}}\hat{j} - \dfrac{1}{\sqrt{2}}\hat{k}\right)$ N

(d) $\dfrac{G}{8}\left(0\hat{i} - \dfrac{1}{\sqrt{2}}\hat{j} + \dfrac{1}{\sqrt{2}}\hat{k}\right)$ N

19. In 2006, the International Astronomical Union redefined "planet." As a result, Pluto is no longer considered to be one of the major planets.
   (a) Use print and/or Internet resources to research and summarize the criteria for an object to be considered a planet.
   (b) One of these criteria is that the object pull itself into an approximately spherical shape. Using concepts from this chapter, explain why a molten body in space will assume a spherical shape.

20. Which of the following is conserved in the orbital motion of a mass in orbit around a much larger second mass (such as a star)?
   (a) only gravitational potential energy
   (b) only total orbital energy (gravitational potential plus kinetic)
   (c) only angular momentum
   (d) angular momentum plus total orbital energy

21. In a gravitational orbital situation, the kinetic energy has value C and the gravitational potential energy has value $-C/2$. What type of orbit is this?
   (a) circular
   (b) elliptic (but not circular)
   (c) hyperbolic
   (d) parabolic

22. The aphelion distance is 3.0 AU and the perihelion distance is 1.0 AU. What must be the eccentricity of the orbit?
   (a) 0.33
   (b) 0.50
   (c) 0.67
   (d) 3.0

PROBLEMS BY SECTION

For problems, star ratings will be used, (★, ★★, or ★★★), with more stars meaning more challenging problems. The following codes will indicate if $\frac{d}{dx}$ differentiation, $\int$ integration, ⬛ numerical approximation, or ⤳ graphical analysis will be required to solve the problem.

Section 11-1   Universal Gravitation

23. ★★ What force does a person of mass 50 kg exert on the Andromeda Galaxy, M31? The Andromeda Galaxy is believed to have a total mass of approximately one trillion times the mass of the Sun and is about 2.3 million light years away. (A light year is the distance that light travels in one year.)

24. ★★ Assume that Earth's orbit is circular with a radius of 1.50×10^{11} m. From first principles, calculate the approximate speed of Earth in its orbit.

25. ★ Calculate the difference in force per kilogram for water on the side of Earth nearest the Moon compared to water on the side farthest from the Moon. Use 6378 km as Earth's equatorial radius, and assume that the centre of the Moon is 3.84×10^8 m from the centre of Earth.

26. ★★ Assume that you are 25.0000 km from the centre of a neutron star that has a radius of 10.0 km and a mass of 4.4×10^{30} kg. Your feet are pointed at the neutron star, and your head is 1.5 m above your feet. What is the *difference* in gravitational force per kilogram on your head compared to your feet?

27. ★★ In an (x, y, z) coordinate system, mass 1 is at (1.00, 0.00, 0.00) (in units of metres) and has a mass of 1.00 kg. Mass 2 is at (2.00, 2.00, −1.00) and has a mass of 3.00 kg. Write the gravitational force of mass 1 on mass 2 in vector form.

Section 11-2   Acceleration Due to Gravity

28. ★ Calculate the value of the acceleration due to gravity on the surface of Mars, given that the mass of Mars is 6.42×10^{23} kg. Assume that Mars is a homogeneous sphere with a radius of 3.40×10^6 m.

29. ★★ Find the radius of a spherical asteroid that has a density of 2900 kg/m³ and an acceleration due to gravity of 0.10 m/s² at the surface.

30. ★★★ Assume that the normal value of g in a region is 9.800 m/s². However, one part of the region has an ore deposit directly below the surface. This deposit has a density of 7800 kg/m³, and the surrounding material has a mean density of 3000 kg/m³. Assume that the ore deposit is spherical with a radius of 500 m. Calculate the value of g above the ore deposit.

31. ★★ The acceleration due to gravity at the surface of a fictional planet is 3.11 m/s², and the mean density of the planet is 880 kg/m³. Find the radius and mass of the planet. Assume that it is spherical.

Section 11-3   Orbits and Weightlessness

32. ★★ Futuristic space colonies might use rotation to simulate gravity. If a space platform is in the shape of a cylinder of diameter 2360 m (with occupants living on the inner face of the surface of the cylinder), what must be the period of revolution if the simulated gravitational acceleration is to have the same value as on the surface of Earth?

33. ★★ A satellite is in a circular orbit 3000. km above the surface of a fictional planet. The satellite has a speed of 8000. m/s and is 8000. km from the centre of the planet.
   (a) Find g at the altitude of the satellite.
   (b) Find g at the surface of the planet.
   (c) Find the mass of the planet.

34. ★★ If a parabolic flight aircraft starts the parabolic portion of a flight at an angle of 40.° above the horizontal, and this portion of the flight lasts 20. s, what was the speed of the aircraft when it started the parabolic portion? Assume that the air resistance is negligible.

Section 11-4 Gravitational Potential Energy

35. ★★ Two spherically symmetric, homogeneous masses are of mass M and are located a distance d apart (see Figure 11-26). A third mass, m, is located a distance c away from the midpoint between the M masses (all three masses lie in a single plane). Derive an expression for the minimum work that must be done to move the mass m from its position to a point infinitely far away.

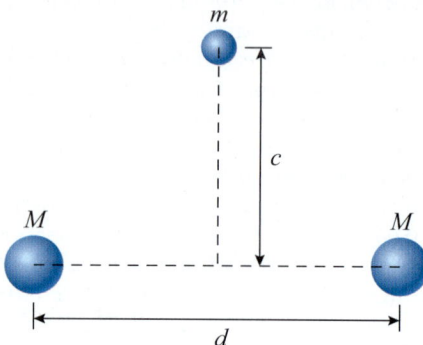

Figure 11-26   Problem 35.

36. ★★ A homogeneous sphere has mass M_1 and is located with its centre at the origin of a Cartesian coordinate system. A second homogeneous sphere is of mass M_2 with a centre at $(p, -q, -r)$, where p, q, and r are positive quantities. Determine an expression for the gravitational potential energy of this configuration.

37. ★★ Olympus Mons is an extinct volcano on Mars that is approximately 22 km high (almost three times the height of Mount Everest). Calculate the work required to raise a 25 kg object from the surface of Mars to the top of Olympus Mons. Assume that Mars is effectively spherical with a radius of 3400 km. The mass of Mars is 6.42×10^{23} kg.

38. ★★ An object is fired vertically from the surface of the Moon with a speed of 1000 m/s. What is the maximum height that it will reach?

Section 11-5 Force from Potential Energy

39. ★ $\frac{d}{dx}$ Find the components of the force in a region where the potential energy function is $U = ky^2$.

40. ★★ $\frac{d}{dx}$ If gravitational potential energy had the form $U(r) = -\dfrac{GMm}{\sqrt{r}}$, what would the gravitational force be?

41. ★★ $\int$ The acceleration due to gravity in a certain region is given by

$$\vec{g} = -4.00\hat{k} \text{ m/s}^2$$

Find an expression for the gravitational potential energy $U(x, y, z)$ for a 1.00 kg mass assuming that $U = 0$ J at the Cartesian point (0, 5.00, 10.00) (displacements in m).

Section 11-6 Escape Speed

42. ★ Calculate the escape speed from the surface of the Moon.

43. ★★ As mentioned above, we can classically calculate the "radius" of the event horizon for a black hole by considering it as an escape problem with the escape speed set equal to the speed of light. Use this approach to find the radius of a black hole that has a mass six times that of the Sun.

44. ★★ The dwarf planet Ceres (Figure 11-27) is approximately spherical with a radius of 488 km. The density of Ceres is 2078 kg/m³.
- (a) What is the escape velocity from the surface of Ceres?
- (b) If an object was launched vertically from the surface of Ceres with a speed that is exactly one-half the escape speed, what height would it reach above the surface of Ceres?

Figure 11-27   Problem 44. Dwarf planet Ceres as pictured by the Dawn space mission in 2015.

45. ★★ Derive an expression for the maximum radius of a spherical, homogeneous asteroid with uniform density ρ if it has an escape speed of v_{esc} from the surface. Assume that the asteroid has no atmosphere.

Section 11-7 Kepler's Laws

46. ★★ An ellipse is specified by the following equation, expressed relative to a coordinate system at the centre of the ellipse:

$$\frac{x^2}{(5.0 \times 10^{11} \text{ m})^2} + \frac{y^2}{(2.0 \times 10^{11} \text{ m})^2} = 1$$

What is the equation of the ellipse expressed in polar coordinate form, with the coordinate system centred at one focus of the ellipse?

47. ★ The semi-minor axis of an ellipse is 3.0 AU, while the semi-major axis is 4.0 AU. Calculate the following:
- (a) eccentricity
- (b) perihelion distance
- (c) aphelion distance

48. ★★ An ellipse is plotted in Figure 11-28.
- (a) What is the perihelion distance?
- (b) What is the aphelion distance?
- (c) Use only your results from (a) and (b) to find values for the semi-major axis a and the eccentricity e (i.e., do *not* use the semi-minor axis).
- (d) What is the total energy per unit mass for a planet in this orbit about the Sun?

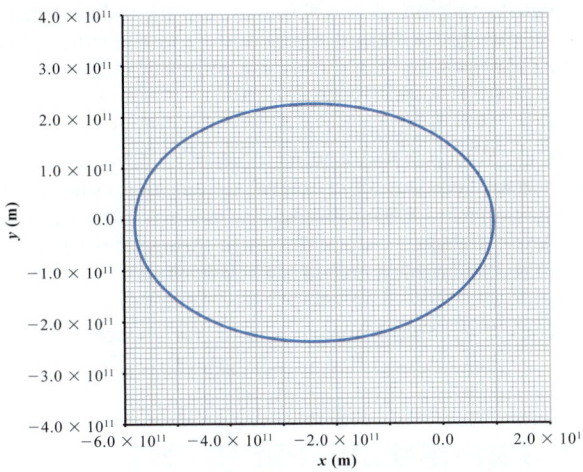

Figure 11-28 Problem 48.

49. ★★ An ellipse is plotted in Figure 11-29.
    (a) What is the perihelion distance?
    (b) What is the aphelion distance?
    (c) What is the semi-major axis?
    (d) Use the answers from (a) and (c) to estimate the eccentricity.
    (e) What is the semi-minor axis?
    (f) Use the results from (c) and (e) to estimate the eccentricity. Your answer should be nearly the same as for (d) but may be slightly different due to the precision of the graphical estimates.

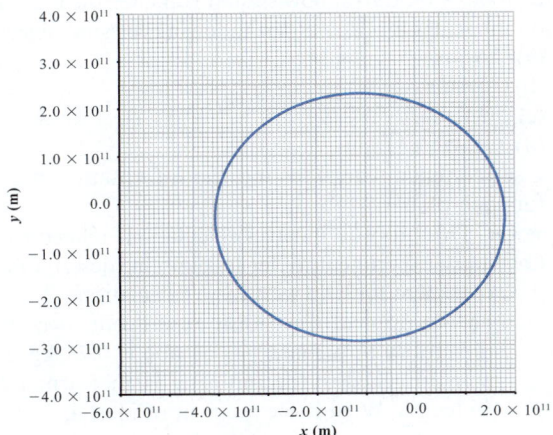

Figure 11-29 Problem 49.

50. ★★★ In this problem you will prove that Kepler's first law can be derived from angular momentum and energy conservation principles. We follow a derivation first proposed by the renowned Canadian physicist Dr. Erich Vogt (1929–2014), who was also a founding member and early director of TRIUMF. While the equations for the ellipse given in the chapter are the most common, a lesser-known equation for the ellipse is

$$\frac{b^2}{h^2} - \frac{2a}{r} = -1 \qquad (1)$$

This is based on a polar coordinate system centred on a focus such as the Sun. As before, r is the radial distance, and a and b are the semi-major and semi-minor axes, respectively. The new parameter, h, is the perpendicular distance from the focus to the backward extension of the instantaneous velocity vector, $\vec{v}$, as shown in Figure 11-30.

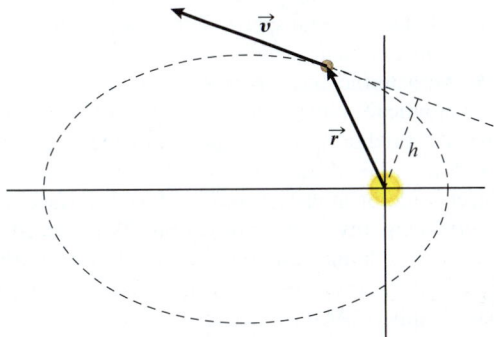

Figure 11-30 Problem 50. Geometry used in proof of Kepler's first law.

(a) Express the angular momentum, L, in terms of the mass of the planet, m; the instantaneous speed, v; and the parameter h. Call this result (2). Although both v and h vary around the orbit, the overall angular momentum must be conserved because there are no external torques.

(b) What is the total orbital energy, E, of a planet of mass m in orbit about a much larger mass M? Express this in terms of the masses, r, v, and constants. The total orbital energy must be conserved. Call this result (3).

(c) Substitute (2) into (3) to obtain an expression without v explicitly shown. Divide each term by the planet mass, m. Call this result (4). From this point on we will keep our expression with the energy per mass (E/m) and the angular momentum per mass (L/m) together because the mass of the planet does not really matter.

(d) Divide each side of result (4) by $-E/m$, calling this new result (5). Remember that for an elliptical orbit the total orbital energy is negative, so this is a positive quantity. You should now have obtained the relationship in the form of the ellipse equation (1) given at the start and therefore have proven Kepler's first law using only conservation of angular momentum and total orbital energy.

(e) By comparing (5) and (1), write an expression for the semi-major axis a. Call this result (6). Does the value of a depend on the angular momentum per unit mass, or only on the total orbital energy per unit mass?

(f) Compare (5) and (1) to write an expression for the semi-minor axis b, calling the result (7).

If desired, the results from (e) and (f) can be combined to find the eccentricity in terms of the angular momentum and the total energy per unit mass. The steps to obtain (1) from the standard form of the equation for the ellipse involve straightforward but tedious mathematical steps. If you want to see the derivation, see the original paper: E. Vogt, "Elementary Derivation of Kepler's Laws," *American Journal of Physics* 64 (1996), 392–396.

51. ★★ It is believed that there is a black hole with a mass of approximately four million times the mass of the Sun near the centre of the Milky Way galaxy. Some stars are relatively close to this black hole. Assume that the semi-major axis of the orbit of one of these stars is 3.5 light days (i.e., the distance that light would travel in 3.5 days).
(a) Find the orbital period of the star.
(b) Find the orbital speed of the star, assuming that it is in a circular orbit.

52. ★★ An asteroid has a perihelion distance of 0.55 AU (astronomical unit) and an aphelion distance of 1.65 AU. Predict the orbital period of the asteroid.

53. ★★ Since the Moon has virtually no atmosphere, an object could orbit just above the Moon's surface without experiencing any aerodynamic drag. What speed would be needed to launch an object so that it would orbit just slightly above the Moon's surface? Treat the Moon as a smooth sphere for this question.

Section 11-8 Types of Orbits

54. ★ At a certain instant, a celestial object has a speed of 31.0 km/s and it is 4.5×10^{11} m from the Sun. What type of orbit is it in?

55. ★★ A comet is in a parabolic orbit. When the comet is still 8.00 AU from the Sun, what is its speed?

56. ★ A Solar System object has a perihelion distance of 0.400 AU and an aphelion distance of 4.00 AU. What is the ratio of its perihelion speed to its aphelion speed?

57. ★★ A Solar System object has semi-major axis a, mass m, and eccentricity $e = 1/4$. Assume that it is in orbit about the Sun, which has mass M, and that $M \gg m$. Derive an expression for the speed at perihelion in terms of G, M, and a.

58. ★★★ A sensitive Doppler radar instrument determines that a small object at a height of 1800 km above Earth's surface has a radial velocity component of 1.3 km/s directed away from Earth's centre. Night-vision images indicate that the object has a transverse velocity component of 6.7 km/s. Assuming that the transverse and radial components are perpendicular, determine whether the object is in a gravitationally bound orbit with Earth or in a hyperbolic orbit that will cause it to escape Earth's gravitational field. Assume no upper atmosphere drag, and neglect the influence of any other masses, such as the Moon and the Sun.

Section 11-9 Detection of Exoplanets

59. ★★ (a) What is the centre of mass of the Sun–Earth system?
(b) Express your answer from (a) as a fraction of the radius of the Sun.

60. ★★★ Spectroscopy of a star produces radial velocity plots suggesting a peak stellar speed of 34 m/s. The period of the cycle is 25 days. Other measurements give an estimate of 4.0×10^{30} kg for the mass of the star. Calculate the mass of the exoplanet that would account for the radial velocity of the star.

61. ★★★ Consider an exoplanet system with one star and two planets. Doppler spectroscopy shows that one planet causes motion of the star with a peak radial speed of 40 m/s and a period of 10 days, and the other planet causes motion with a peak radial speed of 20 m/s and a period of 50 days.
(a) Which planet is closer to the star?
(b) Which planet is more massive?

62. ★★★ Consider an exoplanet system that consists of a star of mass 1.8×10^{30} kg and a single planet of mass 4.5×10^{26} kg in a circular orbit around their centre of mass. The orbital period is 12 days. Find
(a) the maximum orbital speed of the star
(b) the range of observed values for a spectral line that has an unshifted wavelength of 396.781 00 nm on Earth

COMPREHENSIVE PROBLEMS

63. ★★★ Mars has two small satellites called Phobos and Deimos. Assume that these satellites are spherical each with a mean density of 1880 kg/m³ and that the diameter of Phobos is 21 km.
(a) If you dropped an object from a height of 1.5 m above the surface of Phobos, how long would it take the object to fall to the surface?
(b) Phobos orbits Mars with a period of 0.319 days at a mean distance of 9400 km from the centre of Mars. Calculate the mass of Mars.

64. ★ For a museum exhibit, you want to make a set of scales that show the weight a person would have on different planets. If you start with scales that measure 0 N to 1000 N on Earth, how should you alter the markings on the scales for each of the following celestial objects?
(a) Mars
(b) Venus
(c) Mercury
(d) Earth's Moon
Use reference works to find the information needed for each planet.

65. ★★★ This problem involves calculations related to the Cavendish experiment. Refer to an online description of the experiment for help with these calculations.
(a) In the original Cavendish experiment, two small spheres with a mass, m, of about 0.73 kg each were attached to the end of a balance arm 1.8 m in length. What is the moment of inertia, I, of the two spheres as they rotate about the centre of the balance? Ignore the mass of the balance arm.
(b) The period of a torsion balance is given by the relationship

$$T = 2\pi\sqrt{\frac{I}{\kappa}}$$

where T is the period, I is the moment of inertia, and κ is the torsion coefficient of the suspending wire. The observed period in the Cavendish experiment was about 420 s. Calculate the torsion coefficient for the apparatus.
(c) The large spheres, M, each had a mass of 158 kg. Calculate the gravitational force between a small sphere and the adjacent large sphere when the distance between their centres is 230 mm.

(d) The moment arm for each force is half the length of the balance arm, 0.90 m. Calculate the torque associated with one of the large spheres using the force from part (c). The total torque will be double this value (since there is an equal torque at each end of the arm).

(e) The rotational analogue to Hooke's law is given by $\tau = -\kappa\theta$, where τ is the torque, κ is the torsion coefficient, and θ is the deflection angle. Predict the deflection angle for the apparatus described above. (Remember that the total torque acting on the balance arm is double the value calculated in part (d).)

66. ★★ On Earth, a person has a weight of 520 N. What weight would the person have on Mars? The radius of Mars is 53% of the radius of Earth, and its mass is 11% of the mass of Earth.

67. ★★★ Consider a planet that has a gravitational acceleration of 1.8 m/s² at its surface. Assume that the planet is spherical with a radius of 4300 km.
(a) Calculate the mass of the planet.
(b) Find the escape speed from the surface of this planet, assuming no air resistance.

68. ★★★ Consider a geosynchronous satellite with a mass of 140 kg. Geosynchronous satellites orbit Earth with a period of approximately 24 h. The mass of Earth is 5.97×10^{24} kg. Assume that Earth is exactly spherical with a radius of 6378 km.
(a) At what height above Earth's surface will the geosynchronous satellite orbit?
(b) Calculate the angular velocity for the orbit.
(c) Calculate the angular momentum of the satellite in orbit.
(d) How much additional gravitational potential energy does the satellite have when it is in orbit, compared to when it is on Earth's surface?
(e) The orbital period of a geosynchronous satellite is not precisely 24 h. Explain why.

69. ★★ What is the fastest speed at which an object 1.0 AU from the Sun can move and still be gravitationally bound to the Sun? The object is positioned such that the gravitational forces exerted by Earth and other planets are negligible.

70. ★★ Europa is one of the large "moons" of Jupiter and is believed to have a major ocean under a thick frozen surface. Europa has a radius of 1570 km and a mass of 4.87×10^{22} kg. Assume that Europa is completely spherical.
(a) Find the escape speed for an object at the surface of Europa.
(b) Find the acceleration due to gravity, g, at the surface of Europa.
(c) What would the weight of a 64 kg person be on the surface of Europa?

71. ★★★ The Sun subtends an angle of about 0.53° when photographed from Earth. (This angle varies slightly because Earth follows a slightly elliptical orbit.) The Sun is a distance of 1.50×10^{11} m from Earth. Calculate the mean density of the Sun, assuming that it is spherical. Do not assume a value for the mass of the Sun. (Hint: The period of Earth's orbit is one year.)

72. ★★ Pluto has at least four satellites, the largest of which is Charon. Charon orbits Pluto once every 6.387 days, and the semi-major axis of the orbit is 19 570 km. Find the sum of the masses of Pluto and Charon.

73. ★★ Derive the relationship $M = \dfrac{rv^2}{G}$ for a small mass, m, in a circular orbit at a distance r from the centre of a much larger mass, M. The speed of mass m is v.

74. ★★★ The Solar System orbits the centre of our Milky Way galaxy with a speed of approximately 230 km/s at a distance from the centre of the galaxy of about 26 000 light years.
(a) Use the relationship from problem 73 to estimate the mass within the orbit of the Solar System around the centre of our galaxy.
(b) Express the mass calculated in part (a) in units of solar mass, that is, as a multiple of the mass of the Sun.
(c) How long will it take our Solar System to make one orbit around the galaxy? Express your answer in years.

75. ★★ Imagine that a space colony living in a large vessel is in a circular orbit at a distance of 55 km from the centre of a neutron star. The neutron star has a radius of 10.0 km and a mass of 4.4×10^{30} kg.
(a) Find the velocity of the vessel in its orbit.
(b) Find the period of the orbit.

76. ★★ A comet discovered in an orbit around the Sun has a semi-major axis of 36 AU (astronomical units).
(a) Use Kepler's third law to estimate the period for the orbit of this comet. Express your answer in years.
(b) If the same comet were in orbit around an exoplanet of mass 8.6×10^{29} kg with the same semi-major axis of 36 AU, what would be the period of the orbit? Convert the final answer to years.
(c) Would the orbit about the exoplanet be circular? Justify your answer.

77. ★★ Stars lose mass from gradual processes, such as the solar wind, and from the ejection of material in eruptions and explosions. Consider a star with several planets in circular orbits.
(a) Describe qualitatively what would happen to the orbits of the planets if the star lost a significant part of its mass (say 20%).
(b) How much of its original mass would a star have to lose to make the orbits of the planets hyperbolic?

78. ★★★ Consider an exoplanet system with a star of mass 2.5×10^{30} kg and a single planet of mass 1.5×10^{28} kg in a circular orbit around their centre of mass. The orbital period is 25 days.
(a) Calculate the semi-major axis of the planet's orbit.
(b) Where is the centre of mass of the system located?
(c) Find the approximate orbital speed of the planet.
(d) Find the approximate orbital speed of the star.
(e) What is the range of observed wavelengths for a spectral line with a wavelength of 656.280 nm emitted by the star in this system?

79. ★★★ Some meteorites come from the Moon. Calculate the minimum ejection speed needed to move a stone from the surface of the Moon to Earth, ignoring the orbital motion of the Moon and Earth. (Hint: First find the point on the line from the Moon to Earth where the gravitational forces balance.)

80. ★★ ∫ Using work–energy principles, derive the expression for gravitational potential energy at a distance r from a symmetric spherical mass M.

81. ★★ d/dx Suppose that the potential energy between two masses, M and m, has the form $U(r) = -\dfrac{kMm}{r^3}$, where k is a constant. Use calculus to find an expression for the magnitude of the related force.

82. ★★ A visible star with an unseen companion star often emits X-rays. The visible star in one such pair has spectral characteristics that indicate it has a mass of about 7.6 solar masses (i.e., 7.6 times the mass of the Sun). After careful observations you have determined that the semi-major axis of the orbit of the visible star is 26 million km. The orbital period of that star is 8.5 Earth days.
    (a) Use the general version of Kepler's third law to determine the mass of the unseen star.
    (b) Theory suggests that a stellar remnant will not form a black hole with a mass less than 3.0 solar masses. Is the unseen object a black hole?

83. ★★ Assume that the Sun started as a thin, spherical shell of material with the same mass that it has now and with a radius of one light year.
    (a) If we take three-quarters of the current radius of the Sun as the average distance of its mass from the centre of mass, how much gravitational potential energy would have been released as the Sun contracted from the large shell to its current size?
    (b) For what amount of time could the released energy power the Sun at its current output of about 3.85×10^{26} W?

84. ★★ Assume that an object is perfectly spherical and has a uniform density equal to the density of water (1000 kg/m³).
    (a) Find the radius and mass of the object that would make the escape speed at the surface equal to the speed of light.
    (b) What does the answer to part (a) tell you about black holes?

85. ★★ Consider four masses at the corners of a square of side d, as shown in Figure 11-31. Three of the corners have a mass M, and one corner has a mass $2M$. A small mass, m, is placed at the centre of the square as shown. What is the net gravitational force on the small mass m? Assume that the configuration is in a region of deep space with no other significant gravitational forces.

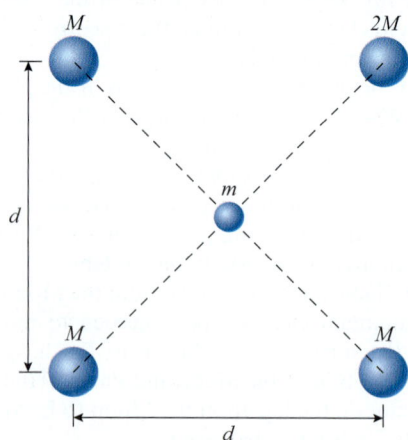

Figure 11-31   Problem 85.

DATA-RICH PROBLEM

86. ★★★ A space probe to a fictitious planet accurately measures the radius of the (assumed perfectly spherical) planet as 5038 km. The space probe measures the g-value at a number of heights above the surface of the planet as given in Table 11-1. From these data determine
    (a) the mass of the planet
    (b) the mean density of the planet
    (c) the value of g at the surface of the planet

Table 11-1   Problem 86

Height (m)	g (m/s²)
10 000	3.92583
15 000	3.91807
20 000	3.91033
25 000	3.90261
30 000	3.89491
35 000	3.88724
40 000	3.87959
45 000	3.87196
50 000	3.86435
55 000	3.85677
60 000	3.84921
65 000	3.84167
70 000	3.83415
75 000	3.82666

OPEN PROBLEMS

87. ★★★ At the beginning of the chapter, it is suggested that the event 65 million years ago that resulted in the extinction of the dinosaurs and perhaps 75% of all species was probably caused by a comet or an asteroid impact.
    (a) Under a worst-case scenario, what would be the approximate maximum speed with which a comet could impact Earth? Show calculations to support your answer, and assume that the comet is gravitationally bound to the Solar System.
    (b) Consult sources to suggest a reasonable radius and density for a comet, and use this information to calculate the comet's mass.
    (c) Using your results from parts (a) and (b), find the kinetic energy of the comet just before impact. To put this in perspective, convert the result into the energy equivalent of a number of tonnes of TNT.

88. ★★★ In 2014, the European Space Agency Rosetta mission went into orbit about Comet 67P/Churyumov-Gerasimenko (see Figure 11-32). As part of this mission, a small lander, Philae, was to perform observations from the surface. It turns out that the lander ended up in a poor orientation to receive solar power, and the mission was much shorter than planned. This was probably due to a much higher bounce than expected. In this problem, we want you to estimate how high Philae might reasonably bounce off the comet. Assume that Philae approached with a speed of about 1.5 m/s. Assuming that the surface was a fairly rigid dust-ice matrix, estimate the rebound speed from the ice considering similar collisions here on Earth (do an experiment if you wish). Use gravitational principles to estimate how far away from the surface of the comet the lander would probably go after the bounce. The comet is irregularly shaped (see Figure 11-32), about 4.3 by 4.1 km.

Figure 11-32 Problem 88. Comet 67P/Churyumov-Gerasimenko as imaged by the Rosetta mission.

CHAPTER 12 | Fluids

The human cardiovascular system consists of a pump (the heart), a fluid (blood), and blood vessels of various cross-sectional areas and structure (Figure 12-1). Oxygen-rich blood leaves the heart at a pressure greater than atmospheric pressure. As blood flows through smaller arteries, frictional forces result in a loss of pressure. By the time the blood enters the capillaries, it is almost at atmospheric pressure. Clogged arteries restrict blood from flowing freely, and the heart has to pump harder to maintain circulation, resulting in higher blood pressure. Thus, high blood pressure is an indicator of clogged arteries. To understand how the pressure and flow of fluids are related, we will analyze some basic properties of static and moving fluids.

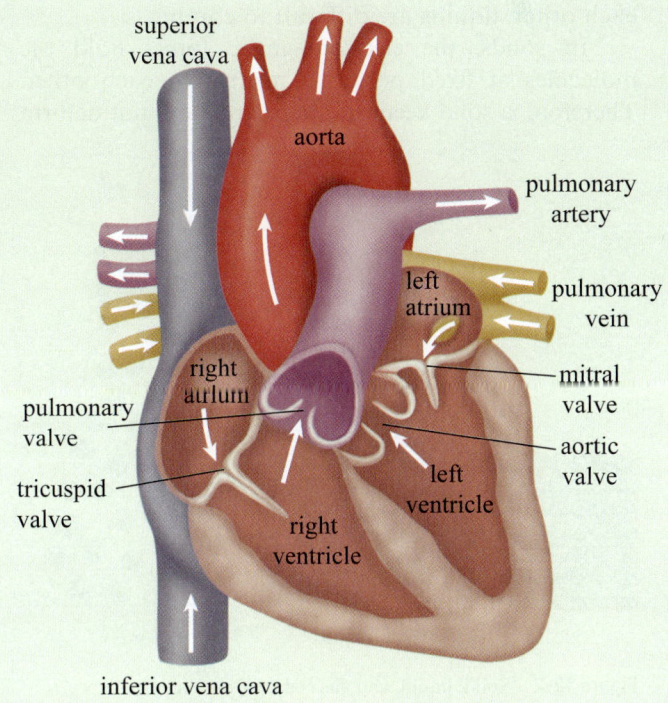

Figure 12-1 The human heart is a powerful muscle that controls the blood flow through the body.

12-1 Phases of Matter

Most matter at everyday temperatures and pressures exists in three **phases** (or **states**), classified as gases, liquids, and solids (Figure 12-2). The physical properties of a given phase are determined by the average distance between its constituents (atoms and molecules) and the strength of the electromagnetic forces between them.

In gases, the average distance between the molecules is large compared to the size of the atoms; hence, the electromagnetic forces between the molecules are so weak that the molecules do not bind with each other and can move freely. At 20 °C and sea level, the average distance between air molecules is 33×10^{-10} m, which is approximately 10 times the diameter of a nitrogen molecule. Only when two molecules are very close to each other do they experience an appreciable force, and we say that a collision has occurred between the two molecules. Such collisions obey Newton's laws of motion. Because of the large distances between gas molecules, gases are easily compressed.

In liquids, the molecules are very close to each other, and the electromagnetic forces between them are sufficiently strong to weakly bind the molecules but not strong enough to hold them at fixed positions. The molecules are still free to slide around each other without leaving the body of the liquid, enabling a liquid to flow under the influence of an external force. The average distance between water molecules is approximately 3×10^{-10} m, which is about the same as the diameter of a water molecule. Because the molecules in liquids are very close to each other, liquids are difficult to compress.

In solids, the electromagnetic forces hold the molecules at fixed positions relative to each other. Therefore, a solid keeps its shape and does not deform

easily. However, solid molecules are not completely stationary: they vibrate rapidly about their equilibrium positions.

There is a fourth phase of matter, called **plasma**. Plasmas form at very high temperatures and in strong electric fields. In this phase, collisions between molecules are so strong that the electrons can be knocked from the outer atomic orbits. Consequently, plasmas are a combination of ions and electrons. This state of matter exists in lightning, fluorescent lamps, neon signs, the solar wind, and stars.

Gases and liquids are collectively called **fluids** (from the Latin word *fluidus*, meaning "to flow") because of their ability to flow. Fluids do not retain a specific shape, so we will define several new variables to describe their physical properties. Although the mathematical equations describing the behaviour and motion of fluids look different from what we have studied so far, the underlying physics is still based on Newton's laws and principles of conservation of energy momentum, and angular momentum.

Solids and Fluids under Stress

A stress consists of opposing forces acting on an object. In Section 10-5 we learned that solid objects either deform or break under the influence of external stresses. Consider a solid cylinder being compressed by opposing forces applied to its opposite ends, pointing toward its centre. The resulting stress, called a compressive stress, tends to compress the cylinder. If the opposing forces are applied away from the centre of the cylinder, the resulting stress, called a tensile stress, tends to stretch the cylinder. In both of these cases the atoms of the cylinders are either slightly pushed together (compressive stress) or pulled apart (tensile stress) and the volume of the cylinder changes, although this change may be negligible. If the opposing forces are applied tangential to its surface (see Figure 12-3), the cylinder twists slightly around its centre but its volume does not change. In this case, instead of being pulled apart (or pushed together), the layers of atoms are forced to slide over one another. This type of stress is called a shear stress. For solids the amount of deformation, called strain, is proportional to applied stress as long as the elastic limit is not reached. When the stress is removed, the object returns to its original shape as atoms move to their original positions.

Unlike a solid, an open body of fluid cannot support compressive, tensile, or shear stresses. Since the atoms of a fluid are freer to move, under the influence of an external stress they move toward the region of lower stress. Try compressing air by slowly compressing your hands together. Air molecules will simply slide through the spaces between and around

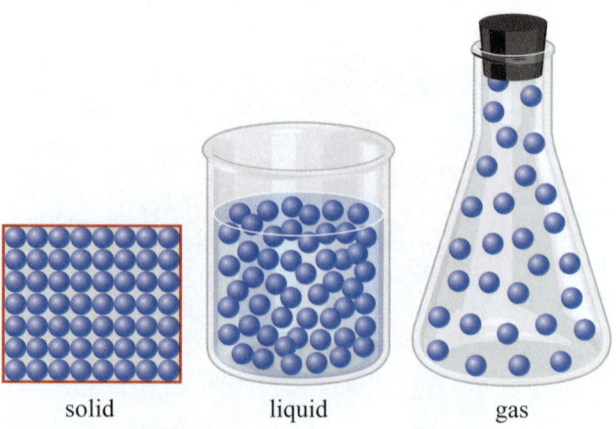

solid   liquid   gas

Figure 12-2   Solid, liquid, and gas states of matter.

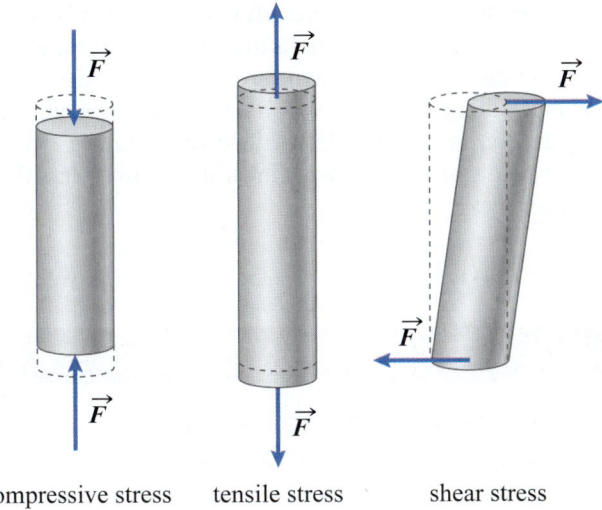

compressive stress   tensile stress   shear stress

Figure 12-3 Compressive, tensile, and shear stresses on a solid cylinder.

your hands. If you move your hands quickly, the air pressure (and hence the density) adjacent to your hands does increase slightly, but when the motion stops the air molecules redistribute and the pressure equalizes. Similarly, under the influence of a shear stress, a fluid begins to flow and continues to do so as long as the stress exists. This is strictly true only if the frictional forces between the atoms of a fluid can be ignored. Such fluids are called non-viscous. Gases and water are good examples of non-viscous fluids. Honey and tar are examples of fluids where frictional forces are not negligible. Such fluids can sustain shear and tensile stresses, though not ones as large as solids can.

If a stress is applied uniformly across the whole surface of a fluid, then its atoms do not have a region of lower stress to move to. Such a stress is called a bulk stress (see Figure 12-4). The atoms of a fluid under a

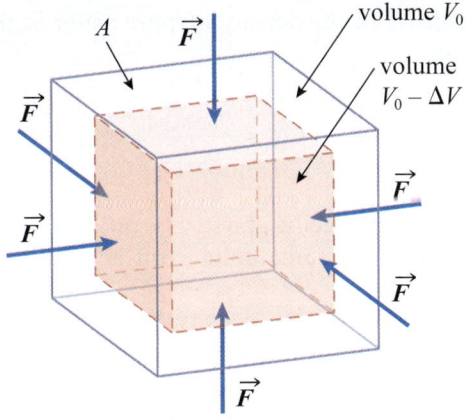

Figure 12-4 A cubical object, solid or fluid, is compressed under a bulk stress.

bulk stress are forced to move closer together, resulting in a decrease in its volume and hence an increase in its density.

Interatomic distances in liquids are about the same as in solids, so a liquid is difficult to compress. Atoms in a gas, at normal pressure, are far apart. Therefore, gases easily compress in response to a bulk stress. So we see that a fluid can only sustain a stress as long as the forces are perpendicular to the surface of the fluid and the stress is applied uniformly along its entire surface.

It is this different behaviour between solids and fluids under the influence of external stresses that requires the introduction of variables such as density and pressure (compared to mass and force for solids) to describe static and moving fluids.

12-2 Density and Pressure

Mass Density

The mass density, ρ, of a substance is the mass per unit volume of the substance. For an object of mass M that occupies a volume V,

KEY EQUATION

$$\rho = \frac{M}{V} \qquad (12\text{-}1)$$

Mass density is a scalar quantity. Note that Equation 12-1 gives the **average mass density** of an object. When we use the word *density* by itself, we are referring to the mass density. (The concept of density can be applied to other physical quantities, such as charge and current.) The SI units for density are kilograms per cubic metre (kg/m^3). Densities are also often expressed in grams per cubic centimetre (g/cm^3):

$$1\,\frac{kg}{m^3} = \frac{10^3\,g}{(10^2\,cm)^3} = 10^{-3}\,\frac{g}{cm^3}$$

$$1\,\frac{g}{cm^3} = 1000\,\frac{kg}{m^3}$$

The density of a substance depends on the mass of its constituents (atoms) and the average spacing between them. Table 12-1 lists the densities of various substances. The densities of liquids are typically thousands of times the densities of gases because the molecules are much farther apart in a gas than in a liquid. By definition, the density of a pure substance is independent of its volume. A steel nail has exactly the same density as a large block of steel.

Table 12-1 Mass Densities of Common Substances

Substance		Mass Density, ρ (kg/m³)
Solids		
Aluminum		2 700
Copper		8 890
Diamond		3 520
Gold		19 300
Ice		917
Lead		11 300
Silver		10 500
Wood (various types)	Balsa	160
	Maple	600–750
Liquids		
Alcohol		800
Blood (at 37 °C)		1060
Mercury		13 600
Water (at 4 °C)		1 000
Gases		
Air		1.29
Helium		0.179
Hydrogen		0.090
Nitrogen		1.25
Oxygen		1.43

EXAMPLE 12-1

The Density of the Sun

The mass of the Sun can be accurately determined from the length of the year on Earth, the distance between Earth and the Sun, and the value of the gravitational constant, G. The mass of the Sun is 1.99×10^{30} kg, and its radius is approximately 6.96×10^8 m. What is the average density of the Sun? Assume that the Sun is spherical.

Solution

The Sun is assumed to be a sphere of radius 6.96×10^8 m. Therefore, the volume of the Sun is

$$V_{sun} = \frac{4\pi}{3}(6.96 \times 10^8 \text{ m})^3 = 1.41 \times 10^{27} \text{ m}^3$$

The mass density of the Sun, ρ_{sun}, is now calculated from Equation 12-1:

$$\rho_{sun} = \frac{M_{sun}}{V_{sun}} = \frac{1.99 \times 10^{30} \text{ kg}}{1.41 \times 10^{27} \text{ m}^3} \approx 1400 \text{ kg/m}^3$$

Making sense of the result

The *average* density of the Sun is approximately 40% greater than the density of water. The inner core of the Sun has a density of approximately 1.5×10^5 kg/m³; the outer corona consists of hot gases with an average density of approximately 10^{-11} kg/m³. For intermediate layers, the density varies between these values. The value of 1400 kg/m³ therefore represents an average density for the entire Sun.

The density of an unconfined gas depends on its temperature. As the temperature of a gas increases, its volume increases and, therefore, its density decreases. When quoting the density of a gas, we will assume that the density is given at 0 °C. The temperature dependence of density for liquids and solids is much less than that for gases.

Specific Gravity

The **specific gravity** (s.g.) of a material is the ratio of the mass of the material to the mass of an equal volume of water. Since mass = density × volume, the specific gravity of a material can also be defined as the ratio of its density to the density of pure water at the same temperature:

$$\text{s.g.} = \frac{\rho_{material}}{\rho_{water}} \qquad (12\text{-}2)$$

As you can see, specific gravity is a ratio of two quantities that have the same units, so it is a dimensionless quantity.

In Table 12-1, the density of blood is listed as 1060 kg/m³. Therefore, its specific gravity is

$$\text{s.g.} = \frac{\rho_{material}}{\rho_{water}} = \frac{1060 \text{ kg/m}^3}{1000 \text{ kg/m}^3} = 1.06$$

Pressure

When you push against a wall with your hand, the force exerted by your hand on the wall is distributed over the area of your hand that is touching the wall. **Pressure**, P, is defined as the magnitude of the component of the force perpendicular to a surface divided by the area of the surface on which the force acts.

Consider a force $\vec{F}$ acting on surface of area A. Let us denote by F_n the component of $\vec{F}$ that is perpendicular to the surface. Then, the pressure exerted by the force on the surface is given by

KEY EQUATION
$$P = \frac{F_n}{A} \qquad (12\text{-}3)$$

Since the component of a force and the area are both scalars, pressure is a scalar quantity. The SI unit of pressure is the **pascal** (symbol Pa), which is equal to one newton per square metre (N/m²):

$$1 \text{ Pa} = 1 \text{ N/m}^2$$

The name "pascal" honours the French mathematician, philosopher, and physicist Blaise Pascal (1623–1662), who, among many other things in his short life, did some of the pioneering work on the properties of fluids.

For a given magnitude of force, pressure can be increased by applying the force over a smaller surface area. When you stand on one foot, the pressure that your foot exerts on the ground is twice the pressure that it exerts when you stand on both feet.

EXAMPLE 12-2

Calculating Pressure

A 0.500 kg book is resting on a table. Its cover has a surface area of 0.060 m². You push down on the book with your hand with a force of magnitude 50.0 N, with your arm inclined at an angle of 40.0° with respect to the normal to the book's surface (Figure 12-6). The area of your hand in contact with the book, A_{hand}, is 0.012 m².

(a) Find the pressure exerted by your hand on the book.
(b) What is the pressure exerted on the table by the book when your hand is pressing on the book?

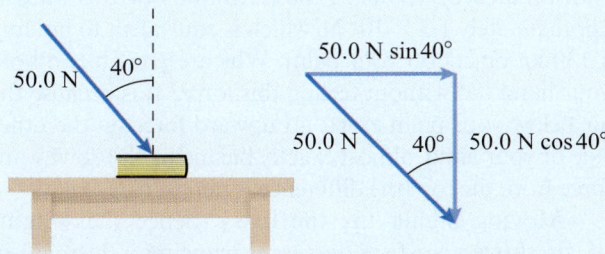

Figure 12-6 The force exerted by the hand on the book is represented by an arrow that makes an angle of 40° with respect to the normal to the book.

Solution

(a) The pressure exerted by your hand on the book is equal to the normal component of your hand's force on the book, $\vec{F}_{\text{hand-book}}$, divided by the area of the hand in contact with the book.

From Figure 12-6, the normal component of the force exerted by the hand on the book $F_{n,\text{hand-book}}$, is given by

$$F_{n,\text{hand-book}} = 50.0 \text{ N} \times \cos 40° = 38.3 \text{ N}$$

Therefore, the pressure exerted by the hand on the book is

$$P_{\text{hand-book}} = \frac{F_{n,\text{hand-book}}}{A_{\text{hand}}} = \frac{38.3 \text{ N}}{0.012 \text{ m}^2} = 3.2 \times 10^3 \text{ N/m}^2$$
$$= 3.2 \text{ kPa}$$

(b) The net force exerted on the table by the book is the vector sum of the force exerted by the hand on the book and the weight of the book:

$$\vec{F}_{\text{book-table}} = \vec{F}_{\text{hand-book}} + \vec{W}_{\text{book}}$$

The normal component of $\vec{F}_{\text{book-table}}$ is the sum of the normal components of the two forces on the right-hand side of the above equation:

$$F_{n,\text{book-table}} = F_{n,\text{hand-book}} + m_{\text{book}}g$$
$$= 50.0 \text{ N} \times \cos 40° + 0.500 \text{ kg} \times 9.81 \text{ m/s}^2$$
$$F_{n,\text{book-table}} = 43.2 \text{ N}$$

The pressure exerted by the book on the table is equal to the normal component of the force exerted by the book on the table, divided by the cross-sectional area of the book:

$$P_{\text{book-table}} = \frac{F_{n,\text{book-table}}}{A_{\text{book}}} = \frac{43.2 \text{ N}}{0.060 \text{ m}^2} = 7.2 \times 10^2 \text{ Pa}$$

Making sense of the result

The pressure calculated in part (b) is considerably smaller than the pressure calculated in part (a), even though the normal force exerted on the table by the book is greater than the normal force exerted by the hand on the book. The reason is the following. The contact area between the book and the table is larger than the contact area between the hand and the book. Therefore, the normal force exerted by the book is distributed over a larger area, so the pressure exerted by the book on the table decreases.

12-3 Pressure in Fluids

A fluid exerts pressure at every point within it and at every point where it contacts a surface, such as the wall of a container. This pressure is called **fluid pressure**. Fluid pressure results from the weight of the fluid and from the constant collisions of the molecules with each other and with the walls of the containing vessel. For liquids, the major contribution to fluid pressure is the weight of the liquid. In contrast, the pressure exerted by a gas usually arises mainly from molecular collisions because the molecules of a gas are free to move and the densities of gases are much less than the densities of liquids (Figure 12-8). However, due to the enormous size of the atmosphere, the weight of air is the dominant contributor to atmospheric pressure.

Atmospheric Pressure

Earth's atmosphere consists mostly of nitrogen (78.1%), oxygen (20.9%), and argon (0.9%). Although there is no well-defined boundary between the atmosphere and space, it is generally agreed that Earth's atmosphere extends to about 120 km above sea level. The weight of the air molecules exerts a pressure at every point in the atmosphere and on Earth's surface. This pressure is called **atmospheric pressure**. The atmospheric pressure at sea level at 20 °C is experimentally determined to be 1.013×10^5 Pa ($\approx$101 kPa).

Objects on Earth are subject to atmospheric pressure. When you hold your hand flat, the air above your hand exerts a downward force on top of your hand. For a hand with an area of 130 cm², the magnitude of this force is approximately 1.3×10^3 N, which is equivalent to holding a 130 kg object on your palm. Why are you able to hold your hand flat without feeling this force? It is because the air below your palm exerts an upward force on the other side of your hand, almost exactly balancing the downward force from the top (the difference is minuscule).

Moving higher up, the mass (hence the weight) of air above a surface decreases, causing a decrease in atmospheric pressure with increasing height. Table 12-2 shows how atmospheric pressure varies with height above sea level.

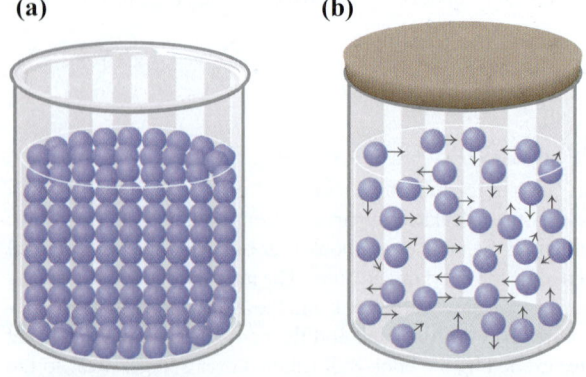

Figure 12-8   (a) In a liquid, the major source of pressure is the weight of the liquid. (b) In a gas, where densities are approximately 1/1000 the densities of the liquids, pressure exists because of collisions between gas particles.

Table 12-2   Variation of Atmospheric Pressure with Altitude

Height above Sea Level (m)	Atmospheric Pressure* (kPa)
0	101
10 000	32
20 000	10
30 000	3
40 000	1
50 000	0.30
60 000	0.01

* These values were calculated assuming an exponential decrease in air density with increasing altitude and are approximate.

EXAMPLE 12-3

Measuring the Mass of Earth's Atmosphere

Given that atmospheric pressure at Earth's surface is 1.01×10^5 Pa and the mean radius of Earth is about 6.370×10^6 m, calculate the approximate mass of the air surrounding Earth. What assumptions do you need to make to calculate the mass of the air?

Solution

The atmospheric pressure at Earth's surface results from the weight of the air. If we assume that Earth is spherical, we can calculate Earth's surface area from its radius. The total downward force exerted by the air, which is equal to the weight of the entire atmosphere, is the product of Earth's surface area and the atmospheric pressure at sea level.

Assuming that the acceleration due to gravity from sea level to the top of the atmosphere remains constant, the magnitude of the downward force on Earth's surface due to the weight of the air is

$$F_{\text{air, down}} = M_{\text{air}} g \tag{12-4}$$

This force is equal to the product of the atmospheric pressure P_0 at sea level and Earth's surface area:

$$F_{\text{air, down}} = P_0 \times 4\pi R_{\text{earth}}^2 \tag{12-5}$$

From Equations (12-4) and (12-5), we get

$$M_{\text{air}} = \frac{P_0 \times 4\pi R_{\text{earth}}^2}{g}$$

$$= \frac{1.01 \times 10^5 \frac{\text{N}}{\text{m}^2} \times 4\pi \times (6.370 \times 10^6 \,\text{m})^2}{9.81 \,\text{m/s}^2} \tag{12-6}$$

$$\approx 5.25 \times 10^{18} \,\text{kg}$$

Making sense of the result

The mass of the air surrounding Earth should be a very large number.

Hydrostatic Pressure

Consider a container that is at rest and is open to the atmosphere and filled with a fluid of density ρ_f (Figure 12-9). Imagine a cylindrical element of this fluid with cross-sectional area A, with its top end at the surface and bottom end at a depth d below the surface. The following external forces are acting on this element in the vertical direction:

- The weight $(\vec{W})$ of the fluid element acting vertically downward: The magnitude of the weight is

  $W = $ mass of the fluid element $\times g$
  $= $ volume of the fluid $\times$ density of the fluid $\times g$
  $= (Ad)\rho_f g$

- The downward force $(\vec{F}_{\text{down}})$ on the top surface, exerted by the atmospheric pressure: The magnitude of the downward force due to atmospheric pressure is

  $F_{\text{down}} = $ atmospheric pressure
  $\times$ cross-sectional area of the cylinder
  $= P_0 A$

- The upward force $(\vec{F}_{\text{up}})$ that the rest of the fluid exerts on the bottom end of the fluid element: Let P be the pressure in the container at a depth d below its surface. Then the magnitude of the upward force is

  $F_{\text{up}} = $ Pressure at the bottom of fluid element
  $\times$ cross-sectional area of the fluid element
  $= PA$

There is no net horizontal flow of fluid because the *horizontal* force exerted on any point of the cylinder is exactly balanced by an equal horizontal force

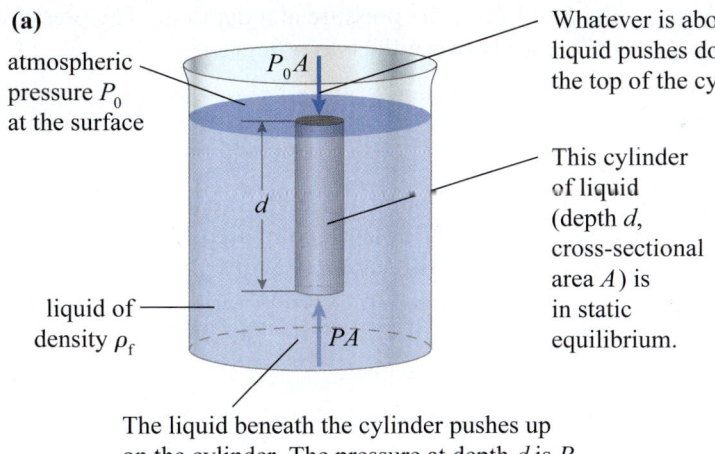

(a)

atmospheric pressure P_0 at the surface

$P_0 A$

Whatever is above the liquid pushes down on the top of the cylinder.

This cylinder of liquid (depth d, cross-sectional area A) is in static equilibrium.

liquid of density ρ_f

PA

The liquid beneath the cylinder pushes up on the cylinder. The pressure at depth d is P.

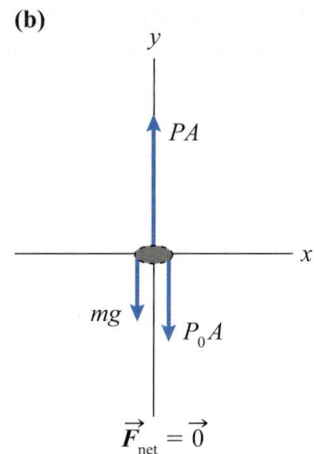

(b)

y

PA

x

mg

$P_0 A$

$\vec{F}_{\text{net}} = \vec{0}$

Figure 12-9 (a) Forces acting on a cylindrical element of a fluid in a container. (b) The FBD of the column of liquid. Here m is the mass of the liquid element under consideration.

on the opposite side of the point. Taking the y-axis along the vertical direction with the positive y-axis upward, the net vertical force on the cylindrical volume is given by

$$\Sigma F_y = F_{up} - F_{down} - W$$
$$= PA - P_0 A - \rho_f g(Ad) \qquad (12\text{-}7)$$

Since the fluid element is stationary, the downward forces due to the weight of the fluid and the atmospheric pressure must be balanced by an upward force exerted by the rest of the fluid. Therefore, $\Sigma F_y = 0$, and we obtain the following relationship for the fluid pressure at a depth d below the surface of a fluid that is open to the atmosphere:

KEY EQUATION $\qquad P = P_0 + \rho_f g d \qquad (12\text{-}8)$

The second term on the right-hand side of Equation 12-8 arises due to the weight of the fluid. Notice two important consequences of this simple equation:

- The pressure within a fluid increases with depth.

- All points at a given depth within a connected body of fluid are at the same pressure.

The second point means that the pressure along a horizontal line through a connected body of a fluid is the same, no matter how the container is shaped. The pressure depends only on the depth below the surface. The pressure at a depth of 5 m inside a narrow tube filled with water is exactly the same as the pressure at a depth of 5 m in a large swimming pool, for example.

Equation 12-8 assumes that the fluid density does not change with depth. Therefore, it is not applicable for the atmosphere, where the density of air changes with elevation.

Water Keeps Its Level

The water level in two arms of a U-shaped tube always rises to the same level (Figure 12-10). Why? Suppose that the water level is higher in the left arm. According to Equation 12-8, the pressure at the bottom of the left arm is higher than the pressure on the right side. The higher pressure at the left forces the water to flow from left to right until the level is the same in the two arms.

Suppose that the cross-sectional area of one arm of the tube is twice the cross-sectional area of the other (Figure 12-11). Is it possible to have an equilibrium situation where the fluid height in the wider arm is half the fluid height in the narrower arm? The volume and, therefore, the weight of the fluid in the two arms are then equal. Therefore, a force of equal magnitude is exerted at the bottom of

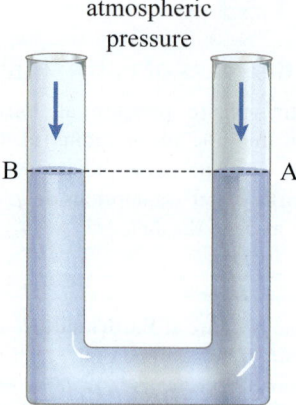

Figure 12-10 The pressure at the bottom of the two arms of the tube is the same, so the height of the fluid in the two arms must be equal.

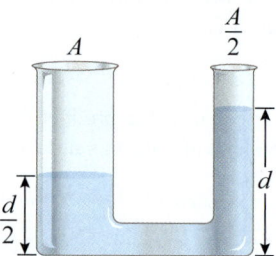

Figure 12-11 Is it possible for a fluid at rest in a tube with two arms of unequal cross-sectional area to be at two different levels?

the two arms of the tube. You might think that the fluid is in static equilibrium. However, with unequal levels of fluid in the two arms, the pressure at the bottom of the narrower arm is greater, causing the fluid to flow toward the wider arm until the fluid level in the two arms is the same.

Equation 12-8 can be used to find the pressure difference between two points within a fluid. Let P_1 be the pressure at a point that is located at a depth d_1 within a fluid and P_2 be the pressure at a depth d_2. The pressure difference between the two points is

$$P_2 - P_1 = (P_0 + \rho_f g d_2) - (P_0 + \rho_f g d_1)$$
$$= \rho_f g (d_2 - d_1) \qquad (12\text{-}9)$$

In a room that is 5 m high, the increase in the air pressure at the floor level, compared to the pressure at the ceiling, is $(1.3 \, \text{kg/m}^3) \times (9.8 \, \text{m/s}^2) \times (5.0 \, \text{m}) \approx 64 \, \text{Pa}$. This increase is a tiny fraction of the average atmospheric pressure of $1.01 \times 10^5 \, \text{Pa}$.

Gauge Pressure

Pressure is usually measured with respect to a reference pressure, usually taken to be atmospheric pressure.

The absolute pressure (P), the reference pressure (P_0), and the measured or **gauge pressure** (P_G) are related as follows:

$$\text{absolute pressure} = \text{measured pressure}$$
$$+ \text{reference pressure}$$
$$P = P_G + P_0$$

A tire gauge measures the gauge pressure inside a tire with respect to the atmospheric pressure. So, a reading of zero on a tire gauge indicates that the pressure inside the tire is equal to the atmospheric pressure, indicating that the tire is flat. Similarly, blood pressure measured by a sphygmomanometer is the gauge pressure measured relative to atmospheric pressure.

CHECKPOINT 12-3

Pressure and Force

Four cylinders of different cross-sectional areas but the same height are completely filled with water, as shown in Figure 12-12.

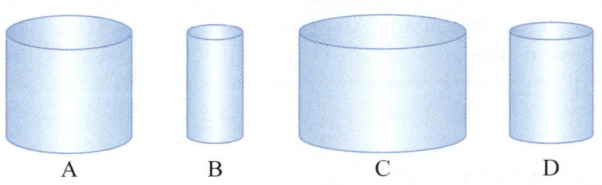

Figure 12-12 Four cylinders of equal height and different cross-sectional areas.

(a) Order the cylinders, from highest to lowest, in terms of the pressure exerted by the water at the bottom of each cylinder.
Highest ___ ___ ___ ___ Lowest
(b) Order the cylinders, from highest to lowest, in terms of the force exerted by the water at the bottom of each cylinder.
Highest ___ ___ ___ ___ Lowest

ANSWER: (a) A = B = C = D; (b) C, A, D, B

Measuring Pressure

A common instrument used to measure gauge pressure is a manometer. A manometer consists of a U-shaped glass tube of uniform cross-sectional area, partially filled with a liquid of known density (Figure 12-13). If the two arms of the manometer tube are at different pressures, then the fluid in the arm at the higher pressure is pressed downward, and the fluid level in the other arm rises.

Consider a manometer of uniform cross-sectional area partially filled with a liquid of density ρ_f (Figure 12-14). One arm (A) of the manometer is connected to a gas-filled cylinder that is maintained at a pressure P. The other arm (B) is open to the atmosphere and under atmospheric pressure P_0,

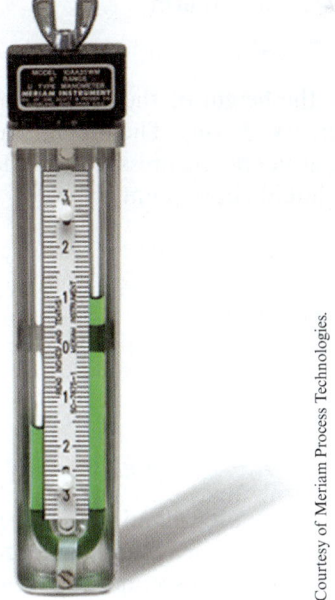

Courtesy of Meriam Process Technologies.

Figure 12-13 A U-shaped manometer. The height difference in the column of fluid between the two arms is proportional to the pressure difference between the two sides.

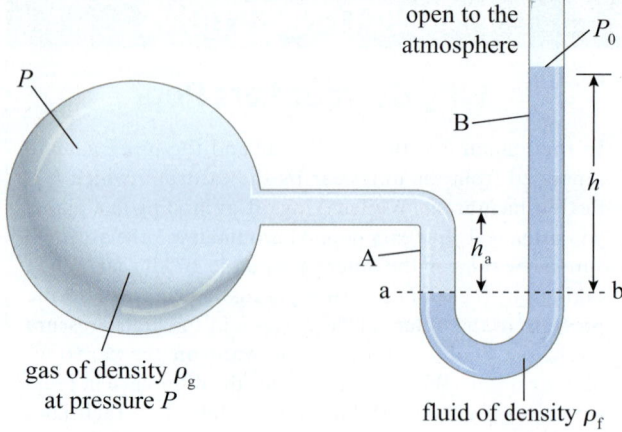

open to the
atmosphere

P_0

B

P

h

h_a

A

a - - - - - - - - - - - - b

gas of density ρ_g
at pressure P

fluid of density ρ_f

Figure 12-14 A U-shaped manometer connected to a pressurized device.

which is assumed to be less than P. Let h be the difference in the heights of the columns of fluids in the two arms. The pressure difference $P - P_0$, between the two arms of the manometer is proportional to h. To see this, draw a straight horizontal line that passes through a point (a) at the gas–fluid interface in arm A and a corresponding point (b) in arm B. Since the pressure in a static fluid is the same at all points at given level,

pressure at point a (P_a) = pressure at point b (P_b)

(12-10)

The pressure at the top of arm A is P; therefore,

$P_a = P +$ pressure because of the weight of the
gas in the left arm

$P_a = P + \rho_g g h_a$

where h_a is the height of the column of gas in arm A and ρ_g is the gas density. The pressure at point b is the sum of the atmospheric pressure and the weight of the column of liquid above point b:

$$P_b = P_0 + \rho_f g h$$

Since $P_a = P_b$, we get

$$P + \rho_g g h_a = P_0 + \rho_f g h$$

Therefore,

$$P - P_0 = \rho_f g h - \rho_g g h_a \qquad (12\text{-}11)$$

The quantity on the left-hand side is the gauge pressure. Since liquid densities are generally approximately 1000 times gas densities, the second term on the right-hand

side of Equation 12-11 can be ignored. With this approximation,

$$P - P_0 = \rho_f g h \qquad (12\text{-}12)$$

This simple equation can be used to determine the pressure in a gas or a liquid pipeline by using a manometer.

EXAMPLE 12-4

Pipeline Pressure

The total pressure in a gas pipeline is to be maintained at three times atmospheric pressure. Figure 12-14 shows the cross-sectional area of the pipeline that is connected to a manometer filled with mercury. An inspector measures a difference of 1.25 m between the two columns of mercury. Is the gas pressure being maintained at the desired value? The density of mercury is 13 600 kg/m^3.

Solution

The open end (B) of the manometer is at atmospheric pressure. We do not know the height of the gas column in arm A, but we can ignore the pressure from the weight of the gas because the gas is much less dense than mercury.

Consider a horizontal line that passes through the gas–fluid interface in arm A at point a and through point b on arm B. The pressures at these two points are equal. If the pipeline is at the desired pressure, then

$$3P_0 = P_0 + \rho_f g h$$

Solving for h, we get

$$h = \frac{2P_0}{\rho_f g} = \frac{2 \times 1.01 \times 10^5 \text{ N/m}^2}{(13.6 \times 10^3 \text{ kg/m}^3)(9.81 \text{ m/s}^2)} = 1.51 \text{ m}$$

The measured height is only 1.25 m, so the gas pressure is below the desired value of $3P_0$.

Making sense of the result

The pressure exerted by a column of mercury that is 1 m high is about $(13.6 \times 10^3 \times 10)$ Pa, which is about 1.3 times the atmospheric pressure. So a 1.5 m high column of mercury exerts about two times the atmospheric pressure.

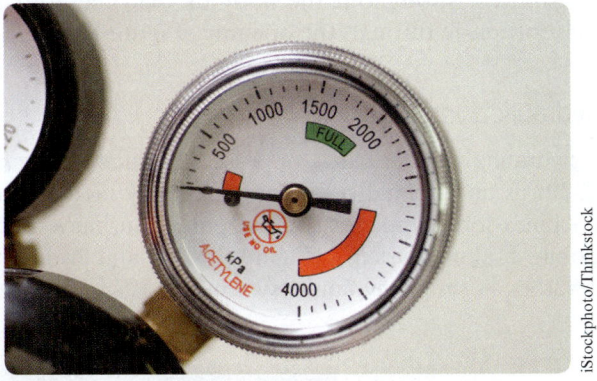
CHECKPOINT 12-4

Pressure in a Closed Tube

In Figure 12-16, the tube is open at one end, closed at the other end, and filled with water. The pressure at the closed end is

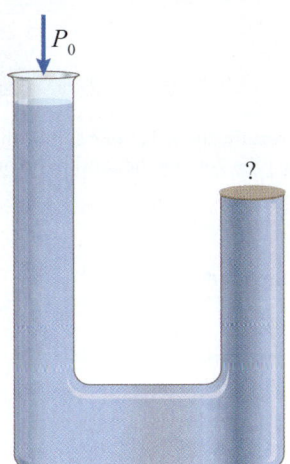

Figure 12-16 Checkpoint 12-4.

(a) greater than the atmospheric pressure
(b) equal to the atmospheric pressure
(c) less than the atmospheric pressure but non-zero
(d) zero

ANSWER: (a)

A Simple Barometer

An interesting application of variation of fluid pressure with depth is a **barometer**. In its simplest form it consists of a long glass tube open at one end and closed at the other, completely filled with a fluid of density ρ_f, and a bowl that is partially filled with the same fluid. The tube is inverted and its open end is carefully placed (so no air can enter the tube) below the fluid surface in the bowl, as shown in Figure 12-17. If the tube is long enough, the fluid from the tube flows into the bowl, creating vacuum at the top. However, not all of the fluid empties into the bowl, and the height, h, of the fluid remaining in the tube is related to the value of the atmospheric pressure, P_0.

Some of the fluid does not empty into the bowl because the atmosphere exerts a pressure on the fluid in the bowl. Consider a point a inside the tube that is at the level of the fluid in the bowl. Since the pressure at the top of the tube is zero, the total pressure at point a due to the weight of the fluid is $\rho_f gh$. Because the pressure everywhere along the level of the fluid in the bowl is equal to the atmospheric pressure P_0, it follows that

$$P_0 = \rho_f gh \tag{12-13}$$

If the pressure at point a were not equal to the atmospheric pressure, there would be a net force on the fluid inside the tube, causing it to flow. Therefore,

$$h = \frac{P_0}{\rho_f g} \tag{12-14}$$

and by measuring h we can calculate the atmospheric pressure. A fluid that is commonly used in a barometer is mercury (Hg), with a density of $1.360 \times 10^4 \text{ kg/m}^3$.

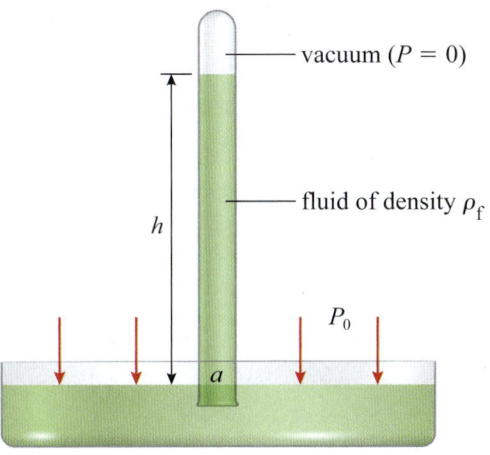

Figure 12-17 The atmospheric pressure, P_0, is related to the height of the fluid in the tube by the relation $P_0 = \rho_f gh$.

The height of a column of mercury that is supported by an atmospheric pressure of 1.013×10^5 Pa is

$$h = \frac{P_0}{\rho_f g}$$

$$= \frac{1.013 \times 10^5 \, \text{Pa}}{(1.360 \times 10^4 \, \text{kg/m}^3)(9.81 \, \text{m/s}^2)} \quad (12\text{-}15)$$

$$= 0.759 \, \text{m} = 759 \, \text{mm}$$

In fact, atmospheric pressure is defined in terms of millimetres of mercury (mmHg) as

$$1 \text{ atmosphere} = 760 \text{ mmHg}$$

From Equation 12-15 we notice that the length of the tube for a mercury barometer must be more than 760 mm. How long would the glass tube have to be if we used water instead of mercury? Using $\rho = 1000 \, \text{kg/m}^3$ in Equation 12-15, we get $h = 10.3$ m. So we see why mercury is the liquid of choice to construct a simple barometer.

Strictly speaking, the pressure at the top of the tube where there is no fluid is not zero and is equal to the vapour pressure of the fluid in the rest of the tube. For mercury, the vapour pressure is negligible, and for water it is about 2300 Pa, still very low compared to the atmospheric pressure.

12-4 Pascal's Principle

Consider a container that is fitted with a movable piston and filled with a liquid (Figure 12-18). When the piston is pushed downward, the layer of liquid in contact with it experiences an increase in pressure. The molecules of a liquid interact with each other and are free to move, so the increase in the pressure is quickly transferred to every part of the liquid and to the walls of the container. This simple fact

was first realized and stated by Blaise Pascal, and is called **Pascal's principle**:

A change in pressure at any point of an enclosed fluid that is at rest is transmitted, without any loss, to every point of the fluid, including the walls of the container.

Notice that for Pascal's principle to hold true, the fluid must be confined. We encounter Pascal's principle in action all around us. When you squeeze the end of a toothpaste tube, the resulting increase in pressure is transmitted throughout the body of the tube, forcing the toothpaste through the opening (Figure 12-19).

Hydraulic Systems

A hydraulic system, based on Pascal's principle (Figure 12-20), is an essential component of many modern-day machines, for example, excavators, pumps, turbines, automobile braking systems (Figure 12-21), and hydraulic elevators. A simple hydraulic system consists of a sealed

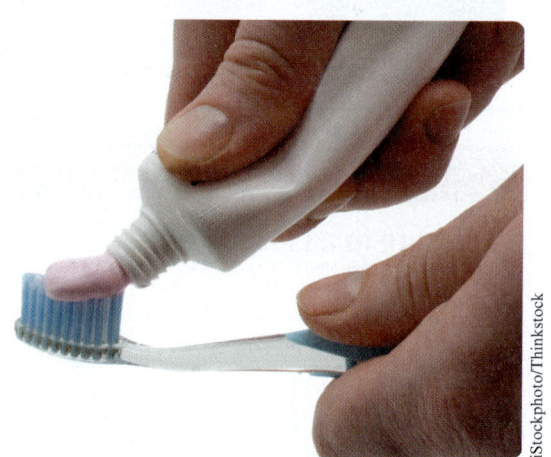

Figure 12-19 Pressure applied at one end of the tube is transmitted to all parts of the tube, forcing the toothpaste out at the open end.

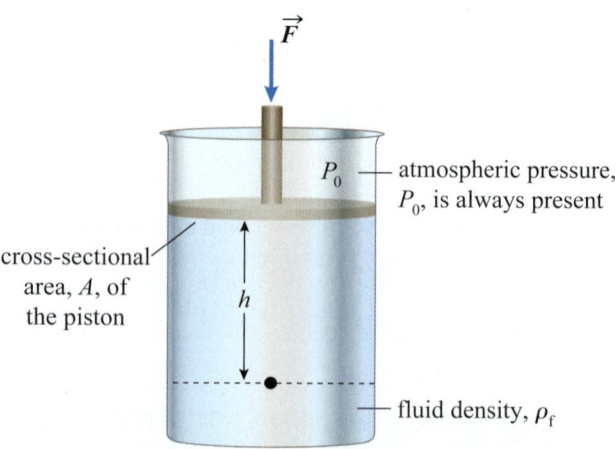

Figure 12-18 The total pressure at any point within the fluid is equal to the sum of the pressures due to the atmosphere, the weight of the fluid, and the external force.

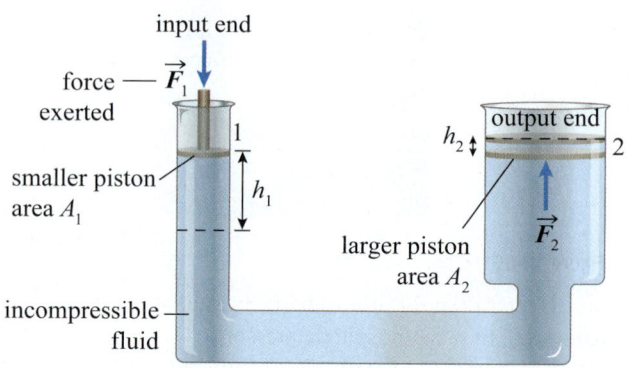

Figure 12-20 A simple hydraulic lift. A force exerted on the input piston results in an increase in pressure in the fluid that is transmitted to the piston at the output end.

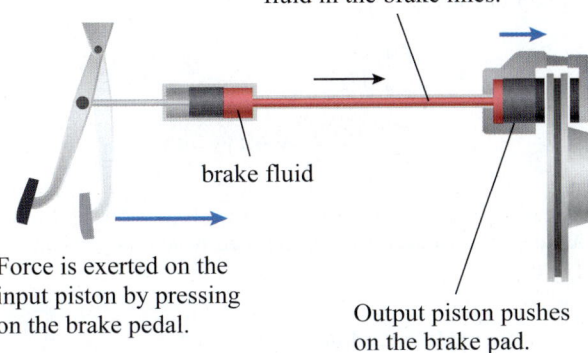

The increase in pressure is transformed through brake fluid in the brake lines.

brake fluid

Force is exerted on the input piston by pressing on the brake pedal.

Output piston pushes on the brake pad.

Figure 12-21 Hydraulic brakes in a car. When the brake pedal is depressed, it pushes on the input piston in the hydraulic cylinder, causing the output pistons to move pads at each wheel.

tube filled with a liquid and fitted with two pistons, usually called an input piston and an output piston. A force exerted on the input piston results in an increase in pressure within the fluid and at the output piston. Suppose the cross-sectional area of the input cylinder is A_1 and that of output cylinder is A_2 and the density of the fluid is ρ_f.

To analyze this system, we will assume that the pistons are massless and that the frictional forces are negligible.

First, consider the case when the only force exerted on the two pistons is due to atmospheric pressure. The total pressure (atmospheric pressure plus pressure due to the weight of the liquid) is the same for all points in the cylinder that are on the same horizontal line. Therefore, the height of the liquid in the two cylinders is the same, and the bottom surfaces of the two pistons are level with each other. Now, assume that the input piston is pushed

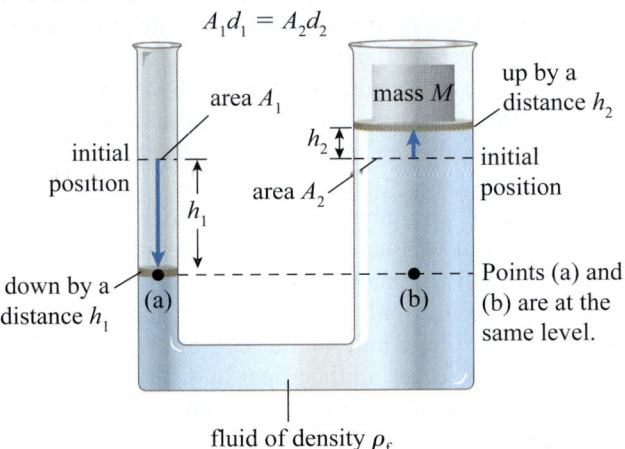

$$A_1 d_1 = A_2 d_2$$

area A_1

mass M

up by a distance h_2

initial position

h_2

initial position

area A_2

h_1

down by a distance h_1 (a)

(b)

Points (a) and (b) are at the same level.

fluid of density ρ_f

Figure 12-22 Applying a downward force to the input piston raises the mass on the output arm of a hydraulic lift.

down by applying a force of magnitude F_1. The increase in the pressure, ΔP, on the input piston is

$$\Delta P = \frac{F_1}{A_1} \tag{12-16}$$

By Pascal's principle, the pressure within the fluid increases by the same amount. This increase in the pressure results in an upward force exerted on the output piston by the fluid. The magnitude of the upward force, F_2, is

$$\begin{align} F_2 &= \Delta P \times A_2 \\ &= \left(\frac{A_2}{A_1}\right) \times F_1 \end{align} \tag{12-17}$$

Note the amplification factor A_2/A_1 on the right-hand side of Equation 12-17. By making the cross-sectional area of the output piston larger than the cross-sectional area of the input piston, a small amount of force can be amplified into a large force. This amplification is one of the most useful features of hydraulic systems.

Consider the situation shown in Figure 12-22. An object (e.g., an automobile) of mass M is placed on the output piston. To hold the two pistons level, a force of magnitude $F_1 = \left(\frac{A_1}{A_2}\right) Mg$ must be exerted on the input piston. Suppose we need to elevate the object by a distance h_2. For this, the input piston needs to be pushed downward to force the liquid into the output end. Since the liquid is neither leaving nor entering the system, the decrease in the volume of the liquid in the input cylinder equals the increase in the volume in the output cylinder.

Let h_1 be the distance by which the input piston needs to be moved downward to raise the output piston by a distance h_2. Then,

$$A_1 \times h_1 = A_2 \times h_2$$

and

$$h_1 = \left(\frac{A_2}{A_1}\right) h_2 \tag{12-18}$$

If $A_2 > A_1$, then $h_1 > h_2$. For example, if A_2 is 20 times A_1 and the output piston is to be raised 10 cm, then the input piston must be pushed down 200 cm. The work done by the external force on the input cylinder is equal to the sum of work done to elevate the mass plus the work done in elevating the volume $A_2 \times h_2$ of the fluid by a height h_2.

As the elevation of the output piston increases, the force at the input end must also increase. Let F_1' be the magnitude of the force needed to hold the mass M at height h_2 with respect to the initial equilibrium position. In this configuration, the difference between the levels of the two pistons is $h_1 + h_2$. Consider points (a) and (b) on a horizontal line that passes through the fluid-piston interface in the input cylinder, as shown in

Figure 12-22. The two points are at the same pressure; therefore,

$$P_0 + \frac{F_1'}{A_1} = P_0 + \frac{Mg}{A_2} + \rho_f g(h_1 + h_2) \quad (12\text{-}19)$$

and

$$F_1' = \left(\frac{A_1}{A_2}\right) Mg + \rho_f g A_1 (h_1 + h_2) \quad (12\text{-}20)$$

The first term on the right-hand side of Equation 12-20 is the force needed to keep the two pistons at the same level. The second term is the additional force needed to balance the weight of the volume $A_1(h_1 + h_2)$ of the liquid above the horizontal line that is level with the input arm. The force exerted at the input arm must balance the combined weight of the object and the lifted liquid.

CHECKPOINT 12-5

Hydrostatic Pressure

A cylindrical U tube, with both arms open to the atmosphere, is partially filled with water and oil (Figure 12-23). The density of oil is less than the density of water. Consider the pressure within the two fluids at points A, B, C, D, E, F, G, and H. Rank the points, from highest to lowest, in terms of the total pressure at these points. Use an equal sign where needed.

Highest ___ ___ ___ ___ ___ ___ ___ Lowest

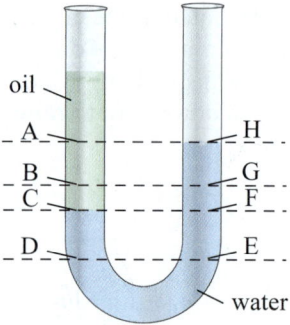

Figure 12-23 A U-shaped manometer.

ANSWER: D = E, C = F, B, G, A, H

EXAMPLE 12-5

Hydraulic Car Jack

A hydraulic car jack has an output cylinder 30.0 cm in diameter and an input cylinder 5.00 cm in diameter and is filled with oil of density 900 kg/m³. An automobile of mass 1100 kg needs to be elevated 40.0 cm.

(a) How far does the piston of the input cylinder need to move?

(b) What is the magnitude of the force on the input piston required to keep the car elevated?

Solution

From Equation 12-18,

$$h_1 = \left(\frac{A_2}{A_1}\right) h_2$$

$$h_1 = \frac{\pi (0.15 \text{ m})^2}{\pi (0.025 \text{ m})^2} \times 0.4 \text{ m} = 14.4 \text{ m}$$

The magnitude of the force that needs to be applied on the input piston to keep the car elevated is given by Equation 12-20:

$$F_1' = \left(\frac{A_1}{A_2}\right) Mg + \rho_f g A_1 (h_1 + h_2)$$

$$F_1' = \left(\frac{1}{36}\right) \times (1100 \text{ kg}) \times (9.81 \text{ m/s}^2)$$

$$+ (900 \text{ kg/m}^3) \times (9.81 \text{ m/s}^2) \times \pi (0.025 \text{ m})^2 \times (14.8 \text{ m})$$

$$= 3.00 \times 10^2 \text{ N} + 2.57 \times 10^2 \text{ N} = 5.57 \times 10^2 \text{ N}$$

Note that the second term in Equation 12-20 is usually negligible in cases where the fluid is a compressed gas.

Making sense of the result

To lift a 1000 kg mass against the force of gravity, one needs a to exert a force of about 10^4 N. Since the ratio of the diameters of the two pistons is 6, the force required to lift the car is about 10^4 N/36 $\approx$ 280 N. This is approximately the value of the first term. The volume of the fluid displaced from the input to the output end is about 0.03 m³ and the weight of this volume of fluid is about 270 N. This is approximately the value of the second term.

12-5 Buoyancy and Archimedes' Principle

If you hold a table tennis ball underwater and then release it, the ball accelerates upward and shoots above the surface. Clearly, the surrounding water must exert an upward force on the ball.

What causes this upward force? The answer lies in the facts that pressure exists at all points within a fluid and that this pressure increases with depth. Therefore, a fluid exerts forces that are normal to every point on the surface of an immersed object. The forces exerted by the water at various points on the surface of a fully immersed ball are shown in Figure 12-24. Since the lower half of the ball is at a greater depth, upward forces on the lower half of the ball are stronger than the downward forces on the upper half. The vector sum of all these forces is a single force that acts upward on the ball.

We can conclude that the horizontal components of forces all cancel out because there is no horizontal motion when the ball is released.

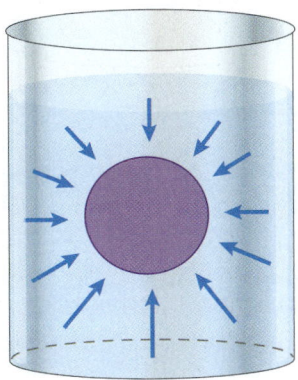

Figure 12-24 The forces due to fluid pressure on a sphere that is fully immersed in a liquid.

The upward force exerted on an object that is fully or partially immersed in a fluid is called the **buoyant force**. We will denote the buoyant force by the symbol $\vec{F}_B$ and its magnitude by the symbol F_B. The direction and the magnitude of the buoyant force are stated by **Archimedes' principle**:

> When an object is fully or partially immersed in a fluid, the fluid exerts an upward force (called the buoyant force) on the object. The magnitude of this upward force is equal to the weight of the fluid displaced by the object.

The genius of Archimedes was in realizing that the magnitude of the buoyant force is equal to the weight of the fluid displaced by the object. The more you immerse a table tennis ball into the water, the greater the volume of water displaced by the ball and the greater the upward force that the water exerts on it. Once the ball is completely submerged, it has displaced the maximum amount of water. If the ball is then moved anywhere within the body of the water, while remaining below the surface, it experiences exactly the same buoyant force (assuming that the density of the water does not change with depth and the ball does not deform).

Imagine a vertical cylinder of height h and cross-sectional area A that is fully immersed in an open container of fluid of density ρ_f (Figure 12-25). The top of the cylinder is at a depth h_1 below the fluid surface, and the bottom of the cylinder is at a depth $h_2 = h_1 + h$ below the surface.

The pressure at the top end of the cylinder, P_1, is

$$P_1 = P_0 + \rho_f g h_1 \tag{12-21}$$

Therefore, the magnitude of the downward force, F_1, on the top of the cylinder is

$$F_1 = (P_0 + \rho_f g h_1)A \tag{12-22}$$

The pressure at the bottom end of the cylinder, P_2, is

$$P_2 = P_0 + \rho_f g h_2 \tag{12-23}$$

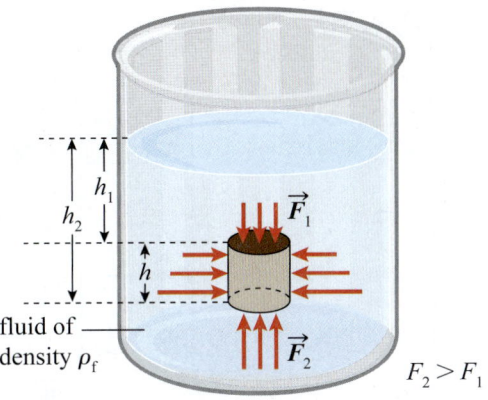

Figure 12-25 The forces exerted by the fluid on a fully submerged cylinder.

The magnitude of the upward force, F_2, on the bottom end of the cylinder is

$$F_2 = (P_0 + \rho_f g h_2)A \tag{12-24}$$

The horizontal force at any given point on the side of the cylinder is cancelled by the corresponding force on the diametrically opposite point. Since $h_2 > h_1$, the magnitude of the upward force is greater than the magnitude of the downward force. The upward force exerted on the cylinder by the fluid, F_B, is given by

$$
\begin{aligned}
F_B = F_2 - F_1 &= (P_0 + \rho_f g h_2)A - (P_0 + \rho_f g h_1)A \\
&= (\rho_f g h_2 - \rho_f g h_1)A \tag{12-25} \\
&= \rho_f g (h_2 - h_1)A = \rho_f g V_0
\end{aligned}
$$

The factor $V_0 = (h_2 - h_1)A$ is the volume of the cylinder and is equal to the volume of the fluid displaced by the cylinder. The quantity $\rho_f V_0$ is the mass of the fluid displaced by the cylinder, and $\rho_f V_0 g$ is the weight of the fluid displaced by the cylinder. Therefore, the magnitude of the buoyant force exerted by the fluid on the cylinder is equal to the weight of the fluid displaced by the cylinder.

KEY EQUATION
$$
\begin{aligned}
F_B &= \text{weight of the displaced fluid} \\
&= (\rho_f V_0)g \tag{12-26}
\end{aligned}
$$

Although we have derived this result for a vertically immersed cylinder, the general result holds for any orientation or shape of an object. Consider a container filled with a fluid, and imagine that a three-dimensional segment of this fluid is suspended within it (Figure 12-26). The fluid is stationary, so this segment is in static equilibrium with the rest of the fluid. Therefore, the buoyant force exerted on this segment by the rest of the fluid must be equal to the weight of the fluid within the segment. Now, suppose that we replace this segment by a solid object of exactly the same shape. The fluid around the object would exert exactly the same buoyant force on the object. Therefore, the magnitude of the buoyant force on a fully or partially immersed object is equal to the weight of the fluid displaced by that object.

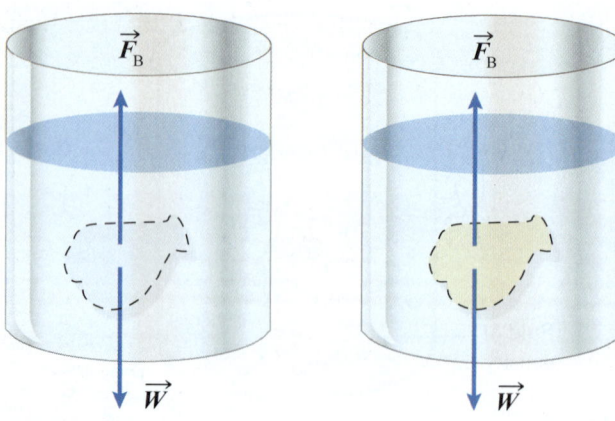

Figure 12-26 The buoyant force on an object immersed in a fluid is equal to the weight of the fluid displaced by the object.

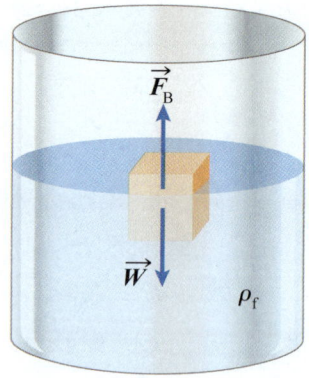

Figure 12-28 For a floating object, the magnitude of the upward buoyant force exerted on the object by the fluid is equal to the magnitude of the weight of the object.

CHECKPOINT 12-6

The Buoyant Force on Spheres

Three identical spheres of different materials are held fully immersed in a container of water, as shown in Figure 12-27. Which of the following statements concerning the magnitude of the buoyant force on each sphere is correct?

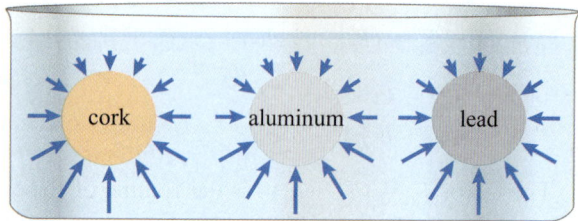

Figure 12-27 Checkpoint 12-6.

(a) $F_{B,cork} < F_{B,aluminum} < F_{B,lead}$
(b) $F_{B,cork} = F_{B,aluminum} = F_{B,lead}$
(c) $F_{B,cork} < F_{B,aluminum} = F_{B,lead}$
(d) $F_{B,cork} > F_{B,aluminum} > F_{B,lead}$

ANSWER: (b)

Flotation

The net vertical force on an object floating in a fluid is zero:

$$\vec{F}_{net} = \vec{F}_B + \vec{W} = 0 \text{ (for a floating object)} \quad (12\text{-}27)$$

Since the buoyant force and the weight act in opposite directions, the magnitude of the buoyant force on a floating object must be equal to the magnitude of the object's weight, W (Figure 12-28):

$$F_B = W \text{ (for a floating object)} \quad (12\text{-}28)$$

Note that Equation 12-28 is derived from Newton's second law of motion and is not a statement of Archimedes'

PEER TO PEER

I have to remind myself that the statement that the magnitude of the buoyant force exerted on an object by a fluid is equal to the magnitude of the weight of the floating object (Equation 12-28) is *only true for objects that float* in that fluid. It is *not true for objects that sink* in that fluid, and it is not a statement of Archimedes' principle.

principle. Only for floating objects is the magnitude of buoyant force on the object equal to its weight.

Consider an object of volume V_0 and density ρ_0 that is floating in a fluid of density ρ_f. Assume that a volume V_{sub} of the object is submerged in the fluid. Then

$$W = \text{weight of the object} = (\rho_0 V_0)g$$
$$F_B = \text{buoyant force on the object}$$
$$= \text{weight of the fluid displaced by the object}$$
$$= (\text{mass of fluid displaced by the object})g$$
$$= (\rho_f V_{sub})g$$

Substituting into Equation 12-28, we get

$$\rho_0 V_0 g = \rho_f V_{sub} g$$

and

$$\frac{V_{sub}}{V_0} = \frac{\rho_0}{\rho_f} \quad (12\text{-}29)$$

Therefore, the fraction of the volume of a floating object below a fluid's surface is equal to the ratio of the density of the object to the density of the fluid.

The density of ocean water is approximately 1020 kg/m^3, and the density of sea ice is 920 kg/m^3. Therefore for ice floating in ocean,

$$\frac{V_{sub}}{V_0} = \frac{920 \text{ kg/m}^3}{1020 \text{ kg/m}^3} = 0.902$$

so approximately 90% of an iceberg is beneath the ocean's surface (Figure 12-29). If the average density

Figure 12-29 Approximately 90% of the volume of an iceberg is beneath the ocean's surface.

of an object is equal to the density of a fluid in which it is fully immersed the object will neither sink or rise. This condition is called neutral buoyancy.

Apparent Weight in a Fluid

If the average density of an object is greater than the density of a fluid, the object will sink in that fluid. The weight of the object is greater than the weight of the displaced fluid. The **apparent weight**, $W_{apparent}$, of the fully submerged object is defined as the difference between the weight of the object when not immersed and the buoyant force exerted on the object when it is immersed in the fluid. The magnitude of the apparent weight is given by

$$W_{apparent} = W - F_B \qquad (12\text{-}33)$$

A convenient method to determine the buoyant force on an object is to weigh the object first when it is outside the fluid and then when it is completely immersed in a fluid. The difference between these two weights equals the magnitude of the buoyant force on the object. This method can also be used to determine the average density of an object.

EXAMPLE 12-6

How Large Is That Helium Balloon?

A balloon is carrying two people in a basket. The total mass of the people, the basket, and the material of the balloon, but excluding the mass of helium gas, is 280 kg. What volume of helium is needed to lift the balloon (Figure 12-30)? The density of the helium is 0.18 kg/m^3, and the density of the air is 1.30 kg/m^3.

Figure 12-30 A helium (or hot air) filled balloon floats when the buoyant force exerted on the balloon by the surrounding air is equal to the weight of the balloon plus the enclosed gas.

Solution

When a balloon is filled with helium gas it displaces air and, the surrounding air exerts a buoyant force on the balloon and the gas inside it. As more helium is added, the volume of the balloon increases, so more air is displaced and the buoyant force on the balloon increases. When the buoyant force equals the total weight of the balloon (including the helium), the balloon will begin to float; adding any more helium will cause the balloon to accelerate upward.

The total weight of the balloon is

W = weight of balloon and its contents
  + weight of the helium gas filling the balloon

$$W = (280 \text{ kg})g + m_{He}g = (280 \text{ kg})g + (V_{He}\rho_{He})g \quad (12\text{-}30)$$

In Equation 12-30, V_{He} is the volume of the helium-filled balloon, and ρ_{He} is the density of helium gas. The buoyant force exerted by the air on the balloon is

F_B = weight of the air displaced by the balloon
$$= m_{air}g = (V_{He}\rho_{air})g \qquad (12\text{-}31)$$

At liftoff, the buoyant force equals the total weight of the balloon, so

$$(280 \text{ kg})g + V_{He}\rho_{He}g = (V_{He}\rho_{air})g$$

and

$$V_{He} = \frac{280 \text{ kg}}{(\rho_{air} - \rho_{He})} = \frac{280 \text{ kg}}{(1.30 - 0.18) \text{ kg/m}^3} = 2.5 \times 10^2 \text{ m}^3 \qquad (12\text{-}32)$$

With 250 m^3 of helium, the balloon is ready to lift off. To initiate an upward velocity, the amount of helium pumped into the balloon must be greater than the above-calculated value.

Making sense of the result

A volume of 250 m^3 of air will exert a buoyant force of approximately $(250 \times 1.3 \times 10)$ N = 3250 N, which is sufficient to lift a mass of about 320 kg against the force of gravity. Since the mass of the people plus the balloon is 280 kg and we have to add to it the mass of helium gas, the above-calculated value of the required volume of helium gas makes sense.

EXAMPLE 12-7

Is the Ring Made of Pure Gold?

Your friend asks you to determine whether the ring that she recently bought is made of pure gold. (The density of gold is 1.93×10^4 kg/m³.) Using a thread of negligible mass and volume, you tie the ring to a very accurate spring balance and determine that in air it weighs 6.40×10^{-3} N. You then fully immerse the ring in a beaker full of water and find that its apparent weight is 6.00×10^{-3} N. Is the ring made of pure gold?

Solution

To solve this problem, we apply the key points of this section as follows:

1. The difference between the weight of the ring in air and the weight of the ring in water is equal to the magnitude of the buoyant force on the ring.
2. The magnitude of the buoyant force is equal to the weight of the water displaced by the ring.
3. Knowing the density of the water and the value of g, the volume of water displaced by the ring can be calculated.
4. The volume of the ring is equal to the volume of the displaced water.
5. The weight of the ring is known; therefore, we can calculate the density of the ring.

Let us now do these calculations step by step:

1. $F_B = W_{ring, air} - W_{ring, water}$
    $= 6.40 \times 10^{-3}$ N $- 6.00 \times 10^{-3}$ N $= 0.40 \times 10^{-3}$ N
    $W_{displaced} =$ Weight of displaced water
    $= \rho_{water} V_{water} g = 0.40 \times 10^{-3}$ N

2. $V_{water} = \dfrac{0.40 \times 10^{-3} \text{ N}}{\rho_{water} g} = \dfrac{0.40 \times 10^{-3} \text{ N}}{1000 \text{ kg/m}^3 \times 9.81 \text{ m/s}^2}$
    $= 4.08 \times 10^{-8}$ m³

3. $V_{ring} = V_{water} = 4.08 \times 10^{-8}$ m³
4. $W_{ring, air} = m_{ring, air} g = (V_{ring} \rho_{ring}) g$

5. $\rho_{ring} = \dfrac{W_{ring, air}}{V_{ring} g} = \dfrac{6.4 \times 10^{-3} \text{ N}}{(4.08 \times 10^{-8} \text{ m}^3) \times 9.81 \text{ m/s}^2}$
    $= 1.6 \times 10^4$ kg/m³

Since the density of the ring is considerably less than the density of pure gold (1.93×10^4 kg/m³), the ring is not made of pure gold.

Making sense of the result

The calculated density of the ring is 16 times the density of water, which is close to the density of gold. So our calculation makes sense.

EXAMPLE 12-8

Weighing an Object Immersed in a Fluid

A solid sphere of density 3.00×10^3 kg/m³ hangs from a spring scale (Scale 1) that reads 200.0 N. A large cylindrical beaker with inside diameter 40.0 cm and partially filled with water sits on Scale 2, which reads 1000 N. The sphere is then lowered into the water until it is fully immersed but not touching the beaker (Figure 12-31). Assume that no water is spilled from the beaker. What are the new readings on the two scales?

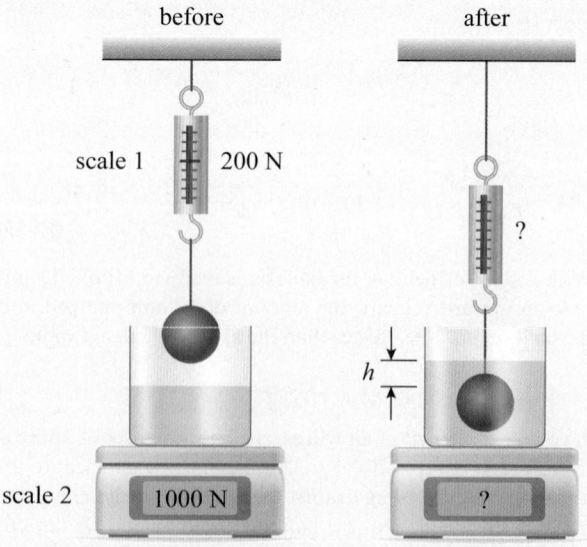

Figure 12-31 Example 12-8.

Solution

When the sphere is hanging freely from a spring scale, the tension in the spring is equal to the weight of the sphere. When the sphere is immersed in water, it will experience a buoyant force exerted by the water. Therefore, the tension in the spring and hence the reading on Scale 1 will decrease. The decrease in the reading is equal to the magnitude of the buoyant force. We therefore need to calculate the buoyant force exerted on the immersed sphere. For this we need to determine the volume of the sphere, which is equal to the volume of the water displaced by the sphere. Since we know the density and the weight of the sphere, we can calculate its volume as follows:

weight of sphere = density of sphere × volume of sphere × g

200.0 N $= (3.00 \times 10^3$ kg/m³$) \times V_{sphere} \times (9.81$ m/s²$)$

Therefore,

$V_{sphere} = \dfrac{200.0 \text{ N}}{(3.00 \times 10^3 \text{ kg/m}^3) \times (9.81 \text{ m/s}^2)} = 6.80 \times 10^{-3}$ m³

The magnitude of the buoyant force, F_B, on the sphere is equal to the weight of the water displace by the sphere. Therefore,

$F_B =$ density of water × V_{sphere} × 9.81 m/s²
$= (1000$ kg/m³$) \times (6.80 \times 10^{-3}$ m³$) \times (9.81$ m/s²$)$
$= 6.7 \times 10^1$ N

The tension, T_s, in the spring is therefore

T_s = weight of sphere − buoyant force on sphere
= 200.0 N − 67.0 N = 133 N

Therefore, the new reading on Scale 1 is 1.33×10^2 N.

Let us now consider the reading on Scale 2 when the sphere is immersed in water. As the sphere will displace water equal to it its volume, the level of water in the beaker will rise and hence the pressure at the bottom of the beaker will increase. This results in greater force exerted by the beaker on the scale, so the reading on the scale will increase. Let h be the increase in the level of water when the sphere is immersed. Since $R = 20$ cm is the inside radius of the beaker, we can calculate h as follows:

volume of water displaced = volume of sphere

$$\pi R^2 h = 6.80 \times 10^{-3} \text{ m}^3$$

Therefore,

$$h = \frac{6.80 \times 10^{-3} \text{ m}^3}{\pi \times (0.20 \text{ m})^2} = 5.4 \times 10^{-2} \text{ m}$$

As the water level rises, the hydrostatic pressure at the bottom of the beaker increases. The increase in the pressure, ΔP (here Δ implies an increase), at the bottom of the container is

$$\Delta P = \text{density of the water} \times g \times h$$
$$= (1000 \text{ kg/m}^3) \times (9.81 \text{ m/s}^2) \times (5.4 \times 10^{-2} \text{ m})$$
$$= 5.3 \times 10^2 \text{ Pa}$$

The increase in the pressure results in an increase in the downward force exerted at the bottom of the container by the water. The magnitude of this downward force, ΔF, is given by

$$\Delta F = \Delta P \times \text{cross-sectional area of interior of container}$$
$$= (5.3 \times 10^2 \text{ Pa}) \times (\pi \times (0.20 \text{ m})^2)$$
$$= 6.7 \times 10^1 \text{ N}$$

Thus the new reading on Scale 2, with the sphere immersed, is

$$1000 \text{ N} + 67 \text{ N} = 1067 \text{ N}$$

Making sense of the result

Note that the increase in the force exerted at the bottom of the beaker by the water due to an increase in the hydrostatic pressure is equal to the buoyant force exerted on the sphere by the water (within rounding). This is not an accident. If we changed the input parameters, the increase in the downward force due to increased hydrostatic pressure would still be equal to the magnitude of the buoyant force on the sphere. Why is that so? Well, you can think of it this way. The water exerts an upward force on the sphere. According to Newton's third law of motion, the sphere must exert an equal and opposite (downward) force on the water, which is transmitted to the bottom of the sphere and hence to the scale.

There is another way of looking at this. The total weight measured by the two scales is equal to g times the total mass of the sphere, the water, and the beaker. The total mass of the system does not change when the sphere is immersed in water. Therefore, when the reading on Scale 1 decreases by a certain amount, the reading on Scale 2 must increase by the same amount.

CHECKPOINT 12-7

Pressure on the Scale

A container, partially filled with water, is resting on a scale that measures its weight. If you immerse your hand in the water, without touching any part of the container or spilling any water, what happens to the reading on the scale? Explain your reasoning.

(a) It stays the same.
(b) It increases.
(c) It decreases.
(d) Insufficient information is provided to answer this question.

ANSWER: (b)

12-6 Fluids in Motion

There is an obvious difference between the energy of water in a large wave and the energy of stationary water in a cup. The water in a wave has significant kinetic energy. Water in a cup is at rest and has no net kinetic energy. Any physical description of a moving fluid must include its kinetic energy. We start our discussion of moving fluids by first introducing the concepts of kinetic energy per unit volume and potential energy per unit volume for a fluid.

Kinetic and Potential Energy per Unit Volume

Consider a fluid of uniform density ρ moving with a velocity $\vec{v}$. The mass of a small volume ΔV of this fluid is $\Delta m = \rho \Delta V$, and its kinetic energy is

$$\Delta K = \frac{1}{2}(\Delta m)v^2 = \frac{1}{2}\rho v^2 (\Delta V) \tag{12-34}$$

The kinetic energy per unit volume of the fluid is

$$\frac{\Delta K}{\Delta V} = \frac{1}{2}\rho v^2 \tag{12-35}$$

The SI unit for kinetic energy per unit volume is the pascal (Pa), the same as pressure:

$$\frac{\text{kg}}{\text{m}^3} \times \frac{\text{m}^2}{\text{s}^2} = \frac{\text{kg}}{\text{m} \cdot \text{s}^2} = \text{Pa}$$

We can similarly define potential energy per unit volume for a fluid. The gravitational potential energy of a small element of fluid that has volume ΔV and is located at a height h from the ground is

$$\Delta U = (\Delta m)gh = \rho gh(\Delta V) \qquad \text{(12-36)}$$

The potential energy per unit volume of the fluid is then

$$\frac{\Delta U}{\Delta V} = \rho gh \qquad \text{(12-37)}$$

The SI unit for potential energy per unit volume is also the pascal.

Ideal Fluids

A general description of moving fluids is complicated because fluids exhibit a wide variety of behaviour that depends on the type of fluid and its speed. The motion of water smoothly emerging from a tap is very different from the chaotic flow of water over Niagara Falls. Similarly, it is easier to suck water through a straw than it is to suck honey through the same straw. However, we can develop a fairly reasonable description of moving fluids by making some key assumptions about the physical properties of the fluids and how the fluids flow. These assumptions are as follows.

A fluid is incompressible We assume that the density of a fluid remains constant and does not change with pressure. This assumption closely approximates the behaviour of liquids. For example, the density of water increases by approximately 1% at a depth of 1000 m below the ocean's surface, where the absolute pressure is 100 times the atmospheric pressure. The assumption that a gas is incompressible is only valid if the change in the gas's pressure is small.

A fluid is inviscid Viscosity is the measure of a fluid's resistance to flow and to any deformation of its shape. For example, water flows much more readily than honey. The assumption that a fluid is inviscid implies that frictional forces within the fluid, as well as between the fluid and the walls of the containing vessel, are negligible. A frictionless flow is an idealization because all real fluids have non-zero friction. We will discuss viscous flow later in the chapter. A fluid that is incompressible and inviscid is called an **ideal fluid**.

The flow of the fluid is steady A **steady flow** is smooth and uniform. In a steady flow, the velocity of the fluid at any *given* point within the region of flow has a fixed magnitude and direction and remains constant with time. The velocities at *different* points in the region of

Figure 12-32 Water gently flowing from a tap is an example of a steady flow.

flow may differ from each other. Water flowing slowly from a tap (Figure 12-32) is an example of a steady flow. In this case, the velocity of water at a fixed distance below the tap is well defined and does not change with time as long as the flow rate through the tap remains the same.

A flow that is not steady is called **turbulent**. In a turbulent flow, the velocity at a given point in the region of flow varies with time. The flow of gases and ash from the eruption of Mount St. Helens is an example of turbulent flow (Figure 12-33). The mathematical description of turbulent flows is quite complex.

The flow of the fluid is irrotational In an irrotational flow, a fluid element does not move in a circular path. An example of irrotational flow is water moving in a

Turbulent flow from the mouth of Mount St. Helens (Washington, U.S.A.). Compare this flow to the flow of water slowly emerging from a tap.

Figure 12-33 The eruption of Mount St. Helens. The flow from the volcano is chaotic and turbulent.

straight pipe or falling straight downward under the influence of gravity.

Streamlines and Flow Tubes

When the flow is steady, fluid particles move along well-defined paths called streamlines. A **streamline** is a curve drawn in the body of a moving fluid such that the velocity of a particle at any point along the streamline is tangent to the streamline at that point. Figure 12-34 shows velocity vectors at various points along a streamline. The instantaneous velocity of a particle must be unique, and for steady flow the velocity at any point in the region of the flow does not change with time. *Therefore, in a steady flow, streamlines do not cross each other.* A steady flow is also called a **streamline flow**, or **laminar flow**.

Imagine a set of neighbouring streamlines that form a closed path in a given region of a moving fluid. The particles of the fluid that are either on the boundary of this closed path or enclosed within it follow their respective streamlines as they move along the fluid. These neighbouring streamlines form a tube, called a **flow tube** (Figure 12-34). The fluid within a flow tube must remain within the tube. There is no mixing of fluid from different flow tubes. The cross-sectional area of a given flow tube may change along its length, but the amount of the fluid that enters at one end of a flow tube must be equal to the amount that leaves through another end down the stream. You can envision a flow tube as an invisible pipe that keeps the fluid within it from mixing with the fluid outside.

There are several ways to visualize streamlines and flow tubes. Smoke in gases and dyes in liquids are often used by engineers and scientists to study flow patterns of fluids around objects (Figure 12-35). With computers, scientists can calculate the pattern of streamlines around objects. Studying such patterns provides information about the resistance (or drag) experienced by an object as it moves through a fluid. Smooth and

Figure 12-35 Streamlines of smoke passing over an automobile in a wind tunnel.

continuous streamlines indicate a steady flow with minimum drag. Closely spaced streamlines indicate high speed.

Keep in mind that a description of an object moving through a stationary fluid is the same as that of a fluid that is moving past a stationary object. That is why streamlines created by moving air around a stationary automobile in a wind tunnel are exactly the same as the streamlines of an automobile moving through stationary air.

Flow Rate

The **flow rate** (Q) is the volume of fluid that flows through a cross-sectional area per unit time:

$$Q = \frac{\text{volume of fluid that flows through a surface}}{\text{time taken}}$$

$$(12\text{-}38)$$

The SI units for flow rate are cubic metres per second (m^3/s). Other convenient units are litres per second (L/s) and cubic centimetres per second (cm^3/s):

$$1 \text{ m}^3/\text{s} = 10^6 \text{ cm}^3/\text{s} = 1000 \text{ L/s}$$

For a fluid moving with a speed v through a cross-sectional area A that is perpendicular to the direction of flow, the volume flow rate is the product of A and v:

KEY EQUATION
$$Q = Av \qquad (12\text{-}39)$$

12-7 The Continuity Equation: Conservation of Fluid Mass

Consider an ideal fluid flowing through a pipe of variable cross-sectional area (Figure 12-36). If there are no

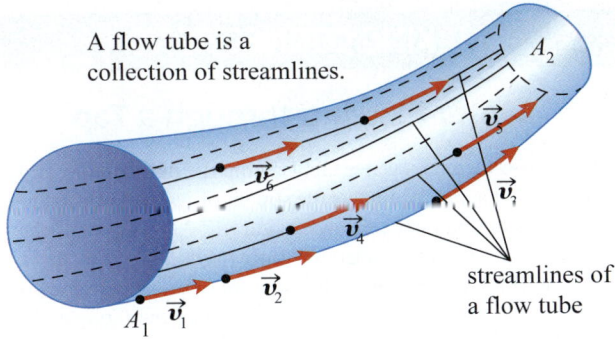

A flow tube is a collection of streamlines.

streamlines of a flow tube

Figure 12-34 A flow tube. Solid lines represent streamlines. The magnitude and direction of the flow velocity at any point along the flow line is given by the length and direction of the arrow at that point. The direction of the arrow is tangential to the streamlines. The group of streamlines forms a stream tube (or flow tube).

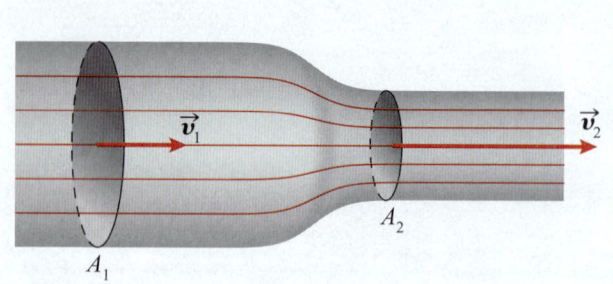

Figure 12-36 Fluid flowing through a horizontal pipe of variable cross-sectional area.

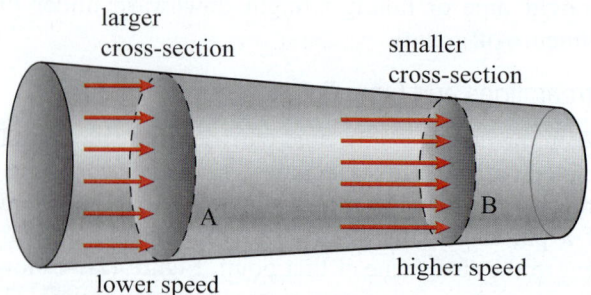

Figure 12-37 When a fluid moves through a horizontal pipe that gradually narrows, the fluid speed increases as the cross-sectional area of the pipe decreases.

other sources of fluid and no holes in the pipe, then the amount of fluid that enters the pipe at one end in some time Δt must be equal to the amount that leaves at the other end in the same time. In other words, what comes in must go out. The mathematical equation that describes this "law of conservation of fluid mass" is called the **equation of continuity**.

Consider a section of a pipe (or a flow tube) where the cross-sectional area is A_1, the speed of the fluid is v_1, and the density of the fluid is ρ_1. In time Δt, a length $\Delta x_1 = v_1 \Delta t$ of the fluid passes through A_1. The volume, ΔV_1, of the fluid passing through A_1 is $\Delta V_1 = A_1(v_1 \Delta t)$. The mass of this volume of the fluid is

$$\Delta M_1 = \rho_1 \Delta V_1 = (\rho_1 A_1 v_1)\Delta t \qquad (12\text{-}40)$$

In the same time interval, a mass ΔM_2 of the fluid must pass through another section of the pipe where the cross-sectional area is A_2, the fluid speed is v_2, and the fluid density is ρ_2:

$$\Delta M_2 = \rho_2 \Delta V_2 = (\rho_2 A_2 v_2)\Delta t \qquad (12\text{-}41)$$

Since the mass of the fluid entering through A_1 must be equal to the mass leaving through A_2, we must have

$$\rho_1 A_1 v_1 = \rho_2 A_2 v_2 \qquad (12\text{-}42)$$

The quantity $\rho A v$ has the dimensions of kg/s and is called the **mass flow rate**. For incompressible fluids ($\rho_1 = \rho_2$), the above equation reduces to a simpler relationship:

KEY EQUATION

$$A_1 v_1 = A_2 v_2 \text{ (for incompressible fluids)} \quad (12\text{-}43)$$

The quantity Av is the volume flow rate, Q. The continuity equation states that for incompressible fluids, the volume flow rate remains constant through all sections of a flow tube. An important consequence of the equation of continuity is that the fluid moves faster through a narrower section of a flow tube and slows when moving through a wider section (Figure 12-37).

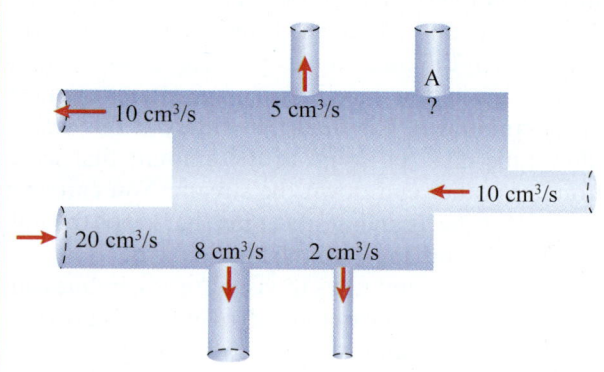
MAKING CONNECTIONS

Water Flowing through a Tap

When you turn on a water tap, keeping the flow smooth and slow, you may notice that as the stream of water flows downward, it continuously narrows (Figure 12-32). Why does the stream narrow as it accelerates downward? The water flows out of the tap with an initial flow rate that is determined by the cross-sectional area of the mouth of the tap and the speed with which the water leaves the tap. Water accelerates due to gravity, so its speed increases. Since the flow rate must remain constant, the cross-sectional area of the stream must decrease with increasing speed of water.

The same phenomenon occurs when you gently tilt a spoonful of honey and watch the stream of honey as it accelerates downward.

12-8 Conservation of Energy for Moving Fluids

According to the equation of continuity, as a fluid passes through a narrower section of a pipe (or a flow tube), its speed increases. For an element of fluid to speed up, the fluid behind it must exert a force on it. Therefore, the pressure in the wider section of the pipe must be higher than the pressure in the narrower section. We thus reach an interesting and non-intuitive conclusion: *In the regions of flow where the fluid moves faster, the pressure within the fluid is lower than the pressure in the regions where the fluid speed is slower.* In this section, we will show that this relationship is a consequence of the law of conservation of energy as applied to moving fluids.

Consider an ideal fluid flowing through a flow tube of variable cross-sectional area and variable height (measured with respect to an arbitrary ground level). The fluid is moving from left to right (Figure 12-39). Let us focus on the body of the fluid that is contained between two sections (1 and 2) of the flow tube. We will calculate the work done on this "test volume" of fluid by the surrounding fluid (Figure 12-40). At section 1, the pressure in the fluid is P_1, the speed of the fluid is v_1, the cross-sectional area of the tube is A_1, and the height of the tube is y_1. Corresponding quantities at section 2 are denoted by P_2, v_2, A_2, and y_2. According to the work-energy theorem, in the absence of frictional forces, the net work done by external forces on an object is equal to the change in the object's total energy:

$$W_{net} = \Delta K + \Delta U \qquad (12\text{-}44)$$

Here, W_{net} is the net work done on the object by external forces, ΔK is the change in the kinetic energy of the object, and ΔU is the change in the object's potential energy. Note that only external forces are considered when calculating the work done on the object because internal forces always come in action-reaction

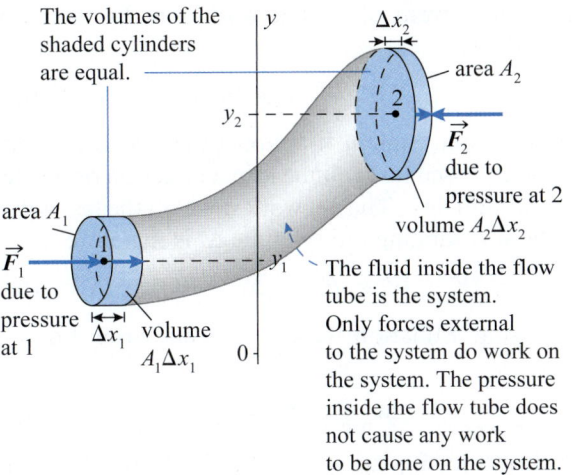

Figure 12-40 Fluid flow through a flow tube of variable cross-sectional area and height.

The volumes of the shaded cylinders are equal. — area A_2 — $\vec{F}_2$ due to pressure at 2 volume $A_2\Delta x_2$

area A_1 — $\vec{F}_1$ due to pressure at 1 volume $A_1\Delta x_1$

The fluid inside the flow tube is the system. Only forces external to the system do work on the system. The pressure inside the flow tube does not cause any work to be done on the system.

pairs and cancel out. Although potential energy could be of several forms, in dealing with fluids we will only consider gravitational potential energy.

In time Δt, a length $\Delta x_1 = v_1\Delta t$ of the fluid enters through section 1. The volume of the entering fluid is

$$\Delta V = A_1(v_1\Delta t) \qquad (12\text{-}45)$$

In the same time, the volume of fluid that leaves through section 2 is $A_2(v_2\Delta t)$. By the equation of continuity,

$$\Delta V = A_1 v_1 \Delta t = A_2 v_2 \Delta t \qquad (12\text{-}46)$$

The entering volume has speed v_1, and the speed of the leaving volume is v_2. Therefore, the change in the kinetic energy (ΔK) of the test volume is

$\Delta K =$ (kinetic energy of leaving fluid)
　　　$-$ (kinetic energy of entering fluid)

$$\Delta K = \frac{1}{2}(\rho\Delta V)v_2^2 - \frac{1}{2}(\rho\Delta V)v_1^2 \qquad (12\text{-}47)$$

The fluid enters at a height y_1 and leaves at a height y_2. Therefore, in time Δt, the change in the potential energy (ΔU) of the test volume is

$\Delta U =$ (potential energy of leaving fluid)
　　　$-$ (potential energy of entering fluid)

$$\Delta U = (\rho\Delta V)gy_2 - (\rho\Delta V)gy_1 \qquad (12\text{-}48)$$

The fluid entering through section 1 exerts a force on the test volume, displacing it toward the right by a distance $v_1\Delta t$ in time Δt. The magnitude of the force exerted by the entering fluid is P_1A_1. Therefore, the **work done on the test volume** by the entering fluid, W_1, is

$$W_1 = P_1A_1(v_1\Delta t) \qquad (12\text{-}49)$$

At section 2, the magnitude of the force exerted by the external fluid on the enclosed test volume is P_2A_2. As the test volume moves to the right, it does work on

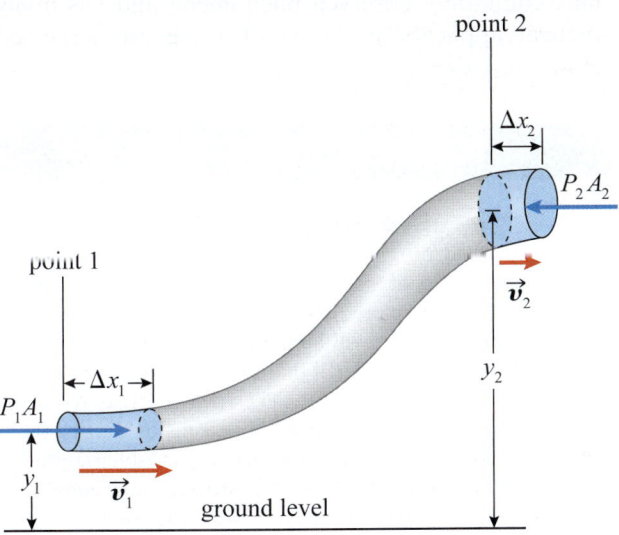

Figure 12-39 An ideal fluid flowing through a pipe of variable height and cross-sectional area.

point 2

Δx_2

P_2A_2

$\vec{v}_2$

point 1

$\vec{v}_2$

Δx_1

y_2

P_1A_1

y_1

$\vec{v}_1$

ground level

the fluid in front of it. The **work done by the leaving fluid**, W_2, is

$$W_2 = -P_2 A_2 (v_2 \Delta t) \qquad (12\text{-}50)$$

The minus sign in W_2 implies that the force exerted on the enclosed test volume is in a direction opposite to that of fluid flow. The net work done on the test volume by the surrounding fluid in time Δt is therefore

$$W_{net} = W_1 + W_2 = P_1 A_1 (v_1 \Delta t) - P_2 A_2 (v_2 \Delta t) \qquad (12\text{-}51)$$

Using Equation 12-46, we can write Equation 12-51 as

$$W_{net} = (P_1 - P_2) \Delta V \qquad (12\text{-}52)$$

Inserting Equations 12-50, 12-51, and 12-52 into Equation 12-44, we obtain

$$(P_1 - P_2) \Delta V = \frac{1}{2}(\rho \Delta V) v_2^2 - \frac{1}{2}(\rho \Delta V) v_1^2 \\ + (\rho \Delta V) g y_2 - (\rho \Delta V) g y_1 \qquad (12\text{-}53)$$

Dividing by ΔV gives

$$P_1 - P_2 = \frac{1}{2}\rho v_2^2 - \frac{1}{2}\rho v_1^2 + \rho g y_2 - \rho g y_1 \qquad (12\text{-}54)$$

Rearranging so that all terms corresponding to a given section are on the same side, we obtain

KEY EQUATION

$$P_1 + \frac{1}{2}\rho v_1^2 + \rho g y_1 = P_2 + \frac{1}{2}\rho v_2^2 + \rho g y_2 \qquad (12\text{-}55)$$

Equation 12-55 is called **Bernoulli's equation**. Bernoulli's equation is a statement of the law of conservation of energy for ideal fluids in the absence of frictional forces. It states that during the steady flow of an ideal fluid, the sum of the fluid pressure, the kinetic energy per unit volume, and the potential energy per unit volume remains constant throughout the body of the fluid. Since points 1 and 2 can be located anywhere along the flow tube, we can also write Bernoulli's equation in the following form:

$$P + \frac{1}{2}\rho v^2 + \rho g y = \text{constant} \qquad (12\text{-}56)$$

Equations (12-43) and (12-55) are the two fundamental equations describing the flow of ideal fluids. In many situations, a description of the flow requires using both of these equations. Let us examine two special cases.

Fluid flow through a horizontal pipe Consider an ideal fluid flowing through a horizontal pipe of variable cross-sectional area, as shown in Figure 12-41. Imagine a streamline that passes through the centre of the pipe, with point 1 located in the wider section and point 2 in the narrower section of the pipe. Since the pipe is

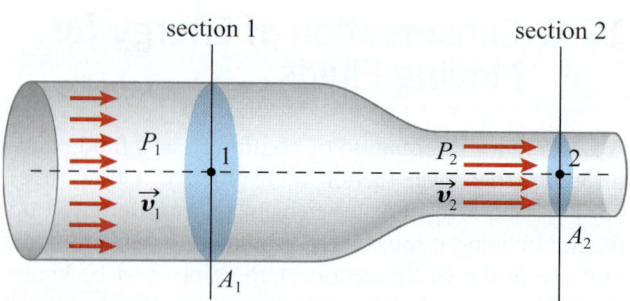

Figure 12-41 Fluid flow along a horizontal pipe of variable cross-sectional area.

horizontal, $y_1 = y_2$, and the gravitational potential energy of the fluid does not change. Bernoulli's equation simplifies to the following:

$$P_1 + \frac{1}{2}\rho v_1^2 = P_2 + \frac{1}{2}\rho v_2^2 \qquad (12\text{-}57)$$

Equation 12-57 implies that the sum of pressure and kinetic energy per unit volume remains constant in a horizontal flow. When passing through the narrow section of the pipe, the fluid speeds up (because of the continuity equation), and, hence, the pressure within the fluid must decrease. Conversely, when a fluid slows down, the pressure within the fluid must increase. You might think that as a fluid speeds up, the pressure within the fluid should also increase, but the opposite is true.

Applying the equation of continuity at points 1 and 2, we get $v_2 = (A_1/A_2)v_1$, and the pressure difference between the two sections of the pipe is

$$P_1 - P_2 = \frac{1}{2}\rho v_1^2 \left(\frac{A_1^2}{A_2^2} - 1 \right) \qquad (12\text{-}58)$$

Thus, $P_1 > P_2$ if $A_1 > A_2$. Bernoulli's equation explains many commonly observed phenomena and has many practical applications. Some of these are discussed below.

MAKING CONNECTIONS

Blowing over a Paper Strip

Cut a small strip of paper (about 2 cm × 16 cm), and place the short end close to your lower lip, holding it with your fingers. The strip will hang downward because of its weight. Now blow hard such that the air moves horizontal to the top surface of the strip. The strip will rise. As you blow, the air above the top surface moves faster, causing the pressure above the strip to drop below atmospheric pressure. Since the pressure beneath the strip remains equal to atmospheric pressure, an upward force acts on the strip, lifting it up.

EXAMPLE 12-9

Blood Flow through a Blocked Artery

Arteriosclerosis, the most common type of cardiovascular disease, is caused by the buildup of cholesterol plaque in an arterial wall resulting in narrowing (called *stenosis*) of an artery. It is the major cause of heart attack and stroke. An 80% blockage of an artery is considered serious and can lead to collapse of that artery under external pressure, a consequence of Bernoulli's equation. This example discusses blood flow through a horizontal artery that is partially blocked. The flow is assumed to be ideal.

Consider blood (density 1060 kg/m³) flow through a horizontal artery that is 80% blocked along a certain section (Figure 12-42). The cross-sectional area of the healthy (i.e., unblocked) part of the artery is 1.0 cm², and the speed of blood through this section is 50 cm/s. Determine the pressure drop in the blood as it passes through the blocked section of the artery.

blocked section

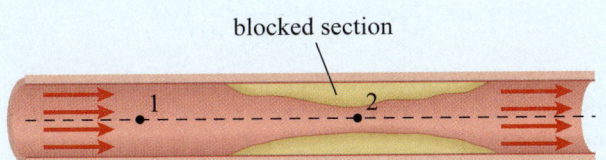

Figure 12-42 Blood flow through a blocked artery. The dashed line shows a horizontal streamline through the section of flow. Points 1 and 2 lie in unblocked and blocked sections of the artery, respectively.

Solution

As the blood passes through the narrower section of the artery, its speed increases. Therefore, the kinetic energy per unit volume of the blood increases and by conservation of energy the blood pressure in the narrower section must decrease. Consider a horizontal streamline that passes from the healthy into the blocked section of the artery. Along this streamline, consider a point 1 in the healthy section and a point 2 in the blocked section. Applying the continuity equation at 1 and 2, we get

$$A_1 v_1 = A_2 v_2$$

Since 80% of the artery is blocked through the narrower section, the area of the narrower section is 20% of that of the healthy section. Therefore, $A_2 = 0.2 A_1$. Since $v_1 = 50$ cm/s = 0.50 m/s, we can calculate v_2, the speed in the narrower section:

$$v_2 = \left(\frac{A_1}{A_2}\right) v_1 = \left(\frac{A_1}{0.2 A_1}\right) v_1 = 5 v_1 = 2.5 \text{ m/s}$$

We can now calculate the pressure difference between 1 and 2 by applying Bernoulli's equation for a horizontal flow:

$$P_1 + \frac{1}{2}\rho v_1^2 = P_2 + \frac{1}{2}\rho v_2^2$$

Here ρ is the density of blood, which we take to be 1060 kg/m³. The change in the blood pressure between the two sections is

$$P_1 - P_2 = \frac{1}{2}\rho v_2^2 - \frac{1}{2}\rho v_1^2$$

$$= \frac{1}{2}\rho (v_2^2 - v_1^2)$$

$$= \frac{1}{2} \times (1060 \text{ kg/m}^3) \times ((2.5 \text{ m/s})^2 - (0.5 \text{ m/s})^2)$$

$$= 3.2 \times 10^3 \text{ Pa}$$

Therefore, the pressure drops by about 3.2×10^3 Pa in the narrower section. The typical blood pressure in a large unblocked artery is about 13.6×10^3 Pa. Because of the low pressure in the blocked section, the external pressure can collapse the artery, cutting off the blood flow.

Making sense of the result

In a blocked section of an artery, the blood speed increases. According to Bernoulli's equation for a horizontal flow, the pressure drops in the section of faster flow. Therefore, the blood pressure in the blocked section drops.

CHECKPOINT 12-9

Blood Pressure in a Clogged Artery

Blood passes through a section of human artery that is partially blocked due to cholesterol buildup. As the blood flows from the unblocked (healthy) part of the artery into the blocked part, the blood pressure in the blocked artery

(a) decreases
(b) remains the same
(c) increases
(d) may increase or decrease, depending on the pressure in the unblocked section

ANSWER: (a)

Constant-velocity flow For flow through a pipe of uniform cross section, the equation of continuity requires that the fluid speed remain constant. For this case, we obtain a simplified form of Bernoulli's equation:

$$P_1 + \rho g y_1 = P_2 + \rho g y_2 \qquad (12\text{-}59)$$

Here, the sum of the pressure and the gravitational potential energy per unit volume remains constant along a streamline. As the height of the fluid increases, the pressure within the fluid decreases. Conversely, a decrease in the height results in an increase in the fluid pressure. Water pressure is lower on higher floors of a building than on the ground floor.

The Venturi Meter

As a fluid passes through a constricted region, its speed increases (equation of continuity) and pressure within the fluid drops (Bernoulli's equation). A Venturi meter uses these principles to measure the speed of a fluid. One type of Venturi meter (Figure 12-43) consists of a horizontal pipe of area A_1

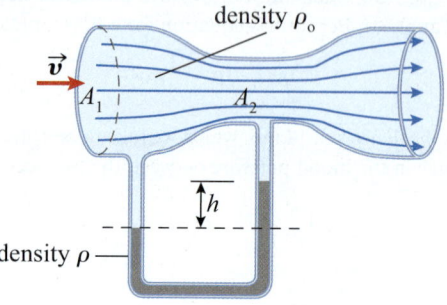

density ρ_0

$\vec{v}$

A_1 A_2

h

density ρ

Figure 12-43   A Venturi meter.

that is gradually constricted at the centre, where the area is A_2. A fluid of density ρ_0, such as air, enters the horizontal tube at one end and leaves through the other. A U-tube of uniform area and partially filled with a liquid of density ρ is attached to the horizontal tube.

When air in the horizontal tube is stationary, the height of the liquid in the two arms of the U-tube is the same. Suppose the air flows through A_1 with speed v. Its speed will increase as it passes through the constricted region, resulting in a pressure drop at the constriction, and the liquid in that arm of the U-tube will rise. The following equation relates the air speed, v, and the difference between the levels of the liquid in the two arms, h:

$$v = \sqrt{2g\left(\frac{\rho - \rho_0}{\rho_0}\right)\left(\frac{A_2^2}{A_1^2 - A_2^2}\right)}\sqrt{h}$$

For a given Venturi meter, all quantities in the square root of the first factor are known, so the air speed is determined by measuring h.

EXAMPLE 12-10

Water Pressure in a Home

Water needs to be pumped to the third floor of a building that is 10.0 m above ground level. The required pressure must maintain a flow rate of 0.25 L/s from a pipe of 2.0 cm diameter. The mechanical pump is located at ground level and pumps the water through a 4.0 cm diameter pipe (see Figure 12.44). Ignore frictional forces.

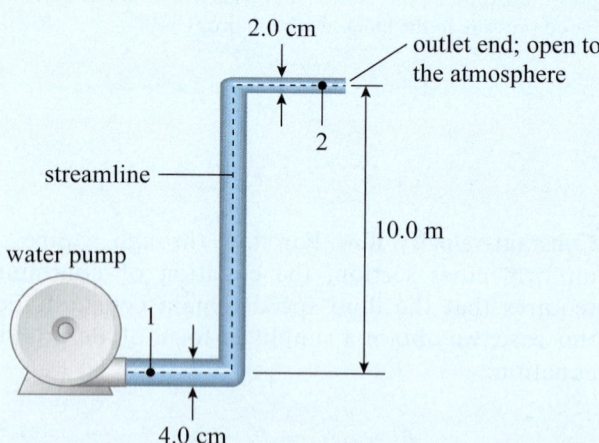

2.0 cm

outlet end; open to the atmosphere

2

streamline

10.0 m

water pump

1

4.0 cm

Figure 12-44   Pumping water.

(a) Calculate the speed of the water as it emerges from the 2.0 cm diameter pipe.
(b) Calculate the speed with which the water flows through the pipe at ground level.

(c) Determine the water pressure that the mechanical pump needs to generate (above the atmospheric pressure) to deliver the water at the required flow rate to the third floor.

Solution

Since the water is moving, its behaviour is described (assuming a steady flow) by both the continuity equation and Bernoulli's equation. Consider a streamline that starts from the pump and ends at the outlet. Choose two points along this streamline. The first point (designated as 1) is along the pipe at ground level, and the other point (2) is at the outlet end. Let P_1, v_1, A_1, and y_1 be the pressure, the fluid speed, the cross-sectional area of the pipe, and its height from ground level at point 1. The corresponding quantities at point 2 of the streamline are denoted by P_2, v_2, A_2, and y_2.

(a) We can determine the required flow speed from the given data. At the output end, the required flow rate, Q_2, is 0.25 L/s. Since $Q_2 = A_2 v_2$, the flow speed at the output must be

$$v_2 = \frac{Q_2}{A_2}$$

The diameter of the elevated pipe is 2.0 cm, so

$$A_2 = \pi R_2^2 = \pi(1.0 \text{ cm})^2 = \pi(1.0 \times 10^{-2}\text{m})^2 = \pi \times 10^{-4}\text{m}^2$$

Converting L/s to m³/s,

$$Q_2 = 0.25 \text{ L/s} = 0.25 \times 10^3 \text{cm}^3/\text{s} = 0.25 \times 10^3 (10^{-2}\text{m})^3/\text{s}$$
$$= 2.5 \times 10^{-4} \text{ m}^3/\text{s}$$

Therefore,

$$v_2 = \frac{Q_2}{A_2} = \frac{2.5 \times 10^{-4} \text{ m}^3/\text{s}}{\pi \times 10^{-4} \text{ m}^2} = 0.80 \text{ m/s}$$

(b) As the water moves from the pump, through the pipe, and out the output end, the flow rate must be constant. Therefore, the flow rate at point 1 is also 0.25 L/s and the water speed in the lower pipe, which has a diameter of 4.0 cm, is given by

$$v_1 = \frac{Q_1}{A_1} = \frac{2.5 \times 10^{-4} \text{ m}^3/\text{s}}{\pi \times 4 \times 10^{-4} \text{ m}^2} = 0.20 \text{ m/s}$$

(c) To determine the pressure, P_1, with which the water must leave the pump to have a speed of 0.80 m/s at the output end, we apply Bernoulli's equation at points 1 and 2 of the streamline. The water emerging from the output is open to the atmosphere; therefore, the pressure at this point is equal to atmospheric pressure, i.e.,

$$P_2 = P_0 = 1.01 \times 10^5 \text{ Pa}$$

It is convenient to measure heights with respect to ground level. Therefore,

$$y_1 = 0, \quad y_2 = 10.0 \text{ m}$$

Applying Bernoulli's equation between points 1 and 2,

$$P_1 + \frac{1}{2}\rho v_1^2 + \rho g y_1 = P_2 + \frac{1}{2}\rho v_2^2 + \rho g y_2$$

Using the given values for y_1, y_2, and P_2, we get

$$P_1 = P_0 + \frac{1}{2}\rho v_2^2 - \frac{1}{2}\rho v_1^2 + \rho g y_2$$

All quantities on the right-hand side are known and we can calculate P_1. It is instructive to calculate the contribution of each term individually:

$$P_1 = P_0 + \frac{1}{2} \times (1000 \text{ kg/m}^3) \times (0.80 \text{ m/s})^2$$

$$- \frac{1}{2}(1000 \text{ kg/m}^3) \times (0.20 \text{ m/s})^2$$

$$+ (1000 \text{ kg/m}^3) \times (9.81 \text{ m/s}^2) \times (10.0 \text{ m})$$

$$P_1 = P_0 + 320 \text{ Pa} - 20 \text{ Pa} + 98\,000 \text{ Pa}$$

The first term on the right is the contribution from atmospheric pressure. The second and third terms are the contributions from the kinetic energy per unit volume in the lower and the upper pipes. Notice that it is the difference of the two that contributes to the required pressure. If the pipe connecting the ground and the upper floors were of uniform cross-sectional area, then these two terms would cancel each other because the water speed would remain constant through the flow. The last term is the contribution from the potential energy per unit volume. It is equal to the energy required to lift a unit volume of water to a height of 10.0 m.

Since atmospheric pressure is present at the ground floor as well as at the third floor, the additional pressure that the water pump needs to generate is equal to $P_1 - P_0$, which is

$$P_1 - P_0 = 9.8 \times 10^4 \text{ Pa}$$

Most of this pressure is needed to lift the water up to the third floor.

Making sense of the result

The work done in moving a unit volume of water to a height of 10.0 m is equal to the gain in the potential energy of the water. This increase in the potential energy is

$$(1000 \text{ kg/m}^3) \times (9.81 \text{ m/s}^2) \times (10.0 \text{ m}) = 9.8 \times 10^4 \text{ J/m}^3$$
$$= 9.8 \times 10^4 \text{ Pa}$$

So our result makes sense.

EXAMPLE 12-11

Water Flow through a Pressurized Tank

Water flows from a large enclosed tank through a horizontal pipe of variable cross-sectional area, as shown in Figure 12-45. The upper end of the tank is maintained at an absolute pressure of $1.20 \times 10^5 \text{ N/m}^2$ by pumping compressed air into it. The horizontal outlet of the pipe has a cross-sectional area of $4.00 \times 10^{-2} \text{ m}^2$ in the larger section and $2.00 \times 10^{-2} \text{ m}^2$ in the smaller section. A vertical tube of cross-sectional area $4.00 \times 10^{-2} \text{ m}^2$ is connected to the horizontal tube and open to the atmosphere.

(a) What is the speed of the fluid leaving the outlet when the height of the water in the large tank is 2.00 m?

(b) What is the speed of the water in the horizontal pipe of cross-sectional area $4.00 \times 10^{-2} \text{ m}^2$?

(c) To what height does the water stand in the vertical tube when the water height in the large tank is 2.00 m?

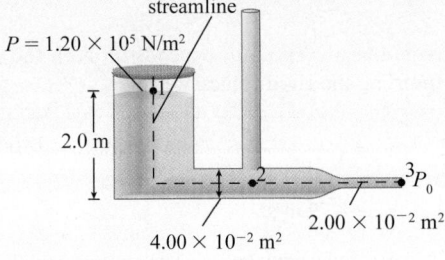

Figure 12-45 Example 12-10.

(*continued*)

Solution

To determine the unknown quantities for parts (a), (b), and (c), we consider a streamline that starts at the top surface of the water in the large tank, passes through the horizontal pipe, and leaves through the outlet. We label three points on this streamline as follows: Point 1 is at the top of the water surface in the large tank because we know the height and the pressure at this point. Point 2 is at the centre of the portion of the vertical tube that is at the bottom of the vertical tube, because we need to determine the height of the column of water in the vertical tube. Point 3 is at the outlet where the water leaves the horizontal pipe, because we know that the pressure at the outlet must be equal to atmospheric pressure.

The variables in Bernoulli's equation and the continuity equation are pressure, speed, height, and the cross-sectional area of the surface perpendicular to the fluid flow. The values of these quantities at the points of interest are as follows:

point 1:

$$P_1 = 1.20 \times 10^5 \text{ Pa}, v_1 = \text{unknown},$$
$$y_1 = 2.00 \text{ m}, A = \text{unknown}$$

point 2:

$$P_2 = \text{unknown}, v_2 = \text{unknown},$$
$$y_2 = 0.00 \text{ m}, A_2 = 4.00 \times 10^{-2} \text{ m}^2$$

point 3:

$$P_3 = P_0, v_3 = \text{unknown},$$
$$y_3 = 0.00 \text{ m}, A_3 = 2.00 \times 10^{-2} \text{ m}^2$$

Here, P_0 is atmospheric pressure, and we have chosen to measure all heights with respect to the streamline that passes through points 2 and 3. There are five unknowns: v_1, A_1, P_2, v_2, and v_3. We can use Bernoulli's equation between points 1 and 2 (or 3) and between points 2 and 3, with two corresponding continuity equations, giving a total of four equations, so there is not enough information to solve for five unknown quantities and we have to make an approximation.

Since the tank is large, the rate at which the level of the water in the tank decreases is negligible. Therefore, we can make the approximation that the speed at which the water is falling in the tank is negligible:

$$v_1 \approx 0 \text{ (large cross-sectional area approximation)}$$

Notice that once we assume that $v_1 \approx 0$, we cannot use the continuity equation between points 1 and 2 (or between points 1 and 3) to determine v_2 or A_2 as both v_1 and A_1 have dropped out of the problem.

Since all the variables except v_3 are known at point 3, we apply Bernoulli's equation between points 1 and 3:

$$P_1 + \frac{1}{2}\rho v_1^2 + \rho g y_1 = P_3 + \frac{1}{2}\rho v_3^2 + \rho g y_3 \tag{12-60}$$

Substituting $P_3 = P_0$ and $y_3 = 0$, using the approximation $v_1 \approx 0$, and solving for v_3, we get

$$v_3 = \sqrt{\frac{2}{\rho}(P_1 - P_0 + \rho g y_1)} \tag{12-61}$$

Note that the speed v_3 depends on both the pressure difference $(P_1 - P_0)$ and the height difference $(y_1 - 0)$. Inserting the given values, we get

$$v_3 = \sqrt{\frac{2}{1000 \text{ kg/m}^3}(1.20 \times 10^5 \text{ Pa} - 1.01 \times 10^5 \text{ Pa} + (1000 \text{ kg/m}^3) \times (9.81 \text{ m/s}^2) \times (2.00 \text{ m}))}$$
$$\tag{12-62}$$
$$v_3 = 8.79 \text{ m/s}$$

(b) Having calculated v_3, we can now calculate the speed and the pressure at point 2. To calculate v_2, we apply the continuity equation between points 2 and 3:

$$A_2 v_2 = A_3 v_3$$

$$v_2 = \frac{A_3}{A_2}v_3 = \frac{2.00 \times 10^{-2} \text{ m}^2}{4.00 \times 10^{-2} \text{ m}^2} \times 8.79 \text{ m/s} = 4.40 \text{ m/s}$$

(c) To find the height, h, of the column of water in the vertical tube, we need to calculate the pressure at a point that is directly underneath the vertical tube. That is why we chose point 2 to be under the vertical tube. The water in the vertical tube is stationary, so the pressure, P_2, at the bottom of the tube and the height, h, are related by

$$P_2 = P_0 + \rho g h$$

Pressure P_2 can be calculated by applying Bernoulli's equation between points 2 and 3:

$$P_2 + \frac{1}{2}\rho v_2^2 + \rho g y_2 = P_3 + \frac{1}{2}\rho v_3^2 + \rho g y_3$$

$$P_2 = P_3 + \left(\frac{1}{2}\rho v_3^2 - \frac{1}{2}\rho v_2^2\right) + (\rho g y_3 - \rho g y_2)$$

Inserting $P_3 = P_0$, $y_2 = y_3 = 0$, $v_2 = 4.40$ m/s, and $v_3 = 8.79$ m/s, we get

$$P_2 = 1.01 \times 10^5\,\text{Pa} + \frac{1}{2} \times (1000\,\text{kg/m}^3) \times ((8.79\text{m/s})^2 - (4.40\text{m/s})^2)$$

$$P_2 = 1.30 \times 10^5\,\text{Pa}$$

The height, h, of the water in the vertical tube can now be calculated:

$$h = \frac{P_2 - P_0}{\rho g} = \frac{1.30 \times 10^5\,\text{Pa} - 1.01 \times 10^5\,\text{Pa}}{(1000\,\text{kg/m}^3) \times (9.81\,\text{m/s}^2)}$$

$$h = 2.96\,\text{m}$$

INTERACTIVE ACTIVITY 12-2

Non-viscous Flow through a Horizontal Pipe

In this activity, you will use the PhET simulation "Non-Viscous Flow through a Horizontal Pipe of Variable Cross-Sectional Area." Work through the simulation and accompanying questions to gain an understanding of the flow of a non-viscous fluid through a horizontal pipe of variable cross-sectional area.

CHECKPOINT 12-10

Pressure and Area

If the cross-sectional area of the vertical tube in Example 12-11 is doubled, the height of the water in the tube

(a) increases by a factor of 4
(b) increases by a factor of 2
(c) remains the same
(d) decreases by a factor of 2

ANSWER: (c)

12-9 Conservation of Fluid Momentum

Consider a fluid moving through a straight horizontal tube that gradually narrows (Figure 12-41). As the cross-sectional area of the tube decreases, the speed of the fluid increases. According to Newton's second law of motion, any change in a fluid's velocity indicates that an external force is acting on the fluid. Similarly, there must be a net force on a fluid that passes through a bend in a tube because its direction changes. To study situations in which a fluid's momentum changes, we need to understand how Newton's second law of motion is applied to moving fluids.

Consider an element of fluid flowing between two sections of a tube of variable cross-sectional area (Figure 12-41). At section 1, the fluid pressure is P_1, the velocity is $\vec{v}_1$, and the cross-sectional area of the tube is A_1. In time Δt, the mass of the fluid that enters section 1 from the left is

$$\Delta M_1 = \rho(A_1 v_1 \Delta t)$$

The momentum of the entering fluid is

$$\Delta \vec{p}_1 = \Delta M_1 \vec{v}_1 = (\rho A_1 v_1 \Delta t)\vec{v}_1$$

In the same time, the mass of fluid leaving the tube at section 2 is

$$\Delta M_2 = \rho(A_2 v_2 \Delta t)$$

and the momentum of the leaving fluid is

$$\Delta \vec{p}_2 = \Delta M_2 \vec{v}_2 = (\rho A_2 v_2 \Delta t)\vec{v}_2$$

The rate of change of momentum of the fluid between the two sections of the tube is, therefore,

$$\frac{\Delta \vec{p}_2 - \Delta \vec{p}_1}{\Delta t} = (\rho A_2 v_2)\vec{v}_2 - (\rho A_1 v_1)\vec{v}_1 \quad (12\text{-}63)$$

By the equation of continuity, $A_1 v_1 = A_2 v_2 \equiv Q$, therefore, the rate of change of the fluid's momentum is

$$\frac{\Delta \vec{p}_2 - \Delta \vec{p}_1}{\Delta t} = \rho Q(\vec{v}_2 - \vec{v}_1) \qquad (12\text{-}64)$$

The rate of change of momentum of a physical object is equal to the sum of all external forces acting on that object. We can then write Equation 12-64 in the form of Newton's second law of motion:

$$\sum \vec{F} = \rho Q(\vec{v}_2 - \vec{v}_1) \qquad (12\text{-}65)$$

Here, $\sum \vec{F}$ denotes the vector sum of all external forces acting on the fluid element. Equation 12-65 is Newton's second law of motion as applied to moving fluids. This equation indicates that a fluid element accelerates or decelerates in response to external forces. By Newton's third law of motion, the fluid exerts an equal and opposite force on the surrounding fluid or the containing vessel.

EXAMPLE 12-12

Keeping a Fire Hose Stationary

Consider a fire hose that has an inner diameter of 4.40 cm. The diameter of the nozzle connected to the end of the hose tapers from 4.40 cm to 2.00 cm. Water is flowing out of the nozzle at a rate of 10 L/s, and the gauge pressure at the wide end of the nozzle is 6.80×10^5 Pa. What force must a firefighter exert to keep the nozzle stationary? (Ignore the weight of the nozzle.)

Solution

To hold the nozzle steady, the firefighter has to exert a force to counter the force that the water exerts on the nozzle. Imagine a flow tube that consists of the fluid inside the nozzle. The forces acting on the element of water within the flow tube are as follows (see Figure 12-46):

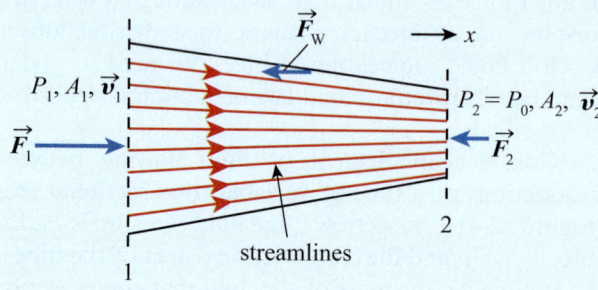

Figure 12-46 Example 12-12.

- $\vec{F}_1$ is the force exerted from the left by the water entering the nozzle;
- $\vec{F}_2$ is the force exerted from the right by the atmospheric pressure P_0 on the water that is leaving the nozzle; and
- $\vec{F}_w$ is the force that the nozzle exerts on the water.
  For this situation, Equation 12-65 gives

$$\vec{F}_1 + \vec{F}_2 + \vec{F}_w = \rho Q(\vec{v}_2 - \vec{v}_1) \qquad (1)$$

Taking the flow direction as the positive x-axis,

$$\vec{F}_1 = P_1 A_1 \hat{x}$$

$$\vec{F}_2 = -P_2 A_2 \hat{x} = -P_0 A_2 \hat{x}$$

$$\vec{F}_w = F_w \hat{x}$$

$$\vec{v}_1 = v_1 \hat{x}, \quad \vec{v}_2 = v_2 \hat{x}$$

Substituting into Equation 1 gives

$$P_1 A_1 - P_0 A_2 + F_w = \rho Q(v_2 - v_1)$$

and the magnitude of the force exerted by the nozzle on the water is

$$F_w = -P_1 A_1 + P_0 A_2 + \rho Q(v_2 - v_1) \qquad (2)$$

All the quantities on the right-hand side of Equation 2 are known:

$$P_0 = 1.01 \times 10^5 \text{ Pa}$$

$$P_1 = P_0 + 6.80 \times 10^5 \text{ Pa} = 7.81 \times 10^5 \text{ Pa}$$

$$A_1 = \frac{\pi D_1^2}{4} = \frac{\pi (4.40 \times 10^{-2} \text{ m})^2}{4} = 1.52 \times 10^{-3} \text{ m}^2$$

$$A_2 = \frac{\pi D_2^2}{4} = \frac{\pi (2.00 \times 10^{-2} \text{ m})^2}{4} = 3.14 \times 10^{-4} \text{ m}^2$$

$$Q = 10.0 \times 10^{-3} \text{ m}^3/\text{s}$$

$$\rho = 1000 \text{ kg/m}^3$$

$$v_1 = \frac{Q}{A_1} = \frac{1.00 \times 10^{-2} \text{ m}^3/\text{s}}{1.52 \times 10^{-3} \text{ m}^2} = 6.58 \text{ m/s}$$

$$v_2 = \frac{Q}{A_2} = \frac{1.00 \times 10^{-2} \text{ m}^3/\text{s}}{3.14 \times 10^{-4} \text{ m}^2} = 31.8 \text{ m/s}$$

Substituting these quantities into Equation 2 gives

$$F_w = -(7.81 \times 10^5 \text{ Pa}) \times (1.52 \times 10^{-3} \text{ m}^2) + (1.01 \times 10^5 \text{ Pa})$$
$$\times (3.14 \times 10^{-4} \text{ m}^2) + (1000 \text{ kg/m}^3)$$
$$\times (1.00 \times 10^{-2} \text{ m}^3/\text{s}) \times (31.8 \text{ m/s} - 6.58 \text{ m/s})$$

$$F_w = -11.87 \times 10^2 \text{ N} + 0.32 \times 10^2 \text{ N} + 2.52 \times 10^2 \text{ N}$$
$$= -9.03 \times 10^2 \text{ N}$$

Making sense of the result

The sign of F_w is negative; therefore, the force exerted by the nozzle on the water is toward the negative x-axis, opposite to the direction of water flow, as it should be. How large is this force? It is equivalent to lifting a 92 kg weight. The firefighters need to be strong and fit.

12-10 Viscous Flow

It takes less force to stir a spoon in a cupful of water than to stir a spoon in a cupful of honey. We say that the honey is more *viscous* than water. Recall from Section 12-6 that viscosity is a measure of the resistance of a fluid to flow or change in its shape. A fluid with non-zero viscosity is called a **viscous fluid**. Frictional forces are always present (except for a few liquids at very low temperatures), so ignoring the viscosity of a moving fluid can give inaccurate results in many situations.

Viscosity can be measured in different ways. Consider the following scenario. Two horizontal flat plates, each of cross-sectional area A, are a distance d apart, and the region between the two plates is filled with a fluid of unknown viscosity (Figure 12-47). The bottom plate is held stationary, and the top plate is moved with a constant velocity by exerting a force on it.

Imagine that the enclosed fluid consists of a very large number of layers of small thickness. Fluid speed decreases from the top plate, where the fluid in contact with the plate moves with speed v, to zero for the fluid that is in contact with the stationary bottom plate. This variation in the fluid speed is a consequence of the frictional forces between the fluid and the plates and between adjacent layers of the fluid.

The speed of a given layer increases with its vertical distance from the bottom plate. Experiments have shown that the magnitude of the force, F, required to move the top plate with a speed v is

- proportional to the cross-sectional area, A, of the plates
- inversely proportional to the distance between the plates
- proportional to the viscosity of the fluid between the plates

Because the frictional (or the viscous) force is directed opposite to the direction of motion of the upper plate, for a plate moving in the positive x-direction the frictional force can be written as

$$F_{vf,x} = -\mu A \frac{v_x}{d} \qquad (12\text{-}66)$$

Here the subscript vf indicates the force due to a viscous flow. The constant of proportionality, μ, is called the viscosity (or the coefficient of viscosity) of the fluid. From Equation 12-66, we see that the units of viscosity are

$$\frac{N \cdot m}{m^2 \cdot (m/s)} = \left(\frac{N}{m^2}\right)s = Pa \cdot s = kg \cdot m^{-1} \cdot s^{-1}$$

If a fluid with a viscosity of 1 Pa·s is placed between the two plates, and the top plate is pushed by applying a stress of 1 Pa, then in 1 s the plate will move a distance equal to the separation between the two plates.

A commonly used unit for viscosity is the poise (P), named after French physicist Jean Poiseuille:

$$1 \text{ P} = 1 \text{ g} \cdot cm^{-1} \cdot s^{-1} = 0.1 \text{ kg} \cdot m^{-1} \cdot s^{-1} = 0.1 \text{ Pa} \cdot s$$

Therefore,

$$10 \text{ P} = 1 \text{ Pa} \cdot s$$

Another commonly used measurement unit is the centipoise (cP):

$$1 \text{ cP} = 10^{-2} \text{ P} = 10^{-3} \text{ Pa} \cdot s$$

In general, viscosities of liquids are greater than viscosities of gases. In addition, the viscosity of liquids decreases with increasing temperature. In gases, viscosity arises from the diffusion of gas molecules within the gas. Therefore, the viscosity of a gas increases with temperature. Table 12-3 lists the viscosities of several fluids.

We can write Equation 12-66 in a more general form. We see that the magnitude of the viscous force changes with the distance between the plates. Taking the y-direction perpendicular to the direction of motion of the fluid, we replace v_x/d by dv_x/dy. The derivative dv_x/dy is called the velocity gradient and is a measure of how the fluid speed changes between adjacent neighbouring layers of fluid. With this change, the viscous force between adjacent layers of a laminar and

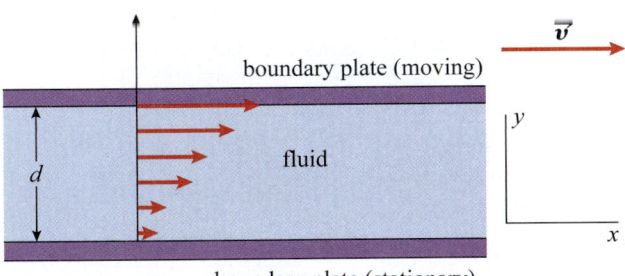

Figure 12-47 Measuring the viscosity of a viscous fluid between two parallel plates.

Table 12-3 Table of Viscosities

Material	Viscosity (Pa·s)	Material	Viscosity (Pa·s)
Air (0 °C)	17.4×10^{-6}	Honey	$2-10$
Hydrogen (0 °C)	8.4×10^{-6}	Corn syrup	1.38
Xenon (0 °C)	2.12×10^{-5}	Olive oil	0.081
Water (20 °C)	8.94×10^{-4}	Peanut butter	~250
Blood	$3 \text{ to } 4 \times 10^{-3}$	Ketchup	$50-100$

CRC Handbook of Chemistry and Physics: A Ready-Reference Book of Chemical and Physical Data by LIDE, DAVID R., JR Copyright 1992. Reproduced with permission of TAYLOR & FRANCIS GROUP LLC-BOOKS in the format Textbook via Copyright Clearance Center.

steady flow along the positive x-direction of a viscous fluid is given by

$$F_{vf,x} = -\mu A \frac{dv_x}{dy} \qquad (12\text{-}67)$$

where the y-direction is perpendicular to the direction of flow.

Poiseuille's Law for Viscous Flow

Consider streamlined flow of a viscous fluid flow through a pipe of length L and radius R. As discussed above, the layer in contact with the walls of the pipe is almost at rest. The layer adjacent to it moves with a slightly higher speed. The farther a layer is from the walls, the faster it moves, with the fluid at the centre of the pipe moving with the highest speed. The speed profile of a fluid is shown in Figure 12-48.

Because of viscosity, the kinetic energy and, hence, the speed of the fluid decrease as the fluid moves through the pipe. Therefore, work must be done on the fluid by applying an external pressure to maintain a constant

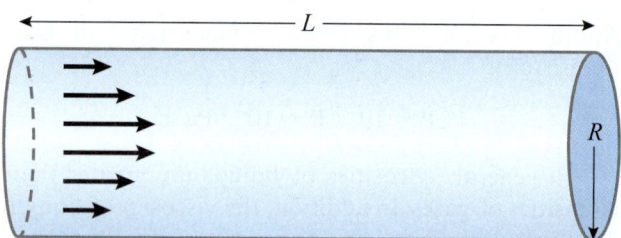

Figure 12-48 The speed profile of a fluid moving through a horizontal tube. The length of the arrows is proportional to the speed of the fluid. The layers of the fluid that are close to the walls of the pipe move slowly. The farther a layer is from the walls, the faster its speed. Fluid moving along the central axis of the pipe has the highest speed.

flow rate. Bernoulli's equation contains no information about frictional forces, so it is not applicable in this case. Poiseuille showed (see the derivation at the end of this section) that for a steady and laminar flow of a viscous fluid through a horizontal and cylindrical pipe of uniform cross-sectional area, the volume flow rate, Q, is related to the pressure difference between the two ends of the pipe by the following equation:

KEY EQUATION
$$P_1 - P_2 = 8\mu \frac{LQ}{\pi R^4} \qquad (12\text{-}68)$$

Here, $P_1 - P_2$ is the pressure difference between the two ends of the pipe. The quantity $\frac{8}{\pi}\left(\frac{\mu L}{R^4}\right)$ can be interpreted as the resistance offered to fluid flow by a pipe of length L and radius R. This quantity is called the total peripheral resistance (TPR). Equation 12-68 shows that to maintain a constant volume flow rate, a non-zero pressure difference needs to be maintained between the ends of a pipe. The required pressure difference is proportional to the length of the pipe and inversely proportional to its radius. The longer the pipe, the greater the energy loss; hence, a greater pressure difference is required to maintain a constant flow rate. For smaller radii, a large fraction of the fluid is close to the inner surface of the pipe, where fluid speed is small. Poiseuille's equation can be derived from Newton's laws of motion.

Note that the pressure difference required to maintain a constant flow rate is inversely proportional to the fourth power of the radius of the pipe. As radius increases, the required pressure difference decreases rapidly. Consequently, Bernoulli's equation can be used for flow through pipes of large cross-sectional areas with negligible error.

EXAMPLE 12-13

Pumping Blood to an Ostrich's Head

An ostrich's neck is approximately 1.20 m long. Assume that the artery supplying blood to the ostrich's head has a radius of 2.00 mm and that the blood flows through it with an average speed of 2.00 cm/s. If the blood arrives at the ostrich's head when it is upright (Figure 12-49) at one atmospheric pressure, with what pressure should the ostrich's heart pump the blood? Assume that the density of the ostrich's blood is 1.06×10^3 kg/m^3 and its viscosity is 4.00×10^{-3} Pa·s.

Solution

The radius of the artery is small, so the viscosity of the blood cannot be ignored. We must use Poiseuille's equation to determine the pressure needed to force the blood through this artery to a height of 1.20 m. The pressure difference between the heart and the head should be large enough to lift the blood up as well as to overcome viscosity. Let P_{head} be the pressure at the ostrich's head and P_{heart} be the pressure with which the heart pumps the blood. We know that a heart contracts and relaxes continuously, so here P_{heart} stands for the average pressure generated by the heart.

Figure 12-49 Example 12-13.

Known quantities:

Radius of the artery: $R = 2.00$ mm $= 2.00 \times 10^{-3}$ m

Length of the artery: $L = 1.20$ m

Average speed of the blood flow: $v = 2.00$ cm/s
$$= 2.00 \times 10^{-2} \text{ m/s}$$

Blood viscosity: $\mu = 4.00 \times 10^{-3}$ Pa·s

Blood density: $\rho_B = 1.06 \times 10^3$ kg/m³

Pressure at ostrich's head: $P_{head} = 1.01 \times 10^5$ Pa

The pressure at the ostrich's heart can be calculated from Equation 12-68, which for the present situation can be written as

$$P_{heart} - P_{head} = \rho_B g L + 8\mu \frac{LQ}{\pi R^4}$$

The first term on the right is the pressure required to lift the blood by a height L, and the second term is the pressure difference needed to overcome viscosity. The flow rate, Q, is

$$Q = (\pi R^2)v = \pi \times (2.00 \times 10^{-3} \text{ m})^2 \times (2.00 \times 10^{-2} \text{ m/s}) = 2.50 \times 10^{-7} \text{m}^3/\text{s}$$

Therefore,

$$P_{heart} - P_{head} = (1.06 \times 10^3 \text{ kg/m}^3) \times (9.81 \text{ m/s}^2) \times (1.20 \text{ m})$$
$$+ \frac{8 \times (4.00 \times 10^{-3}\text{Pa·s}) \times (1.20 \text{ m}) \times (2.50 \times 10^{-7}\text{m}^3/\text{s})}{\pi(2.00 \times 10^{-3}\text{m})^4}$$

$$P_{heart} - P_{head} = 12.4 \times 10^3 \text{ Pa} + 0.20 \times 10^3 \text{ Pa} = 12.6 \times 10^3 \text{ Pa} = 12.6 \text{ kPa}$$

Making sense of the result

Work done in moving a unit volume of blood to a height of 1.2 m is equal to the gain in the potential energy of the blood. This increase in the potential energy is

$$(1.06 \times 10^3 \text{ kg/m}^3) \times (9.81 \text{ m/s}^2) \times (1.20 \text{ m}) = 12.4 \times 10^3 \text{ Pa}$$

The calculated pressure difference between the heart and the head is slightly greater than this value due to the extra pressure difference needed to overcome viscosity. So our result makes sense.

CHECKPOINT 12-11

Viscous Flow through a Pipe

Figure 12-50 shows viscous fluid in a horizontal pipe and two connected vertical tubes that are open to the atmosphere. Which of the following statements correctly describes the motion of the fluid in the pipe? Explain your reasoning.

(a) The fluid is not moving.
(b) The fluid is moving from left to right.
(c) The fluid is moving from right to left.
(d) The motion cannot be determined from the diagram.

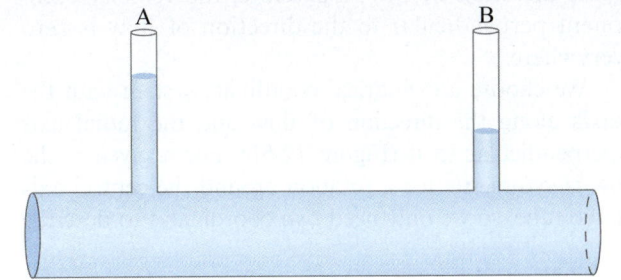

Figure 12-50 Checkpoint 12-11.

ANSWER: (b)

Blood Pressure and Blood Flow

The laws of fluid dynamics apply to the flow of blood through the human body. The human heart has two ventricles, which are chambers that contract to pump blood to the rest of the body (Figure 12-1). With each contraction (heartbeat), the pressure in the arteries—which carry blood from the heart—cycles between a minimum *diastolic* pressure (when the ventricles relax) and a maximum *systolic* pressure (when the ventricles contract). In a healthy adult, the systolic pressure is approximately 16 kPa (120 mm Hg), and the diastolic pressure is approximately 10 kPa (80 mm Hg).

The heart of a healthy adult at rest pumps blood at a rate of approximately 5 L/min, giving an average flow rate of approximately 80 cm^3/s. During strenuous physical activity, the flow rate can increase by a factor of 5. Since the cross-sectional area of the major arteries does not increase significantly during exercise, the blood speed in the major arteries must increase by approximately the same factor.

When a person is standing, the weight of the blood has a substantial effect on the blood pressure throughout the body. Let h_{head} be the vertical distance between the head and the heart, and let h_{feet} be the distance between the feet and the heart. From Equation 12-8, the blood pressure at the head and the feet level is related to the pressure at the heart level as follows:

$$P_{head} = P_{heart} - \rho_{blood}gh_{head}$$

$$P_{feet} = P_{heart} + \rho_{blood}gh_{feet}$$

Taking typical values of $P_{heart} = 13$ kPa, $\rho_{blood} = 1060$ kg/m^3, $h_{head} = 50$ cm, and $h_{feet} = 130$ cm, we find that $P_{head} = 7.8$ kPa and $P_{feet} = 26.5$ kPa. This large difference in relative pressure explains why lying down increases blood flow to the brain and can help relieve swelling in the feet.

Because of viscosity, the blood pressure drops as the blood moves through the arteries. For example, taking the viscosity of blood to be 4.0×10^{-3} Pa·s and using a flow rate of 80 cm^3/s, the pressure drop per unit length in an artery of radius 1.0 cm is

$$\frac{P_1 - P_2}{L} = \frac{8\mu}{\pi}\frac{Q}{R^4} = \frac{8 \times 4.0 \times 10^{-3}\,\text{Pa·s}}{\pi}$$
$$\times \frac{80 \times 10^{-6}\,\text{m}^3/\text{s}}{(1.0 \times 10^{-2}\,\text{m})^4} \approx 80\ \text{Pa/m}$$

By the time the blood reaches the small capillaries, it is at atmospheric pressure. Since changing R greatly affects the total peripheral resistance, a buildup of fats in and around the walls of arteries greatly reduces the blood flow, and the heart has to work harder to maintain the needed flow rate.

Derivation of Poiseuille's Equation

Consider a viscous fluid flowing from left to right through a horizontal cylindrical tube of length L and uniform cross-sectional area with radius R. The flow is assumed to be steady, streamline, and laminar. The flow velocity remains constant along a streamline, and there is no movement of fluid particles perpendicular to the direction of flow. Therefore, the velocity component perpendicular to the direction of flow is zero everywhere.

We choose a cylindrical coordinate system with the x-axis along the direction of flow and the radial axis r perpendicular to it (Figure 12-51). For a given r, the flow is symmetric for a rotation around the central axis of the tube, so we only need two coordinates to describe

three-dimensional flow. As we will see, in this case the laminar sheets (surfaces in the region of flow that have the same fluid velocity) are the hollow cylindrical surfaces of a fixed value of r, centred around the central axis.

As discussed in Section 12-10, the flow is fastest through the middle of the tube, and the fluid layer adjacent to the side of the tube does not move at all. Since viscous forces oppose the flow, an external force must be applied to keep the flow steady. Therefore, to balance the viscous forces, the pressure at the left end of the tube (the fluid is moving from left to right) must be greater than the pressure to its right. Imagine a cylindrical element of the fluid of radius $r < R$ and length L centred around the central axis as it moves through the tube (Figure 12-51). Since its velocity is constant, the sum of all external forces on this element of fluid must be zero. The only forces exerted on it are those due to the pressure on its left and right ends and the viscous forces. Therefore,

$$\vec{F}_{pressure} + \vec{F}_{vf} = 0 \qquad (12\text{-}69)$$

Let P_1 be the pressure on the left side and P_2 the pressure on the right side. Since the flow is along the positive x-axis,

$$F_{pressure,\,x} = P_1(\pi r^2) - P_2(\pi r^2) = (P_1 - P_1)\pi r^2 \quad (12\text{-}70)$$

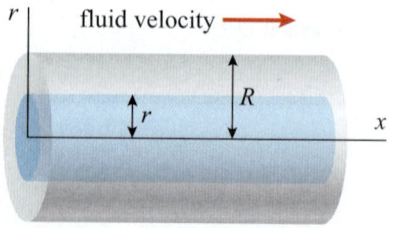

Figure 12-51   Geometry for fluid flow through a cylinder.

If the right end of the tube is open to the atmosphere, then P_2 is atmospheric pressure. The viscous force is given by Equation 12-67. In this case, the layer of the fluid that surrounds the surface of this cylinder of radius r exerts a drag force on it that opposes its flow. The surface area of a cylinder of radius r and length L is $2\pi rL$, and the fluid velocity is a function of r. Therefore,

$$F_{vs,x} = +\mu(2\pi rL)\frac{dv_x}{dr} \qquad (12\text{-}71)$$

Notice the change of variable from y to r in this equation and the minus sign difference between Equations 12-67 and 12-71. In the current configuration, the fluid velocity is maximum at $r = 0$ and vanishes at $r = R$. Therefore, the variable y in Figure 12-47 is related to r by $y = R - r$. Hence, $dv_x/dy = -dv_x/dr$. Inserting Equations 12-70 and 12-71 into Equation 12-69, we get

$$(P_1 - P_2)\pi r^2 + \mu(2\pi rL)\frac{dv_x}{dr} = 0 \qquad (12\text{-}72)$$

This first-order differential equation involves one unknown variable, v_x, and can be solved by integration with the boundary condition that v_x must vanish at $r = R$ (the fluid layer adjacent to the side of the tube does not move). We can write Equation 12-72 as

$$\frac{dv_x}{dr} = -\left(\frac{P_1 - P_2}{2\mu L}\right)r \qquad (12\text{-}73)$$

To determine v_x at r, we must add the contributions of viscous drag from all layers from r to the edge of the tube at R. We therefore integrate both sides of Equation 12-73 from r to R:

$$\int_r^R \frac{dv_x}{dr}\,dr = -\int_r^R \left(\frac{P_1 - P_2}{2\mu L}\right)r\,dr \qquad (12\text{-}74)$$

This gives

$$v_x(R) - v_x(r) = -\left(\frac{P_1 - P_2}{4\mu L}\right)(R^2 - r^2) \qquad (12\text{-}75)$$

Because $v_x(R) = 0$ at the boundary, the flow velocity at a distance r from the central axis is

$$v_x(r) = \left(\frac{P_1 - P_2}{4\mu L}\right)(R^2 - r^2)$$

(flow velocity in a pipe, $r \le R$) $\qquad (12\text{-}76)$

Notice that fluid at the same distance from the central axis moves with the same velocity. The velocity profile is parabolic, with its maximum along the central axis and its zero at the tube wall.

Let us now calculate the flow rate through this tube. Since the flow velocity is a function of r, we cannot use $Q = (\pi R^2)v_x(r)$ to calculate the flow rate. There are two ways to calculate Q in this situation. We can calculate the flow rate for each hollow spherical shell for a fixed r and then add contributions from all shells to calculate the total flow rate, or we can calculate the average velocity from Equation 12-76 and use this velocity to calculate Q. Let us take the latter approach. The average flow velocity, $\bar{v}_x$, is given by

$$\bar{v}_x = \frac{2\pi \int_0^R v_x(r)r\,dr}{2\pi \int_0^R r\,dr} = \frac{\int_0^R \left(\frac{P_1 - P_2}{4\mu L}\right)(R^2 - r^2)r\,dr}{\int_0^R r\,dr} \qquad (12\text{-}77)$$

Using

$$\int_0^R r\,dr = \frac{R^2}{2}$$

$$\int_0^R (R^2 - r^2)r\,dr = \frac{R^4}{4}$$

we get

$$\bar{v}_x = \frac{1}{2}\left(\frac{P_1 - P_2}{4\mu L}\right)R^2 \qquad (12\text{-}78)$$

Notice that the average velocity is half the maximum velocity at the centre ($r = 0$). The flow rate through the pipe is then given by

$$Q = (\pi R^2)\bar{v}_x = (\pi R^2)\left(\frac{P_1 - P_2}{8\mu L}\right)R^2 = \left(\frac{P_1 - P_2}{8\mu L}\right)(\pi R^4) \qquad (12\text{-}79)$$

For a given flow rate, the pressure difference across two ends of the tube must be

$$P_1 - P_2 = 8\mu \frac{LQ}{\pi R^4} \qquad (12\text{-}80)$$

Equation 12-80 is Poiseuille's equation. Notice the R^4 dependence of the flow rate for a given pressure difference. If the radius of a section of artery reduces by half due to blockage, the flow rate through the artery decreases by a factor of 16.

KEY CONCEPTS AND RELATIONSHIPS

Fluids are substances that deform and flow under an applied stress. Liquids and gases are collectively called fluids. Physical laws describing fluid motion are derived using Newton's laws of motion along with the concepts of momentum and energy conservation. The concepts of density and pressure are useful for describing the properties of fluids.

Density

The density of a material is defined as its mass per unit volume. The average density is given by

$$\rho = \frac{M}{V} \qquad (12\text{-}1)$$

The SI units of density are kg/m^3.

Pressure

The pressure exerted by a force on a surface element of area A is defined as the component of the force perpendicular to the surface (F_n) divided by the area of the surface:

$$P = \frac{F_n}{A} \qquad (12\text{-}3)$$

The SI unit of pressure is the pascal (1 Pa = 1 N/m^2).

Pressure in Fluids

A fluid exerts pressure within itself and on the walls of the containing vessel. This pressure arises due to the weight of the fluid and due to collisions between the constituents of the fluids. In liquids, the primary source of pressure is the weight of the liquid. Gas pressure arises mainly due to collisions between gas molecules and between the gas and the walls of the containing vessel.

Atmospheric Pressure

Atmospheric pressure is due to the weight of the air above Earth's surface.

Gauge Pressure

Gauge pressure is the difference between the actual pressure and a reference pressure, usually taken to be atmospheric pressure.

Pressure Variation with Depth

Pressure within the body of a fluid increases with depth. The fluid pressure at a depth d below the surface of a fluid that is open to the atmosphere is

$$P = P_0 + \rho_f g d \qquad (12\text{-}8)$$

If P_a denotes the pressure at a point a within a fluid of density ρ, and P_b is the pressure at a point b that is at a depth d below point a, then

$$P_b = P_a + \rho g d$$

When point a is at the surface of the fluid where the pressure is equal to atmospheric pressure, P_0, then

$$P_b = P_0 + \rho g d$$

Pascal's Principle

Pascal's principle states that any change in the pressure of an enclosed fluid is transmitted undiminished to every part of the fluid and to the walls of the containing vessel.

Archimedes' Principle

Archimedes' principle states that, when an object is fully or partially immersed in a fluid, the fluid exerts a net upward force on the object (called the buoyant force); the magnitude of the buoyant force is equal to the weight of the fluid displaced by the object.

Flotation

An object will float in a fluid when the average density of the object is less than or equal to the density of the fluid.

Ideal Fluids

An ideal fluid is an incompressible fluid that has no viscosity.

Ideal Fluid Flow

A flow that is steady and non-turbulent is called an ideal flow.

Volume Flow Rate (Q)

The volume flow rate of a fluid through a surface that is normal to the direction of flow is equal to the volume of the fluid that passes through the surface per unit time:

$$Q = Av \qquad (12\text{-}39)$$

Here, A is the cross-sectional area of the surface perpendicular to the direction of flow and v is the speed of the fluid.

Streamlines

A streamline is the trajectory taken by an infinitesimal element of fluid in the region of flow. In a steady flow, streamlines never cross.

Continuity Equation

The amount of fluid that passes through a flow tube remains constant:

$$A_1 v_1 = A_2 v_2 \qquad (12\text{-}43)$$

Bernoulli's Equation

For the flow of an ideal fluid, at any point along a streamline through the fluid, the sum of pressure, kinetic energy per unit volume, and potential energy per unit volume remains constant:

$$P_1 + \frac{1}{2}\rho v_1^2 + \rho g y_1 = P_2 + \frac{1}{2}\rho v_2^2 + \rho g y_2 = \text{constant} \qquad (12\text{-}55)$$

Poiseuille's Law for Viscous Flow

When viscosity is taken into account, the volume flow rate, Q, of a fluid of viscosity μ from a pipe of length L and radius R is related to the pressure difference between the two ends of the pipe by the following relationship:

$$P_1 - P_2 = 8\mu \frac{LQ}{\pi R^4} \qquad (12\text{-}68)$$

APPLICATIONS

airplanes and hot-air balloons, spray bottles, hydrometers, hydraulic lifts and brakes, water distribution and irrigation systems, sphygmomanometers, Venturi flow meters, submarines

KEY TERMS

apparent weight, Archimedes' principle, atmospheric pressure, average mass density, barometer, Bernoulli's equation, buoyant force, equation of continuity, flow rate, flow tube, fluid pressure, fluids, gauge pressure, hydraulic lift, ideal fluid, mass flow rate, pascal, Pascal's principle, phases (states), plasma, pressure, specific gravity, steady flow, streamline, streamline flow (laminar flow), turbulent, viscosity, viscous fluid, work done by the leaving fluid, work done on the test volume

QUESTIONS

1. Helium gas contracts when cooled. A rubber balloon is filled with helium gas at room temperature and closed tightly so that the gas cannot escape. When the balloon is cooled, the density of helium inside the balloon
   (a) decreases
   (b) remains the same
   (c) increases

2. A cylindrical glass is filled with water. The gauge pressure at the bottom of the glass is P. A second cylindrical glass with twice the diameter is filled with water to the same height. The gauge pressure at the bottom of the second glass is
   (a) $P/2$
   (b) P
   (c) $2P$
   (d) $4P$

3. A cylindrical glass is filled with water. The magnitude of the net force at the bottom of the glass due to the weight of the water is F. A second cylindrical glass with twice the diameter is filled with water to the same height. The magnitude of the net force at the bottom of the second glass due to the weight of the water is
   (a) $F/2$
   (b) F
   (c) $2F$
   (d) $4F$

4. Is the following statement true or false? Explain your reasoning: The magnitude of the buoyant force on an object that may sink or float in a fluid is equal to the weight of the object.

5. Figure 12-52 shows mass versus volume plots for three fluids, A, B, and C. Order the fluids from highest to lowest density.

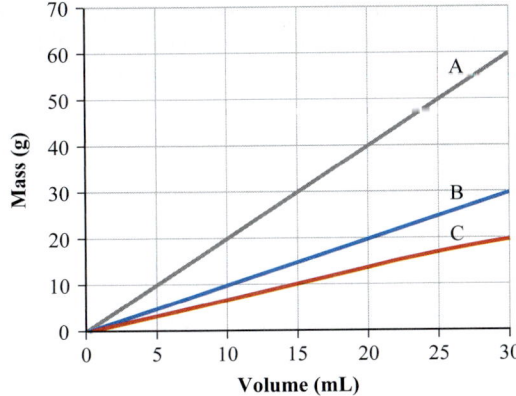

Figure 12-52 Question 5: Plots of mass versus volume for three different fluids.

6. Four objects with identical volume are submerged in a fluid, as shown in Figure 12-53. Each object is in a state of equilibrium. Which of the following statements is true about their average densities?

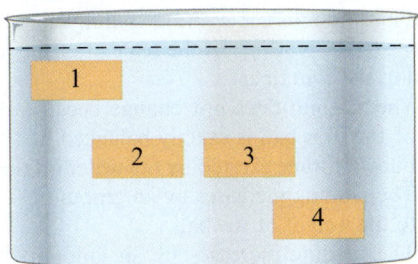

Figure 12-53 Question 6.

   (a) $\rho_1 < \rho_2 < \rho_3 < \rho_4$
   (b) $\rho_1 = \rho_2 = \rho_3 = \rho_4$
   (c) $\rho_1 > \rho_2 > \rho_3 > \rho_4$
   (d) $\rho_1 < \rho_2 = \rho_3 < \rho_4$

7. A plastic toy boat carrying a small iron cube is floating in a glass container. The cube is then gently dropped into the water. The level of the water in the container
   (a) goes down
   (b) goes up
   (c) stays the same
   Explain your reasoning. Now perform this experiment to determine whether or not your prediction agrees with your observations.

8. A rubber ball with average density 900 kg/m³ is floating in a container of water. Oil with a density less than the density of water and that of the ball is slowly poured in the container until it completely covers the ball. When completely immersed in the oil, what fraction of the volume of the ball is submerged in the water? Explain your reasoning.
   (a) the same as before the oil was poured
   (b) less than before the oil was poured
   (c) more than before the oil was poured

9. An ice cube is floating in a partially filled glass of water. After the ice cube melts, the level of the water in the glass
   (a) goes up
   (b) goes down
   (c) remains the same

10. Three identical rectangular pieces of metal are held submerged under water in different orientations, as shown in Figure 12-54. The magnitudes of the buoyant forces on these pieces are $F_{B,1}$, $F_{B,2}$, and $F_{B,3}$. Which of the following statements about the magnitudes of the buoyant forces is correct?

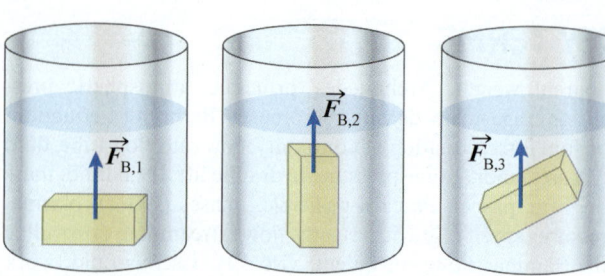

Figure 12-54 Question 10.

(a) $F_{B,1} = F_{B,2} = F_{B,3}$
(b) $F_{B,1} < F_{B,2} = F_{B,3}$
(c) $F_{B,1} = F_{B,2} < F_{B,3}$
(d) $F_{B,1} < F_{B,2} < F_{B,3}$

11. A container, partially filled with water, is resting on a scale that measures its initial weight. A solid iron ball is placed in the container, without spilling any water. What happens to the reading on the scale after the iron ball is placed in the container?
    (a) The reading does not change because the weight of the iron ball is exactly balanced by the upward buoyant force exerted by the water on the ball.
    (b) The reading increases by an amount exactly equal to the weight of the ball.
    (c) The reading increases by an amount that is less than the weight of the ball because the buoyant force cancels a fraction of the weight of the ball.
    (d) The reading increases by an amount that is greater than the weight of the ball because the water on top of the ball further pushes the ball down, thus increasing its weight.

12. A gas-filled balloon of mass 2.25 kg (including the mass of the gas) and volume 5.00 m³ is released from a tall building. The density of air is 1.30 kg/m³. Which of the following statements is correct? Explain your reasoning.
    (a) The balloon will remain stationary.
    (b) The balloon will fall.
    (c) The balloon will rise.

13. Three containers of different shapes but equal base area are filled with water to the same height (Figure 12-55). Is the hydrostatic pressure at the base of each container the same? Explain your reasoning.

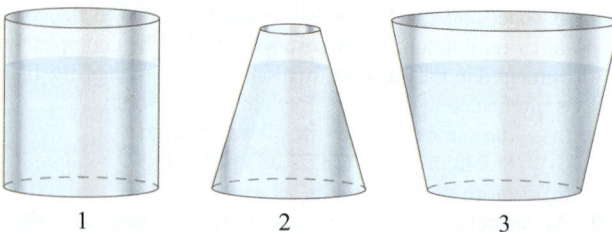

Figure 12-55 Question 13.

14. A container filled with a liquid has a hole in its side through which the liquid is coming out. If the container (with the liquid in it) is dropped from a height and allowed to fall freely, what will happen to the liquid emerging from the container?

(a) It will keep coming out following a similar trajectory as before.
(b) It will start flowing upward.
(c) It will stop flowing.

15. A piece of solid metal of mass M has a density four times the density of water. It is attached to a string and suspended in water, fully immersed. The tension in the string is
    (a) $\dfrac{3}{4} Mg$
    (b) Mg
    (c) $3Mg$
    (d) $4Mg$

16. Scuba divers control buoyancy by wearing a vest that can be inflated or deflated by adding or removing air. The density of water slightly increases with depth. A scuba diver is in a state of neutral buoyancy (neither sinking or rising) at a certain depth. To establish neutral buoyancy at a greater depth, the diver should
    (a) inflate the vest
    (b) deflate the vest
    (c) do nothing because he or she is already in a state of neutral buoyancy

PROBLEMS BY SECTION

For problems, star ratings will be used (★, ★★, or ★★★), with more stars meaning more challenging problems. The following codes will indicate if $\frac{d}{dx}$ differentiation, $\int$ integration, ▢ numerical approximation, or 〰 graphical analysis will be required to solve the problem.

Section 12-1 Phases of Matter

17. ★★ Water has a density of 1.0 g/cm³, and 18 g of water contains 6.02×10^{23} molecules. From this information, estimate the average distance between water molecules.

18. ★★ Given that the molar mass of dry air is 29 g/mol and at STP the density of air is 1.30 kg/m³, estimate the average separation between air molecules.

Section 12-2 Density and Pressure

19. ★ Your mass is 70 kg. The area of the parts of your shoes that are in contact with the ground is 0.010 m². While standing stationary on your two feet, what is the pressure that you exert on the ground?

20. ★★ Two rectangular blocks of masses M_1 and M_2 and cross-sectional areas A_1 and A_2, respectively, are lying on a smooth surface, as shown in Figure 12-56. What is the pressure exerted by the blocks on the surface?
    (a) $\dfrac{M_1 g}{A_1} + \dfrac{M_2 g}{A_2}$

    (b) $\dfrac{M_1 g}{A_1} + \dfrac{M_2 g}{A_1}$

    (c) $\dfrac{M_1 g}{A_2} + \dfrac{M_2 g}{A_2}$

    (d) $\dfrac{M_1 g}{A_1} - \dfrac{M_2 g}{A_2}$

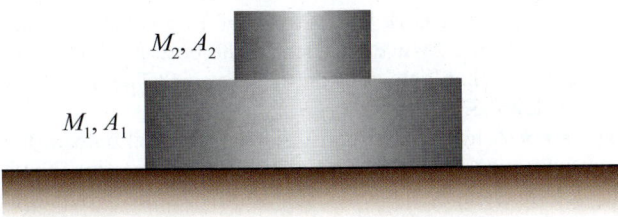

Figure 12-56 Problem 20.

21. ★ A neutron star results when a star in its final stages collapses due to gravitational pressure, forcing the electrons to combine with the protons in the nucleus and converting them into neutrons.
    (a) Assuming that a neutron star has a mass of 3.00×10^{30} kg and a radius of 1.20×10^3 m, determine the density of a neutron star.
    (b) How much would 1.0 cm^3 (the size of a sugar cube) of this material weigh at Earth's surface? Do you think you would be able to lift it?

22. ★ When four tires of an automobile are inflated to a gauge pressure of 240 kPa, the contact area between a tire and a flat road surface is measured to be 0.015 m^2. Determine the mass of the automobile (assume $g = 9.8$ m/s^2).

23. ★★ A spherical metal shell has an outer radius of 15.0 cm and an inner radius of 14.0 cm. The density of the metal is 6000 kg/m^3. What is the average density of the shell? Will this shell float when immersed in water? If yes, what fraction of the volume of the shell will be beneath the water's surface?

Section 12-3 Pressure in Fluids

24. ★ The air provided by a diver's scuba equipment is at the same pressure as the surrounding water. At absolute pressures greater than approximately 1.0×10^6 Pa, the nitrogen in the air becomes dangerously poisonous. At what depth in water does nitrogen become dangerous? The density of the ocean water is 1030 kg/m^3.

25. ★ A submarine is at a depth of 45.0 m under the ocean surface. The inside of the submarine is kept at atmospheric pressure. What is the net force being exerted on a circular window of the submarine that has a radius of 20 cm? The density of the ocean water is 1030 kg/m^3.

26. ★★ A glass tube of cross-sectional area 1.0×10^{-4} m^2 is partially filled with water. An oil with a density of 800 kg/m^3 is slowly poured into the tube so that it does not mix with the water and floats on top of it. The height of the oil above the water surface is 8 cm.
    (a) What is the change in pressure at a depth of 10 cm below the water surface?
    (b) Would the pressure at this depth increase or decrease if you added more oil to the tube? Why?

27. ★ A cylindrical tube that is closed at one end and open at the other is filled with water, as shown in Figure 12-57. Determine the pressure exerted by the water on the closed end of the tube.

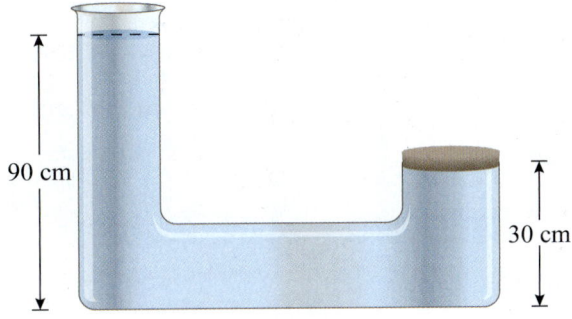

Figure 12-57 Problem 27.

28. ★★ A container of fluid is accelerated vertically downward with acceleration a. Show that the pressure in the fluid at a depth h below the surface is given by

$$P = P_0 + \rho h(g - a),$$

where P_0 is atmospheric pressure and ρ is the fluid density. If the container is allowed to fall freely ($a = g$), why does the pressure not change with depth?

29. ★★ A U-tube of uniform cross section is partly filled with water (density ρ_w). Another liquid of density ρ that does not mix with water is poured into one side until it stands a distance d above the original water level. The water level rises by a distance l in the other arm (Figure 12-58). Find the density of the liquid relative to that of water in terms of d and l.

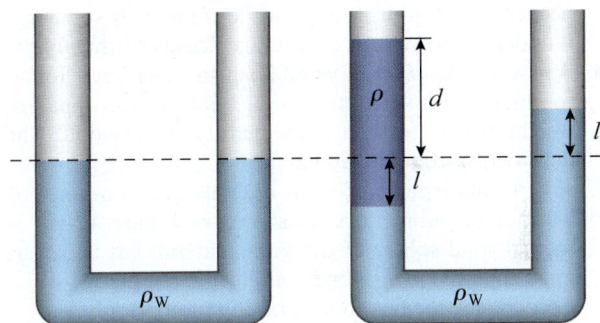

Figure 12-58 Problem 29.

Section 12-4 Pascal's Principle

30. Two pistons of a hydraulic lift have radii of 2.67 cm and 20.0 cm. A mass of 2.00×10^3 kg is placed on the larger piston. Calculate the minimum downward force needed to be exerted on the smaller piston to hold the larger piston level with the smaller piston.

31. Consider the two hydraulic presses shown in Figure 12-59. A force of 1000 N needs to be maintained at the output arm of each press. What is the relationship between the magnitudes of the forces that need to be exerted on the two input arms? Explain your result.
    (a) $F_1 = F_2$
    (b) $F_1 < F_2$
    (c) $F_1 > F_2$

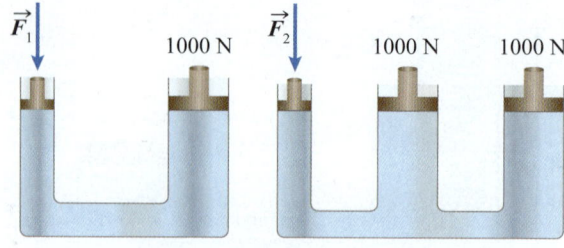

Figure 12-59 Problem 31.

Section 12-5 Buoyancy and Archimedes' Principle

32. ★★ The density of air at sea level is 1.3 kg/m³. Assume that the average density of a human body is 1050 kg/m³.
    (a) Estimate your volume.
    (b) Estimate air's buoyant force on your body.
    (c) What is your apparent weight and what is your true weight in vacuum ($g = 9.8$ m/s²)?

33. ★★ An object with specific gravity 0.85 is immersed in a liquid with specific gravity 1.2. What is the fraction of the volume of the object that submerges in the liquid?

34. ★★ A typical supertanker has a mass of 2.0×10^6 kg and carries oil of mass 4.0×10^6 kg. When empty, 9.0 m of the tanker is submerged in water. What is the minimum water depth needed for it to float when full of oil? Assume the sides of the supertanker are vertical and its bottom is flat.

35. ★★ An object weighs 500.0 N in air. A force of 30.0 N is needed to push the object down to completely submerge it under water. What is the average density of the object?

36. ★★ A 0.48 kg piece of wood floats in water but is found to sink in alcohol. When in the alcohol it has an apparent weight of 0.46 N. What is the density of the wood? The density of alcohol is 790 kg/m³.

37. ★★ A solid sphere of mass m floats in a container of water, half submerged, as shown in Figure 12-60. A second solid sphere of the same material but twice the mass is placed in a second container of water.
    (a) Sketch the final position of the second sphere.
    (b) What is the magnitude of the buoyant force on the first sphere?
    (c) What is the magnitude of the buoyant force on the second sphere?
    (d) Is the buoyant force acting on the second sphere larger than, smaller than, or equal to the buoyant force on the first sphere?
    (e) Is the volume of water displaced by the second sphere larger than, smaller than, or equal to the volume displaced by the first sphere?

Figure 12-60 Problem 37.

38. ★ A piece of cork with a mass of 10 g is held in place under water by a string tied to the bottom of the container. What is the tension in the string? The density of cork is 0.33 g/cm³.

39. ★★★ A hydrometer consists of a spherical bulb with a cylindrical stem (Figure 12-61). The cross-sectional area of the stem is 0.40 cm². The total volume of bulb and stem is 13.2 cm³. When immersed in water, the hydrometer floats with 8.0 cm of the stem above the water surface. In alcohol, 1.0 cm of the stem is above the surface. Find the density of alcohol.

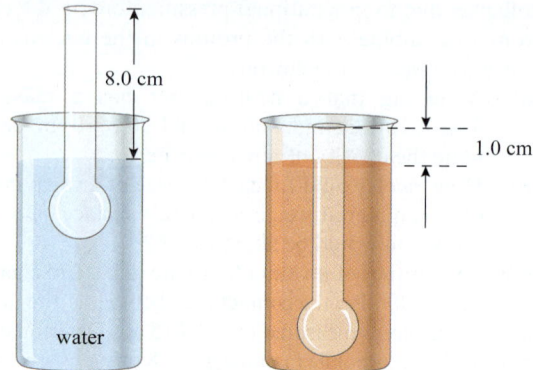

Figure 12-61 Problem 39.

40. ★★★ A glass beaker is placed in a large container of water. The beaker itself has a mass of 390.0 g and an interior volume of 500.0 cm³. A student starts filling the beaker with water and finds that when the beaker is less than half full, it floats (Figure 12-62). But when it is more than half full, it sinks to the bottom. What is the density of the glass making up the beaker?

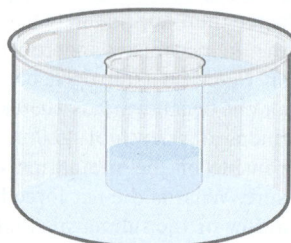

Figure 12-62 Problem 40.

41. ★★ You need to construct a hollow cubical box of aluminum that has an outside length of 20 cm and an average density of 1000 kg/m³. What must be the thickness of the aluminum sheet used to construct this box? The density of aluminum is 2700 kg/m³. Did you have to make any approximations to solve this problem?

42. ★★★ A wooden boat has a density of 700 kg/m³ and a volume of 5.5 m³.
    (a) What fraction of the volume of the boat is submerged in water when empty?
    (b) The boat needs to carry a number of survivors from a desert island. The average mass of each survivor is 65 kg. How many survivors can the boat carry without sinking?
    (c) If some of the survivors sat at the edge of the boat with their legs partially submerged in the water, would the boat be able to carry more people? Why?

43. ★★ An evacuated hollow spherical shell of aluminum floats almost completely submerged under water. The outer diameter of the shell is 100.0 cm, and the density of aluminum is 2.7 g/cm³. What is the inner diameter of the shell? Would your answer increase or decrease if you assumed that the inside of the shell was filled with air?

44. ★★ A solid sphere of aluminum (density 2.7 g/cm³) is gently dropped into a deep ocean. (The density of ocean water is approximately 1.03 g/cm³.) Calculate the sphere's acceleration at the point where it is completely submerged in the ocean. As the sphere drops deeper into the ocean, does the acceleration of the sphere increase or decrease compared to its acceleration just beneath the surface? Explain your reasoning.

45. ★★★ A rectangular block of wood weighs 4.0 N in air and has a density of 600 kg/m³. The block is immersed in water.
    (a) What fraction of the volume of the block is immersed in the water?
    (b) A piece of iron (density 6000 kg/m³) is placed on top of the block so that 85% of the block is now immersed in the water. What is the weight of the iron placed on the block?
    (c) The same piece of iron is now attached to the bottom of the wood that is immersed in the water. What fraction of the volume of the block is now immersed in the water?

46. ★★ Most balloons use hot air to fly. At 50 °C the density of hot air is approximately 0.85 kg/m³. What would be the volume of the balloon if hot air was used in Example 12-6? Assume the density of the ambient air to be 1.28 kg/m³.

47. ★★ A block of solid wood has length 30 cm, width 10 cm, and height 5 cm. The density of wood is 800 kg/m³.
    (a) When placed in water, with its height upright, what fraction of its volume would be immersed in the water?
    You now need to attach a piece of solid aluminum (density 2700 kg/m³) to the block so that it floats with its top just at the surface of the water. What should be the mass of the aluminum piece if you
    (b) place it on the top of the block?
    (c) glue it to the bottom of the block?
    If the two masses are different, explain why.

48. ★★ A cylindrical glass beaker has an inside diameter of 8.0 cm and a mass of 200 g. It is filled with water to a height of 5.0 cm. The water-filled beaker is placed on a weigh scale. What is the reading (in newtons) on the weigh scale?

49. ★★ A solid cylinder of aluminum that is 8.0 cm tall and has a radius of 2.0 cm is tied to a string. The cylinder is lowered into a beaker (that in Problem 48) such that it is half-immersed in the water. What is the reading on the weigh scale now? What is the tension in the string?

50. ★★ There are 12 survivors on an island. They construct a raft out of the trunks of five palm trees. Each tree is cylindrical, with a length of 6.00 m and a radius of 0.10 m. It is found that when the raft (without the survivors) is in the sea it is half-submerged. The density of the seawater is 1030 kg/m³, and the volume of a cylinder of radius R and length L is $\pi R^2 L$.

(a) What is the weight of the raft (without any survivors on it)?
(b) The survivors now stand on the raft. If each survivor has a mass of 60 kg, what is the maximum number of survivors that the raft can carry without completely submerging it in the water?

Section 12-7 The Continuity Equation: Conservation of Fluid Mass

51. ★ Water is travelling at a speed of 0.1 m/s in a portion of pipe that has a radius of 2.0 cm. The water enters another section of the pipe that has a radius of 1.0 cm. What is the speed of the water in this section of the pipe?

52. ★★ $\frac{d}{dx}$ The cross-sectional area of the lake behind a dam is approximately 1.00×10^8 m². Water due to heavy rains enters the lake at a rate of 2.5×10^4 m³/s and leaves through the dam at a rate of 1.9×10^4 m³/s. By how much will the water level increase in the lake in a period of 6.0 h?

53. ★★ $\frac{d}{dx}$ A large spherical balloon made of flexible rubber is filled with water at the rate of 0.1 L/s. What is the rate of increase of the radius of the balloon at the instant when its radius is 20.0 cm?

54. ★★ A water faucet has an inner area of 3.0 cm². The flow of water through the faucet is such that it fills a 500 mL container in 15 s.
    (a) What is the flow rate of the water as it comes out of the faucet?
    (b) What is the velocity with which the water emerges from the faucet?
    (c) What is the velocity of the water 20 cm below the faucet?
    (d) What is the area of the water stream 20 cm below the faucet?

Section 12-8 Conservation of Energy for Moving Fluids

55. ★★ Water in a main pipe at street level is at a gauge pressure of 170 kPa and is moving with negligible speed. A pipe connected to the main pipe is used to deliver water to a kitchen located on the fifth floor of a building, at 15.0 m above street level (Figure 12-63). What is the maximum possible speed with which the water can emerge from an open kitchen faucet?

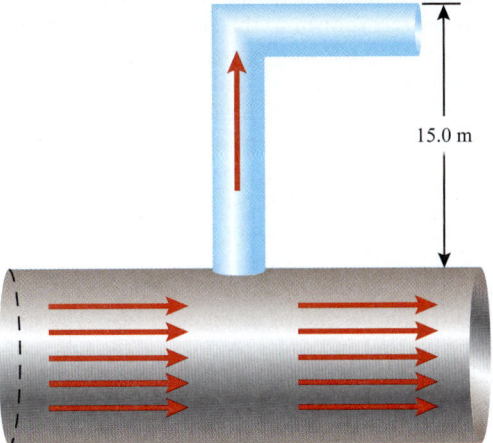

15.0 m

Figure 12-63 Problem 55.

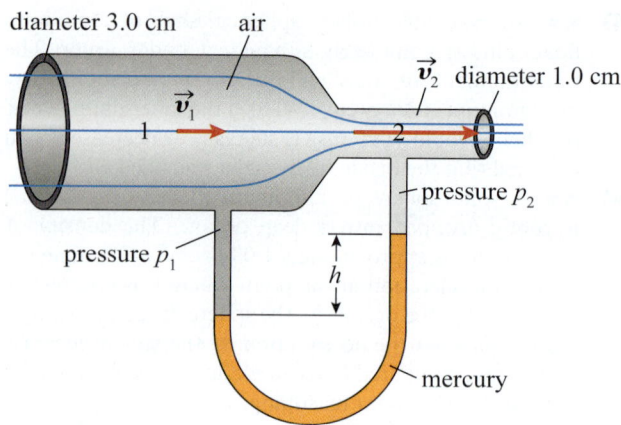

Figure 12-64 Problem 63.

56. ★ Water flows through a horizontal pipe that gradually narrows so that the final inside diameter of the pipe is one-half the original diameter. In the wide section, the flow speed is 2.0 m/s, and the water pressure is two times atmospheric pressure. Determine
   (a) the flow speed of the water in the narrower section
   (b) the water pressure in the narrower section

57. ★ A horizontal pipe, which carries water with a speed of 5.0 cm/s, is smoothly connected to a pipe with a smaller cross-sectional area. The pressure in the small pipe is 7.0 kPa less than the pressure in the large pipe. What is the velocity of the water in the small pipe?

58. ★ Normal blood pressure in a human artery is 13.6 kPa; the blood flows with a speed of 0.12 m/s. Suppose that part of the artery is clogged such that only 20% of the artery's area is open (assume the artery is horizontal).
   (a) What is the blood pressure in the clogged part of the artery?
   (b) What is the percentage change in the blood pressure in the clogged part?

59. ★★ A garden hose has an inside cross-sectional area of 4.00 cm², and the opening in the nozzle has a cross-sectional area of 0.50 cm². The water speed is 0.50 m/s in a segment of the hose that lies on the ground.
   (a) With what speed does the water leave the nozzle when it is held 1.00 m above the ground?
   (b) What is the water pressure in the hose on the ground?

60. ★★★ A 12.0 m long outlet pipe with a diameter of 4.0 cm stands vertically in a water tank that has a diameter of 4.0 m and is open to the atmosphere. The section of the pipe above the water surface is 5.0 m long. Water is pumped from the tank by lowering the pressure at the top end of the pipe. The desired flow rate is 4.0 L/s.
   (a) By how much should the pressure at the top end of the pipe differ from atmospheric pressure?
   (b) Assuming that the pressure at the top of the outlet pipe remains constant, what is the lowest level from which the water can be pumped from the tank?

61. ★★ Water flows through a horizontal pipe of radius 40.0 cm that is located 20.0 m below the surface of a large lake next to a dam. The end of the pipe is connected to a nozzle of radius 10.0 cm, and the outlet end of the nozzle is open to the atmosphere. Assuming that lake is rectangular with verticals walls and that the water behaves as an ideal fluid, find
   (a) the speed with which the water leaves the nozzle
   (b) the water pressure inside the pipe

62. ★★ In problem 61, if the nozzle is closed so that there is no water flow through it, would the pressure in the pipe be less than, greater than, or equal to the pressure calculated in part (b)? Ignore viscosity.

63. ★★ Air is blown through a circular and horizontal Venturi tube, as shown in Figure 12-64. The diameters of the narrow and wider sections of the tube are given, and the height, h, of the mercury column is measured to be 1.0 mm. What are the air speeds in the wider and the narrower sections of the tube? The density of mercury is 13 600 kg/m³.

64. ★★ A tank of cross-sectional area A_2 is filled to a height y_2 with a fluid of density ρ. The tank has a small hole of cross-sectional area A_1 in its side at y_1 from the base. The pressure at the top of the fluid is kept at P_t. Determine an expression for the speed with which the fluid leaves the hole.

65. ★★ A very large tank, closed from above and not open to the atmosphere, is partially filled with water. The pressure at the top surface of the water is kept at three times atmospheric pressure by constantly injecting pressurized air from above. Water emerges from the tank through a circular nozzle that is located close to the base of the tank and is open to the atmosphere. At the instant when the water stands 3.0 m high above the nozzle, what is the water flow rate from the nozzle? The cross-sectional area of the nozzle is 2.0×10^{-4} m².

66. ★★ Water is forced out of a fire extinguisher by pressurized air (Figure 12-65). When the valve is open and the water level in the tank is 0.50 m below the nozzle, the water jet is leaving the nozzle with a speed of 30.0 m/s.
   (a) What is the air pressure inside the tank? (Ignore the speed with which the water level drops in the tank.)
   (b) If the valve is closed, what happens to the water pressure inside the pipe leading to the nozzle?
   (i) increases (ii) decreases (iii) does not change

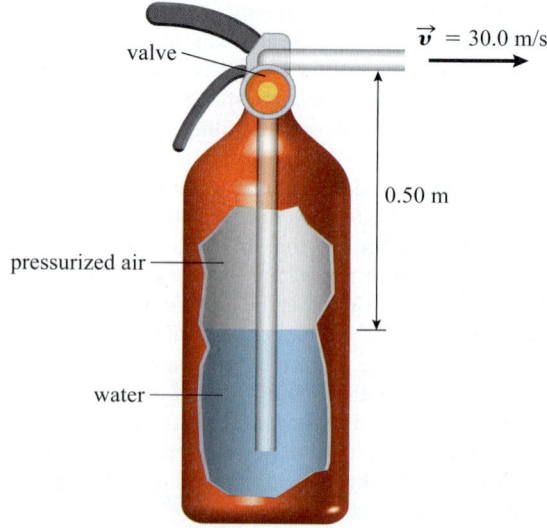

Figure 12-65 Problem 66.

67. ★★ A Venturi flowmeter is used to measure the flow rate of water (Figure 12-66). The flowmeter is inserted in a water pipe that has a diameter of 2.0 cm. At the narrow section, the diameter of the pipe is 0.80 cm. The U-tube contains oil with density 0.80 g/cm³. If the difference in the oil level on the two sides of the flowmeter is 1.5 cm, what is the volume flow rate of water through the pipe? The density of water is 1.0 g/cm³.

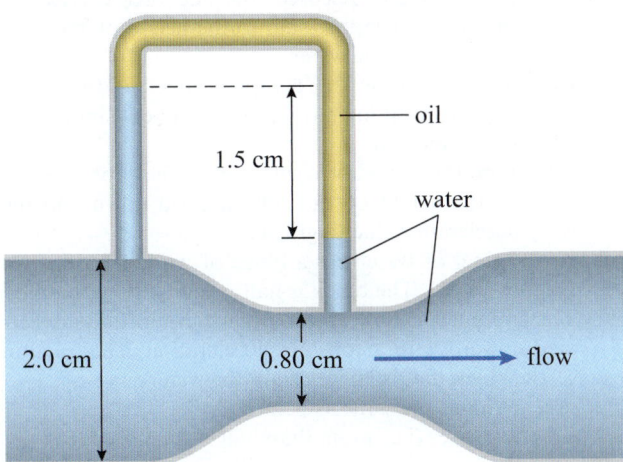

Figure 12-66 Problem 67.

68. ★★ Water emerges from a tap of diameter 2 cm with a flow rate of 0.5 L/s. Determine the average speed of the water as it leaves the tap. Would the average speed of the water be the same 20 cm below the tap as it falls under the influence of gravity? How do the cross-sectional areas of the stream compare at the mouth of the tap and 20 cm below it?

Section 12-9   Conservation of Fluid Momentum

69. ★★★ $\frac{d}{dx}$ A fluid of density ρ is placed in a cylindrical container that is rotated about its symmetry axis with a constant angular velocity ω.
(a) Show that in the radial direction, the variation of pressure within the fluid is given by

$$\frac{dP}{dr} = \rho\omega^2 r,$$

where r is the radial distance from the axis of rotation.
(b) Suppose the pressure along the axis of rotation is P_0. Show that the pressure at a radial distance R from the central axis is

$$P_R = P_0 + \frac{1}{2}\rho\omega^2 R^2$$

(c) Show that the surface of the rotating fluid forms a paraboloid.

70. ★★★ Water passes through a horizontal pipe of uniform cross-sectional area that has a 90° bend. The diameter of the pipe is 10.0 cm, and the volume flow rate is 50.0 L/s. Find the magnitude and direction of the net radial force exerted on the bend as the water passes through it.

Section 12-10   Viscous Flow

71. ★ A cylindrical pipe of length 5.0 m and cross-sectional area 1.0×10^{-4} m² needs to deliver oil at a rate of 5.0×10^{-4} m³/s. What must be the pressure difference

between the two ends of the pipe if the viscosity of the oil is 1.00×10^{-3} Pa·s?

72. ★ Blood flows through an artery at a rate of 200 cm³/s. The cross-sectional area of the artery is 2.0 cm². Determine the pressure drop per unit length of the artery. The coefficient of viscosity of the blood is 4×10^{-2} P (poise).

73. ★ To keep the oil flowing at a rate of 200 L/s through a pipe of length 200 m and radius 5 cm, a pressure difference of 5×10^4 Pa needs to be maintained at both ends of the pipe. Determine the viscosity of the oil.

74. ★ A 1.0 m long pipe, placed vertically, is to pump a liquid of density 1500 kg/m³ and viscosity 2.0 P (poise) such that the flow rate is 1 L/s. If the radius of the pipe is 2 cm, what pressure difference needs to be maintained between the lower and the upper ends of the pipe to maintain the flow?

75. ★★ A needle of inner diameter 0.50 mm is used to draw blood from a patient's arm for a medical test. The average blood pressure in the patient's arm is 85 mm Hg, and the viscosity of the blood is 3.5×10^{-4} Pa·s. The length of the needle is 6.0 cm, and the blood needs to be drawn at a constant flow rate of 0.10 cm³/s. What pressure must be applied to the plunger?

COMPREHENSIVE PROBLEMS

76. A pitot-static tube is a device used to measure the speed of gases. Its uses include measuring the speed of air flow in a wind tunnel and the speed of an airplane in air. In its simplest form, a pitot-static tube consists of a U-shaped tube that is open at both ends and is partially filled with a liquid, as shown in Figure 12-67. One end of the tube (B) is perpendicular to the incoming flow, and the other end (A) is horizontal to the direction of flow. The tube is closed in the middle (it contains liquid), so the air impinging on end A is brought to rest. The pressure at this end is then the total pressure (also called dynamic pressure) of the flowing air. Air is not flowing right at the edge of end B, and the pressure at this end is equal to atmospheric pressure (also called static pressure). Because of the pressure difference at the two ends, the liquid rises in the arm that is at the lower pressure. The height difference, h, between the two arms of the tube is related to the pressure difference between the two ends.
(a) Consider the pitot-static tube in Figure 12-67, which is attached to a pipe through which air is flowing with speed v. The tube is partially filled with a liquid of density ρ_f. Show that, if the difference between the heights of the fluid column in the two arms of the tube is h, then the air speed is related to h by the following relationship:

$$v = \sqrt{2gh\left(\frac{\rho_f}{\rho_{air}}\right)}$$

Thus, by measuring h and knowing the densities, flow speed can be calculated. When pitot-static tubes are used to measure the speed of an airplane, ρ_{air} is the density of the air at the flight altitude and needs to be measured or calculated.

(b) In modern pitot-static tubes, instead of using a liquid, an electronic pressure-measuring instrument is placed between the two ends of the tube. Show that the speed, v, of airflow is related to the measured pressure difference ΔP by the following relationship:

$$v = \sqrt{\frac{2(\Delta P)}{\rho_{air}}}$$

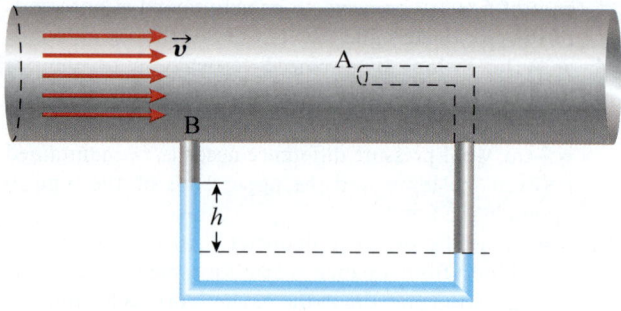

Figure 12-67 Problem 76: A simple pitot-static tube. The height difference of the column of liquid in the two arms of the tube is proportional to the speed of the air flowing through the tube.

77. The density of body fat in humans is approximately 900 kg/m³, and the density of lean muscle is 1100 kg/m³. Suppose a person weighs 790.0 N in air and 50.0 N when completely immersed in water (density 1000 kg/m³).
   (a) Find the person's volume.
   (b) What is the person's density?
   (c) Show that the fraction, x_F, of a person's body mass that is fat is given by

$$x_F = \frac{\rho_F}{\rho_P}\left(\frac{\rho_L - \rho_P}{\rho_L - \rho_F}\right),$$

   where ρ_F is the density of body fat, ρ_L is the density of lean muscle, and ρ_P is the average density of a person.
   (d) What approximations did you have to make to derive the above result?
   (e) What is x_F for the given data?
78. A rotating garden sprinkler consists of a pipe of diameter 2.0 cm that is open and slightly bent at both ends so that the water rushes out from the ends. The pipe is free to rotate around its centre (Figure 12-68). The flow rate from each opening of the pipe is 2.0 L/s. What is the net torque generated by the sprinkler?

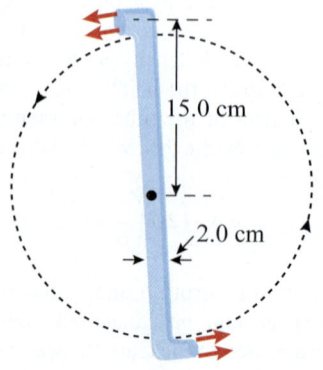

Figure 12-68 Problem 78.

79. A wooden cube is floating at rest in a glass of water with a certain fraction of the cube's volume immersed. In this equilibrium position, the upward buoyant force exerted on the cube by the water is balanced by the weight of the cube. The cube is now gently pressed farther into the water, held stationary, and then released. It bobs up and down before coming to rest in its original configuration.
   (a) Show that when the cube is gently pressed into the water and released, the net force exerted on the cube is proportional to the height of the cube immersed in the water.
   (b) Show, ignoring the damping effect of the water, that the resulting motion of the cube is simple harmonic motion.
   (c) Find the period of oscillation of the cube in terms of the height of the cube pressed down and the acceleration due to gravity.
80. A container of water has a block of wood floating in it (Figure 12-69). The block is half-immersed in the water. The container is now placed in an elevator and accelerated upward with a constant acceleration. The fraction of the block that is immersed in the water
   (a) remains the same as before
   (b) increases (i.e., more than half is immersed)
   (c) decreases (i.e., less than half is immersed)
   (d) There is not enough information to answer this question.

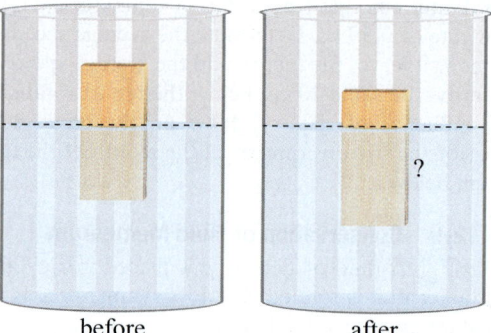

before   after

Figure 12-69 Problem 80.

CHAPTER

13 | Oscillations

Located between New Brunswick and Nova Scotia, the Bay of Fundy is home to some of the highest tides in the world. The average tide height worldwide is about 1 m, and Bay of Fundy tides rise as high as 16 m.

The high tides at the Bay of Fundy result from the shape of the bay (Figure 13-1). The mouth of the bay is 100 km wide and between 120 m and 215 m deep. At its end near Hopewell Rocks, the bay is about 2.5 km wide and 14 m deep, at low tide. This shape funnels the incoming tidal water to create spectacular high tides at the end of the bay. The time it takes for the tidal water to travel the 290 km length of the bay is approximately equal to the time it takes for a new tide to come from the ocean. As a result, the previous tide reinforces the incoming tide in a phenomenon called tidal resonance.

The rise and fall of the water height at inlets like the Bay of Fundy follow a pattern called periodic motion. In this chapter, we will examine the nature of this type of motion.

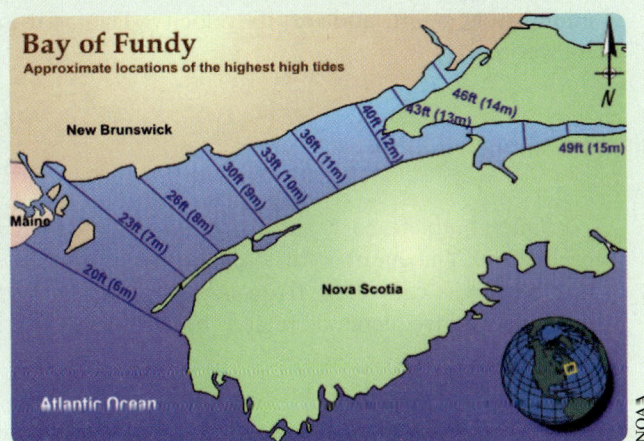

Figure 13-1 The width of the Bay of Fundy at various locations.

13-1 Periodic Motion

A motion that repeats after a finite amount of time is called **periodic motion**, or **harmonic motion**. The amount of time after which the motion repeats is called the **period** of the motion and is denoted by the symbol T. Thus, if a motion is repeated after 20.0 s, then $T = 20.0$ s.

From very small to extremely large scales, examples of periodic motion are all around us. The motion of atoms in molecules, of a mass attached to a spring, of Earth around the Sun, and of the Sun around the centre of the Milky Way are all examples of periodic motion. Although the forces underlying these periodic motions may be different, the mathematical equations describing them are similar. The mathematical descriptions developed in this chapter apply to many types of periodic motion. In later chapters, we will use the descriptions to understand sound, electromagnetic waves, and atomic structure.

We will investigate periodic motion in one dimension. Although motion in one dimension can be discussed using vector notation, the underlying physics is more transparent by avoiding vector notation and distinguishing motion in the positive and negative directions by using plus and minus signs. We will use vector notation where necessary.

The statement that an object repeats its motion periodically means that both the displacement and the velocity of the object will be the same at the beginning and at the end of any interval of T seconds. Let $x(t)$ denote the position of the object, as measured from an equilibrium point, and $v(t)$ its velocity. Then, for any integer n,

$$x(t + nT) = x(t) \tag{13-1}$$

$$v(t + nT) = v(t) \tag{13-2}$$

One complete back-and-forth motion is called an **oscillation**. The number of oscillations completed in one second is called the **frequency**, f, of the periodic motion. The frequency is related to the period of the motion:

$$\text{frequency} = \frac{1}{\text{period}} \tag{13-3}$$

KEY EQUATION

$$f = \frac{1}{T}$$

The SI unit for frequency is the hertz (Hz):

$1\ \text{Hz} = 1\ \text{s}^{-1} =$ one oscillation per second

For example, Earth completes one rotation around the Sun in 365.25 days. The time period of this motion

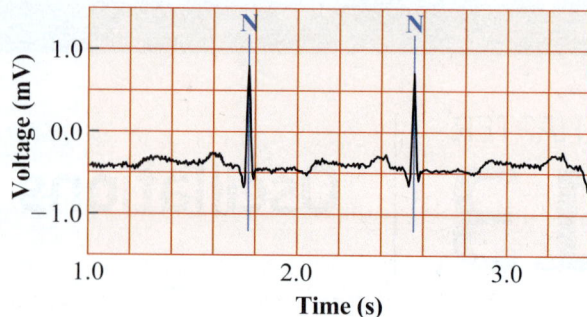

Figure 13-2 An ECG. A plot of voltage as a function of time.

is 365.25 days × 24 h/day × 3600 s/h = 315.57×10^5 s. The corresponding frequency of this periodic motion is $1/(315.57 \times 10^5\ \text{s}) = 3.1688 \times 10^{-8}$ Hz.

For periodic motion, a plot of the displacement, $x(t)$, versus time, t, shows a curve whose shape repeats after every T seconds. Thus, the period and the frequency of a periodic motion can be determined if we are given a graph of its displacement versus time. Two examples of periodic motion are shown in Figures 13-2 and 13-3: Figure 13-2 shows an electrocardiogram (ECG) as a function of time, and Figure 13-3 shows a graph of the displacement of a mass that is attached to a vertical spring and is oscillating about an equilibrium position (the other end of the spring is attached to a rigid support). The graph in Figure 13-3 ignores frictional forces.

Notice two things from these plots. First, both motions are periodic, so we can determine their periods from the graphs. Second, the ECG graph is more complicated than the mass–spring graph. It would be difficult to write an equation that describes the shape of the ECG graph as a function of time, whereas the displacement of the mass–spring system can be described by a cosine or a sine function. In this chapter, we consider a particular type of periodic motion that can be described by a sine or a cosine function, called simple harmonic motion.

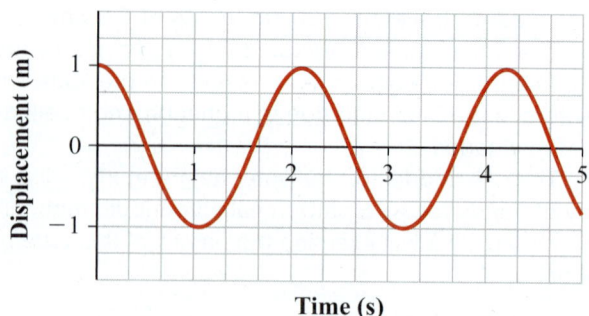

Figure 13-3 A plot of the displacement of an oscillating mass, attached to a spring, as a function of time.

MAKING CONNECTIONS

What Is an Electrocardiogram?

The heart generates rhythmic electrical signals that cause the heart's muscle fibres to contract, pumping blood through the body. An electrocardiogram (ECG or EKG) is a plot of these electric signals (voltage versus time) made while the heart is beating (Figure 13-4). Many heart problems can be diagnosed by comparing the shape of a patient's electric pulses on an ECG to an ECG of a healthy heart.

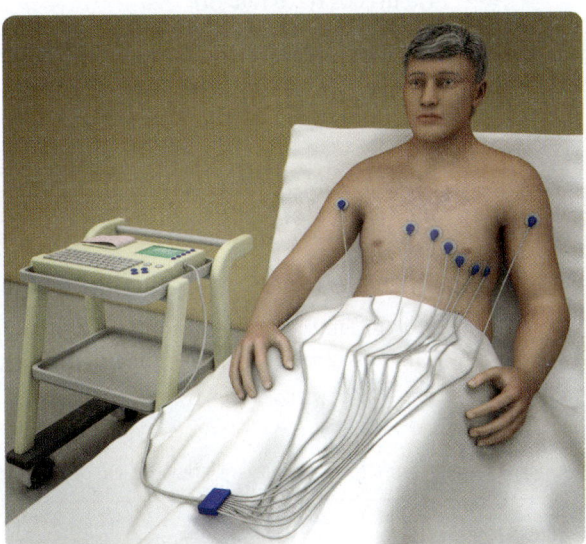

Figure 13-4 An electrocardiogram test being administered. Several electrodes positioned at various locations around the heart monitor the heart's electrical signal, which is then plotted as a function of time.

13-2 Simple Harmonic Motion

When the displacement of an oscillating object from an equilibrium position varies with time as a cosine (or a sine) function, the periodic motion is called **simple harmonic motion**. Therefore, for simple harmonic motion,

KEY EQUATION
$$x(t) = A\cos(\omega t + \phi) \qquad (13\text{-}4)$$

Note that we have chosen to describe the simple harmonic motion using a cosine function. A description using a sine function is equivalent since the two functions are related by $\cos\left(\theta - \dfrac{\pi}{2}\right) = \sin\theta$. The cosine and sine functions are periodic and have a period of 2π rad. An object that is undergoing simple harmonic motion is called a **simple harmonic oscillator**. The argument of a cosine function is a dimensionless quantity and is measured in radians. The quantities in Equation 13-4 are defined as follows.

A is called the **amplitude**. It is a positive number and describes the maximum displacement of the oscillator from its equilibrium position. As the object oscillates, its displacement from the equilibrium position continuously varies between the values $-A$ and $+A$ because the cosine function oscillates between the values -1 and $+1$. The SI unit for amplitude is the metre (m).

ω is called the **angular frequency**. Angular frequency is related to the frequency, f, of the simple harmonic motion as follows:

$$\omega = 2\pi f = \frac{2\pi}{T} \qquad (13\text{-}5)$$

Angular frequency is measured in radians per second (rad/s); since radians are dimensionless, we can simply write per second (s^{-1}).

ϕ is called the **phase constant**. The phase constant is an angle, measured in radians, and its value is determined by the position and the velocity of the motion at a given time, usually taken to be $t = 0$.

We will now explore the role of the amplitude, angular frequency, and phase constant in describing simple harmonic motion. Figure 13-5 shows the displacements of two oscillators described by $x_1(t) = (1.0 \text{ m})\cos(\omega t)$ and $x_2(t) = (2.0 \text{ m})\cos(\omega t)$. For these motions, we have chosen $\omega = 1$ rad/s and $\phi = 0$ rad. Comparing with Equation 13-4, the first oscillator has amplitude $A = 1.0$ m, and the second oscillator has amplitude $A = 2.0$ m. The displacement of the first oscillates between the values $x = +1.0$ m and $x = -1.0$ m, and the displacement of the second oscillates between $x = +2.0$ m and $x = -2.0$ m. Both oscillators reach the maximum displacement, pass through the equilibrium position, and then reach the maximum displacement at the opposite side of the equilibrium position at the same time.

For a given simple harmonic motion, the amplitude is a *constant* and does not change with time. We will see later that the amplitude of a simple harmonic oscillator is related to the energy of the oscillator: an oscillator with greater amplitude has more energy.

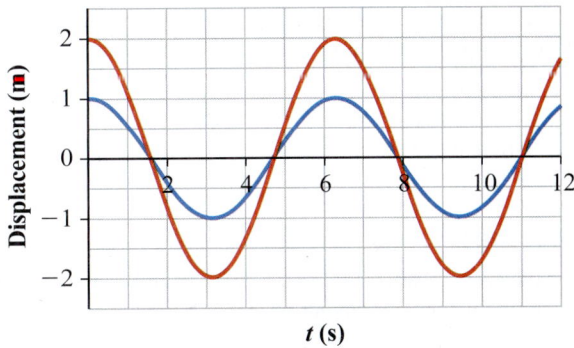

Figure 13-5 Plots of $(1.0 \text{ m})\cos(\omega t)$ and $(2.0 \text{ m})\cos(\omega t)$ as functions of time. For these plots, $\omega = 1$ rad/s.

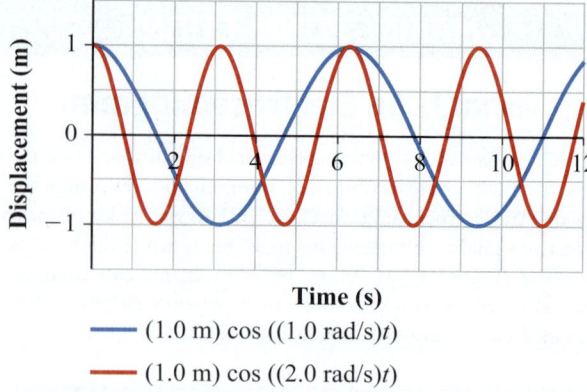

Figure 13-6 Plots of (1.0 m) cos((1.0 rad/s)t) and (1.0 m) cos((2.0 rad/s)t) as functions of time.

Now, we derive the relationship in Equation 13-5 between the angular frequency, ω, and the frequency, f. By the definition of the period,

$$x(t + T) = x(t)$$

Inserting Equation 13-4 in the above relationship, we get

$$A\cos(\omega(t + T) + \phi) = A\cos(\omega t + \phi) \quad (13\text{-}6)$$

$$\cos(\omega t + \omega T + \phi) = \cos(\omega t + \phi) \quad (13\text{-}7)$$

The cosine function repeats after 2π rad, so Equation 13-7 can only be satisfied when

$$\omega T = 2\pi \text{ rad} \quad (13\text{-}8)$$

This condition implies that

$$\omega = \frac{2\pi}{T} = 2\pi f \text{ rad/s} \quad (13\text{-}9)$$

The angular frequency is 2π times the frequency. By adjusting ω, we can describe fast and slow oscillations. Figure 13-6 shows graphs of two simple harmonic motions, with $\omega = 1.0$ rad/s and $\omega = 2.0$ rad/s, and $A = 1.0$ m and $\phi = 0$ rad for both. For $\omega = 1.0$ rad/s, $T = 6.28$ s; for $\omega = 2.0$ rad/s, $T = 3.14$ s. The values of the periods can be approximated from the graphs.

We now examine the meaning of the phase constant, ϕ. Setting $t = 0$ in Equation 13-4, we get

$$x(0) = A\cos\phi$$

Therefore, the phase constant is related to the position of the oscillatory motion at $t = 0$. Different values of ϕ describe different possible initial locations of the oscillator, as shown in the following example.

EXAMPLE 13-1

Calculating the Phase Constant

A simple harmonic oscillator has $T = 2$ s and $A = 0.5$ m. Calculate the values of the phase constant, ϕ, for the following starting ($t = 0$) positions of the oscillatory motion:

(a) $x(0) = 0.5$ m
(b) $x(0) = -0.5$ m
(c) $x(0) = 0.3$ m
(d) $x(0) = 0.0$ m

Solution

For the given values, $T = 2$ s, $\omega = \pi$ rad/s, and $A = 0.5$ m, the equation for simple harmonic motion is given by

$$x(t) = (0.5 \text{ m}) \cos(\pi t + \phi)$$

At $t = 0$,

$$x(0) = (0.5 \text{ m}) \cos\phi$$

For each starting position, the corresponding value of the phase constant can be determined by setting the above equation to the given value.

(a)
$$x(0) = 0.5 \text{ m, then}$$
$$0.5 \text{ m} = (0.5 \text{ m}) \cos\phi$$

$$\cos\phi = 1 \Rightarrow \phi = \cos^{-1}(1) = 0 \text{ rad (or } 2\pi \text{ rad)}$$

The oscillatory motion of the particle is described by the equation

$$x(t) = (0.5 \text{ m}) \cos(\pi t)$$

The displacement versus time graph for the above equation is shown in Figure 13-7.

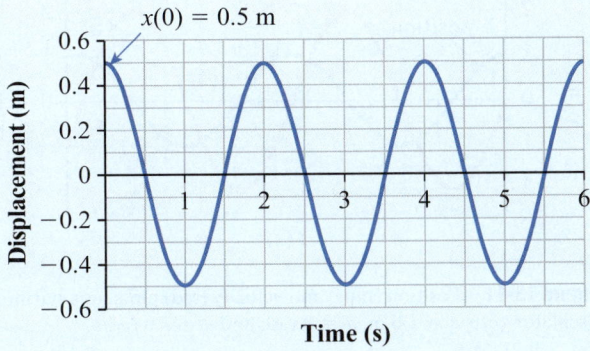

$x(0) = 0.5$ m

Figure 13-7   A plot of $(0.5 \text{ m}) \cos \pi t$ as a function of time.

(b)
$$x(0) = -0.5 \text{ m}$$
$$-0.5 \text{ m} = (0.5 \text{ m}) \cos \phi$$
$$\cos \phi = -1 \Rightarrow \phi = \cos^{-1}(-1) = \pi \text{ rad}$$

The oscillatory motion of the particle is described by the equation

$$x(t) = (0.5 \text{ m}) \cos(\pi t + \pi) \qquad (13\text{-}10)$$

A graph of this equation is shown in Figure 13-8. Note that the displacements described by (a) and (b) are equal in magnitude and opposite in direction for all times.

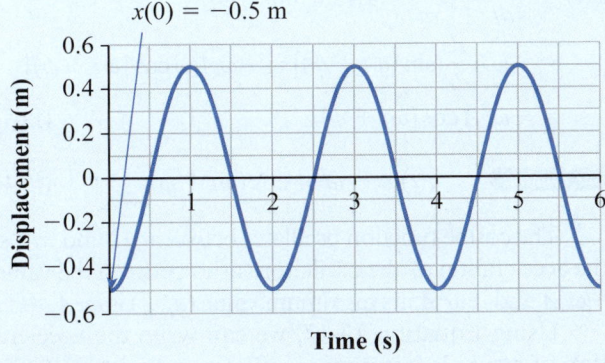

$x(0) = -0.5$ m

Figure 13-8   A plot of $(0.5 \text{ m}) \cos(\pi t + \pi)$ as a function of time.

(c)
$$x(0) = +0.3 \text{ m}$$
$$0.3 \text{ m} = (0.5 \text{ m}) \cos \phi$$
$$\cos \phi = \frac{0.3}{0.5} = 0.6$$
$$\phi = \cos^{-1}(0.6) = 0.93 \text{ rad or } -0.93 \text{ rad}$$

Since $\cos \theta = \cos(-\theta)$, there are two possible values for the phase constant. To distinguish between the two values, additional information is needed, for example, the oscillator's velocity at $t = 0$. In general, a unique determination of the phase constant requires information about the oscillator's position and velocity at $t = 0$.

(d)
$$x(0) = 0.0 \text{ m}$$
$$0.0 \text{ m} = (0.5 \text{ m}) \cos \phi$$
$$\cos \phi = \frac{0.0}{0.5} = 0$$
$$\phi = \cos^{-1}(0) = \frac{\pi}{2} \text{ or } -\frac{\pi}{2} \text{ rad}$$

Again, there are two possible values for ϕ, and we need additional information to determine a unique value. Figure 13-9 shows a plot of $x(t)$ versus t for $\phi = \pi/2$ and $\phi = -\pi/2$.

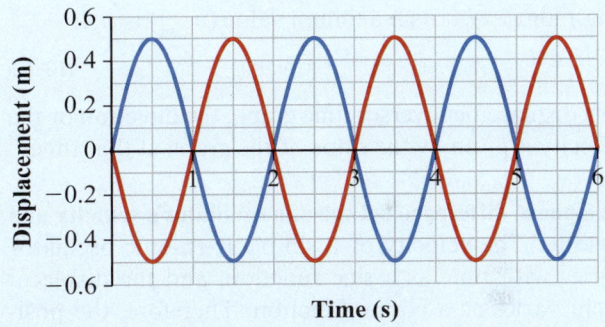

Figure 13-9   A plot of $x(t) = (0.5 \text{ m}) \cos(\pi t + \phi)$ for $\phi = \pi/2$ (blue) and $\phi = -\pi/2$ (red) as a function of time.

CHECKPOINT 13-3

Displacement versus Time Graph

What are the amplitude (A) and angular frequency (ω) of the displacement versus time graph shown in Figure 13-10?

(a) $A = 8$ m, $\omega = 4$ rad/s
(b) $A = 4$ m, $\omega = 2\pi/3$ rad/s
(c) $A = 0$ m, $\omega = 2\pi$ rad/s
(d) $A = -4$ m, $\omega = 2\pi/3$ rad/s

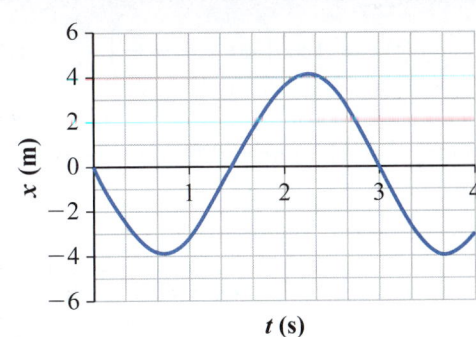

Figure 13-10   Checkpoint 13-3.

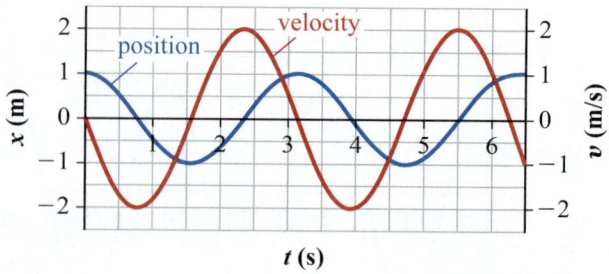

Figure 13-11 Position (blue) and velocity (red) plots for a harmonic oscillator with $A = 1.0$ m, $\phi = 0$ rad, and $\omega = 2.0$ rad/s.

The Velocity of a Simple Harmonic Oscillator

The velocity, $v(t)$, of a simple harmonic oscillator is obtained by taking the derivative of the displacement, $x(t)$, with respect to time, t:

$$v(t) = \frac{dx(t)}{dt} = \frac{d}{dt}[A\cos(\omega t + \phi)]$$

$$= A\frac{d}{dt}[\cos(\omega t + \phi)] = A[-\omega\sin(\omega t + \phi)]$$

$$= -\omega A\sin(\omega t + \phi) \tag{13-11}$$

KEY EQUATION $\qquad v(t) = -\omega A\sin(\omega t + \phi) \tag{13-12}$

The sine function oscillates between -1 and $+1$, so the velocity of the oscillator varies between the values $+\omega A$ and $-\omega A$; its maximum value (v_{max}) is ωA:

$$v_{max} = \omega A \tag{13-13}$$

In a displacement versus time graph, the direction of the velocity is given by the slope of the graph at that time.

The phase difference between an oscillator's velocity and position The velocity of a simple harmonic oscillator varies with time as a sine function, and the displacement varies as a cosine function. Therefore, the position and velocity of the oscillator are $\pi/2$ rad (90°) out of phase. We can derive this result by using the trigonometric identity

$$-\sin\theta = \cos\left(\theta + \frac{\pi}{2}\right)$$

Equation 13-12 for $v(t)$ can now be written as

$$v(t) = \omega A\cos\left(\omega t + \phi + \frac{\pi}{2}\right) \tag{13-14}$$

Both $x(t)$ and $v(t)$ are now written in terms of the cosine function, so the phase difference between $x(t)$ and $v(t)$ is

$$\text{phase of } v(t) - \text{phase of } x(t)$$

$$= \left(\omega t + \phi + \frac{\pi}{2}\right) - (\omega t + \phi) = \frac{\pi}{2}\text{rad} \tag{13-15}$$

Graphs of $x(t)$ and $v(t)$ for an oscillator with $A = 1.0$ m, $\phi = 0$ rad, and $\omega = 2.0$ rad/s are shown in Figure 13-11. Note that when an oscillator's displacement from equilibrium is at a maximum ($x = \pm A$), its velocity is zero, and when an oscillator passes through the equilibrium position ($x = 0$), its velocity is at a maximum ($v = \pm\omega A$). We will see later that this relationship between the position and the velocity is a result of conservation of the oscillator's energy.

The Acceleration of a Simple Harmonic Oscillator

The acceleration, $a(t)$, of a simple harmonic oscillator is obtained by taking the derivative of the velocity, $v(t)$, with respect to time, t:

$$a(t) = \frac{dv(t)}{dt} = \frac{d}{dt}[-\omega A\sin(\omega t + \phi)]$$

$$= -\omega A\frac{d}{dt}[\sin(\omega t + \phi)] = -\omega A[\omega\cos(\omega t + \phi)]$$

$$= -\omega^2 A\cos(\omega t + \phi) \tag{13-16}$$

KEY EQUATION $\qquad a(t) = -\omega^2 A\cos(\omega t + \phi) \tag{13-17}$

The cosine function oscillates between -1 and $+1$, so the acceleration of the oscillator varies between the values $-\omega^2 A$ and $+\omega^2 A$; its maximum value (a_{max}) is $\omega^2 A$.

Using Equation 13-17, we can write the acceleration of a simple harmonic oscillator in terms of its displacement from equilibrium as

KEY EQUATION $\qquad a(t) = -\omega^2(A\cos(\omega t + \phi))$

$$= -\omega^2 x(t) \tag{13-18}$$

This relationship between the acceleration and the displacement is a fundamental property of all simple harmonic oscillators: *For a motion to be simple harmonic motion, the acceleration of the object must be proportional to its displacement from the equilibrium position and opposite in direction, at all times.*

The constant of proportionality between the acceleration and the displacement of a simple harmonic oscillator is equal to the square of its angular frequency.

The phase difference between the acceleration and position of an oscillator The minus sign in Equation 13-17 means that the acceleration and the displacement are π rad (180°) out of phase. Using the trigonometric identity

$$-\cos\theta = \cos(\theta + \pi)$$

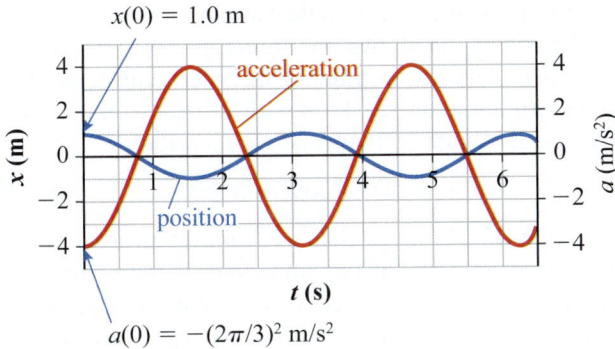

$x(0) = 1.0$ m

acceleration

position

$a(0) = -(2\pi/3)^2$ m/s²

Figure 13-12 Position (blue) and acceleration (red) plots for a harmonic oscillator with $A = 1.0$ m, $\phi = 0$ rad, and $\omega = 2.0$ rad/s.

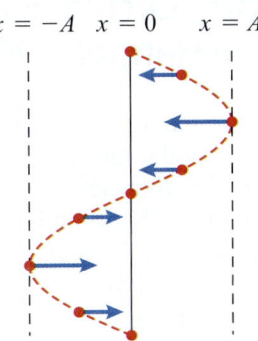

$x = -A \quad x = 0 \quad x = A$

Figure 13-13 The direction of the restoring force for various displacements of a simple harmonic oscillator. The restoring force is always directed toward the equilibrium position.

we can write

$$a(t) = \omega^2 A \cos(\omega t + \phi + \pi) \qquad (13\text{-}19)$$

Therefore, the phase difference between $a(t)$ and $x(t)$ is

$$\begin{aligned} &\text{phase of } a(t) - \text{phase of } x(t) \\ &= (\omega t + \phi + \pi) - (\omega t + \phi) = \pi \text{ rad} \end{aligned} \qquad (13\text{-}20)$$

Graphs of $x(t)$ and $a(t)$ for an oscillator with $A = 1.0$ m, $\phi = 0$ rad, and $\omega = 2.0$ rad/s are shown in Figure 13-12.

The Restoring Force and Simple Harmonic Motion

As an object undergoes simple harmonic motion, its acceleration changes with time, so there must be a time-dependent force acting on the object. What is the nature of this force? Using Equation 13-18 for the acceleration of a simple harmonic oscillator, a mass m that undergoes simple harmonic motion experiences a force

$$F(t) = ma(t) = -m\omega^2 x(t) \qquad (13\text{-}21)$$

Since both the mass, m, and the angular frequency, ω, of the motion are fixed, the quantity $m\omega^2$ is constant for a given oscillator. Therefore,

$$F(t) \propto -x(t) \qquad (13\text{-}22)$$

Figure 13-13 shows the relationship between the direction of the force on the oscillator and its displacement from equilibrium. When the displacement is positive (along the positive x-axis), the force is negative, pointing toward the origin. When the displacement is negative, the force is positive and again points toward the origin. The relationship described by Equation 13-22 is called **Hooke's law**. Elastic objects (e.g., springs) obey Hooke's law when they are deformed, as long as the deformation is small. A force that causes an object to return to its equilibrium position is called a **restoring force**. *Simple harmonic motion is motion by an object that is subjected to a restoring force.*

To determine whether an object, under the influence of an external force, will undergo simple harmonic motion, we can use the following procedure:

1. Displace the object from the equilibrium position by a small displacement, x.
2. Calculate the net force, F, on the object at the displaced position.
3. Determine whether F can be written as

$$F = -\text{constant} \times x \qquad (13\text{-}23)$$

If the displacement from the equilibrium position of the object is proportional to the net force exerted on the object but opposite in sign, the object will undergo simple harmonic motion. The angular frequency of oscillation is determined by comparing Equations 13-21 and 13-23:

$$\begin{aligned} \text{constant} &= m\omega^2 \\ \omega &= \sqrt{\frac{\text{constant}}{m}} \end{aligned} \qquad (13\text{-}24)$$

This procedure also applies for oscillations in two and three dimensions.

CHECKPOINT 13-4

A Ball Bouncing on a Floor

Consider an ideal situation where a ball is bouncing elastically on a perfectly elastic floor without losing any energy. The motion of the ball is an example of
(a) periodic motion that is also simple harmonic motion
(b) periodic motion that is not simple harmonic motion
(c) simple harmonic motion that is not periodic motion
(d) motion that is neither simple harmonic motion nor periodic motion
Explain your reasoning.

ANSWER: (b) A ball bouncing on a perfectly elastic floor experiences the gravitational force when not in contact with the floor and both the gravitational and the reaction force of the floor when in contact with the floor. Neither of these forces depends on the ball's displacement from the equilibrium position (for example, a point halfway between the floor and the maximum height of the ball from the floor). Therefore, although the motion is periodic (because it repeats after a finite time), it is not simple harmonic motion.

13-3 Uniform Circular Motion and Simple Harmonic Motion

A point moving around a circle with uniform angular speed undergoes periodic motion. When this motion is projected onto the diameter of a circle, the point appears to be moving back and forth in periodic motion. As the point completes one revolution around the circle, its projection onto the diameter completes one oscillation. This analogy between the two motions provides a geometric meaning of the phase constant, ϕ, and the angular frequency, ω.

Consider a point P moving counter clockwise in a circle of radius R with an angular speed of ω rad/s. Let Q be the projection of P onto the x-axis. The time taken by P to complete one revolution (2π rad) is equal to $\frac{2\pi}{\omega}$ s, which is equal to the period of a simple harmonic oscillator of angular frequency ω rad/s. Therefore, the angular frequency of a simple harmonic oscillator is the same as the angular speed of a point moving around a circle.

Suppose at $t = 0$, P makes an angle ϕ with respect to the x-axis, as shown in Figure 13-14(a). A time t later, it undergoes an angular displacement of ωt and makes an angle of ($\omega t + \phi$) with respect to the x-axis. The x-coordinate of P at time t is therefore

$$x(t) = R\cos(\omega t + \phi) \qquad (13\text{-}25)$$

Equation 13-25 shows that as P moves in a circle, its projection onto the x-axis undergoes simple harmonic motion. This can be seen from Figure 13-14, where locations of P at various times t_1, t_2, t_3, … are shown. Figure 13-14(b) shows a plot of the projection of P along the x-axis. We see that $x(t)$ describes simple harmonic motion. Similarly, the projection of P along the y-axis (Figure 13-14(c)) also describes simple harmonic motion. Notice that the two graphs are $\pi/2$ rad out phase because the two axes are perpendicular to each other. The phase constant, ϕ, for simple harmonic motion is analogous to the starting angle of circular motion.

Table 13-1 compares uniform circular motion and simple harmonic motion.

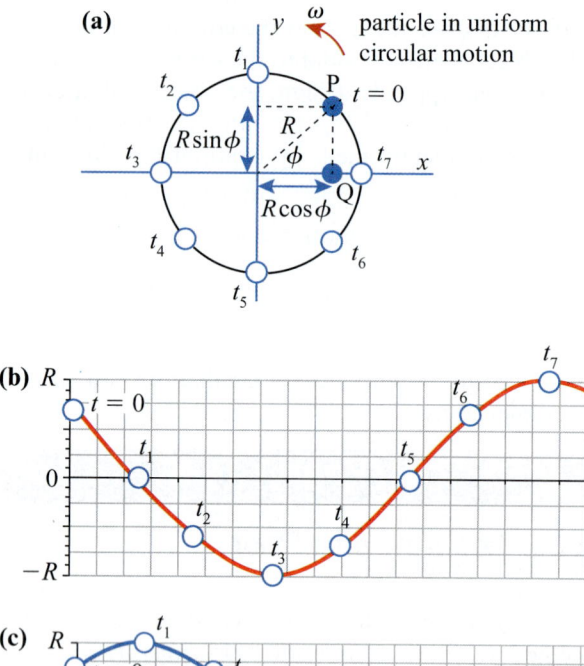

Figure 13-14 (a) A particle moving with angular speed ω in a circle of radius R. At $t = 0$, the particle makes an angle ϕ with respect to the x-axis. (b) projection of particle's position along the x-axis. (c) projection of particle's position along the y-axis.

CHECKPOINT 13-5

Circular Motion and the Phase Constant

Starting positions ($t = 0$) for four particles moving counter-clockwise around a circle of radius R are shown in Figure 13-15. Rank the starting positions in order of increasing phase constant of the corresponding simple harmonic motion.

Lowest ___ ___ ___ ___ Highest

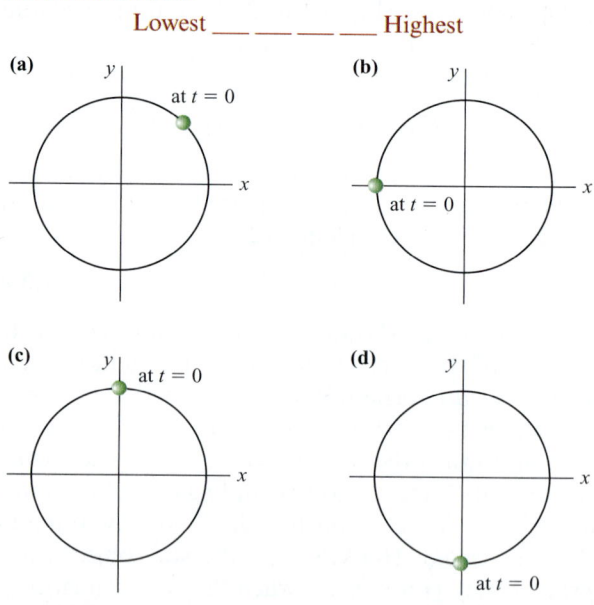

Figure 13-15 Checkpoint 13-5. Starting positions for four particles moving around a circle of radius R.

Table 13-1 Comparison between Uniform Circular Motion and Simple Harmonic Motion

Uniform circular motion	Simple harmonic motion
Radius: R	Amplitude: A
Angular speed: ω	Angular frequency: ω
One complete revolution	One complete oscillation
Time taken for one complete revolution = $2\pi/\omega$	Time period for one oscillation = $2\pi/\omega$
Starting angle: ϕ	Phase constant: ϕ

13-4 Mass-Spring Systems

A Horizontal Mass-Spring System

Figure 13-16 shows a mass, m, attached to one end of a spring of unstretched length L_0 and resting on a horizontal, frictionless surface. The other end of the spring is attached to a rigid support. The mass is restricted to move along the horizontal direction, so gravity does not play any role in its motion. We choose a coordinate system with the origin located at the equilibrium position of the mass (where the spring is neither stretched or compressed) and the positive x-axis toward the right. Suppose a force is exerted on the mass (by pulling on the mass) to displace it toward the positive x-axis by an amount x that does not exceed the elastic limit of the spring. At the displaced position, the change in the length of the spring is x, and the spring exerts an elastic restoring force on the mass, trying to return the mass to the equilibrium position. The elastic restoring force of a spring follows Hooke's law:

$$F_{spring,x} = -kx \qquad (13\text{-}26)$$

The minus sign in Equation 13-26 indicates that the direction of the force exerted by the spring on the mass is opposite to the direction of the displacement of the mass. As long as the mass is held stationary by the external force, the net force on the mass is zero:

$$F_{net,x} = F_{spring,x} + F_{external,x} = 0$$

When the external force is removed, the only force acting on the mass is the elastic restoring force of the spring. Applying Newton's second law of motion,

$$F_{net,x} = F_{spring,x} = -kx = ma_x \qquad (13\text{-}27)$$

$$a_x = -\left(\frac{k}{m}\right)x \qquad (13\text{-}28)$$

The acceleration of the mass is proportional to the displacement of the mass from the equilibrium position and opposite in sign. The mass–spring system will therefore undergo simple harmonic motion, with the displacement of the mass given by the standard equation

$$x(t) = A\cos(\omega t + \phi) \qquad (13\text{-}4)$$

The square of the angular frequency of the motion is obtained by comparing Equations 13-18 and 13-28:

$$\omega^2 = \frac{k}{m}$$

Therefore, for a horizontal mass–spring system,

KEY EQUATION
$$\omega = \sqrt{\frac{k}{m}} \qquad (13\text{-}29)$$

KEY EQUATION
$$T = 2\pi\sqrt{\frac{m}{k}} \qquad f = \frac{1}{2\pi}\sqrt{\frac{k}{m}} \qquad (13\text{-}30)$$

Equation 13-30 tells us that the frequency of oscillation is directly proportional to the stiffness of the spring (k) and is inversely proportional to the mass of the oscillating object. A light mass attached to a stiff spring has a large frequency and, hence, a small period. Note that the frequency of oscillation is independent of the amplitude of motion.

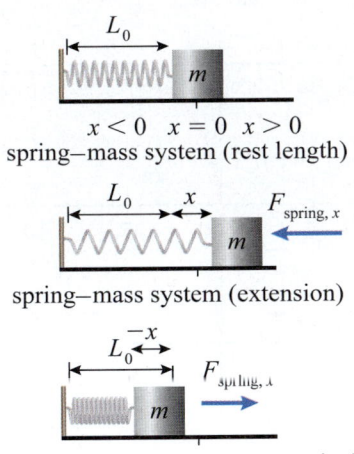

$x < 0 \quad x = 0 \quad x > 0$
spring–mass system (rest length)

spring–mass system (extension)

spring–mass system (compression)

Figure 13-16 A mass attached to a horizontal spring on a frictionless surface.

EXAMPLE 13-2

A Horizontal Mass-Spring System

A 2.00 kg mass, resting on a frictionless table, is attached to a spring of spring constant 20.0 N/m. The spring is stretched 20.0 cm by pulling on the mass, which is held stationary and then released.

(a) Calculate the angular frequency, frequency, and period.
(b) What are the amplitude and the phase constant of oscillations?
(c) What are the maximum speed and maximum acceleration of the mass?
(d) What is the displacement of the mass from equilibrium as a function of time?

Solution

The mass–spring system forms a simple harmonic oscillator, and when the mass is displaced from its equilibrium, it will undergo simple harmonic motion.

We choose a coordinate system with the origin at the equilibrium position of the mass and the positive x-axis toward the right. The initial conditions for this problem are

$$x(0) = 0.200 \text{ m} \quad \text{and} \quad v(0) = 0 \text{ m/s}$$

(a) The angular frequency, frequency, and period of the system are

$$\omega = \sqrt{\frac{20.0 \text{ N/m}}{2.00 \text{ kg}}} = 3.16 \text{ rad/s}$$

$$f = \frac{\omega}{2\pi} = \frac{3.16 \text{ rad/s}}{2\pi \text{ rad}} = 0.503 \text{ Hz}$$

$$T = \frac{1}{f} = 1.99 \text{ s}$$

(b) As the system oscillates, the position and velocity of the mass at any time t are given by

$$x(t) = A \cos(\omega t + \phi)$$
$$v(t) = -\omega A \sin(\omega t + \phi)$$

To determine A and ϕ, we use the given initial position and velocity of the mass. Evaluating $x(t)$ and $v(t)$ at $t = 0$, we obtain

$$x(0) = A \cos\phi \quad \text{(1)}$$
$$v(0) = -\omega A \sin\phi \quad \text{(2)}$$

Dividing both sides of Equation (2) by $-\omega$, squaring and adding it to the square of Equation (1), and using the trigonometric identity $\sin^2\phi + \cos^2\phi = 1$, we can determine A:

$$A = \sqrt{(x(0))^2 + \frac{(v(0))^2}{\omega^2}} \quad \text{(3)}$$

To determine ϕ, divide Equation (2) by Equation (1):

$$\tan\phi = \frac{\sin\phi}{\cos\phi} = -\frac{v(0)}{\omega x(0)} \quad \text{(4)}$$

$$\phi = \tan^{-1}\left(-\frac{v(0)}{\omega x(0)}\right) \quad \text{(5)}$$

Equations (3) and (5) tell us that A and ϕ can be calculated when we know the position and the velocity of an oscillator at $t = 0$.

Inserting the values for $x(0)$ and $v(0)$ into Equations (3) and (5), we get

$$A = \sqrt{(0.200 \text{ m})^2 + \frac{(0 \text{ m/s})^2}{(3.16 \text{ rad/s})^2}} = 0.200 \text{ m}$$

$$\phi = \tan^{-1}\left(-\frac{0 \text{ m/s}}{(3.16 \text{ rad/s}) \times (0.20 \text{ m})}\right) = \tan^{-1} 0 = 0 \text{ rad}$$

(c) The maximum speed of the mass is

$$v_{max} = \omega A = 3.16 \text{ rad/s} \times 0.200 \text{ m} = 0.632 \text{ m/s} \quad \text{(6)}$$

and its maximum acceleration is

$$a_{max} = \omega^2 A = (3.16 \text{ rad/s})^2 \times 0.200 \text{ m} = 2.00 \text{ m/s}^2$$

(d) The displacement of the mass from the equilibrium position at any time t is given by

$$x(t) = (0.200 \text{ m})\cos(3.16t) \quad \text{(7)}$$

Notice that a unique determination of the phase constant requires information about the oscillator's displacement and velocity at the start. A plot of the function $\tan^{-1}\theta$ is shown in Figure 13-17. For any θ there is a single value between $-\pi/2$ and $\pi/2$.

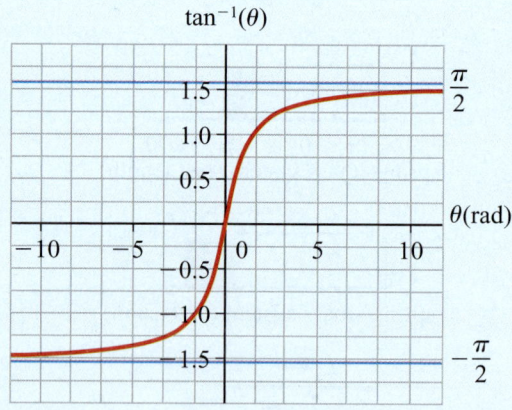

Figure 13-17 A plot of the function $\tan^{-1}\theta$ versus θ, where θ is a real number.

A Vertical Mass-Spring System

Consider a mass, m, suspended vertically from a spring of unstretched length L_0. The hanging mass experiences the elastic restoring forces of the spring and gravity. We take the positive y-axis vertically upward along the spring with $y = 0$ at the equilibrium position of the mass, as shown in Figure 13-18. At the equilibrium position, the spring stretches by the length ΔL such that the elastic restoring force on the mass acting upward balances the force of gravity on the mass acting downward:

$$k(\Delta L) - mg = 0 \qquad (13\text{-}31)$$

Therefore, at the equilibrium position,

$$\Delta L = \frac{mg}{k} \qquad (13\text{-}32)$$

Suppose we displace the mass from the equilibrium position and release it. Would the mass–spring system undergo simple harmonic motion? Let the position of the displaced mass relative to the equilibrium position be y. Compared to the unstretched length L_0, the change in the length of the spring is $\Delta L - y$. Note that y is positive when the mass is above the equilibrium position, that is, when the spring has compressed. When y is negative, the spring has stretched. The net force on the displaced mass is the sum of the force of

gravity and elastic restoring force that could be acting either upward or downward:

$$F_{net, y} = k(\Delta L - y) - mg = k\Delta L - ky - mg$$

Using Equation 13-31 to substitute for ΔL, we get

$$F_{net, y} = -ky \qquad (13\text{-}33)$$

When displaced from the equilibrium position, the net force on the mass is proportional to its displacement and in a direction that is opposite to the displacement. The motion of the mass is therefore simple harmonic motion. Comparing Equation 13-33 to Equation 13-21, we see that

$$m\omega^2 = k$$

and that the angular frequency of oscillation is therefore the same as the angular frequency of a horizontal mass–spring oscillator:

$$\omega = \sqrt{\frac{k}{m}} \qquad (13\text{-}34)$$

$$T = 2\pi\sqrt{\frac{m}{k}} \quad f = \frac{1}{2\pi}\sqrt{\frac{k}{m}} \qquad (13\text{-}35)$$

The displacement of the mass from its equilibrium is given by the standard equation for a simple harmonic oscillator:

$$y(t) = A\cos(\omega t + \phi) \qquad (13\text{-}36)$$

Notice that by choosing to measure the change in the length of the spring with respect to the equilibrium position, we have removed the effect of gravity from this problem. While solving a problem, always think whether a particular choice of coordinate system will make the problem easier to solve.

Equation 13-32 provides a convenient way of measuring the spring constant. Attach one end of a spring to a rigid support, and hang a known mass to its free end. Let the freely hanging mass come to a rest, and measure the change in the length (ΔL) of the spring in this equilibrium position. The spring constant is equal to mg divided by ΔL.

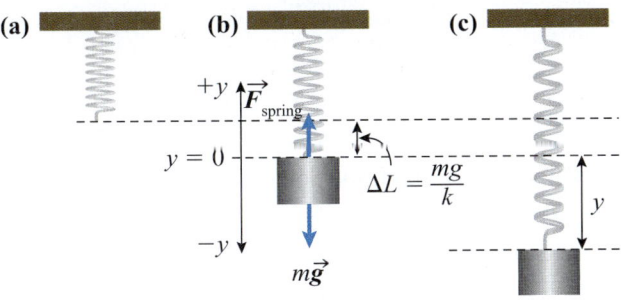

Figure 13-18 A mass attached to a vertically hanging spring. (a) An unstretched spring. (b) A mass, m, hanging free and motionless with the spring, with $y = 0$ taken as the equilibrium position of the mass. (c) The mass displaced from the equilibrium position by an external force.

EXAMPLE 13-3

A Vertical Mass-Spring System

A mass attached to a vertical spring is undergoing simple harmonic motion. A plot of the height of the mass measured with respect to the ground as a function of time is shown in Figure 13-19. The height of the mass at $t = 0$ is 6.3 m.

(a) Find the height of the mass from the ground when the mass is at the equilibrium position.
(b) Find the amplitude and the period.
(c) Write an equation that describes the height of the mass with respect to the ground as a function of time.

Solution

(a) This situation is slightly different in that the height, $h(t)$, of the mass is measured with respect to the ground and not with respect to the equilibrium position of the mass. From the graph, $h(t)$ varies between 6.5 m and 5.5 m. The equilibrium position of the mass is halfway between these two heights. Therefore, the height of the mass from the ground when the mass is in its equilibrium position is 6.0 m.

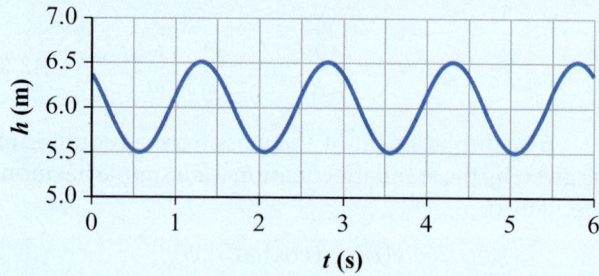

Figure 13-19 Position versus time plot of an oscillating vertical mass–spring system.

(b) From the plot, the period and the amplitude of the oscillations are

$$T = 1.5 \text{ s and } A = 0.50 \text{ m}$$

(c) To write a general equation for $h(t)$, we need to determine the phase constant, ϕ, from the graph. The equation for $h(t)$ can be written as follows:

$$h(t) = (0.50 \text{ m})\cos\left(\frac{2\pi}{1.5}t + \phi\right) + 6.0 \text{ m} \qquad (1)$$

Note that we have added the equilibrium height of the mass to the standard equation for a simple harmonic oscillator. Equation (1) describes a **displaced harmonic oscillator**, where the origin of the coordinate system is not located at the equilibrium position of the oscillator.

To determine ϕ, we note from the graph that $h(0)$ is approximately 6.3 m. Inserting this value and $t = 0$ into the equation for $h(t)$ gives

$$h(0) = (0.50 \text{ m}) \cos\phi + 6.0 \text{ m} = 6.3 \text{ m}$$

Therefore,

$$0.50\cos\phi = 0.30$$
$$\cos\phi = 0.60$$
$$\phi = \cos^{-1}0.60 \approx +0.93 \text{ rad}$$
$$\text{or } -0.93 \text{ rad}$$

The correct value of the phase constant can be determined from Figure 13-19. Choose a point, for example, at $t = 0.5$ s. From the plot, we note that $h(0.50) = 5.5$ m. Now, evaluate $h(0.50)$ from Equation (1) using both values of ϕ:

For $\phi = +0.93$ rad:

$$h(0.50) = (0.50 \text{ m})\cos\left(\frac{2\pi}{1.5} \times 0.50 + 0.93\right) + 6.0 \text{ m}$$
$$= 5.5 \text{ m}$$

For $\phi = -0.93$ rad:

$$h(0.50) = (0.50 \text{ m})\cos\left(\frac{2\pi}{1.5} \times 0.50 - 0.93\right) + 6.0 \text{ m}$$
$$= 6.2 \text{ m}$$

Thus, $\phi = 0.93$ rad is the correct value for the phase constant because it reproduces the given value for $h(0.50)$. The general equation for $h(t)$ is

$$h(t) = (0.50 \text{ m})\cos\left(\frac{2\pi}{1.5}t + 0.93\right) + 6.0 \text{ m}$$

Making sense of the result

Consider a point on the graph in Figure 13-19, for example, at $t = 2.0$ s. Using the above equation for $h(t)$ at $t = 2.0$ s, we get

$$h(2.0) = (0.50 \text{ m})\cos\left(\frac{2\pi}{1.5} \times 2 + 0.93\right) + 6.0 \text{ m} = 5.5 \text{ m}$$

which agrees with the value from the graph.

13-5 Energy Conservation in Simple Harmonic Motion

In the absence of frictional forces, the total energy (E) of an oscillating horizontal mass–spring system is the sum of the kinetic energy (K) of the mass and the elastic potential energy (U) of the spring (Figure 13-20).

Suppose that the mass is pulled from its equilibrium position to $x = A$ and held stationary. In doing so, work has been done to stretch the spring by a length A. This

work is stored as the elastic potential energy of the spring. The total energy of the mass–spring system at $x = A$ is

$$E = K + U$$
$$= 0 + \frac{1}{2}kA^2 \text{ (total energy at } x = A) \qquad (13\text{-}37)$$

where we have used the fact that the work done in stretching (or compressing) a spring by a length x is equal to $\frac{1}{2}kx^2$. When the mass is released, it accelerates toward the equilibrium position due to the elastic restoring force

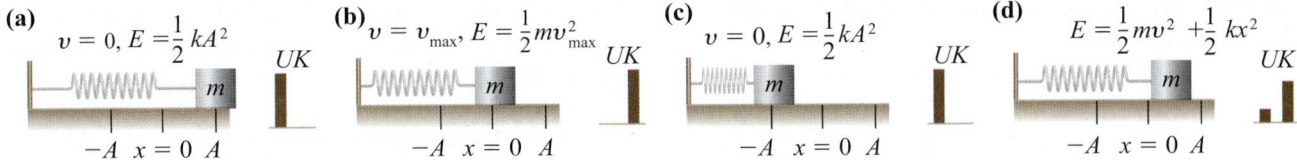

(a) $v = 0$, $E = \frac{1}{2}kA^2$

(b) $v = v_{max}$, $E = \frac{1}{2}mv_{max}^2$

(c) $v = 0$, $E = \frac{1}{2}kA^2$

(d) $E = \frac{1}{2}mv^2 + \frac{1}{2}kx^2$

Figure 13-20 A mass–spring system oscillating in a straight line on a horizontal, frictionless surface. (a) The spring is fully stretched, and the mass is at rest. (b) The spring is unstretched, and the mass is passing through the equilibrium position with maximum speed. (c) The spring is fully compressed, and the mass is at rest. (d) The spring is not fully stretched, and the mass has a nonzero speed.

of the spring. Therefore, its velocity increases, and the length of the spring decreases. Let $x(t)$ be the position of the mass and $v(t)$ its velocity at some time t. At this instant, the kinetic energy of the mass is $\frac{1}{2}mv^2(t)$, and the elastic potential energy of the spring is $\frac{1}{2}kx^2(t)$. The total energy of the mass–spring system is

$$E = \frac{1}{2}mv^2(t) + \frac{1}{2}kx^2(t) \qquad (13\text{-}38)$$

Equating Equations 13-38 and 13-37, we get

$$\frac{1}{2}kA^2 = \frac{1}{2}mv^2(t) + \frac{1}{2}kx^2(t) \qquad (13\text{-}39)$$

At the equilibrium position, the spring is unstretched and the total energy of the system is stored as the kinetic energy of the mass. So the mass must move with maximum speed as it passes through the equilibrium position. Let v_m denote the maximum speed of the mass. Then, at equilibrium,

$$E = \frac{1}{2}mv_m^2 \quad \text{(total energy at } x = 0) \qquad (13\text{-}40)$$

We can express v_m in terms of the maximum displacement A by using Equations 13-37 and 13-40:

$$E = \frac{1}{2}kA^2 = \frac{1}{2}mv_m^2$$

$$v_m = \sqrt{\frac{k}{m}}A = \omega A \qquad (13\text{-}41)$$

This expression for maximum speed is the same as Equation 13-13, which was derived previously. We see that the relationship between the amplitude and the maximum speed of a simple harmonic oscillator is a consequence of the conservation of energy.

Because of inertia, the mass overshoots the equilibrium position and continues to move toward the left, compressing the spring. The resulting restoring force slows the mass until it comes to rest at $x = -A$, where all the energy is again stored as elastic potential energy in the spring:

$$E = K + U = 0 + \frac{1}{2}k(-A)^2$$

$$= \frac{1}{2}kA^2 \quad \text{(total energy at } x = -A) \qquad (13\text{-}42)$$

This continuous transformation of energy between different modes (kinetic energy and elastic potential energy) is a fundamental property of all harmonic motions. In the absence of frictional forces, a harmonic oscillator, once set in motion, will oscillate forever. However, in reality, frictional forces are always present and oscillations eventually stop.

Figure 13-21 shows a plot of the potential, kinetic, and total energy of the mass–spring oscillator. For any position between $-A$ and $+A$, the values of U and K can be read from the corresponding curves. Note that the sum of U and K is always equal to $\frac{1}{2}kA^2$. Using Equation 13-39, we can find the velocity of the mass for a given location x:

$$v = \pm\sqrt{\frac{k}{m}(A^2 - x^2)} \qquad (13\text{-}43)$$

(a)

(b)

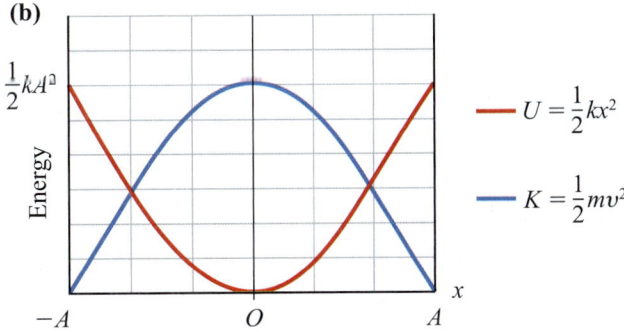

$U = \frac{1}{2}kx^2$

$K = \frac{1}{2}mv^2$

Figure 13-21 (a) A plot of the kinetic energy (K) and elastic potential energy (U) of a simple harmonic oscillator as a function of time. (b) A plot of K and U as a function of displacement.

Similarly, for a given value of velocity, the location of the mass is given by

$$x = \pm\sqrt{A^2 - \frac{m}{k}v^2} \qquad (13\text{-}44)$$

The variation of kinetic and potential energy with time can be determined by using

$$x(t) = A\cos(\omega t + \phi)$$
$$v(t) = -\omega A\sin(\omega t + \phi)$$

Then,

$$U = \frac{1}{2}kx^2(t) = \frac{1}{2}kA^2\cos^2(\omega t + \phi)$$

$$K = \frac{1}{2}mv^2(t) = \frac{1}{2}m\omega^2 A^2\sin^2(\omega t + \phi) \qquad (13\text{-}45)$$
$$= \frac{1}{2}kA^2\sin^2(\omega t + \phi)$$

Note that, individually, both U and K vary with time. However, the sum of the two is always equal to the total energy, E.

$$U + K = \frac{1}{2}kA^2\cos^2(\omega t + \phi) + \frac{1}{2}kA^2\sin^2(\omega t + \phi)$$
$$= \frac{1}{2}kA^2(\cos^2(\omega t + \phi) + \sin^2(\omega t + \phi)) \qquad (13\text{-}46)$$
$$= \frac{1}{2}kA^2$$

The total energy of a mass–spring oscillator is proportional to the square of the amplitude. Doubling the amplitude requires increasing the energy of the oscillator by a factor of 4.

EXAMPLE 13-4

Energy of a Mass-Spring System

A mass–spring oscillator consists of a 1.00 kg mass and a spring with a spring constant of 200 N/m. The oscillator is set in motion by stretching the spring 0.300 m and releasing the mass from rest.

(a) What is the total energy of the mass–spring system?
(b) What is the maximum speed of the mass as it oscillates?
(c) What is the speed of the mass when it is 0.100 m from the equilibrium position?
(d) For which displacements of the mass is the kinetic energy half its total energy?

Solution

We choose a coordinate system with the equilibrium position at the origin and the positive x-axis toward the right. The mass is released from rest at $t = 0$. Therefore,

$$x(0) = 0.300 \text{ m} \quad \text{and} \quad v(0) = 0.00 \text{ m/s}$$

(a) The total energy (E) is the sum of the kinetic energy (K) of the mass and the elastic potential energy (U) of the spring. Using the values of displacement and velocity at $t = 0$,

$$K = \frac{1}{2}mv(0)^2 = 0.00 \text{ J}$$

$$U = \frac{1}{2}kx(0)^2 = \frac{1}{2} \times (200 \text{ N/m}) \times (0.300 \text{ m})^2 = 9.00 \text{ J}$$

Therefore, $E = K + U = 9.00$ J.

(b) When the mass is moving with its maximum speed, the total energy of the oscillator is in the form of the kinetic energy of the mass. From Equation 13-41:

$$v_m = \sqrt{\frac{2E}{m}} = \sqrt{\frac{2 \times 9.00 \text{ J}}{1.00 \text{ kg}}} = 4.24 \text{ m/s}$$

(c) The total energy of the mass–spring oscillator is given by Equation 13-38. Rearranging this equation to solve for v and then evaluating for $x = 0.100$ m, we get

$$E = \frac{1}{2}mv^2 + \frac{1}{2}kx^2$$

$$v = \sqrt{\frac{2E}{m} - \frac{k}{m}x^2}$$

$$v = \sqrt{\frac{2 \times 9.00 \text{ J}}{1.00 \text{ kg}} - \frac{200 \text{ N/m}}{1.00 \text{ kg}}(0.100 \text{ m})^2}$$

$$= 4.00 \text{ m/s}$$

(d) When the kinetic energy of the mass is 4.50 J, the potential energy of the spring must also equal 4.50 J, since the sum of the two must equal 9.00 J. Therefore,

$$4.50 \text{ J} = \frac{1}{2}kx^2$$

$$x = \pm\sqrt{\frac{2 \times 4.50 \text{ J}}{200 \text{ N/m}}} = \pm 0.212 \text{ m}$$

Making sense of the result

The speed calculated in part (c) when the mass is not at the equilibrium position is less than the maximum speed of 4.24 m/s, and the displacement calculated in part (d) is less than the amplitude of oscillation.

13-6 The Simple Pendulum

A **simple pendulum** consists of a point mass attached to a string of negligible mass that does not stretch. The other end of the string is tied to a rigid and frictionless support. When displaced from the vertical equilibrium position and released, the pendulum swings back and forth in a plane under the influence of gravity, which acts as the restoring force. The motion of the pendulum is periodic. Is the motion simple harmonic?

Figure 13-22 shows a mass, m, hanging from a string of length L and making an angle of θ with respect to the equilibrium position ($\theta = 0$). Angular displacement is taken to be positive to the right of the

equilibrium position and negative to the left. The displacement, s, along the arc of the circle is related to the angular displacement as

$$s = L\theta \qquad (13\text{-}47)$$

The forces acting on the mass are $m\vec{g}$, the force of gravity, acting vertically downward, and $\vec{T}_s$, the tension in the string. To analyze the pendulum's motion, we choose a coordinate system with two axes. One is called the radial axis, along the length of the string. The other, called the tangential axis, is tangent to the circular motion of the mass. Note that the radial and tangential axes are perpendicular to each other and their directions change as the mass oscillates. This choice of coordinate system simplifies the analysis because the mass does not move along the radial axis.

The radial axis The force of gravity, $m\vec{g}$, makes an angle of θ with the radial axis. Therefore, it's component along the radial axis is $mg\cos\theta$, and it points away from the suspension point. The tension, $\vec{T}_s$, is directed toward the suspension point. The radial components of the two forces supply the required centripetal acceleration to keep the mass moving in a circular arc of radius L, Therefore,

$$T_s - mg\cos\theta = \frac{mv^2}{L}$$

$$T_s = mg\cos\theta + \frac{mv^2}{L} \qquad (13\text{-}48)$$

Here v is the speed of the oscillating mass. As we will see later in this section, v is zero at the turning points of the motion and is maximum when the mass passes through the equilibrium position.

The tangential axis The tangential component of the force of gravity on the mass is $mg\sin\theta$, and it provides the restoring force that tends to pull the mass toward the equilibrium position. Newton's second law of motion for the tangential component is given by

$$F_{\text{net},t} = -mg\sin\theta = ma_t \qquad (13\text{-}49)$$

where a_t denotes the tangential component of the acceleration. Since $\theta = s/L$,

$$a_t = -g\sin\theta = -g\sin\left(\frac{s}{L}\right) \qquad (13\text{-}50)$$

Note that the acceleration is not proportional to the displacement, s, but to sin (s/L). Therefore, in general, pendulum motion is not simple harmonic motion.

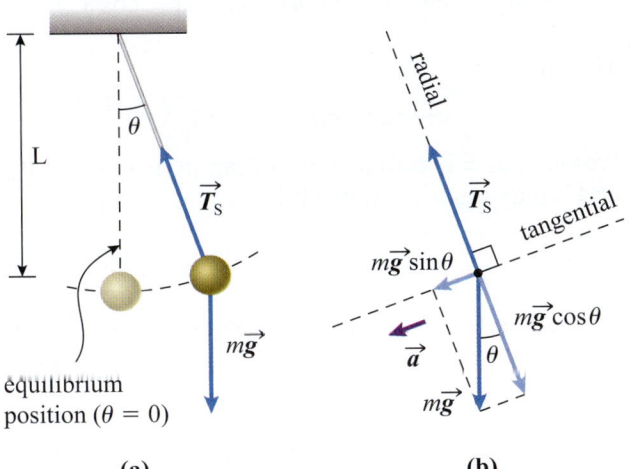

(a)   (b)

Figure 13-22 (a) The simple pendulum. The pendulum swings about the equilibrium position ($\theta = 0$) between angular displacements $\pm\theta_m$. (b) Trigonometric relations. (c) Resolution of force of gravity and tension forces along the radial and tangential directions.

However, if the angle is small, we can use the small-angle approximation for the sine function:

$$\sin x = x \quad \text{(small-angle approximation)} \quad (13\text{-}51)$$

Figure 13-23 shows a plot of $\sin x$ versus x. We see that the small-angle approximation is reasonable for $x \approx 0.2$ rad (approximately 12°).

Using the small-angle approximation,

$$\sin(s/L) = s/L \quad (13\text{-}52)$$

Therefore,

$$a_t = -\left(\frac{g}{L}\right)s \quad \text{(small-angle approximation)} \quad (13\text{-}53)$$

For small angular displacements, the acceleration of the mass is proportional to the displacement and opposite in sign. Therefore, the motion of the pendulum is simple harmonic motion. The angular frequency is obtained by comparing Equation 13-53 with the standard equation for simple harmonic motion:

KEY EQUATION

$$\omega = \sqrt{\frac{g}{L}} \quad (13\text{-}54)$$

The period and the frequency of oscillation of a pendulum are given by

$$T = \frac{2\pi}{\omega} = 2\pi\sqrt{\frac{L}{g}}$$

$$f = \frac{1}{T} = \frac{1}{2\pi}\sqrt{\frac{g}{L}} \quad (13\text{-}55)$$

Notice that the period of a simple pendulum depends on the length of the pendulum and the acceleration due to gravity. The value of g varies slightly across Earth's surface and with height from sea level, so the period of a pendulum of fixed length will vary slightly with the

value of g at a given location. Time can be measured fairly accurately with a simple stopwatch, so we can use a pendulum to measure the variation in the value of g across Earth's surface.

Energy Conservation for a Simple Pendulum

As a pendulum swings, its energy continuously converts between kinetic energy and gravitational potential energy. At the maximum angular displacements ($\pm\theta_m$), the pendulum is momentarily stationary and its energy is entirely in the form of gravitational potential energy. At this displacement the mass is lifted by a height $L(1 - \cos\theta_m)$ with respect to the equilibrium position. Therefore, its total energy is

$$E = mgL(1 - \cos\theta_m) \quad (13\text{-}56)$$

As the angular position decreases from its maximum value, some of the gravitational potential energy is converted into kinetic energy. When the pendulum passes through the equilibrium position, all of its energy is in the form of the kinetic energy and its maximum speed (v_{max}) is given by

$$\frac{1}{2}mv_{max}^2 = mgL(1 - \cos\theta_m) \quad (13\text{-}57)$$

Note that the maximum speed of a simple pendulum is independent of its mass.

Let v be the speed of the mass when the string extends an angle θ with respect to the equilibrium position. Then, by conservation of energy, the sum of the kinetic energy and the gravitational potential energy of the mass must be equal to its total energy, E:

$$\frac{1}{2}mv^2 + mgL(1 - \cos\theta) = mgL(1 - \cos\theta_m)$$

Therefore,

$$v^2 = 2gL(\cos\theta - \cos\theta_m)$$

We see that v is zero at the turning points ($\theta = \pm\,\theta_m$) and is maximum when $\theta = 0$.

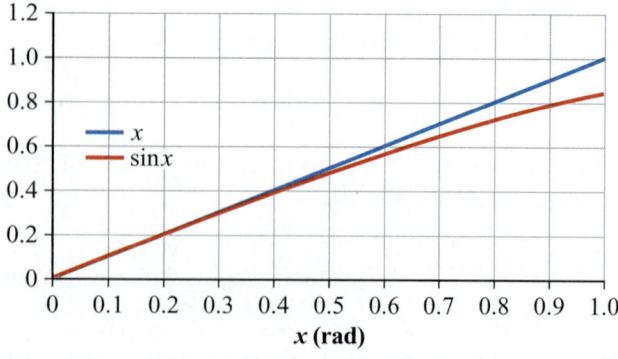

1.2, 1.0, 0.8, 0.6, 0.4, 0.2, 0 (y-axis)

— x
— $\sin x$

0, 0.1, 0.2, 0.3, 0.4, 0.5, 0.6, 0.7, 0.8, 0.9, 1.0

x **(rad)**

Figure 13-23 A plot of $\sin x$ and x versus x (in rad).

WAVES AND OSCILLATIONS

EXAMPLE 13-5

Period and Length of a Simple Pendulum

You need to design a simple pendulum that has a period of 1.00 s. The acceleration due to gravity in your lab is 9.81 m/s². What should the length of the string be?

Solution

The period of a simple pendulum is given by

$$T = 2\pi\sqrt{\frac{L}{g}}$$

Solving for L, we get

$$L = \frac{gT^2}{4\pi^2}$$

$$L = \frac{9.81 \text{ m/s}^2 \times (1.00 \text{ s})^2}{4\pi^2} = 2.48 \times 10^{-1} \text{ m}$$

CHECKPOINT 13-8

Decreasing a Pendulum's Length

A pendulum of length L is undergoing simple harmonic motion. When the length of the string is halved, the period of the pendulum

(a) increases by a factor of $\sqrt{2}$
(b) increases by a factor of 2
(c) decreases by a factor of $\sqrt{2}$
(d) decreases by a factor of 2

ANSWER: (c)

13-7 The Physical Pendulum

An extended object that, when displaced from its equilibrium position, oscillates about an axis is called a **physical pendulum**. This includes objects that oscillate about a pivot point, such as a sphere attached to the end of a rod or a string. Swings and baseball bats can also be considered as physical pendulums.

Consider a sphere of uniform density, radius R, and mass M attached to a rod of length L and negligible mass. The system is free to rotate about an axis that is perpendicular to the plane of the paper and passes through the other end of the rod at the pivot point P (Figure 13-24). Since the rod is assumed to be massless, the centre of mass, C, of the rod–sphere system is located at the centre of the sphere. In the equilibrium

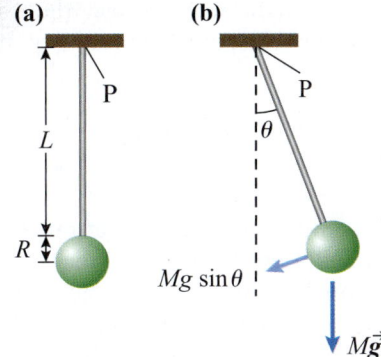

Figure 13-24 A sphere of mass M and radius R is attached to a massless rod of length L, which is hanging from a fixed, frictionless pivot point P. The rod is free to rotate about P. (a) The equilibrium position. (b) The rod is displaced by an angle θ from the equilibrium position.

position, C is directly below the pivot point P, and there is no torque on the sphere due to the force of gravity.

The sphere is then displaced by an angle θ. In the displaced position (Figure 13-24(b)), the gravitational force exerts a torque $\vec{\tau}$ on the sphere about the pivot point, P. The direction of the torque is perpendicular to the plane of the paper, and the magnitude of the torque is given by

$$\tau = -(Mg)(L_{cm})\sin\theta \tag{13-58}$$

Here, $L_{cm} = L + R$ is the distance between the pivot point and the centre of mass of the rod–sphere pendulum. The minus sign indicates that the direction of the torque is opposite to the direction of the angular displacement, θ. Therefore, the torque tends to rotate the sphere toward the equilibrium position. According to Newton's second law of motion for rotational motion, the magnitude of the torque on the sphere is equal to the product of the magnitude of the angular acceleration, α, of the sphere and its moment of inertia, I_P, about the axis passing through the pivot point, P. Therefore,

$$I_P\alpha = -MgL_{cm}\sin\theta$$

$$\alpha = -\left(\frac{MgL_{cm}}{I_P}\right)\sin\theta \tag{13-59}$$

By the parallel-axis theorem (see Section 8-5), the sphere's moment of inertia about the axis through P is

$$I_P = I_{cm} + M(L_{cm})^2$$

where $I_{cm} = \frac{2}{5}MR^2$ is the moment of inertia of the sphere about its centre of mass. When the angular displacement is small, we can use the small-angle approximation $\sin\theta = \theta$, in which case Equation 13-59 becomes

$$\alpha = -\left(\frac{MgL_{cm}}{I_P}\right)\theta \qquad (13\text{-}60)$$

This relationship between the angular acceleration, α, and the angular displacement, θ, is similar to the relationship between the acceleration and displacement of a simple harmonic oscillator. The angular acceleration is proportional to the angular displacement and opposite in sign. Therefore, for small angular displacement, the sphere will undergo simple harmonic motion about the pivot point, P. The angular frequency of oscillation is the square root of the constant of proportionality between α and θ:

KEY EQUATION

$$\omega = \sqrt{\frac{MgL_{cm}}{I_P}} \qquad (13\text{-}61)$$

The period is

$$T = 2\pi\sqrt{\frac{I_P}{MgL_{cm}}} \qquad (13\text{-}62)$$

Inserting the expressions for I_P and I_{cm}, the period for a rod–sphere physical pendulum is

$$T = 2\pi\sqrt{\frac{\frac{2}{5}MR^2 + M(L+R)^2}{Mg(L+R)}}$$

Notice that for $R = 0$, the above expression reduces to $T = 2\pi\sqrt{\dfrac{L}{g}}$, the period for a simple pendulum. The angular displacement of the oscillating sphere is described by an equation analogous to that for linear oscillation:

$$\theta(t) = \theta_m\cos(\omega t + \phi) \qquad (13\text{-}63)$$

Here, θ_m is the maximum angular displacement of the sphere and is the analogue of the amplitude A for linear oscillations.

The moment of inertia, I_P, in Equation 13-62 is calculated with respect to the pivot point, P, and a particular axis of rotation. If either or both of these change, I_P and hence ω and T will change. By comparing Equations 13-62 and 13-55, we notice that a physical pendulum and a simple pendulum have the same time period when

$$L = \frac{I_P}{ML_{cm}} \qquad (13\text{-}64)$$

EXAMPLE 13-6

A Cardboard Sheet as a Physical Pendulum

Consider a square sheet of cardboard with a mass of 20.0 g and a length of 28.0 cm. The sheet is free to rotate about a pivot point, P, that is located directly above the centre of mass of the sheet and 1.00 cm from the top edge. Calculate the period for small angular displacement about the pivot point, P.

Solution

A sheet of cardboard that freely rotates about a frictionless pivot point forms a physical pendulum. The period of the sheet is given by Equation 13-62. The sheet's moment of inertia about the point P is

$$I_p = I_{cm} + ML_{cm}^2$$

The moment of inertia of a square sheet of mass M and length L about its centre of mass is

$$I_{cm} = \frac{1}{12}ML^2$$

The distance L_{cm} is the distance from the centre of the mass of the sheet to the pivot point, so

$$L_{cm} = 14.0\ \text{cm} - 1.00\ \text{cm} = 13.0\ \text{cm}$$

Using the given values for the mass and the length of the sheet, we get

$$\begin{aligned}I_P = &\frac{1}{12}\times(0.0200\ \text{kg})\times(0.280\ \text{m})^2 + (0.0200\ \text{kg})\\ &\times(0.130\ \text{m})^2 \\ = &\ 4.69\times10^{-4}\ \text{kg}\cdot\text{m}^2\end{aligned}$$

Substituting into Equation 13-62 gives

$$T = 2\pi\sqrt{\frac{4.69\times10^{-4}\ \text{kg m}^2}{(0.0200\ \text{kg})\times(9.81\ \text{m/s}^2)\times(0.130\ \text{m})}} = 0.852\ \text{s}$$

Making sense of the result

For a simple pendulum of length 28 cm, the period is

$$T = 2\pi\sqrt{\frac{0.28\ \text{m}}{9.81\ \text{m/s}^2}} = 1.1\ \text{s}$$

which is of the same order of magnitude as the above value.

Walking Motion and the Physical Pendulum

Walking motion can be compared to pendulum motion. The stance foot stays on the ground, and the free leg swings through from behind ($\theta = -\theta_0$) to the front of the stance foot ($\theta = +\theta_0$). For most animals, the maximum deflection angle (Figure 13-25) is typically 20°. The motion of the free leg can be approximated by a regular physical pendulum and that of the stance leg by an inverted physical pendulum. An inverted pendulum has its pivot at the bottom and the oscillating mass distributed at the top. An example of an inverted pendulum is an athlete attempting a pole vault jump. The motions of the free and the stance legs are coupled because the stance leg needs the free leg to complete its motion before the roles of the two legs are switched. So walking motion can be considered as the motion of two coupled pendula (one the regular pendulum and the other an inverted pendulum). In the laboratory frame of reference, these coupled pendula move forward with a speed v_{walk}.

This description of walking motion is a bit complicated, but we can study its main features (the walking speed and the frequency of motion) by modelling it as the motion of a regular physical pendulum. After all, physics is about constructing simple models of natural phenomena.

The period of a physical pendulum of length L, mass M, and moment of inertia I is (see Equation 13-62)

$$T = 2\pi\sqrt{\frac{I}{MgL_{\text{cm}}}}$$

where L_{cm} is the distance of the centre of mass of the pendulum from the pivot point, and I is calculated about an axis that is perpendicular to the plane of rotation and passes through the pivot point. Let us consider a simple model, where a leg is approximated as a rod, and the mass of the leg is uniformly distributed along the length of the rod. This model is reasonable for animals with legs that do not taper much, for example, humans, elephants, and ostriches. The moment of inertia for a rod of mass M and length L that is free to rotate about an axis passing through an end point of the rod is $\frac{1}{3}ML^2$. The centre of mass of the rod is at its centre, so $L_{\text{cm}} = L/2$. Therefore, the period of the rod is

$$T = 2\pi\sqrt{\frac{\frac{1}{3}ML^2}{Mg\left(\frac{L}{2}\right)}} = \sqrt{\frac{2}{3}} \times \left(2\pi\sqrt{\frac{L}{g}}\right) \quad (13\text{-}65)$$

The amplitude of oscillation of the rod is given by

$$A = L\sin\theta_0, \quad (13\text{-}66)$$

where θ_0 is the maximum angle of deflection. The maximum walking speed is, therefore,

$$v_{\text{walk}} = \omega A = \frac{2\pi}{T} \times L\sin\theta_0 = \sqrt{\frac{3}{2}}\sin\theta_0 \times \sqrt{gL} \quad (13\text{-}67)$$

Taking $L = 1.0$ m as the average length of an adult human leg and $\theta_0 = 20°$, we obtain $v_{\text{walk}} = 1.3$ m/s ≈ 4.7 km/h, which is a reasonable result for the maximum walking speed of an adult.

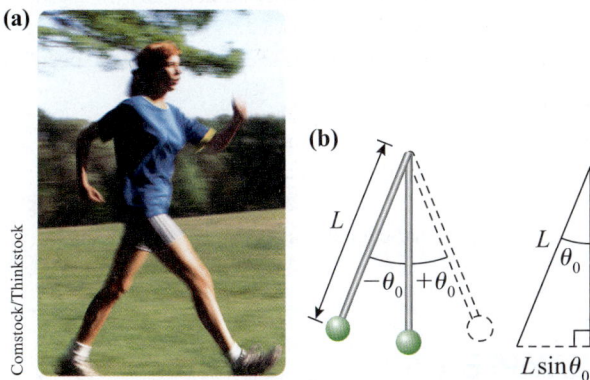

(a)

(b)

Figure 13-25 (a) Free-walking motion of a person. (b) A schematic representation of free-walking motion, where a physical pendulum of length L oscillates between $\theta = -\theta_0$ and $\theta = +\theta_0$.

Several interesting points emerge from this simple model:

- The walking speed is proportional to $\sqrt{g}$. On the Moon, $g = 1.6$ m/s². Using the numbers for leg length and displacement angle as above, we obtain the value $v_{\text{walk}} = 0.5$ m/s $\approx$ 1.8 km/h, which is approximately one-third the walking speed on Earth. Have you ever watched a video of astronauts on the Moon and wondered why they were walking so slowly?

- The walking speed depends on the moment of inertia of the leg. Fast animals (e.g., horses and lions) have legs that are thick at the top and slender at the bottom. A greater fraction of the leg's mass for these animals is located closer to the hip joint, thus decreasing the moment of inertia of the leg about that joint. This distribution decreases the period of the leg's oscillations, resulting in quicker steps and a faster walk.

- As a leopard chases its prey, it hunches down and pounds hard on the ground with its feet (Figure 13-26). By hunching down, the leopard decreases its moment of inertia, and by pounding hard, it increases its effective g as it accelerates upward. Both effects decrease the period of the legs, thus helping it to run faster.

- The walking speed is proportional to the square root of the length of the leg. Animals with longer legs have a faster natural walking speed. The walking speed of a child with legs half the length of an adult is reduced by a factor of $\sqrt{1/2}$, or approximately 30%, compared to that of the adult.

Figure 13-26 A sprinting leopard.

Two animals have similarly shaped legs of the same length. One animal's legs are twice as thick as the other animal's legs. Which animal has the faster walking speed?

(a) the animal with thicker legs
(b) the animal with thinner legs
(c) neither; they both have the same walking speed
(d) This question cannot be answered without knowing the leg mass for each animal.

ANSWER: (b)

13-8 Time Plots for Simple Harmonic Motion

The period, amplitude, and phase constant of a simple harmonic motion can also be determined from the graph of the displacement, the velocity, or the acceleration as a function of time. In this section, we show how this is done through two examples.

EXAMPLE 13-7

Determining an Equation for Simple Harmonic Motion from a Displacement Graph

Figure 13-27 shows a graph of displacement $x(t)$ versus time t for a particle. From this graph, determine an equation for $x(t)$.

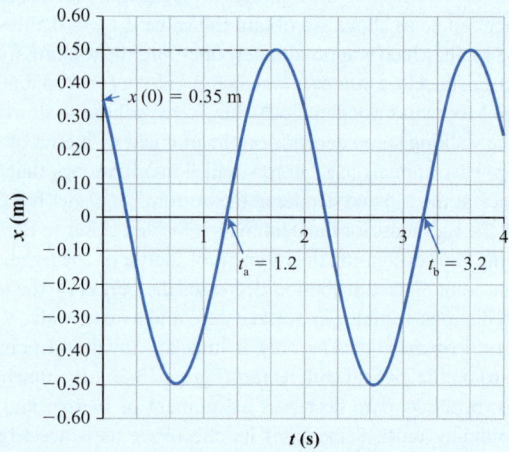

Figure 13-27 The displacement versus time graph for a particle.

Solution

We can see from Figure 13-27 that the graph is a sinusoidal function. Therefore, the motion is simple harmonic motion. We can also see that the maximum displacement of the particle is ± 0.50 m. Therefore, the amplitude of the oscillation is 0.50 m.

We can determine the period by calculating the time taken to complete one oscillation. While any point on the curve can be taken as the starting point, using points where the displacement is zero is often convenient. Using t_a and t_b as the start and end times of an oscillation, the period is

$$T = t_b - t_a = 3.2 \text{ s} - 1.2 \text{ s} = 2.0 \text{ s}$$

The angular frequency is

$$\omega = \frac{2\pi}{T} = \frac{2\pi \text{ rad}}{2.0 \text{ s}} = 3.1 \text{ rad/s}$$

To determine the phase constant, ϕ, we write the standard equation for $x(t)$ with $A = 0.5$ m and $\omega = 3.1$ rad/s:

$$x(t) = (0.50 \text{ m})\cos(3.1t + \phi) \qquad (1)$$

From the graph, $x(0)$ is approximately 0.35 m. Substituting this value into Equation (1) gives

$$x(0) = 0.35 \text{ m} = (0.50 \text{ m})\cos\phi$$

$$\cos\phi = \frac{0.35 \text{ m}}{0.50 \text{ m}} = 0.70$$

$$\phi = \cos^{-1}(0.70) = 0.80 \text{ rad} \quad \text{or} \quad -0.80 \text{ rad}$$

To determine the correct sign of ϕ, we need another piece of information, such as the velocity of the particle at $t = 0$ or its position at a time near $t = 0$. Let us choose $t = 0.33$ s. From the graph, $x(0.33) = -0.10$ m, approximately. Calculating $x(0.33)$ from Equation (1) for $\phi = 0.80$ rad and $\phi = -0.80$ rad,

$\phi = +0.80$ rad:
$$x(0.33) = (0.50 \text{ m})\cos(3.1 \times 0.33 + 0.80) = -0.12 \text{ m}$$

$\phi = -0.80$ rad:
$$x(0.33) = (0.50 \text{ m})\cos(3.1 \times 0.33 - 0.80) = 0.48 \text{ m}$$

Comparing these two values with the value from the graph, we see that $+0.80$ rad is the correct value for the phase constant. Thus, the equation for the displacement shown in Figure 13-27 is

$$x(t) = (0.50 \text{ m})\cos(3.1t + 0.80) \qquad (2)$$

Making sense of the result

With the calculated constants, at $t = 2.0$ s the displacement of the oscillator is

$$x(2.0) = (0.50 \text{ m})\cos(3.1 \times 2.0 + 0.80) = 0.38 \text{ m}$$

which approximately agrees with the value from the graph.

EXAMPLE 13-8

Determining an Equation for Simple Harmonic Motion from Position and Velocity Graphs

Position and velocity graphs of a simple harmonic oscillator are shown in Figure 13-28. Determine the equation of motion for the oscillator from these graphs.

Solution

From the position graph, the amplitude of the oscillations is $A = 0.40$ m, and the period is approximately $T = 4.2$ s, which gives the following value for angular frequency:

$$\omega = \frac{2\pi \text{ rad}}{4.2 \text{ s}} = 1.5 \text{ rad/s}$$

The equations for $x(t)$ and $v(t)$ are

$$x(t) = (0.40 \text{ m}) \cos(1.5t + \phi)$$
$$v(t) = -(1.5 \times 0.40 \text{ m/s}) \sin(1.5t + \phi) \quad \text{(1)}$$

We can use Equation (1) together with the graphs to determine ϕ. From the plots, $x(0) = 0.28$ m and $v(0) = -0.42$ m/s. Evaluating $x(t)$ and $v(t)$ for $t = 0$, we get

$$x(0) = (0.40 \text{ m}) \cos\phi = 0.28 \text{ m}$$
$$v(0) = -(0.60 \text{ m/s}) \sin\phi = -0.42 \text{ m/s}$$

Thus,

$$\cos\phi = 0.70 \text{ and } \sin\phi = 0.70$$

and

$$\tan\phi = \frac{\sin\phi}{\cos\phi} = 1.0 \Rightarrow \phi = \tan^{-1}(1.0) = 0.79 \text{ rad}$$

Thus, the equation of motion for the oscillator is

$$x(t) = (0.40 \text{ m}) \cos(1.5t + 0.79)$$

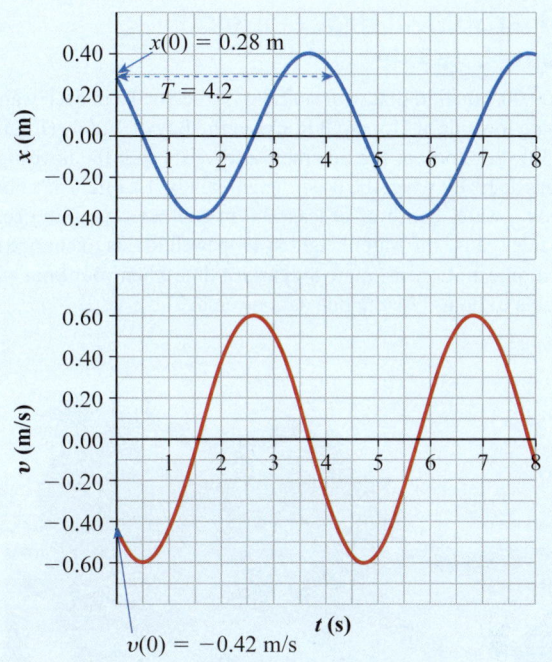

Figure 13-28 Position (blue) and velocity (red) plots for a harmonic oscillator.

Making sense of the result

With the calculated constants, at $t = 3.0$ s the displacement of the oscillator is

$$x(3.0) = (0.40 \text{ m}) \cos(1.5 \times 3.0 + 0.79) = 0.22 \text{ m}$$

which approximately agrees with the value from the graph.

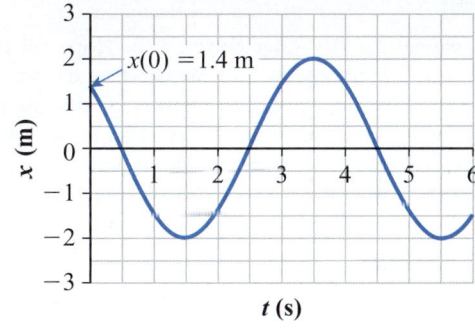

EXAMPLE 13-9

A Boat Stuck at Hopewell Rocks during Low Tide

At 09:00, you find that your sailboat is stuck in 0.3 m deep water during low tide at Hopewell Rocks in the Bay of Fundy (Figure 13-30). You look at the tide table and verify that the next high tide is at 15:15, when the water height will be 13.1 m. Your boat needs a water depth of at least 3.0 m to float. Assuming that the tides cause the water height at Hopewell Rocks to change in harmonic motion, estimate the earliest time when your boat will be able to float. Use the tide data given in Table 13-2.

Table 13-2 Tide Table for a Particular Day at Hopewell Rocks

Time (ADT)	Height (m)
02:48	13.7
09:23	0.3
15:15	13.1
21:44	0.8

Fisheries and Oceans Canada, http://www.lau.chs-shc.gc.ca/english/Canada.shtml. Reproduced with the permission of the Canadian Hydrographic Service. Not to be used for navigation.

Figure 13-30 Hopewell Rocks, Bay of Fundy, New Brunswick, at low tide.

Solution

Let us assume that the water height, h, at Hopewell Rocks varies in simple harmonic motion. We can then use the given data to determine the amplitude, the period, and the phase constant for this motion. From the equation for $h(t)$ we can determine the time when the height will be 3.0 m. The period of the tides is the time between two successive high tides or two successive low tides. Taking the time between two successive high tides, we get

$$T = 15:15 - 02:48 = 12 \text{ h and } 27 \text{ min} = 44\,820 \text{ s}$$

Let us determine the amplitude of the tides. Note that there is a variation in the height of successive high and low tides—the generation of tides is a complicated phenomenon that depends on the relative positioning of the Sun and the Moon. We will take the maximum tide height to be 13.1 m, corresponding to 15:15 ADT (Atlantic Daylight Time). The amplitude of the tides is, therefore,

$$A = \frac{\text{water height at high tide} - \text{water height at low tide}}{2}$$

$$= \frac{13.1 \text{ m} - 0.30 \text{ m}}{2} = 6.4 \text{ m}$$

Since the difference between the highest and the lowest positions in simple harmonic motion is equal to twice the amplitude, we divided the difference between the high and low tides by 2.

In this case, the equilibrium height, denoted by $h(0)$, is the average of the heights at high and low tides:

$$h_0 = \frac{\text{height at high tide} + \text{height at low tide}}{2}$$

$$= \frac{13.1 \text{ m} + 0.30 \text{ m}}{2} = 6.7 \text{ m}$$

Given these quantities, we can now write an equation that describes the height of the water at Hopewell Rocks:

$$h(t) = 6.7 \text{ m} + (6.4 \text{ m})\cos\left(\frac{2\pi}{T}t + \phi\right) \qquad (1)$$

The first term in Equation (1) is the equilibrium height, h_0. *The variation in the water height is measured with respect to h_0.* This term is absent when $x = 0$ is taken as the equilibrium point.

From the previous examples, we know that we can determine ϕ if we know $h(t)$ at a given time. For ease of calculating, we take the morning low tide time, 09:23, as the start time, $t = 0$. At this time, the sea is 0.30 m high. Inserting this information in the above equation yields

$$h(0) = 6.7 \text{ m} + (6.4 \text{ m}) \cos\phi = 0.30 \text{ m}$$

Therefore,

$$\cos\phi = \frac{0.30 \text{ m} - 6.7 \text{ m}}{6.4 \text{ m}} = -1.0$$

$$\phi = \cos^{-1}(-1.0) = \pi \text{ rad}$$

Now we can write Equation (1) for the water height:

$$h(t) = 6.7 \text{ m} + (6.4 \text{ m})\cos\left(\frac{2\pi}{T}t + \pi\right) \qquad (2)$$

WAVES AND OSCILLATIONS

The last task is to determine the time when the water height is 3.0 m. Setting $h(t) = 3.0$ m in Equation (2) and using the trigonometric identity $\cos(\theta + \pi) = -\cos\theta$, we get

$$3.0 \text{ m} = 6.7 \text{ m} + (6.4 \text{ m})\cos\left(\frac{2\pi}{T}t + \pi\right)$$

$$= 6.7 \text{ m} - (6.4 \text{ m})\cos\left(\frac{2\pi}{T}t\right)$$

Solving for t gives

$$\frac{2\pi}{T}t = \cos^{-1}\left(\frac{6.7 \text{ m} - 3.0 \text{ m}}{6.4 \text{ m}}\right) = 0.95$$

$$t = \frac{T}{2\pi} \times 0.95 = \frac{44820 \text{ s} \times 0.95}{2\pi} \approx 6.8 \times 10^3 \text{ s} = 1 \text{ h } 53 \text{ min}$$

Since $t = 0$ was at 09:23, your boat should be ready to float again at about 11:16. Happy sailing, and don't leave your tide table at home!

13-9 Damped Oscillations

So far in our study of simple harmonic motion, we have ignored frictional forces. As a result, our model predicts that once a simple harmonic oscillator is set in motion, it will oscillate indefinitely. However, frictional forces are always present. These forces continuously transform an oscillator's kinetic energy into thermal energy, thus decreasing the amplitude of the oscillation. Eventually, all the energy of the oscillator is transformed into thermal energy, and it comes to rest.

Frictional forces can be quite complicated. For a vertical mass–spring oscillator in air, the frictional force between the mass and the surrounding air depends on the shape of the mass and the speed of oscillation. There is also friction at the contact points where the mass is attached to the spring and where the spring is connected to the rigid support. Furthermore, there are internal frictional forces between the molecules of the spring as it stretches and compresses. A spring that is continuously stretched and compressed gets hot. (Try stretching and compressing a metal spring.) Although an exact description of frictional forces is complicated, two features are common for frictional forces experienced by objects in motion:

1. The magnitude of the frictional force is proportional to the speed of the object.

2. The direction of the frictional force is opposite to the direction of motion of the object.

Oscillations in the presence of frictional forces are called **damped oscillations**. In the study of fluids, such a frictional force is called a **drag force**, or **fluid resistance**. The simplest form of the drag force, $\vec{F}_D$, that an object experiences when moving with a velocity $\vec{v}$ in a medium can be written as

$$\vec{F}_D = -b\vec{v} \tag{13-68}$$

The minus sign indicates that the direction of the drag force is opposite to the direction of the velocity of the object. The proportionality constant, b, is a measure of the strength of the drag force, and for oscillating objects it is called the **damping constant**, or **drag constant**. Its value is positive and depends on the properties of the medium in which the object is moving, as well as the shape of the object (Figure 13-31).

Since the dimensions on both sides of an equation must be same, we can use Equation 13-68 and dimensional analysis to find the dimension of b, denoted $[b]$:

$$\text{kg} \cdot \text{m/s}^2 = [b] \times \text{m/s}$$

$$[b] = \frac{\text{kg}}{\text{s}}$$

Equation 13-68 is only one of many forms for drag forces and is valid only for special conditions, such as the air resistance on small, slowly moving objects. A fast-moving baseball experiences air resistance that is proportional to the square of the speed of the ball.

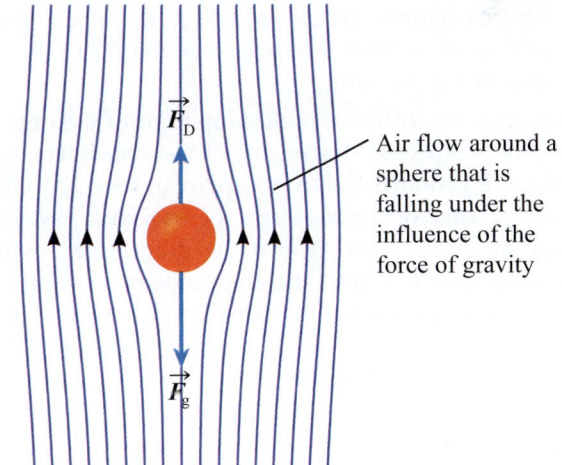

Figure 13-31 The drag force ($\vec{F}_D$) on a sphere falling through the air under the influence of gravity. Note that the direction of the drag force is opposite to the direction of motion of the sphere.

Let us analyze the motion of a mass–spring system, oscillating in the air, in the presence of the drag force of Equation 13-68. The net force, $\vec{F}_{net}$, on the oscillating mass is the sum of the elastic restoring force exerted by the spring and the drag force exerted by the air:

$$\vec{F}_{net} = \vec{F}_{spring} + \vec{F}_D \qquad (13\text{-}69)$$

Taking the x-axis along the direction of motion,

$$F_{net,x} = F_{spring,x} + F_{D,x} = -kx + (-bv_x) \quad (13\text{-}70)$$

Using Newton's second law of motion, the acceleration of the mass is

$$ma_x = -kx - bv_x$$
$$a_x = -\left(\frac{k}{m}\right)x - \left(\frac{b}{m}\right)v_x \qquad (13\text{-}71)$$

Will the mass–spring system undergo simple harmonic motion in the presence of the damping force? Comparing Equation 13-71 with Equation 13-18, we see that in this case the acceleration has an additional term that depends on the velocity of the oscillator. So, in general, the motion is not simple harmonic motion. However, if the velocity-dependent term in Equation 13-71 is small compared to the displacement term, then the acceleration of the mass is approximately proportional to the displacement, and the motion will be nearly simple harmonic motion. Using calculus, it can be shown (see Section 13-11) that the damping force in Equation 13-71 modifies the motion of a simple harmonic oscillator as follows:

1. The amplitude of a damped oscillator decreases exponentially with time.
2. The oscillation frequency of a damped oscillator is lower than the oscillation frequency of an undamped oscillator.

A decrease with time in the amplitude of a damped oscillator is expected. The energy of a simple harmonic oscillator is proportional to the square of its amplitude. In the presence of a drag force, the oscillator loses its energy by doing work against the drag force. As its energy decreases, so does the amplitude of oscillation. The oscillation frequency decreases because, as it loses energy, a damped oscillator slows down and therefore takes more time to complete a cycle.

As shown in Section 13-11, the displacement, $x(t)$, of the damped oscillator described by Equation 13-71 is given by

KEY EQUATION $\quad x(t) = Ae^{-bt/(2m)}\cos(\omega_D t + \phi) \quad (13\text{-}72)$

where

$$\omega_D = \sqrt{\frac{k}{m} - \frac{b^2}{4m^2}} = \sqrt{\omega_0^2 - \left(\frac{b}{2m}\right)^2} \quad (13\text{-}73)$$

and $\omega_0 = \sqrt{k/m}$.

In Equation 13-73, ω_D is the angular frequency in the presence of the drag force, and ω_0 is the angular frequency when $b = 0$. The angular frequency ω_0 is called the **natural frequency** of the oscillation. In Equation 13-72, the constant e is the base of the natural logarithm, and $e^{-bt/(2m)}$ denotes an exponential function that decreases with time. Note that $\omega_D < \omega_0$, so a damped oscillator oscillates at a lower frequency than a corresponding undamped oscillator. We can combine the amplitude, A, and the exponential term to define a time-dependent amplitude $A(t)$ for a damped oscillator and write Equation 13-72 in a form that is similar to the form for an undamped oscillator:

KEY EQUATION $\quad A(t) = Ae^{-bt/(2m)} \qquad (13\text{-}74)$

KEY EQUATION $\quad x(t) = A(t)\cos(\omega_D t + \phi) \qquad (13\text{-}75)$

We now examine the motion of a damped oscillator over the range of possible values for the damping constant, b.

An Underdamped Oscillator $\left(\omega_0 > \dfrac{b}{2m}\right)$

In this case ω_D is positive, and the argument of the cosine function is a real number. The system oscillates with the angular frequency ω_D, and the oscillations are called **underdamped**. For example, when $k = 4.00$ N/m, $m = 0.200$ kg, and $b = 0.200$ kg/s,

$\omega_0 = 4.47$ rad/s, $b/2m = 0.50$ rad/s, and $\omega_D = 4.44$ rad/s

A plot of $x(t)$ for $A = 0.2$ m and $\phi = 0$ rad is shown in Figure 13-32.

For $\omega_0 \gg b/2m$, the main effect of the drag force is to decrease the amplitude exponentially with time. The oscillation frequency, ω_D, is close to the natural frequency, ω_0. The dashed curve in Figure 13-32 is called the **envelope** of the damped oscillations. As the damping constant is increased, the amplitude decreases more rapidly with time, and the difference between ω_D and ω_0 increases. Figure 13-33 shows a plot of $x(t)$

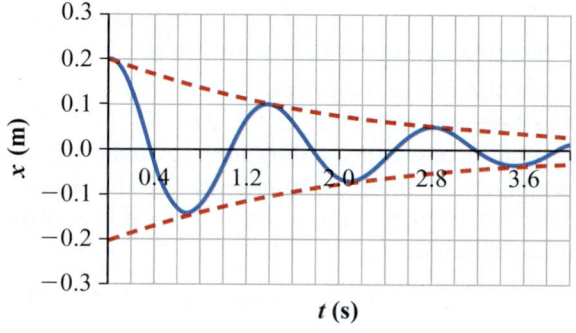

Figure 13-32 Displacement versus time graph for a damped harmonic oscillator, with $k = 4.00$ N/m, $m = 0.200$ kg, and $b = 0.200$ kg/s. The dashed lines are called the envelope of the oscillation and are shown to emphasize the damping.

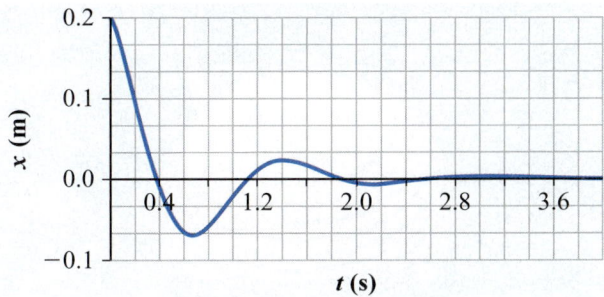

Figure 13-33 A displacement versus time graph for a damped harmonic oscillator, with $k = 4.0$ N/m, $m = 0.20$ kg, and $b = 0.60$ kg/s.

with $b = 1.50$ kg/s, with other parameters kept the same as above. Increasing the value of b by a factor of 3 causes only a small decrease in ω_D (4.21 rad/s), but the effect on the amplitude is quite pronounced. The displacement of the oscillator is unnoticeable after two oscillations.

Most everyday oscillations that you observe—for example, a swinging pendulum and a child on a swing—are situations with low damping, where $\omega_0 \ll b/(2m)$. These systems lose a fraction of their energy with each oscillation, until finally all the energy has been lost. The greater the damping, the quicker the system comes to a stop.

The quantity b/m that appears in Equation 13-72 has dimensions of inverse time. We therefore define the **time constant** (or the mean life) of a damped oscillator by

$$\tau = \frac{m}{b} \quad (13\text{-}76)$$

In terms of τ, the exponential factor in Equation 13-74 is written as $e^{-t/(2\tau)}$. So in a time interval $t = 2\tau$ the amplitude of a damped oscillator decreases to $1/e \approx 0.37$, about a third of its initial value. The larger the time constant, the slower the exponential fall-off.

A Critically Damped Oscillator $\left(\omega_0 = \dfrac{b}{2m}\right)$

As b increases, the frequency of the damped oscillator decreases, and the oscillator takes more and more time to complete an oscillation. The resulting motion is called **critically damped** when $b = 2\sqrt{mk}$ (i.e., $b/(2m) = \omega_0$). In this case, $\omega_D = 0$; therefore, the argument of the cosine function does not change with time, and there is no back-and-forth motion of the oscillator. Using calculus, it can be shown that for a critically damped oscillator the displacement from the equilibrium position as a function of time is given by

$$x(t) = (a_1 + a_2 t)e^{-\omega_0 t} \quad (13\text{-}77)$$

Constants a_1 and a_2 are determined by the initial conditions, $x(0)$ and $v(0)$. When a critically damped oscillator is displaced, it returns to its equilibrium position without completing an oscillation. The time taken for the oscillator to return to the equilibrium position is related to the angular frequency of oscillations, ω_0. *A critically damped oscillator reaches its equilibrium position, without oscillating, faster than for any other value of the damping constant, b.* Figure 13-35 shows a displacement graph for a critically damped mass–spring system. For the graph, we chose $x(0) = 1.0$ m, $v(0) = 0.0$ m/s, $m = 0.200$ kg, and $k = 4.00$ N/m. With these values, $\omega_0 = \sqrt{20}$ rad/s, and critical damping occurs for $b = 2\sqrt{(0.20 \text{ kg}) \times (4.0 \text{ N/m})} = 1.8$ kg/s.

Shock absorbers in cars are a good example of a critically damped system (Figure 13-36). When a car goes over a bump in the road, properly damped shock absorbers should settle the car back to its normal state in a minimal amount of time.

CHECKPOINT 13-11

Damped Mass-Spring Oscillator

Four identical mass–spring systems are subjected to damping forces of the form $F_x = -bv$, with a different value of b for each system. Each system is displaced by 0.10 m from equilibrium and released. The amplitude, $A(t)$, for each is shown in Figure 13-34. Rank in order, from smallest to largest, the damping constants for these systems.

Oscillators from smallest to largest b:

Smallest ___ ___ ___ ___ Largest

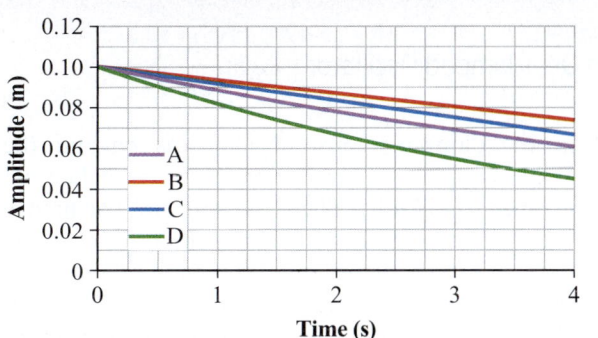

Figure 13-34 Amplitude as a function of time for four different values of damping constant b.

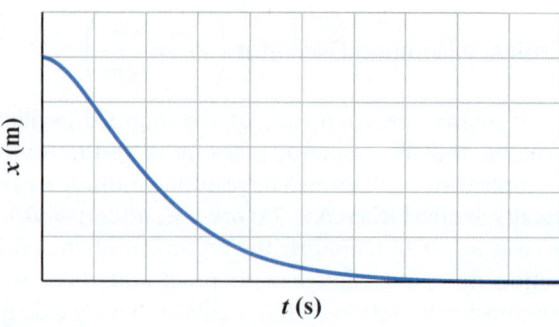

Figure 13-35 A displacement versus time graph for a critically damped oscillator with the initial conditions $x(0) = 1.0$ m, $v(0) = 0.0$ m/s, $\omega_0 = \sqrt{20}$ rad/s, and $b = 1.8$ kg/s.

Figure 13-36 A typical spring-based shock absorber. The inside hydraulic piston contains oil that acts as a damping medium, which absorbs the vibrational energy of the spring and dissipates vibrations.

Another example of a critically damped oscillator is a **tuned-mass damper** (Figure 13-37), which is used in skyscrapers. Tuned-mass dampers are huge blocks of concrete or metal, suspended inside buildings to damp strong vibrations of the building caused by the wind and earthquakes. A detailed explanation is complicated, but basically, the damper absorbs the vibrational energy from the structure when the oscillation frequency of the damper matches the oscillation frequency of the structure.

An Overdamped Oscillator $\left(\omega_0 < \dfrac{b}{2m} \right)$

When the damping constant increases beyond the critically damped value, the oscillator is called an **overdamped** oscillator. There is still no oscillatory motion, but as b increases beyond the critical value, the oscillator takes more and more time to return to its equilibrium position. Heavy doors in public places have hydraulic devices that are overdamped. This keeps the doors from slamming quickly.

Figure 13-37 A tuned-mass damper.

Energy in a Damped Harmonic Oscillator

The energy of an oscillator is proportional to the square of its amplitude, so the energy of a damped harmonic oscillator also decays exponentially with time. Consider an underdamped mass–spring oscillator with an amplitude given by Equation 13-74. The energy of the oscillator at a given instant is

$$E(t) = \frac{1}{2}kA^2(t)$$

$$= \frac{1}{2}kA^2(0)(e^{-bt/(2m)})^2 \qquad (13\text{-}78)$$

$$E(t) = \frac{1}{2}kA^2(0)e^{-bt/m} = \frac{1}{2}kA^2e^{-bt/m}$$

Here, at $t = 0$, we have defined $A(0) = Ae^{-0} = A$. Therefore, the energy at time $t = 0$ is

$$E(0) = \frac{1}{2}kA^2 \qquad (13\text{-}79)$$

Therefore,

$$E(t) = E(0)e^{-bt/m} \qquad (13\text{-}80)$$

After one period, $T_D = 2\pi/\omega_D$, the energy of the oscillator is

$$E(T_D) = \left(\frac{1}{2}kA^2 \right)e^{-bT_D/m} = E(0)e^{-bT_D/m} \quad (13\text{-}81)$$

The **fractional loss of energy** in one oscillation is therefore

$$\frac{E(0) - E(T_D)}{E(0)} \equiv \frac{\Delta E}{E(0)} = 1 - e^{-bT_D/m} \qquad (13\text{-}82)$$

Note two limiting cases of Equation 13-82:

1. When $b = 0$, $\dfrac{\Delta E}{E(0)} = 1 - 1 = 0$, indicating that there is no loss of energy as the system oscillates.

2. At the critical damping limit, $\omega_D = 0$ and therefore $T_D \to \infty$. Since $e^{-\infty} \to 0$, we obtain the result $\Delta E = E(0)$.

PEER TO PEER

If I make an approximation to solve a problem, after solving the problem, I have to remember to evaluate my initial approximation to ensure that it is valid.

INTERACTIVE ACTIVITY 13-3

Determining the Damping Constant of a Damped Oscillator

In this activity, you will use the PhET simulation "Determining the Damping Constant of a Damped Oscillator." Work through the simulation and accompanying questions to gain an understanding of a damped harmonic oscillator.

EXAMPLE 13-10

A Damped Harmonic Oscillator

A damped harmonic oscillator consisting of a block of mass 2.00 kg and a spring of spring constant 10.0 N/m has an amplitude of 25.0 cm at $t = 0$ s. The amplitude falls to 75.0% of its initial value after four oscillations. Assuming that the damping force is of the form $F_{D,x} = -bv_x$, calculate

(a) the value of the damping constant, b
(b) the amount of energy lost during four oscillations

Solution

(a) We are given that the oscillations are damped. Since we know the spring constant and the amplitude at $t = 0$, we can calculate the initial total energy of the oscillator:

$$E(0) = \frac{1}{2}kA^2 = \frac{1}{2} \times (10.0 \text{ N/m}) \times (0.250 \text{ m})^2 = 0.312 \text{ J}$$

We need to determine the period. Since we do not know the damping constant, ω_D is not known. Let us assume that ω_D is approximately equal to ω_0. This is a reasonable assumption because the oscillator loses only 25% of its initial energy in four oscillations, so the damping must be weak. Therefore, the time period of the damped oscillator can be approximated by using the natural frequency, ω_0. Once we have determined the damping constant, we will check the validity of this approximation. Calculating the period, we get

$$T = \frac{2\pi}{\omega_0} = 2\pi\sqrt{\frac{m}{k}} = 2\pi\sqrt{\frac{2.00 \text{ kg}}{10.0 \text{ N/m}}} = 2.81 \text{ s}$$

To determine the damping constant, we use the information that the amplitude falls to 75% of its initial value in four oscillations. Substituting into Equation 13-74 gives

$$A(4T) = Ae^{-4bT/(2m)} = \frac{3}{4}A$$

$$e^{-2bT/m} = \frac{3}{4}$$

Taking the natural logarithm (ln) of both sides of the above equation, we get

$$-\frac{2bT}{m} = \ln\left(\frac{3}{4}\right)$$

$$b = -\frac{m}{2T} \times \ln\left(\frac{3}{4}\right)$$

$$= -\frac{2.00 \text{ kg}}{2 \times 2.81 \text{ s}} \times \ln\left(\frac{3}{4}\right) = 0.102 \text{ kg/s}$$

Let us check whether our original assumption that ω_D is approximately equal to ω_0 was valid. Using Equation 13-73,

$$\omega_D = \sqrt{\frac{k}{m} - \frac{b^2}{4m^2}}$$

$$= \sqrt{\frac{10.0 \text{ N/m}}{2.00 \text{ kg}} - \frac{(0.102 \text{ kg/s})^2}{4 \times (2.00 \text{ kg})^2}} = 2.24 \text{ rad/s}$$

This value is very close to the undamped value, $\omega_0 = \sqrt{5}$ rad/s, so our approximation was excellent!

(b) Using Equation 13-81, the energy of the oscillator after four cycles is related to its energy at $t = 0$:

$$E(4T) = E(0)e^{-4bT/m}$$

Using values of b, T, and m:

$$E(4T) = (0.312 \text{ J})e^{-4 \times 0.102 \text{ kg/s} \times 2.81 \text{ s}/2.00 \text{ kg}} = 0.176 \text{ J}$$

This is the energy of the oscillator after four cycles. Therefore, the energy lost in four cycles is 0.312 J − 0.176 J = 0.136 J.

Making sense of the result

The energy lost in four cycles is less than the starting energy of the oscillator.

The Quality Factor or the Q-Value

From Equation 13-80 we see that the energy of a damped harmonic oscillator decreases exponentially with time. Taking the time derivative of Equation 13-80, we get

$$\frac{dE(t)}{dt} = \frac{d}{dt}\left(E(0)e^{-\frac{bt}{m}}\right) = E(0)\frac{d}{dt}\left(e^{-\frac{bt}{m}}\right)$$

$$= \left(\frac{-b}{m}\right)E(0)e^{-\frac{bt}{m}}$$

Therefore, the rate of energy loss, or the power radiated by a damped oscillator, is proportional to the ratio of its damping constant to its mass (b/m). Two oscillators with different damping constants and masses but the same b/m ratio will lose energy at the same rate. It is always convenient, whenever possible, to work with dimensionless quantities. Therefore, for an underdamped oscillator we define a dimensionless quantity, called the quality factor or the Q-value, denoted by Q, as

$$Q = 2\pi \frac{\text{Energy stored in an oscillator}}{\text{Energy lost in one cycle}}$$

In a weakly damped oscillator, the energy lost per cycle is small and oscillations continue for a long time. Such an oscillator has a high Q-value. In a strongly damped oscillator, the energy lost per cycle is large and the oscillations are quickly damped. Such an oscillator has a low Q-value. Lightly damped piano strings or tuning forks may have Q-values of a few thousand and would oscillate for a few thousand cycles before losing an appreciable fraction of initial energy. The factor of 2π is introduced for the sake of convenience. Notice that the above definition of the Q-value is quite general and is not restricted to a mechanical oscillator.

To see the relationship between the Q-value and the damping constant of a mass–spring oscillator, consider a lightly damped oscillator. If $E(t)$ is the energy of the oscillator at time t and its time period is T_D, then the energy lost in one cycle is defined as $\Delta E = E(t) - E(t + T_D)$, and

$$Q = 2\pi \frac{E(t)}{E(t) - E(t + T_D)}$$

Using Equation 13-80,

$$E(t + T_D) = E(0)e^{-\frac{b(t + T_D)}{m}} = E(0)e^{-\frac{bt}{m}} \times e^{-\frac{bT_D}{m}}$$

$$= E(t)e^{-\frac{bT_D}{m}}$$

Therefore,

$$Q = 2\pi \frac{E(t)}{E(t)\left(1 - e^{-\frac{bT_D}{m}}\right)} = 2\pi \frac{1}{1 - e^{-\frac{bT_D}{m}}}$$

For a lightly damped oscillator, the exponent of the exponential term is small and we can use the approximation $e^{-x} \approx 1 - x$ for $x \ll 1$. The denominator of the above equation can then be written as

$$1 - e^{-\frac{bT_D}{m}} \approx 1 - \left(1 - \frac{bT_D}{m}\right) = \frac{bT_D}{m}$$

and the Q-value can be written as

$$Q = 2\pi \frac{1}{(bT_D/m)} = \left(\frac{2\pi}{T_D}\right)\frac{1}{(b/m)} = 2\pi \frac{\tau}{T_D}$$

Notice that the Q-value is inversely proportional to the ratio b/m. For $\omega_0 \gg b/m$ we can replace T_D by T, where T is the time period of the oscillator, ignoring the damping, and the above equation can be further approximated as

$$Q = 2\pi \frac{\tau}{T}$$

The advantage of specifying an underdamped oscillator by its Q-value is that all oscillators with the same Q-value lose energy at the same rate.

13-10 Resonance and Driven Harmonic Oscillators

Consider a simple mechanical system, for example, a swing, with a natural frequency of oscillation f_0. Left on its own, an oscillating swing loses energy due to frictional forces and comes to rest after a while. The motion can be sustained if energy is transferred to the swing by an external mechanism, for example, by pushing the swing periodically. If the swing is pushed after each oscillation, then the time between successive pushes (t_{push}) is the same as the period of the swing

$(t_{\text{push}} = T = 1/f_0)$. In this case, the frequency of the externally applied force, called the **driving frequency**, is the same as the natural frequency of the swing.

The swing can also absorb energy at other driving frequencies. If the swing is pushed after every other oscillation, the period of successive pushes is twice the period of the swing, and the driving frequency is $f_0/2$. Other possible driving frequencies are $f_0/3, f_0/4, f_0/5, \ldots$.

A mechanical system (such as a swing, a wine glass, a bridge, or a building) is said to be in **resonance** with an externally applied force when the frequency of the externally applied force (f_{driving}) matches the natural frequency (f_0) of the system:

$$f_{\text{driving}} = f_0 \quad \text{(resonance condition)} \quad (13\text{-}84)$$

When the driving and the resonance frequencies match, there is an efficient transfer of energy from the external force to the mechanical system.

There are numerous examples of mechanical resonances: resonating strings in a string instrument such as a violin, resonating membranes of a drum, resonance of the basilar membrane in the ear, and breaking a wine glass using sound waves. Resonance phenomena also play a key role in electromagnetism: when the input circuit of a radio receiver is tuned to match the frequency broadcast by a particular radio station, the amplitude of the received signal is greatly increased.

Consider the damped mass–spring system of Section 13-9 that is subjected to an external harmonic force. Such an oscillator is called a **driven oscillator**, and the resulting oscillations are called **forced oscillations**, or **driven oscillations**. The harmonic driving force has the form

$$F_{\text{driven},x} = F_0 \cos(\omega_{\text{driven}} t) \quad (13\text{-}85)$$

Here, F_0 is the magnitude of the applied force, and ω_{driven} is its angular frequency. The net force on the mass is the sum of the elastic restoring force exerted by the spring, the damping force, and the external driving force:

$$\begin{aligned} \Sigma F_x &= F_{\text{spring},x} + F_{\text{damping},x} + F_{\text{driven},x} \\ &= -kx + (-bv_x) + F_0 \cos(\omega_{\text{driven}} t) \end{aligned} \quad (13\text{-}86)$$

The acceleration of the mass is therefore given by

$$a_x = -\left(\frac{k}{m}\right)x - \left(\frac{b}{m}\right)v_x + \left(\frac{F_0}{m}\right)\cos(\omega_{\text{driving}} t) \quad (13\text{-}87)$$

where $\omega_0 = \sqrt{k/m}$ is the natural frequency of the oscillator.

What is the equation for the displacement of the oscillator as a function of time? Using calculus, it can be shown that the displacement, $x(t)$, of the driven oscillator described by Equation 13-87 is given by

$$x(t) = A_{\text{driven}} \cos(\omega_{\text{driven}} t + \phi_{\text{driven}}) \quad (13\text{-}88)$$

where

$$A_{\text{driven}} = \frac{F_0}{\sqrt{m^2(\omega_{\text{driven}}^2 - \omega_0^2)^2 + b^2\omega_{\text{driven}}^2}}$$

$$\phi_{\text{driven}} = \tan^{-1}\left(\frac{b\omega_{\text{driven}}}{m(\omega_{\text{driven}}^2 - \omega_0^2)}\right) \quad (13\text{-}89)$$

A_{driven} and ϕ_{driven} are, respectively, the amplitude and the phase constant of the driven oscillator. Equation 13-88 is valid after transient effects have died down. (Transient effects are temporary and depend on how the initial motion was started.) Note several properties of driven oscillations:

1. An oscillator subjected to a sinusoidal driving force oscillates at the frequency of the driving force, *not* at the system's natural frequency. For example, an adult human ear resonates at approximately 3000 Hz. However, when an ear is subjected to a sound, for example, at a frequency of 500 Hz, the eardrum vibrates at 500 Hz rather than at its natural frequency.

2. The amplitude and the phase constant of the driven oscillations are determined not by the initial conditions, but by the parameters of the system: the natural frequency of the oscillator, the driving frequency of the external force, the magnitude of the driving force, the damping constant, and the mass of the oscillator. This situation is different from a simple harmonic oscillator, where the amplitude and the phase constant depend on the initial position and velocity of the oscillator.

3. As ω_{driven} approaches ω_0, the amplitude increases and reaches a maximum at $\omega_{\text{driven}} = \omega_0$. As ω_{driven} is increased past the value ω_0, the amplitude decreases again. If the damping constant, b, is small, the amplitude becomes quite large as ω_{driven} approaches ω_0. For $b = 0$, A_{driven} goes to infinity as $\omega_{\text{driven}} \to \omega_0$. In real life, damping is never zero, and the amplitude of driven oscillations can become large but not infinite. This dependence of A_{driven} on ω_{driven} forms the basis of resonance phenomena. For example, a wine glass can shatter when subjected to sound waves whose frequency is very close to the natural frequency of oscillation of the wine glass. Constructing earthquake-resistant buildings requires incorporating enough damping in the structures so that when a building is subjected to oscillations of frequency close to its natural oscillation frequency, the amplitude never becomes large. A plot of the amplitude as a function of b clearly shows the effect of resonance near the natural frequency of the oscillator. For the plot in Figure 13-38, we chose $m = 1.0$ kg, $\omega_0 = 20.0$ rad/s, and $F_0 = 1.0$ N and then plotted A_{driven} for $b = 0.01$ kg/s, 0.1 kg/s, 0.5 kg/s, and 0.8 kg/s. Near the resonance frequency, the amplitude is strongly dependent on the damping constant. Notice

Sympathetic Vibrations–Shattering a Wine Glass with Sound

You can do the following experiment in your physics lab. Take two tuning forks of the same frequency. Set one to vibrate by striking its prongs. If you place the two tuning forks close to each other, but not touching, the second will also begin to vibrate. This happens because the vibrating tuning fork causes the surrounding air to vibrate, which creates pressure waves (i.e., sound) of the same frequency. These pressure waves exert a harmonic force on the second tuning fork. Since the frequency of the pressure waves is the same as that of the tuning fork, a resonance is set up that causes the tuning fork to oscillate with increasing amplitude. Vibrations produced in an oscillator by pressure waves of the same frequency (or an integer multiple of that frequency) are called **sympathetic vibrations**.

Another example of this phenomenon is the shattering of a wine glass by sound waves. If you gently tap a wine glass near its rim (or continuously rub a wet finger around its rim), you will excite the frequency associated with its natural oscillations. An accurate value of this frequency can be determined in a lab by a spectrum analyzer. If the wine glass is now subjected to high-intensity sound waves of this frequency for a sustained duration (usually for a few seconds) from a speaker connected to a wave generator, the wine glass will begin to vibrate at its

natural frequency. A resonance is set up and the amplitude of oscillations quickly increases until the strain becomes so large that the glass shatters into pieces.

Why does a glass shatter? Glass is made by heating a mixture of sand (silicon dioxide) and modifiers (e.g., aluminum oxide, calcium oxide, lead oxide, etc.) until it melts and then cooling it quickly so that its particles do not have time to form a crystalline pattern. A solid that lacks a crystal structure is called an amorphous (meaning "without shape") solid. Unlike a crystalline solid, where particles are arranged in an ordered structure along well-defined planes, an amorphous solid has no long-range pattern. So when the glass is subjected to a stress, because of its irregular structure, its particles cannot move past each other to relieve stress. As a result, small cracks form along its surface. As the amplitude of vibrations increases, the resulting stress in the glass increases and the cracks grow in size until it shatters.

The phenomenon of sympathetic vibrations also occurs when window panes in a building begin to rattle when a large vehicle passes nearby. The rattling occurs when the vibrational frequency of the engine noise matches the natural frequency of the window panes.

---

that the position of the peak amplitude shifts slightly away from the natural frequency as b increases. From Equation 13-89 it can be shown (by finding the maximum of A_{driven} as a function of ω_{driven}) that the amplitude is maximum at

$$\omega_{driven} = \sqrt{\omega_0^2 - \frac{b^2}{2m^2}} \qquad (13\text{-}90)$$

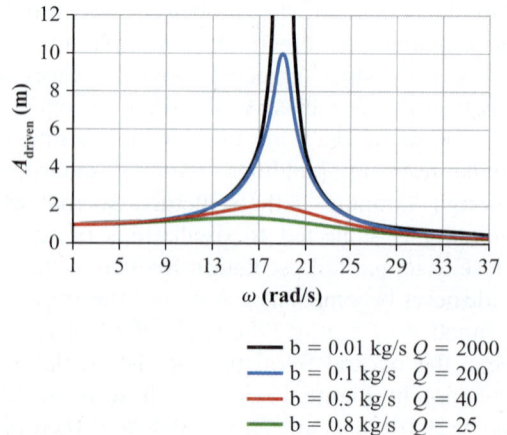

— b = 0.01 kg/s Q = 2000
— b = 0.1 kg/s Q = 200
— b = 0.5 kg/s Q = 40
— b = 0.8 kg/s Q = 25

Figure 13-38 The amplitude for a forced harmonic oscillator for various values of the damping constant, b, and $\omega_0 = 20.0$ rad/s. The Q-value for each b is calculated using $Q = \omega_0/(b/m)$.

13-11 Simple Harmonic Motion and Differential Equations (Optional Section)

An ordinary differential equation is an equation that involves a single unknown function of a single variable and a finite number of its derivatives. Some examples of differential equations are

$$\frac{df(x)}{dx} + xf(x) = 6x$$

$$\frac{d^2x(t)}{dt^2} + e^{-t}\frac{dx(t)}{dt} - x(t) = 0$$

$$\frac{d^2y(x)}{dx^2} - 4\frac{dy(x)}{dx} + 3y(x) = 4e^x + \sin x$$

A linear ordinary differential equation is a differential equation of the form

$$a_n(x)\frac{d^nf}{dx^n} + a_{n-1}(x)\frac{d^{n-1}f}{dx^{n-1}} + a_{n-2}(x)\frac{d^{n-2}f}{dx^{n-2}}$$

$$+ \cdots + a_1(x)\frac{df}{dx} + a_0(x)f = g(x)$$

The coefficients $a_i(x)$ and $g(x)$ are known functions of the independent variable. When $g(x) = 0$, the equation is said to be homogeneous. For non-zero $g(x)$ it is called non-homogeneous. The number of highest derivative in a differential equation is called the order of the differential equation.

We will limit the following discussion to motion in one dimension. Since velocity is the rate of change of position, $x(t)$, and acceleration is the rate of change of velocity, we can write velocity and acceleration as the first and the second derivatives of position as a function of time (our discussion is limited to motion in one dimension):

$$v_x(t) = \frac{dx(t)}{dt}$$

$$a_x(t) = \frac{dv_x(t)}{dt} = \frac{d}{dt}\left(\frac{dx(t)}{dt}\right) = \frac{d^2x(t)}{dt^2} \quad (13\text{-}91)$$

Consider Equation 13-18, which relates the acceleration of a simple harmonic oscillator to its position (here we have replaced the angular frequency ω by ω_0):

$$a_x(t) = -\omega_0^2 x(t)$$

Using Equation 13-91, we can write the above equation as

$$\frac{d^2x}{dt^2} + \omega_0^2 x(t) = 0 \quad (13\text{-}92)$$

Here time, t, is the independent variable and position, x, is the dependent variable. Both coefficients (1 and ω^2) are constants, so Equation 13-92 is a second-order differential equation with constant coefficients. We can now provide still another definition of simple harmonic motion: *The motion of an object or a system whose displacement from the equilibrium position satisfies the second-order differential equation of the form given by Equation 13-92 is called a simple harmonic motion.* The general solution for Equation 13-92 is given in terms of sine and cosine functions and can be written in the following two forms:

$$x(t) = A\cos(\omega_0 t + \phi)$$
$$= a\cos\omega_0 t + b\sin\omega_0 t \quad (13\text{-}93)$$

Here constants A and ϕ (or a and b) are determined from the initial conditions of the motion, usually by knowing the displacement and velocity at $t = 0$. The two forms of the solutions are equivalent. If A and ϕ are known, then we can determine a and b. Alternatively, if a and b are known, then A and ϕ can determined. The two sets of constants are related by the following equations:

$$a = A\cos\phi$$
$$b = -A\sin\phi \quad (13\text{-}94)$$

The elegance of dealing with differential equations is that the dependent variable x (and also the independent variable t) can be replaced by any other variable.

If the angular displacement θ for an object satisfies the differential equation

$$\frac{d^2\theta}{dt^2} + q\theta(t) = 0, \text{ where } q > 0 \quad (13\text{-}95)$$

then we immediately know that the object executes simple harmonic motion and the time dependence of the dependent variable θ is given by

$$\theta(t) = A\cos(\sqrt{q}\,t + \phi) \quad (13\text{-}96)$$

In this case, $\sqrt{q}$ is identified as the angular frequency of oscillations and A represents the maximum angular displacement of the oscillator.

For a damped harmonic oscillator, replacing v_x and a_x in Equation 13-71 with the derivatives, we obtain the following differential equation:

$$\frac{d^2x}{dt^2} + \frac{b}{m}\frac{dx}{dt} + \omega_0^2 x(t) = 0 \quad (13\text{-}97)$$

This differential equation contains both a first- and a second-order derivative. Now we ask ourselves the following question. What is the form of $x(t)$ that satisfies the above equation? Obviously, for $b = 0$, we must get Equation 13-93 for $x(t)$. Let us try a solution of the following form, which is permitted by a well-defined procedure in the study of differential equations:

$$x(t) = Ae^{-at}\cos(dt + \phi) \quad (13\text{-}98)$$

Here A and ϕ represent the amplitude and the phase constant of oscillations, and constants a and d need to be determined. Using the chain rule of differentiation, we get

$$\frac{dx}{dt} = A[-a \times e^{-at}\cos(dt + \phi) + (-d)$$
$$\times e^{-at}\sin(dt + \phi)]$$

$$\frac{d^2x}{dt^2} = A[a^2 \times e^{-at}\cos(dt + \phi) + a \times d \quad (13\text{-}99)$$
$$\times e^{-at}\sin(dt + \phi) + a \times d \times e^{-at}\sin(dt + \phi)$$
$$- d^2 \times e^{-at}\cos(dt + \phi)]$$

Substituting Equation 13-99 into Equation 13-98 and collecting terms for cosine and sine functions, we get

$$\left[a^2 - d^2 - \frac{ab}{m} + \omega_0^2\right]e^{-dt}\cos(dt + \phi)$$

$$+ \left[2ad - \frac{db}{m}\right]e^{-dt}\sin(dt + \phi) = 0 \quad (13\text{-}100)$$

For the above equation to be valid for any time t, the terms in each both square brackets must individually vanish. Therefore,

$$\left[a^2 - d^2 - \frac{ab}{m} + \omega_0^2\right] = 0$$

$$\left[2ad - \frac{db}{m}\right] = 0 \tag{13-101}$$

From the second equation we get

$$a = \frac{b}{2m} \tag{13-102}$$

Substituting this value of a into the first part of Equation 13-101, we get

$$d^2 = \omega_0^2 - \frac{b^2}{4m^2} \quad \Rightarrow \quad d = \pm\sqrt{\omega_0^2 - \frac{b^2}{4m^2}} \tag{13-103}$$

Taking the positive value for d, because the coefficient multiplying t in the argument of the cosine function has the dimensions of angular frequency, which is positive, $x(t)$ is given by

$$x(t) = A\, e^{-\left(\frac{bt}{2m}\right)} \cos\left[\left(\sqrt{\omega_0^2 - \frac{b^2}{4m^2}}\right)t + \phi\right] \tag{13-104}$$

$\sqrt{\omega_0^2 - \frac{b^2}{4m^2}}$ is identified as the angular frequency in the presence of a damping force, and we denote it by

the symbol ω_D. The motion of the damped oscillator is called underdamped for $\omega_0 > \frac{b}{2m}$, critically damped for $\omega_0 = \frac{b}{2m}$, and overdamped for $\omega_0 < \frac{b}{2m}$. Notice that since $e^{\pm 0} = 1$, the above equation reduces to Equation 13-93 when $b = 0$.

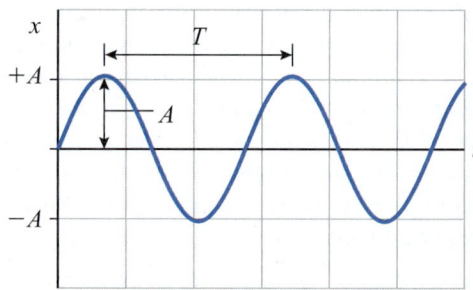

CHECKPOINT 13-13

The Correct Equation for a Simple Harmonic Oscillator

Which of the following differential equations do NOT represent the motion of a simple harmonic oscillator? Here x denotes the position and t the time.

(a) $2\dfrac{d^2x}{dt^2} + 4x(t) = 0$

(b) $\dfrac{d^2x}{dt^2} - 3x(t) = 0$

(c) $-\dfrac{d^2x}{dt^2} - 7x(t) = 0$

(d) $\dfrac{dx}{dt} + 6x(t) = 0$

ANSWER: (b) and (d)

KEY CONCEPTS AND RELATIONSHIPS

Simple Harmonic Motion

A motion that repeats after a finite amount of time is called periodic motion. The time that it takes for the motion to repeat is called the period (T). One complete back-and-forth movement is called an oscillation. The number of oscillations per second completed by an object is the oscillation frequency (f) of the object and is related to the period:

$$f = \frac{1}{T} \tag{13-3}$$

Periodic motion is called simple harmonic motion when, as a function of time, the displacement, $x(t)$, of an object from the equilibrium position is given by a cosine (or a sine) function:

$$x(t) = A\cos(\omega t + \phi) \tag{13-4}$$

The amplitude, A, and the phase constant, ϕ, are determined by the position and the velocity of the oscillator at $t = 0$. The angular velocity, ω, depends on the physical properties of the oscillating object and is related to the period:

$$\omega = 2\pi/T \tag{13-5}$$

The velocity and acceleration of a simple harmonic oscillator are determined by taking the first and second derivatives, respectively, of Equation 13-4 as a function of time:

Figure 13-39 Simple harmonic motion: displacement as a function of time for simple harmonic motion.

$$v(t) = -\omega A \sin(\omega t + \phi) \tag{13-12}$$
$$a(t) = -\omega^2 A \cos(\omega t + \phi) \tag{13-17}$$

An object undergoes simple harmonic motion when the net force acting on the object is proportional to its displacement from the equilibrium position and is opposite in sign to the displacement. Such a force is called a restoring force.

For horizontal and vertical mass–spring systems,

$$\omega = \sqrt{\frac{k}{m}}; \quad T = 2\pi\sqrt{\frac{m}{k}} \tag{13-34, 13-35}$$

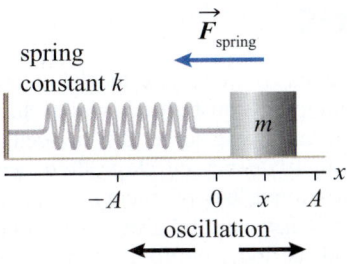

Figure 13-40   A horizontal mass–spring simple harmonic oscillator.

For a pendulum,

$$\omega = \sqrt{\frac{g}{L}}; \quad T = 2\pi\sqrt{\frac{L}{g}} \qquad \text{(13-54), (13-55)}$$

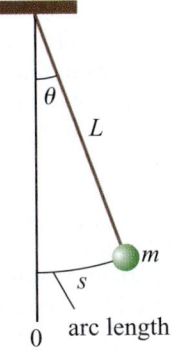

Figure 13-41   A simple pendulum.

Simple Harmonic Motion and Uniform Circular Motion

The projection of a motion with a constant angular velocity along the circumference of a circle is mathematically equivalent to simple harmonic motion. The radius of the circle corresponds to the amplitude of the simple harmonic motion, and the angular velocity along the circle corresponds to the angular frequency of simple harmonic motion.

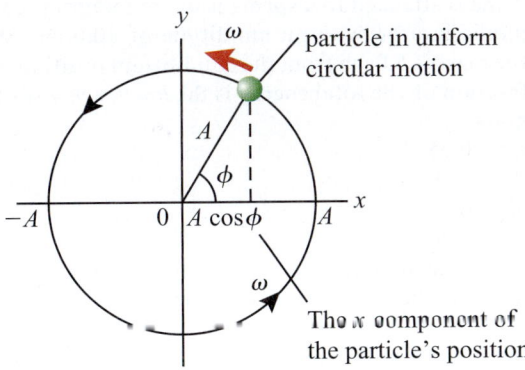

Figure 13-42   Comparing simple harmonic motion and circular motion.

Damped Harmonic Oscillator

When frictional forces are present, the oscillations are damped and the oscillator is called a damped harmonic oscillator. For a velocity-dependent damping force ($\vec{F}_D = -b\vec{v}$), the displacement of a damped harmonic oscillator is given by

$$x(t) = Ae^{-bt/(2m)}\cos(\omega_D t + \phi), \qquad \text{(13-72)}$$

where b is the damping constant, and the oscillation frequency ω_D is given by

$$\omega_D = \sqrt{\frac{k}{m} - \frac{b^2}{4m^2}} \qquad \text{(13-73)}$$

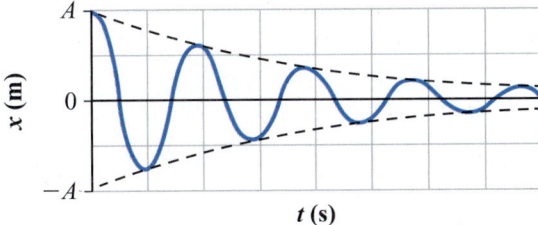

Figure 13-43   Displacement versus time plot for a damped harmonic oscillator.

Energy of a Harmonic Oscillator

The total energy, E, of a simple harmonic oscillator is proportional to the square of its amplitude. In the absence of damping forces, the total energy of a harmonic oscillator remains constant and continuously transforms between kinetic energy, K, and potential energy, U:

$$E = K + U = \frac{1}{2}kA^2 \qquad \text{(13-37)}$$

Energy for a damped harmonic oscillator decreases exponentially with time and is given by

$$E(t) = \frac{1}{2}kA^2e^{-bt/m} \qquad \text{(13-78)}$$

where A is the amplitude of the damped oscillator at $t = 0$.

Quality Factor

The quality factor (Q-factor) of a damped oscillator is defined as the ratio of energy stored in an oscillator to energy lost per cycle due to damping effects.

Resonance

A harmonic oscillator subjected to a sinusoidal external force oscillates with the frequency of the force. A resonance occurs when the frequency of the external force equals the natural frequency of the oscillator:

$$f_{\text{driving}} = f_0 \qquad \text{(13-84)}$$

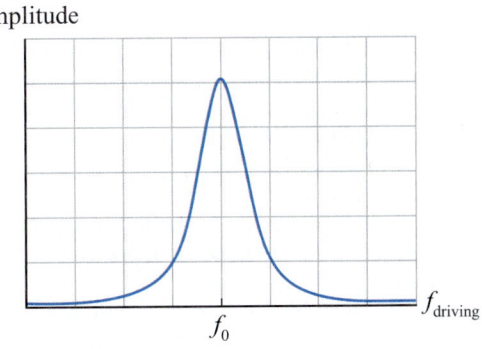

Figure 13-44   The amplitude of a driven oscillator as a function of driving frequency.

APPLICATIONS

pendulum clocks, shock absorbers, tune-mass dampers, electrocardiograms

KEY TERMS

amplitude, angular frequency, critically damped, damped oscillations, damping constant, displaced harmonic oscillator, drag constant, drag force, driven oscillations, driven oscillator, driving frequency, envelope, fluid resistance, forced oscillations, fractional loss of energy, frequency, harmonic motion, Hooke's law, natural frequency, oscillation, over-damped, period, periodic motion, phase constant, physical pendulum, resonance, restoring force, simple harmonic motion, simple harmonic oscillator, simple pendulum, sympathetic vibrations, time constant, tuned-mass damper, underdamped

QUESTIONS

1. What is the total distance travelled in one cycle by a simple harmonic oscillator of amplitude A? What is the net displacement of the oscillator after one cycle?

2. If a spring is cut in half, what happens to its spring constant?

3. Why is the period of a mass–spring oscillator independent of the amplitude?

4. Two masses, m and $4m$, are separately attached to two identical springs. The period of the smaller mass is T. What is the period of the larger mass?
   (a) $T/2$
   (b) T
   (c) $2T$
   (d) $4T$

5. What happens to the period of a mass–spring oscillator if the mass of the spring is not ignored?

6. As a pendulum oscillates, is there a point along its path where both its velocity and its tangential acceleration are zero? Is there a point where both its displacement and its tangential acceleration are zero?

7. What happens to the period of a simple pendulum if the mass of the string is not ignored?

8. If the maximum angular displacement of a pendulum is increased, what happens to its speed as it passes through the equilibrium?

9. A pendulum is suspended from the ceiling of an elevator. When the elevator is at rest, the period of the pendulum is T. When the elevator accelerates upward, the period of the pendulum
   (a) decreases
   (b) remains T
   (c) increases

10. You displace the mass of a mass–spring oscillator from equilibrium by 0.10 m, hold it stationary, and then release it. If you gave the mass a kick as you released it, would the resulting amplitude be less than, equal to, or greater than 0.1 m? Why? Does it matter whether the kick is toward the equilibrium or away from it? Explain.

11. A child is swinging on a playground swing while sitting. When the child stands up, what happens to the period of the swing?

12. A child is swinging on a playground swing while sitting. Another child jumps onto the same swing while it is swinging. Assuming that the children and the swing form a simple pendulum, what happens to the period of the swing? What happens to the speed of the swing as it passes through the equilibrium position?

13. The period of a pendulum is 2.00 s in a vacuum. Its period in air is
   (a) less than 2.00 s
   (b) 2.00 s
   (c) greater than 2.00 s

14. In the following equations, x is displacement and t is time. Which equation describes simple harmonic motion?
   (a) $t(x) = 2.0 \cos(3.0x + \pi/4)$
   (b) $x(t) = 1.0/\sin(2t + \pi)$
   (c) $x(t) = 1.0 \cos^2(4t)$
   (d) $x(t) = 3.0 + 5.0 \cos(2t)$

15. Which of the following forces, acting on a particle of mass m, will *not* result in simple harmonic motion? The displacement of the particle from the equilibrium position is denoted by $x(t)$.
   (a) $F(t) = -2x^2(t)$
   (b) $F(t) = +2x^2(t)$
   (c) $F(t) = +2x(t)$
   (d) $F(t) = -2x(t)$

16. A mass attached to a spring is undergoing simple harmonic motion with an amplitude of 10.0 cm. When the mass is 5.0 cm from the equilibrium position, what fraction of the total energy is the *kinetic energy* of the mass?
   (a) 0.25
   (b) 0.50
   (c) 0.75
   (d) 1.00

17. When the amplitude of a simple harmonic oscillator is doubled, the total energy of the oscillator
   (a) remains the same
   (b) increases by a factor of 2
   (c) decreases by a factor of 2
   (d) increases by a factor of 4
   (e) decreases by a factor of 4

18. The graph in Figure 13-45 shows the displacement versus time plot of an object that is undergoing simple harmonic motion. For the displacement at $t = 3$ s, which of the following statements is correct?
   (a) The velocity of the particle is positive, and its acceleration is positive.

(b) The velocity of the particle is positive, and its acceleration is negative.

(c) The velocity of the particle is negative, and its acceleration is positive.

(d) The velocity of the particle is negative, and its acceleration is negative.

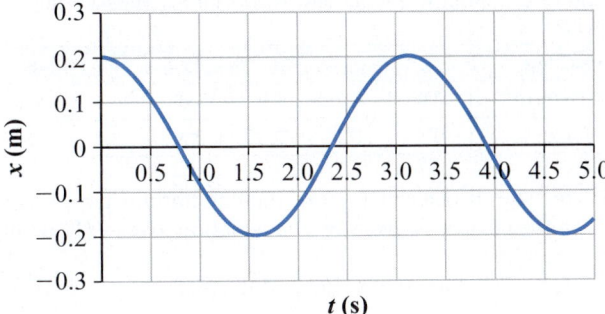

Figure 13-45 Question 18.

PROBLEMS BY SECTION

For problems, star ratings will be used (★, ★★, or ★★★), with more stars meaning more challenging problems. The following codes will indicate if $\frac{d}{dx}$ differentiation, $\int$ integration, ⬚ numerical approximation, or ⤳ graphical analysis will be required to solve the problem.

Section 13-2 Simple Harmonic Motion

19. ★ A mass of 2.0 kg is connected to a spring with a spring constant of 5.0 N/m. The mass is oscillating on a horizontal, frictionless surface. At time $t = 0$, the mass is 0.30 m from the equilibrium position and has zero velocity.
 (a) What is the amplitude?
 (b) What is the maximum speed of the mass?
 (c) What is the maximum acceleration of the mass?
 (d) Write an equation that describes the displacement of the mass from the equilibrium position as a function of time.

20. ★ The displacement of a particle from equilibrium is given by

$$x(t) = (10 \text{ cm}) \cos(6\pi t)$$

 (a) Determine the amplitude, frequency, and phase constant of this oscillator.
 (b) Starting from $t = 0$, what is the first time when the particle passes through the equilibrium position?
 (c) What is the particle's speed at that time?
 (d) What is its maximum acceleration?

21. ★ The displacement of a particle from equilibrium is given by

$$x(t) = (5 \text{ cm}) \cos(2\pi t - \pi/4)$$

 (a) Determine the amplitude, period, and phase constant of this oscillator

(b) Starting from $t = 0$, what is the first time when the particle passes through the equilibrium position?
 (c) What is the particle's speed at that time?
 (d) In what direction is it moving at that time?
 (e) What is its maximum acceleration?

22. ★ A mass–spring system has $k = 18.0$ N/m and $m = 0.71$ kg. The system oscillates with an amplitude of 0.54 m.
 (a) Determine the angular frequency of the oscillations.
 (b) What is the speed of the mass at $x = 0.034$ m?
 (c) What is the displacement of the mass from the equilibrium position when its speed is 0.18 m/s?

23. ★★★ A mass attached to a horizontal spring is pulled to the left and released at $t = 0$ s. The mass passes through $x = +10.0$ cm at $t = 0.3$ s with a velocity of 85.7 cm/s.
 (a) What is the lowest frequency of the system?
 (b) What is the amplitude?
 (c) What is the phase constant?
 (d) Write an equation that describes the position of the mass from the equilibrium position as a function of time.

24. ★ Consider a particle undergoing a simple harmonic motion. Determine its phase constant given that its position at time $t = 0$ is at
 (a) $x = -A$
 (b) $x = 0$ and it is moving from right to left
 (c) $x = +A$
 (d) $x = 0$ and it is moving from left to right

25. ★★★ Two identical mass–spring oscillators are $\pi/3$ rad out of phase and are undergoing simple harmonic motion with amplitude A about the same origin. One of the oscillators is at $x = +A$ at $t = 2.0$ s. When will the second oscillator next reach $x = +A$?

26. ★★ A mass–spring oscillator undergoing simple harmonic motion about $x = 0$ has a period of 2.0 s. At $t = 0$ s, the mass is at $x = 0$; at $t = 0.5$ s, the mass is at $x = -0.40$ m.
 (a) Write an equation that describes the position of the mass as a function of time.
 (b) What are the maximum velocity and acceleration of the mass?
 (c) The mass is 0.10 kg. What is the total energy of the oscillator?

27. ★★ The velocity of a particle is given by

$$v(t) = -(10 \text{ cm/s}) \sin(4t + \pi/2)$$

 (a) Determine the amplitude, period, and phase constant of this oscillator.
 (b) Starting from $t = 0$, what is the first time when the particle passes through the equilibrium position?
 (c) Starting from $t = 0$, what is the first time when the particle has maximum positive displacement from the equilibrium position?
 (d) What is the total distance covered by the particle in one cycle?
 (e) What is its maximum acceleration?
 (f) Write an equation that gives its acceleration as a function of time.

28. ★★★ Two particles oscillate in simple harmonic motion with amplitude A about the centre of a common straight line of length $2A$. Each particle has a period of 1.5 s, and their phase constants differ by $\pi/6$ rad.
   (a) How far apart are the particles (in terms of A) 0.50 s after the lagging particle leaves one end of the path?
   (b) What are their velocities at that time?

29. ★★★ Can we determine the amplitude, A, and the phase constant, ϕ, for a simple harmonic motion given the values of displacement at $t = 0$ and $t = T/2$, where T is the period? Explain your reasoning.

30. ★★ An object, oscillating with simple harmonic motion, has a period of 4.0 s and an amplitude of 0.20 m.
   (a) How long does the object take to move from $x = 0.0$ m to $x = 0.07$ m?
   (b) Would the object take the same time to move from $x = 0.07$ m to $x = 0.14$ m? Explain your answer without any calculations.

31. ★★ Equations for five simple harmonic oscillators are given below:

$$x_1(t) = 0.2\cos(5t + \pi/4)$$
$$x_2(t) = 0.1\cos(2t + \pi)$$
$$x_3(t) = 0.3\cos(t/5)$$
$$x_4(t) = 0.2\cos(7t/2 + \pi/2)$$
$$x_5(t) = 0.6\cos(15t)$$

   (a) Rank the oscillators, from high to low, in order of period:
   High ___ ___ ___ ___ ___ Low
   (b) Rank the oscillators, from high to low, in order of maximum velocity:
   High ___ ___ ___ ___ ___ Low
   (c) Rank the oscillators, from large to small, in order of the phase constant:
   Large ___ ___ ___ ___ ___ Small
   (d) Rank the oscillators, from high to low, in order of maximum acceleration:
   High ___ ___ ___ ___ ___ Low

32. ★★★ Two particles are undergoing simple harmonic motion with the same amplitude and frequency about the same equilibrium position. They pass each other moving in opposite directions, and each time their displacement is half their amplitude. What is the difference between their phase constants?

Section 13-3   Uniform Circular Motion and Simple Harmonic Motion

33. ★ Consider a point located on the equator of Earth. Find the angular speed (in rad/s) and the period of this point.

34. ★ A particle initially at rest on the x-axis, at $x = 3.0$ m, starts moving with a constant angular velocity of 2.0 rad/s in a counterclockwise direction in a circle of radius 3.0 m around the origin. The x-component of the particle's position at any time t is given by which of the following equations?
   (a) $x(t) = 2.0\sin 3.0t$
   (b) $x(t) = 3.0\cos 2.0t$
   (c) $x(t) = 3.0\sin 2.0t$
   (d) $x(t) = 3.0 + \cos 2.0t$

Section 13-4   Mass–Spring Systems

35. ★ The spring constants and masses for six mass–spring oscillators are given in Table 13-3. Rank the oscillators in the order of their periods, from small to large.
   Small ___ ___ ___ ___ ___ ___ Large

Table 13-3 Spring Constants and Masses for Six Mass–Spring Oscillators

Oscillator	A	B	C	D	E	F
k (N/m)	2.0	4.5	1.0	7.0	11.6	4.6
m (kg)	0.5	1.2	2.0	0.3	3.2	3.1

36. ★ The displacement from the equilibrium position of an oscillator is given by the equation $x(t) = (11.0$ cm$)\cos(3.0\,t + 5.0)$.
   (a) What are the period and the phase constant of the oscillation?
   (b) What is the phase of the oscillator at $t = 0$ s and $t = 2.0$ s?
   (c) What are the position and the velocity of the oscillator at $t = 0$ s?
   (d) At what time after $t = 0$ s is the oscillator at $x = 0$ cm?
   (e) What is the total distance covered by the oscillator between $t = 1.3$ s and $t = 1.5$ s?

37. ★★ A simple harmonic oscillator consists of a 0.10 kg mass attached to a spring with a spring constant of 100 N/m. The mass is displaced 0.20 m from the equilibrium position, held motionless, and then released. Calculate the following quantities for this system:
   (a) the angular frequency and the period
   (b) the maximum speed and the maximum acceleration
   (c) the total energy of the mass–spring system

38. ★★ In problem 37, if the mass is released at time $t = 0$ s, at what time will the kinetic energy of the mass be equal to the potential energy of the spring?

39. ★★ Two vertically hanging mass–spring oscillators (one end of the spring is tied to a rigid support and the other to a freely hanging mass) have identical springs, but the mass of one oscillator is twice that of the other. Both masses are pulled from their equilibrium positions by the same amount and then released. Ignore damping.
   (a) Do both oscillators have the same amplitude? Explain your reasoning.
   (b) Do both oscillators have the same total energy? Explain your reasoning.
   (c) Do both oscillators have the same speed while passing through the equilibrium position? Explain your reasoning.

40. ★★ A 2.0 kg mass rests on a smooth, horizontal surface and is attached to a spring, with the other end of the spring attached to a rigid support. An applied force of 10.0 N causes a displacement of 5.0 cm from the equilibrium position.
   (a) Find the spring constant of the spring.
   (b) What is the frequency of the mass–spring system?
   (c) When oscillating, the mass is at a maximum displacement of 5.0 cm at $t = 0.50$ s. Write an equation that describes the position of the mass from the equilibrium position as a function of time.
   (d) What is the total distance travelled by the mass in the interval $t = 0.5$ s to $t = 2.0$ s?

41. ★★ Consider a vertical mass–spring oscillator. The mass is pulled down by 0.20 m from its equilibrium position and given an initial velocity of 1.0 m/s directed toward the equilibrium position. It then oscillates with a period of 2 s.
    (a) Determine the amplitude and the phase constant of the resulting simple harmonic motion.
    (b) What is the speed of the mass when it passes through the equilibrium position?
    (c) Would the amplitude and the phase constant be the same if the initial velocity was directed away from the equilibrium position? Explain your reasoning.

42. ★★★ Figure 13-46 shows a 0.20 kg mass resting on top of a 2.0 kg mass. The lower mass is attached to a spring with a spring constant of 100 N/m. The two masses are oscillating on a frictionless surface. The coefficient of static friction between the two masses is 0.60. What is the maximum amplitude for which the upper block does not slip?

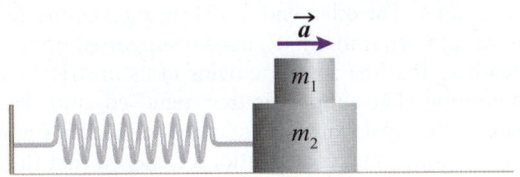

Figure 13-46 Question 42.

43. ★★★ A 500 g block slides along a frictionless surface at a speed of 0.35 m/s. It runs into a horizontal massless spring that extends outward from a wall; the spring constant is 50 N/m. The block compresses the spring and is then pushed back in the opposite direction by the spring, eventually losing contact with the spring.
    (a) How long does the block remain in contact with the spring?
    (b) How would your answer to part (a) change if the block's initial speed were doubled?

44. ★★★ A 0.20 kg block attached to a horizontal spring is oscillating with an amplitude of 5.0 cm and a frequency of 2.0 Hz. Just as it passes through the equilibrium point, moving to the right, a sharp blow directed to the left exerts a 10 N force for 1.0 ms. Calculate the new frequency and amplitude of oscillations.

45. ★★ A mass is attached to a vertical spring and then set to oscillate. At the high point of the oscillations the spring is in its original unstretched position before the mass was attached to it. The low point of the oscillations is 6.0 cm below the high point. What is the period of the oscillations?

46. ★★ An empty compact car has a mass of 1200 kg. Assume that the car has one spring on each wheel, the springs are identical, and the mass is equally distributed over the four springs.
    (a) What is the spring constant of each spring if the car bounces up and down 2.0 times per second?
    (b) What will be the car's oscillation frequency while carrying four passengers each of mass 70 kg?

47. ★★★ Two masses, m_1 and m_2, are connected to two ends of a spring with spring constant k. The masses are free to slide on a horizontal, frictionless surface. The spring is compressed and then released. Find an expression for the frequency of the system.

48. ★★★ A block of mass m, resting on a frictionless surface, is connected to two springs, 1 and 2, with spring constants k_1 and k_2, respectively, as shown in Figure 13-47. The block is displaced from the equilibrium position and then released.
    (a) Show that the resulting motion of the block is simple harmonic motion.
    (b) Show that the frequency of oscillations of the block is given by
    $$f = \sqrt{\frac{f_1^2 f_2^2}{f_1^2 + f_2^2}}$$
    where f_1 is the frequency of the mass when only spring 1 is attached to the mass, and f_2 is the frequency of the mass when only spring 2 is attached to the mass.
    (c) Let U_1 and U_2 denote the elastic potential energies stored in the two springs as the mass oscillates. Show that
    $$\frac{U_1}{U_2} = \frac{k_2}{k_1}$$
    (d) As the mass oscillates, is there a position of the mass where only one of the springs is stretched (or compressed) and the other spring is in its equilibrium position? Explain your reasoning.

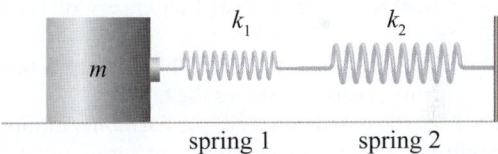

Figure 13-47 Question 48.

49. ★★★ A block of mass m, resting on a frictionless surface, is connected to two springs, 1 and 2, with spring constants k_1 and k_2, respectively, as shown in Figure 13-48. The block is displaced from the equilibrium position and then released.
    (a) Show that the resulting motion of the block is simple harmonic motion.
    (b) Show that the frequency of the block is given by
    $$f = \sqrt{f_1^2 + f_2^2}$$
    where f_1 is the frequency of the mass when only spring 1 is attached to the mass, and f_2 is the frequency when only spring 2 is attached to the mass.
    (c) Let U_1 and U_2 denote the elastic potential energies stored in the two springs as the mass oscillates. Show that
    $$\frac{U_1}{U_2} = \frac{k_1}{k_2}$$

(d) As the mass oscillates, is there a position of the mass where only one of the springs is stretched (or compressed) and the other spring is in its equilibrium position? Explain your reasoning.

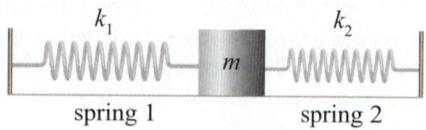

Figure 13-48 Question 49.

50. ★★ A spring hangs vertically from a fixed support. A mass is then attached to the lower end of the spring. When this system undergoes simple harmonic motion, it has a period of 0.50 s. By how much is the spring stretched from its initial length when the mass and spring are hanging motionless?

Section 13-5 Energy Conservation in Simple Harmonic Motion

51. ★ A 2.0 kg object, attached to a spring, oscillates with simple harmonic motion. The spring constant is 200 N/m, and the maximum force acting on the object is measured to be 3.5 N.
   (a) What is the amplitude of the oscillation?
   (b) What is the total energy of the mass–spring system as it oscillates?
   (c) What is the maximum speed of the mass?

52. ★★ An archer pulls her bow string back by 0.50 m before releasing the arrow. The bow–string system has an effective spring constant of 500 N/m.
   (a) What is the elastic potential energy of the drawn bow?
   (b) The arrow has a mass of 30.0 g. What is the speed of the arrow as it leaves the bow?

53. ★★ A particle undergoes simple harmonic motion with amplitude A. At what position (in terms of A) is the kinetic energy of the particle equal to its potential energy?

54. ★★ A mass–spring system is oscillating with an amplitude of 10.0 cm. What is the speed of the mass at a location where the kinetic energy and the potential energy of the mass are equal?

55. ★ A mass of 200 g is hanging motionless from a spring of spring constant 10 N/m. The mass is pulled down by 10 cm from this position and then released.
   (a) Taking the displacement below the equilibrium position to be negative, determine the amplitude, the angular frequency, and the phase constant of the resulting harmonic motion.
   (b) What is the total energy of this oscillator?

56. ★★ A 0.50 kg mass attached to a spring undergoes simple harmonic motion with an amplitude of 0.40 m and a period of 3.0 s. Determine
   (a) the total energy of this oscillator
   (b) the maximum speed of the mass
   (c) the speed when the mass is at $x = \pm 0.20$ m from the equilibrium position
   (d) the elastic potential energy stored in the spring when the mass is moving with half its maximum speed

57. ★★ A student observes that when a mass of 2.0 kg is attached to one end of a vertical spring, the other end of which is connected to a fixed support, the spring stretches 10.0 cm. The student then pulls the mass farther down by 5.0 cm from the equilibrium position and lets it go. As expected, he observes that the mass–spring system undergoes simple harmonic motion.
   (a) What are the amplitude and frequency of oscillations?
   (b) What is the kinetic energy of the mass when it is 3.00 cm below the equilibrium position?
   (c) Assuming that the upward direction is positive, write an equation for the displacement, y, from the equilibrium position, assuming that the mass was released at $t = 0$. Evaluate all parameters.
   (d) If the student had connected two springs identical to the spring above, and then hung a mass of 2.0 kg on them, by how much would the combined springs stretch?

58. ★★★ A mass, m, is attached to a spring with a spring constant k. The other end of the spring is connected to a fixed support. Initially, the mass is supported from underneath so that the spring remains in its unstretched configuration. The support is then removed, and the mass falls vertically downward under the influence of gravity.
   (a) Show, using conservation of energy, that the maximum distance, ΔL, that the mass falls from its initial position is given by $\Delta L = \dfrac{2mg}{k}$.
   (b) Once dropped, the mass will undergo simple harmonic motion about an equilibrium position. Show that the equilibrium position is at a distance $\dfrac{mg}{k}$ below the unstretched position of the spring, the amplitude of oscillations is $\dfrac{mg}{k}$, and the period is $T = 2\pi\sqrt{\dfrac{\Delta L/2}{g}}$.

59. ★★ A 1.0 kg mass is hanging motionless from a spring that is attached to a fixed support. The mass is pulled down 5.0 cm, held motionless, and then released. The mass–spring system begins to oscillate with a period of 2.0 s.
   (a) What is the spring constant of the spring?
   (b) What is the total energy of the mass–spring system?
   (c) What fraction of the system's total energy is the kinetic energy of the mass at $x = 3.0$ cm from the equilibrium position?
   (d) Write an equation that describes the position of the mass from the equilibrium position as a function of time.

Section 13-6 The Simple Pendulum

60. ★★ The lengths and masses for six pendulums are listed in Table 13-4. Rank the pendulums in order of their periods, from small to large.

Small ___ ___ ___ ___ ___ ___ Large

Table 13-4 Lengths and Masses for Six Pendulums

Pendulum	A	B	C	D	E	F
L (m)	0.8	1.1	1.7	0.9	2.1	3.2
m (kg)	0.5	1.2	2.0	0.3	3.2	3.1

61. ★★ A simple pendulum of length 1.00 m is measured to have a period of 2.00 s.
(a) What is the value of g at the location of the pendulum?
(b) The pendulum is now taken to the top of Mount Everest. Does the period increase or decrease? Justify your answer.
(c) What would be the period of this pendulum on the Moon?
(d) What would be the period of this pendulum on the International Space Station?

62. ★★ A pendulum with a length of precisely 1.000 m can be used to measure the acceleration due to gravity, g. Such a device is called a *gravimeter*.
(a) How long do 100 oscillations take at sea level, where $g = 9.81$ m/s^2?
(b) Suppose you take a gravimeter to the top of Mount Everest, where the value of g is 0.28% less than the value at sea level. How long do 100 oscillations take at the top of Mount Everest? Would you be able to measure the difference with a stopwatch that can measure time differences of 0.1 s?

63. ★★ A simple pendulum, consisting of a 0.30 kg mass attached to a string of length 1.0 m, is undergoing simple harmonic motion with a maximum angular displacement of 15°.
(a) What is the total energy of the pendulum?
(b) What is the speed of the mass as it passes through the equilibrium position?

64. ★★ A pendulum consists of a 0.20 kg mass attached to a string of length 1.5 m. The mass is displaced from its equilibrium hanging position by 15° and released. Determine the speed with which the mass swings through the equilibrium position.

65. ★★ (a) Consider a simple pendulum of length L. Show that a change in the length by an amount dL changes the pendulum's period by an amount dT, where

$$dT = \left(\frac{T}{2L}\right)dL$$

(b) If the pendulum clock runs faster by 5 s/h, how should the length be adjusted?

Section 13-7 The Physical Pendulum

66. ★★★ A thin circular ring of mass m and radius R hangs freely on a small peg at its rim. The ring is gently displaced from its equilibrium position and released. Derive an expression for the frequency of small oscillations of the ring.

67. ★★★ A physical pendulum consists of a sphere of mass m and radius R that is attached to a massless string. The other end of the string is attached to a fixed support, and the distance between the centre of the bob and the fixed support is L. Show that the period of small oscillations for this pendulum is given by

$$T = T_0\sqrt{1 + \frac{2R^2}{5L^2}}$$

where $T_0 = 2\pi\sqrt{L/g}$ is the period of a simple pendulum of length L.

68. ★★★ A physical pendulum consists of a sphere of mass m_s and radius R that is attached to a uniform rod of mass m_r and length L. The other end of the rod is attached to a fixed support, and the rod is free to rotate in a plane about an axis that is perpendicular to the plane of rotation and passes through the fixed support.
(a) Show that the period of small oscillations for this physical pendulum is given by

$$T = 2\pi\sqrt{\frac{\frac{2}{5}m_s R^2 + m_r\frac{L^2}{3} + m_s(L+R)^2}{g\left(m_s(L+R) + m_r\frac{L}{2}\right)}}$$

(b) Calculate the period for the following parameters: $m_s = 1.00$ kg, $m_r = 0.20$ kg, $R = 5.0$ cm, and $L = 1.0$ m. Compare this period to the period of a simple pendulum with $L = 1.05$ m. What is the percentage difference between the two periods? Why is the physical pendulum faster than an equivalent simple pendulum?

Section 13-8 Time Plots for Simple Harmonic Motion

69. ★ The position versus time graph for a simple harmonic oscillator is shown in Figure 13-49.
(a) What are the amplitude and the period?
(b) What is the phase constant?
(c) What is the phase at $t = 1$ s and $t = 2$ s?
(d) Write an equation for the oscillator that describes its position, $x(t)$, as a function of time.
(e) What is the velocity of the oscillator at $t = 1.0$ s?
(f) What is the acceleration of the oscillator at $t = 2$ s?

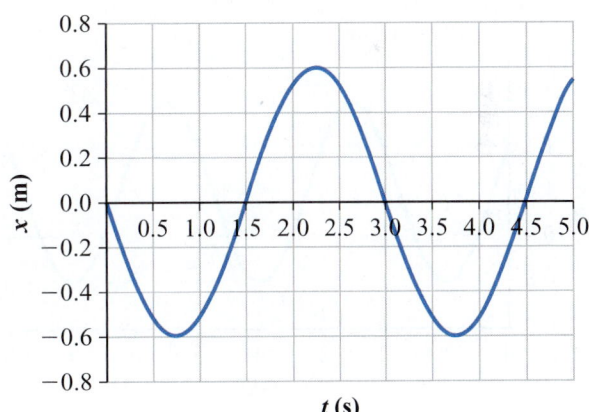

Figure 13-49 Question 69.

70. ★★ The velocity versus time graph for a simple harmonic oscillator is shown in Figure 13-50.
(a) What are the maximum speed and the period?
(b) What is the amplitude?
(c) What is the phase constant?
(d) Write an equation for the oscillator that describes its velocity, $v(t)$, as a function of time.
(e) What is the velocity of the oscillator at $t = 1.0$ s?
(f) What is the position of the oscillator at $t = 3.0$ s?

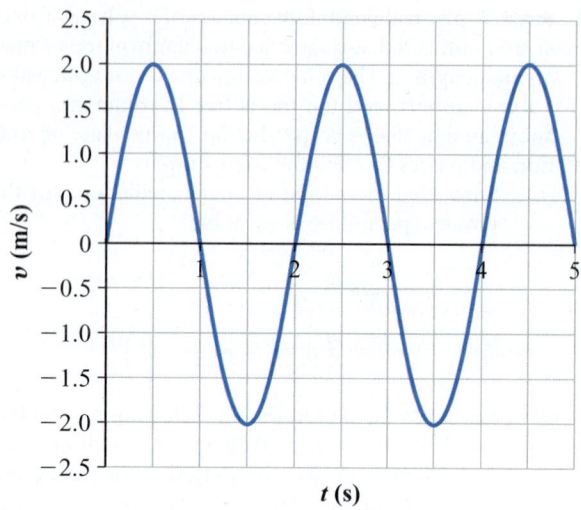

Figure 13-50 Question 70.

71. ★★ The position versus time graph for a simple harmonic oscillator is shown in Figure 13-51.
    (a) What are the amplitude and the period?
    (b) What is the equilibrium position?
    (c) What is the phase constant?
    (d) Write an equation for the oscillator that describes its position, $x(t)$, as a function of time with respect to the point $x = 0$.
    (e) Write an equation for a simple harmonic oscillator that is $\pi/2$ rad out of phase with this oscillator.
    (f) What is the acceleration of the oscillator at $t = 1.0$ s?

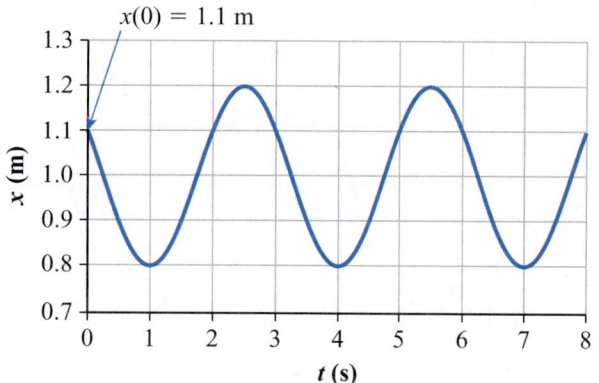

Figure 13-51 Question 71.

72. ★★ The position versus time plots of two oscillators are shown in Figure 13-52.
    (a) What is the phase difference between the two oscillators at $t = 1.0$ s?
    (b) At what times, closest to $t = 0$ s, do the two oscillators have zero velocities?
    (c) At what times, closest to $t = 0$ s, are the two oscillators moving with maximum velocity toward the negative x-axis?

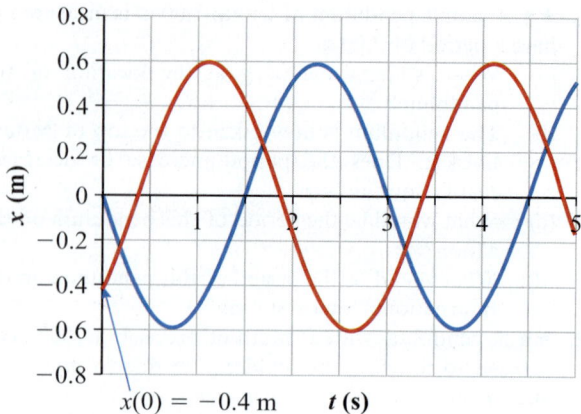

$x(0) = -0.4$ m t (s)

Figure 13-52 Question 72.

73. ★ Figure 13-53 shows the displacement curves, $x(t)$, for three experiments involving three different masses attached to identical springs oscillating with simple harmonic motion. Rank the curves as indicated.
    (a) the system's period
        Greatest ____ ____ ____ Least
    (b) the total energy of the mass–spring system
        Greatest ____ ____ ____ Least
    (c) the potential energy of the spring at $t = 0$
        Greatest ____ ____ ____ Least
    (d) the kinetic energy of the spring at $t = 0$
        Greatest ____ ____ ____ Least

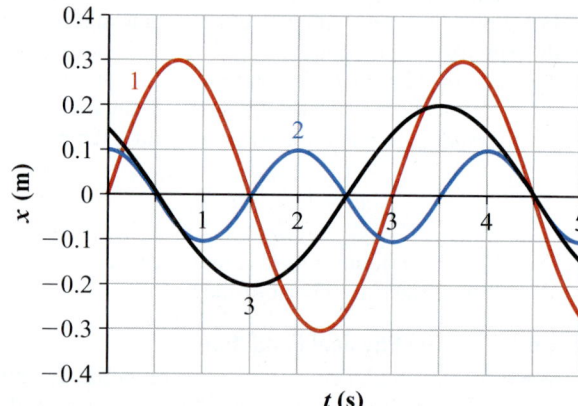

Figure 13-53 Question 73.

Section 13-9   Damped Oscillations

74. ★★ A damped harmonic oscillator consists of a block of mass 2.0 kg and a spring with spring constant $k = 10.0$ N/m. Initially, the system oscillates with an amplitude of 25 cm. Because of the damping, the amplitude decreases by 75% of its initial value at the end of four oscillations.
    (a) What is the value of the damping constant, b?
    (b) What percentage of initial energy has been lost during these four oscillations?

75. ★★ A damped mass–spring oscillator loses 5% of its energy during each cycle.
   (a) The period is 2.0 s, and the mass is 0.25 kg. What is the value of the damping constant?
   (b) If the motion is started with an initial amplitude of 0.30 m, how many cycles elapse before the amplitude reduces to 0.20 m?
   (c) Would it take the same number of cycles to reduce the amplitude from 0.20 m to 0.10 m? No calculations are needed for this part of the problem.

76. ★★ A car and its shock absorbers behave like a damped mass–spring system, with $m = 1500$ kg, $k = 60$ N/m, and $b = 1500$ kg/s. The car hits a pothole and begins to oscillate. After how much time will the displacement drop to half its initial value?

77. ★★ The initial amplitudes and b/m values for three damped oscillators are given in Table 13-5. Rank their amplitudes after 100 s, from smallest to largest, indicating any equivalencies by using the equality sign (=).

   Smallest ___ ___ ___ Largest

Table 13-5 For Problem 77

Damped harmonic oscillator	Amplitude (cm)	b/m (s^{-1})
A	10	0.010
B	20	0.020
C	30	0.030

78. ★ (a) In Problem 77, if the amplitude decreases to 5 cm after 10 oscillations, what is the Q-value of the oscillator?
   (b) If a second mass–spring system starts oscillating with an amplitude of 10 cm and has a Q-value that is twice that of this oscillator, what is the amplitude of the second oscillator after 10 oscillations?

79. ★ A simple pendulum loses 0.5% of its energy in one period. What is its Q-value?

80. ★★ A damped harmonic oscillator loses 5% of its energy each cycle.
   (a) After how many cycles would it lose 50% of its energy?
   (b) What is the Q-value of this oscillator?

81. ★★ (a) A large pendulum consists of a 50.0 kg mass suspended from a 15.0 m wire. Assuming no damping, how many oscillations are completed by the pendulum in one week?
   (b) Slight damping, of course, exists. At a certain time, the amplitude is measured to be 100.0 cm. Five days later, the amplitude has decreased to 80.0 cm. After how many days would the amplitude drop from 100.0 cm to 20.0 cm?

COMPREHENSIVE PROBLEMS

82. ★★ The position of an oscillator is known at two times, $t = 0$ and $t = T/4$, where T is the period. Use this information to determine the amplitude and the phase constant of the oscillator.

83. ★★★ A rectangular block of mass M is attached to a vertical spring and is undergoing simple harmonic motion (ignore damping) with amplitude A and period T. When the spring is fully compressed and the block momentarily comes to rest before changing its direction of motion, a small mass m ($m \ll M$) is gently placed on the block. Let the new amplitude and period be A' and T', respectively.
   (a) Is T' less than, equal to, or greater than T? Explain your reasoning.
   (b) Is A' less than, equal to, or greater than A? Explain your reasoning.
   (c) Does the total energy of the mass–spring oscillator change after the mass m is added? Explain your reasoning.

84. ★★★ A spherical object of mass m and radius r is placed in a large, hemispherical bowl of radius R. The mass rests in the equilibrium position at the bottom of the bowl. The mass is then displaced from its equilibrium position by a small angle, still touching the inner surface of the bowl, and released from rest. The mass rolls down the inner surface of the bowl without slipping. The mass of the bowl is large enough that the bowl remains stationary as the mass moves back to the equilibrium position.
   (a) Show that, when the friction between the mass and the bowl is ignored, the mass will undergo simple harmonic motion about its equilibrium position.
   (b) What is the frequency of the oscillation in terms of the variables m, r, and R?

85. ★★★ The gravitational force acting on a particle that is located inside a solid sphere of uniform density is directly proportional to the distance of the particle from the centre of the sphere. Imagine a narrow, evacuated cylindrical hole along the polar diameter of Earth, passing through its centre. A small object is dropped into this hole at one pole. How much time would the object take to reach the other pole? Does this time depend on the mass of the object? Explain why or why not.

86. ★★★ A simple pendulum consists of a mass m attached to a string of linear mass density p and length L. Determine the period for the pendulum.

87. ★★ An object of uniform cross-sectional area A floats in a liquid. When at rest, a volume, V_0, of the object is immersed in the liquid. Show that when pushed slightly farther into the liquid and released, the object will undergo simple harmonic motion with period $2\pi\sqrt{\dfrac{V_0}{gA}}$, where g is the acceleration due to gravity.

88. ★★ A mass, m, is attached to the free end of an unstretched spring whose other end is attached to a fixed support. The mass is held stationary and then allowed to drop freely. It is noted that the mass drops by a distance h before reversing its direction of motion. Show that the period of oscillations of this mass–spring system is the same as that of a simple pendulum of length $h/2$.

89. ★★ A mass attached to a spring has a period of T. The mass is displaced from its equilibrium position and then released from rest. Determine the time (in terms of T) when
   (a) the kinetic energy of the mass equals the potential energy of the spring
   (b) the kinetic energy of the mass is twice the potential energy of the spring

90. ★★ A block of mass M, resting on a frictionless surface, is attached to a spring with a spring constant k (Figure 13-54). The other end of the spring is tied to a rigid support. The mass is initially at rest. A bullet of mass m, moving horizontally with speed v, strikes the block and bounces off it, straight back. Determine expressions for
   (a) the speed of the block immediately after the collision
   (b) the amplitude of the block–spring oscillations as a result of the collision

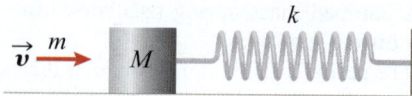

Figure 13-54   Question 90.

CHAPTER 14 | Waves

Earthquakes release an enormous amount of energy (Figure 14-1). This energy is carried from the origin of the earthquake, through Earth, by longitudinal and transverse waves over large distances. What are waves? How do waves transport energy? How does a longitudinal wave differ from a transverse wave? What factors determine the speed with which waves travel? Is the motion of a wave different from the motion of the constituents of the medium through which it moves?

Figure 14-1 Damage caused by the 2005 earthquake in Islamabad, Pakistan.

Maciej Dakowicz / Alamy Stock Photo

14-1 The Nature, Properties, and Classification of Waves

Wave motion is very different from the motion of objects. Some waves, such as waves in an ocean, need a medium in which to travel, but other waves, such as light, do not. A wave passes through a medium unaffected by the presence of other waves in that medium. In this chapter, we will explore the laws governing wave motion and derive mathematical relationships for some properties of waves.

What is a wave? Consider a long rope stretched taut between two rigid supports. If you pluck the rope near one end and then release it, a disturbance originates from where you plucked the rope and travels along the length of the rope. The disturbance moves along the rope, but it does not take the rope with it. As the disturbance passes through a given section of the rope, that section is displaced from its equilibrium position and then returns to it after the disturbance has passed through. Therefore, we can define a **wave** as the *motion of a disturbance*. Waves are classified into two major categories: mechanical waves and electromagnetic waves.

Mechanical waves A **mechanical wave** travels through a physical material, or **medium**, and the speed of the wave depends on the properties of the medium. The particles of the medium must interact so that a disturbance of particles at any given point is transmitted to adjacent particles. The vibration of a violin string, ripples on the surface of water (Figure 14-2), and seismic waves in Earth's crust are all examples of mechanical waves.

Electromagnetic waves Light, microwaves, radio waves, and X-rays are all manifestations of **electromagnetic waves**, or **electromagnetic radiation**.

Electromagnetic waves are generated by charged particles that accelerate, for example, electrons moving back and forth in the antenna of a radio transmitter. Unlike mechanical waves, electromagnetic waves do not need a medium in which to travel. The light that we receive from the Sun travels through the vacuum between the Sun and Earth. This chapter deals with mechanical waves; electromagnetic waves are discussed in Chapter 27.

Waveform The shape or pattern of a wave is called its **waveform**. Waveforms can be represented by plotting some property of the medium as the wave moves through it. For example, a sound wave in air creates regions of higher and lower pressure relative to undisturbed air. If we graph the pressure as a function of time at a given location, the resulting pressure versus time graph shows the shape of the sound wave at that location. A waveform can also be displayed by plotting the shape of the disturbance at a fixed time as a function of position in the medium through which the wave is travelling. Figure 14-3 shows the waveform of a sinusoidal wave on a long string, and Figure 14-4 shows the more complex waveform of a note played on a violin

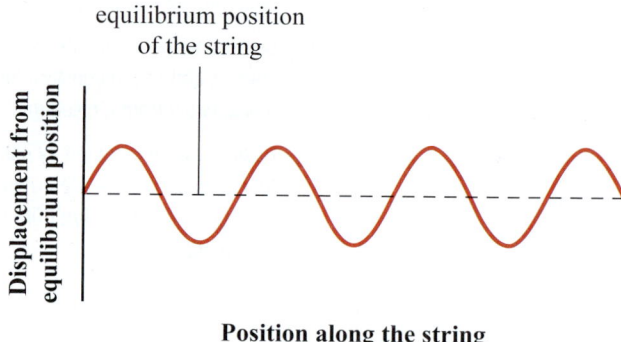

Figure 14-3 The waveform (at a fixed time) generated on a long string by shaking one end of the string in simple harmonic motion.

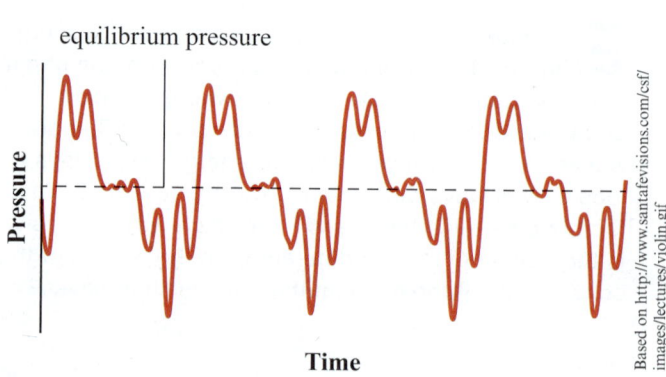

Based on http://www.santafevisions.com/csf/images/lectures/violin.gif

Figure 14-4 A waveform of sound waves generated when playing a note on a violin. The plot shows the change in air pressure with respect to the equilibrium pressure (dashed line, corresponding to no sound) at a given distance from the violin.

Figure 14-2 Concentric waves on a water surface.

string. Figure 14-5 shows an image of a complex waveform for waves on the surface of an ocean at a given instant.

The waveform for a sound wave in the air can be obtained by graphing either the variation in the air pressure or the displacement from the equilibrium position of a small element of air as a function of time. These are all different ways of visualizing the same wave.

Pulse A **pulse** is a disturbance of short duration. The disturbance generated on a stretched string by a quick flip of the hand, up and back to the original position, is a pulse (Figure 14-6). Similarly, a single handclap generates a sound pulse. The shape of a pulse depends on the motion of the source that generates the pulse.

Continuous wave A **continuous wave** (also called a wave) is a continuous series of pulses. Continuously flipping one end of a taut string up and down generates a continuous wave on the string. A vibrating tuning fork generates continuous sound waves as its tines oscillate in repetitive simple harmonic motion.

Wave cycle A continuous wave has a waveform that repeats over and over again. The fundamental unit of the repeating pattern is called a **wave cycle**. The duration of the wave cycle is the period of the wave. Figure 14-7 shows the wave cycles for a sine wave, and Figure 14-8 shows the wave cycles for a more complex wave.

Figure 14-5 Complex waveforms on the ocean surface.

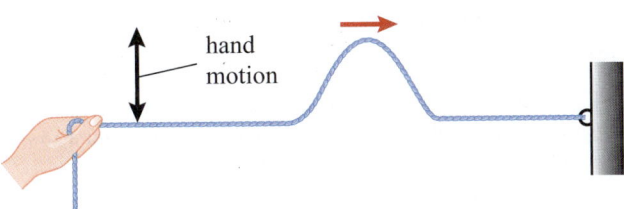

Figure 14-6 A quick flick on one end of a stretched string creates a pulse that travels along the string.

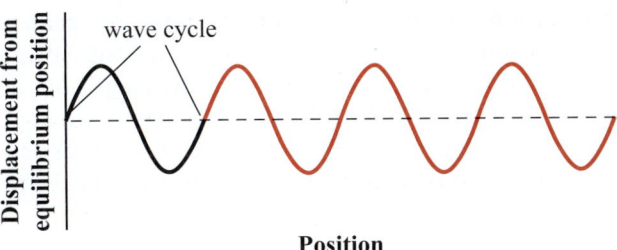

Figure 14-7 The black portion of the wave represents a wave cycle for a sinusoidal wave.

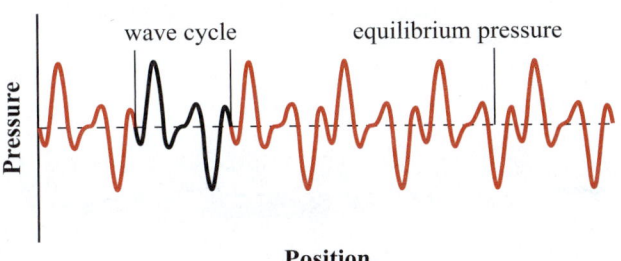

Figure 14-8 The black portion of the wave represents a wave cycle for a more complex wave.

Wave speed The speed at which a disturbance travels through a medium is called the **wave speed**, usually denoted by the letter v. For example, the speed of sound in dry air at sea level at 20 °C is approximately 343 m/s.

Transverse waves As a wave passes through a medium, the constituents (atoms and molecules) of the medium are temporarily displaced from their equilibrium positions in response to the forces exerted by the wave. In a **transverse wave**, the constituents of the medium move *perpendicular* to the direction of propagation of the wave (Figure 14-9). For example, a pulse generated by moving one end of a taut horizontal string up and down travels horizontally along the length of the string, and the segment of the string that the pulse is passing through is displaced vertically from the rest position.

Longitudinal waves In a **longitudinal wave**, the constituents of the medium move *parallel* to the direction

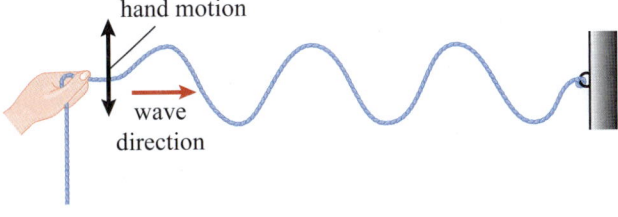

Figure 14-9 A transverse wave on a string.

of motion of the wave. For example, alternately pushing and pulling on one end of a long spring generates a wave that moves along the length of the spring (Figure 14-10(a)). As the wave passes through a given section of the spring, the coils of that section move parallel to the direction of motion, being alternately compressed and stretched. Sound waves are examples of three-dimensional longitudinal waves. As a sound wave passes through air, the air molecules oscillate in the direction of wave motion, alternately creating regions of higher and lower pressure (Figure 14-10(b)).

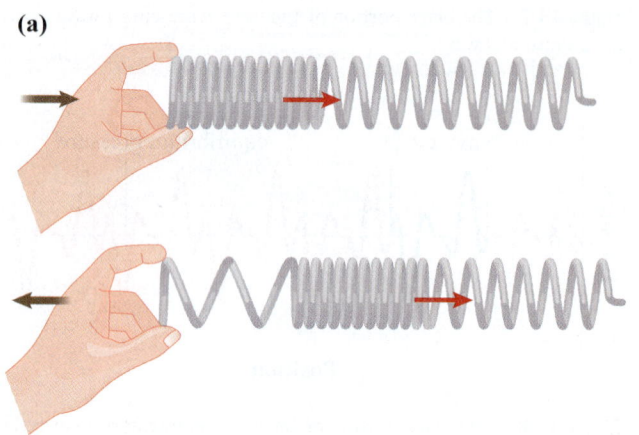

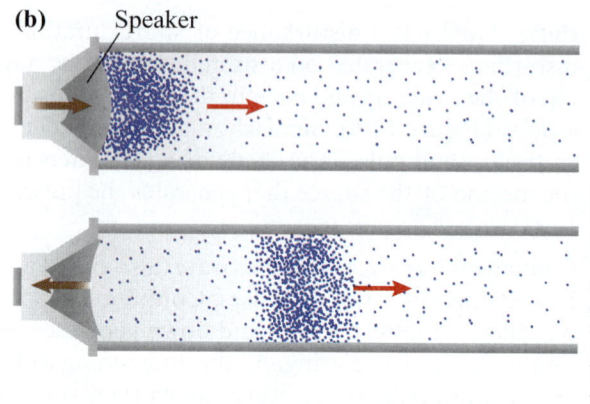

Based on http://sol.sci.uop.edu/~jfalward/wavessound/soundpulseslijnkypulse.jpg

Figure 14-10   (a) A longitudinal wave moving along a stretched spring. The movement of the coils is in the direction of wave motion, creating regions of higher and lower pressure (compressions and rarefactions, respectively). (b) A sound wave travelling along the length of an air-filled tube. As the diaphragm of the speaker oscillates, it creates a periodic pattern of compression and expansion of air.

MAKING CONNECTIONS

Earthquake Waves

During an earthquake, both transverse and longitudinal waves are generated in Earth's crust. The longitudinal waves, called P waves (primary waves), alternately compress and stretch the ground in the direction of wave motion. Transverse waves, called S waves (secondary waves), displace the ground up and down, perpendicular to the direction of wave motion (Figure 14-11). In solids, P waves move approximately twice as fast as S waves. Therefore, during an earthquake, we usually feel two jolts because the P waves arrive ahead of the S waves. The major damage in an earthquake is caused by the up-and-down motion of the ground, that is, by the S waves. Thus, the P waves can give some warning that the more destructive S waves are coming.

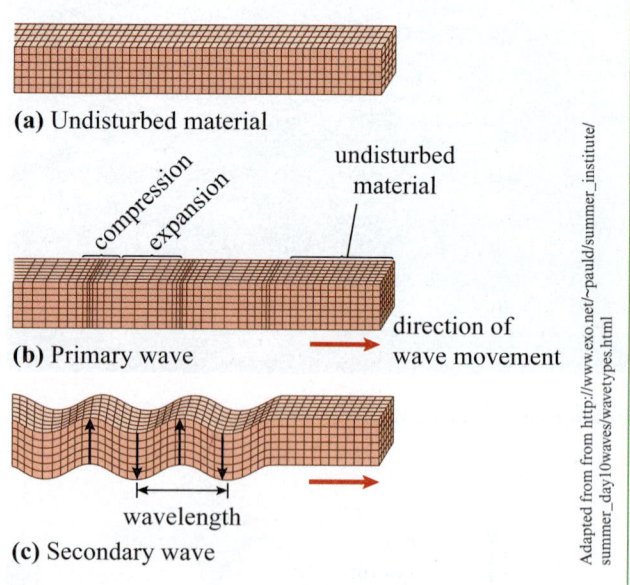

Adapted from from http://www.exo.net/~pauld/summer_institute/summer_day10waves/wavetypes.html

Figure 14-11   Waves generated in an earthquake consist of both longitudinal (P) and transverse (S) waves. The horizontal arrows show the direction of wave propagation.

14-2 The Motion of a Disturbance in a String

Consider a long, stretched string with one end tied to a rigid support and the other pulled taut by your hand. If you quickly flip your hand up and down, you will generate a pulse. The pulse travels along the length of string with a constant speed. This example demonstrates two important properties of waves:

- **Waves carry energy.** In moving the string up and down, your hand did work on the string, thus imparting energy to the string. The pulse carries this energy as it travels along the string. *It takes energy to generate a wave, and the wave carries this energy as it propagates through the medium. A wave can therefore be described as the motion of energy through a medium.*

- **The medium does not travel with the wave.** A given segment of the string is displaced from its equilibrium position while the pulse passes through it and then returns to the equilibrium position once the pulse has gone by. There is no net movement of the string in the direction of motion of the pulse. Similarly, a sound wave does not carry air along with it; the air molecules only oscillate back and forth in the direction of the motion of the sound wave. *Do not confuse the motion of the wave with the motion of the particles of the medium.*

What causes a pulse to move through a string? In Figure 14-12, a pulse is moving from left to right along a string that is held under a constant tension. Let us analyze the motion of a segment of the string located at point P as the pulse passes through it. Point E is at the leading edge of the pulse, and point A is at the trailing edge.

Before the pulse arrives, point P is at rest and is being pulled with tension T_S by the string to its left and its right. When the leading edge of the pulse reaches P, a slightly increased tension is exerted on P from the string to its left, causing it to accelerate upward. As P accelerates upward, the string to its right is also pulled upward. Therefore, P begins to experience a downward force due to increased tension from the string to its right. This causes P to decelerate as it moves farther from the equilibrium position. When P reaches the point of maximum displacement (C), it momentarily comes to a stop. At this displacement,

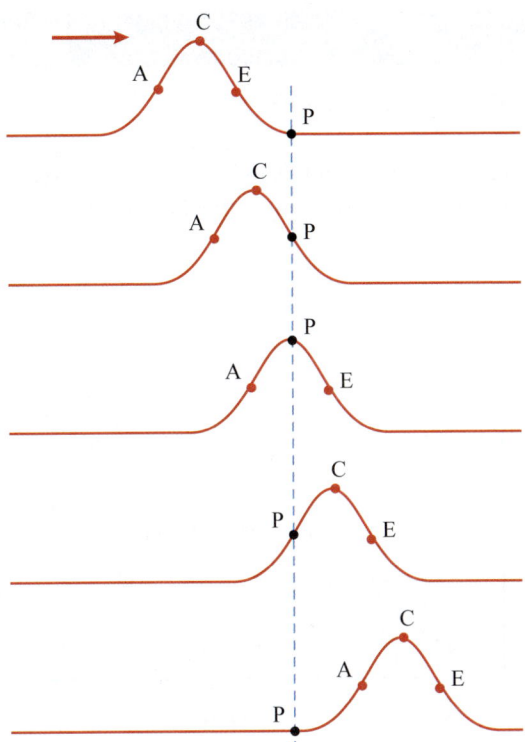

Figure 14-12 The motion of a point (P) on a string as a pulse passes through it. The pulse exerts a net force at P, causing it to be displaced from its equilibrium position.

P is at rest and is now being pulled downward by the string from both sides. Therefore, it experiences maximum downward acceleration at its farthest displacement. As P accelerates toward the equilibrium position, in addition to a downward force from the string to its left, it experiences an upward force from the string to its right, causing it to decelerate. It finally comes to rest after the pulse has passed through.

So, a pulse (or a wave) moves through a medium due to the forces exerted on a given segment by the elements of the medium around that segment.

Figure 14-13 shows a plot of displacement versus time for point P (Figure 14-12). Time $t = t_1$ corresponds to the arrival of the leading edge (E) of the pulse, and $t = t_2$ corresponds to the departure of the trailing edge (A).

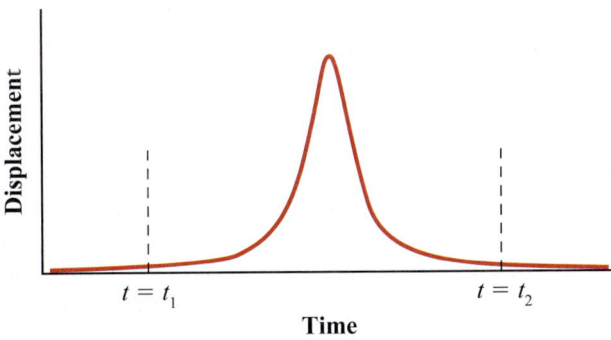

Figure 14-13 A displacement versus time plot for point P (Figure 14-12) as a pulse passes through it.

Displacement Graph

The pulse in Figure 14-14 is about to pass through point P. Which of the graphs in Figure 14-15 correctly describes the displacement of point P, as a function of time, when the pulse passes through P?

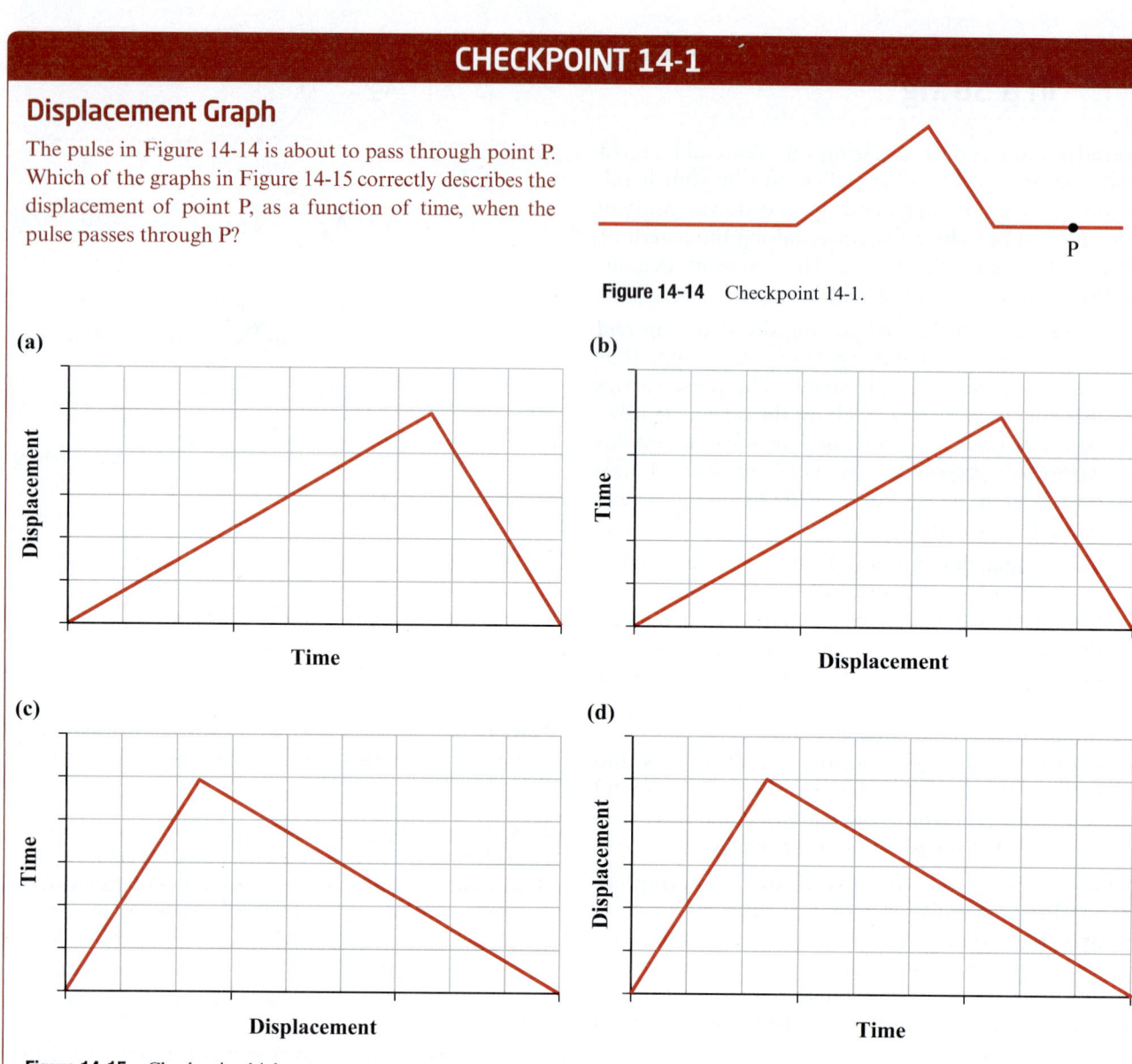

Figure 14-14   Checkpoint 14-1.

(a)

(b)

(c)

(d)

Figure 14-15   Checkpoint 14-1.

ANSWER: (d)

14-3 Equation for a Pulse Moving in One Dimension

Consider a pulse travelling along a stretched string along the x-axis. The transverse displacement of an element of the string from its equilibrium position is denoted by $D(x, t)$ and is plotted along the y-axis. $D(x, t)$ is a function of position along the string (x) and time (t). For example, $D(1, 0)$ represents the displacement from the equilibrium position at time $t = 0$ of an element of the string that is positioned at $x = 1$ m. In the absence of any disturbance, every element of the string is at its equilibrium position, and

$$D(x, t) = 0 \quad \text{for all } x \text{ and } t \tag{14-1}$$

A waveform can be described by giving the displacement of each element of the medium at a fixed time. Figure 14-16(a) shows a pulse with a waveform given by the equation (x and D are in metres and t is in seconds):

$$D(x, 0) = \frac{4}{x^4 + 2}$$

The units of 2 and 4 terms are such that D is in metres. Let us assume that the pulse is moving in the positive x-direction with a speed of 3.0 m/s. After 1.0 s, the pulse has moved 3.0 m toward the right, and the peak of the pulse is at $x = 3.0$ m, as shown in Figure 14-16(b). The waveform of the pulse at $t = 1.0$ s is given by the equation

$$D(x, 1.0) = \frac{4}{(x - 3.0)^4 + 2}$$

After 2.0 s, the pulse has moved 3.0 m farther and its peak is centred 6.0 m from the origin. Its waveform at $t = 2.0$ s (Figure 14-16(c)) is described by the equation

$$D(x, 2.0) = \frac{4}{(x - 6.0)^4 + 2}$$

For any given time t, the peak of the pulse is at $x = 3.0t$, and its waveform is given by

$$D(x, t) = \frac{4}{(x - 3.0t)^4 + 2}$$

If the pulse is moving to the right at an arbitrary speed v (instead of the speed of 3.0 m/s assumed above), we can describe the waveform of the pulse by substituting v for the speed in the preceding equation:

$$D(x, t) = \frac{4}{(x - vt)^4 + 2}$$

The above analysis can be generalized to write an equation for the waveform of any travelling pulse, given its waveform at $t = 0$. If at $t = 0$ the waveform of a pulse is described by a function $D(x)$ and the pulse is moving in the direction of increasing x with speed v, then the waveform at any time t is obtained by the substitution $x \rightarrow x - vt$ in $D(x)$:

$$D(x, t) = D(x - vt) \qquad (14\text{-}2)$$
(a disturbance moving toward increasing x)

Similarly, if the pulse is moving in the direction of decreasing x with speed v, the waveform at any time t is obtained by substituting $x \rightarrow x + vt$ in $D(x)$:

$$D(x, t) = D(x + vt) \qquad (14\text{-}3)$$
(a disturbance moving toward decreasing x)

A function of the form $D(x \mp vt)$ is called a **wave function** or a **displacement function** because it describes a travelling wave by giving the displacement of each point of the medium at any given time. In a wave function for a travelling wave, the variables x and t always occur in the combination $x \mp vt$.

(a)

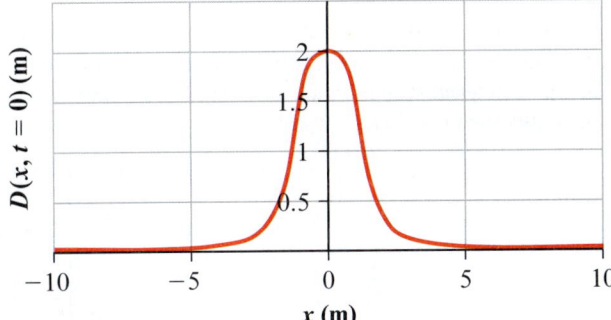

(b)

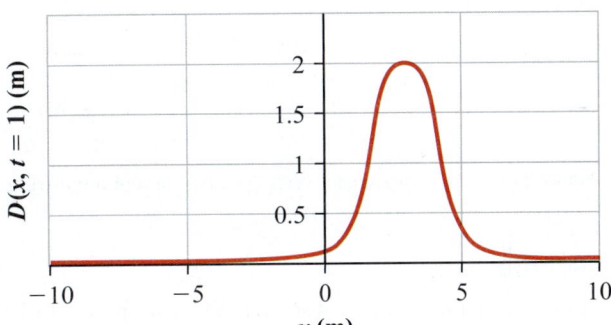

(c)

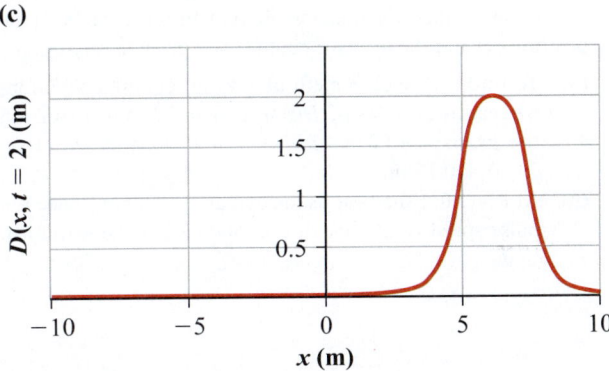

Figure 14-16 Displacement versus position plots of a pulse travelling in the positive x-direction with a speed of 3.0 m/s: (a) at $t = 0$, (b) at $t = 1.0$ s, and (c) at $t = 2.0$ s.

CHECKPOINT 14-2

A Travelling Pulse

Which of the following equations do *not* describe a pulse moving toward the direction of increasing x? (Here x denotes position and t time.)

(a) $D(x, t) = \dfrac{2}{(5.0t - x)^2 + 5}$

(b) $D(x, t) = \dfrac{2}{(x + 0.5t)^2 + 5}$

(c) $D(x, t) = \dfrac{2}{(x^2 - 5.0t) + 5}$

(d) $D(x, t) = \dfrac{3}{(2.0x - 4.0t)^4 + 6}$

ANSWER: (b) and (c)

EXAMPLE 14-1

Pulse on a String

A pulse is travelling from left to right with a speed of 2.0 m/s along a stretched string. Figure 14-17 shows the waveform of the pulse when $t = 1.0$ s.

(a) Find the displacements of the string elements at $x = 0.0$ m and $x = 5.0$ m at this time.
(b) Where was the peak of the pulse at $t = 0.0$ s?
(c) Where will the peak of the pulse be 5.0 s later?

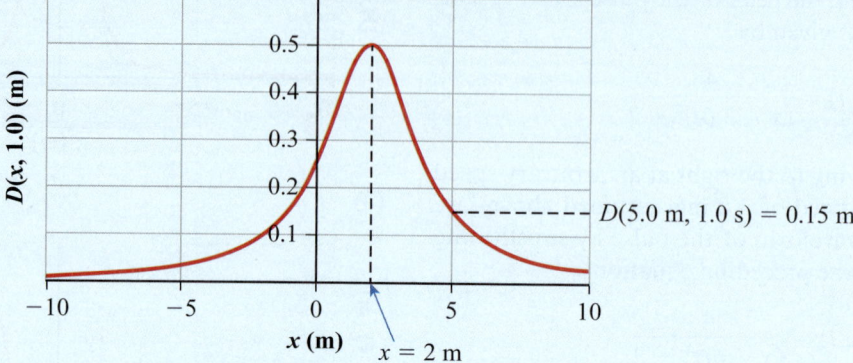

Figure 14-17   A displacement versus position plot of a pulse at $t = 1.0$ s, travelling in the positive x-direction.

Solution

In this example, a plot of displacement versus position of a pulse at $t = 1.0$ s is given. The required information can be read from the graph. When reading a graph, it is very important to understand the units on all axes. In this case, both the position (x) and the displacement (D) are given in metres.

(a) To find the displacement at $x = 0.0$ m and $x = 5.0$ m, we read the values of $D(0.0, 1.0)$ and $D(5.0, 1.0)$ from the graph: $D(0.0$ m, 1.0 s$) = 0.25$ m and $D(5.0$ m, 1.0 s$) = 0.15$ m.

(b) At $t = 1.0$ s, the peak is located at $x = 2.0$ m. Since the pulse speed is 2.0 m/s, one second earlier the waveform

was shifted to the left by 2.0 m. The peak was therefore located at $x = 0.0$ m.

(c) In 5.0 s, the pulse will travel a distance of 10.0 m farther right from its location at $t = 1.0$ s, so the peak will be at $x = 12.0$ m.

Making sense of the result

The displacements calculated in part (a) are between 0.0 m and 0.5 m (the maximum displacement of the pulse). The units are in metres. Also, since the wave speed is 2.0 m/s, one second before the position shown, the peak will be 2.0 m to the left, and 5 seconds later it will be 10 m to the right.

EXAMPLE 14-2

Using Waveform Equations

The equation for a pulse travelling through a medium is given by

$$D(x, t) = \frac{3.0 \text{ m}^3}{(x - (2.0 \text{ m/s})t)^2 + 5.0 \text{ m}^2}$$

where x and D are in metres and t is in seconds.

(a) What is the speed of the pulse, and in which direction is it travelling?
(b) What is the displacement of a point of the medium located at $x = 0.30$ m at $t = 1.0$ s?
(c) What is the maximum displacement of a point as the pulse passes through it? Plot the displacement as a function of time of the point at $x = 15.0$ m as the pulse passes through it.
(d) What is the equation of a pulse that has the same velocity and shape as the above waveform but is inverted relative to it?

Solution

We are given a displacement function describing a travelling pulse. The required information about this pulse is contained in this function. The units of the 3.0 and 5.0 terms in the equation are such that D is in metres.

(a) Since the x- and t-variables appear in the form $x - 2.0t$, the pulse speed is 2.0 m/s. The minus sign between x and t indicates that the pulse is moving in the direction of increasing x.

(b) Inserting $x = 0.30$ m and $t = 1.0$ s in the given waveform, we get

$$D(0.3 \text{ m}, 1.0 \text{ s}) = \frac{3.0 \text{ m}^3}{(0.3 \text{ m} - ((2.0 \text{ m/s})(1.0 \text{ s}))^2 + 5.0 \text{ m}^2}$$
$$= 0.38 \text{ m}$$

(c) The numerator of the given waveform is a constant. Therefore, the displacement is maximum when the denominator is minimum. The term $(x - 2.0t)^2$ has a minimum value of zero. Therefore, the minimum value of the denominator is 5.0. The maximum displacement of a point of the medium is therefore $3.0 \text{ m}^3/5.0 \text{ m}^2 = 0.60 \text{ m}$. A plot of $D(15, t)$ as function of t is shown in Figure 14-18.

(d) The displacement of the inverted pulse is the negative of the displacement of the given pulse. We therefore multiply the numerator of the given wave function by -1 to obtain the wave function of the inverted pulse. The wave function of the inverted pulse is

$$D(x, t) = \frac{-3.0 \text{ m}^3}{(x - (2.0 \text{ m/s})t)^2 + 5.0 \text{ m}^2}$$

Making sense of the result

Since the wave speed is 2.0 m/s, starting from $t = 0$ the peak of the pulse reaches the point at $x = 15.0$ m in $t = (15.0 \text{ m})/(2.0 \text{ m/s}) = 7.5$ s.

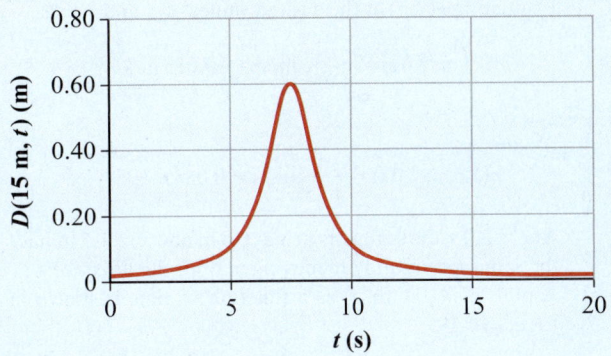

Figure 14-18 The displacement versus time graph at $x = 15.0$ m for the wave function of Example 14-2.

14-4 Transverse Speed and Wave Speed

In the presence of a wave, the instantaneous speed, denoted by $u(x, t)$, of a string element is equal to the rate of change of its displacement, $D(x, t)$, with time. The displacement function depends on both position and time, so the instantaneous speed of an element located at x_0 is obtained by taking the derivative of $D(x, t)$ with respect to time while holding the position fixed at $x = x_0$. When a function depends on several variables, its derivative with respect to one variable, with other variables held constant, is called the partial derivative (see Appendix D). The instantaneous speed of an element of the medium located at $x = x_0$ is therefore obtained by taking the partial derivative of displacement, $D(x, t)$, with respect to time, t, while keeping x fixed at x_0:

$$u(x_0, t) = \frac{\partial D(x, t)}{\partial t}\bigg|_{x = x_0} \tag{14-4}$$

The symbol ∂ indicates a partial derivative, and the term $\partial/\partial t$ indicates a partial derivative with respect to time. The instantaneous speed can also be obtained by graphing $D(x, t)$ as a function of time for a fixed x. The slope of the graph at a given time is equal to the transverse speed of the segment at that time, and the sign of the slope gives the direction of motion.

EXAMPLE 14-3

Transverse Displacement

A pulse travelling through a string is described by the wave function

$$D(x, t) = \frac{2.0 \text{ m}^3}{(x - (0.50 \text{ m/s})t)^2 + 4.0 \text{ m}^2}$$

where x and D are in metres and t is in seconds.

(a) Find the velocity of the pulse.
(b) At $t = 2.0$ s, what are the speeds of the string elements located at $x = 0.50$ m and $x = 1.5$ m?
(c) Graph the speed of the particle located at $x = 10.0$ m between $t = 0.0$ s and $t = 40.0$ s.

Solution

We are given a displacement function describing a travelling pulse. Information about pulse speed can be read from this equation. Also, the velocity of a given segment of the string can be obtained by taking the derivative of the displacement function with respect to time for a fixed x.

(a) The coefficient of the t-term is 0.50, so the pulse speed is 0.50 m/s. The sign between the position, x, and the time, t, is negative, so the pulse is travelling in the direction of increasing x.
(b) To determine the speed of a string element, we take the derivative of the wave function with respect to t, while holding x constant (units in intermediate steps are not shown to keep the calculation easy to follow):

$$u(x, t) = \frac{\partial}{\partial t}\left(\frac{2.0}{(x - 0.50t)^2 + 4.0}\right)$$

$$= 2.0 \times (-1) \frac{2.0 \times (x - 0.50t) \times (-0.50)}{((x - 0.50t)^2 + 4.0)^2}$$

$$= \frac{2.0(x - 0.50t)}{((x - 0.50t)^2 + 4.0)^2}$$

(continued)

Evaluating $u(x, t)$ at the desired values of x and t gives

$$u(0.5\,\text{m}, 2.0\,\text{s}) = \frac{-1}{4.25^2}\,\text{m/s} = -0.055\,\text{m/s}$$

and

$$u(1.5\,\text{m}, 2.0\,\text{s}) = \frac{1}{4.25^2}\,\text{m/s} = 0.055\,\text{m/s}$$

At $t = 2.0$ s, the segments at $x = 0.5$ m and $x = 1.5$ m have the same speeds and are moving in opposite directions.

(c) A plot of $u(10.0\ \text{m}, t)$ as a function of time is shown in Figure 14-19.

Making sense of the result

From the plot, we see that the string element at $x = 10.0$ m begins to move up (positive speed) slowly but increases its speed with time until it reaches a maximum speed. The element then starts to slow down and momentarily comes to a stop. The instant where the velocity changes sign corresponds to the centre of the pulse passing through the element. The element then moves downward, back to the equilibrium position, first gaining speed and then slowing down until it returns to its equilibrium position.

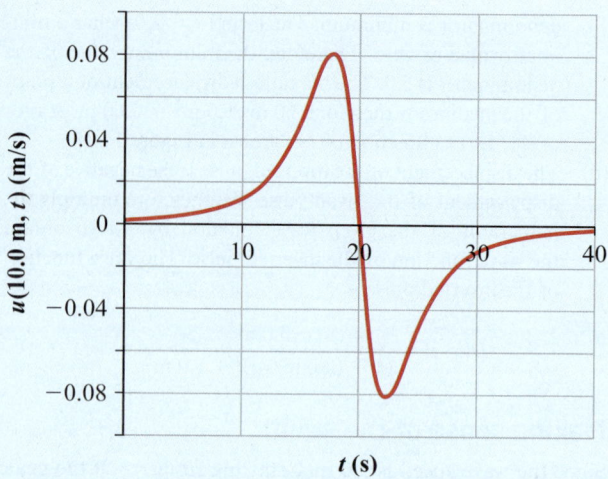

Figure 14-19 The velocity versus time plot of a segment located at $x = 10.0$ m as the pulse described in Example 14-3 passes through the segment.

CHECKPOINT 14-3

Transverse Speed of a String

A waveform is passing through a given segment of a string with speed v. What is the speed of this segment as a function of time?

(a) v
(b) $-v$
(c) zero
(d) changes with time

Wave Speed on a String

The speed of a mechanical wave in a medium depends on the elastic and inertial properties of the medium. For a string, its elastic property is related to the tension in the string, and the inertial property depends on the linear mass density (mass per unit length), denoted by the Greek letter mu, μ. If M is the total mass of a string of length L, then

$$\mu = \frac{M}{L} \qquad (14\text{-}5)$$

The units of linear mass density are kilograms per metre.

As we discussed earlier, in the presence of a wave, a given element of a string is displaced from its equilibrium position by the forces exerted on it by the neighbouring string elements. The greater the tension (the restoring force), the faster a string element moves in response. However, a greater linear mass density means greater inertia and, hence, slower movement. Since the wave speed is determined by the response time of the medium as a wave passes through it, we should expect the wave speed in a string to be proportional to the tension in the string and inversely proportional to its linear mass density.

Consider a pulse moving from left to right with speed v along a string that has linear mass density μ and is under tension of magnitude T_s (Figure 14-20). The pulse speed is measured with respect to a reference frame in which the string is at rest, which we will call the string frame. We assume that the height of the pulse is small compared to the length of the string so that the tension and the linear mass density do not change appreciably in the presence of the pulse. To analyze the motion of the pulse, we choose a frame of reference that is moving with the pulse (call it the pulse frame). In this frame, the pulse is stationary and the string is moving with speed v from right to left. The string and pulse frames are moving with a constant speed relative to each other, so Newton's laws of motion are identical in both frames. Let us focus on a very small element of string at the top of the pulse at some instant, as shown in Figure 14-21. The string element has length ΔL; hence mass $\Delta m = \mu \Delta L$. Since ΔL is very small, this element

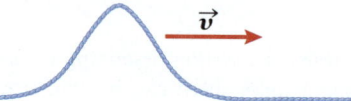

Figure 14-20 A pulse moving with speed v toward increasing x in a string that is under a tension T_s.

516   SECTION 2 | **WAVES AND OSCILLATIONS**

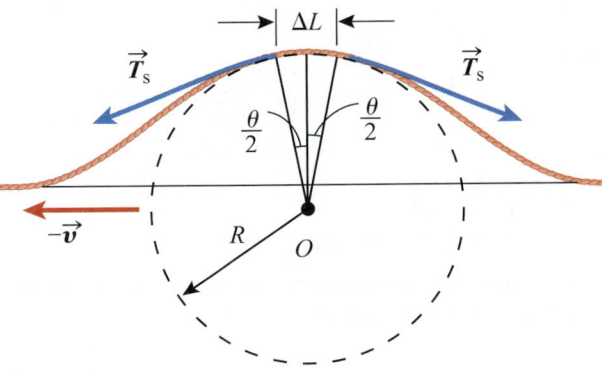

Figure 14-21 The same pulse, viewed from a reference frame in which the pulse is at rest and the string is moving toward the left with speed v.

can be considered to form an arc of a circle of radius R subtending an angle θ with respect to the centre of the circle such that $\Delta L = R\theta$. In the pulse frame, the mass Δm is moving in a circle of radius R with speed v, so it has a centripetal acceleration of magnitude $a_c = \dfrac{v^2}{R}$.

The centripetal acceleration is provided by the tension force, $\vec{T}_s$, in the string, acting tangentially at each end of the element.

The two horizontal components of the tension force each have magnitude $T_s \cos(\theta/2)$ and cancel because they are pointing in opposite directions. Each radial component has a magnitude $T_s \sin(\theta/2)$ pointing toward the centre of the circle. Therefore, the net restoring force on the string element has a magnitude $2T_s \sin \theta/2$ and points toward the centre, O. According to Newton's second law, the net restoring force on this element is equal to its mass times the radial acceleration:

$$F_{net} = \Delta m\left(\frac{v^2}{R}\right) = 2T_s \sin\left(\frac{\theta}{2}\right) \tag{14-6}$$

Since ΔL and, hence, angle θ are small, we can use the small-angle approximation, $\sin\theta \approx \theta$:

$$\Delta m\left(\frac{v^2}{R}\right) \approx 2T_s\left(\frac{\theta}{2}\right)$$
$$= T_s\left(\frac{\Delta L}{R}\right)$$

Solving for the wave speed yields

KEY EQUATION

$$v^2 = T_s\left(\frac{\Delta L}{\Delta m}\right) = \frac{T_s}{\mu}$$
$$v = \sqrt{\frac{T_s}{\mu}} \tag{14-7}$$

This analysis confirms our assumption that the wave speed increases with greater tension in the string and decreases with greater linear mass density. Since we did not specify a particular waveform, Equation 14-7 is valid for a disturbance of any shape. Note that the expression for the wave speed is of the form

$$\text{wave speed} = \sqrt{\frac{\text{elastic property of the string}}{\text{inertial property of the string}}}$$

CHECKPOINT 14-4

Wave Speed in a String

A long string of non-zero linear mass density is hanging from a ceiling. A pulse is generated by shaking the string at the bottom end. As this pulse moves upward toward the ceiling, what happens to its speed? Explain your answer.
(a) The speed remains the same.
(b) The speed increases.
(c) The speed decreases.
(d) The speed increases at first, but then decreases closer to the ceiling.

ANSWER: (b) The tension in the string increases with height from the bottom. Therefore, the wave speed increases.

EXAMPLE 14-4

Pulse Speed on a Hanging String

A string of linear mass density 20.0 g/m and length 10.0 m is hanging vertically from a high ceiling. A mass of 2.00 kg hangs freely from the lower end of the string. A pulse is generated at the lower end of the string and travels upward. What is the speed of the pulse at the lower end of the string, at the middle of the string, and at the top of the string?

Solution

The tension at any given point along the string is due to the weight of the hanging mass and the weight of the portion of the string below the point. Let x denote the height of a segment of the string, measured from the lower end. The total weight

hanging below the point x is $Mg + (\mu x)g$, which is equal to the tension in the string at a height x.

Therefore, the wave speed, $v(x)$, at a height x from the lower end of the string is

$$v(x) = \sqrt{\frac{Mg + \mu xg}{\mu}}$$

We know the following quantities:
linear mass density of the string: $\mu = 20.0$ g/m
$\qquad\qquad\qquad\qquad\qquad\qquad\qquad = 20.0 \times 10^{-3}$ kg/m

mass of the block: $\qquad\qquad M = 2.00$ kg
length of the string: $\qquad\qquad L = 10.0$ m

(continued)

Evaluating the wave speed at $x = 0$, $x = 5.0\,\text{m}$, and $x = 10.0\,\text{m}$, we get

$$v(0) = \sqrt{\frac{Mg}{\mu}}$$

$$= \sqrt{\frac{2.00\ \text{kg} \times 9.81\,\text{m/s}^2}{20.0 \times 10^{-3}\text{kg/m}}} = 31.3\ \text{m/s}$$

$$v(5.0\ \text{m}) = \sqrt{\frac{Mg + 5.0\,\mu g}{\mu}}$$

$$= \sqrt{\frac{(2.00\ \text{kg} + 5.0\ \text{m} \times 20.0 \times 10^{-3}\ \text{kg/m}) \times 9.81\,\text{m/s}^2}{20.0 \times 10^{-3}\ \text{kg/m}}}$$

$$= 32.1\ \text{m/s}$$

$$v(10.0\ \text{m}) = \sqrt{\frac{Mg + 10.0\,\mu g}{\mu}}$$

$$= \sqrt{\frac{(2.00\ \text{kg} + 10.0\ \text{m} \times 20.0 \times 10^{-3}\ \text{kg/m}) \times 9.81\,\text{m/s}^2}{20.0 \times 10^{-3}\ \text{kg/m}}}$$

$$= 32.8\ \text{m/s}$$

Making sense of the result

The tension in the string increases with height, so as the pulse moves toward the ceiling, the pulse speeds up.

14-5 Harmonic Waves

A wave generated by a source that is undergoing simple harmonic motion is called a **harmonic wave** or a **sinusoidal wave**. Imagine a long string with one end attached to a source. As the source oscillates, it generates a continuous wave that travels along the string in simple harmonic motion (Figure 14-22).

A snapshot of the string at, say, $t = 0$ corresponds to a sine (or a cosine) function (Figure 14-22) and can be represented in the following form:

$$D(x) = A \sin(kx) \text{ at } t = 0 \qquad (14\text{-}8)$$

Here, $D(x)$ denotes the transverse displacement of the element of the string located at x. The parameters A and k are described below.

Amplitude As the sine function oscillates between $+1$ and -1, the displacement, $D(x)$, oscillates between the values $\pm A$. A is called the **amplitude** of the wave and corresponds to the maximum displacement from the equilibrium position of any element of the string. Amplitude is a positive quantity. For waves on a string, amplitude has the dimension of length. Figure 14-23 shows two harmonic waves with different amplitudes.

The point of maximum displacement ($D(x) = +A$) in a cycle is called a **crest**, and the point of minimum displacement ($D(x) = -A$) is called a **trough**.

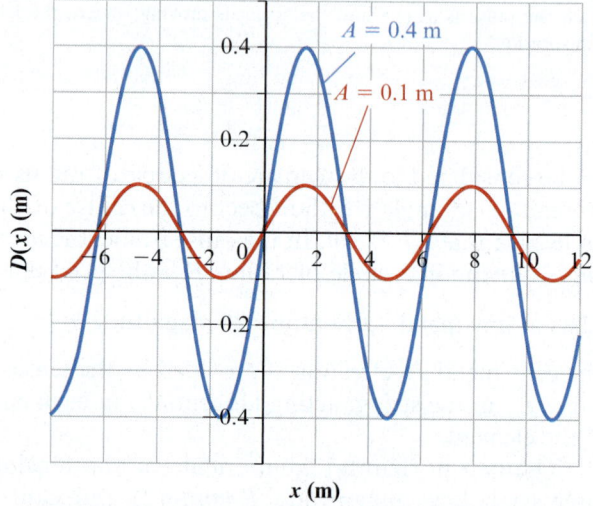

Figure 14-23 A snapshot at a fixed time of two harmonic waves with amplitudes 0.1 m and 0.4 m.

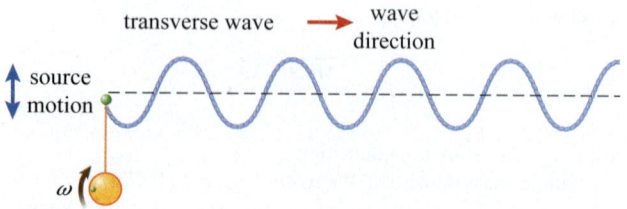

Figure 14-22 A source executing simple harmonic motion generates a simple harmonic wave.

Source: This drawing was adapted from PhET Interactive Simulations, University of Colorado, http://phet.colorado.edu.

Wavelength The shortest distance over which a wave shape repeats is called the **wavelength**. The usual symbol for wavelength is the Greek letter lambda, λ. Wavelength has dimensions of length. In Figure 14-25, the points at x_1 and x_2 are one wavelength apart and the points at x_1 and x_3 are two wavelengths apart. Figure 14-26 shows two waves with wavelengths of 2.0 m and 4.0 m.

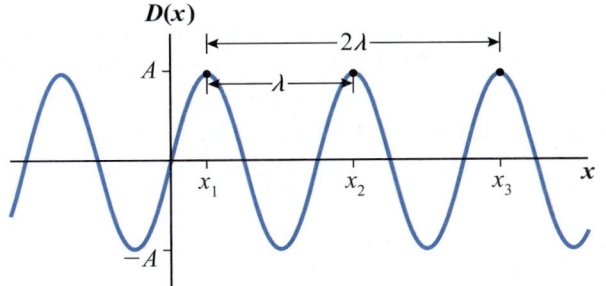

Figure 14-25 The points at $x = x_1$ and $x = x_2$ are one wavelength apart. The points at $x = x_1$ and $x = x_3$ are two wavelengths apart.

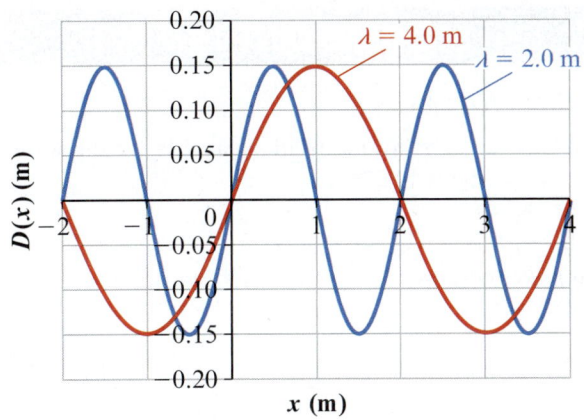

Figure 14-26 Harmonic waves with wavelengths of 2.0 m and 4.0 m.

INTERACTIVE ACTIVITY 14-2

Harmonic Waves

In this activity, you will use the PhET simulation "Harmonic Waves." Work through the simulation and accompanying questions to gain an understanding of harmonic waves.

CHECKPOINT 14-6

Wavelengths

Figure 14-27 shows a plot of three waves with wavelengths λ_1, λ_2, and λ_3. Which of the following wavelength comparisons is correct?

(a) $\lambda_1 < \lambda_2 < \lambda_3$
(b) $\lambda_1 > \lambda_2 > \lambda_3$
(c) $\lambda_1 < \lambda_3 < \lambda_2$
(d) $\lambda_1 > \lambda_2 < \lambda_3$

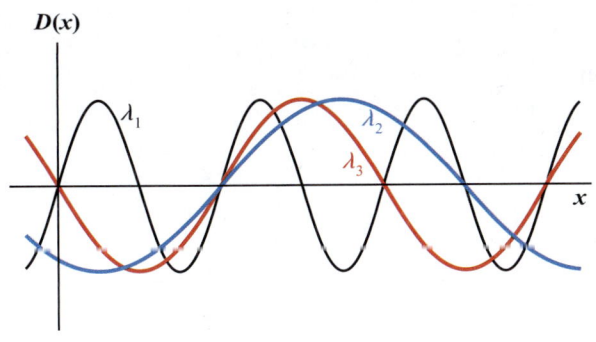

Figure 14-27 Checkpoint 14-6.

ANSWER: (c)

---

between k and λ, we use the fact that a waveform repeats over a length λ. Therefore,

$$D(x) = D(x + \lambda) \text{ for any given location } x \qquad (14\text{-}9)$$

Applying this relationship to the wave function for harmonic waves (Equation 14-8) gives

$$A \sin kx = A \sin k(x + \lambda) \qquad (14\text{-}10)$$

$$\sin kx = \sin(kx + k\lambda) \qquad (14\text{-}11)$$

The sine function repeats after 2π rad, so Equation 14-11 is satisfied when $k\lambda = 2\pi$ rad. Therefore,

KEY EQUATION
$$k = \frac{2\pi}{\lambda}$$

Thus, the wave number is 2π times the reciprocal of the wavelength. Its units are radians/metre. So, the wave number is a measure of the change of phase per unit length. For $\lambda = 1$ m, $k = 2\pi$ rad/m, and for $\lambda = 0.5$ m, $k = 4\pi$ rad/m. As the wavelength increases, the wave number decreases.

PEER TO PEER

I have to remember that the symbol k is used to describe the spring constant as well as the wave number. These are two very different quantities: I must be careful not to confuse one with the other.

Wave number The parameter k in Equation 14-8 is called the **wave number**. To determine the relationship

CHECKPOINT 14-7

Wave Numbers

Which of the following relations applies for the wave-forms in Figure 14-27?

(a) $k_1 < k_2 < k_3$
(b) $k_1 > k_3 > k_2$
(c) $k_1 < k_2 > k_3$
(d) $k_1 > k_2 < k_3$

ANSWER: (b)

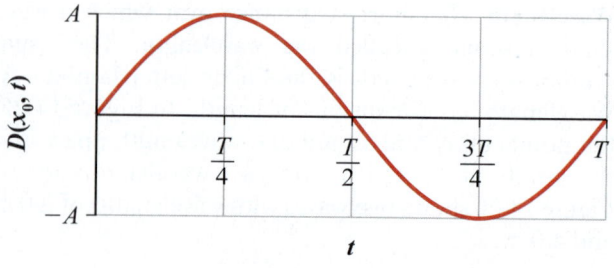

Figure 14-28 The displacement as a function of time as one wave cycle passes through a given point, x_0, of the string.

Period and wave frequency A continuous wave repeats itself in space and time (Figure 14-28). If the period of the source generating a continuous wave is T seconds, one wave cycle is produced in T seconds. As the wave moves through the medium, it takes T seconds for a wave cycle to pass a given point in the medium. The frequency, f, of a wave is equal to the number of wave cycles passing a fixed point in space in 1 s. Therefore, in 1 s, an element of the medium will undergo f oscillations. The relationship between the period and the frequency is given by Equation (13-3)

$$f = 1/T$$

CHECKPOINT 14-8

Wave Frequency

Figure 14-29 shows four plots of displacement versus time for four waves passing through a string element at $x = 0$. Order the frequencies of these waves from lowest to highest.

Lowest ___ ___ ___ ___ Highest

(a)

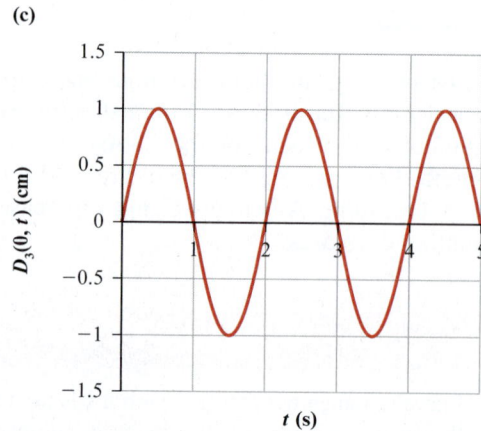

(b)

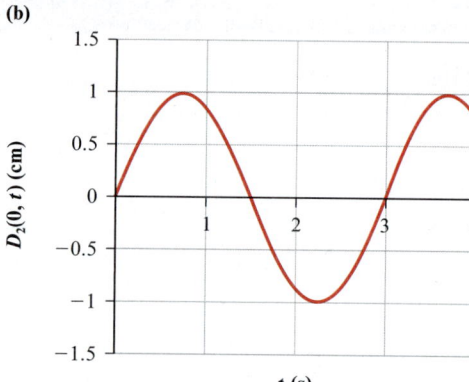

(c)

(d)

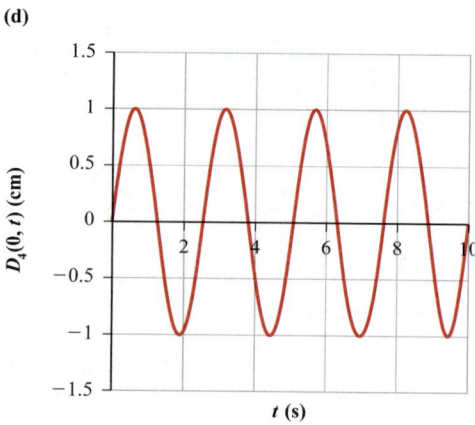

Figure 14-29 Checkpoint 14-8.

ANSWER: (b), (d), (c), (a)

Wave Speed Relationships

Imagine a train moving with a constant speed with respect to a stationary observer standing beside the tracks. Each car of the train has length L, and N cars pass by the observer in 1 s. Therefore, the speed, v, of the train with respect to the observer is $v = LN$. The same reasoning applies to the speed of a mechanical wave with respect to a medium's frame of reference. Consider a harmonic wave of wavelength λ and frequency f. In one second, f wave cycles, each of length λ, pass through a fixed point of the medium. Therefore, the wave speed with respect to the medium's frame of reference is

KEY EQUATION
$$v = \lambda f \qquad (14\text{-}12)$$

Equation 14-12 is valid for all periodic waves. Using $k = 2\pi/\lambda$ and $f = \omega/(2\pi)$, we can write the wave speed in terms of the wave number and the angular frequency of a harmonic wave:

$$
\begin{aligned}
v &= \lambda f \\
&= \left(\frac{2\pi}{k}\right)\left(\frac{\omega}{2\pi}\right) = \frac{\omega}{k}
\end{aligned} \qquad (14\text{-}13)
$$

EXAMPLE 14-5

Wave Speed and Wavelength

The speed of sound waves is approximately 340 m/s in air and 1400 m/s in fresh water. What is the change in the wavelength of a sound wave of frequency 400 Hz when it crosses from air into water?

Solution

The frequency of a wave is determined by the frequency of the source and does not change when the wave moves from one medium into another. Using Equation 14-12, the wavelengths of the sound wave in air and in water are as follows:

$$\lambda_{air} = \frac{v_{air}}{f} = \frac{340 \text{ m/s}}{400 \text{ Hz}} = 0.85 \text{ m}$$

$$\lambda_{water} = \frac{v_{water}}{f} = \frac{1400 \text{ m/s}}{400 \text{ Hz}} = 3.5 \text{ m}$$

Making sense of the result

Since the speed of sound in water is about four times its speed in air, for the same number of wave cycles to pass through a given point in water, the wavelength of the sound wave in water must be about four times the wavelength in air.

Travelling Harmonic Waves

The displacement of any element of a medium from its equilibrium position changes with time as a wave passes through it. Figure 14-30 shows the displacement of two string elements as a wave passes through the string from

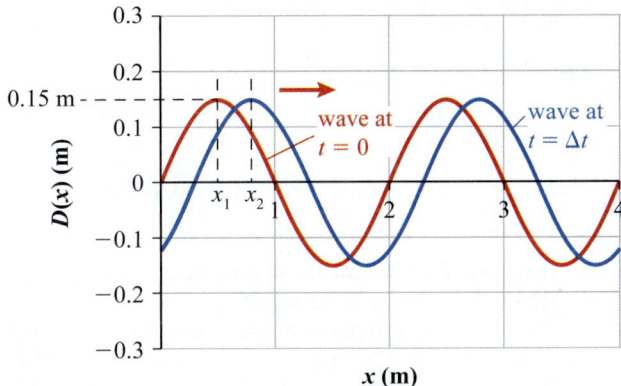

Figure 14-30 The graphs of displacement versus position at time $t = 0$ and $t = \Delta t$ for a sinusoidal wave travelling from left to right.

left to right. At $t = 0$, the point at x_1 is at the crest of the wave and is displaced by 0.15 m, and the point at x_2 is displaced by $+0.10$ m. A short time later, the wave has moved to the right, and the point at x_1 is displaced by $+0.10$ m and the point at x_2 by $+0.15$ m. We can write an equation describing a travelling harmonic wave using the same method that we used to describe a moving pulse. If the equation describing the shape of a wave with a speed of v is known for some fixed time, say $t = 0$, then the wave equation for any arbitrary time t is obtained by

- replacing x by $x - vt$ for a wave travelling in the direction of increasing x
- replacing x by $x + vt$ for a wave travelling in the direction of decreasing x

For a harmonic wave travelling in the direction of increasing x, replacing x by $x - vt$ in Equation 14-8 gives

$$D(x, t) = A \sin (k(x - vt)) \qquad (14\text{-}14)$$

Since

$$kv = \left(\frac{2\pi}{\lambda}\right)v = 2\pi f = \omega$$

Equation 14-14 can be written as

$$D(x, t) = A \sin(kx - \omega t) \qquad (14\text{-}15)$$

Similarly, the wave function for a harmonic wave travelling in the direction of decreasing x is given by

$$D(x, t) = A \sin(kx + \omega t) \qquad (14\text{-}16)$$

We can also write Equation 14-15 in terms of the wavelength and the time period by using the relationships $k = 2\pi/\lambda$ and $\omega = 2\pi f = 2\pi/T$:

$$D(x, t) = A \sin\left(\frac{2\pi}{\lambda}x - \frac{2\pi}{T}t\right) \qquad (14\text{-}17)$$

Since a harmonic wave is described by a sine (or a cosine) function, its properties are determined by its amplitude (A), its wavelength (λ), and its period (T).

EXAMPLE 14-6

Travelling Harmonic Wave

A harmonic wave travelling along a string is described by the wave function

$$D(x, t) = (0.10 \text{ m}) \sin[(2.0 \text{ rad/m})x - (3.0 \text{ rad/s})t]$$

where x is in metres and t is in seconds.

(a) Determine the wavelength and the frequency of the wave.
(b) What is the velocity of the wave?
(c) What is the displacement of the segment of the string located at $x = 0.30$ m at $t = 1.0$ s and at $t = 1.2$ s? Determine the average speed of the segment during this time, and compare it to the wave speed.
(d) Plot the displacement of the segment at $x = 0.3$ m between $t = 1.0$ s and $t = 5.0$ s.
(e) Plot the shape of the section of the string between $x = 0.50$ m and $x = 3.0$ m at $t = 0.50$ s and $t = 0.60$ s.

Solution

The given wave function should be compared with the standard equation for a travelling harmonic wave to determine various constants.

(a) By comparing the given equation with the standard equation for a travelling harmonic wave, Equation 14-15, we find that

$$A = 0.10 \text{ m}, k = 2.0 \text{ rad/m, and } \omega = 3.0 \text{ rad/s}$$

Therefore, the wavelength and frequency of the wave are

$$\lambda = \frac{2\pi}{k} = \frac{2\pi \text{ rad}}{2.0 \text{ rad/m}} = 3.1 \text{ m}$$

$$f = \frac{\omega}{2\pi} = \frac{3.0 \text{ rad/s}}{2\pi \text{ rad}} = \frac{1.5}{\pi} \text{ Hz} = 0.48 \text{ Hz}$$

(b) Because we know λ and f, the wave speed can be calculated from Equation 14-13:

$$v = \frac{\omega}{k} = \frac{3.0 \text{ rad/s}}{2.0 \text{ rad/m}} = 1.5 \text{ m/s}$$

Since the argument of the sine function is of the form $kx - \omega t$, the wave is travelling in the direction of increasing x.

(c) Substituting the given values for x and t in the equation for $D(x, t)$, we get

$$D(0.30 \text{ m}, 1.0 \text{ s}) = (0.10 \text{ m}) \sin((2.0 \text{ rad/m}) \times (0.30 \text{ m})$$
$$- (3.0 \text{ rad/s}) \times (1.0 \text{ s}))$$
$$= (0.10 \text{ m}) \times (-0.68) = -0.068 \text{ m}$$

$$D(0.30 \text{ m}, 1.2 \text{ s}) = (0.10 \text{ m}) \sin((2.0 \text{ rad/m}) \times (0.30 \text{ m})$$
$$- (3.0 \text{ rad/s}) \times (1.2 \text{ s}))$$
$$= (0.10 \text{ m}) \times (-0.14) = -0.014 \text{ m}$$

In 0.20 s, this segment moved a distance of 0.054 m, so the average speed was

$$0.054 \text{ m}/0.20 \text{ s} = 0.27 \text{ m/s}$$

This speed is not the same as the wave speed.

(d) A plot of the displacement of the segment located at $x = 0.30$ m between $t = 1.0$ s and $t = 5.0$ s is obtained by setting $x = 0.30$ m in the wave equation and plotting the

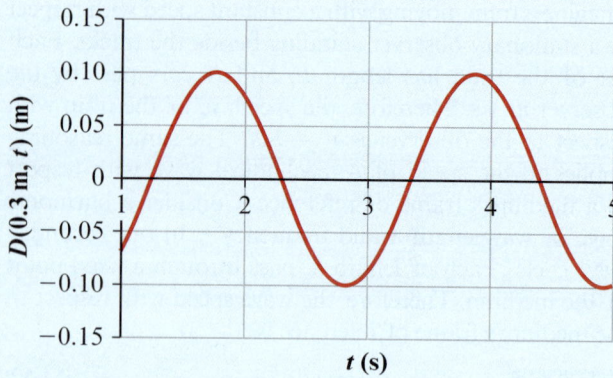

Figure 14-31   The graph of the wave function for Example 14-6 at $x = 0.3$ m for $1.0 \text{ s} \leq t \leq 5.0$ s.

resulting equation between $t = 1.0$ s and $t = 5.0$ s. We thus need to plot

$$D(0.3 \text{ m}, t) = (0.10 \text{ m}) \sin(0.60 - 3.0t), 1.0 \leq t \leq 5.0$$

The resulting plot is shown in Figure 14-31. Note that the above equation is similar to the equation for a simple harmonic oscillator (described by a cosine function) with angular frequency 3.0 rad/s, amplitude 0.10 m, and phase constant $\left(\frac{\pi}{2} - 0.60\right)$ rad.

(e) A plot of the shape of the segment of the string between $x = 0.50$ m and $x = 3.0$ m at $t = 0.50$ s and $t = 0.60$ s is obtained by setting $t = 0.50$ s and $t = 0.60$ s in the wave equation and plotting the resulting equations between $x = 0.50$ m and $x = 3.0$ m:

$$D(x, 0.50 \text{ s}) = (0.10 \text{ m}) \sin(2.0x - 1.5), 0.5 \leq x \leq 3.0$$
$$D(x, 0.60 \text{ s}) = (0.10 \text{ m}) \sin(2.0x - 1.8), 0.5 \leq x \leq 3.0$$

The resulting graphs are shown in Figure 14-32.

Making sense of the result

Since $f = 0.48$ Hz, the time period of the wave is approximately 2.1 s. The time period determined from Figure 14-31 is also about 2.1 s. Also, the wave is moving from left to right. From Figure 14-32, we see that this indeed is the case because the wave at $t = 0.60$ s is to the right of the wave at $t = 0.50$ s.

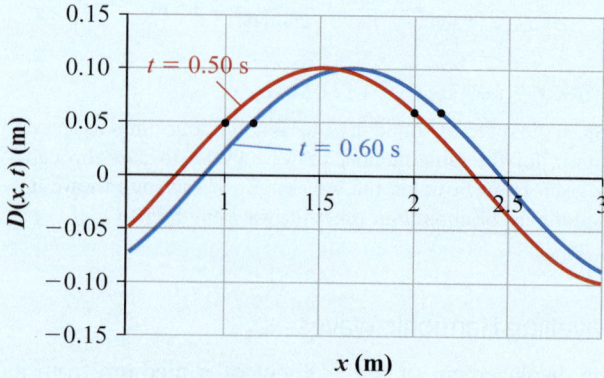

Figure 14-32   Graphs of the wave function for Example 14-6 at $t = 0.50$ s and $t = 0.60$ s.

The Phase Constant, ϕ

Equation 14-15 is not the most general equation for a travelling wave. It restricts the point at $x = 0$ to have zero displacement at $t = 0$. To generalize Equation 14-15, we add a phase constant, ϕ, to the argument of the sine function, as we did for simple harmonic oscillators in Chapter 13. The most general form of the wave function for a harmonic wave travelling in the direction of *increasing* x is

$$D(x, t) = A \sin(kx - \omega t + \phi)$$

KEY EQUATION

$$= A \sin\left(\frac{2\pi}{\lambda}x - \frac{2\pi}{T}t + \phi\right) \quad (14\text{-}18)$$

Similarly, a wave travelling in the direction of *decreasing* x is described by the wave function

$$D(x, t) = A \sin(kx + \omega t + \phi)$$

KEY EQUATION

$$= A \sin\left(\frac{2\pi}{\lambda}x + \frac{2\pi}{T}t + \phi\right) \quad (14\text{-}19)$$

The phase constant, ϕ, can be positive or negative. The sine function is periodic with a period of 2π rad, so it is sufficient to specify ϕ within a range of 0 rad to 2π rad. The phase constant is determined by the initial conditions. If the segment at $x = 0$ and $t = 0$ has a displacement D_0, then from Equation 14-18,

$$D_0 = D(0,0) = A \sin \phi$$

Therefore,

$$\phi = \sin^{-1}\left(\frac{D_0}{A}\right)$$

As the sine function is multivalued, in general a determination of the phase constant from a single initial condition is not possible. We will later see that the phase constant can be uniquely determined from the position or time plot of a harmonic wave.

EXAMPLE 14-7

Phase Constant of a Harmonic Wave

A sinusoidal wave travelling in the positive x-direction has amplitude 0.15 m, wavelength 0.40 m, and frequency 8.0 Hz. The displacement at $t = 0$ of a point located at $x = 0$ is 0.15 m. What is the phase constant of the wave? Write an equation for the wave function of this wave.

Solution

The given wave is travelling along the positive x-axis. For this wave,

$$A = 0.15 \text{ m}, \quad \lambda = 0.40 \text{ m}, \quad \text{and} \quad f = 8.0 \text{ Hz}$$

Since Equation 14-18 describes a wave travelling in the direction of increasing x, we insert these values into this equation, giving

$$D(x, t) = (0.15 \text{ m}) \sin\left(\frac{2\pi}{0.40 \text{ m}}x - 2\pi \times 8.0t + \phi\right)$$

The phase constant is determined by the initial conditions. We are given that the segment at $x = 0$ is displaced by 0.15 m from equilibrium at $t = 0$. Therefore,

$$D(0 \text{ m}, 0 \text{ s}) = 0.15 \text{ m} = (0.15 \text{ m}) \sin \phi$$
$$\sin \phi = 1$$
$$\phi = \frac{\pi}{2} \text{ rad}$$

Therefore, the equation for the wave function is

$$D(x, t) = (0.15 \text{ m}) \sin\left((5.0\pi \text{ rad/m})x - (16.0\pi \text{ rad/s})t + \frac{\pi}{2} \text{ rad}\right)$$

In this case, ϕ is uniquely determined from a single initial condition because the sine function has value $+1$ only once for its argument between 0 rad and 2π rad.

Making sense of the result

The above equation describes a wave moving toward the direction of increasing x, and at $x = 0$, $t = 0$, the displacement $D(0, 0) = 0.15 \text{ m} \sin(\pi/2) = 0.15 \text{ m}$ is the same as given in the problem.

Harmonic Wave Functions

Figure 14-33 shows waveforms (Equation 14-18) of four waves with the same amplitude and wavelength at $t = 0$.

Rank these waveforms in the order of their phase constants, *from small to large*. Assume that the phase constant is between 0 rad and 2π rad.

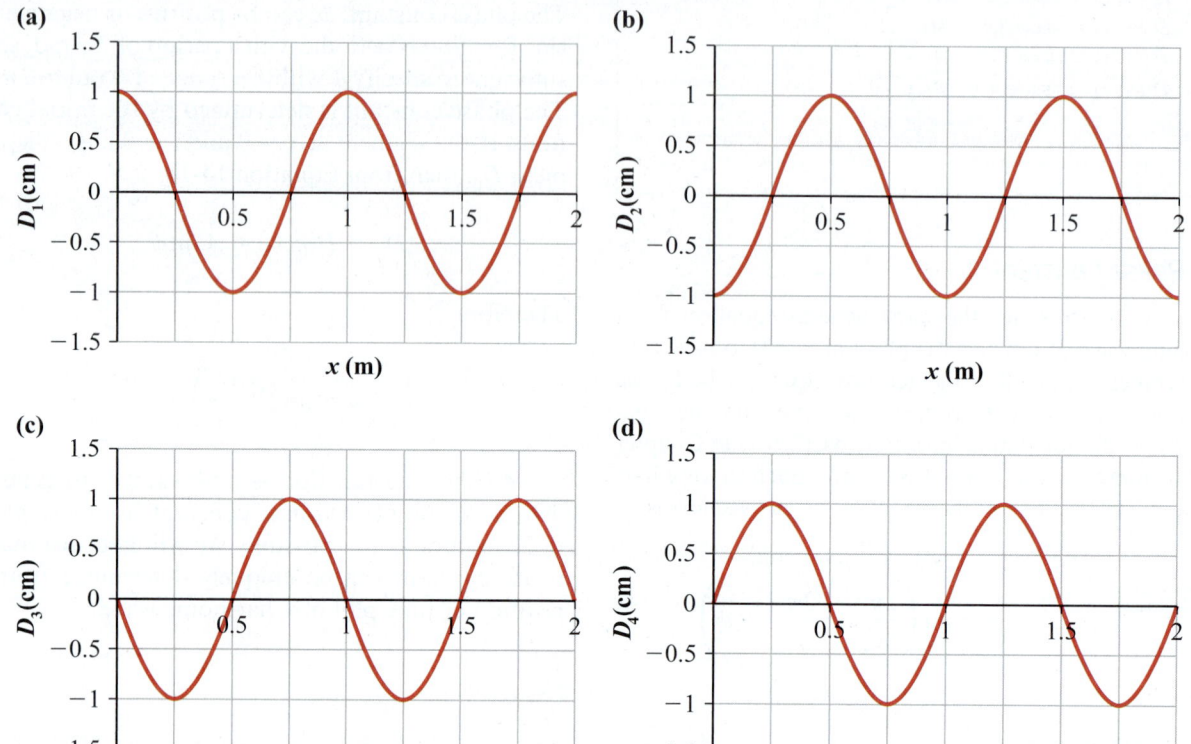

Figure 14-33   Checkpoint 14-10.

ANSWER: (d), (a), (c), (b).

Transverse Velocity and Acceleration for Harmonic Waves

As a harmonic wave travels through a medium, each segment of the medium undergoes periodic motion about its equilibrium position. The instantaneous speed of a segment located at $x = x_0$, denoted by $u(x_0, t)$, is equal to the rate of change of the displacement of the segment. Inserting Equation 14-15 for a harmonic wave into Equation 14-4 and setting $x = x_0$, we obtain

$$u(x_0, t) = \frac{\partial}{\partial t}\left(A\sin(kx - \omega t + \phi)\right)\Big|_{x=x_0}$$

$$= -\omega A\cos(kx_0 - \omega t + \phi) \tag{14-20}$$

Equation 14-20 shows that the velocity of the segment changes with time, oscillating in the range of $\pm\omega A$. This result is the same as the result for the velocity of a simple harmonic oscillator. Notice that, the wave speed and the velocity of a segment of the medium are quite different: a wave moves through a medium with

a constant speed, while each segment of the medium oscillates about its equilibrium position with a continuously changing velocity.

The instantaneous acceleration of a segment of a string located at $x = x_0$, $a(x_0, t)$, is equal to the rate of change of the instantaneous velocity of the segment:

$$a(x_0, t) = \frac{\partial u(x, t)}{\partial t}\bigg|_{x=x_0}$$

$$= \frac{\partial}{\partial t}\left((-\omega A)\cos(kx_0 - \omega t + \phi)\right)$$

$$= -\omega^2 A\sin(kx_0 - \omega t + \phi) \tag{14-21}$$

$$= -\omega^2 D(x_0, t)$$

Notice that the acceleration of a segment is proportional to its displacement and is opposite in sign, a result that also applies to the motion of simple harmonic oscillators. Notice also that each segment of the medium has instantaneous acceleration, but the wave itself has no acceleration since the wave velocity does not change with time.

EXAMPLE 14-8

A Harmonic Wave

A harmonic wave on a string is described by the wave function $D(x, t) = (0.20 \text{ m}) \sin((2.0 \text{ rad/m})x - (3.0 \text{ rad/s})t)$.

(a) What is the velocity at $t = 0.0$ s of a segment of the string located at $x = 0.50$ m?

(b) What is the maximum positive velocity of this segment? When is the first time after $t = 0.0$ that the segment attains this velocity?

(c) Plot the segment's velocity for $0.0 \text{ s} \leq t \leq 10.0 \text{ s}$.

Solution

The velocity of a point at $x = x_0$ of the medium, as a wave passes through it, is obtained by taking the partial derivative of the displacement, $D(x_0, t)$, with respect to time. We take a partial derivative because the position variable x is held constant and only t changes.

(a) We take the partial derivative of the displacement to find an expression for the velocity:

$$u(x, t) = \frac{\partial}{\partial t}\left[(0.20 \text{ m}) \sin((2.0 \text{ rad/m})x - (3.0 \text{ rad/s})t)\right]$$

$$= (-0.60 \text{ m/s}) \cos((2.0 \text{ rad/m})x - (3.0 \text{ rad/s})t)$$

Substituting $x = 0.50$ m and $t = 0.0$ s into this expression, we get

$$u(0.50 \text{ m}, 0.0 \text{ s}) = (-0.60 \text{ m/s}) \cos 1.0$$
$$= -0.32 \text{ m/s}$$

(b) The maximum positive velocity of the segment is $+0.60$ m/s (when the cosine function is -1). To find the time when the segment first attains this velocity, we substitute $+0.60$ m/s in the velocity equation and solve for t:

$$u(0.5 \text{ m}, t) = (-0.60 \text{ m/s}) \cos(1.0 - 3.0t) = +0.60 \text{ m/s}$$

Therefore, $\cos(1.0 - 3.0t) = -1.0$

$$1.0 - 3.0t = \cos^{-1}(-1.0) = \pm\pi$$

Here, we have retained both signs for π since $\cos(\pm\pi) = -1$. Solving for t, we get

$$t = \frac{1.0 \mp \pi}{3.0}$$

Since $t > 0$, choosing the positive sign, we get

$$t = \frac{1.0 + \pi}{3.0} = 1.4 \text{ s}$$

(c) A plot of the velocity of the segment is shown in Figure 14-34.

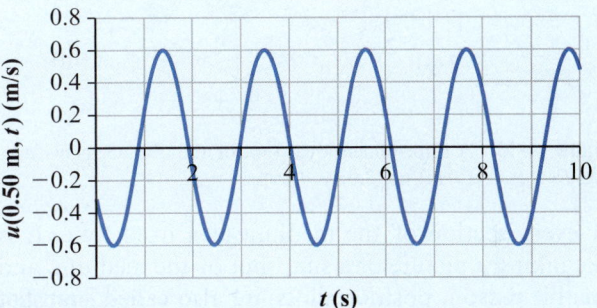

Figure 14-34 A graph of the velocity of the string segment in Example 14-8 at $x = 0.50$ m for $0.0 \text{ s} \leq t \leq 10.0 \text{ s}$.

Making sense of the result

From the velocity graph (Figure 14-34) at $t = 0$ the value of the velocity is about -0.3 m/s, which is consistent with the value calculated in part (a). Also, from the graph we see that the velocity has a maximum value around $t = 1.5$ s, which is consistent with the value that we calculated in part (b).

CHECKPOINT 14-11

Transverse Velocity

A displacement versus time graph for a particular segment of a medium is shown in Figure 14-35. List the labelled points on the displacement curve where the segment's velocity is

(a) positive

(b) negative

(c) zero

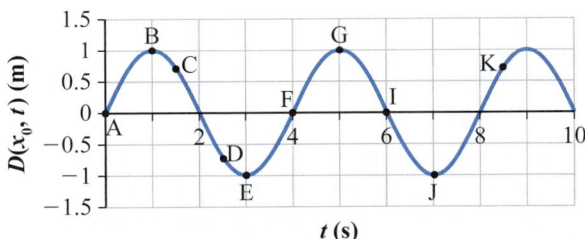

Figure 14-35 Checkpoint 14-11.

14-6 Position Plots and Time Plots

A plot of $D(x, t)$ as a function of position, x, and time, t, results in a three-dimensional graph. Figure 14-36 shows a plot of $D(x, t) = (1.0 \text{ m}) \sin(2.0\pi x - \pi t)$, with position along the x-axis, time along the y-axis, and the displacement, $D(x, t)$, along the z-axis. Although such a plot is useful in visualizing wave motion, it can be difficult to determine the parameters of a wave (wavelength, frequency, amplitude, and phase constant) from a three-dimensional plot. It is easier to use plots in which one of the variables is kept constant and the wave function is plotted against the other variable.

Position Plots

A **position plot** for a travelling wave is obtained by keeping the time fixed and plotting the wave function as a function of position (x). If the time is fixed at $t = t_0$, then the position plot is a graph of $D(x, t = t_0)$ as a function of x. A position plot shows the displacement

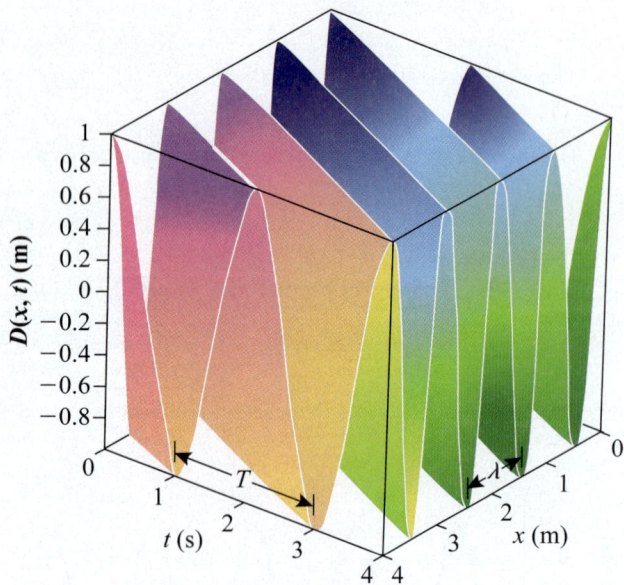

Figure 14-36 A three-dimensional plot of the displacement function $D(x, t) = (1.0\text{ m}) \sin (2.0\pi x - \pi t)$.

of every section of the medium at a fixed time. It is like taking a picture or a snapshot of the medium, and for this reason, position plots are also called **snapshot graphs**. We can determine the amplitude and the wavelength of the wave from its position plot. Figure 14-37 shows a displacement versus position plot of a harmonic wave. From the graph, notice the following:

■ The maximum displacement of a point is ±1.0 m. Therefore, $A = 1.0$ m.

■ The points at $x = 3.0$ m and $x = 6.0$ m are located at two successive crests. Therefore, the wavelength of the wave is $\lambda = 6.0\text{ m} - 3.0\text{ m} = 3.0\text{ m}$.

EXAMPLE 14-9

The Position Plot of a Harmonic Wave

Suppose that the position plot in Figure 14-37 was obtained at $t = 1.0$ s and the wave is moving at a speed of 1.5 m/s toward increasing x. Determine the frequency and the phase constant for this wave, and write an equation for the wave function of this wave.

Solution

The general equation for a sinusoidal wave travelling toward increasing x is given by

$$D(x, t) = A \sin\left(\frac{2\pi}{\lambda}x - \frac{2\pi}{T}t + \phi\right)$$

Substituting $A = 1.0$ m, $\lambda = 3.0$ m, and $t = 1.0$ s, we get

$$D(x, 1.0\text{ s}) = (1.0\text{ m}) \sin\left(\frac{2\pi}{3.0}x - \frac{2\pi}{T} + \phi\right)$$

The wave speed is 1.5 m/s, and the wavelength is known; therefore, we can determine the period:

$$T = \frac{1}{f} = \frac{\lambda}{v} = \frac{3.0\text{ m}}{1.5\text{ m/s}} = 2.0\text{ s}$$

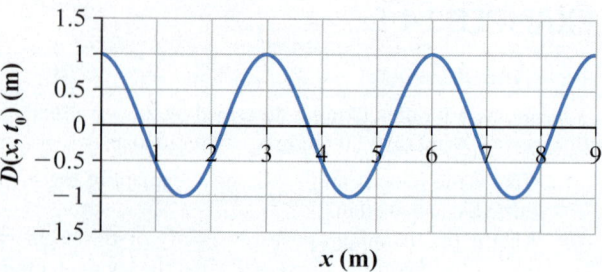

Figure 14-37 The displacement versus position plot of a harmonic wave at a fixed time.

Insert this value for T into the above equation:

$$D(x, 1.0\text{ s}) = (1.0\text{ m}) \sin\left(\frac{2\pi}{3.0}x - \pi + \phi\right)$$

We can determine ϕ by determining the displacement for a given x from the graph and then inserting that information in the wave equation. At $x = 0.0$ m, $D(0.0\text{ m}, 1.0\text{ s}) = 1.0$ m. Inserting these values into the equation yields

$$D(0.0\text{ m}, 1.0\text{ s}) = (1.0\text{ m}) \sin(-\pi + \phi) = 1.0\text{ m}$$
$$\sin(-\pi + \phi) = 1$$
$$-\pi + \phi = \sin^{-1}(1) = \frac{\pi}{2}$$

Therefore,

$$\phi = \frac{3\pi}{2}\text{ rad}$$

Therefore, the wave function for the wave is

$$D(x, t) = (1.0\text{ m}) \sin\left(\frac{2\pi}{3.0}x - \pi t + \frac{3\pi}{2}\right) \tag{14-22}$$

Making sense of the result

Evaluating Equation 14-22 for $x = 0.0$ m and $t = 1.0$ s, we get

$$D(0.0\text{ m}, 1.0\text{ s}) = (1.0\text{ m}) \sin\left(-\pi + \frac{3\pi}{2}\right)$$
$$= (1.0\text{ m}) \sin\left(\frac{\pi}{2}\right) = 1.0\text{ m},$$

which is consistent with the value obtained from the graph.

Time Plots

A time plot for a travelling wave is obtained by plotting the wave function, $D(x, t)$, as a function of time for a fixed position. If the position is fixed at $x = x_0$, then the time plot is a graph of $D(x = x_0, t)$ as a function of t. A time plot shows how the displacement of a given point of the medium varies with time in the presence of a wave. Therefore, the time plots are also called **history graphs**. We can determine the amplitude and the frequency of a wave from its time plot.

EXAMPLE 14-10

The Time Plot of a Harmonic Wave

A sinusoidal wave is travelling with a speed of 2.0 m/s along the positive x-axis on a string. The time plot of the displacement of the section of the string located at $x = 2.0$ m is shown in Figure 14-38. Using the given data and the plot,

(a) determine the wavelength and angular frequency of the wave
(b) determine the phase constant of the wave
(c) write an equation for the wave function, $D(x, t)$, for the wave

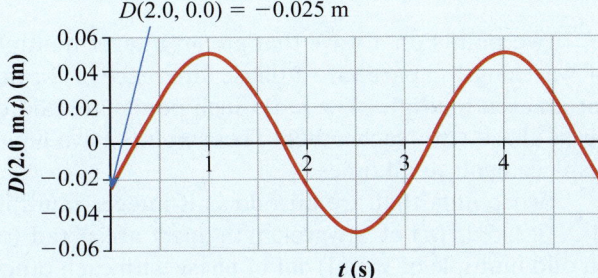

Figure 14-38 The displacement versus time plot of a segment of string located at $x = 2.0$ m as a harmonic wave passes through the point.

Solution

(a) The information about the amplitude and the time period of the wave can be obtained from the given graph.

From the graph, the maximum displacement of the string element is 0.050 m, so the amplitude is $A = 0.050$ m. The period can be calculated from the graph by finding the time difference between two consecutive crests, which we can see occur at $t = 1.0$ s and $t = 4.0$ s. Therefore, the period is

$$T = 4.0 \text{ s} - 1.0 \text{ s} = 3.0 \text{ s}$$

We can now determine the wavelength:

$$\lambda = \frac{v}{f} = vT$$
$$= (2.0 \text{ m/s})(3.0 \text{ s}) = 6.0 \text{ m}$$

The angular frequency is

$$\omega = \frac{2\pi}{T}$$
$$= \frac{2\pi \text{ rad}}{3.0 \text{ s}} = \frac{2\pi}{3.0} \text{ rad/s}$$

We can leave ω as a fraction of π to make the calculation of the phase constant easier.

(b) Since the wave speed is given and we now know the angular frequency, we can determine k:

$$k = \frac{\omega}{v} = \frac{(2\pi/3.0) \text{ rad/s}}{2.0 \text{ m/s}} = \frac{\pi}{3.0} \text{ rad/m}$$

The general equation for a harmonic wave travelling toward increasing x is

$$D(x, t) = A \sin(kx - \omega t + \phi)$$

Inserting $x = 2.0$ m, we get

$$D(2.0, t) = (0.050 \text{ m}) \sin(2.0k - \omega t + \phi)$$

Now, inserting the values of A, ω, and k into the above equation, we get

$$D(2.0, t) = (0.050 \text{ m}) \sin\left(\frac{\pi}{3.0} \times 2.0 - \frac{2\pi}{3.0}t + \phi\right)$$

We need to determine the phase constant, ϕ. For this we can determine the displacement at a given time on the graph, and then use this information to determine ϕ. The graph shows that at $t = 0.0$ s, $D(2.0, 0.0) = -0.025$ m. Therefore,

$$D(2.0, 0.0) = (0.050 \text{ m}) \sin\left(\frac{2\pi}{3.0} + \phi\right) = -0.025 \text{ m}$$

where the right side of the above equation is the value determined from the graph. This gives

$$\sin\left(\frac{2\pi}{3.0} + \phi\right) = -\frac{0.025 \text{ m}}{(0.050 \text{ m})} = -0.50$$

The sine function has value -0.5 when its argument is either $-\frac{\pi}{6}$ rad or $\pi - \left(-\frac{\pi}{6}\right) = \frac{7\pi}{6}$ rad. If we choose the first value, then

$$\frac{2\pi}{3.0} + \phi = -\frac{\pi}{6}$$
$$\phi = -\frac{\pi}{6} - \frac{2\pi}{3.0} = -\frac{5\pi}{6} \text{ rad}$$

If we choose the second value, then

$$\frac{2\pi}{3.0} + \phi = \frac{7\pi}{6}$$
$$\phi = \frac{7\pi}{6} - \frac{2\pi}{3.0} = \frac{\pi}{2} \text{ rad}$$

Only one of these values of ϕ can be correct. Which one?

(c) To determine the correct value of ϕ, use the location of the crest at $t = 1.0$ s. If we choose $\phi = -\frac{5\pi}{6}$ rad, then

$$D(2.0, 1.0) = (0.050 \text{ m}) \sin\left(\frac{2\pi}{3} - \frac{2\pi}{3} - \frac{5\pi}{6}\right)$$
$$= (0.050 \text{ m}) \times (-0.5) = 0.025 \text{ m}$$

which does not agree with the value from the graph. If we choose $\phi = \frac{\pi}{2}$ rad, then the displacement, $D(2.0, 1.0)$, is

$$D(2.0, 1.0) = (0.050 \text{ m}) \sin\left(\frac{2\pi}{3} - \frac{2\pi}{3} + \frac{\pi}{2}\right)$$
$$= (0.050 \text{ m}) \times 1.0 = +0.05 \text{ m}$$

(continued)

which agrees with the graph because $D(2.0, 1.0)$ is on the crest. Therefore, $\phi = \dfrac{\pi}{2}$.

We can now write the general equation for the travelling wave:

$$D(x, t) = (0.050 \text{ m}) \sin\left(\frac{\pi}{3.0} x - \frac{2\pi}{3.0} t + \frac{\pi}{2}\right)$$

A common error while doing similar problems is to omit the phase constant, ϕ, in the wave function.

Making sense of the result

From the given time plot, we see that the displacement is maximum at $t = 4.0$ s. From the wave function calculated in part (c), we find that

$$D(2.0, 4.0) = (0.050 \text{ m}) \sin\left(\frac{\pi}{3.0} 2.0 - \frac{2\pi}{3.0} 4.0 + \frac{\pi}{2}\right)$$

$$= (0.050 \text{ m}) \sin\left(-2.0\pi + \frac{\pi}{2}\right) = 0.050 \text{ m}$$

which agrees with the graph.

14-7 Phase and Phase Difference

The argument of the sine (or the cosine) function of a harmonic wave is called the **phase** of the wave. It is measured in radians and is usually denoted by the Greek letter phi, Φ. For the harmonic waves of Equations 14-18 and 14-19, the phase is

$$\Phi(x, t) = kx \mp \omega t + \phi$$
$$= \left(\frac{2\pi}{\lambda}\right)x \mp (2\pi f)t + \phi \tag{14-23}$$

Do not confuse the phase, $\Phi(x, t)$, with the phase constant, ϕ. The phase constant is a constant determined by the initial conditions, whereas the phase of a wave is a variable that depends on position and time. Consider two points, x_1 and x_2, on a wave that are a distance Δx apart. The difference between the phase at two points is called the **phase difference** and is denoted as $\Delta\Phi$. For harmonic waves,

$\Delta\Phi = $ phase of the wave at x_2 − phase of the wave at x_1
$= (kx_2 - \omega t + \phi) - (kx_1 - \omega t + \phi)$
$= k(x_2 - x_1)$
$= k\Delta x$

Thus,

$$\Delta\Phi = 2\pi\left(\frac{\Delta x}{\lambda}\right) \tag{14-24}$$

Table 14-1 shows the phase differences for various distances between two points on a wave.

Table 14-1 Phase Differences for a Periodic Wave

Distance between Points, Δx (in Multiples of the Wavelength)	Phase Difference, $\Delta\Phi$ (rad)
0	0
$\lambda/4$	$\pi/2$
$\lambda/2$	π
$3\lambda/4$	$3\pi/2$
λ	2π

Two points on a wave that are an integer multiple of wavelengths apart have a phase difference of 2π rad (or an even multiple of π rad). Such points are said to be **in phase** with each other. These points have equal displacements at all times.

Two points that are an odd half-integer multiple (1/2, 3/2, 5/2, …) of a wavelength apart are π rad (or an odd multiple of π rad) **out of phase** with each other. These points have equal but opposite displacements from the equilibrium position at all times. These relationships follow from Equation 14-24. For example, let $\Delta\Phi = \pi$ rad. Then the phase of the wave at x_2 is related to the phase at x_1 as

$$(kx_2 - \omega t + \phi) = (kx_1 - \omega t + \phi) + \pi$$

and

$$D(x_2, t) = A \sin(kx_2 - \omega t + \phi)$$
$$= A \sin(kx_1 - \omega t + \phi + \pi)$$
$$= -A \sin(kx_1 - \omega t + \phi) \tag{14-25}$$
$$= -D(x_1, t)$$

Here, we have used the trigonometric identity $\sin(\theta + \pi) = -\sin\theta$.

CHECKPOINT 14-12

Phase Difference

A harmonic wave is travelling along a string. Two points that were 0.30 m apart when the string was at rest oscillate such that their displacements are always equal and opposite. What is the longest possible wavelength of the wave?

(a) 0.3 m
(b) 0.6 m
(c) 0.9 m
(d) 1.2 m

ANSWER: (b)

14-8 Energy and Power in a Travelling Wave

It takes energy to generate a wave, and the wave carries that energy with it as it travels along a medium. In this section, we will investigate how the power (energy per unit time) carried by a mechanical wave depends on the properties of the wave (frequency, speed, and amplitude) and that of the medium (density) through which it travels. The main results of this section are as follows:

(a) A travelling wave carries energy.

(b) The average power, P_{avg}, carried by a wave is proportional to the wave speed (v), the square of the wave frequency (f^2), the square of the wave amplitude (A^2), and the density of the medium (μ). For a one-dimensional wave in a string,

$$P_{avg} = 2\pi^2 \mu v f^2 A^2$$

where μ is the linear mass density of the string. Notice that this equation is similar to that of the total energy of a mass–spring oscillator (Equation 13-41):

$$E = \frac{1}{2} k A^2 = \frac{1}{2}(m\omega^2)A^2 = (2\pi^2 m)f^2 A^2$$

Intuitively, the dependence of the average power carried by a wave on the properties of the wave and that of the medium is easy to understand. A faster wave delivers more power. It takes more power to generate a rapidly oscillating wave (try generating a wave on a string by flipping your hand up and down first slowly and then rapidly). Therefore, the higher the frequency, the greater the energy needed to generate that wave and hence the greater the energy carried by that wave. Similarly, it takes more energy to generate a wave of large amplitude, so the larger the amplitude, the greater the wave energy. Since a heavier string takes more energy to move, it takes more energy to generate a wave in a heavier string. Thus, waves in a denser medium carry more energy.

Next we derive the above expression for P_{avg} for one-dimensional waves. This result can be generalized for waves in two and three dimensions. The derivation is a bit lengthy and involves slightly advanced calculus, so it may be omitted.

Consider a harmonic wave passing through a string of linear mass density μ. A segment of this string has a mass of $\Delta m = \mu \Delta x$, where Δx is the undisturbed length of the segment (i.e., the length of the segment when the string is at rest). In the presence of the wave, the instantaneous kinetic energy of the oscillating segment is

$$\Delta K = \frac{1}{2} \Delta m (u(x, t))^2 \tag{14-26}$$

where $u(x, t)$ is the instantaneous velocity of the segment.

Using Equation 14-20 for the velocity of the string segment, we can write ΔK as

$$\Delta K = \frac{1}{2}(\mu \Delta x)(A^2 \omega^2 \cos^2 (kx - \omega t))$$

$$= \frac{1}{2}(\mu \Delta x)(A^2 \omega^2 (1 - \sin^2(kx - \omega t))) \tag{14-27}$$

$$= \frac{1}{2}(\mu \Delta x)\omega^2(A^2 - D^2(x, t))$$

Notice that the vibrating segment has zero kinetic energy at $D(x, t) = \pm A$, the maximum displacement from the equilibrium position, where the segment momentarily comes to rest. The segment has a maximum kinetic energy as it passes through its equilibrium position, where $D(x, t) = 0$.

As the wave passes through the segment, the length of the segment changes as it is continually stretched and compressed by the wave. To determine how the change in the length of a segment depends on the shape of the wave, consider a curve in one dimension that is described by a continuous function, $f(x)$. The length, L, of the curve between $x = a$ and $x = b$ is given by the integral

$$L = \int_a^b \sqrt{1 + \left(\frac{df}{dx}\right)^2} \, dx$$

If $\dfrac{df(x)}{dx}$ does not change appreciably in the interval $[a, b]$, then L can be approximated as

$$L = \sqrt{1 + \left(\frac{df}{dx}\right)^2}(b - a)$$

In the presence of the wave the string is deformed— let ΔL be the length of the deformed segment (Figure 14-39). Using the above equation, ΔL can be approximated as

$$\Delta L = \left(\sqrt{1 + \left(\frac{\partial D(x, t)}{\partial x}\right)^2}\right)\Delta x$$

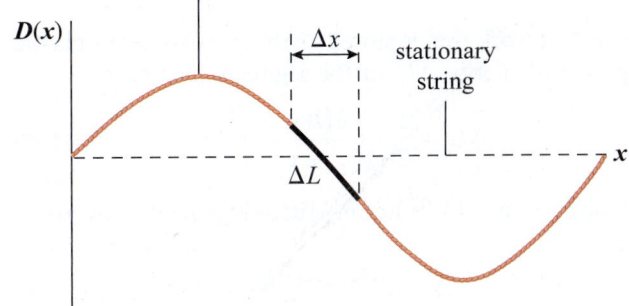

snapshot of the string in the presence of a harmonic wave

Figure 14-39 The change in the length of an infinitesimal segment of a string in the presence of a harmonic wave. The change is maximum when the segment passes through the origin and is zero at the maximum displacement.

where we have assumed that the derivative of $D(x, t)$ does not change appreciably within the segment Δx. A general description of wave propagation is quite complicated, so we will assume that the amplitude of the wave is small compared to its wavelength. In this approximation, the length of the segment in the presence of a wave is not very different from its undisturbed length. Therefore, the second term under the square root is much smaller than 1, and we can expand the square root to write $((1+x)^{(1/2)} = 1 + x/2$ for $x \ll 1)$

$$\Delta L = \left(1 + \frac{1}{2}\left(\frac{\partial D(x, t)}{\partial x}\right)^2\right)\Delta x$$

If Δl is the change in the length of the segment due to the presence of the wave, then $\Delta L = \Delta x + \Delta l$ and Δl is given by

$$\Delta l = \Delta L - \Delta x = \frac{1}{2}\left(\frac{\partial D(x, t)}{\partial x}\right)^2\Delta x \qquad (14\text{-}28)$$

For a harmonic wave, $D(x, t) = A\sin(kx - \omega t)$. Therefore,

$$\frac{1}{2}\left(\frac{\partial D(x, t)}{\partial x}\right)^2 = \frac{1}{2}(kA)^2\cos^2(kx - \omega t)$$

$$= \frac{1}{2}\left(\frac{2\pi A}{\lambda}\right)^2\cos^2(kx - \omega t) \qquad (14\text{-}29)$$

and we can see that in the approximation $2\pi A \ll \lambda$,

$$\frac{1}{2}\left(\frac{\partial D(x, t)}{\partial x}\right)^2 \ll 1.$$

In the absence of a wave, a string segment is under a tension, T_s. As the wave passes through the segment, the tension in the segment varies continually. We now assume that for waves with $A \ll \lambda$ the change in the tension in the presence of a wave is negligible. In this approximation, the work done by the wave in changing the length of the segment, W, is given by

$$W = \text{Tension in the string}$$
$$\times \text{change in the length of the segment}$$

$$= T_s(\Delta l) = T_s\left(\frac{1}{2}\left(\frac{\partial D(x, t)}{\partial x}\right)^2\Delta x\right)$$

It is this work that is stored as the instantaneous elastic potential energy, ΔU, of the segment. Therefore,

$$\Delta U = \frac{1}{2}T_s\left(\frac{\partial D(x, t)}{\partial x}\right)^2\Delta x \qquad (14\text{-}30)$$

Using Equation 14-29, we can write (using $\sin^2\theta + \cos^2\theta = 1$)

$$\Delta U = \frac{1}{2}(T_s\Delta x)(A^2 k^2\cos^2(kx - \omega t))$$

$$= \frac{1}{2}(T_s k^2\Delta x)(A^2 - A^2\sin^2(kx - \omega t)) \qquad (14\text{-}31)$$

$$= \frac{1}{2}(T_s k^2\Delta x)(A^2 - D^2(x, t))$$

Using $v = \sqrt{\dfrac{T_s}{\mu}}$ for the wave speed along a string and $v = \dfrac{\omega}{k}$, we get $T_s k^2 = \mu\omega^2$, and Equation 14-31 can then be written as

$$\Delta U = \frac{1}{2}(\mu\Delta x)\omega^2(A^2 - D^2(x, t)) \qquad (14\text{-}32)$$

Notice that the equation for stored instantaneous potential energy is exactly the same as the equation for the instantaneous kinetic energy of the segment (Equation 14-27). *As a harmonic wave travels through a string, the instantaneous kinetic energy and the potential energy of a given segment of the string are equal.* When the segment is at its maximum displacement ($D = \pm A$), both the kinetic energy and the potential energy of the segment are zero, so the segment has zero total energy. When the segment is passing through the equilibrium position ($D = 0$), it has maximum potential energy as well as maximum kinetic energy. Thus, the behaviour of a string element in the presence of a travelling wave is different from the behaviour of a simple harmonic oscillator. The kinetic and potential energies of a single oscillator continually transform back and forth into each other. When the kinetic energy of a simple harmonic oscillator is maximum, its potential energy is minimum, and vice versa.

The total energy of the segment (ΔE) is equal to the sum of its instantaneous kinetic and potential energies:

$$\Delta E = \Delta K + \Delta U$$
$$= (\mu\Delta x)\omega^2 A^2\cos^2(kx - \omega t) \qquad (14\text{-}33)$$

This energy flows in the direction of the wave motion. The segment receives energy at one end, stores it briefly, and then passes it to the neighbouring segment at the other end. If energy, ΔE, is transmitted through a segment of length Δx in time Δt, then the energy transmitted across the segment per unit time is

$$\frac{\Delta E}{\Delta t} = \mu\left(\frac{\Delta x}{\Delta t}\right)\omega^2 A^2\cos^2(kx - \omega t)$$

The ratio $\Delta x/\Delta t$ is the wave speed. Therefore, the instantaneous power, P, delivered by the wave is

$$P = \frac{\Delta E}{\Delta t} = \mu v\omega^2 A^2\cos^2(kx - \omega t) \qquad (14\text{-}34)$$

The average rate at which the power is transmitted by the wave is obtained by integrating Equation 14-34 over a complete wave cycle. The average value of the square of a cosine (or a sine) function over an integer number of cycles is 1/2. Thus, the average power transmitted by a harmonic wave, P_{avg}, is

KEY EQUATION
$$P_{\text{avg}} = \frac{1}{2}\mu v\omega^2 A^2 = 2\pi^2\mu v f^2 A^2 \qquad (14\text{-}35)$$

The average power delivered by a wave is proportional to the linear mass density of the string, the wave speed, the square of the amplitude of the wave, and the square of its frequency. The dependence of the average power of a wave on the square of its amplitude and frequency is valid for other types of waves as well, including ocean waves, seismic waves, and electromagnetic waves. In real-life situations, frictional forces cause a wave to lose its energy as it moves in a medium. As a wave gradually loses its energy,

it finally disappears. We will neglect effects of frictional forces on wave motion. This means that a wave travels through a uniform medium without a change in its energy.

Note that Equations 14-26 and 14-30 are strictly valid only in the limit when $\Delta x \to 0$. For a segment of finite length, we must integrate these equations over the length of the segment.

CHECKPOINT 14-13

Wave Power

Figure 14-40 shows time plots of four waves passing through four identical strings held under the same tension.

Rank these waves, from highest to lowest, on the basis of the power carried by the waves.

Highest ___ ___ ___ ___ Lowest

(a)

(b)

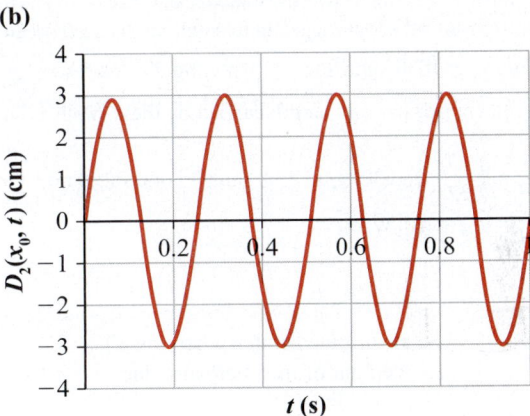

(c)

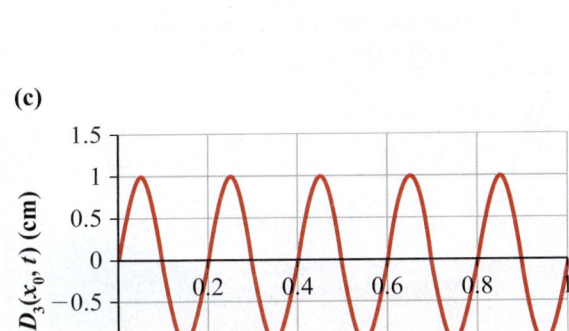

(d)

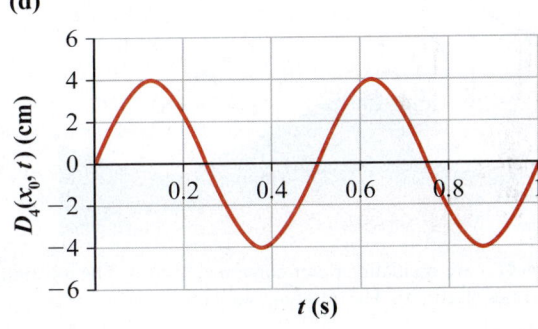

Figure 14-40   Checkpoint 14-13.

MAKING CONNECTIONS

Energy from Ocean Waves

Waves in an ocean are generated by the interaction of the wind with the water's surface. If the wind speed just above the water surface is greater than the speed of the water waves, then energy is transferred from the wind to the water, causing the amplitude of the waves to increase. When the following three conditions are satisfied (this happens during ocean storms), the waves are transformed into a series of repetitive swells:

■ The wind speed is large (several tens of kilometres per hour).

■ The water surface over which the wind is blowing is large (thousands of square kilometres).

■ The wind blows in one direction for a sustained period of time (several hours).

A swell is a long, thick wave that has a height of several metres. Swells are characterized by the wave height, H, which is the crest-to-trough height of the swell, and the period, T, between successive swells. A phrase such as "a 3 m swell at 12 s" means 3 m high swells that are 12 s apart.

(continued)

Swells carry a large amount of energy and travel over large distances from the point of origin. An analysis similar to the analysis that led to the expression for the average power for one-dimensional waves (Equation 14-35) shows that the average power per unit length carried by swells is given by

$$P_{avg} = \left(\frac{1}{32\pi}\right)\rho_{ocean}\, g^2 T H^2$$

Here, g is the acceleration due to gravity, ρ_{ocean} is the density of ocean water, T is the period of the swells, and H is the wave height. Note that just as for one-dimensional waves, the power carried by swells is proportional to the square of the swell height and the density of the medium.

Daily swell heights for all oceans are published by the NOAA (National Oceanic and Atmospheric Administration) in the United States. The period between the swells can range from 12 s to 25 s. Let us assume the following data for swells in the deep ocean:

$$\rho_{ocean} = 1030 \text{ kg/m}^3,\ H = 3.0 \text{ m, and } T = 14.0 \text{ s}$$

The average power per unit length carried by these swells is

$$P_{avg} = \left(\frac{1}{32\pi}\right)(1030 \text{ kg/m}^3)(9.81 \text{ m/s}^2)^2(14.0 \text{ s})(3.0 \text{ m})^2$$
$$= 1.2 \times 10^5 \text{ W/m}$$

This is a large amount of freely available, environmentally friendly, and renewable power, which if efficiently harnessed could greatly help in solving the world's energy problems.

There are a number of efforts underway to harvest this power. One such method is called the oscillating water column (OWC) generator (Figure 14-41). It consists of an air-filled enclosure that has two openings, one underneath and the other at the top. An air turbine (a device that generates electricity when rotated) is connected to the top opening. The enclosure is partially submerged in the water such that the bottom opening is below the waterline and the top opening is above the maximum swell height. When a swell reaches the enclosure, the column of water in the enclosure rises. The air inside the enclosure is compressed and is forced out from the top opening past the turbine. This causes the turbine to rotate and generate electricity. As the swell recedes, the air pressure within the enclosure decreases. This forces the outside air into the enclosure past the turbine, again rotating the turbine and generating electricity.

Located on the island of Islay, off Scotland's west coast, LIMPET (Land Installed Marine Powered Energy Transformer) is the first commercial OWC generator. It produces 500 kW of power (Figure 14-42).

(a)

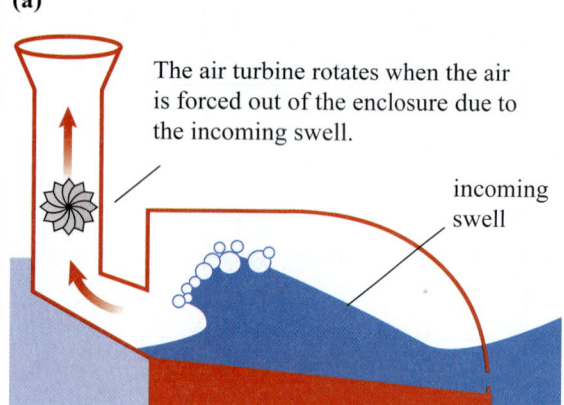

The air turbine rotates when the air is forced out of the enclosure due to the incoming swell.

incoming swell

(b)

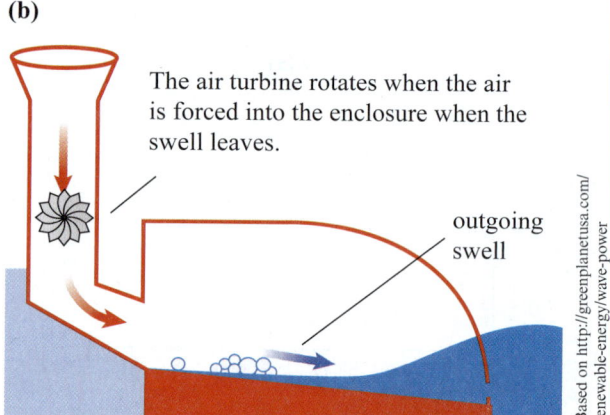

The air turbine rotates when the air is forced into the enclosure when the swell leaves.

outgoing swell

Figure 14-41 An oscillating water column generator. The incoming swell forces the air out of the enclosure, which rotates the air turbine and generates electricity. The outgoing swell forces the air into the enclosure, again generating electricity.

Figure 14-42 A depiction of LIMPET, the first commercial OWC generator, located off the west coast of Scotland.

14-9 Superposition of Waves

The space around us is filled with electromagnetic waves of various frequencies from radio and TV stations, cellphones, satellites, and other sources. The electric current induced by this multitude of waves in the antennas of a radio receiver is quite complicated. However, if you tune a radio to a particular frequency, say 106.9 MHz, you clearly hear the broadcast at that frequency. The electromagnetic waves of a particular frequency are completely unaffected by the presence of other waves. The same is true of sound waves in air. While listening to a concert, you can often identify the sound generated by a particular instrument or singer, even though the sound waves from a number of sources reach your ears simultaneously.

Principle of superposition of waves When more than one wave is present in a medium at the same time, the resultant wave at any point in the medium is equal to the algebraic sum of the individual waves at that point.

For mechanical waves, the superposition principle states that a medium responds to the effects of each wave individually. The presence of other waves does not change in any way the effect of a particular wave on a medium. When multiple waves are travelling along a string, the net displacement of a segment of the string at any time is equal to the sum of the displacements produced by the individual waves.

For a one-dimensional medium carrying a set of waves with individual waveforms given by $D_1(x, t)$, $D_2(x, t)$, $D_3(x, t)$, ..., the resultant waveform, $D(x, t)$, of the medium (i.e., the shape of the resultant wave) is equal to the sum of the individual waveforms:

$$D(x, t) = D_1(x, t) + D_2(x, t) + D_3(x, t) + \cdots \quad (14\text{-}36)$$

The wave obtained by adding the component waves is called the **resultant wave**. The process of combining the waves to produce a resultant wave is called the **superposition** of waves. The physical phenomenon of two or more waves combining to produce a resultant wave is called the **interference** of waves.

EXAMPLE 14-11

Superposition of Two Harmonic Waves

Two harmonic waves with different frequencies, moving in opposite directions, are present in a string at the same time. The wave functions of these waves are given by

$$D_1(x, t) = (0.10 \text{ m}) \sin(2.0x - 10.0t)$$
$$D_2(x, t) = (0.15 \text{ m}) \sin(4.0x + 20.0t)$$

Find the displacement of a segment of the string located at $x = 1.5$ m at $t = 2.0$ s. Plot the displacement of the string between $x = 0.0$ and $x = 5.0$ m at $t = 1.0$ s.

Solution

We apply the principle of superposition to obtain the resultant displacement at $x = 1.5$ m and $t = 2.0$ s:

$$\begin{aligned} D(1.5, 2.0) &= D_1(1.5, 2.0) + D_2(1.5, 2.0) \\ &= (0.10 \text{ m}) \sin(2.0 \times 1.5 - 10.0 \times 2.0) \\ &\quad + (0.15 \text{ m}) \sin(4.0 \times 1.5 + 20.0 \times 2.0) \\ &= 9.6 \text{ cm} + 13.5 \text{ cm} \\ &= 23.1 \text{ cm} \end{aligned}$$

Figure 14-43 shows the position plot of the two waves and the resultant wave at $t = 1$ s. Note that the displacement of each point on the resultant wave is equal to the sum of the displacements due to the component waves.

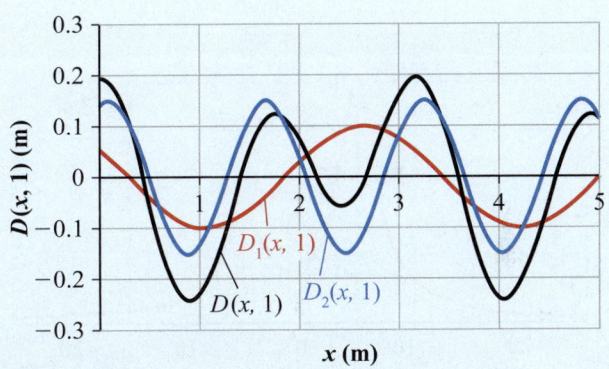

Figure 14-43 Position plots of $D_1(x, 1.0 \text{ s})$ (red), $D_2(x, 1.0 \text{ s})$ (blue), and the resultant wave (black).

Making sense of the result

From Figure 14-43, we see that the resultant displacement (black curve) is the sum of the individual displacements (red and blue curves).

The principle of superposition of waves is easy to understand for mechanical waves. A wave exerts a tension force at the points of the medium through which it is passing. When two or more waves are present, the net force at a point is equal to the sum of all the forces. Therefore, the resultant displacement of a point is equal to the sum of the displacements due to the individual waves. The fact that two or more waves can

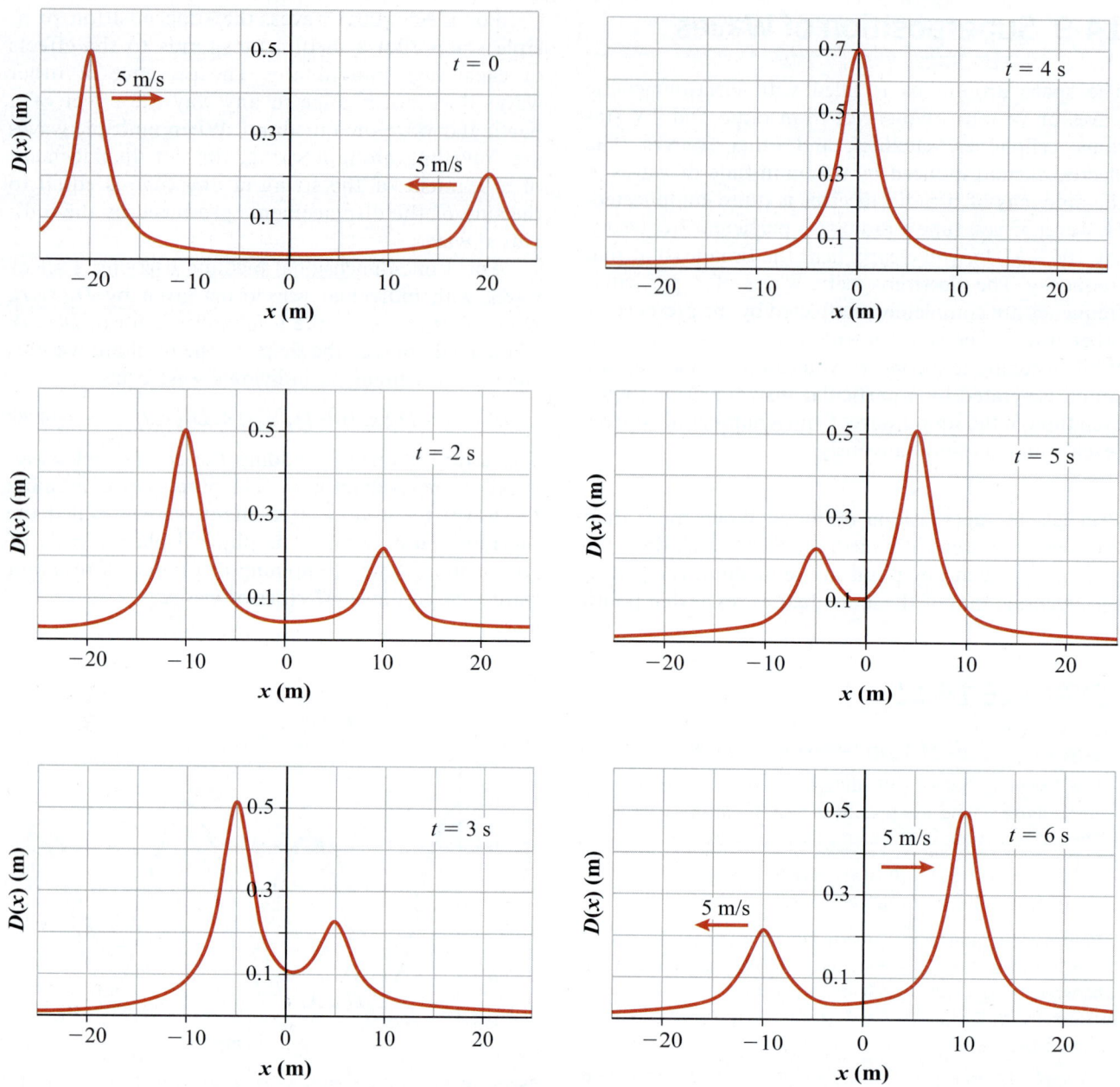

Figure 14-44 Interference of two pulses travelling in opposite directions.

simultaneously exist in a medium without changing each other's shape is quite remarkable. When particles collide, they scatter off each other and their velocities change in accordance with Newton's laws of motion. Waves, on the other hand, pass through each other unaffected.

Figure 14-44 shows a series of position plots of two pulses moving in opposite directions along a stretched string. When the pulses are far apart, the segment of the string where each pulse happens to be conforms to the shape of the pulse. When the two pulses overlap,

the shape of the string at the overlapping position corresponds to the combined effect of the two pulses, in accordance with the principle of superposition.

Now consider two pulses moving in opposite directions with identical shapes but opposite displacements. Figure 14-45 shows a series of position plots of two such pulses at different times. At the instant their centres overlap, the individual displacements cancel each other completely and the string is fully straight. Then each pulse reappears, continuing to travel in its original direction of motion with its original shape.

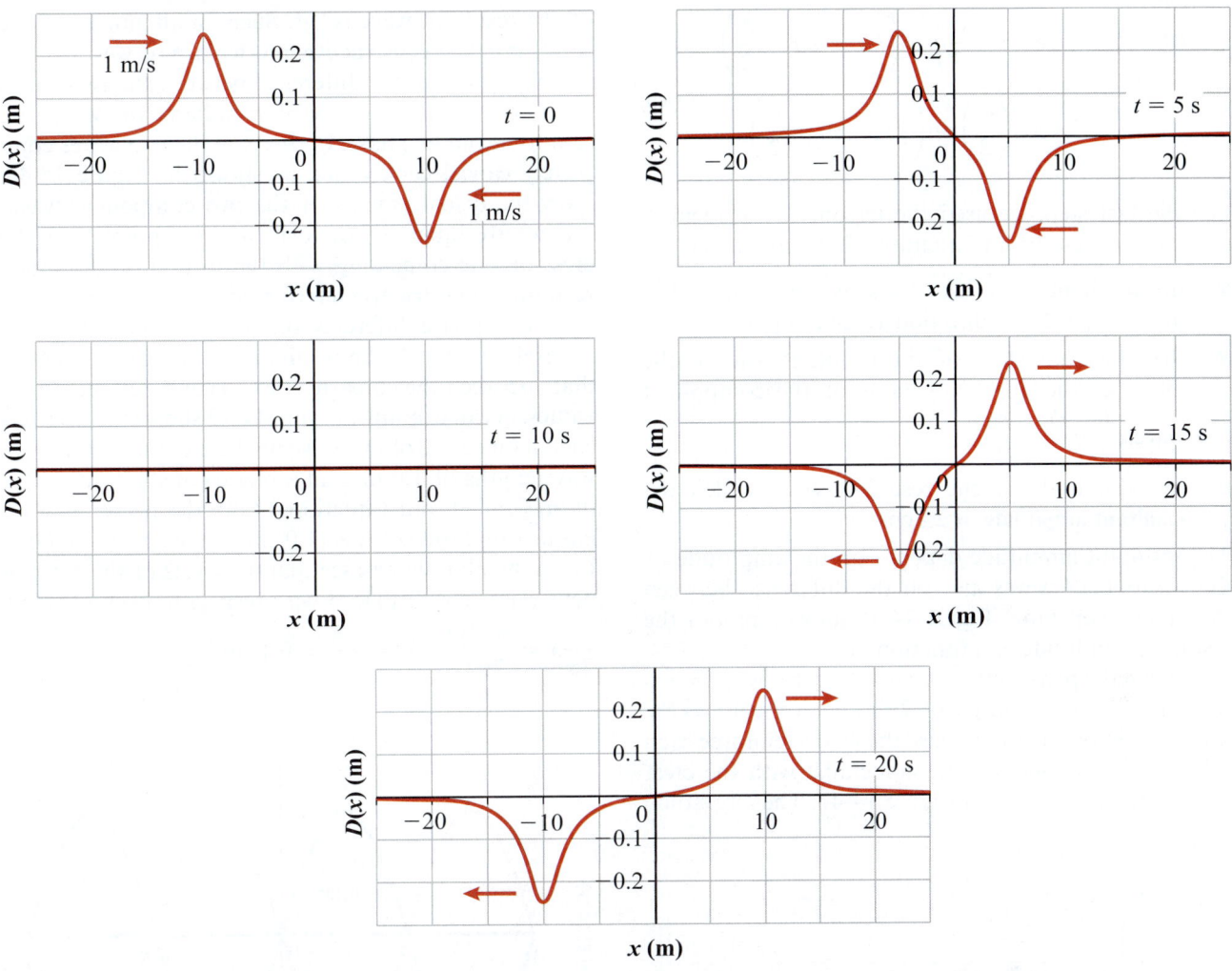

Figure 14-45 Interference of two pulses with identical shapes but opposite amplitudes travelling in opposite directions.

14-10 Interference of Waves Travelling in the Same Direction

We now derive a mathematical expression for the resultant wave from the superposition of two waves with the same wavelength, frequency, amplitude, and direction but different phase constants. To simplify the algebra, we will assume that the phase constant of one of the waves is zero. Then, the wave functions of the two waves are

$$D_1(x, t) = A\sin(kx - \omega t)$$
$$D_2(x, t) = A\sin(kx - \omega t + \phi) \qquad (14\text{-}37)$$

The phase difference between these two waves is equal to the phase constant difference between the two waves:

$$\Delta\Phi = \text{phase of wave 2} - \text{phase of wave 1}$$
$$= (kx - \omega t + \phi) - (kx - \omega t) = \phi \qquad (14\text{-}38)$$

The wave function of the second wave can be rewritten as

$$D_2(x, t) = A\sin\left(k\left(x + \frac{\phi}{k}\right) - \omega t\right)$$
$$= A\sin\left(k\left(x + \frac{\phi}{2\pi}\lambda\right) - \omega t\right) \qquad (14\text{-}39)$$

This form shows that the effect of the phase constant is to shift the second wave by a distance of $\dfrac{\lambda}{2\pi}\phi$ relative to the first wave. The resultant wave function is equal to the sum of the two wave functions:

$$D(x, t) = A\sin(kx - \omega t) + A\sin(kx - \omega t + \phi)$$

Using the trigonometric identity

$$\sin a + \sin b = 2\cos\frac{a - b}{2}\sin\frac{a + b}{2}$$

with $a = kx - \omega t$ and $b = kx - \omega t + \phi$, the equation for the resultant wave function becomes

$$D(x, t) = 2A\cos\left(-\frac{\phi}{2}\right)\sin\left(kx - \omega t + \frac{\phi}{2}\right)$$

$$= 2A\cos\left(\frac{\phi}{2}\right)\sin\left(kx - \omega t + \frac{\phi}{2}\right) \quad (14\text{-}40)$$

For the last step, we used the trigonometric identity $\cos(-\theta) = \cos\theta$. From Equation 14-40, notice that

- the resultant wave has the same wavelength, frequency, speed, and direction as the component waves
- the phase constant of the resultant wave is the mean of the phase constants of the component waves, $\dfrac{0 + \phi}{2} = \dfrac{\phi}{2}$
- the amplitude of the resultant wave, called the **resultant amplitude**, is $2A\cos(\phi/2)$

The resultant amplitude depends on the amplitude of the component waves and on the difference between their phase constants. Figure 14-46 shows a plot of the resultant amplitude as a function of ϕ.

When the phase difference between the two waves is zero or an integer multiple of 2π radians, the waves are said to be in phase. In this case the waves reinforce each other, with the crest of one coinciding with the crest of the other, as shown in Figure 14-47. The amplitude

of the resultant wave is $2A$. Such an alignment of the waves is called **constructive interference** of waves.

When the phase difference between the two waves is an odd integer multiple of π radians, the waves are said to be out of phase. In this case the two waves completely cancel each other, as shown in Figure 14-48. Now the displacements of the two component waves are exactly equal and opposite at every point, with the crest of one coinciding with the trough of the other, resulting in **destructive interference** of the waves.

For a phase difference that is other than an integer multiple of π rad, the resultant amplitude is intermediate between the constructive ($2A$) and destructive (0) values. As an example, Figure 14-49 shows the interference at time $t = 0$ of two waves with amplitudes of 1.0 cm, wavelengths of 1.0 m, and a phase constant difference of $\pi/3$ rad. From Equation 14-40, the resultant wave has an amplitude of $2 \times (1.0\,\text{cm}) \times \cos(\pi/6) = 1.7\,\text{cm}$. From the plot, we can see that the crests of the component waves are displaced with respect to each other by

$$\frac{\phi}{2\pi}\lambda = \frac{\pi/3\,\text{rad}}{2\pi\,\text{rad}} \times \lambda = \frac{\lambda}{6} = 0.17\,\text{m}.$$

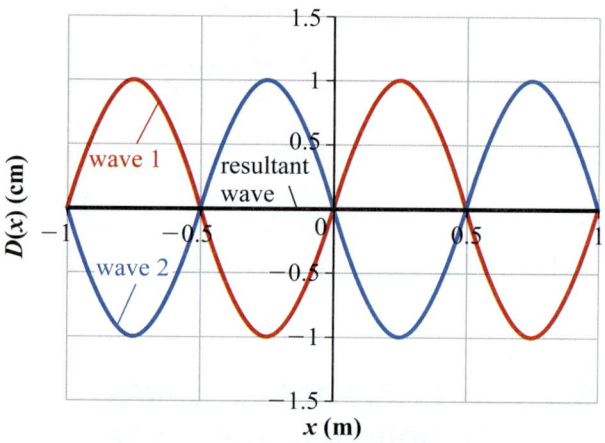

Figure 14-48 Destructive interference of two waves moving in the same direction with equal amplitudes and wavelengths and a phase constant difference of π rad.

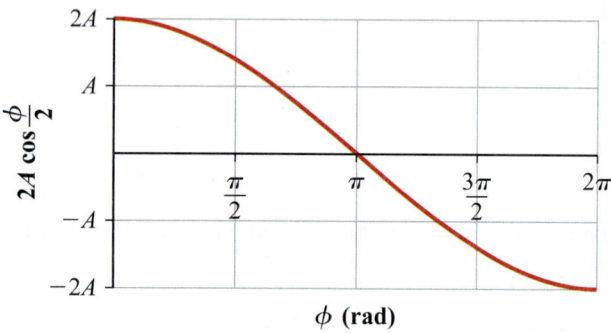

Figure 14-46 The amplitude of the resultant wave as a function of the difference between the phase constants of two waves that have the same wavelength, frequency, amplitude, and direction.

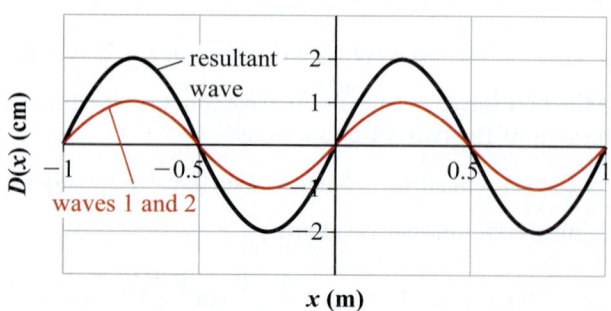

Figure 14-47 Constructive interference of two waves moving in the same direction with equal amplitudes and wavelengths and a phase constant difference of 0 rad.

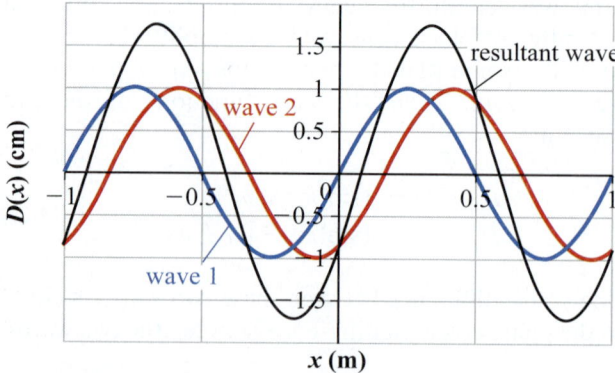

Figure 14-49 The interference of two waves of equal amplitudes and wavelengths and a phase constant difference of $\pi/3$ rad.

WAVES AND OSCILLATIONS

14-11 Reflection and Transmission of Mechanical Waves

When you listen to music in a concert hall, you may notice reverberation, an important factor in the sound quality. Reverberation results from sound waves reflecting from the walls of the hall. However, someone standing in the lobby can also hear the music, albeit considerably muffled. So, a portion of the sound waves has gone through the walls. In this section, we introduce the concepts of reflection and transmission of mechanical waves at boundaries.

Reflection at a fixed end Consider a transverse pulse travelling along a string toward an end that is tied to a rigid support, as shown in Figure 14-50. The transverse displacement of the **incident pulse** shown is upward, so the string exerts an upward force on the support when the pulse reaches the fixed end. The rigid support exerts an equal and opposite reaction force on the string, in accordance with Newton's third law of motion. The reaction force generates an inverted pulse that moves in the opposite direction. A reflection from a rigid boundary is called a **hard reflection**.

The same is true for waves. When a wave reflects from a fixed end, the reflected wave is inverted (compared to the incident wave) and moves in the opposite direction. Since the rigid end has zero displacement, the incident and the reflected waves always cancel each other at this end. Therefore, they are out of phase by π rad (180°). When a wave is reflected from a *fixed* end, the phase constant of the reflected wave changes by π rad compared to the phase constant of the incident wave.

Reflection at a free end Now consider the reflection of the same pulse from an end that is free to move. In Figure 14-51, the end of the string is connected to a light ring that slides freely along a rod. The incident pulse exerts an upward force on the ring, causing it to accelerate upward. Due to inertia, the ring keeps moving upward,

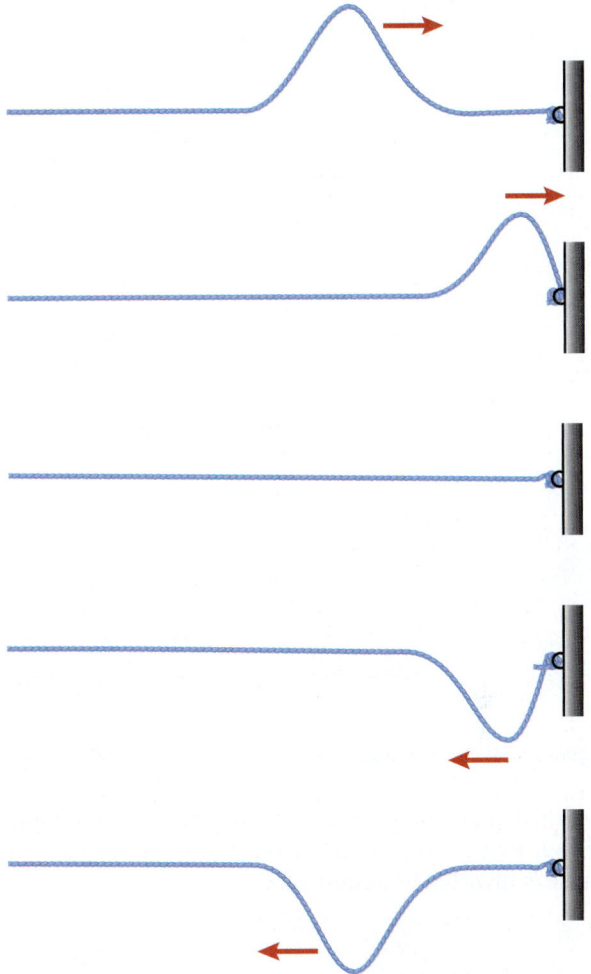

Figure 14-50 Hard reflection: A pulse reflected from a fixed end is inverted. The phase of the reflected pulse is changed by π rad.

overshooting the maximum height of the pulse and pulling the string with it. The reaction force exerted on the string by the ring generates a *backward-moving pulse that is not inverted*. Such a reflection is called a **soft reflection**. If the amplitude of the pulse is A, the maximum displacement of the ring from its equilibrium position is $2A$. Therefore, for a soft reflection, the incident and the reflected pulses are in phase with each other.

Similarly, when a wave reflects from a free end, the reflected wave is not inverted (compared to the incident wave) and moves in the opposite direction. In this case, the incident and the reflected waves are in phase with each other.

Often, an end is neither completely rigid nor fully free. Consider a string of linear mass density μ_1 attached to a string of linear mass density μ_2, with $\mu_1 < \mu_2$. When a pulse travelling in the lighter string reaches the junction between the two strings, part of the incident pulse is reflected and part is transmitted into the thicker string. The reflected pulse is inverted, but the transmitted pulse is not (Figure 14-52). For the reverse situation, when an incident pulse travelling in

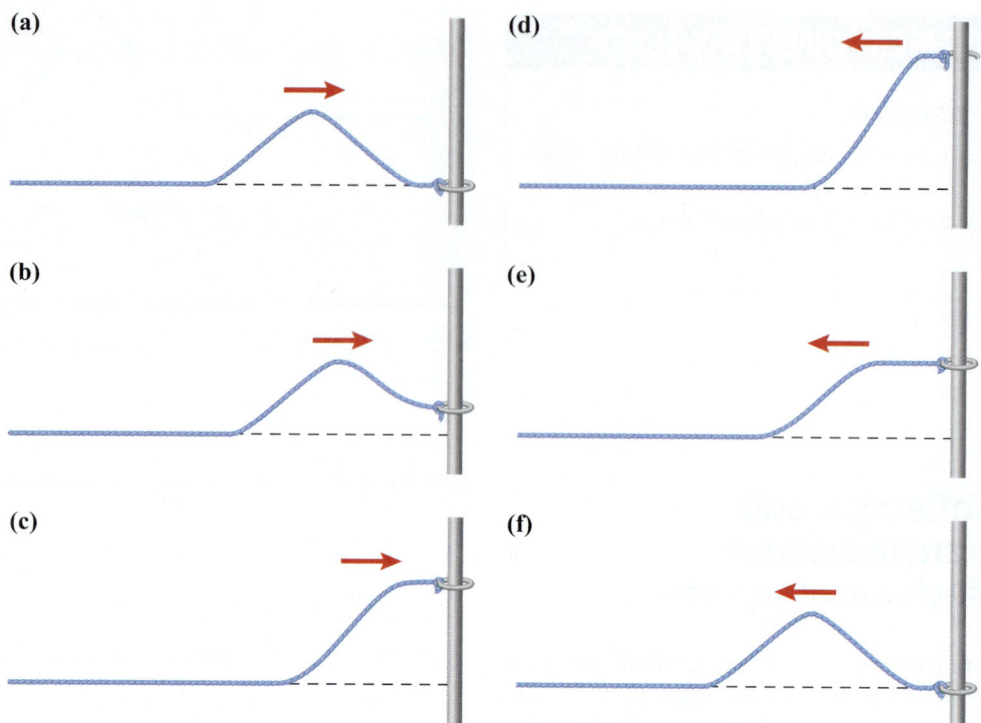

Figure 14-51 Soft reflection: A pulse reflected from a free end is not inverted. The phase of the reflected pulse does not change.

the thicker string reaches the boundary with the lighter string, neither the reflected pulse nor the transmitted pulse is inverted (Figure 14-53).

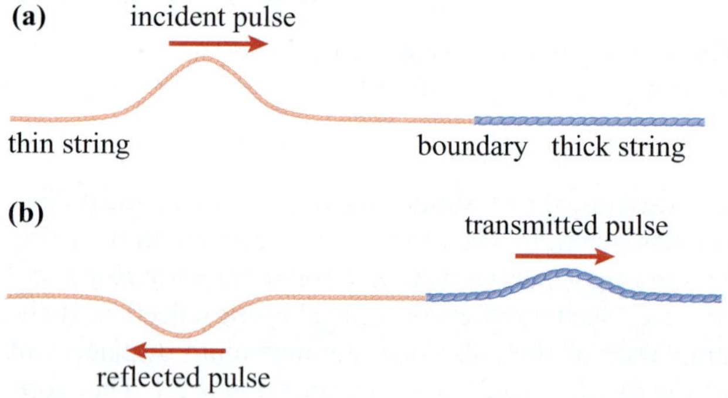

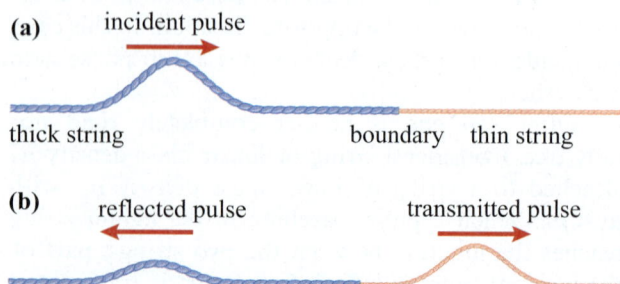

Figure 14-52 When a pulse travels from a less dense medium to a denser medium ($\mu_1 < \mu_2$), the reflected pulse is inverted, but the transmitted pulse is not.

(a) incident pulse

thick string   boundary   thin string

(b) reflected pulse   transmitted pulse

Figure 14-53 When a pulse travels from a dense medium to a less dense medium ($\mu_1 > \mu_2$), neither the reflected pulse nor the transmitted pulse is inverted.

The total energy of the incident wave is equal to the sum of the energies of the reflected and transmitted waves.

Although our discussion here has dealt only with one-dimensional waves on strings, two- and three-dimensional mechanical waves and electromagnetic waves show the same behaviour at boundaries between media.

14-12 Standing Waves

In Section 14-10, we discussed the interference of two waves that are moving in the same direction. Now we consider two harmonic waves of equal amplitude, wavelength, and frequency that are moving in *opposite* directions. For mathematical simplicity, we assume that the phase constants of both waves are zero; however, in general, they may be unequal. The wave functions for the two waves are

$$D_1(x, t) = A \sin(kx - \omega t)$$

(wave moving in the direction of increasing x)

(14-41)

$$D_2(x, t) = A \sin(kx + \omega t)$$

(wave moving in the direction of decreasing x)

According to the principle of superposition, the resultant wave function is

$$
\begin{aligned}
D(x, t) &= D_1(x, t) + D_2(x, t) \\
&= A \sin(kx - \omega t) + A \sin(kx + \omega t) \quad (14\text{-}42) \\
&= A(\sin(kx - \omega t) + \sin(kx + \omega t))
\end{aligned}
$$

Using the trigonometric identity

$$\sin(a - b) + \sin(a + b) = 2 \sin(a) \cos(b)$$

with $a = kx$ and $b = \omega t$, we obtain

KEY EQUATION $\quad D(x, t) = 2A \sin(kx) \cos(\omega t) \quad$ (14-43)

What kind of wave does this wave function describe? A travelling wave must have the position, x, and time, t, in the combination $x \mp vt$, where v is the wave speed. In Equation 14-43, the x- and t-variables appear separately, with x in the argument of the sine function and t in the cosine function. Therefore, the wave described by Equation 14-43 is not a travelling wave. We call it a **standing wave**. To understand the nature of such a wave, let us define a position-dependent amplitude, $A(x)$, as

$$A(x) = 2A \sin(kx) = 2A \sin\left(2\pi \frac{x}{\lambda}\right) \quad (14\text{-}44)$$

Using this definition, we can rewrite Equation 14-43 as

$$D(x, t) = A(x) \cos(\omega t) \quad (14\text{-}45)$$

Therefore, when two waves of equal wavelength, frequency, and amplitude but moving in opposite directions combine, each segment of the string oscillates in simple harmonic motion with the frequency of the individual waves and an amplitude that depends on the location of the segment along the string.

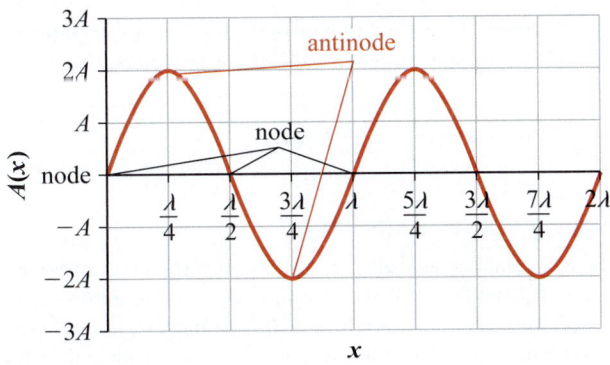

Figure 14-54   The amplitude, $A(x)$, of a standing wave as a function of position along a string.

Figure 14-54 shows a plot of $A(x)$ as a function of position. Amplitude is a sine function, so certain points on the string have zero amplitude and remain at rest at all times. These points are called **nodes**. The points that move with the maximum possible amplitude of $2A$ are called **antinodes**. All the other points have amplitudes between zero and $2A$. Any two points that are one wavelength apart have the same amplitude because

$$
\begin{aligned}
A(x_0 + \lambda) &= 2A \sin\left(2\pi \frac{x_0 + \lambda}{\lambda}\right) \\
&= 2A \sin\left(2\pi \frac{x_0}{\lambda} + 2\pi\right) \\
&\qquad\qquad\qquad\qquad\qquad (14\text{-}46) \\
&= 2A \sin\left(2\pi \frac{x_0}{\lambda}\right) \\
&= A(x_0)
\end{aligned}
$$

Since nodes do not move, no energy flows along a standing wave. As each segment of the medium between two consecutive nodes oscillates in simple harmonic motion, its energy continually transforms between kinetic energy and elastic potential energy, just as for a single harmonic oscillator.

Location of nodes and antinodes At the nodes of a standing wave, $A(x) = 0$. Therefore, nodes occur when

$$\sin\left(\frac{2\pi}{\lambda} x\right) = 0$$

$$\frac{2\pi}{\lambda} x = m\pi, \quad m = 0, \pm 1, \pm 2, \ldots$$

$$x = m\frac{\lambda}{2}, \quad m = 0, \pm 1, \pm 2, \ldots \quad (14\text{-}47)$$

$$x = 0, \pm\frac{\lambda}{2}, \pm\lambda, \pm\frac{3\lambda}{2}, \pm 2\lambda, \ldots$$

Thus, the distance between two consecutive nodes is half a wavelength.

At the antinodes of a standing wave, $A(x) = \pm 2A$, which occurs when

$$\sin\left(\frac{2\pi}{\lambda} x\right) = \pm 1$$

$$\frac{2\pi}{\lambda} x = \left(m + \frac{1}{2}\right)\pi, \quad m = 0, \pm 1, \pm 2, \ldots$$

$$x = \left(m + \frac{1}{2}\right)\frac{\lambda}{2}, \quad m = 0, \pm 1, \pm 2, \ldots \quad (14\text{-}48)$$

$$x = \pm\frac{\lambda}{4}, \pm\frac{3\lambda}{4}, \pm\frac{5\lambda}{4}, \ldots$$

The distance between consecutive antinodes is also half a wavelength. An adjacent node and an antinode are a quarter of a wavelength apart.

To see how a string oscillates in a standing wave pattern, we write the wave function in terms of the period ($T = 1/f$):

$$D(x, t) = 2A \sin\left(\frac{2\pi}{\lambda}x\right) \cos\left(\frac{2\pi}{T}t\right) \quad (14\text{-}49)$$

Figure 14-55 shows the displacement of a section of the oscillating string at intervals of $T/8$ from $T = 0$ to $t = T/2$. Table 14-2 compares the displacements of the section of the string between the first two nodes during the first half of the period. The motion for the next half of the period is in the opposite direction. This oscillatory motion repeats every period.

All points between two consecutive nodes oscillate in phase with each other. The mean speed is greatest for the antinode because it has to cover the longest distance ($8A$) in one period. The speed decreases away from the antinode and is zero at the nodes. Note that the motion of the section between the next two nodes is π rad out of phase with the first section. The sine term in the wave function accounts for this property:

$$A\left(x_0 + \frac{\lambda}{2}\right) = 2A \sin\left(\frac{2\pi}{\lambda}\left(x_0 + \frac{\lambda}{2}\right)\right)$$

$$= 2A \sin\left(\frac{2\pi}{\lambda}x_0 + \pi\right)$$

$$\quad (14\text{-}50)$$

$$= -2A \sin\left(2\pi \frac{x_0}{\lambda}\right)$$

$$= -A(x_0)$$

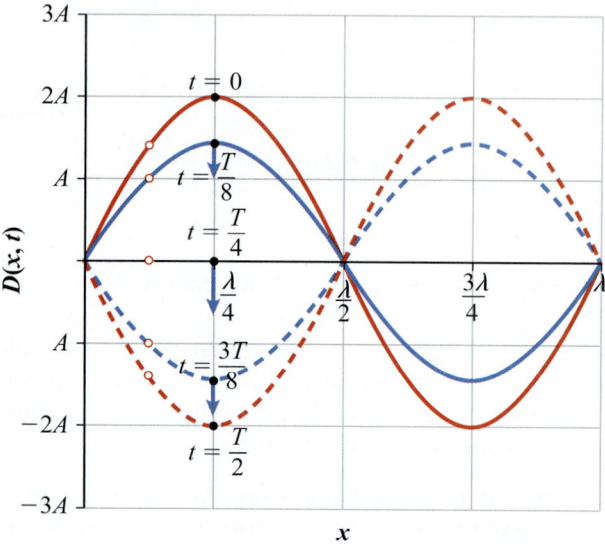

Figure 14-55 Displacement versus position for a standing wave on a string during the first half of a period. The position, x, is expressed in multiples of the wavelength, λ.

EXAMPLE 14-12

Standing Wave

The amplitude of a standing wave is given by

$$A(x) = (25.0 \text{ cm}) \sin (2.00x)$$

where x is in metres. Determine

(a) the amplitude and the wavelength of the constituent travelling waves
(b) the location of the first three nodes and the first three antinodes along the positive x-axis
(c) the location of the first point from the origin where the amplitude is 0.20 m

Solution

(a) Comparing the given amplitude to Equation 14-44, we find that

$$2A = 25.0 \text{ cm} \quad \text{and} \quad \frac{2\pi}{\lambda} = 2.00 \text{ rad/m}$$

Therefore,

$$A = 12.5 \text{ cm and } \lambda = \pi \text{ m} = 3.14 \text{ m}$$

(b) The positions of nodes and antinodes can be calculated from Equations 14-47 and 14-45. The first three nodes occur at

$$x = 0.00 \text{ m}, \ x = \frac{\lambda}{2} = \frac{3.14 \text{ m}}{2} = 1.57 \text{ m}, \ x = \lambda = 3.14 \text{ m}$$

The first three antinodes occur at

$$x = \frac{\lambda}{4} = \frac{3.14 \text{ m}}{4} = 0.785 \text{ m}, \quad x = \frac{3\lambda}{4} = 2.36 \text{ m},$$

$$x = \frac{5\lambda}{4} = 3.92 \text{ m}$$

(c) To determine the location of the point that has amplitude 0.20 m, we substitute this value in the amplitude equation and solve for x:

$$0.20 \text{ m} = (0.25 \text{ m}) \sin 2.00x$$
$$\sin (2.00x) = 0.80 \Rightarrow 2.00x = \sin^{-1}(0.80) \Rightarrow x = 0.46 \text{ m}$$

Making sense of the result

The amplitude is zero at $x = 0.00$ m, and the first antinode at $x = 0.78$ m has a displacement of 0.25 m. Therefore, the first occurrence of an amplitude of 0.20 m must be somewhere between 0.00 m and 0.78 m from the origin, in agreement with the location calculated.

Table 14-2 Displacements of String Elements between Two Nodes of a Standing Wave during Half a Period

Time	Wave Function	Displacement
0	$D(x, 0) = 2A\sin\left(\dfrac{2\pi}{\lambda}x\right)\cos(0) = 2A\sin\left(\dfrac{2\pi}{\lambda}x\right)$	Each segment is at its maximum displacement.
$T/8$	$D\left(x, \dfrac{T}{8}\right) = 2A\sin\left(\dfrac{2\pi}{\lambda}x\right)\cos\left(\dfrac{2\pi}{T}\dfrac{T}{8}\right) = \sqrt{2}A\sin\left(\dfrac{2\pi}{\lambda}x\right)$	The displacement of each segment decreases by a factor of $\sqrt{2}$.
$T/4$	$D(x, T/4) = 2A\sin\left(\dfrac{2\pi}{\lambda}x\right)\cos\left(\dfrac{2\pi}{T}\dfrac{T}{4}\right) = 0$	Each segment reaches the equilibrium position.
$3T/8$	$D\left(x, \dfrac{3T}{8}\right) = 2A\sin\left(\dfrac{2\pi}{\lambda}x\right)\cos\left(\dfrac{2\pi}{T}\dfrac{3T}{8}\right) = -\sqrt{2}A\sin\left(\dfrac{2\pi}{\lambda}x\right)$	Each segment reaches $1/\sqrt{2}$ of the maximum displacement at the opposite side of the equilibrium position.
$T/2$	$D(x, T/2) = 2A\sin\left(\dfrac{2\pi}{\lambda}x\right)\cos\left(\dfrac{2\pi}{T}\dfrac{T}{2}\right) = -2A\sin\left(\dfrac{2\pi}{\lambda}x\right)$	Each segment is at its maximum displacement at the opposite of the equilibrium position.

14-13 Standing Waves on Strings

Consider a string of length L with both ends fixed. When the string is plucked, waves travel back and forth along the string, reflecting from the fixed ends. We then have travelling waves moving in opposite directions, creating standing waves on the string.

We take one end of the string to be the origin and the other end to be at $x = L$. Because the string is clamped at both ends, the amplitude must be zero at $x = 0$ and at $x = L$. The amplitude in Equation 14-44 is zero at $x = 0$, and for it to be zero at $x = L$ we must have

$$\sin\left(\frac{2\pi}{\lambda}L\right) = 0 \qquad (14\text{-}51)$$

and

$$\frac{2\pi}{\lambda}L = m\pi \qquad (14\text{-}52)$$

where m is a positive, non-zero integer (1, 2, 3, …).

Rearranging Equation 14-52, we find that a string with both ends fixed can oscillate in a standing wave pattern with only the following wavelengths:

KEY EQUATION $\qquad \lambda_m = \dfrac{2L}{m} \quad m = 1, 2, 3, 4, \ldots \qquad (14\text{-}53)$

The longest of these wavelengths is $\lambda_1 = 2L$. The series of wavelengths continues with $\lambda_2 = L$, $\lambda_3 = \dfrac{2L}{3}$,

$\lambda_4 = \dfrac{L}{2}$, and so on. Notice that each of these wavelengths is such that an integral number of half wavelengths fit within length L of the string (see Figures 14-56 and 14-57). These standing waves are called the **normal modes** of vibration of the string.

The frequencies corresponding to the normal modes of vibration are

$$f_m = \frac{v}{\lambda_m}$$

KEY EQUATION
$$= m\frac{v}{2L} \qquad (14\text{-}54)$$

$$= \frac{m}{2L}\sqrt{\frac{T_s}{\mu}}$$

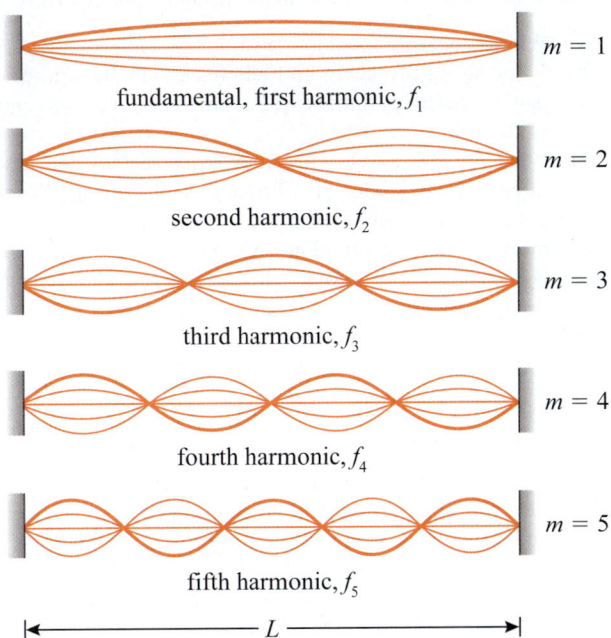

$m = 1$
fundamental, first harmonic, f_1

$m = 2$
second harmonic, f_2

$m = 3$
third harmonic, f_3

$m = 4$
fourth harmonic, f_4

$m = 5$
fifth harmonic, f_5

L

Figure 14-56 Standing waves on a string: the first five harmonics.

INTERACTIVE ACTIVITY 14-4

Standing Waves on a String

In this activity, you will use the PhET simulation "Standing Waves on a String" to gain an understanding of how standing waves arise from the superposition of two travelling waves.

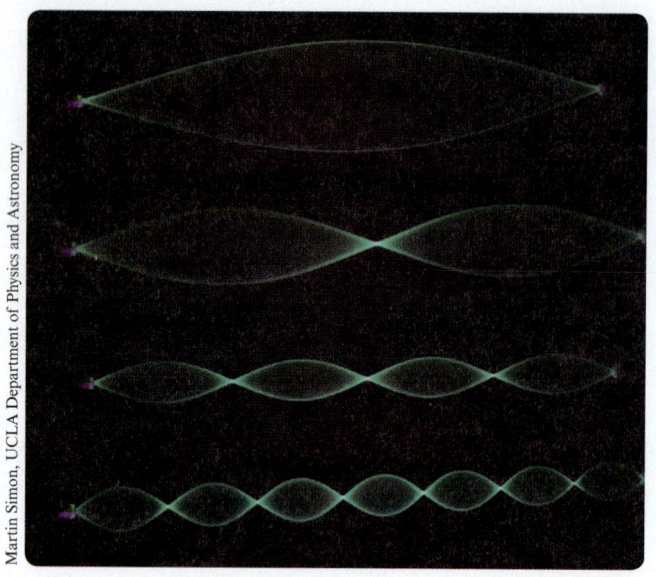

Figure 14-57 The envelopes of a fluorescent string vibrating in various normal modes. The images were taken by shining a strobe light on the string.

Notice that the fundamental frequency of a string is proportional to the square root of the tension in the string and is inversely proportional to its length and to the square root of the linear mass density. Higher frequencies of vibration correspond to higher values of m and are integer multiples of the fundamental frequency, since

$$f_m = mf_1 \qquad m = 1, 2, 3, 4, \ldots \qquad (14\text{-}56)$$

The allowed frequencies are called **harmonics** or **resonant frequencies**. The first few harmonics are

$$f_1 = \frac{1}{2L}\sqrt{\frac{T_s}{\mu}} \qquad \text{fundamental or first harmonic}$$
$$f_2 = 2f_1 \qquad \text{second harmonic}$$
$$f_3 = 3f_1 \qquad \text{third harmonic} \qquad (14\text{-}57)$$
$$f_4 = 4f_1 \qquad \text{fourth harmonic}$$

The lowest frequency corresponds to the longest wavelength, $\lambda_1 = 2L$, and is called the **fundamental frequency** or the **first harmonic**:

$$f_1 = \frac{v}{\lambda_1} = \frac{v}{2L} = \frac{1}{2L}\sqrt{\frac{T_s}{\mu}} \qquad (14\text{-}55)$$

When plucked, a string can vibrate in a single normal mode or a combination of several modes. Notice that m is equal to the number of antinodes between the fixed ends of the string.

As we will see in Chapter 15, sound waves in a wind instrument also form standing wave patterns.

EXAMPLE 14-13

String Vibrations

A violin string is 0.320 m long and has a linear mass density of 3.83×10^{-4} kg/m. The tension in the string is kept at 70.0 N.

(a) What is the wave speed in the string?
(b) If the string is plucked and allowed to vibrate, approximately how many times would a wave reflect from one end of the string in 1 s?
(c) What are the wavelengths and frequencies of the first three normal modes of vibration of the string? What is the spacing between two consecutive nodes for each mode? Sketch the vibration pattern of these modes.
(d) By what amount should the tension be changed to decrease the fundamental frequency by 7 Hz?

Solution

When plucked, the string will vibrate in a standing wave pattern with the wavelengths and frequencies of normal modes given by Equations 14-53 and 14-56.

(a) The wave speed in the string is determined by the tension and linear mass density of the string:

$$v = \sqrt{\frac{T_s}{\mu}} = \sqrt{\frac{70.0\,\text{N}}{3.83 \times 10^{-4}\,\text{kg/m}}} = 4.27 \times 10^2\,\text{m/s}$$

(b) The time that the waves take to travel from one end of the string to the other is

$$\Delta t = \frac{\text{length of the string}}{\text{wave speed}} = \frac{0.320\,\text{m}}{4.27 \times 10^2\,\text{m/s}} = 7.48 \times 10^{-4}\,\text{s}$$

Each reflection from a given end requires a back-and-forth trip. Therefore, the number of times a wave reflects from a given fixed end in 1 s is

$$\frac{1.0\,\text{s}}{2 \times \Delta t} = \frac{1.0\,\text{s}}{2 \times 7.48 \times 10^{-4}\,\text{s}} = 6.68 \times 10^2$$

(c) For the first harmonic:
   wavelength: $\lambda_1 = 2L = 2 \times 0.320\,\text{m} = 0.640\,\text{m}$

   frequency: $f_1 = \frac{v}{\lambda_1} = \frac{427\,\text{m/s}}{0.640\,\text{m}} = 6.68 \times 10^2$

   distance between consecutive nodes: $= \lambda_1/2 = 0.320\,\text{m}$

For the second harmonic:

   wavelength: $\lambda_2 = 2L/2 = 0.320\,\text{m}$
   frequency: $f_2 = 2f_1 = 1.34 \times 10^3\,\text{Hz}$
   distance between consecutive nodes: $= \lambda_2/2 = 0.160\,\text{m}$

WAVES AND OSCILLATIONS

Martin Simon, UCLA Department of Physics and Astronomy

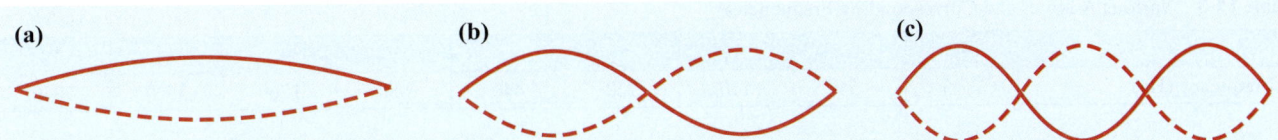

(a) **(b)** **(c)**

Figure 14-58 Example 14-13. Envelopes of the displacements for (a) the first harmonic, (b) the second harmonic, and (c) the third harmonic.

For the third harmonic:

$$\text{wavelength: } \lambda_3 = 2L/3 = 0.213 \text{ m}$$
$$\text{frequency: } f_3 = 3f_1 = 2.00 \times 10^3 \text{ Hz}$$
$$\text{distance between consecutive nodes: } = \lambda_3/2 = 0.106 \text{ m}$$

The vibration patterns of these modes are shown in Figure 14-58.

(d) The new fundamental frequency will be $f_1' = 668 \text{ Hz} - 7 \text{ Hz} = 661 \text{ Hz}$. To decrease the fundamental frequency, the tension in the string must be decreased. If T_s' is the new tension, then

$$f_1' = \frac{1}{2L}\sqrt{\frac{T_s'}{\mu}}$$

Squaring the above equation, we can write

$$T_s' = 4L^2\mu(f_1')^2$$

The fractional decrease in tension needed to lower the frequency by 7.000 Hz is therefore

$$\frac{T_s - T_s'}{T_s} = \frac{(f_1)^2 - (f_1')^2}{(f_1)^2} = \frac{(668 \text{ Hz})^2 - (661 \text{ Hz})^2}{(668 \text{ Hz})^2} = 0.0208$$

Therefore, the tension in the string needs to be decreased by $0.0208 \times 70.0 \text{ N} = 1.46 \text{ N}$.

Making sense of the result

The fundamental frequency calculated in part (c) is the same as the number of times the wave reflects from a fixed end of the string (calculated in part (b)).

CHECKPOINT 14-15

Standing Wave Nodes

How many nodes (excluding the nodes at both ends) are present between the fixed ends of a string vibrating at its seventh harmonic?

(a) 5
(b) 6
(c) 7
(d) 8

ANSWER: (b)

14-14 String Musical Instruments (Optional Section)

A musical instrument that produces sound by vibration of strings is called a string instrument. Such instruments come in many shapes and sizes and include (from smallest to largest string lengths) violin, cello, viola, and bass. Other type of stringed instruments include banjo, various types of guitars, harp, sitar, and more.

Musical Scale

Human ears are sensitive to sounds of frequency between 20 Hz and 20 000 Hz. As we will study in Chapter 15, sound is a pressure wave. A vibrating source creates regions of low and high pressure that travel through the surrounding medium. The faster a source vibrates, the higher the frequency at which the pressure varies above the normal pressure (the pressure when there is no sound wave). Repetitive pressure fluctuation produces an auditory sensation called a **tone** that has a pitch associated with it. For our purposes, we define the **pitch** of a sound to be the frequency of the sound. So high pitch corresponds to high-frequency sound and low pitch corresponds to low-frequency sound.

From the continuum of frequencies within the audible range, a set of frequencies has been chosen to categorize music. This set is called the **musical scale**. How does the musical scale work? It is a common experience that the combined sound from two sources sounds pleasant if the pitch of the two sources differs by a certain interval. This interval is called an **octave**. Pythagoras determined that a pitch difference of one octave implies that the pitch (i.e., the frequency) of the two sounds differs by a factor of two. So if a certain note corresponds to a frequency of f, then a note that is one octave higher corresponds to a frequency of $2f$ and a note that is one octave lower corresponds to a frequency of $f/2$.

In Western music, the interval of an octave is divided into 12 equally spaced subintervals, called **semitones**. The ratio that when multiplied by itself 12 times gives 2 is $2^{1/12} \approx 1.0595$. Therefore, if a note corresponds to frequency f, then the note that is one octave higher has frequency $2f$ and the frequencies of the intervening semitones are $2^{1/12}f, 2^{2/12}f, 2^{3/12}f, \ldots, 2^{11/12}f$. For example, if a

Table 14-3 Various A Notes and Corresponding Frequencies

Note	A_0	A_1	A_2	A_3	A_4	A_5	A_6	A_7	A_8
Frequency (Hz)	27.5	55	110	220	440	880	1760	3520	7040

WAVES AND OSCILLATIONS

note corresponds to a frequency of 300 Hz, then the note that is one semitone higher has frequency $2^{1/12} \times 300$ Hz $\approx 1.0595 \times 300$ Hz = 317.85 Hz, and a note that is 6 semitones higher has a frequency of $2^{6/12} \times 300$ Hz $\approx 1.4142 \times 300$ Hz = 424.26 Hz.

This still leaves the starting frequency undefined. The current standard in Western music is the frequency of 440 Hz, which is given the name A_4. The same letter is used for notes that appear an octave above or an octave below this note, with the number in the subscript increasing or decreasing sequentially. The note that is one octave higher than A_4 has frequency 880 Hz and is named A_5. A note one octave below A_4 has frequency 220 Hz and is called A_3. Table 14-3 shows A notes and the corresponding frequencies.

The semitones between two successive As are also named (this naming is historical). Starting from A_4, the semitones' names and corresponding frequencies are given in Table 14-4. The notes starting from A_4 go up in 12 steps as A_4, $A_4\sharp$, B_4, C_5, $C_5\sharp$, D_5, $D_5\sharp$, E_5, F_5, $F_5\sharp$, G_5, $G_5\sharp$, A_5. The symbol $\sharp$ (called sharp) means "go one semitone above." So $A_4\sharp$ means one semitone above A_4. Notice that the turnover point for the subscript is not A, but C. So in going up from A_4 the subscript increases at the C note, i.e., C_5. To go one semitone below a note we use the symbol $\flat$ (called flat). So $A_4\sharp = B_4\flat$, which means that going one semitone higher from A_4 is the same as going one semitone lower from B_4.

Table 14-4 Semitones and Their Frequencies between A_4 and A_5 Notes

Note	Semitone	Frequency
A_4	0	440 Hz
$A_4\sharp$	1	$2^{1/12} \times 440 = 466.16$ Hz
B_4	2	$2^{2/12} \times 440 = 493.88$ Hz
C_5	3	$2^{3/12} \times 440 = 523.25$ Hz
$C_5\sharp = D_5\flat$	4	$2^{4/12} \times 440 = 554.36$ Hz
D_5	5	$2^{5/12} \times 440 = 587.33$ Hz
$D_5\sharp = E_5\flat$	6	$2^{6/12} \times 440 = 622.25$ Hz
E_5	7	$2^{7/12} \times 440 = 659.26$ Hz
F_5	8	$2^{8/12} \times 440 = 698.46$ Hz
$F_5\sharp = G_5\flat$	9	$2^{9/12} \times 440 = 739.99$ Hz
G_5	10	$2^{10/12} \times 440 = 783.99$ Hz
$G_5\sharp$	11	$2^{11/12} \times 440 = 839.60$ Hz
A_5	0	$2^{12/12} \times 440 = 2 \times 440 = 880$ Hz

An Acoustic Guitar

Let us see how notes are produced on a guitar. A guitar consists of six stretched strings, each of different thickness and hence different linear mass density. In the order of highest to lowest thickness, the strings are named as E, A, D, G, B, e (called high E). This is true for strings made of metal. Steel strings may come in the following gauges (a gauge is diameter measured in thousandths of an inch): 54, 42, 32, 25, 16, and 12. At the lower end, each string is tied to a bridge pin and passes over a saddle on the bridge (see Figure 14-59). At the upper end, the string passes over the nut and then wraps around a tuning post. The distance between the saddle and the nut is between 25 and 26 inches. This is then the free vibrating length of the string.

A guitar's body consists of a hollow chamber with a sound hole. The upper plate of the chamber is tied to the bridge and vibrates in response to the vibrations of the bridge. It is usually made of wood. Inside the chamber there are braces that strengthen the body and help keep the upper plate flat as it vibrates. The vibrations of the lower plate are usually not that important, because it rests against the musician. The guitar's body plays a crucial role in producing sound. The strings are very thin and, hence, when plucked, are not efficient at moving the air around them. The upper plate has a large cross-sectional area and is therefore much more efficient at moving air. As the front plate vibrates, the air inside the body continually expands and compresses. The sound hole couples these vibrations to the outside air. It is a common misconception that a guitar's body amplifies the sound. It just efficiently transforms the energy that was used in plucking the string into sound waves.

Between the nut and the saddle there are a series of thin metal bars, called frets, placed perpendicular to the length of the strings. By pressing on a fret, the length of the vibrating section of a string can be changed, which then changes the frequency at which the string vibrates. The distance between two successive frets decreases in going from the nut to the saddle.

In Section 14-13 we discussed how the resonant frequencies of a string, fixed at both ends, depend on its length, linear mass density, tension, and mode of vibration. The formulas describing these resonant frequencies are (Equations 14-56 and 14-57)

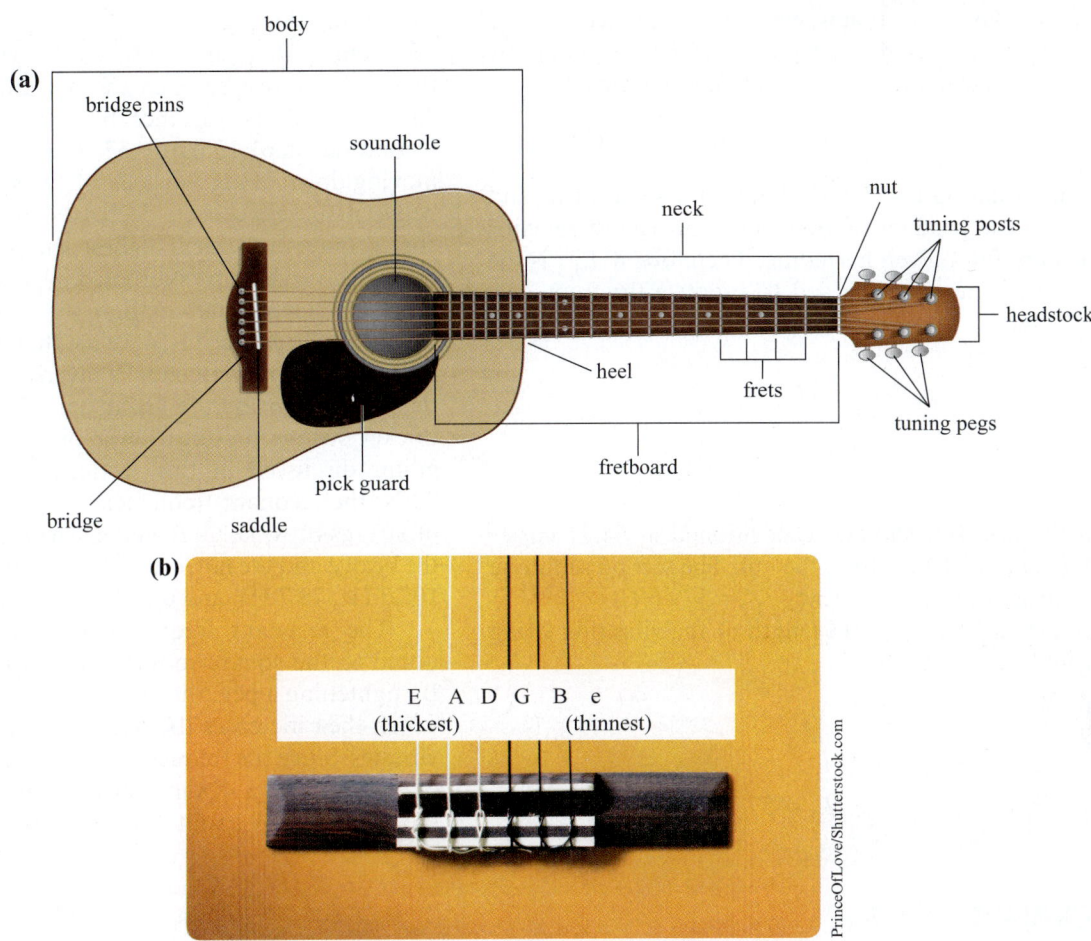

Figure 14-59 (a) Anatomy of an acoustic guitar. (b) Guitar strings.

PrinceOfLove/Shutterstock.com

$$f_m = mf_1, \quad m = 1, 2, 3, 4, \ldots$$

$$f_1 = \frac{1}{2L_1}\sqrt{\frac{T_s}{\mu}}$$

Here L_1 is the length of the vibrating section of the string, μ is its linear mass density, T_s is the tension in the string, and m is the harmonic number. The frequency f_1 corresponds to the fundamental (or the first) harmonic, and $f_2, f_3, \ldots$ correspond to the second, third, … harmonics, respectively.

Let us see how various notes are generated. Consider the E string, and let us determine the tension required to play the E_4 note at 329.6 Hz as the string vibrates in the first harmonic between the nut and the saddle. We will use the following data for the string:

$$L_1 = 25.50 \text{ inches} = 25.50 \text{ inches} \times \frac{2.54 \text{ cm}}{\text{inch}} = 64.77 \text{ cm}$$

$$= 0.6477 \text{ m}$$

$$D = 1.00 \times 10^{-2} \text{ inch} = 2.54 \times 10^{-2} \text{ cm}$$

$$= 2.54 \times 10^{-4} \text{ m}$$

Here D is the string's diameter. Let us assume that the string is made of steel of density 7750 kg/m³. Its linear mass density is

μ = cross-sectional area of the wire

$\quad \times$ density of material

$= \pi R^2 \times \rho_{steel}$

$= \pi \left(\dfrac{2.54 \times 10^{-4} \text{ m}}{2}\right)^2 \times 7750 \text{ kg/m}^3$

$= 3.93 \times 10^{-4} \text{ kg/m}$

Squaring Equation 14-57, we can now write tension in terms of the known quantities:

$T_s = (2f_1 L_1)^2 \times \mu$

$\quad = (2 \times 329.6 \text{ s}^{-1} \times 0.6477 \text{ m})^2 \times (3.93 \times 10^{-4} \text{ kg/m})$

$\quad = 71.6 \text{ N}$

So, to play the E_4 when the string vibrates freely between the nut and the saddle, the tension in the string should be 71.6 N.

A guitar player plays different notes by pressing on frets and therefore shortening the length of the vibrating string. Why does the spacing between successive frets decrease in going from top to bottom? Suppose we want to play a note that is one semitone higher than E_4. This note is F_4, and its frequency, which we denote by $f_{1,1}$ (first harmonic, first semitone), is given by

$$f_{1,1} = 2^{1/12}f_1 = 2^{1/2} \times 329.6 \text{ Hz} = 349.2 \text{ Hz}$$

From Equation 14-57, note that for the same string and tension the product of frequency and length remains constant for a given harmonic. Therefore, if $L_{1,1}$ is the length of the string needed to produce the frequency $f_{1,1}$, then

$$L_{1,1} = \left(\frac{f_1}{f_{1,1}}\right)L_1$$
$$= \frac{1}{2^{1/2}}L_1 = 61.13 \text{ cm}$$

So the first fret should be positioned at 64.77 cm − 61.13 cm = 3.64 cm from the nut. The second semitone (the $F_4\sharp$ note) has frequency $f_{1,2} = 329.6 \text{ Hz} \times 2^{2/12} = 370 \text{ Hz}$ and therefore the length of the vibrating string should be

$$L_{1,2} = \left(\frac{f_1}{f_{1,2}}\right)L_1$$
$$= \frac{1}{2^{2/12}}L_1 = 57.70 \text{ cm}$$

Therefore, the second fret should be positioned at 64.77 cm − 57.70 cm = 7.07 cm below the nut. Notice that the distance between the nut and the first fret is 3.64 cm and the distance between the first and the second fret is 61.13 cm − 57.70 cm = 3.43 cm. So, in moving down along the neck of the guitar, the spacing between successive frets decreases. To play a note that corresponds to the next highest semitone, the length of the string needs to be shortened by a factor of $1/2^{1/12} \approx 0.944$ of the previous length.

From Equation 14-54 we see that for a given length and tension the frequency is inversely proportional to the square root of the linear mass density. Therefore, thicker strings produce lower pitch. For the acoustic guitar discussed in this section, using a tension of 72 N, the resonant frequencies for the first harmonics of strings E, A, D, G, B, and e, when vibrating between the bridge and the nut, are 275.4 Hz, 206.6 Hz, 132.2 Hz, 103.3 Hz, 78.7 Hz, and 61.2 Hz.

The resonant frequency of a string is proportional to the square root of the tension in the string. By tightening a peg, the tension in a string is increased, which then increases the frequency at which the string vibrates. Since the frequency changes as the square root of the tension, a 4% increase in the tension increases the frequency by about 2%.

EXAMPLE 14-14

An Acoustic Guitar

Consider an acoustic guitar with strings made of steel of density 7750 kg/m³. The six strings, named E, A, D, G, B, and e (called high E), have diameters of 0.0120, 0.0160, 0.0250, 0.0320, 0.0420, and 0.0540 inches, respectively. If the tension in each string is kept at 72.0 N, determine the resonant frequency of each string while it vibrates in the first harmonic between the nut and the bridge. The length of the string is 25.5 inches.

Solution

We have six guitar strings made of steel. The strings have the same length and are under the same tension but have different diameters. So each string has a different linear mass density. The resonant frequencies of a string that is fixed at both ends are given by Equation 14-56. Also, we are given that the strings vibrate in the first harmonic, so we will use $m = 1$. To calculate resonant frequency, we will need to convert information about the diameter into the linear mass density of a string. The linear mass density is related to the density of the material and the diameter of the wire:

> Linear mass density = volume of a unit length of string × density of material

Let us calculate the linear mass density of the E string:

$$D = \text{diameter} = 0.0120 \text{ inch} = 0.0120 \text{ inch} \times \frac{2.54 \text{ cm}}{\text{inch}}$$
$$= 0.0305 \text{ cm} = 3.05 \times 10^{-4} \text{ m}$$

The linear mass density of the string can now be calculated:

$$\mu = \text{cross-sectional area of wire} \times \text{density of material}$$
$$= \pi(D/2)^2 \times \rho_{\text{steel}}$$
$$= \pi(3.05 \times 10^{-4} \text{ m}/2)^2 \times 7750 \text{ kg/m}^3$$
$$= 5.66 \times 10^{-4} \text{ kg/m}$$

The other known quantities are the length of the vibrating section of the string, $L = 25.5$ inches $= 0.6477$ m, and the tension in the string, $T_s = 72.0$ N. The frequency of the first harmonic can now be calculated:

$$f_1 = \frac{1}{2L}\sqrt{\frac{T_s}{\mu}}$$
$$= \frac{1}{2 \times 0.6477 \text{ m}}\sqrt{\frac{72.0 \text{ N}}{5.66 \times 10^{-4} \text{ kg/m}}} = 2.75 \times 10^2 \text{ Hz}$$

Therefore, the resonant frequency of the E string is 275 Hz. Performing similar calculations for the other strings, we get the following frequencies:

String A: $f_1 = 2.07 \times 10^2$ Hz
String D: $f_1 = 1.32 \times 10^2$ Hz
String G: $f_1 = 1.03 \times 10^2$ Hz
String B: $f_1 = 7.87 \times 10^1$ Hz
String e: $f_1 = 6.12 \times 10^1$ Hz

Making sense of the result

First let us look at the resonant frequency of string E. The factor inside the brackets is of the order of 12×10^4, the square root of which is $\sqrt{12} \times 100$, which is about $3.5 \times 100 = 350$. Since the factor outside the square root is about $1/1.3$, the answer should be about $350/1.3 \approx 270$. Our actual answer of 275 Hz is very close to this number. Second, the linear mass density of strings increases as we go from the E to the e string. The resonant frequency is inversely proportional to the diameter of the string. Therefore, the resonant frequencies should decrease with increasing thickness. Our answers show such a decrease. The ratio of the diameters of strings e and E is $0.054/0.012 = 4.5$, and we see that $275/61 \approx 4.5$.

CHECKPOINT 14-16

String Musical Instruments

A guitar string vibrates between the bridge and the nut at 512 Hz, forming the vibrational pattern shown. What is the frequency of the second harmonic for this string (see Figure 14-60)?

Figure 14-60 Checkpoint 14-16.

(a) 64 Hz   (b) 128 Hz   (c) 256 Hz
(d) 512 Hz   (e) 1024 Hz

ANSWER: (c)

14-15 The Wave Equation in One Dimension (Optional Section)

As a wave passes through a medium, any given segment of the medium executes harmonic motion about its equilibrium position. By applying Newton's second law to a segment, we can derive a mathematical equation (which is a second-order differential equation) called the **wave equation** that describes the segment's motion as a function of time in the presence of the wave. We will derive a wave equation for a transverse sinusoidal wave passing through a string. Even for this rather simple situation, the resulting wave equation is complicated, so we need to make a few assumptions to keep it simple. In particular, we will make the following assumptions:

(a) The string is of uniform linear mass density.

(b) The motion of the string is perpendicular to the direction of motion of the wave.

(c) There is no motion of the string along the direction of wave propagation.

(d) The amplitude of the wave is much smaller than its wavelength.

(e) There are no frictional forces, and the effect of gravity is ignored.

Consider a very small segment of length Δx of this string located between points 1 and 2. Point 1 of this segment is located at x and point 2 is at $x + \Delta x$ (Figure 14-61). In the absence of a wave, this segment is horizontal and under a tension $\vec{T}_{\text{s}}$. In the presence of a wave, the segment continually deforms as it is pulled from both ends by the rest of the string.

Let us apply Newton's second law to this vibrating segment. The string makes an angle of θ_2 with respect to the horizontal direction at point 2 and an angle of θ_1 at point 1. The tension at any point of the string is along the tangent to that point. Let the tension at point 2 be $\vec{T}_{\text{s},2}$ and at point 1 be $\vec{T}_{\text{s},1}$. The net force along the horizontal direction on this segment is

$$\sum F_x = T_{\text{s},2}\cos\theta_2 - T_{\text{s},1}\cos\theta_1 \qquad (14\text{-}58)$$

Because the string does not move along the horizontal direction, the net force along this direction must vanish. Therefore,

$$T_{\text{s},2}\cos\theta_2 - T_{\text{s},1}\cos\theta_1 = 0 \qquad (14\text{-}59)$$

As points 1 and 2 can be located anywhere along the string, Equation 14-59 states that the x-component of

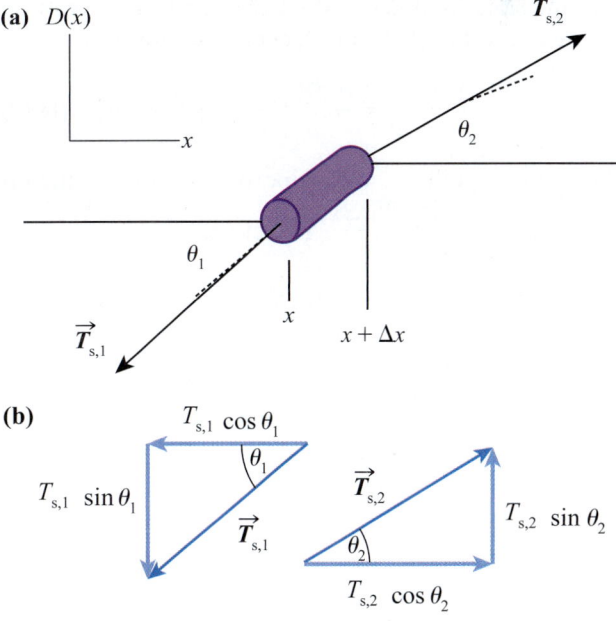

(a) $D(x)$

(b)

Figure 14-61 (a) A string segment as a wave passes through it. Tensions $\vec{T}_{\text{s},1}$ and $\vec{T}_{\text{s},2}$ act at the left and the right ends of this segment, producing an upward acceleration. (b) Resolving $\vec{T}_{\text{s},1}$ and $\vec{T}_{\text{s},2}$ along their horizontal and vertical components.

tension at every point of the string has the same value, which we denote by T_s:

$$T_{s,2} \cos \theta_2 = T_{s,1} \cos \theta_1 = T_s \qquad (14\text{-}60)$$

Notice that in the absence of a wave, $\theta_1 = \theta_2 = 0$, so T_s is therefore the magnitude of the tension in the undisturbed string. The net force along the vertical direction on this segment is

$$\sum F_y = T_{s,2} \sin \theta_2 - T_{s,1} \sin \theta_1 \qquad (14\text{-}61)$$

This net force must be equal to the mass of the segment times its acceleration along the y-axis, a_y. Therefore,

$$(\mu \Delta x) a_y = T_{s,2} \sin \theta_2 - T_{s,1} \sin \theta_1 \qquad (14\text{-}62)$$

Dividing both sides of the above equation by $T_s = T_{s,2} \cos \theta_2 = T_{s,1} \cos \theta_1$, we get

$$\frac{(\mu \Delta x)}{T_s} a_y = \frac{T_{s,2} \sin \theta_2}{T_{s,2} \cos \theta_2} - \frac{T_{s,1} \sin \theta_1}{T_{s,1} \cos \theta_1}$$

Therefore,

$$\frac{(\mu \Delta x)}{T_s} a_y = \tan \theta_2 - \tan \theta_1 \qquad (14\text{-}63)$$

Note that $\tan \theta_1$ and $\tan \theta_2$ describe the slope of the segment in the presence of the wave at points x and $x + \Delta x$, respectively. Remember that the wave function $D(x,t)$ describes the displacement, from the equilibrium position, of an element of string located at position x at time t. To see how a differential equation emerges from the above equation, let us write a_y in terms of $D(x,t)$. Since acceleration is equal to the second derivative of displacement with respect to time, the acceleration of an element of string located at point x is given by (we will take the limit $\Delta x \to 0$ so the acceleration can be written in terms of the displacement at point x)

$$a_y = \frac{d^2 D(x,t)}{dt^2} \qquad (14\text{-}64)$$

At a given point x, the slope of the wave function $D(x,t)$ is given by $\dfrac{dD(x,t)}{dx}$. Therefore,

$$\tan \theta_2 = \frac{dD(x,t)}{dx}\Big|_{x+\Delta x}$$

Here the symbol $\big|_{x+\Delta x}$ means "evaluate the derivative of $D(x,t)$ with respect to x and then replace x by $x + \Delta x$ in the result." Then the right-hand side of Equation 14-63 is

$$\tan \theta_2 - \tan \theta_1 = \left(\frac{dD(x,t)}{dx}\Big|_{x+\Delta x} - \frac{dD(x,t)}{dx}\Big|_{x} \right) \qquad (14\text{-}65)$$

Dividing both sides of Equation 14-63 by Δx, we get

$$\left(\frac{\mu}{T_s} \right) \frac{d^2 D(x,t)}{dt^2} = \frac{1}{\Delta x} \left(\frac{dD(x,t)}{dx}\Big|_{x+\Delta x} - \frac{dD(x,t)}{dx}\Big|_{x} \right)$$

$$(14\text{-}66)$$

In the limit when Δx goes to zero, the right-hand side of the above equation is equal to the second derivative of $D(x,t)$ with respect to x:

$$\lim_{\Delta x \to 0} \frac{1}{\Delta x} \left(\frac{dD(x,t)}{dx}\Big|_{x+\Delta x} - \frac{dD(x,t)}{dx}\Big|_{x} \right)$$

$$= \frac{d}{dx} \left(\frac{dD(x,t)}{dx} \right) = \frac{d^2 D(x,t)}{dx^2}$$

Therefore, Equation 14-66 can be written as

$$\frac{d^2 D(x,t)}{dt^2} = \left(\frac{T_s}{\mu} \right) \frac{d^2 D(x,t)}{dx^2}$$

Since the wave speed, the tension in the string, and the linear mass density are related by $v = \sqrt{\dfrac{T_s}{\mu}}$, we can write the above equation as

$$\frac{d^2 D(x,t)}{dt^2} = v^2 \frac{d^2 D(x,t)}{dx^2} \qquad (14\text{-}67)$$

The above equation is called the **wave equation in one dimension**. Every wave function describing a travelling harmonic wave in one dimension and satisfying the assumptions stated earlier must satisfy this equation. The wave equation is satisfied by waves in a string, sound waves in long tubes, and other types of one-dimensional waves. In the case of sound waves, v represents the speed of sound in that medium.

We have been a bit careless in dealing with derivatives. The function $D(x, t)$ is a function of two variables, x and t. Because acceleration is equal to the second derivative of position with respect to time, to calculate acceleration at a point x from $D(x,t)$ we must keep x constant and take derivatives of $D(x,t)$ with respect to t. We should therefore have take the partial derivative of $D(x,t)$ with respect to t. Similarly, when calculating the slope of $D(x,t)$ at given position x we must keep t fixed. In terms of partial derivatives, the wave equation is written as

KEY EQUATION

$$\frac{\partial^2 D(x,t)}{\partial t^2} = v^2 \frac{\partial^2 D(x,t)}{\partial x^2} \qquad (14\text{-}68)$$

It is easy to show that the wave function for a travelling harmonic wave, $D(x,t) = A \sin(kx - \omega t)$, satisfies the above equation:

$$\frac{\partial^2}{\partial t^2} A \sin(kx - \omega t) = \frac{\partial}{\partial t} \left(\frac{\partial}{\partial t} A \sin(kx - \omega t) \right)$$

$$= \frac{\partial}{\partial t} (-A \omega \cos(kx - \omega t))$$

$$= -\omega^2 A \sin(kx - \omega t)$$

And

$$\frac{\partial^2}{\partial x^2} A\sin(kx - \omega t) = \frac{\partial}{\partial x}\left(\frac{\partial}{\partial x} A\sin(kx - \omega t)\right)$$

$$= \frac{\partial}{\partial x}\left(kA\cos(kx - \omega t)\right)$$

$$= -k^2 A\sin(kx - \omega t)$$

Therefore,

$$-\omega^2 A\sin(kx - \omega t) = -(v^2 k^2)A\sin(kx - \omega t)$$

Since $v = \omega/k$, the above equality holds.

KEY CONCEPTS AND RELATIONSHIPS

The Nature, Properties, and Classification of Waves

A mechanical wave is the propagation of a disturbance through a medium. It takes energy to generate a wave, and the wave carries this energy with it as it passes through a medium. In the presence of a wave, the particles of the medium oscillate about their equilibrium positions but are not carried along with the wave. In longitudinal waves, the motion of the particles of the medium is in the direction of propagation of the wave. In transverse waves, the motion of the particles of the medium is perpendicular to the direction of wave propagation.

Equation for a Pulse Moving in One Dimension

If the shape of a wave is described by a function $D(x)$ at $t = 0$, then its shape at time x is obtained by the substitution $x \rightarrow x \mp vt$, where v is the speed of the wave. $D(x, t)$ is called the wave function. The combination $x - vt$ describes a wave moving in the direction of increasing x, and the combination $x + vt$ describes a wave moving in the direction of decreasing x.

Transverse Speed and Wave Speed

The wave speed in a medium depends on the properties of the medium. For waves on a string with linear mass density μ and under a tension T_s the wave speed is given by,

$$v = \sqrt{\frac{T_s}{\mu}} \qquad (14\text{-}7)$$

Harmonic Waves

For continuous waves the wave speed is related to the wavelength (λ) and the frequency (f) of the waves by

$$v = \lambda f \qquad (14\text{-}12)$$

A sinusoidal wave is generated by a source that is undergoing simple harmonic motion. The wave function of a sinusoidal wave is given by

$$D(x,t) = A\sin(kx - \omega t + \phi)$$

$$= A\sin\left(\frac{2\pi}{\lambda}x - 2\pi ft + \phi\right) \qquad (14\text{-}18)$$

where A is the amplitude, λ is the wavelength, f is the frequency, $k = \dfrac{2\pi}{\lambda}$ is the wave number, $\omega = 2\pi f = \dfrac{2\pi}{T}$ is the angular frequency, and ϕ is the phase constant.

Position Plots and Time Plots

A position plot (also called a snapshot plot) displays the shape of the wave as a function of position for a fixed time. A time plot (also called a time plot) displays the displacement of a point of the medium as a function of time. The value of the phase constant can be determined from a position or a time plot.

Phase and Phase Difference

The argument of the sine (or cosine) function of a harmonic wave is called the phase of the wave. It is measured in radians and is denoted by the Greek letter phi, Φ. For the harmonic waves of Equations 14-18 and 14-19, the phase is

$$\Phi(x,t) = kx \mp \omega t + \phi \qquad (14\text{-}23)$$

The phase difference between two points on a harmonic wave that are separated by a length Δx is denoted as $\Delta\Phi$ and is given by,

$$\Delta\Phi = 2\pi\left(\frac{\Delta x}{\lambda}\right) \qquad (14\text{-}24)$$

Energy and Power in a Travelling Wave

The average power transported by a wave is

$$P_{\text{avg}} = 2\pi^2\mu v f^2 A^2 \qquad (14\text{-}35)$$

Superposition of Waves

When more than one wave is present in a medium at the same time, the resultant wave at any point in the medium is equal to the algebraic sum of the individual waves at that point.

Interference of Waves Travelling in the Same Direction

When two waves with the same wavelength and frequency and travelling in the same direction combine, the amplitude of the resultant wave is maximum when the phase difference between the two waves is an integer multiple of 2π rad and zero when the difference is an odd multiple of π rad.

Reflection and Transmission of Mechanical Waves

When a wave reflects from a fixed end, the reflected wave is inverted (compared to the incident wave) and moves in the opposite direction. When a wave reflects from a free end, the reflected wave is not inverted (compared to the incident wave) and moves in the opposite direction. A transmitted wave is not inverted.

Standing Waves

Two waves with the same frequency and amplitude travelling in opposite directions produce standing waves. The wave function of a standing wave is

$$D(x, t) = 2A\sin(kx)\cos(\omega t) \qquad (14\text{-}43)$$

where A is the amplitude of each wave. In a standing wave, energy is not transported through the medium.

Standing Waves on Strings

The harmonics for a string that is fixed at both ends are

$$\lambda_m = \frac{2L}{m} \tag{14-53}$$

$$f_m = m\frac{v}{2L} \tag{14-54}$$

where L is the length of the string, v is the wave speed, and $m = 1, 2, 3, \dots$.

The Wave Equation in One Dimension

Sinusoidal waves in one dimension satisfy the following wave equation:

$$\frac{\partial^2 D(x,t)}{\partial t^2} = v^2 \frac{\partial^2 D(x,t)}{\partial x^2} \tag{14-68}$$

where $D(x,t)$ is the wave function.

APPLICATIONS

energy generation (from tidal waves), energy transfer, oil and mineral detection, sonar, dental cleaning and food preservation using ultrasonic waves, medical imaging, industrial cleaning, musical instruments.

KEY TERMS

amplitude, antinodes, constructive interference, continuous wave, crest, destructive interference, displacement function, electromagnetic radiation, electromagnetic waves, first harmonic, fundamental frequency, hard reflection, harmonic wave, harmonics, history graphs, in phase, incident pulse, interference, longitudinal wave, mechanical wave, medium, musical scale, nodes, normal modes, octave, out of phase, phase, phase difference, pitch, position plot, principle of superposition, pulse, resonant frequencies, resultant amplitude, resultant wave, semitones, sinusoidal wave, snapshot graphs, soft reflection, standing wave, superposition, tone, transverse wave, trough, wave, wave cycle, wave equation, wave equation in one dimension, wave function, wave number, wave speed, waveform, wavelength

QUESTIONS

1. Indicate whether each of the following statements is true (T) or false (F). Explain your answer.
   (a) As a wave passes through a medium, it carries the particles of the medium with it.
   (b) The wave speed of a mechanical wave depends on the frequency of the source.
   (c) Waves with different wavelengths, travelling in the same medium, can have different speeds.
   (d) Mechanical waves do not require a medium to travel.
   (e) A wave, travelling in a uniform medium, can have a non-zero acceleration.
   (f) A travelling wave can have zero energy.
   (g) When two waves of amplitudes 1.0 m and 2.0 m interfere, the amplitude of the resultant wave can be greater than 3.0 m.
   (h) Standing waves can be produced by combining two waves that are moving in the same direction.
2. Waves carry energy. Do they carry momentum?
3. The speed of sound waves in air is 340 m/s, and the speed of light is 3×10^8 m/s. Consider a sound wave and light both of frequency 20 000 Hz. Which has the longer wavelength?
4. In the presence of a travelling wave, are there any particles of the medium that are always at rest?
5. Suppose a longitudinal harmonic wave is moving through a spring with a speed of 10.0 m/s. Does each coil of the spring oscillate about its equilibrium position with a speed of 10.0 m/s?

6. A pulse with total energy E_1 is travelling along a string that is connected to another string. At the boundary of the two strings, the pulse is partly transmitted into the second string and partly reflected into the first string. The energy of the reflected pulse is E_2, and the energy of the transmitted pulse is E_3. How are E_1, E_2, and E_3 related? What assumptions did you make to reach your conclusion?
7. When you throw a stone on the surface of a pond, the wave travels away from the stone and its amplitude decreases. Why?
8. Two wires, stretched under the same tension, have linear mass densities that differ by a factor of 4. The wave speeds in the wires differ by a factor of
   (a) $\sqrt{2}$
   (b) 2
   (c) 4
   (d) 16
9. When the frequency of a wave is doubled, the power carried by the wave
   (a) remains the same
   (b) increases by a factor of 2
   (c) increases by a factor of 4
   (d) decreases by a factor of 2
   (e) decreases by a factor of 4
10. A source is generating waves at a certain frequency. When the power of the source is doubled, the amplitude of the wave
    (a) remains the same
    (b) increases by a factor of $\sqrt{2}$

WAVES AND OSCILLATIONS

(c) increases by a factor of 2
(d) decreases by a factor of $\sqrt{2}$
(e) decreases by a factor of 2

11. Why do two travelling waves (moving in opposite directions) with the same amplitude and frequency generate a standing wave? Can we produce a standing wave by combining two travelling waves of
   (a) the same amplitude but different frequencies?
   (b) different amplitudes but the same frequency?

12. A travelling wave transports energy between two points. Can a standing wave transport energy between two points?

13. Standing waves with a frequency of 440 Hz are generated on two strings of the same length. Must the two strings have exactly the same linear mass density?

14. Two waves of amplitude A are travelling on a string. The waves interfere to produce a resultant wave of amplitude $A/2$. Is the total energy of the resultant wave equal to the sum of the energies of the constituent waves? Explain your reasoning.

PROBLEMS BY SECTION

For problems, star ratings will be used (★, ★★, or ★★★), with more stars meaning more challenging problems. The following codes will indicate if $\frac{d}{dx}$ differentiation, $\int$ integration, ▢ numerical approximation, or ∿ graphical analysis will be required to solve the problem.

Section 14-1   The Nature, Properties, and Classification of Waves

15. ★ A wave has a wavelength of 2.00 m and a frequency of 10 Hz. Determine
   (a) the wave number
   (b) the angular frequency
   (c) the wave speed

16. ★ The speed of sound in air is 340.0 m/s. The angular frequency of a sound wave is 600.0 rad/s. Determine the wave's
   (a) frequency
   (b) period
   (c) wavelength

17. ★ A wave is travelling along a stretched string with a speed of 150.0 m/s. The wave number of the wave is 6.00 rad/m. Determine the wave's
   (a) frequency
   (b) period
   (c) wavelength

18. ★ Electromagnetic waves in a microwave oven have a frequency of 4.0×10^9 Hz. What is the wavelength of the waves?

19. ★ Human ears can detect sounds with frequencies from 20 Hz to 20 000 Hz. The speed of sound in air is approximately 340 m/s. What is the corresponding range of wavelengths that the human ear can detect?

20. ★ A transverse wave takes 2.0 s to travel the full length of a 3.0 m string. What is the frequency of the wave when its wavelength is 0.50 m?

21. ★ The speed of sound waves in ocean water is approximately 1500 m/s. Dolphins produce sound waves with frequencies in the range of 250 Hz to 150 kHz. What is the range of wavelengths of the waves in ocean water and in air?

22. ★ The speed of sound in air is 340 m/s, and the speed of light is 3.0×10^8 m/s. What frequency of light has the same wavelength as a 200.0 Hz sound wave?

Sections 14-2 and 14-3   The Motion of a Disturbance in a String and Equation for a Pulse Moving in One Dimension

23. ★★ At time $t = 0$, the transverse displacement of a string due to a pulse travelling through it is given by the equation

$$D(x,0) = \frac{0.3}{x^2 + 1.2}$$

where x and D are measured in metres. When the pulse is moving at 4.0 m/s toward the positive x-axis, what is the displacement of a segment of the string that is located at $x = 1.5$ m at $t = 2.0$ s? What are the maximum and minimum displacements of this segment?

24. ★ Rank the following pulses in order of wave speed and amplitude, from lowest to highest, using an equality sign where needed. Also, indicate the direction of motion of each pulse.
   (a) $D(x,t) = \dfrac{3}{4 + (x - 0.5t)^2}$
   (b) $D(x,t) = \dfrac{2}{3 + (x + t)^2}$
   (c) $D(x,t) = \dfrac{11}{15 + (x - 5t)^2}$
   (d) $D(x,t) = \dfrac{7}{5 + (x - 4t)^2}$

25. ★★ $\frac{d}{dx}$ The displacement function for a pulse passing through a string is given by

$$D(x,t) = \frac{-3.0}{6.0 + (x + 3.0t)^2}$$

where x and D are in metres and t is in seconds.
   (a) What is the speed of the pulse, and in which direction is it travelling?
   (b) What is the displacement of a point located at $x = 2.0$ m at $t = 3.0$ s?
   (c) What is the maximum displacement of a point on the string?
   (d) What is the transverse speed of a point that is located at $x = 2.0$ m at $t = 3.0$ s?
   (e) Draw a displacement versus position (x) graph for $t = 0$ s, 1 s, 2 s, and 3 s for this waveform.

26. ★ $\frac{d}{dx}$ At time $t = 0$, the transverse displacement of a string due to a pulse travelling at a speed of 2.0 m/s is given by

$$D(x,0) = \frac{3.0}{x^4 + 10.0}$$

where x and D are in metres.
   (a) What is the amplitude of the pulse?
   (b) What is the displacement of a point located at $x = 1.0$ m at $t = 0.0$ s?
   (c) The pulse is moving toward increasing x. Write an equation that describes the displacement of the string as a function of x and t.
   (d) What is the transverse speed of a point located at $x = 2.0$ m at $t = 1.0$ s?

27. ★★ The displacement function for a travelling pulse is given by the equation

$$D(x,t) = \begin{cases} +2\,\text{m} & \text{if } |x - t| \leq 1 \\ -2\,\text{m} & \text{if } |x - t| > 1 \end{cases}$$

where x is in metres and t is in seconds.

(a) Draw a displacement versus position graph for the pulse for $t = 1$ s, 2 s, and 3 s.
(b) Determine the wave speed from the graph. In which direction is the disturbance travelling?
(c) What is the displacement of a point located at $x = 0.5$ m at $t = 1.0$ s?
(d) Write the equation for a pulse that has the same shape but is moving in the opposite direction.

28. ★ $\frac{d}{dx}$ The displacement function for a pulse travelling along a string is given by the equation

$$D(x, t) = \frac{4.0}{(x - 2.0t - 10.0)^2 + 6.0},$$

where x is in metres and t is in seconds.

(a) What is the speed of the pulse, and in which direction is it travelling?
(b) What is the displacement of a point located at $x = 4.0$ m at $t = 1.0$ s?
(c) What is the speed of a point that is located at $x = 5.0$ m at $t = 1.0$ s?
(d) What is the maximum displacement of a point as the pulse passes through it?

29. ★★ Two pulses travelling on a string are described by the displacement functions

$$D_1(x, t) = \frac{2}{(x - 5t + 10)^2 + 4}$$

and

$$D_2(x, t) = \frac{-2}{(x + 5t - 10)^2 + 4}$$

(a) Sketch the shape of the pulses at $t = 0$.
(b) Where is the peak of each pulse located when $t = 0$?
(c) At what time do the two pulses cancel each other?
(d) Is there a point along the string that has zero displacement for all t? If yes, where is it located?

Section 14-4 Transverse Speed and Wave Speed

30. ★ The E string on a violin has a diameter of 0.039 cm and is made of steel of density (ρ) 7.86 g/cm³. What is the linear mass density of the string?

31. ★ The density of steel is 7.86 g/cm³. It is observed that transverse waves in a 0.40 mm diameter steel wire propagate at 160 m/s. What is the tension in the wire?

32. ★★ A steel wire of cross-sectional area 1.00×10^{-4} m² is under a tension of 1.00×10^3 N. At what speed does a transverse wave move along the wire? The density of steel is 7860 kg/m³.

33. ★ A 50.0 m string of uniform density has a mass of 0.10 kg. When the tension in the string is 100.0 N, what is the speed of a pulse travelling along the length of the string?

34. ★★ The string in problem 33 is hanging from a tall ceiling, and a weight is tied to its lower end to produce a tension of 10.0 N. A pulse generated at the lower end travels upward toward the ceiling. The pulse passes through the midpoint of the string.

(a) What is the speed of the pulse at the midpoint if the mass of the string is ignored?
(b) What is the speed of the pulse at the midpoint if the mass of the string is *not* ignored?

35. ★ String A is 10.0 m long and has a linear mass density of 2.0 g/m. String B is 20.0 m long and has a linear mass density of 5.0 g/m. The two strings are tied together and kept under a constant tension of 50.0 N. Two identical pulses are generated at the two opposite ends of the strings. Where will the two pulses first meet along the string?

36. ★★★ ∫ A thick rope of uniform density hangs vertically from a fixed support.

(a) Show that the wave speed at a distance y from the lower end is given by $v = \sqrt{yg}$.
(b) Show that the time taken by a pulse to travel the entire length, L, of the rope is given by $t = 2\sqrt{L/g}$.

37. ★★ The speed of waves on the surface of a deep ocean is given by $v = \sqrt{\dfrac{g\lambda}{2\pi}}$, where λ is the wavelength of the waves and g is the acceleration due to gravity.

(a) Show that the expression on the right side of this equation has the dimensions of speed.
(b) What is the period of the waves?
(c) Successive crests pass a stationary oil platform every 12 s. What is the speed of the waves?

Section 14-5 Harmonic Waves

38. ★ The displacement versus position graphs at a fixed time for four waves, travelling in the same medium, are shown in Figure 14-62. Rank these waves in order of frequency, from highest to lowest.

(a)

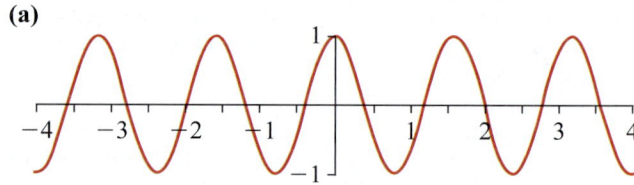

(b)

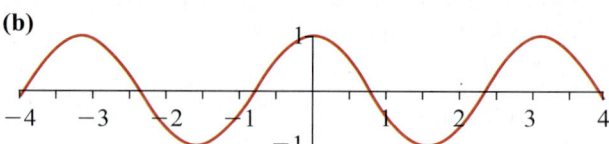

(c)

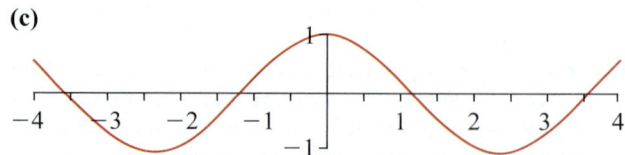

(d)

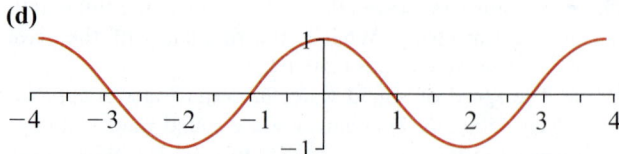

Figure 14-62 Problem 38.

39. ★ Rank the following waves in order of wavelength and frequency, from lowest to highest. Also, indicate the direction of motion, toward increasing or decreasing x, for each wave. Use an equality sign where needed. Here x is in metres and t is in seconds.
(a) $D(x, t) = (0.25 \text{ m}) \sin(x + t)$
(b) $D(x, t) = (1.50 \text{ m}) \sin(2x - t)$
(c) $D(x, t) = (1.25 \text{ m}) \sin(x + 2t)$
(d) $D(x, t) = (0.75 \text{ m}) \sin\left(\dfrac{x}{2} - t\right)$

40. ★ Wave functions for five waves are given below. Here, x is in metres and t is in seconds.
(a) $D(x, t) = (0.3 \text{ m}) \sin(x - 3t)$
(b) $D(x, t) = (0.5 \text{ m}) \sin(2x - 6t)$
(c) $D(x, t) = (0.6 \text{ m}) \sin(2x - 3t)$
(d) $D(x, t) = (0.1 \text{ m}) \sin(4x - 7t + \pi)$
(e) $D(x, t) = (0.5 \text{ m}) \sin(7x - 13t + \pi/2)$
Using equality signs where needed, rank the waves from lowest to highest in terms of
(i) amplitude
(ii) wavelength
(iii) frequency
(iv) wave speed

41. ★★ Write a displacement equation for each wave shown in Figure 14-63. The wavelength of D_2 is 1.0 m. Assume that both waves are moving in the same direction in the same medium.

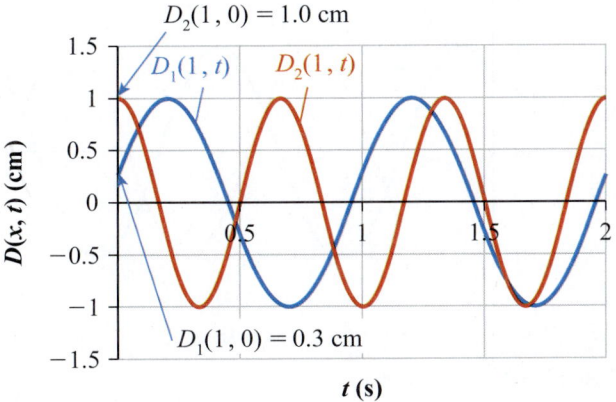

$D_2(1, 0) = 1.0$ cm

$D_1(1, t)$ $D_2(1, t)$

$D(x, t)$ (cm)

t (s)

$D_1(1, 0) = 0.3$ cm

Figure 14-63   Problem 41.

42. ★★ The displacement function at $x = 2.00$ m for a travelling wave is given by

$$D(2.00, t) = (0.75 \text{ m}) \sin(10.0 - 3.50t),$$

where x is in metres and t is in seconds.
(a) If the phase constant is zero, what are the wavelength, frequency, and speed of the wave?
(b) Write the displacement function for an arbitrary x and t, that is, an equation for $D(x, t)$.

43. ★★ The displacement function at $t = 2.00$ s for a travelling wave is given by

$$D(x, 2.00) = (0.05 \text{ m}) \sin(10.0x - 15.0)$$

where x is in metres and t is in seconds.
(a) If the phase constant is zero, what are the wavelength, frequency, and speed of the wave?
(b) Write a displacement equation, that is, an equation for $D(x, t)$, for the wave.

44. ★★ The displacement function for a travelling wave is given by

$$D(x, t) = (0.75 \text{ m}) \sin(10.0x - 3.5t + 0.25)$$

where x is in metres and t is in seconds.
(a) What are the wavelength, frequency, phase constant, and speed of the wave?
(b) What is the displacement of a point located at $x = 0.10$ m at $t = 2.0$ s?
(c) What is the displacement of a point located at $x = 0.15$ m at $t = 2.0$ s?
(d) At $t = 0.5$ s, what is the phase difference between points located at $x = 1.0$ m and $x = 1.5$ m along the wave?
(e) What is the phase difference between the displacements of a point that is located at $x = 1.0$ m at $t = 1.0$ s and $t = 1.5$ s?
(f) Plot $D(x, 1.0)$ and $D(x, 1.5)$ for the wave. Determine the wave speed from these graphs. Does this speed agree with the speed calculated in part (a)?

45. ★★ $\frac{d}{dx}$ A wave on a string is described by the displacement equation

$$D(x, t) = (0.02 \text{ m}) \sin\left(\frac{2\pi}{3.0}x + \frac{2\pi}{6.0}t - \frac{\pi}{3}\right)$$

where x is in metres and t is in seconds.
(a) What are the wavelength, frequency, and speed of the wave?
(b) In which direction is the wave travelling?
(c) Draw a displacement versus time graph for a point located at $x = 0.5$ m for $t = 0$ s to $t = 2.0$ s.
(d) What is the velocity of the segment of the string that is located at $x = 0.5$ m at $t = 2.0$ s?

46. ★★★ $\frac{d}{dx}$ A transverse wave on a string is described by the displacement function

$$D(x, t) = (0.20 \text{ m}) \sin(\pi x + 2\pi t)$$

where x is in metres and t is in seconds.
(a) What is the wave speed, and in which direction is the wave travelling?
(b) What is the transverse velocity of the segment at $x = 0.5$ m at $t = 3.0$ s?
(c) What is the transverse acceleration of the segment at $x = 0.5$ m at $t = 3.0$ s?

Section 14-6   Position Plots and Time Plots

47. ★ A position plot of a 5.0 Hz wave, moving toward the right, is shown in Figure 14-64.
- (a) What is the wavelength of the wave?
- (b) What is the speed of the wave?
- (c) What is the phase constant of the wave?
- (d) Write a displacement equation for the wave.

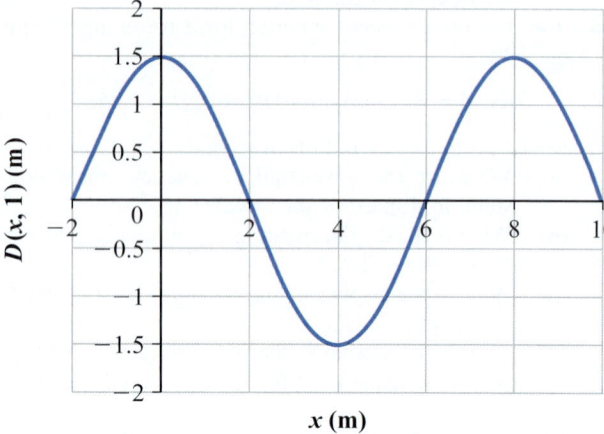

Figure 14-64   Problem 47.

48. ★ The time plot for a wave of wavelength 0.50 m, moving toward the right, is shown at $x = 1.0$ m in Figure 14-65.
- (a) What is the angular frequency of the wave?
- (b) What is the speed of the wave?
- (c) What is the phase constant of the wave?
- (d) Write a displacement equation for the wave.

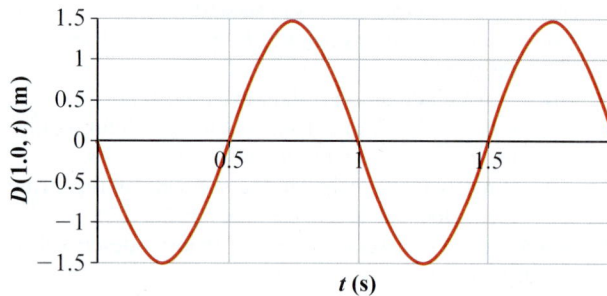

Figure 14-65   Problem 48.

49. ★ A time plot of a wave at $x = 2.5$ m, moving toward the right with a speed of 4.0 m/s, is shown in Figure 14-66.
- (a) What is the frequency of the wave?
- (b) What is the wavelength of the wave?

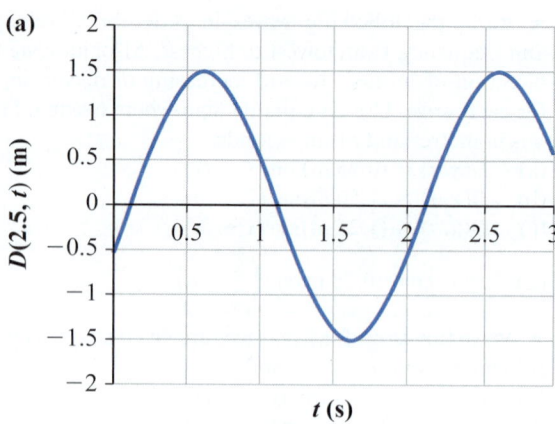

Figure 14-66   Problem 49.

- (c) What is the phase constant of the wave?
- (d) Write a displacement equation for the wave.

50. ★★ The position plots for two 50 Hz waves at $t = 1.0$ s are shown in Figure 14-67.
- (a) What is the phase difference between the waves if the waves are moving toward the right?
- (b) What is the phase difference between the waves if the waves are moving toward the left?

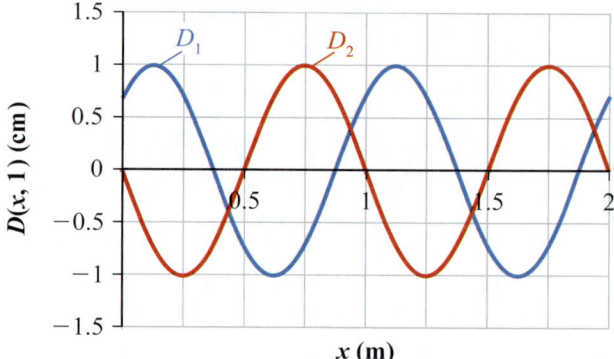

Figure 14-67   Problem 50.

51. ★★ Figure 14-68(a) shows a position plot at $t = 1.0$ s and Figure 14-68(b) shows a time plot at $x = 1.2$ m for a travelling wave.
- (a) Write a general displacement equation for the travelling wave.
- (b) Write an equation for a travelling wave that, when combined with the wave in part (a), produces a standing wave.

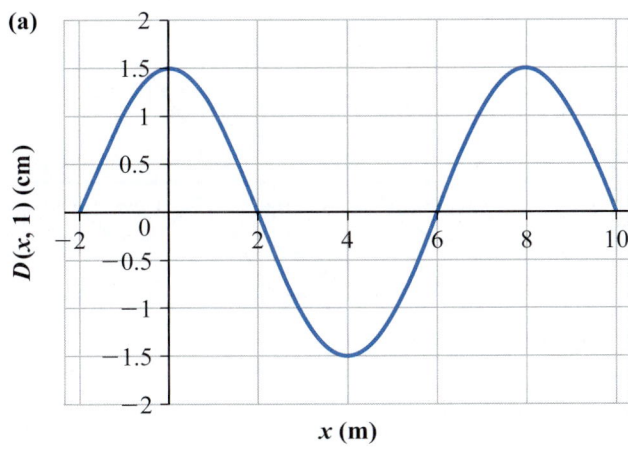

(a)

(b)

Figure 14-68 Problem 51.

52. ★★ A one-dimensional wave is travelling along the x-axis in the positive direction. Figure 14-69(a) shows the displacement versus position graph for the wave at $t = 1.0$ s. Figure 14-69(b) shows the displacement versus time graph of the wave at $x = 1.0$ m.
   (a) Use the graphs to determine the displacement equation, that is, $D(x, t)$, for the wave.
   (b) What is the speed of the wave, and in which direction is it travelling?

53. ★★ A one-dimensional wave is travelling along the x-axis in the positive direction. Figure 14-70(a) shows the displacement versus position graph for the wave at $t = 1.0$ s. Figure 14-70(b) shows the displacement versus time graph of the wave at $x = 1.0$ m.
   (a) Use the graphs to determine the displacement equation, that is, $D(x, t)$, for the wave.
   (b) What is the speed of the wave?

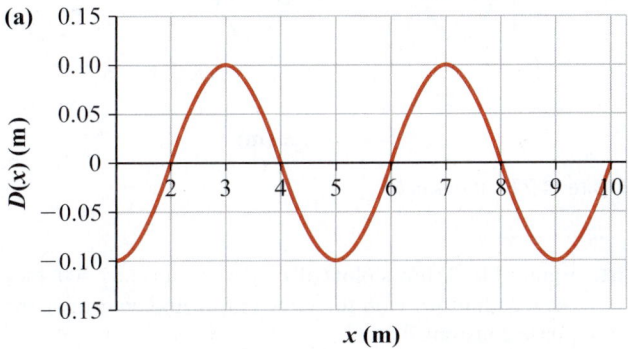

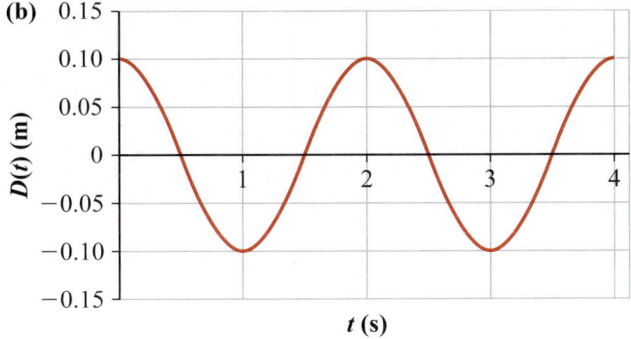

Figure 14-69 Problem 52.

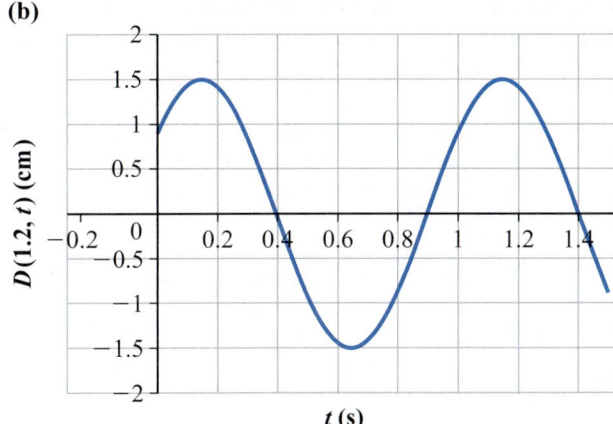

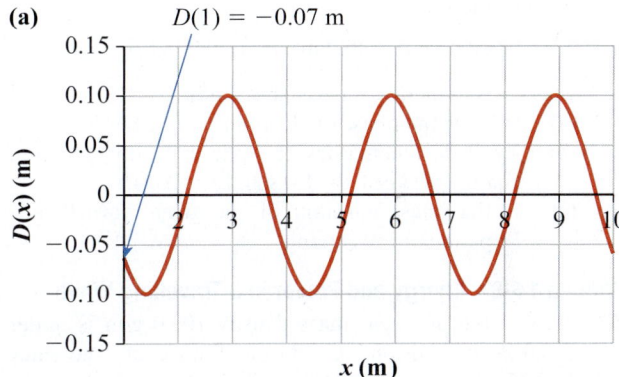

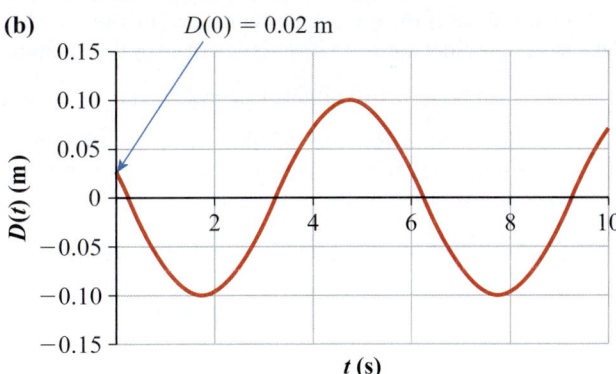

Figure 14-70 Problem 53.

Section 14-7 Phase and Phase Difference

54. ★ The wave function of a wave is given by the equation

$$D(x, t) = (0.2 \text{ m}) \sin(2.0x - 4.0t + \pi)$$

where x is in metres and t is in seconds.
(a) What is the phase constant of the wave?
(b) What is the phase of the wave at $t = 1.0$ s and $x = 0.5$ m?
(c) At a given instant, what is the phase difference between two points that are 0.5 m apart?
(d) At what speed does a crest of the wave move?

55. ★★ The wave function of a wave is given by the equation

$$D(x, t) = (0.1 \text{ m}) \sin(4.0x - 0.5t + 0.2)$$

where x is in metres and t is in seconds.
(a) What is the phase constant of the wave?
(b) What is the phase of a point located at $x = 0.5$ m at $t = 2.0$ s?
(c) Does the phase of the point $x = 0.5$ m change with time? Explain your reasoning.
(d) At $t = 1.0$ s, what is the phase difference between two points that are 0.1 m apart?
(e) Is the phase difference between these points the same at $t = 2.0$ s?
(f) What is the phase difference between two points that are one wavelength apart?

56. ★★ The wave function of a wave at $x = 2.0$ m is given by the equation

$$D(2.0 \text{ m}, t) = (0.3 \text{ m}) \sin(12.0 + 5.0t)$$

where x is in metres and t is in seconds.
(a) What is the phase of the point at $t = 1.0$ s?
(b) By how much does the phase at this location change between $t = 1.0$ s and $t = 2.0$ s?
(c) Is the phase constant of the given wave 0 rad? Explain your reasoning.

Section 14-8 Energy and Power in a Travelling Wave

57. ★ A string of linear mass density 100.0 g/m is under 200.0 N of tension. A sinusoidal wave of frequency 10.0 Hz and amplitude 1.0 cm is propagating along the string. What is the average power carried by the wave?

58. ★ A sinusoidal wave on a string is described by the equation

$$D(x, t) = (0.10 \text{ m}) \sin(4\pi x - 200\pi t)$$

The linear mass density of the string is 10.0 g/m. Determine the direction of motion and the average power transmitted by the wave.

Sections 14-9 and 14-10 Superposition of Waves and Interference of Waves Travelling in the Same Direction

59. ★ Figure 14-71 shows position versus displacement plots, at a common time, of two harmonic waves that are travelling in the same medium in the same direction.
(a) Plot the resultant wave.
(b) What are the amplitude and the wavelength of the resultant wave?
(c) Is the speed of the resultant wave the same as the speed of the component waves? Explain your reasoning.
(d) Is the resultant wave a standing wave or a travelling wave?

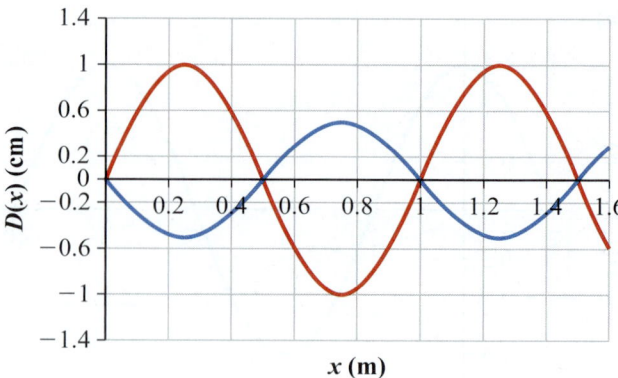

Figure 14-71 Problem 59.

60. ★ Figure 14-72 shows displacement versus position plots for two harmonic waves. Draw the resultant wave obtained by adding these two waves.

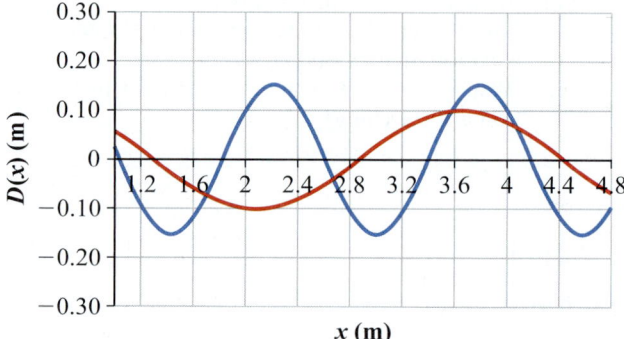

Figure 14-72 Problem 60.

61. Figure 14-73 shows plots of two waves at time $t = 0$. Plot the resultant wave on the same graph, and determine the phase constant.

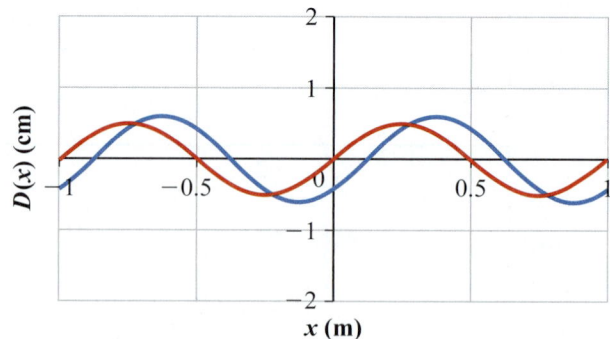

Figure 14-73 Problem 61.

62. ★ Two waves with the same amplitude and frequency arrive at a point P. The waves are $\pi/2$ rad out of phase at this point. What is the resulting amplitude of the waves at P?

63. ★★ Two harmonic waves with the same amplitude, wavelength, and frequency are travelling in the same direction. The amplitude of the resultant wave is half the amplitude of the component waves. Find the difference between the phase constants of the two component waves, and the difference between the phase constant of the resultant wave and that of either component wave.

64. ★ The phase constants of two harmonic waves of the same frequency and wavelength, travelling in the same direction, differ by π rad. The amplitudes of the waves are A_1 and A_2. Determine the amplitude of the resultant wave.

65. ★ Two waves of amplitudes A_1 and A_2 are simultaneously present in the same medium.
    (a) What is the minimum possible amplitude of the resultant wave?
    (b) What is the maximum possible amplitude of the resultant wave?

66. ★★ The wave functions for two harmonic waves are given by

$$D_1(x, t) = (0.2\,\text{m})\sin(2.0x - 3.0t),$$
$$D_2(x, t) = (0.2\,\text{m})\sin(2.0x - 3.0t + \pi/4)$$

where x is in metres and t is in seconds. Write the wave function of the resultant wave when the two waves interfere. What is the amplitude of the resultant wave?

67. ★★ The wave functions for two harmonic waves that interfere with each other are given by

$$D_1(x, t) = A\sin(kx - \omega t + \phi_1)$$
$$D_2(x, t) = A\sin(kx - \omega t + \phi_2)$$

Show that the wave function $D(x, t)$ of the resultant wave is given by

$$D(x, t) = 2A\cos\left(\frac{\phi_1 - \phi_2}{2}\right)\sin\left(kx - \omega t + \frac{\phi_1 + \phi_2}{2}\right)$$

Section 14-11   Reflection and Transmission of Mechanical Waves

68. ★ A pulse travelling along a string is reflected from a boundary. The displacement function of the pulse is given by

$$D(x, t) = \frac{1}{(x - 3t)^2 + 4}$$

    (a) Write the displacement function of the reflected pulse for a hard reflection.
    (b) Write the displacement function of the reflected pulse for a soft reflection.

69. ★ A harmonic wave travelling along a string is reflected from a boundary. The displacement function of the wave is given by

$$D(x, t) = 0.2\sin(3x + 4t)$$

    (a) Write the displacement function of the reflected wave for a hard reflection.
    (b) Write the displacement function of the reflected wave for a soft reflection.

Section 14-12   Standing Waves

70. ★ A travelling wave is described by

$$D(x, t) = (0.5\,\text{m})\sin(x - 4t)$$

where x is in metres and t is in seconds. Which of the following wave(s), when added to the above wave, will result in a standing wave?
    (a) $D(x, t) = (-0.5\,\text{m})\sin(x - 4t)$
    (b) $D(x, t) = (0.5\,\text{m})\sin(x - 4t)$
    (c) $D(x, t) = (-0.5\,\text{m})\sin x\cos 4t$
    (d) $D(x, t) = (0.5\,\text{m})\sin(x + 4t)$

71. ★ A string oscillates in a standing wave pattern given by the equation (x is in metres and t in seconds)

$$D(x, t) = (1.50\,\text{cm})\sin 0.6x\cos 20\pi t$$

    (a) Determine the wavelength and frequency of the standing wave.
    (b) What is the distance between two consecutive nodes, and between a node and an adjacent antinode?
    (c) Write equations for two travelling waves that produce the above standing wave when superimposed.
    (d) What is the displacement of a particle located at $x = 0.20$ m when $t = 3.0$ s?

72. ★ The equation for a sinusoidal wave is given by

$$D(x, t) = (0.50\,\text{cm})\sin\left(2.0x - 20.0t + \frac{\pi}{4}\right)$$

where x is in metres and t is in seconds.
    (a) Write the equation of a wave that produces a standing wave when added to the given wave.
    (b) Find the frequency of the resulting standing wave.
    (c) What is the amplitude of a point located at an antinode of the standing wave?
    (d) Find the distance between two consecutive nodes of the standing wave.
    (e) Find the distance between a node and an adjacent antinode in the standing wave.

73. ★★ Two sinusoidal waves, simultaneously travelling along a string, are described by

$$D_1(x, t) = (1.0\,\text{mm})\sin(\pi x - 0.5\pi t)$$
$$D_2(x, t) = (1.0\,\text{mm})\sin(\pi x + 0.5\pi t)$$

where x is in metres and t is in seconds.
    (a) Determine the equation for the resultant wave.
    (b) Starting from $x = 0$, what are the values of x corresponding to the first three nodes and the first three antinodes?
    (c) What is the distance between two consecutive nodes?

74. ★★ Two sinusoidal waves with frequencies of 100 Hz, wavelengths of 0.500 m, amplitudes of 2.0 cm, and phase constants of 0 moving in opposite directions along a string interfere to produce a standing wave.
    (a) Starting from $x = 0$, what are the values of x corresponding to the first three nodes and the first three antinodes of the standing wave?
    (b) What is the amplitude of the standing wave at $x = 2.0$ m?

75. ★★ The wave functions for two interfering harmonic waves are given by

$$D_1(x, t) = A \sin(kx - \omega t + \phi_1)$$
$$D_2(x, t) = A \sin(kx - \omega t + \phi_2)$$

Show that the wave function, $D(x, t)$, of the resultant wave is given by

$$D(x, t) = 2A \sin\left(kx + \frac{\phi_1 + \phi_2}{2}\right) \cos\left(\omega t - \frac{\phi_1 - \phi_2}{2}\right)$$

Section 14-13 Standing Waves on Strings

76. ★ A stretched string of length L that is fixed at both ends is vibrating in its third harmonic. How far from the end of the string can the blade of a screwdriver be placed against the string and not disturb the amplitude of the vibrations?

77. ★ A string fixed at both ends is 0.500 m long and has a tension that causes the frequency of the first harmonic (fundamental frequency) to be 260.0 Hz. The tension is increased by 4%. What is the new fundamental frequency of the string?

78. ★★ A 1.0 m long string is fixed at both ends.
 (a) What is the longest-wavelength standing wave that can exist on this string?
 (b) The wave speed on the string is 300.0 m/s. What is the frequency of the longest standing wave?
 (c) How should the length of the string be changed to increase the lowest frequency of the string by 5% while keeping the tension the same?

79. ★★ A 2.0 m long string is clamped at both ends and is vibrating so that there are two nodes along the length of the string, in addition to the nodes at the fixed ends.
 (a) What is the longest-wavelength standing wave possible on this string?
 (b) When the wave speed is 200.0 m/s, what frequency corresponds to the longest wavelength?
 (c) Can a standing wave of frequency 1.5 times the frequency calculated in part (b) be generated on this string without changing its length or tension?
 (d) Determine the frequencies of the second and fourth harmonics, and sketch the corresponding vibrational pattern of the string.

80. ★★ A string is clamped at both ends. The string vibrates at 440 Hz and 660 Hz with no other vibrational frequencies in between. What is the lowest frequency at which this string can vibrate in a standing wave pattern?

81. ★★ A string of length 100.0 cm is fixed at both ends and is kept under a tension of 25.00 N. The linear mass density of the string is 0.650 g/m.
 (a) What is the lowest resonant frequency of the string?
 (b) What are the frequencies of the second and third harmonics?
 (c) What is the wave speed for this string?
 (d) What is the lowest resonant frequency when the tension is increased to 35.00 N?

82. ★★ One end of a string with a linear mass density of 7.60×10^{-4} kg/m is connected to an oscillator with a frequency of 50.0 Hz. The other end is connected to a hanging variable mass, as shown in Figure 14-74. The string passes over a pulley, and the string between the oscillator and the pulley can vibrate freely.
 (a) The length of the vibrating section of the string is 0.500 m. What mass must be attached so that the string vibrates with
  (i) one antinode between the pulley and the oscillator?
  (ii) three antinodes between the pulley and the oscillator?
 (b) What is the oscillation frequency of the string in each case in part (a)?

Figure 14-74 Problem 82.

Section 14-14 String Musical Instruments

83. ★ Suppose you are asked to design a new musical scale where the interval within an octave is divided into 10 equally spaced semitones. Determine the frequencies of all semitones between the A_4 (440 Hz) and the A_5 (880 Hz) notes.

84. ★ The G_2 note corresponds to a 98 Hz frequency.
 (a) What are the names and the frequencies of the notes that are an octave lower and an octave higher than G_2?
 (b) Determine the frequencies of all semitones between G_1 and G_2. What are the names of the corresponding notes?

85. ★★ A cello's A-string vibrates in its first harmonic with a frequency of 220 Hz. The vibrating segment of the string is 69.0 cm long and has a mass of 1.90 g.
 (a) What should be the tension in the string?
 (b) What should be the vibrating length of this string if a note is to be played that is two semitones above 220 Hz? Assume that the tension in the string does not change.
 (c) Determine the frequency of vibration when the string vibrates in the third harmonic.

86. ★★ A cello string is to be tuned to play a D_3 note at 146.8 Hz. The musician determines that the string is playing a 152 Hz note.
   (a) Would she have to decrease or increase the tension in the string to play the D_3 note? Explain your reasoning.
   (b) By what fraction does the tension need to be changed?

87. ★★ The thickest string on a guitar has a diameter of 0.054 inches and is stretched under a tension of 220 N (Figure 14-75).
   (a) If it vibrates along the entire length of 65 cm between the bridge and the nut, what will be the frequency of the fundamental harmonic?
   (b) The guitar player wants the string to vibrate at twice the frequency calculated in part (a). For this she presses the string at a fret that is located a distance L from the bridge before plucking the string. Determine L.

Figure 14-75   Problem 87.

88. ★★ A 69 cm long wire vibrates in its third harmonic with a frequency of 330 Hz. The wire is made of copper ($\rho = 8600$ kg/m³) and has a radius of 0.02 cm. Determine the linear mass density and the tension in the wire. By what fraction does the tension need to be changed for the wire to start vibrating in the first harmonic?

89. ★★ Consider four strings of the same linear mass density, vibrating as shown in Figure 14-76. The length and the tension for each string are given. For the modes shown, rank the frequencies of these strings from lowest to highest. Use an equality sign where needed.

Lowest _____ _____ _____ _____ Highest
Frequency   Frequency

(a)   **(b)**

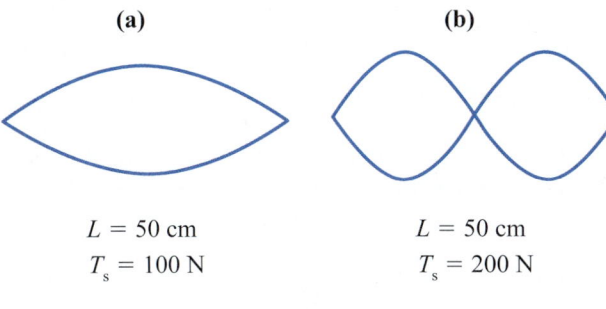

$L = 50$ cm $L = 50$ cm
$T_s = 100$ N $T_s = 200$ N

(c)

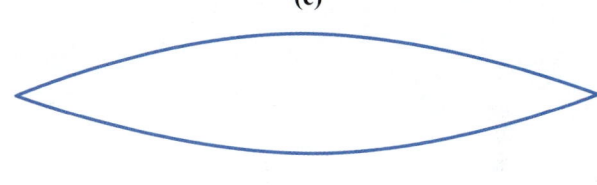

$L = 100$ cm
$T_s = 100$ N

(d)

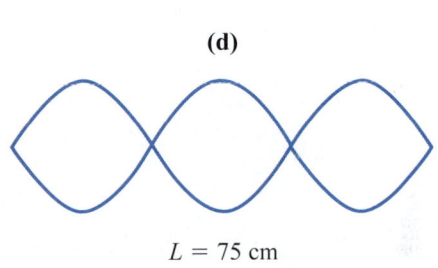

$L = 75$ cm
$T_s = 200$ N

Figure 14-76   Problem 89.

90. ★ A string, fixed at both ends, vibrates at 220 Hz. How long does it take for a wave to travel along the length of the string?

Sound and Interference

Rock bands use stacks of powerful amplifiers to amplify their music for concerts in large venues (Figure 15-1). Such sound systems typically have a total power of approximately 50 kW. How much of this power reaches your ears when you attend a rock concert? Can such power levels be harmful? In 2009, Ottawa bylaw officers measured a sound level of 136 dB (decibels) during an outdoor performance by KISS. The officers asked the band to turn the volume down. What is a decibel? What are safe sound levels? We will explore the answers to these questions in this chapter.

Kevin Winter / Staff / Getty Images Entertainment / Getty Images

Figure 15-1 Arcade Fire in concert.

15-1 Sound Waves

In Chapter 14, we restricted our discussion of wave motion to one dimension. In this chapter, we will examine the properties of sound waves. Sound waves can propagate in one (in a pipe), two (on a drum head), and three (in air) dimensions, and we will examine some of the consequences of that. Although our discussion will focus on sound waves, the way they behave is characteristic of many other types of waves. For example, the principles of wave interference described in Section 14-10 apply to all types of waves, including sound waves. Sound waves are a type of mechanical wave. Unlike light waves, which can propagate in a vacuum, mechanical waves need a medium in which to propagate.

Sound Waves Are Longitudinal Waves

Recall from Chapter 14 that sound is a longitudinal wave: As a sound wave passes through a medium, the molecules of the medium oscillate, alternately creating regions of higher and lower pressure. The molecules are displaced in directions that are either parallel or antiparallel to the direction of propagation of the wave. Figure 15-2 illustrates the direction of propagation, as well as the alternate compressions (darker regions) and rarefactions (lighter regions) in a longitudinal wave.

Accompanying these particle displacements are pressure variations. In regions in which the medium is compressed, the pressure is elevated above the normal or ambient pressure. Regions in which the medium is stretched (rarefaction) correspond to a lowering of the pressure. These pressure variations are illustrated in Figure 15-3. Note that pressure variations are scalar quantities and are not to be confused with the particle displacements, which are either parallel or antiparallel to the direction of propagation.

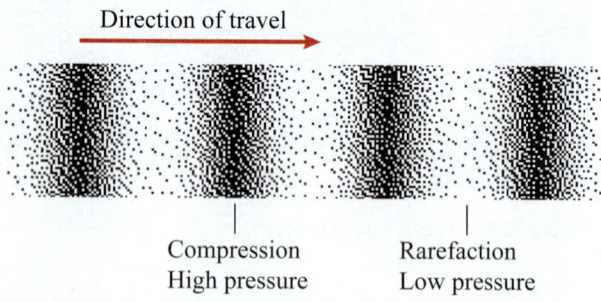

Figure 15-2 Particle positions at an instant of time in a longitudinal wave. The wave is travelling from left to right. The dark regions correspond to areas of compression, while the light regions are areas of rarefaction. For these regions to exist, the particles in the medium through which the wave is travelling must be displaced from their equilibrium positions, and these displacements must be either parallel or antiparallel to the direction of travel of the wave.

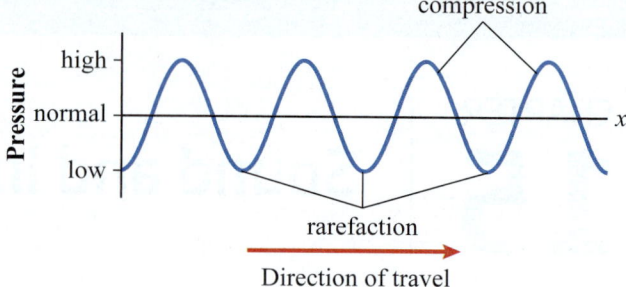

Figure 15-3 Pressure variations in the sound wave illustrated in Figure 15-2. The wave travels from left to right. The vertical axis shows the pressure variations associated with the wave, not to be confused with the particle displacements.

The Speed of Sound

The speed of a sound wave depends on the properties of the medium through which it is propagating. As you learned in Chapter 14, the speed of waves on strings depends on the linear mass density of the string, μ, and the tension in the string, T_s:

$$v = \sqrt{\frac{T_s}{\mu}} \qquad (14\text{-}7)$$

Equation 14-7 uses a measure of the stiffness of the medium (T_s) and a measure of how much mass is oscillating (μ). The appropriate measure of the string's mass is the linear mass density. For air, and any other medium through which sound can pass, we have to use the density, ρ. What quantity can we use to describe how "stiff" air is?

We know that a spring constant, a measure of the stiffness of a spring, is defined in terms of how much the length of the spring changes when we exert a force on it. For a three-dimensional material, we instead measure by what fraction the volume changes when we change the pressure exerted on the material. The ratio of the change in pressure, Δp, divided by the fractional change in the volume, $\Delta V/V$, is called the **bulk modulus**, B:

$$B = -\frac{\Delta p}{\left(\dfrac{\Delta V}{V}\right)} = -V\left(\frac{\Delta p}{\Delta V}\right) \qquad (15\text{-}1)$$

The negative sign is included in the definition because the sign of the fractional change in volume is always the opposite of the sign of the pressure change—if the pressure on a system increases, its volume decreases. The bulk modulus has units of pressure, the pascal. Analogous to Equation 14-7, the speed of sound is given by

KEY EQUATION

$$v = \sqrt{\frac{B}{\rho}} \qquad (15\text{-}2)$$

Table 15-1 Bulk Modulus, Density, and Speed of Sound for Various Media

Medium	Speed of Sound (m/s)	Bulk Modulus (Pa)	Density (kg/m³)
Air (20 °C) (gas)	343	1.41×10^5	1.2
Hydrogen (gas)	1 284	1.51×10^5	0.089
Water (20 °C) (liquid)	1 482	2.2×10^9	1 000
Mercury (liquid)	1 450	2.85×10^{10}	13 534
Gold (solid)	3 054	1.8×10^{11}	19 300
Diamond (solid)	11 200	4.42×10^{11}	3 520

This result is quite general and is not restricted to air or, for that matter, even gases. Table 15-1 gives the speed of sound, the bulk modulus, and the density for a variety of media. Note the general trend of increasing sound speed as we move from gases to solids, even though solids tend to be denser than gases. The differences in bulk modulus more than offset the effects of the differences in density. The bulk modulus is a measure of the stiffness of the system. The molecules in a gas are quite far apart and tend to interact very weakly, while in solids and liquids they are much closer together and interact more strongly. The bulk modulus of water is about 10^4 times that of air but the density is only about 10^3 times as great—hence the higher speed of sound in water.

INTERACTIVE ACTIVITY 15-1

Speed of Sound

Go outside with a friend and stand far enough away from them so that when they clap their hands, you can measure the time between seeing the clap and hearing the sound. Record your data and use it to estimate the speed of sound in air.

Mathematical Description of the Displacement Amplitude

We can write a mathematical description of a travelling sinusoidal sound wave in one dimension. We will assume that the wave is travelling in the positive x-direction; therefore, the displacement of the particles in the medium is longitudinal in the x-direction. As in Chapter 14, we use D to denote the displacement of an element of the medium from its equilibrium position in the presence of a travelling sound wave:

KEY EQUATION $D(x, t) = A \cos(kx - \omega t + \phi)$ (15-3)

In Equation 15-3, A is the maximum displacement from equilibrium, $k = \dfrac{2\pi}{\lambda}$ is the wave number, $\omega = 2\pi f = \dfrac{2\pi}{T}$ is the angular frequency, and ϕ is the phase constant. Note that the form of Equation 15-3 is similar to Equation 14-18 for a travelling wave, except

we have used a cosine instead of a sine function. Sound waves are travelling waves.

As shown in Chapter 14, the relationship between the speed of a sound wave, v, the frequency, f, and the wavelength, λ, is

$$v = f\lambda = \frac{\omega}{k} \qquad (14\text{-}13)$$

EXAMPLE 15-1

Equation for Longitudinal Waves

In Equation 15-3 we have written the expression to describe the displacement of the particles in a medium for a sound wave travelling in the positive x-direction. Sound waves can also propagate in other directions. Write a general expression for the wave function of a longitudinal wave that travels in the negative z-direction.

Solution

We will still use the variable D to denote the displacement. Recall that for a wave travelling in the positive x-direction the argument of the function has the characteristic form

(continued)

$D(x,t) = D(kx - \omega t)$, while a wave travelling in the negative direction has the form $D(x,t) = D(kx + \omega t)$. However, we wish to describe a wave that travels in the z-direction. Thus, the displacement will depend upon the position, z, and the time, t:

$$D(z,t) = A\cos(kz + \omega t + \phi)$$

Making sense of the result

We have just changed x to z to get the wave traveling in the z-direction. Notice however that the \omega t term comes in with a plus sign since the wave is travelling in the negative z-direction.

Relationships between Displacement, Pressure, and Intensity

In addition to having a function that describes the displacements of the particles from their equilibrium positions, a sound wave also has pressure variations associated with these particle displacements. The gas that a wave travels through has an equilibrium pressure when there is no sound wave present. The pressure variations of the sound wave correspond to increases and decreases in pressure from the equilibrium pressure. The pressure variations have the same periodicity, in both time and space, as the displacement variations. For a sound wave with a single frequency, the pressure varies sinusoidally. However, the phase relationship between the pressure variations and the displacement is subtle.

We will make our argument in one dimension for clarity and simplicity. One might at first think that the pressure variation is greatest where the displacement is greatest, that is, the two variations are in phase. They are not. If we consider a region of high pressure, it is a region that must have had particles pushed into it from the right and the left. On the left side of a high-pressure region, the displacement is positive, while on the right side, it is negative. It must be the case then that the displacement goes through zero right at the point of maximum pressure. A similar argument can be made for the low-pressure regions: the point of low pressure is also a point of zero displacement.

We begin by considering an element of a medium centred at the point x, with width Δx and cross-sectional area A. The displacement, D, of this element is in the range $0 \leq D \leq A$. The volume of this element of the medium will change with x because the displacement of the right side of the element will generally be different from the displacement of the left side. The change in displacement is given by

$$\Delta D = \left(\frac{\partial D}{\partial x}\right)\Delta x \tag{15-4}$$

We have used the partial derivative because, in general, D depends on both x and t. To find the change

in volume, we multiply ΔD by the cross-sectional area, A:

$$\Delta V = A\Delta D = A\left(\frac{\partial D}{\partial x}\right)\Delta x \tag{15-5}$$

However, the volume of the element is $V = A\Delta x$, to first order, so Equation 15-5 can be rearranged to give

$$\frac{\Delta V}{A\Delta x} = \frac{\Delta V}{V} = \frac{\partial D}{\partial x} \tag{15-6}$$

Since we are interested in how the pressure changes, we use the bulk modulus to make the connection between the volume change and the pressure change. Solving Equation 15-1 for $\Delta V/V$ and substituting into Equation 15-6, we obtain

$$\frac{\Delta V}{V} = -\frac{\Delta p}{B} = \frac{\partial D}{\partial x} \tag{15-7}$$

and

$$\Delta p = -B\frac{\partial D}{\partial x} \tag{15-8}$$

We can use Equation 15-3 to calculate the partial derivative of D with respect to x:

$$\frac{\partial D}{\partial x} = \frac{\partial}{\partial x}\left[A\cos(kx - \omega t + \phi)\right] = -kA\sin(kx - \omega t + \phi) \tag{15-9}$$

Finally, we substitute this result into Equation 15-8 to obtain

KEY EQUATION
$$\Delta p = BkA\sin(kx - \omega t + \phi) \tag{15-10}$$

We can identify the amplitude of the pressure variations as the coefficient of the sine factor in Equation 15-10:

$$\Delta p_m = BkA \tag{15-11}$$

Comparing Equations 15-3 and 15-10, we see that although the wave displacement has a cosine function and the pressure is a sine function, the arguments are the same in both cases. Thus, they have the same wavelength, period, and wave speed but are $\pi/2$ rad out of phase with each other. These relationships are plotted in Figure 15-4 for a 1000 Hz sound wave with an amplitude that is at the threshold of human hearing—if this sound wave were any quieter, we would not be able to hear it. At first glance it seems quite incredible that our ear is able to detect sounds with displacements on the order of 10^{-11} m, less than the diameter of a hydrogen atom. However, we must keep in mind that what we really are sensitive to are the changes in pressure produces by these displacements. It is still quite remarkable, though, that we are able, although only just, to detect pressure variations on the order of 10^{-5} Pa when atmospheric pressure is roughly 10^5 Pa—better than one part in a billion! If our ears were slightly more sensitive, we would hear a constant background noise due to the thermal motion of the air molecules. Our ears have

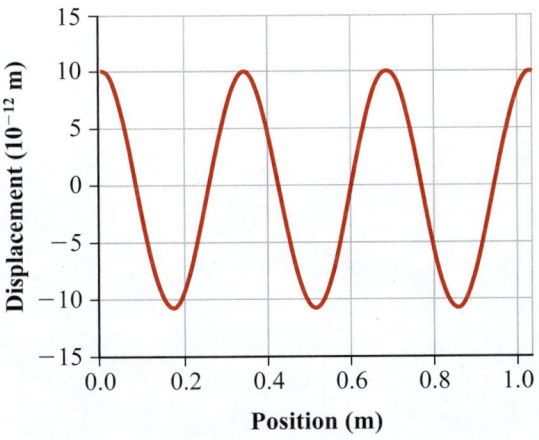

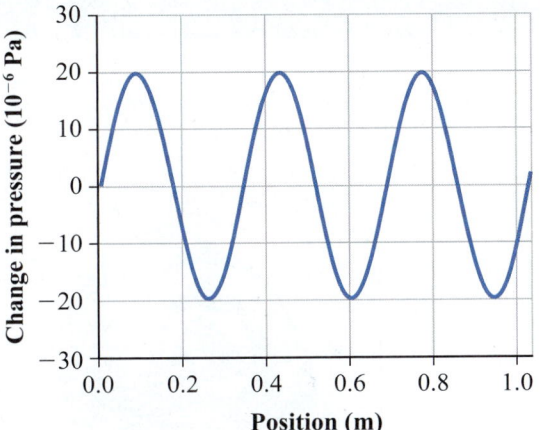

Figure 15-4 Displacement and pressure variations for a 1000 Hz sound wave in air.

evolved to be as sensitive as possible without subjecting us to constant background noise.

EXAMPLE 15-2

Wave Amplitude

Calculate the displacement amplitude of a 1000 Hz sound wave whose pressure amplitude is 100.0 μPa.

Solution

This is a simple application of the relationship expressed in Equation 15-11. Using the bulk modulus for air found in Table 15-1, we rearrange Equation 15-11 to find

$$A = \frac{\Delta p}{Bk} = \frac{\Delta p_m}{B\frac{2\pi f}{v}} = \frac{0.000\,1000\ \text{Pa}}{1.41 \times 10^5\ \text{Pa}\ \frac{2\pi(1000\ \text{s}^{-1})}{343\ \text{m}\cdot\text{s}^{-1}}}$$

$$= 3.87 \times 10^{-11}\ \text{m}$$

Making sense of the result

We have a pressure amplitude that is about four times that shown in Figure 15-4. Since the pressure amplitude is proportional to the displacement amplitude, we should find a displacement amplitude that is four times that shown in Figure 15-4.

We will now examine the energy that a sound wave delivers. We define the **intensity**, I, of a wave as the power delivered per unit area:

$$I = \frac{P}{A} \tag{15-12}$$

where P is the rate at which the wave delivers energy and A is the area that the wave is impinging upon. We need to be careful that we do not confuse pressure (p) with power (P).

As was shown in Chapter 14, the power of a mechanical wave in one dimension is

$$P_{\text{avg}} = \frac{1}{2}\mu v \omega^2 A^2 \tag{14-35}$$

where μ is the linear mass density, v is the wave speed, ω is the angular frequency, and A is the amplitude of the wave.

For a sound wave, we replace the linear mass density, which has units of kg/m, with the mass density, ρ, which has units of kg/m^3. This substitution gives a quantity with units of W/m^2. Thus, the intensity of a sound wave is given by

$$I = \frac{1}{2}\rho v \omega^2 A^2 \tag{15-13}$$

where ρ is the density of the medium and v is the wave speed.

EXAMPLE 15-3

Intensity and Power

Determine the intensity of the 1000 Hz sound wave in Example 15-2. Assuming that a human eardrum has a radius of approximately 0.75 cm, determine how much power is delivered to the ear by such a wave.

(continued)

Solution

The intensity of a sound wave is expressed in $W \cdot m^{-2}$. To find the power (in W) delivered to the eardrum, we need to multiply the intensity by the area of the eardrum.

We use Equation 15-13 to determine the intensity:

$$I = \frac{1}{2}(1.2 \text{ kg} \cdot m^{-3})(343 \text{ m} \cdot s^{-1})(2\pi \cdot 1000 \text{ s}^{-1})^2$$
$$\times (3.87 \times 10^{-11} \text{ m})^2$$
$$= 1.22 \times 10^{-11} \text{ W} \cdot m^{-2}$$

To find the power, we multiply this result by the area of the eardrum to get

$$P = IA = (2.38 \times 10^{-11} \text{ W} \cdot m^{-2})(\pi)(0.0075 \text{ m})^2$$
$$= 2.15 \times 10^{-15} \text{ W}$$

Making sense of the result

This is a very small number, and we really do not have anything else to compare it to in order to see if it reasonable. However, as we shall see later in the chapter, the human ear is a remarkably sensitive device that is able to operate over many orders of magnitude of intensity. The particular sound wave we considered here would be barely audible.

15-2 Wave Propagation and Huygens' Principle

Because in general sound waves are not restricted to one dimension, we need to consider how they propagate in higher dimensions.

Spherical Waves

We begin by considering a single point source of sound. This situation is analogous to dropping a stone in a pool of water. In the water, the wave spreads out from the location of the disturbance but is restricted to travel along the surface of the water (Figure 15-5(a)). In open air, a sound wave can propagate in three dimensions and it does so uniformly in all directions (Figure 15-5(b)). Each crest of a sound wave forms a **spherical wave** front centred on the point source.

The mathematical description of such a wave is much more complex than the description given in Equation 15-3 for a one-dimensional wave. Rather than being a function of just x and t, a spherical wave is a function of x, y, z, and t because the wave propagates in three dimensions. In addition, the amplitude of a spherical wave decreases as it moves outward from the source. We will consider these amplitude changes in more detail in Section 15-6.

Christiaan Huygens, a seventeenth-century Dutch scientist, recognized that he could describe the propagation of a wave by considering every point on a wave front to be a point source of spherical waves. The resulting wave is then determined by adding all the waves from the point sources. His description is called **Huygens' principle**. Here is a procedure for applying Huygens' principle:

(a)

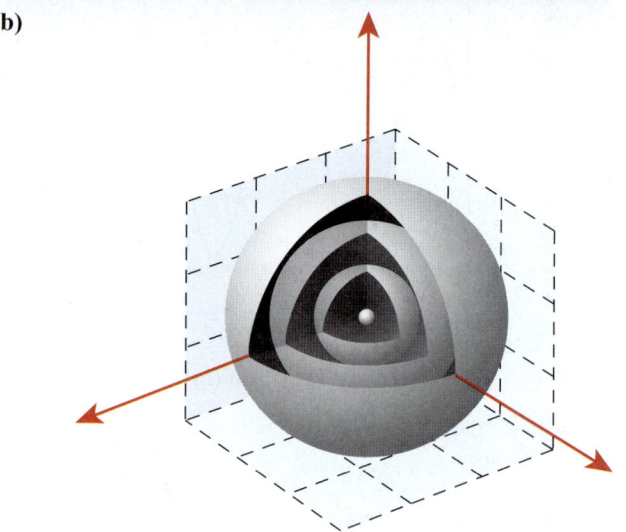

(b)

Figure 15-5 (a) The waves created by dropping a stone into a pond propagate uniformly on the water's surface. (b) A spherical wave propagates in three dimensions from a point source. The arrows indicate the direction of propagation of the wave.

1. Select a set of equally spaced points on a wave front.
2. Draw a set of circles, all of the same radius, each centred at a different point.
3. Construct the resulting wave front by drawing tangents to the circles.

Figure 15-6 illustrates this procedure for a spherical wave.

Plane Waves

We will now consider **plane waves**, which have wave fronts that lie in planes parallel to each other (Figure 15-7). You can create a plane wave in a pond by dropping a long, straight stick horizontally into the pond. You can create a plane sound wave by vibrating a large, flat plate in a direction perpendicular to the surface of the plate. We can use Huygens' principle to show that a plane wave, once created, continues to propagate as a plane wave. In Figure 15-7, notice that the crests and troughs of the wave are parallel to each other.

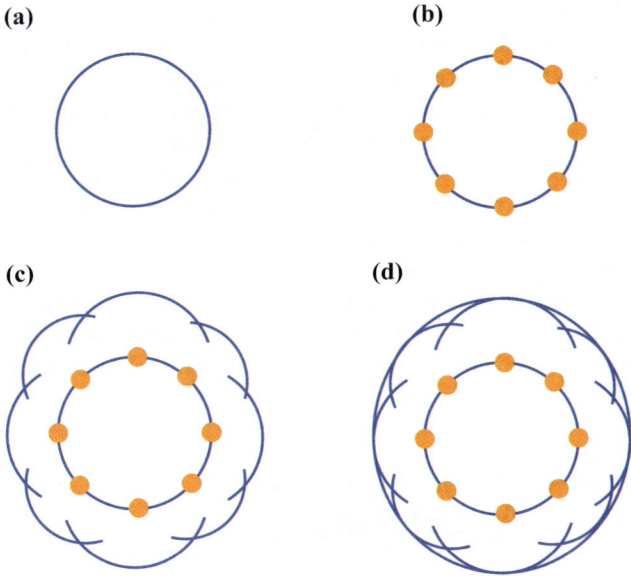

Figure 15-6 Applying Huygens' principle. (a) Draw a spherical wave. (b) Draw a set of points along the wave front. (c) Draw a spherical wave from each point. (d) Draw a curve tangent to the individual circular wave fronts.

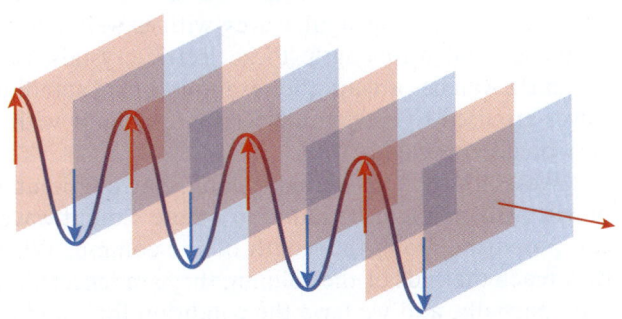

Figure 15-7 A plane wave propagating in the direction of the black arrow. In the figure, the red and blue planes depict planes of maximum (red) and minimum (blue) pressure. The wave fronts extend in both directions perpendicular to the direction of propagation.

Source: Constant314/Public Domain

15-3 Reflection and Refraction

We can use Huygens' principle to help understand what happens to a sound wave when it encounters a boundary between two media where the wave speed changes. At such a boundary, some of the wave can be reflected from the boundary and some transmitted across the boundary into the second medium. The portion of the wave that is reflected travels back into the first medium. As we saw in Chapter 14, the reflected wave may have its phase changed upon reflection. The angle of reflection is the same as the angle of incidence.

The portion of the wave that crosses the boundary into the second medium is called a refracted (or transmitted) wave. In general, the direction of propagation of the refracted wave is different from the direction of the incident wave; we can use Huygens' principle to understand these effects.

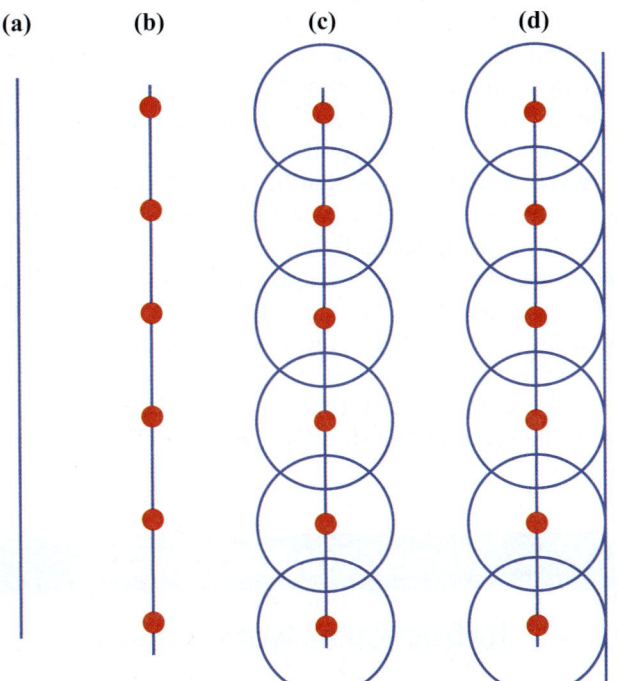

Figure 15-8 Huygens' principle applied to a plane wave that is travelling from left to right.

Consider a plane wave encountering a planar boundary. To apply Huygens' principle, we place a line of equally spaced dots at the boundary between the two media, as shown in Figure 15-9. In the figure, the speed

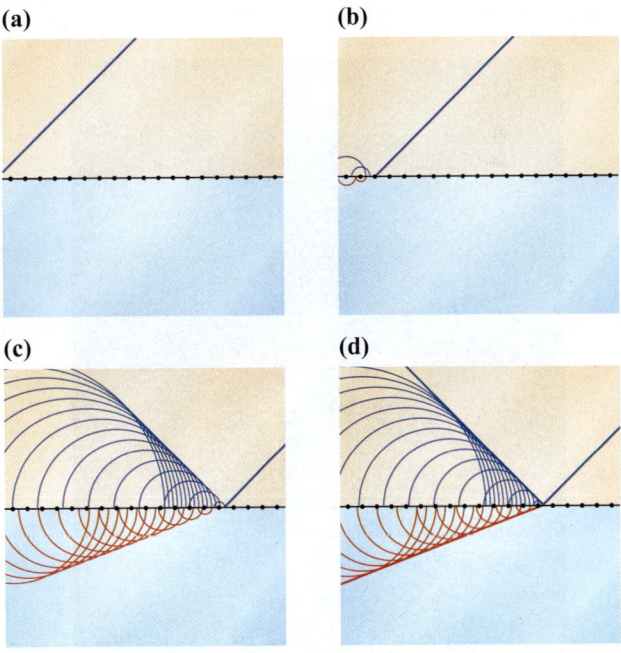

Figure 15-9 Reflection and refraction at a boundary. In (a), we see an incoming plane wave front that has not struck the boundary. In (b), the wave has just struck the boundary. In (c), we have allowed the wave front to continue striking the barrier and we have simply repeated the process we started in (b). In (d), we have added to the sketch in (c) lines that are tangential to all of the individual wave fronts from each dot to see the resultant wave fronts.

Source: (a–d) Re-drawn screenshots from an Applet available at http://www.walter-fendt. de/ph14e/huygenspr.htm

of sound in medium 2 is less than the speed of sound in medium 1. Each time the wave front encounters one of the dots on the boundary, we draw a semicircle around the dot. The radius of each semicircle is directly proportional to the speed in the medium in which we are drawing it and the time that has elapsed since the wave front hit the dot. Thus, for each dot we draw larger circles in the upper half than the lower half, since the wave speed is larger in the upper half. And, as we move from left to right in the drawing, the size of the circles decreases, since the left-most dots were struck by the wave front first.

We can see that the shape of the wave does not change, but the direction of propagation does.

MAKING CONNECTIONS

Refraction of Water Waves

Have you ever wondered why waves reaching a beach are always parallel to the beach? For water waves, the speed of the wave depends on the depth of the water when the depth is comparable to the wavelength. As the depth decreases, so does the wave speed. As we saw in Figure 15-9, waves tend to bend toward the region of slower wave speed. Thus, the water waves tend to bend until they are parallel to the beach, as the image in Figure 15-10 illustrates. Despite the complexity of the coastline, the waves are still coming in parallel to the beach.

Chris Cheadle / Alamy Stock Photo

Figure 15-10   Ocean waves coming in parallel to the beach at Pacific Rim National Park Reserve in British Columbia.

The reflected wave emerges from the surface at the same angle as the incoming wave came in—in this case 45°. The refracted wave (in the lower medium) has had its direction of propagation changed. We measure the direction of propagation relative to a surface normal—in this case, the angle between the direction of propagation and the surface normal has decreased. The wave is bent toward the medium with the lower speed, medium 2.

However, if the angle of incidence is 0°, measured with respect to a boundary normal, then both the transmitted wave and the reflected wave leave normal to the boundary.

15-4 Standing Waves in Air Columns

The formation of standing waves in brass, woodwind, and stringed instruments is responsible for producing musical notes. Standing waves are also important in other applications not involving sound. The resonant cavity of a laser reflects the light waves back and forth and creates standing light waves with a well-defined frequency and wavelength. Piezoelectric crystals used in quartz watches are another example where standing waves are used to create an electrical oscillator with a very precise frequency.

We start by considering a long, narrow column of air, such as might be found in a flute. Sound waves can propagate along the length of the column. When they reach the ends of the column, they are reflected, at least partially, and we have the condition for standing waves that we saw in Chapter 14: waves with the same frequency travelling in opposite directions in the same medium.

Similar to strings, air columns can have open (free) and closed (fixed) ends. The closed end of an air column is a displacement node (and hence a pressure antinode) because the air molecules cannot oscillate across such a fixed boundary. If we have an air column that is closed at both ends, the displacement function must have nodes at both ends. The possible standing waves in such a column are then exactly what we would see for a string with fixed ends.

With an air column that is open at both ends, the ends are displacement antinodes, and the standing waves are exactly what are expected in a string with open ends. These results are summarized in Table 15-2. Because the pressure is $\pi/2$ rad out of phase with the displacement, the displacement for a column with both ends open has the same shape as the pressure for a column with both ends closed.

For a column that is open at one end and closed at the other, one end is a displacement node and the other is a displacement antinode; see Table 15-3.

Table 15-2 Standing Waves in an Air Column with Either Both Ends Closed or Both Ends Open

Displacement Amplitude		Mode	Wavelength	Frequency
Both ends closed	Both ends open			
(a)	(b)	$n = 1$	$\lambda_1 = 2L$	$f_1 = \dfrac{v}{2L}$
(c)	(d)	$n = 2$	$\lambda_2 = \dfrac{2L}{2}$	$f_2 = \dfrac{2v}{2L}$
(e)	(f)	$n = 3$	$\lambda_3 = \dfrac{2L}{3}$	$f_3 = \dfrac{3v}{2L}$
(g)	(h)	$n = 4$	$\lambda_4 = \dfrac{2L}{4}$	$f_4 = \dfrac{4v}{2L}$
General case		n	$\lambda_n = \dfrac{2L}{n}$	$f_n = \dfrac{nv}{2L}$

Table 15-3 Displacement Amplitudes for a Tube That Is Open at One End and Closed at the Other End

Displacement Amplitude	Mode	Wavelength	Frequency
One end open, one end closed			
(a) L	$n = 1$	$\lambda_1 = 4L$	$f_1 = \dfrac{v}{4L}$
(b) L	$n = 3$	$\lambda_3 = \dfrac{4L}{3}$	$f_3 = \dfrac{3v}{4L}$
(c) L	$n = 5$	$\lambda_5 = \dfrac{4L}{5}$	$f_5 = \dfrac{5v}{4L}$
(d) L	$n = 7$	$\lambda_7 = \dfrac{4L}{7}$	$f_7 = \dfrac{7v}{4L}$
General case	only odd n	$\lambda_n = \dfrac{4L}{n}$	$f_n = \dfrac{nv}{4L}$

Standing waves contribute to the rich variety of sounds that musical instruments produce. In Table 15-2, we listed the different types of standing waves that are possible for an air column with both ends open or both ends closed. The rich sound of a musical instrument results from the fact that a string or an air column can simultaneously sustain combinations of the possible allowed standing waves. In fact, it is very difficult to avoid having multiple standing waves. The lowest-frequency standing wave is called the fundamental, and the higher frequencies are called the harmonics.

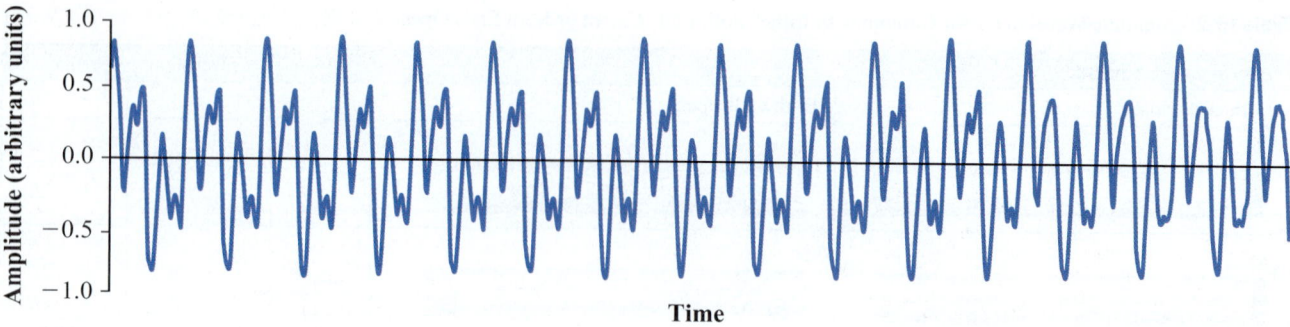

Figure 15-11 Sound wave amplitude recorded by blowing into one end of a pipe that was open at the other end, recorded over an interval of approximately 0.1 s.

Figure 15-11 shows the amplitude as recorded on a computer by blowing into one end of a pipe that is open at the other end. The variations in amplitude are periodic but are clearly not simple sine waves. We can create complex periodic waveforms by adding together sine waves with frequencies that are all integer multiples (harmonics) of a fundamental frequency. For example, the wave in Figure 15-12 is the sum of sine waves with

five different frequencies, where we have chosen $x = 0$ for simplicity:

$$D(0, t) = 0.56\cos(t) + 0.75\cos(3t) + 0.33\cos(5t)$$
$$- 0.16\cos(7t) - 0.05\cos(9t)$$

Notice that the waveform in Figure 15-12 is quite similar to the wave in Figure 15-11. You can also see that all the waves have a frequency that is an integer multiple of the fundamental, $f_0 = \dfrac{1}{2\pi}$ s^{-1}. For this particular wave, the frequencies are all odd multiples of the fundamental, and some of the terms have a negative sign, which means they are 180° out of phase from a positive term with the same frequency.

Musical instruments produce a sound composed of many different frequencies. For instruments that rely on vibrating air columns (e.g., trumpets, flutes, and clarinets), vibrating strings (e.g., violins, pianos, and banjos), and vibrating bars (e.g., xylophones, marimbas, and gamelans), the component frequencies in the sound are integer multiples of the fundamental frequency. However, the sound from instruments based on a vibrating membrane (such as drums) contains multiple frequencies that are not all integer multiples of the fundamental. This is why a drum does not have as distinct a note as, say, a trumpet.

CHECKPOINT 15-2

Fundamental Frequencies

Consider the following pipes:

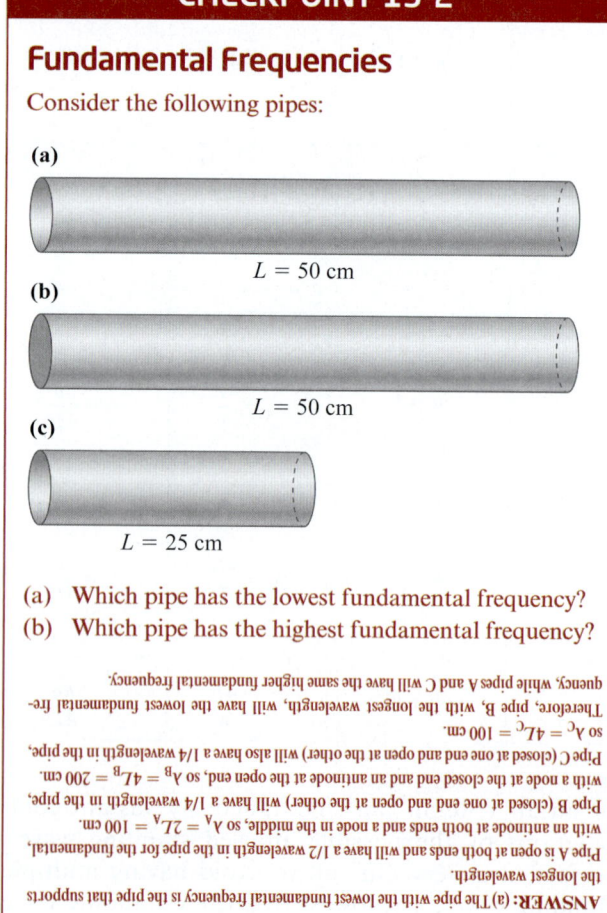

(a)

$L = 50$ cm

(b)

$L = 50$ cm

(c)

$L = 25$ cm

(a) Which pipe has the lowest fundamental frequency?
(b) Which pipe has the highest fundamental frequency?

ANSWER: (a) The pipe with the lowest fundamental frequency is the pipe that supports the longest wavelength.

Pipe A is open at both ends and will have a 1/2 wavelength in the pipe for the fundamental, with an antinode at both ends and a node in the middle, so $\lambda_A = 2L_A = 100$ cm.

Pipe B (closed at one end and open at the other) will have a 1/4 wavelength in the pipe, with a node at the closed end and an antinode at the open end, so $\lambda_B = 4L_B = 200$ cm.

Pipe C (closed at one end and open at the other) will also have a 1/4 wavelength in the pipe, so $\lambda_C = 4L_C = 100$ cm.

Therefore, pipe B, with the longest wavelength, will have the lowest fundamental frequency, while pipes A and C will have the same higher fundamental frequency.

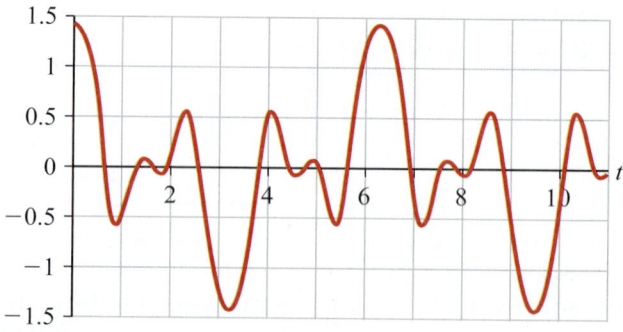

Figure 15-12 A standing wave composed of multiple sine waves with different frequencies and amplitudes.

Fourier's Theorem

Figure 15-12 demonstrates that we can generate complicated waveforms by combining sinusoidal waveforms of different wavelengths. In fact, a theorem developed by the French mathematician Jean-Baptiste Joseph Fourier (1768–1830) states that any periodic waveform can be written as a sum of sine and cosine waves with appropriately chosen frequencies and amplitudes.

According to Fourier's theorem, if a function has a period, T, such that $D(t) = D(t + T)$, then

$$D(t) = a_0 + \sum_{n=1}^{\infty} \left[a_n \cos \left(\frac{2\pi nt}{T} \right) + b_n \sin \left(\frac{2\pi nt}{T} \right) \right]$$

(15-14)

Note that the arguments of the sine and cosine functions are all integer multiples of a fundamental frequency that is related to the period by

$$f_0 = \frac{1}{T}$$

(15-15)

In stating the theorem, Fourier also described how to determine the constants a_n and b_n when given $D(t)$. The technique, which involves integral calculus, is beyond the scope of this book.

INTERACTIVE ACTIVITY 15-2

Making Waves

In this activity, you will use the PhET simulation "Making Waves" to synthesize waves by combining sine functions whose frequencies are all integer multiples of one another. You can also listen to the sound each function you create makes. See if you can manipulate the harmonics to create a square wave. Do you need only odd, only even, or both odd and even harmonics? What is the ratio of the amplitude of each harmonic to the amplitude of the fundamental?

Wind Instruments

Wind instruments employ resonant air columns. To make a sound, the player blows air either into or over a mouthpiece that sets up standing waves in the column.

For the brass instruments, the player vibrates their lips inside a mouthpiece and adjusts the tension in their lips to assist in producing the desired note. Woodwinds, on the other hand, employ various techniques for exciting the air in the column. In some, the player causes a reed to vibrate, which then sets up the vibrations in the air column. Recorders and flutes use a different mechanism, where the player blows air past a sharp edge or over a hole.

As we saw previously, there is a connection between the allowed frequencies for standing waves in an air column and the length of the column. Many of the wind instruments employ mechanisms that allow the musician to alter the length of the air column and hence the notes that are available to be played. For the brass instruments, the two principal mechanisms that are employed are valves (trumpet, French horn, tuba) and the slide (trombone).

The modern trumpet has three valves that can be used either singly or in combination. When a valve is pressed, it lengthens the resonating tube and lowers the frequency of the note that is produced. The player also has to adjust the tension in their lips at the same time. The trombone uses a slide that allows the player to continuously vary the length of the air column.

Woodwinds employ a different strategy for changing the note produced by the instrument. They have holes in the side of the column that are closed or opened either with the tips of the musician's fingers (recorder) or a series of mechanically operated pads (saxophone (Figure 15-13), flute).

One of the simplest brass instruments is the bugle (Figure 15-14), which has no mechanisms for adjusting the length of the instrument.

Courtesy of Mike Murley

Figure 15-13   Saxophone player using the keys on the sax to select notes.

UltraOrto, S.A. / Shutterstock.com

Figure 15-14   The bugle is a brass instrument with a fixed length.

The bugle player still has a range of notes available to them because of the harmonics that an air column can sustain. You would expect that a bugle would behave like a pipe that is open at one end (the bell) and closed at the other—the mouthpiece. As a consequence, only the odd harmonics would be available. In fact, such a set of harmonics is considered too sparse to play music.

The detailed physics of a bugle are rather complex. The bell and the mouthpiece are designed in such a way to pull the lower harmonics up in frequency and the higher harmonics down so that the resulting tones are very close to both the even and odd harmonics. The player still has a limited range of notes available, and the tunes we are accustomed to hearing from the bugle—such as Reveille and Taps—reflect that.

15-5 Interference

Interference of three-dimensional waves is more complex than interference of one-dimensional waves. In one dimension, waves with the same frequency and wavelength have a fixed phase difference that depends only on the difference between the phase constants of the two waves. The phase difference does not vary with time or position. However, when waves propagate in different directions, their relative phase varies with position.

Figure 15-15 shows two waves with the same wavelength and frequency meeting each other at right angles. There is a doubling of the amplitude due to constructive interference.

Interference in Space

We will start our exploration of interference in more than one dimension by considering the interference pattern that is produced by two point sources. The two sources are in phase (they both have the same phase constant) and produce waves with the same frequency and wavelength. Each source produces spherical wave fronts. We can look at the resulting interference pattern in a plane that contains the two sources, as shown in Figure 15-16. This image represents the interference pattern at a particular instant in time.

We can mathematically determine the points where the two waves interfere constructively. At such points, the two waves must be in phase regardless of the elapsed time. Let us consider an instant when both sources are at the peak positive amplitude. If we then locate a point where the waves are interfering constructively, there are two ways this can happen—both waves must be at either peak positive amplitude at that point or peak negative amplitude at that point. If they are at peak positive amplitude, there must be an integer number of wavelengths between each source and the point we are considering. Figure 15-17 shows one such point.

Alternatively, if the two waves are at peak negative amplitude at the point of interest, then there must be a half-integer number of wavelengths between each source and that point.

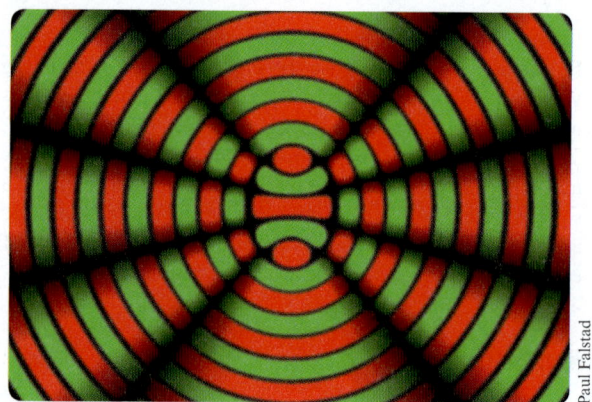

Figure 15-16 Interference from two identical point sources. The red regions represent large positive amplitude, and the green regions represent large negative amplitude.

Figure 15-15 Two waves with the same wavelength meet at right angles. Where the waves overlap, there is a doubling of the amplitude due to constructive interference.

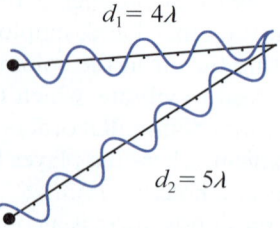

Figure 15-17 Constructive interference of two waves from two point sources occurs when the path difference from the two sources is an integer multiple of the wavelength—in this case one wavelength.

INTERACTIVE ACTIVITY 15-3

Wave Interference

In this activity, you will use the PhET simulation "Wave Interference" to explore interference.

If we consider the instant at which the two sources are at large negative amplitude, the same argument applies and there is either an integer number or a half-integer number of wavelengths between each source and the point. However, this interference now produces large negative amplitude. In general, any point that is either an integer number of wavelengths from both sources or a half-integer number of wavelengths will be a point of constructive interference. Defining d_1 as the path length from source 1, and d_2 as the path length from source 2, the condition for constructive interference is

$$d_1 = m\lambda \text{ and } d_2 = n\lambda$$

or

$$d_1 = \left(m + \frac{1}{2}\right)\lambda \text{ and } d_2 = \left(n + \frac{1}{2}\right)\lambda \quad (15\text{-}16)$$

$$m = 0,1,2,3, \ldots$$
$$n = 0,1,2,3, \ldots$$

These conditions can be made more general by considering the difference between the distances from the two sources to the point of constructive interference:

$$\Delta d = d_2 - d_1 = (n - m)\lambda = p\lambda$$

or

$$\Delta d = d_2 - d_1 = \left(n + \frac{1}{2} - m - \frac{1}{2}\right)\lambda = p\lambda \quad (15\text{-}17)$$

$$p = \pm 1, \pm 2, \pm 3, \ldots$$

Thus, the condition for constructive interference is that the path difference between the two sources that are in phase must be an integer multiple of the wavelength. If the sources are π radians out of phase, then the path difference must be a half-integer multiple of the wavelength.

The approach we used to arrive at Equation 15-17 demanded that both paths individually be integer multiples of the wavelength. However, that does not necessarily have to be true for Equation 15-17 to be satisfied. Let us see if the path difference condition is sufficient to produce constructive interference.

As a spherical wave travels away from its source, it oscillates in space and time. However, the amplitude is constant over any spherical surface centred on the source. Thus, the spatial variations can be described as a function of r, the distance from the source. We can write the wave function of a spherical wave as

$$D(r,t) = A(r)\cos(kr - \omega t + \phi) \quad (15\text{-}18)$$

This expression looks very similar to the expression for a one-dimensional wave in Equation 15-3. In the argument of the cosine, we have simply replaced x with r. However, the amplitude, $A(r)$, for a spherical wave depends on r because, unlike a wave in one dimension, the wave front spreads out over a larger area as it propagates outward; consequently, the amplitude decreases

as r increases. We will consider this in more detail in Section 15-6.

For two waves to be in phase, the arguments of the cosine function for each wave must differ by an integer multiple of 2π. Since we have assumed that both sources are in phase, we can neglect the phase constant and write

$$(kd_2 - \omega t) - (kd_1 - \omega t) = k(d_2 - d_1) = p2\pi$$

$$d_2 - d_1 = p\frac{2\pi}{k} = p\lambda, \quad (15\text{-}19)$$

$$p = 0, \pm 1, \pm 2, \pm 3, \ldots$$

From this result, we can see that constructive interference occurs whenever the path difference is an integer multiple of the wavelength. For the special case of $d_2 = d_1 = d$, we can easily add the two waves together to find the resultant wave:

$$D_{\text{Total}}(d,t) = A(d)\cos(kd - \omega t + \phi)$$
$$+ A(d)\cos(kd - \omega t + \phi) \quad (15\text{-}20)$$
$$= 2A(d)\cos(kd - \omega t + \phi)$$

We again consider a situation in which both sources are in phase but will turn our attention to determining the conditions for destructive interference. Consider an instant when both sources are at large positive displacement. If we select a point that is an integer number of wavelengths away from the first source, the wave from that source will exhibit a large positive displacement at that point. If we choose the point carefully so that it is also a half-integer number of wavelengths from the second source, the wave from the second source will exhibit a large negative displacement at that point.

One such point is shown in Figure 15-18. At this point, the wave from source 1 has travelled 3 wavelengths, and the wave from source 2 has travelled 3.5 wavelengths. Consequently, the waves are completely out of phase and interfere destructively. The resulting displacement at this point is quite small.

The general condition for destructive interference is that the path difference is a half-integer multiple of the wavelength (i.e., an odd number of half wavelengths):

$$\Delta d = d_2 - d_1 = \left(n + \frac{1}{2}\right)\lambda, \, n = 0, \pm 1, \pm 2, \pm 3, \ldots \quad (15\text{-}21)$$

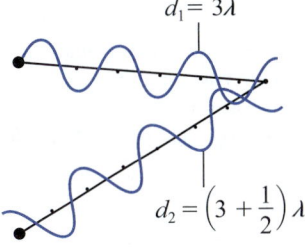

Figure 15-18 The destructive interference of two waves from two point sources.

EXAMPLE 15-4

Interference Pattern

Two speakers (treated as point sources) are placed 2.0 m apart and are emitting sound waves with the same amplitude and phase; the wavelength is 1.0 m. Sketch the interference pattern, and identify points of constructive and destructive interference.

Solution

We make a sketch with concentric circles around the two sources, with each circle having a radius that is one wavelength larger than the previous circle (Figure 15-19).

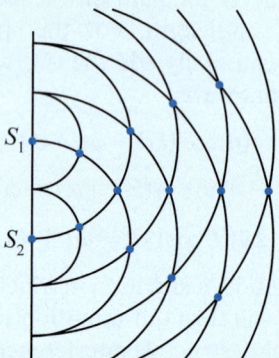

Figure 15-19   Wave fronts for the speakers in Example 15-4.

We can clearly see that there are points where the wave crests overlap—points of constructive interference. However, recognizing that the waves are travelling waves, we should also consider what happens for all times, not just the particular instant shown here.

For all points on the horizontal line that passes midway between the two sources, the path difference is zero. Therefore, that entire line will undergo constructive interference. The other points of constructive interference also appear to lie on lines, and we ask whether these lines are also lines of constructive interference.

Consider the situation shown in Figure 15-20. To have constructive interference, the path difference must be an integer multiple of the wavelength:

$$d_2 - d_1 = \sqrt{x^2 + (y+1)^2} - \sqrt{x^2 - (y-1)^2} = n\lambda$$

The general solution to this equation is complex. However, for $n = 1$, $\lambda = 1.0$ m, we find that

$$y = \pm\frac{\sqrt{12x^2 + 9}}{6}$$

Figure 15-21 shows the two curves defined by this equation. We can see that the curves pass through the points of intersection of the circles representing wave fronts 1.0 m apart. Because every point on the two curves satisfies the condition that the path length

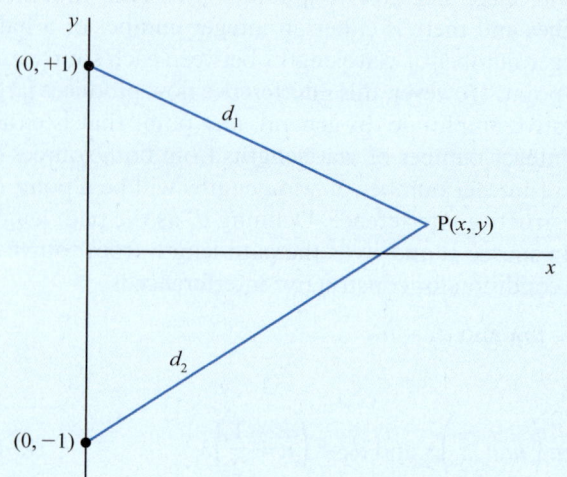

Figure 15-20   The path difference between the two sources and a point P(x, y).

difference is one wavelength, constructive interference occurs all along the curves. We can infer that there is a curved line of destructive interference somewhere between each of the plotted curves and the x-axis, where the path difference is zero.

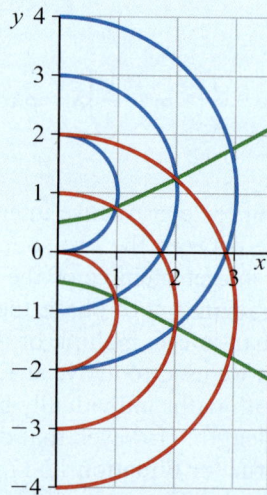

Figure 15-21   The two lines show the solutions corresponding to a path difference of one wavelength.

Making sense of the result

Along the lines of constructive interference, we will have travelling waves whose amplitude is twice that of an individual source. Therefore, if you sit at one place on such a line, you will see waves that alternately have large positive and negative amplitude placements. Along the lines of destructive interference, the two waves always cancel each other out and there will be no amplitude displacement.

Interference Patterns

Two sources that are in phase produce the interference pattern shown in Figure 15-22.

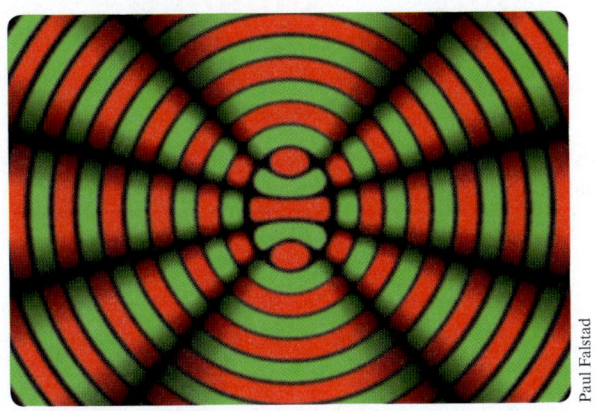

Figure 15-22   Checkpoint 15-3.

If the two sources are out of phase by π, we will observe

(a)   no interference pattern

(b)   the same interference pattern

(c)   an interference pattern similar to the figure here but with the destructive interference located where the constructive interference was observed, and vice versa

(d)   none of the above

ANSWER: (c) One source will emit a crest, and $T/2$ s later, the other source will emit a crest. The crest from the first source will have travelled half a wavelength in that time. All the places where the in-phase sources had an integer wavelength path difference will now have a half-integer difference in the path and will be the locations of destructive rather than constructive interference. A similar argument applies to the locations that experienced destructive interference with the sources in phase—they will be the locations of constructive interference.

Beats

In the patterns described above, regions of constructive and destructive interference are separated by intervals of space. Constructive and destructive interference can also be separated by intervals of time.

When we listen to two waves that have slightly different frequencies, we hear the amplitude of the resultant wave increasing and decreasing as a function of time. This type of variation in amplitude is called a **beat**, and the rate at which the beat occurs is called the **beat frequency**. The beat frequency is proportional to the frequency difference. However, when the frequency difference is large, we hear the sounds as two distinct tones rather than one tone that varies in intensity.

We can see why this occurs by considering two sources that differ slightly in frequency and imagining we are equidistant from the two. The amplitudes of the two sources and the resultant total amplitude are

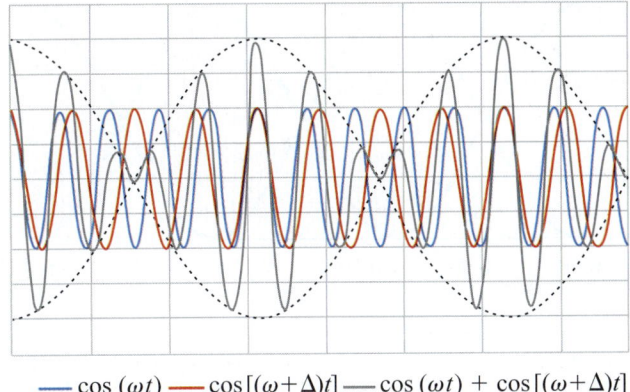

$$\text{—} \cos(\omega t) \text{—} \cos[(\omega+\Delta)t] \text{—} \cos(\omega t) + \cos[(\omega+\Delta)t]$$

Figure 15-23   Two waves that differ by Δ Hz (red and blue) and the resultant wave (grey). The dashed lines outline the amplitude envelope of the resultant wave.

shown in Figure 15-23—we are at a fixed point in space and plot the amplitude we observe at that point as a function of time.

The sound waves start off interfering constructively. At the point on the graph where the dashed lines first cross zero, the red wave has completed two full cycles, and the blue wave has only completed one and one half cycles. Consequently, they are out of phase and interfere destructively. Two red cycles later, they are back in phase. We can see from the dashed lines that the amplitude of the resultant wave (grey) strongly and slowly varies in amplitude (is modulated), and we hear this as a beating effect.

For simplicity, we will develop the mathematics of beats in one dimension. Consider two waves that have the same amplitude but different angular frequencies, ω_1 and ω_2, with different wavelengths and thus different wave numbers. We will set both phases to be zero and consider the resultant amplitude at a fixed position in space, $x = 0$:

$$D_1(0,t) = A\cos(-\omega_1 t) = A\cos(\omega_1 t)$$
$$D_2(0,t) = A\cos(-\omega_2 t) = A\cos(\omega_2 t) \tag{15-22}$$

To find the resulting wave, we apply the principle of superposition and add the two waves:

$$\begin{aligned}
D_{\text{Total}}(0,t) &= D_1(0,t) + D_2(0,t) \\
&= A\cos(\omega_1 t) + A\cos(\omega_2 t) \\
&= 2A\cos\left(\frac{\omega_1 + \omega_2}{2}t\right)\cos\left(\frac{\omega_1 - \omega_2}{2}t\right)
\end{aligned} \tag{15-23}$$

To simplify this expression, we define the following quantities:

$$\overline{\omega} = \frac{\omega_1 + \omega_2}{2} \quad \text{Mean angular frequency}$$
$$\Delta\omega = \frac{\omega_1 - \omega_2}{2} \quad \text{Angular frequency difference} \tag{15-24}$$

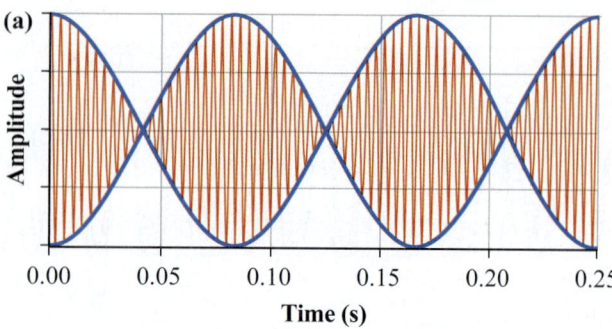

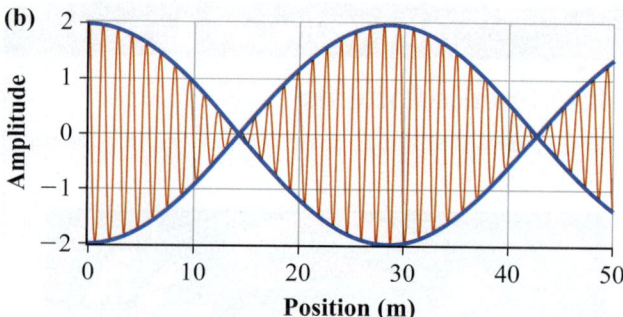

Figure 15-24 (a) The amplitude versus time graph for the sum of a 244 Hz wave and a 256 Hz wave. (b) The amplitude versus position graph for the sum of a 244 Hz wave and a 256 Hz wave. Notice that the spatial modulation of the amplitude has a wavelength of approximately 30 m, which corresponds to a frequency of about 12 Hz.

With these definitions, we can rewrite Equation 15-23 as

$$D_{\text{Total}}(0,t) = 2A\cos(\overline{\omega}t)\cos(\Delta\omega t) \qquad (15\text{-}25)$$

We have a product of two cosines—one oscillates at the mean angular frequency, $\overline{\omega}$, and the other oscillates at the angular frequency difference, $\Delta\omega$. When the two waves are close to each other in frequency, we hear a tone that has the mean frequency $\overline{f} = \overline{\omega}/(2\pi)$, whose amplitude is modulated.

Figure 15-24 shows the result of adding waves with frequencies of 244 Hz and 256 Hz. The red curve shows the resultant amplitude, and the blue curves show the envelope of the amplitude modulation. From this figure, we can see that the amplitude modulation has a period of approximately 0.08 s, which corresponds to a frequency of approximately 12 Hz. We hear this modulation as a beat with a frequency of 12 Hz.

The frequency of the beat is the difference in the frequencies of the two tones, $f_{\text{beat}} = f_2 - f_1$, which is not the same as the difference frequency, $\Delta f = (f_2 - f_1)/2$. The amplitude is modulated at the difference frequency, which means there are two maxima and two minima in one modulation period. Thus, $f_{\text{beat}} = 2\Delta f$.

However, when the two waves are not close in frequency, we perceive them as two separate tones (Figure 15-25).

Interestingly, if we listen to two separate tones through headphones, with one frequency in one ear

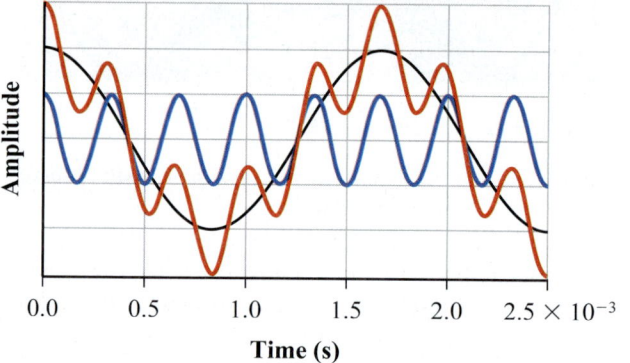

Figure 15-25 Two sound waves with very different frequencies. The black wave has a frequency of 600 Hz, and the blue wave has a frequency of 3000 Hz. In the resultant wave, shown in red, we can clearly see both waves.

and a different but close frequency in the other, we still hear beats even though the waves do not interfere outside our head. It is believed that this effect, called binaural beating, results from the way our brain processes signals from our ears rather than from any interference of the sound waves travelling through our skull.

CHECKPOINT 15-4

Beat Frequency

A 100 Hz tone and a 105 Hz tone are played simultaneously. What is the beat frequency?
(a) 2.5 Hz
(b) 5 Hz
(c) 10 Hz
(d) 205 Hz

ANSWER: (b) Even though the difference frequency is defined as $\Delta\omega = (\omega_1 - \omega_2)/2$, the effect of this is to modulate the amplitude. The amplitude will be large for both the positive and negative crests; therefore, the amplitude modulations have the same frequency as the difference between the two frequencies.

Using Beat Frequencies

Beats can indicate that an instrument is out of tune. Musicians can tune an instrument by adjusting the frequency of its notes until the beating stops (Figure 15-26). The beating effect is part of the reason why instruments that are badly out of tune sound so disagreeable. However, a Balinese gamelan orchestra uses pairs of instruments that are intentionally slightly out of tune from each other to add a "shimmering" effect to the music (Figure 15-27).

The advantage of the beat tuning technique is that we are easily able to detect increases and decreases in amplitude at quite low frequencies—easily below 1 Hz. However, when we compare one frequency to another when we play them sequentially, we can only distinguish tones that differ by about 3 Hz (for notes in the range of 440 Hz).

Figure 15-26   Orchestral musician tuning a violin.

Figure 15-27   A Balinese gamelan orchestra.

15-6 Measuring Sound Levels

As mentioned previously, the quietest sound that a human ear can detect corresponds to an intensity on the order of 10^{-12} W/m^2. At the other extreme, the threshold of pain at 1000 Hz corresponds to an intensity of about 10 W/m^2. This intensity is what you would experience if you were standing about 8 m from a jetliner during takeoff. The human ear has a dynamic range (ratio of loudest to quietest sound intensity) of approximately 10^{13}. For such large ranges, it is often convenient to work with a logarithmic scale.

Decibels

The **decibel** (dB) is a commonly used unit for logarithmic comparisons of power or intensity. Since logarithmic scales are relative scales, we must choose a reference level. For sound, we use the quietest intensity we can hear at 1000 Hz as our reference intensity. We define the zero of our logarithmic sound level scale to correspond to this reference level. The reference sound level of 0 dB corresponds to an intensity of

$$I_0 = 10^{-12} \text{ W/m}^2 \qquad (15\text{-}26)$$

To express any other **sound level** (β) in decibels, we use

KEY EQUATION $\qquad \beta(I) = 10\log_{10}\left(\dfrac{I}{I_0}\right) \qquad (15\text{-}27)$

The decibel is a convenient unit for β because it is about the smallest change in sound level that a human can typically detect. Table 15-4 lists typical β-values for some common sources. Since decibels indicate the logarithm of the ratio of two levels, small differences in decibel measurements indicate much larger differences in the intensity, I. The decibel is one tenth of a bel, named in honour of Alexander Graham Bell (1847–1922).

Table 15-4   Typical Sound Levels for Some Common Sources

Sound Source	Sound Level (dB)
Chainsaw, 1 m away	110
Lawnmower, 1 m away	107
Dance club speaker, 1 m away	100
Portable stereo, half-volume	94
Transport truck, 10 m away	90
Side of busy road	80
Vacuum cleaner, 1 m away	70
Normal speech, 1 m away	60
Average home	50
Quiet library	40
Quiet bedroom at night	30
Quiet sound studio	20

EXAMPLE 15-5

Adding Sound Levels

In this example we wish to explore how to determine the sound level due to multiple sound sources. We can see in Table 15-4 that if we are 1 m away from a lawnmower, the sound level is 107 dB, while if we are 1 m away from a chainsaw, the sound level is 110 dB. Suppose you are 1 m away from both a lawnmower and a chainsaw. What are the resultant sound intensity and sound level?

Solution

We need to be careful here—while the sound intensities just add, the sound levels, which are on a logarithmic scale, do not. Our strategy will be to determine the sound intensities for each source, add them together, and use that sound intensity to find the resultant sound level. From Table 15-4, we can see that the sound level for the chainsaw is 110 dB, while the sound level for the lawnmower is 107 dB.

We will first do the intensity level calculation for the chainsaw. We use Equation 15-27 and recall that $I_0 = 10^{-12}$ W/m^2:

$$110 = 10\log_{10}\left(\frac{I_{\text{chainsaw}}}{I_0}\right)$$

$$11.0 = \log_{10}\left(\frac{I_{\text{chainsaw}}}{10^{-12}\,\text{W}\cdot\text{m}^{-2}}\right)$$

$$10^{11.0} = \frac{I_{\text{chainsaw}}}{10^{-12}\,\text{W}\cdot\text{m}^{-2}}$$

$$I_{\text{chainsaw}} = 10^{11.0} \times 10^{-12}\,\text{W}\cdot\text{m}^{-2}$$

We can perform a similar calculation to find that

$$I_{\text{lawnmower}} = 10^{10.7} \times 10^{-12}\,\text{W}\cdot\text{m}^{-2}$$

We can find the total sound intensity by adding the two intensities found above:

$$I = I_{\text{chainsaw}} + I_{\text{lawnmower}} = (10^{10.7} + 10^{11.0}) \times 10^{-12}\,\text{W}\cdot\text{m}^{-2}$$

We put this result back in to Equation 15-27 to get the sound level:

$$\beta = 10\log_{10}\left(\frac{I_{\text{chainsaw}} + I_{\text{lawnmower}}}{I_0}\right)$$

$$= 10\log_{10}\left(\frac{(10^{10.7} + 10^{11.0}) \times 10^{-12}\,\text{W}\cdot\text{m}^{-2}}{10^{-12}\,\text{W}\cdot\text{m}^{-2}}\right)$$

$$= 10\log_{10}(10^{10.7} + 10^{11.0}) = 112\,\text{dB}$$

Making sense of the result

This result might seem a bit puzzling. One might have guessed that the result would be 207 dB, obtained by simply adding the two sound levels together. We need to recall that while the intensities add directly, the sound levels (dB) do not, which is a consequence of the logarithmic scale we use to calculate the decibel levels. Let's do a very simple calculation that illustrates exactly how this works. Suppose we have some source that has intensity I_1. We then double the intensity to $I_2 = 2I_1$. Let's see how the sound level changes.

We again use Equation 15-27 to find

$$\beta_1 = 10\log_{10}\left(\frac{I_1}{I_0}\right)$$

$$\beta_2 = 10\log_{10}\left(\frac{I_2}{I_0}\right)$$

$$= 10\log_{10}\left(\frac{2I_1}{I_0}\right)$$

$$= 10\log_{10}\left(\frac{I_1}{I_0}\right) + 10\log 2$$

$$= \beta_1 + 3\,\text{dB}$$

(Note that we used the property of logarithms that the log of a product is the sum of the logs.) With that result in mind, we can see that we did not quite double the intensity in our example and we had slightly less than a 3 dB increase in the sound level.

PEER TO PEER

To help me remember about the logarithmic nature of decibels, I always try to recall that if we double the intensity, that corresponds to an increase of 3 dB, not a doubling of the sound level in decibels.

INTERACTIVE ACTIVITY 15-4

Moving away from a Sound Source

See if you can find a steady source of sound and observe how loud it sounds as you walk around the source at a constant distance from the source. Then move closer to the source. Does the loudness increase or decrease? Move away from the source. Does the loudness increase or decrease? Many smartphones have free sound level apps. See if you can quantify the relationship between distance and sound level.

EXAMPLE 15-6

Concert Sound Level

You are at a concert where the sound intensity measures 3.00 W/m^2. What is the sound level in decibels? Should you be concerned about your hearing?

Solution

We use Equation 15-27 and recall that $I_0 = 10^{-12} \text{ W/m}^2$.

$$\beta = 10\log_{10}\left(\frac{I}{I_0}\right)$$

$$= 10\log_{10}\left(\frac{3.00 \text{ W} \cdot \text{m}^{-2}}{10^{-12} \text{ W} \cdot \text{m}^{-2}}\right)$$

$$= 125 \text{ dB}$$

This level is close to the threshold of pain and is substantially above the level permitted in a workplace. There is cause for concern about hearing loss, especially since hearing damage is cumulative and permanent.

Making sense of the result

This sound level is not uncommon at rock concerts.

In Example 15-6, the sound intensity was a modest 3 W/m^2 at what most of us would consider a loud concert. What happens to all the power from the amplifiers? The loudness of a point sound source diminishes as you move away from the source. However, as you move around a point source at a constant distance from the source, the loudness does not vary. In practice, it is very difficult to produce an actual point source for sound waves, but the mathematics of such a source is relatively straightforward.

If we think about conservation of energy, it makes sense that the intensity of a source diminishes as we move away from the source. If we enclose the source with a spherical surface of radius r, all the power radiated by the source must pass through the surface of the sphere. If the power is radiated **isotropically** (uniformly in all directions) by the source, then the intensity over the surface of the sphere will be uniform. Consequently, the total power radiated, P, is equal to the intensity, I, multiplied by the surface area of the sphere:

$$P = I \times 4\pi r^2 \tag{15-28}$$

We can rearrange this equation to find

KEY EQUATION

$$I = \frac{P}{4\pi r^2} \tag{15-29}$$

The intensity of a sound source that radiates equally in all directions falls off as the inverse square of the distance from the source.

MAKING CONNECTIONS

Hearing Safety

Exposure to loud noise can cause permanent damage to your hearing. In Canada, provincial Occupational Health and Safety legislation restricts the maximum β in a work environment for an 8 h shift to between 85 dB and 90 dB. If the β is above this limit, either hearing protection must be worn or the exposure time must be reduced to avoid damage to the workers' hearing.

Musicians are particularly concerned about hearing damage. Interestingly, it appears that the incidence of hearing loss is higher among classical musicians than among rock musicians. High-quality earplugs that reduce the sound level while still allowing a musician to hear the music adequately have been developed to help prevent such hearing loss (Figure 15-28). Ideally, the earplugs should reduce all frequencies equally.

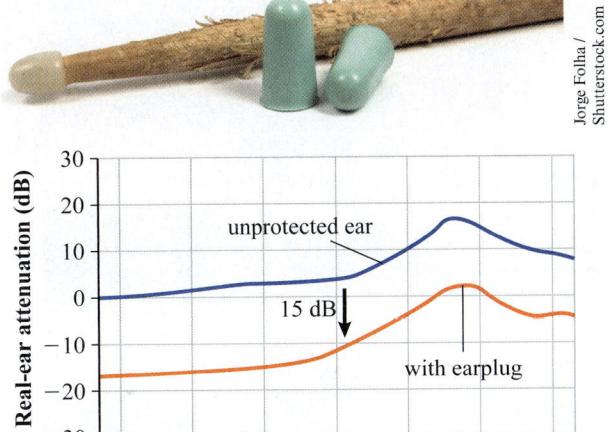

Jorge Folha / Shutterstock.com

Figure 15-28 High-quality earplugs provide uniform attenuation at all frequencies.

EXAMPLE 15-7

Speaker Power

Assume that you are 50. m from the stage at a concert where the sound intensity is 0.30 W/m². How much power are the speakers radiating, assuming that they direct their sound output toward the audience such that the sound intensity in that direction is 8 times as great as if the speakers were isotropic point sources?

Solution

The sound intensity is 0.30 W/m². To find the power for an isotropic source, we use Equation 15-28:

$$P = I \times 4\pi r^2 = (0.30 \text{ W/m}^2)(4\pi)(50. \text{ m})^2 = 9.4 \times 10^3 \text{ W}$$

Since the speakers direct most of their sound toward the audience, their total sound output is about one eighth that of an isotropic source that produces the same sound intensity in all directions:

$$P = \frac{9.4 \times 10^3 \text{ W}}{8} = 1.2 \text{ kW}$$

Making sense of the result

We can see that the point source must be quite powerful to produce this level of sound at 50. m. However, as we noted at the beginning of the chapter, amplifier systems at rock concerts are typically capable of delivering approximately 50 kW of electrical power to the speakers. Speakers themselves typically only have an efficiency of about 1%. Thus, if we connect a 100 W amplifier to a speaker, it only radiates about 1 W in sound energy. The rest of the energy is lost as heat in the amplifiers.

MAKING CONNECTIONS

Refraction of Sound Waves

If you have ever been out in a canoe on a large lake on a calm day, you may have noticed that it is possible to hear sounds clearly over long distances. The same effect is rarely observed on land. The air near the surface of the lake is cooler than the air well above the lake (Figure 15-29). The speed of sound in air increases as the temperature increases. As we saw in Section 15-3, sound waves tend to bend toward regions where the wave speed is slower. Thus, sound waves tend to be refracted down to the surface of the lake. Consequently, the power from the sound source is not radiated isotropically.

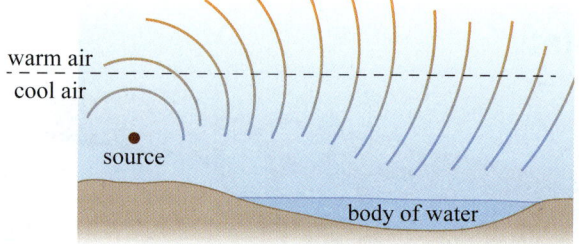

Figure 15-29 On a calm warm day, cool air over the surface of a lake causes sound waves to refract down toward the lake surface, greatly increasing the range of the sound.

Response of the Ear

When we listen to sound, a complex process takes place that converts the sound pressure in the air into our perception of the sound. The outer ear captures the sound waves and directs them to the eardrum, located in the middle ear (Figure 15-30). The sound pressure makes the eardrum vibrate, and three delicate bones connected to the eardrum transmit these vibrations to the inner ear. The inner ear contains the cochlea (shown as a blue coiled object), which is filled with fluid. The vibrations in this fluid stimulate delicate "hairs" on the basilar membrane. Nerve cells in the basilar membrane convert the vibrations of the hairs into electrical impulses that travel to the brain over the auditory nerve.

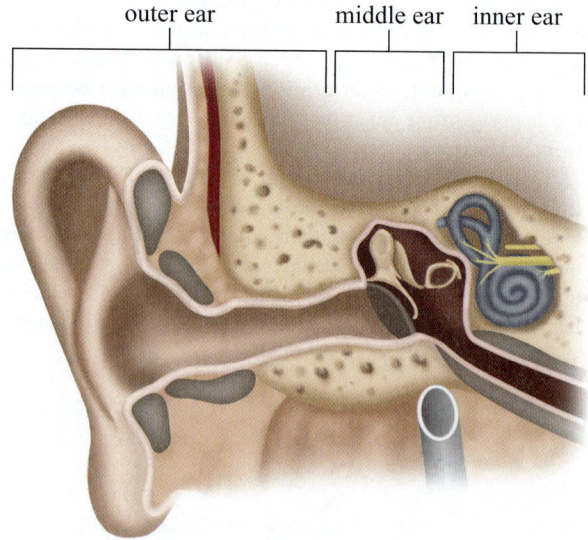

Figure 15-30 A cross section of the human ear.

15-7 The Doppler Effect

When we stand on the side of a highway and a car goes by, the sound that we hear has a higher frequency when the car is approaching and shifts to a lower frequency as the car passes by. Similarly, if we generate sound as we stand on the side of the road, the people in the car will hear a higher frequency as they approach and a lower frequency as they move away. In general, if there is relative motion between the source of the sound and the receiver of the sound, the frequency at the receiver is different from the frequency that is transmitted by the source. If the two are moving toward each other, the received frequency is higher, and if they are moving apart, the received frequency is lower. This phenomenon is called the **Doppler effect**, named for the Austrian physicist Christian Doppler (1803–1853), who first described it.

It is possible to write a single equation for the Doppler effect that covers the three possible scenarios:

- stationary receiver, moving source
- stationary source, moving receiver
- moving source and receiver

We denote the frequency of the source as f_s and the frequency at the receiver as f_r. If the source and receiver are moving along the line that connects them, the general relationship between the two frequencies is

KEY EQUATION

$$f_r = \frac{v \pm v_r}{v \mp v_s} f_s \qquad (15\text{-}30)$$

where v is the speed of sound in air and v_r and v_s are the speed of the receiver and the source, respectively, relative to the air.

Note that Equation 15-30 has $\pm$ in the numerator but $\mp$ in the denominator. If the receiver is moving toward the source, we observe an increased frequency and use the top sign in the numerator. If the source is moving toward the receiver, we also observe an increase in the frequency and use the top sign in the denominator. In both cases, if the motion is toward, we use the top sign.

We now consider the common special cases, where either $v_r = 0$ or $v_s = 0$.

Moving Source, Stationary Receiver

Figure 15-32(a) shows a stationary source emitting spherical waves that travel outward at the same speed in all directions. A receiver at any point around this source will detect the same frequency because the wave fronts are all equally spaced and all travel at the same speed. However, if the source is moving to the right, the wave fronts to the right of the source are closer together, and the wave fronts to the left of the source are farther apart, as shown in Figure 15-32(b). Since the wave fronts are still all travelling at the same speed, a receiver to the right of the source will detect more wave fronts per second than a receiver to the left of the source will. More wave fronts per second correspond

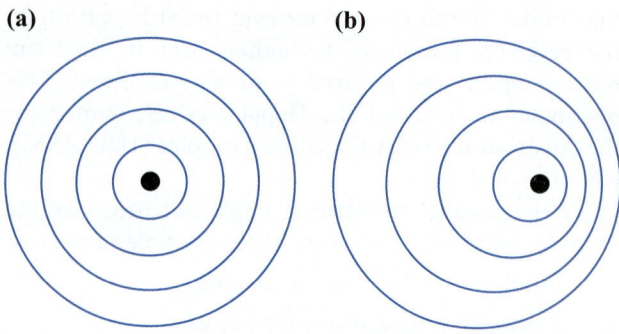

Figure 15-32 (a) A stationary source emitting waves. (b) The source moving to the right.

to a higher frequency, and fewer wave fronts per second correspond to a lower frequency.

Another way to understand the Doppler effect is this: recall that the speed of sound in air at a given temperature is constant. Clearly, the wavelength to the right of the source in Figure 15-32(b) is shorter, and since $f = v/\lambda$, a shorter wavelength implies a higher frequency.

We can calculate the change in the wavelength. Consider the shortened wavelengths observed to the right of the source. The source emits a wave front every

T seconds, where T is the period of the source. In each period, the wave front travels a distance $D = vT$, where v is the speed of the wave. In the same time, the source travels a distance $d = v_s T$, where v_s is the speed of the source. Thus, the distance between the source and the last wave front it emitted is

$$\lambda_r = \frac{v}{f_r} = D - d = (v - v_s)T = \frac{v - v_s}{f_s} \qquad (15\text{-}31)$$

Solving for f_r, we find that for a source moving toward a stationary receiver:

$$f_r = \frac{v}{v - v_s} f_s \qquad (15\text{-}32a)$$

(source moving toward stationary receiver)

Had we considered the situation with the receiver to the left of the source (i.e., the source moving away from a stationary receiver), we would have a plus sign in the denominator to get

$$f_r = \frac{v}{v + v_s} f_s \qquad (15\text{-}32b)$$

(source moving away from stationary receiver)

EXAMPLE 15-8

The Doppler Effect

A high-speed train travelling at 250 km/h blows its whistle as it approaches a level crossing. The whistle has a frequency of 660 Hz. What frequency would be perceived by someone standing at the crossing as the train

(a) approaches the crossing?
(b) moves away from the crossing?

Solution

(a) The speed of the source (the train) is 250 km/h = 69.4 m/s, and the speed of the receiver is zero. The speed of sound in air is approximately 343 m/s. Substituting these values into Equation 15-30 gives

$$f_r = \left(\frac{343 \text{ m/s} + 0 \text{ m/s}}{343 \text{ m/s} - 69.4 \text{ m/s}} \right) 660 \text{ Hz} = 827 \text{ Hz}$$

Notice that we chose a minus sign in the denominator: since the train was moving toward the receiver, we use the top sign.

(b) When the train is leaving the crossing, we use the bottom sign:

$$f_r = \left(\frac{343 \text{ m/s} - 0 \text{ m/s}}{343 \text{ m/s} + 69.4 \text{ m/s}} \right) 660 \text{ Hz} = 549 \text{ Hz}$$

Making sense of the result

As the train approaches, we hear a higher pitch, and as it moves away, we hear a lower pitch. This is similar to what we observe when we stand near the side of a highway.

Moving Receiver, Stationary Source

When the source is stationary and the receiver is moving away from it, the receiver detects a reduction in the wave speed, and a receiver moving toward the source detects an increase in the wave speed. First, consider a receiver approaching the stationary source. When the receiver is positioned at a wave crest, the next wave crest is a distance λ away and moving toward the receiver at a speed relative to the

receiver of $v + v_r$. The time this second wave crest takes to reach the receiver is

$$T_r = \frac{\lambda}{v + v_r} = \frac{v}{v + v_r}\left(\frac{1}{f_s}\right) = \frac{1}{f_r} \qquad (15\text{-}33)$$

Solving for f_r, we find

$$f_r = \frac{v + v_r}{v} f_s \qquad (15\text{-}34a)$$

(receiver moving toward a stationary source)

The Doppler Effect and Electromagnetic Radiation

The Doppler effect also occurs with electromagnetic radiation (light waves). According to the basic postulate of the theory of relativity (Chapter 30), as measured in an inertial frame of reference, the speed of light (c) in vacuum is independent of the motion of the source. Therefore, the Doppler effect for electromagnetic waves has only two scenarios. The relative motion of the source and the observer is either toward or away from each other. When the relative speed, u, is such that $u \ll c$, then the observed frequency, f', and the emitted frequency, f, for these two scenarios are related as follows:

Source and observer moving away from each other:

$$f' = f\left(\frac{c - u}{c}\right) \tag{1}$$

Source and observer moving toward each other:

$$f' = f\left(\frac{c + u}{c}\right) \tag{2}$$

It is usual to write the above equations in terms of the emitted wavelength ($\lambda = c/f$) and the observed wavelength ($\lambda' = c/f'$) as follows:

Source and observer moving away from each other:

$$\lambda' = \frac{\lambda}{\left(1 - \dfrac{u}{c}\right)} \tag{3}$$

Source and observer moving toward each other:

$$\lambda' = \frac{\lambda}{\left(1 + \dfrac{u}{c}\right)} \tag{4}$$

Therefore, the observed light from a source that is moving away from an observer is shifted toward longer wavelengths (this phenomenon is called the **redshift**), while the light emitted from a source that is moving toward an observer is shifted toward smaller wavelengths (this phenomenon is called the **blueshift**).

This Doppler shift of light has a very interesting implication in astronomy. If we are living in an expanding universe, then the light emitted from a distant galaxy that is moving away from Earth's frame of reference will be redshifted when it reaches Earth. Consider, for example, the $H_\alpha = 6562.8$ Å (1 Å $= 10^{-10}$ m $= 0.1$ nm) line in the hydrogen spectrum. A photon of this wavelength is emitted when an electron in the hydrogen atom jumps from an $n = 3$ orbital to an $n = 2$ orbital. Figure 15-33 shows the observed hydrogen spectrum from galaxy NGC 2276, which is estimated to be about 90 Mpc (i.e., 90 megaparsec or 90×3.26 million light years) from Earth. In this spectrum, the wavelength of the

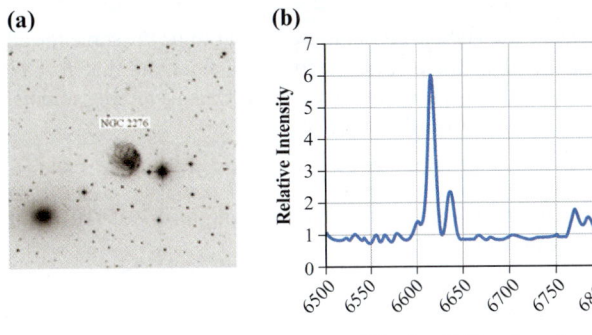

(a)   **(b)**

Figure 15-33 (a) An image of galaxy NGC 2276 and (b) the spectrum of light from this galaxy. Notice that the H_α line has shifted to 6617 Å.

Sources: (a) Space Telescope Science Institute, http://www.astro.washington.edu/courses/labs/clearinghouse/labs/HubbleLaw/galaxies.html; (b) Based on data from http://cdsarc.u-strasbg.fr/viz-bin/nph-Plot/Vgraph/txt?VII%2f141%2f.%2fsp%2fNGC2276

H_α line is observed to be 6617 Å. This means that NGC 2276 is moving away from Earth. Its recession speed can be calculated using (3) as follows:

$$u = c\left(\frac{\lambda' - \lambda}{\lambda'}\right)$$

Using $\lambda = 6562.8$ Å and $\lambda' = 6617$ Å, we obtain

$$u = 0.0082c \approx 2500 \text{ km/s}$$

So NGC 2276 is receding from Earth (and hence our galaxy) with a speed of 2500 km/s. Note that an observer at NGC 2276 would also determine that our galaxy, the Milky Way, is receding with the same speed.

American astronomer Edwin Hubble (1889–1953) made a plot of the Doppler shift of distant galaxies versus their distance from Earth and noted that the light from more or less all of the galaxies was redshifted and that the magnitude of the redshift was roughly proportional to the distance, D, of the galaxy from Earth. His original data are shown in Figure 15-34. The relationship between the recession speed and the distance is linear, so that

$$u = HD \tag{5}$$

This is **Hubble's law**, and H is called the **Hubble constant**. It has dimensions of speed/distance or (time)$^{-1}$. It is measured in (km/s)/Mpc and is the same for all galaxies. Since farther galaxies are receding from Earth with higher speeds, Hubble concluded that the universe is expanding. There is another way to understand the meaning of the Hubble constant. If we assume that a galaxy is at a distance D and has been moving

(continued)

WAVES AND OSCILLATIONS

away from us with a constant speed u, then that galaxy was at zero distance from us at time

$$t_H = \frac{D}{u} = \frac{1}{H} \tag{6}$$

Because the Hubble constant is assumed to be the same for all galaxies, this time is the same for all galaxies and therefore t_H is a measure of the age of the universe. Using a recent estimate, $H = 72.6 \pm 3.1$ (km/s)/Mpc, we get

$$t_H = (13.5 \pm 0.6) \times 10^9 \text{ years}$$

This idea forms the basis of the Big Bang theory of the origin of the universe.

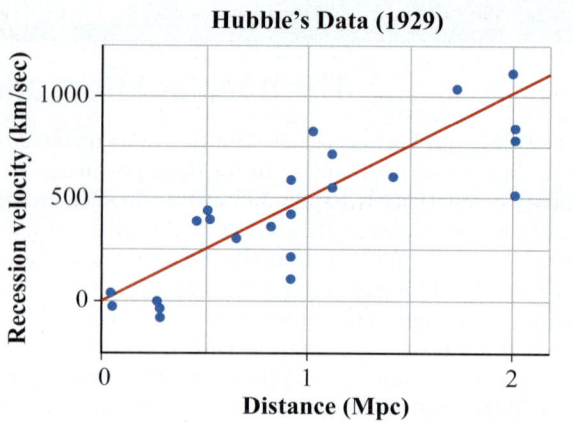

Hubble's Data (1929)

Figure 15-34 The slope of this line is known as Hubble's constant, recently estimated to be $H = 72.6 \pm 3.1$ (km/s)/Mpc. These are the standard units astronomers use to express the number, where Mpc is a megaparsec. A parsec is about 3.09×10^{16} m ≈ 3.26 light years.

Source: Data from Edwin Hubble, "A relation between distance and radial velocity among extra-galactic nebulae," PNAS March 15, 1929 vol. 15 no. 3 168–173.

This expression is the special case of Equation 15-30 with $v_s = 0$ and the receiver moving toward the source. For a receiver moving away from a stationary source, we use a minus sign in the numerator instead to obtain

$$f_r = \frac{v - v_r}{v} f_s \tag{15-34b}$$

(receiver moving away from a stationary source)

With Equations 15-32 and 15-34, we can readily derive the general result we stated in Equation 15-30. If both the source and the receiver are moving, we replace the stationary source frequency in Equation 15-34 with the stationary receiver frequency from Equation 15-32 to obtain Equation 15-30:

$$f_r = \frac{v \pm v_r}{v} \frac{v}{v \mp v_s} f_s$$

$$f_r = \frac{v \pm v_r}{v \mp v_s} f_s \tag{15-30}$$

The sign conventions use the upper sign when the motion is toward and the lower sign when the motion is away.

EXAMPLE 15-9

Suppose a bat is flying at 6.00 m/s chasing a moth that is flying at 4.00 m/s directly away from the bat. If the bat is emitting echolocation cries at 82.0 kHz, at what frequency will the moth hear the cries?

Solution

This is a Doppler effect problem and we clearly have a moving source (bat) and receiver (moth) in this case. We will use Equation 15-30 to find

$$f_{receiver} = \frac{v - v_r}{v - v_s} f_{source}$$

The numerator deals with the motion of the receiver (moth), which is moving in the same direction as the bat, so we use a negative sign, as we would expect the moving moth to hear a lower frequency than a stationary moth. The denominator deals with the motion of the source and so we also use a negative sign, because we would expect to perceive the frequency of a source moving toward us to be Doppler-shifted up. When we put in the speeds and frequencies, we find

$$f_{receiver} = \frac{343 \text{ m/s} - 4.00 \text{ m/s}}{343 \text{ m/s} - 6.00 \text{ m/s}} (82.0 \text{ kHz}) = 82.5 \text{ kHz}$$

Making sense of the result

The moth perceives the bat to be crying at 82.5 kHz, which is higher than the 82 kHz that the bat is emitting. This makes sense as the bat is getting closer to the moth, so the perceived frequency should be Doppler-shifted up.

We can think about this problem in another way. The moth is unable to know whether it perceives the sound of a stationary or a moving source. So, we can first determine what frequency the bat would appear to emit to a stationary receiver and then consider the bat to be a stationary source that emits at that frequency.

We first consider what frequency a stationary receiver that the bat is flying toward would perceive:

$$f_{\text{stationary receiver}} = \frac{v}{v - v_s} f_{\text{source}}$$

$$= \frac{343 \text{ m/s}}{343 \text{ m/s} - 6.00 \text{ m/s}} (82.0 \text{ kHz}) = 83.5 \text{ kHz}$$

We now consider the moth to be a receiver that is moving away from a stationary source that is emitting a frequency of 83.5 kHz. It then receives a frequency of

$$f_{\text{moving receiver}} = \frac{v - v_r}{v} f_{\text{source}}$$

$$= \frac{343 \text{ m/s} - 4.00 \text{ m/s}}{343 \text{ m/s}} (83.5 \text{ kHz}) = 82.5 \text{ kHz}$$

This clearly illustrates the effect of the motion of both sources. The forward motion of the bat produces an upward shift in frequency that is somewhat offset by the forward motion of the moth. The net effect is that the moth perceives a frequency that is shifted upward because the bat is closing in on the moth.

MAKING CONNECTIONS

Bats and the Doppler Effect

Most of us are aware that bats use echolocation to detect objects and prey as they fly around at night. They emit pulses of sound and "listen" for the echoes. The distance to the target can be determined from the time between emitting the pulse and the detection of the echo. Naturally, bats need to have incredibly sensitive hearing—many bats have a dip in their threshold of hearing near the frequency at which they emit their calls. Figure 15-35 shows the threshold of hearing measured on a greater horseshoe bat, which is found in Europe, Africa, South Asia, and Australia.

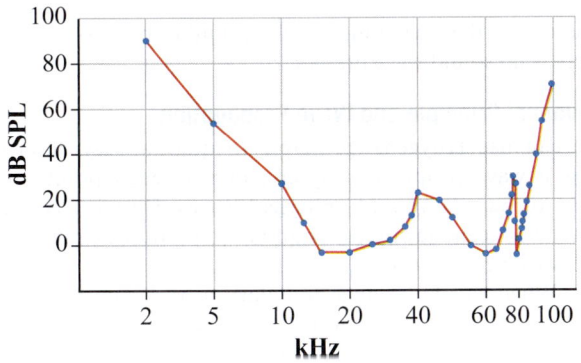

Figure 15-35 Threshold sound pressure level in dB SPL as a function of frequency in *Rhinolophus ferrumequinum*. This is the quietest sound that the bat is able to perceive. These bats have three frequency regions where their hearing is very sensitive. There is a fairly broad band between 15 kHz and 30 kHz, a somewhat narrower band around 60 kHz, and a very sharp band with a dip at about 85 kHz.

Source: Journal of Comparative Physiology, "Behavioural audiograms from the bat, Rhinolophus ferrumequinum Glenis," Volume 100, Issue 3, 1975, pp. 211–219, Glenis R. Long, Hans-Ulrich Schnitzler, Copyright © 1975, Springer-Verlag. With permission of Springer.

When hunting, the horseshoe bat emits cries with a frequency of about 82 kHz, which is below the frequency at which their hearing is most sensitive. Incredibly, the difference between these two frequencies is related to the speed at which the bat flies. The bat itself is a moving source that causes an upward Doppler shift of the cry for any target in front of it:

$$f_{\text{target}} = \frac{v}{v - v_{\text{bat}}} f_{\text{cry}}$$

The target reflects this higher frequency and the bat is then a moving receiver, and a second shift to higher frequency occurs for the bat:

$$f_{\text{echo}} = \left(1 + \frac{v_{\text{bat}}}{v}\right) f_{\text{target}} = \frac{1 + \dfrac{v_{\text{bat}}}{v}}{1 - \dfrac{v_{\text{bat}}}{v}} f_{\text{cry}}$$

We can rearrange this equation to solve for the speed of the bat, and we find

$$v_{\text{bat}} = \frac{f_{\text{echo}} - f_{\text{cry}}}{f_{\text{echo}} + f_{\text{cry}}} = \frac{85 \text{ kHz} - 82 \text{ kHz}}{85 \text{ kHz} + 82 \text{ kHz}} (343 \text{ m/s}) = 6.2 \text{ m/s}$$

which is well within the range of observed bat flight speeds. Even more incredible is the fact that bats adjust the frequency of their cries, as the relative speed of the target varies, to ensure that the frequency of the echo lies where their hearing is most sensitive.

Not surprisingly, faced with such a sophisticated hunter, bat prey have evolved sophisticated defence mechanisms, which include enhanced hearing sensitivity in the frequency range of the bat hunting cry and producing their own sounds that serve to confuse the bats.

18. ★★ A sound wave is described by $D(x, t) = 2 \times 10^{-9} \cos(37x - 12566t)$. Find the wave's
    (a) wavelength
    (b) frequency
    (c) speed

19. ★ Write an expression for the displacement of a 60 Hz sound wave that travels in the positive y-direction in air at 20 °C.

20. ★★ Medical ultrasound machines typically use frequencies in the range of 2 MHz to 18 MHz.
    (a) What are the corresponding wavelengths in air?
    (b) Given that the human body contains primarily water, estimate the wavelength range in the body.

21. ★ A 1000 Hz sound wave with a pressure amplitude of 20 N/m² in air is at the threshold of pain for human hearing. It has a pressure amplitude of $\Delta p = 20$ N/m². What is the corresponding displacement amplitude?

22. ★ What is the intensity of a 2000 Hz sound wave in air that has a pressure amplitude of 1.5×10^{-3} N/m²?

23. ★ When the displacement amplitude of a sound wave doubles, by what factor does the pressure amplitude increase?

Section 15-2 Wave Propagation and Huygens' Principle

24. ★★ For the wave front in Figure 15-37, use Huygens' principle to determine the shape of the resulting wave front
    (a) close to the original wave front
    (b) far from the original wave front

Figure 15-37 Problem 24.

Section 15-3 Reflection and Refraction

25. ★★ In Figure 15-38, a plane wave is approaching a rectangular corner. Draw the reflected wave using Huygens' principle.

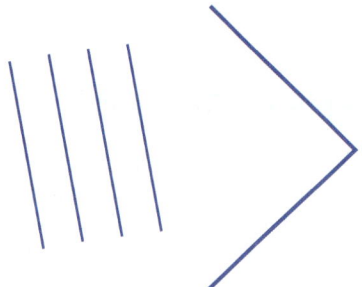

Figure 15-38 Problem 25.

26. ★★ Repeat the sequence of drawings shown in Figure 15-9 for the case where the index of refraction is 1.5 in the lower medium and 2 in the upper medium.

Section 15-4 Standing Waves in Air Columns

27. ★ An air column has a length of 0.75 m and is open at both ends. What is the lowest frequency of a standing wave that can be found in such a column?

28. ★★ Standing waves can form in a box. A speaker enclosure has dimensions 10 cm × 15 cm × 20 cm. What are the lowest-frequency standing waves (at 20 °C) that can form in the three directions?

29. ★★★ A student places a speaker at one end of a tube and a microphone at the other end. The student connects the speaker to a function generator and sweeps the frequency of the generator from 100 Hz to 500 Hz. The student measures the amplitude of the signal from the microphone while doing this and observes maxima in the microphone signal at frequencies of 160 Hz, 240 Hz, 320 Hz, 400 Hz, and 480 Hz. What is the length of the tube?

Section 15-5 Interference

30. ★★ Two point sources are 25 cm from each other, and both emit a 2000 Hz tone. Sketch the resulting interference pattern.

31. ★ Two violins produce frequencies of 294 Hz and 296 Hz. What beat frequency is heard?

32. ★★ Two point sources are in phase and emit a 500 Hz tone. What is the phase difference at a point that is 5.0 m from one source and 6.0 m from the other?

33. ★ In North America, the telephone dial tone (the sound you hear when you pick up a land-line phone) consists of a 350 Hz tone and a 440 Hz tone. What is the resulting beat frequency?

Section 15-6 Measuring Sound Levels

34. ★ A food processor produces a sound level of 87 dB at a distance of 1.0 m. How many watts per square metre does this sound level correspond to?

35. ★ A single table saw produces a sound level of 95 dB at a distance of 3.0 m. What would the sound level be for two table saws, both 3.0 m away?

36. ★ What is the displacement amplitude for an 800 Hz sound wave that has an intensity of 20 mW/m²?

37. ★★ A sound level of 60 dB is measured at a distance of 40 m from a lion when it roars. What sound level would be measured 10 m from the lion?

Section 15-7 The Doppler Effect

38. ★ A police siren emits a frequency of 700.0 Hz. When the police car approaches you at a speed of 80.0 km/h, what frequency do you perceive?

39. ★ The Doppler effect can be used to measure flow rates in blood. A sound wave is transmitted into a vein or an artery, where it is reflected off the moving red blood cells and then detected by a receiver. A transmitter uses a frequency of 10.0 MHz, and the blood flow rate is 0.50 m/s. What is the expected frequency shift?

40. ★ Some bats use the Doppler shift to detect the direction of motion of prey. The bats use a frequency of about 61 kHz and can detect shifts as small as 35 Hz. To what velocity does this correspond?

41. ★★ A bat is flying at 7.00 m/s and is emitting cries at a frequency of 82.0 kHz. It is chasing a moth and is detecting the echoes from the moth. Moths fly somewhat erratically. Calculate the frequency detected by the bat when the moth is
    (a) stationary
    (b) flying at 1.00 m/s in the same direction as the bat
    (c) flying at 2.00 m/s toward the bat

COMPREHENSIVE PROBLEMS

42. ★ At large concerts, it is sometimes disconcerting to observe the musicians moving apparently out of sync with the music. This results from the time it takes the sound to travel from the stage to you. When the musicians are playing at 100 beats/min, at what distance from the stage will they appear to be one full beat behind?

43. ★ While camping, you observe a lightning bolt, and 4.0 s later you hear the associated thunder. How far away is the strike?

44. ★ Two sound sources differ in sound level by 20 dB. Find the ratio of their
  (a) displacement amplitudes
  (b) pressure amplitudes
  (c) intensities

45. ★★★ Two sound waves of frequency 200 Hz travel in the x-direction. One has a displacement amplitude of 10 nm, and the other has a displacement amplitude of 20 nm. The phase difference between them is $\pi/4$ rad. What is the resultant amplitude?

46. ★★ If we allow the phase difference between the two waves in problem 45 to vary, what are the smallest and largest possible displacement amplitudes?

47. ★ The sound level a distance of 23 m from an isotropic point source is 65 dB. What is the intensity, and what is the total power radiated by the source?

48. ★★★ B flat played on a tuba has a fundamental frequency of approximately 117 Hz.
  (a) Write equations for this fundamental and for the second and third harmonics with half the amplitude of the fundamental.
  (b) On the same graph, plot the waveform of the fundamental, the two harmonics, and the resultant of these three tones.
  (c) Compare the period and amplitude of the resultant to those of the fundamental.

49. ★★ Police radar uses the Doppler shift to determine whether you are speeding. The radar unit transmits a wave, which is reflected off your car and Doppler-shifted as a result. Rather than detecting the frequency shift directly, the radar receiver mixes the transmitted and received beams and measures the resulting beat frequency. For a radar unit that operates at 10 GHz and for a driving speed of 130 km/h, what beat frequency would result?

50. ★ An interesting demonstration of standing waves can be done by twirling one end of a flexible, corrugated hose that is open at both ends. The twirling motion causes air to be drawn through the hose, and air rushing over the corrugations creates turbulence in the hose. The turbulence excites waves. It is usually possible to get the three lowest tones from the hose. For a 1.50 m hose, what frequencies will be heard?

51. ★★ A bugle is an instrument that has an air column with a fixed length. Different notes can be achieved by exciting different harmonics in the column, not an easy task. The bugle actually acts like a tube with both ends open. For a certain length, L, a bugle will have a sequence of four consecutive harmonics with frequencies very nearly equal to the frequencies associated with the notes G_4, C_5, E_5, and G_5. Determine the length of the bugle.

52. ★★ A plane wave is normally incident on a boundary where the speed of sound changes. The wave travels from a medium in which the speed of sound is 1200 m/s to a medium in which the speed of sound is 400 m/s. The incident wave has a frequency of 440 Hz.
  (a) What is the period of a 440 Hz wave?
  (b) How far will a wave front travel in the second medium during the time from part (a)?
  (c) What is the wavelength in the second medium?
  (d) What is the frequency in the second medium? (Hint: Use the speed and the wavelength from part (c).)

53. ★★ The Bay of Fundy has some of the highest tides in the world, with a tidal range as large as 15 m. The bay behaves like a resonator that is open at one end and closed at the other. The Bay of Fundy is approximately 270 km long, and the tidal period is approximately 12.5 h. Assuming that the tide corresponds to the lowest-frequency standing wave, what is the wave speed in the Bay of Fundy?

54. ★ The air cavity inside stringed instruments is designed to have resonant frequencies that are close to the resonant frequencies that are produced by the instrument. For a 12 cm deep guitar body cavity, what are the frequencies of the standing waves that could form between the top and back of the guitar?

55. ★★ The reverberation time in a performance space is a measure of how long it takes a sound to fade away once it has stopped being produced. It is defined as the amount of time it takes the sound to drop by 60 dB. For an initial sound intensity of 0.2 W/m², what is the corresponding reduction in intensity after one reverberation time?

56. ★★★ A sound wave loses 10% of its intensity each time it reflects from a particular surface. How many reflections will be required to reduce the sound level by 3 dB?

57. ★★ To determine the depth of a well, you drop a stone into the well and hear the splash 2.8 s later. How deep is the well?

58. ★★ A jet produces a sound level of approximately 140 dB at a distance of 30 m. Assuming that your ear has an effective radius of about 2.0 cm, what is the total power received by your ear?

59. ★★★ One way of measuring the frequency of a sound is to make measurements of the interference pattern that is created when the sound is produced by two sources. You find that for two sources, the sound is loudest when you are equidistant from the sources. You then determine that there is a minimum in intensity when you are 2.8 m farther from one source than the other. What is the frequency of the source?

60. ★★ Standing waves in a music room can cause certain spots in the room to have lower volume levels. For a room with dimensions 4.00 m × 6.00 m × 2.40 m, what are the possible standing wave frequencies?

61. ★★ Standing waves can be set up in solid rods, and this is the basis for a popular demonstration. The rod, clamped in the middle, is stroked (sometimes a cloth is used) and begins to emit a tone. For a 120 cm aluminum rod in which the speed of sound is 6420 m/s, find
  (a) the fundamental wavelength on the rod
  (b) the fundamental frequency

62. ★★ A typical audio amplifier does not have the same gain at all frequencies. An amplifier is rated 100.0 W at 1.0 kHz. The manufacturer specifies that the output falls by 3.0 dB when the frequency output is 20 kHz. What power can the amplifier produce at 20 kHz?

63. ★★★ Show that we still get beats if we add a phase constant to each of the two waves in Equation 15-22.

64. ★★ You have been asked to design a set of pipes for an organ. Assuming that the pipes have one open end and one closed end, calculate the lengths you will need to go from 440 Hz (A_4) to 880 Hz (A_5).

65. ★ The H_α line for a particular galaxy is observed to occur at $\lambda' = 6800.0 \times 10^{-10}$ m. How fast is this galaxy receding from us, and, using Hubble's Law, how far away is it?

66. ★★★ Lasers are used to "cool" atoms down in atomic traps. The lasers are tuned to have a frequency that is slightly below an absorption frequency of the atom. If the atom is drifting toward the laser, the laser light appears to be Doppler-shifted up, and the atom can absorb a photon. The photon carries momentum in the direction of the beam and thus the atom slows down when it absorbs the photon, but it will be in an excited state. The atom will then emit a photon to get back down to its lowest-energy state, but this photon will be emitted in a random direction. Hence, 50% of the time the atom will slow further in the direction parallel to the beam and 50% of the time it will speed up again but, on average, it will slow down, since it always slows when it absorbs the photon.

    An Na atom is in such a trap and we hope to cool it using a transition that occurs at 589.6 nm. The atom is moving at 560 m/s. How much below the 589.6 nm wavelength should we tune the laser?

67. ★★★ You may have noticed that you can hear traffic on a busy highway from a very long distance away. Another example is thunder, which is the sound generated by a long column of lightning. In both these cases, the sound is not a point source but is best modelled as a line source.

In this chapter we argued that for a point source, the intensity drops off as $1/r^2$. The basis for this argument is that the sound energy is radiated isotropically. If you surround a point source with a sphere that is centred on the source, the radiated energy will be uniformly spread out over the surface of the sphere, which has an area of $\sim r^2$. Extend this reasoning to predict how the intensity would drop off versus distance from a very long line sound source such as a busy highway. Hint: Consider what type of surface you could surround a line with so that the sound energy is uniform over the surface.

DATA-RICH PROBLEM

68. ★★★ You have been hired by an environmental consulting firm to do a noise analysis for a quarry operation. The operation uses two trucks, a drill, and a crusher. The manufacturers of the equipment provided the specifications in Table 15-5 for the sound level of the various pieces of equipment. Local bylaws specify that the maximum sound level at the perimeter of the quarry property not exceed 85 dB. How close to the perimeter can the quarry operate all these devices simultaneously?

Table 15-5   Data for Problem 68

Equipment	Sound Level at 10 m
Truck	85 dB
Drill	110 dB
Crusher	110 dB

OPEN PROBLEM

69. ★★★ Many of us have heard the effect that can be produced by inhaling helium and speaking. The speaker's voice is shifted to higher frequencies. Discuss the physics behind this effect.

CHAPTER
16

Temperature and the Zeroth Law of Thermodynamics

When you have completed this chapter, you should be able to

LO1 Explain why we use macroscopic statistical measures to describe the bulk behaviour of matter.

LO2 List the three main states of matter and some of their associated physical characteristics.

LO3 Discuss the state variables for some systems.

LO4 Describe the relationship between the average speed of a particle in a gas and the pressure that the gas exerts on the walls of its container.

LO5 Define what is meant by the thermometric property and do calculations related to the linear expansion of solids.

LO6 Be familiar with some common temperature scales.

LO7 State the zeroth law of thermodynamics, and describe the concept of thermal equilibrium.

LO8 Use the ideal gas law to determine various properties of a gas.

LO9 Describe how a universal temperature scale can be based on an ideal gas thermometer.

LO10 Discuss the relationship between the temperature of a gas and the average mechanical energy of the atoms in that gas.

LO11 Qualitatively interpret a probability distribution.

LO12 Read a phase diagram.

The morning landscape in Figure 16-1 contains all the familiar phases of matter: solids in the trees and bushes, liquid in the tiny droplets of water in the fog, and invisible gases in the air.

On such a morning, you would find the grass and bushes covered with heavy dew. In addition to liquid water, fog also contains a lot of water vapour. So, water is present in two phases: liquid and gas. While this is a relatively common occurrence for water, we might wonder whether other substances exhibit this behaviour, and whether water can be present in all three states simultaneously.

Figure 16-1 Early morning fog in the Annapolis Valley, Nova Scotia.

Courtesy of Rebecca Dimock

16-1 The Need for a Macroscopic Description

In this chapter and the two that follow, we will examine the behaviour of large collections of particles, primarily from an energy perspective. As we shall see, these energies are related to the types of energies we have already encountered in our study of mechanics—kinetic and potential energy. However, unlike our study of mechanics, where we treated objects as either point particles or rigid bodies, the systems we will be considering contain large collections of particles, and we will no longer treat them as rigid bodies.

In mechanics, we studied single point particles and described their motion by specifying their positions and velocities as a function of time. For the systems we now wish to consider, this would be a formidable task. For example, a mole of gas contains on the order of 10^{23} particles. For each particle, assuming we can treat them as point particles, we would need to specify six numbers (the three components of both their position and velocity vectors) at every instant of time. Assuming we use 16 bits (2 bytes) to store each number, we would require 2×10^{23} bytes of storage. The storage capacity of a typical computer drive is on the order of 10^{12} bytes. To simply store this much data, we would need about 10^{11} such drives. This would correspond to about 30 drives for every single person on earth. This is clearly not practical, as that is what would be required to simply store the information at a single instant in time.

Rather than describing systems from such a microscopic point of view, we instead use a macroscopic point of view using variables such as temperature and **pressure**. For example, we can describe the state of a pure gas in terms of the number of particles in the system, the **volume** of space occupied, the pressure that the gas exerts on the walls of the container, and the temperature of the gas. As we shall see later in this chapter, the pressure and the temperature are both related to the microscopic details of the system. However, they are related to averages over all the particles in the system.

16-2 Solids, Liquids, and Gases

On Earth, matter is usually found in **solid**, **liquid**, and **gas** phases, as described in Chapter 12. Under suitable conditions, most materials can exist in all three phases. For example, water is commonly found as a solid (ice), a liquid, and a gas (steam). There is also a fourth phase, called plasma. Plasma is a highly ionized gas and is the predominant phase in the universe. The Sun, which contains over 99% of the mass in the solar system, is in a plasma phase.

Solids tend to be very rigid and do not compress or stretch readily. The atoms in solids tend to remain at fixed positions, but the atoms in a liquid and in a gas can drift around. However, the atoms in solids do vibrate around their equilibrium positions. In crystalline solids, the atoms are arranged in regular periodic arrays, and each atom is bonded to a set of adjacent atoms (Figure 16-2). Crystalline solids are usually composed of a collection of small crystals. Even microscopic individual crystals still contain a huge number of atoms and extend over a very large number of interatomic distances. Such materials are described as having long-range order. In some materials, such as diamond, the individual crystals are often large enough to be seen without magnification (Figure 16-3).

Liquid phases are usually somewhat less dense and more compressible than the solid phase of the same material. Liquids have no long-range order (Figure 16-4). Over short distances, the positions—and sometimes the relative orientations—of adjacent particles in liquids are correlated. As a liquid cools, the extent of this ordering increases, becoming long-range order at the freezing point. Liquids consist of either atoms or molecules.

solid

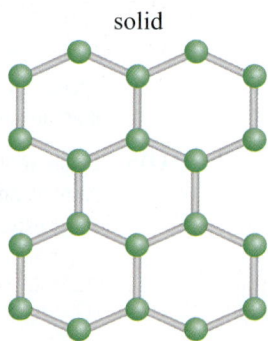

Figure 16-2 A periodic lattice from a solid. The atoms are closely spaced and form a periodic array.

Figure 16-3 A large single crystal of diamond. The faces of the diamond correspond to planes in the crystal structure.

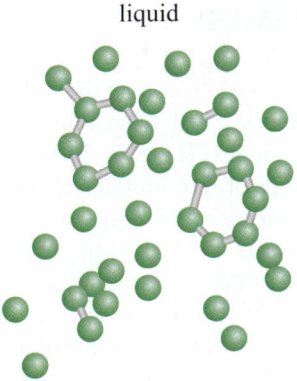

liquid

Figure 16-4 A liquid phase does not have long-range order, and the atoms are somewhat more loosely packed than atoms in the solid phase.

Gases are much less dense and much more compressible than the liquid and solid phases of the same material. Gases have no long- or short-range order, so the particles are completely free to move around (Figure 16-5). Like liquids, gases consist of either atoms or molecules. In the case of a multispecies gas, they can contain both atoms and molecules.

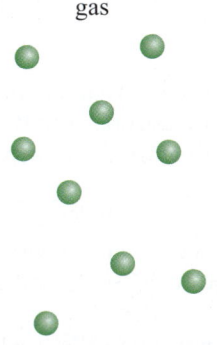

gas

Figure 16-5 The atoms in a gas are relatively far apart, and their positions are not related to each other.

CHECKPOINT 16-1

Importance of Interactions?

True or false: Because the particles in a gas have much larger interparticle spacing (due to the lower density) than a solid, the strength of the interaction between them is much weaker.

ANSWER: True. Interatomic forces depend upon the distance between particles and get weaker the farther apart they are.

16-3 State Variables

A key concept in thermodynamics is the **state** of a system. If we know all the forces that act on a particle, and we know the position and velocity of the particle at some instant, then we can, at least in principle, describe the motion of the particle for all time. This description would then specify the state of the particle. However, as we noted previously, this approach is not practical for our present purposes.

We can describe the state of a pure gas in terms of the number of particles in the system, the volume of space occupied, the pressure that the gas exerts on the walls of the container, and the temperature of the gas. We will discuss these **state variables** for gases in Section 16-8. Other types of systems use different variables. Table 16-1 gives a few examples. Notice that all the variables in Table 16-1 refer to macroscopic properties of the systems. Although in some instances these macroscopic properties are related to average values of the microscopic properties of individual particles, none of the state variables allow us to know the properties of specific individual particles.

CHECKPOINT 16-2

Are the Details in the State Variables?

Do state variables depend on all the details of the particles in a system?

ANSWER: Yes. For example, we will see that the pressure in a gas depends on the square of the average speed of the particles in the gas. To determine the average, you need to know the speeds of all the particles.

CHECKPOINT 16-3

Can the Details Be Extracted from the State Variables?

Can information about specific particles in a system be determined from the state variables?

ANSWER: No. The state variables usually allow us to determine some average values. However, it is not possible to extract information about single particles from the average values.

Table 16-1 State Variables for Various Types of Systems

System	State Variables			
Gas	Temperature	Volume	Pressure	Number of particles
Magnet	Temperature	Magnetization	Applied magnetic field	Number of magnetic dipoles*
Dielectric	Temperature	Polarization	Applied electric field	Number of electric dipoles*
*Chapters 19 and 24 describe electric and magnetic dipoles, respectively.				

CHAPTER 16 | **TEMPERATURE AND THE ZEROTH LAW OF THERMODYNAMICS**

The fundamentals of thermodynamics were developed during the Industrial Revolution in the late eighteenth and early nineteenth centuries. The original focus of thermodynamics was on the behaviour of gases (steam in particular) with the goal of improving the efficiency of steam engines (Figure 16-6).

The four thermodynamic state variables for a gas are the number of gas particles, the volume, the pressure, and the temperature. The number of particles in a gas is a straightforward quantity. Similarly, the volume occupied by a gas is relatively straightforward to understand and is usually simply determined by the size of the container holding the gas. However, the remaining two variables—pressure and temperature—are somewhat less obvious. We will first consider pressure.

Figure 16-6 Improving the efficiency of steam engines provided much of the impetus for the early work in thermodynamics.

16-4 Pressure

The pressure that a gas exerts on the walls of its container results from collisions of individual particles with the wall of the container (Figure 16-7). When a particle collides with the walls of its container, it exerts a force on the wall, and the wall exerts an equal but opposite force on the particle. The pressure is the sum of all of the forces on the wall of the container resulting from each individual collision, divided by the area of the wall of the container.

Consider a gas particle of mass m that is moving with velocity $\vec{v}_{\text{initial}} = v_x\hat{i} + v_y\hat{j} + v_z\hat{k}$ in a cubic container with sides of length L and volume $V = L^3$. When the particle collides with the wall of the container, we assume that the collision is elastic. Since energy is conserved in an elastic collision, the particle has the same speed before and after the collision, but one component of the particle's velocity has been reversed. For example, if a particle collides with a vertical wall that is perpendicular to the x-axis, the x-component of the velocity is reversed, and $\vec{v}_{\text{final}} = -v_x\hat{i} + v_y\hat{j} + v_z\hat{k}$. The change in momentum of the particle is then

$$\Delta\vec{p}_{\text{particle}} = \vec{p}_{\text{f}} - \vec{p}_{\text{i}} = m\vec{v}_{\text{final}} - m\vec{v}_{\text{initial}} = -2mv_x\hat{i}$$

(16-1)

Since momentum is conserved, the change in momentum of the container wall must be exactly equal but opposite to the change in momentum of the particle:

$$\Delta\vec{p}_{\text{wall}} = -\Delta\vec{p}_{\text{particle}} = 2mv_x\hat{i}$$

(16-2)

Equation 16-2 gives us the momentum imparted to the container wall due to a single collision with a single particle. To determine the pressure exerted by the gas, we also need to know the rate at which collisions are occurring, since we can relate the force to the rate of change of momentum.

The time between collisions between a given particle and the vertical wall is the time it takes the particle to cross the length of the container twice:

$$\Delta t = \frac{2L}{v_x}$$

(16-3)

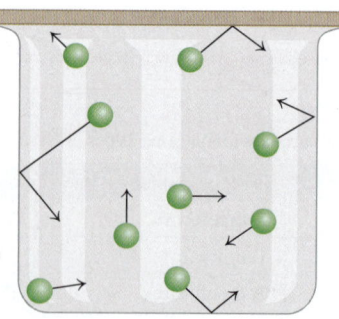

Figure 16-7 Gas atoms collide with the walls of a container.

PEER TO PEER

For an ideal gas system, I only need to know three state variables to know everything about the system. This means that I can calculate the fourth state variable if I know the three other state variables. For example, if I know the volume, temperature, and number of particles in a system, I can use the ideal gas law to determine the pressure. More complex systems are more difficult to describe and therefore require more state variables to do so.

Dividing Equation 16-2 by Equation 16-3, we obtain the rate of change of momentum of the wall:

$$\frac{\Delta p_{\text{wall}}}{\Delta t} = \frac{2mv_x}{\frac{2L}{v_x}} = \frac{mv_x^2}{L} \qquad (16\text{-}4)$$

As we saw in Chapter 7, we can write Newton's second law as $\vec{F} = \frac{\Delta \vec{p}}{\Delta t}$, so

$$F_x = \frac{\Delta p_x}{\Delta t} = \frac{mv_x^2}{L} \qquad (16\text{-}5)$$

To determine the total force on the wall, we sum this expression over all the particles to get

$$\begin{aligned}
F_{x,\text{Total}} &= \sum_i F_{x,i} \\
&= \sum_i \frac{mv_{x,i}^2}{L} \\
&= \left(\frac{mv_{x,1}^2}{L} + \frac{mv_{x,2}^2}{L} + \frac{mv_{x,3}^2}{L} + \cdots + \frac{mv_{x,N}^2}{L} \right) \\
&= \frac{m}{L}(v_{x,1}^2 + v_{x,2}^2 + v_{x,3}^2 + \cdots + v_{x,N}^2)
\end{aligned} \qquad (16\text{-}6)$$

where N is the total number of particles in the box.

We note that the average value of the square of the x-components of the velocities is given by

$$(v_x^2)_{\text{avg}} = \frac{1}{N}\left(v_{x,1}^2 + v_{x,2}^2 + v_{x,3}^2 + \cdots + v_{x,N}^2 \right) \qquad (16\text{-}7)$$

Rearranging this equation gives

$$N(v_x^2)_{\text{avg}} = (v_{x,1}^2 + v_{x,2}^2 + v_{x,3}^2 + \cdots + v_{x,N}^2) \qquad (16\text{-}8)$$

Thus, the total force exerted on the wall of the container is

$$F_{x,\text{Total}} = \frac{m}{L}N(v_x^2)_{\text{avg}} \qquad (16\text{-}9)$$

To determine the pressure that the gas exerts on the wall, we divide the total force by the area of the wall to obtain

$$P = \frac{\frac{m}{L}N(v_x^2)_{\text{avg}}}{L^2} = \frac{mN(v_x^2)_{\text{avg}}}{L^3} \qquad (16\text{-}10)$$

We can generalize this expression to three dimensions. For any set of velocities,

$$(v^2)_{\text{avg}} = (v_x^2)_{\text{avg}} + (v_y^2)_{\text{avg}} + (v_z^2)_{\text{avg}} \qquad (16\text{-}11)$$

Because the container is cubical and there is nothing special about the x-, y-, or z-directions (if we assume there are no external forces acting on the gas), it must be true that

$$(v_x^2)_{\text{avg}} = (v_y^2)_{\text{avg}} = (v_z^2)_{\text{avg}} = \frac{(v^2)_{\text{avg}}}{3} \qquad (16\text{-}12)$$

Finally, recalling that $V = L^3$, we can write

KEY EQUATION
$$P = \frac{1}{3}\frac{mN(v^2)_{\text{avg}}}{V} \qquad (16\text{-}13)$$

CHECKPOINT 16-4

Pressure-Speed Connection in a Gas

If we observe that the pressure in a sample of gas changes, can we conclude that the square of the average speed has changed correspondingly?

ANSWER: No. Either the number of particles or the volume that the gas occupies could have changed instead.

You may have questioned the assumption that particles colliding with the container emerge from the collision with their speeds unchanged. Implicit in this assumption is that the velocity of the container did not change as a result of the collision. This situation might appear to violate conservation of momentum. However, gas molecules collide with all the walls of the container. For every particle that collides with the wall on the right, there is a particle with the same speed that collides with the wall on the left. If this were not the case, the total momentum of the gas would be changing, and the centre of mass of the gas would move. In other words, the net momentum transfer resulting from the random motion of the huge number of gas particles averages to zero.

You might also wonder about the assumption implicit in Equation 16-3, which states that the time for the particle to go from one side of the container to the other is equal to the length of the container divided by the speed of the particle. This relationship is only strictly true for a single particle that does not change its velocity during this time. In fact, a particle in a gas collides with other particles, and its velocity does change. However, our assumption is still valid overall because the gas is in a state of equilibrium, with constant volume, pressure, and temperature. To maintain the state of equilibrium, it must be true that for every collision that causes a particle with speed v_x to change its speed to v_x', there must be a corresponding collision that causes a particle with speed v_x' to change its speed to v_x. Otherwise, the total momentum of the gas would change, and the gas would not be in a state of equilibrium. *On average*, there is a constant number of particles with a given speed v_x, and that number of particles will, *on average*, cross the container in the time L/v_x.

CHAPTER 16 | TEMPERATURE AND THE ZEROTH LAW OF THERMODYNAMICS

THERMODYNAMICS

EXAMPLE 16-1

How Fast Are the Molecules Going?

At standard temperature (0 °C) and pressure (101.325 kPa), or STP, a mole of gas occupies a volume of 22.4 L. The molar mass of an atom of nitrogen is approximately 14 g/mol. Nitrogen is diatomic, that is, its atoms bond in pairs to form N_2 molecules. Estimate the average speed of a nitrogen molecule at STP.

Solution

Since each nitrogen molecule contains two atoms, a mole of nitrogen gas has a mass of approximately 28 g. Solving Equation 16-13 for the speed, we get

$$(v^2)_{avg} = \frac{3PV}{mN}$$

$$\sqrt{(v^2)_{avg}} = v_{rms} = \sqrt{\frac{3PV}{mN}}$$

where v_{rms} is the root-mean-square velocity, that is, the square root of the average (or mean) of the squares of the individual velocities.

Substituting in the known values, and converting litres to cubic metres (1 L = 0.001 m³), we get

$$v_{rms} = \sqrt{\frac{3 \times 101.325 \times 10^3 \, \text{kg} \cdot \text{m}^{-1}\text{s}^{-2} \times 0.0224 \, \text{m}^3}{0.028 \, \text{kg}}}$$

$$= \sqrt{2.43 \times 10^5 \, \text{m}^2\text{s}^{-2}} = 490 \, \text{m/s}$$

Making sense of the result

We can use other data to make a second estimate. It takes 200 kJ of energy to convert 1.0 kg of liquid nitrogen to gaseous nitrogen. If we assume that all of this energy goes into kinetic energy, then

$$v = \sqrt{\frac{2E}{m}} = \sqrt{\frac{2 \times 2 \times 10^5 \, \text{J/kg}}{1 \, \text{kg}}} = 630 \, \text{m/s}$$

Thus, our result seems reasonable.

16-5 Temperature and Thermal Expansion

Temperature is something that we all discuss on a daily basis: What is the temperature today? At what temperature should the oven be set to cook a particular dish? In fact, thermometry (the measurement of temperature) was so taken for granted that the law of thermodynamics that deals with it was only developed after the formulation of the other laws. Since these had already been assigned to the first and second laws, thermometry was assigned the zeroth law. Before we attempt to relate temperature to the microscopic properties of a system, we will examine how temperature is measured.

To measure temperature, we use a thermometric property of some object. One common thermometer is the alcohol thermometer, which relies on the thermal expansion of alcohol: as we warm alcohol, it expands. Many other materials with thermometric properties are used in thermometers; some of them are listed in Table 16-2.

Thermal Expansion of Solids

Most solids expand when warmed. For a long, thin piece of solid material, the expansion occurs principally along the long axis of the material.

If we do not change the temperature of the material too much, we can assume that the expansion is linear, that is, the expansion is directly proportional to the temperature change. Since the molecular bonds in the material expand equally, the total amount of expansion is also proportional to the length of the material. For a given temperature change, the increase in the length of a piece of metal that is 100 cm long is twice the increase in a 50 cm length of the same material.

For linear thermal expansion, the coefficient of thermal expansion, α, is defined as

KEY EQUATION

$$\alpha = \frac{1}{L}\frac{\Delta L}{\Delta T} \tag{16-14}$$

where ΔL is the observed increase in length, L is the length of the object at the initial temperature, and ΔT is the temperature change.

To calculate the change in length of a sample, we rearrange Equation 16-14:

$$\Delta L = L\alpha\Delta T \tag{16-15}$$

Because the expansion depends on the details of the bonds between the atoms in the solid, different materials expand to a greater or lesser extent. Table 16-3 lists some coefficients of linear thermal expansion for various materials.

When designing devices that are made with different materials, it is important to keep in mind these differing coefficients to ensure that the devices function properly across the range of temperatures in which they are expected to operate.

Table 16-2 Various Common Temperature-Measuring Devices and Their Associated Thermometric Properties

Temperature-Measuring Device	Thermometric Property
Galinstan*	Thermal expansion
Bimetallic strip	Differential thermal expansion
Thermocouple	Thermoelectric effect
* Galinstan is a liquid alloy used as a substitute for highly toxic mercury.	

MAKING CONNECTIONS

Using Physics in the Kitchen

When we have a jar that is difficult to open, it can help to run hot water over the lid of the jar. The metal in the lid expands, making it easier to unscrew.

Table 16-3 Coefficients of Linear Thermal Expansion for Various Materials

Material at 20 °C	Coefficient of Thermal Expansion (10^{-6}/°C)
Aluminum	23
Brass	19
Carbon steel	10.8
Concrete	12
Copper	17
Diamond	1
Glass	8.5
Gold	14
Iron	11.1
Lead	29
Pine (perpendicular to the grain)	34
Stainless steel	17.3
Water	~69

EXAMPLE 16-2

Thermal Expansion of a Rod

A long, thin rod of aluminum is measured to have a length of 0.250 m at 20.0 °C. By how much will the length of the rod increase if it is heated to 30.0 °C?

Solution

This is a simple application of the linear expansion Equation 16-15. According to Table 16-3, the coefficient of linear thermal expansion for aluminum is 23.0×10^{-6}/°C. We use Equation 16-15 to find the change in length:

$$\Delta L = L\alpha\Delta T$$
$$= (0.250 \text{ m})(23.0 \times 10^{-6}/°C)(30.0 °C - 20.0 °C)$$
$$\Delta L = 5.75 \times 10^{-5} \text{ m}$$

Thus, at 30.0 °C, the change in length of the rod will be 0.0575 mm.

Note that if we had cooled the rod by 10 °C, the change in length would have been negative.

MAKING CONNECTIONS

Sea Level Rise: Melting Ice or Thermal Expansion?

As the temperature of the atmosphere increases, so too does the temperature of the ocean, although it takes some time for the lower levels of the ocean to warm up (Figure 16-8). As the oceans warm, they also expand. Some researchers predict that this thermal expansion will make the largest contribution to sea level rise caused by global warming.

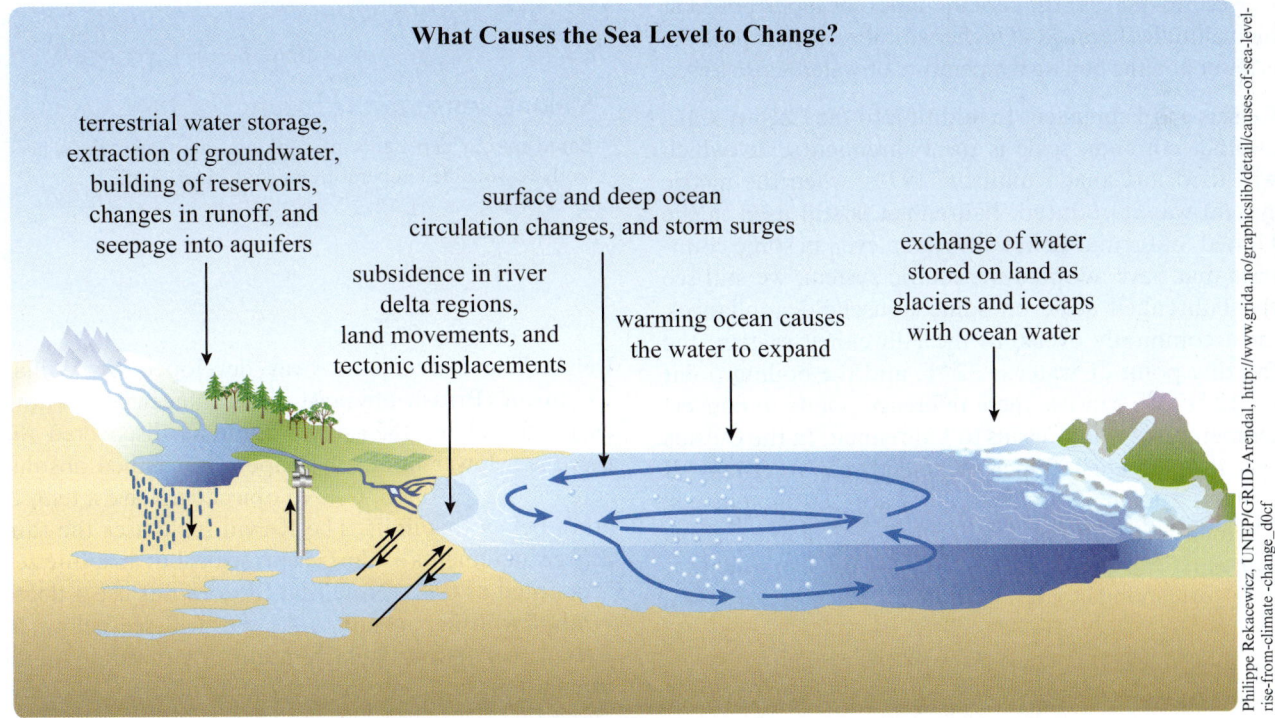

What Causes the Sea Level to Change?

terrestrial water storage, extraction of groundwater, building of reservoirs, changes in runoff, and seepage into aquifers

surface and deep ocean circulation changes, and storm surges

subsidence in river delta regions, land movements, and tectonic displacements

warming ocean causes the water to expand

exchange of water stored on land as glaciers and icecaps with ocean water

Philippe Rekacewicz, UNEP/GRID-Arendal. http://www.grida.no/graphicslib/detail/causes-of-sea-level-rise-from-climate-change_d0cf

Figure 16-8 Thermal expansion and global warming.

(continued)

CHAPTER 16 | **TEMPERATURE AND THE ZEROTH LAW OF THERMODYNAMICS**

THERMODYNAMICS

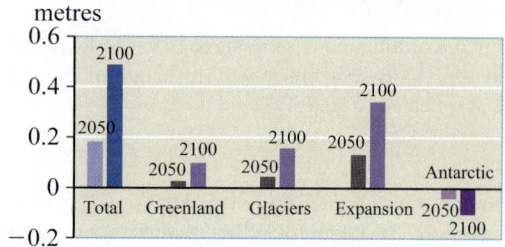

Components of Mean Sea Level Rise

Figure 16-8 Continued

Various assumptions need to be made to model sea level rises. The model depicted assumes rapid economic growth, a global population that peaks in the twenty-first century and then declines, and the rapid introduction of new and more efficient technologies. The model also assumes a more global economy with a substantial reduction in regional differences in per capita incomes. Another key assumption is what energy sources we rely on, and this model assumes a fossil fuel–intensive energy supply.

16-6 Thermometers and Temperature Scales

Temperature-sensitive properties of materials are the basis for various types of thermometers, which convert measurements of physical quantities such as volume and resistance into temperature readings. To quantify the measurements, we must assign a temperature scale to the readings of our thermometer. The most common scale in the world today is the Celsius scale, which assigns a temperature of 0 °C to the freezing point of water and 100 °C to the boiling point. In practice, we also need to specify the pressure for these calibration points because many of them also depend on the pressure. For example, we need to increase the cooking times for foods boiled at high altitudes because, at higher altitudes, the air pressure is lower and the boiling temperature of water is also lower.

Celsius and Fahrenheit In addition to the Celsius scale, another common scale is the Fahrenheit scale, which was used in Canada until the 1970s, when the metric system was introduced. Fahrenheit is still used in the United States and Belize. However, even in some countries that have adopted the metric system, we still see the Fahrenheit scale on some household appliances, most commonly ovens. In the Fahrenheit system, the freezing point of water is 32 °F, and the boiling point is 212 °F. We can use these reference points to convert temperatures from Celsius to Fahrenheit. In the Celsius system there are 100 increments between freezing and boiling, and in Fahrenheit there are 180 increments. Therefore, the ratio of a Fahrenheit degree to a Celsius degree is 180 °F/100 °C = 9 °F/5 °C. We also note that 0 °C corresponds to 32 °F. Thus,

$$°F = \frac{9\ °F}{5\ °C} \times °C + 32\ °F$$

$$°C = \frac{5\ °C}{9\ °F} \times (°F - 32\ °F)$$

(16-16)

Kelvin The Kelvin scale was developed by William Thomson (British physicist and engineer, also known as Lord Kelvin, 1824–1907) after he discovered that there is a lower limit for temperature, called absolute zero (about −273.15 °C). No object can have a temperature below this limit. The Kelvin scale uses the same degree increments as the Celsius scale, but absolute zero is denoted as 0 K. (Note that SI does not use a degree symbol with kelvin units.) As you will see below, the Kelvin scale is particularly useful for thermodynamic calculations and predicting the behaviour of gases. Table 16-4 lists some common temperatures in Celsius, Kelvin, and Fahrenheit.

Table 16-4 Some Common Temperatures Listed in Celsius, Kelvin, and Fahrenheit

	Celsius	Kelvin	Fahrenheit
Absolute zero	−273.15 °C	0 K	−459.67 °F
Freezing point of water	0 °C	273.15 K	32 °F
Typical normal body temperature	37 °C	310.15 K	98.6 °F
Room temperature	20 °C	293.15 K	68 °F
Surface temperature of the Sun	~6000 °C	~6273 K	~10 832 °F

16-7 The Zeroth Law of Thermodynamics

Suppose we place a thermometer in thermal contact with an object that is at some steady temperature. We will assume that our thermometer is so small that it does not affect the temperature of the object. The thermometer reading will change when we first establish contact with the object, but the reading will eventually reach a steady state. At this point, we say that the thermometer is in thermal equilibrium with the object. Two objects are said to be in **thermal equilibrium** when placing them in thermal contact with each other causes no change in the temperature of either object—they are at the same temperature.

If we now place the thermometer in thermal contact with a second object and find that the thermometer stabilizes at the same reading, we can say that the thermometer is in thermal equilibrium with the second object.

If we then place the first and second objects in thermal contact with each other, the reading on our thermometer will not change, regardless of which object it contacts. This observation demonstrates the **zeroth law of thermodynamics**: If object A is in thermal equilibrium with object B, and object A is also in thermal equilibrium with object C, then object C is also in thermal equilibrium with object B.

In other words, if we get the same temperature reading on a thermometer from two different objects, then those objects are at the same temperature, and if we put them together, their temperatures will not change.

CHECKPOINT 16-5

Wood or Metal: Which Is "Warmer"?

Consider a metal object and a wooden object sitting on a desk. Typically, the metal object feels cool to the touch, but the wooden object does not. Yet, the objects have been sitting side by side in the same room and should be at the same temperature. Why do they not feel equally warm?

ANSWER: When we touch the wood or the metal, thermal energy is transferred from our hand to the object, because our hand is warmer than the object. Wood has a very poor ability to conduct thermal energy, so the spot we are touching quickly warms up to about the same temperature as our hand. Metal, however, is a very efficient thermal conductor, and the energy from our finger rapidly spreads into the entire object. The spot we are touching does not warm up much and therefore feels cooler than the wood.

CHECKPOINT 16-6

Holding Hands

When we hold hands with another person, we notice that the other person's hand feels warm, even if the temperature of the other person is exactly the same as ours. Are we in thermal equilibrium with the other person? If so, why does the person's hand feel warm to us?

ANSWER: Our hand is in thermal equilibrium with the other person's hand if that person is at the same temperature as us. Our skin is usually in contact with air that is cooler than our body temperature. As a result, thermal energy is constantly flowing out of our bodies into our surroundings, and our skin is at some temperature that is lower than our body's core temperature. Further, we have evolved to feel comfortable (i.e., neither hot nor cold) at "room temperature," so when we hold hands, the temperature of the surface of our skin is elevated above its normal temperature, and that feels warm to us.

16-8 Ideal Gases

One difficulty with the system of thermometry we have described so far is getting two thermometers to agree with each other. While identical thermometers are consistent with each other, two different thermometers typically agree only at common calibration points. Much experimentation has been done to develop thermometers that are accurate over their full range. The discrepancies between various types of thermometers are largely due to differences in the materials used to construct them. The particles in solids and liquids are relatively close together and interact strongly with each other. These interactions differ in the various materials and affect how the materials respond to changes in temperature. Consequently, solids and liquids differ both in the linearity of their response and in the temperature range over which the nonlinear response becomes significant.

Gases, however, have greater interatomic distances and, consequently, much weaker interactions between particles. The behaviour of gases is much more weakly dependent on the type of gas being studied than is the case for solids and liquids. Table 16-5 compares the densities and interatomic distances for ice, liquid water, and water vapour.

Much experimentation has been done with gases to determine the relationships between their pressure, temperature, and volume. The motivation for some of these studies was not to develop better thermometers, but to understand and improve the hot-air balloon!

Table 16-5 Densities and Interatomic Distances for Phases of Water

Material	Density (g/cm³)	Typical Interatomic Distance (nm)
Ice at −10 °C	0.998	~0.3–0.4
Water at 40 °C	0.992	~0.3–0.4
Water vapour at 100 °C and 1 atm	0.000 59	~3–4

Gas Properties

The ideal gas law states that $P^aV^b \propto N^cT^d$, where P is the pressure, V is the volume, N is the number of particles in the gas, T is the temperature in kelvins, and a, b, c, and d are all integers. In this activity, you will use the PhET simulation "Gas Properties" to determine a, b, c, and d.

It was found that all gases exhibit the same behaviour at low densities, regardless of the chemical composition of the gas. This behaviour is described by the **ideal gas law**:

KEY EQUATION
$$PV = NkT \qquad (16\text{-}17)$$

where P is the pressure, V is the volume, N is the number of particles in the gas, T is the temperature in kelvins, and k is Boltzmann's constant, $1.38 \times 10^{-23}\ \text{J}\cdot\text{K}^{-1}$ (named for the Austrian physicist Ludwig Boltzmann, 1844–1906).

EXAMPLE 16-4

Using the Gas Law

The pressure in a 5.00 L flask of gas is 200. kPa, and the temperature is 300. K. Assuming that the gas is ideal, how many gas particles are in the flask?

Solution

We can use the ideal gas law and solve for N:

$$N = \frac{PV}{kT}$$

We can now substitute in the known quantities. However, we must be careful to convert the units properly:

$$1\ \text{Pa} = 1\ \text{N/m}^2 = 1\ \text{kg}\cdot\text{m}^{-1}\cdot\text{s}^{-2}$$

$$1\ \text{J} = 1\ \text{N}\cdot\text{m} = 1\ \text{kg}\cdot\text{m}^2\cdot\text{s}^{-2};\ 1\ \text{L} = 1000\ \text{cm}^3 = 10^{-3}\ \text{m}^3$$

Using these conversions, we find that

$$N = \frac{PV}{kT} = \frac{(2.00 \times 10^5\text{kg}\cdot\text{m}^{-1}\text{s}^{-2})(5.00 \times 10^{-3}\text{m}^3)}{(1.38 \times 10^{-23}\text{kg}\cdot\text{m}^2\text{s}^{-2}\text{K}^{-1})(300.\,\text{K})}$$

$$= 2.42 \times 10^{23}$$

Making sense of the result

At first glance, the result seems like a huge number. However, one mole of gas at standard temperature (0 °C) and pressure (101.325 kPa) occupies 22.4 L. We have arrived at a result of 2.42×10^{23} particles, which is about one third of a mole. Our sample is at approximately twice standard pressure but somewhat less than one quarter of the volume, so we would expect to have approximately 0.5 mol of gas. However, our temperature is higher than STP, so we should have slightly less than 0.5 mol. Thus, our result seems quite reasonable.

Moles versus Number of Particles

Chemists often write the ideal gas law in terms of the number of moles of gas that are present. The amount (number of moles in the gas) is the number of particles in the gas expressed as a multiple of Avogadro's number, 6.02×10^{23}. Avogadro's number, N_A, is defined as the number of atoms in exactly 12 g of carbon-12. Denoting the amount as n, we can then write

$$PV = nN_AkT = nRT$$

where

$$R = N_Ak = 8.314\,472\ \text{J}\cdot\text{K}^{-1}\cdot\text{mol}^{-1}$$

Note that real gases behave like ideal gases only at very low densities. Nonetheless, it is somewhat remarkable that all gases do behave the same at low densities, and we will devote some time to understanding why that is so.

Recall that gases are much less dense than liquids and solids and that the typical distance between particles in a gas is large. Consequently, gas particles in a container move about with relatively few interactions with the other gas particles. As we reduce the density of the gas, the fraction of the time that a particle spends interacting with other particles becomes smaller and smaller. As a result, the details of those interactions become less and less significant in determining the behaviour of the gas.

Figure 16-9 shows a generic interatomic potential. At large separations, the interaction energy goes to zero. At small separations, the interaction energy is very large and positive: the particles strongly repel each other. The minimum in the potential corresponds to the equilibrium bond length, which is what we observe in a molecule or a solid. The particles in a low-density

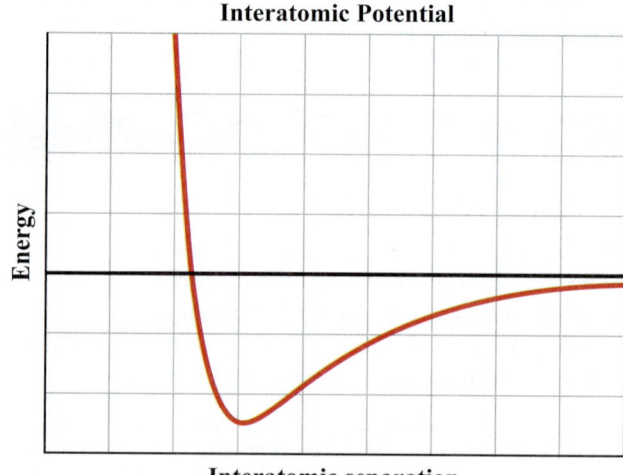

Interatomic Potential

Energy (vertical axis) · **Interatomic separation** (horizontal axis)

Figure 16-9 A typical interatomic potential.

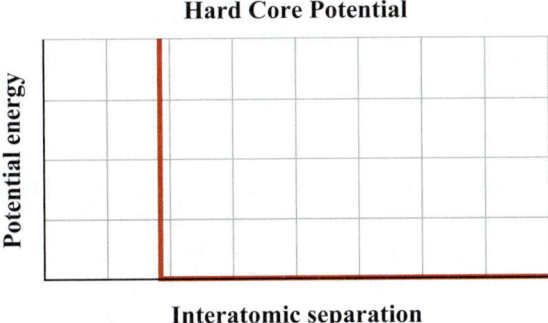

Figure 16-10 The hard core approximation for the interatomic potential for a dilute gas. The particles experience an infinitely strong repulsive force when they collide but do not interact otherwise.

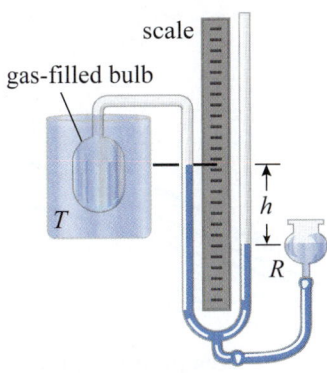

Figure 16-11 A constant-volume gas thermometer.

gas spend most of their time at large interparticle distances. When they do collide, the large repulsive force dominates the interaction. Consequently, a reasonable approximation to this potential, called the hard core approximation, is shown in Figure 16-10.

Even though the approximate potential in Figure 16-10 does not have a minimum, this is not significant for a dilute gas, because the minimum is responsible for the attractive forces that ultimately produce the liquid and solid phases, which are not observed to occur in a dilute gas. Another way of thinking about it is that in a dilute gas, the particles do not spend a lot of time close to each other; thus, the short-range details of the potential are not important.

CHECKPOINT 16-7

Gases

True or false: Since the interactions between gas particles get weaker as the gas becomes more dilute, we expect all gases to behave in the same way in the dilute limit.

ANSWER: True. At large interatomic distances, all of the potentials approach zero.

It is possible to calculate the equation of state for a system of particles that interact through the hard sphere potential, although the details are beyond the scope of this book. The result is the ideal gas law, $PV = NkT$.

16-9 The Constant-Volume Gas Thermometer

The discovery that all gases have the same behaviour in the low-density limit provided the basis for a universal thermometer against which all other thermometers could be calibrated: the **constant-volume gas thermometer**. This thermometer measures the temperature of a gas-filled bulb. In Figure 16-11, this bulb is

shown submerged in a liquid of unknown temperature. The bulb on the right is filled with mercury, and its height can be adjusted so that the mercury in the left-hand tube is brought to the zero mark, keeping the volume of gas constant. The difference in height between the left- and right-hand tubes can then be used to determine the pressure in the gas-filled bulb.

CHECKPOINT 16-8

What Is Held Constant?

In addition to the volume, what other state variable is held constant in the constant-volume gas thermometer?
(a) temperature
(b) pressure
(c) number of gas particles

ANSWER: (c) Since the gas is sealed in the thermometer, the number of gas particles also remains constant.

If the gas in the bulb is behaving as an ideal gas, then the temperature of the liquid will be proportional to the pressure because we are keeping the volume and quantity of the gas constant:

$$T = CP \qquad (16\text{-}18)$$

where C is a constant that must be determined.

The pressure in the bulb is given by

$$P = P_{atm} - \rho_{Hg}gh \qquad (16\text{-}19)$$

where P_{atm} is the atmospheric pressure, ρ_{Hg} is the density of mercury, g is the acceleration due to gravity, and h is the height difference between the two mercury columns.

To determine C, we place the bulb at some known temperature, say the triple point of water (see Section 16-12), and we measure the pressure. At this point we find that

$$T_3 = CP_3 \qquad (16\text{-}20)$$

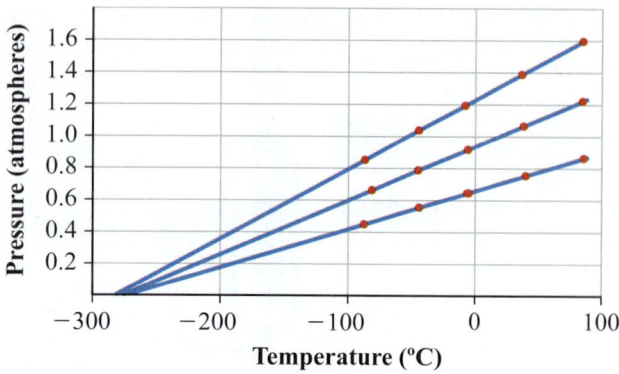

Figure 16-12 The three lines represent running the experiment with successively smaller quantities of gas in the system. The lines all go to zero at approximately −273 °C.

Eliminating C from Equations 16-18 and 16-20, we find

KEY EQUATION
$$T = T_3 \frac{P}{P_3} = (273.16 \text{ K}) \frac{P}{P_3} \tag{16-21}$$

Carrying out this procedure with different gases, we find that we get consistent results only at very low densities. By making a series of measurements with successively smaller densities, we can find a limit that gives a precise value for the temperature:

$$T = (273.16 \text{ K}) \lim_{\substack{\text{gas} \\ \text{density} \to 0}} \frac{P}{P_3} \tag{16-22}$$

It was discovered that for all gases at the low-density limit, the pressure goes to zero at the same temperature: 0 K. Figure 16-12 shows data confirming this discovery.

16-10 Temperature and Mechanical Energy

We now return to the question of the relationship between the temperature—a macroscopic thermodynamic variable—and the molecular-level details of the system under consideration.

Recall that for a gas we showed that

$$P = \frac{1}{3} \frac{mN(v^2)_{\text{avg}}}{V} \tag{16-13}$$

However, the ideal gas law can be solved for the pressure:

$$P = \frac{NkT}{V} \tag{16-23}$$

Eliminating P from Equations 16-13 and 16-23 and solving for kT, we obtain

$$kT = \frac{VmN(v^2)_{\text{avg}}}{3NV} = \frac{1}{3}m(v^2)_{\text{avg}} = \frac{2}{3}\left(\frac{1}{2}m(v^2)_{\text{avg}}\right) \tag{16-24}$$

Note that the product kT has units of energy. The last term in parentheses in Equation 16-24 is the average kinetic energy of a particle in the gas. Rearranging Equation 16-24 yields

KEY EQUATION
$$\frac{3}{2}kT = \frac{1}{2}m(v^2)_{\text{avg}} \tag{16-25}$$

Equation 16-25 provides us with an elegant interpretation of the temperature of a gas: the **temperature** of a gas is a measure of the average kinetic energy of the gas particles.

CHECKPOINT 16-9

Temperature and Energy?

Suppose we have two samples of the same gas. One sample has twice the number of particles and occupies twice the volume but is at the same pressure as the smaller sample. If the total kinetic energy of the larger sample is twice the total kinetic energy of the smaller sample, will the temperature also be twice that of the smaller sample?

ANSWER: No. The temperature is a measure of the average kinetic energy. The larger sample has twice the total kinetic energy but that energy is shared among twice the number of particles. Therefore, the average kinetic energy, and hence the temperature, are the same.

Equipartition of Energy

We now consider the factor of 3 that appears on the left-hand side of Equation 16-25. It originated in Equation 16-13, which we derived by arguing that

$$(v_x^2)_{\text{avg}} = (v_y^2)_{\text{avg}} = (v_y^2)_{\text{avg}} = \frac{(v^2)_{\text{avg}}}{3} \tag{16-12}$$

The factor of 3 results from the fact that there are three directions in which the gas particles can move. If the gas were confined to the xy-plane, we would have concluded that

$$(v_x^2)_{\text{avg}} = (v_y^2)_{\text{avg}} = \frac{(v^2)_{\text{avg}}}{2} \tag{16-26}$$

In both cases, the energy in the system is equally divided among the degrees of freedom that the system has. The number of ways in which a system can absorb energy equals the number of degrees of freedom for the system. For an ideal gas in a container, there are three directions in which the gas particles can move, and the kinetic energy associated with those three directions is equal. The energy is equally partitioned between the three directions.

Energy considerations are more complex for a gas in which the particles are molecules instead of single atoms. For example, the atmosphere is composed of approximately 78% nitrogen gas, which consists of nitrogen molecules, N_2. A gas molecule has three degrees of freedom associated with its translational kinetic energy. However, the molecule can also absorb energy in other ways, which gives it more degrees of freedom than a single atom.

A nitrogen molecule, or any diatomic molecule, can rotate and vibrate. In Chapter 8, you learned that the kinetic energy associated with a rotating object is

$$K_{rot} = \frac{1}{2}I\omega^2 \qquad (16\text{-}27)$$

where I is the moment of inertia about the axis of rotation and ω is the angular velocity.

For a diatomic molecule, the rotational inertia about the axis that joins the two atoms is negligible. However, there is a significant moment of inertia about the two axes that are perpendicular to the axis that joins the two atoms. Since neither of these two rotational axes is preferred, the energy is evenly distributed between them. Thus, a diatomic gas molecule has two additional degrees of freedom associated with the rotational kinetic energy.

There is also energy associated with vibrations in a diatomic gas. The chemical bond that joins the two atoms is not infinitely rigid, and it has a potential as shown in Figure 16-13. This potential curve is called the Lennard–Jones potential. The equilibrium separation corresponds to the minimum in the potential. However, the minimum in the potential is a stable equilibrium point. For reasonably small excursions from the equilibrium point, the potential can be modelled as a parabolic well.

As we know from Chapter 14, a parabolic potential is characteristic of a simple harmonic oscillator. Associated with the oscillator are two energies: the kinetic energy and the potential energy of vibration. Thus, a diatomic gas molecule has two additional degrees of freedom associated with its vibrational motion.

The effects of these additional degrees of freedom are observed in experimental data, but only at sufficiently high temperatures. A full understanding of molecular energy requires quantum theory, which is introduced in Chapter 32.

MAKING CONNECTIONS

Lower Dimensions

Are there always three degrees of translational kinetic energy in a gas? Can atoms and molecules be confined to an xy-plane? In fact, there are various techniques for making such systems. One involves creating layered structures of semiconducting materials. At the interface between the two appropriately chosen materials, a large potential well can be created that confines the electrons in a very narrow region at the interface, as illustrated in Figure 16-14.

The electrons are free to move parallel to the interface. In the direction perpendicular to the interface, the confinement is such that the laws of classical physics no longer apply, and we must use quantum mechanics. (Strictly speaking, we need to use quantum mechanics in the other direction as well, but the quantization effects are very small.) The result is that the energy for motion in the z-direction becomes quantized, with only certain discrete values allowed. If we cool such a device to very low temperatures, the electrons are all forced into the lowest-energy level for motion perpendicular to the interface while remaining relatively free to move in the xy-plane parallel to the interface. Studies of such a system in strong magnetic fields led to the discovery of the integer quantum Hall effect. In 1985, the Nobel Prize in Physics was awarded to German physicist Klaus von Klitzing (b. 1943) for this discovery.

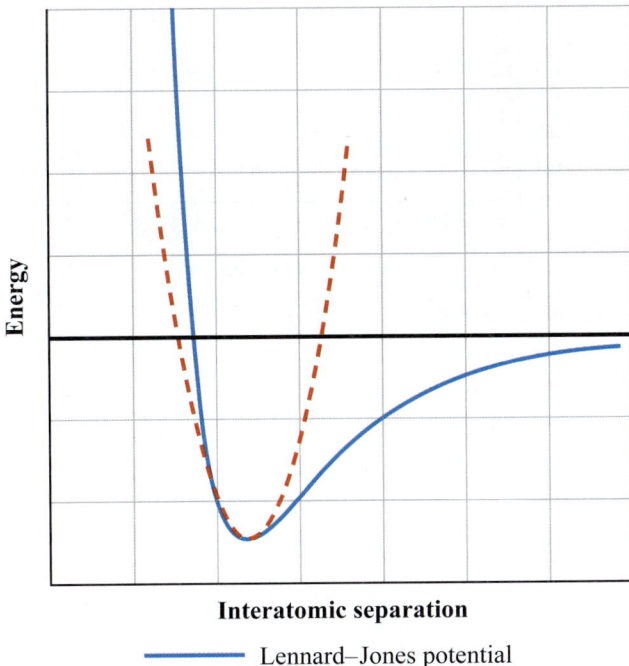

Interatomic separation

— Lennard–Jones potential
- - - Parabolic approximation

Figure 16-13 The interatomic potential in a diatomic molecule. The solid curve represents the Lennard–Jones potential, and the dashed curve shows a parabolic approximation.

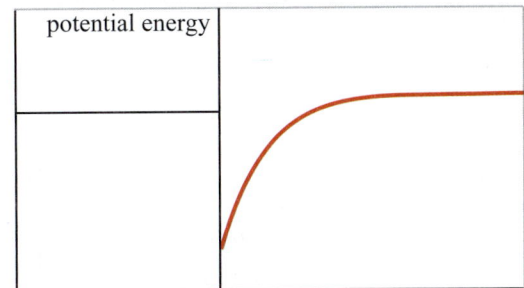

Figure 16-14 Deep potential well near the sample surface.

16-11 Statistical Measures

When we calculated the average speed for particles in a gas, we were using a statistic to describe the state of the gas. How accurately does a statistic such as average speed represent the actual behaviour of the gas? In other words, what is the probability that the theoretical average we have calculated matches the actual behaviour of a sample of gas particles in a laboratory experiment?

Suppose we want to determine whether a coin is fair, that is, balanced such that there is an equal likelihood of getting either heads or tails when we toss the coin. To quantify the results of the coin toss, we count heads as 1 and tails as 0. If the coin is fair, the average value of a large number of coin tosses should be 0.5. The results of such an experiment are shown in Figure 16-15. We find that the running average gives a reliable description of the coin only when the number of coin tosses is quite large.

Similarly, there must be a large number of particles in a system for a statistical thermodynamic description to apply. As the number of particles in the sample increases, so does the probability that the statistics accurately describe the sample. As the number

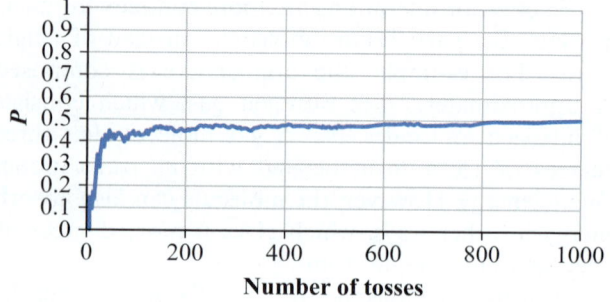

Figure 16-15 Running average value in a coin toss experiment.

of particles approaches infinity, the system approaches the **thermodynamic limit**, where the statistical measures are completely accurate. For practical purposes, when the number of particles in a system is of the order of Avogadro's number, the system behaves as though it were at the thermodynamic limit.

However, statistical measures do not give any details about the behaviour of an individual particle. For example, the average speed gives no indication of the proportion of particles that are moving faster or slower than the average at any given moment.

EXAMPLE 16-5

Road Trip Statistics

Suppose that a family drives from Nova Scotia to Manitoba and back for their summer holiday. The total time for the trip is three weeks, with lots of stops along the way to visit and sightsee. The distance from Halifax to Winnipeg is approximately 3700 km. When they are in the car, they drive at a speed of 120 km/h. What is the average speed? Plot the speed probability distribution.

Solution

To determine the average speed, we divide the total distance travelled, 7400 km, by the total time, 504 h (3 weeks × 7 days/week × 24 h/day):

$$\text{speed}_{\text{avg}} = \frac{\text{total distance travelled}}{\text{total time for trip}} = \frac{7400 \text{ km}}{504 \text{ h}} = 15 \text{ km/h}$$

On this basis alone, you might conclude that this family has very cautious drivers, not to mention being fairly annoying to fellow motorists.

To construct the speed probability distribution, we need to calculate the probability that the car is travelling at a particular speed. We already know that the total time for the trip is 504 h. The time spent driving is

$$\frac{7400 \text{ km}}{120 \text{ km/h}} = 62 \text{ h}$$

Therefore, the probability of finding the speed to be 120 km/h at any time on the trip is

$$\text{Prob}(120 \text{ km/h}) = \frac{\text{time at 120 km/h}}{\text{total time for trip}} = \frac{62 \text{ h}}{504 \text{ h}} = 0.12$$

The remainder of the trip, 504 h − 62 h = 442 h, was spent at a speed of 0 km/h. Thus, the probability of the speed being 0 km/h is

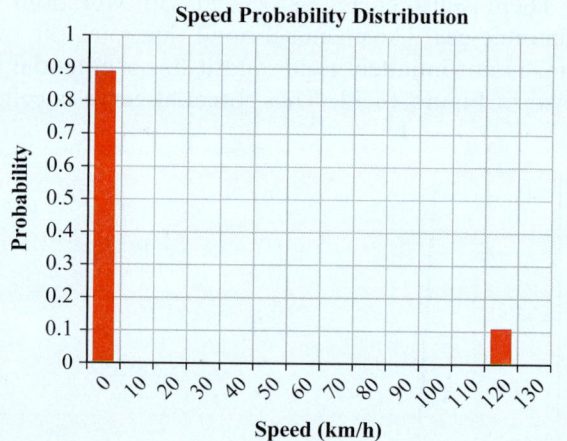

Figure 16-16 Speed probability distribution.

$$\text{Prob}(0 \text{ km/h}) = \frac{442}{504} = 0.877$$

The probability of finding any other speed is zero, if we neglect the short times it takes to get up to speed and slow to a stop. Figure 16-16 shows a plot of these probabilities.

Making sense of the result

This example highlights the fact that much information is masked when we use statistical measures. The average value does not tell us very much about the driving habits of the family. The speed probability distribution tells us much more—when they drive they travel pretty quickly, but they like to stop a lot. However, a lot of details are still unknown—how many stops were made and where, for example.

What is the probability that an individual particle in a gas will have a speed within a certain range? An apparatus for measuring the distribution of particle speeds is shown in Figure 16-17(a). The source oven is filled with a gas. The collimating slits ensure that only those gas molecules with horizontal velocity enter the apparatus. The shutter on the left is opened for a brief time, Δt. At some time τ later, the right-hand shutter is also open for a time Δt. The detector produces a signal proportional to the number of particles that impinge upon it. To determine the speeds of the particles that can make it through the apparatus, we construct a space-time diagram as shown in Figure 16-17(b).

As we can see from Figure 16-17(b), the slowest particles detected travel the distance D in a time $\tau + \Delta t$, and the fastest particles can make it in a time $\tau - \Delta t$. Thus, particles that make it through to the detector while the right-hand shutter is open have a speed, s, that falls in the range of

$$\frac{D}{\tau + \Delta t} \leq s \leq \frac{D}{\tau - \Delta t} \quad (16\text{-}28)$$

where D is the distance between the shutters of the apparatus, τ is the delay between the opening of the left- and right-hand shutters, and Δt is the length of time that the shutters are open.

Only those particles that have a speed in this range will be detected. Particles with faster speeds will hit the right-hand shutter before it opens, and particles with slower speeds will arrive after the shutter has closed. By varying τ, we can select the speed of the particles that reach the detector; thus, we can get data that tell us how many particles there are in a particular speed range. By making Δt small, we can improve the resolution of the measurement, at the cost of increasing the time needed to complete the experiment.

The results of such experiments show that the particle speeds have a well-defined distribution that depends on the temperature of the gas, as shown in Figure 16-18. The experimental results are well modelled by a distribution predicted by a statistical model developed by James Clerk Maxwell (Scottish physicist, 1831–1879) and Ludwig Boltzmann (Figure 16-19). The equation they derived for this distribution is

KEY EQUATION $\quad f(v) = 4\pi \left(\frac{m}{2\pi kT}\right)^{3/2} v^2 e^{-mv^2/(2kT)} \quad (16\text{-}29)$

The fact that we get a well-defined velocity distribution indicates that the system is near the thermodynamic limit and is in thermal equilibrium. There are many possible ways to distribute the speeds such that the total kinetic energy is the same. However, as the particles collide with each other, they quickly equilibrate, and the speed distribution approaches the Maxwell–Boltzmann distribution.

(a)

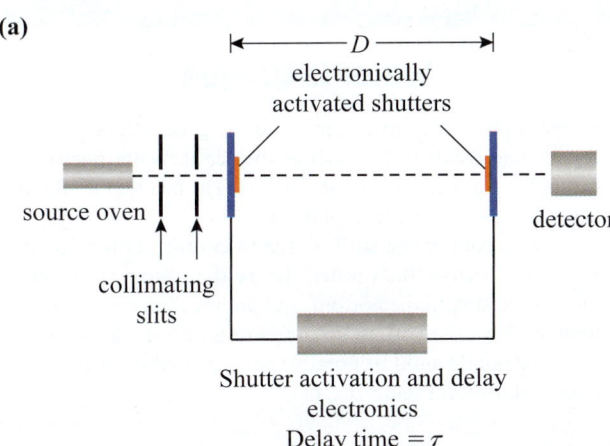

(b)

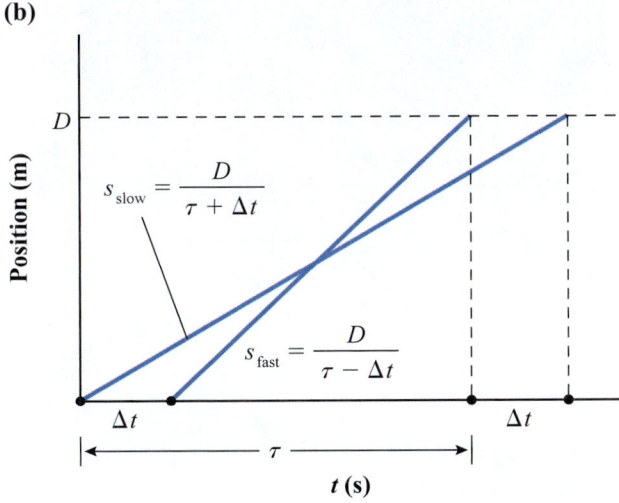

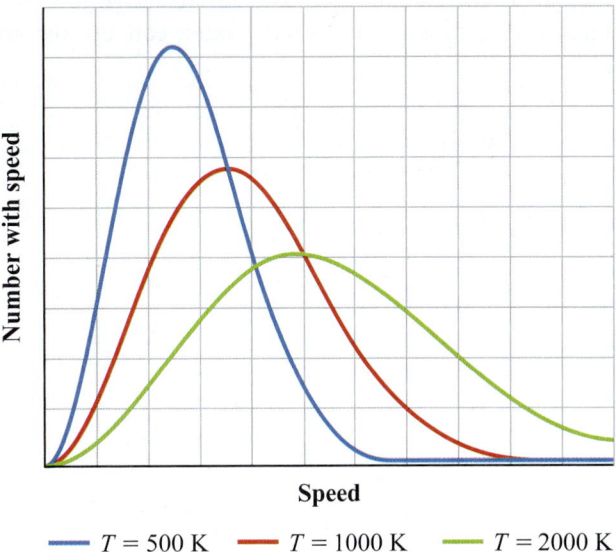

Figure 16-18 The Maxwell–Boltzmann distribution for three different temperatures. It has been shown to be in excellent agreement with experimental data.

Figure 16-17 (a) An apparatus for measuring molecular speed distributions. (b) A space-time diagram for a molecular speed apparatus.

Figure 16-19 Ludwig Boltzmann's grave in Austria. At the top of the marker you can see Boltzmann's equation for entropy, one of the fundamental concepts of thermodynamics.

16-12 Phase Diagrams

Now that we have a set of variables to describe a thermodynamic system, we can begin to discuss the behaviour of a particular system as a function of those variables. We can use a **phase diagram**, which is a diagram that shows the phases of a substance at various temperatures and pressures. A phase diagram for water is shown in Figure 16-20.

We can see that there are distinct regions for each phase. Along the boundaries of each region, different phases can coexist at the same temperature and pressure. In Figure 16-20, there is a broad range of pressures and temperatures where water can coexist in

liquid and gas phases along the line extending to the right of the triple point. The conditions for the fog in the image at the beginning of this chapter lie on this line. The triple point, where the three lines intersect in a phase diagram, indicates the pressure and temperature at which the solid, liquid, and gas can all coexist. For water, the triple point occurs at a pressure of 0.006 atm and a temperature of 0.01 °C.

Also note that the line separating the solid and liquid phases for water slants upward to the left from the triple point. This slope is a consequence of the fact that water expands when it freezes. If we increase the pressure on solid ice while keeping the temperature constant, we move upward vertically in the phase diagram. Moving upward in the solid region, we cross the boundary into the liquid region, indicating that the ice will melt. However, it is also the case that when we increase the pressure on a substance, its volume always decreases. Thus, the volume of the liquid water is less than that of the solid water.

MAKING CONNECTIONS

Water Is Unique

Water is an important exception to the general tendency of liquids and solids to contract as they cool. Water begins to expand as we cool it below about 4 °C. The fact that ice floats is a consequence of this expansion.

Lakes cool at the surface due to contact with cold air. If ice were denser than water, the ice that forms at the surface would sink to the bottom, and bodies of water would fill with ice from the bottom up in freezing weather. Skating on a lake or pond would be possible only after all the water had frozen, unlike in Figure 16-21.

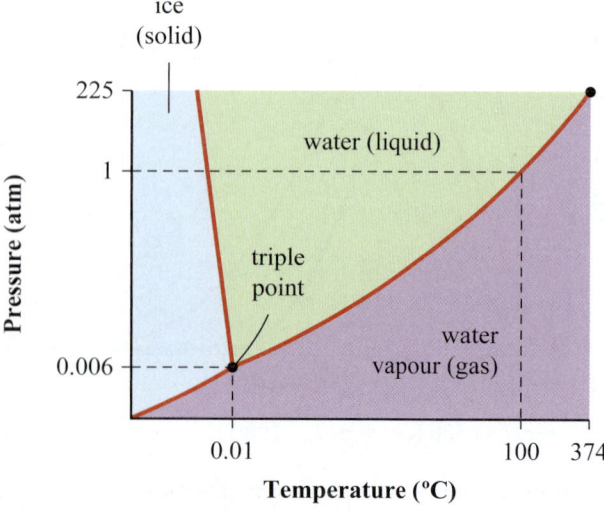

Figure 16-20 A phase diagram for water.

Figure 16-21 Outdoor skating on the Rideau Canal in Ottawa.

KEY CONCEPTS AND RELATIONSHIPS

A Macroscopic Description

In this chapter, we used state variables, such as volume, pressure, and temperature, to describe a macroscopic assembly of particles. We also established a firm basis for the science of thermometry with the zeroth law of thermodynamics.

We used average values to relate state variables to the properties of the individual particles, recognizing that statistical measures are fully valid only for a system with a very large number of particles.

State Variables

We can relate the pressure that a gas exerts on the walls of its container to the square of the average speed of the gas particles:

$$P = \frac{1}{3} \frac{mN(v^2)_{avg}}{V} \qquad (16\text{-}13)$$

At the low-density limit, all gases exhibit the same behaviour, which is described by the ideal gas law:

$$PV = NkT \qquad (16\text{-}17)$$

Temperature

We saw that temperature can be measured using thermometric properties of materials. Most materials expand when heated, and the coefficient of linear expansion is defined as

$$\alpha = \frac{1}{L} \frac{\Delta L}{\Delta T} \qquad (16\text{-}14)$$

APPLICATIONS

thermometry, gas handling, material response to temperature and pressure changes

We can use the universal low-density behaviour of gases to create an ideal gas thermometer:

$$T = T_3 \frac{P}{P_3} = (273.16\,\text{K})\frac{P}{P_3} \qquad (16\text{-}21)$$

The temperature of a gas is a measure of the average kinetic energy of the molecules in the gas:

$$\frac{3}{2}kT = \frac{1}{2}m(v^2)_{avg} \qquad (16\text{-}25)$$

The Zeroth Law

The zeroth law of thermodynamics is the foundation for the measurement of temperature: If object A is in thermal equilibrium with object B, and object A is also in thermal equilibrium with object C, then object C is also in thermal equilibrium with object B.

Statistical Measures

For a gas that is at a constant temperature, pressure, and volume, the probability of finding a gas particle with speed v is given by the Maxwell–Boltzmann distribution:

$$f(v) = 4\pi \left(\frac{m}{2\pi kT}\right)^{3/2} v^2 e^{-mv^2/(2kT)} \qquad (16\text{-}29)$$

Phase Diagrams

A phase diagram captures the complex behaviour of a material as its temperature and pressure are varied.

KEY TERMS

constant-volume gas thermometer, gas, ideal gas law, liquid, phase diagram, pressure, solid, state, state variables, temperature, thermal equilibrium, thermodynamic limit, volume, zeroth law of thermodynamics

QUESTIONS

1. Two different thermometers are calibrated by placing them in the same ice-water bath and then in the same pot of boiling water. The respective readings of the two thermometers are assigned to be 0 °C and 100 °C. The scale on each thermometer is divided into 100 equal increments. If we then use these thermometers to measure the temperature of an object that is at some intermediate temperature, will they agree?

2. A piece of sodium metal and a bath of water are at the same temperature. If we place them in thermal contact, their temperatures do not change. However, if we place them in physical contact, we observe a rather spectacular reaction. Discuss.

3. The maximum in a probability distribution indicates the value that is most likely to be observed. Is the most likely value the same as the average value? Use sketches to illustrate your answer.

4. Most materials expand as they are heated. Discuss why this occurs using the shape of the Lennard–Jones potential. Explain why this expansion would be greatly reduced if the potential were parabolic, as shown in Figure 16-13.

5. Using the phase diagram of water in Figure 16-20, discuss whether there are limits on the temperatures at which the three phases can be observed. How would your answers change if liquid water did not expand when cooled?

6. A mercury thermometer uses the fact that mercury expands when heated. The coefficient of thermal expansion for mercury is $6 \times 10^{-5}/°\text{C}$. This means that if we heat a column of mercury that is 1 m long and cause its temperature to change by 1 °C, the column will increase in length by 6×10^{-5} m. However, in practice, the length of the mercury column changes by several centimetres with small changes in temperature. How do mercury thermometers work?

7. Two gases are at the same temperature. Do they necessarily exert the same pressure on the walls of their respective containers?

8. Measuring very high temperatures is a challenge. If we try to use a gas, we must measure its pressure and, hence, use a container. However, imagine trying to insert a container of gas into a lava flow. Considering your own experience with the light that is emitted by hot objects, suggest a method of determining the relative temperatures of such objects.

9. Figure 16-22 shows the pressure in a gas sample as a function of temperature, with the volume held constant. Does this gas behave as an ideal gas?

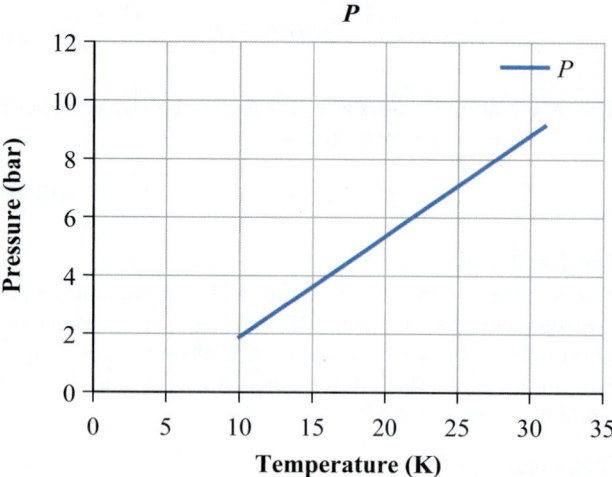

Figure 16-22   Question 9.

10. When we derived an expression for the pressure that a gas exerts on the walls of its container, we assumed that the collisions that the gas particles made with the walls were elastic, and the particles emerged from the collisions with the same speed. If that assumption were invalid, what would happen to the gas in the container?

11. Many of us have experienced the phenomenon of a sea breeze. We wake up on a summer morning near the water and there is no wind. As the Sun rises and the air begins to warm up, we notice that a breeze develops that blows from the water onto the land. Keeping in mind that winds blow from areas of high pressure to areas of low pressure, explain why sea breezes occur.

12. The ideal gas model assumes that there are no attractive interactions between the molecules of a gas, and this is a good approximation at low densities. At densities where the approximation is not valid, would you expect the pressure of a real gas to be higher or lower than the pressure of an ideal gas? Assume the same quantity, volume, and temperature for the real and ideal gases.

13. Many of us have had the experience of pumping up a bicycle tire with a manual pump. You may have observed while doing so that the hose from the pump to the tire becomes warm to the touch, indicating that the pumping process causes the temperature of the air to increase. Clearly, compressing the gas in the pump causes the volume to decrease and the pressure to increase. We might assume, for example, that if we halve the volume we double the pressure. However, if the temperature also increases, is this assumption true? Explain your answer.

PROBLEMS BY SECTION

For problems, star ratings will be used (★, ★★, or ★★★), with more stars meaning more challenging problems. The following codes will indicate if $\frac{d}{dx}$ differentiation, $\int$ integration, ⌨ numerical approximation, or ∿ graphical analysis will be required to solve the problem.

Section 16-1   The Need for a Macroscopic Description

14. ★ Assuming that a 1 TB (terabyte) hard drive is 2 cm thick, how high would a stack be if you stacked up enough drives to store the data for a mole of gas particles?

15. ★ Assuming it takes 1 ps (picosecond) per particle to perform the calculations required to advance a mole of gas particles through 1 μs of a simulation, how long would it take to do the calculations for a 2 s simulation?

Section 16-2   Solids, Liquids, and Gases

16. ★★ The density of solid copper is approximately 9.00 g/cm^3, and the atomic mass of copper is 63.5 amu.
   (a) How many atoms of copper are in 1.00 cm^3 of copper?
   (b) What is the average volume occupied by one copper atom?
   (c) What is the average distance between adjacent copper atoms?

17. ★ One mole of gas at standard temperature and pressure occupies 22.4 L. What is the average volume occupied by each gas particle? What is the average spacing between particles?

Section 16-3   State Variables

18. ★ If we know the temperature of a system in Celsius, is the temperature in kelvins an additional state variable?

19. ★ The reading of a thermometer in some system fluctuates. Is the system in thermodynamic equilibrium?

Section 16-4   Pressure

20. ★ The SI unit of pressure is the pascal. Other common units of pressure are the atmosphere, the bar, the torr, and pounds per square inch (psi). Look up the conversions between these units, and express the standard pressure of 101.325 kPa in units of
   (a) atmospheres
   (b) torr
   (c) pounds per square inch

21. ★★ Racing bicycles have very narrow tires, which reduce the rolling friction. Assume that such a tire has a diameter of 70.0 cm with a cross-sectional area of 4.00 cm^2, and that it is inflated to 110. psi.
   (a) Convert the pressure to pascals.
   (b) Determine the volume of the tire.
   (c) Assuming $mv^2 = 1.24 \times 10^{-20}$ J, how many moles of gas are required to inflate the tire?
   (d) What would the mass of this gas be if it were nitrogen? Helium?

Section 16-5   Temperature and Thermal Expansion

22. ★ Bridges are usually constructed with expansion joints so that they do not buckle when heated (Figure 16-23). Assume that the expansion joint of a bridge is dominated by structural steel with a linear thermal expansion coefficient of approximately 1×10^{-5}/°C. Suppose we wish to design expansion joints in a bridge that will be exposed to temperatures ranging from −40 °C to +40 °C.
   (a) At what temperature will the gap in the joint be the largest?
   (b) How long can a section be if it can expand no more than 3.0 cm?

Figure 16-23 Problem 22.

23. ★ A 2.00 m copper rod has its temperature increased from 20.0 °C to 40.0 °C. By how much does the length of the rod increase?

24. ★★ A thin aluminum ring has a circumference of 20.0 cm. The ring is heated from 10.0 °C to 50.0 °C. By how much does the diameter of the ring change?

Section 16-6 Thermometers and Temperature Scales

25. ★ Convert −40 °F to Celsius.

26. ★ On February 3, 1947, Weather Service of Canada workers at Snag, Yukon, filed a notch into the glass casing of an alcohol thermometer because the indicator within fell below the lowest number, −80 °F. When they later sent the thermometer to Toronto, officials there determined that the temperature had dropped to −81.4 °F, the lowest official temperature ever recorded in North America. Convert this temperature to
    (a) Celsius
    (b) kelvin

27. ★ The highest temperature ever officially recorded is 58 °C at Azizia, Libya, in the Sahara desert, recorded in 1922. Convert this temperature to
    (a) Fahrenheit
    (b) kelvin

Sections 16-8 and 16-9 Ideal Gases and The Constant-Volume Gas Thermometer

28. ★★ A hydrogen gas thermometer is found to have a volume of 100.0 cm^3 when placed in an ice-water bath at 0 °C. When the same thermometer is immersed in boiling liquid bromine, the volume of hydrogen at the same pressure is found to be 121.6 cm^3. Assume that the hydrogen gas behaves as an ideal gas.
    (a) Write the ideal gas law for both circumstances.
    (b) Recognizing that the pressure is the same for both cases, solve each equation in part (a) for pressure.

(c) Set the two equations in part (b) equal to each other, and find the temperature of the bromine.

29. ★★ A high-altitude balloon contains helium, whose molar mass is 4.00 g/mol. At the balloon's maximum altitude, its volume is 792 m^3 and the outside temperature and pressure are −53.0 °C and 5.00 kPa, respectively. Assume that the helium in the balloon is in equilibrium with the outside air temperature and pressure.
    (a) Use the ideal gas law to find the amount of helium in the balloon at maximum altitude.
    (b) Assuming no loss of helium, what would be the volume of the balloon when it was launched from the ground, where the air temperature and pressure are 20.0 °C and 101 kPa, respectively?
    (c) If the volume of the balloon when it was launched was 65.0 m^3, how many moles of helium were lost during the balloon's ascent?

30. ★ An 18.0 L container holds 16.0 g of oxygen gas (O_2) at 45.0 °C.
    (a) Look up the molar mass of oxygen gas, and use that to determine the amount of gas.
    (b) Use the ideal gas law to find the pressure in the container.

31. ★ All gases behave as ideal gases at sufficiently low densities. Consider one mole of gas at standard temperature and pressure, and assume that it behaves as an ideal gas.
    (a) Calculate the volume occupied by the gas.
    (b) If the gas is nitrogen, what is the average speed of a nitrogen molecule?
    (c) If the gas is oxygen, what is the average speed of an oxygen molecule?
    (d) What is the ratio of the average speeds from parts (b) and (c)?

Section 16-10 Temperature and Mechanical Energy

32. ★★ Massive amounts of energy are transferred when the atmosphere close to Earth's surface shifts between daytime and nighttime temperatures.
    (a) Calculate the number of atoms in 1.0 km^3 of nitrogen gas at standard temperature and pressure, assuming that nitrogen behaves as an ideal gas.
    (b) Calculate the average kinetic energy of a nitrogen molecule at 25 °C.
    (c) Calculate the average kinetic energy of a nitrogen molecule at 15 °C.
    (d) How much energy transfers out of 1.0 km^3 of nitrogen gas when it cools by 10 °C?

33. ★ What would the average speed of a nitrogen molecule be at the temperatures in problems 26 and 27?

Section 16-11 Statistical Measures

34. ★ Consider the function $f(t) = at^2$. What is the average value of f between 0 s and T s? Hint: You will need to do some integration.

35. ★★ Use Equation 16-29 to compute the average speed of a molecule in an ideal gas.

Section 16-12 Phase Diagrams

36. Solid carbon dioxide is sometimes called *dry ice* because it goes directly from the solid to the gas phase when warmed at a pressure of 1 atm. What can you infer about the triple point of carbon dioxide?

37. On a newly discovered planet, the atmospheric pressure is 500 kPa and the temperature is 200 K. What phase(s) will water, oxygen, nitrogen, and carbon dioxide be found in? You will need to look up the phase diagrams for these substances.

COMPREHENSIVE PROBLEMS

38. ★ We have an ideal gas, and we double the quantity of gas. What other quantities would we need to change, and how, for the temperature of the gas to remain constant?

39. ★ Your body contains approximately 60% water by mass. Estimate the amount of water in your body.

40. ★★ Very cold temperatures cannot be measured with a simple mercury thermometer since mercury freezes at $-38.72\,°C$. However, solid-state diodes (described in Chapter 33) can be used as sensors for low temperatures. The relationship between the current, I, in a diode and the voltage, V, across the diode is $I = I_0(e^{q_e Vk/T} - 1)$, where I_0 is the saturation current, q_e is the fundamental unit of charge (1.602×10^{-19} C), k is Boltzmann's constant, and T is the temperature in kelvins.
  (a) Rearrange this equation to find the relationship between the voltage across the diode and the temperature, assuming a constant current through the diode.
  (b) Assuming $I_0 = 1.0 \times 10^{-12}$ A, what voltage will appear across the diode at a temperature of $-190\,°C$ if a current of 5.0 mA flows through the diode?

41. ★ The lowest temperatures ever reached in a laboratory are on the order of nanokelvins, 10^{-9} K. Calculate the average speed of a cesium atom in a gas that has been cooled to 1×10^{-9} K.

42. ★★ You work for a major theatre chain and have been assigned the job of determining the optimum conditions for running the popcorn machines to get the largest, most uniformly sized popcorn. The popcorn size distributions from two runs of your experiment are shown in Figure 16-24.
  (a) Which series has the largest average kernel size?
  (b) Which series has the most uniform kernel size?
  (c) Which series has the largest number of kernels?
  (d) Which series is the more reliable measurement?

43. ★ A 30 L container at a pressure of 150 kPa contains three moles of an ideal gas. What is the temperature?

44. ★★ A sample of ideal gas is compressed until its volume is one third of its original volume.
  (a) If the compression is carried out at constant temperature, what is the resulting pressure?
  (b) If the compression is carried out at constant pressure, what is the resulting temperature?

45. ★★★ We showed that the pressure in a gas is given by $P = \dfrac{1}{3}\dfrac{mNv_{avg}^2}{V}$. Estimate the number of collisions per second per unit area on the walls of the container. Use data for nitrogen at standard temperature and pressure.

46. ★★ You are designing storage tanks that will be filled with helium. The tanks will be filled at 20 °C to a pressure of 100.0 atm. What maximum and minimum pressures would you expect to observe in a full tank if the ambient temperature will range from $-20\,°C$ to 30 °C?

47. ★ Gravity pulls the matter in a star toward the centre. When the star eventually collapses, the volume is reduced, and the temperature and pressure increase. The increase in pressure is eventually sufficient to offset the gravity, and the star is said to be in hydrostatic balance. The increase in temperature is enough to allow nuclear fusion to occur, which further increases the temperature, leading to a new equilibrium. The net result is that in the core of a star we observe very high densities, pressures, and temperatures. At the core of the Sun, the density is approximately 150 000 kg·m^{-3}, and the temperature is as high as 13.6×10^6 K. In this question, we naively treat the Sun as an ideal gas. Calculate
  (a) the pressure at the core of the Sun
  (b) the average speed of a hydrogen atom at the core of the Sun

48. ★★★ An alternative method of measuring the velocity distribution of particles in a gas is to use a rotating cylinder with a helical groove cut in the side, as shown in Figure 16-25. If the cylinder is rotating at an angular speed of ω, what speed of particle will it be selecting for? Express your result in terms of the length and radius of the cylinder, the helix angle, and the angular speed. Discuss how the width of the groove affects measurements made with this device.

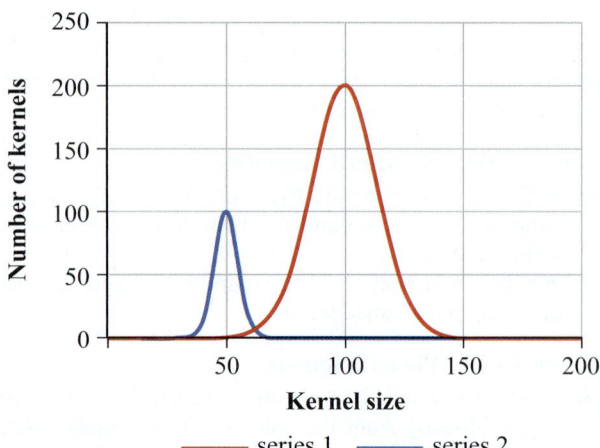

Figure 16-24 Problem 42.

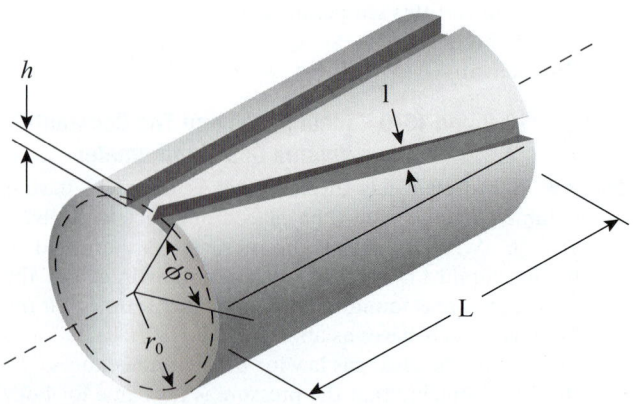

Figure 16-25 Problem 48.

49. ★★ Table 16-6 contains some observations of the phase that a particular substance is in at various pressures and temperatures. Use the data to construct a rough phase diagram.

Table 16-6 Data for Problem 49

Temperature (°C)	Pressure (atm)	Observed Phase
−78	0.5	Gas
−78	1.1	Solid
−78	10	Solid
−57	4	Gas
−57	6	Solid
−57	19	Solid
−50	6	Gas
−50	7	Liquid
−50	15	Solid
−30	7	Gas
−30	8	Liquid
−30	100	Solid

50. ★★ Roll a six-sided die, and record the numbers that you roll in a spreadsheet. Calculate the running average value of all of your rolls as you go, and plot your data.

51. ★★ To construct a thermometer, a physicist decides to use a resistor whose resistance varies as a function of temperature. The physicist and her team construct a circuit that passes a constant current of 1 mA through the resistor; they then measure the voltage across the resistor. Table 16-7 lists the calibration data. Fit the data with a suitable polynomial, and predict the resistance value at a temperature of 400 K.

Table 16-7 Data for Problem 51

T (K)	77.0	97.0	117	137	157	177	197
R (Ω)	132.2	134.7	139.5	143.8	146.5	152.0	153.6

T (K)	217	237	257	277	297	317
R (Ω)	157.8	163.7	167.8	168.9	172.2	177.9

OPEN PROBLEM

52. ★★★ In problem 45, we found that there is an enormous number of collisions per second with the walls of a container holding a gas. Suppose the pressure sensor in our system has a response time of 1 ms. To what value would we have to reduce the concentration of particles in our system to see fluctuations in the reading of the pressure sensor? State any assumptions you make.

CHAPTER 17

Heat, Work, and the First Law of Thermodynamics

Learning Objectives

When you have completed this chapter, you should be able to

LO1 Discuss what is meant by the term *heat*.

LO2 Define specific heat, and perform calculations using heat capacities.

LO3 Calculate the final temperature of a system when heat flows from one part of the system to another.

LO4 Define heats of fusion and vaporization, and use them in calculations.

LO5 Calculate the work done by a gas during an expansion.

LO6 State the first law of thermodynamics.

LO7 Define what is meant by adiabatic, isothermal, isobaric, and isochoric processes.

LO8 Describe various energy transfer processes, and perform calculations for radiative and conductive processes.

The first step in brewing beer is mixing malted barley and water together to form a "mash". The mash is then held at a temperature that allows enzymes in the barley to convert starches into fermentable sugars (Figure 17-1). How can the brewer determine what initial temperature of the water will produce the desired temperature for the mixture?

Figure 17-1 Malted barley and water in the mash tun at a brewery.

Cephas Picture Library/Alamy Stock Photo

17-1 What Is Heat?

The nature of heat puzzled scientists for centuries. Early theories suggested that there was a "caloric fluid" that flowed from a hot object to a cold object. Anglo-American physicist, inventor, and military officer Count Rumford (Benjamin Thompson, 1753–1814) was among the first to dispel this notion and make a connection between motion and heat. He performed an experiment in which he bored out the barrel of a brass cannon with a dull cutting tool in a water bath. He monitored the temperature of the water to determine the heat produced during the boring. He showed that this mechanical process could produce significant heat continuously for a virtually unlimited time. He also noted that the physical properties of the shavings produced as a result of the boring appeared to be unchanged. On the basis of these findings, Rumford claimed that there could not be a caloric fluid because surely it would be depleted and would limit the amount of heat that could be extracted from an object. In addition, if the heat were some type of fluid, you would expect there to be some observable change in the physical properties of a material if the fluid were removed. He therefore suggested that there is a connection between motion and heat.

In Chapter 16, we noted that the temperature of an ideal gas is a measure of the average kinetic energy of the particles that make up the gas. If we place a sample of gas into contact with a hot object, we observe that the temperature of the gas increases. Based on these observations, it would seem that heating an object somehow transfers energy into the object and, at least in the case of an ideal gas, the energy is stored in the form of kinetic energy. For more complex systems, in addition to kinetic energy, the energy may also be stored as potential energy. We call the energy stored in the system the **internal energy** of the system.

When we place a hot object in contact with a cooler object, energy is transferred from the hot object to the cooler object and we say we are heating the cooler object. **Heat** is the energy that is transferred from one object to another when a temperature difference exists between the two objects. Because heat is a form of energy it can be measured in joules. We can gain some insight into how this energy transfer occurs by considering what happens when we place a hot object and a cold object in contact with each other. The particles in the hot object have greater kinetic energy than the particles in the cold object. When they collide, energy is transferred from the more energetic particles to the less energetic particles. Consequently, the average kinetic energy decreases in the hot object and increases in the cold object, and the temperatures of the objects change correspondingly.

We need to be careful to distinguish between internal energy and heat. An object possesses a well-defined quantity of internal energy in the form of kinetic and potential energy. However, heating a system is not the only way to add energy to a system. As we shall see later in the chapter, it is also possible to change the internal energy of a system by doing work on the system. We could have two identical systems that start in the same state and end in the same state, but have moved from the initial state to the final state by different processes. They start and end with the same internal energy, and thus the change in the internal energy of each system is also the same. However, because the processes differ, the heat required in each process is not the same, nor is the work. By simply examining the systems in their end states, it is not possible to distinguish between them and we cannot say how much heat or work is in each system. We can only specify the total internal energy.

17-2 Temperature Changes Due to Heat Transfer

We now consider quantitatively how the temperature of an object changes when it is heated. We can make an object absorb heat by placing it in contact with an object that is at a higher temperature, such as when we put a pot of food on a stove element.

Since temperature is a measure of the average kinetic energy of the particles (at least for an ideal gas), we would expect that the temperature change due to heating would be proportional to the amount of heat put into the system:

$$\Delta T \propto Q \tag{17-1}$$

where ΔT is the temperature change and Q is the heat flow into the system. Notice that when $Q > 0$, $\Delta T > 0$, and when $Q < 0$, $\Delta T < 0$.

Two other factors affect the change in temperature: the mass of the material present and the type of material. The fact that the temperature change depends on the amount of material makes sense. For example, if we double the number of particles, the kinetic energy added per particle is halved, and we would expect the change in temperature to be halved as well.

The atoms of a simple monatomic gas, such as helium, can only have kinetic energy from linear motion. A diatomic gas, such as oxygen, can also have rotational kinetic energy, vibrational kinetic energy, and vibrational potential energy. These different modes are illustrated in Figure 17-2.

As we discussed in Chapter 16, these different ways of absorbing energy are called degrees of freedom, and in a classical system, the absorbed energy is shared

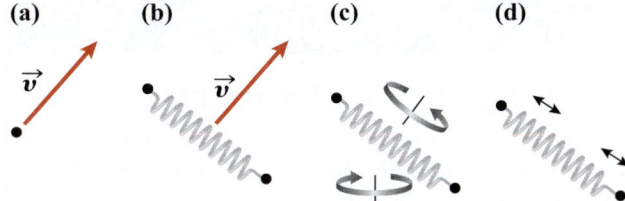

Figure 17-2 (a) A monatomic particle with translational kinetic energy. (b), (c), and (d) A diatomic particle with translational, rotational, and vibrational energy.

Table 17-1 Specific Heat Capacities of Some Common Materials at Constant Pressure (101 kPa)

Material	Specific Heat Capacity, c (J/(kg·K))
Copper (20 °C)	390
Lead (20 °C)	129
Glass (20 °C)	840
Air (20 °C)	1000
Animal tissue (20 °C)	3500
Water, ice (−5 °C)	2100
Water, liquid (15 °C)	4186
Water, steam (110 °C)	2010

equally among all the degrees of freedom. Quantum effects complicate this situation somewhat at low temperatures, but at higher temperatures the energy is shared equally. As a result, at high temperatures, the absorption of a given amount of heat causes a much smaller increase in translational kinetic energy for a diatomic gas than for a monatomic gas.

We can quantify these differences by defining the **heat capacity**, C, of an object as the ratio of the heat added to the object to the resulting temperature change:

$$C = \frac{Q}{\Delta T} \tag{17-2}$$

An object with a large heat capacity undergoes a small temperature change for a given heat input compared to an object with a small heat capacity. The units of heat capacity are J/K.

CHECKPOINT 17-1

Internal Energy and Temperature

Does an object at a higher temperature always contain more internal energy than an object at a lower temperature?

ANSWER: No. Objects with different heat capacities can absorb the same amount of heat and exhibit significantly different temperature changes.

The temperature change observed when a sample absorbs heat depends on both the type and quantity of material in the sample. We define the **specific heat capacity**, c, as $c = C/m$, which is the quantity of heat required to raise the temperature of a unit mass by a unit degree. This allows us to relate heat flows and temperature changes:

KEY EQUATION
$$Q = mc\Delta T \tag{17-3}$$

where Q is the amount of heat absorbed, m is the mass of the sample, c is the specific heat capacity, and ΔT is the temperature change, $T_{\text{final}} - T_{\text{initial}}$.

Table 17-1 lists the specific heat capacities of some common substances.

EXAMPLE 17-1

Temperature Change Due to an Input of Heat

Find the resultant temperature change when 100. g of water absorbs 1000. J of heat.

Solution

We simply rearrange Equation 17-3 to solve for the temperature change:

$$\Delta T = \frac{Q}{mc} = \frac{1000.\ \text{J}}{(0.100\ \text{kg})(4186\ \text{J/(kg} \cdot \text{K))}} = 2.39\ \text{K}$$

Making sense of the result

A cup of tea is made with about 100 g of water, and it typically takes about 5 min to boil a kettle of water (say 1000 g) on a 1500 W stove element. 1000 J is the energy consumed by a 1500 W element in less than second. So, this small temperature change seems reasonable.

MAKING CONNECTIONS

Water Helps Regulate Temperature

You will note that the specific heat capacity of liquid water is the highest value listed in Table 17-1. This property of water helps us regulate our body temperature and plays a significant role in moderating the climate of coastal areas. For example, in Vancouver, British Columbia, average maximum daily temperatures are about 6 °C in winter and 22 °C in summer. In contrast, the corresponding averages for Winnipeg, Manitoba, which is at about the same latitude, are −13 °C and 26 °C, respectively. We can see that Winnipeg has both a lower winter temperature and a higher summer temperature. The climate in Vancouver is moderated by the Pacific Ocean.

THERMODYNAMICS

17-3 The Flow of Heat between Objects

We now consider what happens when a hot object is in thermal contact with a cold object. For simplicity, we will assume that the two objects are completely thermally insulated from their surroundings. Thus, there is only heat flow between the two objects and any heat that leaves the hot object, Q_{hot}, will go into the cold object. Thus,

$$Q_{hot} = -Q_{cold} \qquad (17\text{-}4)$$

where Q_{cold} is the heat flow into the cold object. We can replace the heat flows for each object with the mass, specific heat, and temperature change from Equation 17-3 to get

$$m_{hot}c_{hot}\Delta T_{hot} = -m_{cold}c_{cold}\Delta T_{cold} \qquad (17\text{-}5)$$

EXAMPLE 17-2

Mashing

A particular recipe for home-brewed beer calls for 3.00 kg of crushed barley to be mixed with 14.0 L of water. The desired temperature of the mixture is 66.0 °C. The crushed barley is at room temperature, and malted barley has a specific heat of 1600 J/(kg·K). What should the initial temperature of the water be?

Solution

When we mix the hot water with the room temperature barley, heat from the hot water will flow into the barley. This flow will continue until the barley and the water are at the same temperature. Table 17-1 lists a specific heat of 4186 J/(kg·K) for water. Since 1 L of water has a mass of 1 kg, we have 14.0 kg of water. Applying Equation 17-5, we get

$$m_{water}c_{water}(T_{f,water} - T_{i,water}) = -m_{malt}c_{malt}(T_{f,malt} - T_{i,malt})$$

$$\Rightarrow T_{i,water} = T_{f,water} + \frac{m_{malt}c_{malt}}{m_{water}c_{water}}(T_{f,malt} - T_{i,malt})$$

Since $T_{f,water} = T_{f,malt} = 66\,°C$,

$$T_{i,water} = 66\,°C + \frac{3.00\ kg\cdot 1600\ J/(kg\cdot°C)}{14.0\ kg\cdot 4186\ J/(kg\cdot°C)}(66\,°C - 20\,°C)$$

$$= 69.8\,°C$$

Note that we have kept our temperatures in Celsius and have expressed the specific heats in Celsius. We can do this because the unit in the specific heat is a change in temperature, and a change of 1 K is the same as a change of 1 °C.

Making sense of the result

The initial water temperature is only a few degrees higher than the temperature of the final mixture both because there is much more water than malt and because the specific heat of water is greater than the specific heat of malted barley.

EXAMPLE 17-3

Using Specific Heat

A 20.0 g block of copper at a temperature of 100. °C is placed in good thermal contact with a 40.0 g block of copper that is at 10.0 °C. What is the final temperature?

Solution

We once again apply Equation 17-5:

$$m_{hot}c_{hot}\Delta T_{hot} = -m_{cold}c_{cold}\Delta T_{cold}$$

In this case, both blocks are made of the same material, so the specific heat cancels. We also know that $T_{f,hot} = T_{f,cold} = T_f$.

Therefore,

$$T_f = \frac{m_{hot}T_{i,hot} + m_{cold}T_{i,cold}}{m_{hot} + m_{cold}}$$

$$= \frac{20.0\ g \times 100\,°C + 40.0\ g \times 10.0\,°C}{20.0\ g + 40.0\ g} = 40\,°C$$

Making sense of the result

The final temperature is closer to the temperature of the larger block than to the temperature of the smaller block because the larger block has a larger heat capacity.

17-4 Phase Changes and Latent Heat

When you boil water in a pot on a stove, the burner transfers significant energy to the pot, yet the temperature of the liquid water sits at 100 °C. The energy being absorbed by the water is being used to break the bonds that exist between the water molecules in the liquid phase so they can move to the gas phase. In some sense, the energy that is being added to the system is "hidden" in that it does not manifest itself as a resulting temperature change. We call these "hidden" heats **latent heats**.

A latent heat called the **heat of vaporization** is associated with the transition from the liquid to the gaseous state. Similarly, a latent heat called the **heat of fusion** is associated with the change from the solid to the liquid state. When a material goes from the solid to the liquid state, the heat responsible for the phase change must be supplied from the surroundings. Conversely, when the material goes from the liquid to the solid state, heat is liberated to the surroundings. Table 17-2 lists heats of fusion and vaporization for a few common materials. Note that all the values are in J/kg. To determine the total amount of heat required to make a transition, we need to multiply these quantities by the mass of the sample. Thus, the heat required for the phase change is

KEY EQUATION

$$Q_{\text{phase}} = mL \qquad (17\text{-}6)$$

where m is the mass of the sample and L is the latent heat.

MAKING CONNECTIONS

Reusable Hand Warmers

Reusable hand warmers rely on heat of fusion (Figure 17-3). They contain a supersaturated solution of a salt, typically sodium acetate, which is in a liquid state. You liquefy the solution by placing the pack in boiling water for a few minutes. Snapping a small metal circle in the pack triggers the crystallization of the solution from the liquid to the solid state. You can see that the pack in the image is crystallizing from right to left. As it crystallizes, the energy that was put into the pack to liquefy the solution is liberated, and the pack feels quite warm to the touch. Once the pack has liberated all the energy, the pack contents can be returned to the liquid state by again immersing the pack in boiling water.

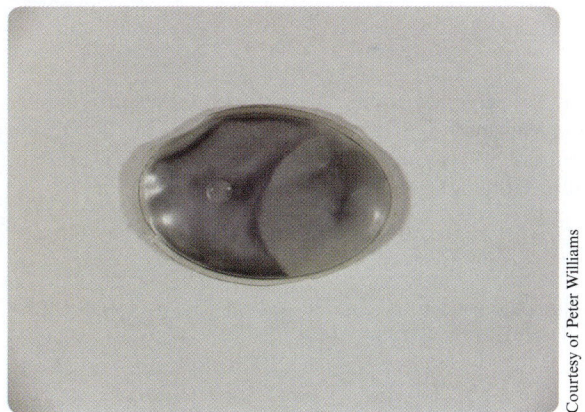

Courtesy of Peter Williams

Figure 17-3 A reusable hand warmer.

Table 17-2 Melting Points, Heats of Fusion, Boiling Points, and Heats of Vaporization for Some Common Materials

Substance	Melting Point (°C)	Heat of Fusion (J/kg)	Boiling Point (°C)	Heat of Vaporization (J/kg)
Water	0	3.33×10^5	100	2.26×10^6
Nitrogen	−210	2.6×10^4	−195.8	2×10^5
Lead	327	2.5×10^4	1750	8.7×10^5
Tungsten	3410	1.84×10^5	5900	48×10^5

THERMODYNAMICS

When doing heat flow (calorimetric) calculations, we consider whether a substance goes through a phase change. If so, we modify Equation 17-3 as follows:

$$Q = mc_1(T_{\text{phasechange}} - T_i) + mL$$
$$+ mc_2(T_f - T_{\text{phasechange}}) \tag{17-7}$$

where c_1 and c_2 are the specific heats of the two phases and L is the latent heat per unit mass.

EXAMPLE 17-4

Temperature Changes Due to Heating When There Is Also a Phase Change

How much heat is required to raise the temperature of 250. g of water from 80.0 °C to 120. °C?

Solution

For clarity, we will calculate each term in Equation 17-7 separately. We will first calculate how much heat is needed to raise the temperature of the water to 100. °C. We assume that the specific heat of water in this temperature range is a constant 4186 J·kg^{-1}·°C^{-1}.

$$Q_{80.0-100.} = (0.250 \text{ kg})(4186 \text{ J/(kg·K)})(100. \text{ °C} - 80.0 \text{ °C})$$
$$= 20\,930 \text{ J}$$

Next, we will determine how much heat is required to convert the liquid water to steam. We refer to Table 17-2 to find the heat of vaporization for water to write

$$Q_{\text{water–steam}} = (0.250 \text{ kg})(2.26 \times 10^6 \text{ J/kg})$$
$$= 5.65 \times 10^5 \text{ J}$$

Finally, we will determine how much heat is required to raise the temperature of the steam from 100. °C to 120. °C. We will assume that the specific heat of the steam is a constant 2010 J/(kg·K):

$$Q_{100.-120.} = (0.250 \text{ kg})(2010 \text{ J/(kg·K)})(120. \text{ °C} - 100. \text{ °C})$$
$$= 10\,050 \text{ J}$$

Thus, the total heat required to heat 250. g of water from 80.0 °C to 120. °C is

$$Q_{\text{Total}} = Q_{80.0-100.} + Q_{\text{water–steam}} + Q_{100-120}$$
$$= 20\,930 \text{ J} + 565\,000 \text{ J} + 10\,050 \text{ J}$$
$$= 596\,000 \text{ J}$$

Making sense of the result

Since the heat of vaporization is quite large compared to the heat required to warm the water and subsequently the steam, most of the energy required goes into the phase change.

EXAMPLE 17-5

What Is the Final Phase?

Calculate the final temperature and phase when 100. g of ice at −5.00 °C is mixed with 200. g of water at 20.0 °C, as well as how much ice will be melted. Assume that no heat is transferred between the mixture and its surroundings.

Solution

There are three possible outcomes for a mixture of ice and water. First, all the ice could melt, resulting in water at some temperature above 0 °C. Second, all the liquid could freeze, producing ice at some temperature below 0 °C. Finally, the resultant mixture could be part liquid and part solid at a temperature of 0 °C.

To determine what actually happens, we will proceed iteratively through the process.

First, we assume that the mixture will come to 0 °C and calculate the heat transfers required to bring both the water and the ice to that temperature:

$$Q_{\text{cool water}} = (0.200 \text{ kg})(4186 \text{ J/(kg·K)})(0 \text{ °C} - 20.0 \text{ °C})$$
$$= -16\,744 \text{ J}$$

$$Q_{\text{warm ice}} = (0.00 \text{ kg})(2100 \text{ J/(kg·K)})(0 \text{ °C} - (-5.00 \text{ °C}))$$
$$= 1050 \text{ J}$$

The minus sign indicates that cooling the water transfers heat *from* the water to the ice. The net transfer of heat to the ice is 16 744 J − 1050 J = 15 694 J.

To get to this temperature, we have taken 16 744 J from the water and used 1050 J to warm the ice. We will now calculate how much ice will be melted by the remaining 15 694 J of heat. If the amount melted is less than the amount of ice we have, the final mixture will consist of ice and water at 0 °C. Using the heat of fusion from Table 17-2:

$$m = \frac{Q}{L} = \frac{15\,694 \text{ J}}{3.33 \times 10^5 \text{ J/kg}}$$
$$= 4.71 \times 10^{-2} \text{ kg}$$
$$= 47.1 \text{ g}$$

Thus, 47.1 g of ice at 0 °C will be converted to liquid at 0 °C. So, the final mixture will be 247.1 g of liquid water and 52.9 g of solid water, all at a temperature of 0 °C.

Making sense of the result

In this case, the final outcome was a solid–liquid mixture at 0 °C. This is because the heat that was transferred from the water to the ice was only enough to warm the ice to the melting point and then melt some of it. If the ice melted had exceeded 100 g, we would have concluded that the final mixture would have been all water. Conversely, if the energy required to warm the ice had been greater than the energy required to cool the water, some or perhaps all of the water would have frozen.

MAKING CONNECTIONS

Home Ice Cream Makers

A home ice cream maker (like the one in Figure 17-4) utilizes a mixture of ice and salt water to achieve temperatures below 0 °C. Crushed ice and a saturated salt solution are placed in the outer part of the ice cream maker. The saturated salt solution has a freezing temperature of about −21 °C. Heat flows from the salt solution into the ice. The solution cools and the ice warms up. However, because the ice dominates the mixture, the final temperature of the mixture is closer to the initial temperature of the ice than that of the salt solution.

Susanne Kischnick/Alamy Stock Photo

Figure 17-4 A home ice cream maker uses a mixture of ice and salt water to freeze the ice cream.

17-5 Changing the Internal Energy Via Work

In the previous sections, we examined energy going into and out of a system by placing it into contact with another object at a different temperature. There is another way to change the internal energy of a system. We saw in Chapter 6 that doing work on an object changes its kinetic energy. We now consider the effects of doing work on a system and having a system do work on its surroundings. Recall that when a force $\vec{F}$ is exerted on a particle that moves from $\vec{r}_1$ to $\vec{r}_2$, the work done is

$$W_F = \int_{\vec{r}_1}^{\vec{r}_2} \vec{F} \cdot d\vec{r} \qquad (6\text{-}15)$$

We also recall the work–energy theorem, which for a single particle states that

$$W_T = \Delta K \qquad (6\text{-}30)$$

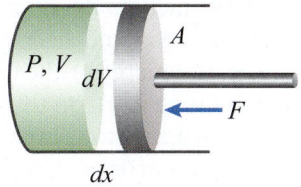

Figure 17-5 A piston compressing a gas.

where ΔK is the change in kinetic energy of the particle and W_T is the net work done on the particle by the net force.

How can we do work on an assembly of particles? Let us consider a gas in a cylinder, as shown in Figure 17-5. A force is applied to the piston, causing it to move to the left and compress the gas. As the piston moves, the pressure in the gas increases. To determine the work done on the piston, we consider an infinitesimal motion of the piston, denoted as dx in the figure. The work that the applied force does on the piston is given by

$$dW = \vec{F} \cdot d\vec{r} = F\,dx \qquad (17\text{-}8)$$

which is positive because the force and the displacement are parallel.

If we assume that the piston either starts and ends at rest, or travels at a constant speed during the process, the kinetic energy of the piston does not change and therefore the net work done on the piston must be zero. Therefore, the gas must do some negative work on the piston to balance off the positive work done by the external force, and it does so through pressure. The pressure of the gas exerts a force to the right on the piston and has a magnitude of $F_{\text{pressure}} = PA$. The work done on the piston by the gas is then

$$dW_{\text{gas on piston}} = -PA\,dx \qquad (17\text{-}9)$$

The minus sign is included because the force and the displacement are in opposite directions. Since the net work done on the piston is zero,

$$F\,dx - PA\,dx = 0 \qquad (17\text{-}10)$$

Thus, the force applied on the piston by the gas has magnitude PA and is equal but opposite to the applied external force. Newton's third law requires that the piston exert an equal but opposite force on the gas. Consequently, the piston does work on the gas:

$$dW_{\text{piston on gas}} = PA\,dx = -P\,dV \qquad (17\text{-}11)$$

The last step in this equation follows from the fact that as we move the piston we change the volume of the gas: $dV = Adx$. We include the minus sign because we are compressing the gas and hence the volume is decreasing. However, the energy of the gas is increasing because the piston is doing work on the gas.

How does this moving piston actually increase the kinetic energy of the gas particles? Consider what happens when the gas particles collide with the walls of the cylinder. If we assume that the collisions are elastic, then a particle emerges from the collision with the same speed and, therefore, the same kinetic energy that it had going into the collision. The only change that occurs is that the component of its velocity vector perpendicular to the face of the piston changes sign. However, in a collision with the moving piston, a gas particle emerges with that component of its velocity increased in addition to having its sign changed. As we saw in Chapter 7, for a perfectly elastic collision in one dimension,

$$u' = \frac{2m}{m + M} v + \frac{M - m}{m + M} u$$
$$v' = \frac{m - M}{m + M} v + \frac{2M}{m + M} u$$
(7-26)

where M and m are the masses of the two objects, and u and v are the initial velocities, with the primes denoting final velocities. In our case, the mass, M, of the piston is much larger than the mass, m, of the gas particle. Let v be the component of the velocity of the gas particle that is perpendicular to the face of the piston, and $-U$ be the velocity of the piston. In the limit of $M \gg m$, Equation 7-26 reduces to

$$u' = -U \text{ and } v' = -v - 2U$$
(17-12)

Thus, the speed in that direction is increased by twice the speed of the piston.

As you can see from the above discussion, we need to be very careful when defining the work done to specify whether we mean the work done on the system by an external agent or the work done by the system on an external agent. You can choose either convention, but then you must be consistent with your choice. For our purposes, we will define the work to be that done by the system on an external agent. With this choice, if the gas increases in volume, a positive amount of work is done on the external agent and therefore negative work is done on the system.

With this convention, the general expression for calculating the work done by the system in a finite volume change is

KEY EQUATION
$$W = \int_{V_1}^{V_2} P dV$$
(17-13)

The magnitude of this quantity corresponds to the area under the curve on a **P-V diagram**, as shown in Figure 17-6. Note that the amount of work done in going from one point to another depends on the path taken on the P-V diagram.

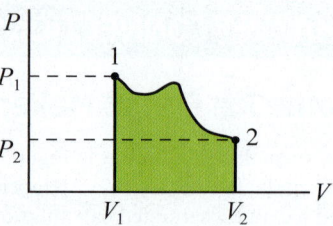

Figure 17-6 A P-V diagram. The shaded area represents the work done in the process of going from point 1 to point 2.

17-6 The First Law of Thermodynamics

We saw in the previous two sections that we can add energy to a thermodynamic system in two ways: heating the system and doing work on the system. Applying conservation of energy to a thermodynamic system suggests that the change in internal energy of the system is simply the sum of the heat put into the system and the work done by the system:

KEY EQUATION
$$\Delta U = Q - W$$
(17-14)

where U is the internal energy of the system, Q is the heat flow *into* the system, and W is the work done by the system. The minus sign in front of the W is a consequence of the fact that we defined W to be the work done by the system on an external agent. Thus, if W is positive, the energy of the external agent has been increased and therefore the energy of the gas has decreased. When heat flows into the system, which raises the internal energy, we consider that to be a positive flow and Q is positive. Conversely, when heat flows out of the system, the internal energy decreases and Q is negative.

Equation 17-14 is the **first law of thermodynamics**, which states that the change in the internal energy of a system is equal to the heat that flows into the system less the work done by the system.

What do we mean by "a system" in this context? In general, a system is simply the set of objects we wish to consider, and we are usually free to define our system in any way that is convenient for us. For example, if we wished to consider what happens when we place a quantity of ice cubes in a glass of water, we could define our system as the water and the ice. This choice is a **closed system** because the total amount of material in the system does not change—no mass enters or leaves the system. Alternatively, we could have chosen just the glass of water as our system. This choice defines an **open system**, which is a system in which mass is added or removed.

Another important distinction is deciding whether or not the system is isolated. In an isolated system, we assume that no heat flows into or out of the system. In general, it is very difficult to make an isolated system. A reasonable approximation might be an insulated cup, which can be treated as an isolated system over short periods of time.

CHECKPOINT 17-5

Sign of the Heat Flow

Ice cubes are placed in a glass of water. When we consider the ice and the liquid to be separate systems, which of the following statements is (are) correct?
(a) The heat flow from the ice is negative.
(b) The heat flow from the liquid is negative.
(c) The heat flow from the ice is equal in magnitude but opposite in sign to the heat flow from the liquid.

ANSWER: (b) and (c) Heat flows from the warm liquid to the cooler ice. Thus, for the liquid system, the heat flow is negative (out of the liquid); for the ice system, the heat flow is into the liquid and is positive. The magnitude of the two flows is the same, assuming that the two systems are isolated from all other systems except each other. All the heat that flows from the liquid goes into the ice.

The variables in Equation 17-14 can have different properties. For example, in the case of an ideal gas, the internal energy is a function of the temperature of the gas and has a well-defined value at a given temperature. The internal energy is a state variable—once we know the temperature, we know the internal energy, or vice versa. However, as we noted in Section 17-5, the work done in a process that carries a system from one state to another depends on how the process is carried out. The work is the area under the P-V curve, and since it is possible to connect two points on a P-V diagram via different paths, different amounts of work can be done. Consequently, we cannot associate a quantity of work with the state of a system. Work is not a state variable. In addition, the heat that flows into or out of a system must also depend on the way the process is carried out, such that variations in the work done are offset by equal but opposite variations in the heat flow. Otherwise, the first law would not hold for all processes. Therefore, heat is also not a state variable. We will discuss different types of processes in the next section.

EXAMPLE 17-6

Internal Energy and Heating

Calculate the change in internal energy of an isolated closed system when

(a) 3000 J of heat is added to it, and the system does 4000 J of work
(b) 3000 J of heat is added to it, and 4000 J of work is done on the system

Solution

The calculation is a straightforward application of Equation 17-14, but we need to pay careful attention to the sign conventions.

(a) When heat is added to the system, it increases the internal energy of the system and is therefore positive. When the system does work, energy is transferred out of the system. Thus, the net change in the internal energy is

$$\Delta U = Q - W = 3000 \text{ J} - 4000 \text{ J} = -1000 \text{ J}$$

(b) When the system has 4000 J of work done on it, the work done by the system is -4000 J. Then the net change in internal energy is

$$\Delta U = Q - W = 3000 \text{ J} - (-4000 \text{ J}) = 7000 \text{ J}$$

Making sense of the result

In the first case, the system loses more energy by doing work than it received via heating, so the net change in internal energy is negative. In the second case, work is done *on* the system, which raises its internal energy.

CHECKPOINT 17-6

Heat and Work

True or false? The heat flow and the work that is done in a process depend on the details of the process, but the difference between the heat flow and the work, $Q - W$, does not, and only depends on the start and end points of the process.

ANSWER: True. By the first law of thermodynamics, $Q - W = \Delta U$, and the internal energy is a state variable. Therefore, the change in the internal energy in any process that has the same start and end points must be the same, regardless of the details of the process.

17-7 Different Types of Processes

There are many different ways a system can be taken from an initial state to a final state. It is useful to focus on processes where one of the variables is held constant. A process carried out at constant temperature is called an **isothermal** process. Similarly, an **isobaric** process is carried out at constant pressure, and an **isochoric** process is carried out at constant volume.

When an isolated system undergoes a process, there is no heat flow into or out of the system. We call such a process **adiabatic**.

Isothermal Processes

For an ideal gas,

$$PV = nRT \qquad (17\text{-}15)$$

where P is the pressure, V is the volume, n is the number of moles of the gas, R is the gas constant, and T is the temperature in kelvin. When we carry out an isothermal process on a closed system, the product PV must be a constant, since n and T are constant. Since the internal energy for an ideal gas depends only on the temperature, in an isothermal process there is no change in the internal energy of the system. Thus,

$$\Delta U = W - Q = 0$$
$$\therefore W = Q \qquad (17\text{-}16)$$

Consider the work done in an isothermal expansion of an ideal gas in a cylinder with a movable piston. We immerse the cylinder in a bath of water large enough that any heat flow between the cylinder and the bath has a negligible effect on the temperature of the bath. We also carry out the process slowly enough that the temperature of the gas in the piston is always the same as the temperature of the surrounding water bath. In fact, this process is only **quasi-static** because the piston would have to move infinitesimally slowly to keep the temperature completely constant. We can apply Equation 17-13 to calculate the work done by the gas:

$$W = \int_{V_1}^{V_2} P\,dV = \int_{V_1}^{V_2} \frac{nRT}{V}\,dV = nRT \ln\!\left(\frac{V_2}{V_1}\right) \qquad (17\text{-}17)$$

Here, we have used the ideal gas law to replace P in the integral with $\dfrac{nRT}{V}$. The work in such a process is illustrated in Figure 17-7. The solid black curve is an **isotherm** (line of constant temperature), and the work done is the shaded area under the curve.

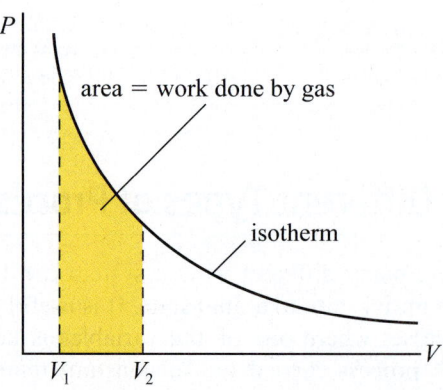

Figure 17-7 A P-V diagram for an isothermal process in an ideal gas.

Isobaric Processes and the Constant-Pressure Heat Capacity

When we hold the pressure constant while compressing an ideal gas,

$$P = \frac{nRT}{V} = \text{constant} = P_c$$
$$\therefore T = \frac{P_c}{nR}V \qquad (17\text{-}18)$$

Thus, if we change the volume while holding the pressure constant, the temperature must also change by heat either flowing into or out of the system. In Chapter 16, we showed that the average kinetic energy of the particles in an ideal monatomic gas is related to the temperature of the gas. For a monatomic ideal gas, all of the energy is the translational kinetic energy of the gas particles. It therefore follows that the internal energy of an ideal gas is given by

$$U = \frac{3}{2}nRT \qquad (17\text{-}19)$$

Therefore, this isobaric process involves heat flow, work, and a change in the internal energy of the system.

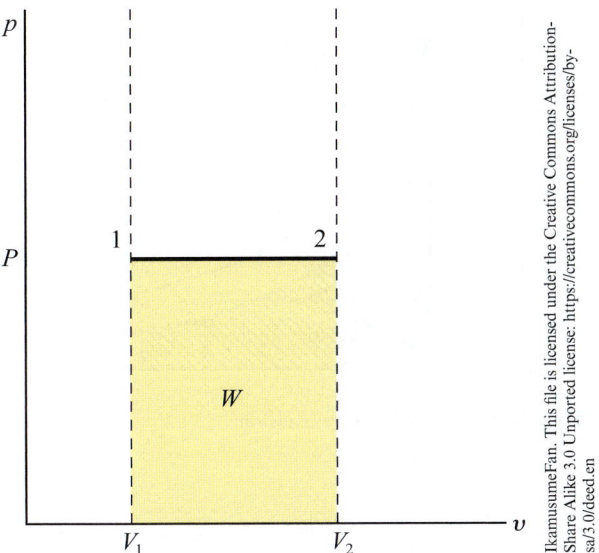

Figure 17-8 An isobaric process.

Since the pressure is constant, the integral for work in Equation 17-13 becomes

$$W = \int_{V_1}^{V_2} PdV = P_c \int_{V_1}^{V_2} dV = P_c(V_2 - V_1) \quad (17\text{-}20)$$

The shaded area in Figure 17-8 corresponds to the work done in this process.

The temperature change can be found using Equation 17-15:

$$\Delta T = T_2 - T_1 = \frac{P_c V_2}{nR} - \frac{P_c V_1}{nR} = \frac{P_c}{nR}(V_2 - V_1) \quad (17\text{-}21)$$

Using this expression for the temperature change with Equation 17-19 gives an equation for the change in internal energy:

$$\Delta U = \frac{3}{2}nR\Delta T = \frac{3}{2}P_c(V_2 - V_1) \quad (17\text{-}22)$$

Rearranging Equation 17-14 to solve for the heat flow, we get

$$Q = \Delta U + W = \frac{3}{2}P_c(V_2 - V_1) + P_c(V_2 - V_1)$$
$$= \frac{5}{2}P_c(V_2 - V_1) \quad (17\text{-}23)$$

When the ideal gas is compressed, all the terms on the right of Equation 17-23 are negative, and heat must flow out of the system. However, if the ideal gas expands, all the terms on the right of Equation 17-23 are positive, and heat must flow into the system. From Equation 17-21, we can see that in the case of an expansion ($V_2 > V_1$), the work done by the gas is positive and, hence, comes at a cost to the internal energy of the gas; for a compression the converse is true.

EXAMPLE 17-8

Work, Heat Flow, and Internal Energy

1.00 mol of an ideal gas is at a temperature of 293 K and a volume of 22.0 L. It is compressed in an isobaric process to a volume of 10.0 L. Find the work done, the heat flow, and the change in internal energy.

Solution

We can use Equation 17-20 to calculate the work, but first we need to determine the pressure. To do that, we use the ideal gas law and solve for P:

$$P = \frac{nRT}{V} = \frac{1.00 \text{ mol} \times 8.314 \text{ J/(mol}\cdot\text{K)} \times 293 \text{ K}}{0.0220 \text{ m}^3}$$
$$= 110\,727 \text{ Pa}$$

We can then determine the work:

$$W = P(V_2 - V_1) = (110\,727 \text{ Pa})(0.0100 \text{ m}^3 - 0.0220 \text{ m}^3)$$
$$= -1330 \text{ J}$$

The change in internal energy is

$$\Delta U = \frac{3}{2}P(V_2 - V_1) = \frac{3}{2}W = -1990 \text{ J}$$

and the heat flow is

$$Q = \frac{5}{2}P(V_2 - V_1) = \frac{5}{2}W = -3320 \text{ J}$$

Making sense of the result

The work done by the gas is negative, as it always is in a compression.

We can use Equation 17-23 to determine the heat capacity for an ideal gas in an isobaric process. The heat capacity is the ratio of the heat that flows into a system to the resulting temperature change, so

$$C_p = \frac{Q}{\Delta T} = \frac{\frac{5}{2}P(V_2 - V_1)}{T_2 - T_1} \quad (17\text{-}24)$$
$$= \frac{5}{2}\frac{nR(T_2 - T_1)}{T_2 - T_1} = \frac{5}{2}nR$$

The subscript p denotes that this is a constant-pressure heat capacity.

Isochoric Processes

In an isochoric process, we keep the volume constant and vary the pressure. We can carry out such a process by placing the gas in thermal contact with a heat reservoir and allowing heat to flow into or out of the gas. Because the volume is constant, no work is done in such a process, and the first law becomes

$$\Delta U = Q \quad (17\text{-}25)$$

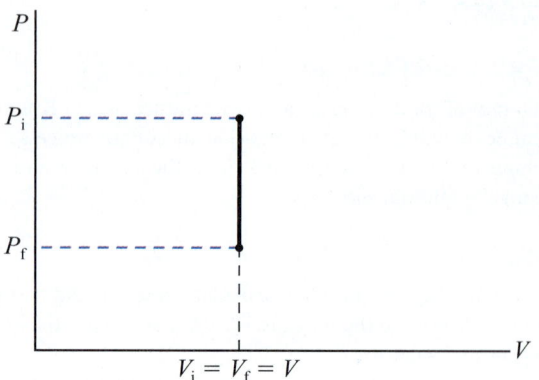

Figure 17-9 In an isochoric process, the pressure and temperature are varied and the volume remains constant.

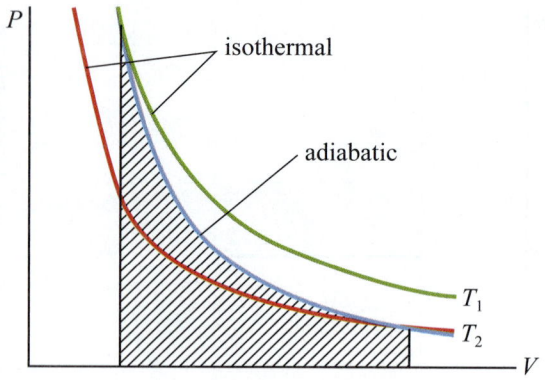

Figure 17-10 An adiabatic expansion

Figure 17-9 shows a *P-V* graph for an isochoric process. Note that since the volume does not change, there is no area under the curve and no work is done.

There is a change in the temperature of the gas during this process, and the change in internal energy is

$$\Delta U = \frac{3}{2}nR(T_2 - T_1) = Q \qquad (17\text{-}26)$$

The heat capacity for an isochoric process can be found from Equation 17-26:

$$C_V = \frac{Q}{\Delta T} = \frac{3}{2}nR \qquad (17\text{-}27)$$

The heat capacity for an isochoric process is less than the heat capacity for an isobaric process. For a given input of heat, we get a larger temperature change in an isochoric process than we get in an isobaric process. This is because in an isochoric process, no work is done by the gas, whereas in an isobaric process, the gas does work and some of the internal energy is converted into work done by the gas.

Adiabatic Processes

In an adiabatic process, there is no heat flow into or out of the system. Such a process can be accomplished by carefully insulating the system from the surroundings. Carrying out a process very quickly also reduces heat flow to and from the system. In an adiabatic process, the first law of thermodynamics becomes

$$\Delta U = -W \qquad (17\text{-}28)$$

In an adiabatic expansion, work is done and the internal energy of the system changes. For an ideal gas, this change means there is also a change in the temperature of the system and, hence, changes in the pressure and volume of the gas. Since the gas does work as it expands, the change in the internal energy is negative. If the process started at some temperature T_1 and

ended at a lower temperature, T_2, the *P-V* diagram for the process would look like Figure 17-10.

The online supplement for this textbook shows that for adiabatic processes in ideal gases,

$$PV^\gamma = \text{constant} \qquad (17\text{-}29)$$

where $\gamma = \frac{f + 2}{f}$ and f is the number of degrees of freedom, as discussed in Chapter 16.

A monatomic gas only has three translation degrees of freedom and therefore $f = 3$. Therefore, the work done in an adiabatic expansion of a monatomic ideal gas is

$$W = \int_{V_1}^{V_2} P dV = \int_{V_1}^{V_2} \frac{C}{V^\gamma} dV = \frac{1}{1 - \gamma} \frac{C}{V^{\gamma - 1}} \bigg|_{V_1}^{V_2} \qquad (17\text{-}30)$$

The constant C can be determined from the starting point using the ideal gas law.

State Variables

We now further investigate the idea that the internal energy is a state variable, but neither the work nor the heat is. We will compare two different ways of getting from an initial state to a final state for an ideal gas: an isothermal process that takes an ideal gas from (P_1, V_1, T) to (P_2, V_2, T) versus a two-step process with an isochoric stage from (P_1, V_1, T) to (P_2, V_1, T_i), followed by an isobaric stage from (P_2, V_1, T_i) to (P_2, V_2, T). These two different processes are illustrated in Figure 17-11.

For the one-step isothermal process, there is no change in the temperature and, therefore, no change in the internal energy. We know from Equation 17-17 that the work done is

$$W_{\text{isothermal}} = nRT \ln\left(\frac{V_2}{V_1}\right) = Q \qquad (17\text{-}31)$$

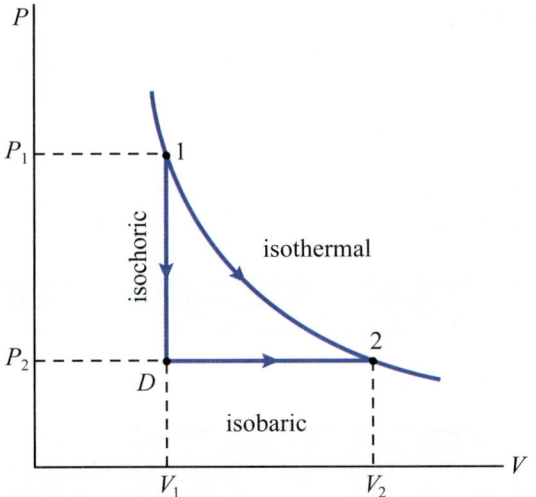

Figure 17-11 An isothermal process and an isochoric/isobaric process.

For the isochoric part of the two-step process, there is no work done but there is a change in temperature and, hence, a change in the internal energy:

$$\Delta U_{\text{isochoric}} = \frac{3}{2}P_2(V_2 - V_1) = Q_{\text{isochoric}} \qquad (17\text{-}32)$$

For the isobaric part of the two-step process,

$$W_{\text{isobaric}} = P_2(V_2 - V_1) \qquad (17\text{-}33)$$

$$Q_{\text{isobaric}} = \frac{5}{2}P_2(V_2 - V_1) \qquad (17\text{-}34)$$

and

$$\Delta U_{\text{isobaric}} = Q_{\text{isobaric}} - W_{\text{isobaric}}$$
$$= \frac{3}{2}P_2(V_2 - V_1) \qquad (17\text{-}22)$$

The net change in internal energy for the two-step isobaric–isochoric process is

$$\Delta U_{\text{isochoric+isobaric}} = \frac{3}{2}P_2(V_2 - V_1) + \frac{3}{2}P_2(V_2 - V_1)$$
$$= 0 = \Delta U_{\text{isothermal}} \qquad (17\text{-}35)$$

However, we can see that the net work in the isothermal process and the two-step process is not the same, and the net heat flows also differ.

17-8 Energy Transfer Mechanisms

We will conclude this chapter with a discussion of three different processes by which energy can move from one body to another.

Radiation

We are all familiar with the fact that a hot stove element radiates energy. In fact, all objects whose temperature is not zero kelvin emit electromagnetic radiation, which carries energy. An object radiates energy at a rate given by

KEY EQUATION $$P_{\text{radiate}} = \sigma \varepsilon A T^4_{\text{object}} \qquad (17\text{-}36)$$

where $\sigma = 5.67 \times 10^{-8}$ W/(m$^2 \cdot$K^4) is the **Stefan-Boltzmann constant**, ε is the emissivity coefficient of the object, A is the surface area of the object, and T is the temperature of the objects in kelvin. The emissivity coefficient is a characteristic of the surface of the object and has a value between zero and one. A surface with an emissivity of one is called a **blackbody radiator**. The details of blackbody radiation are discussed in Chapters 31 and 32. Table 17-3 lists emissivity coefficients for some common materials.

If we put some typical values into Equation 17-36, such as an emissivity on the order of one and a temperature, T, on the order of 300 K, we would see that many everyday objects at room temperature radiate energy at a rate of approximately 400 W/m^2. You might think that these objects would cool down rapidly. However, each object is also receiving radiation from its surroundings at a rate that depends on the ambient temperature:

$$P_{\text{absorbed}} = \sigma \varepsilon A T^4_{\text{ambient}} \qquad (17\text{-}37)$$

The net power radiated by an object is

KEY EQUATION $$P_{\text{net}} = P_{\text{radiate}} - P_{\text{absorbed}}$$
$$= \sigma \varepsilon A (T^4_{\text{object}} - T^4_{\text{ambient}}) \qquad (17\text{-}38)$$

Table 17-3 Emissivity Coefficients for Various Materials

Surface Material	Emissivity Coefficient
Aluminum, heavily oxidized	0.2–0.31
Aluminum, highly polished	0.039–0.057
Asphalt	0.93
Brick, red, rough	0.93
Concrete	0.85
Cotton, cloth	0.77
Glass, smooth	0.92–0.94
Gold, polished	0.025
Ice, rough	0.985
Oil paints, all colours	0.92–0.96
Porcelain, glazed	0.92
Paper	0.93
Plastics	0.91
Water	0.95–0.963

We have used the same constants and parameters in Equations 17-37 and 17-38. If these values were not the same, there would be a net flow of energy between an object and its surroundings when they are at the same temperature. Since such flows do not occur, we know that objects are equally efficient at radiating and at absorbing.

Conduction

When two objects are in contact with each other, the molecules at the contact surface can collide and transfer energy between the objects. Within a single object that has one end at a higher temperature than the other, the particles that make up the object itself can transfer energy between each other, which leads to a flow of heat from the hot end to the cold end.

MAKING CONNECTIONS

Where Does the Energy Go?

Staying warm in the winter in northern climates can be a challenge. In 2008, a total of 920.8 PJ (1 PJ $= 10^{15}$ J) was used for domestic space heating in Canada. This number represents 55% of total domestic energy consumption, excluding energy used in our domestic transportation. The false-colour thermal image in Figure 17-12 was produced by measuring infrared radiation from houses. The red areas have the greatest rate of energy loss. The house in the centre of the image is poorly insulated and is radiating significantly through the walls, as indicated by the orange-red colour of large areas of the walls. The house to the right of the image has well-insulated walls (mostly blue) but is radiating from the windows and under the eaves. Insulating drapes on the windows can reduce energy loss. More dramatic improvements in the windows can be achieved by using double-paned windows with a low thermal conductivity gas between them. It is also possible to use thin-film coatings on the glass to lower the emissivity.

Figure 17-12   A false-colour thermal image of a poorly insulated house showing hot spots in red and cool spots in blue.

EXAMPLE 17-9

Hot Coffee

A ceramic cup of hot coffee is placed on the counter. The liquid is at a temperature of 80. °C. Estimate the rate at which the cup and coffee are radiating energy.

Solution

We will use Equation 17-38, but we will have to make some assumptions about the surface area of the cup, the emissivity of the cup, and the ambient temperature. We can model the cup as a cylinder of radius 5.0 cm and height 10. cm. Then we get a surface area of $A = \pi r^2 + \pi r^2 + 2\pi r h = 0.047$ m². From Table 17-3, we see that that the emissivity of glazed porcelain is 0.92 and the emissivity of water is between 0.95 and 0.963. So,

we can use 0.92 as a reasonable approximation for the cup and coffee system. Assuming that the ambient temperature is 21 °C, and converting to kelvin, we get

$$P = (5.67 \times 10^{-8}\,\text{W} \cdot \text{m}^{-2} \cdot \text{K}^4)(0.92)(0.047\,\text{m}^2)$$
$$\times [(353.16\,\text{K})^4 - (294.16\,\text{K})^4]$$
$$= 20\,\text{W}$$

Making sense of the result

Putting your hand near a 20 W light bulb is similar to grabbing a hot cup of coffee. Putting a lid on the cup and insulating it can significantly reduce the surface temperature of the container and reduce the cooling rate.

Thermal Conduction

It is found experimentally that when we place a layer of material between a hot reservoir and a cold reservoir, the rate at which energy is transported from the hot side of the material to the cold side is given by

KEY EQUATION
$$P_{\text{conduction}} = \frac{\kappa A(T_{\text{hot}} - T_{\text{cold}})}{x} = \frac{\kappa A \Delta T}{x}$$

$$(17\text{-}39)$$

where κ is the thermal conductivity of the material in units of W/(m·K), A is the area, and x is the thickness of the material.

Table 17-4 lists the thermal conductivities of various materials. In general, gases tend to have low thermal conductivities, insulating solids have somewhat higher conductivities, and metals have the highest thermal conductivities (metals are also good conductors of electricity). In metals, both the atoms and some of the electrons contribute to the thermal

Table 17-4 Thermal Conductivities of Various Materials

Material/Substance	Thermal Conductivity at 25 °C (W/(m · K))
Argon (gas)	0.016
Brass	109
Brickwork, common	0.6–1.0
Carbon dioxide (gas)	0.0146
Concrete, stone	1.7
Copper	401
Glass	1.05
Gold	310
Hardwoods (oak, maple, etc.)	0.16
Insulation materials	0.035–0.16
Mica	0.71
Nitrogen (gas)	0.024
Paper	0.05
Snow (temperature < 0 °C)	0.05–0.25
Water	0.58

conductivity. For materials that do not conduct electricity, only the atoms in the material can transport the energy.

EXAMPLE 17-10

Efficient Windows

A 2.0 m² glass window is at a temperature of 20. °C on the inside surface and −20. °C on the outside. The glass is 5.0 mm thick. At what rate does energy flow out through the window glass? Compare this rate to the energy flow through a double-paned window that has a 2.0 cm gap between the panes filled with argon gas.

Solution

For the single-paned window, we use Equation 17-39 and the thermal conductivity of glass from Table 17-4:

$$P_{conduction} = \frac{1.05 \text{ W/(m·K)}(2.0 \text{ m}^2)(+20.\,°C - (-20.\,°C))}{0.0050 \text{ m}}$$

$$= 17\,000 \text{ W}$$

For the double-paned window, we will assume that both surfaces of the inner pane are at 20. °C and both surfaces of the outer pane are at −20. °C. Then, the rate of energy conduction through the argon between the panes is

$$P_{conduction} = \frac{0.016 \text{ W/(m·K)}(2.0 \text{ m}^2)(+20.\,°C - (-20.\,°C))}{0.020 \text{ m}}$$

$$= 64 \text{ W}$$

Making sense of the result

Clearly, investing in double-paned windows in your house can result in a significant reduction in energy consumption. We should also remember that windows allow energy to flow into a house in the form of sunlight and are important elements in a passive solar design.

Another measure of energy flow through a material that is commonly used when discussing insulation is the *R*-value, a measure of the thermal resistance of a material. The *R*-value combines the thermal conductivity with the thickness of the material, *l*, and is defined as

$$R = \frac{l}{\kappa} \tag{17-40}$$

In the SI system, the units for *R* are $m^2/(W \cdot K)$, often called RSI values. *R*-values simplify calculations for building insulation. If you use a multi-layered insulation system, you simply add the *R*-values for the individual layers to find the total *R*-value. To calculate the energy loss of a particular building element, you divide the area of the element by the total *R*-value and multiply by the temperature difference across the element.

EXAMPLE 17-11

Insulate the Roof

A cathedral (sloped) ceiling in a house has a total area of 100. m². The ceiling is covered with drywall (RSI 0.45), and the rafters are filled with insulation (RSI 40.). The roof is sheathed with 1/2 in. plywood (RSI 0.63) and asphalt shingles (RSI 0.44). Calculate the rate at which energy is lost, assuming that the interior temperature is 20. °C and the exterior temperature is 2.0 °C. Assume that energy losses from conduction through the wood rafters and from ventilation of the roof cavity are negligible.

Solution

$$P_{conduction} = \frac{A}{R}(T_{inside} - T_{outside})$$

$$= \frac{100 \text{ m}^2}{(0.45 + 40. + 0.63 + 0.44) \text{ m}^2/(W \cdot K)}$$
$$\times (20.\,°C - 2.0\,°C)$$

$$= 43 \text{ W}$$

Making sense of the result

While this may seem like a modest amount of power, we must remember that this is lost continuously and over time can add up to a significant amount of energy. Also note that we did not convert the temperatures to kelvin since they appear as a difference and temperature differences are the same in both scales.

Convection

The third energy transport mechanism is convection, which occurs in gases and liquids that are in a gravitational field. When we heat a portion of a gas or liquid, it expands and the density decreases. If there is a cooler, higher-density region above it, there is a buoyant force on the warmer portion, so the warmer portion rises and the cooler region falls.

You may have experienced getting into a swimming pool or a lake on a hot summer day and noticing that the top layers of the water are nice and warm, but the water farther down is much cooler. Sunlight striking the surface heats the water from the top. The less dense warm water stays at the top, and the cooler water stays below it. There is no circulation; thus, no energy is transported by the water other than by conduction from the top to the bottom of the lake. Figure 17-13 shows temperature measurements in a lake on a sunny summer day. Near the surface, the temperature approaches 25 °C, and at the deepest point it is only about 6 °C. The lake is **thermally stratified**.

However, when a fluid is heated from the bottom, the heated material rises and the cooler material at the top falls. When the cooler material gets to the bottom, it is heated and rises. As long as the rising material loses energy by some process, there will be a continuous **convection cycle**.

Hot water heating systems use both radiation and convection to distribute the energy throughout a room. The radiator through which the hot water is circulated is usually placed on or near the floor close to a wall. The radiator is built to have a large surface area to radiate energy, which is absorbed by the air surrounding the radiator. The heated air rises, and the cooler air falls. Because the radiator is placed against one wall, the hot air tends to spread out across the ceiling away from that wall, pushing the cooler air toward the opposite wall, where it falls, as shown in Figure 17-14. This cooler air is then drawn across the floor to the radiator, where it is heated. It then rises to the ceiling, continuing the convection cycle.

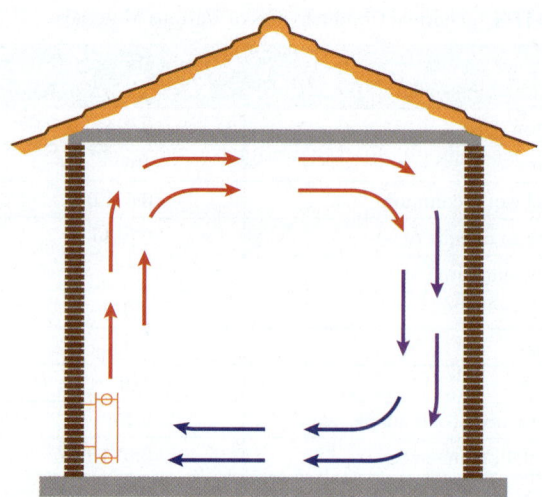

Figure 17-14   Thermal convection in a home heating system.

MAKING CONNECTIONS

Interesting Convection Patterns

A Rayleigh-Bénard cell consists of two parallel plates that are sealed together around the edges, with the space between the plates filled with liquid. The bottom plate is maintained at a higher temperature than the top plate. When the temperature difference between the top and bottom plates is quite small, there is only thermal conduction through the liquid. As the temperature difference increases, however, convection occurs, creating beautiful patterns like that shown in Figure 17-15. Different patterns emerge as the difference in temperature between the plates changes. When the temperature difference is great enough, the convection becomes completely turbulent. This sequence of transitions from a well-ordered system to chaotic turbulence is characteristic of nonlinear dynamical systems.

Convection is also very important in global climate. The Gulf Stream is a large convection current that transports warm water from the Gulf of Mexico to the North Atlantic. The Gulf Stream is responsible for keeping Europe warm. Palm trees can grow in southern England despite it being at about the same latitude as Winnipeg, Manitoba.

Figure 17-15   A time exposure of convection in hexagonal Rayleigh-Bénard cells. The warm fluid is upwelling from the centre of each cell and sinking around the edges. The flow has been made visible by adding small aluminum flakes to the liquid.

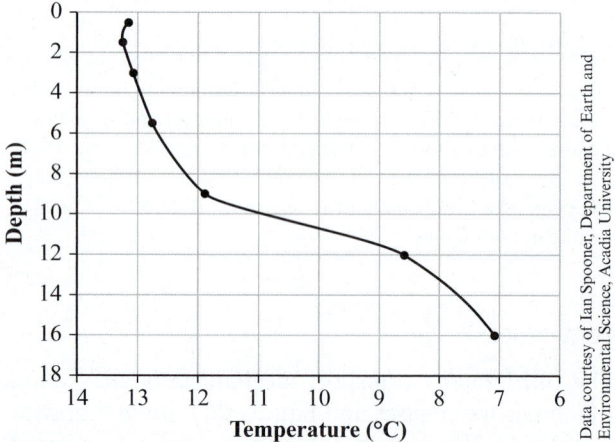

Data courtesy of Ian Spooner, Department of Earth and Environmental Science, Acadia University

Figure 17-13   Thermal stratification in a lake.

KEY CONCEPTS AND RELATIONSHIPS

What Is Heat?

When two objects of different temperature are placed in good thermal contact, energy flows from the hotter object to the cooler object. We say that we are heating the cooler object and we call the energy that is transferred in such a process heat.

It is important to distinguish between heat and the internal energy of a system. The internal energy of a system can be changed by adding heat to the system and by the system doing work on its surroundings. In various processes with the same start and end points, the energy added/removed via heating/cooling and the work done may differ, but the net energy change of the system is always the same, which is the basis of the first law of thermodynamics. As a result, heat and work are not state variables.

Temperature Changes Due to Heat Transfer

The ability of a material to absorb heat is characterized by the material's specific heat capacity. When we heat a system and there is no phase change, the relationship between the heat added and the resulting temperature change is

$$Q = mc\Delta T \tag{17-3}$$

where Q is the heat flow, m is the mass, c is the specific heat, and ΔT is the temperature change.

Flow of Heat between Objects

When two objects are placed in good thermal contact, the heat that flows out of the hotter object flows into the cooler object. The final temperature of such a system can be found using

$$m_{hot}c_{hot}\Delta T_{hot} = -m_{cold}c_{cold}\Delta T_{cold} \tag{17-5}$$

Phase Changes and Latent Heat

When we heat a system and it undergoes a phase change, the temperature of the system does not change. Instead, the heat is used to change the phase, and the heat required is given by

$$Q_{phase} = mL \tag{17-6}$$

where m is the mass of the system and L is the latent heat of the transition.

Work

When a system undergoes a volume change, the system does work that is given by

$$W = \int_{V_1}^{V_2} P\,dV \tag{17-13}$$

The work done depends on the path taken between the start and end points.

First Law of Thermodynamics

The first law of thermodynamics states that the difference between the heat that is added to a system and the work done by the system is independent of the path taken to carry out the process. This difference is called the internal energy, U:

$$\Delta U = Q - W \tag{17-14}$$

Different Types of Processes

There are countless ways that a system can be taken from one point on a P-V diagram to another. Some common processes are the following:

- Isobaric: constant pressure

- Isochoric: constant volume

- Isothermal: constant temperature

- Adiabatic: no heat flow

Energy Transfer Mechanisms

Energy transfer mechanisms include radiation, conduction, and convection.

All objects radiate energy as given by

$$P_{radiate} = \sigma \varepsilon A T_{object}^4 \tag{17-36}$$

They also absorb energy from the other objects in their surroundings. The net power due to these two processes is given by

$$P_{net} = \sigma \varepsilon A (T_{object}^4 - T_{ambient}^4) \tag{17-38}$$

where σ is the Stefan-Boltzmann constant, ε is the emissivity, and A is the surface area.

For conduction,

$$P_{conduction} = \frac{\kappa A (T_{hot} - T_{cold})}{x} = \frac{\kappa A \Delta T}{x} \tag{17-39}$$

where κ is the thermal conductivity, A is the surface area, x is the thickness of the material, and ΔT is the temperature difference.

In convection, a gas or fluid that is heated tends to expand and rise.

APPLICATIONS

heating, cooling, moving energy, energy-efficient windows and buildings, temperature control of mixtures

KEY TERMS

adiabatic, blackbody radiator, closed system, convection cycle, first law of thermodynamics, heat, heat capacity, heat of fusion, heat of vaporization, internal energy isobaric, isochoric, isotherm, isothermal, latent heats, open system, P-V diagram, quasi-static, specific heat capacity, Stefan-Boltzmann constant, thermally stratified

QUESTIONS

1. You are asked to heat a sample of gas using the least heat possible. Should you carry out a constant-volume process or a constant-pressure process?
2. Two boards of insulation are made of the same material. One board is twice as thick as the other.
   (a) Which board has the greater thermal conductivity?
   (b) Which board has the greater R-value?
3. Figure 17-16 shows a temperature versus time graph for a sample that is heated from the solid phase to the gas phase. What phases are present in each section?

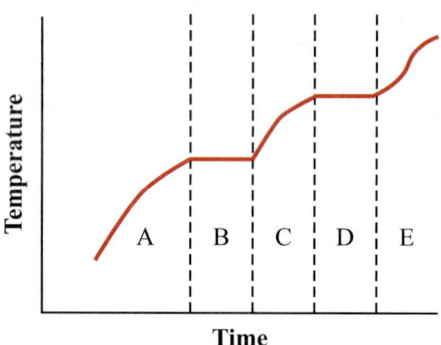

Figure 17-16 Question 3.

4. Two objects with the same mass are heated to the same temperature. They are placed into identical insulated water baths that are initially at room temperature. The final temperatures of the water baths with the objects in them are different. Would the bath with the higher final temperature contain the object with the greater or the lesser specific heat?
5. In the winter, even though the air temperature in our homes is 20 °C, we usually wear warm clothes. Conversely, on a summer evening, when the inside temperature of the house drops to 20 °C, we tend to be quite comfortable in shorts and a light shirt. Why? (Hint: Consider differences in radiant energy absorption by your skin.)
6. Figure 17-17 shows a temperature versus time graph for a sample of ice and a sample of water that are mixed together. What is the final phase?

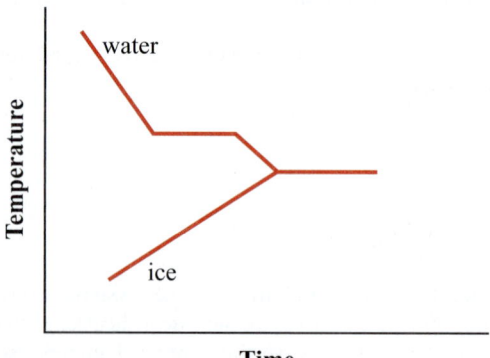

Figure 17-17 Question 6.

7. For the P-V process for an ideal gas shown in Figure 17-18, the system starts at A and goes to B. Is the work done positive or negative?

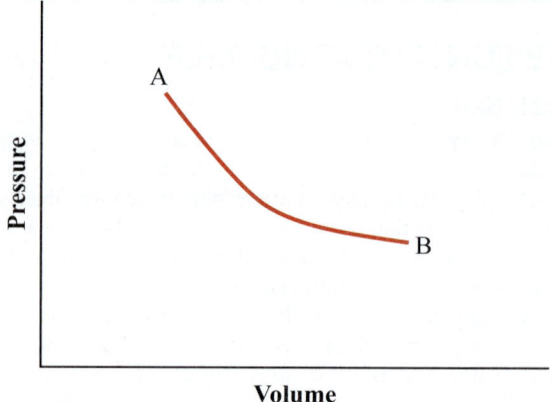

Figure 17-18 Question 7.

8. When we double the temperature of a hot object, by what factor does the rate at which it radiates energy increase?
9. When we double the temperature difference across an insulator, by what factor does the energy flow through the insulator increase?
10. Many animals sweat or perspire to keep cool in hot weather. How does perspiring help remove energy from your body?
11. In the cooking technique called *sous-vide*, raw food is sealed in an airtight plastic bag and then immersed in a hot water bath; the water temperature is maintained at the desired final temperature for the food. The types of changes that occur in the food as it cooks are determined primarily by the temperature of the water bath, so the flavour and texture of food cooked *sous-vide* differ from those of the same food cooked by other methods that use higher temperatures. Many types of food can be held in a *sous-vide* water bath for extended periods without overcooking. Explain why this is so.
12. Pressure cookers are used to shorten cooking times for foods that are boiled. How do pressure cookers work?
13. A hot object and a cold object are placed into thermal contact. Heat flows from the hot object to the cold object, and eventually they reach the same temperature. Under what circumstances is the final temperature halfway between the two initial temperatures?

PROBLEMS BY SECTION

For problems, star ratings will be used (★, ★★, or ★★★), with more stars meaning more challenging problems. The following codes will indicate if $\frac{d}{dx}$ differentiation, $\int$ integration, ▱ numerical approximation, or ⌤ graphical analysis will be required to solve the problem.

Section 17-2 Temperature Changes Due to Heat Transfer

14. ★ A 50.0 g block of copper is at a temperature of 20.0 °C. When 2.00 kJ of heat is added to the block, what temperature will it reach?
15. ★★ You wish to make tea, and you put 1.0 kg of water at 5.0 °C into your 500.0 W kettle. How long will it take for the water to boil? Assume no heat loss.

16. ★★ Alberta experiences chinooks in the winter, which are winds that can cause the air temperature to rise dramatically. In one instance, the temperature rose from $-30.$ °C to 5.0 °C. We assume that the atmosphere is heated by this amount to an altitude of 5.0 km. How much heat per square kilometre is required?

Section 17-3   The Flow of Heat between Objects

17. ★ A 100. g block of aluminum at a temperature of 100. °C is placed in a well-insulated water bath at a temperature of 10.0 °C. If the bath holds 500. g of water, what is the temperature of the system when it reaches equilibrium?

18. ★ 20.0 g of milk at a temperature of 2.00 °C is added to 300. g of coffee at a temperature of 170. °C. The coffee cup has a heat capacity of 109 J/K. What is the equilibrium temperature of the mixture?

19. ★★ You determine in an experiment that a 100. g sample of an unknown metal experiences a 5.0 °C temperature increase when 65 J of heat is added to the sample. What is the specific heat of the metal?

Section 17-4   Phase Changes and Latent Heat

20. ★★ A 500. W kettle has 1.00 kg of water in it at a temperature of 100. °C. How long will it take for the kettle to boil dry?

21. ★★ 20.0 g of ice at a temperature of -4.00 °C is placed in 100. g of water at a temperature of 30.0 °C. Find the final temperature of the mixture.

Section 17-5   Changing the Internal Energy Via Work

22. ★★ 1.00 mol of an ideal gas is compressed isobarically from a volume of 40.0 L to a volume of 30.0 L at a pressure of 110. kPa. How much work is done?

23. ★ 3.00 mol of an ideal gas is expanded isothermally at a temperature of 100. °C. During the expansion, the volume increases by 30.0%. How much work is done?

24. ★ A gas undergoes an isochoric process in which the pressure increases from 50.0 kPa to 100. kPa. How much work is done?

Section 17-6   The First Law of Thermodynamics

25. ★ An ideal gas undergoes an isothermal process in which the gas does 100. J of work. How much heat flows into the gas?

26. ★★ A gas undergoes an isothermal expansion from a certain pressure, temperature, and volume. The volume of the gas doubles. The same gas undergoes an adiabatic expansion from the same starting point, and again the gas doubles in volume. In which process is more work done? In which process is there a heat flow?

Section 17-8   Energy Transfer Mechanisms

27. ★ Comet Ikeya-Seki was observed to reach a temperature of approximately 650 °C at its closest approach to the Sun. At what rate does the comet radiate energy per square metre? Assume an emissivity of 0.50.

28. ★ A particular insulating material is 10.0 cm thick and has an RSI value of 40.0. What is the thermal conductivity of the material?

29. ★★ A 0.50 cm thick piece of copper has an area of 4.0 cm^2. One side of the copper is at a temperature of 100. °C, and the other side is at 400. °C. At what rate does energy flow through the copper?

COMPREHENSIVE PROBLEMS

30. ★★ The Sun delivers radiant energy to Earth at a rate that averages approximately 1.3 kW per square metre of the cross-sectional area of Earth. If we assume that all of this energy is absorbed by Earth (i.e., that Earth has an emissivity of one), at what rate must Earth radiate energy if Earth's temperature is to remain constant? What would that constant temperature be? (Note: The actual temperature of Earth is somewhat higher due to the greenhouse effect of the atmosphere.)

31. ★ $\frac{d}{dx}$ A researcher heats a 100. g sample using a 400. W resistive heater. The temperature of the sample increases at a rate of 3.0 °C/min. Find the heat capacity and the specific heat of the sample.

32. ★★ (a) The Sun has a surface temperature of approximately 6.0×10^3 K and a diameter of 1.4×10^6 km. What is the total power radiated by the Sun?
    (b) The Sun derives its energy from fusion reactions in which $E = mc^2$, where E is the energy released, m is the mass converted to energy, and c is the speed of light. At what rate does the Sun lose mass as radiated energy?

33. ★★ Athletes can lose significant amounts of water from their bodies as a result of perspiring during a competition. During an intense 2.0 h tennis match, a player lost 4.0 kg. Estimate the rate at which her body was using energy to evaporate all of that perspiration. List any assumptions you make for the estimate.

34. ★★ 3.00 mol of an ideal gas is at a temperature of 273 K and a pressure of 100. kPa. The gas is compressed isothermally to half its original volume. It is then expanded adiabatically to its original state. How much work is done in the compression? In the expansion? What is the highest temperature reached by the gas? The lowest?

35. ★★ 1.5 mol of an ideal gas is at standard temperature and pressure. The gas is isothermally compressed to one third its original volume. How much work is done?

36. ★★ The compression in problem 35 is carried out adiabatically. How much work is done in that case?

37. ★★ The Etruscan shrew is a small rodent that has a very high metabolic rate: its heart beats over 1000. times/min. The animal is approximately 4.0 cm long and has a body temperature of approximately 36 °C. Estimate the rate at which it loses energy due to radiation.

38. $\frac{d}{dx}$ ★★★ The ice on the surface of a lake in winter helps insulate the lake. For these questions, assume that the air temperature is $-10.$ °C, the lower surface of the ice is at 0.0 °C, and the ice is 10. cm thick.
    (a) How much energy per square metre is lost from the water to the atmosphere?
    (b) Using the rate in part (a), determine how long it will take another centimetre of ice to form.

39. ★★ An ideal gas is in a cylinder that has a light frictionless piston. The pressure outside the cylinder is 101 kPa. You slowly add 1000. kJ of heat to the gas and observe the volume to increase from 10.0 m^3 to 15.0 m^3. Determine the work done by the gas and the change in the internal energy of the gas.

40. ★★ An ideal gas is at an initial pressure of 3000. kPa and a volume of 7.00 L. The gas is expanded at constant pressure to a volume of 11.0 L. It is then cooled at constant volume until the temperature returns to its original value. Calculate the work done, the change in internal energy, and the heat flow into or out of the gas.

41. ★★ Consider an ideal gas that initially has a pressure of 1909 kPa and a volume of 11.0 L. The gas is compressed at constant pressure to a volume of 7.00 L. Heat is then added at constant volume until the pressure reaches 3000. kPa. Calculate the work done, the change in internal energy, and the heat flow into or out of the gas.

42. ★★ An ideal monatomic gas undergoes the increases in pressure and volume shown in the graph in Figure 17-19. The initial temperature of the gas is 297 K.
  (a) Calculate the work done.
  (b) Calculate the temperature of the gas at the end of the process.
  (c) Calculate the change in internal energy in this process.
  (d) Calculate the quantity of heat added to or removed from the gas during this process.

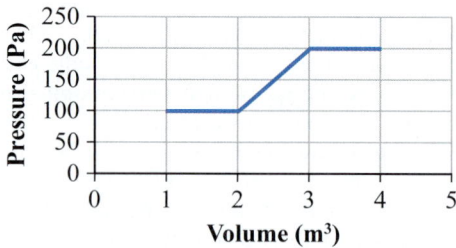

Figure 17-19 Problem 42.

43. An ideal gas is carried through a thermodynamic cycle consisting of two isobaric and two isothermal processes, as shown in Figure 17-20. Calculate the total work done in the entire cycle when $V_1 = 0.500$ m³, $V_2 = 2.00$ m³, $P_1 = 1.00 \times 10^5$ Pa, and $P_2 = 1.50 \times 10^5$ Pa.

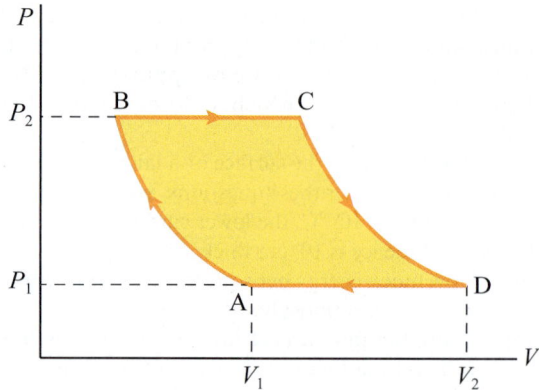

Figure 17-20 Problem 43.

44. ★★ 1.00 mol of an ideal gas is at a temperature of 50.0 °C and a pressure of 100. kPa. The gas is expanded isothermally to three times its initial volume. Find the final temperature, the heat input to the gas, the work done by the gas, and the change in the internal energy of the gas. Draw the process on a P-V diagram.

45. ★★ 1.00 mol of an ideal gas is at a temperature of 50.0 °C and a pressure of 100. kPa. The gas is expanded isobarically to three times its initial volume. Find the final temperature, the heat input to the gas, the work done by the gas, and the change in the internal energy of the gas. Draw the process on a P-V diagram.

46. ★★ 1.00 mol of an ideal gas is at a temperature of 50.0 °C and a pressure of 100. kPa. The gas is expanded adiabatically to three times its initial volume. Find the final temperature, the heat input to the gas, the work done by the gas, and the change in the internal energy of the gas. Draw the process on a P-V diagram.

DATA-RICH PROBLEM

47. $\frac{d}{dx}$ Table 17-5 lists temperatures recorded as a sample of water was heated from about 5.00 °C to 30.0 °C. The ambient temperature in the room was 20.0 °C. The heat was supplied by a 10.0 W heater. Determine the mass of the water being heated.

Table 17-5 Data for Problem 47

Time (s)	Temperature (°C)
0	4.79
40	6.53
80	8.04
120	9.86
160	11.05
200	12.50
240	13.56
280	14.94
320	15.80
360	17.04
400	17.83
440	18.90
480	19.98
520	20.90
560	21.92
600	22.51
640	23.63
680	24.28
720	24.94
760	26.00
800	26.31
840	27.03
880	27.71
920	28.62
960	28.98
1000	29.88
1030	30.06

OPEN PROBLEM

48. $\frac{d}{dx}$ ⌣ ★★★ We can reduce the cost of heating our homes by increasing the amount of insulation we put into them. Of course, there is a cost to installing the insulation. Suppose that energy costs $\$Y$ per joule and that insulation costs $\$X$ per square metre per centimetre. This seems like a peculiar unit, but the variable when installing installation is the thickness of the insulation and the cost of insulating is roughly proportional to the thickness of the insulation. Interpret the unit as cost per square metre of area per centimetre of thickness.

Determine the thickness of insulation one should use when building a house that minimizes the total cost of insulating and heating the house. You will have to make some assumptions about how long the house will be lived in, the thermal conductivity of the insulation, and the average inside–outside temperature difference.

CHAPTER 18

Heat Engines and the Second Law of Thermodynamics

When you have completed this chapter, you should be able to

LO1 Describe the operation of a refrigerator, a heat pump and a heat engine.

LO2 State the definitions of the efficiencies of various thermal engines and calculate them in terms of a Carnot cycle.

LO3 Describe entropy in terms of macroscopic variables.

LO4 State the second law of thermodynamics in several ways.

LO5 Discuss the circumstances under which thermodynamics is applicable.

LO6 Discuss some of the consequences of the second law of thermodynamics.

LO7 Describe entropy from a microscopic perspective.

What do the refrigerator and car engine in Figure 18-1 have in common? In the refrigerator, we supply electrical energy to power a motor, which does work on a gas. The gas is circulated so heat is transferred from the interior of the refrigerator to the coils on the back. Work is done on the system, and heat is moved from a lower to a higher temperature.

In the car engine, fuel is burned at a high temperature inside cylinders with pistons. The hot gases do work on the pistons, pushing them outward. The gases cool somewhat during this expansion and are then ejected through the exhaust pipes. The moving pistons rotate the crankshaft, which is connected to a gearing system that drives the wheels. In this engine, heat is moved from a higher temperature to a lower temperature, and the system does work on the car.

One of the goals of this chapter is to understand how heat pumps and heat engines work. As you will see, the study of these devices was a key step in the formulation of the second law of thermodynamics. The second law places some fundamental limits on the efficiency of these devices. As we shall also see, it places some restrictions on the types of processes that naturally occur in nature—it sets the direction of the "arrow of time."

The second law of thermodynamics introduces the concept of entropy. We will first describe entropy in terms

Figure 18-1 A refrigerator and an internal combustion engine

of macroscopic quantities: heat and temperature. Later in the chapter, we will look at a microscopic description. We begin our approach to the second law by examining in more detail how heat engines and refrigerators (and heat pumps) work.

18-1 Heat Engines and Heat Pumps

Heat Engines

A **heat engine** is a device that performs work while allowing heat to flow from a high-temperature reservoir to a low-temperature reservoir, as shown in Figure 18-2. A **reservoir** is an object whose heat capacity is so large that it can be considered to be at a constant temperature. The reservoir from which we are removing heat is called the source, and the reservoir into which we are putting heat is called the **sink**.

A high-temperature source provides heat to the engine. For a heat engine, the temperature of the source of heat (hot reservoir) is always greater than the temperature of the heat sink (cold reservoir). The engine does some work and also delivers a quantity of heat to the low-temperature sink. From the perspective of the engine, there is a positive flow of heat in from the hot reservoir, a negative flow of heat to the cold reservoir, and some energy lost by doing work on the surroundings. Since energy must be conserved,

KEY EQUATION
$$+|Q_H| - |Q_C| - |W| = 0 \qquad (18\text{-}1)$$

In Equation 18-1, we have used absolute value signs to avoid any ambiguity that could arise from the fact that heat flows can be either positive or negative, depending on our reference point. For example, Q_H is a negative heat flow from the perspective of the high-temperature source but a positive heat flow from the perspective of the engine.

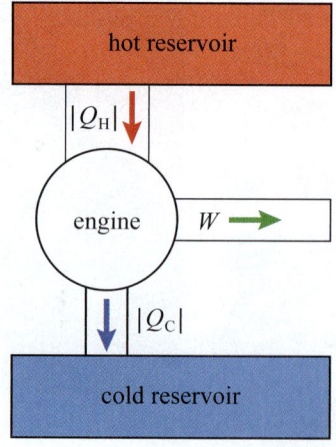

Figure 18-2 Schematic illustration of a heat engine.

Steam engines are a common type of heat engine. While obsolete for railway locomotives, a steam engine is the mechanism that converts heat into mechanical energy in thermal power plants (both fossil fuel–fired and nuclear). In a steam engine, water is heated to produce steam at a high temperature. In a fossil fuel–fired plant, hot combustion gases from burning fuel heat the water flowing through tubes in the boiler. In a nuclear plant, the reactor core has coolant pumped through it to capture energy from the fission reaction, and that energy is used to generate the steam. Such plants use a steam turbine to drive an electrical generator, as shown in Figure 18-3.

Often, pressurized water is circulated through the reactor core to both control the temperature of the core and capture the energy that is generated by the nuclear reactions in the core. Since the water is under high pressure, it can be heated to a high temperature (well above 100 °C) without boiling. The pressurized water circulates through pipes in the steam generator, where heat from these pipes vaporizes water in the turbine loop to form high-pressure steam. The reactor coolant, which has been exposed to intense radiation in the core, circulates in a closed loop of piping and does not mix with the water in the turbine loop. The steam is fed to the turbine, causing it to turn. The turbine drives the generator, producing electrical power. The steam expands and cools as it passes through the turbine into the condenser. However, not all the heat in the high-temperature steam is converted into work in the turbine. To increase the efficiency of the engine, the steam in the condenser is cooled so that the heat

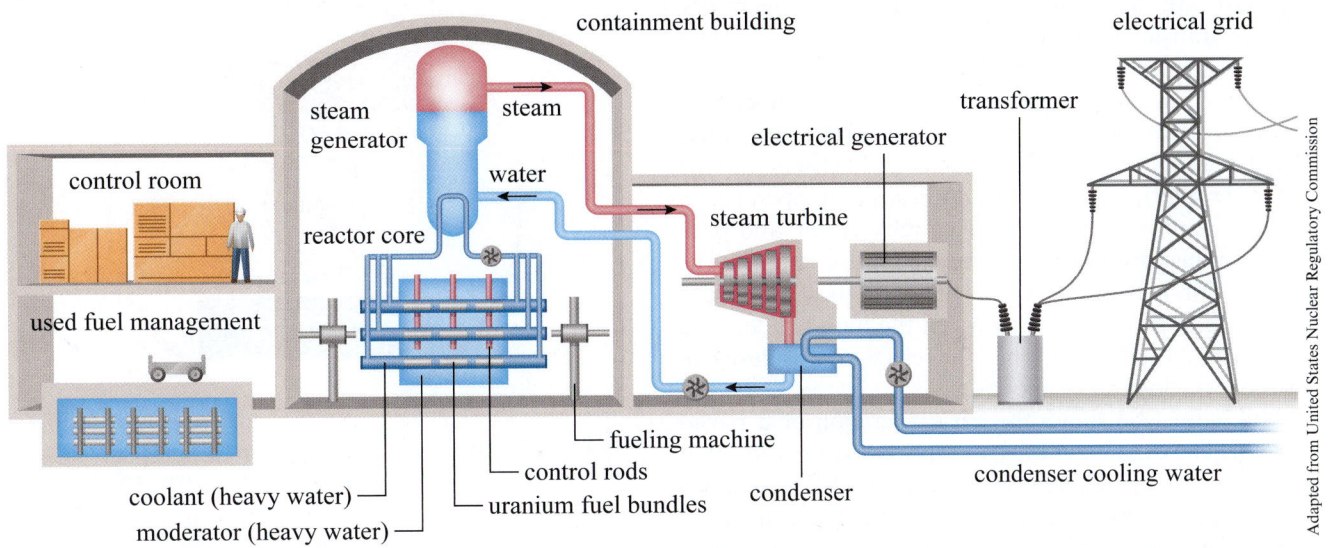

Figure 18-3 A steam turbine in a CANDU nuclear power plant. Considered one of Canada's top 10 engineering achievements, CANDU reactors provide about half the electrical energy consumed in Ontario.

ejected from the engine is at as low a temperature as possible. We will examine heat engine efficiencies in Section 18-2.

A heat engine always generates some waste heat. In a car, that heat is dissipated through the radiator, although some of this heat can be used to warm the passenger compartment on cold days. The fact that waste heat is unavoidable leads to the **Kelvin–Planck statement** of the **second law of thermodynamics**: no heat engine can convert heat completely into work.

The device shown in Figure 18-4 is not possible because the sole effect of the device is to convert heat completely into work.

Heat Pumps and Refrigerators

Heat pumps and refrigerators are both devices that use energy to transfer heat from a low-temperature source to a high-temperature sink, as shown in Figure 18-5.

When we apply conservation of energy to a heat pump, we find that the heat expelled to the high-temperature reservoir must equal the heat extracted from the low-temperature reservoir plus the work input. From the perspective of the pump, we can then write

KEY EQUATION
$$+ |Q_C| + W - |Q_H| = 0 \qquad (18\text{-}2)$$

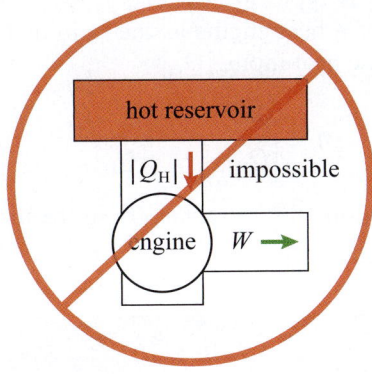

Figure 18-4 This heat engine process is impossible, a fact summarized by the Kelvin–Planck statement of the second law of thermodynamics.

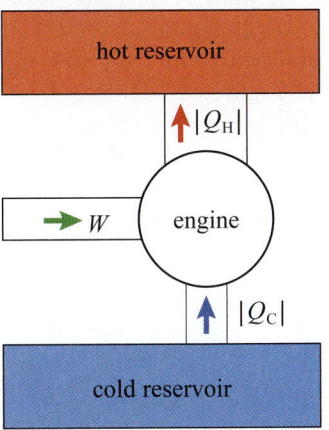

Figure 18-5 A heat pump and a refrigerator.

We are all familiar with household refrigerators, which transfer heat from a cool interior compartment to the warmer surroundings. With heat pumps, the focus is on the heat that is delivered to the higher-temperature area. For example, a heat pump for home heating extracts heat from the exterior of the house and delivers it to the warmer interior of the house. A heat pump operates on the same physical principles as a refrigerator, and both transfer heat from a cooler body to a warmer body. The distinction between the two is based on their purpose—a heat pump is intended to warm the high-temperature reservoir (your house), while the refrigerator is designed to cool the interior of the fridge.

Figure 18-6 shows the basic operation of a refrigerator. The expansion valve is a spring-loaded valve that requires a high pressure in the external heat exchange coil to open. This pressure is created by the compressor. When the pressure in that coil becomes sufficiently high, the expansion valve opens. Because it is a spring-loaded valve, the pressure on the external heat exchanger side of the valve is always higher than the pressure on the internal heat exchanger side. Thus, when the high-pressure gas passes through the expansion valve, the gas expands and cools to a temperature that is lower than the temperature of the interior of the refrigerator. Heat in the air in the refrigerator is then absorbed by the cool gas. The gas is then compressed as it is pumped into the external heat exchange coil. This compression raises the temperature of the gas above the ambient temperature in the room, so heat transfers from the coil on the back of the refrigerator to the room. Thus, the refrigerator transfers heat from its cool interior to the warmer room. To accomplish this transfer, energy must be supplied to the compressor, which does work on the gas.

We find that heat does not flow from a cooler body to a warmer body unless we do work on the system. This observation leads to the **Clausius statement** of the second law of thermodynamics: no device can transfer heat from a low-temperature reservoir to a high-temperature reservoir without work input. The heat pump shown in Figure 18-7 violates the Clausius statement because it has no work input.

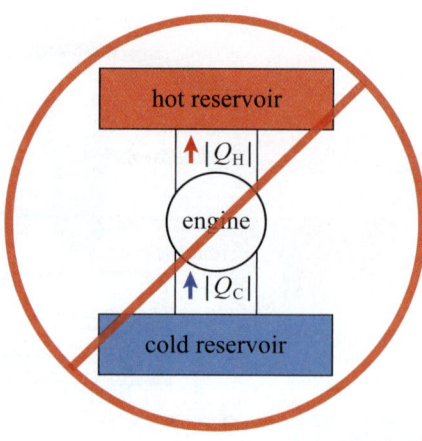

Figure 18-7 This heat pump process is impossible, a fact summarized by the Clausius statement of the second law of thermodynamics.

18-2 Efficiency and the Carnot Cycle

We have seen that heat engines cannot completely convert heat into work and that heat pumps (e.g., refrigerators) cannot simply move heat from a low temperature to a high temperature without work being done. Of course, when designing a heat engine, we would like to extract as much work as possible from a given quantity of heat. Similarly, we would like to minimize the energy that a heat pump requires to transfer a given quantity of heat. Now, we consider the efficiencies of these devices.

Heat Engine Efficiency

For a heat engine, we are primarily interested in maximizing the amount of work done for a given heat input. In so doing, we are also minimizing the heat output, which is typically lost energy. We define the **efficiency**, η, of a heat engine as the ratio of the work done, $|W|$, to the heat input, $|Q_H|$:

KEY EQUATION
$$\eta \equiv \frac{|W|}{|Q_H|} = \frac{|Q_H| - |Q_C|}{|Q_H|} \qquad (18\text{-}3)$$

We used Equation 18-1 to replace W in the numerator with $|W| = |Q_H| - |Q_C|$.

CHECKPOINT 18-2

Limits of Efficiency

A manufacturer of a heat engine claims that the engine has an efficiency of 110%. Is this claim reasonable?

ANSWER: No. Getting more work out than the energy you put in would violate conservation of energy. In addition, we can see from Equation 18-3 that the only way to get an efficiency greater than one is to have $|Q_C| < 0$, which is not possible.

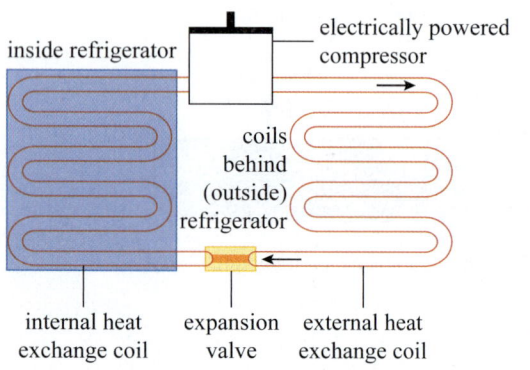

inside refrigerator

electrically powered compressor

coils behind (outside) refrigerator

internal heat exchange coil

expansion valve

external heat exchange coil

Washington University in St Louis. Chemistry 152. Refrigeration, Figure 5. http://www.chemistry.wustl.edu/~courses/genchem/Tutorials/Fridge/refrigeration.htm. © 2004, Washington University. Used by permission.

Figure 18-6 Schematic of a refrigerator.

EXAMPLE 18-1

Work, Heat, and Efficiency

An automobile engine has an efficiency of 20.0%. When the engine radiates heat at a rate of 100. kJ/s, how much mechanical work is being extracted?

Solution

This is a straightforward application of Equation 18-3, by which we can see that to determine W, we need to know Q_H. We are given Q_C, so we start by solving for Q_H:

$$|Q_C| = |Q_H| - \eta |Q_H|$$

$$\therefore |Q_H| = \frac{|Q_C|}{1 - \eta} = \frac{100.\ kJ/s}{1 - 0.200} = 125.\ kJ/s$$

We can now determine the work to be

$$W = |Q_H| - |Q_C| = 125.\ kJ/s - 100.\ kJ/s = 25.0\ kJ/s$$

Making sense of the result

The work done is certainly less than the total heat input, so we know that the engine has not violated the Kelvin–Planck statement of the second law of thermodynamics.

Efficiency of Heat Pumps and Refrigerators

For heat pumps and refrigerators, the definition of efficiency focuses on the amount of work done per unit of heat transferred. For a refrigerator, we are primarily interested in the heat transferred from the cool interior. Therefore, we define the **coefficient of performance of a refrigerator**, CP_R, as

KEY EQUATION
$$CP_R = \frac{|Q_C|}{|W|} \qquad (18\text{-}4)$$

In contrast, our primary focus with a heat pump is the heat delivered to the higher temperature. We define the **coefficient of performance of a heat pump**, CP_H, as

KEY EQUATION
$$CP_H = \frac{|Q_H|}{|W|} \qquad (18\text{-}5)$$

CHECKPOINT 18-3

Too Good to Be True?

A manufacturer of home-heating systems claims that its heat pump can deliver up to 3 J of heat for every joule of electrical energy it consumes. Is this even possible?

ANSWER: Yes. An examination of Figure 18-5 shows that the heat delivered to the high-temperature reservoir is, in general, greater than the energy supplied in the form of work. As we will see later in this section, the actual ratio depends on the difference between the inside and outside temperatures.

The Carnot Cycle

French scientist Nicolas Léonard Sadi Carnot (1796–1832) devoted considerable attention to maximizing the efficiencies of heat engines. His studies introduced some of the fundamental concepts of thermodynamics.

As is often done in physics, Carnot chose to study a simple model system under ideal conditions. He considered a heat engine that used an ideal gas in a

THERMODYNAMICS

reversible process. A **reversible process** is a process that is carried out so slowly that the system is always in equilibrium. A system is in equilibrium when none of the state variables vary with time or location within the system. An ideal gas is in equilibrium when the pressure and temperature are the same everywhere in the gas, the number of particles in the gas is constant, and the volume of the gas is constant.

All real processes are *irreversible*. For example, if we suddenly compress a gas, we create a pressure wave that travels throughout the gas. The wave is a region of high pressure, and the compression reduces the volume of the gas, so clearly the gas is no longer in equilibrium. As the wave propagates, the molecules in the gas collide with each other. The wave spreads out, and eventually the pressure becomes uniform. At this point, the gas has returned to equilibrium.

However, we could imagine breaking up the sudden compression into smaller, sudden steps. At each step, the departure from equilibrium is smaller than for the entire compression, and the time taken to return to equilibrium is shorter. If we break up the process into a series of infinitesimal steps, the departure from equilibrium becomes infinitesimally small, and the time away from equilibrium becomes infinitesimally short. In the limit, the process maintains equilibrium throughout. However, it will take an infinite amount of time to carry out the process. Nonetheless, this infinite process can be a useful theoretical construct.

Figure 18-9 illustrates the **Carnot cycle** for a heat engine. As indicated by the arrows, we proceed clockwise around the cycle. We could also proceed counter-clockwise to describe a Carnot cycle for a heat pump or refrigerator.

The Carnot cycle consists of four reversible processes. Starting in the upper left-hand corner of the cycle, we first carry out an isothermal expansion at T_H. During this process, heat $|Q_H|$ is absorbed by the gas. We then carry out an adiabatic expansion where

no heat flows into or out of the gas, but the temperature is lowered to T_C. Once we reach that temperature, we carry out an isothermal compression, and heat $|Q_C|$ is removed from the system. The final step is an adiabatic compression that raises the temperature back to T_H. During the two expansions, the gas does work on its surroundings. During the two compressions, the surroundings do work on the gas. The net work done on the surroundings is represented by the pink shaded area in Figure 18-9.

We can determine how much work is done in this cycle by calculating all the heat flows. In Chapter 17, we showed that for isothermal processes,

$$W = \int_{V_1}^{V_2} P dV = \int_{V_1}^{V_2} \frac{nRT}{V} dV = nRT \ln\left(\frac{V_2}{V_1}\right)$$

(17-17)

Thus, for the Carnot cycle we have

$$W_{1-2} = nRT_H \ln\left(\frac{V_2}{V_1}\right)$$

(18-6)

Since this is an isothermal process, there is no change in the internal energy of the gas. The first law of thermodynamics then reduces to $W = Q$ (Equation 17-16), and we can write

$$|Q_H| = nRT_H \ln\left(\frac{V_2}{V_1}\right)$$

(18-7)

All the same arguments apply to the isothermal compression. Therefore,

$$|Q_C| = nRT_C \ln\left(\frac{V_3}{V_4}\right)$$

(18-8)

Recall that $|Q_C|$ is the magnitude of the heat absorbed by the low-temperature reservoir. To make that quantity positive, we used V_3/V_4 as the argument of the logarithm.

For the adiabatic portions of the cycle, we can eliminate the volumes and express the heat flows simply in terms of the temperatures. In Chapter 17, we showed that for an ideal gas undergoing an adiabatic process,

$$PV^\gamma = \text{constant}$$

(17-29)

where $\gamma = \dfrac{f + 2}{f}$ and f is the number of degrees of freedom.

Thus, we can write

$$P_2 V_2^\gamma = P_3 V_3^\gamma$$
$$P_4 V_4^\gamma = P_1 V_1^\gamma$$

(18-9)

Since we are using an ideal gas, we can apply the ideal gas law to give

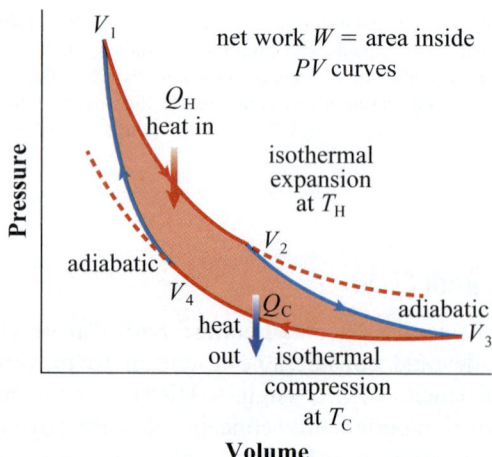

Figure 18-9 A *P-V* diagram for a heat engine operating on a Carnot cycle.

$$\frac{P_2 V_2}{T_H} = \frac{P_3 V_3}{T_C}$$

$$\frac{P_4 V_4}{T_C} = \frac{P_1 V_1}{T_H} \qquad (18\text{-}10)$$

In Equation 18-10, all four ratios are equal, but we have written the pairings to mirror those in Equation 18-9. Dividing the corresponding pairs in Equation 18-9 by those in Equation 18-10, we obtain

$$T_H V_2^{\gamma-1} = T_C V_3^{\gamma-1}$$
$$T_H V_1^{\gamma-1} = T_C V_4^{\gamma-1} \qquad (18\text{-}11)$$

Dividing the upper row by the lower row in Equation 18-11, we find

$$\left(\frac{V_2}{V_1}\right)^{\gamma-1} = \left(\frac{V_3}{V_4}\right)^{\gamma-1} \qquad (18\text{-}12)$$

Therefore,

$$\ln\left(\frac{V_2}{V_1}\right) = \ln\left(\frac{V_3}{V_4}\right) \qquad (18\text{-}13)$$

Using this result in Equations 18-7 and 18-8, and combining these two equations, we find (finally!)

$$\frac{|Q_H|}{|Q_C|} = \frac{T_H}{T_C} \qquad (18\text{-}14)$$

Rearranging gives

$$\frac{|Q_H|}{T_H} = \frac{|Q_C|}{T_C} \qquad (18\text{-}15)$$

We can use this result to determine the efficiency of a heat engine operating on a Carnot cycle:

KEY EQUATION

$$\eta = \frac{|Q_H| - |Q_C|}{|Q_H|} = 1 - \frac{|Q_C|}{|Q_H|} = 1 - \frac{T_C}{T_H} \qquad (18\text{-}16)$$

It is important to remember that these temperatures are expressed using the Kelvin scale. Note that the efficiency of such an engine is always less than one for an engine operated at finite temperatures. Since the Kelvin–Planck statement of the second law indicates that a heat engine cannot have an efficiency of 100%, T_C must be greater than 0 K. We will explore this lower bound to the Kelvin scale more fully in Section 18-6.

Carnot was able to prove that this efficiency applies for any reversible engine operating between two given temperatures. He was also able to show that any irreversible engine has a lesser efficiency. This result is called **Carnot's theorem** and can be stated as follows: the efficiency of any reversible engine operating between two fixed temperatures is the same, and any irreversible engine operating between the same two temperatures is not as efficient.

Thus, Carnot's theorem establishes the benchmark against which all real (irreversible) engines are measured.

As mentioned earlier, we can run the Carnot cycle in reverse. The coefficients of performance for a heat pump and for a refrigerator operating on a Carnot cycle are as follows:

$$\text{Heat pump: } CP_H = \frac{|Q_H|}{|W|} = \frac{|Q_H|}{|Q_H| - |Q_C|}$$

$$= \frac{|Q_H|}{|Q_H| - |Q_H|\dfrac{T_C}{T_H}} = \frac{1}{1 - \dfrac{T_C}{T_H}}$$

$$= \frac{T_H}{T_H - T_C}$$

Therefore,

KEY EQUATION

$$CP_H = \frac{T_H}{T_H - T_C} \qquad (18\text{-}17)$$

$$\text{Refrigerator: } CP_R = \frac{|Q_C|}{|W|} = \frac{|Q_C|}{|Q_H| - |Q_C|}$$

$$= \frac{|Q_C|}{|Q_C|\dfrac{T_H}{T_C} - |Q_C|} = \frac{1}{\dfrac{T_H}{T_C} - 1}$$

$$= \frac{T_C}{T_H - T_C}$$

Therefore,

KEY EQUATION

$$CP_R = \frac{T_C}{T_H - T_C} \qquad (18\text{-}18)$$

Note that in both cases the temperature difference $T_H - T_C$ appears in the denominator. As a consequence, the coefficients of performance of these devices decrease as the temperature difference increases. Thus, we can reduce the energy consumption of a refrigerator by placing it in a cool place. Similarly, heat pumps work more effectively when the low-temperature reservoir is not too cold.

CHECKPOINT 18-4

Practical Limits to Efficiency

Can a heat engine operating between 400. K and 300. K have an efficiency of 24%?

ANSWER: The maximum possible efficiency for a reversible engine operating between the given temperatures is

$$\eta = 1 - \frac{T_C}{T_H} = 1 - \frac{300. \text{ K}}{400. \text{ K}} = 0.250 = 25.0\%$$

Thus, an efficiency of 24% is theoretically possible. However, an efficiency this high is very unlikely because the best practical heat engines only approach approximately 75% of the maximum possible efficiency.

THERMODYNAMICS

18-3 Entropy

Now, we consider the direction of the heat flow during a Carnot cycle in a heat engine. Q_H is the heat absorbed by the engine at the high temperature, and our convention is that heat flow into the engine is a positive quantity. Therefore, $|Q_H| = Q_H$. However, Q_C is the heat transferred from the engine to the low-temperature reservoir and is therefore a negative quantity. Thus, $|Q_C| = -Q_C$. Therefore, we can write Equation 18-15 as

$$\frac{Q_H}{T_H} = -\frac{Q_C}{T_C} \tag{18-19a}$$

or

KEY EQUATION

$$\frac{Q_H}{T_H} + \frac{Q_C}{T_C} = 0 \tag{18-19b}$$

CHECKPOINT 18-5

Sign Conventions

How would Equation 18-19a change if we considered a heat pump instead of a heat engine?

ANSWER: There would be no change. We would replace $|Q_H|$ with $-Q_H$ and $|Q_C|$ with Q_C. Thus, Equation 18-15 would become $-\frac{Q_H}{T_H} = \frac{Q_C}{T_C}$, and multiplying both sides by -1 gives an equation identical to Equation 18-19a.

So far, we have demonstrated only that Equation 18-19a applies to a Carnot cycle. We now consider whether this relationship applies to any reversible process for an ideal gas. Any such process takes us from one point on a P-V diagram to another and can be accomplished with a combination of a reversible adiabatic process and a reversible isothermal process. The adiabatic process gets us from the starting temperature to the isotherm that the end point lies on. Then the isothermal process takes us to the end point, as shown in Figure 18-10. We could also start with an isothermal process and then carry out an adiabatic process.

We could break up any arbitrary reversible process that takes the system from A to B into a large number of reversible adiabatic and isothermal steps. The greater the number of steps, the better the approximation. Similarly, we could approximate any arbitrary reversible process that returns the system to point A from point B with a large number of reversible adiabatic and isothermal steps. Thus, any reversible process over a closed path (a path that returns the system to its starting point) can be approximated as a series of small adiabatic and isothermal steps. In the limit of an infinite number of steps, we would have a perfect approximation. Thus, any closed path

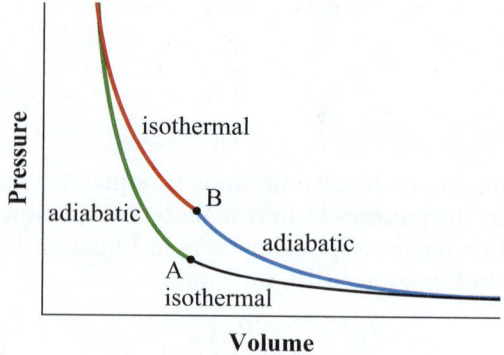

Figure 18-10 We can get from point A to point B via either the path at the upper left, which consists of an adiabatic process (green) followed by an isothermal process (red), or the path at the lower right, which consists of an isothermal process (black) followed by an adiabatic process (blue).

on a P-V diagram can be approximated with arbitrary accuracy with a series of Carnot cycles. Such a process is illustrated in Figure 18-11.

We now turn our attention to the heat flows that occur in such a stepwise process. We can generalize Equation 18-19b to

$$\sum_{\text{reversible}} \frac{Q_i}{T_i} = 0 \tag{18-20}$$

where Q_i denotes the heat flow for each successive step, and the temperature at which it occurs is T_i.

In the limit of an infinite number of steps, the summation becomes an integral:

$$\oint_{\text{reversible}} \frac{dQ}{T} = 0 \tag{18-21}$$

Since the integral sums to zero over a closed path, regardless of the shape of the path, there must be a state variable associated with this integral. Recall that a state variable has a well-defined value that is unique to the state of the system. For example, when we specify the temperature, pressure, and volume for an ideal gas with a fixed number of particles, we have specified the state of the gas. We saw in Chapter 17 that for an ideal

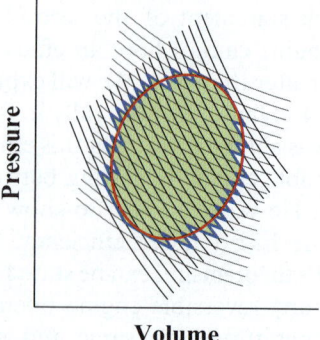

Figure 18-11 Approximating an arbitrary process with a series of Carnot cycles.

gas the internal energy is a state variable, but the work and the heat are not.

We call this new state variable **entropy** and denote it with the symbol S. For a reversible process going from a to b,

$$S_b - S_a = \Delta S = \int_a^b \frac{dQ}{T} \qquad (18\text{-}22)$$

From Equation 18-22 we can identify

KEY EQUATION
$$dS = \frac{dQ}{T} \qquad (18\text{-}23)$$

Entropy has units of J/K.

18-4 Entropy and the Second Law of Thermodynamics

We have seen several statements of the second law of thermodynamics so far, and it is not immediately apparent how they are related. As you will see, these statements are all linked by the concept of entropy. To show this, we will consider how the entropy of the system will change if each particular statement of the law is violated. Our approach is called proof by contradiction. If we see that the entropy changes in a certain way when the statement is violated, we conclude that the entropy cannot change in that way.

Clausius Statement Violated

The Clausius statement says that heat cannot flow from a cold temperature to a hot temperature without work being done. If such a process were to occur, the net effect would be a flow of heat out of the low-temperature reservoir while the same amount of heat flows into the high-temperature reservoir. The net change in entropy of the system comprising the two reservoirs is

$$\Delta S = -\frac{Q}{T_C} + \frac{Q}{T_H} < 0 \qquad (18\text{-}24a)$$

Thus, if the Clausius statement of the second law is violated, the entropy of the system will decrease.

On the other hand, if we operate a heat pump on a Carnot cycle, which is the most efficient way to operate a heat pump, Equation 18-15 tells us that $\frac{|Q_C|}{T_C} = \frac{|Q_H|}{T_H}$ and, therefore,

$$\Delta S = -\frac{|Q_C|}{T_C} + \frac{|Q_H|}{T_H} = 0 \qquad (18\text{-}24b)$$

We conclude that for an ideal heat pump, if we respect the Clausius statement of the second law, the entropy remains constant, but, if we violate the statement, the entropy decreases.

Kelvin–Planck Statement Violated

The Kelvin–Planck statement of the second law says that no process can completely convert heat into work. If such a process were possible, the net effect would be a removal of heat from a high-temperature reservoir. Thus, the change in entropy would be

$$\Delta S = \frac{-Q}{T_H} < 0 \qquad (18\text{-}25a)$$

Thus, if the Kelvin–Planck statement of the second law is violated, the entropy will decrease.

However, if we operate a heat engine on a Carnot cycle, we find that

$$\Delta S = \frac{|Q_C|}{T_C} - \frac{|Q_H|}{T_H} = 0 \qquad (18\text{-}25b)$$

where we have again used Equation 18-15. We reach the same conclusion here—for an ideal heat engine, the entropy remains constant. However, if we violate the Kelvin–Planck statement of the second law, the entropy decreases.

Carnot Theorem Violated

Suppose we have a heat engine that is more efficient than a Carnot engine. For example, the efficiency would be greater if the heat flow at the lower temperature were somewhat reduced with a corresponding increase in the amount of work done by the engine. We denote the reduced low-temperature heat flow as Q_C'. We are assuming that $Q_C' < Q_C$, where Q_C is the heat flow for an engine that does not violate Carnot's theorem. It must also be true that $\frac{Q_C'}{T_C} < \frac{Q_C}{T_C} = \frac{Q_H}{T_H}$. The change in entropy will then be

$$\Delta S = \frac{Q_C'}{T_C} - \frac{Q_H}{T_H} < \frac{Q_C}{T_C} - \frac{Q_H}{T_H} = 0 \qquad (18\text{-}26)$$

Thus, there would also be a decrease in entropy if a heat engine were more efficient than a Carnot engine.

Summary

We can see that in each of these processes that violate one of the statements of the second law, the entropy of the system *decreases*. We also saw that for a Carnot cycle, operated in either direction, the entropy of the system remained constant. Based on these observations, we might be tempted to conclude that the second law of thermodynamics requires that the entropy always remain constant. However, we have restricted ourselves to considering reversible processes so far, and all real processes are irreversible.

For a real heat engine, we find that we are able to extract less work than with a reversible process. Since energy is conserved, the heat delivered to the low-temperature reservoir must be greater than for a reversible process. The change in entropy for an irreversible process is

$$\Delta S = \frac{-Q_H}{T_H} + \frac{Q''_C}{T_C} > 0 \qquad (18\text{-}27)$$

for efficiency less than Carnot efficiency, where $Q''_C > Q_C$ is the increased amount of heat.

Thus, the entropy either remains constant (reversible processes) or increases (all real processes). The most general statement of the second law of thermodynamics is as follows: *The total entropy never decreases. The result of all real processes is that the total entropy increases.*

EXAMPLE 18-2

Entropy Change in a Reversible Process

Find the change in total entropy when a sample of liquid water at 0 °C is placed in thermal contact with a large reservoir that has a temperature infinitesimally lower than 0 °C.

Solution

Heat will slowly flow out of the liquid into the reservoir, and the liquid will freeze. The change in entropy of the liquid is

$$\Delta S_{liquid} = \frac{-mL}{273.15\,\text{K}}$$

where m is the mass of the liquid and L is the latent heat associated with the liquid–solid transition.

We can see that the entropy of the liquid decreases. However, we must also consider the change in the entropy of the reservoir. The heat that comes out of the liquid transfers to the reservoir. Since the reservoir is large and its temperature is only infinitesimally lower than the temperature of the liquid water, the amount of heat transferred to the reservoir is so small that the temperature of the reservoir does not change. The change in the entropy of the reservoir is

$$\Delta S_{reservoir} = \frac{mL}{273.15\,\text{K}}$$

Thus, the total change in the entropy of the system consisting of the liquid sample and the reservoir is zero.

Making sense of the result

The process is "reversible," so we would expect the change in the entropy to be zero. In fact, because the reservoir was at a temperature that was infinitesimally lower than 0 °C to get the sample to cool, the increase in entropy of the reservoir would be infinitesimally larger than the decrease of the water sample.

EXAMPLE 18-3

Entropy Change in an Irreversible Process

Find the change in the total entropy when 0.100 kg of liquid water at 10.00 °C is placed in thermal contact with a large thermal reservoir at exactly 0 °C.

Solution

Heat flows from the warmer sample into the cooler reservoir until the sample cools to 0.000 °C. The heat that flows out of the water sample is

$$\begin{aligned}
Q_{water} &= mc\Delta T \\
&= (0.100\,\text{kg})(4185.5\,\text{J}\cdot\text{kg}^{-1}\cdot\text{K}^{-1}) \\
&\quad (273.15\,\text{K} - 283.15\,\text{K}) \\
&= 4185.5\,\text{J}
\end{aligned}$$

The same amount of heat flows into the reservoir. Since the reservoir is large, its temperature does not change, and the change in entropy of the reservoir is

$$\Delta S_{reservoir} = \frac{Q_{reservoir}}{T_{reservoir}} = \frac{4185.5\,\text{J}}{273.15\,\text{K}} = 15.32\,\text{J}\cdot\text{K}^{-1}$$

However, the temperature of the liquid is not constant during the heat transfer. Recall from Chapter 17 that the heat flow in a process where there is a finite temperature change is

$$Q = mc\Delta T \qquad (17\text{-}3)$$

For an infinitesimal temperature change, Equation 17-3 becomes

$$dQ = mc\,dT$$

We can use this result to determine the change in entropy for the liquid:

$$\begin{aligned}
\Delta S_{liquid} &= \int \frac{dQ}{T} = \int_{283.15\,\text{K}}^{273.15\,\text{K}} \frac{mc\,dT}{T} \\
&= mc\ln T \Big|_{283.15\,\text{K}}^{273.15\,\text{K}} \\
&= (0.100\,\text{kg})(4185.5\,\text{J}\cdot\text{kg}^{-1}\cdot\text{K}^{-1})\ln\left(\frac{273.15\,\text{K}}{283.15\,\text{K}}\right) \\
&= -15.04\,\text{J}\cdot\text{K}^{-1}
\end{aligned}$$

The total change in entropy of the system is then

$$\begin{aligned}
\Delta S_{Total} &= \Delta S_{liquid} + \Delta S_{reservoir} \\
&= -15.04\,\text{J}\cdot\text{K}^{-1} + 15.32\,\text{J}\cdot\text{K}^{-1} \\
&= 0.280\,\text{J}\cdot\text{K}^{-1}
\end{aligned}$$

Making sense of the result

The change in the total entropy is positive, as we would expect for an irreversible process.

18-5 The Domain of the Second Law of Thermodynamics

The second law of thermodynamics establishes a direction for all processes. Only those processes that lead to a net increase in entropy are observed to occur.

In Example 18-3, we examined a situation in which an object was placed in contact with a cooler object. Heat flowed from the warmer object to the cooler object, and the total entropy increased. Imagine two bodies at the same temperature—placed in contact—and having heat flow from one to the other so that a temperature difference is created. Such a process would not violate conservation of energy because the exact amount of heat that left one body would move to the other body. However, this flow of heat would result in a net decrease in entropy and, hence, violate the second law. Such processes are never observed.

Let us consider some other natural processes for which the reverse process never occurs. Place a full glass of water on the counter in a sealed room. Over a few days, the water evaporates from the glass, and the room becomes somewhat more humid. However, placing an empty glass on the counter never results in the room becoming less humid and the glass filling with water. Similarly, when we open a bottle of perfume in a closed room, the scent spreads throughout the room and remains uniformly distributed after all the perfume has evaporated from the bottle. We do not observe the perfume refilling the bottle, nor does the scent concentrate in any one part of the room. In both of these examples, the water in the glass and the perfume in the bottle are at the same temperature as the room. It is not obvious how heat has flowed from the room into the perfume or the water to cause it to evaporate. However, we can see that the degree of organization or "order" of the system has changed. Having all the water in the glass is a state that is more ordered than having the water molecules dispersed throughout the room. Conversely, there is more disorder when the water or perfume molecules are dispersed throughout the room than there is when they are all in a container. Such examples lead us to another statement of the second law: all natural processes tend toward states of greater disorder.

So, we can interpret the entropy of a system as a measure of the disorder in the system. Increasing disorder equates to increasing entropy.

We can think of many examples of things that we see happen on a regular basis that are never observed to occur in reverse. If a glass falls off a counter and shatters on the floor, we would be startled to see the shards come back together and the reformed glass then leap back up onto the counter. In fact, we usually find it amusing to watch motion pictures run backward because of the absurdity of seeing such processes occur in reverse.

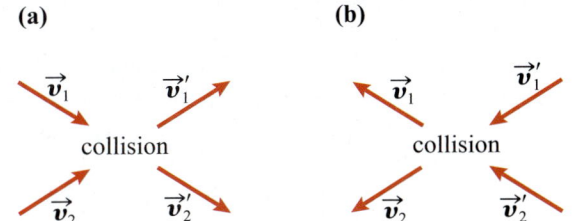

Figure 18-12 Two perfectly elastic collisions between two point particles.

However, we can think of certain events that are readily reversible. For example, consider a perfectly elastic collision between two particles, as shown in Figure 18-12. In the collision, both momentum and energy are conserved. We can see that the collision in Figure 18-12(b) is simply the reverse of the collision in Figure 18-12(a). If we were to observe either of these collisions, we would not in any way be able to tell if the "clock" was running backward or forward.

What about entropy? Doesn't the second law of thermodynamics apply to all processes? One important difference between the processes illustrated in Figure 18-12 and the processes that we discussed previously is the number of particles involved.

INTERACTIVE ACTIVITY 18-1

How Many Particles?

In this activity, you will use the PhET simluation "Reversible Reactions" to explore how changing the number of particles in a system affects some of the thermodynamic properties.

Consider a vessel full of some ideal gas and equipped with pressure and temperature sensors. As we saw in Chapter 16, the pressure is created by collisions between the gas molecules and the walls of the container. Similarly, the temperature is measured by allowing the thermometer to experience collisions with the gas molecules. These collisions transfer energy back and forth between the gas molecules and the material the thermometer is made of. Eventually, the net transfer of energy back and forth is zero, and the gas and the thermometer are in thermal equilibrium. At this point, the thermometric property of the thermometer stops changing and we can read the temperature. For nitrogen gas at standard temperature and pressure, there are approximately 10^{25} molecular collisions with the walls of the container per square metre per second. As a result, the readings on the pressure and temperature sensors appear constant.

Now consider what happens when we remove some of the gas molecules. We would expect the reading on the pressure sensor to decrease because there would be fewer gas molecules colliding with the sensor. If we were careful about how we removed molecules, it would be possible to keep the temperature constant. We would need to ensure that the average kinetic energy of the molecules we removed was equal to the average kinetic energy of the molecules in the gas. However, as we reduced the number of molecules further and further, we would find that the frequency of collisions with the pressure sensor would become so low that instead of a steady reading, we would see spikes in the readings as individual molecules hit the sensor.

Our thermometer readings would also start to become peculiar as the number of gas molecules got very low. As we saw in Chapter 16, not all of the molecules have the same speed in a gas; hence, there is a distribution of kinetic energies as well. As the frequency of collisions became very small, we would expect fluctuations in the temperature reading, assuming we had an "ideal" thermometer that responded instantaneously to each collision. When a very energetic molecule hit the sensor, it would deposit a lot of energy and make the sensor reading spike up. Conversely, when a low-speed molecule collided with the sensor, we would detect a transfer of energy from the sensor to the molecule, and the reading would spike down.

Once the number of molecules got so low that the readings of our pressure sensor and thermometer were no longer constant, these variables would no longer be well-defined, and we would begin to question whether a thermodynamic description of the system is appropriate.

How would we approach thinking about the entropy of a system that contained a single molecule? It would be very difficult to discuss, for example, the degree of disorder of a single gas molecule. Suppose we had two molecules. As they moved around the container, we would not be terribly surprised if at some point they passed very close to each other. However, if we had several moles of gas in the container, we would be completely surprised if all of those molecules congregated in the same place at the same time.

All of these state variables—temperature, pressure, and entropy—are based on averages, and these averages are meaningful only when taken over a large number of particles. The number of particles needed to make a thermodynamic approach valid is not precisely known. Strictly speaking, thermodynamics only applies perfectly in the limit of an infinite number of particles—the **thermodynamic limit**. As we reduce the number of particles, we begin to see fluctuations in the readings of thermometers and pressure sensors. Once the fluctuations become comparable to the readings themselves, the variables are no longer well defined. Then we are no longer in the thermodynamic limit, and the concept of entropy no longer applies.

When Does Thermodynamics Apply?

Which of the following systems can we describe using thermodynamics?
(a) a gas sample containing several moles of gas
(b) a hydrogen atom
(c) a transistor
(d) a galaxy

ANSWER: All but (b) should be amenable to a thermodynamic analysis, because these systems contain a very large number of particles.

18-6 Consequences of the Second Law of Thermodynamics

Absolute Zero

In Section 18-2, we found that the efficiency of a heat engine operating on a Carnot cycle is

$$\eta = \frac{|Q_H| - |Q_C|}{|Q_H|} = 1 - \frac{|Q_C|}{|Q_H|} = 1 - \frac{T_C}{T_H} \quad (18\text{-}16)$$

The Kelvin–Planck statement of the second law asserts that no heat engine can convert all the heat into work. Consequently, $|Q_C| > 0$ and $\eta < 1$. It follows that, for any finite T_H, $T_C > 0$. Thus, there is a lower limit on temperature below which one cannot go. That limit temperature is called **absolute zero**. Absolute zero is 0 K on the Kelvin scale and $-273.15\,°C$ on the Celsius scale. The lowest temperatures that have been produced in laboratories are on the order of 10^{-10} K. We note that this lower limit rests on the assumption that T_H is finite. Since temperature is a measure of the average kinetic energy of the particles in a system, we could have an infinite temperature only if we had an infinite amount of energy.

MAKING CONNECTIONS

How Cold Is That?

When we read that the lowest temperature recorded in a lab is on the order of 10^{-10} K, it sounds pretty impressive but it is difficult to understand just how cold that is. Suppose we had a thermometer that spanned Canada, with the 0 K mark at the clock tower on Citadel Hill in Halifax, and the 1 K mark at the top of Grouse Mountain in Vancouver. On such a thermometer, the 10^{-10} K would be located slightly less than half a millimetre away from the 0 K mark! The 2001 Nobel Prize in physics was jointly awarded to Eric A. Cornell, Wolfgang Ketterle, and Carl E. Wieman for their work with Bose–Einstein condensates at low temperatures.

You Are a Thermodynamic Fluctuation

The fact that the universe is not in heat death tells us something about the structure of the universe at its beginning. Because there is structure in the universe today, there must have been structure in the universe when it began. This structure is responsible for our existence. Evidence of this structure can be seen in two ways. The way that matter is distributed in the universe on a very large scale—how galaxies are distributed, for example—reflects the structure in the early universe. Other signs of structure are the fluctuations in the cosmic microwave background (CMB) radiation throughout the universe.

As we saw in Chapter 15, the Doppler shift of the light emitted from distant galaxies is evidence that the universe is expanding. Observations suggest that all matter in the early universe was primarily hydrogen, and most of the other elements were formed by fusion processes in stars that coalesced from the hydrogen gas. If we extrapolate the expansion backward to before the formation of stars, the universe becomes denser, and it becomes hotter if we model the hydrogen as a simple ideal gas (a reasonable approximation). At some earlier time, the temperature would have been hot enough that all the hydrogen atoms would have been ionized into electrons and protons. Light scatters from charged particles, and any radiation in the universe at that point would have been interacting significantly with the matter in the universe.

When we have light that is in thermal equilibrium with matter at a fixed temperature, the light has a characteristic pattern of frequencies called a blackbody spectrum. Analysis of such a spectrum reveals the temperature of the object. The CMB radiation is microwave radiation that appears almost uniformly throughout the sky. A detailed analysis of this radiation shows that it is consistent with an object that was in thermodynamic equilibrium with the radiation, and the characteristic temperature of the radiation is now 2.725 K. The "photon gas" that makes up the CMB radiation has also expanded and has been shifted to longer wavelengths and lower frequencies as a result. Consequently, it appears to have reached this low temperature from the temperature of approximately 3000 K at which the electrons and protons combined. Although the radiation is remarkably uniform, it is not perfectly so, and these fluctuations are believed to result from small deviations from equilibrium that existed very early in the universe (Figure 18-13).

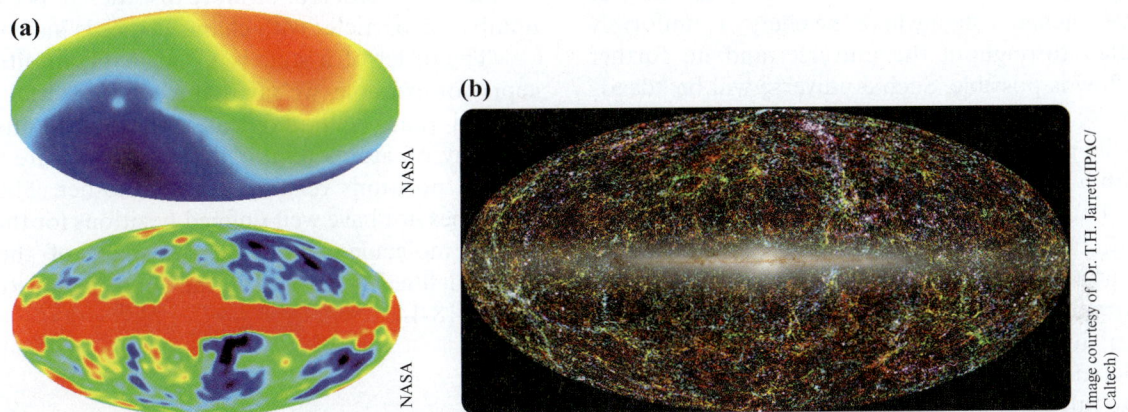

Figure 18-13 (a) Fluctuations in the CMB. False-colour images of the temperature inferred from CMB radiation measurements made by the COBE satellite. In the upper image, red corresponds to 2.729 K, and blue corresponds to 2.721 K. The dominant structure is due to the Doppler shift that results from Earth's motion. When that is removed, we are left with the bottom image, where the difference between red and blue is only 0.002 K. (b) Large-scale structure in the universe. A near-infrared image of the sky, which reveals the distribution of galaxies beyond our own. Galaxies that are coloured blue are closest, and galaxies that are coloured red are farthest away. The band across the centre is the Milky Way galaxy.

Heat Death

The second law of thermodynamics provides all real processes with a direction, which some call the "arrow of time." We now consider in which direction the arrow points and the ultimate fate of the universe.

All processes either move heat from a high temperature to a lower temperature and do some work, or require some work input to move heat from a lower temperature to a higher temperature. The second law establishes that to perform work, we must have temperature differences because there is no heat engine that can solely convert heat into work. Consider a simple model universe that consists of a high-temperature reservoir and a low-temperature reservoir. If we run a heat engine between these two reservoirs, the engine will perform work *and* transfer heat from the high-temperature reservoir to the low-temperature reservoir. The temperature of the high-temperature reservoir decreases as we remove heat from it, and the temperature of the low-temperature reservoir increases as heat is deposited into it. At some point, the two reservoirs reach the same temperature, making it impossible for the heat engine to produce any more work. This universe will have reached a state where the energy is uniformly distributed throughout the universe, and no further energy flow is possible. Such a universe will be "dead" in the sense that all matter will be uniformly distributed and no life possible.

Such a universe would not necessarily have zero energy. The energy would simply be uniformly distributed. Fortunately, this process will take a very long time to occur. The universe is estimated to be approximately 14 billion years old, and there is still plenty of heat flowing around. In fact, there is even some debate over whether this heat death will be the ultimate fate of our universe, because it depends on some things that we do not know, such as whether the universe will continue to expand or whether the expansion will stop and the universe will collapse back in on itself.

18-7 A Microscopic Look at Entropy

We begin our examination of how the microscopic details of a system are related to entropy by examining the concepts of macrostate and microstate.

A *macrostate* of an ideal gas is specified by knowing the pressure, volume, temperature, and number of particles in the system. We have no knowledge of the positions or speeds of any of the individual particles in the system. We do, however, have knowledge of the average speeds of the particles because, as we saw in Chapter 16, we can relate the temperature and the pressure of an ideal gas to those quantities. We also know that the average velocity of all the particles is zero if the gas has no net momentum.

The *microstate* of an ideal gas is specified by stating how many particles have a particular speed and the volume of the gas. Knowing this, for every speed that is observed in the gas, we can then calculate the total number of particles; the average speed; and, hence, the temperature and pressure.

Let us consider how many different microstates correspond to a given macrostate. The two quantities that must be conserved are the number of particles and the average speed of the particles. One way to create the macrostate is to use a microstate in which each particle has a speed equal to the average speed. There is only one way to do this, however, and we say that such a microstate is highly ordered.

Suppose we increase the speed of one particle and decrease the speed of another particle so we do not change the average speed. There are many ways to do this simple modification because we can change the speed by an arbitrary amount. If we extend this to changing the speeds of three particles, we get an even larger number of ways to do it. These microstates would be less ordered, or more disordered, because the number of particles with changed speeds increases.

The order of a state is a reasonably intuitive concept. For example, we say that a solid material is more ordered than its liquid state. The solid consists of an array of atoms and/or molecules that are in well-defined positions relative to one another. The liquid state does not have well-defined positions for the atoms and/or molecules, and even the shape of the liquid is less defined than that of the solid, as illustrated in Figure 18-14.

Figure 18-14 A melting ice cube. The solid is in a more ordered state than the liquid.

Similarly, the gas is less ordered than the liquid, and we expect the water from the melted ice cube to eventually evaporate into the air. The natural tendency is toward states of greater disorder.

It appears as though systems naturally evolve toward states of higher disorder, and from our consideration of the microstates of a gas, there are many more disordered microstates corresponding to a particular macrostate than there are ordered microstates.

A fundamental concept in statistics is that the probability of an event occurring is proportional to the number of different ways that the event can occur. For example, if we roll a die, there are six possible outcomes, each of which is equally likely. The probability of getting any number on the roll is then 1/6. Suppose we roll two dice, one red and one blue. There are now 36 possible outcomes for the roll because, for each number on the first die, there are 6 possible outcomes for the second die.

If we categorize the rolls based on the sum of the two dice, we get the results in Table 18-1. The most probable outcome is that the dice sum to 7 because that result can be achieved in the largest number of ways.

Table 18-1 Outcomes for Rolling Two Dice

Sum of Two Dice (Macrostate)	Possible Outcomes (Microstates)						Number of Outcomes (Microstates)
2	1, 1						1
3	1, 2	2, 1					2
4	1, 3	2, 2	3, 1				3
5	1, 4	2, 3	3, 2	4, 1			4
6	1, 5	2, 4	3, 3	4, 2	5, 1		5
7	1, 6	2, 5	3, 4	4, 3	5, 2	6, 1	6
8	2, 6	3, 5	4, 4	5, 3	6, 2		5
9	3, 6	4, 5	5, 4	6, 3			4
10	4, 6	5, 5	6, 4				3
11	5, 6	5, 6					2
12	6, 6						1

We make the analogy between the dice and the thermodynamics systems as follows. Knowing the sum of the two dice is equivalent to specifying the macrostate of the system. Specifying the result of each die is specifying the microstate.

The important point to note is that, in general, there is more than one way to achieve the same outcome. The same is true of a thermodynamic system. In general, there are many microstates that correspond to the same macrostate.

EXAMPLE 18-4

States for Coin Tosses

Enumerate the possible macrostates (number of heads and tails) and their corresponding microstates (head and tail for each coin) when four coins are tossed.

Solution

The possible macrostates are 4 tails, 3 tails and 1 head, 2 tails and 2 heads, 1 tail and 3 heads, and 4 heads.

There is only one microstate each corresponding to all heads (H, H, H, H) and all tails (T, T, T, T).

For the case of 3 tails and 1 head, we can have (T, T, T, H), (T, T, H, T), (T, H, T, T), and (H, T, T, T). The case

of 1 tail and 3 heads can be arrived at by switching all heads for tails, and vice versa. We then obtain (H, H, H, T), (H, H, T, H), (H, T, H, H), and (T, H, H, H).

The case of 2 tails and 2 heads can be arrived at with (H, H, T, T) and (T, T, H, H), (H, T, T, H) and (T, H, H, T), and (H, T, H, T) and (T, H, T, H). Note that each pair is the complement of each other, obtained by swapping all of the heads for tails and vice versa. We summarize our results in Table 18-2.

Making sense of the result

We find that the greatest number of microstates corresponds to the most "disordered" state: 2 heads and 2 tails.

Table 18-2 Results for Example 18-4

Macrostate	Microstates					Number of Microstates
4 tails	T, T, T, T					1
3 tails, 1 head	H, T, T, T	T, H, T, T	T, T, H, T	T, T, T, H		4
2 tails, 2 heads	T, T, H, H	H, H, T, T	T, H, H, T	H, T, T, H	T, H, T, H	H, T, H, T	6
1 tail, 3 heads	T, H, H, H	H, T, H, H	H, H, T, H	H, H, H, T		4
4 heads	H, H, H, H					1

THERMODYNAMICS

The results of a coin toss can be generalized to a large number of coins. For example, if we toss 100 coins, there are on the order of 10^{30} possible microstates. On the one hand, the probability of getting all heads or all tails is thus about 10^{-30}, a very small number. On the other hand, approximately 8% of the states are states in which there are the same number of heads and tails. Seventy-three percent of the states lie between 45% and 55% being all heads. As we increase the number of coins, more and more of the probability shifts toward the same number of heads and tails, and in the limit of an infinite number of coins, the probability of getting 50% tails and 50% heads is one.

The same thing happens in a thermodynamic system. All microstates are occupied with equal probability, but the macrostate that is observed is the one that has the largest number of microstates associated with it. In the thermodynamic limit of an infinite number of particles, this corresponds to a single macrostate. If we prepare a system in another macrostate that has fewer microstates, it will evolve toward the macrostate with the largest number of microstates, and this is also the most disordered state.

KEY CONCEPTS AND RELATIONSHIPS

Refrigerators, Heat Pumps, and Heat Engines

A heat engine is a device that moves heat from a high-temperature reservoir to a low-temperature reservoir and performs some work:

$$+|Q_H| - |Q_C| - |W| = 0 \qquad (18\text{-}1)$$

The Kelvin–Planck statement of the second law of thermodynamics is made in terms of a heat engine: no heat engine can convert heat completely into work.

A heat pump is a device that uses work input to move heat from a low-temperature reservoir to a high-temperature reservoir:

$$+|Q_C| + W - |Q_H| = 0 \qquad (18\text{-}2)$$

The Clausius statement of the second law of thermodynamics is made in terms of a heat pump: no heat pump can move heat from a low-temperature reservoir to a higher temperature reservoir without work input.

Efficiencies and the Carnot Cycle

The efficiency of a heat engine is defined as the ratio of the work done by the engine to the heat input:

$$\eta \equiv \frac{|W|}{|Q_H|} = \frac{|Q_H| - |Q_C|}{|Q_H|} \qquad (18\text{-}3)$$

We define the coefficient of performance of a refrigerator, CP_R, as the ratio of the heat removed from the low-temperature reservoir to the work required to do so:

$$CP_R = \frac{|Q_C|}{|W|} \qquad (18\text{-}4)$$

We define the coefficient of performance of a heat pump, CP_H, as the ratio of the heat delivered to the high-temperature reservoir to the work required to do so:

$$CP_H = \frac{|Q_H|}{|W|} \qquad (18\text{-}5)$$

A reversible process is an idealized process that occurs so slowly that the system remains in equilibrium throughout

the process. The most efficient engine possible is one that uses reversible processes. Adiabatic and isothermal processes are used, and such an engine is said to operate on a Carnot cycle. For all Carnot cycles operating between a low and a high temperature, we find that

$$\frac{Q_H}{T_H} + \frac{Q_C}{T_C} = 0 \qquad (18\text{-}19b)$$

The efficiency of a heat engine operating on a Carnot cycle is given by

$$\eta = \frac{|Q_H| - |Q_C|}{|Q_H|} = 1 - \frac{|Q_C|}{|Q_H|} = 1 - \frac{T_C}{T_H} \qquad (18\text{-}16)$$

The coefficients of performance for a heat pump and for a refrigerator operating on a Carnot cycle are as follows:

$$\text{Heat pump: } CP_H = \frac{T_H}{T_H - T_C} \qquad (18\text{-}17)$$

$$\text{Refrigerator: } CP_R = \frac{T_C}{T_H - T_C} \qquad (18\text{-}18)$$

Entropy

Equation 18-19b for a Carnot cycle suggests the existence of a state variable called entropy, defined by

$$dS = \frac{dQ}{T} \qquad (18\text{-}23)$$

Entropy and the Second Law of Thermodynamics

For reversible processes, the entropy remains constant; for all other processes, the entropy increases. This led to the most general statement of the second law of thermodynamics: entropy always increases or stays the same.

The Domain of Thermodynamics

Thermodynamics only applies to systems where the number of particles is large enough that statistical measures of their behaviour are valid.

Some Consequences of the Second Law of Thermodynamics

The idea of entropy also leads to an understanding of the "arrow of time." Processes that are always observed to occur in one direction but never in reverse are processes in which the entropy increases.

Entropy from a Microscopic Perspective

Entropy is a measure of the disorder of a system. The second law of thermodynamics can be stated as follows: all natural processes tend toward states of greater disorder.

APPLICATIONS

engine design, home heating, refrigeration, power generation, improving energy efficiency of devices

KEY TERMS

absolute zero, Carnot cycle, Carnot's theorem, Clausius statement, coefficient of performance of a heat pump, coefficient of performance of a refrigerator, efficiency, entropy, heat engine, heat pumps, Kelvin–Planck statement, reservoir, reversible process, second law of thermodynamics, sink, thermodynamic limit

QUESTIONS

1. Two physics students are trying to stay cool on a hot day. They read in their textbook that a refrigerator and an air conditioner are both heat pumps. They come up with the idea of opening the door of their refrigerator to try to cool their kitchen. Will this idea work? Explain.
2. Discuss how the order of a system changes as it is cooled from the gas phase to the solid phase.
3. An inventor claims to use the waste heat from a heat engine as heat input to the engine and achieve an efficiency greater than the Carnot efficiency. What is wrong with this claim?
4. When a liquid evaporates from a glass into a room, it is not completely clear how the latent heat of evaporation is supplied to the water for it to evaporate, because, if the water is in equilibrium with the room, the water is at the same temperature as the room and heat only flows spontaneously from high to low temperatures. Can this truly be an equilibrium process?
5. An inventor has designed a heat pump to recover waste heat from a heat engine and use it to resupply the engine. The inventor claims that adding this heat pump will improve the efficiency of any heat engine. Is this claim valid? Explain.
6. Which is the most effective way to increase the efficiency of a Carnot heat engine: increase the temperature of the high-temperature reservoir or decrease the temperature of the low-temperature reservoir by the same amount?
7. Two keen physics students tell their parents they do not want to clean their rooms because they feel they will be violating the second law of thermodynamics. Are they correct? Explain.
8. Describe a process that respects conservation of energy but violates the second law.
9. In Section 18-5, we considered an elastic collision between two particles and claimed that we were unable to distinguish between the collisions depicted in Figure 18-12 on thermodynamic grounds. Suggest another example.

PROBLEMS BY SECTION

For problems, star ratings will be used (★, ★★, or ★★★), with more stars meaning more challenging problems. The following codes will indicate if $\frac{d}{dx}$ differentiation, $\int$ integration, ▭ numerical approximation, or ∿ graphical analysis will be required to solve the problem.

Section 18-1   Heat Engines and Heat Pumps

10. ★ A heat engine exhausts 1100 J of heat in the process of doing 400 J of work. What is the heat input to the engine?
11. ★ A car has a fuel consumption of 5.8 L/100 km while travelling at 100. km/h. The power output of the engine is 40. kW. When we burn gasoline, the heat released is approximately 33 MJ/L.
    (a) How many litres of gasoline does the engine burn per second?
    (b) How much heat is released per second as a result?
12. ★ A heat engine absorbs 20. kJ of energy from a high-temperature reservoir and ejects 15 kJ of energy to a low-temperature reservoir. How much work does the engine do?
13. ★★ A camper has a propane-powered refrigerator. The refrigerator consumes approximately 0.50 kg of propane per day. When propane is burned completely, it releases approximately 50. MJ/kg. The interior of the refrigerator is maintained at an average temperature of 2.0 °C, and the exterior temperature is 22 °C. Assume that the refrigerator has a surface area of 2.0 m² and that the insulation has an R-value of 3.2 K·m²/W.
    (a) At what rate does heat flow into the refrigerator? (Hint: See Example 17-9.)
    (b) At what rate must the refrigerator remove heat from its interior?
    (c) How much energy is released per second from the propane combustion?
    (d) How much heat is delivered to the room?

CHAPTER 18 | HEAT ENGINES AND THE SECOND LAW OF THERMODYNAMICS

14. ★ A heat pump operates between the outside and the inside of a house. The pump absorbs 10. kJ of energy from the cool exterior and delivers 30. kJ to the interior. How much work is required to accomplish this?

Section 18-2 Efficiency and the Carnot Cycle

15. ★ A heat engine operates between a high temperature of 1000. °C and a low temperature of 500. °C. Assume that it operates on a Carnot cycle.
(a) What is the efficiency of the engine?
(b) When the heat engine absorbs 100. kJ from the high-temperature reservoir, how much work does it do?

16. ★ A heat pump operates on a Carnot cycle between the cool exterior of a house and the warm interior. When the exterior temperature is 2.0 °C and the interior temperature is 20. °C, what is the coefficient of performance for the heat pump?

17. ★★ The heat pump in problem 16 delivers 63 MJ of energy to the interior of the house. How much energy input is required in the form of work?

Section 18-3 Entropy

18. ★★★ ∫ Suppose we have two identical samples, one at a temperature of T_2 K and the other at a temperature of $-T_1$ K, where T_1 and T_2 are both positive. The two samples are placed in thermal contact.
(a) Determine the final temperature of the two samples.
(b) What is the change in entropy of the sample that was at temperature T_2?
(c) What is the change in entropy of the sample that was at temperature $-T_1$?
(d) What is the net change in entropy?
(e) Are negative temperatures possible on the Kelvin scale?

Section 18-4 Entropy and the Second Law of Thermodynamics

19. ★★ ∫ An ideal gas is placed in good thermal contact with a reservoir that is held at a constant temperature of 200. K. The temperature of the gas is initially 210. K. The volume of the gas is held constant.
(a) How much heat, per mole of gas, flows from the gas to the reservoir to cool the gas to 200. K?
(b) What is the change in the entropy of the reservoir?
(c) What is the change in the entropy of the gas? (Hint: You will need to do an integral.)
(d) What is the overall change in the entropy of the reservoir–gas system?
(e) Is this a reversible process?

20. ★★ A 20.0 g piece of copper at an initial temperature of 50.0 °C is placed in an insulated container that holds 300. g of water at an initial temperature of 20.0 °C.
(a) What is the final temperature of the system?
(b) How much heat flows out of the copper?
(c) How much heat flows into the water?
(d) What is the change in the entropy of the copper?
(e) What is the change in the entropy of the water?
(f) What is the total change in the entropy of the copper–water system?

21. ★★★ $\frac{d}{dx}$ A heavy-duty restaurant pot is placed on a hot burner on a stove. The temperature of the burner is 500. °C, and the pot is filled with tap water at a temperature of 10.0 °C. The bottom of the pot is steel with a thickness of 1.00 cm and an area of 0.0300 m².
(a) At what rate does heat flow through the bottom of the pot?
(b) What is the change in entropy per second of the burner?
(c) What is the change in entropy per second of the water?
(d) What is the net rate of change of entropy?

Section 18-5 The Domain of the Second Law of Thermodynamics

22. ★★ 〰 Suppose you toss six coins, which can land either heads or tails. If we treat each coin as being unique, there are 64 unique results for the toss. However, in terms of the number of heads and tails possible, there are only 7 possible outcomes for a toss, ranging from all heads and no tails to all tails and no heads.
(a) In how many ways can you toss
(i) all heads?
(ii) all tails?
(iii) only one head?
(iv) only two heads?
(v) three heads?
(vi) four heads?
(vii) five heads?
(b) Plot a histogram showing the number of ways each result can be achieved.

23. ★ There is a seemingly never-ending reduction in the size of electronic devices. However, at some point, transistors will become small enough that they will no longer be in the thermodynamic limit. How many atoms would be in a cube of silicon that is 20. nm on an edge?

24. ★★ 〰 In a coin toss experiment, there are only two possible outcomes, and the probability of getting a certain outcome is given by the binomial probability distribution. When we toss N coins, the probability of getting n heads is given by

$$P(n, N)\frac{N!}{n!(N-n)!}\,0.5^N$$

(a) Use software (such as a spreadsheet) to calculate $P(n, N)$ for tossing 100 coins for $0 \le n \le 100$.
(b) What is the probability of getting 50 heads?
(c) What is the probability of getting between 45 and 55 heads?

Section 18-6 Consequences of the Second Law of Thermodynamics

25. ★★ ∫ A 200. g block of copper is cooled to very low temperatures. Calculate the reduction in the entropy of the block in going from
(a) 100. K to 99.0 K
(b) 10.0 K to 9.00 K
(c) 2.00 K to 1.00 K
(d) 0.0200 K to 0.0100 K
For simplicity, assume that the specific heat of copper does not change as the block is cooled. (In fact, the specific heat is proportional to T^3.)

26. ★ Helium atoms interact very weakly with each other—helium only liquefies at 4.2 K at standard pressure. What is the kinetic energy of a helium atom
    (a) at 100. K?
    (b) at 10.0 K?

27. ★★ A suggested strategy to avoid heat death is to use the work from a heat engine to drive a heat pump to restore the original temperature difference. Is this possible?

COMPREHENSIVE PROBLEMS

28. ★★ Human beings are highly ordered arrangements of molecules. Do we violate the second law of thermodynamics?

29. ★ You have been asked to review a patent application for a new design for a heat engine. The inventor has supplied the following data on the application:

- The engine takes in 4.50 kW of heat at a temperature of 400. K.

- The engine expels 2.00 kW of heat at a temperature of 200. K.

- The engine performs work at a rate of 2.50 kW.
  Do you approve or reject the application? Explain.

30. ★★★ A reasonable approximation to the cycle in an internal combustion engine is the Otto cycle, which is depicted in Figure 18-15. At point 1, the piston is at the bottom of its stroke, and the cylinder is filled with exhaust gases. The exhaust valve was opened at point 4. The exhaust valve remains open, and the piston moves up in the cylinder, pushing the exhaust gases out. This is depicted by the blue horizontal line in the figure. Once the piston reaches the top of the stroke, the exhaust valve closes and the intake valve opens. The piston moves down and draws the fuel–air mixture into the cylinder. This is represented by the green horizontal line, and we have returned to point 1. The intake valve closes and the piston moves up, compressing and raising the temperature of the fuel–air mixture along the yellow line, which we approximate as an adiabatic process ending at point 2. The spark plug then ignites the fuel–air mixture, leading to a sharp increase in temperature and pressure, as indicated by the red segment going from 2 to 3, which we approximate as an isochoric process. This increase in pressure forces the piston downward along the red line from 3 to 4, and the pressure and temperature both decrease in this adiabatic process. At point 4, the exhaust valve opens, and the pressure and temperature drop along the blue line from 4 to 1 in another isochoric process. Compare the work done in this process with the work that would be done by a Carnot cycle operating between temperatures T_1 and T_3.

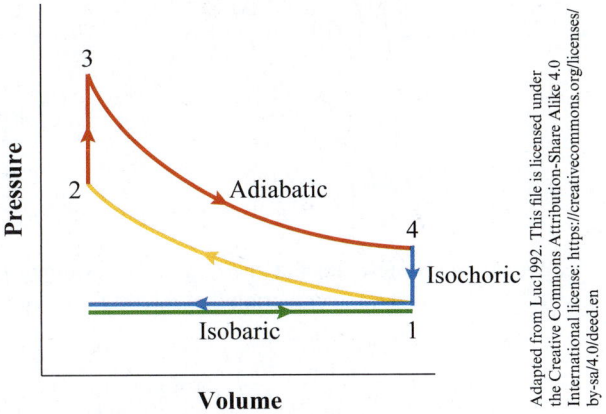

Figure 18-15 Problem 30. The Otto cycle.

31. ★★★ ∫ We want to investigate how the reading on a sensor in a system becomes more stable as we add more particles to the system. We will consider a very simple one-dimensional system of length L. All the particles in the system travel with the same speed, v m/s. A detector that has width $\Delta L \ll L$ is placed in the middle of the system; the detector measures some property of the particle. When a particle is in the detector, the detector reads a value Y_0, and when there is no particle in the detector, it reads zero. If we have a single particle in the system, it will travel from one end to the other in time $T = L/v$. During that time, it will pass through the detector once and spend time $\Delta T = L/v$ in the detector. As a result, the reading on our sensor will look something like that shown in Figure 18-16. Clearly, the sensor reading is not very stable.

Assuming that the particle enters the sensor at time t_0, the average value of the sensor reading is

$$\overline{Y} = \frac{1}{T}\int_0^T Y(t)dt = \frac{1}{T}\int_{t_0}^{t_0+\frac{\Delta L}{v}} Y_0 dt = \frac{1}{T}\frac{Y_0\Delta L}{v}$$
$$= \frac{v}{L}\frac{Y_0\Delta L}{v} = \frac{Y_0\Delta L}{L}$$

We will characterize the "wobble" in the sensor with a statistical measure called the standard deviation, σ_Y. It is defined as the positive square root of

$$\sigma_Y^2 = \frac{1}{T}\int_0^T (Y(t) - \overline{Y})^2 dt$$

Notice that this measure adds up the square of the deviations from the average value and is zero when the sensor reading is constant. For our sensor, with a single particle, we calculate this as follows. We will simplify the integrations by assuming $t_0 = 0$:

$$\sigma_Y^2 = \frac{1}{T}\int_0^T \left(Y(t) - \frac{Y_0 \Delta L}{L}\right)^2$$

$$= \frac{1}{T}\left[\int_0^{\frac{\Delta L}{v}}\left(Y_0 - \frac{Y_0 \Delta L}{L}\right)^2 dt + \int_{\frac{\Delta L}{v}}^0 \left(\frac{Y_0 \Delta L}{L}\right)^2 dt\right]$$

$$= \frac{Y_0 \Delta L}{L}\left(1 - \frac{\Delta L}{L}\right)$$

We skipped a few steps at the end. If we examine the ratio of σ_Y to $\overline{Y}$, remembering that $\Delta L \ll L$, we find

$$\frac{\sigma_Y}{\overline{Y}} = \frac{\sqrt{\frac{Y_0^2 \Delta L}{L}\left(1 - \frac{\Delta L}{L}\right)}}{\frac{Y_0^2 \Delta L}{L}} \approx \sqrt{\frac{L}{\Delta L}} \gg 1$$

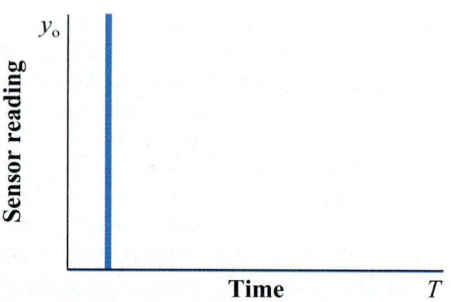

Figure 18-16 Problem 31.

Generalize this result to N particles, and show that the ratio gets small in the limit of large N.

32. ★★ A block of wood is sliding across a rough surface. The block has a mass of 10.0 kg and is moving with an initial speed of 10.0 m/s. It slides across the surface and comes to rest. The kinetic energy of the block has been converted into heat. Assume that the block and the surface are at a constant temperature of 293 K.
    (a) What is the kinetic energy of the block?
    (b) How much heat has been deposited in the block and the surface?
    (c) What is the change in entropy?
    (d) Is this a reversible process?

33. ★★ $\int$ A kettle is put on the stove and brought to a boil. 1.0 L of liquid water is converted to steam. What is the change in the entropy of the water if the initial temperature of the water was 22 °C?

34. ★★ $\int$ A typical shower uses approximately 6.0 L of water per minute. If the initial temperature of the water is 40. °C and the water ends up in the sewer at 5.0 °C, how much does the entropy of the water change during a 5.0 min shower?

35. ★★ Wood burns at a temperature of approximately 500. °C. One cord of seasoned (dry) firewood releases approximately 2.0×10^{10} J of energy when it is burned. All the energy is delivered to the room at approximately 20. °C. What is the change in the entropy of the wood? The change in the entropy of the air in the room? The total change in entropy of the wood–air system?

36. ★★ An air conditioner removes heat from the interior of a house at 20. °C and releases it to the exterior of the house at 35 °C. Assume that the system operates on a Carnot cycle. How much energy is removed from the interior per 1000. J of energy supplied to run the device?

37. ★★ $\int$ The temperature of 10.0 mol of an ideal monatomic gas is increased reversibly from 300. to 400. K. During the process, the volume is held constant. What is the change in the entropy of the gas?

38. ★★ Two constant-temperature reservoirs are kept in thermal contact with a 1.00 kg aluminum bar. One reservoir is at 100. °C, and the other is at 50.0 °C. After 1.00 kJ of energy has flowed from the hotter reservoir to the cooler reservoir, determine the change in entropy of
    (a) the hotter reservoir
    (b) the cooler reservoir
    (c) the bar

39. ★★ Heat pumps for home heating commonly have a coefficient of performance, CP_H, of about 3.5 when the temperature inside the home is 20 °C and the temperature outside is 10. °C. Estimate the coefficient of performance for such a heat pump when the outside temperature is −10. °C, assuming that the change in the coefficient is determined primarily by the difference between the inside and outside temperatures.

40. ★ Coal releases approximately 30. MJ/kg of heat when it is burned. A power plant produces approximately 1500 MW h while burning 7.0×10^5 kg of coal. How efficient is the plant?

41. ★★★ d/dx, $\int$ Ice on the surface of a lake forms an insulating barrier that slows further freezing. Assume that the temperature profile shown in Figure 18-17 is valid for this question.
    (a) For a given thickness of ice, at what rate will heat go from the water to the air?
    (b) At what rate will the entropy of the air change?
    (c) At what rate will the entropy of the water change?
    (d) At what rate will the thickness of the ice grow?
    (e) Put your answers to (a) and (d) together to show that the thickness of the ice is proportional to the square root of time.

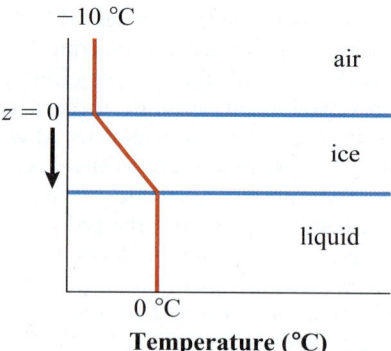

Figure 18-17 Problem 41.

42. ★★ When dealing four cards from a full deck of cards, determine the probability of dealing
    (a) four cards of the same face value (e.g., four threes or four queens)

(b) four completely different cards (i.e., four cards all with different face values and different suits)

(Hint: Consider how many different ways (microstates) each outcome (macrostate) can be generated.)

43. ★★ An object of mass m falls from height h and lands on the ground, coming to rest at temperature T. What is the change in the entropy of the system? Discuss the change in entropy in terms of order and disorder.

44. ★★ You clean your room and put all of your clothes and books away in an orderly arrangement. The disorder of your room has decreased. Have you violated the second law of thermodynamics? Note: This is *not* a good excuse for refusing to clean your room.

45. ★★ As we mentioned in the Making Connections feature called "You Are a Thermodynamic Fluctuation," there was some structure in the early universe. The fluctuations have been amplified by gravity over time to result in the large-scale structure we observe in the universe today: galaxies, etc. Thus, it seems that the universe has evolved from a very nearly uniform state (high disorder) to one of significant structure (low disorder). Does this violate the second law of thermodynamics?

46. ★★ The coefficient of performance of a heat pump decreases as the temperature difference increases, so heat pumps become less efficient on very cold days. A clever scientist came up with the idea of using two heat pumps in a home heating system. One transfers heat from the surroundings of the house to a heat reservoir and runs only when the exterior temperature is high enough for the process to be reasonably efficient. The second transfers heat from the reservoir to the interior of the house, and runs when the house needs to be heated. Discuss the practicality of such an arrangement. What would be a good material to use for the heat reservoir? How large would the reservoir need to be? Clearly state any assumptions.

47. ★★★ Assume that a Carnot heat engine operates between reservoirs T_2 and T_1. The work derived from the heat engine is used to drive an irreversible heat pump that operates between the same two reservoirs. Consider the following three cases:

(a) The efficiency of the irreversible heat pump is less than the efficiency of the Carnot engine.

(b) The efficiency of the irreversible heat pump is the same as the efficiency of the Carnot engine.

(c) The efficiency of the irreversible heat pump is greater than the efficiency of the Carnot engine.

Prove Carnot's theorem by showing that only one of these cases is possible.

48. ★★ ∫ 5.0 mol of an ideal monatomic gas is at standard temperature and pressure. Calculate the work done and the change in the entropy of the gas when the gas doubles in volume during

(a) an adiabatic expansion

(b) an isothermal expansion

OPEN PROBLEM

49. ★★ People have long looked to natural temperature gradients to exploit for driving heat engines. Examples include differences in sea water temperature between the surface and the depths, and differences in the temperature between Earth's surface and core. Estimate the efficiencies of these two examples, and identify any adverse environmental effects that a scheme using these differences may pose.

CHAPTER

19 | Electric Fields and Forces

During a spectacular lightning discharge, there is a flow of electric charge (Figure 19-1). How does the prior atmospheric charge separation occur, and what initiates the lightning strike? When we have separation of electric charges, we also have electrical forces. In this chapter, we will examine the electrical forces exerted on charged objects and define the new concept of electric field as electrical force per unit charge. We will apply electric fields to various natural and technological phenomena. Electric and magnetic fields are among the most important concepts in physics.

Figure 19-1 A lightning strike results in the rapid flow of a very large electric charge.

19-1 Electric Charge

What force supports you when you walk on the floor? While the obvious answer is the contact force of the floor on your body, we saw in Chapter 5 that this was not a fundamental force. The real source of the force is the electrical force that we consider in this chapter. The forces that muscles in your body exert are fundamentally electrical in nature, and indeed electrical signals play critical roles in almost all aspects of your body. Essentially all modern technology depends on electrical interactions. An electric car operates via electric motors powered by charge stored in batteries, electric energy is generated in power plants and transmitted on the electric grid to your home, and your smartphone is not only powered by electricity but depends critically on signals transmitted in the form of electromagnetic waves (see Chapter 27). Try this for yourself: over one day make a list of every interaction that you have that depends in some way on electromagnetism.

The electromagnetic interactions represent one of the four fundamental interactions and are among the most important topics in all of physics. Sensors and measurements in virtually all areas of science and engineering are electrical in nature, so no matter what your career plans, understanding electricity is important.

To have electrical forces, we must have electric charges. Just like mass, **electric charge** is a fundamental property. However, while mass comes in one form, electric charge comes in two forms (positive and negative). Electric charge is a conserved quantity: the net charge in a closed system does not change. As far as we know, the universe is electrically neutral overall; that is, it has equal amounts of positive and negative charges. Electric charges interact through exerting forces on each other: like electric charges (e.g., two negative charges) repel each other, and unlike charges (e.g., one negative and one positive) attract each other.

The SI unit for electric charge is the coulomb (C), named in honour of the French physicist Charles Augustin de Coulomb (1736–1806). Many (but not all) elementary particles have charge. For example, protons are positively charged, electrons are negatively charged, and neutrons have no charge at all. The magnitude of the electric charge of a proton or an electron is called the **elementary charge** and is usually represented by the symbol e. The coulomb is vastly larger than the elementary charge: $e = 1.60 \times 10^{-19}$ C, so 1 C contains 6.25×10^{18} elementary charges. The electron carries a negative charge, $-e$, while the proton carries a positive charge, $+e$. Later you will consider the charges on other particles and on ions.

While the charge transferred in a lightning strike is large, it is not as huge as you might have thought. As well as how much charge is moved, it is important

to know how quickly that process happens. **Current** is defined as the rate of flow of electric charge:

$$I = \frac{\Delta q}{\Delta t} \qquad (19\text{-}1)$$

where I is the current (in amperes) and Δq is the amount of charge (in coulombs) that flows past a point in a time Δt (in seconds). In terms of the SI base units, one coulomb is equal to one ampere times one second: $1 \text{ C} = 1 \text{ A} \cdot \text{s}$. When we take the limit $\Delta t \to 0$ we obtain the differential form of Equation 19-1:

$$I = \frac{dq}{dt} \qquad (19\text{-}2)$$

Sometimes the symbol i is used instead of I when we are talking about instantaneous values of a changing current. While the lightning discharge did not involve a particularly huge charge, it did involve a very large current flow because of the short time for the discharge (as we will see in Chapter 21, it also involved a very large energy).

CHECKPOINT 19-1

Charge and Current

Suppose that some dust particles have become charged such that each particle has a charge of 0.0200 mC. What current is flowing if 100 of these dust particles moving in the same direction pass a point in 0.100 s?

(a) 0.500 mA

(b) 2.00 mA

(c) 20.0 mA

(d) 200. mA

ANSWER: (c) With each charge being 0.0200 mC, if we have 100 of the charges, we will have a total charge of 2.00 mC flowing in 0.100 s. Thus, the current is $I = 2.00 \text{ mC}/0.100 \text{ s} = 20.0 \text{ mA}$.

EXAMPLE 19-1

How Many Charges per Second?

A current of 0.250 μA is flowing in an electrical circuit. How many elementary charge units (electrons in this case) are passing a point in this circuit each second?

Solution

First, we convert the current into SI units of amperes:

$$\frac{0.250\,\mu\text{A}}{1} \times \frac{1 \times 10^{-6}\,\text{A}}{1\,\mu\text{A}} = 2.50 \times 10^{-7}\,\text{A}$$

A current of 1 A means that the amount of charge flowing through the cross-sectional area of the circuit in 1 s is 1 C. Since the current we have is 2.50×10^{-7} A, we can calculate the amount of charge associated with it:

$$\Delta q = I\Delta t \Rightarrow \Delta q = (2.50 \times 10^{-7}\,\text{A})(1\,\text{s}) = 2.50 \times 10^{-7}\,\text{C}$$

To find the number of electrons flowing through the circuit per second, we can use the charge on each electron:

$$N_e = \frac{I}{e} = \frac{2.50 \times 10^{-7}\,\text{C/s}}{1.60 \times 10^{-19}\,\text{C/electron}} = 1.56 \times 10^{12}\,\text{electrons/s}$$

Therefore, we would have 1.56×10^{12} elementary charge units (in this case, they would be negatively charged electrons) flowing past a point in the circuit every second.

Making sense of the result

We see that even a small dust particle contains a large number of elementary charges. The elementary unit of charge, e, is very small, and most everyday charging situations contain large numbers of them.

Not all materials readily conduct current. We call a material that can readily conduct current a conductor, while one with zero or very low current conduction is an insulator. We will formally define these terms, and give examples, in the next section, but before that we suggest that you make two lists, one of substances that you believe are good electrical conductors, and another of substances that are insulators.

Static electricity refers to a lack of balance of positive and negative charges on an object and is one of the oldest recorded physics phenomena. Thales of Miletus (circa 600 BCE) made observations of static electricity. Indeed, the word "electricity" comes from the Greek word ἤλεκτρον for amber, one of the first materials found to be easily charged with static electricity. In ancient times, the terms *charge* and *electricity* were used interchangeably, but now the term *electricity* is a broad term for electrical phenomena.

Let us now consider how objects are charged. If on a dry day you scuff your feet across a carpeted floor, your body will become charged. When you then reach for a metal doorknob, current will flow to discharge your body. Complete Interactive Activity 19-1 to investigate this phenomenon.

Although each atom has positive and negative charges, most of the time objects are electrically **neutral** (no net electrical charge), or very nearly so. When we rub two different materials together, points on the surface of one object repeatedly contact and separate from points on the surface of the other object, causing positive charge to build up on one of the objects and negative charge on the other. This process is called **triboelectric charging**.

However, rubbing two materials together will result in significant charging only when the two materials have different electrical properties, with one holding on to electrons weakly compared to the other. Qualitatively, we can indicate the relative ability of materials to retain electrons by placing the materials in a **triboelectric series**, like the one in Table 19-1. Materials in the positive column acquire a positive charge, and materials in the negative column acquire a negative charge in triboelectric interactions.

We get significant triboelectric charging when we rub materials that are in different positions in the triboelectric series. For example, if we rub a rubber balloon

Table 19-1 The Triboelectric Series for Some Common Materials

Positive	Negative
Leather	
Glass	
Hair	
Nylon	
Wool	
Silk	
Paper	
Cotton	
	Wood
	Acrylic
	Rubber
	Polyester
	Polystyrene foam
	Cellophane tape
	Plastic wrap
	Polyethylene
	Polyvinyl chloride (PVC)
	Silicon

with hair, the table indicates that the hair will become positively charged and the balloon will become negatively charged. No significant charging will result from rubbing together two materials that are in the centre of the table, or two materials that are close together in the table and are both in the positive or negative column. For example, rubbing cotton with wood will leave the wood with a tiny amount of positive charge at most. While triboelectric charging often does involve the movement of electrons, at other times large macromolecules break, leaving charged fragments that produce the charging.

Because rubbing is the normal process to produce static charging, it is sometimes thought, incorrectly, to be necessary. All that we require is contact and breaking of contact between two appropriate surfaces (or indeed molecular collisions in some cases). This is still triboelectric charging, even though no rubbing is involved. Problem 96 provides an example of this in an experiment you can do at home.

There are other ways that objects become charged. When objects with different amounts of charging come in contact, conduction of charge can happen (see Section 19-2). Metals can be charged by the **photoelectric effect** (see Section 31-8), in which incident photons provide the energy needed to strip some electrons from the surface of the metal. Heating can also cause charging in some materials.

19-2 Charging by Electrical Induction

In some materials, a flow of electric charges can readily occur; these materials are called **conductors**. In the preceding section, we asked you to make up a list of good conductors. Examples of good conductors that might be on your list are metals such as copper, steel, gold, and aluminum.

In most situations, current is due to the flow of negatively charged electrons, and the flow of electrons is in one direction, with the defined direction of current flow in the opposite direction. You will learn more details about the processes of electrical conduction in Chapter 23.

In other substances, called **insulators**, there is almost no flow of electrical charges. Examples of electrical insulators are glass, porcelain, and most plastics. Materials called **semiconductors** are intermediate in conductivity and are very important in various technologies. You will learn more about these in Chapter 33.

When two oppositely charged conductors are placed in contact, a flow of charges results. Even if two objects are not in direct contact, one charged object can influence the distribution of charges on other objects—we call this charging by **induction**. We examine electrical forces later in this chapter, but to understand induction we need to know that unlike charges attract each other, while like charges repel each other.

It is normal to talk about distributions of positive and negative charges. Most of the time the mobile charge carriers are negatively charged electrons, but we can treat an absence of electrons as a positive charge.

As an example of induction, assume that we move a positively charged object near (but not touching) a second object. That second object is overall electrically neutral. The negative charges (electrons) in the second object are attracted by the positive charge and move closer to the first object. Similarly, positive charges on the second object could be (if mobile) repelled and shift away from the positively charged object.

Let's look step by step at the situation shown in Figure 19-2.

- We have two conducting spheres, on insulating bases. Initially (a) they are in contact. At this time, both are neutral and there is no separation of charge on either.

- In part (b) of Figure 19-2, they are still in contact, but now a positively charged third object is brought near to, but not touching, the left-hand sphere. The electrons in the left sphere are attracted to the left side, leaving a net positive charge on the right side of the right sphere, as shown.

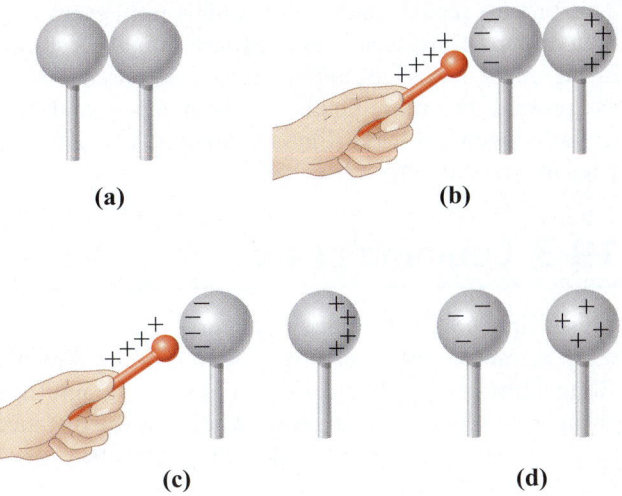

(a)   **(b)**

(c)   **(d)**

Figure 19-2 The temporary presence of a nearby (but not touching) third (red) positively charged object can cause charge separation on two initially neutral conducting spheres by induction. See the text for the details of each step. In this example the stands are insulating so negligible charge flows to or from the ground.

- While keeping the positively charged third object close to the left sphere, we separate the two identical spheres from each other. This is shown in part (c) of Figure 19-2. Remember that the bases are insulating, so no charges can flow to or from the ground.

- Finally, in part (d), we take the third object away. Note that now we have separated charge on the two conductors, with the left one negative and the right one positive. Each will be charged with the same magnitude of charge, but of opposite sign.

Courtesy of Marc Nantel.

Figure 19-3 The child became charged by triboelectricity as she slid down the plastic tube. The like charges on her hair cause it to repel and stand on end.

When a non-conducting object is charged, electrical charges with the same sign in the different parts of the object repel each other. This repulsion makes people's hair stand on end (Figure 19-3). When it is humid, there is enough conduction through the air that usually the amount of charging that remains is low.

In summary, objects can be charged by conduction (when charges actually flow between objects) and by induction (no actual contact). When some charge flows by conduction, both objects will have charge of the same sign. Some situations involve both processes. When analyzing a situation, it is important to be clear on whether flow from the ground is allowed or not.

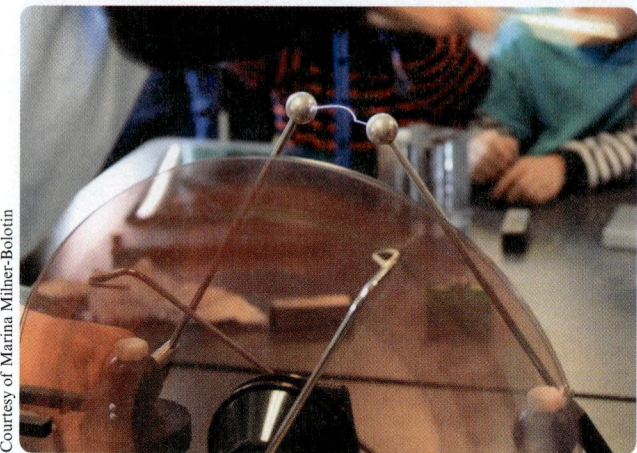

Courtesy of Marina Milner-Bolotin

Figure 19-4 A Wimshurst machine is an electrostatic generator that works by induction. The hand-cranked machine has two disks rotating in opposite directions. At a certain point the breakdown of air results in the jump of a spark, as shown.

An elegant device to produce impressive levels of electrostatic charging from manual cranking is the Wimshurst machine (see Figure 19-4). This device is named after the British engineer James Wimshurst (1832–1903), who perfected it in about 1883 (cruder forms had been constructed by others approximately 20 years earlier). The machine operates by electrical induction. It has two insulating disks (often of plastic) that rotate rapidly in opposite directions, each with a series of regular metal (often aluminum) plates. On each side are double-ended brushes, which are perpendicular on the two sides of the machine. The key idea is that, say, a negative charge on one of the metal plates on one side rotates until it lines up opposite the brush from the other side, when it will induce a negative charge on that brush.

This process repeats and rapidly builds until eventually breakdown of air occurs, and a spark jumps, as shown in Figure 19-4. A slight initial charge is required to start the process. For more details on the machine and how it works, view the MIT Physics video at https://www.youtube.com/watch?v=Zilvl9tS0Og.

19-3 Coulomb's Law

We saw in Chapter 11 that every point mass in the universe attracts every other mass with a force directed along a line joining their centres, and with a magnitude given by the following equation, where r is the distance between the centres of mass of the two objects:

$$F_G = \frac{Gm_1m_2}{r^2} \tag{11-1}$$

In an analogous way, every point electric charge exerts a force on every other point electric charge, with the force being attractive when the charges are of opposite sign (i.e., one positive and one negative charge) and repulsive when the two charges are of the same sign. We could express the magnitude of the electrical force by the following, where k is a universal constant sometimes called the **Coulomb constant** (note that although we use the same symbol this has nothing to do with the spring constant k), the two charges are q_1 and q_2, and r is the distance between the point charges:

$$F_E = k\frac{q_1q_2}{r^2} \tag{19-3}$$

This relationship is called **Coulomb's law**. Since we quite frequently deal with electrical forces along with some other force, such as gravitational, we will generally designate the force as $\vec{F}_E$, but at times when it is clear from context that we are dealing only with electrical forces, the notation is simplified to just $\vec{F}$.

To get the direction of the force, we use the fact that like charges (e.g., two positive or two negative charges) repel, while unlike charges attract. We show this in Figure 19-5. So while there is a similarity to the gravitational force, in the case of gravitation only attraction takes place, while both possibilities exist in electrical situations. Note that each charge exerts an

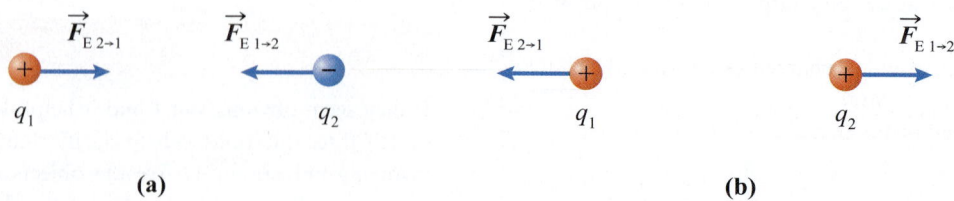

(a)   (b)

Figure 19-5 We show the direction of the electrical forces on each charge for the case of (a) unlike charges (attract) and (b) two positive charges (repel).

Whether I write it as $F_{E\ 1\rightarrow2}$ or $F_{E\ 12}$, it means the force that charge 1 exerts on charge 2. If we are talking about the force of charge 2 on charge 1, I write it as $F_{E\ 2\rightarrow1}$. In drawing an FBD for a charge, I include only the forces acting *on that charge*.

After I got used to it, I found the arrow notation used in this book helped me to avoid direction errors. Note that when dealing with a *displacement*, $\hat{r}_{1\rightarrow2}$ means a unit vector pointing in direction from object 1 toward object 2, but when dealing with *forces*, $\vec{F}_{1\rightarrow2}$ means the force that object 1 exerts on object 2.

electrical force of equal magnitude on the other, but the two forces are in opposite directions, as shown. Newton's third law requires that the force that charge 1 exerts on charge 2 (we will write this as $F_{E\ 1\rightarrow2}$) will be equal in magnitude but opposite in direction to the force exerted by charge 2 on charge 1 ($F_{E\ 2\rightarrow1}$).

While some treatments do use the Coulomb constant k, it is more normal to write this in terms of another constant of nature, ε_0:

$$k = \frac{1}{4\pi\varepsilon_0} \qquad (19\text{-}4)$$

Here ε_0 is a constant of nature called the **permittivity** of free space and has a value of 8.85×10^{-12} $C^2/(N \cdot m^2)$. Strictly speaking, this is the value in a vacuum, but the value in air is almost identical. In Chapter 20 we will show why we prefer to express the constant in terms of the permittivity. Coulomb's law is only applicable when the charges are not moving. You will see in later chapters how to handle situations in which the charges are moving.

While it is possible to solve electrical force problems using Equations 19-3 and 19-4 for the magnitude of the force, along with the fact that like charges repel and unlike charges attract to determine the direction, many physicists prefer to write Coulomb's law in vector form. To do this we first define a unit displacement vector pointing in the direction from charge 1 toward charge 2 as $\hat{r}_{1\rightarrow2}$. Recall that a unit vector is a dimensionless vector of magnitude 1. As we saw, the electrical force that point electric charge q_1 exerts on a second point charge q_2 is called $\vec{F}_{E\ 1\rightarrow2}$. Using this notation, Coulomb's law in vector form can be expressed as

KEY EQUATION
$$\vec{F}_{E\ 1\rightarrow2} = \frac{1}{4\pi\varepsilon_0}\frac{q_1 q_2}{r^2}\hat{r}_{1\rightarrow2} \qquad (19\text{-}5)$$

Here, the electric charges q_1 and q_2 must be expressed using the appropriate positive or negative sign. When the two charges have the same sign, then $\vec{F}_{E\ 1\rightarrow2}$ and $\hat{r}_{1\rightarrow2}$ will be in the same direction (Figure 19-6a), but when one charge is positive and one is negative they will be in opposite directions (Figure 19-6b). The distance between the point charges is r. To obtain a force in SI units (N), we must also use SI units for the charges (C), the separation distance (m), and the permittivity of free space ($C^2/(N \cdot m^2)$).

In Equation 19-5 and Figure 19-6 we show only the force that charge 1 exerts on charge 2. We can use Newton's third law to find the force that charge 2 exerts on charge 1. By Newton's third law the two forces must be equal in magnitude but opposite in direction.

CHECKPOINT 19-3

Electrical Force of Charge 2 on Charge 1

Consider the situation shown in Figure 19-6. What would be the direction of the force that charge 2 exerts on charge 1 (i.e., $\vec{F}_{E\ 2\rightarrow1}$)?

(a) In situation (a), $\vec{F}_{E\ 2\rightarrow1}$ will point to the left, and in situation (b) to the right.

(b) In situation (a), $\vec{F}_{E\ 2\rightarrow1}$ will point to the right, and in situation (b) to the left.

(c) In both cases, $\vec{F}_{E\ 2\rightarrow1}$ will point to the right.

(d) In both cases, $\vec{F}_{E\ 2\rightarrow1}$ will point to the left.

ANSWER: (a) By Newton's third law, the force that charge 2 exerts on charge 1 must be equal in magnitude but opposite in direction to the force that charge 1 exerts on charge 2.

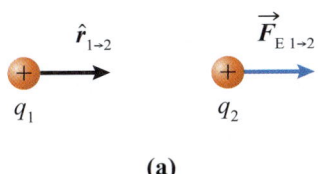

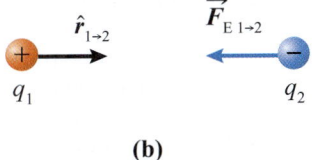

(a) (b)

Figure 19-6 When both charges have the same sign (a), the electrical force that charge q_1 exerts on charge q_2 is in the same direction as the unit displacement vector $\hat{r}_{1\rightarrow2}$ pointing from charge 1 to charge 2. When the charges have opposite signs, (b) the two vectors point in opposite directions.

ELECTRICITY, MAGNETISM, AND OPTICS

Coulomb's law applies to point charges. Of course, in the real world, we essentially always have not just single point charges (remember how tiny the unit of fundamental charge is), but a large number of electric charges in some configuration. In the next section, we will see how to apply Coulomb's law to situations with multiple point charges, and then in the following section to situations with continuous distributions of charges.

CHECKPOINT 19-4

Electrical Force Directions

Charge q_1 is a negative charge (say $-Q$) placed in the xy-plane at position ($x = 0$ m, $y = -1$ m), and charge q_2 is a positive charge (say $+3Q$) placed at ($x = 0$ m, $y = +1$ m). In terms of the ijk unit vector notation, what is the direction of the force acting on charge q_2?

(a) the $+\hat{i}$-direction
(b) the $-\hat{i}$-direction
(c) the $+\hat{j}$-direction
(d) the $-\hat{j}$-direction

ANSWER: (d) First, find the direction of the unit vector pointing from q_1 toward q_2, which must be in the $+\hat{j}$-direction. Since one charge is positive and the other is negative, there is a net negative sign in the equation for Coulomb's law. Thus, the direction of the force on q_2 must be opposite to the direction of the unit vector. This makes sense because unlike charges attract, so the top charge (2) will be pulled downward, in the $-y$-direction.

EXAMPLE 19-2

Charge on Hanging Masses

Two 100. g balls with equal but unknown charge q are hanging from massless strings, each making an angle of 15.0° with the vertical. The strings are 40.0 cm long. Find the magnitude of charge q. Can you deduce the sign of the charges from the information provided?

Solution

As shown in Figure 19-7, we label the charges q_1 and q_2. There are three forces acting on each ball: a repulsive electrical force, a downward gravitational force, and a tension force, directed along the string. Since the masses are equal, the two gravitational forces have the same magnitude and direction. We can break the tension force into x- and y-components so all forces are in one of those directions.

We will select the horizontal and vertical directions for our coordinate axes, and define positive as to the right and upward. As shown in Figure 19-7, we can break the tension forces into x- and y-components so all forces are in one of those directions. The system is in static equilibrium, so the forces must add to zero in each of the x- and y-directions. In the vertical direction, we have the following. Note that when we write a force without the vector sign we refer to the magnitude of that force:

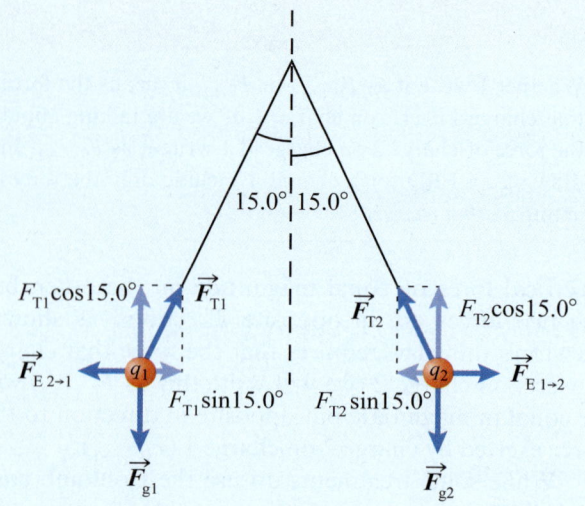

Figure 19-7 The charged balls repel each other. There are three forces: tension, $\vec{F}_T$; electrical, $\vec{F}_E$; and the gravitational force, $\vec{F}_g$, that together keep the system in static equilibrium.

$$F_{T1} \cos 15.0° - F_{g1} = 0$$
$$F_{T1} \cos 15.0° - mg = 0$$

If we substitute $m = 0.100$ kg, we obtain for the magnitude of the tension force

$$F_{T1} = \frac{mg}{\cos 15.0°} = 1.02 \text{ N}$$

Similarly, if we set the sum of the horizontal force components to zero for static equilibrium, we have the following for the forces on charge 1:

$$F_{T1} \sin 15.0° - F_{E2 \to 1} = 0$$
$$F_{E2 \to 1} = 0.264 \text{ N}$$

Now that we know the electrical force, we can apply Coulomb's law. We let q represent the value of each charge (they are both the same):

$$\vec{F}_{E2 \to 1} = \frac{1}{4\pi\varepsilon_0} \frac{q_1 q_2}{r^2} \hat{r}_{2 \to 1}$$
$$F_{E2 \to 1} = \frac{1}{4\pi\varepsilon_0} \frac{q^2}{r^2}$$

We can use trigonometry to find the distance, r, between the charges:

$$r = 2 \times 0.400 \text{ m} \times \sin 15.0° = 0.207 \text{ m}$$

Now we solve for the unknown charge q:

$$q^2 = 4\pi \times 8.85 \times 10^{-12} \frac{\text{C}^2}{\text{N} \cdot \text{m}^2} (0.207 \text{ m})^2 \times 0.264 \text{ N}$$
$$q = 1.12 \times 10^{-6} \text{ C} = 1.12 \text{ } \mu\text{C}$$

We cannot determine the sign of the charges from the information provided. The charges could both be negative or both be positive because the electrical force would be repulsive in either case.

Making sense of the result

This configuration is used in an electroscope to measure electric charges. If the balls had been separated at a greater angle, then we would find a greater value for q. Note that in most everyday situations, static charges are a tiny fraction of a coulomb.

19-4 Multiple Point Charges and the Superposition Principle

When there are more than two point charges, we find the net force on any one point charge by using Coulomb's law to find the force on that charge due to each of the other charges, and then add these forces as vectors. Calculating the net force in this way is an application of the superposition principle.

The **superposition principle** states that we can calculate the response due to individual stimuli (in this case, the electrical force due to one charge) and then add the individual contributions to find the total response. As we will see, we can often use symmetry to simplify the situation because some components will cancel and do not need to actually be calculated. The superposition principle requires that the system be **linear**, meaning if you double the stimulus the response will also be doubled (you will learn a more complete definition in later mathematics or physics courses). Many physical situations are approximately linear, and therefore the superposition principle is important and widely used. In situations where there are many point charges in a complex arrangement, we can use numerical software to implement the superposition principle to find the net electrical force.

Example 19-3 illustrates the process of combining electrical forces for multiple point charge situations. The technique is the same as the technique you used for net forces in Chapter 5, except that you now use Coulomb's law to calculate the forces.

PEER TO PEER

I find I make fewer mistakes when applying Coulomb's law in situations involving multiple charges if I draw the directions of the arrows for each electrical force using the fact that like charges repel and unlike charges attract. I then just use Coulomb's law to find the magnitude of each force, putting in the directions myself, rather than using the vector form of the law. If you do use the vector form, still check your answer this way.

CHECKPOINT 19-5

Zero Net Electrical Force

As shown in Figure 19-8, the charge $-2Q$ is located on the x-axis at a distance D to the right of another charge, $+Q$. Is there a point anywhere on the x-axis where we could place a third positive charge, $+q$, such that it would experience zero net electrical force?

(a) No, there is no such point.
(b) Yes, and it would be between the charges.
(c) Yes, and it would be to the left of the $+Q$ charge.
(d) Yes, and it would be to the right of the $-2Q$ charge.

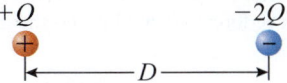

Figure 19-8 Checkpoint 19-5.

ANSWER: (c) Nowhere between the two charges will we be able to find a point with zero net force. This is because the $+q$ charge is attracted to the $-2Q$ and repelled by the $+Q$ charge, in both cases resulting in a force in the same direction and therefore no balance. If we are to the left of the $+Q$ charge, then we would feel a repulsion from the $+Q$ charge (a force to the left) and an attraction toward the $-2Q$ charge (a force to the right). At the right distance, these two forces can add to produce zero net force. In the region to the right of the $-2Q$ charge, the force from that charge is always stronger because the charge is greater and the distance to it is smaller.

EXAMPLE 19-3

Net Force from Three Charges

Charge q_1 is $+1.00$ μC. Located 0.500 m to the right of q_1 is charge q_2 ($+2.00$ μC). At a distance of 1.50 m in the $+y$-direction from charge q_1 is charge q_3 (-3.00 μC). Find the magnitude and direction of the net electrical force on charge q_1.

Solution

The physics principle here is Coulomb's law (we implicitly assume that only electrical forces are important). We will find the electrical force due to each interaction and then apply the superposition principle to combine them into one net force. Let us define a coordinate system with charge q_1 at the origin, as shown in Figure 19-9. We will apply the vector form of Coulomb's law. Since we want to find the net force acting on q_1, we draw unit vectors for direction from q_2 and q_3 to q_1. When we draw the directions of the corresponding forces, we need to keep in mind that the charge product $q_2 q_1$ is positive and the charge product $q_3 q_1$ is negative. Consequently, the direction of the force that q_3 exerts on q_1 will be opposite to that of $\hat{r}_{3 \to 1}$. This is the same result we would get from observing that q_1 and q_3 have opposite signs and attract, while q_1 and q_2 repel since they have the same sign.

(continued)

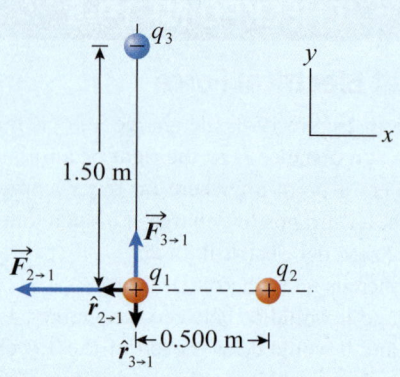

Figure 19-9 FBD for the electrical forces acting on q_1 in Example 19-3. The forces are in blue and the unit displacement vectors in black.

We will first calculate the force that q_2 exerts on q_1:

$$\vec{F}_{E\,2\to1} = \frac{1}{4\pi\varepsilon_0}\frac{q_2 q_1}{r_{21}^2}\hat{r}_{2\to1}$$

$$= \frac{1}{4\pi\varepsilon_0}\frac{(+2.00\times10^{-6}\,\text{C})(+1.00\times10^{-6}\,\text{C})}{(0.500\,\text{m})^2}\hat{r}_{2\to1}$$

$$= 0.0719\,\text{N}\,\hat{r}_{2\to1}$$

It is important to realize that since the unit vector points in the $-x$-direction, we express the force as $-0.0719\,\text{N}\,\hat{i}$.

Similarly, for the force that q_3 exerts on q_1, we have

$$\vec{F}_{E\,3\to1} = \frac{1}{4\pi\varepsilon_0}\frac{q_3 q_1}{r_{31}^2}\hat{r}_{3\to1}$$

$$= \frac{1}{4\pi\varepsilon_0}\frac{(-3.00\times10^{-6}\,\text{C})(+1.00\times10^{-6}\,\text{C})}{(1.50\,\text{m})^2}\hat{r}_{3\to1}$$

$$= -0.0120\,\text{N}\,\hat{r}_{3\to1}$$

If we use a conventional axis orientation (x-axis is horizontal with positive to right and y-axis is vertical with positive upward), then $\hat{r}_{3\to1}$ is in the $-y$-direction, and the force $\vec{F}_{E\,3\to1}$ is in the $+y$-direction:

$$\vec{F}_{E\,3\to1} = +0.0120\,\text{N}\,\hat{j}$$

Thus, the net force acting on q_1 can be written as

$$\vec{F}_{\text{net}} = -0.0719\,\text{N}\,\hat{i} + 0.0120\,\text{N}\,\hat{j}$$

The magnitude of the net force is

$$F_{\text{net}} = \sqrt{(-0.0719)^2 + (+0.0120)^2}\,\text{N} = 0.0729\,\text{N}$$

The direction will be above the $-x$-axis at an angle ϕ given by

$$\tan\phi = \frac{0.0120}{0.0719};\ \phi = 9.47°$$

Making sense of the result

Charge q_2 will repel charge q_1, so charge q_1 will experience a force directed to the left. Charge q_3 will attract charge q_1, so charge q_1 will experience an upward force. These directions are consistent with what we found. The separation of charges q_1 and q_3 is three times the separation for charges q_1 and q_2, and the same magnitude of charge q_3 is 3/2 times the magnitude of charge q_2. Thus, the inverse square distance

factor, 1/9, will more than offset the greater charge magnitude. When we combine the two forces, we should get a net force that is to the left and upward, with the leftward component dominating, which is consistent with our result. We have used the vector form of Coulomb's law but we could have equally well used it just for the magnitudes of the forces, and used like charges repel and unlike attract to get the directions for each.

Before we leave this section, let us summarize a best practice procedure for dealing with Coulomb's law problems:

- Start by drawing a diagram showing all charges (including the sign of each, if known).

- Be clear which charge you are calculating the force on. For each of the other charges, draw a vector representing the electrical force due to that charge. You can either use common sense and like charges repel and unlike attract, or the vector form of Coulomb's law to find the direction for each force, combining them on an FBD.

- Now you will use the superposition principle to find the net force. Use symmetry here to reduce the calculations. You will probably need to break forces into components in finding the net force (see Chapters 2 and 5 to review this procedure if necessary). Remember to clearly indicate your choice of axes and the positive direction, and be consistent with your choice.

- After you have found the net force, make sure that you decide if the final answer makes sense, keeping in mind which forces cancel, the direction of each force, and the approximate magnitude of each force.

19-5 Electrical Forces for Continuous Charge Distributions

In this section, we consider situations in which we have a single point charge along with a continuous distribution of charge. The physics principle here is similar to the many point charges we considered in the previous section, where we used Coulomb's law for each charge pair and combined the results using the superposition principle. We imagine that the continuous distribution is made up of many tiny differential elements of charge, dq, and then integrate over the charge distribution. We need to take into account the vector nature of Coulomb's law, although symmetry can often simplify calculations.

In doing problems involving continuous charge distributions, it is helpful to define charge densities (depending on the situation, these can be linear, area, or volume charge densities). The idea is similar to mass densities, except instead of mass per unit volume (or per unit length or area), we are talking about the electrical charge per unit volume.

First, we will define **linear charge density**. When a total charge, Q, is uniformly distributed along a thin rod of length L, we can define the average linear charge density, λ, as follows:

KEY EQUATION
$$\lambda \equiv \frac{Q}{L} \qquad (19\text{-}6)$$

Next we define the average **surface charge density**, σ, as the charge, Q, per unit area, A, of a surface.

KEY EQUATION
$$\sigma \equiv \frac{Q}{A} \qquad (19\text{-}7)$$

Finally we define the **volume charge density**, ρ, as the charge per unit volume. Volume charge density is analogous to mass density:

KEY EQUATION
$$\rho \equiv \frac{Q}{V} \qquad (19\text{-}8)$$

CHECKPOINT 19-6

Linear Charge Density

When a charge of 25 μC is uniformly distributed on a thin rod that is 2.5 cm long, the linear charge density is

(a) 1.0 mC/m
(b) 10. mC/m
(c) 10. C/m
(d) 1.0 kC/m

ANSWER: (a) Divide the charge expressed in coulombs by the length in metres to get the linear charge density of 0.0010 C/m.

To find the electrical force on a point charge, Q, from a continuous distribution of charge, apply the following procedure:

■ Find the linear, area, or surface charge density according to the type of object the charge is distributed on.

■ Divide that continuous distribution of charge into differential elements, dq, and find the electrical force, dF_E, on the point charge, Q, due to that differential element of charge using the following modified form of Coulomb's law:

$$dF_E = \frac{1}{4\pi\varepsilon_0}\frac{Qdq}{r^2} \qquad (19\text{-}9)$$

■ Use the fact that like charges repel and unlike charges attract to interpret the direction of the force given by Equation 19-9.

■ Integrate the differential force given by Equation 19-9 over all the differential charge elements of the object. If the forces are not in the same direction, you must take this into account. The symmetry of the problem will often allow you to simplify these calculations (see Examples 19-4, 19-5, and 19-6).

This process is equivalent to considering the object as many tiny point charges (the differential charge elements); using Coulomb's law to find the force on the point charge, Q, from each of those differential elements; and then adding all the forces using the superposition principle.

It is usually obvious whether you should employ a line, surface, or volume charge density. If the charges occur one after the other in sequence, even if along a curved arc (see Example 19-5), then we use a linear charge density, and the resulting integration will be over a single variable. If the charge is distributed over a two-dimensional surface, then we use a surface charge density and the integral will be over two variables. If the charge is over a three-dimensional object, then we use the volume charge density and integrate over three variables.

EXAMPLE 19-4

Electrical Force from Finite Line of Charge

A total positive charge of 12.00 μC is evenly distributed on a straight thin rod of length 6.00 cm. A positive point charge, $Q = 4.00$ nC, is located a distance of 5.00 cm above the midpoint of the rod. What will be the electrical force on the point charge?

Solution

Assume that the rod is aligned with the x-axis, as shown in Figure 19-10, and define the origin as the midpoint of the rod. Consider that the rod is divided up into differential elements of charge dq and length dx. Since we are dealing only with electrical forces, we have simplified the notation by dropping the subscript E from forces. The differential force, $d\vec{F}$, on the point charge Q

from that differential element of charge will be in the direction shown (since both are positive, the force is repulsive). Now we need to integrate over all elements of charge from the entire rod. We can use symmetry to simplify our calculations. If we divide the differential force, $d\vec{F}$, into x- and y-components, the summation over the x-components must add to zero by symmetry, and therefore we need only consider the y-components, dF_y.

We use the Pythagorean theorem to find the distance from the differential element to the point charge Q as $\sqrt{d^2 + x^2}$. The magnitude of the force due to the differential element of charge is

$$dF = \frac{1}{4\pi\varepsilon_0}\frac{Qdq}{d^2 + x^2} \qquad (1)$$

(continued)

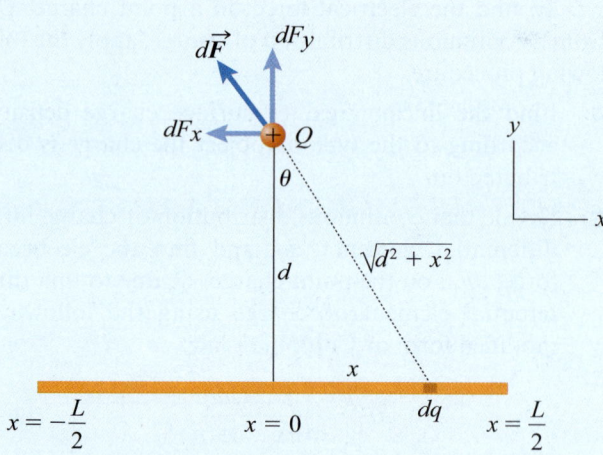

Figure 19-10 Example 19-4. A positive point charge, Q, is above the midpoint of a positively charged thin rod. We show the force and its y-component due to one differential charge element.

We will call the uniform linear charge density on the rod λ, which has numerical value

$$\lambda = \frac{12.00\,\mu\text{C}}{0.0600\,\text{m}} = 2.00 \times 10^{-4}\,\text{C/m} \quad (2)$$

The differential charge can be expressed in terms of the differential dx:

$$dq = \lambda dx \quad (3)$$

We are interested in only the y-component of the force, which is given by the following (see Figure 19-10 for the definition of angle θ):

$$dF_y = dF\cos\theta \quad (4)$$

$$\cos\theta = \frac{d}{\sqrt{d^2 + x^2}} \quad (19\text{-}14)$$

When we substitute (1), (3), and (5) into (4), we obtain

$$dF_y = \frac{Q\lambda d}{4\pi\varepsilon_0}\frac{dx}{(d^2 + x^2)^{3/2}} \quad (6)$$

We now integrate over all charge elements to get the net force:

$$
\begin{aligned}
F_{\text{net}} &= \int dF_y = \frac{Q\lambda d}{4\pi\varepsilon_0}\int_{-L/2}^{+L/2}\frac{dx}{(d^2 + x^2)^{3/2}} \\
&= \frac{Q\lambda d}{4\pi\varepsilon_0}\left[\frac{x}{d^2\sqrt{d^2 + x^2}}\right]_{-L/2}^{+L/2} \\
&= \frac{Q\lambda d}{4\pi\varepsilon_0}\left[\frac{L/2}{d^2\sqrt{d^2 + (L/2)^2}} - \frac{-L/2}{d^2\sqrt{d^2 + (-L/2)^2}}\right] \\
&= \frac{Q\lambda d}{4\pi\varepsilon_0}\frac{2L}{d^2\sqrt{4d^2 + L^2}} = \frac{Q\lambda}{4\pi\varepsilon_0}\frac{2L}{d\sqrt{4d^2 + L^2}}
\end{aligned}
$$

When we substitute in the linear charge density from (2) and the values $L = 0.0600$ m, $d = 0.0500$ m, and $Q = 4.00 \times 10^{-9}$ C, we obtain the magnitude of the net force as $F_{\text{net}} = 0.148$ N. The direction is directly away from the rod (in our configuration it would be in the $+\hat{j}$-direction).

Making sense of the result

One way to see if this answer is of the right order of magnitude is to calculate the repulsive force between two point charges separated by 5.00 cm. One of the point charges is the charge on Q and the other is the total charge on the rod. We expect the value to be a bit larger than when the charge is spread over the rod. We find a value of 0.17 N, which confirms the reasonableness of our answer.

EXAMPLE 19-5

Electrical Force from an Arc of Charge

A total charge, $+Q$, is uniformly distributed along an arc that is exactly 1/4 of a circle of radius R, as shown in Figure 19-11. A positive point charge, also $+Q$, is placed at the centre of the circle. Find the magnitude and direction of the net electrical force on this point charge.

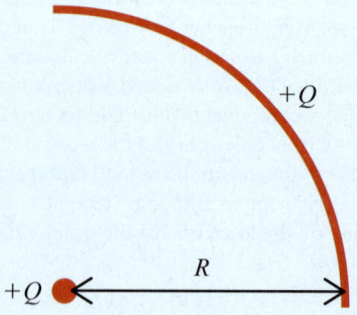

Figure 19-11 Charge $+Q$ is uniformly distributed over the arc of radius R.

Solution

The physics principle here is Coulomb's law. We are assuming implicitly that only electrical forces need be considered, so we have dropped the subscript E, since all forces are electrical. We will find the force exerted on the point charge from a differential unit of charge on the arc and then integrate over the entire arc, keeping in mind the vector nature of the force. The situation is pictured in Figure 19-12.

First we calculate the linear charge density along the arc. Note that we can have a linear charge density even when the object is not straight, provided that it is very small in the other two dimensions:

$$\lambda = \frac{Q}{L} = \frac{2Q}{\pi R}$$

This means that the differential charge element can be represented by

$$dq = \lambda R d\theta$$

The force components on the point charge from the charge in this differential element are the following (with x and y defined in the traditional sense with positive to the right and upward):

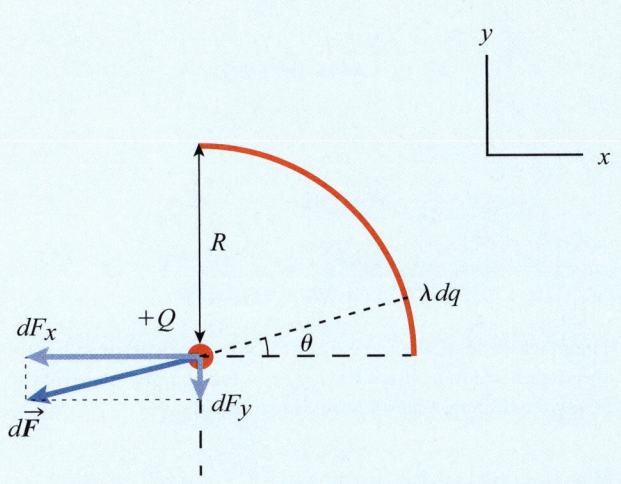

Figure 19-12 Example 19-5. Force from a differential element of charge on the arc.

$$dF_x = -\frac{1}{4\pi\varepsilon_0}\frac{Qdq}{R^2}\cos\theta \quad dF_y = -\frac{1}{4\pi\varepsilon_0}\frac{Qdq}{R^2}\sin\theta$$

When we substitute in the expressions for dq and λ, we have

$$dF_x = -\frac{1}{4\pi\varepsilon_0}\frac{2Q^2}{\pi R^2}\cos\theta\, d\theta \quad dF_y = -\frac{1}{4\pi\varepsilon_0}\frac{2Q^2}{\pi R^2}\sin\theta\, d\theta$$

We now integrate these over the range of angles to obtain the force components for the effects of all elements of charge along the arc:

$$F_x = -\frac{1}{4\pi\varepsilon_0}\frac{2Q^2}{\pi R^2}\int_0^{\pi/2}\cos\theta\, d\theta = -\frac{1}{4\pi\varepsilon_0}\frac{2Q^2}{\pi R^2}$$

$$F_y = -\frac{1}{4\pi\varepsilon_0}\frac{2Q^2}{\pi R^2}\int_0^{\pi/2}\sin\theta\, d\theta = -\frac{1}{4\pi\varepsilon_0}\frac{2Q^2}{\pi R^2}$$

Therefore, the net force is given by

$$F_{net} = -\frac{1}{4\pi\varepsilon_0}\frac{2Q^2}{\pi R^2}\hat{i} - \frac{1}{4\pi\varepsilon_0}\frac{2Q^2}{\pi R^2}\hat{j}$$

If desired, these could be combined and expressed in magnitude and direction notation.

Making sense of the result

As expected from the symmetry of the situation, our force has equal magnitude components in the x- and y-directions, and these are both in the negative direction because positive charges repel each other. We would expect the magnitude of the force to be a bit less than that of a second point charge, Q, at a distance R since the various charge elements exert forces in somewhat different directions. That is what we find. Note that this analysis depends on the charge being uniformly spread over the arc.

When we have a flat two-dimensional object, such as a circular plate or a rectangular sheet, the surface charge density is used. We divide the surface into small differential elements, find an expression for the charge on the differential element, and then integrate in two dimensions to allow for the contributions from all parts of the object. Since force is a vector quantity, direction must be taken into account. Usually symmetry considerations are needed to readily obtain a solution. We illustrate this technique in Example 19-6.

EXAMPLE 19-6

Force on Point Charge from a Charged Circular Plate

A positive point charge, Q, is located at a distance h directly above the centre of a charged thin non-conducting circular plate of radius R (see Figure 19-13). The plate carries a total positive charge, Q, spread uniformly over its surface area. What will be the electrical force on the point charge?

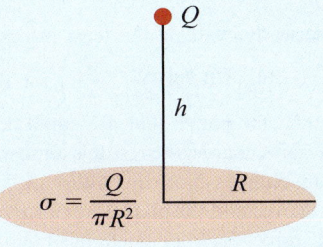

Figure 19-13 Example 19-6.

Solution

We first calculate the surface charge density by dividing the charge, Q, on the plate by the area of the circle:

$$\sigma = \frac{Q}{\pi R^2} \quad (1)$$

Next we divide the surface into differential area elements, dq. Given that it is a circular plate, it is easiest to use plane polar coordinates. The radial distance from the centre is r, and the angle is ϕ. As shown in the inset of Figure 19-14, the differential area is approximated by a rectangle with one side dr and the other side $rd\phi$.

Therefore, the charge on the differential area element is

$$dq = \sigma r dr d\phi \quad (2)$$

The magnitude of the contribution to the electrical force due to just this element (see Figure 19-14) will be given by the following, where we have used the Pythagorean theorem to find

(continued)

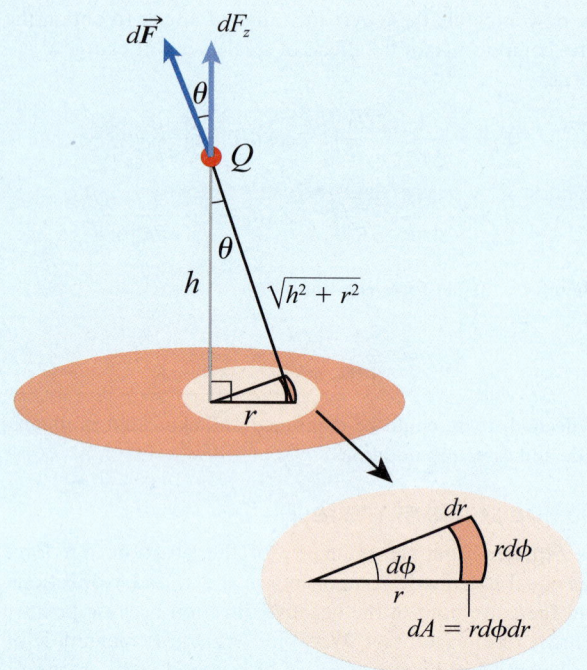

Figure 19-14 Differential area element $dA = rdrd\phi$ will have charge $dq = \sigma rdrd\phi$ and will contribute $d\vec{F}$ to the electrical force on point charge Q.

the distance from the differential charge element to the point charge:

$$dF = \frac{1}{4\pi\varepsilon_0}\frac{Qdq}{(r^2 + h^2)} \tag{3}$$

Since we are dealing only with electrical forces in this question, we have dropped the subscript E to simplify the notation.

At this point, we invoke the symmetry of the situation to see that the components of the electrical force in the z-direction will add, while those in the xy-plane will cancel as we consider all differential elements in the plate. The z-component of the electrical force on the point charge, Q, due to the differential charge element dq will be the following, where we write the cosine in terms of h and r in the second line:

$$dF_z = \frac{1}{4\pi\varepsilon_0}\frac{Qdq}{r^2 + h^2}\cos\theta$$
$$= \frac{1}{4\pi\varepsilon_0}\frac{Qdq}{r^2 + h^2}\frac{h}{\sqrt{r^2 + h^2}} = \frac{1}{4\pi\varepsilon_0}\frac{Qhdq}{(r^2 + h^2)^{3/2}} \tag{4}$$

If we substitute (1) and (2) into (4), we have

$$dF_z = \frac{1}{4\pi\varepsilon_0}\frac{Q^2h}{\pi R^2}\frac{rdrd\phi}{(r^2 + h^2)^{3/2}} \tag{5}$$

To find the total electrical force in the z-direction, we then integrate this over the area of the circular plate:

$$F_z = \frac{1}{4\pi\varepsilon_0}\frac{Q^2h}{\pi R^2}\int_0^R\frac{rdr}{(r^2 + h^2)^{3/2}}\int_0^{2\pi}d\phi$$
$$= \frac{1}{4\pi\varepsilon_0}\frac{Q^2h}{\pi R^2}2\pi\int_0^R\frac{rdr}{(r^2 + h^2)^{3/2}}$$
$$= \frac{1}{4\pi\varepsilon_0}\frac{Q^2h}{\pi R^2}2\pi\left[-\frac{1}{\sqrt{r^2 + h^2}}\right]_0^R$$
$$= \frac{1}{4\pi\varepsilon_0}\frac{2Q^2h}{R^2}\left(\frac{1}{h} - \frac{1}{\sqrt{R^2 + h^2}}\right) \tag{6}$$

The magnitude of the force on the point charge is given by (6), and the direction is directly away from the charged plate (in the positive z-direction as we have drawn the plate).

Making sense of the result

As expected, since we were considering a two-dimensional flat plate, we have a double integral. We can see that the result given in (6) is dimensionally correct, since the dimension units of length for h in the numerator cancel the units of length in the denominator in the term in brackets, and the rest of the expression has the same dimensions as for the force between two point charges. We would expect the electrical force to be less than for two point charges, Q, at a separation distance h. To see if this is so, let us consider the result if $h = R$. When we substitute that into (6) and simplify, we find

$$F_z = \frac{1}{4\pi\varepsilon_0}\frac{2Q^2h}{h^2}\left(\frac{1}{h} - \frac{1}{\sqrt{h^2 + h^2}}\right)$$
$$= \frac{1}{4\pi\varepsilon_0}\frac{2Q^2h}{h^2}\left(\frac{1}{h} - \frac{1}{h\sqrt{2}}\right)$$
$$= \frac{1}{4\pi\varepsilon_0}\frac{Q^2}{h^2}\left[\frac{2(\sqrt{2} - 1)}{\sqrt{2}}\right]$$

This gives us a result that is about 0.59 that of the force on two point charges, Q, which seems reasonable. Now let us consider the special case $R = h/10$, where we would expect the result to be much closer to that for two point charges, since the dimension of the plate is small compared to the distance to the point charge. For this approximation, result (6) reduces to

$$F_z = \frac{1}{4\pi\varepsilon_0}\frac{2Q^2h}{R^2}\left[\frac{1}{h} - \frac{1}{\sqrt{R^2 + h^2}}\right]$$
$$= \frac{1}{4\pi\varepsilon_0}\frac{2Q^2h}{(h/10)^2}\left[\frac{1}{h} - \frac{1}{\sqrt{(h/10)^2 + h^2}}\right]$$
$$= \frac{1}{4\pi\varepsilon_0}\frac{Q^2}{h^2}\frac{200}{1}\left[1 - \frac{1}{\sqrt{101/100}}\right]$$

When we numerically evaluate the term in brackets and multiply by 200, we obtain $0.99\left(\frac{1}{4\pi\varepsilon_0}\frac{Q^2}{h^2}\right)$ for the force in this case. As expected, for plates that are small compared to the distance to the point charge, the result is almost identical to the case of two point charges. Note that although we were told that both the point charge and the plate were positively charged, the direction of the force would be unchanged had they both been negatively charged.

We could use the volume charge density, and a triple integral, to calculate the electrical force for a three-dimensional situation in an analogous fashion. Sometimes we can simplify the situation, however. If we have a uniformly charged sphere that is small compared to the distance a point charge is away from the centre of the sphere, to good accuracy we can consider the sphere as a point charge for the purposes of calculating the force on the point charge.

In all of these cases, it is important to realize that we have assumed a uniform charge distribution. If the surface charge density, σ, or the volume charge density, ρ, were not uniform, the technique could still be applied, but in this case the charge density could not be taken outside the integral sign, and the mathematics of the situation would be more complex.

19-6 Electric Field

In this section, we define a new and very important concept: the electric field. When a charge experiences an electrical force, you could consider that the force is caused by a so-called action-at-a-distance, due to effects from various other charges located at different positions. Another way to explain the force is that the other charges create an electric field, and the charge responds to that electric field. Before reading further, think about these two ways to view the situation, and decide which you prefer, and why.

Both approaches are valid. When dealing with a few charges, it is often easiest to consider a vector superposition of individual electrical forces. However, as we will see in Chapter 22, we can define an energy density in an electric field, which is helpful in situations involving distributions of many charges. Most physicists highly value this field view.

Before we consider electric fields in detail, it is helpful to think about gravitational fields. Recall that we can think of the acceleration due to gravity, $\vec{g}$, as being a gravitational field, as described in Chapter 11. If we place a mass, m, in this field, we observe a gravitational force, $m\vec{g}$. If we want to calculate the gravitational field in a region, we can find the gravitational force on some test mass, divide by that test mass, and the result would be the gravitational field, $\vec{g}$.

We can define an electric field in an analogous way: place a small positive test charge, q, in a region;

measure the electrical force on it, $\vec{F}_E$; and divide that force by the test charge to obtain the field, $\vec{E}$. In the same way that current is defined in terms of a flow of positive charges by convention only, again, electric field direction is by convention always defined in terms of the force on a positive test charge. Therefore, **electric field** is defined as the electrical force per unit charge (and has SI units of N/C):

$$\vec{E} \equiv \frac{\vec{F}_E}{q} \qquad (19\text{-}10)$$

The magnitude of the electric field is referred to as the **electric field strength**.

English physicist Michael Faraday (1791–1867) developed early ideas of what we now call the electric field. In his view, the presence of charges created a field in the region near the charges. That field would then interact with additional charges placed in the region, causing what we now call the electrical force. We can place a tiny test charge in a region to determine if an electric field is present by measuring the electrical force on the test charge, and then applying Equation 19-10.

For an electric field at a distance r from a single point charge Q, we can combine Coulomb's law with Equation 19-10:

KEY EQUATION

$$\vec{E} = \frac{1}{4\pi\varepsilon_0} \frac{Q}{r^2}\hat{r} \qquad (19\text{-}11)$$

where $\hat{r}$ is a unit displacement vector pointing away from charge Q (i.e., in the direction of increasing r). Note that charge Q can be either positive or negative, so the electric field can be in either the same or the opposite direction as unit vector $\hat{r}$.

From Equation 19-11, we can see that the electric field points *away* from a positive charge and *toward* a negative charge. The electric field is strongest near the charge, and decreases as the inverse square of the distance from the centre of the charge. These characteristics are illustrated in Figure 19-15. In Section 19-8 we will introduce two ways to graphically represent the electric field at various points.

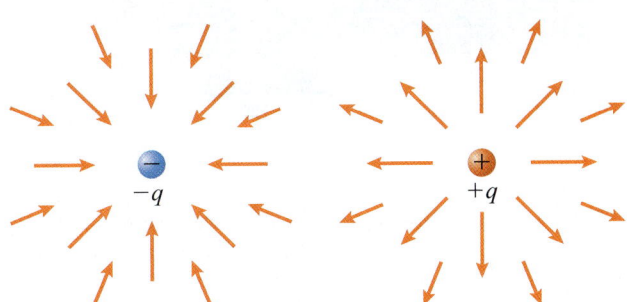

Figure 19-15 The electric field at various points around a negative charge (left) and a positive charge (right). Note that the electric field is stronger nearer the charge.

I have to be careful with wording! If the question asks for an electric field, that is a vector quantity, and I must give the direction. If it asks for electric field strength, it is just asking for the magnitude of the electric field.

CHAPTER 19 | ELECTRIC FIELDS AND FORCES   671

ELECTRICITY, MAGNETISM, AND OPTICS

MAKING CONNECTIONS

Inkjet and Laser Printers

Inkjet printers use electric fields to deflect tiny charged drops of ink to form the desired character or image (Figure 19-16). The drops are formed by rapidly heating and vaporizing ink in a chamber or by using a piezoelectric crystal to compress the ink chamber. In both cases, the drops carry charges, and their vertical motion is precisely controlled by electric fields. Modern inkjet printers have ink drops as small as a picolitre (10^{-12} L), and the ink is typically moving at 10 m/s to 50 m/s when it leaves the print head. The electric fields used for deflection are typically of the order of 100 kN/C. The concept of inkjet printing was first proposed by Lord Kelvin in 1867, but the technology did not become commercially viable until the mid-1970s.

Laser printers use electric charge as well but in a different way. Laser light temporarily changes the conductivity of a photosensitive coating on a drum inside the printer, creating a pattern of static charge on the drum. Toner particles are electrostatically attracted to the charged areas of the drum. When a sheet of paper contacts the drum, most of the toner particles adhere to the paper, transferring the image. A heater then fuses the powdery toner to the paper, making the image indelible.

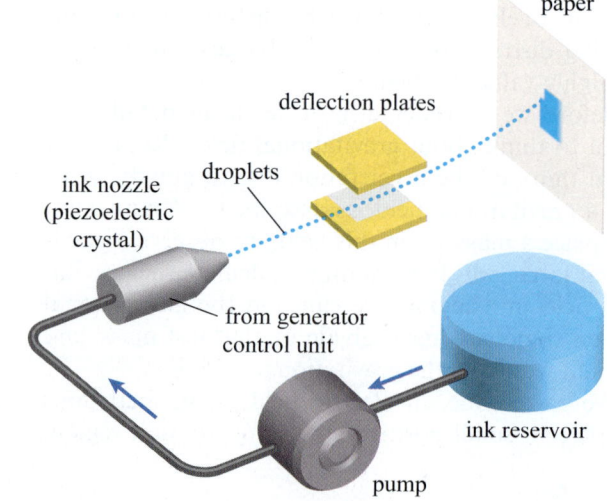

Figure 19-16 Inkjet printers use electric fields to deflect charged ink drops.

EXAMPLE 19-7

Electric Field and Unknown Charge

When a small positive test charge, q (of magnitude 2.00 μC), is placed at a distance of 2.50 m from an unknown charge, it is observed that a force of 34.0 N is exerted on q and it is in the direction toward the unknown charge.

(a) What is the electric field at this point (i.e., 2.50 m away from the unknown charge)?
(b) What must be the sign and magnitude of the unknown charge?

Solution

In this problem, we can use the definition of the electric field (i.e., force per unit test charge). In finding the unknown charge, we can use the expression for the electric field for a single point charge.

(a) We can use the defining relationship for electric field to find the magnitude of the field:

$$E = \frac{F}{q} = \frac{34.0\,\text{N}}{2.00 \times 10^{-6}\,\text{C}} = 1.70 \times 10^7\,\text{N/C}$$

Since our test charge is positive, the direction of the force on it will also be the direction of the electric field, i.e., toward the unknown charge.

(b) Because a positive test charge is attracted toward the unknown charge, we know that the unknown charge must be of opposite sign and therefore negative. The electric field at a distance r from a charge Q is given by

$$\vec{E} = \frac{1}{4\pi\varepsilon_0}\frac{Q}{r^2}\hat{r}$$

This can be rearranged to solve for the magnitude of the charge Q (using magnitudes only of all quantities):

$$|Q| = 4\pi\varepsilon_0 E r^2 = 0.0118\,\text{C}$$

Therefore, the unknown charge is -11.8 mC.

Making sense of the result

In questions like this, it is always good to check that your answer is consistent with common sense. You know that charges of different sign attract, so the directions of the force and your result are consistent. You could also perform a check that the numbers are consistent by using Coulomb's law to calculate the force between the charges. If you do that, you will obtain 34.0 N, as expected.

While we can think of Equation 19-10 as a way to obtain the electric field from the measured electrical force on a known point charge, in many situations we know the electric field and use the relationship to solve for the force on some charge.

$$\vec{F} = q\vec{E} \tag{19-12}$$

There is no need for the charge to be positive or to be small. This is one advantage of the field approach. Once you have found the electric field at a position, you can use that field to find the electrical force on *any* point charge that you place in that position. We demonstrate this technique in Example 19-8. Interactive Activity 19-4 gives you an opportunity to direct charges using the electric fields generated by other charges.

MAKING CONNECTIONS

North America's First Electron Microscope

The resolution of an optical microscope is limited according to the wavelength of the light that it uses (see Chapter 29). The effective wavelength of electrons—the de Broglie wavelength (Chapter 32)—can be much shorter than the wavelength of visible light. In 1938, two University of Toronto graduate students, James Hillier (1915–2007) and Albert Prebus (1913–1997), built the first electron microscope in North America. The instrument, which was capable of magnifications of 20000×, was a major advance over earlier European attempts and is sometimes credited as the first successful electron microscope anywhere. Electron microscopes use electric fields to accelerate the electrons that form the image and to position the beam of electrons when scanning an object. Scanning electron microscopes (SEMs) have many applications in science, engineering, and medicine. Figure 19-17 is an image taken with a modern electron microscope.

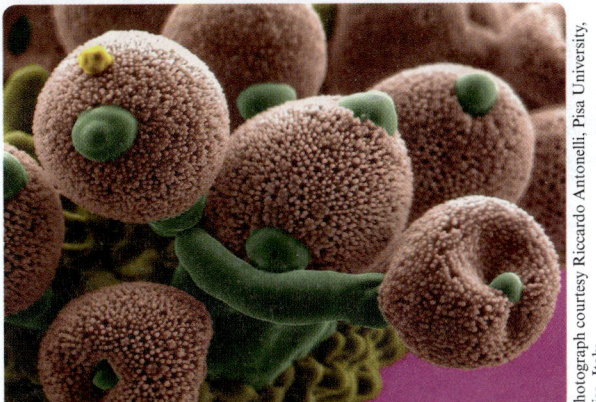

Figure 19-17 An electron microscope image of pollen from a geranium leaf.

Photograph courtesy Riccardo Antonelli, Pisa University, Pisa, Italy

EXAMPLE 19-8

Acceleration of Gold Ions

Gold has a very high electrical conductivity, and as a result it is widely used as a thin conductive coating in a variety of space, industrial, and consumer applications. In an experimental setup, Au^{3+} ions are accelerated in a vacuum from rest in a region with uniform electric field strength of 2750 N/C. What will be the speed and the distance travelled by the ion after 1.50 μs? Assume that the mass of one gold ion is 197 atomic mass units.

Solution

Let us first map out the strategy we will use. Since we know the ion charge and the electric field strength, we can find the force. From the mass of the ion, we can then calculate the acceleration. Since the electric field strength is constant in the region, the acceleration will be constant, and we can use the relationships for constant-acceleration motion from Chapters 3 and 4 to find the speed and distance travelled.

The ion is Au^{3+}, which means that three electrons have been removed, so the charge on the ion is $q = 3 \times 1.60 \times 10^{-19}$ C $= 4.80 \times 10^{-19}$ C. Therefore, the magnitude of the electrical force on the ion in the uniform electric field is

$$F = qE = (4.80 \times 10^{-19}\,\text{C})(2750\,\text{N/C}) = 1.32 \times 10^{-15}\,\text{N}$$

The atomic mass of gold is 197 u, or in SI units of kg,

$$m = (197\,\text{u})(1.66 \times 10^{-27}\,\text{kg/u}) = 3.27 \times 10^{-25}\,\text{kg}$$

Now we can calculate the magnitude of the acceleration, which is constant while within the uniform electric field:

$$a = \frac{F}{m} = \frac{1.32 \times 10^{-15}\,\text{N}}{3.27 \times 10^{-25}\,\text{kg}} = 4.04 \times 10^{9}\,\text{m/s}^2$$

Since the ion starts from rest, $v_0 = 0$, and the speed after 1.50 μs is given by the following constant-acceleration relationship:

$$v = v_0 + at = 0 + (4.04 \times 10^{9}\,\text{m/s}^2)(1.50 \times 10^{-6}\,\text{s})$$
$$= 6060\,\text{m/s}$$

The distance travelled is given by the following constant-acceleration relationship:

$$\Delta x = v_0 t + \frac{1}{2}at^2 = 0 + \frac{1}{2}(4.04 \times 10^{9}\,\text{m/s}^2)(1.50 \times 10^{-6}\,\text{s})^2$$
$$= 4.54 \times 10^{-3}\,\text{m}$$

Making sense of the result

Even though the force is a small value by macroscopic standards, relative to the mass of a single ion it produces an acceleration of more than a billion metres per square second. Starting from rest, over the time of 1.50 μs, the ion travels several millimetres. Remember that the analysis given here is only valid if the electric field strength is uniform.

19-7 Electric Fields and the Superposition Principle

Since electric fields are electrical forces per unit charge, it is not surprising that we can calculate the net electric field for an assembly of point charges by adding the electric field due to each using the superposition principle. When combining the electric fields, we must keep in mind that electric fields are *vectors*, and they must be added as vectors. Sometimes the symmetry of a situation can help us reduce the required calculations because the fields in some dimensions will cancel each other (see Example 19-9).

CHECKPOINT 19-8

Electric Field from an Array of Charges

Shown in Figure 19-18 is an array of six charges. The top charges are all positive and of the same magnitude, $+Q$, and there is a distance a between the charges. At a distance $2a$ below this line of charges are three other charges, with the outer two being $+Q$ and the central one $-Q$. What is the net electric field at a point P at the centre of the charge array?

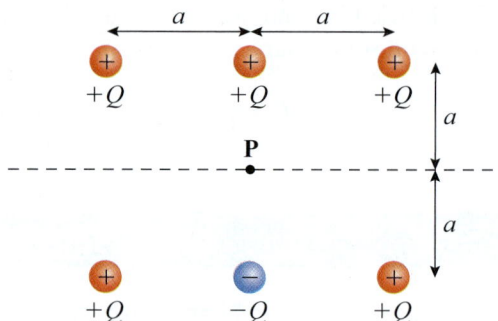

Figure 19-18   Charge array for Checkpoint 19-8.

(a) upward direction and magnitude $\dfrac{1}{4\pi\varepsilon_0}\dfrac{Q}{a^2}$

(b) downward direction and magnitude $\dfrac{1}{4\pi\varepsilon_0}\dfrac{2Q}{a^2}$

(c) downward direction and magnitude $\dfrac{1}{4\pi\varepsilon_0}\dfrac{2\sqrt{2}Q}{a^2}$

(d) downward direction and magnitude $\dfrac{1}{4\pi\varepsilon_0}\dfrac{Q}{\sqrt{2a^2}}$

ANSWER: (b) We can use symmetry to significantly reduce the number of actual calculations. If we place a small test charge at point P, the forces from the top left and the bottom right will be of equal magnitude but opposite direction and will cancel. The same is true for the charges at the top right and the bottom left. Therefore, we only need to consider the two central charges. Since the top one is positive and the bottom one is negative, a positive test charge placed at P would experience downward forces from both charges. We simply calculate the equal electric fields using the point charge relationship and then use the superposition principle to combine.

EXAMPLE 19-9

Electric Field from Multiple Point Charges

The situation for this problem is shown in Figure 19-19. Find the electric field at point P due to the combined effect of the four charges shown.

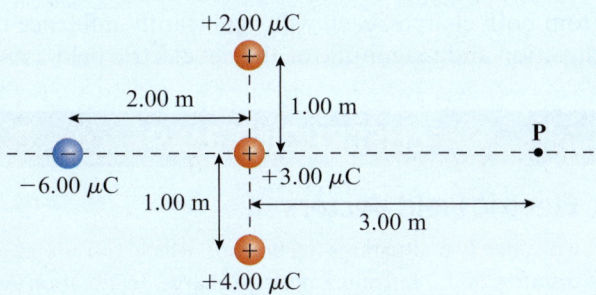

Figure 19-19   Example 19-9.

Solution

In this problem, we will find the electric field at P for each of the four charges using the expression for electric field due to a point charge. We will then use the superposition principle to find the total electric field, adding the electric fields as vectors. Let us first draw in the directions of the electric fields due to each of the four charges, remembering that the electric field is in the direction of the electrical force on a positive test charge at P (and that like charges repel and unlike charges attract). In Figure 19-20 we show the electric field vectors at point P for each of the charges (we have numbered the charges and used the same subscript on the electric field vectors).

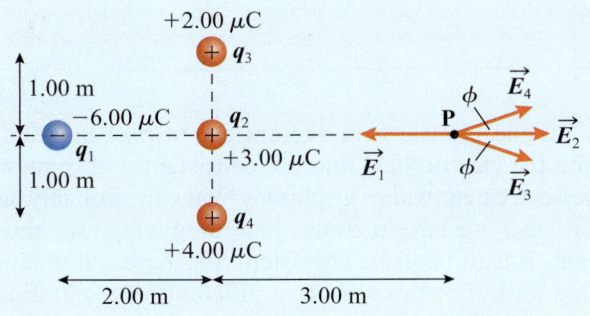

Figure 19-20   The electric field vectors for the four charges of Example 19-9.

We can use the Pythagorean theorem to find the distance between P and charges 3 and 4, which is $\sqrt{10.0}$ m. We can find the angle ϕ using trigonometry:

$$\tan\phi = \frac{1.00 \text{ m}}{3.00 \text{ m}} = 0.333$$

$$\phi = 18.43°$$

Now we are ready to use the expression for the electric field for a single point charge:

$$\vec{E} = \frac{1}{4\pi\varepsilon_0}\frac{Q}{r^2}\hat{r}$$

The electric fields for all four charges are given below using a coordinate system centred at P with positive x to the right and positive y upward:

$$\vec{E}_1 = -\frac{1}{4\pi\varepsilon_0}\frac{6.00 \times 10^{-6}\text{ C}}{(5.00 \text{ m})^2}\hat{i}$$

$$= -2158 \text{ N/C }\hat{i}$$

$$\vec{E}_2 = \frac{1}{4\pi\varepsilon_0}\frac{3.00 \times 10^{-6}\text{ C}}{(3.00 \text{ m})^2}\hat{i}$$

$$= 2997 \text{ N/C }\hat{i}$$

$$\vec{E}_3 = \frac{1}{4\pi\varepsilon_0}\frac{2.00 \times 10^{-6}\text{ C}}{10.00 \text{ m}^2}(\cos\phi\,\hat{i} - \sin\phi\,\hat{j})$$

$$= 1706 \text{ N/C }\hat{i} - 568.7 \text{ N/C }\hat{j}$$

$$\vec{E}_4 = \frac{1}{4\pi\varepsilon_0}\frac{4.00 \times 10^{-6}\text{ C}}{10.00 \text{ m}^2}(\cos\phi\,\hat{i} + \sin\phi\,\hat{j})$$

$$= 3412 \text{ N/C }\hat{i} + 1137 \text{ N/C }\hat{j}$$

We used 4 significant digits above, but in final answer below express the final answer to the correct 3 significant digits. We can then add these four electric fields to find the net electric field:

$$\vec{E}_{\text{net}} = \vec{E}_1 + \vec{E}_2 + \vec{E}_3 + \vec{E}_4$$

$$= 5960 \text{ N/C }\hat{i} + 569 \text{ N/C }\hat{j}$$

If desired, these could be converted into magnitude and direction format.

Making sense of the result

As our vector diagram for the four electric fields (Figure 19-20) suggested qualitatively, we are not surprised that the net electric field is to the right and upward, with the x-component much larger than the y-component.

We will extend the treatment to computation of electric fields from continuous distributions of charge in Section 19-9. In this case, we divide the charged object into many small differential charge elements and then integrate over the entire object to find the net electric field according to the superposition principle.

19-8 Electric Field Vectors and Lines

There are two widely used ways to graphically represent electric fields. One of these is **electric field vectors**. These are drawn at various locations, with the length representing the strength of the electric field at that point and the direction of course being the direction of the electric field at that point. We demonstrate electric field vectors for a single positive point charge in Figure 19-21. Before reading further in this chapter, examine this figure carefully, considering the following questions. How would you describe to a friend who could not see the figure the orientation of the electric field arrows? If your friend asked you about the length of the different vectors, what response could you give?

The electric field is in the direction of electrical force on a positive test charge and therefore points radially outward from the charge. Since the strength of the electric field for a point charge varies inversely with the square of the distance from the charge, the closer we go to the charge, the longer the electric field vector arrow. If the charge is negative, then the arrows point in toward the charge.

In Figure 19-22, we show a collection of electric field vectors for a system with equal-magnitude positive and negative charges. Here it is a bit more complicated to compute the electric field vectors since we need to use superposition of the independent electric fields from each charge, like we did in the previous section. When you are quite near a charge, the electric field due to that charge will dominate over the one from the other charge, whereas when you are about equally far from both charges, each will significantly influence the direction and magnitude of the net electric field.

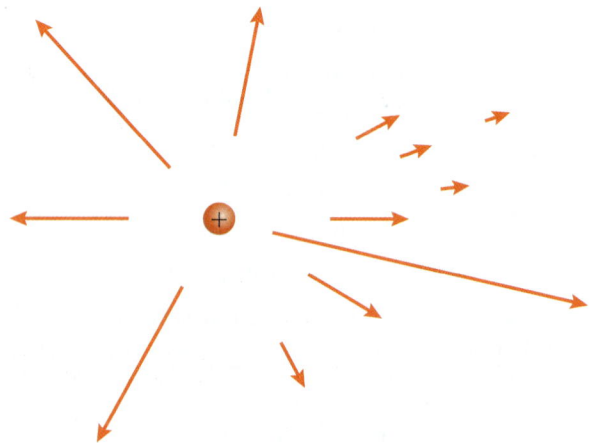

Figure 19-21 Electric field vectors for a single positive charge. The closer the point to the charge, the stronger the electric field and therefore the longer the electric field vector.

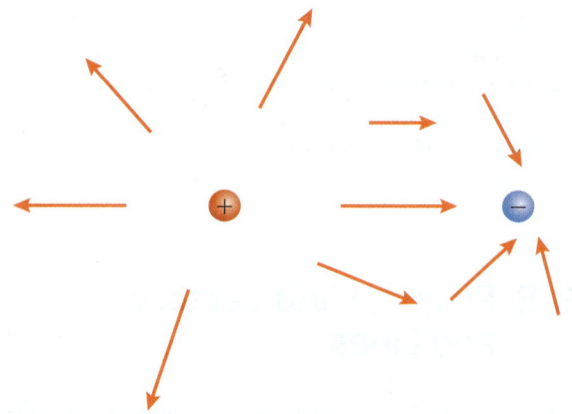

Figure 19-22 Electric field vectors for one positive and one negative charge of equal magnitude. The electric field at any point is obtained by superposition of the electric field due to each charge.

CHECKPOINT 19-9

Electric Field Vectors

Consider two situations, in both of which you are at a distance of 1.0 m from a positive charge. In situation A, there is only that charge. In situation B, there is also a negative charge of equal magnitude an equal distance away on the opposite side (i.e., the point is midway between the two charges in B). Which statement correctly relates the electric field vectors in the two cases?

(a) The electric field vector will be of equal length in the two cases and point away from the positive charge.

(b) The electric field vector will be of zero length in B and point away from the positive charge in A (and toward the negative charge).

(c) The electric field vector in situation B will be twice as strong as in A, but in both cases it will point away from the positive and toward the negative charge.

(d) The electric field vector in situation B will be half as strong as in A, but in both cases it will point away from the positive and toward the negative charge.

ANSWER: (c) The electric field in B is repulsive away from the positive charge and attractive toward the negative charge, so both are in the same direction and double the field strength in situation A.

While electric field vectors are helpful in visualizing the electric field and are consistent with how we use vectors elsewhere in physics, you can probably see that since we have a choice regarding where we show them, it is difficult be consistent and have a non-cluttered look. For this reason, a different representation, **electric field lines**, is commonly used instead to represent electric fields. We show in Figure 19-23 the electric field lines for configurations of two charges.

A useful way to think about how electric field lines are drawn is to start the first line somewhere on the surface of a positive charge. At that point the charge will start radially outward. You then move slightly, and compute the direction of the electric field at that point, and then move in the direction of the electric field to a new point, and repeat the process. Software automates this, so it is of course not done with manual calculations. The electric field line eventually ends on a negative charge (or at infinity if there are no negative charges present).

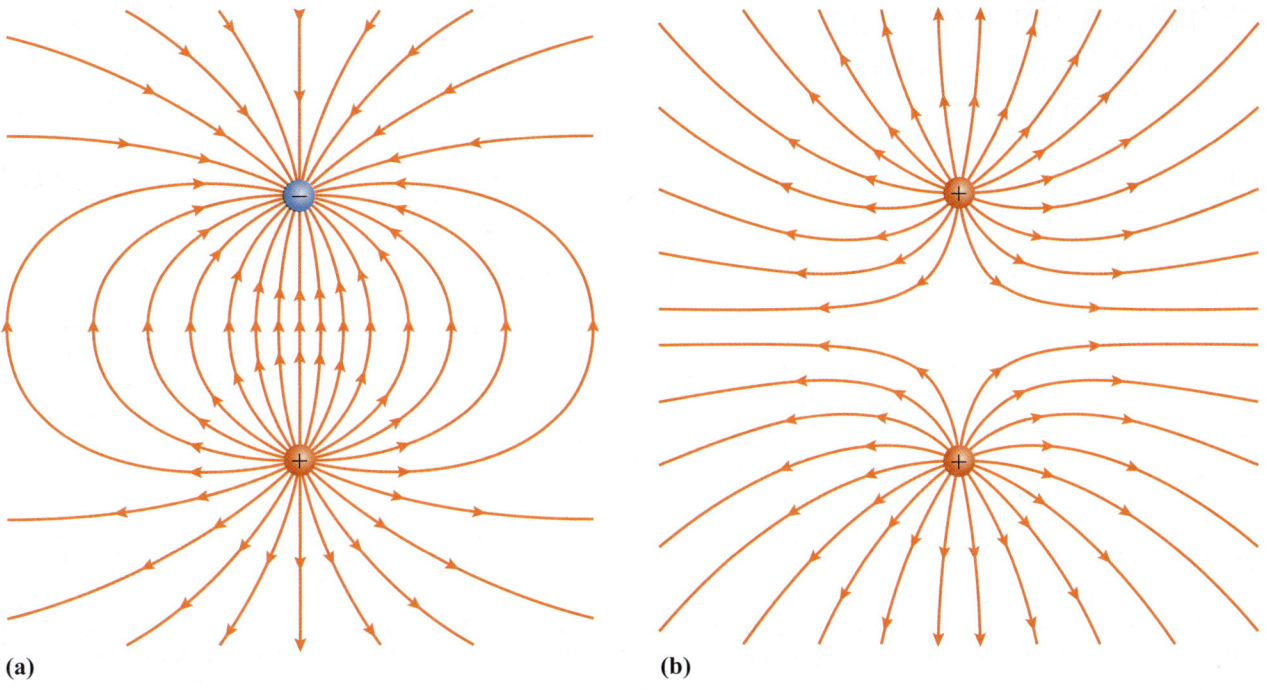

(a) (b)

Figure 19-23   Electric field lines for (a) one positive and one negative charge of equal magnitude and (b) two equal-magnitude positive charges.

In Figure 19-23(a) we see that the electric field lines start on the positive and end on the negative charge. In Figure 19-23(b) there are no negative charges, so the electric field lines go off to infinity. We show in Figure 19-24 a more complex configuration with two positive and two negative charges, all of equal magnitude.

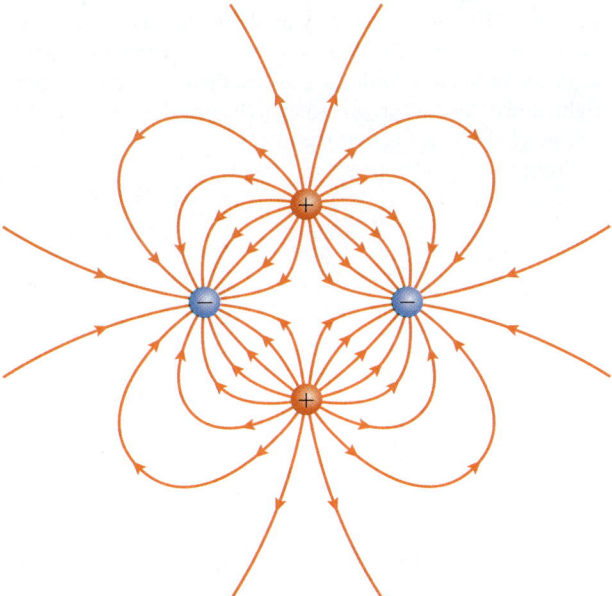

Figure 19-24   Electric field lines for a configuration of two positive and two negative charges, all of equal magnitude. Note that the field lines always start on a positive charge and end on a negative charge when there are equal numbers of both charges.

Recall that the electric field strength is highest when we are very near the charge (since it goes as the inverse square of the distance from the charge). Also, you will note in Figure 19-23 that the electric field lines are closer together in this region. That is a general feature: the closer the electric field lines, the higher the electric field strength. Also it is not possible to have two electric field lines cross, since that would imply that the electric field had two different values at exactly the same point.

We summarize the main characteristics of electric field lines below:

- At any point, the electric field line is in the direction of the electric field at that point.

- Electric field lines start on positive charges and end on negative charges. If there are more positive charges than negative charges, some of the electric field lines that start on the positive charges will go to infinity (or some of the electric field lines that end on the negative charges will come from infinity if there are excess negative charges in the configuration).

- Electric field lines never cross.

- Electric field lines being closer together implies a stronger electric field.

- Electric field lines on conductors are perpendicular to the local surface (we will see why this is in the next chapter).

Now that you have seen both electric field vectors and electric field lines, summarize in your own words what you see as the key advantages and limitations of each approach.

19-9 Electric Fields from Continuous Charge Distributions

So far, we have calculated electric fields produced by a finite number of discrete point charges. Now, we will consider electric field due to a charge spread uniformly over a linear element (and later over a surface). We will use the concept of linear, surface, and volume charge density that we introduced earlier in the chapter.

To find the net electric field around a continuously distributed charge, we need to use calculus. We write an expression for the field produced by each infinitesimal charge element, dq, and then integrate over all of the charge. Often, symmetry allows us to simplify the calculation because components in one or more directions cancel out.

Consider a thin, uniformly charged rod with linear charge density λ. If we define an x-axis along the length of the rod, a length element, dx, of the rod will have a charge dq given by

$$dq = \lambda dx \qquad (19\text{-}13)$$

The contribution from this element of charge to the electric field at a point a distance r from the element is

$$d\vec{E} = \frac{1}{4\pi\varepsilon_0}\frac{dq}{r^2}\hat{r} \qquad (19\text{-}14)$$

Note that the unit vector $\hat{r}$ points away from the charge element. Therefore, the electric field element points away from the charge element when dq is positive and toward the charge element when dq is negative.

We then integrate over the entire charged object to find the total electric field:

KEY EQUATION
$$\vec{E}_{net} = \int d\vec{E} = \int_{object} \frac{1}{4\pi\varepsilon_0}\frac{dq}{r^2}\hat{r} \qquad (19\text{-}15)$$

CHECKPOINT 19-10

Electric Field inside a Charged Spherical Shell

Consider a thin spherical shell with equally distributed positive surface charge. In case A, the shell is of radius R, and in case B it is of radius $2R$. How does the electric field at exactly the centre of the shell compare in the two cases? The total charge on the shell is the same in both cases.

(a)   Case A will be double that of case B.
(b)   Case A will be four times that of case B.
(c)   Case A will be half that of case B.
(d)   The electric field is exactly zero in both cases.

ANSWER: (d) From the symmetry of the situation, the electric field vectors from any one element of charge will be exactly cancelled by the vector from an element on the opposite side of the sphere, so the net electric field is zero.

EXAMPLE 19-10

Electrical Force from a Charged Rod

A thin, horizontal rod of length 2.00 m carries a total positive charge of 98.0 μC distributed uniformly along its length. A single point charge, Q, of $+2.00$ μC, is 1.00 m from the right end of the charged rod (see Figure 19-25). Find the total electrical force acting on the charge Q.

Solution

We will divide the rod into very small differential charge elements, write an expression for the electric field due to each element, and then integrate to find the net field. As we can see from Figure 19-25, every element of the charge on the rod produces a contribution to the electric field that is in the $+x$-direction, so these contributions all simply add together.

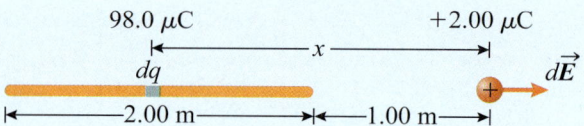

Figure 19-25 To find the net electric field from a charged rod, we integrate the contributions from the various infinitesimal elements of the rod.

First, we will express the charge element dq in terms of the linear charge density of the rod and a length element, dx:

$$\lambda = \frac{charge}{length} = \frac{98.0 \times 10^{-6}\text{ C}}{2.00\text{ m}} = 4.90 \times 10^{-5}\text{ C/m}$$

$$dq = \lambda dx = (4.90 \times 10^{-5}\text{ C/m})dx$$

Since all elements of charge on the rod produce force elements with the same direction, we can simplify Equation 19-15 to solve for just the magnitude of the electrical force. We integrate over the length of the rod. We define the x-axis to lie along the rod, with the positive direction being to the right, and define the origin as being the position of the point charge Q. This means that the integration over the rod will go from $x = -3.00$ m to $x = -1.00$ m:

$$E_{net} = \frac{1}{4\pi\varepsilon_0}\int_{-3}^{-1}\frac{\lambda dx}{x^2}$$

$$= \frac{\lambda}{4\pi\varepsilon_0}\int_{-3}^{-1}\frac{dx}{x^2}$$

$$= \frac{\lambda}{4\pi\varepsilon_0}\left[-\frac{1}{x}\right]_{-3}^{-1}$$

$$= \frac{\lambda}{4\pi\varepsilon_0}\left[-\frac{1}{-1}-\left(-\frac{1}{-3}\right)\right]$$

$$= \frac{2}{3}\frac{\lambda}{4\pi\varepsilon_0} = 2.94 \times 10^5\text{ N/C} = 294\text{ kN/C}$$

To find the magnitude of the net force on the charge Q, we use the relationship between electric field and force:

$$F_{net} = qE_{net} = (2.00 \times 10^{-6}\text{ C})(2.94 \times 10^5\text{ N/C}) = 0.588\text{ N}$$

Since Q is a positive charge, the direction of the force is the same as for the electric field, that is, in the $+x$-direction. Therefore, $\vec{F}_{net} = 0.588$ N $\hat{i}$.

Making sense of the result

To check the order of magnitude of the force, we can replace the charge along the rod with a point charge of 98 μC at a distance of 2 m (average position) from the other charge. Applying Coulomb's law then gives a force of 0.4 N, which indicates that the answer we obtained is reasonable. (Note that this approximation underestimates the actual force because the force depends on the inverse square of the distance between the charges.)

In Example 19-11, we demonstrate the use of superposition, integration, symmetry, and surface charge density in finding the electric field of a charged ring and an infinite plane.

EXAMPLE 19-11

Electric Field from a Charged Ring and from an Infinite Plane

(a) A horizontal ring of radius r carries a uniform charge and has a total charge of $+Q$. What is the electric field at a point a distance y vertically above the plane of the ring?

(b) Use your result from part (a) to find the electric field a distance y above an infinite plane with surface charge density σ.

Solution

(a) Again we will use differential charge elements and integrate over the electric field contributions from each. The situation is shown in Figure 19-26, where we have shown the electric field $d\vec{E}$ from an infinitesimal charge element dq. We can consider the infinitesimal element as approximating a point charge dq, and we obtain the following relationship for the magnitude of the contribution of the electric field from dq:

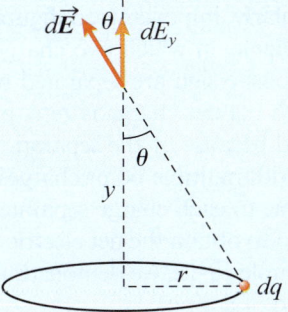

Figure 19-26 Example 19-11. A narrow ring carries a uniform charge.

$$dE = \frac{1}{4\pi\varepsilon_0}\frac{dq}{(y^2 + r^2)}$$

At this point, it is important to consider the symmetry of the situation. Since the ring is uniformly charged, the components of the electric field in the xz-plane will all cancel, and the net electric field must be vertical. The magnitude of the vertical component of the electric field is given by

$$dE_y = dE\cos\theta = \frac{1}{4\pi\varepsilon_0}\frac{y\,dq}{(y^2 + r^2)^{3/2}},$$

where we have used

$$\cos\theta = \frac{y}{(y^2 + r^2)^{1/2}}$$

Calculating the integral, and using the facts that both y and r are constant in this situation and that the integral of the charge element around the ring is simply the total charge on the ring Q, we have the following for the magnitude of the net electric field:

$$E_{net} = \int dE_y = \frac{1}{4\pi\varepsilon_0}\frac{yQ}{(y^2 + r^2)^{3/2}}$$

This expression gives the magnitude of the electric field. Its direction is vertically upward when the ring charge Q is positive and downward when it is negative.

(b) In this situation, we want to consider the electric field at a distance y above an infinite plane that is uniformly charged with surface charge density σ. The approach we will take is to imagine the plane as being made up of an infinite number of rings of charge of increasing radius. We then integrate over those rings. We could have done the differential elements as we did in Example 19-6, but we see that the same result can be obtained more readily by this approach of integrating over rings, rather than area elements.

If we consider a ring of radius r and width dr, then the charge on just that ring (the entire ring, not just a little piece of it as we did for dq), which we will call dQ, is given by $dQ = 2\pi\sigma r dr$.

As in part (a), the symmetry of the situation implies that the net electric field must be in a vertical direction. We can use the result from part (a) to find the magnitude of the contribution to the electric field due to just this ring:

$$dE = \frac{1}{4\pi\varepsilon_0}\frac{y\,dQ}{(y^2 + r^2)^{3/2}} = \frac{1}{4\pi\varepsilon_0}\frac{y\,2\pi rdr\sigma}{(y^2 + r^2)^{3/2}}$$

To find the total electric field (due to all the rings), we need to integrate over all r values to infinity:

$$E_{net} = \int dE = \frac{2\pi\sigma}{4\pi\varepsilon_0}\int_{r=0}^{r=\infty}\frac{y\,rdr}{(y^2 + r^2)^{3/2}}$$

In calculating the integral, we can make the following substitution:

$$u = y^2 + r^2 \quad \text{and} \quad du = 2rdr$$

(continued)

This allows us to calculate the integral and to find the magnitude of the net electric field:

$$E_{\text{net}} = \frac{\pi \sigma y}{4\pi\varepsilon_0} \int_{u=y^2}^{u=\infty} \frac{du}{u^{3/2}} = \left[-2\frac{\pi \sigma y}{4\pi\varepsilon_0} u^{-1/2} \right]_{y^2}^{\infty}$$

$$= -2\frac{\sigma y}{4\varepsilon_0}\left(0 - \frac{1}{y}\right) = \frac{\sigma}{2\varepsilon_0}$$

The electric field is vertical (upward when the charge is positive) and has the magnitude given above.

Making sense of the result

At first glance, it may seem surprising that the strength of the electric field does not depend on the distance, y, above the charged plane. Qualitatively, what is happening here is that, although the element of the charged plane nearest to you does produce a stronger electric field, the contributions from more distant parts of the plane are at a larger angle when you are close and tend to cancel as vectors. It turns out that the net result is independent of the distance from the charged infinite plane. In Chapter 20, you will see an alternative way to obtain the same result.

19-10 Dielectrics and Dipoles

We can classify materials according to how readily charge moves in response to an electric field. Some of the electrons in a conductor, such as copper, move readily when we apply an electric field, resulting in significant current flow (see Chapter 23 for a more detailed discussion). In contrast, the electrons in insulators are so tightly bound that applying an electric field produces very little, if any, current. Examples of good insulators are glass, ceramic, and paper, although many other materials, such as various plastics, are relatively good insulators.

Some molecules are polarized, although they have no net charge. In polar molecules, electrons are shared unequally in bonds between the atoms. For example, in a water molecule, the oxygen atom attracts the shared electrons more strongly, so it is negatively charged, and the hydrogen atoms are positively charged.

Consider an external electric field produced in a region between two charged plates, with the top plate positive and the bottom plate negative. Therefore, what we will refer to as the *external* electric field points from top to bottom. Now if we place some polar molecules in the space between the plates, they will preferentially align, with the more negative ends of the molecules closer to the positive plate and the more positive ends of the molecules closer to the negative plate (since

unlike charges attract). This means that the *internal* electric field of the aligned polar molecules is *opposite* to the direction of the external electric field. As a result, the net electric field, from combining the external and internal fields, is less than the value of the original external electric field.

Materials that respond to electric fields in this way are called **dielectrics**. In Chapter 22, we will examine an important application of dielectric materials in electronic capacitors. Capacitors are pairs of charged plates that can be used to store charge (and energy). Since the net electric field has been reduced by the addition of the dielectric, it turns out that it is more efficient at storing charge than a similar capacitor without a dielectric.

While polar molecules lead to the strongest dielectric materials, it is possible to have polarization in an electric field for non-polar molecules and for ionic crystals. The term *dielectric* was first proposed by the historian and philosopher of science William Whewell (1794–1866). He also introduced a number of other terms you study in physics, including anode, cathode, physicist, and scientist.

A particularly important configuration of charges is an **electric dipole**, in which two charges of equal magnitude but opposite sign are separated by some distance, d. Even though the net charge is zero, a dipole produces an electric field because of the separation of the charges. In situations with multiple point charges, we can find the electric field due to each charge separately and then do a vector addition to obtain the net electric field by superposition. In Example 19-12, we demonstrate this process for the electric field of a dipole.

EXAMPLE 19-12

Electric Field of a Dipole

Consider a dipole lying along the x-axis and centred at the origin. Derive an expression for the electric field at a point on the y-axis.

Solution

Let r represent the distance from the point on the y-axis to the origin (the centre of the dipole). Draw the dipole and the electric field vectors for each charge, as shown in Figure 19-27. The magnitudes of these two electric field vectors are the same because the magnitudes of the charges are identical, as are the distances to the point. Therefore, by symmetry, the downward component of one field vector exactly cancels the upward component of the other, and the net electric field is the sum of the two x-components (both of which point in the $-x$-direction). Symmetry can often be used to simplify applications of the superposition principle for electric forces and fields.

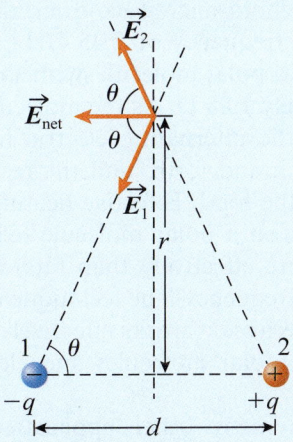

Figure 19-27 Example 19-12. An electric dipole with charges $-q$ and $+q$ separated by distance d.

The magnitude of each of the two electric field vectors is

$$E_1 = E_2 = \frac{1}{4\pi\varepsilon_0} \frac{q}{(r^2 + (d/2)^2)} = \frac{1}{4\pi\varepsilon_0} \frac{q}{(r^2 + d^2/4)}$$

The net electric field is to the left ($-\hat{i}$-direction), and its magnitude is

$$E_{net} = 2E_1 \cos\theta$$

We can express the angle θ in terms of d and r:

$$\cos\theta = \frac{d/2}{\sqrt{r^2 + d^2/4}}$$

Combining these results, we have

$$\vec{E}_{net} = \left(-\frac{1}{4\pi\varepsilon_0}\right) \frac{qd}{(r^2 + d^2/4)^{3/2}} \hat{i}$$

If we take the limit $r \gg d$, we obtain the following approximate result:

$$\vec{E}_{net} \approx -\frac{1}{4\pi\varepsilon_0} \frac{qd}{r^3} \hat{i} \quad \text{when } r \gg d$$

Making sense of the result

The direction of the field is the same as the direction from the positive to the negative charge, which makes sense. At large distances from the dipole ($r \gg d$), the electric field is approximately proportional to $1/r^3$, and the field from a single charge is proportional to $1/r^2$. Thus, the electric field from a dipole weakens faster with increasing distance, which makes sense because the fields from the two opposite charges partially cancel each other.

We saw in Example 19-12 that the electric field for a dipole goes down as the inverse cube of the distance when you are far away from the dipole. It is common to define an **electric dipole moment**, $\vec{p}$, for charges $+q$ and $-q$ separated by a displacement vector $\vec{d}$ that points from the negative charge to the positive charge:

KEY EQUATION
$$\vec{p} = q\vec{d} \quad \text{(19-16)}$$

The SI units for electric dipole moments are coulomb-metres. We can write the result from Example 19-12 in terms of the electric dipole moment:

KEY EQUATION
$$\vec{E}_{dipole} \approx -\frac{1}{4\pi\varepsilon_0} \frac{\vec{p}}{r^3} \quad \text{when } r \gg d \quad \text{(19-17)}$$

Electric dipoles are of crucial importance in molecular physics. The 1936 Nobel Prize in Chemistry was awarded to Petrus Josephus Wilhelmus Debye (1884–1966) for his work on dipole moments. A convenient unit for dipole moments of molecules, the debye (D), was named in his honour (1 D = 3.34 × 10^{-30} C·m).

Polar substances can be characterized by a dipole moment. For example, water (H_2O) has a dipole moment of about 1.85 D, while that of ammonia (NH_3) is 1.47. Molecules of diatomic elements, such as chlorine (Cl_2), have no separation of charge and, hence, zero dipole moment. A number of other non-ionic molecules, including CO and CO_2, have near-zero dipole moments. Sodium chloride (NaCl) does not exist as individual molecules, but rather as a crystal ionic compound. It can still be characterized by a dipole moment and has a high value of 9.0 D.

When an external electric field is applied to a substance with a dipole moment, the molecules will align if the molecules are free to rotate. A torque is present until this alignment is achieved (see Example 19-13).

EXAMPLE 19-13

Torque on a Dipole from an Electric Field

As shown in Figure 19-28, an electric dipole consists of $+q$ and $-q$ charges separated by a distance d. The dipole is oriented at an angle θ with respect to an external electric field that points horizontally from right to left. What will be the torque on the dipole due to the electric field?

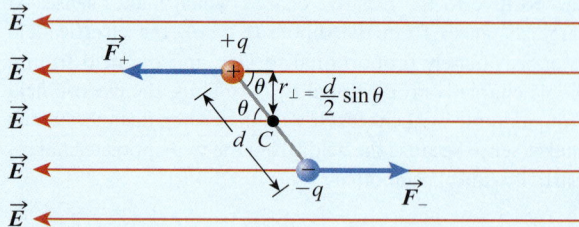

Figure 19-28 Example 19-13. An electric field exerts a torque on an electric dipole unless it is oriented along the field.

Solution

The upper (positive) charge will experience an electrical force in the direction of the electric field, in this case to the left. We label this force $\vec{F}_+$ and it will have a magnitude of qE. The bottom charge is negative, and therefore the direction of the electrical force will be opposite to that of the electric field, to the right in this case. We label this force $\vec{F}_-$. Since these forces have equal magnitudes, and point in opposite directions, there is no net force on the electric dipole from the external electric field.

There is, however, a torque on the electric dipole due to these forces since they are applied at opposite ends of the dipole. One way to calculate the magnitude of the torque (see Chapter 8) is $\tau = r_\perp F$, where $r_\perp$ is the perpendicular moment arm:

$$r_\perp = \frac{d}{2}\sin\theta$$

The contribution to the torque from just the positive charge is $qE\frac{d}{2}\sin\theta$. Both electrical forces tend to make the dipole rotate in a counter-clockwise fashion (as viewed from above) and therefore add to give a net torque of $qEd\sin\theta$. The direction of the torque can be obtained by curling the fingers of the right hand in the direction that the dipole tends to rotate, with the thumb giving the direction. In our case, this is upward out of the plane of the electric field and dipole.

Making sense of the result

The force and moment arm give the reasonable result that the torque is proportional to the electric field, the charge magnitude, and the separation distance of the charges in the dipole. As illustrated in the text of the section, this result can be expressed in terms of the dipole moment. Note that if the dipole is aligned (or anti-aligned) with the electric field with $\theta = 0°$ (or $\theta = 180°$), the torque will be zero. It will have a maximum value when the dipole is perpendicular to the electric field.

We can express the result obtained in Example 19-13 in terms of the vector cross product of the dipole moment and the electric field (see Figure 19-29). Recall that the direction of $\vec{p}$ points from the negative charge to the positive charge.

KEY EQUATION

$$\vec{\tau} = \vec{p} \times \vec{E} \qquad (19\text{-}18)$$

The direction of the torque obtained by the right-hand rule for the vector cross product will be out of the plane, as obtained earlier in Example 19-13. We can see that the magnitude of the torque obtained by Equation 19-18 is also consistent with that from the example problem:

$$\tau = pE\sin\theta = qdE\sin\theta$$

If we rapidly change the orientation of the electric field, we can use the resultant torques to constantly change the orientation of the molecules. Such movement is the basis for heating in microwave ovens. In a typical household microwave oven, the electric field is applied at a frequency of 2.45 GHz. As mentioned above, water is a polar molecule, with a dipole moment of approximately 1.85 D. As the molecules try to align according to the alternating electric field, they rapidly rotate back and forth, and the resulting thermal motion heats the food. Effective heating by this technique depends on a polar molecule (e.g., microwaves heat water more effectively than fats) and an appropriate applied frequency. The technique is called dielectric heating (because it works effectively on dielectrics, materials with polar molecules and electric insulator properties).

There are many other applications of electric dipoles in science and technology. While not as well known as the radiation and chemical therapies, in hyperthermia cancer cells are heated slightly using high-frequency dielectric heating to damage cancer cells. Some muscle therapies also use dielectric heating. Dielectric heating is used to control pests in some food crops, especially nuts.

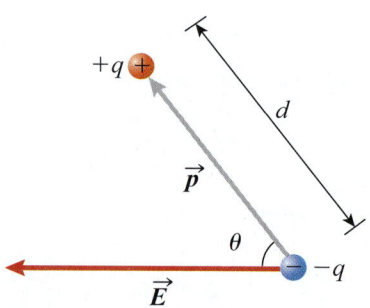

Figure 19-29 The torque on an electric dipole is obtained by taking the vector cross product of the dipole moment and the electric field.

19-11 Electric Field Essentials

Electricity and magnetism are among the most important and pervasive topics in physics. While this chapter has concentrated on electrical force, you will in the next series of chapters expand your understanding of electricity and magnetism. The electric field is not only important in its own right, but electrical field ideas played a key role in how a lot of later physics was developed.

In this final section, we will consider again the nature of the electric field, as well as reflecting on techniques and potential pitfalls for computing electric fields. While from its definition the electric field is the electrical force per unit charge if a small positive test charge is placed at that location, the concept of a field is more essential than this definition might imply. Electric fields exist around charge configurations in the same sense that gravitational fields exist around a planet, and *the electric field exists whether we place a small test charge there or not* (in the same way a planet's gravitational field exists whether we place a small mass to test it or not). For example, energy can be considered to be stored in electric fields. We have considered static electric fields in this chapter, but in Chapter 27 we will encounter electromagnetic fields for changing electric and magnetic fields. Broadcast radio and television signals and wireless telephone networks depend on the transmission of power by these electromagnetic fields.

You might have wondered why we specified that the test charges for the electric field must be small and positive. The idea of a small test charge is simply that we do not want the electrical influence of the charge to be enough that it will change the configuration of external charges that produce the electric field being studied. The requirement that it be positive is not essential if we consider $\vec{E} = \vec{F}/q$ and take into account the signs on both the force and the charge, but it is simpler to say that the electric field is in the direction of the force on a positive test charge.

While we first defined electric fields in terms of electrical force per charge, often we use this relationship the other way around. We know, or compute, the electric field in a region, and we use that to calculate the electrical force. *If we know the electric field at all positions, then we know the electrical force on any charge we place in that region.* Remember that *the direction of the force on a charge will be the same as the direction of the electric field for positive charges, and in the opposite direction for negative charges.*

You learned techniques for calculating the electric field in this chapter for single point charges, configurations of point charges, and continuous charge distributions. It is critical when doing these computations to remember that *electric fields are vector fields* and that *when the superposition principle is employed for*

electrical fields the individual electric field contributions must be added as vectors.

When computing electric fields from point charges, you can get the direction either by using the vector form of the expression for the electric field (Equation 19-11) or by using common sense—like charges repel and unlike attract—to decide the direction of the force on a positive test charge. Even if you use the vector form of the equation, it is always a good idea to *check your results by common-sense consideration of attraction or repulsion on a positive test charge.*

Often the symmetry of the situation permits simplification of the actual computations. When dealing with collections of charges, see if some can be paired to cancel electric field contributions. Even with continuous charge distributions, the electric field due to one part may be cancelled by that due to another part of the charged object.

CHECKPOINT 19-12

Electric Field Misconceptions

Consider the following statements regarding the electric field. For each, indicate whether the statement is true or false.

(a) Electric field and electric field strength mean the same thing.

(b) Electric fields only exist when we place a positive test charge in the location.

(c) If you are precisely halfway between two equal-magnitude negative charges, the electric field will be zero.

(d) The electric field for a single point charge depends only on the distance from the charge and magnitude of the charge, and not the sign of the charge.

ANSWER: (a) False: The electric field is a vector quantity and electric field strength refers only to the magnitude of the electric field and is not a vector. (b) False: The electric field exists independent of whether an actual test charge is present. (c) True: The electric field contributions from the charges will be exactly equal but in opposite directions and will add to zero. (d) False: The electric field is a vector quantity. The direction of the electric field depends on whether the charge is positive or negative.

The point charges that played such a central role in this chapter are seldom present exactly in situations that we consider. As you develop your skills and intuition as a physicist, you will gradually come to determine when a spherical distribution of charge can be approximated by a point charge of the same magnitude.

Most of the next chapter is concerned with an additional technique for computation of the electric field. This technique, called Gauss's law, can be used in certain situations where there is sufficient symmetry in the charge configuration. In such cases, this technique is usually computationally easier than adding up vector electric fields by superposition (or integration for continuous charge distributions). Nevertheless, the

ELECTRICITY, MAGNETISM, AND OPTICS

superposition techniques of this chapter, when coupled with modern numerical computing, can be used to find the electric field to desired precision for essentially any distribution of charges.

We want to end the chapter as we started it, by pointing out how pervasive electrical forces and signals are in all aspects of our lives. We communicate with electromagnetic waves, convert energy using electromagnetism, use electromagnetic motors to do work for us, and measure electrical signals for monitoring of environmental, medical, and industrial situations. Our quality of life would be much poorer were it not for the electromagnetic principles that you will learn in this section of the textbook.

KEY CONCEPTS AND RELATIONSHIPS

Electric charge is a fundamental property, and electric charges are either positive or negative. Objects are electrically charged if they feel an electrical force in the presence of other charges. Like charges (i.e., two positive or two negative charges) repel, and unlike charges (one positive and one negative) attract. Most everyday forces are electromagnetic in nature.

Some materials hold onto their electrons more firmly than other materials, and contact followed by separation causes triboelectric charging. The triboelectric series is used to predict which materials take on positive charges and which take on negative charges when rubbed.

Conductors are materials in which a current flows in response to an applied electric field. Insulators are materials in which virtually no current flows. Dielectrics do not have a net current flow, but they do have charges that can be displaced when an external electric field is applied.

Charge and Current

The SI unit of charge is the coulomb, C. Current is the amount of charge that flows past a point in a given time:

$$I = \frac{\Delta q}{\Delta t} \qquad (19\text{-}1)$$

Conductors, Insulators, and Charging by Induction

In conductors, electric charges readily move, while they do not in insulators. When a charged object is placed near a conductor, charges of opposite sign are attracted to that side of the conductor. These can come from another object if the conductor is not isolated. Objects can be charged in this way, which is called induction.

Coulomb's Law

The electrical force exerted on one point charge by another is given by Coulomb's law:

$$\vec{F}_{E\,1\to2} = \frac{1}{4\pi\varepsilon_0}\frac{q_1 q_2}{r^2}\hat{r}_{1\to2}, \qquad (19\text{-}5)$$

where q_1 and q_2 are the two charges, r is the distance between the centres of the charges, $\hat{r}_{1\to2}$ is a unit vector that points from q_1 to q_2, ε_0 is the permittivity of free space, and $\vec{F}_{E\,1\to2}$ is the electrical force that charge q_1 exerts on charge q_2. According to Newton's third law, the force of charge 2 on charge 1 has the same magnitude but opposite direction.

Superposition Principle

For systems of point charges, we can find the electrical force or electric field separately for each charge and then vectorially add the results to find the net electrical force or electric field.

Charge Densities

In dealing with situations involving continuous charge distributions, we divide the charged object into infinitesimal differential charge elements and then integrate to find the total electrical force (or field), keeping in mind the vector nature. In these situations, we define the linear, surface, or volume charge density:

$$\lambda \equiv \frac{Q}{L} \qquad (19\text{-}6)$$

$$\sigma \equiv \frac{Q}{A} \qquad (19\text{-}7)$$

$$\rho \equiv \frac{Q}{V} \qquad (19\text{-}8)$$

A form of Coulomb's law in which one charge is replaced by a differential charge element is then used in computing the net force.

Electric Field Definition

The electric field, $\vec{E}$, is defined as the net electrical force on a small positive point charge, q, divided by the magnitude of the point charge:

$$\vec{E} \equiv \frac{\vec{F}_E}{q} \qquad (19\text{-}10)$$

Electric Field for Point Charges

For a single point charge, the electric field at a distance r from the charge, Q, has the following form, where $\hat{r}$ is a unit vector pointed radially outward from the charge. Note that the correct sign must be used for the charge, Q:

$$\vec{E} = \frac{1}{4\pi\varepsilon_0}\frac{Q}{r^2}\hat{r} \qquad (19\text{-}11)$$

When we have a collection of a number of point charges, we can compute the electric field due to each and then combine the results as vectors to find the total electric field according to the superposition principle.

ELECTRICITY, MAGNETISM, AND OPTICS

Electric Field Vectors and Lines

We can represent electric fields by electric field vectors, which have a direction corresponding to that of the electric field and a magnitude equal to the strength of the electric field. We can alternatively use electric field lines that are everywhere parallel to the local electric field. Electric field lines start on positive charges and end on negative charges, never cross, and are perpendicular to the local surface just outside a conductor. The density of electric field lines indicates the strength of the electric field in that region.

Electric Fields from Continuous Charge Distributions

For continuous charge distributions, the net electric field is given by

$$\vec{E}_{net} = \int_{object} \frac{1}{4\pi\varepsilon_0} \frac{dq}{r^2}\hat{r} \qquad (19\text{-}15)$$

Often, we can use symmetry to simplify the calculations.

Dipole Moment

For two opposite charges of equal magnitude $(+q, -q)$ separated by displacement $\vec{d}$, the dipole moment is

$$\vec{p} = q\vec{d} \qquad (19\text{-}16)$$

The electric field at a large distance, r, from a dipole is approximately equal to

$$\vec{E}_{dipole} \approx -\frac{1}{4\pi\varepsilon_0}\frac{\vec{p}}{r^3} \quad \text{when } r \gg d \qquad (19\text{-}17)$$

The torque on a dipole in an electric field can be determined from

$$\vec{\tau} = \vec{p} \times \vec{E} \qquad (19\text{-}18)$$

APPLICATIONS

cathode ray tube (CRT) devices, dielectric heating, electrical sensors, electron microscopes, electrostatic damage to electronics, industrial process and quality monitoring, inkjet and laser printers, medical monitoring, microwave heating, sputtering coating, static charging, thermal therapies

KEY TERMS

conductors, Coulomb constant, Coulomb's law, current, dielectrics, dipole moment, electric charge, electric dipole, elementary charge, electric field, electric field lines, electric field strength, electric field vectors, induction, insulators, linear, linear charge density, neutral, permittivity, photoelectric effect, semiconductors, static electricity, superposition principle, surface charge density, triboelectric charging, triboelectric series, volume charge density

QUESTIONS

1. You remove a piece of polyethylene tape from a glass surface. Which statement best describes what will happen? Table 19-1 may be helpful in answering this question.
   (a) You need rubbing for triboelectric charging, so no electrostatic charge results.
   (b) The tape becomes positively charged, and the glass becomes negatively charged.
   (c) The glass becomes positively charged, and the tape becomes negatively charged.
   (d) Both the tape and the glass take on a positive charge.

2. You rub a balloon through your hair, and it becomes charged. You rub a second similar balloon through your hair. When the two balloons are tied to electrically insulating strings, what happens when you hold the ends of the strings away from the balloon?
   (a) The balloons move away from each other.
   (b) The balloons come together and discharge.
   (c) The balloons come together, but each retains its charge.
   (d) The balloons come together, and each charges to a higher value.

3. Use Table 19-1 to determine which of the following combinations will have the greatest triboelectric charge when rubbed together.
   (a) paper and polystyrene
   (b) plastic wrap and glass
   (c) hair and wool
   (d) PVC and polyethylene tape

4. Refer to Table 19-1 when answering this question. Which of the following statements applies when you rub plastic wrap and glass together?
   (a) There is very little charging.
   (b) The glass becomes positively charged, and the plastic wrap becomes negatively charged.
   (c) The glass becomes negatively charged, and the plastic wrap becomes positively charged.
   (d) Both the glass and the plastic wrap become positively charged.

5. Some polyester shirts and wool socks have been tumbled together in a clothes dryer. Which of the following statements is correct? Refer to Table 19-1.
   (a) The socks are negatively charged, and the shirts are positively charged. The socks stick to each other, and the shirts stick to each other.
   (b) The socks are positively charged, and the shirts are negatively charged. The socks stick to each other, and the shirts stick to each other.
   (c) The socks are negatively charged, and the shirts are positively charged. The socks stick to the shirts but not to the other socks.
   (d) The socks are positively charged, and the shirts are negatively charged. The socks stick to the shirts but not to the other socks.

6. A charge of 50 μC flows past a point in 0.05 s. What is the resulting current?
  - (a) 10 μA
  - (b) 10 mA
  - (c) 1 mA
  - (d) 10 A

7. A positive charge, q, is placed near a larger, stationary positive charge, $5q$. What happens when you let go of charge q?
  - (a) It accelerates toward the larger charge.
  - (b) It moves away from the larger charge with constant velocity.
  - (c) It moves away from the larger charge with constant acceleration.
  - (d) It moves away from the larger charge but with an acceleration that decreases the farther it moves away.

8. You have two charges, q and $4q$. Charge q exerts a force of magnitude F on charge $4q$. What is the magnitude of the force that $4q$ exerts on q?
  - (a) $F/4$
  - (b) F
  - (c) $2F$
  - (d) $4F$

9. You have two fixed positive charges, $+Q$ on the left and $+4Q$ on the right. Where could you place a third small positive charge, $+q$, along a line joining the two charges such that it would experience zero net electrostatic force?
  - (a) at a point beyond the $+Q$ charge to the left
  - (b) at a point 1/4 of the way between the charges, closer to the $+Q$ charge
  - (c) at a point 1/3 of the way between the charges, closer to the $+Q$ charge
  - (d) There is no location where it would experience zero electrostatic force.

10. Two positive charges, $+Q$ and $+Q$, are located a distance L apart. A second pair of charges, $+2Q$ and $-2Q$, is placed at a distance $2L$ apart. Which situation has the larger electrostatic force?
  - (a) The force is the same in the two situations.
  - (b) The magnitude of the force is greater in the case with the $+Q$ charges.
  - (c) The magnitude of the force is greater in the case with the $+2Q$ charges.
  - (d) The force is zero in both cases.

11. A proton is moving in the $+x$-direction, and its speed is increasing with a constant acceleration. What can you conclude about the external electric field?
  - (a) The electric field is in the $-x$-direction and constant.
  - (b) The electric field is in the $-x$-direction and increasing.
  - (c) The electric field is in the $+x$-direction and constant.
  - (d) The electric field is in the $+x$-direction and increasing.

12. A positive charge is located at the origin of an xyz-coordinate system. What is the electric field direction at the point ($x = 0$, $y = -1$, $z = 0$)?
  - (a) $-\hat{i}$
  - (b) $+\hat{i}$
  - (c) $-\hat{j}$
  - (d) $+\hat{j}$

13. An electron is located at the origin of an xyz-coordinate system. What is the direction of the electric field at the point ($x = 1$, $y = 0$, $z = 0$)?
  - (a) $-\hat{i}$
  - (b) $+\hat{i}$
  - (c) $-\hat{j}$
  - (d) $+\hat{j}$

14. You are midway between two point charges, a positive charge $+Q$ to the left ($-x$-direction) and a negative charge $-Q$ to the right ($+x$-direction). What is the direction of the electric field at your location?
  - (a) $-\hat{i}$
  - (b) $+\hat{i}$
  - (c) There is zero net electric field at your location.
  - (d) $+\hat{j}$

15. If you slightly increase the distance between a proton and an electron, what happens to the dipole moment?
  - (a) The dipole moment is slightly larger.
  - (b) The dipole moment is slightly smaller.
  - (c) There is no dipole moment before or after.
  - (d) The dipole moment has not changed because the charges are still the same.

16. Consider two oppositely charged vertical parallel plates with equal magnitudes of charge. The left plate is positively charged. The space between the conductors is filled with a dielectric. How are the molecules in the dielectric material aligned?
  - (a) The positive ends of the molecules are toward the left.
  - (b) The positive ends of the molecules are toward the right.
  - (c) The positive ends of the molecules are toward the bottom.
  - (d) There is no net alignment of the molecules.

17. A radio wave has an electric field of approximately 0.15 N/C. What is the approximate electrical force on an electron that is in the path of the radio wave?
  - (a) 0.15 N in a direction opposite to the direction of the electric field
  - (b) 0.15 N in the same direction as the direction of the radio wave electric field
  - (c) 2.4×10^{-20} N in a direction opposite to the direction of the electric field
  - (d) 2.4×10^{-20} N in the same direction as the direction of the radio wave electric field

18. Which of the following is NOT a property of electric field lines?
  - (a) Lines closer together imply a stronger electric field at that point.
  - (b) Electric field lines never cross.
  - (c) Electric field lines always start on negative charges and end on positive charges.
  - (d) Electric field lines just outside a conductor are perpendicular to the surface.

19. A proton and an electron are separated by 10.0 nm. What is the dipole moment?
  - (a) 1.60×10^{-27} C m directed toward the proton
  - (b) 1.60×10^{-27} C m directed toward the electron
  - (c) 3.20×10^{-27} C m directed toward the proton
  - (d) 3.20×10^{-27} C m directed toward the electron

20. Polar molecules experience the most heating in an alternating electric field when their dipole moments are
    (a) 0.0 D
    (b) 5.0 D
    (c) 8.0 D
    (d) 11 D
21. When at a distance of 16 cm from a point charge, an electric field vector is 2 cm long. If you are at a distance of 8 cm from the same charge, what is the length of the electric field vector now?
    (a) 1 cm
    (b) 2.8 cm
    (c) 4 cm
    (d) 8 cm

PROBLEMS BY SECTION

For problems, star ratings will be used (★, ★★, or ★★★), with more stars meaning more challenging problems. The following codes will indicate if $\frac{d}{dx}$ differentiation, $\int$ integration, ▢ numerical approximation, or ⤴ graphical analysis will be required to solve the problem.

Section 19-1 Electric Charge

22. ★ If 250. billion elementary charges pass a point in 1.00 s, what is the current?
23. ★★ When you rub a gold sphere with rabbit's fur, the gold takes on a negative charge (and the rabbit fur positive). Suppose the gold sphere has a mass of 100.0 g and it obtains a net charge of 0.500 μC. Estimate the ratio of the number of electrons added to the gold sphere to the number originally there.
24. ★ There are a number of different types of lightning with different characteristics. One estimate for a large bolt of positive-polarity lightning is that it transfers 350 C of charge and has an average current of 120 kA. Estimate how long the lightning current flows.

Section 19-2 Charging by Electrical Induction

25. ★ (a) As shown in Figure 19-30, when an external object of unknown charge is brought near to the right-hand side of two conducting spheres (on insulated stands) that are touching, the induced charge is as shown. What must be the sign of the charge on the unknown charged object?

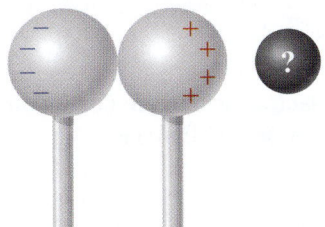

Figure 19-30 Problem 25.

(b) The two conducting spheres are then separated (while the charged object is still close), and then the charged object is removed. Sketch the distribution of charge on each of the spheres.

Section 19-3 Coulomb's Law

26. ★ Many first-year university textbooks write Coulomb's law with a simplified form for the constant, calling the whole expression k. What is the value of k? Specify its units.
27. ★★ One estimate of the mean separation between the proton and the electron in a hydrogen atom is 53.0 pm (if we picture them as classical particles).
    (a) What is the Coulomb force between the proton and the electron?
    (b) Set this force equal to ma, and for the acceleration use the classical centripetal acceleration. What is the speed of the electron?
28. ★★ A +35.0 nC charge is placed at the origin of a coordinate system, and a −25.0 nC charge is placed in the xy-plane at the point ($x = 0.100$ m, $y = -0.0500$ m).
    (a) Determine the magnitude and the direction of the force that the positive charge exerts on the negative charge. Use ijk notation for your result.
    (b) Determine the magnitude and direction of the force of the negative charge on the positive charge.

Section 19-4 Multiple Point Charges and the Superposition Principle

29. ★★ A −2.00 μC charge is 3.00 cm to the left of a +1.00 μC charge. Where must a +4.00 μC charge be placed to make the net electrical force on the +1.00 μC charge zero?
30. ★★★ A square of side length $\sqrt{2}D$ has charge $+Q$ on three of the corners and charge $-Q$ on the bottom right corner. A charge $+q$ is placed at the exact centre of the square. What is the net force on this charge?
31. ★★ Three equal charges of +0.250 μC are at the vertices of an equilateral triangle with side length 3.25 cm. What is the magnitude of the net electrical force on each charge? What is the direction of each force?

Section 19-5 Electrical Forces for Continuous Charge Distributions

32. ★★ (a) What would be the surface charge density in SI units on a sphere of radius 2.50 cm if a total charge of 75.0 μC is spread evenly over the surface?
    (b) What would be the net electric field at a point precisely in the centre of the sphere?
33. ★★★ $\int$ As shown in Figure 19-31, a point charge of +15.0 nC is located $d = 6.50$ cm above the midpoint of a uniformly charged fixed thin rod of length 4.00 cm and carrying a total charge of +80.0 μC. Calculate the net electrical force on the +15.0 nC charge.

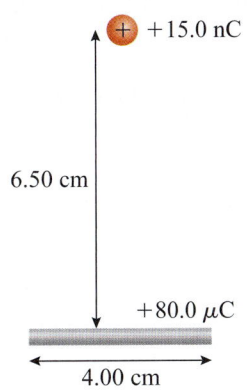

Figure 19-31 Problem 33.

Section 19-6   Electric Field

34. ★★ A proton has an instantaneous acceleration of 4.0×10^8 m/s^2 in the $+x$-direction. What electric field could account for this acceleration?

35. ★ Strong electric fields can be present inside biological cell membranes. The field in a particular cell at a certain location is 5.00 N/μC (note the units). What magnitude of force would result on an ion that has a charge of $+2e$?

36. ★★ (a) What is the ratio of the electrical force divided by the gravitational force for two protons at a distance of 350 pm apart?
(b) Would the answer in (a) be different if the distance had a different value? Explain.

37. ★★ An unknown charge is at the origin of an xy-coordinate system. When a temporary test charge of -4.00 nC is placed at ($x = 0.00$ m, $y = 1.00$ m) in SI units, the resultant electrical force on it is given by

$$\vec{F} = -6.00 \mu N \hat{j}$$

What must be the unknown charge?

38. ★★ For the situation described in problem 37, what would be the electrical field from the central charge only at the point ($x = 1.00$ m, $y = 2.00$ m)?

Section 19-7   Electric Fields and the Superposition Principle

39. ★★ Consider a point midway between two point charges, a $+24.0$ pC charge on the right and a -12.0 pC charge on the left. The charges are separated by 0.240 mm. What is the electric field at this midway point?

40. ★★ (a) A $+8.00$ μC charge is placed at the point ($x = 0.00$ m, $y = 16.00$ m). What charge must be placed at the origin for the electric field to be zero at the point ($x = 0.00$ m, $y = 4.00$ m)?
(b) Is there any charge that could be placed at the origin for the electric field to be zero at the point ($x = 4.00$ m, $y = 4.00$ m)?

41. ★★ You have an infinite line of closely spaced positive point charges at $x = 2.00$ m and for all values of y. You have another infinite line of point charges, also positive, at $x = 14.00$ m and for all values of y. The spacing of both sets of point charges is the same, but the magnitude of each point charge in the second set is double that of the first. Describe where, if anywhere, in the space the electric field will be zero.

42. ★★ Charges Q and q are stationary on corners of a square with side a, as shown in Figure 19-32. Assume that Q is a positive charge. Derive an expression for the value of charge q in terms of the variables of the question so that there is no net force on charge Q. As part of your answer, clearly indicate whether charge q is positive or negative.

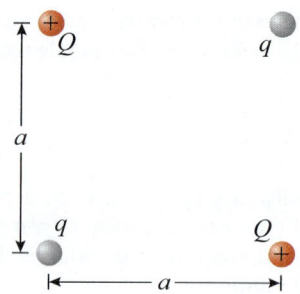

Figure 19-32   Problem 42.

Section 19-8   Electric Field Vectors and Lines

43. ★★ If you agree on a convention that electric field vectors will have a length of 1.00 cm for an electric field of 25.0 N/C, what should be the length and orientation for the electric field vector at the point ($x = 3.00$ m, $y = -2.00$ m) from a point charge of $+45.0$ nC placed at the origin of the coordinate system?

44. ★ Summarize the advantages and disadvantages of electric field vector representations compared to electric field line representations.

45. ★★ A negative charge lies in the middle between two positive charges on a horizontal line. Sketch the electric field lines diagram.

46. ★ Consider the electric field line diagram shown in Figure 19-33. What charge configuration is implied? Only part of the charge configuration is shown in the figure.

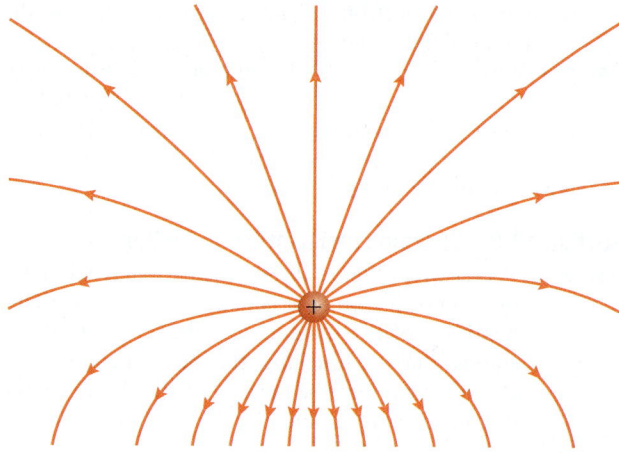

Figure 19-33   Problem 46.

Section 19-9   Electric Fields from Continuous Charge Distributions

47. ★★ ∫ A thin, fixed rod is 25.0 mm long and carries a total positive charge of 75.0 nC distributed uniformly along its length. A single small positive point charge, $Q = 1.00$ μC, is placed 100.0 mm from the left end of the charged rod in a direction along the line of the rod. What is the total electrical force on the charge Q?

48. ★★ ∫ A thin, fixed rod is 40.0 mm long and carries a total positive charge of $+40.0 \mu C$. The rod is located on the x-axis, with its centre at the origin of the coordinate system. What is the electric field at a point on the y-axis directly above the centre of the rod and a distance of 120. mm from the rod?

Section 19-10 Dielectrics and Dipoles

49. ★★ An electric dipole consists of one positive and one negative elementary charge. The magnitude of the dipole moment is 1.75 D. What is the separation of the two charges? Express your answer in nanometres.

50. ★★ In Example 19-12, we found an expression for the electric field for a dipole when we were at a perpendicular distance from the dipole. Derive a similar expression for the case when you are at a distance in line with the dipole direction. What is the approximate expression for this case when the distance from the dipole is much greater than the dimensions of the dipole?

Section 19-11 Electric Field Essentials

51. ★★ Write a comparison of gravitational to electrical forces and fields based on this chapter and Chapter 11.

52. ★ It was said in this section that electric field lines start on positive charges and end on negative charges. If we consider gravitational field "lines," where do they start and end?

53. ★ You are trying to convince a friend that electric fields are real and not simply a notation for representing something. Write a concise argument.

54. ★★ (a) Draw electric field lines for a single positive test charge.
(b) Now draw electric field lines for a positively charged conducting spherical shell considering what we said about directions of field lines near conductors in this section.
(c) Repeat part (b) but for a negatively charged conducting spherical shell.
(d) If you have both a positive point charge $+Q$ and a spherical conducting shell with exactly the same charge magnitude but negatively charged $-Q$, and if the point charge is exactly at the centre of the spherical shell, draw the electric field lines both inside and outside the spherical shell.

COMPREHENSIVE PROBLEMS

55. ★★ You have probably heard the story of the electrical kite experiment (see Figure 19-34) possibly conducted by Benjamin Franklin.
(a) Use reliable sources and write a concise account of the experiment.
(b) What was he trying to prove with the proposed experiment?
(c) Was the experiment really conducted by Benjamin Franklin?
(d) Why would the experiment be highly dangerous to really carry out? Note: Never experiment with this—people have been killed doing it.

Figure 19-34 Problem 55.

56. ★★ Suppose that Earth and the Moon had opposite charges of equal magnitude. What would be the charge on each if the electrostatic force was to be equal to the gravitational attraction between Earth and the Moon?

57. ★ (a) In a triboelectric process, an object gets a net charge of $+34.5 \mu C$. Assuming that this is by gain or loss of electrons, how many electrons have moved?
(b) If the charging takes place over 2.00 s, what is the magnitude of the average current flow?

58. ★★ In Figure 19-35, a small charge with a mass of 1.00 g and a charge of $-0.800 \mu C$ is attached to the end of a massless string. The charge hangs at an angle of $30.0°$ when a uniform electric field, E, is applied in the horizontal direction.
(a) What is the direction of the electric field?
(b) Draw a free body diagram for the small mass.
(c) Calculate the magnitude of the electric field.

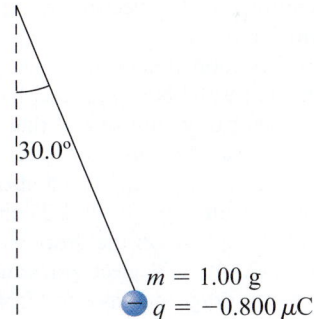

Figure 19-35 Problem 58.

59. ★★ Four electric charges are arranged as shown in Figure 19-36. Find the electric field at point P.

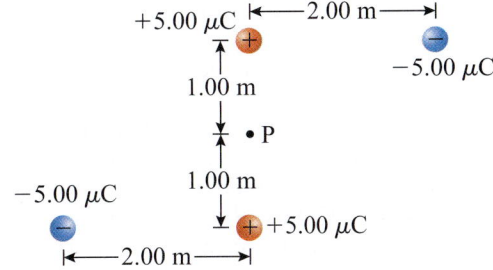

Figure 19-36 Problem 59.

ELECTRICITY, MAGNETISM, AND OPTICS

60. ★★★ The charges in Figure 19-37 are held rigidly in their positions. Determine the direction and magnitude of the net electric field at point P. Assume the charges and other quantities are known to two significant figures (e.g., 2.0 μC). (Hint: Use symmetry to simplify the calculation.)

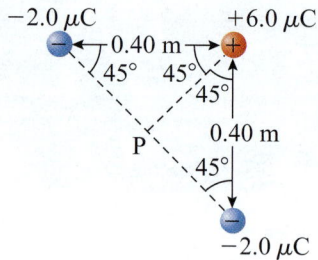

Figure 19-37   Problem 60.

61. ★ The electric field near Earth's surface points downward and has a magnitude of approximately 150 N/C. What is the force on an electron due to this electric field? Find the ratio of this force to the gravitational force on the electron.

62. ★★ A positive ion with a charge of $+3e$ is fixed at the origin of a coordinate system. An electron is located at $x = 0.20\ \mu$m on the x-axis. Where is the electric field zero?

63. ★★ A cathode ray tube (CRT) display accelerates an electron from rest to 2.50×10^7 m/s within a distance of 0.950 cm in the acceleration region.
   (a) What is the electric field in the CRT acceleration chamber?
   (b) Does the electric field point in the direction of acceleration of the electron or in the opposite direction? Explain.

64. ★★★ In the deflection area of an inkjet printer, there is an electric field of 95.0 kN/C. Assume that one drop has a volume of 2.00 pL (picolitre) and that the density of the ink is the same as the density of water. A drop moves horizontally at a speed of 25.0 m/s and travels between deflecting plates with a length of 1.25 cm. What is the magnitude of the charge on the drop when the vertical deflection is 0.225 mm? Neglect gravitational acceleration, aerodynamic drag, and any effect from Earth's electric field.

65. ★★ Some barbecues and camping stoves (among other devices) use a technique called piezo ignition. Explain what is happening in these devices.

66. ★ Charge has been separated, with positive charge on one rectangular plate and negative charge on a nearby parallel rectangular plate (in Chapter 22 you will see that this is called a capacitor). You place a test charge in the region between the plates to determine the electric field. What is the problem if too large a test charge is used?

67. ★★★ Consider two dipoles, each with the same dipole moment p (with $+q$ and $-q$ charges separated by distance d). The dipoles are both aligned with the x-axis and are located a distance D apart. The negative end of the right dipole faces the positive end of the left dipole.
   (a) Derive an expression for the net force between the dipoles.

(b) What does this expression reduce to for the large-distance approximation of $D \gg d$?

68. ★★★ Dipoles placed in an electric field experience a torque unless they are aligned with the field. For an electric field of 30.0 kN/C directed in the $+x$-direction, find the torque exerted on a dipole that is aligned in the y-axis direction, with a charge of $-e$ at $y = -25.0$ nm and $+e$ at $y = +25.0$ nm.

69. ★★★ In 1909, Robert Millikan (1868–1953) used his famous oil-drop experiment to estimate the elementary charge. In this experiment, small drops of oil were charged, and an electric field was adjusted to help the drops stay suspended. From the required field, the charge on various drops was calculated—they were all whole-number ratios of the elementary charge. The actual analysis has to take into account buoyant forces and other details, but we will simplify it for this problem. Assume that the density of the oil used in the experiment is 820. kg/m^3, a drop has a radius of 1.10 μm, and a drop has a charge of $+2e$. What applied electric field is necessary to keep the drop suspended against gravity?

70. ★★ ∫ A charge of 25.0 μC is uniformly distributed around a thin ring of radius 4.00 cm that is in the xy-plane and centred at the origin. What is the electric field at a point 10.0 cm directly above the centre of this ring? (Hint: Use symmetry to simplify the situation.)

71. ★★★ ∫ A charge, α, is uniformly distributed over a circular plate of radius R. Derive an expression for the electric field at a distance d above the centre of the plate.

72. ★★ You rub a leather rod with some plastic wrap.
   (a) What is the sign of the charge on each object?
   (b) Describe how you could use the leather rod to induce a positive charge on a conducting sphere using the rod (assume you have two conducting spheres on insulated stands).

73. ★★ An electric field vector representation uses vectors of 1.00 cm length for every 40 N/C. When the vector is located 0.500 m from an unknown charge, the correctly drawn electric field vector is 2.50 cm long and points toward the charge. What are the sign and magnitude of the unknown charge?

74. ★★ A proton accelerates in a uniform electric field, going from rest to 4.00×10^6 m/s in a time of 45.0 ms. What must be the magnitude of the electric field?

75. ★★ (a) An unknown point charge is placed at an unknown location in a planar xy-coordinate system. It is observed that the electric field vector at ($x = 1.00$ m, $y = -1.00$ m) is $\vec{E} = -12.0$ kN/C $\hat{j}$.
   At another point ($x = 3.00$ m, $y = 1.00$ m) the electric field vector is given by $\vec{E} = +M$ kN/C $\hat{i}$, where M is an unknown positive magnitude. What must be the location and magnitude of the unknown point charge?
   (b) What is the value of M?

76. ★★ (a) Look up in a reliable resource what is meant by electrical breakdown.
   (b) What is the approximate breakdown electric field in air? (Note: You may find the units expressed in V/m, which you will encounter in Chapter 21, but 1 V/m = 1 N/C).
   (c) If breakdown in air was occurring in a certain situation, would placing wax paper in the region instead of air help?

(d) Sometimes after electrical breakdown there is a distinctive smell. What is this smell due to?

77. ★ In a certain room, static charging and discharge (e.g., by touching a doorknob) is a problem. Suggest some ways that this could be alleviated.

78. ★ When you scuff across a carpeted floor, what sign of charge do you and the floor get?

79. A single positive point charge, $+Q$, is located at the origin of a coordinate system. When a second positive point charge, $+q$, of mass m is placed at $x = r_1$ along the positive x-axis, it feels an acceleration of 4.80×10^5 m/s^2. When it is at r_2, a distance 1.00 m farther from Q than r_1, it feels an acceleration of 3.00×10^4 m/s^2. What are the values of the distances r_1 and r_2?

80. ★★ Draw an electric field line diagram for the case of two positive charges, separated by a horizontal distance, but with the charge on the right having a higher magnitude than the charge on the left.

81. ★★ Draw an electric field line diagram for a higher-magnitude positive charge located to the left of a lower-magnitude negative charge (on a horizontal line).

82. ★ A cathode ray tube has a beam of electrons; since the electrons are negatively charged, it would seem that they should immediately repel and make the beam very wide, and not useful. Can you hypothesize why this does not occur?

83. ★★ You observe that in a certain region of space the electric field points in the direction of the $+x$-axis and that it has no significant y-component, even at points well off the x-axis. Also, the electric field has approximately constant magnitude.
(a) Is this situation possible with one or two single point charges?
(b) Can you think of a charge configuration that might lead to this situation? If so, describe it.

84. ★★ Three charges are placed on the x-axis. There is a charge of $+8.00\ \mu$C placed at $x = -2.00$ m; an unknown charge, Q, placed at $x = +2.00$ m; and a charge of $+16.00\ \mu$C placed at $x = +5.00$ m. If the electric field at the point $x = +1.00$ m is to be exactly zero, what must be the sign and magnitude of the unknown charge, Q?

85. ★★ An ion has 27 protons, 26 neutrons, and 25 electrons. What would be the electric field at a distance of 1400 pm from the ion (assuming that it is sufficiently distant that we can consider the other charged particles as though they are at the centre of the ion)?

86. ★★ ∫ Two concentric rings of charge are centred on the origin and lie in the xy-plane. One ring has radius r_1 and a total charge (equally distributed) of Q_1, while the other has radius r_2 and total charge Q_2. Assume that both charges are positive. What will be the electric field at a point a distance d above the origin of the xy-plane?

87. ★★ (a) If you rub a plastic pipe with wool, what sign charge do you expect on the pipe?
(b) Now do an experiment in which you let a tiny stream of water flow from a faucet (or you can achieve this by putting a small hole in the bottom of a plastic cup). Once the water is flowing smoothly, bring the charged plastic pipe near to (but not touching) the water stream. Describe what happens.
(c) Explain what is happening in terms of the nature of the water molecule.

88. ★★ ∫ A positive point charge $(+Q)$ is located at the origin of a coordinate system. Centred on that origin and lying in the xy-plane is a thin ring of charge of radius R and total charge (equally distributed) of $-Q$. What will be the net electric field at a point a distance d directly above the origin of the coordinate system?

DATA-RICH PROBLEMS

89. ★★★ (a) Table 19-2 lists the electric field at various locations along the $+x$-axis. The negative sign means that the electric field points in the $-x$-direction. Assuming that the field was created by a single point charge, determine the value and location of this charge.
(b) What is the electric field for the two values that are missing in the table?

Table 19-2 Data for Problem 89

Position, x (m)	Electric Field, E (kN/C)
2.00	−281.0
3.00	?
4.00	−179.8
5.00	−148.6
6.00	?
7.00	−106.4
8.00	−91.8
9.00	−79.9

90. ★★ ◠ A point charge of magnitude $+750.\ \mu$C is located at $x = 1.50$ m on the x-axis. Calculate and plot the electric field for points along the x-axis from $x = -2.00$ m to $x = +1.00$ m and from $x = +2.00$ m to $x = +5.00$ m.

91. ★★ ◠ The magnitude of the electric field is plotted in Figure 19-38 for a single unknown point charge. Use this information to estimate the location, sign, and magnitude of the charge.

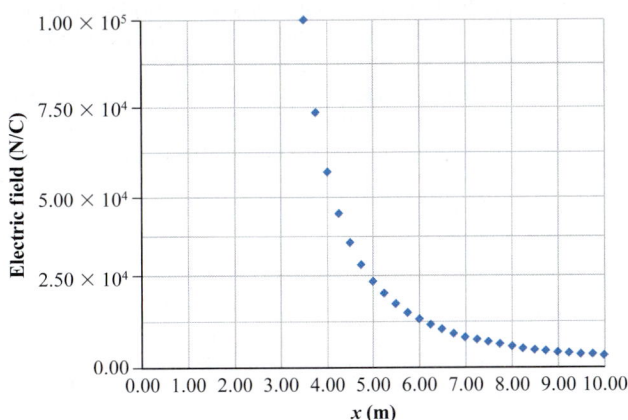

Figure 19-38 Problem 91.

92. ★★ In an experiment designed to determine the permittivity of air, measurements of electric field strength are made at different distances from a $+2.60\ \mu C$ charge (Table 19-3). Use these data to find a mean value for the permittivity of air as well as the standard deviation of the mean (see Chapter 1).

Table 19-3 Problem 92

Distance (m)	Electric Field (MN/C)
0.250	0.38271
0.500	0.09513
1.000	0.02420
2.000	0.00577
3.000	0.00255
4.000	0.00144
5.000	0.00094

OPEN PROBLEMS

93. ★★ (a) Walk in a room with a synthetic fibre carpet to charge your body. Do this in a controlled way, with the same distance walked and the same number of similar steps each time. Approach a metal doorknob holding a small bolt in your fingers and carefully approach it until a spark jumps. Have a friend record the distance the spark jumped. Repeat this several times and average the results.

(b) Now repeat with a pointed object, such as a nail, instead of the rounded bolt, but keeping everything else constant. Again perform a number of trials and average them.

(c) Which type of object must have the larger electric field (for the same amount of charging)? You will see why this is true in the next two chapters.

94. ★★★ In this problem, we will consider several aspects of atmospheric electricity and lightning discharges. Look up information or make reasonable estimates as required.

(a) First, consider how charging might take place. Lightning-producing clouds are typically very high (10 km or more above Earth's surface), and during the active phase of these storms, there are extensive updrafts and downdrafts. In addition, the clouds usually have a mix of hail and water droplets. Explain a possible production mechanism for charging by triboelectricity. Do you expect the top of the clouds to be positively or negatively charged?

(b) As shown in the opening image of this chapter, the lightning strike does not take place in one direct line but in a series of near paths. Explain what you think might be happening.

(c) Weather services register lightning strikes. Look up information online to find out where the highest number of lightning strikes per day occurs.

(d) List some guidelines for protecting human life during a lightning storm. Certain animals are more frequently killed by lightning. Which animals do you think might be at greater risk? Why?

(e) What are some concerns regarding airplanes and lightning discharges?

95. ★★ If the negative charge on electrons and the positive charge on protons were not equal, then atoms would not be electrically neutral. Let's assume that the two varied by one part in a billion. Estimate the resulting force between two iron spheres each of radius 1.0 cm when held 1 m apart (at this distance you can assume them to be point charges).

96. ★★ In this experiment you can do at home, you will prove that triboelectric charging is possible without rubbing. Obtain two pieces of clear tape (ideally the wider, less glossy type), each about 12 cm long. Attach the tape to the surface of a table, and then carefully peel each away, making sure they do not touch each other.

(a) Now bring the two pieces of tape close together, holding each at one end only, and record what happens.

(b) Explain the cause of what you see in part (a).

(c) From this experiment, is rubbing required for triboelectric charging?

(d) Considering the triboelectric series, do you expect the tape to be negatively or positively charged?

(e) Now go into a darkened room, and watch carefully as you peel some tape off the roll. What do you observe?

(f) Explain what is happening in this case.

You can find a description of a similar experiment at http://amasci.com/emotor/sticky.html.

Gauss's Law

Australian Peter Terren, also known as Dr. Electric, does spectacular demonstrations involving electrical discharge, such as the one shown in Figure 20-1. Past a certain critical electric field strength, we have electrical breakdown in air, as shown. In this chapter, you will learn a new and powerful technique called Gauss's law to calculate the electric field in different regions. You will also learn about the electric field just outside objects of different shapes, and that the electric field is higher near sharply edged conductors. Predicting the electric field can help protect sensitive electronic devices and assess human safety in high electric field situations. Gauss's law is one of the most important fundamental relationships in physics, and we will also see that an analogous law can be used to find the gravitational field strength in symmetrical cases.

Figure 20-1 An impressive Tesla coil demonstration by Peter Terren in Australia.

Courtesy of Peter Terren

20-1 Gauss's Law and Electric Field Lines

This chapter introduces a single new idea: Gauss's law. While we will not formally state Gauss's law until Section 20-3, in this section we qualitatively consider the essential idea of the law, as well as why it is important in physics. In Chapter 19 you learned several techniques for calculating the electric field. For a few point charges, it is easy to find the electric field due to each and then add them as vectors using the superposition principle. For a continuous distribution of charge, calculus can be used to find the electric field by integrating over the electric field due to each charge element. Because of the vector nature of the electric field, these methods can become challenging to apply. When there is symmetry in the electric field, a rather simple and elegant alternative technique involving Gauss's law can be used.

We saw in Section 19-8 that the number of field lines per unit area is a measure of the electric field strength. Figure 20-2 shows electric field lines for two uniformly charged spheres carrying different charges.

Imagine now that we draw a concentric sphere outside each of the charged spheres and count the number of field lines that pass through the surface of the concentric sphere (see Figure 20-3). Clearly, more field lines pass through the sphere around the object with the larger charge. In fact, when charge is twice as much, the number of field lines that pass through the surface is doubled.

The number of electric field lines intersecting a surface is related to a quantity called electric flux that we consider in detail in the next section. The key idea of Gauss's law is that the electric flux through a surface is proportional to the enclosed electric charge.

Now consider the situation shown in Figure 20-4. We again have a positively charged sphere, but this time we have a concentric spherical shell that has a total charge exactly equal in magnitude to the charge on the smaller sphere, but it is negatively charged. You will recall from Section 19-8 that electric field lines start on positive charges and end on negative charges if there are equal numbers of charges (when unequal some will start or end at infinity). In Figure 20-4 we have an equal number of positive and negative charges, and therefore there must be equal numbers of electric field lines leaving the positively charged sphere and arriving on the negatively charged sphere. The number of field lines is constant through any spherical surface that we draw between the positively charged sphere and the inside of the negatively charged spherical shell, and it turns out the flux (interpreted here as the number of field lines we "catch") is also constant.

Electric field lines start on positive charges and end on negative charges. The number of electric field lines we draw is arbitrary, but we must be consistent throughout an electric field line diagram, with the number of field lines per area proportional to the electric field at that point. For point charges, we must draw the same number of field lines starting (or ending) on each point charge if the values of the charges are the same.

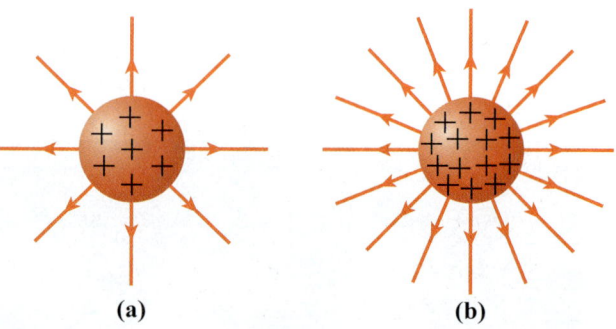

Figure 20-2 Electric field lines for two symmetric positively charged spheres, with double the charge in (b) compared to (a).

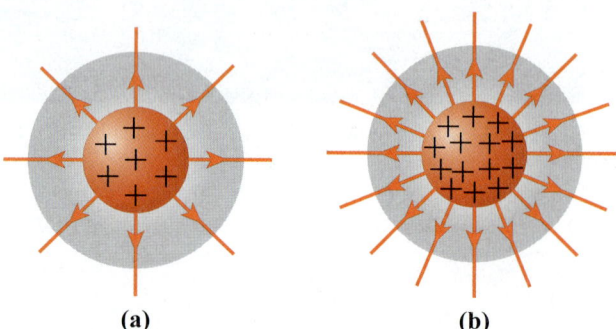

Figure 20-3 The spherical surface in (b) will collect twice as many electric field lines as the corresponding surface in (a) since the magnitude of the charge is double.

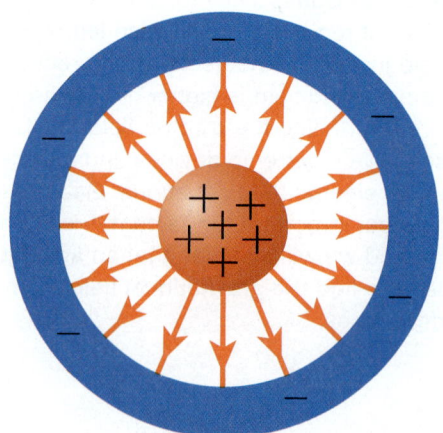

Figure 20-4 A negatively charged spherical shell is concentric with a positively charged smaller sphere. The number of charges is the same on both, and they are symmetrically distributed.

Therefore, in the case of Figure 20-4, the same number of electric field lines that leave the positive charges on the inner sphere must end on an equal number of negative charges on the outer spherical shell. Since every electric field line that starts on the inner sphere ends on the negatively charged outer shell, no electric field lines are outside the spherical shell and the electric field must be zero there. This demonstrates another aspect of Gauss's law: it is the *net* enclosed charge that is important, and we must take into account the sign of enclosed charges.

We can now summarize Gauss's law qualitatively. We draw an imaginary surface (this is called a Gaussian surface) consistent with the symmetry of the charge distribution. The strength of the electric flux (the electric field strength times the area it intersects or, equivalently, a measure of how many electric field lines it catches) is proportional to the net enclosed charge. Therefore, we can calculate the electric field strength provided that there is enough symmetry that we can show that the electric field strength is constant over the Gaussian surface.

Gauss's law does more than provide an easier way to calculate the electric field strength in symmetrical situations. Gauss's law constitutes one of the four relationships that make up Maxwell's equations (see Chapter 27), an elegant, powerful, and concise way to summarize all of electricity and magnetism, not just for static fields but also for changing fields such as those of electromagnetic waves.

CHECKPOINT 20-2

Electric Fields and Charges

Concentric with a $-2.0\ \mu C$ point charge is a thin spherical shell that carries a uniformly distributed $-7.0\ \mu C$ charge. A larger thin concentric spherical shell carries a charge of $+9.0\ \mu C$. What is the electric field in each of the regions?

(a) Outside the larger shell the field is zero, between the two shells it is radially outward, and between the point charge and the inner shell it is also radially outward.

(b) Outside the larger shell the field is zero, between the two shells it is radially inward, and between the point charge and the inner shell it is also radially inward.

(c) Outside the larger shell the field is radially outward, between the two shells it is radially inward, and between the point charge and the inner shell it is also radially inward.

(d) It is radially inward in all three regions.

ANSWER: (b) Electric field lines start on positive charges and end on negative charges. Since the magnitude of the positive charge on the outer shell equals the sum of the magnitudes of the negative charges on the inner shell and the point charge, every electric field line that leaves the outer shell ends on either the inner shell or the point charge. This means the field outside the larger shell is zero. In both other regions the field lines go from outside to inside, so the electric field points radially inward.

CHECKPOINT 20-1

Field Lines

Consider a situation like the one shown in Figure 20-4, but assume that the outer spherical shell has 25 units of negative charge and the inner sphere has 20 units of positive charge. Which statement would correctly describe the electric field in both the region between the sphere and the spherical shell and the region outside the spherical shell?

(a) There is a radially outward electric field in the region between the inner charge and the spherical shell, and a radially inward electric field in the region outside the shell.

(b) There is a radially inward electric field in the region between the inner charge and the spherical shell, and a radially outward electric field in the region outside the shell.

(c) There is a radially outward electric field in the region between the inner charge and the spherical shell, and zero electric field in the region outside the shell.

(d) There is a radially inward electric field in the region between the inner charge and the spherical shell, and zero electric field in the region outside the shell.

ANSWER: (a) There is less positive charge on the inner sphere than negative charge on the spherical shell. This means that all of the electric field lines from the inner charged sphere end up on the outer spherical shell, but also some additional lines of charge must end up on the outer shell. These must come from infinitely far away, so the field outside the shell must be radially inward (toward the negative shell). Between the sphere and the spherical shell the electric field must point from the positively charged object toward the negatively charged object.

20-2 Electric Flux

We saw in the preceding section that an important idea was the number of electric field lines intersecting a surface. How many electric field lines we catch depends both on how dense the electric field lines are in that region (which as we saw in Section 19-8 depends on the electric field strength) and on the area of the surface. It will also depend on a third parameter, the angle between the collecting surface and the electric field lines. In this section, we formally introduce a number of terms and concepts related to this flux that are needed to fully understand Gauss's law.

Vector Field

A **vector field** refers to a physical situation where at each point in space we can assign a vector to some physical quantity. An example is the flow of a fluid

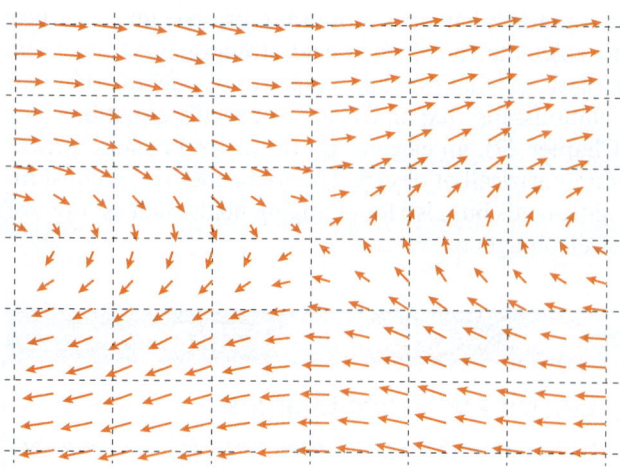

Figure 20-5 A vector field plot, perhaps corresponding to the flow of a fluid.

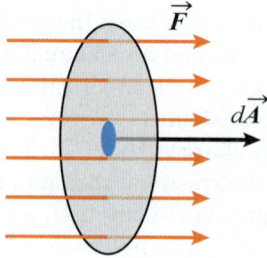

Figure 20-6 A vector field, $\vec{F}$, leading to a flux through a circular area.

in a region. At every point, we could in theory measure the velocity of the fluid, and all of these measurements taken together would constitute a vector field. We show an example of a vector field in Figure 20-5. It might, for example, represent a fluid that flows left to right at the top, right to left at the bottom, and somewhat chaotically in the transition between the two. The electric field, as we saw in Chapter 19, is a vector quantity, and therefore the set of electric field values at every point in a space constitutes a vector field. In fact, the electric field vector plots of Section 19-8 were vector field plots. There are many other examples of vector fields in physics and other sciences, so the techniques you learn in this chapter will find broad application.

Flux for Open Surfaces

Now consider placing an imaginary surface in a vector field like that of Figure 20-5 (except defined in three-dimensional space). If the field is a fluid, for example, we want to measure how much fluid passes through that surface per unit time. We can define a **flux**, Φ, through the following relationship:

$$\Phi \equiv \iint_{surface} \vec{F} \cdot d\vec{A} \qquad (20\text{-}1)$$

Here $\vec{F}$ is the series of vectors representing the vector field, $d\vec{A}$ is a small differential element of the surface being used for the flux (e.g., the circle in Figure 20-6), and the two vectors have a dot product operation as shown. The integral represented by the symbol $\iint_{surface}$ is called a **surface integral**. This means that at every point on the surface we find the value of the vector quantity $\vec{F}$; take the vector dot product with the differential area of the surface, $d\vec{A}$; and then integrate over

the surface. We demonstrate the process for calculating a flux in Example 20-1.

A surface like that shown in Figure 20-6 is called an **open surface** (we will soon consider surfaces such as spheres that curve around and make a closed surface). The direction of $d\vec{A}$ is always taken as perpendicular to the surface. For an open surface, it could be defined to point in either direction, that is to the right or to the left in Figure 20-6. It is conventional to select a direction for $d\vec{A}$ so that the dot product will be positive for flow into the surface and negative for flow out of the surface, as viewed from the perspective of the fluid, and that is what we have done. This is a convention, though, and you may well encounter the opposite choice for the direction of $d\vec{A}$ for open surfaces. Note that although the field is vector in nature, the flux is a scalar quantity (although it can have positive and negative values).

EXAMPLE 20-1

Flux through a Circle

As shown in Figure 20-7, the momentum flow in a fluid vector field has a magnitude of 12.5 kg·m/s oriented at an angle of 65.0° to the horizontal. What is the flux through a horizontal circle of radius 25.0 cm?

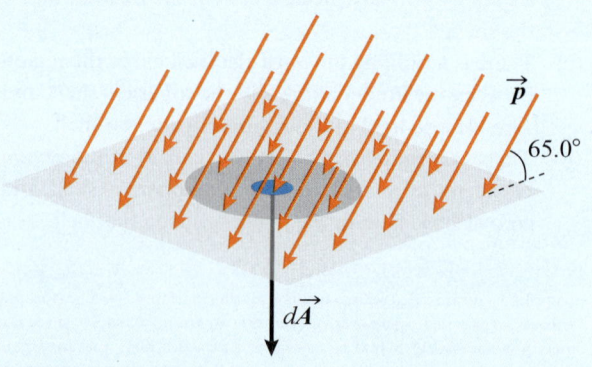

Figure 20-7 Example 20-1. The flux of vector field $\vec{p}$ through the circular area.

Solution

We use the symbol $\vec{p}$ for the vector field. It is a matter of convention which perpendicular direction we take as positive for open surface $d\vec{A}$, but we will use the convention of taking the direction consistent with a positive value for flow into the surface as viewed from the fluid entry. Although $\vec{p}$ makes an angle of 65.0° with respect to the horizontal circle, it makes an angle of 25.0° with respect to the direction of the differential area vector $d\vec{A}$. Therefore, the solution for the flux gives us the following, using the expression for dot product from Chapter 2:

$$\Phi = \iint_{\text{circle}} \vec{p} \cdot d\vec{A}$$

$$= \iint_{\text{circle}} p \, dA \cos 25.0°$$

The vector field $\vec{F}$ is uniform and does not depend on position and can therefore be taken outside the surface integral:

$$\Phi = p \cos 25.0° \iint_{\text{circle}} dA$$

The surface integral over the circle is simply the area of the circle:

$$\Phi = p \cos 25.0° \, \pi r^2$$
$$= 2.22 \text{ kg} \cdot \text{m}^3/\text{s}$$

Making sense of the result

The value we get is a bit less than if the field had been vertical (which would have yielded $\Phi = 2.45$ kg·m³/s), which makes sense. Note that although the units have been expressed in a combined fashion, they represent vector field units times area units. The flux is a scalar, and in this case it is positive by the convention we have adopted, since the flow represents an influx.

Electric Flux for Open Surfaces

Since the electric field is a vector field, we apply Equation 20-1 to find the **electric flux** Φ_E through an open surface as

$$\Phi_E \equiv \iint_{\text{open surface}} \vec{E} \cdot d\vec{A} \qquad (20\text{-}2)$$

Here $d\vec{A}$ is as before a differential area element oriented perpendicularly to the surface. Since this is an open surface, it is a matter of convention which direction is considered positive, with the usual choice being the direction that makes an influx of electric field lines positive.

The electric flux is a scalar quantity that depends on the following three parameters:

- the electric vector field
- the surface area through which the flux is calculated

- the angle between the perpendicular to the surface and the electric field

If the electric field, $\vec{E}$, is perpendicular to $d\vec{A}$, then the electric flux will be zero. It is only when the electric field is constant in magnitude that we can take it outside the integral of Equation 20-2 and simply multiply by the area. We show in Example 20-2 a case where the electric field has two different values.

EXAMPLE 20-2

Electric Flux with Different Electric Field in Two Parts

As shown in Figure 20-8, a uniform electric field, $\vec{E} = -95.0$ N/C $\hat{j}$, is present when $x < 0$, and $\vec{E} = 0$ N/C $\hat{j}$ when $x > 0$. A 10.0 cm by 24.0 cm rectangle lies in the xz-plane with half of it in the region $x < 0$. What is the electric flux through the rectangle?

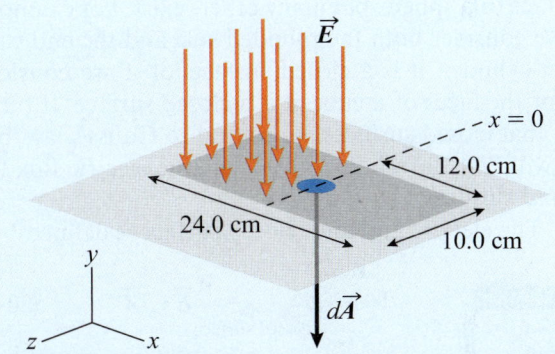

Figure 20-8 Example 20-2. A uniform electric field is perpendicular to one half of a rectangular area (blue), with the electric field being zero in the other half of the rectangle.

Solution

For open surface integrals, either perpendicular direction can be used for the surface area. We will define downward as the direction for $d\vec{A}$, and therefore $\vec{E}$ and $d\vec{A}$ are in the same (downward) direction with an angle of 0° between them. We will use the expression for computation of the electric flux, shown below, but will divide the surface integral over the rectangle into two parts, since the electric field is zero in one half:

$$\Phi_E = \iint \vec{E} \cdot d\vec{A}$$

$$\Phi_E = E(\cos 0°)(0.100 \text{ m})(0.120 \text{ m}) + (0)(0.100 \text{ m})(0.120 \text{ m})$$
$$= E(\cos 0°)(0.100 \text{ m})(0.120 \text{ m})$$
$$= (95.0 \text{ N/C})(1.00)(0.100 \text{ m})(0.120 \text{ m})$$
$$= 1.14 \text{ N} \cdot \text{m}^2/\text{C}$$

(continued)

Note that E refers to the magnitude of the electric field $\vec{E}$ and is always positive. The direction of the area relative to the field is taken into account by the angle that we use in the cosine term.

Making sense of the result

Remember that when doing problems with different values for the electric field in different regions, you must do each part separately. The units are electric field units (N/C) multiplied by area units (m²), but some write it as $\Phi_E = 1.14$ (N/C)·m² to show this more clearly. Both ways are correct. Note that since the surface is open, either perpendicular direction could have been defined as positive, with the flux taking on a negative value if the other choice had been used.

Electric Flux for Closed Surfaces

Now we will extend this to the electric flux through a **closed surface**, which for our purposes we can define as a three-dimensional surface without a boundary (i.e., it closes back on itself). The simplest closed surface is the surface of a sphere, but many others exist. For example, if we consider both the rounded side and the end caps of a cylinder, it is a closed surface, or if we consider all of the faces of a cube, it is a closed surface. It turns out that closed surfaces are central to Gauss's law, but we will first learn how to calculate the electric flux for closed surfaces.

The electric flux for a closed surface is defined by

KEY EQUATION

$$\Phi_E \equiv \oiint_{\text{closed surface}} \vec{E} \cdot d\vec{A} \qquad (20\text{-}3)$$

This is similar to the situation for an open surface, but in the case of closed surfaces the orientation of $d\vec{A}$ is no longer a matter of convention and must be perpendicular to the closed surface and pointing in an outward sense. Note that the circle in the integral notation tells us that it is a closed surface.

For a closed surface, we can find the total electric flux by first calculating the flux through each part and then adding the contributions to find the total flux. In the case of closed surfaces, the direction of $d\vec{A}$ is still perpendicular to the surface, but the orientation

is no longer arbitrary. The direction of $d\vec{A}$ for a closed surface is perpendicular to the local surface and is always outward. We demonstrate the technique for calculating the electric flux for a closed surface in Example 20-3.

EXAMPLE 20-3

Electric Flux for a Closed Cylinder

The uniform electric field in a certain region is given by

$$\vec{E} = -2.00 \text{ N/C } \hat{k}$$

If we draw a cylinder of radius R and length L aligned along the z-axis and define the centre of a coordinate system at the centre of the cylinder (see Figure 20-9), what is the electric flux for the surface of the cylinder?

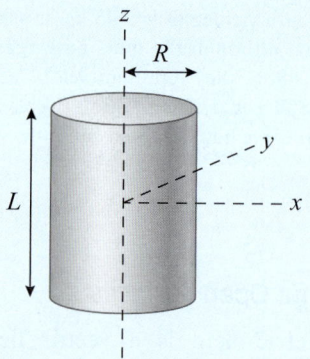

Figure 20-9 Example 20-3.

Solution

We will start with the definition of electric flux for a closed surface:

$$\Phi_E \equiv \oiint_{\text{closed surface}} \vec{E} \cdot d\vec{A}$$

The electric field points in the negative z-direction. Since the orientation of the electric field relative to $d\vec{A}$ is different in different parts of the cylinder, we divide the surface into different parts: the curved vertical wall and the two end caps.

In Figure 20-10 we have drawn the cylinder with each of these three parts in a different colour.

The electric field $\vec{E}$ is constant in both direction and magnitude at all points. The differential surface area element $d\vec{A}$ for the top cap points upward (i.e., outward) while for the bottom cap it points downward (remember that by definition $d\vec{A}$ is always outward). For the curved surface, it points everywhere outward; for example, on the right it points to the right, while on the left it points to the left. We have only shown it for two locations, but it could similarly be drawn for other locations.

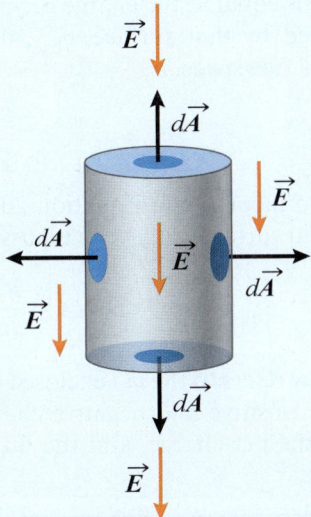

Figure 20-10 The electric field and area directions for Example 20-3.

Now if we consider the dot product of the electric field and the surface area vector for the top end cap, the result will be negative (since $\vec{E}$ and $d\vec{A}$ point in opposite directions), for the bottom it will be positive (since $\vec{E}$ and $d\vec{A}$ point in the same direction), and for the curved surface it will be zero, since $\vec{E}$ and $d\vec{A}$ are always perpendicular so their dot product will equal zero.

Therefore, the net electric field is given by

$$\Phi_{E\,net} = \Phi_{E\,top\,cap} + \Phi_{E\,curved\,wall} + \Phi_{E\,bottom\,cap}$$

$$= -E\pi R^2 + 0 + E\pi R^2$$

$$= 0$$

where E represents the magnitude of the electric field.

Making sense of the result

From the symmetry of the situation, it is clear that the amount of flux on the two end caps must cancel for this uniform electric field. If we imagine the electric field as a smooth flow of a fluid, as much flows into the can as flows out of it, and none will flow through the side walls. In the next section, we will see why the electrical flux had to be zero in this case in terms of Gauss's law.

CHECKPOINT 20-3

Electric Flux

As shown in Figure 20-11, a cube is aligned with the axes of an xyz Cartesian coordinate system. A uniform electric field points in the negative y-direction. Which statement best describes the electric flux?

(a) The only surfaces with non-zero flux are the faces parallel to the yz-plane, and the net flux for the entire cube is zero.

(b) The only surfaces with non-zero flux are the faces parallel to the xy-plane, and the net flux for the entire cube is negative.

(c) The only surfaces with non-zero flux are the faces parallel to the xz-plane, and the net flux for the entire cube is zero.

(d) The only surfaces with non-zero flux are the faces parallel to the xz-plane, and the net flux for the entire cube is negative.

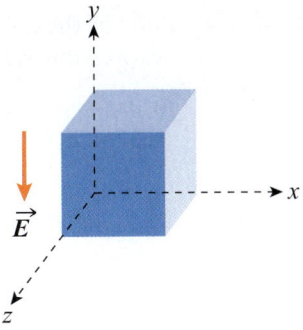

Figure 20-11 Checkpoint 20-3.

ANSWER: (c) The electric flux uses a dot product between the vectors $d\vec{A}$ and $\vec{E}$. Therefore, for the four faces parallel the xy- and yz-planes, the flux is zero. Just the top and the bottom faces of the cube have non-zero electric flux. Since the direction of the $d\vec{A}$ vector is perpendicular to the surface and pointing outward, the electric flux is negative for the top face of the cube (where the vectors $\vec{E}$ and $d\vec{A}$ are anti-parallel) and positive for the bottom face (where $\vec{E}$ and $d\vec{A}$ point in the same direction). Since the magnitudes of these two fluxes are equal, the net flux through the cube is zero.

20-3 Gauss's Law

We argued in the first section that the number of electric field lines captured was proportional to the amount of charge surrounded by a surface. Now we will develop a formal statement of Gauss's law. Consider a single positive charge Q, with a concentric sphere of radius r (see Figure 20-12).

From Chapter 19, we know that the electric field points radially outward for a point charge, given by the following expression:

$$\vec{E} = \frac{1}{4\pi\varepsilon_0}\frac{Q}{r^2}\hat{r} \qquad (19\text{-}11)$$

The electric flux for a closed surface is given by

$$\Phi_E \equiv \oiint_{\text{surface}} \vec{E} \cdot d\vec{A} \qquad (20\text{-}3)$$

Our sphere of radius r is a closed surface, with the differential area elements $d\vec{A}$ perpendicular to the surface and radially outward. This means that everywhere $\vec{E}$ and $d\vec{A}$ point in the same direction, so the dot product in the expression for electric flux will just be the product of the magnitudes. Also from Equation 19-11, the magnitude of $\vec{E}$ is constant over the sphere, allowing us to take the electric field strength outside the integral:

$$\Phi_E = E \oiint_{\text{sphere}} dA$$

But the integration of the area elements over a sphere is simply the surface area of the sphere, so we have the following (where E is the magnitude of the electric field at radius r):

$$\Phi_E = 4\pi r^2 E$$

If we substitute the magnitude of the electric field from Equation 19-11 into this expression we have the following result for the electric flux:

$$\Phi_E = 4\pi r^2 \frac{1}{4\pi\varepsilon_0}\frac{Q}{r^2}$$

$$= \frac{Q}{\varepsilon_0}$$

So at least for this one case we have shown that the electric flux through a closed surface surrounding the charge is the enclosed charge divided by the fundamental physical constant, the permittivity of free space, ε_0. While we have only proven this for a single charge, the result is in fact true in general and is called Gauss's law.

Gauss's law states that the electric flux through any closed surface is equal to the charge *enclosed* within the region bounded by that surface, q_{enc}, divided by the permittivity of free space, ε_0:

KEY EQUATION
$$\Phi_E = \frac{q_{\text{enc}}}{\varepsilon_0} \qquad (20\text{-}4)$$

Some prefer to incorporate Equation 20-3 for electric flux for a closed surface, and write Gauss's law as

KEY EQUATION
$$\Phi_E = \oiint_{\text{surface}} \vec{E} \cdot d\vec{A} = \frac{q_{\text{enc}}}{\varepsilon_0} \qquad (20\text{-}5)$$

Gauss's law refers to the *net* enclosed charge. If you have a mix of positive and negative charges, you add them to find the net charge, and the flux depends on that net charge.

CHECKPOINT 20-4

Largest Electric Flux

Four situations involving point charges (all of equal magnitude) and closed surfaces of various sizes are shown in Figure 20-13. Which of them results in the largest electric flux through the closed surface?

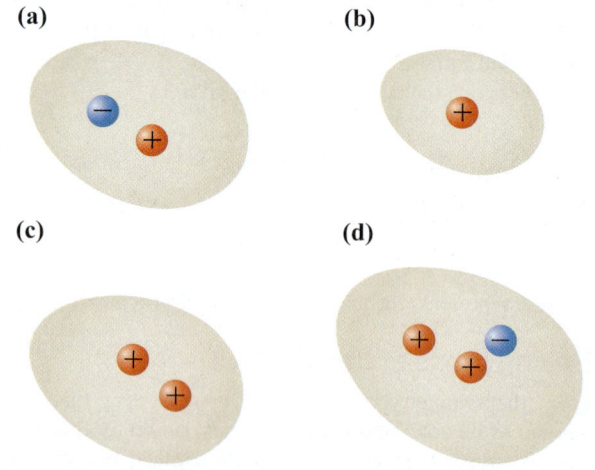

(a)　　　　　　　　　　**(b)**

(c)　　　　　　　　　　**(d)**

Figure 20-13　Checkpoint 20-4.

ANSWER: (c) By Gauss's law, the electric flux depends only on the net enclosed charge, not on the size or shape of the closed surface. The net enclosed charge is zero for (a), $+q$ for (b), $+2q$ for (c), and $+q$ for (d), so (c) has the highest flux.

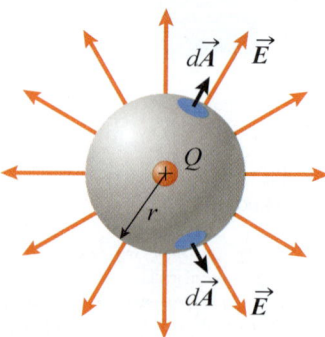

Figure 20-12　A sphere of radius r concentric with a single positive point charge, Q. The surface area elements, $d\vec{A}$, point radially outward everywhere and are therefore parallel to the electric field at the surface.

Note that Gauss's law says that we can use *any* closed surface. Skill in applying Gauss's law involves knowing what sort of surface to draw, and we call the closed surface that we use a **Gaussian surface**. As we will see later, the symmetry of the charge distribution determines the most appropriate Gaussian surface. While Gauss's law is true for any closed surface, it is only useful for calculating the electric field in certain symmetrical cases.

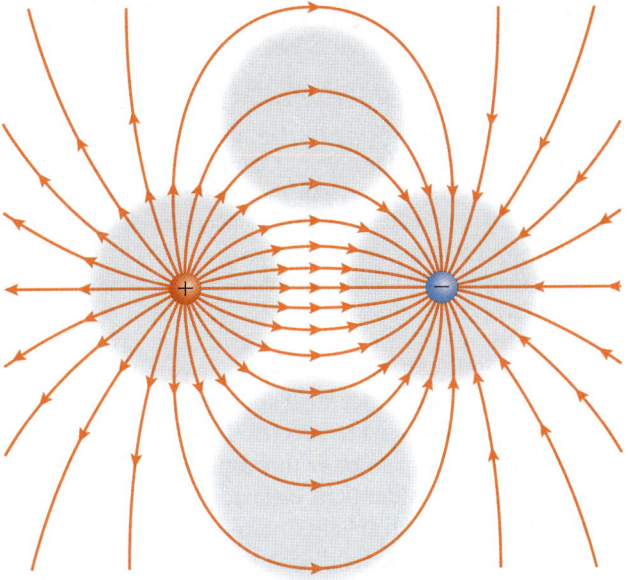

Figure 20-14 If you count electric field lines in and out of each of the Gaussian spheres, the result is positive when a positive charge is enclosed (more lines out) and is net zero in regions where no charge is enclosed.

Next we will consider a Gaussian sphere that does not enclose any charges. The upper sphere has three electric field lines entering, but three electric field lines also leave, so the net flux is zero. We would expect this from Gauss's law since that surface has no enclosed charge. The bottom sphere has four electric field lines entering and four electric field lines leaving, so the net flux is again zero. When there is zero net enclosed charge, the electric flux through that closed surface is also zero. This semi-quantitative example does not prove Gauss's law, but it is nice to see that it is consistent with it.

CHECKPOINT 20-5

Electric Flux through the Face of Cube

A positive charge, $2q$, is placed exactly in the centre of a cube of side length L (all side lengths are equal). What is the electric flux through *one* face of the cube?

(a) $\dfrac{q}{3\varepsilon_0}$

(b) $\dfrac{2q}{\varepsilon_0}$

(c) $\dfrac{q}{3\varepsilon_0 L^2}$

(d) $\dfrac{2q}{\varepsilon_0 L^2}$

ANSWER: (a) By Gauss's law, the electric flux depends on the enclosed charge, and in this case the combined flux through all the faces is $\dfrac{2q}{\varepsilon_0}$. There are six faces of the cube, and by symmetry they each have 1/6 of the total flux, or $\dfrac{q}{3\varepsilon_0}$. Note that the size does not come into this. A larger cube has a larger area, but weaker electric field, with the flux being the same.

We can understand Gauss's law by examining electric field diagrams. Electric field lines entering a closed surface correspond to a negative electric flux, and electric field lines leaving a closed surface correspond to a positive electric flux. We can relate the magnitude of the electric flux to the net number of electric field lines that enter or leave a closed surface. Let us consider the situation shown in Figure 20-14. We have drawn four different (in grey) spherical Gaussian surfaces, one around a positive charge, one around a negative charge, and two in regions with no enclosed charges.

The Gaussian sphere around the positive charge has 24 electric field lines leaving the sphere, so there is a +24 electric flux (we are just doing relative qualitative fluxes here; for example, +20 would be twice +10). Now consider the Gaussian sphere around the negative charge. Even though we have offset it from the centre of the charge, there are still 24 electric field lines entering the sphere, so the net flux is −24. So when positive charge is enclosed, the flux is positive, and negative enclosed charges lead to negative flux, just as we would expect from Gauss's law.

CHECKPOINT 20-6

Electric Flux for Different Shapes

Which of the following closed surfaces has the greatest net electric flux passing through it if they all enclose the same net electric charge?
(a) a small sphere
(b) a large sphere
(c) a cube
(d) The flux is the same for all three surfaces.

ANSWER: (d) By Gauss's law, the net electric flux through any of the closed surfaces depends only on the enclosed charge, which is the same for all three surfaces.

PEER TO PEER

In applying Gauss's law, I must remember that the charge is the total *net* charge *inside* the closed surface. I must not count charges that are outside the closed surface.

INTERACTIVE ACTIVITY 20-1

Gauss's Law for a Point Charge

In this activity, you will use the OpenSource Physics simulation "Electric Field for a Point Charge" to draw different Gaussian surfaces and explore the electric field and flux values at different radial distances from a point charge. You can also vary the magnitude and sign of the charge and observe electric field line configurations.

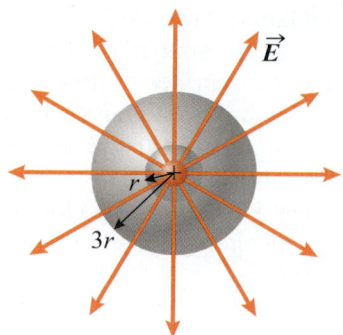

Figure 20-15 shows a single positive point charge, Q, enclosed by two concentric spheres, one of radius r and one of radius $3r$. The electric field lines that start on the positive charge pass through both spheres, so both spheres have identical electric flux. The smaller sphere has more electric field lines per unit area, related to the stronger electric field.

Coulomb's law is an inverse square law—the strength of the electric field due to a point charge is inversely proportional to the square of the distance from the charge. In Chapter 19, we saw that the electric field depended on radial distance according to

$$\vec{E} = \frac{1}{4\pi\varepsilon_0}\frac{Q}{r^2}\hat{r} \qquad (19\text{-}11)$$

Figure 20-15 Gaussian spheres with different radii do not have the same surface area or electric field strengths, but the electric flux through them is the same.

Therefore, the electric field strength at the larger sphere will be 1/9 the electric field strength at the smaller sphere, since the radius is three times as great.

But the larger sphere has nine times the surface area of the smaller sphere (surface area of a sphere of radius r is $4\pi r^2$). Therefore, this exactly cancels the amount by which the electric field is weaker, and the electric flux for each sphere is exactly the same. This is consistent with Gauss's law, since the enclosed charge is identical for the two spheres.

MAKING CONNECTIONS

Carl Friedrich Gauss

Many consider Carl Friedrich Gauss (1777–1855) to be the greatest mathematician of the last 2000 years (Figure 20-16). He made major contributions to both physics and mathematics, including advances in vector calculus, statistics, electrical theory, astronomy, optics, and geophysics. He showed great promise even as a child and performed his most important work in number theory when he was only 19. Gauss completed his doctorate by the age of 22. He was a perfectionist, often waiting years before publishing results, and then releasing only what he thought was truly important and complete. For example, he developed what we now call Gauss's law in 1835 but did not publish it until 1867. Starting in 1831, Gauss collaborated with the physicist Wilhelm Weber; this collaboration resulted in the application of Gauss's mathematical insights to many branches of physics. One of his first contributions to physics was an accurate calculation of the orbit of the "lost" dwarf planet Ceres by developing new methods for extrapolating the scant data available. The Dawn space mission provided spectacular images of Ceres in 2015.

While tremendously successful in his professional life, Gauss faced many personal challenges, including the early death of his first wife and a baby son, the lengthy illness and death of his second wife, and disagreements with his surviving sons. It is said that he struggled with periodic depression,

Carl Friedrich Gauss (1777–1855), painted by Christian Albrecht Jensen

Figure 20-16 A painting of Carl Friedrich Gauss.

which makes his contributions all the more remarkable. A Canadian national school mathematics prize competition is named in his honour.

We can obtain all aspects of Coulomb's law through the combination of the definition of electric field (force per charge) and Gauss's law. In one of the most brilliant contributions to physics, James Clerk Maxwell (1831–1879) described all aspects of static and changing electromagnetic fields (and, therefore, electromagnetic waves such as light) in only four powerful equations. Gauss's law is one of these relationships. You will see the complete set of equations in Chapter 27.

Although Gauss's law is true for *any* closed surface, it is only *useful for calculating the electric field* when there is *sufficient symmetry* so that we know the electric field has a constant magnitude and a consistent direction relative to the surface over some Gaussian surface. We show this for a spherically symmetric situation in Example 20-4 and consider other possibilities in the next few sections.

An analogy with light is helpful in understanding the interplay of electric flux and the nature of the closed surface. Imagine a tiny point source that emits light uniformly in all directions. The light source is analogous to a charge, and the light rays it emits are analogous to electric field lines. Now imagine that you surround that light with an opaque plastic sphere, a closed surface. All of the light emitted is absorbed by that sphere whether we have a large or a small sphere, or even some different shape including an irregular one, as long as it is a closed surface and the light source is inside. The light absorbed by the surface is like the electric flux, and we can see that the flux depends on the source (light here, or charge in the Gauss's law case) but *not* the size or type of closed surface, as long as it totally encloses the point light source. If our closed surface is deformed in shape, as long as no holes are created in the surface, it does not matter for the total light absorbed (the flux in the electrical case).

If we are not satisfied with the total light collected by the surface, but also want to find the illumination at any point, then we need symmetry. For example, if we have a 10. W point source light, and the closed surface is a concentric sphere of 50. cm radius, the illumination is given by

$$\frac{10.\ \text{W}}{4\pi (0.50\ \text{m})^2} = 3.2\ \text{W/m}^2$$

But this is only true if the light is emitted symmetrically and the closed surface is a concentric sphere. In a corresponding way, we can only use the enclosed charge and Gauss's law to find the electric field when there is symmetry in the situation.

EXAMPLE 20-4

Electric Field inside and outside a Charged Sphere

A solid non-conducting sphere of radius R carries a total positive charge of Q. The charge is uniformly spread through the sphere.

(a) Derive an expression for the electric field at a radial distance $r < R$.
(b) What is the electric field at the surface of the sphere?
(c) What is the electric field for points outside the charged sphere, $r > R$?
(d) Plot the electric field strength as a function of radial distance.

Solution

In this case, we have spherical symmetry about the centre of the charged sphere, with the electric field strength being constant in all directions at any given radial distance. Therefore, we draw a spherical Gaussian surface concentric with the charged sphere. From the symmetry of the problem, and the positive charge, we know that the electric fields must point radially outward and be constant at any particular radial distance.

(a) Let us first calculate the electric flux through a Gaussian sphere of radius r where $r < R$. We show the situation

in Figure 20-17, with the smaller darker sphere being the Gaussian surface. We have only shown it in one place, but the electric field, $\vec{E}$, points radially outward, as does the differential area element, $d\vec{A}$:

$$\Phi_E = \oiint_{\text{sphere}} \vec{E} \cdot d\vec{A} \qquad (1)$$

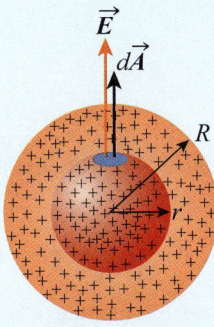

Figure 20-17 Example 20-4. A Gaussian sphere of radius r inside a uniformly charged non-conducting sphere of radius R.

(continued)

Since $\vec{E}$ and $d\vec{A}$ are parallel, the vector dot product can be written in terms of the magnitudes only:

$$\vec{E} \cdot d\vec{A} = EdA\cos 0° = EdA(1) = EdA \qquad (2)$$

If we insert result (2) into (1), we have

$$\Phi_E = \oiint_{\text{sphere}} EdA \qquad (3)$$

where E and dA are now magnitudes.

Since the electric field strength is by symmetry constant at any particular radius, we can take it outside the integral of (3):

$$\Phi_E = E\oiint_{\text{sphere}} dA \qquad (4)$$

In the same way that $\int dx = x$, the integral over a surface of just an area differential is simply the total area of that surface (the surface area of the sphere in this case):

$$\oiint_{\text{sphere}} dA = A = 4\pi r^2 \qquad (5)$$

When we substitute (5) into (4), we obtain the electric flux through the Gaussian sphere of radius r:

$$\Phi_E = 4\pi r^2 E \qquad (6)$$

Now we need to calculate the charge enclosed within the Gaussian sphere. First we calculate the volume charge density:

$$\rho = \frac{Q}{V} = \frac{Q}{\frac{4}{3}\pi R^3} \qquad (7)$$

The charge enclosed by the Gaussian surface is the volume charge density from (7) multiplied by the volume of the Gaussian sphere of radius r:

$$q_{\text{enc}} = \frac{4}{3}\pi r^3 \rho = \frac{4}{3}\pi r^3 \frac{Q}{\frac{4}{3}\pi R^3} = \frac{Qr^3}{R^3} \qquad (8)$$

Finally, we can substitute results (6) and (8) into Gauss's law to find the magnitude of the electric field at radial distance r, where $r < R$:

$$\Phi_E = \frac{q_{\text{enc}}}{\varepsilon_0}$$

$$4\pi r^2 E = \frac{Qr^3/R^3}{\varepsilon_0}$$

$$E = \frac{1}{4\pi\varepsilon_0}\frac{Qr}{R^3}$$

As we argued earlier, from symmetry and the positive charge, the direction of the electric field is radially outward, so we can write it as

$$\vec{E} = \frac{1}{4\pi\varepsilon_0}\frac{Qr}{R^3}\hat{r} \qquad (9)$$

where $\hat{r}$ is a unit vector pointing radially outward.

(b) We have two approaches to finding the electric field at the surface of the charged sphere. We can either substitute $r = R$ into result (9) from part (a), or we can draw a Gaussian sphere of radius $r = R$ with the enclosed charge

being the total charge on the sphere now: $q_{\text{enc}} = Q$. Of course both methods yield identical results:

$$\vec{E} = \frac{1}{4\pi\varepsilon_0}\frac{Qr}{R^3}\hat{r}$$

$$= \frac{1}{4\pi\varepsilon_0}\frac{QR}{R^3}\hat{r}$$

$$= \frac{1}{4\pi\varepsilon_0}\frac{Q}{R^2}\hat{r}$$

Again the direction of the field is radially outward.

(c) For the space outside the sphere, we proceed similarly but this time draw a Gaussian sphere with $r > R$. As in part (a), we calculate the electric flux as follows:

$$\Phi_E = \oiint_{\text{sphere}} \vec{E} \cdot d\vec{A}$$

$$\vec{E} \cdot d\vec{A} = EdA\cos 0° = EdA(1.00) = EdA$$

$$\Phi_E = \oiint_{\text{sphere}} EdA = E\oiint_{\text{sphere}} dA = 4\pi r^2 E$$

Next we apply Gauss's law, but this time the enclosed charge is simply the total charge Q:

$$\Phi_E = \frac{q_{\text{enc}}}{\varepsilon_0}$$

$$4\pi r^2 E = \frac{Q}{\varepsilon_0}$$

$$E = \frac{1}{4\pi\varepsilon_0}\frac{Q}{r^2}$$

As before, the direction of the electric field is radially outward, so we can write the electric field in vector form:

$$\vec{E} = \frac{1}{4\pi\varepsilon_0}\frac{Q}{r^2}\hat{r}$$

(d) We show a plot of the electric field strength, E, versus radial distance, r, in Figure 20-18. The electric field strength increases linearly inside the uniformly charged sphere, reaching a maximum value at the surface. Outside it decreases according to the inverse square of the radial distance. Note that at the exact centre, the electric field strength is zero.

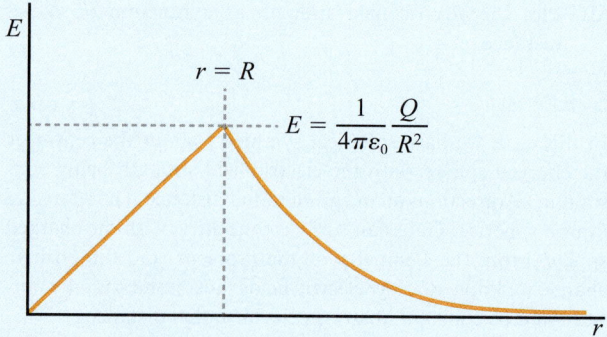

Figure 20-18 Plot of electric field strength versus radial distance for Example 20-4.

Note that the results of parts (b) and (c) agree when $r = R$, as expected. Outside the charged sphere the electric field strength is the same as for a point charge Q. The electric field strength increases with radial distance inside the charged sphere because the volume and enclosed charge vary with the cube of the radius of the Gaussian sphere, while the surface area of the Gaussian sphere increases only as the square of the radius. Note that this solution is for a non-conducting, uniformly charged sphere. Later in the chapter we will consider conducting objects.

In the next section, we develop a strategy for applying Gauss's law to calculate the electric field and then apply it to a variety of types of symmetry later.

20-4 Strategy for Using Gauss's Law

Before we go on to apply Gauss's law to other situations, we summarize a strategy for applying Gauss's law to calculate the electric field. Note that some questions use Gauss's law just to find the electric flux, rather than the electric field. This procedure is for calculating the electric field. The steps are best considered in the context of an example, so you may wish to go on to a problem, and then make reference to this as you work through the solution.

1. *Sketch the charge distribution*, as well as the position(s) where the electric field is to be calculated.

2. *Determine the type of symmetry.* Gauss's law is useful when we have **spherical symmetry, cylindrical symmetry**, or **planar symmetry**. Ask yourself what shape of object would have constant electric field strength over its surface. With spherical symmetry, at any given radial distance, the electric field strength is constant. The types of symmetry are illustrated in Figure 20-19 and summarized in Table 20-1. If the object does not have any of these three types of symmetry, Gauss's law, even though still true, is unlikely to be helpful in calculating the electric field strength.

Table 20-1 Type of Symmetry and the Corresponding Gaussian Surface That Should Be Used

Symmetry	Type of Gaussian Surface	Examples
spherical	sphere concentric with point	point charge charged sphere spherical shell
cylindrical	cylinder coaxial with line	line of charge charged cylinder coaxial cylindrical shells
planar	cylinder or box perpendicular to plane	charged parallel plane(s) large flat object

3. *Draw the appropriate closed Gaussian surface* (see Table 20-1) based on the symmetry, choosing the size of the surface according to where the electric field is to be determined. For example, in the case of a charged sphere, a Gaussian sphere is drawn concentric with the centre of the object and is of radius equal to the radial distance where the electric field is to be computed. Sketch in the directions of $\vec{E}$ and $d\vec{A}$ on your diagram.

4. *If necessary, divide the closed surface into parts.* In some cases, the electric field is not constant over the entire surface but is known over each part. In this case, divide the Gaussian surface into different parts, ensuring that the entire closed surface is included. For example, we earlier divided the cylinder into two end caps and the curved side.

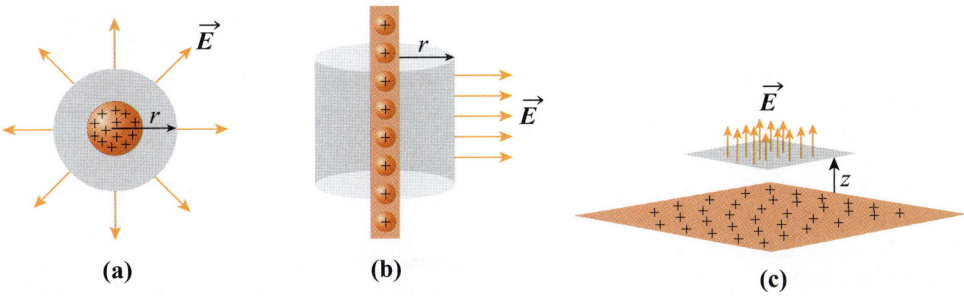

(a)   **(b)**   **(c)**

Figure 20-19 In spherical symmetry (a), the points on a sphere have constant electric field strength. For cylindrical symmetry (b), points at a common distance from a line have constant electric field strength. In planar symmetry (c), points on a plane parallel to the charged plane have constant electric field strength.

ELECTRICITY, MAGNETISM, AND OPTICS

5. For each part, *determine the orientation of $\vec{E}$ relative to $d\vec{A}$* and use that to find the dot product for that part.

6. Formally *integrate over the Gaussian surface to compute the electric flux*. However, if you have a surface with uniform electric field strength (and constant direction relative to the surface normal) you can take E out of the integral and simply multiply the constant E by the surface area multiplied by the $\cos\theta$ value.

7. *Calculate the net charge enclosed by the Gaussian surface*. Remember that only the charge inside the surface needs to be considered, and when there are positive and negative charges inside the surface it is the net charge that is used.

8. *Use Gauss's law* with the expression for the electric flux (from step 6) and the enclosed charge (step 7) *to solve for the electric field strength*.

9. The charge sign and symmetry will make the direction of the electric field obvious. Remember that it points away from positive charges and toward negative charges. *Write the result (step 8) in vector form.*

10. *Check that your answer makes sense* with limiting cases (i.e., does your relationship make sense just outside a surface, or at infinite distance).

PEER TO PEER

In deciding the type of symmetry, I ask myself, "Where I could walk and still have a constant electric field?" If I can walk anywhere on a flat plane, then my Gaussian surface should be a plane. For spherical symmetry, I can wander around the surface of a sphere (all a constant distance from the symmetry point) and the electric field strength is constant.

EXAMPLE 20-5

Two Charged Spherical Shells

Two thin concentric spherical shells carry positive charges of Q_1 and Q_2 and have radii R_1 and R_2, where $R_1 < R_2$. Find expressions for the electric field in all three regions:

(a) outside both shells, $r > R_2$
(b) between the two shells, $R_1 < r < R_2$
(c) inside the smaller shell, $r < R_1$

Solution

(a) Outside both shells ($r > R_2$): We use the numbering of steps given in the strategy.
   (1) We sketch the situation in Figure 20-20.
   (2) We have spherical symmetry about the centre point of the concentric shells.
   (3) Since we have spherical symmetry, the Gaussian surface is a sphere, and we have drawn this in Figure 20-20. The radius of our sphere is r, since that is where we want to calculate the electric field.
   (4) In this case, dividing into parts is not needed, since the electric field strength is uniform over the Gaussian sphere.
   (5) The shells are both positively charged and the charge is symmetrically distributed, so we know that the

electric field must point radially outward. By definition, the differential area element $d\vec{A}$ points outward from the local surface. We have added both $\vec{E}$ and $d\vec{A}$

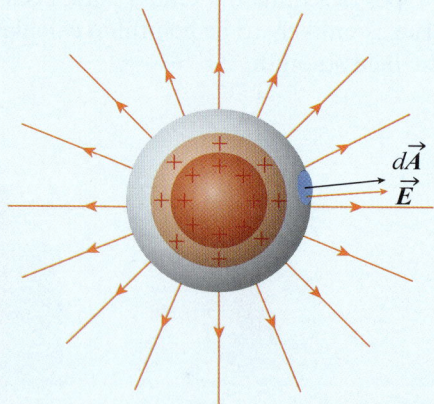

Figure 20-20 Example 20-5 with two positively charged concentric spherical shells. We have drawn the Gaussian sphere (grey) for the case outside both charged shells. Note that we only show field lines for the situation outside the Gaussian surface where we calculate them. See the example for details in other regions.

in Figure 20-20. They both point radially outward and are parallel to each other everywhere.

Since over the Gaussian surface, $\vec{E}$ and $d\vec{A}$ are parallel, the dot product simply becomes the product of the two magnitudes:

$$\vec{E} \cdot d\vec{A} = E \, dA \cos 0° = E \, dA (1.00) = E \, dA$$

where E and dA are the magnitudes of the electric field and the surface differential area, respectively.

(6) Now we are ready to calculate the electric flux. The magnitude of the electric field (and its direction relative to the surface) are everywhere the same over the Gaussian sphere, so we can take E outside the flux calculation:

$$\Phi_E = \oiint_{\text{sphere}} \vec{E} \cdot d\vec{A}$$

$$= \oiint_{\text{sphere}} E \, dA$$

$$= E \oiint_{\text{sphere}} dA$$

$$= EA$$

$$= 4\pi r^2 E$$

We have used the fact that the surface area of a sphere of radius r is $4\pi r^2$.

(7) Since we are dealing with the region outside both shells, the net enclosed charge is $q_{\text{enc}} = Q_1 + Q_2$.

(8) Now we are ready to use Gauss's law to solve for the magnitude of the electric field:

$$\Phi_E = \frac{q_{\text{enc}}}{\varepsilon_0}$$

$$4\pi r^2 E = \frac{Q_1 + Q_2}{\varepsilon_0}$$

$$E = \frac{1}{4\pi\varepsilon_0} \frac{Q_1 + Q_2}{r^2}$$

(9) The electric field is radially outward (since positively charged), which we can express using the radially outward unit vector $\hat{r}$:

$$\vec{E} = \frac{1}{4\pi\varepsilon_0} \frac{Q_1 + Q_2}{r^2} \hat{r}$$

(10) We can look at some limiting cases. If one of the charges goes to zero, we are left with the expression for electric field outside a point charge, which makes sense. The electric field goes to zero at infinite distance.

We will do the other two cases in less detail but will still follow the key parts of the procedure.

(b) Between the two shells ($R_1 < r < R_2$): We again have symmetry about a point and will draw a spherical Gaussian surface, but this time with a radius between that of the two shells: $R_1 < r < R_2$. This time the enclosed charge is simply Q_1. Otherwise, everything is essentially the same as in part (a) and we obtain

$$\vec{E} = \frac{Q_1}{4\pi r^2 \varepsilon_0} \hat{r}$$

(c) Inside the inner shell ($r < R_1$): This time we draw a spherical Gaussian surface with radius $r < R_1$. The electric flux is given by

$$\Phi_E = \oiint_{\text{sphere}} E \, dA = E \oiint_{\text{sphere}} dA = EA = 4\pi r^2 E$$

Here the enclosed charge is zero, so from Gauss's law the electric flux must also be zero:

$$\Phi_E = \frac{q_{\text{enc}}}{\varepsilon_0} = 0$$

$$4\pi r^2 E = 0$$

$$E = 0$$

A flux of zero does not always imply an electric field of zero (since we might have a case with the electric field and area vectors perpendicular, but that is not the situation here).

Making sense of the result

We showed in case (a) that the answer worked for the limiting case of one shell with zero charge. Also, the electric field goes to zero at infinite distance, which makes sense. In part (b), the result that we obtained is the same as for a point charge, so the shell acts as though all of the charge is concentrated at the centre. While it may seem strange that we have nearby charge but no electric field in part (c), as long as there is a symmetric distribution of that charge, the vector sum at any interior point adds to zero.

We apply this strategy in Example 20-5, and for different types of symmetry in the later sections.

While Gauss's law may seem mathematically complex at first, in most symmetrical cases the surface integral is easily computed. In the next two sections, this technique is applied to cases with the other two types of symmetry. As with other topics in physics, at first it will be helpful to follow the strategy closely, but soon the key elements and concepts will become natural as you approach problems.

PEER TO PEER

At first I wasn't sure about when I could calculate electric field and when I could only use Gauss's law to find the electric flux. I can *always* use Gauss's law to find electric flux over any closed surface as long as I know the enclosed charge. But I can find the electric field *only* if I know the enclosed charge and there is enough symmetry that I can find a surface that has constant electric field strength.

20-5 Gauss's Law for Cylindrical Symmetry

Using symmetry, we can readily apply Gauss's law to calculate electric fields around charge distributions that have symmetry about a point, a line, or a plane. In this section, we consider cases of symmetry about a line, also called cylindrical symmetry. This type of symmetry applies to cases such as long lines of charge, charged cylinders, and charged cylindrical shells. When dealing with objects with symmetry about a line, we usually use the linear charge density that you encountered in Chapter 19, the charge per unit length. While the symmetry strictly requires infinite lengths, it will be a good approximation, except near the ends, for cases of finite-length objects.

Consider the very long straight line of positive charge with uniform linear charge density λ shown in Figure 20-21. We want to find the electric field at a distance r from the line. If we assume that the line of charge is essentially infinite, we know from symmetry that the electric field must point radially away (since it is positively charged) from the line of charge and have the same magnitude, E, at every point that is at the same radial distance r from the line of charge (see the end view from Figure 20-21).

It is clear that we have cylindrical symmetry in this case, so the correct Gaussian surface will be a cylinder coaxial with the line of charge. The radius of the Gaussian cylinder needs to be r, but the length, ℓ, is arbitrary. We will see that ℓ cancels out in the final results.

This is a situation when we must divide the closed surface of the cylinder into different parts. We will find the total electric flux through the cylinder by dividing it into three parts: the two end caps and the curved surface (see Figure 20-22). We show the orientation of $\vec{E}$ and $d\vec{A}$ in the different parts. For the two end caps, $\vec{E}$ and $d\vec{A}$ are perpendicular to each other, and therefore the dot product is zero for each of these parts. For the curved walls of the cylinder, the directions of

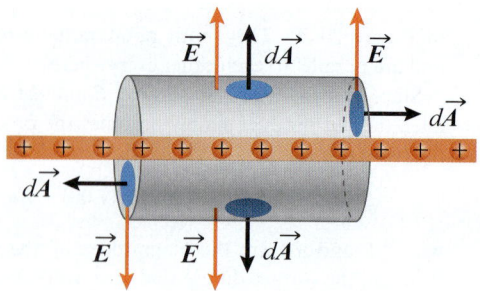

Figure 20-22 The orientation of $\vec{E}$ and $d\vec{A}$ for the three parts (two end caps plus curved side) of the Gaussian cylinder around a long line of charge.

both $\vec{E}$ and $d\vec{A}$ change with position, but they are both always radially outward and parallel to each other. The dot product used in the electric flux calculation is summarized below for each of the three parts, with E and dA representing the magnitude of the corresponding vector quantities:

left end cap: $\vec{E} \cdot d\vec{A} = EdA \cos 90° = EdA(0) = 0$
curved wall: $\vec{E} \cdot d\vec{A} = EdA \cos 0° = EdA(1.00) = EdA$
right end cap: $\vec{E} \cdot d\vec{A} = EdA \cos 90° = EdA(0) = 0$

$$(20\text{-}6)$$

The contributions of electric flux from the two end caps are zero, and we only have the contribution from the curved surface. We obtain the net electric flux by adding the flux from the various parts, as shown:

$$\Phi_{E\,net} = \Phi_{E\,left\,cap} + \Phi_{E\,curved\,surface} + \Phi_{E\,right\,cap}$$

$$= 0 + \oiint_{curved\,surface} \vec{E} \cdot d\vec{A} + 0$$

$$= \oiint_{curved\,surface} EdA$$

$$= E \oiint_{curved\,surface} dA \qquad (20\text{-}7)$$

$$= EA_{curved\,surface}$$

$$= E(2\pi r)\ell$$

$$= 2\pi r \ell E$$

Note that we have used the dot product relationships from above, and also the fact that to find the surface area of the curved surface we could imagine it flattened out to become a rectangle with length ℓ and width $2\pi r$, the circumference of the curved surface.

We can find the charge enclosed within the cylindrical Gaussian surface of length ℓ by using the linear charge density:

$$q_{enc} = \lambda \ell \qquad (20\text{-}8)$$

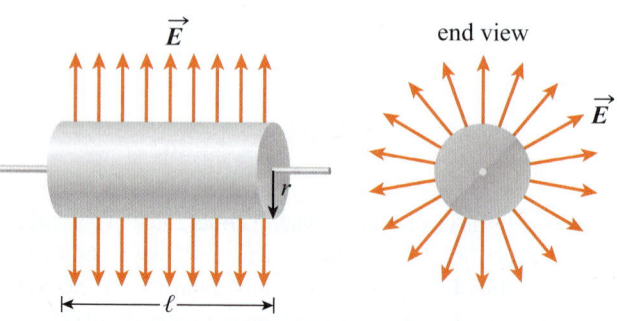

Figure 20-21 A long line of charge with a Gaussian cylinder of radius r and arbitrary length ℓ. The electric field points radially outward from the wire in all directions but is shown only at the horizontal surfaces of the cylinder here.

If we substitute results from Equations 20-7 and 20-8 into Gauss's law, we can solve for the magnitude of the electric field:

$$\Phi_E = \frac{q_{enc}}{\varepsilon_0}$$

$$2\pi r \ell E = \frac{\lambda \ell}{\varepsilon_0} \qquad (20\text{-}9)$$

$$E = \frac{\lambda}{2\pi\varepsilon_0 r}$$

Equation 20-9 gives the electric field strength, E, at a distance r from an infinite line with charge density λ. The direction of the electric field is radially away from the line of charge.

It makes sense that the electric field is proportional to the charge per unit length since twice as many charges enclosed would yield twice the electric flux and field. Note that the electric field is inversely proportional to the radial distance, r, not the square of the distance, as it is for a point charge. If the charge were negative, the results would be exactly the same except that the electric field would point radially inward toward the line of charge.

PEER TO PEER

I don't need to memorize results such as the electric field around a line of charge, but rather I must be sure I know how to use Gauss's law in any applicable situation.

INTERACTIVE ACTIVITY 20-2

Linear Charge and Gauss's Law

In this activity, you will use the OpenSource Physics simulation "Linear Charge Gauss's Law" to investigate electric field and flux for a very long solid charged cylinder. You can draw Gaussian surfaces of different sizes. You do not know how the charge is distributed in the solid cylinder, but you will infer that from the observations made.

CHECKPOINT 20-8

Coaxial Cable

Coaxial cables (see Figure 20-23) are commonly used to transmit signals, both in consumer applications like television and in scientific equipment. A coaxial cable consists of a central wire, a layer of insulation, a concentric cylindrical second conductor, and an outer layer of insulation.

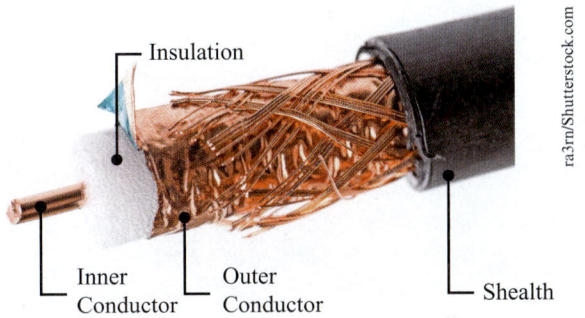

Figure 20-23 A coaxial cable (Checkpoint 20-8).

Which of the following statements best describes the electric field between the conductors and outside a length of coaxial cable that has a positive charge of q on the inner wire and a negative charge of the same magnitude on the outer conductor?

(a) The electric field is directed away from the wire in both regions.
(b) The electric field is directed toward the wire in both regions.
(c) The electric field is zero in the region between the conductors and directed outward outside the cable.
(d) The electric field is directed radially away from the central wire in the space between the conductors and is zero outside the cable.

ANSWER: (d) By Gauss's law, the net electric flux outside the cable must be zero because the total enclosed charge is zero inside a Gaussian box that encloses both the inner wire and the outer conductor. In the space between the conductors, the electric field is directed outward because the inner wire has a positive charge.

EXAMPLE 20-6

Electric Field for a Charged Cylinder

A non-conducting very long cylinder of radius R has a positive uniform volume charge density ρ throughout. Derive expressions for the electric field both (a) inside ($r < R$) and (b) outside ($r > R$) the cylinder.

Solution

We have cylindrical symmetry, so the appropriate closed surface is a coaxial Gaussian cylinder of radius r. The length of

the Gaussian cylinder, L, is arbitrary and will cancel out in the final answer.

(a) Inside the cylinder ($r < R$): For this part, we draw the Gaussian cylinder with radius r less than the radius, R, of the charged cylinder (i.e., the Gaussian surface is inside the physical cylinder, but coaxial with it). The situation is illustrated in Figure 20-24, with the Gaussian cylinder in grey. From the symmetry of the situation, the positively charged cylinder results in an electric field that points

(continued)

radially outward and has constant electric field strength at any particular radial distance.

This is a situation in which we need to divide the closed Gaussian surface into three parts, the two end caps and the curved side of the cylinder. In all cases, $d\vec{A}$ points outward from the closed Gaussian surface. This is pictured for the grey Gaussian surface in Figure 20-24. For the two end caps, $d\vec{A}$ is parallel to the axis of the cylinder, pointing in different directions at the two ends. For the curved surface of the cylinder, $d\vec{A}$ is everywhere radially outward (we only show $\vec{E}$ and $d\vec{A}$ at one point on the curved Gaussian surface, but both are similarly radially outward for all other points on the curved surface).

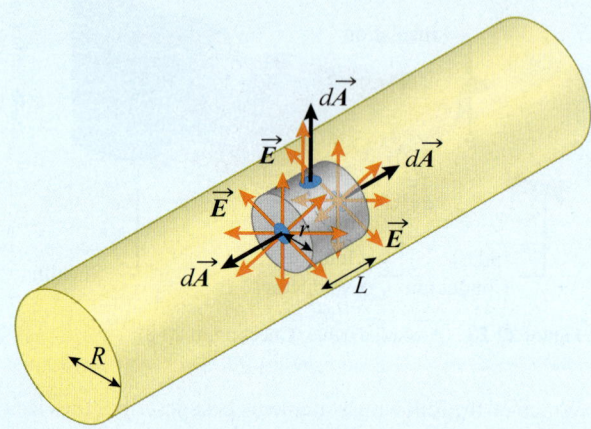

Figure 20-24 Example 20-6. The very long charged cylinder (yellow) has a uniform positive charge density. In grey is drawn the smaller Gaussian cylinder for finding the electric field inside the charged cylinder.

We find the electric flux by combining the contributions from the three parts of the surface:

$$\Phi_{\text{E net}} = \Phi_{\text{E left cap}} + \Phi_{\text{E curved surface}} + \Phi_{\text{E right cap}}$$

Everywhere along the curved surface, $\vec{E}$ and $d\vec{A}$ are parallel to each other, while on both end caps they are perpendicular to each other (see Figure 20-24). This means that the vector dot product of $\vec{E}$ with $d\vec{A}$ yields

left end cap: $\vec{E} \cdot d\vec{A} = EdA\cos 90° = EdA(0) = 0$
curved wall: $\vec{E} \cdot d\vec{A} = EdA\cos 0° = EdA(1.00) = EdA$
right end cap: $\vec{E} \cdot d\vec{A} = EdA\cos 90° = EdA(0) = 0$

Here E represents the magnitude of the electric field, and dA is the differential area element without a vector direction. Therefore, substituting into the relationship for electric flux, we have

$$\Phi_{\text{E left cap}} = \oint_{\text{left cap}} \vec{E} \cdot d\vec{A} = 0$$

$$\Phi_{\text{E curved wall}} = \oint_{\text{curved wall}} \vec{E} \cdot d\vec{A} = \oint_{\text{curved wall}} Ed A$$

$$= E\oint_{\text{curved wall}} dA = EA_{\text{curved wall}}$$

$$\Phi_{\text{E right cap}} = \oint_{\text{right cap}} \vec{E} \cdot d\vec{A} = 0$$

The area of the curved wall can be determined if we imagine it has been unwrapped into a corresponding rectangle. One side of the rectangle is L and the other is the circumference of the surface, $2\pi r$:

$$\Phi_{\text{E curved wall}} = EA_{\text{curved wall}} = EL(2\pi r)$$
$$= 2\pi rLE$$

Since the two end caps have zero electric flux, this is also the net electric flux:

$$\Phi_{\text{net}} = 2\pi rLE$$

Now we need to find the net enclosed charge. We multiply the volume charge density, ρ, by the volume of the Gaussian cylinder. The volume of a cylinder is the cross-sectional area times the length of the cylinder:

$$q_{\text{enc}} = \pi r^2 L\rho$$

Finally, we invoke Gauss's law and solve for the electric field strength:

$$\Phi_{\text{E}} = \frac{q_{\text{enc}}}{\varepsilon_0}$$

$$2\pi rLE = \frac{\pi r^2 L\rho}{\varepsilon_0} \qquad (20\text{-}10)$$

$$E = \frac{r\rho}{2\varepsilon_0}$$

The direction of the electric field points radially outward from the axis of the cylinder.

(b) Outside the charged cylinder ($r > R$): In this case, we make a Gaussian cylinder of radius larger than that of the charged cylinder, but otherwise the solution is similar. The electric flux process is identical to that given in part (a), with the same result:

$$\Phi_{\text{net}} = 2\pi rLE$$

When computing the enclosed charge, we now use the volume of the charged cylinder, rather than the larger Gaussian cylinder:

$$q_{\text{enc}} = \pi R^2 L\rho$$

We invoke Gauss's law and solve for the electric field strength.

$$\Phi_{\text{E}} = \frac{q_{\text{enc}}}{\varepsilon_0}$$

$$2\pi rLE = \frac{\pi R^2 L\rho}{\varepsilon_0} \qquad (20\text{-}11)$$

$$E = \frac{R^2 \rho}{2r\varepsilon_0}$$

Again, the direction of the electric field is radially outward from the axis of the cylinder.

Making sense of the result

Inside the charged cylinder, the electric field increases linearly with distance from the axis. At the exact centre of the cylinder, the electric field is zero. Outside the charged cylinder, the electric field is proportional to $1/r$, approaching zero at infinite distance. When $r = R$, we require that the results from the two parts be consistent, which they are. Note that this solution is for the case of a non-conducting cylinder with the charge uniformly distributed.

20-6 Gauss's Law for Planar Symmetry

In this section, we consider applications of Gauss's law to situations with planar symmetry. For exact results, we need infinite planes, but as long as the distance to the edge is very large compared to the distance you are away from the surface, the techniques of this section will yield sufficiently accurate results.

In most planar problems, the surface charge density, σ, the charge per unit area (see Chapter 19), will be used. In Chapter 22, you will encounter planar objects of great practical importance, capacitors, made up of two flat planes with opposite charge. We will use the techniques of this chapter to find the electric fields in capacitors in that chapter.

The type of Gaussian surface that we draw for an object with planar symmetry can be a Gaussian cylinder or a Gaussian rectangular cube, in both cases oriented perpendicular to the charged plane. We demonstrate this technique in Example 20-7.

EXAMPLE 20-7

Electric Field from an Infinite Slab of Charge

Derive an expression for the electric field a distance d from a very large (assumed infinite) non-conducting flat charged slab with uniform charge density σ. The precise thickness of the slab does not matter, as long as it is thin. We assume that σ represents the total charge through the slab per unit area.

Solution

We will assume that the charge is positive, although other than reversing the direction of the electric field, the analysis is the same for either sign. The approach is similar to that done earlier for other types of symmetry. Since we have planar symmetry, we draw a cylindrical Gaussian surface (or we could have used a cube) that extends an equal distance d above and below the infinite charged plane, as shown in Figure 20-25. We will call the area of each end cap A, although this variable will cancel out of our final expression.

Here we must divide the closed cylindrical Gaussian surface into three parts, the top end cap, the curved wall of the cylinder, and the bottom end cap. We will combine the individual electric flux values:

$$\Phi_{E\text{ net}} = \Phi_{E\text{ top cap}} + \Phi_{E\text{ curved wall}} + \Phi_{E\text{ bottom cap}} \quad (20\text{-}12)$$

Considering the assumed positive charge, and the symmetry of the situation, the electric field, $\vec{E}$, points perpendicularly away

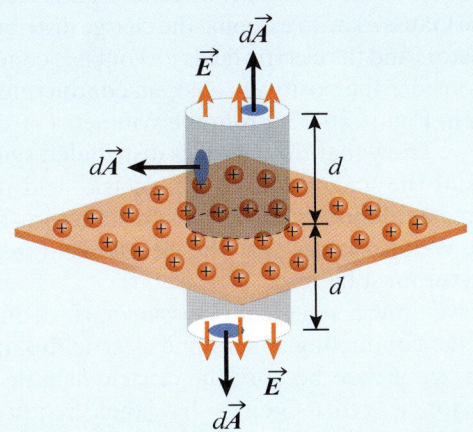

Figure 20-25 A thin, infinite slab of charge with a Gaussian surface extending equal distances above and below the surface. For clarity, the electric field vectors are shown only where they pass through the surface of the Gaussian cylinder.

from the infinite charged plane, as shown in Figure 20-25. The direction of the differential area elements, $d\vec{A}$, is always perpendicular to the local surface in the outward direction. We show several of these in Figure 20-25. Note that for both end caps, $\vec{E}$ and $d\vec{A}$ point in the same direction, while along the curved wall, $d\vec{A}$ is perpendicular to $\vec{E}$ everywhere. By symmetry, the electric field strength, E, is the same on the top and the bottom caps. If we call dA the magnitude of the differential area element, we must have that $\vec{E} \cdot d\vec{A} = E dA$ for both end caps, and $\vec{E} \cdot d\vec{A} = 0$ all along the curved wall. Therefore, we have the following for the electric flux in each part:

$$\Phi_{E\text{ top cap}} = \oiint_{\text{top cap}} E dA = E \oiint_{\text{top cap}} dA = EA$$

$$\Phi_{E\text{ curved wall}} = 0$$

$$\Phi_{E\text{ bottom cap}} = \oiint_{\text{bottom cap}} E dA = E \oiint_{\text{bottom cap}} dA = EA$$

We use Equation 20-12 to combine these into one electric flux for the entire closed surface:

$$\Phi_{E\text{ net}} = \Phi_{E\text{ top cap}} + \Phi_{E\text{ curved wall}} + \Phi_{E\text{ bottom cap}}$$
$$= EA + 0 + EA = 2EA \quad (20\text{-}13)$$

Next we calculate the enclosed charge, the charge bounded by the Gaussian surface, which is simply the charge density, σ, multiplied by the cross-sectional area, A, of the Gaussian cylinder:

$$q_{\text{enc}} = \sigma A \quad (20\text{-}14)$$

Now we are ready to substitute our result for the electric flux (Equation 20-13) and the enclosed charge (Equation 20-14) into Gauss's law (Equation 20-4), and then solve for the electric field strength:

$$\Phi_E = \frac{q_{\text{enc}}}{\varepsilon_0}$$

$$2EA = \frac{\sigma A}{\varepsilon_0} \quad (20\text{-}4)$$

$$E = \frac{\sigma}{2\varepsilon_0}$$

The direction is away from the plane for a positive surface charge and toward the plane for a negative surface charge.

(continued)

Making sense of the result

You may be surprised that the strength of the electric field does not depend on the distance from the infinite uniform plane of charge. The contributions to the electric field from distant points almost entirely cancel at points near the plane of charge, but for greater distances from the plane, these contributions provide a more significant vertical resultant (see Figure 20-26). It turns out that this effect exactly offsets the decrease in strength from the point on the plane nearest the measurement point, so the net effect is that the electric field does not depend on the distance from the plane.

The answer we obtained in this example is identical to the result that we obtained in Chapter 19 (Example 19-11) using superposition and direct integration. Gauss's law provides a shorter, less complex way to obtain the result.

(a)

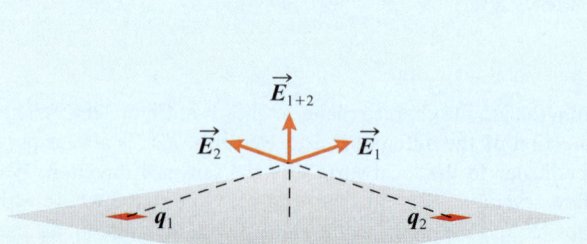

(b)

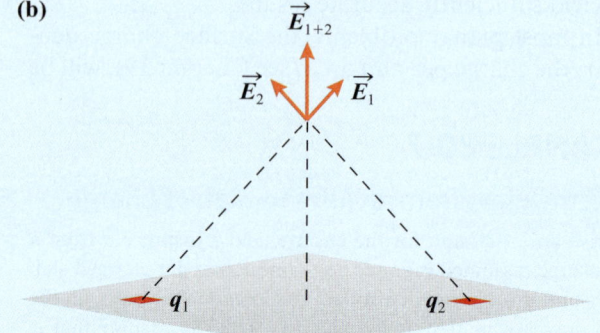

Figure 20-26 At a point near a charged plane (a), contributions to the electric field from distant parts of the plane more nearly cancel than for a more distant point (b). Even though $\vec{E}_1$ and $\vec{E}_2$ have larger magnitudes in (a) because of smaller distances to the charges, the resultant of adding the two is less than it was in case (b).

Example 20-7 derived the following important expression for the electric field for a very large flat non-conducting charged object with charge per unit area σ:

$$E = \frac{\sigma}{2\varepsilon_0} \qquad (20\text{-}15)$$

The direction is perpendicular to the charged slab and points toward it for a negative surface charge and away from it for a positive surface charge.

Of the situations considered so far, this is the only one in which the field did not become weaker as we moved farther away. As mentioned at the outset, this result is only exact for an infinite plane, but it will be a good approximation for large flat objects.

CHECKPOINT 20-9

Surface Charge Density Doubled

If you double the charge density on large thin flat slab, how does the electric flux through a circle parallel to the plane of the slab, but displaced from it, change?

(a) There is zero flux in both cases.
(b) The electric flux is one half its previous value.
(c) The electric flux is double its previous value.
(d) The electric flux is four times its previous value.

ANSWER: (c) By the technique of Example 20-7, we know that the electric field strength is proportional to the charge density, so the electric field strength will double. The electric flux through the circle parallel to the slab, and therefore perpendicular to the electric field, is simply the area of the circle times the electric field strength. Therefore, we expect the electric flux to also double.

20-7 Conductors and Electric Fields

In conductors, the free electrons are mobile (see Chapter 23 for a discussion of the microscopic nature of current flow). This means that if we have a static situation (i.e., enough time has elapsed that charges have flowed to reach a final configuration), there must by definition be no further charge flow. This is only possible if there is no electric force on the charges. But since the electric force is simply the product of the electric field and the charge, this requires that in a static situation there must be zero electric field *inside* an ideal conductor. In this section, we will use Gauss's law to examine the charge distribution in conductors and the electric fields just outside conductors.

Consider the positively charged conducting sphere shown in Figure 20-27. From the symmetry of the situation, we know that the charge is distributed symmetrically and the magnitude of the electric field must be the same at all points at the same distance, r, from the centre. We also know that the electric field inside the conductor must be zero.

If we draw a spherical Gaussian surface anywhere inside the conducting sphere, the electric flux through the surface is zero because the electric field inside the conductor is zero. Gauss's law then requires that the enclosed charge be zero. Since we can increase the radius of this Gaussian sphere right up to just below the radius of the conducting sphere and the argument still holds, the charge on the conducting sphere must all reside on the *outer surface* of the conductor.

If we draw a Gaussian sphere just slightly outside the conducting sphere, however, then the enclosed charge is the total charge on the spherical conductor and there is an electric field. In fact, the electric field outside the conductor will be the same as at that at the same radial distance for single point charge with the same charge. The exterior electric field points radially outward for a charged spherical conductor.

In fact, charges on conductors of any shape always lie on the outer surfaces in static conditions. We could prove this by drawing a Gaussian surface the same shape as the object, but slightly smaller. Since the Gaussian surface is inside a conductor, the electric field must be zero, and therefore the electric flux must also be zero. We show this in Figure 20-28. Now, however, the charges are not uniformly spread over the surface—there are more at sharply curved parts. We will prove why this is so in Chapter 21, after we have introduced the concept of electric potential.

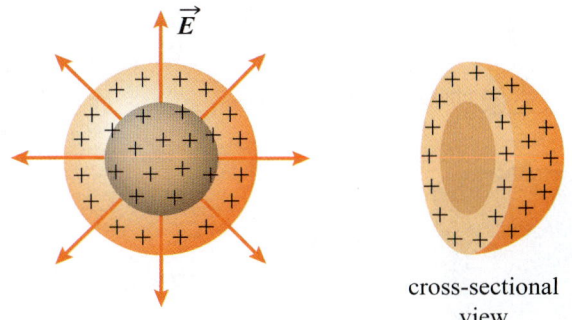

Figure 20-27 Since the electric flux through any spherical Gaussian surface inside a spherical conductor is zero, the enclosed charge must be zero. We can conclude that all charge must reside on the outer surface of a conductor.

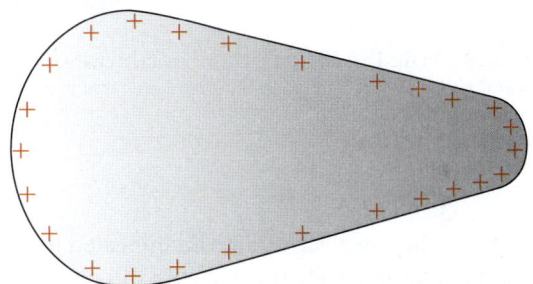

Figure 20-28 When the conductor is irregularly shaped, the charges are still on the outside, but now they are not uniformly spread.

CHECKPOINT 20-10

Electric Field for a Charged Conducting Shell

Which statement correctly describes the charge distribution and the electric field in the vicinity of a negatively charged spherical conducting shell?

(a) Charges are uniformly distributed over the outer surface. There will be zero electric field inside the conductor, while just outside the conductor the electric field will be perpendicular to the surface and point inward.

(b) Charges are uniformly distributed over the outer surface. There will be zero electric field inside the conductor, while just outside the conductor the electric field will be perpendicular to the surface and point outward.

(c) Charges are uniformly distributed over the outer surface. Both inside and just outside the conductor the electric field points radially inward.

(d) Charges are uniformly distributed over the inner surface. Both inside and just outside the conductor the electric field points radially inward.

ANSWER: (a) As we argued in this section, the electric field inside a conductor is zero and all charge goes to the outer surface of a conductor. Just outside a conductor, the field, if any, must be perpendicular. Since the object is negatively charged, the direction is radially inward.

In this section, we have argued that all charges on an ideal conductor under static conditions are on the outer surface and the electric field is zero at all interior points. Therefore, we can obtain protection from high electric fields by being inside a conducting enclosure. Such shielding enclosures are called Faraday cages,

after Michael Faraday (see Making Connections: Faraday Cage Protection). In 1816, Faraday used them to demonstrate that no charge was present inside a conducting shell. Actually, Benjamin Franklin had made the same discovery earlier while experimenting with charge distributions in 1755.

Although the absence of charges inside a closed conducting surface is strictly true only in electrostatic situations, such surfaces are often effective shields that block electromagnetic waves from penetrating the surface, provided that any holes in the surface are small compared to the wavelength of the electromagnetic waves (see Chapter 27).

Arguably the first experiment that studied static charges in a quantitative fashion was conducted by Michael Faraday in 1843 and is called the ice pail experiment. The key aspects of the experiment are illustrated in Figure 20-29. A metal pail (the "ice pail," explaining the name for the experiment) is placed on an insulating stand so charge cannot flow to or from the ground. A charged object is lowered into the pail, without actually touching the sides, and once inside the pail a charge will be measured on the electroscope. As we saw in Section 19-2, the nearby charged object will induce charge separation in the pail. The question is: Will there be charge on the outside surface of the

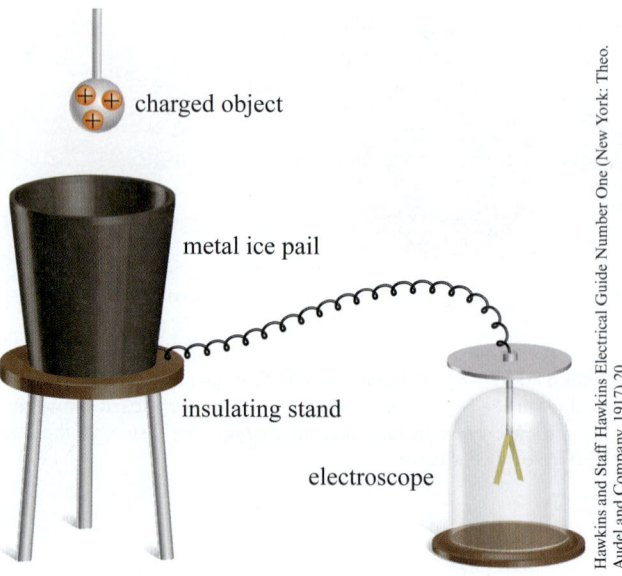

charged object

metal ice pail

insulating stand

electroscope

Hawkins and Staff Hawkins Electrical Guide Number One (New York: Theo. Audel and Company, 1917) 20

Figure 20-29 In the Faraday ice pail experiment, a charge is induced on a metal pail, and it is demonstrated that the charge is present on the outside of the pail.

pail? To test this, an electroscope is connected to the ice pail, and it does illustrate the presence of charge on the outer surface of the ice pail, with the same polarity as the lowered charged object. The modern lecture demonstration version of the experiment usually replaces the pail with a spherical metal shell that has a hole the charged object can be inserted through. The location of the charged object inside the sphere does not change the charge reading, as long as it does not touch the metal.

Suppose we place an uncharged conducting sphere in an external electric field directed downward, as shown in Figure 20-31. The conductor has no net charge, but it does have mobile electrons that are free to move within the conductor. We must still have zero net electric field inside the conductor—the mobile charge carriers in the conductor move to positions such that their own electric fields exactly cancel the external electric field within the conductor. Free electrons move to the top of the sphere, producing a negative charge there and leaving a net positive charge at the bottom of the sphere, a phenomenon called induced charge separation. This is to produce the electric field required to cancel the external field inside the conductor. Note that the electric field lines in Figure 20-31 are distorted by the presence of the conductor: some electric field lines end at the negative induced charges on the conductor, and an equal number of electric field lines start at the positive induced charges. The total electric field in the conductor is exactly zero in electrostatic equilibrium. These lines still approach and leave perpendicular to the surface, as we have seen for the electric field in other conductors in this section.

MAKING CONNECTIONS

Faraday Cage Protection

At the Boston Museum of Science is located the world's largest Van de Graaff static electricity generator, which was invented in 1931 by physicist Robert J. Van de Graaff (1901–1967). Even though there are huge electric fields outside the charged spheres, the operator is protected from harm inside the metal mesh of the Faraday cage pictured (Figure 20-30) at the bottom of the photo.

Peter Menzel/Science Source

Figure 20-30 Boston Museum of Science display operator is protected in a Faraday Cage from the high fields around the world's largest Van de Graaff static electricity generator.

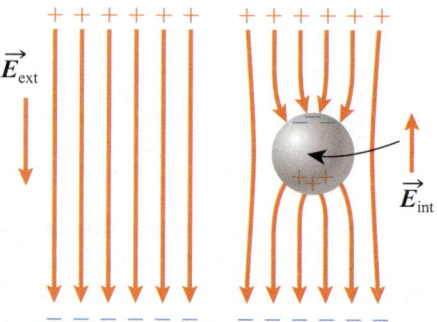

Figure 20-31 Charge separation induced in a conductor by an external electric field.

The induced charge can also be explained in terms of electric forces. The external electric field in Figure 20-31 exerts an upward force on negative charges and within the conductor a downward force on positive charges.

In the previous section (Example 20-7), we found an expression for the electric field outside a non-conducting near-infinite flat charged slab. We obtained the interesting result that the electric field was independent of the distance away from the charged flat slab. Let us now calculate the electric field outside a near-infinite charged conducting plane (Example 20-8).

EXAMPLE 20-8

Electric Field from a Large Conducting Plane

Derive an expression for the electric field a distance d from a very large (assume near-infinite) charged conducting plane with a uniform surface charge density σ (assume a positive charge). Note that the surface charge density refers to the charge on *one* side only for conducting planes.

Solution

There is planar symmetry, so we will draw a cylindrical Gaussian surface perpendicular to the plane. This time, however, we will make the Gaussian cylinder start *inside* the conductor and extend out to the point of computation of the electric field. We call the cross-sectional area of the Gaussian surface A, although that variable will not appear in our final answer and the cross section of the cylindrical Gaussian surface does not matter. The key difference from our approach for the non-conducting plane is that we place one end of the Gaussian surface inside the conductor, since we know that the electric field is zero everywhere inside a conductor.

For this cylindrical Gaussian surface, we must break the closed cylindrical surface into parts since the relationship between $\vec{E}$ and $d\vec{A}$ varies in the different parts. As seen in earlier examples, we can consider the electric flux as made up of three parts, on the top and bottom caps and on the curved side surface (see Figure 20-32):

$$\Phi_{E\ net} = \Phi_{E\ top\ cap} + \Phi_{E\ curved\ wall} + \Phi_{E\ bottom\ cap}$$

In Figure 20-32, we indicate the directions of $\vec{E}$ and $d\vec{A}$ for the three parts of the Gaussian cylinder. For the top cap, $\vec{E}$ and $d\vec{A}$ are parallel, and the magnitude of the electric flux is the product of their magnitudes. Along the curved side of the cylinder, $\vec{E}$ and $d\vec{A}$ are at right angles, so the electric flux through this curved surface is zero. Since the bottom cap is in a region with zero electric field, there is zero flux through the bottom cap. At the top cap, the symmetry of the situation requires that the electric field strength, E, remain constant over the cap. Therefore, we have

$$\Phi_{E\ top\ cap} = \oiint_{top\ cap} \vec{E} \cdot d\vec{A} = \oiint_{top\ cap} E\,dA$$

$$= E \oiint_{top\ cap} dA = EA$$

$$\Phi_{E\ curved\ wall} = \oiint_{curved\ wall} \vec{E} \cdot d\vec{A} = 0$$

$$\Phi_{E\ bottom\ cap} = \oiint_{bottom\ cap} \vec{E} \cdot d\vec{A} = \oiint_{bottom\ cap} 0\,dA = 0$$

$$\Phi_{E\ net} = \Phi_{E\ top\ cap} + \Phi_{E\ curved\ wall} + \Phi_{E\ bottom\ cap}$$

$$= EA + 0 + 0 = EA$$

The charge per unit surface area is σ, so the charge enclosed within the area of the plane bounded by the Gaussian surface, A, is

$$q_{enc} = \sigma A$$

Finally, we use Gauss's law, with the electric flux and the enclosed charge results, solving for the electric field strength:

$$\Phi_E = \frac{q_{enc}}{\varepsilon_0}$$

$$EA = \frac{\sigma A}{\varepsilon_0}$$

$$E = \frac{\sigma}{\varepsilon_0}$$

The direction is away from the plane for a positive surface charge.

Making sense of the result

Again we find that for a very large flat plane, the strength of the electric field does not depend on the distance away from the plane. You may be surprised that we obtained a different result by a factor of 2 compared to what we found in Example 20-7 for a non-conducting plane. If the surface charge is the same, shouldn't the electric field be the same? We have given a hint at the resolution of this paradox in Figure 20-32. When we place charge on a conducting plane, it spreads along all the outer surfaces, both top and bottom, and therefore we essentially have double the charge for the same surface charge density (on one side of the surface) and hence the factor of 2. If we had drawn a Gaussian cylinder that extended symmetrically on both sides of the plane, like we did for the non-conducting plane in the previous section, the enclosed charge would be double, as would the electric field strength, and we would have again obtained the same result.

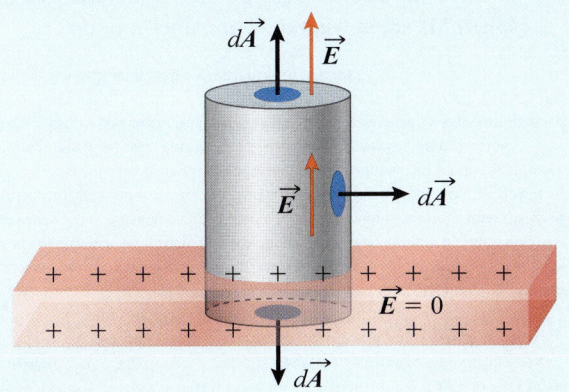

Figure 20-32 Example 20-8.

We have derived the important result that the electric field strength for any distance from an infinite charged conducting plane depends only on the surface charge density, σ (and the permittivity of free space):

$$E = \frac{\sigma}{\varepsilon_0} \tag{20-16}$$

Note that σ is the surface charge density on *one* side of the conductor. The direction of the electric field is perpendicular to the infinite charged plane and points away from the plane for a positive surface charge and toward it for a negative charge.

Charges on a conductor go to the outer surface, even for irregularly shaped objects. However, the charges are not equally distributed, and that is why we cannot readily use Gauss's law to find the electric field just outside the surface in these cases. Although the electric field is still perpendicular to the local surface, there is not constant electric field strength, as shown in Figure 20-33. After we have introduced electric potential in the next chapter, we will return to consider situations such as this in more detail.

Here is one final challenge for you before we leave the topic of conductors. Consider the situation illustrated in Figure 20-34. There is a hole in a spherical conductor, and in the exact centre of that empty space

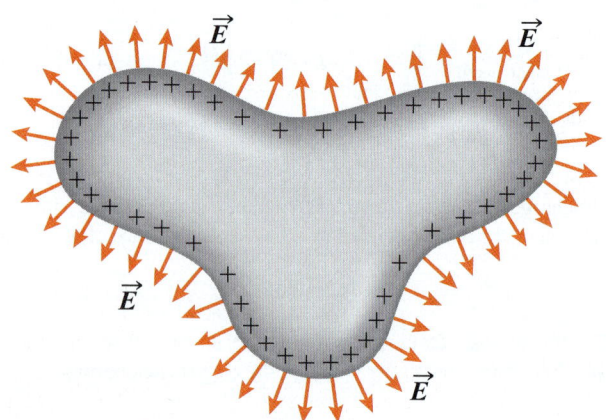

Figure 20-33 Even on irregular conductors, the charges are on the outside surface, and the electric field is perpendicular to the local surface. For irregularly shaped objects, the electric field strength varies with position, however.

CHECKPOINT 20-11

Which Are True for Conductors?

For charged conductors, mark each as true or false.
(a) The electric field is zero inside the conductor.
(b) The charges will spread into a uniform charge distribution.
(c) The charges are distributed only on the surface of the conductor.
(d) If there is negative charge on a conductor, the electric field will point toward it.
(e) The electric field strength is constant at all points just outside the surface of the conductor.

ANSWER: (a), (c), and (d) are true, while (b) and (e) are not always true. A uniform distribution of the charges on the surface only happens for objects with symmetry, such as infinite planes and conducting spheres. It is not true when the object is irregular in shape (see, e.g., Figure 20-28). Again, electric field strength is uniform for a symmetrical object, but not in cases such as that shown in Figure 20-33.

CHECKPOINT 20-12

Conductor with Empty Space

As shown in Figure 20-34, a conducting sphere of radius r_2 carries no net charge. It has a cavity out to radius r_1, and a positive point charge, $+Q$, is placed at the exact centre of the cavity. You do not need to calculate the electric field, just in each case indicate if there is one and the direction.

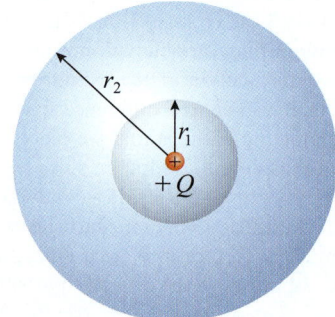

Figure 20-34 Cross-sectional view of a positive point charge, $+Q$, placed in the centre of a cavity in a conducting sphere.

(a) Is there an electric field inside the cavity? If so, in what direction does it point?
(b) Is there an electric field inside the conductor? If so, what is the direction?
(c) There is no net charge on the conducting sphere, but is there a separation of charges to different regions? If so, how are they distributed?
(d) Is there an electric field in the region outside the sphere? If so, in what direction does it point?

ANSWER: (a) Yes. There is an electric field and it points radially outward. If we drew a Gaussian spherical surface in this region, it would contain the $+Q$ charge, so there would be an electric flux. Everything is spherically symmetric, so the electric flux implies an electric field. Since the charge is positive, the electric field points radially outward. (b) No. Under static conditions, there cannot be an electric field in a perfect conductor. (c) Yes. There has to be a separation of charge, with negative charge induced on the inner surface of the conductor. For a spherical Gaussian surface of radius $r_1 < r < r_2$, we know that the electric field is zero, and therefore the electric flux is also zero. By Gauss's law, this is only possible if the total enclosed charge is zero. Therefore, we need a total charge of $-Q$ induced on the inner surface, which, when added to the $+Q$ point charge, gives a net enclosed charge of zero. Since the overall spherical conductor has no net charge, if $-Q$ is induced on the inner surface, then $+Q$ must be on the outer edge of the conductor. (d) Yes. There is an electric field and it points radially outward. A Gaussian spherical surface of radius r_2 has an enclosed charge of $+Q$. The electric flux must be positive, implying a radially outward direction for the electric field.

a positive point charge $+Q$ is placed. See if you can use the expertise you have developed to answer the questions posed in Checkpoint 20-12 for this novel situation.

20-8 When Can Gauss's Law Be Used to Find the Electric Field?

We have seen that Gauss's law provides a powerful technique for calculating the electric field in cases with sufficient symmetry. In this section, we stress key points with respect to Gauss's law and use the superposition principle in conjunction with Gauss's law to show how it may be applied in more complex situations.

Gauss's law is always true for any closed surface you choose to draw. This does not mean, however, that it is always useful for computing the electric field. *It is only helpful for finding the electric field when the electric field symmetry allows the unknown electric field strength to be moved outside the surface integral.*

Electric flux plays a key role in Gauss's law. While it is proper to talk about electric flux through open two-dimensional surfaces (and indeed easiest to first explore flux in this way), we must keep in mind that *the electric flux of Gauss's law is the total over a closed three-dimensional surface.* Often the trick in successfully employing Gauss's law is to break up this surface into different parts, as we saw for cylindrical Gaussian surfaces.

The most frequent mistake made in applying Gauss's law is to assume that the electric field is constant along a surface when it is in fact not. Always ask yourself if the symmetry of the situation supports your assumption regarding the direction and magnitude of the electric field.

We have developed three techniques for finding the electric field: direct computation for point charges (including vector superposition for multiple point charges), integration of the electric field from a continuous distribution of charge, and now Gauss's law. You might think that we have exhausted the different ways to compute electric field. However, in the next chapter we will introduce the powerful concept of electric potential and see that it can be used to calculate electric field.

We can use the superposition principle with Gauss's law to find the electric field in situations that at first seem to lack sufficient symmetry. We itemize the steps below and apply the procedure in Example 20-9:

- Divide the charge distribution into several parts, each of which has symmetry that permits application of Gauss's law.

- Use Gauss's law to find the electric field at the desired point due to each of the charge parts.

- Use vector superposition to combine the electric fields into one net result.

CHECKPOINT 20-13

Point Charge in a Cube

A single positive point charge, $+Q$, is placed at the exact centre of a cube of side length L (see Figure 20-35).

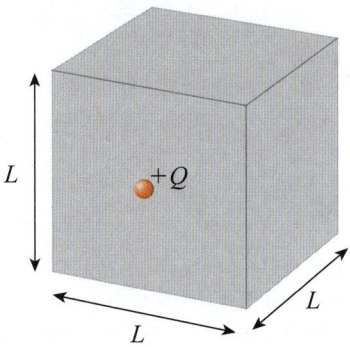

Figure 20-35 A single point charge, $+Q$, is at the exact centre of a cube. The length of each side of the cube is L.

(a) Is it possible to calculate the electric flux through one face of the cube? If so, what is the value?

(b) Can Gauss's law be used to find the electric field on the face of the cube? If so, what is the value?

ANSWER: (a) Yes. The enclosed charge is $+Q$, so by Gauss's law the *total electric flux* through all six faces of the cube is $\dfrac{Q}{\varepsilon_0}$. By symmetry, all the faces are equal, so the electric flux through any one face is just 1/6 of this value, $\dfrac{Q}{6\varepsilon_0}$.

(b) No. You might think we could use Gauss's law by drawing a cubical Gaussian surface, but we do not know the distribution of electric field strength across the faces, so we cannot take E out of the integral.

EXAMPLE 20-9

Electric Field from Plane Plus Point Charge

A very large (and thin) non-conducting plane slab carries a uniform surface charge density σ (assume positively charged). A single positive point charge, $+Q$, is located a distance d above the plane. What is the electric field at a point between the point charge and the plane and at a distance b from the plane?

Solution

The situation is drawn in Figure 20-36. We will use the superposition principle along with Gauss's law to solve this problem. If we consider only the charged plane, we can find the electric field due to the plane. Similarly, we can also find the result for only the point charge. We will refer to the electric field due to the point charge as $\vec{E}_Q$ and that due to the charged plane as $\vec{E}_\sigma$. We then use superposition to combine these two electric fields as vectors.

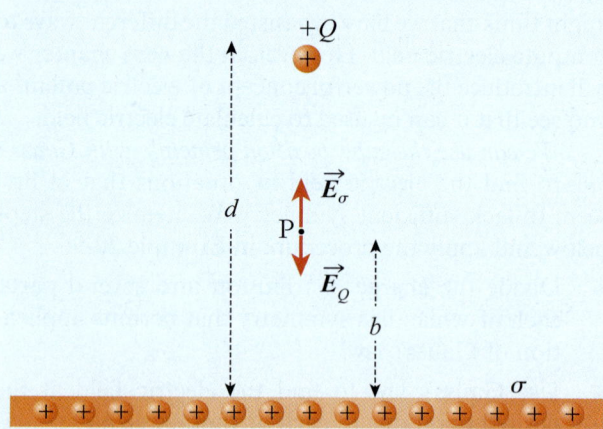

Figure 20-36 Situation of Example 20-9 with the electric field at point P shown separately due to the non-conducting plane of charge $\vec{E}_\sigma$ and the point charge $\vec{E}_Q$.

First consider only the thin slab of charge. There is planar symmetry, and we will draw a cylindrical Gaussian surface perpendicular to the plane. The cylinder length will be $2b$ (to make it symmetrical about the charged plane) and the cross-sectional area is A (which will cancel out in the end). The situation is exactly the one we considered in Example 20-7, so we will not repeat the details here:

$$\oiint_{\text{cylinder}} \vec{E} \cdot d\vec{A} = \frac{q_{\text{enc}}}{\varepsilon_0}$$

$$E_\sigma A + 0 + E_\sigma A = \frac{\sigma A}{\varepsilon_0}$$

$$E_\sigma = \frac{\sigma}{2\varepsilon_0}$$

The direction of this electric field is perpendicular to the plane and away from it (due to the positive charge), as shown in Figure 20-36.

For the electrical field due to only the point charge, we have symmetry about a point and draw a spherical Gaussian surface centred on the point charge and of radius $d - b$. We show this in Figure 20-37. This is similar to the case of Example 20-4(c).

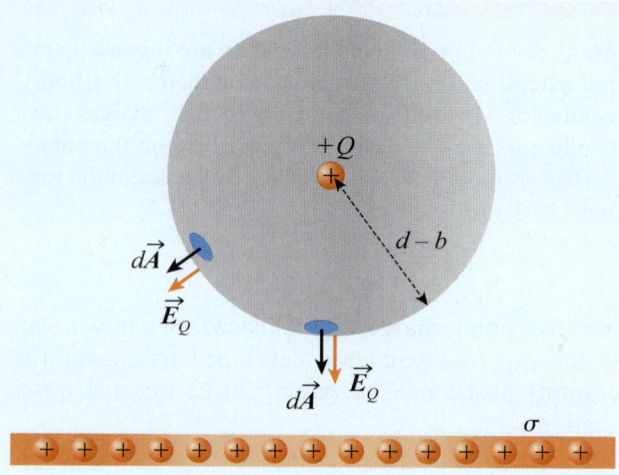

Figure 20-37 Example 20-9. Spherical Gaussian surface for the point charge. Note that we have shown just two of the electric field and differential area vectors, but they are everywhere parallel to each other.

At all points on this sphere, the electric field and the differential area vector are parallel, so the dot product simply becomes the product of the magnitudes. Also by symmetry, the electric field strength must be constant over the sphere. Therefore, we have the following by Gauss's law:

$$\oiint_{\text{sphere}} \vec{E} \cdot d\vec{A} = \frac{q_{\text{enc}}}{\varepsilon_0}$$

$$E_Q \oiint_{\text{sphere}} dA = \frac{q_{\text{enc}}}{\varepsilon_0}$$

$$E_Q 4\pi (d - b)^2 = \frac{Q}{\varepsilon_0}$$

$$E_Q = \frac{Q}{4\pi\varepsilon_0 (d - b)^2}$$

The direction of this electric field is away from the point charge, and therefore down toward the plane. We can use vector superposition to combine the two electric fields. If we define the z-axis as toward the top of the page, we can write the net electrical field as

$$\vec{E}_{\text{net}} = \vec{E}_\sigma + \vec{E}_Q$$

$$= \left(\frac{\sigma}{2\varepsilon_0} - \frac{Q}{4\pi\varepsilon_0 (d - b)^2} \right) \hat{k}$$

The net electric field could be either upward or downward, depending on the relative magnitudes of the point charge and the surface charge density.

Making sense of the result

If we can split up a charge distribution into two (or more) parts (each with sufficient symmetry), we can apply Gauss's law to each and then combine the electric field vectors. The technique shown here would not have been valid if the plane had been a conductor, since the charges on the plane would then have been free to move to a non-symmetrical arrangement due to the presence of the point charge.

While the superposition technique can be powerful, you have to be careful not to use it when it is not valid. If you have a conductor, the presence of another charged object will change the distribution of charges on the conductor, so we could not assume that the plane is still uniformly charged.

A clever application of the technique of using multiple parts involves situations involving voids in charged objects. For example, consider a uniformly charged solid sphere with a hole in one part of it. You could picture this as a combination of two objects, a uniformly charged sphere with no hole, and a second sphere with the size and location of the hole, but assumed to have a charge density of the same magnitude but opposite sign. Again, remember that we need to have non-conductors for this to be a valid approach.

In working with Gauss's law, always ask yourself what implicit assumptions you are making about charge distribution and symmetry of the electric field, and be sure that they are valid. Examine the "solutions" offered in Checkpoints 20-15 to 20-17, and try to identify what is wrong in each case.

CHECKPOINT 20-16

What Is Wrong? Long Triangular Object

As shown in Figure 20-39, a very long (assume near infinite) charged object has an equilateral triangle cross section. It is a non-conductor with uniform surface charge density σ on all surfaces (assume it is positively charged). What is the electric field just outside the surface?

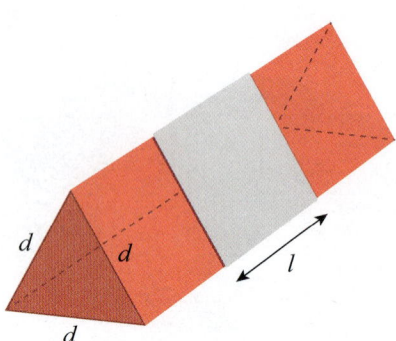

Figure 20-39 Checkpoint 20-16. A Gaussian surface (grey) is just barely outside the surface and of identical shape.

What is wrong with this solution? As shown in Figure 20-39, we draw a Gaussian surface of length ℓ also with an equilateral triangle cross section that is just slightly outside the charged object. By symmetry, we assume that the electric field strength, E, is uniform over all the plates and that it points outward (since positive charge) and perpendicular to the surface. We can use Gauss's law to solve for the electric field strength. Note that the surface area of each face of the Gaussian surface is $d\ell$, so the total surface area is three times this amount:

$$\oiint_{\text{surface}} \vec{E} \cdot d\vec{A} = \frac{q_{\text{enc}}}{\varepsilon_0}$$

$$E \oiint_{\text{surface}} dA = \frac{3\ell d\sigma}{\varepsilon_0}$$

$$E 3\ell d = \frac{3\ell d\sigma}{\varepsilon_0}$$

$$E = \frac{\sigma}{\varepsilon_0}$$

So the electric field is of this strength and is outward perpendicular to the surfaces. *What error have we made in this solution?*

ANSWER: Even though the charge density is constant on this non-conductor, that does not mean that the strength of the electric field is uniform here, and therefore we are not allowed to take E outside the surface integral. Integration of lines of charge (as we did in Chapter 19) would prove that the direction and strength of the electric field vary with position on the surface. This does not have symmetry matching the assumption we made.

CHECKPOINT 20-17

What Is Wrong? Point Charge and Charged Conducting Plane

As shown in Figure 20-40, we have a very large positively charged conducting plane. It has a surface area (one side) of A and carries total positive charge of $+Q$. At a distance d above the conducting plane, is a negative point charge, $-Q$. What is the net electric field at a point a distance $2d$ above the plane?

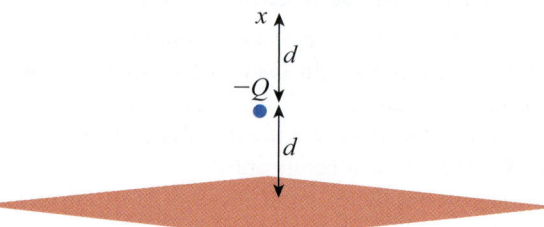

Figure 20-40 Checkpoint 20-17.

What is wrong with this solution? The combination of the point charge and the charged plane taken together does not have adequate symmetry to directly apply Gauss's law. However, we will split them into two parts, finding the electric field first due to just the charged plane, and then due to just the point charge. We then will combine the two results using the superposition principle. Let us first consider the charged conducting plane. It carries a total charge of $+Q$ over the entire plate. A single conducting plane by itself will spread the charge uniformly, including on both sides of the conducting plane. Therefore, the charge density on *one side* of the conducting plane is

$$\sigma = \frac{Q}{2A} \tag{1}$$

We derived in Example 20-8 the following result (note that σ is the surface charge density on one side) for the electric field strength at any distance from a conducting plane:

$$E = \frac{\sigma}{\varepsilon_0} \tag{20-16}$$

If we substitute (1) into Equation 20-16, we have the following result for the electric field strength:

$$E_{\text{plane}} = \frac{Q}{2A\varepsilon_0}$$

Because of the positive charge on the plane, the direction of this electric field at the indicated point will be away from the plane, i.e., upward. If we draw a y-axis in the vertical direction, calling positive upward, we can express the electric field due to the charged plane alone as

$$\vec{E}_{\text{plane}} = \frac{Q}{2A\varepsilon_0}\hat{j} \tag{2}$$

For the point charge, we can either use the expression from Chapter 19 for the electric field due to a point charge or draw a Gaussian spherical surface of radius d centred on the charge and apply Gauss's law. Either will yield the following result for the electric field at the point $2d$ above the plane:

$$\vec{E}_{\text{point}} = -\frac{1}{4\pi\varepsilon_0}\frac{Q}{d^2}\hat{j} \tag{3}$$

We can find the net electric field by combining results (2) and (3) using the superposition principle:

$$\vec{E}_{\text{net}} = \vec{E}_{\text{plane}} + \vec{E}_{\text{point}}$$

$$= \left(\frac{Q}{2A\varepsilon_0} - \frac{1}{4\pi\varepsilon_0}\frac{Q}{d^2}\right)\hat{j}$$

According to the relative size of A and d, the direction of the net electric field may be either up or down according to this relationship. *What error have we made in this solution?*

ANSWER: When we apply superposition, the presence of one charged object must not influence the charge distribution of the other. In this case, the presence of the negative point charge above the conducting plane will change the distribution of charges on the plane. Therefore, we cannot use superposition. If the plane had been non-conducting, then we could have used superposition and obtained the answer derived here.

Gauss's law forms the first of Maxwell's four equations of electromagnetism. With Gauss's law, we do not need Coulomb's law as well, as it can be deduced from Gauss's law and the definition of electric field as force per unit charge. As well as its central role in electromagnetism, Gauss's law and the associated idea of electric flux formed a framework that has been applied to a number of other areas of physics.

In the previous chapter, we stressed that the electric field is real, and not simply a way for humans to picture the force per unit charge. For example, electromagnetic waves transmit energy that is stored in the electric and magnetic fields and, as you will see in Chapters 27 and 30, transfer momentum as well. You will see in Chapter 22 that energy storage in a capacitor can be viewed similarly. Electric flux is essentially a way to measure the "amount" of electric field captured by a surface, so it is also a quantity with a real meaning. The fact that flux enters directly and prominently in Maxwell's equations, which specify all of electromagnetism (see Chapter 27), is one measure of the reality of electric flux.

20-9 Gauss's Law for Gravity

We saw in Chapter 19 that the acceleration due to gravity, $\vec{g}$, can be thought of as a gravitational field analogous to the electric field, $\vec{E}$. The gravitational field is the gravitational force per unit mass, while the electric field is the electrical force per unit charge. While we often use the gravitational acceleration at the surface of Earth, we can calculate the magnitude of $\vec{g}$ at

any location. We showed in Section 11-2 that it is given by the following expression

$$g = \frac{GM}{r^2} \tag{11-4}$$

where G is the universal gravitational constant and r is the radial distance from the centre of a radially symmetric large mass, M.

For this chapter, we will need the direction as well as the magnitude. This can be written in terms of the unit vector $\hat{r}$ that points radially outward from the centre of the large mass, M:

$$\vec{g} = -\frac{GM}{r^2}\hat{r} \tag{20-17}$$

We define a **gravitational flux**, Φ_g, in an analogous way to the electric flux of Equation 20-2. The integral is over a closed surface, and it is a vector dot product between the gravitational field $\vec{g}$ and an outward-pointing differential surface element $d\vec{A}$:

KEY EQUATION

$$\Phi_g = \oiint_{\text{surface}} \vec{g} \cdot d\vec{A} \tag{20-18}$$

The gravitational flux is a scalar quantity that can take on positive and negative values. You should think of $\vec{g}$ as a gravitational field here, rather than an acceleration, although they have the same numerical value. We call the magnitude of $\vec{g}$ the **gravitational field strength**, g.

We will now calculate the gravitational flux around Earth. For a radially symmetric planet, we have spherical symmetry, and our closed surface is a sphere. In Figure 20-41 we illustrate Earth and a larger concentric Gaussian sphere. As long as the mass of Earth is distributed in a radially symmetric way, the gravitational field must have constant magnitude (but not direction) all around the closed Gaussian surface. Note that

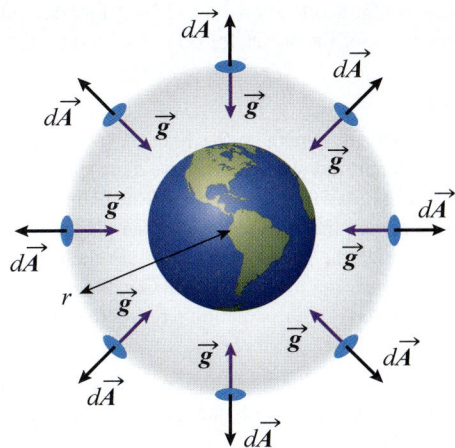

Figure 20-41 For the case of a Gaussian sphere around a radially symmetric Earth, the gravitational field $\vec{g}$ and the differential area vectors $d\vec{A}$ are directly opposite each other.

everywhere the gravitational field $\vec{g}$ and the differential surface vector $d\vec{A}$ are in exactly opposite directions.

The gravitational flux can be calculated as shown, where g is the scalar gravitational field strength at distance r from the centre of Earth:

$$\Phi_g = \oiint_{\text{sphere}} \vec{g} \cdot d\vec{A}$$

$$\Phi_g = \oiint_{\text{sphere}} g\,dA\cos 180°$$

$$\Phi_g = -g \oiint_{\text{sphere}} dA$$

$$\Phi_g = -g4\pi r^2$$

Now we can substitute the value of the gravitational field strength, g, at distance r using Equation 11-4:

$$\Phi_g = -\left(\frac{GM}{r^2}\right)4\pi r^2$$

$$\Phi_g = -4\pi GM$$

(20-19)

We have only developed this for the case of a spherically symmetric mass, but it turns out we can generalize the result from Equation 20-19 to obtain **Gauss's law for gravity**: the gravitational flux through any closed

EXAMPLE 20-10

Gravitational Field inside and outside a Homogeneous Planet

Differentiated planets such as Earth, with an iron core, have higher mass densities near the core. However, in this problem, assume a spherical planet of radius R_p and total mass M_p that has uniform mass density throughout. Use Gauss's law for gravity to find expressions for the acceleration due to gravity for points at distance r from the centre of the planet, for (a) $r > R_p$ and (b) $r < R_p$.

Solution

We have spherical symmetry. To find the acceleration due to gravity at different radial distances, r, we draw spherical Gaussian surfaces at distance r from the centre of the planet. When we are outside the planet, the enclosed mass is simply the total mass of the planet. Inside the planet, we have to use the mass density to find the mass enclosed by the spherical Gaussian surface. In both cases, the symmetry of the situation requires that the direction of the acceleration due to gravity be radially inward.

(a) $r > R_p$: Here we have a situation similar to that of Figure 20-41 and draw a Gaussian sphere concentric with the planet, but larger:

$$\oiint_{\text{surface}} \vec{g} \cdot d\vec{A} = -4\pi GM_{\text{enc}}$$

Since $\vec{g}$ and $d\vec{A}$ are in opposite directions, $\vec{g} \cdot d\vec{A} = g\,dA\cos 180° = -g\,dA$. From the symmetry of the situation, the gravitational field strength is constant at any particular radial distance and can therefore be taken outside the integral:

$$\oiint_{\text{surface}} \vec{g} \cdot d\vec{A} = -4\pi GM_p$$

$$-\oiint_{\text{surface}} g\,dA = -4\pi GM_p$$

$$-g\oiint_{\text{surface}} dA = -4\pi GM_p$$

$$-4\pi r^2 g = -4\pi GM_p$$

Therefore, $g = -\dfrac{GM_p}{r^2}$, or, in vector form since the gravitational field is directed radially inward,

$$\vec{g} = -\frac{GM_p}{r^2}\hat{r}$$

(b) $r < R_p$: In this case, we draw a concentric spherical Gaussian sphere of radius r inside the planet (see Figure 20-42).

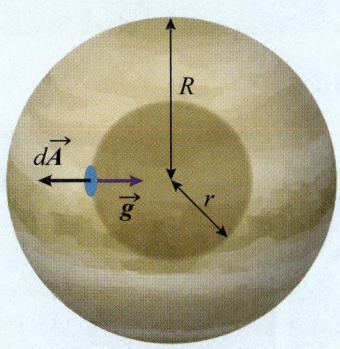

Figure 20-42 Example 20-10. To calculate the gravitational field at a point inside the planet, we draw a concentric spherical Gaussian surface (the darker smaller sphere) of radius r less than the radius of the planet. Everywhere on this Gaussian surface, $\vec{g}$ will have a constant magnitude and will be directly opposite in direction to $d\vec{A}$.

First we calculate the mass density of the spherical planet:

$$\rho = \frac{M_p}{V} = \frac{M_p}{\frac{4}{3}\pi R^3} = \frac{3M_p}{4\pi R^3} \qquad (1)$$

The gravitational field $\vec{g}$ points radially inward and the differential area element $d\vec{A}$ radially outward:

$$\vec{g} \cdot d\vec{A} = gdA\cos 180° = -gdA \qquad (2)$$

We know from the symmetry of the situation that the gravitational field strength is constant at any particular radial distance. Therefore, we compute the gravitational flux using result (2). The integral over the differential area elements for the Gaussian surface simply yields the surface area of a sphere of radius r:

$$\Phi_g = \oiint_{\text{sphere}} \vec{g} \cdot d\vec{A}$$

$$= -g \oiint_{\text{sphere}} dA \qquad (3)$$

$$\Phi_g = -g4\pi r^2$$

For the Gaussian surface inside the planet, M_{enc} is not the entire mass but rather just the part of the mass inside the Gaussian surface. We can calculate it using the density we found earlier in (1):

$$M_{\text{enc}} = \frac{4}{3}\pi r^3 \rho$$

$$= \frac{4}{3}\pi r^3 \frac{3M_p}{4\pi R^3} \qquad (4)$$

$$M_{\text{enc}} = M_p \frac{r^3}{R^3}$$

Therefore, Gauss's law for gravity in this case becomes the following, using results (3) and (4):

$$\Phi_g = -4\pi GM_{\text{enc}}$$

$$-g4\pi r^2 = -4\pi GM_p \frac{r^3}{R^3}$$

$$g = \frac{GM_p r}{R^3}$$

Since the acceleration due to gravity is radially inward, we can write this in vector form, using the outward-pointing radial unit vector, as

$$\vec{g} = -\frac{GMr}{R^3}\hat{r}$$

Making sense of the result

The result we obtained in part (b) reduces to the result for gravitational acceleration at the surface of a planet when we set $r = R$. When we are inside the surface of the planet, we get the interesting result that (for this homogeneous-density planet) the strength of the gravitational field is directly proportional to the distance from the centre of the planet. Right at the centre, as expected, the relationship yields a value of zero.

surface is proportional to the mass enclosed by the closed surface (with an opposite sign):

KEY EQUATION
$$\Phi_g = -4\pi GM_{\text{enc}} \qquad (20\text{-}20)$$

Gauss's law for gravity is frequently written in a form that includes the gravitational flux relationship:

$$\oiint_{\text{surface}} \vec{g} \cdot d\vec{A} = -4\pi GM_{\text{enc}} \qquad (20\text{-}21)$$

Just as we saw for the electrical case, while Gauss's law for gravity applies no matter the shape of the closed surface, it is only helpful in calculating the gravitational

field strength when there is sufficient symmetry that we can choose a surface with constant field strength. The same considerations we used in the electrical case for the type of symmetry and the corresponding shape of Gaussian surface to draw apply here.

In Example 20-10, we show the application of Gauss's law for a situation that would not have been easy to find directly using integration and the law of universal gravitation.

Gauss's law for gravity can also be applied to cases with cylindrical and planar symmetry, and some of the end-of-chapter problems will give you opportunities to apply it with these symmetries. With superposition it can be used in more complex cases, as required in one of the open problems.

Evidence for Dark Matter

It appears that ordinary matter makes up less than 5% of the universe (see Figure 20-43). A much larger amount of the material in the universe is so-called **dark matter**, which we do not directly observe. In addition, a large part of the mass-energy of the universe is something else, called dark energy, that we will explore in Chapters 30 and 35.

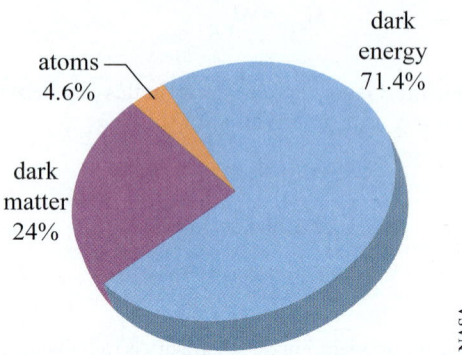

Figure 20-43 The composition of the universe shows that there is far more dark matter than ordinary matter.

While there are a number of pieces of evidence for dark matter, strong early evidence was obtained in the 1970s by American astrophysicist Vera Rubin Cooper (1928–2016). She measured rotation rates of stars at different radial distances for nearby galaxies. As we show on the right of Figure 20-44, we would expect the rotational speed to decrease with radial distance, but it is observed to be almost constant at large distances. This is evidence for the presence of large amounts of dark matter concentrated in the outer regions of galaxies.

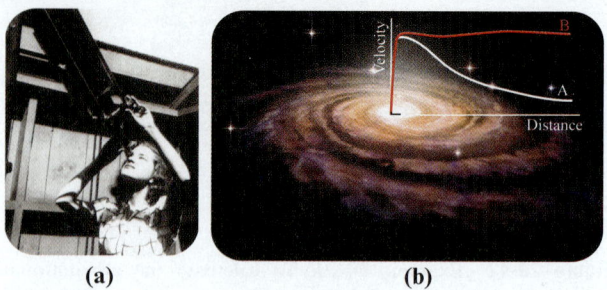

Figure 20-44 (a) Vera Rubin Cooper at the Kitt Peak telescope used for her observations. (b) The simplified plot shows the observed and predicted motion of stars at different radial distances.

Sources: (a) Emilio Segrè Visual Archives / American Institute of Physics / Science Source; (b) Alex Mit/Shutterstock.com. Data from PhilHibbs, https://en.wikipedia.org/wiki/Gravity

If we assume that the mass in a galaxy is spherically symmetric, we can at any particular radius r draw a spherical Gaussian surface and apply Gauss's law for gravity to get the result $-4\pi r^2 g = -4\pi G M_{\text{enc}}$. If we relate the gravitational field to that required to provide the centripetal acceleration, we have $g = v^2/r$. Combining these results, we find the speed as a function of radial distance and enclosed mass:

$$v = \sqrt{\frac{GM_{\text{enc}}}{r}}$$

The only way that the speed can stay almost constant at large radial distances is if the enclosed mass keeps getting larger roughly proportional to r. Since there is not significant visible matter at large radial distances, the implication is that there must be large amounts of dark matter in the outer regions of galaxies.

KEY CONCEPTS AND RELATIONSHIPS

Electric Flux

Electric flux is proportional to the number of electric field lines captured by a surface. It is computed over a closed surface by the relationship

$$\Phi_{\text{E}} \equiv \oiint_{\text{surface}} \vec{E} \cdot d\vec{A}, \tag{20-3}$$

where $\vec{E}$ is the electric field and $d\vec{A}$ is a differential surface element perpendicular to the surface. For a closed surface, differential element $d\vec{A}$ is defined as outward pointing and perpendicular to the surface.

Gauss's Law

Gauss's law states that the electrical flux over any closed surface equals the net charge enclosed by that surface divided by the constant permittivity of free space:

$$\Phi_{\text{E}} = \frac{q_{\text{enc}}}{\varepsilon_0} \tag{20-4}$$

Most of the time we combine the definition of electric flux in Gauss's law:

$$\Phi_{\text{E}} \equiv \oiint_{\text{surface}} \vec{E} \cdot d\vec{A} = \frac{q_{\text{enc}}}{\varepsilon_0} \tag{20-5}$$

The electric flux is integrated over some closed surface called a Gaussian surface. The vector dot product is taken between the electric field and a differential surface element, $d\vec{A}$, perpendicular to the surface and pointing outward. The total net electric charge enclosed by the closed surface is q_{enc}, and ε_0 is a constant of nature called the permittivity of free space.

Symmetry Types and Gaussian Surfaces

Gauss's law is useful for calculating the electric field in symmetrical situations. We show the three types of symmetry, and the Gaussian closed surface that we use for each, in Table 20-2.

Table 20-2 Type of Symmetry and the Corresponding Gaussian Surface That Should Be Used

Symmetry	Type of Gaussian Surface	Examples
spherical (about a point)	sphere concentric with point	point charge charged sphere spherical shell
cylindrical (about a line)	cylinder coaxial with line	line of charge charged cylinder coaxial cylindrical shell
planar (about a plane)	cylinder or box perpendicular to plane	charged parallel plane(s) large flat object

Strategy for Application of Gauss's Law

The following strategy can be used to employ Gauss's law to find the electric field.

- First sketch the situation, including the direction of the anticipated electric field.
- From this sketch, determine the type of symmetry, which will set the shape of the Gaussian surface.
- The size of the Gaussian surface is determined by the need for the point where the electric field strength is desired to lie on the Gaussian surface.
- If the electric field strength is not uniform over the entire Gaussian surface, divide the surface into parts, with each part having a constant electric field.
- Indicate the directions of the outward-pointing area elements perpendicular to the Gaussian surface, and from this compute the electric flux over the closed surface.
- Find the net electric charge enclosed by the Gaussian surface, and relate that and the electric flux using Gauss's law.
- Express your final answer for the electric field in vector form, and make sure that the result is reasonable.

When Gauss's Law Is Useful

While Gauss's law holds for any situation and any choice of closed surface, it is useful for computing the electric field only when the electric field strength is constant over the chosen surface and the direction of the electric field relative to the area vector is uniform. When necessary, we can break the Gaussian surface up into different parts, each with constant electric field strength.

Electric Fields and Conductors

In static situations, the charges on conductors reside on the outer surface, and the electric field inside the conductor is zero. The electric field at the surface of the conductor is perpendicular to the conductor. Charges on irregularly shaped conductors are not uniformly distributed, and the electric field strength just outside irregularly shaped conductors is not uniform.

Superposition

Sometimes you can divide a charge distribution into two or more parts, each of which has a symmetry permitting application of Gauss's law. Then vector superposition is used to combine the two electric fields. In general, this cannot be done with conductors, however, since the charges on one can move in response to the presence of the other.

Gauss's Law for Gravity

Gauss's law can be applied in gravitational situations where the enclosed mass, rather than the enclosed charge, is used. The acceleration due to gravity can be considered a gravitational field, and analogous to the electric flux we can calculate the gravitational flux over a closed surface:

$$\Phi_g = \oiint_{\text{surface}} \vec{g} \cdot d\vec{A} \tag{20-18}$$

Differential surface element $d\vec{A}$ is perpendicular to the closed surface and outward pointing. Gauss's law for gravity is used to calculate the gravitational field, $\vec{g}$, in certain symmetrical situations:

$$\Phi_g = -4\pi G M_{\text{enc}} \tag{20-20}$$

Here M_{enc} is the total mass enclosed by the Gaussian surface.

APPLICATIONS

atmospheric and planetary electricity, charging by induction, electric field measurement, electric flux sensing, Faraday cage, gravitation, nanotechnology electric fields, producing electric fields of controlled strength, shielding from electromagnetic fields

KEY TERMS

closed surface, cylindrical symmetry, dark matter, electric flux, flux, Gauss's law, Gauss's law for gravity, Gaussian surface, gravitational field strength, gravitational flux, open surface, planar symmetry, spherical symmetry, surface integral, vector field

QUESTIONS

1. Which one of the following statements about Gauss's law is *not* true?
   (a) Gauss's law is only true in situations with a lot of symmetry.
   (b) We can use Gauss's law to calculate the electric field in different regions.
   (c) Gauss's law (in different forms) can be applied in both electrical and gravitational situations.
   (d) You still need Coulomb's law to find forces, even with Gauss's law.

2. You have three charges in a line: the left charge is $+12\ \mu C$, the centre is $-20\ \mu C$, and the right is $+8\ \mu C$. An electric field line diagram is drawn. Which statement is true for that diagram?
   (a) Field lines go from each of the left and right positive charges toward the negative charge in the middle, but some also go out to infinite distance.
   (b) Field lines end on the two positive charges and start on the one negative charge, with the number starting on the negative charge equal to the sum of the two ending on the positive charges.
   (c) More field lines start on the $+12\ \mu C$ charge than on the $+8\ \mu C$ charge (one and a half times as many), and all the lines that start on the positive charges end on the negative charge.
   (d) An equal number of lines start or end on each charge, with the lines starting on positive charges and ending on negative, with the extra lines going off to infinity.

3. Consider positive charge uniformly spread over a sphere. In the region outside this sphere, which statement correctly describes the electric field and electric flux on Gaussian spheres of different radii?
   (a) The electric field decreases in magnitude at larger distances, and the electric flux is zero everywhere.
   (b) The electric flux is constant (but not zero) for all radial distances, but the electric field decreases with increasing distance.
   (c) Both the electric flux and the electric field are constant at all distances outside the sphere.
   (d) The electric flux decreases with increasing distance, but the electric field is constant everywhere.

4. There are two concentric charged conducting spherical shells, the inner one with a total charge of $+80\ \mu C$ and the outer one with a total charge of $+20\ \mu C$. What is the direction of the electric field inside the inner shell and in the region between the two shells?
   (a) Inside the inner shell the electric field is zero, and in the region between the two shells it is radially outward.
   (b) Inside the inner shell the electric field is zero, and in the region between the two shells it is radially inward.
   (c) It is radially inward in both regions.
   (d) It is radially outward in both regions.

5. A spherical shell with total charge $+2Q$ is concentric with a smaller sphere with total charge $-Q$. In both cases, the charge is uniformly spread. What is the direction of the electric field in the region between the inner sphere and the outer spherical shell, and in the region outside the spherical shell?
   (a) Inside the shell (but outside the sphere) the electric field is radially outward, and in the region outside the shell it is radially inward.
   (b) Inside the shell (but outside the sphere) the electric field is radially inward, and in the region outside the shell it is radially outward.
   (c) It is radially inward in both regions.
   (d) It is radially outward in both regions.

6. Which of the following statements is correct?
   (a) Electric field is a vector, and electric flux is a scalar.
   (b) Electric field is a scalar, and electric flux is a vector.
   (c) Electric field and electric flux are both scalars.
   (d) Electric field and electric flux are both vectors.

7. A small positively charged sphere is at the origin of a coordinate system. You centre two Gaussian surfaces on the charge, one a sphere of radius R and the other a sphere of radius $2R$. The charged sphere has a radius smaller than R. How does the flux compare in the two cases?
   (a) It is exactly the same in both cases.
   (b) It is double in the case of the larger Gaussian surface.
   (c) It is about 1.4 times as much in the case of the larger Gaussian surface.
   (d) It is 4 times as much in the case of the larger Gaussian surface.

8. Your friend says: "My prof claims that Gauss's law is true no matter what the shape of the surface used, so I am going to always draw heart-shaped Gaussian surfaces when I use Gauss's law to calculate electric fields." What is wrong with that reasoning?

9. You try to use Gauss's law for the electric field in the space between the two charges of an electric dipole. It seems you could draw a spherical Gaussian surface that contains only one of the charges, and use that to calculate the electric field at P (see Figure 20-45), and the other charge does not come into it at all. What is wrong, if anything, with the method? *Assume that we desire an answer with a single integration, not invoking superposition techniques.*
   (a) Nothing is wrong.
   (b) Gauss's law can only be used with continuous distributions of charge, not with point charges.
   (c) The situation does not have spherical symmetry, and the electric field is not constant all along that spherical Gaussian surface.
   (d) A dipole refers to a positive and negative charge at the same point, so it makes no sense talking about the electric field in the space between the charges.

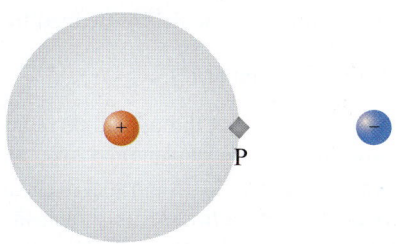

Figure 20-45 Question 9.

10. If the electric flux on some surface is zero, what can we conclude?
    (a) The electric field must also be zero.
    (b) The electric field may be zero but is not necessarily zero.
    (c) The electric field is constant in magnitude but is not necessarily zero.
    (d) The electric field must be perpendicular to the surface.

11. A single positive charge is placed inside a small cube. What can be said about the net electric flux over the surfaces of the cube?
    (a) The net electric flux is negative, and the value depends on where in the cube the charge is placed.
    (b) The net electric flux is positive, and the value depends on where in the cube the charge is placed.
    (c) The net electric flux is negative, and the value is independent of the position of the charge inside the cube.
    (d) The net electric flux is positive, and the value is independent of the position of the charge inside the cube.

12. A positive ion contains a nuclear charge of $+5e$ and has an electronic charge of $-4e$. What is the electric field at a distance r well outside the ion? Assume spherical symmetry.
    (a) The electric field is $\dfrac{1}{4\pi\varepsilon_0}\dfrac{e}{r^2}$ in a direction radially inward toward the ion.
    (b) The electric field is $\dfrac{1}{4\pi\varepsilon_0}\dfrac{e}{r^2}$ in a direction radially outward away from the ion.
    (c) The electric field is $\dfrac{1}{4\pi\varepsilon_0}\dfrac{5e}{r^2}$ in a direction radially inward toward the ion.
    (d) The electric field is $\dfrac{1}{4\pi\varepsilon_0}\dfrac{9e}{r^2}$ in a direction radially outward away from the ion.

13. You draw a spherical Gaussian surface just outside an electric dipole (equal positive and negative charges separated by a small distance). What can you conclude about the electric flux through that surface?
    (a) As long as the dipole is completely inside the surface, the flux is zero.
    (b) The flux is only zero when the surface is centred on the midpoint of the dipole.
    (c) The flux is approximately but not exactly zero, even when the surface is centred on the midpoint of the dipole.
    (d) None of the above are true.

14. Some charge is placed on a closed (and hollow) tin can. Where is the charge located?
    (a) The charge is located equally throughout the metal of the can.
    (b) All the charge is on the inside surface.
    (c) All the charge is on the outside surface.
    (d) All the charge is on the two ends (none on the curved surface).

15. Consider two solid spheres, one conducting and one non-conducting. Both spheres carry the same total charge and have equal radii. The non-conducting sphere has a uniform volume charge density. Which statement correctly describes the electric field just inside and just outside the surface of each sphere?
    (a) Both have equal electric fields inside and just outside.
    (b) The conducting sphere has zero electric field inside and a larger electric field outside than the non-conducting sphere.
    (c) The conducting sphere has zero electric field inside and an electric field just outside equal to that in the non-conducting sphere.
    (d) The conducting sphere has zero electric field inside and a smaller electric field outside (compared to the non-conducting sphere).

16. When the charge on a conducting sphere is doubled, how does the electric field *inside* the sphere change?
    (a) The electric field is now four times as large.
    (b) The electric field is now two times as large.
    (c) The electric field is now approximately $\sqrt{2}$ times as large.
    (d) The electric field is zero in both cases.

17. Two planets have the same radius and the same overall density. However, planet A has a higher density in the central core, while planet B has a uniform density. Considering Gauss's law for gravity, how would the acceleration due to gravity compare? Assume that a deep hole has been drilled in each planet to perform measurements of $\vec{g}$ at various interior points.
    (a) Surface gravity would have the same value for both planets, but at points deep in the planet $\vec{g}$ would be higher in B.
    (b) Surface gravity would have the same value for both planets, but at points deep in the planet $\vec{g}$ would be higher in A.
    (c) Surface gravity would be higher in A, but at points deep in the planet $\vec{g}$ would be higher in B.
    (d) Surface gravity would be higher in B, but at points deep in the planet $\vec{g}$ would be higher in A.

18. Three concentric spherical shells carry charges of -1 mC, $+2$ mC, and -3 mC. The radius of the outer shell is 25 cm. What is the electric field just outside the outer shell ($\hat{r}$ is an outward-pointing radial unit vector)? The answer is expressed in SI units.
    (a) $-\dfrac{1}{500\pi\varepsilon_0}\hat{r}$
    (b) $+\dfrac{1}{500\pi\varepsilon_0}\hat{r}$
    (c) $-\dfrac{1}{125\pi\varepsilon_0}\hat{r}$
    (d) $+\dfrac{1}{125\pi\varepsilon_0}\hat{r}$

19. Three concentric spherical shells carry charges of $+4$ mC, $+3$ mC, and $+1$ mC. The radius of the outer shell is 10. cm. What is the electric field (in SI units) just outside the outer shell ($\hat{r}$ is an outward-pointing radial unit vector)?

 (a) $-\dfrac{1}{5\pi\varepsilon_0}\hat{r}$

 (b) $+\dfrac{1}{5\pi\varepsilon_0}\hat{r}$

 (c) $-\dfrac{1}{25\pi\varepsilon_0}\hat{r}$

 (d) $+\dfrac{1}{25\pi\varepsilon_0}\hat{r}$

20. A solid long conducting cylinder carries a net negative charge. Which statement correctly describes the electric field?

 (a) The electric field is radially outward inside the conductor and is zero outside.

 (b) The electric field is radially inward inside the conductor and is zero outside.

 (c) The electric field is zero inside the conductor and radially outward outside the conductor.

 (d) The electric field is zero inside the conductor, and radially inward outside the conductor.

PROBLEMS BY SECTION

For problems, star ratings will be used (★, ★★, or ★★★), with more stars meaning more challenging problems. The following codes will indicate if $^d/_{dx}$ differentiation, ∫ integration, ▢ numerical approximation, or ∿ graphical analysis will be required to solve the problem.

Section 20-1   Gauss's Law and Electric Field Lines

21. ★ Consider two very long cylindrical shells. The inner shell has radius R and carries a positive charge of $+Q$, while the outer cylindrical shell is of radius $2R$ and is negatively charged with $-Q$. Draw an electric field diagram for the situation (in a cross-sectional view).

22. ★★ In an electric field line diagram, 40 lines leave object A, with 20 of them ending on object B and the other 20 going off to infinite distance. What can you conclude about the charge on objects A and B?

23. ★★ An inner sphere is surrounded by a larger-radius spherical shell. In an electric field diagram, you note that 30 lines leave the inner sphere and end on the inside of the spherical shell. There also are 90 electric field lines that come in from an infinite distance and end on the outside of the shell.

 (a) What can you conclude about the charges (both signs and relative magnitudes) on the sphere and the shell?

 (b) What is the direction of the electric field, if any, in the region between the sphere and the shell and in the region outside the shell?

Section 20-2   Electric Flux

24. ★★ In a certain region, the electric field points in the $+y$-direction and has a magnitude of 35.0 N/C. What is the electric flux through a circle of radius 1.00 cm that lies in the xz-plane?

25. ★★ A constant electric field is defined by the following vector:

$$\vec{E} = 4.00 \text{ N/C}\,\hat{i} + 25.0 \text{ N/C}\,\hat{k}$$

What is the electric flux through a 1.00 m by 1.00 m rectangular area that is in the xy-plane and is centred on the origin?

26. ★★ An electric field, $\vec{E}$, is parallel to the axis of a hemispherical shell of radius R, as shown in Figure 20-46. What is the electric flux through the surface?

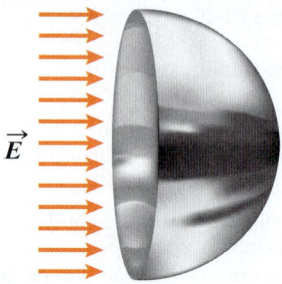

Figure 20-46   Problem 26.

27. ★★ Consider a cube with equal sides of length $L = 1.00$ m oriented as shown in Figure 20-47. A uniform electric field of $\vec{E} = 3.50$ N/C $\hat{i}$ is acting on the cube. Consider the surface of the cube as a closed surface. Calculate the electric flux for each face of the cube, as well as the total electric flux for the entire closed surface of the cube.

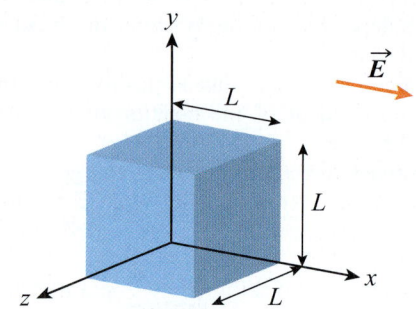

Figure 20-47   Problem 27.

28. ★★ ∫ The electric field in a certain region, when both x and E are expressed in SI units, is given by

$$\vec{E} = \frac{2}{x^2}\hat{k}$$

What is the electric flux through a flat plate in the xy-plane extending from ($x = 1.00$ m, $y = 1.00$ m) to ($x = 2.00$ m, $y = 3.00$ m)? Assume that $d\vec{A}$ is oriented in the $+\hat{k}$ direction.

29. ★★ ∫ The electric field in a certain region, when both r and E are expressed in SI units, is given by

$$\vec{E} = 4r^2\hat{k}$$

What is the electric flux through a flat circular plate of radius 1.00 m oriented in the xy-plane? Assume that $d\vec{A}$ is oriented in the $+\hat{k}$ direction.

Section 20-3   Gauss's Law

30. ★★ ∫ A $+25.0$ nC point charge is surrounded by a concentric conducting spherical shell of radius 2.40 cm that carries a charge of -20.0 nC. Derive expressions for the electric field as a function of radial distance both inside and outside the shell.

31. ★★ ∫ A solid, non-conducting sphere of radius r_0 has a uniform volume charge density ρ (assume a positive charge). Derive expressions for the electric field as a function of radial distance, r, both inside and outside the sphere, and provide a plot of the results.

32. ★ An unknown point charge has a Gaussian sphere of radius $R = 0.200$ m drawn around it, and the total flux through that closed surface is $+2500$ N·m²/C.
    (a) Is the charge enclosed by the surface positive or negative? What is its magnitude?
    (b) If we replace the spherical Gaussian surface with a cube of side length R, what is the total electric flux through the surfaces?

33. ★ A non-conducting sphere has a radius of 2.75 cm and a total charge of $+3.00\ \mu$C uniformly spread throughout the sphere. What is the volume charge density?

34. ★★ ∫ A -5.00 nC point charge is located at the centre of a charged spherical shell of radius 45.0 cm. You know that the charge on the shell is uniformly distributed. An outward-pointing electric field of 1.65 kN/C is measured just outside the surface of the shell. What must be the charge on the shell?

35. ★★ ∫ The electric field near Earth's surface points downward and has a magnitude of approximately 150 N/C. If the field were due to a thin layer of charge spread evenly over Earth's surface, what would the magnitude and sign of the electric charge be?

Section 20-4 Strategy for Using Gauss's Law

36. ★ A non-conducting cube (all sides are equal) has a uniform charge density on all surfaces. Can you use Gauss's law to find the electric field by drawing a Gaussian surface that is cube shaped and just slightly larger than the cube? If not, explain why not.

37. ★ For each of the following, state if a Gaussian surface can be drawn that would be useful in calculating the electric field just outside the object, and if so what shape that would be.
    (a) a plus sign–shaped charged conductor
    (b) a very long flat ribbon type conductor
    (c) an insulating sphere with a second spherical shell concentric with it
    (d) a dumbbell-shaped conductor
    (e) a uniformly charged insulating ellipsoid (3D oval)

38. ★★ ∫ You have three concentric charged (but non-conducting) spherical shells. They are of radii 10.0 cm, 25.0 cm, and 40.0 cm. The charges on them are $+2.00$ nC, -3.00 nC, and $+6.00$ nC, respectively. Find the electric field just outside each of the three shells.

Section 20-5 Gauss's Law for Cylindrical Symmetry

39. ★ A charge of $+165\ \mu$C is uniformly distributed around a very thin ring of radius 4.00 cm. What is the linear charge density along the ring?

40. ★★ ∫ A long charged line is of length 25.0 m and radius 2.40 mm and carries a total charge of $+36.0\ \mu$C. What is the electric field at a point near the middle of the wire and offset by 5.00 cm from the wire?

41. ★★ ∫ A very long solid, non-conducting cylinder of radius r_0 has uniform charge density ρ. Derive expressions for the electric field as a function of radial distance, r, both inside and outside the cylinder. Plot the electric field strength as a function of radial distance, including the results for both inside and outside the cylinder.

42. ★★★ ∫ Two long, concentric conducting cylinders have radii 4.50 cm and 8.50 cm (Figure 20-48). The inner cylinder carries a charge per unit length of $+15.0\ \mu$C/m, and the outer cylinder carries a charge per unit length of $-10.0\ \mu$C/m. Derive and plot expressions for the electric field as a function of radial distance, r, in all three regions: inside the inner cylinder, the space between the two cylinders, and outside the outer cylinder.

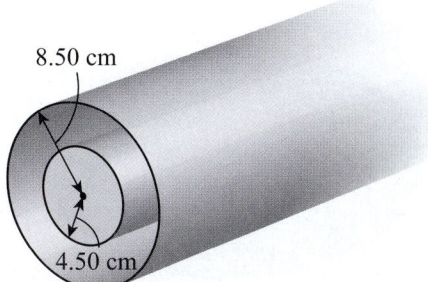

8.50 cm

4.50 cm

Figure 20-48 Problem 42.

Section 20-6 Gauss's Law for Planar Symmetry

43. ★ A thin rectangular plate has dimensions 3.50 cm × 12.0 cm and carries a charge density of 45.0 μC/m². How many elementary charge units (i.e., 1.60×10^{-19} C) are on the entire plate?

44. ★★ ∫ (a) You measure an electric field of 350 N/C directed toward a very large flat non-conducting charged sheet from a point 15.0 cm away from the sheet. What is the surface charge density on the sheet?
    (b) What is the electric field at a point 30.0 cm from the sheet?

45. ★★ ∫ A non-conducting rectangular plate carries a uniformly distributed charge of 600.0 nC. The plate is 18.00 cm long, 40.00 cm wide, and 0.500 cm thick.
    (a) Use Gauss's law to approximate the electric field at a point 1.00 cm above the surface of the plate.
    (b) What is the approximate electric field at a point 2.00 cm above the plate?

46. ★★ ∫ For an experiment, you want the electric field to point perpendicularly away from a charged rectangular plate and to have a magnitude of 75.0 kN/C at a point only 2.50 mm above the plate. Assume that the plate is non-conducting and has dimensions 30.0 cm by 50.0 cm. How much charge needs to be added to the plate? Do your solution from first principles (rather than using results derived in the chapter).

47. ★ For symmetry about a plane, we stated that the Gaussian surface could either be a cylinder perpendicular to the surface or a cube. Why does it not matter which you use?

48. ★ Explain with an electric field line diagram why it is necessary for the electric field from an infinite plane of charge to not depend on the distance away from that plane.

Section 20-7 Conductors and Electric Fields

49. ★ What is one example of a Faraday cage present in your kitchen?

50. ★★ ∫ You have a solid conducting sphere of radius R and with total positive charge Q.
   (a) What is an expression for the electric field at radial distances $r < R$?
   (b) What is an expression for the electric field at radial distances $r > R$?
   (c) What is the surface charge density on the sphere?

51. ★ A conducting solid oval carries a negative charge (Figure 20-49). Sketch the electric field vectors, if any, (a) inside, (b) on the surface, and (c) slightly away from the surface.

Figure 20-49 Problem 51.

52. ★★ ∫ A conducting spherical shell extends from a radius of R to $2R$. It has positive surface charge density σ. Derive expressions for the electric field as a function of radial distance for $r < R$, $R < r < 2R$, and $r > 2R$.

53. ★★ ∫ A spherical conductor of radius R has no *net* charge on it. However, there is a cavity inside it of radius a and concentric with the conductor and there is a point charge, $+Q$, in the centre of that cavity. The situation is shown in Figure 20-50.

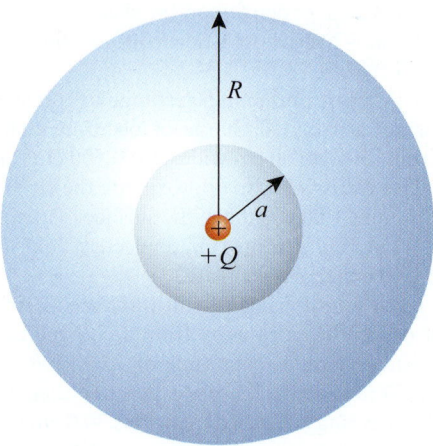

Figure 20-50 Problem 53.

   (a) Is there any induced charge on the inner surface of the conductor? If so, work out the surface charge density.
   (b) Is there a surface charge on the outside surface of the conductor? If so, what is the surface charge density there?
   (c) If you were at a point outside the conductor, what electric field, if any, would you feel?

Section 20-8 When Can Gauss's Law Be Used to Find the Electric Field?

54. ★★ ∫ A total net electric flux of 8600. $N \cdot m^2/C$ is measured over the entire surface area of a box of dimensions 5.00 cm by 4.00 cm by 2.50 cm. There is a point charge of unknown charge located somewhere within that box. What is the electric field at a distance of 25.0 cm from the point charge?

55. ★ If you were asked to justify why essentially a single law warrants an entire chapter in a physics textbook, how would you explain the importance of Gauss's law?

56. ★★★ ∫ As shown in Figure 20-51, a non-conducting sphere of radius $5R$ has a uniform positive volume charge density of ρ. However, there is a void in the sphere (a vacuum with no charge in that part) of radius R and located at a distance of $2R$ from the centre of the sphere. What is the electric field at point P located a distance of $10R$ from the centre of the sphere?

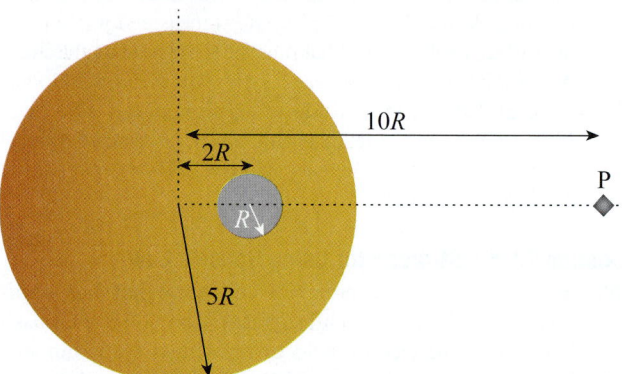

Figure 20-51 Problem 56. A void in a non-conducting sphere.

Section 20-9 Gauss's Law for Gravity

57. ★★ ∫ Consider the situation of Figure 20-52 with two concentric spherical mass shells, one of radius R and one of radius $2R$. The total mass of the inner shell is M and the total mass of the outer shell is $3M$. What is the magnitude of the acceleration due to gravity at a distance r ($r > R$ and $r < 2R$) from the centre in the space between the two shells?

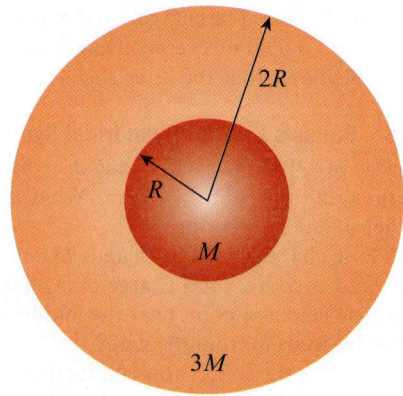

Figure 20-52 Problem 57.

58. ★★ ∫ A spherical planet of radius 2100. km has a uniform density of 3400. kg/m^3 over most of the planet. However, at the core out to a radial distance of 700. km, it has a density of 7800. kg/m^3. Use Gauss's law for gravity to determine the acceleration due to gravity at a height 500. km above the surface.

59. ★★★ ∫ A spherical dust cloud of radius R has a mass density ρ (in kg/m^3) as a function of radial distance, r, given by

$$\rho = \rho_0 \frac{R}{r}$$

where ρ_0 is a constant. Use Gauss's law for gravity to find an expression for the gravitational field, $\vec{g}$, as a function of radial distance r for (a) $r > R$ and (b) $r \le R$.

COMPREHENSIVE PROBLEMS

60. ★★ ∫ A conducting sphere of radius 12.0 cm has a spherical cavity of radius 5.00 cm at its centre. The sphere carries a total charge of +240. nC. Derive expressions for the electric field as a function of radial distance, r. Plot the electric field strength as a function of radial distance.

61. ★ Are you related to Gauss, academically speaking? The Mathematics Genealogy Project (genealogy.math. ndsu.nodak.edu/) allows you to track the academic genealogy of mathematicians through graduate supervisors. For example, Euler was Lagrange's advisor, who in turn was Fourier's advisor, and so on. Why not try tracing some of your mathematics professors to see if they are related to Gauss? Even if you do not find a link to Gauss, you may find ancestry to some famous mathematical names, such as Euler, Lagrange, Poisson, and Dirichlet. There is not a similarly complete academic genealogy project for physicists, but phdtree.org can be used to trace academic backgrounds.

62. ★★★ ∫ The diameter of the central wire in a very long triple conductor coaxial cable is 4.00 mm. Concentric around the wire is a middle cylindrical conductor with a diameter of 2.60 cm and an outer conductor with a diameter of 5.00 cm. At some instant, the wire carries a charge density of +15.0 μC/m, the middle conductor carries a charge density of −7.00 μC/m, and the outer conductor carries a charge density of −8.00 μC/m. Derive expressions for the electric field as a function of radial distance in each of the following regions:
(a) between the wire and the central conductor
(b) between the middle and outer conductors
(c) outside the outer conductor

63. ★★★ ∫ Assume that a non-conducting solid sphere has an isotropic positive charge distribution that depends on radial distance according to some function $\rho(r)$. If the electric field strength is to be constant at different radial distances, how must $\rho(r)$ depend on radial distance?

64. ★★ ∫ A relatively large Van de Graaff generator has a conducting sphere of radius 0.600 m. Assume that a total charge of 94.0 μC is present on the sphere.
(a) What is the electric field inside the conducting sphere?
(b) What is the electric field just outside the conductor?

65. ★★★ ∫ (a) We assume that dry air breaks down electrically when the electric field exceeds 2.50×10^6 N/C. Determine the maximum charge that can be placed on a conducting sphere with each of the following radii: (i) 10.0 cm and (ii) 20.0 cm.
(b) For each case in part (a), determine the surface charge density.

66. ★★ (a) Faraday cages also shield electromagnetic waves, provided that the openings in the cage are much smaller than the wavelength of the radio waves. Get a portable radio and tune in AM and FM channels.
(b) Then go inside some sort of metal (or metal wire) enclosed space (you can use a vehicle, for example) and see if the reception is reduced.
(c) Try stations of various frequencies. Do high or low frequencies seem to be more reduced?
(d) Would you expect shielding to be most complete for high-frequency or low-frequency stations? You can use the relationship $c = f\lambda$, where c is the speed of light, f is the frequency of the station, and λ is the wavelength, to help you answer this question.

67. ★★ ∫ Imagine that the attractive force between the Sun and Earth is electrostatic rather than gravitational. Assume that Earth has a negative charge and the Sun has a positive charge of equal magnitude.
(a) What charge would produce an electric force equal to the actual gravitational force?
(b) Suppose that the charge is distributed evenly over only the surface of Earth. What would the surface charge density be?
(c) What would be the electric field strength just above Earth's surface?

68. ★★ ∫ A series of concentric conducting spherical shells have radii R, $2R$, $3R$, $4R$, ..., with respective positive charge on each shell of $+Q$, $+2Q$, $+3Q$, $+4Q$, ... Where does the electric field have the highest magnitude, and what is the electric field strength at this point?

69. ★ You overhear students talking about Gauss's law and hear mention of something about seasons on Earth. Can you explain what the connection might be?

70. ★★ Why do you think they say that inside a car is a relatively safe place to stay when there is a lightning storm in the area?

71. ★★ A solid non-conducting sphere has a uniform charge density except for one region inside it where there is a spherical hole. Can you nevertheless use Gauss's law to find the electric field inside and outside the sphere under each of the following possibilities?
(a) The hole is centred on the centre of the sphere.
(b) The hole is totally inside the sphere but offset from the centre of the sphere.
(c) The spherical hole is only partly inside the charged sphere.

72. ★★★ ∫ Three drops each have a radius of 250. μm and a charge of +50.0 nC. Assume that the drops are perfectly spherical (this is a good approximation because of surface tension).
(a) What is the electric field just outside the surface of one drop?
(b) The three drops coalesce, with the total charge now spread evenly over the resultant larger drop. What is the new electric field just outside the surface of the drop?

73. ★★ ∫ (a) Derive an expression for the electric field at a distance r from a very long line of charge that carries a constant positive linear charge density λ.
    (b) If the electric field is 1200 N/C at a distance of 1.00 m, how far do we need to move away from the wire to have a field of 120 N/C?

74. ★★ ∫ If in an otherwise empty universe you had a near infinite length cylindrical mass of radius 250. m and of uniform mass density 3000. kg/m³, what would be the gravitational acceleration at a distance of 400. m from the centre of the cylinder?

75. ★★ ∫ If you lived on an infinite-dimensional flat planet with a thickness of 900. km and a uniform mass density of 2500. kg/m³, what would be the acceleration due to gravity at a height of 1.00 m above the surface?

76. ★★★ ∫ A long non-conducting charged cylinder has radius R and a positive charge density given by
    $\rho = \rho_0\left(1 - \dfrac{r}{R}\right)$, where r is the radial distance and ρ_0 is a constant density.
    (a) Describe in words how the density varies.
    (b) Derive an expression for the electric field for points inside the cylinder.
    (c) What is the electric field at the surface of the charged cylinder?
    (d) Derive an expression for the electric field for points outside the cylinder.

DATA-RICH PROBLEMS

77. ★★ The electric field as a function of radial distance is given in Table 20-3. A negative sign means that it points radially inward. Describe precisely what object is consistent with these data (i.e., give the charge and dimension of the object).

Table 20-3 Problem 77

Radial Distance (m)	Electric Field (N/C)
0.100	0.000
0.200	0.000
0.300	0.000
0.400	−140.5
0.500	−89.9
0.600	−62.4
0.700	−45.9
0.800	−35.1
0.900	−27.8
1.000	−22.5

78. ★★ 〰 Plotted in Figure 20-53 is the electric field magnitude at different radial distances from a conducting sphere. The electric field is directed radially outward.
    (a) How large is the sphere?
    (b) What is the total charge on the sphere?

(c) What is the corresponding surface charge density?
(d) From these data, could you differentiate a solid conducting sphere from a conducting spherical shell?

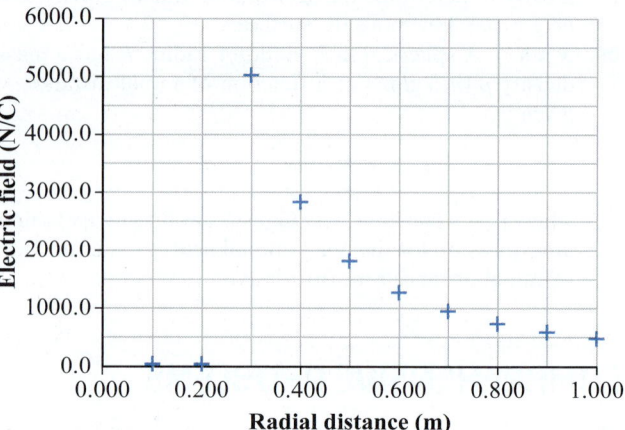

Figure 20-53 Problem 78.

79. ★★★ ∫, 〰 Derive expressions for the electric field in the different regions for the following situation, and then plot the electric field as a function of radial distance. A solid conducting sphere has radius 10.0 cm and carries a charge of +0.500 nC. Concentric with that sphere are two thin conducting spherical shells. One shell extends from 30.0 cm to 40.0 cm and has a charge of −6.00 nC. The other extends from 80.0 cm to 90.0 cm and carries a charge of +16.00 nC. Plot your data from 0 cm to 150 cm radial distance.

80. ★★★ 〰, ∫ Consider the plot in Figure 20-54 of electric field versus radial distance for a set of charged spherically symmetric objects. Describe in detail the characteristics of the objects (e.g., size and charge of different parts). Clearly outline your reasoning. A positive electric field in the graph means a radially outward electric field, while a negative value means one that points radially inward.

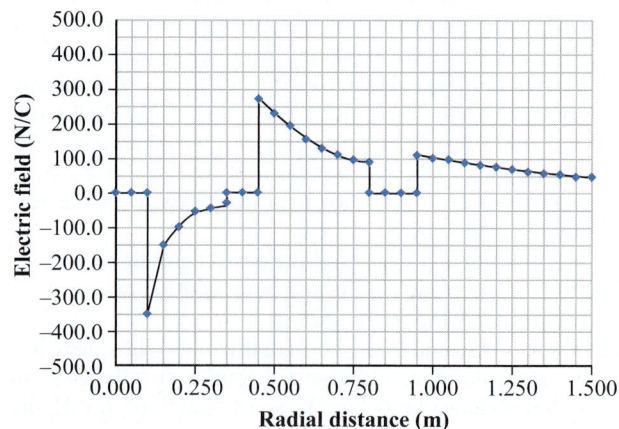

Figure 20-54 Problem 80.

OPEN PROBLEMS

81. ★★★ If you scuff along a carpeted floor on a dry day and then approach a metal object, such as a doorknob, a spark can jump from your finger. If you hold a conducting object, such as a metal ball, the spark will jump from that. In this problem, you will estimate the amount of charge on a metal ball that you hold in performing this experiment.

(a) When the electric field reaches a certain value, breakdown will occur. That is what happens when the spark jumps. Use a reliable source to determine a value for breakdown in dry air. Note it may be expressed in V/m, but 1 V/m = 1 N/C, as you will see in Chapter 21.

(b) Find a room with a metal doorknob that is approximately spherical. Hold a conducting ball of similar size in your hands as you scuff on carpet on a dry day. Very slowly approach the doorknob, and have a friend observe and measure the separation at the instant the spark jumps.

(c) If you have access to a very high input impedance voltmeter, measure the voltage (see Chapter 21 for details on the meaning of electric potential and voltage) prior to having the spark jump (your friend will have to do this using the insulated probe on a suitable voltmeter, touching one probe to the sphere you hold and measuring the peak transient voltage value).

(d) Combine your measurements from parts (b) and (c) to estimate the electric field just prior to discharge.

(e) Use Gauss's law to estimate the electric charge on the sphere that you hold in your hands, assuming it is isolated. This technique overestimates the actual charge, since the electric field is influenced by the presence of the metal doorknob.

(f) Estimate the surface charge density on the conducting sphere.

82. ★★★ ∫ (a) Use some method to estimate the charge on two vinyl balloons (e.g., from force of repulsion between two similarly charged balloons).

(b) How many electrons must have been added or subtracted to produce the charge in part (a)?

(c) If we compare to the number of electrons originally present in the atoms of the neutral balloon before charging, what percentage charging occurred?

(d) Use Gauss's law to estimate the electric field just outside each balloon.

83. ★★ Devices that measure electric fields are usually called electrometers, and they must have very high input impedances (i.e., they draw very small currents).

(a) Borrow a hand-held electrometer, and measure the electric field in various everyday situations. Record your observations (even when you get null results), and try to find locations with significant electric fields.

(b) From an averaged series of outside measurements, estimate the magnitude of the electric field near Earth's surface.

84. ★★★ ∫ Some astronomical objects have complex shapes, such as Comet 67P/Churyumov–Gerasimenko, pictured in Figure 20-55. Look up reasonable estimates for the dimensions and density of the comet online, and then use Gauss's law for gravity to find the acceleration due to gravity at point P just above the surface in the neck region of the comet. Assume that one can approximate the comet as two spheres.

Figure 20-55 Problem 84. Comet 67P/Churyumov–Gerasimenko.

Electrical Potential Energy and Electric Potential

Learning Objectives

When you have completed this chapter, you should be able to

LO1 Calculate the work done by electrical forces to prove that the electric field is conservative.

LO2 Calculate electrical potential energy for configurations of point charges.

LO3 Define and calculate electric potential for groups of point charges.

LO4 Draw and interpret the meaning of equipotential and electric field line diagrams.

LO5 Calculate electric potential from continuous distributions of charge.

LO6 Define and use the electron volt as a unit of energy.

LO7 Use calculus to determine electric fields from electric potentials and vice versa.

LO8 Apply electric potential and field ideas to situations involving conductors.

LO9 Employ a systematic strategy for solving electric problems using electric potential and electric field.

The Large Hadron Collider (LHC), operated by the European Organization for Nuclear Research (CERN), is the world's largest and most powerful particle accelerator. A small portion of the accelerator is shown in Figure 21-1. Protons are accelerated in packets that travel in opposite directions in a huge oval—the entire accelerator, which was built underground near Geneva, Switzerland, has a circumference of about 27 km. The main purpose of the experiment is to perform proton–proton collisions at extremely high relative speeds and to study the fundamental particle physics that only becomes obvious under those conditions. After the upgrade and restart of LHC in 2015, two protons collided with a total relativistic kinetic energy of about 13 TeV (tera electron volts, or 10^{12} eV). At this energy, the protons are moving very close to the speed of light. The electron volt (eV) is a unit of energy commonly used in particle and atomic physics, and it is one of the ideas you will learn about in this chapter.

Just as we found it helpful in Chapters 19 and 20 to introduce the electric field, defined as the electrical force per unit charge, in this chapter we will introduce the electric potential, defined as the electrical potential energy per unit charge. Electric potential is measured in units of volts (V).

Figure 21-1 A portion of the LHC, the largest and most powerful particle accelerator in the world.

21-1 Work and Electric Fields

You first encountered conservative fields in Chapter 6. A **conservative field** implies that we can take any **closed path** (any path that returns at the end to the starting point), and the net amount of work done while we traverse that path will be zero. We implicitly assume that the motion around the path is at infinitesimal speed, so no work goes into change of kinetic energy. For conservative fields, potential energy functions exist—recall that potential energy differences depend on the end points and not the path taken between these points. It is often much easier to use energy techniques since they are scalars rather than the vector electric fields and forces.

In Example 21-1, we compute the work done in moving a small test charge around a closed path in the electric field created by a larger fixed charge. We show that the total work is zero for the closed path. While this single case does not prove that all other closed paths would also lead to zero work, that is indeed true.

EXAMPLE 21-1

Work over a Closed Path in an Electric Field

As shown in Figure 21-2, a positive point charge (Q) produces an electric field. This electric field points radially outward, with an electric field strength that varies inversely with the square of radial distance, as we saw in Chapter 19. Now consider a closed path ABCDA (Figure 21-2). Show that the total work done by an *external agent* (henceforth we will simply call it an agent) in moving a small positive test charge q around the closed path is zero. By **external agent** we mean someone exerting the required forces to move the charge with negligible acceleration along the indicated path in the electric field. We are assuming that we can neglect forces other than electrical. We need the assumption that the speed is infinitesimal, so no work goes into a change in kinetic energy. Assume that charge Q is fixed in position.

Solution

In this problem, we consider the electrical force on charge q due to the electric field created by charge Q. In calculating the work, we implicitly assume that the external agent moves the charge at infinitesimal speed so no work is done to increase kinetic energy, and the charge starts and ends at rest. You saw in Section 6-4 that the work done in moving from point a to point b over a path is given by

$$W = \int_a^b \vec{F} \cdot d\vec{r} \qquad (6\text{-}15)$$

In this chapter, and in the later chapters in electromagnetism, we will use the notation $d\vec{\ell}$ instead of $d\vec{r}$ as the differential element along the path.

Let us consider the direction of the force required by the external agent. The agent must balance the electrical force so that there will be negligible acceleration. Charge q is repelled by Q, so in moving charge q from A to B, to keep the charge from accelerating, the external agent must apply a force toward Q; that is, the direction of the applied force is opposite to the direction of the displacement $d\vec{\ell}$. Therefore, the work done by the agent is negative along this segment of the path. From B to C, the force and the displacement are everywhere perpendicular ($\vec{F}$ and $d\vec{\ell}$ are perpendicular), so the vector dot product results in zero contribution to the work for this part of the path. From C to D, there is repulsion between the two positive charges, and the agent must push against this repulsion. Therefore, the agent will do a positive amount of work in moving charge q from point C to point D ($\vec{F}$ and $d\vec{\ell}$ are in the same direction). From point D to point A, $\vec{F}$ and $d\vec{\ell}$ are again perpendicular, so the dot product leads to zero work for that segment of the path. We show the vectors for the various segments of the closed loop in Figure 21-3.

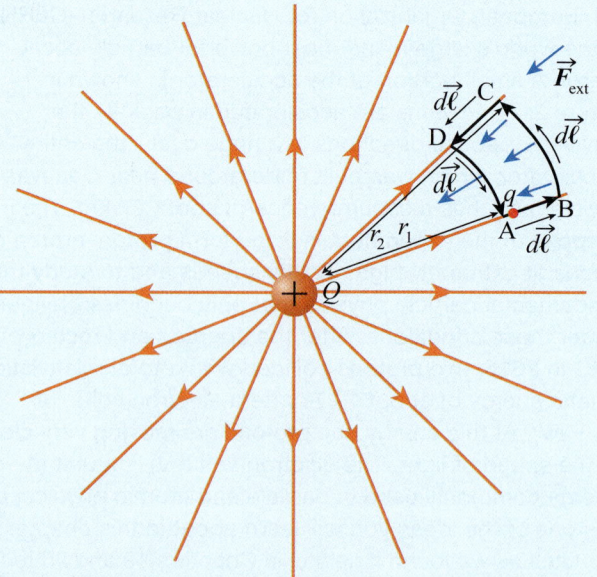

Figure 21-3 Example 21-1. The forces (blue vectors) required of an *external agent* to move a positive test charge q around the different segments of the closed path (in black).

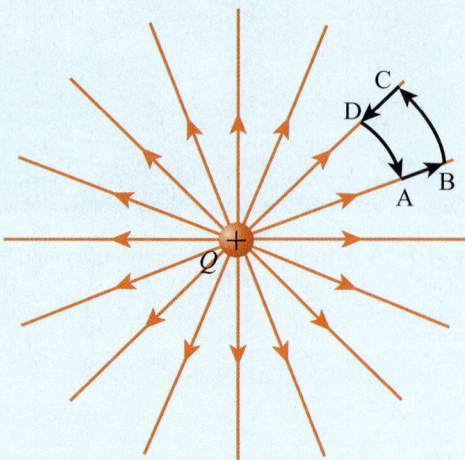

Figure 21-2 Example 21-1.

It is important to be clear on the meaning of the vectors $d\vec{r}$, $d\vec{\ell}$, and $\hat{r}$ that we use in computing work along different paths. Vector $d\vec{\ell}$ represents a differential step along the closed path we are using. As shown in Figure 21-3, it will point in different directions in different segments of the path. The radial vector $d\vec{r}$ points in the direction of increasing radial distance r (i.e., radially outward). The unit-length (but dimensionless) vector in this direction is called $\hat{r}$. In the same way that $\hat{i}$ points in the direction of increasing positive x, $\hat{r}$ points in the direction of increasing r.

Now we are ready to do the actual computation. From Chapter 19 we have the following result for the electric field of a point charge, Q:

$$\vec{E} = \frac{1}{4\pi\varepsilon_0}\frac{Q}{r^2}\hat{r} \qquad (19\text{-}11)$$

Recall that $\hat{r}$ is a radially outward unit vector, and r is the radial distance from point charge Q. The electrical force on the small test charge is simply the product of this electric field and the charge, q, that is being moved. The force that must be exerted by the agent moving charge q is always *opposite* in direction to the electrical force on the particle q.

First we compute the work done by the agent in moving charge q from point A to point B (we use the notation $W_{A \to B}$ for the work done by the agent in moving the charge from A to B). We will also use the notation r_1 for the radial distance to point A from charge Q, and r_2 for the radial distance to point B:

$$W_{A\to B} = \int_A^B \vec{F}\cdot d\vec{\ell}$$
$$= \int_{r_1}^{r_2} -\frac{1}{4\pi\varepsilon_0}\frac{Qq}{r^2}\hat{r}\cdot dr\,\hat{r}$$
$$= -\frac{Qq}{4\pi\varepsilon_0}\int_{r_1}^{r_2}\frac{dr}{r^2}$$
$$= -\frac{Qq}{4\pi\varepsilon_0}\left[-\frac{1}{r}\right]_{r_1}^{r_2}$$
$$= -\frac{Qq}{4\pi\varepsilon_0}\left[\frac{1}{r_1}-\frac{1}{r_2}\right]$$

Since distance r_1 is less than r_2, $1/r_1$ is greater than $1/r_2$ and $W_{A\to B}$ is negative, as expected (if we assume that both charges have the same polarity).

For path segment B $\to$ C (and D $\to$ A), the force and path step are perpendicular, so $\vec{F}\cdot d\vec{\ell}=0$ and therefore $W_{B\to C}=W_{D\to A}=0$.

In path segment C $\to$ D, the agent must exert a radially inward force to overcome the outward electrical force on charge q, so we expect the work to be positive:

$$W_{C\to D} = \int_C^D \vec{F}\cdot d\vec{\ell}$$
$$= \int_{r_2}^{r_1}\left(-\frac{1}{4\pi\varepsilon_0}\frac{Qq}{r^2}\right)\hat{r}\cdot(-d\ell)\hat{r}$$
$$= \frac{Qq}{4\pi\varepsilon_0}\int_{r_2}^{r_1}\frac{d\ell}{r^2}$$

To evaluate the integral, we need to have the differential in terms of dr. Note that as $d\ell$ (the step along the path) increases in this segment, dr (the step in the increasing r-direction) decreases, so $d\ell = -dr$:

$$W_{C\to D} = -\frac{Qq}{4\pi\varepsilon_0}\int_{r_2}^{r_1}\frac{dr}{r^2}$$
$$= -\frac{Qq}{4\pi\varepsilon_0}\left[-\frac{1}{r}\right]_{r_2}^{r_1}$$
$$= -\frac{Qq}{4\pi\varepsilon_0}\left[-\frac{1}{r_1}+\frac{1}{r_2}\right]$$
$$= -\frac{Qq}{4\pi\varepsilon_0}\left[\frac{1}{r_2}-\frac{1}{r_1}\right]$$
$$= \frac{Qq}{4\pi\varepsilon_0}\left[\frac{1}{r_1}-\frac{1}{r_2}\right]$$

As expected, we obtain a positive result for the work done by the external agent over the C $\to$ D segment. We now combine results to find the total work done by the external agent as it moves the charge q around the closed path A $\to$ B $\to$ C $\to$ D $\to$ A. We have the following expression for the path:

$$W_{\text{external agent}} = W_{A\to B} + W_{B\to C} + W_{C\to D} + W_{D\to A}$$
$$= -\frac{Qq}{4\pi\varepsilon_0}\left[\frac{1}{r_1}-\frac{1}{r_2}\right] + 0 + \frac{Qq}{4\pi\varepsilon_0}\left[\frac{1}{r_1}-\frac{1}{r_2}\right] + 0$$
$$= 0$$

Making sense of the result

As expected, we found that the work done by the external agent in moving the charge along the closed path is zero, which implies a conservative force. While this does not prove the result in general, in fact *any* closed path, even a very complicated one, will yield a value of zero for the total work done. We have taken the viewpoint of work done by an external agent, but we could equally well use the viewpoint of work done by the electric field. From that viewpoint, the work in each segment would have the opposite sign, but the total for the path would still add to zero.

While Example 21-1 does not constitute a general proof, in fact, static electric fields are always conservative. By static we mean that the charges that create the electric field are not in motion. We stressed in Chapters 19 and 20 that electric fields are real, and not only a construct to help us understand the physics. Since it is the field that is conservative, many argue that it is better to calculate the work done by the force due to the electric field, rather than that of an external agent. We adopted this approach in Chapter 6.

If we repeated Example 21-1 from that perspective, the force on small charge q exerted by the electric field produced by charge Q would have (for positive charges) the direction shown in Figure 21-4. Since these are just the opposite of the force directions for the external agent of Figure 21-3, the work from A to B will now be positive, while the work for the segment from C to D will be negative (the other two segments will still be zero). The sum around the closed path will still add to zero.

The argument in favour of the external agent approach is based on the idea that potential energy represents the energy stored in a configuration (e.g., a stretched spring stores energy, as does a mass lifted in a gravitational field or a positive charge held separated from a negative charge). Energy is conserved in conservative fields, so the work done by the external agent will equal the increase of potential energy of the system (assuming, as we have, that none goes into a change in kinetic energy).

However, the majority of physicists prefer to consider the work done by the associated field. One argument in favour of the field approach is that it keeps forces in the direction associated with the field. Also, many physicists argue that it is the *field* that is conservative, so we should calculate the work done by the *field*, rather than that done by an external agent. With the external agent approach, we needed to specify that the motion was made without significant acceleration, but that requirement is avoided in the field approach.

Mathematically the two approaches are almost identical, with the sign simply being different in each segment of the path. It is not that one approach is right and one is wrong, but rather a matter of preference. You should be familiar with both approaches, but make sure you use one or the other consistently in a problem. The situation is similar to the gravitational case, where we can calculate the positive work done by a person lifting a mass near Earth, or the negative work done by the gravitational field over the same lift.

21-2 Electrical Potential Energy

Potential energy functions exist for conservative fields (static electric and gravitational fields are always conservative). The work required to move a charge from one point to another depends only on the end points (and the charge being moved), and not on the path taken between the points. Since potential energies are

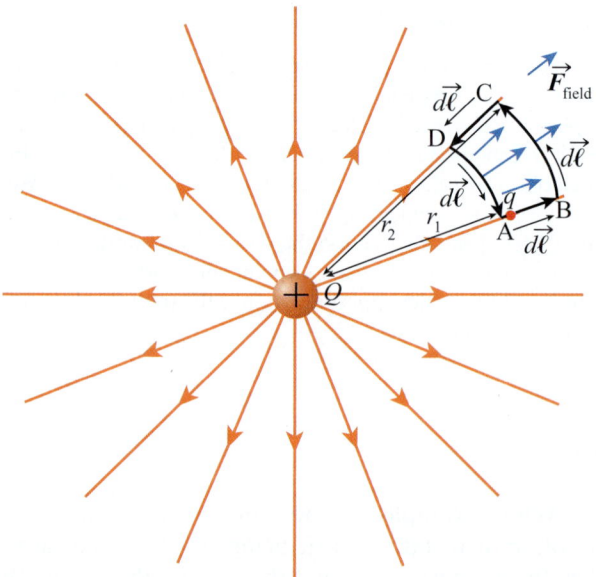

Figure 21-4 The same situation as Figure 21-3, but the force vectors are now the forces the *electric field* of fixed central charge Q exerts on the moving test charge q.

scalars, this provides a mathematically simpler method to solve many problems than using vector superposition of electrical forces.

The potential energy difference between two configurations is equal to the work required to change one configuration into the other. Most of the time we talk about potential energy differences, but if we define one configuration as having zero potential energy, then we can discuss absolute potential energies relative to that zero reference. Sometimes the choice of the zero reference is obvious—for example, we normally define the unstretched state of a spring mass system as the zero. At other times it is not—for example, when dealing with gravitational energies near the surface of the Earth, we could call sea level the zero reference, but could just as well choose any other horizontal level. As we saw in Chapter 11, we are free to choose the configuration that we will define as the zero of the potential energy.

Let us first consider the situation when we move a charge between two positions, a and b, in a region with an electric field. Call U_b and U_a the **electrical potential energy** when the charge is in the two configurations, b and a. If $\vec{F}_{agent}$ is the force an external agent would need to apply in moving (with negligible acceleration) an electric charge from a to b, and $d\vec{\ell}$ is a vector along the path taken, we have the following relationship for the difference in electrical potential energies between the two configurations:

$$U_b - U_a = \int_a^b \vec{F}_{agent} \cdot d\vec{\ell} \qquad (21\text{-}1)$$

We calculate the work done by the agent moving the object from a to b, and the result is the change in potential energy. Note that the result can be positive or negative, so we can have an increase or decrease in potential energy of the configuration. If the agent did positive work in moving to configuration b, that configuration has a higher potential energy. It is important to remember that Equation 21-1 gives the *change* in electrical potential energy between the two charge configurations. Sometimes it is abbreviated to ΔU, but we prefer the form given above, which guarantees the correct sign for the potential energy difference.

Most physicists prefer to use the field, rather than the agent, approach, in which we integrate the work done by the field as the charge is moved from point a to point b. Now, however, we take the *negative* of that work done by the force associated with the field to find the change in potential energy as the charge is moved from a to b:

KEY EQUATION
$$U_b - U_a = -\int_a^b \vec{F}_{field} \cdot d\vec{\ell} \qquad (21\text{-}2)$$

Since $\vec{F}_{agent} = -\vec{F}_{field}$, the two relationships always give the same result. We recommend that you use the field approach represented by Equation 21-2. However, it is

often helpful to consider the agent view as a check that the sign you obtain is reasonable. For example, if your field calculation shows that the new configuration has a higher potential energy, ask yourself if an agent would have done positive work to create that configuration.

The above relationships are true for any conservative force field. Let us now derive the change in electrical potential energy when a small charge, q, is moved from distance r_1 to a larger distance r_2 away from fixed charge, Q, (see Figure 21-5). In showing force directions, we assume both charges are positive, but the result we obtain will be true whether the charges are positive or negative as long as we use the appropriate signs. We will assume that charge Q is fixed relative to the coordinate system and only q moves.

We will employ the electric field viewpoint, rather than that of an external agent, in this derivation. To make this obvious, we label the electrical force on charge q with $\vec{F}_{field}$. Since both charges are assumed positive, the force on q due to the electric field of Q will be in the direction shown in Figure 21-5. We start with Equation 21-2 to compute the difference in electrical potential energy as we move charge q from separation r_1 to r_2:

$$U_b - U_a = -\int_a^b \vec{F}_{field} \cdot d\vec{\ell} \qquad (21\text{-}2)$$

Since both charges are positive, the force $\vec{F}_{field}$ is to the right and in the same direction as $d\vec{\ell}$. We could have used a Cartesian coordinate system, but we will employ a radial unit vector $\hat{r}$ to the right, as shown in Figure 21-5:

$$U_2 - U_1 = -\int_{r_1}^{r_2} \frac{1}{4\pi\varepsilon_0} \frac{Qq}{r^2} \hat{r} \cdot d\vec{\ell} \hat{r}$$

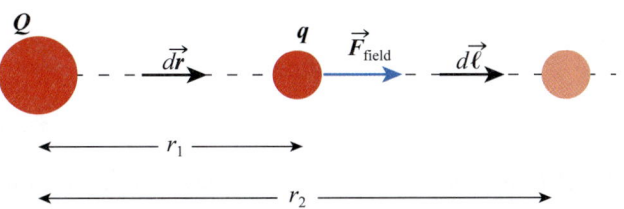

Figure 21-5 We calculate the change in electrical potential energy as charge q moves from r_1 to r_2 away from fixed charge Q. The differential vector $d\vec{\ell}$ points along the path between the two positions, while $d\vec{r}$ is a differential displacement vector in the direction of increasing value of r. They both point in the same direction in this case.

Here the vectors $d\vec{r}$ and $d\vec{\ell}$ both point in the same direction, and $dr = d\ell$. We also have that the dot product of a unit vector with itself produces a unit result: $\hat{i} \cdot \hat{i} = 1$:

$$U_2 - U_1 = -\frac{Qq}{4\pi\varepsilon_0}\int_{r_1}^{r_2}\frac{dr}{r^2}$$

$$= -\frac{Qq}{4\pi\varepsilon_0}\left[-\frac{1}{r}\right]_{r_1}^{r_2}$$

$$= -\frac{Qq}{4\pi\varepsilon_0}\left[\left(-\frac{1}{r_2}\right) - \left(-\frac{1}{r_1}\right)\right]$$

Therefore, the electrical potential energy *difference* when two charges, Q and q, are moved from a separation of r_1 to a different separation of r_2 is

KEY EQUATION

$$U_2 - U_1 = \frac{Qq}{4\pi\varepsilon_0}\left[\frac{1}{r_2} - \frac{1}{r_1}\right] \qquad (21\text{-}3)$$

Both charges are positive; therefore, since $1/r_2 < 1/r_1$, $U_2 - U_1$ is negative. This makes sense, since positive charges repel. Therefore, the potential energy decreases as separation increases for two positive (or two negative) charges. Equation 21-3 also applies when one charge is positive and the other negative, and the charges attract. In that case, moving the charges farther apart corresponds to an increase in potential energy.

As mentioned earlier, all potential energies are defined relative to some assumption of a zero point. For example, when a mass is on a table, we can use the surface of the table as the zero reference for gravitational potential energy, resulting in the mass having zero potential energy. However, if we use the floor as the zero reference, the mass has positive potential energy. We saw in Chapter 11 that, generally, the most convenient way to define the zero of gravitational potential energy is for the zero level to be when the masses are infinitely far apart. A similar convention is often adopted for electrical potential energy—zero when the charges are infinitely far apart. When $r_1 \to \infty$, Equation 21-3 yields the following expression for the **electrostatic potential energy** when charges q and Q are a distance r apart *with an assumption of U = 0 when the charges are infinitely separated*:

KEY EQUATION

$$U = \frac{1}{4\pi\varepsilon_0}\frac{Qq}{r} \qquad \text{assuming } U \to 0 \text{ when } r \to \infty$$

$$(21\text{-}4)$$

By **electrostatic** we mean that the charge configuration does not change with time. Usually we will just refer to this as electrical potential energy, rather than the more specific term electrostatic potential energy, but strictly speaking Equation 21-4 is only valid for static charge

situations. Sometimes the symbol U_E is used to differentiate electrical potential energy from other forms of potential energy, but in this chapter when we refer to potential energy it will always mean electrical, so we will not use the subscript. If we insert an infinite value for r in Equation 21-4, we get the result $U = 0$, consistent with our definition of zero level.

PEER TO PEER

Most of the time the zero point for electrical potential energy is defined as the potential energy when the charges are infinitely separated, but I need to make sure that is the indeed the definition of zero point being used before I use an equation based on that!

CHECKPOINT 21-2

Electrical Potential Energy

Consider a situation with a negative charge moved from a great distance to a position near a positive charge. If the potential energy was defined as zero when the charges were infinitely separated, then

(a) positive work must be done by an external agent to move the charge to that position, and the potential energy of the system is positive

(b) positive work must be done by an external agent to move the charge to that position, and the potential energy of the system is negative

(c) negative work must be done by an external agent to move the charge to that position, and the potential energy of the system is positive

(d) negative work must be done by an external agent to move the charge to that position, and the potential energy of the system is negative

ANSWER: (d) The charges are of opposite sign and therefore attract each other. The applied external force counteracts this force and is in the opposite direction to the displacement that brings the particles closer together. Therefore, the net amount of work is negative. This results in a negative potential energy.

Equation 21-4 gives the electrical potential energy for a pair of charges (assuming the $U = 0$ at $r = \infty$ definition of the zero point for potential energy). With three charges, there is a potential energy contribution from each pairing of the charges. This is expressed as follows:

$$U = \frac{1}{4\pi\varepsilon_0}\left(\frac{q_1q_2}{r_{12}} + \frac{q_1q_3}{r_{13}} + \frac{q_2q_3}{r_{23}}\right) \qquad (21\text{-}5)$$

where, for example, r_{12} represents the separation between charge 1 and charge 2. We can readily extend this idea to larger numbers of charges, keeping in mind that we must have a term in the potential energy equation for each pairing, but we do not duplicate any pair (i.e., we would not have both a 12 term and a 21 term). Example 21-2 shows how to apply the technique to a configuration of four charges.

If no net work is done by an agent, total energy is conserved in electrostatic situations. Note here we refer to the situation as electrostatic since the charges are not moving at the end configurations, even though we of course must move them at infinitesimal speeds to change from one configuration to the other. We could write this in either of the following equivalent forms:

$$U_i + K_i = U_f + K_f \qquad (21\text{-}6)$$

$$\Delta U + \Delta K = 0 \qquad (21\text{-}7)$$

Here, U represents potential energy, K represents kinetic energy, and i and f refer to the initial and final values. Equation 21-7 is obtained from Equation 21-6 by subtracting $U_i + K_i$ from each side of the equation and then expressing differences using Δ notation. Some prefer to express Equation 21-7 in the following more explicit form:

$$(U_f - U_i) + (K_f - K_i) = 0 \qquad (21\text{-}7a)$$

We demonstrate how to employ these ideas in Example 21-3.

EXAMPLE 21-2

Electrical Potential Energy for Four Charges

(a) Find the total work required to move four charges (+2.00 nC, +3.00 nC, +4.00 nC, +5.00 nC) from initial positions infinitely far apart to the corners of a square with sides of 0.150 m, as shown in Figure 21-6.

(b) Define the zero of electrical potential energy to be when the charges are infinitely far apart. What is the electrical potential energy of the charge configuration shown in Figure 21-6?

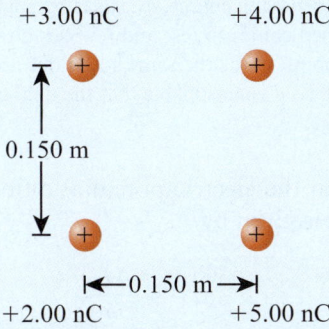

+3.00 nC   +4.00 nC

0.150 m

+2.00 nC   +5.00 nC
|←— 0.150 m —→|

Figure 21-6 Four charges are located at the corners of a square.

Solution

(a) You might at first consider calculating work using force and displacement. However, the force between two charges depends on the distance between them. Therefore, you would need to integrate, which becomes much more cumbersome once we add the third and fourth charges because the force vectors have different directions. One strength of energy approaches to solving problems is that the potential energy for conservative forces depends only on the actual configuration of the system and is not affected by how the system got to that configuration. Therefore, we will use electrical potential energy, expanding Equation 21-5 to the situation for four charges. We will number the charges 1 to 4 in same order as the charge magnitudes (i.e., the 2.00 nC charge

is number 1). Since energy values are scalars, the terms simply add:

$$U = \frac{1}{4\pi\varepsilon_0}\left(\frac{q_1 q_2}{r_{12}} + \frac{q_1 q_3}{r_{13}} + \frac{q_1 q_4}{r_{14}} + \frac{q_2 q_3}{r_{23}} + \frac{q_2 q_4}{r_{24}} + \frac{q_3 q_4}{r_{34}}\right)$$

Note that we have included a term for each charge pair combination, using the notation, for example, r_{34} to represent the distance between charges 3 and 4. We use the Pythagorean theorem to obtain the distance between charges on opposite corners of the square. All the terms being summed have identical units, so we can move these units into the common factor for clarity:

$$U = \frac{1.00 \times 10^{-18}\ \text{C}^2/\text{m}}{4\pi \times 8.85 \times 10^{-12}\ \text{C}^2/(\text{N} \cdot \text{m}^2)}$$

$$\times \left(\frac{2.00 \times 3.00}{0.150} + \frac{2.00 \times 4.00}{0.212} + \frac{2.00 \times 5.00}{0.150}\right.$$

$$\left. + \frac{3.00 \times 4.00}{0.150} + \frac{3.00 \times 5.00}{0.212} + \frac{4.00 \times 5.00}{0.150}\right)$$

$$= 3.85 \times 10^{-6}\ \text{N} \cdot \text{m}$$

$$= 3.85\ \mu\text{J}$$

Thus, we must do 3.85×10^{-6} J of positive work to bring the charges into this configuration from an initial position of infinite separation.

(b) The electrical potential energy of the configuration is simply the work required by an external agent to create the configuration, which is 3.85×10^{-6} J, or 3.85 μJ.

Making sense of the result

We expect the work required to be positive because the positive charges repel and it takes positive work to push them closer together. Compared to energies we encountered in mechanics, this is a very tiny value. The charges in this question are relatively small—static electricity from rubbing objects typically creates nC to μC range charged objects.

EXAMPLE 21-3

Charged Particle Speed

Consider the situation shown in Figure 21-7. Initially, a negative charge q_2 of mass m is at a distance r_2 from a *fixed in position* positive charge q_1. At this position q_2 was given a speed of v_0 in a direction directly toward q_1.

(a) What is the speed of q_2 when it reaches a smaller distance, r_1, from q_1?
(b) Is the speed of charge q_2 more or less when it moves to r_1?
(c) If the charge had the same initial speed v_0 but in the opposite direction (away from q_1), how would the answer to (a) be different? Justify your response.

You can assume that changes in gravitational potential energy are negligible.

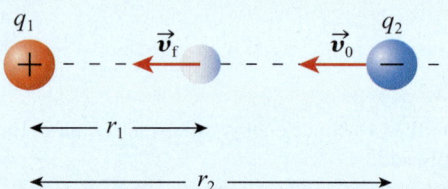

Figure 21-7 Example 21-3. Charge q_1 is fixed in position and positively charged, while charge q_2 is negatively charged and has velocity $\vec{v}_0$ initially (on the right) and $\vec{v}_f$ when it has moved closer to the positive charge.

Solution

(a) The computationally simplest way to approach this problem is using conservation of energy. Since energies are scalars, no vectors need to be employed. While an external force provided the initial speed, from that point on we can assume there is no external work done

and energy is conserved. It is assumed that gravitational potential energy changes are negligible and we only need to consider electrical potential energy:

$$U_i + K_i = U_f + K_f$$

$$\frac{1}{4\pi\varepsilon_0}\frac{q_1 q_2}{r_2} + \frac{1}{2}mv_0^2 = \frac{1}{4\pi\varepsilon_0}\frac{q_1 q_2}{r_1} + \frac{1}{2}mv_f^2$$

$$v_f^2 = v_0^2 + \frac{1}{4\pi\varepsilon_0}\frac{2q_1 q_2}{m}\left(\frac{1}{r_2} - \frac{1}{r_1}\right)$$

$$v_f = \sqrt{v_0^2 + \frac{q_1 q_2}{2\pi\varepsilon_0 m}\left(\frac{1}{r_2} - \frac{1}{r_1}\right)}$$

(b) Since $r_1 < r_2$, then $1/r_1 > 1/r_2$ and therefore $(1/r_2 - 1/r_1) < 0$. However, one charge is positive and one is negative, so $q_1 q_2 < 0$. This means that the term added to v_0^2 is positive, and charge q_2 will be moving faster when closer to q_1. This makes sense since the charges are of opposite sign and will attract each other.

(c) The answer will be exactly the same, since the kinetic energy and potential energy at the initial position are identical for this case.

Making sense of the result

As expected, the charge moves faster when closer because unlike charges attract and it is moving to a configuration with a lower potential energy (a higher-magnitude negative potential energy). In part (c) of the problem, we get the same answer since kinetic energy depends on the speed but not the direction of that speed. In this case, the moving charge, q_2, starts by going farther away from q_1 and then comes to rest and reverses direction. When it passes the r_2 point, it has the same speed but now in the direction toward q_1, so it is reasonable that the final position answer does not change.

In most problems in this chapter, the term *potential energy* will refer to the electrical potential energy, but if gravitational or other forms of potential energy are relevant, they must be considered as well. In doing problems like this, make sure you note whether one charge is fixed in position, or if both are allowed to move. In problem 31, we consider a situation in which both charges are free to move and therefore have a share of the kinetic energy.

The work done by the electric field is taken into account in the electrical potential energy and is not treated as external work. We adopted a similar approach when dealing with the gravitational field in Chapter 11.

21-3 Electric Potential

In the same way that we defined electric field as the electrical force per unit charge, a powerful idea is that of **electric potential**, V, which is defined as electrical potential energy per unit charge. Since electrical potential energies are relative, so are electric potentials. Therefore, we usually speak of **electric potential differences**, ΔV. If ΔU is the change in potential energy of the configuration when charge q is moved from one point to

another, then the electric potential difference between these points is given by

KEY EQUATION

$$\Delta V = \frac{\Delta U}{q} \qquad (21\text{-}8)$$

Just as we defined a zero point for electrical potential energy, we also define a zero point for electric potential. Normally, we define electrical potential energy to be zero for infinitely separated charges, and similarly define the zero point of electric potential as a point infinitely far from any charges. You should keep in mind that you are free to define the point for zero electric potential as long as you stay consistent with that definition throughout the question. If a definition of the zero point has been made, we can drop the delta (meaning change in) symbols from Equation 21-8:

KEY EQUATION

$$V = \frac{U}{q} \qquad (21\text{-}9)$$

Equations 21-8 and 21-9 look similar, but Equation 21-8 is used when we are comparing the potential difference between two different points, whereas Equation 21-9 is used for an absolute electric potential after we have defined the zero point of electric potential at infinite distance from the charges. Strictly speaking, the term *electric potential difference* should be used when we are talking about the difference in electric potential between two points, while electric potential is used only after a zero reference has been defined. Note also that the terms *electric potential* and *electrical potential* are both used by physicists and mean the same thing (and similarly *electric potential difference* and *electrical potential difference*).

The SI unit for electric potential is the **volt** (V). We can see from Equation 21-9 that 1 V = 1 J/C. At first, it might seem confusing to use the same symbol, V, for the quantity electric potential and the unit used to measure it. However, there is a subtle difference in how we write the two symbols: unit symbols should be written as roman (upright) text, and symbols for variables are written as italic (slanted) text. The term **voltage** is often used interchangeably with *electric potential difference*, although the formal term is *electric potential difference*.

It is critical to always keep in mind that electric potentials are relative and depend on a defined zero point. In electric circuits, this zero point is called **ground** because in an electrical system it may well be directly connected to a conducting plane placed in the ground. For example, consider two points, A at +2 V and B at +5 V. Both points are positive with respect to the ground potential, but point A has a negative electric potential if compared to the electric potential at point B.

We can combine Equation 21-9 with Equation 21-4 to obtain an expression for the electric potential at a distance r from a single point charge, Q, keeping in mind that we have defined the zero point as the electric potential at infinite distance from the charge:

KEY EQUATION
$$V = \frac{Q}{4\pi\varepsilon_0 r} \tag{21-10}$$

We see from Equation 21-10 that the electric potential is positive in the region near a positive charge, negative near a negative charge, and zero at an infinite distance from the charge Q. It is important to realize that electric potential normally varies with position, and it is often written explicitly as $V(r)$ to show this dependence; for example, we could write Equation 21-10 as

$$V(r) = \frac{Q}{4\pi\varepsilon_0 r} \tag{21-10a}$$

We apply the superposition principle to situations involving multiple charges, using Equation 21-10 to find the contributions to the electric potential from each charge. Since these contributions are scalar quantities, it is easy to add them to obtain a net electric potential, as demonstrated in Example 21-4.

EXAMPLE 21-4

Electric Potential for Three Charges

(a) Find the electric potential at point P in Figure 21-8.
(b) How much work is required to move a +1.00 μC charge to point P from an infinite distance?

Solution

(a) According to the superposition principle, we can find the electric potential at point P due to each of the charges separately, and then simply add the three individual potentials algebraically. We use Equation 21-10 with the appropriate charge and the distance from each charge to point P. Distances r_1 and r_3 are both $\sqrt{2.00}$ m:

$$V = \frac{q_1}{4\pi\varepsilon_0 r_1} + \frac{q_2}{4\pi\varepsilon_0 r_2} + \frac{q_3}{4\pi\varepsilon_0 r_3}$$

$$= \frac{1}{4\pi\varepsilon_0}\left(\frac{-4.00 \times 10^{-6}\,\text{C}}{\sqrt{2.00}\,\text{m}} + \frac{+8.00 \times 10^{-6}\,\text{C}}{1.00\,\text{m}}\right.$$

$$\left. + \frac{-4.00 \times 10^{-6}\,\text{C}}{\sqrt{2.00}\,\text{m}}\right)$$

$$= \frac{1\,\text{C/m}}{4\pi \times 8.85 \times 10^{-12}\,\text{C}^2/(\text{N}\cdot\text{m}^2)}$$

$$\times (-2.828 \times 10^{-6} + 8.00 \times 10^{-6} - 2.828 \times 10^{-6})$$

$$= +2.11 \times 10^4\,\text{V}$$

(b) When the +1.00 μC charge is infinitely far away, it will have an electric potential of zero, so the potential difference between that location and point P is +21.1 kV. Since

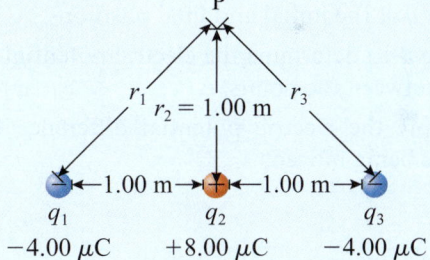

Figure 21-8 Three charges are in a line; the middle charge is positive.

the electric potential is the electrical potential energy per unit charge, the electrical potential energy when the 1.00 μC charge is at point P is

$$(+1.00 \times 10^{-6}\,\text{C})(+2.11 \times 10^4\,\text{V}) = +0.0211\,\text{J}$$

Therefore, the work required to move the charge from infinitely far away to point P is 0.0211 J, or 21.1 mJ.

Making sense of the result

Although there are equal amounts of positive and negative charge in the initial configuration, the net electric potential is positive because point P is closer to the positive charge than to the negative charges. You may be surprised by how large a potential difference is produced by a small fraction of a coulomb of charge. Remember from Chapter 19 that a coulomb is a huge amount of charge, with most charging situations involving μC or less. In part (b), we are moving a positive charge to a region with a more positive electric potential; therefore, the work required is positive.

PEER TO PEER

I must admit that at first I found the whole idea of electric potentials pretty abstract and confusing. My "aha" moment was when I realized electric potentials are telling me where a positive test charge would want to go. If the potential is positive compared to where I am, then a positive charge is repelled and does not want to go there. It will only go there if something does work to move it there. On the other hand, if the potential is more negative, a positive charge will just love to go to that region and will accelerate as it goes there.

CHECKPOINT 21-3

Work Required to Move Charge q

Two electric point charges of equal magnitude but opposite signs, $+Q$ and $-Q$, are located at fixed positions along the x-axis, with $+Q$ on the left. A small positive point charge, q, is moved along the line joining the charges, from a point near the $+Q$ charge to a point near the $-Q$ charge. The work required to move the charge (assuming that these are the only significant electric charges, and no forces other than electrical are important) is

(a) positive

(b) negative

(c) either positive or negative, depending on where the zero point is for the electric potential

ANSWER: (b) We urge you to draw a diagram to aid in your understanding of the situation. Since the positive charge, q, is repelled by the $+Q$ charge and attracted to the $-Q$ charge, a force must be applied in the opposite direction to the motion to keep the positive charge from accelerating. This suggests negative work. Another way to think about it is in terms of electric potential difference. The electric potential is positive near the $+Q$ charge and negative near the $-Q$ charge because the closer charge dominates in each case. Thus, the electric potential difference as we move from near $+Q$ to near $-Q$ is negative, and when multiplied by a positive test charge, q, yields a negative value for the work required.

Example 21-4 demonstrates a useful strategy for finding the work required to move a charge q (we of course are assuming only electrical forces):

(1) Find the electric potentials due to all the charges *except q* at the initial and final positions.

(2) Subtract to determine the electric potential difference between the points, ΔV.

(3) Multiply the electric potential difference by the charge being moved.

We can represent this as follows:

KEY EQUATION
$$W = q\Delta V \qquad (21\text{-}11)$$

where W is the work required to move charge q through a potential difference ΔV. Note that any of q, W, and ΔV can be positive or negative, and that we have assumed that only electrical forces are present. Check your understanding of this relationship with Checkpoint 21-3. We demonstrate how this relationship can be used in Example 21-5.

EXAMPLE 21-5

Work to Move Charge

A particle has a positive charge of 3.00 nC and a mass of 3.07×10^{-23} kg. Initially it is at rest at a point at an electric potential of -12.0 V. How much work is required to move it to another point that is at an electric potential of $+36.0$ V? Assume that it moves in a horizontal plane so that there is no change in gravitational potential energy, and no other forces are present.

Solution

An agent must do positive work to move the positive charge to a point with a higher potential. What is important is the potential difference between the two points, which is $\Delta V = 48.0$ V. We can use Equation 21-11 to find the required work:

$$
\begin{aligned}
W &= q\Delta V \\
&= 3.00 \times 10^{-9}\,\text{C} \times 48.0\,\text{V} \\
&= 1.44 \times 10^{-7}\,\text{J}
\end{aligned}
$$

Making sense of the result

Note that the work we need to do to move the charge only depends on the charge and the electric potential difference. The mass does not enter into the calculation. If the charge had been negative, a negative amount of work would have been required to move it to a point at a higher potential. We considered this from the point of view of an external agent. Had we instead considered the work done by the electric field, that would have been negative, since the charge is moved opposite to the direction of the electrical force on the charge.

Sometimes we use a symbol such as Q in a sense where it can take on any value, positive or negative, and is a true variable. For example, the Q in Equations 21-9 and 21-10 (and many other equations) is to be interpreted this way. Sometimes, in a situation such as Checkpoint 21-3, we have specific charges, one of which is positive and one is negative, but instead of stating specific numerical values, such as $+4.0\ \mu C$ and $-4.0\ \mu C$, we call them $+Q$ and $-Q$. It should always be clear from the context whether a symbol is meant to be a variable that can take on any value or one that is positive.

21-4 Equipotential Lines and Electric Field Lines

It is useful to draw lines, called **equipotential lines**, through points of equal electric potential. We normally draw a series of these at different electric potentials; for example, a series of equipotential lines might be drawn for points at -20 V, -10 V, 0 V, $+10$ V,

$+20$ V, $+30$ V, etc. We see from Equation 21-10 that for a single point charge, equipotential lines are concentric circles centred on the point charge. When the charge is positive, the electric potential values are positive, with increasing values as one approaches the charge. Figure 21-9 shows these equipotential lines along with electric field lines that point outward from a positive charge. (Recall from Chapter 19 that electric fields point in the direction of the electrical force on a positive test charge, and therefore away from a positive charge.)

Note that the electric field lines are perpendicular to the equipotential lines. In fact, we can easily show that electric field lines are always perpendicular to lines or surfaces made up of points that are at equal electric potential. Assume for a moment that the electric field, $\vec{E}$, was not perpendicular to the equipotential lines. Then there would be a component of the electric field along the equipotential line and, hence, a force acting on any charge moving along the line. But this force would do work as the charge moved along the line, which would make the electric potential different at each point, so the line could not be an equipotential line. This contradiction shows that *electric field lines must always be perpendicular to equipotential lines* when they cross.

We show another situation in Figure 21-10 where electric field lines and equipotential lines are plotted for a set of two charges of equal magnitude, one positive and one negative. The field lines are closest together near the charges, indicating that the electric field is strongest there. Very near each charge, the effect of that charge dominates on the equipotential

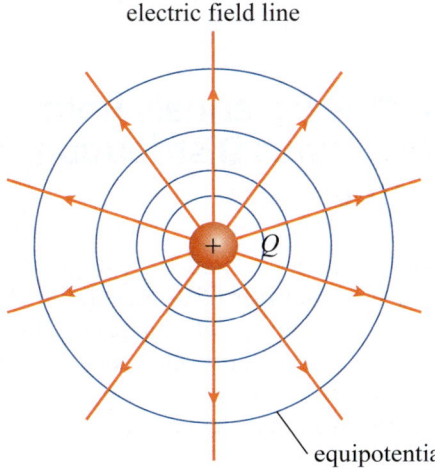

Figure 21-9 Equipotential lines and electric field lines around a single positive charge. Since electric potential is proportional to $1/r$, the equipotential lines will not be equally spaced for constant-voltage intervals, being closer together near the charge.

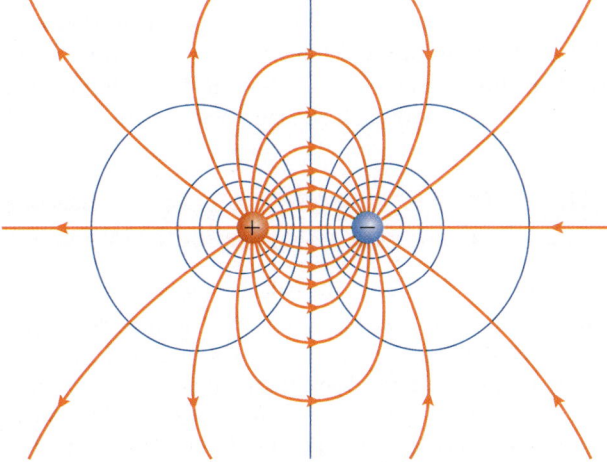

Figure 21-10 Equipotential lines and electric field lines for a pair of charges of equal magnitude but opposite sign.

plots, which are not strongly different from circles, but farther away the superposition of the two electric potentials produces more noticeable effects. Note that electric field lines always cross equipotential lines perpendicularly.

Figure 21-11 shows a fixed positive central charge with one negative point charge of lesser magnitude symmetrically placed on each side. The positive charge is +4 nC, and the negative charges are −1 nC. We have plotted 13 equipotential lines at 10 V intervals. There would be higher and higher potentials near each charge. Note that very near each charge, the contribution from

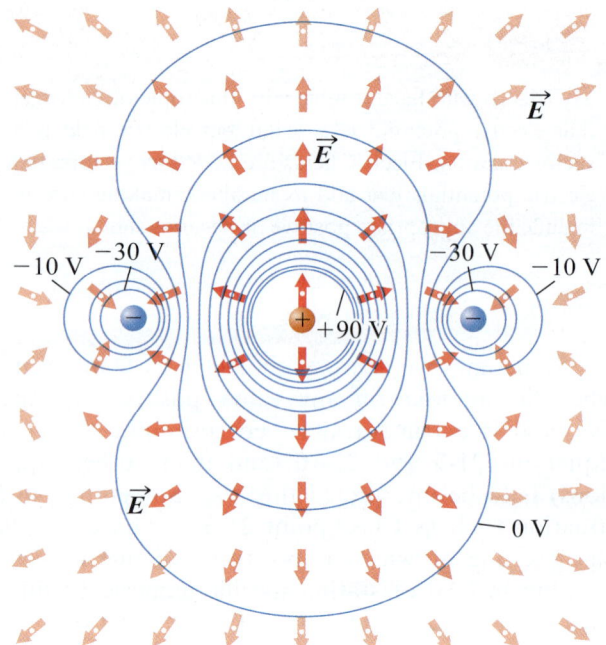

Figure 21-11 Equipotential and electric field vectors for a situation with a positive charge (+4 nC) in the centre and one equal negative (−1 nC) charge symmetrically on each side. The plotted equipotential lines are at intervals of 10 V, starting with the equipotential line closest to the positive charge at +90 V. The orange arrows represent the approximate direction of the electric field at each point and the depth of colour the field strength. You can produce your own plots like this using the PhET simulation "Charges and Fields."

that charge dominates, and the equipotential lines approximate concentric circles. Farther away, the influence of other charges on the net potential becomes apparent. The PhET "Charges and Fields" plots electric vectors instead of electric field lines, in a manner similar to Figure 21-11. In the figure and in the PhET, the depth of colour on the vectors is used to represent the strength of the electric field.

21-5 Electric Potentials from Continuous Distributions of Charge

So far, we have only considered the electric potential produced by a small number of discrete point charges. For continuous distributions of charge, we divide the charge into infinitesimal charge elements, dq, and use the following equation to find the electric potential contribution, dV, from that element:

KEY EQUATION

$$dV = \frac{dq}{4\pi\varepsilon_0 r} \tag{21-12}$$

We integrate over the entire object to find the net electric potential, in a similar manner to what we did in

Chapter 19 for electric fields. Note that r represents the distance from dq to the point where electric potential is being measured. Since electric potentials are scalar quantities, the integrations are considerably easier than the electric field calculations for charge distributions in Chapter 19, where direction information must be considered. We demonstrate the technique in Example 21-6. We will frequently use linear, surface, or volume charge densities in solving problems with continuous distributions of charge, as we did in Chapters 19 and 20.

EXAMPLE 21-6

Electric Potential for a Charged Plastic Washer

A thin plastic washer (non-conducting) has an inner radius of 2.00 cm and an outer radius of 3.00 cm. It is uniformly charged with surface charge density $\sigma = 240$ nC/m². What is the electric potential at a point $h = 4.00$ cm directly above the centre of the washer (point P in Figure 21-12)?

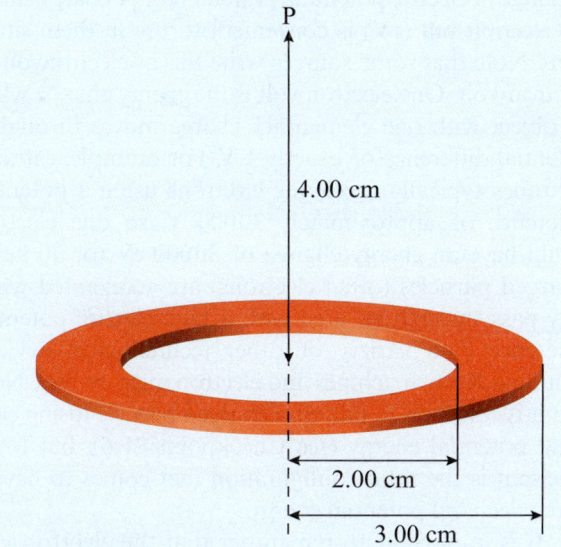

Figure 21-12   Example 21-7.

Solution

We divide the washer into differential charge elements and use Equation 21-12 to find the contribution of each to the electric potential. We then integrate over the entire washer to get the net electric potential.

We divide the washer into small thin flat rings of radius r and thickness dr, as shown in Figure 21-13. The differential area of one ring is approximated by a rectangle of length $2\pi r$, the circumference of the ring, and of width dr, and will therefore have area $dA = 2\pi r dr$. Therefore, the charge on that ring will be

$$dq = \sigma 2\pi r dr$$

The distance to point P from all points on the ring of radius r under consideration is $\sqrt{h^2 + r^2}$. Therefore, the contribution to the electric potential from that differential ring of charge is

$$dV = \frac{\sigma 2\pi r dr}{4\pi \varepsilon_0 \sqrt{h^2 + r^2}} = \frac{\sigma r dr}{2\varepsilon_0 \sqrt{h^2 + r^2}}$$

As the Peer to Peer box above points out, the symbol r is sometimes (as here) used for a radial distance to an element of charge, but in Equation 21-12 it means the distance from the

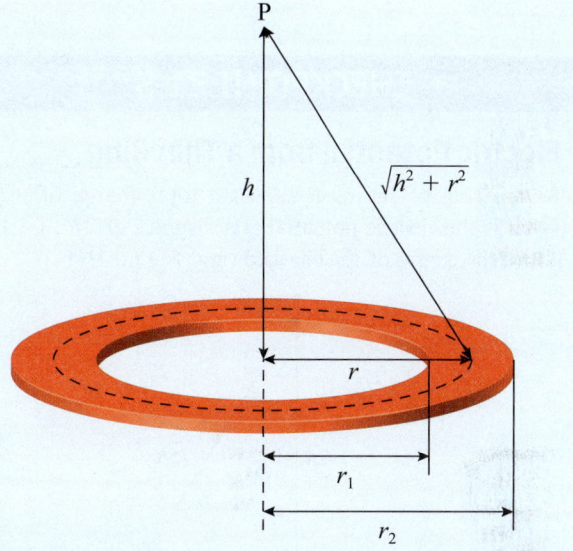

Figure 21-13   Divide the washer of Example 21-7 into thin charged rings of radius r and width dr.

charge element to the point above the washer where the electric potential is calculated.

We integrate over all rings within the washer to find the total electric potential:

$$V = \int_{\text{annulus}} dV = \frac{\sigma}{2\varepsilon_0} \int_{r_1}^{r_2} \frac{r dr}{\sqrt{h^2 + r^2}}$$

$$= \frac{\sigma}{2\varepsilon_0} \left[\sqrt{h^2 + r^2} \right]_{r_1}^{r_2}$$

$$= \frac{\sigma}{2\varepsilon_0} \left(\sqrt{h^2 + r_2^2} - \sqrt{h^2 + r_1^2} \right)$$

If we substitute in the values for h, r_1, and r_2, we obtain the final result, $V = 71.6$ V.

Making sense of the result

We can estimate if the result is (at least) approximately correct by seeing if it lies within the range of the value for a point charge (with the entire charge on the washer) and a distance corresponding to the inner edge of the washer and the value when we use the outer edge of the washer. The distance to the point P from the inner edge of the washer is 0.0447 m and for the outer edge is 0.0500 m. We find the total charge on the washer by multiplying the charge density by the area of the washer: $\pi(r_2^2 - r_1^2)\sigma = 3.78 \times 10^{-10}$ C. The two estimates for the electric potential obtained in this way are 67.8 V and 75.8 V. As expected, the value we obtained in the solution falls between these minimum and maximum estimates.

CHECKPOINT 21-5

Electric Potential from a Thin Ring

A thin ring of radius R carries a total charge of $+Q$. What is the electric potential at a distance of $2R$ directly above the centre of the charged ring (Figure 21-14)?

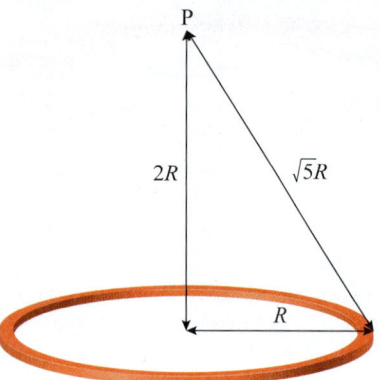

Figure 21-14   Checkpoint 21-5.

(a) $\dfrac{Q}{4\sqrt{5}\pi\varepsilon_0 R}$ (c) $\dfrac{Q}{8\pi\varepsilon_0 R}$

(b) $\dfrac{Q}{4\sqrt{5}\pi\varepsilon_0 R^2}$ (d) $\dfrac{Q}{8\pi\varepsilon_0 R^2}$

ANSWER: (a) The critical point to realize here is that all differential elements are at the *same* distance from the point ($\sqrt{5}R$). Because they are all at the same distance, the electric potential will be identical to the expression for the electric potential at that distance from a similarly charged point charge.

21-6 The Electron Volt

Rearranging Equation 21-8, we see that, when we move a charge q through an electric potential difference ΔV, the change in electrical potential energy, ΔU, is

$$\Delta U = q\Delta V \qquad (21\text{-}13)$$

We have used the notations ΔV and ΔU because we are considering the *changes* in electric potential and electrical potential energy. When we move a small positive charge to a region with higher electric potential, the electrical potential energy of the configuration increases. However, the electrical potential energy of

the system decreases when a negative charge moves to a region of higher electric potential.

When zero work is done by external forces, a loss in electrical potential energy must correspond to a gain in some other form of energy, usually kinetic energy. Ultimately, the accelerations produced in particle accelerators come from this effect, although the details are complex and normally involve alternating electric fields (Chapters 26 and 27) and magnetic forces (Chapter 24) as well. While you could describe the acceleration in terms of forces exerted on the charged particles in electric fields, in many situations it is easier to use an electrical potential energy approach.

In many devices, electrons, protons, and ions undergo a change in electric potential. A small unit of energy called the **electron volt** (eV) is convenient to use in these situations. Note that some sources write this as electronvolt or electron-volt. One electron volt is the energy change when an object with one elementary charge moves through a potential difference of exactly 1 V. For example, cathode ray tubes typically accelerate electrons using a potential difference of approximately 30000 V, so one electron would have an energy change of 30000 eV (or 30 keV). Charged particles (often electrons) are accelerated when they pass through an area with a high electric potential difference in a variety of other technological devices, including X-ray machines and electron microscopes. Note that it is common to talk of a charged particle losing electrical potential energy (see Checkpoint 21-6), but to be precise it is the total configuration that comes to have a lower electrical potential energy.

It is important to remember that the electron volt is a unit of energy, and the volt is a unit of electric potential difference (energy per unit charge). To convert electron volts to joules, we use the magnitude of the elementary charge, 1.60×10^{-19} C, for q:

$$1 \text{ eV} = 1.60 \times 10^{-19} \text{ C} \times 1 \text{ V} = 1.60 \times 10^{-19} \text{ J}$$

The electron volt is a tiny amount of energy by macroscopic standards, but it is a convenient amount on the atomic scale. For example, it takes approximately 13.6 eV to ionize hydrogen and approximately 5.2 eV to remove the first electron from sodium. Those working with atomic spectra usually state energy level differences in electron volts.

Proton beams, such as those produced at Canada's TRIUMF facility, are used to treat certain types of cancer. Proton beams are preferable to X-rays for cancer treatment for a number of reasons, the main one being that the beam width and the effective penetration depth in the body are better controlled. The proton beam damages the DNA of the cancer cells, thereby reducing the ability of the cancerous cells to reproduce. The proton beam for this type of treatment usually needs to be in the energy range of approximately 75 MeV to 225 MeV; that is, each proton has an energy of between 75 million electron volts and 225 million electron volts.

EXAMPLE 21-7

Proton Beam Therapy

TRIUMF can produce protons with energies between 5 MeV and 120 MeV in the lower energy beam (called BL2C). Consider a beam of 100 MeV protons used for treating cancer.

(a) What is the energy of one of these protons, in joules?

(b) The energy of the protons is built up after many passes through an accelerating potential of 94.0 kV. How many passes are needed to produce 100 MeV protons?

(c) The protons are accelerated by an electric potential difference across a gap between two halves of the cyclotron chamber. Which side of this acceleration region has the higher electric potential?

Solution

(a) Since one electron volt equals 1.60×10^{-19} J,

$$100 \text{ MeV} \times \frac{1.00 \times 10^6 \text{ eV}}{1 \text{ MeV}} \times \frac{1.60 \times 10^{-19} \text{ J}}{1 \text{ eV}}$$

$$= 1.60 \times 10^{-11} \text{ J}$$

(b) The proton has a charge of one elementary charge (1.60×10^{-19} C). Therefore, the energy gained when it passes through an accelerating potential of 94.0 kV is 94.0 keV. Therefore, the number of passes required to produce protons with the indicated energy is

$$100 \text{ MeV} \times \frac{1.00 \times 10^6 \text{ eV}}{1.00 \text{ MeV}} \times \frac{1 \text{ pass}}{94.0 \text{ keV}} \times \frac{1.00 \text{ keV}}{1.00 \times 10^3 \text{ eV}}$$

$$= 1064$$

To three significant digits, 1060 passes through the accelerating region are needed.

(c) Protons are positively charged; therefore, they are accelerated toward a more negative region. The end of the acceleration region must be at a lower electric potential.

Making sense of the result

This example demonstrates how convenient the electron volt unit is when dealing with charged particles. Since higher positive electric potentials are in more positive regions, it makes sense that positively charged particles lose electrical potential energy and gain kinetic energy as they move toward more negative regions.

MAKING CONNECTIONS

Ion Propulsion Rockets

The underlying principle of any rocket engine is conservation of linear momentum: fuel is expelled in one direction with a high relative velocity, and the rocket moves in the opposite direction (due to Newton's third law), with the total momentum of the rocket-plus-fuel system unchanged. Most rocket engines, such as those used in space launches, burn fuel for exhaust propulsion. Ion propulsion systems are more energy efficient than chemical propulsion systems. These systems are relatively simple: the engine ionizes a material, such as xenon gas, using electrical energy from solar cells on the spacecraft. The positive ions are then accelerated through a large electric potential difference. High ejection velocities are possible, making ion propulsion quite efficient. Ion propulsion is also safer and more reliable than chemical rockets, which burn fuel in violent oxidation reactions. Ion propulsion systems are in use, but the thrusts available are still tiny compared to chemical propellants. At present, ion propulsion technology is best suited for long-duration voyages where a spacecraft can maintain a tiny thrust for a long time. NASA's *Dawn* mission (Figure 21-15) used ion propulsion to move the space vehicle into the orbits required to provide close-up views of the two largest asteroids, Ceres and Vesta, from 2007 to 2015.

Figure 21-15 NASA's *Dawn* spacecraft employed ion propulsion engines for flights to the asteroids Ceres and Vesta. An image of part of the surface of Ceres is shown here.

ELECTRICITY, MAGNETISM, AND OPTICS

21-7 Calculating Electric Field from Electric Potential

This section, which involves partial derivatives, could be regarded as optional, and the rest of the chapter does not depend on it. In Chapter 6, you saw how one could obtain a conservative force from a potential energy function using partial derivatives. In Chapter 11, we applied this to obtain the gravitational force from the change in gravitational potential energy. This technique is valid for any conservative force field, and in this section we apply it here to find electric fields.

The qualitative idea is that the magnitude of the force depends on how the potential energy is changing over an area. A two-dimensional analogy is a model of the surface height in a region. Where the height, corresponding to potential energy, changes sharply, there are hills (the slope corresponds to the force). If there is no potential energy change when you go in a certain direction, then the force component in that direction is zero. When the potential energy changes strongly as you go in a certain direction, the force component is relatively large in that direction.

We can express this mathematically as follows, where the force components can be calculated from any potential:

$$F_x = -\frac{\partial U}{\partial x}; \ F_y = -\frac{\partial U}{\partial y}; \ F_z = -\frac{\partial U}{\partial z} \qquad (21\text{-}14)$$

Here U is the potential energy function (although we have not shown it explicitly as $U(x, y, z)$, remember that the important point is how it varies with position). Keep in mind that these partial derivatives are taken with respect to one coordinate variable while holding the others constant. For example, for the potential energy given by $U = 20y - 10$ (in J), the x- and z-components of the corresponding force are zero, and the component in the y-direction is -20 N.

Equation 21-14 applies for Cartesian coordinates. If we have radial symmetry so that the potential energy only depends on the radial distance from some point, and not the direction, the radial force is given by

$$F_r = -\frac{\partial U(r)}{\partial r} \qquad (21\text{-}15)$$

Electric potentials are commonly used in place of electrical potential energy when dealing with electric fields. Recall that electric potential is the electrical potential energy per unit charge, and electric field is electrical force per unit charge. If we divide both sides

of Equation 21-14 by charge, we get the following important relationships:

$$E_x = -\frac{\partial V}{\partial x}; \ E_y = -\frac{\partial V}{\partial y}; \ E_z = -\frac{\partial V}{\partial z}$$

$$(21\text{-}16)$$

The electric field components can always be calculated when we know the electric potential at all points. Remember that electric potential, V, is a scalar quantity, while the electric field, $\vec{E} = E_x\hat{i} + E_y\hat{j} + E_z\hat{k}$, is a vector quantity. If the electric potential does not change at all along a certain direction, there is no electric field component in that direction. When the electric potential changes sharply as you move in a direction, there is a strong electric field.

The above relationship between the electric field and the electric potential suggests that the SI units for electric field are V/m. But from the definition of electric field (Chapter 19) as electrical force per unit charge, we have units of N/C. Clearly these must be equivalent, so we have the following:

$$1\frac{V}{m} = 1\frac{N}{C} \qquad (21\text{-}17)$$

CHECKPOINT 21-7

What Is a V · C?

What common unit is equal to a volt times a coulomb?

ANSWER: J (joule, a unit of energy). From Equation 21-17, we can see that $V \cdot C = N \cdot m$, which is equal to a joule and is the unit of energy. We could alternatively have come up with this result using the definition of electric potential as electrical potential energy per charge, so when we multiply the volt by the coulomb it must give units of energy.

PEER TO PEER

It's so easy to forget the negative sign in the expression for electric field from electric potential. I must not forget that it is not the partial derivative, but the *negative* of the partial derivative. I find it helpful to think about the gravity case: when I go up in gravitational potential energy, the gravitational force points down.

If you express the volt, newton, and coulomb in terms of primary SI units (see Chapter 1), you can independently show that this relationship is valid.

If you know the electric potential at all points, then you can use these relationships to find the electric field at all points. Since electric potentials are scalar quantities, and we can calculate them using the superposition principle, this approach is often the easiest way to calculate electric fields, even in complex situations.

CHECKPOINT 21-8

Average Electric Field

The electric potential at a point on the x-axis, $x = 1.0$ m, is $+50$. V, assuming some definition of the zero point for electric potential. At the point $x = 3.0$ m, also on the x-axis, the electric potential is -30. V. What must be the average value of the x-component of the electric field between the two points?

other components in this case.
that we don't have information about points off the x-axis, so we don't know if there are
The x-component of the electric field points in the positive direction in this case. Note

$$\overline{E}_x = -\frac{\Delta V}{\Delta x} = -\left(\frac{-30 \text{ V} - 50 \text{ V}}{3.0 \text{ m} - 1.0 \text{ m}}\right) = +40 \text{ V/m}$$

with changes in quantities:
express the average value. We therefore rewrite Equation 21-16 by replacing derivatives
ANSWER: $\overline{E}_x = +40$ V/m. Since we do not have the full functional form, we can only

---

Often we have spherical symmetry, where the electric potential only depends on the radial distance from some point. In this case, we can calculate electric field from the following relationship, where $\hat{r}$ represents a radially outward unit vector:

KEY EQUATION
$$\vec{E} = -\frac{\partial V(r)}{\partial r}\hat{r} \qquad (21\text{-}18)$$

Note that if the electric potential is decreasing at greater radial distance, then the electric field direction will be radially outward due to the negative sign. Don't forget that this is only valid when we have radial symmetry—the general case for spherical coordinates could have three components for the electric field.

One can elegantly express the operation of Equations 21-16 and 21-18 in terms of a mathematical operation called the **gradient** that is independent of a particular coordinate representation:

$$\vec{E} = -\vec{\nabla}V \qquad (21\text{-}19)$$

The gradient does not change the physics, just how we write the operation. If you prefer not to use this notation, it is not essential. It is still true that the electric field component in each direction depends on the negative of the partial derivative of the electric potential in that direction. In directions with sharply changing

electric potential, the electric field strength (magnitude of the electric field) is high.

CHECKPOINT 21-9

Electric Potential and Electric Field

The electric potential (in V) in a certain region is given by $V = 3x^2 - 5$. What is the electric field at the origin of the coordinate system?
(a) All components are zero.
(b) The y- and z-components are zero, and the x-component is in the $+x$-direction.
(c) The y- and z-components are zero, and the x-component is in the $-x$ direction.
(d) The x-component is zero, and the y- and z-components are both infinite.

origin.
derivative with respect to the x-coordinate gives $E_x = -6x$, which has a zero value at the
tial, clearly, those components of the electric field are zero everywhere. Taking the partial
ANSWER: (a) Since y and z do not appear explicitly in the expression for electric poten-

---

You may wonder whether we can reverse the process used above and determine the resultant electric potential from the electric field. Indeed, you can perform an integral along a path between two points to find the potential difference between the points. In a conservative field, the path taken does not matter, only the end points. We let V_a and V_b represent the electric potentials at points a and b, and $d\vec{\ell}$ represent a small element of the path from point a to point b. The electric potential *difference* between the two points is given by the following integral:

KEY EQUATION
$$V_b - V_a = -\int_a^b \vec{E} \cdot d\vec{\ell} \qquad (21\text{-}20)$$

Of course, *absolute* electric potential can have an arbitrary constant value added; this constant is determined

ELECTRICITY, MAGNETISM, AND OPTICS

EXAMPLE 21-8

Electric Field from Electric Potential

In a region of space, the electric potential, $V(x, y)$, as a function of position is given by $V(x, y) = \left(\dfrac{25.0}{x^2 + y^2} + 80.0\right)$ V.

Note that the V on the left refers to the electric potential, while the V on the right means units of volts.

(a) Is the potential defined and finite everywhere? Qualitatively describe the electric potential.
(b) Derive expressions for the components of the electric field.
(c) Find the electric field at the point $(1.5, 0.0, 0.0)$ m.
(d) Sketch the magnitude of the x-component of the electric field along the x-axis as a function of distance from the origin.

Solution

(a) The potential is defined everywhere except at the origin, where it becomes infinite. The electric potential is positive everywhere and decreases as the distance from the origin in the xy-plane increases. The potential approaches a value of $+80.0$ V for large absolute values of either the x- or y-coordinate.

(b) To obtain the components of the electric field, we use Equation 21-16. Rewriting the electric potential as $V(x, y) = (25(x^2 + y^2)^{-1} + 80)$ V makes it easier to take the partial derivatives. For the x-component of the electric field, we have the following, where for clarity we have suppressed units until the final expression:

$$E_x = -\frac{\partial V}{\partial x} = -25(-1)(x^2 + y^2)^{-2}(2x) + 0$$

$$= (50x(x^2 + y^2)^{-2}) \text{ V/m}$$

Similarly,

$$E_y = -\frac{\partial V}{\partial y} = (50y(x^2 + y^2)^{-2}) \text{ V/m}$$

Since z does not appear in the expression for the electric potential, the partial derivative with respect to z is zero, so $E_z = 0$.

(c) Substituting $x = +1.5$ m, $y = 0$ m, and $z = 0$ m into the expressions from part (b), we obtain

$$E_x = 50x(x^2 + y^2)^{-2} = 50(+1.5)(1.5^2 + 0^2)^{-2} = +14.8 \text{ V/m}$$
$$E_y = 0 \text{ V/m}$$
$$E_z = 0 \text{ V/m}$$

Therefore, the electric field at the point $(1.5$ m, 0 m, 0 m) has a magnitude of 14.8 V/m and points in the $+x$-direction.

(d) Figure 21-16 shows plots of the magnitude of the electric potential and the magnitude of the electric field along the x-axis. Note that the magnitudes of both V and E_x are infinite at $x = 0$.

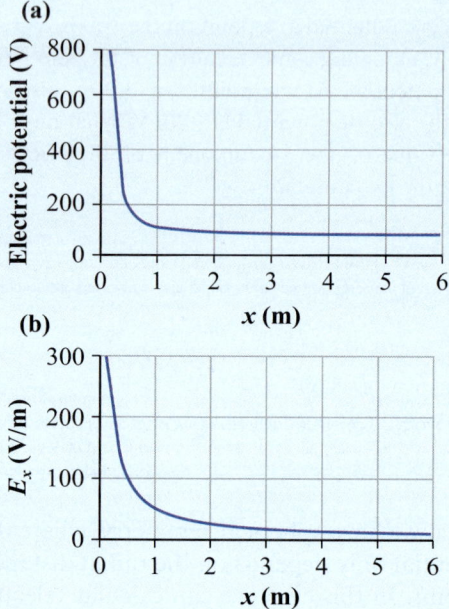

Figure 21-16 (a) A plot of the electric potential (V) as a function of position along the $+x$-axis. (b) A plot of the x-component of the electric field along the $+x$-axis.

Making sense of the result

The first term in the potential function decreases as the square of the distance in the xy-plane from the origin. The constant term does not affect the electric field because the derivative of a constant is zero. Since the electric potential is positive along the x-axis and increases closer to the origin, we expect a positive test charge on the positive x-axis to be repelled with the x-component of the electric field in the positive direction, as obtained above. Careful examination of the two plots shows that the values of E_x correspond to the negative of the slope of the V plot, as expected, because the slope corresponds to the derivative of the potential function.

---

by the user-defined zero point for electric potential. Normally, however, we talk about the electric potential difference between two points. We urge you to use Equation 21-20 in the form shown, rather than replacing the left side with the ambiguous ΔV, since it is important to specify which electric potential is more positive.

Pay careful attention to the signs of the various quantities. You should always check whether the result that you obtain is consistent with what you know about electric fields and potentials. For example, for the situation in Figure 21-17, an electric field points straight down. By definition, the path element $d\vec{\ell}$ points along the path going from a to b. Since $\vec{E}$ and $d\vec{\ell}$ have almost

ELECTRICITY, MAGNETISM, AND OPTICS

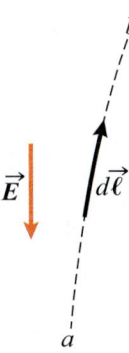

Figure 21-17 In this situation, the electric field points from the top to the bottom. The element $d\vec{\ell}$ always points along the integration path going from point a to point b.

opposite directions, the dot product of the vectors is negative. The right side of Equation 21-20 also has a negative sign, so the net sign is positive, indicating that V_b is more positive than V_a.

Remember that you can take any path in a conservative field, so you are free to choose a path for which the dot product and integral are easy to calculate. We demonstrate this in Example 21-9.

21-8 Electric Potentials and Fields for Conductors

In this section, we return to the consideration of charges and electric fields on conductors started in Section 20-7 Now we will also consider electric potentials, and prove that the electric field is stronger near sharply curved edges of irregular conductors. This has many practical applications, and is visually demonstrated in the opening image of the previous chapter (Figure 20-1).

Consider the conducting sphere shown in Figure 21-18. An ideal conductor allows charge to flow without resistance. An infinitesimally small electrical force would move a charge within this conductor, and if the charges are moving, the situation is not static. Consequently, for static (no flow of charge) situations, there must be zero electric field *within* the conductor (but very definitely one outside the conductor, as shown).

We used Gauss's law in Section 20-7 to prove that charges must reside only on the outer surface, since if we draw any spherical Gaussian surface, even one just barely inside the surface, it will have no electric

EXAMPLE 21-9

Calculating Electric Potential Difference

In a region of space, the electric field has a constant value of $\vec{E} = -5.0 \text{ V/m}\hat{j}$.

(a) What is the electric potential difference between the origin of a Cartesian coordinate system and the (x, y, z) point $(0, 1.0, 0)$ m?

(b) What is the potential difference between $(0, 1.0, 0)$ m and $(1.0, 1.0, 0)$ m?

(c) What is the potential difference between the origin and $(1.0, 1.0, 0)$ m?

Solution

(a) We have not drawn a diagram for you, but urge you to draw one to aid your understanding as you go through the steps of the solution. The electric field points in the $-\hat{j}$ direction and is of constant magnitude, while our path from the origin to the indicated point is in the $+\hat{j}$ direction. Therefore, if we apply Equation 21-20 and define a as the origin and b as $(0, 1.0, 0)$ m, we obtain

$$V_b - V_a = -\int_a^b \vec{E} \cdot d\vec{\ell} = -\int_a^b (-5.0 \text{ V/m}\hat{j}) \cdot (+dy \text{ m}\hat{j})$$

$$= (5.0 \text{ V/m}) \int_{0\text{ m}}^{1.00\text{ m}} dy = 5.0 \text{ V}$$

Therefore, the point $(0, 1.0, 0)$ m is at 5.0 V higher electric potential than the origin.

(b) In this case, the electric field still points in the $-\hat{j}$ direction, but the path we are moving along is in the $+\hat{i}$ direction, so the two are perpendicular and the vector dot product yields a zero result. Therefore, the two points are at the same electric potential.

(c) While we could directly apply Equation 21-20 with a path that has both $\hat{i}$- and $\hat{j}$-components, we can make our work easier by realizing that we can take any path between end points in a conservative field and the result will be the same. Therefore, we can first go from the origin to $(0, 1.0, 0)$ m, and then from there to $(1.0, 1.0, 0)$ m point, using the two results we obtained earlier. Therefore, the potential difference between the origin and the $(1.0, 1.0, 0)$ m point is 5.0 V, with $(1.0, 1.0, 0)$ m being at the higher potential.

Making sense of the result

This example demonstrates three important ideas. First, it is only the component of the electric field along the path you are taking between positions that matters. Second, when the electric field strength is constant, we simply multiply the electric field strength by the path difference, along with the appropriate sign from the dot product, to find the electric potential difference. Finally, realizing that we can use any path often makes the computation simpler. It makes sense that we get a higher electric potential at the second point, since the electric field points toward the origin from that point and we would have to do positive work to move a small positive test charge to the point.

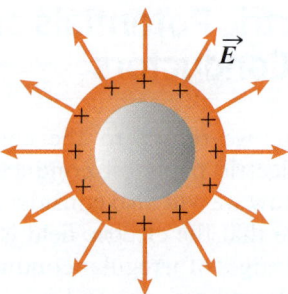

Figure 21-18 For a conducting sphere, since the electric field inside the conductor is zero, Gauss's law requires that all charge reside on the outer surface. The direction of the electric field outside the conductor is that for a positively charged conductor.

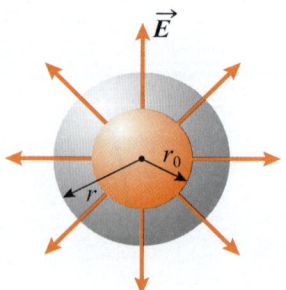

Figure 21-19 A conducting sphere (red) of radius r_0 carries a positive charge Q. The charge is on the outer surface, and the electric field lines start there. We can draw a Gaussian sphere (grey) of radius r to find the electric field at that distance.

flux (since the electric field must be zero) and therefore must have no enclosed electric charge. We also argued in Section 20-7 that the electric field just outside the conductor must be perpendicular to the surface. Note that this orientation applies in static situations, but not necessarily when there are moving charges or changing electric fields, topics considered in Chapter 27. As we saw in Chapter 20, the direction of the electric field (outward or inward) depends on the sign of the charge on the conductor, with positively charged conductors having electric field lines directed outward.

Now we extend our discussion and consider the electric potential for the conducting sphere of Figure 21-18. The electric potential must be exactly the same everywhere within an ideal conductor in the static case (or else the potential difference would result in charge flow, which would be in conflict with our static assumption). Superconductors are the only known materials with exactly zero resistance, but some metals, such as copper and silver, are very close to being ideal conductors, and the potential difference across them can be considered negligible for most situations.

Consider the case of a conducting sphere of radius r_0 that carries charge Q. We draw in Figure 21-19 a Gaussian sphere outside the conducting sphere at some distance r, with $r > r_0$. By the symmetry of the situation, the electric field has a constant magnitude at any particular radial distance, and $\vec{E}$ and $d\vec{A}$ are everywhere directed outward and everywhere parallel to each other. All the charge is enclosed, so we can apply Gauss's law to the Gaussian sphere to find the magnitude of the electric field at a distance r:

$$\oiint_{\text{sphere}} \vec{E} \cdot d\vec{A} = \frac{q_{\text{enc}}}{\varepsilon_0} \tag{21-21}$$

We can express the radially outward directions of $\vec{E}$ and $d\vec{A}$ in terms of the outward-pointing unit radial vector $\hat{r}$. Also, the symmetry of the question requires that the electric field strength be constant over the Gaussian sphere, so it can be taken outside the integration:

$$\oiint_{\text{sphere}} E\hat{r} \cdot dA\hat{r} = \frac{Q}{\varepsilon_0}$$

$$\hat{r} \cdot \hat{r} = 1$$

$$E \oiint_{\text{sphere}} dA = \frac{Q}{\varepsilon_0}$$

The integration over dA is simply the surface area of the Gaussian sphere:

$$E(4\pi r^2) = \frac{Q}{\varepsilon_0}$$

$$E = \frac{1}{4\pi\varepsilon_0} \frac{Q}{r^2} \tag{21-22}$$

Equation 21-22 is consistent with the expression we obtained in Chapter 19 for the electric field around a *point* charge Q. We obtain the electric potential, V, by integrating the electric field over a path from the reference point of zero potential (normally at infinite distance). Since the electric field is the same for points at the same radial distance from a point charge and for points at the same radial distance outside a concentric uniformly charged sphere with the same electric charge Q, the expression we obtained for the electric potential at a distance r from a point charge also applies for the potential from a charged conducting sphere:

$$V(r) = \frac{Q}{4\pi\varepsilon_0 r} \tag{21-23}$$

We have used the notation $V(r)$ to show explicitly that the electric potential is a function of the radial distance, r. It is important to remember that Equation 21-23 only applies for points *outside* the charged conducting sphere. For these exterior points we have now proven that the electric field and the electric potential are the same as for a point charge of the same magnitude. This can help us simplify computations in cases with spherical symmetry.

EXAMPLE 21-10

Electric Potential on a Spherical Shell with an Enclosed Charge

As shown in Figure 21-20, a conducting spherical shell has an inner radius R_1 and an outer radius R_2. The shell does not carry any net charge, but a positive point charge, Q, is placed at its centre. What are the electric potentials on the inner and outer edges of the conducting spherical shell? Assume a definition of the zero point for electric potential of $V = 0$ at $r = \infty$.

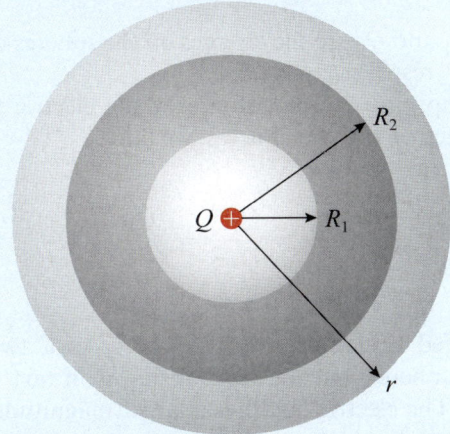

Figure 21-20 Example 21-10. A positive point charge Q surrounded by a concentric conducting spherical shell that carries no net charge.

Solution

There is spherical symmetry, so we draw a Gaussian sphere. For points outside the conducting spherical shell, we use a Gaussian sphere of radius r, where $r > R_2$:

$$\oiint_{\text{sphere}} \vec{E} \cdot d\vec{A} = \frac{q_{\text{enc}}}{\varepsilon_0}$$

The symmetry of the situation requires that the electric field strength, E, be constant over the Gaussian sphere, so it can be taken outside the surface integral. Also, the direction of $\vec{E}$ and $d\vec{A}$ is radially outward everywhere (and they are parallel to each other at each point):

$$\oiint_{\text{sphere}} E\hat{r} \cdot dA\hat{r} = \frac{Q}{\varepsilon_0}$$

$$\hat{r} \cdot \hat{r} = 1$$

$$E \oiint_{\text{sphere}} dA = \frac{Q}{\varepsilon_0}$$

The integration is simply the surface area of the sphere:

$$E(4\pi r^2) = \frac{Q}{\varepsilon_0}$$

$$E = \frac{1}{4\pi\varepsilon_0}\frac{Q}{r^2}$$

We can express the result vectorially using the outward-pointing radial unit vector:

$$\vec{E} = \frac{1}{4\pi r^2}\frac{Q}{\varepsilon_0}\hat{r}$$

Now, using the expression relating the change in electric potential and the electric field, we obtain

$$V_b - V_a = -\int_a^b \vec{E} \cdot d\vec{\ell}$$

$$V_\infty - V_{R_2} = -\int_{R_2}^\infty \frac{1}{4\pi\varepsilon_0}\frac{Q}{r^2}\hat{r} \cdot dr\hat{r}$$

$$V_\infty - V_{R_2} = -\frac{Q}{4\pi\varepsilon_0}\int_{R_2}^\infty \frac{1}{r^2}dr$$

$$V_\infty - V_{R_2} = -\frac{Q}{4\pi\varepsilon_0}\left[-\frac{1}{r}\right]_{R_2}^\infty$$

$$0 - V_{R_2} = \frac{Q}{4\pi\varepsilon_0}\left[0 - \frac{1}{R_2}\right]$$

$$V_{R_2} = \frac{Q}{4\pi\varepsilon_0 R_2}$$

This is the electric potential on the outer edge of the conducting shell. Since it is a conductor, the electric potential must be constant throughout, so the electric potential on the inner edge will have the same electric potential:

$$V_{R_1} = \frac{Q}{4\pi\varepsilon_0 R_2}$$

We show in Figure 21-21 a plot of how the potential would vary with radial distance for distances R_1 and greater.

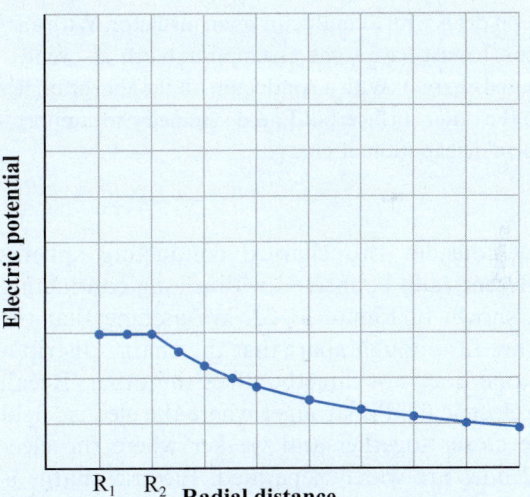

Figure 21-21 A plot of the electric potential for radial distances greater than R_1. Note that the electric potential is constant within the conductor.

Making sense of the result

Since outside the conductor the electric field is the same as for the case of a similarly charged concentric point charge, we are not surprised that we obtained the expression for the electric potential of a point charge. Since the charge is positive, the potential will also be positive (with the standard definition of zero level). Note that expressions for electric potential depend on the zero point chosen, which we have assumed here as $V = 0$ when $r \to \infty$. As we saw in Chapter 20, negative charge will be induced on the inner surface of the spherical shell, leaving positive charge on the outer edge, but the net charge on the shell is zero.

Now consider two charged conducting spheres with different radii connected with a long conducting wire, as shown in Figure 21-22. We assume that the spheres are far enough apart that the charge distribution on one does not directly affect the other. Recall that the electric field is stronger where the electric field lines are closer together and weaker where the electric field lines are widely separated. Electric charge is placed on one of the spheres. Some of that charge will then flow through the conducting wire so both spheres are charged. In electric equilibrium, the charge must be distributed so the electric potential is the same on the

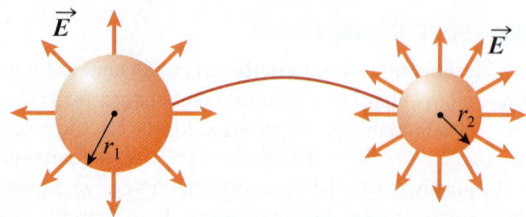

Figure 21-22 Two positively charged conducting spheres with different radii connected with a conducting wire. Note that the spheres are actually farther apart than could be indicated in the diagram.

two spheres (since otherwise charges would flow in the conducting wire and the system would not be in electrical equilibrium as assumed).

The expression we obtained (Equation 21-23) must apply for each of the spheres, and they must both have the same electric potential (since they are connected conductors). Therefore, in electrostatic equilibrium the following must be true:

$$\frac{Q_1}{r_1} = \frac{Q_2}{r_2} \quad (21\text{-}24)$$

where Q_1 and Q_2 are the charges on the spheres of radii r_1 and r_2, respectively.

Comparing the magnitude of the electric field at the surface of the two spheres, we find

$$\frac{E_1}{E_2} = \frac{\dfrac{1}{4\pi\varepsilon_0}\dfrac{Q_1}{r_1^2}}{\dfrac{1}{4\pi\varepsilon_0}\dfrac{Q_2}{r_2^2}} = \frac{Q_1 r_2^2}{Q_2 r_1^2} = \frac{Q_1 r_2^2}{Q_1 r_2 \dfrac{r_1^2}{r_1}} = \frac{r_2}{r_1} \quad (21\text{-}25)$$

We derived Equation 21-25 for the case of two conducting spheres, but the relationship is in fact true in general. The electric field has a larger magnitude when the radius of curvature of a charged conducting surface is smaller. That is, the electric field is stronger near points and has low values beside flat or nearly flat regions of a surface, as illustrated in Figure 21-23. We argued this result qualitatively in Section 20-7, but now we have established a quantitative expression for exactly how the electric field strength varies on surfaces with different curvature.

Because the electric field is much greater in sharply pointed regions, it may become large enough for the air or other gas to break down and conduct a current because we have ionized material near those points. **Corona discharge** is electric discharge due to ionization of material in a fluid (such as air) surrounding a conductor. This discharge happens near pointed regions on conductors. The minimum electric field needed for breakdown is called the breakdown field and is typically 3 MV/m for fairly dry air. Corona discharge is used in ozone generators, in electrical precipitators that remove particulates in

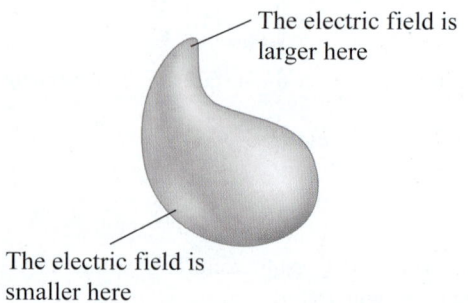

The electric field is larger here

The electric field is smaller here

Figure 21-23 The electric field is stronger near sharply curved points on a conductor.

MAKING CONNECTIONS

Electrostatic Precipitators

Undesired by-products of power plant and industrial combustion processes are particulates in the exhaust gases. Electrostatic precipitators can reduce particulate concentrations. The gas containing particulates passes through a chamber region with high electric potential differences applied to a grid with many sharp points (like a screen with spikes or barbed wire points). Figure 21-24 shows the configuration of a typical electrostatic precipitator. The electric field is high enough near these points to cause corona discharge, which makes some of the particulates electrically charged. When the exhaust then passes near electrode plates with the opposite electric polarity, the charged particles precipitate onto the plates, thus cleaning the exhaust gases. Mechanical or fluid cleaning periodically removes the buildup of particulates from the plates. A similar process can be used in electrostatic cleaners as part of hot-air heating systems in the home and in free-standing air-cleaning units. The idea of charging dust to remove it was devised in the nineteenth century, but the first modern electrostatic precipitator was not patented until 1907, by American scientist and engineer Dr. Frederick G. Cottrell (1877–1948). The research corporation that Cottrell set up using revenues from these patents continues to fund university scientific research.

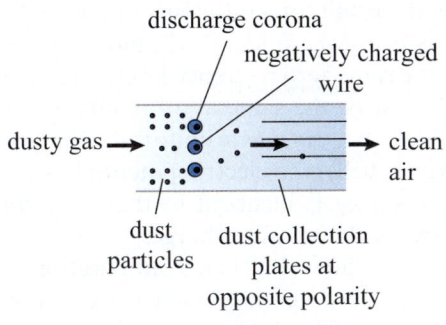

Electrostatic precipitator

Figure 21-24   An electrostatic precipitator.

smokestacks, and in forced-air heating systems (see the Making Connections box). It is also important in many other applications, including lightning protection rods.

21-9 Electric Potential: Powerful Ideas

Electric potential is a very important idea in physics, and you will encounter it repeatedly in the later electricity and magnetism chapters, as well as in modern physics. In this section, we highlight concepts related to electric potential applications, including common mistakes to avoid. Also, we will introduce a new method

for solving electric potential and electric field problems called the method of images.

Electric potential is defined as electrical potential energy per charge. Since electrical potential energy must be specified relative to a user-defined zero point, electric potential is also defined relative to a defined zero point (commonly called the ground). The definition of zero point will not matter if we use *changes* in potential energy between two configurations, or *changes* in electric potential, and this is often done:

$$\Delta V = \frac{\Delta U}{q} \tag{21-8}$$

While Equation 21-8 is a defining relationship for electric potential difference, it is most frequently applied to solve for ΔU and find the loss or gain in potential energy as a charge is moved from one position to another. We calculate the difference in electric potential between the points and then multiply by the charge that is moved between the positions to find the change in electrical potential energy.

If we do define a zero point for electric potential, then it is possible to talk about electric potential in absolute terms (relative to that reference). While any user definition of zero point is allowed, the one that is most frequently used is that the electric potential is zero at infinite distance from the charges. If we do this, then the electric potential at a point r away from a point charge Q is given by

$$V = \frac{Q}{4\pi\varepsilon_0 r} \tag{21-10}$$

Note that Equation 21-10 implies that (for this definition of zero level) the electric potential is positive in regions around positive charges and negative near negative charges. This makes sense in terms of Equation 21-8 since it would require work to move a positive test charge from an infinite distance to near a positive charge. This is helpful to keep in mind when you are making sure the results you obtain make sense.

Electric potentials follow the superposition principle, which means that we can find the electric potential due to each point charge separately and then simply sum the results. Since electric potentials are scalar quantities, they simply add as numbers. Therefore, superposition of electric potentials is much easier than for the case of vector superposition for electric fields. When we have continuous distributions of charge, we divide the charge into differential elements and integrate over the object.

When using superposition, remember that for conductors, the presence of one charged object can influence the charge distribution on the other conductor. Therefore, superposition of electric potential is most often used for non-conducting objects and for collections of fixed point charges.

Electric Potential between Two Charged Spheres

As shown in Figure 21-25, two charged non-conducting spheres each carry uniformly distributed charges of Q. One sphere is of radius R and the other is of radius $2R$. What is the electric potential at the point, P, where the two spheres almost touch?

(a) $\dfrac{Q}{8\pi\varepsilon_0 R}$

(b) $\dfrac{3Q}{8\pi\varepsilon_0 R}$

(c) $\dfrac{3Q}{4\pi\varepsilon_0 R}$

(d) $\dfrac{3Q}{\pi\varepsilon_0 R}$

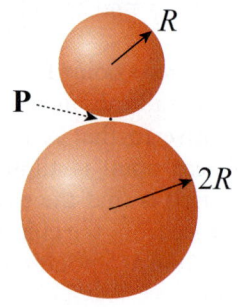

Figure 21-25   Checkpoint 21-11.

ANSWER: (b) We add the electric potential due to each sphere using superposition. Since the electric field everywhere outside the sphere is by Gauss's law the same as for a point charge of the same charge (assuming the uniform charge distribution on the non-conducting spheres), we have $\dfrac{1}{4\pi\varepsilon_0}\left\{\dfrac{Q}{R} + \dfrac{Q}{2R}\right\} = \dfrac{3Q}{8\pi\varepsilon_0 R}$. Note that we could not have done this if the spheres had been conducting, since then we would have had to know how the charges were distributed on the surfaces of the spheres.

---

scalar superposition to find the electric potential, and then taking partial derivatives to find the electric field. This approach can be implemented using numerical computation to find approximate electric fields. In Chapter 20, you encountered several ways to find the electric field. Now we have an additional one: find the electric potential at different points, and use the gradient to find the electric field.

Sometimes you can use one result to simplify obtaining the result for a different situation. If the electric field is everywhere identical (and both configurations use the same zero reference level), then the electric potential must be the same at all points. We used this idea to show that *outside* a uniformly charged sphere (or spherical shell), the electric potential as a function of radial distance is identical to that of a concentric point charge with the same charge.

If we can show that two configurations produce identical electric potentials, then we can use one to model the other. This is the basis of a powerful technique called the **method of images** (see Example 21-11), in which we replace a single point charge and a grounded conducting plane with two point charges. It can also be used in more complicated configurations. This technique is normally covered in higher-level courses and can be considered optional.

In Chapter 22 we extend the ideas of this chapter and Chapter 20 to a particularly important technological case of nearby charged plates called capacitors. Capacitors play important roles in almost every electronic device.

You can determine electric fields from electric potential distributions (or vice versa). We can express this relationship in Cartesian coordinates as

$$\vec{E} = -\frac{\partial V}{\partial x}\hat{i} - \frac{\partial V}{\partial y}\hat{j} - \frac{\partial V}{\partial z}\hat{k} \qquad (21\text{-}26)$$

Note that electric potential is a scalar, but the derivatives with respect to spatial coordinates result in a vector electric field. For many complex problems, the easiest way to compute the electric field is by first using

ELECTRICITY, MAGNETISM, AND OPTICS

EXAMPLE 21-11

Method of Images

(a) A positive charge, Q, is placed at some distance, d, to the left of an infinite conducting plate that is grounded. What sign of charge is induced in the conducting plate at the point nearest Q?

(b) Sketch electric field lines for the situation.

(c) In more advanced electromagnetism courses, a technique called the method of images is used. In this case, we can imagine a charge of the opposite polarity (an "image charge") placed behind the plate at an equal distance, d, from the plate. We know that this is the correct placement because it will lead to a zero potential along the conducting plate, as required. If two configurations produce the same electric potential at all points in a region, then the electric field must also be the same everywhere in that region. We can replace the conducting plate with the image charge and consider the two charges when calculating electric potentials. Use this idea to sketch electric equipotential lines.

(d) If $Q = 2.50$ nC and $d = 0.150$ m, what is the electric potential at a point halfway between charge Q and the plate?

Solution

(a) Negative charge is induced in the parts of the plate nearer the charge.

(b)

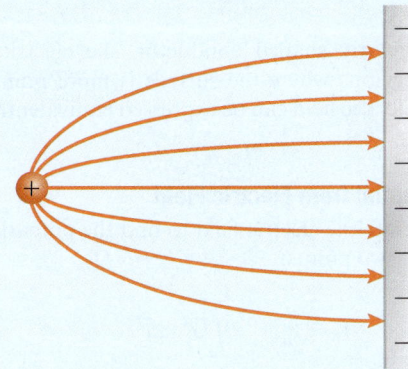

Figure 21-26 Electric field lines for Example 21-11.

(c) The equipotential lines are shown dashed in Figure 21-27. They have positive potentials when around the positive charge

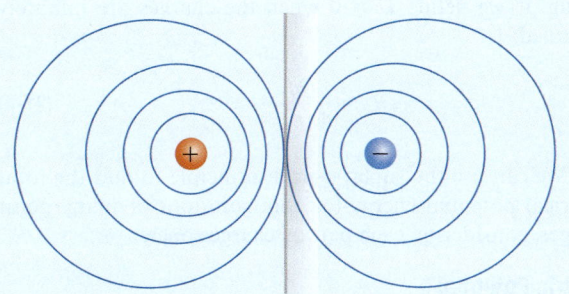

Figure 21-27 Equipotential lines for Example 21-11.

and negative potentials when around the image charge. The grounded conducting plane has zero potential.

(d) We can simply use the superposition principle along with the expression for the electric potential for a point charge. Note that we assume the usual zero point definition of zero potential infinitely far from point charges and use r_1, r_2 for the distances from the point to the positive, negative charges:

$$V = \frac{Q}{4\pi\varepsilon_0 r_1} - \frac{Q}{4\pi\varepsilon_0 r_2}$$

$$= \frac{2.50 \times 10^{-9}\,\text{C}}{4\pi\varepsilon_0 (0.075\ \text{m})} - \frac{2.50 \times 10^{-9}\,\text{C}}{4\pi\varepsilon_0 (0.225\ \text{m})}$$

$$= \frac{2.50 \times 10^{-9}\,\text{C}}{4\pi\varepsilon_0}\left(\frac{1}{0.075\ \text{m}} - \frac{1}{0.225\ \text{m}}\right)$$

$$= 200.\ \text{V}$$

Making sense of the result

We expect the potential to be positive since the point is closer to the positive charge than to the negative image charge. We expect it to be less than the potential at that point due to only the positive charge. Note that the method of images offers an easy solution to what would otherwise be a very complex computational problem if we had to determine the induced charge distribution in the conductor and then integrate over that distribution to find the potential at the point. It can be challenging to apply the method of images to more complex problems. Remember that the charges and image charges you incorporate must make the electric potential zero where any grounded plates are present. Problems 73 and 93 deal with using the method of images in more complicated situations.

KEY CONCEPTS AND RELATIONSHIPS

Electrical Potential Energy

The difference in electrical potential energy can be calculated from the negative of the work done by the force associated with the electric field in moving the charges between the two configurations:

$$U_b - U_a = -\int_a^b \vec{F}_{\text{field}} \cdot d\vec{\ell} \qquad (21\text{-}2)$$

U_a and U_b are the electrical potential energies of configurations a and b, respectively.

When two charges, Q and q, are moved from a separation distance of r_1 to r_2, the change in electrical potential energy is given by

$$U_2 - U_1 = \frac{Qq}{4\pi\varepsilon_0}\left[\frac{1}{r_2} - \frac{1}{r_1}\right] \qquad (21\text{-}3)$$

The electrical potential energy of a configuration of two charges, Q and q, a distance r apart is given by the following, if we define $U = 0$ when the charges are infinitely separated:

$$U = \frac{Qq}{4\pi\varepsilon_0 r} \tag{21-4}$$

We can use the superposition principle to find the total electrical potential energy for configurations of many point charges, considering each pair of charges once.

Electric Potential

Electric potential (V) is measured in volts and is defined as electrical potential energy per unit charge:

$$V = \frac{U}{q} \tag{21-9}$$

The SI unit for electric potential is the volt (V), with one volt equal to one joule per coulomb.

Since electrical potential energy and electric potential are relative to some defined zero reference, we often write ΔV and ΔU to remind ourselves that it is differences that are physically important:

$$\Delta V = \frac{\Delta U}{q} \tag{21-8}$$

The electric potential at a distance r from a single point charge Q is given by the following, if we define the zero of electric potential at a point infinitely far from the charge:

$$V = \frac{Q}{4\pi\varepsilon_0 r} \tag{21-10}$$

The superposition principle can be used to find the electric potential due to collections of point charges. Since electric potentials are scalars, superposition is much easier for electric potentials than for electric fields.

The work required to move a charge q through an electric potential difference ΔV is

$$W = q\Delta V \tag{21-11}$$

The work is positive when moving a positive charge to a position with a more positive electric potential.

We can integrate over a continuous distribution of charge to find the net electric potential using the following differential relationship:

$$dV = \frac{dq}{4\pi\varepsilon_0 r} \tag{21-12}$$

Equipotential Lines

Equipotential lines and surfaces join points of equal electric potential. Usually they are drawn at some equal intervals, such as every 25 V. Electric field lines must always be perpendicular to equipotential lines. The surface of an ideal electrical conductor must be at one common electric potential. There is no electric field inside a conductor in electrical equilibrium. Electric fields just outside a conductor are perpendicular to the surface.

Electron Volt

The electron volt (eV) is a unit of energy that corresponds to the change in electrical potential energy when one elementary charge unit moves through a potential difference of one volt.

Electric Field from Electric Potential

Electric fields correspond to the negative of the rate of change of electric potential with spatial coordinates. In a Cartesian coordinate system, this is expressed in terms of the components of the electric field as

$$E_x = -\frac{\partial V}{\partial x}; \ E_y = -\frac{\partial V}{\partial y}; \ E_z = -\frac{\partial V}{\partial z} \tag{21-16}$$

If we have spherical symmetry, we can compute the electric field from

$$\vec{E} = -\frac{\partial V(r)}{\partial r}\hat{r} \tag{21-18}$$

For an irregularly shaped conductor, the electric field is stronger in regions where the surface is more pointed. The units of the electric field can be expressed equivalently as V/m or N/C.

Electric Potential from Electric Field

We can integrate the electric field to find the potential difference between two points:

$$V_b - V_a = -\int_a^b \vec{E} \cdot d\vec{\ell} \tag{21-20}$$

Conductors

All parts of ideal conductors must be at the same electric potential, and they must have zero electric field at interior points. Just outside an irregularly shaped conductor, the electric field is stronger in regions where the surface is more pointed and can be shown to depend inversely on the local radius of curvature of the surface.

APPLICATIONS

atmospheric electricity, cathode ray tube devices, electron microscopes, electrostatic precipitators for dust control, ion propulsion engines, ozone production by corona discharge, particle accelerators, protection from electrical breakdown, radiation treatment

KEY TERMS

closed path, conservative field, corona discharge, electric potential, electric potential differences, electrical potential energy, electrical equilibrium, electron volt, electrostatic, electrostatic potential energy, equipotential lines, external agent, gradient, ground, method of images, volt, voltage

QUESTIONS

1. An electron ($-e$ charge) is in orbit about a hydrogen nucleus ($+e$ charge) at a distance of r_0. What is the electrical potential energy of this system?

   (a) $-\dfrac{1}{4\pi\varepsilon_0}\dfrac{e^2}{r_0}$

   (b) $+\dfrac{1}{4\pi\varepsilon_0}\dfrac{e^2}{r_0}$

   (c) $-\dfrac{1}{4\pi\varepsilon_0}\dfrac{e^2}{r_0^2}$

   (d) $+\dfrac{1}{4\pi\varepsilon_0}\dfrac{e^2}{r_0^2}$

2. You have two unknown point charges, Q_1 and Q_2. You observe that the electric potential is zero exactly halfway between the two charges. What conclusion can you make?

   (a) The two charges have the same magnitude and the same sign.
   (b) The two charges have the same magnitude and the opposite sign.
   (c) There is no way that you can have zero potential at the midpoint.
   (d) The magnitudes of both charges are zero.

3. An electron is moving past a positive ion on the path shown in Figure 21-28. At which of the following points does the electron–ion system have the highest speed? Assume that there are no other charges present.

   (a) A
   (b) B
   (c) C
   (d) D

Figure 21-28 Question 3.

4. An electron goes from a region at -10 kV to a region at $+30$ kV. What is the electron's change in kinetic energy?

   (a) decrease by 40 keV
   (b) increase by 20 keV
   (c) increase by 30 keV
   (d) increase by 40 keV

5. You have two fixed charges, a positive charge, $+Q$, on the left and a charge, $-4Q$, on the right, separated by a distance of 3 m. Where is one location with zero electric potential?

   (a) at a point 1 m to the left of the $+Q$ charge
   (b) at a point 1 m to the right of the $+Q$ charge
   (c) at a point 1 m to the left of the $-4Q$ charge
   (d) at a point 1 m to the right of the $-4Q$ charge

6. Which of the following statements is correct?

   (a) Electric field and electric potential are both vectors.
   (b) Electric field and electric potential are both scalars.
   (c) Electric field is a vector, and electric potential is a scalar.
   (d) Electric field is a scalar, and electric potential is a vector.

7. A proton is moving in the $+x$-direction, and its speed is increasing but its acceleration is constant. Which of the following statements is correct?

   (a) The proton is moving into a region with higher electric potential.
   (b) The proton is moving into a region with lower electric potential.
   (c) The acceleration is constant, so there is no change in electric potential.
   (d) The scenario is impossible.

8. A positive charge is located at the origin of an xyz-coordinate system (and no other charges are present). What can you conclude about the electric potential at the points A(0, -1, 0) and B(0, $+1$, 0) in SI units?

   (a) Points A and B are both at a negative electric potential.
   (b) Points A and B are both at a positive electric potential.
   (c) Point A is at a negative electric potential, and point B is at a positive electric potential.
   (d) Point A is at a positive electric potential, and point B is at a negative electric potential.

9. When an electric potential increases with distance in the x-direction according to $V = 3x^2 + 20$, what is the electric field at the point $x = y = z = 1$ m?

   (a) The electric field components are $+20$ N/C in the y- and z-directions and -6 N/C in the x-direction.
   (b) The electric field components are $+20$ N/C in the y- and z-directions and $+6$ N/C in the x-direction.
   (c) The electric field components are zero in the y- and z-directions and $+14$ N/C in the x-direction.
   (d) The electric field components are zero in the y- and z-directions and -6 N/C in the x-direction.

10. When you slightly increase the distance between a proton and an electron in a hydrogen atom, what happens to the electrical potential energy of the system?

    (a) The electrical potential energy increases.
    (b) The electrical potential energy decreases.
    (c) The electrical potential energy is negative both before and after but does not change in magnitude.
    (d) The electrical potential energy is zero both before and after.

11. The eV is a unit of

    (a) electric flux
    (b) electric potential
    (c) energy
    (d) power

12. An ion contains a nuclear charge of $+5e$ and has a total electronic charge of $-4e$. What is the electric potential at a distance r well outside the atom? Assume spherical symmetry. Assume that the electric potential is zero at infinite distance.

    (a) $-\dfrac{1}{4\pi\varepsilon_0}\dfrac{e}{r^2}$

    (b) $+\dfrac{1}{4\pi\varepsilon_0}\dfrac{e}{r^2}$

    (c) $+\dfrac{1}{4\pi\varepsilon_0}\dfrac{e}{r}$

    (d) $+\dfrac{1}{4\pi\varepsilon_0}\dfrac{9e}{r}$

13. The work done by an electric field in moving a $+200\ \mu C$ charge from point A to point B is $+40$ mJ. If point A is at $+500$ V potential, what is the electric potential at point B?
   (a) -200 V
   (b)   300 V
   (c)   700 V
   (d)   900 V

14. For some configuration of point charges, a small test charge q is moved clockwise around a closed circle of radius R. A second test charge, also q, is moved around a closed circle of radius $2R$ but going counterclockwise. Which statement correctly relates the work done in the two cases?
   (a) The work is zero in both cases.
   (b) The work for the larger loop will have the opposite sign and double the magnitude.
   (c) The work for the larger loop will have the opposite sign and half the magnitude.
   (d) The work will be positive in both cases, but there will be four times as much work for the larger loop.

15. Two conducting spheres with different radii are joined by a wire, and a negative charge is placed on the system. Which statement correctly describes the electric field and electric potential just outside the surface of each sphere?
   (a) The electric field points toward the surface in both cases and is stronger on the smaller sphere. The electric potential is negative and the same for both spheres.
   (b) The electric field points away from the surface in both cases and is stronger on the smaller sphere. The electric potential is zero for both spheres.
   (c) The electric field points toward the surface in both cases and is stronger on the larger sphere. The electric potential is negative and the same for both spheres.
   (d) The electric field points toward the surface in both cases and is of the same strength. The potential is negative in both cases and higher for the smaller sphere.

16. When the charge on a conducting sphere is doubled, how does the electric potential *inside* the sphere change?
   (a) The electric potential is now four times as large.
   (b) The electric potential is now two times as large.
   (c) The electric potential is now approximately $\sqrt{2}$ times as large.
   (d) The electric potential is zero in both cases.

17. An external agent does $+24.0$ mJ of work on a $+6.00\ \mu C$ charge in moving it to a new position (assume at infinitesimal speed). If it started from a position at ground potential (0 V), what is the electric potential at the new position?
   (a) -144 V
   (b) $+144$ V
   (c) -4.00 kV
   (d) $+4.00$ kV

18. Three $-q$ charges are at the corners of an equilateral triangle of side r. What is the electrical potential energy of the configuration?
   (a) $\dfrac{3q^2}{4\pi\varepsilon_0 r}$
   (b) $-\dfrac{3q^2}{4\pi\varepsilon_0 r}$
   (c) $\dfrac{2q^2}{4\pi\varepsilon_0 r^2}$
   (d) $-\dfrac{2q^2}{4\pi\varepsilon_0 r^2}$

19. Three charges lie along the x-axis: $+Q$ at the origin, $+Q$ at $x = 2$, and $+3Q$ at $x = 3$. What is the electrical potential energy of the configuration (in J)?
   (a) $\dfrac{9Q^2}{4\pi\varepsilon_0}$
   (b) $\dfrac{3Q^2}{4\pi\varepsilon_0}$
   (c) $\dfrac{3Q^2}{2\pi\varepsilon_0}$
   (d) $\dfrac{9Q^2}{8\pi\varepsilon_0}$

20. Equipotential lines are shown in Figure 21-29. A $+1.00$ mC point charge is moved between points A and B as shown. What *net* amount of work needs to be done by the external agent moving that charge?

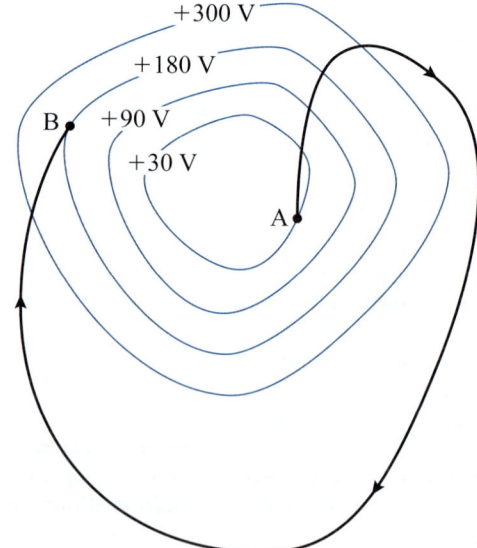

Figure 21-29   Problem 20.

   (a) 150 eV
   (b) 150 mJ
   (c) 300 mJ
   (d) 600 mJ

PROBLEMS BY SECTION

For problems, star ratings will be used (*, **, or ***), with more stars meaning more challenging problems. The following codes will indicate if $\frac{d}{dx}$ differentiation, $\int$ integration, ▢ numerical approximation, or ⌁ graphical analysis will be required to solve the problem.

Section 21-1 Work and Electric Fields

21. ★ If you had only a wire, an ammeter, and a region with an electric field, describe a method to prove that the electric field is conservative.

22. ★★ If in a certain region of space the electric field is constant with value $\vec{E} = -25.0 \text{ N/C}\hat{j}$, calculate the work done by the electric field in moving a 0.200 mC charge from the (x, y, z) position $(0, 0, 0)$ m to $(4.00, 4.00, 0)$ m on each of the following two paths.
    (i)   two steps: from $(0, 0, 0)$ m to $(4.00, 0, 0)$ m, and then on to $(4.00, 4.00, 0)$ m
    (ii) one step: diagonally from $(0, 0, 0)$ m to $(4.00, 4.00, 0)$ m

23. ★★ $\int$ An electric dipole has ± 2.00 nC charges separated by 50.0 μm. How much work is needed to double the distance between the charges (assuming the positive charge is rigidly attached to a point)?

24. ★ What is wrong with the following reasoning? Since the field force and the agent force are of equal magnitude but in opposite direction, you can cancel them out and ignore them both.

25. ★★ An agent does $+50.0$ mJ of work in moving a -25.0 μC charge from point A to point B. Points A and B are 50.0 cm apart. If the electric field is of constant magnitude, what can you conclude about its direction and magnitude (expressing direction in terms of the positions A and B)?

Section 21-2 Electrical Potential Energy

26. ★ We assume that an electron and a proton are on average separated by 53.0 pm. What is the electrical potential energy of a hydrogen atom?

27. ★★ How much work is required to move four charges from infinite initial separation into the following configuration along the x-axis: $+1.00$ μC at $+2.00$ m, $+2.00$ μC at $+3.00$ m, $+3.00$ μC at $+4.00$ m, and $+4.00$ μC at $+5.00$ m?

28. ★ Two protons are 3.00 fm apart. What is the electrical potential energy of this configuration?

29. ★ An electric dipole has two charges, $+1.00$ nC and -1.00 nC, separated by a distance of 0.250 μm. What is the electrical potential energy of the configuration?

30. ★★ A positive charge, $+q$, of mass m is moving at speed v_0 toward a second positive charge, $+Q$, when it is at an initial distance, R, away. The second charge, $+Q$, is assumed to be fixed in position. Also assume that any changes in gravitational potential energy are negligible. Derive an expression for the closest distance (call this r) charge $+q$ gets to $+Q$ before it changes position.

31. ★★★ Two positive charges ($+30.0$ μC) have equal masses (25.0 g). They are initially held at rest with a separation of 4.00 cm. The charges are then released and begin to move apart. What speed is one charge moving *relative to the other charge* when they are at a separation of 12.0 cm?

Section 21-3 Electric Potential

32. ★ What is the electric potential at a point 0.250 cm away from a -1.50 nC charge? Assume the standard reference of $V = 0$ infinitely far from the charge.

33. ★ How far would you need to be from a $+3.00$ nC charge for the electric potential to be $+50.0$ V? Assume the standard reference of $V = 0$ infinitely far from the charge.

34. ★★ What is the electric potential at a point midway between two charges if one is $+0.340$ μC and the other is -0.180 μC? The charges are a distance of 96.0 cm apart.

35. ★★ Four $+2.00$ nC charges are located at the corners of a square that is 10.0 cm on each side. What is the electric potential at the exact centre of the square? Assume the standard reference of $V = 0$ infinitely far from the charges.

36. ★★ In Figure 21-30, a -3.00 μC charge is 0.500 m to the right of a $+6.00$ μC charge. Point P_1 is located 0.250 m to the right of the negative charge, and point P_2 is located a further 0.250 m to the right. Assume the standard reference of $V = 0$ infinitely far from the charges.
    (a) What is the electric potential at point P_1?
    (b) What is the electric potential at point P_2?
    (c) How much work is required to move a -1.00 nC charge from point P_2 to point P_1?

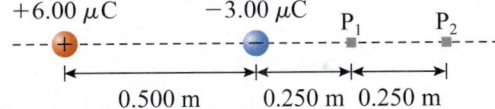

Figure 21-30 Problem 36.

37. ★ How much work is required to move a $+1.00$ μC charge from a point at an electric potential of $+400$. V to another at $+550$. V?

38. ★★ Two charges ($+20.0$ nC and $+75.0$ nC) are separated by 3.00 m. How much work is required to move a third charge, $+1.00$ nC, from a point between the charges and 1.00 m from the $+20.0$ nC charge to a point between the charges but 1.00 m from the $+75.0$ nC charge?

Section 21-4 Equipotential Lines and Electric Field Lines

39. ★ Consider the sketch of equipotential and electric field lines shown in Figure 21-31.
    (a) Identify the orange and the blue lines according to which are equipotential and which are electric field lines.
    (b) The charges have been hidden. For each, indicate if it is positive or negative.
    (c) Of the equipotential lines drawn, which one has the highest positive value?

ELECTRICITY, MAGNETISM, AND OPTICS

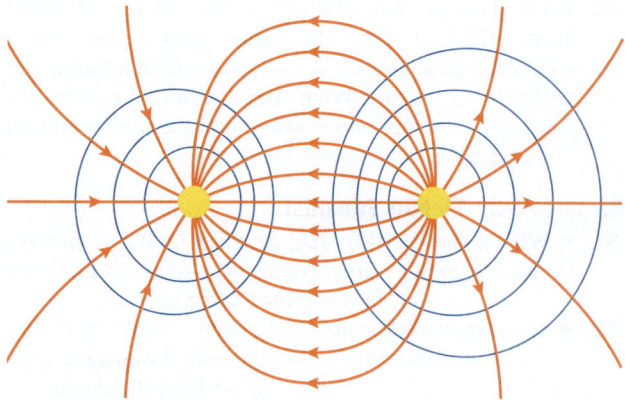

Figure 21-31 Problem 39.

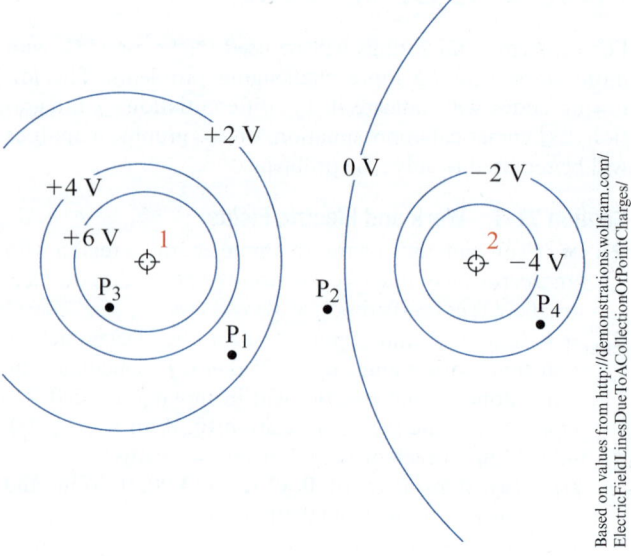

Figure 21-34 Problem 42.

40. ★ Figure 21-32 shows plots of some equipotential lines. Sketch in the corresponding electric field lines, and indicate the relative strength of the electric field vectors in different regions.

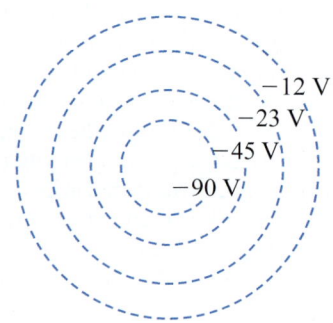

Figure 21-32 Problem 40.

41. ★★ The central $+2$ nC charge in Figure 21-33 is flanked on the left and right (at equal distances) by -1 nC charges. Qualitatively sketch the electric field lines and equipotential lines for the system.

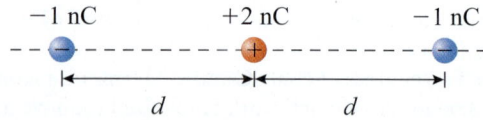

Figure 21-33 Problem 41.

42. ★★ A set of equipotential lines are drawn in Figure 21-34 for two point charges along with the electric potential on each (assuming a reference of zero at infinite distance).
   (a) Is charge 1 positive or negative?
   (b) Is charge 2 positive or negative?
   (c) Which charge must be of higher magnitude?
   (d) For each of the points P_1 through P_4, indicate the approximate orientation and relative strength of the electric field vectors.
   (e) If we placed positive point charges of equal magnitudes at P_1 and P_2 and negative charges at P_3 and P_4, draw arrows to represent the relative forces on these particles.

Section 21-5 Electric Potential from Continuous Distributions of Charge

43. ★ ∫ A total charge of -5.60 nC is equally spread around a thin ring of radius 7.50 cm. What is the electric potential at the centre? Assume a reference with $V = 0$ at infinite distance.

44. ★ A non-conducting sphere has a radius of 2.75 cm and a total charge of $+3.00$ μC uniformly spread throughout the sphere. What is the volume charge density?

45. ★★ ∫ For the situation of problem 44, what is the electric potential at the exact centre of the sphere (assuming a reference with $V = 0$ at infinite distance)?

46. ★★ ∫ A thin ring of charge of radius a is in the xy-plane, centred on the origin. The ring has linear charge density λ. Derive an expression for the electric potential at the point $(0, 0, z)$. In your final expression, include a constant of integration, representing the zero definition of electric potential.

47. ★★★ ∫ Using calculus on differential elements of charge, derive an expression for the electric potential at a point located a distance z above the centre of a disk of radius a. Assume that the disk has a surface charge density of σ (measured in coulombs per square metre).

48. ★★ ∫ A non-conducting sphere of radius $4R$ has a concentric spherical cavity of radius $3R$. The material of the sphere carries a uniform charge of volume charge density ρ. Derive an expression for the electric potential at the centre, assuming a reference with $V = 0$ at infinite distance.

49. ★★★ ∫ What is the electric potential at a point a height h above one end of a uniformly charged line of charge of length L and with linear charge density λ (see Figure 21-35)?

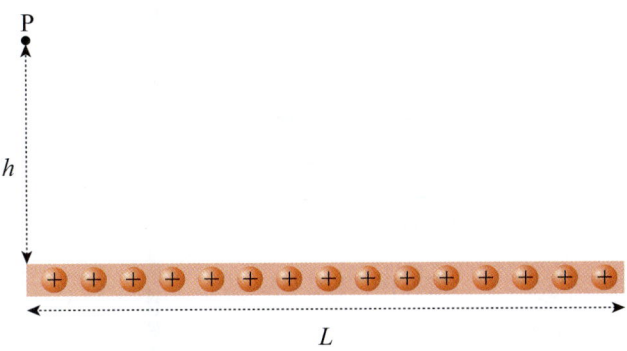

P

h

L

Figure 21-35 Problem 49.

Section 21-6 The Electron Volt

50. ★ A particle has kinetic energy of 1.03×10^{-15} J. What is that in eV units?

51. ★★ A mosquito has a mass of 2.50 mg and is flying at 2.35 km/h. What is its kinetic energy, expressed in units of eV?

52. ★★ An LHC proton beam contains 2808 bunches, with each bunch containing 1.15×10^{11} protons. If the energy of one proton is 7.00 TeV, what is the nominal energy of one beam expressed in joules?

53. ★★ (a) A proton is travelling at a speed of 2.50×10^7 m/s. What is the kinetic energy, in eV? Ignore relativistic effects.
    (b) Repeat part (a) with an electron.

54. ★★ The electron beam in a typical scanning electron microscope produces kinetic energies of 25.0 keV.
    (a) Assuming that relativistic effects can be ignored, what is the speed of the electrons?
    (b) What potential difference is required to produce electrons of this energy?
    (c) This accelerating potential difference is applied over a distance of 2.50 cm. What is the electric field in that region?

55. ★★ A particular chemical reaction requires 2.45 eV to create a new molecule.
    (a) What is this energy per molecule expressed in units of joules?
    (b) What is this energy in thermochemical calories (1 calorie = 4.184 J)?

Section 21-7 Calculating Electric Field from Electric Potential

56. ★ The magnitude of an electric field at a certain point is 4.50 kV/m. What is this in N/C?

57. ★★ At a certain point, the electrical force on a $+3.00$ nC charge is $\vec{F} = -2.75 \times 10^{-7}$ N $\hat{j}$. What is the electric field in V/m?

58. ★★ $\frac{d}{dx}$ An electric potential in SI units has the form $V = -100.0 + 25.0y$.
    (a) Write expressions for the three components of the electric field.
    (b) What is the electric field at the point (1.00 m, 2.00 m, 0.00)?

59. ★ High electric fields can be present inside biological cell membranes. The field in a particular cell over a small region is assumed to have a constant electric field strength of 5.00 N/μC. What is the magnitude of the potential difference over a 2.00 μm distance?

60. ★★ An electric field vector is $\vec{E} = -20.0$ N/C$\hat{i}$.
    (a) Describe the equipotential surfaces.
    (b) If $V = +25.0$ V at the origin, where is $V = +75.0$ V?

61. ★★★ ∫ The following expression represents the electric field (in SI units) as a function of radial distance, r (assume that it is valid for r-values greater than 0.050 m):

$$\vec{E} = -\frac{12.0}{r^2}\hat{r}$$

    (a) If the electric potential is $+20.0$ V at $r = 2.00$ m, what is the value at $r = 5.00$ m?
    (b) What is the electric potential at infinite distance?
    (c) Where does the electric potential equal zero?

62. ★★★ ∫ The electric field (in SI units) as a function of radial distance r is given by

$$\vec{E} = +\frac{20.0}{r^2}\hat{r}$$

    (a) If the electric potential is $+10.0$ V at $r = 1.00$ m, what is the value at $r = 9.00$ m?
    (b) What is the electric potential at infinite distance?

Section 21-8 Electric Potentials and Fields for Conductors

63. ★★ What is the total charge on a spherical conductor of radius 2.50 cm if the electric potential of the conductor is 48.0 V (assuming a zero electric potential reference at infinite distance)?

64. ★★ (a) What is the surface charge density on a spherical conductor of radius 4.65 cm if the electric potential of the conductor is 160.0 V (assuming a zero electric potential reference at infinite distance)?
    (b) Would the answer be different had it been a very thin spherical conducting shell of the same radius? Justify your answer.

65. ★★ Two isolated identical conducting spheres each have a radius of 8.50 cm. One sphere initially carries a charge of -14.00 nC, while the other is uncharged. What is the electric potential on each sphere if they are connected with a long conducting wire?

66. ★★ Two isolated conducting spheres have radii of 5.00 cm and 7.50 cm. The 5.00 cm radius sphere initially has a charge of -16.00 nC, while the other sphere is initially uncharged. What is the electric charge on each sphere if they are electrically connected with a long conducting wire?

67. ★★ A long wire connects two well-separated conducting spheres. One sphere has a radius of 10.0 cm and an electric potential of $+250.$ V (using a reference of the zero point at infinite distance). The other conducting sphere has a radius of 16.0 cm. What is the charge on that sphere?

68. ★★ In this problem, use a definition of zero point such that at infinite distance from the charges we define the electric potential to be +200 V. A conducting sphere of radius 7.75 cm has a surface charge density of 4.25 nC/m². What is the electric potential of the conducting sphere?

69. ★★★ ∫ A conductor has two approximately spherical extensions, separated by a long conducting "neck." One extension has a radius of 2.00 cm, while the other is 0.50 cm. It is observed that the electric potential of the conductor is 3.50 kV. In doing this problem, assume that the extensions are sufficiently spaced both from each other and from the rest of the conductor that we can treat the extension as isolated.
    (a) What is the electric field just outside the surface of the 2.00 cm spherical extension?
    (b) Answer the same question for the 0.50 cm spherical extension.

Section 21-9 Electric Potential: Powerful Ideas

70. ★★ Point A is at an electric potential of +900. V and point B is at an electric potential of +800. V. If a particle of mass 2.00 ng carrying a charge of +3.50 nC starts at rest at point A, what is its speed by the time it reaches point B? Assume that the motion is not relativistic.

71. ★★ Two identical non-conducting spherical shells each have a radius of 8.40 cm and each carry uniform charge distributions. One sphere has surface charge density $\sigma_1 = +26.0$ nC/m², while the other has surface charge density $\sigma_2 = -12.0$ nC/m². The two spheres almost touch. What is the electric potential at the point between the spheres where they almost touch (assume a zero level for electric potential at infinite distance)?

72. ★★ $\frac{d}{dx}$ You will see in the next chapter that a set of two closely spaced parallel plates is called a capacitor. Assume that the plates are parallel to the yz-plane, with one plate at the origin and the other at $x = 0.00800$ m. The electric potential (in SI units for both V and x) is given by $V = 200 + 3800x$. Assume three significant digits in each term.
    (a) What must be the units for each of the numerical terms in the expression?
    (b) What is the electric potential on each of the two plates?
    (c) Draw an equipotential plot.
    (d) What is the electric field at a point halfway between the plates?
    (e) What is the electric field at the point $x = 0.00100$ m?
    (f) What is the electric potential difference between the two plates?
    (g) If one plate carries positive charge and the other negative charge, which plate is charged with which sign?

73. ★★★ As shown in Figure 21-36, a large grounded plate has a +2.00 nC charge at a distance of 2.00 cm to the left of the plate and a −1.00 nC charge at a distance of 3.00 cm from the plate. Use the method of images to find the electric potential at a point halfway between the two charges.

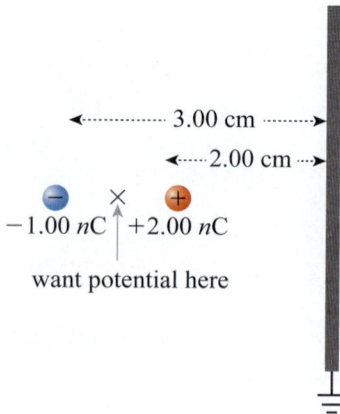

Figure 21-36 Problem 73.

COMPREHENSIVE PROBLEMS

74. ★★ Three +2.00 nC charges are placed at the vertices of an equilateral triangle with 25.0 cm sides.
    (a) What is the electrical potential energy of this charge distribution?
    (b) How much work is needed to bring the charges from an infinite distance into this configuration?

75. ★★ Two positive charges in space, each with a charge of +25.0 nC and a mass of 5.00 mg, are initially held at a fixed separation of 35.0 mm. They are released and allowed to separate under mutual repulsion. What speed will they each attain when they are very far apart? (Hint: Use an energy approach. By conservation of linear momentum, you know that the situation must result in symmetric motions relative to the centre of mass.)

76. ★★ An electron is accelerated through a constant electric field until it reaches a speed of 3.00×10^5 m/s. What is the potential difference between the starting and ending positions?

77. ★★ Consider the distribution of the four point charges shown in Figure 21-37. What is the electric potential, V, at a point halfway between charges A and B?

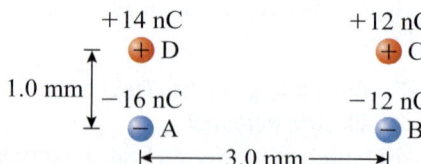

Figure 21-37 Problem 77.

78. ★★ Consider the arrangement of four electric charges shown in Figure 21-38. Find the electric potential at point P, assuming that the potential has a zero point at infinite distance.

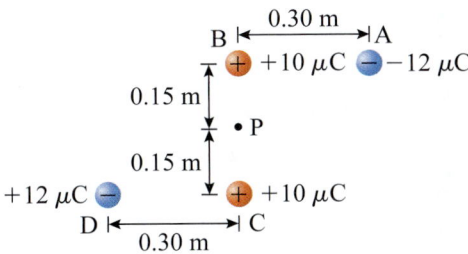

Figure 21-38 Problem 78.

79. ★★★ Three identical positive charges, $+q$, are placed at three of the four corners of a square with side lengths d.
  (a) What is the electric potential at the corner that is missing a charge?
  (b) What is the work required to move a fourth charge, $+q$, from an infinite distance to the corner that is missing a charge?
  (c) Check your answer to part (b) by calculating the difference between the electrical potential energy in the cases of three charges and four charges.

80. ★★ In the famous Rutherford scattering experiment, α-particles were fired at a thin gold foil. When the α-particles passed near a nucleus, they were deflected by large angles. For the purpose of this question, we will consider the situation of an α-particle directed exactly at a gold nucleus. Calculate how close an α-particle could get before it "bounced back" due to electric repulsion. Use energy considerations to solve this problem. Gold has 79 protons in the nucleus. An α-particle is made up of 2 protons and 2 neutrons. The α-particle had a speed of 1.50×10^7 m/s when it was a large distance away. Assume that the gold nucleus is sufficiently massive that it does not move. Treat the gold nucleus as though it were a point charge of the same total charge for the purpose of solving this problem.

81. ★★★ A Van de Graaff generator employs a conducting sphere of radius 0.600 m. Assume that a total charge of 94.0 μC is present on the sphere. Assume a reference with zero electric potential at infinite distance.
  (a) What is the electric potential at the surface of the conductor?
  (b) What is the electric potential inside the conducting sphere?
  (c) Calculate and plot the electric potential as a function of distance.

82. ★★ (a) Determine the sign and direction of the electric field required for an electron to have an upward electrical force that just balances the downward force of gravity.
  (b) With this field, we define the electric potential as 0 V at the surface. What is the potential at a height of 3.00 m? Assume a constant field.

83. ★★ A cathode ray tube display accelerates an electron from rest to 2.25×10^7 m/s within a distance of 1.25 cm in the acceleration region.

(a) What is the magnitude of the electric potential difference in this accelerating region?
(b) What is the increase in the kinetic energy of the electrons, in eV?

84. ★★★ A conducting sphere of radius 12.0 cm has a spherical cavity of radius 5.00 cm at its centre. The sphere carries a total charge of $+240.$ nC. Derive expressions for the electric field magnitude and the electric potential as a function of radial distance, r.

85. ★★ 〰 Figure 21-39 shows a plot of an electric potential as a function of distance along the x-axis. From this information, determine the x-component of the electric field at each of the following displacements.
  (a) $x = 0.50$ m
  (b) $x = 1.50$ m
  (c) $x = 3.00$ m
  (d) $x = 4.50$ m

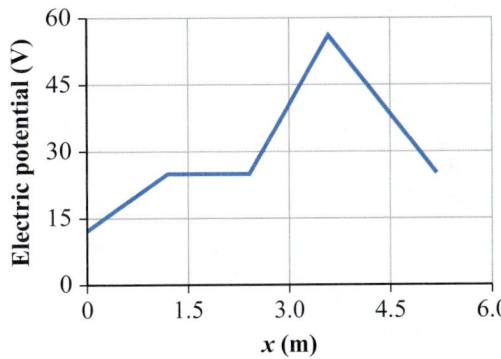

Figure 21-39 Problem 85.

86. ★★ As mentioned earlier in the chapter, electrostatic dust precipitators have a section where there are spikes to make high electric fields to ionize the dust grains (followed by plates at high electric potential difference to collect the charged dust). Assume that the spikes can be approximated by a hemisphere, with a diameter of 3.00 mm, at the top of a rod, as shown in Figure 21-40. When the plate is held at a potential of 800. V, what is the electric field near a spike surface? (Assume for this purpose that it is the same field as for an entire sphere of the same radius, and that the "spikes" are sufficiently isolated from each other that the field from one does not affect the other.)

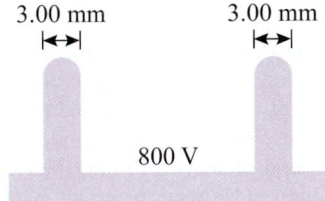

Figure 21-40 Problem 86.

87. ★★★ Imagine that the attractive force between the Sun and Earth is electrical rather than gravitational. Assume that Earth has a negative charge and the Sun has a positive charge of equal magnitude.
    (a) What charge would produce an electrical force equal the actual gravitational force?
    (b) Suppose that the charge is distributed evenly over the surface of Earth. What would the surface charge density be?
    (c) What would the electric field just above Earth's surface be?
    (d) What would the electric potential on Earth's surface be?

88. ★★ A drop is perfectly spherical (this is a good approximation because of surface tension). Assume that the liquid of the drops is electrically conducting, and the charge spreads uniformly on the surface of the drop. The drop has a radius of 250. μm and a charge of 50.0 nC. Assume a reference with zero electric potential at infinite distance.
    (a) What is the electric potential just at the surface of one drop?
    (b) If three of these identical drops coalesce, and the liquid of the new drop is also a perfect sphere, what is the new electric potential? Again assume the charge is uniformly spread over the surface of the drop.

89. ★★★ ∫ Use calculus to find the electric potential at distances $a/2$ and $2a$ from an infinite line of charge having a linear charge density λ (measured in coulombs per metre). Assume that zero electric potential is defined at distance a from the line of charge.

90. ★★★ In Chapter 11 you encountered the idea of open orbits. If we can consider an electron as a classical particle, what is the highest speed it could have and still be electrically bound to an ion with a charge of $+4e$, assuming that it is at a distance of 180. pm?

91. ★★★ ∫ Consider two concentric conducting spherical shells. The inner shell has radius R and carries a total charge of $-Q$. The outer shell has radius $2R$ and carries a total charge of $+2Q$. Find expressions for the electric potential in all three regions (i.e., $r < R$, $R < r < 2R$, and $r > 2R$). Assume a reference with zero electric potential at infinite distance.

92. ★★★ ∫ A conducting spherical shell has inner radius R and outer radius $2R$. It carries a total charge of $+2Q$. Concentric with the spherical shell is a point charge, $-Q$, located at the centre. Find expressions for the electric potential in all three regions (i.e., $r < R$, $R < r < 2R$, and $r > 2R$). Assume a reference with zero electric potential at infinite distance.

93. ★★★ An L-shaped grounded conductor (of large dimension) is oriented on the x- and y-axes. A single positive point charge of $+5.00$ nC is located at a point 2.00 cm from both the x- and y-axes, as shown in Figure 21-41. Use the method of images to find the electric potential at a point midway between the point charge and the part of the conductor along the y-axis (see Figure 21-41). Assume a zero level for electric potential at infinite distance. Hint: You will need three image charges to provide zero potential along both the x- and the y-axes.

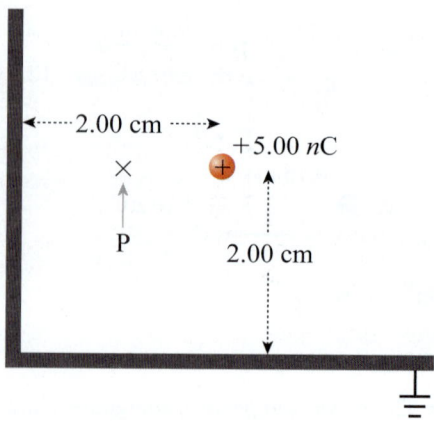

Figure 21-41   Problem 93.

DATA-RICH PROBLEMS

94. ★★★ 〰 Table 21-1 lists data on electric potential as a function of radial distance for a spherically symmetric object. The object is some sort of sphere.
    (a) Plot the electric potential data.
    (b) From the data, determine the approximate electric field, and plot the electric field data. Indicate the direction of the electric field.
    (c) Is the object conducting or non-conducting? Justify your answer.
    (d) What is the radius of the object?
    (e) Calculate the approximate charge distribution for the object: If the charge is all on a surface, calculate the surface charge density. If the charge is uniformly spread throughout the object, calculate the volume charge density, assuming that it is constant. Is the charge positive or negative?
    (f) How far from the object would you need to be for the electric potential to be 500 V?

Table 21-1   Data for Problem 94

R (m)	Electric Potential (V)
0.0	7000
0.1	7000
0.2	7000
0.3	7000
0.4	7000
0.5	6300
0.6	5250
0.7	4500
0.8	3940
0.9	3500
1.0	3150
1.1	2865
1.2	2625
1.3	2425
1.4	2250
1.5	2100
1.6	1970
1.7	1855
1.8	1750

95. ★★ 🖵 The grid in Table 21-2 shows electric potential values (in V) at different x- and y-locations expressed in metres (assume there is no z-dependence of electric potential).

(a) Use these data to estimate the approximate electric field at each of the following locations in (x, y) format.

(i)   (1.00, 1.00)
(ii) (7.00, 7.00)
(iii) (4.00, 8.00)

(b) Determine an expression for the electric potential that fits the data and extrapolate that (assuming it is valid outside this range) to give what the electric potential would be at the (x, y) points (12.00 m, 12.00 m) and (8.00 m, 14.00 m).

Table 21-2   Problem 95

x / y	0.00	1.00	2.00	3.00	4.00	5.00	6.00	7.00	8.00	9.00	10.00
0.00	0	10	20	30	40	50	60	70	80	90	100
1.00	−30	−20	−10	0	10	20	30	40	50	60	70
2.00	−60	−50	−40	−30	−20	−10	0	10	20	30	40
3.00	−90	−80	−70	−60	−50	−40	−30	−20	−10	0	10
4.00	−120	−110	−100	−90	−80	−70	−60	50	−40	−30	−20
5.00	−150	−140	−130	−120	−110	−100	−90	−80	−70	−60	−50
6.00	−180	−170	−160	−150	−140	−130	−120	−110	−100	−90	−80
7.00	−210	−200	−190	−180	−170	−160	−150	−140	−130	−120	−110
8.00	−240	−230	−220	−210	−200	−190	−180	−170	−160	−150	−140
9.00	−270	−260	−250	−240	−230	−220	−210	−200	−190	−180	−170
10.00	−300	−290	−280	−270	−260	−250	−240	−230	−220	−210	−200

96. ★★★ 🖵, $\frac{d}{dx}$ The grid in Table 21-3 shows electric potential values (in V) at different x- and y-locations expressed in metres (assume there is no z-dependence of electric potential).

(a) Use these data to estimate the approximate electric field at each of the following locations in (x, y) format.

(i)   (2.00, 4.00)
(ii) (3.00, 6.00)
(iii) (5.00, 9.00)

(b) Write an electric potential function to fit the data.

(c) Use your result from (b) to confirm your answer from (a).

Table 21-3   Problem 96

x / y	0.00	1.00	2.00	3.00	4.00	5.00	6.00	7.00	8.00	9.00	10.00
0.00	100	75	50	25	0	−25	−50	−75	−100	−125	−150
1.00	105	80	55	30	5	−20	−45	−70	−95	−120	−145
2.00	120	95	70	45	20	−5	−30	−55	−80	−105	−130
3.00	145	120	95	70	45	20	−5	−30	−55	−80	−105
4.00	180	155	130	105	80	55	30	5	−20	−45	−70
5.00	225	200	175	150	125	100	75	50	25	0	−25
6.00	280	255	230	205	180	155	130	105	80	55	30
7.00	345	320	295	270	245	220	195	170	145	120	95
8.00	420	395	370	345	320	295	270	245	220	195	170
9.00	505	480	455	430	405	380	355	330	305	280	255
10.00	600	575	550	525	500	475	450	425	400	375	350

97. ★★★ 〰️, $\frac{d}{dx}$ Shown in Figure 21-42 is a plot of electric potential as a function of radial distance. Note that you can*not* assume a reference of zero electric potential at infinite distance.

(a) What is the electric potential reference, that is, the electric potential at infinite distance?

(b) Write down a function for electric potential as a function of radial distance to approximate the data.

(c) Use your answer in part (b) to write an expression for the electric field.

(d) What type of object or objects are consistent with these observations?

(e) What must be the approximate surface charge density on the object?

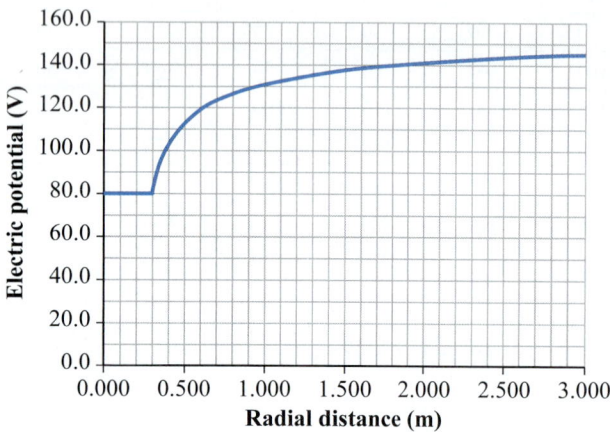

Figure 21-42 Problem 97.

98. ★★★ 〰️, $\frac{d}{dx}$ Figure 21-43 is a plot of electric potential as a function of radial distance for some spherically symmetric object. Do *not* assume a reference of zero electric potential at infinite distance.

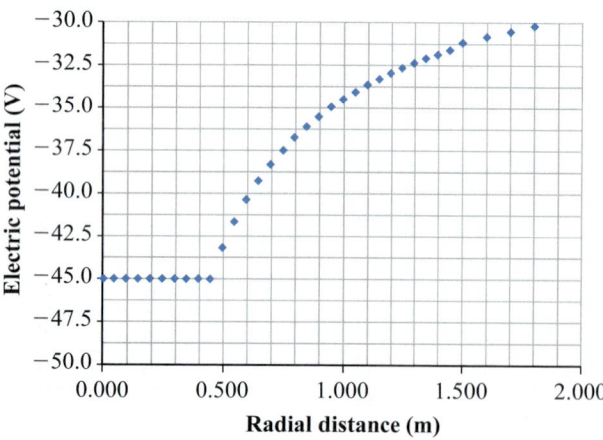

Figure 21-43 Problem 98.

(a) Write down a function for electric potential as a function of radial distance to approximate the data.

(b) Use your answer to determine an expression for the electric field.

(c) What type of object or objects are consistent with these observations?

(d) What must be the approximate surface charge density on the object?

OPEN PROBLEMS

99. ★ Why is it particularly dangerous to keep playing golf when there is a threat of a lightning storm?

100. ★★ ∫ Estimate the maximum voltage possible on a typical demonstration Van de Graaff generator (Figure 21-44) before electrical breakdown in thin air will occur.

Figure 21-44 Problem 100.

101. ★ Using reasonable numbers, determine how much electrical potential energy is "lost" in a single lightning strike (Figure 21-45).

Figure 21-45 Problem 101.

102. ★★★ When you rub a balloon with your hair, you can charge it by triboelectric effects. On a dry day, charge two roughly spherical balloons and attach them with threads so their repulsion causes them to be suspended roughly as shown in Figure 21-46. Measure the distance between the balloons, along with the balloon parameters, and use the data to estimate the electric potential on each balloon.

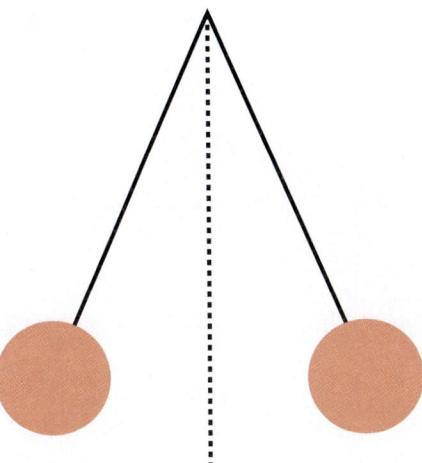

Figure 21-46   Problem 102.

103. ★★ Sharks are very sensitive to electric fields.
(a) Look up in a reliable source the smallest electric field a shark can detect.
(b) A television producer wants you to express the field from part (a) as the field produced by one plate hooked up to the positive terminal of a 1.5 V dry cell and the other plate connected to the negative terminal. How far apart would the plates be?
(c) Give two reasons why the simplification of part (b) is not realistic to model the sensitivity of electric field detections by sharks. If you want to read more on this topic, see the article in *Wired* magazine by physicist Rhett Allain: www.wired.com/2013/08/how-sensitive-are-sharks-to-electric-fields/.

Figure 21-47   Problem 103.

CHAPTER 22 | Capacitance

Every day, defibrillators save many lives. Fibrillation is a series of rapid, irregular muscle contractions. Fibrillation of the heart can be fatal because it prevents proper blood flow and can seriously damage heart tissue. A defibrillator uses an electric shock to momentarily stop the heart and give it a chance to resume beating normally. Figure 22-1 shows a common type of external defibrillator. A miniature internal defibrillator can be implanted in patients who have chronic problems with heart fibrillation. A key part of the defibrillator circuit is the capacitor, which is a device that can store charge and electrical energy, so it can deliver a controlled pulse of electric current to the patient. In this chapter, you will learn how capacitors are made, how they operate, and how these relatively simple devices have innumerable applications in modern technology.

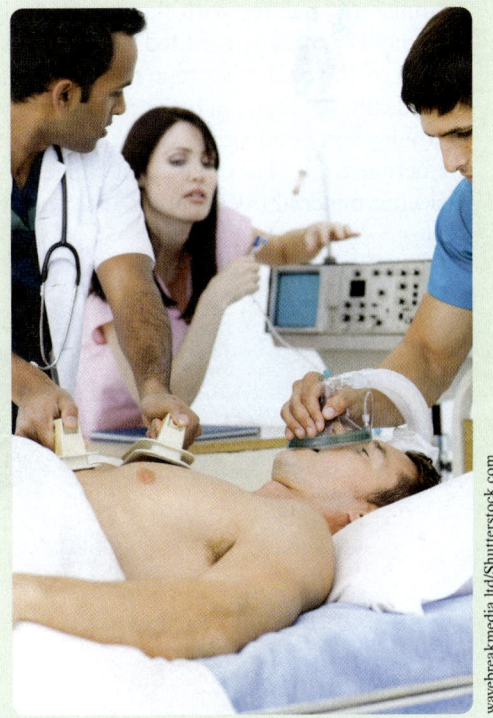

wavebreakmedia ltd/Shutterstock.com

Figure 22-1   A paddle-type defibrillator.

22-1 Capacitors and Capacitance

A **capacitor** is a device that can store charge and electrical energy. You have probably used capacitors numerous times without realizing it. For example, an electronic camera flash uses a capacitor to store the energy that is released during the flash. Turn indicators on some automobiles use a circuit with capacitors and resistors to control the rate at which the signals flash. Digital cameras form images by storing charge on an array of tiny photosensitive capacitors.

Capacitors come in many shapes and sizes and can be made with a variety of materials (Figure 22-2); however, most capacitors consist of two closely spaced parallel plates (Figure 22-3). When the capacitor is charged, one of these plates carries a positive charge, and the other plate carries a negative charge of the same magnitude. The words *capacitor* and *capacity* have the same root, which reflects the key feature of capacitors: their ability to store charge.

The first capacitors were invented independently around 1745 by the German scientist Ewald Georg von Kleist (1700–1748) and the Dutch professor Pieter van Musschenbroek (1692–1761). The very first models consisted simply of a glass jar of water with a metal wire or chain suspended from an insulating stopper. The water acted as one plate and the experimenter's hand as the other plate. Researchers soon found that coating the inside and outside of the jar with metal foil increased the amount of charge that it could store. These glass jar capacitors were called Leyden jars after the university where Musschenbroek conducted his research. Benjamin Franklin was possibly the first to make a capacitor with flat, parallel plates. Michael Faraday is generally credited with the first clear explanation of the charge-storage process in capacitors.

Ideally, the material between the plates of a capacitor is totally non-conducting. Some capacitors simply have air between the plates, but most capacitors have a solid dielectric material (see Chapter 19). Since unlike charges attract, electrical forces hold the charge on the plates of a capacitor when the charge source is disconnected from the capacitor. An ideal capacitor would stay charged forever, but an actual capacitor discharges because charge gradually leaks through the dielectric between its plates or through the air between its terminals.

You may wonder about our statement that the vast majority of capacitors are of the parallel-plate type, since the capacitors in Figure 22-2 appear cylindrical in shape. These are actually parallel-plate capacitors, but they have been rolled up into a cylindrical form to make them smaller. To make a capacitor, start with two very long rectangular strips of flexible metal foil and a thin strip of flexible insulator, such as polyethylene plastic wrap, of the same dimensions. Then wrap the plates and insulator tightly into a cylindrical shape, making sure the different metal layers are insulated from each other.

Let us consider what we mean by **charging** a capacitor. What really happens is that the energy of a power source, such as a battery, moves mobile charge carriers, normally negatively charged electrons, to one plate, with a similar-magnitude positive charge remaining on the opposite plate. We call this process charging a capacitor. Therefore, a charged capacitor has an equal amount of charge on each plate, but with opposite polarity on the two plates. We refer to this magnitude as the charge on the capacitor and use the symbol Q for this charge. If we consider both plates of a capacitor together, there is no net charge.

If we measure the electric potential difference between the charged plates of an ideal capacitor, it is equal to the potential difference of the power source that was used to charge the capacitor. Although we do not typically use the ΔV notation with capacitors, but rather simply V, it is important to realize that V does represent the potential *difference* between the capacitor plates. The plate with the positive charge is the one at the higher electric potential. Sometimes the electric potential difference across the plates of a capacitor is simply referred to as the voltage of the capacitor.

The electric field between the plates of the capacitor in Figure 22-3 is directed from the positively charged

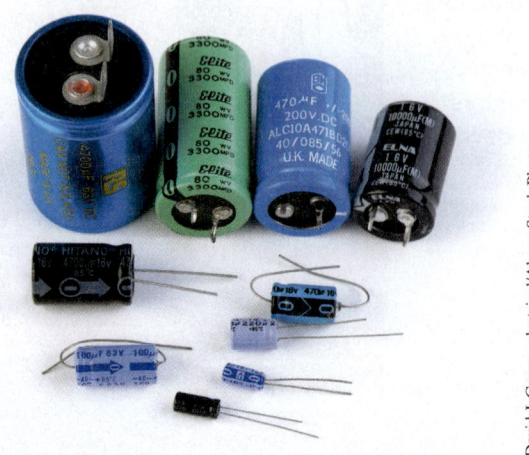

Figure 22-2 An assortment of modern capacitors.

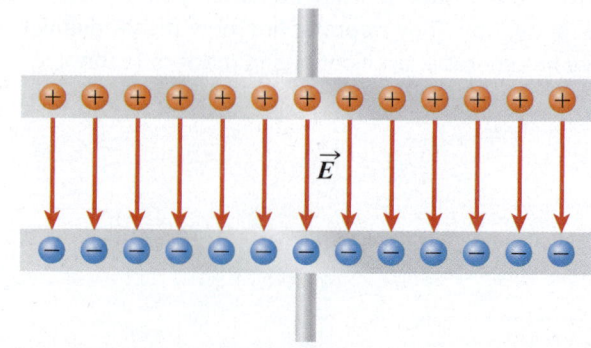

Figure 22-3 An idealized parallel-plate capacitor.

David J. Green - electrical/Alamy Stock Photo

plate to the negatively charged plate because that would be the direction of the force on a small positive test charge placed between the plates. As we will prove in the next section, this field is essentially constant: it is the same everywhere between the plates except very near the edge of a plate.

Recall from the previous chapter that the electrical potential difference between two points can be found by integrating the vector dot product of the electric field along a path joining the points:

$$V_b - V_a = -\int_a^b \vec{E} \cdot d\vec{\ell} \qquad \text{(21-20)}$$

Since the electric field is constant in both direction and magnitude between the plates of a capacitor (proved in the next section), we can take E outside the integral, giving a simple expression for the potential difference, V, between two plates a distance d apart:

KEY EQUATION
$$V = Ed \qquad \text{(22-1)}$$

As we mentioned earlier, a capacitor carries charge of equal magnitude, but opposite signs, on the two plates. If, for example, we double the charge on each plate, the electric field must also double due to this extra charge. By Equation 22-1, since the plate spacing does not change, the potential difference, V, between the capacitor plates also doubles. In other words, we have argued that the magnitude of charge, Q, and the electric potential difference, V, are proportional to each other. This is written as follows:

KEY EQUATION
$$Q = CV \qquad \text{(22-2)}$$

where the proportionality constant, C, is called the **capacitance**. As we will see, the capacitance depends only on the physical dimensions and characteristics of the capacitor.

INTERACTIVE ACTIVITY 22-1

Capacitor Basics

In this activity, you will use Capacitor Basics in the PhET simulation "Capacitor Lab" to investigate how capacitance and charge for a parallel-plate capacitor depend on the applied electric potential difference (voltage), the plate area, and the separation between the plates. You vary one parameter at a time to see how the capacitance changes. Ideally, complete this activity before reading further in this chapter.

It is important to realize that the capacitance does not change when you change the potential difference. Capacitance is determined by the dimensions and physical properties of the materials in the capacitor. A useful way to think about capacitance is that it is a measure of how much charge can be stored for a certain potential

difference. High capacitance means that you can store more charge at the same potential difference than in a capacitor with less capacitance.

The SI unit of capacitance is the **farad** (F), named in honour of Michael Faraday. As you can see from Equation 22-2, one farad is equal to one coulomb per volt. A farad is a huge amount of capacitance for everyday purposes; most capacitors that you will use in the lab or encounter in household devices are measured in picofarads (pF) or microfarads (μF). By convention, capacitance is rarely measured in nanofarads.

Capacitors are charged by connecting them to a source of electric potential difference, such as the battery in the circuit shown in Figure 22-4. The symbol for a capacitor is simply two parallel lines, representing the plates. Some textbooks show capacitors with one thick or curved line, but in all cases the lines are the same length. The end connected to the positive terminal of the battery is positively charged. We have indicated the polarity in Figure 22-4, but it is not usual to show this on the symbol.

The battery provides electrical energy per charge, or electric potential difference, and therefore will be measured in units of volts (see Chapter 21). The formal name (developed in the early days of understanding the nature of electricity) of these circuit elements is **electromotive force (emf)**, and the symbol for an emf is ε (Figure 22-4). The emf converts some other form of energy, such as chemical in the case of a battery, into electrical potential energy. The emf does not create charge but rather provides the energy needed to separate the charges onto the two plates of a capacitor.

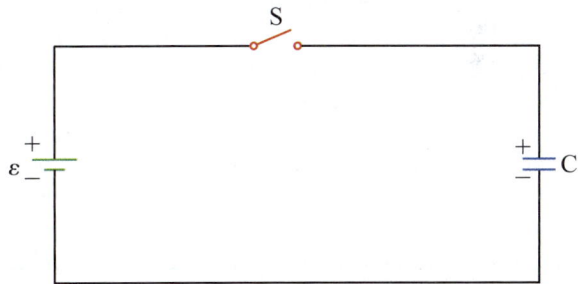

Figure 22-4 When the switch (S) is closed, the capacitor (C) will charge to the same potential difference as the ideal battery (ε). The end connected to the positive terminal of the battery is the positively charged capacitor side.

INTERACTIVE ACTIVITY 22-2

Capacitor Electric Fields

In this activity, you will use Capacitor Electric Fields in the PhET simulation "Capacitor Lab" to investigate how the electric field varies in the region between the plates of a parallel plate capacitor and as one goes outside that region. In addition, you will see how the strength of the electric field depends on the potential difference between the plates and the spacing of the capacitor plates.

22-2 Electric Fields in Parallel-Plate Capacitors

In this section, we will develop a relationship for the capacitance of an ideal parallel-plate capacitor. By an ideal capacitor we mean one that has plates of essentially infinite area. The results we obtain will be valid for actual capacitors as long as the plate area is large.

We will show that the capacitance depends only on the geometry of the capacitor (and the material between the plates). We first need to find how the electric field between the capacitor plates is related to the charge stored on the plates. We will use Gauss's law in this proof. We stated without proof in the previous section that the electric field between the plates of an ideal capacitor is uniform, and the electric field outside that region is zero. We will prove that this is the case. The analysis will also show us where the charge is located on the plates of the capacitor.

If you have omitted Chapter 20 on Gauss's law, and if you accept that the electric field between the plates of an ideal capacitor is uniform and is given by Equation 22-6 ($E = Q/(\varepsilon_0 A)$), below, you can continue to the section "Capacitance of a Parallel-Plate Capacitor."

The Electric Field between Parallel Plates Using Superposition

We assume a parallel-plate capacitor like that of Figure 22-3, with the top plate positively charged. We will let A represent the area of each plate and d the distance between the plates. We require the dimensions of the plate to be much larger than the plate separation for the results to be valid. We will call the magnitude of charge on each plate Q.

Initially, we will not make any assumption about whether the charge is distributed on both sides of the plate, or only on one side. We know from Chapter 20 that charge on an isolated metal plate is distributed equally on both sides, and only on the surfaces, but this is not an isolated single plate. Although we will use the same symbol as used previously for surface charge density, here it will refer to the total charge on one plate (no matter which side of the plate the charge is on) divided by the plate area:

$$\sigma = \frac{Q}{A} \qquad (22\text{-}3)$$

Since one plate of the capacitor is positively charged and one is negatively charged, we lack direct symmetry about a single plane. We have planar symmetry, however, if we consider only one charged plate at a time and then use superposition to combine the results from the two plates. The situation for only the positive plate is shown in Figure 22-5. By the symmetry of the situation, the electric field lines need to be vertical, pointing in both directions, and since the plate is positively charged, the electric field lines point away from the plate.

Although this is similar to a situation we considered using Gauss's law in Chapter 20, because of the subtle difference in meaning of the charge density, we will repeat the analysis here, stressing assumptions made for the solution. By the symmetry of the situation, we consider the charge uniformly spread but make no assumption about whether it is on one side or both sides of the plate. The symmetry only holds if the plates are very large compared to the separation between them.

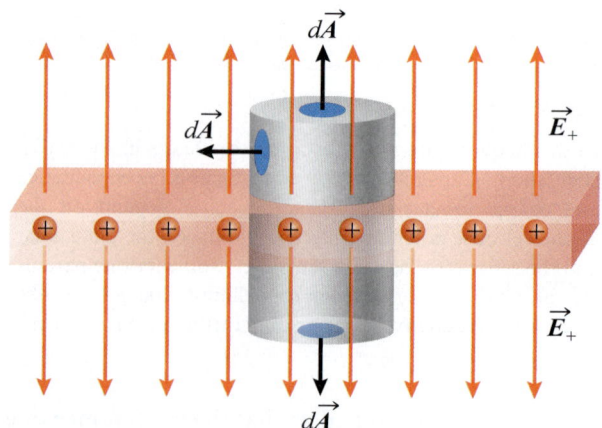

Figure 22-5 The electric field if only the positively charged plate of the capacitor is considered. We use the notation $\vec{E}_+$ to indicate that this is the electric field when only the positively charged plate is considered. A Gaussian cylinder is drawn perpendicular to the plate, with an equal extension on each side.

Since the situation has planar symmetry, we draw a Gaussian cylinder perpendicular to the surface. The cylinder is symmetrically placed relative to the plate, with equal extensions on each side. The size of the cylinder will not matter in the final solution, but for now we will assume it has radius r and total height $2d$ (i.e., d on each side of the plate). Note that d here is not the entire distance between the plates as used earlier.

The electric flux can be considered to be made up of three parts, through the two end caps and through the curved surface:

$$\Phi_{E\ net} = \Phi_{E\ top\ cap} + \Phi_{E\ curved\ wall} + \Phi_{E\ bottom\ cap}$$

As shown in Figure 22-5, for the two end caps, $\vec{E}_+$ and $d\vec{A}$ point in the same direction, so

$$\vec{E}_+ \cdot d\vec{A} = E_+\,dA \cos 0° = E_+\,dA \qquad \text{(end caps)}$$

For the curved surface, $\vec{E}_+$ and $d\vec{A}$ are perpendicular, so there is no contribution to the electric flux:

$$\vec{E}_+ \cdot d\vec{A} = E_+\,dA \cos 90° = 0 \qquad \text{(curved surface)}$$

By symmetry, the electric field strength must be uniform over each end cap, and since we have placed the Gaussian cylinder symmetrically relative to the plate, we expect each end cap to have the same electric field strength. We therefore have for the electric flux

$$\Phi_{E\ net} = \Phi_{E\ top\ cap} + \Phi_{E\ curved\ wall} + \Phi_{E\ bottom\ cap}$$

$$= \iint_{top\ cap} E_+\,dA + 0 + \iint_{bottom\ cap} E_+\,dA$$

$$= E_+ \iint_{top\ cap} dA + 0 + E_+ \iint_{bottom\ cap} dA$$

$$= E_+\pi r^2 + 0 + E_+\pi r^2$$

$$= 2E_+\pi r^2$$

The enclosed charge is obtained by multiplying the enclosed area by the charge density defined earlier:

$$q_{enc} = \pi r^2 \sigma = \pi r^2 \frac{Q}{A}$$

Now we apply Gauss's law to obtain the electric field strength due to only the positively charged plate, E_+:

$$\Phi_E = \frac{q_{enc}}{\varepsilon_0}$$

$$2E_+\pi r^2 = \frac{\pi r^2 Q}{\varepsilon_0 A} \qquad (22\text{-}4)$$

$$E_+ = \frac{Q}{2\varepsilon_0 A}$$

Note that the strength of the electric field is independent of the distance away from the plate.

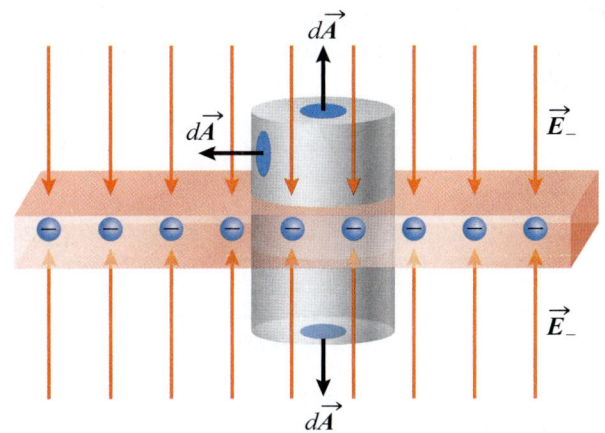

Figure 22-6 The electric field if only the negatively charged plate of the capacitor is considered. We use the notation $\vec{E}_-$ to indicate that this is the electric field when only the negatively charged plate is considered. A Gaussian cylinder is drawn perpendicular to the plate, with an equal extension on each side.

The analysis is similar when we consider only the negatively charged plate. In this case (see Figure 22-6), the electric field lines end on the plate, and therefore point inward.

We again have planar symmetry and draw a Gaussian cylinder perpendicular to the plate (Figure 22-6) and placed symmetrically with equal extensions on the two sides. Now, however, $\vec{E}_-$ and $d\vec{A}$ point in opposite directions for the two end caps and we have

$$\vec{E}_- \cdot d\vec{A} = E_-\,dA \cos 180° = -E_-\,dA \qquad \text{(end caps)}$$

The electric field and area element vectors are still perpendicular for the curved surface, so we have no contributions to the electric flux from the curved surface:

$$\Phi_{E\ net} = \Phi_{E\ top\ cap} + \Phi_{E\ curved\ wall} + \Phi_{E\ bottom\ cap}$$

$$= -\iint_{top\ cap} E_-\,dA + 0 - \iint_{bottom\ cap} E_-\,dA$$

$$= -E_- \iint_{top\ cap} dA + 0 - E_- \iint_{bottom\ cap} dA$$

$$= -E_-\pi r^2 - E_-\pi r^2$$

$$= -2E_-\pi r^2$$

From the direction of the electric field lines, we expect a negative electric flux.

The charge on this plate is $-Q$, so the charge density has a negative value, and the enclosed charge for this plate is

$$q_{enc} = -\pi r^2 \frac{Q}{A}$$

When we apply Gauss's law, we obtain the following result for the electric field strength due to only the negative plate:

$$\Phi_E = \frac{q_{enc}}{\varepsilon_0}$$

$$-2E_-\pi r^2 = -\frac{\pi r^2 Q}{\varepsilon_0 A} \qquad (22\text{-}5)$$

$$E_- = \frac{Q}{2\varepsilon_0 A}$$

Again, the electric field strength is independent of the distance from the plate.

Finally, we use the principle of superposition to combine the individual results and obtain an electric field for the entire capacitor. Remember that these are electric fields, so it is vector superposition that is required. We show the situation in Figure 22-7. Remember that the electric fields due to the individual plates have the same strength, and that strength does not depend on the distance from the plate.

The electric field directions are everywhere uniform and away from the plate for the positively charged plate, and uniform and toward the plate for the negatively charged plate. Note that in the space outside the capacitor, they point in opposite directions, so we have proven that the electric field outside the capacitor is zero. In the space between the plates of the capacitor, however, they add and we obtain the following expression for the electric field strength between the plates of the capacitor:

$$E_{net} = E_+ + E_- = \frac{Q}{2\varepsilon_0 A} + \frac{Q}{2\varepsilon_0 A} = \frac{Q}{\varepsilon_0 A}$$

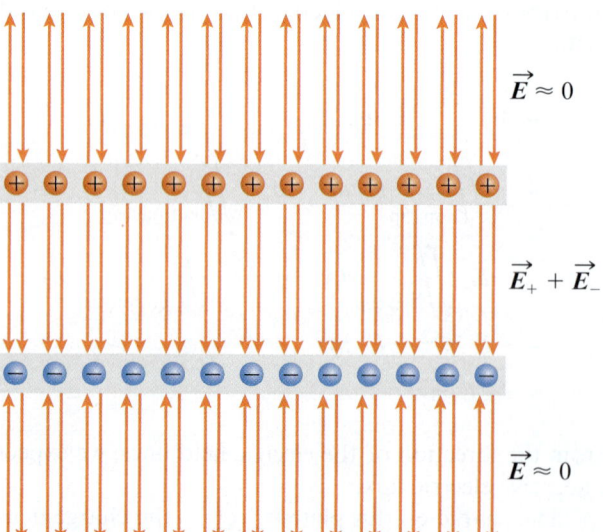

Figure 22-7 The result of applying superposition to the individual results for each plate of the capacitor. In the space between the plates, the electric fields are in the same direction and add, while outside the plates, they are in opposite directions and cancel.

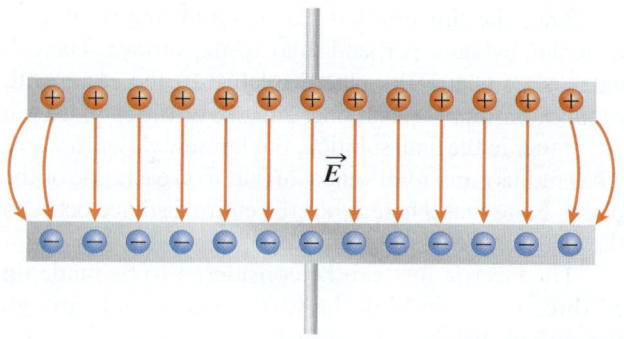

Figure 22-8 Electric field lines for an actual parallel-plate capacitor. Note that near the ends of the capacitor the lines are curved and extend into the space outside the region between the plates.

Therefore, we have proven that the electric field has a constant strength between the two plates, as follows:

$$E = \frac{Q}{\varepsilon_0 A} \qquad (22\text{-}6)$$

You may be concerned that we applied the principle of superposition, since we argued in Chapter 20 that you had to be very careful in applying superposition to situations involving conductors, since the charges are free to move position in response to the other charged object. The fact that the plates are very large, in fact almost infinite, argues that the charge distribution on the plates must be uniform by symmetry, even in the presence of the plate with the opposite charge. This assumption breaks down if the plates are not sufficiently large, and near the end of the capacitor the electric field will not be confined to the space between the plates, as shown in Figure 22-8. The lines also gently curve, and the charge distribution there is not quite uniform, with more charges near the ends.

The Electric Field between Parallel Plates without Superposition

We will now see if we can obtain the same result for the electric field between the plates of a capacitor, but without needing to invoke superposition. In Chapters 19 and 20, we pointed out that electric field lines start on positive charges and end on negative charges (if there are unequal numbers of charges, some will start or end at infinite distance). Since in the standard capacitor configuration the positive plate and negative plate have the same number of charges, just opposite sign, each electric field line that starts on a charge on the positive plate ends on a charge on the negative plate (see Figure 22-9). This requires that no field lines go to (or come from) infinity, and this qualitative argument suggests that the electric field outside the capacitor should be zero.

Since charges in conductors are free to move, they will move to a position that has a minimum electrical potential energy. This is analogous to the mechanical

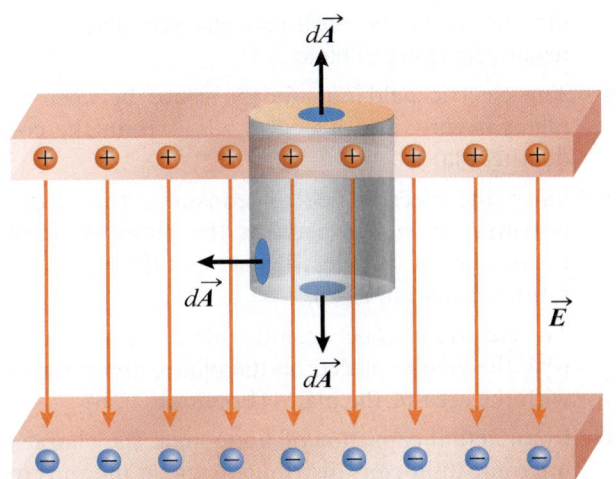

Figure 22-9 The charges on a capacitor are only on the inner surfaces. Every electric field line starts on a positive charge and ends on a negative charge, so none are outside the capacitor. The Gaussian cylinder has its top end inside the upper plate, where the electric field is zero because it is a conductor.

situation in which a marble in a bowl will seek the position of lowest gravitational potential energy at the bottom of the bowl. Here that position of minimum electrical potential energy is when the charges are as close as the conductor will allow, which is on the inner surfaces only. Let us assume that this is so and see if we obtain a result consistent with our earlier analysis.

We again draw a Gaussian cylindrical surface perpendicular to the positive plate, but this time we make the top cap of the plate end *inside* the metal of the plate, as shown in Figure 22-9. Recall from Chapter 20 that the electric field inside a conductor must be zero, and therefore the electric flux must also be zero through the top cap. The bottom cap of the Gaussian cylinder ends somewhere in the space between the plates.

Now we will calculate the electric flux for the closed surface of the cylinder, dividing the closed surface into three parts:

$$\Phi_{E\ net} = \Phi_{E\ top\ cap} + \Phi_{E\ curved\ wall} + \Phi_{E\ bottom\ cap}$$

We argued that the flux through the top cap is zero because that cap is totally within a conductor, and the electric field inside a perfect conductor must be zero. For the curved surface, the electric flux is also zero, this time because $\vec{E}$ and $d\vec{A}$ are perpendicular for this segment:

$$\vec{E} \cdot d\vec{A} = EdA \cos 90° = 0 \quad \text{(curved surface)}$$

For the bottom cap, $\vec{E}$ and $d\vec{A}$ are in the same direction, so we have

$$\vec{E} \cdot d\vec{A} = EdA \cos 0° = EdA \quad \text{(bottom cap)}$$

Note that the symmetry of the situation suggests that the electric field must be uniform over the bottom cap. Therefore, the total electric flux over the closed surface of the cylinder can be readily calculated as

$$\Phi_{E\ net} = \Phi_{E\ top\ cap} + \Phi_{E\ curved\ wall} + \Phi_{E\ bottom\ cap}$$

$$= 0 + 0 + \iint_{bottom\ cap} EdA$$

$$= E \iint_{bottom\ cap} dA$$

$$= E\pi r^2$$

Here we again use r for the radius of the Gaussian cylinder. The charge enclosed by the Gaussian cylinder is

$$q_{enc} = \pi r^2 \frac{Q}{A}$$

where we assume that all of the charge is on the inner surfaces of the plates so we can still use Equation 22-3 for the charge density.

Finally, we apply Gauss's law to find the electric field strength in the space between the plates:

$$\Phi_E = \frac{q_{enc}}{\varepsilon_0}$$

$$E\pi r^2 = \frac{\pi r^2 Q}{\varepsilon_0 A}$$

$$E = \frac{Q}{\varepsilon_0 A}$$

This is the same result that we obtained in Equation 22-6, but this time we derived it without using superposition, but we did need the argument that the charge is all on the inner surfaces. We could argue that by obtaining the same result both ways, this is a proof that the charge must all reside only on the inner surfaces of a capacitor.

You could have obtained the same result by using a Gaussian cylinder that has its bottom cap in the metal of the negatively charged plate and its top cap in the space between the plates. We would have a negative sign on both sides when we used Gauss's law, so the result would be the same.

Capacitance of a Parallel-Plate Capacitor

Now that we have proven that the electric field is constant in the space between the plates of an ideal parallel-plate capacitor, and found Equation 22-6 for that electric field strength, we are ready to find an expression for the capacitance of a parallel-plate capacitor. If we substitute Equation 22-6 into Equation 22-1, we get the following result for the electric potential difference between the plates of the capacitor, where d is the plate separation:

$$V = Ed = \frac{Qd}{\varepsilon_0 A} \qquad (22\text{-}7)$$

This can be rearranged as

$$Q = \frac{\varepsilon_0 A}{d} V \qquad (22\text{-}7a)$$

If we compare Equation 22-7a to Equation 22-2, we see that the capacitance for an air-filled parallel-plate capacitor is given by

KEY EQUATION
$$C = \frac{\varepsilon_0 A}{d} \qquad (22\text{-}8)$$

The capacitance increases with plate area, which makes sense since a capacitor with a larger area could hold more charge for the same potential difference. The capacitance also increases as we *decrease* the separation distance, d, between the plates. We can see why this is, since from Equation 22-1 a smaller plate separation results in a larger electric field for the same potential difference. From Equation 22-6, this allows the capacitor to store more charge, and therefore the smaller plate separation increases the capacitance.

Important Results for the Ideal Parallel-Plate Capacitor

We summarize the important results from this section below. These require a parallel-plate capacitor with large plates compared to the separation between the plates. We further assume that the plates are constructed of ideal conductors.

- The electric field is uniform in the space between the plates. It is perpendicular to the plates and directed from the positively charged plate to the negatively charged plate.

- The electric field outside the space between the plates is zero, and it is also zero within the conducting plates.

- Since the electric field is constant, the electric potential difference between the plates is simply the product of the electric field and the plate separation (Equation 22-1).

- The electric field between the plates varies linearly with the charge placed on the plates, and inversely with the area of the plates (Equation 22-6).

- In an actual capacitor of finite area, the charge is not quite uniformly distributed, and at the ends the electric field lines curve outside the space between the plates.

- The electric charge is located entirely on the inner surfaces of the capacitor plates.

- The capacitance of an air-filled parallel-plate capacitor depends only on the plate area and the plate separation (and on the permittivity of free space). It varies linearly with area and inversely with plate separation (Equation 22-8).

PEER TO PEER

I can solve many capacitor-related problems by applying two simple relationships: $Q = CV$ for all capacitors and $E = V/d$ for the constant electric field between parallel plates, where d is the plate spacing. I need to remember to make my units consistent (e.g., express the capacitance in farads and the plate spacing in metres).

EXAMPLE 22-1

A Parallel-Plate Capacitor

A capacitor that has two parallel plates of dimensions 8.00 cm × 10.00 cm separated by 2.00 mm of air is connected to a 24.0 V battery. Find

(a) the capacitance
(b) the charge on one plate
(c) the electric field in the capacitor

Solution

(a) We can find the capacitance from the relationship for a parallel-plate capacitor:

$$C = \frac{\varepsilon_0 A}{d} = \frac{8.85 \times 10^{-12}\,\text{C}^2 \cdot \text{N}^{-1} \cdot \text{m}^{-2} \times 0.0800\,\text{m} \times 0.100\,\text{m}}{0.002\,00\,\text{m}}$$

$$= 3.54 \times 10^{-11}\,\text{F} = 35.4\,\text{pF}$$

Remember that electrical potential is electrical potential energy per charge; therefore, $1\,\text{V} = 1\,\text{J/C} = 1\,(\text{N} \cdot \text{m})/\text{C}$. Hence, $1\,\text{C}^2/(\text{N} \cdot \text{m}) = 1\,\text{C/V} = 1\,\text{F}$.

(b) To determine the charge on one plate, we need the capacitance and the applied potential difference. The potential difference is given, and we now know the capacitance:

$$Q = CV = 3.54 \times 10^{-11}\,\text{F} \times 24.0\,\text{V}$$
$$= 8.50 \times 10^{-10}\,\text{C} = 0.850\,\text{nC}$$

(c) The electric field is essentially constant between the plates of a parallel-plate capacitor. Therefore, we can find the field by dividing the potential difference by the plate spacing:

$$E = \frac{V}{d} = \frac{24.0\,\text{V}}{0.002\,00\,\text{m}} = 1.20 \times 10^4\,\text{V/m} = 12.0\,\text{kV/m}$$

Making sense of the result

These results demonstrate that typical plate dimensions and separations for parallel plates in air lead to modest values for the capacitance, usually in the picofarad range. You may be surprised by how large the electric field is, but do not confuse a field of thousands of volts per metre with the potential difference (just 24 V here). Remember that in this capacitor, the field extends only a few millimetres.

22-3 Calculating Capacitance

By far, the most common capacitance configuration is that of two parallel plates. We saw in the last section how the capacitance depends on the area and spacing of the plates for a parallel-plate configuration.

It is important to realize that capacitance is a feature of various configurations of conductors, and not

just parallel plates. Capacitance is defined as the ratio of the stored charge, Q, to the potential difference, V, between the plates:

$$C = \frac{Q}{V} \qquad (22\text{-}2a)$$

This definition of capacitance is equivalent to the definition in Section 22-1.

Here is a procedure for finding the capacitance of any configuration of two plates:

- Assume some arbitrary charge, Q, on the plates, and use Gauss's law to determine the electric field in the space between the plates.

- Integrate the electric field between the two plates to obtain the potential difference, V.

- Take the ratio of the charge, Q, to the potential difference, V, to obtain the capacitance. We take the magnitude, since capacitance is always positive.

Example 22-2 demonstrates the application of this procedure.

EXAMPLE 22-2

Capacitance for Coaxial Cylinders

Consider a system of two long coaxial cylinders. The inner cylinder is solid and has radius r_1, and the outer cylindrical conductor is thin and has radius r_2. The inner cylinder carries a charge of $+Q$ over a length L, and the outer cylinder carries a charge of $-Q$ over the same length.

(a) What is the capacitance of length L this configuration?

(b) When $r_1 = 4.00$ mm, and $r_2 = 5.00$ mm, what is the capacitance in a 0.500 m length of the capacitor?

Solution

(a) As we saw in Chapter 20, all the charge on the inner solid conductor is on its outer surface. The outer conductor is thin, and its charge is on its outer surface as well, but we are assuming that the radii of its inner and outer surfaces are essentially the same. As shown in Figure 22-10, the electric field reflects the symmetry of the cylinders and is directed radially outward (from positive to negative charge).

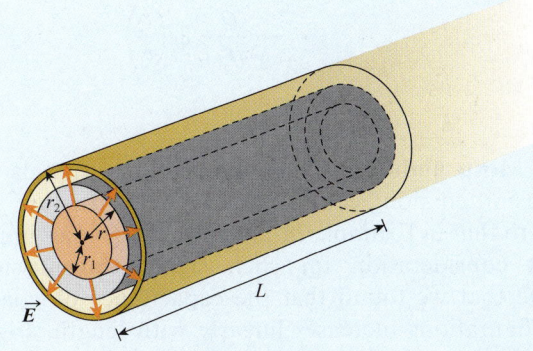

Figure 22-10 Example 22-2. Gaussian cylinder of radius r drawn on coaxial cylindrical conductors. The actual cylinder is much longer than the Gaussian surface section.

To find the capacitance, we apply the three-stage process of first finding the electric field by Gauss's law, then the

(continued)

potential difference between the surfaces, and finally the charge–to–potential difference ratio.

The system has linear symmetry, so we draw a concentric Gaussian cylinder with length L and radius r intermediate between r_1 and r_2. We can ignore flux through the ends of the Gaussian cylinder because the surface area vector and the electric field are perpendicular, making the vector dot product vanish. Over the curved side of the Gaussian cylinder, the electric field is of constant magnitude (at the same r) and is everywhere parallel to the area vector element, so we simplify the integral as shown below. Note that the enclosed charge is only the charge on the inner conductor, $+Q$:

$$\oint \vec{E} \cdot d\vec{A} = E \oint dA = \frac{q_{enc}}{\varepsilon_0} = \frac{Q}{\varepsilon_0}$$

The area of the curved part of the cylinder is simply the circumference times the length. Thus,

$$E(2\pi r L) = \frac{Q}{\varepsilon_0}$$

$$E = \frac{Q}{\varepsilon_0 2\pi r L}$$

Now we have an expression for the magnitude of the electric field between the two conductors as a function of radial distance. Although this electric field is not constant, we can still integrate it over a path between the cylindrical conductors to find the potential difference. Since $\vec{E}$ and $d\vec{A}$ both point radially outward, their dot product becomes a simple multiplication. Therefore, the potential difference between the plates is

$$V_{r_2} - V_{r_1} = -\int_{r_1}^{r_2} \vec{E} \cdot d\vec{r}$$

$$= -\int_{r_1}^{r_2} \frac{Q}{\varepsilon_0 2\pi r L} dr = -\frac{Q}{2\pi \varepsilon_0 L} \int_{r_1}^{r_2} \frac{1}{r} dr$$

Since $\int \frac{1}{r} dr = \ln r$,

$$V_{r_2} - V_{r_1} = -\frac{Q}{2\pi \varepsilon_0 L} (\ln r_2 - \ln r_1)$$

$$= -\frac{Q}{2\pi \varepsilon_0 L} \ln\left(\frac{r_2}{r_1}\right)$$

To find the capacitance, we simply take the ratio of the magnitude of the charge to magnitude of the potential difference. Note that we ignore the signs of the charge and the potential difference:

$$C = \frac{Q}{V} = \frac{Q}{\dfrac{Q}{2\pi \varepsilon_0 L} \ln\left(\dfrac{r_2}{r_1}\right)} = \frac{2\pi \varepsilon_0 L}{\ln\left(\dfrac{r_2}{r_1}\right)}$$

(b) Substituting the given quantities yields

$$C = \frac{2\pi \varepsilon_0 L}{\ln\left(\dfrac{r_2}{r_1}\right)} = \frac{2\pi \times 8.85 \times 10^{-12}\ \text{C}^2 \cdot \text{N}^{-1} \cdot \text{m}^{-2} \times 0.500\ \text{m}}{\ln\left(\dfrac{0.005\ 00\ \text{m}}{0.004\ 00\ \text{m}}\right)}$$

$$= 1.25 \times 10^{-10}\ \text{F} = 125\ \text{pF}$$

Making sense of the result

Since the electric field in part (a) is directly proportional to the charge and inversely proportional to the length, this field really depends only on the linear charge density (the charge per unit length). It makes sense that the electric field decreases in magnitude as the radial distance increases (to conserve electric flux). In the expression for the potential difference, we found that the potential at the outer conductor is less than the potential at the inner conductor. This makes sense because the inner conductor carries the positive charge. To determine whether the capacitance in part (b) is reasonable, we could approximate the coaxial conductors as a parallel-plate capacitor of area $2\pi r L$ with spacing equal to $r_2 - r_1$:

$$C \approx \frac{\varepsilon_0 A}{d} \approx \frac{\varepsilon_0 2\pi r L}{d}$$

$$\approx \frac{8.85 \times 10^{-12}\ \text{C}^2 \cdot \text{N}^{-1} \cdot \text{m}^{-2} \times 2\pi \times 0.00450\ \text{m} \times 0.500\ \text{m}}{0.001\ \text{m}}$$

$$= 1.25 \times 10^{-10}\ \text{F}$$

In this case, the parallel-plate approximation yields the correct result for the capacitor. In fact, when two surfaces are close together compared to their radii, you can often reasonably approximate them as a parallel-plate capacitor. We have neglected end effects. The separation of the cylinders is small compared to the length, so this should be a fairly good approximation.

Most audio and communications signals are transmitted along coaxial cables with a structure similar to that in Example 22-2. Capacitance is an important consideration for such signal transmissions. Note that we found that the capacitance of coaxial configurations increases linearly with length. Figure 22-11 shows a common type of flexible coaxial cable. You will learn more about uses of coaxial cables in Chapter 24.

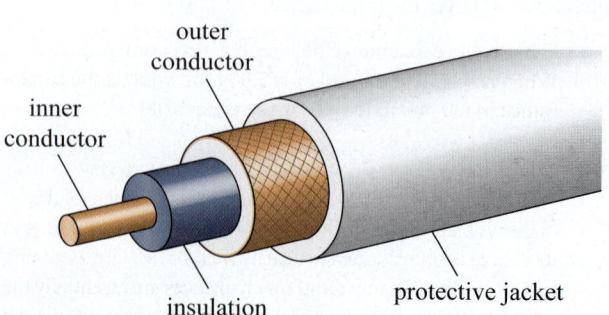

Figure 22-11 The structure of the coaxial cables that are frequently used for signal transmission.

22-4 Combining Capacitors

If you had only, say, 1.0 μF capacitors and wanted to make a 0.25 μF capacitance, could you do it? What if you needed a 5.0 μF capacitor? In fact, you can indeed combine capacitors to produce different equivalent capacitances. Combinations of capacitors are often used in circuits in which a required value is not available as a standard part.

Components in **series** are connected such that the current must flow through one component followed by each of the others, and the electric potential differences across the individual components sum to the total applied potential difference. When components are connected in **parallel**, they all have the same electric potential difference; one end of each individual component is connected to the same common terminal, and the other end of each parallel component is connected to a second common terminal.

When you connect a capacitor in parallel to another capacitor, you *increase* the net capacitance:

KEY EQUATION
$$C_{parallel} = C_1 + C_2 \qquad (22\text{-}9)$$

When you connect a capacitor in series with another capacitor, you *decrease* the net capacitance:

KEY EQUATION
$$\frac{1}{C_{series}} = \frac{1}{C_1} + \frac{1}{C_2} \qquad (22\text{-}10)$$

INTERACTIVE ACTIVITY 22-3

Capacitors in Parallel

In this activity, you will use Capacitors in Parallel in the PhET simulation "Capacitor Lab" to simulate a situation with three capacitors in parallel with a battery. You will determine the potential difference and charge on each of the individual capacitors and gain an understanding of how capacitors combine in parallel.

Consider two capacitors, C_1 and C_2, connected in parallel to a battery, as shown on the left in Figure 22-12. On the right is an equivalent circuit with C_{eq} charged by the same battery. If the circuit is equivalent,

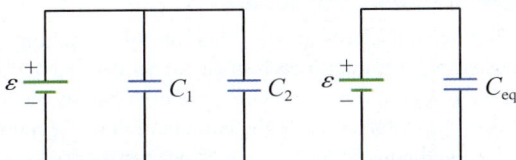

Figure 22-12 A parallel combination of two capacitors charged by battery ε (left) and the equivalent capacitance form of the circuit (right).

then the same total charge must be redistributed by the battery in both cases:

$$Q_1 + Q_2 = Q_{eq} \qquad (22\text{-}11)$$

The charge on each capacitor is given by $Q = CV$. Since the capacitors are in parallel, the potential difference across each capacitor is equal to the potential difference of the battery ε, as is the potential difference across the equivalent capacitor. Therefore,

$$C_1\varepsilon + C_2\varepsilon = C_{eq}\varepsilon \qquad (22\text{-}12)$$

We simply divide by the battery potential difference ε to get the relationship for capacitors in parallel. Another way to think about capacitors in parallel is this: you are connecting their plates together, thus increasing the total area of the plates. It makes sense that the capacitances will add together in the same way.

INTERACTIVE ACTIVITY 22-4

Capacitors in Series

In this activity, you will use Capacitors in Series in the PhET simulation "Capacitor Lab" to simulate three capacitors in series and investigate how the sum of the potential differences on the individual capacitors compares to the battery voltage. Also, you will determine the charge on each of the capacitors in series and determine the rule regarding charges on capacitors in series.

When capacitors are combined in series, as in Figure 22-13, all the capacitors have the same charge, and that charge must be the same as on the equivalent single capacitor. The potential differences across the individual capacitors must sum to the applied potential difference.

To understand why the charges on capacitors in series must be the same, consider what a capacitor is—two conducting plates separated by an insulating

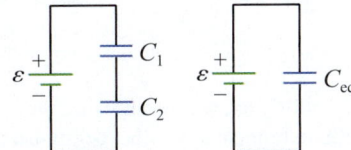

Figure 22-13 A series combination of two capacitors charged by a battery, ε (left), and the equivalent capacitance form of the circuit (right).

PEER TO PEER

Capacitors in series each have the same charge, even when they have different capacitances. The really important point, though, is that they have the same charge as would be on the single equivalent capacitance in the same circuit.

region. The energy of the battery will cause a flow of charges, charging the left plate of C_1 positively and the right plate of C_3 negatively, as shown in Figure 22-14. The insulating regions of each capacitor prevent the flow of current through the capacitor, so overall, the

regions marked A and B on the figure must remain electrically neutral since the total charge is conserved. The right plate of capacitor C_1 will get some negative charge (electrons) attracted to the positive charges on the other plate of that capacitor. Since region A stays neutral overall, the left end of C_2 must develop an equal magnitude of positive charge. Similar arguments show that all the series capacitors have the same charge magnitudes.

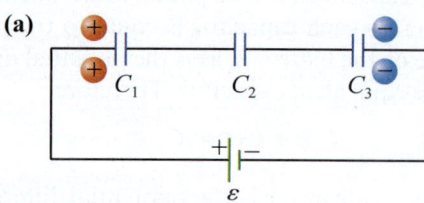

(a)

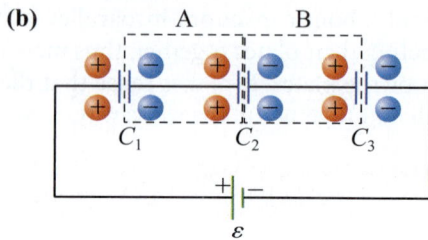

(b)

Figure 22-14   Charge on capacitors in series.

EXAMPLE 22-3

Capacitor Combination

What is the net capacitance between terminals A and B of the circuit in Figure 22-15 when $C_1 = 12.0\,\mu F$, $C_2 = 4.00\,\mu F$, $C_3 = 3.00\,\mu F$, and $C_4 = 6.00\,\mu F$?

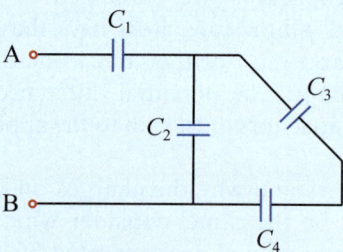

Figure 22-15   Example 22-3.

Solution

Capacitors C_3 and C_4 are in series because they are connected end to end with no branching to other components at the junction. We first find the equivalent capacitance for these two capacitors. For simplicity of notation, we leave out the units until the final step, noting that all the capacitances in this circuit are measured in microfarads:

$$\frac{1}{C_{34}} = \frac{1}{C_3} + \frac{1}{C_4} = \frac{1}{3.00} + \frac{1}{6.00} = \frac{3}{6.00}$$

$$C_{34} = 2.00\,\mu F$$

It is helpful when simplifying combinations of components to redraw the circuit diagram after each step. Figure 22-16 shows the circuit with C_3 and C_4 replaced by the equivalent of $2.00\,\mu F$.

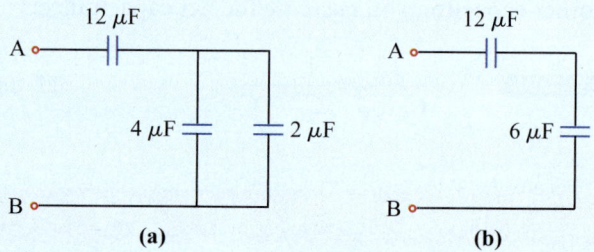

Figure 22-16   In (a), C_3 and C_4 from the original circuit have been combined in series to produce the $2\,\mu F$ capacitor. In (b), we further reduce the circuit by combining the $4\,\mu F$ and $2\,\mu F$ capacitors of (a) in parallel.

From the revised circuit diagram in Figure 22-16(a), it is clear that the next step is to combine $2.00\,\mu F$ and $4.00\,\mu F$, which are in parallel:

$$C_{234} = 2.00\,\mu F + 4.00\,\mu F = 6.00\,\mu F$$

Finally, this equivalent $6.00\,\mu F$ capacitor (shown in Figure 22-16(b)) is in series with the $12\,\mu F$ capacitor, so the net equivalent capacitance is

$$\frac{1}{C_{1234}} = \frac{1}{12.0} + \frac{1}{6.00} = \frac{3}{12.0}$$

$$C_{1234} = 4.00\,\mu F$$

Making sense of the result

When finding equivalents for combinations of capacitors, you should check your results at each stage. Remember that parallel capacitors always add to a greater value than any of the individual capacitors, for example the combination of C_{34} and C_2 is the sum of the individual values. Series capacitors yield an equivalent that is less than any of the individual capacitors. The calculations satisfy these criteria, for example C_{34} is somewhat smaller than C_3 or C_4 individually.

EXAMPLE 22-4

Charges on Capacitors in a Network

A capacitor circuit is shown in Figure 22-17. Assume all capacitors and the battery are ideal. Find the charge on each individual capacitor if $C_1 = 6.00~\mu F$, $C_2 = 2.00~\mu F$, and $C_3 = 16.00~\mu F$.

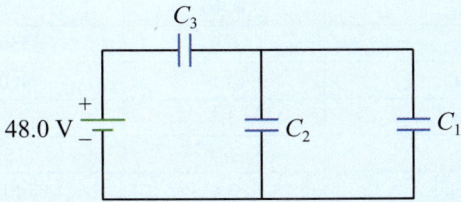

Figure 22-17   Example 22-4.

Solution

In cases with multiple capacitors, the strategy to use is to first reduce to an equivalent circuit with a single capacitor and then find the charge on that equivalent capacitor. Then we work from that to find the individual charges, as illustrated below. In doing this problem, we use the notation where the subscript indicates which individual capacitors have been combined.

First we combine C_1 and C_2 in parallel:

$$C_{12} = C_1 + C_2 = 8.00~\mu F$$

Next we combine this capacitance with C_3 in series:

$$\frac{1}{C_{123}} = \frac{1}{C_{12}} + \frac{1}{C_3} = \frac{1}{8.00~\mu F} + \frac{1}{16.00~\mu F} = \frac{3}{16.00~\mu F}$$

$$C_{123} = \frac{16.00~\mu F}{3} = 5.33~\mu F$$

Now we calculate the charge on this equivalent capacitance:

$$Q_{123} = C_{123}V_{123} = 5.33~\mu F \times 48.0~V = 256~\mu C$$

The charge on capacitors in series must be the same, and they must be the same as that series combination. That means this must also represent the charge on C_3:

$$Q_3 = 256~\mu C$$

The result of the combination for capacitors 1 and 2 is in series with this capacitor, so we know that the charge on that combination must also take on this value:

$$Q_{12} = 256~\mu C$$

We can use this to find the electric potential difference across these parallel capacitors:

$$V_{12} = \frac{Q_{12}}{C_{12}} = \frac{256~\mu C}{8.00~\mu F} = 32.0~V$$

This must be the electric potential difference across each of the capacitors because they are in parallel. We can use this to calculate the charge on each:

$$Q_2 = C_2 V_2 = 2.00~\mu F \times 32.0~V = 64.0~\mu C$$

$$Q_1 = C_1 V_1 = 6.00~\mu F \times 32.0~V = 192.0~\mu C$$

We have worked in μC and μF, but if desired we could have converted to C and F.

Making sense of the result

Remember that when we combine capacitors in parallel, the net capacitance is larger than that of individual capacitors, and when we combine them in series, the equivalent capacitance is less. This is confirmed with the results obtained here. Other checks are possible. We would expect the combined charges on C_1 and C_2 to equal the 256 μC on C_3, which is confirmed. If we use $V_3 = Q_3/C_3 = 256~\mu C/16.0~\mu F = 16.0~V$, we can see that the sum of the potential differences on C_3 and on the parallel of C_1 and C_2 equals the battery voltage of 48.0 V, as expected.

PEER TO PEER

As everywhere in physics, it is always okay to convert to SI units of farads and coulombs. However, it is frequently handy to leave capacitances in μF, in which case the charge I get when using $Q = CV$ will be in μC if V is in volts..

22-5 Dielectrics and Capacitors

So far we have considered capacitors with air between the plates. We can increase the capacitance by using a dielectric material between the plates. Most commercial capacitors employ some type of dielectric. As we saw in Chapter 19, many dielectrics have molecules that are either polar or can be polarized. The polarized molecules can be aligned in the presence of an external electric field.

Figure 22-18 shows a capacitor with a dielectric material inserted between its plates. The top plate is positively charged, so the electric field due to the charge on the plates (the solid arrows) points from top to bottom. The molecules of the dielectric (represented by ovals) orient themselves so that their negative end (shaded blue) is slightly closer to the positively charged top plate on the capacitor. Therefore, the electric field from the polarized dielectric molecules points upward (dashed arrows). The net electric field is the vector sum of these two electric fields. Thus, the presence of the dielectric reduces the net electric field produced between the plates of a capacitor by a given amount of charge on the plates.

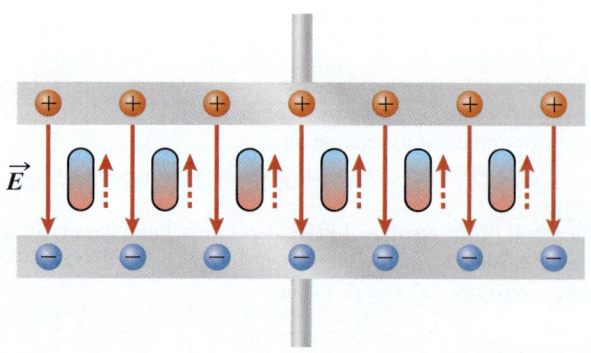

Figure 22-18 A capacitor with a dielectric material between the parallel plates. The internal electric field of the dielectric (dashed red arrows) opposes the field produced by the charges on the capacitor plates (solid red arrows). While a single layer is shown schematically, really there are many layers of charges.

If we let E_0 represent the electric field strength between the plates of the capacitor in the absence of a dielectric, and E the electric field strength with a dielectric completely filling the space between the capacitor plates, the two are related by

$$E = \frac{E_0}{\kappa} \tag{22-13}$$

where κ is the **dielectric constant** and depends on the dielectric material. The dielectric constant, κ, is dimensionless and always has a value of one or more. For a vacuum, $\kappa = 1$, and it is nearly the same when air fills the space between the plates. Table 22-1 shows values of the dielectric constant for some materials commonly used in capacitors. Equation 22-13 is consistent with the earlier statement that the presence of the dielectric material makes the net electric field strength, E, in the presence of the dielectric weaker than it was in the absence of a dielectric.

CHECKPOINT 22-5

A Dielectric in a Capacitor

An air-filled parallel-plate capacitor with capacitance C is charged using a battery with potential difference ε, which is then disconnected. When a dielectric with dielectric constant 2 is inserted to totally fill the space between the plates,
(a) the capacitance is C, and the potential difference is 2ε
(b) the capacitance is $2C$, and the potential difference is ε
(c) the capacitance is $2C$, and the potential difference is $\varepsilon/2$
(d) the capacitance is $2C$, and the potential difference is 2ε

ANSWER: (c) The dielectric will increase the capacitance by a factor equal to the dielectric constant, so the capacitance becomes $2C$. Since the battery is disconnected, the charge must stay the same on the capacitor. Since $Q = CV$, Q is constant, and C is doubled, the new potential difference must be $\varepsilon/2$.

It is helpful to think of capacitors like this: the capacitance depends on how much charge can be stored for a given electric potential difference between the plates. Equivalently, a higher capacitance corresponds

Table 22-1 Dielectric Constants and Maximum Electric Field Strength before Electrical Breakdown for Various Materials

Dielectric Material	Dielectric Constant (unitless)	Typical Breakdown Field Strength (MV/m)
Dry air	1.0006	3
Aluminum oxide	7–9	700
Ceramic	10–100	10
Glass	4–10	10
Mica	6.9	150
Mylar	3.1	400
Nylon	3.5	15
Paper	3–4	15
Plexiglas	3.4	40
Polyethylene	2.3	20
Rubber	2–7	10
Skin	30–40	
Tantalum oxide	20–30	500
Water	80	30 (pure water)
Waxed paper	4	50

to a lower electric potential difference for the same stored charge. Since the capacitor plate spacing is constant, and the electric potential difference is simply the net electric field times that plate spacing, we expect that the presence of a dielectric will reduce the electric potential difference (for the same capacitor charge) and therefore increase the capacitance.

According to Equation 22-13, the electric field is reduced by the dielectric constant, κ. For a fixed amount of charge on the capacitor, the electric potential difference is reduced by the same factor. Therefore, the capacitance when a dielectric fills the space between the plates is increased by this factor. In the presence of a dielectric that completely fills the space between the capacitor plates, the capacitance of a parallel-plate capacitor is given by

KEY EQUATION

$$C = \frac{\kappa \varepsilon_0 A}{d} \tag{22-14}$$

Equation 22-8 can be considered a special case of Equation 22-14 when $\kappa = 1$.

Sometimes the relationship is written in terms of the permittivity of the dielectric filling the space between the plates, ε:

$$C = \frac{\varepsilon A}{d} \tag{22-15}$$

The permittivity of the material, the dielectric constant, and the permittivity of free space are related by

$$\varepsilon = \kappa \varepsilon_0 \tag{22-16}$$

Good dielectrics have higher κ-values (see Table 22-1). In addition to the dielectric constant, it is important to consider how strong the field can be before electric breakdown occurs and conduction takes place through the dielectric—this is given in the right-hand column of Table 22-1.

If the dielectric does not completely fill the space between the plates, we must take that into account in calculating the capacitance. When a portion is completely filled and a portion not filled at all, we can view the configuration as two capacitors in parallel.

When a dielectric is inserted, or removed, from a capacitor, the result is different depending on whether the charging source (e.g., battery) is still connected to the capacitor. When the charging source remains connected, it holds the electric potential difference between the capacitor plates at a fixed value, but the charge on the plates can change. If the battery has been disconnected before adding the dielectric, however, it is the charge that must stay constant, while the electric potential difference will change.

In Interactive Activity 22-5, you will explore for yourself how the capacitance, the electric field, and the energy stored in the capacitor vary when a dielectric is added. We formally consider the topic of energy storage in capacitors in the next section.

INTERACTIVE ACTIVITY 22-5

Capacitors and Dielectrics

In this activity, you will use Capacitors and Dielectrics in the PhET simulation "Capacitor Lab" to gain an understanding of how capacitance, electric field, and stored energy change as a dielectric is inserted between the plates of a parallel-plate capacitor while a battery is connected and how the situation changes when the dielectric is pulled out after the battery has been disconnected from the capacitor.

22-6 Energy Storage in Capacitors

As mentioned at the beginning of this chapter, one of the main applications of capacitors is storing electrical energy. We now develop the key relationships for energy storage in capacitors. The electric potential, V, is defined as the electrical potential energy per unit charge, so you might expect that the energy stored in a capacitor carrying a charge Q would be QV, where V is the potential difference across the fully charged capacitor. However, the potential difference across the capacitor is zero when charging starts and reaches V only as the last of the charge moves onto the plates.

Consider the electrical potential energy of a capacitor of capacitance C as it is charged with Q coulombs

of charge. Let V represent the changing electric potential difference on the capacitor. As a small element of charge, dq, is added, V increases, as does the electric potential energy, U. From the definition of electric potential, we have

$$\int dU = \int_0^Q V dq$$

We can express V in terms of the charge, q, at any given moment:

$$V = \frac{q}{C}$$

Therefore, the electrical energy, U, stored in the capacitor when it carries charge Q is given by

$$U = \int_0^U dU = \int_0^Q \frac{q}{C} dq = \frac{1}{C} \frac{Q^2}{2}$$

We can use the basic capacitor relationship $Q = CV$ to substitute for either of the variables on the right side of this equation.

The electrical potential energy of a charged capacitor is given by

KEY EQUATION
$$U = \frac{1}{2} \frac{Q^2}{C} = \frac{1}{2} CV^2 = \frac{1}{2} QV \qquad (22\text{-}17)$$

A potential energy value is relative to an assumed zero reference. For capacitors, the potential energy is zero when the capacitor is completely uncharged.

CHECKPOINT 22-6

Potential Energy in a Capacitor

A capacitor with air between the plates is charged with a battery, which is then disconnected. If a dielectric (with a dielectric constant greater than that of air) is then slid into the space between the plates, the electrical potential energy stored in the capacitor

(a) increases

(b) decreases

(c) does not change

(d) may increase or decrease depending on the sign of the dielectric constant

ANSWER: (b) Dielectric constants are always positive and have values of one and above, so answer (d) does not make sense. When the dielectric is inserted, the capacitance increases. Because the capacitor had previously been disconnected from the battery, no charge can flow, so charge Q is unchanged. From the equation for the potential energy of a charged capacitor, this means that the energy must have decreased. How is this change of energy possible? The dielectric is pulled into the space; therefore, negative work is done by the person who is guiding the dielectric into the space, and the potential energy of the capacitor decreases. Note that the answer would have been different if the capacitor had remained connected to the battery.

ELECTRICITY, MAGNETISM, AND OPTICS

The National Ignition Facility

The National Ignition Facility, operated by the Lawrence Livermore National Laboratory in California, seeks to produce energy by nuclear fusion. In nuclear fusion, light atoms combine into heavier atoms; for example, hydrogen atoms fuse to form helium. Nuclear fusion is the source of energy for the Sun and other stars and could be the basis for power reactors. Fusion reactors are inherently safer than current nuclear fission reactors. However, a key challenge in producing controlled nuclear fusion is creating the incredible pressures and temperatures needed to initiate nuclear fusion. The National Ignition Facility creates these conditions by directing 192 powerful laser beams simultaneously onto a tiny target (Figure 22-19). The target contains deuterium and tritium (two isotopes of hydrogen) fuel enclosed in a plastic sheath. In a test conducted in 2012, the lasers delivered 5×10^{14} W of power to the tiny target, almost one thousand times the electrical energy consumption rate for the entire United States. This rapidly delivered energy initiates intense X-rays in the enclosing container, followed by rapid heating of the hydrogen fuel. This process, called inertial confinement, can produce the high temperature and pressure conditions required for nuclear fusion. While other approaches are also being considered, inertial confinement is believed to be the most feasible method of achieving controlled nuclear fusion. The goal of nuclear fusion experiments is to produce more energy by fusion than the energy initially applied to the target. You may wonder how it is possible for one experiment

Figure 22-19 The target area at the National Ignition Facility. The target is positioned to a precision of less than the width of a human hair (less than 100 μm).

to use power at a higher rate than the entire U.S. national electrical grid can supply, and that is how capacitors come into the picture. The National Ignition Facility uses banks of high energy density capacitors to rapidly supply the power for the laser flash lamps that start the laser pulse process.

EXAMPLE 22-5

Defibrillator

A defibrillator uses a 5.00 kV power supply to charge a 32.0 μF capacitor. How much energy is stored in this capacitor when fully charged?

Solution

Since we know the potential difference and the capacitance, we choose the form of Equation (22-17) that expresses the electrical potential energy in terms of these two quantities:

$$U = \frac{1}{2} CV^2 = \frac{1}{2} (32.0 \times 10^{-6} \text{ F}) (5.00 \times 10^3 \text{ V})^2 = 400. \text{ J}$$

Making sense of the result

The stored energy is significant—it is approximately equivalent to the energy needed to lift a 40 kg person 0.5 m. Since the energy is not required continuously, storing energy in a capacitor makes it possible to use a power supply that is smaller, lighter, and less expensive than a power supply that supplies energy directly to the defibrillator paddles.

PEER TO PEER

When calculating energy, I must convert capacitances to units of farads and charges to units of coulombs (the result for energy will be in joules).

We can think of the energy in a capacitor as being stored in the electric field in the space between the plates of a capacitor. Consider an air-filled parallel-plate capacitor with plate area A and spacing d. For this capacitor, the electric field is related to the potential difference by $V = Ed$. Therefore, the electrical potential energy can be written as

$$U = \frac{1}{2} CV^2 = \frac{1}{2} C(Ed)^2 \qquad (22\text{-}18)$$

As we saw earlier in this chapter, the capacitance of this type of capacitor is given by

$$C = \frac{\varepsilon_0 A}{d} \qquad (22\text{-}8)$$

Substituting for C in Equation (22-18) gives

$$U = \frac{1}{2}\left(\frac{\varepsilon_0 A}{d}\right)(Ed)^2 = \frac{1}{2}\varepsilon_0(AdE)^2 \qquad (22\text{-}19)$$

The area of the plates times the spacing of the plates, Ad, equals the volume between the plates, which is the volume containing the electric field. So, Equation (22-19) shows that the electrical energy stored in the capacitor depends on the volume between the plates and the square of the electric field.

It is common, especially in more advanced physics courses, to work in terms of energy density, u, which is defined as energy per unit volume (see Section 27-5). Dividing Equation (22-19) by the volume gives us an expression for the energy density between the capacitor plates:

KEY EQUATION
$$u = \frac{1}{2}\varepsilon_0 E^2 \qquad (22\text{-}20)$$

This expression for energy density is generally true for electric fields in a vacuum and can be applied, for example, to the energy density of electromagnetic waves.

The energy stored in electromagnetic fields is ultimately the source of chemical energy in batteries and biochemical energy storage in biological systems. Molecular lipid bilayers act as insulators between conducting regions in living systems, so these act as

EXAMPLE 22-6

Energy Density in a Capacitor

An air-filled parallel-plate capacitor has plate dimensions of 9.00 mm by 7.50 mm and a plate separation of 1.50 mm. It carries a charge of 1.25 nC. What is the energy density in the region between the plates?

Solution

We can first find the capacitance from the relationship for a parallel-plate capacitor. After we know the capacitance (and the charge, which is given in the problem) we can use $Q = CV$ to find the electric potential difference across the capacitor. Once we know the electric potential difference and the plate separation, the electric field between the plates can be determined. Finally, we can use this to find the energy density in the electric field:

$$C = \frac{\varepsilon_0 A}{d} = \frac{\varepsilon_0 \times 0.00900 \text{ m} \times 0.00750 \text{ m}}{0.00150 \text{ m}} = 3.98 \times 10^{-13} \text{ F}$$

$$V = \frac{Q}{C} = \frac{1.25 \times 10^{-9} \text{ C}}{3.98 \times 10^{-13} \text{ F}} = 3140 \text{ V}$$

$$E = \frac{V}{d} = \frac{3140 \text{ V}}{0.00150 \text{ m}} = 2.09 \times 10^6 \text{ V/m}$$

$$u = \frac{1}{2}\varepsilon_0 E^2 = 19.4 \text{ J/m}^3$$

Making sense of the result

We can check this result by multiplying the energy density by the volume of the space between the plates to find the total stored energy:

$$U = uAd = 19.3 \text{ J/m}^3 \times 0.00900 \text{ m} \times 0.00750 \text{ m}$$
$$\times 0.00150 \text{ m} = 1.95 \times 10^{-6} \text{ J}$$

This should be equal, within rounding, to the energy stored in the capacitor:

$$U = \frac{1}{2}CV^2 = \frac{1}{2}(3.98 \times 10^{-13} \text{ F}) \times (3140 \text{ V})^2 = 1.96 \times 10^{-6} \text{ J}$$

We might also point out that the field is just less than that for electrical breakdown in air (see Table 22-1), so the capacitor is near its maximum charge and potential difference.

capacitors. The axons of nerve cells responsible for transmission of electrical impulses in the body are cylindrical capacitors. An understanding of capacitors has an importance that goes well beyond the use of electrical capacitors in circuits.

22-7 Applications of Capacitors

Most capacitor applications involve charging or discharging. As mentioned earlier in the chapter, the battery provides the energy to move existing charges; it does not create or destroy electric charge. Once the capacitor starts to charge, the flow of more electrons onto the negatively charged plate is opposed by electrical repulsion. The battery provides the energy needed to overcome this repulsion. This energy will then be stored as electrical potential energy in the capacitor. When a capacitor is discharged by connecting the two plates with an external conductor, electrons flow from the negative plate to the positive plate, eventually resulting in zero net charge on each plate.

Storing Charge

Probably the most direct application of capacitors is the storage of charge. Dynamic random access memory (DRAM), a common type of electronic memory, has arrays of microscopic capacitors, with a high bit or low bit indicated by whether the stored charge is above or below some critical value. In the charge coupled device (CCD) image sensor, tiny capacitors each store charge corresponding to the brightness of a pixel of the image (Figure 22-20).

Storing Energy

A charged capacitor can be used to store electrical potential energy, energy that we can get back by allowing partial or full discharge of the capacitor. Capacitors store energy for many electrical devices,

Figure 22-20 The CCD image sensor is an array of photosensitive capacitive cells.

including the defibrillator mentioned at the beginning of the chapter. Similarly, camera flash units store the energy needed for a bright flash in capacitors. A wide variety of "instant on" electronic devices store enough energy in a capacitor to start up even when they are switched off.

Sensing

Sensors that measure capacitance are used in a variety of devices around the home and in industry. For example, when hanging a heavy painting or a shelf, you can use a stud sensor to locate optimum attachment points. Mechanical stud sensors use compass-type magnetic sensors to detect the presence of either metal studs (common in commercial and apartment buildings) or nails along studs in wood construction found in individual homes. Now these mechanical stud sensors have largely been replaced with electronic stud sensors that work by detecting the change in dielectric constant, and hence measured capacitance, due to the wood of the stud. While the wood is outside the volume directly enclosed by the capacitor plates, the small plates in the sensor have extensive end effects (see Figure 22-8), and a change in dielectric constant in that region can be readily detected. Can you think of more applications of capacitive sensing in everyday life? See problem 80 for another example.

MAKING CONNECTIONS

The Nd:YAG Laser

Nd:YAG lasers, like the one in Figure 22-21, are widely used in medicine, dentistry, industrial machining, and cosmetic surgery. Nd:YAG lasers use a neodymium (Nd) doped yttrium aluminum garnet (YAG) crystal as the lasing medium. An optical flash lamp close to the crystal adds energy to the crystal atoms, populating excited (higher-energy) states. A number of other types of lasers use similar optical pumping systems. The energy to fire the flash lamp is built up in a capacitor, which can then quickly deliver a pulse of energy to the flash tube.

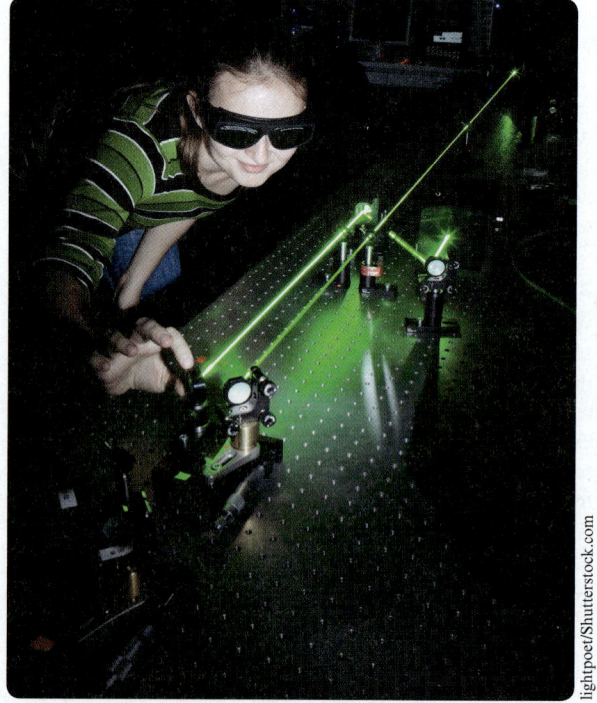

Figure 22-21 The Nd:YAG laser uses a flash tube to energize the crystal that produces the laser light.

Filtering

In scientific instrumentation, as well as in everyday applications such as audio amplifiers, one needs to filter signals. A filter circuit allows certain frequencies through while blocking others. For example, when you use an audio equalizer or even simple bass and treble controls, you are adjusting the settings of various filters.

Most filters use capacitors, often in combination with resistors. Resistors oppose the flow of current in a circuit: the greater the resistance, the less the current that flows for a given applied potential difference, as you will see in the discussion of current and resistance in Chapter 23. A low-pass filter lets only low frequencies through. The simplest low-pass filter consists of just a single resistor and a single capacitor. You will learn more about filters when you study the response of circuits to sinusoidal signals in Chapter 26.

DC Power Supplies

Some electrical circuits work on **direct current (DC)**, in which the current always flows in the same direction. Sources of potential difference, such as batteries, lead to DC current in the circuit. Other electrical devices operate on **alternating current (AC)**, in which the current has a sinusoidal form and flows half the time in one direction and the other half in the opposite direction. Chapter 26 provides a detailed consideration of AC circuits, including circuits that combine capacitors with other circuit elements.

For efficiency and ease of conversion to different potential difference values, most power systems transmit electricity as AC with a sinusoidal waveform. In North America, most household circuits have a potential difference of 120 V AC. However, most electronic devices require a lower DC potential difference. The power supplies for these devices convert household AC into lower potential difference DC. The essential parts of a power supply are a transformer, which steps down the AC potential difference; diodes, which allow the current to flow in only one direction; and a capacitor, which alternately stores and releases charge so that the DC output is approximately constant. Most DC power supplies use electrolytic capacitors, a type of capacitor that allows relatively large capacitances to fit in compact packages (Figure 22-22). Most high-quality power supplies also have a regulator section to help maintain constant potential difference under varying conditions.

Uniform Electric Fields

Since the electric field inside a parallel-plate capacitor is essentially constant, this configuration is often used in applications that require uniform electric fields. Such fields are used to select, direct, and accelerate electrons and other subatomic particles in particle accelerators at major scientific facilities, X-ray machines, and mass spectrometers.

Figure 22-22 Electrolytic capacitors of the type used in a DC power supply.

CHECKPOINT 22-8

Particle Physics Experiment

A parallel-plate capacitor in a particle physics experiment provides a uniform electric field of 14 000 V/m directed upward. The plate separation is 1.0 cm. When the bottom plate is at −100. V relative to the ground, what is the potential of the top plate?

(a) −240 V
(b) −140 V
(c) +140 V
(d) +800 V

ANSWER: (a) The potential difference, ΔV, equals Ed, where d is the plate spacing. Substituting the given values, $d = 0.01$ m and $E = 14\,000$ V/m, we get $\Delta V = 140$ V. Since the electric field points upward, the bottom plate must be more positive. Therefore, the potential at the top plate must be $-100.\ \text{V} - 140\ \text{V} = -240\ \text{V}$.

Timing Circuits

Many general-purpose timing circuits use the time required to charge, or discharge, a capacitor to a certain level to control the timing. In these devices, a comparison circuit detects when the electric potential difference across a capacitor reaches the threshold value. Although high-precision timing requires other approaches (such as the use of piezoelectric crystals in resonant circuits), for many technological and scientific applications, capacitor timers are sufficient and cost-effective. See Section 23-6 for a more detailed treatment.

Transducers

A transducer is a device that provides an electrical signal that depends on some physical parameter. For example, a microphone is a transducer that transforms sound pressure waves into electrical signals. Many transducers, including some types of microphones,

ELECTRICITY, MAGNETISM, AND OPTICS

Touch-Sensitive Screens

Touch-sensitive screens on devices such as smartphones and tablets have revolutionized portable electronics. Although several different mechanisms are used for touch screens, including thermal, light, and resistive sensing, the majority of modern touch screens use capacitive effects. Transparent conducting grids lie above the display and the glass layer. In one design, lines for an electrically driven grid are separated by an insulator from sensing lines, forming, in essence, an array of capacitors. When you touch the screen, essentially grounding that point, you alter the charge stored at that point, and the electronics detect which capacitor has been grounded (Figure 22-23).

Roman Pyshchyk/Shutterstock.com

Figure 22-23 The iPad and the iPhone use touch screens that depend on capacitive effects.

pressure sensors, and fluid-level sensors, make use of capacitive effects. Recall that the capacitance depends on the area of the plates, the separation of the plates, and the dielectric properties of any material between the plates:

$$C = \frac{\varepsilon A}{d} \qquad (22\text{-}15)$$

A capacitive-based transducer may work by changing any of these three parameters. For example, in most capacitive pressure sensors, the distance between two plates changes according to the pressure on the transducer. In some position sensors, the effective area of the plates changes as one plate slides relative to the other. The capacitance of a fluid-level sensor can change as the level of a dielectric fluid between

the plates of the capacitor changes. Capacitive-based condenser microphones are used in many smartphones.

Transistor Capacitors

While passive elements such as resistors and capacitors are fabricated as part of integrated circuits, it is the active semiconductor devices—diodes and transistors—that play the central role.

Transistors are essentially control devices that allow control of a much larger electric potential difference or current by a small potential difference or current. They are key elements of amplifiers but are also used in most modern electronic devices. Huge numbers of transistors can be integrated into a single semiconductor device called an **integrated circuit**. Computer integrated circuits contain millions to billions of transistors. Transistors are constructed from semiconductor layers where the various regions have had a small amount of impurity added (e.g., a silicon region has some of the atoms replaced with phosphorus). There are two main types of transistors, called BJT (bipolar junction transistor) and FET (field effect transistor), with many variants on each.

The most important type of transistor for integrated circuits is the metal oxide semiconductor field effect transistor (MOSFET), which is essentially a capacitor with an insulating layer separating two charge regions. Physicist Julius Edgar Lilienfeld (1882–1963) sought the first ever patent for an FET in 1925; it was awarded by the Canadian government in 1927. When compared to BJT, the other major type of transistor, FET and especially MOSFET devices have significantly lower power requirements. However, it would be many decades before mass production of MOSFETs became practical. Lilienfeld also invented the electrolytic capacitor.

Biological Measurements

Natural systems show capacitive effects, which are widely used in biological research and development. For example, the first accurate measurement of the thickness of the cell wall was done using a capacitive measurement. Samples of different types of cells have capacitances that vary according to the DNA content.

You will learn additional applications of capacitors in Section 23-6, in which you will study the details of charge/discharge circuits involving capacitors and resistors.

KEY CONCEPTS AND RELATIONSHIPS

A common type of capacitor consists of two parallel conducting plates separated by a small gap, which may be filled with air or a dielectric. Capacitors are used to store charge and energy and have many applications, for example, timing and measurement circuits.

Capacitance is a measure of how much charge can be stored for a given applied potential difference and is measured in farads (F).

Capacitance

The relationship between the stored charge, Q; the applied electric potential difference, V; and the capacitance, C, is

$$Q = CV \qquad (22\text{-}2)$$

Electric Fields

Except for small differences near the ends of the capacitor, the electric field is essentially constant in the area between the plates. The relationship between the potential difference across the plates, V; the plate spacing, d; and the magnitude of the electric field, E, is

$$V = Ed \qquad (22\text{-}1)$$

Parallel-Plate Capacitors

The capacitance for vacuurm or air-filled parallel-plate capacitors depends on the plate area, A; the plate spacing, d; and the permittivity of free space, ε_0:

$$C = \frac{\varepsilon_0 A}{d} \qquad (22\text{-}8)$$

Calculating Capacitance

The term *capacitance* can be applied to configurations other than parallel plates. For a general configuration, capacitance can be calculated by finding the ratio of charge to potential difference between the conductors. Gauss's law is used to find the electric field between the conductors, which is then integrated over a path between the conductors to find the potential difference.

Combining Capacitors

When capacitors are connected in parallel, the capacitances add:

$$C_{\text{parallel}} = C_1 + C_2 \qquad (22\text{-}9)$$

When capacitors are connected in series, each of the capacitors has the same charge and

$$\frac{1}{C_{\text{series}}} = \frac{1}{C_1} + \frac{1}{C_2} \qquad (22\text{-}10)$$

Dielectrics and Capacitors

Adding a dielectric between the plates of a parallel plate capacitor decreases the net electric field (for the same applied potential difference) and increases the capacitance.

$$C = \frac{\kappa \varepsilon_0 A}{d} \qquad (22\text{-}14)$$

The dielectric constant, κ, is a dimensionless ratio of the permittivity of the material filling the space between the plates to the permittivity of free space.

Energy Storage

The energy stored in a capacitor (capacitance C) carrying a charge, Q, with a potential difference, V, is

$$U = \frac{1}{2}\frac{Q^2}{C} = \frac{1}{2}CV^2 = \frac{1}{2}QV \qquad (22\text{-}17)$$

Energy is stored in the electric field between the plates. The energy density in this field is

$$u = \frac{1}{2}\varepsilon_0 E^2 \qquad (22\text{-}20)$$

APPLICATIONS

capacitive measurements of biological cells, energy storage in capacitors (e.g., camera flash units and defibrillators), capacitance-based measurements (e.g., fluid levels), touch screens, charge coupled devices (CCDs) for images, timing, filtering of signals

KEY TERMS

alternating current (AC) capacitance, capacitor, charging, dielectric constant, direct current (DC), electromotive force (emf), farad, integrated circuit, parallel, series, transistor

QUESTIONS

1. A capacitor is connected to a 40 V battery. It is observed that the charge on one of the plates is 160 μC. What is the capacitance?
   (a) 0.25 μF
   (b) 4.0 μF
   (c) 640 μF
   (d) 4.0 F

2. A capacitor has capacitance C and applied potential difference V. What is the net total charge on the capacitor?
   (a) 0
   (b) CV
   (c) C/V
   (d) $2CV$

3. The spacing between the plates of a 10. μF capacitor is 1.0 mm. It is connected to a 10. V battery for charging. What is the electric field in the capacitor?
   (a) 0.010 kV/m
   (b) 0.10 kV/m
   (c) 1.0 kV/m
   (d) 10. kV/m

4. Two parallel plates have been charged with equal magnitude and opposite polarity charges. The plates are then disconnected from the charging source, and the spacing between the plates is decreased. Which of the following statements correctly describes what happens?
   (a) The capacitance and the electric potential difference both decrease.
   (b) The electric potential difference increases, and the capacitance decreases.
   (c) The capacitance increases, and the electric potential difference decreases.
   (d) The capacitance and the electric potential difference both increase.

5. Consider two cylindrical capacitors: A has radii 2.5 cm and 2.6 cm and length 2.0 m; B has radii 1.0 cm and 3.0 cm and length 1.0 m. Rather than using the exact relationship for cylindrical capacitors, you want to approximate them as parallel-plate capacitors. Which of the following statements best describes the situation?
   (a) You cannot approximate them at all as a parallel-plate configuration.
   (b) You can approximate A very well, but B will probably be a poor approximation.
   (c) You can approximate B very well, but A will probably be a poor approximation.

6. A coaxial cable is made up of two concentric cylinders. You have two cables, A and B, with the same distance between the cylinders (i.e., from the outside of the inner cylinder to the outer cylinder, which is considered to be thin). Both cables also use the same dielectric material. However, the diameter of cable B is larger than that of cable A. How does the capacitance per unit length compare for the two cables?
   (a) They both have exactly the same capacitance per unit length.
   (b) A has a larger capacitance per unit length.
   (c) B has a larger capacitance per unit length.
   (d) It cannot be determined from the data supplied.

7. You need a slightly smaller capacitance than the value of a capacitor that you are using. How should you add a capacitor to accomplish this?
   (a) Add a much smaller capacitor in series.
   (b) Add a much smaller capacitor in parallel.
   (c) Add a much larger capacitor in series.
   (d) Add a much larger capacitor in parallel.

8. You have a single 1.0 μF capacitor in a circuit. When you connect a 10. μF capacitor in parallel with the 1.0 μF capacitor, what is the new effective capacitance?
   (a) 0.10 μF
   (b) 0.90 μF
   (c) 1.1 μF
   (d) 11 μF

9. An air-filled parallel-plate capacitor initially has some capacitance, C. It is charged using a battery of potential difference X, which is left connected. A dielectric with dielectric constant 2 is then inserted, totally filling the space between the plates. Which of the following statements describes the new condition?
   (a) The capacitance is C, and the potential difference is 2X.
   (b) The capacitance is 2C, and the potential difference is X.
   (c) The capacitance is 2C, and the potential difference is X/2.
   (d) The capacitance is 2C, and the potential difference is 2X.

10. An air-filled parallel-plate capacitor is charged with a battery, which is kept connected. A dielectric (with a dielectric constant greater than that of air) is then inserted into the space between the plates. What change, if any, is there for the electrical potential energy stored in the capacitor?
    (a) The electrical potential energy increases.
    (b) The electrical potential energy decreases.
    (c) The electrical potential energy does not change.
    (d) It depends on whether the dielectric constant is positive or negative.

11. When the potential difference applied to the plates of a capacitor is doubled, what happens to the energy stored in the device?
    (a) The energy is one-quarter the previous value.
    (b) The energy is one-half the previous value.
    (c) The energy is double the previous value.
    (d) The energy is four times the previous value.

12. Two capacitors have identical dielectrics and plate spacing, and they are connected to the same battery. However, capacitor A has a plate area that is four times the plate area of capacitor B. How do the electrical potential energy values stored in the capacitors compare?
    (a) The electrical potential energy of A is 16 times the electrical potential energy of B.
    (b) The electrical potential energy of A is 4 times the electrical potential energy of B.
    (c) The electric potential energy of A is 2 times the electrical potential energy of B.
    (d) The electrical potential energy of B is 2 times the electrical potential energy of A.

13. The power supply you have constructed has too much "ripple," which means that the potential difference goes down too much from a constant value during the 1/60 s period until it is refreshed from the transformer–rectifier network. Which statement correctly describes how the situation might be improved?
    (a) You could add a capacitor in series with the existing capacitor.
    (b) You could add a capacitor in parallel with the existing capacitor.
    (c) Changing the capacitor value will not help the situation.
    (d) Replacing the capacitor with a large resistor will help most.

ELECTRICITY, MAGNETISM, AND OPTICS

14. In a particle physics experiment, you use the area between capacitor plates to create an electric field for exerting a force on electrons. You find that the force is too strong. What could you do to help the situation?
  (a) Increase the plate area and leave everything else the same.
  (b) Decrease the plate spacing and leave the applied potential difference constant.
  (c) Decrease the applied potential difference and leave the plate spacing the same.
  (d) Add a second capacitor in parallel, and keep the applied potential difference the same.

15. Four capacitors (with different capacitances) are connected in series, the combination being charged by a battery. Which statement is correct?
  (a) The four capacitors have the same potential difference and the same charge.
  (b) All four capacitors have the same charge, and the potential differences add to equal the battery voltage.
  (c) All four capacitors have the same potential difference, but the charges are different on each.
  (d) Both the charges and the potential differences are different for each capacitor.

16. Three capacitors (with different capacitances) are connected in parallel, the combination being charged by a battery. Which statement is correct?
  (a) All three have the same potential difference and the same charge.
  (b) All three have the same charge, and the potential differences add to equal the battery voltage.
  (c) All three have the same potential difference, but the charges are different on each.
  (d) Both the charges and the potential differences are different for each.

17. Two capacitors have equal plate separations. Capacitor A has capacitance C and applied potential difference V. Capacitor B has capacitance $4C$ and applied potential difference $V/2$. Which of the following statements is true?
  (a) Both have the same stored energy and the same energy density.
  (b) B has the higher stored energy and the higher energy density.
  (c) Both have the same stored energy, but A has the higher energy density.
  (d) Both have the same stored energy, but B has a higher energy density.

18. Two air-filled capacitors have the same capacitance and the same applied potential difference, but one capacitor has a smaller plate separation. Which of the following statements is true?
  (a) Both have the same energy density.
  (b) The capacitor with the smaller plate separation has the higher energy density.
  (c) The capacitor with the larger plate separation has the higher energy density.
  (d) It is impossible to decide the one with the higher energy density from the information provided.

19. Two capacitors, $2.0\ \mu F$ and $3.0\ \mu F$, are connected in parallel to a 15 V battery. What is the combined charge on the two capacitors?
  (a) $3.0\ \mu C$
  (b) $25\ \mu C$
  (c) $75\ \mu C$
  (d) $90.\ \mu C$

20. Two capacitors, one $4.0\ \mu F$ and the other $8.0\ \mu F$, are connected in series. A 12 V battery is connected to that series combination. What is the charge on the $4.0\ \mu F$ capacitor?
  (a) $4.0\ \mu C$
  (b) $8.0\ \mu C$
  (c) $16\ \mu C$
  (d) $32\ \mu C$

PROBLEMS BY SECTION

For problems, star ratings will be used (★, ★★, or ★★★), with more stars meaning more challenging problems. The following codes will indicate if $\frac{d}{dx}$ differentiation, $\int$ integration, $\square$ numerical approximation, or $\sim$ graphical analysis will be required to solve the problem.

Section 22-1   Capacitors and Capacitance

21. ★ You need to store a charge of 6.00 mC when a 12.0 V battery is applied to a capacitor. What must be the capacitance?

22. ★★ A capacitor has a charge of $30.0\ \mu C$ when a potential difference X is applied to it. When the potential difference is increased to $X + 10.0$ V, the new charge on the capacitor is $40.0\ \mu C$.
  (a) What is the potential difference, X?
  (b) What is the capacitance?

23. ★ When a battery is attached to a $6.00\ \mu F$ capacitor, it is observed that $+3.00\ \mu C$ of charge flows to plate A of the capacitor and $-3.00\ \mu C$ flows to plate B of the capacitor. What is the potential difference between the plates, and which plate will be at the more positive electric potential?

24. ★ When a potential difference of 12.0 V is applied to a capacitor, a charge of $\pm 6.00\ \mu C$ flows to the two plates. If the electric potential is increased to 48.0 V, what *additional* magnitude of electric charge flows from the source to the plates?

Section 22-2   Electric Fields in Parallel-Plate Capacitors

25. ★★ A capacitor has plates separated by a 2.00 mm air gap. A charge of magnitude $56.0\ \mu C$ is on each plate. The electric field between the plates is 2400. V/m.
  (a) What must be the electrical potential difference between the plates?
  (b) What is the capacitance?

26. ★★ An air-filled parallel-plate capacitor has a plate spacing of 1.50 mm and an electrical potential difference of 65.0 V.
  (a) What is the electric field between the plates?
  (b) Calculate the surface charge density on each plate.
  (c) The effective dimensions of the plate are 4.00 cm $\times$ 5.00 cm. What is the total charge on each plate?

27. ★★ Two air-filled capacitors have plates with the same area. The spacing between the plates is x in the first capacitor. The second capacitor has double the applied potential difference of the first capacitor, but it is observed that both capacitors carry the same charge. What must be the plate spacing in the second capacitor (expressed in terms of x)?

28. ★★ The electric field strength in an air-filled (effectively a vacuum) parallel-plate capacitor is 450. kV/m. The plate separation distance is 2.00 mm. Each plate has dimensions 4.00 cm by 4.00 cm.
  (a) What is the electric potential difference between the plates?
  (b) Calculate the capacitance.
  (c) Find the magnitude of the charge on each plate.

29. ★★ A 2.50 μF capacitor has a charge of 50.0 μC. The spacing between the plates is 0.0320 mm and the effective area of each plate is 820.0 cm^2. What must be the magnitude of the electric field in the space between the plates?

Section 22-3 Calculating Capacitance

30. ★★ An air-filled parallel-plate capacitor consists of two circular plates of radius 7.50 cm. To achieve a capacitance of 115 pF, what must be the separation of the plates?

31. ★★ An air-filled (effectively a vacuum) cylindrical capacitor has a length of 7.50 cm. The inner conductor has a radius of 9.50 mm, while that of the outer conductor is 10.50 mm. What is the capacitance?

32. ★★ The capacitance per unit length of a coaxial line is 64.0 pF/m. The outside diameter is 5.00 mm. What is the radius of the inner conductor? Assume air (effectively a vacuum) fills the space between the conductors.

33. ★★★ ∫ Consider a spherical capacitor that is made up of two spheres of radii 8.00 cm and 9.00 cm. The inner sphere is negatively charged with 450.0 nC, and the outer sphere is positively charged to the same magnitude. Air fills the region between the conductors.
  (a) Use Gauss's law to find an expression (as a function of radial distance r) for the electric field in the region between the two spheres. Include the direction of the field.
  (b) What is the electric field in the region outside both spheres?
  (c) What is the potential difference between the two spheres?
  (d) Calculate the capacitance.
  (e) Estimate the capacitance by approximating the configuration as a set of parallel plates.

Section 22-4 Combining Capacitors

34. ★★ Explain how you could combine an unlimited supply of 500. pF capacitors to obtain each of the following.
  (a) 0.002 μF
  (b) 0.0001 μF
  (c) 0.001 25 μF

35. ★★ If you had only four capacitors, of values 1.00 μF, 2.00 μF, 3.00 μF, and 4.00 μF, what would be the smallest and the largest capacitances that you could obtain by combinations of these capacitors? In each case, draw the circuit and show the calculation for the equivalent capacitance.

36. ★ If 1.00 μF, 2.00 μF, 3.00 μF, 4.00 μF, and 5.00 μF capacitors are placed in series, what is the total equivalent capacitance?

37. ★ You have a 4.70 μF capacitor but need an effective capacitance of 5.60 μF. What capacitor needs to be placed in parallel with the 4.70 μF capacitor to achieve this?

38. ★★ What is the effective capacitance for the network of capacitors shown in Figure 22-24?

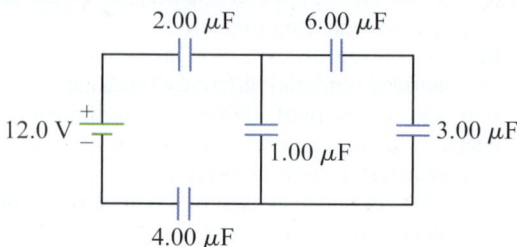

Figure 22-24 Problem 38.

Section 22-5 Dielectrics and Capacitors

39. ★★ A parallel-plate capacitor has rectangular plates 6.50 cm × 8.50 cm, spaced 2.00 mm apart. It has a dielectric material totally filling the space between the plates. The capacitor has a capacitance of 169 pF. Which of the materials in Table 22-1 could be the dielectric?

40. ★★ A parallel-plate capacitor has a capacitance of 0.002 00 μF when the space between the plates is filled with air. The plate spacing is 0.200 mm. It is connected throughout to an ideal 24.0 V battery.
  (a) What is the charge on each plate?
  (b) What is the electric field strength in the space between the plates?
  (c) A dielectric with $\kappa = 3.50$ is inserted, totally filling the space between the plates. What is the charge on each plate now?
  (d) What is the net electric field between the plates now?

41. ★ A charged capacitor (now disconnected from the charging source) has a potential difference of 8.00 V when no dielectric is inserted and 4.00 V when the dielectric is inserted. What must be the dielectric constant?

42. ★★ A capacitor has a plate spacing of 5.00 mm. It remains connected throughout to a 36.0 V charging source with the positive lead connected to the top plate of the capacitor. Initially it has no dielectric between the plates.
  (a) What is the electric field between the plates at this time (no dielectric)?
  (b) At this point, a dielectric with $\kappa = 4.00$ is inserted to fill the space between the plates. The capacitor remains connected to the charging source. Using the terminology of Interactive Activity 22-5, what are the plate, dielectric, and net (or sum) electric fields now?

43. ★★ With a dielectric present, it is observed that the electric field of the *dielectric* alone is $\vec{E}_{dielectric} = +800.$ V/m$\hat{j}$. The *net* electric field for the same situation is given by $\vec{E}_{net} = -200.$ V/m$\hat{j}$. The spacing between the plates is 3.50 mm.
  (a) What is the plate electric field, $\vec{E}_{plate}$ (see Interactive Activity 22-5 for terminology if necessary)?

(b) What is the potential difference between the plates? Is the upper or lower plate more positive?

(c) If the dielectric is removed (while keeping the charging source connected), what will the electric field be between the plates?

Section 22-6 Energy Storage in Capacitors

44. ★ A 200. μF capacitor is charged by connecting it to an ideal 24.0 V source. How much electrical energy is stored?

45. ★★ Three capacitors (1.00 μF, 2.00 μF, and 3.00 μF) are connected in series, and a 12.0 V battery is connected to the combination, as shown in Figure 22-25. Calculate the electrical energy stored in each capacitor.

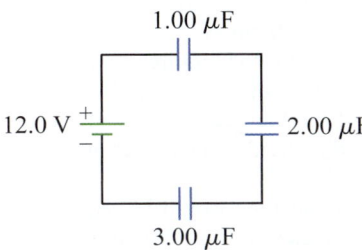

1.00 μF

12.0 V

2.00 μF

3.00 μF

Figure 22-25 Problem 45.

46. ★★ An air-filled (effectively a vacuum) parallel-plate capacitor is charged by connection to a 18.0 V source. The rectangular plates are 5.00 cm × 3.00 cm with a spacing of 3.50 mm. Calculate the electrical energy stored in the capacitor using the following two techniques.

(a) Calculate the capacitance and use the expression for the electrical energy stored in a capacitor.

(b) Find the electrical energy density in the electric field between the plates and multiply that by the volume of the space between the plates.

47. ★★ You have two identical 5.00 μF capacitors and a single 12.0 V ideal battery. What is the total electrical energy stored in the system under each of the conditions below?

(a) The two capacitors are placed in series, and that combination is charged with the 12.0 V battery.

(b) The two capacitors are placed in parallel, and that combination is charged with the 12.0 V battery.

48. ★★ An air-filled capacitor charged with an ideal battery with potential difference V has a total stored electrical energy U.

(a) The space between the plates is now filled with a dielectric of dielectric constant κ. The battery is left connected throughout. What is the amount of stored electrical energy now?

(b) Answer part (a) for the case in which the battery has been disconnected prior to insertion of the dielectric material.

Section 22-7 Applications of Capacitors

49. ★★ A pulsed laser that uses optical flash pumping produces 25.0 mJ per output pulse. Assume that the system is only 20% efficient, so five times as much energy must be stored in a capacitor to fire the flash each time. The

capacitance of the capacitor is 50.0 μF. What is the minimum applied potential difference needed to charge this capacitor?

50. ★★ A pressure sensor (see Figure 22-26) consists of a flexible membrane that moves the rectangular plates of a capacitor closer together when the pressure is higher. Assume that the plates are 1.00 cm × 5.00 cm. At a low pressure, the plate separation is 2.15 mm, while at a high pressure the plates are 1.85 mm apart. Assume that the space between the plates contains air. What is the change in capacitance between the two pressure values? Is the capacitance higher or lower when the pressure is higher?

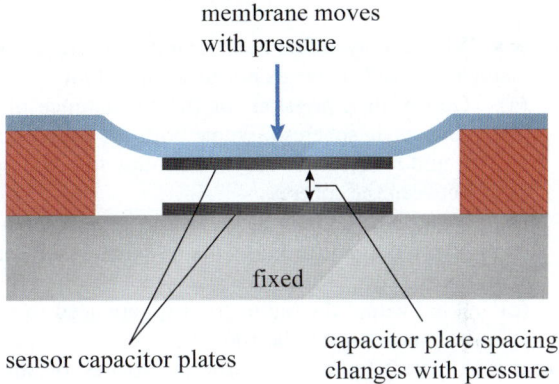

membrane moves with pressure

fixed

sensor capacitor plates

capacitor plate spacing changes with pressure

Figure 22-26 Problem 50.

51. ★★ It is desired to have the following force on an electron when it enters a region between two parallel plates (the plates are parallel to the xz-plane):

$$\vec{F} = -4.50 \times 10^{-16} \, \text{N} \hat{j}$$

The plates are separated by 15.0 mm.

(a) What should be the potential difference between the two plates?

(b) Should the top or bottom plate be at the higher electrical potential?

COMPREHENSIVE PROBLEMS

52. ★ Prove that the permittivity of free space, ε_0, can be expressed in units of F/m.

53. ★★ Capacitors can be made on semiconductor integrated circuits essentially as parallel-plate capacitors, although it is difficult to achieve high capacitance values (and also challenging to create precise values). Assume that you want a 13.0 pF capacitor and that the layers of the integrated circuit allow you to have a spacing of 1.50 μm. Assume that the material has a dielectric constant of 3.10. What is the necessary area of the plates on this integrated circuit capacitor?

54. ★★ A single cell in a computer memory chip has a capacitance of 90.0 fF (femtofarads) and operates at a potential difference of 3.30 V when charged.

(a) What is the charge on the cell?

(b) Express this charge in terms of the number of electrons.

55. ★★ An air-filled capacitor is to use circular plates of radius 5.00 cm that are separated by 3.00 mm. It is to have a maximum electric field of 1.00 kV/m in the space between the plates. Calculate
   (a) the capacitance
   (b) the maximum potential difference that should be applied
   (c) the maximum charge that can be stored
   (d) the maximum energy that can be stored

56. ★★ ∫ A capacitor consists of two thin coaxial cylinders. The inner radius is 1.00 cm, the outer radius is 2.50 cm, and the length of both is 2.00 m. They carry equal amounts of charge per unit length but have opposite polarities. What is the capacitance per metre for the system?

57. ★★ While it may at first seem strange, you can consider a single charged sphere as having a capacitance.
   (a) Derive an expression for the capacitance of two concentric spheres (assume both are thin) of radii r_1 and r_2. Assume that a vacuum fills the space between the spheres.
   (b) Now, find the limit as we let the second sphere be of infinite radius. This will give you a result for the capacitance of a single sphere.
   (c) What would the radius for a sphere need to be for its capacitance to be 100. pF?

58. ★★ A parallel-plate capacitor has plates of area A and separation d. Two different dielectrics (κ_1, κ_2) are inserted (Figure 22-27), each filling half the space. Derive an expression for the capacitance. (Hint: Treat the setup as two capacitors in parallel.)

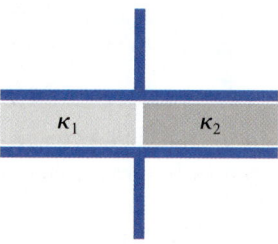

Figure 22-27 Problem 58.

59. ★★★ ∫ One form of transmission line consists of two long conducting wires (of radius r) separated by a distance D. Assume that D is much larger than r and that one wire has a linear charge density of $+\lambda$ (measured in coulombs per metre) and the other wire has a linear charge density of $-\lambda$.
   (a) Calculate the capacitance per unit length of the transmission line. In the early days of television, the signal from the antenna to the television was carried on this type of transmission line.
   (b) Find a numerical value for the capacitance per unit length for $D = 7.50$ mm and $r = 0.320$ mm.

60. ★★ Five 2.00 μF capacitors are connected in series, and that combination is charged with a 9.00 V ideal battery.
   (a) What is the charge on one plate of each capacitor?
   (b) What is the potential difference across each capacitor?
   (c) How much energy is stored in one capacitor?

61. ★★ What is the effective capacitance between points A and B in the circuit shown in Figure 22-28?

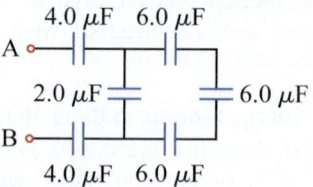

Figure 22-28 Problem 61.

62. ★★★ An air-filled parallel-plate capacitor is to be used to create an approximately constant electric field that will accelerate electrons with an acceleration of 7.00×10^{15} m/s^2. Assume that the spacing between the plates is 4.00 mm.
   (a) What is the required electric field strength?
   (b) What must be the applied potential difference?
   (c) Will the acceleration of the electrons be in the direction of the higher or lower electrical potential plate?
   (d) What time would be required for the electrons to accelerate from rest to a speed of 2.50×10^7 m/s, assuming that there is sufficient space between plates to reach this speed?
   (e) What distance would the electrons travel in reaching this speed? Is it possible to reach this speed with the given plate spacing?

63. ★★ A 0.100 μF capacitor is charged to 15.0 V. It is then connected in parallel to a second 0.100 μF capacitor, which was originally uncharged. Calculate the charge and potential difference on each capacitor afterward.

64. ★★ Consider the circuit shown in Figure 22-29. The potential difference is shown at a variety of points in the circuit. You are given the capacitance of the leftmost capacitor. Find the value of each of the other capacitors.

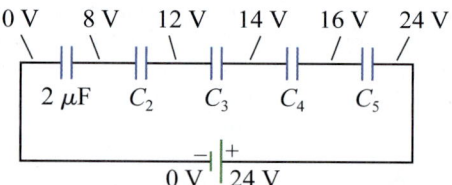

Figure 22-29 Problem 64.

65. ★★ A 2.00 μF capacitor is charged to 12.0 V. It is then disconnected from the battery and connected across an uncharged 4.00 μF capacitor. Calculate the resulting charge and potential difference on each capacitor.

66. ★★★ A 4.00 cm by 5.00 cm parallel-plate capacitor uses a dielectric with a dielectric constant of 4.00. The capacitor is designed so that when a potential difference of 60.0 V is applied, the charge stored on each plate has a magnitude of 150. nC. Calculate
   (a) the necessary capacitance
   (b) the spacing between the plates
   (c) the energy that will be stored in the capacitor under these conditions
   (d) the electric field due to charge on the plates (not the net field considering the dielectric)

67. ★★ An air-filled parallel-plate capacitor has a plate area of 0.100 m², a plate spacing of 2.50 mm, and an applied potential difference of 120. V. What is the energy density in the space between the plates of the capacitor?

68. ★★ What is the energy density just outside the surface of a conducting sphere of radius 3.50 cm that contains 3.50 nC of charge?

69. ★★★ A charge coupled device (CCD) chip has an active area of 12.0 mm × 9.00 mm and a resolution of 9.00 megapixels; the pixels are square. Note that here we are using the binary definition of megapixel: 1 megapixel = 2^{20} pixels = 1 048 576 pixels.
   (a) What are the dimensions of the device, expressed in pixels (i.e., x pixels × y pixels)? Round to the nearest integer number of pixels.
   (b) What are the dimensions of each pixel, in micrometres?
   (c) Using your answer in (b) to find the plate area, and assuming a dielectric constant of 3.10 and a plate spacing of 1.00 μm, what is the capacitance of each pixel cell? (Note: This is a very simplified model of a CCD.)

70. ★★ The outer layers of biological cells (lipid bilayers) act as capacitors, with the two sides holding opposite charge (the inside of the cell is negatively charged). Cells come in a variety of sizes, but assume a spherical cell of radius 5.00 μm and a dielectric constant of 4.00. Assume an electrical potential difference between the layers of 75.0 mV.
   (a) The thickness of the bilayer is 20.4 nm. What is the capacitance of the cell? (Hint: Model it as two thin spherical shells.)
   (b) What is the charge on each side?
   (c) How many elementary charges does this represent?

71. ★★ You are camping in a remote location and would like to have power at night. However, you do not want to carry a heavy battery to store the energy from the light-weight solar collectors that you are planning to carry in. The solar array that you have can generate 5.50 W in full sunlight. The array produces a potential difference of 12.0 V.
   (a) How much capacitance is needed to store the energy produced by solar array in 6.00 h of full sunlight?
   (b) The largest electrolytic capacitors you have available are 1000. μF each. How many would be needed in parallel?
   (c) How long could the stored energy power a tablet that uses 8.00 W?

72. ★★ Current high-energy capacitor designs (e.g., like those used in the National Ignition Facility) achieve energies of about 2.00 J for every cubic centimetre of the capacitor. If one of these capacitors were in the shape of a cylinder of radius 5.00 cm and length 10.0 cm, and had a maximum voltage rating of 750. V, what would its capacitance be?

73. ★★★ Consider the circuit shown in Figure 22-30. The values of the capacitors are $C_1 = 10.0\ \mu$F, $C_2 = 30.0\ \mu$F, $C_3 = 20.0\ \mu$F, and $C_4 = 10.0\ \mu$F. Assume that the

battery and all capacitors are ideal. Find the potential difference and charge on each of the four capacitors.

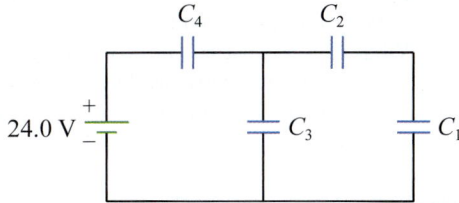

Figure 22-30 Problem 73.

74. ★★ An ideal 12.0 μF capacitor is charged with 240. μC of charge. It is then disconnected from the charging source and its leads are connected in parallel with an originally uncharged 24.0 μF capacitor.
   (a) What is the charge on each capacitor after they are connected together?
   (b) What is the voltage (electric potential difference) on the capacitors in the final state?

75. ★★★ Consider the circuit shown in Figure 22-31. Assume that the battery and all capacitors are ideal. The values of the capacitors are $C_1 = 8.00\ \mu$F, $C_2 = 4.00\ \mu$F, $C_3 = 6.00\ \mu$F, and $C_4 = 10.00\ \mu$F. Find the potential difference for each of the four capacitors.

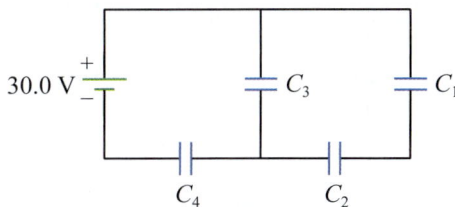

Figure 22-31 Problem 75.

76. ★★★ In the circuit shown in Figure 22-32, the battery is ideal, as are all capacitors. The values of the capacitors are $C_1 = 12.0\ \mu$F, $C_2 = 24.0\ \mu$F, $C_3 = 36.0\ \mu$F, and $C_4 = 8.0\ \mu$F. Find the electrical energy stored for each of the four capacitors.

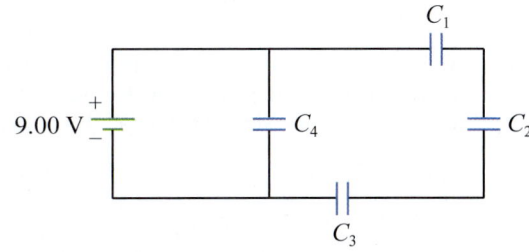

Figure 22-32 Problem 76.

DATA-RICH PROBLEM

77. ★★★ Table 22-2 shows a relationship between electrical potential (in V) and radial distance, r (in cm).
   (a) Describe a situation that could correspond to these data.
   (b) What is the electric field at $r = 6$ cm?

(c) What is the electric field at $r = 12$ cm?

(d) What is the electric field just outside $r = 10$ cm?

(e) If this is some sort of capacitor, what is the potential difference between the two plates?

(f) What charge must be on the inner plate to be consistent with the electric field in part (d)?

(g) Use your data from parts (e) and (f) to estimate the capacitance.

Table 22-2 Data for Problem 77

Radial Distance (cm)	Electrical Potential Difference (V)
0	27 000
2	27 000
4	27 000
6	27 000
8	27 000
10	27 000
12	22 500
14	19 300
16	16 900
18	15 000
20	13 500
22	13 500
24	13 500
26	13 500
28	13 500

OPEN PROBLEMS

78. ★★★ In this question, you will make reasonable estimates to determine whether it is feasible to power an electric vehicle with charged capacitors instead of batteries.

(a) Estimate (or look up specifications for) the energy required for an electric vehicle to drive 50 km. List any assumptions that you make for your estimate.

(b) Design an optimally efficient (in terms of energy storage per volume) capacitor. Include the design (e.g., shape and size), materials (e.g., dielectric used), and estimates of the working electric potential and energy storage per capacitor unit. Provide comments on all aspects of your design.

(c) To provide the energy estimated in part (a), what volume and mass of capacitors are required, considering your capacitor design in part (b)? Comment on how reasonable these requirements are.

(d) Current technology uses lithium ion batteries for electric automobile energy storage. Look up, or estimate, the volume and mass requirements using that technology to provide the same amount of energy.

(e) Comment on the general advantages and disadvantages (within an electric automobile context) of capacitor-based energy storage compared to lithium ion batteries or fuel cells.

79. ★★ You have been asked to design a capacitor-based portable device that can store sufficient energy to start an automobile that has had its battery discharged.

(a) Using realistic numbers for the current draw on the starter motor of an automobile, and the length of time this current needs to be supplied, estimate the energy needed to be stored in the capacitor.

(b) Considering the 12 V supply, what total capacitance value is needed?

(c) Assume that you will use commercial electrolytic capacitors in the application. Look up on the web typical sizes and capacitance values for these, and estimate the minimum volume and weight of the device.

(d) Also, estimate the cost, assuming commercial prices for the capacitors and reasonable costs for additional parts of the device.

(e) Assess the feasibility of the device that you have designed. Look up on the web to see if such devices are in use, and include descriptions of any commercially available devices that you find.

80. ★★ When there is a flood in a condominium or apartment building, it is important to make sure that moisture does not stay in the walls, where it would promote the growth of mould. Professionals who do this sort of work carry hand-held moisture meters that can detect moisture in walls by simply holding them against the wall. Speculate on the principle of operation of these devices based on the material of this chapter.

CHAPTER 23

Electric Current and Fundamentals of DC Circuits

Learning Objectives

When you have completed this chapter, you should be able to

LO1 Derive and use the microscopic model of electric current to describe its flow through a conductor.

LO2 Describe the conductivity and resistivity of a metal quantitatively and qualitatively using the microscopic model of electric current.

LO3 Describe the relationship between electric current, voltage, and resistance in metals using the macroscopic model of electric current: Ohm's law.

LO4 Derive relationships for equivalent resistance in series and parallel electric circuits.

LO5 Analyze simple and compound DC circuits, and use Kirchhoff's laws to predict the circuit power output, current flow, and potential difference across different circuit components.

LO6 Analyze series RC circuits.

In 1791, Italian physician and physicist Luigi Galvani (1737–1798) discovered bioelectricity. Shortly thereafter, his associate, Alessandro Volta (1745–1827), invented the first electrical battery (the voltaic pile). Less than 250 years later, we rely heavily on ubiquitous access to electric power. Terms such as *electricity*, *electric current*, *electric resistance*, *electric power*, *electricity generation*, and *transmission* have entered our everyday vocabulary. Sometimes, we use them interchangeably, even though they mean different things. Electrical phenomena are the basis of our lives on both microscopic (e.g., physiological phenomena and electron motion) and macroscopic (e.g., everyday electric appliances) levels. To appreciate the importance of electrical phenomena in our lives, we need only recall the havoc created in northeastern Canada and the United States on the hot summer day of August 14, 2003. On that day, more than 55 million people in Ontario and in the northeastern United States were left without electric power due to the failure of a transmission grid as a result of a 3500 MW (3.5×10^9 W) power surge (Figure 23-1). In this chapter, we examine the nature of electric current and its applications in direct current (DC) circuits.

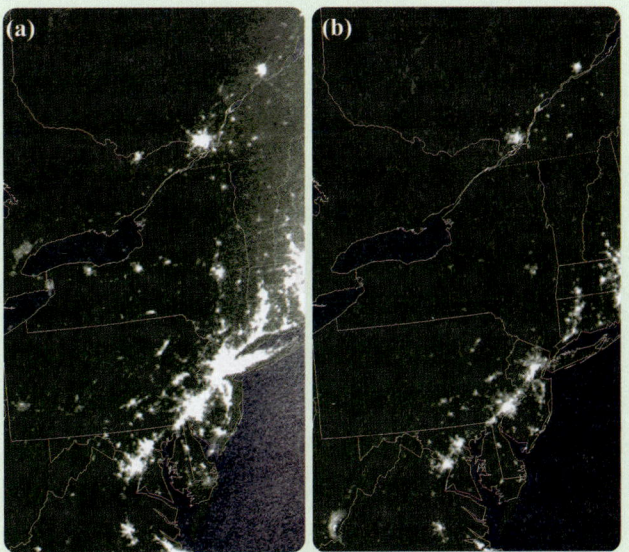

Figure 23-1 The "Northeast Blackout" of 2003 affected more than 55 million people in North America and was the second most widespread blackout in history. A view of the 2003 blackout from space: Ontario and the northeastern United States before (left) and during (right) the blackout.

23-1 Electric Current: The Microscopic Model

In Chapter 22, we saw that electric charges can move through a metal conductor to charge or discharge a capacitor. Various experiments, such as studies of the Hall effect (described in Chapter 24) and the Tolman–Stewart experiment, conducted in 1916, have shown that the moving charges in metals are electrons. Thus, we can describe the electric current in metals as an **electron current** or an **electron flow**. In some materials, such as plasmas (ionized gases) and ionic solutions, the charge carriers may include positive and negative ions.

We model a metal as a **crystal lattice** (a symmetrical three-dimensional ordered arrangement of atoms, ions, or molecules in a crystalline material) of positively charged ions immersed in a cloud, or sea, of **free (delocalized) electrons**. These electrons are dislodged from the outer shell of the metal atoms. As a result, they are not attached to any specific atom and are free to move inside the metal under the influence of electric or magnetic fields. We will discuss these phenomena in more detail in Chapter 33. In most metals, such as silver, copper, gold, and aluminum, each atom loses one outermost (valence) electron, thus turning the atom into a positive ion (Figure 23-2). The solid metal is held together by the electrostatic attraction forces between the positive lattice and a negative free electron cloud. These forces are called **metallic bonds**. The presence of free electrons in metals explains why they are good conductors of both thermal energy (heat) and electricity. It also explains why insulators (materials that do not have free electrons) are poor conductors.

Metal ions are composed of the metal nuclei and bound electrons. For example, an aluminum ($^{13}\text{Al}_{27}$) ion consists of the nucleus with 13 protons, 14 neutrons, and 12 bound electrons. The 13th electron becomes a free electron in the metal lattice. The nuclei of metal atoms consist of protons and neutrons that have approximately 2000 times the mass of electrons ($m_\text{p} \approx m_\text{n} \approx 2000 m_\text{e}$), so the thermal motion of the metal ions (Chapter 16) is negligible compared to the thermal motion of the much lighter free electrons. Therefore, we consider the heavy lattice ions almost static compared to the fast but randomly moving free electrons. The **free** (or **conduction**) **electron density** (n_e) of a material is defined as the number of free electrons per cubic metre of the material. In Example 23-1, we will see how this density can be estimated. Table 23-1 lists the free electron densities for some familiar metals. We will calculate the average speed for the thermal motion of free electrons in Chapter 33.

Table 23-1 Conduction Electron Density in Metals

Metal	Conduction Electron Density, n_e (m^{-3})
Aluminum	6.0×10^{28}
Copper	8.4×10^{28}
Gold	5.9×10^{28}
Iron	8.4×10^{28}
Nickel	9.0×10^{28}
Silver	5.8×10^{28}
Zinc	6.0×10^{28}

EXAMPLE 23-1

Estimating the Free Electron Density in Aluminum

Using the free electron model described above, estimate the free electron density in aluminum, given that the density of aluminum is $\rho_\text{Al} = 2.7 \times 10^3$ kg/m^3.

Solution

Assuming that every atom of aluminum contributes one free electron to the free electron cloud that can be described using the orbital notation you have used in your chemistry courses (Al: $1s^2$, $2s^2$, $2p^2$, $3s^2$, $3p^1$), the free electron density is equal to the number of aluminum atoms per cubic metre. That number can be calculated from the molar mass of aluminum (M_Al), its density (ρ_Al), and the number of atoms in one mole of aluminum (Avogadro's number, N_A):

$$n_\text{e} = \frac{\rho_\text{Al}}{M_\text{Al}} \cdot N_\text{A} = \frac{2.7 \times 10^6 \dfrac{\text{g}}{\text{m}^3}}{27 \dfrac{\text{g}}{\text{mol}}} (6.022 \times 10^{23} \text{ mol}^{-1})$$

$$= 6.0 \times 10^{28} \text{ m}^{-3}$$

Making sense of the result

Our estimated value for the free electron density in aluminum agrees with the experimentally observed value listed in Table 23-1. Thus, the free electron cloud model produces a prediction supported by an experiment.

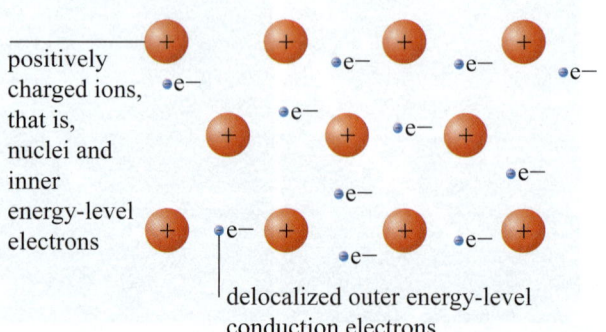

positively charged ions, that is, nuclei and inner energy-level electrons

delocalized outer energy-level conduction electrons

Figure 23-2 The sea-of-electrons model in a metal. Delocalized conduction electrons are immersed in the ionic lattice of a metal. These electrons are free to move inside the metal and can often go past hundreds of atoms before bumping into them, consequently changing the course of the electrons' motion. These electrons make metals good conductors of both heat and electricity.

As long as no external electric fields are applied to conduction electrons, they move in random directions as a result of collisions with ions of the metal lattice, so that their average velocity is zero (Chapter 16): on average, there are just as many electrons moving in one direction as in the other direction. However, when a potential difference is applied across a conductor, it creates an external electric field, $\vec{E}$. This electric field results in a net force acting on the electron cloud directed opposite to the direction of the electric field (Figure 23-3). If the electrons were located in a vacuum, the electric field would continuously accelerate them to very high speeds. However, since the electrons are located in a metal, they have to move through the crystal lattice, undergoing multiple collisions, thus deflecting from their original path (**electron scattering**). This process slows them down significantly, making their average velocity quite small. This average velocity of an electron cloud is called the **drift velocity**, $\vec{v}_d$. The drift velocity depends on the properties of the metal and on the applied external potential difference (external electric field). The higher the applied potential difference, the higher the drift velocity. This drift velocity can be measured experimentally, and we will also estimate its magnitude later on in this chapter.

Let us estimate the number of electrons, N_e, passing through the cross-sectional area, A, of a cylindrical aluminum wire in a time interval, Δt, as a result of an applied external electric field, $\vec{E}$ (Figure 23-4). Note that due to this external electric field, there is an electric field inside a wire carrying an electric current, in contrast to the electrostatic case, in which the electric field inside a perfect conductor always settles to zero in electrostatic equilibrium, as described in Chapters 20 and 21.

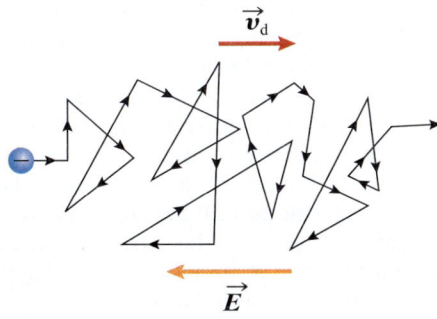

Figure 23-3 An external electric field applies an electric force on electrons in a metal, so they start moving with non-zero average velocity.

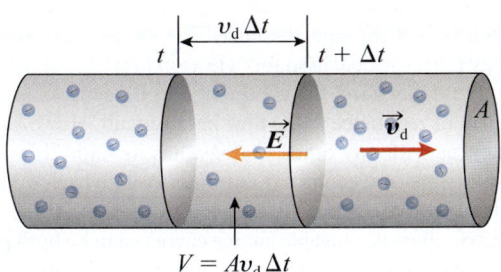

Figure 23-4 The movement of electrons in a conductor under the influence of an external electric field.

Since the electron cloud moves a distance $\Delta x = v_d \Delta t$ during the time interval Δt, the number of electrons passing through the cross-sectional area A over this time can be calculated as

$$N_e = n_e V = n_e A \Delta x = n_e A v_d \Delta t \qquad (23\text{-}1)$$

Notice, in Equation 23-1, that V stands for volume: $V = A \Delta x$. In Chapters 23 to 25, a special symbol for volume (V) is chosen to distinguish it from the potential difference V.

Electron current is defined as the number of electrons passing through a conductor per unit time. Thus,

$$I_e = \frac{N_e}{\Delta t} = n_e v_d A \qquad (23\text{-}2)$$

Let us estimate the value of the electron current in an aluminum wire with a cross-sectional area of 1 mm² when an external electric field of 1 V/m is applied to it. The value for the conduction electron density of aluminum, n_e, was found in Example 23-1 (see also Table 23-1). The value of the drift velocity of electrons in an aluminum wire when an electric field of 1 V/m is applied to it was experimentally found to be 10^{-4} m/s. Thus, we can estimate the electron current through this aluminum wire:

$$I_e = n_e v_d A = (6 \times 10^{28}\ \text{m}^{-3})\left(10^{-4}\ \frac{\text{m}}{\text{s}}\right)(10^{-6}\ \text{m}^2)$$

$$= 6 \times 10^{18}\ \text{s}^{-1} \qquad (23\text{-}3)$$

The number of electrons flowing through this aluminum wire is very large. To get a sense of how large this number is, it is almost a billion times the population of planet Earth at the beginning of the 21st century!

It is important to note that while moving through the wire, these electrons carry electric charge with them. Thus, an electron flow through the wire is a flow of electric charge through it. Therefore, it is generally more practical to measure the current in terms of the amount of electric charge rather than the number of electrons carried through the wire per unit time. The flow of charge is called the **electric current**, I. Historically, it was thought that positive charges and not electrons moved in a wire. For this reason, by convention, electric current is defined as the motion of positively charged particles. Therefore, the directions of electric and electron currents are opposite. Since the electric charge carried by one electron is $q_e = -e = -1.6 \times 10^{-19}$ C, where e is the elementary charge, the relationship between electric current and electron current is as follows:

KEY EQUATION

$$I = -I_e \cdot e = -(n_e v_d A)e$$

$$= -(n_e v_d A)(1.6 \times 10^{-19})\frac{\text{C}}{\text{s}} \qquad (23\text{-}4)$$

Electric current is a scalar quantity describing the rate of flow of electric charge in the direction that a positive moving charge would take. This explains why electron and electric currents have opposite signs.

We will discuss the meaning of the sign of the electric current in Section 23-4. The magnitude of the electric current represents the amount of electric charge flowing through the surface per unit time.

KEY EQUATION
$$I = \frac{\Delta Q}{\Delta t} \tag{23-5}$$

The unit of current, the ampere (a base unit of SI), is named in honour of French scientist Andre-Marie Ampère (1775–1836). It is often abbreviated as A or amp:

$$1\,\text{A} = \frac{1\,\text{C}}{1\,\text{s}}$$

We can use Equations 23-3 and 23-4 to estimate the value of the electric current in the aluminum wire when an electric field of 1 V/m is applied across it (Example 23-1):

$$I = (6.0 \times 10^{18}\,\text{s}^{-1})(1.6 \times 10^{-19}\,\text{C}) \approx 1.0\,\frac{\text{C}}{\text{s}} = 1.0\,\text{A}$$

The **current density**, J, in a conductor represents the amount of electric current, I, flowing per unit of cross-sectional area, A:

KEY EQUATION
$$J = \frac{I}{A} = \frac{n_e v_d A e}{A} = n_e v_d e \tag{23-6}$$

The unit of current density is defined as

$$[J] = \frac{\text{C}}{\text{s}\cdot\text{m}^2} = \frac{\text{A}}{\text{m}^2}$$

As we will see later in this chapter, it is easy to measure electric current flowing through a wire. Knowing the physical properties of the wire (its dimensions and the metal it is made of) allows us to estimate the drift velocity of the electrons in it. For example, the drift velocity of the electrons in the aluminum wire in Example 23-1, carrying electric current of 1 A, can be estimated using Equation 23-6 as

$$v_d = \frac{I}{A n_e e}$$

$$= \frac{1\,\text{A}}{(10^{-6}\,\text{m}^2)(6.0 \times 10^{28}\,\text{m}^{-3})(1.6 \times 10^{-19}\,\text{C})} \tag{23-7}$$

$$= 10^{-4}\,\frac{\text{m}}{\text{s}}$$

This means that it takes an electron 1000 s (almost 17 min) to move a distance of 1 m! Notice that if the potential difference across the wire increases, the drift velocity of electrons increases as well. For example, if in Example 23-1 the potential difference across the wire and consequently the electric field in it doubled (to 2 V and 2 V/m, respectively), the drift velocity of the electrons would double as well to $2 \times 10^{-4}\,\frac{\text{m}}{\text{s}}$.

MAKING CONNECTIONS

Turn on the Light: A Microscopic View of Electric Current

As we discussed above, the drift velocity of electrons in metals when a potential difference of about 1 V is applied across them has order of magnitude 10^{-4} m/s (which is only 3.6 m/h, which means it would take them about an hour to move across your bedroom). Why, then, do lights turn on almost instantly when you flip the switch? The answer is that a metal wire is full of conduction electrons even before you turn the light on, and what propagates through the wire is a vibrational wave in the free electron cloud (the transfer of energy between electrons), not the cloud itself (the cloud would have moved with the drift velocity). This vibrational wave propagates through the metal lattice at a speed close to the speed of light ($c = 3 \times 10^8$ m/s), thus making the light turn on almost instantaneously. We will discuss this in more detail in Chapters 26 and 27. You can think of it as a hose full of water. The water is almost incompressible, and when you turn on the tap on one end of the hose, the water starts flowing from the other end relatively fast, as the wave of compression moves through the hose. The electronic–hydraulic analogy is very widely used for "electron flow" in metals. It is helpful to understand the conservation of electric charge or the effect of resistance to the flow in the circuit. However, the analogy, as with any other analogy, has its limitations. For example, water molecules cannot interact with each other at a distance, while electric charges can. Moreover, in the case of electric charges, movable charge carriers can be both positive and negative, while in the case of water, we have only one kind of charge. Thus, you always have to be aware when an analogy is valid and when it breaks down.

23-2 Electric Conductivity and Resistivity: The Microscopic Model

When a copper wire is connected to a battery or to a charged capacitor, there is an excess of positive charge at the end of the wire connected to the positive terminal of the battery or to the positive plate of the capacitor (high potential) and an excess of negative charge (electrons) at the end of the wire connected to the negative terminal or plate (low potential) (Chapter 21). The non-uniform charge distribution at the ends of the wire creates an internal electric field, $\vec{E}$, inside it. This electric field is directed from the end of the wire that has an excess of positive charges (high external potential) to the end of the wire that has an excess of negative charges (low external potential). As we discussed in Chapters 20 and 21, in the case of electrostatics, when the charges in a conductor are in a state of static equilibrium, the electric field inside the conductor is zero. But in this case, we have electric current flowing through the wire, so this is not a case of static equilibrium. The electric field in the current-carrying wire is non-zero. Since electrons are negatively charged, this internal electric field accelerates them in the direction opposite to $\vec{E}$:

$$\vec{a} = \frac{\vec{F}}{m_e} = -\frac{e\vec{E}}{m_e} = -(1.6 \times 10^{-19}\,\text{C})\frac{\vec{E}}{m_e} \quad (23\text{-}8)$$

Thus, the collective motion of electrons in the direction opposite to the internal electric field, $\vec{E}$, is superimposed on the random motion of electrons caused by their collisions with the lattice ions (Figure 23-2). The result is the drifting of the electrons in the direction opposite to $\vec{E}$ (Figure 23-3). We can estimate the drift velocity of electrons from the microscopic model of electric current if we assume that, on average, electrons spend τ seconds between collisions with the positive ions of the metal. The value of τ, which is called the **mean free time** of electrons, can be measured experimentally and is a property of the metal. As discussed earlier, in the absence of an electric field, the average velocity of the free electrons is zero. However, when an electric potential difference is applied across the metal, there is an electric field inside it. Therefore, during the time between collisions, electrons experience an electrostatic force $\vec{F} = -e\vec{E}$ that causes them to accelerate, gaining drift speed:

$$v_d = \frac{eE}{m}\tau \quad (23\text{-}9)$$

Substituting the above expression into the equations for electron and electric currents, respectively (Equations 23-3, 23-4), we find the following:

$$I_e = n_e \frac{eE}{m}\tau A = \frac{n_e e \tau A}{m}E$$

$$I = I_e e = \frac{n_e e^2 \tau A}{m}E \quad (23\text{-}10)$$

Equation 23-10 shows that the values of both the electron current, I_e, and the conventional electric current, I, are proportional to the electric field, E. All the other variables in Equation 23-10 are fully determined by the electrons and the physical properties of the metal wire. Let us rewrite the above equation in terms of the current density, J, to exclude the specific dimensions of the wire:

$$J = \frac{I}{A} = \frac{n_e e^2 \tau}{m}E = \sigma E, \text{ where } \sigma = \frac{n_e e^2 \tau}{m} \quad (23\text{-}11)$$

The physical quantity σ has a special name: **electric conductivity**. The units of electric conductivity are

$$[\sigma] = \frac{C^2 \cdot s}{m^3 \cdot kg} = \frac{A}{V \cdot m}$$

Equation 23-11 describes a causal relationship between the electric field and the current density inside the wire: the electric field in the wire causes the electric current inside it. In this equation, σ represents how much electric current flows through the wire in response to a given electric field. Electric conductivity is an internal property of a material. The greater the electric conductivity, the better the material conducts electric current. It is important to notice that historically the same letter, σ, has been used to denote two very different physical concepts: electric conductivity and surface charge density (Chapters 19 and 20). Thus, you have to pay extra attention when you see this symbol.

Multiplying Equation 23-11 by the cross-sectional area of the wire, A, gives an expression for the electric current:

$$I = JA = \frac{n_e e^2 \tau}{m}AE = \sigma EA = \frac{1}{\rho}EA \quad (23\text{-}12)$$

The **electric resistivity**, ρ, of a material describes how strongly the material opposes the electric current. It is defined as follows:

KEY EQUATION
$$\rho = \frac{1}{\sigma} = \frac{EA}{I} = \frac{m}{n_e e^2 \tau} \quad (23\text{-}13)$$

The units of electric resistivity are defined as

$$[\rho] = \frac{(V/m)m^2}{A} = \frac{V \cdot m}{A}$$

Once again, be careful: the same letter, ρ, is used to denote both the density of a material and the electric resistivity. Note that it is also sometimes used to denote a charge density, but we will use the index e in that case, ρ_e, to reduce confusion.

The shorter the mean free time (τ) between collisions of electrons within the crystal lattice, the more collisions will take place and, consequently, the more kinetic energy of the electrons will be converted into thermal energy. This will take place during the collisions of electrons with the ions in the metal's crystal lattice. During this process, part of the kinetic energy of the electrons will be converted into the vibrational energy of the ions of the lattice, thus increasing the temperature of the metal. This energy transfer explains why wires heat up when electric current flows through them.

This also explains why the resistivity of a metal also increases with temperature, and the conductivity decreases. As the metal heats up, the time between collisions, τ, decreases, which affects both ρ and σ, as can be seen from Equation 23-13. For most metals, the dependence of resistivity on temperature is linear for a limited range of temperatures:

KEY EQUATION

$$\rho = \rho_0 + \alpha\rho_0(T - T_0) = \rho_0[1 + \alpha(T - T_0)]$$
$$= \rho_0(1 + \alpha\Delta T) \tag{23-14}$$

where ρ is the resistivity at temperature T, measured in degrees Celsius; ρ_0 is the resistivity at the baseline temperature T_0 (often chosen to be 20 °C); and α is the temperature coefficient of resistivity, which has units of 1/K.

Table 23-2 Resistivities, Conductivities, and Temperature Coefficients for Different Metals

Metal	ρ ($\Omega \cdot$m) at 20 °C	σ ($\Omega \cdot$m)$^{-1}$ at 20 °C	Temperature Coefficient α (K^{-1})
Aluminum	2.65×10^{-8}	3.77×10^7	3.80×10^{-3}
Copper	1.72×10^{-8}	5.95×10^7	4.30×10^{-3}
Gold	2.24×10^{-8}	4.10×10^7	3.40×10^{-3}
Iron	9.7×10^{-8}	1.00×10^7	5.00×10^{-3}
Lead	2.06×10^{-7}	4.55×10^6	3.90×10^{-3}
Nichrome	1.10×10^{-6}	9.09×10^5	4.00×10^{-3}
Nickel	6.85×10^{-8}	1.43×10^7	6.41×10^{-3}
Platinum	1.06×10^{-7}	9.43×10^6	3.92×10^{-3}
Silver	1.59×10^{-8}	6.30×10^7	3.80×10^{-3}
Tin	1.09×10^{-7}	9.17×10^6	4.50×10^{-3}
Tungsten	5.60×10^{-8}	1.79×10^7	4.50×10^{-3}
Zinc	5.90×10^{-8}	1.69×10^7	3.70×10^{-3}

Source: Based on http://en.wikipedia.org/wiki/Resistivity#Table_of_resistivities and http://www.engineeringtoolbox.com/resistivity-conductivity-d_418.html

The resistivities, conductivities, and temperature coefficients for various metals are given in Table 23-2.

You have probably heard about materials that can conduct electric current without it experiencing any resistance. A **superconductor** is a material with a resistivity that abruptly drops to zero below a critical temperature (less than 100 K for most currently known superconductors).

23-3 Ohm's Law

Electric charges flowing through the conducting wires in our homes and through many of the electrical appliances we use in everyday life allow us to transfer electrical energy and use it to power electrical appliances, generate light, and do mechanical work. To be able to convert electrical energy into mechanical work and other forms of energy, we need to be able to analyze electrical circuits that contain various combinations of electrical elements. First, we derive a relationship between electric current, potential difference, and electrical resistance.

Consider a conducting wire of length ℓ. In Chapter 22, we saw that in the case of a constant and uniform electric field in a wire of length ℓ, the relationship between electric field strength and electric potential is given by $E = V/\ell$, where V represents the potential difference across the wire (Equation 22-1, with $d = \ell$). Substituting for E in Equation 23-12, we obtain

$$I = \sigma E A = \frac{1}{\rho}EA = \frac{1}{\rho}\frac{V}{\ell}A = \frac{A}{\rho\ell}V = \frac{V}{R} \tag{23-15}$$

CHECKPOINT 23-2

Dimensional Analysis in Action

Which of the following equations has incorrect dimensions? Square brackets indicate the units of the enclosed physical quantity.

(a) $[I] = \dfrac{C^2 \cdot s \cdot V}{kg \cdot m^2}$

(b) $[J] = \dfrac{C^2 \cdot s \cdot V}{kg \cdot m^4}$

(c) $[\rho] = \dfrac{kg \cdot m^3}{C^2 \cdot s}$

(d) $[\sigma] = \dfrac{C^2 \cdot s}{kg \cdot m^3}$

(e) $[I_e] = \dfrac{C \cdot s \cdot V}{m^3 \cdot kg}$

ANSWER: (e) $[I_e] = \dfrac{q_e}{I} = \dfrac{C}{C/s} = \dfrac{C}{1/s} = \dfrac{s}{V} \cdot \dfrac{A}{C} \neq \dfrac{C \cdot s \cdot V}{m^3 \cdot kg}$

In this equation, we introduce a new variable, R, to denote **electric resistance**:

$$R = \rho \frac{\ell}{A} \qquad (23\text{-}16)$$

where A is the cross-sectional area of the material, ℓ is its length, and ρ is its resistivity. We can see that the resistivity of a wire represents the resistance of a wire that is 1 m long and has a cross-sectional area of 1 m². Once again, be careful: the same letter, R, depending on the context, may denote electrical resistance or a radius. To avoid confusion, whenever possible in this chapter we will use a lowercase d to represent the diameter of a wire instead of using the radius notation.

Equation 23-16 allows us to calculate the resistance of a wire if we know the material it is made of (which gives us information about its resistivity) and its dimensions. Let us assume the wire has the shape of a cylinder with cross-sectional area A and length ℓ. From Equation 23-16, we deduce that a wire with a shorter length or a larger cross-sectional area has a lower resistance than a longer and narrower wire. Here the water analogy might be helpful. A longer and narrower water hose provides a higher resistance to the water flow than a shorter water hose that has a higher cross-sectional area (see Interactive Activity 23-1).

Equation 23-15, which showed the relationship between electric current through a wire, its resistance, and the potential difference across it ($I = V/R$) was derived for a special case—a cylindrical electrical wire with a uniform cross-sectional area—but it can be generalized for a large variety of electrical components, such as **resistors**—circuit elements with constant specified resistance. As we discussed earlier (Chapter 22), resistors limit or regulate the flow of electric current in an electric circuit. The equation shows a very important relationship between electric current flowing through a wire and the potential difference across it: the amount of electric current through the wire is directly proportional to the potential difference across it. Because

potential difference is expressed in units of volts, it is also often referred as voltage. Voltage, as a potential difference, is always measured between two points, as discussed in Chapter 21. For example, the voltage between points A and B is denoted as $V_{A \to B} = V_B - V_A$. However, when the points between which the potential difference is measured can be unequivocally deduced from the context, these indices can be dropped, and the notation V can be used to denote the potential difference or voltage across a specific circuit element, such as a wire, a resistor, a battery, or a capacitor. While some books use ΔV to denote potential difference, we will omit the delta to simplify the notation and to avoid the use of the incorrect terms *voltage drop* and *voltage difference*. Introducing the concept of electric resistance allows us to describe the relationship between electric current through the wire, voltage across the wire, and the physical properties of the wire using a simple relationship called **Ohm's law**. Ohm's law states that for some circuit elements, the electric current through them is directly proportional to the potential difference, V, applied across it and inversely proportional to the resistance of this element (Figure 23-5(a)):

$$I = \frac{V}{R} \qquad (23\text{-}17)$$

The units of electric current are defined as

$$1\ \text{A} = \frac{1\ \text{V}}{1\ \Omega}$$

Devices for which this relationship holds are called **ohmic conductors** and include resistors and wires, such as various heating elements, carbon-composition resistors, and film or cermet resistors (Figure 23-6). This relationship was discovered by Georg Simon Ohm (1787–1854), a German scientist. The unit of electric resistance, the ohm (Ω), is named in his honour. (We recommend that you explore the PhET "Ohm's Law" at phet.colorado.edu/en/simulation/ohms-law.) Ohm's law is widely used in designing

ELECTRICITY, MAGNETISM, AND OPTICS

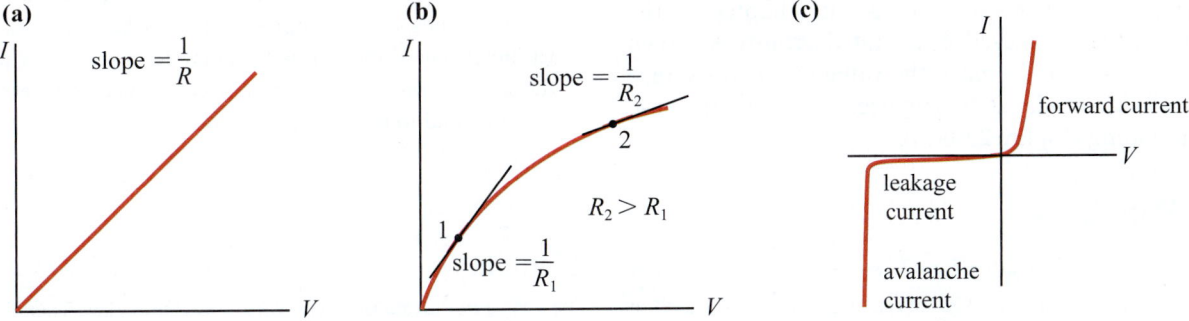

Figure 23-5 The dependence of current on applied voltage (a current–voltage characteristic or an *I-V* curve) across (a) an ohmic circuit element and two non-ohmic circuit elements, (b) an incandescent light bulb, and (c) a diode in the area of forward current. R_1 and R_2 denote the resistance of the element when different voltages are applied across it.

CHAPTER 23 | **ELECTRIC CURRENT AND FUNDAMENTALS OF DC CIRCUITS**

and analyzing electric circuits. However, you have to remember that the electric resistivity, ρ, and the electric resistance, R, for many materials depend on temperature. Thus, the slope of the I-V curve (the common name for the $I(V)$ graph), which is represented by $1/R$, might change if the material heats up as the current increases (Figure 23-5(b)). This heating would make the relationship nonlinear. In cases when it is applicable, Ohm's law accurately describes how current through an element of a circuit depends on the voltage across it while the temperature of the element remains constant. This limits the applicability of Ohm's law. For example, the resistance of the filament of an incandescent light bulb increases markedly when current heats it enough to produce visible light; therefore, the filament of an incandescent light bulb is a non-ohmic circuit element. Thus, the resistance of the filament is relatively low when we turn the light bulb on (it is cold at the beginning), which means a higher amount of current flows through it at the beginning, slowly decreasing as the filament heats up and increases its resistance. It also explains why light bulbs usually burn out just as they are turned on: when you turn the light on, the filament is still cold, so its resistance is low and, as a result, the current is high. This high current can damage the filament and thus burn out the light bulb. An **electric diode**, a semiconductor device that allows the flow of current in one direction only, is another example of a non-ohmic circuit element (Figure 23-5(c)).

The dependence of electric current on temperature can be derived from Equations 23-14 and 23-16:

$$I = \frac{V}{R} = \frac{VA}{\rho_0[1 + \alpha(T - T_0)]\ell} = \frac{V}{R_0[1 + \alpha(T - T_0)]}$$

(23-18)

Electric circuit components with resistance, such as the ones shown in Figure 23-6(a), are found in all electrical and electronic devices. Often an electric circuit for a particular application will need many circuit elements that have constant specified resistances—electrical resistors. As we discussed in Chapter 22, the symbol for resistor used in circuit diagrams is shown in Figure 23-6(b). To mark the value of the resistance on small circuit elements, engineers and scientists use colour coding (Figure 23-6(c)).

(a)

edography/Shutterstock.com

(b)

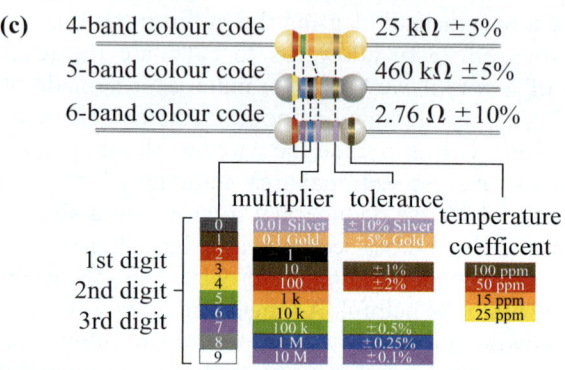

Figure 23-6 (a) Miniature electrical resistors inside a digital clock. (b) The symbol for a resistor in a circuit diagram. (c) Colour-coding scheme of electrical resistors. To find the value of the resistor, first look at the left bands (25, 460, 276 in our examples) and then multiply this value by the corresponding power of 10 expressed by the following band. This value can be found in the second column of the diagram above (1000, 1000, and 0.01 in our examples). The value of the resistance is the product of the two: 25 kΩ, 460 kΩ, and 2.76 Ω for the resistors in the example.

PEER TO PEER

At first I was confused about why we plot an I-V curve this way, since the slope of this curve is $1/R$. I thought it would be easier to plot it as an I-V graph with slope R. Then I realized that we do this because the I-V plot shows how the electric current in the circuit (the dependent variable) depends on the applied voltage (the independent variable). The slope of this I-V curve represents how well the circuit conducts electric current, which is inversely proportional to the resistance, $1/R$.

EXAMPLE 23-2

Applying Ohm's Law to a Current-Carrying Wire

A potential difference of 100. mV is applied across a 5.00 m long conducting wire that has a diameter of 0.50 mm. The resulting electric current is 66.7 mA.

(a) Find the resistance of the wire.
(b) Find the resistivity of the wire, and suggest what metal it could be made of.

Solution

(a) We assume that the metal wire obeys Ohm's law. Therefore, we can find its resistance, R, using Equation 23-17:

$$R = \frac{V}{I} = \frac{100.\text{ mV}}{66.7 \text{ mA}} = \frac{100. \times 10^{-3}\text{ V}}{66.7 \times 10^{-3}\text{ A}} = 1.50 \ \Omega$$

(b) Rearranging Equation 23-16, we can find the resistivity of the wire:

$$\rho = \frac{RA}{\ell} = \frac{(1.50\ \Omega) \cdot \pi (0.25 \times 10^{-3}\ \text{m})^2}{5.0\ \text{m}} = 5.9 \times 10^{-8}\ \Omega \cdot \text{m}$$

Comparing this value to the resistivities listed in Table 23-2, we find that the wire could be made of zinc.

Making sense of the result

Resistance, R, and resistivity, ρ, are different physical quantities. The resistance, R, describes how a particular piece of wire resists the flow of electric current. In our example, it is $R = 1.50\ \Omega$. The resistivity is the resistance of a hypothetical wire that is 1 m long and has a cross-sectional area of 1 m²: $\rho = 5.9 \times 10^{-8}\ \Omega \cdot \text{m}$. Therefore, knowing the resistance of the wire without knowing its dimensions is insufficient to deduce what material the wire is made of. However, knowing its resistivity can help answer this question.

CHECKPOINT 23-3

Ranking Resistances of Metal Wires

Which of the following statements correctly represents the ranking of the resistances of the five copper wires shown in Figure 23-7? Notice that D in Figure 23-7 represents the diameter of the wire. How would you rank the resistivities of these wires?

(a) $R_b > R_a = R_e > R_c > R_d$
(b) $R_b > R_a > R_e > R_c > R_d$
(c) $R_c > R_a = R_e > R_b > R_d$
(d) $R_d > R_c > R_e = R_a > R_b$
(e) $R_d > R_a = R_e > R_c > R_b$

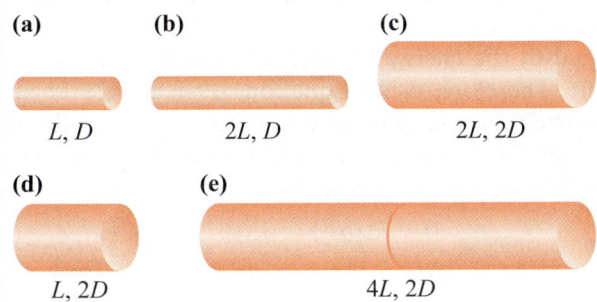

(a) L, D **(b)** $2L, D$ **(c)** $2L, 2D$

(d) $L, 2D$ **(e)** $4L, 2D$

Figure 23-7 Checkpoint 23-3.

23-4 Series and Parallel Electric Circuits

Now we are ready to explore how Ohm's law can be applied to a combination of resistors. Let us model a resistor as a wire of length ℓ, cross-sectional area A, and resistivity ρ. Imagine that we have two identical resistors, each with resistance R. What is the effective resistance of a combination of these resistors? The answer depends on how the resistors are connected.

Resistors connected in series When two or more resistors are connected in series (Figure 23-8(a)), the circuit is called

a **series circuit**. In that case, the same electric current flows through each resistor because the current that exits the first resistor must enter the second resistor. Connecting two identical resistors in series affects their total length while leaving the other parameters unchanged. It is convenient to define the equivalent resistance of electric circuits that are composed of two or more resistors. The **equivalent resistance** is the resistance of a resistor that, when connected in series to the circuit's battery, will produce an electric current equal to the electric current in the original circuit. To find the value of the equivalent resistance in a series circuit, R_{series}, we use Equation 23-16:

$$R_{series} = \rho \frac{2L}{A} = 2\rho \frac{L}{A} = 2R \qquad (23\text{-}19)$$

Although this expression was derived for the special case of two identical resistors, we will show later that it applies to any number of different resistors connected in series:

KEY EQUATION
$$R_{series} = \sum_{i=1}^{N} R_i \qquad (23\text{-}20)$$

Equation 23-20 indicates that connecting resistors in series increases their total or equivalent resistance. The term *equivalent* here indicates that resistors connected in series behave as one equivalent resistor that has a resistance equal to the sum of all the individual resistances.

(a)

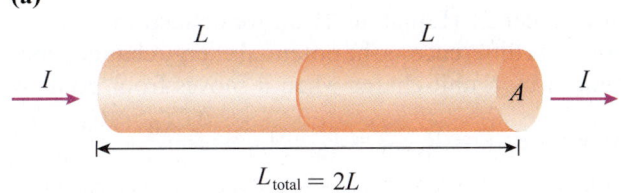

$L_{total} = 2L$

(b)

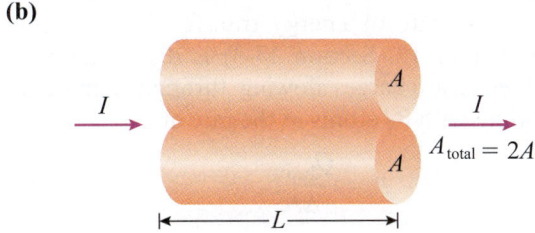

$A_{total} = 2A$

Figure 23-8 Electrical resistors connected (a) in series and (b) in parallel.

Resistors connected in parallel A circuit in which the electric current has more than one continuous path for electrons to flow (**electric circuit branch**) is called a **parallel circuit**. A parallel circuit has **junction points**, which are points where electric current can split up and flow through different branches of the circuit. When two resistors are connected in parallel (Figure 23-8(b)), some of the current will flow through the upper resistor, and the rest will flow through the lower resistor. Connecting identical resistors in parallel increases the effective cross-sectional area while leaving the length of the path for electrons unchanged. To find the value of the equivalent resistance for two identical resistors connected in parallel, $R_{parallel}$, we again use Equation 23-16:

$$R_{parallel} = \rho \frac{L}{2A} = \frac{1}{2} \rho \frac{L}{A} = \frac{R}{2} \qquad (23\text{-}21)$$

Although we used the special case of two identical resistors to derive this equation, we will show later that the equivalent resistance for any number of different resistors connected in parallel is given by

KEY EQUATION

$$\frac{1}{R_{parallel}} = \sum_{i=1}^{N} \frac{1}{R_i} \qquad (23\text{-}22)$$

Equation 23-22 indicates that connecting resistors in parallel reduces the equivalent resistance.

Power in DC Circuits

In Chapter 21 (Equation 21-8), we defined the electric potential difference as the potential energy change experienced by a unit charge when it moves from point A to point B: $V \equiv V_{A \to B} = V_B - V_A = \dfrac{\Delta U_{AB}}{q} = \dfrac{U_B - U_A}{q}$

In Section 23-2, we saw that a current of free electrons in a material heats the material through collisions with its atoms. As we discussed in Chapter 6, **power**, P, is defined as the rate of energy transfer. The thermal energy (Chapter 17, Section 17-3) is dissipated as a result of electric charges flowing through a material and colliding with its atoms at the rate of

$$P = \frac{Q_{heat}}{\Delta t} \qquad (23\text{-}23)$$

where Q_{heat} is the thermal energy dissipated during the time interval Δt.

When current flows in a material, the electric potential energy of the electrons is converted first to kinetic energy (moving particles) and then to thermal energy (for example, through collisions with the atoms of the crystal lattice). As a result, the material initially increases its temperature, which causes the dissipation of thermal energy. Eventually, an equilibrium state is reached where the energy radiated by the material is equal to the energy transferred from the electric current. Therefore, using Equations 19-1 and 21-11 and realizing that the generated heat, Q_{heat}, is a manifestation of the change of electrical potential energy due to the flow of electric charge, we write

$$P = \frac{Q_{heat}}{\Delta t} = \frac{W}{\Delta t} = \frac{\Delta(qV)}{\Delta t} = \frac{q\Delta V}{\Delta t} + \frac{\Delta q}{\Delta t} V = IV \quad (23\text{-}24)$$

Note that since the electric potential is constant, one of the terms in Equation 23-24 disappears. Using Ohm's law, Equation 23-24 can be rewritten as

KEY EQUATION

$$P = IV = I^2 R = \frac{V^2}{R} \qquad (23\text{-}25)$$

$$[P] = \frac{J}{s} = \text{watt (W)}$$

$$1\,\text{W} = 1\,\text{A} \cdot 1\,\text{V}$$

Consequently, the energy dissipated by a conductor of resistance R carrying electric current I over the time period Δt can be expressed as

KEY EQUATION

$$Q_{heat} = I^2 R \Delta t \qquad (23\text{-}26)$$

This relationship was discovered independently by two scientists—British physicist James Prescott Joule (1818–1889) and Russian-German-Estonian physicist Heinrich Lenz (1804–1865)—and is often called the Joule–Lenz law.

Household Electrical Wiring: How Much Current Is Too Much?

All your household appliances are connected in parallel to the external source of electric energy through the electrical outlets in the walls (Figure 23-9). By connecting more appliances in a parallel circuit, you are reducing the equivalent resistance of the combined load, thus increasing the current drawn from the electrical outlet:

$$I_{outlet} = I_1 + I_2 + I_3 + \cdots = \frac{V}{R_1} + \frac{V}{R_2} + \frac{V}{R_3} + \cdots$$

$$= V\left(\frac{1}{R_1} + \frac{1}{R_2} + \frac{1}{R_3} + \cdots\right) = \frac{V}{R_{parallel}}$$

In North America, common electrical outlets are designed to supply currents of up to 15 A. Larger currents in such circuits can overheat the wires enough to damage the insulation and possibly cause a fire. For this reason, strict safety regulations are in place that require all household circuits to be protected by a fuse or a circuit breaker, which will cut off the current when the circuit is overloaded.

In an electric circuit, the total power output of the circuit equals the sum of the power outputs of all the appliances. Therefore, to avoid reaching the maximum allowable current, the sum of all the power values produced by the appliances in a 15 A circuit should not exceed 1800 W. For a parallel circuit in your home,

$$P_{max} = IV = (I_1 + I_2 + I_3 + \cdots)V$$

$$= I_1 V + I_2 V + I_3 V + \cdots = \frac{V^2}{R_1} + \frac{V^2}{R_2} + \frac{V^2}{R_3} + \cdots$$

$$= P_1 + P_2 + P_3 + \cdots$$

$$= P_{1,nominal} + P_{2,nominal} + P_{3,nominal} + \cdots$$

Let us do the calculations for a parallel electric circuit in your home, where the AC voltage in your outlets is 120 V:

$$P_{max} = IV = (15\ A)(120\ V) = 1800\ W$$

$$P_1 + P_2 + P_3 + \cdots \le P_{max} = 1800\ W$$

In a series circuit connected to a voltage source V, the power output of the circuit will also be equal to the sum of the power outputs of its elements. However, the power output of each of the elements will be less than the **nominal power output** they were designed for, as the actual power output of a circuit element depends not only on its resistance, but also on the potential difference across it (Equation 23-25). The potential difference across each one of the circuit elements connected in series to a voltage source V will be less than V, and since this was the potential difference used to calculate the nominal power output, the actual power output of the appliance will also be less than its nominal power output. For example, if the nominal power output of a light bulb sold in Canada is 60 W, this means it was rated for a voltage of 110 V. However, if

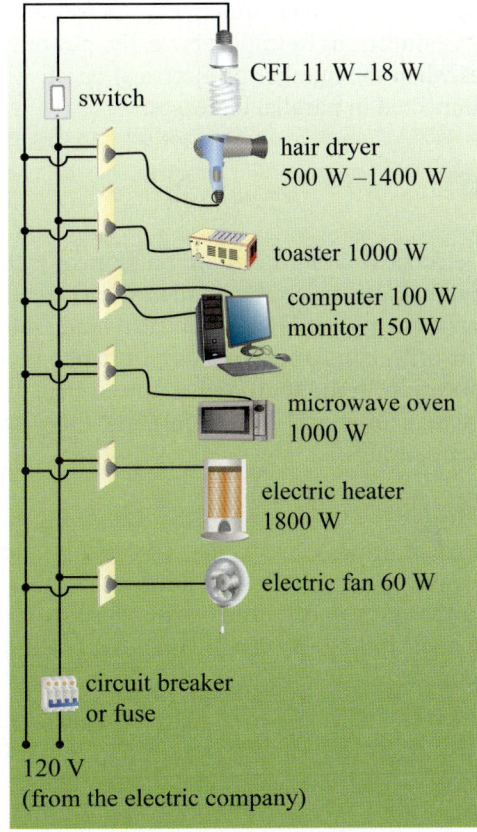

this light bulb is plugged into voltage source of 55 V (half the voltage it was designed for), the power produced by the light bulb will only be a quarter of its nominal power (Equation 23-25):

$$P_{actual} = \frac{V_{actual}^2}{R} = \frac{(0.5V_{nominal})^2}{R} = \frac{1}{4}P_{nominal}$$

It is also important to mention that at home we use alternating current (AC), while the calculations above were done for direct current (DC) circuits, yet this principle is also valid for AC circuits and thus can be applied to household appliances. You might also want to calculate the maximum power output in European households, where the outlet voltage is 220 V. You can assume the same value of maximum current as in Canada.

Figure 23-9 All your household appliances are connected in parallel to the external sources of electric energy (AC power sources) through the electrical outlets in the walls. To prevent fires, the electrical appliances at your home are connected to different circuits. In addition, circuit breakers and fuses are used to "break" the circuit if too many electrical appliances are used simultaneously and a dangerously high current (more than 15 A in North America) is drawn through the power outlet.

In the figure:
- switch
- CFL 11 W–18 W
- hair dryer 500 W–1400 W
- toaster 1000 W
- computer 100 W monitor 150 W
- microwave oven 1000 W
- electric heater 1800 W
- electric fan 60 W
- circuit breaker or fuse
- 120 V (from the electric company)

ELECTRICITY, MAGNETISM, AND OPTICS

One of the implications of the Joule-Lenz's law and the analysis of a parallel circuit described above is that in a parallel electric circuit, the power output of the circuit equals the sum of the nominal power outputs (the power output an electrical component was designed for) of its components. As we saw in the Making Connections example above, the power output increases when the number of electrical resistive appliances connected in parallel increases:

KEY EQUATION

$$P_{\text{parallel}} = \sum_{i=1}^{N} P_i \qquad (23\text{-}27)$$

An electrical component or portion of a circuit that consumes electric power is called an electrical load. The electric energy used by an electrical load (such as a toaster or other appliance) in a circuit depends on the power drawn from the source when the electrical appliances are turned on and on the length of time, Δt, that the circuit operates:

$$E_{\text{el}} = P\Delta t \qquad (23\text{-}28)$$

Electrical utility companies commonly measure energy consumption in kilowatt hours (kW·h) (Figure 23-10), although some companies now use megajoules (MJ) for their billing. By converting hours to seconds, we can convert a kilowatt hour into joules (Chapter 6):

$$1 \text{ kW·h} = (1000 \text{ W})(3600 \text{ s})$$
$$= 3.6 \times 10^6 \text{ J} = 3.6 \text{ MJ} \qquad (23\text{-}29)$$

BC Hydro Electric Charges

Apr 07 to Jun 04 (Residential Conservation Rate 1101)

Basic charge: 59 days @ $0.13410/day	7.91*
Usage charge:[1]	
Step 1: 935 kW·h @ $0·06270/kW·h	58.62*
Step 2: 0 kW·h @ $0·08780/kW·h	0.00
Rate Rider at 4.0%	2.66*
Innovative Clean Energy Fund Levy at 0.4%	0.28
Regional transit levy: 59 days @ $0.06240/day	3.68*
* GST	3.64
	$76.79

Your total consumption for the billing period is 935 kW·h

Figure 23-10 The residential electricity bill of one of the authors. Notice the dot in the energy units.

PEER TO PEER

I knew that on an electrical bill the energy is written as kW·h, but only recently did I realize that this stands for the product of the two, not the ratio (i.e., it is *not* kW per hour). When measuring energy consumption in everyday life, the unit J is inconvenient because it is too small. So we use kW·h instead: $1 \text{ kW·h} = 3.6 \times 10^6$ J (Equation 23-29). To show the ratio of two numbers, we use the term *per* (e.g., kilometres per hour), but when we want to show the product of two numbers, we just write them side by side (e.g., kilowatt hours).

MAKING CONNECTIONS

The Invention of the Incandescent Light Bulb

Humphry Davy (1778–1829) built the first electric light bulb in 1809. Over the next century, a number of inventors developed a variety of designs, which led to the mass production of affordable and reasonably reliable incandescent light bulbs. Among these inventors were Sir Joseph Wilson Swan (1828–1914) in England, who invented a bulb using a carbon filament, and Henry Woodward and Mathew Evans in Canada, who ended up selling their incandescent light bulb patent to an American, Thomas Alva Edison (1847–1931). Edison had the financial backing needed to improve and commercialize their design. Today, the filaments in incandescent lamps are made of tungsten,

which can withstand higher temperatures and therefore produce more light than carbon filaments. The energy efficiency of tungsten bulbs, in this case, the amount of energy that is converted into visible light as opposed to heat, is 2% to 10%. This shows that only a tiny fraction of the light they emit is in the range of visible light, while most is converted into heat. Therefore, traditional incandescent light bulbs are gradually being replaced by more efficient bulbs, such as halogen incandescent light bulbs, fluorescent lamps, and light-emitting diodes (LEDs). This is why some incandescent bulbs have been banned in many jurisdictions and are gradually being replaced by LEDs.

EXAMPLE 23-3

Power Rating of a Light Bulb

The antique 30.0 W light bulb in Figure 23-11(a) was designed to be connected to a 200. V battery. What is the power output of this light bulb when it is operated by a 100. V battery (Figure 23-11(b))?

Solution

The power output of the light bulb depends on the resistance of its filament (its internal property) and the potential

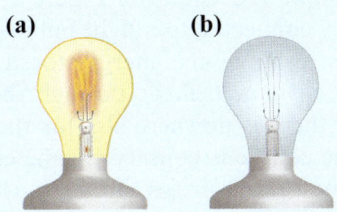

(a)   **(b)**

Figure 23-11 This antique incandescent light bulb uses a carbon filament. The grey colour results from sublimated carbon that is gradually deposited on the inner surface of the bulb. This light bulb is powered by (a) a 200 V source and (b) a 100 V source.

difference across the energy source. Since the light bulb was designed for 200 V, the **nominal resistance** (the resistance of the filament when the bulb is connected to a 200 V source) of the carbon filament of this light bulb is

$$R = \frac{V_1^2}{P_1} = \frac{(200.\,\text{V})^2}{30.0\,\text{W}} = \frac{4.00 \times 10^4 \text{V}^2}{30.0\,\text{W}} = 1.33\,\text{k}\Omega$$

When the light bulb is connected to a source with a lower potential difference, the power output is also reduced:

$$P_2 = \frac{V_2^2}{R} = \frac{(100.\,\text{V})^2}{1.33\,\text{k}\Omega} = \frac{1.00 \times 10^4 \text{V}^2}{1.33 \times 10^3\,\Omega} = 7.50\,\text{W}$$

Making sense of the result

Since the power depends on the square of the potential difference (voltage), decreasing the voltage across the light bulb by half reduces the power output of the bulb by a factor of 4.

MAKING CONNECTIONS

You Are Grounded!

You have probably heard the term *electrical grounding* and may have wondered what it meant. **Electrical grounding** was invented as a safety measure to protect people who accidentally touch a faulty electrical appliance, becoming a path for the electric current and getting shocked. For example, consider the heating element inside your electric toaster. When the toaster is operating properly, electric current flows through the heating element, thus heating your Montréal bagel. If for any reason the heating element comes into contact with the metal case of the toaster, the case will become a part of the circuit. If you accidentally touch the case, your body will become part of the circuit as well, providing a path for electric current from a high-voltage power source (high potential energy) to the ground (low potential energy): the current will flow through the device, through the metal case, and then through you into the ground. This will expose you to high voltage.

While the current through your body might be low due to your high internal resistance (1000 Ω if your skin is dry and 100 000 Ω if it is wet), it is still very dangerous. A person will feel a shock if only 5 mA of current flows through the body; 100 mA of current could be lethal. The low threshold of dangerous electric currents for humans is very important

because a circuit breaker is built to break the circuit when the current levels reach much higher values to prevent fires. For Canadian households, lighting and small appliance circuits are designed for 15 A, while the circuits for major appliances are designed for much higher currents of 20 A, 50 A, or even 60 A. However, a circuit breaker will not prevent a human from being shocked by a faulty appliance. By connecting the toaster case to the ground using a low-resistance copper wire, the current will have an alternative (parallel low-resistance) path to flow from the toaster case (should the toaster become faulty) to the ground. The connection made with Earth (ground) is vital, as it uses Earth as an endless reservoir for excess electrical charges. Thus, next time you use your powerful electrical appliances, notice the three-prong electrical plugs that provide safe electrical grounding and protect you from being shocked by faulty appliances. You might also have encountered four-prong plugs. These require four-hole receptacles that are still rare in homes but can be often found in industrial applications. These four-prong plugs correspond to new regulations that require separating the neutral wire and the ground wire. If you buy an appliance with a four-prong plug, you will need to have an electrician update your home's outlets.

23-5 Analysis of DC Circuits and Kirchhoff's Laws

A **direct current (DC)** circuit is a circuit in which electric current flows in one direction. A simple DC electric circuit consists of circuit elements, such as batteries, resistors, switches, and capacitors, connected by conducting wires that allow the flow of electric current. We will use the concepts of electric current, I; resistance, R; power, P; and voltage (potential difference), V, to describe the operation of these circuits. First, we consider the source of electric potential energy.

Source of electromotive force Electric batteries and generators are sources of electric energy. Historically, they were called **sources of electromotive force** (emf). However, as you will see below, electromotive force is not a real force but is a measure of electric potential energy per unit charge. To avoid confusion while adhering to the tradition, we will use the abbreviation *emf* to refer to a constant voltage produced by a source. An example of a constant-voltage source is an ideal battery—a battery that has zero internal resistance, so the voltage across it does not depend on the load of the circuit the battery is connected to (Figure 23-12(a)). Since any real (non-ideal) emf source has some internal resistance, we model it by connecting an internal resistance, r, in series to an ideal source that has zero internal resistance (Figure 23-12(b)). The internal resistance of the real battery (for example, your smartphone battery) slowly increases, decreasing the output voltage across it. This happens for a number of reasons. One of them is, as the battery is discharged, the electrolyte concentration in it is reduced, as compared to how it was when the battery was fully charged. The change in the electrolyte concentration increases the internal resistance of the battery. Thus, not surprisingly, real batteries age.

Figure 23-12 shows representations of an ideal battery and a real battery.

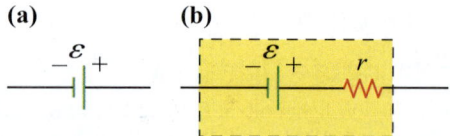

Figure 23-12 Schematic representations of ideal and real batteries. (a) An ideal battery (b) A real battery that has a non-zero internal resistance r.

The simple DC circuit in Figure 23-13(a) consists of a battery connected to a light bulb with conducting wires. As a result of the energy stored in the battery, an electric current, I, flows through the circuit. Note that, since there is nowhere else for the current to go (the electric charge is conserved), the current through any element of this series circuit must be the same. We assume that the resistance of the wires compared to the resistance of the light bulb, R, is negligible; the light bulb is in its steady state ($R = $ const); but the emf source (the battery) is real, which means it has an internal resistance, r. We apply Ohm's law (Equation 23-17) to each segment of this circuit and summarize the results in Table 23-3.

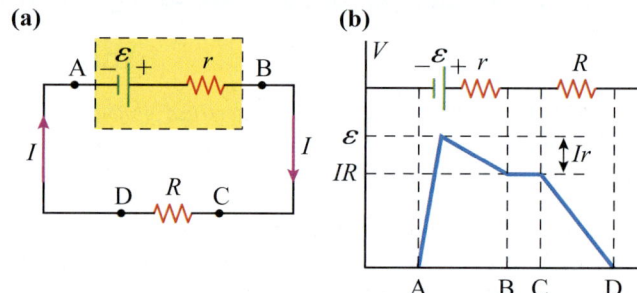

Figure 23-13 Analysis of a simple DC circuit consisting of a non-ideal emf source (a battery), a light bulb (in a steady state, so its resistance R is not changing), and connecting wires. (a) A schematic representation of a real-life circuit. (b) A schematic representation of the energy use in the circuit.

Table 23-3 Analysis of a Simple DC Circuit Containing a Non-ideal emf Source (a Battery), a Light Bulb, and Connecting Wires for Figure 23-13

Segment	I (A)	R (Ω)	V (V)	Meaning
AB	I	r	$V_{AB} = V_B - V_A = (\varepsilon - Ir)$	The positive terminal, B, has a higher potential than the negative terminal, A; therefore V_{AB} is positive. However, to move charged particles through the non-ideal battery, positive work equal to Ir needs to be done. Thus, the voltage output between the terminals of the emf source, V_{AB}, will be reduced by Ir as compared to the emf, ε.
BC	I	0	$V_{BC} = V_C - V_B = I(0\,\Omega) = 0$ V	No work is required to drive electric current through this segment, so there is zero potential difference.
CD	I	R	$V_{CD} = V_D - V_C = IR$	A light bulb converts electric potential energy into thermal energy, thus dissipating electric energy as heat.
DA	I	0	$V_{DA} = V_A - V_D = I(0\,\Omega) = 0$ V	No work is required to drive electric current through this segment, so there is zero potential difference.
Total: ABCDA	I	$R + r$	$IR - (\varepsilon - Ir) = 0$ $IR = \varepsilon - Ir$ $\varepsilon = I(R + r)$	Since electric charges complete a circuit, their total change in electric potential must be zero: $\Delta V_{total} = 0$.

The analysis presented in Table 23-3 is an application of the law of conservation of energy to a simple DC circuit. As with gravitational potential energy, if one returns to the starting point, the sum of all the gains and losses of potential energy along the way must be zero.

Figure 23-13(b) shows potential differences across different elements of the circuit shown in Figure 23-13(a). This diagram clearly identifies the segments where the sources of energy are located. When a positive electric charge moves from point A to point B through the non-ideal emf source (a real battery), it gains energy. Since the internal resistance of the source is non-zero, the energy gain per unit charge is $\varepsilon - IR$. On the other hand, there is no change in energy when the charge moves along segments BC and DA, and the electric potential energy of the charge is converted into thermal energy over segment CD.

As discussed above,

$$V_{AB} = \varepsilon - Ir \qquad (23\text{-}30)$$

Since the light bulb is connected to the emf source with conducting wires that have negligible resistance, the potential difference across the light bulb equals the potential difference across the terminals of the emf source:

$$IR = \varepsilon - Ir$$
$$I = \frac{\varepsilon}{(R + r)} \qquad (23\text{-}31)$$

The thermal energy dissipated by this circuit includes the energy dissipated by the light bulb and by the battery:

$$I\varepsilon = I^2R + I^2r = I^2(R + r) \qquad (23\text{-}32)$$

This analysis illustrates how the law of conservation of energy can be applied to electric charges moving through electric circuits.

Meters in an electric circuit To confirm that our analysis of a circuit is correct, we can use an **ammeter** to measure the currents through various circuit elements

and a **voltmeter** to measure the potential differences across them. An ammeter has to be connected in series with the circuit elements so that the same current flows through the ammeter and through the element, as shown in Figure 23-14(a). An ammeter must have a very low internal resistance to minimize its effect on the circuit, as the equivalent resistance of the series circuit equals the sum of the resistances of its elements (Equation 23-20). However, a voltmeter has to be connected in parallel to a circuit element to measure the voltage across it. Therefore, a voltmeter (Figure 23-14(b)) must have a very high internal resistance to produce a minimal effect on the circuit. The higher the resistance of the voltmeter, the less current will flow through it and the smaller will be its impact on the circuit (Equation 23-22). We will consider the impact of a voltmeter and an ammeter on a circuit in Example 23-4.

Both ammeter and voltmeter measurements have to be made on an operating circuit.

An **ohmmeter** (Figure 23-14(c)) measures the electrical resistance of a circuit element when the element is disconnected from a circuit. A device that can be set to measure current, voltage, or resistance is called a **multimeter**.

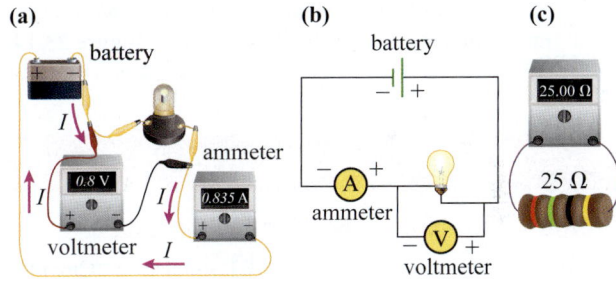

Figure 23-14 (a) An ammeter and a voltmeter connected to an electric circuit. (b) A circuit diagram showing the ammeter and the voltmeter. (c) An ohmmeter connected to a resistor.

EXAMPLE 23-4

Evaluating the Impact of an Ammeter and a Voltmeter on an Electric Circuit

Figure 23-14 shows an electric circuit consisting of a 12.00 V voltage source and 40.00 Ω light bulb. To measure the current through the light bulb and the potential difference across it, you decide to use an ammeter and a voltmeter. The internal resistance of the ammeter you are using is 0.1000 Ω and that of the voltmeter is 1000. Ω. Calculate the impact of the meters on your circuit.

Solution

To solve the problem we will compare the theoretical and the measured values of electric current through and potential difference across the light bulb. First of all, let us find the current through the light bulb and the voltage across it theoretically,

when the meters are not part of the circuit. Since the light bulb is the only load in the circuit (we neglect the resistance of the connecting wires), the potential difference across it is also 12 V. Then, using Ohm's law (Equation 23-17), we find

$$V_1 = 12.00 \text{ V}$$

$$I_1 = \frac{V_1}{R_1} = \frac{12.00 \text{ V}}{40.00 \Omega} = 0.3000 \text{ A} = 300.0 \text{ mA}$$

Now let us consider what happens when an ammeter is connected to the circuit. The ammeter is connected in series with the light bulb. Therefore, using Equation 23-20, we can find the equivalent resistance of this circuit:

$$R_{2\text{eq}} = \sum_{i=1}^{N} R_i = 40.00 \Omega + 0.1000 \Omega = 40.10 \Omega$$

(continued)

Using Ohm's law, we can find the electric current through the circuit (in a series circuit the same current flows through all its elements):

$$I_2 = \frac{V_{\text{battery}}}{R_{2\text{eq}}} = \frac{12.00 \text{ V}}{40.10 \ \Omega} = 0.2993 \text{ A} = 299.3 \text{ mA}$$

Then the potential difference across the bulb is

$$V_2 = I_2 R_2 = (0.2993 \text{ A})(40.00 \ \Omega) = 11.97 \text{ V}$$

Now let us consider when only a voltmeter is connected to the circuit. The voltmeter is connected in parallel to the light bulb. Therefore, using Equation 23-22, we can find the equivalent resistance of this parallel circuit:

$$\frac{1}{R_{3\text{eq}}} = \sum_{i=1}^{N} \frac{1}{R_i} = \frac{1}{40.00 \ \Omega} + \frac{1}{1000. \ \Omega}$$

$$R_{3\text{eq}} = 38.46 \ \Omega$$

Using Ohm's law, we can find the current in this circuit (the current flowing through the battery):

$$I_3 = \frac{V_{\text{battery}}}{R_{3\text{eq}}} = \frac{12.00 \text{ V}}{38.46 \ \Omega} = 0.3120 \text{ A} = 312 \text{ mA}.$$

However, not all of the current will flow through the light bulb; some of it will flow through the voltmeter. Since the light bulb is connected in parallel to the battery, the potential difference across it is exactly 12 V. Therefore, the current through it is 300 mA.

Now let us consider what happens when both the ammeter and the voltmeter are connected as shown in Figure 23-14.

Then the equivalent resistance of the circuit is

$$R_{4\text{eq}} = 38.46 \ \Omega + 0.100 \ \Omega = 38.56 \ \Omega$$

The current in the circuit is

$$I_{4\text{battery}} = \frac{V_{\text{battery}}}{R_{4\text{eq}}} = \frac{12.00 \text{ V}}{39.56 \ \Omega} = 0.3112 \text{ A} = 31.12 \text{ mA}$$

The potential difference across the ammeter according to Ohm's law is then

$$V_{4\text{ammeter}} = I_4 R_{\text{ammeter}} = (0.3112 \text{ A})(0.1000 \ \Omega) = 0.03112 \text{ V}$$

Therefore, the potential difference across the light bulb is 12.00 V − 0.03112 V = 11.97 V. Consequently, the current through the light bulb is

$$I_{4\text{light bulb}} = \frac{V_{\text{light bulb}}}{R_{\text{light bulb}}} = \frac{11.97 \text{ V}}{40.00 \ \Omega} = 0.2992 \text{ A} = 299.2 \text{ mA}$$

Making sense of the result

This problem illustrates how using meters in a circuit might alter it. It also explains why the voltmeter must have as large a resistance as possible, while an ammeter might have the smallest possible resistance. An ideal ammeter must have a zero resistance, while an ideal voltmeter will have an infinitely large resistance. As we will see in the following chapters, electric currents produce magnetic fields. This allows the design of non-contact ammeters that will measure the value of the electric current by measuring the magnetic field it creates without altering the circuit itself.

INTERACTIVE ACTIVITY 23-2

Simple and Not So Simple DC Circuits

In this activity, you will use the PhET simulation "Circuit Construction Kit (AC and DC)" to explore how DC circuits work and to understand the functions of different circuit components and circuit configurations, as well as the operation of multimeters.

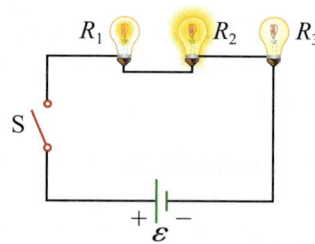

Figure 23-15 Three non-identical light bulbs connected in series.

Kirchhoff's Laws

Series circuits Figure 23-15 shows three different light bulbs connected in series with an ideal battery (zero internal resistance; thus $V_{\text{battery}} \equiv \varepsilon$) and a switch. When the switch is closed, it has negligible resistance and the potential difference across it is zero. As a consequence of the law of energy conservation, the potential difference across the battery, ε (energy source), must be equal to the sum of the potential differences across all circuit elements that consume this energy. Therefore,

$$\varepsilon = V_1 + V_2 + V_3 \tag{23-33}$$

Equation 23-33 can be generalized to state that the algebraic sum of all the potential differences (voltages)

across all circuit elements around a closed circuit loop must be zero:

KEY EQUATION

$$\sum_{\text{closed loop}} \Delta V = 0 \tag{23-34}$$

This loop rule, called **Kirchhoff's voltage law** (or Kirchhoff's second law), after German physicist Gustav Kirchhoff (1824–1887), is a consequence of the law of energy conservation. To apply it properly, we must agree on sign conventions for potential differences across different circuit elements. This is shown in Table 23-4. Notice that the sign of the potential difference depends on the nature of the element (a battery, a resistor, etc.), the direction of the traverse, and the direction of the electric current.

Table 23-4 Sign Conventions for Applying Kirchhoff's Laws

Sign Convention for Potential Difference across Circuit Elements	Potential Difference across the Source	Potential Difference across the Resistor	Potential Difference across the Capacitor
Positive:	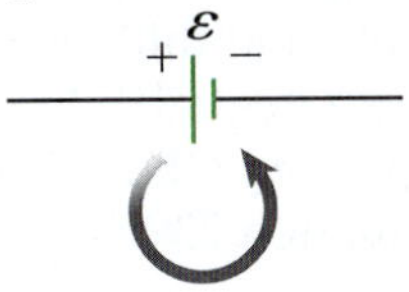 ε The direction of the traverse around the loop is from the positive terminal to the negative terminal of the emf source.	IR 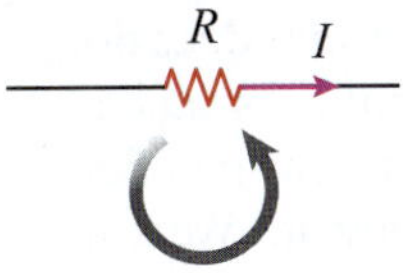 The direction of the traverse around the loop is opposite to the direction of the current through the resistor (R).	VC 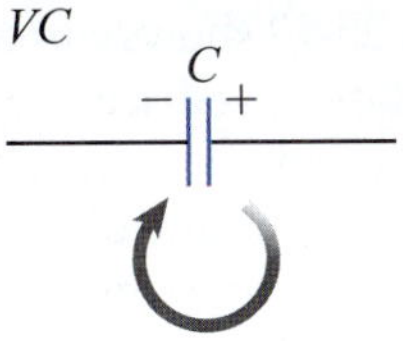 The direction of the traverse around the loop is from the positive to the negative plate of the capacitor.
Negative:	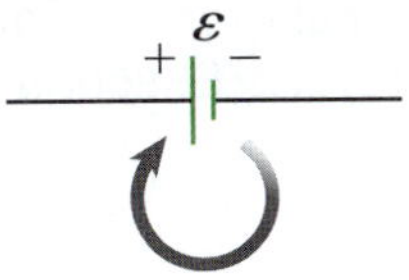 $-\varepsilon$ The direction of the traverse around the loop is from the negative to the positive terminal of the emf source.	$-IR$ 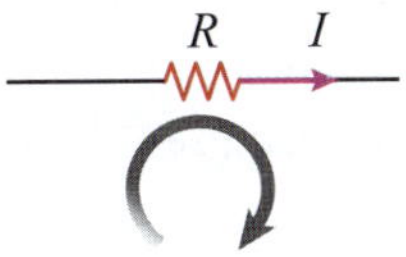 The direction of the traverse around the loop is in the same direction as the current through the resistor (R).	$-VC$ 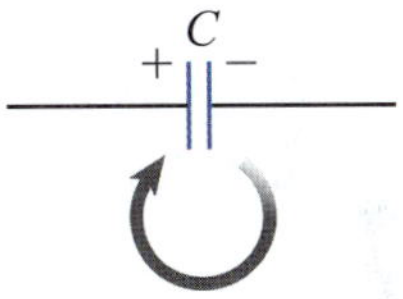 The direction of the traverse around the loop is from the negative to the positive plate of the capacitor.

Next, we note that the same current, I, flows through all the elements of our series circuit. Applying Ohm's law to each light bulb and once again using R_{series} for the equivalent resistance of the entire series circuit (Figure 23-15), we obtain

$$\varepsilon = IR_1 + IR_2 + IR_3 = I(R_1 + R_2 + R_3)$$

Since

$$\varepsilon = IR_{series}$$

we have

$$IR_{series} = I(R_1 + R_2 + R_3)$$

$$R_{series} = R_1 + R_2 + R_3$$

We can apply the same analysis to any number of different resistors connected in series, thus proving the general formula for resistors connected in series we discussed earlier (Equation 23-20):

$$R_{series} = \sum_{i=1}^{N} R_i$$

CHECKPOINT 23-4

Three Light Bulbs in a Series Electric Circuit

The three light bulbs in Figure 23-15 (above) are connected in series. The middle light bulb is the brightest, and the bulb on the left is the dimmest. For these light bulbs,

(a) $R_1 > R_3 > R_2$; $I_1 = I_2 = I_3$
(b) $R_1 = R_2 = R_3$; $I_1 < I_3 < I_2$
(c) $R_1 > R_3 > R_2$; $I_1 < I_3 < I_2$
(d) $R_1 = R_2 = R_3$; $I_2 < I_3 < I_1$
(e) $R_2 > R_3 > R_1$; $I_1 = I_2 = I_3$

ANSWER: (e) The light bulbs are connected in series; therefore, the same amount of current flows through them. The brightness of the light bulb is a consequence of its power output. As we discussed earlier, in the case of a light bulb, most of the electrical energy is dissipated as thermal energy, but some of it is radiated as visible light. The power output is $P = I^2R$ (Equation 23-25): the more power dissipated (thermal energy per unit time), the brighter the light bulb. Therefore, the greater the resistance of the light bulb, the greater the potential difference across it and the higher the power it generates, thus dissipating more thermal energy, some of which is released in the form of visible light.

Parallel circuits Figure 23-16 shows three light bulbs connected in parallel to an ideal battery. Each light bulb is connected directly to the battery and thus forms a separate branch of the circuit. A point where different branches join together is called a **junction**, or a **node**.

Figure 23-17 shows currents flowing through two different junctions. As discussed earlier, the current does not necessarily divide equally among the parallel branches. However, the amount of current entering any junction must equal the amount of current exiting it. We can paraphrase this by saying that the algebraic sum of these currents equals zero, provided that the currents entering the junction are assigned a positive sign and the currents exiting it are assigned a negative sign. We can express this mathematically using the following equation:

KEY EQUATION
$$\sum_{k=1}^{N_1} I_{\text{entering junction}} = \sum_{k=1}^{N_2} I_{\text{exiting junction}}$$

or

$$\sum_{k=1}^{N} I_{\text{junction}} = \sum_{k=1}^{N_1} I_{\text{entering junction}} - \sum_{k=1}^{N_2} I_{\text{exiting junction}} = 0$$

(23-35)

where N represents the total number of currents, and N_1 and N_2 represent the currents entering and exiting the junction, respectively: $N = N_1 + N_2$.

Equation 23-35 is a direct consequence of the **law of electric charge conservation**, which states that electric change can be neither created nor destroyed. The equation is called **Kirchhoff's current (first) law** (or the **junction rule**).

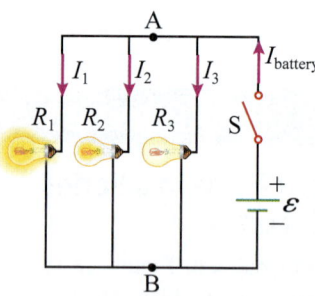

Figure 23-16 Three light bulbs connected in parallel: $I_{\text{battery}} = I_1 + I_2 + I_3$.

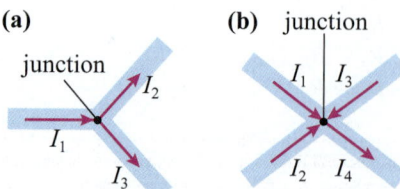

Figure 23-17 The algebraic sum of all the currents flowing through the junction equals zero: $\sum_{k=1}^{N} I_k = 0$, provided we follow the sign convention. (a) $I_1 = I_2 + I_3$. (b) $I_1 + I_2 = I_3 + I_4$.

Applying Kirchhoff's current law to junction A in the parallel circuit (Figure 23-16), we obtain what is often called the junction rule:

$$I_{\text{battery}} = I_1 + I_2 + I_3$$

(23-36)

Since each light bulb is connected directly to the battery, the potential difference across each bulb must be equal to the potential difference across the battery's terminals. Applying Ohm's law to substitute for the currents in Equation 23-36, we find

$$\frac{V_{\text{battery}}}{R_{\text{parallel}}} = \frac{V_{\text{battery}}}{R_1} + \frac{V_{\text{battery}}}{R_2} + \frac{V_{\text{battery}}}{R_3}$$

$$\frac{1}{R_{\text{parallel}}} = \frac{1}{R_1} + \frac{1}{R_2} + \frac{1}{R_3}$$

We can apply the same argument to any number of different resistors connected in parallel, thus proving the general formula for resistors connected in parallel we discussed earlier (Equation 23-22):

$$\frac{1}{R_{\text{parallel}}} = \sum_{i=1}^{N} \frac{1}{R_i}$$

CHECKPOINT 23-5

Three Light Bulbs in a Parallel Electric Circuit

The three light bulbs in Figure 23-16 (above) are connected in parallel to a battery. The left light bulb is the brightest, and the light bulb in the middle is the dimmest. For these light bulbs,

(a) $R_1 > R_3 > R_2$; $I_1 = I_2 = I_3$
(b) $R_1 = R_2 = R_3$; $I_1 > I_3 > I_2$
(c) $R_1 > R_3 > R_2$; $I_1 < I_3 < I_2$
(d) $R_1 = R_2 = R_3$; $I_2 < I_3 < I_1$
(e) $R_2 > R_3 > R_1$; $I_1 > I_3 > I_2$

ANSWER: (e) The light bulbs are connected in parallel; therefore, they have the same potential difference across them. Since $P = V^2/R$ (Equation 23-25), the lower the resistance of the light bulb, the more current will flow through it and the more thermal energy per unit of time (power) it will produce, thus dissipating more heat.

We can use the properties of series and parallel circuits to find the simplest way to determine the power dissipated in the components of circuits. In a series circuit, the same current flows through each component, so it is convenient to use the equation $P = I^2 R$ to determine the power dissipated by each component. However, the branches in a parallel circuit are all connected to the same battery; thus, the potential differences across them must be the same and equal to the potential difference across the battery. Therefore, it is more convenient to use $P = V^2/R$ to determine the electric power dissipated in each branch.

branch with less resistance has greater current. In the example above,

$$\frac{I_{top2}}{I_{bottom2}} = \frac{R_{bottom2}}{R_{top2}}$$

$$\frac{0.168 \text{ A}}{0.0168 \text{ A}} = \frac{550. \ \Omega}{55.0 \ \Omega}$$

■ The equivalent resistance in a parallel circuit is less than the lowest resistance of each of its branches. In the example above, the equivalent resistance of the parallel components of the second circuit is 50.0 Ω, which is less than either 55.0 Ω or 549 Ω. (See problems 33, 43, and 44 at the end of this chapter.)

Example 23-5 demonstrates two important properties of parallel circuits:

■ In a parallel circuit, the ratio of the currents through parallel branches is equal to the inverse ratio of the resistances of these branches. Consequently, a

EXAMPLE 23-5

Resistors in Series and in Parallel

Two different combinations of resistors are shown in Figure 23-18. Each combination is then connected to a 12.0 V battery, which is connected to points A and B, respectively. The current through the battery in the first circuit is twice the current in the second circuit.

(a) What is the value of R in the second circuit?

(b) Find the current in each branch of the two circuits.

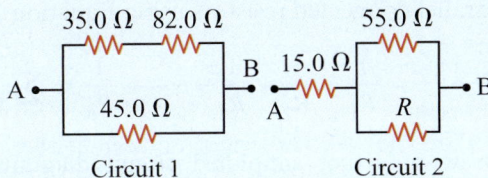

Figure 23-18 Example 23-5.

Solution

(a) In the first circuit diagram, the two resistors in the top branch are connected in series, so their equivalent resistance is

$$(R_{eq})_{top1} = 35.0 \ \Omega + 82.0 \ \Omega = 117.0 \ \Omega$$

Since the top and bottom branches are connected in parallel, their equivalent resistance is

$$\frac{1}{R_{eq1}} = \frac{1}{117.0 \ \Omega} + \frac{1}{45.0 \ \Omega}$$

$$\frac{1}{R_{eq1}} = 0.030769 \ \Omega^{-1}$$

$$R_{eq1} = 32.499 \ \Omega$$

We now use Ohm's law to find the current flowing through the battery:

$$I_{battery} = \frac{V_{battery}}{R_{eq1}} = \frac{12.0 \text{ V}}{32.499 \ \Omega} = 0.36924 \text{ A}$$

Since the current in the second circuit is half the current in the first one, $I_2 = 0.18462$ A. To produce this current by connecting the combination of resistors to a 12.0 V battery, the equivalent resistance in the second circuit must be

$$R_{eq2} = \frac{12.0 \text{ V}}{0.18462 \text{ A}} = 64.998 \ \Omega$$

(continued)

ELECTRICITY, MAGNETISM, AND OPTICS

To find R, we write an equation for the equivalent resistance in the second circuit:

$$R_{eq2} = 64.998\ \Omega$$

$$64.998\ \Omega = 15.0\ \Omega + \left[\frac{1}{55.0\ \Omega} + \frac{1}{R}\right]^{-1}$$

$$49.998\ \Omega = \frac{(55.0\ \Omega)R}{R + 55.0\ \Omega}$$

$$(49.998\ \Omega)R + 2749.89\ \Omega^2 = (55.0\ \Omega)R$$

$$R = 549.758\ \Omega \approx 550.\ \Omega$$

Note that we kept extra significant digits in the calculations, but rounded to three digits for the final answer.

(b) To find the current flowing through each branch, we apply Ohm's law to each one. For diagram (a), the two parallel branches have a potential difference of 12.0 V across them. Therefore, the currents through the branches are as follows:

$$I_{top1} = \frac{12.0\ \text{V}}{117\ \Omega} = 0.103\ \text{A}$$

$$I_{bottom1} = \frac{12.0\ \text{V}}{45.0\ \Omega} = 0.267\ \text{A}$$

For diagram (b), the potential difference across each resistor is less than 12.0 V because the parallel combination of resistors is connected in series with the 15.0 Ω resistor. Therefore,

$$I_{top2} = \frac{12.0\ \text{V} - (0.184\,62\,\text{A})(15.0\ \Omega)}{55.0\ \Omega} = 0.168\,\text{A}$$

$$I_{bottom2} = \frac{12.0\ \text{V} - (0.184\,62\,\text{A})(15.0\ \Omega)}{549.758\ \Omega} = 0.0168\,\text{A}$$

Making sense of the result

To verify our calculations, we can check that the sum of the currents in each circuit equals the current flowing through the battery:

$$I_{battery1} = I_{top1} + I_{bottom1} = 0.103\ \text{A} + 0.267\ \text{A} = 0.370\ \text{A and}$$
$$I_{battery2} = I_{top2} + I_{bottom2} = 0.168\ \text{A} + 0.0168\ \text{A} = 0.185\ \text{A}$$

These results correspond to the electric current values found earlier within rounding error.

In summary, in a series circuit, the same current flows through all the resistors connected in series. The equivalent resistance of the circuit grows as more resistors are connected in series. As the result, the current in the series circuit decreases when the number of resistors connected in series increases. The sum of all potential differences across the resistors connected in series is equal to the potential difference across the battery (the energy source). On the other hand, in a parallel circuit, the potential differences across all the branches connected in parallel are equal, while the currents flowing through different branches might be different, depending on the resistances of the branches. Only when the resistances of parallel branches are equal will the currents flowing through them be equal. Increasing the number of resistors connected in parallel (the number of parallel branches) decreases the equivalent resistance of the circuit, thus increasing the current flowing through the battery. Lastly, for any one of the junctions in a parallel circuit, the amount of current entering the junction equals the amount of current exiting it.

Resolving compound electric circuits Most practical circuits that you will deal with have both series and parallel connections of multiple resistors. In this section, we describe a strategy for finding a single equivalent resistance for these circuits. This strategy works for many, but not all, circuits. For circuits that cannot be analyzed in this way, you can use Kirchhoff's rules, as described in the next section.

A Strategy for Analyzing Compound DC Circuits

1. Draw the circuit diagram, label all the circuit elements, and record all the known quantities.

2. Identify all the resistors connected in series. Calculate the equivalent resistance of each set of series-connected resistors using Equation 23-20:

$$R_{series} = R_1 + R_2 + R_3 + \cdots + R_N = \sum_{i=1}^{N} R_i$$

3. Draw a simplified circuit diagram, and replace each series combination of resistors with its equivalent resistance.

4. Identify all the resistors connected in parallel. Calculate the equivalent resistance of each set of parallel-connected resistors using Equation 23-22:

$$\frac{1}{R_{parallel}} = \frac{1}{R_1} + \frac{1}{R_2} + \frac{1}{R_3} + \cdots + \frac{1}{R_N} = \sum_{i=1}^{N} \frac{1}{R_i}$$

5. Draw a further simplified circuit diagram, and replace each parallel combination of resistors with its equivalent resistance.

6. Repeat steps 4 and 5, if necessary, for series and parallel combinations that include the equivalent resistances until you have simplified the DC circuit to one final equivalent resistance.

EXAMPLE 23-6

Analyzing a Compound Electric Circuit

Figure 23-19(a) shows a compound electric circuit that has nine identical 2.0 Ω resistors connected to a 12 V ideal battery.

(a) Find the equivalent resistance for the circuit.
(b) Find the total electric current flowing from the battery.
(c) Identify the resistors that have the greatest potential difference across them.

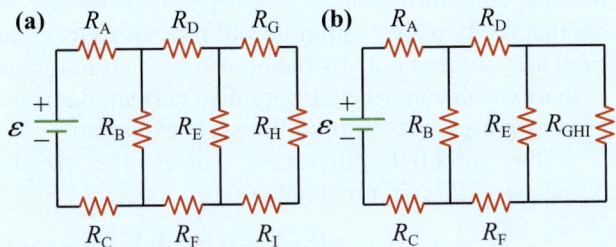

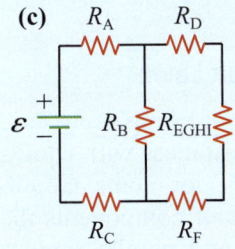

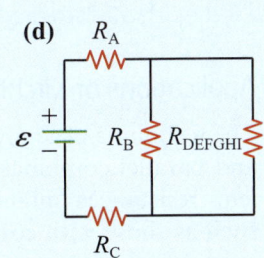

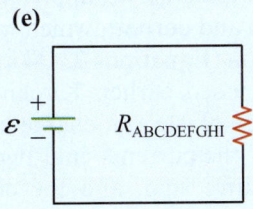

Figure 23-19   Example 23-6.

Solution

(a) Although the resistors are identical, the currents flowing through them likely differ. Therefore, we give each resistor a separate label: R_A, R_B, R_C, and so on. Then, we identify the resistors connected in series: R_G, R_H, and R_I. In this case, these are the resistors in the most inward branch of the circuit. Their equivalent resistance is

$$R_{GHI} = R_G + R_H + R_I = 6.0\ \Omega$$

We can simplify the circuit using this equivalent resistance (Figure 23-19(b)). Since there are no more resistors connected only in series, we now find the equivalent resistance of the two resistors connected in parallel, R_E and R_{GHI}:

$$\frac{1}{R_{EGHI}} = \frac{1}{R_E} + \frac{1}{R_{GHI}} = \frac{1}{2.0\ \Omega} + \frac{1}{6.0\ \Omega} \Rightarrow \frac{1}{R_{EGHI}} = \frac{2}{3.0\ \Omega}$$

$$\Rightarrow R_{EGHI} = 1.5\ \Omega$$

We can simplify the circuit further using the equivalent resistance R_{EGHI} (Figure 23-19(c)). In this simplified circuit, R_D, R_F, and R_{EGHI} are connected in series. Their equivalent resistance is

$$R_{DEFGHI} = 2.0\ \Omega + 2.0\ \Omega + 1.5\ \Omega = 5.5\ \Omega$$

Continuing this procedure, we eventually get a circuit that has only one equivalent resistance (Figure 23-19(e)):

$$R_{ABCDEFGHI} = 5\frac{7}{15}\ \Omega \approx 5.5\ \Omega$$

(b) We apply Ohm's law to find the current output of the battery:

$$I_{battery} = \frac{\varepsilon}{R_{ABCDEFGHI}} = \frac{12\ V}{5\frac{7}{15}\ \Omega} = \frac{90}{41}\ A \approx 2.2\ A$$

(c) To identify the resistors that have maximum potential difference across them, we use Ohm's law once again. Since all the resistors have the same resistance, the one with the largest current flowing through it will have the greatest potential difference across it ($V = IR$). Based on the diagram, you can see that all the current from the battery flows through resistors R_A and R_C, and only part of this current flows through the other resistors. Therefore, the potential difference across resistors R_A and R_C is the highest among them all:

$$V_A = V_C = I_{battery}(2.0\ \Omega) = \left(\frac{90}{41}\ A\right)(2.0\ \Omega) = 4\frac{16}{41}\ V \approx 4.4\ V$$

Making sense of the result

Notice the problem-solving strategy: we identify the part of a circuit that can be simplified and continue simplifying step by step until we arrive at a final circuit that contains a battery and an equivalent resistance. You can check your calculations using PhET or other circuit simulation software by constructing and measuring the circuit.

CHECKPOINT 23-7

Ranking the Relative Brightness of Light Bulbs

The circuit in Figure 23-20 has six identical light bulbs and an ideal battery. Initially, switch S is open. Which of the following statements correctly describes what happens to the brightness of bulbs A and D when the switch is closed?

(a) Bulb A gets brighter, and bulb D turns off.
(b) Bulb A gets dimmer, and bulb D gets brighter.
(c) Bulb A turns off, and bulb D gets brighter.
(d) The brightness of bulbs A and D remains the same.
(e) Both bulb A and bulb D get dimmer.

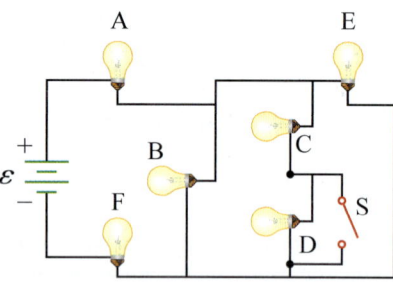

Figure 23-20 Checkpoint 23-7.

ANSWER: (a) By closing switch S, you will create a zero-resistance path for the electric current that will bypass light bulb D. Thus, the equivalent resistance of the circuit will decrease. Assuming that all light bulbs have equal resistance, R, the equivalent resistance initially equals $R_{eq1} = R + \left(\dfrac{1}{3} + \dfrac{1}{2R}\right)^{-1} + R = R + \dfrac{2}{5}R + R = 2\dfrac{2}{5}R$. After the switch is closed, the equivalent resistance is $R_{eq2} = R + \left(\dfrac{1}{R} + \dfrac{2}{R}\right)^{-1} + R = 2\dfrac{1}{3}R$. Decreasing the equivalent resistance means increasing the current through bulb A and decreasing the current to zero through bulb D (we say this bulb was shorted out).

Consider a circuit with two parallel branches having resistances R_1 and R_2 respectively, one of which has negligibly small (almost zero) resistance, $R_1 \rightarrow 0$. The equivalent resistance of such a circuit also approaches zero:

$$\frac{1}{R_{parallel}} = \frac{1}{R_1} + \frac{1}{R_2}$$

$$R_{parallel} = \frac{R_1 \cdot R_2}{R_1 + R_2} = \frac{0 \cdot R_2}{R_2} = 0 \qquad (23\text{-}37)$$

According to Ohm's law, a zero-equivalent resistance electric circuit connected to a real battery produces a very high (infinite) electric current. The current will travel through the zero-resistance branch, causing it to overheat. We say that the branches with non-zero resistance (R_2 in our case) are "shorted out" by the branch with no resistance. **A short circuit** can generate very high current, damaging power supplies and causing a fire or shock hazard.

The potential differences across the parallel branches of this circuit are also zero:

$$V_{\text{across branch 1}} = R_1 \cdot I_1 = 0 \cdot I = 0 \text{ V}$$
$$V_{\text{across branch 2}} = R_2 \cdot I_2 = 0 \cdot I = 0 \text{ V} \qquad (23\text{-}38)$$

Applications of Kirchhoff's Circuit Laws

The technique of analyzing a circuit by replacing series and parallel combinations of resistances with equivalent resistances might not work for some circuits, such as those with configurations of components that cannot be separated into series and parallel connections. For such circuits, we apply Kirchhoff's circuit laws for voltage and current, which we introduced earlier in this section (Equations 23-34 and 23-35).

As we discussed earlier, Kirchhoff's current law is a consequence of electric charge conservation, and it states that, if the currents entering the junction are assigned a positive sign, and the currents exiting it are assigned a negative sign, the algebraic sum of all the currents equals zero. Kirchhoff's voltage law is a consequence of energy conservation, and it states that the algebraic sum of the potential differences (voltages) across the circuit elements around a closed circuit loop is zero. To demonstrate a step-by-step procedure for applying Kirchhoff's laws, we analyze the circuit shown in Figure 23-21. Our goal is to find the current flowing through each circuit branch and the potential difference across each circuit element.

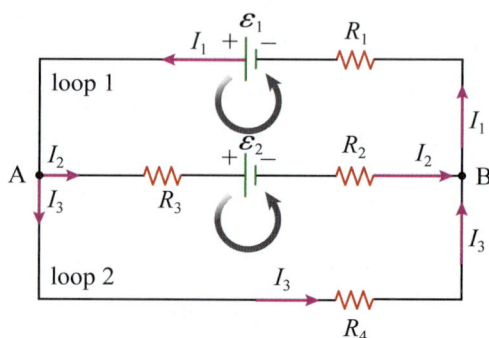

Figure 23-21 Circuit analysis using Kirchhoff's laws.

Circuit Analysis Using Kirchhoff's Laws

1. Identify and label all the junctions and branches. In Figure 23-21, there are three parallel branches (top, middle, and bottom) connecting at the two junctions, A and B.

2. Identify and label the currents in each branch. In our problem, we have three unknown currents: I_1, I_2, and I_3. Do not worry about the currents' directions. You can choose them arbitrarily and then stick with your choice during the entire solution. If your choice happens to be opposite to the actual direction of the current, the sign of the current you calculate will indicate this. For example, if in our example you calculate that the value of I_2 is negative, it will mean that I_2 enters junction A and exits junction B. Remember that the current through each series component in a branch is the same. For example, the current leaving battery ε_1 in Figure 23-21 is labelled I_1 because it is the same current as the current entering R_1 from junction B.

3. Apply Kirchhoff's current law (junction rule) to each junction. In general, the number of independent equations you obtain from applying the current law is one less than the number of junctions. In our example, the equations for junctions A and B are the same: $I_1 = I_2 + I_3$. Therefore, so far we have three unknowns and only one equation.

4. Identify the loops. Our circuit has three loops: loop 1, loop 2, and the external loop that consists of the elements of loop 1 and loop 2. Notice that all of these loops include batteries in addition to resistors. We need to have two additional equations to solve the problem (the number of independent equations has to be equal to the number of unknown variables). Therefore, analyzing two independent loops, for example, loops 1 and 2, will suffice to solve it. These two loops include different elements, and together they comprise all the branches of the circuit.

5. Apply Kirchhoff's voltage law (loop rule) to as many loops as needed to include each branch of the circuit at least once. You can choose a starting point in each loop arbitrarily, but you have to make sure that you go around the entire loop. Follow the same direction (clockwise or counter-clockwise) in each loop, and use the sign convention listed in Table 23-4. For example, in loop 1, we have chosen the positive terminal of the battery as the starting and ending point and counter-clockwise as the direction of the traverse.

For our example, applying Kirchhoff's voltage law produces the following equations:

Loop 1: $-I_2R_3 - \varepsilon_2 - I_2R_2 - I_1R_1 + \varepsilon_1 = 0$

Loop 2: $I_2R_3 - I_3R_4 + I_2R_2 + \varepsilon_2 = 0$

6. Identify the number of unknowns, and choose all the independent junction equations and enough additional independent loop equations to match the number of unknown currents. For the example circuit, we have three unknowns (currents I_1, I_2, and I_3), so we need three independent equations. Solve the equations in general form. Do not substitute known values right away.

$$\begin{cases} -I_2R_3 - \varepsilon_2 - I_2R_2 - I_1R_1 + \varepsilon_1 = 0 \\ I_2R_3 - I_3R_4 + I_2R_2 + \varepsilon_2 = 0 \\ I_1 = I_2 + I_3 \end{cases}$$

$$\begin{cases} -I_2(R_3 + R_2) - \varepsilon_2 - I_1R_1 + \varepsilon_1 = 0 \\ I_2(R_3 + R_2) - I_3R_4 + \varepsilon_2 = 0 \\ I_1 = I_2 + I_3 \end{cases}$$

$$\begin{cases} -I_2(R_3 + R_2) - \varepsilon_2 - (I_2 + I_3)R_1 + \varepsilon_1 = 0 \\ I_2(R_3 + R_2) - I_3R_4 + \varepsilon_2 = 0 \\ I_1 = I_2 + I_3 \end{cases}$$

$$\begin{cases} I_2 = \dfrac{\varepsilon_1 - I_3(R_1 + R_4)}{R_1} & \text{(add the first two equations of the previous set)} \\ I_2 = \dfrac{I_3R_4 - \varepsilon_2}{R_2 + R_3} & \text{(solve the second equation above for } I_2) \\ I_1 = I_2 + I_3 \end{cases}$$

$$\begin{cases} I_1 = \dfrac{\left(\dfrac{\varepsilon_1(R_2 + R_3) + \varepsilon_2R_1}{R_1R_4 + R_1R_2 + R_1R_3 + R_2R_4 + R_3R_4}\right)R_4 - \varepsilon_2}{R_2 + R_3} \\ \qquad + \dfrac{\varepsilon_1(R_2 + R_3) + \varepsilon_2R_1}{R_1R_4 + R_1R_2 + R_1R_3 + R_2R_4 + R_3R_4} \\ I_2 = \dfrac{I_3R_4 - \varepsilon_2}{R_2 + R_3} \\ \quad = \dfrac{\left(\dfrac{\varepsilon_1(R_2 + R_3) + \varepsilon_2R_1}{R_1R_4 + R_1R_2 + R_1R_3 + R_2R_4 + R_3R_4}\right)R_4 - \varepsilon_2}{R_2 + R_3} \\ I_3 = \dfrac{\varepsilon_1(R_2 + R_3) + \varepsilon_2R_1}{R_1R_4 + R_1R_2 + R_1R_3 + R_2R_4 + R_3R_4} \end{cases}$$

7. Use dimensional analysis to check whether the expressions for the unknowns make sense. In our example, all the final variables (I) should have units of current (amperes or volts/ohm). Example 23-7 demonstrates this technique.

8. Use limiting cases to check your final answers. Example 23-7 demonstrates this technique.

9. Use Ohm's law and the calculated values for the currents to find the potential differences across the circuit elements.

EXAMPLE 23-7

Analyzing a Multi-loop and Multi-battery Electric Circuit

Assume that the components in Figure 23-21 have the following values:

$$\varepsilon_1 = 10.\ \text{V}; \varepsilon_2 = 15\ \text{V (ideal batteries)}$$
$$R_1 = 5.0\ \Omega; R_2 = 10.\ \Omega; R_3 = 15\ \Omega; R_4 = 20.\ \Omega$$

(a) Find the current through each branch of the circuit.
(b) Find the potential difference across each circuit element.
(c) How will your answers to parts (a) and (b) differ if the batteries are real batteries?

Solution

The detailed solution to this problem was explained in the earlier discussion. Now, we can substitute the values and make sense of the result.

(a) Substituting the given values into the expressions derived for the currents in the three branches, we obtain the following:

$$I_2 = \frac{I_3 R_4 - \varepsilon_2}{R_2 + R_3}$$

$$= \frac{\left(\dfrac{\varepsilon_1(R_2 + R_3) + \varepsilon_2 R_1}{R_1 R_4 + R_1 R_2 + R_1 R_3 + R_2 R_4 + R_3 R_4}\right) R_4 - \varepsilon_2}{R_2 + R_3}$$

$$= -\frac{7}{29}\ \text{A} \approx -0.24\ \text{A}$$

$$I_3 = \frac{\varepsilon_1(R_2 + R_3) + \varepsilon_2 R_1}{R_1 R_4 + R_1 R_2 + R_1 R_3 + R_2 R_4 + R_3 R_4} = \frac{13}{29}\ \text{A} \approx 0.45\ \text{A}$$

$$I_1 = I_2 + I_3 = \frac{6}{29}\ \text{A} \approx 0.21\ \text{A}$$

These results indicate that our choice of current directions for I_1 and I_3 was correct, and current I_2 flows in the opposite direction to the one originally chosen in the diagram.

(b) To find the potential difference across each resistor, we use Ohm's law:

$$V_{R_1} = I_1 R_1 = \left(\frac{6}{29}\ \text{A}\right)(5.0\ \Omega) = 1\frac{1}{29}\ \text{V} \approx 1.0\ \text{V}$$

$$V_{R_2} = I_2 R_2 = \left|-\frac{7}{29}\ \text{A}\right|(10\ \Omega) = \frac{70}{29}\ \text{V} = 2\frac{12}{29}\ \text{V} \approx 2.4\ \text{V}$$

$$V_{R_3} = I_2 R_3 = \left|-\frac{7}{29}\ \text{A}\right|(15\ \Omega) = \frac{105}{29}\ \text{V} = 3\frac{18}{29}\ \text{V} \approx 3.6\ \text{V}$$

$$V_{R_4} = I_3 R_4 = \left(\frac{13}{29}\ \text{A}\right)(20\ \Omega) = \frac{260}{29}\text{V} = 8\frac{28}{29}\ \text{V} \approx 9.0\ \text{V}$$

Notice that we use fractional representation here until we get to the final answer to avoid rounding errors.

(c) If the batteries are real, we have to account for their internal resistance. This can be done by adding their internal resistances in series to each battery. For the circuit, this is equivalent to replacing resistances R_1 and R_2 with R'_1 and R'_2:

$$\begin{cases} R'_1 = R_1 + r_1 \\ R'_2 = R_2 + r_2 \end{cases}$$

Making sense of the result

To check your calculations, you can use Wolfram Alpha (www.wolframalpha.com) or other software for solving simultaneous equations. However, checking the calculations does not tell you if your initial equations are correct. To check these equations, we can apply dimensional analysis, analysis of limiting cases, and the law of conservation of energy.

Dimensional Analysis

The currents should all have dimensions of amperes:

$$[I_3] = \left[\frac{\varepsilon_1(R_2 + R_3) + \varepsilon_2 R_1}{R_1 R_4 + R_1 R_2 + R_1 R_3 + R_2 R_4 + R_3 R_4}\right]$$

$$= \frac{\text{V} \cdot \Omega + \text{V} \cdot \Omega}{\Omega^2} = \frac{\text{V}}{\Omega} = \text{A}$$

$$[I_2] = \left[\frac{I_3 R_4 - \varepsilon_2}{R_2 + R_3}\right] = \frac{\text{A} \cdot \Omega - \text{V}}{\Omega} = \frac{\text{V}}{\Omega} = \text{A}$$

$$[I_1] = [I_2 + I_3] = \text{A} + \text{A} = \text{A}$$

Remember, dimensional analysis helps identify errors but does not guarantee that the solution is correct.

ELECTRICITY, MAGNETISM, AND OPTICS

Limiting Case Analysis

Let us consider the limiting case, where no source of energy is present in the circuit. If we replace the two batteries with conductors, all the currents become zero. The expressions derived above equal zero when the values of two emfs are zero. So, these equations are valid for this limiting case. If only the first battery, ε_1, is removed from the circuit, then the equivalent resistance of the circuit becomes

$$R_{eq1} = 10\ \Omega + 15\ \Omega + \cfrac{1}{\cfrac{1}{5.0\ \Omega} + \cfrac{1}{20\ \Omega}} = 29\ \Omega$$

and the current flowing through the battery is

$$I_2 = \frac{\varepsilon_2}{R_{eq1}} = \frac{15\ \text{V}}{29\ \Omega} = \frac{15}{29}\ \text{A} \approx 0.52\ \text{A}$$

Let us check this by substituting $\varepsilon_1 = 0$ in the general solution, making sure we get the same answer:

$$(I_2)_{\varepsilon_1 = 0} = \frac{\left(\dfrac{\varepsilon_2 R_1}{R_1 R_4 + R_1 R_2 + R_1 R_3 + R_2 R_4 + R_3 R_4}\right)R_4 - \varepsilon_2}{R_2 + R_3}$$

$$= -\frac{15}{29}\ \text{A}$$

Analysis of Electrical Potentials

You can check that the potential difference across all the branches between junctions A and B are equal, as we would expect because the branches are parallel.

Top branch:

$$V_{AB} = I_1 R_1 - \varepsilon_1 = \left(\frac{6}{29}\ \text{A}\right)(5.0\ \Omega) - 10\ \text{V}$$

$$= \left(\frac{30}{29} - 10\right)\text{V} = -8\frac{28}{29}\ \text{V}$$

Middle branch:

$$V_{AB} = -I_2 R_3 - \varepsilon_2 - I_2 R_2$$

$$= -\left(-\frac{7}{29}\ \text{A}\right)(15\ \Omega) - 15\ \text{V} - \left(-\frac{7}{29}\ \text{A}\right)(10\ \Omega)$$

$$= -8\frac{28}{29}\ \text{V}$$

Bottom branch:

$$V_{AB} = -I_3 R_1 = -\left(\frac{13}{29}\ \text{A}\right)(20\ \Omega) = -8\frac{28}{29}\ \text{V}$$

The values for V_{AB} are identical across the branches, which tells us that we solved the problem correctly. Note that when finding the potential difference V_{AB} along a particular branch, we traverse the branch from point A to point B.

23-6 RC Circuits

This section will help you learn how to analyze electric circuits that in addition to batteries and resistors also include capacitors. Capacitors in series RC circuits exhibit very interesting behaviour when the current in the circuit changes in direction or in magnitude. Therefore, we will consider the response of these circuits to sudden changes in DC voltage (called **transient voltage**). These sudden changes happen in a circuit connected to a DC power supply during the time when a switch is being opened or closed. In Chapters 25 and 26, we will continue to analyze series circuits that include inductors in addition to resistors and capacitors. We will begin the analysis of series RC circuits with an Interactive Activity that will help us make important observations.

Interactive Activity 23-4 examines a circuit with a capacitor, a battery, a light bulb (as a resistor), and a switch, like the one shown in Figure 23-22. The circuit uses an ammeter (a device that measures electric current), indicated by an A inside a circle, and two voltmeters (devices that measure electric potential difference, or voltage), indicated by a V inside a circle. The brightness of the light bulb in this circuit changes with

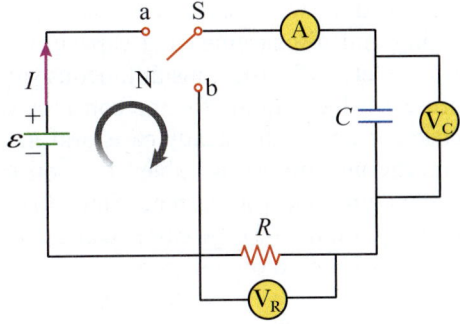

Figure 23-22 A series RC circuit consists of a resistor (R) and a capacitor (C). Such a circuit might (when the switch is in position a) or might not (when the switch is in position b) include a battery. The ammeter registers electric current flowing through the resistor and a capacitor, while the voltmeters (V_C and V_R) measure potential differences across the capacitor and the resistor, respectively.

time, indicating that the current through the circuit is time dependent. Since the potential differences across the capacitor and across the light bulb are both related to the current in the circuit, they also vary with time. We will now use Kirchhoff's laws to derive mathematical expressions describing how the current through and potential differences across the circuit's elements change as a capacitor charges or discharges. While the direct derivations of the mathematical expressions will require some knowledge of differential equations, you can always guess a solution and then differentiate it to show that it works.

Charging a Capacitor

Consider the circuit in Figure 23-22. Initially, the capacitor (C) is uncharged. Let us assume that the switch is flipped to position a (from the neutral position, N) at time $t = 0$. By moving this switch to position a, we complete a series circuit that includes an initially uncharged capacitor (C), a resistor (R), and an emf source (ε). As a result, the current in the circuit will start flowing in a clockwise direction and the capacitor will begin charging. Its top plate will become positively charged and its bottom plate will become negatively charged. Applying Kirchhoff's loop rule to the circuit (Table 23-4), while traversing it in a clockwise direction, which is also the chosen direction of the electric current I, we obtain

$$\varepsilon - IR - V_C = 0 \qquad (23\text{-}39)$$

Equation 23-39 indicates that the capacitor is charging, thus storing some of the energy from the battery. In Chapter 22, we discussed how the potential difference across a capacitor, V_C, is related to the charge on each of its plates (q) and to its capacitance (C):

$$V_C = \frac{Q}{C} \qquad (22\text{-}2)$$

In this section, we consider what happens during the charging and discharging of a capacitor. We will use lowercase letters for these instantaneous values. Do not confuse I, which stands for the constant value of the electric current in the steady case, and $i(t)$, which represents the instantaneous value of a continuously varying (changing) electric current. Since $i(t) = dq/dt$ (similar to Equation 23-5), we can substitute for I and V_C in Equation 23-39 to obtain

$$\varepsilon - \frac{dq}{dt}R - \frac{q}{C} = 0 \qquad (23\text{-}40)$$

This is a differential equation. One way of solving it is to guess the solution and then show that it works. We can also solve it by reasoning through it, as discussed

below. Because the electric charge on the capacitor's plates varies with time, we denote it as $q(t)$. Then we can rearrange the above equation as follows:

$$\varepsilon - \frac{dq(t)}{dt}R - \frac{q(t)}{C} = 0$$

$$\frac{dq(t)}{dt} = \frac{\varepsilon}{R} - \frac{q(t)}{RC}$$

$$\frac{dq(t)}{dt} = -\frac{q(t) - \varepsilon C}{RC}$$

$$\frac{dq(t)}{q(t) - \varepsilon C} = -\frac{dt}{RC}$$

Noticing that the quantity εC is constant (so $dq(t) \equiv d(q(t) - \varepsilon C)$), we can rewrite this as

$$\frac{d(q(t) - \varepsilon C)}{q(t) - \varepsilon C} = -\frac{dt}{RC} \qquad (23\text{-}41)$$

Now we can use the following substitution:

$$x(t) \equiv q(t) - \varepsilon C \Rightarrow dx(t) \equiv dq(t) \qquad (23\text{-}42)$$

As a result, Equation 23-41 can be rewritten as

$$\frac{dx(t)}{x} = -\frac{dt}{RC} \qquad (23\text{-}43)$$

Moreover, at time $t = 0$, the capacitor had no charge on its plates, so $q(t = 0) = 0 \Rightarrow x(t = 0) = -\varepsilon C$, and at time t, the plates had a charge of $\pm q(t)$. Integrating both sides of Equation 23-43 gives

$$\int_{-\varepsilon C}^{q - \varepsilon C} \frac{dx(t)}{x} = -\int_{0}^{t} \frac{dt}{RC}$$

$$\ln(x(t))\Big|_{-\varepsilon C}^{q(t)-\varepsilon C} = -\frac{t}{RC}$$

$$\ln(q(t) - \varepsilon C) - \ln(-\varepsilon C) = -\frac{t}{RC}$$

$$\ln\left(\frac{q(t) - \varepsilon C}{-\varepsilon C}\right) = -\frac{t}{RC}$$

$$\frac{q(t) - \varepsilon C}{-\varepsilon C} = e^{-\frac{t}{RC}}$$

$$q(t) = (-\varepsilon C)e^{-\frac{t}{RC}} + \varepsilon C = \varepsilon C(1 - e^{-\frac{t}{RC}}) \quad (23\text{-}44)$$

In Chapter 22, we showed that the charge on the capacitor's plates is related to the voltage as $q = VC$ (see Equation 22-2). Therefore, the maximum charge on the capacitor, q_{max}, is related to the maximum voltage across its plates, V_{max}, as $q_{max} = V_{max}C$. In our circuit, the maximum potential difference across the plates equals the potential difference across the battery, so $V_{max} = q_{max}/C = \varepsilon$, and $q_{max} = \varepsilon C$. Therefore, while a capacitor is charging, the charge on the capacitor's plates and the potential difference across it can be expressed as

KEY EQUATION
$$q(t) = q_{max}\left(1 - e^{-\frac{t}{RC}}\right) = \varepsilon C\left(1 - e^{-\frac{t}{\tau}}\right) \quad (23\text{-}45)$$

$$V_C = \frac{q(t)}{C} = \frac{q_{max}}{C}\left(1 - e^{-\frac{t}{RC}}\right) = \varepsilon\left(1 - e^{-\frac{t}{\tau}}\right) \quad (23\text{-}46)$$

Let us check the dimensions of the product RC:

$$[RC] = \Omega \cdot F = \frac{V}{A} \cdot \frac{C}{V} = \frac{C}{A} = \frac{C}{C/s} = s \quad (23\text{-}47)$$

The product of the capacitance and the resistance of a series RC circuit has dimensions of time. It is called the **time constant** of the circuit, $\tau = RC$. Therefore, as we should have expected, the exponent in Equation 23-45 is dimensionless: $[t/\tau] = 1$.

Differentiating Equation 23-45, we find the dependence of electric current on time:

$$i(t) = \frac{dq(t)}{dt} = \frac{d\left(q_{max}\left(1 - e^{-\frac{t}{RC}}\right)\right)}{dt}$$

$$= \frac{d\left(\varepsilon C\left(1 - e^{-\frac{t}{RC}}\right)\right)}{dt}$$

$$= \frac{\varepsilon C}{RC}e^{-\frac{t}{RC}} = \frac{\varepsilon}{R}e^{-\frac{t}{\tau}} \quad (23\text{-}48)$$

The ratio ε/R represents the maximum current in this RC circuit (the current in the absence of a capacitor). This current flows in the circuit only at the initial moment when the switch is being closed ($t = 0$) and the capacitor begins charging. You should have noticed this in Interactive Activity 23-4. Thus,

KEY EQUATION
$$i(t) = I_0 e^{-\frac{t}{RC}} = I_0 e^{-\frac{t}{\tau}} = \frac{\varepsilon}{R}e^{-\frac{t}{\tau}} \quad (23\text{-}49)$$

(charging)

Equation 23-49 describes the dependence of electric current on time during the charging of a capacitor in an RC circuit. In Chapter 22, we found that the energy stored in a charging capacitor depends on the square of the charge stored by it. Substituting for $q(t)$ in Equation 22-17, $U = \frac{1}{2}\frac{Q^2}{C}$, gives an expression for the energy stored in a capacitor as a function of time:

$$U_C = \frac{(q(t))^2}{2C} = \frac{\left(q_{max}\left(1 - e^{-\frac{t}{RC}}\right)\right)^2}{2C}$$

$$= \frac{q_{max}^2\left(1 - e^{-\frac{t}{RC}}\right)^2}{2C} \quad (23\text{-}50)$$

$$= \frac{C\varepsilon^2}{2}\left(1 - e^{-\frac{t}{RC}}\right)^2$$

Graphs of $q(t)$ and $i(t)$ for this process are shown in Figure 23-23.

The potential difference across the capacitor asymptotically approaches its maximum value, $V_{max} = \varepsilon C$. At the same time, the current in this RC circuit decreases exponentially but never reaches zero. As we discussed in Chapter 21, for exponential decay or growth, it is common to describe the relationship using the time constant. In the case of charging a capacitor, it is $\tau = RC$. The time constant is the time that it takes the capacitor to charge to about 63% of its maximum:

$$q(\tau) = q_{max}(1 - e^{-\frac{\tau}{\tau}}) = q_{max}\left(1 - \frac{1}{e}\right) \approx 0.632\,q_{max} \quad (23\text{-}51)$$

At the same time ($t = \tau$), the current across the capacitor is a factor of e less than the maximum current, I_0:

$$i(\tau) = I_0 e^{-\frac{\tau}{\tau}} = \frac{I_0}{e} \approx 0.368\,I_0 \quad (23\text{-}52)$$

For most practical purposes, a capacitor can be considered fully charged after 5 to 10 time constants have elapsed. The relationships described above explain the behaviour of a light bulb in a series RC circuit consisting of a battery, a capacitor, and a light bulb (such as the circuit shown in Figure 23-22, where the resistor is replaced with a light bulb). When the capacitor begins charging, the current is near its maximum and the light bulb is bright. Then, as the current decreases, the light bulb gets progressively dimmer and eventually stops glowing.

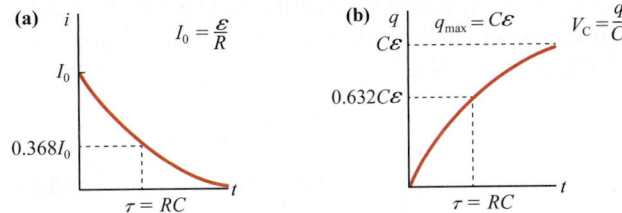

Figure 23-23 When the switch in a series RC circuit is flipped from position N to position a, the capacitor begins charging: (a) The dependence of the electric current through the series RC circuit on time. (b) The dependence of the electric charge on the capacitor's plates on time. Since the voltage across the capacitor is proportional to the charge on its plates, this graph also represents the voltage across the capacitor.

EXAMPLE 23-8

Charging a Capacitor

A 500. μF capacitor is connected to a 9.00 V battery through a 2.00 kΩ resistor, thus completing a simple RC series circuit.

(a) How long does it take for the capacitor to reach 99% of q_{max}? Express your answer in terms of the time constant, τ. Assume that 99% is an exact quantity.

(b) How much charge will have accumulated on the capacitor's plates when $t_1 = \tau$, $t_2 = 2\tau$, and $t_3 = 3\tau$?

(c) Find the current in the circuit at the times in part (b).

Solution

(a) For this circuit,

$$\tau = RC = (2.00 \times 10^3\,\Omega)(5.00 \times 10^{-4}\,F) = 1.00\,s$$

$$0.99\,q_{max} = q_{max}\left(1 - e^{-\frac{t}{RC}}\right)$$

$$0.99 = 1 - e^{-\frac{t}{RC}}$$

$$e^{-\frac{t}{RC}} = 0.01$$

$$t = -RC \ln 0.01 = -\tau \ln 0.01$$

$$t = -(1.00\,s)\ln 0.01$$

$$= 4.61\,s$$

(b) $q_{max} = \varepsilon C = (9.00\,V)(5.00 \times 10^{-4}\,F) = 4.50 \times 10^{-3}\,C$

$$q(t) = q_{max}\left(1 - e^{-\frac{t}{RC}}\right)$$

$$q(\tau) = q_{max}\left(1 - e^{-\frac{\tau}{RC}}\right) = q_{max}\left(1 - e^{-\frac{RC}{RC}}\right)$$

$$= (4.50 \times 10^{-3}\,C)\left(1 - \frac{1}{e}\right) = 2.84 \times 10^{-3}\,C$$

$$q(2\tau) = q_{max}\left(1 - e^{-\frac{2\tau}{RC}}\right) = q_{max}\left(1 - e^{-\frac{2RC}{RC}}\right)$$

$$= (4.50 \times 10^{-3}\,C)\left(1 - \frac{1}{e^2}\right) = 3.89 \times 10^{-3}\,C$$

$$q(3\tau) = q_{max}\left(1 - e^{-\frac{3\tau}{RC}}\right) = q_{max}\left(1 - e^{-\frac{3RC}{RC}}\right)$$

$$= (4.50 \times 10^{-3}\,C)\left(1 - \frac{1}{e^3}\right) = 4.28 \times 10^{-3}\,C$$

(c) $I_0 = \dfrac{\varepsilon}{R} = \dfrac{9.00\,V}{2.00\,k\Omega} = 4.50 \times 10^{-3}\,A$

$$I(t) = I_0 e^{-\frac{t}{RC}}$$

$$I(\tau) = I_0 e^{-\frac{\tau}{RC}} = I_0 e^{-\frac{RC}{RC}} = (4.50 \times 10^{-3}\,A)\left(\frac{1}{e}\right)$$

$$= 1.66 \times 10^{-3}\,A$$

$$I(2\tau) = I_0 e^{-\frac{2\tau}{RC}} = I_0 e^{-\frac{2RC}{RC}} = (4.50 \times 10^{-3}\,A)\left(\frac{1}{e^2}\right)$$

$$= 0.609 \times 10^{-3}\,A$$

$$I(3\tau) = I_0 e^{-\frac{3\tau}{RC}} = I_0 e^{-\frac{3RC}{RC}} = (4.50 \times 10^{-3}\,A)\left(\frac{1}{e^3}\right)$$

$$= 0.224 \times 10^{-3}\,A$$

Making sense of the result

It is often helpful to calculate the time constant of the circuit first ($\tau = 1.00$ s in this example). You can see that the capacitor acquired 99% of its maximum charge before $t = 5\tau = 5$ s.

Discharging a Capacitor

If we wait long enough ($t \gg \tau$) when the switch is in position a, the capacitor in Figure 23-22 will be fully charged. Let us move the switch to position b. The capacitor and resistor are now connected in series, leaving out the emf source (ε). With these connections, the capacitor will start discharging. Applying Kirchhoff's voltage law (loop rule) while paying careful attention to the sign conventions (Table 23-4), we obtain

$$-V_C - IR = 0$$

$$-\frac{q(t)}{C} - IR = 0 \tag{23-53}$$

We can derive equations for the charge on the capacitor's plates and the current through the circuit by using the same method as for the example above. For the discharging capacitor,

$$-\frac{q(t)}{C} - \frac{dq(t)}{dt}R = 0 \tag{23-54}$$

At time $t = 0$, the capacitor was fully charged: $q(0) = C\varepsilon = q_{max}$. Again, we find the solution by rearranging the differential equation and integrating both sides:

$$\frac{dq(t)}{q(t)} = \frac{dt}{RC}$$

$$\int_{\varepsilon C}^{q(t)} \frac{dq(t)}{q(t)} = -\int_0^t \frac{dt}{RC}$$

$$\ln q(t) - \ln(\varepsilon C) = -\frac{t}{RC}$$

$$\ln \frac{q(t)}{\varepsilon C} = -\frac{t}{RC}$$

$$q(t) = \varepsilon C e^{-\frac{t}{RC}} = q_{max} e^{-\frac{t}{RC}} = q_{max} e^{-\frac{t}{\tau}}$$

Therefore, the charge on the plates of a capacitor and the potential difference across it when it is discharging can be expressed as follows:

KEY EQUATION

$$q(t) = q_{max} e^{-\frac{t}{RC}} = \varepsilon C e^{-\frac{t}{\tau}} \tag{23-55}$$

$$V_C = \frac{q(t)}{C} = \frac{q_{max}}{C}e^{-\frac{t}{RC}} = \varepsilon e^{-\frac{t}{\tau}} \qquad (23\text{-}56)$$

Consequently, the current is

$$i(t) = \frac{dq(t)}{dt} = \frac{d\left(q_{max}e^{\frac{-t}{RC}}\right)}{dt}$$

$$= -\frac{\varepsilon C}{RC}e^{-\frac{t}{RC}} = -\frac{\varepsilon}{R}e^{-\frac{t}{RC}} = -\frac{\varepsilon}{R}e^{-\frac{t}{\tau}}$$

The ratio ε/R once again represents the maximum current in this RC circuit. This current flows at the instant that switch S_2 is closed ($t = 0$). Thus,

KEY EQUATION
$$i(t) = -I_0 e^{-\frac{t}{RC}} = -I_0 e^{-\frac{t}{\tau}} = -\frac{\varepsilon}{R}e^{-\frac{t}{\tau}}$$

(discharging)   (23-57)

The negative value for current in this equation indicates that it flows in the opposite direction to the current when the capacitor was charging.

Equations 23-55, 23-56, and 23-57 describe the time dependence of the charge, potential, and current during the discharge of a capacitor in a series RC circuit. As above, we use these equations to find an expression for the electric potential energy stored in the capacitor at any given moment:

$$U_C = \frac{(q(t))^2}{2C} = \frac{\left(q_{max}e^{-\frac{t}{RC}}\right)^2}{2C} = \frac{(q_{max})^2 e^{-\frac{2t}{RC}}}{2C}$$

$$= \frac{C\varepsilon^2}{2}e^{-\frac{2t}{RC}} \qquad (23\text{-}58)$$

Figure 23-24 shows $i(t)$ and $q(t)$ graphs for a capacitor discharging through a resistor. Comparing the graphs in Figures 23-23 and 23-24, we see that the

$i(t)$ graphs are the same: the current starts with its maximum value and then diminishes exponentially with time. However, the $q(t)$ and correspondingly the $V_C(t)$ graphs for charging and discharging a capacitor have opposite trends.

The analysis above can be extended to circuits that have a combination of resistors and capacitors by calculating the equivalent resistance and the equivalent capacitance of the combinations.

CHECKPOINT 23-8

Discharging a Capacitor

The five circuits in Figure 23-25 consist of identical capacitors and resistors. At $t = 0$ s, the switch is closed and the capacitors start discharging. In which circuit will the capacitors discharge to 5% of their maximum charge in the shortest time? Explain.

(a) circuit (a)
(b) circuit (b)
(c) circuit (c)
(d) circuit (d)
(e) circuit (e)

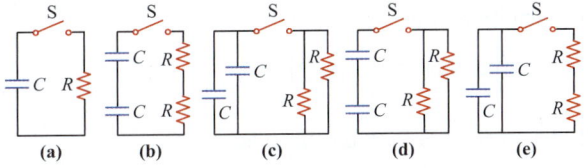

Figure 23-25   Checkpoint 23-8.

ANSWER: (d) The circuit time constant $\tau = RC$ shows how fast the capacitor in the circuit will charge or discharge. The time constants of the circuits are $\tau_1 = RC$; $\tau_2 = 2R\frac{C}{2} = RC$; $\tau_3 = RC$; $\tau_4 = \frac{R}{2}2C = RC$; $\tau_5 = \frac{R}{2}\frac{C}{2} = \frac{2}{4} = 2R2C = 4RC$.

MAKING CONNECTIONS

How Do Touch-Sensitive Lamps Work?

Touch-sensitive lamps commonly have a power-level switch controlled by an RC circuit. Your body is an electrically conductive object, so it can store electric charge. Thus, it can act as a capacitor. The typical capacitance of the human body is in the tens to low hundreds of picofarads. When you touch the metal body of a touch-sensitive lamp, you add capacitance to the control circuit and usually some static charge from your body, as well. These changes in the capacitance trigger the control circuit so it changes the voltage supplied to the lamp's bulb. Often, touch-sensitive switches cycle through three power levels: off-low-medium-high, then back to off. The same principle is used for the operation of touch-sensitive kitchen faucets.

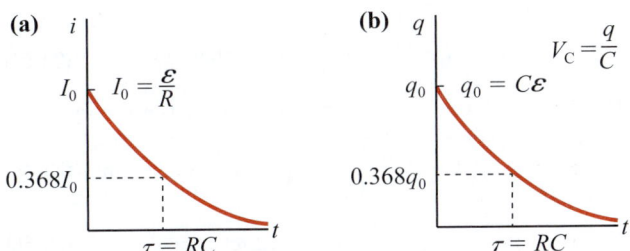

Figure 23-24   When the switch in a series RC circuit is flipped from position a to position b, the capacitor begins discharging: (a) The dependence of the electric current through the series RC circuit on time. (b) The dependence of the electric charge on the capacitor's plates on time. Since the voltage across the capacitor is proportional to the charge on its plates, this graph also represents the voltage across the capacitor.

KEY CONCEPTS AND RELATIONSHIPS

Applications of electric circuits have become so ubiquitous in our lives that only a "shocking" event such as the Northeast Blackout of 2003 could remind us how much we depend on them. Metals have free electrons that can drift in the direction opposite to the external electric field. Moving electric charges create electric current. Electric current flowing through a conductor carries energy that can be converted into heat, which is dissipated by the conductor. There is a fundamental relationship between electric current through a resistor, its resistance, and the voltage across it. DC circuits have currents that flow in only one direction through the components. Ohm's law, equivalent resistances, and Kirchhoff's laws can be used to analyze circuits that contain a number of components such as resistors, capacitors, and emf sources.

Electric Current: The Microscopic Model

The electron current in a metal wire, i, is related to its electric current and its electric current density, J. These depend on the electron density, n_e; the drift speed, v_d; and the cross-sectional area, A:

$$I = -I_e \cdot e = -(n_e v_d A)e = -n_e v_d A(1.6 \times 10^{-19})\frac{C}{s} \quad (23\text{-}4)$$

The definition of electric current is

$$I = \frac{\Delta Q}{\Delta t} \quad (23\text{-}5)$$

Electric current density is defined as

$$J = \frac{I}{A} = \frac{n_e v_d A e}{A} = n_e v_d e \quad (23\text{-}6)$$

The electric conductivity, σ, and the resistivity, ρ, describe how well a specific metal conducts current; these properties are temperature dependent:

$$\rho = \frac{1}{\sigma} = \frac{m}{n_e e^2 \tau} \quad (23\text{-}13)$$

$$\rho = \rho_0[1 + \alpha \Delta T] \quad (23\text{-}14)$$

where α is the temperature coefficient of resistivity, T is temperature, and ρ_0 is the resistivity at a baseline temperature.

Electric Current, Resistance, and Voltage: The Macroscopic Model

The electric current, I, through a resistor with resistance R and voltage across it V are related by Ohm's law:

$$I = \frac{V}{R} \text{ (for ohmic resistors only)} \quad (23\text{-}17)$$

The resistance of a conductor depends on its resistivity, ρ; length, l; and cross-sectional area, A:

$$R = \rho \frac{\ell}{A} \quad (23\text{-}16)$$

A resistor with resistance R that carries electric current I dissipates thermal energy Q_{heat} at the rate described by the Joule–Lenz law:

$$P = IV = I^2 R = \frac{V^2}{R} \quad (23\text{-}25)$$

The energy dissipated by a resistor R carrying current I over time period Δt is given by

$$Q_{heat} = I^2 R \Delta t \quad (23\text{-}26)$$

The electrical power dissipated by a parallel circuit is given by

$$P_{parallel} = \sum_{i=1}^{N} P_i \quad (23\text{-}27)$$

Analysis of Compound Electric Circuits: Equivalent Resistance

For resistors connected in series,

$$R_{series} = \sum_{i=1}^{N} R_i \quad (23\text{-}20)$$

For resistors connected in parallel,

$$\frac{1}{R_{parallel}} = \sum_{i=1}^{N} \frac{1}{R_i} \quad (23\text{-}22)$$

Kirchhoff's Laws for the Analysis of Multi-loop DC Circuits

Kirchhoff's voltage law (loop rule or second law) is given by

$$\sum_{closed\ loop} \Delta V = 0 \quad (23\text{-}34)$$

Kirchhoff's current law (junction rule or first law) is given by

$$\sum_{k=1}^{N_1} I_{entering\ junction} = \sum_{k=1}^{N_2} I_{exiting\ junction}$$

$$\sum_{k=1}^{N} I_{junction} = \sum_{k=1}^{N_1} I_{entering\ junction} - \sum_{k=1}^{N_2} I_{exiting\ junction} = 0$$

$$(23\text{-}35)$$

RC Circuits

In a series RC circuit, the electric current across a capacitor as it is charging or discharging changes as

$$i(t) = I_0 e^{-\frac{t}{RC}} = \frac{\varepsilon}{R} e^{-\frac{t}{\tau}} \text{ (charging)} \quad (23\text{-}49)$$

$$i(t) = -I_0 e^{-\frac{t}{RC}} = -\frac{\varepsilon}{R} e^{-\frac{t}{\tau}} \text{ (discharging)} \quad (23\text{-}57)$$

The electric charge on the capacitor's plates is given by

$$q(t) = q_{max}\left(1 - e^{-\frac{t}{RC}}\right) = \varepsilon C\left(1 - e^{-\frac{t}{\tau}}\right) \text{ (charging)} \quad (23\text{-}45)$$

$$q(t) = q_{max} e^{-\frac{t}{RC}} = \varepsilon C e^{-\frac{t}{\tau}} \text{ (discharging)} \quad (23\text{-}55)$$

The electric potential across the capacitor's plates is given by

$$V(t) = \frac{q_{max}}{C}\left(1 - e^{-\frac{t}{RC}}\right) = \varepsilon\left(1 - e^{-\frac{t}{\tau}}\right) \text{ (charging)} \quad (23\text{-}46)$$

$$V(t) = \frac{q_{max}}{C} e^{-\frac{t}{RC}} = \varepsilon e^{-\frac{t}{\tau}} \text{ (discharging)} \quad (23\text{-}56)$$

APPLICATIONS

electrical appliances, electric power transmission, residential and commercial electrical wiring, electric or medical equipment (e.g., defibrillators), electronic equipment, physiological applications

KEY TERMS

ammeter, conduction electron density, crystal lattice, current density, direct current (DC), drift velocity, electric circuit branch, electric conductivity, electric current, electric diode, electric resistance, electric resistivity, electrical grounding, electron current, electron flow, electron scattering, equivalent resistance, free (delocalized) electrons, free electron density, junction, junction points, junction rule, Kirchhoff's current (first) law, Kirchhoff's voltage law, law of electric charge conservation, mean free time, metallic bonds, multimeter, node, nominal power output, nominal resistance, ohmic conductors, ohmmeter, Ohm's law, parallel circuit, power, resistors, series circuit, short circuit, sources of electromotive force, superconductor, time constant, transient voltage, voltmeter

QUESTIONS

1. Which of the following statements is true? Explain your answer.
   (a) The directions of electron and electric currents in metals are always the same.
   (b) If the resistivities of two wires are the same, then their resistances must also be the same.
   (c) The free electron density in a metal depends on its temperature.
   (d) The resistivity and conductivity of a metal depend on the metal's temperature.
   (e) The electron drift speed depends on the resistivity of the metal.

2. In a box labelled "spare 40 W and 60 W bulbs," there are two clear incandescent light bulbs. One bulb has a thicker filament than the other. Which bulb is likely to be rated at 40 W?
   (a) thicker filament
   (b) thinner filament
   (c) It is impossible to know without trying each one.

3. In circuit I, a light bulb is connected to a battery. In circuit II, 10 light bulbs, all identical to the light bulb in circuit I, are connected in series to an identical battery. In circuit III, 10 light bulbs are connected in parallel to an identical battery. Which statement correctly describes the relative power outputs of these circuits? Explain your answer.
   (a) circuit I > circuit II > circuit III
   (b) circuit III > circuit I > circuit II
   (c) circuit I = circuit II = circuit III
   (d) circuit II > circuit I > circuit III
   (e) circuit III > circuit II > circuit I

4. Which circuit in question 3 will produce the most light? Explain.
   (a) circuit I
   (b) circuit II
   (c) circuit III
   (d) Circuits II and III will both produce 10 times the light of circuit I.

5. The resistance of a long copper wire is R. You cut the wire into N equal pieces and connect the pieces in parallel. What is the resistance of this combination?
   (a) N^2R
   (b) NR
   (c) R
   (d) R/N
   (e) R/N^2

6. How much energy is consumed by a 60 W bulb if you leave it on for 10 h?
   (a) $60 \text{ kW} \cdot \text{h}$
   (b) 60 kJ
   (c) 600 J
   (d) $3600 \text{ kW} \cdot \text{h}$
   (e) 2.16 MJ

7. How much energy is stored in a fully charged 12 V car battery rated at $100 \text{ A} \cdot \text{h}$?
   (a) $12 \text{ MW} \cdot \text{h}$
   (b) $1.2 \text{ MW} \cdot \text{h}$
   (c) $1.0 \text{ MW} \cdot \text{h}$
   (d) 4320 kJ
   (e) $432 \text{ kW} \cdot \text{h}$

8. What is the ratio of the filament resistance of a 70 W light bulb designed for North America (110 V) and a 140 W light bulb designed for Europe (220 V)?
   (a) 2:1
   (b) 1:2
   (c) 2:2
   (d) 4:1
   (e) 1:4
   (f) 1:1

9. You have six identical $4.0 \ \Omega$ resistors and one 3.0 V battery. How can you connect the resistors to the battery such that the current flowing from the battery is 4.5 A?
   (a) Attach all the resistors in series, and connect them to the battery.
   (b) Attach all the resistors in parallel, and connect them to the battery.
   (c) Attach two resistors in series and the rest in parallel, and then connect them to the battery.
   (d) Attach three resistors in series and the rest in parallel, and then connect them to the battery.
   (e) It is not possible.

10. Which of the following statements is false?
    (a) According to Ohm's law, an increase in a circuit's equivalent resistance means a decrease in the circuit's power output.
    (b) A voltmeter is always connected in parallel to the circuit element across which we want to measure a voltage.

(c) An ammeter is always connected in series to the circuit element through which we want to measure a current.

(d) A short circuit means a circuit has a very large electric resistance, thus becoming a fire hazard.

(e) If a piece of metal wire is heated, its resistivity will increase and its conductivity will decrease.

11. You need a 6.0 Ω resistor, but you only have four 4.0 Ω resistors. How many resistors should you use and how should you connect them to produce a total resistance of 6.0 Ω?

(a) three 4.0 Ω resistors connected in series

(b) three 4.0 Ω resistors connected in parallel

(c) two 4.0 Ω resistors connected in parallel and then connected in series to a 4.0 Ω resistor

(d) two 4.0 Ω resistors connected in series and then connected in parallel to a 4.0 Ω resistor

(e) two 4.0 Ω resistors connected in parallel, connected in series to two 4.0 Ω resistors connected in series

12. What is the operating resistance of a 100. W (110. V) light bulb?

(a) 1.11 Ω

(b) 12.1 Ω

(c) 121 Ω

(d) 12.1 kΩ

(e) 121 kΩ

13. Figure 23-26 shows an electric circuit consisting of six identical light bulbs, each with an operating resistance of 10.0 Ω, connected to a 20.0 V battery. What is the equivalent resistance of this circuit?

(a) 60.0 Ω

(b) 22.5 Ω

(c) 21.7 Ω

(d) 18.8 Ω

(e) 18.4 Ω

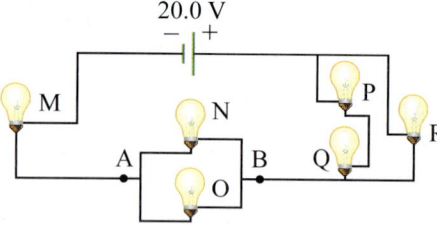

Figure 23-26 Questions 13–21.

14. What is the electric current flowing through the battery in Figure 23-26?

(a) 1.07 A

(b) 1.00 A

(c) 0.923 A

(d) 0.890 A

(e) 1.09 A

15. In the circuit shown in Figure 23-26, what is the value of the potential difference between points A and B?

(a) 9.24 V

(b) 10.0 V

(c) 18.5 V

(d) 5.45 V

(e) 4.62 V

16. In the circuit shown in Figure 23-26, what is the value of the potential difference across resistor M?

(a) 3.08 V

(b) 6.16 V

(c) 4.62 V

(d) 9.23 V

(e) 20.0 V

17. In the circuit shown in Figure 23-26, what is the value of the potential difference across resistor R?

(a) 3.08 V

(b) 6.15 V

(c) 4.62 V

(d) 9.23 V

(e) 20.0 V

18. In the circuit shown in Figure 23-26, what is the value of the potential difference across resistor P?

(a) 3.08 V

(b) 6.16 V

(c) 4.62 V

(d) 9.23 V

(e) 20.0 V

19. In the circuit shown in Figure 23-26, what is the value of the potential difference across resistor N?

(a) 3.08 V

(b) 6.16 V

(c) 4.62 V

(d) 9.23 V

(e) 20.0 V

20. Electric current through resistor N is the same as electric current through resistor O. However, electric current through R is twice larger than electric current through O. Why does this happen?

(a) Electric current always splits equally between branches, so the statement above is not true.

(b) Branches N and O have equal resistances, but branches R and PQ also have equal but larger resistances.

(c) Branches N and O have equal resistance, but the resistance of branch R is twice the resistance of branch PQ.

(d) Branches PQ and R have equal resistance, but the resistance of branch N is twice the resistance of branch O.

(e) Branches N and O have equal resistance, but the resistance of branch PQ is twice the resistance of branch R.

21. Rank the brightness of the light bulbs in Figure 23-26. Use your answers to questions 13–20.

(a) M > N > O > P = Q > R

(b) M = R > N = O > P = Q

(c) M > R > N = O > P = Q

(d) M = N = O > P = Q < R

(e) M > N > O > R > P = Q

22. A 12.0 V battery is connected to the combination of nine identical 400. Ω resistors shown in Figure 23-27. What is the current flowing through this battery?

(a) 3.33 mA

(b) 1.33 A

(c) 8.92 mA
(d) 11.0 mA
(e) 533 A

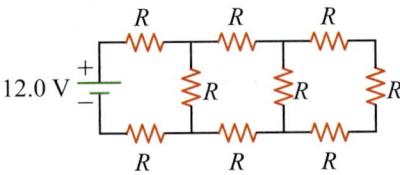

Figure 23-27 Question 22.

23. Six identical 5.0 Ω light bulbs—A, B, C, D, E, and F—are connected as shown in Figure 23-28; $\varepsilon = 24$ V. Which of the following ranks the amount of current flowing through the light bulbs from the largest to the smallest?
(a) $I_D, I_A, I_B, I_F, I_C = I_E$
(b) $I_A, I_D, I_B = I_F, I_C = I_E$
(c) $I_A, I_D, I_B, I_F, I_C = I_E$
(d) $I_A, I_D, I_B, I_F = I_C = I_E$
(e) $I_D, I_A, I_B, I_F = I_C = I_E$

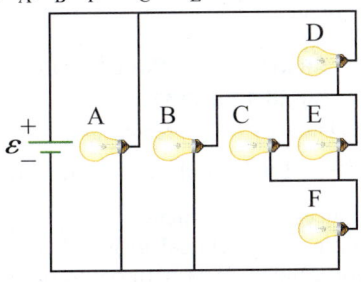

Figure 23-28 Question 23.

24. A 5 μF capacitor is connected through a 10 kΩ resistor to a 12 V battery. It takes this capacitor 0.002 s to be charged to 95% of its maximum charge in this series circuit. How long would it take the same capacitor to become 95% charged (if initially uncharged) when the 12 V battery is replaced by a 6 V battery?
(a) 0.004 s
(b) 0.002 s
(c) 0.001 s
(d) 0.0005 s
(e) 0.0001 s

25. An RC electric circuit is shown in Figure 23-29. Initially, switch S is open and the capacitors are completely uncharged. After the switch has been closed for 3.0 min, how much energy will be stored in the 5.0 μF capacitor?
(a) 0.30 mJ
(b) 0.60 mJ
(c) 1.5 mJ
(d) 36 mJ
(e) 72 mJ

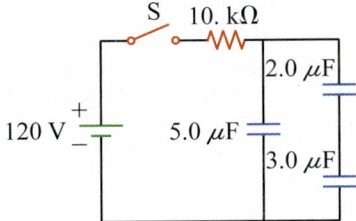

Figure 23-29 Question 25.

26. A series RC circuit consists of a resistor, a battery, wires, and a capacitor. You are worried that the wires might not withstand the current flowing through them as the capacitor charges. When are the wires more likely to burn?
(a) right at the beginning of the experiment, when the capacitor just starts charging
(b) in the middle of the experiment, when the capacitor is about 32% charged
(c) in the middle of the experiment, when the capacitor is about 67% charged
(d) at the end of the experiment, when the capacitor is almost fully charged

PROBLEMS BY SECTION

For problems, star ratings will be used (★, ★★, or ★★★), with more stars meaning more challenging problems. The following codes will indicate if $\frac{d}{dx}$ differentiation, $\int$ integration, 🖥 numerical approximation, or 〰 graphical analysis will be required to solve the problem.

Section 23-1 Electric Current: The Microscopic Model

27. ★ Estimate the conduction (free) electron density in copper given that the atomic mass of copper is 63.546 amu and the density of copper at 293 K is 8.96 g/cm³. Compare your estimate with an experimentally established value.

28. ★ An electric current of 2.0 A is flowing through a copper wire that has a diameter of 2.0 mm.
(a) At what rate are electrons flowing through the wire?
(b) Find the electric current density in the wire.
(c) How much electric charge will be carried through the cross section of the wire in 2.0 s?

29. ★ Your classmate hears the professor mention that the average drift speed of electrons in metals is about 0.1 mm/s. This does not make sense to your classmate: "If this were true," she says, "I would have to keep my key in the ignition switch of my car for more than 4 h for the electrons to flow from the negative terminals of the battery to the starter motor and back to the positive terminal of the battery." How will you respond to her?

Section 23-2 Electric Conductivity and Resistivity: The Microscopic Model

30. ★ An electric current of 2.0 A flows through a copper wire at 15 °C. The wire has a circular cross section and is 2.0 mm in diameter.
(a) Using the information in Table 23-2, find the conductivity and the resistivity of the wire.
(b) Using Table 23-1, find the drift velocity of the conduction electrons in the wire.
(c) Estimate the mean time, τ, that free electrons in the wire spend between collisions.
(d) Find the electric field inside the wire.
(e) How would your answers to parts (a) to (d) change if the electric current through the wire doubled (i.e., 4.0 A instead of 2.0 A)?

(f) Suppose the cross section of the wire were changed from circular to square with the same area. What effect would this change have on the quantities in parts (a) to (d)?

(g) How would the quantities in parts (a) to (d) change if the temperature of the wire were raised from 15 °C to 35 °C while the current remained constant?

Sections 23-3 and 23-4 Ohm's Law and Series and Parallel Combinations of Resistors

31. ★ The resistance of the tungsten filament in a common size of incandescent light bulb was 15 Ω. The diameter of the filament was 0.10 mm. What was the length of the filament wire?

32. ★ The mass of the copper wire in a coil is 1.78 g, and its resistance is 34.0 Ω. Estimate the length and the cross-sectional area of the wire in the coil.

33. ★ (a) Compare the relative change of the resistances and resistivities of an aluminum wire and a Nichrome wire that are both heated to 500 °C. Neglect the thermal expansion of the wires.

(b) How would your answers change if you took the thermal expansion of the wires into account?

(c) How would your answers change if the lengths of the wires were doubled?

34. ★ How much current flows through a 1.0 kW electric toaster connected to a 120 V outlet? (The voltage is AC, not DC, but the physics is the same.) How many toasters can you operate at the same time before you blow a fuse or circuit breaker? (Assume that the maximum allowed current is 15 A.)

35. ★ You have two conductors: Conductor 1 is a 10 m long tin wire that has a 1 mm^2 square cross-sectional area. Conductor 2 is a strip of tin foil with dimensions 10 mm × 200 mm and thickness 100 μm. Compare the resistance of both conductors, assuming that the current flows along the longest dimension of the tin strip.

36. ★ A wire made of a manganese alloy has a length of 56 cm and a diameter of 0.40 mm. The resistivity of the alloy is $\rho = 0.45 \ \mu\Omega \cdot m$.
(a) Find the resistance of the wire.
(b) Find the current through the wire when the potential difference across it is 3.0 V.

37. ★★ For your science experiment, you require a resistor that has a resistance of 10.0 Ω. You have a spool of copper wire ($\rho_{Cu} = 172 \ \mu\Omega \cdot m$) that has a diameter of 1.00 mm.
(a) How long a wire you should cut for it to have the required resistance?
(b) After you realize the wire is too long, you decide to wrap it around a plastic spool that has a diameter of 10.0 cm. How many loops will you have when you wrap it around the spool?
(c) If instead of copper wire, you had a nickel wire ($\rho_{Ni} = 685 \ \mu\Omega \cdot m$), how would you answers to questions (a) and (b) have changed?

38. ★★ Will a 3 m long aluminum wire melt if you heat it enough to triple its resistance? Explain your answer.

39. ★★ You have 10 resistors that have resistances of 1.00 Ω, 2.00 Ω, 3.00 Ω, 4.00 Ω, 5.00 Ω, 6.00 Ω, 7.00 Ω, 8.00 Ω, 9.00 Ω, and 10.0 Ω. You first connect them in series and then connect them in parallel.
(a) What is their equivalent resistance when they are connected in series?

(b) What is their equivalent resistance when they are connected in parallel?

(c) What is the ratio of the equivalent resistance in (a) to the equivalent resistance in (b)?

40. ★★ Prove that the equivalent resistance of any number of parallel resistors will never exceed the resistance of the branch that has the least resistance among all parallel branches:

$$R_{parallel} < (R_{min})_{in \ parallel \ branch}$$

41. ★ Use Figure 23-9 to determine three possible different combinations of household electrical appliances that can be used simultaneously without exceeding the maximum current of 15 A.

42. ★★ You decide to decrease your electricity bill by reducing the usage of your toaster from 5.0 min/day to 3.0 min/day. Your toaster has a wattage of 1.0 kW, and you pay 6.5¢ per kW·h.
(a) How much energy will you save per year?
(b) How much less will you pay for your electricity bill per year than before?
(c) Suggest more effective ways of reducing your electricity bill.

Section 23-5 Analysis of DC Circuits and Kirchhoff's Laws

43. ★ Ohm's law can be expressed as $I = V/R$, $I = \varepsilon/R$, or $I = \dfrac{\varepsilon}{R + r}$. Explain when each equation applies and the differences between them.

44. ★ You have four identical ideal 1.5 V batteries. How should you connect these batteries to produce
(a) a 1.5 V emf source that will outlast a single battery?
(b) a 3.0 V emf source that will outlast a single battery?
(c) a 4.5 V emf source?
(d) a 6.0 V emf source?
Check your answer by using the PhET "Circuit Construction Kit (AC and DC)."

45. ★★ A series electric circuit consists of a real battery ($\varepsilon = 12.0$ V; $r = 3.00 \ \Omega$) and an ohmic resistor with $R = 20.0 \ \Omega$.
(a) Draw a schematic diagram for the circuit.
(b) Derive an expression to determine how much current will flow through the battery in the circuit, and calculate the current.
(c) Derive an expression to determine the potential difference across the battery, and calculate its value.
(d) Without doing the calculations, describe what would happen to the answers above if another identical resistor ($R = 20. \ \Omega$) were connected
  (i) in parallel with the original resistor
  (ii) in series with the original resistor
(e) Modify the expressions derived in parts (b) and (c) to determine whether your predictions in part (d) were correct. Calculate the current through the circuit and the potential difference across the battery.

46. ★★ Figure 23-30 shows an electric circuit that consists of nine resistors (all multiples of $R = 5.0 \ \Omega$) connected to a 12 V ideal emf source.
(a) Calculate the equivalent resistance of the circuit and the electric power dissipated by it.
(b) Find the electric current that flows through each resistor.

(c) Calculate the potential difference across each resistor.
(d) Calculate the power dissipated by each resistor.
(e) Compare your answer in question (d) to the electric power dissipated by the circuit you found in part (a). What does your comparison tell you?

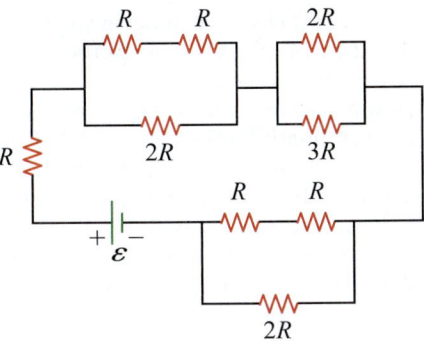

Figure 23-30 Problem 46.

47. ★★★ An ammeter is always connected in series (it has a very small internal resistance), and a voltmeter is always connected in parallel (it has a very large internal resistance). Assume that the resistance of an ammeter is 1.000 $\mu\Omega$ and the resistance of a voltmeter is 10.00 kΩ. If a series circuit consists of a 12.00 V battery and a 5.000 Ω resistor, what will be the effect of measuring the electric current in this circuit and the electric voltage across the resistor using the ammeter and the voltmeter? Calculate the percentage change of electric current and voltage before and after the meters are connected.

48. ★★ Seven identical resistors, resistance R, are connected as shown in Figure 23-31.
(a) What is the equivalent resistance between points A and B?
(b) How would your answers to the previous question change, if each side of the equilateral triangle had a resistance of 2R (one resistors is removed from the horizontal branch)?
(c) In the configuration depicted in Figure 23-31, is it possible to make the equivalent resistance of this combination of resistors to be less than R while still having seven resistors?

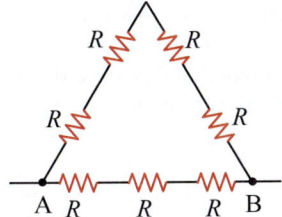

Figure 23-31 Problem 48.

49. ★★★ Twelve identical resistors, resistance R, are connected as shown in Figure 23-32. What is the equivalent resistance between points A and B?

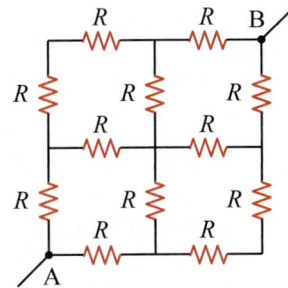

Figure 23-32 Problem 49.

50. ★★★ Twelve identical 3.0 Ω resistors are connected as shown in Figure 23-33. What is the equivalent resistance between points A and B?

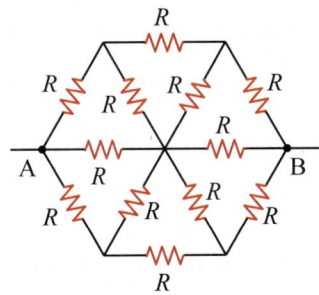

Figure 23-33 Problem 50.

51. ★★★ Figure 23-34 shows a multi-loop electric circuit. The components have the following values:

$$\varepsilon_1 = 10.0 \text{ V}; \varepsilon_2 = 20.0 \text{ V (ideal batteries)}$$
$$R_1 = 10.0 \text{ }\Omega; R_2 = 15.0 \text{ }\Omega; R_3 = 15.0 \text{ }\Omega; R_4 = 20.0 \text{ }\Omega$$

(a) Find the currents through all the branches of the circuit.
(b) Find the potential difference across each circuit element.
(c) How will your answers to parts (a) and (b) differ if the batteries are real batteries?

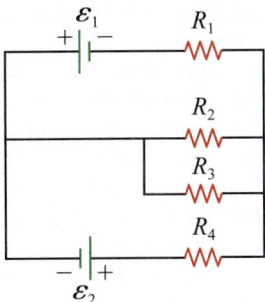

Figure 23-34 Problem 51.

Section 23-6 RC Circuits

52. ★ A series RC circuit consists of a 300.0 μF capacitor, a 12.00 V battery, and a 5.000 kΩ resistor.
    (a) How long does it take for the capacitor to reach 99.9% of q_{max}? Express your answer in terms of the time constant, τ.
    (b) How much charge will be accumulated on the capacitor's plates after $t_1 = \tau$, $t_2 = 2\tau$, and $t_3 = 3\tau$?
    (c) What will be the value of the electric current in the circuit in each of the above cases?

53. ★ A series RC circuit consists of a 450. μF capacitor, a 6.00 V battery, and a 5.00 kΩ resistor.
    (a) What is the time constant of this circuit?
    (b) You connect a very sensitive ammeter to this series RC circuit. The ammeter has a precision of 1.00 μA. How long will it take since you started charging your capacitor for the ammeter to show zero current?

54. ★★ An initially uncharged 500.0 μF capacitor is being connected in series to a 2500 V emf source with a 10.0 kΩ resistor.
    (a) What is the time constant of this capacitor?
    (b) How long will it take for the voltage across this capacitor to reach 95% of its maximum value?
    (c) What is the maximum value of the voltage across the capacitor?
    (d) When the capacitor reaches 95% of its maximum voltage, the charging circuit is removed and the capacitor is left charged but can slowly discharge through the moist air with a resistance of 1000.0 MΩ. What will be the time constant of this discharging RC circuit?
    (e) How long will it take this moist-air RC circuit to discharge this capacitor to 20.0 V through the air? What does your answer tell you?

55. ★★★ You have four identical 400.0 μF capacitors, a 10.0 kΩ resistor, and a 20.0 V battery.
    (a) You first connect the capacitors in series with each other and then connect them in series with the resistor and the battery. What is the time constant of this circuit?
    (b) What is the voltage across each one of these capacitors when they are fully charged?
    (c) How long will it take each one of these capacitors to charge to 90% of their maximum charge value?
    (d) What is this maximum charge value on each one of these capacitors? Answer this question using two different methods.
    (e) Now you connect the four identical capacitors in parallel with each other and then you connect them in series with the resistor and to the battery. What is the time constant of this circuit?
    (f) What is the voltage across each one of these capacitors when they are fully charged?
    (g) How long will it take each one of these capacitors to charge to 90% of their maximum charge value?
    (h) What is this maximum charge value on each one of these capacitors? What do your results tell you?

COMPREHENSIVE PROBLEMS

56. ★ A high-quality dry-cell D battery (1.5 V) can provide 0.67 A for about 5.0 h before dying.
    (a) Compare the cost of the energy you get from buying a dry-cell D battery to the cost of the energy you get from the electrical outlet in your wall. (Assume that the cost of 1 kW·h is $0.100 and one D dry-cell battery costs $1.00.)
    (b) How much would an average Canadian household pay per month for their electric bill if they were to use dry cells to satisfy all their electricity needs?

57. ★ When operating, the resistance of the filament of a 60.0 W incandescent light bulb is 240. Ω. However, if you were to measure the resistance of the filament using a regular ohmmeter, you would find it to be about one twelfth the size (20.0 Ω).
    (a) How can you explain this phenomenon? (Hint: Think what happens to the light bulb when it is plugged in.)
    (b) Do appropriate calculations to show that when operating, the tungsten filament does have the appropriate resistance.
    (c) Based on your calculations, estimate the temperature of the incandescent light bulb when operating.

58. ★ Four resistors connected to a battery have the resistances indicated in Figure 23-35; $R = 5.0\ \Omega$. The battery is an ideal emf source that maintains a constant potential difference across it of 10. V. A voltmeter is used to measure the following potential differences: V_{AB}, V_{AG}, V_{BH}, V_{HF}, V_{FG}, V_{HD}, V_{EF}, V_{CD}, V_{DE}, V_{BE}, and V_{CF}. Determine the readings of the voltmeter.

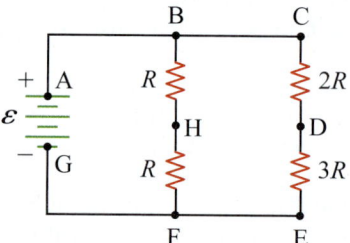

Figure 23-35 Problem 58.

59. ★★ An electric circuit consists of six resistors and a real battery with $\varepsilon = 9.00$ V and an internal resistance of $r = 0.500\ \Omega$ (Figure 23-36). Find the battery current, the voltage across each resistor, and the terminal voltage of the battery.

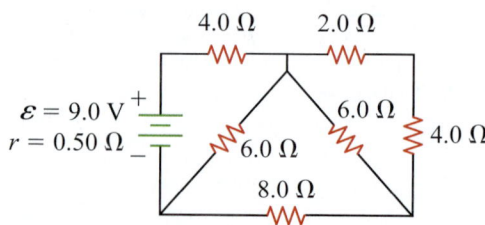

Figure 23-36 Problem 59.

60. ★ Figure 23-37 shows two different electric circuits.
  (a) Compare the current flowing through the batteries in both circuits.
  (b) Compare the potential differences across points A and B in each circuit. What does your comparison tell you?
  (c) Compare the brightness of the light bulbs in both circuits.

(a)

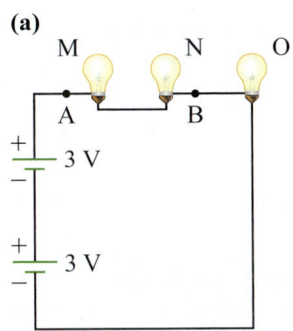

(b)

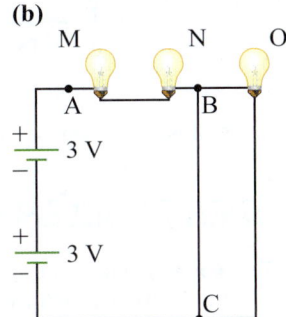

Figure 23-37 Problem 60.

61. ★★ In Figure 23-38, a metal net is connected to an emf source through points A and B. The conductivity of the external frame (solid lines) of the net is very high, and the resistance of each internal segment (dashed) is r. What is the resistance of the net between points A and B?

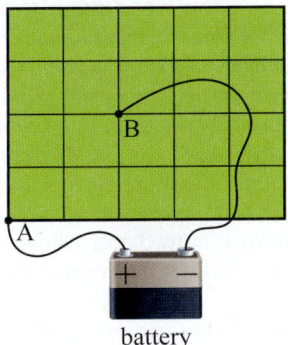

Figure 23-38 Problem 61.

62. ★★★ You have an aluminum wire and a tungsten wire that have the same resistance at room temperature. The temperature resistivity coefficient of aluminum is approximately half the temperature resistivity coefficient of tungsten (see Table 23-2). When answering the questions below, assume that when their temperatures are equal, the wires lose thermal energy to the surroundings at the same rate. For each part, provide a detailed explanation.
  (a) When the wires are connected in parallel to the battery, which wire will have a higher temperature when the system reaches equilibrium?
  (b) When the wires are connected in series to the battery, which wire will have a higher temperature when the system reaches equilibrium?

63. ★★ Twelve identical 4.0 Ω resistors are connected to form a cube, as shown in Figure 23-39. What is the equivalent resistance between points A and B?

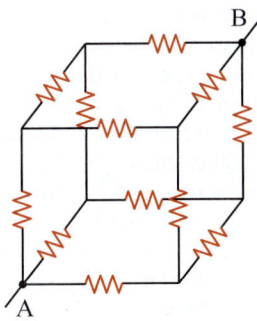

Figure 23-39 Problem 63.

64. ★★★ Figure 23-40 shows a multi-loop electric circuit. The values of the components are as follows:

$$\varepsilon_1 = 15.0 \text{ V}; \varepsilon_2 = 20.0 \text{ V (ideal batteries)}$$
$$R_1 = 10.0 \text{ Ω}; R_2 = 10.0 \text{ Ω}; R_3 = 15.0 \text{ Ω}; R_4 = 20.0 \text{ Ω};$$
$$R_5 = 20.0 \text{ Ω}; R_6 = 10.0 \text{ Ω}$$

  (a) Find the current through each branch of the circuit.
  (b) Find the potential difference across each circuit element.
  (c) How will your answers to parts (a) and (b) differ if the batteries are real batteries?

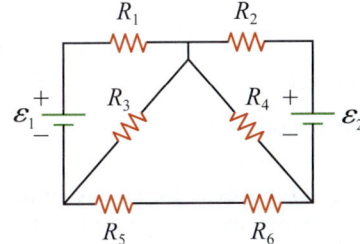

Figure 23-40 Problem 64.

65. ★★ Design an electric circuit that can be described by the following system of equations:

$$\begin{cases} I_1 + I_2 + I_3 = 0 \\ -V_1 + I_1 R_1 - I_3 R_3 = 0 \\ I_3 R_3 - I_2 R_2 + V_2 = 0 \end{cases}$$

66. ★★★ Analyze the DC circuit shown in Figure 23-41 by finding the current flowing through each light bulb and the potential difference across each light bulb. All the emf sources in this circuit are ideal. The values of the emf sources and resistances are as follows:

$$\varepsilon_1 = 25.0 \text{ V}; \varepsilon_2 = 10.0 \text{ V}; \varepsilon_3 = 12.0 \text{ V}$$
$$R_1 = 10.0 \text{ Ω}; R_2 = 5.0 \text{ Ω}; R_3 = 15.0 \text{ Ω}; R_4 = 20.0 \text{ Ω};$$
$$R_5 = 10.0 \text{ Ω}$$

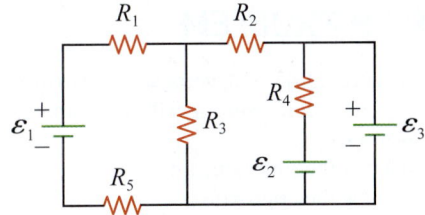

Figure 23-41 Problem 66.

67. ★★ A capacitor is connected to a battery through a resistor and allowed to charge fully so the voltage across the capacitor is V. Then the distance between the plates of the capacitor is quadrupled, while the capacitor remains connected to the battery. Describe what happens to the charge (Q) of the capacitor, its electric field (E), and the voltage across it (V) as a result of the increase in distance. Explain why this happens. If needed, include appropriate diagrams to support your explanation.

68. ★★ The switch in the circuit shown in Figure 23-42 has been closed for a long time. Assume that all the values are given to three significant figures.
(a) Find the potential difference across each circuit element.
(b) When the switch is reopened, what is the time constant of the circuit?

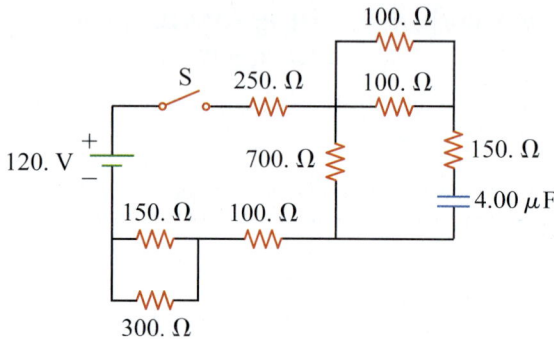

Figure 23-42   Problem 68.

69. ★★ A pacemaker uses an RC circuit for timing.
(a) If the pacemaker is to operate at 72 beats/min, and if the circuit "fires" when the capacitor is 65% charged, what must the time constant be?
(b) When $C = 7.50\ \mu F$, what value of resistance should be used?

70. ★★★ Assume that a capacitor used for smoothing in a DC power supply is to discharge by no more than 0.100 V (from an initial value of 10.0 V) during 1/60.0 s. The power supply has specifications of output voltage of 10.0 V and a supplied current of 250.0 mA.
(a) What is the effective "resistance" of the circuit connected to the power supply?
(b) What must the minimum value for the smoothing capacitor be?
(c) How much energy is stored in the capacitor when it is fully charged?
(d) How much energy has been drawn from the capacitor when it has had its voltage reduced by 0.100 V?

DATA-RICH PROBLEM

71. ★★★ Figure 23-43 is a voltage versus time graph for a RC series circuit using a 5.00 μF capacitor and an unknown resistor.
(a) What is the time constant?
(b) What is the approximate value of the resistor?
(c) What is the approximate voltage of the charging source?

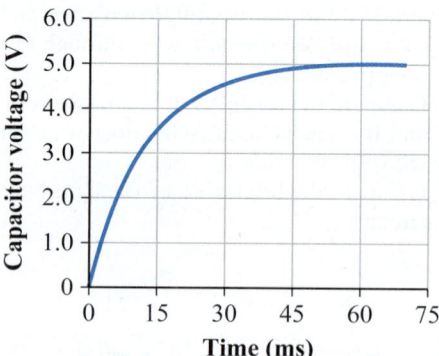

Figure 23-43   Problem 71.

OPEN PROBLEMS

72. ★★★ Estimate the annual energy use of (a) your household; (b) your educational institution; (c) the town or city you live in. Clearly indicate the assumptions you are making in your estimate.

73. ★★★ Design your own low-cost smoke detection and alarm system. On April 12, 2015, the bodies of two people were found in a fire on Hastings Street in East Vancouver. According to CBC News, more than 50 firefighters were called out to fight this three-alarm fire in an old two-storey building. The ready-for-demolition building apparently didn't have smoke detectors that could have saved lives. According to recent statistics, more than 65% of fire deaths happen in homes that do not have fire alarms. Most households in Canada today have smoke alarms like the one shown in Figure 23-44. These are especially important in wood-framed structures. There are a number of circuit designs for common fire alarm systems. Most fire alarms have a slot between the infrared (LED) light source and a receiver (such as a phototransistor) to detect smoke. As soon as the signal from the LED is interrupted by the smoke, the phototransistor switches the module output from "on" to "off," thus alerting the occupants of the problem. However, fire alarms have various designs. Work with a partner and suggest at least three possible working designs for a fire alarm and build them using available electronics. Compare these designs in terms of different parameters that you deem important for the operation of the fire alarm system.

Figure 23-44   An example of a common smoke detector.

Magnetic Fields and Magnetic Forces

When you have completed this chapter, you should be able to

LO1 Describe magnetic fields and forces using the concept of magnetic field lines.

LO2 Calculate the magnitude and direction of the magnetic force acting on a charged particle moving inside a magnetic field, and describe the particle's trajectory.

LO3 Describe applications of the motion of a charged particle inside crossed electric and magnetic fields.

LO4 Calculate the magnitude and direction of a magnetic force applied by a magnetic field on current-carrying wires.

LO5 Calculate the magnitude and direction of a magnetic torque applied by a magnetic field on current-carrying loops.

LO6 Calculate the magnitude and direction of a magnetic field created by moving electric charges using the Biot–Savart law.

LO7 Calculate the magnitude and direction of a magnetic field created by moving electric charges using Ampère's law.

LO8 Describe the magnetic forces between two current-carrying wires quantitatively and qualitatively.

LO9 Describe the magnetic properties of materials and some everyday life applications of magnetic phenomena.

You have probably used a plastic card with a magnetic stripe, such as a credit card, a debit card, a student identification card, a driver's licence, or a transit pass (Figure 24-1). Have you ever wondered how these cards work? Magnetic stripes are made of iron-based magnetic particles approximately 5 μm across embedded in a thin film. To record information on the magnetic stripe, groups of these particles are magnetized in a specific direction. The card is read by measuring the magnetic field along the stripe.

The process of attaching a magnetic stripe to a plastic card was developed in the late 1960s by IBM engineer Forrest Parry, who was working on a security system for the U.S. government. Parry found that it was quite difficult to attach a magnetic stripe to a plastic card without warping the stripe or having its surface damaged or contaminated by the adhesives used to glue it in place. The solution came from Parry's wife, Dorothea Tillia, who suggested using an iron to bond the stripe to the card.

Frank McNamara came up with the concept of the modern credit card. It was intended to be used by salespeople who travelled a lot and did not want to carry large amounts of cash to pay for their meals and entertainment. Hence, the first credit card was the Diners' Club card. Today, credit cards increasingly use integrated circuits (ICs), although for now they still have magnetic stripes for backward compatibility.

Figure 24-1 The magnetic stripe on the back of a credit card.

24-1 Magnetic Field and Magnetic Force

Earlier chapters describe electric and gravitational fields and illustrate how field lines can be used to represent the direction and strength of a field. The orientation of the field at any given point is always tangential to the field line passing through that point, and the strength of the field is represented by the density of the field lines. The concept of fields and field lines was originally developed by British scientist and engineer Michael Faraday (1791–1867) to represent a magnetic field. In 1845, Faraday used **magnetic field lines** (which he called "lines of force") to describe the space surrounding a permanent magnet.

The needle of the compass is itself a small magnet, and it deflects differently depending on its location relative to the magnet that generates the magnetic field. Magnetic field lines are always drawn so that at any point, the needle of the compass points along the tangent to the magnetic field line. Therefore, you can map the **magnetic field** created by a small permanent magnet by moving a compass around it and tracing the direction of the compass needle (Figure 24-2). Note that, like gravitational and electric field lines, magnetic field lines never cross. This happens because there can only be exactly one value and direction of the magnetic field at every location in space. If two magnetic field lines crossed each other at a point, it would mean that there were two distinctly different values of the magnetic field at this point. We can conduct an experiment and see that this is not the case. The same applies to gravitational and electric fields. However, unlike gravitational and electric field lines, which begin and end on gravitational masses and electric charges (also referred to as monopoles), respectively, magnetic field lines do not originate or terminate at specific points (monopoles)—they form closed loops. Magnetic field lines generated by a permanent bar magnet are represented by closed loops, with the field direction from the north pole of the magnet toward the south pole in the vicinity of the magnet. Inside the magnet, the direction of the magnetic field lines is from the south pole to the north pole of the magnet.

The end of the compass needle pointing toward Earth's magnetic north pole is called a north-seeking

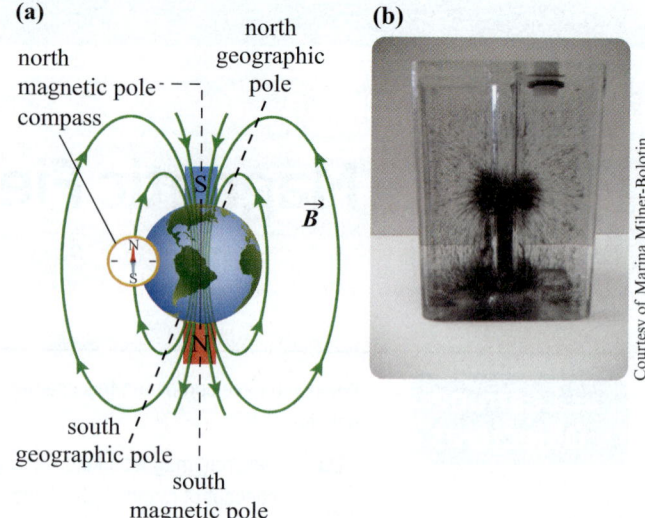

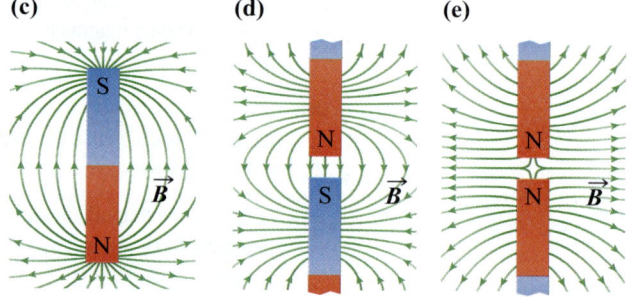

Figure 24-2 (a) Earth's magnetic field, $\vec{B}$, can be visualized as the magnetic field of a giant bar magnet. (b) Magnetic field lines created by iron filings suspended in oil. (c), (d), and (e) Magnetic field patterns for various magnet configurations.

pole, or simply a north pole. Similarly, the end of the compass needle pointing toward Earth's magnetic south pole is called a south-seeking pole, or a south pole (Figure 24-2(a)). Earth's magnetic field is likely caused by electric currents flowing in its molten outer core. This magnetic field is similar to the field that would be produced if a giant bar magnet were located inside Earth's core. Earth's magnetic north pole would be located near the south pole of this imaginary bar magnet, and Earth's magnetic south pole would be near the magnet's north pole. It is important to mention that Earth's magnetic and geographical poles do not coincide. In fact, the magnetic poles are gradually moving relative to the geographic poles.

Another way of displaying the magnetic field patterns is to use iron filings. In a magnetic field, suspended iron filings behave as tiny magnets and orient themselves in the direction of the magnetic field lines. To visualize a three-dimensional magnetic field, we can use iron filings suspended in oil (Figure 24-2(b)). Figure 24-2(c)–(e) shows magnetic field lines created by different magnet configurations.

Electric and Magnetic Fields and Forces Acting on Charged Particles

Like electric fields, magnetic fields are vector physical quantities that have direction, magnitude, and units and are represented by vectors. The magnetic field is denoted as $\vec{B}$, also referred to as the **magnetic flux density**. Recall that we defined the direction and magnitude of an electric field, $\vec{E}$, in terms of the electrostatic force that it exerts on a test electric charge. The electric field and the electrostatic force are always collinear, and $\vec{F}_E = q\vec{E}$, where $\vec{F}_E$ is the electrostatic force, q is the electric charge, and $\vec{E}$ is the electric field. This means that the electrostatic force is directed parallel to the electric field when the particle is positively charged, and the electrostatic force is directed anti-parallel to the electric field when the particle is negatively charged. The electrostatic force, $\vec{F}_E$, acting on a charged particle depends only on the particle charge and on the strength and direction of the electric field. However, unlike the electrostatic force, the **magnetic force**, $\vec{F}_B$, acting on a charged particle depends not only on the particle's charge and the strength and direction of the magnetic field, but also on the velocity of the particle and its orientation relative to the magnetic field. Moreover, the magnetic force is *not* collinear (parallel or antiparallel) with the magnetic field. The forces exerted on charged particles by electric and magnetic fields differ significantly:

- The magnitude, F_B, of the magnetic force, $\vec{F}_B$, on a charged particle is proportional to the particle's speed. A faster-moving particle experiences a greater magnetic force, and a particle at rest does not experience any magnetic force.

- Unlike the electrostatic force, the magnetic force depends on the direction of the motion of the charged particle relative to the direction of the magnetic field. For a given speed, a charged particle moving perpendicular to the magnetic field lines experiences the greatest magnetic force, and a charged particle moving parallel to the magnetic field lines experiences no magnetic force. When the particle moves at an angle θ to the magnetic field lines, the force exerted by the magnetic field is proportional to $\sin\theta$ (Figure 24-3).

- The magnetic force is always perpendicular to the particle's velocity, $\vec{v}$, and to the magnetic field, $\vec{B}$: $\vec{F}_B \perp \vec{v}$ and $\vec{F}_B \perp \vec{B}$. In Figure 24-3, it is directed in or out of the page. Consequently, the magnetic force does zero work on the particle:

$$dW = (\vec{F}_B \cdot d\vec{r}) = (\vec{F}_B \cdot \vec{v}dt) = (\vec{F}_B \cdot \vec{v})dt = 0$$

Thus, the magnetic force cannot change the particle's kinetic energy, just as the gravitational force acting on an object moving along a circular orbit around a planet does not change the object's

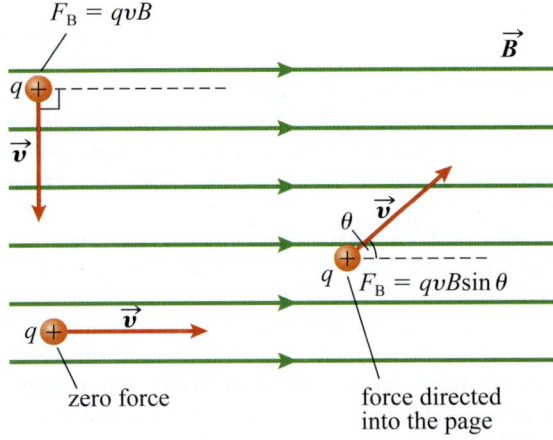

kinetic energy (see Chapter 11). However, the magnetic force can affect the *direction* of the particle's motion.

Can you think of any other case in which a force applied on an object does not change its speed, but only affects the direction of its motion? (See Section 4-3.)

Calculating the magnitude and direction of the magnetic force requires knowing not only the magnetic field, $\vec{B}$, and the particle's charge, q, but also its velocity, $\vec{v}$. Moreover, the three vectors $\vec{v}$, $\vec{B}$, and $\vec{F}_B$ are never all located in the same plane. Therefore, problems involving magnetic forces are always three-dimensional. Based on the above observations, the magnetic force acting on a charged particle can be described succinctly using the cross product discussed earlier (Section 2-5) and illustrated in Figure 24-4:

KEY EQUATION

$$\vec{F}_B = q\vec{v} \times \vec{B} \Rightarrow F_B = qvB\sin\theta \text{ and } \vec{F}_B \perp \vec{v}, \vec{B} \quad (24\text{-}1)$$

Figure 24-3 The magnetic force acting on a charged particle moving inside a uniform magnetic field at different angles relative to the magnetic field lines.

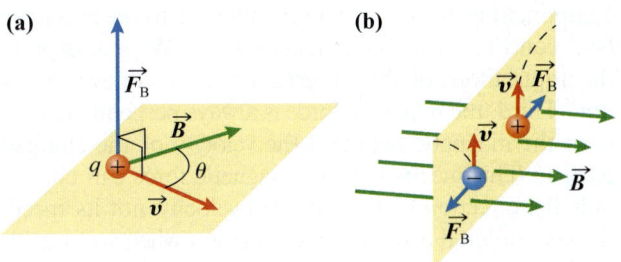

Figure 24-4 (a) The direction of the magnetic force, $\vec{F}_B$, on a positively charged particle q moving with velocity $\vec{v}$ in a magnetic field $\vec{B}$. (b) Magnetic forces on positively and negatively charged particles moving inside a uniform magnetic field $\vec{B}$. (We indicate magnetic field vectors with green arrows, magnetic forces with blue arrows, and velocities with red arrows.)

Equation 24-1 has three important implications. First, the magnetic forces acting on oppositely charged particles moving in the same direction in a uniform magnetic field are indeed opposite. Therefore, when computing the value of the magnetic force, remember that the order of factor multiplication in the cross product matters—the cross product is anticommutative (Section 2-5): $\vec{v} \times \vec{B} = -\vec{B} \times \vec{v}$. Second, since the vector product of two collinear vectors is always zero, the component of the particle's velocity parallel or antiparallel to the magnetic field has no impact on the magnetic force. Only the component of the velocity perpendicular to the magnetic field contributes to the magnetic force. We will explore the implications of this observation in the following section. Third, the magnetic force is always perpendicular to both the magnetic field and the velocity of the charged particle. This means that the magnetic force can change only the direction of the particle's velocity, not its speed. We encountered a similar phenomenon when we considered circular motion (Sections 4-3 and 5-9). For example, the gravitational force acting on a satellite moving in an almost circular orbit around Earth changes the direction of the satellite's motion, but not its speed.

When a charged particle is moving in crossed magnetic and electric fields, it experiences both electrostatic and magnetic forces:

KEY EQUATION
$$\vec{F} = \vec{F}_E + \vec{F}_B = q\vec{E} + q\vec{v} \times \vec{B} \qquad (24\text{-}2)$$

The expression for the combined force is called the **Lorentz force**, in honour of Dutch physicist Hendrik Lorentz (1853–1928), who formulated it in 1892. However, Equation 24-2, although not in the present form, appeared 31 years earlier in the works of Scottish mathematician and theoretical physicist James Clerk Maxwell (1831–1879).

The SI unit of magnetic field is the **tesla** (T), named after Serbian-American inventor, physicist, and engineer Nikola Tesla (1856–1943). A particle carrying a charge of 1 C moving with a speed of 1 m/s perpendicular to a magnetic field of 1 T experiences a magnetic force of 1 N:

$$1\,\text{N} = 1\,\text{C} \cdot 1\,\frac{\text{m}}{\text{s}} \cdot 1\,\text{T} \quad \text{or} \quad 1\,\text{T} = 1\,\frac{\text{N} \cdot \text{s}}{\text{C} \cdot \text{m}}$$

The tesla can also be expressed as follows:

$$1\,\text{T} = 1\,\frac{\text{N} \cdot \text{s}}{\text{C} \cdot \text{m}} = 1\,\frac{\text{N}}{\text{C/s} \cdot \text{m}} = 1\,\frac{\text{N}}{\text{A} \cdot \text{m}} \qquad (24\text{-}3)$$

As we will discover shortly, a magnetic field of 1 T is quite strong. The magnetic fields that we typically encounter have strengths of less than 1 mT.

Another commonly used unit for magnetic fields is the gauss (G), named after German mathematician and physical scientist Carl Friedrich Gauss (1777–1855): 10^4 G = 1 T or 1 G = 10^{-4} T.

CHECKPOINT 24-1

Ranking the Magnitudes of Magnetic Forces

Four charged particles, $q_1 = q$, $q_2 = -q$, $q_3 = 2q$, and $q_4 = -2q$, moving with velocities $\vec{v}_1 = 3v\hat{i}$, $\vec{v}_2 = 3v\hat{j}$, $\vec{v}_3 = -2v\hat{i}$, and $\vec{v}_4 = 3v\hat{j}$, enter a uniform magnetic field $\vec{B} = -2B_0\hat{i}$. Which of the following statements correctly ranks the magnitudes of the magnetic forces acting on these particles?

(a) $F_{B_1} < F_{B_2} < F_{B_3} < F_{B_4}$

(b) $F_{B_1} > F_{B_2} > F_{B_3} > F_{B_4}$

(c) $F_{B_1} = F_{B_2} > F_{B_3} = F_{B_4}$

(d) $F_{B_1} > F_{B_2} = F_{B_3} > F_{B_4}$

(e) $F_{B_1} = F_{B_3} < F_{B_2} < F_{B_4}$

ANSWER: (e) The magnetic force can be calculated using Equation 24-1 and the rules for cross product calculations described in Section 2-5.

Right-Hand Rules for Finding the Direction of the Magnetic Force

Below we will learn how we can determine the direction of the magnetic force acting on a moving charged particle in the case where the magnetic field and the velocity of the charged particle are perpendicular to one another: $\vec{v} \perp \vec{B}$. In the following section, we will extend our discussion to the more general case.

As we discussed in Chapter 2, right-hand rules are useful tools for finding the direction of the cross product of two given vectors. Since the magnetic force acting on a charged particle moving in a magnetic field can be expressed as the cross product of the particle's velocity and the magnetic field (Equation 24-1), we can determine the direction of this force using the right-hand rules discussed in Section 2-5. Below we describe three commonly used versions of the **right-hand rule** applied to finding the direction of the magnetic force acting on a charged particle moving inside a magnetic field (Figure 24-6(a)–(c)). These are three alternatives among others (see Chapters 2 and 8), and you should use the one that is more appealing to you, but it is good to be conversant in more than one version.

Right-hand rule—version 1 (curled fingers): Point your four right-hand fingers in the direction of motion of a positively charged particle, and curl them 90° in the direction of the magnetic field, as shown in Figure 24-6(a). Your stretched-out thumb points in the direction of the magnetic force acting on the positively charged particle (Equation 24-1). If the particle is negatively charged, the direction of the force is opposite.

Right-hand rule—version 2 (extended palm): Position your right hand so that your fingers point in the direction of the magnetic field and your thumb points in the direction of the velocity of a positively charged particle (Figure 24-6(b)). Then, when you curl your fingers 90°, they point in the direction of the magnetic force acting on the positively charged particle (Figure 24-6(b)). When the particle is negatively charged, the direction of the force is opposite.

Right-hand rule—version 3 (three fingers): Extend your thumb, your index finger, and your middle finger along three mutually perpendicular directions, (think of the x-, y- and z-axes of a Cartesian coordinate system) (Figure 24-6(c)). If you line up your thumb with the velocity vector of the particle and your index finger with the magnetic field, then your middle finger will give the direction of the magnetic force acting on this positively charged particle. The magnetic force points in this direction if the charge is positive and in the opposite direction if the charge is negative.

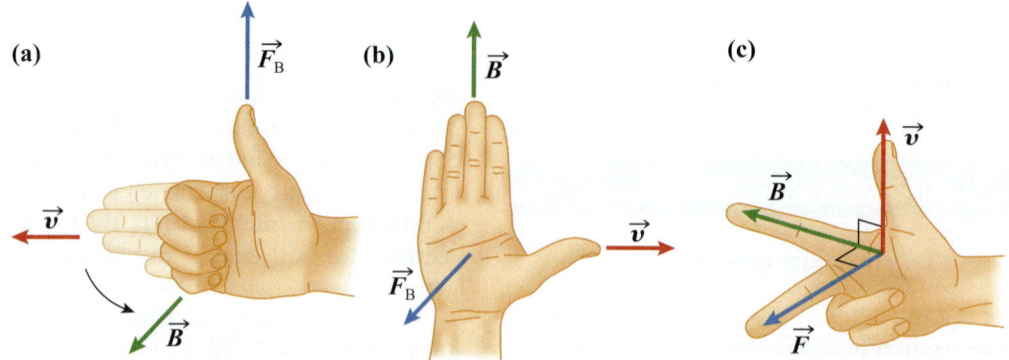

Figure 24-6 Three versions of a right-hand rule for determining the direction of a magnetic force acting on a positively charged particle moving inside a magnetic field. (a) Right-hand rule—version 1. (b) Right-hand rule—version 2. (c) Right-hand rule—version 3.

Check for yourself that the three versions of the right-hand rule described above are equivalent. Remember that the direction of the cross product in Equation 24-1 always represents the direction of the magnetic force on a positively charged particle. For a negatively charged particle, the direction of the magnetic force is opposite to the direction indicated by the right-hand rule.

Since problems involving the calculation of magnetic forces involve three dimensions, it is useful to have conventions to represent three-dimensional situations as two-dimensional diagrams. We represent a vector directed away from us into the page as a cross (like the end view of the tail of an arrow). Similarly, we represent a vector coming out of the page toward us as a dot (the front view of the tip of an arrow) (see Figure 24-7).

A **uniform magnetic field** is a magnetic field that does not change in space (Figure 24-7(a) and (b)), and a **constant magnetic field** is a magnetic field that does not change in time. A uniform field is not necessarily constant, nor is a constant field necessarily uniform.

(a) $\vec{B}$ out of the page:

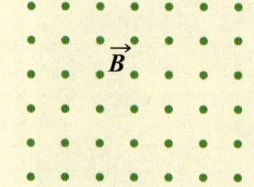

(c) $\vec{B}$ out of the page:

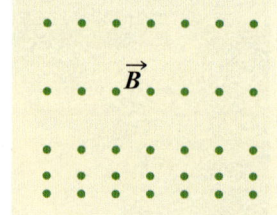

(b) $\vec{B}$ into the page:

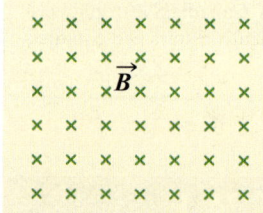

(d) $\vec{B}$ into the page:

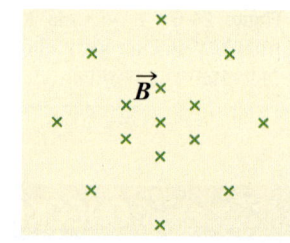

Figure 24-7 Representations of magnetic fields perpendicular to the page. (a) A uniform magnetic field directed out of the page. (b) A uniform magnetic field directed into the page. (c) A non-uniform magnetic field directed out of the page. (d) A non-uniform magnetic field directed into the page.

EXAMPLE 24-1

The Right-Hand Rule and the Cross Product

A positively charged particle, q, enters a region with a uniform magnetic field, $\vec{B} = B_0\hat{i}$. The particle's initial velocity is $\vec{v} = 3v_0\hat{j}$ (Figure 24-8). Determine the magnetic force acting on the particle just as it enters the magnetic field.

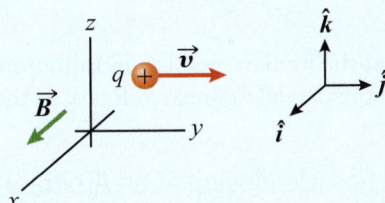

Figure 24-8 A positively charged particle moves inside a uniform magnetic field.

Solution

Method 1: Using the Right-Hand Rule

Applying the right-hand rule, we find that the magnetic force points in the negative z-direction. Its magnitude is given by

$$F_B = qvB\sin\theta = q(3v_0)B_0\sin 90° = 3qv_0B_0$$

In vector form, the magnetic force can be represented as $\vec{F}_B = -3qv_0B_0\hat{k}$.

Method 2: Using the Cross Product

$$\vec{F}_B = q\vec{v} \times \vec{B} = q(3v_0)\hat{j} \times B_0\hat{i} = 3qv_0B_0\begin{vmatrix} \hat{i} & \hat{j} & \hat{k} \\ 0 & 1 & 0 \\ 1 & 0 & 0 \end{vmatrix}$$

$$= 3qv_0B_0[\hat{i}(1\cdot 0 - 0\cdot 0) - \hat{j}(0\cdot 0 - 1\cdot 0) + \hat{k}(0\cdot 0 - 1\cdot 1)]$$

$$= -3qv_0B_0\hat{k}$$

Making sense of the result

As expected, both solutions produce the same result. In this case, Method 1 appears to be more straightforward; however, in general, an algebraic approach is more versatile.

CHECKPOINT 24-2

Determining the Direction of the Magnetic Force

An electron is moving at speed v in the positive x-direction. The magnetic field, $\vec{B}$, points in the negative x-direction. What is the magnetic force experienced by the electron?

(a) $F_B = qvB$ pointing in the positive y-direction
(b) $F_B = qvB$ pointing in the negative y-direction
(c) $F_B = qvB$ pointing in the positive z-direction
(d) $F_B = qvB$ pointing in the negative z-direction
(e) $F_B = 0$

ANSWER: (e) A magnetic field that is parallel or antiparallel to the charged particle's velocity exerts no force on the particle (Equation 24-1).

So far we have learned how to find the magnetic force acting on a charged particle when the velocity of the particle is perpendicular to the magnetic field. In the next section, we will consider the more general case of when the particle is moving at an arbitrary angle to the magnetic field. We will also explore the trajectory of a charged particle moving inside a magnetic field.

24-2 The Motion of a Charged Particle in a Uniform Magnetic Field

As noted earlier, the magnetic force acting on a charged particle can change the *direction* of its motion but not its speed. In this section, we use Newton's laws and the description of the magnetic force (Equation 24-1) to investigate the trajectory of a charged particle moving in a uniform magnetic field.

A Charged Particle Moving Perpendicular to a Uniform Magnetic Field

Consider an electron moving with velocity $\vec{v}$ directed to the right entering a uniform magnetic field, $\vec{B}$, directed into the page *perpendicular* to the electron's velocity (Figure 24-9). Applying the right-hand rule, we find that the direction of the magnetic force acting on the electron at the moment when it enters the magnetic field region is downward (Equation 24-1). Since the magnetic force is always perpendicular to the electron's velocity, the electron will start moving along a circular path with radius r (Section 4-3), accelerating at a rate of $a_r = v^2/r$. As we discussed in Section 4-3, the direction of this acceleration is toward the centre of the circle (centripetal acceleration). Applying Newton's second law to this circular motion and considering the x-axis to be directed toward the centre of the circle along its radius, we get

$$\sum \vec{F} = m\vec{a}$$

$$qvB\sin\theta = ma_R$$

$$qvB\sin 90° = m\frac{v^2}{r}$$

$$qvB = m\frac{v^2}{r}$$

Therefore, the radius of the electron's trajectory can be expressed as

$$r = \frac{mv}{qB} = \frac{p}{qB} \qquad (24-4)$$

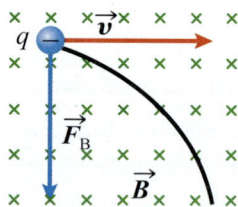

Figure 24-9 An electron moving in a uniform magnetic field directed perpendicular to its velocity has a circular trajectory.

While we derived Equation 24-4 for the case of an electron's motion, it is applicable to describing the trajectory of any charged particle moving inside a uniform magnetic field directed perpendicular to its velocity. This equation indicates that the radius of the particle's trajectory is proportional to the magnitude, p, of its momentum, $\vec{p}$, and is inversely proportional to its charge and the magnitude of the magnetic field.

We now consider a stream of identical particles all moving in the same direction but with different speeds. Let us consider the case where these particles enter a uniform magnetic field, $\vec{B}$, perpendicular to the direction of their motion. For these identical particles, the mass-to-charge ratio is the same, and since the magnetic field is uniform, $\dfrac{m}{qB} = $ const. Therefore, we can rewrite Equation 24-4 as

$$r = \frac{mv}{qB} = \frac{m}{qB} \cdot v$$

Since the particles have different speeds, they move along circles of different radii.

For any circular trajectory of a charged particle, we can define the period, T, as the time required for a particle to complete one revolution:

$$T = \frac{2\pi r}{v} \tag{24-5}$$

Combining Equations 24-4 and 24-5 yields

$$T = \frac{2\pi r}{v} = \frac{2\pi mv}{vqB} = 2\pi \frac{m}{qB} \tag{24-6}$$

We note that the period, T, depends only on the ratio $\dfrac{m}{qB}$ and is not affected by the particle's speed. Since the relationship between the magnitudes of the angular and linear velocity is $v = \omega r$,

$$\omega = \frac{v}{r} = \frac{qBR}{mr} = \frac{qB}{m}$$

Thus, the angular velocity, ω, of a charged particle moving in a uniform magnetic field $\vec{B}$ directed perpendicular to its velocity depends only on the particle's charge and mass and the magnitude of the magnetic field. The angular velocity of the particle, ω, divided by 2π has units of 1/s or Hz, $[\omega] = \dfrac{1}{s} \equiv$ Hz, and is

called the **cyclotron frequency**, f_{cyc}. A **cyclotron** is a type of particle accelerator that accelerates charged particles moving along circular trajectories (see Section 24-3). We can connect the cyclotron frequency and the angular cyclotron frequency using Equation 13-5: $\omega_{cyc} = 2\pi f_{cyc}$:

$$f_{cyc} = \frac{\omega}{2\pi} = \frac{qB}{2\pi m} \tag{24-7}$$

We note that the final expression in Equation 24-7 does not include the particle's linear velocity or the radius of its trajectory.

A Charged Particle Moving at an Arbitrary Angle to a Magnetic Field

Let us consider what happens when a charged particle enters a uniform magnetic field at an arbitrary angle θ to the field lines. We can resolve the particle's velocity into two components perpendicular ($\vec{v}_\perp$) and parallel ($\vec{v}_\parallel$) to the magnetic field (Figure 24-10):

$$\vec{v} = \vec{v}_\parallel + \vec{v}_\perp \qquad v = \sqrt{v_\parallel^2 + v_\perp^2} \tag{24-8}$$

where $v_\parallel = v\cos\theta$ and $v_\perp = v\sin\theta$.

As we discussed above, the magnetic force acting on the particle due to the perpendicular component of the particle's velocity causes the particle to move along a circular path with a radius given by Equation 24-4. The velocity component parallel to the magnetic field lines is not affected by the magnetic field. The parallel velocity component remains constant along the particle motion. In a similar fashion to our discussion of torque in Chapter 8, the magnetic force can be expressed as

$$\vec{F}_B = q\vec{v} \times \vec{B} = q(\vec{v}_\parallel + \vec{v}_\perp) \times \vec{B}$$
$$= q\vec{v}_\parallel \times \vec{B} + q\vec{v}_\perp \times \vec{B} = q\vec{v}_\perp \times \vec{B} \tag{24-9}$$

Had the particle's total velocity vector been perpendicular to the field, the magnetic force would cause the particle to move along a circular path with a radius given by Equation 24-4. However, the particle has a velocity component that is parallel to the magnetic field lines and is not affected by the magnetic field. Hence, the charged particle follows a helical path, a combination of circular motion in a plane perpendicular to the magnetic field and uniform linear motion parallel to it:

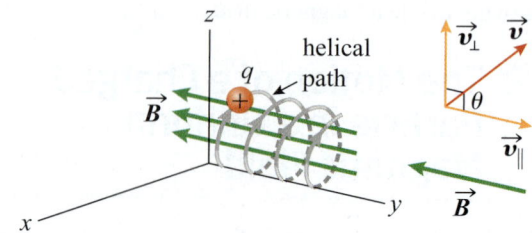

Figure 24-10 A positively charged particle moving at angle θ to a uniform magnetic field follows a helical trajectory.

Auroras

Spectacular auroras occur in regions 10° to 20° of latitude from the magnetic poles: the aurora borealis (Northern Lights) in the northern hemisphere (Figure 24-11(a)) and the aurora australis in the southern hemisphere. On rare occasions, auroras can be visible as far south as New Brunswick and even Vancouver, British Columbia. Auroras happen when fast-moving electrons and protons enter the magnetosphere, which is the region of strong magnetic fields with a boundary at approximately 15 to 25 Earth radii above Earth's surface. These charged particles follow magnetic field lines moving along helical paths (Figure 24-11(b)), as described above. When they collide with oxygen and nitrogen atoms at altitudes of 30 km to 300 km above Earth's surface, they transfer part of their energy to atmospheric atoms and molecules, which then emit visible light. The colour of the aurora depends on the atoms that were struck. Auroras occur on other planets as well. For example, auroras occur on Saturn, as was proven by the ultraviolet images of Saturn's southern polar region taken by the Hubble Space Telescope.

(a)

Pi-Lens/Shutterstock.com

(b) charged particle approaching Earth

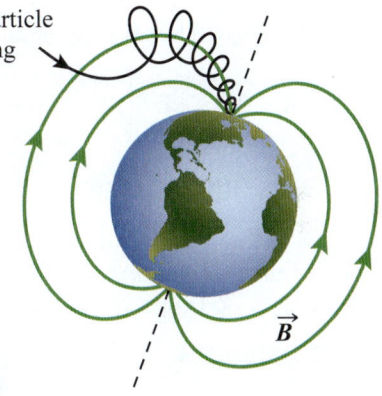

$\vec{B}$

(c)

NASA/Alamy Stock Photo

Figure 24-11   (a) The aurora borealis in northern Canada. (b) Schematic representation of auroras. (c) Saturn aurora.

In the plane perpendicular to the magnetic field—circular motion:

$$r = \frac{mv\sin\theta}{qB}$$

In the direction parallel to the magnetic field—motion with constant velocity:

$$\vec{v}_\parallel = \text{const} \qquad (24\text{-}10)$$

The distance along the magnetic field travelled by a charged particle during one revolution (pitch):

$$d = v_\parallel t = v\cos\theta\frac{2\pi m}{qB}$$

$$= \frac{2\pi mv\cos\theta}{qB}$$

ELECTRICITY, MAGNETISM, AND OPTICS

EXAMPLE 24-2

A Charged Particle Moving in a Uniform Magnetic Field

A particle with a charge of 1.3 nC and a mass of 1.4×10^{-9} kg moves with velocity $\vec{v} = (1.5 \times 10^4 \text{ m/s})\hat{i} + (1.2 \times 10^4 \text{ m/s})\hat{j}$ as it enters a uniform magnetic field of $\vec{B} = (1.4 \text{ T})\hat{i}$.

(a) Determine the magnetic force acting on the particle as it enters the field.
(b) Describe the path of the particle, and determine its radius.
(c) Find the time it takes for the particle to complete one revolution.
(d) Find the distance the particle travels along the magnetic field lines as it completes one revolution.

Solution

(a) This part can be solved in a way similar to Example 24-1:

$$\vec{F}_B = q\vec{v} \times \vec{B} = (1.3 \times 10^{-9}) \begin{vmatrix} \hat{i} & \hat{j} & \hat{k} \\ 1.5 \times 10^4 \text{ N} & 1.2 \times 10^4 \text{ N} & 0 \\ 1.4 & 0 & 0 \end{vmatrix}$$

$$= (1.3 \times 10^{-9})(10^4 \text{ N})[\hat{i}(1.2 \cdot 0 - 0 \cdot 0) - \hat{j}(1.5 \cdot 0 - 0 \cdot 1.4)$$
$$+ \hat{k}(1.5 \cdot 0 - 1.4 \cdot 1.2)]$$

$$= (1.3 \times 10^{-5} \text{ N})(-1.68\hat{k})$$

$$= (-2.184 \times 10^{-5} \text{ N})\hat{k}$$

$$= (-2.2 \times 10^{-5} \text{ N})\hat{k}$$

(b) The particle moves along a helical path because the x-component of the particle's velocity is parallel to the magnetic field and is therefore unaffected by the field. So, the particle moves along a helical trajectory whose projection onto the yz-plane is a circle, while in the x-direction the velocity of the particle is constant. The radius of the helix can be found using Equation 24-4:

$$r = \frac{mv}{qB} = \frac{(1.4 \times 10^{-9} \text{ kg})(1.2 \times 10^4 \text{ m/s})}{(1.3 \times 10^{-9} \text{C})(1.4 \text{ T})} = 9.2 \times 10^3 \text{ m}$$

(c) We can use Equation 24-6 to find the period of the particle's motion:

$$T = \frac{2\pi m}{qB} = \frac{2\pi(1.4 \times 10^{-9} \text{ kg})}{(1.3 \times 10^{-9} \text{C})(1.4 \text{ T})} = 4.8 \text{ s}$$

(d) It will take the particle 4.8 s to complete one revolution. During this time, the particle will move a distance d along the x-axis. This distance is called the **pitch of the helix**:

$$d = Tv_x = (4.8 \text{ s})(1.5 \times 10^4 \text{ m/s}) = 7.2 \times 10^4 \text{ m}$$

CHECKPOINT 24-3

Location of Auroras

Auroras do not occur near the equator because
(a) charged particles never reach the equator
(b) the magnetic field lines are strongest near the equator
(c) the magnetic field lines near the equator are directed toward Earth
(d) there is no magnetosphere near the equator
(e) none of the above

24-3 Applications: Charged Particles Moving in a Uniform Magnetic Field

The behaviour of charged particles in a magnetic field has a number of applications in medicine, industry, and scientific research. In this section, we see how the physics of this behaviour is used in several important devices.

Velocity Selector

Charged particles produced in a lab or arriving from the Sun often have a range of velocities. A **velocity selector** uses perpendicular electric and magnetic fields to filter a beam of charged particles, leaving only those with a particular desired velocity.

The velocity selector in Figure 24-12 has a chamber with crossed uniform electric and magnetic

(a)

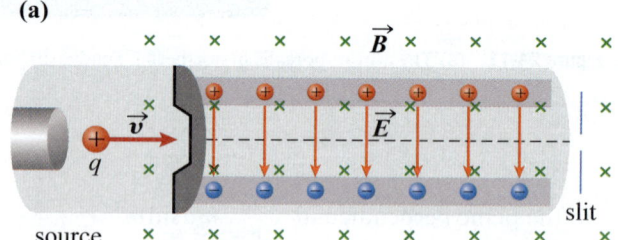

(b)

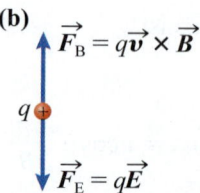

Figure 24-12 (a) The operation of a velocity selector. (b) An FBD for a positively charged particle moving inside a velocity selector.

fields. When a positively charged particle enters this chamber, the electric force acting on the particle is directed along the electric field lines (downward). However, the magnetic force is directed upward, perpendicular to the magnetic field lines. Particles with a velocity such that the magnetic force balances the electric force pass through the chamber without changing their velocity, as indicated by the dashed line. All other charged particles are deflected either upward or downward, depending on which force is stronger for each particle. The magnetic force can be opposite to the electric force for only the particles entering the region with velocity perpendicular to both $\vec{B}$ and $\vec{E}$. When the electric and magnetic forces balance,

$$F_{E} = F_{B}$$

$$qE = qvB \qquad (24\text{-}11)$$

$$E = vB$$

and

$$v = \frac{E}{B} \qquad (24\text{-}12)$$

Thus, only the particles whose velocity satisfies Equation 24-12 pass straight through the velocity selector.

CHECKPOINT 24-4

Operation of the Velocity Selector

Which of the following statements represents the operation of a velocity selector?

(a) The polarity of the electric and magnetic fields in the velocity selector depends on the charge of the selected particles.

(b) The ratio $\frac{E}{B}$ of the velocity selector depends on the mass of the charged particles to be selected.

(c) If in the velocity selector $E \neq vB$, the particle moves in a circular trajectory.

(d) If a positive particle is moving faster than the speed chosen in the velocity selector, the particle is deflected in the direction of the electric field.

(e) If a negative particle is moving faster than the speed chosen in the velocity selector, the particle is deflected in the direction of the electric field.

ANSWER: (e) The speed of the particle has no effect on the electric force acting on it, but it has an effect on the magnetic force. Since the particle in the velocity selector is negatively charged, the magnetic force acting on it will be larger than the electric force. The magnetic force in that case is directed in the direction of the electric field.

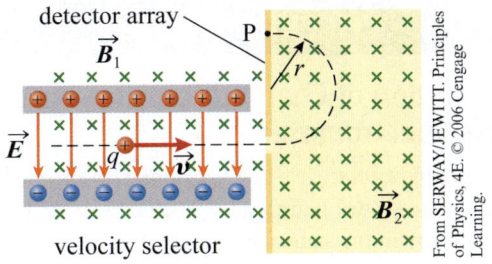

Figure 24-13   The operation of a mass spectrometer.

From SERWAY/JEWITT. Principles of Physics, 4E. © 2006 Cengage Learning.

Mass Spectrometers

The charged particles that pass through a velocity selector all have the same velocity, but they do not necessarily have the same masses or charges. A **mass spectrometer** sorts particles according to their mass-to-charge ratios, making it a powerful tool for chemical analysis.

As shown in Figure 24-13, a mass spectrometer has a velocity selector connected to a chamber with a uniform magnetic field. Charged particles leaving a velocity selector have a known speed. When they enter the second chamber, the uniform magnetic field causes them to move along semicircular trajectories with radii that depend on their mass and charge. Combining Equations 24-4 and 24-12, we obtain

$$r = \frac{mv}{qB_2} = \frac{mE}{qB_1B_2} \qquad (24\text{-}13)$$

where B_1 represents the magnetic field inside velocity selector chamber 1, and B_2 represents the magnetic field inside chamber 2.

Thus, by measuring the radius of a particle's trajectory, we can determine its mass-to-charge ratio:

$$\frac{m}{q} = \frac{rB_1B_2}{E} \qquad (24\text{-}14)$$

If the magnetic fields in the two chambers are equal, which is not always the case, we have $B_1 = B_2 = B$, and

$$\frac{m}{q} = \frac{rB^2}{E} \qquad (24\text{-}15)$$

Some mass spectrometers might not have a separate velocity selector chamber, as shown in Figure 24-13. These devices (see Figure 24-13 and Figure 24-86 (in problem 97)) have the advantage of being lighter and lower in cost. We will consider this model of a mass spectrometer in problem 97.

Determining the Electron's Mass-to-Charge Ratio

Experiments by British scientist J.J. Thomson (1856–1940) culminated in 1897 with the discovery of the electron. Thomson measured the electron's mass-to-charge ratio using a cathode ray tube (CRT) like the one shown in Figure 24-14. Thomson's apparatus consisted of a vacuum tube containing an electrode that emits electrons (a cathode), a fluorescent screen, plates to apply an electric field, and coils to apply a magnetic field. Electrons emitted by the cathode passed through a slit into a region of crossed magnetic and electric fields; then they struck a fluorescent screen, causing it to glow at the point of impact. Thomson determined the radius of the path of the electrons from the position of the glowing point. Then he calculated the $\dfrac{m_e}{e}$ ratio for the electron:

$$\frac{m_e}{e} = \frac{rB}{v} \tag{24-16}$$

Although Thomson's equipment produced only an approximate value for the ratio $\dfrac{m_e}{e}$, it was nonetheless clear

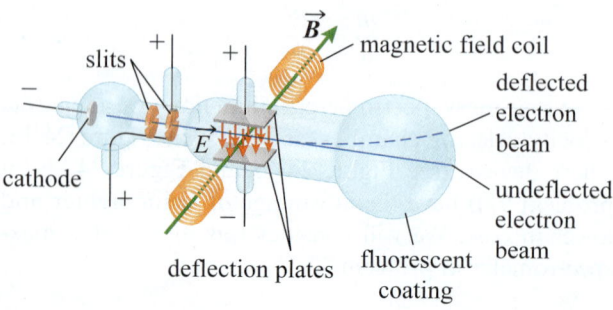

Figure 24-14 Thomson used a CRT, similar to the one shown here, to determine the mass-to-charge ratio for the electron.

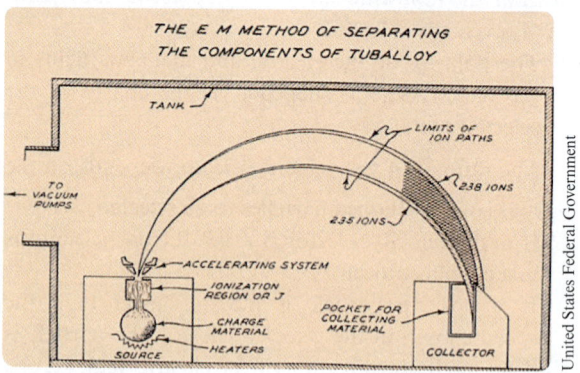
evidence that "cathode rays" are tiny negatively charged particles. Almost 15 years later, American physicist Robert A. Millikan (1868–1953) devised an ingenious method to extend the results of Thomson's experiments: he used electric fields to determine the mass and charge of the electron. Millikan's experiment is described in Chapter 31.

CHECKPOINT 24-6

The CRT

Which particle *cannot* be measured using the crossed magnetic and electric field method suggested by Thomson? Explain.

(a) electron
(b) proton
(c) α-particle
(d) neutron
(e) hydrogen ion

ANSWER: (d) Neutrons have no electric charge, so they will move along the CRT undeflected.

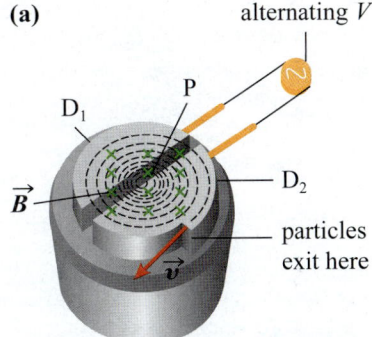

(a) alternating V

D_1 P

$\vec{B}$ D_2

$\vec{v}$ particles exit here

(b)

LAWRENCE BERKELEY LABORATORY/SCIENCE PHOTO LIBRARY

Figure 24-16 (a) A schematic representation of a cyclotron. (b) The first cyclotron, invented by E.O. Lawrence and M.S. Livingston in 1934.

Cyclotrons

Cyclotrons (Figure 24-16) use electric and magnetic fields to accelerate charged particles to very high kinetic energies. A uniform electric field acts on a charged particle that moves a distance d along the electric field lines and changes the particle's kinetic energy:

$$\Delta K = W$$

$$W = (\vec{F}_E \cdot \vec{d}) = qEd = q\Delta V \qquad (24\text{-}17)$$

$$\Delta K = qEd = q\Delta V$$

Increasing the distance d and strengthening the electric field $\vec{E}$ increases the gain in the particle's kinetic energy. However, building very long accelerators and using extremely strong electric fields is costly and technologically difficult. The cyclotron reduces some of these challenges by using crossed electric and magnetic fields to accelerate charged particles as they follow a spiral path (Figure 24-16(a)).

A cyclotron consists of two hollow semicircular chambers called "dees," labelled D_1 and D_2 in Figure 24-16(a). A high-frequency voltage is applied across the gap between the dees, and a uniform magnetic field $\vec{B}$ is directed perpendicular to the plane of the dees. An ion source feeds ions to the centre of the cyclotron (point P, Figure 24-16(a)). When an ion enters a region of the uniform magnetic field, it starts moving along a circular path, repeatedly crossing the gap between the dees. The alternating voltage source across the gap ensures that the ion experiences an electric field in the direction of its motion, thus increasing its kinetic energy. Example 24-3 discusses how the frequency of the alternating voltage across the gap is calculated. After exiting the gap between the dees, the ion is subjected to a uniform magnetic field only, perpendicular to its velocity. As a result, the ion continues to move along a circular path. However, since the speed of the ion is higher than it was before it entered the gap, the radius of its trajectory increases. When the ion enters the second gap between the dees, the potential difference between the dees reverses, so the electric field is once again

pointing in the direction of the ion's motion and the ion gains additional kinetic energy. After the ion exits the gap between the dees, it is moving even faster, causing an additional increase in the radius of its trajectory. Eventually, the trajectory radius increases enough that the ion travels along the edge of the dee and passes through the exit slit.

The exit kinetic energy of the ion is

$$K = \frac{mv^2}{2} = \frac{m}{2}\left(\frac{rqB}{m}\right)^2 = \frac{r^2q^2B^2}{2m} \qquad (24\text{-}18)$$

Cyclotrons can accelerate charged particles to speeds approaching the speed of light. For example, the giant TRIUMF cyclotron (www.triumf.ca), located in Vancouver, British Columbia, accelerates negatively charged hydrogen ions (called H⁻ ions) to 224 000 km/s, approximately 75% of the speed of light (Figure 24-17). These high-energy particles can be used to produce beams of neutrons and short-lived particles called mesons. TRIUMF is so efficient at creating pi-mesons that it is often called a meson factory. The high-energy particles are also used to treat several forms of cancer.

ELECTRICITY, MAGNETISM, AND OPTICS

Figure 24-17 The cyclotron at TRIUMF is the largest cyclotron in the world. It accelerates 10^{15} protons per second to speeds of 75% of the speed of light. TRIUMF's cyclotron accelerates H^- ions: it can extract multiple beams with varying energy levels at very high efficiency, making TRIUMF's accelerator one of the most versatile accelerators in the world.

MAKING CONNECTIONS

Cyclotron Use in Medicine

Canada is one of the world leaders in the production of medical isotopes. For example, Nordion, a company founded in Canada, produces radioisotopes for cancer treatment, such as iodine-123 and palladium-103. Palladium-103 is used to treat prostate cancer, and iodine-123 is primarily used in the diagnosis of thyroid cancer and neurological disorders (such as Alzheimer's and Parkinson's disease). Nordion supplies over two thirds of the world's reactor-produced isotopes. It is estimated that nearly 20 million nuclear medicine procedures are conducted worldwide each year. Nordion currently operates two commercial cyclotrons at the TRIUMF site located on the University of British Columbia campus, and a third in Fleurus, Belgium.

EXAMPLE 24-3

The Cyclotron Frequency of the TRIUMF Cyclotron

The magnetic field used in the cyclotron at TRIUMF to accelerate H^- ions for medical research has a strength of 0.300 T.

(a) Calculate the cyclotron frequency of the TRIUMF cyclotron.
(b) Calculate the period of one revolution.
(c) How will increasing the magnetic field by a factor of approximately 1.5, to 0.460 T, affect the cyclotron frequency?

Solution

(a) By Equation 24-7, the cyclotron frequency is defined as the inverse of the time it takes a charged particle to complete one revolution inside a cyclotron (a circular trajectory of radius R). In the case of the TRIUMF cyclotron, the charged particles are H^- ions. Therefore, their charge is -1.6×10^{-19} C, and their mass is approximately equal to the mass of one proton (the mass of two electrons is negligible):

$$m_p \sim 1840 m_e \rightarrow m_{H^-} = m_p + 2m_e \sim 1.67 \times 10^{-27} \text{ kg}$$

Therefore, the cyclotron frequency, f_{cyc}, is

$$f_{cyc} = \frac{qB}{2\pi m} = \frac{|-1.602 \times 10^{-19}\text{C}|(0.300 \text{ T})}{2\pi(1.67 \times 10^{-27} \text{ kg})} = 4.58 \times 10^6 \text{s}^{-1}$$

$$= 4.58 \times 10^6 \text{ Hz}$$

The cyclotron frequency is always positive, so we use the absolute value of the charge in calculations.

(b) The period of one revolution is the inverse of the cyclotron frequency:

$$T = \frac{1}{f_{cyc}} = (4.58 \times 10^6 \text{ s}^{-1})^{-1} = 2.18 \times 10^{-7} \text{ s}$$

(c) As discussed in part (a), the cyclotron frequency is proportional to the magnetic field and to the charge-to-mass ratio of the accelerated particles (Equation 24-7). Therefore, if the magnetic field increased by a factor of 1.5, the cyclotron frequency would increase in the same proportion.

Making sense of the result

The cyclotron frequency does not depend on the radius of the particle's motion. Thus, the oscillating voltage between the dees can have a constant frequency, equal to the cyclotron frequency.

The Hall Effect

Early studies of electricity could not determine if the charge carried in electric circuits was positive charge, negative charge, or maybe both. Magnetic fields can be used to answer this question by determining the sign of charge carriers. Consider a wide conducting strip carrying electric current placed perpendicular to and inside a uniform magnetic field, $\vec{B}$ (Figure 24-18).

As discussed in Section 23-1, charge carriers can be considered to move with an average drift speed of v_d. Let us assume that an electric current flows clockwise, which can be represented by negative charges moving counter-clockwise (Figure 24-18(b)) or by positive charges moving clockwise (Figure 24-18(a)). Let us assume first that the charge carriers in this case are positive ions. When positive charge carriers are moving clockwise, the magnetic field pointing perpendicular to the conducting strip (directed into the strip) exerts an upward magnetic force on the positive charges, so they move toward the top of the conducting plate. The drift of positive charge carriers leaves an excess of negative charges behind, and the bottom of the conducting plate becomes negatively charged. This creates an electric potential difference across the conductor perpendicular to the direction of the current. The top of the conductor will thus have a *higher* electric potential than the bottom. However, if the charge carriers in this case are electrons (negative particles), the direction of their motion will be opposite to the direction of the electric current in the wire (Figure 24-18(b)). When negative particles are moving counter-clockwise, the same magnetic field exerts an upward magnetic force on them, shifting them toward the top of the conductor, leaving an excess of positive charges behind. Then the top of the conductor becomes negatively charged and has a *lower* potential than the bottom (Figure 24-18(b)). Thus, depending on the sign of the charge carriers, an electric potential difference will be generated across the conductor in the direction perpendicular to the direction of the electric current and of the magnetic field. This effect, called the **Hall effect**, is named after Edwin Hall (1855–1938), who discovered it in 1879.

A potential difference across a conductor placed in a magnetic field is called the **Hall potential difference**. A metal conductor has only one type of charge carrier—either positive or negative. Therefore, the Hall potential difference can be calculated using the fact that in a steady state, the magnetic force acting on moving charge carriers generates an electric field across the conductor. This electric field produces an electric force equal and opposite to the magnetic force acting on the charge carriers. In a metal conductor with cross-sectional area $w \times h$ (Figure 24-18) and with charge carriers moving with a drift velocity of $\vec{v}_d$ along the conductor,

$$F_B = qv_d B$$

$$F_E = qE_{Hall} = q\frac{\Delta V_{Hall}}{w}$$

Since $F_E = F_B$,

$$qv_d B = q\frac{\Delta V_{Hall}}{w}$$

$$v_d B = \frac{\Delta V_{Hall}}{w}$$

$$\Delta V_{Hall} = wv_d B$$

According to Equation 23-6,

$$v_d = \frac{J}{n_q q}$$

so

$$\Delta V_{Hall} = w\frac{J}{n_q q}B = w\frac{I/A}{n_q q}B = w\frac{I}{whn_q q}B = \frac{I}{hn_q q}B$$

KEY EQUATION $$\Delta V_{Hall} = wv_d B = \frac{IB}{hn_q q} \qquad (24\text{-}19)$$

where w is the width of the conductor, v_d is the drift speed, I is the electric current, B is the magnetic field, h is the dimension of the conductor parallel to the magnetic field, n_q is the density of the charge carriers (the number of charge carriers per cubic metre; Section 23-1), and q is the charge of a charge carrier (note that charge carriers might or might not be electrons). Observations of the Hall effect in metals allowed scientists to determine that charge carriers in metals are in most cases negatively charged (electrons), while in semiconductors charge carriers can be both negative (electrons) and positive (electron holes), as we will discuss in Chapter 33.

The Hall effect is used in magnetic field sensors, proximity switches, position and speed detectors, and current sensors. A Hall effect sensor is a transducer

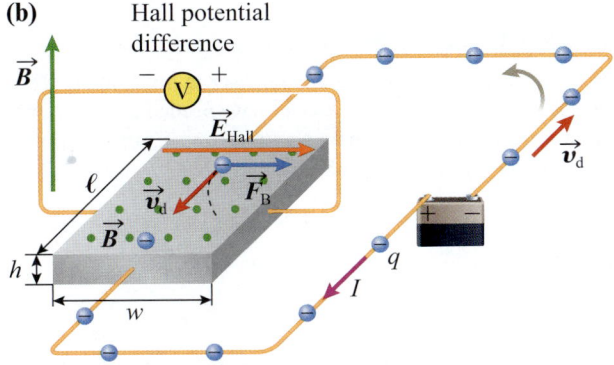

(a) Hall potential difference

(b) Hall potential difference

Figure 24-18 The Hall effect. A magnetic field perpendicular to the direction of an electric current creates a potential difference in a direction perpendicular to both the electric current and the magnetic field. (a) The Hall effect in the case of positively charged carriers. (b) The Hall effect in the case of negatively charged carriers. The Hall potential difference can be measured using a voltmeter.

that varies its output voltage in response to changes in a magnetic field. To determine the strength of a magnetic field, a Hall effect sensor measures voltage across a conductor in a direction perpendicular to the current flowing in it and then uses this voltage, the current, and the width of the conductor to calculate the magnetic field.

However, almost 50 years before Hall, British scientist Michael Faraday realized that moving electric charges in a uniform magnetic field create a potential difference in a direction perpendicular to the magnetic field and the charge flow. To investigate this phenomenon, in the 1830s, Faraday unsuccessfully attempted to measure the electric voltage between the sides of the Waterloo Bridge across the Thames River in London, England (see Example 24-4). Faraday expected that salt water flowing through Earth's magnetic field as the tide ebbed would produce a potential difference across the river (Figure 24-19).

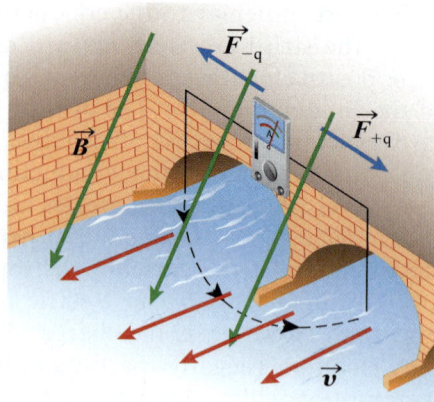

Figure 24-19 In 1831, Faraday tried to measure the potential difference across the Waterloo Bridge over the Thames River due to the motion of ions in Earth's magnetic field. This schematic representation of Faraday's original experiment shows the magnetic field, velocity, and magnetic forces acting on positive and negative ions.

EXAMPLE 24-4

The Thames River as a Natural Dynamo

Estimate the maximum potential difference across the Waterloo Bridge in London, England (Figure 24-19). The width of the Thames River at the bridge is approximately 265 m. The average flow rate of the Thames at London is approximately 65 m^3/s. Assume that the average depth of the river near the bridge is 2.5 m. London is close enough to the sea that the tides affect the flow of the river at Waterloo Bridge and the salinity of the water—the river here is brackish—it is saltier than fresh water.

Solution

Let us consider a charged particle moving in brackish Thames River waters, and assume that the vertical component of Earth's magnetic field in this area equals 45 μT. The magnitude of the magnetic force acting on such a particle is given by the magnetic component of the Lorentz force, Equation 24-1: $\vec{F}_B = q\vec{v} \times \vec{B}$.

Since the magnetic force is perpendicular to both the magnetic field and the direction of water flow, the force is parallel to the bridge. This magnetic force acts in opposite directions on positive and negative ions, creating a potential difference of $\Delta V = Ed$. In a state of equilibrium, the electrostatic force attracts oppositely charged ions and is equal to the magnetic force pulling them apart:

$$qE = qvB$$

Therefore,

$$\Delta V = vBd \qquad (1)$$

The average speed of the water flowing past the bridge equals the flow rate divided by the cross-sectional area of the river:

$$v \approx \frac{65\ m^3/s}{(2.5\ m)(265\ m)} \approx 0.10\ m/s$$

The flow rate is greatest when the tide is ebbing. As an approximation, we assume that the maximum speed of the water is approximately 0.30 m/s (three times the average value). Then the potential across the bridge, according to (1), is

$$\Delta V = vBd \approx \left(0.30\ \frac{m}{s}\right)(0.045\ mT)(265\ m)$$

$$= 0.0036\ V = 3.6\ mV$$

Notice that the ion concentration in the Thames does not affect the potential difference caused by the Hall effect.

Making sense of the result

The potential difference across the river is quite small. To detect this voltage, Faraday would have needed a voltmeter sensitive enough to measure a few millivolts. Since he didn't have one, he wasn't able to observe this effect.

EXAMPLE 24-5

Using the Hall Effect to Describe Electric Properties

A Hall probe containing a strip of copper 1.50 mm thick and 1.00 cm wide is placed in a uniform magnetic field of 1.50 T. The probe measures a Hall potential difference of 0.108 μV when the current through the probe is 10.0 A. Determine

(a) the drift speed of the electrons in the copper strip
(b) the strength of the electric field across the strip
(c) the density of the charge carriers in the strip

Solution

(a) Equation 24-19 can be used to find the drift speed of the electrons inside the probe:

$$v_d = \frac{\Delta V_{Hall}}{wB} = \frac{0.108 \times 10^{-6}\,V}{(1.50 \times 10^{-3}\,m)(1.50\,T)} = 4.80 \times 10^{-5}\,m/s$$

(b) The strength of the electric field inside the wire due to the Hall effect is

$$E = \frac{\Delta V_{Hall}}{w} = \frac{(0.108\,\mu V)}{(1.50 \times 10^{-3}\,m)} = 72.0\,\mu V/m$$

(c) Then the density of the charge carriers (electrons in this case) can be expressed as

$$n_q = \frac{I}{e v_d A}$$

$$= \frac{10.0\,A}{(1.60 \times 10^{-19}\,C)(4.80 \times 10^{-5}\,m/s)(1.50 \times 10^{-3}\,m)(1.00 \times 10^{-2}\,m)}$$

$$= 8.68 \times 10^{28}\,m^{-3}$$

Making sense of the result

The values of the drift speed (4.80×10^{-5} m/s) of the charge carriers in copper (electrons) and of the charge density in copper (8.68×10^{28} m^{-3}) agree with the experimentally measured values, where the former is discussed in Section 23-2 and the latter is listed in Table 23-1 in Chapter 23.

24-4 The Magnetic Force on a Current-Carrying Wire

As discussed earlier, a charged particle moving inside a magnetic field experiences a magnetic force. Therefore, a magnetic field exerts a force on each charged particle moving inside a current-carrying wire. The net magnetic force acting on the wire depends on the net motion of the electrons and, hence, on their drift velocity.

Figure 24-20 shows a series of experiments to study the effect of a magnetic field on a current-carrying wire placed between the poles of a permanent magnet. As expected, a wire carrying no current is unaffected by the magnetic field. Since free electrons inside the wire move randomly ($v_d = 0$), the effect of the magnetic field on the electrons averages to zero. However, when electric current flows in the wire, the effect of the magnetic field becomes pronounced. The direction of conventional current is defined in terms of the flow of positively charged carriers. Thus, to find the direction of the magnetic force on the wire, we use the same right-hand rule that we used to find the direction of a magnetic force acting on a positively charged carrier. Note that a magnetic field directed parallel or antiparallel to the current does not exert any force on the wire.

We now derive an expression for the magnitude of the magnetic force on a current-carrying wire of length ℓ and cross-sectional area A (Figure 24-21). Let us assume that all charge carriers have the same charge, e (we are considering the magnitude of the charge in this case), and are moving with the same drift velocity, $\vec{v}_d$.

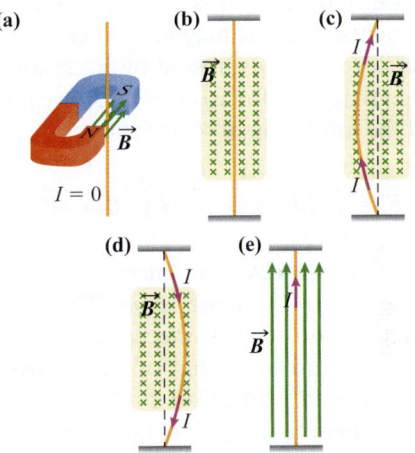

Figure 24-20 The effect of a magnetic field on a current-carrying wire. (a) A wire suspended vertically between the poles of a permanent magnet. (b) A wire carrying no current ($I = 0$) in a uniform magnetic field perpendicular to the wire. (c) A wire carrying an upward non-zero current in a uniform magnetic field perpendicular to the wire. (d) A wire carrying a downward non-zero current in a uniform magnetic field perpendicular to the wire. (e) A wire carrying a non-zero current in a uniform magnetic field parallel to the wire.

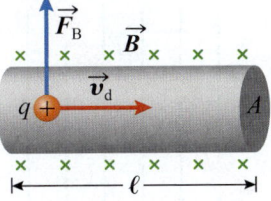

Figure 24-21 Calculating the effect of the magnetic field on a straight current-carrying wire. To calculate the net force on a segment of the wire, we calculate the magnetic force on each moving charge (considering that the charges move with an average drift velocity $\vec{v}_d$) and then add the forces acting on all the charges in this segment.

Then the magnetic force acting on each charge carrier is

$$\vec{F}_B = e\vec{v}_d \times \vec{B}$$

Since the volume of the wire is $A\ell$, the total charge on the charge carriers in the wire is $q = nA\ell$, where n is the charge density in the wire.

The net magnetic force acting on the wire is

$$\vec{F} = n_q \ell A \vec{v}_d \times \vec{B} \qquad (24\text{-}20)$$

Substituting $I = n_q A v_d$ and recalling that the direction of the drift velocity of positive charge carriers coincides with the direction of the electric current, we get

$$\vec{F} = n_q A v_d \vec{\ell} \times \vec{B} = I\vec{\ell} \times \vec{B}$$

KEY EQUATION
$$\vec{F} = I\vec{\ell} \times \vec{B} \qquad (24\text{-}21)$$

where $\vec{\ell}$ has a magnitude equal to the length of the wire and it is oriented in the direction of the electric current, I, flowing through the wire.

For the above calculation, we assumed that the wire is straight. We now consider a wire of arbitrary shape (Figure 24-22), which consists of an infinite number of small segments, $d\vec{\ell}$. The magnetic force acting on each individual segment is expressed by Equation 24-21. To find the net magnetic force on the wire, we integrate the forces on the infinitesimal segments:

$$\vec{F} = \int_a^b I d\vec{\ell} \times \vec{B} = -I \int_a^b \vec{B} \times d\vec{\ell} \qquad (24\text{-}22)$$

Note that *the electric current flowing through all the wire segments is the same.* The negative sign on the right-hand side of the equation results from the anticommutative nature of the cross product: $d\vec{\ell} \times \vec{B} = -\vec{B} \times d\vec{\ell}$. Equation 24-22 provides a way to define the ampere: a uniform magnetic field of 1 T perpendicular to the wire carrying an electric current of 1 A exerts a force of 1 N on a 1 m length of wire:

$$1\,A = \frac{1\,N}{1\,m \cdot 1\,T} \qquad (24\text{-}23)$$

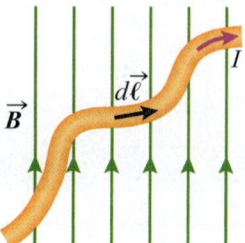

Figure 24-22 Calculating the effect of the magnetic field on a current-carrying wire of an arbitrary shape. The wire can be split into small straight segments $d\vec{\ell}$, and the effects of the magnetic field on all the segments can be added to find the net force on the wire.

24-5 The Torque on a Current-Carrying Loop in a Magnetic Field

In many everyday applications, such as electric motors, we use current-carrying loops submerged in magnetic fields. In this section, we will explore the behaviour of these loops and understand why they have such a wide range of applications. We will begin with the case of a rectangular loop. Let us consider a rectangular loop of wire, with dimensions a and b, carrying a current in a clockwise direction (Figure 24-23(a)). This loop is placed in a uniform magnetic field, $\vec{B}$, directed to the right. The magnetic field is parallel to sides BC and DA of the loop, so $\vec{F}_{BC} = \vec{F}_{DA} = 0$. Sides AB and CD are perpendicular to the magnetic field. Since the electric current in AB flows in the opposite direction to the electric current in CD, the magnetic forces on the two opposite sides of the loop are opposite to each other: the magnetic force on AB is directed into the page, and the magnetic force on CD is directed out of the page (Figure 24-23(b)). These forces create a torque on the loop (Section 8-3). Mathematically, this result can be expressed as

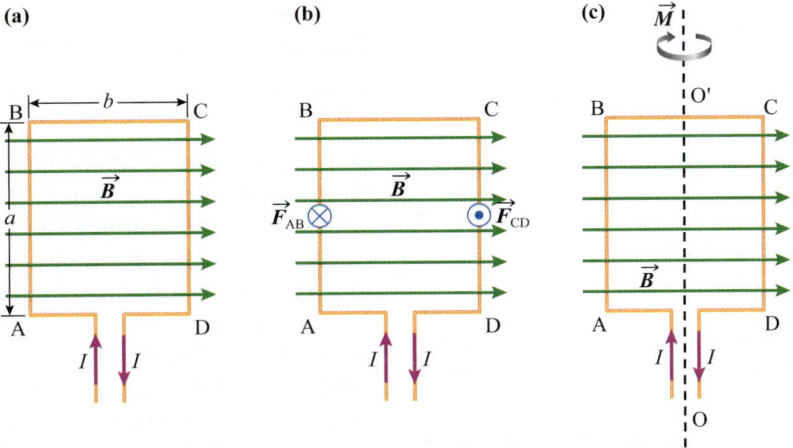

Figure 24-23 (a) A rectangular current-carrying loop in a uniform magnetic field. (b) Magnetic forces acting on opposite sides of the current-carrying loop. (c) A couple moment M (torque) acting on a current-carrying loop in a uniform magnetic field.

$$\vec{F}_{ABCD} = \vec{F}_{AB} + \vec{F}_{BC} + \vec{F}_{CD} + \vec{F}_{DA}$$
$$= I\vec{\ell}_{AB} \times \vec{B} + I\vec{\ell}_{BC} \times \vec{B} + I\vec{\ell}_{CD} \times \vec{B} + I\vec{\ell}_{DA} \times \vec{B}$$
$$= \vec{F}_{AB} + \vec{F}_{CD}$$

Forces $\vec{F}_{AB}$ and $\vec{F}_{CD}$ are equal in magnitude and opposite in direction. Their magnitudes can be expressed as

$$F_{AB} = F_{CD} = IaB \qquad (24\text{-}24)$$

Since forces $\vec{F}_{BC}$ and $\vec{F}_{DA}$ are applied to opposite sides of the loop and are perpendicular to its plane, they create a torque on the loop, causing it to rotate about the OO'-axis (Figure 24-23(c)). The magnitude of this torque is

$$\tau_{ABCD} = \tau_{AB} + \tau_{CD}$$
$$= \frac{b}{2}F_{AB} + \frac{b}{2}F_{CD} = \frac{b}{2}IaB + \frac{b}{2}IaB = abIB = AIB$$

or

$$\tau_{ABCD} = IAB \qquad (24\text{-}25)$$

where A is the area of the loop: $A = ab$.

The torque exerted on the loop by the magnetic field is proportional to the area of the loop, the electric current, and the magnitude of the magnetic field. It is common to describe the orientation of the loop by a vector $\vec{A}$ whose magnitude represents the area of the loop and whose direction is defined by a right-hand rule: when the fingers of your right hand are curled in the direction of the electric current, the direction of your thumb indicates the direction of vector $\vec{A}$ (Figure 24-24(a)), as well the direction of the magnetic field created by this loop, $\vec{B}_{loop}$ (Figure 24-24(b)). When applying this rule, remember that the direction of the electric current is defined as the direction of motion of positive charges. Note that, not surprisingly, the structure of the magnetic field created by a current-carrying loop is similar to the structure of the magnetic field created by a permanent magnet (Figure 24-2(a)). This also

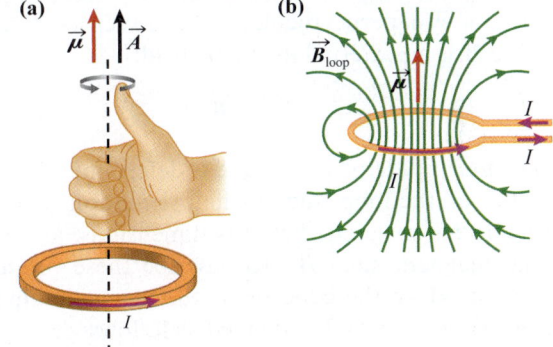

Figure 24-24 (a) A version of the right-hand rule for determining the orientation of a current-carrying loop using vector $\vec{A}$ and the magnetic dipole moment $\vec{\mu}$. (b) Magnetic field lines created by the current-carrying loop.

hints at the nature of the magnetic field and its relationship to electric current, which we will explore in detail in the following section.

As we just discussed, current-carrying conducting elements, including wires and loops, interact with the external magnetic fields they are immersed in. At the same time, these current-carrying elements also produce their own magnetic fields. When considering current-carrying elements submerged in external magnetic fields, it is important to distinguish between the magnetic field created by the element, $\vec{B}_{element}$, which does not affect the element (Figure 24-24(b)), and the external magnetic field, $\vec{B}$. When computing the force or torque on a current-carrying element, we consider the impact of the external field on that element. This external field can also be generated by a different element within the same component. An example of this phenomenon is the attractive force between the loops of a solenoid when current runs through it. Can you think of why the loops of a solenoid exert an attractive force on one another? We will explore the answer to this question in the following sections.

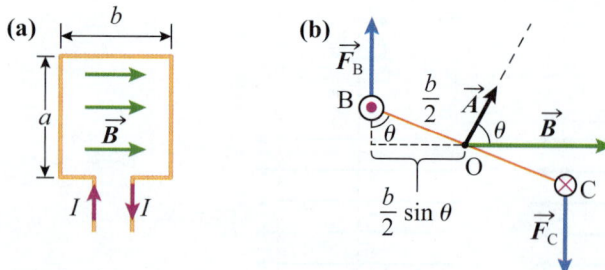

Figure 24-25 (a) The side view and (b) the top view of a vertical rectangular loop of width b, carrying current I submerged in a uniform horizontal magnetic field $\vec{B}$. The loop is free to rotate about its vertical axis of symmetry.

As discussed in Chapter 8, torque is a vector quantity that points along the object's axis of rotation, indicating clockwise or counter-clockwise rotation. The direction of rotation is called the "sense of the torque vector." For the loop in Figure 24-25, for example, the direction of $\vec{\tau}$ is perpendicular to both vectors $\vec{A}$ and $\vec{B}$, where $\vec{B}$ is an external magnetic field:

$$\vec{\tau}_{ABCD} = I\vec{A} \times \vec{B} \tag{24-26}$$

The magnitude of the torque is determined by the properties of the loop, such as its orientation relative to the external magnetic field, its dimensions, and the external magnetic field $\vec{B}$. To describe these properties, we introduce the concept of the **magnetic dipole moment**, $\vec{\mu}$ (Figure 24-24), defined as follows:

KEY EQUATION
$$\vec{\mu} = I\vec{A} \tag{24-27}$$

The concept of the magnetic dipole helps clarify a key difference between electric and magnetic fields. The magnetic field profile from a magnetic dipole, such as a bar magnet, is similar to the electric field profile from an electric dipole. An electric dipole is a combination of two equal but opposite charges. Electric field lines created by the electric dipole originate at the positive charge and end at the negative charge. Positive and negative charges are electric monopoles. In a uniform electric field, an electric dipole experiences a torque that acts to align the dipole with the electric field lines: $\vec{\tau}_E = \vec{p} \times \vec{E}$ (Chapter 19, Equation 19-18), where $\vec{p}$ is the electric dipole moment. However, there is no evidence that magnetic charge or magnetic monopoles exist. A **magnetic dipole** is a current-carrying loop or a permanent magnet, not a combination of two magnetic monopoles. One cannot say that there is a north pole separable from a south pole. This is due to the electric current nature of the source of the magnetic field.

For example, the current-carrying loop in Figure 24-24 creates a magnetic field $\vec{B}_{loop}$. As a result, the current-carrying loop behaves as a permanent magnet (Figure 24-2(a)), with the polarity of the magnet depending on the direction of the current in the loop. Thus, a current-carrying loop becomes a magnetic dipole. An external magnetic field $\vec{B}$ exerts a torque on this magnetic dipole, directed such that the magnetic

dipole moment aligns with the external magnetic field. Equation 24-26 can be expressed as

KEY EQUATION
$$\vec{\tau}_B = \vec{\mu} \times \vec{B} \tag{24-28}$$

Only the external magnetic field component parallel to the plane of the loop results in a torque on it. When a loop positioned parallel to a uniform magnetic field starts rotating, the effect of the magnetic field on the loop changes as it rotates (Figure 24-25).

The component of the magnetic field $\vec{B}$ parallel to the surface of the rotating loop can be expressed as $B\sin\theta$, where θ is the angle between the normal to the loop (or the area vector, or the dipole moment) and the magnetic field (Figure 24-26). Then the net torque (a couple moment) produced by the magnetic force is

$$\tau = bIaB\sin\theta = AIB\sin\theta \tag{24-29}$$

Consequently, the net torque on the loop varies as the loop rotates, reaching a maximum value when the magnetic field is parallel to the plane of the loop ($\theta = 90°$) and a minimum value when the magnetic field is perpendicular to it ($\theta = 0°$):

$$\begin{aligned} \tau_{max} &= bIaB\sin 90° = bIaB & \text{when } \theta = 90° \\ \tau_{min} &= bIaB\sin 0° = 0 & \text{when } \theta = 0° \end{aligned} \tag{24-30}$$

For example, the maximum torque on the current-carrying loop shown in Figure 24-26, which can rotate around its vertical axis, is $0.15\ N\cdot m$:

$$\tau_{max} = (0.10\ m)(0.25\ m)(5.0\ A)(1.2\ T) = 0.15\ N\cdot m$$

Although we derived this expression for a rectangular loop, we can show that it holds for any loop.

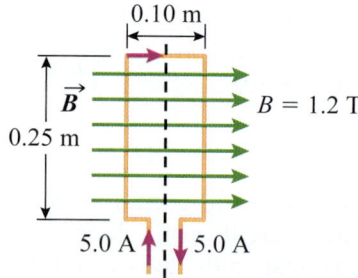

Figure 24-26 A rectangular vertical loop with dimensions $0.10\ m \times 0.25\ m$ carrying current of $5.0\ A$ in a horizontal and uniform magnetic field of $1.2\ T$. The maximum torque on this current-carrying loop is $\tau_{max} = 0.15\ N\cdot m$ (Equation 24-30).

PEER TO PEER

At first I didn't realize how important it is to distinguish between the magnetic field created by a current-carrying wire, $\vec{B}_{wire}$, and an external magnetic field, $\vec{B}$. If I have both in one problem, I use indices to label them differently. If I have only one field, I can label it $\vec{B}$ without worrying about getting confused. For example, while using the Biot–Savart law to calculate magnetic fields created by different current-carrying wires, I just use $\vec{B}$ (see Section 24-6).

Interaction of a Current-Carrying Loop and a Magnetic Field

A vertical rectangular current-carrying loop with area A hangs from a light thread, as shown in Figure 24-27. The loop is submerged in a uniform magnetic field, $\vec{B}$, oriented perpendicular to the loop's surface. Which statement best describes the magnitude of the torque generated by this magnetic field on the loop?

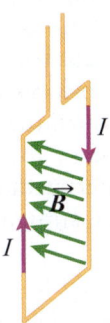

Figure 24-27   Checkpoint 24-8.

(a) The loop experiences torque of magnitude $\tau = AIB$ because the surface of the loop is perpendicular to $\vec{B}$.

(b) The loop experiences torque of magnitude $\tau = -AIB$ because the surface of the loop is parallel to $\vec{B}$.

(c) The loop experiences zero torque because the magnetic field is parallel to the loop's surface.

(d) The loop experiences zero torque because the magnetic field is perpendicular to the loop's surface.

(e) There is not enough information to determine the torque.

ANSWER: (d) This current-carrying loop experiences zero magnetic torque (Equation 24-30) because of its orientation relative to the magnetic field. Note that the angle should be measured between the normal to the surface of the loop and the magnetic field. In this case, the angle between A and B is $0°$, so $\sin 0° = 0$ and the torque is zero.

EXAMPLE 24-6

The Magnetic Torque on a Simple Electric Motor

Estimate the maximum value of the magnetic torque in the simple electric motor shown in Figure 24-28, given that the magnetic field created by the magnets is 0.40 T, the battery has a voltage of 3.0 V, the number of copper wire loops is 2, and the area of each loop is 5.0 cm². Assume that the resistance of all the wire used in the motor is 1.0 Ω.

Solution

The magnetic torque acting on a current-carrying loop in a magnetic field can be found using Equation 24-25. The factor of 2 comes from the fact that there are two loops, hence double the area:

$$\tau_{max} = 2IAB = 2\frac{V}{R}AB = 2\frac{3.0\ V}{1.0\ \Omega}(5.0 \times 10^{-4}\ m^2)(0.40\ T)$$

$$= 1.2 \times 10^{-3}\frac{V \cdot m^2 \cdot T}{\Omega} = 1.2 \times 10^{-3}\ N \cdot m$$

Making sense of the result

This simple experiment helps us make two interesting observations. First, the low resistance of the wire allows a relatively large current to flow from the battery. Nevertheless, the magnetic torque is very small. Therefore, for this motor to work, the loop has to be very light. The second observation is more profound. We discussed earlier that magnetic force cannot do work on a moving charge because it is directed perpendicular to its velocity (Equation 24-1). In the case of an electric motor, it appears to do work on a moving current-carrying wire. How can this happen? The solution to this apparent paradox is that the physics involved is really more complex than finding the magnetic force on the electrons moving in the wire, although that standard derivation does give the correct answer. In reality, the magnetic force on the electrons exerts a transverse force on them as in the Hall effect (Equation 24-19). This charge separation creates an electric force across the wire, which exerts a force on the positively charged lattice atoms, doing the work.

MAKING CONNECTIONS

Can You Build an Electric Motor?

Many everyday appliances use electric motors, for example, microwave ovens, food processors, fans, DVD players, and washing machines. These motors all apply the concept of a magnetic torque on a current-carrying loop. Although some motors, such as those in DVD players and computer hard drives, involve complex precision technology, you can build a simple DC motor with a battery, a few pieces of wire, a small permanent magnet, a pair of scissors, a few paper clips, tape, and any convenient cylinder, such as a thick marker. We wrap a piece of wire around the cylinder to create a solenoid, as shown in Figure 24-28.

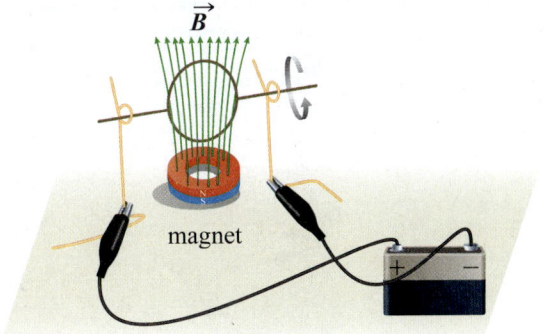

Figure 24-28   A simple DC motor.

24-6 The Biot-Savart Law

Electric charges create electric fields, and we have seen some experimental evidence in the behaviour of current-carrying loops in external magnetic fields that electric currents might be the sources of magnetic fields. In this section, we will explore the sources of magnetic fields in more detail. The nature of the sources of magnetic fields was a mystery for centuries. The equations we have derived so far show how magnetic fields affect moving charges, but not vice versa. Is it reasonable to assume that a magnetic field can be affected by an electric current? Could electric currents be the source of magnetic fields? Although the answers to these questions might seem obvious to some today, two centuries ago they were at the forefront of scientific inquiry. The first step in solving this mystery was taken by Danish scientist Hans Christian Oersted (1777–1851).

In 1820, Oersted showed that a compass needle deflects from magnetic north when an electric current is switched on or off in a nearby wire. This discovery motivated French scientists Jean-Baptiste Biot (1774–1862) and Félix Savart (1791–1841) to conduct experiments not only to confirm that electric current is the source of magnetic fields but to find a mathematical expression for the relationship between them. The magnetic field generated by an infinitesimally small element of a steady electric current is given by the Biot–Savart law (Figure 24-29):

KEY EQUATION
$$d\vec{B} = \frac{\mu_0}{4\pi} \frac{Id\vec{\ell} \times \hat{r}}{r^2} = \frac{\mu_0}{4\pi} \frac{Id\vec{\ell} \times \vec{r}}{r^3} \qquad (24\text{-}31)$$

In this equation, $d\vec{\ell}$ is an infinitesimally small element of the wire carrying current I, so $d\vec{\ell}$ points in the direction of the conventional current, as shown in Figure 24-29; r is the distance from the element $d\vec{\ell}$ to an arbitrary point at which the magnetic field is calculated; $\hat{r}$ is the unit vector in the direction of $\vec{r}$, so $\hat{r} = \frac{\vec{r}}{r}$; and μ_0 is a constant called the **permeability constant of free space** and has a value of $4\pi \times 10^{-7}$ T·m/A. The term 4π was introduced in the definition to make Maxwell's

equations of electromagnetism more convenient for calculations, as you will see later in this chapter. This bears a resemblance to the introduction of $k = \frac{1}{4\pi\varepsilon_0}$ in Coulomb's law (Equations 19-3 and 19-4) to simplify the applications of Gauss's law (Equation 20-4). Note that the position vector, $\vec{r}$, points from the given current element, $Id\vec{\ell}$, to the point where the magnetic field is being observed.

Let us examine the Biot–Savart law more closely:

■ Since the expression for $d\vec{B}$ includes the cross product of $d\vec{\ell}$ and $\vec{r}$, $d\vec{B}$ is always perpendicular to the element of electric current that created it, as well as to the position vector $\vec{r}$ connecting $d\vec{\ell}$ and the point at which the magnetic field is calculated. The direction of the magnetic field can also be established experimentally by moving a compass around a long straight current-carrying wire (Figure 24-30).

■ The magnitude of $d\vec{B}$ is inversely proportional to the square of the distance between the point at which the field is calculated and the infinitesimal element of electric current creating the field.

■ Both $d\vec{B}$ and the resultant magnetic field are proportional to the current.

■ Since $d\vec{B}$ is proportional to the cross product of $d\vec{\ell}$ and $\vec{r}$, the magnetic field depends on the sine of the angle between the infinitesimally small element of the wire, $d\vec{\ell}$, and the position vector, $\vec{r}$.

The Direction of the Magnetic Field and the Right-Hand Rule

The direction of the magnetic field due to a current element can be found by using one or more versions of the right-hand rule introduced earlier. To find the direction of the magnetic field due to a current element, keeping your thumb, index finger, and middle

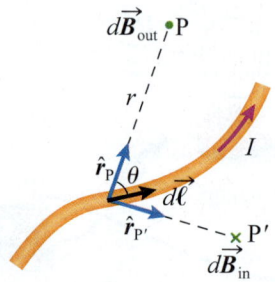

Figure 24-29 Using the Biot–Savart law to find the magnetic field created by a long current-carrying wire.

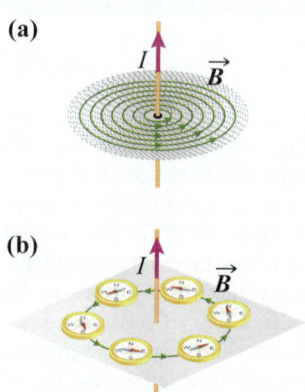

Figure 24-30 (a) A magnetic field created by a long straight current-carrying wire. (b) Compass needles indicate the direction of the magnetic field surrounding the current-carrying wire.

finger mutually perpendicular, if you line up your thumb with the current and your index finger with the position vector (as measured from the current element), then the direction of the magnetic field is given by your middle finger. For example, if the current element runs straight up in the plane of this page, and you wish to determine the direction of the magnetic field directly to the right of the current element, by the right-hand rule, the magnetic field points directly into the page. The discussion on how to use the right-hand rule and how to predict the direction of the cross product between vectors that are not perpendicular, as presented earlier in the chapter, applies in this case as well.

EXAMPLE 24-7

Establishing the Direction of the Magnetic Field Due to a Long Current-Carrying Wire

Current I runs through a straight conducting horizontal wire to the right, as shown in Figure 24-31. Point A is to the right of and above the current element, as shown.

Use a right-hand rule to establish the direction of the magnetic field due to the current element shown at point A.

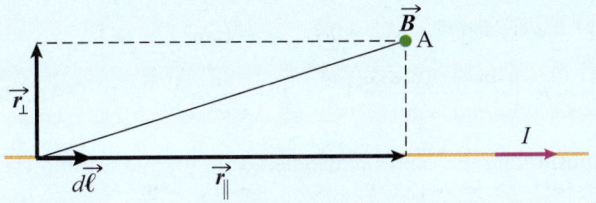

Figure 24-31 The direction of the magnetic field at point A is given by the cross product of $d\vec{\ell}$ and $\vec{r}_\perp$.

Solution

The direction of the magnetic field, $d\vec{B}$, created by the current element of length $d\vec{\ell}$, is established by the right-hand rule, as discussed earlier; the magnetic field is given by the cross product of $d\vec{\ell}$ and $\vec{r}$, the position vector of the point of observation as measured from the current element. Since these two vectors, as shown in Figure 24-31, are not perpendicular, we need only consider the component of one of the vectors that is perpendicular to the other vector, as we saw earlier in the chapter. We will consider the component of $\vec{r}$ that is perpendicular to $d\vec{\ell}$, $\vec{r}_\perp$. Notice that $\vec{r}_\parallel$, the component of $\vec{r}$ that is parallel to $d\vec{\ell}$, does not contribute to the cross product.

Using the "three-finger" right-hand rule, keeping the thumb, the index finger, and the middle finger in mutually perpendicular positions, we line up our thumb with the current element (to the right) and our index finger with $\vec{r}_\perp$; our middle finger will point out of the page and give us the direction of the magnetic field at A.

Making sense of the result

When taking a cross product, only perpendicular components matter. You can also use the right-hand curl rule to establish the direction of the magnetic field, or any of the other versions of the right-hand rule.

The Biot–Savart law for magnetism is analogous in some respects to Coulomb's law for electric fields. Both laws show that the strength of the field is proportional to its source: q for an electric field and the electric current element $d\vec{I} = Id\hat{\ell}$, where $\hat{\ell}$ is a unit vector in the $\vec{\ell}$-direction (the direction of the conventional current), for a magnetic field. In both laws, the field strength is inversely proportional to the square of the distance from the source of the field.

However, there are also some important differences: electric field lines always emanate from positive charges and terminate at negative charges. Sources of electric fields (positive and negative charges) can be separated, with a positive charge defined as a source (electric field lines are distributed isotopically around the charge, and they point directly away from the charge extending to infinity) and a negative charge defined as a sink (electric field lines originate infinitely far in space, they are distributed isotopically around the charge, and they point directly toward it) (Figure 19-15). As we discussed earlier, such a positive or a negative charge is referred to as an electric monopole. We cannot, however, separate the south and north poles of a magnet because, as far as we know, there are no magnetic monopoles. Since there is no magnetic monopole, magnetic field lines are continuous and closed: they have neither a beginning nor an end, and they loop around the electric current that created them. Due to the closed-loop nature of magnetic field lines, they do not follow straight lines, and it becomes very difficult to produce magnetic fields that are spatially uniform (we will discuss how one can produce uniform magnetic fields in a small region in the following sections). However, reasonably uniform electric fields can, in principle, be produced in the vicinity of large plates of evenly distributed electric charge.

EXAMPLE 24-8

The Magnetic Field Created by an Infinitely Long Straight Current-Carrying Wire

(a) Find the magnetic field created by a straight current-carrying wire of length L with current I running through it, at point P located a distance R from the wire along its perpendicular bisector (Figure 24-32).

(b) Using your result in part (a), find the magnetic field due to an infinitely long current-carrying wire.

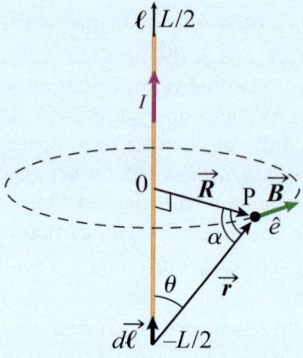

Figure 24-32 Calculating the magnetic field created by a long straight current-carrying wire of length L at point P located a distance R from the wire along its perpendicular bisector.

Solution

(a) We use the Biot–Savart law to calculate the magnetic field at point P. Let us denote the current flowing through the wire as I, and the shortest distance from the wire to the point where the magnetic field is to be calculated as R. It is important to note that point P is located exactly in the middle of the wire, so that the problem is symmetrical along its perpendicular bisector. If we consider the vertical axis directed along the wire to have its origin exactly in the middle of the wire, then the top point of the wire segment is $L/2$ and the bottom one is $-(-L/2)$. To use the Biot–Savart law (Equation 24-31), we have to calculate the contribution of the current in the wire segment $d\vec{I}$, where $d\vec{I} = Id\hat{e}$, to the magnetic field at point P. Then we add the effects of all the segments along the wire. The magnetic field produced by the wire segment $d\vec{\ell}$ has direction $d\vec{\ell} \times \vec{r}$. The resultant vector is perpendicular to both $d\vec{\ell}$ and $\vec{r}$, and we denote its direction using a unit vector, $\hat{e}$:

$$d\vec{\ell} \times \vec{r} = \hat{e}\, r\sin\theta\, d\ell = \hat{e}Rd\ell$$

Now we are ready to apply the Biot–Savart law:

$$d\vec{B} = \frac{\mu_0}{4\pi} \frac{Id\vec{\ell} \times \vec{r}}{r^3} = \frac{\mu_0 I}{4\pi} \frac{\hat{e}Rd\ell}{(\ell^2 + R^2)^{3/2}} = \hat{e}\frac{\mu_0 IR}{4\pi} \frac{d\ell}{(\ell^2 + R^2)^{3/2}}$$

where $r = \sqrt{\ell^2 + R^2}$.

The value of the magnetic field at point P can be expressed as the sum (integral) of all the infinitesimally small contributions $d\vec{B}$. Since the wire has length L, we can choose the centre

of the wire segment as $\ell = 0$. Therefore, the value of the magnetic field can be expressed as follows:

$$\vec{B} = \int_{-L/2}^{L/2} \hat{e}\frac{\mu_0 IR}{4\pi} \frac{d\ell}{(\ell^2 + R^2)^{3/2}} = \hat{e}\frac{\mu_0 IR}{4\pi} \int_{-L/2}^{L/2} \frac{d\ell}{(\ell^2 + R^2)^{3/2}} \quad (1)$$

Note that since the unit vector $\hat{e}$ in the calculation has a constant magnitude and since we are looking for the magnitude of the magnetic field, we can move it outside the integral sign. This integral can be calculated using a trigonometric substitution: $\ell = R\tan\alpha$. Then we can express $d\ell$ as

$$d\ell = d(R\tan\alpha) = Rd\tan\alpha = R\left(\frac{\sin^2\alpha + \cos^2\alpha}{\cos^2\alpha}\right)d\alpha$$

$$= R\frac{1}{\cos^2\alpha}d\alpha = R(\sec^2\alpha)d\alpha \quad (2)$$

The denominator of (1) under the integral sign can be expressed as

$$(\ell^2 + R^2)^{3/2} = (R^2\tan^2\alpha + R^2)^{3/2} = R^3(\tan^2\alpha + 1)^{3/2}$$

$$= R^3\left(\frac{\sin^2\alpha + \cos^2\alpha}{\cos^2\alpha}\right)^{3/2} = R^3\left(\frac{1}{\cos^2\alpha}\right)^{3/2} = R^3\sec^3\alpha \quad (3)$$

Substituting the expressions obtained in (2) and (3) into (1), we get

$$\vec{B} = \int_{-L/2}^{L/2} \hat{e}\frac{\mu_0 IR}{4\pi} \frac{d\ell}{(\ell^2 + R^2)^{3/2}} = \hat{e}\frac{\mu_0 IR}{4\pi} \int_{\text{Wire } L} \frac{R\sec^2\alpha\, d\alpha}{R^3\sec^3\alpha}$$

$$= \hat{e}\frac{\mu_0 I}{4\pi R} \int_{\text{Wire } L} \frac{d\alpha}{\sec\alpha} = \hat{e}\frac{\mu_0 I}{4\pi R} \int_{\text{Wire } L} \cos\alpha\, d\alpha \quad (4)$$

When performing integration by substitution, we usually have to recalculate the integration limits. However, in our case, we will revert to integrating over the original variable, ℓ. Therefore, we do not need to compute the new limits of integration for the new variable α. However, to remember that we are integrating over a wire of length L, we write "Wire L" next to the integration sign.

Examining (4), we recognize that the antiderivative of cosine is sine:

$$\int \cos\alpha\, d\alpha = \sin\alpha + \text{const} \quad (5)$$

We can also express $\sin\alpha$ in terms of the original variables as

$$\sin\alpha = \frac{\ell}{\sqrt{(\ell^2 + R^2)}}$$

As a result, we obtain

$$\vec{B} = \hat{e}\frac{\mu_0 I}{4\pi R}\sin\alpha\bigg|_{\text{Wire } L} = \hat{e}\frac{\mu_0 I}{4\pi R}\frac{\ell}{\sqrt{\ell^2 + R^2}}\bigg|_{-L/2}^{L/2}$$

$$= \hat{e}\frac{\mu_0 I}{4\pi R}\frac{L}{\sqrt{\frac{L^2}{4} + R^2}} \quad (6)$$

(b) To find the expression for an infinitely long wire, we assume that in (6) L goes to infinity. Since R remains the same, $L \gg R$. In the limiting case of $L \to \infty$, we obtain

$$\vec{B} = \lim_{L \to \infty} \left(\hat{e} \frac{\mu_0 I}{4\pi R} \frac{L}{\sqrt{\frac{L^2}{4} + R^2}} \right) - \hat{e} \frac{\mu_0 I}{4\pi R} \lim_{L \to \infty} \frac{L}{\sqrt{\frac{L^2}{4} + R^2}}$$

$$= \hat{e} \frac{\mu_0 I}{4\pi R} 2 = \hat{e} \frac{\mu_0 I}{2\pi R}$$

(7)

Making sense of the result

The derivation above might look a little cumbersome. So it is always important to check the units. In both (6) and (7), the units of magnetic field are T, as should be expected. We can also check the limiting cases. For instance, when the electric current in the wire diminishes to zero, we would expect that no magnetic field would be created. This is seen from both (6) and (7). Finally, the results indicate that the magnetic field created by a straight infinitely long current-carrying wire is inversely proportional to the distance from the wire and directly proportional to the current flowing in it. It is directed along concentric circles surrounding the wire.

From the above example, we conclude that the magnitude of the magnetic field due to a very long or infinitely long conductor a distance R away is given by

KEY EQUATION

$$B_{\substack{\text{outside long} \\ \text{straight wire}}} = \frac{\mu_0 I}{2\pi R}$$

(24-32)

The Right-Hand Curl Rule and the Direction of the Magnetic Field Due to a Long Conductor

The direction of the magnetic field due to the electric current can be conveniently established using the following right-hand curl rule for the magnetic field (Figure 24-33): When you point your right thumb in the direction of the electric current and curl your fingers, the direction of your curled fingers is the direction of the magnetic field lines that form concentric circles around the current-carrying wire. Note that this is a different right-hand rule from

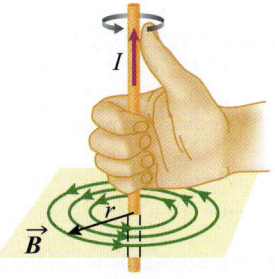

Figure 24-33 The right-hand curl rule for determining the direction of the magnetic field surrounding a long straight current-carrying wire. The magnetic field lines create closed circular concentric loops. The magnitude of the magnetic field can be found using Equation 24-32.

the one used to determine the Lorentz force acting on a charged particle moving in a magnetic field (Equation 24-2). To distinguish between these two different right-hand rules, we will call the latter the **right-hand curl rule**.

EXAMPLE 24-9

The Magnetic Field Created by a Circular Current-Carrying Loop

In the previous section, we discussed the fact that a current-carrying circular loop creates a magnetic field. (a) Derive an expression for the magnetic field along the axis of symmetry of a circular loop of radius R that carries an electric current I (Figure 24-34). (b) Find the magnetic field at the centre of the ring. (c) Find the magnetic field a very large distance away from the ring's centre along its axis of symmetry. (d) Express the answer in (c) in terms of the magnetic dipole moment of the loop, and comment on the distance dependence of the magnetic field at large distances.

Solution

(a) For convenience, we choose a coordinate system with the current-carrying loop in the yz-plane (Figure 24-34). We divide the loop into an infinitely large number of infinitesimally small current elements: $d\vec{I} = I d\vec{\ell}$. The Biot–Savart law can be used to calculate the contribution of each element to the magnetic field at point P on the axis of

symmetry of the loop. Let x be the distance of this point from the centre of the loop, O. Then

$$d\vec{B} = \frac{\mu_0}{4\pi} \frac{I d\vec{\ell} \times \hat{r}}{r^2} = \frac{\mu_0 I}{4\pi} \frac{d\vec{\ell} \times \hat{r}}{(x^2 + R^2)}$$

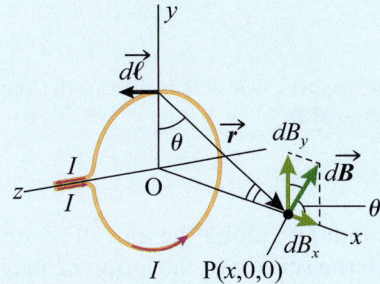

Figure 24-34 Example 24-9.

First, consider the cross product of $d\vec{\ell}$ and $\vec{r}$. Since point P is on the loop's axis of symmetry, the unit vector $\hat{r}$ is perpendicular to any loop element $d\vec{\ell}$. Therefore,

(continued)

ELECTRICITY, MAGNETISM, AND OPTICS

$|d\vec{\ell} \times \hat{r}| = d\ell$. The direction of the cross product, shown in Figure 24-34, is perpendicular to both $d\vec{\ell}$ and $\vec{r}$. The magnetic field created by this loop element has two components, B_y and B_z. Now consider another element of the loop, $d\vec{\ell}\,'$, diametrically opposite to the element $d\vec{\ell}$. Since the elements $d\vec{\ell}$ and $d\vec{\ell}\,'$ are antiparallel, the magnetic fields created by them have x-components that are equal and y-components that are equal in magnitude but opposite in direction. The same consideration applies to any pair of symmetrical loop elements: the magnetic field components perpendicular to the loop's axis of symmetry cancel out, and the components parallel to the axis add up. The dB_x-component can be calculated as follows:

$$dB_x = \frac{\mu_0}{4\pi} \frac{I\cos\theta \, d\ell}{(x^2 + R^2)} = \frac{\mu_0}{4\pi} \frac{I d\ell}{(x^2 + R^2)}\cos\theta$$

$$= \frac{\mu_0}{4\pi} \frac{I d\ell}{(x^2 + R^2)} \frac{R}{(x^2 + R^2)^{1/2}}$$

$$= \frac{\mu_0}{4\pi} \frac{IR d\ell}{(x^2 + R^2)^{3/2}}$$

To calculate the net magnetic field created by the entire loop, we integrate this expression over the entire loop:

$$B_{\text{loop}} = \int_0^{2\pi R} \frac{\mu_0}{4\pi} \frac{IR d\ell}{(x^2 + R^2)^{3/2}}$$

$$= \frac{\mu_0 IR}{4\pi(x^2 + R^2)^{3/2}} \int_0^{2\pi R} d\ell$$

$$= \frac{\mu_0 IR 2\pi R}{4\pi(x^2 + R^2)^{3/2}}$$

$$= \frac{\mu_0}{2} \frac{IR^2}{(x^2 + R^2)^{3/2}}$$

Thus, the magnetic field created by a circular current-carrying loop along its axis of symmetry is collinear with this axis. The direction of this magnetic field can be determined using the right-hand curl rule for current. The magnitude of this magnetic field is

$$B_{\text{loop}} = \frac{\mu_0}{2} \frac{IR^2}{(x^2 + R^2)^{3/2}} \qquad (1)$$

(b) The magnetic field at the centre of the loop ($x = 0$) can be expressed as

$$B_{\text{loop-centre}} = \frac{\mu_0}{2} \frac{IR^2}{R^3} = \frac{\mu_0}{2} \frac{I}{R} \qquad (2)$$

(c) To find the magnetic field along the axis of the loop a large distance away from its centre, we take the limit of expression (1) for the case when $x \gg R$. In the limiting case, we see that the denominator of (1) approaches x^3. Therefore, the expression we obtained in (a) becomes

$$B_{\text{loop}} = \frac{\mu_0}{2} \frac{IR^2}{x^3} \qquad (3)$$

which can be rewritten as

$$B_{\text{loop}} = \frac{\mu_0}{2} \frac{IR^2}{x^3} = \frac{\mu_0 I}{2\pi} \frac{\pi R^2}{x^3} = \frac{\mu_0}{2\pi} \frac{IA}{x^3} \qquad (4)$$

where A is the area of the loop.

As we saw earlier in the chapter, the product IA represents the magnetic dipole moment of the loop, $\vec{\mu} = I\vec{A}$ (Equation 24-27), and (4) can be rewritten as

$$B_{\text{loop}} = \frac{\mu_0}{2\pi} \frac{\mu}{x^3} \qquad (5)$$

In vector form, this can be expressed as

$$\vec{B}_{\text{loop}} = \frac{\mu_0}{2\pi} \frac{\vec{\mu}}{x^3} \qquad (6)$$

The magnetic dipole moment of the loop here is expressed as a vector whose magnitude is equal to the product of the current in the loop and its area vector, as established earlier in this chapter. We recall here that the direction of the magnetic dipole moment of a current loop points in the same direction as the magnetic field due to the current at the centre of the loop.

We also note here another important result. The magnetic field along the loop's axis of symmetry decreases as the cube of the distance from the centre of the loop when this distance becomes large as compared to the size of the loop ($x \gg R$).

Making sense of the result

This result supports our earlier observation that a current-carrying circular loop behaves as a magnetic dipole. This is a very important observation that has profound implications, which we will discuss when we analyze the structure of an atom and the magnetic properties of different materials in the following sections.

In summary, along the axis of symmetry perpendicular to the surface of the loop, the magnitude of the magnetic field can be calculated as

KEY EQUATION

$$B_{\text{loop}} = \frac{\mu_0}{2} \frac{IR^2}{(x^2 + R^2)^{3/2}} \qquad (24\text{-}33)$$

In the centre of the current-carrying loop, the magnitude of the magnetic field can be calculated as

KEY EQUATION

$$B_{\text{loop-centre}} = \frac{\mu_0}{2} \frac{IR^2}{R^3} = \frac{\mu_0}{2} \frac{I}{R} \qquad (24\text{-}34)$$

And, lastly, a current-carrying circular loop behaves as a magnetic dipole whose magnetic field at large distances decreases as the cube of the distance from the loop. This behaviour is analogous to that of the electric field generated by an electric dipole along its axis at a large distance from the dipole. This is an important analogy in electromagnetism.

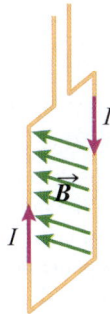
24-7 Ampère's Law

Although the Biot–Savart law provides a technique for determining the magnetic field around a wire of any arbitrary shape, the calculations can be somewhat daunting even for relatively simple shapes, as we saw in Examples 24-8 and 24-9. In this respect, the Biot–Savart law resembles Coulomb's law, which can in principle be used to determine the electric field around any charge distribution but involves difficult calculations unless the charge distribution is highly symmetric. When a continuous charge distribution has certain symmetries, Gauss's law significantly simplifies the electric field calculations (Chapter 20). Similarly, Ampère's law allows much easier calculations of magnetic fields when the path of the current is symmetric. Before we define Ampère's law, we need to introduce line integrals.

Overview of Line Integrals

In previous chapters, we saw that the integral of a function, $\int_a^b f(x)dx$, is an infinite sum of the products of the values of the function, $f(x)$, and an infinitesimally small element, dx, along a path of integration from $x = a$ to $x = b$, where a and b are the limits of integration. Thus, dx is the lower limit of small steps, Δx, along the path of integration, and

$$\lim_{N\to\infty}\sum_{i=1}^{N}f(x)\Delta x_i = \int_a^b f(x)dx \qquad (24\text{-}35)$$

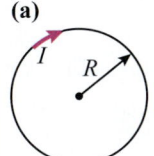 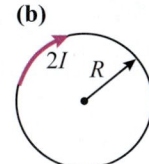 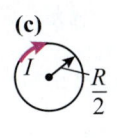 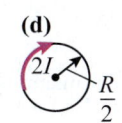 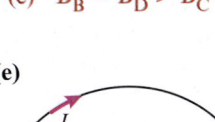 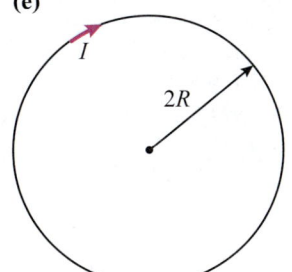 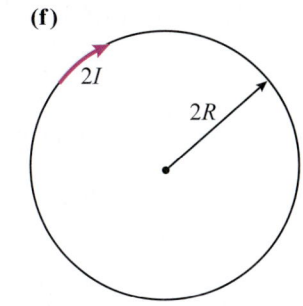

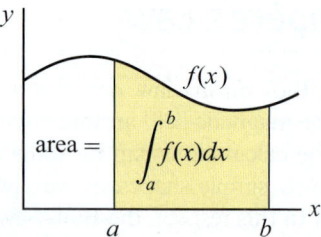

Figure 24-37 The area under the graph is a geometric representation of integration.

Consequently, the integral represents the area under the graph of $f(x)$ (Figure 24-37).

Sometimes, instead of varying an independent variable along a linear coordinate axis, it is useful to vary it along a two- or three-dimensional curve (Figure 24-38). Such line integrals are defined as

$$\lim_{N\to\infty}\sum_{i=1}^{N} f(\vec{r}) \cdot \Delta\vec{\ell}_i = \int_a^b f(\vec{r}) \cdot d\vec{\ell} \qquad \text{(24-36)}$$

The integral defined by Equation 24-36 is the general case of a line integral along a line, such as the x-axis. The function $f(x)$ that we want to integrate can be either a scalar or a vector function. Each infinitesimally small element of the curve is considered an infinitesimally small vector element $d\vec{\ell}$ because it has both magnitude and direction. Consider a line integral where the magnetic field has a constant magnitude and is parallel to the line of integration. For example, consider the case when the magnetic field (e.g., due to a long straight current-carrying wire) follows concentric circles, and where the line of integration is along one of these circles (Figure 24-38). In this case, the integral of the scalar product of the magnetic field and the curve element can be simplified. Notice that although

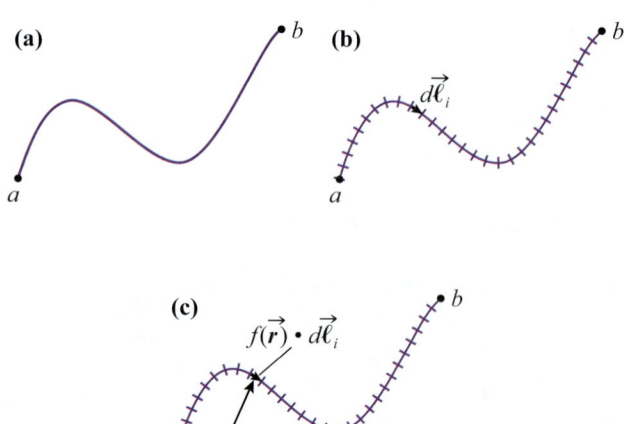

Figure 24-38 (a) An arbitrary line connecting two points a and b along the path of integration. (b) The line can be divided into infinitesimally small segments, $d\vec{\ell}_i$. (c) The integral along the path of integration can be approximated by the sum of the products of $f(\vec{r})$ and $d\vec{\ell}_i$.

the magnetic field depends on $\vec{r}$, for simplicity, we will omit $\vec{r}$ from the notation of the magnetic field and will denote it as $\vec{B}$:

$$\int_a^b \vec{B} \cdot d\vec{\ell} = \int_a^b B \cdot d\ell = B \int_a^b d\ell = BL \qquad \text{(24-37)}$$

where L is the total displacement along the curve of integration.

Closed-loop path integrals are integrals around a closed loop, so the initial and final points of integration coincide. Such integrals are denoted by a small circle over the conventional integral sign, for example, $\oint \vec{B} \cdot d\vec{\ell}$. Closed-loop path integrals are used to calculate the magnetic fields created by electric currents.

EXAMPLE 24-10

Calculate a Line Integral for a Current-Carrying Wire

Consider an infinitely long wire with current I running through it. Calculate the line integral of the scalar product of the field and the infinitesimal path length along a closed circle of radius r, whose plane is perpendicular to the wire and whose centre coincides with it (Figure 24-33).

Solution

The scenario here is very much like the scenario described earlier in Figure 24-31:

$$\oint \vec{B} \cdot d\vec{\ell} = \oint |\vec{B}\| d\vec{\ell}| \cos\theta$$

In that example, we expressed the scalar product explicitly as the product of the magnitudes of the two vectors and the cosine of the angle between them. Later in the chapter, we will drop the full formalism and simply express the integral as

$$\oint \vec{B} \cdot d\vec{\ell} = \oint B d\ell \cos\theta$$

The magnitude of the magnetic field is constant over the loop, so it can come outside the integral sign, and the angle between the field and the line element is zero, so the expression becomes

$$\oint \vec{B} \cdot d\vec{\ell} = B \oint d\ell = BL_{loop}$$

The magnitude of a field due to a long conductor a distance r away is given by $B = \dfrac{\mu_0 I}{2\pi r}$, and the length of the loop is $L = 2\pi r$, so the integral reduces to

$$\oint \vec{B} \cdot d\vec{\ell} = \mu_0 I$$

Making sense of the result

The field is parallel to the infinitesimal element of length around the circle, and hence the line integral should reduce to the product of the magnitude of the field and the circumference of the circle.

The result of the previous example has significant implications, as we shall see below. The line integral of the field and the curve element around the closed loop is equal to the product of the permeability of free space and the enclosed current. This is in fact true not only for the simple case of the circular path we have taken, but for any arbitrary closed path. We shall expand on this in what follows.

Ampère's Law

French physicist and mathematician André-Marie Ampère (1775–1836), after whom the unit of electric current is named, studied magnetic fields created by current-carrying wires. He noticed that the direction of a magnetic field is always perpendicular to the direction of the wire. He also realized that magnetic field lines should have a symmetry corresponding to the symmetry of the wire. The only way this symmetry can occur is if the magnetic field lines form concentric circles around the wire in the planes perpendicular to it (Figure 24-39).

Ampère's law states the following:

1. The magnetic field in the space around an electric current is proportional to the electric current that serves as its source, just as an electric field is proportional to the charge that serves as its source.

2. For any closed-loop path (Figure 24-39), the integral of the scalar products of the length element $d\vec{\ell}$ and the magnetic field $\vec{B}$ along the closed path (which can be approximated by the sum of the scalar products of the length element $d\vec{\ell}$ and the magnetic field $\vec{B}$ along the closed path) is equal to the permeability of free space (or magnetic constant) μ_0 multiplied by the electric current enclosed by the loop:

KEY EQUATION
$$\oint \vec{B} \cdot d\vec{\ell} = \mu_0 \sum I_{enc} \qquad (24\text{-}38)$$

where the directions of the enclosed current are taken into account.

It is customary to refer to the closed path or loop in the context of Ampère's law applications as

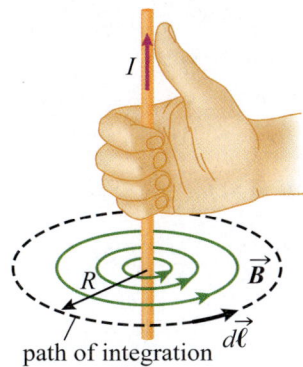

Figure 24-39 The magnetic field created by a straight current-carrying wire. The direction of the field can be determined using the right-hand curl rule: point your right-hand thumb in the direction of the electric current. Your curled fingers will show the direction of the magnetic field created by this current.

an Amperian loop, analogous to the concept of the Gaussian surface in Gauss's law.

When electric currents enclosed by a path of integration flow in opposite directions, they have opposite signs. The direction of the magnetic field can be determined using the right-hand curl rule: when you orient your right hand so that your thumb points in the direction of the electric current, your curled fingers point in the direction of the magnetic field (see Figure 24-39). It is especially convenient to apply Ampère's law when you can choose a path of integration that is either parallel to the magnetic field lines or perpendicular to them. Notice that only electric currents *enclosed* by the path of integration contribute to the line integral (similar to Gauss's law, described by Equation 20-5), where only electric charges enclosed by the surface, accounting for their sign, are used to calculate the electric flux contribution to the electric field).

Symmetry Considerations in the Application of Ampère's Law: Guidelines for Choosing the Path of Integration Just like Gauss's law, Ampère's law is universal—it holds everywhere. However, the applicability of Ampère's law to calculating magnetic fields due to current distributions depends to a large degree on the symmetry of the current distributions. Much like in the case of Gauss's law, the calculation of the line integral becomes easier when the magnitude of the magnetic field along the path of integration is a constant, and when the angle between the field and the path of integration is either 0° or 90°.

Applications of Ampère's Law

We now look at three common applications of Ampère's law. We first apply Ampère's law to the case of long current-carrying wire.

An infinitely long straight current-carrying wire Let us calculate the magnetic field generated by the long current-carrying wire using Ampère's law. While we already know the result from the application of the Biot–Savart law discussed earlier (Equation 24-31), we will see that Ampère's law can help us simplify the derivation of this result significantly. From the symmetry of the problem, we can infer that the magnetic field should have a profile of concentric circles around the wire. We can also see that the magnitude of the magnetic field at a given distance from the wire should be constant. We need a path of integration along which the magnitude of the magnetic field is constant, and at the same time we need the direction of the magnetic field to be either parallel or perpendicular to the path. This will allow us to simplify the calculations.

For the path of integration, we choose a circle centred on the wire and whose plane is perpendicular to it (Figure 24-39). Thus, we can apply Ampère's law (Equation 24-38) as follows:

$$\oint \vec{B}(\vec{r}) \cdot d\vec{\ell} = \mu_0 \sum I_{enc}$$

The scalar product inside the line integral can be expanded, noticing that due to our choice of the path of integration, the magnetic field lines are always parallel to the line of integration: $\vec{B} \parallel d\vec{\ell}$. Therefore, the scalar product can be expressed as

$$\vec{B}(\vec{r}) \cdot d\vec{\ell} = B(\vec{r}) \cdot d\ell = B \cdot d\ell$$

We also notice that the enclosed current is the current flowing in the wire, I, and the magnitude of the magnetic field at all locations at a distance R from the wire is constant; we shall call it $B(R)$. Therefore, we can move the magnetic field $B(R)$ out of the integral. At the same time, the integral of the length element over the circular path of radius R is the length of the circular path: $\oint dl = 2\pi R$. Consequently, Ampère's law for this case can be expressed as

$$\oint B dl = \mu_0 I$$

$$B(R) \oint dl = \mu_0 I$$

$$B(R) \cdot 2\pi R = \mu_0 I$$

$$B(R) = \frac{\mu_0 I}{2\pi R}$$

The same result was obtained earlier using the Biot–Savart law (Equation 24-31). However, applying Ampère's law for this highly symmetrical case greatly simplifies the derivation.

EXAMPLE 24-11

The Magnetic Field around a Long Straight Current-Carrying Wire

Find the magnetic field at a distance of 1.0 m from a long straight wire carrying a current of 1.0 A.

Solution

The magnitude of the magnetic field created by a long straight wire is described by Equation 24-32:

$$B = \frac{\mu_0 I}{2\pi R} = \frac{\left(4\pi \times 10^{-7}\, \frac{\text{T} \cdot \text{m}}{\text{A}}\right)(1.0\,\text{A})}{2\pi(1.0\,\text{m})}$$

$$= 2.0 \times 10^{-7}\,\text{T} = 2.0 \times 10^{-1}\,\mu\text{T}$$

Making sense of the result

It is worth noting that Earth's magnetic field is in the range of 30 μT to 60 μT, which is approximately 100 times the magnetic field 1.0 m from a long straight wire carrying 1.0 A of electric current.

CHECKPOINT 24-11

Ranking Line Integrals

Figure 24-40 shows six different configurations of electric currents going into or coming out of the plane of the page. The closed-loop path of integration is indicated by a dashed line in each diagram. Which of the following correctly ranks the magnitudes of the $(\vec{B}(\vec{r}) \cdot d\vec{\ell})$ integrals for these configurations?

(a) $C > E > A = B = D > F$
(b) $F > A = B = D = C > E$
(c) $C > E > A > B > F > D$
(d) $F > E > D = B = A > C$
(e) $A > D > B = C = E > F$

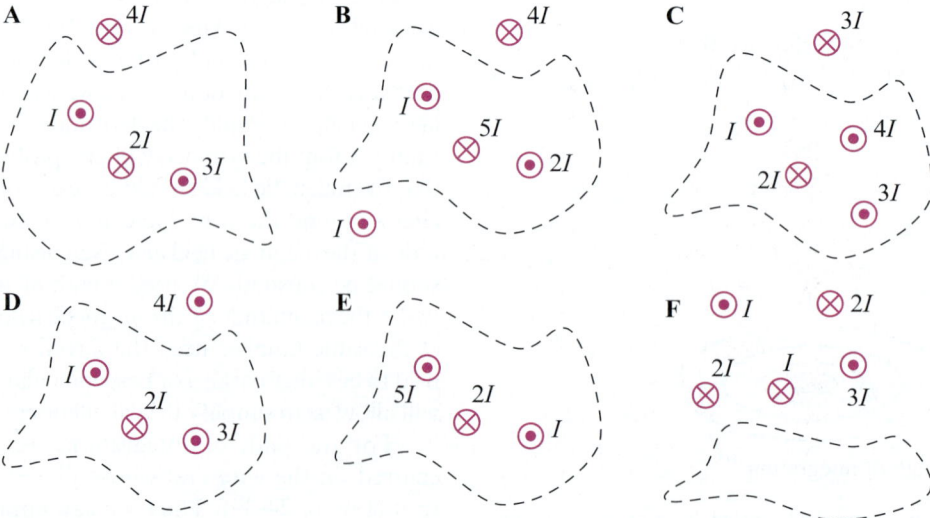

Figure 24-40   Checkpoint 24-11.

EXAMPLE 24-12

The Magnetic Field inside and outside a Long Straight Current-Carrying Wire

Determine the magnetic field inside and outside a long straight wire of radius R carrying an electric current I. Plot how this magnetic field depends on the distance from the wire. Clearly indicate the assumptions you are making to solve the problem.

Solution

Figure 24-41(a) shows an electric wire of radius R carrying an electric current I. To find the magnetic field inside the wire, let us assume a constant current density across the wire. Let us choose an arbitrary point P a distance r from the centre of the wire, such that $r < R$. Now let us draw a circular loop of radius r with the wire at its centre. This loop goes through point P. To find how much current is enclosed by the loop ($I_{enclosed}$), we calculate the current density inside the wire and multiply it by the cross-sectional area of the wire enclosed by it:

$$\text{Current density in the wire:} \quad J = \frac{I}{\pi R^2} \quad (1)$$

The enclosed current when $r < R$ can be calculated as the product of the cross-sectional area of the enclosed region (πr^2) and the current density J:

$$I_{enclosed} = J\pi r^2 = \frac{I}{\pi R^2}\pi r^2 = I\frac{r^2}{R^2} \quad (2)$$

Then the magnetic field inside the wire at a distance r from its centre, when $r < R$, can be calculated as

$$B_{\text{inside wire}} = \frac{\mu_0 I_{enclosed}}{2\pi r} = \frac{\mu_0}{2\pi r}I\frac{r^2}{R^2} = \frac{\mu_0 I}{2\pi R^2}r, \text{ when } r < R \quad (3)$$

The magnetic field *outside* the current-carrying wire was found earlier—Equation 24-32. This magnetic field is inversely proportional to the distance from the centre of the wire:

$$B_{\text{outside wire}} = \frac{\mu_0 I}{2\pi r}, r > R \quad (4)$$

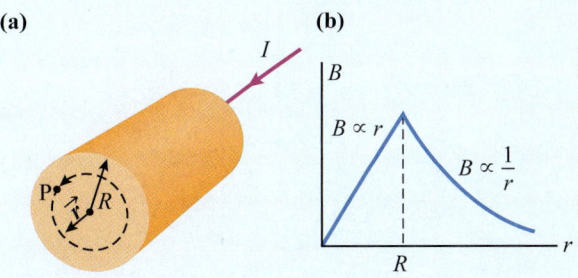

(a) **(b)**

Figure 24-41 (a) The cross section of a current-carrying wire. (b) A plot of $B(r)$, the dependence of the magnitude of the magnetic field created by the current-carrying wire on the distance from the centre of the wire.

Making sense of the result

The magnetic field *inside* the long straight current-carrying wire is proportional to the distance from the centre of the wire (in (3) above). The magnetic field *outside* the wire is inversely proportional to the distance from its centre, as shown in (4). These relationships are plotted in Figure 24-41(b). Notice that since the magnetic field must be continuous, the two expressions (3) and (4) should have the same value at the surface of the wire: $B_{\text{wire surface}} = \frac{\mu_0 I}{2\pi R}, r = R$.

Coaxial cables A **coaxial cable** is an electrical cable that consists of an inner conductor surrounded by an outer cylindrical conductor. The two conductors are separated by an insulating spacer of constant thickness, and the cable usually has an insulating jacket over the outer conductor (Figure 24-42). The term *coaxial* is used because the inner and outer conductors have the same axis of symmetry. Coaxial cables have many applications, in particular, for carrying high-frequency voltages for TV and radio signals. Coaxial cables have electrical characteristics that help reduce signal losses in long cable runs and shield against electromagnetic interference.

Ampère's law can be used to show that the electromagnetic field carrying the signal exists primarily in the space between the inner and outer conductors of the coaxial cable when the currents passing through the two conductors have equal magnitude but opposite direction. Outside the coaxial cable,

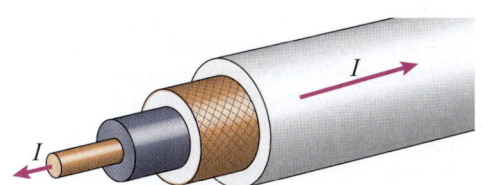

Figure 24-42 A cross section of a coaxial cable.

the magnetic field created by the electric current flowing in one direction is equal and opposite to the magnetic field created by the opposite current (see problem 91). Lastly, it is important to mention that Ampère's law in the form discussed above allows us to calculate the magnetic field for steady (unchanging) currents. In Chapter 27, we will consider how to calculate magnetic fields created by electric currents changing in time.

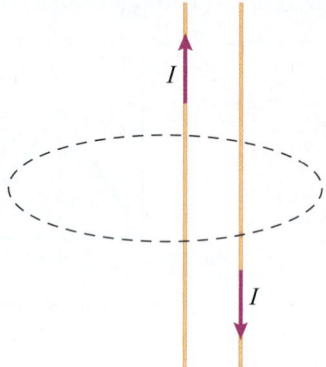
The Magnetic Field inside an Ideal Solenoid In Chapter 22, we showed that a capacitor consisting of two large oppositely charged parallel metal plates creates a uniform electric field in the space between the plates. A solenoid can be regarded as the magnetic counterpart of a capacitor. A solenoid is a device that produces a nearly uniform magnetic field (Figure 24-44). Solenoids are widely used because the uniform magnetic field inside them is determined solely by their spatial properties and the current flowing through them.

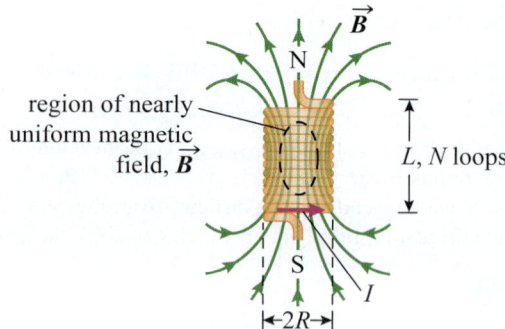

Figure 24-44 A solenoid of length L and radius R consists of N current-carrying loops.

Recall the expression for the magnetic field created in the centre ($x = 0$) of a current-carrying loop (Equation 24-34):

$$B_{\text{loop-centre}} = \frac{\mu_0}{2}\frac{I}{R}$$

Now we draw magnetic field lines created by a single loop, by three neighbouring loops, and by four neighbouring loops (Figure 24-45).

The magnetic field lines inside the loops (Figure 24-45) reinforce each other, and the magnetic field lines outside adjacent loops have opposite directions, so they nearly cancel each other. The more loops that are placed next to each other, the more their magnetic fields are reinforced inside the loops and cancel each other outside the loops. Let us consider a solenoid that has 10 loops (Figure 24-46(a)).

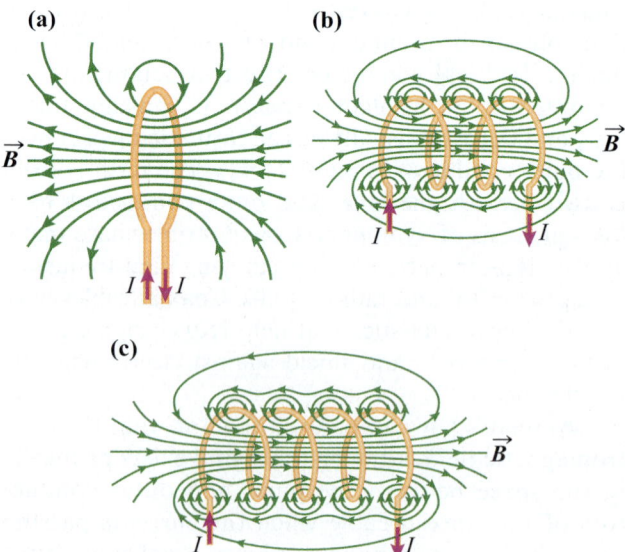

Figure 24-45 Calculating the magnetic field of a solenoid. (a) The magnetic field created by a single loop. (b) The magnetic field created by three adjacent loops. (c) The magnetic field created by four adjacent loops.

ELECTRICITY, MAGNETISM, AND OPTICS

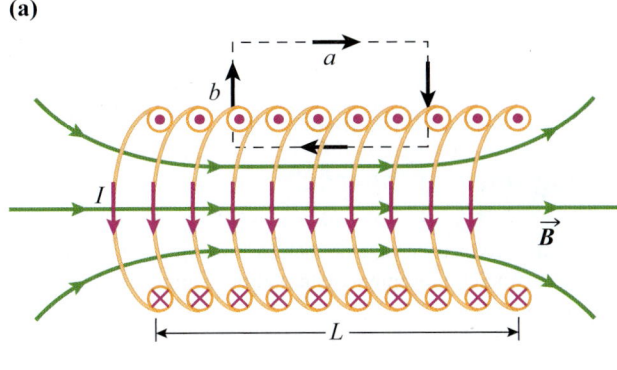

(a)

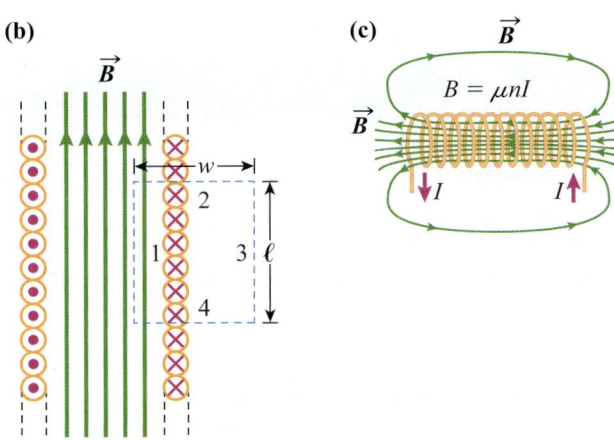

(b)   **(c)**

Figure 24-46   (a) The magnetic field created by a 10-loop solenoid, (b) the cross section of a very long solenoid, and (c) a very long solenoid consisting of N loops.

The cross section of this solenoid is shown in Figure 24-46(b). The magnetic field is almost uniform in the centre of the solenoid and is relatively weak outside it. Increasing the number of loops increases the magnetic field inside the solenoid and makes it more uniform (Figure 24-46(c)). A solenoid is also called an **electromagnet** because it uses an electric current to produce a magnetic field. However, unlike permanent magnets, the magnetic field of a solenoid depends on its physical properties and on the amount and direction of the electric current that flows through it.

To apply Ampère's law in this case, we will choose an integration path that follows the guidelines we highlighted earlier: one that is either parallel or perpendicular to the magnetic field, and one along which the magnitude of the magnetic field is uniform. As we discussed earlier, this is done to simplify the calculation of the line integral.

The magnetic field is parallel to the axis of the solenoid, so we choose a rectangular path with sides of length a and b (Figure 24-46(a)). The sides of length a are parallel to the axis of the solenoid. For the side that is inside the solenoid, the magnitude of the magnetic field along it is uniform and the angle between the length element of the path and the magnetic field

is zero: $\vec{B}_a \| d\vec{\ell}_a \Rightarrow \vec{B}_a \cdot d\vec{\ell}_a = Bd\ell$. The vertical sides of the rectangle (sides b) are oriented at right angles to the field inside the solenoid. Thus, the resulting scalar product between the length element and the field for these vertical sides inside the solenoid is zero: $\vec{B}_b \perp d\vec{\ell}_b \Rightarrow \vec{B}_b \cdot d\vec{\ell}_b = 0$. Since the magnetic field outside the solenoid is zero, there is no contribution to the line integral from the side of the path located outside the solenoid.

Now we are ready to apply Ampère's law:

$$\oint \vec{B} \cdot d\vec{\ell} = \mu_0 \sum I_{enc}$$

$$\int_{inside\, a} Bd\ell = \mu_0 \sum I_{enc}$$

$$Ba = \mu_0 \sum I_{enc}$$

Next we find an expression for the current enclosed in the Amperian loop. The amount of current running though the loop is equal to the number of turns going through the loop of length a, N_a, multiplied by the current, I: $I_{enc} = N_a I$. The number of turns going through the loop is the number of turns per unit length multiplied by the length of the side of the loop. Since the number of turns per unit length, n, is equal to the total number of turns of the solenoid divided by the length of the solenoid, we obtain $\sum I_{enc} = N_a I = \dfrac{N}{L} aI$. Equating this to the result of the line integral above, we have

$$Ba = \frac{N}{L} aI$$

or

KEY EQUATION
$$B_{solenoid} = \mu_0 \frac{IN}{L} = \mu_0 In \qquad (24\text{-}39)$$

Therefore, the magnetic field inside the solenoid depends on the density of the solenoid's loops ($n = N/L$) and the value of the electric current, I. Unlike the magnetic field created by a current-carrying loop, the magnetic field of a solenoid does not depend on the radius of the coils.

INTERACTIVE ACTIVITY 24-2

The Magnetic Field of a Solenoid

In this activity, you will use the PhET simulation "Magnets and Electromagnets" to gain a conceptual understanding of how the current and the number of loops affect the magnetic field lines created by a solenoid.

EXAMPLE 24-13

Coils in an MRI Magnet

Some MRI scanners need a magnetic field of 3.0 T. How many loops per metre does an electromagnet require to produce this magnetic field with a current of 15 A?

Solution

We can use Equation 24-39 to calculate the turn density of the coil:

$$B = \mu_0 \frac{IN}{L} = \mu_0 In$$

Therefore,

$$n = \frac{B}{\mu_0 I} = \frac{3.0 \text{ T}}{(4\pi \times 10^{-7} \text{ T·m/A})(15 \text{ A})} = 1.6 \times 10^5 \text{ loops/m}$$

Making sense of the result

The number of loops per metre is very high. To have fewer turns and to reduce the size of the electromagnets, MRI scanners use a superconducting wire that allows significantly higher electric currents.

MAKING CONNECTIONS

Magnetic Resonance Imaging

Although magnetic resonance imaging (MRI) was invented less than 50 years ago, it is difficult to imagine a modern hospital without it. MRI technology produces images of body tissue and can show details of soft tissue that are not visible using other imaging technologies, such as X-rays and CT scans (Figure 24-47). In addition to the advantage of having good resolution for soft tissue, we should mention that MRI does not involve ionizing radiation, so it does not carry the potential risks of X-rays. MRI machines generate these images by measuring changes in the magnetic fields of hydrogen nuclei in water molecules in the body (the human body is 70% water). These changes are caused by the nuclear magnetic resonance of the hydrogen nuclei: when a person is inside the scanner, the hydrogen nuclei (i.e., protons) align with a strong magnetic field in the scanning chamber. A radio wave at just the right frequency (the resonance frequency) for the protons to absorb energy pushes some of the protons out of alignment. The protons then snap back to alignment, producing a detectable rotating magnetic field as they do so. Since protons in different types of tissue (e.g., fat, muscle, or tumours) realign at different speeds, these tissues can be distinguished in the MRI images. Most MRI systems use a superconducting magnet to generate the very strong magnetic fields required, typically 0.5 T to 3.0 T (approximately 10 000 times Earth's magnetic field). The 2003 Nobel Prize in Physiology or Medicine was awarded to Paul C. Lauterbur (a chemist) and Peter Mansfield (a physicist) for the discovery of MRI. In their press release, the Nobel Committee stated: "Imaging of human internal organs with exact and non-invasive methods is very important for medical diagnosis, treatment and follow-up. This year's Nobel Laureates in Physiology or Medicine have made seminal discoveries concerning the use of magnetic resonance to visualize different structures. These discoveries have led to the development of modern magnetic resonance imaging, MRI, which represents a breakthrough in medical diagnostics and research."

(a)

REUTERS/Alamy Stock Photo

(b)

epa european pressphoto agency b.v./Alamy Stock Photo

(c)

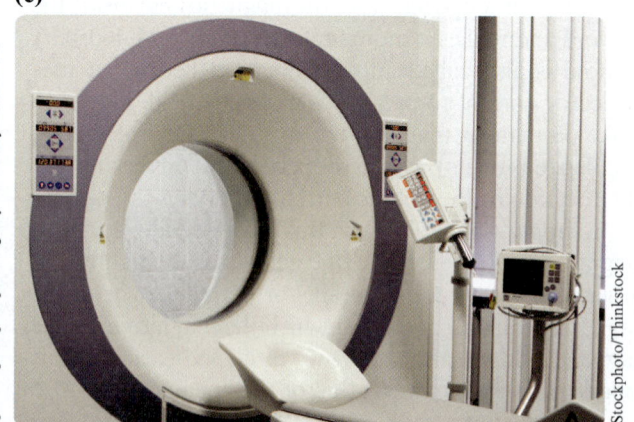
iStockphoto/Thinkstock

Figure 24-47   (a) British physicist Peter Mansfield and (b) American chemist Paul C. Lauterbur discovered magnetic resonance imaging. (c) An MRI scanner.

24-8 The Magnetic Force between Two Parallel Current-Carrying Conductors

Earlier we discovered that electric currents generate magnetic fields. At the same time, wires carrying electric currents are affected by external magnetic fields. Therefore, it is reasonable to assume that two current-carrying wires will interact with each other via these generated magnetic fields. Imagine two straight parallel wires that are separated by a distance R and carry electric currents I_1 and I_2 in the same direction, as shown in Figure 24-48(b). We first consider the magnetic field created by the top wire. Using the right-hand curl rule for determining the direction of the magnetic field created by a long straight current-carrying wire and illustrated in Figure 24-39, we see that the magnetic field created by I_1 is directed into the page directly below the wire and out of the page directly above the wire (Figure 24-48(a)). The magnitude of this field is given by Equation 24-32: $B_1 = \frac{\mu_0 I_1}{2\pi R}$.

This magnetic field exerts a magnetic force on the second wire carrying current I_2. Using Equation 24-21 for the magnetic force exerted on a current-carrying wire, we can see that the magnetic field from the top wire exerts an upward force on the bottom current-carrying wire (Figure 24-48(b)). Consequently, the magnetic force experienced by I_2 due to the magnetic field created by I_1 is a force of attraction. The magnitude of this force can be calculated using Equation 24-21:

$$\vec{F}_{1 \text{ on } 2} = I_2 \vec{\ell} \times \vec{B}$$

$$= I_2 \ell \frac{\mu_0 I_1}{2\pi R} = \frac{\mu_0 I_1 I_2}{2\pi R} \ell$$

Similarly, the magnetic force experienced by I_1 due to the magnetic field created by I_2 is directed down (it is also a force of attraction) and has a magnitude of

$$F_{2 \text{ on } 1} = I_1 \ell \frac{\mu_0 I_2}{2\pi R} = \frac{\mu_0 I_1 I_2}{2\pi R} \ell$$

The fact that these two forces are equal in magnitude and opposite in direction is not surprising because the interacting wires should obey Newton's third law. The magnitude of force per unit length is also equal for the two long straight wires:

KEY EQUATION
$$\frac{F_{1 \text{ on } 2}}{\ell} = \frac{F_{2 \text{ on } 1}}{\ell} = \frac{\mu_0 I_1 I_2}{2\pi R} \qquad (24\text{-}40)$$

Following the same steps, we can show that whenever two parallel wires carry electric currents in opposite directions, they repel each other (Figure 24-48(c)).

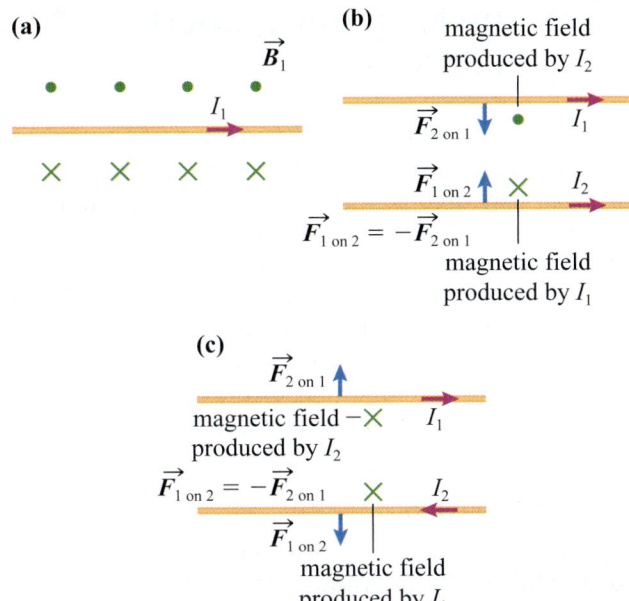

Figure 24-48 (a) The magnetic field produced by I_1. (b) Two parallel straight long wires carrying currents that flow in the same direction attract each other. (c) Two parallel straight long current-carrying wires carrying opposite currents repel each other.

In summary, two parallel wires carrying currents in the same direction attract each other; when they are carrying currents in opposite directions, they repel each other.

The force between two straight, parallel wires provides a way to define the SI unit of electric current, the ampere. When two parallel wires each carry an electric current of 1 A and are a distance of 1 m from each other, the force between the wires per metre of wire is

$$\frac{F_{1 \text{ on } 2}}{\ell} = \frac{F_{2 \text{ on } 1}}{\ell} = \frac{\mu_0 I_1 I_2}{2\pi R} = \frac{\mu_0 (1 \text{ A})(1 \text{ A})}{2\pi (1 \text{ m})}$$

$$= \frac{4\pi \times 10^{-7}}{2\pi} \text{ N/m} = 2 \times 10^{-7} \text{ N/m}$$

CHECKPOINT 24-13

The Magnetic Force between Two Current-Carrying Wires

Two straight parallel wires carry electric currents of $I_1 = I$ and $I_2 = 2I$. Which of the following equations correctly describes the magnetic forces that the wires exert on each other?

(a) $F_{1 \text{ on } 2} = 4 F_{2 \text{ on } 1}$
(b) $F_{1 \text{ on } 2} = 2 F_{2 \text{ on } 1}$
(c) $F_{1 \text{ on } 2} = F_{2 \text{ on } 1}$
(d) $2 F_{1 \text{ on } 2} = F_{2 \text{ on } 1}$
(e) $4 F_{1 \text{ on } 2} = F_{2 \text{ on } 1}$

ANSWER: (c) Think of Newton's third law.

ELECTRICITY, MAGNETISM, AND OPTICS

24-9 The Magnetic Properties of Materials

We began this chapter with a discussion of how credit cards work. To store information on a credit card, an array of microscopic magnetic particles is arranged. Credit cards, magnetic tapes, computer hard drives, and refrigerator magnets all depend on the magnetic properties of matter. In general, magnetic materials can be classified into three broad groups based on their magnetic properties: paramagnetic, diamagnetic, and ferromagnetic. The behaviour of these three classes of materials arises from interactions at the atomic level, which we look at first.

The Bohr Magneton

At the beginning of the 20th century, Danish physicist Niels Bohr (1885–1962) proposed a model for the structure of atoms (which will be discussed in more detail in Chapter 31). In Bohr's model, atoms consist of a nucleus made of protons and neutrons, with electrons orbiting around it. These orbiting electrons behave like microscopic current loops inside the atom and give the atom a magnetic moment. In some materials, the atomic magnetic moments orient themselves along the external magnetic field, increasing its strength. We call this process magnetization and say that some materials can be magnetized by an external magnetic field.

To understand how atomic magnetic moments contribute to the magnetic properties of matter, we look at hydrogen, the simplest atom. A semi-classical model of the hydrogen atom consists of one electron orbiting a nucleus that consists of just one proton. As discussed in Section 24-5, the magnetic dipole moment of the current loop can be expressed as $\vec{\mu} = I\vec{A}$, where I is the electric current and A is the area of the loop (Equation 24-27). Applying this definition to an electron orbiting the nucleus in an atom of hydrogen, we obtain

$$\mu = IA = \frac{e}{T}A = \frac{e}{2\pi r/v}\pi r^2 = \frac{evr}{2} \qquad (24\text{-}41)$$

It is often convenient to express the dipole moment of the hydrogen atom in terms of the angular momentum of an electron moving in a circular orbit (Section 8-7), $L = rp = rmv$:

$$\mu = \frac{evr}{2} = \frac{e}{2m}L \qquad (24\text{-}42)$$

Expression 24-42 shows that the magnetic dipole moment of a hydrogen atom is proportional to the angular momentum of its electron. Bohr suggested that this angular momentum must always be an integer multiple of $h/(2\pi)$, where h is Planck's constant, a fundamental

physical constant with a value of 6.626×10^{-34} J·s. Bohr's postulate of the quantization of angular momentum is a cornerstone of quantum mechanics (Chapter 32). Until now, we have discussed angular momentum as being a continuous physical quantity. However, due to the extremely small value of Planck's constant, the quantum properties of matter only become relevant at the atomic scale.

Using the quantization of angular momentum, we can rewrite Equation 24-42, describing the dipole moment of the hydrogen atom as

KEY EQUATION

$$\mu_n = \frac{e}{2m}L_n = \frac{e}{2m}\frac{h}{2\pi}n = \frac{eh}{4\pi m}n = \mu_B n \qquad (24\text{-}43)$$

where $n = 1, 2, 3, \ldots$

The quantity $\mu_B = \dfrac{eh}{4\pi m} = 9.274 \times 10^{-24}\,\text{J·T}^{-1} = 9.274 \times 10^{-24}\,\text{A·m}^2$ represents the fundamental unit of magnetic moment and is called the **Bohr magneton**. It is convenient to express magnetic moments of charged particles in terms of the Bohr magneton. Equation 24-43 can also be used to calculate the magnetic potential energy for a magnetic moment in a magnetic field:

$$U = -\vec{\mu} \cdot \vec{B} \qquad (24\text{-}44)$$

Electron Spin

In 1925, three young doctoral students—George Uhlenbeck, Samuel Goudsmit, and Ralph Krönig—independently came up with the then radical idea that an electron has an inherent property (similar to its mass or charge) that interacts with a magnetic field. This inherent property of an electron is called **electron spin**, and its magnitude is slightly greater than the Bohr magneton, $1.001\mu_B$. The net magnetic moment of an atom is the sum of the magnetic moments of the electrons orbiting the nucleus and the spins of these electrons.

Paramagnetism

In most materials, the net magnetic moments in the atoms are zero. However, in certain materials, some atoms have non-zero magnetic moments on the order of magnitude of the Bohr magneton, μ_B. When these materials—called paramagnetic materials—are placed in an external magnetic field, the unbalanced atomic magnetic moments experience a magnetic torque of $\vec{\tau} = \vec{\mu} \times \vec{B}$. The unbalanced torques will try to align with the magnetic field to decrease the potential energy of the system, thus increasing the net magnetic field. This phenomenon is called **paramagnetism**.

Since the net magnetic dipole moment of the material, $\vec{\mu}_{net}$, depends on the amount of material, it is more meaningful to discuss the net magnetic dipole moment of the material per unit volume. This quantity is called the **magnetization** of the material and is denoted by $\vec{M}$:

$$\vec{M} = \frac{\vec{\mu}_{net}}{V} \qquad (24\text{-}45)$$

This additional magnetic dipole moment produces an additional magnetic field, which can be calculated in the same way as the magnetic field produced by a current-carrying loop: $\vec{B}_{add} = \mu_0\vec{M}$. Then the total magnetic field inside such a material is given by

$$\vec{B} = \vec{B}_0 + \mu_0\vec{M}$$

where $\vec{B}_0$ is the external magnetic field.

It is common to describe the behaviour of paramagnetic materials using the **relative permeability** of the material, K_m, which indicates how much greater the magnetic field inside a paramagnetic material in an external magnetic field is than in the external magnetic field:

KEY EQUATION $\qquad \vec{B} = \vec{B}_0 + \mu_0\vec{M} = K_m\vec{B}_0 \qquad (24\text{-}46)$

Table 24-1 shows the relative magnetic permeability, K_m, of different paramagnetic and diamagnetic materials at room temperature ($20\,°C$). The relative magnetic permeabilities of these materials typically range from 1.000 01 to 1.007 20. Another way of thinking of the relative magnetic permeability is to use a coefficient that compares the magnetic field inside the paramagnetic material to the magnetic field in a vacuum immersed in the same external magnetic field. This view of relative permeability allows us to extend all the magnetic field derivations to current-carrying conductors embedded in paramagnetic materials. For such derivations, μ_0 should be replaced by the **magnetic permeability**, μ:

$$\mu = \mu_0 K_m \qquad (24\text{-}47)$$

It is important to mention that we have to be careful with Equation 24-47, as it has limited applicability. For example, when we use ferromagnetic materials as the iron core of a solenoid, we magnify the magnetic field inside it by aligning the magnetic domains inside the iron with the external magnetic field created by the solenoid. However, the effect of magnification will only work up to the point when the magnetic field in the core reaches its **magnetic saturation level** (when all the magnetic domains have been aligned). Then increasing the external magnetic field will not increase the magnetic field inside this ferromagnetic material. Core saturation in a transformer can increase its temperature. Due to magnetic saturation, the magnetic permeability of ferromagnetic materials reaches its

Table 24-1 Relative Magnetic Permeability K_m and Magnetic Susceptibility of Paramagnetic and Diamagnetic Materials, χ_m, at Room Temperature (20 °C)

Material	Relative Magnetic Permeability, K_m	Magnetic Susceptibility, $\chi_m = K_m - 1$
Paramagnetic materials		
Iron oxide	1.007 20	0.007 20
Ammonium iron (III) sulfate (iron alum)	1.000 66	0.000 66
Uranium	1.000 40	0.000 40
Platinum	1.000 26	0.000 26
Tungsten	1.000 068	0.000 068
Cesium	1.000 051	0.000 051
Aluminum	1.000 022	0.000 022
Lithium	1.000 014	0.000 014
Magnesium	1.000 012	0.000 012
Sodium	1.000 007 2	0.000 007 2
Oxygen gas	1.000 001 9	0.000 001 9
Diamagnetic materials		
Bismuth	0.999 834	−0.000 166
Mercury	0.999 971	−0.000 029
Silver	0.999 974	−0.000 026
Carbon (diamond)	0.999 979	−0.000 021
Carbon (graphite)	0.999 984	−0.000 016
Lead	0.999 982	−0.000 018
Sodium chloride	0.999 986	−0.000 014
Copper	0.999 99	−0.000 010
Water	0.999 991	−0.000 009 1

Source: Adapted from HyperPhysics, http://hyperphysics.phy-astr.gsu.edu/hbase/Tables/magprop.html.

maximum and then begins to decline. The magnetic saturation phenomenon has wide technical applications. It is also discussed in problem 56.

Be careful not to confuse the magnetic permeability of a material with the magnetic dipole moment, which is also denoted by μ.

Another convenient way of describing paramagnetic materials is to use the **magnetic susceptibility**, χ_m, which shows by how much the magnetic permeability differs from one. For paramagnetic materials, the magnetic susceptibility is positive (Table 24-1):

$$\chi_m = K_m - 1 \qquad (24\text{-}48)$$

For paramagnetic materials, the magnetization of the material depends on the absolute temperature. In many cases, this dependence can be described by Curie's law, named after renowned French scientist Pierre Curie (1859–1906):

$$M = C\frac{B}{T} \qquad (24\text{-}49)$$

CHAPTER 24 | **MAGNETIC FIELDS AND MAGNETIC FORCES**

where C is the Curie constant, T is the absolute temperature, and B is the magnetic field. Above the Curie temperature, paramagnetic materials lose their permanent magnetic properties, which are replaced by induced magnetism.

Different materials have different Curie constants. Curie's law can be understood when we consider how the variables affect the atomic magnetic moments: thermal motion tends to randomize atomic magnetic moments, and an external magnetic field tends to align the atomic magnetic moments along its direction.

Diamagnetism

Unlike paramagnetic materials, the net magnetic moments of atoms in diamagnetic materials are zero when there is no external magnetic field present. However, an external magnetic field affects the motion of the electrons within the atoms, creating magnetic dipole moments within the atoms. These induced magnetic dipole moments are always directed opposite to the external magnetic field. We explain the reason for this orientation when we talk about Faraday's law of electromagnetic induction in Chapter 25. **Diamagnetism** can be compared to the effect of polarization by an electric field (Section 19-6). For example, a neutral water molecule can become polarized in the presence of an electric field and act as an electric dipole. However, unlike electrical polarization, the polarization of diamagnetic materials is always opposite to the external magnetic field.

Diamagnetic materials always have a negative susceptibility, as shown in Table 24-1. The relative permeability of diamagnetic materials is slightly less than one. In contrast to paramagnetic materials, the susceptibility of diamagnetic materials is almost independent of temperature.

Ferromagnetism

Note that Table 24-1 does not include the metals that come to mind when we talk about magnetic materials: iron, nickel, cobalt, and their alloys. These materials constitute a third group of magnetic materials, called ferromagnetic materials. In ferromagnetic materials, strong magnetic interactions between atomic magnetic dipole moments make them align parallel to each other within small regions even when there is no external magnetic field, a phenomenon called **ferromagnetism**. These regions of aligned dipole moments are called **magnetic domains**. The magnetic moment of a magnetic domain can be of the order of magnitude of $1000\mu_B$ because most of the individual atomic magnetic moments within the domain are parallel.

When there is no external magnetic field, $\vec{B}_0$, as shown in Figure 24-49, the magnetic domains are randomly oriented, and the net magnetic moment of the

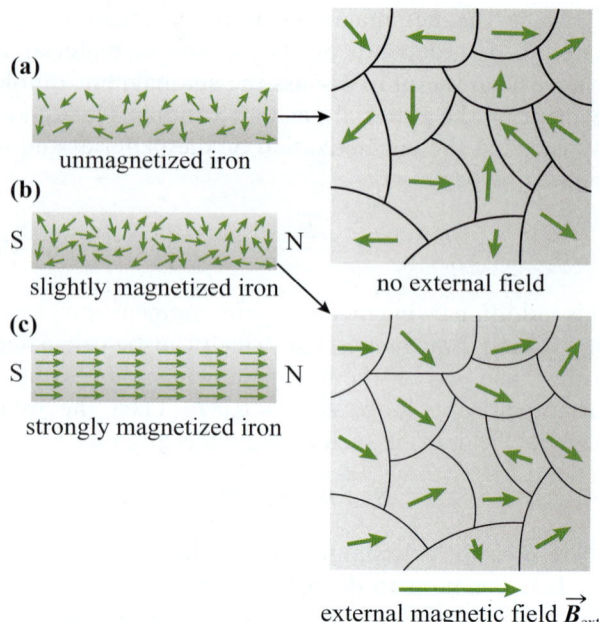

Figure 24-49 (a) Magnetic domains inside a ferromagnetic material when there is no external magnetic field present. (b) Magnetic domains in the presence of a weak external magnetic field. (c) Magnetic domains in the presence of a strong magnetic field.

material is almost zero. However, in the presence of an external magnetic field, $\vec{B}_0$, the domains orient themselves in the direction of the field, creating a strong magnetic field within the material.

The boundaries of magnetic domains in ferromagnetic materials are also affected by the external magnetic field: domains oriented in the direction of an external magnetic field grow as the external magnetic field increases. As a result, the relative magnetic permeability of a ferromagnetic material, K_m, is 1000 to 1 000 000 times the relative permeability of a paramagnetic material (Table 24-2). Consequently, ferromagnetic materials, such as iron, are strongly attracted to permanent magnets, and paramagnetic materials, such as aluminum, are not. Aluminum does experience some magnetic attraction, but the magnetic attraction of iron is thousands of times stronger in the same magnetic field.

When an external magnetic field increases, more and more domains in the ferromagnetic material align with the external magnetic field. As a result, the magnetization and the relative magnetic permeability of the ferromagnetic material increase. When all the magnetic domains have aligned with the external magnetic field, increasing the external field will not produce any further increase in relative permeability. At this point, the ferromagnetic sample has reached **saturation of magnetization**.

Figure 24-51 shows how the magnetization of a ferromagnetic material varies as the external

Table 24-2 Initial and Maximum Relative Magnetic Permeabilities, K_m, of Some Ferromagnetic Materials at Room Temperature (20 °C)

Material	Treatment	Initial Relative Magnetic Permeability, K_m	Maximum Relative Magnetic Permeability, K_m
Iron, 99.8% pure	Annealed	150	5 000
Iron, 99.95% pure	Annealed in hydrogen	10 000	200 000
78 permalloy	Annealed, quenched	8 000	100 000
Super-permalloy	Annealed in hydrogen, controlled cooling	100 000	1 000 000
Cobalt, 99% pure	Annealed	70	250
Nickel, 99% pure	Annealed	110	600

Source: Adapted from HyperPhysics, http://hyperphysics.phy-astr.gsu.edu/hbase/Tables/magprop.html.

MAKING CONNECTIONS

Animal Magnetism: How Do Animals Navigate?

Many of us have wondered how animals as diverse as sea turtles, whales, birds, worms, wolves, and butterflies can navigate hundreds and even thousands of kilometres without getting lost along the way. For example, monarch butterflies, arguably the most beautiful of all butterflies, migrate a thousand kilometres from Canada to Mexico every year (Figure 24-50). What helps their navigation? How can whales undergo months-long migrations and arrive exactly at their final destination?

Just recently, scientists have discovered that animal magnetic sense (the seventh sense) comes from a special protein that acts as a compass. This protein always points north, thus directing the animals in their journey. This is a groundbreaking discovery that shows how magnetic phenomena shape animals' behaviour.

Figure 24-50 Monarch butterflies can navigate thousands of kilometres during their annual migration thanks to their magnetosensor.

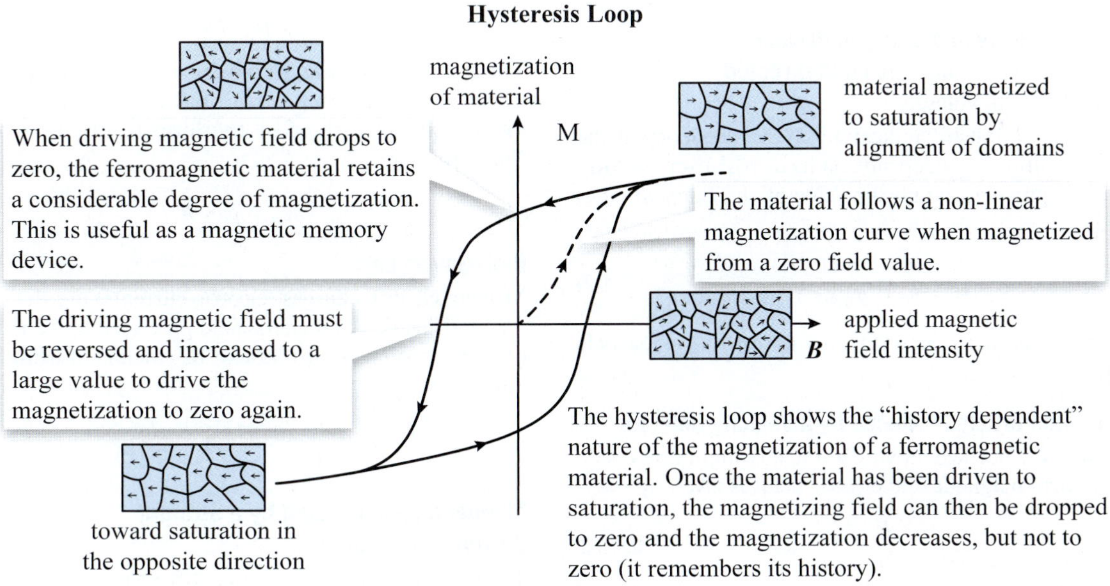

Hysteresis Loop

When driving magnetic field drops to zero, the ferromagnetic material retains a considerable degree of magnetization. This is useful as a magnetic memory device.

The driving magnetic field must be reversed and increased to a large value to drive the magnetization to zero again.

toward saturation in the opposite direction

material magnetized to saturation by alignment of domains

The material follows a non-linear magnetization curve when magnetized from a zero field value.

The hysteresis loop shows the "history dependent" nature of the magnetization of a ferromagnetic material. Once the material has been driven to saturation, the magnetizing field can then be dropped to zero and the magnetization decreases, but not to zero (it remembers its history).

Figure 24-51 Magnetic hysteresis loop for a ferromagnetic material.

Source: Ferromagnetic materials, http://electrons.wikidot.com/ferromagnetic-materials, licensed under Creative Commons Attribution-ShareAlike 3.0 License: http://creativecommons.org/licenses/by-sa/3.0/

magnetic field changes. This nonlinear behaviour is called magnetic **hysteresis**. As the external magnetic field increases, the magnetization of a ferromagnetic sample increases until it reaches its maximum value (saturation). However, when the external magnetic field is removed, the magnetization of the sample does not completely disappear. To decrease the magnetization of the sample to zero, we must apply a magnetic field in the opposite direction. The area inside the hysteresis curve indicates the relative amount of energy required to magnetize and demagnetize a sample: a ferromagnetic material that has a large area inside the hysteresis loop requires more energy to re-magnetize it than a ferromagnetic sample with a smaller area. The more energy required to demagnetize a material, the better this material can serve as a permanent magnet.

We finish this chapter by returning to the credit card stripe we began with (Figure 24-1). Magnetic stripes are made of ferromagnetic materials, usually iron oxide or a barium ferrite, that can be magnetized in particular ways to store the information on the card. As we understand now, these materials retain their properties unless you expose them for a prolonged time to very strong magnetic fields (this is why you should not place your credit cards near strong permanent magnets). Reading the information off the magnetic stripe requires the laws of electromagnetic induction, which we will explore in the following chapter.

KEY CONCEPTS AND RELATIONSHIPS

Magnetic fields affect *moving* electric charges, and, at the same time, moving electric charges (electric currents) are the sources of magnetic fields.

The Magnetic Force

Charged particles *moving* inside a magnetic field, $\vec{B}$, experience a magnetic force, $\vec{F}_B$, that is perpendicular to both the particle's velocity, $\vec{v}$, and the direction of the magnetic field:

$$\vec{F}_B = q\vec{v} \times \vec{B} \Rightarrow F_B = qvB\sin\theta \quad \text{and} \quad \vec{F}_B \perp \vec{v}, \vec{B} \quad (24\text{-}1)$$

The Lorentz Force

Charged particles moving in electric and magnetic fields experience a Lorentz force:

$$\vec{F} = q\vec{E} + q\vec{v} \times \vec{B} \quad (24\text{-}2)$$

Cyclotron Frequency of a Charged Particle Moving in a Uniform Magnetic Field Directed Perpendicular to Its Velocity

The following relationship indicates that the frequency of the particle's revolution only depends on its charge-to-mass ratio (the particle's properties) and the value of the external magnetic field:

$$f_{cyc} = \frac{\omega}{2\pi} = \frac{qB}{2\pi m} \quad (24\text{-}7)$$

It is the basis for the operation of a cyclotron—a particle accelerator.

Velocity of the Particle Produced by a Velocity Selector

The velocity of a particle moving along a straight line inside crossed uniform magnetic and electric fields is given by

$$v = \frac{E}{B} \quad (24\text{-}12)$$

Hall Effect

The Hall effect is the generation of a potential difference (Hall potential difference) across a current-carrying conductor submerged in a magnetic field perpendicular to the direction of the current. The potential difference is generated in the conductor along the direction perpendicular to the directions of an electric current in the conductor and the external magnetic field:

$$\Delta V_{Hall} = wv_d B = \frac{IB}{hn_q q} \quad (24\text{-}19)$$

Magnetic Force on a Current-Carrying Wire

A current-carrying wire submerged in a magnetic field experiences a force given by

$$\vec{F} = I\vec{\ell} \times \vec{B} \quad (24\text{-}21)$$

It interacts with the magnetic field.

Magnetic Torque Exerted on a Current-Carrying Loop

The magnetic dipole moment is defined as

$$\vec{\mu} = I\vec{A} \quad (24\text{-}27)$$

The current-carrying loop behaves like a magnet in a magnetic field. The component of the magnetic field perpendicular to the surface of the loop results in a torque on the loop:

$$\vec{\tau}_B = \vec{\mu} \times \vec{B} \quad (24\text{-}28)$$

Biot–Savart Law

Moving electric charges (electric currents) are the sources of magnetic fields. A magnetic field produced by a small element of wire carrying a current can be calculated using the Biot–Savart law:

$$d\vec{B} = \frac{\mu_0}{4\pi}\frac{Id\vec{\ell} \times \hat{r}}{r^2} = \frac{\mu_0}{4\pi}\frac{Id\vec{\ell} \times \vec{r}}{r^3} \quad (24\text{-}31)$$

Magnetic Field Created by a Straight Current-Carrying Wire

The magnetic field has a concentric shape, and its magnitude can be described as

$$B_{\text{outside straight long wire}} = \frac{\mu_0 I}{2\pi R} \quad (24\text{-}32)$$

Magnetic Field Created by a Circular Current-Carrying Loop

Along the axis of symmetry perpendicular to the surface of the loop, the magnitude of the magnetic field can be calculated as

$$B_{\text{loop}} = \frac{\mu_0}{2} \frac{IR^2}{(x^2 + R^2)^{3/2}} \qquad (24\text{-}33)$$

In the centre of the current-carrying loop, the magnitude of the magnetic field can be calculated as

$$B_{\text{loop-centre}} = \frac{\mu_0}{2} \frac{IR^2}{R^3} = \frac{\mu_0}{2} \frac{I}{R} \qquad (24\text{-}34)$$

Ampère's Law

Ampère's law provides another way of calculating the magnetic field created by a wire carrying a current:

$$\oint \vec{B} \cdot d\vec{\ell} = \mu_0 \sum I_{\text{enc}} \qquad (24\text{-}38)$$

Magnetic Field inside a Solenoid

The magnetic field inside a solenoid is given by

$$B_{\text{solenoid}} = \mu_0 \frac{IN}{L} = \mu_0 In \qquad (24\text{-}39)$$

Magnetic Force between Two Parallel Current-Carrying Wires

The magnetic force between two parallel current-carrying wires is given by

$$\frac{F_{1\text{ on }2}}{\ell} = \frac{F_{2\text{ on }1}}{\ell} = \frac{\mu_0 I_1 I_2}{2\pi R} \qquad (24\text{-}40)$$

Two parallel wires carrying currents flowing in the same direction attract, and two wires carrying currents flowing in opposite directions repel.

Magnetic Properties of Matter

The quantization of the dipole moment of the hydrogen atom is given by

$$\mu_n = \frac{e}{2m}L_n = \frac{e}{2m}\frac{h}{2\pi}n = \frac{eh}{4\pi m}n = \mu_{\text{B}}n \qquad (24\text{-}43)$$

where $n = 1,2,3,\dots$

We describe the magnetic field inside a paramagnetic material as a function of the external magnetic field, $\vec{B}_0$, and the relative permeability of the material, K_{m}:

$$\vec{B} = \vec{B}_0 + \mu_0 \vec{M} = K_{\text{m}} \vec{B}_0 \qquad (24\text{-}46)$$

APPLICATIONS

mass spectrometers, electromotors, magnetic field sensors, proximity switches, position and speed detectors, current sensors, coaxial cables, solenoids

KEY TERMS

Ampère's law, Bohr magneton, coaxial cable, constant magnetic field, cyclotron, cyclotron frequency, diamagnetism, electromagnet, electron spin, ferromagnetism, Hall effect, Hall potential difference, hysteresis, Lorentz force, magnetars, magnetic dipole, magnetic dipole moment, magnetic domains, magnetic field, magnetic field lines, magnetic flux density, magnetic force, magnetic permeability, magnetic saturation level, magnetic susceptibility, magnetization, mass spectrometer, paramagnetism, permeability constant of free space, pitch of the helix, relative permeability, right-hand curl rule, right-hand rule, saturation of magnetization, tesla, uniform magnetic field, velocity selector

QUESTIONS

1. A charged particle moves in a straight line through a particular region of space. Could there be a non-zero magnetic field in this region? Explain.

2. An electron is travelling above and parallel to a conducting wire when a current in the wire is turned on. The electron curves upward and away from the wire. What is the direction of the current in the wire?
   (a) the same direction as the electron's motion
   (b) the opposite direction to the electron's motion
   (c) Both answers are possible.
   (d) The current in the wire will have no effect on the electron.

3. Find the direction of the force on the negative charge for each diagram in Figure 24-52. Copy the diagram into your notebook, and indicate the direction of the force on your diagram.

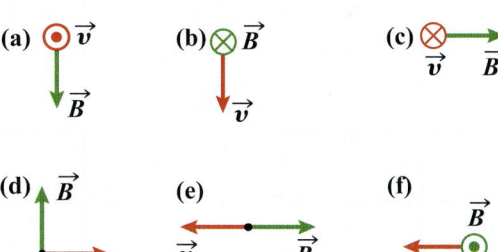

Figure 24-52 Question 3.

4. A charged particle is moving along a circular path under the influence of a uniform magnetic field. An electric field that points in the same direction as the magnetic field is turned on. Describe the path that the charged particle will take.

5. A rectangular piece of a semiconductor is inserted in a magnetic field, and a battery is connected to its ends, as shown in Figure 24-53. When a sensitive voltmeter is connected between points D (back face) and C (front face), it is found that point D is at a higher potential than point C. What is the sign of the charge carriers in this semiconductor material? Explain.

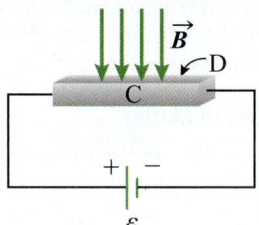

Figure 24-53 Question 5.

6. Figure 24-54 shows several situations with two wires with currents of 2 A flowing out of or into the page. Rank the magnetic field at point P from greatest to least. Denote the downward-pointing field as negative and the upward-pointing field as positive. Assume that all the adjacent lines on the grid are separated by the same distance and that each situation is independent of the others.
Greatest 1___ 2___ 3___ 4___ 5___ 6___ 7___ Least
Or, the magnetic field is the same in all cases. ___
Or, the magnetic field is zero in all cases. ___
If the magnetic field is the same for two or more cases, clearly indicate it on the ranking scheme. Explain your reasoning.

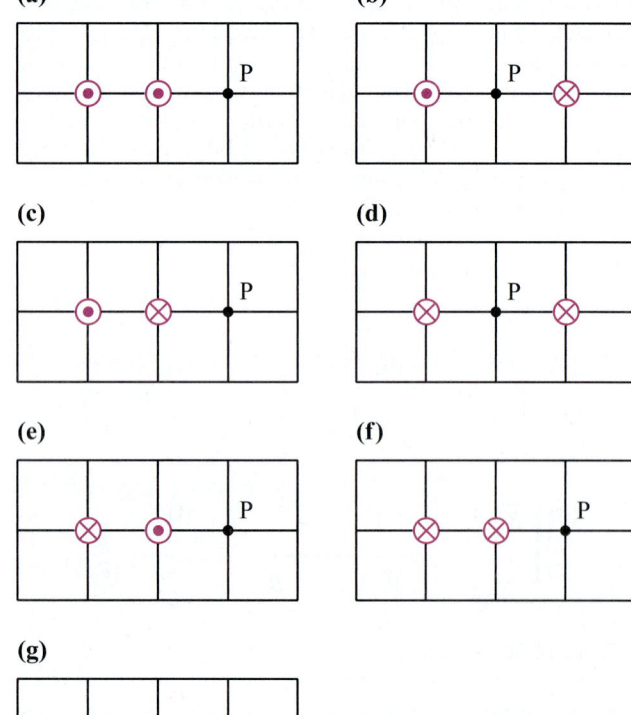

Figure 24-54 Question 6.

7. The coil shown in Figure 24-55 attracts a permanent bar magnet. Which of the following statements correctly describes the polarity of the permanent bar magnet? Draw a diagram for this problem, and show your solution on the diagram.
(a) The end of the bar magnet that is close to the coil is north; the opposite end is south.
(b) The end of the bar magnet that is close to the coil is south; the opposite end is north.
(c) The upper side of the bar magnet is north; the bottom side is south.
(d) The upper side of the bar magnet is south; the bottom side is north.
(e) Since we have a permanent magnet, we cannot determine its polarity in this example.
(f) A permanent magnet (unlike a temporary magnet) has zero polarity; it is neutral.

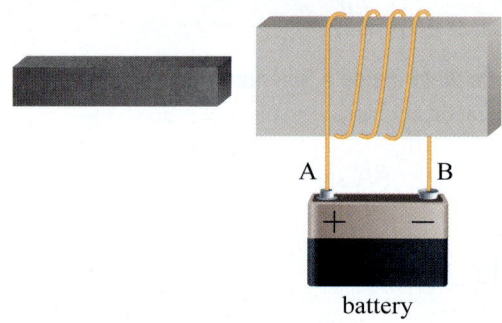

Figure 24-55 Question 7.

8. The coil in Figure 24-56 repels the iron bar on the left by causing it to become a temporary magnet. Which of the following statements correctly describes the polarity of the temporary bar magnet in this situation? Draw a diagram for this problem, and show your solution on the diagram.
(a) The end of the bar magnet that is close to the coil is north; the opposite end is south.
(b) The end of the bar magnet that is close to the coil is south; the opposite end is north.
(c) The upper side of the bar magnet is north; the bottom side is south.
(d) The upper side of the bar magnet is south; the bottom side is north.
(e) A temporary magnet cannot repel the coil, so this example does not make sense.

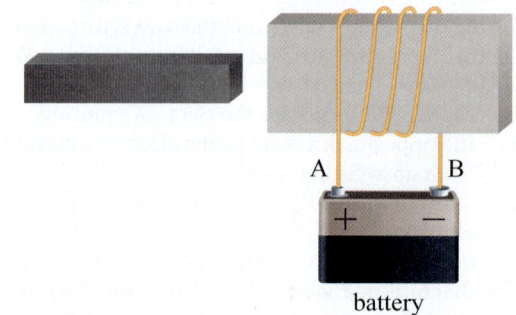

Figure 24-56 Question 8.

9. A proton moving upward enters the magnetic field created by two bar magnets, as shown in Figure 24-57. The direction of the force exerted on the proton by the magnetic field is
  (a) in the direction of the proton motion (upward)
  (b) in the direction opposite to the proton motion (downward)
  (c) left (i.e., toward the south pole of the left magnet)
  (d) right (i.e., toward the north pole of the right magnet)
  (e) into the page
  (f) out of the page

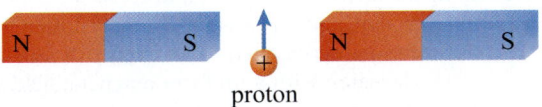

Figure 24-57 Question 9.

10. What would your answer be to question 9 if there was an electron instead of a proton between the magnets?

11. An electric current in a long vertical straight wire is flowing straight up (Figure 24-58). What is the direction of the magnetic field at point P?
  (a) up (vertically)
  (b) down (vertically)
  (c) into the page
  (d) out of the page
  (e) to the right (horizontally)
  (f) to the left (horizontally)

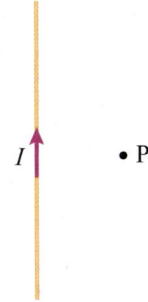

Figure 24-58 Question 11.

12. Two long parallel vertical wires carry electric currents of I and $3I$ in the same direction, as shown in Figure 24-59. The wires are separated by a distance d. At what point will the net magnetic field created by these wires equal zero?
  (a) between the wires, a distance $d/3$ from the wire carrying current I and $2d/3$ from the wire carrying current $3I$
  (b) between the wires, a distance $2d/3$ from the wire carrying current I and $d/3$ from the wire carrying current $3I$
  (c) outside the wire carrying current I, a distance $d/3$ from this wire and $4d/3$ from the wire carrying current $3I$
  (d) between the wires, a distance $d/4$ from the wire carrying current I and $3d/4$ from the wire carrying current $3I$

(e) between the wires, a distance $3d/4$ from the wire carrying current I and $d/4$ from the wire carrying current $3I$

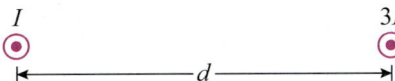

Figure 24-59 Question 12.

13. A negatively charged ion moving with speed v (Figure 24-60) enters three adjacent regions of uniform magnetic fields but of varying magnitudes. The magnetic field in each region is directed perpendicular to the plane of the paper. Which of the following correctly ranks the magnitudes of these magnetic fields?
  (a) $B_3 > B_1 > B_2$
  (b) $B_3 > B_2 > B_1$
  (c) $B_2 > B_1 > B_3$
  (d) $B_2 > B_3 > B_1$
  (e) $B_1 > B_3 > B_2$
  (f) $B_1 > B_2 > B_3$

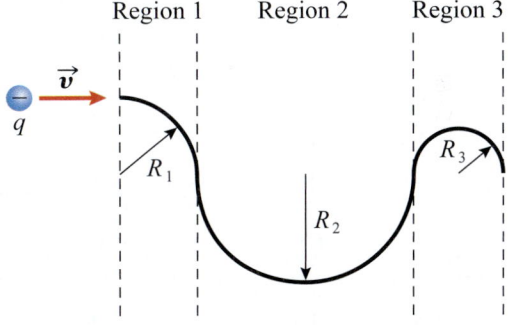

Figure 24-60 Question 13.

14. What is true about the directions of the magnetic fields in question 13?
  (a) Magnetic fields $\vec{B}_1$ and $\vec{B}_3$ are into the page, and magnetic field $\vec{B}_2$ is out of the page.
  (b) Magnetic fields $\vec{B}_1$ and $\vec{B}_3$ are out of the page, and magnetic field $\vec{B}_2$ is into the page.
  (c) All of the magnetic fields are directed into the page.
  (d) All of the magnetic fields are directed out of the page.
  (e) There is not enough information to determine the directions.

15. An electron is moving inside the cathode ray tube shown in Figure 24-14. What happens to the deflection of an electron when the magnetic field decreases by a factor of 2?
  (a) The deflection remains the same.
  (b) The deflection decreases by a factor of 2.
  (c) The deflection decreases but not by a factor of 2.
  (d) The deflection increases by a factor of 2.
  (e) The deflection increases but not by a factor of 2.

16. Rank the magnitudes of the magnetic fields produced at the centres of curvature (points O) of the wires shown in Figure 24-61 from highest to lowest. All the wires carry an identical electric current. The radii of the circular segments are shown in the figure.

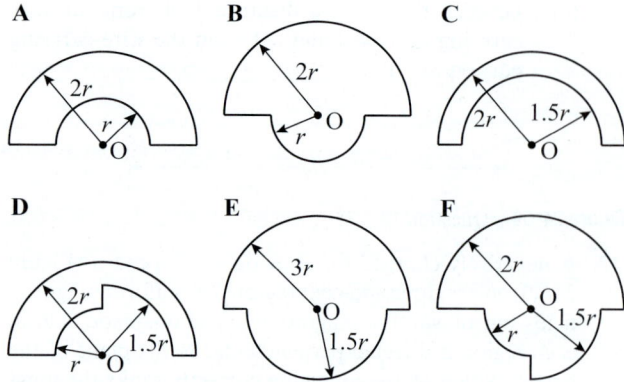

Figure 24-61 Question 16.

17. Rank the magnitudes of the magnetic fields in the centres of the six squares shown in Figure 24-62. The magnetic fields are created by four different parallel or antiparallel currents, I, flowing through the vertices of each square.

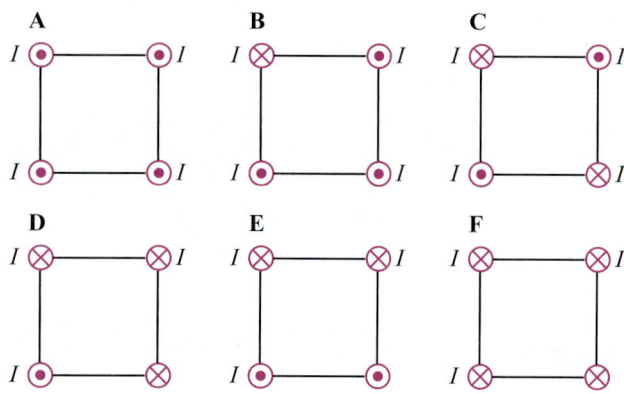

Figure 24-62 Question 17.

18. Each wire in Figure 24-63 carries an electric current of 1 A into or out of the page. Each diagram indicates the path of a line integral $\oint \vec{B} \cdot d\vec{\ell}$ to be used while applying Ampère's law for calculating the magnitude of the resultant magnetic field. Rank the values of the line integrals.

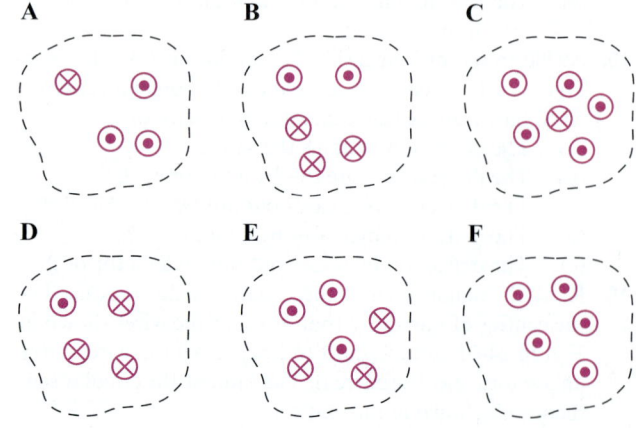

Figure 24-63 Question 18.

PROBLEMS BY SECTION

For problems, star ratings will be used (★, ★★, or ★★★), with more stars meaning more challenging problems. The following codes will indicate if $\frac{d}{dx}$ differentiation, $\int$ integration, ▭ numerical approximation, or ◠ graphical analysis will be required to solve the problem.

Section 24-1   Magnetic Field and Magnetic Force

19. ★ An electron moving with velocity $\vec{v} = (4.00 \times 10^4 \text{ m/s})\hat{i} - (8.00 \times 10^4 \text{ m/s})\hat{j}$ enters a region where a uniform magnetic field of $\vec{B} = (0.800 \text{ T})\hat{i} - (0.600 \text{ T})\hat{j}$ is present. Determine the force on the electron in terms of its magnitude and direction, and express it in algebraic notation.

20. ★ (a) An electron, accelerated to an energy of 100.0 eV, enters a region with a uniform magnetic field with a magnitude of 1.500×10^{-2} T. When a magnetic field is directed in the positive x-direction and an electron starts moving in the positive z-direction, what are the direction and magnitude of the force acting on the electron?
    (b) How would your answer change if instead of an electron, a proton with the same energy entered this region of the magnetic field? Explain.

21. ★ Figure 24-64 shows a proton entering a horizontal magnetic field. The proton's velocity is in a vertical plane at an angle to the magnetic field as shown.

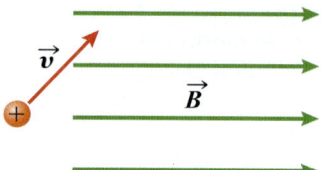

Figure 24-64 Question 21.

22. ★ A particle carrying a charge of 10. nC is moving at $0.001c$ along the positive x-direction in the region of a uniform magnetic field directed in the negative z-direction. The magnitude of the magnetic field is 1.5 T. Calculate the magnetic force on the particle, indicating both the magnitude and the direction.

23. ★★ A droplet has a mass of 2.0×10^{-4} g, a charge of 35nC, and an initial horizontal velocity of $(6.0 \times 10^5 \text{ m/s})\hat{j}$. The electric field in the vicinity of Earth is approximately 100. N/C directed downward. Find the magnitude and direction of the magnetic field that will keep the droplet moving in this direction. What assumptions did you make to solve the problem? (Hint: Do you need to take the gravitational force into account?)

Section 24-2   The Motion of a Charged Particle in a Uniform Magnetic Field

24. ★ An electron, a proton, and an α-particle are moving with the same velocity. At some instant, they enter the region of a uniform magnetic field, $\vec{B}$, pointing perpendicular to the direction of their motion.
    (a) Compare the radii of the orbits of the particles upon entering this region.
    (b) Compare the time it takes for each particle to complete one revolution (period of motion).

25. ★ An electron moving perpendicular to a magnetic field of magnitude 2.4×10^{-2} T has a circular trajectory of radius r. The speed of the electron is 5% of the speed of light.
   (a) Determine the radius of the electron's trajectory.
   (b) Determine the period of revolution.

26. ★ A proton enters a uniform magnetic field that causes it to follow a circular path of radius 0.100 mm. The field strength is 0.900 T.
   (a) Calculate the proton's speed.
   (b) Calculate the proton's angular frequency.

27. ★★ An electron of mass m and charge $-e$ moving with an initial velocity of $\vec{v} = (v_{0x} \text{ m/s})\hat{i} + (v_{0y} \text{ m/s})\hat{j}$ enters a uniform magnetic field, $\vec{B} = B\hat{j}$. Derive an expression, in algebraic notation, for the velocity of the electron at any later instant, t.

28. ★★ A proton (mass m_p), a deuteron ($m_d = 2m_p$, $Q = e$), and an α-particle ($m\alpha = 4m_p$, $Q = 2e$) are accelerated by the same potential difference, V, and then enter a uniform magnetic field, B, where they move in circular paths perpendicular to B. Express the radii of the paths for the deuteron and α-particle in terms of the proton's radius.

29. ★★ An electron is travelling at 100. km/s parallel to a long straight horizontal conductor a distance of 3.00 cm from the conductor. A current of 12.0 A runs through the wire as the electron travels parallel to it, in the same direction as the electron's velocity. Find the strength of the external electric field that will prevent the electron from deviating from its original path. Express your answer in V/m.

Section 24-3 Applications: Charged Particles Moving in a Uniform Magnetic Field

30. ★ A magnetic field at the surface of a neutron star can reach 3.00×10^7 T. What would be the radii of curvature for the paths of three electrons moving in the vicinity of a neutron star with speeds of 5%, 10%, and 20% of the speed of light? What are the magnitudes of the magnetic forces acting on these electrons?

31. ★★ Charged particles, such as protons, found in cosmic rays, enter the Van Allen belt as they approach Earth. The average strength of the magnetic field of the lower Van Allen belt at a height of 3.00×10^3 km is approximately 1.00×10^{-5} T. Assume that the protons are moving with a speed of 3.00×10^7 m/s (10% of the speed of light) and you can apply classical mechanics to describe their motion. What is the radius of curvature and the cyclotron frequency of a proton entering the lower Van Allen belt at an angle of 30.0°? Describe the trajectory of the proton. Why do scientists say that a "proton was captured by the Van Allen belt"?

32. ★★ Cathode ray tube TVs used crossed electric and magnetic fields to deflect electrons before they struck a fluorescent screen. A beam of electrons moving with speed 6.0×10^7 m/s strikes a fluorescent screen located 0.50 m away. However, even when the magnetic field of the cathode ray is turned off, Earth's magnetic field is still present in that region. When an electron beam is moving along the x-axis, and Earth's magnetic field in the region has components $(15\ \mu\text{T}, 20.\ \mu\text{T}, 18\ \mu\text{T})$, what is the deflection, in mm, of the electron beam due to Earth's magnetic field?

33. ★★ The core of any microwave oven is a high-voltage system called a magnetron tube. The magnetron is a diode-type electron tube, which is used to produce electromagnetic waves with a frequency of 2450 MHz that are strongly absorbed by water. These waves are microwaves and have a wavelength of approximately 0.120 m. Determine the strength of the magnet needed to move the electrons in the magnetron tube at that frequency. Compare the strength of this magnetic field to the strength of Earth's magnetic field.

34. ★★★ Figure 24-65 shows a simple mass spectrometer, designed to analyze and separate atomic and molecular ions with different charge-to-mass ratios. In the design shown, ions are accelerated through a potential difference, V, after which they enter a region containing a uniform magnetic field. They describe semicircular paths in the magnetic field and land on a detector a lateral distance x from where they entered the field region, as shown. Assume that there is no magnetic field in the accelerating region of the initial electric field, and no electric field in the semicircular path region.
   (a) With the situation shown in Figure 24-65, are the ions positively or negatively charged? Clearly explain your reasoning.
   (b) Derive an expression for x in terms of the other quantities: V, B, and the charge-to-mass ratio q/m. Assume that the ions start from rest at the beginning of the acceleration region.
   (c) If the mass spectrometer used has a potential difference of 11.5 kV in the accelerating region, and the ion being accelerated has a net charge of $+2e$ and a mass of 95.0 atomic mass units, what must the magnetic field strength be to produce $x = 8.25$ cm?

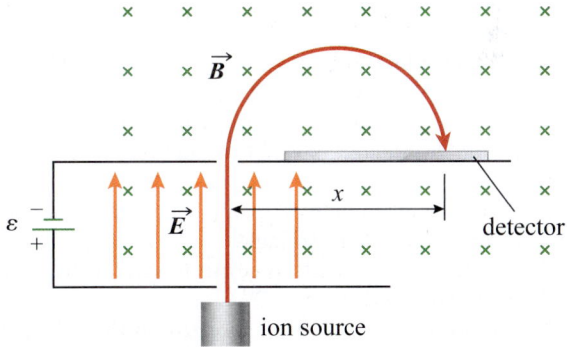

Figure 24-65 Problem 34.

Section 24-4 The Magnetic Force on a Current-Carrying Wire

35. ★ (a) Find the magnetic force per unit length on a long straight wire carrying 10. A of electric current in the positive x-direction in a uniform magnetic field of 1.8 T directed in the negative y-direction.
   (b) What happens to this force when the magnitude of the magnetic field is halved and the electric current in the wire doubles?

36. ★★ You have a long thick copper wire carrying 10.00 A of electric current. The wire has a diameter of 1.024 mm. Its density is 8.900 g/cm³. Compare the magnitudes of the magnetic and gravitational forces per unit length acting on the wire in each case.

    (a)   The wire is submerged in a horizontal magnetic field of 1.000 T perpendicular to the direction of the current in the field.

    (b)   The wire is submerged in the magnetic field of Earth. Assume Earth's magnetic field, with a magnitude of 50.00 μT, is directed horizontally and is perpendicular to the direction of the current in the wire.

37. ★★ Suppose you have a very long straight horizontal copper current-carrying wire with a linear density of 0.065 kg/m. The wire is suspended in a horizontal magnetic field perpendicular to it with a magnitude of 2.55 T.

    (a)   What should the direction of the magnetic field relative to the direction of the current in the wire be for the wire to be suspended in air? Assume the current in the wire flows from west to east.

    (b)   What amount of current must the wire carry for the wire to be suspended?

Section 24-5   The Torque on a Current-Carrying Loop in a Magnetic Field

38. ★ A rectangular copper loop has an area of 10.0 cm². It is carrying 2.00 A of electric current.

    (a)   The loop is submerged in a magnetic field of 1.50 T. What should the direction of this magnetic field be for the magnetic torque acting on the loop to reach its maximum value?

    (b)   Calculate the absolute value of the maximum magnetic torque acting on this loop.

    (c)   What would the maximum torque value be if instead of a single loop, you had a solenoid with 100 loops, with each one having the same area as the single loop in part (a)?

39. ★★ A copper loop in the shape of an equilateral triangle ABC (AB = AC = BC = a), carrying an electric current, I, is placed in a uniform magnetic field of magnitude B directed parallel to side AC of the triangle (Figure 24-66).

    (a)   Find the magnitude of the magnetic force exerted on each side of the triangle.

    (b)   Find the net magnetic force exerted on this triangle by the magnetic field.

    (c)   Find the magnetic torque exerted on this triangle by this magnetic field.

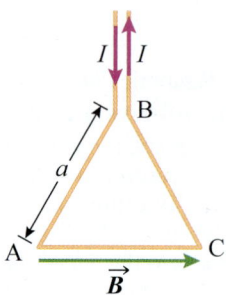

Figure 24-66   Problem 39.

Section 24-6   The Biot–Savart Law

40. ★ A long straight conductor carrying a current, I, splits into two identical semicircular arcs of 5 cm radius (Figure 24-67). Calculate the magnitude and the direction of the magnetic field at the centre of the circle (point O).

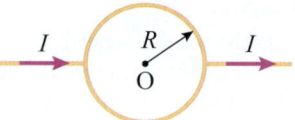

Figure 24-67   Problem 40.

41. ★ The magnitude of a magnetic field created by a long straight wire carrying an electric current 25.0 cm from the wire is 10.0 μT.

    (a)   What is the current through the wire?

    (b)   What will happen to the magnitude of the magnetic field when the current in the wire doubles?

42. ★★★ $^{d}/_{dx}$, $\int$ An infinitely long thin conducting sheet of width w positioned along the x-axis lies in the xy-plane as shown. The sheet carries a uniform current with a linear current density of J_s measured in amperes per metre. Electric current flows in the positive y-direction, as shown in Figure 24-68.

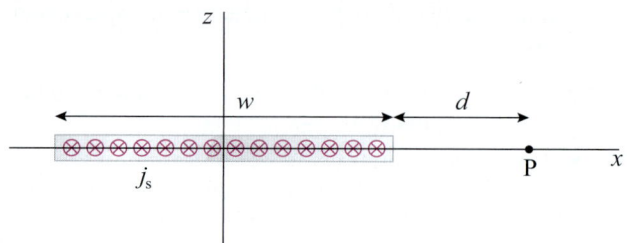

Figure 24-68   Problem 42.

    (a)   Find the total current flowing in the sheet.

    (b)   Find the current flowing in the sheet over an infinitesimally small width dx.

    (c)   Find the magnetic field at point P located a distance d away from the sheet in the plane of the sheet due to the current in it.

43. ★★★ $^{d}/_{dx}$, $\int$ Show that the magnetic field vector at a point P in the xy-plane and in the vicinity of the straight current-carrying wire shown in Figure 24-69 is given by

$$\vec{B} = \frac{\mu_0 I}{4\pi R}(\cos\theta_1 - \cos\theta_2)\hat{k}$$

(Note that while this wire must be part of a closed circuit, in this problem we are only concerned with calculating the magnetic field due to the part of the wire of length $2L$ that carries electric current I, as shown).

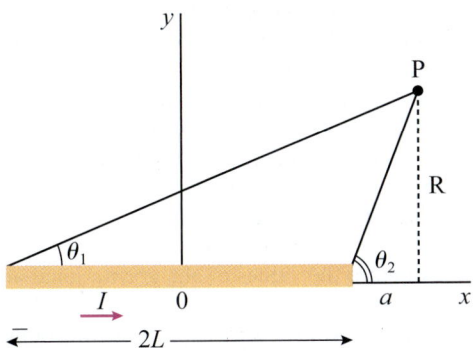

Figure 24-69 Problem 43.

Section 24-7 Ampère's Law

44. ★ A friend of yours claims that if the line integral of the magnetic field along the path around a closed loop is zero, the magnetic field around that loop must also be zero. Do you agree or disagree with your friend? Justify your position.

45. ★ A lightning bolt can generate electric currents of magnitude on the order of 10^4 A. What are the magnitudes of the magnetic fields generated (a) 10.00 m, (b) 100.0 m, and (c) 1.000 km from a lightning bolt that generated a 1.000×10^4 A current? Compare these magnitudes to the magnitude of Earth's magnetic field.

46. ★ A straight vertical wire carries 10.0 A of electric current. Find the direction and magnitude of the magnetic fields created by this wire at a distance of (a) 10.0 cm, (b) 20.0 cm, and (c) 50.0 cm from the wire.

47. ★★ ∫, ∿ A straight long cylindrical conductor of radius R carries current I along its length *with uniform current density*.
    (a) Find the magnetic field for $r < R$.
    (b) Find the magnetic field for $r > R$.
    (c) Sketch the plot representing the dependence of the magnetic field on the radius of the conductor, $B(r)$, for all r.

48. ★★ ∿, d/dx An electric cable has a diameter of 10.0 mm and carries an electric current of 25.0 A.
    (a) Find the strength of the magnetic field at distances of 2.00 mm and 7.00 mm from the centre of the cable.
    (b) Plot a graph of the magnitude of the magnetic field as a function of the distance from the centre of the cable for each distance.
    (c) In Examples 24-11 and 24-12, we found expressions for the magnitude of the magnetic field inside and outside a long current-carrying wire. Compare these expressions with the expressions for the magnitude of the electric field inside and outside a uniformly charged insulating sphere. What does this comparison tell you?

49. ★★ ∿, d/dx, ∫ A long cylindrical coaxial cable has a solid inner conductor of radius a with current I flowing along its length, and an outer cylindrical shell of inner radius b and outer radius c, also with current I flowing along the length of the shell in the opposite direction to that of the inner conductor. Assume that the current densities within each conductor are uniform. Find an expression for the magnitude of the magnetic field as a function of r, $B(r)$. Assume that the relative permeability of the conductor material is essentially unity.

50. ★★★ d/dx, ∫ A Helmholtz coil (Figure 24-70) is a structure used for creating uniform B-fields at the centre. Consider two thin circular coaxial coils each of radius R, having N turns, carrying current I in the same direction and separated by distance d.
    (a) Find B on the axis of symmetry, the x-axis of the Helmholtz coil.
    (b) Show that $\dfrac{dB}{dx} = 0$ at a point midway between the two coils.
    (c) Show that if $d = R$, $\dfrac{d^2B}{dx^2} = 0$ at the midway point between the coils.
    (d) Explain what your results in (b) and (c) mean.

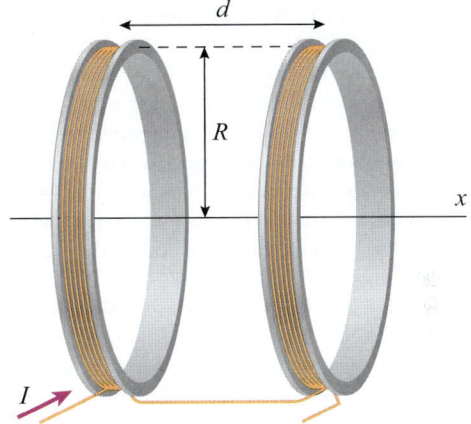

Figure 24-70 Problem 50.

Section 24-8 The Magnetic Force between Two Parallel Current-Carrying Conductors

51. ★ Two straight parallel superconducting cables can withstand a maximum tension of 10.0 kN/m without breaking. An engineer wants to use these cables in a device in which each cable is securely fastened and carries 20.0 kA of electric current. The currents will be flowing in opposite directions. What is the minimum separation required to keep the cables from breaking? Explain.

52. ★ Two very long straight parallel wires carry 10.0 A (wire A) and 15.0 A (wire B) of electric current, respectively. The currents in the wires flow in the same direction. The wires are located 1.0 m from each other.
    (a) What is the magnitude of the magnetic force that wire A exerts on wire B?
    (b) What is the magnitude of the magnetic force that wire B exerts on wire A?
    (c) Compare your answers to parts (a) and (b) and explain them.

53. ★★ A rectangular loop of dimensions of 100. cm × 1.00 cm carries 10.0 A of electric current.
    (a) Find the magnitudes and directions of the magnetic forces experienced by the long sides of the loop.
    (b) Find the magnitudes of these forces per unit length of the wire.
    (c) How would your answers to parts (a) and (b) change if the direction of the current in the loop was reversed?
    (d) Would you be able to use the same approach to find the forces acting on the short sides of the rectangle? Explain what assumptions you are making to solve the problem.

Section 24-9   The Magnetic Properties of Materials

54. ★ The magnetic field inside an air-filled solenoid is measured to be 1.20 T. Will the field inside the solenoid change if the air is pumped out of the core of the solenoid? If so, by how much will the magnetic field change? Explain.

55. ★ A 15.0 A current flows through a 50.0 cm long solenoid that has 1000 turns. Will the magnetic field inside this solenoid change if the air in its core is replaced by the following materials (justify your answers)?
    (a)   water at 20.0 °C
    (b)   oxygen at 20.0 °C
    (c)   99% pure annealed cobalt

56. ★ You have a 2.00 m long solenoid of radius $r = 0.250$ m that has 1000 turns per metre and 0.100 A of current flowing through it.
    (a)   What is the magnetic field inside this solenoid?
    (b)   How would your answer to the previous question change if the length of the solenoid was halved but the number of loops remained the same?
    (c)   What if you insert an iron core with $\chi_{iron} = 4.00 \times 10^3$ into the solenoid in part (a)? Can you use Equation 24-39 to estimate the magnetic field inside the solenoid? Explain your answer.

57. ★★ The magnetic field inside an air-filled solenoid that has N turns and carries current I is measured to be B. Explain how the field inside the solenoid will change if
    (a)   the number of turns in the solenoid doubles but its length remains the same
    (b)   both the number of turns and the length of the solenoid double
    (c)   an iron core is inserted inside the solenoid but the number of turns and its length remain the same
    (d)   a plastic core is inserted inside the solenoid but the number of turns and the length of the solenoid do not change

COMPREHENSIVE PROBLEMS

58. ★★ A proton moving at $t = 0$ with velocity $v_x = 2.0 \times 10^5$ m/s, $v_y = 0$, and $v_z = 1.5 \times 10^5$ m/s enters a uniform magnetic field of magnitude 0.45 T directed along the z-axis. Assume that the only force acting on the proton is the magnetic force.
    (a)   Find the magnetic force acting on the proton as it enters the region of the magnetic field, and compare its magnitude with the gravitational force. Use your findings to justify the above assumption.
    (b)   Describe the proton's trajectory as it travels inside the magnetic field region.
    (c)   Find the proton's angular speed and its period of circular motion.
    (d)   Find the radius of the proton's trajectory and the pitch of the helix.
    (e)   How would your answers to parts (a) to (d) change if, instead of a proton, an electron entered this region at the same velocity?

59. ★★ At a certain instant, an electron with kinetic energy 20. eV is moving in a westward direction. The horizontal component of Earth's magnetic field in this region is 18 μT north, and its vertical component is 48 μT down.
    (a)   What is the net magnetic field in this region in terms of its magnitude and direction as compared to the north direction?
    (b)   What is the trajectory of the electron?
    (c)   What is the radius of curvature of the electron's path?
    (d)   Compare the values of the gravitational and magnetic forces on the electron and decide if the gravitational force should be taken into account. Justify your answer.

60. ★★★ An α-particle travels in a circular path of radius 15 cm in a uniform magnetic field of 2.0 T.
    (a)   Find the speed of the α-particle. Express the speed of the α-particle as a fraction of the speed of light. Is it justified to use classical physics to express the kinetic energy of this particle, or should relativity be used?
    (b)   What is the kinetic energy of the α-particle in eV?
    (c)   What potential difference must the α-particle be accelerated through to achieve this speed?

61. ★★★ A proton is moving in a circular orbit of radius 14 cm in a uniform 0.35 T magnetic field perpendicular to the velocity of the proton.
    (a)   Find the linear speed of the proton. Do you need to take relativistic effects into account based on this speed? Justify your answer.
    (b)   Find the time it takes for the proton to complete one revolution.
    (c)   Suddenly, a uniform electric field is turned on in the direction opposite to the direction of the magnetic field. The magnitude of the electric field is 400. kV/m.
        (i)   Draw an FBD for the proton.
        (ii)   Draw the trajectory of the proton when both fields are on. Explain your drawing.
        (iii) How would your answer to part (i) be different if instead of a proton an electron were moving in a uniform magnetic field of 0.35 T and the electric field mentioned above were turned on?
        (iv)   How would your answer to part (ii) be different if instead of a proton an electron were moving in a uniform magnetic field of 0.35 T and the electric field mentioned above were turned on?

62. ★★ A straight vertical wire carries a downward current of 1.5 A. The wire is placed in the area of a uniform magnetic field created by a strong superconducting electromagnet. The magnitude of this magnetic field reaches 0.65 T. Calculate the magnetic force per metre of the wire applied on it by the magnetic field in each scenario. Draw a diagram for each scenario.
    (a)   The direction of the magnetic field is perpendicular to the direction of the wire.
    (b)   The direction of the magnetic field is parallel to the wire.
    (c)   The magnetic field is directed at 45° to the direction of the wire.

63. ★ Due to health regulations, the magnetic field produced by a current-carrying wire at a distance of 30.0 cm is not allowed to exceed Earth's magnetic field (5.50×10^{-5} T).
   (a) What is the maximum current that this wire can carry?
   (b) What would the maximum current be if the regulations required the same limit at half the original distance (15.0 cm)?

64. ★ A jumper cable used to start a vehicle that has a dead battery carries a 65.0 A current. What is the magnitude of the magnetic field created by the cable (a) 10.0 cm away? (b) 25.0 cm away? Compare the values of the magnetic field to the magnitude of Earth's magnetic field.

65. ★★ Figure 24-71 shows six sets of five equally spaced long parallel wires carrying electric currents of equal magnitudes into or out of the page. Rank the magnitude of the magnetic forces acting on the central wire in each scenario, due to the currents in the other four wires, from greatest to smallest. Support your ranking with the appropriate calculations.

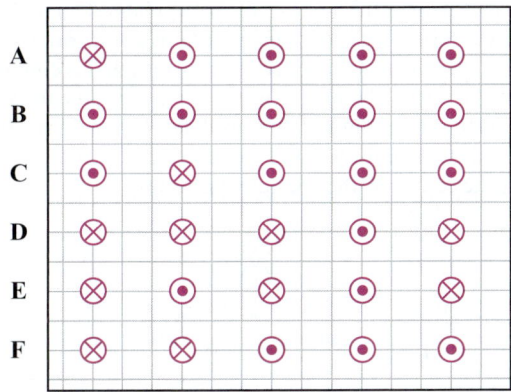

Figure 24-71   Problem 65.

66. ★★ In Example 24-9 of this chapter, we proved that in Cartesian coordinates, the magnetic field due to a current-carrying loop along its axis of symmetry can be expressed as (Equation 24-33) $B_{\text{loop}} = \dfrac{\mu_0}{2} \dfrac{IR^2}{(x^2 + R^2)^{3/2}}$.

In this expression, I represents electric current, R is the radius of the loop, and x is the distance from the centre of the loop along its axis of symmetry. Show that this expression can be rewritten as $B_{\text{loop}} = \dfrac{\mu_0 I}{2R} \sin^3 \phi$, where $\tan \phi = \dfrac{R}{x}$. What does angle ϕ represent here?

67. ★★★ $\frac{d}{dx}$, $\int$ Consider a short solenoid of length L, radius R, and number of turns N (Figure 24-72).
   (a) Find the magnetic field strength on the axis of the solenoid at point P, located outside the solenoid, a distance x away from its centre. Give the answer in terms of the angles θ_1 and θ_2, where the angles are given by $\tan \theta_1 = \dfrac{x - L/2}{R}$ and $\tan \theta_2 = \dfrac{x + L/2}{R}$.
   (b) Find the magnetic field strength at the centre of the short solenoid when the length of the solenoid is much larger than the diameter.
   (c) Do your results confirm what you expected? Explain.

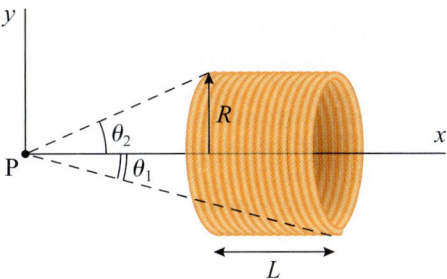

Figure 24-72   Problem 67.

68. ★★ 〰, $\frac{d}{dx}$, $\int$ Consider current I flowing along an infinitely long copper pipe with inner radius R_1 and outer radius R_2.
   (a) Find the expression for the magnitude of the current density in the pipe, assuming it is uniform.
   (b) Calculate the magnitude of the magnetic field B produced by the current at distance r from the axis (centre) of the pipe for
      (i) $r < R_1$
      (ii) $R_1 < r < R_2$
      (iii) $R_2 < r$
   (c) Sketch the magnetic field vectors, assuming that the pipe is perpendicular to the page and the current is flowing into the page.

69. ★★ A portion of a magnetic lens is being considered for use in an electron microscope. An electron beam with electrons initially moving at speed v_0 to the right is fired into a region where the magnetic field points out of the plane of the paper, as shown in Figure 24-73. The electron beam is seen to exit the field region 0.0300 ns later deflected at an angle of $\phi = 4.50°$ compared to the initial direction. Ignore other forces, such as gravitational or those due to other magnetic fields.
   (a) Find the strength of the magnetic field.
   (b) The horizontal distance between the point of entrance and the point of exit is measured to be 0.250 mm. Find the speed of the electrons.

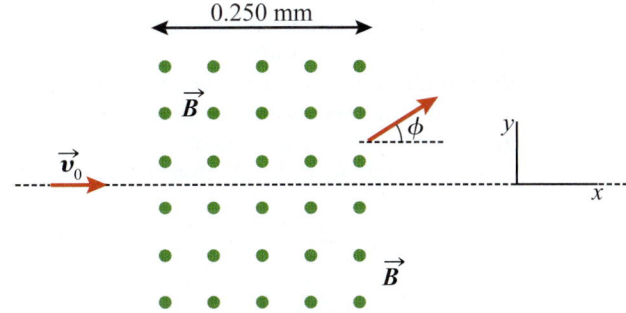

Figure 24-73   Problem 69.

70. ★★ A circular 100-loop coil with a radius of 0.100 m is located in the horizontal xy- plane, as shown in Figure 24-74. A 10.0 A current flows clockwise through the coil (as seen from above). The coil is in a uniform magnetic field of 1.00 T, directed in the positive y-direction with $\vec{B} = (1.00 \text{ T})\hat{j}$.
   (a) Calculate the magnetic moment of each loop of the coil.

(b) Calculate the magnetic moment of the entire coil.

(c) Calculate the magnetic torque experienced by the coil. Clearly identify the direction of the torque.

(d) How will your answers to parts (a) to (c) change if the direction of the current in the coil is reversed?

(e) How will your answers to parts (a) to (c) change if the direction of the magnetic field is changed to $\vec{B} = (1.00\ \text{T})\hat{i}$ or $\vec{B} = (1.00\ \text{T})\hat{k}$? Explain.

(f) The coil rotates from the position where its magnetic moment is parallel to the direction of the magnetic field to the position where its magnetic moment is perpendicular to it. What is the change in the potential energy of the coil? Explain.

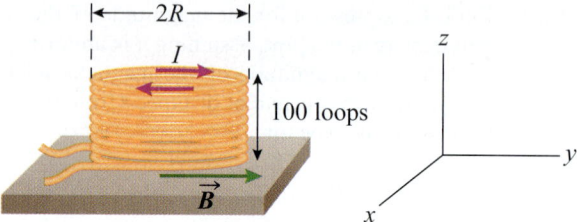

Figure 24-74 Problem 70.

71. ★★ A current-carrying wire is located along the y-axis. An electric current of 5.00 A flows through this wire in the positive y-direction. Calculate the magnetic force per metre experienced by this wire when it is placed in the following magnetic fields:

(a) $\vec{B} = (0.550\ \text{T})\hat{i}$

(b) $\vec{B} = (0.550\ \text{T})\hat{j}$

(c) $\vec{B} = (0.550\ \text{T})\hat{k}$

(d) $\vec{B} = (0.550\ \text{T})\hat{i} - (0.550\ \text{T})\hat{k}$

72. ★★★ ⌇ A long wire carries a 5.00 A current toward the top of the page, as shown in Figure 24-75. An electron is travelling parallel to the wire with an initial separation of 2.00 cm and an initial velocity of 1.50×10^7 m/s toward the top of the page.

(a) What are the magnitude and direction of the magnetic force on the electron?

(b) What would your answers to part (a) be if instead of an electron, there was a proton moving in the same direction at the same speed?

(c) Draw the trajectories of the electron and the proton.

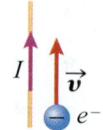

Figure 24-75 Problem 72.

73. ★★★ A charged particle carrying charge q and moving with speed v enters a region of uniform magnetic field $\vec{B}$ directed perpendicular to the particle's velocity. The particle spends time Δt in this region.

(a) Assume that Δt is small. Estimate the angle of deflection, θ, of the particle. When we assume that Δt is small, to what are we comparing it? Explain your answer.

(b) Suppose the charged particle is an electron moving with a speed of 1.5×10^7 m/s. What should be the size (length along the particle's original path) of the region of a uniform magnetic field of 2.0×10^{-2} T to produce a deflection of 0.15 rad?

74. ★★★ A long straight conductor carrying current I splits into two semicircular arcs. The resistance of the upper arc is twice the resistance of the bottom arc (Figure 24-76). Calculate the magnitude and the direction of the magnetic field at the centre of the circle (point O). Assume the arcs are made of wires with the same resistivity and cross-sectional area.

Figure 24-76 Problem 74.

75. ★★ In Figure 24-77, three parallel current-carrying wires pass through the vertices of an equilateral triangle. Two wires carry an electric current that is going into the page, and one wire carries an electric current that is coming out of the page. The magnitudes of the currents are 5.00 A, and the side length of the equilateral triangle is 5.00 cm. Find the magnitude and direction of the magnetic field at the centre of the triangle (point O).

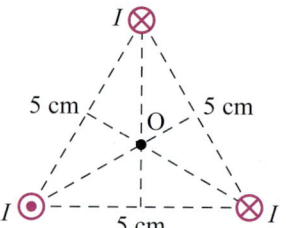

Figure 24-77 Problem 75.

76. ★★★ $^{d}/_{dx}, \int, ⌇$ A 40. cm solenoid has a radius of 0.050 m and consists of 50 circular loops. An electric current of 1.0 A flows through the solenoid.

(a) Determine the magnetic field in the centre of the solenoid.

(b) Derive the formula for the magnetic field along the axis of symmetry of the solenoid (Hint: You can check if your answer makes sense by using it to answer part (a)). Use this formula to find the magnetic field along the axis of the solenoid 15 cm off its centre.

(c) Plot the graph of the magnetic field along the axis of symmetry of the solenoid as a function of the distance from its centre (use graphing software).

(d) From the plot in (c), estimate at what distance from the centre of the solenoid the magnetic field will decrease by 5%.

77. ★★ Find an expression for the magnetic field inside a toroidal coil that has a total of N turns wound around an iron doughnut (thick ring) of inner radius a and outer radius b and thickness w (see Figure 24-78). The relative permeability of the iron core is μ_r.

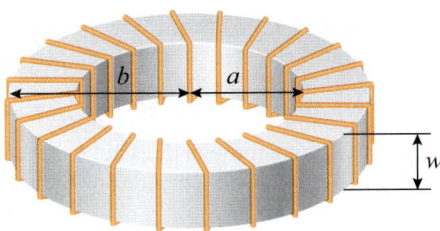

Figure 24-78 Problem 77.

78. ★★ Two parallel current-carrying wires are separated by a distance d. The wires carry currents of equal magnitude. In Figure 24-79(a), both currents are going into the page, and in Figure 24-79(b), one current is going into the page and the other is coming out of the page. At what location(s) does the resultant magnetic field equal zero for each case? Explain.

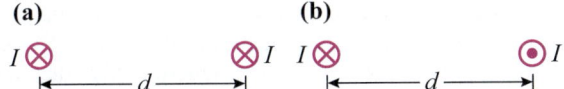

Figure 24-79 Problem 78.

79. ★★ Two long parallel vertical wires carry electric currents of I and $2I$ in opposite directions (Figure 24-80). Find the magnitude and direction of the magnetic fields at points A, B, C, and D when $I = 10$ A and the distance between the wires is 1.0 m.

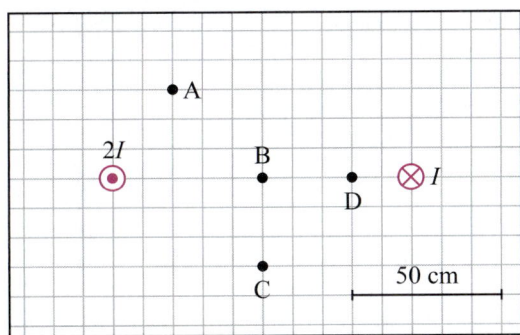

Figure 24-80 Problem 79.

80. ★★ Two parallel wires separated by a distance of 50.0 cm carry currents in opposite directions (Figure 24-81). The left wire carries a current of 10.0 A. Point A is the midpoint between the wires, and point B is 5.00 cm to the left of the 10.0 A current. The current, I, is adjusted so that the magnetic field at B is zero.
(a) Find the strength of the magnetic field 10.0 cm to the left of A.
(b) Find the direction and magnitude of the force per unit length on the left wire. Ignore gravity.

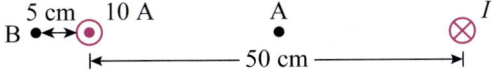

Figure 24-81 Problem 80.

81. ★★★ $^d/_{dx}$, ∫ The magnetic field inside a long solid cylindrical conductor with radius R whose long axis of symmetry lies along the z-axis is measured to be $\vec{B} = \dfrac{B_0 r}{6R}\hat{k}$, where r represents the distance from the axis of symmetry of the conductor. Suggest a current density inside the conductor that would give this field. Assume that the relative permeability of the conductor material is essentially unity.

82. ★★★ $^d/_{dx}$, ∫ The magnetic field in a long solid cylindrical conductor that is parallel to the z-axis is measured to be

$$B_0\left[\frac{r}{4R} + \frac{1}{6}\left(\frac{r}{R}\right)^2\right]\hat{e}$$

What current density inside the conductor would give this field? Assume that the relative permeability of the conductor material is essentially unity.

83. ★★★ $^d/_{dx}$, ∫ A long straight hollow cylindrical conductor of inner radius a and outer radius b carries current I along its length. The current density varies as $J = J_0/r^2$ within the conductor, where r is the radial distance from the axis of symmetry of the conductor. Assume that the relative permeability of the conductor material is essentially unity.
(a) Find the constant J_0 in terms of the given quantities.
(b) Find the magnetic field strength for $a < r < b$.
(c) Find the magnetic field strength for $r > b$.

84. ★★★ $^d/_{dx}$, ∫ A long cylindrical coaxial cable has a solid inner conductor of a radius a with current I flowing along its length, and an outer cylindrical shell of inner radius b and outer radius c, also with current I flowing along the length of the shell in the opposite direction to that of the inner conductor. The current densities vary linearly with r: $J = \alpha r$ in the inner conductor, and $J = \beta r$ in the outer conductor. Assume a relative permeability of one everywhere.
(a) Find α and β.
(b) Find the strength of the magnetic field as a function of r for all r.

85. ★★★ $^d/_{dx}$, ∫ A conducting disk of radius a is located in the xy-plane with its centre at the origin. Electric current of density J flows in the disk in a clockwise direction around its centre. The density of this current is given by $J = Cr$, where C is a constant and r represents the radius of the disk. Find the magnetic field generated by this current at a point a distance z away from the centre of the disk along the z-axis.

86. ★★ A horizontally oriented metal ring of radius R has total charge Q distributed uniformly on it. The ring is spun in the xy-plane about its vertical axis of symmetry (which coincides with the z-axis) with an angular speed of ω. When viewed from above, the current in the ring runs counter-clockwise. Answer the following questions, giving all your answers in full vector notation.
(a) Find the magnetic field along the z-axis a distance z away from the centre of the ring.
(b) Find the magnetic field in the centre of the ring.
(c) Calculate the magnetic dipole moment of the ring.
(d) Show that the field you obtained in part (a) decreases as the cube of the distance for large distances from the centre of the ring, and express your answer in terms of the dipole moment of the ring.

87. ★★★ A rectangular loop of dimensions 30.0 cm × 20.0 cm carries 10.0 A of electric current (Figure 24-82). The loop is oriented in the *xz*-plane. An infinitely long straight vertical wire carrying 20.0 A of electric current in the positive *z*-direction is located 60.0 cm from the centre of the loop. Find the resultant force and torque acting on the loop.

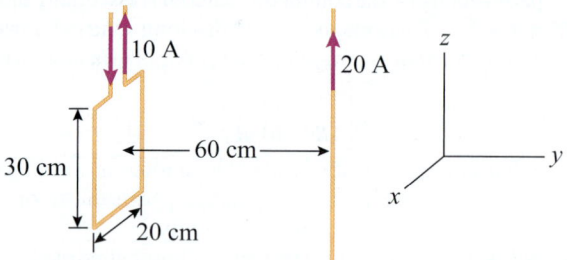

Figure 24-82   Problem 87.

88. ★★ A long copper wire of length *L* carries an electric current, *I*. This wire can be formed into a coil consisting of multiple loops while preserving the total length of the wire. Find the optimal number of loops in the coil to maximize the torque on the coil when it carries a current, *I*, in a uniform magnetic field, *B*. Prove that the maximum value of this torque is $\tau = (1/4\pi)L^2IB$.

89. ★★ A long straight wire carrying an electric current of 50.0 A is placed in a uniform magnetic field of 5.00 mT directed toward the north magnetic pole. The wire is oriented such that the electric current flows from east to west. Find the points at which the net magnetic field is zero. (Hint: Think of the influence of Earth's magnetic field.)

90. ★★★ $\frac{d}{dx}$, ∫ Two charged particles moving with the same speed but in opposite directions pass each other at a separation distance *d*. The speed of the particles, *v*, is much less than the speed of light, $v \ll c$. Compare the magnetic and electric forces that the particles exert on each other at the instant when the particles are closest to each other. Perform the calculations for the three cases listed below.
(a)   The charged particles are two protons.
(b)   The charged particles are two electrons.
(c)   One charged particle is a proton, and the other charged particle is an electron.

91. ★★ A coaxial cable is oriented along the *x*-axis. It has an external radius of 5.0 mm and an internal radius (the radius of the inner conductor) of 2.0 mm, and it carries an electric current of 5.0 A in each direction. Derive an equation to determine the magnitude and direction of the magnetic field generated by this cable at an arbitrary distance from its axis. Assume that the electric current in the inner conductor flows in the positive *x*-direction, while the current in the outer conductor flows in the negative *x*-direction. Then use your equation to find the magnetic fields at the following distances from its axis:
(a)   1.0 mm
(b)   3.0 mm
(c)   5.0 mm

92. ★★★ $\frac{d}{dx}$, ∫ Use the Biot–Savart law to prove that the magnetic field at the centre of a wire shaped into a circular arc (Figure 24-83) is given by $B = \dfrac{\mu_0 I \phi}{4\pi R}$, where *I* is the electric current flowing through the wire, *R* is the radius of curvature, and ϕ is the angle subtended by the circular arc.

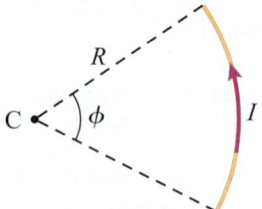

Figure 24-83   Problem 92.

93. ★★ Each wire in Figure 24-84 carries an electric current of 5.00 A. (Hint: Use the results from problem 92.)
(a)   In Figure 24-84(a), the radius of the arc is 10.0 cm. Find the strength of the magnetic field at the centre of the wire.
(b)   In Figure 24-84(b), the radius of the arc is 20.0 cm. Find the strength of the magnetic field at the centre of the wire. (Hint: Use your result from part (a).)

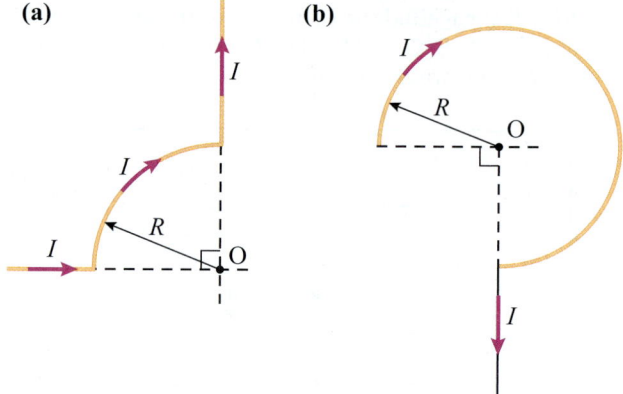

Figure 24-84   Problem 93.

94. ★★★ A straight wire of length *L* carries an electric current of magnitude *I*.
(a)   Use the Biot–Savart law to prove that the magnetic field at point P_1, located a distance *R* from the centre of the wire along the line perpendicular to it, can be expressed as

$$B_{P_1} = \frac{\mu_0 I}{2\pi R} \frac{L}{(L^2 + 4R^2)^{1/2}}$$

(b)   What result would you expect if the length of the wire became infinitely long? Show that the result you proved in part (a) satisfies the limiting case of an infinitely long wire.

(c) Use the Biot–Savart law to prove that the magnetic field at point P_2, located a distance R from the end of the wire along the line perpendicular to it, can be expressed as

$$B_{P_2} = \frac{\mu_0 I}{4\pi R} \frac{L}{(L^2 + 4R^2)^{1/2}}$$

(d) Show that the results you proved in parts (a) and (c) are consistent.

95. ★★ A copper wire is bent to create a square loop with sides of length d. An electric current of magnitude I flows counter-clockwise through this loop. Find the magnitude and direction of the magnetic field created at the centre of this loop.

96. ★★★ A toroidal (doughnut-shaped) coil is closely wound with one continuous wire, as shown in Figure 24-85. Assume that the current in the wire is I, the number of turns in the coil is N, and the radius of the toroid (the average of the inner and outer radii of the toroid) is R.
(a) Derive an expression for the magnetic field both inside and outside the toroid.
(b) Compare your answer for the magnetic field of a toroid to the magnetic field of a solenoid.
(c) What does your comparison in part (b) tell you?
(d) Calculate the magnetic field of a 1000-turn toroid of radius 0.500 m when 15.0 A of electric current flows through it.
(e) Compare your result in part (d) to the magnetic field inside a 1.00 m long solenoid that has 1000 turns and a 15.0 A current flowing through it. What does your comparison tell you?

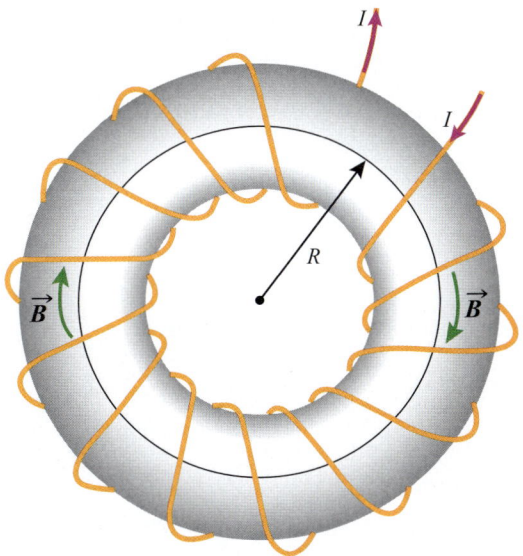

Figure 24-85 Problem 96.

97. ★★★ As discussed earlier, there are different versions of mass spectrometers. Figure 24-86 shows a mass spectrometer that does not have a velocity selection stage (compare it to the mass spectrometer shown in Figure 24-13). Derive an expression that shows how the radius of the charged particle in the mass spectrometer depends on the mass of the particle; its charge; its magnetic field, $\vec{B}$; and the potential difference, V. What assumption have you made for your derivation? Compare the operation of the two mass spectrometer models (Figures 24-13 and 24-86), and discuss their advantages and disadvantages.

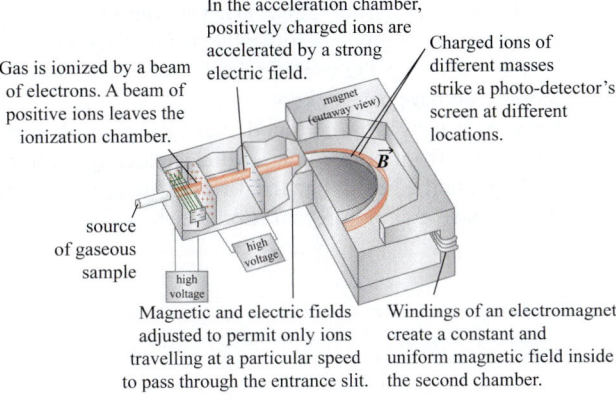

In the acceleration chamber, positively charged ions are accelerated by a strong electric field.

Gas is ionized by a beam of electrons. A beam of positive ions leaves the ionization chamber.

Charged ions of different masses strike a photo-detector's screen at different locations.

source of gaseous sample

Magnetic and electric fields adjusted to permit only ions travelling at a particular speed to pass through the entrance slit.

Windings of an electromagnet create a constant and uniform magnetic field inside the second chamber.

Figure 24-86 Problem 97.

98. ★★★ An infinitely large thin metal sheet is located in the horizontal xy-plane at $z = 0$. An electric current with a uniform current density of J flows through this sheet, as shown in Figure 24-87. The current is flowing in the positive y-direction.
(a) What assumptions do you need to make to calculate the magnetic field at a point located a distance z above the sheet?
(b) Calculate the magnetic field at a point located a distance z above the sheet. Indicate the direction of the field.
(c) Describe the differences and the similarities between this problem and how you calculated the value of the electric field created by a uniformly charged infinitely large sheet of metal.

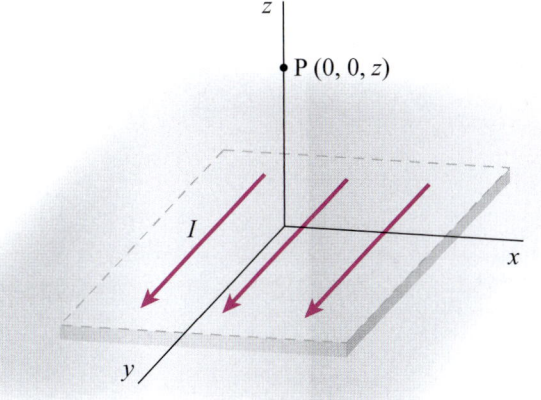

Figure 24-87 Problem 98.

CHAPTER 24 | MAGNETIC FIELDS AND MAGNETIC FORCES

DATA-RICH PROBLEMS

99. ★★★ Could using a device such as an iPod be hazardous for a person who has a pacemaker? A paper called "Low frequency magnetic emissions and resulting induced voltages in a pacemaker by iPod portable music players," by Howard Bassen, attempts to provide an experimental answer to this question (www.pubmedcentral.nih.gov/articlerender.fcgi?artid=2265271). Use the information in the paper to give scientifically justified recommendations regarding the use of portable music devices by people who have pacemakers.

100. ★★★ You hear from a friend that electric and magnetic fields from power lines and electrical appliances might present a health hazard. Investigate the problem and prepare a short presentation on the topic for the general public. Remember that your goal is to use the physics knowledge you have gained to distinguish reliable from unreliable sources and explanations. You might want to explore relevant websites from the World Health Organization or the Government of Canada, e.g., www.canada.ca/en/health-canada/services/home-garden-safety/electric-magnetic-fields-power-lines-electrical-appliances.html, or other reliable sources.

101. ★★★ On March 9, 2009, Ed Young, a research blogger, posted an article titled "Power lines disrupt the magnetic alignment of cows and deer" (blogs.discovermagazine.com/notrocketscience/2009/03/16/power-lines-disrupt-the-magnetic-alignment-of-cows-and-deer/#.WLn4ODvyuUk). Read the article and write a review commenting on the validity of its argument.

Electromagnetic Induction

anada is the second-largest country in the world, spanning over 9300 km from east to west. Because the country is so large, transportation is a big challenge for the economy and for the environment. About a quarter of Canadian CO_2 emissions come from transportation, mostly from driving. But airline flights also produce a great deal of CO_2. For example, a round-trip flight from Vancouver to Toronto (6717 km) produces 1.536 tonnes of CO_2 emissions.[1] Consequently, the development of energy-efficient transportation is important for Canadians. While oil-fuelled automobiles, trucks, and airplanes dominated 20th-century transportation, electrically powered vehicles will become the transportation of the future. One promising technology is magnetic levitation (maglev) trains, which are suspended over conductive rails by electromagnetic forces (Figure 25-1). Maglev trains are powered by electric energy, which can be produced with minimal damage to the environment using hydroelectric power plants. In addition to reducing CO_2 emissions (assuming the electricity is not fossil-fuel generated), maglev trains are much more energy efficient than other transportation modes. They are also fast—they reach speeds comparable to modern jets: up to 590 km/h. They can travel at high speeds because levitating the train eliminates friction and vibrations from contact between the wheels and the rails, thus reducing waste energy.

The first maglev vehicle patent was obtained in 1902 by an American, A. Zehden. However, this technology required very strong magnets, so practical maglev vehicles were not developed until the early 1960s. The world's first commercial maglev train was introduced in Birmingham, England, in 1984, but it was discontinued in 1995 due to technical difficulties. In 2002, German scientists and engineers

Figure 25-1 Magnetic levitation (maglev) trains are high-speed trains that operate using the law of electromagnetic induction.

from Transrapid International developed the first maglev system still in operation today: a maglev line between Shanghai's Pudong airport and its financial district (30.5 km). Other commercial maglev trains operate in Japan. Currently, there are a few maglev projects under construction in Europe, China, Japan, and the United States. These maglev trains use the law of electromagnetic induction discovered by Michael Faraday almost two centuries ago, and they might be part of the solution to the energy-efficient transportation problem.

1. http://www.less.ca.

25-1 In Faraday's Lab: Science in the Making

As described in Section 24-6 in the previous chapter, the first solid evidence for a connection between electric and magnetic phenomena was uncovered by Danish physicist and chemist Hans Christian Oersted (1777–1851). In 1821, while showing an experiment during a lecture, Oersted noticed that an electric current deflected the needle of a nearby compass, suggesting that the current generates a magnetic field. Oersted's findings intrigued British scientist Michael Faraday (1791–1867), who set out in the late 1820s to answer one of the most important questions in the history of physics: *If electric current is capable of generating a magnetic field, can a magnetic field generate an electric current?*

You can recreate Faraday's experiments either with available equipment or with the computer simulation in Interactive Activity 25-1 described below.

INTERACTIVE ACTIVITY 25-1

In Faraday's Lab

In this activity, you will use the PhET simulation "Faraday's Electromagnetic Lab: Pickup Coil" to gain an understanding of how Michael Faraday generated electric current in the coil by moving a magnet toward and away from it.

Faraday observed that moving a magnet toward a coil generated an electric current in one direction, and moving the same magnet away from the coil generated an electric current in the opposite direction. When the magnet was not moving relative to the coil, no current flowed in the coil. Faraday made several important observations from this series of experiments:

- A magnetic field can induce (generate) an electric current in a coil; thus, the phenomenon is called **electromagnetic induction** (Figure 25-2).

- The magnitude and the direction of the induced current depend on the rate of change of the number of magnetic field lines passing through the coil. As soon as the change ceases, the generation of electric current stops.

Another series of Faraday's original experiments (Figure 25-3) used two adjacent electric circuits called the primary and secondary circuits. The primary circuit consisted of a coil connected to a battery, while the secondary circuit consisted of a coil connected to an ammeter. The coils were made of insulated wire wound around opposite sides of an iron ring, which kept most of the magnetic field inside the coils. Faraday noticed that, when a steady electric current flowed through the primary circuit, the ammeter in the secondary circuit indicated zero current. However, when the circuit was

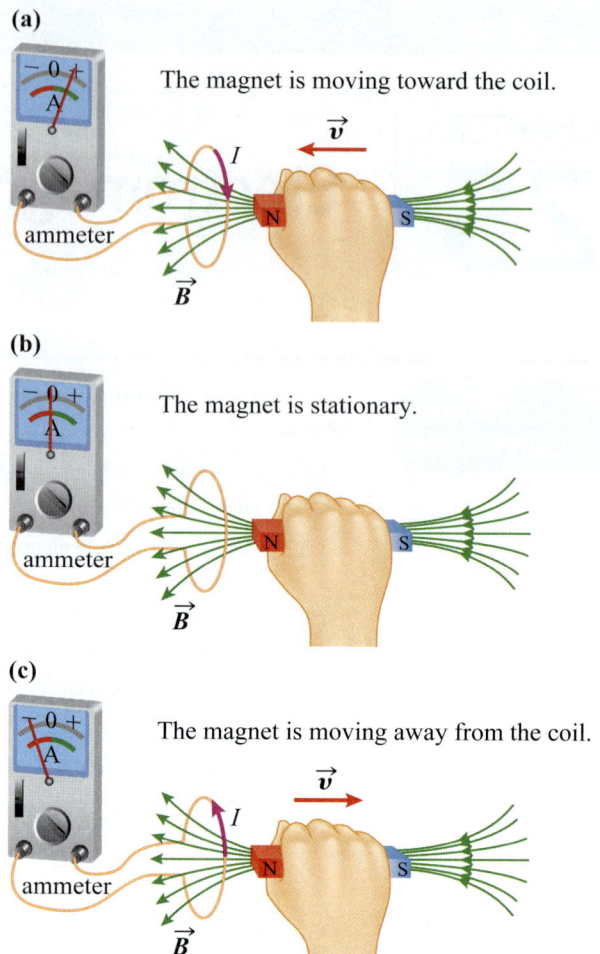

(a) The magnet is moving toward the coil.

(b) The magnet is stationary.

(c) The magnet is moving away from the coil.

Figure 25-2 The effects of moving a magnet through a conducting loop. (a) When the magnet moves toward the loop, an electric current is generated in the loop that is registered by an ammeter. (b) When the magnet is stationary relative to the loop, no current in the loop is detected. (c) When the magnet moves away from the loop, a current is generated again. The direction of the current depends on the relative motion of the magnet and the loop.

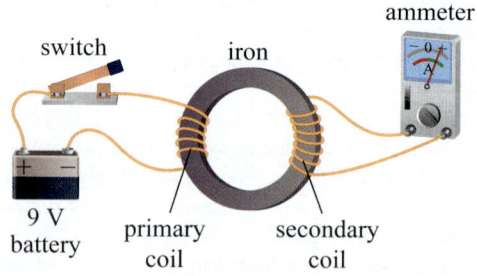

Figure 25-3 A modern version of Faraday's original experiment. Notice that there is no electrical conduction connection between the two coils—they are not connected with electrical wires. These coils are part of two different circuits: the primary and the secondary circuits.

switched on or off, the needle of the ammeter deflected for an instant, indicating the generation of an electric current in the secondary circuit. Faraday hypothesized that the changing electric current in the primary circuit induced a changing magnetic field that, in turn, induced a changing electric current in the secondary circuit. The secondary coil is sometimes called a pickup coil because it is picking up the signal of the varying magnetic field generated by the primary coil.

25-2 Magnetic Flux and Its Rate of Change

To explain how a varying magnetic field generates electric current, Faraday introduced the concept of **magnetic flux**. You learned about electric flux in Chapter 20 (Equation 20-1). To distinguish between electric and magnetic flux, we will use the subscripts E and B, respectively. In Section 24-5, we defined the orientation of a flat two-dimensional loop with area vector $\vec{A}$, where the magnitude of the vector represents the area of the loop and its direction is perpendicular to the plane of the loop. If a uniform magnetic field is directed at an arbitrary angle θ to the surface of the loop (Figure 25-4(a)), the magnetic flux through the loop is defined as

$$\Phi_B = BA\cos\theta \tag{25-1}$$

The magnetic flux through the loop is a scalar quantity. It represents the number of magnetic field lines passing through the loop. In analogy with electric flux, it can be described more concisely using a scalar product:

$$\Phi_B = BA\cos\theta = \vec{B} \cdot \vec{A} \tag{25-2}$$

When a uniform magnetic field is directed perpendicular to the plane of the loop (Figure 25-4(b)), the angle between the vectors $\vec{A}$ and $\vec{B}$ is 0°; thus, the magnetic flux Equation 25-2 becomes

$$\Phi_{B(\vec{B}\,\|\,\vec{A})} = BA\cos 0° = BA \quad \text{when } \vec{B}\,\|\,\vec{A} \tag{25-3}$$

When the magnetic field lines are parallel to the plane of the loop, the angle between $\vec{A}$ and $\vec{B}$ is 90°; thus the magnetic flux through the loop becomes zero (Figure 25-4(c)):

$$\Phi_{B(\vec{B}\perp\vec{A})} = BA\cos 90° = 0 \quad \text{when } \vec{B}\perp\vec{A} \tag{25-4}$$

The SI unit for magnetic flux is the weber (Wb):

$$[\Phi_B] = 1\ \text{T} \cdot 1\ \text{m}^2 \equiv 1\ \text{Wb} \tag{25-5}$$

Thus, 1 Wb of magnetic flux passes through each square metre of a surface perpendicular to a magnetic field of 1 T.

For a non-uniform magnetic field, magnetic flux is defined using the concept of the average magnetic field in that region, $\vec{B}_{avg}$:

$$\Phi_B = \vec{B}_{avg} \cdot \vec{A} = B_{avg}A\cos\theta \tag{25-6}$$

We can divide an arbitrary surface into infinitesimally small area elements, $d\vec{A}$, calculate the magnetic flux $d\Phi_B$ through each element, and then integrate over the surface to determine the net flux:

KEY EQUATION

$$\Phi_B = \int \vec{B} \cdot d\vec{A} \tag{25-7}$$

For a three-dimensional surface (Figure 25-5), the direction of the normal to the surface element is always chosen to point outward. Therefore, a vector field directed inward through the surface generates a negative flux, and a field directed outward generates a positive flux. For a flat surface, the direction of the normal to the surface (and, hence, the sign of the magnetic flux) is chosen arbitrarily.

The breakthrough idea in Faraday's discovery of electromagnetic induction was his realization of the importance of the **rate of change of the magnetic flux in time,** $\dfrac{d\Phi_B(t)}{dt}$.

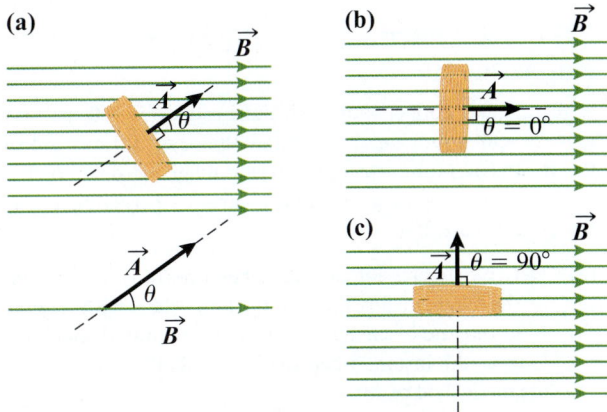

Figure 25-4 (a) A magnetic field is directed at an angle θ to the normal to the loop: $\Phi_B = BA\cos\theta$. (b) The magnetic field is perpendicular to the plane of the loop: $\Phi_B = BA\cos 0° = BA$. (c) The magnetic field is parallel to the plane of the loop: $\Phi_B = BA\cos 90° = 0$.

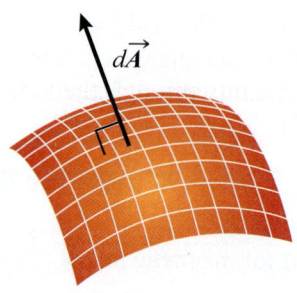

Figure 25-5 A three-dimensional surface can be divided into small surface elements $d\vec{A}$. The direction of each $d\vec{A}$ element is normal to the surface and points outward.

Let us calculate the rate of change of the magnetic flux through a coil. Since the coil can be represented as a stack of closed loops, we will first learn how to calculate the rate of change of the magnetic flux through a single loop:

$$\frac{d\Phi_B(t)}{dt} = \frac{d(\vec{B}(t) \cdot \vec{A}(t))}{dt}$$

$$\frac{d\Phi_B}{dt} = \frac{d[B(t)A(t)\cos\theta(t)]}{dt}$$

$$\frac{d\Phi_B}{dt} = \underbrace{A(t)\cos\theta(t)\frac{dB(t)}{dt}}_{B \neq \text{const}} + \underbrace{B(t)\cos\theta(t)\frac{dA(t)}{dt}}_{A \neq \text{const}}$$

$$+ \underbrace{B(t)A(t)\frac{d\cos\theta(t)}{dt}}_{\theta \neq \text{const}} \qquad (25\text{-}8)$$

In general, when the number of loops in a coil is N,

$$\left(\frac{d\Phi_B}{dt}\right)_{\text{coil}} = N\left(\frac{d\Phi_B}{dt}\right)_{\text{loop}} \qquad (25\text{-}9)$$

PEER TO PEER

Don't be put off by calculus when you need to calculate the rate of change of a magnetic flux or of a different physical quantity. I often try to do these problems step by step, and I can also use the WolframAlpha website (www.wolframalpha.com/) or the app on my smartphone to check if my calculations are accurate. Checking the units in the final expression before plugging in the numbers is very important, as it helps me identify problems with my solution before I put in any additional effort.

CHECKPOINT 25-2

Magnetic Flux through the Surface of a Closed Loop

The magnetic flux through the plane of a closed loop is zero ($\Phi_B = 0$). What must be true?

(a) There is no magnetic field in that space.
(b) The magnetic field lines are perpendicular to the plane of the loop.
(c) The magnetic field lines are parallel to the plane of the loop.
(d) Either statement (a) or statement (b) must be true.
(e) Either statement (a) or statement (c) must be true.

ANSWER: (e) The key word in this question is *must*. There can be a magnetic field in the area, but if the magnetic field lines are parallel to the surface, the flux is zero. On the other hand, it is possible that there is no magnetic field, so the magnetic flux is zero. Therefore, either (a) or (c) must be true.

EXAMPLE 25-1

Calculating the Rate of Change of a Magnetic Flux

A rectangular copper loop with dimensions $\ell = 60.$ cm and $w = 20.$ cm is moving at a constant speed of 1.0 m/s out of the region of a uniform time-varying magnetic field, $B(t) = (t + 5.0)$ mT. The magnetic field is directed perpendicular to the loop (Figure 25-6).

(a) Derive an expression that describes how the rate of change of the magnetic flux through the loop varies with time.
(b) Find the rate of change of the magnetic flux through the loop at the instant when the loop is halfway out of the magnetic field region.
(c) Find the rate of change of the magnetic flux through the loop when the loop has moved completely out of the magnetic field region.

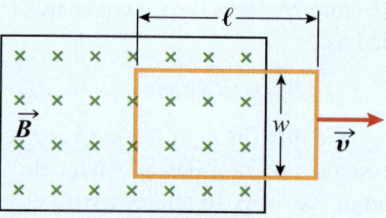

Figure 25-6 A rectangular copper loop is moving at a constant velocity out of a uniform magnetic field region. Magnetic field is directed into the page.

(d) Find the rate of change of the magnetic flux when the loop is stationary, while still located within the magnetic field region.
(e) Recalculate parts (a) to (d) for a coil consisting of three loops, all identical to the loop described above.

Solution

(a) We use Equation 25-8 to calculate the rate of change of the magnetic flux. In this problem, the plane of the loop is always perpendicular to the magnetic field, which means that the magnetic field is parallel to the loop area vector: $\vec{A} \parallel \vec{B} \Rightarrow \theta = 0°$. Therefore, $\cos \theta = \cos 0° = 1$, so Equation 25-8 simplifies to

$$\frac{d\Phi_B}{dt} = \frac{dB(t)}{dt}A(t) + \frac{dA(t)}{dt}B(t)$$

Since the loop is pulled out of the magnetic field region with constant speed v, the area of the loop intercepted by the magnetic field decreases with time. Therefore, the rate of change of the magnetic flux should be negative. If the loop was entirely inside the magnetic field region at time $t = 0$ s, at time t, the area of the loop inside the magnetic field is

$A(t) = w(\ell - vt)$, provided $t \le \dfrac{\ell}{v} = 0.60$ s. Therefore,

$$\frac{dA(t)}{dt} = \frac{d[w(\ell - vt)]}{dt} = -wv$$

Now we can combine all this information and finally calculate the rate of change of the magnetic flux:

$$B(t) = (t + 5.0) \times 10^{-3}\ \text{T}$$

$$\frac{dB(t)}{dt} = 10^{-3}\ \text{T·s}^{-1}$$

$$A(t) = w(\ell - vt)\ \text{m}^2 = (0.12 - 0.20t)\ \text{m}^2$$

$$\frac{dA(t)}{dt} = -wv\ \text{m}^2 \cdot \text{s}^{-1} = -0.20\ \text{m}^2 \cdot \text{s}^{-1}$$

Therefore, the rate of change of the magnetic flux is

$$\frac{d\Phi_B}{dt} = B(t)\frac{dA(t)}{dt} + \frac{dB(t)}{dt}A(t)$$

$$= (-8.8 - 4.0t) \times 10^{-4}\ \text{T·m}^2 \cdot \text{s}^{-1}$$

$$= -(0.88 + 0.40t)\ \text{mWb·s}^{-1},$$

where we have used $w = 0.20$ m, $\ell = 0.60$ m, and $v = 1.0$ m/s.

This result applies only during the time when the loop is partially submerged in the magnetic field region. After 0.60 s, the loop is clear of the magnetic field and is no longer affected by it. As expected, the rate of change of the magnetic field is time dependent.

(b) The loop is halfway out of the magnetic field at $t = 0.30$ s, and

$$\left(\frac{d\Phi_B}{dt}\right)_{t=0.30\ \text{s}} = -(0.88 + 0.40(0.30))\ \text{mWb·s}^{-1}$$

$$= -1.0\ \text{mWb·s}^{-1}$$

(c) When the loop has moved completely outside the magnetic field region, the magnetic flux through the loop is zero: $\Phi_B \equiv \text{const} \equiv 0$, and its rate of change is also zero:

$$\frac{d\Phi_B}{dt} = 0.$$

(d) When the entire loop is held stationary in the uniform, but changing with time, magnetic field, the area of the loop submerged in the magnetic field remains constant. However, the magnetic flux through the loop is changing because the magnetic field itself changes with time. Consequently, Equation 25-8 simplifies to

$$\frac{d\Phi_B}{dt} = \frac{dB(t)}{dt}A(t)$$

Since in this case $A(t)$ is constant, $A(t) = w\ell = 0.12$ m². Therefore,

$$\frac{d\Phi_B}{dt} = (10^{-3}\ \text{T·s}^{-1})(0.12\ \text{m}^2)$$

$$= 1.2 \times 10^{-4}\ \text{T·m}^2 \cdot \text{s}^{-1} = 0.12\ \text{mWb·s}^{-1}$$

(e) When the number of loops is tripled, the effective area of the loop and thus the rate of change of the magnetic flux also triple. Therefore, the answers to the previous parts triple as well.

Making sense of the result

Since the solution to the problem depends on the expression derived in part (a), it is worth checking it using dimensional analysis:

$$\frac{d\Phi_B}{dt} = B(t)\frac{dA(t)}{dt} + \frac{dB(t)}{dt}A(t)$$

$$\left[\frac{d\Phi_B}{dt}\right] = \text{T} \cdot \frac{\text{m}^2}{\text{s}} + \frac{\text{T}}{\text{s}} \cdot \text{m}^2 = \frac{\text{T·m}^2}{\text{s}} = \frac{\text{Wb}}{\text{s}}$$

$$\left[\frac{d\Phi_B}{dt}\right] = \frac{\text{Wb}}{\text{s}}$$

The answer to part (c) is zero because the loop in (c) is unaffected by the magnetic field. The answers to parts (b) and (d) are non-zero. Moreover, the rate of change of the magnetic flux in (b) is larger than that in part (d). This happens because, in part (d), the only cause of the change of the magnetic flux is the change of the magnetic field with time. Yet in part (b) the changes in both the magnetic field and in the area of the loop submerged in the magnetic field contribute to the change of the magnetic flux. The negative sign in part (b) indicates that the magnetic flux is decreasing—the loop is coming out of the magnetic field region.

25-3 Faraday's Law of Electromagnetic Induction

In 1831, Michael Faraday formulated a theory of electromagnetic induction to explain the results of his experiments. **Faraday's law of electromagnetic induction** states that a time-varying magnetic flux through a conducting loop induces an emf ε_{ind} in the loop, which is equal to the negative of the time rate of change of the magnetic flux enclosed by the loop. Thus, Faraday's law provides a way to calculate the magnitude and direction of the induced emf and consequently of the induced current in the loop.

Faraday's law of electromagnetic induction can be expressed mathematically as

KEY EQUATION
$$\varepsilon_{\text{ind}} = -\left(\frac{d\Phi_B}{dt}\right) \tag{25-10}$$

ELECTRICITY, MAGNETISM, AND OPTICS

The minus sign in the law of electromagnetic induction emphasizes that the direction of the induced current is such that the magnetic field induced by it ($\vec{B}_{ind}$) *opposes* the change in the magnetic flux that produced it.

To determine the units for the rate of the change of the magnetic flux, we first link the unit of magnetic field, the tesla, to other SI units. We can do this using the Lorentz force expression $\vec{F}_B = q\vec{v} \times \vec{B}$ (Equation 24-1):

$$[B] = T = \frac{N}{C \cdot \frac{m}{s}} = \frac{kg \cdot \frac{m}{s^2}}{C \cdot \frac{m}{s}} = \frac{kg}{s \cdot C} \Rightarrow T \equiv \frac{kg}{s \cdot C}$$

Therefore, the unit of the rate of change of the magnetic flux (Equation 25-7) can be expressed as

$$\left[\frac{d\Phi_B}{dt}\right] = \frac{Wb}{s} = \frac{T \cdot m^2}{s} = \frac{kg}{s \cdot C} \cdot \frac{m^2}{s}$$

$$= \frac{kg\left(\frac{m}{s}\right)^2}{C} = \frac{J}{C} = V \qquad (25\text{-}11)$$

Thus, dimensional analysis confirms that the rate of change of magnetic flux relates to the electric potential difference. A conducting loop in a magnetic field acts as an emf source, as long as the magnetic flux keeps changing in time.

When a coil consists of N identical loops, the law of electromagnetic induction (Equation 25-10) can be rewritten as follows:

KEY EQUATION $\quad \varepsilon_{ind} = -\left(\dfrac{d\Phi_B}{dt}\right)_{coil} = -N\left(\dfrac{d\Phi_B}{dt}\right)_{loop}$ $\quad$ (25-12)

Equations 25-8 and 25-12 can be combined to obtain

$$\varepsilon_{ind} = -N\left(\underbrace{\frac{dB(t)}{dt}A\cos\theta}_{B \neq const} + \underbrace{\frac{dA(t)}{dt}B\cos\theta}_{A \neq const} + \underbrace{\frac{d\cos\theta(t)}{dt}BA}_{\theta \neq const}\right)$$

$$(25\text{-}13)$$

As discussed earlier, Equation 25-13 has three distinct parts. Each one reflects the generation of an induced emf in the circuit due to the changing magnetic field, the changing area of the loop, and the changing angle between the magnetic field and the plane of the loop. Table 25-1 shows how you can visualize each part of Equation 25-13.

PEER TO PEER

When I understood that the magnetic flux though a loop can change for three different reasons, I started seeing magnetic flux problems in a different light. Now while reading a problem, I pay attention to three key variables: $\vec{B}$, $\vec{A}$, and θ. I check if the magnitude of the magnetic field is changing with time, if the area of the conducting loop submerged in the magnetic field region is changing with time, and if the angle the magnetic field forms with the plane of the loop is changing with time. As soon as I see that at least one of these variables is changing, I know that an emf is induced in the loop.

Table 25-1 Three Causes of Induced emf

The magnitude of the magnetic field varies with time ($B \neq const$).	
The area of the loop changes with time ($A \neq const$).	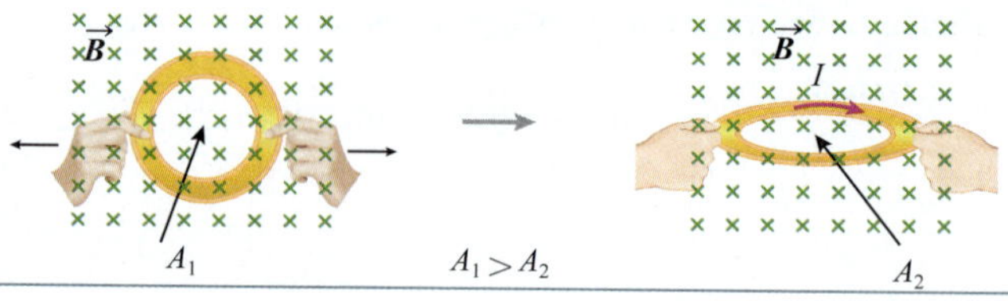
The angle between vectors $\vec{B}$ and $\vec{A}$, represented as θ, changes with time.	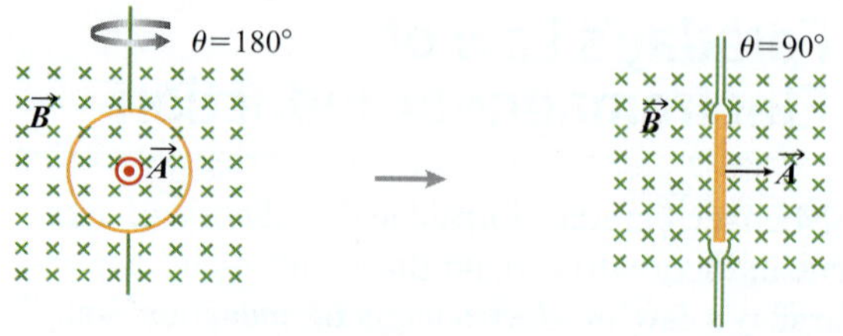

As pointed out earlier, the direction of the induced emf (Equation 25-12) in Faraday's law of electromagnetic induction always opposes the change that caused it in the first place. The direction or sign of the induced emf has very significant practical and theoretical implications, and because this part of Faraday's law was rediscovered independently in 1834 by Russian physicist Heinrich Lenz (1804–1865), today it is called **Lenz's law**. We will explore Lenz's law and its implications in more detail below.

Lenz's Law

Lenz's law states that *the direction of the induced current in a circuit always opposes the change that produced it.* Let us consider what this means and how Lenz's law can help us understand how electromagnetic circuits obey Newton's third law and the conservation of energy law. Let us modify and re-examine Example 25-1. In this case, we simplify it by considering a region of constant magnetic field.

EXAMPLE 25-2

Exploring the Implications of Lenz's Law: The Case of a Rectangular Loop

A rectangular copper loop with dimensions $\ell = 60.0$ cm and $w = 20.0$ cm is moving at a constant speed of 1.00 m/s out of the region of a uniform and constant magnetic field $B(t) = 0.500$ T. The magnetic field is directed perpendicular to the loop, and the resistance of the copper in the loop is 26.5 mΩ (Figure 25-7). When answering the questions, assume the loop is still partially submerged in the magnetic field region.

(a) Describe qualitatively the direction and causes of all the forces acting on each side of the loop as it is moving out of the region of constant magnetic field.

(b) Find the magnitude and direction of the force acting on side CD of the loop, $\vec{F}_{CD}$, and the magnitude and direction of the induced electric current in the loop.

(c) Derive an expression for the external force, $\vec{F}_{ext}$, that must be applied on the loop to move it with this constant speed outside the magnetic field region.

(d) Derive an expression that describes the rate of change of the magnetic flux through the loop, and find the direction and magnitude of the induced current in the loop.

(e) Explain how your answers to parts (a) to (d) are related.

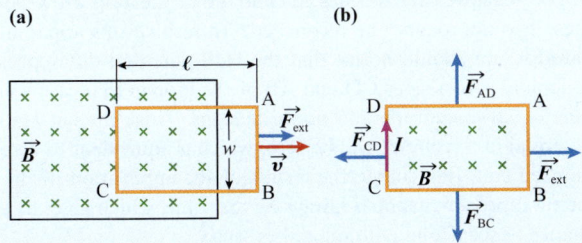

Figure 25-7 (a) A rectangular copper loop is moving at a constant velocity out of a uniform and constant magnetic field region. (b) Forces exerted on the sides of the moving rectangular loop.

Solution

(a) Let us consider each side of the loop. When the loop is stationary and there is no electric current flowing in it, the charged particles inside the loop have zero average velocity (their drift velocity is equal to zero). However, when the loop is moving to the right with a constant velocity of 1.00 m/s the charges inside the loop start moving to the right as well. These charges are also moving with an

average velocity of 1.00 m/s (Figure 25-7(a)). Since they are moving inside the region of a constant magnetic field, these charges will experience a magnetic force described by Equation 24-1. As a result, positive charges will experience a magnetic force directed upward, while negative charges will experience a downward magnetic force. We have already considered a similar case when discussing the Hall effect in Section 24-3. This charge separation will be the most pronounced inside side BC, which is directed perpendicular to both the magnetic field, $\vec{B}$, and the velocity, $\vec{v}$. This charge separation will induce an electric current in the loop in the clockwise direction. Now we can consider the effect of this induced current on each side of the loop. We will use Equation 24-21, which describes the magnetic force on a current-carrying wire, to find the forces acting on each side of the loop.

Side BC: As a result of the induced electric current flowing in the clockwise direction, side BC experiences a downward magnetic force of $\vec{F}_{BC}$.

Side CD: As a result of the induced electric current flowing in the clockwise direction (due to the Hall effect), side CD experiences a magnetic force directed to the left of $\vec{F}_{CD}$.

Side AD: As a result of the induced electric current flowing in the clockwise direction, side AD experiences an upward magnetic force $\vec{F}_{AD}$.

Side AB: While there is electric current flowing through AB, it is located outside the region of the magnetic field, so it will not experience any magnetic force. The only force applied on it is the external force $\vec{F}_{ext}$ pulling the loop outside of the magnetic field region.

(b) The force acting on side CD can be found using Equation 24-19 describing the Hall potential difference ($\Delta V_{Hall} = vBw$) and the magnetic force on the current-carrying wire (Equation 24-21: $\vec{F}_B = I\vec{\ell} \times \vec{B}$):

$$F_{CD} = I\ell B = \frac{\Delta V_{Hall}}{R} wB = \frac{vBw}{R} wB = \frac{v}{R}(wB)^2$$

$$= \frac{1.00 \text{ m/s}}{26.5 \text{ m}\Omega}((0.200 \text{ m})(0.500 \text{ T}))^2$$

$$= 0.377 \text{ N}$$

While in this example, the force on side CD is small, it is important to notice that the force is proportional to the square of the magnetic field. Moreover, it is inversely

(continued)

proportional to the resistance of the loop. By decreasing the resistance, we can increase the force. This relationship is important for many applications, such as magnetic brakes, discussed later in this chapter.

The direction of the induced electric current, as discussed above, is clockwise, and its magnitude is equal to

$$I_{ind} = \frac{\Delta V_{Hall}}{R} = \frac{vBw}{R}$$

$$= \frac{(1.00 \text{ m/s})(0.200 \text{ m})(0.500 \text{ T})}{26.5 \text{ m}\Omega} = 3.77 \text{ A}$$

(c) According to Newton's first law, for the loop to move with a constant velocity, the sum of the forces acting on it must balance out: $\sum \vec{F} = 0$. Therefore, the external force pulling on the loop must be equal and opposite to the force on side CD (Figure 25-7(b)): $\vec{F}_{CD} = -\vec{F}_{ext} \Rightarrow F_{ext} = 3.77 \text{ mN}$. Notice that the forces on the horizontal sides of the loop pull in opposite directions and cancel each other.

(d) To find the rate of change of the magnetic flux in the loop, we use the results from Example 25-1, noting that the change of magnetic flux in this example is caused by the changing area of the loop that is immersed in the magnetic field that is perpendicular to the plane of the loop:

$$\frac{d\Phi_B}{dt} = \underbrace{\frac{dA(t)}{dt}}_{A \neq const} B\cos\theta = \frac{w\Delta x}{\Delta t}B = wvB = 0.100 \text{ V}$$

Using Faraday's law of electromagnetic induction, we can find the emf induced in the loop (Equation 25-12):

$$\varepsilon_{ind} = -\frac{d\Phi_B}{dt} = -0.100 \text{ V}$$

To interpret the meaning of the negative sign, we can apply Lenz's law: the direction of the induced emf should oppose the changes that caused it. Since the loop is moving out of the region of a constant magnetic field, the magnetic flux through it is decreasing. To prevent the decrease, the current induced in the loop should be directed to generate an induced magnetic field in the direction of the original magnetic field. This means that the induced current in the loop will be flowing in the clockwise direction.

(e) We have considered the case of closed loop moving out of the region of uniform and constant magnetic field from the perspective of magnetic forces acting on moving charges (parts (a) to (c)) and from the perspective of Faraday's law and magnetic flux. As expected, the two approaches give the same results. The Hall potential difference generated across side BC is precisely equal to the induced emf, according to Faraday's law: $\varepsilon_{ind} = V_{Hall} = vwB$. Moreover, this induced potential difference prevents the loops from moving out of the magnetic field. The negative sign in Faraday's law, which is reflected in Lenz's law, emphasizes that one needs to do work on the loop (invest energy into the system) to pull the loop out of the magnetic field region. We can understand this because, during the process, electric current is generated and the mechanical energy we invested into moving the loop out is transformed into electromagnetic energy. If not for the opposite sign in Faraday's law, we could have generated an infinite amount of electromagnetic energy (significant current) from investing a very small amount of mechanical energy into the system (which did not have energy stored in it) by just giving the loop a small initial kick—an initial velocity. Notice that similar reasoning can be applied to explain why the minus sign in Hooke's law is crucial from the energy conservation standpoint. In summary, if the direction of the induced emf were not opposed to the change of the magnetic flux, we would get a perpetual motion machine arising from the positive feedback. This "runaway" scenario contradicts the law of energy conservation.

Making sense of the result

As mentioned above, we can use different approaches to solve problems where emf and consequently electric current are induced in a loop as a result of changing the magnetic flux through it. Whatever our approach is, our solution should obey Newton's laws and the law of energy conservation. Moreover, different approaches should lead to the same results. For example, if we imagine a loop moving in a uniform, constant, and infinitely large magnetic field, the magnetic flux through the loop will remain constant. Therefore, no emf will be induced in it. However, the Hall potential difference would still be generated across sides CD and AB of the loop. How can these two approaches be reconciled? To resolve this apparent paradox, we should notice that the Hall potential differences generated across sides CD and AB of the loop in that case will have equal magnitude and opposite signs. Thus, the net Hall potential in the circuit will be zero, which is equivalent to zero induced emf. Remember, the results of the application of different approaches should always be the same, which gives us a chance to see if our solution makes sense.

Now you are ready to explore Interactive Activity 25-2. We recommend you pay careful attention to both the magnitude and the direction of the induced emf and induced current. You can do it by using the meters in the simulation.

INTERACTIVE ACTIVITY 25-2

Exploring Faraday's Law of Electromagnetic Induction

In this activity, you will use the PhET simulations "Faraday's Electromagnetic Lab: Pickup Coil" and "Faraday's Electromagnetic Lab: Electromagnet" to gain an understanding of Faraday's law of electromagnetic induction and Lenz's law.

(a)

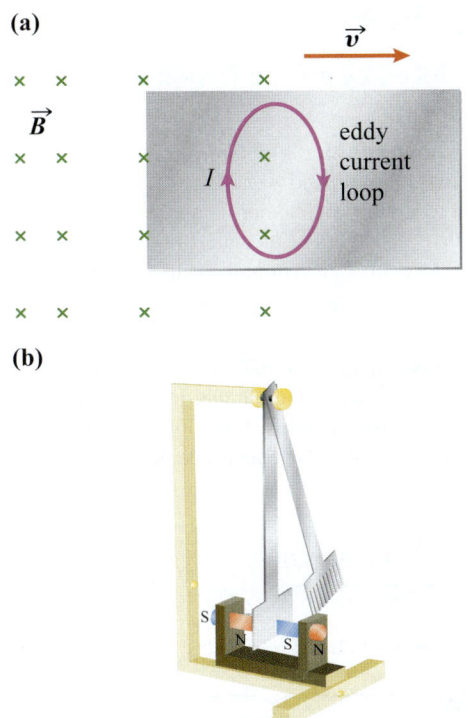

(b)

(c)

Figure 25-9 Eddy currents. (a) One of the eddy current loops inside a metal sheet moving to the right through a non-uniform magnetic field. (b) Two metal pendulums are placed between the poles of two permanent magnets. As the pendulums swing, eddy currents generated in the pendulum that consists of a solid metal sheet (no slits) slow the motion of the pendulum (a consequence of Lenz's law). On the other hand, the pendulum that has slits in it does not slow down, since the slits prevent the generation of eddy currents. (c) A magnet levitates above a superconductor as a result of the Meissner effect.

CHECKPOINT 25-3

Induced Current in a Metal Ring

In which direction should the bar magnet in Figure 25-8 move to induce the electric current in the ring shown in the diagram (clockwise direction as seen from left)?

(a) horizontally toward the ring
(b) horizontally away from the ring
(c) upward
(d) downward
(e) none of the above

Figure 25-8 Checkpoint 25-3.

ANSWER: (b) Apply Lenz's law.

Eddy Currents

In 1851, almost 20 years after Faraday's discovery, French physicist Léon Foucault (1819–1868) noticed that when the magnetic flux through a large piece of conductor varies with time, the flux induces electric current loops inside the conductor (Figure 25-9(a)). Such currents are called **eddies** or **eddy currents**. According to Faraday's law, these induced currents oppose the change of the original magnetic flux. The faster the magnetic flux changes, the stronger the induced eddy currents. The direction of eddies can be determined using Lenz's law. Since all conductors (except for superconductors) have some resistance, eddy currents generated in them by the changing magnetic flux will heat the conductors up. This heating means generation of thermal energy. Therefore, eddy currents can convert mechanical energy (e.g., motion of a conductor through the non-uniform magnetic field) into thermal energy (heat dissipated by the current-carrying conductors). In some applications, such as a hand-crank generator flashlight, the electrical energy generated by the changing magnetic flux can be stored. However, in other applications, the energy dissipation from eddy currents is undesirable. To reduce energy losses, conductors are made with slits (Figure 25-9(b)) or are constructed with thin layers (laminations) that are insulated from each other by a non-conductive varnish. These techniques reduce undesirable eddy currents by making the area of the closed loops much smaller. Eddy currents are used in magnetic brakes and magnetic dampers, as discussed in Section 25-6.

Since superconductors below the critical temperature have zero resistance (and, hence, no thermal energy dissipation), eddy currents generated on the surface of a superconductor will flow indefinitely. Consequently, such eddy currents induce a magnetic field that *completely* opposes the external magnetic flux that induced them. Therefore, the superconductor expels the magnetic field from its interior. This phenomenon is called the Meissner effect (Figure 25-9(c)), named after German physicist Walther Meissner (1882–1974), who discovered it in 1933.

EXAMPLE 25-3

Current Induced by a Time-Dependent Magnetic Field

A rectangular stationary loop with dimensions $\ell = 60.$ cm and $w = 20.$ cm (Figure 25-10) and a resistance of 0.10 Ω is placed perpendicular to a uniform time-dependent magnetic field with $B(t) = (0.040t + 0.0020)$ T. Calculate the direction and the magnitude of the electric current induced in the loop.

Solution

It is often convenient to separately find the magnitude and direction of the induced emf or induced current due to the changing magnetic flux. Let us use Equation 25-13, taking into account that $N = 1$, to find the magnitude of the emf induced in the loop:

$$|\varepsilon_{ind}| = N \underbrace{\left| \frac{dB(t)}{dt} A(t) \cos \theta(t) \right|}_{B \neq const}$$

$$= \underbrace{\left| \frac{d(0.040t + 0.020)\text{ T}}{dt} (0.20\text{ m} \cdot 0.60\text{ m}) \cos 0° \right|}_{B \neq const}$$

$$= 0.040\text{ T/s} \cdot 0.12\text{ m}^2 = 0.0048\text{ V} = 4.8\text{ mV}$$

To calculate the magnitude of the induced current, we use Ohm's law:

$$I_{ind} = \frac{\varepsilon_{ind}}{R} = \frac{4.8\text{ mV}}{0.10\text{ Ω}} = 48\text{ mA}$$

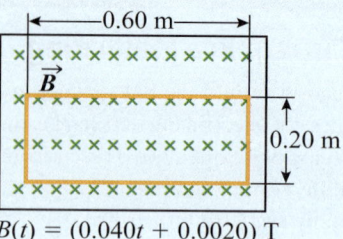

$$B(t) = (0.040t + 0.0020)\text{ T}$$

Figure 25-10 Example 25-3. A rectangular loop in a uniform magnetic field that changes with time.

Now we are ready to find the direction of the induced current. We can use Lenz's law to do that. Since the magnetic field increases with time, the magnetic flux increases with time. According to Lenz's law, the direction of the induced current is such that the induced magnetic field, $\vec{B}_{ind}$, opposes the magnetic flux change. The direction of the original magnetic field, $\vec{B}(t)$, is into the page in Figure 25-10. The field is increasing, causing the magnetic flux to increase as well. For the induced magnetic field $\vec{B}_{ind}$ to oppose this change in magnetic flux, its direction should be opposite to the direction of the original magnetic field. Therefore, $\vec{B}_{ind}$ is directed out of the page. As a result, the induced current flows counter-clockwise.

Making sense of the result

Since the original magnetic field, $\vec{B}(t)$, changes with time linearly, its rate of change is constant. Consequently, the induced emf and induced current are also constant.

Going back to Faraday's law of electromagnetic induction (Equation 25-13), we recall that an induced emf is generated when the magnetic flux through a loop (or through a coil) is changing with time. Thus, an induced emf can be caused by a time-dependent magnetic field (as discussed in Examples 25-1 and 25-3); by the movement of the loop inside a non-uniform magnetic field, thus by changing the area of the loop submerged in the field (Example 25-2); by varying the angle between the magnetic field and the loop; or by all of the above. Whenever an emf is generated as a result of the motion of the loop relative to the magnetic field, we call it **motional emf**. Generating motional emf is very common, as it allows the transfer of mechanical energy (for example, potential energy of water) into electromagnetic energy, as is done in hydroelectric power stations. For simplicity, we will consider a coil consisting of identical rectangular loops. Let us first assume that each rectangular loop has

width w and length ℓ. Let us calculate the induced emf in this rectangular coil when it enters or leaves a region of a uniform and constant magnetic field, $\vec{B} = $ const, that is directed perpendicular to the coil, $\theta = 0°$ (see Figure 25-11(a) in Example 25-4 below). Moreover, the coil is moving at a constant velocity $\vec{v}$ such that its plane is perpendicular to the direction of the magnetic field. Taking into account that the magnetic field is constant and using Equation 25-13, we obtain the expression for the induced emf in the coil:

KEY EQUATION
$$\varepsilon_{ind} = -NB\underbrace{\frac{dA(t)}{dt}}_{\substack{\text{area changes} \\ \text{with time}}} = -NB v \ell \qquad (25\text{-}14)$$

where N is the number of loops in the coil and ℓ is the length of the side of the coil that is inside the constant magnetic field and perpendicular to both the field and the velocity.

EXAMPLE 25-4

Motional emf in a Rectangular Coil (Revisiting Example 25-2)

A horizontal, rectangular (5.00 cm × 10.0 cm) coil has 100 turns and a total resistance of 10.0 Ω. Initially, it is sitting with the right side of the coil at the edge of a vertical uniform magnetic field with $B = 0.400$ T, as shown in Figure 25-11(a). The coil is then pulled to the right, out of the field.

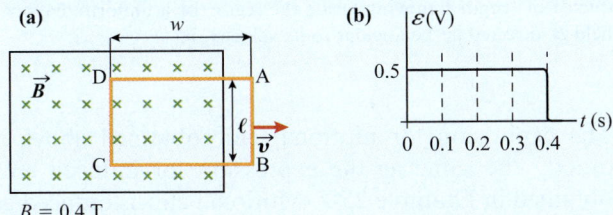

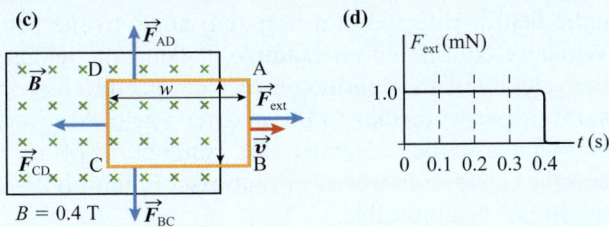

Figure 25-11 (a) A rectangular coil is moving with constant velocity out of a region of uniform and constant magnetic field. (b) Induced emf in the coil as a function of time during its motion. (c) Forces acting on the coil during its motion. (d) The dependence of the external force on time.

(a) Determine the velocity of the coil needed to generate an emf of 0.500 V.

(b) Draw a graph that shows the dependence of the induced emf on time while the coil is leaving the magnetic field at the constant velocity $\vec{v}$ found in part (a).

(c) Find the direction and the magnitude of the electric current induced in the coil during this process.

(d) Find the magnitude of the external force, $\vec{F}_{ext}$, needed to pull the coil with the constant velocity $\vec{v}$. Assume that friction is negligible.

(e) Sketch a graph that represents the dependence of the external force $\vec{F}_{ext}$ on time. Compare this graph with the emf graph in part (b).

Solution

This example should remind you of the problem discussed in Example 25-2.

(a) Applying Equation 25-14, we find the speed of the coil that will generate 0.500 V of induced emf:

$$v = \frac{\varepsilon_{ind}}{NB\ell} = \frac{0.500 \text{ V}}{(100)(0.400 \text{ T})(0.0500 \text{ m})}$$

$$= 0.250 \text{ m/s} = 25.0 \text{ cm/s}$$

(b) Moving at a constant speed of 25.0 cm/s, it takes 0.400 s for the coil to move 10.0 cm to get completely outside the magnetic field region. During this time, the induced emf is constant (Equation 25-14). However, as soon as the coil is clear of the magnetic field, the induced emf in it drops to zero, as the magnetic flux through the coil stops changing (Figure 25-11(b)).

(c) To find the magnitude of the electric current induced in the coil while it is being pulled out of the field, we use Ohm's law:

$$I_{ind} = \frac{\varepsilon_{ind}}{R} = \frac{NBv\ell}{R} = \frac{0.500 \text{ V}}{10.0 \text{ Ω}} = 0.0500 \text{ A} = 50.0 \text{ mA}$$

Since the magnetic flux through the coil decreases as the coil leaves the field, Lenz's law requires the induced current to move in a clockwise direction to induce a magnetic field $\vec{B}_{ind}$ in the same direction as the original field (into the page).

(d) Since the coil is moving with a constant velocity, the net force acting on it must be zero. It might seem that there is no need to apply an external force, $\vec{F}_{ext}$, to pull the coil out of the magnetic field. However, let us consider each side of the coil more carefully, as we did in Example 25-2 (Figure 25-11(c)). Side AB is never inside the magnetic field, so there is no magnetic force acting on it. However, there are magnetic forces acting on the other three sides. Using the right-hand rule for determining the magnetic force on the current-carrying wire (Section 24-4) and the direction of the current in the coil, we find that the magnetic forces acting on sides AD and BC are directed upward and downward, respectively (Figure 25-11(c)), and the magnetic force acting on side CD is directed to the left, resisting the coil's motion. The magnitude of the resultant horizontal magnetic force can be found using Equation 24-21:

$$F_B = N(I\ell B) = N\frac{NBv\ell}{R}\ell B = \frac{N^2 B^2 \ell^2 v}{R} \quad (1)$$

Since the net force acting on the coil is zero, the external force pulling it must be directed to the right with a magnitude of

$$F_{ext} = \frac{N^2 B^2 \ell^2 v}{R}$$

$$= \frac{(100)^2 (0.400 \text{ T})^2 (0.0500 \text{ m})^2 (0.250 \text{ m/s})}{10.0 \text{ Ω}}$$

$$= 0.100 \text{ N}$$

(e) Figure 25-11(d) shows how the magnitude of the pulling force, $\vec{F}_{ext}$, depends on time.

Making sense of the result

The induced emf is generated as long as the magnetic flux through the coil is changing, that is, while the coil is moving out of the region of a constant and uniform magnetic field. During this process, the magnetic force acting on the coil resists its motion. Notice that we obtained a similar result in Example 25-2.

Let us consider the energy transfer in Example 25-4. We generated an electric current by pulling the coil out of the region of a constant and uniform magnetic field. To pull the coil with constant velocity $\vec{v}$, we had to apply an external force $\vec{F}_{\text{ext}}$ (Figure 25-12), which did work on the coil (Equation (1) in Example 25-4). The rate at which this work was done equals the generated power, P:

$$P = F_{\text{ext}}v = \frac{N^2 B^2 \ell^2 v}{R} v = \left(\frac{NB\ell v}{R}\right)^2$$

$$R = \left(\frac{\varepsilon_{\text{ind}}}{R}\right)^2 R = I_{\text{ind}}^2 R \qquad (25\text{-}15)$$

Equation 25-15 shows that the power delivered by the external force dissipates as thermal energy.

We now consider motional emf at the microscopic level. When a conductor is moving with a constant velocity $\vec{v}$ inside a uniform magnetic field $\vec{B}$ directed perpendicular to its velocity, the charged particles inside the conductor move with the same average velocity $\vec{v}$. Therefore, positive and negative charges inside this moving conductor experience magnetic forces acting in opposite directions (Figure 25-13). Consequently, the charges separate. This separation, as we discussed earlier, is a manifestation of the Hall effect. Electrons are free to move inside a conductor, so they will be affected by this magnetic force. This will generate a potential difference across the conductor (the Hall potential difference). The separation of charges continues until the magnitudes of the electric and magnetic forces equalize:

$$F_{\text{E}} = F_{\text{B}} \Rightarrow E_{\text{ind}}q = qvB_\perp \Rightarrow E_{\text{ind}} = vB_\perp$$

In a conductor of length ℓ, an induced electric field $\vec{E}_{\text{ind}}$ generates an induced potential difference across the conductor:

$$\Delta V_{\text{ind}} = E_{\text{ind}}\ell = vB_\perp \ell \qquad (25\text{-}16)$$

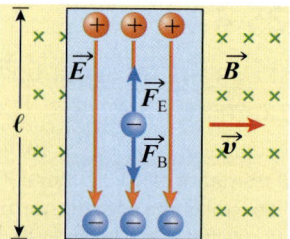

Figure 25-13 The separation of charges inside a rectangular conductor of length ℓ moving inside the region of a uniform magnetic field $\vec{B}$, directed perpendicular to its velocity, $\vec{v}$.

The expression for motional emf obtained above is exactly the same as the expression for induced emf obtained in Example 25-2. Motional emf is induced as long as the velocity of a conductor moving in a magnetic field is directed at a non-zero angle to the field. While we considered an example of constant velocity, the velocity of the conductor does not need to be constant or perpendicular to the magnetic field. Motional emf can also be calculated for conductors of more complex shapes; however, in that case, Equation 25-15 might not be applicable.

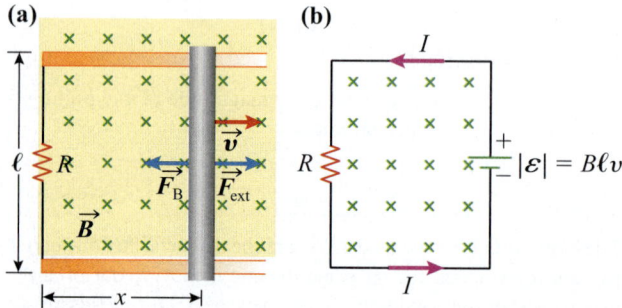

Figure 25-12 (a) A conducting bar is in a constant and uniform magnetic field. An applied external force $\vec{F}_{\text{ext}}$ slides the bar with constant velocity along the frictionless rails. (b) The equivalent circuit diagram for the conducting bar sliding with constant velocity.

CHECKPOINT 25-4

Charge Distribution on a Plate Moving Inside a Magnetic Field

A rectangular metal plate initially located outside a constant and uniform magnetic field is given an initial push toward the magnetic field region and then let go. Which diagram in Figure 25-14 correctly represents the charge distribution on the plate as it moves into the magnetic field?

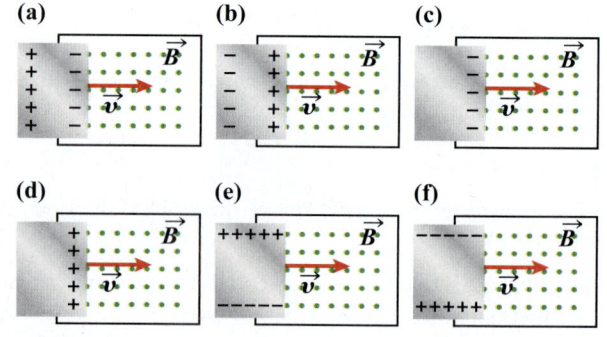

Figure 25-14 Checkpoint 25-4.

ANSWER: Diagram (f). Use the right-hand rule to determine the direction of the magnetic force on charged particles moving inside a magnetic field. Notice that the right-hand rule for magnetic force (Equation 24-1) applies to positive charges, so the effect on negative charges will be opposite.

25-4 Induced emf and Induced Electric Fields

We now are ready to take a closer look at the current induced in a circular conducting loop by a time-varying magnetic flux (Figure 25-15). As the magnetic field in the shaded area changes with time, it generates a changing magnetic flux through the loop. This changing magnetic flux induces an emf and consequently an electric current in the loop. A **galvanometer** (G) (an instrument for detecting and measuring very small electric currents; a galvanometer is a very sensitive ammeter) connected in series with the loop can measure this current. The presence of an electric current indicates the presence of an electric field that applies an electric force on charge carriers. Since the loop is circular, the direction of the electric field must be tangent to the loop to force the charged particles to move along the circular loop (Figure 25-15(a)). An electric field is induced in the space surrounding the time-varying magnetic flux (Figure 25-15(b)). Note that this **induced electric field**, $\vec{E}_{\text{ind}}$, is independent of the presence of free electric charges in the space surrounding the loop.

The properties of an induced electric field differ substantially from the properties of an electrostatic electric field.

Table 25-2 compares the properties of the electrostatic field $\vec{E}$ created by a stationary electric charge with the properties of the induced electric field $\vec{E}_{\text{ind}}$.

Consider the positively charged particle q moving along a circular loop in the counter-clockwise direction in Figure 25-15(a). Let us calculate the work done by the induced electric field on the particle. Both the force acting on the charged particle and the induced electric field lines are always tangent to the loop. From symmetry considerations, we deduce that at any given instant the magnitude of the induced electric field $\vec{E}_{\text{ind}}$ is the same at every point along the circular loop. When the electric charge moves a distance ds along the loop, the work done on it is $dW = \vec{F} \cdot d\vec{s} = q\vec{E} \cdot d\vec{s}$. The energy acquired by a charged particle as a result of the work done on it by the electric field, as the particle moves across the potential difference ε_{ind}, can be expressed as $W = q\varepsilon_{\text{ind}}$. Therefore, we can calculate this work as

$$
\begin{cases}
W_{\text{along the loop}} = q\varepsilon_{\text{ind}} \\
W_{\text{along the loop}} = \oint dW = \oint q\vec{E}_{\text{ind}} \cdot d\vec{s} \quad (25\text{-}17) \\
\qquad\qquad = qE_{\text{ind}} \oint ds = qE_{\text{ind}} 2\pi r
\end{cases}
$$

Since both expressions represent the work done on a charged particle moving along a closed circular loop, they must be equal, and

$$q\varepsilon_{\text{ind}} = qE_{\text{ind}} 2\pi r \Rightarrow \varepsilon_{\text{ind}} = 2\pi r E_{\text{ind}}$$

We can now calculate the induced electric field:

$$E_{\text{ind}} = \frac{\varepsilon_{\text{ind}}}{2\pi r} = \frac{1}{2\pi r}\left(-\frac{d\Phi_B}{dt}\right)$$

$$E_{\text{ind}} = -\frac{1}{2\pi r}\left(\frac{dB(t)}{dt}\right)\pi r^2 = -\frac{r}{2}\left(\frac{dB(t)}{dt}\right) \quad (25\text{-}18)$$

In this derivation, we expressed the magnetic flux through the loop as $\Phi_B = B(t)\pi r^2$. Equation 25-18 indicates that as long as the magnetic field varies with time, an induced electric field, E_{ind}, is created. The minus sign indicates that the induced electric field opposes the change in the magnetic flux. Since the change in the magnetic flux is proportional to the area of the loop (provided the loop is situated within the magnetic field region), a bigger loop means a greater induced emf and induced electric field. Therefore, the intensity of the induced field increases as one moves away from the centre of the loop, as long as the loop is located inside the region of the changing magnetic field. (Notice how the density of electric field lines changes in Figure 25-15(b).)

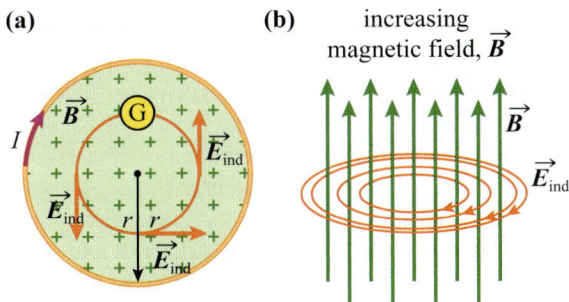

(a)

(b) increasing magnetic field, $\vec{B}$

Figure 25-15 (a) A galvanometer, G, records the presence of electric current in the coil, induced by an increasing magnetic flux through the coil: $\vec{B}$ increases with time. (b) Induced electric field lines surrounding the area of the increasing magnetic field.

Table 25-2 Comparison of Electrostatic Field, $\vec{E}$, and Induced Electric Field $\vec{E}_{ind}$

Electrostatic Field $\vec{E}$ (Figure 19-15)	Induced Electric Field $\vec{E}_{ind}$ (Figure 25-15(b))
	(b) increasing magnetic field, $\vec{B}$

	Electrostatic Field $\vec{E}$ (Figure 19-15)	Induced Electric Field $\vec{E}_{ind}$ (Figure 25-15(b))
Definition	The electrostatic field, $\vec{E}$, at a point represents the **electrostatic force** per unit charge acting on a test charge at that point: $\vec{E} = \dfrac{\vec{F}_{q_0}}{q_0}$. The electrostatic field is a vector field directed parallel to the electrostatic force acting on a positively charged test particle.	The induced electric field, $\vec{E}_{ind}$, at a certain point in space represents the **electromagnetic force** per unit charge acting on a test charge placed at that point: $\vec{E}_{ind} = \dfrac{\vec{F}_{q_0}}{q_0}$. The induced electric field is a vector field directed parallel to the electromagnetic force acting on a positively charged test particle.
Source	Electric charges	Time-varying magnetic fields
Field lines	Emanate from positive charges and point toward negative charges	Are closed loops without a beginning or an end point
Work done on a charged particle	Depends on initial and final position of the particle, but not on the path between them; hence, electrostatic fields are **conservative fields**.	Done on a charged particle moving along a closed loop and is non-zero; hence, induced electric fields are **non-conservative fields**.
Potential energy	The electrostatic potential energy is defined as $U_E = \dfrac{Qq}{4\pi\varepsilon_0 r}$.	We cannot associate potential energy with an induced electric field because it is a non-conservative field. We will discuss this in more detail in Chapter 27.

Equation 25-17 can be extended to the general case of a closed path of an arbitrary shape:

$$\begin{cases} W_{\text{along the loop}} = \oint \vec{F} \cdot d\vec{s} = \oint q\vec{E}_{ind} \cdot d\vec{s} \\ q\varepsilon_{ind} = q\oint \vec{E}_{ind} \cdot d\vec{s} \Rightarrow \varepsilon_{ind} = \oint \vec{E}_{ind} \cdot d\vec{s} \\ \text{Since } \varepsilon_{ind} = -\dfrac{d\Phi_B}{dt} \Rightarrow \oint \vec{E}_{ind} \cdot d\vec{s} = -\dfrac{d\Phi_B}{dt} \end{cases}$$

The last equation leads to a reformulation of Faraday's law of electromagnetic induction: a time-varying magnetic flux in a certain region of space always induces a *non-conservative* electric field $\vec{E}_{ind}$ in that region (see Table 25-2), even in the absence of electric charges. The magnitude of this induced electric field can be expressed as

KEY EQUATION　　$\left| \varepsilon_{ind} \right| = \left| \oint \vec{E}_{ind} \cdot d\vec{s} \right| = \left| -\dfrac{d\Phi_B}{dt} \right|$　　(25-19)

The direction of the induced electric field is such that the electric current generated by it induces a magnetic field that *opposes* the change in the magnetic flux.

CHECKPOINT 25-5

Exploring Induced Electric Fields

A uniform but time-varying magnetic field $\vec{B}(t)$ induces an electric field that is constant in magnitude but has a varying direction. What must be true about the dependency of the magnetic field $\vec{B}$ on time?

(a) The magnetic field does not depend on time; it is constant.

(b) The magnetic field changes with time exponentially: $B(t) = Be^{-Ct}$, where C is a constant.

(c) The magnetic field changes with time at a constant rate $B(t) = Bt + C$, where B and C are constants.

(d) The magnetic field B changes with time as $B(t) = At^2 + Bt + C$, where A, B, and C are constants.

ANSWER: (c) Use Faraday's law of electromagnetic induction: The induced electric field is proportional to the rate of change of the magnetic field. Therefore, a magnetic field that changes with time at a constant rate induces an electric field that has a constant magnitude. As you can see in Figure 25-15, the direction of this induced electric field will be changing.

25-5 Self-Inductance and Mutual Inductance

Understanding the concepts of induced emf and induced electric field leads us to another question: What effect does the presence of a loop have on a circuit consisting of a battery, a resistor, a coil, and a switch? To distinguish between the emf from a battery and an induced emf, we denote the battery emf as ε and the emf induced in a coil as ε_{ind}. Induced currents can be generated in any closed loop. However, they become especially important in circuits containing various coils, significantly affecting the behaviour of the circuit. Such elements of electric circuits are called **inductors** and are denoted in circuit diagrams by a helical symbol and the letter L (in honour of Heinrich Lenz), as shown in Figure 25-16. Let us consider an inductor that has a very low internal resistance ($R \approx 0\ \Omega$) and is connected in series to a resistor, R, and a battery, ε. This circuit is called a **series RL circuit**.

First, we consider the circuit when the switch has been in position a for a long time, and a steady electric current flows through the RL circuit (Figure 25-17). Let us use a voltmeter (V_L) to measure the potential difference across the inductor, ε_{ind}, and an ammeter to measure the current I through it (since it is a series

circuit, the same current flows through all of the circuit elements). Since a constant current flows through the inductor, it generates a constant magnetic field, so there is no change in the magnetic flux through the inductor. In that case, the inductor behaves like an ordinary wire with a very low resistance, R. Therefore, the potential difference across this inductor is zero: $\varepsilon_{ind} \approx 0$.

Now let us examine what happens immediately after the switch has been flipped from position a to position b. Initially, the current in the circuit was I_0. However, because the battery was effectively removed from the circuit, the current in the circuit will start rapidly decreasing. We will denote the instantaneous current in an RL circuit as i (similar to what we did with RC circuits, discussed in Chapter 23) to distinguish it from a constant current. This **transient** (changing) **electric current** in the circuit induces an emf ε_{ind} in the coil. We apply Faraday's law (Equation 25-19) to find the direction of the induced emf. Since the induced emf is directed such that it *opposes* the change in the magnetic flux through the coil, this emf induces electric current i_{ind} in the direction of the original current I_0. Thus, the magnetic field, $\vec{B}_{ind}$, induced by this current has the same direction as the original magnetic field created by the decreasing current. If we flip the switch once again from position b to position a, so the battery becomes a part of the circuit again, the induced emf will have the opposite direction. In this case, the induced electric current, i_{ind}, will be in the opposite direction to the current I in the circuit. Interactive Activity 25-3 provides an opportunity to take a closer look at the behaviour of an RL circuit and build a deeper understanding of this phenomenon.

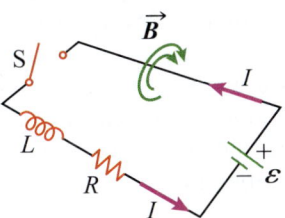

Figure 25-16 A series RL circuit that consists of a battery, a switch, a resistor, and a coil (also called an inductor and denoted as L).

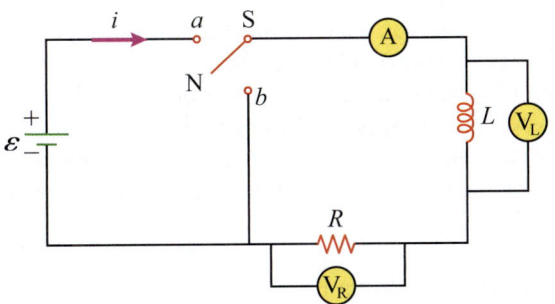

Figure 25-17 The RL circuit in this figure is designed such that when the switch is in position a, the battery, the resistor, and the inductor are part of the series RL circuit. When the switch is in position b, only the resistor and the inductor are part of the series RL circuit. When the switch has been in position a for a long time, the potential difference across the inductor is zero: $\varepsilon_{ind} = 0$. When the current across the inductor is changing (the switch is being flipped from a to b or vice versa), the changing current (called transient current) in the circuit that follows the flipping of the switch induces a non-zero emf across the inductor: $\varepsilon_{ind} \neq 0$.

An induced emf across an inductor, ε_{ind}, is often called a **back emf** because it *opposes* the change in the magnetic flux. Notice that the physical nature of the back emf is very different from the emf created by the battery. In a coil consisting of many loops, the magnetic field of each loop creates a back emf, which affects each of the other loops, multiplying the effect. Every loop affects itself as well as its neighbours. Therefore, the process of creating a back emf is called **self-induction**.

Recall from Chapter 24 that a solenoid is a coil with evenly spaced loops. Consider a solenoid of radius r and length ℓ, consisting of N closely spaced loops. Using Faraday's law of electromagnetic induction

(Equation 25-10) and the expression for the magnetic field of a solenoid (Equation 24-39), we can derive an expression for the back emf of a solenoid:

$$\varepsilon_{\text{ind}} = -N\left(\frac{d\Phi_B}{dt}\right) = -N\frac{d}{dt}(BA) = -NA\frac{dB(t)}{dt}$$

$$B(t) = \mu_0 nI = \mu_0 \frac{N}{\ell}I \Rightarrow \frac{dB(t)}{dt} = \mu_0 \frac{N}{\ell}\frac{dI}{dt}$$

Therefore, the back emf induced in a solenoid can be expressed as

$$\varepsilon_{\text{ind}} = -N\pi r^2 \mu_0 \frac{N}{\ell}\frac{dI}{dt} = -\mu_0 \frac{N^2\pi r^2}{\ell}\frac{dI}{dt}$$

$$= -L\frac{dI}{dt} \qquad (25\text{-}20)$$

where L stands for the **inductance** of the solenoid. In the case of a solenoid (Equation 24-39),

$$L = L_{\text{solenoid}} = \mu_0 \frac{N^2\pi r^2}{\ell}$$

The back emf depends on the rate of change of the electric current through the solenoid and on its inductance, L:

KEY EQUATION
$$\varepsilon_{\text{ind}} \equiv \varepsilon_L = -L\frac{dI}{dt} \qquad (25\text{-}21)$$

While Equation 25-21 is true for any inductor, the following derivation for inductance, L, applies only to a solenoid with circular coils of radius r:

KEY EQUATION
$$L_{\text{solenoid}} = \mu_0 \frac{N^2\pi r^2}{\ell} = \mu_0 \frac{N^2\ell\pi r^2}{\ell^2}$$

$$= \mu_0 n^2 \ell A = \mu_0 n^2 V \qquad (25\text{-}22)$$

where $V = \ell\pi r^2$ is the volume of a solenoid.

The inductance of a solenoid depends solely on its geometric and physical properties. The inductance is proportional to the volume V (to distinguish between volume and voltage in this chapter, we will use V for voltage and V for the solenoid's volume) and to the square of the density

of the loops. A solenoid with an iron core often has an inductance hundreds of times that of an air-core solenoid of the same size. Using the expression for magnetic permeability (Equation 24-47), the inductance of a circular solenoid with an inserted iron core can be expressed as

$$L_{\text{solenoid}} = K_{\text{m}}\mu_0 \frac{N^2\ell\pi r^2}{\ell^2} = K_{\text{m}}\mu_0 n^2 \ell A$$

$$= \mu n^2 V \qquad (25\text{-}23)$$

where μ is the magnetic permeability of the solenoid's core.

We can also calculate the inductance of a solenoid that has a core with permeability μ using the expression for its magnetic field (Equation 24-39) and the law of electromagnetic induction:

$$\varepsilon_L = -N\frac{d\Phi_B}{dt} = -N\frac{d(\mu_0 K_{\text{m}} nIA)}{dt} = -N\mu nA\frac{dI}{dt} = -L\frac{dI}{dt}$$

$$L_{\text{solenoid}} \equiv NK_{\text{m}}\mu_0 nA = \mu n^2 \ell A = \mu n^2 V \qquad (25\text{-}24)$$

Therefore, considering that for a solenoid $\Phi_B = \mu nAI$, the inductance of a solenoid that has a core with permeability μ can be expressed as

$$L = N\mu nA = \frac{N\Phi_B}{I} \qquad (25\text{-}25)$$

As we should have expected, Equation 25-24 is a general version of Equation 25-22 when a solenoid has a core with permeability μ.

Now we are ready to compare the functions of resistors and inductors in electric circuits. We will denote the potential difference across the resistor as V_R and the potential difference across the conductor as V_L. According to Ohm's law $I = \frac{V_R}{R}$ (Equation 23-17), increasing the resistance in a circuit reduces the current that converts electric energy to thermal energy. Thus, the potential difference across a resistor, V_R, is proportional to both the current and the resistance: $V_R = IR$ (Figure 25-18(a)). Current flows through a resistor from a higher potential to a lower potential. Inductors, however, act to oppose or impede the *change* in electric current. The potential difference across an inductor is proportional to its inductance and the rate of change of the electric current: $V_L = -L\frac{dI}{dt}$. Notice the direction of the induced potential difference in the inductor (Figure 25-18(b)). Thus, inductors behave somewhat like resistors to AC current. However, when a steady (DC) current flows through the inductor, the potential difference across the inductor is zero, because $\frac{dI}{dt} = 0$. The inductor, unlike the resistor, provides no resistance in a DC circuit. This explains why a transformer can burn out if you try to use DC instead of AC current.

ELECTRICITY, MAGNETISM, AND OPTICS

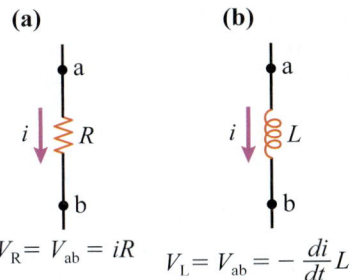

(a) **(b)**

$V_R = V_{ab} = iR$ $\quad V_L = V_{ab} = -\dfrac{di}{dt}L$

Figure 25-18 Comparison of the resistive and inductive elements of an electric circuit. (a) In the case of a resistor, electric current flows from a higher potential to a lower potential: $V_a > V_b$. (b) In the case of an inductor, the induced current in the coil is what generates the potential difference V_{ab}, such that $V_b > V_a$.

The SI unit for inductance is the henry (symbol H), named after American physicist Joseph Henry (1797–1878). An inductance of 1 H corresponds to an inductor that self-induces an emf of 1 V when the electric current in it changes at a rate of 1 A/s:

$$1\ \text{H} = \frac{1\ \text{V}}{1\ \text{A/s}} \Rightarrow [L] = \text{H} = \frac{\text{V}}{\text{A/s}} \qquad (25\text{-}26)$$

CHECKPOINT 25-6

The Direction of an Induced emf in an Ideal Inductor

Figure 25-19 shows an ideal inductor ($R = 0$). The potential at point a of the inductor is measured to be higher than the potential at point b. Which of the following current flow(s) could account for this potential difference?

(a) A steady current flows from a to b.
(b) A steady current flows from b to a.
(c) A decreasing current flows from a to b.
(d) An increasing current flows from a to b.
(e) A decreasing current flows from b to a.
(f) An increasing current flows from b to a.

$$a \quad \underset{\text{000}}{\overline{\qquad V_a > V_b \qquad}} \quad b$$

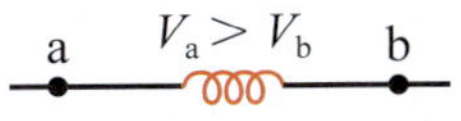

Figure 25-19 Checkpoint 25-6.

ANSWER: (d) and (e), according to Faraday's or Lenz's law.

Our daily lives are impossible to imagine without a simple device called a **transformer**. It is often called the heart of the AC system because it helps transfer electrical energy, convert AC power to DC power, bring the voltage across electrical devices up or down, and much more. You are using a transformer when you charge your computer or smartphone. We will discuss some of the applications of transformers in the following sections in more detail. A transformer consists of two inductively **coupled inductors**. Consider two solenoids wound around an iron core such that the magnetic fluxes through them are equal, as shown in Figure 25-20. Let us assume that these solenoids have N_1 and N_2 loops, respectively; the average circumference of the iron core is ℓ; and its cross-sectional area is A. Faraday's law of electromagnetic induction states that a time-varying electric current in the first solenoid, $I_1(t)$, generates a time-varying magnetic field $\vec{B}(t)$ through the second solenoid, which induces an electric current $I_2(t)$. Let us calculate this induced current.

To calculate the magnetic field induced by the first solenoid, we apply Ampère's law (Equation 24-38) to a closed contour ℓ located inside the iron core:

$$\oint_L \vec{B} \cdot d\vec{\ell} = \mu \sum I_{\text{enc}}$$

Since the magnetic field has the same magnitude at all points inside the iron core and is always directed along the tangent to the contour ℓ, we rewrite the above equation as

$$B\ell = \mu N_1 I_1 \Rightarrow B = \frac{\mu N_1 I_1}{\ell}$$

where μ is the magnetic permeability of iron.

Not surprisingly, this is the equation of the magnetic field created by a solenoid, discussed in Chapter 24 (Equation 24-39).

Using Faraday's law, the induced emf in the second solenoid can be expressed as

$$(\varepsilon_{\text{ind}})_2 = -N_2 \frac{d\Phi_B}{dt} = -N_2 \frac{d(AB)}{dt} = -N_2 A \frac{dB}{dt}$$

$$= -N_2 A \frac{d}{dt} \frac{\mu N_1 I_1}{\ell} = -\frac{\mu A N_2 N_1}{\ell} \frac{dI_1}{dt}$$

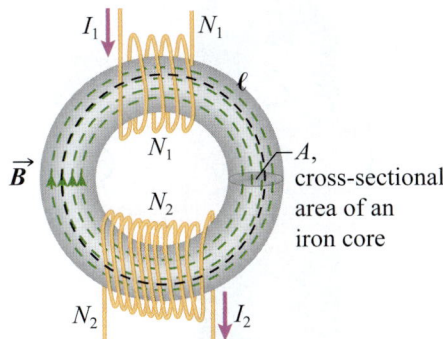

Figure 25-20 Two coils coupled through an iron core of average circumference ℓ and cross-sectional area A.

Then, the emf induced in the second solenoid can be expressed as

$$(\varepsilon_{ind})_2 = -M_{12}\frac{dI_1}{dt} \qquad (25\text{-}27)$$

where

$$M_{12} = \mu A\frac{N_1 N_2}{\ell}$$

The coefficient M_{12}, called the **coefficient of mutual inductance**, indicates the strength of coupling between the two solenoids. Similarly, if we consider the emf induced in the first solenoid by the varying current in the second, we find that

$$(\varepsilon_{ind})_1 = -M_{21}\frac{dI_2}{dt}; \quad M_{21} = \mu A\frac{N_2 N_1}{\ell} \qquad (25\text{-}28)$$

Comparing Equations 25-27 and 25-28, we see that

$$M_{12} = M_{21} = M \qquad (25\text{-}29)$$

Thus, the coefficient of mutual inductance between two coils is denoted as M.

Now we consider two long, thin solenoids of length ℓ wound over top of each other so their cross-sectional areas are approximately equal (Figure 25-21). The inner and outer solenoids have N_1 and N_2 coils, respectively, and their cross-sectional area is A. Once again, if a time-varying electric current, I_1, flows through the inner solenoid, it will induce an emf, ε_2, in the outer coil:

$$(\varepsilon_{ind})_2 = -N_2\frac{d\Phi_B}{dt} = -N_2\frac{d(AB)}{dt} = -N_2 A\frac{dB}{dt}$$

$$= -N_2 A\frac{d}{dt}\frac{\mu N_1 I_1}{\ell} = -\frac{\mu A N_2 N_1}{\ell}\frac{dI_1}{dt}$$

and

$$(\varepsilon_{ind})_2 = -M\frac{dI_1}{dt}$$

where

$$M = \frac{\mu A N_2 N_1}{\ell} \qquad (25\text{-}30)$$

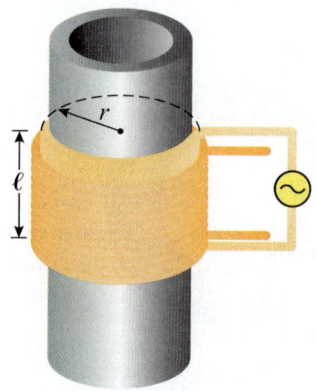

Figure 25-21 Two long, thin concentric solenoids wound on top of each other.

Once again, we can show that the mutual inductance $M_{12} = M_{21} = M$. Comparing Equation 25-30 with the expression for the self-inductance of a solenoid (Equation 25-22), we find

$$M = \frac{\mu A N_1 N_2}{\ell} = \sqrt{L_1 L_2} \qquad (25\text{-}31)$$

In the derivation above, we assumed that all the magnetic flux created by the first solenoid passes through the second solenoid. This is rarely true; therefore, Equation 25-31 should be adjusted to include the coefficient of coupling between the inductors, k, which varies from zero to one. Perfectly coupled coils have a coefficient of coupling of one, and perfectly decoupled coils have a coefficient of coupling of zero:

KEY EQUATION

$$M = k\sqrt{L_1 L_2} \qquad (25\text{-}32)$$

MAKING CONNECTIONS

How Does an Electric Toothbrush Get Charged?

Have you ever wondered how an electric toothbrush can be recharged when the charging stand and the battery-powered toothbrush are both sealed and have no exposed metal contacts to make a circuit (Figure 25-22)? The answer is **inductive charging**. The stand and toothbrush essentially operate like a transformer. The base connected to the AC power outlet serves as the primary coil, and a coil inside the toothbrush acts as the secondary coil. Every time you slide your toothbrush into the base, you couple these coils, allowing current to be induced in the toothbrush. The induced current recharges the toothbrush battery and also makes the handle of the toothbrush feel warm. So, the next time you brush your teeth, think of Michael Faraday, whose discovery of electromagnetic induction made the operation of your toothbrush possible.

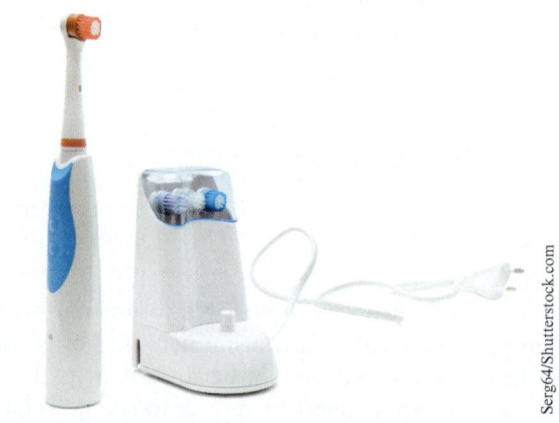

Figure 25-22 An electric toothbrush uses an inductive charger, which operates like a transformer.

EXAMPLE 25-5

Calculating the Mutual Inductance of Two Coupled Solenoids

Two long, thin solenoids of equal length ℓ and equal radius r are wound over top of each other (Figure 25-21). The inner and outer solenoids have N_1 and N_2 turns, respectively. An alternating electric current $I_1 = (I_1)_{max} \sin(\omega t)$ flows through the inner solenoid, where $(I_1)_{max}$ represents the maximum value of the current in the inner solenoid.

(a) Calculate the mutual inductance of the solenoids, and prove that it satisfies Equations 25-30 and 25-32.
(b) Find an expression for the emf induced in the outer solenoid.

Solution

(a) Using Equation 25-23 for the inductance of a solenoid and substituting it into Equation 25-30 for the mutual inductance of two solenoids gives

$$M = \mu_0 A \frac{N_1 N_2}{\ell} = \sqrt{\left(\mu_0 A N_1 \frac{N_1}{\ell}\right)\left(\mu_0 N_2 A \frac{N_2}{\ell}\right)} = \sqrt{L_1 L_2}$$

In this derivation, we assumed that the cross-sectional area of each solenoid is A. Since the outer solenoid surrounds the inner solenoid, all the magnetic flux produced by the inner solenoid passes through the outer solenoid. Therefore, $k = 1$, and the solution also satisfies Equation 25-32.

(b) To find the emf induced in the outer solenoid, we use Faraday's law (Equation 25-27) for a solenoid, taking into account that the cross-sectional area of a solenoid is πr^2:

$$(\varepsilon_{ind})_2 = -M \frac{dI_1}{dt} = -M \frac{d}{dt}(I_1)_{max} \cos(\omega t)$$

$$= (I_1)_{max} \omega \mu_0 \pi r^2 \frac{N_1 N_2}{\ell} \sin(\omega t)$$

Making sense of the result

As expected, a sinusoidal emf is generated in the outer coil because it is being induced by a sinusoidal current, $I = (I_1)_{max} \sin(\omega t)$, in the inner coil. We will discuss alternating currents (AC) and their use in more detail in Chapter 26.

In the following section, we will consider some of the most common applications of Faraday's law of electromagnetic induction.

25-6 Applications of Faraday's Law of Electromagnetic Induction

As we pointed out earlier, few scientific discoveries have had a more profound impact on our lives than Faraday's law of electromagnetic induction. In this section, we look at a few examples of how this law is applied in the multitude of devices used in homes, businesses, and industry every day.

AC Power Generators

Most of the electric power generated in the world is produced using generators that convert mechanical energy into electric energy using the same principle as the simple generator shown in Figure 25-23.

The core of an electric generator is a system of conductive loops that rotate in a uniform and constant magnetic field. As each loop rotates, the magnetic flux through it changes, inducing an emf in it. Consider a single loop of area A and resistance R rotating at an angular velocity ω that corresponds to frequency f ($\omega = 2\pi f$) in a uniform magnetic field $\vec{B}$ (Figure 25-23(a)). The angle between the magnetic field and the normal to the plane of the loop can be described as $\theta = \omega t$. Using Equation 25-8, we can calculate the rate of change of the magnetic flux through the loop as

$$\frac{d\Phi_B}{dt} = \frac{d(BA\cos\theta)}{dt} = \frac{BAd\cos(\omega t)}{dt}$$

$$= -BA\omega\sin(\omega t) \qquad (25\text{-}33)$$

(a)

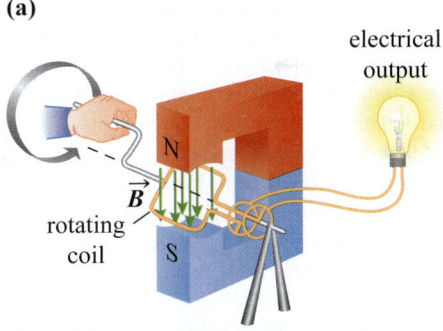

electrical output

rotating coil

(b)

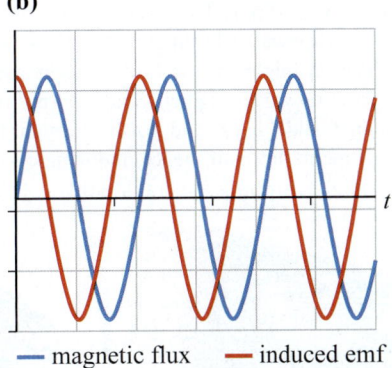

— magnetic flux — induced emf

Figure 25-23 (a) The operation of an electric generator. (b) The dependence of the magnetic flux through the rotating coil and of the induced emf on time. The graphs are offset by $\pi/2$, so the max or min of the magnetic flux corresponds to the zero of the induced emf, while the zero of the magnetic flux corresponds to the max or min of the induced emf.

We use Faraday's law to find the induced emf and current in the loop:

$$\varepsilon_{ind}(t) = -\frac{d\Phi_B}{dt} = BA\omega\sin(\omega t) = V_{max}\sin(\omega t)$$

where $V_{max} = Ba\omega = 2\pi fBA$. Since according to Ohm's law, $I = V/R$,

$$I_{ind}(t) = \frac{V(t)}{R} = \frac{V_{max}\sin(\omega t)}{R} = I_{max}\sin(\omega t)$$

where $I_{max} = V_{max}/R = 2\pi fBA/R$.

To summarize,

$$\varepsilon_{ind}(t) = V_{max}\sin(\omega t) \text{ where } V_{max} = 2\pi fBA$$

$$I_{ind}(t) = I_{max}\sin(\omega t) \text{ where } I_{max} = \frac{2\pi fBA}{R} \quad (25\text{-}34)$$

Thus, such a generator produces an emf that varies sinusoidally with time, which is the most common type of **alternating emf**. An electric current generated by such a generator is called **alternating current** (AC). For a rotating coil with N loops that rotates with frequency f, related to angular frequency ω via $\omega = 2\pi f$, the expressions for emf and current become

$$\varepsilon_{ind}(t) = NBA\omega\sin(\omega t) = V_{max}\sin(\omega t)$$

where $V_{max} = NBa\omega = 2\pi fNBA$. Since according to Ohm's law, $I = V/R$,

$$I_{ind}(t) = \frac{V(t)}{R} = \frac{V_{max}\sin(\omega t)}{R} = I_{max}\sin(\omega t)$$

where $I_{max} = V_{max}/R = 2\pi fNBA/R$.

To summarize:

$$\varepsilon_{ind}(t) = V_{max}\sin(2\pi ft) \text{ where } V_{max} = 2\pi fNBA$$

$$I_{ind}(t) = I_{max}\sin(2\pi ft) \text{ where } I_{max} = \frac{2\pi fNBA}{R} \quad (25\text{-}35)$$

MAKING CONNECTIONS

Sources of Electrical Energy

In North America, the standard frequency for AC current is 60 Hz; in Europe and some other countries, it is 50 Hz. The most common source of mechanical energy for electric generators worldwide is steam from burning fossil fuels or from nuclear fission. However, in Canada, the most common source of this mechanical energy is hydroelectric—kinetic energy from falling water (approximately 63%). Coal supplies slightly less than 10%, and nuclear power supplies approximately 13%. Canada is the third-largest hydroelectric power producer in the world following China and Brazil. It is interesting that Paraguay produces 100% of its electricity from hydroelectric dams, while Norway produces about 99% of its electricity from hydroelectric sources. In 2015, renewable energy sources accounted for about 10% of total U.S. energy consumption and about 13% of energy generation.

The same principles of electromagnetic induction that underlie giant generators (Figure 25-24) feeding national power grids also apply to diesel-powered backup generators for hospitals, gas-powered portable generators on construction sites, and even hand-powered flashlights, like those shown in Figure 25-25. The flashlight on the bottom right uses a stationary coil and a movable magnet. Shaking this flashlight makes the magnet slide back and forth through the coil, changing the magnetic flux and, hence, inducing current. On hand-crank flashlights, turning the crank either rotates a coil inside a stationary magnet directly or winds a spring, which powers some clockwork, which rotates the coil until the spring runs down. All of these flashlights use electromagnetic induction to convert human mechanical energy into electrical energy.

Figure 25-24 The Churchill River Underground Power Plant in Newfoundland and Labrador is the second largest underground power plant in the world.

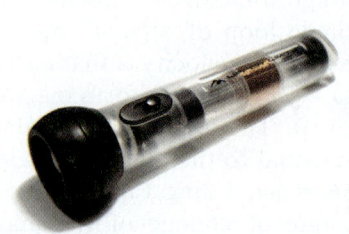

Figure 25-25 Several types of hand-powered generator flashlights.

EXAMPLE 25-6

Exploring How an Electric Generator Works

An electric generator has a circular coil with 20 turns, each 0.60 m in diameter. The coil makes 60. revolutions per second. What strength of magnetic field is required to produce a peak output voltage of 170 V?

Solution

Using Equation 25-35, we find

$$V_{max} = NBA\omega = 2\pi NBAf$$

$$B = \frac{V_{max}}{2\pi NAf} = \frac{V_{max}}{2\pi N\left(\pi \dfrac{d^2}{4}\right)f}$$

$$= \frac{170 \text{ V}}{2\pi(20 \text{ turns})\left(\pi \dfrac{(0.60 \text{ m})^2}{4}\right)(60. \text{ Hz})} = 0.080 \text{ T}$$

Making sense of the result

A strong permanent magnet can produce a magnetic field with strength about 0.1 T, so the value of 0.080 T seems reasonable.

Transformers

Have you ever wondered why electric power delivered to our homes comes in the form of alternating current (AC) rather than direct current (DC)? At the end of the 19th century, the struggle between the proponents of AC and DC power generation was so fierce that it was dubbed the "war of currents." Thomas Alva Edison (1847–1931) advocated for DC power, and Nikola Tesla (1856–1943; Figure 25-26) and George Westinghouse (1846–1914) promoted AC power generation. The key issue that swayed the argument in favour of AC power was power transmission. For various practical reasons,

Figure 25-26 The Nikola Tesla monument in Queen Victoria Park at Niagara Falls, Ontario, was unveiled on July 9, 2006. Tesla is standing on top of an AC motor, one of the 700 inventions he patented.

large electric power plants are often located away from densely populated areas, so the electric current has to travel through wires over long distances to reach consumers. The amount of power dissipated as thermal energy in the transmission lines depends on the current and the resistance of the wires (Equation 23-25):

$$P = I^2R$$

The longer the transmission line, the higher its resistance and the greater the proportion of electrical energy lost as thermal energy. Edison tried to minimize such losses by building electric power plants very close to consumers, for example, in New York City. However, this approach turned out to be impractical.

Tesla realized that AC voltages can be stepped up or down using a transformer. By using high voltages and much smaller currents, thermal energy dissipation in transmission lines can be greatly reduced. In Edison's time, there was no practical and economically viable way of changing the voltage of large amounts of DC. As we discussed in Section 25-5, transformers use inductively coupled coils, so that changing magnetic flux generated by alternating current in the primary coil induces alternating emf in the secondary coil. Since the principle of operation of transformers is based on changing magnetic flux, they need alternating and not direct current. This was one of the main reasons why the war of currents was won by AC and Tesla over DC and Edison.

INTERACTIVE ACTIVITY 25-4

Operation of a Transformer

In this activity, you will use the PhET simulation "Faraday's Electromagnetic Lab: Transformer" to explore the operation of a transformer.

As we discussed earlier, a transformer consists of two adjacent coils: a primary coil, which is connected to the power supply, and a secondary coil, which is connected to the load (Figure 25-27). Usually, the coils are wound on the same iron core. A transformer is based on the principle of mutual inductance discussed in Section 25-5. We say that the primary and secondary coils of a transformer are mutually coupled, since the changing magnetic field through the primary coil generates a time-varying magnetic field through the secondary coil. Let us denote the potential difference across the primary coil of a transformer as V_1 and the potential difference across the secondary coil as V_2. Sinusoidal alternating currents continuously change their magnitude and direction, but due to symmetry considerations, the average values of V_1 and V_2 are zero since the average of sine and cosine functions is also zero $(\sin x = \cos x = 0)$. For power calculations, we use the expression $P = I^2R = V^2/R$ (Equation 23-25), so we

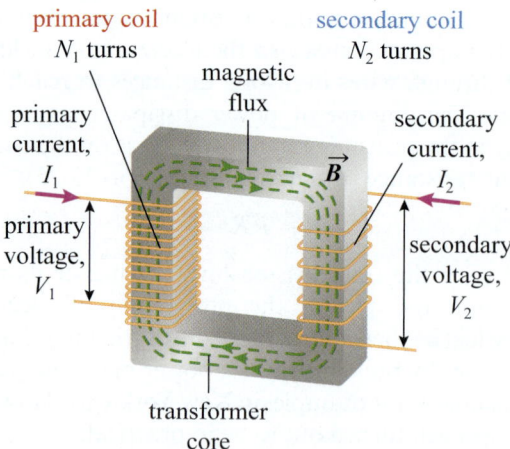

primary coil
N_1 turns

secondary coil
N_2 turns

magnetic flux

primary current, I_1

secondary current, I_2

$\vec{B}$

primary voltage, V_1

secondary voltage, V_2

transformer core

Figure 25-27 A step-down transformer.

need to average the square of the current or voltage. We will show in Chapter 26 that we can do this using root mean square (rms) values for AC voltages and currents, which for sine and cosine functions will be different from zero: the rms values for any sine function, such as a function describing AC current or AC voltage, equal $I_{rms} = I_{max}/\sqrt{2}$ and $V_{rms} = V_{max}/\sqrt{2}$, respectively, where I_{max} and V_{max} represent the maximum values of the electric current and the electric voltage, respectively. Therefore, for AC current, Equation 23-25 can be expressed as $P = \dfrac{I_{max}^2}{2}R = \dfrac{V_{max}^2}{2R}$. This will be discussed in more detail in Chapter 26.

Transformers transform lower voltages into higher voltages and back by converting electric energy into magnetic energy and vice versa. The law of energy conservation states that if there are no energy losses (such as dissipated thermal energy), the power delivered to the primary coil equals the power generated in the secondary coil. Actual transformers are not ideal due to a number of loss mechanisms; they have power efficiencies of about 95%. Using Equation 23-24,

$$P_1 = P_2$$
$$I_1 V_1 = I_2 V_2$$

(25-36)

Since power dissipated through thermal energy is proportional to the square of the current ($P = I^2R$), to reduce energy transmission losses, the current transmitted through the wires should be as low as possible. Therefore, power stations have transformers that step up the voltage (transforming low voltage into high voltage) and, consequently, lower the current in the transmission lines. Transmission voltages generally range from 138 kV to 765 kV. At the other end of the transmission line, the voltage is stepped down (transformed from high voltage to lower voltage) in several stages to the standard 120 V/240 V for most residential customers.

Let us prove that the function of a transformer (step-up or step-down) is defined by the ratio of the

number of loops in its primary and secondary coils. As discussed above, we assume the transformer is ideal, so the magnetic flux through both coils is the same and will be denoted as Φ_B. According to Faraday's law of electromagnetic induction, the emf generated in the secondary coil, as a result of the AC through the primary coil, is related to the magnetic flux as

$$V_2 = (\varepsilon_{ind})_2 = -N_2 \frac{d\Phi_B}{dt}$$

At the same time, the magnetic flux through the primary coil is related to the alternating potential difference across the primary coil as

$$V_1 = -N_1 \frac{d\Phi_B}{dt}$$

Therefore, the ratio of potential differences across the primary and secondary coils of a transformer can be expressed as

KEY EQUATION

$$\frac{V_2}{V_1} = \frac{N_2}{N_1}$$

(25-37)

CHECKPOINT 25-7

The Operation of a Transformer

Figure 25-28 shows a simple transformer: two coils wound around an iron core. The primary coil is connected to a DC power supply. The voltmeter shows the voltage induced in the secondary coil of the transformer. When was the voltmeter reading taken?

(a) The voltmeter reading was taken during the opening of the circuit, when the current in the circuit was decreasing.

(b) The voltmeter reading was taken during the closing of the circuit, when the current in the circuit was increasing.

(c) The voltmeter reading was taken when the current in the circuit was steady: the circuit has been closed for a while.

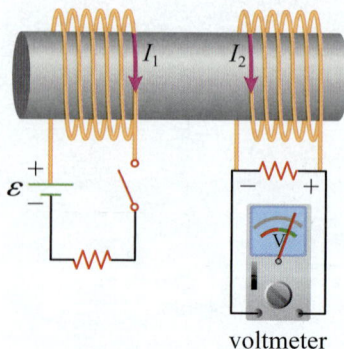

I_1

I_2

ε

voltmeter

Figure 25-28 Checkpoint 25-7.

ANSWER: Use Lenz's law to verify that (a) is correct. Notice that connecting a transformer to a DC power source might burn it.

EXAMPLE 25-7

Transformer Ratio

A transformer is used to change a 120. V, 3.00 A current to 12.0 V in a power supply for a laptop computer.

(a) Is this a step-up or a step-down transformer? Explain.
(b) What is the ratio of the number of secondary coil turns to the number of primary coil turns in this transformer?
(c) What current would be induced in the secondary coil?

Solution

(a) This is a step-down transformer because it transforms higher voltage into lower voltage.
(b) To find the ratio of the number of secondary coil turns to the number of primary coil turns, we use Equation 25-37:

$$\frac{V_2}{V_1} = \frac{N_2}{N_1} \Rightarrow \frac{N_2}{N_1} = \frac{12.0\,\text{V}}{120\,\text{V}} = 1.00{:}10.0$$

(c) The law of energy conservation can be used to find the current in the secondary coil (Equation 25-36):

$$I_1 V_1 = I_2 V_2 \Rightarrow I_2 = \frac{I_1 V_1}{V_2} = \frac{(3.00\,\text{A})(120.\,\text{V})}{12.0\,\text{V}} = 30.0\,\text{A}$$

Making sense of the result

The secondary current is 10 times the primary current, which is reasonable because the secondary voltage is one tenth the primary voltage.

Electromagnetic Damping and Electromagnetic Braking

Electromagnetic dampers consist of a strong electromagnet or permanent magnet and a moving non-magnetic metal part whose vibrations are to be damped. In the presence of a strong magnetic field, eddy currents are created inside the moving part. These eddy currents oppose the motion that induces them and thus act to damp vibration in the moving part. Electromagnetic dampers are clean, easy to adjust, and less sensitive to temperature than hydraulic dampers. For example, when sheet metal is fed along a production line for coating, transverse vibrations in the sheets can make the coating uneven. Placing a powerful electromagnet near the sheets induces eddy currents that dampen the vibrations significantly.

Electromagnetic braking uses the same principle, slowing an object by inducing eddy currents that convert mechanical energy into thermal energy or electrical energy. The first electromagnetic brakes were used on mountain trains. Today, electromagnetic brakes are used on vehicles such as roller coasters, conventional and magnetic levitation trains, and hybrid vehicles. To slow the vehicle, the electric drive motor is switched to act as a generator. The motor then converts kinetic energy into electric energy and thermal energy. As a coil on the shaft of the motor rotates in a magnetic field inside the motor, a back emf is induced on the coil and fed to the battery, recharging it while slowing the rotation of the motor shaft.

The Electric Guitar

In both acoustic and electric guitars, sound is produced by the vibration of the strings. In an acoustic guitar, the sound of the strings is boosted by a soundboard and a resonant cavity (Chapter 15). In an electric guitar, electromagnetic induction produces an electric current that corresponds to the vibration of the strings. The signal is then amplified electronically to drive a speaker. The strings of an electric guitar are made of a ferromagnetic material (often steel or nickel). Electric guitars typically have two or three sets of pickups positioned along the strings (Figure 25-29(a)). Each set has six coils, one aligned directly underneath each of the six strings on the guitar. The design of the coils and the position of the pickups affect the signal they produce, so the guitarist can change the sound of the guitar by choosing which pickups are connected to the amplifier. The vibrating ferromagnetic strings create time-varying magnetic fields through the pickup coils (Figure 25-29(b)), which then induce a changing emf and a changing electric current, which control the sound created by the speakers.

(a)

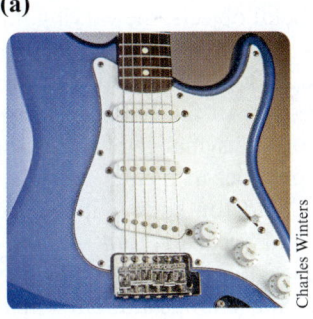

Charles Winters

(b)

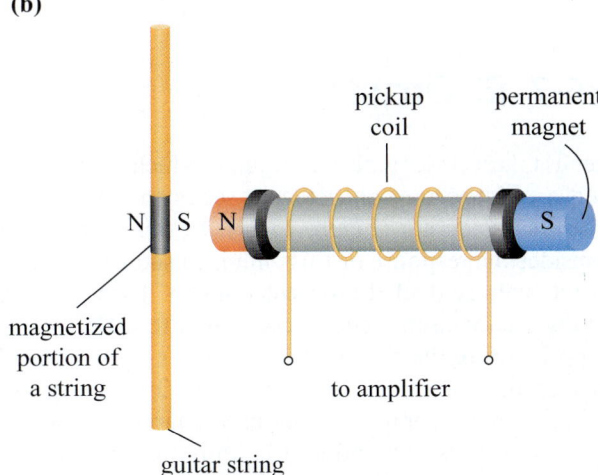

Figure 25-29 The operation of an electric guitar. (a) Pickup coils (metal circles) are located under the strings of an electric guitar. (b) The moving steel string changes the flux from the permanent magnet, inducing a current in the pickup coil.

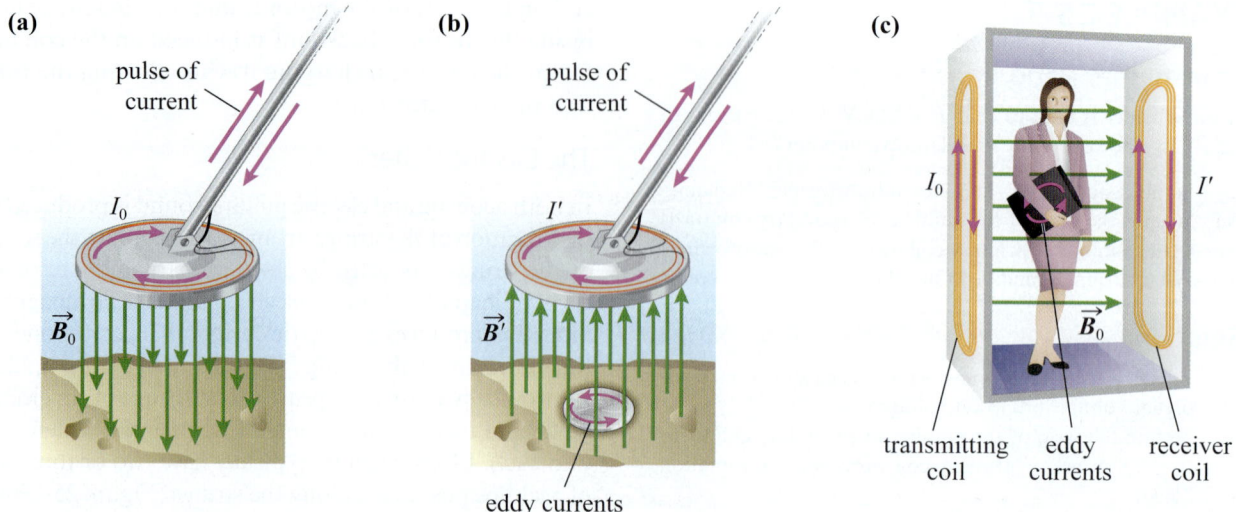

(a) pulse of current — I_0 — $\vec{B}_0$

(b) pulse of current — I' — $\vec{B}'$ — eddy currents

(c) I_0 — $\vec{B}_0$ — I' — transmitting coil — eddy currents — receiver coil

Figure 25-30 Metal detectors. (a) A single-coil metal detector consists of a transmitter coil that generates a pulsing magnetic field. (b) In the presence of a coin, for example, in the vicinity of the coil, the eddy currents induced in the coin affect the current in the transmitter coil and indicate the presence of metal. (c) A two-coil metal detector.

Metal Detectors

There are several types of metal detectors, and they all use electromagnetic induction. In the single-coil metal detector shown in Figure 25-30(a), a pulsing current is applied to the coil. This current induces a time-varying magnetic field and, hence, a time-varying magnetic flux through nearby objects. In metal objects, the varying flux induces eddy currents, which generate their own magnetic fields, shown in red in Figure 25-30(b). As a result, an electric current opposite in direction to the original electric current is generated in the coil. This change of current in the coil is used to detect the presence of metal.

In the two-coil metal detector shown in Figure 25-30(c), the transmitter and receiver coils act as a transformer. The presence of metal near the receiver coil induces eddy currents that change the current in the receiver coil.

25-7 RL Circuits

Let us take a closer look at a circuit in which an inductor and a resistor are connected in series to a battery—a series RL circuit (Figure 25-31). In this section, we consider the response of this circuit to sudden changes in DC voltage (called **transient voltage**). These sudden changes happen in a circuit connected to a DC power supply during the time when a switch is being opened or closed. In Chapter 26, we will consider AC power sources that supply continuously changing voltage. Since an emf is only induced in an inductor when the magnetic flux through it changes with time, an inductor influences the circuit behaviour only when the current through it changes. Therefore, the inductor shown in

Figure 25-31 affects the current only for a relatively short time immediately after the switch is turned on or off. We will consider each of these two cases separately. (You started exploring the effect of such transient currents in Interactive Activity 25-3).

Connecting a Battery to a Series RL Circuit

Let us assume that the switch is being flipped to position a (from the neutral position, N) at time $t = 0$. Since all the circuit components are connected in series, the law of energy conservation and Kirchhoff's second law require that

$$\varepsilon + V_R + V_L = 0 \Rightarrow \varepsilon - iR + \varepsilon_{ind} = 0 \quad (25\text{-}38)$$

where i is the instantaneous current at time t, $V_R = -iR$, and

$$V_L = \varepsilon_{ind} = -L\frac{di}{dt}$$

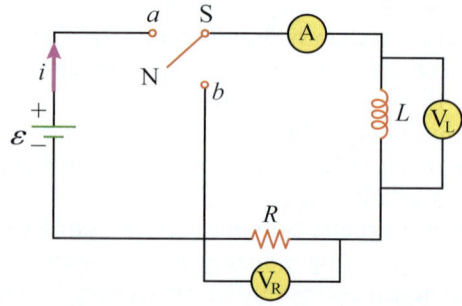

Figure 25-31 A series RL circuit consists of a resistor (R) and an inductor (L). Such a circuit might (when the switch is in position a) or might not (when the switch is in position b) include a battery. The ammeter (A) registers electric current flowing through the resistor and the inductor, while the voltmeters (V_L and V_R) measure potential differences across the inductor and the resistor, respectively.

These expressions can be combined:

$$\varepsilon - iR - L\frac{di}{dt} = 0$$

$$\frac{di}{dt} = \frac{\varepsilon - iR}{L}$$ (25-39)

To solve this differential equation, we set $x = \varepsilon - iR$. Then $dx = d(\varepsilon - iR) = -Rdi$, and Equation 25-39 can be rewritten as

$$x + \frac{L}{R}\frac{dx}{dr} = 0 \Rightarrow \frac{dx}{x} = -\frac{R}{L}dt$$ (25-40)

We solve this equation by integrating both sides. Since we are interested in knowing the behaviour of the current from the initial time $t_0 = 0$ to the final time $t_f = t$, we choose these values as the limits of integration. The corresponding limits of integration for the variable x are denoted as x_0 and x:

$$\int_{x_0}^{x}\frac{dx}{x} = -\int_0^t\frac{R}{L}dt \Rightarrow \int_{x_0}^{x}\frac{dx}{x} = -\frac{R}{L}\int_0^t dt$$

$$\ln\frac{x}{x_0} = -\frac{R}{L}t \Rightarrow \frac{x}{x_0} = e^{-\frac{R}{L}t}$$ (25-41)

Let us verify that Equation 25-41 has the proper units. The exponent should be a pure number, so the units of the ratio L/R should be time (seconds). Substituting for the inductance gives

$$\left[\frac{L}{R}\right] = \frac{\left(\frac{V \cdot s}{A}\right)}{\Omega} = \frac{\Omega \cdot s}{\Omega} = s$$

Using $x = \varepsilon - iR$ and taking into account that the current at $t = 0$ is zero, $I_0 = 0$, Equation 25-41 can be rewritten as

$$x = x_0 e^{-\frac{R}{L}t} \Rightarrow \varepsilon - iR = (\varepsilon - I_0 R)e^{-\frac{R}{L}t}$$

$$\varepsilon - iR = \varepsilon e^{-\frac{R}{L}t}$$

$$\varepsilon\left(1 - e^{-\frac{R}{L}t}\right) = iR$$

$$i(t) = \frac{\varepsilon}{R}\left(1 - e^{-\frac{R}{L}t}\right)$$

The ratio L/R has units of time and is called the **time constant**, τ, of the RL circuit. Substituting for L/R in the above expression gives

$$i(t) = \frac{\varepsilon}{R}\left(1 - e^{-\frac{t}{L/R}}\right) = \frac{\varepsilon}{R}\left(1 - e^{-\frac{t}{\tau}}\right)$$

where $\tau = L/R$.

Figure 25-32(a) shows a graph of $i(t)$ versus time. The $i(t)$ graph starts from zero, and then the current increases exponentially with time, asymptotically approaching its maximum value of ε/R. This is the value of the current that would have been observed if there were no inductor in the circuit. This makes sense because as time goes on, the rate of change of the current diminishes (the current approaches the steady state as the $i(t)$ slope becomes more horizontal); as a result, ε_L approaches zero. Another important observation is that at time $t = \tau$, the current reaches approximately 63% of its maximum value:

$$i(\tau) = \frac{\varepsilon}{R}\left(1 - e^{-\frac{\tau}{\tau}}\right) = \frac{\varepsilon}{R}(1 - e^{-1})$$

$$\approx \frac{\varepsilon}{R}(1 - 0.368) = 0.632\frac{\varepsilon}{R}$$ (25-42)

To examine how the potential difference across the inductor, $V_L \equiv \varepsilon_L$, depends on time, we find the derivative of current with respect to time, $di(t)/dt$, and again use Equation 25-21:

$$V_L \equiv \varepsilon_L = -L\frac{di}{dt} = -L\frac{d}{dt}\left(\frac{\varepsilon}{R}\left(1 - e^{-\frac{t}{\tau}}\right)\right) = -\frac{L\varepsilon}{R\tau}e^{-\frac{t}{\tau}}$$

$$= -\frac{L\varepsilon}{R\frac{L}{R}}e^{-\frac{t}{\tau}} = -\varepsilon e^{-\frac{t}{\tau}}$$

Combining the results of the above equations, we obtain

KEY EQUATION
$$\begin{cases} i(t) = \frac{\varepsilon}{R}\left(1 - e^{-\frac{t}{\tau}}\right) \\ V_L \equiv \varepsilon_L = -\varepsilon e^{-\frac{t}{\tau}} \end{cases}$$ (25-43)

where $\tau = L/R$.

From the graph of $\varepsilon_L(t)$, we see that at $t = 0$, $\varepsilon_L(0) = \varepsilon$, and $\varepsilon_L(t)$ diminishes exponentially with time (Figure 25-32(b)). Therefore, at $t = 0$, the emf across the inductor equals the value of the emf supplied by

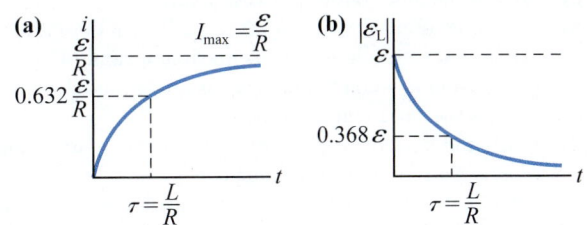

Figure 25-32 When the switch in a series RL circuit is flipped from position N to position a, the current starts flowing through the circuit. (a) The dependence of the electric current through the series RL circuit on time. (b) The dependence of the induced emf across the inductor on time.

the battery. Then, after τ seconds have passed, the value of ε_L is approximately equal to 37% of ε:

$$\varepsilon(\tau) = \varepsilon e^{-\frac{t}{\tau}} = \varepsilon e^{-1} \approx 0.368\varepsilon \qquad (25\text{-}44)$$

Both graphs are asymptotic: it takes an infinitely long time for the current to reach its steady-state value of ε/R and for the voltage across the inductor ε_L to reach zero. However, the time constants for most practical inductive circuits are short enough that the difference between the actual value and the final steady-state value is negligible after a few seconds.

CHECKPOINT 25-8

Exploring RL Circuits

How long does it take for the current in the RL circuit in Figure 25-31 to reach 74% of its maximum value?

(a) τ
(b) 1.34τ
(c) 1.50τ
(d) 2.00τ
(e) 2.67τ

ANSWER: (b) Use Equation 25-43 to check the answer.

Disconnecting a Battery from a Series RL Circuit

We can use the same method to determine what happens to the circuit when the switch is moved to position b after it has been in position a for a long time. Taking the instant when the switch is opened as $t = 0$, we have $I_0 = i(0) = I_{max} = \varepsilon/R$ and $i(\infty) = 0$. Then the

expressions for the current in the circuit and the emf across the inductor can be rewritten as

KEY EQUATION
$$\begin{cases} i(t) = \dfrac{\varepsilon}{R}e^{-\frac{t}{\tau}} \\ V_L \equiv \varepsilon_L = \varepsilon e^{-\frac{t}{\tau}} \end{cases} \quad \text{where } \tau = L/R \qquad (25\text{-}45)$$

Figure 25-33 shows how $i(t)$ and $\varepsilon_L(t)$ change immediately after the battery is removed from the series RL circuit.

The time constant, τ, of the circuit determines how quickly the current in the circuit reaches its steady state when the circuit is connected to a battery and how quickly it diminishes to zero when the battery is switched off. The time constant is directly proportional to L and inversely proportional to $R(\tau = L/R)$, so, when $L/R \gg 1$, the inductor dominates the circuit behaviour (inductive circuit), and when $L/R \ll 1$, the resistance dominates the circuit behaviour (resistive circuit).

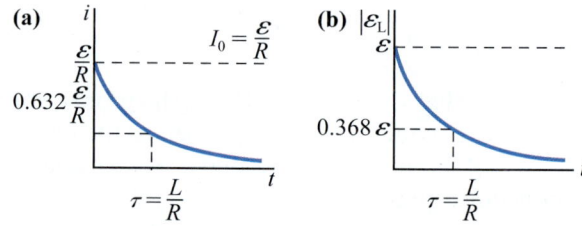

Figure 25-33 When the switch in a series RL circuit is flipped from position a to position b, the circuit is disconnected from the battery. (a) The dependence of the electric current through this circuit on time, $i(t)$. (b) The dependence of the potential difference across the inductor on time, $\varepsilon_L(t)$.

EXAMPLE 25-8

Exploring RL Circuits

In the RL circuit shown in Figure 25-34, the battery has an emf of 20.0 V, the resistor has a resistance of 10.0 Ω, and the inductor has an inductance of 40.0 mH.

(a) What is the time constant of this circuit?
(b) How long after the switch is set to position a does it take the current to reach 99% of its maximum value? Express your answer in seconds and in terms of the time constant.
(c) What is the current at this time?
(d) When the current reaches 99% of its maximum value, what is the emf across the inductor?
(e) After 20.0 s, the switch is moved to position b. How long will it take for the current to decrease to 10% of its maximum value?

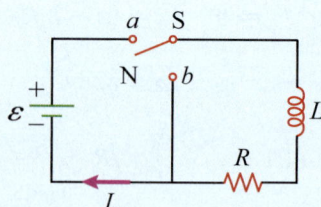

Figure 25-34 An RL circuit.

Solution

(a) The time constant of this circuit, according to Equation 25-45, is $\tau = L/R = \dfrac{40.0 \text{ mH}}{10.0\,\Omega} = 4.00 \text{ ms}$

(b) To find how long it will take the current to reach 99% of its maximum value, we use Equation 25-43:

$$i(t) = \frac{\varepsilon}{R}\left(1 - e^{-\frac{t}{\tau}}\right)$$

$$0.99\frac{\varepsilon}{R} = \frac{\varepsilon}{R}\left(1 - e^{-\frac{t}{\tau}}\right)$$

$$e^{-\frac{t}{\tau}} = 0.01 \Rightarrow t = -\tau \ln 0.01$$

Since $\ln 0.01 = -4.61 \Rightarrow$

$$t = 4.61\tau = 4.61\frac{L}{R} = 4.61\left(\frac{40.0 \times 10^{-3}\text{ H}}{10.0\ \Omega}\right)$$

$$= 4.61(4.00\text{ ms}) = 18.4\text{ ms}$$

The current will reach 99% of its maximum value after 4.61τ or 18.4 ms.

(c) The maximum value of the current is

$$I_{max} = \frac{\varepsilon}{R} = \frac{20.0\text{ V}}{10.0\ \Omega} = 2.00\text{ A}$$

and 99% of this value is 1.98 A.

(d) When the current reaches 99% of its maximum value, the emf across the inductor can be calculated using Equation 25-45:

$$\varepsilon_L = \varepsilon e^{-\frac{t}{\tau}} = (20.0\text{ V})e^{-\frac{18.4\text{ ms}}{4.00\text{ ms}}} = (20.0\text{ V})e^{-4.6} = 0.201\text{ V}$$

(e) The switch is moved to position *b* after 20.0 s. Since this time interval is over 4000 times the circuit time constant (4.00 ms), we can assume that at the time the switch was moved to position *b*, the current had already reached its maximum value. Let t_{10} represent the time when the current reaches 10% of its initial value. Then, $i(t_{10}) = 0.1I_{max} = 0.1\varepsilon/R$. Substituting into Equation 25-45 gives

$$0.1I_{max} = \frac{\varepsilon}{R}e^{-\frac{t}{\tau}} \Rightarrow 0.1\frac{\varepsilon}{R} = \frac{\varepsilon}{R}e^{-\frac{t}{\tau}}$$

$$0.1 = e^{-\frac{t}{\tau}} \Rightarrow t = -\tau \ln 0.1 = 2.3\tau$$

$$= 2.3(4.00\text{ ms}) = 9.2\text{ ms}$$

Making sense of the result

When the current is changing rapidly, the back emf across an inductor is large, and when the current is steady, the back emf across the inductor becomes negligible. Thus, the presence of an inductor opposes the rapid changes in the circuit. Therefore, we conclude that due to the presence of an inductor in an electric circuit, the current cannot reach its maximum value or diminish to zero instantaneously.

INTERACTIVE ACTIVITY 25-5

Induced emf across an Inductor in an RL Circuit (Part II)

In this activity, you will use the PhET simulation "Circuit Construction Kit (AC + DC)" to explore the behaviour of RL circuits. Think how RL circuits might be different from and similar to RC circuits you encountered in Chapter 23.

25-8 Energy Stored in a Magnetic Field

Example 25-8 demonstrated that the current in an RL circuit continues to flow for some time after the battery is disconnected. This tells us that an inductor is supplying energy to the circuit, somewhat like a battery. In fact, the energy is stored in the magnetic field of the inductor. This is analogous to the energy stored in the electric field of a capacitor (see Equation 22-17).

Let us once again consider the RL circuit shown in Figure 25-31. When the switch is in position *a*, the law of energy conservation demands that Equation 25-39 hold true. Multiplying both sides of this equation by *i*, we obtain

$$\varepsilon i - i^2 R - Li\frac{di}{dt} = 0$$

$$\varepsilon i = i^2 R + Li\frac{di}{dt} \tag{25-46}$$

This equation describes the law of energy conservation for an RL circuit. The εi-component represents the rate at which the energy is supplied by the battery to the circuit, that is, the electric power delivered to the circuit. Some of this energy is delivered to the resistor and is dissipated by it at a rate of i^2R, and the rest is delivered to the inductor at a rate of $Li\frac{di}{dt}$. Since the latter expression represents the rate of energy transfer to the inductor, we define the energy stored in the magnetic field, U_B, as

$$\frac{dU_B}{dt} = Li\frac{di}{dt} \tag{25-47}$$

$$dU_B = (Li)\,di$$

Knowing the rate of energy transfer from the inductor, we calculate the energy stored in it as the current in the circuit changes from zero to its maximum (steady-state) value:

$$\int_0^{U_B} dU_B = \int_0^I (Li)\,di$$

KEY EQUATION

$$U_B = \frac{LI^2}{2} \tag{25-48}$$

This important relationship indicates that the energy stored in the inductor is proportional to the inductance, *L*, and to the square of the current flowing through the inductor. This energy is stored in the magnetic field of the inductor when current flows through it. Notice that U_B is present only when electric charge is moving through the inductor. This magnetic energy expression is somewhat analogous to the kinetic energy expression, as it is related to the motion of electrical charges: $K = \dfrac{mv^2}{2}$. On the other hand, the energy stored in the electric field of a capacitor is somewhat analogous to the potential energy, as it is related to the configuration of electric charges and not to their motion, $U_E = \dfrac{CV^2}{2}$ (Equation 22-17).

Let us now examine Faraday's law of electromagnetic induction from the energy perspective. When electric current starts flowing through a solenoid, the current generates a magnetic field inside the solenoid. We determined (Equation 25-48) that magnetic energy is stored within this magnetic field. This energy must have come from somewhere. As the current builds up in the circuit, the inductor acts to oppose that current increase by converting kinetic energy carried by the moving electric charges into magnetic energy stored in the inductor's magnetic field. We have to remember that an ideal inductor has zero resistance, so all of the energy carried by moving electric charges is converted into magnetic energy. In real inductors that have non-negligible resistance, some of this kinetic energy is converted into heat through thermal energy losses.

When the switch disconnects the battery from the circuit, moving into position b (Figure 25-31), the current in the solenoid, already at its maximum value, $I_{max} = \dfrac{\varepsilon}{R}$, starts to drop off. The energy stored in the inductor also begins to drop off according to Equation 25-48. Since the back emf induced in the inductor opposes the resulting decrease in current, the energy stored in the magnetic field of the inductor reduces the rate of decrease of the electric current in the circuit. Therefore, the inductor can prevent the abrupt decrease of the current in the circuit. Therefore, the current induced by the inductor is directed so the induced emf has the same direction as the emf from the original battery. As when the circuit was first switched on, the inductor opposes the *change* in electric current.

Let us calculate the energy stored in the magnetic field of a long solenoid with loop density n, radius R, and length ℓ:

$$U_B = \dfrac{LI^2}{2} = \dfrac{1}{2}(\mu_0 n^2 \ell A)I^2 = \dfrac{1}{2}\mu_0 n^2 \ell (\pi R^2)\left(\dfrac{B}{\mu_0 n}\right)^2$$

$$= \dfrac{B^2}{2\mu_0}\ell(\pi R^2) = \dfrac{B^2}{2\mu_0}V$$

$$U_B = \dfrac{B^2}{2\mu_0}V \qquad (25\text{-}49)$$

where $V = \pi R^2 \ell$ is the volume of a solenoid.

Equation 25-49 shows that the energy stored in a solenoid is determined by its volume and the strength of the magnetic field inside it. For many practical applications, it is useful to know the energy density (amount of energy per unit volume):

KEY EQUATION
$$u_B = \dfrac{B^2}{2\mu_0} \qquad (25\text{-}50)$$

Although we derived Equation 25-50 for a particular type of solenoid, the equation is valid for any region of space that contains a magnetic field. Compare this equation to the equation describing the energy density of the electric field, $u = \dfrac{1}{2}\varepsilon_0 E^2$ (Equation 22-20). Both types of energy density depend solely on the square of the magnitude of the respective field. We will apply this crucial observation when discussing the transfer of energy by electromagnetic waves in Chapter 26.

It is instructive to compare what you have learned about RL circuits in this chapter with what you learned about RC circuits in Chapter 23, focusing on the functions of the resistors, capacitors, and inductors in the circuits. While we will discuss circuits combining resistors, capacitors, and inductors in Chapter 26 in detail, we will summarize what we have learned so far in Table 25-3 on the next page.

Comparing the Response of RC and RL Circuits to Transient Voltage

While analyzing RL circuits and experimenting with the PhET simulations, you probably noticed that the response of RL circuits to sudden changes of voltage across the power supply is in many ways similar to the behaviour of RC circuits, which we discussed in Chapter 23. We will finish this section by comparing and contrasting the responses of RC and RL circuits to transient voltage.

CHECKPOINT 25-9

Energy Stored in a Magnetic Field

What action performed alone will quadruple the energy stored in the magnetic field of a solenoid?
(a) quadrupling the current through the solenoid
(b) doubling the length of the solenoid
(c) doubling the density of coils of the solenoid
(d) doubling the diameter of a solenoid
(e) quadrupling the number of coils of the solenoid

ANSWER: (c) The energy stored in the magnetic field of a solenoid can be expressed as $U_B = LI^2/2$. Therefore, according to Equation 25-48, we should either change the current through the solenoid or its inductance. However, quadrupling the current will increase the energy sixteen-fold. Therefore, we have to quadruple its inductance. Since, according to Equation 25-22, the inductance of a solenoid is $L_{solenoid} = \mu_0 n^2 \ell A$, we can double the density of the coils of the solenoid. If the rest of the parameters (its length, its cross-sectional area) remain the same, L will quadruple and the energy stored in the solenoid will quadruple as well.

ELECTRICITY, MAGNETISM, AND OPTICS

Table 25-3 Comparison of the Response of Series RC and RL Circuits to Transient Voltage

	Transient RC Circuit	Transient RL Circuit
Circuit diagram	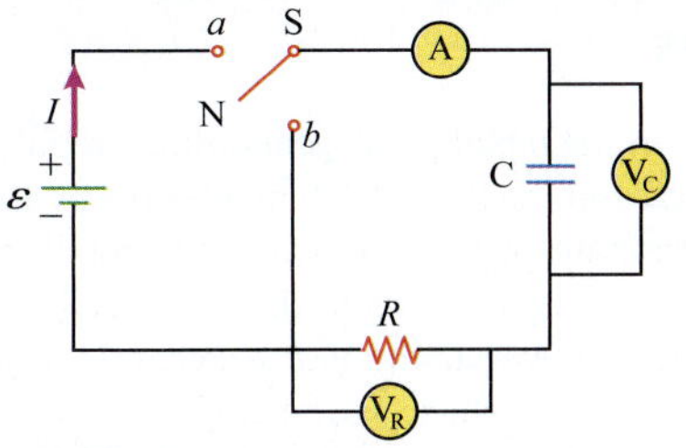 **Figure 25-35** A series RC circuit.	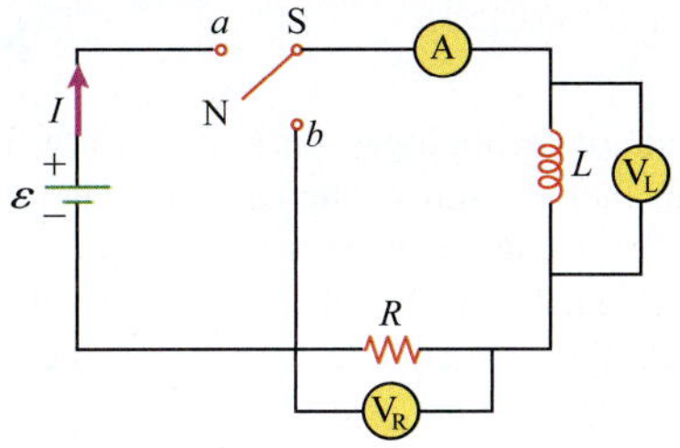 **Figure 25-36** A series RL circuit.
Dependence of current and voltage across the capacitor or an inductor on time	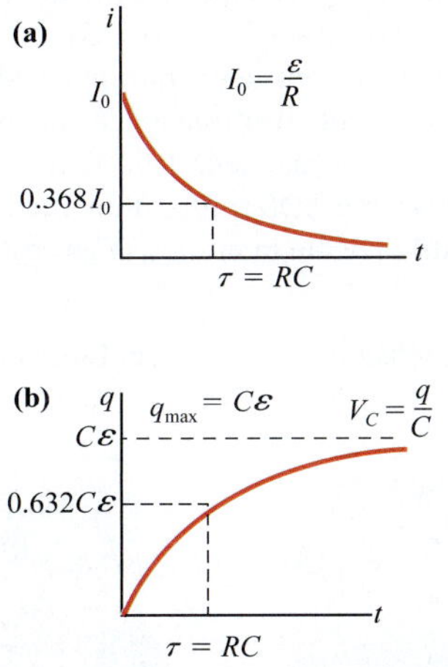 **Figure 25-37** Charging capacitor in a series RC circuit. (a) The dependence of the electric current on time. (b) The dependence of the charge stored on the capacitor (potential difference across it) on time. 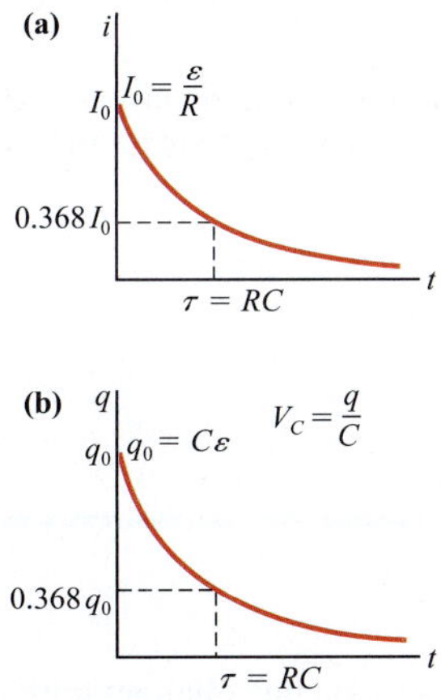 **Figure 25-39** A series RC circuit is being disconnected from a battery. (a) The dependence of the electric current on time. (b) The dependence of the charge stored on the capacitor (potential difference across it) on time.	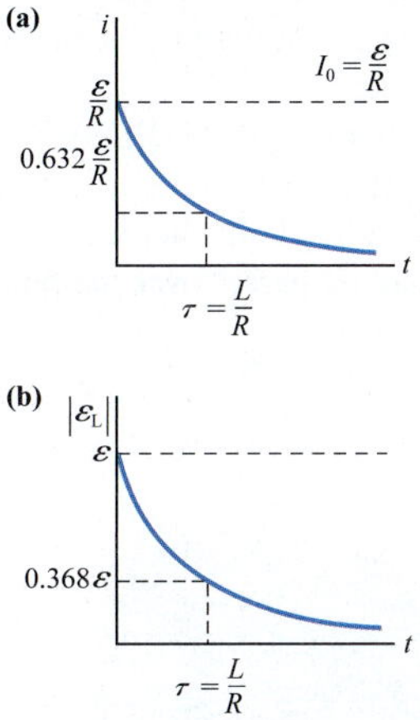 **Figure 25-38** A series RL circuit is being connected to a battery. (a) The dependence of the electric current on time. (b) The dependence of the induced emf across the inductor on time. 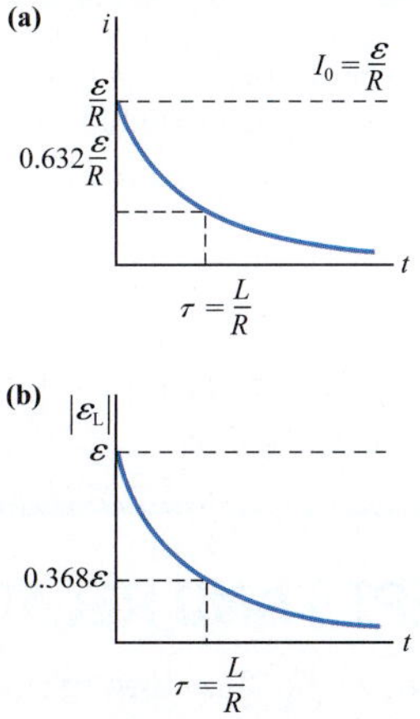 **Figure 25-40** A series RL circuit is being disconnected from a battery. (a) The dependence of the electric current on time. (b) The dependence of the induced emf across the inductor on time.
Time constant τ	Represents the time it takes an RC or an RL series circuit to reach a steady state. The bigger it is, the longer it takes the circuit to reach a steady state.	
	$\tau = RC$	$\tau = L/R$
Energy stored	Energy stored in the magnetic field of a solenoid: $U_B = LI^2/2$	Energy stored in the electric field between the plates of a capacitor: $U_E = CV^2/2$

Maglev Trains

There are three major technologies currently in use for magnetic levitation in maglev trains. The electromagnetic system (EMS) uses the attractive force between a support magnet located under the vehicle and a rail that is made of a ferromagnetic material. The attractive force between the steel rail and the supporting magnet (Figure 25-35(a)) lifts the vehicle upward. EMS technology is used in maglev trains in Germany. The distance between the electromagnets and the rail is critical, so this system requires a control mechanism that continuously measures this distance and adjusts it by varying the current in the guidance magnet.

The electrodynamic system (EDS) is used in maglev trains in Japan (Figure 25-35(b)). These trains carry magnets, and a large metal plate runs along the centre of the maglev track. When a train's magnet passes over the plate, eddy currents are created in the metal plate, generating a repulsive force between the track and the train. EDS maglev trains are naturally stable: if the train drops toward the rail, the repulsive force increases; if the train rises too high above the rail, the repulsive force decreases. However, the electromagnetic repulsion is present only when the vehicle is moving, and the vehicle must have wheels for takeoff and landing (unlike EMS trains). Moreover, the electric currents induced in the metal plates dissipate energy and create a significant drag force on the vehicle.

Inductrack, a variation of EDS technology, uses permanent room-temperature magnets attached to the vehicle (instead of electromagnets or superconducting magnets) to produce a magnetic field that levitates the train over passive coils. Like EDS trains, Inductrack maglev trains do not levitate until they are in motion (Figure 25-41(c)).

(a) Inductrack **(b)** Electromagnetic suspension **(c)** Electrodynamic suspension

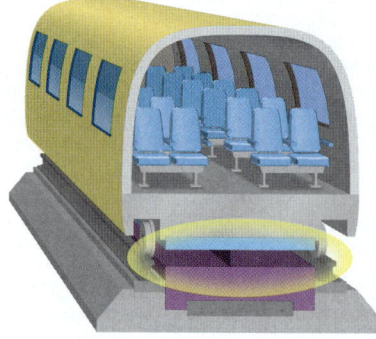 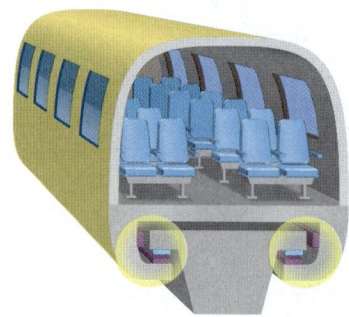 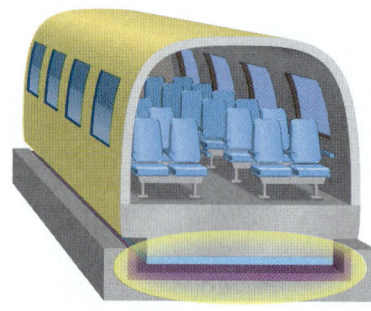

Figure 25-41 (a) The electromagnetic suspension system is used in maglev trains in Germany. (b) The electrodynamic suspension system is used in maglev trains in Japan. (c) Inductrack uses permanent magnets to levitate the train over passive coils.

KEY CONCEPTS AND RELATIONSHIPS

A time-varying magnetic flux through a conducting loop induces an emf in it. To produce a magnetic flux that changes with time, we must have either a magnetic field that changes with time or a constant magnetic field but with a loop whose effective area changes with time. The latter can be achieved, for example, by rotating the loop in the constant magnetic field.

Magnetic Flux through a Surface

Magnetic flux is a scalar quantity (can be positive, negative, or zero) and is defined as an integral over the surface:

$$\Phi_B = \int \vec{B} \cdot d\vec{A} \qquad (25\text{-}7)$$

Magnetic flux changes with time when the magnetic field, the area of the loop, and/or the angle between the normal to the loop and the magnetic field vary in time (Figure 25-4).

Faraday's Law of Electromagnetic Induction

A time-varying magnetic flux through a conducting loop induces an emf in it:

$$\varepsilon_{\text{ind}} = -\left(\frac{d\Phi_B}{dt}\right) \qquad (25\text{-}10)$$

A time-varying magnetic flux through a coil that has N loops induces an emf in it:

$$\varepsilon_{ind} = -\left(\frac{d\Phi_B}{dt}\right)_{coil} = -N\left(\frac{d\Phi_B}{dt}\right)_{loop} \qquad (25\text{-}12)$$

Lenz's law states that the directions of the induced emf and induced current always oppose (reduce) the change in the magnetic flux that caused it.

The motional emf generated in a rectangular coil of width ℓ, having N loops, and moving out of a magnetic field region with speed v is given by

$$\varepsilon_{ind} = -NBv\ell \qquad (25\text{-}14)$$

The magnitude of an induced electric field by a changing magnetic field is given by

$$|\varepsilon_{ind}| = |\oint \vec{E}_{ind} \cdot d\vec{s}| = \left|-\frac{d\Phi_B}{dt}\right| \qquad (25\text{-}19)$$

The direction of the induced electric field is such that the electric current generated by it induces a magnetic field that *opposes* the change in the magnetic flux.

Inductance and Mutual Inductance

The back emf induced in a solenoid is

$$\varepsilon_{ind} \equiv \varepsilon_L = -L\frac{dI}{dt} \qquad (25\text{-}21)$$

The inductance of a solenoid with circular loops is given by

$$L_{solenoid} = \mu_0 n^2 \ell A = \mu_0 n^2 V \qquad (25\text{-}22)$$

The mutual inductance of two coupled solenoids is

$$M = k\sqrt{L_1 L_2} \qquad (25\text{-}32)$$

The ratio of voltages to the number of loops in the coils of a transformer is given by

$$\frac{V_2}{V_1} = \frac{N_2}{N_1} \qquad (25\text{-}37)$$

Response of Series RL Circuits to Transient Voltage

The time constant of a series RL circuit is $\tau = L/R$.

The dependence of the current through and the potential difference across an inductor on time when a series RL circuit is being connected to a battery is given by

$$\begin{cases} i(t) = \dfrac{\varepsilon}{R}\left(1 - e^{-\frac{t}{\tau}}\right) \\[2mm] V_L \equiv \varepsilon_L = -\varepsilon e^{-\frac{t}{\tau}} \end{cases} \quad \text{where } \tau = L/R \qquad (25\text{-}43)$$

The dependence of the current through and the potential difference across an inductor on time in a series RL circuit that is being disconnected from a battery is

$$\begin{cases} i(t) = \dfrac{\varepsilon}{R}e^{-\frac{t}{\tau}} \\[2mm] V_L \equiv \varepsilon_L = -\varepsilon e^{-\frac{t}{\tau}} \end{cases} \quad \text{where } \tau = L/R \qquad (25\text{-}45)$$

Energy Stored in a Magnetic Field

The energy stored in a magnetic field can be expressed as

$$U_B = \frac{LI^2}{2} \qquad (25\text{-}48)$$

The energy stored in a magnetic field per unit volume (energy density) is

$$u_B = \frac{B^2}{2\mu_0} \qquad (25\text{-}50)$$

APPLICATIONS

AC generator, AC circuits (RL, RC, and RLC circuits), electromagnetic brakes and dampers, transformers, metal detectors, electric guitars, magnetic levitation trains

KEY TERMS

alternating current, alternating emf, back emf, coefficient of mutual inductance, conservative fields, coupled inductors, eddies, eddy currents, electromagnetic force, electromagnetic induction, electrostatic force, Faraday's law of electromagnetic induction, galvanometer, induced electric field, inductance, inductive charging, inductors, Lenz's law, magnetic flux, motional emf, mutual inductance, non-conservative fields, rate of change of the magnetic flux in time, self-induction, series RL circuit, time constant, transformer, transient electric current, transient voltage

QUESTIONS

1. A rectangular metal loop is located in the region of a uniform magnetic field directed perpendicular to the plane of the loop (Figure 25-42). An induced electric current in the clockwise direction is detected in the loop. What must be true about this magnetic field?
   (a) The magnetic field is directed into the page and is steady.
   (b) The magnetic field is directed into the page and is increasing.
   (c) The magnetic field is directed into the page and is decreasing.
   (d) The magnetic field is directed out of the page and is steady.
   (e) The magnetic field is directed out of the page and is increasing.
   (f) The magnetic field is directed out of the page and is decreasing.
   (g) Only (a) and (b) are possible.
   (h) Only (c) and (e) are possible.
   (i) Only (d) and (f) are possible.
   (j) All the answers can be correct.

Figure 25-42 Question 1.

2. Which of the following equations accurately describe a magnetic field that can produce a constantly increasing emf in a loop oriented perpendicular to its direction? (Choose all that apply.)
   (a) $B(t) = (3t + 278)$ mT
   (b) $B(t) = (5\sin(2\pi t))$ mT
   (c) $B(t) = (5\sin(3\pi t) + 10)$ mT
   (d) $B(t) = (5\cos(3\pi t) + 30)$ mT
   (e) $B(t) = (5t^2 + 15)$ mT

3. Which of the following scenarios describes a magnetic flux through a loop that does not change with time?
   (a) A uniform magnetic field changes with time, and the area of the loop and its orientation remain constant.
   (b) A loop is moving with constant velocity through a non-uniform and constant magnetic field.
   (c) A square loop is rotating about its diagonal in a uniform and constant magnetic field directed perpendicular to the loop's diagonal.
   (d) A loop is moving at a constant velocity (its orientation is not changing) in a uniform and constant magnetic field.
   (e) The area of a loop is increasing at a constant rate while it is in a uniform magnetic field whose magnitude is also increasing and whose direction is always parallel to the surface of the loop.

4. In which one of the following cases will there be an electric current induced in the ring in Figure 25-43?
   (a) The ring is moving in a horizontal direction in a uniform and constant magnetic field.
   (b) The ring is moving vertically in a uniform and constant magnetic field.
   (c) The ring is stationary, but the magnetic field is increasing.
   (d) The ring is stationary, but the magnetic field is decreasing.
   (e) The ring is spinning about a horizontal axis parallel to the B lines and going through its centre.
   (f) The ring is spinning about a horizontal axis perpendicular to the B lines and going through its centre.
   (g) The ring is spinning about a vertical axis perpendicular to the B lines and going through its centre.
   (h) The ring is moving diagonally (without spinning) at a constant velocity.
   (i) The ring is moving diagonally (without spinning) but at an increasing speed.
   (j) Parts (b), (c), (d), (e), (f), (h), and (i) are true.
   (k) Parts (c), (d), (f), (g), and (i) are true.
   (l) Parts (c), (d), (f), and (g) are true.
   (m) Parts (a), (c), (d), (f), (g), and (i) are true.
   (n) All of parts (a) through (i) are true.

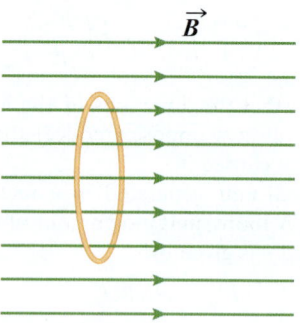

Figure 25-43 Questions 4 and 5.

5. What will the direction of the current in the metal ring in Figure 25-43 be if the magnetic field starts increasing?
   (a) counter-clockwise (if seen from the left)
   (b) clockwise (if seen from the left)
   (c) There will be no current generated in the ring, and the ring is stationary.

6. A metal ring is dropped between the poles of a horseshoe magnet (Figure 25-44). What is true about the electric current in the ring as observed from the right?
   (a) The electric current is clockwise while entering the loop and counter-clockwise while exiting.
   (b) The electric current is counter-clockwise when entering the loop and clockwise while exiting.
   (c) The electric current is counter-clockwise all the time.
   (d) The electric current is clockwise all the time.
   (e) There is no current in the ring.

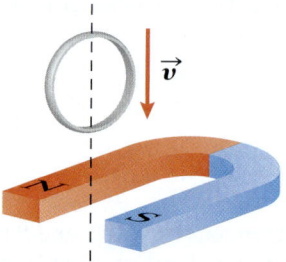

Figure 25-44 Question 6.

7. A very light, 1 m long metal rod is placed in a uniform and constant horizontal magnetic field $B = 1$ T, directed as shown in Figure 25-45. The rod is free to move along frictionless metal rails that have negligibly small resistance. A light bulb, requiring a potential difference of 10 V to light up, is attached to the metal rails as shown. What should the minimal speed of the rod be for the bulb to light up?
   (a) 0.01 m/s
   (b) 0.1 m/s
   (c) 1 m/s
   (d) 10 m/s
   (e) 100 m/s

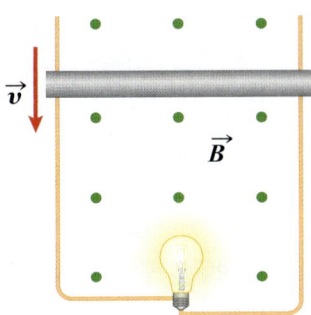

Figure 25-45   Question 7.

8. Three light rectangular blocks with the same shape and approximately the same mass are placed on top of an aluminum inclined plane relatively far from each other (Figure 25-46). Block A is made of wood, block B is made of a magnetic material, and block C is made of copper. The coefficients of friction between the blocks and the incline are negligible. The blocks, initially at rest, are let go from the top of the incline. Which of the following statements most accurately describes what happens to the blocks?
   (a) The copper block will come down last because copper is attracted to aluminum.
   (b) The magnetic block will come down last because, during its motion, eddy currents will be created in the incline, which will slow it down significantly.
   (c) The magnetic block will come down first because, during its motion, eddy currents will be created in the incline, speeding it up.
   (d) The wooden block will come down last because wood will experience significant friction that will slow it down.
   (e) Since friction is negligible, all the blocks will have the same acceleration. As a result, they will reach the bottom simultaneously.

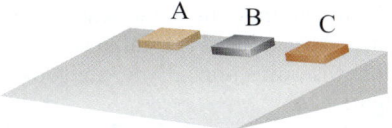

Figure 25-46   Question 8.

9. A step-up transformer has N_1 = 25 loops and N_2 = 25 000 loops. The output voltage, V_2, has to be 120 000 V, and the output current, I_2, has to be 10. mA. What are the input voltage, V_1, and current, I_1?
   (a) V_1 = 12 V; I_1 = 10. mA
   (b) V_1 = 12 V; I_1 = 1.0 A
   (c) V_1 = 12 V; I_1 = 10. A
   (d) V_1 = 120 V; I_1 = 10. A
   (e) V_1 = 120 V; I_1 = 1.0 A
   (f) V_1 = 120 V; I_1 = 10. mA
   (g) V_1 = 120 V; I_1 = 0.010 mA

10. A current through an inductor starts increasing at a steady rate, as shown in Figure 25-47. Which of the following is true about the potentials at points a and b?
   (a) $V_a = V_b$
   (b) $V_a < V_b$
   (c) $V_a > V_b$

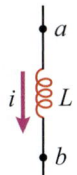

Figure 25-47   Question 10.

11. A circular copper loop is moving away from a long, straight current-carrying wire, as shown in Figure 25-48. Which of the following statements correctly describes what is happening to the loop as a result of its motion?
   (a) The circular loop is attracted to the current-carrying wire because copper is magnetic and it is attracted to the current.
   (b) A current in the clockwise direction is induced in the loop, which attracts the loop to the current-carrying wire.
   (c) A current in the counter-clockwise direction is induced in the loop, which attracts the loop to the current-carrying wire.
   (d) A current in the clockwise direction is induced in the loop, which repels the loop from the current-carrying wire.
   (e) A current in the counter-clockwise direction is induced in the loop, which repels the loop from the current-carrying wire.
   (f) There are no forces of attraction or repulsion between the loop and the wire because copper is a non-magnetic material.

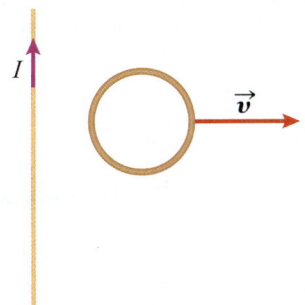

Figure 25-48   Question 11.

12. A circular copper loop is moving along a long straight current-carrying wire, as shown in Figure 25-49. The speed of the loop is steadily increasing. Which of the following statements correctly describes what is happening to the loop as a result of its motion?
   (a) A counter-clockwise current is induced in the loop, which attracts the loop to the current-carrying wire.
   (b) A clockwise current is induced in the loop, which repels the loop from the current-carrying wire.
   (c) A current in the counter-clockwise direction is induced in the loop, which repels the loop from the current-carrying wire.
   (d) A current in the clockwise direction is induced in the loop, which attracts the loop to the current-carrying wire.
   (e) There are no forces of attraction or repulsion between the loop and the wire because the magnetic flux is constant.

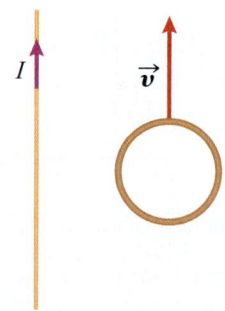

Figure 25-49   Question 12.

13. A metal ring is moved over a permanent magnet, as shown in Figure 25-50(a). Which of the following statements correctly describes what happens to the ring and the direction of the current as seen from above?
    (a) A clockwise current is induced in the ring before it reaches the middle of the magnet, and then the induced current flows in the counter-clockwise direction.
    (b) A counter-clockwise current is induced in the ring before it reaches the middle of the magnet, and then the current is induced in the clockwise direction.
    (c) A clockwise current is induced in the ring only when the magnet is inside the ring.
    (d) A counter-clockwise current is induced in the ring before it reaches the magnet, and then the current stops.
    (e) A current is induced in the ring in the clockwise direction only when the ring leaves the magnet. (The magnet is no longer inside the ring.)

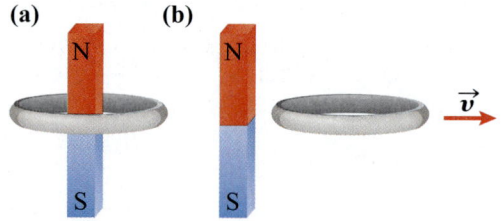

Figure 25-50   Questions 13 and 14.

14. A metal ring is located near a permanent magnet, as shown in Figure 25-50(b). The ring is moved away from the magnet. Which of the following statements correctly describes the direction of the current in the ring as seen from above?
    (a) There is an induced current in the clockwise direction.
    (b) There is an induced current in the counter-clockwise direction.
    (c) There is no current induced in the ring.
15. Rank the inductances of the five solenoids shown in Figure 25-51 from the largest to the smallest.
    Largest 1___ 2___ 3___ 4___ 5___ Smallest
    Or, the inductances are the same in all cases. ___
    If the inductance is the same for two or more cases, clearly indicate it on the ranking scheme. Explain your reasoning.

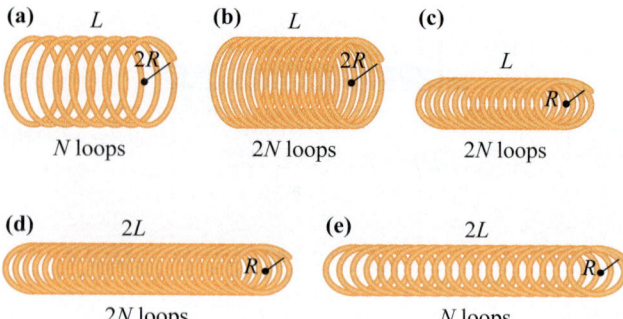

Figure 25-51   Questions 15, 16, and 17.

16. The same electric current flows through the five solenoids shown in Figure 25-51. Rank the energy stored inside the solenoids from largest to smallest if a 1 A electric current flows through each solenoid.
    Largest 1___ 2___ 3___ 4___ 5___ Smallest
    Or, the stored energy is the same in all cases. ___
    If the energy is the same for two or more cases, clearly indicate it on the ranking scheme. Explain your reasoning.
17. Rank the energy density inside the five solenoids in Figure 25-51 from largest to smallest when a 1 A electric current flows through each solenoid.
    Largest 1___ 2___ 3___ 4___ 5___ Smallest
    Or, the energy density is the same in all cases. ___
    If the energy density is the same for two or more cases, clearly indicate it on the ranking scheme. Explain your reasoning.
18. Two copper coils are facing each other, as shown in Figure 25-52. Coil 1 is connected to a battery, so a constant current, I_1, flows through it. At time $t = 0$, coil 2 starts moving to the right with velocity $\vec{v}$. Which of the following statements correctly describes what happens as soon as coil 2 starts moving away from coil 1?
    (a) There is an induced current in coil 2 in the clockwise direction if you are looking from the right.
    (b) There is an induced current in coil 2 in the counter-clockwise direction if you are looking from the right.
    (c) There is no current induced in coil 2 because current I_1 is constant.
    (d) All of the above are possible, depending on the velocity of coil 2 relative to coil 1.

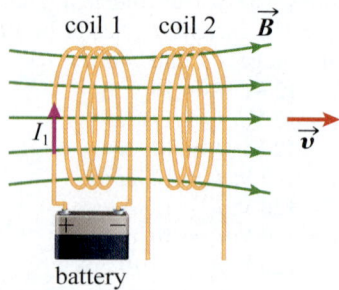

Figure 25-52   Question 18.

19. Five series RL circuits (like the RL circuit in Figure 25-31) have different resistors, inductors, and batteries. The information for each circuit is shown in Table 25-4. The circuits were initially in position N; at time $t = 0$ s, the switch was flipped to position a. Rank the times it takes for the current in each circuit to reach 75% of its maximum value from largest to smallest.

Largest 1___ 2___ 3___ 4___ 5___ Smallest

Or, the time is the same for each circuit. ___

If the time is the same for two or more circuits, clearly indicate it on the ranking scheme. Explain your reasoning.

Table 25-4 Data for Question 19

	R	L	ε
Circuit 1	10. Ω	10. mH	10. V
Circuit 2	20. Ω	20. mH	10. V
Circuit 3	20. Ω	20. mH	20. V
Circuit 4	50. Ω	30. mH	25 V
Circuit 5	50. Ω	30. mH	30. V

20. Five different series RL circuits (like the RL circuit in Figure 25-31) have the same batteries (20 V each), the same resistors (20 Ω each), but different inductors: $L_1 = 10$ mH, $L_2 = 20$ mH, $L_3 = 30$ mH, $L_4 = 40$ mH, and $L_5 = 50$ mH. The circuits are initially in position N; at time $t = 0$ s, the switch was flipped to position a. What is true about the values of the maximum current in these circuits?
(a) $I_1 > I_2 > I_3 > I_4 > I_5$
(b) $I_1 < I_2 < I_3 < I_4 < I_5$
(c) $I_1 = I_2 = I_3 = I_4 = I_5$

21. The light bulb in Figure 25-53 is a part of a series circuit powered by an AC source and containing an air-filled inductor. When an iron bar is moved inside the inductor's core, what happens to the brightness of the light bulb? Explain.
(a) The brightness remains the same.
(b) The brightness increases.
(c) The brightness decreases.

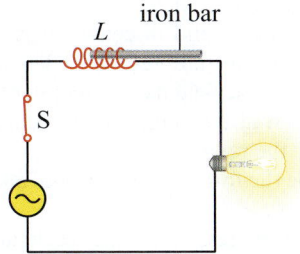

Figure 25-53 Question 21.

PROBLEMS BY SECTION

For problems, star ratings will be used (★, ★★, or ★★★), with more stars meaning more challenging problems. The following codes will indicate if $\frac{d}{dx}$ differentiation, $\int$ integration, ▭ numerical approximation, or ∿ graphical analysis will be required to solve the problem.

Sections 25-1 and 25-2 In Faraday's Lab and Magnetic Flux and Its Rate of Change

22. ★★ $\frac{d}{dx}$ A circular loop of area $A = 3.0$ cm^2 is in a region where there is a uniform and steadily decreasing magnetic field. The direction of the magnetic field is at a 30° angle to the loop's axis of symmetry. At time $t = 2.0$ s, the magnetic field has a magnitude of 10. mT, and at time $t = 6.0$ s, the magnetic field has a magnitude of 2.0 mT. Find the rate of change of the magnetic flux through the loop.

23. ★★ $\frac{d}{dx}$ A rectangular copper loop with dimensions $\ell = 50.0$ cm and $w = 25.0$ cm is moving at a constant velocity of 1.50 m/s out of a uniform magnetic field directed into the page. The magnetic field varies with time as $B = (0.00200t + 0.00100)$ T.
(a) Find the rate of change of the magnetic flux through this loop at the instant when the loop is three quarters of its length out of the magnetic field.
(b) Answer part (a) when the loop has moved completely out of the magnetic field.
(c) How will your answers to parts (a) and (b) change if, instead of a single loop, there is a coil consisting of 20 loops?

24. ★★ $\frac{d}{dx}$ A square copper loop of dimensions 5.0 cm × 5.0 cm is located in the area of a uniform, time-varying magnetic field. The loop is directed perpendicular to the magnetic field. The magnetic field varies with time, as shown in Figure 25-54. Find the emf induced in the loop during the time intervals I, II, and III. Assume all the values in the graphs are given to two significant figures.

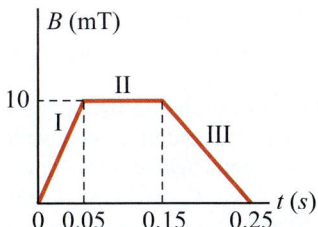

Figure 25-54 Problem 24.

25. ★★ $\frac{d}{dx}$ Two long, concentric (coaxial) copper solenoids have the same length, $\ell = 50$. cm. Their radii are $R_1 = 2.0$ cm and $R_2 = 3.0$ cm, and their numbers of turns are 100 and 200, respectively. When the electric current in the inner solenoid changes at a rate of 0.50 A/s, what is the rate of change of the magnetic field through the outer solenoid?

26. ★★★ d/dx A circular coil with 10 loops and a radius of 5.0 cm rotates about its diameter in a uniform and constant magnetic field with a magnitude of 10. mT. The axis of rotation of the coil is always perpendicular to the magnetic field. When the angular velocity of rotation is 4.0 rpm, what is the maximum rate of change of the magnetic flux through the coil?

Section 25-3 Faraday's Law of Electromagnetic Induction

27. ★ The magnetic flux through a solenoid with 100 turns changes from 6.0×10^{-2} Wb to 8.0×10^{-4} Wb in 0.10 s. Find the induced emf in the coil.

28. ★★ d/dx A rectangular metal loop with dimensions 5.0 cm × 10. cm is inserted in a uniform but changing magnetic field. The resistance of the loop is 3.0 Ω. Find the rate of change of the magnetic field so that 5.0 A of electric current is induced in the loop when
  (a) the magnetic field is perpendicular to the surface of the loop
  (b) the magnetic field is directed at a 30° angle to the loop's normal
  (c) the magnetic field is directed parallel to the loop's surface

29. ★ To find the magnitude of the magnetic field between the poles of a large horseshoe magnet, a coil of wire consisting of 20 turns and with a cross-sectional area of 6.0 cm² is inserted between the poles of the magnet perpendicular to the magnetic field lines. When the coil is pulled out of the magnetic field in 0.020 s, an induced emf of 45 mV is measured in the coil. What is the magnitude of the magnetic field between the poles of the magnet?

30. ★★ d/dx, ⤳ A rectangular 4.0 cm × 8.0 cm metal coil has 50 turns and a resistance of 10. Ω. The coil is placed on a horizontal frictionless table and is completely submerged in a vertical uniform magnetic field, $B = 0.50$ T.
  (a) How fast should the coil be pulled out of this uniform magnetic field so that the emf generated in the coil reaches 0.50 V? The velocity of the coil, $\vec{v}$, is directed perpendicular to one of its sides (Figure 25-55).
  (b) Draw a graph that shows the dependence of the induced emf on time while the coil is pulled out of the magnetic field at the constant velocity $\vec{v}$ found in part (a).
  (c) Find the direction and the magnitude of the electric current induced in the coil while it is pulled out of the magnetic field at a constant velocity $\vec{v}$.
  (d) Find the force, $\vec{F}$, that needs to be applied to pull the coil with this constant velocity $\vec{v}$.
  (e) Draw the graph representing the dependence of F on time. Draw this graph under the graph you made for part (b) to see the correspondence.

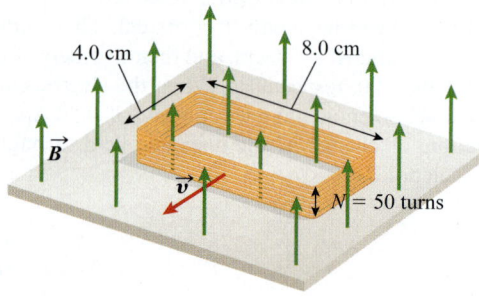

Figure 25-55 Problem 30.

31. ★ Another way of looking at Faraday's law of electromagnetic induction is to consider the induced charge in a wire placed in a magnetic field (due to the Lorentz force acting on moving electric charges submerged in a magnetic field). The induced charge in a coil with N loops and resistance R can be expressed as

$$\Delta Q = \frac{-N\Delta \Phi_B}{R}$$

Show that this relationship is a direct consequence of the law of electromagnetic induction.

32. ★★★ d/dx, ∫ A conducting bar of length $\ell = 0.20$ m and mass $m = 0.10$ kg is submerged in a uniform and constant magnetic field $B = 0.50$ T (Figure 25-56). The bar is free to slide along two frictionless metal rails. At time $t = 0$ s, the bar is given an initial velocity of $\vec{v}_i = 2.0$ m/s directed to the right. The resistance $R = 4.0$ Ω is the resistance of the entire circuit.
  (a) Find the velocity of the bar 1.0 s after it was pushed.
  (b) Find the acceleration of the bar at $t = 1.0$ s.
  (c) Determine the distance travelled by the bar during the first second.
  (d) Calculate the distance travelled by the bar before its speed decreases to $v_i/4$.

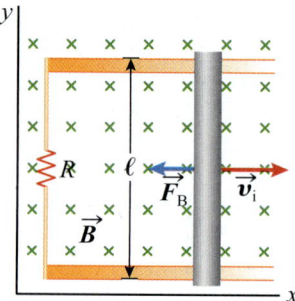

Figure 25-56 Problem 32.

33. ★★★ d/dx, ∫ A metal rod of mass $m = 250.$ g and length $\ell = 50.0$ cm is free to slide on two parallel, frictionless horizontal metal rails (Figure 25-57). The rails are connected at one end so that they and the rod form a closed circuit. The rod and the rails have a resistance of 2.00 Ω. A uniform magnetic field B, perpendicular to the plane of this circuit and directed vertically upward, is decreasing at a constant rate of 0.300 mT/s. At time $t = 0$ s, the magnetic field has a strength of 15.0 mT, and the rod is at rest at a distance of 150. cm from the connected end of the rails.
  (a) Describe the forces acting on the rod at $t = 0$ s and thereafter.
  (b) Derive an expression for the acceleration of the rod at $t = 0$ s in terms of the given quantities.

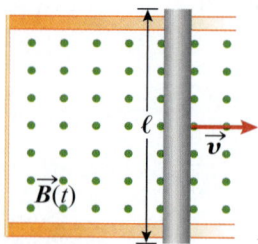

Figure 25-57 Problem 33.

Section 25-4 Induced emf and Induced Electric Fields

34. ★★ $\frac{d}{dx}$ A circular loop of radius $R = 10.0$ cm is coaxial with a solenoid ($r = 0.0500$ m, $\ell = 1.00$ m, and 750 turns), as shown in Figure 25-58. The current in the circuit increases linearly from 1.00 A to 4.60 A over 0.500 s.
    (a) Calculate the induced emf in the loop.
    (b) Calculate the induced electric field in the loop.
    (c) When the resistance of the loop is 20.0 Ω, how much energy will be dissipated by the loop in 10.0 s?

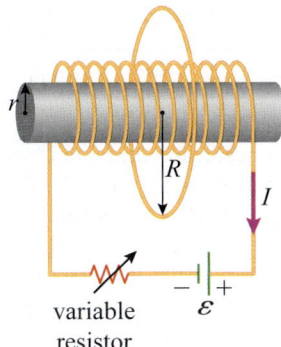

variable resistor

Figure 25-58 Problem 34.

35. ★★★ $\frac{d}{dx}$, ∿ A solenoid with length ℓ and radius R has N turns. The solenoid carries an electric current that varies with time as $i(t) = I_{max} \sin (2\pi ft)$, where I_{max} represents the maximum current and f is the frequency of the AC source.
    (a) Derive an expression that describes the magnitude of the electric field outside the solenoid ($r > R$).
    (b) Derive an expression that describes the magnitude of the electric field inside the solenoid ($r < R$).
    (c) For parts (a) and (b), draw graphs showing the dependence of the magnitude of the electric field on r, $E(r)$, at a certain instant of time t. Explain your results.

Section 25-5 Self-Inductance and Mutual Inductance

36. ★ The current through an air-filled coil is changing at a rate of 200. A/s. As a result of this change, the induced emf in it is measured to be 250. V.
    (a) What is the self-inductance of the coil?
    (b) How will the value of the self-inductance change if an iron core is inserted in this coil?
    (c) What is the rate of change of the magnetic flux through this coil?

37. ★ Two electric circuits are located near each other. When the electric current in the first circuit changes at a rate of 200. A/s, an induced emf of 20.0 V is measured in the second circuit. What is the mutual inductance of the circuits?

38. ★ Two coaxial air-filled solenoids of equal length ($\ell = 20.0$ cm) and equal radius ($r = 5.00$ cm) have 100 turns and 400 turns, respectively.
    (a) Find the mutual inductance of these solenoids.
    (b) How will your answer to part (a) change if iron cores are inserted in the solenoids?

39. ★★ $\frac{d}{dx}$ A 2.00 m long air-filled solenoid has a radius of 5.00 cm and 25 000 turns. A 500-turn coil is tightly wound over the outside of the solenoid, as shown in Figure 25-59.
    (a) Find the mutual inductance of the solenoid and the coil.
    (b) The current in the solenoid is changing according to $i_{solenoid}(t) = \left(10.0 \frac{A}{S}\right)t$. Find the emf induced in the coil.

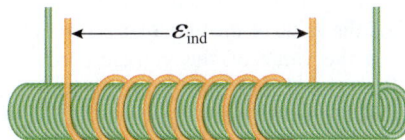

Figure 25-59 Problem 39.

Section 25-6 Applications of Faraday's Law of Electromagnetic Induction

40. ★★ $\frac{d}{dx}$ An engineering student proposes to use the law of electromagnetic induction to power aircraft equipment. The proposal uses the Boeing 757-200 as an example. The aircraft has a maximum cruising speed of 914.0 km/h and a wing span of 38.05 m. Suppose the aircraft is flying in a region where the vertical component of the magnetic field is 3.500×10^{-5} T. What is the potential difference generated across the wings? Is it sufficient to power the onboard equipment?

41. ★★ $\frac{d}{dx}$ A simple AC generator, like the one described in Figure 25-23, consists of a single loop with an area of 100. cm² that rotates with a frequency of 60.0 Hz in a uniform magnetic field of 3.00×10^{-2} T.
    (a) Find the maximum emf generated by this generator.
    (b) Find the maximum emf that would be generated by this generator if instead of a single loop it had 100 loops.
    (c) Describe mathematically how the output voltage depends on time. The output voltage is maximum when $t = 0$ s.

42. ★★★ $\int$, $\frac{d}{dx}$ A horizontal copper disk 40.0 cm in diameter is placed in a constant and uniform magnetic field of 8.00×10^{-2} T directed vertically upward. The disk spins about a vertical axis through its centre, with an angular velocity of 10.0 rad/s.
    (a) What is the magnitude of the emf induced between the centre of the disk and its rim?
    (b) What is the magnitude of the induced emf between diametrically opposite points on the disk?

43. ★★ To reduce the 120. V AC voltage to 12.0 V AC to operate a standard doorbell, a step-down transformer is used. (Assume that 120. V is an rms voltage).
    (a) This transformer has 1000 turns in the primary coil. How many turns does it have in the secondary coil?
    (b) When the doorbell button is pushed, 15.0 Ω of resistance is introduced in the secondary circuit. What are the values of the maximum and rms currents in the primary and secondary coils? Assume that the transformer used in the doorbell has 100% efficiency.
    (c) How would your answers to part (b) change if the efficiency of the transformer were 90%?

Section 25-7 RL Circuits

44. ★★ A series electric circuit consists of a 24 V battery, a switch, a 20. Ω resistor, and an inductor. The inductor is a 30. cm long solenoid that has 1000 turns and a radius of 2.5 cm. After the switch is closed, it takes 0.28 ms for the current in the circuit to reach half its maximum value.
    (a) Find the inductance of the solenoid using two different methods.
    (b) Derive an expression for the induced emf in the solenoid during the time when the current is changing from zero to its maximum value.
    (c) Find the value of the maximum current in the circuit.
    (d) Find the value of the voltage across the resistor when the induced emf in the solenoid is at one half its maximum value.
    (e) The switch is left on for a long time. What is the value of the induced emf in the solenoid at $t = 0.10$ s? What is the value of the voltage across the resistor at that time?

Section 25-8 Energy Stored in a Magnetic Field

45. ★ Find the energy stored in the solenoid described in problem 44 when the current in the circuit
    (a) is at its maximum value
    (b) is at one half its maximum value
    (c) is zero

COMPREHENSIVE PROBLEMS

46. ★★ $\frac{d}{dx}$ A square loop of wire with a side length of 5.0 cm is bent along its diagonal such that its two triangular halves are located in perpendicular planes (Figure 25-60). The loop is then submerged in a uniform magnetic field B, directed perpendicular to the diagonal of the square loop and at a 45° angle to the planes of the bent square. The resistance of the loop is 3.0 Ω, and the magnetic field in the region changes with time according to $B(t) = (50.t + 3.0)$ mT. Find the magnitude and the direction of the induced emf and the induced electric current in the loop during the time interval 0 s $< t <$ 5.0 s.

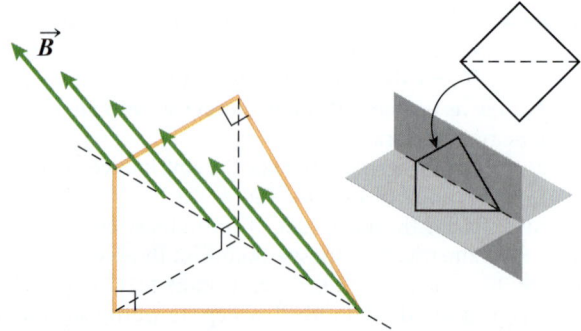

Figure 25-60 Problem 46.

47. ★★ ∫ A horizontal, 10. cm long aluminum rod is located inside a vertical, uniform, and constant magnetic field of 35 mT. One end of the rod is secured to a hinge, such that the rod can rotate around it in a horizontal plane (Figure 24-61). The rod makes five revolutions per second. What is the induced emf along its length?

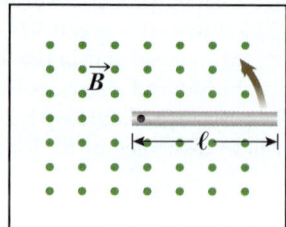

Figure 25-61 Problem 47.

48. ★★ Two concentric aluminum loops have radii $R_1 = 2.5$ cm and $R_2 = 5.0$ cm.
    (a) The current in the outside loop is changing according to $i(t) = [15 \sin (30\pi t)]$ A, and the resistance of the inside loop is 10. Ω. Derive expressions for the induced emf and the induced electric current in the inside loop.
    (b) Compare the frequency of the induced emf and the induced current with the frequency of the current in the outside loop.

49. ★★ $\frac{d}{dx}$ A copper loop in the shape of an equilateral triangle with sides of 10.0 cm is placed in a uniform and constant magnetic field, $B = 45.0$ mT, directed perpendicular to the surface of the loop (Figure 25-62). The resistance of the loop is 1.00 Ω. The loop is pulled out of the magnetic field with a constant velocity of 2.00 m/s.
    (a) What is the magnitude of the emf induced in the loop as it moves out of the magnetic field?
    (b) What is the value of induced emf at $t = 0$ s?
    (c) What is the direction of the induced current?
    (d) Find the force that needs to be exerted on the loop to pull it out with a speed of 4.00 m/s.

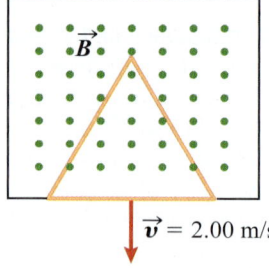

Figure 25-62 Problem 49.

50. ★★★ $\frac{d}{dx}$, ∫ A rectangular metal loop is located a distance d from a long, straight wire carrying electric current I (Figure 25-63). The short side of the loop is parallel to the wire, and the loop and the wire are located in the same plane. The electric current in the wire changes according to $i(t) = [10.0 \sin (120\pi t)]$ A. Derive an expression that describes the rate of change of the magnetic flux through the loop.

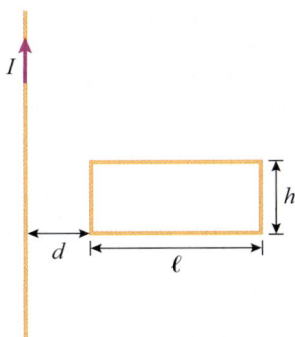

Figure 25-63 Problem 50.

51. ★★ $\frac{d}{dx}$ A circular elastic band is made out of a conducting material. Its initial radius is 5.00 cm, and it is stretched in a horizontal plane. The band is located in a uniform and constant magnetic field of $B = 50.0$ mT. The band begins expanding uniformly in all directions such that its radius changes at a rate of 0.0250 cm/s. What is the expression for induced emf in the band as a function of time?

52. ★★ $\frac{d}{dx}$ A square metal loop with dimensions 5.00 cm × 5.00 cm is located in the region of a non-uniform but constant magnetic field directed perpendicular to the surface of the loop (Figure 25-64). The magnetic field is uniformly decreasing along the direction of motion of the loop at a rate of 5.00 mT/cm. When the loop is pulled with a velocity of 1.00 m/s, as shown in the figure, what is the emf induced in it?

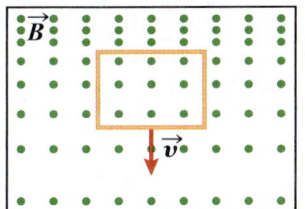

Figure 25-64 Problem 52.

53. ★★★ $\frac{d}{dx}$ A metal rod with length $\ell = 0.200$ m, mass 300. g, and cross-sectional area $A = 0.010$ m² slides down very long frictionless, tilted aluminum rails (Figure 25-65). The rod starts from rest at the top of the rails tilted at a 30.0° angle to the horizontal. The apparatus is placed in a uniform and constant vertical magnetic field $B = 0.500$ T, and the resistance of the rails and the rod is 2.00 Ω.
   (a) Describe the motion of the rod as it slides down the incline. Indicate all the forces acting on the rod.
   (b) Derive an expression for the induced emf across the rod as a function of the rod's velocity.
   (c) Derive an expression for the velocity of the rod as a function of time.
   (d) Derive an expression for the emf induced across the rod as a function of time.
   (e) Find the terminal velocity of the rod as it slides down the incline.

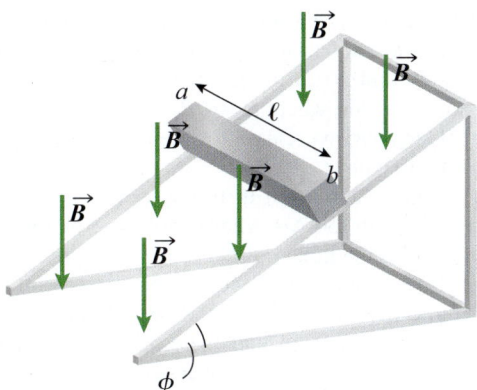

Figure 25-65 Problem 53.

54. ★★ Two coaxial air-filled solenoids, A and B, of equal length ($\ell = 20.0$ cm) and radii $R_A = 2.50$ cm and $R_B = 5.00$ cm have $N_A = 100$ turns and $N_B = 400$ turns.
   (a) The current in the inner solenoid increases from 0 A to 15.0 A in 0.0100 s. What is the induced emf in the outer solenoid during that time?
   (b) After 0.0100 s, the current in the inner solenoid becomes constant. What is the induced emf in the outer solenoid at that time?
   (c) How will your answer to part (b) change if an iron core is inserted in the solenoids?

55. ★★ Toroidal inductors (Figure 25-66) and transformers are electronic components that consist of a ring-shaped magnetic core around which a wire is wound to make an inductor. A toroidal inductor is built by winding 1000 turns of wire around a soft iron ring with a rectangular cross section of 10. mm × 15 mm. The mean radius of the inductor is 30. cm, and the magnetic permeability of the iron used in this particular inductor is $\mu = 1.2 \times 10^{-3}$ H/m.
   (a) Calculate the magnetic field inside the toroid when the electric current through it is 0.10 A.
   (b) Calculate the inductance of the toroid.
   (c) How would your answers to parts (a) and (b) change if the electric current through the inductor doubled?

 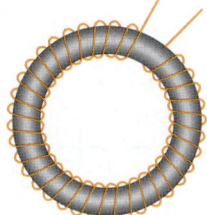

Figure 25-66 Problem 55.

56. ★★ $\frac{d}{dx}$ Two coaxial coils oriented in the same plane are arranged as shown in Figure 24-67. The larger coil has radius R_1 and N_1 turns, and the smaller coil has radius R_2 and N_2 turns, where $R_2 \ll R_1$.
  (a) Derive an expression for the mutual inductance of these coils.
  (b) When the current in the outer coil is changing at a rate of 150 A/s, what is the value of the induced emf in the inner coil?
  (c) If the inner coil is connected to the emf source, and the outer coil is not, how will the value of the mutual inductance of these coils change? How can you make sense of your answer?

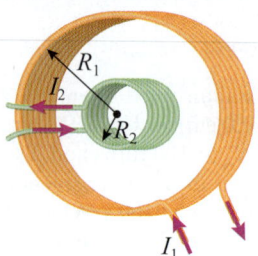

Figure 25-67   Problem 56.

57. ★ The starter motor in your car has a resistance of 0.40 Ω in its armature windings. (An armature is a conductive coil used in a motor.) The car battery has a voltage of 12 V. When the motor operates at its normal speed, it has a back emf of approximately 10. V. Estimate the amount of current this car motor draws
  (a) when running at its operating speed
  (b) during start-up
58. ★★ The switch in the circuit shown in Figure 25-68 has been open for a long time. Then, at $t = 0$ s, the switch is closed.
  (a) Derive a relationship that shows how the current through the inductor depends on time.
  (b) Derive an expression that shows how the induced emf across the inductor depends on time.
  (c) Plot the graphs of the quantities found in parts (a) and (b) as function of time. Assume the inductor is ideal.

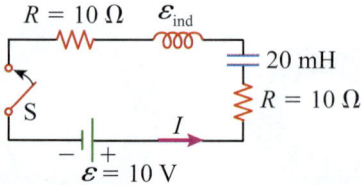

Figure 25-68   Problem 58.

59. ★★★ 〜 Faraday's law of electromagnetic induction can be used to measure the speed of a projectile. Imagine a small magnet imbedded into a projectile, as shown in Figure 25-69. The apparatus consists of two coils separated by a distance d; each coil is connected to an oscilloscope. When the magnet-carrying projectile passes through a coil, it induces an emf in the coil that is recorded by the oscilloscope. Since the time of the emf pulses can be recorded accurately, the speed of the projectile can be calculated.
  (a) Draw what you would see in an oscilloscope connected across the inductor when the projectile is approaching the coils from the left. Denote a current flowing in the clockwise direction as seen by the observer launching the projectile as positive. Clearly indicate the pulse induced in coil 1 and in coil 2.
  (b) How would your graphs change if the projectile moved twice as fast?
  (c) How would your graphs change if the number of loops in the left coil doubled, and the number of loops in the right coil remained the same?
  (d) The time separation between the pulses is 3.5 ms, and the distance between the coils is 2.0 m. What is the speed of the projectile?
  (e) The method described above is used to measure the speed of a projectile up to tens of kilometres per second. What are some limitations of this method? Why do you think it is not used to measure speeds of faster projectiles?

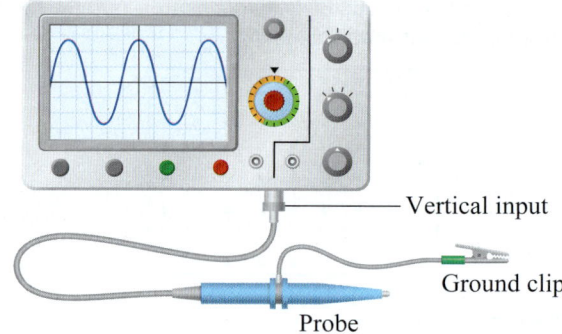

Figure 25-69   Problem 59.

60. ★★ A loop of wire, shown in Figure 25-70, is placed in a uniform but non-constant magnetic field. The magnitude and the direction of the magnetic field change continuously with time, such that at times t_1 and t_2 the magnetic flux values are $\Phi_B(t_1)$ and $\Phi_B(t_2)$, respectively.
  (a) Prove that the net electric charge that flows through the resistor during the time $t = t_2 - t_1$ can be expressed as

$$q(t) = \frac{\Phi(t_2) - \Phi(t_2)}{R}$$

  (b) Analyze the expression for electric charge for the following limiting cases: constant magnetic field, very low resistance, and very high resistance. What does the analysis of these limiting cases tell you?

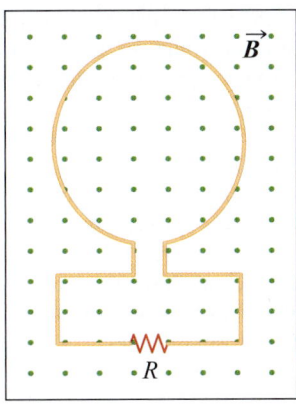

Figure 25-70 Problem 60.

61. ★★ A piece of insulated conducting wire is shaped as shown in Figure 25-71. The radius of the large circular loop is twice the radius of the small loop (4.00 cm and 2.00 cm, respectively), and the resistance of the small loop is 0.500 Ω. The wire is placed in a uniform but time-varying magnetic field. The rate of change of the magnetic field is 0.500 mT/s. Find the direction and magnitude of the electric current induced in each one of the loops.

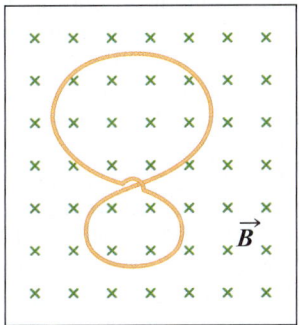

Figure 25-71 Problem 61.

62. ★★ A series RL circuit consists of a 9.0 V battery, a 5.0 Ω resistor, and a 4.0 H inductor. After the circuit has been turned on, how long will it take for the current in this circuit to reach
(a) 50% of its maximum value?
(b) 75% of its maximum value?
(c) 90% of its maximum value?
(d) 99% of its maximum value?
Supplement your algebraic solution by drawing an $I(t)$ graph for this circuit.

63. ★★ A series RL circuit consists of a 12.0 V battery, a 15.0 Ω resistor, and a 3.00 mH inductor.
(a) Complete Table 25-5 by calculating the values of the voltage across the resistor, the voltage across the inductor V_L, the current through the battery, and the voltage across the battery.
(b) How do you know if the values you found are reasonable? Assume the battery is ideal.
(c) How would the answers to the problem change if the battery were not ideal?

Table 25-5 Table for Problem 63

Time	V_R (V)	V_L (V)	I (A)	ε Battery (V)
0.00100 s				
0.0100 s				
0.100 s				
0.200 s				
0.400 s				

64. ★ You and your friend decide to design a simple magnetometer. Your device is mounted on your car and consists of a 1.00 m long metal pole (a radio antenna, e.g.) and a voltmeter that measures the potential difference across it. When the car is moving due north along a horizontal road at a speed of 60.0 km/h, your magnetometer measures the potential difference across the antenna to be 1.00 mV. You know that in the area where you were driving the direction of the magnetic field is approximately 63.0° below the horizontal. What is the magnitude of the magnetic field there?

65. ★★ A 60.0 cm long bar can slide without friction along the long horizontal rails shown in Figure 24-72. The bar and the rails have a resistance of 5.00 Ω, and the entire setup is placed in a uniform magnetic field of 3.00 mT.
(a) Derive an expression for the force needed to move the bar to the left at a constant speed.
(b) Use the expression you derived in part (a) to find the force required to pull the bar with a constant speed of 1.5 m/s.
(c) Derive the expression for the energy delivered to the resistor in t seconds when the bar is moving to the right with constant speed $\vec{v}$.
(d) How much energy was delivered to the resistor when the bar was moving with a speed of 1.50 m/s during 3.00 s?

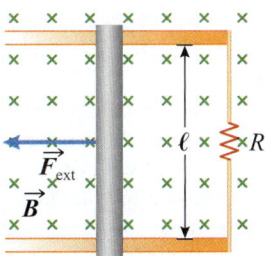

Figure 25-72 Problem 65.

66. ★★★ $\frac{d}{dx}$, $\int$ The 60.0 cm long metal bar in Figure 25-65 (see problem 53) is given an initial velocity of 2.50 m/s up the incline (up and right) and then let go.
(a) How long will it take the bar to stop moving, and how far will the bar move along the metal rails provided the friction between the rails and the bar is negligible?
(b) Provide at least two ways of solving the problem: one solution based on Newton's laws and another solution based on energy considerations.

67. ★★ The RL circuit shown in Figure 25-34 is designed such that the switch S is used to choose between having the battery as part of the circuit (the switch is in position *a*) or having the battery removed from the circuit (the switch is in position *b*). The battery has an emf of 12.0 V, the resistance of the resistor is $R = 5.00 \ \Omega$, and the inductance of the inductor is $L = 40.0$ mH. At first, the switch is moved to position *a*.

(a) How long does it take the current to reach 99% of its maximum value? What is this value?

(b) When the current reaches 99% of its maximum value, what is the potential difference across the inductor?

(c) After 25.0 s, the switch is moved to position *b*. How long will it take for the current to decrease to 50% of its maximum value?

68. ★★ A medium-sized electric generating plant produces electric energy at 50.0 A and 20.0 kV. The electricity produced by the power plant has to be transmitted 50.0 km over transmission lines, which have a resistance of 10.0 Ω/km.

(a) Compare the power output of this electric generating plant with the electric power output of the Sir Adam Beck #2 Power Plant at Niagara Falls, Ontario (1.29 GW).

(b) What is the power loss during the transmission when the energy is transmitted at 20.0 kV?

(c) To what value should the voltage output value of the generator be stepped up to reduce the energy loss by a factor of 20?

(d) How many average families can this power plant support? List and justify your assumptions.

69. ★★ An electric generator consists of a 1000-loop coil formed into a rectangle with sides of 30. cm and 60. cm. The coil is placed in a uniform magnetic field of magnitude $B = 2.5$ T.

(a) Derive an expression for the emf induced by this generator, provided the coil is spun about the axis perpendicular to the magnetic field with angular velocity ω.

(b) Find the maximum value of the emf produced by this generator when the coil is rotated about an axis perpendicular to $\vec{B}$ with $f = 1200$ rev/min.

(c) Use the expression for the induced emf to explain what can be done to the generator to change the frequency of the induced emf and/or its maximum value.

70. ★★ The RL circuit in Figure 25-73 consists of an ideal battery; a resistor, R; an inductor, L; and a switch, S, that has three different positions: *a*, *b*, and N. Initially, the switch is in position N (the circuit is open). After the circuit switch is moved to position *a* and time Δt elapses, the current in the circuit reaches 50% of its maximum value. The switch remains in position *a* for the time interval 10τ. After that, the switch is moved into position *b*.

(a) What is the time constant of this RL circuit?

(b) What are the values of the electric current in the circuit, the energy density in the coil, and the emf across the coil

  (i) when the switch has been at position *a* for a long time?

  (ii) after the switch has moved to position *b* and τ seconds have elapsed?

  (iii) after the switch has moved to position *b* and 3τ seconds have elapsed?

Support your answers with a graph of the electric current in the circuit versus time.

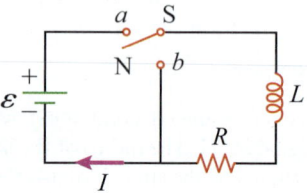

Figure 25-73 Problem 70.

71. ★★★ The value of the electric field at Earth's surface depends on the location on Earth, the time of the day, and the weather, and ranges from 40.0 V/m to 150. V/m. The magnetic field values at Earth's surface vary from 2.50×10^{-5} T to 5.70×10^{-5} T (being the highest at the poles). Compare the energy densities for the electric and magnetic fields near Earth's surface. What does your comparison tell you?

72. ★★ Magnetic resonance imaging (MRI) is a medical technique used widely in modern hospitals to produce images of the interior of the body. During an MRI scan, a patient is placed in the region of a constant magnetic field of up to 5.0 T. While operating properly, the magnetic field is kept constant. However, in case of equipment failure, the magnetic field might be shut off suddenly. This rapidly decreasing magnetic field will generate an induced emf, potentially affecting the charged particles in the body fluids and producing life-threatening electric currents. To prevent this from happening, the engineers make sure that it takes at least time τ to shut the MRI magnetic field off.

(a) Taking into account that for safety reasons, the maximum induced emf should be kept below 0.010 V and the cross-sectional area of the part of the human body exposed to the magnetic field does not exceed 0.040 m², calculate the value of τ.

(b) One of the important safety regulations for an MRI scan is that a patient or a technician never wear any metal objects. Explain why.

73. ★★ Baby breathing monitors became popular in the mid-1990s and today are widely used in homes and in hospitals. Although not a life-saving device by itself, the monitor is sensitive to the slightest movement and will sound an alarm if the baby stops breathing for more than 20.0 s. There are two main types of baby monitor sensors: the mattress pad sensor, placed under the baby's mattress, and the body sensor, which is a thin belt wrapped around the baby's chest. Both detect the changes in Earth's magnetic field flux through a coil as a result of the movement of a patient. The simplest baby body sensor monitor consists of a 300-turn coil wrapped around the baby's chest and connected to a sensitive voltmeter. Estimate the sensitivity of this voltmeter (how accurately it should be able to measure an induced emf) to measure the emf induced by a baby's breathing. Assume that the magnetic field component perpendicular to the coils is 30.0 μT, the baby's normal breathing rate is approximately 40 times a minute (much higher than the adult rate), and the smallest increase in the area of the belt around the baby's chest is 5.00 cm^2.

74. ★★ $\frac{d}{dx}$ An electric guitar uses a circular pickup coil that detects the vibrations of a steel guitar string. The pickup coil consists of a magnet and a coil with as many as 7000 turns of wire wound around it. A vibrating steel string modulates a changing magnetic flux through the pickup coil. The component of the magnetic field perpendicular to the pickup coil can be described as $B(t) = [45.00 + 3.500 \sin (2\pi 622.0t)]$ mT. The diameter of the pickup coil is 5.400 mm. Find the time dependence of the emf induced in the coil.

75. ★★★ Engineers proposed using superconducting materials in the coaxial cables for power transmission lines. For the purpose of this problem, you can think of it as the coaxial cable shown in Figure 25-74. The inner and outer wires of such a superconducting cable carry currents in opposite directions. Such a superconducting coaxial cable can carry up to 1.00 GW (1 × 10^9 W) of power at 200. kV DC over a distance of 1500 km, almost without a loss in power strength. The radius of the inner wire is 2.50 cm, and the radius of the outer wire is 5.00 cm. Calculate the value of
(a) the magnetic field at the surface of the inner conductor
(b) the magnetic field at the inner surface of the outer conductor
(c) the magnetic field at the outer surface of the outer conductor
(d) the energy density between the conductors
(e) the energy stored in the space between the conductors over the 1500 km distance
(f) the pressure exerted on the outer conductor as a result of the power transmission.

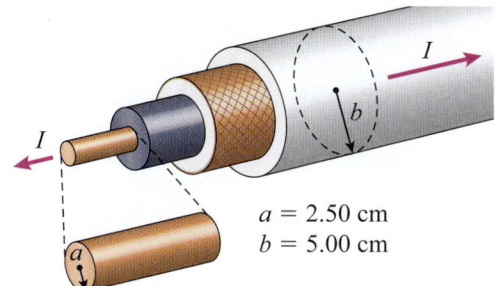

Figure 25-74 Problem 75.

76. ★★★ ▭, $\frac{d}{dx}$ The switch in the circuit shown in Figure 25-75 has been open for a long time. Then, at $t = 0$ s, the switch is closed.
(a) Derive an expression that shows how the current in the inductor, the current in the battery, and the current in the switch depend on time.
(b) Derive an expression that shows how the induced emf across the inductor depends on time.
(c) Plot the graphs for the quantities found in parts (a) and (b) as a function of time. Assume it is an ideal inductor.

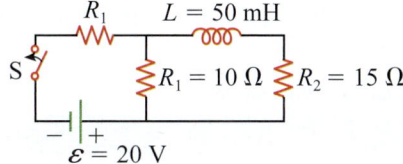

Figure 25-75 Problem 76.

77. ★★★ (a) Show that the equivalent inductance of two or more inductors can be calculated as follows:
(i) For inductors connected in series:
$$L_{eq} = L_1 + L_2 + \cdots$$
(ii) For inductors connected in parallel:
$$\frac{1}{L_{eq}} = \frac{1}{L_1} + \frac{1}{L_2} + \cdots$$
(b) Use the PhET simulations from the online activities to verify whether these formulas hold.
(c) Design more complex RL circuits, calculate their parameters, and then test your prediction using the PhET simulations incorporated in the Interactive Activities in this chapter.

78. ★★ A student in a physics lab investigates the operation of an AC generator. She collects data on the emf of the generator as a function of time (Figure 25-76). The generator consists of a 200-turn coil that is spun in a strong magnetic field. The coil has a cross-sectional area of 25.0 cm^2.
(a) Determine the frequency, f (Hz), of the generated current.
(b) Determine the frequency, f (Hz), and the angular frequency, ω (rad/s), of the rotation of the coil.
(c) Find the magnitude of the magnetic field.

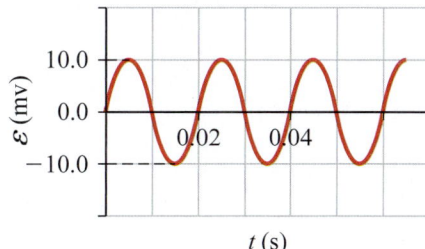

Figure 25-76 Problem 78.

DATA-RICH PROBLEM

79. ★★★ 🖥, 〰 A student in a physics lab investigates the operation of a series RL circuit. She uses a 2.000 Ω resistor, a battery, an inductor with an unknown value, and a voltmeter. After the inductor is connected to the battery (the switch is in position *a*), she measures the voltage across the inductor at different times. Her observations are recorded in Table 25-6.

   (a) Find the inductance value of the inductor.
   (b) Find the time constant of the circuit.
   (c) Find the values of the electric current at each time. (Hint: Use a spreadsheet program to answer the questions.)

Table 25-6 Data for Problem 79

Time (ms)	V_L (V)	Time (ms)	V_L (V)	Time (ms)	V_L (V)
1.000	8.598	3.500	3.737	6.000	1.624
1.500	7.278	4.000	3.163	6.500	1.375
2.000	6.161	4.500	2.678	7.000	1.164
2.500	5.215	5.000	2.267	7.500	0.985
3.000	4.414	5.500	1.919	8.000	0.834

OPEN PROBLEM

80. ★★★ $\frac{d}{dx}$ An 8.00 cm long bar magnet (Figure 25-77) is dropped into a copper coil from a height of 50.0 cm above the coil. The coil is 5.00 cm in radius, consists of 400 turns, and is 10.0 cm long, and the bar magnet has a strength of 0.200 T.

   (a) Estimate the value of the maximum emf induced in the coil. Clearly list your assumptions.
   (b) Draw the graphs of the induced emf in the coil as a function of time and distance.

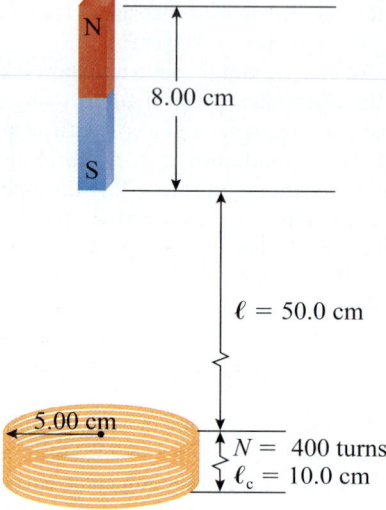

Figure 25-77 Problem 80. Not to scale.

Alternating Current Circuits

Learning Objectives

When you have completed this chapter, you should be able to

LO1 Derive expressions for the current in alternating current (AC) circuits with resistive, inductive, or capacitive loads.

LO2 Analyze LC circuits.

LO3 Use phasors and phasor diagrams to represent electric current and voltage in AC circuits.

LO4 Analyze series RLC circuits.

LO5 Determine the resonant frequency of a series RLC circuit, and describe how the current and component voltages vary with the frequency of the emf.

LO6 Calculate power dissipated in AC circuits.

According to Natural Resources Canada, the annual energy consumption of an average Canadian household in 2007, including all energy sources except fuel for transportation, was 1.06×10^{11} J $\approx$ 30 000 kW · h (distributed as shown in Figure 26-1). This annual consumption is equivalent to 3.35 kJ of energy per second year-round. Approximately 38% of this is electrical energy supplied in the form of alternating current (AC) through a nation wide power grid, a gigantic network of AC circuits. This chapter introduces the basic principles of AC circuits.

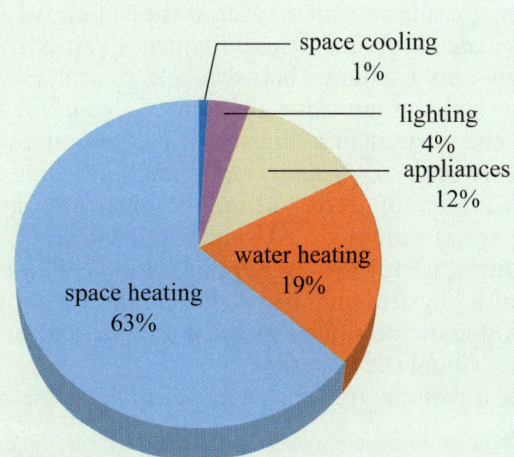

space cooling 1%
lighting 4%
appliances 12%
water heating 19%
space heating 63%

Energy Efficiency Trends in Canada 1990–2013, figure 3.3, https://www.nrcan.gc.ca/sites/www.nrcan.gc.ca/files/energy/pdf/trends2013.pdf, Office of Energy Efficiency, Natural Resources Canada. Reproduced with the permission of the Minister of Public Works and Government Services Canada, 2016.

Figure 26-1 Domestic energy usage in Canada.

26-1 Simple Loads in AC Circuits

In Chapter 25, we discussed the generation of electrical energy. Most modern power plants generate **alternating current** (AC), which is electric current that changes direction and magnitude many times a second. In North America, the frequency of AC oscillation is 60 Hz; in the European Union and many other countries, the AC frequency is 50 Hz.

In this chapter, we will use the term *voltage* for the electric potential difference across the elements of a circuit. Voltage is the more common term in electronics and engineering and has the virtue of being a more compact expression.

Due to the dependence of AC current and voltage on time, we cannot describe these quantities the same way we did for DC circuits. However, we can apply the major concepts and relationships of DC circuits, such as potential difference, electric current, resistance, and Ohm's law, to AC circuits by using some additional techniques and terminology. Unlike a DC circuit, an AC power supply produces an oscillating emf (voltage) that continually changes its magnitude and direction. Therefore, the voltage across the power supply changes with time. For AC circuits, we denote instantaneous values of current and voltage using lowercase letters: $i(t)$ and $v(t)$, respectively.

In this chapter, we discuss AC power supplies with an emf that can be described using a sine function:

KEY EQUATION $\varepsilon(t) = \varepsilon_{max}\sin(\omega t) = \varepsilon_{max}\sin(2\pi f t)$ (26-1)

where ε_{max} is the maximum value of the emf and ω and f are the angular frequency and frequency, respectively.

This emf oscillates between $+\varepsilon_{max}$ and $-\varepsilon_{max}$ and has an average value of zero. As you will see below, the current in a sinusoidal AC circuit oscillates between $+i_{max}$ and $-i_{max}$ and also has an average value of zero. Generally, the **root mean square** (rms) values of AC emfs and currents are much more useful for circuit calculations. The rms value of a physical quantity is the square root of the mean value of the square of the mathematical function describing the quantity.

For a periodic function, $f(t)$, we compute the rms value as

$$f_{rms} = \sqrt{\frac{1}{T}\int_0^T f^2(t)dt}$$ (26-2)

where T is the period. For sinusoidal quantities, we can take the rms over one period, or cycle. For emf $\varepsilon(t) = \varepsilon_{max}\sin(\omega t)$, the period of oscillation is $T = \dfrac{1}{f} = \dfrac{1}{\omega/(2\pi)} = \dfrac{2\pi}{\omega}$. So, we can substitute the sine

function into the expression for the rms value and then integrate over one period:

$$\varepsilon_{rms} = \sqrt{(\varepsilon^2)_{avg}}$$

$$= \sqrt{(\varepsilon_{max}^2\sin^2(\omega t))_{avg}}$$ (26-3)

$$= \varepsilon_{max}\sqrt{(\sin^2(\omega t))_{avg}}$$

We will compute the average of the square of the sine function separately:

$$(\sin^2(\omega t))_{avg} = \frac{\displaystyle\int_0^T \sin^2(\omega t)dt}{T}$$

$$= \frac{1}{T}\int_0^T \frac{1}{2}(1 - \cos(2\omega t))dt$$ (26-4)

$$= \frac{1}{T}\frac{T}{2}$$

$$= \frac{1}{2}$$

Thus, we put the result of Equation 26-4 into Equation 26-3 to obtain

$$\varepsilon_{rms} = \varepsilon_{max}\sqrt{\frac{1}{2}} = \frac{\varepsilon_{max}}{2}$$ (26-5)

The same method can be used to find the rms value of an alternating current or any other sine or cosine quantity.

Therefore, the rms values for the emf and current in an AC circuit, with a sinusoidal emf, can be expressed as

KEY EQUATION $$\varepsilon_{rms} = \frac{\varepsilon_{max}}{\sqrt{2}}$$ (26-6)

KEY EQUATION $$i_{rms} = \frac{i_{max}}{\sqrt{2}}$$ (26-7)

CHECKPOINT 26-1

The Root Mean Square Value of a Square Wave

Is the rms value of a square wave that goes between $+V_{max}$ and $-V_{max}$ greater than, less than, or equal to the rms value of a sine wave with the same amplitude?

ANSWER: Greater. The rms is defined as the square root of the area under one full waveform squared, divided by the period of the waveform. A square wave has an amplitude that is either $+\varepsilon_{max}$ or $-\varepsilon_{max}$, whose square is ε_{max}^2, so the area is $\varepsilon_{max}^2 T$. When we divide by the period and take the square root, we obtain $\varepsilon_{rms} = \varepsilon_{max}$ for a square wave.

Circuit Construction

In this activity, you will use the PhET simulation "Circuit Construction Kit (AC+DC)" to explore the relationships between current and voltage in simple AC circuits that contain just one of a resistor, a capacitor, and an inductor.

Figure 26-2 An AC circuit with a resistive load.

Resistive Load

Let us start with an AC circuit that consists simply of an AC power supply connected to a resistor, as shown in Figure 26-2.

Just as with DC circuits, we can apply Kirchhoff's loop rule to obtain

$$\varepsilon(t) - iR = 0 \qquad (26\text{-}8)$$

Solving for i and using Equation 26-1 for the emf, we find

$$i(t) = \frac{\varepsilon(t)}{R} = \frac{\varepsilon_{max}}{R}\sin(\omega t) = i_{max}\sin(\omega t) \qquad (26\text{-}9)$$

Not surprisingly, the current in this circuit is also sinusoidal, and the amplitude of the current is related to the amplitude of the emf:

$$i_{max} = \frac{\varepsilon_{max}}{R} \qquad (26\text{-}10)$$

Equation 26-10 has the same form as Ohm's law, $I = \dfrac{V}{R}$: current equals voltage divided by resistance. Figure 26-3 shows a plot of the emf and the resulting current versus time.

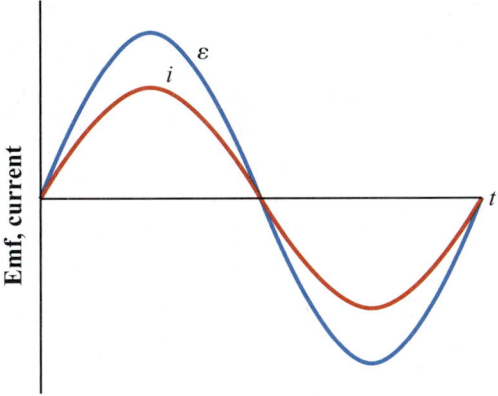

Figure 26-3 The emf and current in an AC circuit with only a resistor. Note that they are in phase.

The voltage that appears across the resistor is given by

$$v_R(t) = i(t)R = i_{max}R\sin(\omega t) \qquad (26\text{-}11)$$

From this equation, we can identify the maximum value of the voltage across the resistor:

KEY EQUATION $$V_{R,max} = i_{max}R \qquad (26\text{-}12)$$

EXAMPLE 26-1

Maximum and Root Mean Square Current through a Resistor

A 1.00 kΩ resistor is connected across an emf that is given by $\varepsilon(t) = 20.0\sin(50.0t)$. Find an expression for the current as a function of time and the rms value of the current.

Solution

In a purely resistive AC circuit, the current is in phase with the emf and has the same angular frequency. Therefore, to write an expression for the current, we need to find the amplitude of the current. We can use Equation 26-10 for i_{max}:

$$i_{max} = \frac{\varepsilon_{max}}{R} = \frac{20.0\text{ V}}{1.00\text{ k}\Omega} = 0.020\text{ A}$$

We can then use Equation 26-9 to write

$$i(t) = 0.020\sin(50.0t)$$

To find the rms value is a straightforward application of Equation 26-7:

$$i_{rms} = \frac{i_{max}}{\sqrt{2}} = \frac{0.020\text{ A}}{\sqrt{2}} = 0.014\text{ A}$$

Making sense of the result

The magnitude of the current is given by Ohm's law. We have a 20.0 V source across a 1.00 kΩ resistor, so we would expect 20.0 mA or 0.020 A.

Inductive Load

We now consider an AC emf connected to just an inductor, as shown in Figure 26-4. We call such a circuit an AC circuit with an inductive load.

Figure 26-4 An AC circuit with an inductive load.

Again applying Kirchhoff's loop rule, we obtain

$$\varepsilon(t) - L\frac{di}{dt} = 0 \qquad (26\text{-}13)$$

Solving for the derivative of the current, we find

$$\frac{di}{dt} = \frac{\varepsilon(t)}{L} = \frac{\varepsilon_{max}}{L}\sin(\omega t) \qquad (26\text{-}14)$$

A solution to this differential equation is

$$i(t) = -\frac{\varepsilon_{max}}{\omega L}\cos(\omega t) = -i_{max}\cos(\omega t) \qquad (26\text{-}15)$$

We note that $i_{max} = \dfrac{\varepsilon_{max}}{\omega L}$ has the form of Ohm's law—a current is given by a voltage divided by another quantity. The quantity ωL does indeed have units of ohms, and we define the **inductive reactance** as

KEY EQUATION

$$X_L = \omega L \qquad (26\text{-}16)$$

If we plot the emf and the current through the inductor, as shown in Figure 26-5, we can see that the current peaks a quarter cycle after the emf. Thus, the current *lags* the emf.

Since $\sin\left(\omega t - \dfrac{\pi}{2}\right) = -\cos(\omega t)$, we can rewrite the current as

$$i(t) = i_{max}\sin\left(\omega t - \frac{\pi}{2}\right) \qquad (26\text{-}17)$$

The maximum voltage that appears across the inductor is given by

KEY EQUATION

$$V_{L,max} = i_{max}X_L \qquad (26\text{-}18)$$

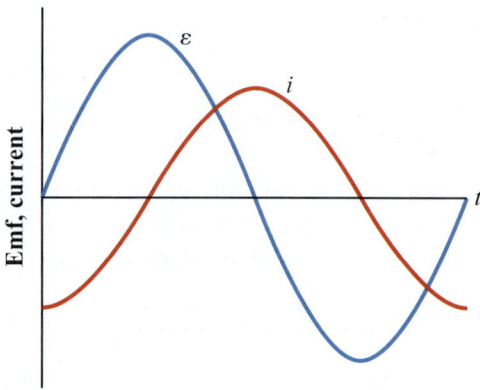

Figure 26-5 Plot of the emf and the current for an AC circuit with an inductive load. Note that the voltage leads the current. This means that the maximum value of the emf occurs earlier in the cycle than the maximum value of the current.

Capacitive Load

We now consider an AC voltage source connected to just a capacitor, as shown in Figure 26-6.

Applying Kirchhoff's loop rule, we obtain

$$\varepsilon(t) - \frac{q}{C} = 0 \qquad (26\text{-}19)$$

Solving for q, the charge on the capacitor, gives

$$q(t) = C\varepsilon(t) = C\varepsilon_{max}\sin(\omega t) \qquad (26\text{-}20)$$

Since $i(t) = dq/dt$,

$$
\begin{aligned}
i(t) &= \frac{d}{dt}\left(C\varepsilon_{max}\sin(\omega t)\right) \\[4pt]
&= C\varepsilon_{max}\frac{d\sin(\omega t)}{dt} \qquad (26\text{-}21)\\[4pt]
&= \omega C\varepsilon_{max}\cos(\omega t)
\end{aligned}
$$

In Equation 26-21, we see that the maximum value of the current is given by $i_{max} = \omega C\varepsilon_{max}$, which is not quite in the form of Ohm's law, where we would expect a voltage divided by a resistance. However, the quantity ωC does have units of Ω^{-1}. If we define the **capacitive reactance** as

KEY EQUATION

$$X_C = \frac{1}{\omega C} \qquad (26\text{-}22)$$

we can write

$$i_{max} = \frac{\varepsilon_{max}}{X_C} \qquad (26\text{-}23)$$

Figure 26-6 An AC circuit with a capacitive load.

ELECTRICITY, MAGNETISM, AND OPTICS

EXAMPLE 26-2

Current in an Inductor

A 10.0 mH inductor is connected across an emf $\varepsilon(t) = 50.0 \sin(20.0\,t)$. Find

(a) the inductive reactance
(b) the maximum current through the inductor

Solution

(a) We will use Equation 26-16 for the inductive reactance. From the given emf, we see that $\omega = 20.0$ rad/s; thus,

$$X_L = \omega L = (20.0 \text{ rad/s})(1.0 \times 10^{-2} \text{ H}) = 0.200 \ \Omega$$

(b) Noting that $\varepsilon_{max} = 50.0$ V, we can find the maximum current:

$$i_{max} = \frac{\varepsilon_{max}}{X_L} = \frac{50.0 \text{ V}}{0.200 \ \Omega}$$

$$= 0.250 \text{ kA}$$

Making sense of the result

The inductive reactance is very small, so we have a very large current. In practice, an inductor also has some resistance due to the wire it is wound from, which will limit the current somewhat.

The expression for AC current again has the same form as Ohm's law, and the maximum voltage that appears across the capacitor is given by

$$V_{C,max} = i_{max} X_C \qquad (26\text{-}24)$$

When we plot the emf and the current in this circuit, as shown in Figure 26-7, we again see that the maximum values of the emf and the current do not occur at the same time—they are not in phase. However, in this case, the maximum value of the current is reached before the maximum value of the emf, and we say that the current leads the emf by a quarter cycle.

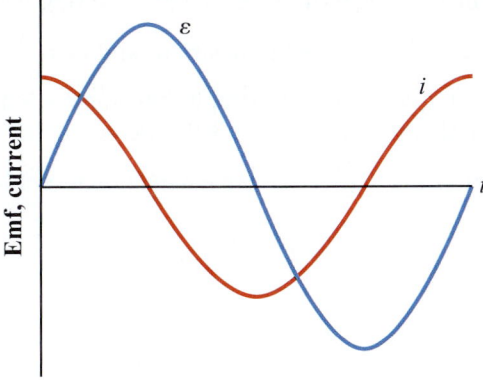

Figure 26-7 Plot of the emf and the current for an AC circuit with a capacitive load. Note that the current leads the emf.

Again, we can rewrite the current in terms of a sine function. Since $\sin\left(\omega t + \dfrac{\pi}{2}\right) = \cos(\omega t)$,

$$i(t) = i_{max}\sin\left(\omega t + \frac{\pi}{2}\right) \qquad (26\text{-}25)$$

In contrast to the inductor current, the phase difference in the argument of the sine function of the current is positive.

Talking about current "in" or "through" capacitors requires some clarification. We know that a capacitor is two conductors separated by an insulator—typically a dielectric. No current can flow through the dielectric. However, current can flow into one side of a capacitor, and the charge builds up on the plate. At the same time, current is flowing out of the opposite plate, leaving behind a buildup of negative charge. In a DC circuit, this buildup of charge is what eventually causes the current to stop in the circuit. In the AC case, the current flows back and forth, repeatedly charging and discharging the capacitor. So, although the charge does not cross over between the plates, whatever current flows onto the positively charged plate is exactly matched by a current flowing off the negatively charged plate, so in that sense, the current flows "through" the capacitor.

EXAMPLE 26-3

Current in a Capacitor

A 1.00 nF capacitor is connected across an emf $\varepsilon(t) = 30.0 \sin(40.0\,t)$. Find

(a) the capacitive reactance
(b) the maximum current through the capacitor

Solution

(a) We will use Equation 26-22 for the capacitive reactance. From the emf, we can see that $\omega = 40.0$ rad/s; thus,

$$X_C = \frac{1}{\omega C} = \frac{1}{(40.0 \text{ rad/s})(1.00 \times 10^{-9} \text{ F})}$$

$$= 0.250 \text{ G}\Omega$$

(b) Noting from the emf that $\varepsilon_{max} = 30.0$ V, we can find the maximum current:

$$i_{max} = \frac{\varepsilon_{max}}{X_C} = \frac{30.0 \text{ V}}{0.250 \text{ G}\Omega}$$

$$= 1.20 \ \mu\text{A}$$

Making sense of the result

In this case, the capacitive reactance is quite large, so we have a small current. The capacitive reactance is large because both the capacitance and the angular frequency are small.

The introduction of either an inductor or a capacitor into an AC circuit produces the interesting result that the current is not in phase with the emf. Such components are called **reactive circuit elements**, and the phase difference between the emf and the current is called the **phase shift**.

The magnitude of the current observed in an AC circuit depends on the angular frequency and the values of the reactive circuit components. Inductors have large reactances at high angular frequency, while capacitors have large reactances at low angular frequency.

The large reactance of an inductor at high angular frequency can be understood in terms of Faraday's law. At higher angular frequencies, the rate of change of the magnetic flux, and hence the induced voltage across the inductor, is larger than at lower angular frequencies.

At low angular frequency, a capacitor has more time to charge, and the voltage across it is proportional to the charge on the capacitor.

PEER TO PEER

I use the phrase "ELI the ICE man" to help me remember that the Emf comes before the I (current) in an L (inductor), while the I (current) comes before the Emf in a Capacitor.

26-2 The LC Circuit

We will now consider the simple combination of a capacitor and an inductor, called an **LC circuit**, shown in Figure 26-8. Instead of using an AC emf, we will charge the capacitor with a battery. We will remove it from the battery and connect it to an inductor. We will assume that the conductors and components in this circuit have no resistance. Once the capacitor has been fully charged to some initial charge q_0 by the battery, we open switch S_1 to disconnect the battery. At time $t = 0$, we then close switch S_2 to connect the inductor across the capacitor.

We apply Kirchhoff's law around the loop containing the capacitor and the inductor to find

$$\frac{q}{C} - L\frac{di}{dt} = 0 \qquad (26\text{-}26)$$

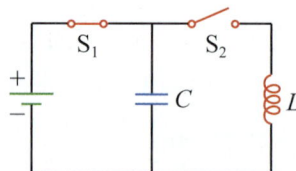

Figure 26-8 An LC circuit with the capacitor fully charged.

An LC Circuit

In this activity, you will use the PhET simulation "Circuit Construction Kit (AC+DC)" to explore the relationship between L, C, and the oscillation frequency in an LC circuit.

Since $i = -\dfrac{dq}{dt}$, $\dfrac{di}{dt} = -\dfrac{d^2q}{dt^2}$, which we substitute into Equation 26-26:

$$\frac{q}{C} + L\frac{d^2q}{dt^2} = 0 \qquad (26\text{-}27)$$

$$\frac{d^2q}{dt^2} = -\frac{1}{LC}q$$

This equation has the same form as a simple harmonic oscillator:

$$a(t) = -\omega^2(A\cos(\omega t + \phi)) = -\omega^2 x(t) \qquad (13\text{-}18)$$

Just as for simple harmonic oscillators (Chapter 13), Equation 26-27 has a solution of the form

$$q(t) = q_0\cos(\omega_0 t) \qquad (26\text{-}28a)$$

where

KEY EQUATION $$\omega_0 = \frac{1}{\sqrt{LC}} \qquad (26\text{-}28b)$$

So, the charge on the capacitor, and hence the current in the circuit, oscillates with a frequency that is determined by the values of the inductor and the capacitor.

The LC circuit is analogous to the oscillating mass–spring system in Chapter 13. Inductance plays the role of the mass, and the reciprocal of the capacitance plays the role of the spring constant. We can see the similarity between the two systems by comparing Equations 26-27 and 13-18.

It is useful to look at the two equations together. Equation 13-18 is written as $a(t) = -\omega^2 x(t)$, where a is the acceleration, x is the position, and $\omega = \dfrac{k}{m}$ is the angular frequency of the simple harmonic oscillator. Recalling that the acceleration is the second derivative of the position, $a(t) = \dfrac{d^2x}{dt^2}$, we can rewrite Equation 13-18 as

$$\frac{d^2x}{dt^2} = -\frac{k}{m}x$$

Compare this to Equation 26-27:

$$\frac{d^2q}{dt^2} = -\frac{1}{LC}q \qquad (26\text{-}27)$$

EXAMPLE 26-4

An LC Circuit

A 1.0 μF capacitor charged with 10. μC is connected across a 10. mH inductor.

(a) What is the initial voltage across the capacitor?
(b) What is the angular frequency of the oscillation?
(c) Plot the voltage across the capacitor and the voltage across the inductor versus time.

Solution

Finding the initial voltage across the capacitor is a straightforward application of Equation 22-2. Similarly, the angular frequency can be found simply from Equation 26-28b.

Plotting the voltages is a little trickier. The voltage across the capacitor is related to the charge on the capacitor, which is given by Equation 26-28a.

The voltage across the inductor is proportional to the negative of the derivative of the current. The current is proportional to the negative of the derivative of the charge on the capacitor. Thus, the voltage across the inductor is proportional to the second derivative of the charge on the capacitor.

This problem is a relatively straightforward calculation. We can determine the initial voltage across the capacitor, and we know it will then be a cosine function with angular frequency given by Equations 26-28. We also know that the voltage across the inductor will be exactly equal but opposite to that of the capacitor. However, it is useful to do the calculation in detail.

(a) We use $CV = q$ for a capacitor to solve for V:

$$V = \frac{q}{C} = \frac{10. \ \mu\text{C}}{1.0 \ \mu\text{F}} = 10. \text{ V}$$

(b) We calculate the angular frequency of oscillation using Equation 26-28b:

$$\omega_0 = \frac{1}{\sqrt{LC}} = \frac{1}{\sqrt{10. \text{ mH} \times 1.0 \ \mu\text{F}}} = 1.0 \times 10^4 \text{ s}^{-1}$$

(c) The voltage across the capacitor is found from

$$v_C(t) = \frac{q(t)}{C} = \frac{10. \ \mu\text{C} \cos(\omega_0 t)}{1.0 \ \mu\text{F}} = 10. \cos\left[(1.0 \times 10^4 \text{ s}^{-1}) \cdot t\right] \text{ V}$$

The voltage across the inductor is given by

$$v_L(t) = L\frac{d^2q}{dt^2}$$

$$= -Lq_0\omega_0^2 \cos(\omega_0 t)$$

$$= -10. \text{ mH} \cdot 10. \ \mu\text{C} \cdot (1.0 \times 10^4 \text{ s}^{-1})^2 \cos\left[(1.0 \times 10^4 \text{ s}^{-1})t\right]$$

$$= -10. \cos\left[(1.0 \times 10^4 \text{ s}^{-1})t\right]$$

The plot of these voltages in Figure 26-9 shows that they are exactly out of phase.

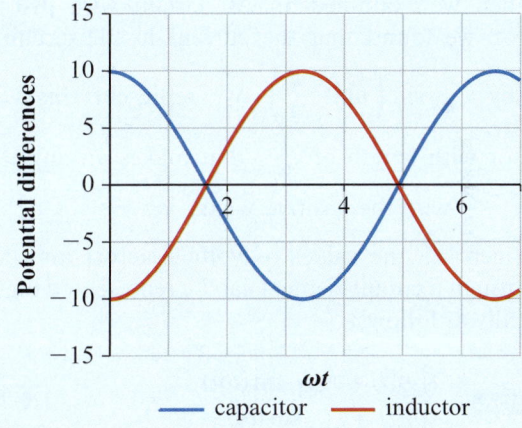

Figure 26-9 Capacitor and inductor voltages versus time.

Making sense of the result

At any instant in time, the voltage across the capacitor is exactly opposite to the voltage across the inductor. This is consistent with Kirchoff's loop rule.

From a mathematical perspective, these two equations are identical—the second derivative of a function is given by the negative of the function itself multiplied by a constant. The solution to equations of this form is always that of a simple harmonic oscillator.

26-3 Phasors

As we saw in Section 26-1, the phases of the AC current and voltage sine functions can shift relative to each other depending on the nature of the load in the circuit. A useful way to represent the various quantities in AC circuits is by using something called a phasor.

CHECKPOINT 26-3

How to Change the Oscillation Frequency

When we increase the capacitance, does the oscillation frequency of an LC circuit
(a) increase?
(b) decrease?
(c) stay the same?

ANSWER: The oscillation frequency decreases when we increase the capacitance because the capacitance is in the denominator of the expression.

A **phasor** is a two-dimensional vector. The tail of the vector is located at the origin of the coordinate system, and the vector has a length that is equal to the amplitude of the quantity we wish to represent. The angle that the phasor makes with the horizontal axis is time dependent, and, as a consequence, the phasor rotates about the origin. We will choose counter-clockwise to be positive. The rate at which the phasor rotates is determined by the angular frequency, ω. We can determine the instantaneous value of the quantity the phasor is being used to represent by taking the projection of the phasor onto the vertical axis. For example, the phasor representing the emf in an AC circuit has a length of ε_{max} and makes an angle of $\theta = \omega t$ with the x-axis. The projection of the phasor on the vertical axis is then $\varepsilon_{max} \sin(\omega t)$, which is $\varepsilon(t)$, as given by Equation 26-1. This is shown in Figure 26-10.

When we examined an AC circuit with just an inductor, we found that the current in the circuit is given by $i_{max} \sin\left(\omega t - \dfrac{\pi}{2}\right)$. We represent this with a phasor with length of i_{max} that makes an angle of $\theta = \omega t - \dfrac{\pi}{2}$ with the positive x-axis.

In general, the values of voltage across and current through a circuit element can be represented mathematically as follows:

KEY EQUATION
$$\begin{cases} \varepsilon(t) = \varepsilon_{max} \sin(\omega t) \\ i(t) = i_{max} \sin(\omega t - \phi) \end{cases} \quad (26\text{-}29)$$

where ϕ is the phase shift or phase angle. The phase angle is the angle the emf phasor makes relative to the current phasor.

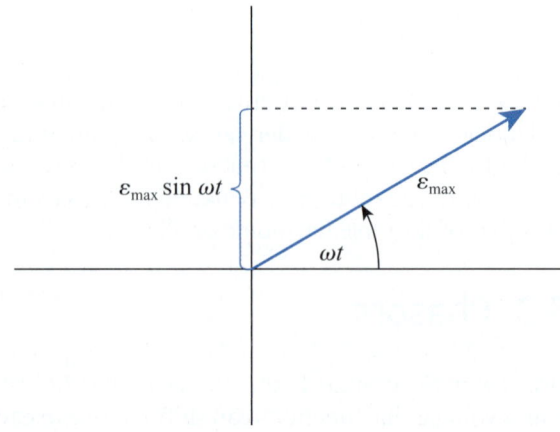

Figure 26-10 A phasor representing the voltage associated with an emf given by $\varepsilon(t) = \varepsilon_{max} \sin(\omega t)$. The projection of the phasor onto the vertical axis is what we would measure at any instant in time.

Phasors and Phase Shifts

(a) Match the phasor diagrams on the left to the appropriate voltage and current versus time diagrams on the right in Figure 26-11.
(b) Which of the phasor diagrams corresponds to a capacitive circuit?

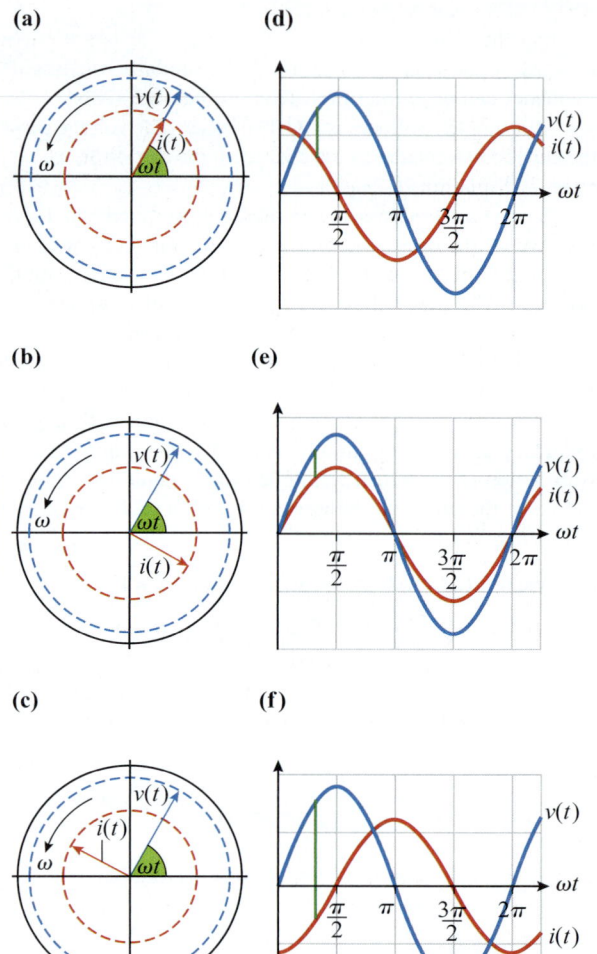

Figure 26-11 Match the phasor diagrams (a), (b) and (c), to their corresponding current-voltage plots (d), (e) and (f).

ANSWER: (a) In diagram A, the current and voltage phasor are parallel, indicating that they are in phase. The voltage and current are in phase in diagram E on the right. In diagram B, the voltage phasor leads the current phasor by $\dfrac{\pi}{2}$ rad. We can see that the instantaneous value of the current is negative and increasing toward zero, while the voltage is positive and increasing toward a maximum value. In figure F, for $0 \le \omega t \le \dfrac{\pi}{2}$, the current is negative and increasing, while the voltage is positive and increasing. Therefore, B matches with F. We could pair C and D simply by elimination, but let's repeat our analysis. The current phasor leads the voltage phasor by $\dfrac{\pi}{2}$ rad, and, when $0 \le \omega t \le \dfrac{\pi}{2}$, the current is positive and decreasing, while the voltage is positive and increasing, as in D.

(b) For a purely capacitive circuit, the current leads the voltage, so phasor diagram C represents a capacitive circuit.

26-4 Series RLC Circuits

We now consider a circuit with an AC emf, a resistor, an inductor, and a capacitor in series, as shown in Figure 26-12. This type of circuit is called a **series RLC circuit**.

Kirchhoff's loop rule (Chapter 23, Section 23-4) requires that the instantaneous voltages around the circuit loop add to zero. We can use phasors in this process to determine the magnitude and phase of the current in the circuit. We know the phase of the voltages across the resistor, inductor, and capacitor relative to the current. Specifically, the resistor voltage is in phase with the current, the inductor voltage leads the current, and the capacitor voltage lags the current. Because all the circuit elements are connected in series, the same current flows through them. Consequently, we can draw a phasor diagram for the circuit by first placing a phasor that represents the current on the circuit. This phasor makes an angle of $\omega t - \phi$ with respect to the positive x-axis, as shown in Figure 26-13.

Next, we use the voltage–current phase relationships to place phasors for the resistor, inductor, and capacitor voltages, as shown in Figure 26-14.

By Kirchhoff's loop rule, the vector sum of the voltage phasors must equal the AC emf phasor, as shown in Figure 26-15.

In the right-angled triangle ABC in Figure 26-15, the emf phasor is the hypotenuse. The other two sides are given by the magnitude of the resistor voltage phasor and the difference in magnitude of the inductor voltage phasor and the capacitor voltage phasor. Applying the Pythagorean theorem gives

$$\varepsilon^2_{max} = v^2_{R,\,max} + (v_{L,max} - v_{C,max})^2 \qquad (26\text{-}30)$$

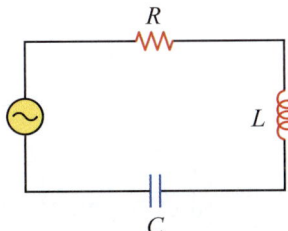

Figure 26-12   A series RLC circuit.

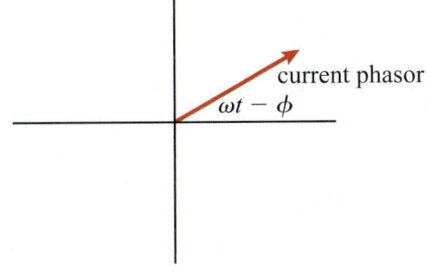

Figure 26-13   A current phasor for the series RLC circuit.

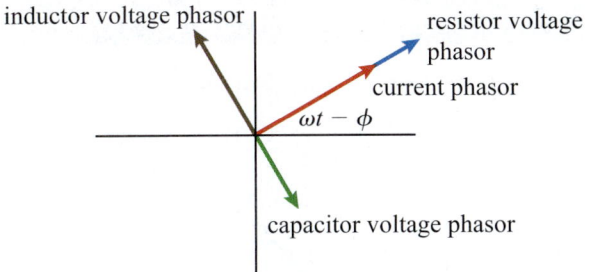

Figure 26-14   Current and voltage phasors diagrams for the RLC circuit.

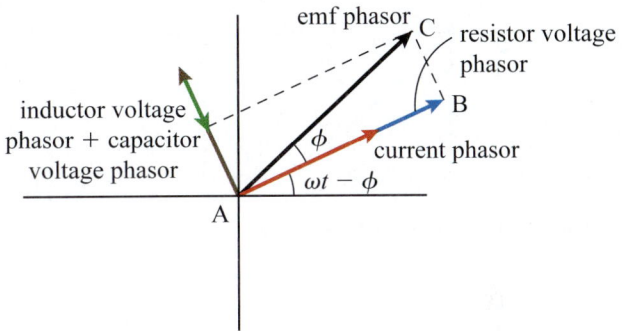

Figure 26-15   The sum of the phasors for the resistor, inductor, and capacitor voltages gives the AC emf phasor.

Substituting from Equations 26-12, 26-18, and 26-24, we get

$$\varepsilon^2_{max} = (i_{max} R)^2 + (i_{max} X_L - i_{max} X_C)^2 \qquad (26\text{-}31)$$

Solving for i_{max} gives

$$i_{max} = \frac{\varepsilon_{max}}{\sqrt{R^2 + (X_L - X_C)^2}} \qquad (26\text{-}32)$$

Equation 26-32 gives us the magnitude of the maximum current in the circuit as a function of the magnitude of the maximum AC emf, the resistance, the capacitive reactance, and the inductive reactance. Note that both the capacitive and inductive reactance depend on the frequency of the AC emf; consequently, so does the magnitude of the maximum current.

We can see that Equation 26-32 looks similar to Ohm's law in that a current is given by a ratio of a voltage to a quantity that has units of ohms. We define **impedance**, Z, as

KEY EQUATION $$Z = \sqrt{R^2 + (X_L - X_C)^2} \qquad (26\text{-}33)$$

Now that we have found an expression for the magnitude of the maximum current, we turn our attention to the phase angle. From Figure 26-15, we can see that the tangent of the angle ϕ is given by

KEY EQUATION

$$\tan \phi = \frac{i_{max}(X_L - X_C)}{i_{max}R} = \frac{X_L - X_C}{R} \qquad (26\text{-}34)$$

The Sign of the Phase Angle

In Figure 26-16, is the phase angle positive or negative? Explain.

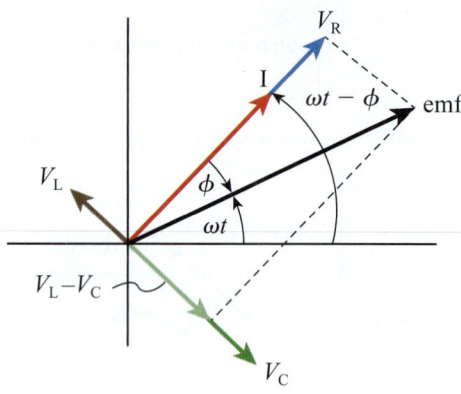

Figure 26-16 Checkpoint 26-5.

ANSWER: Negative. Recall that the phase angle is defined to be the angle between the current and the applied emf, measured relative to the current. We can see from the figure that the current is ahead (more counter-clockwise) of the emf, so the phase angle must be negative. Another way of looking at it is to recall that the emf makes an angle of ωt with the x-axis, and the current makes an angle of $\omega t - \phi$ with the x-axis. The only way for $\omega t - \phi > \omega t$ is if $\phi < 0$. Or, since $V_C > V_L$, we have $X_C > X_L$. Then, from Equation 26-34, the phase angle must be negative.

The maximum voltages that appear across the components in an AC circuit reflect their reactances. The maximum voltage across the inductor in Example 26-5 is quite small, and the maximum voltage that appears across the capacitor is relatively large. When we add the maximum voltages, we get

$$v_{R,max} + v_{L,max} + v_{C,max} = 65.0 \text{ V} + 5.20 \text{ V} + 81.1 \text{ V}$$

$$= 151.3 \text{ V}$$

This total is much greater than the maximum applied voltage of 100. V. How can this be? Remember that the voltages that appear across these elements are time dependent and are not in phase. The three voltages do not all reach their maximum voltage simultaneously, as we can see from a plot of the voltages versus time. The time-dependent voltages across the RLC circuit elements are given by

$$v_R(t) = Ri(t) = Ri_{max}\sin(\omega t - \phi)$$

$$v_L(t) = L\frac{di}{dt} = \omega L i_{max}\cos(\omega t - \phi)$$

$$= X_L i_{max}\cos(\omega t - \phi) \qquad (26\text{-}35)$$

$$v_C(t) = \frac{q(t)}{C} = -\frac{1}{\omega C}i_{max}\cos(\omega t - \phi)$$

$$= -X_C i_{max}\cos(\omega t - \phi)$$

EXAMPLE 26-5

Analyzing a Series RLC Circuit

A series RLC circuit consists of an emf $\varepsilon = (1.00 \times 10^2 \text{ V})\sin((8.00 \times 10^3 \text{ rad} \cdot \text{s}^{-1})t)$, a $1.00 \times 10^2 \ \Omega$ resistor, a 1.00 mH inductor, and a 1.00 μF capacitor. Determine

(a) the maximum current flow
(b) the phase shift
(c) the maximum voltages across the three components

Solution

(a) We will first determine the impedance using Equation 26-33:

$$Z = \sqrt{R^2 + (X_L - X_C)^2}$$

$$= \sqrt{(1.00 \times 10^2 \ \Omega)^2 + \left(8.00 \times 10^3 \text{ rad} \cdot \text{s}^{-1} \cdot 1.00 \text{ mH} - \frac{1}{(8.00 \times 10^3 \text{ rad} \cdot \text{s}^{-1}) \cdot 1.00 \ \mu\text{F}}\right)^2}$$

$$= 1.54 \times 10^2 \ \Omega$$

Using this result, we can determine i_{max}:

$$i_{max} = \frac{\varepsilon_{max}}{Z} = \frac{1.00 \times 10^2 \text{ V}}{1.54 \times 10^2 \ \Omega} = 0.650 \text{ A}$$

(b) The phase shift can be determined using Equation 26-34:

$$\tan\phi = \frac{X_L - X_C}{R}$$

$$= \frac{(8.00 \times 10^3 \text{ rad} \cdot \text{s}^{-1})(1.00 \text{ mH}) - \frac{1}{(8.00 \times 10^3 \text{ rad} \cdot \text{s}^{-1})(1.00 \ \mu\text{F})}}{1.00 \times 10^2 \ \Omega} = -1.17$$

Therefore, $\phi = \tan^{-1}(-1.17) = -49.5°$.

(c) We find the maximum voltages across the resistor, the inductor, and the capacitor from Equations 26-14, 26-18, 26-24, respectively:

$$v_{R,max} = i_{max}R = (0.650 \text{ A})(1.00 \times 10^2 \text{ } \Omega) = 65.0 \text{ V}$$

$$v_{L,max} = i_{max}X_L = (0.650 \text{ A})(8.00 \times 10^3 \text{ rad} \cdot \text{s}^{-1})(1.00 \text{ mH}) = 5.20 \text{ V}$$

$$v_{C,max} = i_{max}X_C = (0.649 \text{ A}) \frac{1}{(8.00 \times 10^3 \text{ rad} \cdot \text{s}^{-1})(1.00 \text{ } \mu\text{F})} = 81.2 \text{ V}$$

Making sense of the result

We can get a sense of what is going on in the circuit from the reactances:

$$X_L = \omega L = (8.00 \times 10^3 \text{ rad} \cdot \text{s}^{-1})(1.00 \text{ mH}) = 8.00 \text{ } \Omega$$

$$X_C = \frac{1}{\omega C} = \frac{1}{(8.00 \times 10^3 \text{ rad} \cdot \text{s}^{-1})(1.00 \text{ } \mu\text{F})} = 1.25 \times 10^2 \text{ } \Omega$$

The capacitive reactance is much larger than the inductive reactance but comparable to the resistance. As such, we would expect the circuit to behave somewhere in between a purely resistive circuit and a purely capacitive circuit. In a purely resistive circuit, the current is in phase with the applied emf. In a purely capacitive circuit, the current leads the applied emf. For this particular circuit, we found a phase angle of $-49.5°$. The minus sign indicates that the current is leading the emf, and the 49.5° angle indicates that the circuit is somewhere between a purely resistive circuit ($\phi = 0$) and a purely capacitive circuit ($\phi = -90°$), as we predicted from the reactances.

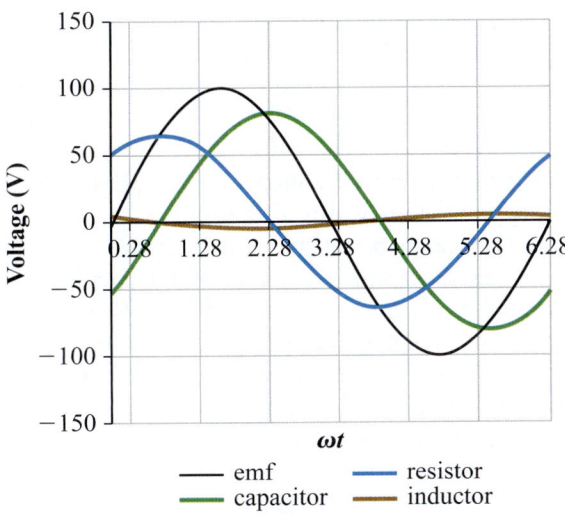

Figure 26-17 Voltages versus time for the RLC circuit of Example 26-5.

In Figure 26-17, we plot the voltages across the various elements in the circuit versus time along with the AC emf. The voltages are indeed out of phase, and the component voltages do sum to the AC emf at any given instant.

26-5 Resonance

We will now consider how the maximum current and the phase shift vary in a series RLC circuit as we change the frequency of the emf. The inductive reactance increases as we increase the frequency. On the other hand, the capacitive reactance increases when we decrease the frequency. At low frequencies, we would expect the capacitor to dominate the behaviour of the circuit, while at high frequencies, we would expect the

inductor to dominate. In Figure 26-18, we plot i_{max} and ϕ versus angular frequency for a series RLC circuit with $R = 100 \text{ } \Omega$, $L = 1 \text{ mH}$, and $C = 1 \text{ } \mu\text{F}$. The applied emf has an amplitude of 100 V.

We can see that the current tends to small values of much less than 1 A at low and high angular frequencies and peaks at 1 A at some intermediate angular frequency. The phase starts off at $-\pi/2$ at low angular frequency and approaches $\pi/2$ at high angular frequency. The phase goes through zero at the angular frequency where the peak in the current occurs. The point at which the phase goes to zero is determined by the numerator in Equation 26-34, that is, when $X_L = X_C$. The angular frequency at which the inductive and capacitive reactances are equal is the **resonance frequency**, ω_0, for this circuit.

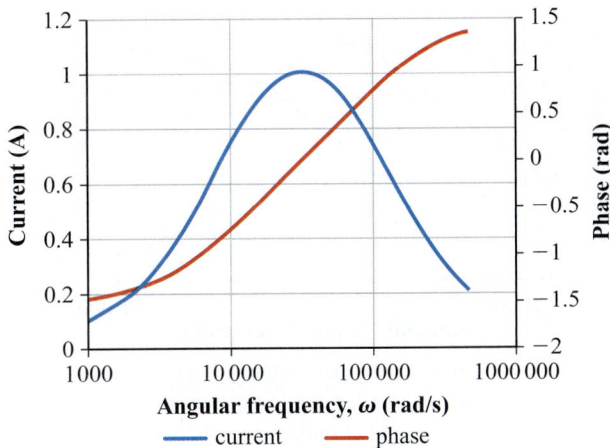

Figure 26-18 Current (left axis) and phase (right axis) versus angular frequency for an RLC circuit.

We can equate the expressions for these reactances and solve for the resonant frequency:

$$\omega_0 = \frac{1}{\sqrt{LC}} \qquad (26\text{-}36)$$

Note that the resonant frequency is the same as the oscillation frequency of the LC circuit with no resistance in Section 26-2. The driven RLC circuit is analogous to a driven damped harmonic oscillator, with the resistance acting as the damping term. In Figure 26-19, we plot the current in a driven RLC circuit with various values of resistance while holding the inductance and the capacitance constant. As the resistance decreases, the resonance peak gets taller and narrower.

Now consider what happens to the voltages that appear across the various circuit elements as we vary the frequency. We plot the maximum resistor, capacitor, and inductor voltages versus frequency for $R = 50\ \Omega$, $L = 1$ mH, and $C = 1\ \mu$F in Figure 26-20. We can see that the voltage across the resistor has the same shape as the current, which we would expect from Ohm's law. At low frequencies, the voltage across the capacitor becomes large,

which is reasonable because the capacitive reactance gets large at low frequencies. Conversely, at high frequencies, the voltage across the inductor becomes relatively large because the inductive reactance gets large at high frequencies.

Remember that Figure 26-20 shows voltage *amplitudes*, which do not generally add to the applied emf of 100 V. However, at resonance, the voltage across the resistor is exactly 100 V, and the voltages across the inductor and the capacitor add to zero because they have equal magnitudes and opposite phases.

If we reduce the value of the resistor, the resonance sharpens. We plot the amplitude of the voltage across the resistor, capacitor, and inductor versus $R = 10\ \Omega$, $L = 1$ mH, and $C = 1\ \mu$F in Figure 26-21.

We can see that the highest the voltage across the resistor can get is still 100 V. At low angular frequencies, the voltage across the capacitor is approximately 100 V, while the inductor voltage is small. Conversely, at high angular frequencies, the voltage across the inductor is about 100 V and the voltage across the capacitor is small. In the resonance region, the voltages across both the capacitor and the inductor get large and substantially exceed the applied voltage. We must again remind ourselves that the graphs show amplitudes and the voltages are not in phase.

We characterize the width of a resonance by a **quality factor**, or ***Q*-factor**, defined as

$$Q = \frac{f_0}{\Delta f} \qquad (26\text{-}37)$$

where f_0 is the centre frequency of the peak and Δf is the half-power bandwidth of the peak.

The **half-power bandwidth** is the difference between the two frequencies at which the power in the circuit is equal to half the power at the centre frequency (Figure 26-22). A circuit with a high *Q*-factor has a narrow, tall resonance peak. The *Q*-factor can be determined from the power dissipated in the resistor of the circuit, which we will examine more closely in the next section.

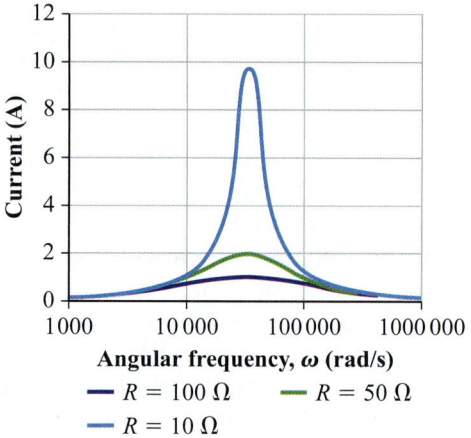

Figure 26-19 Current versus angular frequency in a driven RLC circuit. The different curves correspond to different values of the resistance.

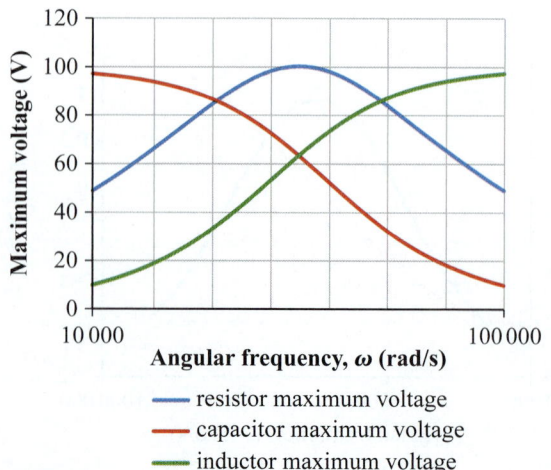

Figure 26-20 Voltage amplitudes versus angular frequency in an RLC circuit with $R = 50\ \Omega$, $L = 1$ mH, and $C = 1\ \mu$F.

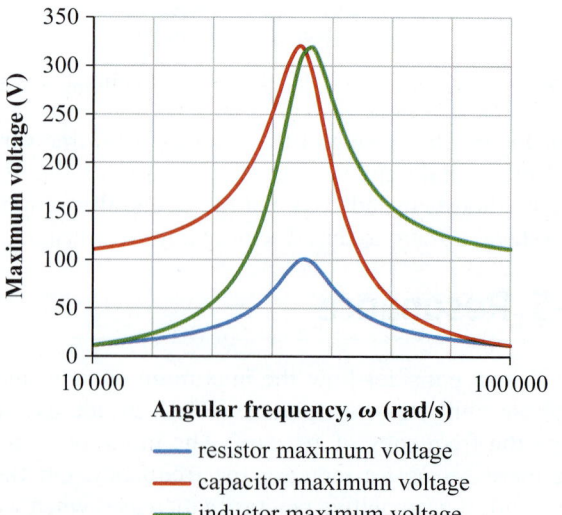

Figure 26-21 Maximum voltages in an RLC circuit for $R = 10\ \Omega$, $L = 1$ mH, and $C = 1\ \mu$F.

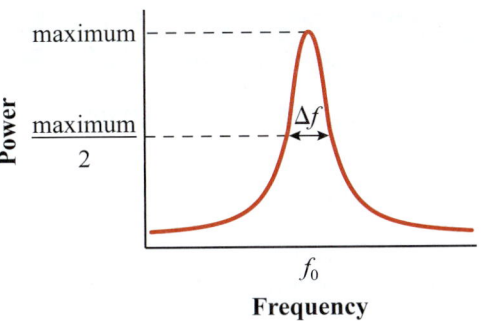

Figure 26-22 f_0 and Δf on a resonance peak.

MAKING CONNECTIONS

Radio Resonance

A tuner in a radio or a TV is an RLC circuit. Typically, the resistance in the circuit is made as small as possible to give as sharp a resonance as possible. Figure 26-23 shows a type of tuning capacitor commonly used in radios with manual tuning. Turning the knob on the capacitor changes the overlap between the interleaved stationary bottom plates and the moving top plates, changing the effective area of the capacitor plates.

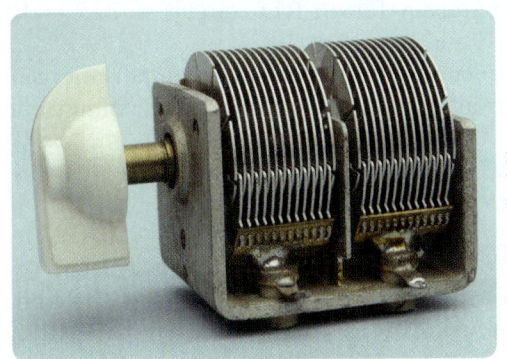

Figure 26-23 A variable tuning capacitor.

26-6 Power in AC Circuits

The individual voltages and currents in an AC circuit all have an average value of zero, because they oscillate symmetrically about zero. However, the power dissipated in a circuit element is the time average of the product of the voltage and the current, and, as we will see, this product generally does not average to zero.

We first consider the power dissipated by a resistor in an RLC circuit. The current through the resistor is given by

$$i(t) = i_{max}\sin(\omega t - \phi) \tag{26-38}$$

where i_{max} is given by Equation 26-32 and ϕ is given by Equation 26-34.

The time-dependent voltage across the resistor is given by

$$v_R(t) = Ri(t) = Ri_{max}\sin(\omega t - \phi) \tag{26-39}$$

The instantaneous power dissipated by the resistor is thus given by

$$P_R(t) = v_R(t)i(t) = Ri_{max}\sin(\omega t - \phi)i_{max}\sin(\omega t - \phi)$$

$$= i_{max}^2 R \sin^2(\omega t - \phi) \tag{26-40}$$

Recall from Section 26-1 that $(\sin^2 x)_{avg} = \dfrac{1}{2}$. So, averaging the instantaneous power over one cycle gives

KEY EQUATION

$$P_{R,avg} = \frac{i_{max}^2}{2}R = \left(\frac{i_{max}}{\sqrt{2}}\right)^2 R = i_{rms}^2 R \tag{26-41}$$

For the inductor, we also calculate the power as the product of the voltage and the current. We get the voltage across the inductor from Equation 26-35:

$$P_L(t) = v_L(t)i(t) = X_L i_{max}\cos(\omega t - \phi)i_{max}\sin(\omega t - \phi)$$

$$= X_L i_{max}^2 \cos(\omega t - \phi)\sin(\omega t - \phi) \tag{26-42}$$

When we average this power over one complete cycle, we can use the trigonometric identity $\sin(2\theta) = 2\sin(\theta)\cos(\theta)$ to get

$$P_{L,avg} = \frac{1}{T}\int_0^T X_L i_{max}^2\cos(\omega t - \phi)\sin(\omega t - \phi)dt$$

$$= \frac{X_L i_{max}^2}{T}\int_0^T \cos(\omega t - \phi)\sin(\omega t - \phi)dt$$

$$= \frac{X_L i_{max}^2}{T}\int_0^T \frac{1}{2}\sin(2(\omega t - \phi))dt$$

$$= 0$$

A plot of the instantaneous power versus time verifies that the average power is zero. For each complete cycle, the area enclosed by the curve below the x-axis

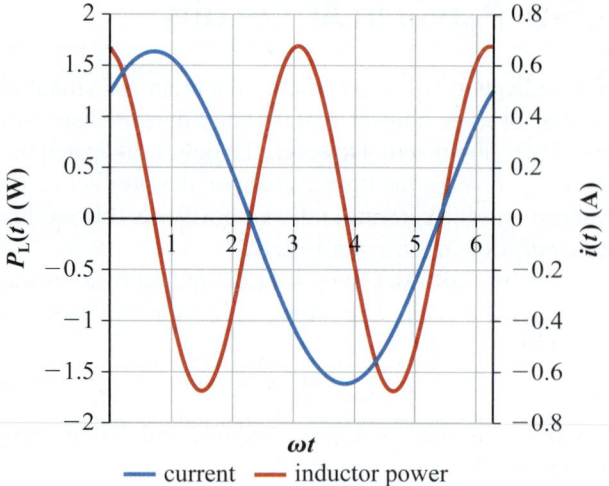

— current — inductor power

Figure 26-25 Power (red) and current (blue) in an inductor in an RLC circuit with $R = 100\ \Omega$, $L = 1$ mH, $C = 1\ \mu$F, and $\varepsilon_{max} = 100$ V. Note that as the magnitude of the current increases, the inductor absorbs energy. As the current returns to zero, the inductor delivers energy back to the circuit.

exactly cancels the area enclosed by the curve above the x-axis, as can be seen in Figure 26-25.

When the power is positive, the inductor is absorbing energy. Unlike in the resistor, this energy is not dissipated but is stored in the magnetic field of the inductor, as was shown in Chapter 25. While the current is increasing, the inductor absorbs energy. When the current starts to decrease, the inductor returns the stored energy to the circuit, and the power is negative.

EXAMPLE 26-6

Power and the Capacitor

Find the average power in the capacitor in an RLC circuit.

Solution

The instantaneous power in the capacitor is given by

$$P_C(t) = v_C(t)i(t) = -X_C i_{max}\cos(\omega t - \phi)i_{max}\sin(\omega t - \phi)$$
$$= -X_C i^2_{max}\cos(\omega t - \phi)\sin(\omega t - \phi)$$

Averaging this power over one complete cycle, we find

$$P_{C,avg} = \frac{1}{T}\int_0^T -X_C i^2_{max}\cos(\omega t - \phi)\sin(\omega t - \phi)dt$$

$$= \frac{-X_C i^2_{max}}{T}\int_0^T \cos(\omega t - \phi)\sin(\omega t - \phi)dt = 0$$

Making sense of the result

The instantaneous power in the capacitor is exactly out of phase with the instantaneous power in the inductor. When the inductor is absorbing energy, the capacitor is releasing energy, and vice versa.

Electrical Heating

A heater element toaster (Figure 26-24) is a simple example of resistive power dissipation in an AC circuit. Similar elements are found in electric water heaters, electric clothes dryers, blow dryers, and electric baseboard heaters. Heating elements are made of metals with low resistance and high melting points.

Figure 26-24 Heater elements in a toaster.

Thus, we find that the only element in the driven RLC circuit that dissipates power is the resistor. Total energy must be conserved, so all the power dissipated by the resistor comes from the emf source. The amount of power that the resistor dissipates depends on the frequency. Figure 26-26 shows a plot of power transfer versus frequency for an RLC circuit with $R = 10\ \Omega$, $C = 1\ \mu$F, and $L = 10$ mH.

Frequency dependence is an important general result for all oscillating systems. The maximum energy transfer between the driving emf and the oscillator occurs when the frequency of the driver matches the resonance frequency of the oscillator.

Now, we will examine how the quality factor is related to the power curve and the parameters of the RLC circuit. We can use Equation 26-41 to find the power in the resistor:

$$P_{R,avg} = i^2_{rms}R = \frac{\varepsilon^2_{rms}R}{Z^2} = \frac{\varepsilon^2_{rms}R}{R^2 + (X_L - X_C)^2} \quad (26\text{-}43a)$$

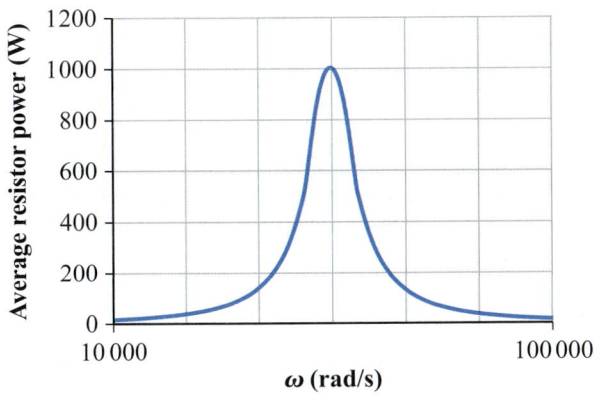

Figure 26-26 Frequency dependence of power dissipated by the resistor of an RLC circuit.

CHECKPOINT 26-7

Power in an RLC Circuit

True or false? At any instant of time, the power being delivered by the emf in an RLC circuit is equal to the power being dissipated by the resistor.

ANSWER: False. The only condition under which this statement is true is when the circuit is being driven at the resonant frequency. At that frequency, the voltages across the inductor and the capacitor are equal but opposite. However, the current through both elements is the same because they are in series. Therefore, the power delivered to one of the elements is exactly equal to the power delivered by the other, and the emf is simply providing the power dissipated in the resistor. For any other drive frequency, the emf must deliver (or absorb) the excess power absorbed/delivered by the reactive elements. It is important to recall that the power in the reactive elements, averaged over a single cycle, is equal to zero, but it is not zero at every instant in time.

Since the Q-factor is defined by the half-power points, we need to find the frequencies at which the power is half the peak value. The peak occurs at resonance when $Z = R$; thus, $P_{R,avg,max} = \dfrac{\varepsilon_{rms}^2}{R}$. At the half-power points, the denominator in Equation 26-43a must be $2R^2$, which requires $(X_L - X_C)^2 = R^2$. Substituting expressions for the reactances, we find

$$\left(\omega L - \frac{1}{\omega C}\right)^2 = R^2$$

$$\text{or } \left(\omega L - \frac{1}{\omega C}\right) = \pm R \tag{26-43b}$$

Multiplying through by ω and rearranging terms yields

$$L\omega^2 \pm R\omega - \frac{L}{C} = 0 \tag{26-43c}$$

This equation is quadratic in ω. Quadratic equations have two solutions, but in this case, because of the $\pm$

in front of the linear term, there are four solutions. They are

$$\omega = \frac{-R + \sqrt{R^2 + \dfrac{4L^2}{C}}}{2L}$$

$$\omega = \frac{R + \sqrt{R^2 + \dfrac{4L^2}{C}}}{2L}$$

$$\omega = \frac{-R - \sqrt{R^2 + \dfrac{4L^2}{C}}}{2L} \tag{26-43d}$$

$$\omega = \frac{R - \sqrt{R^2 + \dfrac{4L^2}{C}}}{2L}$$

We can see that the third and fourth values of ω in Equation 26-43d are both negative, so we reject these roots. The second value of ω is the larger of the two angular frequencies, so the half-power bandwidth is given by

$$\Delta\omega = \frac{R + \sqrt{R^2 + \dfrac{4L^2}{C}}}{2L} - \frac{-R + \sqrt{R^2 + \dfrac{4L^2}{C}}}{2L}$$

$$= \frac{R}{L} \tag{26-43e}$$

Therefore, for a series RCL circuit, the quality factor can be expressed as

KEY EQUATION
$$Q = \frac{f_0}{\Delta f} = \frac{\omega_0}{\Delta\omega} = \frac{\sqrt{\dfrac{1}{LC}}}{\dfrac{R}{L}} = \frac{1}{R}\sqrt{\frac{L}{C}} \tag{26-44}$$

EXAMPLE 26-7

The Q-factor from the Resistor Power

Find the quality factor for the RLC circuit with the power curve shown in Figure 26-26.

Solution

The component values are $R = 10.0\ \Omega$, $C = 1.00\ \mu F$, and $L = 10.0\ mH$. Therefore,

$$Q = \frac{1}{R}\sqrt{\frac{L}{C}} = \frac{1}{10.0\ \Omega}\sqrt{\frac{10.0\ mH}{1.00\ \mu F}} = 10.0$$

Making sense of the result

We can compare this value to what we would expect for a radio receiver. FM radios receive signals with a frequency on the order of 100 MHz, and the channels are separated by 0.2 MHz. The Q-factor for the tuner must therefore be on the order of 100 MHz/0.2 MHz = 500. So, a Q-factor of 10 is not particularly high.

KEY CONCEPTS AND RELATIONSHIPS

Simple AC Circuits

The AC circuits studied in this chapter are driven by an emf, $\varepsilon(t)$, that varies sinusoidally with time:

$$\varepsilon(t) = \varepsilon_{max}\sin(\omega t) = \varepsilon_{max}\sin(2\pi f t) \tag{26-1}$$

We frequently find it convenient to express voltages and currents in AC circuits in terms of their root mean square values. For a sinusoidal emf, or current, the rms values are given by

$$\varepsilon_{rms} = \frac{\varepsilon_{max}}{\sqrt{2}} \tag{26-6}$$

$$i_{rms} = \frac{i_{max}}{\sqrt{2}} \tag{26-7}$$

For an AC circuit that contains only a resistor, the current is in phase with the emf and is given by $i(t) = i_{max}\sin(\omega t)$, where $i_{max} = \dfrac{\varepsilon_{max}}{R}$. The resulting voltage across the resistor is given by $v_R(t) = i(t)R = i_{max}R\sin(\omega t)$. Thus, the maximum voltage that appears across the resistor, $V_{R,max}$, is

$$V_{R,max} = i_{max}R \tag{26-12}$$

For an AC circuit that only contains an inductor, we find that the resulting current is sinusoidal but no longer in phase with the emf and can be written as $i(t) = i_{max}\sin\left(\omega t - \dfrac{\pi}{2}\right)$. We say the current lags the voltage across an inductor. The amplitude of the current is $i_{max} = \dfrac{\varepsilon_{max}}{\omega L}$. We note the similarity of this expression to Ohm's law. We define the inductive reactance, X_L, as

$$X_L = \omega L \tag{26-16}$$

With this definition, we can write the amplitude of the voltage across the inductor as

$$V_{L,max} = i_{max}X_L \tag{26-18}$$

For an AC circuit that contains only a capacitor, we find that the resulting current leads the voltage across the capacitor and can be written as $i(t) = i_{max}\sin\left(\omega t + \dfrac{\pi}{2}\right)$, where $i_{max} = \dfrac{\varepsilon_{max}}{X_C}$ and

$$X_C = \frac{1}{\omega C} \tag{26-22}$$

The LC Circuit

If we connect an inductor across a charged capacitor, we find that the resulting current is sinusoidal and has angular frequency

$$\omega_0 = \frac{1}{\sqrt{LC}} \tag{26-28b}$$

Phasors

Phasors are a graphical device that allows us to represent time-varying voltages and currents that are not in phase with one another. For example, for a purely inductive circuit, where the voltage leads the current, we can represent these quantities with the phasor diagram in Figure 26-27.

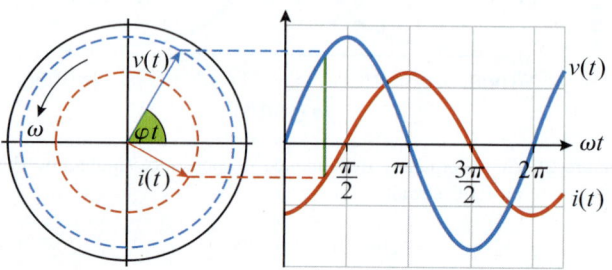

Figure 26-27 Phasor diagram and corresponding time plots representing the voltage and current in a purely inductive circuit. The instantaneous value is obtained from the phasor diagram by looking at the projection of the phasor onto the vertical axis.

In general, the current can be out of phase with the voltage by some angle, ϕ, and we write for the instantaneous values of the emf and the current

$$\varepsilon(t) = \varepsilon_{max}\sin(\omega t)$$
$$i(t) = i_{max}\sin(\omega t - \phi) \tag{26-29}$$

Series RLC Circuits

We can use a phasor diagram to add the voltages across the elements in a series RLC circuit and require that they add up to the voltage of the applied emf, $\varepsilon(t) = \varepsilon_{max}\sin(\omega t)$. When we do this, we find that the current in the circuit is given by $i(t) = i_{max}\sin(\omega t - \phi)$, where $i_{max} = \dfrac{\varepsilon_{max}}{\sqrt{R^2 + (X_L - X_C)^2}}$. We define the impedance, Z, of an RLC circuit as

$$Z = \sqrt{R^2 + (X_L - X_C)^2} \tag{26-33}$$

The phase angle, ϕ, is given by

$$\tan\phi = \frac{X_L - X_C}{R} \tag{26-34}$$

The time-dependent voltages across the resistor, inductor, and capacitor are then found from $v_R(t) = Ri_{max}\sin(\omega t - \phi)$, $v_L(t) = X_L i_{max}\cos(\omega t - \phi)$, and $v_C(t) = -X_C i_{max}\cos(\omega t - \phi)$, respectively.

Resonance

If we drive an RLC circuit with an emf with angular frequency

$$\omega_0 = \frac{1}{\sqrt{LC}} \tag{26-36}$$

we are said to be driving the circuit at resonance. At this angular frequency, we observe the current amplitude to be at a maximum.

Power in AC Circuits

The resistor is the only element in an AC circuit that dissipates power:

$$P_{R,avg} = i_{rms}^2 R \qquad (26\text{-}41)$$

If capacitors or inductors are present in the circuit, they dissipate no power on average.

By examining how the resistor power varies with angular frequency, we can determine the quality factor, or Q-factor, for an RLC circuit as

$$Q = \frac{\omega_0}{\Delta\omega} = \frac{1}{R}\sqrt{\frac{L}{C}} \qquad (26\text{-}44)$$

Here, $\Delta\omega$ is the width of the power resonance curve measured between points of half maximum power.

APPLICATIONS

energy consumption, power plants, electric circuits, radio and TV tuners, home heater elements

KEY TERMS

alternating current, capacitive reactance, half-power bandwidth, impedance, inductive reactance, LC circuit, phase shift, phasors, quality factor, Q-factor, reactive circuit elements, resonance frequency, root mean square, series RLC circuit

QUESTIONS

1. An RLC circuit is driven at a certain frequency. When the frequency is increased, the maximum current decreases. Is the circuit above or below resonance?

2. What is the phase angle of a series RLC circuit at resonance?

3. At very low frequencies, an inductor has little effect on a series RLC circuit. Why?

4. Two circuits with identical inductors and capacitors have quality factors of 10 and 100. Which has the smaller resistor?

5. Does capacitive reactance increase or decrease with frequency?

6. When we add the maximum voltages that appear across the elements in a series RLC circuit, we usually get a voltage that is greater than the applied emf. Does this violate Kirchhoff's loop rule?

7. Can the rms values of a voltage ever exceed the maximum value? Equal it?

8. What is the limit on the maximum voltage that can appear across the capacitor (or inductor) in a series RLC circuit?

9. What does the phase angle in a series RLC circuit tend toward as the frequency increases significantly?

10. For fixed L and C, how does one increase the quality factor of a series RLC circuit?

11. The maximum voltage across a particular element in an AC circuit occurs when the current is zero. Is the element the
    (a) resistor?
    (b) capacitor?
    (c) inductor?
    (d) capacitor or inductor?
    (e) There is not enough information to answer the question.

12. The inductor and the capacitor in an LC circuit are both halved. The oscillation frequency will increase by a factor of
    (a) 2?
    (b) 4?
    (c) 8?
    (d) $\sqrt{2}$?

13. Consider the square wave shown in Figure 26-7. Instead of going between $+\varepsilon_{max}$ and $-\varepsilon_{max}$, it is shifted up by ε_{max} and thus goes between 0 and $+2\varepsilon_{max}$. The rms value of an unshifted square wave is ε_{max}. Will the rms value of the shifted wave be
    (a) $2\varepsilon_{max}$?
    (b) $\sqrt{2}_{max}$?
    (c) $4\varepsilon_{max}$?
    (d) none of the above?

14. An inductor is connected across the terminals of a fully charged capacitor at $t = 0$. Which of the following statements is true?
    (a) The current is zero and the voltage across the inductor is zero.
    (b) The current is at its maximum value and the voltage across the inductor is at its maximum negative value.
    (c) The current is zero and the voltage across the inductor is at its maximum negative value.
    (d) The answer will depend upon the values of the capacitor and inductor.

15. A sinusoidal emf has a maximum amplitude of 200. V. What is its average value?
    (a) 200. V
    (b) 141 V
    (c) 0.00 V
    (d) need to know the frequency to answer this question

PROBLEMS BY SECTION

For problems, star ratings will be used (★, ★★, or ★★★), with more stars meaning more challenging problems. The following codes will indicate if $\frac{d}{dx}$ differentiation, $\int$ integration, ▢ numerical approximation, or ◠ graphical analysis will be required to solve the problem.

Section 26-1   Simple Loads in AC Circuits

16. ★★ $\int$ Show that the rms value for a triangular wave is
$$\frac{1}{\sqrt{3}}\varepsilon_{max}.$$

17. ★ The wall outlets in many North American homes provide 110 V rms. What is the maximum amplitude of this emf? In Europe, it is 220 V rms. What is this maximum amplitude?

18. ★ A 60. W light bulb is connected to 120 V rms.
  (a) What is the rms current?
  (b) What is the resistance of the bulb filament?

19. ★ A 200. Ω resistor is connected across a 20. V emf with a frequency of 440 Hz.
  (a) What is the amplitude of the current?
  (b) What is the angular frequency of the emf?
  (c) Write an expression for the current through the resistor as a function of time.

20. ★ A 1.0 mH inductor is connected across a 20. V rms emf with an angular frequency of 1000. rad/s.
  (a) What is the inductive reactance?
  (b) What is the amplitude of the resulting current?
  (c) Write an expression for the current through the inductor as a function of time.

21. ★ An inductor is connected across a 100 V emf (RMS) with an angular frequency of 2000 rad/s. The resulting current is given by $i(t) = 0.001 \sin\left(\omega t - \frac{\pi}{2}\right)$.
  (a) What is the amplitude of the current?
  (b) What is the inductive reactance?
  (c) What is the inductance?

22. ★ A 1.0 nF capacitor is connected across a 20. V rms emf with an angular frequency of 100. rad/s.
  (a) What is the capacitive reactance?
  (b) What is the amplitude of the current in the circuit?
  (c) Write an expression for the current in the circuit as a function of time.

23. ★ A 1.0 nF capacitor is connected across an emf with a frequency of 200. Hz. A current with an amplitude of 10. μA results.
  (a) What is the capacitive reactance?
  (b) What is the amplitude of the emf?

Section 26-2   The LC Circuit

24. ★★ A 10. μF capacitor is charged and connected across an inductor. The resulting current has a frequency of 10. kHz. What is the value of the inductor?

25. ★ A 1.0 nF capacitor is charged and connected across a 0.10 mH inductor. What is the angular frequency of the resulting oscillation?

26. ★ By what factor does the resonance frequency increase when the capacitance of a capacitor is doubled in value?

Section 26-3   Phasors

27. ★ The voltage across a circuit element is represented by a phasor with an amplitude of 100. V that makes an angle of $\theta = 200. t$ with respect to the positive x-axis. What is the value of the voltage across the element at $t = 10.$ s?

28. ★★ The current in a circuit is represented by a phasor with an amplitude of 20. mA that makes an angle of $\theta = 5000. t$ with respect to the positive x-axis.
  (a) At what time(s) is the current zero?
  (b) At what times is the current at its maximum positive value?
  (c) At what times is the current at its maximum negative value?

29. ★★ A circuit has three elements and an emf. The phasors representing the voltages across the elements are given by $v_1 = 20 \sin(\omega t)$, $v_2 = 40 \cos(\omega t)$, and $v_3 = -60 \cos(\omega t)$. Add the three phasors to find
  (a) the amplitude of the emf
  (b) the angle between the emf and v_1

Section 26-4   Series RLC Circuits

30. ★★ A 10. Ω resistor, a 5.0 nF capacitor, and a 20. mH inductor are connected in series across an emf $\varepsilon(t) = 25 \sin(33\,000t)$.
  (a) Find the capacitive reactance.
  (b) Find the inductive reactance.
  (c) Find the impedance.
  (d) Find the phase angle.
  (e) Find the amplitude of the current, i_{max}.

31. ★★ A 25 nF capacitor and a 30. mH inductor are connected in series with an unknown resistor across an emf $\varepsilon(t) = 25 \sin(33\,000t)$. The current has amplitude of 0.10 A.
  (a) What is the value of the resistor?
  (b) What is the phase angle?

32. ★★ A 100. Ω resistor and a 5.0 nF capacitor are connected in series with an unknown inductor across an emf $\varepsilon(t) = 250 \sin(9000. t)$. The phase angle is -0.79 rad.
  (a) What is the inductive reactance?
  (b) What is the inductance?
  (c) What is the amplitude of the current?

Section 26-5   Resonance

33. ★ An RLC circuit has an inductor of 23 mH, a capacitor of 1.0 μF, and a resistor of 20. Ω. What is the resonance frequency?

34. ★ An RLC circuit has a quality factor of 15, a capacitor of 0.10 μF, and a resistor of 100. Ω. What is the value of the inductor?

35. ★★★ A circuit has a quality factor of 200. By what factor, compared to the value at resonance, does the maximum current change when the frequency is changed to 90% of the resonance frequency?

Section 26-6 Power in AC Circuits

36. ★ A 110 V rms house circuit is fused for 15 A. How many 100. W light bulbs can go on the circuit?

37. ★ How much power is dissipated in a 100. Ω resistor connected across an emf with an amplitude of 10. V?

38. ★★ A series LC circuit is connected across an emf. At $t = 0$, the capacitor is uncharged and there is no current flowing in the circuit. The emf is then turned on. Does the emf deliver any energy to the circuit?

COMPREHENSIVE PROBLEMS

39. ★★ A 20. Ω resistor, a 1.0 nF capacitor, and a 20. mH inductor are in series with an emf $\varepsilon(t) = 50.\sin(6000.)t$. Find the maximum current and the phase angle.

40. ★★ A 40. Ω resistor, a 200. μF capacitor, and an inductor are in series with an emf $\varepsilon(t) = 25.\sin(4000.)t$. The current is observed to have an amplitude of 45 mA. Find the inductance and the phase angle.

41. ★★★ 〰 RLC circuits can be used as filters. Plot the amplitude of the maximum voltage versus frequency across the resistor, capacitor, and inductor for a series RLC circuit with a 500. Ω resistor, a 1.0 μF capacitor, and a 100. mH inductor. Assume that the emf has an amplitude of 1.0 V, and plot in the range $0 < \omega < 20\,000$ rad/s.

42. ★ At an angular frequency of 1000. rad/s, a capacitor has a reactance of 300. Ω. What is its capacitance value?

43. ★★ A series RLC circuit is observed to have a phase angle of −45°. The amplitude of the voltage across the capacitor is twice the amplitude of the voltage across the resistor. Relative to the capacitor, what is the amplitude of the voltage across the inductor?

44. ★★ The windings in electric motors have an electrical resistance from the wire as well as an inductance from the fact that they are wound in coils. A motor that is operated at 110 V rms and 60. Hz draws 2.0 A of current. The windings have a resistance of 30. Ω. What is the inductance of the motor?

45. ★★ A 0.10 μF capacitor and a 10. mH inductor are connected. The current in the circuit is observed to have a maximum amplitude of 10. mA. What is the frequency of oscillation, and what is the maximum charge on the capacitor?

46. ★★ A 0.10 μF capacitor is in an LC circuit that oscillates at 200. kHz. The capacitor is initially charged to 200. nC. What is the value of the inductor, and what is the amplitude of the current?

47. ★★ In an RLC circuit, the resistor, inductor, and capacitor have voltage amplitudes of 10. V, 20. V, and 30. V, respectively. What are the phase angle and the amplitude of the emf generator?

48. ★★ In an RLC circuit, the voltage amplitude across the resistor is 20. V, the frequency is 2000. Hz, and the phase angle is −45°. What is the instantaneous voltage across the resistor at $t = 10.$ s?

49. ★★ A series circuit contains a resistor and an inductor. The inductance is 10. mH, and the emf has a frequency of 2.0 kHz. The phase angle is 45°. What is the value of the resistance?

50. ★ A series circuit contains a resistor and a capacitor. The emf is $\varepsilon(t) = 45.\sin(5000.t)$, and the current is $i(t) = 0.0010\sin(5000.t + 0.70 \text{ rad})$. Find the resistance and the capacitance.

51. ★ A 20. Ω resistor, a 1.0 nF capacitor, and a 20. mH inductor are in series with an emf $\varepsilon(t) = 50.\sin(6000.t)$. How much power does the resistor dissipate?

52. ★★ What should the phase angle be in a circuit to deliver maximum power from the emf (source) to the circuit (load)?

53. ★★ The emf and current in an RLC circuit are shown in Figure 26-28. What are the phase angle and resistance?

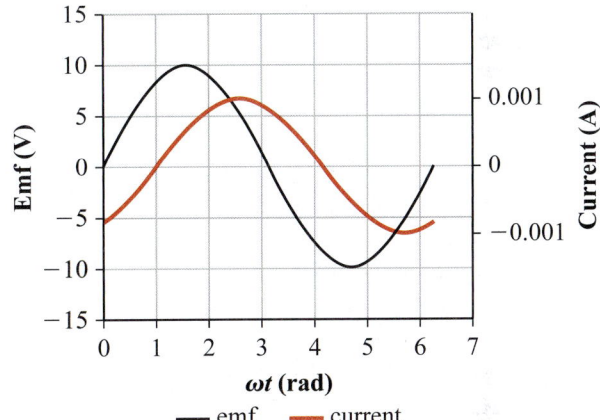

Figure 26-28 Problem 53.

54. ★★ A filter motor on a swimming pool draws 5.0 A from a 110 V, 60. Hz wall socket and dissipates 500. W rms of power. Find the resistance of the motor and the phase angle.

55. ★ The receiver in an FM radio has to be tuned from a low frequency of 88 MHz up to 108 MHz. A fixed inductor of 3.0 nH is used. What range must the variable tuning capacitor have to tune across this frequency band?

56. ★★★ FM radio stations are typically separated by 200. kHz in frequency. Each individual station transmits energy in a frequency range from approximately 20. kHz below to 20. kHz above its centre frequency. Design a tuner circuit for an FM radio receiver, using 101.5 MHz as the centre frequency. Your tuning circuit should have a half-power bandwidth of 40. kHz and use a 10. nH inductor.

DATA-RICH PROBLEM

57. ★★★ ⟋, ▭ The markings have worn off the resistor in a series RLC circuit. The circuit has a 1.0 μF capacitor and a 100. mH inductor. You measure the voltage across the resistor as you vary the frequency of the emf and obtain the results in Table 26-1. Use some fitting software to determine the amplitude of the emf and the unknown resistance.

OPEN PROBLEM

58. ★★★ You wish to design a very sensitive thermometer that is based on a capacitor whose capacitance changes with temperature. Consider using a driven RLC circuit, and outline the strategy you would take.

Table 26-1   Voltage versus Frequency Data for Problem 57

Angular Frequency (rad/s)	Resistor Voltage (V)
1 000	11.0
2 000	31.6
3 000	94.9
4 000	55.5
5 000	31.6
6 000	22.5
7 000	17.7
8 000	14.7
9 000	12.6
10 000	11.0
11 000	9.9
12 000	8.9
13 000	8.1
14 000	7.5
15 000	7.0
16 000	6.5
17 000	6.5

Electromagnetic Waves and Maxwell's Equations

Learning Objectives

When you have completed this chapter, you should be able to

LO1 Explain how a changing electric field induces a magnetic field and how a changing magnetic field induces an electric field.

LO2 Explain why Maxwell needed to add a displacement current to Ampère's law.

LO3 Write Maxwell's equations in integral form, and explain the terms in these equations.

LO4 Classify the electromagnetic spectrum according to the range of wavelengths (or frequencies).

LO5 Describe how electromagnetic waves carry energy and momentum, and calculate the intensity and momentum of a plane electromagnetic wave.

LO6 Calculate the pressure that electromagnetic radiation exerts on a surface.

LO7 Explain how electromagnetic waves are generated by accelerating charges.

LO8 Describe the polarization of electromagnetic waves, explain how a polarizer works, and calculate the intensity of light that passes through two polarizers.

The radio waves transmitted by local radio stations, the red laser light used to scan a product's bar code, the radiation that cooks food in a microwave oven, the sunlight that makes life possible on Earth, the X-rays used to detect hidden cavities in your teeth, and the gamma rays emitted by quasars are all electromagnetic waves with different frequencies.

Along with Newton's laws of motion, relativity theory, and quantum mechanics, the theory of electromagnetic waves forms the backbone of modern science and our understanding of the universe. In this chapter, we explore some of the important properties of electromagnetic waves.

NASA, ESA, ESO, CXC & D. Coe (STScI)/J. Merten (Heidelberg/Bologna)

Figure 27-1 The galaxy cluster Abell 2744 emits visible light and X-rays. In this colour configuration, the red colour corresponds to regions of the galaxy that are emitting X-rays, and the blue colours correspond to regions containing dark matter.

27-1 The Laws of Electric and Magnetic Fields

So far, we have used electric and magnetic fields to describe the forces that act on stationary and moving charges. In 1862, Scottish physicist James Clerk Maxwell (1831–1879) showed that electric and magnetic fields are two components of a single field, which he called the **electromagnetic field**. Maxwell noted the strong similarities between the key laws that describe electric and magnetic fields, leading him to develop a unified model of the fields. Maxwell's theory also predicts that oscillating electromagnetic waves can exist in free space and can travel without any medium.

Chapters 20, 24, and 25 discuss two laws that govern electric fields and two similar laws for magnetic fields. Table 27-1 lists the equations for these four laws, and the laws themselves are summarized below.

Gauss's Law for Electric Fields

Gauss's law for electric fields relates the net flux of an electric field, $\vec{E}$, through a closed surface, S, to the net electric charge, q_{enc}, enclosed within that surface. The quantity $\oint \vec{E} \cdot d\vec{A}$ is the electric flux through the surface S and is denoted by Φ_E. The area vector element $d\vec{A}$ is normal to the surface and directed outward (Figure 27-2).

Table 27-1 Four Laws of Electric and Magnetic Fields (for Steady Currents)

Gauss's Law for Electric Fields		Faraday's Law for Electric Fields	
$\oint \vec{E} \cdot d\vec{A} = \dfrac{q_{enc}}{\varepsilon_0}$	(20-5)	$\oint \vec{E} \cdot d\vec{\ell} = -\dfrac{d\Phi_B}{dt}$	(25-20)
Gauss's Law for Magnetic Fields		**Ampère's Law for Magnetic Fields**	
$\oint \vec{B} \cdot d\vec{A} = 0$	(27-3)	$\oint \vec{B} \cdot d\vec{\ell} = \mu_0 I_{enc}$	(24-38)

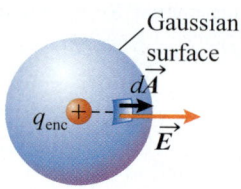

Figure 27-2 Gauss's law for electric fields. The total electric flux through a closed surface is proportional to the total charge enclosed within the surface.

Gauss's Law for Magnetic Fields

Gauss's law for magnetic fields states that the net flux of a magnetic field, $\vec{B}$, through a closed surface, S, is zero. The quantity $\oint \vec{B} \cdot d\vec{A}$ is the magnetic flux through the surface S and is denoted by the symbol Φ_B.

Faraday's Law for Electric Fields

Faraday's law for electric fields states that a changing magnetic flux produces an electric field and therefore induces an electromotive force (emf) in a closed circuit. The magnitude of the induced emf is equal to the rate of change of the magnetic flux through the circuit bounded by a closed loop, C.

Ampère's Law for Magnetic Fields

A steady electric current generates a magnetic field that encircles the current. Ampère's law for magnetic fields states that the line integral of a magnetic field, $\vec{B}$, around a closed loop that encloses a current (denoted by I_{enc} in Equation 24-38) passing through the loop is equal to the permeability times the electric current.

As you can see from Table 27-1, the equations for these four laws are quite similar. The left sides of the equations for both of Gauss's laws describe the flux of the field through a closed surface. For the electric field, the flux depends on the quantity of the charge enclosed by that surface. When the net charge enclosed within a surface is positive, a net electric flux *emerges* through that surface. When the net charge is negative, a net flux *enters* the surface. However, the net magnetic flux through a closed surface is zero because sources that produce magnetic fields always have both a north pole and a south pole. An isolated magnetic monopole (a single north pole or a single south pole) has never been observed. Therefore, magnetic field lines that originate from the north pole of a magnet eventually have to re-enter the magnet's south pole. Consequently, the magnetic flux leaving a closed surface equals the flux re-entering the surface. If magnetic monopoles are ever discovered, a non-zero term will have to be added to the right side of the equation for Gauss's law for magnetic fields, making it completely symmetric with Gauss's law for electric fields.

The left sides of the equations for Faraday's law for electric fields and Ampère's law for magnetic fields are both line integrals of a field around a closed loop. However, the right sides of these two equations look quite different. The equation for Faraday's law has no term proportional to a current because there are no magnetic monopoles. (If monopoles did exist, then the right side of Faraday's law would have a term describing a magnetic current due to the flow of magnetic monopoles.) The right side of the equation for Faraday's law indicates that the induced electric field is proportional to the rate of change of the magnetic flux, but Ampère's law has no corresponding term. Does a changing electric flux produce a magnetic field?

Maxwell realized that Ampère's law as stated in Table 27-1 holds only when the electric current, I, is constant. When a current changes with time, the equation for Ampère's law needs an additional term.

CHECKPOINT 27-1

Gauss's Law for Electric Fields

Rank each charge configuration in Figure 27-3 in decreasing order of the amount of net electric flux passing through each closed spherical surface. Use the equality sign if needed.

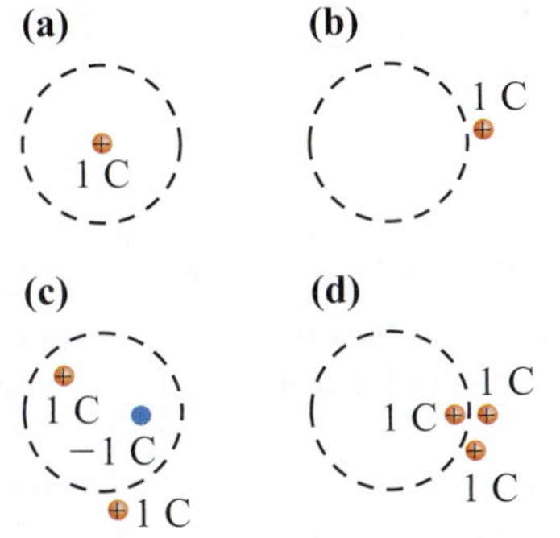

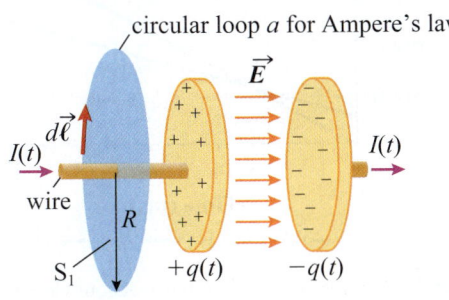

Figure 27-3 Checkpoint 27-1.

ANSWER: (a) = (d) < (b) = (c)

27-2 Displacement Current and Maxwell's Equations

Consider the process of charging a parallel-plate capacitor by passing a current, $I(t)$, through a connecting wire. As the capacitor charges, a charge, $q(t)$, accumulates on its plates and an instantaneous electric field, $\vec{E}(t)$, begins to build up between the two plates of the capacitor. We know that a magnetic field exists around a current-carrying wire. Let us apply Ampère's law to the circular loop, a, of radius R that is centred on the wire, as shown in Figure 27-4. The loop a encloses the planar surface S_1 that is being penetrated

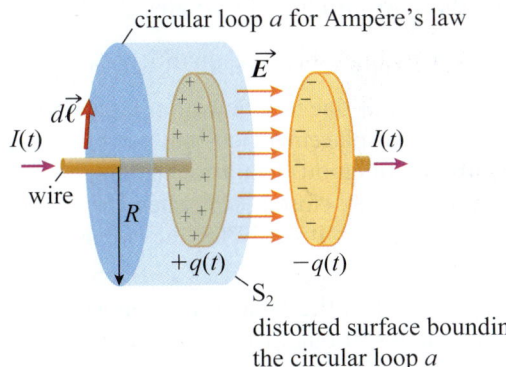

circular loop a for Ampère's law

distorted surface bounding the circular loop a

Figure 27-5 There is no conduction current through the surface S_2; however, there is a displacement current through this surface due to a changing electric flux as the charge builds up on the capacitor plates.

by the current $I(t)$. So the magnitude of the magnetic field around the path is $B(t) = [\mu_0/(2\pi R)]\, I(t)$. Now let us distort the surface S_1 (still bounded by the same circular loop a) to a surface S_2 such that the surface S_2 passes through the space between the plates of the capacitor, as shown in Figure 27-5. In this case, no current is penetrating through the surface S_2 (because there is no flow of current between the two plates) and hence the line integral of $\vec{B}$ around the circular path vanishes. So Ampère's law as given by Equation 24-38 is ambiguous. It gives different results depending on how the surface bounding the circular loop is chosen. Notice that this ambiguity arises only when the capacitor is being charged (or discharged). When the capacitor is fully charged, current flow through the wire stops and the line integral of $\vec{B}$ is zero for any choice of surface bounding the circular path.

Maxwell noted that as the capacitor is being charged, the flux of the electric field between the capacitor plates increases with time. So the *changing* electric field must generate a magnetic field that has exactly the same magnitude as the magnetic field generated when the electric current passes through surface S_1. Using this insight, Maxwell extended Ampère's law:

$$\oint \vec{B} \cdot d\vec{\ell} = \mu_0 I + \mu_0\left(\varepsilon_0 \frac{d\Phi_E}{dt}\right) \qquad (27\text{-}1)$$

where Φ_E denotes electric flux, $\dfrac{d\Phi_E}{dt}$ is the rate of change of electric flux with time, and $I = I_{enc}$ is the current due to the flow of the charged particles, called the conduction current. Equation 27-1 is called the Ampère–Maxwell equation. The quantity $\varepsilon_0 \dfrac{d\Phi_E}{dt}$ has units of current, as shown below, and is called the **displacement current**, I_{disp}:

$$[I_{disp}] = \frac{C^2}{N \cdot m^2} \times \frac{1}{s} \times \frac{N \cdot m^2}{C} = \frac{C}{s}$$

circular loop a for Ampère's law

Figure 27-4 A parallel-plate capacitor being charged. The conduction current, $I(t)$, induces a magnetic field around the wire, which can be calculated by applying Ampère's law to the flat surface, S_1.

Note that no current is actually displaced; the name simply distinguishes the quantity $\varepsilon_0 \dfrac{d\Phi_E}{dt}$ from the current, I, that is due to a flow of charges. With the addition of the displacement current term, Ampère's law gives an unambiguous result for the magnetic field generated around any closed loop when the current is not steady. This is how it works for the two surfaces of Figures 27-4 and 27-5.

- Surface S_1: When the capacitor is being charged, a current I flows through the surface, but there is no electric flux through S_1. Hence, the displacement current is zero, and the magnetic field is generated by the conduction current, I.

- Surface S_2: There is no conduction current through the surface, but there is a changing electric flux through it. The magnetic field is generated by the displacement current (i.e., by the changing electric flux). As will be shown below, this magnetic field is exactly the same as the magnetic field generated by the conduction current, I.

When the capacitor is fully charged, the conduction current stops. There is an electric flux through the surface S_2, but it is now constant. So the displacement current is also zero. Hence, there is no magnetic field in the circuit.

We showed that the dimension of the displacement current is the same as that of the conduction current (due to the flow of electrons). Now we will show that it has the same magnitude. Let $q(t)$ be the charge on a capacitor plate at time t. The charge density (charge per unit area) on a plate of cross-sectional area A is

$$\sigma(t) = \frac{q(t)}{A}$$

Therefore, the magnitude of the electric field between the two plates of the capacitor, ignoring edge effects, is

$$E(t) = \frac{\sigma(t)}{\varepsilon_0} = \frac{q(t)}{A\varepsilon_0}$$

Assuming that the electric field between the plates is uniform, the electric flux, Φ_E, between the plates is the product of the electric field and the area of a single plate:

$$\Phi_E(t) = E(t) \times A = \frac{q(t)}{\varepsilon_0}$$

The rate of change of the electric flux is

$$\frac{d\Phi_E(t)}{dt} = \frac{1}{\varepsilon_0} \frac{dq(t)}{dt}$$

Since the rate of change of charge on the plates is equal to the conduction current,

$$I_{\text{disp}}(t) = \varepsilon_0 \frac{d\Phi_E(t)}{dt} = \frac{dq(t)}{dt} = I(t)$$

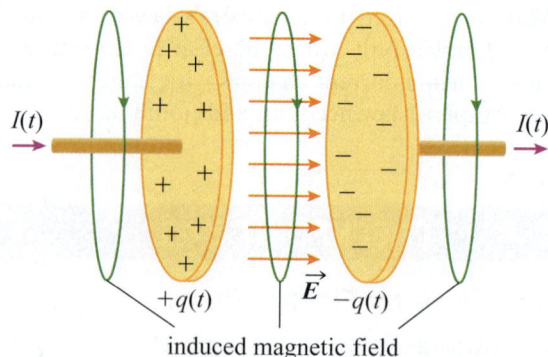

Figure 27-6 As a capacitor is charged, the induced magnetic field around the wire is due to the conduction current, and the field induced between the plates is due to the displacement current.

Thus, *the magnitude of the displacement current is the same as the magnitude of the conduction current.* We emphasize that no electrons are flowing across the gap between the capacitor plates. The changing electric flux between the plates generates a magnetic field that is exactly the same as the magnetic field generated by the conduction current in the wires (Figure 27-6). If the circuit were enclosed in a sealed box, you would detect a magnetic field around the box, but you would not be able to do an experiment to determine whether the magnetic field is due to a conduction current or a displacement current.

CHECKPOINT 27-2

Displacement Currents

The graph in Figure 27-7 shows the magnitudes of four changing uniform electric fields, all with the same direction. Consider a square loop of area 5 cm^2 perpendicular to the electric fields. Rank the displacement currents from the changing fields from least to greatest.

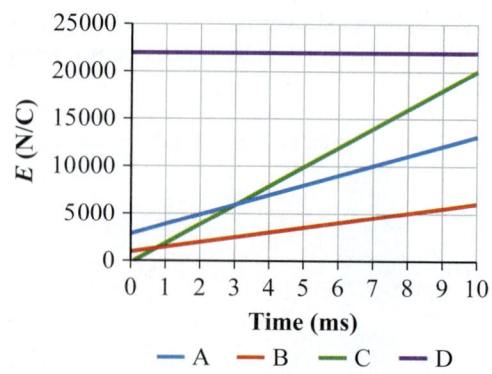

Figure 27-7 A graph of the magnitudes of four electric fields as a function of time.

ANSWER: C < A < B < D

EXAMPLE 27-1

Magnetic Field Due to the Displacement Current

D.F. Bartlett and T.R. Cole measured the induced magnetic field inside a charging capacitor in 1984 using a very sensitive magnetometer. They used a capacitor with two circular plates that were 1.22 cm apart. Instead of using direct current to charge the capacitor, they charged the capacitor using a generator that produced a 1.25 kHz sinusoidal voltage with an amplitude of 340 V. The arrangement is similar to the one shown in Figure 27-6. Calculate the peak strength of the magnetic field around a loop of radius 2.00 cm centred on the central axis of the capacitor. Assume that the radius of the capacitor plates is 6.00 cm so that the edge effects can be ignored.

Solution

As the capacitor is being charged, the electric flux between the plates increases and generates a magnetic field. The magnitude and direction of the induced magnetic field are given by the Ampère–Maxwell Equation 27-1. Since there is no conduction current in the region between the plates:

$$\oint \vec{B} \cdot d\vec{\ell} = \mu_0 \left(\varepsilon_0 \frac{d\Phi_E}{dt} \right)$$

We now evaluate each side of this equation.

Right Side

The electric field is perpendicular to the surface of the plates and has the same magnitude, $E(t)$, anywhere between the plates at a given time, t. The electric flux, Φ_E, through a disk of radius r centred on the central axis of the plates is given by

$$\Phi_E(t) = E(t) \times \pi r^2$$

The rate of change of this flux is

$$\frac{d\Phi_E(t)}{dt} = \pi r^2 \frac{dE(t)}{dt} \qquad (1)$$

The potential difference, V, between the two plates has a maximum value of 340 V and varies sinusoidally with a frequency of 1.25 kHz. Therefore,

$$V(t) = V_{max} \sin(2\pi f t)$$
$$= (340 \, \text{V}) \sin(2\pi \times (1250 \, \text{s}^{-1})t)$$

The electric field between the plates is related to the potential difference between the plates and their separation, L:

$$E(t) = \frac{V(t)}{L} = \frac{(340 \, \text{V}) \sin(2\pi \times (1250 \, \text{s}^{-1})t)}{L}$$

Inserting this expression for $E(t)$ into the right side of (1), we get

$$\mu_0 \varepsilon_0 \frac{d\Phi_E(t)}{dt} = \mu_0 \varepsilon_0 \frac{\pi r^2}{L} (340 \, \text{V}) \frac{d \sin((2.50 \times 10^3 \, \text{s}^{-1})\pi t)}{dt}$$
$$= 8.50 \times 10^5 \left(\frac{\pi^2 r^2 \mu_0 \varepsilon_0}{L} \right) \cos(2.50 \times 10^3 \pi t) \frac{\text{C}}{\text{s}}$$

Left Side

We need to calculate the line integral around the circular amperian loop of radius r shown in Figure 27-6. The magnetic field is tangent to the loop, and the loop is centred symmetrically on the central axis. Therefore, the magnetic field has the same magnitude at every point around the loop. Taking $\vec{B}$ and the path element $d\vec{\ell}$ parallel to each other, we get

$$\oint \vec{B} \cdot d\vec{\ell} = \oint B \cos d\ell = B \oint d\ell = B(2\pi r)$$

Equating the left and the right sides, we get the magnitude of the induced magnetic field around a circular loop of radius r that lies inside the two plates:

$$B(t) = 8.50 \times 10^5 \left(\frac{\pi r \mu_0 \varepsilon_0}{2L} \right) \cos((2.50 \times 10^3 \, \text{s}^{-1})\pi t) \, \text{T}$$

Notice that both the electric field and the induced magnetic field oscillate with the same frequency and we have shown the explicit time dependence of the magnetic field. The amplitude of the magnetic field is equal to the factor multiplying the cosine term. Substituting the known values, $r = 2.00 \times 10^{-2}$ m, $L = 1.22 \times 10^{-2}$ m, $\mu_0 = 4\pi \times 10^{-7}$ T·m/A, and $\varepsilon_0 = 8.85 \times 10^{-12}$ C^2/(N·m^2), we get

$$B(t) = (2.43 \times 10^{-11}) \cos((2.50 \times 10^3 \, \text{s}^{-1})\pi t) \, \text{T}$$

The peak strength of the magnetic field occurs when $\cos((2.50 \times 10^3 \, \text{s}^{-1})\pi t) = \pm 1$ and is therefore $\pm 2.43 \times 10^{-11}$ T.

Making sense of the result

The magnitude of Earth's magnetic field ranges between 3×10^{-5} T and 6×10^{-5} T, which is much larger than the induced magnetic field in even a large capacitor. Physicists in the time of Faraday and Ampère were unable to detect such small magnetic fields due to changing electric flux in their experiments.

The four equations in Table 27-1, with Ampère's law replaced by Equation 27-1, are collectively called Maxwell's equations. **Maxwell's equations** describe the physics of all phenomena that involve electric and magnetic fields, including the attraction between nuclei and electrons, van der Waals forces between molecules, friction, viscosity, and the propagation of electromagnetic waves. Maxwell's equations are to electromagnetism what Newton's laws are to mechanics.

Maxwell's Equations in a Vacuum

There is no matter in a vacuum, so there are no charges or currents. Consequently, Maxwell's equations become much simpler in a vacuum:

$$\oint \vec{E} \cdot d\vec{A} = 0 \quad \text{(Gauss's law for electric fields)} \qquad (27\text{-}2)$$

$$\oint \vec{B} \cdot d\vec{A} = 0 \quad \text{(Gauss's law for magnetic fields)} \qquad (27\text{-}3)$$

$$\oint \vec{E} \cdot d\vec{\ell} = -\frac{d\Phi_B}{dt} \quad \text{(Faraday's law)} \qquad (27\text{-}4)$$

$$\oint \vec{B} \cdot d\vec{\ell} = \mu_0 \varepsilon_0 \frac{d\Phi_E}{dt}, \quad \text{(Ampère's law)} \qquad (27\text{-}5)$$

In a vacuum, Gauss's laws for electric and magnetic fields are identical in form. There are no charges in a vacuum, so the net flux of the electric field through any closed surface is zero. Similarly, the net magnetic flux through a closed surface is zero because there are no magnetic monopoles.

Other than a multiplying factor, Faraday's and Ampère's laws also have identical forms in a vacuum. A changing magnetic flux produces an electric field, and any change in electric flux produces a magnetic field. The minus sign on the right side of Faraday's law is equivalent to Lenz's law, which states that the direction of the induced electric field is such that the induced current produces a magnetic field that opposes the original change in the magnetic flux.

Equations 27-4 and 27-5 indicate that the electric and magnetic fields in a vacuum are coupled. If the strength of the electric field changes with time, the changing electric flux produces a magnetic field that also changes with time. A changing magnetic field means that magnetic flux is changing; hence, a changing electric field is produced. This electric field, in turn, produces a changing magnetic field, and so on, as shown in Figure 27-8. The coupling of the time-varying electric and magnetic fields continuously produces electric and magnetic fields and sets up an electromagnetic wave that propagates in the vacuum. We will study such propagation in the next section.

Equations 27-2 to 27-5 contain integrals of electric and magnetic fields and are called Maxwell's equations in integral form. We can also write Maxwell's equations in terms of derivatives of electric and magnetic fields, and these equations are called Maxwell's equations in differential form.

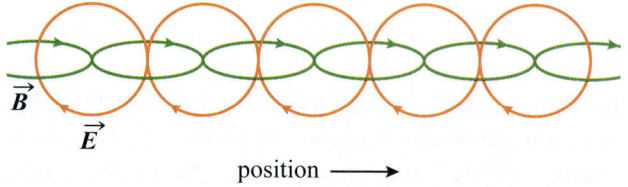

Figure 27-8 An electric field that changes with time produces a magnetic field that also changes with time, and vice versa.

Maxwell published his work in 1873. Although relativity and quantum theory were developed in the next century, Maxwell's equations are consistent with both of these theories. Maxwell's careful synthesis of a large amount of seemingly unrelated experimental data produced one of the most elegant theories in physics.

27-3 Electromagnetic Waves

An **electromagnetic wave** consists of coupled electric and magnetic fields that vary with time. Electromagnetic waves are generated by accelerating charges and by quantum-mechanical processes (see Section 27-7). Once an electromagnetic wave is generated, it propagates on its own and does not require any physical medium to continue its propagation. The wave generates itself because a time-varying electric field generates a time-varying magnetic field, which in turn generates a time-varying electric field. This process is continually repeated.

Electromagnetic waves have the following properties:

1. In a vacuum, all electromagnetic waves travel at the speed of light.

2. At any instant, the electric and magnetic fields of an electromagnetic wave are perpendicular to each other.

3. The direction of motion of the wave is perpendicular to the direction of oscillation of the wave's electric and magnetic fields. Therefore, electromagnetic waves are transverse.

4. Electromagnetic waves are three-dimensional because the electric field, the magnetic field, and the propagation direction of the wave are all perpendicular to each other.

The existence of electromagnetic waves can be derived from Maxwell's equations using vector calculus. However, here we will simply show that a sinusoidal plane electromagnetic wave in a vacuum satisfies Maxwell's equations.

Recall from Chapter 15 that a plane wave is a wave with a fixed frequency whose wave fronts are parallel planes of constant amplitude and propagate perpendicular to the direction of motion of the wave. Therefore, the properties of a plane wave are the same when observed at any point on a given plane perpendicular to the direction of motion of the wave (Figure 27-9). A plane wave is an approximation of a spherical wave that propagates outward from a source. At distances that are much larger than the wavelength of the emitted wave, a spherical wave can be accurately approximated as a plane wave.

Figure 27-10 shows a sinusoidal plane electromagnetic wave travelling in the positive x-direction. The directions of oscillation of the electric field and the magnetic field are chosen as the y-axis and z-axis, respectively. Recall from Chapter 14 that a sinusoidal wave travelling in the positive x-direction is described by the function $\sin(kx - \omega t)$. Here, $k = \dfrac{2\pi}{\lambda}$ is the wave number and

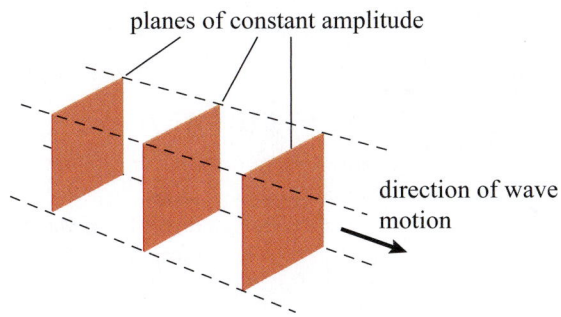

Figure 27-9   A three-dimensional plane wave.

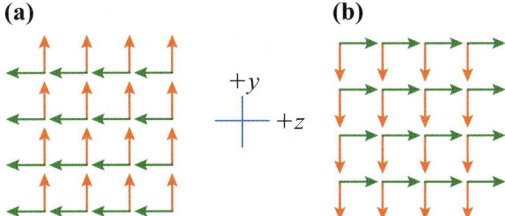

Figure 27-11 (a) A sinusoidal plane electromagnetic wave at an instant when the electric fields (orange arrows) and the magnetic fields (green arrows) reach their maximum strengths in the positive *y*- and *z*-directions, respectively. The electric and magnetic fields are in the plane of the paper, and the wave is travelling toward the reader. (b) The electric and magnetic fields of the wave half a time period later.

$\omega = 2\pi f$ is the angular frequency of the wave. The components of the electric and magnetic fields of the plane electromagnetic wave in Figure 27-10 are as follows:

$$E_x(x,t) = 0, E_y(x,t) = E_0 \sin(kx - \omega t), E_z(x,t) = 0 \tag{27-6}$$

$$B_x(x,t) = 0, B_y(x,t) = 0, B_z(x,t) = B_0 \sin(kx - \omega t) \tag{27-7}$$

where E_0 is the amplitude of the electric field (measured in V/m) and B_0 is the magnitude of the magnetic field (measured in T).

Note that the electric and magnetic fields have the same wavelength, frequency, and phase constant (which we have chosen to be zero). According to Equation 27-6, the magnitude of the electric field at $x = 0$ varies with time as $E_0 \sin(-\omega t) = -E_0 \sin(\omega t)$, and the direction of the electric field parallels the *y*-axis. A charge of $+q$ at $x = 0$ would experience a force of magnitude $qE_0 \sin(-\omega t)$ directed along the *y*-axis. Similarly, the magnitude of the magnetic field at $x = 0$ varies with time as $B_0 \sin(-\omega t)$, and the direction of the magnetic field parallels the *z*-axis.

For a plane electromagnetic wave travelling along the *x*-axis, the magnitudes of the electric and magnetic fields are the same everywhere on a given *yz*-plane.

If the *yz*-plane is the plane of the page with the *x*-axis pointing toward the reader, then at any given instant the electric field has the same magnitude and direction at all points in the plane of the page of the paper. The same is true for the magnetic field. The magnitudes of the electric and magnetic fields oscillate with time, reaching maximum and minimum values in unison (Figure 27-11).

CHECKPOINT 27-3

Field Orientation of Electromagnetic Waves

Figure 27-12 shows four orientations of electric (orange) and magnetic (green) fields in a vacuum at a given instant of time. Which orientations of these fields could represent an electromagnetic wave? Explain why, and determine the direction in which the wave is travelling.

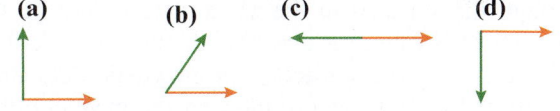

Figure 27-12   Checkpoint 27-3.

ANSWER: (a) and (d) The electric and magnetic fields are perpendicular to each other. These waves are travelling in a direction that is perpendicular to the plane of the page.

We will now consider each of Maxwell's equations as they apply to a plane electromagnetic wave travelling in a vacuum.

Gauss's Law for Electric Fields

Since there are no charges in a vacuum, the net flux of the electric field through any closed surface must be zero. Imagine a cubical Gaussian surface of length *h*, like the one in Figure 27-13(a), in the presence of a plane electromagnetic wave described by Equations 27-6 and 27-7. Since the electric field does not depend on the *y*-coordinate, the magnitude of the electric field entering the cube through the bottom surface is equal

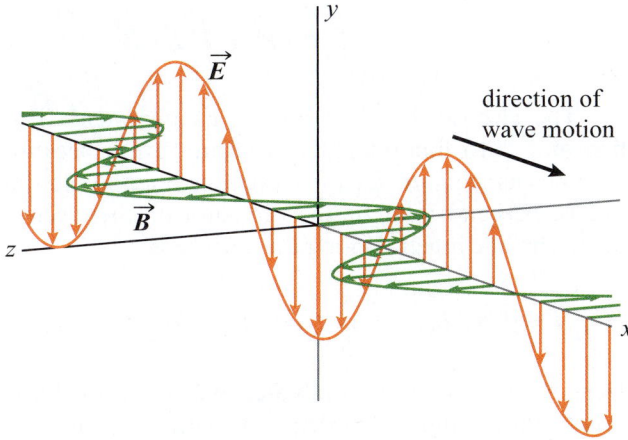

Figure 27-10   A sinusoidal travelling electromagnetic wave at a moment when the wave crests are at the origin of the coordinate system.

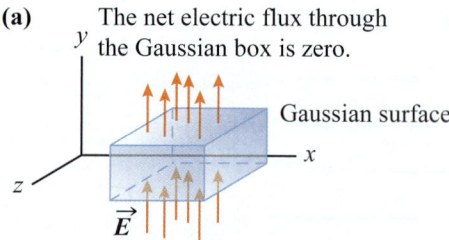

(a) The net electric flux through the Gaussian box is zero.

Gaussian surface

$\vec{E}$

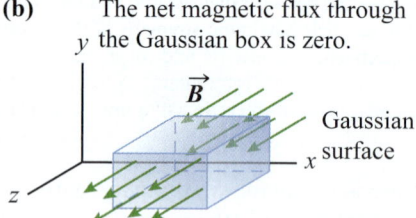

(b) The net magnetic flux through the Gaussian box is zero.

$\vec{B}$

Gaussian surface

Figure 27-13 (a) The electric flux entering the bottom surface in the *xz*-plane of the cube is equal to the flux leaving through the corresponding top surface. (b) Similarly, the magnetic flux entering from the back *xy*-surface of the cube is equal to the flux leaving from the corresponding front surface.

to the magnitude of the field that leaves the top surface of the cube. Therefore, the electric flux entering the bottom surface is equal to the flux leaving the top surface. The electric flux through the surfaces in the *xy*- and *yz*-planes of the cube is zero because the direction of the electric field is parallel to these surfaces. Therefore, the net electric flux through the box is zero, and Gauss's law for electric fields is satisfied.

Gauss's Law for Magnetic Fields

Consider the magnetic flux through the cubical Gaussian surface in Figure 27-13(b). The magnetic flux through the surfaces in the *xz*- and *yz*-planes of the cube is zero because the direction of the magnetic field is parallel to these surfaces. The magnetic field does not depend on the *z*-coordinate, so the magnetic flux entering the back surface in the *xy*-plane is equal to the flux leaving through the parallel front surface. Thus, the net magnetic flux through the Gaussian surface is zero, and Gauss's law for magnetic fields is satisfied.

Faraday's Law

According to Faraday's law, a changing magnetic flux through a closed loop induces an electric field around that loop. The loop may be an imaginary surface or a part of a physical circuit, like a loop of wire. Consider an infinitesimally narrow, rectangular strip in the *xy*-plane, with height *h* and width Δx (Figure 27-14), in the presence of a plane electromagnetic wave described by Equations 27-6 and 27-7. The left edge of the strip is at *x*, and the right edge is at $x + \Delta x$. The width is sufficiently small that the magnitude of the magnetic field passing through the strip can be assumed to remain constant over the entire width. Both the magnetic field

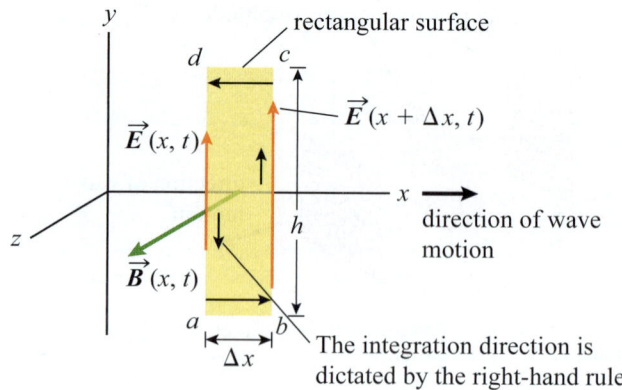

The integration direction is dictated by the right-hand rule.

Figure 27-14 The closed rectangular surface in the *xy*-plane has an infinitesimally small width. The surface is placed in the path of an electromagnetic wave that has $\vec{E}$ along the *y*-axis and $\vec{B}$ along the *z*-axis.

and the normal to the strip are pointing along the *z*-axis. Therefore, the magnetic flux through the strip is the product of the magnitude of the magnetic field, $B_z(x, t)$, and the area of the strip:

$$\Phi_B(x, t) = B_z(x, t)(h\Delta x)$$

The magnetic field is a function of time, so the magnetic flux passing through the strip changes with time, and

$$\frac{\partial}{\partial t}\Phi_B(x, t) = \frac{\partial}{\partial t}[B_z(x, t)(h\Delta x)] = (h\Delta x)\frac{\partial B_z(x, t)}{\partial t} \quad (27\text{-}8)$$

The changing magnetic flux induces an electric field around the strip. *This induced electric field is the electric component of the electromagnetic wave.* The magnetic field points along the positive *z*-axis; therefore, according to the right-hand rule, the circulation of the electric field is in the counter-clockwise direction, as shown in Figure 27-14:

$$\oint \vec{E} \cdot d\vec{\ell} = \int_a^b \vec{E} \cdot d\vec{\ell} + \int_b^c \vec{E} \cdot d\vec{\ell}$$

$$+ \int_c^d \vec{E} \cdot d\vec{\ell} + \int_d^a \vec{E} \cdot d\vec{\ell} \quad (27\text{-}9)$$

The electric field has only a *y*-component. The first and third line integrals therefore vanish because the line elements $d\vec{\ell}$ for these sections point along the *x*-axis, perpendicular to the direction of the electric field. The line integral along the path *bc* is

$$\int_b^c \vec{E} \cdot d\vec{\ell} = (E_y(x + \Delta x, t)\hat{j}) \cdot (h\hat{j}) = hE_y(x + \Delta x, t)$$

Here, the electric field is evaluated at $x + \Delta x$, the location of the *bc* edge of the strip. The path *da* is traversed in the negative *y*-direction, from top to bottom (counter clockwise). Therefore,

$$\int_d^a \vec{E} \cdot d\vec{\ell} = (E_y(x,t)\hat{j}) \bullet ((h)(-\hat{j})) = -hE_y(x,t)$$

Combining contributions from all the sides, the line integral of the electric field around the rectangular strip is

$$\oint \vec{E} \cdot d\vec{\ell} = E_y(x + \Delta x, t)h - E_y(x,t)h \qquad (27\text{-}10)$$

The expression on the right side is related to the derivative of $E_y(x, t)$ with respect to x. By definition, the derivative of a function $f(x)$ of a variable x is related to its values at two nearby points by

$$\frac{df(x)}{dx} = \lim_{\Delta x \to 0} \left(\frac{f(x + \Delta x) - f(x)}{\Delta x} \right)$$

Therefore, as the width $\Delta x \to 0$,

$$E_y(x + \Delta x, t) - E_y(x, t) \Rightarrow \Delta x \frac{dE_y(x, t)}{dx}$$

and

$$\oint \vec{E} \cdot d\vec{\ell} = (h\Delta x) \frac{dE_y(x, t)}{dx} \qquad (27\text{-}11)$$

Inserting Equations 27-11 and 27-8 into Equation 27-4, we get

$$\frac{\partial E_y(x, t)}{\partial x} = -\frac{\partial B_z(x, t)}{\partial t} \qquad (27\text{-}12)$$

In Equation 27-12, we have replaced the total derivative of the electric field by the partial derivative because the electric field depends on both position and time. Equation 27-12 emphasizes that while taking the derivative of $E_y(x, t)$ with respect to x, we must keep t constant. Similarly, while taking the derivative of $B_z(x, t)$ with respect to t, we must keep x constant.

When applied to plane electromagnetic waves, Faraday's law requires that *the rate of change of the electric field with position in the neighbourhood of a given point be equal and opposite to the rate at which the magnetic field changes with time at that point.* Therefore, if the electric field changes with position, then the magnetic field must change with time.

So far, we have not used the fact that the plane electromagnetic wave is a sinusoidal travelling wave. We can evaluate both sides of Equation 27-12 using Equations 27-6 and 27-7:

$$\frac{\partial E_y(x, t)}{\partial x} = \frac{\partial}{\partial x} (E_0 \sin(kx - \omega t)) = kE_0 \cos(kx - \omega t)$$

and

$$-\frac{\partial B_z(x, t)}{\partial t} = -\frac{\partial}{\partial t} (B_0 \sin(kx - \omega t)) = \omega B_0 \cos(kx - \omega t)$$

Equating the expressions and dividing out the cosine term, we get

$$kE_0 = \omega B_0 \qquad (27\text{-}13)$$

Thus, Faraday's law requires that the amplitude of the magnetic field be equal to the amplitude of the electric field divided by the speed of the electromagnetic waves:

$$B_0 = \frac{k}{\omega} E_0 = \frac{E_0}{v_{em}} \qquad (27\text{-}14)$$

where $v_{em} = \omega/k$ is the speed of the electromagnetic waves in a vacuum.

Ampère's Law

The magnetic field of the plane electromagnetic wave points along the z-axis (Equation 27-7). We therefore choose a closed rectangular strip in the xz-plane, as shown in Figure 27-15.

The electric field is normal to the plane of the strip, and the electric flux passing through the strip is the product of the magnitude of the electric field, $E_y(x, t)$, and the area of the strip:

$$\Phi_E(x, t) = E_y(x, t)(h\Delta x)$$

The rate at which the electric flux passing through the strip varies with time is given by

$$\frac{\partial}{\partial t} \Phi_E(x, t) = \frac{\partial}{\partial t} [E_y(x, t)(h\Delta x)] = (h\Delta x) \frac{\partial E_y(x, t)}{\partial t} \qquad (27\text{-}15)$$

The changing electric flux induces a magnetic field around the rectangular strip. *The induced magnetic field forms the magnetic field component of the electromagnetic wave.* The calculation of the line integral of the magnetic field around the strip is similar to the calculation for the electric field. The result is

$$\oint \vec{B} \cdot d\vec{\ell} = (h\Delta x) \frac{\partial B_z(x, t)}{\partial x} \qquad (27\text{-}16)$$

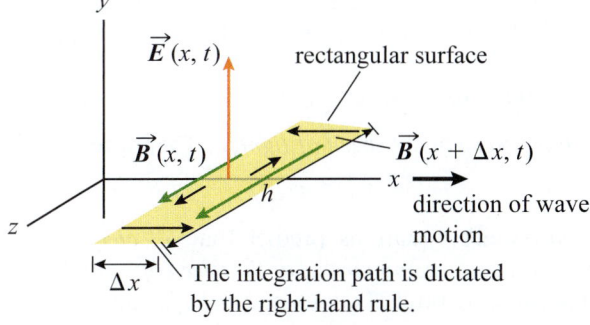

Figure 27-15 The closed rectangular surface has an infinitesimally small width in the xz-plane. The electromagnetic wave has $\vec{E}$ along the y-axis and $\vec{B}$ along the z-axis.

Inserting Equations 27-15 and 27-16 into Equation 27-5 for Ampère's law, we get

$$\frac{\partial B_z(x, t)}{\partial x} = -\varepsilon_0 \mu_0 \frac{\partial E_y(x, t)}{\partial t} \quad (27\text{-}17)$$

When applied to plane electromagnetic waves, Ampère's law requires *that the rate of change of the magnetic field with position in the neighbourhood of a given point be proportional and opposite to the rate at which the electric field changes with time at that point.*

Inserting Equations 27-6 and 27-7 and evaluating the partial derivatives with respect to x and t, we get the following constraint imposed by Ampère's law on electric and magnetic fields of a plane electromagnetic wave:

$$k B_0 = \varepsilon_0 \mu_0 (\omega E_0) \quad (27\text{-}18)$$

The Speed of Electromagnetic Waves

When we combine the constraints of Equations 27-14 and 27-18, we get a remarkable result. Equation 27-18 can be written as

$$B_0 = \varepsilon_0 \mu_0 \left(\frac{\omega}{k}\right) E_0 = \varepsilon_0 \mu_0 v_{em} E_0 \quad (27\text{-}19)$$

Combining Equations 27-19 and 27-14, we get

$$v_{em} = \frac{1}{\sqrt{\varepsilon_0 \mu_0}} \quad (27\text{-}20)$$

This equation shows that the speed of electromagnetic waves in a vacuum depends *only* on the electric permittivity and the magnetic permeability of the vacuum. The speed does not depend on the frequency, wavelength, or amplitude of the waves. Substituting the known values for ε_0 and μ_0 into Equation 27-20 gives

$$v_{em} = \frac{1}{\sqrt{\varepsilon_0 \mu_0}}$$

$$= \frac{1}{\sqrt{(8.85 \times 10^{-12} \text{ C}^2/\text{N} \cdot \text{m}^2)(4\pi \times 10^{-7} \text{ T} \cdot \text{m}/\text{A})}}$$

$$= 3.00 \times 10^8 \text{ m/s}, \quad (27\text{-}21)$$

which is precisely the speed of light, c. Therefore,

$$v_{em} = c$$

So, Maxwell's equations predict that *all electromagnetic waves in a vacuum travel at the speed of light.* From Equations 27-19 and 27-21, using $k = 2\pi/\lambda$ and $\omega = 2\pi f$, we get

$$c = \frac{\omega}{k} = \lambda f \quad (27\text{-}22)$$

To summarize, Maxwell's equations predict the existence of electromagnetic waves with the following properties:

1. Electromagnetic waves travel in a vacuum at the speed of light. The wavelength and frequency of an electromagnetic wave are related to its speed by $\lambda f = c$.
2. The electric and magnetic fields are perpendicular to the direction of propagation of the wave. Electromagnetic waves are therefore transverse waves.
3. The electric and magnetic fields are perpendicular to each other.
4. The amplitudes of electric and magnetic fields of an electromagnetic wave are related such that

KEY EQUATION

$$B_0 = \frac{E_0}{v_{em}} = \frac{E_0}{c}$$

We have shown that sinusoidal plane electromagnetic waves satisfy Maxwell's equations. Electromagnetic waves with complex waveforms can be produced by the superposition of sinusoidal waves of various frequencies and amplitudes, similar to the way that complex sound waves can be formed by a combination of harmonics. Since Maxwell's equations are linear in electric and magnetic fields, such complex electromagnetic waves also satisfy Maxwell's equations.

CHECKPOINT 27-4

Plane Electromagnetic Waves

The four plots in Figure 27-16 show the directions of electric (orange) and magnetic (green) fields at a given instant. Determine whether the electric and magnetic fields in each plot correspond to those of a plane electromagnetic wave. If yes, determine the direction of motion of the wave.

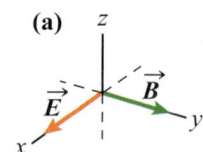 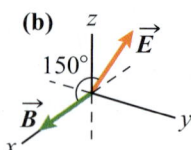

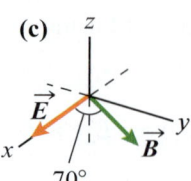 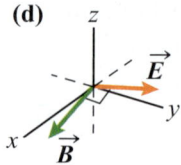

Figure 27-16 Checkpoint 27-4.

ANSWER: (a) Yes, z; (b) no; (c) no; (d) yes, z.

EXAMPLE 27-2

A Helium–Neon Laser

A helium–neon laser produces a beam of light with a wavelength of 633 nm and an electric field strength of 1.00 k V/m. The direction of motion of the beam is taken to be along the z-axis, and the electric field is parallel to the x-axis. A laser beam is an electromagnetic plane wave.

(a) Find the frequency of the laser beam.
(b) What are the amplitude and the direction of the magnetic field of the beam?

Solution

(a) The wavelength and the frequency of this wave are related by Equation 27-22. Rearranging and solving for f, we get

$$f = \frac{c}{\lambda} = \frac{3.00 \times 10^8 \, \text{m/s}}{633 \times 10^{-9} \, \text{m}} = 4.74 \times 10^{14} \, \text{Hz}$$

(b) The amplitude of the magnetic field is related to the amplitude of the electric field by Equation 27-14. Solving for B_0 gives

$$B_0 = \frac{E_0}{c} = \frac{1.00 \times 10^3 \, \text{V/m}}{3.00 \times 10^8 \, \text{m/s}} = 3.33 \times 10^{-6} \, \text{T}$$

The direction of the magnetic field is perpendicular to the direction of motion of the laser beam and that of the electric field; therefore, the direction of the magnetic field must be parallel to the y-axis.

Making sense of the result

(a) Since the wavelength is of the order of 10^{-7} m and the speed of light is approximately 3×10^8 m/s, the frequency of the light should be of the order of 10^{15} Hz.
(b) The magnitude of the electric field is 1000 V/m and the speed of light is approximately 3×10^8 m/s, so the magnitude of the electric field should be of the order of 10^{-6} T.

27-4 The Electromagnetic Spectrum

The wavelength (and, hence, the frequency) of an electromagnetic wave can have any positive value. The largest possible wavelength is limited only by the size of the universe. The shortest wavelength can tend to zero. This range of allowed wavelengths is called the **electromagnetic spectrum**. As shown in Figure 27-17, regions of the electromagnetic spectrum have their own names. We outline below the properties and some applications of electromagnetic waves in each region. Note, however, that the boundaries of these regions are not precisely defined.

Radio Waves

Radio waves ($\lambda \approx 100$ km to 0.3 m, $f \approx 3 \times 10^3$ Hz to 1×10^9 Hz) have the longest wavelengths in the electromagnetic spectrum. Radio waves are used for radio and TV signals, radar, and numerous other applications. Some properties of radio waves depend on the wavelength. For example, shorter-wavelength radio waves can reflect off the ionosphere and thus travel beyond the horizon. Consequently, shortwave radio broadcasts (1.8 MHz to 30 MHz) can be received at long distances from the transmitter.

Many galaxies, including the Milky Way, emit radio waves as well as visible light. The radio emissions from galaxies are generated by fast-moving electrons that

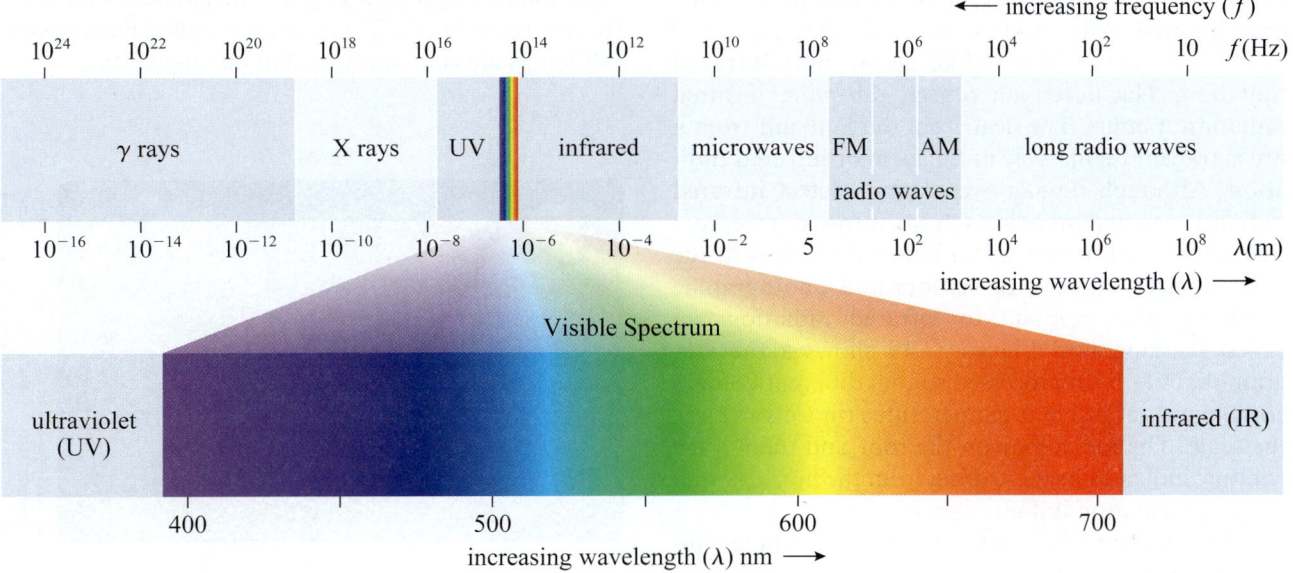

Figure 27-17 The electromagnetic spectrum.

Figure 27-18 A radio frequency image of the central region of the Milky Way.

Figure 27-19 A thermogram of a house at night. The scale at the right shows the temperatures corresponding to the various colours.

spiral around the magnetic field lines of the galaxy. The branch of astronomy that studies radio waves from space is called **radio astronomy**. The study of radio wave emission from galaxies provides extremely useful information that cannot be obtained by observing only visible light. Figure 27-18 shows a radio image of the central region of the Milky Way.

Microwaves

Applications of microwaves ($\lambda \approx 0.3$ m to 0.3 mm, $f \approx 1 \times 10^9$ Hz to 1×10^{12} Hz) include wireless communications (Bluetooth and IEEE 02.11 devices use 2.4 GHz microwaves), cellphone networks, satellite radio (2.3 GHz), garage door openers, the Global Positioning System (GPS), and microwave ovens. Microwaves are also generated by stars, including the Sun.

Infrared Radiation

The infrared region ($\lambda \approx 0.3$ mm to 700 nm, $f \approx 1 \times 10^{12}$ Hz to 4.3×10^{14} Hz) of the electromagnetic spectrum lies between microwaves and visible light. The part of the region closest to microwaves is called **far infrared**, and the part closest to visible light is called **near infrared**. All objects emit some infrared radiation. Even cold objects, like snow, emit infrared radiation. The hotter an object, the more infrared radiation it emits. The heat from the Sun and from a fire is transmitted largely in the form of infrared radiation. Although human eyes cannot detect infrared radiation, many organisms have infrared receptors that allow them to see warm-blooded animals in the dark. Images taken with sensors and photographic emulsions that respond to infrared radiation are called **thermograms**. Figure 27-19 shows a thermogram that has been processed so that different colours correspond to different temperatures (in Celsius); see the scale. The red colour on the roof and the middle window indicates heat escaping from the house, a sign of poor insulation in these areas.

Near-infrared radiation is used in TV remote controls and similar devices. A near-infrared wave causes much less heating than a far-infrared wave.

Visible Light

The narrow band of the electromagnetic spectrum that our eyes can detect is called the **visible spectrum** ($\lambda \approx 750$ nm to 400 nm, $f \approx 4.0 \times 10^{14}$ Hz to 7.5×10^{14} Hz). Although we divide the visible spectrum into colours (often listed as red, orange, yellow, green, blue, indigo, and violet), the wavelength changes continuously within the region, as shown in the enlarged section of Figure 27-17. Human eyes are most sensitive to light with a wavelength of 555 nm, which has a green colour.

Ultraviolet Light

The ultraviolet (UV) region ($\lambda \approx 400$ nm to 10 nm, $f \approx 7.5 \times 10^{14}$ Hz to 3×10^{16} Hz) lies between visible light and X-rays. Ultraviolet radiation has sufficient energy to ionize atoms and cause chemical reactions. Many minerals absorb ultraviolet light and then emit radiation of longer wavelengths, often in the visible spectrum (Figure 27-20). This process is called **fluorescence**. UV lamps are commonly used in tanning salons.

Figure 27-20 Various-coloured light emitted by different minerals when exposed to ultraviolet radiation.

The Sun is a major source of UV radiation, which is divided into three bands: UV-A, UV-B, and UV-C.

- The UV-A band contains wavelengths between 320 nm and 400 nm. Radiation in the UV-A band is not absorbed by the atmosphere and is not particularly harmful to living organisms.

- The UV-B band contains wavelengths between 320 nm and 280 nm. Radiation in this band is partly absorbed by the atmosphere. UV-B radiation is harmful to human skin, and prolonged exposure can cause skin cancer.

- The UV-C band contains wavelengths between 280 nm and 100 nm. Radiation in this band is mostly absorbed by the atmosphere. UV-C radiation is more energetic and harmful to human skin. Prolonged exposure can cause sunburn and eventually skin cancer.

MAKING CONNECTIONS

What Is the UV Index?

The UV index is a numbering system that indicates the level of harm to human skin due to the intensity of UV radiation (Table 27-2). This index was first developed by Environment Canada, and Canada was the first country to issue UV-level forecasts for the public. The UV index is now standardized by the World Health Organization and used worldwide. The higher the category, the more harmful the effect of the UV radiation. A colour is associated with each risk category.

Table 27-2   The UV Index

UV Index	Risk Category	Associated Colour
0–2	Low	Green
3–5	Moderate	Yellow
6–7	High	Orange
8–10	Very high	Red
11 and higher	Extreme	Violet

X-rays

Electromagnetic waves with wavelengths between 3 nm and 0.003 nm are called X-rays ($f \approx 1 \times 10^{17}$ Hz to 1×10^{20} Hz). Wavelengths between 10 nm and 0.10 nm are classified as **soft X-rays**, and wavelengths between 0.10 nm and 0.010 nm are classified as **hard X-rays**. Hard X-rays easily penetrate skin and soft body tissue but are absorbed by dense material such as bones and teeth. Hard X-rays are used for making diagnostic images of the interior of the human body, studying the structure of crystals, and nondestructively testing objects such as structural materials. The X-ray image is either recorded on a film (a **radiograph**) or stored as an electronic image. Different

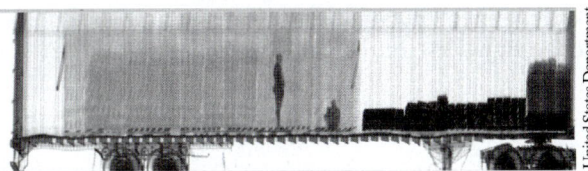

Figure 27-21   A gamma ray image of the inside of a truck. The image clearly shows the boxes and human figures (dummies) inside the truck.

materials appear light or dark depending on the proportion of the X-rays the materials absorb. Calcium in bones absorbs large amounts of X-rays; therefore, bones look white on the radiograph, which is viewed as a negative image. Fat and other soft tissue absorbs less and look grey. Air absorbs very few X-rays, so lungs look black. X-rays are high-energy radiation, and overexposure to them can cause severe damage to human tissue.

Gamma Rays

Gamma rays ($\lambda \approx 0.003$ nm and smaller, $f \approx 1 \times 10^{20}$ Hz and greater) are the most energetic electromagnetic waves. Gamma rays are emitted by nuclei and can also be produced by collisions of particles and antiparticles. Gamma rays are highly penetrating radiation, and in recent years they have been used in different industries, for example, cancer therapy (destroying cancerous cells), food preservation (killing bacteria and other microorganisms in food), sterilizing medical equipment (killing bacteria), and testing structural materials. Gamma ray detectors are also used at seaports and border crossings to image the insides of containers (Figure 27-21).

27-5 The Energy and Momentum of Electromagnetic Waves

Like all waves, electromagnetic waves carry energy and momentum. The energy and momentum flow in the direction of motion of the wave. In this section, we will explore how the energy and momentum of an electromagnetic wave depend on the amplitudes of the wave's electric and magnetic fields.

We know from electrostatics that electric and magnetic fields store energy. Recall that the magnitude of the *electric energy density* (the electric potential energy per unit volume) stored in any electric field, $\vec{E}$, is

$$u_{\mathrm{E}} = \frac{1}{2}\,\varepsilon_0 E^2 \tag{22-20}$$

Similarly, the magnitude of the *magnetic energy density* of a magnetic field, $\vec{B}$, is

$$u_{\mathrm{B}} = \frac{1}{2\mu_0}\,B^2 \tag{25-51}$$

Consider a sinusoidal electromagnetic wave described by Equations 27-6 and 27-7 travelling in a vacuum in the

ELECTRICITY, MAGNETISM, AND OPTICS

positive x-direction. Substituting Equations 27-6 and 27-7 into the equations for the energy densities gives

$$u_E(x, t) = \frac{1}{2}\varepsilon_0 E_y^2 = \frac{1}{2}\varepsilon_0 E_0^2 \sin^2(kx - \omega t)$$

$$u_B(x, t) = \frac{1}{2\mu_0} B_z^2 = \frac{1}{2\mu_0} B_0^2 \sin^2(kx - \omega t) \tag{27-23}$$

Note that the energy densities for a travelling wave are position and time dependent. Since $B_0 = E_0/c$ for an electromagnetic wave and $c = \dfrac{1}{\sqrt{\varepsilon_0 \mu_0}}$, the magnetic energy density can be written as follows:

$$u_B(x, t) = \frac{1}{2\mu_0}\left(\frac{E_0}{c}\right)^2 \sin^2(kx - \omega t)$$

$$= \frac{1}{2}\varepsilon_0 E_0^2 \sin^2(kx - \omega t) \tag{27-24}$$

Thus, *in a travelling electromagnetic wave, the energy density of the magnetic field is equal to the energy density of the electric field.* In any given region of space, the total energy of the electromagnetic field is equally shared between the electric and the magnetic field. This result is a consequence of the symmetry of Maxwell's equations with respect to electric and magnetic fields in a vacuum.

The total energy density of the wave, $u(x, t)$, is the sum of the electric and magnetic energy densities and can be written in several equivalent ways:

$$u(x, t) = u_E(x, t) + u_B(x, t)$$

$$= 2u_E(x, t) = \varepsilon_0 E_0^2 \sin^2(kx - \omega t)$$

$$= \varepsilon_0 c E_0 B_0 \sin^2(kx - \omega t) \tag{27-25}$$

The amount of energy flowing per unit time per unit area perpendicular to the direction of the wave propagation is called the **energy flux** (S) or the **intensity** of the wave. Imagine a 1.0 m^2 rectangular area perpendicular to the path of an electromagnetic wave in a vacuum. Since an electromagnetic wave travels with the speed of light, c, the energy that passes through the rectangular area in one second is equal to the energy contained in a cube of volume 1.0 $m^2 \times c \times 1.0$ s (Figure 27-22). Therefore,

$$S(x, t) = u(x, t)c = \varepsilon_0 c^2 E_0 B_0 \sin^2(kx - \omega t)$$

$$= \frac{E_0 B_0}{\mu_0} \sin^2(kx - \omega t) \tag{27-26}$$

The units of energy flux are W/m^2, which we can confirm by considering the dimensions of the quantities in Equation 27-26:

$$[\mu_0] = \frac{kg \cdot m}{C^2} \qquad [E_0] = \frac{N}{C} \qquad [B_0] = T = \frac{kg}{C \cdot s}$$

So,

$$[S] = \frac{C^2}{m \cdot kg} \times \frac{N}{C} \times \frac{kg}{C \cdot s} = \frac{N}{m \cdot s} = \frac{N \cdot m}{m^2 \cdot s} = \frac{W}{m^2}$$

The energy flow varies with the frequency of the electric and magnetic fields. We do not notice these fluctuations in visible light because the frequency ($\sim 10^{14}$ Hz) is far too fast for us to perceive. The time and position average of the function $\sin^2(kx - \omega t)$ over one oscillation is $\dfrac{1}{2}$. Therefore, we can define the **average intensity** ($\bar{S}$), or **irradiance**, as

KEY EQUATION

$$\bar{S} = \frac{E_0 B_0}{2\mu_0} = \frac{E_0^2}{2\mu_0 c} = \frac{\varepsilon_0 c E_0^2}{2} \tag{27-27}$$

From Equations 27-26 and 27-27, the total average energy density, $\bar{u}$, of an electromagnetic wave is related to the average intensity of the wave by

$$\bar{S} = c\bar{u} \tag{27-28}$$

Notice that as for mechanical waves, the energy carried by an electromagnetic wave is proportional to the square of its amplitude.

Consider a source of power that is emitting spherical electromagnetic waves, for example, a light bulb or an omnidirectional radio antenna. The waves emerging from the source spread outward at the speed of light. At a distance r from the source, the power, P, is distributed over a surface area, $4\pi r^2$. Therefore, the average intensity of the waves, $\bar{S}$, at a distance r from the source is

KEY EQUATION

$$\bar{S} = \frac{P}{4\pi r^2} \tag{27-29}$$

The intensity of electromagnetic waves falls off as the square of the distance from the source (Figure 27-23).

incident electromagnetic wave

1.0 m^2

3.00 $\times$ 10^8 m

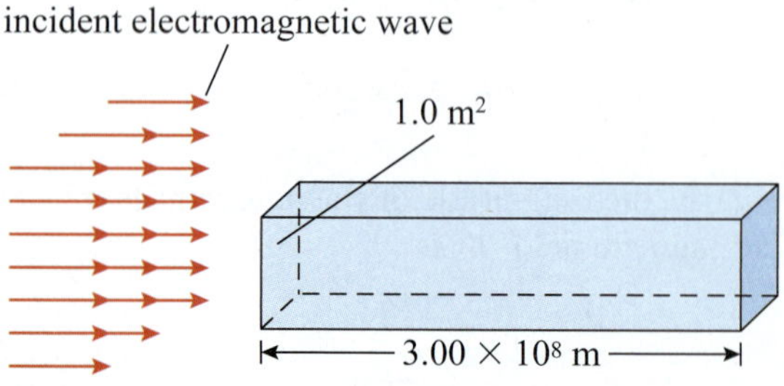

Figure 27-22 The energy flux, S, is defined as the amount of electromagnetic energy that passes through a unit cross-sectional area perpendicular to the path of the incident wave, in one second.

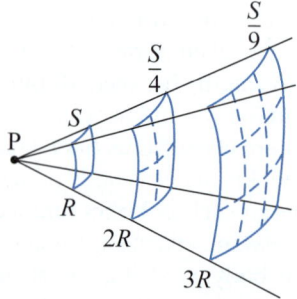

Figure 27-23 The intensity of the three-dimensional spherical waves falls off as the square of the distance from the point source producing the waves.

EXAMPLE 27-3

Solar Power Reaching Earth

The average intensity of sunlight, $\bar{S}$, at the upper atmosphere is approximately $1400\,\text{W/m}^2$. The average distance between Earth and the Sun is 1.50×10^{11} km, and the mean radius of Earth is 6370 km.

(a) What are the amplitudes of the electric and magnetic fields in the sunlight?
(b) What is the average power generated by the Sun?
(c) What is the total power intercepted by Earth?

Solution

(a) From Equation 27-27, the amplitude of the electric field is related to the average intensity by

$$E_0 = \sqrt{2\mu_0 c \bar{S}}$$

$$= [2(4\pi \times 10^{-7}\,\text{T·m/A})(3.00 \times 10^8\,\text{m/s})(1400\,\text{W/m}^2)]^{1/2}$$

$$= 1.03 \times 10^3\,\text{V/m}$$

The amplitude of the magnetic field is

$$B_0 = \frac{E_0}{c} = \frac{1.03 \times 10^3\,\text{V/m}}{3.00 \times 10^8\,\text{m/s}}$$

$$= 3.43 \times 10^{-6}\,\text{T}$$

(b) The average power of the Sun can be calculated from Equation 27-29:

$$P = 4\pi r^2 \bar{S}$$

$$= 4\pi (150 \times 10^9\,\text{m})^2 (1400\,\text{W/m}^2)$$

$$= 3.96 \times 10^{26}\,\text{W}$$

(c) The total power intercepted by Earth is the product of the incident intensity and the cross-sectional area of Earth. At such a large distance from the Sun, electromagnetic waves reaching Earth are plane waves. Earth projects a disk with a radius of 6370 km to these waves. Therefore, the power intercepted by Earth is

$$P_{\text{Earth}} = (\pi r_{\text{Earth}}^2)\bar{S}$$

$$= \pi (6370 \times 10^3\,\text{m})^2 \times (1400\,\text{W/m}^2)$$

$$= 1.78 \times 10^{17}\,\text{W}$$

Making sense of the result

In 2008, the average power consumption of the world was approximately 1.5×10^{13} W. So, the Sun could provide all of this power if sufficiently efficient and cost-effective ways to harness solar power can be developed.

MAKING CONNECTIONS

How Do Microwave Ovens Heat Food?

A microwave oven produces electromagnetic radiation with a frequency of approximately 2.45 GHz ($\lambda \approx 0.12$ m). This radiation is absorbed by water, fats, and sugar molecules contained in the food, thus heating and cooking the food. A water molecule consists of one oxygen and two hydrogen atoms and is electrically neutral overall. However, within the water molecule the oxygen atom is slightly negatively charged, and the hydrogen atoms are slightly positively charged. Therefore, the water molecules act like tiny dipoles. In the presence of an electric field, the positive end of a dipole experiences a force in the direction of the electric field, and the negative end experiences a force in the opposite direction, thus twisting the dipole. When an alternating electromagnetic field is applied to a water molecule, as in a microwave oven, the molecule continuously twists about its centre of charge (Figure 27-24). Thus the water molecules absorb electromagnetic energy from the microwaves and convert it to thermal energy (molecular motion). This thermal energy is then transferred to adjacent non-polar molecules through collisions and thus heating the food. Microwave radiation heats food quickly and effectively because it is absorbed throughout the food. Longer wavelengths would pass through the food, and shorter wavelengths would be absorbed, mainly at the surface. Microwave radiation does not have sufficient energy to remove electrons from the atoms, so it cannot chemically change the food.

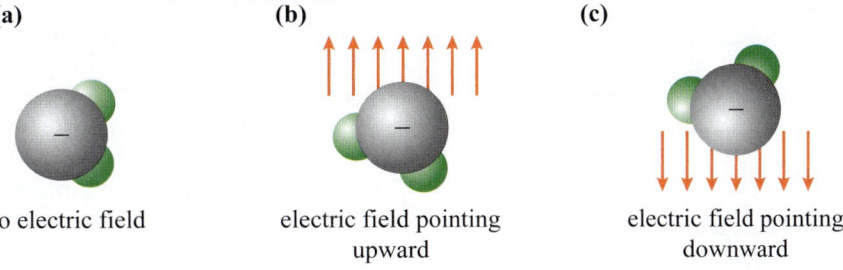

(a) no electric field   (b) electric field pointing upward   (c) electric field pointing downward

Figure 27-24 Orientations of a water molecule with (a) no electric field, (b) the electric field directed upward, and (c) the electric field directed downward.

The Poynting Vector and Wave Momentum

Recall that the cross product of vectors $\vec{a}$ and $\vec{b}$ is a vector that is perpendicular to the plane containing the two vectors. Since the direction of energy flow of an electromagnetic wave is perpendicular to the direction of the electric and magnetic fields, we define a vector, $\vec{S}$, called the **Poynting vector**, as

$$\vec{S} = \frac{1}{\mu_0} \vec{E} \times \vec{B} \qquad (27\text{-}30)$$

The magnitude of $\vec{S}$ is the energy flux S as defined by Equation 27-26, and the direction of $\vec{S}$ is that of the wave propagation, as shown in Figure 27-25. The magnitude of the Poynting vector oscillates with the frequency of the wave. The Poynting vector was introduced by English physicist John Henry Poynting (1852–1914).

Electromagnetic Wave Momentum

Consider an electromagnetic wave incident on a collection of charges. Even if all the charges are initially at rest, the electromagnetic wave exerts a force on the charges, causing them to move and hence gain momentum. By the law of conservation of momentum, an increase in the momentum of the charges is equal to the loss in the wave's momentum. Therefore, in addition to energy, an electromagnetic wave must also carry momentum.

According to quantum theory, electromagnetic waves are made of massless particles called photons (see Chapter 31). The relationship between the energy and the momentum of a fast-moving particle is given by Einstein's relativistic energy-momentum relationship (Chapter 30, Equation 30-38):

$$E = \sqrt{m^2 c^4 + p^2 c^2}$$

where E is the energy, p the momentum, m is the rest mass of the particle, and c is the speed of light in vacuum. The

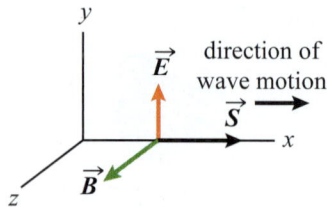

The electric and magnetic fields are in the yz-plane.

Figure 27-25 The Poynting vector $\vec{S}$ of an electromagnetic wave.

energy, E, and the magnitude of the momentum, p, of a photon ($m = 0$) are therefore related by

$$p = \frac{E}{c} \qquad (27\text{-}31)$$

Because electromagnetic waves are made of photons, the relationship 27-31 also applies to these waves. The momentum carried by an electromagnetic wave is equal to the energy of the wave divided by the speed of light. It follows from the above equation that the magnitude of the time-averaged **radiation momentum density** (momentum per unit volume), $\bar{g}_{rad}$, of an electromagnetic wave equals its time-averaged energy density, $\bar{u}$ (Equation 27-28), divided by c:

$$\bar{g}_{rad} = \frac{\bar{u}}{c} = \frac{1}{c}\left(\frac{\varepsilon_0 E_0^2}{2}\right) \qquad (27\text{-}32)$$

Because the direction of energy and momentum flow of a particle is along its direction of motion, the momentum of an electromagnetic wave points along the direction of the energy flow. As defined in Equation 27-30, the Poynting vector, $\vec{S}$, describes the direction of energy flow of an electromagnetic wave. Therefore, the radiation momentum density, $\vec{g}_{rad}$ (a vector quantity that describes both the magnitude of the momentum density and the direction of the momentum flow), of an electromagnetic wave is related to $\vec{S}$ by

$$\vec{g}_{rad} = \frac{\vec{S}}{c^2} \qquad (27\text{-}33)$$

The momentum density equals the energy flux divided by c^2. This relationship between the energy flux of a wave and its momentum density is also true for other types of waves, for example, sound waves.

The relationship of Equation 27-33 can also be derived using Maxwell's equations and was known before the advent of quantum theory. However, that derivation involves the use of vector calculus, a branch of mathematics that you may not have studied yet.

27-6 Radiation Pressure

Radiation pressure is defined as the force per unit area exerted by electromagnetic waves on an object. Light falling on the surface of an object, such as black paper, is absorbed by the surface. As the light is absorbed, it transfers its momentum to the object, thus exerting a force on the object. By definition, the force per unit area on a surface is the pressure exerted on that surface. Therefore, light exerts pressure on an absorbing surface.

Now consider electromagnetic radiation with average intensity $\bar{S}$ incident on a surface of area A

that absorbs all the incident radiation. The energy Δu absorbed by the surface in time Δt is given by

energy absorbed = average intensity × surface area × Δt

$$\Delta \bar{u} = \bar{S}A\Delta t$$

Since all the radiation is absorbed, the total momentum transferred to the surface, Δp, is equal to the time-averaged **radiation momentum**, $\bar{g}_{rad}$, given by Equation 27-32:

$$\Delta p = \frac{\Delta \bar{u}}{c} = \frac{1}{c}(\bar{S}A\Delta t)$$

The rate of change of momentum per unit area of the surface is, therefore,

$$\frac{1}{A}\left(\frac{\Delta p}{\Delta t}\right) = \frac{\bar{S}}{c}$$

The quantity on the left side is the pressure exerted on the surface by the radiation, P_{rad}. Therefore, radiation pressure on a perfectly absorbing surface is

KEY EQUATION
$$P_{rad} = \frac{\bar{S}}{c} \qquad (27\text{-}34)$$

If the surface is a perfect reflector, then the change in the momentum of the reflected waves is $2\Delta p$ for normal incidence (just like the change in the momentum of a ball when it rebounds off a perfectly elastic surface is twice the incident momentum). The pressure exerted on a perfectly reflecting surface is therefore,

KEY EQUATION
$$P_{rad} = 2\frac{\bar{S}}{c} \qquad (27\text{-}35)$$

EXAMPLE 27-4

Radiation Pressure at Earth's Surface

Estimate the radiation pressure at Earth's surface. Is this pressure uniformly distributed over Earth's entire surface at any given time?

Solution

The intensity of solar radiation at Earth's surface is approximately 1.00×10^3 W/m^2. It is less than the intensity at the top of Earth's atmosphere due to absorption in the atmosphere. Assuming that all of the sunlight is reflected from Earth's surface, the radiation pressure exerted on Earth's surface is approximately

$$P_{rad} = 2\frac{1.00 \times 10^3 \text{ W/m}^2}{3.00 \times 10^8 \text{ m/s}} = 6.67 \times 10^{-6} \text{ Pa}$$

The amount of sunlight reflected by Earth's surface varies considerably across the surface and depends on the presence of cloud cover, snow, and other variables. On the other hand, if we assume that all of the sunlight is absorbed by Earth, then the radiation pressure exerted on Earth's surface will be half the calculated value. The actual value is therefore between 3.33×10^{-6} Pa and 6.67×10^{-6} Pa. The radiation pressure is only exerted on the side of Earth that is facing the Sun.

Making sense of the result

The atmospheric pressure at sea level is 1.01×10^5 Pa, so the radiation pressure is approximately 7×10^{-11} times the atmospheric pressure. That is why we do not feel radiation pressure.

EXAMPLE 27-5

Solar Sailing: IKAROS

On May 21, 2010, the Japan Aerospace Exploration Agency launched a spacecraft called IKAROS (Interplanetary Kite-craft Accelerated by Radiation of the Sun) toward Venus to demonstrate propulsion by a solar sail. The sail of the spacecraft (Figure 27-26(a)) consists of 200. m^2 of a reflective polyimide resin membrane as thin as 0.0075 mm. The mass of the solar sail is 15 kg, and the total mass of the spacecraft is 315 kg. Once clear of Earth, the spacecraft is propelled toward Venus by pressure from solar radiation. Estimate the magnitude of acceleration generated by radiation pressure on the sail.

Solution

Let $\bar{S}_E$ be the solar radiation intensity just above Earth's atmosphere. When the spacecraft is at a distance r from the Sun, the solar intensity at this distance, $\bar{S}(r)$, can be related to $\bar{S}_E$ using Equation 27-29:

$$\bar{S}(r) = \bar{S}_E\left(\frac{r_E^2}{r^2}\right)$$

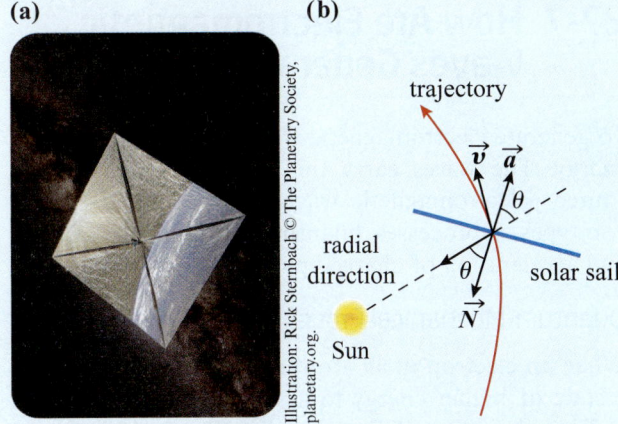

(a) (b)

Figure 27-26 (a) The IKAROS solar sail. (b) The orientation of the solar sail with respect to the Sun. The normal to the sail ($\vec{N}$) makes an angle of θ with respect to the radial direction. The red curve shows the trajectory of the spacecraft.

(*continued*)

where r_E is the distance between Earth and the Sun. Assuming that all incident radiation is reflected, we can use Equation 27-35, so the radiation pressure at a distance r from a Sun is

$$P_{rad}(r) = \frac{2\bar{S}(r)}{c}$$

For a solar sail with cross-sectional area A perpendicular to the incident radiation, the magnitude of the force exerted by the radiation is

$$F_{rad}(r) = A \times P_{rad}(r) = \frac{2A\bar{S}(r)}{c}$$

For a general trajectory (Figure 27-26(b)), the sail makes an angle of θ with respect to the radial direction between the Sun and the spacecraft. So, the effective area perpendicular to the direction of the incident radiation is $A\cos\theta$ (normal incidence corresponds to $\theta = 0°$). The magnitude of acceleration generated by the radiation pressure is

$$a(r) = \frac{F_{rad}(r)}{m} = \bar{S}_E\left(\frac{r_E^2}{r^2}\right)\frac{2(A\cos\theta)}{mc}\cos\theta$$

where m is the total mass of the spacecraft.

The extra factor of $\cos\theta$ occurs because the acceleration is directed perpendicular to the plane of the sail, at an angle of θ with respect to the radial direction.

Let us assume that $r = r_E$, $\theta = 0°$ (normal incidence), and $\bar{S}_E = 1400$ W/m^2. Then the acceleration due to radiation pressure when the spacecraft is close to Earth is approximately

$$a(r_E) = \frac{(1400 \text{ W/m}^2)2(200. \text{ m}^2)}{(315 \text{ kg})(3.00 \times 10^8 \text{ m/s})} = 5.9 \times 10^{-6} \text{ m/s}^2$$

Making sense of the result

The acceleration from solar radiation is very small. The effects of gravity from Earth and the Sun would have to be included when calculating the trajectory of the spacecraft.

CHECKPOINT 27-5

Solar Sail

Assume that the speed of the IKAROS spacecraft when it leaves the atmosphere is 7 km/s and that the acceleration due to solar radiation pressure remains constant at 6×10^{-6} m/s^2. How long would it take the spacecraft to double its speed?

(a) approximately 6 months
(b) between 5 and 10 years
(c) between 10 and 50 years
(d) over 100 years

ANSWER: (c)

27-7 How Are Electromagnetic Waves Generated?

To generate electromagnetic waves, a source must lose energy. The waves carry this energy away from the source. Electromagnetic waves are generated during two types of processes: quantum-mechanical processes and acceleration of charged particles.

Quantum-Mechanical Processes

When an electron in an atom makes a transition from a state of higher energy to a state of lower energy, it radiates the energy difference as electromagnetic waves. Infrared light, visible light, ultraviolet light, and X-rays can be generated by this process. We will discuss the process of emission of light from atoms in Chapter 32.

Protons and neutrons in a nucleus can also make transitions from a state of higher energy to a state of lower energy, emitting gamma rays in the process. We will discuss the process of emission of gamma rays from a nucleus in Chapter 34.

In matter–antimatter annihilation, a charged particle can combine with its antiparticle (which has an equal but opposite charge), and the mass of both particles can be converted to energy in the form of electromagnetic radiation. Conversion of mass to energy is discussed in Chapter 30, but the details of annihilation processes are beyond the scope of this book.

Acceleration of Charged Particles

Electromagnetic waves cannot be generated by stationary charges or charges that are moving with a constant velocity. A stationary charge has an electric field but no magnetic field. Therefore, it cannot generate an electromagnetic wave. The electric field of a stationary charge remains *attached* to the charge, never leaving it.

Now consider a charge moving with a constant velocity. In a frame of reference in which the charge is moving with a uniform velocity, it generates both electric and magnetic fields. However, in the frame of reference of the charge, the charge is at rest and therefore has only an electric field. So, in its own frame of reference, a charge cannot emit electromagnetic radiation. The laws of physics must be the same in all frames that move with a uniform velocity with respect to each other. So, if a charge cannot emit electromagnetic radiation in its own reference frame, it cannot emit electromagnetic radiation in any frame moving with uniform velocity with respect to the charge.

Thus, a charge can emit electromagnetic radiation in free space *only when accelerating*. Note the qualifier "in free space." A charge moving with a uniform velocity through a dielectric medium can emit electromagnetic radiation. Such radiation is called **Cerenkov radiation**.

We briefly discuss a few examples of electromagnetic radiation emitted during accelerated motion of charged particles.

A charge (or an electric current) undergoing simple harmonic motion Simple harmonic motion is accelerated motion. Therefore, a charged particle or an electric current undergoing oscillatory motion emits electromagnetic waves. The frequency of the radiated wave is equal to the oscillation frequency of the motion. Harmonic motion of charges underlies the generation of electromagnetic waves for radio, TV, and other types of communication signals.

A charged particle moving in a curved orbit Motion along a curved path is accelerated motion. Therefore, a charged particle moving in a curved trajectory radiates electromagnetic waves, called **synchrotron radiation** or **cyclotron radiation**. The term *synchrotron radiation* is generally used for radiation emitted by a charged particle moving rapidly in a magnetic field. In the presence of a magnetic field, a charged particle experiences a force that is perpendicular to both the particle's velocity and the direction of the magnetic field. This force causes the particle to spiral in the direction of the magnetic field (Figure 27-27), continuously emitting electromagnetic radiation. Synchrotron radiation is also generated when fast-moving electrons spiral through the intense magnetic fields of astronomical objects.

A charged particle stopping abruptly When a fast-moving charged particle stops abruptly, it undergoes large deceleration and emits electromagnetic radiation. Such radiation is called **bremsstrahlung**, a German word meaning "braking radiation." The X-ray machine in a dentist's office generates X-rays by firing a beam of fast-moving electrons at a metal plate. When electrons collide with the plate, they come to an abrupt stop, emitting X-rays. The

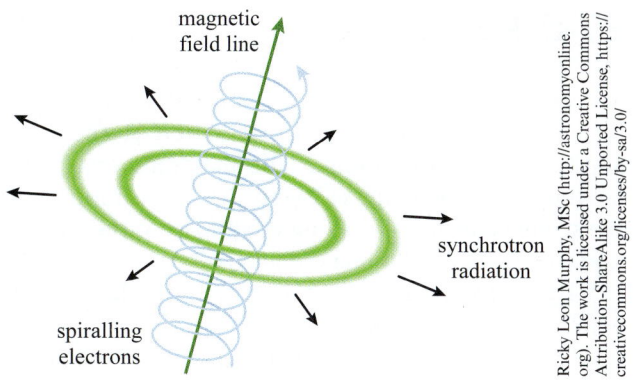

Figure 27-27   The spiral motion of a charged particle in a magnetic field generates synchrotron radiation.

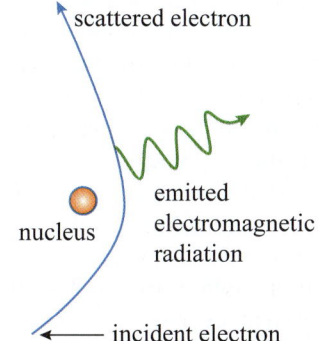

Figure 27-28   An electron moving within the vicinity of a nucleus is deflected by the electric and magnetic forces between the two.

electrons actually follow a curved path as they decelerate, as shown in Figure 27-28. The electric fields of the nuclei of the metal atoms deflect the electrons.

MAKING CONNECTIONS

Half-Wave Dipole Antennas

A dipole antenna consists of two conducting metal wires (or rods) of equal length, separated by a narrow insulating gap. Dipole antennas can be used for both generating and receiving electromagnetic waves. To generate waves, the two wires are powered, often by connecting them to an RLC circuit, as shown in Figure 27-29. (An RLC circuit is an electrical circuit consisting of a resistor, an inductor, and a capacitor; see Chapter 26.) The current and voltage in the wires vary sinusoidally with frequency, f_0, the resonant frequency of the RLC circuit. The sinusoidal output of the RLC circuit forces electrons in the antenna to undergo harmonic motion, and the accelerated electrons emit electromagnetic waves at frequency f_0, which then propagate away from the antenna. The frequency of the emitted waves can be changed by changing the inductance or the capacitance in the RLC circuit.

For maximum efficiency, the total length of the antenna should be approximately half the wavelength of the electromagnetic waves emitted. So, for example, if the frequency of the RLC circuit is 98 MHz (which is approximately in the middle of the FM radio band in North America), the wavelength of the emitted waves is

$$\lambda_0 = \frac{c}{f_0} = \frac{3.00 \times 10^8 \text{ m/s}}{98.0 \times 10^6 \text{ Hz}} = 3.06 \times 10^2 \text{ cm}$$

Therefore, the optimal total length of the antenna is approximately 1.5 m and each wire would be approximately 75 cm long (depending on the length of the gap between the two wires). The exact length of the antenna is also affected by other factors, such as the thickness of the wires and the height of the antenna above the ground.

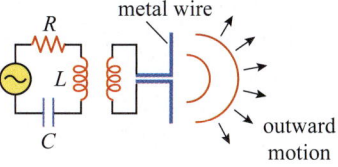

Figure 27-29   A dipole antenna connected to an RLC circuit emits electromagnetic waves at the resonant frequency of the RLC circuit.

27-8 Polarization

The **polarization** of a transverse wave at a given point is defined as the direction of oscillation of the wave at that point. For example, for the transverse wave moving along the string in Figure 27-30, the motion of the string is taken to be along the y-axis, so the wave is said to be polarized along the y-axis.

By convention, the polarization of electromagnetic waves is described by specifying the orientation of the wave's *electric* field at a given point over one time period. The electric field of the plane electromagnetic wave in Figure 27-31(a) is oriented along the y-axis, so the wave is polarized along the y-axis. When the electric field is oriented along a single line, the polarization is called **linear polarization**, and the wave is **linearly polarized**. The plane containing the electric field and the direction of motion of the wave is called the **plane of polarization**. The wave shown in Figure 27-31(a) is linearly polarized along the y-axis, and its plane of polarization is the yz-plane.

Since the electric and magnetic fields of electromagnetic waves are always perpendicular to the direction of motion, these waves are always polarized perpendicular to their direction of motion. Figure 27-31(b) shows a wave linearly polarized at an angle of $45°$ in the xy-plane.

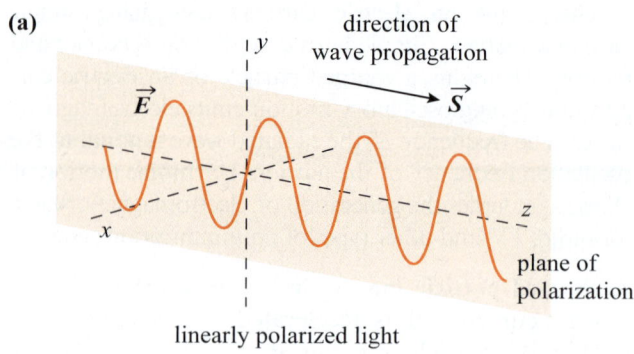

(a)

linearly polarized light

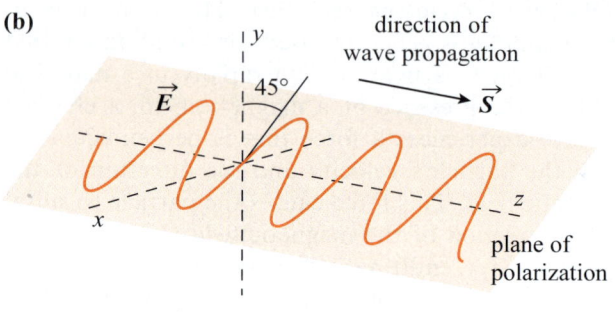

(b)

linearly polarized light

Figure 27-31 (a) A linearly polarized electromagnetic wave with its electric field along the y-axis. (b) An electromagnetic wave linearly polarized at an angle of $45°$ in the xy-plane. The shaded plane is the plane of polarization.

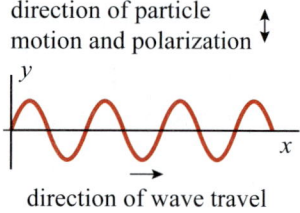

Figure 27-30 A wave on a string is polarized perpendicular to its direction of motion.

CHECKPOINT 27-6

Polarization

A linearly polarized electromagnetic wave is travelling along the x-axis, and its magnetic field is oscillating along the y-axis. Which of the following statements is correct?

(a) The wave is polarized along the y-axis, and its plane of polarization is the xy-plane.

(b) The wave is polarized along the x-axis, and its plane of polarization is the xy-plane.

(c) The wave is polarized along the z-axis, and its plane of polarization is the xz-plane.

(d) The wave is polarized along the y-axis, and its plane of polarization is the xz-plane.

ANSWER: (c) Because the wave is moving along the x-axis, the electric and magnetic fields are in the yz-plane. Since the magnetic field is along the y-axis, the electric field must be along the z-axis. Therefore, the wave is polarized along the z-axis and its plane of polarization is the xz-plane.

Unpolarized Light

Light from the Sun and from a common light bulb consists of a large number of waves emitted by individual atoms. In general, an electromagnetic wave emitted by one atom has no correlation with the waves emitted by other atoms, and the electric fields of individual waves point in random directions. For example, in a 50 W light bulb, approximately 10^{19} atoms emit individual electromagnetic waves with randomly oriented electric fields every second. Such light is **unpolarized**. The electric field vector of unpolarized light consists of a random superposition of a very large number of linearly polarized waves (Figure 27-32).

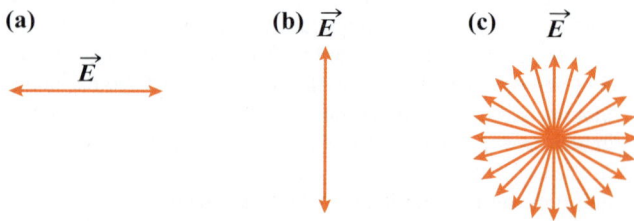

(a)

(b)

(c)

Figure 27-32 (a) An electromagnetic wave polarized along the horizontal direction. (b) An electromagnetic wave polarized along the vertical direction. (c) An unpolarized electromagnetic wave consisting of random superposition of linearly polarized waves. The waves are travelling perpendicular to the page.

Polarization by Absorption

We can produce polarized light from a beam of unpolarized light by filtering out all but one of the polarizations. **Polarizers** are materials that only allow electric fields with a particular orientation to pass through. The electric field of the transmitted light is oriented in that direction and is therefore polarized.

A commonly used polarizer is Polaroid film, which is a flexible, plastic sheet made of polyvinyl alcohol (PVA). The sheet is heated and stretched to align the PVA molecules into long, parallel rows. The sheet is then dipped into an iodine dye. The iodine atoms, which have loosely bound outer electrons, bond to the long chains of PVA molecules, making them conductive. The PVA chains can then act as long, extremely thin, closely spaced metal wires. When unpolarized light falls on the plastic film, the electric field of the light exerts a force on the free-moving iodine electrons. The electrons can move along the length of the stretched chains of molecules, so the component of the electric field that is parallel to the direction of these chains is easily absorbed by the electrons. However, the iodine electrons cannot readily move perpendicular to the stretched chains and therefore cannot absorb the component of the electric field in this direction. Consequently, this perpendicular component of the electric field passes through the material. The direction along which the electric field is transmitted is called the **transmission axis**, or **polarizing axis**. Note that the transmission axis of Polaroid film is perpendicular to the axis of long-chained molecules.

Figure 27-33 shows unpolarized light incident on a polarizer that lies in the plane of the page. The red arrow shows the direction of the electric field, $\vec{E}$, of one of the randomly polarized light waves. The transmission axis of the polarizer is along the y-axis, and the electric field makes an angle of θ with respect to the transmission axis. The electric field can be resolved into a component along the transmission axis ($E_y = E \cos \theta$) and a component perpendicular to the transmission axis ($E_x = E \sin \theta$). Only the E_y-component passes through the polarizer.

When a polarized beam of light passes through a polarizer, the intensity of the transmitted light depends on the angle between the light's initial polarization direction and the transmission axis of the polarizer. Let I_0 be the intensity of a polarized beam of light that falls on a polarizer. We denote the angle between the polarization direction of the light and the transmission axis of the polarizer as θ and the intensity of light that is transmitted through the polarizer as $I(\theta)$. Then

KEY EQUATION
$$I(\theta) = I_0 \cos^2 \theta \qquad (27\text{-}36)$$

This relationship was discovered experimentally by Étienne-Louis Malus in 1809 and is called **Malus's law**. We can derive Malus's law from the properties of electromagnetic waves.

Consider the arrangement shown in Figure 27-34. A beam of unpolarized light is incident on polarizer P_1, which polarizes the light along its transmission axis. Let the amplitude of the electric field that is transmitted through P_1 be E_0, and the intensity of the transmitted light be I_0. A second polarizer, P_2 (called an **analyzer**), is placed in front of the polarized beam with its polarizing axis at an angle of θ with respect to the transmission axis of P_1. The amplitude of the electric field transmitted through P_2 is $E_0 \cos \theta$. The intensity of light is proportional to the square of the amplitude of the electric field (Equation 27-27); therefore, the intensity of the light transmitted through P_2 is

$$I(\theta) = \frac{1}{2} c \varepsilon_0 (E_0 \cos \theta)^2 = \left(\frac{1}{2} c \varepsilon_0 E_0^2 \right) \cos^2 \theta$$
$$= I_0 \cos^2 \theta \qquad (27\text{-}37)$$

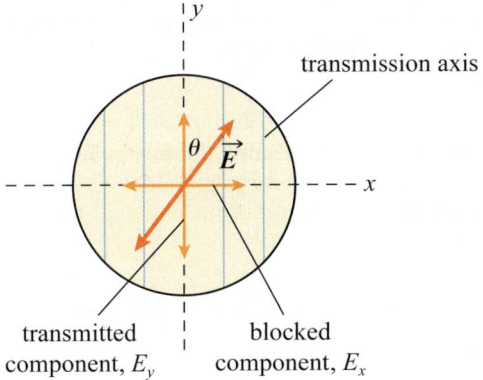

Figure 27-33 Only the component of the electric field that is parallel to the transmission axis of the polarizer is transmitted. The component that is perpendicular to the transmission axis is blocked.

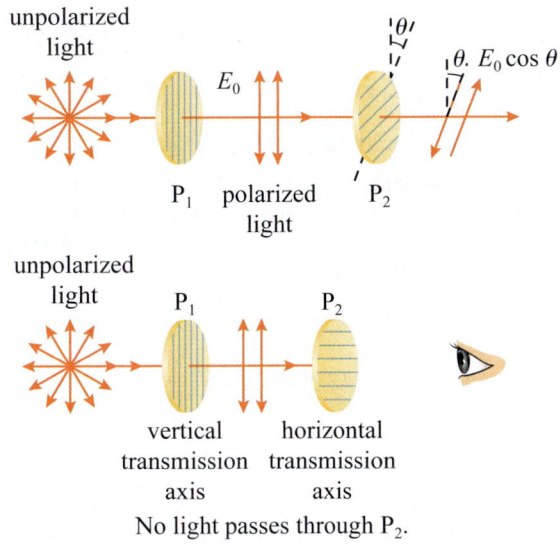

Figure 27-34 When an electromagnetic wave passes through a polarizer, only the component of the electric field that is parallel to the transmission axis of the polarizer is transmitted. The component that is perpendicular to the transmission axis is blocked.

Note these consequences of Malus's law (see Figure 27-34):

- When the polarizing axes of two polarizers are parallel to each other ($\theta = 0°$ or $\theta = 180°$), the intensity of the light passing through the analyzer remains I_0. The analyzer does not block any light that is polarized along its transmission axis.

- When the polarizing axes of two polarizers are perpendicular to each other ($\theta = 90°$), the intensity of the light passing through the analyzer is zero. The analyzer completely blocks the light that is polarized perpendicular to its transmission axis.

CHECKPOINT 27-7

Malus's Law

A beam of light is polarized along the vertical axis and passes through two polarizers. The transmission axis of the first polarizer (P_1) is inclined at 45° with respect to the light's polarization direction. The transmission axis of the second polarizer (P_2) is inclined at 45° with respect to P_1 so that it is perpendicular to the polarization direction of the incident beam. The intensity of the initial beam is I_0. What is the intensity of the beam after it passes through P_2?

(a) zero

(b) $\dfrac{I_0}{4}$

(c) $\dfrac{I_0}{2}$

(d) I_0

ANSWER: (b) The electric field is reduced by a factor of $\sqrt{2}$ after passing through the first analyzer; therefore, the intensity is halved after emerging from the first analyzer. The electric field is again reduced by $\sqrt{2}$ after passing through the second analyzer, so the intensity is halved again.

Polarization by Reflection

A wave travelling in a medium divides into two parts when it encounters the boundary of another medium. The part that is reflected into the original medium is called the **reflected wave**. The part that is transmitted into the second medium is the **transmitted wave**. The plane containing the direction of the incident wave and the normal to the surface between the two media is called the **plane of incidence** (Figure 27-35).

Consider a beam of unpolarized light incident at an angle to the surface of a flat glass plate in air. The electric field of the incident beam can be resolved into two components: one in the plane of incidence and the other perpendicular to it, as shown in Figure 27-36. For unpolarized light, both components have the same amplitude.

The reflected and transmitted waves are generated by electrons on the surface of the glass plate. The incident

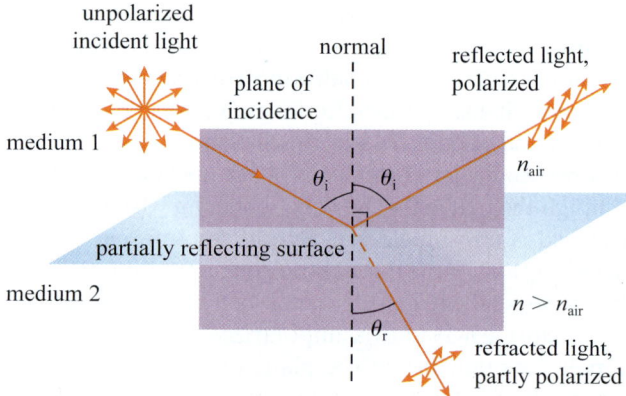

Figure 27-35 The incident, reflected, and transmitted waves at the boundary between two media. Here θ_i is the angle of incidence with respect to the normal to the reflecting surface and θ_r is the angle of refraction.

EXAMPLE 27-6

Beam Intensity through Two Polarizers

A beam of unpolarized light with intensity I_0 is incident on a polarizer with a vertical transmission axis. The beam then passes through a second polarizer, the transmission axis of which is at an angle of 60° to the vertical.

(a) What is the intensity of the light after it passes through the first polarizer?

(b) What is the intensity of the beam that emerges from the second polarizer in terms of I_0?

Solution

(a) We can resolve the electric field of the unpolarized light into two perpendicular components. We choose one component to be in the vertical direction and the other component to be in the horizontal direction. For unpolarized light, these two components have equal magnitudes.

When the light passes through the first polarizer, the vertical component is transmitted and the horizontal component is absorbed. Therefore, half the incident intensity passes through the first polarizer:

$$I_1 = \frac{I_0}{2}$$

(b) The light that passes through the first polarizer is linearly polarized. We can now apply Malus's law to find its intensity after transmission through the second polarizer:

$$I_2 = I_1 (\cos 60°)^2 = \frac{I_0}{2} \times \left(\frac{1}{2}\right)^2 = \frac{I_0}{8}$$

Making sense of the result

Only 12.5% of the initial intensity emerges from the second polarizer.

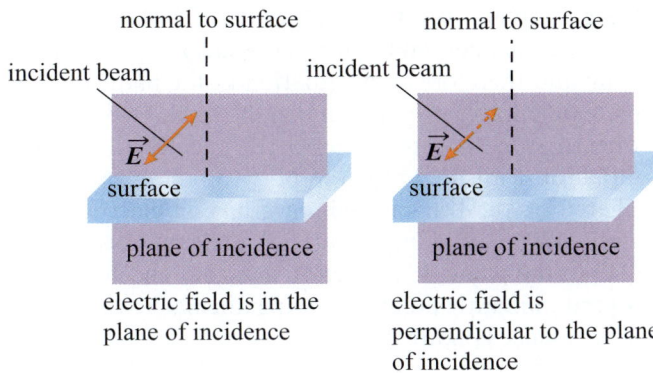

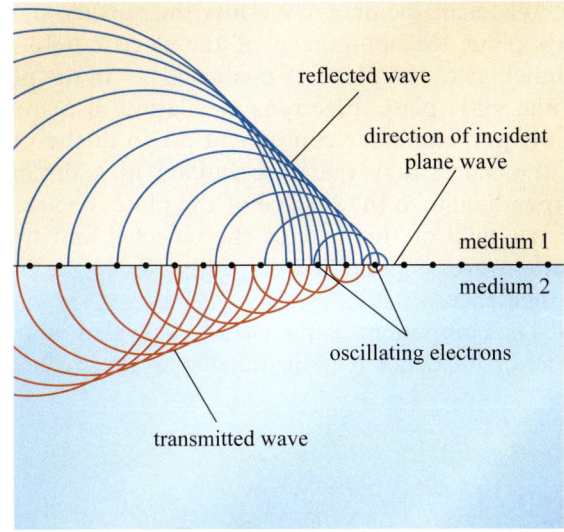

Figure 27-36 The electric field of the incident beam of light can be resolved into a component in the plane of incidence (the plane of the page) and a component perpendicular to the plane of incidence.

electric field exerts an oscillatory force on the electrons, causing them to oscillate. Each oscillating electron becomes a source of electromagnetic waves that propagate outward, forming the reflected and transmitted waves (Figure 27-37).

Figure 27-37 When subjected to an incident electromagnetic wave, the atomic electrons at the interface of two surfaces act as sources of spherical electromagnetic waves that form the reflected and the transmitted waves.

MAKING CONNECTIONS

Polarization and 3D Movies

Hold a finger about 30 cm in front of your nose and look at it by alternately closing your left and right eye every couple of seconds. You will see a slightly different view of your finger each time. The location of the image slightly shifts from left (when the right eye is shut) to right (when the left eye is shut). Your left eye sees more of the left side of the finger, and your right eye sees more of its right side. When viewing with both eyes open, your brain superimposes the two images together, allowing you to see in three dimensions. This is known as *stereoscopic vision*.

Three-dimensional (3D) movies work in the same way. 3D movies are filmed by using two lenses that are as far apart from each other as the average distance between two human eyes, about 6.5 cm (Figure 27-38(a)). Each lens films a scene from a slightly different perspective. This difference in perspective is called parallax. The film for 3D movies therefore contains more information than film shot with a single lens.

In a movie theatre, the two reels of the film (one filmed by each lens) are projected onto a screen through a special projector with two lenses. Each lens has a polarizer filter in front of it. The transmission axes of the two filters are perpendicular to each other. So if one filter transmits horizontally polarized light, then the other transmits vertically polarized light. These horizontally and vertically polarized images reflect from the screen and enter the audience's eyes.

People in the audience wear special glasses made of polarizing filters (Figure 27-38(b)), with the transmission axes of the two filters perpendicular to each other. The polarizer for the left eye allows only the horizontally polarized light to pass through, and that for the right eye allows only the vertically polarized light to pass. So each eye sees a different image with a slightly different perspective, and when these two images superimpose on the retina, the brain is "fooled" into perceiving a 3D image.

(a)

© Mike K. | Dreamstime.com

(b)

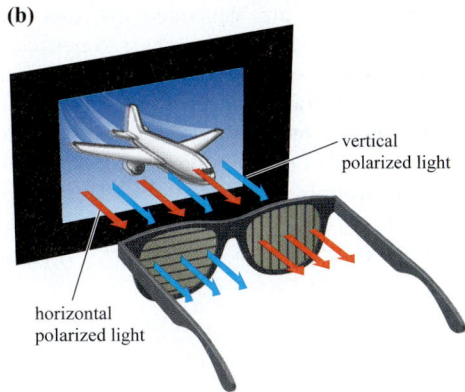

Figure 27-38 (a) A 3D movie camera with two lenses. Each lens films a slightly different perspective of a scene. (b) The horizontally and vertically polarized images pass through different eyes and form two slightly different images on the retina. The superimposed image is perceived as a 3D image by the brain.

When an incident wave hits the surface of the glass plate, the component of the electric field perpendicular to the plane of incidence lies in the plane of the glass plate. Electrons oscillating in response to the perpendicular component do so in the plane of the plate; they therefore radiate in a direction perpendicular to the surface of the plate. Hence, the electric field of this part of the reflected and transmitted waves is polarized perpendicular to the plane of incidence.

The component of the electric field that is in the plane of incidence is perpendicular to the surface of the glass plate. Electrons responding to this component oscillate perpendicular to the surface of the glass plate and therefore emit radiation in the plane of the plate (like waves propagating on the surface of a pond when a stone is thrown into it). Virtually all of this radiation transmits into the body of the glass plate. Therefore, the transmitted wave has an electric field with substantial components parallel and perpendicular to the plane of incidence, and the reflected wave is predominantly polarized perpendicular to the plane of incidence (Figure 27-36).

KEY CONCEPTS AND RELATIONSHIPS

Maxwell's Equations

Maxwell discovered that a varying electric flux produces a magnetic field. To incorporate this phenomenon, he added a displacement current term, $\varepsilon_0 \dfrac{d\Phi_E}{dt}$, to Ampère's law.

Maxwell's equations describe how electric charges, electric currents, electric fields, and magnetic fields interact with each other; the equations also predict the existence of travelling electromagnetic waves.

Properties of Electromagnetic Waves

Electromagnetic waves sustain themselves once generated, travel in a vacuum at the speed of light, and have electric and magnetic fields in phase with each other with amplitudes related by $E_0 = cB_0$.

In a travelling electromagnetic wave, the electric field, the magnetic field, and the direction of motion of the wave are all perpendicular to each other.

Electromagnetic waves span a broad spectrum, from very large to extremely small wavelengths. Radio waves, TV waves, microwaves, infrared radiation, visible light, ultraviolet radiation, X-rays, and gamma rays are all electromagnetic waves.

Electromagnetic waves are generated by quantum-mechanical processes and by accelerating charged particles.

Electromagnetic waves carry energy and momentum. The rate at which energy is transported per unit area per unit time is called the energy flux or intensity. The Poynting vector, $\vec{S}$, describes the magnitude of energy flux and the direction of the energy flow in electromagnetic waves:

$$\vec{S} = \frac{1}{\mu_0}\vec{E} \times \vec{B} \qquad (27\text{-}29)$$

The average intensity flow in one cycle is given by the time-averaged energy flux, $\bar{S}$:

$$\bar{S} = \frac{E_0 B_0}{2\mu_0} \qquad (27\text{-}27)$$

In an electromagnetic wave, the energy density of the electric field is equal to the energy density of the magnetic field.

The intensity of a three-dimensional spherical wave varies inversely as the square of the distance from the source.

Electromagnetic waves exert radiation pressure on objects:

$$P_{rad} = \frac{\bar{S}}{c} \text{ (when absorbed)} \qquad (27\text{-}34)$$

$$P_{rad} = 2\frac{\bar{S}}{c} \text{ (when reflected)} \qquad (27\text{-}35)$$

Polarization describes the orientation of the electric field of an electromagnetic wave. The plane of polarization is a plane that contains the direction of motion of the wave and the direction of the electric field of the wave.

When polarized light with intensity I_0 falls on a polarizer, which has its transmission axis inclined at an angle of θ with respect to the direction of polarization of the incident light, the intensity of the emitted light, I_1, is related to I_0 by Malus's law:

$$I_1 = I_0\cos^2\theta \qquad (27\text{-}36)$$

APPLICATIONS

wireless communications, astronomical research, food preservation, security, GPS, UV index, medical applications

KEY TERMS

analyzer, average intensity, bremsstrahlung, Cerenkov radiation, cyclotron radiation, displacement current, electromagnetic field, electromagnetic spectrum, electromagnetic wave, energy flux, far infrared, fluorescence, hard X-rays, intensity, irradiance, linear polarization, linearly polarized, Malus's law, Maxwell's equations, near infrared, plane of incidence, plane of polarization, polarization, polarizers, polarizing axis, Poynting vector, radiation momentum, radiation momentum density, radiation pressure, radio astronomy, radiograph, reflected wave, soft X-rays, synchrotron radiation, thermograms, transmission axis, transmitted wave, unpolarized, visible spectrum

QUESTIONS

1. Explain why a displacement current is needed to explain the propagation of electromagnetic waves.
2. An electromagnetic wave is travelling along the positive z-axis. If at a given instant and position the magnetic field is pointing along the negative y-axis, the electric field at that instant and position points along the
   (a) negative y-axis
   (b) positive y-axis
   (c) negative x-axis
   (d) positive x-axis
3. Which of the following sets of equations for electric and magnetic fields does *not* represent a travelling electromagnetic wave? Why not?

   (a) $\vec{E}(x, t) = E_0 \sin(x - ct)\hat{k}$; $\vec{B}(x, t) = \dfrac{E_0}{c} \sin(x - ct)\hat{k}$

   (b) $\vec{E}(x, t) = E_0 \sin(x - ct)\hat{k}$; $\vec{B}(x, t) = \dfrac{E_0}{c} \sin(x - ct)\hat{j}$

   (c) $\vec{E}(x, t) = E_0 \sin(x - ct)\hat{j}$; $\vec{B}(x, t) = \dfrac{E_0}{c} \cos(x - ct)\hat{k}$

   (d) $\vec{E}(x, t) = E_0 \sin(x - ct)\hat{j}$; $\vec{B}(x, t) = E_0 \sin(x - ct)\hat{k}$

4. Give several examples of electromagnetic waves from very large to very small frequencies. Compare and contrast your examples.
5. Compare and contrast sound and electromagnetic waves.
6. (a) A dipole antenna transmits electromagnetic waves at a frequency of 1130 kHz. Determine the length of the antenna.
   (b) A dipole antenna transmits radio waves at a frequency of 98.8 MHz. Determine the length of the antenna.
7. Describe various ways that a charged particle can emit electromagnetic radiation.
8. Unpolarized light is incident on two polarizers (Figure 27-34). Their transmission axes are oriented so that no light is transmitted through the second polarizer. A third polarizer with its transmission axis oriented at 45° with respect to that of the first polarizer is inserted between the two. Can the light now pass through the second polarizer? Explain your reasoning.
9. Can an isolated charged particle move in a circular orbit of constant radius while emitting electromagnetic radiation? Explain your reasoning.
10. Can a laser beam be used to levitate a small and perfectly reflecting object? Explain your reasoning.

11. Is the radiation pressure greater on a surface that absorbs all incident radiation or a surface that is a perfect reflector? Explain your reasoning.

PROBLEMS BY SECTION

For problems, star ratings will be used (★, ★★, or ★★★), with more stars meaning more challenging problems. The following codes will indicate if $\frac{d}{dx}$ differentiation, $\int$ integration, ▢ numerical approximation, or 〰 graphical analysis will be required to solve the problem.

Section 27-2 Displacement Current and Maxwell's Equations

12. ★ A uniform electric field is increasing at a rate of 2.0×10^6 V/(m·s). What is the displacement current through an area of 5.0 3 10^{-4} m² that is perpendicular to the direction of the field?
13. ★ A parallel-plate capacitor has circular plates of radius 0.010 m each that are 0.10 m apart. The capacitor is being charged at a rate of 100. V/(m·s). What is the displacement current in the capacitor?
14. ★★ Circular plates of a capacitor are 5.0 cm in radius and 2.0 mm apart and have air between them. The voltage across the plates is changing at a rate of 60.0 V/s. Determine
   (a) the rate of change of the electric field between the plates
   (b) the displacement current between the plates
15. ★ A uniform displacement current of magnitude 1.00 A, directed out of the plane of the paper, passes through a circular region of radius 5.00 cm. Determine the magnitude of the magnetic field due to the displacement current at the radial distances (a) 2.00 cm and (b) 10.00 cm.
16. ★★ A parallel-plate capacitor with circular plates of radius 10. cm is being charged. Consider a circular loop of radius 20. cm centred at the central axis of the plates. The displacement current through the loop is 1.0 A. What is the rate of change of the magnitude of the electric field between the plates?
17. ★★ A 100. pF capacitor has circular plates of 10.0 cm radius that are 5.0 mm apart and have air between them. The capacitor is charged by connecting it to a 12.0 V battery through a 1.0 Ω resistor.

(a) Determine the current through the plates at $t = 0$ (when the battery is connected).

(b) What is the current through the plates at $t = 60.$ s?

(c) Determine the rate at which the electric field between the plates changes at $t = 0$ and at $t = 60.$ s.

(d) Determine the magnetic field between the plates at $t = 0$ and at $t = 60.$ s.

18. ★★ The rate of change of a uniform electric field through a region of space is shown in Figure 27-39. Determine the displacement current through a 0.25 m² area perpendicular to the direction of the electric field during the time intervals $0 < t < 12$, $12 < t < 22$, and $22 < t < 30$, where t is in milliseconds.

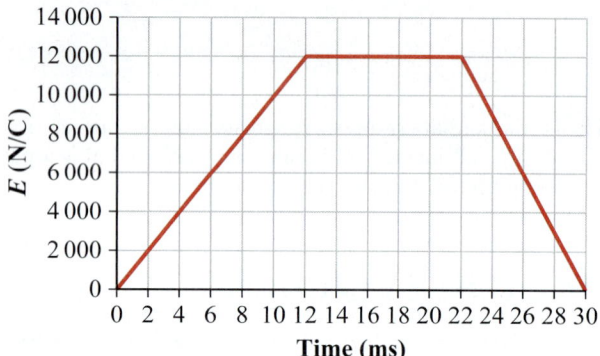

Figure 27-39 Problem 18.

Section 27-3   Electromagnetic Waves

19. ★ Determine the direction of propagation of the electromagnetic wave for each of the following cases.

(a) $\vec{E}$ is in the z-direction; $\vec{B}$ is in the x-direction.

(b) $\vec{E}$ is in the y-direction; $\vec{B}$ is in the x-direction.

(c) $\vec{E}$ is in the x-direction; $\vec{B}$ is in the y-direction.

(d) $\vec{E}$ is in the z-direction; $\vec{B}$ is in the y-direction.

20. ★ The amplitude of the magnetic field of an electromagnetic wave is 2.0×10^{-2} T. What is the amplitude of the electric field of the electromagnetic wave? If the magnetic field is pointing along the positive z-axis, what are the possible directions of the electric field?

21. ★ What is the amplitude of the magnetic field of an electromagnetic wave that has an electric field of amplitude 5.0×10^{-3} V/m? When the magnetic field is pointing along the positive x-axis, what are the possible directions of the electric field?

22. ★ The electric field of an electromagnetic wave in a vacuum is given by

$$\vec{E}(x, t) = (100 \text{ V/m}) \sin((2.00 \times 10^9 \text{m}^{-1})x - \omega t)\hat{k}$$

(a) What are the wavelength and frequency of this wave?

(b) Write equation for $\vec{B}(x, t)$.

(c) In which direction is the wave travelling?

23. ★ The magnetic field of an electromagnetic wave is given by

$$\vec{B}(x, t) = (5.0 \times 10^{-6} \text{ T}) \sin((4.00 \times 10^9 \text{m}^{-1}x - \omega t)\hat{k}$$

(a) Determine the wavelength and the frequency of the wave.

(b) What is the amplitude of the wave's electric field?

(c) Write the equation for $\vec{E}(x, t)$.

(d) What is the direction of propagation of the wave?

24. ★ A plane electromagnetic wave with a wavelength of 500. nm is travelling in the positive x-direction, and its electric field points along the z-axis. The amplitude of the electric field is 2.0 V/m. Write the equations for the electric and magnetic fields of this wave as a function of position and time.

25. ★ A sinusoidal electromagnetic field is travelling in the $+y$-direction. At a certain position and time, its electric field has a magnitude of 1.0 V/m and is pointing along the $-x$-direction. What are the magnitude and direction of the wave's magnetic field at that position and location?

26. ★★ Show that the electric and magnetic fields of Equations 27-6 and 27-7 satisfy the following wave equations:

$$\frac{\partial^2 E_y(x,t)}{\partial t^2} = c^2 \frac{\partial^2 E_y(x,t)}{\partial x^2}$$
$$\frac{\partial^2 B_z(x,t)}{\partial t^2} = c^2 \frac{\partial^2 B_z(x,t)}{\partial x^2}$$

27. ★ The electric field of an electromagnetic wave is given by

$$\vec{E}(y, t) = (5.0 \text{ V/m}) \sin\left(\frac{2\pi}{10^6} \text{ m}^{-1}y - 2\pi ft\right)\hat{i}$$

(a) Determine the frequency of the wave.

(b) In which direction is the wave travelling?

(c) Write the equation for $\vec{B}(y, t)$.

Section 27-4   The Electromagnetic Spectrum

28. ★ Rank the following electromagnetic waves in the order of their frequencies from smallest to largest: infrared light, gamma rays, yellow light, blue light, microwaves, and X-rays.

29. ★ What is the wavelength of gamma rays that have a frequency of 10^{20} Hz?

30. ★ Determine the frequency of light of wavelengths (a) 450 nm, (b) 533 nm, and (c) 700 nm.

31. ★ Determine the wavelength of radio waves of each frequency.

(a) 530 kHz

(b) 1200 kHz

(c) 88.0 MHz

(d) 104.1 MHz

32. ★ Which of the following combinations of wavelength and frequency are not allowed for electromagnetic waves in a vacuum? Why? For the speed of light, use $c = 3.0 \times 10^8$ m/s.

(a) $\lambda = 100.$ m; $f = 3.0 \times 10^6$ Hz

(b) $\lambda = 0.1$ m; $f = 3.0 \times 10^8$ Hz

(c) $\lambda = 6.0 \times 10^{-7}$ m; $f = 5.0 \times 10^{14}$ Hz

(d) $\lambda = 2.0 \times 10^{-15}$ m; $f = 3.0 \times 10^{-23}$ Hz

33. ★ What should the frequency of a sound wave be so that its wavelength is the same as that of radio waves of frequency 600. kHz? Take the speed of sound in air to be 340 m/s and the index of refraction of air to be one.

Sections 27-5 and 27-6 The Energy and Momentum of Electromagnetic Waves and Radiation Pressure

34. ★ A 100. W point source produces light with a wavelength of 550 nm. Determine the amplitudes of the electric and magnetic fields of the light at (a) 10. m, (b) 100. m, and (c) 1.0×10^5 m from the source. What is the intensity of the electromagnetic wave at the three distances?

35. ★ The intensity of solar radiation at the upper atmosphere of Earth is about 1300 W/m². What is the intensity at half the distance between the Sun and Earth? The distance between the Sun and Earth is not needed for this problem.

36. ★ The intensity of the sunlight on top of Earth's atmosphere is approximately 1400 W/m². Calculate the amplitudes of the electric and magnetic fields corresponding to this intensity.

37. ★ A laser beam delivers 0.10 J/m² of energy in 5.0 s to a surface that is placed normal to the direction of motion of the beam. What is the irradiance of the beam?

38. ★★ A laser pointer has a radius of 2.0 mm and delivers 3.2×10^{-3} W of power.
   (a) What is the intensity of the laser light?
   (b) For a beam length of 5.0 m, how much energy is contained in the beam?

39. ★ The cosmic microwave background (CMB) radiation is the electromagnetic radiation that is left over from the Big Bang and fills the universe. It has an average energy density of 4×10^{-14} J/m³. What are the peak values (i.e., the amplitudes) of the electric and magnetic fields of the CMB radiation?

40. ★★ The National Ignition Facility in California fired the world's most powerful laser in March 2012. The pulse from the laser lasted for 23×10^{-9} s and delivered 4.11×10^{14} W of power. Assume that the laser beam was focused on an area of 1.0 mm².
   (a) What was the total energy delivered by the laser?
   (b) What were the amplitudes of the electric and magnetic fields?
   (c) What was the radiation pressure on the 1.0 mm² area?

41. ★ What is the radiation pressure 2.0 m away from a 100 W light bulb that emits uniformly in all directions on a surface that (a) totally reflects the incident radiation and (b) totally absorbs the incident radiation? How far away does the surface need to be moved away from the bulb if the radiation pressure is to be decreased by a factor of 20?

42. ★ The irradiance at a distance of 1.0 m from a light bulb is 2.0 W/m². What is the irradiance from this bulb at a distance of 5.0 m?

43. ★ A point source emits lights uniformly in all directions. At 10.0 m from the source, the electric field strength of the light is 50. N/C. Determine the average power emitted by the source.

44. ★ The average solar power that reaches Earth's surface is 1.0×103 W/m². What is the magnitude of the force exerted by the sunlight on a free electron that is moving with a speed of 1.1×10^5 m/s?

45. ★★★ The radiation pressure of a laser beam can be used to levitate a small particle. What should be the minimum power of a laser beam to levitate a particle of mass 1.0×10^{-6} kg? Assume that the beam is totally reflected from the particle and that the cross-sectional area of the particle is greater than the cross-sectional area of the beam.

46. ★★ Photovoltaic cells convert electromagnetic radiation into electricity. A solar panel consists of an assembly of photovoltaic cells (Figure 27-40). A typical efficiency for these cells is approximately 15%, which means that 15% of the incident radiation is converted into electricity. How many kilowatt hours of energy will solar panels that cover an area of 16 m² on the roof of a house generate in one year? Assume that the average solar intensity at Earth's surface is approximately 200. W/m². This takes into account the fact that Earth represents a disk of radius 6370 km to the incoming sunlight and that a fraction of the incident energy is either absorbed or reflected by the atmosphere.

Figure 27-40 Problem 46. Solar panels on a rooftop.

47. ★★ A sinusoidal electromagnetic wave from a satellite passes perpendicularly through a window with a cross-sectional area of 0.75 m². The amplitude of the electric field of the wave entering the window is 4.0×10^{-3} V/m. How much energy passes through the window in one hour?

48. ★★ A laser has a power of 5.00 mW and produces a spot with a radius of 1.00 mm on a white screen.
   (a) What is the intensity of the laser light on the spot?
   (b) When all the light reflects from the white screen, what are the radiation pressure and the magnitude of the force exerted by the light on the screen?

49. ★★ The power emitted by the Sun is approximately 3.8×10^{26} W.
   (a) What is the average intensity of the Sun's electromagnetic waves at Earth's location? (The average distance between the Sun and Earth is approximately 1.5×10^{11} m.)
   (b) What is the radiation pressure exerted on Earth by the sunlight?

Section 27-8 Polarization

50. ★ An electromagnetic wave is moving in the direction of the positive x-axis. At a given instant, the magnetic field of the wave is pointing toward the negative z-axis. What is the direction of polarization of the wave?

51. ★ In a certain setup, light is polarized in the vertical direction, and only 75% of the incident light is transmitted through the polarizer. What angle does the transmission axis of the polarizer make with the vertical?

52. ★ In a certain setup, light is polarized in the horizontal direction, and only 25% of the incident light is transmitted through the polarizer. What angle does the transmission axis of the polarizer make with the vertical?

53. ★ Linearly polarized light is incident on a polarizer (see Figure 27-33). The intensity of the incident light is 1.0 W/m². Calculate the intensity of the transmitted light when the angle between the transmission axis and the direction of the electric field is (a) 30.°, (b) 60.°, and (c) 90.°.

54. ★ Unpolarized light passes through two polarizers (see Figure 27-34), and the transmission axes of the two polarizers are inclined at an angle of 45° with respect to each other. The initial light has an intensity of 1.0 W/m². What is the intensity of the light that emerges from the second polarizer?

55. ★ A polarized beam of light of intensity 10. W/m² is incident normal to the surface of a polarizer. If the angle between the polarizing axis and the transmission axis is 20°, what is the intensity of the transmitted beam?

56. ★★ Unpolarized light of intensity 100 W/m² is normally incident on a pair of closely placed polarizers. The angle between the transmission axes of the two polarizers is 60°. What is the intensity of the light that transmits through the two polarizers?

57. ★★ Linearly polarized light with an intensity of 5.0 W/m² passes through two polarizers. The transmission axis of the first polarizer is inclined at 45°, and the transmission axis of the second polarizer is inclined at 90° with respect to the polarization direction. What is the intensity of the transmitted light?

58. ★ A polarized beam of light of intensity 20 W/m² is incident normal to the surface of a polarizer. The intensity of the transmitted beam is measured to be 14 W/m². What is the angle between the polarizing axis and the transmission axis?

59. ★★ The transmission axes of two polarizers are inclined at 90° with respect to each other. For the following cases, determine the ratio of incident to transmitted intensity of the light that passes through the two polarizers.
(a) The initial light is unpolarized.
(b) The initial light is linearly polarized with its polarizing axis along the transmission axis of the first polarizer it encounters.
(c) The initial light is linearly polarized with its polarizing axis perpendicular to the transmission axis of the first polarizer it encounters.
(d) The initial light is linearly polarized with its polarizing axis inclined at 45° with respect to the transmission axis of the first polarizer it encounters.

60. ★★ Vertically polarized light is incident on two polarizers, as shown in Figure 27-41. The transmission axis is indicated by the solid black line. The incident intensity is 5.0 W/m². Calculate the transmitted intensity for each configuration.

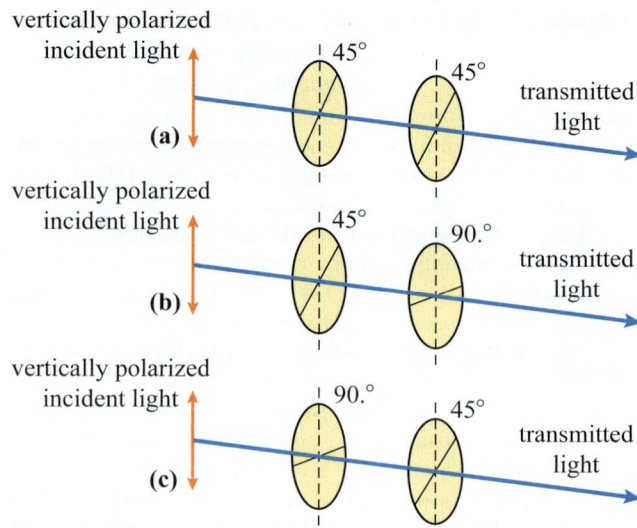

Figure 27-41   Problem 60.

61. ★★ Unpolarized light from a laser is incident on three polarizers, as shown in Figure 27-42. The intensity of the incident light is I_0. Determine the intensity of the light at points A, B, and C.

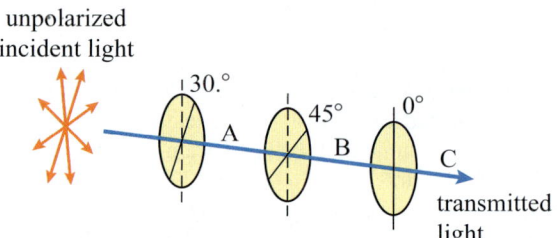

Figure 27-42   Problems 61 and 62.

62. ★★ Consider the transmission of unpolarized light through three polarizers (Figure 27-42). Determine the ratio of incident to transmitted intensity for the following combinations of the directions of the transmission axes of the polarizers. The angles are with respect to the horizontal direction.
(a) 90.°; 45°; 0°
(b) 90.°; 45°; 90.°
(c) 90.°; 0°; 45°
(d) 0°; 45°; 45°

63. ★★★ Five identical polarizers are placed in a straight line next to each other with their transmission axes inclined at angles of θ, 2θ, 3θ, 4θ, and 5θ from left to right, with respect to the vertical direction. Unpolarized light is incident from the left. What must the angle θ be so that only 10.% of the incident light is transmitted through the polarizers?

64. ★★ Unpolarized light is incident on two closely spaced polarizers that have their transmission axes inclined at an angle of θ with respect to each other. After the light passes through the second analyzer, its intensity is determined to be 20.% of the incident intensity. Determine the angle θ.

65. ★★ Unpolarized light is incident on two polarizers with their transmission axes initially parallel to each other. The transmission axis of the second polarizer is first rotated by 20° and then by another 20° in the same direction. What is the ratio of the intensities of transmitted light at these two positions?

COMPREHENSIVE PROBLEMS

66. ★★★ Two identical circular disks of radius 1.00 cm and mass 2.00 g each are attached to the opposite ends of a thin wire of length 10.0 cm and mass 1.00 g. The wire is free to rotate about an axis that passes through its centre (Figure 27-43). One of the disks is coloured black to absorb all the incident radiation, and the other disk is a perfect reflector. The arrangement is placed in the sunlight, which falls perpendicular on one face of each disk.
(a) Will the wire rotate in the presence of the sunlight? Explain your reasoning.

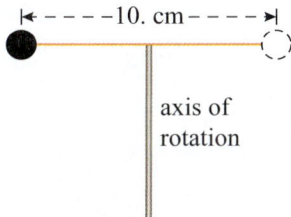

Figure 27-43 Problem 66.

(b) If yes, determine the angular acceleration of the wire.
(c) Will the wire rotate if both disks are perfect reflectors or coloured black? Ignore the effect of air drag in this problem. The intensity of the sunlight is 1000. W/m².

67. ★★★ As a comet approaches the Sun, the solar radiation causes some materials within the comet to vaporize and form an atmosphere around the comet. The force due to solar radiation pressure pushes lighter particles in the atmosphere away from the Sun, forming a tail (Figure 27-44). Assume that the particles in the comet's atmosphere are spherical and have a density of 2500. kg/m³.
(a) For what size would the force due to radiation pressure be equal to the gravitational force of the Sun on a particle? Assume that a particle reflects all incident radiation.
(b) What would happen to particles smaller than the size you calculated in part (a)? Would you expect the comet's tail to become longer as the comet approached the Sun?

Figure 27-44 Problem 67. Comet Hyakutake.

28

Geometric Optics

When you have completed this chapter, you should be able to

LO1 Provide evidence for the geometric optics approach to describing optical phenomena and solving problems.

LO2 Derive the law of reflection from Fermat's principle, and use the law of reflection to describe image formation in a plane mirror.

LO3 Predict the properties of optical images formed by plane and spherical mirrors.

LO4 Derive the law of refraction from Fermat's principle, and use the law of refraction to describe light propagation through various media.

LO5 Predict the properties of optical images formed by thin lenses.

LO6 Describe how the human eye and corrective lenses function.

LO7 Use the concept of Brewster's angle to describe the polarization of refracted and reflected light.

The 110 tonne elliptical sculpture shown in Figure 28-1 stands in Chicago's Millennium Park. This sculpture was designed by the British artist Anish Kapoor, who forged highly polished curved steel plates to form a huge bean-shaped sculpture that reflects Chicago's famous skyline. The sculpture is called *Cloud Gate* because the shape of the sculpture resembles the shape of a cloud. Visitors are invited to walk through the sculpture and view their images on the curved mirror surfaces. The sculpture is warmly referred to by the locals as *The Bean*. This unforgettable sculpture, a convergence of science and art, makes us wonder about the formation of optical images and prompts us to ask scientific questions that can be answered using the concepts of geometric optics, the topic of this chapter.

Figure 28-1 The *Cloud Gate* sculpture in Chicago's Millennium Park.

28-1 Evidence for the Geometric Optics Approach

Although light is an electromagnetic wave, it is often possible to describe the propagation of light using a simplified model in which light is treated as a collection of geometric rays. The geometric optics approach to describing light is applicable only when the dimensions of the objects that light interacts with are much larger than the wavelength of the light (400 nm–750 nm for visible light, as described in Chapter 27). The branch of physics that uses ray approximations is called **geometric** or **ray optics**. The ray approximation is a straightforward geometric way of describing optical phenomena such as reflection, refraction, eclipses, rainbows, mirages, and images created by mirrors and lenses. However, some light phenomena, including interference and diffraction, can be explained only by considering the wave nature of light, as you will learn in Chapter 29.

In geometric optics, we treat light as a collection of rays that propagate in straight lines through a homogeneous medium. In physics, a **medium** (plural form media) is any substance through which something (such as waves or energy) is transmitted. For example, water, air, and metals can be media for mechanical waves. On the other hand, electromagnetic waves, as you saw in Chapter 27, do not require a medium for their propagation—they can propagate through a vacuum.

Let us consider a few of the natural phenomena that led to the development of geometric optics. A shadow is created when an opaque object is located between a light source and a screen. An **opaque** material does not transmit any light. In contrast, **transparent** materials transmit all light, and **translucent** materials transmit some light. The location and size of the shadows can be predicted by noting that light in a homogeneous medium propagates in straight lines, a property called

the rectilinear propagation of light. If a light source produces a beam of parallel rays (also called collimated light), it produces a shadow, as shown in Figure 28-2(a). A beam of parallel rays results when a light source is located very far from an observer, such as the Sun's rays observed from Earth. Some light sources, such as lasers, can produce a highly collimated beam of light with rays that are almost parallel. A non-collimated light source, such as a light bulb, produces light rays propagating in many different directions. When such a source is relatively close to the object casting a shadow, or when a light source is much bigger than an object casting a shadow, both a **full shadow (umbra)** and a **partial shadow (penumbra)** result (Figure 28-2(b)).

Observations of shadows cast by terrestrial and celestial bodies support the geometric optics approach: in a homogeneous medium light appears to propagate in straight lines and can therefore be considered as a collection of rectilinear rays. As shown in Figure 28-3, we can use light rays to help us understand how eclipses occur.

CHECKPOINT 28-1

Partial Shadows

To produce a partial shadow,
(a) the light rays illuminating an opaque object must diverge (spread apart) or converge (come together)
(b) the light rays illuminating an opaque object must come from a distant light source
(c) the light rays illuminating an opaque object must be parallel
(d) both the light source and the opaque object must be of comparable size
(e) Both (a) and (d) are true.

ANSWER: (e) See Figure 28-3.

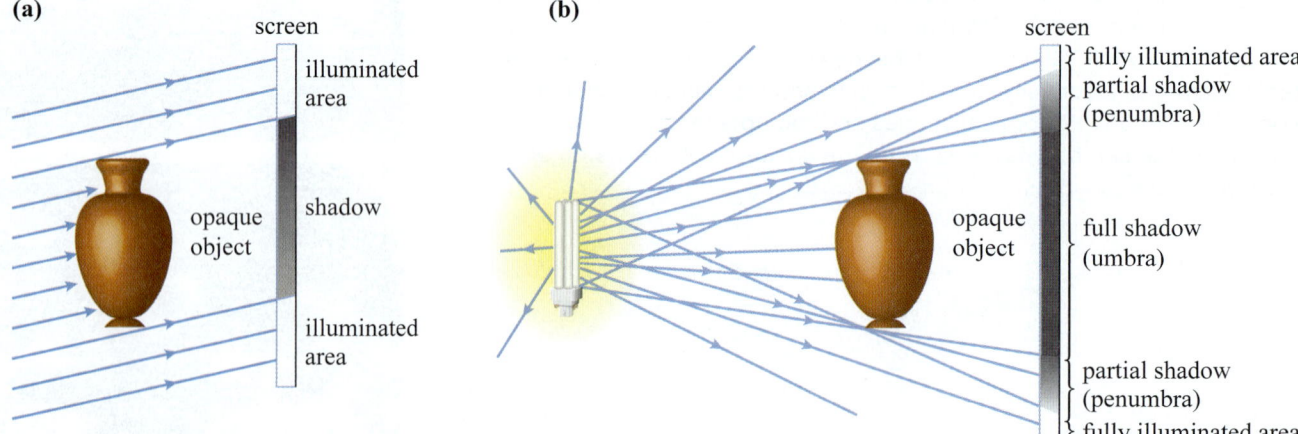

(a) The light source is very far away from an opaque object (a vase).

(b) An extended light source is close to an opaque object (a vase).

Figure 28-2 Opaque objects casting shadows while illuminated by (a) a collimated beam of light and (b) a non-collimated beam of light.

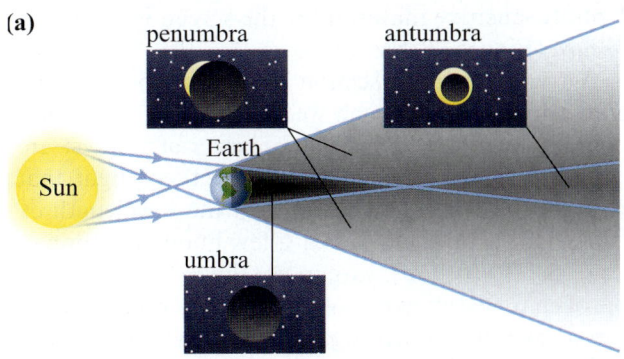

(a)
penumbra
antumbra
Earth
Sun
umbra

(b)
Sun
penumbra
Earth
umbra
moon's orbit

Figure 28-3 (a) Shadows cast by Earth on the Moon. (b) A total lunar eclipse takes place when the Sun, the Moon, and Earth are oriented along a straight line (neither figure is drawn to scale).

Eratosthenes Measures Earth

If you are able to travel to Egypt during the summer solstice, you can recreate the original experiments performed by Eratosthenes more than 2000 years ago. However, if a trip to Egypt is a far-fetched dream for you, you can do it virtually—using a computer simulation. In this activity, you will use the Open Source Physics simulation "Eratosthenes Measures Earth," designed by Todd K. Timberlake, to gain an understanding of how Eratosthenes was able to measure the size of our planet without making a round-the-world trip.

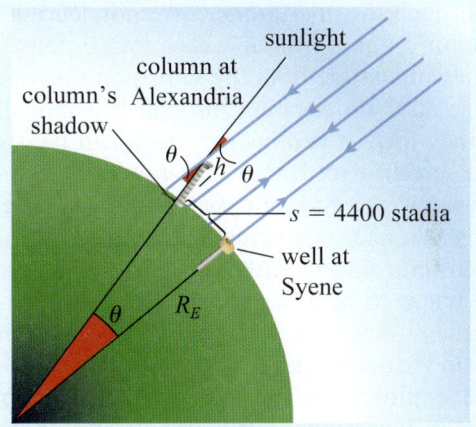

sunlight
column at
column's Alexandria
shadow
θ h θ
$s = 4400$ stadia
well at
Syene
θ R_E

Figure 28-4 Example 28-1.

EXAMPLE 28-1

How Eratosthenes Estimated Earth's Radius

Ancient Greek mathematician, geographer, and astronomer Eratosthenes (276–194 BCE) noticed that at noon during the summer solstice the Sun's rays were exactly perpendicular to the water surface of a local well in the Egyptian city of Syene (today's Aswan). He realized that if he measured the angle of inclination of the Sun's rays in another Egyptian city, Alexandria, at noon on the same day, he would be able to estimate the radius of Earth (Figure 28-4). He estimated the distance between Syene and Alexandria to be approximately 4400 stadia (singular form stadium) by averaging the speed of camel caravans. The stadium is an ancient unit of length, but it is not known whether Eratosthenes' calculations used the Greek stadium (157 m) or the Egyptian stadium (166 m). Eratosthenes used a shadow cast by a vertical column located in Alexandria to measure the angle between the two cities to be 7°12', as shown in Figure 28-4. Use Eratosthenes' data to estimate Earth's radius in stadia and metres (using both conversion factors), and compare the results to the current accepted value of 6370 km, which is often rounded to 6400 km, which we will use for this problem.

Solution

From the diagram in Figure 28-4, we can see that the angle between radii passing through Syene and Alexandria equals the angle between the Sun's rays and the vertical column in Alexandria. Therefore, the ratio of this angle to 360°

(a full circle) equals the ratio of the distance between the cities (length of the arc) to Earth's circumference:

$$\frac{7°12'}{360°} = \frac{d_{Alexandria-Syene}}{2\pi R_E}$$

$$R_E = \frac{360°}{7°12'} \frac{d_{Alexandria-Syene}}{2\pi} = \left(\frac{50.0}{2\pi}\right)(4400 \text{ st})$$

$$= 35\,014 \text{ st}$$

Assuming Greek stadia:

$$R_{E1} = 35\,014 \text{ st} \cdot 157 \frac{m}{st} = 5.50 \times 10^6 \text{ m} = 5500 \text{ km}$$

Assuming Egyptian stadia:

$$R_{E2} = 35\,014 \text{ st} \cdot 166 \frac{m}{st} = 5.81 \times 10^6 \text{ m} = 5810 \text{ km}$$

The current accepted value for the mean radius of Earth is 6370 km. Depending on which stadium unit Eratosthenes used, his estimate was either 9% or 14% off the actual value, $R_E = 6370$ km, which is very impressive!

Making sense of the result

Eratosthenes' estimate was based on basic geometry, the observation that the Sun's rays are straight and parallel, and the assumption that Earth is spherical (or nearly so). Since the actual angle between Syene and Alexandria is approximately 7°5', his estimate of 7°12' gives a slightly low value for Earth's radius.

We finish this section with a brief discussion of optical image formation. In geometric optics, an **object** can be any material that either emits or reflects light. Whatever the object is, light rays emanate from it. Although all real objects have some physical size, an object can often be modelled as a point. For example, a small light-emitting diode viewed from a distance of 5 m can be considered a point object (a point light source), but when viewed from a distance of 5 cm, we usually need to treat it as an extended object (an extended light source). Extended objects (like the Sun) can be visualized as a combination of an infinite number of point objects.

An **image** is an apparent reproduction of an object formed by a lens, mirror, and/or other optical apparatus through the one-to-one correspondence between each point on an object and each image point. If the light rays emanating from the same point on an object converge after propagating through the optical apparatus, they form a **real image**, which can be observed using a screen placed where the rays converge. If the light rays diverge and only their imaginary extensions converge, they form a **virtual image**, which cannot be captured on a screen. Although our brains perceive a virtual image to be in a particular location, light does not actually emanate from that location. Light rays that form virtual images diverge.

We explore the formation of images by different configurations of optical devices in the following sections. However, an image can be formed without mirrors or lenses using a simple technique first recorded approximately 2400 years ago by both Greek and Chinese scholars. A **pinhole camera** (or *camera obscura*) is a closed box, or a darkened room, with a screen on one side and a small hole (a pinhole aperture) in the middle of the opposite side. Some of the light emitted by an object passes through the pinhole to reach the screen. Since the pinhole is very small and the light propagates in straight lines, one can imagine that only one ray emitted by every point of an object will be able to reach the screen, thus creating a real image that corresponds point by point to the object (Figure 28-5(a)). The images created by pinhole cameras are reversed left to right and inverted (upside down). In the mid-1820s, the first photographs were made by substituting a plate of photosensitive material for the screen in a *camera obscura* (Figure 28-5).

We will finish this section by pointing out that the geometric optics approach we consider in this chapter is applicable when the sizes of objects of interest are significantly larger than the wavelength of light that produces the images. Recalling that the wavelength of visible light is on the order of a few hundred nanometres ($\sim 500 \times 10^{-9}$ m), we can use the geometric optics approach to investigate images created by objects larger than a micron. To investigate smaller objects, we will need to use either electromagnetic waves that have a shorter wavelength than visible light or a different approach—wave optics. We will consider wave optics in Chapter 29.

(a)

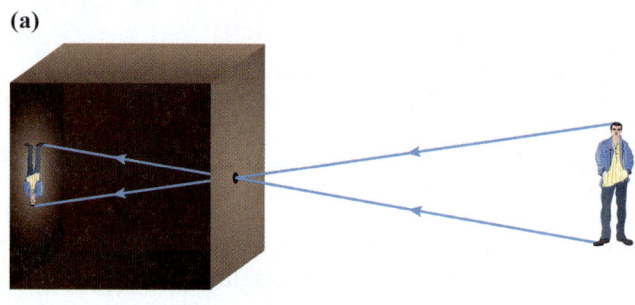

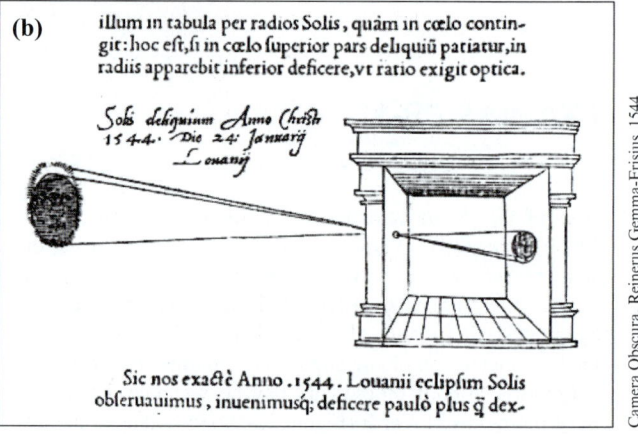

Camera Obscura, Reinerus Gemma-Frisius, 1544

Figure 28-5 (a) The formation of an image of a person in a pinhole camera. (b) On January 24, 1544, Reinerus Gemma-Frisius observed an eclipse of the Sun at Louvain, Belgium. Later, he used this illustration of the event in his book *De Radio Astronomica et Geométrico*, 1545. It is thought to be the first published illustration of a *camera obscura*.

Pinhole Cameras in Nature, Art, and Science

Walking through the shady trees of the beautiful University of British Columbia campus, one sometimes notices projected images of the Sun on the ground under the trees. These images are formed by the small openings in the leaf canopy. Thousands of years ago, ancient Chinese, Greeks, and Arabs used pinhole cameras to study the nature of light. By the Renaissance, many people were fitting darkened rooms with an outside wall with a pinhole so that the outdoor scene could be projected onto an opposite interior wall. Newton conducted optics experiments with such a room. Some famous Renaissance painters used images from pinhole cameras to help them draw small details in their paintings, and the military used pinhole cameras for surveillance. Later on, rich nobility used darkened carriages for entertainment, like the one sculpted by Rodney Graham and exhibited on the University of British Columbia campus and shown in Figure 28-6. Pinhole cameras are still used in artists' studios and in science labs today. Scientists use pinhole cameras to make images using types of radiation that do not pass through lenses readily, such as α-particles and gamma rays.

Source: Rodney Graham Millennial Time Machine, 2003 Landau carriage converted to a mobile camera obscura Collection of the Morris and Helen Belkin Art Gallery, The University of British Columbia. Gift of the artist with support from the Canada Council for the Arts Millennium Fund, the Morris and Helen Belkin Foundation, British Columbia 2000 Recognition Plan, and The University of British Columbia, 2003. Photo: Tim Newton

Figure 28-6 *The Millennial Time Machine* is a sculpture by Rodney Graham. The sculpture is an artistic rendition of a pinhole camera used for entertainment during the Renaissance.

28-2 Reflection of Light

The rectilinear propagation of light in a homogeneous medium follows from the "principle of least time" postulated by the French mathematician Pierre de Fermat (1601–1665). The principle of least time, or **Fermat's principle**, states that a light ray propagating between any two points always takes the path that it can traverse (travel across) in the shortest time. During the 20th century, Fermat's principle was modified to describe newly discovered phenomena, such as the curved paths of light near supermassive objects, a result of gravitational lensing. For our purposes in this chapter we will use Fermat's original principle.

When light rays encounter a surface or a boundary between two media, some of them change their direction and bounce back. This process is called **reflection**. A very smooth surface (such as a mirror) produces **specular reflection**, and a rough surface (such as a painted wall) produces **diffuse reflection** (Figure 28-7). Note that the roughness or the smoothness of the surface is measured compared to the wavelength of the light reflecting from the surface.

(a) Specular Reflection

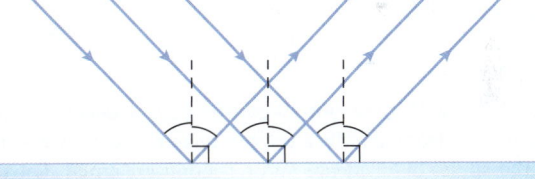

(b) Diffuse Reflection

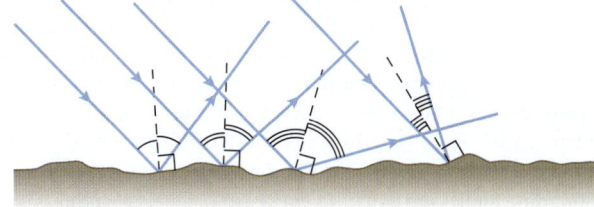

Figure 28-7 Specular (a) and diffuse (b) reflection.

A ray travelling toward a surface is called an **incident ray**, and a ray leaving the surface is called a **reflected ray**. By convention, the **angle of incidence** and the **angle of reflection** are measured relative to a normal to the surface at the point of incidence (Figure 28-8).

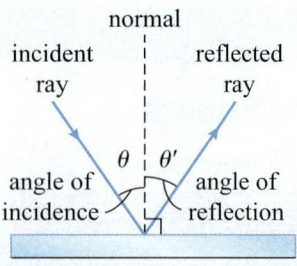

Figure 28-8 The law of reflection.

Therefore, a light ray perpendicular to the surface has an angle of incidence of 0°. All reflected rays have a direction determined by the **law of reflection**: the incident ray, the reflected ray, and the normal to the surface at the point of incidence lie in the same plane, and the angles of incidence and reflection are always equal:

KEY EQUATION
$$\theta = \theta'$$ (28-1)

In Example 28-2, the speed of light along both paths is the same, so the fastest optical path coincides with the shortest geometric path. However, when light travels through different media, the fastest optical path and the shortest geometric paths could differ. Moreover, if the

light travelled in the opposite direction along the path (from point B to point A), our conclusions would not change: thus, the optical path between two points is the shortest path, independent of the direction of the light travel. In other words, the *optical path is reversible*. We will rely on the **reversibility of light** when solving problems and analyzing more complex optical apparatus.

CHECKPOINT 28-2

Photographing a Mirror Image

You are photographing your friend looking at herself in a wall mirror. You want to make sure your photo captures the reflection of her face. While you are photographing her, what must she see in the mirror?

(a) Her face
(b) Her entire body
(c) Your entire body
(d) Your camera

ANSWER: (d) Due to the reversibility of light, she must see your camera for the camera to "see" her eyes and her face. However, she might not be able to see herself in the mirror.

EXAMPLE 28-2

The Law of Reflection

Show that the law of reflection is a direct consequence of Fermat's principle.

Solution

We need to prove that the shortest path for a ray that leaves point A and reflects from a surface to pass through point B occurs when the angles of incidence and reflection are equal. Let us choose the two possible optical paths shown in Figure 28-9. For path AOB, the angles of incidence and reflection are equal ($\theta_1 = \theta_1'$), but they are not equal for path AO'B ($\theta_2 \neq \theta_2'$).

We draw a normal to the mirror from point B and extend it an equal distance beyond the mirror to point B'. To emphasize that the rays through the mirror are imaginary, we show them as dashed lines. We now extend ray AO beyond the mirror to point B'. Right-angled triangles BOC and B'OC are congruent because they have a common side (OC) and their corresponding angles are equal. Therefore, their corresponding sides must be equal: OB = OB'. By the same reasoning, O'B = O'B'. Now we compare the optical paths AOB and AO'B:

$$AOB = AO + OB = AO + OB' = AOB'$$
$$AO'B = AO' + O'B = AO' + O'B' = AO'B'$$
$$AOB' < AO'B' \Rightarrow AOB < AO'B$$

The same argument applies for any location of point O' on the mirror. Therefore, the path with $\theta_1 = \theta_1'$ is shorter than any

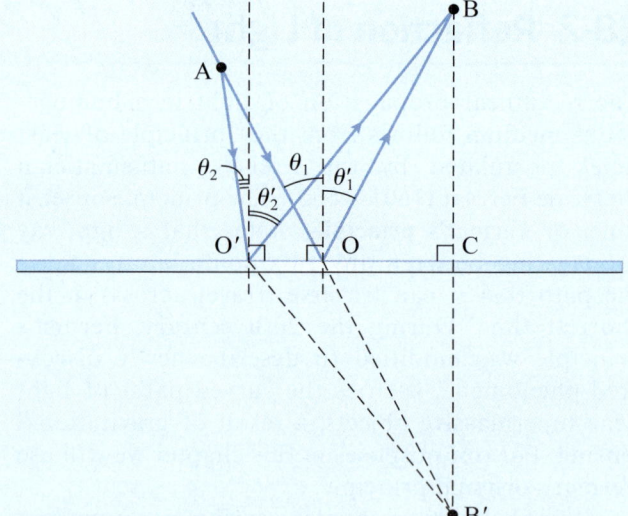

Figure 28-9 Proof of the law of reflection using Fermat's principle.

path where the angles of incidence and reflection differ, such as path AO'B.

Making sense of the result

Fermat's principle can be used to prove the laws of geometric optics. We will also use it when discussing the law of refraction of light.

28-3 Mirrors

In this section, we use the law of reflection to understand the formation of images by plane, concave, and convex mirrors.

Images in Plane Mirrors

A **plane mirror** is a highly reflective flat surface, such as a flat sheet of glass with a reflective coating or the surface of a calm lake. Consider an arbitrary point P on a coffee mug sitting in front of a mirror. When illuminated by incident rays with a range of angles of incidence, which is the usual case when the source of illumination is an extended source (e.g., lighting in a room), the optically rough surface of the mug reflects incident light in all outward directions (diffuse reflection). Some of the light rays emanating from point P reach the mirror and reflect off it in accordance with the law of reflection, as shown in Figure 28-10(a).

To an observer standing in front of the mirror, it appears as if the light rays originated from behind the mirror, from point P′, which is a virtual image of point P. The image is a virtual image because the light rays diverge after reflecting from the mirror (no light rays actually penetrate the mirror's reflective surface). Rays from other points on the surface of the mug reflect in the same way. The region where the image can be seen is called the **field of view**. This region is bounded by rays connecting the end points of an extended object to the adjacent ends of the mirror, as shown in Figure 28-10(b).

Figure 28-11 is an example of a ray diagram. In a **ray diagram**, an object is represented by an arrow to make it clear when the image is inverted. We call the distances between the object and the mirror and the image and the mirror d_o and d_i, respectively. We also call an imaginary line perpendicular to the mirror and passing through its centre the **principal optical axis** of the mirror. In the case of a plane mirror, there is no special point that is considered to be the centre of the mirror. Thus, a plane mirror can have an infinite number of principal optical axes. However, the concept of a principal optical axis will be particularly significant for curved mirrors, as you will see later in the chapter.

To find the location of the image of the top point of the arrow, we need to use at least two rays emanating from it. It is convenient to use a ray parallel to the principal optical axis of the mirror, such as PQ in Figure 28-11. For the second ray, we use a light ray reflected off the mirror at the point where the principal axis meets the mirror, ray PR. We can easily draw its reflection, remembering that the angles of incidence and reflection are equal. The intersection of the continuations of these two rays gives the location of the image of point P (point P′). Extensions of any additional reflected rays from point P pass through the same image point, P′. The same procedure can be repeated for any other point of the object. To help distinguish between the object and the optical image in the diagrams, we will use orange colour to denote the arrows representing optical images.

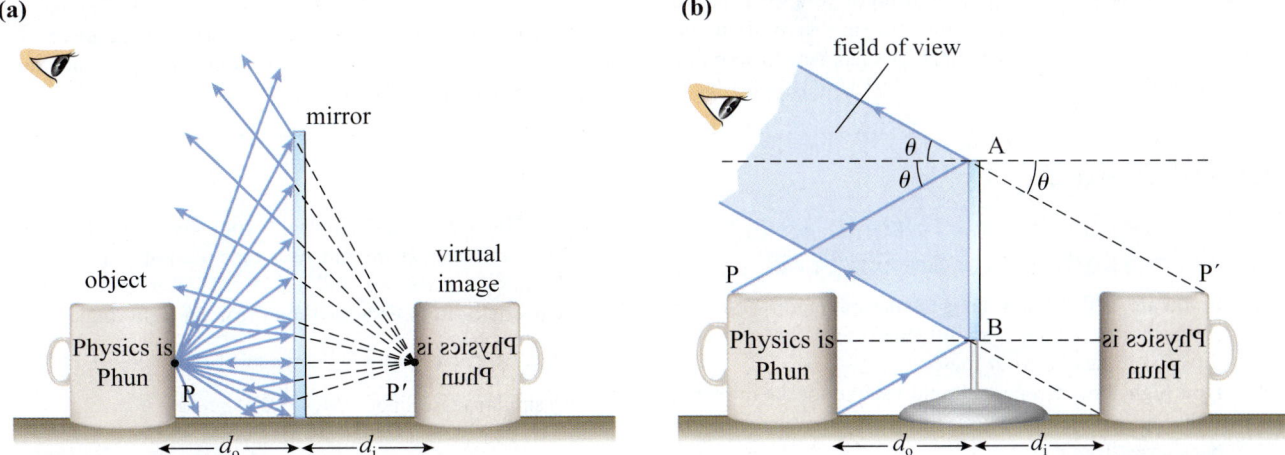

Figure 28-10 (a) Some of the light rays from point P reflect off the mirror in accordance with the law of reflection. (b) Only observers located within the shaded region can see the reflection of the entire coffee mug in the mirror.

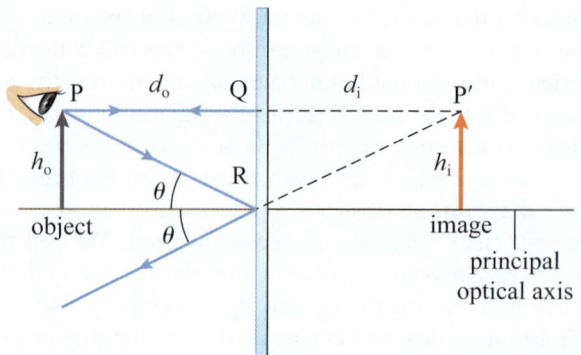

Figure 28-11 A ray diagram representing the formation of an image in a plane mirror. An object (represented by an upright grey arrow) is located along the mirror's principal axis. Ray diagrams often show a number of light rays emanating from the object and reflected off a mirror. It is convenient to use special rays: a ray parallel to the principal axis and a ray reflected off the mirror at the location where the principal axis meets the mirror (point R). The intersection of these rays helps identify the location of the image.

Right-angled triangles PQR and P'QR are congruent because they have a common side and equal angles. As a result, their corresponding sides are equal: PQ = P'Q. Therefore, the distances from the object to the mirror and from the mirror to the image are equal: $d_o = d_i$. Also, the sizes of the image and the object are equal, $h_i = h_o$, and the **linear magnification**, M, of a plane mirror is 1:

$$M = \frac{h_i}{h_o} = 1 \qquad (28\text{-}2)$$

To summarize, a plane mirror forms a virtual, upright image that is the same size as the object ($M = 1$).

MAKING CONNECTIONS

Are Images in a Plane Mirror Laterally Inverted?

Some people believe that a mirror exchanges left and right. However, this is not so.

You can test this by standing in front of a mirror. Move your right arm about, and it seems that the left arm in the mirror is moving. Is this really so? If your identical twin were to stand in place of the image, and then the mirror were removed, it would indeed be the twin's left arm moving. If you hold something in your right hand, you will see that the mirror image arm that is moving also holds something; that is, it is the mirror image of the right hand moving, not a left hand. So why does it happen?

The moving hand in the mirror appears to be a left hand, because we are used to seeing real people standing in front of us, and if it were a real person, then it would be a left hand moving, not a right hand. Imagine your identical twin initially stood by your side. Then to stand in front of you, she needs to rotate by 180°. The mirror does not perform this rotation and it fools us into thinking of the image as a real person standing in front of us.

What the mirror actually does it inverts in a direction perpendicular to the mirror; that is, it inverts "depth." You can see this by having several people stand in a line that is perpendicular to the mirror, in the order A, B, C. The mirror images are in the order C, B, A.

A further illustration results from placing a mirror on a floor and standing on it, or looking up at a mirror on a ceiling. This should clarify that a mirror inverts "depth" and not left and right.

Someone may counter-argue by saying, "Suppose that we write the word PHYSICS on translucent paper and then turn it over and place it in front of a mirror. It is clear that the mirror has inverted the word laterally, that is, exchanged left and right."

ƧƆISYHԳ

Note that it is the person that has inverted the piece of paper laterally, not the mirror! Mirrors invert depth; they do not produce lateral inversion.

EXAMPLE 28-3

Mirror, Mirror on the Wall, I Want to See My Full-Length Reflection or Nothing at All!

(a) If you are 1.80 m tall, what is the minimum mirror length in which you can see your full-length reflection from the top of your head to your toes?

(b) How high above the floor should you mount the mirror?

(c) Will you see more of yourself in a smaller mirror if you back away from it?

Solution

(a) First, we draw a ray diagram of your reflection in the mirror, as shown in Figure 28-12. Point S represents the location of your shoes, and point E is at the height of your eyes. For you to see the image of your shoes, ray OE must enter your eyes. If the mirror is the minimum length that allows you to see your full-length reflection, point O will be at the bottom edge of the mirror. In right-angled triangle EE'S', the point N bisects side EE', so NO = $\frac{1}{2}$E'S'. By similar reasoning, NM = $\frac{1}{2}$E'H', and

$$OM = (ON + NM) = \frac{1}{2}S'E' + \frac{1}{2}E'H'$$

$$= \frac{1}{2}(S'E' + E'H') = \frac{1}{2}S'H' = \frac{1}{2}SH$$

Since SH is your height (1.80 m), the minimum length for the mirror is 90 cm (half your height).

(b) Assuming that the height of your eyes (point E) is approximately 1.66 m, then EH = 14 cm and NM = 7 cm. Therefore, the top edge of the mirror should be 173 cm above the floor.

(c) The calculation of the minimum mirror length for a complete image did not involve the object–mirror distance, d_o, at all. Therefore, you cannot see a full image of yourself in a mirror less than half your height, regardless of your distance from the mirror. You can confirm this result by redrawing Figure 28-12 with the object distance doubled.

Making sense of the result

The reasoning in part (a) does not depend on how far you are from the mirror, but only on your height. Therefore, everyone who wants to see their full-length image needs a mirror no shorter than half their height. In addition, it is important how high off the floor you mount the mirror. The height of mounting of the mirror must be exactly right for the person for this to work. Since people who have different heights will be using the same mirror, full-length mirrors are in practice

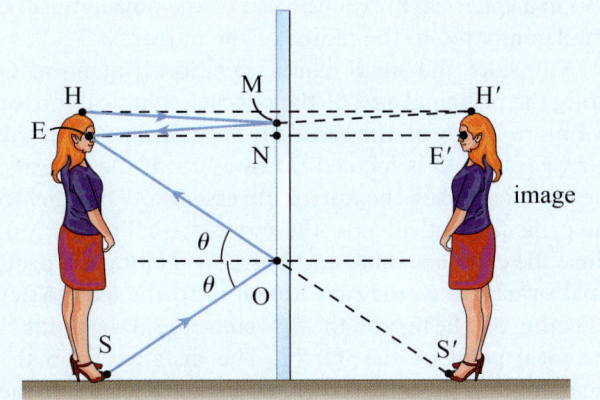

Figure 28-12 To see your entire body in the mirror, the length of the mirror must be at least half your height.

much longer than the minimum. Contrary to what one might assume, moving away from a mirror that is shorter than half your height or coming closer to it does not let you see a complete image of yourself.

Images Produced by Spherical Mirrors

Spherical mirrors are segments of a spherical reflective surface (Figure 28-13). The inner face of the segment is a **concave mirror** because it curves away from objects reflected in it, and the outer face of the segment is a **convex mirror** because it curves toward objects reflected in it.

Images produced by concave spherical mirrors The **centre of curvature** of the mirror shown in Figure 28-14 is the centre of the sphere. We denote the centre of curvature as C and the **radius of curvature** as R. Note that light rays passing through the centre of curvature of the mirror (point C) are perpendicular to the mirror. The line passing through the centre point of the mirror, V, and the centre of curvature is the principal optical axis. If we place a light source beyond the centre of curvature of this concave mirror (Figure 28-14(b)), some of the light rays will reach the mirror, reflect from it, and converge at point O′, which denotes the location of a real image of the light source. To an observer, the light rays appear to originate from this point.

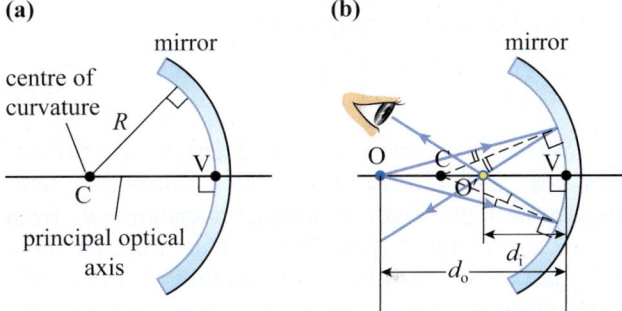

Figure 28-14 (a) A concave spherical mirror. (b) Image formation by a concave mirror.

The light rays from a point on an object do not all pass through a single point when they reflect from a spherical mirror. Spherical mirrors produce sharp images only for **paraxial rays**, which are incident rays that are close to the principal axis and form small angles with it (Figure 28-15). To form a sharp image

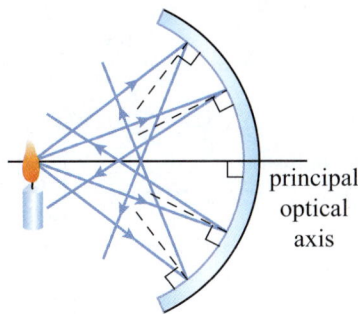

Figure 28-15 Spherical mirrors create sharp images only when the rays coming from an object are close to the principal optical axis and are nearly parallel to it. In this case, no sharp image will be created as reflected rays will not converge in one point.

Figure 28-13 Spherical mirrors.

(O') in a spherical mirror, the size of the object must be small compared to the radius of the mirror.

Consider the small candle positioned at point O along the principal axis of the concave spherical mirror in Figure 28-16. If the candle is very distant from the mirror (point O is located far away from the mirror), the rays that reach the mirror are essentially parallel to the principal optical axis (these rays are called P-rays). Since the candle is small and positioned along the principal optical axis, the rays are close to the axis. After reflecting off the mirror, the rays converge at the point F, the **focal point** of the mirror. The distance from the focal point to the mirror is called the **focal length** of the mirror, f. Concave mirrors have a *real* focal point and a *positive* focal length.

In Figure 28-16, the top P-ray has angle of incidence θ. Since the angles of reflection and incidence are equal, triangle CFA is an isosceles triangle with $\sphericalangle ACF = \sphericalangle CAF = \theta$ and CF = AF. Moreover, AF $\approx$ FV because we are dealing with paraxial rays. Therefore,

$$CF = VF \approx \frac{R}{2}$$

and, for concave spherical mirrors,

$$f = \frac{R}{2} \tag{28-3}$$

We can use a ray diagram to determine the properties of an image formed by a mirror. Consider a luminous object of height h_o located a distance d_o from a concave mirror (Figure 28-17). The image of point H at the top of the object is formed by the rays emanating from this point and converging to point H' after reflecting off the mirror. The location of this image point, H', is given by the intersection of any two of these rays. We can confirm this location by drawing a third ray. As shown in Figure 28-17, a ray that reflects from the centre of the mirror (V) has a path that is symmetrical about the principal axis. It is often convenient to use three special rays:

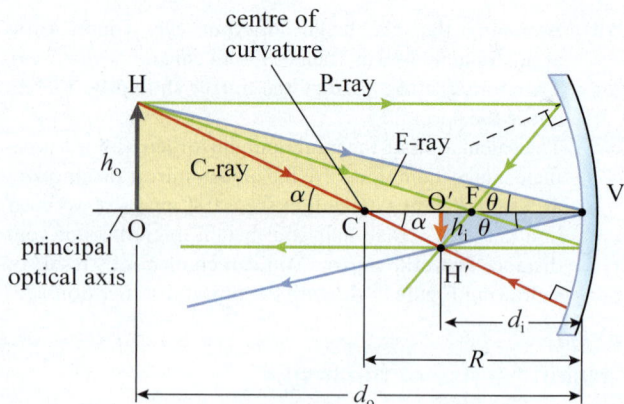

Figure 28-17 We can use a ray diagram to help determine the properties of an image formed by a spherical mirror.

- P-ray: A ray parallel to the principal axis. After being reflected, P-rays pass through the focal point of the mirror, point F.

- C-ray: A ray that passes through the mirror's centre of curvature, point C. Since incident C-rays are perpendicular to the mirror's surface ($\theta = 0°$), the reflected C-ray coincides with the incident ray.

- F-ray: A ray that passes through the focal point. After reflecting from the mirror, F-rays are parallel to the principal optical axis. The path of an F-ray is the reverse of the path of a P-ray due to the reversibility of light.

We follow the same procedure to locate the image of any point of the object. Since the image is inverted, its height, h_i, is represented by a negative quantity. We define the magnification of a concave spherical mirror by considering the similar right-angled triangles COH and CO'H'. Since in similar triangles the ratios of corresponding sides are equal, we write

$$M = \frac{h_i}{h_o} = -\frac{R - d_i}{d_o - R}$$

Since triangles HOV and H'O'V are also similar, the ratios of their corresponding sides are equal. This allows us to derive the magnification equation for a spherical mirror:

$$M = \frac{h_i}{h_o} = -\frac{d_i}{d_o}$$

Equating the two previous expressions yields

$$\frac{R - d_i}{d_o - R} = \frac{d_i}{d_o}$$

$$\frac{1}{d_o} + \frac{1}{d_i} = \frac{2}{R} \tag{28-4}$$

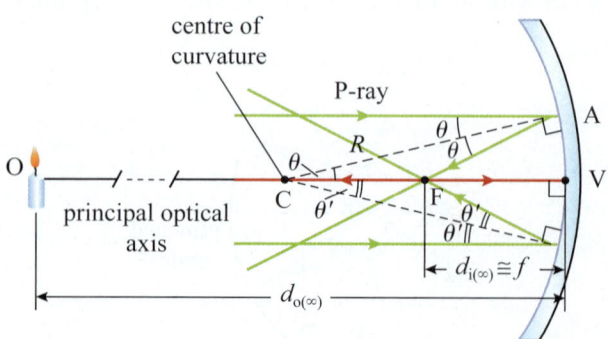

Figure 28-16 The formation of an image of a distant object by a concave spherical mirror.

Using Equation 28-3 to substitute for the focal length of a concave mirror allows us to derive a complete algebraic description of image formation by spherical mirrors:

KEY EQUATION
$$\frac{1}{d_o} + \frac{1}{d_i} = \frac{1}{f} \qquad (28\text{-}5a)$$

KEY EQUATION
$$M = \frac{h_i}{h_o} = -\frac{d_i}{d_o} \qquad (28\text{-}5b)$$

Equations 28-5 are used to describe images formed by concave spherical mirrors. A plane mirror can be considered to be a part of an infinitely large concave mirror—a concave mirror with an infinitely large focal length, $f = \infty$. Substituting this information into Equations 28-5, we find that this description also applies to plane mirrors. We will soon discover that these equations apply to convex mirrors, as well. Thus, Equations 28-5 are called the mirror equations.

You can use these expressions to determine the location of an image when a ray diagram approach might be too cumbersome or imprecise. You can also verify that the ray diagrams you construct represent the images correctly.

PEER TO PEER

At first I was confused by the negative sign of magnification. Later I realized that the negative sign only indicates that the image is inverted compared to the object. Thus, a magnification of 2 means that the orientation of the image and of the object is the same and the image is twice as tall as the object. However, a magnification of -2 means that the image is twice as tall as the object but is inverted (upside down) compared to it.

CHECKPOINT 28-3

Concave Spherical Mirrors

Which of the following statements correctly describes images formed by a concave spherical mirror?
(a) When $d_o \gg f$, $d_i = 2f$.
(b) It is impossible to have a negative image–mirror distance ($d_i < 0$) in a spherical mirror.
(c) Spherical mirrors always produce virtual, inverted, and magnified images.
(d) When $0 < d_o < f$, the image is virtual, magnified, and upright.
(e) When $d_o = f$, the image is virtual and located at $d_i = -2f$.
(f) None of the above are correct.

ANSWER: (d) Draw a ray diagram or use Equation 28-5a to verify that this answer is correct.

EXAMPLE 28-4

Touching an Image

While waving your finger in front of the concave surface of a shiny metal soup ladle, you notice that at one point your finger appears to touch its own reflection. The ladle has a spherical shape with a diameter of 10 cm. How far is your finger from the ladle when this happens?

Solution

We consider the ladle to be a spherical concave mirror, so its focal length is equal to half its radius (Equation 28-3):

$$f = \frac{R}{2} = \frac{D/2}{2} = 2.5 \text{ cm}$$

Since the locations of the image and the object (your finger) coincide, $d_i = d_o$. Using the spherical mirror Equation 28-5a and remembering that a concave mirror has a positive focal length,

$$\frac{1}{d_o} + \frac{1}{d_i} = \frac{1}{f} \Rightarrow \frac{2}{d_o} = \frac{1}{f} \Rightarrow d_o = 2f = 5.0 \text{ cm}$$

$$d_o = d_i = 5.0 \text{ cm} \Rightarrow M = \frac{h_i}{h_o} = -\frac{d_i}{d_o} = -1$$

Making sense of the result

Looking at the calculations in this example, we can see that the result can be generalized to any concave spherical mirror: when the object–mirror distance equals twice the focal length ($2f$), the formed image is a real inverted image that is the same size as the object ($M = -1$).

Images in convex spherical mirrors Convex mirrors are diverging mirrors. A collimated light beam parallel to the mirror's principal optical axis is reflected so the rays appear to diverge from a point behind the mirror (Figure 28-18(a)). This point is called a **virtual focal point** and is also denoted as F. Since convex mirrors always form virtual images, their focal length, f, is always negative. Using similar reasoning as for concave mirrors, we can show that Equation 28-3 holds for convex mirrors if we assign the radius of a convex mirror a negative value. Thus, $f_{\text{convex}} = \frac{R}{2} < 0$. This sign convention, $+R$ for concave surfaces and $-R$ for convex surfaces, is accepted in mathematics and science. The mirror equations discussed earlier (Equations 28-5) also hold for convex mirrors. For a convex mirror, both the focal point F and the centre of curvature C are located behind the mirror, as shown in Figure 28-18.

In conclusion, Equations 28-5 can be applied to any plane or spherical mirror provided we only deal with paraxial rays. This is because a spherical mirror, unlike a parabolic mirror, only focuses paraxial rays into a point. If we shine a non-paraxial parallel beam onto a spherical mirror, the reflected rays will not converge onto one point. Therefore, large mirrors used in optical instruments are parabolic and not spherical.

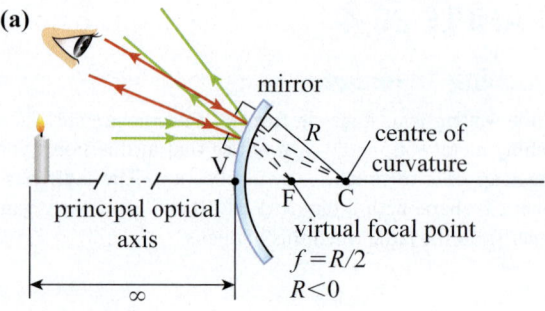

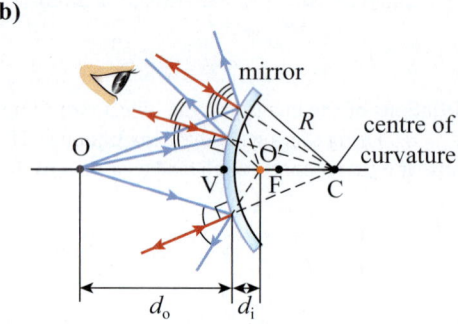

Figure 28-18 (a) A convex spherical mirror has a virtual focal point. (b) Image formation by a convex mirror.

EXAMPLE 28-5

Comparing Convex and Concave Mirrors

A shiny metal hemispherical bowl has a diameter of 32 cm. You notice that when you place a small object along its axis of symmetry on the concave side, the image of the object appears in front of the mirror, inverted and approximately half the height of the object. Describe the properties of the image formed if you place the object at the same distance from the bowl on the axis of symmetry on the convex side.

Solution

This shiny bowl can be considered as a sphere silvered on both sides. Since both reflective surfaces are spherical, their focal length values are

$$f_{concave} = \frac{R}{2} = \frac{16 \text{ cm}}{2} = 8.0 \text{ cm}$$

and

$$f_{convex} = \frac{R}{2} = \frac{-16 \text{ cm}}{2} = -8.0 \text{ cm}$$

We use the spherical mirror Equations 28-5 to find the object–mirror distance for the real image produced by the concave surface:

$$M = \frac{h_i}{h_o} = -\frac{d_i}{d_o} = -\frac{1}{2} \Rightarrow d_i = \frac{d_o}{2}$$

$$\frac{1}{d_o} + \frac{1}{d_i} = \frac{1}{f} \Rightarrow \frac{1}{d_o} + \frac{1}{d_o/2} = \frac{1}{f} \Rightarrow \frac{3}{d_o} = \frac{1}{f}$$

$$d_o = 3f = 3(8.0 \text{ cm}) = 24 \text{ cm}$$

Now we apply the same spherical mirror equations to find the properties of the image produced by the convex surface:

$$\frac{1}{d_o} + \frac{1}{d_i} = \frac{1}{f} \Rightarrow \frac{1}{24 \text{ cm}} + \frac{1}{d_i} = \frac{1}{-8.0 \text{ cm}} \Rightarrow \frac{1}{d_i} = -\frac{1}{6.0 \text{ cm}}$$

$$d_i = -6.0 \text{ cm}$$

$$M = \frac{h_i}{h_o} = -\frac{d_i}{d_o} = -\frac{-6.0 \text{ cm}}{24 \text{ cm}} = \frac{1}{4} \Rightarrow h_i = \frac{1}{4}h_o$$

Since d_i is negative and M is positive, the image produced by the convex surface is virtual, one quarter the height of the object, and upright.

Making sense of the result

For an object located $3f$ from a convex mirror, the image is virtual, upright, and one quarter the height of the object ($M = 0.25$). Concave and convex mirrors with the same curvature produce markedly different images of an object located at the same distance from them.

Variation of images with the object's location The ray diagrams and images in Figure 28-19 illustrate that the properties of the image produced by a spherical mirror change as the location of the object relative to the focal point changes. For example, the object in Figure 28-19(a) is to the far left of C, and the image is real and inverted. In Figure 28-19(b), the same object is located between F and the mirror, and the image is virtual and upright. The object in Figure 28-19(c) is located in front of the convex mirror. Its image is virtual (located behind the mirror) and upright.

Table 28-1 summarizes the properties of images in spherical mirrors. The absolute value of the magnification, M, shows whether the image is magnified ($|M| > 1$) or reduced ($|M| < 1$), and its sign indicates whether the image is upright (positive) or inverted (negative). Convex mirrors always produce reduced images.

Remember that $|f| = \dfrac{R}{2}$ for spherical mirrors.

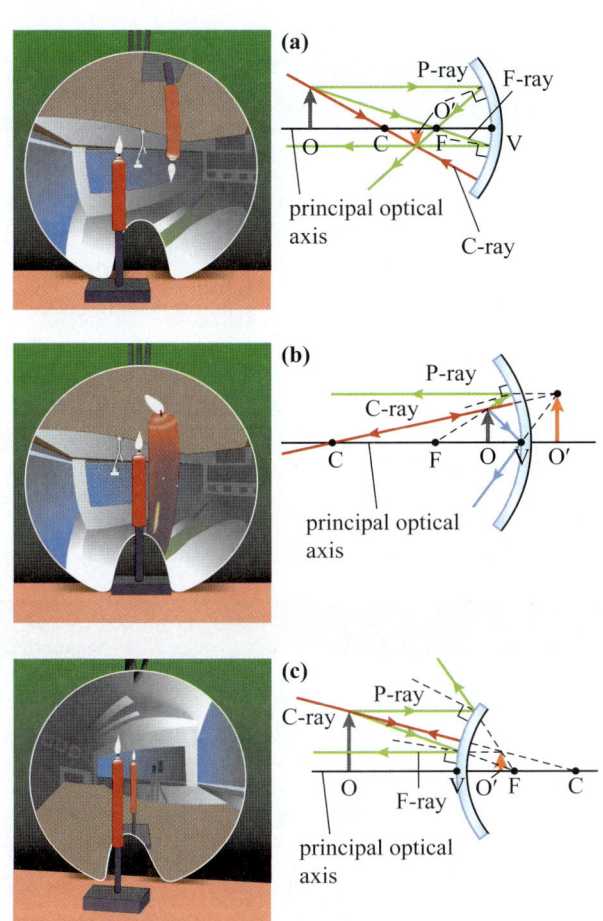

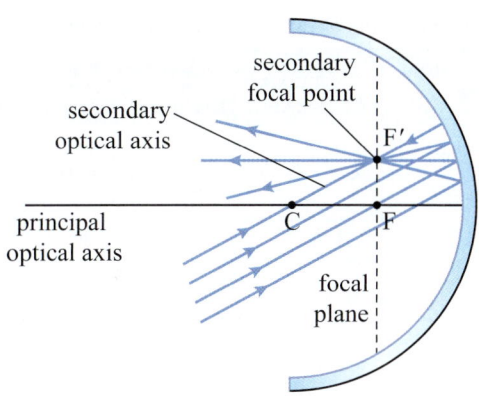

Figure 28-20 The focal plane of a spherical mirror passes through the focal point and is perpendicular to the principal optical axis.

Focal plane The focal plane of a spherical mirror passes through its focal point and is perpendicular to the principal optical axis. A parallel beam of light incident on a spherical concave mirror always converges at a point located in its focal plane. Only rays parallel to the principal axis converge at the mirror's focal point. A beam of parallel rays that are incident at an angle to the principal axis (Figure 28-20) converge at a **secondary focal point** that is located at the intersection of the focal plane of the mirror and the secondary optical axis. A secondary optical axis passes through the centre of the mirror and is parallel to the beam. Thus, a concave mirror has one primary focal point and an infinite number of secondary focal points. All the focal points of a concave mirror are real. Similar reasoning can be applied to convex mirrors, but all its focal points are virtual.

Figure 28-19 Ray diagrams for image formation for an object at various locations near (a, b) concave and (c) convex spherical mirrors.

Table 28-1 Properties of Images Produced by Spherical Mirrors

Object–Mirror Distance	Concave Mirror: $R > 0, f = R/2 > 0$		Convex Mirror: $R < 0, f = R/2 < 0$	
	Mathematical Description	**Image Properties**	**Mathematical Description**	**Image Properties**
$0 < d_o < |f|$	$d_i < 0$	Virtual, appears behind the mirror	$d_i < 0$	Virtual, appears behind the mirror
	$h_i > 0$	Upright	$h_i > 0$	Upright
	$M = \dfrac{h_i}{h_o} = -\dfrac{d_i}{d_o} > 1$	Magnified	$M = \dfrac{h_i}{h_o} = -\dfrac{d_i}{d_o} < 1$	Reduced
	$|d_i| > |d_o|$	Located farther from the mirror than the object	$(0 < M < 1)$ $|d_i| < |d_o|$	Located closer to the mirror than the object
$d_o = |f|$	No image is created		$d_i = \dfrac{f}{2} < 0$	Virtual, appears behind the mirror
			$h_i > 0$	Upright
			$M = -\dfrac{d_i}{d_o} = \dfrac{1}{2}$	Reduced by half
			$|d_i| = \dfrac{|f|}{2} < |d_o|$	Closer to the mirror than the object
$f < d_o < 2|f|$	$d_i > 0$	Real, appears in front of the mirror	$d_i < 0$ $|d_i| < |f| < |d_o|$	Virtual, appears behind the mirror
	$h_i < 0$	Inverted	$h_i > 0$	Upright
	$M = \dfrac{h_i}{h_o} = -\dfrac{d_i}{d_o} < -1$	Magnified	$M = \dfrac{h_i}{h_o} = -\dfrac{d_i}{d_o} < 1$	Reduced
	$|d_i| > |d_o|$	Farther from the mirror than the object	$|d_i| < |d_o|$	Closer to the mirror than the object

(continued)

Table 28-1 Properties of Images Produced by Spherical Mirrors (*Continued*)

Object–Mirror Distance	Concave Mirror: $R > 0, f = R/2 > 0$		Convex Mirror: $R < 0, f = R/2 < 0$	
	Mathematical Description	**Image Properties**	**Mathematical Description**	**Image Properties**
$d_o = 2|f|$	$d_i > 0$ $d_i = -d_o$	Real, appears in front of the mirror	$d_i = \dfrac{2}{3}f < 0$ $0 < |d_i| < |f|$	Virtual, appears behind the mirror
	$h_i < 0$	Inverted	$h_i > 0$	Upright
	$M = \dfrac{h_i}{h_o} = -\dfrac{d_i}{d_o} = -1$	Same size as the object	$M = \dfrac{h_i}{h_o} = -\dfrac{d_i}{d_o} < 1$	Reduced, closer to the mirror than the object
$d_o > 2|f|$	$d_i > 0$	Real ($d_i > 0$), appears in front of the mirror	$d_i < 0$	Virtual, appears behind the mirror
	$h_i < 0$	Inverted	$h_i > 0$	Upright
	$-1 < M < 0$	Reduced, closer to the mirror than the object	$M = \dfrac{h_i}{h_o} = -\dfrac{d_i}{d_o} < 1$ $|d_i| < |d_o|$	Reduced, closer to the mirror than the object

CHECKPOINT 28-5

Focal Points

A parallel beam of light strikes a concave spherical mirror. Which ray diagram in Figure 28-21 correctly represents the reflected beam?

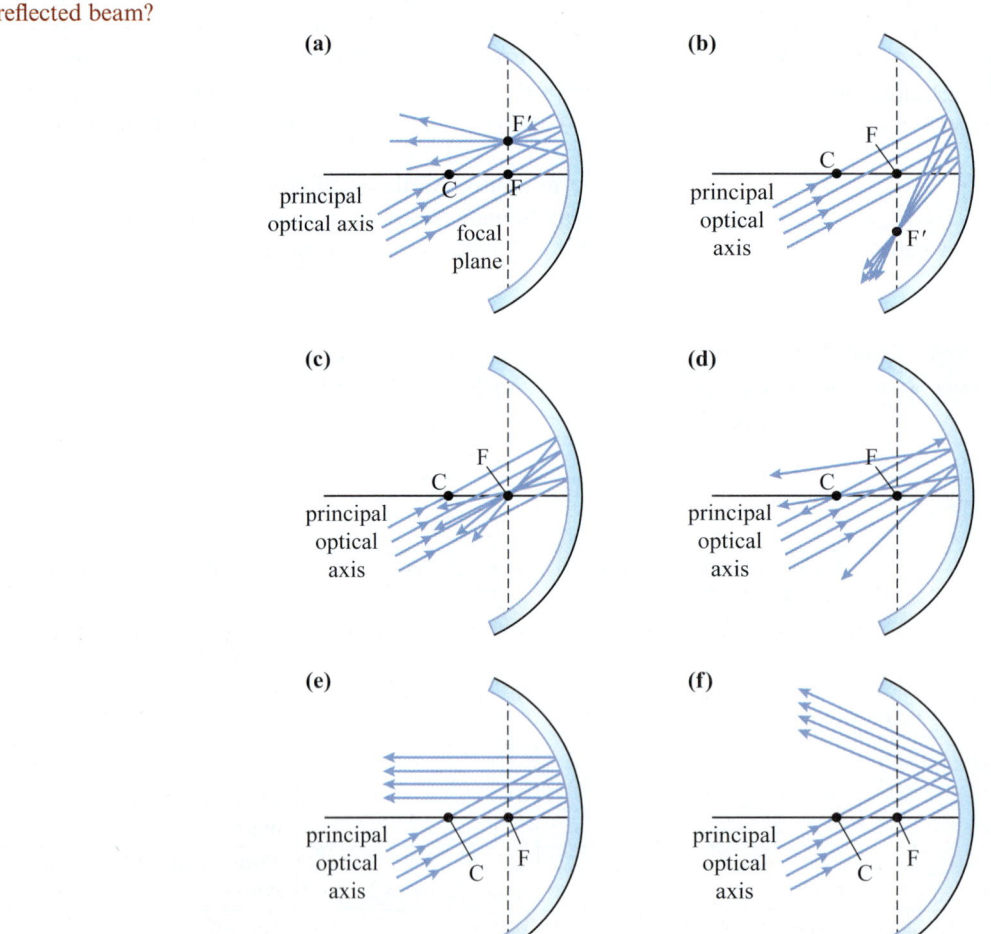

Figure 28-21 Checkpoint 28-5.

ANSWER: (a) The parallel beam reflects off the mirror and the reflected rays are at the secondary focal point F′. A secondary focal point for a spherical mirror is located at the intersection of the mirror's focal plane and its secondary optical axis, which is parallel to the beam and passes through the centre of the mirror, point C (Figure 28-20).

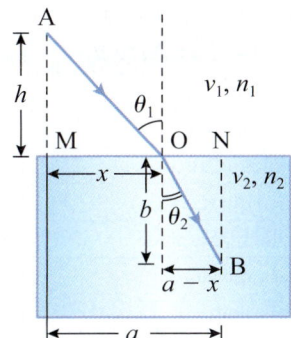

Figure 28-22   Fermat's principle applied to the refraction of light.

MAKING CONNECTIONS

Linear and Angular Magnification

If you buy a telescope or a pair of binoculars, you will see that the magnification provided in the specifications of the optical device is not linear but **angular magnification**. You might wonder why this is. Linear magnification, as described in this section, is used when we are projecting an image on a screen, as it is defined as the ratio of the size of the image to the actual size of an object (Equation 28-2). However, when we view an object with an eyepiece, we cannot easily measure the size of the image or even of the object. In that case, it is more convenient to use angular magnification, which is defined as the ratio of the tangent of the angle subtended at the focal point of the eyepiece to the tangent of the angle subtended at the focal point of the objective. Therefore, if you are viewing something with the naked eye and it covers 0.035 rad (2.0°) with a 5× magnification it will appear to cover 0.17 rad (10°). We can ignore tangents in the definition of angular magnification because when dealing with small angles, provided we measure angles in radians, we can substitute the tangent of an angle with its value: $\tan\theta \approx \theta$. In this chapter, we will use linear magnification (unless indicated otherwise); thus, when we use the word *magnification*, we mean linear magnification.

28-4 Refraction of Light

Let us consider what happens when a ray of light encounters a boundary between two transparent media, for example, when light travelling through air enters glass. As shown in Figure 28-22, the direction of the light changes. This process is called **refraction**. In the figure, θ_1 denotes the angle of incidence and θ_2 denotes the **angle of refraction**.

We can use Fermat's principle to explain why refraction occurs. The speed of light depends on the medium, so we denote the speed of light in the medium

of the incident ray as v_1 and in the medium where the light is refracted as v_2. Consider the ray of light that is travelling from point A to point B in Figure 28-22. The time, t, the light takes to travel between the points is the sum of the time taken along segment AO in the first medium and segment OB in the second medium. Applying the Pythagorean theorem to the two right-angled triangles AMO and BNO, we obtain

$$t = \frac{\sqrt{h^2 + x^2}}{v_1} + \frac{\sqrt{b^2 + (a - x)^2}}{v_2}$$

According to Fermat's principle, light follows the path for which this expression has a minimum value. Therefore,

$$\frac{dt(x)}{dx} = 0$$

$$\frac{d}{dt}\left(\frac{\sqrt{h^2 + x^2}}{v_1} + \frac{\sqrt{b^2 + (a - x)^2}}{v_2}\right) = 0$$

$$\frac{x}{v_1\sqrt{h^2 + x^2}} + \frac{-(a - x)}{v_2\sqrt{b^2 + (a - x)^2}} = 0$$

$$\frac{\sin\theta_1}{v_1} = \frac{\sin\theta_2}{v_2} \qquad (28\text{-}6)$$

The **index of refraction** of a medium is defined as

KEY EQUATION $$n = \frac{c}{v} \qquad (28\text{-}7)$$

where c is the speed of light in a vacuum and v is the speed of light in the medium.

Therefore, the index of refraction of a vacuum is 1, and $n > 1$ for all other media (Table 28-2). The term **optical density** is often used to describe the refractive properties of materials; the greater a medium's index of refraction, the greater its optical density.

ELECTRICITY, MAGNETISM, AND OPTICS

Table 28-2 Indexes of Refraction for Various Substances

Substance (Solid at 20 °C)	Index of Refraction	Substance (Liquid at 20 °C)	Index of Refraction
Cubic zirconium	2.20	Benzene	1.501
Diamond (C)	2.419	Carbon disulfide	1.628
Fluorite (CaF_2)	1.434	Carbon tetrachloride	1.461
Fused quartz (SiO_2)	1.458	Corn syrup	2.210
Gallium phosphide	3.50	Ethyl alcohol	1.361
Glass, crown	1.52	Glycerine	1.473
Glass, flint	1.66	Vegetable oil	1.470
Ice (H_2O)	1.309	Water	1.333
Gases at 0 °C, 1 atm	**Index of Refraction**		
Carbon dioxide	1.000 45		
Air	1.000 293		

Equation 28-6 is called the **law of refraction**, and it can be rewritten in terms of the indexes of refraction of the two media:

KEY EQUATION
$$n_1\sin\theta_1 = n_2\sin\theta_2 \qquad (28\text{-}8)$$

This formula was discovered experimentally by the Dutch mathematician Willebrord Snellius (1580–1626) and is often called **Snell's law**. The word *refraction* stems from the Latin word *refractionem* meaning "breaking up." It is derived from past participle stem of Latin *refringere*, "to break up." This phenomenon is called light refraction because a light ray "breaks" at the border between two media. Consequently, the objects observed at the interface of different media appear broken or fractured. For example, a pencil placed in a glass filled with water appears to an observer to be broken because the image of the bottom part of the pencil is closer to the water–air boundary than the real pencil is (Figure 28-23).

Applying Snell's law, we conclude the following:

■ When an incident light ray is normal to the boundary between two media ($\theta_1 = 0°$), the **refracted ray** is also normal to the boundary, $\theta_1 = \theta_2 = 0°$.

■ When light propagates from a less optically dense medium to a more optically dense medium ($n_1 < n_2$), the light ray bends toward the normal: $\theta_2 < \theta_1$.

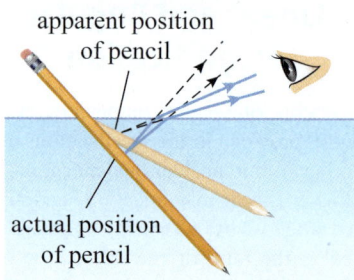

Figure 28-23 A pencil partially submerged in water appears to be broken at the air–water border.

■ When light propagates from a more optically dense medium to a less optically dense medium ($n_1 > n_2$), the light ray bends away from the normal: $\theta_2 > \theta_1$.

■ A beam of polychromatic light (light with more than one frequency) disperses in space when it crosses the boundary between media because the speed of light varies with the frequency of the light. The dependence of the speed of light (**phase velocity** of light waves), and hence the index of refraction, on the frequency of light, known as **dispersion**, is the reason we see a range of colours when a beam of white light enters a prism at an angle (Figure 28-24).

(a)

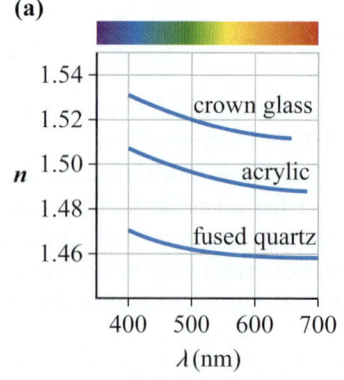

(b)

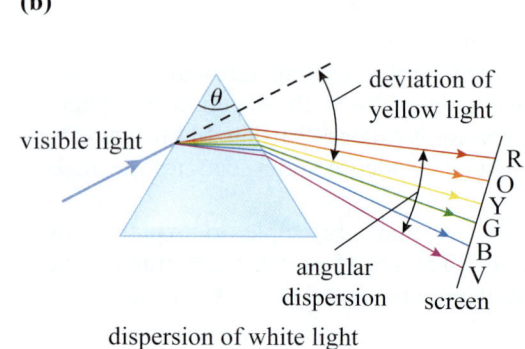

Figure 28-24 (a) Variation of the index of refraction with frequency. (b) Dispersion of sunlight in a glass prism.

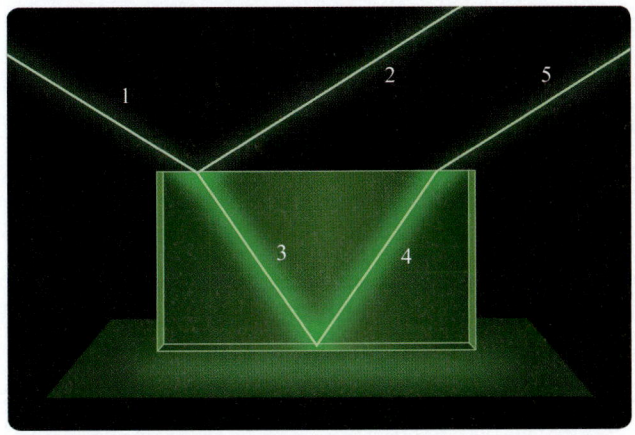

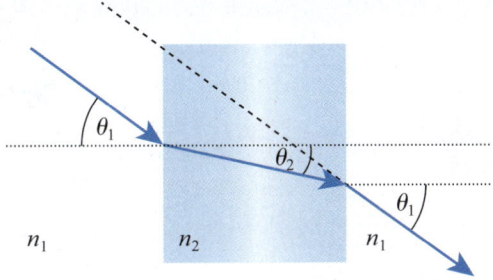

Figure 28-25 (a) A beam of light incident on a glass plate in air. Some of the beam undergoes reflection (path 1–2), some undergoes total internal reflection (path 3–4), and some undergoes refraction (paths 1–3 and 4–5). (b) A beam of light travelling through a plane-parallel glass plate.

In summary, when a beam of light propagating through air is incident on a plane-parallel glass plate (Figure 28-25(a)), some of the light reflects (paths 1–2 and 3–4), and some of the light refracts (paths 1–3 and 4–5). When a beam of light travels through this plane (Figure 28-25(b)), the incident beam and the beam that exits the glass slab are parallel to each other.

Total Internal Reflection

We can rewrite Snell's law (Equation 28-8) as

$$\sin\theta_2 = \frac{n_1}{n_2}\sin\theta_1 \qquad (28\text{-}9)$$

In Figure 28-28, light is propagating from a more optically dense medium to a less optically dense medium ($n_1 > n_2$) (e.g., from water to air). At some angle of incidence, the right side of Equation 28-9 will equal one, making the angle of refraction, θ_2, equal 90° (Figure 28-28(c)). This angle of incidence is called the **critical angle**, θ_{cr}, and

$$\sin 90° = \frac{n_1}{n_2}\sin\theta_{cr}$$

$$\frac{n_1}{n_2}\sin\theta_{cr} = 1$$

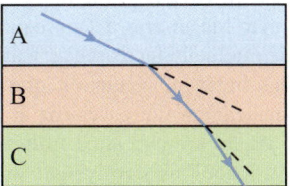
Therefore,

KEY EQUATION
$$\theta_{cr} = \sin^{-1}\left(\frac{n_2}{n_1}\right) \qquad (28\text{-}10)$$

For example, at a water–air boundary,

$$\theta_{cr} = \sin^{-1}\left(\frac{1}{1.33}\right) \approx 48.8°$$

When the angle of incidence exceeds the critical angle, the light cannot propagate through the boundary between the media and is reflected off it (Figure 28-28(d) and Figure 28-25 (path 3–4)). This phenomenon is called **total internal reflection**.

Total internal reflection is essential for the operation of the optical fibres that are widely used in communications and in medicine. The light travelling inside a fibre

ELECTRICITY, MAGNETISM, AND OPTICS

Mirages

The indexes of refraction of many optical media (e.g., air) vary with temperature. Such variations in the index of refraction of air can cause refractions that produce mirages. For example, when driving on a hot summer day, you may have observed what appear to be puddles on the road that disappear as you approach them. Similarly, in a hot desert, people sometimes see what appear to be pools of water (Figure 28-27). Mirages occur because the air closest to the hot ground or pavement is much hotter than the air above. Consequently, the index of refraction of air gradually increases as you move away from the ground. This index of refraction gradient refracts Sun rays that are incident on the pavement at very large angles (remember, the angles are measured from the normal to the surface), making it look as if this light were reflected from water in a pool ahead of the observer. Invoking Fermat's principle once again, we can say that as the speed of light in the air increases with rising air temperature, the light rays bend down where they can travel faster, to make the time from a faraway object, such as the tree in the desert shown in the figure to your eye a minimum.

Although less common, mirages also occur when there is cooler air near the ground, with a layer of warmer air above it.

If there is a sufficient difference in temperature, light from objects below the horizon is refracted downward enough to offset the curvature of Earth and make the objects visible to a distant observer. This type of mirage can make a ship that is beyond the horizon appear to be on the horizon or even floating above it.

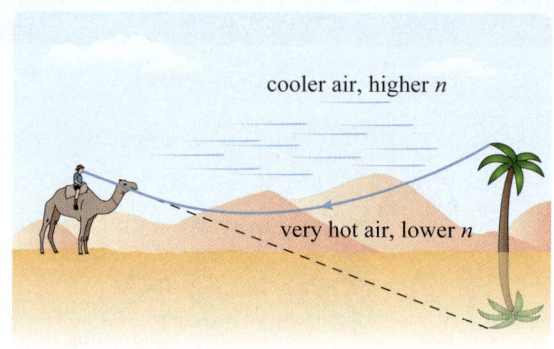

Figure 28-27 The formation of a mirage.

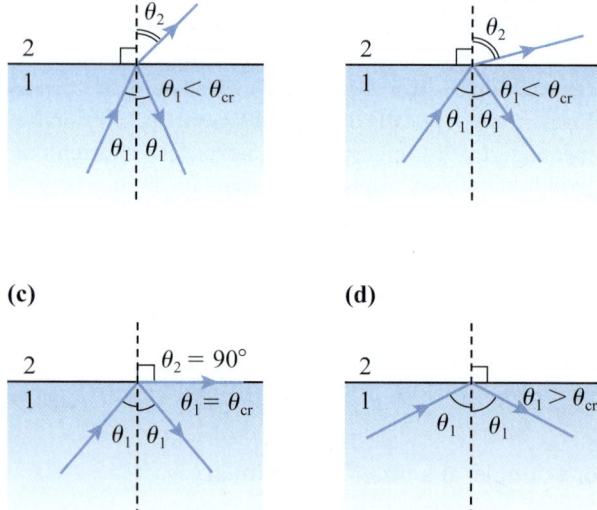

Figure 28-28 (a), (b) The propagation of light from a more optically dense medium to a less optically dense medium, $n_1 > n_2$. (c) When the angle of refraction is 90°, the refracted beam propagates along the border between the media. (d) Total internal reflection.

optic cable undergoes multiple total internal reflections, so the light does not leave the fibre. Consequently, by bending the optical fibre, the beam of light can be forced to bend. Optical fibres play such an important role in our lives that one half of the 2009 Nobel Prize in physics was awarded to the "father of fiber optics"—Charles Kuen Kao—"for groundbreaking achievements concerning the transmission of light in fibers for optical communication." Optical fibres are also widely used in medical applications, such as biomedical sensing, endoscopic imaging, micro-miniature fibre optics pressure sensors, and many others. Total internal reflection is also used in many other devices.

Dispersion and total internal reflection are also common in nature. For example, the total internal reflection of light and light dispersion in raindrops produce rainbows. As shown in Figure 28-29, we can approximate raindrops as small spheres. Sunlight entering a raindrop is refracted at the air–water boundary as it enters the raindrop. At the water–air boundary, some of the Sun's rays then undergo total internal reflection.

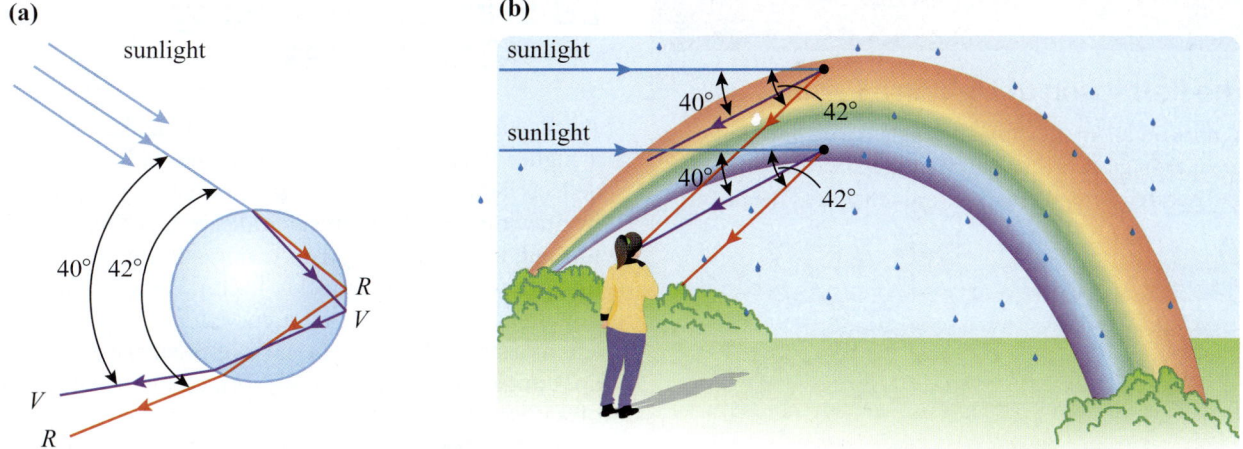

Figure 28-29 (a) Dispersion and total internal reflection of light in a spherical raindrop. (b) The formation of a rainbow. Notice that the Sun is located behind the person.

Refraction of Light in a Triangular Prism

We now consider the refraction of monochromatic light in a triangular prism. Let the index of refraction of the prism, n_p, be higher than the index of refraction of the ambient medium, n_{amb}, as, for example, in a glass prism surrounded by air. In the prism, the refracted ray is closer to the normal at the air–glass boundary than the incident ray is. The process is reversed at the glass–air boundary. As a result, the light ray exiting the prism is deflected toward its base by an angle δ (Figure 28-30(a)). When white light enters the prism, the deflection of rays of different colours varies with the wavelength of the light. Consequently, the light exiting the prism spreads out with the highest frequencies (shortest wavelengths) being deflected the most (Figure 28-30(b)).

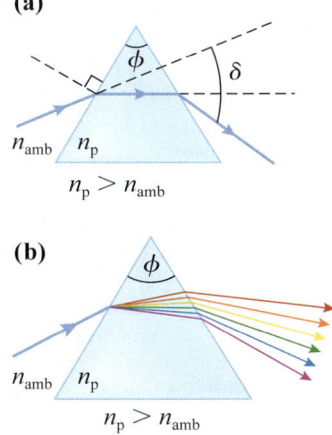

Figure 28-30 (a) The refraction of monochromic light in a triangular prism. (b) The dispersion of white light by a triangular prism.

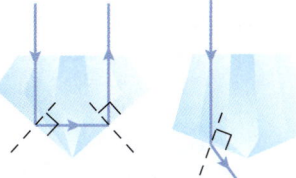
ELECTRICITY, MAGNETISM, AND OPTICS

A horizontal, monochromatic light beam travelling in air enters a glass prism. Which of the diagrams in Figure 28-32 correctly represents the beam that exits the prism?

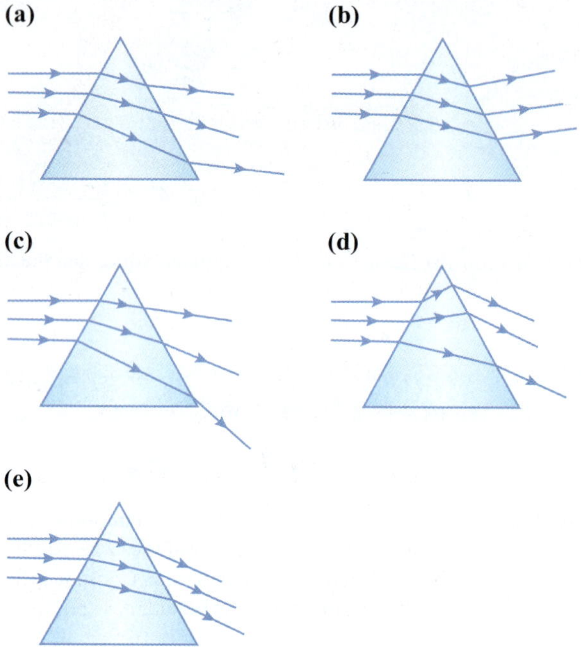

(a)

(b)

(c)

(d)

(e)

Figure 28-32 Checkpoint 28-7.

ANSWER: (e) See Figure 28-30.

28-5 Images Formed by Thin Lenses

As shown in the previous section, a triangular prism with a higher index of refraction than the ambient medium deflects light toward its base. In contrast, as you can check by drawing a ray diagram, a parallel beam of light passing through a rectangular prism is not deflected (it will shift up or down a prism parallel to itself). We can understand how images are formed by a curved lens by treating the lens as a stack of prisms (or parts of prisms) with different apex angles, which is the angle formed by the pointed tip of a prism (Figure 28-33).

A **thin lens** is a lens whose thickness is small relative to its other dimensions. Figure 28-34 shows six basic shapes of thin lenses. Provided that the index of

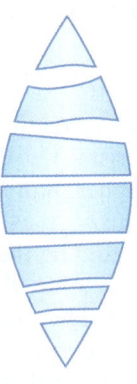

Figure 28-33 A representation of a biconvex lens (a lens that has convex faces on both sides) as a stack of prisms or parts of prisms with different apex angles.

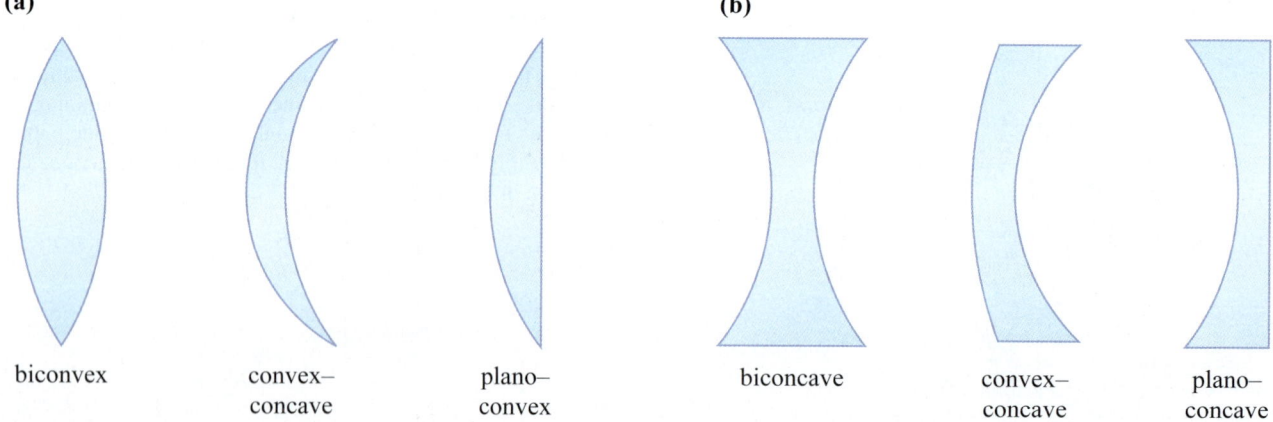

(a)

biconvex   convex–concave   plano–convex

(b)

biconcave   convex–concave   plano–concave

Figure 28-34 Various shapes of lenses whose index of refraction is higher than the ambient index of refraction. (a) Converging lenses have a positive focal length. (b) Diverging lenses have a negative focal length.

Introducing Lenses

You most certainly have played with a magnifying glass, a pair of glasses, a camera, or other optical instruments that contain lenses. Moreover, each one of us relies on one lens in each of our own eyes. In this activity, you will use the PhET simulation "Introducing Lenses" to gain intuitive understanding of how these fascinating devices work through virtual experience with lenses.

PEER TO PEER

To remember what to call each lens, I use a simple mnemonic: the word *concave* has "cave" in it, so the concave side of a lens should look like a cave (Figure 28-34). The convex side of a lens, then, should look like a *belly*—definitely not a cave. I always describe each side of the lens while looking at it from its own direction: I am looking at the left side of the lens from the left and at the right side of the lens from the right. A lens has two sides, so I name it by describing each one of the sides. The order of the name doesn't matter, so a concave–convex lens can also be called a convex–concave lens.

refraction of the lens is higher than the ambient index of refraction, the **converging lenses** are biconvex, convex–concave, and plano–convex. The **diverging lenses** are biconcave, convex–concave, and plano–concave.

The distance from an object to a lens, d_o, is measured from the centre plane of the lens. Unlike the case of two stacked inverted prisms (Figure 28-35(a), the left image), a beam of light parallel to the major optical axis of a converging thin lens (Figure 28-35(b),

(a)

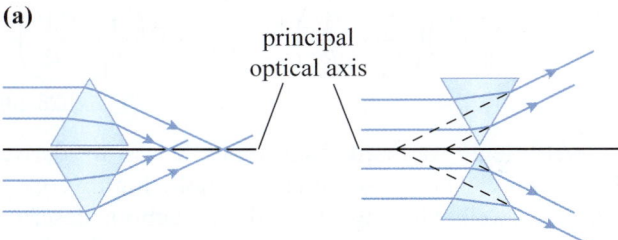

(b)

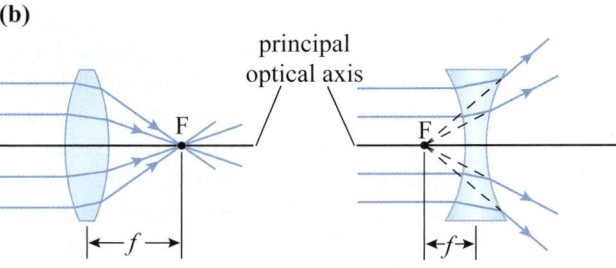

Figure 28-35 Parallel beams of light passing through (a) two stacked prisms, and (b) converging (left image) and diverging (right image) lenses.

the left image) converges at the focal point F behind the lens. Such a lens has a positive focal length, $f > 0$. A beam of light parallel to the principal optical axis of a diverging thin lens (Figure 28-35(b), the right image) diverges so the rays appear to have originated from an imaginary focal point F in front of the lens. Notice that if this diverging lens were replaced with two inverted prisms (Figure 28-35(a), the right image), the light rays would not appear to originate from one imaginary focal point. The focal length of a diverging lens is negative, $f < 0$. Every lens has two equidistant focal points, one located in front of the lens, and the other behind it. We can easily see that symmetrical lenses, like those in Figure 28-35(b), have this property. Using ray diagrams (or more advanced methods of analysis), we can show that asymmetrical lenses also have two equidistant focal points, one on each side. Consequently, all lenses have two focal planes, each perpendicular to the principal optical axis crossing it at one of the two focal points. Since the two focal points are always located the same distance from the lens, even when the lens is asymmetrical, this explains why you can call a lens either a convex–concave or a concave–convex lens.

Images Produced by Thin Lenses

We denote the centre point of a converging (biconvex) lens (Figure 28-36) as C. The axis of symmetry perpendicular to the curved surfaces of a lens is the principal optical axis of the lens. Consider a small light source (object) at point O, a distance d_o from a lens. Every ray of light that emanates from point A of the source and reaches the lens refracts according to the law of refraction (Equation 28-8). If the light source is small and the lens is thin, refracted rays converge at point A′ behind the lens, forming a real image of point A. Just as all the rays from point A on the object converge at point A′ in the image, each point on the object has a corresponding point on the image at which all rays emanating from it converge. To an observer standing behind the lens, the light rays appear as if they originated from point A′. The same applies for any other point of the light source. Like spherical mirrors, thin lenses create sharp images when the incident rays are paraxial.

PEER TO PEER

Until I was shown a computer simulation that traced rays with different types of lenses, I wasn't able to convince myself that asymmetrical lenses (e.g., concave–convex and plano–convex lenses) have two symmetrical focal points. In addition, it is important to remember that you can always reverse the direction of light. So, if a light ray travels from point A to point B along a certain path, it will travel along the same path when it travels from point B to point A. This also follows from the principle of the reversibility of light.

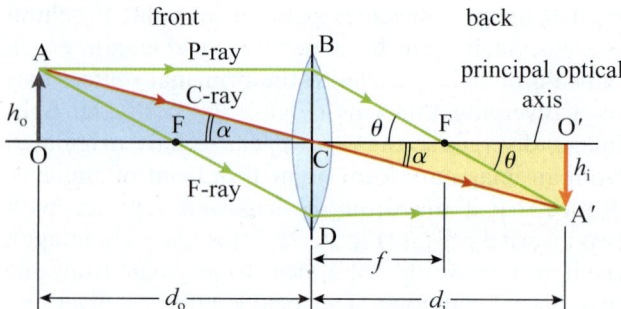

Figure 28-36 A ray diagram for image formation by a converging lens.

Now consider an object on the principal axis of a converging lens, as shown in Figure 28-36. Incident light ray AB (a P-ray) passes through the focal point F after refracting through the lens. Ray AD, an F-ray passing through the front focal point of the lens, exits parallel to the principal axis behind the lens. Incident ray AC, a C-ray passing through the centre of the lens, however, does not deflect noticeably because the lens is thin. These rays intersect at point A', which is the location of the image of point A.

Since the right-angled triangles AOC and A'O'C are similar, the ratios of the corresponding sides must be equal. This leads to a lens magnification equation, which is identical to the magnification equation for a spherical mirror (Equation 28-5b):

$$M = \frac{h_i}{h_o} = -\frac{d_i}{d_o}$$

As with mirror magnification, the minus sign indicates that the image is inverted. Right-angled triangles BCF and A'O'F are also similar, and CB = h_o. Therefore,

$$\begin{cases} \tan\theta = \dfrac{h_o}{f} \text{ from } \Delta\text{BCF} \\ \Rightarrow \\ \tan\theta = \dfrac{-h_i}{d_i - f} \text{ from } \Delta\text{A}'\text{O}'\text{F} \end{cases}$$

$$\frac{h_o}{f} = \frac{-h_i}{d_i - f} \Rightarrow \frac{h_o}{f} = \frac{h_i}{f - d_i} \Rightarrow \frac{h_o}{h_i} = \frac{f}{f - d_i}$$

Combining the two previous equations gives

$$-\frac{d_o}{d_i} = \frac{f}{f - d_i}$$

Rearranging the above equation leads to the thin lens equation, which, in combination with the lens magnification equation, provides an algebraic description of image formation by thin lenses:

$$\frac{1}{d_o} + \frac{1}{d_i} = \frac{1}{f} \qquad (28\text{-}11\text{a})$$

$$M = \frac{h_i}{h_o} = -\frac{d_i}{d_o} \qquad (28\text{-}11\text{b})$$

You can use Equations 28-11 to predict the properties of an image formed by a thin lens. You can also use it to check that your ray diagrams correctly represent the images formed by lenses. Noticing the similarities in the derivations of the thin lens and the mirror equations, it should not be surprising that the thin lens equations are the same as the spherical mirror equations, Equations 28-5. However, the relationship between the focal length and the geometric and physical characteristics of the lens is far from obvious. The relationship can be derived theoretically by examining image formation at both refracting surfaces. Since it was first discovered experimentally by lens makers it is called the lens maker's equation for a thin lens:

$$\frac{1}{f} = \left(\frac{n_{lens} - n_{amb}}{n_{amb}}\right)\left(\frac{1}{R_1} - \frac{1}{R_2}\right) \qquad (28\text{-}12)$$

where n_{lens} is the index of refraction of the lens, n_{amb} is the index of refraction of the ambient medium, and R_1 and R_2 are the radii of curvature of the front (closest to the light source) and back (farthest from the light source) surfaces of the lens, respectively.

The radius of curvature of a lens is considered to be positive when the surface is convex as observed from the side of the light source, and negative when the surface is concave. For example, R_1 is positive and R_2 is negative for a biconvex lens. The bigger the difference in the indexes of refraction, the shorter the focal length and the more the lens refracts light. Unless indicated otherwise, we assume that the ambient medium is air. Since n_{air} is very close to one, the lens maker's equation for lenses in air reduces to

$$\frac{1}{f} = \left(\frac{n_{lens} - 1}{1}\right)\left(\frac{1}{R_1} - \frac{1}{R_2}\right) = (n_{lens} - 1)\left(\frac{1}{R_1} - \frac{1}{R_2}\right) \qquad (28\text{-}13)$$

Although we used converging lenses to derive Equations 28-11, these equations hold for diverging lenses, as well. Thus, we derived an algebraic description of images produced by thin lenses and stated the formula for describing the focal length of a lens based on its physical properties (Equation 28-13). Notice that since we most often deal with glass or plastic lenses surrounded by air, the index of refraction of a lens is often larger than the index of refraction of the ambient medium: $n_{lens} > n_{ambient}$. Therefore, the value of $(n_{lens} - 1)$ in Equation 28-13 is positive. This explains why we often assume that the lenses shown in Figure 28-34 are converging and diverging lenses. However, if we had the reversed case, where the index of refraction of a lens was less than the ambient index of refraction ($n_{lens} < n_{ambient}$), the descriptions of converging and diverging lenses in Figure 28-34 would be reversed. In that case, the lenses in Figure 28-34(a) would be diverging lenses and the lenses shown in Figure 28-34(b) would be diverging lenses. Therefore, always make sure

you take into account not only the index of refraction of a lens but also the index of refraction of the ambient medium. Table 28-3 summarizes the sign conventions for the variables used in these equations. These conventions match the conventions used for spherical mirrors. Notice that the phrase *in front of the lens* indicates that it is the side of the lens that faces the object. Thus, the light from the object enters the lens from its *front* side and exits from the *back* side of the lens.

PEER TO PEER

I have to remind myself that no light rays pass through either virtual images or virtual objects: it only appears to us that the light rays come from or go to virtual images or objects! This is why we call them virtual. Whenever the object is a real object (the light comes from the object to the lens), the distance between the object and the lens is positive: $d_o > 0$. It doesn't matter if the object is placed to the right or to the left of the lens. However, in the case of multiple lenses, a virtual image formed by one lens can become an object for the second lens. In that case, we have a virtual object, and $d_o < 0$.

Table 28-3 Sign Conventions for Lenses

Quantity	Positive	Negative	Comments
d_o, object–lens distance	The object is in front of the lens (real object).	The object is behind the lens (virtual object).	A negative object–lens distance is often found in systems of multiple lenses.
d_i, object–image distance	The image is behind the lens (real image).	The image is in front of the lens (virtual image).	
f, the focal length	Converging lens	Diverging lens	The focal length depends on the geometry of the lens, the properties of the materials it is made of and the surrounding medium.
h_o, the height of an object	Most cases: when an object is located above the principal optical axis	Special cases: when an object is located above the principal optical axis	Whatever the object orientation is, it is most often considered to be positive.
h_i, the height of an image	The image is upright compared to the object.	The image is inverted compared to the object.	
n_{lens}, the index of refraction of the lens	Always positive		Important for calculating the focal length of a lens using the lens maker's equation, Equation 28-12
n_{amb}, the index of refraction of the surrounding medium	Always positive		
R_1, the radius of curvature of the lens surface closest to the light source	The surface is convex (from the side of the light source).	The surface is concave (from the side of the light source).	Important for calculating the focal length of a lens using the lens maker's equation, Equation 28-12
R_2, the radius of curvature of the lens surface farthest from the light source	The surface is convex (from the side of the light source).	The surface is concave (from the side of the light source).	

ELECTRICITY, MAGNETISM, AND OPTICS

EXAMPLE 28-6

Is a Biconvex Lens a Converging Lens after All?

A biconvex lens shell made of a thin transparent plastic is filled with water. A 5.00 cm tall plastic toy placed 25.0 cm in front of the lens produces a sharp image that is 10.0 cm tall on a screen. The lens shell is then emptied and filled with vegetable oil (n_{oil} = 1.47). Ignore the small amount of refraction in the plastic holder in solving the problem.

(a) What is the focal length of the lens when it is filled with water?
(b) What are the characteristics of the image of the toy when the lens is filled with water?
(c) What is the focal length of the lens when it is filled with oil?
(d) What are the characteristics of the image of the toy on the screen when the lens is filled with oil?
(e) The same demonstrations were performed when the lens was filled with air and immersed in water. What would the properties of the toy's image be in that case?

Solution

(a) From Table 28-2, we have n_{water} = 1.33 (to three significant digits). Since the image is real and it is a one-lens system, it must be inverted. Therefore, h_i = −10.0 cm, and the magnification is $M = \dfrac{h_i}{h_o} = \dfrac{-10.0 \text{ cm}}{5.00 \text{ cm}} = -2.00$

From Equation 28-11a, we can see that the image–lens distance must be twice as long as the object–lens distance: d_i = 2(25.0 cm) = 50.0 cm.

Now we use the thin lens equation, Equation 28-11a, to find the focal length of the water-filled lens in air:

$$\frac{1}{d_o} + \frac{1}{d_i} = \frac{1}{f_{w-a}}$$

$$\frac{1}{25.0 \text{ cm}} + \frac{1}{50.0 \text{ cm}} = \frac{1}{f_{w-a}} \Rightarrow f_{w-a} = 16.7 \text{ cm}$$

(b) The height and magnification found in part (a) indicate that the image is real, inverted, and magnified (M = −2).

(c) When the lens is filled with oil instead of water, the index of refraction of the lens changes but R_1 and R_2 do not change. We can use the lens maker's equation, Equation 28-13, to find the focal length of the lens filled with oil and surrounded by air, f_{o-a}:

$$\frac{1}{f_{w-a}} = (n_w - 1)\left(\frac{1}{R_1} - \frac{1}{R_2}\right)$$

$$\left(\frac{1}{R_1} - \frac{1}{R_2}\right) = \frac{1}{f_{w-a}(n_w - 1)} = \frac{1}{(16.67 \text{ cm}) \cdot 0.33}$$

$$= 0.182 \text{ cm}^{-1}$$

$$\frac{1}{f_{o-a}} = (n_o - 1)\left(\frac{1}{R_1} - \frac{1}{R_2}\right) = (1.47 - 1)(0.182 \text{ cm}^{-1})$$

$$= 0.086$$

$$f_{o-a} = 11.628 \text{ cm} \approx 11.6 \text{ cm}$$

The focal length of the oil-filled lens is shorter than the focal length of the water-filled lens.

(d) The object is 25.0 cm from the oil-filled lens:

$$\frac{1}{d_o} + \frac{1}{d_i} = \frac{1}{f_{w-a}}$$

$$\frac{1}{25.0 \text{ cm}} + \frac{1}{d_i} = \frac{1}{11.628 \text{ cm}} \Rightarrow \frac{1}{d_i} = 0.046 \text{ cm}^{-1}$$

$$d_i = 21.7 \text{ cm}$$

Since the image–lens distance is positive, the image is real. We find the image height and the magnification using Equation 28-11b:

$$M = \frac{d_i}{d_o} = -\frac{21.7 \text{ cm}}{25.0 \text{ cm}} = -0.87$$

$$M = \frac{h_i}{h_o} \Rightarrow h_i = h_o \cdot M = -8.7 \text{ cm}$$

Thus, the image is real, inverted, and reduced.

(e) From Equation 28-12, we see that when the lens is filled with air and surrounded by water, it becomes a diverging lens. Thus, the lens produces images that are virtual and upright. These images cannot be viewed on a screen.

Making sense of the result

This example illustrates that the focal length of a lens depends on its geometry, the material it is made of, and the optical properties of the ambient medium. As we discussed above, a lens can be either converging or diverging, depending on the relative indexes of refraction of the lens and the ambient medium.

Ray Diagrams for Thin Lenses

In Figure 28-36, we constructed a basic ray diagram for a thin lens. We now look at ray diagrams for thin lenses in more detail. As with spherical mirrors, we find the image location first by finding the location of the image of a point at the top of the object. Again, it is convenient to use three special rays:

■ The P-ray is parallel to the principal optical axis. For a *converging* lens, the P-ray passes through the focal point *behind* the lens (Figure 28-37(a), (b)). However, for a *diverging* lens, the P-ray refracts so it appears to originate from a focal point *in front of* the lens (Figure 28-37(c)).

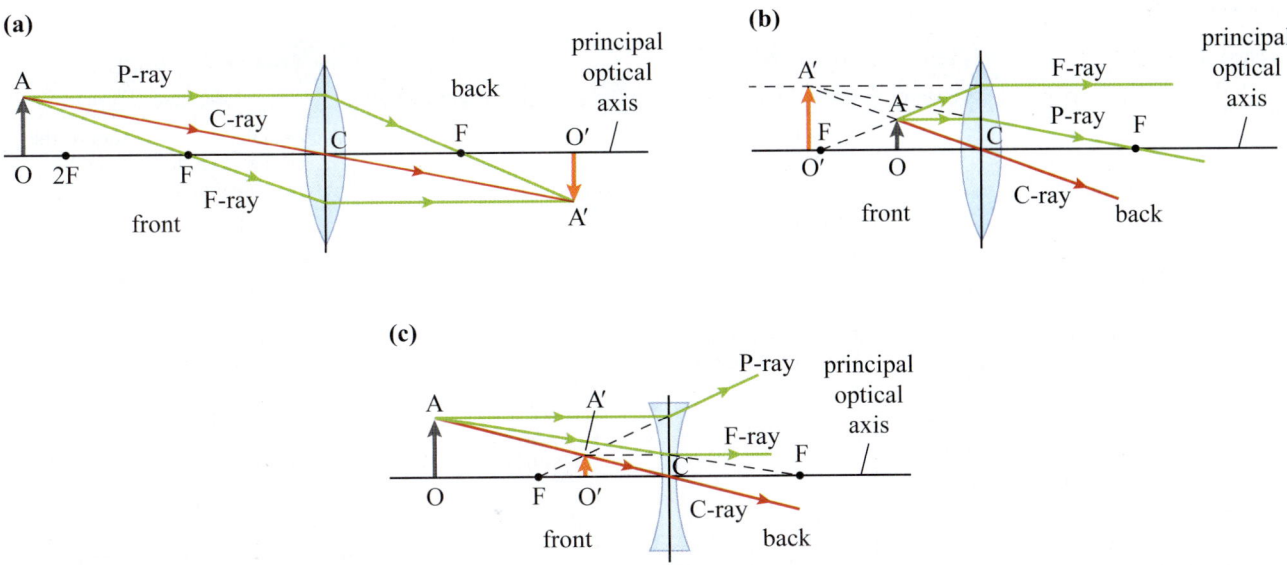

Figure 28-37 Ray diagrams for locating images created by thin lenses. (a) An object that is farther than twice the focal length from a converging lens creates a real, inverted, and reduced image. (b) An object between the mirror and the focal length of a converging lens creates a virtual, upright, and reduced image. (c) An object in front of a diverging lens creates a virtual, upright, and reduced image.

- The C-ray passes through the centre of the lens. The displacement of this ray is negligible for thin lenses.

- The F-ray passes through the focal point in front of the lens. This ray refracts so the ray emerges from the lens parallel to its principal axis.

The point where the rays actually intersect indicates the location of a real image, and the point from which the diverging rays appear to originate indicates the location of a virtual image. With thin lenses and small objects, every point of an object is located approximately the same distance from the lens, and every point of the image is also located approximately the same distance from the lens. Thus, the image location is determined by drawing a normal to the principal axis from the point where the image of point A is located.

Table 28-4 summarizes the properties of images formed by thin lenses. The absolute value of the magnification indicates whether the image is magnified ($|M| > 1$) or reduced ($|M| < 1$), and its sign indicates whether the image is upright (positive) or inverted (negative). Note that diverging lenses *always* produce reduced virtual images.

Very often we deal with images created by a combination of lenses. For example, twin-lens reflex cameras, dual-lens smartphone cameras, telescopes, and compound microscopes have two or even more lenses. Your eye has a lens as well and if you wear

CHECKPOINT 28-9

Image Formed by a Converging Lens

A lit candle is 20 cm from a converging lens ($f = 10$ cm). What happens to the candle's image if you cover the bottom part of the lens?
(a) Only the bottom half of the original image is visible; the image is dimmer.
(b) Only the top half of the original image is visible; the image is dimmer.
(c) The image is dimmer and half the size as before, but the entire candle is visible.
(d) The image is the same size as before but dimmer; the entire candle is visible.
(e) The image does not change in brightness but is twice as big; the entire candle is visible.

ANSWER: (d) Draw a ray diagram to verify the answer. Every point of an object emits light in all directions, so rays from every point of an object go through different parts of the lens. When some of the lens is blocked, fewer rays converge at the location of an image, thus creating a dimmer image.

ELECTRICITY, MAGNETISM, AND OPTICS

Table 28-4 Properties of Images Created by Thin Lenses

Object–Lens Distance	Converging Lens: $f > 0$		Diverging Lens: $f < 0$	
	Mathematical Description	**Image Properties**	**Mathematical Description**	**Image Properties**
$0 < d_o < |f|$	$d_i < 0$	Virtual, appears in front of the lens	$d_i < 0$	Virtual, appears in front of the lens between the lens and its front focal point
	$h_i > 0$	Upright	$h_i > 0$	Upright
	$M = \dfrac{h_i}{h_o} = -\dfrac{d_i}{d_o} > 1$	Magnified	$M = \dfrac{h_i}{h_o} = -\dfrac{d_i}{d_o} < 1$	Reduced
	$|d_i| > |d_o|$	Located farther from the lens than the object	$(0 < M < 1)$ $|d_i| < |d_o|$	Located closer to the lens than the object
$d_o = |f|$	No image is created.		$d_i = \dfrac{f}{2} < 0$	Virtual, appears behind the lens
			$h_i > 0$	Upright
			$M = -\dfrac{d_i}{d_o} = \dfrac{1}{2}$	Reduced by half
			$|d_i| = \dfrac{|f|}{2} < |d_o|$	Closer to the lens than the object
$|f| < d_o < 2|f|$	$d_i > 0$	Real, appears on the other side of the lens	$d_i < 0$ $|d_i| < |f| < |d_o|$	Virtual, appears in front of the lens between the lens and its front focal point
	$h_i < 0$	Inverted	$h_i > 0$	Upright
	$M = \dfrac{h_i}{h_o} = -\dfrac{d_i}{d_o} = -1$	Magnified	$M = \dfrac{h_i}{h_o} = -\dfrac{d_i}{d_o} < 1$	Reduced
	$|d_i| > |d_o|$	Farther from the lens than the object	$|d_i| < |d_o|$	Closer to the lens than the object
$d_o = 2|f|$	$d_i > 0$ $d_i = -d_o$	Real, appears on the other side of the lens	$d_i = \dfrac{2}{3}f < 0$ $0 < |d_i| < |f|$	Virtual, appears in front of the lens between the lens and its front focal point
	$h_i < 0$	Inverted	$h_i > 0$	Upright
	$M = \dfrac{h_i}{h_o} = -\dfrac{d_i}{d_o} = -1$	Same size as the object	$M = \dfrac{h_i}{h_o} = -\dfrac{d_i}{d_o} < 1$	Reduced
$d_o > 2|f|$	$d_i > 0$	Real, appears on the other side of the lens	$d_i < 0$	Virtual, appears in front of the lens between the lens and its front focal point
	$h_i < 0$	Inverted	$h_i > 0$	Upright
	$-1 < M < 0$	Reduced	$M = \dfrac{h_i}{h_o} = -\dfrac{d_i}{d_o} < 1$	Reduced
	$|d_i| < |d_o|$	Closer to the mirror than the object	$|d_i| < |d_o|$	Closer to the lens than the object

glasses you have two lenses per eye. Example 28-7 illustrates how to find an image formed by a combination of thin lenses.

In general, a combination of thin lenses can be replaced by a single equivalent lens. The optical properties of the lenses and the distance between them define the properties of the image. Lens combinations are used in telescopes, microscopes, cameras, and other optical devices. Interestingly, when you wear prescription eyeglasses, each of your eyes sees an image formed by a combination of the lens in the glasses and the lens in your eye.

EXAMPLE 28-7

Combination of Two Thin Lenses

Two converging lenses with focal lengths of $f_1 = 20.0$ cm and $f_2 = 25.0$ cm are placed parallel to each other so their centres are aligned along the principal optical axis (Figure 28-38). The lenses are separated by a distance of $d = 10.0$ cm. A 30.0 cm tall object is located 100 cm to the left of lens 1 (L_1). Describe the properties of the image produced by this combination of lenses.

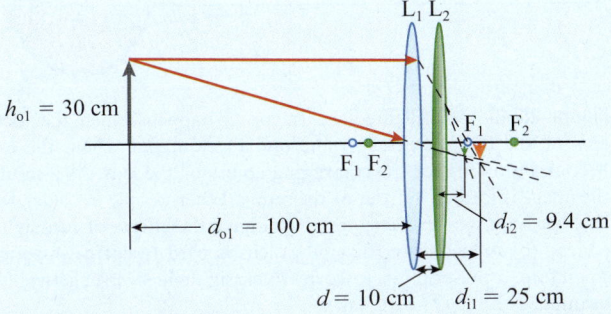

$h_{o1} = 30$ cm

$d_{o1} = 100$ cm

$d_{i2} = 9.4$ cm

$d = 10$ cm $d_{i1} = 25$ cm

Figure 28-38 Image formed by a combination of two lenses. The image formed by lens L_1 is shown in orange; the final image, formed by L_2, is shown in green. Dashed lines show the location of an image that would have been formed by L_1 if L_2 were not present.

Solution

Let us consider the image formed by the first lens, L_1. The distance between the object and L_1 is 100 cm. We can use the thin lens Equation 28-11a to find the lens–image distance:

$$\frac{1}{d_{o1}} + \frac{1}{d_{i1}} = \frac{1}{f_1}$$

$$\frac{1}{100 \text{ cm}} + \frac{1}{d_{i1}} = \frac{1}{20.0 \text{ cm}} \Rightarrow d_{i1} = 25.0 \text{ cm}$$

This distance is shown in red in the diagram. Since the lens–image distance is positive, the image is real and inverted. We find its height by using the lens magnification Equation 28-11b:

$$M_1 = \frac{h_{i1}}{h_{o1}} = -\frac{d_{i1}}{d_{o1}} = -\frac{25.0 \text{ cm}}{100 \text{ cm}} = -\frac{1}{4}$$

$$h_{i1} = -\frac{1}{4}h_{o1} = -\frac{30.0 \text{ cm}}{4} = -7.50 \text{ cm}$$

The negative height of the image indicates that it is inverted. We now consider the second lens, L_2. The image formed by L_1 becomes an object for L_2. However, since L_2 is located to the left of the image created by L_1, we are dealing with a virtual object and $d_{o2} < 0$. From Figure 28-38 we see that the distance between the second lens and the virtual object is 15.0 cm. Now we use the thin lens Equation 28-11a to find the location of the image formed by L_2:

$$\frac{1}{d_{o2}} + \frac{1}{d_{i2}} = \frac{1}{f_2}$$

$$\frac{1}{-15.0 \text{ cm}} + \frac{1}{d_{i2}} = \frac{1}{25.0 \text{ cm}} \Rightarrow d_{i2} = 9.38 \text{ cm}$$

To find the height of the final image, we use Equation 28-11b:

$$M_2 = \frac{h_{i2}}{h_{o2}} = -\frac{d_{i2}}{d_{o2}} = -\frac{9.38 \text{ cm}}{-15.0 \text{ cm}} = 0.625$$

$$h_{i2} = 0.625h_{o2} = 0.625h_{i1} = 0.625(-7.50 \text{ cm}) = -4.69 \text{ cm}$$

The overall magnification is

$$M_{\text{final}} = M_1 \cdot M_2 = \left(-\frac{1}{4}\right)(0.625) = -0.156$$

$$h_{i2} = -M_{\text{final}}h_{o1} = (-0.156)(30.0 \text{ cm}) = -4.69 \text{ cm}$$

Thus, the final image is inverted (relative to the object) and reduced such that the height of the final image are 16% of the original object's height.

28-6 The Human Eye and Vision Correction

Our fascination with vision, image formation, light, and colour is millennia old. It also took us centuries to understand the science behind the process of seeing and to confirm this understanding experimentally. There were two major ancient Greek schools of thought, offering basic explanations of how vision is carried out. The first theory (called the **extramission** or **emission theory of vision**) was put forward by the ancient Greek scholar Euclid (about 300 BCE) in his book *Optics* and was championed by many scholars, including Ptolemy (CE 100–c. 170). It stated that visual perception is accomplished by rays of light emitted by the eyes. The proponents of a competing theory, called the **intromission theory of vision**, such as Aristotle (384–322 BCE), claimed that vision happens when light rays emitted by objects enter our eyes, representing the object. In the 11th century, the medieval Arab scholar Ibn al-Haytham, known in the west as Alhazen (CE 965–1040), composed a seven-volume treatise called *The Book of Optics*. In his study, among other things, Alhazen investigated human vision. Alhazen presented *experimentally founded arguments* against the extramission theory of vision, arguing in favour of the intromission theory. What is even more interesting is that long before Isaac Newton (1643–1727) and Gottfried Wilhelm Leibniz (1646–1716) and their work on calculus, Alhazen suggested considering an object as being composed of infinitely many infinitely small points, such that each point on the object's surface emits light rays that travel to the viewer's eyes, creating its representation. The compilation of all the representations of these points creates a representation of the object. More than 600 years later, Pierre Gassendi (1592–1655), Newton, John Locke (1632–1704), and others firmly held not only that vision was intromissionist but that the rays emitted by seen objects were composed of actual matter, or corpuscles, that entered the seer's mind by way of the eye. Thus, the **corpuscular theory of light** was born. The debate about the nature of light continued into the 20th century, as you will see in Chapters 29 to 31.

We can use the principles of geometric optics to understand how the human eye functions. Indeed, we can use the same principles to understand the functioning of the eyes of insects, fish, birds, and other animals, even though the structure of their eyes is often radically different from the structure of human eyes.

The human eye is a complex optical device (Figure 28-39). Optically, the key parts are as follows. The iris adjusts the size of the opening (the pupil) through which light passes to the lens, which focuses the light onto the retina. The cornea and a chamber of fluid (called the aqueous humour) are in front of the lens, and a thicker fluid (called the vitreous humour) is in a chamber between the lens and the retina. The retina (the screen the images we see are formed on) contains rod cells, which provide black and white vision, and cone cells, which provide colour vision. The rod cells are not sensitive to colour (they are only sensitive to black and white), but they are more sensitive and more numerous than the cone cells. Thus, in dim light we cannot see in colour, we can only see in black and white. The optic nerve carries electrochemical signals from the retina to the brain.

For us to perceive an object clearly, the light emitted by the object has to be focused on the retina. Since the lens–retina distance is essentially constant, muscles in the human eye change the shape and, hence, the focal length of the lens. This allows the lens to form sharp images of the objects located at different distances from the eye on the retina. (In contrast, a fish eye can adjust the distance between its lens and retina, as shown in Interactive Activity 28-4, below.) If the eyeball is too long or the combination of cornea and lens is too powerful optically, the eye muscles may not be able to bring distant objects completely into focus, resulting in **nearsightedness** or **short-sightedness (myopia)** (Figure 28-40). Similarly, a short eyeball or a weaker lens can cause **farsightedness** or **long-sightedness (hyperopia)** (Figure 28-41). It is also important to note that the images formed on the retina are inverted images. This might be very surprising to you. Why, then, doesn't the world look upside down to us? The answer lies in the power of the human brain to adapt the sensory information it receives

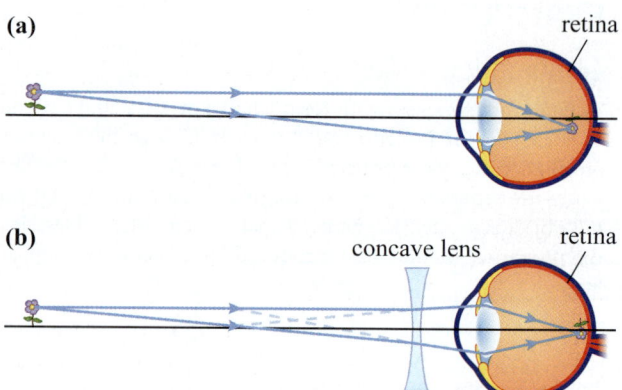

Figure 28-40 Nearsightedness (myopia) happens when the focal lens of the eye decreases or the eyeball elongates. Then, the eye becomes too long for the converging ability of the lens. As a result, the image is formed in front of the retina. When people get older, the lens becomes weaker, which might mitigate the effects of nearsightedness. (b) A biconcave (diverging) lens is used to correct myopia. Therefore, a prescription to correct nearsightedness is negative, for example, -2.5 D ($f = -0.4$ m).

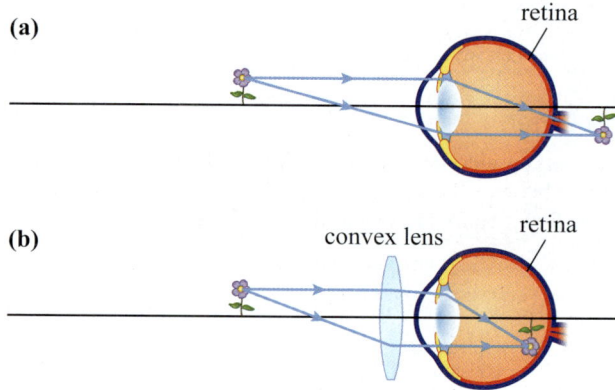

Figure 28-41 Farsightedness (hyperopia) happens when the focal lens of the eye increases or the eyeball shortens. Then, the eye becomes too short for the converging ability of the lens. As a result, the image is formed behind the retina. (b) A biconvex (converging) lens is used to correct hyperopia. Therefore, a prescription to correct farsightedness is positive, for example, $+2.5$ D ($f = 0.4$ m).

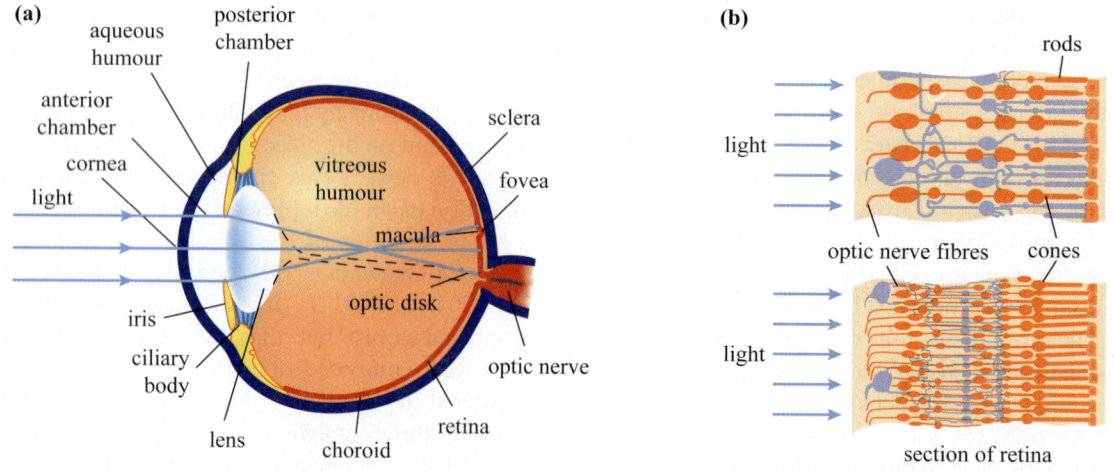

Figure 28-39 (a) A schematic representation of a human eye. (b) Details of the retina.

and make it fit with what it already knows. Essentially, your brain takes the raw, inverted data and turns it into a coherent, right-side-up image. If you doubt the truth of it, you can follow in the footsteps of the American psychologist George M. Stratton (1865–1957), who pioneered the study of visual perception. In the early 20th century, he conducted a series of experiments with special glasses that inverted images. He found that after wearing these glasses for only a few days, his brain was able to adjust to the new environment and see everything correctly.

Optometrists use a unit called a dioptre (symbol D) to measure the optical power of a lens. **Optical power** is the reciprocal of focal length, when the latter is measured in metres:

$$P = \frac{1}{f}$$

$$[P] = \frac{1}{m} \equiv D \qquad (28\text{-}14)$$

For example, a lens that has a focal length of 5.0 cm (0.050 m) has an optical power of 20 D. Note that "optical power" has nothing to do with the rate at which energy is transferred, so the word *power* here has a different meaning than the concept of power used earlier. Lenses that have shorter focal distance have higher optical power.

The reason it is convenient to use optical power, rather than the focal length of a lens, is that most of the lens equations (Equations 28-11, 28-12, and 28-13) use the reciprocal of the focal length. Moreover, the optical power of a combination of two or more thin lenses that are close together (essentially at the same location) is

approximately equal to the sum of the optical powers of the individual lenses:

$$P_{net} = \sum_{i=1}^{n} P_i \qquad (28\text{-}15)$$

MAKING CONNECTIONS

Early Applications of Lenses

At the end of the 13th century, people started wearing spectacles, a precursor of modern eyeglasses (Figure 28-42). Then a magnifying glass (a converging lens) began to be used to observe small objects. At the end of the 16th century, Dutch spectacle makers started experimenting with combinations of thin lenses to create more powerful magnifying glasses, resulting in the invention of the microscope. The telescope, originally developed for maritime use, was also invented in the Netherlands. When the telescope was brought to Italy at the beginning of the 17th century, Galileo used it to study the sky. He made several ground-breaking discoveries, including sunspots, the moons of Jupiter, the phases of Venus, features of the Moon, and the nature of the Milky Way. In 1784, Benjamin Franklin invented bifocal lenses.

R.Sekulovich/Public Domain

Figure 28-42 People started wearing spectacles in the 13th century. Early spectacles were the precursors of modern eyeglasses.

Exploring Human Vision

In this activity, you will use a number of simulations and animations to explore the functioning of human eye. They will help you understand the ways to correct eye defects using lenses.

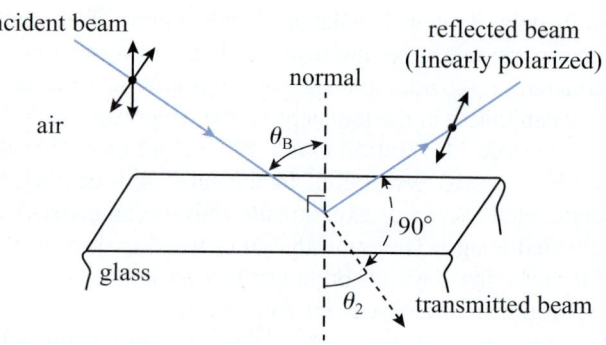

Figure 28-43 When the incident angle is equal to Brewster's angle, θ_B, the reflected light is completely polarized perpendicular to the plane of incidence. The transmitted light is partially polarized parallel to the plane of incidence.

28-7 Brewster's Angle

In Chapter 27, we considered the physical properties of light as an electromagnetic wave. We also discussed the concept of polarization (Section 27-8). Using Maxwell's equations, it can be shown that reflected light is completely polarized when the reflected and the transmitted waves form a right angle (Figure 28-43). This fact was first discovered by Scottish physicist David Brewster (1781–1868) in 1811, long before Maxwell's equations were developed. The angle of incidence for this complete polarization is called **Brewster's angle**, θ_B.

Let the index of refraction of air be n_1 and the index of refraction of the glass in Figure 28-43 be n_2. Let θ_B be the angle that the incident beam makes with respect to the normal to the plate and θ_2 be the angle that the transmitted beam makes with respect to the normal (the angle formed by the refracted beam).

(a) Due to the law of reflection, the angle between the reflected beam and the normal is θ_B.

(b) The angle of incidence and the angle of transmission are related by Snell's law:

$$n_1 \sin \theta_B = n_2 \sin \theta_2$$

We require that the reflected and the transmitted (refracted) beams make a right angle. From the geometry of Figure 28-43, we have

$$\theta_B + 90° + \theta_2 = 180°$$
$$\theta_2 = 90° - \theta_B$$

Inserting θ_2 into Snell's law, we get

$$n_1 \sin \theta_B = n_2 \sin (90° - \theta_B) \Rightarrow$$
$$n_1 \sin \theta_B = n_2 \cos \theta_B,$$

which gives

$$\tan \theta_B = \frac{n_2}{n_1}$$

MAKING CONNECTIONS

Polarizing Sunglasses

Sunlight is unpolarized. Light reflected from a horizontal surface is partially polarized in the horizontal plane. Polarizing sunglasses contain a coating that absorbs light that is polarized in the horizontal direction (Figure 28-44), greatly reducing the glare from road and water surfaces.

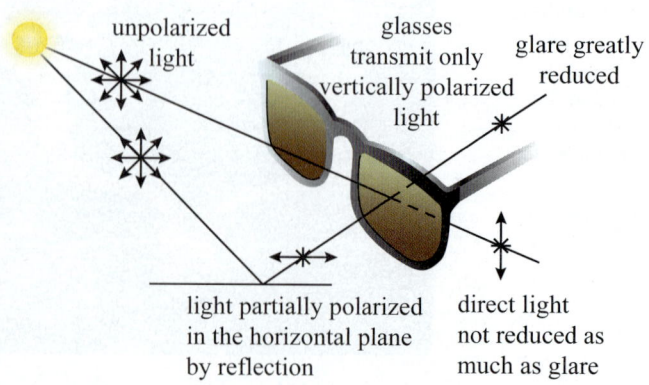

vision without polarized sunglasses

vision with polarized sunglasses

Courtesy of Marina Milner-Bolotin

Figure 28-44 Polarized sunglasses reduce glare by absorbing horizontally polarized sunlight.

$$\theta_B = \tan^{-1}\left(\frac{n_2}{n_1}\right) \quad \text{(Brewster's angle)} \quad (28\text{-}16)$$

For the air–glass interface, $n_1 = 1$ (air) and $n_2 = 1.52$ (glass). The value of Brewster's angle in this case is

$$\theta_B = \tan^{-1}\left(\frac{1.52}{1.00}\right) = 56.7°$$

(1) When unpolarized light from air strikes a glass surface at 57° (with respect to the normal to the surface), the reflected light is completely polarized perpendicular to the plane of incidence.

(2) When the incident light is polarized in the plane of incidence and strikes the glass surface at 57°, there is no reflected beam. The incident beam is completely transmitted into the glass.

As we discussed earlier, the refractive index for a given medium depends on the wavelength of light propagating through it (dispersion of light), so Brewster's angle also depends on the wavelength.

CHECKPOINT 28-11

Exploring Polarized Sunglasses

You are offered five pairs of sunglasses, as shown in Figure 28-45. Which pair of sunglasses would you choose to eliminate the glare resulting from sunlight reflecting off the calm waters of a lake? Explain your reasoning. The polarization axes, when relevant, are shown by the straight lines with the arrows. The sunglasses in (a) are not polarized.

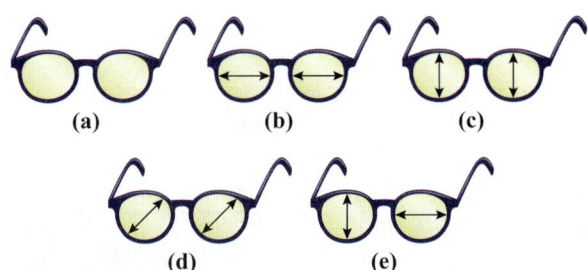

Figure 28-45 Checkpoint 28-11.

ANSWER: (c) Light reflected from smooth horizontal surfaces, such as water or a flat road, is generally horizontally polarized (its polarization is aligned to the surface the light is reflected from). Therefore, to block it, you need to choose a pair of vertically polarized sunglasses.

KEY CONCEPTS AND RELATIONSHIPS

Fermat's Principle

Light always travels between any two points along the path that requires the least amount of time, as compared to other possible paths. In a homogeneous medium, light always propagates along a straight line. In the heterogeneous medium, light can change its direction through undergoing reflection or refraction.

Optical Images

In order for us to see an object the light emitted by or reflected off the object must enter our eyes. The objects that emit light are called light source. The objects that reflect light partially or fully are light reflectors. A real image of an object is formed in the point of convergence of the light rays originated from an object and passing through different media and undergoing reflection or refraction. Real images can be produced by converging lenses and by concave mirrors when an object is placed beyond the focal point of these lenses or mirrors. If we place a screen in the plane of a real image the image will generally become visible on this screen. Examples of real images include the image on a cinema screen or images formed on your eyeball retina. A virtual image, on the other hand, is formed when the light rays originated from an object diverge as a result of refraction or reflection. A virtual image appears to be located at the point of divergence. Since the light rays diverge they cannot be projected onto a screen, so virtual images cannot be seen on a screen. Both virtual and real images can be either upright or inverted, magnified or reduced. A plane mirror, diverging lenses and convex lenses form virtual images. Converging lenses and concave mirrors also can produce virtual images if the objects are placed in front of the focal point ($d_o < f$).

The Law of Reflection

The angle of incidence equals the angle of reflection:

$$\theta = \theta' \quad (28\text{-}1)$$

The incident ray, the reflected ray, and the normal to the surface at the point of incidence are always located in the same plane.

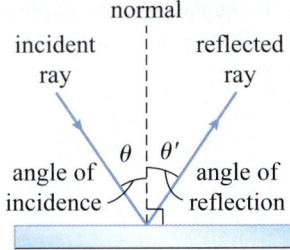

The Law of Refraction

The index of refraction of a medium is

$$n = \frac{c}{v} \quad (28\text{-}7)$$

where c is the speed of light in a vacuum and v is the speed of light in the medium.

When a ray of light propagates through the boundary between two media, it changes direction through undergoing refraction such that

$$n_1\sin\theta_1 = n_2\sin\theta_2 \quad (28\text{-}8)$$

where n_1 and n_2 are the indexes of refraction of the two media and θ_1 and θ_2 are the angles of incidence and refraction, respectively.

(a)

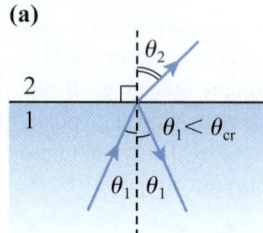

(b)

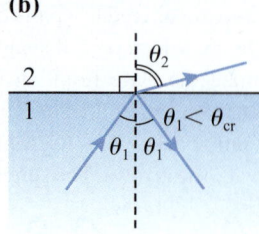

(c)

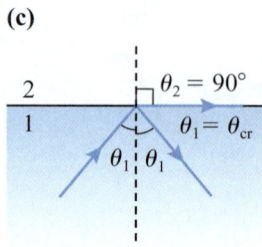

(d)

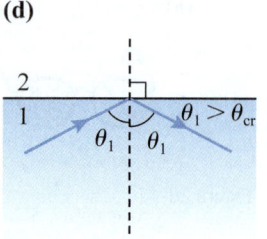

Total Internal Reflection and Critical Angle

When the angle of incidence exceeds the critical angle, the light propagating from a more optically dense medium to a less optically dense medium undergoes total internal reflection:

$$\theta_{cr} = \sin^{-1}\left(\frac{n_2}{n_1}\right) \qquad (28\text{-}10)$$

The Formation of Images by Mirrors and Lenses

The locations of an object and the image formed by a spherical mirror or a thin lens are related as follows:

$$\frac{1}{d_o} + \frac{1}{d_i} = \frac{1}{f} \qquad (28\text{-}5a, 28\text{-}11a)$$

where d_o is the object–mirror/lens distance, d_i is the mirror/lens–image distance, and f is the focal length of the mirror/lens. The focal length is positive for concave mirrors and converging lenses and negative for convex mirrors and diverging lenses.

The magnification of an image is determined using the following equation:

$$M = \frac{h_i}{h_o} = -\frac{d_i}{d_o} \qquad (28\text{-}5b, 28\text{-}11b)$$

where M is the linear magnification and h_o and h_i are the heights of the object and of the image, respectively.

The relationship between the focal length and the geometric and physical characteristics of the lens is called the lens maker's equation for a thin lens:

$$\frac{1}{f} = \left(\frac{n_{lens} - n_{amb}}{n_{amb}}\right)\left(\frac{1}{R_1} - \frac{1}{R_2}\right), \qquad (28\text{-}12)$$

where n_{lens} is the index of refraction of the lens, n_{amb} is the index of refraction of the ambient medium, and R_1 and R_2 are the radii of curvature of the front (closest to the light source) and back (farthest from the light source) surfaces of the lens, respectively.

Brewster's Angle

Brewster's angle is the angle of incidence at which light reflected from a surface is completely polarized:

$$\theta_B = \tan^{-1}\left(\frac{n_2}{n_1}\right) \qquad (28\text{-}16)$$

APPLICATIONS

optical instruments, such as eyeglasses, telescopes, microscopes, and magnifying glasses; natural phenomena, such as shadows, eclipses, rainbows, and mirages; fibre optics cables; polarized sunglasses

KEY TERMS

angle of incidence, angle of reflection, angle of refraction, angular magnification, Brewster's angle, centre of curvature, concave mirror, converging lenses, convex mirror, corpuscular theory of light, critical angle, diffuse reflection, dispersion, diverging lenses, emission theory of vision, extramission theory of vision, farsightedness, Fermat's principle, field of view, focal length, focal point, full shadow, geometric optics, hyperopia, image, incident ray, index of refraction, intromission theory of vision, law of reflection, law of refraction, linear magnification, long-sightedness, medium, myopia, nearsightedness, object, opaque, optical density, optical power, paraxial rays, partial shadow, penumbra, phase velocity, pinhole camera, plane mirror, principal optical axis, radius of curvature, ray diagram, ray optics, real image, reflected ray, reflection, refracted ray, refraction, reversibility of light, secondary focal point, short-sightedness, Snell's law, specular reflection, thin lens, total internal reflection, translucent, transparent, umbra, virtual focal point, virtual image

QUESTIONS

1. You are standing 2.0 m in front of a plane mirror and then move 0.50 m away from the mirror. By how much will the distance between you and your image change?

2. You are moving toward a mirror at a speed of 0.50 m/s. How fast are you moving toward your own image?

3. Why does an image in a plane mirror undergo depth inversion? Is it correct to say that a mirror interchanges left and right? Why is there no up–down inversion? Use a plane mirror, a piece of paper with a word written on it, and a ray diagram to test your explanation. How is this phenomenon used in everyday life? (Hint: Think of ambulances.)

4. Ordinary window glass reflects approximately 4% of incident light and transmits the remaining 96%. Use this information to explain why ordinary window glass works like a mirror for a person inside the house at night but looks transparent during the day. Remember, the light entering the glass encounters two boundaries: an air–glass boundary upon entering the glass pane and a glass–air boundary at the exit.

5. Which of the following statements correctly describes the images formed by a plane mirror?
   (a) The images are always real, inverted, and magnified.
   (b) The images are always virtual, inverted, and reduced.
   (c) The images are always virtual, upright, and the same size as the objects.
   (d) The images are always virtual and upright, and they appear closer than the objects.
   (e) The images in the mirror can be viewed from any location in front of the mirror.

6. A friend of yours claims that the image formed in a plane mirror is formed on the surface of the mirror, so the image does not have depth. What would you do to convince your friend that the location of the image is behind the mirror?

7. On a car's side-view mirror you notice a warning: "Objects in mirror are closer than they appear." Explain the physics behind this statement. What kind of mirror is it, and why is it used as a side-view mirror?

8. While surfing the Internet, you notice an advertisement for a car rear-view mirror that claims to solve the blind spot problem. The problem arises because the driver's field of view in a regular rear-view mirror is limited. Thus, the driver must move her head to be able to see all the cars in the vicinity of her own car. Alternatively, she must use additional mirrors. Draw a ray diagram to illustrate the problem, and propose a solution using optical devices discussed in this chapter. See whether your proposal makes sense the next time you have a chance to take a closer look in a rear-view mirror.

9. Figure 28-46 shows a rose that is located off to one side of a plane mirror. Use a ray diagram to answer the following questions.
   (a) Will an image of the rose be formed?
   (b) If you answered "yes" to part (a), where will the image be formed, and where should an observer be located to be able to view the image? (That is, determine the field of view for this image.)

Figure 28-46   Questions 9 and 10.

10. Replace the plane mirror in Figure 28-46 with a concave mirror and then with a convex mirror, and answer the questions asked in question 9. Compare your answers for the three different mirrors, and draw conclusions.

11. Two plane mirrors form the legs of a 60° angle. An object is placed between the two mirrors. What is the maximum number of images that could be viewed while looking into the mirrors? Explain.

12. If you want to see a full-length image of yourself (from the top of your head to your toes) in a plane mirror, what should the minimum size of the mirror be?
    (a) slightly more than my height
    (b) exactly my height
    (c) about three-quarters of my height
    (d) about one-half of my height
    (e) about one-quarter of my height
    (f) It depends on my distance from the mirror.

13. Explain your reasoning for your answer to question 12. How would your answer change if instead of a plane mirror you used a spherical mirror?

14. The Mirage (see optigone.com) is a classic toy that consists of two parabolic concave mirrors (Figure 28-47). The top mirror has a small circular opening on the top. A small object is placed in the centre of the bottom mirror. When assembled, the mirrors produce a life-like three-dimensional image of the object inside the device. The image is located right above the opening, as shown in Figure 28-47. Draw a ray diagram showing how the image is formed. Why must the object be small for the Mirage to work? How many images will be created by this toy? What is their relative orientation?

Figure 28-47 Question 14.

15. You decide to buy yourself a magnifying makeup mirror. What kind of a mirror is it? Use a ray diagram to explain your answer.

16. Use the mirror equation to plot a graph of the magnification of (a) a concave spherical mirror, (b) a convex spherical mirror, and (c) a plane mirror as a function of the object–mirror distance. Use Tables 28-1 and 28-3 to make sense of your graphs. Do the two representations (the table and your graph) match?

17. Three light rays simultaneously emitted by a candle pass through a converging lens, as shown in Figure 28-48. Which ray arrives at the screen first? Explain.

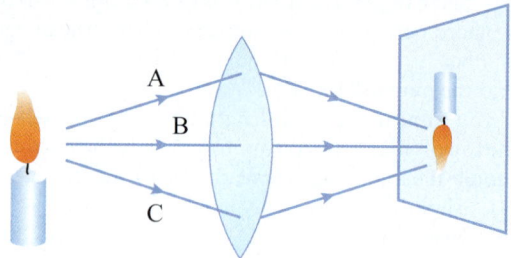

Figure 28-48 Question 17.

18. Use the thin lens equation to plot a graph of the magnification of (a) converging lenses and (b) diverging lenses as a function of the object–lens distance. Use Table 28-4 to make sense of your graphs. Do the two representations (the table and your graph) match? Compare the graphs you plotted for the mirrors in question 16 and for the lenses. What do they have in common, and how do they differ?

19. An object is placed in front of a converging lens such that its image is formed on a screen. What happens to the image when the bottom half of the lens is covered? Explain.

20. Give an example of where one lens can be a converging lens under some conditions and a diverging lens under other conditions. Where might this be important? (Hint: Think of seeing under water.)

21. A parallel beam of light is incident on a converging lens at an angle to the principal optical axis. Draw a ray diagram illustrating the propagation of the beam through the lens. Will the beam converge after exiting the lens? Where will it happen? Compare your result to the case of a spherical concave mirror. (Hint: Think of the lens's focal plane.)

22. Explain the difference between nearsighted vision (myopia) and farsighted vision (hyperopia), and use a ray diagram to show which lenses can be used to correct these problems.

PROBLEMS BY SECTION

For problems, star ratings will be used (★, ★★, or ★★★), with more stars meaning more challenging problems. The following codes will indicate if $\frac{d}{dx}$ differentiation, $\int$ integration, ⬜ numerical approximation, or 〰 graphical analysis will be required to solve the problem.

Section 28-1 Evidence for the Geometric Optics Approach

23. ★ You want to figure out the height of a tree on your street. You use a 1.00 m long vertical stick to cast a shadow (Figure 28-49). You can use the length of its shadow to determine other lengths or distances. On a bright sunny day, you find that at a certain time, the shadow cast by the stick is 3.00 m long. The stick is 100. m behind the tree, and you measure the shadow just before the Sun disappears behind the tree. What is the height of the tree? What assumptions did you make?

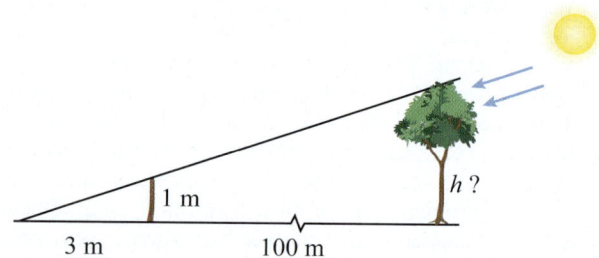

Figure 28-49 Problem 23.

24. ★ A pinhole camera like the one in Figure 28-5(a) is 20.0 cm deep and has a hole 1.00 mm in diameter. A 20.0 m tall tree is 50.0 m from the camera. What is the size of the image of the tree on the camera screen? How does the size of the image change when the diameter of the pinhole is doubled? Does anything else change as a result? Explain.

25. ★★ An eclipse of the Sun can be total or partial, depending on where on Earth one is located, as shown in Figure 28-50.
    (a) Explain what observers A, B, and C will see during the eclipse.
    (b) The figure shows an instant of a solar eclipse. Show how the area of the total and partial solar eclipse moves on Earth's surface during the eclipse.

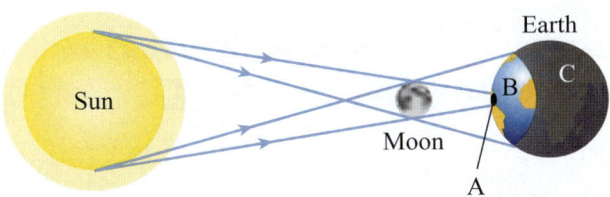

Figure 28-50   Problem 25. A total solar eclipse (not to scale).

26. ★★ The phases of the Moon are shown in Figure 28-51.
    (a) Use Figure 28-51 to explain what phase the Moon is in during a total solar eclipse. Why is a total solar eclipse such a rare phenomenon?
    (b) At some times during the Moon cycle, we can see the Moon and the Sun at the same time. When does this happen in terms of the phases of the Moon?
    (c) The popular phrase "the dark side of the Moon" reflects the fact that we can only see one side of the Moon. Explain why.
    (d) Draw a diagram that shows how lunar eclipses happen, and why at different locations around the globe people can see a total or a partial lunar eclipse.

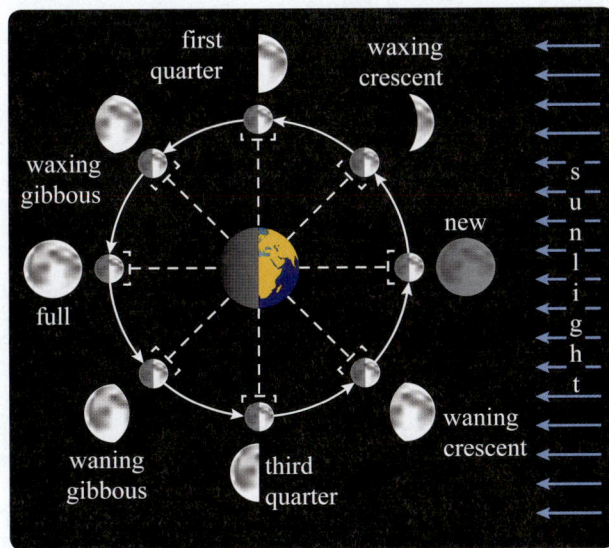

Figure 28-51   Problem 26.

Section 28-2   Reflection of Light

27. ★ A laser beam is reflected by a plane mirror. It is observed that the angle between the incident and reflected beams is 45°. The mirror is then rotated so that the angle of incidence decreases by 7.0°. What is the new angle between the incident and reflected beams?

28. ★ A laser beam enters a 10.0 m long optical fibre at an angle of incidence of 60.0°, which far exceeds the critical angle of the fibre. The core of the fibre is 60.0 mm in diameter, and the cladding is 140 mm (Figure 28-52). How many times will the laser light be reflected inside the fibre prior to exiting it? Assume the fibre is reasonably straight.

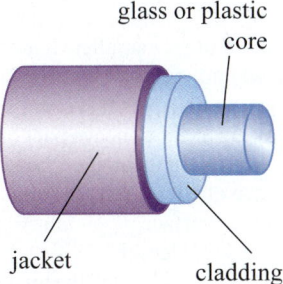

Figure 28-52   Problem 28.

29. ★ Sunlight enters a room at an angle of 30.0° above the horizontal and reflects from a small mirror lying flat on the floor. The reflected light forms a spot on a wall that is 2.50 m behind the mirror (Figure 28-53). You place a notebook under the edge of the mirror nearest to the wall, tilting it upward by 4.00°. How much higher on the wall (Δh) is the spot?

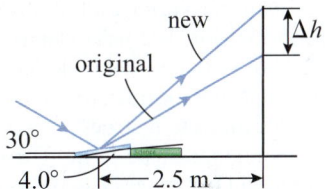

Figure 28-53   Problem 29.

Section 28-3   Mirrors

30. ★ Two plane mirrors meet at an angle θ, such that $\frac{360°}{\theta}$ is a whole number.
    (a) Use ray diagrams to help you derive an equation that shows the number of images that will be created by this configuration of mirrors.
    (b) Apply the equation you derived in part (a) to the case where two mirrors are parallel to each other. What conclusion can you make? How can you verify your conclusion?

31. ★ A periscope is an optical device that allows you to see from a concealed position. The simplest periscope consists of a tube at each end of which are tilted mirrors. Draw a diagram of a periscope, and use your diagram to answer the following questions.
    (a) Is the image of an object seen in your periscope inverted? Explain why or why not.
    (b) Sometimes two simple lenses are added to a periscope to focus the images of the objects observed. Use a ray diagram to show how this works.
    (c) Now suppose that you use curved mirrors in your periscope. Would this be a viable design? Explain why or why not.

32. ★ A 1.50 m tall child is standing in front of a plane mirror. His eyes are 9.00 cm below the top of his head.
  (a) What is the shortest mirror height for the child to see his entire body?
  (b) How high from the floor should the mirror hang so that the child can see his entire body?
  (c) How will your answers to parts (a) and (b) change if instead of a plane mirror a spherical mirror is used?
  (d) If the child moves back from the mirror, will he be able to use a mirror smaller than the size you found in part (a) and still see his entire body?

33. ★ While playing with a metal serving spoon, you notice that when looking at the concave part of the spoon, your finger appears to touch its image when it is 7.00 cm from the inner surface of the spoon. Assume that the spoon is a spherical mirror surface.
  (a) What is its focal length?
  (b) What are the properties of the image of your finger?

34. ★★ A 15.0 cm tall object is placed 0.500 m in front of a concave spherical mirror that is 60.0 cm in diameter. Justify your answers to the following questions.
  (a) Where will the image of the object be formed?
  (b) Describe the properties of the image (type, orientation, and magnification).
  (c) Instead of a concave mirror, you use a convex mirror. How do your answers to parts (a) and (b) change?

35. ★★ To increase security in a store, a storekeeper decides to replace an old plane mirror with a new convex mirror with the same diameter. He finds a 12.0 in. heavy-duty glass indoor spherical mirror. The seller claims that the mirror has a 160.° field of view area that is ideal for the store.
  (a) What should the focal length of the mirror be for the specifications to be true?
  (b) If a person stands 2.00 m away from this mirror, how large will her image be in the mirror? What are the properties of the image?

Section 28-4 Refraction of Light

36. ★ A laser beam is shone on a horizontal slab of 3.0 cm thick glass ($n = 1.5$) at a 30.° angle from the normal. By how much will the light ray be shifted from along the glass slab upon exiting it? (In other words, what is the distance between the point where the laser beam leaves the glass slab and where it would have left it if the slab had had an index of refraction of $n = 1$?)

37. ★ An underwater flashlight is pointed upward at the bottom of a 2.00 m deep pool filled with fresh water. Estimate the area of the light spot at the surface of the pool. List any assumptions you make for your estimate.

38. ★ (a) Find the index of refraction of a substance in which light travels 0.620 m in 3.00 ns.
  (b) How far does light travel in air in 3.00 ns?
  (c) How long does it take light to travel 1.00 m in oil?

39. ★★ A coin is placed at the bottom of an 80.0 cm deep aquarium filled with water. You are looking at the coin from above the aquarium. How deep will the coin appear to you? Explain using ray diagrams and the necessary calculations.

40. ★ While scuba diving, you look up and notice a beautiful sunset. To you, the Sun appears to be 50.° above the horizon, yet it was a sunset. What is the actual angle of the Sun above the horizon? Explain.

41. ★ A ray of light enters the long side of a 45°-90°-45° prism and undergoes two total internal reflections (Figure 28-54). As a result, the ray reverses its direction.
  (a) What is the minimum value of the prism's index of refraction for these internal reflections to be total?
  (b) How will your answer to part (a) change if instead of air the prism is submerged in oil?

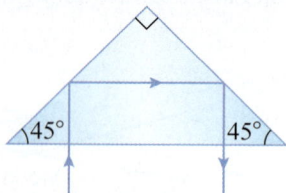

Figure 28-54 Problems 41 and 59.

42. ★★ A ray of He-Ne laser light is incident on a 45°-90°-45° glass prism at an angle of 60°, 5.0 cm above its base, as shown in Figure 28-55. Assume that the dimensions of the lens are given to two significant figures.
  (a) What is the angle of deflection (the angle between the incident and refracted beams) for this ray as it exits the prism?
  (b) At what point will the ray exit the prism?
  (c) How will your answers to parts (a) and (b) change if instead of an He-Ne laser (which produces a monochromatic light), white light were used for this question?

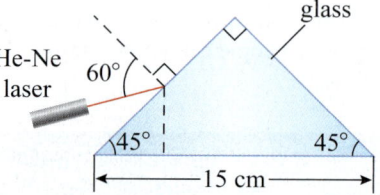

Figure 28-55 Problem 42.

Section 28-5 Images Formed by Thin Lenses

43. ★ A lens for an old-fashioned 35 mm camera has a focal length of 45 mm.
  (a) How close to the film should the lens be placed to form a sharp image of an object that is 6.0 m away?
  (b) What is the magnification of the image on the film?

44. ★★ A plastic magnifying lens ($n = 1.45$) has a magnification of 2.5 when the object–lens distance is 4.0 cm. The lens is placed in a large container filled with water, and its magnification is measured when the object–lens distance is once again 4.0 cm.
  (a) What is the magnification when the ambient medium is water?
  (b) Justify your answer to part (a) using the law of refraction.

45. ★ A thin lens has a focal length of −4.00 cm. Justify your answers to the following questions using both equations and ray diagrams.
  (a) Find the image distance and magnification that result when an object is placed 25.0 cm in front of the lens.
  (b) What type of lens is it?
  (c) What type of image is it?

46. ★★ Two converging thin lenses (f_1 = 15 cm and f_2 = 20. cm) are separated by 25 cm. A candle is placed 40. cm in front of the first lens.
 (a) Where will the image of the candle be created by the two-lens system?
 (b) Describe the properties of the image.
 (c) Describe how the image changes as the lenses are moved closer together.
 (d) The lenses are now moved right next to each other. What is the equivalent optical power of the system? What conclusion can you draw?

47. ★★ A glass concave–convex lens has radii of 12.0 cm and 14.0 cm, respectively. A 5.00 cm tall object is placed 12.0 cm in front of the lens. Justify your answers to the following questions using equations and ray diagrams.
 (a) The lens and the object are submerged in water. What is the magnification of the lens?
 (b) The lens is now surrounded by air. What is the magnification?

Section 28-6 The Human Eye and Vision Correction

48. ★ A classmate of yours who is not taking any physics courses asks you to explain what it means to be nearsighted or farsighted. Explain to her what each one of the conditions means and how each one of these vision problems can be corrected. Draw relevant diagrams to explain your reasoning. Feel free to use relevant computer simulations.

Section 28-7 Brewster's Angle

49. ★ Find Brewster's angle for the water–air interface.

COMPREHENSIVE PROBLEMS

50. ★ An observer is positioned so that the far edge of the bottom of an empty glass is just visible (Figure 28-56). When the glass is filled to the top with water, the centre of the bottom of the glass is just visible to the observer. The width of the glass is 6.8 cm.
 (a) What is the height, h, of the glass?
 (b) What does the observer see when the glass is filled with oil instead of water?

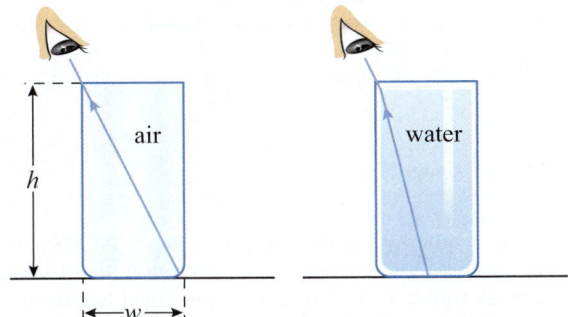

Figure 28-56 Problem 50.

51. ★ A small light source is located at the bottom of a large aquarium filled with water. A circular disk is floating on the surface of the water right above the light source. The height of the water in the aquarium is H. What should the smallest radius of the disk be so no light exits the surface of the water?

52. ★ A light ray is incident on a horizontal glass slab like the one in Figure 28-25(a), so the angle between the reflected and refracted rays is 90°. What is the angle of incidence of this ray?

53. ★ A narrow beam of light is incident on an air–glass boundary at an angle of 30°. Calculate the angle of deflection (the angle between the incident and refracted beams) of this beam. Will the angle of deflection increase or decrease when the angle of incidence increases to 45°? Justify your answer with calculations and a diagram.

54. ★ While visiting the Vancouver Aquarium, you watch beluga whales through the glass wall of their water tank. When a beluga is swimming toward you at a speed of 1.0 m/s, what is the apparent speed of the whale as observed by you? Should these two speeds be different? Explain.

55. ★ A small object is 0.15 m from a diverging lens, which has a focal length of −30. cm. At what distance from the lens is the image of this object located? Will you be able to see this image on a screen? Justify your answer using calculations and a diagram.

56. ★ An object is 4.2 m from a screen.
 (a) Where should you place a converging lens to see a focused image of the object on the screen with 20.× magnification?
 (b) Determine the focal length and the optical power of the lens.

57. ★ A camera lens has a focal length of 5.0 cm and a charge coupled device (CCD) sensor that measures 6.0 cm × 8.0 cm. You want to take an image of an object that is 60. cm × 60. cm.
 (a) How far from the object should the lens of the camera be to make the image as big as possible?
 (b) How far from the lens should the CCD sensor be? The sensor, which is where the image is formed, is hidden inside the camera.

58. ★ A monochromatic parallel beam of light emitted by an He-Ne laser strikes a triangular 30°-60°-90° prism surrounded by air 5 cm above the base of the prism. The incident beam is perpendicular to one of the prism's faces, as shown in Figure 28-57. After undergoing refraction in the prism, the light comes out of the bottom right vertex of the prism. Calculate the index of refraction of the prism for this beam of light. Draw a ray diagram of this scenario.

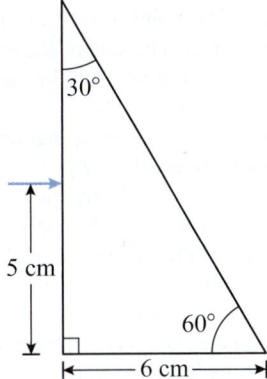

Figure 28-57 Problem 58.

59. ★★ The 45°-90°-45° triangular glass prism in Figure 28-54 (see problem 41) is surrounded by air. A ray of light is incident on the left face at an angle of 60.0°. The point of incidence is high enough that the refracted ray hits the opposite sloping side.
    (a) Through which face of the prism does the light exit?
    (b) What is the angle that the exiting beam of light makes with the horizontal?
    (c) The length of the hypotenuse of the cross-sectional area of the prism is 10. cm. How high from the bottom of the prism will the light ray exit it?

60. ★ A 5.0 cm tall vertical rod is 60. cm from a convex mirror, which has a radius of curvature of 40. cm.
    (a) Where is the image of the rod located?
    (b) What are the properties (i.e., orientation, magnification, real or virtual) of the image?

61. ★★ Two thin lenses have optical powers of 4.0 D and 5.0 D, respectively. They are 0.90 m from each other.
    (a) Find the focal length of the lenses.
    (b) Find the location of the image of an object that is 0.50 m in front of the first lens.
    (c) Can you use the equation for the combined power of the lenses to solve this problem? Justify your answer.

62. ★ During an eye exam, you notice that to be able to read a book, you have to hold it 16 cm in front of your eyes. For a healthy eye, this distance should be approximately 25 cm.
    (a) What vision problem does this indicate?
    (b) What further information do you need to determine the optical power of eyeglasses to correct your vision?

63. ★★ Derive an equation for the angle of deflection of a light beam propagating through an isosceles triangular prism as a function of the angle of the apex of the prism, ϕ; its index of refraction, n; and the angle of incidence, θ_i. Assume that the base of the prism is its largest face and that the light beam encounters the prism high enough that the refracted ray meets the opposite sloping side.

64. ★★★ A horizontal monochromatic circular beam of light (3.00 mm in diameter) is incident on an equilateral triangular prism made of glass ($n = 1.52$), which sits on a wooden table. The point of incidence is 8.00 cm above the base of the prism. The side of the equilateral triangle is 15.0 cm. What is the area of the spot formed by the emergent beam on the table?

65. ★★ A parallel beam of light is incident on a diverging lens that has a focal length of F_1.
    (a) How far from the diverging lens should a converging lens ($F_2 = -3F_1$) be located so that the beam exiting the diverging lens is also a parallel beam of light?
    (b) How does your answer to part (a) change if the light beam enters the converging lens first and then the diverging lens? Explain.

66. ★★ A light source is located at the bottom of a 20.0 cm deep aquarium filled with glycerine ($n = 1.47$), as shown in Figure 28-58. A screen is placed 10.0 cm above the surface of the glycerine. A converging lens with a focal length of 1.60 cm is placed between the surface of the glycerine and the screen. The lens is used to create a focused image of the source on the screen. What is the size of the image of the light source on the screen as compared to the size of the light source?

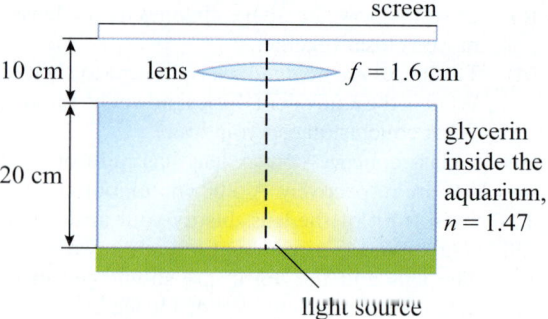

Figure 28-58 Problem 66.

67. ★★ Water is poured inside a concave mirror that has a radius of curvature of 50 cm.
    (a) Draw a diagram of this optical system.
    (b) What are the focal lengths and optical powers of the mirror and the water lens?
    (c) What is the optical power of the system?

68. ★★★ The inside surface of the convex part of a plano–convex plastic lens with radius of curvature $R = 40.0$ cm and index of refraction $n = 1.52$ is coated with silver, so the inside of the convex surface acts as a mirror (Figure 28-59). A collimated beam of light parallel to the principal optical axis of the lens is incident on the mirror surface.
    (a) Describe what happens to the beam of light in the figure.
    (b) Where is the focal point of the system?
    (c) An object is placed 15.0 cm in front of the plane side of the lens. Where is its image located? Describe the properties of the image.

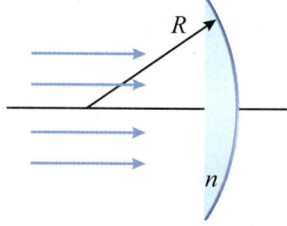

Figure 28-59 Problem 68.

69. ★★★ A small 50. g marble is moving at 6.0 m/s along the principal optical axis of a converging plano–convex lens, as shown in Figure 28-60. The focal length of the lens is 10. cm. The lens is securely mounted on a stand, and the friction between the stand and the table is negligible. The combined mass of the lens and the stand is 0.40 kg. After an elastic collision between the marble and the lens, the marble bounces back. Find the time interval during which the lens produces a virtual image of the marble. Neglect the gravitational force.

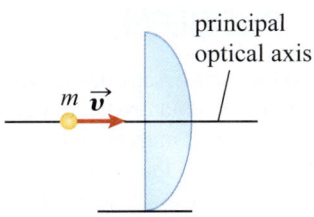

principal
optical axis

$m \vec{v}$

Figure 28-60 Problem 69.

70. ★★ A converging beam of light is incident on a screen, as shown in Figure 28-61. The screen has an opening 10. cm in diameter exactly at the location of the beam. The angle between the envelope of the beam and the axis of symmetry of the opening is 30°. Support your answers to the following questions with ray diagrams.
   (a) How far from the screen does the beam converge after going through the opening?
   (b) A converging lens of $P_1 = 10$. D is placed in the opening. How far from the screen does the beam converge?
   (c) If instead of a converging lens a diverging lens of $P_2 = -10$. D is placed in the opening, how far from the screen will the beam converge?

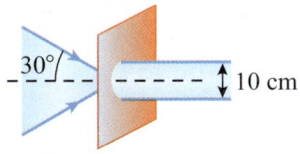

30°

10 cm

Figure 28-61 Problem 70.

71. ★★★ Prove that a pair of thin lenses located very close to each other (essentially at the same location) act as a single lens with a power P_{net} equal to the sum of the powers of the two lenses making up the combination, as shown in Equation 28-15.
72. ★★★ Prove that the combination of a thin lens and a spherical mirror placed very close to each other (essentially at the same location) acts as a single mirror with a power, P_{net}, equal to the sum of powers of the lens and mirror making up the combination, as shown in Equation 28-15.
73. ★★★ A thin converging lens ($f = 0.15$ m) is connected with a string to the ceiling, as shown in Figure 28-62. The lens is moving along a circle such that its optical axis always remains perpendicular to the floor. The lens produces an image of the hook on the ceiling that the string is attached to. At what angular velocity of the lens is the image always located on the floor? The height of the ceiling is 3.00 m.

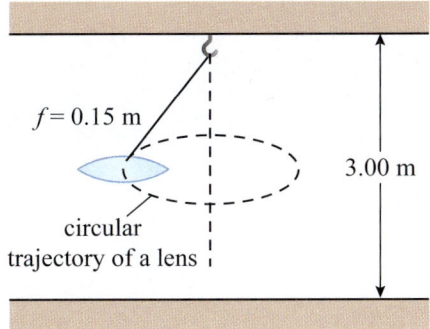

$f = 0.15$ m

3.00 m

circular
trajectory of a lens

Figure 28-62 Problem 73.

74. ★★★ A magnifying glass (either a biconvex lens or a combination of two lenses) is a converging lens that has a short focal length, approximately 1 cm to 10 cm, depending on the required magnification. When an object is located between the lens and its focal point, its virtual magnified image should be located at a distance of approximately 25 cm, which is the optimal distance for a healthy eye.
   (a) Prove that the magnification of a magnifying glass is given by

$$M = \frac{f}{f - d_o}$$

   (b) Prove that the magnification of a magnifying glass with the above focal length, which varies from 1.0 cm to 10 cm, varies between 3.5 and 26.

75. ★★★ A compound microscope is an optical instrument that uses at least two converging lenses (an eyepiece and an objective) separated by a distance L. (Figure 28-63). Two converging lenses for a compound microscope have focal lengths of 5.00 cm and 30.0 cm and are positioned so they are 40.0 cm apart.
   (a) How far from the left lens should an object be located so the final image is located exactly at the midpoint between the two lenses?
   (b) What is the overall linear magnification of the compound microscope?
   (c) The final image in this case is inverted. Is it possible to produce an upright final image with this microscope?
   (d) Use graphing software, the equations you derived, and/or a virtual microscope simulation (such as www1.udel.edu/biology/ketcham/microscope/scope.html) to explore how changing each of the parameters of the compound microscope (distance between the lenses, power of each one of the lenses, and the distance between the object and the objective lens) will affect the magnification.

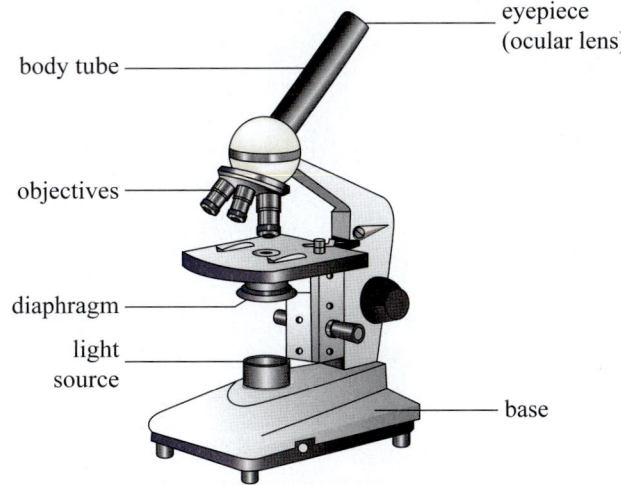

eyepiece
(ocular lens)

body tube

objectives

diaphragm

light
source

base

Figure 28-63 Problem 75.

76. ★★★ Galileo's telescope used two lenses: a diverging lens as an eyepiece and a large converging lens as an objective. Kepler's telescope used two converging lenses. Kepler's telescope produced inverted images, and Galileo's telescope produced upright images (he discovered three of the moons of Jupiter the first time he used it to look at the planet!). However, Galileo's design has a serious disadvantage, even though it is still used today in opera glasses. Draw a ray diagram for each telescope, and explain the problem with Galileo's telescope. (You might find it interesting to explore the Newtonian reflecting telescope as well).

77. ★★★ A friend of yours claims that he can design an optical system that can project all the light from a large spherical light source into a significantly reduced image of the light source.

   (a) Is this optical system possible? Will it violate any law of physics? (Hint: Think thermodynamics.)

   (b) Would a system that focused all the light from a small light source into a magnified image be possible? Explain why or why not.

DATA-RICH PROBLEM

78. ★★★ Use the Interactive Activities described in this chapter to collect quantitative data on the relationship between the parameters of spherical mirrors and lenses. Use the simulations to illustrate how the images are formed and how their parameters change as a function of the object–optical mirror/lens distance. Prove that the same equation can be used to describe spherical mirrors and thin lenses.

OPEN PROBLEM

79. ★★★ In his paper titled "Optics Demonstration with Student Eyeglasses Using the Inquiry Method,"[1] Dr. Mark C. James explains how he can "magically" guess his students' eyeglass prescriptions by looking through their eyeglasses. Explore the paper, and develop a short presentation to share this method with other students. Will this technique work for both positive and negative prescription eyeglasses? Why or why not?

1. Mark C. James, "Optics Demonstration with Student Eyeglasses Using the Inquiry Method," *The Physics Teacher*, 49 (2011), pp. 357–359.

Physical Optics

Figure 29-1 shows a radio telescope near the town of Penticton in southern British Columbia. Why do radio telescopes need such large diameters to have even modest resolution? How do you think the resolution of this telescope compares to the resolution of a small optical telescope or a pair of binoculars? The surface of this radio telescope is not even solid, whereas the surface of a mirror in an optical telescope needs to be very precise. Why the difference?

Figure 29-1   The radio telescope operated by the National Research Council near Penticton, British Columbia.

29-1 Physical and Geometric Optics

In Chapter 28, we saw that light rays can be applied to solve problems in reflection and refraction, including how images are formed. That branch of optics is **called geometric (ray) optics** because we apply the geometry of light rays to solve the associated problems. The geometric optics approach is accurate as long as the wavelength of the light is considerably smaller than the dimensions of the objects being studied. When that condition is not true, and the wavelengths are comparable to the object dimensions, we can no longer accurately use ray optics, and we must consider interference between waves travelling different paths. This branch of optics, which forms the content of this chapter, is called **wave** or **physical optics**.

While we will primarily discuss phenomena involving visible light, the concepts can be applied to any type of electromagnetic radiation.

Recall that the speed of light, c; the wave frequency, f; and the wavelength, λ; for all types of electromagnetic waves are related by the following equation:

$$c = f\lambda \tag{29-1}$$

Physical optics effects are more pronounced for electromagnetic radiation with longer wavelengths because more objects have dimensions similar to the waves. Thus, the wave behaviour of radio waves is more readily evident than the wave behaviour of visible light and much more apparent than the wave behaviour of gamma rays.

Recall from Section 27-4 that radio waves, microwaves, infrared radiation, visible light, ultraviolet light, X-rays, and gamma rays are all electromagnetic waves. The difference is the wavelength, with radio waves having the longest wavelength and gamma rays the shortest. Visible light has wavelengths varying from 400 nm to 700 nm, with violet at the shorter wavelength and red at the longer. See Section 27-4 for more details on the electromagnetic spectrum.

29-2 Interference

At the core of physical optics is **interference** between waves. Recall from Chapters 14 and 15 that two mechanical waves with the same frequency will interfere constructively when they are in phase and destructively when they are out of phase. Electromagnetic waves (see Chapter 27) interfere in exactly the same way. Of course, waves usually do not have precisely the same amplitude and phase difference to completely reinforce or completely cancel each other.

In interference experiments and calculations, we need to consider both the wavelength and the relative phase of the waves. Ordinary light bulbs produce light with mixed wavelengths and random phases. **Lasers**, however, produce light that is both **monochromatic** (with exactly a single wavelength and, hence, colour) and **coherent** (with the same phase). (Gas discharge tubes with appropriate filters can produce monochromatic light, but the light lacks coherence over significant time intervals.)

Complications also arise from the fact that the planes of the electric and magnetic fields are important when we add two electromagnetic waves. However, the addition of two waves with the same frequency and polarization is fairly straightforward. The intensity of the resulting combination is given by

$$I = I_1 + I_2 + 2\sqrt{I_1 I_2} \cos \Delta\phi \tag{29-2}$$

where the intensities of the waves are I_1 and I_2, and the phase difference between them is $\Delta\phi$.

You may have expected that when we combined waves of identical frequency and intensity in phase, we would get two times the intensity of one of the individual waves, but the correct result is four times. The explanation is that wave intensities are proportional to the square of the wave amplitude. When we combine two identical waves in phase, we double the amplitude but have an intensity four times the individual intensities. Also remember that this quadrupling only applies for combinations of individual waves of identical frequency and equal strength. We demonstrate in

---

INTERACTIVE ACTIVITY 29-1

Interference of Light Waves

In this interactive activity, you will use the PhET simulation "Wave Interference" to simulate the interference between two waves. In particular, you will see how the result depends on the colour (wavelength) of the light used, and on the physical spacing of the two sources.

Example 29-2 a case where the intensities of the two waves are not equal.

When the intensities of the two initial waves are equal ($I_1 = I_2$) and the phase difference, $\Delta\phi$, is 0 (or any integer multiple of 2π), the total intensity is four times the intensity of either wave alone. When $\Delta\phi$ is π rad (or 3π, 5π, and so on), the resultant intensity, I, is zero.

We normally express phase differences in radians, although degrees are sometimes used. A phase difference of π radians, corresponding to 180°, results in destructive interference. A phase difference of 2π radians, corresponding to 360°, results in constructive interference. Note that 0 radian and 2π radian phase shifts produce the same result.

Sometimes we express phase differences in terms of wavelengths, with a path difference of $\lambda/2$ corresponding to destructive interference, while a 0 or λ path difference results in constructive interference. At first it may seem strange to express phase differences

in terms of wavelengths, since wavelengths are measured in units of length. It is important to be clear on what we really mean when we say a "phase difference of one-half wavelength." It means a phase difference of $2\pi\dfrac{0.5\lambda}{\lambda} = \pi$ rad. Expressing phase differences in terms of wavelengths is particularly helpful when the phase difference arises from combining two signals that have travelled different distances in some medium.

Thinking about phase differences in terms of distances expressed in wavelengths can be particularly helpful when analyzing situations involving interference. If the wave is travelling in a medium, we must use the wavelength of the electromagnetic wave in that medium.

EXAMPLE 29-1

Combining Waves with Different Phases

Two electromagnetic waves with identical frequencies have individual intensities $I_1 = 4.00$ W/m² and $I_2 = 2.00$ W/m².

(a) What phase difference results in a maximum combined intensity, and what is that intensity?
(b) What phase difference results in a minimum combined intensity, and what is that intensity?
(c) If we want a combined wave with an intensity of $I_T = 7.00$ W/m², what must the phase difference between I_1 and I_2 be?

Solution

(a) We get a maximum combined intensity when the phase difference is 0 (or 2π, 4π, etc.) radians. We use the expression for combining wave intensities to find the result:

$$I = I_1 + I_2 + 2\sqrt{I_1 I_2}\cos\Delta\phi$$
$$I = 4.00\ \text{W/m}^2 + 2.00\ \text{W/m}^2$$
$$+ 2\sqrt{(4.00\ \text{W/m}^2)(2.00\ \text{W/m}^2)}\cos 0$$
$$I = 4.00\ \text{W/m}^2 + 2.00\ \text{W/m}^2 + 2\sqrt{(4.00\ \text{W/m}^2)(2.00\ \text{W/m}^2)}$$
$$I = 11.7\ \text{W/m}^2$$

(b) For a minimum result, we require a phase difference of π (or 3π, 5π, etc.) radians. This gives the following combined intensity:

$$I = I_1 + I_2 + 2\sqrt{I_1 I_2}\cos\Delta\phi$$
$$I = 4.00\ \text{W/m}^2 + 2.00\ \text{W/m}^2$$
$$+ 2\sqrt{(4.00\ \text{W/m}^2)(2.00\ \text{W/m}^2)}\cos\pi$$

$$I = 4.00\ \text{W/m}^2 + 2.00\ \text{W/m}^2 - 2\sqrt{(4.00\ \text{W/m}^2)(2.00\ \text{W/m}^2)}$$
$$I = 0.343\ \text{W/m}^2$$

(c) This time we substitute the desired resultant intensity and solve for the phase difference:

$$I = I_1 + I_2 + 2\sqrt{I_1 I_2}\cos\Delta\phi$$
$$7.00\ \text{W/m}^2 = 4.00\ \text{W/m}^2 + 2.00\ \text{W/m}^2$$
$$+ 2\sqrt{(4.00\ \text{W/m}^2)(2.00\ \text{W/m}^2)}\cos\Delta\phi$$
$$1.00\ \text{W/m}^2 = 5.66\ \text{W/m}^2\cos\Delta\phi$$
$$\cos\Delta\phi = \frac{1.00\ \text{W/m}^2}{5.66\ \text{W/m}^2} = 0.177$$
$$\Delta\phi = 1.39\ \text{rad}$$
$$\Delta\phi = 79.8°$$

Since we used an inverse cosine function, the negative of this angle is also consistent with the problem. We could add any multiple of 2π radians, but it is normal to express the smallest value as the solution.

Making sense of the result

When we have no phase difference and equal intensities, we obtain a combined intensity of four times the individual intensities. We expect a bit less when the intensities are not equal, but we see that the value is roughly four times the average intensity and therefore it seems reasonable. For identical individual intensities with a π phase difference, the result is zero, so here we expect a small but not zero resultant intensity. If the question does not ask you for a specific format, it is acceptable to express the phase difference in either radians or degrees.

ELECTRICITY, MAGNETISM, AND OPTICS

One way to observe interference is to use a **Michelson interferometer**: split a beam of light, send the two parts along different paths of different lengths (causing a phase difference), and then recombine them. The key elements of a Michelson interferometer are shown in Figure 29-2. This instrument can measure tiny shifts in distance or wave speed. The light beam from the source (usually a laser) is split into two beams by a partially silvered mirror. One beam reflects off the silvered edge of the beam splitter (which is assumed to be farther from the source) and bounces off the movable mirror before passing through the splitter again to the detector. The other beam passes through the splitter, bounces off a fixed mirror back to the splitter, and then reflects from the splitter to the detector. The glass compensator plate makes the total distance that each beam travels in glass equal. If the beam splitter is silvered on the side away from the source, the compensator is placed as shown in Figure 29-2.

When a Michelson interferometer is placed in a vacuum, the speed of light in both arms is c. If we start with the two paths equal in length and gradually adjust the position of the movable mirror, the image at the detector goes from bright (zero path difference) to dim (a path difference of one-half wavelength) to bright again (one wavelength difference), to dim again (one and one-half wavelength difference), and so on. Note that the change in the path difference distance is twice the distance, d, that the mirror moves because the light travels to the movable mirror and then back to the splitter. For example, moving the mirror a quarter wavelength changes the path length by a half wavelength. The condition for an intensity maximum at the detector (constructive interference) is

$$m\lambda = 2d \qquad (29\text{-}3)$$

where m is any integer and λ is the wavelength of the light.

When the path difference is 0, constructive interference produces a bright disk of light at the detector; a total path difference of 0.5λ produces complete destructive interference. As we continue to increase the path difference, the centre of the output is bright for integer multiples of the wavelength and dark for $m + \frac{1}{2}$ multiples, but there are also alternating light and dark rings around the centre, as shown in Figure 29-3. These interference patterns result

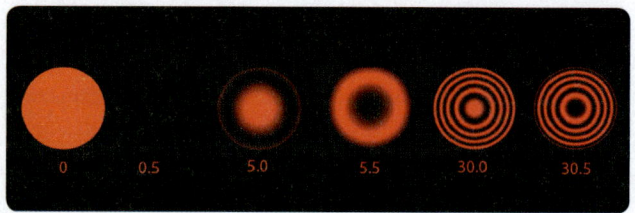

Figure 29-3 The output of a Michelson interferometer for various path length differences (labelled as multiples of the wavelength).

from the fact that the path lengths increase slightly with distance from the centre of the detector.

The Michelson interferometer is named in honour of its inventor, physicist Albert Abraham Michelson (1852–1931). He worked with chemist Edward Williams Morley (1838–1923) to improve the precision of his original model. In 1887, Michelson and Morley used the improved version shown in Figure 29-4 to carry out a key experiment in the development of modern physics. Their results revealed that the speed of light in a vacuum is the same for all observers, regardless of their relative motion. As you will see in Chapter 30, this puzzling observation became one of the postulates of special relativity.

Interferometers continue to play central roles in modern physics. They provide ways to detect tiny changes in distance and to measure indexes of refraction with outstanding precision. An interferometer is a key instrument in the search for evidence of gravitational waves.

Figure 29-4 The original Michelson–Morley interferometer.

© Huntington Library/SuperStock

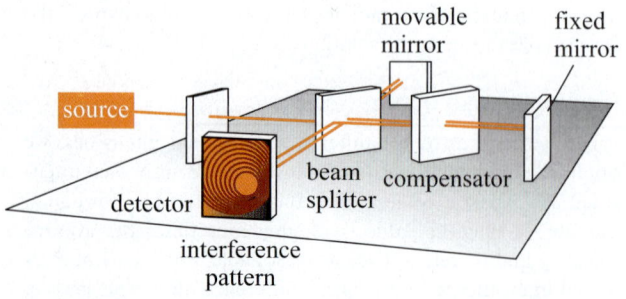

Figure 29-2 In a Michelson interferometer, a light beam is split by a half-silvered mirror, the two parts travel different paths, and then the parts combine to produce an interference pattern.

CHECKPOINT 29-2

Interferometer

A Michelson interferometer is operating in a vacuum chamber using light from a laser with a wavelength of 632 nm. The movable mirror is adjusted to give a maximum bright output at the detector. How far must the movable mirror be moved to produce the next output maximum?

(a) 1264 nm   (c) 632 nm

(b) 948 nm   (d) 316 nm

ANSWER: (d) The path difference must be increased by one wavelength for the next maximum. Since the change in path length is double the distance that the mirror moves, the mirror needs to move by only 316 nm.

Laser Interferometer Gravitational Wave Observatory (LIGO)

LIGO was designed by the California Institute of Technology and the Massachusetts Institute of Technology with the aim of detecting gravitational waves. Early in 2016 the LIGO scientific collaboration announced the first detection (a signal from coalescing black holes)—we will provide more detail on that detection in Chapter 30. Gravitational waves have been referred to as "ripples in spacetime." When a gravitational wave passes, distances between points in one direction are very slightly expanded, while they are similarly contracted in a perpendicular direction. While a mechanical ruler would be similarly stretched and would not show the effect, we can measure with light to show the expansion. Essentially LIGO is a perpendicular-arm Michelson interferometer of incredible precision.

As shown in Figure 29-5, a splitter divides the laser beam, with part going down (and back) each perpendicular arm. The perpendicular interferometer arms are about 4 km in length each and need to be evacuated to a high degree ($\approx 10^{-12}$ atm). The beams that come from the two arms are then recombined, and if a gravitational wave temporarily stretched one arm and contracted the other, then an interference pattern would be observed in the detector. To make it more precise, each arm uses something called a Fabry Perot cavity that enhances the interference through multiple passages of the light beam in the cavity. The laser used in the LIGO detectors operates in the near-infrared region, with a wavelength of 1064 nm.

Advanced LIGO can detect a strain of less than one part in 10^{21}. The signal produced by the coalescing black holes (see Chapter 30) had a peak displacement of ± 0.002 fm (or 2×10^{-18} m) in the interferometer arms. By comparison, the radius of a proton is about 0.9 fm, so LIGO can measure a tiny fraction the size of a single proton over a distance of 4 km. As the LIGO executive director pointed out, this would be equivalent to measuring the distance to the nearest star (other than our Sun), Proxima Centauri, with a precision of about the width of a human hair!

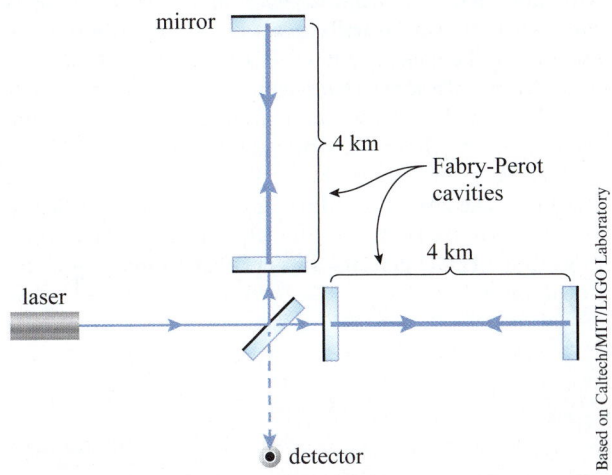

Figure 29-5 In LIGO, a laser beam is split, with part going through each of two parallel arms. They are then combined and interference patterns used to detect slight differences in the lengths of the two arms.

Because of the incredibly small signals it measures, LIGO must be as immune as possible to noise. One of the techniques it employs is coincidence detection. There are two similar LIGO detectors, one at Livingston, Louisiana, and the other at Hanford, Washington (Figure 29-6). True gravitational waves from the distant universe will arrive at both detectors (although with a tiny time separation due to the approximately 3000 km distance between the detectors), while most noise sources will be local and affect only one of the detectors. Coincidence detection is widely used in physics to distinguish true signals.

LIGO has been in design and planning for a very long time, and construction began in 1994. The first observations were in 2001, the first science-grade precision achieved in 2005, and full routine operation began in 2014. The LIGO scientific collaboration currently involves more than one thousand scientists from many different institutions.

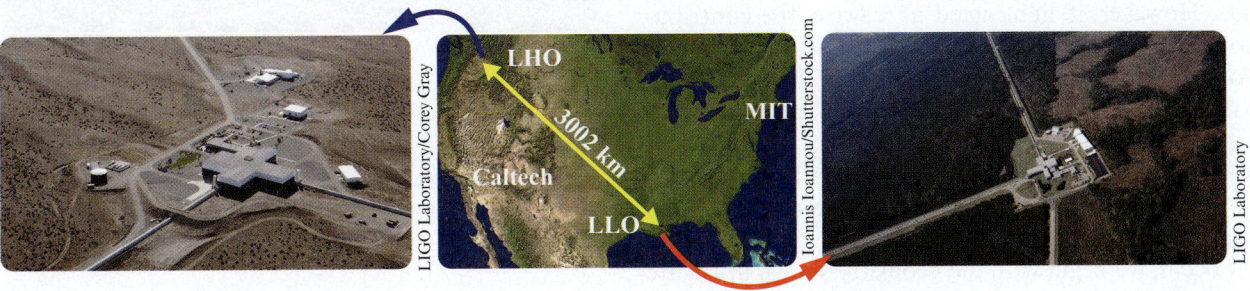

Figure 29-6 The two LIGO detectors (Hanford, Washington, left, and Livingston, Louisiana, right).

(continued)

While others currently direct the mission, LIGO was co-founded in 1984 by American physicist Kip Stephen Thorne (b1940). Kip Thorne (see Figure 29-7) has contributed extensively to the fields of gravitation and cosmology, with more than 150 scientific papers and 50 Ph.D. students supervised. He is also well known for his contributions to science for the general public. His book *Black Holes and Time Warps: Einstein's Outrageous Legacy* was very widely read. He served as the science consultant for the 2014 film *Interstellar*. His character appears in the 2014 film *A Theory of Everything*, an account of the life and work of Stephen Hawking. The 2017 Nobel Prize in Physics was awarded to Rainer Weiss, Barry Barish, and Kip Thorne for "contributions to the LIGO detector and the observation of gravitational waves."

Figure 29-7   Dr. Kip Thorne.

29-3 Double-Slit Interference

At the beginning of the nineteenth century, Thomas Young (1773–1829) performed a crucial experiment that showed that light has wave properties. Young had worked previously with sound waves, and in 1799 he proposed that light is also a wave. This theory contradicted the view expressed by many important scientists, who shared Isaac Newton's firm belief that light was composed only of tiny particles. General acceptance of Young's double-slit experiment as evidence for the wave nature of light did not come for some time.

Figure 29-8 shows a representation of Young's experiment. A source of light illuminates two narrow, closely spaced slits, projecting an image on a screen some distance away. If light were particles, one would expect two maxima on the screen corresponding to straight-line motion of the particles through the two narrow slits. However, if light acted as a wave, the contributions from the two slits would interfere with each other. Young observed an alternating series of light and dark bars on the screen, confirming the wave nature of light. Figure 29-9 shows results from a modern re-enactment of the double-slit experiment using laser light.

To derive a mathematical relationship for the angles at which maxima in light intensity from a double slit occur, we begin with the geometry shown in Figure 29-10. Light passing through slit S_1 travels a shorter distance to point P than does light passing through slit S_2. We draw a right-angled triangle to divide the distance from S_2 to P into two parts, as shown. The distance from S_1 to P is approximately equal to the distance from point Q to P as long as the distance D from the slits to the screen is much greater than the slit spacing d. This restriction is called the **Fraunhofer**

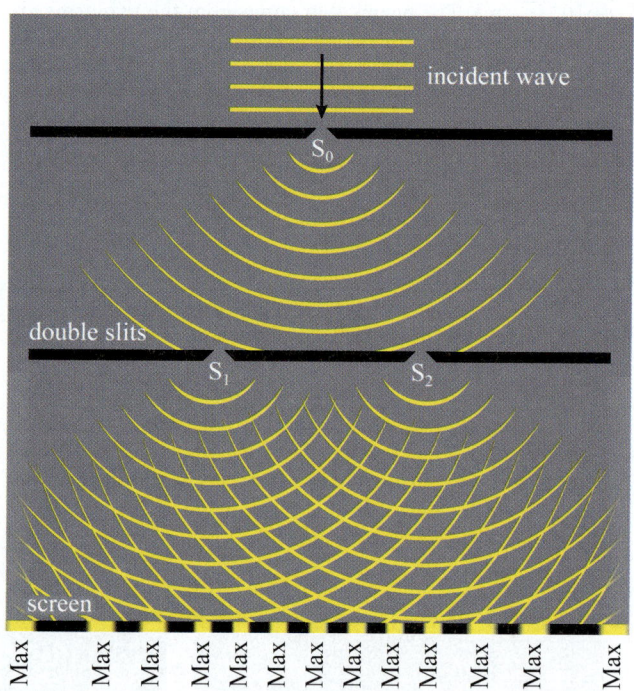

Figure 29-8   Young's double-slit experiment. Light illuminates two narrow slits, producing an interference pattern of alternating bright and dark patches on a distant screen. Note that it is bright when the light from the two slits arrives with the same number of wavelengths, and dark when they differ by one-half wavelength.

condition (named for the German optician and physicist Joseph von Fraunhofer (1787–1826)) and is true in most important cases. We can readily show that the upper angle in the right-angled triangle equals the deflection angle θ to point P (measured from a normal to the slit plate). Therefore, the extra distance that the light from S_2 travels to reach point P is $d \sin \theta$.

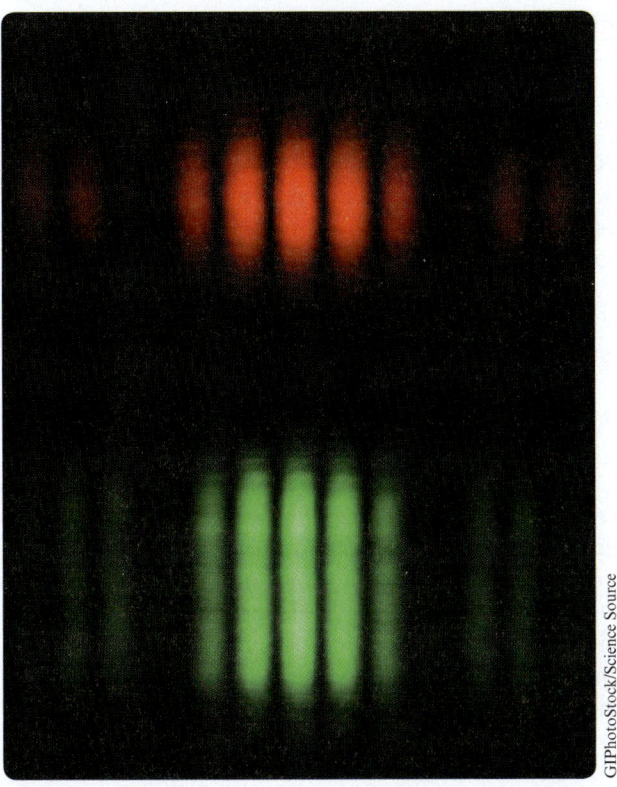

Figure 29-9 Output on a screen from an actual double-slit experiment performed using red (wavelength 632 nm) and green (543 nm) laser light with two slits 75 μm apart. The interference pattern is more spread out for the longer wavelength.

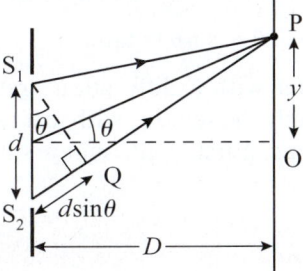

Figure 29-10 Geometry of the double-slit experiment. The distance between the two slits is d, and the distance to the screen is D.

When the extra distance corresponds to exactly a whole number of wavelengths, then constructive interference occurs, and a bright region appears on the screen at angle θ. Thus, constructive interference occurs when

KEY EQUATION $$d\sin\theta = m\lambda \qquad (29\text{-}4)$$

where m is an integer, often called the spectral order.

When $m = 0$, we get an angle of 0° for any wavelength. Therefore, when we shine white light at a pair of slits, the beam in the centre of the interference pattern is white. For all other values of m, the maxima for different wavelengths occur at different angles, with longer wavelengths having larger deflection angles. The direct central maximum is called the zero order, and the spectral orders on one side are normally labelled

$m = +1$, $+2$, and so on, and the spectral orders on the other side are labelled $m = -1$, -2, and so on.

In interferometry experiments, we often want to know the displacement distance, y, and not the angle

CHECKPOINT 29-3

Double-Slit Spacing

When analyzing the interference pattern from two slits, you see that the $m = 1$ and $m = 2$ fringes are too close together to measure easily. Which of the following statements correctly describes the ways that the interference pattern spacing could be increased?

(a) Move the slits farther from the screen, reduce the wavelength of the light, or increase the spacing of the slits.

(b) Move the slits farther from the screen, increase the wavelength of the light, or increase the spacing of the slits.

(c) Move the slits farther from the screen, reduce the wavelength of the light, or decrease the spacing of the slits.

(d) Move the slits farther from the screen, increase the wavelength of the light, or decrease the spacing of the slits.

ANSWER: (d) To increase the spacing between the fringes, we could simply keep the angle the same and move the screen farther away. Otherwise, we have to increase the angle. Rearranging Equation (29-4), we have $\sin\theta = \dfrac{m\lambda}{d}$, which shows that we get a larger angle when d is less or when λ is more.

(see Figure 29-10). The displacement is readily determined using trigonometry (see Example 29-2 below). Other times, we want to determine the maximum order (m-value) that can occur. In that case, we set $\sin\theta$ to 1 (because just under 90° is the maximum angle possible), solve for a value of m, and then truncate this value down to the nearest integer. For example, if setting $\sin\theta = 1$ gives $m = 3.8$, then 3 spectral orders are visible.

Although we are usually primarily concerned with the bright fringes in an interference pattern, we sometimes want to find the angles for the dark fringes. The

EXAMPLE 29-2

Double-Slit Interference Pattern

A laser beam with a wavelength of 532 nm illuminates a double slit and produces an interference pattern on a screen 9.00 cm away. The distance between the $m = 0$ fringe and one of the $m = 1$ fringes is 0.850 cm on the screen. How many bright fringes are on the screen? Assume that the screen extends to a large distance in each direction and that even the faint fringes are visible.

Solution

Here, we have the geometry shown in Figure 29-10 with values of $D = 9.00$ cm, $\lambda = 532$ nm, and $y = 0.850$ cm for $m = 1$. We do not have values for d and θ.

Our strategy will be to first use trigonometry to find the angle θ, and then solve for the slit spacing d. Then we can calculate the maximum order, m_{max}.

We need to use consistent units, so we will express all distances in metres:

$$\tan\theta = \frac{8.50 \times 10^{-3}\,\text{m}}{9.00 \times 10^{-2}\,\text{m}}$$

$$\theta = 5.39°$$

Now we can use Equation (29-4) to solve for the slit spacing d:

$$d = \frac{m\lambda}{\sin\theta} = \frac{1 \times 5.32 \times 10^{-7}\,\text{m}}{\sin 5.39°} = 5.66 \times 10^{-6}\,\text{m}$$

Now we find the maximum possible value for m by setting $\theta = 90°$:

$$m = \frac{d\sin\theta}{\lambda} = \frac{5.66 \times 10^{-6}\,\text{m} \times \sin 90°}{5.32 \times 10^{-7}\,\text{m}} = 10.6$$

Truncating down to the nearest integer gives $m = 10$. Therefore, there are 10 bright fringes on each side of the central zero-order maximum, making a total of 21 bright fringes.

Making sense of the result

The angle of approximately 5° that we found for $m = 1$ seems reasonable for the given dimensions. Since the bright fringes become farther apart with each order, it seems reasonable that there are 10 fringes in the 90° span on each side of the zero-order maximum.

condition for destructive interference is simply a path difference equal to an odd number of half wavelengths:

$$d\sin\theta = \left(m + \frac{1}{2}\right)\lambda \qquad (29\text{-}5)$$

The analysis above holds only if the slits themselves are narrow compared to the distance separating them. We will see later in the chapter that there are diffraction effects for light from a single slit that need to be taken into account for precision work. In addition, we have only considered the angles corresponding to the maxima (or minima) and the intensities of the various fringes. Generally, the higher the order, the dimmer each successive maximum. Finally, we mention again that our derivation depends on the Fraunhofer condition, $D \gg d$.

29-4 Diffraction Gratings

You saw in Chapter 28 that a glass prism can be used to split light into different colours in order by wavelength. Different wavelengths of light have slightly different speeds in glass (and, therefore, different indexes of refraction), resulting in varying angles of refraction at the prism's air–glass interfaces. However, most scientific instruments that divide light into constituent wavelengths use **diffraction gratings**, which are plates with typically a thousand or more evenly spaced slits close together.

The condition for constructive interference with many slits with constant spacing d between adjacent slits is exactly the same as for one pair of slits. This is because if one pair of slits meets the condition for constructive interference, then all pairs will. For example, if distance b,c for the diffraction grating in Figure 29-11 is one wavelength, then distance d,e is two wavelengths, and distance f,g is three wavelengths. Consequently, the angles for constructive interference for light of wavelength λ passing through the diffraction grating are given by

$$d\sin\theta = m\lambda \qquad (29\text{-}6)$$

where m is the spectral order.

For calculations involving first-order angles, it is often possible to use the small-angle approximations: when an angle is small, then the sine, the tangent, and the angle itself (expressed in radians) are all approximately equal:

$$\theta \approx \sin\theta \approx \tan\theta \qquad (29\text{-}7)$$

Why bother with a diffraction grating when the angles are exactly the same as with just two slits? There are two reasons. First, with many slits, more light

PEER TO PEER

It is important to remember that I can only replace $\sin\theta$ with θ when two conditions are met: the angle is very small and the angle is expressed in radians.

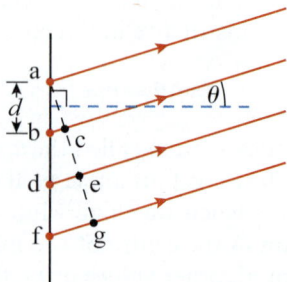

Figure 29-11 Geometry of the diffraction grating. The distance between each pair of slits is d, so the relative distance to the screen is one wavelength longer (for the correct angle) for light from each subsequent slit.

passes through the plate, and the interference pattern is much brighter. Second, with more slits the diffraction grating better resolves different wavelengths. Figure 29-12 illustrates how the constructive interference peaks for monochromatic light become markedly sharper as the number of slits increases. Thus, diffraction gratings can produce separate peaks for wavelengths of light that are only slightly different. Figure 29-13 shows a high-resolution spectrum of light from the Sun, made with a series of diffraction gratings with different slit spacings.

Usually, diffraction grating specifications are given in terms of the number of lines per millimetre, N, rather than the distance, d, between adjacent slits. Clearly,

$$d = \frac{1}{N} \tag{29-8}$$

For example, if there are 200 lines/mm, then $d = 1/200$ of a millimetre, or 5×10^{-6} m.

Throughout this section we have assumed a narrow, parallel beam of light. If the light source is not a laser, a slit should be placed between the source and the diffraction grating so that the incident light that reaches the grating is essentially a narrow, parallel beam.

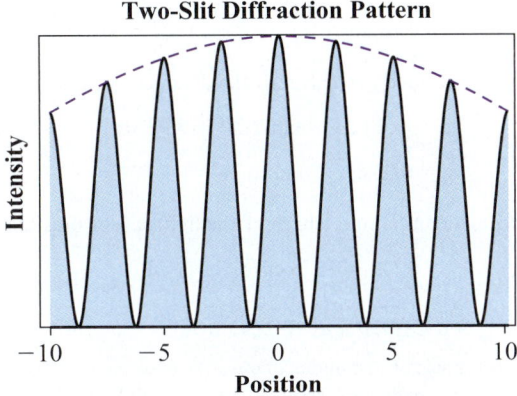

Two-Slit Diffraction Pattern

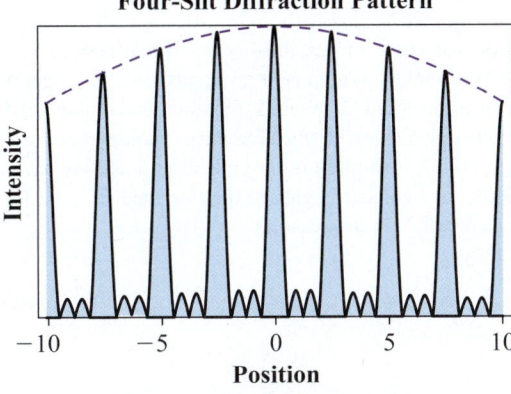

Four-Slit Diffraction Pattern

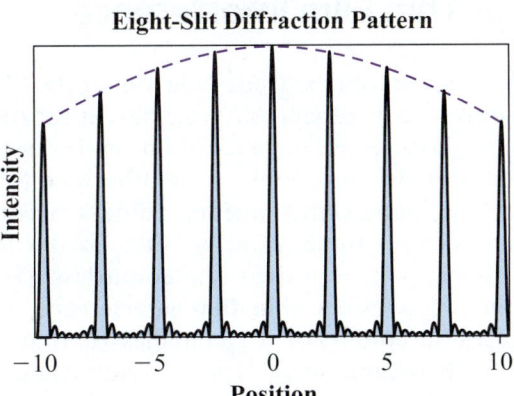

Eight-Slit Diffraction Pattern

Figure 29-12 Interference patterns for monochromatic light passing through gratings with 2, 4, and 8 slits.

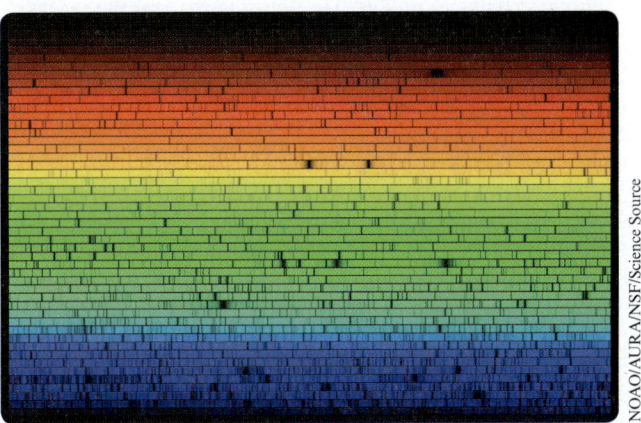

Figure 29-13 Spectrum of visible light from the Sun, compiled from a series of detailed spectra of regions of the visible spectrum. The dark lines represent absorption by different atoms and molecules in the outer regions of the Sun.

CHECKPOINT 29-4

Two Diffraction Gratings

You have two diffraction gratings. Grating A is known to have 100 lines/mm, but the specifications sheet has been lost for grating B. The first-order fringe produced by monochromatic light shone on the two gratings appears at an angle of 4° for grating A and an angle of 8° for grating B. What is the spacing of the slits in grating B?

(a)   25 lines/mm
(b)   50 lines/mm
(c)   200 lines/mm
(d)   400 lines/mm

ANSWER: (c) Rearranging the diffraction grating equation to $d = \frac{m\lambda}{\sin \theta}$, we see that a larger angle corresponds to a smaller slit spacing, d. Evaluating the sines (or using the small-angle approximation) shows that the spacing, d, for grating B is half that of grating A. Therefore, grating B has twice as many lines per millimetre, which is 200 lines/mm.

EXAMPLE 29-3

Design a Spectrometer

You want to design a spectrometer to cover the visible spectrum from violet (400. nm) to red (700. nm) using first-order images from a diffraction grating with 400. lines/mm. The spacing from the grating to the electronic image sensor used to record the spectrum is 6.00 cm. What is the minimum dimension for the image sensor?

Solution

The design of the spectrometer is shown in Figure 29-14. The sensor needs to cover the first-order image on only one side. From the relationship $d \sin\theta = m\lambda$, we know that the longer wavelength corresponds to a larger deflection angle. Therefore, B corresponds to the red light, and A corresponds to the violet light. We will find the angles for these limiting wavelengths and use the angles and the distance to the screen to find the two displacements, y_A and y_B. The sensor must cover the ranges of these two displacements.

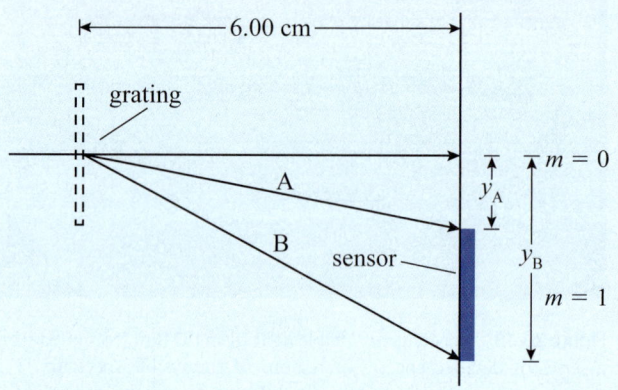

Figure 29-14   Example 29-3.

First, we use the number of lines per millimetre to find the slit spacing d: $d = \dfrac{1}{400}$ mm $= 2.50 \times 10^{-6}$ m.

Next, we solve for the angle to the A line in the diagram (400 nm wavelength):

$$\sin\theta_A = \frac{m\lambda}{d} = \frac{1 \times 4.00 \times 10^{-7} \text{ m}}{2.50 \times 10^{-6} \text{ m}} = 0.160$$
$$\theta_A = 9.21°$$

We use the same method to find the angle to the B line (700 nm wavelength):

$$\sin\theta_B = \frac{m\lambda}{d} = \frac{1 \times 7.00 \times 10^{-7} \text{ m}}{2.50 \times 10^{-6} \text{ m}} = 0.280$$
$$\theta_B = 16.26°$$

We use trigonometry to solve for the distances y_A and y_B:

$$y_A = 6.00 \text{ cm} \times \tan 9.21° = 0.973 \text{ cm}$$
$$y_B = 6.00 \text{ cm} \times \tan 16.26° = 1.750 \text{ cm}$$

Therefore, the minimum length of the light sensor must be

$$1.750 - 0.973 \text{ cm} = 0.777 \text{ cm}$$

Making sense of the result

The detector size of just under 1 cm seems reasonable. Most commercial spectrometers are designed not to include the zero-order maximum because it does not provide any spectral information. If we had wanted to detect the zero-order maximum as well as one set of first-order images, the detector would have to be twice as wide. We should also comment on significant digits here. When you are near the boundary above 9, the simplified significant digits rule breaks down. Clearly, when a number such as 9.98 increases slightly to 10.02, it should not lose precision. Here, we have kept one additional significant figure in the intermediate steps for the angle just over 16° for this reason.

INTERACTIVE ACTIVITY 29-3

Optical Pulse Shaping

In this interactive activity, you will use the PhET simulation "Optical Quantum Control." The simulation allows control of the frequencies present in an optical pulse in order to shape it. In particular, you will seek to create a shape that is well matched to provide just the right electric field to provide electrical forces to tear apart the molecule.

The diffraction gratings we have described so far are all **transmission gratings**, which have evenly spaced slits. There are also **reflection gratings**, which have evenly spaced lines that reflect light, producing interference patterns similar to those from transmission gratings.

29-5 Thin Film Interference

In his book on optics, Isaac Newton wrote, "It has been observed by others that transparent substances such as glass, water, air, etc. when made very thin by being blown into bubbles, or otherwise formed into plates, do exhibit various colours according to their various thinness, altho' at a greater thickness they appear very clear and colourless." In this section, we consider **thin film interference**, which produces the colours of soap bubbles, oil films, peacock feathers, and some types of butterflies. Thin films are also important in antireflection coatings in quality optics and in energy-saving technologies for windows.

Constructive or destructive interference can occur at the surface of a thin film, depending on the thickness of the film and any phase changes that occur during reflection at the surfaces of the film.

Let us review what happens when waves encounter a boundary between media with different optical properties. As we have seen with mechanical waves, some of the incident wave is reflected from the boundary surface, and some is transmitted through it. Light will reflect, at least partially, from the boundary surface when there is a difference in the index of refraction (which corresponds to the speed of light in the medium).

Figure 29-16 shows a thin film designed to reduce reflection from a lens. Here, we have air ($n = 1.00$) and glass (assume $n = 1.52$) separated by a thin coating of uniform thickness, t. For a variety of practical reasons, magnesium fluoride (MgF_2) is one of the most common materials for antireflection coatings on optical lenses and glasses. Magnesium fluoride has an index of refraction of 1.38.

When a light ray (1) strikes the air–coating boundary, it is partly reflected (2) and partly transmitted (3). The transmitted light is refracted because the index of refraction of the coating is greater than the index of refraction of air. When the refracted light (3) encounters the coating–glass boundary, the light is again partly reflected (4) and partly transmitted with further refraction (5). When (4) strikes the coating–air boundary, part will be transmitted back into the air (6). The net reflection depends on the interaction between (2) and (6). If they are in phase, constructive interference will make the net reflected ray relatively bright. If (2) and (6) are out of phase, destructive interference will result and there will be essentially no net reflection.

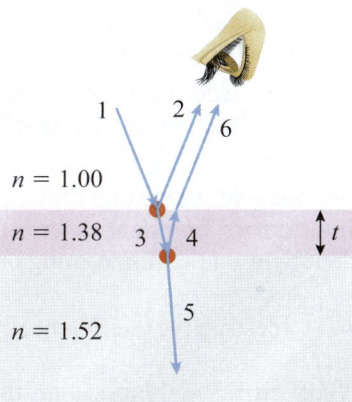

Figure 29-16 Reflection and refraction of light at the boundary surfaces of an antireflection coating. The small circles indicate a hard reflection (phase change).

The function of an antireflection coating is to use destructive interference to minimize the amount of light reflected from the lens.

The relative phase of (2) and (6) depends, in part, on the extra distance travelled by (6). If for now we assume that the initial ray (1) is near vertical, then the distance travelled by light after leaving the coating is essentially equal and constant, and the extra distance travelled by (6) is approximately $2t$, twice the thickness of the coating.

Note that the extra distance is travelled in the coating, not in the air. When an electromagnetic wave is in a medium, the wavelength is shorter than it would be in a vacuum. The frequency does not change when the wave enters a different medium, but the speed decreases compared to the speed in a vacuum; therefore, the wavelength decreases correspondingly. The wavelength, λ_n, of the wave in a medium with an index of refraction n is given by

KEY EQUATION
$$\lambda_n = \frac{\lambda}{n} \qquad (29\text{-}9)$$

Now we must take into account the phase changes that can occur when a wave reflects. Reflected light will experience a 180° phase change (a **hard reflection**) when it reflects from a medium with a greater index of refraction, and no phase change (a **soft reflection**) when the reflection is from a medium with a smaller index of refraction. For the coating we are considering, there is a phase change for incident light reflected from both the outer and inner surfaces.

These **phase changes** are analogous to the reflections of mechanical waves at boundaries considered in Section 14-11. We summarize the conditions for optical waves below:

- Transmitted waves do not undergo a phase change.

- When waves are reflected from a medium with a higher index of refraction, there is a half-wavelength (180° or π radians) phase change (hard reflection).

- When waves are reflected from a medium with a lower index of refraction, there is no phase change (soft reflection).

Combining these ideas, we find that a condition of destructive interference (minimal reflection) for near-vertical incident light on the thin coating on glass is

$$2t = \frac{\lambda_n}{2} \quad \text{(destructive interference)} \qquad (29\text{-}10)$$

Equation (29-10) gives the thinnest film that will produce destructive interference. We will also have destructive interference when the difference in path length is $3\lambda_n/2$, $5\lambda_n/2$, and so on. The general relationship is

$$2t = \left(m + \frac{1}{2}\right)\lambda_n \quad \text{(destructive interference)}, \qquad (29\text{-}11)$$

where m is an integer ≥ 0.

Note that this relationship holds only for media that cause a phase shift at each of the two reflections. Also, a different condition is required for constructive interference.

A procedure for thin film problems is outlined below:

1. For each of the two paths, decide the phase changes by identifying which reflections are hard ones from materials with a higher index of refraction.

2. Add up all the phase changes of the hard reflections on the path. Each hard reflection has a π radian phase change.

3. If the sum of the reflection phase changes is 0 (or 2π) radians, use the top row of Table 29-1. If the reflection phase changes add to π (or 3π) radians, use the second row in the table.

4. Ask yourself whether you are working out a situation with constructive (bright) or destructive (dark) interference. This determines which column of the table you use.

5. For approximately vertical incidence, the extra two-way distance in the thin film ($2t$) is given by the appropriate expression in the table. Note that the wavelength is that of the medium where the extra distance is travelled and is therefore equal to λ/n, where λ is the wavelength in vacuum.

We demonstrate this procedure in Examples 29-4 and 29-5.

Table 29-1 Conditions for Constructive and Destructive Interference between Two Waves in a Thin Film of Thickness t and Index of Refraction n

	Constructive Interference	Destructive Interference
Phase constant difference between waves, not counting extra distance, is 0 radians.	$2t = m\dfrac{\lambda}{n}, \quad m = 1, 2, 3, \ldots$	$2t = \left(m + \dfrac{1}{2}\right)\dfrac{\lambda}{n}, \quad m = 0, 1, 2, 3, \ldots$
Phase constant difference between waves, not counting extra distance, is π radians.	$2t = \left(m + \dfrac{1}{2}\right)\dfrac{\lambda}{n}, \quad m = 0, 1, 2, 3, \ldots$	$2t = m\dfrac{\lambda}{n}, \quad m = 1, 2, 3, \ldots$

ELECTRICITY, MAGNETISM, AND OPTICS

EXAMPLE 29-4

Soap Bubble

When a narrow beam of white light shines almost perpendicular to the surface of a soap bubble, the bubble appears blue (480 nm in air) in reflected light. The index of refraction of the soapy water is approximately 1.33. What are three possible thicknesses for the wall of the soap bubble?

Solution

The situation is pictured in Figure 29-17. Here, we have a 180° phase change at the first reflection but not at the second. To have *constructive* interference (as required for the blue appearance of the outer surface), the extra distance travelled by the

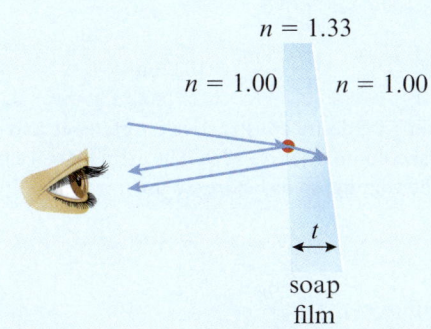

$$n = 1.33$$
$$n = 1.00 \qquad n = 1.00$$

$$t$$

soap film

Figure 29-17 Example 29-4. Reflection and refraction at the surfaces of a soap bubble. The small circle indicates a hard reflection.

light that reflects from the inner surface must correspond to another 180° phase change. Therefore,

$$2t = \frac{\lambda_n}{2}$$

We need to find the wavelength of the blue light in the soapy water:

$$\lambda_n = \frac{\lambda}{n} = \frac{480 \text{ nm}}{1.33} = 361 \text{ nm}$$

Therefore, the thinnest possible wall for a blue colour is

$$t = \frac{\lambda_n}{4} = \frac{361 \text{ nm}}{4} = 90.2 \text{ nm}$$

The next two possible conditions for constructive interference of blue light are

$$2t = \frac{3\lambda_n}{2} \quad \text{therefore, } t = 271 \text{ nm}$$

$$\text{and} \quad 2t = \frac{5\lambda_n}{2} \quad \text{therefore, } t = 451 \text{ nm}$$

Making sense of the result

As expected, the wavelength of the light in the soapy water is less than the wavelength in air. We need an extra 180° phase shift added to the phase shift of the reflected wave to produce constructive interference, so we need wall thicknesses of $\lambda_n/4$, $3\lambda_n/4$, $5\lambda_n/4$, and so on. We have calculated wall thicknesses for the maximum reflection of blue light. However, the incident light is white, so there will also be partial reflection of other colours.

Figure 29-18 shows a soap film exposed to white light. You can see that the film of soapy water has a varying thickness; we would expect the higher parts of the film to be thinner due to the effect of gravity.

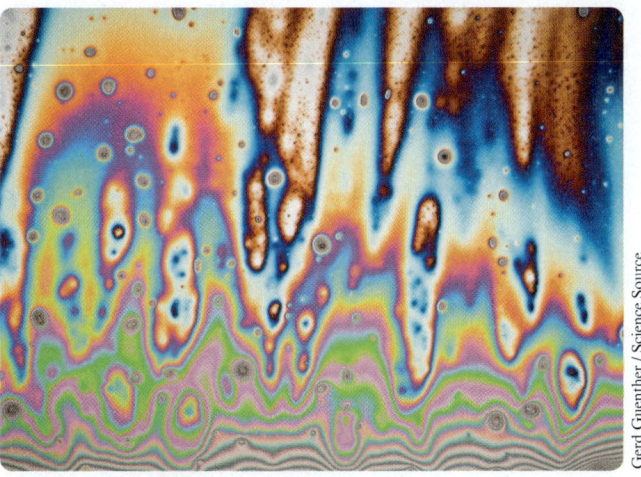

Figure 29-18 The varying thicknesses of the soap bubble result in constructive interference for different colours in different regions.

The variations in thickness result in different colours being dominant in constructive interference in various regions of the soap film.

Thin films are critical in modern science and technology. (See the Making Connections feature on applications to energy-efficient windows, called "Smart Windows.")

You may have noticed that the derivation for thin films did not specifically require that they be thin. Are interference effects also observed when the film is relatively thick? While interference can occasionally be observed in these situations, in general significant interference is restricted to relatively thin films. One reason is that as the film becomes much thicker, it is more difficult to keep the two sides parallel and of constant thickness, so interference effects may partly cancel.

The more important reason, however, is related to the concept of **coherence length**, which refers to the distance over which the electromagnetic waves making up a beam will have a consistent phase. A detailed treatment of coherence length is beyond the scope of this book, but the coherence length is inversely proportional to the variability in wavelength of the source. For

EXAMPLE 29-5

Lens Antireflection Coating

Camera lenses ($n = 1.52$) are often coated with magnesium fluoride (MgF_2; $n = 1.38$). The principle behind these coatings is to use destructive interference to reduce reflections from the surface of the lens.

(a) Calculate the minimum thickness required for an MgF_2 coating that would give destructive interference for reflected light in the blue–green region ($\lambda = 565$ nm).

(b) For this thickness of MgF_2, are there any wavelengths in the visible spectrum (wavelengths of 400 nm to 700 nm) for which the reflected light interferes constructively?

Solution

(a) The situation is shown in Figure 29-19. The numbered waves here refer to that diagram, with (1) being the incident wave. Wave (2) has undergone one hard reflection at the air–MgF_2 interface, producing a π radian phase shift. The other wave (6) undergoes no phase change when it is transmitted through the air–MgF_2 interface, but it does have a π radian phase change due to the hard reflection at the MgF_2–glass interface.

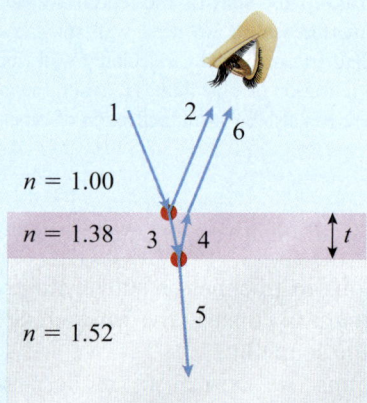

Figure 29-19 Example 29-5. Part of incident ray (1) undergoes a hard reflection at the first interface (2), while part is transmitted (3), and then splits into a reflected (4) and a transmitted ray at the second interface.

Since the paths for both waves have one hard reflection, resulting in one π radian phase shift, the phase difference between the two paths is 0 and we use the first row in Table 29-1. We want destructive interference, so we use the right column in the table. The extra distance travelled is in the MgF_2 coating, so we use $n = 1.38$ in computing the wavelength in the medium:

$$2t = \left(m + \frac{1}{2}\right)\frac{\lambda}{n} \quad m = 0, 1, 2, 3, \ldots$$

We desire the minimum thickness, so we will use $m = 0$. If we leave the wavelength in nanometres, the thickness will be in the same units:

$$t = \frac{\lambda}{4n}$$

$$= \frac{565 \text{ nm}}{4 \times 1.38} = 102 \text{ nm}$$

(b) This time we desire constructive interference and will use the first column of Table 29-1. The extra distance travelled is in the coating, so as before we use $n = 1.38$:

$$2t = m\frac{\lambda}{n} \quad m = 1, 2, 3, \ldots$$

$$\text{For } m = 1: \lambda = \frac{2tn}{m}$$

$$= \frac{2 \times 102 \text{ nm} \times 1.38}{1} = 282 \text{ nm}$$

This is in the ultraviolet region and is not visible light. Higher values of m will lead to even shorter wavelengths, so there is not a visible wavelength for this thickness that results in constructive interference.

Making sense of the result

It is reasonable that for destructive interference we will need to take a one-quarter wavelength (in the medium) each way so the total extra distance is half a wavelength. That is what we obtained. While there is no visible wavelength for this coating thickness resulting in constructive interference (part (b)), there are of course other thicknesses that would result in constructive interference.

---

a helium–neon laser, the coherence length can be up to 30 cm, but for a sodium discharge tube it is only about 0.6 mm. Therefore, we should be able to observe film-type interference effects with laser light on relatively thick films, if they can be constructed with consistent thickness. However, for a gas discharge or other light source, the short coherence length will mask the film interference effects.

We have not formally considered the case where the incident light is not normal to the surface, but exactly the same principles apply. The only difference is that you need to take into account that the difference travelled is not the thickness (in each direction) but rather $\frac{t}{\cos\theta}$, where θ is the angle of the light relative to a perpendicular to the surface.

29-6 Single-Slit Diffraction

Two related but distinct phenomena—interference and diffraction—form the core of physical optics. Both depend on the wave nature of light. Interference allows us to understand how waves that travel different paths in the same medium interact. When analyzing diffraction, we consider each point on a wave front to be the source of new waves, which then allows us to analyze the interference between those waves. Diffraction gives us a model to understand how light spreads out and bends around a small object. Diffraction is also crucial in understanding the resolution limit (Section 29-8) imposed by the wave nature of light.

Young's double-slit experiment was a crucial proof of the wave nature of light, but another experiment, this one by Dominique Arago (see the Making Connections box) involving diffraction, played a critical historical role. To understand diffraction, we will use **Huygens' principle** (sometimes called the Huygens–Fresnel principle), which was described in Section 15-2. According to Huygens' principle, each point along a wave front acts as a new source of waves that spread out spherically from that point. The wave front at a later time is defined by the unique tangent to the spherical waves at that time.

 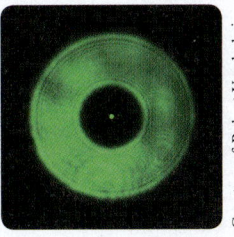

Consider a single slit illuminated by a source of parallel monochromatic light, which is also coherent, at least over a short distance. At first, you might think that a single slit would not produce any interference effects, but when we apply Huygens' principle to the area of the slit, we have a whole series of point sources of waves that can interfere with each other. Figure 29-22 shows a row of such sources spanning the width, w, of a slit. We can pair each source in the upper half of the slit with a corresponding source in the lower half. For the sources shown in Figure 29-22, we would pair the first source (counting from the top) with the fifth source, the second with the sixth, the third with the seventh, and so on. For a distant screen, the path difference for each pair of sources is approximately $\frac{w}{2}\sin\theta$.

If this path difference equals an odd number of half wavelengths, destructive interference will produce a dark region on the screen:

$$\frac{w}{2}\sin\theta = \left(m + \frac{1}{2}\right)\lambda$$

$$w\sin\theta = (2m + 1)\lambda$$

(29-12)

where m is an integer and $m \geq 0$.

Thus, destructive interference will occur when $w\sin\theta = \lambda, 3\lambda, 5\lambda, 7\lambda, \ldots$.

However, we could divide the secondary sources into four regions instead if two, and pair each source in the first (uppermost) region with a corresponding source in the next region, and similarly pair each source in the third region with a corresponding source in the fourth region. Now the condition for a half-wavelength path difference is

$$\frac{w}{4}\sin\theta = \left(m + \frac{1}{2}\right)\lambda$$

$$w\sin\theta = (4m + 2)\lambda$$

(29-13)

where m is an integer and $m \geq 0$.

This will result in destructive interference when $w\sin\theta = \lambda, 2\lambda, 6\lambda, 10\lambda, \ldots$.

We can apply the same reasoning when we divide the secondary sources into any even number of groups. Combining all these conditions for *destructive* interference gives a simple relationship for the destructive interference cases:

KEY EQUATION

$$w\sin\theta = m\lambda$$

(29-14)

where m is an integer > 0.

This relationship is almost the same as the relationship for *constructive* interference for a double slit, where the width, w, of the single slit replaces the spacing, d, between the double slits. However, it is important to remember for the single slit that the equation gives the angle for dark fringes (destructive interference), and the similar equation for the double slit gives the angles for bright fringes (constructive interference). Also, the double-slit equation can be applied with $m = 0$, but the single-slit equation cannot.

Figure 29-23 shows the interference pattern on a screen 1 m away for a single slit 0.40 mm wide illuminated with 633 nm (red) laser light. The pattern has a bright central band, becomes dark at a distance of about 1.6 mm on each side, and then alternates light fringes (with decreasing intensity) and dark fringes as the distance from the centre increases. Note that most of the intensity is contained within the **central maximum**, which we define as the region between the first minimum on one side and the first minimum on the other side.

What will happen if we make the single slit wider? Figure 29-24 shows the results of using the same light source (633 nm) with a slit 0.80 mm wide. Now the central band extends only to approximately 0.8 mm on each side. Making the slit wider has made the image on the

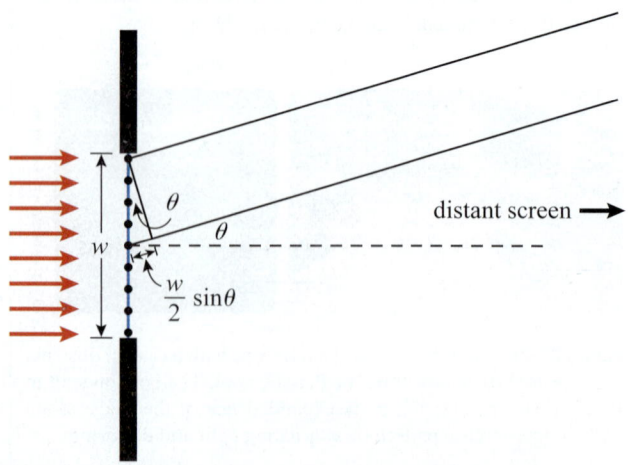

Figure 29-22 Huygens' secondary sources (marked as circles) in a row across the width of a single slit.

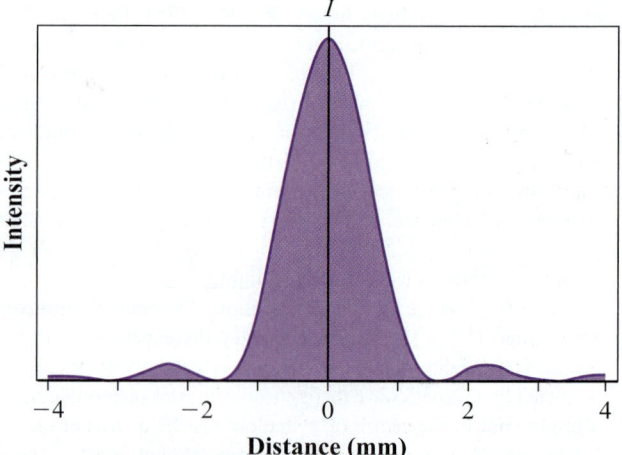

Figure 29-23 The relative intensity plot for coherent light with a wavelength of 633 nm falling on a single slit of width 0.40 mm. The plot gives distance in millimetres on a screen that is 1 m away from the slit.

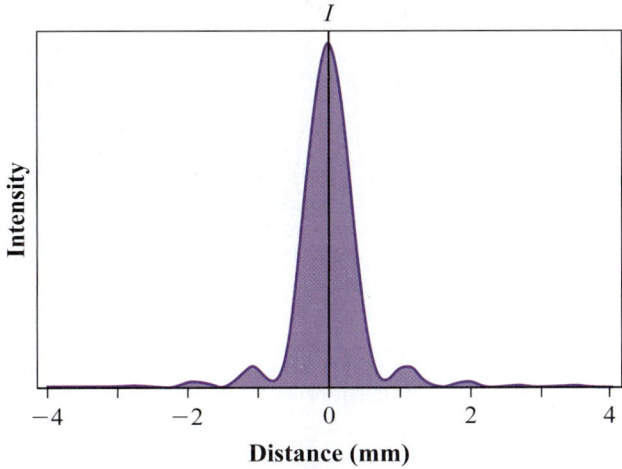

Figure 29-24 The intensity plot for a situation like Figure 29-23 but for a single slit of width 0.80 mm.

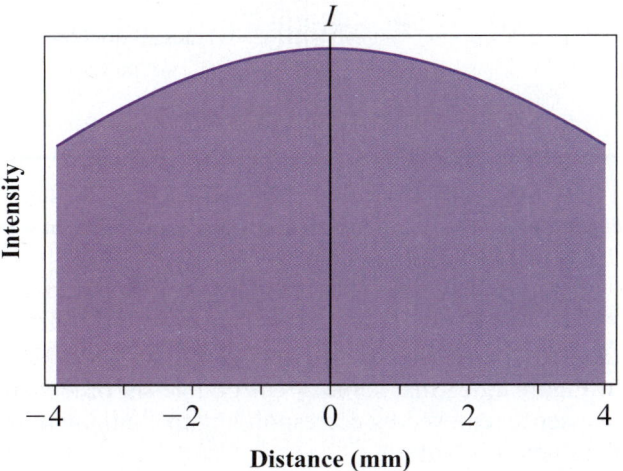

Figure 29-25 The relative intensity plot for a narrower single slit of width 0.05 mm.

screen narrower. Note that in this diffraction pattern the maxima are not located halfway between the minima.

From Equation 29-14, it is clear why the size of the central maximum of the interference pattern for a given wavelength decreases when we make the slit width, w, larger. Equation 29-14 also indicates that a smaller slit width will produce larger angles for the same wavelength, resulting in a broader pattern, as shown in Figure 29-25.

CHECKPOINT 29-6

Single-Slit Diffraction with Blue and Red Filters

White light passes through a blue filter before reaching a narrow single slit. Which statement correctly describes the position of the first bright band away from the centre when a red filter replaces the blue filter?

(a) The position is exactly the same when either filter is used.

(b) The bright band is slightly closer to the centre when the blue filter is used.

(c) The bright band is slightly farther from the centre when the blue filter is used.

(d) There is only the central bright band because you need white light to form a complete interference pattern.

ANSWER: (b) Red light has a longer wavelength than blue light (and passing the white light through a coloured filter leaves light predominantly of that colour). When the wavelength is longer, the angle θ for both dark and light fringes is greater. Therefore, we expect the position of the next bright fringe to be more distant for red and closer for blue.

EXAMPLE 29-6

Single-Slit Image Width

A single slit of width 0.750 mm is illuminated by red laser light with a wavelength of 656 nm. Find the width of the central (bright) band on a screen 1.00 m from the slit.

Solution

The width of the central band is usually defined as the distance from the first dark region on one side of the central band to the first dark region on the other side. Therefore, we will find the angle of the first dark region on either side and then use trigonometry to calculate the distance on the screen.

The first angle at which destructive interference occurs has $m = 1$, so

$$w\sin\theta = \lambda$$

We need to use consistent units, so we will express all distances in metres:

$$\sin\theta = \frac{\lambda}{w} = \frac{6.56 \times 10^{-7}\ \text{m}}{7.50 \times 10^{-4}\ \text{m}} = 8.75 \times 10^{-4}$$

$$\theta = 0.0501°$$

The distance from the centre to this point is

$$y = 1.00\ \text{m} \times \tan 0.0501° = 8.744 \times 10^{-4}\ \text{m}$$

The width of the central band is double this distance: 1.75×10^{-3} m, or 1.75 mm.

Making sense of the result

The width of the central image depends on both the wavelength of the light and the slit width. Here, the slit is three times as wide as in Figure 29-23, but the wavelength is longer. The wider slit makes the pattern narrower, and the longer wavelength makes the pattern slightly wider. Our calculated angle of approximately 0.05° is about twice the angle in the figure, which appears reasonable.

PEER TO PEER

When I'm solving double- and single-slit problems, I find that it is easy to forget to make the units consistent. Usually, widths and spacing of slits are given in millimetres and wavelengths in nanometres, but for calculations, I have to put them in the same units. Similarly, in resolution problems, I need to put the diameter and the wavelength in the same units.

29-7 Actual Intensity Pattern for Double Slits

The equations we derived in Section 29-3 do not, in fact, fully describe double-slit interference. When we want the precise intensity pattern, we must consider both the double-slit interference and the diffraction effects due to the finite width of the slits. The exact mathematics of this combination can be daunting, but we can use a simple approximation that is reasonably accurate.

Let us first briefly compare what we have found for single-slit and double-slit interference patterns (Figures 29-26 and 29-27):

single slit: for destructive interference, $w \sin\theta = m\lambda$

double slit: for constructive interference, $d \sin\theta = m\lambda$

Here, λ is the wavelength of the light, θ is the angle from the centre of the pattern, m is the spectral order, w is the slit width, and d is the spacing between the slits.

Both patterns have most of the intensity in the central bright band, and the intensity of the other bright fringes decreases farther from the central point.

We can approximate the actual interference pattern for a double slit by multiplying the intensity pattern for the idealized double slit by the intensity pattern for single-slit diffraction, giving the intensity pattern shown in Figure 29-28. Compare this prediction with the actual patterns for two slits in Figure 29-9. Note in Figure 29-28 that some of the double-slit orders are in essence removed by corresponding to a minimum of the single-slit pattern.

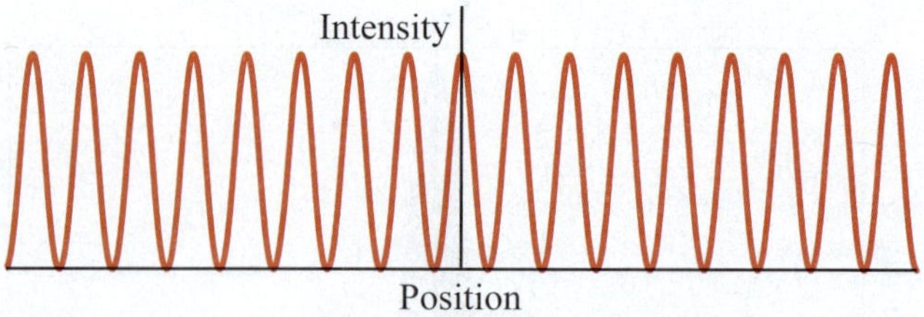

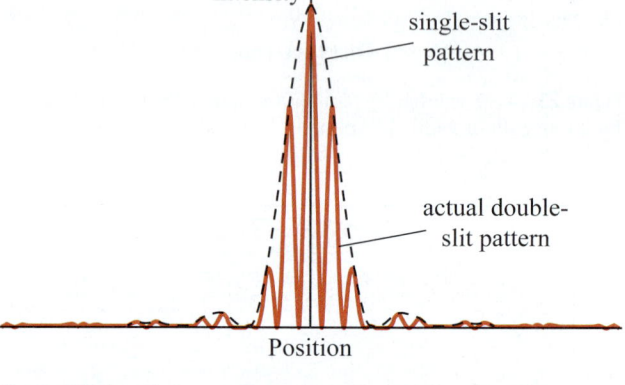

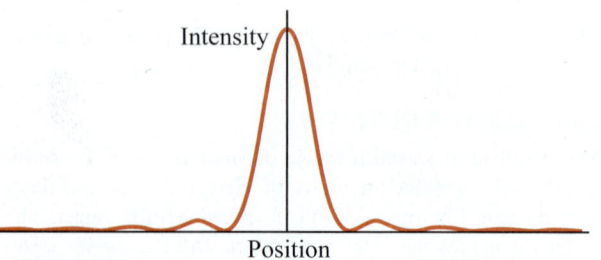

Figure 29-26 A simplified intensity plot for interference from a single slit.

Figure 29-27 A simplified intensity plot for interference from an idealized double slit.

Figure 29-28 The approximate intensity plot for a double slit with diffraction from the individual slits factored in.

CHECKPOINT 29-7

A Real Double Slit with Different Slit Widths

When you shine monochromatic light through two slits, you can distinguish the central bright fringe and three more light fringes on each side. You make the slits narrower and increase the intensity of the incident light to keep the amount of light that passes through the slits constant. What will you observe?

(a) the same number of light fringes in the same positions
(b) the same number of light fringes, but closer to the centre
(c) fewer light fringes
(d) more light fringes

ANSWER: (d) We multiply the single-slit pattern by the theoretical double-slit interference pattern. If we make the slits narrower, the single-slit image becomes broader, and we can see a higher order (i.e., more) of the double-slit images. The narrowing of the slits could make the higher-order fringes too faint to see if we had not increased the incident intensity.

EXAMPLE 29-7

How Many Fringes?

A laser emitting green coherent light with a wavelength of 532 nm illuminates two slits that are 0.150 mm apart. The width of each slit is 0.0200 mm. How many fringes are visible on a screen? Assume that only the fringes within the central maximum are visible.

Solution

We need to determine how many fringes are within the single-slit envelope at the centre of the interference pattern, which has a shape like that shown Figure 29-28. We will write an expression for the angle corresponding to the first zero for the single-slit pattern, set that expression equal to the expression for double-slit interference, and solve for the order, m.

The $m = 1$ zero for single-slit diffraction occurs when $\sin\theta = \dfrac{1\lambda}{w}$.

For double-slit interference, we have $\sin\theta = \dfrac{m\lambda}{d}$.

Equating these two expressions and solving for m, we have

$$m = \frac{d}{w} = \frac{1.50 \times 10^{-4}\,\text{m}}{2.00 \times 10^{-5}\,\text{m}} = 7.5$$

Thus, double-slit fringes up to and including the 7th order are visible, but the 8th- and higher-order fringes are outside the single-slit envelope; they are assumed to be too faint to be seen. Since the double-slit pattern has a bright zero-order central band, a total of 15 fringes are visible (-7, -6, -5, -4, -3, -2, -1, 0, 1, 2, 3, 4, 5, 6, and 7).

Making sense of the result

Although 15 light fringes are visible, they are not equally bright: the higher-order fringes are much fainter, as we would expect from the shape of the combined pattern in Figure 29-28.

29-8 Resolution Limit

We have seen that the wave nature of light can produce diffraction and interference. These effects can mask the true nature of objects. We call the smallest dimensions that we can see the **resolution limit**.

For example, suppose you are on the Prairies on a clear night and see an automobile far away. The automobile has two headlights, but if it is too far away, you will not be able to clearly resolve them as two separate sources of light; instead you see just see one bright light. In this situation, the distance separating the two headlights is beyond the resolution limit. Figure 29-29 shows two adjacent light sources photographed such that they are not resolved (left), marginally resolved (centre), and resolved (right).

The resolution limit depends on the wavelength of the light and the diameter of the light-gathering area of the detection instrument. We can see the origin of this relationship by considering the effect of a single slit: the narrower the slit relative to the wavelength, the wider the central image. Therefore, the smaller the aperture (opening) of the observing instrument, the more spread out the image, making two adjacent point objects more difficult, or even impossible, to resolve.

Figure 29-30 shows the diffraction pattern for coherent light with a wavelength of 532 nm passing through an aperture with a diameter of 0.1 mm. The central image has a radius of approximately 0.19°, surrounded by alternating light and dark rings, called an **Airy pattern** (named for the English astronomer George Biddell Airy, 1801–1892). If we make the aperture much larger, the pattern has essentially the same form, but there is much better resolution. With an aperture diameter of 15 mm, the central image is only approximately 0.0012° across.

Figure 29-29 Improving the resolution reveals that the bright spot in the left photograph is actually two separate sources.

Figure 29-30 With a small aperture, the effects of diffraction are apparent: the image is broad and surrounded by an Airy pattern of light and dark rings.

The Rayleigh criterion is widely used to decide whether two sources are too close to be clearly distinguished. Developed by the English physicist Baron Rayleigh (John William Strutt, 1842–1919), the **Rayleigh criterion** states that two objects can be resolved if they are separated by an angle at least large enough for the central maximum of one object to correspond to the first minimum in the diffraction pattern of the other.

If we apply the equation for single-slit diffraction to the Rayleigh criterion, we have $m = 1$ because the criterion involves the first minimum of the diffraction pattern. It follows that for light with a wavelength of λ and an optical instrument whose aperture has an effective diameter of D, the minimum resolved angle, θ_r, is given by

$$\sin\theta_r = \frac{\lambda}{D} \tag{29-15}$$

When the angular separation of two objects is less than θ_r, they cannot be resolved. However, most apertures in optical instruments are circular. The mathematics of the fringes around images formed from small circular apertures (Figure 29-30) is somewhat more complex. The Rayleigh criterion for circular apertures can be shown to be

KEY EQUATION
$$\sin\theta_r = 1.22\frac{\lambda}{D} \tag{29-16}$$

When applying this criterion, we must ensure that the wavelength and diameter are expressed in the same units. Optical telescopes with a large diameter collect more light, which helps us detect faint objects; the large diameter also improves resolution. Equation (29-16) also applies to the resolution limit for the human eye, which depends on the diameter of the pupil.

The resolution criterion can be applied to any wavelength, not just to the visible region of the electromagnetic spectrum. We now see that radio telescopes (like the one shown at the beginning of the chapter) must have large diameters to have even modest angular resolution because radio wavelengths are relatively long. While radio telescopes must be large in diameter, because the wavelength is so large the requirements on surface smoothness are much less stringent than for optical telescopes.

EXAMPLE 29-8

Resolving Binary Stars

A pair of stars in a binary system is separated by 0.200 light years. The system is 7800 light years from us, and the stars' orbits are oriented perpendicular to our line of sight.

(a) What minimum aperture diameter for your telescope do you need to resolve them? Assume light from the stars has a wavelength of 525 nm.
(b) If the stars are bright enough, can they be resolved by an unaided human eye? Explain.

Solution

(a) We first need to find the angle subtended by the two stars as seen from Earth. Then we will use the Rayleigh criterion with this angular separation to calculate the necessary diameter:

$$\tan\theta = \frac{0.200 \text{ ly}}{7800 \text{ ly}} = 2.56 \times 10^{-5}$$

$$\theta = \tan^{-1}(2.56 \times 10^{-5}) = 1.47 \times 10^{-3°}$$

Now we use the Rayleigh criterion and solve for the diameter, D:

$$D = 1.22\frac{\lambda}{\sin\theta_r} = 1.22\frac{5.25 \times 10^{-7} \text{ m}}{\sin(1.47 \times 10^{-3°})} = 0.025 \text{ m}$$

A telescope with an aperture diameter of 2.5 cm is the minimum size needed.

(b) The diameter of the pupil of the human eye, even under dark-adjusted conditions, is only approximately 5 mm, much less than the required aperture diameter. Therefore, the binary stars cannot be resolved by an unaided human eye.

Making sense of the result

It makes sense that an optical instrument with an aperture diameter of only 2.5 cm can resolve closely spaced objects better than the human eye can because the aperture diameter of the eye is only approximately one-fifth as large. The Rayleigh criterion only gives us the resolution imposed by the wave nature of light. Other effects, such as imperfections in the optics and motion of air with varying temperatures and humidity, can increase the resolution limit.

According to Equation 29-16, there are two ways to improve the resolution for circular apertures: increase the effective diameter of the optical instrument or use light with a shorter wavelength. For example, other things being equal, a system using ultraviolet light will resolve features better than a system that uses infrared or visible light.

Electron Microscopes

In 1938, two students at the University of Toronto, James Hillier (1915–2007) and Albert Prebus (1913–1997), constructed the first workable electron microscope. Their instrument achieved magnifications of 7000×, several times that of the best optical microscopes. Hillier and Prebus based their design on the work of the German physicist Ernst August Friedrich Ruska (1906–1988), who showed that electrons can be used instead of light in a microscope to greatly increase the resolution of the images. As you will see in later chapters, electrons have both wave and particle properties, including an effective wavelength that depends on the momentum of the electrons. Electron wavelengths are tiny compared to even ultraviolet light, so the most modern electron microscopes can have resolutions less than 50 pm. Figure 29-31 is an electron microscope image of part of a butterfly wing.

The wave nature of electrons means that a small aperture for an electron beam will produce an interference pattern very similar to that seen in Figure 29-30 for light. Figure 29-32 shows the diffraction pattern of a beam of electrons passing through a graphite block.

Figure 29-31 An electron microscope image of part of a butterfly wing reveals a pattern in the wing's structure. This pattern acts like a diffraction grating, and the colour of the wing results primarily from interference effects rather than pigments.

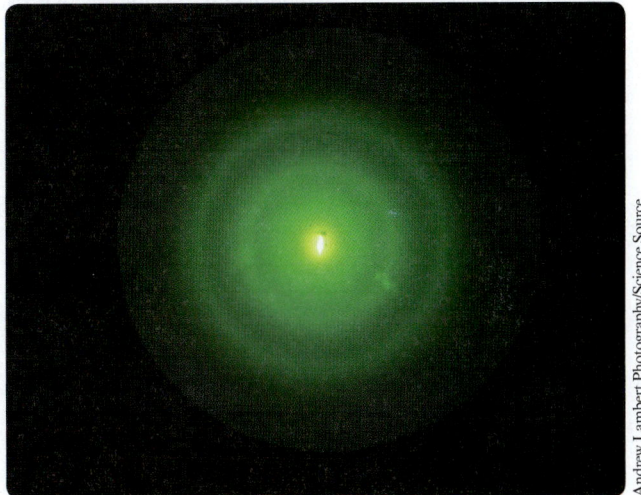

Figure 29-32 Diffraction of an electron beam passing through graphite.

KEY CONCEPTS AND RELATIONSHIPS

Geometric and Physical Optics

When solving problems in which the size of an object is comparable to the wavelength of the electromagnetic radiation illuminating it, we must explicitly consider the wave nature of light, including interference effects. We can consider light to travel in straight lines (rays) in geometric optics only when the wavelength is much shorter than the dimensions of the object being studied.

Conditions for Interference

Electromagnetic waves, including light, interfere constructively when they are in phase and destructively when they are out of phase. Phase differences between two waves are often due to a greater distance travelled by one of the waves.

Ideal Double Slit

Ignoring diffraction from the individual slits, constructive interference of light of wavelength λ passing through two narrow slits separated by a distance d occurs at any angle θ for which

$$d\sin\theta = m\lambda \qquad (29\text{-}4)$$

where m, the spectral order, is an integer ≥ 0.

Diffraction Grating

A diffraction grating consists of many equally spaced slits. The condition for maximum brightness is the same as for a double slit, i.e.,

$$d\sin\theta = m\lambda \qquad (29\text{-}7)$$

where d is the spacing between slits, m is the order of the image, and λ is the wavelength. The images are sharper and brighter for a diffraction grating than for a double slit. The spacing, d, is $1/N$, where N is the number of lines per unit length for the grating.

Thin Film Interference

Constructive interference occurs when the total phase difference between two possible light paths is 0, 2π, 4π, etc., radians, while destructive interference happens for a total phase difference of π, 3π, 5π, etc., radians. The phase difference is due to both extra distance travelled and different types of reflections. When a wave is reflected at an interface with a higher index of refraction material (called a hard reflection), the reflected wave has a π phase shift. When the reflection is at an interface to a material with a lower index of refraction (a soft reflection), there is no phase shift in the reflected wave. Transmitted waves undergo no phase change. When the extra distance travelled is one-half wavelength, there will be a π phase change, but we must take the wavelength in the medium, which is the wavelength in vacuum divided by the index of refraction. The various thin film interference

possibilities are summarized in Table 29-1. We always use the value of n for the material where the extra distance is travelled:

$$\lambda_n = \frac{\lambda}{n} \qquad (29\text{-}9)$$

Single-Slit Diffraction

By Huygens' principle, we can regard the region in a single slit to be composed of a number of secondary sources. *Destructive* interference between waves from these sources in a slit of width w occurs at any angle θ for which

$$w\sin\theta = m\lambda \qquad (29\text{-}14)$$

where m is an integer > 0.

The narrower the slit, the wider the region over which the light is spread. Most of the intensity falls between the $m = 1$ dark fringes, and we define the size of the central maximum as the distance between these fringes.

Actual Intensity Pattern of Double Slits

The interference pattern for a double slit is approximately equal to the product of the ideal intensity pattern for single-slit diffraction and the intensity pattern for an ideal double slit.

Resolution Limit

The wave nature of light limits the detail that can be seen with a given wavelength. The Rayleigh criterion states that two objects can be resolved if they are separated by an angle at least large enough for the central maximum of one object to correspond to the first minimum in the diffraction pattern of the other.

For a circular aperture of diameter D, the smallest subtended angle, θ_r, that can be resolved with light (or any other electromagnetic wave) of wavelength λ in accordance with the Rayleigh criterion is

$$\sin\theta_r = 1.22\frac{\lambda}{D} \qquad (29\text{-}16)$$

APPLICATIONS

interferometry, diffraction gratings, spectrometers, quantum control and ultrafast physics, antireflection optical coatings, energy-efficient windows, LIGO, resolution of optical instruments, electron microscopes

KEY TERMS

airy pattern, central maximum, coherence length, coherent, diffraction gratings, Fraunhofer condition, geometric optics, hard reflection, Huygens' principle, interference, lasers, Michelson interferometer, monochromatic, phase change, physical optics, ray optics, Rayleigh criterion, reflection gratings, resolution limit, soft reflection, thin film interference, transmission gratings, wave optics

QUESTIONS

1. Which of the following electromagnetic waves is most accurately approximated by geometric (ray) optics when used to view print on a page?
   (a) infrared
   (b) radio
   (c) visible light
   (d) X-ray

2. A thermal imaging camera can work in two modes, one with a wavelength of 2 μm and the other with a wavelength of 10 μm. The camera is being used to study emissions from an organism that is approximately spherical with a diameter of 5 mm. We want to solve problems

related to the thermal imaging camera. Which of the following statements is true?
   (a) We can use geometric optics to solve problems, and the approximation is better at the 2 μm wavelength.
   (b) We can use geometric optics to solve problems, and the approximation is better at the 10 μm wavelength.
   (c) We cannot use geometric optics to solve problems in this situation.
   (d) We can use geometric optics to solve problems, and the approach is equally valid at both wavelengths.

3. A Michelson interferometer apparatus is used with two different monochromatic sources, one with a wavelength of 430 nm and the other with a wavelength of 645 nm. When the light with the 645 nm wavelength is used, 64 fringes are counted for a certain movement of the mirror. What is the number of fringes counted for the same mirror movement when the 430 nm wavelength is used?
   (a) 43
   (b) 64
   (c) 96
   (d) 144

4. Monochromatic laser light with a wavelength of 532 nm illuminates two closely spaced slits to form an interference pattern on a screen. Bright interference dots form at a spacing of approximately 2 cm on the screen. If nothing else is changed when an infrared laser of wavelength 1064 nm is substituted, what is the spacing of the bright interference dots on the screen?
   (a) 0.5 cm
   (b) 1 cm
   (c) 4 cm
   (d) 8 cm

5. One of the wavelengths produced by an argon laser is 488 nm. Light from this laser is used to illuminate two slits, and the pattern on a screen has bright fringes approximately 1.2° on each side of the central bright band. With a different two-slit pattern, the first-order bright fringes are now 2.4° from the central image. The original slits were separated by a distance s. What is the spacing of the second pair of slits?
   (a) $s/4$
   (b) $s/2$
   (c) $2s$
   (d) $4s$

6. Spectrometer A has a diffraction grating with 9000 lines/cm, and spectrometer B (otherwise similar) has a diffraction grating with 3000 lines/cm. Which statement correctly describes the images produced by these spectrometers?
   (a) The first-order image for spectrometer B starts at a larger angle and is more spread out for spectrometer A.
   (b) The first-order image for spectrometer A starts at a larger angle and is more spread out for spectrometer B.
   (c) The first-order image starts at the same angle for both spectrometers but is more spread out for spectrometer A.
   (d) The first-order image starts at the same angle for both spectrometers but is more spread out for spectrometer B.

7. A diffraction grating is held in front of the lens of a camera pointed at stars in the night sky. Which of the following statements correctly describes the image produced?
   (a) There is no slit, so there is no interference pattern. Nothing will be seen.
   (b) On each side of each star the light spreads out into colours, with violet toward the centre and red farther out (but no spot directly in the centre for each star).

   (c) There is a white dot for each star, and on each side of that dot, the light spreads out into colours, with red nearer the white dot and violet farther away.
   (d) There is a white dot for each star, and on each side of that dot, the light spreads out into colours, with violet nearer the white dot and red farther away.

8. When a slit is narrow in both directions, there are diffraction patterns in both directions. The image in Figure 29-33 was produced with monochromatic blue light and a slit that is narrow in both directions. Which of the following most accurately describes the dimensions of the slit used to produce the image?
   (a) The slit is small in both dimensions but is narrower along the x-axis.
   (b) The slit is small in both dimensions but is narrower along the y-axis.
   (c) The slit is of equal width in both dimensions.
   (d) The relative size cannot be determined from the information provided.

Figure 29-33   Questions 8 and 9.

9. If instead of monochromatic blue light we use monochromatic red light to produce an image like the one in Figure 29-33, how will the diffraction pattern change?
   (a) The pattern will not change at all.
   (b) The pattern will be more spread out in both directions.
   (c) The pattern will be more compressed in both directions.
   (d) The pattern will be more compressed in the x-direction but more spread out in the y-direction.

10. A circular diffraction pattern is observed by shining monochromatic light through a small circular opening. Which statement correctly describes how the pattern changes with wavelength and the aperture size?
   (a) The size of the diffraction pattern increases as the size of the opening increases or the wavelength increases.
   (b) The size of the diffraction pattern increases as the size of the opening decreases or the wavelength increases.
   (c) The size of the diffraction pattern increases as the size of the opening increases or the wavelength decreases.
   (d) The size of the diffraction pattern increases as the size of the opening decreases or the wavelength decreases.

11. When white light is shining perpendicular to the surface of a soap bubble, the wall of the bubble initially appears to be green. What happens to the colour of that region as the soap film begins to evaporate?
(a) The soap film stays the same colour.
(b) The soap film becomes blue.
(c) The soap film becomes yellow.
(d) The soap film becomes white.

12. A test thickness, t, of an antireflection coating works excellently for red light with a wavelength (in a vacuum) of 640 nm. What thickness should be used to prevent reflection of ultraviolet radiation with a wavelength of 160 nm?
(a) $t/4$
(b) $t/2$
(c) $2t$
(d) $4t$

13. A certain microscope is unable to resolve details of an object when visible light is used. Assuming appropriate detectors, how could you improve the resolution?
(a) Use infrared light instead of visible light.
(b) Use ultraviolet light instead of visible light.
(c) Use microwaves instead of visible light.
(d) None of the above will help because the resolution depends only on the diameter of the objective lens.

14. The radio telescope pictured in Figure 29-1 is used to study both atomic hydrogen at a frequency of 1.42 GHz and methanol at a frequency of 6.67 GHz. Compare the angular resolution at these two frequencies.
(a) The angular resolution is the same for both.
(b) The angular resolution is approximately 4.7 times better for the atomic hydrogen observations.
(c) The angular resolution is approximately 4.7 times better for the methanol observations.
(d) The angular resolution is approximately 22 times better for the methanol observations.

15. You have a double-slit apparatus with the slits separated by a distance x and a distance to the screen of y. It is illuminated with visible yellow light. What is the requirement for the Fraunhofer condition in this situation?
(a) $x \gg 500$ nm
(b) $x \ll 500$ nm
(c) $x \gg y$
(d) $y \gg x$

16. Two electromagnetic waves have the same frequency, polarization, and amplitude. The phase difference between the two waves is $\pi/4$ radians. If I_0 is the intensity of one wave alone, what is the intensity when the two waves are combined?
(a) $\sqrt{2}I_0$
(b) $2I_0$
(c) $4I_0$
(d) $(2 + \sqrt{2})I_0$

17. As shown in Figure 29-34, a light wave goes from water to air, with a portion reflected at the interface. What are the phase differences (in radians) of the reflected and the transmitted waves near the interface relative to the incident wave?

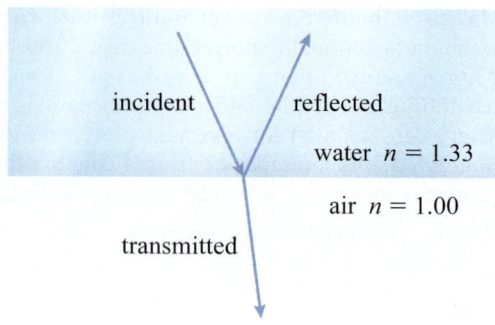

Figure 29-34 Question 17.

(a) reflected: 0; transmitted: 0
(b) reflected: 0; transmitted: π
(c) reflected: π; transmitted: 0
(d) reflected: π; transmitted: $\pi/2$

18. Which of the following equations correctly describes the conditions for *constructive* interference between waves 1 and 2 in the situation of Figure 29-35? Assume that the incidence is near enough to vertical that it can be considered approximately vertical. Here m represents integer values and λ is the wavelength in air.

(a) $2t = m\dfrac{\lambda}{n_2}$

(b) $2t = \left(m + \dfrac{1}{2}\right)\dfrac{\lambda}{n_2}$

(c) $2t = m\dfrac{\lambda}{n_3}$

(d) $2t = \left(m + \dfrac{1}{2}\right)\dfrac{\lambda}{n_3}$

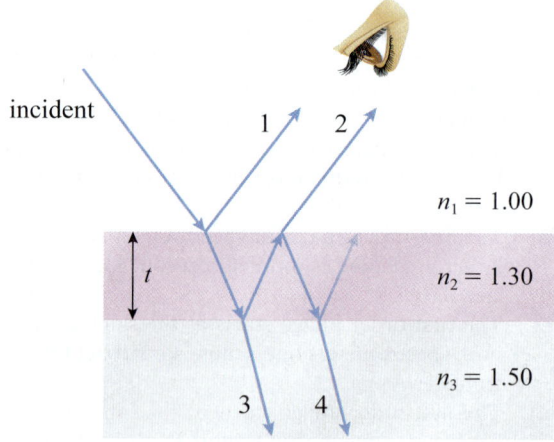

Figure 29-35 Questions 18–21.

19. Which of the following equations correctly describes the conditions for *destructive* interference between waves 1 and 2 in the situation of Figure 29-35? Assume that the incidence is near enough to vertical that it can be considered approximately vertical. Here m represents integer values and λ is the wavelength in air.

(a) $2t = m\dfrac{\lambda}{n_2}$

(b) $2t = \left(m + \dfrac{1}{2}\right)\dfrac{\lambda}{n_2}$

(c) $2t = m\dfrac{\lambda}{n_3}$

(d) $2t = \left(m + \dfrac{1}{2}\right)\dfrac{\lambda}{n_3}$

20. Again making reference to the situation of Figure 29-35, which of the following equations correctly describes the conditions for *constructive* interference between waves 3 and 4? Assume that the incidence is near enough to vertical that it can be considered approximately vertical. Here m represents integer values and λ is the wavelength in air.

(a) $2t = m\dfrac{\lambda}{n_2}$

(b) $2t = \left(m + \dfrac{1}{2}\right)\dfrac{\lambda}{n_2}$

(c) $2t = m\dfrac{\lambda}{n_3}$

(d) $2t = \left(m + \dfrac{1}{2}\right)\dfrac{\lambda}{n_3}$

21. Again making reference to the situation of Figure 29-35, which of the following equations correctly describes the conditions for *destructive* interference between waves 3 and 4? Assume that the incidence is near enough to vertical that it can be considered approximately vertical. Here m represents integer values and λ is the wavelength in air.

(a) $2t = m\dfrac{\lambda}{n_2}$

(b) $2t = \left(m + \dfrac{1}{2}\right)\dfrac{\lambda}{n_2}$

(c) $2t = m\dfrac{\lambda}{n_3}$

(d) $2t = \left(m + \dfrac{1}{2}\right)\dfrac{\lambda}{n_3}$

22. What is the minimum possible intensity when two electromagnetic waves of the same frequency and polarization are combined, assuming that I_0 and $3I_0$ are the intensities of the two waves individually?

(a) $3\sqrt{2}I_0$

(b) $2\sqrt{3}I_0$

(c) $(4 - 2\sqrt{3})I_0$

(d) $(4 + 2\sqrt{3})I_0$

23. What is the maximum possible intensity when two electromagnetic waves of the same frequency and polarization are combined, assuming that I_0 and $5I_0$ are the intensities of the two waves individually?

(a) $6I_0$

(b) $(6 - 2\sqrt{5})I_0$

(c) $(6 + 2\sqrt{5})I_0$

(d) $24I_0$

24. A double-slit optical device is designed for infrared radiation with $\lambda = 800$ nm, producing an $m = 1$ maximum at an angle from the centre of $\theta = 1.6°$. If it is instead illuminated with violet light of wavelength $\lambda = 400$ nm, where will the $m = 1$ maximum now occur (approximately)?

(a) 0.4°

(b) 0.8°

(c) 1.6°

(d) 3.2°

25. If you want the interference pattern from a double-slit apparatus to be spread farther apart (i.e., at greater angles), which of the following changes should be made? Assume that the light used is of a single fixed wavelength.

(a) Illuminate it with light of higher intensity.

(b) Illuminate it with light of lower intensity.

(c) Make the slit spacing closer together.

(d) Make the slit spacing farther apart.

26. You use white light to illuminate a double-slit apparatus and observe the interference pattern on a screen on the other side of the slits. Which statement correctly describes the appearance on the screen?

(a) The central maximum is white, with all of the others separated into colours, with red the closest to the centre.

(b) The central maximum is white, with all of the others separated into colours, with yellow the closest to the centre.

(c) The central maximum is white, with all of the others separated into colours, with violet the closest to the centre.

(d) Each maximum, including the central one, consists of just a single colour, with the colour depending on the spacing of the slits.

PROBLEMS BY SECTION

For problems, star ratings will be used (★, ★★, or ★★★), with more stars meaning more challenging problems. The following codes will indicate if $\frac{d}{dx}$ differentiation, $\int$ integration, ⬜ numerical approximation, or 〰 graphical analysis will be required to solve the problem.

Section 29-1 Physical and Geometric Optics

27. ★ An electromagnetic wave has a frequency of 469 THz. What is the wavelength, expressed in nanometres, and what type of radiation (e.g., X-ray, radio wave, microwave, visible light, ultraviolet, etc.) is this? If visible, identify the colour.

28. ★ The FM radio band extends from about 88 MHz to 108 MHz. What would be the corresponding range of wavelengths?

29. ★ A single-wavelength electromagnetic wave passes through an opening that has dimensions of 1 mm × 10 cm. Assess whether geometric optics approximations can be used in each of the following cases. Justify your answer.
(a) visible green light with a wavelength of 532 nm
(b) infrared radiation with a wavelength of 5.00 μm
(c) ultraviolet radiation with a wavelength of 180 nm
(d) radio waves with a frequency of 88.0 MHz

30. ★★ The brightness of a surface varies in the form of a sine wave with a wavelength of 2.5 mm (i.e., the distance between one brightness maximum and the next is 2.5 mm). Assume that the wavelength of the radiation must be one-tenth as large as the object being viewed for a geometric optics approach to be valid. What are the wavelength and frequency of the maximum wavelength of radiation that meets this condition, and what type of radiation is this?

Section 29-2 Interference

31. ★ Two electromagnetic waves have the same frequency, polarization, and intensity. The phase difference between the waves is $\frac{\pi}{6}$. What will be the intensity of the combined wave compared to the intensity of one of the individual waves?

32. ★ Two electromagnetic waves have the same frequency, polarization, and intensity. What should be the phase difference between the waves if the combined intensity is 3.00 times that of each individual wave?

33. ★★ Two electromagnetic waves have the same wavelength and polarization. The intensity of one of the waves is I_0, while that of the other is $9I_0$. If we allow the full set of possible phase differences between the waves, what is the range of possible intensities for the combined waves (expressed in terms of I_0)?

34. ★★ Two radio antennas are being driven in phase by the same 1500 kHz signal. The antennas are 2.00 km from each other.
(a) Would constructive interference or destructive interference occur at a detector located exactly halfway between the two transmitters?
(b) The detector is moved to a position of destructive interference. Determine the minimum distance the detector needs to be moved to this position.

35. ★★ Laser light with a wavelength of 633 nm is used to illuminate a Michelson interferometer in a vacuum chamber. You count 90 fringes as one mirror is moved. Through what distance must the mirror have moved?

Section 29-3 Double-Slit Interference

36. ★ A double-slit apparatus has a spacing of 3.80 mm between the slits. It is illuminated by a monochromatic sodium discharge tube with a wavelength of 589 nm. What will be the angles for the $m = 1$ and $m = 2$ order maxima?

37. ★★ A screen is located a distance of 24.5 cm from two slits that are spaced 0.450 mm apart. Incident white light with wavelengths from 400 nm to 700 nm is used. What distance on the screen (measured from the centre of the $m = 0$ spot) corresponds to the different colours for the $m = 1$ interference pattern?

38. ★★ A double slit is illuminated with light containing wavelengths from 400 nm to 650 nm. The screen is at a distance of 20.0 cm from the slits. For the 400 nm wavelength violet light, it is observed that the $m = 1$ maximum on the screen is at a distance of 0.120 cm from the $m = 0$ maximum.
(a) What is the spacing between the slits?
(b) What is the distance from the $m = 0$ to $m = 1$ maxima for the red 650 nm wavelength light?

39. ★★ The $m = 4$ maximum is located 2.40 cm from the $m = 0$ image when monochromatic radiation illuminates a double slit. The screen is 18.0 cm from the slits. The slit spacing is $d = 0.150$ mm.
(a) What must be the wavelength of the radiation?
(b) What type of radiation (i.e., visible, infrared, ultraviolet, microwave, etc.) is this?

40. ★★ Two narrow slits are 0.250 mm apart. They are illuminated by light from a helium–neon laser with a wavelength of 633 nm, and the interference pattern is observed on a screen 45.0 cm from the slits.
(a) What is the angle between the zero-order maximum and one of the first-order images?
(b) What is the distance between the two first-order images?

41. ★★★ 🖳 In a double-slit experiment, you want the separation between the $m = 1$ and $m = 2$ images on the same side of the zero-order maximum to be 4.00 mm when the screen is 0.750 m from the slits. Green laser light with a wavelength of 532 nm is being used for illumination. Hint: Use a numerical approach to solve the relationships you develop.
(a) What is the required spacing between the slits?
(b) What would the spacing between the $m = 1$ and $m = 2$ images be if a red laser with a wavelength of 633 nm were used for illumination?

Section 29-4 Diffraction Gratings

42. ★ A diffraction grating has 6.00×10^3 lines/cm. What is the spacing between successive lines expressed in micrometres (μm)?

43. ★★ A diffraction grating has 5000. lines/cm. It is illuminated with monochromatic light of frequency 500. THz. At what angle, in degrees, will the first-order image appear?

44. ★★ Monochromatic radiation of wavelength 398 nm is incident on a diffraction grating. The grating has 2500. lines/cm. What is the *difference* in angles between the $m = 1$ and $m = 2$ diffraction images?

45. ★★ Two hydrogen spectral lines have wavelengths of 486 nm and 656 nm. If a 3000. lines/cm diffraction grating is used with a screen 15.0 cm from the grating, what will be the distance on the screen between the first-order images (maxima) for these two wavelengths?

46. ★★ You are asked to design a spectrometer such that on a screen positioned 0.750 m from the diffraction grating the $m = 1$ image for red light with a wavelength of 700 nm is at a distance (from the centre of the $m = 0$ image) of 2.00 cm.
(a) How many lines per centimetre should the diffraction grating have?
(b) What is the distance to the $m = 1$ image for light with a wavelength of 400 nm?

47. ★★★ 🖳 You want to design a spectrometer to cover the first-order spectrum completely from 360 nm to 675 nm. The image sensor that will detect the first-order image on one side is 22.5 mm across. The spacing from the diffraction grating to the detector is 5.00 cm. What grating spacing (in lines/cm) should be used? This is challenging to solve analytically, so use a numerical approach to the computation.

Section 29-5 Thin Film Interference

48. ★ The index of refraction for a material is 1.50. If light of wavelength 450 nm travels an extra distance of 450 nm in this material, what will be the phase difference for this extra distance?

49. ★★ Making reference to Figure 29-36 (and assuming near vertical incidence), what are the three smallest thickness (t) values that result in constructive interference when viewed from above? Assume monochromatic light of wavelength 589 nm (in vacuum) is used.

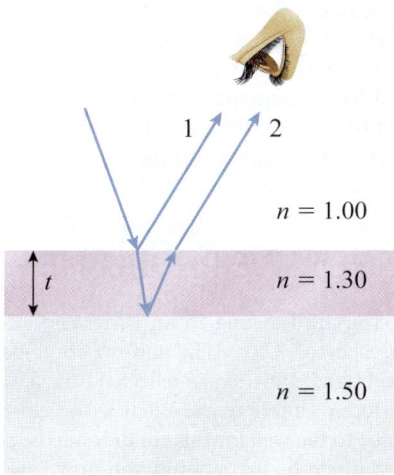

Figure 29-36 Problem 49.

50. ★★ As shown in Figure 29-37, a part of a soap bubble has varying thickness, being thicker lower down. White light (400 nm to 700 nm wavelength in air) enters approximately perpendicular to the soap film, and the bubble is viewed through reflected light. What must be the minimum soap film thickness in order to produce colours of red (assume 650 nm), yellow (589 nm), and violet (400 nm) light, where all wavelengths are expressed in vacuum? Also state which colour would appear higher in the diagram. You can observe effects like these in real soap bubbles and are urged to try this experiment at home.

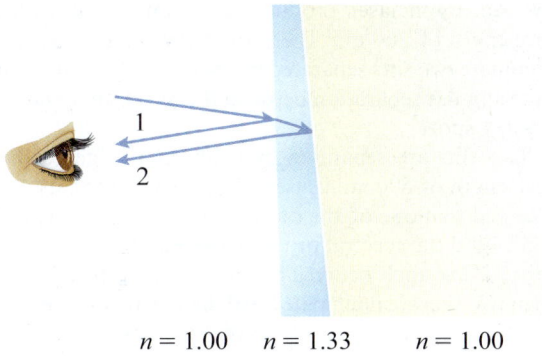

Figure 29-37 Problem 50.

51. ★★ Making reference to Figure 29-38, what is the smallest thickness (t) resulting in destructive interference when viewed from above? Assume monochromatic light of wavelength 550 nm (in vacuum).
   (a) Find the thickness assuming the angle of incidence is perpendicular to the surface.
   (b) Find the thickness if we don't make this assumption and the angle of incidence (to the perpendicular) is 15°.

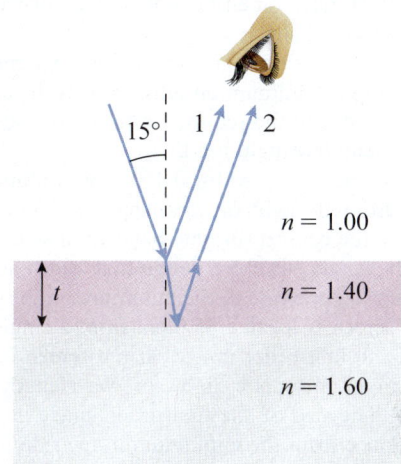

Figure 29-38 Problem 51.

52. ★★ Derive an expression for the phase difference (in radians) for the two light paths shown in Figure 29-39. Assume that $n_1 < n_2 < n_3$ and that near-perpendicular (to surface) incidence can be assumed.

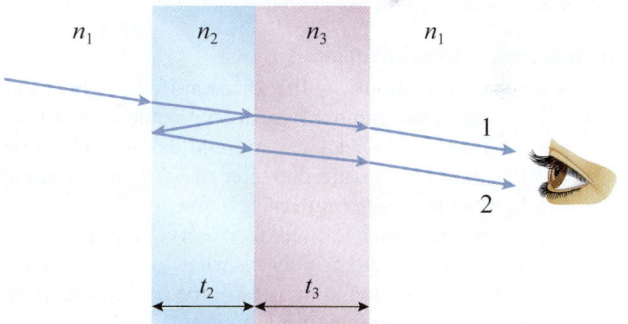

Figure 29-39 Problem 52.

53. ★★ You need to design an antireflection coating for infrared light with a wavelength of 1250 nm. The coating has an index of refraction of 1.38, and it is to be placed on glass with an index of refraction of 1.52.
   (a) What is the thinnest coating that you could use? Assume that the infrared beam is perpendicular to the coating.
   (b) What wavelength shorter than 1250 nm would produce constructive interference with the coating you designed in part (a)?

54. ★★ What is the minimum thickness of a soap bubble wall that will result in constructive interference of yellow light with a wavelength (in a vacuum) of 580 nm. Assume that the beam is approximately perpendicular to the soap bubble and that the index of refraction of the soapy water is $n = 1.33$.

Section 29-6 Single-Slit Diffraction

55. ★ Monochromatic light with a wavelength of 589 nm illuminates a single slit. Destructive interference appears at angles of 1.15° on either side of the normal. What is the width of the slit?

56. ★★ During a single-slit diffraction experiment, you want the first minimum on either side to be at an angle of 60°. Calculate the required ratio of the slit width to the wavelength of light used.

57. ★★ A single slit of width 0.450 mm is illuminated by monochromatic light of wavelength 589 nm. Find the width of the central (bright) band on a screen 50.0 cm from the single slit if we define that width as from the minimum on one side to the minimum on the other.

58. ★★ A screen is located 85.0 cm from a single slit. The size of the first-order image on the screen is 3.65 mm when monochromatic light of wavelength 454 nm is used. The size of the first-order image is defined as the distance from the minimum on one side to the minimum on the other side. What must be the width of the single slit?

Section 29-7 Actual Intensity Pattern for Double Slits

59. ★★★ Laser light with a wavelength of 633 nm is used to illuminate two slits separated by 0.125 mm. The width of each slit is 0.0150 mm. Assuming that only fringes between the first minima in the pattern are counted, how many bright fringes are visible?

Section 29-8 Resolution Limit

60. ★★ (a) If limited only by the wave nature of light, what would be the minimum angle that could be resolved (assume the Rayleigh criterion holds) by a telescope that has an aperture diameter of 15.0 cm? Assume light with a wavelength of 550. nm.

(b) If the telescope is used to observe a binary star system that is 16.5 ly away, what is the minimum separation of the stars (in kilometres) that can be resolved (assuming optimum orientation)?

61. ★★ You are driving at night when the pupil of your eye is 2.85 mm in diameter. A car's headlights are 1.45 m apart. How close must a distant car be for you to see two separate headlights? Assume a wavelength of 550.0 nm.

62. ★ If they had the same quality of optics and diameter, would a microscope that worked in the visible, the infrared, or the ultraviolet have the best resolution? Justify your answer.

63. ★★ A smartphone camera has a focal length of 4.15 mm and a focal ratio of 2.20. What would be the minimum angle, in degrees, resolvable with this camera? Note: The focal ratio of a lens is the ratio of the focal length to the diameter. Assume visible light with a wavelength range from 400 nm to 650 nm.

64. ★★ You want a camera to be able to resolve an object that is 25.0 cm in size from a distance of 850 m. The camera lens has a focal length of 50.0 mm. What should be the focal ratio of the lens? The focal ratio is the ratio of the focal length to the diameter. Assume that the camera works over wavelengths from 400 nm to 650 nm.

65. ★★ The James Webb Space Telescope uses a complex optical design with 18 hexagonal mirror segments, but you can assume that the effective diameter is 6.50 m. The telescope operates over wavelengths from 0.600 μm to 28.5 μm.

(a) What region(s) of the electromagnetic spectrum does the telescope operate in?

(b) If we assume that the telescope is diffraction limited, what is the minimum angle that could be resolved when the telescope is used at a wavelength of 28.5 μm? Express your answer in arcseconds (1 degree = 3600 arcseconds).

(c) If the telescope studied an astronomical object at a distance of 120 ly (light years), what would be the width of the smallest object that could be resolved? Express your answer in AU (astronomical unit, the mean Sun to Earth distance).

COMPREHENSIVE PROBLEMS

66. ★★ (a) You are listening to an FM radio station with a frequency of 92.0 MHz, and the combination of the direct signal and a signal reflected from a wall produces destructive interference. What is the closest that you can be to the wall for this condition to be possible?

(b) How much farther should you move away from the wall to reach a point of constructive interference?

67. ★★★ A Michelson interferometer is constructed such that one of the equal arms can hold gas or be evacuated. Each arm is 0.365 000 m long. An interference pattern is observed as a gas is added slowly, and it is observed that in going from evacuated to containing the gas a total of 205 fringes are observed. The interferometer uses a helium–neon laser with a wavelength of 633.000 nm. Find the index of refraction of the gas.

68. ★★ A sodium discharge tube produces results in a doublet of two relatively closely spaced spectral lines with wavelengths of 588.995 nm and 589.592 nm. You want these lines to be separated by 2.40 mm on a screen that is 50.0 cm from a diffraction grating. How many lines/cm should the diffraction grating have?

69. ★★ An argon laser produces ultraviolet light with a wavelength of 264 nm. Light from that laser is used to illuminate two slits separated by exactly 0.100 mm. What is the angular separation between the $m = 0$ spot and the $m = +1$ spot?

70. ★ Two slits are separated by 0.100 mm. Light with a wavelength of 500 nm is used to illuminate the slits. It is observed that one of the bright fringes is at an angle of 1.43°. Find the spectral order of this band.

71. ★★ ▱ The angle between the two $m = 2$ maxima is 1.30° when two slits are illuminated with light with a wavelength of 480 nm. Find the spacing between the slits.

72. ★★ A laser with a wavelength of 633 nm is used to illuminate a diffraction grating with 7500 lines/cm. What is the total number of bright spots that appear on a screen? (Hint: Remember that the maximum possible value of the angle is 90° and that spots appear on both sides of the zero-order maximum.)

73. ★★★ The Ocean Optics company makes a variety of miniature spectrometers, which are widely used in scientific labs. One version of this particular spectrometer uses a diffraction grating with 600 lines/mm and is capable of detecting light from 350 nm to 1000 nm in a first-order image.
    (a) The grating has a simple design (without mirror stages). What is the deflection angle for light with a wavelength of 350 nm? For a 1000 nm wavelength?
    (b) The detector used in this spectrometer has a row of 2048 pixels, each of which is 14 μm wide. What is the effective width of the detector, in millimetres?
    (c) Considering the angles determined in part (a) and the width of the detector, what is the effective distance from the grating to the sensor if the spectrometer does not include mirrors?

74. ★★ A thin liquid layer with index of refraction $n = 1.44$ floats on top of water ($n = 1.33$). What is the minimum thickness possible when light incident perpendicular to the surface is most strongly reflected at a wavelength of 540 nm?

75. ★★ Two glass plates with very accurate, flat surfaces are slightly inclined relative to each other, as shown in Figure 29-40. The separation, s, between the plates is 75.0 μm at one side and 0 at the opposite side. Monochromatic light with a wavelength of 532 nm in air is vertically incident. Due to interference between the light that reflects at the top and bottom of the varying air gap between the plates, you observe a series of light and dark fringes. How many bright regions do you see when viewing the plates from above?

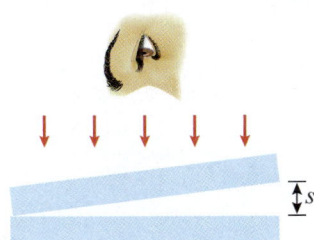

Figure 29-40 Problems 75 and 76.

76. ★★ For the two glass wedged plates in problem 75, the space between the plates is now filled with water ($n = 1.33$). How many bright fringes are there?

77. ★★ The Canada-France-Hawaii Telescope (CFHT) has an objective mirror 3.60 m in diameter. Assume a wavelength of 550.0 nm.
    (a) If only the wave nature of light limited resolution, what would be the limiting angular resolution (assuming the Rayleigh criterion)?
    (b) Astronomers often express angles in units of arcseconds (symbol ″), where $1″ = 1°/3600$. Convert your answer from (a) into arcseconds.

78. ★★ The Penticton radio telescope in Figure 29-1 has a diameter of 26.0 m. One of the main uses is to study electromagnetic emissions from atomic hydrogen at a frequency of 1.42 GHz. Calculate the angular resolution limit according to the Rayleigh criterion of the telescope at this frequency.

79. ★★ To improve effective resolution, separate radio telescopes are connected in interferometry arrays that approximate a single very large instrument. In 1967, this was done for synchronized signals recorded separately at the radio telescope in Penticton, British Columbia, and another radio telescope in Algonquin Park, Ontario, a spacing of approximately 3000 km. What is the approximate angular resolution achieved with this array? Assume the same radio frequency as for problem 78.

80. ★★ The Very Long Baseline Array (VLBA) uses 10 telescopes with a baseline extending 8611 km from Mauna Kea, Hawaii, to St. Croix, U.S. Virgin Islands. Find the angular resolution of this array at the extremes of the range of frequencies used: 1.20 GHz to 96.0 GHz.

81. ★★ Atmospheric scintillation, which makes stars appear to twinkle, limits the resolution of observations made through the atmosphere. A typical resolution limit from atmospheric scintillation is approximately 1.5″. What size of instrument aperture would have an equal resolution limit due to the wave nature of light? Assume a wavelength of 500 nm.

82. ★★ The Canadian NEOSSat mission utilizes a small satellite (about the size of a suitcase) in Earth orbit to detect asteroids near Earth. NEOSSat uses a telescope with a diameter of 15.0 cm. If there were no other limitations, what is the minimum separation of two sources that this telescope can resolve at a wavelength of 560 nm? Express your answer in units of arcseconds.

83. ★★★ It is desired to have an actual double slit with a spacing such that the third-order image does not appear (because it corresponds to a diffraction minimum from the width of the slits). If the separation between the centres of the slits is 0.125 mm and if light with a wavelength of 532 nm is used, what should the width of the slits be?

DATA-RICH PROBLEM

84. ★★★ 〰 In this question you will plot the relative intensity at different angles, θ, due to the illumination of a single slit of width w with monochromatic light of wavelength λ. The intensity function is given by the following:

$$I_0 \left\{ \frac{\sin\left(\dfrac{\pi w \sin\theta}{\lambda}\right)}{\left(\dfrac{\pi w \sin\theta}{\lambda}\right)} \right\}^2$$

where I_0 is the intensity at the centre and the angles are expressed in radians. (The function $\dfrac{\sin \pi x}{\pi x}$ is called the sinc function. It often appears in advanced physics, especially in signal processing.)

(a) Use a computer program or spreadsheet to calculate the intensity for angles from $-60°$ to $+60°$. First, use $w/\lambda = 4$. Plot your output and record the plot.

(b) Determine where the intensity approaches zero. Do these angles correspond to the angles predicted by the single-slit equation derived in Section 29-6?

(c) Now repeat part (a) for $w/\lambda = 8$ and then $w/\lambda = 2$. Compare the resulting plots with the one from part (a). How does the slit width affect the intensity pattern?

(d) Set $w/\lambda = 1$. What do you now observe?

OPEN PROBLEM

85. ★★★ You need to estimate the volume of a kerosene spill in tropical waters by observing it from the air. When the Sun is approximately overhead and you are nearly directly above the spill, you observe that the most strongly reflected light has a wavelength of approximately 500 nm. When viewed from 1200 m above the water, the spill subtends an angle of approximately 20° side to side. Estimate the volume of kerosene, clearly stating any assumptions.

CHAPTER

30 | Relativity

When you have completed this chapter, you should be able to

LO1 Differentiate between special and general relativity, and cite applications of each.

LO2 Describe the Michelson–Morley experiment, and explain its significance.

LO3 Use the postulates of special relativity to derive the time dilation relationship.

LO4 Solve problems related to length contraction and time dilation.

LO5 Use the Lorentz transformation for relativistic coordinate transformations.

LO6 Describe spacetime intervals, and apply them to solve relativity problems.

LO7 Explain mass–energy equivalence, and calculate relativistic momentum and kinetic energy.

LO8 Add relativistic speeds in special relativity situations.

LO9 Derive and apply the Doppler effect for light.

LO10 Explain and apply concepts of gravitational time dilation.

LO11 Explain how special and general relativity considerations affect the GPS system.

Global Positioning System (GPS) receivers are more than just a convenience for drivers (Figure 30-1). These devices are the core of modern navigation systems for planes and ships and of emergency beacons that broadcast their exact location. A GPS receiver system uses the time delays of signals from a series of satellites orbiting Earth to calculate the position of the receiver. Without corrections for relativistic effects, the GPS system would be hopelessly inaccurate. Relativity theory also underlies research and technology ranging from astrophysics to nuclear medicine to power plants.

Figure 30-1 A GPS system requires calculations using relativity theory to determine your precise location.

30-1 Special and General Relativity

Relativity is divided into two branches: special relativity and general relativity. Most of this chapter deals with special relativity, but we will introduce the basic concepts of general relativity near the end.

Special relativity describes how observers moving relative to each other measure different values for the same physical quantities. These differences are substantial when the relative speed is significant compared to the speed of light ($c = 3.00 \times 10^8$ m/s) but negligible for moderate relative speeds. Perhaps the most surprising consequence of special relativity is that time slows for moving observers, an effect called **time dilation**. Length measurements also depend on the frame of reference of the observer. Consequently, the concepts of kinetic energy and linear momentum developed earlier in this book need to be modified. Relativity theory treats space and time as components of a single four-dimensional entity called **spacetime**.

The speed of light appears to be the upper limit for the speeds of objects in our universe. The relativistic effects on time, length, momentum, and kinetic energy become apparent when the speed of an object approaches the speed of light. Another important consequence of relativity is that energy and mass are viewed as different aspects of a unified entity, and objects possess energy by virtue of their mass.

As the name implies, **general relativity** is a more general form of relativity, but it is also a radically new theory of gravitation. General relativity does not treat gravitation as a force but rather proposes that masses bend spacetime and that objects move according to the resulting curvature of spacetime. The curvature caused by Earth's mass is small, but its effects are measurable. The effects of the curvature near very dense objects, such as stellar cores and black holes, are extreme and rather surprising.

The concepts of relativity can be challenging because they seem contrary to common sense. However, countless experiments on Earth and observations of objects in space have validated the ideas of both special and general relativity. For example, measurements of the decay of unstable subatomic particles moving at speeds near to the speed of light have confirmed that time dilation occurs exactly as predicted by special relativity.

MAKING CONNECTIONS

Albert Einstein

The name Albert Einstein (1879–1955) is almost synonymous with scientific creativity and intellectual brilliance. Like Newton and many other famous scientists, Einstein did his most important work as a young scientist. In 1905, Einstein published five works, including key papers on relativity, Brownian motion, and the photoelectric effect. Einstein was then only 26 years old, and his ideas were developed when he was even younger (Figure 30-2).

Einstein's contribution to our understanding of the photoelectric effect, for which he later received a Nobel Prize, helped set the foundation for quantum mechanics. However, Einstein is best known for the development of relativity theory. He published the main elements of special relativity in 1905, including a paper that developed the famous $E = mc^2$ mass–energy equivalence, although the term *special relativity* was not introduced until later. Approximately 10 years later, he finished work on a generalization of relativity that incorporated gravitation. This version is what we now call general relativity.

Figure 30-2 Albert Einstein at about the time he developed special relativity.

Lucien Chavan

CHECKPOINT 30-1

Special and General Relativity

Identify the branch of relativity theory associated with the following concepts:

(a) Gravitational effects result from the curvature of spacetime.

(b) Time passes at a different rate for observers in relative motion at speeds near the speed of light.

(c) Objects cannot travel faster than the speed of light.

(d) Mass has an energy equivalent.

ANSWER: (a): general relativity; (b), (c), and (d): special relativity.

30-2 Reference Frames and the Michelson–Morley Experiment

The story of relativity begins with the speed of light. As you saw in Chapter 27, Maxwell's elegant set of equations summarizing electricity and magnetism led directly to a relationship between the speed of electromagnetic waves in a vacuum (c) and the product of the permeability of free space (μ_0) and the permittivity of free space (ε_0):

$$c = \frac{1}{\sqrt{\varepsilon_0 \mu_0}} \qquad (27\text{-}20)$$

Thus, the fields of electricity, magnetism, and optics had been unified. In this chapter, we will see that the speed of light has a key role in mechanics as well.

In the second half of the 19th century, physicists puzzled over the reference frame for measuring the speed of light. Any speed must be measured relative to something. Since electromagnetic waves propagate in a vacuum (for example, we get energy radiated from the Sun through space, which is largely a vacuum), the reference frame was not obvious. One proposed solution was the theory that electromagnetic waves travelled relative to an **ether**, a massless, invisible medium present even in a vacuum. So the search was on for the reference frame of this hypothetical ether. The most obvious way to establish this reference frame was to carefully measure the speeds of electromagnetic waves transmitted when moving in different directions relative to the ether. Differences in these speeds would indicate the speed of the ether relative to the observer.

Several attempts were made to detect the flow of the ether, the most famous being the interferometer experiments conducted by Albert Michelson and Edward Morley, working at what is now Case Western Reserve University in Cleveland, Ohio. Interferometers can measure very small changes in distance (comparable to the wavelength of light) and, if the distances are held fixed, correspondingly small differences in the speed of light. Section 29-2 described the principles of interferometers and the basics of the design developed by Michelson and Morley.

Figure 30-3 shows a simplified version of the experiment. As described in Chapter 29, the Michelson interferometer uses a half-silvered mirror to split the light from a source into two beams that travel different paths, and then recombines the beams to produce an interference pattern. In this experiment, the paths in the perpendicular arms of the interferometer have exactly the same length. By varying the orientation of the apparatus relative to the motion of Earth around the Sun, Michelson and Morley hoped to determine the ether reference frame. Thus, the **Michelson–Morley experiment** is an interferometer-based attempt to measure the reference frame for electromagnetic waves such as light.

Assume that the ether frame moved at speed v from right to left, as shown in Figure 30-4. Then the beam that travels in the arm *parallel* to the direction of the ether would move at speed $c + v$ when moving to the left and at speed $c - v$ when moving to the right. The time for the two-way trip is then

$$t_{\parallel} = \frac{d}{c - v} + \frac{d}{c + v} \qquad (30\text{-}1)$$

where d is the distance travelled.

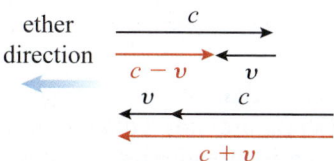

Figure 30-4 Light moving parallel and antiparallel to an assumed ether reference frame.

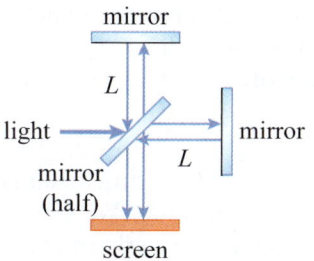

Figure 30-3 Simplified version of the Michelson–Morley experiment.

CHECKPOINT 30-2

Movement in the Ether

If the ether were moving horizontally at a speed of $0.25c$, what would be the total time required to travel a horizontal distance Δx and back?

(a) $\dfrac{7\Delta x}{4c}$

(b) $\dfrac{2\Delta x}{c}$

(c) $\dfrac{32\Delta x}{15c}$

(d) $\dfrac{9\Delta x}{4c}$

ANSWER: (c) When travel is in the direction of the ether flow, the time is $\dfrac{\Delta x}{0.75c}$. For a direction opposite to the ether flow, the time is $\dfrac{\Delta x}{1.25c}$. Adding these times gives a total travel time of $\dfrac{32\Delta x}{15c}$.

MODERN PHYSICS

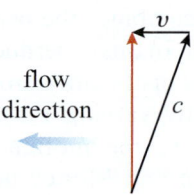

Figure 30-5 Movement perpendicular to an ether reference frame.

Now consider the case when the ether flow is perpendicular (Figure 30-5). Analogous to rowing a boat across a slow-moving river, if you try to row directly across the river, the current will push you slightly upstream or downstream, and you will not end up exactly opposite where you started. Similarly, the beam in the arm perpendicular to the ether flow would have to be aimed at an angle as shown in Figure 30-5 to compensate for the displacement caused by the ether flow. Applying the Pythagorean theorem gives an effective speed of $\sqrt{c^2 - v^2}$ in each direction. Therefore, the total travel time in the perpendicular arm is

$$t_\perp = \frac{2d}{\sqrt{c^2 - v^2}} \qquad (30\text{-}2)$$

We can show that the travel times in Equations (30-1) and (30-2) are not equal (see problem 23). Therefore, a precise enough experiment would show the direction and speed of the flow of ether. Since Earth moves at approximately 30 km/s in its orbit around the Sun, Michelson and Morley hoped that orienting their interferometer with one arm parallel to the direction of this motion would result in a detectable difference between the speeds of the two light beams.

Michelson detected no difference when he first tried the experiment in 1881 while working in Germany. After he moved to the United States, Michelson worked with Morley to substantially increase the precision of the equipment, but their experiments in 1887 still found no difference in light speed for the two paths. Einstein, who became interested in the ether question while still in high school, wanted to try the same experiment when he was an undergraduate at university in Zurich, Switzerland, but his professor would not let him use the university laboratory for the experiment. Modern experiments with extremely high precision have confirmed Michelson and Morley's null result.

At the time, various attempts were made to "explain" the null result. One was that the movement of the ether relative to Earth was simply too little (they could measure a movement of about 2 km/s). Some physicists argued that an ether was present but dragged along with Earth, but that seemed unlikely. Another

possibility proposed was that the ether compressed objects, cancelling out the interference results sought. However, the simplest interpretation is that there is no ether, and therefore all observers would measure the same value for the speed of light. As we will see in the next section, the idea that the speed of light does not depend on the reference frame is the starting point for the special theory of relativity.

30-3 Postulates of Special Relativity and Time Dilation

Special relativity applies for **inertial reference frames**, which are reference frames in which Newton's laws are valid. We can in general consider these to be unaccelerated reference frames. Examples of non-inertial reference frames are rotating and linearly accelerated reference frames.

To derive special relativity, Einstein made two assumptions:

- **Observers in different inertial reference frames agree on the *laws* of physics**, even though they may measure different *values* for physical quantities. Although we cannot prove that physical laws are the same throughout the universe, so far, there is no evidence to contradict this assumption.

- **All inertial observers measure the same value for the speed of light in a vacuum.** For example, if you were on a spaceship passing a planet at half the speed of light and shone a laser beam in front of you, you and an observer on the planet would both measure the same speed for the laser light. If there is no ether, based on the most obvious interpretation of the Michelson–Morley experiment, then there is nothing for different observers to measure the speed of light against, and all should measure the same value for the speed of light in a vacuum.

We will use the term *event* frequently in this chapter. In relativity theory, an event is something that happens independent of the space and time reference frame used to describe it. Different reference frames assign different space and time values to the same event. Some examples of events are an atom emitting a photon, a radioactive nucleus emitting an α-particle, a flare being fired, and two particles colliding with each other.

Time Dilation

Now, we shall see how Einstein's assumptions led to an amazing result: time dilation. We start with a thought experiment. In a **thought experiment**, you use the results of practical experiments to predict how a hypothetical

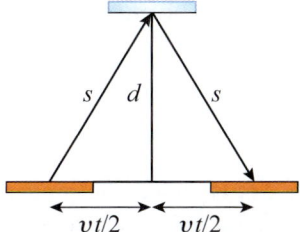

Figure 30-8 A clock moving to the right at speed v relative to an observer.

MAKING CONNECTIONS

Olympia Academy

Einstein developed special relativity while he was working at the patent office in Bern, Switzerland. He became life-long friends with two of his co-workers, Maurice Solovine and Conrad Habicht. The three men spent many hours discussing ideas of physics and philosophy, and they called their discussion group the Olympia Academy. Einstein's first wife, Mileva Marić, sometimes took part in the discussions of the Olympia Academy. Marić and Einstein studied physics together at the Zurich Polytechnic.

Source: (Left): Vollenweider und Sohn (Bern), ca 1903/e-pics; (Right): ETH-Bibliothek Zürich, Bildarchiv / Fotograf: Unbekannt / Portr_03106 / Public Domain Mark

Figure 30-6 Members of the Olympia Academy, from left to right: Conrad Habicht, Maurice Solovine, and Albert Einstein. Shown on the right is Mileva Marić.

experiment would turn out, especially when you cannot actually perform the hypothetical experiment. Einstein liked thought experiments. Newton used thought experiments in some of his work, as did some of the founders of quantum mechanics.

Consider a clock based on the time that light takes to travel some fixed distance, as shown in Figure 30-7. In the figure, we have a light pulse that starts at the bottom, travels a distance d to the top, where it is reflected from a mirror, and then is detected when it reaches the bottom again. We will assume that the light travels in a vacuum inside the clock and that the clock is stationary relative to the reference frame being used for measurements. We can treat the time that the light takes to travel from the bottom to the mirror and back again as a unit of time. For a clock at rest in our reference frame, we will call this time t_0:

$$t_0 = \frac{2d}{c} \qquad (30\text{-}3)$$

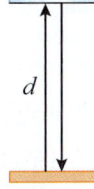

Figure 30-7 A clock based on the time for light to travel a fixed distance. In this case, the clock is not moving relative to an observer.

Now imagine the same clock moving at speed v to the right relative to some observer. Since the entire clock is moving at the same constant speed, the light beam will still reflect from the same point on the mirror, but the distance, s, travelled by the light will be somewhat longer relative to the observer, as shown in Figure 30-8. We will call the time (for one complete round-trip light pulse to the mirror and back) measured by this clock t. Our goal here is to obtain a relationship between the times measured by the two clocks, t_0 and t.

Using the Pythagorean theorem, for the second clock we can obtain the following relationship for the distance s:

$$s = \sqrt{d^2 + \frac{v^2 t^2}{4}} \qquad (30\text{-}4)$$

Remember that the time for one-half of the two-way light trip for this moving clock is $\frac{t}{2}$.

According to the second assumption of special relativity, an observer in the reference frame of the moving clock measures the same value for the speed of light, c. Therefore, the travel time for the light pulse in the moving clock is given by

$$t = \frac{2s}{c} = \frac{2\sqrt{d^2 + v^2 t^2/4}}{c} = \frac{\sqrt{4d^2 + v^2 t^2}}{c} \qquad (30\text{-}5)$$

If we square both sides of this expression and then substitute for d from Equation 30-3, we obtain

$$t^2 = \frac{4d^2 + v^2 t^2}{c^2} = \frac{c^2 t_0^2 + v^2 t^2}{c^2} = t_0^2 + \frac{v^2}{c^2} t^2 \qquad (30\text{-}6)$$

We rearrange this to obtain the following equation relating the times as measured by the two clocks:

$$t = \frac{t_0}{\sqrt{1 - v^2/c^2}} \qquad (30\text{-}7)$$

Usually, it is the intervals between events that are important, for example, the time between a spaceship starting and ending a trip or between the formation and decay of a subatomic particle. Therefore, the time

MODERN PHYSICS

dilation relationship is normally written in terms of time intervals:

KEY EQUATION

$$\Delta t = \frac{\Delta t_0}{\sqrt{1 - v^2/c^2}} \qquad (30\text{-}8)$$

The expression Δt_0 is called the **proper time interval**. The **proper time** is the time measured by the observer who records the two events taking place at the *same* spatial coordinates.

Observers are at rest in their own frame of reference. Therefore, proper time is measured by an observer who is at rest relative to that clock. Said another way, proper time is the elapsed time between two events as measured by a clock that coincides with both events. The proper time of a moving object is the time measured by a clock that moves with that object.

The relationship given by Equation 30-8 is called **special relativity time dilation**, which is relevant when one observer is moving at a constant velocity relative to another observer. The word *dilation* refers to a stretching or shrinking, and we see that time itself is stretched in special relativity. Later in the chapter, we will see that there is also a different time dilation effect related to strong gravitational fields.

Some texts use the term **coordinate time** for Δt. If we considered a tightly spaced grid of clocks at different spatial positions and read the time off the clocks nearest the beginning and the ending spacetime events, that would give us the coordinate time for the interval. Since different inertial coordinate systems (in constant velocity, one relative to the other) would show different values for the interval, it is called coordinate time.

PEER TO PEER

I must remember that the subscript 0 notation in this chapter, such as Δt_0 for the proper time interval, refers to the reference frame used and does *not* mean initial time.

Although an observer in any other reference frame does not measure the proper time, measurements in that observer's reference frame are still self-consistent and equally valid. The light emission and light detection events occur at the same position in our reference frame in the stationary light clock in Figure 30-7, but not for the moving clock in Figure 30-8.

Note that the denominator in Equation (30-8) is less than 1 since $v < c$. Therefore,

$$\Delta t > \Delta t_0 \qquad (30\text{-}9)$$

Thus, the proper time interval for any pair of events is less than the coordinate time interval measured in any other reference frame. Strictly speaking, this conclusion applies only to time intervals measured using the passage of light beams because our derivation is based on a clock that uses that process. However, repeated experiments have shown that time dilation affects every process that physicists have been able to test. It appears that moving clocks of any type—mechanical, electronic, biological, chemical, and atomic—run slower. In other words, time itself runs slower and affects all time-related processes. Consequently, an observer cannot detect this slowing by measuring only objects in his or her own reference frame.

CHECKPOINT 30-3

Proper Time

If a spaceship travels from Earth to a planet in the Alpha Centauri star system, which observer measures the proper time for the trip?
(a) an observer on Earth
(b) an observer on the planet in the Alpha Centauri star system
(c) an observer on the spaceship
(d) an observer deep in space far from the planet and Earth

ANSWER: (c) For an observer on the spaceship, the trip starts and ends at the same spatial coordinates; therefore, the spaceship reference frame gives the proper time measurement.

EXAMPLE 30-1

Time Dilation for the Lifetime of a Muon

A muon is an unstable subatomic particle with the same charge as the electron, but its mass is approximately 207 times the mass of the electron. Muons at rest decay in 2.20 μs on average. Cosmic rays hitting Earth's atmosphere can create muons travelling near the speed of light. Assume that one of these muons has a speed of 0.990c. Note: The reference frames are not strictly inertial due to the rotation of Earth, but we can neglect that in this question since the associated rotational speed is much less than the speed of light.

(a) What lifetime does an observer on the ground measure for the muon?
(b) How far does the muon travel according to the observer?

Solution

(a) The pair of events birth and decay define the lifetime of the muon. Clearly, the beginning and end of this lifetime have the same location only in a reference frame fixed to the muon. Therefore, an observer in the muon's reference frame measures the proper time, and

$$\Delta t_0 = 2.20 \times 10^{-6} \text{ s}$$

We want to find the time interval Δt for the ground-based observer. The speed of the muon relative to the ground-based observer is $v = 0.990c$. Applying the time dilation equation gives

$$\Delta t = \frac{\Delta t_0}{\sqrt{1 - v^2/c^2}} = \frac{2.20 \times 10^{-6} \text{ s}}{\sqrt{1 - (0.990)^2}} = 1.56 \times 10^{-5} \text{ s}$$

(b) The ground-based observer sees a muon with the lifetime calculated in part (a) travelling at a speed of 0.990c. Therefore, the observer measures the distance travelled to be

$$\Delta x = 0.990 \cdot 3.00 \times 10^8 \text{ m/s} \cdot 1.56 \times 10^{-5} \text{ s} = 4630 \text{ m}$$

Making sense of the result

The muon is moving very fast relative to the observer, so the observer sees time for the muon substantially dilated. So it is reasonable that an observer in Earth's reference frame measures a lifetime that is almost an order of magnitude longer than that of the muon reference frame. Observations of high-energy muons in the upper atmosphere do indeed confirm the time dilation predicted by special relativity.

The Twin "Paradox"

Consider a thought experiment involving identical twins who separate: one stays on Earth, and the other leaves in a spaceship travelling at a constant speed near the speed of light, and then reverses direction to return to Earth at the same speed. If we consider the time to make a one-way trip, it is the twin who travels who measures the proper time, since for that twin the start and finish of the trip is at the same coordinate. According to Equation 30-9, $\Delta t > \Delta t_0$, so the twin who stays on Earth must measure a longer time for the trip. The same argument holds for the return trip, since it is only the speed that enters into the relationship, and therefore we would expect the twin on Earth to be older than the one who had been on the trip.

But now let us take a viewpoint from a different reference frame, since motions are relative. If we viewed the situation from the reference frame of the travelling twin, we could view Earth move away at a high speed, then reverse direction, and then approach at the same speed. From this perspective, it is the other twin who measures the proper time. In this point of view, the twin who stays on Earth expects to be younger when they meet. At first glance, it seems paradoxical that both twins expect the other to be younger, depending on the point of view. This is called the **twin paradox**.

The resolution of the paradox lies in the fact that the motion of the twins is not symmetrical. The twin who stays on Earth is always in the same reference frame, and the travelling twin undergoes accelerations: when leaving Earth, when turning around, and when arriving back at Earth. She is on one reference frame during the trip out and a different reference frame on the return, with periods of acceleration in between. Special relativity applies only to inertial reference frames, and the travelling twin is not in an inertial reference frame when accelerating. It is the twin who experiences the accelerations who will be younger. One way to look at the twin paradox is as follows: according to general relativity, we cannot differentiate between an accelerated frame of reference and a frame of reference in a gravitational field. Later in the chapter you will see that time passes more slowly in strong gravitational fields.

Often in relativity problems we define a dimensionless **speed parameter**, β, as the ratio of the speed to the speed of light:

$$\beta \equiv \frac{v}{c} \tag{30-10}$$

With this definition, Equation 30-8 can be rewritten as

$$\Delta t = \frac{\Delta t_0}{\sqrt{1 - \beta^2}} \tag{30-11}$$

You will use this speed parameter in Interactive Activity 30-1 to explore the twin paradox.

INTERACTIVE ACTIVITY 30-1

Twin Paradox

In this activity, you will use the interactive simulation "Twin Paradox," which uses the Java program written by Kiril Simov and Wolfgang Christian, available at www.compadre.org. You will be presented with a spacetime diagram of a trip along with clocks showing time as measured by two observers. You can adjust parameters such as the distance travelled and the speed. Answer the series of questions to increase your understanding of time dilation, proper time, and inertial reference frames.

The predictions of time dilation are tested every day in the more than 30 000 particle accelerators in operation around the world. The most impressive particle accelerator is the Large Hadron Collider (LHC) (see Figure 30-9), which as of 2015 produced proton–proton collisions with a total energy of about 14 TeV. At this energy, the protons move at a speed of about 0.999 999 991c, so relativistic time dilation effects are

Figure 30-9 A portion of the Large Hadron Collider (LHC). This CERN-operated 27 km circumference particle accelerator is at the Switzerland–France border near Geneva.

Testing Time Dilation with Atomic Clocks

The most direct test of time dilation is to use precise clocks in an experiment analogous to the Twin Paradox experiment. This was first done in October 1971 when physicist Joseph C. Hafele (1933–2014) and astronomer Richard E. Keating (1941–2006) took precise cesium beam atomic clocks on commercial aircraft. They compared the time on clocks that stayed at the U.S. Naval Observatory to time on similar clocks that flew twice around the Earth on the commercial aircraft (see Figure 30-10).

They tried flights in both directions around the world. It turns out that both special and general relativity needed to be taken into account, since the clocks on the flight were moving and also at a height with a weaker gravitational field. The results of the experiment agreed with relativity theory within the precision of the experiments. For example, for the westward flights it was predicted that the time difference would be 275 ns, and the observed value was 273 ± 7 ns. More precise repetitions of the experiment were conducted in 1996 and 2010, again showing agreement between experiment and relativity. Interestingly, in 2015 NASA astronaut twins Scott and Mark Kelly underwent a space twin experiment, in which Scott Kelly spent a year in flight on the International Space Station while his identical twin, also an astronaut, was on Earth. While this

Figure 30-10 Joseph C. Hafele (left) and Richard E. Keating on a commercial aircraft with two of the atomic clocks used in the time dilation experiment.

was used to test various biological and psychological aspects of space flight, body clocks are not precise enough to test special relativity.

huge. As outlined in Example 30-1, cosmic rays produce energetic particles that travel at speeds comparable to the speed of light. Special relativity successfully predicts the behaviour of these particles, as well as of those produced in particle accelerators. Atomic clocks have been used several times to directly test time dilation (see Making Connections: Testing Time Dilation with Atomic Clocks).

Time dilation—the idea that time itself passes at a different speed for observers in relative motion—is a truly remarkable result. The fact that we can predict the exact numerical relationship using only a thought experiment and two postulates is impressive. Time dilation is confirmed in particle accelerators every day. As you will see in the final section of this chapter, time dilation is also crucial to a workable GPS system. But it is not just time that depends on the motion of the observer, and in the next section we show that lengths in the direction of motion will be contracted.

30-4 Length Contraction

In this section, we will see that special relativity also predicts that when an object is moving relative to an observer, the length of the object in the direction of motion contracts. To understand length contraction,

we first define **proper length** in a manner analogous to proper time: An observer measures the proper length, L_0, of an object only if the object is *not* in motion relative to the observer's reference frame. An observer in any other inertial frame measures a different length, L.

Consider a hypothetical trip by a spaceship travelling at a constant speed, v, between exoplanets A and B, which are not moving relative to each other. (We will approximate the situation as inertial reference frames, although actual exoplanets do not have exactly inertial frames of reference due to motion about their parent stars, galactic rotation, and expansion of the universe.) An observer on each exoplanet measures the proper length for the distance between the planets, L_0, but does not measure the proper time, because the beginning and the end of the trip occur at different spatial coordinates in the reference frame of the planet (top of Figure 30-11). Instead, a planetary observer measures a dilated time, Δt. An observer on the spaceship (bottom of Figure 30-11) measures the proper time, Δt_0, because the start and end of the trip do occur at the same spatial coordinates in this reference frame. However, this observer does not measure the proper length because the spaceship moves relative to A and B.

According to the planet-based observer, the speed of the spaceship is

$$v = \frac{L_0}{\Delta t} \qquad (30\text{-}12)$$

Similarly, from the point of view of the spaceship observer, the speed for the trip is

$$v = \frac{L}{\Delta t_0} \qquad (30\text{-}13)$$

Consider two inertial reference frames, one moving at speed v relative to the other. For example, if observer A thinks that B is moving at speed v along the $+x$-direction, then the symmetry of the situation requires that observer B think that A must be moving in the $-x$-direction. Although they disagree on direction, they will agree on the speed. This means that we can use Equations 30-12 and 30-13 to obtain the following relationship:

$$L = \frac{L_0 \Delta t_0}{\Delta t} \qquad (30\text{-}14)$$

When we substitute the time dilation result (Equation 30-8) into Equation 30-14, we obtain the following **length contraction** relationship:

KEY EQUATION

$$L = L_0 \sqrt{1 - v^2/c^2} \qquad (30\text{-}15)$$

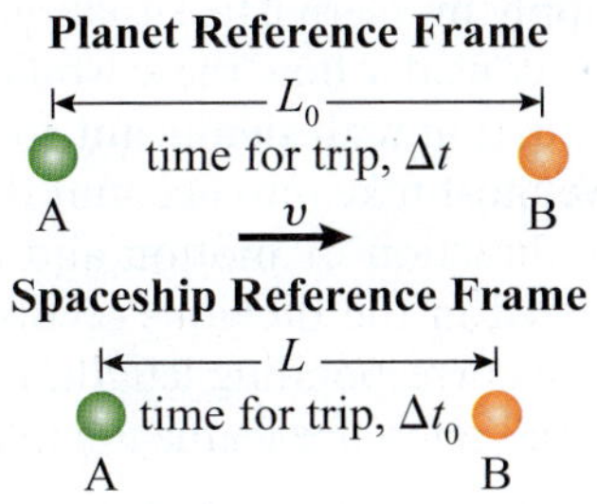

Planet Reference Frame

$$\vdash\!\!\!-\!\!\!-\!\!\!-\,L_0\,-\!\!\!-\!\!\!-\!\!\dashv$$
time for trip, Δt
A $\qquad v \qquad$ B

Spaceship Reference Frame

$$\vdash\!\!\!-\!\!\!-\,L\,-\!\!\!-\!\!\dashv$$
time for trip, Δt_0
A $\qquad\qquad$ B

Figure 30-11 From a reference frame on one planet (top), a trip between the planets will take time Δt and be of distance L_0. A spaceship observer will measure time Δt_0 and distance L.

EXAMPLE 30-2

Spaceship Length Contraction

A spaceship is 25.0 m long as measured in its own reference frame. What length do you measure as it moves past you at 90% of the speed of light?

Solution

First, ask yourself, Who measures the proper length? It is an observer on the spaceship. So, $L_0 = 25.0$ m and $v = 0.900c$. Using the length contraction relationship, we have

$$L = L_0 \sqrt{1 - v^2/c^2} = (25.0 \text{ m})\sqrt{1 - 0.900^2}$$
$$= 10.9 \text{ m}$$

Making sense of the result

As expected, the spaceship moving past you is contracted in length in the direction of motion. The value you measure is less than the proper length of the spaceship.

Remember that length contraction takes place only in the direction corresponding to the motion of the object.

Special relativity makes it possible (theoretically at least) to travel to distant exoplanets within a human lifetime. For example, if you travel at 99% of the speed of light, a trip that is 10 yr (years) in duration for an Earth-based reference frame will only be approximately 1.4 yr to an observer on the spaceship. However, you will learn later that the energy that a spaceship requires to reach speeds close to the speed of light is huge. There is an additional issue related to the physiological effects of acceleration to get to the high relativistic speeds, as well as the time required for reasonable accelerations.

For problems involving space travel, it is often convenient to work in distance units of light years and instead of writing it as ly, to express it as c yr. This allows us to divide out, for example, the speed of light, c. We also express all speeds as a fraction of the speed of light and use years as the time unit. This approach to units is demonstrated in Example 30-3. Of course, you can equally well convert all quantities to SI units.

PEER TO PEER

In doing relativistic trip type problems, I find the most important thing is to keep in mind the definitions of proper length and proper time. The person on the trip measures the proper time (if the time interval is the trip), but a different observer not moving with respect to the end points of the trip measures the proper length.

EXAMPLE 30-3

Space Travel

A spaceship travels at 98.0% of the speed to light to an exoplanet 45.0 ly away.

(a) What distances do observers on Earth and on the spaceship measure for the trip?

(b) What time does an observer on Earth measure for the one-way trip?

(c) What time do the occupants of the spaceship measure for the one-way trip?

Note that we are ignoring the times required for the acceleration stages.

Solution

The spaceship traveller measures the proper time for the trip because the start and end occur at the same place in the reference frame of the spaceship.

(a) The Earth-based observer measures the proper length for the trip, 45.0 ly. We use the length contraction relationship to find the length for the trip as measured by an observer on the spaceship. For convenience, we will use light years as the unit for length:

$$L = L_0\sqrt{1 - v^2/c^2} = (45.0 \text{ ly})\sqrt{1 - 0.980^2} = 8.95 \text{ ly}$$

(b) We now use the notation suggested earlier: an observer on Earth says the distance is 45.0 ly = 45.0c yr, and we know the speed is 0.980c, so the time required is

$$\frac{45.0c \text{ yr}}{0.980c} = 45.9 \text{ yr}$$

(c) To find the time as measured by occupants of the spaceship, we can use the distance result from part (a):

$$\Delta t = \frac{8.95c \text{ yr}}{0.98c} = 9.13 \text{ yr}$$

Alternatively, we can apply the time dilation relationship:

$$\Delta t = \frac{\Delta t_0}{\sqrt{1 - v^2/c^2}}$$

$$\Delta t_0 = \Delta t\sqrt{1 - v^2/c^2} = (45.9 \text{ yr})\sqrt{1 - 0.98^2} = 9.13 \text{ yr}$$

Making sense of the result

The length contraction and time dilation approaches for finding the trip duration in part (c) yield the same value, as expected.

30-5 Lorentz Transformation

While the techniques for relativistic computation of lengths and time intervals covered up to this point are perfectly sufficient for solving problems involving space and time in special relativity, it is often convenient to use a transformation between the space and time measurements of different inertial observers.

Let us first consider the transformation applicable to an everyday non-relativistic situation. As shown in Figure 30-12, a primed reference frame is moving at speed v in the positive x-direction compared to the unprimed coordinate system. The two coordinate systems are coincident at $t = 0$. Assume that some spacetime event occurs (e.g., at the dot in Figure 30-12).

From the original coordinate system, we could give it spacetime coordinates (t, x, y, z) and from the primed system we could use coordinates (t', x', y', z').

Remembering that the two reference frames are coincident at $t = 0$, and that when relativity is not considered all observers see time passing at the same rate, we have the following classical (non-relativistic) transformation coordinates in the two reference frames:

$$\begin{aligned} x' &= x - vt & x &= x' + vt \\ y' &= y & y &= y' \\ z' &= z & z &= z' \\ t' &= t & t &= t' \end{aligned} \qquad (30\text{-}15)$$

These non-relativistic transformation relationships are frequently called the Galilean or Newtonian transformation. Note that the transformations of Equation 30-15 do *not* apply in relativistic situations.

Let us now consider how these would be altered if we were in a situation with significant special relativity corrections. We must take into account the length contraction in the direction of motion and the time dilation. As explained in the previous section, the proper length, L_0, and a corresponding length, L, as measured in another inertial reference frame are related by

$$L = L_0\sqrt{1 - v^2/c^2} \qquad (30\text{-}16)$$

The primed system would measure the length (from its coordinate origin to the spacetime event) as $x - vt$ if there were no relativistic effects, but this will

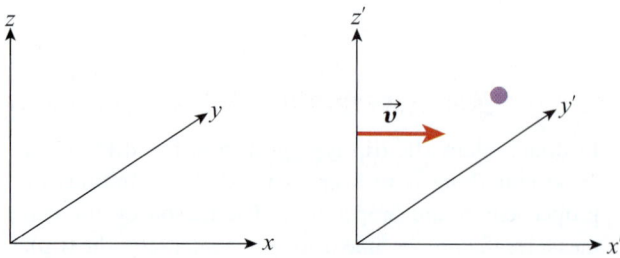

Figure 30-12   A primed coordinate system moves at speed v in the positive x-direction relative to the unprimed coordinate system. At $t = 0$ the reference frames are coincident. The purple dot represents a spacetime event that can be measured using either coordinate system.

be modified by the effects of length contraction, which makes the object appear longer in the moving primed system than observed in the unprimed system, yielding the following transformation relationship:

$$x' = \frac{x - vt}{\sqrt{1 - v^2/c^2}} \tag{30-17}$$

The symmetry of the situation of the two inertial reference frames will yield the relationship for the x' to x transformation by replacing v with $-v$. Since lengths are only contracted in the direction of motion, we expect no change in the y- and z-values between the two reference frames.

The derivation for the time transformation is a bit more complex. As expected, there will be the time dilation term (see Equation 30-8), but we also must take into account the difference in spatial coordinates since the two clocks were synchronized. These are shown in the following equation, where the numerator is due to the change in spatial coordinates and the denominator is due to the time dilation;

$$t' = \frac{t - vx/c^2}{\sqrt{1 - v^2/c^2}} \tag{30-18}$$

We can do several checks on the transformations of Equations 30-17 and 30-18. If $v \ll c$, relativistic effects should not be significant, and the transformations should, and do, reduce to the Galilean relationships.

The complete transformations are given in Equation 30-19, below, with the left set of equations used when going from the unprimed to the primed spacetime coordinates, and the right set when moving from primed to unprimed. These are called the Lorentz transformation (sometimes written plural) after Dutch physicist Hendrik A. Lorentz (1853–1928). Lorentz believed in the ether and developed the transformation prior to relativity in an attempt to reconcile observations on electromagnetic waves. Einstein interpreted the transformations in terms of the integration of space and time, and they became a key part of special relativity:

$$
\begin{array}{ll}
x' = \dfrac{x - vt}{\sqrt{1 - v^2/c^2}} & x = \dfrac{x' + vt'}{\sqrt{1 - v^2/c^2}} \\[2ex]
y' = y & y = y' \\[1ex]
z' = z & z = z' \\[1ex]
t' = \dfrac{t - vx/c^2}{\sqrt{1 - v^2/c^2}} & t = \dfrac{t' + vx'/c^2}{\sqrt{1 - v^2/c^2}}
\end{array}
\tag{30-19}
$$

The transformations can be written in a particularly elegant fashion if we use matrix multiplications on four-dimensional spacetime. Since the four dimensions should have common dimensions, it is natural to use ct instead of just time. It also allows them to

EXAMPLE 30-4

Transforming Spacetime Coordinates

There are two inertial reference frames, with the primed system travelling in the positive x-direction (similar to Figure 30-12) at a speed corresponding to $\beta = \sqrt{3}/2$. In the unprimed coordinate system, a spacetime event is at $t = 0$, $x = 100$. m, $y = 200$. m, $z = 300$. m. What will be the x'-coordinate of this spacetime event?

Solution

From the β-value, we know that the speed of the primed reference frame is $v = \sqrt{3}c/2$. Therefore,

$$x' = \frac{x - vt}{\sqrt{1 - v^2/c^2}} = \frac{100 - v \times 0}{\sqrt{1 - 3/4}}\,\text{m} = \frac{100.}{1/2}\,\text{m} = 200.\,\text{m}$$

Making sense of the result

Note that the denominator term is dimensionless, and the units in the numerator are consistent with units of length as required. Since the primed system is moving in the positive x-direction, we expect x' to be less than x, which it is. The question did not ask for y' and z', but they would be unchanged since the motion is in the x-direction.

be written more concisely if we use the β speed factor defined in Equation 30-10 and also define a **Lorentz factor** according to the following relationship:

KEY EQUATION

$$\gamma \equiv \frac{1}{\sqrt{1 - v^2/c^2}} = \frac{1}{\sqrt{1 - \beta^2}} \tag{30-20}$$

With these definitions, the Lorentz transformation from the unprimed to the primed spacetime coordinates is

MODERN PHYSICS

$$\begin{pmatrix} ct' \\ x' \\ y' \\ z' \end{pmatrix} = \begin{pmatrix} \gamma & -\beta\gamma & 0 & 0 \\ -\beta\gamma & \gamma & 0 & 0 \\ 0 & 0 & 1 & 0 \\ 0 & 0 & 0 & 1 \end{pmatrix} \begin{pmatrix} ct \\ x \\ y \\ z \end{pmatrix} \quad (30\text{-}21)$$

Standard matrix multiplication rules show that this is consistent with the first column of Equation 30-19 (see Example 30-4). If we replace v with $-v$ we obtain the inverse transformation:

$$\begin{pmatrix} ct \\ x \\ y \\ z \end{pmatrix} = \begin{pmatrix} \gamma & +\beta\gamma & 0 & 0 \\ +\beta\gamma & \gamma & 0 & 0 \\ 0 & 0 & 1 & 0 \\ 0 & 0 & 0 & 1 \end{pmatrix} \begin{pmatrix} ct' \\ x' \\ y' \\ z' \end{pmatrix} \quad (30\text{-}22)$$

EXAMPLE 30-5

Matrix Form of the Lorentz Transformation

Prove that the matrix form of the Lorentz transformation in Equation 30-21 is equivalent to the form expressed in Equation 30-19.

Solution

We will start with the matrix form shown below and multiply the matrix by the vector elements as detailed below:

$$\begin{pmatrix} ct' \\ x' \\ y' \\ z' \end{pmatrix} = \begin{pmatrix} \gamma & -\beta\gamma & 0 & 0 \\ -\beta\gamma & \gamma & 0 & 0 \\ 0 & 0 & 1 & 0 \\ 0 & 0 & 0 & 1 \end{pmatrix} \begin{pmatrix} ct \\ x \\ y \\ z \end{pmatrix}$$

The multiplication yields the following four relationships:

$$ct' = \gamma ct - \beta\gamma x$$
$$x' = -\beta\gamma ct + \gamma x$$
$$y' = y$$
$$z' = z$$

If we divide both sides of the top equation by c and then substitute the definitions for β and γ, we obtain the following, which is equivalent to the time transformation in Equation 30-19:

$$t' = \gamma t - \frac{\beta\gamma x}{c}$$
$$= \gamma\left(t - \frac{\beta x}{c}\right)$$
$$= \gamma\left(t - \frac{vx}{c^2}\right)$$
$$= \frac{t - vx/c^2}{\sqrt{1 - v^2/c^2}}$$

Now we will show that the x-transformation is also equivalent to Equation 30-19:

$$x' = -\beta\gamma ct + \gamma x$$
$$= \gamma(x - \beta ct)$$
$$= \gamma\left(x - \frac{v}{c}ct\right) = \gamma(x - vt)$$
$$= \frac{x - vt}{\sqrt{1 - v^2/c^2}}$$

As expected, $y' = y$ and $z' = z$ in both formulations. Therefore, we have proven that the two forms are completely equivalent.

Making sense of the result

As expected, the two forms of the relationship are completely equivalent. While the matrix form provides a more concise way to write the relationships, the physics is identical in the two expressions.

There is a deeper meaning to the Lorentz transformation. It turns out that not just the four-dimensional spacetime coordinates, but also a second set of four momentum energy components transform according to the Lorentz transformation. You will explore this in Section 30-7.

30-6 Spacetime

When solving problems related to relativity, it is important to consider spatial and time coordinates as integrated into one spacetime system. In spacetime, we view each event as being specified by four spacetime coordinates, such as (ct, x, y, z). The time coordinate is multiplied by the speed of light so that all four coordinates have units of distance. Another observer could associate a different set of spacetime coordinates, (ct', x', y', z'), to the same event, similar to the way two observers could assign different coordinates to the same point in three-dimensional space. For example, if two spots are marked on a blackboard, students could draw different x- and y-axes and assign different coordinates to the same spot, as shown in Figure 30-13. However, the students would agree on the length of the line joining the two spots.

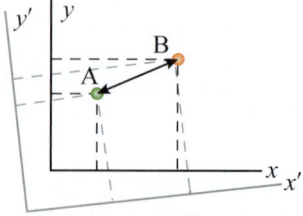

Figure 30-13 In an analogy to spacetime intervals in two-dimensional space, two reference frames do not agree on the coordinates, but they do agree on the length between the points.

A similar situation applies in spacetime, where different observers agree on the **spacetime interval** described in the following equation 30-23:

$$(\Delta s)^2 = (c\Delta t)^2 - (\Delta x)^2 - (\Delta y)^2 - (\Delta z)^2 \qquad (30\text{-}23)$$

We will define $(\Delta s)^2$ as the spacetime interval. Some texts define Δs as the spacetime interval.

We demonstrate in Example 30-6 how the constancy, or invariance, of this spacetime interval can be used to solve special relativity problems. Four-dimensional spacetime diagrams are called Minkowski diagrams, named for the German mathematician Hermann Minkowski (see the Making Connections feature "The Development of Spacetime Ideas").

In special relativity we call a quantity **invariant** when it has the same value for all inertial reference frames. The spacetime interval of Equation 30-23 is an invariant quantity. One way to justify the invariance of the spacetime interval is to consider two light clocks (of the type introduced in Section 30-3) travelling at different constant speeds (Figure 30-14). From one reference frame we have a time interval (for one complete up-and-down cycle of the light pulse) of Δt_1, and that clock moves along the x-axis a distance of Δx_1. From the viewpoint of the other reference frame, the time interval is Δt_2 and the distance travelled is Δx_2.

The two clocks will agree on the direction perpendicular to motion (d) and the speed of light (c). This allows us to write the following expressions for the two clocks using the Pythagorean theorem:

$$d^2 = \left(c\frac{\Delta t_1}{2}\right)^2 - \left(\frac{\Delta x_1}{2}\right)^2$$
$$d^2 = \left(c\frac{\Delta t_2}{2}\right)^2 - \left(\frac{\Delta x_2}{2}\right)^2 \qquad (30\text{-}24)$$

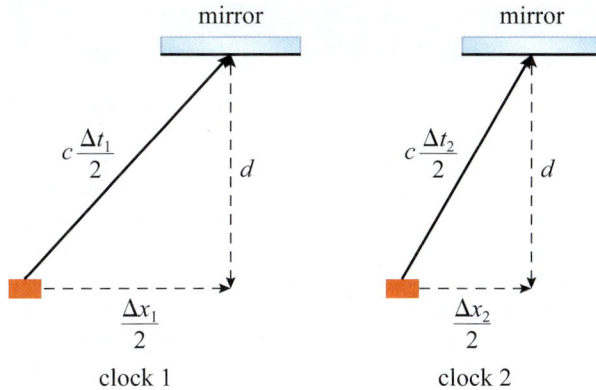

Figure 30-14 Two light clocks moving at constant, but different, speeds along the positive x-axis. We just show the situation for one-half of a clock cycle.

When we multiply each side by 4 and then equate the two expressions, we obtain

$$(c\Delta t_1)^2 - (\Delta x_1)^2 = (c\Delta t_2)^2 - (\Delta x_2) \qquad (30\text{-}25)$$

When we include possible motion in the other spatial dimensions, we obtain the spacetime invariance relationship of Equation 30-23.

EXAMPLE 30-6

Spacetime Intervals

Karen sees two events occur at the same place but separated by 2.50 μs in time. Another observer, Tasha, sees the two events separated by 3.50 μs in time. According to Tasha, what is the spatial separation between the two events?

Solution

For Karen, we have $\Delta x = \Delta y = \Delta z = 0$, so the spacetime interval is

$$(\Delta s)^2 = (c\Delta t)^2 - (\Delta x)^2 - (\Delta y)^2 - (\Delta z)^2$$
$$= (3.0 \times 10^8 \text{ m/s} \cdot 2.5 \times 10^{-6} \text{ s})^2 = 5.62 \times 10^5 \text{ m}^2$$

Now, both observers must have the same value for the spacetime interval. Therefore, we have the following, where we have let $\Delta x'$ represent the spatial separation of the events for Tasha:

$$5.62 \times 10^5 \text{ m}^2 = (c\Delta t')^2 - (\Delta x')^2$$
$$= (3.0 \times 10^8 \cdot 3.5 \times 10^{-6})^2 - (\Delta x')^2$$
$$(\Delta x')^2 = 1.103 \times 10^6 \text{ m}^2 - 5.62 \times 10^5 \text{ m}^2$$

Thus, in Tasha's reference frame the two events are separated by 735 m.

Making sense of the result

In calculations of spacetime intervals, the squared spatial coordinates are subtracted from the squared time coordinate, so if the time coordinate of a given spacetime interval is less, the spatial components must also be less, and vice versa. Therefore, it makes sense that Karen measures a shorter time with the events closer together, while Tasha measures a longer time interval and a greater spatial separation.

CHECKPOINT 30-6

Two Perspectives

John sees two spacetime events occur with a shorter time separation than Jillian, who observes the same two events. What can we conclude about their perspectives on the spatial separation of the events?

(a) John thinks they are closer together spatially.

(b) Jillian thinks they are closer together spatially.

(c) They both agree on exactly the same spatial separation.

(d) The situation of the question is impossible: they cannot measure different time separations between the same events.

ANSWER: (a) If the time separation is less, the spatial separation must also be less for the spacetime interval to be consistent for observers.

MODERN PHYSICS

The Development of Spacetime Ideas

While the core concepts of spacetime were present in Einstein's 1905 relativity papers, it was Hermann Minkowski (1864–1909) who developed much of the mathematics we now associate with spacetime. Minkowski was Einstein's mathematics teacher at Zurich Polytechnic. The class Minkowski taught Einstein had only five students, including only two physics majors, Einstein and Mileva Marić, who would become Einstein's first wife. Einstein liked to concentrate on what interested him and not always on what was being taught. It is said that Minkowski was surprised at the brilliance of Einstein's 1905 paper and the extent to which it was influenced by the mathematics Minkowski had taught.

Spacetime diagrams (also called Minkowski diagrams) normally plot ct on the vertical axis and spatial coordinates on the horizontal axes. Since it is not possible to show all four dimensions of spacetime, the diagrams include either one or two space coordinates. Any point on the diagram represents a **spacetime event**, and a series of linked spacetime points represent the **world line** representing the motion in space and time of some object (the pink line of Figure 30-15). Since the fastest speed possible is c, if the scales on the two axes are the same, a 45° **light cone** represents the possible world line paths of a flash of light initiated at the origin of the spacetime coordinate system.

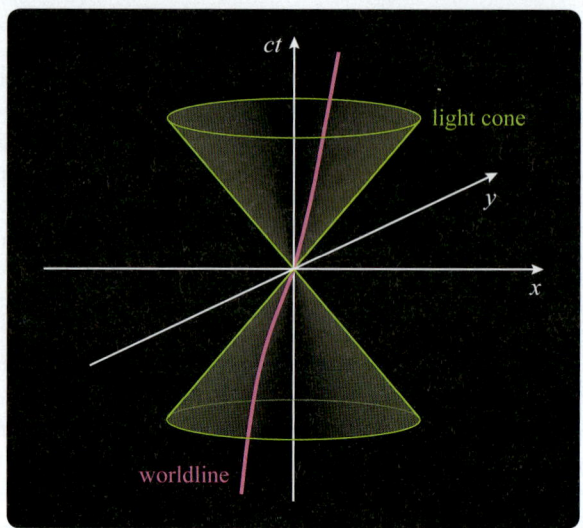

Figure 30-15 A Minkowski spacetime diagram. The pink curve represents the world line of some moving object. The yellow light cone is the path of a light pulse from the origin of the spacetime coordinates. The top half represents the future, and the bottom represents the past.

EXAMPLE 30-7

Lorentz Transformation and Spacetime Intervals

Consider two spacetime events, A and B. From the point of view of observer 1, spacetime event A happens at $t_{1A} = 1.00\ \mu s$, $x_{1A} = 50.0$ m, $y_{1A} = 10.0$ m, $z_{1A} = 20.0$ m, while spacetime event B occurs at $t_{1B} = 2.00\ \mu s$, $x_{1B} = 100.0$ m, $y_{1B} = 50.0$ m, $z_{1B} = 60.0$ m.

(a) From the point of view of reference frame 1, what will be the spacetime interval between the two events?

(b) If a second reference frame 2 is in motion along the positive x-axis with $v = \dfrac{\sqrt{3}}{2}c$, what will be the coordinates of the two spacetime events according to this reference frame? (Assume that v is known precisely enough to justify answers to three significant digits.)

(c) Calculate the spacetime interval from the point of view of the second reference frame using the results from (b).

Solution

(a) The spacetime interval is computed from the relationship

$$(\Delta s)^2 = (c\Delta t)^2 - (\Delta x)^2 - (\Delta y)^2 - (\Delta z)^2$$

If we define $\Delta x = x_B - x_A$ and similarly for the other spacetime coordinates, we have

$$(\Delta s)^2 = (c(t_B - t_A))^2 - (x_B - x_A)^2 - (y_B - y_A)^2 - (z_B - z_A)^2$$

$$(\Delta s)^2 = (c(2.00 \times 10^{-6}\ s - 1.00 \times 10^{-6}\ s))^2$$
$$- (100.0\ m - 50.0\ m)^2 - (50.0\ m - 10.0\ m)^2$$
$$- (60.0\ m - 20.0\ m)^2$$
$$(\Delta s)^2 = 84\,300\ m^2$$

(b) The Lorentz transformation is

$$x' = \frac{x - vt}{\sqrt{1 - v^2/c^2}}$$
$$y' = y$$
$$z' = z$$
$$t' = \frac{t - vx/c^2}{\sqrt{1 - v^2/c^2}}$$

We will define reference frame 1 as the unprimed and 2 as the primed system. The time is converted to SI units of seconds. When we substitute in the values, we obtain the following results for the spacetime coordinates from reference frame 2:

$$t_{2A} = 1.71\ \mu s;\ x_{2A} = -420\ m;\ y_{2A} = 10.0\ m;\ z_{2A} = 20.0\ m$$
$$t_{2B} = 3.42\ \mu s;\ x_{2B} = -839\ m;\ y_{2B} = 50.0\ m;\ z_{2B} = 60.0\ m$$

(c) We now use these values to calculate the spacetime interval from the point of view of this reference frame:

$$(\Delta s)^2 = (c(t_B - t_A))^2 - (x_B - x_A)^2 - (y_B - y_A)^2 - (z_B - z_A)^2$$
$$(\Delta s)^2 = (c(3.4226 \times 10^{-6}\ s - 1.7113 \times 10^{-6}\ s))^2$$
$$- (-839.23\ m - (-419.62\ m))^2 - (50.0\ m - 10.0\ m)^2$$
$$- (60.0\ m - 20.0\ m)^2$$
$$(\Delta s)^2 = 84\,300\ m^2$$

Making sense of the result

As expected, the spacetime interval is the same when viewed from the different inertial reference frames since the spacetime interval is an invariant. While we have only shown this for one particular pair of reference frames, it would hold for any pair of reference frames. Note that from the point of view of the second reference frame, the time interval between the spacetime events is larger, but the spatial separation is also greater, and the spacetime interval is identical. Note that since each term is squared in the spacetime interval calculation, it does not matter whether you define the difference as A − B or B − A on the spacetime coordinates as long as you are consistent.

The important point is that all inertial reference frames agree on the spacetime interval, although they do not agree on the four components that make it up. Another way to justify the relationship is to use the Lorentz transformation to find the spacetime coordinates in a different reference frame, compute the spacetime interval from the point of view of each reference frame, and see that it is constant. We do this in Example 30-7. This can actually be proven in the general case using the Lorentz transformation.

You will notice that in our justification of spacetime intervals we could have inserted a negative sign in each side of the equation. This means that it is possible to define the spacetime interval with the opposite sign (i.e., the spatial components are positive and the time component is negative), but the convention we have used here is the most common, and the one that we will exclusively use.

If the spacetime events defining an interval involve transmission of light in vacuum (e.g., the spacetime event light is emitted first and then that light pulse is received), then the spacetime interval will have a value $(\Delta s)^2 = 0$ and we say that the interval is **light-like**. Two points along the light cone of Figure 30-14 make a light-like spacetime interval.

As we have seen in this chapter, the speed of light seems to impose a maximum possible speed limit. Therefore, any two causally related spacetime events not involving light must have (according to our sign convention on spacetime intervals) $(\Delta s)^2 > 0$, and we call these spacetime intervals **time-like**.

Of course, we can compute a spacetime interval for any two spacetime points. If a computed interval has $(\Delta s)^2 < 0$, we say that it is **space-like**, and because of the maximum speed of light, the first spacetime event of the interval could not have caused the second event.

We call the set of spacetime coordinates a **four vector** (ct, x, y, z). Normally, to get the length of a vector we take the vector dot product with an identical vector and then take the square root of the result. In the case of four vectors it is slightly different, in that we attach a negative sign so that the invariant length of the above spacetime four vector is

$$\text{invariant length} = \sqrt{(ct)^2 - x^2 - y^2 - z^2} \quad (30\text{-}26)$$

A four vector length is invariant under the Lorentz transformation. We will encounter another four vector later in the chapter.

30-7 Relativistic Momentum and Energy

In special relativity, linear momentum can still be expressed in the form $m\vec{v}$, but a velocity-dependent factor appears when we consider the **relativistic momentum**, that is, the momentum in terms of the rest mass, m_0, of an object. **Rest mass** is the mass as measured in a reference frame in which the object is not moving:

KEY EQUATION
$$\vec{p} = \frac{m_0 \vec{v}}{\sqrt{1 - v^2/c^2}} \quad (30\text{-}27)$$

The conservation of linear momentum (e.g., in collisions) is a cornerstone of classical physics. In the following section, you will learn how to transform relativistic velocities for different reference frames. If we define linear momentum classically as simply the rest mass times the velocity, it can be shown that linear momentum is not conserved when viewed in different reference frames. Physicists sought a relativistic definition of linear momentum that would support conservation of momentum in collisions even when viewed from different reference frames. It was found that Equation 30-27 satisfies this requirement.

One way to arrive at this expression for relativistic momentum is to first write the velocity using the particle's own reference frame. It will measure its own rest mass and will measure proper time for any time interval since its own clock travels with the particle. It will not, however, measure the proper length for the distance travelled, so we write the distance travelled as Δx in the following equation:

$$p = m_0 \frac{\Delta x}{\Delta t_0} \quad (30\text{-}28)$$

Now consider the particle as observed by some other reference frame. If we now use the time dilation relationship to express Δt in terms of Δt_0, we obtain the following result, where we have interpreted the speed of the particle as seen by the other reference frame as $v = \Delta x/\Delta t$:

$$p = m_0 \frac{\Delta x}{\Delta t \sqrt{1 - v^2/c^2}} = \frac{m_0}{\sqrt{1 - v^2/c^2}} \frac{\Delta x}{\Delta t}$$

$$= \frac{m_0 v}{\sqrt{1 - v^2/c^2}} = \gamma m_0 v \qquad (30\text{-}29)$$

We see that the value for the linear momentum is consistent with Equation 30-27.

While you might be tempted to think of γm_0 as a speed-dependent relativistic mass, and some physicists do view it in this way (including Einstein in his early papers), most now prefer to think of the mass of a particle as invariant, and the factor of γ as part of the momentum rather than the mass (see problem 40).

The important point is that with linear momentum defined by Equation 30-27, conservation of linear momentum is retained, no matter which inertial reference frame we use. For speeds $v \ll c$, the relativistic momentum reduces to the classical expression for linear momentum. As speeds become large, the relativistic expression gives a larger value for linear momentum than would be found from the classical expression.

Probably the most famous equation in all of science is the mass–energy equivalence formulated by Einstein in 1905 in one of his "miracle year" papers:

$$E = mc^2 \qquad (30\text{-}30)$$

Actually, the equation $E = mc^2$ never appeared in Einstein's paper (see problem 81). He did argue, however, that if a body lost energy, then its mass would change by that energy loss divided by the square of the speed of light. According to this relationship, it is possible to convert mass into energy, or energy into mass (see Example 30-8).

Einstein's justification for his derivation of mass–energy equivalence was based on the fact that electromagnetic waves carry momentum. In the 19th century, physicists established that a packet of electromagnetic waves with energy E carries linear momentum p of value E/c.

Although Einstein's complete derivation is too complex to include here, we can demonstrate the underlying principle. If the energy of a light wave did have a mass equivalent, m, its momentum would be $p = mv = mc$ because the energy is moving at the speed of light. As described above, the momentum of the light wave also equals E/c, so we can readily obtain $E = mc^2$. However, the real proof of $E = mc^2$ and the relativistic momentum relationship consists of the countless experiments that have proven that the universe operates in a way that is consistent with this relationship.

For example, mass-to-energy conversion occurs in nuclear power reactors, where large nuclei are broken into two or more lighter nuclei. During this process, a small

amount of the mass of each original nucleus is converted into energy. The Sun and similar stars obtain their energy by nuclear fusion in their core, where temperatures and pressures are immense. In **nuclear fusion** processes, light elements fuse together to form heavier elements. Each fusion reaction has a small net mass loss; the "lost" mass corresponds to the energy produced. In fact, at the fundamental level, all chemical reactions also derive their energy from slight changes in mass (see Example 30-8).

So we see that in a sense an object has an energy by virtue of having mass. We call this **rest mass energy**,

EXAMPLE 30-8

Mass-Energy Conversion in the Sun

In the dominant reaction that produces energy in the core of the Sun, four hydrogen nuclei (protons) fuse into one helium nucleus (α-particle) and emit two positrons and two neutrinos of negligible mass. The rest mass of a proton is 1.007 276 467 u, and the rest mass of an α-particle is 4.001 506 177 u, where u is the unified atomic mass unit (1 u = $1.660\ 539 \times 10^{-27}$ kg). Material in the interior of stars such as the Sun is essentially totally ionized, so we use the mass of the nuclei, not the neutral atoms, in the calculation. The rest mass of a positron is $9.109\ 38 \times 10^{-31}$ kg.

(a) What is the mass loss in this fusion reaction?
(b) What percentage of the original mass is lost?
(c) How much energy is released by one reaction? Express the energy in joules and electron volts.
(d) The power output of the Sun is 3.85×10^{26} W. How many hydrogen fusion reactions per second are needed to provide this energy output?

Solution

(a) The net reaction is 4 protons → 1 α-particle + 2 positrons + energy. We need to carry several significant figures because the difference in mass is a very small percentage.

Initial mass:

4 protons: $4 \cdot 1.007\ 276\ 467$ u $\cdot\ 1.660\ 539 \times 10^{-27}$ kg/1 u

$= 6.690\ 487\ 428\ 9 \times 10^{-27}$ kg

Products:

α-particle: $1 \cdot 4.001\ 506\ 174\ 7$ u $\cdot\ 1.660\ 539 \times 10^{-27}$ kg/1 u

$= 6.644\ 657\ 062 \times 10^{-27}$ kg

2 positrons: $2 \cdot 9.109\ 38 \times 10^{-31}$ kg $= 1.821\ 876 \times 10^{-30}$ kg

Therefore, the overall mass loss is $4.400\ 849 \times 10^{-29}$ kg.

(b) The percentage of the original mass lost is

$(4.400\ 849 \times 10^{-29}$ kg$/6.690\ 487 \times 10^{-27}$ kg$) \times 100\% = 0.658\%$

While additional significant digits are possible here, the rest of the question simply has three significant digits, so we state this part to that precision as well.

(c) To calculate the energy released, we use the mass loss from (a) for m in $E = mc^2$. Note that only the mass difference is converted into energy, not the entire original mass. Putting in the numbers, we obtain 3.96×10^{-12} J. Using the conversion 1 eV = 1.60×10^{-19} J, we have 2.48×10^7 eV, or 24.8 MeV.

(d) To determine the number of fusion reactions, we divide the power output of the Sun by the energy produced by each fusion reaction:

$$\frac{3.85 \times 10^{26} \text{ J/s}}{3.96 \times 10^{-12} \text{ J/reaction}} = 9.72 \times 10^{37} \text{ reactions/s}$$

Making sense of the result

A tiny amount of mass is lost in each hydrogen fusion reaction. In most nuclear reactions only a small part of the original mass is lost. We see that nuclear processes create, per reaction, much more energy than chemical processes (although all energy does ultimately come from the same mass–energy equivalence). The Sun actually has several different nuclear reactions taking place, but the result for the number of reactions per second is a good first approximation. The number of hydrogen fusion reactions is huge, and each one produces two neutrinos. Therefore, the Sun produces an incredible number of neutrinos each second. It has been estimated that approximately 100 trillion neutrinos from the Sun pass through your body each second.

E_0, and based on the mass–energy equivalence justified earlier it is given by

$$E_0 = m_0 c^2 \tag{30-31}$$

If we do not have any other forms of energy (such as electromagnetic), then we can express the total energy as the sum of this rest mass energy and the kinetic energy K:

$$E = E_0 + K = m_0 c^2 + K \tag{30-32}$$

Let's look a bit more closely at the total energy, E. By analogy with our justification for the relativistic momentum, we might expect **total energy** to be given by

$$E = mc^2 = \gamma m_0 c^2 \tag{30-33}$$

As mentioned earlier, some physicists prefer not to talk about a relativistic mass but rather to assume that there is only a single mass. Those who favour such an approach write the right-hand term directly and use m instead of m_0 everywhere. See problem 40

to investigate the controversy regarding the best way to teach relativistic momentum and energy. In analogy with proper time and proper length, we prefer to use rest mass, but you should be aware of the arguments for both approaches. The important thing is that the right-hand term of Equation 30-33 is valid no matter which approach you favour (although if you do not like the idea of relativistic mass, you can simply call it m).

Relativistic Kinetic Energy

The classical expression for kinetic energy is not valid at relativistic speeds. The **relativistic kinetic energy** is the difference between the total energy E, K, and the rest mass energy:

$$K = E - m_0 c^2 = \gamma m_0 c^2 - m_0 c^2 \tag{30-34}$$

KEY EQUATION
$$K = m_0 c^2 \left(\frac{1}{\sqrt{1 - v^2/c^2}} - 1 \right) \tag{30-35}$$

MODERN PHYSICS

You may wonder how this equation relates to the classical result, $K = \frac{1}{2}mv^2$. If we use a series expansion for the square root function in Equation 30-35, we have

$$K = m_0c^2[(1 - v^2/c^2)^{-1/2} - 1]$$

$$= m_0c^2\left[1 + \frac{1}{2}\frac{v^2}{c^2} - \frac{1\cdot3}{2\cdot4}\left(\frac{v^2}{c^2}\right)^2 + \frac{1\cdot3\cdot5}{2\cdot4\cdot6}\left(\frac{v^2}{c^2}\right)^3 - \cdots - 1\right]$$

$$= m_0c^2\left[\frac{1}{2}\frac{v^2}{c^2} - \frac{3}{8}\left(\frac{v^2}{c^2}\right)^2 + \frac{15}{48}\left(\frac{v^2}{c^2}\right)^3 - \cdots\right] \quad (30\text{-}36)$$

$$K = \frac{1}{2}m_0v^2 - m_0\left(\frac{3}{8}\frac{v^4}{c^2} - \frac{5}{16}\frac{v^6}{c^4} + \cdots\right)$$

CHECKPOINT 30-7

Relativistic Kinetic Energy

If an object has a total relativistic energy that is three times its rest mass energy, what fraction of the total energy is the relativistic kinetic energy?

(a) 1/9
(b) 1/6
(c) 1/3
(d) 2/3

ANSWER: (d) If $E = 3m_0c^2$ and $K = E - m_0c^2$, then K must be 2/3 of the total energy, E.

EXAMPLE 30-9

Relativistic Kinetic Energy and Momentum

A spaceship has a rest mass of 2000 kg and travels at 90.0% of the speed of light.

(a) What is the relativistic kinetic energy of the spaceship?
(b) What is the kinetic energy according to classical physics?
(c) What is the magnitude of the relativistic momentum of the spaceship?

Solution

(a) We can use the relationship for relativistic kinetic energy (Equation 30-35):

$$K_{rel} = m_0c^2\left(\frac{1}{\sqrt{1 - v^2/c^2}} - 1\right)$$

$$= (2000 \text{ kg})(3.00 \times 10^8 \text{ m/s})^2\left(\frac{1}{\sqrt{1 - 0.900^2}} - 1\right)$$

$$K_{rel} = 2.33 \times 10^{20} \text{ J}$$

(b) Using the classical mechanics expression for kinetic energy:

$$K_{cl} = \frac{1}{2}mv^2 = \frac{1}{2}(2000 \text{ kg})((0.900)(3.00 \times 10^8 \text{ m/s}))^2$$

$$= 7.29 \times 10^{19} \text{ J}$$

(c) The relativistic momentum has the same direction as the velocity of the spaceship and has a magnitude of

$$p = \frac{m_0v}{\sqrt{1 - v^2/c^2}} = \frac{(2000 \text{ kg})(2.70 \times 10^8 \text{ m/s})}{\sqrt{1 - 0.900^2}}$$

$$= 1.24 \times 10^{12} \text{ kg}\cdot\text{m/s}$$

Making sense of the result

We see that the relativistic kinetic energy is significantly greater than the classical kinetic energy, as expected at a speed of 90% of the speed of light. In part (c), we see that the relativistic momentum is similarly much greater than the classical momentum.

When $v \ll c$, we can ignore all but the first term in the series, leaving an expression that simplifies to the classical result for kinetic energy.

We can see that relativistic effects become extreme as v approaches c. For example, if an object were moving at the speed of light, the length of the object would contract to zero, and its kinetic energy would become infinite. The speed of light appears to be a universal upper limit for speeds. Some theoretical models suggest that particles that always travel at speeds greater than the speed of light may exist. To date, these particles, called tachyons, have never been observed.

When we introduced the four vector, we mentioned that there were other four vectors, one of which has energy and momentum components (for a Cartesian coordinate system) of (E, p_xc, p_yc, p_zc). As before, E represents the total energy. This four vector can be transformed from one reference frame to another using the Lorentz transformation.

In a similar way that we found the spacetime invariant earlier, we have the following:

$$E^2 - (p_xc)^2 - (p_yc)^2 - (p_zc)^2 = \text{invariant} \quad (30\text{-}37)$$

We can relate the components of p to the square of the magnitude of the total relativistic momentum:

$$p^2 = p_x^2 + p_y^2 + p_z^2$$

We can substitute this into Equation 30-37 to simplify the expression. What is the invariant quantity in Equation 30-37? It has units of energy squared, and it turns out that it is equal to the square of the rest mass energy (you will prove this in problem 83):

KEY EQUATION $$E^2 - (pc)^2 = (m_0c^2)^2 \quad (30\text{-}38)$$

MODERN PHYSICS

30-8 Relativistic Velocity Addition

Consider the situation shown in Figure 30-17, in which the S' reference frame is moving at speed v in the positive x-direction with respect to the S reference frame. We assume that at $t = 0$ the two reference frames are aligned. Assume that some object is moving at speed u' in the positive x-direction with respect to S'. If we did not have to take relativity into account, the speed in S would be related to that in S' by

$$u = u' + v \qquad (30\text{-}39)$$

This can be justified by common sense, or we can differentiate the Galilean (Newtonian) transformation relationship in Section 30-5 with respect to time. If we solve for u' we have

$$u' = u - v \qquad (30\text{-}40)$$

Equation 30-39 must not hold when special relativity is taken into account, since even if both u' and v were near the speed of light, we know that the speed u must be less than the speed of light, c.

We can derive the relativistic velocity addition relationship starting with the Lorentz transformation relationships of Section 30-5:

$$x' = \frac{x - vt}{\sqrt{1 - v^2/c^2}} \qquad (30\text{-}17)$$

$$t' = \frac{t - vx/c^2}{\sqrt{1 - v^2/c^2}} \qquad (30\text{-}18)$$

The velocity in the primed system is obtained from these as shown below, where we have used the transformation to write t' in terms of t, v, and x and to write x' in terms of x, v, and t:

$$u' = \frac{dx'}{dt'} = \frac{d(x - vt)}{d(t - vx/c^2)}$$

This looks pretty formidable, but if we keep in mind that c is constant and also we assume that the speed v is also constant, we can write both numerator and denominator as differentials as shown below:

$$u' = \frac{dx - vdt}{dt - vdx/c^2}$$

We divide all terms in both numerator and denominator by dt, obtaining

$$u' = \frac{\dfrac{dx}{dt} - v\dfrac{dt}{dt}}{\dfrac{dt}{dt} - \dfrac{vdx}{c^2dt}}$$

If we then recognize that $u = dx/dt$, we obtain the following final result for addition of relative velocities, which is valid even when they are near the speed of light:

KEY EQUATION
$$u' = \frac{u - v}{1 - vu/c^2} \qquad (30\text{-}41)$$

We can see that Equation 30-41 behaves as expected in the classical limit. When u and v are both much less than c, it will reduce to Equation 30-40.

If we use the inverse Lorentz transformation in Equation 30-19 with a similar derivation, we obtain

KEY EQUATION
$$u = \frac{u' + v}{1 + vu'/c^2} \qquad (30\text{-}42)$$

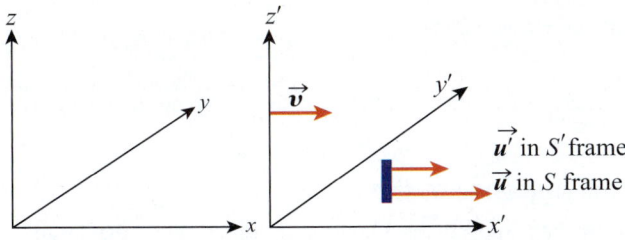

Figure 30-17 The S' coordinate system moves in the positive x-direction at speed v relative to the S coordinate system. At $t = 0$ the reference frames are coincident. An object marked by the purple rectangle moves at speed u when measured with respect to the S reference frame and speed u' relative to the S' reference frame.

CHECKPOINT 30-8

Relative Speeds

An observer on a planet sees two spaceships, each travelling at half the speed of light, approaching from opposite directions. What is the relative speed of one spaceship as measured by the other?

(a) c

(b) $\dfrac{4}{5}c$

(c) $\dfrac{3}{4}c$

(d) $\dfrac{1}{2}c$

ANSWER: (b) If we take the left spaceship as the reference, both u and v are $\frac{1}{2}c$, and

$$v = \frac{c/2 + c/2}{1 + \dfrac{(c/2)(c/2)}{c^2}} = \frac{c}{\dfrac{5}{4}} = \frac{4}{5}c.$$

When u and v are both much less than c, this will reduce to Equation 30-39. When u' and v both approach the speed of light, c, then the result for u' will also approach c and, as required by relativity, not exceed the speed of light.

While we have not used vector notation, these are one-dimensional velocities, where the signs are important. This is similar to the treatment of one-dimensional velocities and accelerations in Chapter 3. One reason not to use vector notation is that the relationships only hold for motion along (or opposite to) the direction of motion of the reference frame, while vector notation might have been interpreted otherwise.

Note that the relationships 30-41 and 30-42 are only used for velocities in the direction (or anti-direction) of the motion of one reference frame relative to the other. Test your understanding of relativistic velocity addition with Checkpoint 30-8. Example 30-10 illustrates relativistic velocity addition.

EXAMPLE 30-10

Galaxy Velocity Addition

Galaxy G1 is moving directly away from Earth at a speed of $0.50c$. Another galaxy, G2, is also moving directly away from Earth but in the opposite direction at a speed of $0.40c$ (see Figure 30-18). What speed would an observer on galaxy G2 measure for the motion of G1?

Figure 30-18 Example 30-10.

Solution

To use the notation of the chapter, we will call galaxy G2 the S reference frame and Earth the S' reference frame. With this definition, and using the terminology of this section, we have the situation shown in Figure 30-19. Note that if G2 moves to the left at $0.40c$ from the perspective of Earth we can say that Earth moves to the right with a speed of $0.40c$.

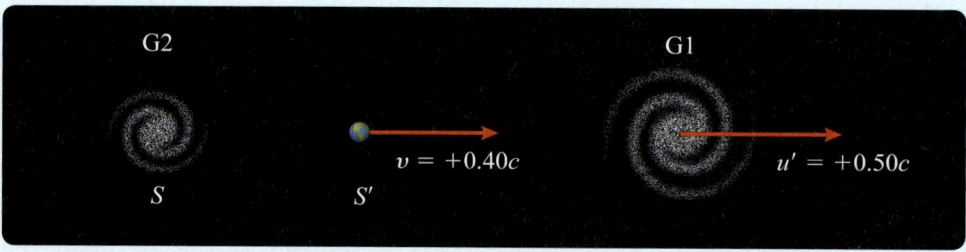

Figure 30-19 From the perspective with G2 as reference frame S and Earth as reference frame S'.

We can now use Equation 30-42 to compute the speed of G1 as measured by a G2-centred reference frame:

$$u = \frac{u' + v}{1 + vu'/c^2}$$

$$u = \frac{0.50c + 0.40c}{1 + \dfrac{0.40c \times 0.50c}{c^2}}$$

$$u = \frac{0.90c}{1 + 0.20}$$

$$u = 0.75c$$

If desired we could insert the value for c to compute the speed in m/s.

Making sense of the result

If we had applied the Galilean transformation, the result would have been $0.90c$, so we expect the relativistic result to be somewhat less than that, which is consistent with our result.

30-9 Relativistic Doppler Shift

When a source is moving away from an observer at speed v_s, if v is the speed of sound in the medium and f_r is the frequency of the received wave when f_s is the frequency of the wave emitted by the source, you saw in Section 15-7 that the Doppler effect for sound waves is given by

$$f_r = \frac{v}{v + v_s} f_s \qquad (15\text{-}33b)$$

If we have light waves, you might think we need only replace v with c, as shown:

$$f_r = \frac{c}{c + v_s} f_s \qquad (30\text{-}43)$$

According to relativity, time intervals are governed by the time dilation relationship, however. Realizing that $f \sim 1/\Delta t$ and using the time dilation, we see that Equation 30-43 should be modified, where γ is the Lorentz factor of Equation 30-20:

$$f_r = \frac{c}{c + v_s} \frac{1}{\gamma} f_s \qquad (30\text{-}44)$$

It is normal to divide numerator and denominator by c to obtain

$$
\begin{aligned}
f_r &= \frac{1}{1 + v_s/c} \frac{1}{\gamma} f_s \\
&= \frac{1}{1 + v_s/c} \sqrt{1 - v_s^2/c^2} f_s \\
&= \frac{1}{1 + v_s/c} \sqrt{(1 - v_s/c)(1 + v_s/c)} f_s \\
&= \frac{\sqrt{1 - v_s/c}}{\sqrt{1 + v_s/c}} f_s
\end{aligned}
$$

KEY EQUATION
$$f_r = \frac{\sqrt{1 - v_s/c}}{\sqrt{1 + v_s/c}} f_s \qquad (30\text{-}45)$$

If the source is moving toward the detector, we simply change the sign in front of v_s in Equation 30-45.

Since the speed of light is constant, by $c = f\lambda$ we must have that $\lambda \sim \dfrac{1}{f}$, and therefore the corresponding relationship expressed in terms of wavelength is given by

KEY EQUATION
$$\frac{\lambda_r}{\lambda_s} = \frac{\sqrt{1 + v_s/c}}{\sqrt{1 - v_s/c}} = \frac{\sqrt{c + v_s}}{\sqrt{c - v_s}} \qquad (30\text{-}46)$$

When we receive light from a source moving away from us at a speed comparable to the speed of light, the received wavelength will be longer than the emitted wavelength, and we call this **redshift**. If the source is moving toward the detector, the wavelength received will be shorter, and this is called **blueshift**. It is important to realize that a redshifted light is not necessarily red, it is simply moved to a longer wavelength.

Astrophysicists define a **redshift parameter z** by

$$z \equiv \frac{\lambda_r}{\lambda_s} - 1 = \frac{\lambda_r - \lambda_s}{\lambda_s} = \frac{\Delta\lambda}{\lambda_s} \qquad (30\text{-}47)$$

CHECKPOINT 30-9

Doppler Shift of Light

A light source of wavelength λ is moving directly toward you at speed $\frac{2}{3}c$. Will you observe it redshifted or blueshifted? What will be the received wavelength?

(a) blueshifted, received wavelength $\dfrac{1}{\sqrt{5}}\lambda$

(b) blueshifted, received wavelength $\dfrac{2}{3}\lambda$

(c) redshifted, received wavelength 1.5λ

(d) redshifted, received wavelength $\sqrt{5}\lambda$

ANSWER: (a) It is moving toward you, so you know it will be blueshifted. To get the received wavelength, we use Equation 30-46, but since the source velocity is toward us, we use $-\frac{2}{3}c$ for v, yielding the result $\lambda_r = \dfrac{1}{\sqrt{5}}\lambda_s$.

If we combine this with Equation 30-46, we have

$$z = \frac{\Delta\lambda}{\lambda_s} = \frac{\sqrt{1 + v_s/c}}{\sqrt{1 - v_s/c}} - 1 \qquad (30\text{-}48)$$

Equation 30-48 is the exact relationship, but we can use a binomial expansion, keeping only the first-order terms, and the result will be valid as long as the recession speed is not too close to the speed of light:

$$z \approx \frac{1 + \frac{1}{2}\frac{v_s}{c}}{1 - \frac{1}{2}\frac{v_s}{c}} - 1$$

$$\frac{\lambda_r - \lambda_s}{\lambda_s} \approx \frac{2c + v_s}{2c - v_s} - 1$$

$$\frac{\lambda_r - \lambda_s}{\lambda_s} \approx \frac{2c + v_s - 2c + v_s}{2c - v_s}$$

$$\frac{\lambda_r - \lambda_s}{\lambda_s} \approx \frac{2v_s}{2c - v_s}$$

KEY EQUATION
$$z = \frac{\Delta\lambda}{\lambda_s} \approx \frac{v_s}{c} \qquad (30\text{-}49)$$

This was the approximate relationship we encountered in Chapter 11 when calculating the masses of exoplanets from the Doppler shift in the light of their stars. As long as the recession speed is less than 10% of the speed of light, we can use this approximate expression.

We saw in Chapters 1 and 15 that the Doppler effect for light was critical in our understanding of the nature of the universe. The recession speed of distant galaxies is measured from their redshifted spectra (see Example 30-11). Hubble and collaborators were able to determine that more distant galaxies moved away from us at higher recession speeds, which is indicative of an expanding universe.

EXAMPLE 30-11

Redshift Galaxy M101

First discovered in 1781, one of the most beautiful galaxies is M101, pictured in Figure 30-20. This galaxy is about 27 million light years from us. It is moving away from us with a speed of 241 km/s.

(a) Calculate the redshift parameter z using both the exact and approximate expressions.
(b) If hydrogen light with spectral line wavelength 486.13 nm is emitted by the galaxy, what will be the observed wavelength when received at Earth?

Solution

(a) We convert the recession speed to SI units of metres per second and then substitute in the speed of light to calculate the redshift parameter z:

$$z = \frac{\sqrt{1 + v_s/c}}{\sqrt{1 - v_s/c}} - 1$$

$$z = \frac{\sqrt{1 + \dfrac{2.41 \times 10^5 \text{ m/s}}{3.00 \times 10^8 \text{ m/s}}}}{\sqrt{1 - \dfrac{2.41 \times 10^5 \text{ m/s}}{3.00 \times 10^8 \text{ m/s}}}} - 1$$

$$z = 8.037 \times 10^{-4}$$

If we use the approximate expression valid for recession speeds much less than the speed of light, then

$$z \approx \frac{v_s}{c}$$

$$z \approx \frac{2.41 \times 10^5 \text{ m/s}}{3.00 \times 10^8 \text{ m/s}}$$

$$z \approx 8.033 \times 10^{-4}$$

(b) Since the galaxy is moving away from us, its light will be redshifted, or moved to a longer wavelength. We can compute the wavelength shift $\Delta\lambda$ using the exact and approximate expressions:

$$\lambda_r = \lambda_s + \Delta\lambda$$
$$= \lambda_s + z\lambda_s$$
$$= 486.13 \text{ nm} + 0.000\,8037 \times 486.13 \text{ nm}$$
$$= 486.13 \text{ nm} + 0.39 \text{ nm} = 486.52 \text{ nm}$$

Figure 30-20 Example 30-11. M101 Galaxy.

Jean-Charles Cuillandre (CFHT), Hawaiian Starlight, CFHT

The approximate expression yields the same result to five significant digits:

$$\lambda_r = \lambda_s + z\lambda_s$$
$$\approx 486.13 \text{ nm} + 0.000\,8033 \times 486.13 \text{ nm}$$
$$\approx 486.13 \text{ nm} + 0.39 \text{ nm}$$
$$\approx 486.52 \text{ nm}$$

Making sense of the solution

Since the speed of recession, while large by terrestrial speeds, is very small compared to the speed of light, we expect the approximate expression to give a similar value to the exact expression, and this is what we found. In fact, to five significant digits there is no difference in the predicted redshifted wavelength.

30-10 Gravitational Time Dilation in General Relativity

Einstein's **mass–energy equivalence** equation, $E = mc^2$, indicates that mass can be created from energy as long as all conservation laws are obeyed. According to classical Newtonian physics, any existing mass would instantly experience the gravitational force exerted by the newly created mass. But this instantaneous effect requires information to be transmitted faster than the speed of light, contrary to special relativity. Einstein saw this problem while developing special relativity and spent about a decade working on a solution: general relativity.

The first step was the equivalence principle. The **equivalence principle** states that the acceleration of a reference frame and the effect of gravity cannot be distinguished. For example, if you were in a small, windowless box, you could not distinguish the presence of a gravitational field from being in an accelerating spaceship far from any masses. In what Einstein later described as his happiest (some prefer the translation "most fortunate") thought, he realized that the way to remove the effect of gravity was to consider reference frames in free fall.

If gravitation is not a force, what does it become in general relativity? The answer is that masses curve spacetime. Imagine first that we had no masses in the universe. The spacetime would be flat and would not vary from location to location. If we then add a mass, say a dense star, the spacetime curves, or warps, in the region near the star, as shown in Figure 30-21. The motion of objects follows this curved spacetime, somewhat like the way your motion is dictated by the contour of the landscape when you drive in hilly country. The renowned American theoretical physicist John Archibald Wheeler (1911–2008) concisely summarized general relativity as this: matter tells spacetime how to curve, and curved spacetime tells matter how to move.

The field equations that describe general relativity are mathematically complex even for symmetrical situations.

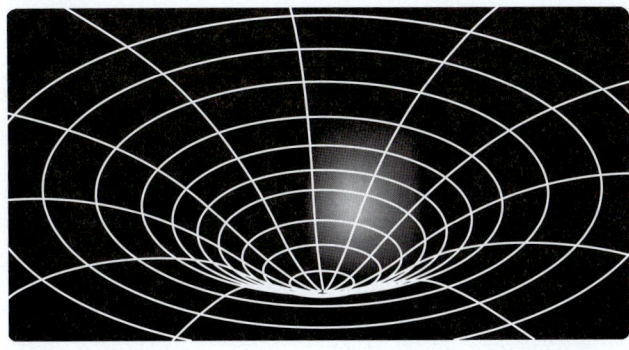

Figure 30-21   Spacetime is curved in the region around a mass.

You might wonder why they are called the **field equations**. An analogy with electromagnetic fields may be helpful. Electromagnetic fields (see Chapters 19 and 24) are caused by charges and movement of charges, and the field at a point governs how a small test charge placed there will move. In general relativity, the distribution of mass–energy and linear momentum governs the spacetime geometry, and therefore the movement of a test mass. The actual field equations are complex and are represented using **tensors**. We saw earlier in the text that you can think of a vector equation as really N coupled equations (in N-dimensional space), one for each vector component. A vector is a tensor of first order, but the main tensors of general relativity field equations are second order, meaning that in N-dimensional spacetime the tensor equation implies $N \times N$ relationships (not all of these are independent). The field equations of general relativity can only be solved exactly in situations involving symmetry, but powerful numerical techniques have recently allowed approximate solutions in more complex cases.

General relativity provides an alternative to Newtonian gravity for explaining the motion of planets and other objects in elliptical orbits around the Sun. The speed of an object in an elliptical orbit varies as its distance from the Sun varies because the curvature of spacetime is greater closer to the Sun.

The presence of masses does not just curve spacetime; it also causes time to slow down. The greater the curvature of spacetime, the more slowly time moves. This effect is called **gravitational time dilation**. To understand why general relativity predicts time dilation, we consider a related effect, gravitational redshift.

It can be shown that gravitational time dilation is described by the following relationship:

KEY EQUATION
$$\frac{\Delta t_0}{\Delta t_\infty} = \left(1 - \frac{2GM}{R_0 c^2}\right)^{1/2} \tag{30-50}$$

where M is the mass (assumed to be spherically symmetric), R_0 is the distance of the observer who measures Δt_0 from the centre of mass M, and Δt_∞ is the time measured by an observer at an infinite distance, where spacetime is assumed to be flat. Since the quantity in brackets is always less than 1, $\Delta t_0 < \Delta t_\infty$, and time moves slower in curved spacetime.

Frequency is proportional to the inverse of the period, and therefore the frequency of an electromagnetic wave must be as shown here:

$$\frac{f_\infty}{f_0} = \left(1 - \frac{2GM}{R_0 c^2}\right)^{1/2} \tag{30-51}$$

An electromagnetic wave emitted at infinite distance (where spacetime is flat) with frequency f_∞ will be observed at distance R_0 from the centre of a symmetric mass M to have frequency f_0, and vice versa. Note that $f_0 > f_\infty$.

All observers agree on the speed of light, so from $c = f\lambda$ the frequency, f, and wavelength, λ, are inversely related. This means that we can express the relationship of Equation 30-51 in terms of wavelengths:

KEY EQUATION
$$\frac{\lambda_0}{\lambda_\infty} = \left(1 - \frac{2GM}{R_0 c^2}\right)^{1/2} \qquad (30\text{-}52)$$

Equation 30-52 is called the **gravitational redshift** relationship. When light (or any other electromagnetic wave) of wavelength λ_0 is emitted from curved spacetime (i.e., at a distance R_0 from the centre of a spherically symmetric dense star of mass M), it will be observed from flat spacetime to have a longer wavelength, $\lambda_\infty > \lambda_0$ (hence the name redshift). It is important to realize that this is a different type of redshift from the Doppler effect for light considered in the previous section. If there is motion of the source relative to the observer, both need to be taken into account.

We will use the thought experiment shown in Figure 30-22 to justify gravitational redshift. An elevator car has an apparatus on the floor that emits a photon of light toward a detector on the ceiling. At exactly the instant that the photon is emitted, the cable on the elevator is released, allowing the elevator car to fall freely toward Earth. (We neglect Earth's rotation and orbital motion in this thought experiment.)

We might expect the light to be blueshifted when it reaches the detector because the detector is falling toward the photon. However, Einstein proposed that any reference frame in gravitational free fall is just as valid as a reference frame at rest for general relativity. Therefore, there should not be any unusual results in the free-falling frame of the elevator car; hence, the photon should have the same wavelength when it is detected as when it was emitted.

We can reconcile the wavelength measurements only if the curved spacetime of Earth's gravitational field causes a redshift that exactly cancels the blueshift in the free-fall reference frame. Numerous observations

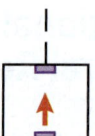

Figure 30-22 A source emits a photon from the bottom at the instant the elevator goes into free fall. A detector is at the top of the elevator box.

have demonstrated that curved spacetime does indeed cause a gravitational redshift. In Example 30-13, you will see how this thought experiment validates Equation 30-52 to first order.

EXAMPLE 30-12

Gravitational Time Dilation Near a Dwarf Star

Say 100.000 s passes on a clock that is in flat spacetime, far away from any masses. What time passes on a clock near the surface of a white dwarf star with a mass of $1.800\,00 \times 10^{30}$ kg and a radius of $6.100\,00 \times 10^6$ m?

Solution

Applying the relationship for gravitational time dilation (Equation 30-50), and suppressing units until the final answer,

$$\frac{\Delta t_0}{\Delta t_\infty} = \left(1 - \frac{2GM}{R_0 c^2}\right)^{1/2}$$

$$= \left(1 - \frac{2(6.674\,08 \times 10^{-11})(1.800\,00 \times 10^{30})}{(6.100\,00 \times 10^6)(2.997\,92 \times 10^8)^2}\right)^{1/2}$$

$$= 0.999\,781$$

$$\Delta t_0 = 0.999\,781 \times 100.000 \text{ s} = 99.978\,1 \text{ s}$$

Making sense of the result

Even near massive high-density objects such as a white dwarf star, the effect of gravitational time dilation is slight.

MODERN PHYSICS

EXAMPLE 30-13

Thought Experiment Justification of Gravitational Redshift

(a) Mathematically develop the falling elevator thought experiment to obtain an expression for gravitational redshift.

(b) Use a series expansion to approximate the gravitational redshift for weakly curved spacetime.

(c) Use a series expansion for Equation 30-52 valid when the square root is only slightly less than one. Show that this is equal to the result you obtained in part (b).

Solution

(a) According to the thought experiment, there must be a redshift in curved spacetime to just cancel the Doppler blueshift of the elevator in free fall (see Figure 30-22). We will first consider the difference in a small radial distance from the planet and then integrate to get the entire gravitational redshift from the surface to an infinite distance where spacetime is flat.

Let us call the height of the elevator box dr (because we are going to later integrate). Therefore, the time, dt, for the photon to go from the emitter to the detector in Figure 30-22 will be given by

$$dt = \frac{dr}{c} \qquad (1)$$

We will call r the radial distance from the centre of a spherically symmetric mass, M. From Chapter 11, the magnitude of the acceleration due to gravity at some radial distance r is given by the following, pointing radially inward toward the centre of the mass:

$$a = \frac{GM}{r^2} \qquad (2)$$

Since we have assumed that the elevator starts its free fall at $t = 0$, its speed at the time the photon reaches the detector will be, using the results (1) and (2),

$$dv = a\,dt = \frac{GM}{r^2}\frac{dr}{c} \qquad (3)$$

We saw in the previous section (Equation 30-49) that as long as the speed is not too near the speed of light, we can approximate the Doppler shift for light by the following, where we have replaced $\Delta\lambda$ with $d\lambda$ and used dv for v under the assumption of the box starting in free fall at the instant the photon is released:

$$\frac{d\lambda}{\lambda} \approx \frac{dv}{c} \qquad (4)$$

Now if we substitute (3) into (4), we obtain

$$\frac{d\lambda}{\lambda} \approx \frac{GM}{c^2}\frac{dr}{r^2} \qquad (5)$$

Next we will integrate both sides from the surface of the spherical mass (which we will call R_0) out to an infinite distance where spacetime is flat. We will use the notation λ_0 and λ_∞ for the wavelength of the light at the surface of

the spherical mass and in the infinitely distant flat spacetime, respectively:

$$\int_{\lambda_0}^{\lambda_\infty}\frac{d\lambda}{\lambda} \approx \int_{r_0}^{\infty}\frac{GM}{c^2}\frac{dr}{r^2} \qquad (6)$$

Performing the integrations and then taking the inverse of the expression within the natural logarithm (to obtain a format to permit comparison with our gravitational redshift expression), we have the following solution:

$$\int_{\lambda_0}^{\lambda_\infty}\frac{d\lambda}{\lambda} \approx \frac{GM}{c^2}\int_{R_0}^{\infty}\frac{dr}{r^2}$$

$$[\ln \lambda]_{\lambda_0}^{\lambda_\infty} \approx \frac{GM}{c^2}\left[-\frac{1}{r}\right]_{R_0}^{\infty}$$

$$\ln \frac{\lambda_\infty}{\lambda_0} \approx \frac{GM}{c^2}\left[0 + \frac{1}{R_0}\right]$$

$$-\ln \frac{\lambda_0}{\lambda_\infty} \approx \frac{GM}{R_0 c^2}$$

This can be written as

$$\frac{\lambda_0}{\lambda_\infty} \approx e^{-\frac{GM}{R_0 c^2}} \qquad (7)$$

You may be disappointed that this is not the same as Equation 30-52, but remember that this is just an approximate expression for relatively weak spacetime. We will show in parts (b) and (c) of this example that to first order it is equivalent.

(b) We will start with the exact expression stated for the gravitational redshift:

$$\frac{\lambda_0}{\lambda_\infty} = \left(1 - \frac{2GM}{R_0 c^2}\right)^{1/2} \qquad (30\text{-}52)$$

Assume that we are in only weakly curved spacetime, where the term $\dfrac{2GM}{R_0 c^2} \ll 1$. This will allow us to do a series expansion using the following with $x = -\dfrac{2GM}{R_0 c^2}$:

$$(1 + x)^{1/2} \approx 1 + \frac{1}{2}x - \frac{1}{8}x^2 + \frac{1}{16}x^3 + \cdots$$

If spacetime is sufficiently weakly curved, we can keep only the first term (after the fixed 1) and obtain

$$\frac{\lambda_0}{\lambda_\infty} = \left(1 - \frac{2GM}{R_0 c^2}\right)^{1/2} \approx 1 - \frac{GM}{R_0 c^2} \qquad (8)$$

This is the result obtained for the gravitational redshift if we use the exact relationship, but assume that M/R_0 is sufficiently small that we need only keep the first two terms in the series expansion.

(c) In this part, we will start from the approximate expression for the gravitational redshift that we obtained in part (a):

$$\frac{\lambda_0}{\lambda_\infty} \approx e^{-\frac{GM}{R_0 c^2}} \qquad (7)$$

Again we are assuming that the spacetime is weakly curved and the expression in the exponent is not much different from

(continued)

zero. The series expansion for an exponential under these conditions is

$$e^{-x} \approx 1 - x + \frac{1}{2}x^2 - \frac{1}{6}x^3 + \cdots$$

This means we can approximate (7) by the following, which is equal to what we found in part (b):

$$\frac{\lambda_0}{\lambda_\infty} \approx e^{-\frac{GM}{R_0 c^2}} \approx 1 - \frac{GM}{R_0 c^2}$$

Making sense of the result

While the thought experiment leads to a result approximately equal to the exact expression, when spacetime is sharply curved, the two will diverge. The true "proof" of the gravitational redshift is that it makes predictions confirmed by observations, including the correction to the GPS system considered in the final section of this chapter.

Black Holes

The right side of Equation 30-50 is zero when

$$R_0 = \frac{2GM}{c^2} \tag{30-53}$$

An observer at distance R_0 would record zero local time change even when a distant observer in flat spacetime notes a very long time interval. For this reason, we call this value of R_0 the **event horizon**. The stopping of time at this distance from the mass justifies the expression, "nothing ever happens at the event horizon." We will see below that Equation 30-53 is also an expression for the dimension of a black hole.

The idea of a **black hole**—a region that does not let even light escape—may seem far-fetched, but there is excellent evidence for the existence of several types of black holes in the universe. First, we will use a classical argument to derive the size of the boundary of the black hole. It is a boundary in the sense that if you pass within, you will no longer be able to communicate with the rest of the universe.

Recall from Chapter 11 that escape from near the surface of a spherical object of mass M and radius R requires the total of the kinetic energy and the gravitational potential energy to be at least zero:

$$\frac{1}{2}mv_{esc}^2 - \frac{GMm}{R^2} \geq 0 \tag{11-17}$$

When we divide both sides by m, set the expression to the minimum value of 0, use c for the escape speed, and solve for this special value of R, which we will call R_S, we get the following result:

KEY EQUATION
$$R_S = \frac{2GM}{c^2} \tag{30-54}$$

While this classical approach to the escape of light is not strictly valid, the field equations of general relativity yield exactly the same result when solved for a single,

MAKING CONNECTIONS

Black Holes

The first to consider the idea of a dense star trapping light was probably the British clergyman Rev. John Michell (1724–1793), who proposed the idea in 1783. In the view of general relativity, a black hole is an object that is completely collapsed in terms of spacetime curvature. The term *black hole* was coined by American physicist John Wheeler because it was awkward to keep saying "completely gravitationally collapsed object." Wheeler was passionate about the importance of teaching and learning physics and continued to teach first-year physics through most of his career. Together with Edward Taylor he wrote one of the best books on relativity, called *Spacetime Physics*. Theoretical calculations of the collapse of large stars led to predictions of the existence of stellar black holes. Observations of an object near the X-ray source Cygnus X-1 by Canadian astrophysicists including

Dr. Tom Bolton of the University of Toronto established the first (and some of the most conclusive) evidence for the existence of a stellar black hole. Canadian rock band Rush wrote and performed a song called "Cygnus X-1" about this discovery and the strange nature of black holes.

More recent data suggest the existence of much larger black holes at the centres of most galaxies, including our own Milky Way galaxy. The LIGO experiment (see Chapter 29 and the Detecting Gravitational Waves box) tests general relativity, including the detection of interacting black holes. The famous physicist Dr. Stephen Hawking (b. 1942) has proposed the possible existence of tiny primordial black holes formed in regions of extremely dense matter during the initial stage of the expansion of the universe.

Solving for Black Holes

Einstein had thought that it would be many years before the field equations for general relativity would be solved, but only months after receiving the equations in a letter from Einstein, the brilliant German physicist and astrophysicist Karl Schwarzschild (Figure 30-23) completed an exact solution for a single, spherical non-rotating mass, a solution that could be applied to black holes. Soon afterward, Schwarzschild produced a more general solution with much wider applications. At that time, Schwarzschild was literally in the trenches of the Russian front during World War I. Tragically, he contracted a rare skin disease and died at age 42, shortly after publishing the second solution.

Schwarzschild published two papers on celestial mechanics before he was 16, and he speculated on the possibility of curved spacetime well before Einstein developed the concept. Although Schwarzschild's best-known work is related to general relativity, he also made significant contributions to astronomical research and quantum mechanics.

Public Domain

Figure 30-23 Dr. Karl Schwarzschild.

radially symmetric mass (like a black hole). The escape limit for light, R_S, is called the **Schwarzschild radius** in honour of the German scientist Karl Schwarzschild (1873–1916), the first person to solve Einstein's field equations for the case of a black hole.

What are black holes in terms of general relativity? The mass of a black hole is so concentrated that it causes the extreme curvature of spacetime. As described above, gravitational time dilation at the Schwarzschild radius (event horizon) essentially slows time to a stop. If one is inside the event horizon of a black hole, it is impossible to ever escape from the black hole, or even to send information past the event horizon of the black hole. Thus, the Schwarzschild radius is a measure of the size of a black hole. If we shrink the physical size of an object to less than this limit, the object will become a black hole. For example, an object with Earth's mass would have to shrink to a radius of 8.8 mm to become a black hole. Some very massive stars become black holes when they exhaust their supply of nuclear fuel and collapse. It appears that natural processes also form much larger black holes at the centres of galaxies.

Detecting Gravitational Waves

In Chapter 29, we looked at how the LIGO detector works. In 2016, the LIGO group announced the detection of gravitational waves. So how are gravitational waves generated, and why was this discovery so important?

We saw earlier in this chapter that according to general relativity, masses warp spacetime. It would seem qualitatively reasonable that if we had moving masses, the warping of spacetime would change location, and perhaps some sort of wave would be generated. It turns out that to generate a gravitational wave, we need a little bit more than simple motion, we need an acceleration of the masses, in the same way that accelerating charges produce electromagnetic waves. Actually, to be precise,

we require even more than just acceleration of a mass: there needs to be a gravitational quadrupole moment. We will not consider quadrupole moments in detail, but in Section 19-10 we considered electric dipole moments, and two of these, oppositely directed, can be used to create an electric quadrupole.

One way for masses to accelerate is when two masses are in orbit around a common centre of mass (this is called a binary system). It turns out that this has a gravitational quadrupole moment and will generate gravitational waves. We illustrate the process in Figure 30-24. The "tracks" behind the masses represent the change in the ripples of spacetime as the masses move. The accelerating masses generate gravitational waves that take away

(continued)

MODERN PHYSICS

some of the orbital energy, so the two masses spiral closer to each other. This in turn causes even more gravitational wave generation, and the masses spiral closer and closer.

It turns out that the gravitational wave power generated in a binary system depends on the inverse fifth power of the separation distance, so the dominant gravitational wave signal will be in the moments just before the eventual coalescence of the two masses.

With this background we show in Figure 30-25 the signals detected by the two LIGO detectors for the event announced in 2016 (but actually detected in September 2015). In the upper half of Figure 30-25 is the relative strain plotted versus time (as we saw in Chapter 29, LIGO can detect signals well under one part in 10^{21}). The entire sequence shown here is only a few tenths of a second. We see that, particularly when a best fit is applied to the signal (middle portion of the figure), the signals detected by the two LIGO sites agree very well. The bottom part of the plot gives the frequency of the signal, in hertz, as a function of time, as well as the relative strength of the signal (plotted in terms of the colour). We see, as expected, that the frequency increases, and the signal gets stronger, as the two black holes spiral closer and closer.

It is only in the last decade that numerical methods of approximately solving difficult general relativity field equation problems have permitted accurate models for events such as the inward spiral of two massive objects. We show in Figure 30-26 the results of such simulations for the inward spiral of the black holes believed to be responsible for the signal detected by LIGO. We see that the segment observed lasts only about 0.2 s, and the amplitude and frequency of the predicted signal, as well as the overall shape, agree very well with the LIGO detection of Figure 30-25. The black holes achieve orbital speeds of about 60% of the speed of light just prior to merger.

So what do we actually mean by merger? As we saw in this chapter, there is an event horizon around a black hole of size determined by the Schwarzschild radius. Once the black holes get close enough to each other, they will merge into a single black hole with a larger Schwarzschild radius. Using calculations that formed the basis of Figure 30-26, physicists were able to determine quite a bit about the merging black holes detected by LIGO. One black hole had a mass between 32 and 41 times that of the Sun, while the other was between 25 and 33 solar masses. This means that they were the two largest stellar black holes detected by any method (much larger black holes do exist at the centres of galaxies). After the merger, the combined mass was about 62 solar masses, compared to approximately 65 solar masses prior to the merger. The difference is emitted as energy, most in the form of gravitational wave energy. The location in the sky of the source can only be confined loosely to an arc in the southern hemisphere. We do know that the black hole merger occurred very far away, somewhere between 0.7 billion light years and 1.9 billion light years.

So why was this so important? It provided a direct confirmation of gravitational waves, and as such another confirmation of the predictions of general relativity. At the same time we have detection of the largest stellar black holes ever confirmed. But even more important, the gravitational wave detectors provide a new observational tool for looking at the universe, just as radio or X-ray telescopes gave us new ways to research astrophysical processes. Estimates suggest that LIGO will be able to detect perhaps 20 black hole mergers per year, and 40 mergers of neutron stars. Therefore, LIGO will soon, probably by the time you are reading this, provide significant statistics on the populations and interactions of the largest mass stellar remnants in our universe.

NASA/Dana Berry, Sky Works Digital

Figure 30-24 This shows a sequence as two masses in rotation about a common centre of mass move closer and closer as they lose energy through generated gravitational waves.

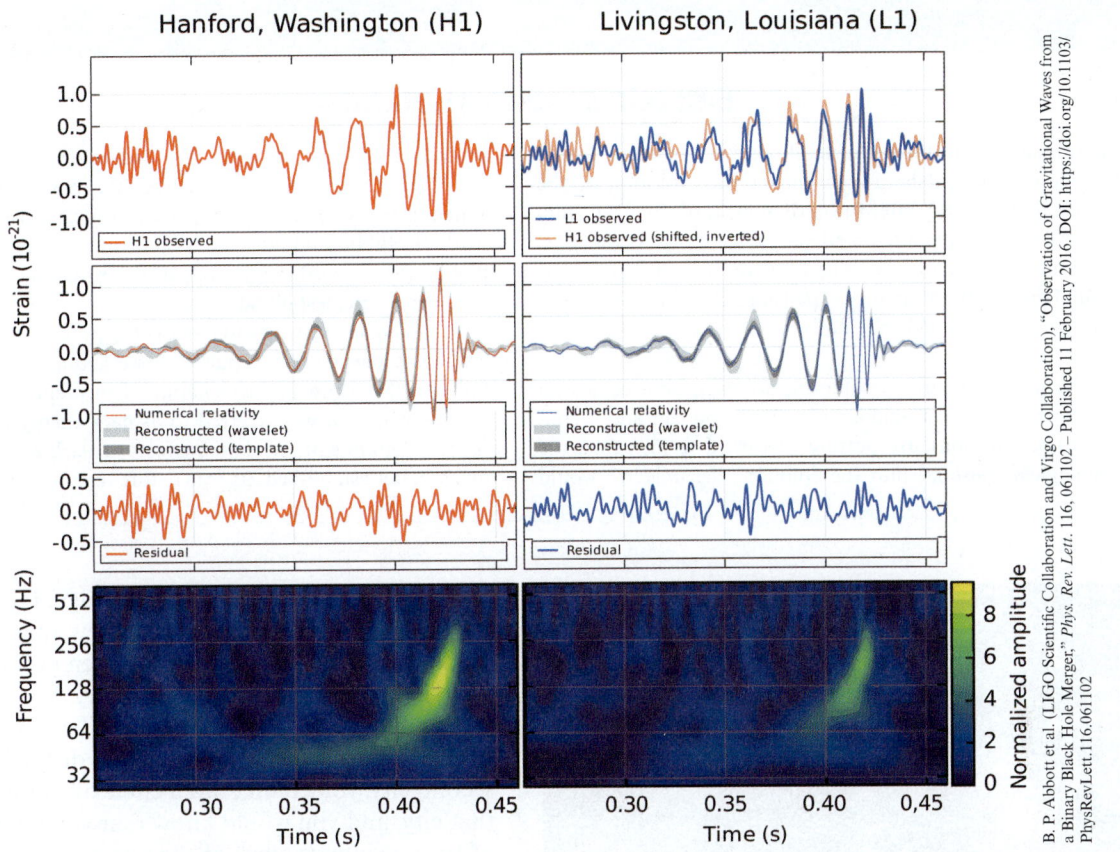

B. P. Abbott et al. (LIGO Scientific Collaboration and Virgo Collaboration), "Observation of Gravitational Waves from a Binary Black Hole Merger," *Phys. Rev. Lett.* 116, 061102 – Published 11 February 2016. DOI: https://doi.org/10.1103/PhysRevLett.116.061102

Figure 30-25 Signals from the coalescing black holes detected by LIGO. The top is the actual signal, and the middle provides an optimum fit to the data. The left is from the Hanford site, and the right from Livingston. The bottom part of the graph shows the frequency of the signal, and the relative strength displayed by colour.

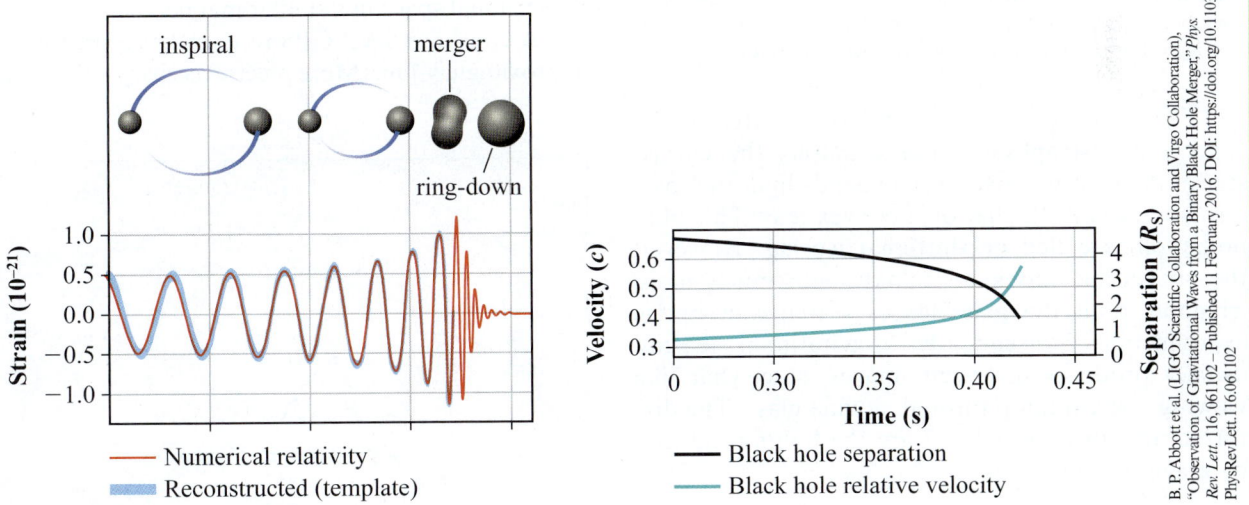

B. P. Abbott et al. (LIGO Scientific Collaboration and Virgo Collaboration), "Observation of Gravitational Waves from a Binary Black Hole Merger," *Phys. Rev. Lett.* 116, 061102 – Published 11 February 2016. DOI: https://doi.org/10.1103/PhysRevLett.116.061102

Figure 30-26 The graph on the left shows the expected strain for the inward spiral of the two black holes. In the graph on the right are the orbital speeds of the black holes expressed in terms of a fraction of the speed of light, and the separation of their centres, expressed in terms of the radius of the Sun.

MODERN PHYSICS

Einstein's Greatest Blunder?

The ultimate application of general relativity as a new theory of gravitation would be a solution for the evolution of the universe itself. Shortly after the development of the field equations, Albert Einstein attempted such a solution. At the time, the expansion of the universe had not yet been discovered by Edwin Hubble. Without this expansion, the solution of the field equations resulted in a universe that would collapse on itself.

The general relativity solution, however, allowed for an extra term (somewhat analogous to how you can add a constant of integration). By putting this extra term, called the **cosmological constant**, into the solution, the universe would no longer collapse on itself. The cosmological constant has the effect of a repulsive force between the mass in the universe. When, a few years after his solution, Einstein learned of the work of Hubble and the expansion of the universe, he realized that he did not need the cosmological constant, and he referred to it as his "greatest blunder."

Starting in 1998, astrophysicists began to realize that there was some unexplained repulsive force acting throughout the universe. The cause and true nature of this repulsive force was not known, so it was referred to as **dark energy** (dark in the sense of unknown). While the effects of dark energy are now much better established, its true nature remains unclear.

Andrew Fruchter (STScI) et al., WFPC2, HST, NASA

Figure 30-27 This astronomical image is proof that mass really does bend spacetime. The dense galaxy cluster called Abell 2218 bends light from more distant galaxies, stretching their images into arcs.

The concepts of general relativity affect many aspects of astrophysics. For example, the curved spacetime near massive objects bends light in a pattern somewhat like that of a convex lens. This phenomenon is called **gravitational lensing**. However, the spacetime lenses of galaxies are much more irregular than the glass lenses you use in optics experiments. Consequently, gravitational lensing usually produces distorted images, somewhat like viewing a street lamp through a wine glass. The distorted arcs in Figure 30-27 are the lights produced by gravitational lenses.

30-11 Relativity and the Global Positioning System

The **Global Positioning System** (GPS) measures time delays in signals received from four different satellites. With this information, we can calculate position coordinates for the receiver and also obtain a precise time.

Consider this two-dimensional analogy: If you were on the surface of a flat Earth and knew that you were a certain distance, d_1, from a city, then you could draw a circle of radius d_1 around that city. You would then know that your position was somewhere on that circle. If you also knew that you were at a distance d_2 from a second city, you could draw a circle of radius d_2 around that city, and you would know that you must be at one of the two intersections of the two circles. For example, Figure 30-28 shows one point, Thompson, Manitoba: the point is 1188 km from Calgary and 2211 km from Montréal. Knowing your distance from a third city lets you uniquely determine which intersection point is correct. The GPS system calculates a position in three-dimensional space in a similar manner.

A consumer-level GPS receiver has a precision of approximately 1 m. (More precise receivers are available

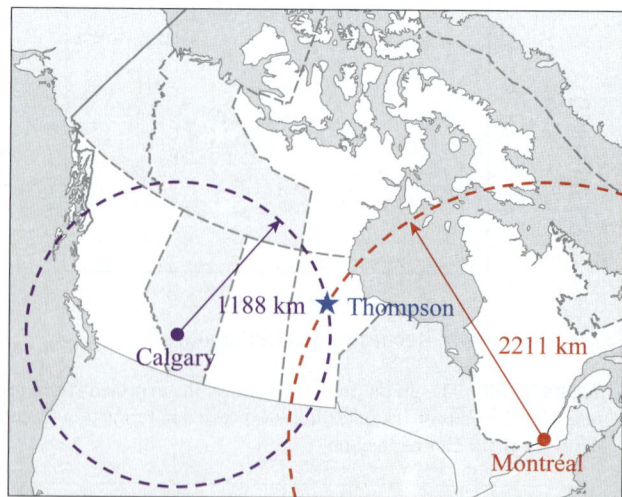

Figure 30-28 You can use known distances from two cities to find two possible locations on a flat Earth (in this case one near Thompson, Manitoba, and the other in the northern United States). The GPS system works in three dimensions with more points.

for military and government operations.) At present, there are more than 30 active GPS satellites. These satellites orbit at a distance of approximately four Earth radii in six planes inclined at about 55° with respect to the equator, as shown in Figure 30-29. Each satellite circles Earth approximately twice every day.

If the desired precision is of the order of 1 m, then the time must be accurate to that distance divided by the speed of light, or approximately 3.3 ns. The atomic (cesium) clocks that are on the satellites have sufficient precision (a typical cesium atomic clock has a precision of 1 ns or better). However, the satellite clocks are all moving relative to the receiver on Earth, and they are several times farther from Earth's centre.

Even though the curvature of spacetime is not severe near Earth, there is still a small effect due to gravitational time dilation. Time runs more slowly in the more sharply curved spacetime near Earth's surface. Therefore, the satellite clocks will, by general relativity, tick faster than one on Earth. The difference is approximately 46 μs a day (see Example 30-14), great enough that the needed 3.3 ns precision would be lost in less than 10 s.

However, we must also consider special relativity time dilation because the satellites in orbit are travelling at a significant speed compared to the ground-based receiver. Even though the speed is not large compared to the speed of light, the precision needed means that relativistic effects are still important. Note that time dilation causes the satellite clocks to run more slowly than clocks on the ground by approximately 7 μs per day (see Example 30-14). Since the two time dilations have opposite effects, the special relativity dilation partly cancels the general relativity dilation, leaving a net time error of approximately 39 μs per day.

The designers of the GPS system understood both special and general relativity effects and built the corrections into the system's calculations. With these corrections, based directly on Einstein's discoveries about a century ago, we have a remarkably precise system for navigation worldwide.

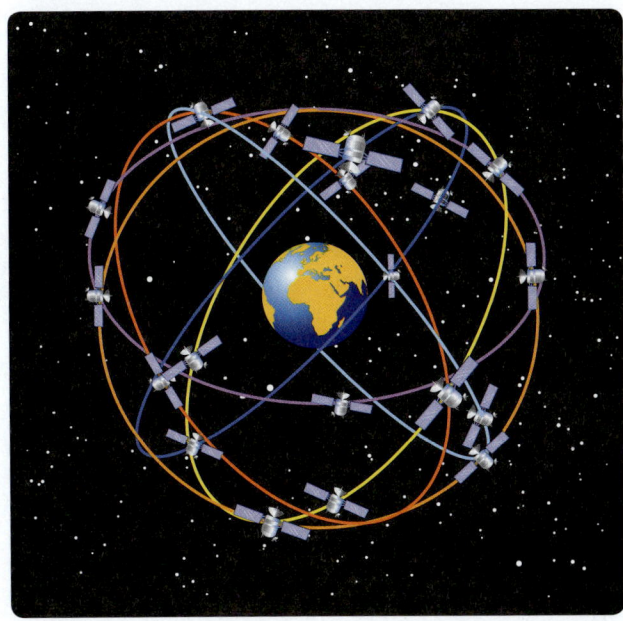

Figure 30-29   The configuration of GPS satellites.

CHECKPOINT 30-11

GPS Corrections

Which statement correctly summarizes the relativistic effects for a clock on a GPS satellite orbiting Earth as compared to an identical clock on the surface of Earth?
(a) The GPS clock runs faster due to special relativity and more slowly due to general relativity.
(b) The GPS clock runs faster due to general relativity and more slowly due to special relativity.
(c) The GPS clock runs more slowly due to both effects.
(d) The GPS clock runs faster due to both effects.

ANSWER: (b) The clock on the GPS satellite is moving relative to the clock on Earth, so the GPS clock runs more slowly due to special relativistic time dilation. However, the spacetime around the satellite is less curved than the spacetime at Earth's surface, so the satellite clock runs faster because the gravitational time dilation is nearer to Earth's centre.

EXAMPLE 30-14

GPS Relativistic Corrections

(a) The GPS satellites make one orbit around Earth in one-half of a sidereal day. A sidereal (star-based) day is approximately 23 h and 56 min. Find the radius of the orbit of a GPS satellite, assuming that the orbit is circular.
(b) What is the orbital speed of a GPS satellite relative to the centre of Earth?
(c) Calculate the amount per solar day by which a GPS satellite clock differs from a clock on Earth due to special relativistic time dilation. State any assumptions you make for this calculation.

(d) Calculate the amount per solar day by which a GPS satellite clock differs from a clock on Earth due to gravitational time dilation.

Solution

(a) We will first use the Newtonian form of Kepler's third law to solve for the length of the semi-major axis, a. Since we are assuming a circular orbit, this axis will be the same as the radius of the orbit:

$$T^2 = \frac{4\pi^2}{G(m_1 + m_2)} a^3$$

(continued)

MODERN PHYSICS

The satellite period, T, is 11 h and 58 min, or 43 080 s. The mass of the satellite is negligible compared to the mass of Earth, so for the mass term in the denominator we use Earth's mass, 5.974×10^{24} kg. Substituting these values and solving for a gives

$$a = 2.656 \times 10^7 \text{ m}$$

(b) For a circular orbit, the orbital speed is given by $v = \dfrac{2\pi a}{T}$. Substituting $a = 2.656 \times 10^7$ m and $T = 43080$ s, we obtain $v = 3.874 \times 10^3$ m/s.

(c) The clock on Earth moves as Earth rotates, so the relative speed of the satellite varies because its orbit is not directly above the equator. We will assume that the mean speed of the satellite relative to the clock on Earth is equal to the orbital speed of the satellite (which is a reasonable approximation). Both clocks are actually accelerating constantly, so neither is in a truly inertial reference frame. We will assume that the effects of these accelerations can be ignored.

We saw in the special relativity time dilation section that

$$\Delta t = \frac{\Delta t_0}{\sqrt{1 - v^2/c^2}}$$

Using the orbital speed from part (b), we obtain

$$\frac{\Delta t}{\Delta t_0} = 1.000\ 000\ 000\ 0834$$

To find the time dilation per day, multiply the difference between this ratio and 1 by the number of seconds in a solar day:

$$0.000\ 000\ 000\ 0834 \times (24 \text{ h} \times 3600 \text{ s/h}) = 7.20 \times 10^{-6} \text{ s}$$
$$= 7.2\ \mu\text{s}$$

(d) The general relativity time dilation relationship relates the time measured by an observer at a distance R_0 from the centre of a spherical mass M to the time measured by an observer in flat spacetime, infinitely far from any masses:

$$\frac{\Delta t_0}{\Delta t_\infty} = \left(1 - \frac{2GM}{R_0 c^2}\right)^{1/2}$$

However, we want to compare two clocks that are both in curved spacetime. We use the subscript E to denote quantities related to the Earth clock and the subscript S for quantities related to the satellite clock:

$$\frac{\Delta t_S}{\Delta t_E} = \frac{\dfrac{\Delta t_S}{\Delta t_\infty}}{\dfrac{\Delta t_E}{\Delta t_\infty}} = \frac{\left(1 - \dfrac{2GM}{R_S c^2}\right)^{1/2}}{\left(1 - \dfrac{2GM}{R_E c^2}\right)^{1/2}}$$

We know from part (a) that $R_S = 2.656 \times 10^7$ m. The radius of Earth is 6.378×10^6 m, and the mass of Earth is 5.974×10^{24} kg. Substituting these values, we obtain

$$\frac{\Delta t_S}{\Delta t_E} = 1.000\ 000\ 000\ 528$$

Again, we multiply the difference between this ratio and 1 by the number of seconds in a solar day to find the time dilation per day:

$$0.000\ 000\ 000\ 528 \times (24 \text{ h} \times 3600 \text{ s/h}) = 4.56 \times 10^{-5} \text{ s}$$
$$= 46\ \mu\text{s}$$

Making sense of the result

In part (a), we found that the GPS satellites were at a radial orbital distance of about four Earth radii, a reasonable result. Because the satellites orbit twice each day, we expect them to be at a lower altitude than the geosynchronous satellites studied in Chapter 11.

In part (b), we obtained an orbital speed of approximately 3.9 km/s, which is somewhat less than the escape speed (11.2 km/s) and therefore a reasonable result. In parts (c) and (d), we found that the correction due to general relativity is greater than the correction due to special relativity, and both corrections are essential for determining accurate locations using the GPS system.

KEY CONCEPTS AND RELATIONSHIPS

SPECIAL RELATIVITY

Special relativity deals with objects moving near the speed of light as seen from two inertial (non-accelerating) reference frames. As we approach speeds that are comparable to the speed of light, there is a slowing of time, an increase in linear momentum and energy, and a contraction of the lengths of objects in the direction of motion.

Time Dilation

Proper time is measured by an observer for whom the interval events begin and end at the same spatial coordinates. The time interval measured by any other observer is

$$\Delta t = \frac{\Delta t_0}{\sqrt{1 - v^2/c^2}} \tag{30-8}$$

where Δt_0 is the proper time interval, v is the relative speed, and c is the speed of light.

Length Contraction

In special relativity, the observer for whom an object is not moving measures the proper length, L_0, for that object. Lengths are contracted in the direction of motion such that

$$L = L_0 \sqrt{1 - v^2/c^2} \tag{30-15}$$

The same observer does not measure both proper time and proper length.

Relativistic Momentum

Relativistic momentum is given by

$$\vec{p} = \frac{m_0 \vec{v}}{\sqrt{1 - v^2/c^2}} \qquad (30\text{-}27)$$

where m_0 is the rest mass of the object.

Relativistic Kinetic Energy

Mass and energy are different forms of a unified quantity, with mass–energy equivalence given by $E = mc^2$.

Relativistic kinetic energy is defined as the difference between the total relativistic energy, E, and the rest mass energy:

$$K = mc^2 - m_0c^2 = m_0c^2 \left(\frac{1}{\sqrt{1 - v^2/c^2}} - 1 \right) \qquad (30\text{-}35)$$

Relativistic momentum, total energy, and rest mass energy are related by

$$E^2 - (pc)^2 = (m_0c^2)^2 \qquad (30\text{-}38)$$

Spacetime Intervals

According to relativity, we must unify space and time coordinates into spacetime. Different observers no longer agree on various physical measurements, although they do agree on spacetime intervals between pairs of events:

$$(\Delta s)^2 = (c\Delta t)^2 - (\Delta x)^2 - (\Delta y)^2 - (\Delta z)^2 \qquad (30\text{-}23)$$

Lorentz Transformation

Consider two spacetime reference systems, S and S', with S' moving at speed v in the $+x$-direction of S. The Lorentz transformation converts between spacetime coordinates measured in the two reference frames. We can define a relativistic spacetime four vector and use the Lorentz transformation on that four vector as follows:

$$\begin{pmatrix} ct' \\ x' \\ y' \\ z' \end{pmatrix} = \begin{pmatrix} \gamma & -\beta\gamma & 0 & 0 \\ -\beta\gamma & \gamma & 0 & 0 \\ 0 & 0 & 1 & 0 \\ 0 & 0 & 0 & 1 \end{pmatrix} \begin{pmatrix} ct \\ x \\ y \\ z \end{pmatrix} \qquad (30\text{-}21)$$

We have used the speed factor, β, and the Lorentz factor, γ, to simplify the expression:

$$\beta \equiv \frac{v}{c} \qquad (30\text{-}10)$$

$$\gamma \equiv \frac{1}{\sqrt{1 - \dfrac{v^2}{c^2}}} = \frac{1}{\sqrt{1 - \beta^2}} \qquad (30\text{-}20)$$

The inverse Lorentz transformation takes us from S' to S spacetime coordinates:

$$\begin{pmatrix} ct \\ x \\ y \\ z \end{pmatrix} = \begin{pmatrix} \gamma & +\beta\gamma & 0 & 0 \\ +\beta\gamma & \gamma & 0 & 0 \\ 0 & 0 & 1 & 0 \\ 0 & 0 & 0 & 1 \end{pmatrix} \begin{pmatrix} ct' \\ x' \\ y' \\ z' \end{pmatrix} \qquad (30\text{-}22)$$

Relativistic Velocity Addition

If reference frame S' is moving at velocity v along the x-axis relative to an S reference frame, we can relate the velocity in the x-direction of some object as measured in the two coordinate systems (u' in S' and u in S):

$$u = \frac{u' + v}{1 + vu'/c^2} \qquad (30\text{-}42)$$

We have not used vector notation, but the correct signs must be included for these one-dimensional velocities. The associated speeds will all be less than c in either coordinate system.

Relativistic Doppler Shift

When a source is moving away from the observer at a speed v_s, the wavelength of the electromagnetic wave received, λ_r, will be related to the wavelength emitted at the source, λ_s, by

$$\frac{\lambda_r}{\lambda_s} = \frac{\sqrt{1 + v_s/c}}{\sqrt{1 - v_s/c}} = \frac{\sqrt{c + v_s}}{\sqrt{c - v_s}} \qquad (30\text{-}46)$$

When the source moves away, there is a redshift (toward longer wavelength) in the received wavelength, and when the source moves toward you, there is a blueshift.

GENERAL RELATIVITY

General relativity expands special relativity to include the gravitational effects of masses by treating gravitation as a curvature of spacetime.

Gravitational Time Dilation

Time passes more slowly in regions of curved spacetime. The time interval, Δt_0, measured by an observer at a distance R_0 from a mass M is

$$\frac{\Delta t_0}{\Delta t_\infty} = \left(1 - \frac{2GM}{R_0c^2} \right)^{1/2} \qquad (30\text{-}50)$$

where Δt is the time interval as measured by an observer far from any masses.

Gravitational Redshift

One consequence of gravitational time dilation is that electromagnetic waves received from regions of sharply curved spacetime will be redshifted:

$$\frac{\lambda_0}{\lambda_\infty} = \left(1 - \frac{2GM}{R_0c^2} \right)^{1/2} \qquad (30\text{-}52)$$

Here λ_0 is the wavelength emitted from curved spacetime (distance R_0 from the centre of a spherically symmetric mass M) and λ_∞ is the wavelength observed from flat spacetime.

Black Holes

Sufficiently dense masses can curve spacetime sufficiently that not even light can escape. The "radius" of the event horizon for a black hole of mass M is given by

$$R_S = \frac{2GM}{c^2} \qquad (30\text{-}54)$$

APPLICATIONS

nuclear fusion power, gravitational lensing, Global Positioning System (GPS), dynamics of particles in accelerators and in cosmic ray physics, black hole physics, space travel

KEY TERMS

black hole, blueshift, coordinate time, cosmological constant, dark energy, equivalence principle, ether, event horizon, field equations, general relativity, Global Positioning System, gravitational lensing, gravitational redshift, gravitational time dilation, inertial reference frames, invariant, length contraction, light-like, light cone, mass–energy equivalence, Michelson–Morley experiment, nuclear fusion, proper length, proper time, proper time interval, radiation pressure, redshift, redshift parameter z, relativistic kinetic energy, relativistic momentum, rest mass, Schwarzschild radius, space-like, spacetime, spacetime diagrams, spacetime event, spacetime interval, special relativity, special relativity time dilation, tensors, thought experiment, time dilation, time-like, total energy, twin paradox, world line

QUESTIONS

1. Which of the following statements is correct with respect to the laws of physics in different inertial reference frames?
   (a) The laws of physics are different in different inertial reference frames.
   (b) The laws of physics vary by a constant in different inertial frames.
   (c) The laws of physics are the same in different inertial frames.
   (d) The laws of physics are scaled by some factor in different inertial reference frames.

2. Parvati moves past you at a speed one-half the speed of light. He measures 100 s for a trip he takes., What time would you obtain for Parvati's trip?
   (a) more than 100 s
   (b) exactly 100 s
   (c) less than 100 s but more than 50 s
   (d) exactly 50 s

3. On what do different inertial observers agree?
   (a) only the laws of physics
   (b) only the laws of physics and the speed of light
   (c) the laws of physics, the speed of light, and the spacetime interval between two events
   (d) the laws of physics, the speed of light, the times of events, and the spatial coordinates of events

4. According to special relativity, for observers moving past you at a high speed, you observe their time running slowly, their lengths contracted in the direction of motion, and an increase in their mass. What will these observers say about you?
   (a) My time runs fast, my mass decreases, and my length increases.
   (b) My time runs fast, but my mass and length do not change.
   (c) My time runs slowly, my mass decreases, and my length increases.
   (d) My time runs slowly, my mass increases, and my length contracts.

5. A black hole has a mass M, and its Schwarzschild radius is R_S. We add mass so that its mass is now $2M$. What is the new Schwarzschild radius?
   (a) It is no longer a black hole.
   (b) $R_S/2$
   (c) $1.4R_S$
   (d) $2R_S$

6. If the Sun were to become a black hole (astrophysics tells us that this will not happen for stars the size of the Sun), what would happen to Earth?
   (a) Earth would fall into the black hole Sun.
   (b) Earth would stay in exactly its current orbit.
   (c) Earth would gradually spiral in until it fell within the Schwarzschild radius of the black hole Sun.
   (d) Earth would break free from the gravity of the Sun and move into interstellar space.

7. On which of the following objects would time be expected to run most slowly?
   (a) a planet of mass M and radius R
   (b) a planet of mass M and radius $2R$
   (c) a planet of mass M and radius $R/2$
   (d) a planet of mass $2M$ and radius $2R$

8. Consider three particles. According to some inertial reference frame,
   • A has a rest mass energy of E_0 and a total energy of $2E_0$.
   • B has a rest mass energy of $2E_0$ and a total energy of $3E_0$.
   • C has a rest mass energy of $3E_0$ and a total energy of $4E_0$.
   Which statement about the relative rest masses is correct?
   (a) A has the least rest mass, and C has the greatest rest mass.
   (b) A, B, and C all have exactly the same rest mass values.
   (c) C has the least rest mass, and A has the greatest rest mass.
   (d) A and C have a larger rest mass than B.

9. Considering the data in question 8, which statement about the relative kinetic energies of the objects is correct?
   (a) A has the least and C has the greatest kinetic energy.
   (b) C has the least and A has the greatest kinetic energy.
   (c) B has the least and A has the greatest kinetic energy.
   (d) All three have exactly the same kinetic energy values.

10. Considering the data from questions 8 and 9, which particle has the greatest speed?
    (a) A
    (b) B
    (c) C
    (d) They all have exactly the same speed.

11. You are travelling in a spaceship at a speed in the $+x$-direction of $0.5c$ relative to an Earth-based reference frame. Within the spaceship, you are conducting an experiment in which subatomic particles are moving at a speed of $0.5c$ relative to you (in the $+x$-direction as well). What is the approximate speed of the particles relative to Earth?
    (a) $0.3c$
    (b) $0.5c$
    (c) $0.8c$
    (d) $1.0c$

12. You observe light produced by atoms located just outside the surface of a very dense white dwarf star. A spectral line observed from these atoms compared to the same spectral line on Earth has
    (a) exactly the same wavelength
    (b) a slightly longer wavelength
    (c) a slightly shorter wavelength
    (d) a much longer wavelength

13. Which of the following statements is *false*?
    (a) The speed of light (in a vacuum) is constant.
    (b) Observers in different inertial reference frames measure the same speed of light.
    (c) The speed of light slows down in the region near (but outside of) a black hole.
    (d) As an object approaches the speed of light, the relativistic kinetic energy of the object becomes very large.

14. If the Lorentz factor is $\gamma = 5$, what is the speed?
    (a) $\sqrt{\dfrac{4}{5}}c$
    (b) $\sqrt{\dfrac{24}{25}}c$
    (c) $\sqrt{\dfrac{25}{24}}c$
    (d) $\sqrt{\dfrac{5}{4}}c$

15. Which statement is true regarding the gravitational redshift?
    (a) It is simply another way of looking at the relativistic Doppler shift.
    (b) It is required to keep relative speeds less than the speed of light.

    (c) It comes directly out of the Lorentz transformation for black holes.
    (d) Gravitational redshift follows from gravitational time dilation.

16. How fast must a source be moving away from you if the received wavelength of an electromagnetic wave is 3/2 that of the original wavelength emitted?
    (a) $\dfrac{5}{13}c$
    (b) $\dfrac{4}{9}c$
    (c) $\dfrac{2}{3}c$
    (d) $\dfrac{12}{13}c$

17. The redshift parameter z is 0.020 for a receding galaxy. If infrared light of wavelength 2000. nm is emitted by the galaxy, what will be the observed wavelength of the received radiation? Assume no significant gravitational redshift.
    (a) 1960. nm
    (b) 2004 nm
    (c) 2040. nm
    (d) 2400. nm

18. A quasar has $z = 3$. What can you conclude about its speed of recession?
    (a) It would imply movement at $3c$, which is impossible.
    (b) The quasar is moving toward you at $\dfrac{1}{3}c$.
    (c) The quasar is moving away at $\dfrac{8}{9}c$.
    (d) The quasar is moving away at $\dfrac{15}{17}c$.

19. If a mass is moving at a speed of $\dfrac{2}{3}c$, how will the magnitude of the relativistic momentum (p_r) compare to what would be calculated using the non-relativistic classical expression (p_c)?
    (a) $p_r = \dfrac{4}{9}p_c$
    (b) $p_r = \dfrac{2}{3}p_c$
    (c) $p_r = \dfrac{3}{\sqrt{5}}p_c$
    (d) $p_r = \dfrac{3}{2}p_c$

20. Two identical precise atomic clocks (called B and T) are available. They are synchronized, and then clock B is placed at the surface of Earth and T on the top of a high tower. Then, after some time has passed, clock T is taken down from the tower, placed beside B, and the times compared. Which of the following statements is true?
    (a) The two clocks will show exactly the same time.
    (b) Clock B will show a smaller time interval.
    (c) Clock T will show a smaller time interval.
    (d) It cannot be determined without knowing the height of the tower.

PROBLEMS BY SECTION

For problems, star ratings will be used, (★, ★★, or ★★★), with more stars meaning more challenging problems. The following codes will indicate if $\frac{d}{dx}$ differentiation, $\int$ integration, ▭ numerical approximation, or ∼ graphical analysis will be required to solve the problem.

Section 30-1 Special and General Relativity

21. ★ You are writing an introduction to relativity for non-specialist readers. Select one of the following as the best description of the key idea of *general* relativity. Support your choice through an analysis of all four statements.
    (a) No object can travel faster than the speed of light.
    (b) When objects travel near the speed of light, time goes more slowly.
    (c) Gravity is not a force but rather a curvature of spacetime near massive objects.
    (d) The laws of physics are the same for observers in relative motion.

Section 30-2 Reference Frames and the Michelson – Morley Experiment

22. ★ You learned in Chapter 20 that the permittivity of free space, ε_0, has the value 8.85×10^{-12} C²/(N·m²). Use the speed of light, $c = 3.00 \times 10^8$ m/s, to find a value for the permeability of free space, μ_0.

23. ★★ In their 1887 experiment, Michelson and Morley used an effective length of 1.10 m for the distance, d, in the interferometer. One arm of an interferometer is parallel to the motion of Earth in its orbit at a speed of 29 800 m/s. Calculate the time *difference* for light travel along the two paths, assuming that the standard value for the speed of light is used for the other arm and that speeds add in the Galilean way expected prior to relativity. Use Michelson and Morley's value for d.

Section 30-3 Postulates of Special Relativity and Time Dilation

24. ★★ (a) How long, from the perspective of an Earth-bound observer, would a flight around Earth's circumference at an altitude of 1000 km above Earth's surface take if you could travel at 50% of the speed of light? For this question, treat the reference frames as inertial.
    (b) What time would be required from the perspective of someone who was in the capsule taking the flight?

25. ★★ A fundamental particle called a muon has a lifetime, in its own reference frame, of 2.20 μs. Muons from a cosmic ray atmospheric burst have an apparent lifetime of 15.0 μs for a ground-based observer. What must be the speed of the muon relative to the ground-based observer?

Section 30-4 Length Contraction

26. ★★ The nearest star similar to the Sun is Alpha Centauri, which is at a distance of 4.40 ly. If one travels at 75% of the speed of light, how long would a one-way trip to the star from Earth require from the viewpoint of
    (a) an Earth-based observer?
    (b) an occupant on the spacecraft?
    Assume that the initial acceleration is sufficient that the entire trip is essentially at full speed.

27. ★★ (a) If you make a round trip to a star that is 400.0 ly away (as measured by an Earth-based observer) in a total time of 40.00 yr (as measured by people on the spacecraft), what speed is needed? (Assume, somewhat unrealistically, that the time that the spacecraft takes to accelerate to its cruising speed is negligible.)
    (b) What time would the Earth-based observer measure for the trip?
    (c) What distance (one way) would the spaceship determine for the trip?

Section 30-5 Lorentz Transformation

28. ★★ Reference frame S' is moving in the direction of the positive x-axis with a speed of $\frac{3}{4}c$ compared to a different reference frame, S. The origins of both reference frames are aligned at $t = 0$. In the S' reference frame the coordinates of a spacetime event are $t' = 1.15$ μm, $x' = 1250$ m, $y' = 300$ m, $z' = 420$ m. Use the Lorentz transform to determine the coordinates in the S reference frame.

29. ★★ Reference frame S' is moving in the direction of the positive x-axis with a speed of $0.900\,c$ compared to a different coordinate system, S. The origins of both reference frames are aligned at $t = 0$. In the S reference frame, the coordinates of a spacetime event are $t = 0.950$ μs, $x = 270$. m, $y = 450$. m, $z = 150$. m. Use the Lorentz transformation to determine the coordinates in the S reference frame.

30. ★ The Lorentz transformation between S and S' reference frames is given by

$$
\begin{pmatrix} ct' \\ x' \\ y' \\ z' \end{pmatrix} = \begin{pmatrix} \frac{2}{\sqrt{3}} & -\frac{1}{\sqrt{3}} & 0 & 0 \\ -\frac{1}{\sqrt{3}} & \frac{2}{\sqrt{3}} & 0 & 0 \\ 0 & 0 & 1 & 0 \\ 0 & 0 & 0 & 1 \end{pmatrix} \begin{pmatrix} ct \\ x \\ y \\ z \end{pmatrix}
$$

Describe the motion of the S' system relative to the S reference frame.

31. ★★ The Lorentz transformation between S and S' reference frames is given by

$$
\begin{pmatrix} ct' \\ x' \\ y' \\ z' \end{pmatrix} = \begin{pmatrix} \frac{5}{3} & -\frac{4}{3} & 0 & 0 \\ -\frac{4}{3} & \frac{5}{3} & 0 & 0 \\ 0 & 0 & 1 & 0 \\ 0 & 0 & 0 & 1 \end{pmatrix} \begin{pmatrix} ct \\ x \\ y \\ z \end{pmatrix}
$$

    (a) Describe the motion of the S' reference frame relative to the S reference frame.
    (b) If a spacetime event in the S reference frame is $t = 1.50$ μs, $x = 120$ m, $y = 150$ m, $z = 250$, what will be the spacetime coordinates of the event in the S' reference frame?
    (c) Write the inverse transformation that takes one from the S' reference frame coordinates to the S reference frame coordinates.
    (d) Use the transformation of part (c) to transform an S' spacetime event $t' = 2.40$ μs, $x' = 420$. m, $y' = 390$. m, $z' = 560$. m to coordinates in the S system.

Section 30-6 Spacetime

32. ★ Consider three spacetime reference frames, S, S', and S''. S' is moving in the direction of the positive x-axis at a speed of $0.75c$, while S'' is moving in the same direction but with a speed of $0.90c$ with respect to S. The three reference frames are coincident at $t = 0$. The spacetime interval, $(\Delta s)^2$, for some pair of spacetime events when measured by reference frame S is 8.40×10^3 m². What will be the spacetime interval as measured in the S' and in the S'' reference frames?

33. ★★ According to an observer, two spacetime events have the following coordinates: $t_1 = 12.0$ μs, $x_1 = 250.$ m, $y_1 = 300.$ m, $z_1 = 400.$ m; $t_2 = 8.00$ μs, and $x_2 = 400.$ m, $y_2 = 500.$ m, $z_2 = 600.$ m.
  (a) What is the spacetime interval between the two events?
  (b) Is this interval time-like, space-like, or light-like?

34. ★★ According to an observer, two spacetime events have the following coordinates: $t_1 = 4.5$ μs, $x_1 = 750$ m, $y_1 = 100$ m, $z_1 = 200$ m; $t_2 = 5.50$ μs, $x_2 = 900$ m, $y_2 = 500$ m, $z_2 = 300$ m
  (a) What is the spacetime interval, $(\Delta s)^2$, between the two events?
  (b) Is this interval time-like, space-like, or light-like?
  (c) Could one event have caused the other?

35. ★ Observer Mei sees two events as being simultaneous in time but separated by 300. m in space. Li sees the two events as taking place with a spatial separation of 500. m. What will Li say is the time separation between the two events?

36. ★★ One observer sees two spacetime events happen separated by 10.0 ns in time and by 3.0 m in space. Another inertial observer sees the same events as happening at the same time. What will the inertial observer say is the spatial separation of the events?

37. ★★ (a) Danielle sees two events occur at the same place but separated by 2.00 μs in time. Another observer, Wei, sees the two events separated by 100.0 m in space. What does Wei say is the time separation between the two events?
  (b) What is the relative speed of their reference frames?

Section 30-7 Relativistic Momentum and Energy

38. ★ A particle is travelling at a speed of $0.960c$. What is the ratio of the relativistic linear momentum to that calculated using the classical expression for linear momentum?

39. ★ An electron has a speed of $0.980c$.
  (a) What is its total energy, E?
  (b) What is its kinetic energy?

40. ★★ (a) Consult a variety of textbooks and popular articles, and determine whether each uses the rest mass or not.
  (b) What are the arguments in favour of using the idea of rest mass in teaching or explaining relativity?
  (c) What are the objections to using the idea of rest mass?
  (d) Do you feel that there is a compelling argument in favour of or against the use of the term *rest mass*?

41. ★ A particle has rest mass m_0 and total energy $E = 2m_0c^2$. What must be its relativistic linear momentum in terms of m_0 and c?

42. ★★ Assume that a high-energy cosmic ray has a relativistic kinetic energy of 75.0 pJ. Assume that the cosmic ray is a proton. Determine the speed at which it must be travelling relative to Earth.

43. ★★ A free neutron can decay into a proton as long as an electron and an electron neutrino are also produced (to conserve electrical charge and lepton characteristics in the reaction). The rest mass of a neutron is $1.674\,927 \times 10^{-27}$ kg, the proton rest mass is $1.672\,622 \times 10^{-27}$ kg, and the rest mass of an electron is $9.109\,38 \times 10^{-31}$ kg. What is the kinetic energy that must be released in the decay? Express your answer in both J and MeV. Assume that no rest mass is associated with the neutrino.

Section 30-8 Relative Velocity Addition

44. ★★ We are being visited by two alien civilizations, each approaching us from opposite directions. Alien civilization A has a spacecraft approaching us at a relative speed of $0.80c$, and civilization B is approaching us at a speed of $0.90c$. What speed does civilization A measure for the spacecraft of civilization B?

45. ★★ Particle A is travelling radially away from the Earth at a speed of $0.900c$. This particle then disintegrates into other particles, one of which, particle B, also moves radially away from Earth at a speed of $0.950c$ relative to the original speed of particle A. What speed is B moving relative to Earth?

46. ★★★ In reference frame S, a particle is observed to have a speed of $0.985c$. In reference frame S', the same particle is observed to have a speed of $0.965c$. What is the speed of reference frame S' relative to reference frame S? Assume that all motions are in the same direction.

47. ★★ In reference frame S, a particle is observed to have a speed of $0.850c$. In reference frame S', the same particle is observed to have a speed of $0.750c$. What is the speed of reference frame S' relative to reference frame S? Assume that all motions are in the same direction.

48. ★★★ $\frac{d}{dx}$ Reference frame S' is moving along the positive x-axis with respect to reference frame S at a speed of $0.600c$. Reference frame S'' is moving along the positive x-axis with respect to reference frame S' at a speed of $0.700c$. An object is moving with the speed $0.800c$, also along the positive x-axis, relative to the S'' reference frame. What is the speed of this object when measured with respect to the S reference frame?

Section 30-9 Relativistic Doppler Shift

In doing these questions, you need only consider the Doppler effect for light and can ignore any gravitational redshift(s).

49. ★ An astrophysical source is moving away from us at a speed of $0.400c$. The source emits electromagnetic waves of frequency 6.35×10^{15} Hz. What will be the frequency when the signal is received near Earth?

50. ★ A galaxy is moving away with a speed of $0.200c$. If hydrogen line emission of wavelength 410.174 nm is emitted by the galaxy, what will be the received wavelength here?

51. ★★ The spin flip transition for neutral hydrogen has a wavelength of 21.106 114 cm. We receive the radiation from an interstellar cloud and measure a wavelength of 21.106 026 cm.
    (a) What region of the electromagnetic spectrum is this radiation in?
    (b) Is the gas cloud moving toward or away from us?
    (c) What is the speed of the cloud relative to us?

52. ★★ One of the Balmer lines of hydrogen emits light of wavelength 656.2852 nm. We receive a shifted wavelength of 658.3453 nm.
    (a) Is the source moving toward or away from us?
    (b) What is the redshift parameter z?
    (c) Assuming that we can use the low-speed relationship, what is the speed of the source relative to us?

53. ★★ A quasar moving away from us at a speed of $0.550c$ emits light of wavelength 486.13 nm.
    (a) What is the redshift parameter z?
    (b) What would be the received wavelength? Assume that all gravitational redshifts can be ignored.

Section 30-10 Gravitational Time Dilation in General Relativity

54. ★★ Consider two locations, one at sea level and one high in the Rockies at an elevation of 3400 m above sea level.
    (a) Would time move more slowly at sea level or high in the Rockies?
    (b) Over a 70 yr lifetime, what would be the difference in time, due to gravitational time dilation, between the two locations?

55. ★ A black hole has a mass that is 12 times the mass of the Sun. What is the radius of the event horizon around the black hole?

Section 30-11 Relativity and the Global Positioning System

56. ★★★ An exoplanet has a mass exactly double Earth's mass and a radius that is 85.0% of Earth's radius. The planet has a sidereal (relative to distant stars) rotation period of 7.45×10^4 s. Assume that the exoplanet has a GPS system similar to ours, and that the periods of the GPS satellites are one-half of the planet's sidereal period.
    (a) What is the radius (from the centre of the exoplanet) of the satellite orbits?
    (b) What is the special relativity time correction between the satellite and the surface of the exoplanet? Express your answer in terms of difference in time over a sidereal "day" on the exoplanet.
    (c) What is the general relativity time correction per sidereal day?
    (d) What is the net correction per day?

COMPREHENSIVE PROBLEMS

57. ★ Calculate the length of one light year in metres. Assume that the duration of the year is 365.25 days.

58. ★★ The Sun is located at a distance of approximately 25 500 ly from the centre of our Milky Way galaxy. If one wanted to travel from Earth to the centre of the galaxy in 1450 yr, at what speed should the spacecraft move?

59. ★★ The bright star Vega is 25.3 ly from Earth. A planet around that star is featured in the movie *Contact*.
    (a) What time is required to send a radio signal from Earth to the planet orbiting Vega and back again?
    (b) If a spacecraft were to travel at a speed of $0.985c$ from Earth to the planet and back again, what time would an Earth-based observer measure for the return trip? (Assume that the acceleration time is negligible.)
    (c) What time would an observer on the spacecraft measure for the round trip?

60. ★★ A friend passes you in her spaceship travelling at a relative speed of $0.80c$. As she passes you, you measure her spaceship to be 5.2 m long and 1.5 m high. It is moving in the direction of the length.
    (a) What are the length and height of the spaceship as measured in a reference frame in which the spaceship is at rest?
    (b) The rest mass of your friend's spaceship is 5.0×10^3 kg. What is the relativistic kinetic energy of the spaceship as it travels at $0.80c$?

61. ★★ A hypothetical high-speed transit line advertises that passengers will be carried from Halifax, Nova Scotia, to Vancouver, British Columbia, a distance of approximately 6.0×10^3 km, at such a high speed that a passenger's watch only registers half the time for the trip that synchronized clocks on the stations will read.
    (a) What is the speed of the transit vehicles?
    (b) What is the time for a one-way trip, as measured by the clocks at the stations?

62. ★ In the electromagnetism chapters, you learned about the electron volt as a unit of energy. What is the speed of an electron that has a total energy, E, of 1.2 MeV? Express your answer in terms of c.

63. ★★★ Older televisions and monitors used CRT (cathode ray tube) displays. Assume that the electrons in a CRT display are accelerated by a potential difference of 32.0 kV.
    (a) Use the classical expression for kinetic energy to determine the speed of the electrons.
    (b) Now use relativistic kinetic energy to determine the speed of the electrons.
    (c) Are relativistic considerations important in the design of CRT electronics?

64. ★★ An electron is accelerated through a potential difference of 8.00 MV (million volts). Calculate the speed of the electron using classical and relativistic kinetic energy equations. Are relativistic effects important in this situation?

65. ★ When a star goes supernova, the light from the star might reach Earth a few hours after a burst of neutrinos. Is this in conflict with special relativity? Explain.

66. ★★ (a) An interstellar probe travelling at $0.85c$ relative to a local planetary system hits a meteoroid in that planetary system (i.e., the relative speed of the meteoroid is $0.85c$). The meteoroid has a mass of 1.6 g (a little bit less than the mass of a dime). Find the relativistic kinetic energy of the meteoroid.
    (b) Find the TNT equivalent to the kinetic energy of the meteoroid, given that 1 kg of TNT explosive releases approximately 4.2 MJ.

67. ★★★ The Sun has a luminosity of 3.85×10^{26} W. This is the total of the power that it radiates in all directions.
    (a) According to $E = mc^2$, how much mass per second must the Sun convert to create this much power?
    (b) The most important nuclear fusion reaction sequence in the Sun is one in which four hydrogen nuclei (protons) combine, forming one helium nucleus (an α-particle), plus some low-mass products such as neutrinos. In this reaction, 0.66% of the initial mass of the hydrogen is converted to energy. What mass of hydrogen must the Sun convert to helium each year?
    (c) The current mass of the Sun is about 2.0×10^{30} kg. Approximately 10% of this mass is hydrogen available for nuclear fusion reactions (not all the hydrogen can be brought to the core for nuclear reactions). For how long can the Sun produce its current power output?

68. ★★ One of the exoplanets recently discovered is around a star 49.8 ly from Earth. You are proposing that a spaceship travel at a speed of $0.980c$ to this exoplanet. Assume this speed is reached sufficiently quickly that we can consider the spaceship to make the entire voyage at this speed.
    (a) How long (in years) will a one-way trip take, according to clocks on the spaceship?
    (b) What is the time (in years) for the one-way trip, according to Earth-based clocks?
    (c) Assume that the spaceship has a mass of 12 million tonnes (1.20×10^{10} kg), which is the minimum suggested by considerations of rocket design and closed systems for long voyages, even with propulsion systems. Find the relativistic kinetic energy of the spaceship when it reaches cruising speed.

69. ★★ In one coordinate system, two events occur with a spatial separation of 4.00 m and separated in time by 20.0 ns. Another inertial observer travels so that both events occur at the same spatial location.
    (a) What approximate time interval does a clock with that observer measure between the two events?
    (b) What is the speed of one reference frame relative to the other?

70. ★★ The Sun has a mass of approximately 2.0×10^{30} kg and a radius of 6.96×10^8 m.
    (a) Assuming that the Sun is spherical, what is its mass density?
    (b) If a star has the same density as the Sun but greater mass, could it ever become a black hole? What would the radius need to be?
    (c) Does an object need to be very dense to become a black hole? (In 1792, the great French scientist Pierre-Simon de Laplace (1749–1827) commented on how a star with such dimensions could trap light.)

71. ★★ An interstellar spacecraft of length 250 m is moving at a speed of $0.75c$ with respect to its home planet reference frame. An observer on the planet observes a tiny meteoroid moving in a direction opposite to the motion of the spacecraft and at a speed of $0.50c$ relative to the planet. What is the time, from the viewpoint of an observer on the spacecraft, for the meteoroid to move past the length of the spacecraft?

72. ★★ Consider three inertial reference frames, A, B, and C. B and C are in motion in the same direction (along the $+x$-axis) with respect to A. Assume that reference frame B moves at a relative speed of $0.50c$ with respect to A, and reference frame C moves at a speed of $0.40c$ with respect to B. An object moves, also along the $+x$-axis, at a speed of $0.30c$ with respect to reference frame C. What is the speed of the object as measured by reference frame A?

73. ★ Look up the lyrics to the song "Cygnus X-1" by the Canadian band Rush. Analyze the lyrics in terms of what we know about black holes and the candidate black hole near the X-ray source Cygnus X-1.

74. ★★ Dr. Stephen Hawking and others have proposed the possibility that small black holes, called primordial black holes, might have formed shortly after the Big Bang. It is probable that these primordial black holes were formed with masses ranging from perhaps 1.0×10^{-8} kg to 1.0×10^{35} kg. Calculate the Schwarzschild radius for the largest of these proposed primordial black holes, and find the mass density.

75. ★★ If an object has the density of water, what would its radius need to be to become a black hole?

76. ★★★ The world's most powerful particle accelerator, the Large Hadron Collider (LHC) at CERN, became operational in 2008–2009. ATLAS, one of the two main experiments, involves the collaboration of more than 1000 scientists, including many from Canada. This experiment sends protons in opposite directions around a circular path with a 27 km circumference. After acceleration to very high speeds, some of the protons collide head-on, liberating enough energy to create a number of different particles.
    (a) The kinetic energy of each proton is up to 7.0 TeV. By what percentage is the speed of a proton less than c?
    (b) How many times per second will a proton go around the 27 km ring?
    (c) What is the relativistic momentum of a proton at the kinetic energy stated in part (a)?

77. ★★★ The Alberta Large-area Time-coincidence Array (ALTA) experiment uses an array of detectors, many on roofs of schools, to detect very high-energy cosmic rays by searching for time-coincident events over large areas. Such arrays sometimes detect cosmic rays with energies as high as 1.0×10^{19} eV or more. Interestingly, the origin of these high-energy cosmic rays is not definitively known.
    (a) A cosmic ray proton has an energy of 1.0×10^{19} eV. Which produces higher-energy particles: the CERN LHC (see problem 76) or cosmic rays? By what factor?
    (b) By what percentage is the speed of a proton less than the speed of light, if it has a relativistic kinetic energy of 1.0×10^{19} eV?
    (c) What is its relativistic momentum?
    (d) What is the time we measure for the proton to cross our galaxy, a distance of about 280 000 ly (if we include the halo region)?
    (e) What is the time as measured by a clock that travels with the proton?

78. ★★★ One observer sees two particles, each of rest mass m_0, collide head-on, with one particle travelling $0.80c$ in the positive x-direction and the other at at the same speed in the opposite direction. A second observer sees the same collision but is in an inertial reference frame in which one of the particles is at rest.

(a) At what speed is the other particle travelling in that reference frame?

(b) Find the kinetic energy of the particles in the first reference frame.

(c) Find the kinetic energy of the moving particle in the second reference frame.

79. ★★★ A particle of rest mass m_0 moves (according to some inertial reference frame) in the $+x$-direction with a speed v. The particle then has a totally inelastic collision with a particle of rest mass $m_0/2$ moving with speed v but in the $-x$-direction. What is the rest mass of the particle produced in the collision? (Hint: Both total relativistic energy and total relativistic momentum must be conserved in the collision.)

80. ★★★ Electromagnetic waves carry momentum. The momentum carried by a wave packet of energy E is E/c. The effect of momentum transfer from electromagnetic waves is called radiation pressure.

(a) Derive an expression for the largest radius of a dust grain for which radiation pressure just balances the gravitational pull of the Sun. Assume that dust grains are completely reflecting spheres of radius r and mass density ρ. Denote the luminosity (power radiated) by the Sun as L and the distance of the grain from the Sun as d. (Hint: Think about the momentum transfer from the wave to the dust grain.)

(b) Does the distance from the Sun really matter in this calculation?

(c) If the dust particle density is 3400 kg/m^3 calculate the dust particle radius for the case of equal magnitude gravitational and radiation pressure forces. The Sun has a luminosity of 3.85×10^{26} W.

81. ★★ The complete works of Albert Einstein are available in the digital archive at Princeton: einsteinpapers.press. princeton.edu. Note that there are copies of the papers in English translation as well as the original German. This question explores the paper on which $E = mc^2$ is based (although it does not appear in that form in the paper).

(a) Find the paper with the title "Does the Inertia of a Body Depend on Its Energy Content?" What year was this published, and in which issue of which journal?

(b) The paper starts with some principles. Summarize these.

(c) The main idea of the experiment (second page) is based on a thought experiment. Qualitatively describe this thought experiment.

(d) Summarize the key result (on last page).

(e) What symbol does Einstein use for energy?

(f) The value of the square of the speed of light is given as 9.0×10^{20}. Einstein does not state the units for the speed, but what must they be?

(g) Energy is expressed in units of ergs. Look up and record the conversion between ergs and joules.

(h) Near the end of the paper, Einstein speculates on a possible way to test the theory. What is that test?

(i) What strikes you about the paper, when you compare it to modern physics literature?

82. ★★ There is an alternative global positioning system to GPS called GLONASS. Look up information on GLONASS to assist with answering the following.

(a) How many active GLONASS satellites are there?

(b) Are GPS and GLONASS at the same orbital radius? If not, which is lower?

(c) Does the answer to part (b) imply that the period for GLONASS satellites will be longer or shorter than that of the GPS satellites? Review material from Chapter 11, if necessary, to answer this question.

(d) What is a sidereal day?

(e) Describe the orbital period of each type of satellite in terms of sidereal days.

(f) Each orbital plane has eight satellites in the GLONASS system. What is the significance of this number?

(g) Do smart phones such as the iPhone use GPS, GLONASS, or neither?

(h) What is the advantage of having both a GPS and a GLONASS system?

83. ★★★ The rest mass energy, E_0; the total energy, E; and the magnitude of the relativistic momentum, p, are related as follows:

$$E^2 = p^2 c^2 + E_0^2$$

Derive this relationship. (Hint: Start with an expression for the relativistic momentum, square that, and multiply by c^2.)

84. ★★ Original estimates for the first black hole pair merger reported by LIGO (see Making Connections in the chapter) suggested an initial total mass of the two black holes of about 65 solar masses, and the final combined mass after merger of about 62 solar masses. For this question, assume that the merger occurred over a period of 0.10 s. The distance to the black holes is rather uncertain, but assume a value of 1.3 billion light years.

(a) How much energy is converted from mass loss in the merger? Express your answer in joules. It is assumed that most of this energy is radiated in the form of gravitational waves.

(b) Assuming that the merger effectively took place over 0.10 s, what is the average power radiated during the black hole merger? Express your answer in watts.

(c) The luminosity of the Sun is about 3.85×10^{26} W. The power radiated during the black hole merger would correspond to the total radiated power of how many stars like the Sun?

(d) If we assume that the energy is radiated equally in all directions from the black hole merger, what is the intensity received at Earth (expressed in W/m^2)?

DATA-RICH PROBLEM

85. ★★★ 🖳 Table 30-1 gives the magnitude of the relativistic momentum, p, and the relativistic kinetic energy, K, for a series of unknown velocities. Use numerical or analytical approaches with the data to answer the following questions.
   (a) What is the rest mass of the object?
   (b) What is the range of velocities represented in the table, expressed in terms of the speed of light?
   (c) What is the value for relativistic linear momentum in the blank cell in the left column?
   (d) What is the value for kinetic energy in the blank cell in the right column?
   (e) What total energy, E, corresponds to the data in the last row of the table?

Table 30-1   Data for Problem 85.

p (kg·m/s)	K (J)
7.54×10^7	1.13×10^{15}
1.53×10^8	4.64×10^{15}
2.36×10^8	1.09×10^{16}
3.27×10^8	2.05×10^{16}
4.33×10^8	3.48×10^{16}
5.63×10^8	5.63×10^{16}
	9.01×10^{16}
1.00×10^9	

OPEN PROBLEM

86. ★★★ You have been asked to plan the first interstellar space mission to carry people to another planetary system.
   (a) Look up recent data on exoplanet discoveries, and select a specific target planet. Explain why you chose that planet (in terms of habitable zone and other features). Note the distance to the exoplanet system.
   (b) What would you consider a reasonable round-trip time for the visit (time as measured by the travellers)?
   (c) Use relativity to calculate the speed required, assuming that most of the trip is done at a constant speed and the time for acceleration and deceleration is not significant compared to the duration of the voyage.
   (d) How many people, minimum, do you feel that such a mission should carry?
   (e) Considering your answers to (b) and (c), what would you suggest as the minimum mass of the spacecraft? Provide the analysis upon which your answer is based (e.g., the amount of mass needed for fuel, food production, protective shell). Canadian scientist Dr. Mike Dixon and his research team at the University of Guelph are working on closed environmental systems to be used to process wastes and grow food for long-duration missions to Mars.
   (f) How much energy would be required to accelerate the spacecraft to the speed you calculated in part (c)? Express this energy in joules and in terms of the annual energy use in Canada.

CHAPTER 31 | Fundamental Discoveries of Modern Physics

The ATLAS detector is part of the Large Hadron Collider at CERN (Figure 31-1). The detector was built as a collaboration among more than 3000 scientists from 174 universities and laboratories and 38 nations—one of the largest collaborations in the physical sciences. Of the 3000 scientists, approximately 1000 are students. The Large Hadron Collider is the most powerful particle accelerator built to date, producing energies on the order of 10^{15} electron volts. In this chapter, we will see how particle collisions helped physicists realize that there is a realm of physics beyond what is now called classical physics.

The genesis of classical physics was the publication of Newton's *Philosophiæ Naturalis Principia Mathematica* in 1687, which included his three laws of motion and his law of universal gravitation. This view of the world formed the basis of a great deal of scientific advance and reigned well into the 1800s. At that time, experimental results began to emerge that could not be reconciled with some aspects of Newtonian physics. This period coincided with the Industrial Revolution, when scientific advances both enabled and were enabled by technological advances. In this chapter, we will examine some of the experimental evidence and early theoretical efforts that ultimately led to the development of quantum physics.

Much of the progress occurred during a time of scientific confusion and controversy. At its emergence, quantum theory was hotly debated because it required discarding some of the concepts that were the basis of extremely successful models for describing nature.

We will not present the material in a strictly chronological order but in a sequence that allows for the logical development of the topics. Keep in mind that this sequence does not reflect the considerable difficulties that physicists faced in developing and accepting radical new theories during the first half of the 20th century.

Figure 31-1 The ATLAS detector at CERN.

MAXIMILIEN BRICE, CERN/SCIENCE PHOTO LIBRARY

31-1 The Connection between Matter and Electricity

In 1834, Faraday published the results of a series of experiments that led to an understanding of the electrical properties of matter. Prior to this, it was thought that electricity was some type of massless fluid, or fluids, that flowed between objects.

Faraday passed electric currents through solutions and observed that gas evolved at the electrodes. This process is called electrolysis. A typical setup for such an experiment is shown in Figure 31-2.

Faraday noted that the amount of gas evolved at each electrode was always proportional to the total charge that passed through the cell. He also observed that the relative amounts of gas produced are in the same proportion as they are known to be in the substance being electrolyzed. For example, in the electrolysis of water, twice as much hydrogen gas (by volume) is evolved as oxygen. Faraday found that when he passed the same amount of electricity through different cells, the mass of each element produced at the electrodes was proportional to the atomic weight of the element divided by its **valence**, which is a measure of the number of bonds formed by each atom of the element. The constant of proportionality, now called Faraday's constant, F, has a value of 96 485 C/mol.

Faraday's observations established a clear link between matter and electricity. Faraday observed that the ratio of the charge passed through the system to the mass of gas evolved is the same from experiment to experiment for a particular substance, but a different ratio is obtained for different substances. It seemed that each substance had a well-defined charge-to-mass ratio. However, his results gave no insight into whether the variations in the ratio from substance to substance were due to differences in mass, charge, or both.

CHECKPOINT 31-1

Charge and Mass

Which of the following statements is true?
(a) Faraday demonstrated that the charge-to-mass ratio is the same for all elements.
(b) Faraday demonstrated that each element has a unique charge-to-mass ratio.
(c) Faraday demonstrated that there is no relationship between mass and charge.
(d) Faraday demonstrated that all elements have the same amount of charge but differ in their mass.

ANSWER: (b) Faraday showed that for a given element, the gas evolved at an electrode is always proportional to the total charge passed through the cell and that each element has a unique constant of proportionality.

EXAMPLE 31-1

The Electrolysis of Water

During a water electrolysis experiment, a current of 36 mA is passed through the cell for 60. s. How many hydrogen molecules are liberated from the cell during that time?

Solution

The total charge that passes through the cell is given by

$$Q = \int_0^{60.} I dt = It \Big|_0^{60.} = 2.16\,\text{C}$$

In Figure 31-2, we see that one hydrogen molecule is liberated for every two electrons that pass through the cell. Therefore, the number of hydrogen molecules produced is

$$N_{H_2} = \frac{Q}{2e} = \frac{2.16\,\text{C}}{(2)(1.602 \times 10^{-19}\,\text{C})} = 6.74 \times 10^{18}$$

In one mole of material, there are 6.022×10^{23} particles. Thus, the quantity liberated in this example is 1.12×10^{-5} mol.

Making sense of the result

One mole of ideal gas at standard temperature and pressure occupies 22.4 L, so the volume of gas produced in 1 min occupies approximately 0.25 mL. The current is relatively small, so this amount seems reasonable.

$$2\,H_2O\,(l) \rightarrow O_2(g) + 4\,H^+\,(aq) + 4e^-$$

$$2\,H_2O\,(l) + 2e^- \rightarrow H_2(g) + 2\,OH^-\,(aq)$$

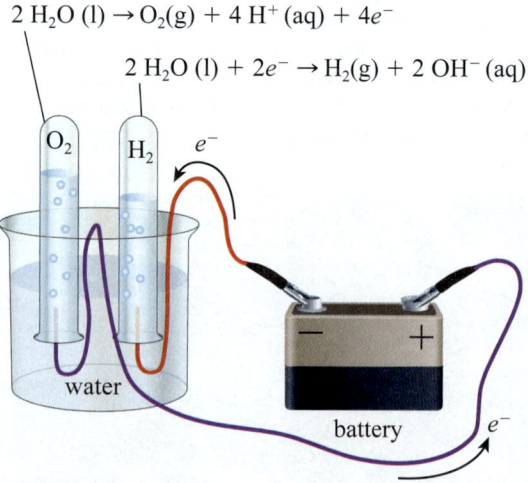

Figure 31-2 An apparatus for the electrolysis of water. A weak acid such as vinegar is added to the water to improve conduction. As current flows through the cell, hydrogen gas and oxygen gas are evolved at the electrodes.

31-2 Temperature and the Emission of Light

Some of the early investigations into how the temperature of an object affects the light that it emits played a fundamental role in the development of modern physics.

If we heat an object to a sufficiently high temperature, the object will visibly glow. Thus, hot objects, such as the stove element in Figure 31-4, emit visible light. In fact, an object can emit all types of electromagnetic radiation, and the spectrum the object emits is characteristic of the temperature of the object. Detailed experiments show that the behaviour of most hot objects approximates the spectrum of a blackbody. A **blackbody** is an object for which the character of the light emitted depends only on the temperature of the body. The spectrum of a blackbody is continuous; that is, a blackbody emits electromagnetic waves at all wavelengths, although the intensity varies considerably with wavelength.

Numerous efforts to describe blackbody radiation have been made. As we saw in Chapter 17 (Equation 17-36), the total power radiated by a hot object is

KEY EQUATION
$$P = \sigma \varepsilon A T^4 \qquad (31\text{-}1)$$

where $\sigma = 5.67 \times 10^{-8}$ W/(m^2·K^4) is the Stefan-Boltzmann constant, ε is the emissivity of the object, A is the surface area, and T is the temperature of the object in kelvin.

Figure 31-4 A stove element glows red-hot.

The wavelength at which the greatest intensity of radiation is emitted has been shown to be given by **Wien's law**:

KEY EQUATION
$$\lambda_{\max} = \frac{2.898 \times 10^{-3}\,\text{m} \cdot \text{K}}{T} \qquad (31\text{-}2)$$

EXAMPLE 31-2

The Power of the Sun

The Sun has a surface temperature of about 6.0×10^3 K and a surface area of about 6.1×10^{18} m^2. How much power does the Sun radiate, and what is the peak wavelength in the spectrum, assuming it radiates as a blackbody with an emissivity of one?

Solution

We can use the two relationships above to find these answers. We can determine the total power radiated using Equation 17-36:

$$P_{\text{Sun}} = 5.67 \times 10^{-8}\frac{\text{W}}{\text{m}^2 \cdot \text{K}^4} \times 1 \times 6.1 \times 10^{18}\,\text{m}^2$$
$$\times (6.0 \times 10^3\,\text{K})^4$$
$$= 4.5 \times 10^{26}\,\text{W}$$

The peak wavelength in the spectrum is obtained from Equation 31-2:

$$\lambda_{\max} = \frac{2.898 \times 10^{-3}\,\text{m} \cdot \text{K}}{6.0 \times 10^3\,\text{K}} = 480\,\text{nm}$$

Making sense of the result

The peak wavelength falls right into the visible spectrum of light. In fact, this is why the visible spectrum falls where it does—our eyes have evolved to be most sensitive in the part of the spectrum where there is the greatest intensity, with the peak sensitivity for human vision falling at 555 nm.

The power result is huge. If we divide this result by the surface area of a sphere whose radius is the orbit of Earth, we get a power density of about 1.6 kW/m^2, which agrees reasonably well with the measured solar flux.

MAKING CONNECTIONS

Spectrometers

How do we characterize the light that an object emits? One method is to measure the intensity of the light as a function of wavelength. In a **spectrometer**, light passes through a dispersive element, such as a prism or a diffraction grating. The intensity of the light at the different wavelengths is then measured at a detector, as shown in Figure 31-3. Adjusting the angle of the grating determines which wavelengths strike the detector, thus allowing the measurement of intensity as a function of wavelength.

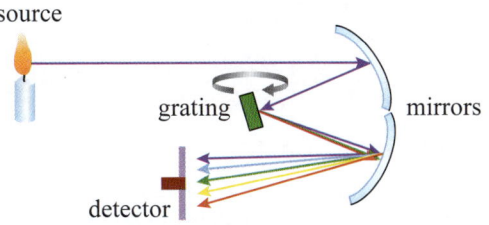

Figure 31-3 A schematic diagram of a spectrometer.

MODERN PHYSICS

Blackbody Spectra

We know from experience that a "white-hot" object is at a higher temperature than a "red-hot" object. In this activity, you will use the PhET simulation "Blackbody Spectrum" to investigate what colours of light a white-hot object emits that a red-hot object does not and whether these colours are at the long wavelength or the short wavelength end of the spectrum.

These two results were based on thermodynamic reasoning and gave some insight into the nature of blackbody radiation. A full description of the spectrum did not come until 1900, when Max Planck (1858–1947) made a bold assumption. He assumed that the atoms in the heated material oscillated with frequency f and emitted radiation when they did so. However, he then postulated that the radiation could only be emitted in discrete amounts with an energy, E, given by

KEY EQUATION

$$E = hf \qquad (31\text{-}3)$$

where $h = 6.626 \times 10^{-34}$ J·s is a constant known as **Planck's constant**. Based on this, he showed that the spectral radiance of the emitted radiation depended upon the frequency and the temperature through the relationship

KEY EQUATION

$$B(f,T) = \frac{8\pi h f^3}{c^2}\frac{1}{e^{\frac{hf}{kT}} - 1} \qquad (31\text{-}4)$$

The units of B are J·m^{-2}. To get the total power radiated by the object, we need to integrate B with respect to f. This result shows excellent agreement with experiment. Interestingly, the key assumption of quantizing the energy was not immediately recognized as being of great significance—Planck himself did not place great emphasis on the assumption in his initial work. However, it turned out to be the birth of quantum physics.

Blackbodies

As we increase the temperature of a blackbody,
(a) the intensity changes
(b) the peak intensity wavelength changes
(c) the general shape of the curve changes
(d) none of the above

ANSWER: (a) and (b) The important feature of blackbody radiators is that they all have the same emission spectra if they are all at the same temperature.

Set the Controls for the Heart of the Sun

The surface temperature of the Sun is approximately 6000 K. Have you ever wondered how this temperature was determined? Obviously, no one has been to the Sun with a thermometer. Astronomers analyze the light emitted by the Sun and compare it to a blackbody spectrum. As shown in Figure 31-5, a blackbody temperature of 5777 K gives a reasonably good fit to the solar spectrum.

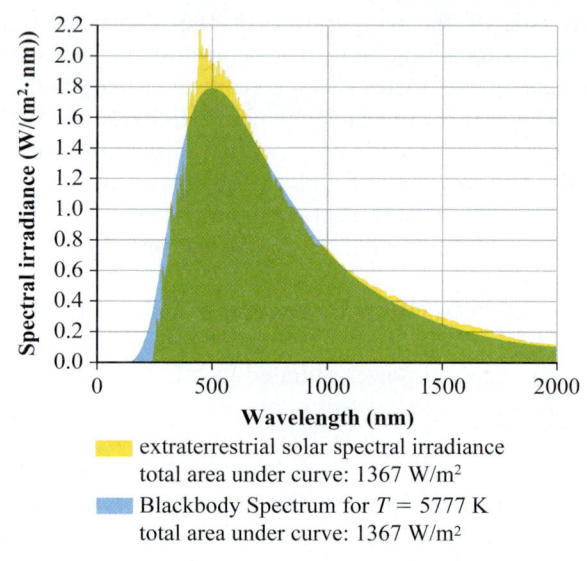

extraterrestrial solar spectral irradiance
total area under curve: 1367 W/m^2

Blackbody Spectrum for $T = 5777$ K
total area under curve: 1367 W/m^2

Figure 31-5 A spectrum of the Sun with a blackbody fit.

31-3 Gas Discharge Spectra

In this section, we will concentrate on the light emitted from excited gases. A familiar source of this type of light is a neon sign, as shown in Figure 31-6. A neon sign is an example of a discharge tube. A **discharge tube** contains a gas at a relatively low pressure. Electrodes at each end of the tube allow an electric potential difference to be applied across the gas. When a sufficiently high potential difference is applied, a current flows through the gas, which then emits light. The spectrum of this light is not continuous—the light is emitted at discrete wavelengths that depend on the type of gas in the tube.

Gas Discharge Spectra

In this activity, you will use the PhET simulation "Neon Lights and Other Discharge Lamps" to explore the types of spectra that are produced when we bombard a gas with electrons and to compare these spectra to those produced by a blackbody radiator.

Figure 31-6 A neon sign emits light with a characteristic reddish-orange colour.

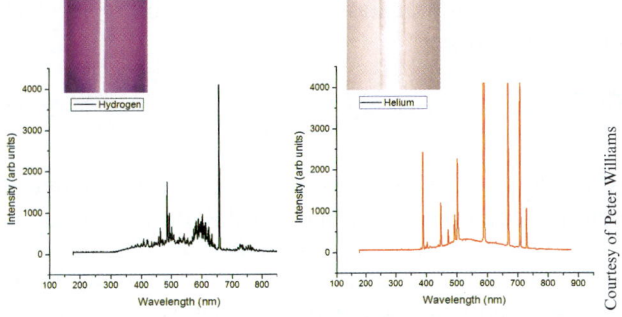

Figure 31-7 Intensity versus wavelength for a hydrogen discharge tube (left) and a helium discharge tube (right). The insets show the tubes themselves.

Figure 31-7 shows spectra obtained with a spectrometer for a hydrogen discharge tube and a helium discharge tube. You can see that there is significant intensity at only certain wavelengths. Further, the two gases have very different spectra.

These observations presented a formidable challenge. Maxwell demonstrated in 1846 that light is a form of electromagnetic radiation. In 1887, Heinrich Hertz (1857–1894) showed that electromagnetic radiation is produced when charged particles are accelerated. The fact that a gas discharge spectrum has discrete frequencies suggests that charged particles in the gas oscillate at only those frequencies. Thus, any theory that tried to account for these spectra would have to include oscillating charged particles that were somehow constrained to oscillate only at certain frequencies.

Much effort was expended to determine whether these spectral lines could be described by mathematical formulas. Swedish physicist Johannes Rydberg (1854–1919) is credited with developing a general formula applicable to the spectra of all elements. His formula describes the spectral lines in terms of wave numbers instead of wavelengths. The **wave number**, $\widetilde{v}$, is simply the reciprocal of the wavelength, λ, and therefore equals the number of wavelengths per unit distance:

$$\widetilde{v} = \frac{1}{\lambda} \tag{31-5}$$

Rydberg noted that the wave numbers of spectral lines are not uniformly distributed and tend to fall into groups or series. If we assign an integer to each line within a series and plot the wave number versus the series number, we see that each series follows a smooth curve with a similar shape, as shown in Figure 31-8 for hydrogen. This similarity suggested to Rydberg that a single formula could be used to describe all the spectral lines.

In 1885, Swiss mathematician Johann Jakob Balmer (1823–1898) derived an equation that matches the pattern of the visible spectral lines of hydrogen. When Rydberg saw Balmer's equation, he generalized

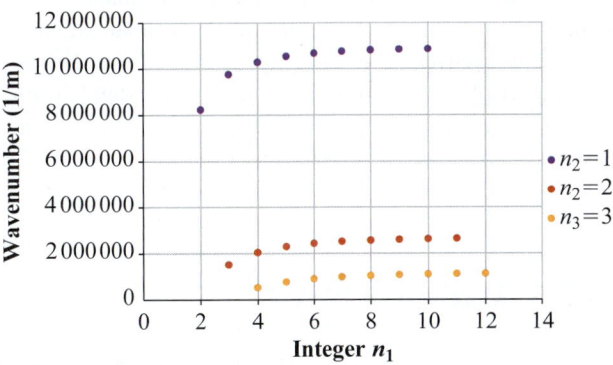

Figure 31-8 Wave number series for hydrogen. These plots are generated by assigning successive integers to each wave number in a series.

it, producing an equation that predicts all the spectral lines of hydrogen:

$$\tilde{v} = R_{\text{H}}\left(\frac{1}{n_2^2} - \frac{1}{n_1^2}\right) \qquad (31\text{-}6)$$

where $R_{\text{H}} = 1.097 \times 10^7/\text{m}$ is a constant that Rydberg determined empirically, so it is called Rydberg's constant.

The Balmer series corresponds to $n_2 = 2$, the only hydrogen series to have been observed at the time. Rydberg's equation accurately predicts other series of infrared and ultraviolet lines. Any successful atomic theory would have to produce the Rydberg formula.

EXAMPLE 31-3

Hydrogen Spectral Lines

Find the wave number, wavelength, and frequency of the hydrogen spectral line that corresponds to $n_2 = 2, n_1 = 3$.

Solution

We can calculate the wave number using Equation 31-6:

$$\tilde{v} = R_{\text{H}}\left(\frac{1}{n_2^2} - \frac{1}{n_1^2}\right) = (1.097 \times 10^7/\text{m})\left(\frac{1}{2^2} - \frac{1}{3^2}\right)$$

$$= 1.524 \times 10^6/\text{m}$$

The wavelength is the reciprocal of the wave number:

$$\lambda = \frac{1}{\tilde{v}} = 6.562 \times 10^{-7}\text{m} = 656.2 \text{ nm}$$

To find the frequency, we recall that $v = f\lambda$; therefore,

$$f = \frac{v}{\lambda} = \frac{3.00 \times 10^8 \text{ m/s}}{6.562 \times 10^{-7} \text{ m}} = 4.57 \times 10^{14} \text{ Hz}$$

Making sense of the result

The wavelength of 656.2 nm is toward the red end of the visible spectrum and corresponds to the tallest line in the hydrogen spectrum in Figure 31-7.

PEER TO PEER

Both the wavelength and the frequency are difficult numbers to understand because they are very small and very large, respectively. It is useful to remember that the visible part of the electromagnetic spectrum extends from approximately 380 nm (violet) to 750 nm (red), or from 8×10^{12} Hz to 4×10^{12} Hz, respectively.

31-4 Cathode Rays

Experimenters working with gas discharge tubes at extremely low pressures noticed that the glass of the tubes would glow at the end opposite to the cathode, the negative electrode. Further experimentation

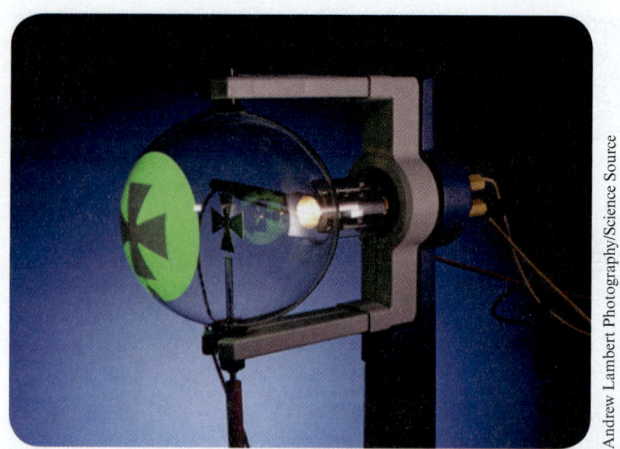

Figure 31-9 Cathode rays in the tube emanate from the cathode, and when they strike the glass walls of the tube, they cause the tube to glow. When the metal cross is placed in the path, it creates a shadow on the glass.

revealed that an object placed in the tube casts a shadow on the end of the tube away from the cathode (Figure 31-9). This result suggested that whatever was causing the tube to glow was emanating from the cathode. Consequently, these rays were called cathode rays.

In a related experiment, a paddle wheel placed between the electrodes rotated when the tube was energized, indicating that whatever was moving through the tube carried both energy and momentum. Further experimentation revealed that cathode rays can be deflected by both electric and magnetic fields, providing evidence that the rays could be moving charged particles.

These experiments culminated in the discovery of the electron by J.J. Thomson (1856–1940) in 1897. As described briefly in Section 23-3, Thomson measured the charge-to-mass ratio of cathode rays, thus showing them to be tiny charged particles with mass. We now look at Thomson's experiment in more detail.

Thomson's apparatus is illustrated schematically in Figure 31-10. The filament serves as the cathode in this apparatus. The rays are attracted to the **anode**,

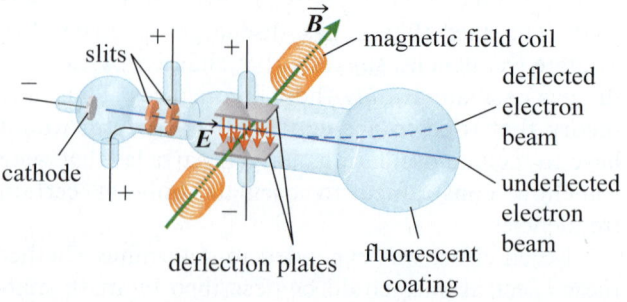

Figure 31-10 A schematic diagram of the apparatus used by J.J. Thomson to measure the charge-to-mass ratio of cathode rays.

the positive electrode of the tube. The anode has a small hole in the middle, which allows a narrow beam of cathode rays to reach the fluorescent zinc sulfide screen at the end of the tube. The beam passes between the deflection plates, which have a potential difference across them that creates an electric field directed downward. In the same region, an electromagnet creates a uniform magnetic field directed into the page.

Thomson made his measurement in two steps. First, he carefully adjusted the electric and magnetic fields until the beam of cathode rays passed straight through the deflection plate area. As described in Chapter 23, Thomson could determine the speed of the particles from the strengths of the electric and magnetic fields. He then turned off the electric field and measured the deflection caused by the magnetic field, which he used to calculate the charge-to-mass ratio of the particles in the cathode rays.

Consider the forces that a charged particle moving with velocity $\vec{v}$ experiences in electric and magnetic fields. These fields, and the associated forces that the particle experiences, are shown in Figure 31-11. The negatively charged particle experiences an upward force due to the electric field and a downward force due to the magnetic field.

When Thomson adjusted the fields so that the beam was not deflected, the force caused by the electric field was equal but opposite to the force caused by the magnetic field. Consequently, as we saw in Chapter 23,

$$qvB = qE$$

and

$$v = \frac{E}{B} \qquad (31\text{-}7)$$

Figure 31-12 shows the geometry of the vertical deflection, δ, when the electric field is switched off. We can see from Figure 31-12 that $r^2 = (r - \delta)^2 + L^2$, where L is the length of the plates. Solving for the path radius r gives

$$r = \frac{\delta^2 + L^2}{2\delta} \qquad (31\text{-}8)$$

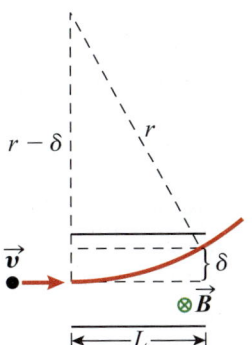

Figure 31-12   Cathode rays deflected by a magnetic field.

Thus, by measuring δ, we can determine the radius of curvature of the particle trajectory in the magnetic field. A charged particle in a magnetic field experiences a force that is perpendicular to the velocity of the particle. This force causes the particle to travel along a curved arc of a circle. Since the particle is travelling along a circular path with constant velocity, the magnitude of the acceleration must be

$$a_c = \frac{v^2}{r} = \frac{qvB}{m} \qquad (31\text{-}9)$$

Solving for $\frac{q}{m}$ and using Equation 31-7, we find

$$\frac{q}{m} = \frac{v}{rB} = \frac{E}{rB^2} \qquad (31\text{-}10)$$

With this result, Thomson determined that the negatively charged particles in the cathode rays have a charge-to-mass ratio that is over 1000 times the ratio determined by Faraday for a hydrogen ion. Thomson repeated his experiments with different cathode materials and found the same charge-to-mass ratio. These results led Thomson to postulate that cathode rays contain a new type of particle: the electron. Because Thomson could measure only the charge-to-mass ratio, it was unclear whether either the mass or the charge of this new particle was comparable to that of a hydrogen ion.

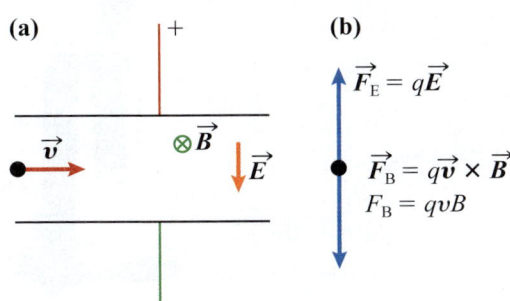

Figure 31-11   Field configuration (a) and free body diagram (FBD) (b) for Thomson's crossed-field apparatus.

EXAMPLE 31-4

Determining the Speed of a Charged Particle

In a crossed-field experiment with a magnetic field of 3.00 mT, a potential difference of 200. V is required for a beam of electrons to pass undeflected through the plates, which are separated by 2.00 cm.

(a) How fast are the electrons moving? (Ignore relativistic effects.)

(b) What potential difference is required for the electrons to be accelerated to the speed you found in part (a)?

(continued)

Solution

(a) We can use Equation 31-7 to calculate the velocity if we know the strength of both fields. Since the electric field is produced by parallel plates, we know that

$$E = \frac{V}{d} = \frac{200.\ \text{V}}{0.0200\ \text{m}} = 1.00 \times 10^4\ \text{V/m}$$

Therefore,

$$v = \frac{E}{B} = \frac{1.00 \times 10^4\ \text{V/m}}{0.003\,00\ \text{T}} = 3.33 \times 10^6\ \text{m/s}$$

(b) If we accelerate an electron at rest through a potential difference of ΔV, the electron will have a kinetic energy of approximately $K = \frac{1}{2}mv^2 = q\Delta V$, because the speed is low enough to ignore relativistic effects. We can solve for ΔV to find

$$\Delta V = \frac{mv^2}{2q} = 31.5\ \text{V}$$

Making sense of the result

One way to quickly check on this is to recall that if we accelerate an electron through 1 V, it acquires an energy of 1 eV, or 1 eV $= 1.602 \times 10^{-19}$ J. The kinetic energy in this case is $K = \frac{1}{2}mv^2 = \frac{1}{2}(9.11 \times 10^{-31}\ \text{kg})(3.33 \times 10^6\ \text{m} \cdot \text{s}^{-1})^2 = 5.05 \times 10^{-19}$ J. We can convert this to electron volts to find $K = \dfrac{5.05 \times 10^{-18}\ \text{J}}{1.602 \times 10^{-19}\ \text{J/eV}} = 31.5$ eV, which agrees with our result.

31-5 The Millikan Oil Drop Experiment

American physicist Robert Andrews Millikan (1868–1953) conducted a series of experiments that established that there is a fundamental unit of charge. The apparatus that Millikan used is illustrated in Figure 31-14. He used a spray device to produce a mist of tiny oil drops. Some of the oil drops fell into the region between the charged plates, where the drops could be observed with a telescope and manipulated by an electric field between the plates. Millikan observed that some of these drops experience a force when placed in an electric field, indicating that they must be charged. The drops acquire a charge through friction with the nozzle, in much the same way that we acquire a static charge when we shuffle our feet across a carpet. (Ionizing radiation can also be used to charge the drops.)

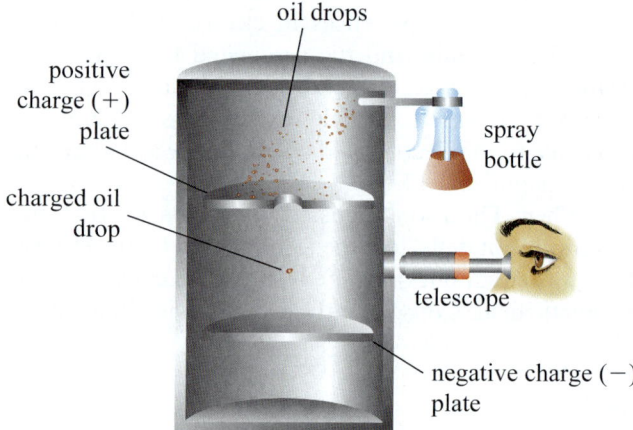

Figure 31-14   The Millikan oil drop experiment.

MAKING CONNECTIONS

Display Technology

The Thomson apparatus is the precursor of the cathode ray tube (CRT) display, which has only recently been supplanted by liquid crystal display (LCD) technology. Figure 31-13 shows a CRT display on the left and an LCD display on the right. In a CRT display, an electron beam is swept across the phosphorescent screen line by line, in what is called a raster pattern. Modulating the intensity of the beam as a function of screen position varies the brightness and thus creates an image.

Figure 31-13   Cathode ray tube technology (left) has been replaced by LCD technology (right).

When there is no field, an oil drop falls under the influence of gravity but quickly reaches a terminal velocity due to the drag force of the surrounding air. Millikan knew that the magnitude of the drag force is given by Stokes' law, an equation derived by mathematician and physicist George Gabriel Stokes (1819–1903):

$$F_D = 6\pi r \eta v_d \tag{31-11}$$

where r is the radius of the oil drop, η is the viscosity of the air, and v_d is the terminal velocity of the falling drop.

The weight of the oil drop is partially offset by the buoyant force of the air surrounding it. So, the magnitude of the effective weight of the drop is

$$W = mg - m_{air}g = \frac{4}{3}\pi r^3(\rho - \rho_{air})g \tag{31-12}$$

where g is the acceleration due to gravity, m is the mass of the drop, m_{air} is the mass of the same volume of air, and ρ and ρ_{air} are the densities of the oil and the air, respectively.

When the drop reaches terminal velocity, the drag force is equal to the effective weight of the drop. Therefore,

$$6\pi r \eta v_d = \frac{4}{3}\pi r^3(\rho - \rho_{air})g$$

and

$$r^2 = \frac{9\eta v_1}{2g(\rho - \rho_{air})} \tag{31-13}$$

Once we have a value for r, we can readily calculate the effective weight, W, using Equation 31-12.

When the electric field, E, is turned on, the drop experiences a force due to the field. Depending on the sign of the charge on the drop and the direction of the field, this force will be either upward or downward. For the purpose of our derivation, we will assume the force is directed upward. Thus, the drop begins to rise and quickly reaches a new terminal velocity, v_2. The forces acting on the drop are now an upward force of magnitude qE, where q is the charge on the drop, and the downward forces are the effective weight and the drag force, F_D. Thus,

$$qE = W + F_D$$
$$= W + 6\pi r \eta v_2$$

KEY EQUATION

$$= W + \frac{W}{v_1}v_2 \tag{31-14}$$

$$\therefore q = \frac{W}{E}\left(1 + \frac{v_2}{v_1}\right)$$

Since the electrodes are parallel plates, the magnitude of the electric field can be readily calculated from the potential difference between the plates and the plate separation. Thus, by measuring the terminal velocities with the field off and the field on, it is possible to calculate the charge on a drop.

Millikan made the necessary measurements on a number of drops and calculated the charge on each drop. He showed that all the drops had charges that are integer multiples of a fundamental unit of charge. In his initial results, published in 1910, he reported a value of 1.63×10^{-19} C for the fundamental charge. After modifying his equipment and measuring a larger number of drops, he published a revised value of 1.59×10^{-19} C in 1913. The currently accepted value for the **fundamental** (or **elementary**) **charge**, e, is approximately 1.602×10^{-19} C.

CHECKPOINT 31-3

The Millikan Experiment

Millikan showed that the charge on an oil drop only comes in discrete units of approximately 1.602×10^{-19} C. True or false?

ANSWER: True. Millikan found that the fundamental unit of charge is 1.59×10^{-19} C, which is quite close to the current accepted value.

EXAMPLE 31-5

An Oil Drop Experiment

Students replicating Millikan's experiment measure a free-fall terminal velocity of $v_1 = 3.30 \times 10^{-5}$ m/s for an oil drop as it falls. When they apply a potential difference of 10.0 V to the plates, which are 5.00 mm apart, the students observe that the drop moves upward with speed $v_2 = 2.48 \times 10^{-5}$ m/s. Find the charge on the drop given that $\rho_{oil} = 1.05 \times 10^3$ g/cm³, $\rho_{air} = 0.129 \times 10^3$ g/cm³, and $\eta = 1.73 \times 10^{-5}$ N·s/m².

Solution

We use Equation 31-13 to determine the radius of the drop from the free-fall velocity, v_1:

$$r = \sqrt{\frac{(9)(1.73 \times 10^{-5}\,\text{N·s/m}^2)(3.30 \times 10^{-5}\,\text{m/s})}{(2)(9.81\,\text{m/s}^2)(1.05 \times 10^6\,\text{kg/m}^3 - 0.129 \times 10^6\,\text{kg/m}^3)}}$$

$$= 1.69 \times 10^{-8}\,\text{m}$$

With this radius, we then calculate the effective weight W using Equation 31-12:

$$W = \frac{4}{3}\pi r^3(\rho - \rho_{air})g = 1.83 \times 10^{-16}\,\text{N}$$

We then use Equation 31-14 and solve for q:

$$q = \frac{W}{E}\left(1 + \frac{v_2}{v_1}\right) = \frac{W}{\frac{V}{d}}\left(1 + \frac{v_2}{v_1}\right) = 1.63 \times 10^{-19}\,\text{C}$$

Dividing this charge by 1.602×10^{-19} C, we obtain, within the precision of the data, e as the charge on the oil drop.

Millikan's results suggest that all particles carry some integer multiple of the same fundamental unit of charge. It appears that positively and negatively charged particles have the same magnitude of the

MODERN PHYSICS

charge. Thus, the charge-to-mass ratio for cathode ray particles indicates that these particles must be much lighter than atomic masses; that is, electrons are subatomic particles. These ideas set the stage for the development of atomic models.

31-6 Thomson's Model of the Atom

Atoms are neutral objects in bulk substances, yet Faraday's experiments clearly established that there is electrical charge associated with atoms. Taken together, Thomson's and Millikan's results show that there exist negatively charged particles that are much less massive than atoms. So, it seemed reasonable to postulate that atoms consist of a collection of rather massive positive charges balanced by the negative charge of the less massive electrons. However, the structure of the atom was unknown.

Thomson proposed that the atom was a sphere of positive charge in which were embedded enough electrons to make the atom electrically neutral. This model explained the incompressibility of solids because there would be strong repulsive forces between the spheres of positive charge if they were forced to overlap. Without any such overlap, the electrostatic forces are quite weak because the atoms are neutral overall. Consequently, one would expect these atoms to have strong repulsive interactions when packed together in solids but very weak long-range interactions in a gas state.

In this model, the electrons experience a linear restoring force inside the sphere of positive charge. If we assume this sphere is a cloud of uniform charge density with a radius of r_0 and a total charge of Q, we can use Gauss's law (Chapter 20) to calculate the electric field, E, inside the sphere:

$$E = \frac{Q}{4\pi\varepsilon_0} \frac{r}{r_0^3} \tag{31-15}$$

The cloud has a positive charge, so the electric field points radially outward. An electron placed in the cloud therefore experiences a force, eE, that is directed radially inward. Due to the spherical symmetry of the positive charge, the net force acting on an electron at any distance r from the centre of the sphere is entirely radial with no transverse components.

Now we consider the motion of an electron inside the sphere of positive charge. We will assume that this motion is radial, so the two angular coordinates do not change with time. Using Newton's second law, we can then write

$$ma_r = -qE = -\frac{qQ}{4\pi\varepsilon_0 r_0^3} r \tag{31-16}$$

where a_r is the acceleration in the radial direction.

MAKING CONNECTIONS

Spherical Coordinates

Spherical coordinates are particularly useful in situations in which a force acts only in the radial direction. Spherical coordinates describe a position in terms of a radial distance and two angles. The radial coordinate, r, is the magnitude of the position vector and is related to the Cartesian coordinates by $r = \sqrt{x^2 + y^2 + z^2}$. The polar angle, θ, is the angle that the position vector makes with the positive z-axis, and the azimuthal angle, ϕ, is the angle between the positive x-axis and the projection of the vector onto the xy-plane, as shown in Figure 31-15. From the figure, we can see that

$$z = r \cos \theta$$
$$x = r \sin \theta \cos \phi$$
$$y = r \sin \theta \sin \phi$$

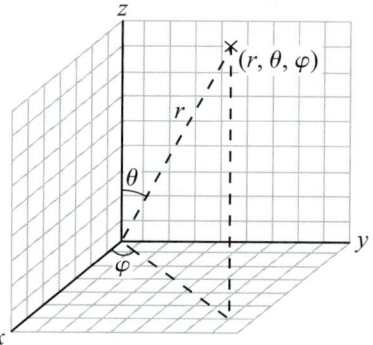

Figure 31-15 Spherical coordinates.

Equation 31-16 is an equation for simple harmonic motion with an angular frequency given by

KEY EQUATION

$$\omega = \sqrt{\frac{qQ}{4\pi\varepsilon_0 r_0^3 m}} \tag{31-17}$$

Therefore, electrons in the Thomson model undergo simple harmonic motion inside the atom. Particles undergoing simple harmonic motion are accelerating, and charged particles emit radiation when they accelerate. Thus, we would expect Thomson atoms to radiate light at the frequency of oscillation given by Equation 31-17. This frequency depends on Q, the charge on the cloud. For $q = Q = e$, $m_e = 9.1 \times 10^{-31}$ kg, $r_0 = 1.0 \times 10^{-9}$ m, and $\varepsilon_0 = 8.854 \times 10^{-12}$ F·m^{-1}, we find

$$\omega = 5.0 \times 10^{14} \text{ s}^{-1}$$
$$f = 2\pi\omega \approx 3.1 \times 10^{15} \text{ s}^{-1} \tag{31-18}$$
$$\lambda = \frac{c}{f} \approx 100 \text{ nm}$$

This wavelength has the correct order of magnitude, but the model does not explain how an atom can produce many different spectral lines for a given value of Q.

31-7 Rutherford Scattering

In 1909, New Zealand physicist Ernest Rutherford (1871–1937), working with Hans Geiger (1882–1945) and Ernest Marsden (1889–1970) at the University of Manchester, performed experiments that revealed the structure of the atom. The researchers directed a beam of α-particles produced by the decay of radium at a thin gold foil. Rutherford had made reasonable estimates for the speed and charge-to-mass ratio of α-particles using a crossed-field apparatus similar to Thomson's. The particles scattered from the gold foil were counted using a detector developed by Rutherford and Geiger. A zinc sulfide screen in the detector emitted a flash of light when struck by an α-particle.

MAKING CONNECTIONS

Rutherford's Influence on the Next Generation

Rutherford was born in New Zealand, where he completed his undergraduate studies and a master's degree. After two years of research in electrical technology, he moved to England for more postgraduate studies at the University of Cambridge, where he worked with J.J. Thomson. In 1898, Rutherford was the Chair of Physics at McGill University in Montréal, Québec. While there, he conducted research on radioactivity and the transmutation of elements, for which he was awarded the Nobel Prize in Chemistry in 1908. Rutherford returned to England in 1907 to take the Chair of Physics at the University of Manchester, where he devised his famous scattering experiment and supported the early work of Niels Bohr (1885–1962), whose quantum model of the atom is described in Section 31-9. In 1919, Rutherford became director of the Cavendish Laboratory at the University of Cambridge. Several researchers working under his direction received Nobel prizes, including James Chadwick (1891–1974; discovery of the neutron), Edward Appleton (1892–1965; discovery of the ionosphere), and John Cockcroft (1897–1967) and Ernest Walton (1903–1995) (splitting the atom).

Based on the Thomson model with electrons embedded in a cloud of positive charge, Rutherford's group thought that the α-particles would be deflected by at most a few degrees. Since α-particles are much heavier than electrons, the researchers expected the electrons to have very little effect on the trajectory of the α-particles, and any measurable deflection would be caused by the positive charge. For an α-particle travelling along a line passing through the centre of the cloud, the electric field is given by

$$E = \begin{cases} \dfrac{1}{4\pi\varepsilon_0}\dfrac{Q}{r^2} & r > r_0 \\[2ex] \dfrac{1}{4\pi\varepsilon_0}\dfrac{Qr}{r_0^3} & r \leq r_0 \end{cases} \quad (31\text{-}19)$$

where r_0 is the radius of the cloud of positive charge. Integrating this electric field gives the electric potential at the centre of the cloud of positive charge:

$$V_{r=0} = \frac{3Q}{8\pi\varepsilon_0 r_0} \quad (31\text{-}20)$$

The radius, r_0, was assumed to be of the order of interatomic spacing. Gold has a molar mass of approximately 197. g/mol and a density of 19.3 g/cm^3. Using these data with Avogadro's number, N_A, we can estimate the volume associated with a single gold atom and, hence, its radius:

$$\frac{4}{3}\pi r_0^3 = \frac{197.\text{ g/mol}}{19.3\text{ g/mol}}\frac{1}{N_A} = 1.69 \times 10^{-23}\,\text{cm}^3 \quad (31\text{-}21)$$

$$r_0 = 1.6 \times 10^{-10}\,\text{m} \quad (31\text{-}22)$$

Rutherford's measurements showed that the speeds of the α-particles emitted by radium are in the range of 2×10^7 m/s to 3×10^7 m/s. Thus, the α-particles have a kinetic energy of, using $m = 6.64 \times 10^{-27}$ kg and $v = 2.5 \times 10^7$ m$\cdot$s^{-1}, approximately

$$K = \frac{1}{2}mv^2 = 2 \times 10^{-12}\,\text{J} \quad (31\text{-}23)$$

An α-particle has a charge of $+2e$, and a gold nucleus has a charge of $+79e$. We can use Equation 31-20 to find the potential energy of an α-particle at the centre of a cloud of charge of $+79e$ and with a radius of 1.6×10^{-10} m:

$$U = qV = \frac{474e^2}{8\pi\varepsilon_0 r_0} = 3.4 \times 10^{-16}\,\text{J} \quad (31\text{-}24)$$

This potential energy is approximately one thousandth the kinetic energy of an α-particle, so we would expect the particle to pass right through the cloud of positive charge with very little deflection.

If the path of the α-particle is offset from the centre of the cloud, the calculation of the trajectory is substantially more complex. Suppose the trajectory of the α-particle is off centre by r_0, the radius of the cloud. We can make a rough estimate of the deflection by assuming that the particle experiences a force that is perpendicular to its trajectory for a time that is equal to the time it takes for the particle to travel the diameter of the atom. The time for the α-particle to travel this distance, using an average speed of 2.50×10^7 m/s, is

$$\Delta t = \frac{2r_0}{v} = 1.28 \times 10^{-17}\,\text{s} \quad (31\text{-}25)$$

The magnitude of the force varies as the distance between the cloud and the α-particle changes, but we will approximate the force as constant with a magnitude equal to the force that the α-particle experiences at the surface of the atom. The force at the surface of the cloud is given by the product of the charge on the α-particle, $+2e$, and the electric field due to the cloud, which we can get from Equation 31-19:

$$F = qE = (2e)\frac{1}{4\pi\varepsilon_0}\frac{79e}{r_0^2} = 1.4 \times 10^{-6}\,\text{N} \qquad (31\text{-}26)$$

If the particle experiences this force for the entire time it takes to cross the diameter of the nucleus, the change in momentum perpendicular to the incident direction is given by

$$\Delta p_\perp = F_{\text{avg}}\Delta t = 1.8 \times 10^{-23}\,\text{kg·m/s} \qquad (31\text{-}27)$$

Since the mass of the α-particle is approximately four times the mass of a proton, the change in velocity perpendicular to the direction of the incident beam is approximately

$$\Delta v_\perp = \frac{\Delta p}{m} = \frac{1.8 \times 10^{-23}\,\text{kg·m·s}^{-1}}{6.64 \times 10^{-27}\,\text{kg}} = 2.7 \times 10^3\,\text{m/s} \qquad (31\text{-}28)$$

Since the incident velocity is of the order of 10^7 m/s, the deflection from the incident direction is quite small.

Rutherford and his co-workers observed that the majority of the α-particles *did* simply pass through the foil, but some were scattered through quite large angles, and a few particles rebounded almost straight back from the foil. Rutherford commented, "It was almost as incredible as if you fired a 15-inch shell at a piece of tissue paper and it came back and hit you."

Rutherford realized that the scattering pattern showed that the positive charge could not be distributed in a cloud throughout the gold atoms. For the α-particles to be scattered backward, the electric fields produced by the positive charges inside the atom must be much greater than the fields predicted by the Thomson model. If we examine Equation 31-19 closely, we can see that the largest value of the electric field that a cloud of charge can produce occurs at the surface of the charge and is given by

$$E_{\text{max}} = \frac{1}{4\pi\varepsilon_0}\frac{Q}{r_0^2} \qquad (31\text{-}29)$$

Thus, if the strength of the electric field is much greater, the radius of the cloud of charge must be very small. Rutherford estimated the maximum possible radius of the cloud of positive charge by assuming that an α-particle travelling straight toward the centre of the cloud of positive charge comes to a stop just at the surface of the positive charge before being accelerated back the way it came. If an α-particle stops in this way, all the kinetic energy of the particle must be converted to potential energy. At the surface of a charged sphere,

the electric potential is equal to the electric potential due to a point charge:

$$V = \frac{1}{4\pi\varepsilon_0}\frac{Q}{r} \qquad (21\text{-}10)$$

Multiplying the electric potential by the charge of the α-particle gives us the potential energy of the α-particle at the surface of the cloud. The α-particle stops just at the surface of the charge if this electric potential energy equals the initial kinetic energy of the α-particle:

$$qV = \frac{1}{2}mv^2$$
$$(2e)\frac{1}{4\pi\varepsilon_0}\frac{(79e)}{r_0} = \frac{1}{2}mv^2 \qquad (31\text{-}30)$$

Solving for r_0, we find

$$r_0 = \frac{79e^2}{\pi\varepsilon_0 mv^2}$$
$$= \frac{79(1.602 \times 10^{-19}\,\text{C})^2}{\pi(8.85 \times 10^{-12}\,\text{F·m}^{-1})(6.64 \times 10^{-27}\,\text{kg})(2.5 \times 10^7\,\text{m·s}^{-1})^2}$$
$$= 1.76 \times 10^{-14}\,\text{m} \qquad (31\text{-}31)$$

This result suggests that for the α-particles to be back-scattered, all the positive charge in the atom must be concentrated in a sphere with a radius no larger than approximately 10^{-14} m.

Rutherford realized the rather startling implications of this result. If all the positive charge is confined to such a small region, then most of the atom is empty space. If atoms are mostly empty space, what prevents them from collapsing in on each other? The small radius of the charge also indicates that the material carrying the charge has a density of at least 10^{15} kg/m^3, many orders of magnitude greater than any previously measured (e.g., lead has a density of approximately 1×10^4 kg/m^3).

Rutherford suggested that his results supported a **planetary model** of the atom, in which the electrons could perhaps be orbiting a central positive charge. However, orbiting electrons would be accelerating and thus should be emitting light continuously. Such electrons would be losing energy and should therefore simply spiral into the positive charge. Nevertheless, the planetary model was a key step in understanding atomic structure.

31-8 The Photoelectric Effect

Some materials can eject electrons when they are exposed to light, a phenomenon called the **photoelectric effect**. Early observations of the photoelectric effect were made by French physicist Antoine Henri Becquerel (1852–1908) with electrodes immersed in conducting solutions. Hertz reported the first observations of a similar effect in air. During his experiments with spark gaps, Hertz found that when he placed a spark gap in the dark to observe the sparks more easily, he had to narrow the gap considerably to get a spark to jump.

MODERN PHYSICS

Systematic investigations of the photoelectric effect revealed the following:

- Every photoelectric material has a **threshold frequency**, which is the minimum frequency of incident radiation below which no electrons are emitted, regardless of the intensity of the incident radiation.

- Above the threshold frequency, the number of electrons emitted is proportional to the intensity of the incident radiation.

- Above the threshold frequency, the energy of the electrons is not proportional to the intensity of the incident radiation.

The relationship between the number of electrons emitted and the radiation intensity made sense because Maxwell's description of electromagnetic radiation indicates that the total energy in an electromagnetic wave is proportional to the intensity of the wave. However, the existence of threshold frequencies was baffling. Why could one not simply increase the intensity to eject electrons?

In 1900, Max Planck developed a theory that describes the blackbody spectra (like those in Figure 31-5). Planck had postulated that electromagnetic radiation comes in discrete parcels, which we now call **photons**. Planck asserted that the energy, E, of a photon is related to its frequency, f, through the relation

KEY EQUATION

$$E = hf \qquad \text{(31-2a)}$$

where $h = 6.626 \times 10^{-34}$ J·s is Planck's constant.

EXAMPLE 31-6

Photon Energy

Calculate the energy associated with a photon that has a wavelength of 665. nm.

Solution

We can use Equation 31-2 in combination with the relation $f = \dfrac{c}{\lambda}$:

$$E = hf = h\frac{c}{\lambda} = \frac{(6.626 \times 10^{-34} \text{ J} \cdot \text{s})(3.00 \times 10^{8} \text{ m/s})}{6.65 \times 10^{-7} \text{ m}}$$
$$= 2.99 \times 10^{-19} \text{ J}$$

Making sense of the result

This amount of energy is very small, but it is on the scale of energies typically associated with atoms.

In 1905, Einstein applied Planck's concept of quantized radiation energy to the photoelectric effect. He assumed that an electron would absorb only one photon at a time (the probability of absorbing two at the same time is very small). The electrons are bound to the material, so a certain minimum amount of energy is required to liberate an electron. This threshold energy is called the **work function** and is usually denoted by E_0. Applying conservation of energy to the process of an electron absorbing a photon, we find that

$$hf - E_0 = K_{\text{max}} \qquad \text{(31-32)}$$

where K_{max} is the maximum kinetic energy of the ejected electron. We measure the kinetic energy of the ejected electrons by slowing them down by applying an electric potential difference between the target and a cathode that collects the electrons. We monitor the electron current as we increase the potential difference, and when the current goes to zero, we call that potential difference the **stopping potential**. At that potential difference, $K_{\text{max}} = e\Delta V_{\text{stopping}}$. If $K_{\text{max}} > 0$, then $hf > E_0$. This condition requires the photon to have a frequency greater than the threshold frequency, f_0, where

$$f_0 = \frac{E_0}{h} \qquad \text{(31-33)}$$

Radiation with frequency less than f_0 will not be able to eject an electron from the material because the individual photons do not have sufficient energy. Increasing the intensity of the radiation simply increases the *number* of photons that hit the material, but each individual photon still has insufficient energy to eject an electron from an atom in the material. Once the frequency threshold is exceeded, increasing the intensity increases the number of electrons ejected because each individual photon now has enough energy to free an electron from the material.

Einstein's explanation of the photoelectric effect was an important analysis because it suggested that light, which had been thought to be a wave, has some particle properties. Further, Einstein extended Planck's idea that physical quantities could be quantized. Einstein's 1921 Nobel Prize in Physics was awarded largely for his work on the photoelectric effect.

31-9 The Bohr Model of the Atom

To develop an atomic model, Danish physicist Niels Henrik David Bohr (1885–1962) first considered the simplest possible atom, hydrogen, which has just one electron and one proton. Hydrogen also has the simplest emission spectrum of all the elements. Bohr's model, published in 1913, accurately predicts the emission spectra for hydrogen.

INTERACTIVE ACTIVITY 31-4

Bohr Model

In this activity, you will use the PhET simulation "Models of the Hydrogen Atom" to help you develop a clear picture of the different spectra produced by the plum pudding model and the Bohr model of the hydrogen atom.

To explain why orbiting electrons do not continuously emit electromagnetic radiation, Bohr postulated that there are stable states for the electrons and that the lowest-energy state did not decay. An electron could go from a high-energy state to a low-energy state and vice versa. For energy to be conserved during such transitions, the atom would either emit or absorb energy equal to the difference in energy between the two states of the atom. The energy that was emitted or absorbed would be in the form of electromagnetic radiation. This radiation should have frequencies that match the observed spectral lines for the atom.

Planck and Einstein had suggested that energy in electromagnetic radiation is quantized in terms of Planck's constant, so Bohr conjectured that some quantity in the atom is correspondingly quantized. As you saw in Chapter 8, an object of mass m travelling with constant speed v in a circular orbit of radius r has an angular momentum of $L = mvr$. Bohr suggested that the electron in a hydrogen atom has a quantized angular momentum given by

$$L = mvr = n\frac{h}{2\pi} = n\hbar \qquad (31\text{-}34)$$

where n is a positive integer, $\hbar = \dfrac{h}{2\pi}$, and m is the mass of the electron.

Each value of the quantum number n is associated with a unique angular momentum and a unique energy, E_n. Bohr assumed that the electron was kept in a circular orbit around the nucleus by the Coulomb force between the positive nucleus and the negative electron. Thus,

$$\frac{ke^2}{r^2} = \frac{mv^2}{r} \qquad (31\text{-}35)$$

where k is Coulomb's constant.

If we solve Equation 31-34 for v, we get

$$v = \frac{n\hbar}{mr} \qquad (31\text{-}36)$$

Substituting this result into Equation 31-35 gives

$$\frac{ke^2}{r^2} = \frac{m}{r}\left(\frac{n\hbar}{mr}\right)^2 \qquad (31\text{-}37)$$

and

$$r = \frac{\hbar^2}{kme^2}n^2 = \frac{L^2}{kme^2} \qquad (31\text{-}38)$$

where we have used Equation 31-34 to substitute L for $n\hbar$. Thus, only certain orbital radii are permitted. The smallest allowed value of r, corresponding to $n = 1$, is called the **Bohr radius** and has a value of

$$r_0 = \frac{\hbar^2}{kme^2} = 5.30 \times 10^{-11}\,\text{m} \qquad (31\text{-}39)$$

Note that the standard notation for the Bohr radius is r_0, even though it is the radius for $n = 1$. The Bohr radius is a reasonable value for the size of a hydrogen atom based on the density of liquid hydrogen. However, the radius of the electron's orbit grows very rapidly with increasing n.

The total energy of an orbiting electron is the sum of its kinetic energy and its electric potential energy. Thus, the energy for any orbit (or state) is given by

$$E_n = \frac{1}{2}mv^2 - \frac{ke^2}{r} \qquad (31\text{-}40)$$

Substituting $mv^2 = \dfrac{ke^2}{r}$ from Equation 31-35, we obtain

$$E_n = \frac{1}{2}\frac{ke^2}{r} - \frac{ke^2}{r} = -\frac{1}{2}\frac{ke^2}{r} \qquad (31\text{-}41)$$

Next, we substitute the expression for r from Equation 31-38:

KEY EQUATION $\quad E_n = -\dfrac{1}{2}\dfrac{ke^2}{\frac{\hbar^2 n^2}{kme^2}} = -\dfrac{1}{2}\dfrac{k^2me^4}{\hbar^2 n^2} = -\dfrac{1}{2}\dfrac{k^2me^4}{L^2}$

$$(31\text{-}42)$$

Equation 31-42 shows that the quantized angular momentum results in quantized energy levels. Thus, the Bohr model allows only quantized values of the energy. Note that these energy levels are negative because we have taken the potential energy to be zero at infinite separation. For the electron to remain bound to the nucleus, the kinetic energy of the electron must be less than the potential energy. The state with $n = 1$ has the lowest energy and is called the **ground state**.

The change in the energy of the electron when it moves from some state n_1 to a different state n_2 is given by

KEY EQUATION

$$\Delta E_{\text{atom}} = E_{n_2} - E_{n_1} = -\frac{1}{2}\frac{k^2me^4}{\hbar^2}\left(\frac{1}{n_2^2} - \frac{1}{n_1^2}\right) \quad (31\text{-}43)$$

The change in the energy of the atom is negative when $n_2 < n_1$, that is, when the electron moves to a lower state. To conserve energy, the photon emitted during this transition must have a positive energy equal to the decrease in the energy of the atom:

$$E_{\text{photon}} = hf = -\Delta E_{\text{atom}} \qquad (31\text{-}44)$$

To facilitate comparison with Rydberg's formula, we now derive an expression for the wave number of the photon. Since $\tilde{v} = \dfrac{1}{\lambda} = \dfrac{v}{c}$,

$$\tilde{v} = \frac{4\pi^2k^2me^4}{2h^3c}\left(\frac{1}{n_2^2} - \frac{1}{n_1^2}\right) \qquad (31\text{-}45)$$

The form of Equation 31-45 is the same as Rydberg's formula (Equation 31-6). When we evaluate the constant in Equation 31-45, we find $\dfrac{4\pi^2k^2me^4}{2h^3c} = 1.09 \times 10^7/\text{m}$, which is in excellent agreement with Rydberg's empirical value.

Although Bohr's model was soon supplanted by a more complete description, it was a significant breakthrough in the development of modern physics. While Planck had previously hypothesized that radiation could be quantized, the Bohr model was the first to predict that matter particles can have quantized physical quantities such as angular momentum and energy. However, the Bohr model still has electrons as particles orbiting a nucleus, and, as such, as we saw in Section 27-7, they should radiate their energy away and thus not be stable.

EXAMPLE 31-7

The Ground State of Hydrogen

Calculate the energy for the ground state of the hydrogen atom using the Bohr model. Express this energy in J and eV.

Solution

Setting $n = 1$ in Equation 31-42 gives the ground-state energy for hydrogen:

$$E_1 = -\frac{1}{2}\frac{k^2 m e^4}{\hbar^2 1^2} = -2.18 \times 10^{-18}\,\text{J}$$

Since $1\,\text{eV} = 1.602 \times 10^{-19}\,\text{J}$,

$$E_1 = -\frac{2.18 \times 10^{-18}\,\text{J}}{1.602 \times 10^{-19}\,\text{J/eV}} = -13.6\,\text{eV}$$

Making sense of the result

You can see why it is common to express atomic energies in electron volts, as numbers on the order of 10 or so are much easier to think about than a number as small as 10^{-18}.

MAKING CONNECTIONS

Stellar Spectra

Have you ever wondered how we know the chemical composition of the Sun? The surface of the Sun is hot and emits a spectrum like a blackbody. That surface is enclosed in an "atmosphere" of hot gases surrounding the Sun. As light from the surface passes through the solar atmosphere, some frequencies are absorbed by the atoms in the atmosphere, causing dark absorption lines in the spectrum, as shown in Figure 31-17. By carefully examining the absorption lines, we can determine which elements are in the atmosphere. Absorption spectroscopy is also a common technique used in chemistry. Light is passed through a sample of interest, and the location and depth of the absorption lines give information about the atoms in the sample and their relative abundances.

Figure 31-17 In this solar spectrum, we see all the characteristic colours from the visible spectrum, but some dark lines also appear in the spectrum. Some of these are due to absorption occurring in the Sun, and some are related to absorption of light as it passes through Earth's atmosphere.

31-10 Compton Scattering

Experimental confirmation that light consists of photons was provided in 1923 by American physicist Arthur Holly Compton (1892–1962). He scattered X-rays of well-defined energy from electrons bound to carbon atoms in a graphite sample.

Compton's experimental setup is shown in Figure 31-18. Compton produced X-rays by bombarding a metal target with electrons and then passed the X-rays through a series of collimating slits to produce a narrow beam with a known wavelength. He directed the beam at a carbon target and measured the intensity and energy (and, hence, the wavelength) of the X-rays scattered at various angles.

According to the classical theory of scattering, electromagnetic waves passing through an atom force the atom's electrons into harmonic motion. The accelerating electrons then emit radiation with the same frequency as the radiation that caused the electrons to accelerate. Let I_{inc} be the intensity of the X-rays incident on the target and $I_{\text{sca}}(\theta)$ be the intensity of the scattered X-rays at an angle of θ relative to the incident beam direction and at a distance R from the scattering target. Then, according to Maxwell's equations,

$$I_{\text{sca}}(\theta) = I_{\text{inc}} \frac{Ne^4}{2m_e^2 c^4 R^2}(1 + \cos^2\theta) \qquad (31\text{-}46)$$

where N is the number of electrons per unit volume of the target, m_e is the electron mass, e is the magnitude of the charge of the electron, and c is the speed of light.

From Equation 31-46, we see that the intensity of the scattered X-rays does not depend on the wavelength of the incident beam. The scattered intensity depends on $\cos^2\theta$ because only a fraction of the incident beam is scattered and the rest of the beam passes through the target undeflected.

Earlier experiments by other researchers showed that Equation 31-46 agrees with observations for incident X-rays of low energy at small scattering

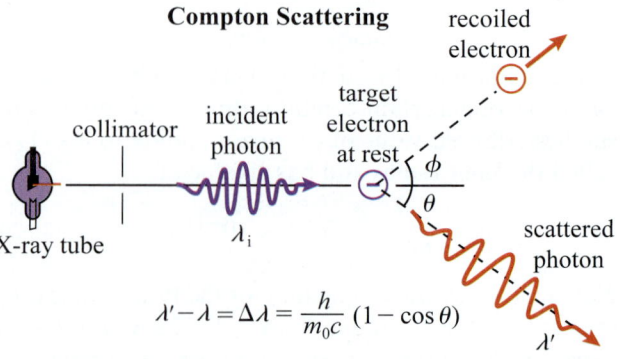

Figure 31-18 A schematic diagram of Compton's scattering experiment.

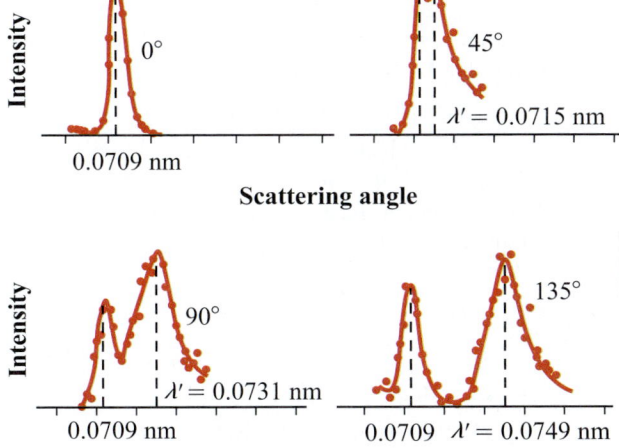

Figure 31-19 The intensity of scattered X-rays for various scattering angles.

angles. However, when Compton used high-energy X-rays (of the order of 10^4 eV) and measured the scattered intensity for large angles, he observed that the scattered X-rays have intensity peaks at two different wavelengths (Figure 31-19). One of the wavelengths is the incident wavelength, λ. The other wavelength, λ', is larger than λ and increases as the scattering angle, θ, increases.

How does the scattering process generate X-rays of different wavelengths? Compton found that all explanations based on classical physics failed to account for the increase in the wavelength of scattered X-rays. For example, the Doppler effect could lower the frequency (and, hence, increase the wavelength) of a scattered X-ray if the target electron were moving away from the incident beam of X-rays. However, a simple calculation showed that for the Doppler effect to account for the observed increase in the wavelength, the electrons in the target would have to be moving away from the direction of incident X-rays at half the speed of light, which is clearly not possible for electrons bound in a carbon atom.

Next, Compton considered the hypothesis proposed by Planck and Einstein that X-rays are photons, each carrying energy $E = hf = \dfrac{hc}{\lambda}$. Then the X-rays are scattered by elastic collisions of an electron and a single photon. The energy of the incident photons is so much greater than the binding energy of an electron in graphite (carbon) that the target electron can be assumed to be initially at rest and free. Figure 31-18 shows the geometry of a collision between a stationary electron and a photon of wavelength λ. During the collision, the photon transfers some of its energy and momentum to the electron and scatters at an angle of θ. The electron, having received energy and momentum,

scatters at an angle of ϕ. Compton applied the laws of energy and momentum conservation to this scattering process.

Let $\vec{p}_\lambda$, $\vec{p}_{\lambda'}$, $\vec{p}_e$, and $\vec{p}_{e'}$ denote the momentum of the incident photon, the scattered photon, the stationary electron, and the scattered electron, respectively, and let E_λ, $E_{\lambda'}$, $E_{e'}$, and E_e denote the corresponding energies. Since momentum must be conserved in the scattering process,

$$\vec{p}_\lambda + \vec{p}_e = \vec{p}_{\lambda'} + \vec{p}_{e'} \tag{31-47}$$

Similarly, energy conservation requires that

$$E_\lambda + E_e = E_{\lambda'} + E_{e'} \tag{31-48}$$

Using relativistic kinematics, the energy and momentum of an electron are related by

$$E = \sqrt{m_e^2 c^4 + p^2 c^2}$$

Assuming that a photon has zero mass, its energy and momentum are related by $E = pc$. Since the target electron is assumed to be at rest, $\vec{p}_e = 0$ and Equation 31-47 can be written as

$$\vec{p}_{e'} = \vec{p}_{\lambda'} - \vec{p}_\lambda \tag{31-49}$$

Using the identity $\vec{A} \cdot \vec{A} = A^2$, where A denotes the magnitude of vector $\vec{A}$, we can write

$$p_{e'}^2 = \vec{p}_{e'} \cdot \vec{p}_{e'}$$
$$= p_\lambda^2 + p_{\lambda'}^2 - 2\vec{p}_\lambda \cdot \vec{p}_{\lambda'}$$
$$= p_\lambda^2 + p_{\lambda'}^2 - 2p_\lambda p_{\lambda'} \cos\theta \tag{31-50}$$

Equation 31-48 for energy conservation can be written as follows:

$$p_\lambda c + m_e c^2 = p_{\lambda'} c + \sqrt{m_e^2 c^4 + p_{e'}^2 c^2}$$

Bringing the $p_{\lambda'}c$ term to the left and squaring both sides gives

$$p_\lambda^2 + p_{\lambda'}^2 - 2p_\lambda p_{\lambda'} + 2m_e c(p_\lambda - p_{\lambda'}) = p_{e'}^2 \tag{31-51}$$

Inserting Equation 31-50 into Equation 31-51, we obtain

$$m_e c(p_\lambda - p_{\lambda'}) = p_\lambda p_{\lambda'} (1 - \cos\theta) \tag{31-52}$$

The energy and the wavelength of a photon are related by

$$E = pc = hf = \frac{hc}{\lambda}$$

Therefore, the momentum of a photon is given by

$$p = \frac{h}{\lambda} \tag{31-53}$$

Writing Equation 31-52 in terms of wavelength, we get

KEY EQUATION $$\lambda' - \lambda = \frac{h}{m_e c}(1 - \cos\theta) \tag{31-54}$$

Equation 31-54 relates the wavelength of the scattered photon to the scattering angle and accurately predicts the change in the wavelength of scattered X-rays for a given scattering angle. Notice that the change in the wavelength ($\lambda' - \lambda$) is independent of the original wavelength. The quantity $h/(m_e c)$ is called the **Compton wavelength**, λ_C, and its value is

$$\lambda_C = \frac{h}{m_e c} = \frac{6.62 \times 10^{-34}\,\text{J} \cdot \text{s}}{(9.11 \times 10^{-31}\,\text{kg})(3.00 \times 10^8\,\text{m/s})}$$

$$= 2.42 \times 10^{-12}\,\text{m} \tag{31-55}$$

The small value of the Compton wavelength explains why the classical description of the scattering experiment (Equation 31-46) agrees with experiments for low-energy (long-wavelength) X-rays. The shift in the wavelength is very small and becomes noticeable only when the incident wavelength is small.

Compton's explanation of his observations proved without a doubt that the scattering of X-rays from electrons is a quantum phenomenon. His experiment showed that single photons transfer energy and momentum to electrons during X-ray scattering; therefore, X-rays and, by extension, other electromagnetic radiation, consist of photons. The scattering of photons from electrons is now called **Compton scattering**, and the increase in the wavelength of the scattered photon is called the **Compton shift**.

EXAMPLE 31-8

Wavelength Change in Compton Scattering

In one of his experiments, Compton scattered 17.4 keV X-rays from a graphite target.

(a) What is the wavelength of photons that are scattered at an angle of 135° with respect to the incident beam?
(b) What is the percentage change in the observed wavelength?

Solution

(a) The wavelength of the incident photons is given by the relationship $E = hc/\lambda$. Therefore,

$$\lambda = \frac{hc}{E} = \frac{(6.62 \times 10^{-34}\,\text{J} \cdot \text{s})(3.00 \times 10^8\,\text{m/s})}{(17.4 \times 10^3\,\text{eV})(1.60 \times 10^{-19}\,\text{J/eV})}$$

$$= 7.13 \times 10^{-11}\,\text{m}$$

The wavelength of the photon scattered at 135° is given by Equation 31-54:

$$\lambda' = \lambda + \frac{h}{m_e c}(1 - \cos\theta)$$

$$= 7.13 \times 10^{-11}\,\text{m} + (2.42 \times 10^{-12}\,\text{m})(1 - \cos 135°)$$

$$= 7.61 \times 10^{-11}\,\text{m}$$

(b) The percentage change in the wavelength is

$$\frac{\lambda' - \lambda}{\lambda} = \frac{0.48 \times 10^{-11}\,\text{m}}{7.13 \times 10^{-11}\,\text{m}} \times 100\% = 6.7\%$$

CHECKPOINT 31-5

Compton Scattering

In a Compton scattering experiment, energies of scattered electrons are measured at angles of $\phi = 0°$, 90°, 135°, and 180°. At which angle does the electron carry maximum energy? Assume that initially the target electron is at rest.

(a) 0°
(b) 90°
(c) 135°
(d) 180°

ANSWER: (a) Referring to Figure 31-18, we can see that if the electron emerges from the collision at an angle of $\phi = 0°$, then the electron will emerge with an angle of $\theta = 180°$ to ensure momentum is conserved. This large scattering angle for the electron produces the largest possible increase in wavelength for the photon and thus the largest possible decrease in energy for the photon. To conserve energy, all of this energy is transferred to the electron.

KEY CONCEPTS AND RELATIONSHIPS

The early developments in modern physics were centred on gaining an understanding of the electrical properties of matter and the generation of light by matter.

The Connection between Matter and Electricity

Faraday's electrolysis experiments established a clear connection between electric charge and matter. Prior to these experiments, electricity had been thought of as a massless fluid that flowed from one object to another. However, Faraday showed that the amount of gas evolved, and hence mass, in an electrolysis experiment is proportional to the amount of charge passed through the cell, and he determined charge-to-mass ratios for various elements. This clearly established a connection between charge and mass and dispelled the massless fluid theories.

Temperature and the Emission of Light

Experiments with light emission established that hot objects emit light at all wavelengths, and the spectrum is characteristic of the temperature of the object. The Stefan–Boltzmann law,

$$P = \sigma \varepsilon A T^4 \tag{31-1}$$

describes the total power radiated by an object at a temperature T, while Wiens' law relates the temperature of an object to the wavelength in the spectrum at which the most power is radiated:

$$\lambda_{max} = \frac{2.898 \times 10^{-3}\,\text{m} \cdot \text{K}}{T} \tag{31-2}$$

Max Planck accurately described the blackbody spectrum by assuming that the light is emitted by the object in discrete quanta with energy

$$E = hf \tag{31-3}$$

Using this assumption, Planck derived the following expression for the intensity of blackbody radiation:

$$B(f,T) = \frac{8\pi h f^3}{c^2} \frac{1}{e^{\frac{hf}{kT}} - 1} \qquad (31\text{-}4)$$

which agrees very well with experimental data. Although Planck did not attach much physical significance to his assumption that the energies of photons were quantized, the idea was subsequently found to be quite profound.

Gas Discharge Spectra

In addition to emission spectra generated by hot objects, gas discharge spectra were also being studied. In contrast with a blackbody, a gas discharge spectrum only contains intensity at certain wavelengths, called spectral lines. In addition, each different gas has a unique set of spectral lines. Balmer analyzed the wave numbers, $\tilde{v} = \frac{1}{\lambda}$, of the light in the hydrogen discharge spectrum and showed that it could be described by

$$\tilde{v} = R_H \left(\frac{1}{n_2^2} - \frac{1}{n_1^2} \right) \qquad (31\text{-}6)$$

Both the blackbody radiation and the gas discharge results suggested that energies are somehow quantized.

Cathode Rays

Experiments with gas discharge tubes at very low pressures led to Thomson's discovery of negatively charged particles (electrons) with a charge-to-mass ratio at least 1000 times any previously observed. This result suggested that there was some type of particle (ultimately determined to be the electron) that existed at the subatomic level, since its mass was so small compared to those determined by Faraday.

The Millikan Oil Drop Experiment

Millikan's oil drop experiments allowed him to measure the charges on small oil drops:

$$q = \frac{W}{E}\left(1 + \frac{v_2}{v_1} \right) \qquad (31\text{-}14)$$

He showed that charges come in integer multiples of a fundamental unit of charge, e, now known to be approximately $e = 1.602 \times 10^{-19}$ C. This result helped determine that an electron carries a single unit of negative charge and, consequently, must be very light, based on the charge-to-mass ratios determined by the cathode ray experiments.

Thomson's Model of the Atom

Thomson postulated a model of an atom that consisted of electrons embedded in a spherical cloud of positive charge, which contained almost all the mass of the atom. This model

predicted that the electrons oscillate, and thus radiate, but only at one frequency for a given element:

$$\omega = \sqrt{\frac{qQ}{4\pi\varepsilon_0 r_0^3 m}} \qquad (31\text{-}17)$$

which did not agree with the experimental results.

Rutherford Scattering

Rutherford and his co-workers scattered α-particles off thin gold foils and found that some of the particles were scattered through very large angles. Based on these results, Rutherford showed that the positive charge in the atom must be concentrated in a very small volume, and he proposed an atomic structure with the electrons in orbits about a positive nucleus.

The Photoelectric Effect

Einstein's analysis of the photoelectric effect, using Plank's hypothesis that the energy of a photon is quantized,

$$E = hf \qquad (31\text{-}3)$$

further bolstered the idea that photons were particles that carried discrete amounts of energy.

The Bohr Model of the Atom

Bohr introduced the notion of stationary orbits that had quantized angular momentum. Bohr's model, which correctly predicts the wavelengths of the spectral lines for hydrogen, was an important first step in the development of quantum theory. He found that the wave numbers could be described by

$$\tilde{v} = \frac{4\pi^2 k^2 m e^4}{2h^3 c} \left(\frac{1}{n_2^2} - \frac{1}{n_1^2} \right) \qquad (31\text{-}45)$$

Compton Scattering

Compton's X-ray scattering experiments provided proof that Einstein and Planck's idea that light consists of discrete particles called photons is correct. He determined that the change in wavelength in a photon when it scatters off a free electron is given by

$$\lambda' - \lambda = \frac{h}{m_e c}(1 - \cos\theta) \qquad (31\text{-}54)$$

Putting all of these results together set the stage for the development of quantum theory. Charge and mass were shown to be connected to one another, dispelling the massless fluid ideas of electricity. Analysis of blackbody radiation, gas discharge spectra, and the photoelectric effect established that photon energies come in discrete amounts and that the energy levels in atoms are also quantized. Cathode ray experiments established that there are very light subatomic particles, now known to be electrons, and the Millikan experiments established that charge was also quantized. The success of the Bohr Model of the atom indicated that quantization of energies in atoms was a key feature, and the Compton experiments established that photons were indeed particles.

APPLICATIONS

particle accelerators, spectrometers

KEY TERMS

anode, blackbody, Bohr radius, Compton scattering, Compton shift, Compton wavelength, discharge tube, elementary charge, fundamental charge, ground state, photoelectric effect, photons, Planck's constant, planetary model, spectrometer, stopping potential, threshold frequency, valence, wave number, Wien's law, work function

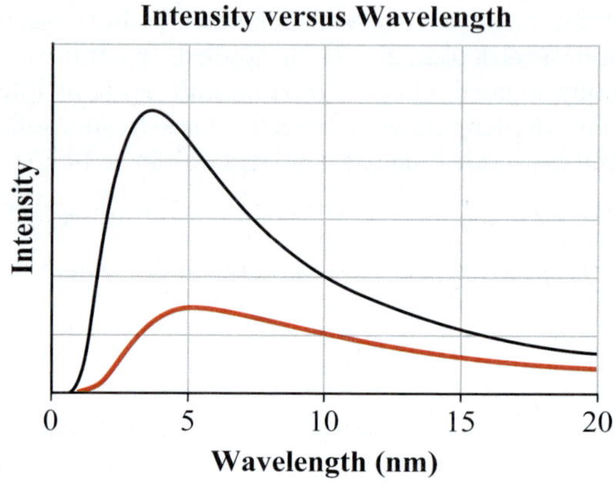

QUESTIONS

1. Which experiments led to the postulation of the existence of the electron?
2. When an atom emits light, does the atom make a transition to a higher or a lower electron energy state?
3. What principal facts need to be explained by a successful atomic model?
4. What is the key property required for a nucleus to back-scatter α-particles?
5. What physical quantity did Bohr quantize to develop his model of the atom?
6. How can you distinguish between different gases by examining their discharge spectra?
7. What are the essential differences between a discharge spectrum and a blackbody spectrum?
8. What important question did Millikan resolve with his oil drop experiment?
9. Some electronic medical thermometers measure body temperature with a probe placed in the ear for a few seconds. How might such a device work?
10. What idea did Planck and Einstein introduce that was key to the development of the quantum theory?

PROBLEMS BY SECTION

For problems, star ratings will be used (★, ★★, or ★★★), with more stars meaning more challenging problems. The following codes will indicate if $\frac{d}{dx}$ differentiation, $\int$ integration, ▱ numerical approximation, or ⟋ graphical analysis will be required to solve the problem.

Section 31-1   The Connection between Matter and Electricity

11. ★ Calculate the charge-to-mass ratio for hydrogen and oxygen.
12. ★ During an electrolysis experiment, 3.0 mol of hydrogen gas is evolved.
    (a)   How much charge has passed through the cell?
    (b)   The current was 10. mA. How long did the experiment take?

Section 31-2   Temperature and the Emission of Light

13. ★ Figure 31-20 shows blackbody spectra for two objects.
    (a)   Which object is at the higher temperature?
    (b)   Is the light emitted from these objects visible to the human eye?

Intensity versus Wavelength

Figure 31-20   Problem 13.

Section 31-3   Gas Discharge Spectra

14. ★★ Calculate the wave numbers for the Balmer series of hydrogen, which corresponds to $n_2 = 2$.
15. ★ A spectral line in a hydrogen discharge spectrum is observed to have a wavelength of 1875 nm.
    (a)   What wave number does this correspond to?
    (b)   What values of n_1 and n_2 does this correspond to?
    (c)   What part of the electromagnetic spectrum does this radiation appear in?

Section 31-4   Cathode Rays

16. ★★ An electron is accelerated from rest through 200. V. It then enters a crossed-field apparatus whose plates are separated by 1.00 cm with a potential difference of 100. V across them.
    (a)   How fast is the electron moving when it enters the crossed fields?
    (b)   What is the electric field between the plates?
    (c)   What magnetic field is required to have zero deflection?
17. ★★ The path of an electron is observed to have a radius of curvature of 20.3 cm in a magnetic field of 1.80 mT.
    (a)   How fast is the electron moving?
    (b)   What electric field is required to cause the electron to move in a straight line?
    (c)   The field is produced by a potential of 150. V. What is the separation of the plates?
18. ★★ An α-particle consists of two protons and two neutrons.
    (a)   What is the mass of an α-particle?
    (b)   What is the charge of an α-particle?
    (c)   We accelerate an α-particle through 200. V. How fast is it moving?

(d) An electric field of 10.0 kV/m causes the α-particle to pass straight through the crossed-field region. What is the magnetic field?

(e) If we turned off the electric field, what would the radius of curvature of the path of the α-particles be?

19. ★★ In our analysis of crossed-field experiments, we did not consider the effects of gravity on the electrons. Suppose we have a plate separation of 2.00 cm and a voltage of 300. V.

(a) What is the electric field between the plates?

(b) What is the force on the electron due to the magnetic field?

(c) What is the force on the electron due to gravity?

(d) Was it reasonable to neglect gravity in our analysis? Explain.

Section 31-5 The Millikan Oil Drop Experiment

20. ★★ A latex sphere with a radius of 1.0 μm is between plates that are separated by 15 mm. A potential difference of 38 V is just enough to keep the sphere suspended.

(a) What is the gravitational force on the sphere?

(b) What is the electric field between the plates?

(c) What is the magnitude of the sphere's charge?

(d) The top plate is positive relative to the bottom plate. What is the sign of the charge on the sphere?

21. ★★★ If you do not account for the buoyancy force on spheres in an oil drop experiment, what percentage error do you make in the weight force for the following? (Hint: You will need to look up the densities of oil and latex.)

(a) oil drops

(b) latex spheres

22. ★★ Calculate the free-fall speed of a 0.800 μm oil drop. Assume that the density of the oil is 821 kg/m^3.

Section 31-6 Thomson's Model of the Atom

23. ★★★ $\int$ In this question, you will calculate the energy involved in assembling the positive charge in a Thomson atom. Treat the positive charge as a sphere with charge uniformly distributed throughout. The potential at a distance r from the centre is given by

$$V(r) = \frac{kQ_{enc}}{r} = \frac{kQr^2}{R^3}$$

where $+Q$ is the atom's total charge and R is its radius. Consider a thin shell of charge with radius r and thickness dr. Such a shell carries the charge

$$dq = 4\pi r^2 dr\rho = \frac{3Qr^2 dr}{R^3}$$

(a) Put the above expressions together to derive $dU = Vdq$.

(b) Integrate the resultant expression from $r = 0$ to $r = R$ to get the total potential energy of the sphere of charge.

(c) Evaluate the resulting expression for $Q = 79e$ and $R = 1.0$ nm.

24. ★★ Use Equation 31-17 to plot ω versus Q for $e < Q < 100e$. Set $q = e$, and use m as the mass of the electron.

Section 31-7 Rutherford Scattering

25. ★★ Equation 31-31 gives the radius of a nucleus required to stop an α-particle with an initial kinetic energy of

$$K = \frac{1}{2}mv^2 = 5.0 \times 10^{-13}\,\text{J}$$ at the surface of a nucleus

with a radius of 1.2×10^{-14} m.

(a) Suppose the α-particle has one-half that initial energy. At what distance from the centre of the nucleus will it stop?

(b) Suppose the α-particle has twice that initial energy. How far into the nucleus would it penetrate?

26. ★★ Evaluate the expression you obtained in problem 23(b) for $Q = 79e$ and $R = 1.0 \times 10^{-14}$ m.

27. ★ We accelerate an electron from rest through a potential difference of 1.00 kV.

(a) What is the kinetic energy of the electron in J?

(b) What is the speed of the electron?

(c) What is the kinetic energy of the electron in eV?

28. ★ Some theorists have speculated that the mass of particles may be due to their "self-energy"; that is, the energy required to assemble them is given by $E = mc^2$. Does the result of problem 26 give a reasonable estimate for the mass of the gold nucleus?

Section 31-8 The Photoelectric Effect

29. ★ Sodium has a work function of 2.28 eV.

(a) Convert this energy to J.

(b) What is the threshold frequency for the photoelectric effect in sodium?

(c) To what wavelength does this frequency correspond?

30. ★ A certain material is observed to have a threshold wavelength of 263 nm.

(a) To what frequency does this wavelength correspond?

(b) What energy is carried by each photon at this frequency?

(c) What is the work function of the material?

(d) Express your result in eV.

31. ★★ 〰 A student collects the data in Table 31-1 during a photoelectric effect experiment. Recall that $hf - E_0 = K_{max}$, and K_{max} is related to the stopping potential, ΔV, through $K_{max} = e\Delta V$, where $e = 1.602 \times 10^{-19}$ C.

(a) Convert the wavelengths in Table 31-1 to frequencies.

(b) Convert the stopping potentials into values of K_{max}.

(c) Plot K_{max} versus frequency, and fit a straight line.

(d) How does the value of your slope compare to Planck's constant?

(e) Divide the intercept by e to get the work function in eV.

Table 31-1 Data for Problem 31

Wavelength (nm)	Stopping Potential (V)
260	0.46
250	0.66
240	0.87
230	1.09

CHAPTER 31 | FUNDAMENTAL DISCOVERIES OF MODERN PHYSICS   1119

Section 31-9 The Bohr Model of the Atom

32. ★ Consider a spectral line corresponding to a transition from $n_1 = 2$ to $n_2 = 1$.
  (a) Calculate the change in energy of the atom.
  (b) Calculate the frequency of the emitted light.
  (c) Calculate the wavelength of the emitted light.
  (d) Calculate the wave number of the emitted light.

33. ★ What is the speed of an electron in a Bohr orbit corresponding to $n = 1$?

34. ★★ What is the radius of the $n = 7$ orbit of a Bohr atom?

35. ★ What is the period of the $n = 1$ Bohr orbit?

36. ★★ In 1924, Louis de Broglie (1892–1987) postulated that particles have a wavelength of $\lambda = \dfrac{h}{p}$, where h is Planck's constant and $p = mv$.
  (a) What is the momentum of an electron in the $n = 1$ Bohr orbit?
  (b) What is the wavelength of an electron in the $n = 1$ orbit?
  (c) What is the circumference of the $n = 1$ orbit?

Section 31-10 Compton Scattering

37. ★ A Compton scattering experiment uses 350 keV X-rays. The detector is placed at an angle of 40.° relative to the incident beam. Find
  (a) the energy of the scattered X-ray
  (b) the energy of the scattered electron

38. ★★ A photon with a wavelength of 750 pm causes an electron to recoil with a speed of 1.35×10^6 m/s in a Compton experiment. Find
  (a) the change in the photon's wavelength
  (b) the change in the photon's frequency
  (c) the angle through which the photon is scattered

COMPREHENSIVE PROBLEMS

39. ★ A 5.0 MeV α-particle approaches a nucleus head-on and stops at a distance of 7.0 fm from the centre of the nucleus. What is the charge on the nucleus?

40. ★ What element is in the sample with the absorption spectrum given in Figure 31-21?

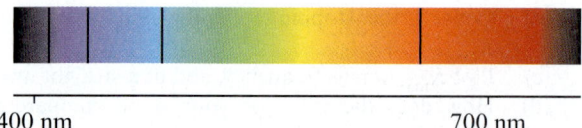

400 nm   700 nm

Figure 31-21 Absorption spectrum for problem 40.

41. ★★★ 〰 In the Sun, two protons can fuse to form deuterium. In this process, a positron, an electron neutrino, and some energy are released. For the reaction to take place, the protons must come close enough together for the weak interaction forces to overcome the repulsive Coulomb interaction. Make a plot of the Coulomb potential as a function of proton–proton separation, assuming that the protons are point charges. It is generally thought that this reaction can proceed at temperatures of approximately 1.5×10^7 K. What separation distance does such a temperature allow?

42. ★ The diameter of a hydrogen nucleus is taken to be 1×10^{-14} m, and we assume that the volume of a hydrogen atom is defined by the Bohr radius. What is the ratio of the volume of the nucleus to the volume of the atom?

43. ★★ In a mass spectrometer, a magnetic field is used to cause charged particles to move in circular paths with radii $r = \dfrac{mv}{qB}$. Determining r then allows us to calculate m, assuming that v, q, and B are known. To simplify the experiment, a velocity selector is used to make sure that all the particles have the same speed, and a crossed-field configuration is used. If the radius of the spectrometer magnet is 0.50 m and it has a field strength of 300. mT, what is the maximum speed that incoming particles can have if we wish to measure masses up to the mass of an argon atom? Assume that all particles are singly ionized.

44. ★★ Suppose that electrons are accelerated through a 1.00×10^3 V potential difference and then strike a 5.00 mg aluminum target. It is observed that the temperature of the target rises by 4.00° C in 8.00 s. Assume that all of the energy of the electrons is converted into thermal energy. How many electrons have struck the target during the 8 s? Find the beam current.

45. ★★ An electron in a cathode ray beam enters a crossed-field region. The plates are 1.0 cm apart and have a potential difference of 200. V across them. The magnetic field has a strength of 0.80 mT when the electrons pass straight through. When the potential difference is removed, what is the radius of curvature of the electron beam? What is the maximum length of the plates that will allow the beam to exit without hitting the plates? Assume that the beam enters the plates halfway between them.

46. ★★ Suppose that an electron orbits a proton and radiates light at the orbital frequency. What does the radius of the orbit need to be for the atom to emit light with a wavelength of 656 nm?

47. ★★★ $\frac{d}{dx}$ In Chapter 32, you will see that a blackbody spectrum can be described by

$$I(f,T) = \frac{2hf^3}{c^2} \frac{1}{e^{\frac{hf}{kT}} - 1}$$

where I is the intensity of the radiation, f is the frequency of the radiation, h is Planck's constant, c is the speed of light, k is Boltzmann's constant, and T is the temperature in K. Differentiate the intensity to find the frequency at which the intensity is maximum.

48. ★★ 〰 The following wave numbers are measured in some other universe that has a different Rydberg constant: 3472/m, 4688/m, 5250/m, 5556/m, 5740/m, and 5859/m. Perform a fit to the data to determine n^2 and the value of the Rydberg constant in that universe.

49. ★★ It is not known whether the electron is a point particle. If you want to do electron–electron scattering to probe the structure of the electron, approximately what energies are required? Assume that the radius of the electron is 10^{-17} m and that the electrons must be able to get at least this close to each other.

50. ★★ An α-particle scatters from an oxygen nucleus. The initial speed of the α-particle is $v_1 = 1.00 \times 10^7$ m/s. The α-particle is deviated by an angle of 90.0°, and it emerges from the collision with speed $v_f = 0.317 \times 10^7$ m/s. Determine the magnitude and direction of the recoil momentum of the nucleus, and discuss whether the collision was elastic.

51. ★★★ Suppose you wanted to use a crossed-field apparatus built for studying electrons to measure the charge-to-mass ratio of the proton. What changes would you need to make to the apparatus to make a successful measurement?

52. ★★★ Another way of doing the Millikan experiment is to reverse the direction of the applied field and observe the terminal velocity reached in both directions. The radii of the drops must be known. A student is doing such an experiment with 1.00 μm latex spheres and a potential difference of 100. V between plates, which are separated by 5.00 mm. What velocities would be observed for a drop with a charge of $+e$? $-2e$?

53. ★★ The radiant flux at Earth due to the Sun is approximately 1.30 kW/m². This energy is spread across the visible spectrum and extends both down into the infrared and up into the ultraviolet. For the purpose of this problem, assume that all of this intensity is at 500. nm. How many photons per square metre does such a flux correspond to?

DATA-RICH PROBLEM

54. ★★ ⌇ Go to the text's online resource and download the hydrogen data file, which contains all the wave numbers that are plotted in Figure 31-8. Download the data file, plot the data, and try to fit the data to the form of Equation 31-6. In the process, you should arrive at a value for the Rydberg constant.

OPEN PROBLEM

55. ★★★ Hydrogen is attractive as an alternative fuel because it is plentiful and its combustion product is simply water. Estimate how much water would have to be electrolyzed to meet your annual energy needs.

Introduction to Quantum Mechanics

When you have completed this chapter, you should be able to

LO1 Explain de Broglie's hypothesis and the Davisson–Germer experiment, and calculate an object's wavelength given its momentum.

LO2 Explain how the width of a wave packet is related to the uncertainty in the position of the particle described by the wave packet, and how uncertainties in position and momentum are related by Heisenberg's uncertainty principle.

LO3 State the four assumptions that Schrödinger made to derive his wave equation, and write the Schrödinger equation in one dimension for a given potential.

LO4 Write solutions for a second-order differential equation of the form $\dfrac{d^2 f(x)}{dx^2} \pm a^2 f(x) = 0$, where a is a real number, for given boundary conditions.

LO5 Write the normalized wave functions for a particle in an infinite square well potential. Calculate the energy of the emitted photon when a charged particle makes a transition between two bound states of an infinite square well potential.

LO6 Qualitatively explain the behaviour of a particle confined in a finite square well potential. Explain how energy eigenvalues are calculated for such a particle.

LO7 Describe what is meant by quantum tunnelling, and calculate the non-normalized probability of tunnelling through a finite square well potential.

LO8 Describe the quantization of angular momentum. Calculate the magnitude and the z-component of the angular momentum in quantum mechanics.

LO9 Explain the similarities and differences between the Bohr model and the Schrödinger equation description of a hydrogen atom.

LO10 Describe the Stern–Gerlach experiment, and explain how it led to the discovery of quantized spin angular momentum.

Quantum mechanics predicts the behaviour of particles and waves at the atomic and subatomic scales. This behaviour can appear strange and even baffling because it often differs radically from behaviour at the scale of the everyday objects we have handled all of our lives. In this chapter, you will learn how to apply the equations that describe the quantum-mechanical behaviour of particles and the hydrogen atom. Many modern instruments and devices, such as laser scanners, smoke detectors, flash memory (Figure 32-1), tunnelling electron microscopes, diode lasers, and nuclear magnetic resonance (NMR) machines, are based on the theory of quantum mechanics. New theoretical and experimental research involving quantum computers and quantum teleportation has the potential for a new generation of practical applications.

Some derivations in this chapter require a basic understanding of complex numbers and partial derivatives.

Figure 32-1 The principles of quantum mechanics underlie the technology of the memory chips in this music player.

32-1 Matter Waves and de Broglie's Hypothesis

Quantum mechanics is the branch of physics that deals with the motion and properties of objects of extremely small masses, such as electrons, protons, neutrons, nuclei, atoms, and molecules. The following comparison will help provide an idea of the scale of subatomic particles: the mass of the smallest known virus is of the order of 10^{-20} kg, which is approximately 10 million times the mass of a proton and 40 000 times the mass of the heaviest nucleus. For subatomic masses, Newton's laws of motion break down—quantum mechanics provides a much more accurate description of the motion of these objects. Numerous experiments have verified the principles of quantum theory and have shown that its strange and far-reaching predictions are correct.

As described in Chapter 31, Planck had to assume that electromagnetic radiation is emitted in discrete parcels, called quanta, to explain the radiation spectrum of a blackbody, and Einstein had to make the same assumption to explain the photoelectric effect. Not only is the radiation *emitted* in quanta, but it also *travels* in quanta. So, by the early 20th century, scientists had realized that electromagnetic waves have particle-like properties. In 1924, French physicist Louis de Broglie (1892–1987) argued that in a symmetric universe particles should also possess wave-like properties. He proposed that *moving objects behave like a wave and therefore have a wavelength and a frequency*. For a particle with energy E and momentum p, the frequency, f, and the wavelength, λ, of the associated wave are given by

$$f = \frac{E}{h} \tag{32-1}$$

KEY EQUATION
$$\lambda = \frac{h}{p} \tag{32-2}$$

Equations 32-1 and 32-2 are the same equations that Planck and Einstein proposed for the energy and momentum of a photon. A wave associated with a moving particle is called a **de Broglie wave** or **matter wave**, and its wavelength is called the **de Broglie wavelength**.

EXAMPLE 32-1

De Broglie Wavelengths

Determine whether the de Broglie effect (wave–particle) is detectable for the following objects:

(a) a space shuttle entering orbit
(b) a fastball thrown by a major-league baseball pitcher
(c) an electron accelerated through the potential difference of 1.5 V

Solution

(a) The approximate mass of a space shuttle is 22 000 kg, and its speed just before it enters its orbit is approximately 8 000 m/s. The de Broglie wavelength of a shuttle at this speed is

$$\lambda = \frac{h}{p} = \frac{6.63 \times 10^{-34} \text{ J} \cdot \text{s}}{(22\,000 \text{ kg})(8\,000 \text{ m/s})} = 3.8 \times 10^{-42} \text{ m}$$

This wavelength is approximately 10^{-29} times the size of a proton. No existing instrument or method can measure a wavelength this small. Note that we can use the non-relativistic formula for momentum ($p = mv$) because the speed of the shuttle is much less than the speed of light.

(b) The mass of a baseball is approximately 0.14 kg, and the speed of the fastest thrown baseball was recorded at 103 mph (46.0 m/s). The de Broglie wavelength of the baseball is

$$\lambda = \frac{6.63 \times 10^{-34} \text{ J} \cdot \text{s}}{(0.14 \text{ kg})(46.0 \text{ m/s})} = 1.0 \times 10^{-34} \text{ m}$$

This wavelength is still beyond the measurement range of any instrument.

(c) An electron has a mass of 9.11×10^{-31} kg. So when it is accelerated through a 1.5 V potential difference, it acquires 1.5 eV of kinetic energy. The speed can be calculated using kinematic relationships:

$$K = \text{kinetic energy of the electron} = 1.5 \text{ eV}$$
$$= 1.5 \times 1.60 \times 10^{-19} \text{ J} = 2.4 \times 10^{-19} \text{ J}$$

$$v = \sqrt{\frac{2K}{m_e}} = \sqrt{\frac{(2)(2.4 \times 10^{-19} \text{ J})}{9.11 \times 10^{-31} \text{ kg}}} = 7.3 \times 10^5 \text{ m/s}$$

Again, the speed is low enough that we can ignore relativistic effects. The de Broglie wavelength of the electron is

$$\lambda = \frac{6.63 \times 10^{-34} \text{ J} \cdot \text{s}}{(9.11 \times 10^{-31} \text{ kg})(7.3 \times 10^5 \text{ m/s})} = 1.0 \times 10^{-9} \text{ m}$$

This wavelength is of the order of the wavelength of high-energy X-rays and the size of atoms. Such wavelengths are easily measured with modern instruments.

CHECKPOINT 32-1

The de Broglie Wavelength

An electron, a proton, and an iron nucleus are moving at speeds such that they have the same momentum. Which has the shortest de Broglie wavelength?
(a) the electron
(b) the proton
(c) the iron nucleus
(d) They all have the same wavelength.

ANSWER: (d) The de Broglie wavelength of a particle depends on its momentum only.

MODERN PHYSICS

Bragg's Law

As described in Chapter 29, the wavelength of light can be measured by observing the diffraction pattern formed when the light passes through a diffraction grating. For accurate measurements, the spacing between the slits of the grating must be of the same order of magnitude as the wavelength of the light. So, the slit separation of an optical diffraction grating is too large to measure wavelengths of X-rays, for example.

The German physicist Max Theodor Felix von Laue (1879–1960) showed that crystals, whose interatomic spacing is of the order of 10^{-10} m, can be used as a diffraction grating for measuring the wavelengths of X-rays. Figure 32-2 shows X-rays scattering from individual atoms of successive layers of a crystal. The path difference between two rays that scatter at an angle θ from two adjacent planes is $2d \sin \theta$. The condition for constructive interference between these two waves is

$$2d \sin \theta = m\lambda \qquad \text{(Bragg's law)} \qquad (32\text{-}3)$$

where m is a non-zero positive integer. This condition for constructive interference is called **Bragg's law**. Figure 32-3 shows the diffraction pattern produced by X-rays scattering from an aluminum oxide foil.

The Davisson–Germer Experiment

In 1926, the German-born American physicist Walter Maurice Elsasser (1904–1991) suggested that if electrons have wavelengths comparable to X-ray wavelengths, then electron scattering from crystals should produce the same type of diffraction pattern as observed in X-ray scattering.

In 1927, American physicists Clinton Davisson (1881–1958) and Lester Germer (1896–1971) tested Elsasser's suggestion and performed electron diffraction experiments using the setup shown in Figure 32-4. Electrons emitted from a heated metal filament are accelerated through a potential difference and collimated into a narrow beam. Davisson and Germer directed the electron beam at the face of a crystal, and they measured the angles at which the reflected beam had maximum intensity. They then used Bragg's law to calculate the wavelength corresponding to these angles—they found that it matched the wavelength predicted by de Broglie. Davisson and Germer varied the potential difference to change the energy of the electrons in the beam, producing maxima and minima at different angles in the diffraction patterns. In each case, the wavelength calculated using the Bragg equation agreed with the de Broglie wavelength. This series of experiments proved conclusively that electrons do have a wavelength. Similar experiments with other particles have confirmed the wave nature of matter.

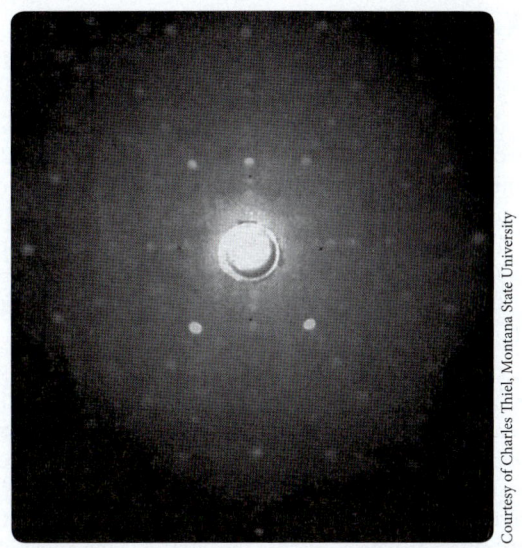

Figure 32-3 A diffraction pattern formed by X-rays scattering from an yttrium aluminum oxide target.

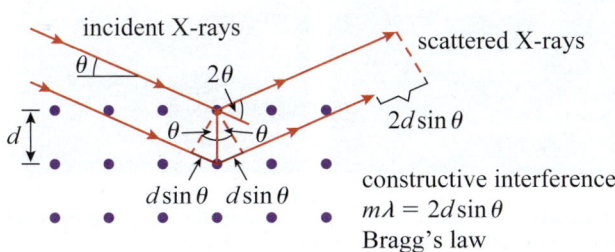

Figure 32-2 The scattering of X-rays from adjacent planes in a crystal. The path difference for waves scattered from two successive planes is $2d \sin \theta$.

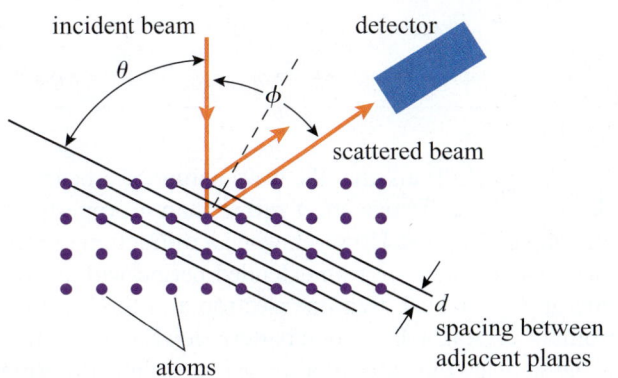

Figure 32-4 The Davisson–Germer experiment, showing the relationship between the incident angle θ and the scattering angle ϕ.

EXAMPLE 32-2

The de Broglie Wavelength of an Electron

In one experiment, Davisson and Germer accelerated electrons through a potential difference of 54 V and scattered the electrons from a nickel crystal. They observed a maximum in the intensity of reflected electrons at an angle of $\phi = 50°$. The spacing between the scattering planes of the crystal was calculated from X-ray diffraction to be $d = 0.091$ nm. Show that these results agree with de Broglie's hypothesis for particle wavelengths.

Solution

According to Bragg's law, when waves of wavelength λ are reflected from the planes of a crystal that are separated by a distance d, the scattering angle at which maximum constructive interference occurs, θ, is given by Equation 32-3. For scattering from adjacent planes, $m = 1$, the angle θ is the angle of incidence with respect to a particular scattering plane of the crystal (Figure 32-4). By conservation of momentum, the reflected beam must also make an angle θ with respect to the same scattering plane. Therefore, the angle between the incident and the reflected beams, ϕ, is

$$\phi = \pi - 2\theta$$

The location of the detector with respect to the incident beam corresponds to this angle, ϕ. For scattering from two adjacent planes, Bragg's law gives

$$\lambda = 2d\cos\left(\frac{\phi}{2}\right)$$

Inserting the given data, we get

$$\lambda = (2)(0.091 \text{ nm})\cos 25° = 0.165 \text{ nm}$$

For non-relativistic speeds, the kinetic energy of an electron accelerated through a potential V is given by

$$\frac{1}{2}m_e v^2 = eV$$

Therefore, the de Broglie wavelength of this electron is

$$\lambda = \frac{h}{m_e v} = \frac{h}{m_e\sqrt{\dfrac{2eV}{m_e}}} = \frac{h}{\sqrt{2eVm_e}}$$

Substituting $V = 54$ V gives a de Broglie wavelength of (1 V $= 1$ J/C)

$$\lambda = \frac{6.63 \times 10^{-34} \text{ J} \cdot \text{s}}{\sqrt{(2)(1.60 \times 10^{-19} \text{ C})(54 \text{ J/C})(9.11 \times 10^{-31} \text{ kg})}}$$

$$= 0.167 \text{ nm}$$

This value is in good agreement with the experimentally determined wavelength.

CHECKPOINT 32-2

Proton Diffraction

If the electron beam in Example 32-2 is replaced by a proton beam ($m_p \approx 1836\,m_e$) of the same de Broglie wavelength, the angle at which the diffraction peak occurs

(a) is the same as for the electron beam
(b) increases
(c) decreases
(d) is too small to measure because protons are much more massive than electrons

ANSWER: (a) The location of the diffraction peak is the same for particles with the same de Broglie wavelength.

You might think that the wave nature of a beam of electrons is a collective effect rather than a property of individual electrons. However, when a diffraction experiment is performed over an extended period with a low-intensity beam such that one electron at a time scatters from the crystal, a diffraction pattern slowly emerges that is the same as the pattern produced by a short-duration high-intensity beam with the same number of electrons. So, each electron independently displays wave properties.

Diffraction experiments have been performed with a variety of particles. Figure 32-5 shows diffraction patterns of X-ray beams and electron beams passing through aluminum foil. The diffraction of neutron beams from crystal surfaces displays similar diffraction effects.

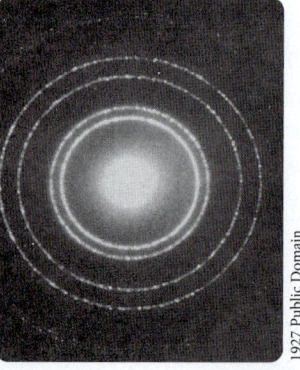

1927 Public Domain

Figure 32-5 Diffraction patterns formed by the scattering of X-ray beams and electron beams from aluminum foil.

MODERN PHYSICS

Electron Diffraction from a Crystal

In this activity, you will use the PhET simulation "Electron Diffraction from a Crystal." Work through the simulation and accompanying questions to learn more about the diffraction of electrons from crystals.

32-2 Heisenberg's Uncertainty Principle

What kind of wave is associated with the de Broglie wavelength? In Chapter 14, you learned that a travelling harmonic wave on a string can be described by a sinusoidal wave function of the form $D(x, t) = A \sin(kx - \omega t)$. Sound and electromagnetic waves can also be represented by harmonic waves. The square of the wave function at a given position, x, and time, t, is proportional to the *intensity* of the wave at that position and time. If a de Broglie wave is represented by a harmonic wave, then the square of the wave function at a given position and time is related to the *probability* of the object being at that position at that time. However, the sinusoidal wave function has an infinite extent with a constant amplitude, so the probability of finding the particle is the same for all locations. Therefore, we cannot represent a de Broglie wave with a sinusoidal wave of the form $\sin(kx - \omega t)$. Such a wave function accurately describes the wavelength of the object and, hence, its momentum (because $p = h/\lambda$) but leaves its location undetermined.

We need to represent a moving particle by a wave that is localized in space. Such a wave is called a **wave packet** or a **wave group**. We can construct a wave group by combining a large number of harmonic waves with slightly different wavelengths. Figure 32-6 shows a single harmonic wave with $k = 10$ rad/m and a wave group made up of 10 harmonic waves with the wave number increasing by 0.1 rad/m, starting at 10.0 rad/m. With a localized wave group, the position of an object is represented more accurately. However, the wavelength (and, hence, momentum) of the object become less certain because the wave group consists of a number of waves with closely spaced wave numbers. In fact, there is a *fundamental* limit to the accuracy with which we can measure both the position and the momentum of a moving object.

Consider a wave group that results from combining two waves with wave numbers k and $k + \Delta k$ and angular frequencies ω and $\omega + \Delta \omega$. The wave function of this wave group is given by

$$D(x, t) = A \sin((k + \Delta k)x - (\omega + \Delta \omega)t) + A \sin(kx - \omega t)$$

Using the trigonometric identity

$$\sin A + \sin B = 2 \cos\left(\frac{A - B}{2}\right) \sin\left(\frac{A + B}{2}\right)$$

we get

$$D(x, t) = 2A \cos\left(\frac{\Delta k}{2}x - \frac{\Delta \omega}{2}t\right) \times \sin\left(\frac{2k + \Delta k}{2}x - \frac{2\omega + \Delta \omega}{2}t\right) \tag{32-4}$$

A graph of this equation at $t = 0$ for $A = 1$ m, $k = 6$ rad/m, and $\Delta k = 1$ rad/m is shown in Figure 32-7. This situation is similar to the situation that you encountered while studying the beat phenomenon for sound waves.

The cosine term in Equation 32-4, called the modulation, has a wave number $\Delta k/2$, and its wavelength is $\frac{2\pi}{\Delta k/2} = \frac{4\pi}{\Delta k}$. From Figure 32-7, we notice that the width,

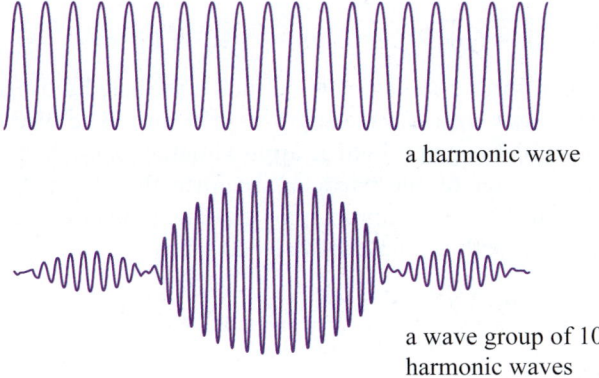

Figure 32-6 A harmonic wave and a wave group consisting of 10 harmonic waves with closely spaced wave numbers.

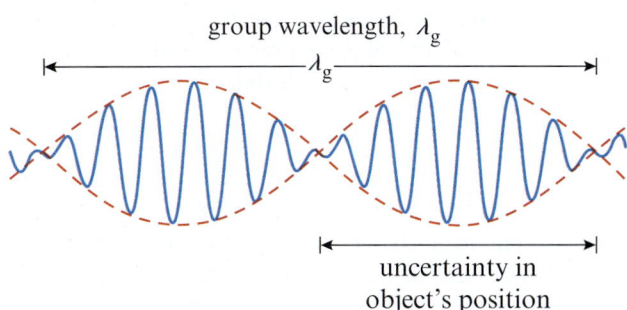

Figure 32-7 A wave group. The dashed curve represents the envelope of the wave group. The uncertainty in the object's position is equal to half the wavelength, λ_g, of the wave group.

Δx, of the wave group is equal to half the wavelength of the modulation. Therefore,

$$\Delta x = \frac{1}{2}\left(\frac{4\pi}{\Delta k}\right) = \frac{2\pi}{\Delta k} \qquad (32\text{-}5)$$

The uncertainty in the position of the wave group is of the same order of magnitude as its width. Therefore, it is reasonable to assume that the uncertainty in the wave number of the wave group is related to the uncertainty in its position by

$$\Delta k = \frac{2\pi}{\Delta x}$$

From Equation 32-2, we know that the wave number, k, of the de Broglie wave that is associated with a particle of momentum p is given by

$$p = \frac{h}{\lambda} = \frac{h}{2\pi/k} = \frac{h}{2\pi}k$$

Therefore, an uncertainty, Δk, in the wave number of the wave group corresponds to an uncertainty, Δp, in the particle's momentum such that

$$\Delta p = \frac{h}{2\pi}\Delta k$$

Using Equation 32-5, we can now relate the uncertainty in the particle's momentum to the uncertainty in its position (i.e., the uncertainty in the position of the wave group that represents the particle) as

$$\Delta p = \frac{h}{2\pi}\left(\frac{2\pi}{\Delta x}\right) = \frac{h}{\Delta x}$$

The product of the uncertainty in the particle's position (Δx) and the uncertainty in its momentum (Δp) satisfies the relationship

$$\Delta x\Delta p = h$$

Note that Δx and Δp are the *minimum* uncertainties due to the wave nature of the object. Any simultaneous measurement of the object's position and momentum will result in uncertainties such that

$$\Delta x\Delta p_x \geq h$$

In the above equation, we have replaced Δp with Δp_x to emphasize that the uncertainty is in that component of the momentum that is related to the uncertainty in a given component of the position vector. Our approach did not rigorously define what we mean by uncertainty. A more careful analysis yields the relationship called **Heisenberg's uncertainty principle**:

KEY EQUATION $$\Delta x\Delta p_x \geq \frac{\hbar}{2} \qquad (32\text{-}6)$$

where $\hbar = h/(2\pi)$. The German physicist Werner Karl Heisenberg (1901–1976) derived this relationship in 1927 from the general principles of quantum physics. He found that the product of the uncertainties in the simultaneous measurement of the position and momentum of a particle cannot be less than $\hbar/2$, and therefore the position and the momentum cannot be simultaneously measured with infinite accuracy. In very precise measurements, any technique that reduces the uncertainty in the position measurement increases the uncertainty in the momentum measurement, and vice versa.

Since $p_x = mv_x$, where m is the mass of the object, Equation 32-6 can be written in terms of uncertainties in position and velocity:

$$\Delta x\Delta v_x \geq \frac{\hbar}{2m} \qquad (32\text{-}7)$$

Let us consider another physical basis of the uncertainty principle. We determine the position of an object by illuminating the object with light. The light scatters from the object, enters the field of view of an eye or a microscope, and forms an image of the object. From Chapter 29 we know that to form a clear image, the wavelength of the light must be less than the dimensions of the object it illuminates. If the wavelength is of the same order of magnitude as the size of the object, the diffraction effects produce a blurred image. Let us use this mechanism to determine the position and the velocity of a small classical object and then an electron.

Consider a carbon particle with a radius of 1 μm (1×10^{-6} m) moving with a speed of 0.1 m/s. The density of carbon is 2.25×10^3 kg/m^3. Therefore, the mass and momentum of this particle are

$$m = \frac{4\pi}{3}R^3 \times \rho = \frac{4\pi}{3}(1 \times 10^{-18}\text{ m}^3)(2.25 \times 10^3\text{ kg/m}^3)$$
$$\approx 10^{-14}\text{ kg}$$

and

$$p = (10^{-14}\text{ kg})(0.1\text{ m/s}) = 10^{-15}\text{ kg·m/s}$$

We can form a clear image of such a particle by using green light with a wavelength of 550 nm. (The wavelength of the green light is approximately one-quarter the diameter of the particle.) To form the image, the light must scatter from the particle. The momentum of a photon with a wavelength of 550 nm is

$$p = \frac{h}{\lambda} = \frac{6.63 \times 10^{-34}\text{kg·m}^2\text{/s}}{550 \times 10^{-9}\text{m}} = 1.2 \times 10^{-27}\text{kg·m/s}$$

This is a factor of 10^{-12} smaller than the momentum of the particle. Therefore, when a photon hits a carbon particle, there is no noticeable change in the momentum of the particle.

Now we replace the carbon particle with an electron. Let us assume that the electron is moving with a speed of 1.0×10^8 m/s. The momentum of the electron is

$(9.11 \times 10^{-31}$ kg$)(1.0 \times 10^8$ m/s$) = 9.1 \times 10^{-23}$ kg$\cdot$m/s

(We have used non-relativistic kinematics because we are interested in order-of-magnitude calculations.) The current experimental data suggest that the radius of an electron must be smaller than 10^{-18} m. Let us assume that the radius of an electron is 10^{-18} m. To form a sharp image of an electron, the wavelength of the photon must be at least 10^{-18} m. Suppose that we scatter photons with a wavelength of 10^{-18} m from the electron. The momentum of such a photon is

$$p = \frac{h}{\lambda} = \frac{6.63 \times 10^{-34} \text{kg} \cdot \text{m}^2/\text{s}}{10^{-18} \text{m}} = 6.63 \times 10^{-16} \text{kg} \cdot \text{m/s}$$

This momentum is a factor of 7×10^6 times the momentum of the electron. Such a photon striking an electron would be like a freight train smashing into a bicycle. If we illuminate an electron with a photon of small enough wavelength to form a sharp image, we knock the electron out of its position and do not know its position or momentum. If we try using a photon with a much longer wavelength (and, hence, much smaller momentum) so that the electron is not disturbed, diffraction effects blur its image and we cannot accurately determine the position of the electron.

Our inability to accurately measure both the position and the momentum of a particle is not simply a matter of needing more precise instruments. The uncertainty limit determined by Heisenberg is part of the nature of the microscopic world. At this scale, we have to deal with probabilities rather than definite values. A statement such as "at time $t = 0$, the electron is at $x = 0.2$ m and moving with a speed of 10^4 m/s" gets replaced by "at time $t = 0$, the probability of finding the electron around $x = 0.2$ m is 0.9, and the probability that it is moving with a speed of 10^4 m/s is 0.6." This lack of certainty is one of the more confusing aspects of quantum theory. Even Einstein was perplexed about why nature would have a built-in mechanism that prevents precise measurements.

EXAMPLE 32-3

Uncertainty in the Momentum of a Particle Confined within a Box

An object with a mass of 1.0×10^{-3} kg is confined within a box of length 1.0 m. What is the minimum uncertainty in the momentum and the speed of the object?

Solution

The object is confined within a box of length 1.0 m, so the maximum uncertainty in the position of the object is 1.0 m. Using $\Delta x = 1.0$ m in Equation 32-6, we get

$$\Delta p_x \geq \frac{\hbar}{2 \times \Delta x} = \frac{1.05 \times 10^{-34} \text{ J} \cdot \text{s}}{2 \times 1.0 \text{ m}} = 0.52 \times 10^{-34} \text{ kg} \cdot \text{m/s}$$

for the minimum uncertainty in the momentum of the particle. Therefore, the speed of the particle cannot be determined to an accuracy better than

$$\Delta v_x = \frac{\Delta p_x}{m} = \frac{0.52 \times 10^{-34} \text{ kg} \cdot \text{m/s}}{1.0 \times 10^{-3} \text{ kg}} = 5.2 \times 10^{-30} \text{ m/s}$$

Making sense of the result

Since the uncertainty in position is very large ($\Delta x \sim 1.0$ m) relative to a microscopic scale, the uncertainty in the speed is very small. Over a billion years, a difference in speed of 5.2×10^{-30} m/s would cause a difference in location of only about 1.6×10^{-13} m.

EXAMPLE 32-4

Uncertainty in the Momentum of a Proton in a Nucleus

Find the minimum uncertainty in the momentum and the speed of a proton in the nucleus of an iron atom. The mass of a proton is 1.67×10^{-27} kg, and the radius of an iron nucleus is approximately 4.6×10^{-15} m.

Solution

The proton is confined within the iron nucleus, so it is reasonable to expect that the uncertainty in the position of the proton is equal to the diameter of the iron nucleus:

$$\Delta x = 2 \times 4.6 \times 10^{-15} \text{ m} = 9.2 \times 10^{-15} \text{ m}$$

The uncertainty in the momentum of the proton is then given by

$$\Delta p_x \geq \frac{\hbar}{2 \times \Delta x} = \frac{1.05 \times 10^{-34} \text{ J} \cdot \text{s}}{2 \times 9.2 \times 10^{-15} \text{ m}} = 5.7 \times 10^{-21} \text{ kg} \cdot \text{m/s}$$

The speed of the proton therefore cannot be determined to an accuracy better than

$$\Delta v_x = \frac{\Delta p_x}{m} = \frac{5.7 \times 10^{-21} \text{ kg} \cdot \text{m/s}}{1.67 \times 10^{-27} \text{ kg}} = 3.4 \times 10^6 \text{ m/s}$$

Making sense of the result

A proton is localized in a very small volume, so the uncertainty in the location of the proton is very small (Δx: 9.2×10^{-15} m). Therefore, the uncertainty in its speed, and hence momentum, is large.

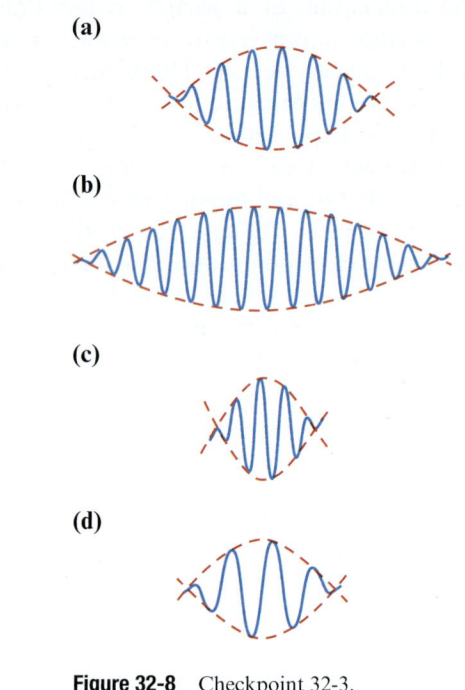
Other Uncertainty Relationships

The position and the momentum of a particle are called **observables** because these quantities can be measured in an experiment. The uncertainty relationship is also valid for other combinations of observables, such as time, energy, angular momentum, and acceleration. The time–energy uncertainty relationship is

KEY EQUATION

$$\Delta t\, \Delta E \geq \frac{\hbar}{2} \qquad (32\text{-}8)$$

where Δt is the time taken to measure the energy of a system (e.g., a hydrogen atom in an excited state) and ΔE is the accuracy to which the energy of the system can be measured in that time. The time–energy uncertainty principle states that the time required to measure the energy of a system to an accuracy of ΔE must be greater than $\frac{\hbar}{2\Delta E}$. Thus, the law of conservation of energy can only be verified to an accuracy of $\frac{\hbar}{2\Delta t}$, where Δt is the time taken to measure the energy.

The Double-Slit Experiment with Electrons

In Chapter 29 we discussed Young's double-slit experiment using waves. In this experiment, waves emerging from the slits interfere with each other, producing a series of dark (destructive interference) and bright (constructive interference) bands on a screen in front of the slits. Double-slit interference is a wave phenomenon, and Young's double-slit experiment, to the dismay of Newton, proved that light is a wave.

Let us replace the source that generates light with a source that generates electrons. This might be a heated filament. A screen with two narrow slits (these need to be much narrower than in Young's experiment) is placed in front of the source and a potential difference is applied between the source and the screen to accelerate the electrons toward the slits. A second screen that detects the electrons is placed in front of the slits. This might be a phosphorescent plate. Every time an electron hits a point on the plate, a tiny flash of light is produced and the location of the flash is recorded. The whole apparatus is placed in a vacuum. We let the experiment run for a sufficiently long time so that a large number of flashes have been recorded. The resulting flash pattern consists of a series of bands (Figure 32-9(a)), similar to that of the double-slit interference pattern with waves. This is a surprising result given the fact that an electron is assumed to be a point-like particle.

If we perform this experiment with one slit closed, we will see the pattern shown in Figure 32-9(b), without alternating bright and dark bands. If we alternately close each slit (so that only one slit is open at a given time), we will get a double-exposure pattern of two bands, as

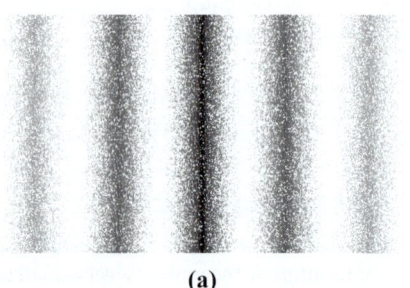

(a)

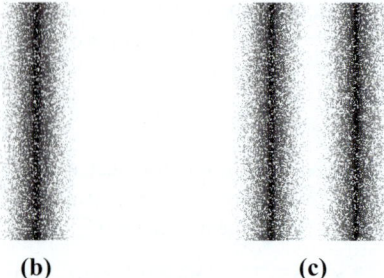

(b)   **(c)**

Figure 32-9 (a) Pattern of flashes on the screen in Young's double-slit experiment using electrons. (b) Pattern observed when one slit is closed and electrons emerge from a single screen. (c) Pattern observed when each slit is opened alternately but only one screen is open at a given time.

MODERN PHYSICS

shown in Figure 32-9(c). Why don't we see the two-band pattern when both slits are simultaneously open? How does an electron know if the other slit is open or closed? Surely each electron passes through only one of the two slits. Or does an electron pass through both slits? You might say that when travelling between the slits and the detecting plate, somehow electrons scatter off each other and that is why the flash distributions in Figure 32-9(a) and 32-9(b) look different. If we decrease the intensity of the source so that a single electron is released and passes through the slits at a given time, we still obtain a flash pattern that is similar to Figure 32-9(a) (although we have to run the experiment for a very long time).

If we treat an electron as an elementary particle, then the results of this experiment are truly bizarre. However, if we assign a wave-like property to an electron, then the results are easily explained. In this case, each electron has an associated wave function (just like the wave functions we discussed for a harmonic wave in Chapter 14) and the *wave describing an electron passes through both slits*. It does not mean that an electron splits into two electrons. That is not possible by charge conservation. The electron waves emerging from the slits interfere with each other, resulting in the interference pattern of Figure 32-9(a) on the screen. When an electron hits the screen, it produces a bright spot at a particular point on the screen, so the electron must arrive as a particle. The probability that an electron arrives is, however, distributed like the intensity of a wave. So an electron behaves like both a particle and a wave. This wave-particle duality is a reflection of how nature works. German physicist Max Born stated this duality in the form of a principle called the **complementarity principle**. It states that objects have complimentary properties that cannot be measured accurately simultaneously. We can describe the particle properties of an object by using Newton's laws of motion, appropriately modified by the theory of relativity where needed. How do we calculate the wave function of an object? We will discuss this question next.

32-3 The Schrödinger Equation

Macroscopic objects obey Newton's laws of motion. Harmonic travelling waves obey the wave equation, and electromagnetic waves are described by Maxwell's equations. What is the nature of the laws that describe matter waves? In 1926, Heisenberg formulated a description of matter waves using matrices. In the same year, German physicist Erwin Schrödinger (1887–1961) used a wave equation to accurately describe the wave behaviour of matter. The two approaches are equivalent. Here, we will follow Schrödinger's approach.

In Chapter 14 we learned that the wave function $D(x, t)$ for a one-dimensional travelling harmonic wave satisfies the differential equation

$$\frac{\partial^2 D(x,t)}{\partial x^2} = \frac{1}{v^2} \frac{\partial^2 D(x,t)}{\partial t^2} \tag{32-9}$$

where v is the wave speed in the medium.

To derive a similar differential equation that describes the motion of a matter wave, Schrödinger made the following four assumptions:

1. Matter waves satisfy the de Broglie relation $\lambda = \dfrac{h}{p}$. Therefore, the wave number, k, of a matter wave is related to the corresponding particle's momentum:

$$k = \frac{2\pi}{\lambda} = 2\pi\left(\frac{p}{h}\right) = \frac{p}{\hbar} \tag{32-10}$$

2. Einstein's equation, $f = \dfrac{E}{h}$, relates a particle's energy to the frequency of the corresponding matter wave. Therefore, the angular frequency of a matter wave is related to the particle's energy:

$$\omega = 2\pi f = 2\pi\left(\frac{E}{h}\right) = \frac{E}{\hbar} \tag{32-11}$$

3. The phase of a travelling matter wave has the same form as the phase of a mechanical wave. Therefore, the relationships in the first two assumptions can be used to derive an expression for the phase:

$$kx - \omega t = \left(\frac{p}{\hbar}\right)x - \left(\frac{E}{\hbar}\right)t = \frac{1}{\hbar}(px - Et) \tag{32-12}$$

4. A wave equation that describes the motion of a particle must satisfy the classical energy–momentum relationship for that particle.

Schrödinger first considered a free particle of mass m that is moving in one dimension (along the x-axis) with momentum p. Assuming that the particle has no potential energy and that relativistic effects are negligible, the total energy, E, of the particle is equal to its kinetic energy ($p^2/2m$); therefore,

$$\frac{p^2}{2m} = E \tag{32-13}$$

Schrödinger realized that to satisfy the energy and momentum relation in Equation 32-13, the wave function of the matter wave must contain a combination of sine and cosine functions of the form

$$\cos\theta + i\sin\theta \tag{32-14}$$

Here, i denotes the imaginary number defined by $i^2 = -1$. Therefore, $i = \sqrt{-1}$. Equation 32-14 can be written as an exponential of an imaginary number (problem 35):

$$e^{i\theta} = \cos\theta + i\sin\theta \tag{32-15}$$

Using the Greek letter ψ (psi, pronounced "sigh") to denote the wave function for matter waves, Schrödinger wrote the wave function for a particle as a sum of a cosine and a sine travelling wave:

$$\psi(x, t) = \psi_M\left[\cos\left(\frac{p}{\hbar}x - \frac{E}{\hbar}t\right) + i\sin\left(\frac{p}{\hbar}x - \frac{E}{\hbar}t\right)\right]$$

$$(32\text{-}16)$$

where the constant ψ_M is the amplitude of the wave function. Using Equation 32-15, $\psi(x, t)$ can be expressed as a compact exponential function:

$$\psi(x, t) = \psi_M e^{(i/\hbar)(px - Et)} \qquad (32\text{-}17)$$

To write an equation for $\psi(x, t)$ that satisfies the energy relationship in Equation 32-13, Schrödinger multiplied both sides by $\psi(x, t)$:

$$\left(\frac{p^2}{2m}\right)\psi(x, t) = E\psi(x, t) \qquad (32\text{-}18)$$

Note that

$$\frac{\partial}{\partial t}\psi(x, t) = \frac{\partial}{\partial t}\left(\psi_M e^{(i/\hbar)(px - Et)}\right) = -\frac{iE}{\hbar}\psi_M e^{(i/\hbar)(px - Et)}$$

$$(32\text{-}19)$$

$$= -\frac{iE}{\hbar}\psi(x, t)$$

Here, we have used the fact that $\frac{d}{dx}e^{ax} = ae^{ax}$. Multiplying both sides by $i\hbar$ and using the identity $i \times i = -1$, we get

$$\left(i\hbar\frac{\partial}{\partial t}\right)\psi(x, t) = E\psi(x, t) \qquad (32\text{-}20)$$

Similarly, taking the partial derivative of $\psi(x, t)$ with respect to position x gives

$$\frac{\partial}{\partial x}\psi(x, t) = \frac{\partial}{\partial x}(\psi_M e^{(i/\hbar)(px - Et)})$$

$$= +\frac{ip}{\hbar}\psi_M e^{(i/\hbar)(px - Et)} = +\frac{ip}{\hbar}\psi(x, t)$$

$$(32\text{-}21)$$

Multiplying both sides by $-i\hbar$, we get

$$\left(-i\hbar\frac{\partial}{\partial x}\right)\psi(x, t) = p\psi(x, t) \qquad (32\text{-}22)$$

Schrödinger then took the second derivative of $\psi(x, t)$ with respect to x (since Equation 32-18 contains a p^2):

$$\frac{\partial}{\partial x}\left(\frac{\partial}{\partial x}\psi(x, t)\right) = \frac{ip}{\hbar}\frac{\partial}{\partial x}(\psi_M e^{(i/\hbar)(px - Et)})$$

$$(32\text{-}23)$$

$$= -\frac{p^2}{\hbar^2}\psi(x, t)$$

Multiplying both sides of Equation 32-23 by $-\frac{\hbar^2}{2m}$ gives

$$\left(-\frac{\hbar^2}{2m}\frac{\partial}{\partial x^2}\right)\psi(x, t) = \frac{p^2}{2m}\psi(x, t) \qquad (32\text{-}24)$$

Inserting Equations 32-24 and 32-20 into Equation 32-18, we get the following differential equation for the wave function:

$$-\frac{\hbar^2}{2m}\frac{\partial^2\psi(x, t)}{\partial x^2} = i\hbar\frac{\partial\psi(x, t)}{\partial t} \qquad (32\text{-}25)$$

Note that in obtaining this differential equation, Schrödinger did the following:

1. He started with the equation that relates the momentum and total energy of the particle.

2. He multiplied the equation by a wave function that uses de Broglie's relation for the wavelength and Einstein's relation for the frequency of the wave.

3. Then he derived a wave equation involving derivatives of the wave function by making the following substitutions:

$$E \rightarrow i\hbar\frac{\partial}{\partial t} \quad \text{and} \quad p \rightarrow -i\hbar\frac{\partial}{\partial x} \qquad (32\text{-}26)$$

Equations 32-20 and 32-22 are called **eigenvalue** equations, and the derivatives $i\hbar\frac{\partial}{\partial t}$ and $-i\hbar\frac{\partial}{\partial x}$ are called operators because they operate on the wave function. When the operator $i\hbar\frac{\partial}{\partial t}$ acts on the wave function $\psi(x, t)$, it gives the energy eigenvalue E. Similarly, the operator $-i\hbar\frac{\partial}{\partial x}$ gives the momentum eigenvalue when it acts on the wave function. Note that operators in equations are placed to the left of the wave function on which they act.

What happens if the particle also has a potential energy, $V(x)$? In classical mechanics, the total energy of the particle is the sum of its kinetic and potential energies:

$$\frac{p^2}{2m} + V(x) = E \qquad (32\text{-}27)$$

Schrödinger applied the procedure outlined above to Equation 32-27 to obtain a differential equation for the wave function when the potential energy is not zero. The resulting equation is called the **time-dependent Schrödinger wave equation**:

KEY EQUATION

$$-\frac{\hbar^2}{2m}\frac{\partial^2\psi(x, t)}{\partial x^2} + V(x)\psi(x, t) = i\hbar\frac{\partial\psi(x, t)}{\partial t} \qquad (32\text{-}28)$$

Equation 32-28 describes the motion of a particle with rest mass m that is moving in one dimension and has potential energy $V(x)$. Note that the potential energy term is accounted for by simply multiplying $V(x)$ by the wave function and adding this term to Equation 32-25.

Schrödinger's equation is to quantum physics what Newton's laws are to classical physics. The above procedure of obtaining the Schrödinger equation is a prescription that cannot be derived from more fundamental principles, just as Newton's laws of motion cannot be derived from more fundamental principles. As with Newton's laws of motion and relativity theory, experimental results have consistently agreed with predictions based on the Schrödinger equation.

The Time-Independent Schrödinger Equation

When a particle is in a state with definite energy E (such as an electron moving in a given energy state around a nucleus), the time variation of the wave function of the particle can be written in the simple form $e^{-iEt/\hbar}$. We can then separate the wave function $\psi(x, t)$ into two parts: $\psi(x)$, which only depends on the position, x, and $e^{-iEt/\hbar}$, the time-dependent part. Thus,

$$\psi(x, t) = \psi(x)\, e^{-iEt/\hbar}$$

Substituting this wave function into the time-dependent equation, we get

$$-\frac{\hbar^2}{2m}\left(\frac{d^2\psi(x)}{dx^2}\right)e^{-iEt/\hbar} + V(x)\psi(x)e^{-iEt/\hbar}$$

$$= i\hbar\psi(x)\left(\frac{de^{-iEt/\hbar}}{dt}\right)$$

$$= E\psi(x)e^{-iEt/\hbar}$$

Here, we have replaced the partial derivatives with total derivatives because $\psi(x)$ is a function of x only, and the $e^{-iEt/\hbar}$ term only depends on t. Dividing out the factor $e^{-iEt/\hbar}$ gives the **time-independent Schrödinger equation** in one dimension:

KEY EQUATION
$$-\frac{\hbar^2}{2m}\frac{d^2\psi(x)}{dx^2} + V(x)\psi(x) = E\psi(x) \quad \text{(32-29)}$$

For a given $V(x)$, Equation 32-29 is satisfied for only certain values of energy E. These values of energy are called eigenvalues. For a given eigenvalue, the corresponding wave function that satisfies the Schrödinger equation is called the **eigen wave function**, or simply the **eigenfunction**. The first successful application of the Schrödinger equation was in calculating energy eigenvalues and eigenfunctions for the electron in a hydrogen atom.

EXAMPLE 32-5

The Schrödinger Equation for a Harmonic Oscillator Potential

Consider a particle of mass m undergoing simple harmonic motion along the x-axis under the influence of an elastic restoring force, $F(x) = -kx$. Write the Schrödinger equation for this particle.

Solution

As discussed in Chapter 14, the elastic potential energy of a particle undergoing simple harmonic motion is $\frac{1}{2}kx^2$ (Figure 32-10), and the total energy of the particle is

$$V(x) = \frac{1}{2}kx^2$$

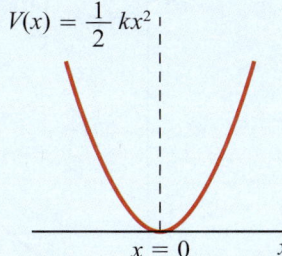

Figure 32-10 A harmonic oscillator potential, $V(x) = \frac{1}{2}kx^2$, in one dimension.

$$\frac{1}{2}mv^2 + \frac{1}{2}kx^2 = \frac{p^2}{2m} + \frac{1}{2}kx^2 = E$$

To obtain the Schrödinger equation for this simple harmonic oscillator, we start with the equation that relates its kinetic energy, potential energy, and total energy. Multiplying the energy equation by the wave function $\psi(x, t)$ gives

$$\left(\frac{p^2}{2m} + \frac{1}{2}kx^2\right)\psi(x, t) = E\psi(x, t)$$

Next, we replace the total energy, E, and the momentum, p, by the derivative operators from Equation 32-26 to obtain the Schrödinger equation:

$$\frac{-\hbar^2}{2m}\frac{\partial^2}{\partial x^2}\psi(x, t) + \frac{1}{2}kx^2\psi(x, t) = i\hbar\frac{\partial}{\partial t}\psi(x, t)$$

Making sense of the result

Here, we have used Schrödinger's procedure to turn the classical energy equation into the corresponding Schrödinger equation. To determine the wave function $\psi(x, t)$, we need to solve this differential equation.

The Schrödinger Equation in Three Dimensions

The total energy, E, of a particle moving in three dimensions is given by

$$\frac{p_x^2}{2m} + \frac{p_y^2}{2m} + \frac{p_z^2}{2m} + V(x, y, z) = E \qquad (32\text{-}30)$$

where p_x, p_y, and p_z are the Cartesian components of the particle's momentum and $V(x, y, z)$ represents the potential energy in three dimensions.

To obtain the corresponding Schrödinger equation, we replace the momentum components and the energy with derivatives that act on a wave function $\psi(x, y, z, t)$ that depends on all three space coordinates and time:

$$p_x \rightarrow -i\hbar\frac{\partial}{\partial x}, p_y \rightarrow -i\hbar\frac{\partial}{\partial y}, p_z \rightarrow -i\hbar\frac{\partial}{\partial z}$$

$$E \rightarrow i\hbar\frac{\partial}{\partial t} \qquad (32\text{-}31)$$

The resulting time-dependent Schrödinger equation is

$$-\frac{\hbar^2}{2m}\left(\frac{\partial^2}{\partial x^2} + \frac{\partial^2}{\partial y^2} + \frac{\partial^2}{\partial z^2}\right)\psi(x, y, z, t)$$

$$+V(x, y, z)\psi(x, y, z, t) = i\hbar\frac{\partial\psi(x, y, z, t)}{\partial t} \qquad (32\text{-}32)$$

As for the one-dimensional case, the right side of Equation 32-32 is equal to $E\psi(x, y, z, t)$ for stationary states. The corresponding time-independent Schrödinger equation is

$$-\frac{\hbar^2}{2m}\left(\frac{\partial^2}{\partial x^2} + \frac{\partial^2}{\partial y^2} + \frac{\partial^2}{\partial z^2}\right)\psi(x, y, z)$$

$$+ V(x, y, z)\psi(x, y, z) = E\psi(x, y, z) \qquad (32\text{-}33)$$

The Physical Meaning of the Wave Function

For mechanical waves, $D(x, t)$ denotes the displacement of a particle from its equilibrium position. However the physical meaning of, the quantity represented by the wave function $\psi(x, y, z, t)$ is not obvious. The physical interpretation of $\psi(x, y, z, t)$ was provided by Max Born (1882–1970), a German British physicist and mathematician. Born noted that the square of the absolute value of the wave function is a real number given by

$$|\psi(x, y, z, t)|^2 = \psi^*(x, y, z, t) \times \psi(x, y, z, t)$$

where $\psi^*(x, y, z, t)$ represents the complex conjugate of the wave function $\psi(x, y, z, t)$. Born argued that this number is equal to the probability per unit volume of finding a particle at position (x, y, z) at time t. For a small volume element $dV = dxdydz$ centred around a point (x, y, z), the quantity $|\psi(x, y, z, t)|^2dV$, called

the probability density, is equal to the probability of finding the particle in that volume at time t. Since the probability of finding the particle somewhere in the universe must be one (a certainty), the integral of probability density over all space must be equal to one.

$$\int_{\text{all space}} |\psi(x, y, z, t)|^2 dV = 1 \qquad (32\text{-}34)$$

The constraint imposed on the wave function by Equation 32-34 is called a **normalization condition**, and a wave function that satisfies this condition is called a **normalized wave function**. For the one-dimensional case, dV is replaced by dx. For wave functions in one dimension, the normalization condition is given by

KEY EQUATION
$$\int_{-\infty}^{\infty} |\psi(x, t)|^2 dx = 1 \qquad (32\text{-}35)$$

Max Born also proposed that for the wave function to be a mathematically well-defined function it must satisfy certain conditions. Let us consider these conditions for a wave function of a one-dimensional time-dependent Schrödinger equation, Equation 32-28. Born's conditions that a wave function $\psi(x, t)$ must satisfy are as follows:

1. **The wave function must be single valued.** A single-valued function is a function that for each point in its domain has a unique value in the range. This means that for any given values of x and t, $\psi(x, t)$ must have a unique value. This guarantees that there is only a single value for the probability of the system being in a given state. The condition that a function is single valued is a requirement of all proper mathematical functions.

2. **The wave function must be continuous everywhere.** If a function has a discontinuity, such as a sharp jump (see Figure 32-11), then the derivative of the function is infinite at that point. Since the momentum of a particle is proportional to the derivative of its wave function with respect to position (Equation 32-21),

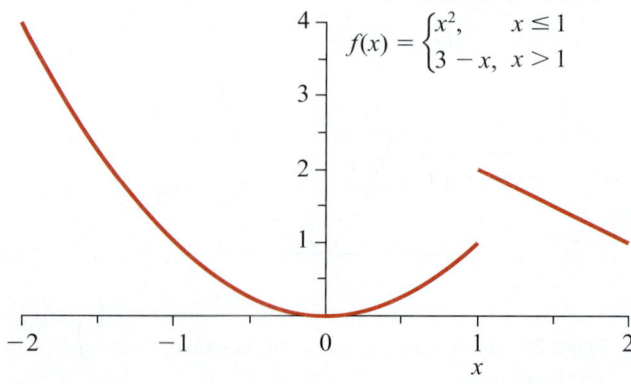

Figure 32-11 The function $f(x)$ is discontinuous at $x = 1$ as it jumps from one branch to the other branch at this point.

a discontinuity in the wave function would imply an infinite momentum, which is not possible. Also, since the square of the absolute value of a wave function, $|\psi(x, t)|^2 dx$, is equal to the probability of finding the particle between x and $x + dx$, the continuity of the wave function ensures that there are no sudden jumps in the probability density when moving through space.

3. **All first-order derivatives of the wave function must be continuous.** Following the same reasoning as in condition 2, a discontinuous first derivative would imply an infinite second derivative, and since the kinetic energy of a system is proportional to the second derivative of its wave function with respect to position (Equation 32-24), a discontinuous second derivative would imply an infinite kinetic energy, which again is not physically realistic.

4. **The wave function must be square-integrable.** This condition requires that the integral of $|\psi(x, t)|^2$ over all space must be finite. Since this integral is equal to the probability of finding the particle over all space, it must be finite. For a properly normalized wave function (see Section 32-5), the value of this integral is equal to one.

Consider the potential $V(x)$ shown in Figure 32-12. $V(x)$ is zero for $x < a$ and is equal to V_0 for $x > a$. Therefore, the potential is discontinuous at $x = a$. If the wave function in the region $x < a$ is $\psi_1(x)$ and for $x > a$ is $\psi_2(x)$, then conditions 2 and 3 require

$$\psi_1(a) = \psi_2(a)$$

$$\left.\frac{d\psi_1(x)}{dx}\right|_{x=a} = \left.\frac{d\psi_2(x)}{dx}\right|_{x=a}$$

(32-36)

Here the symbol $\left.\dfrac{d\psi(x)}{dx}\right|_{x=a}$ means "take the derivative of $\psi(x)$ with respect to x and then set $x = a$ in the result."

It turns out that some systems, like a particle confined in an infinite square well potential, do violate one or more of these conditions. The wave functions for a particle in an infinite square well potential are continuous at the boundary of the potential but the derivative of these wave functions is *not* continuous at the boundary. This happens because an infinite potential is an idealization, and such potentials do not exist in nature.

32-4 Solving the Time-Independent Schrödinger Equation

The Schrödinger equation is a second-order differential equation. Generally, wave functions that satisfy this equation for a given potential energy term $V(x)$ can be complex (i.e., have both a real and an imaginary part). A complete solution of the Schrödinger equation requires that we know either the value of the wave function at two different values of x, or the value of the wave function and its derivative at a given x.

Let us consider solutions of two second-order differential equations. These solutions will help us understand how the Schrödinger equation is solved for simple cases.

Case 1: Consider a second-order differential equation of the form

$$\frac{d^2f(x)}{dx^2} + a^2 f(x) = 0$$

(32-37)

where a is a real number. The two conditions to be satisfied by the function are $f(0) = 0$ and $f(L) = 0$. These conditions require the solution of Equation 32-37 to vanish at $x = 0$ and $x = L$.

A solution of Equation 32-37 can be written in terms of

(a) sine and cosine functions:

$$f(x) = A_1 \sin(ax) + B_1 \cos(ax)$$

where A_1 and B_1 are arbitrary constants that can be determined by applying the given conditions on $f(x)$

(b) exponential functions, e^{iax} and e^{-iax}:

$$f(x) = A_2 e^{iax} + B_2 e^{-iax}$$

Using the Euler identity, $e^{i\theta} = \cos\theta + i\sin\theta$, we can see that these forms are equivalent. Let us choose a solution of Equation 32-37 in terms of sine and cosine functions:

$$f(x) = A_1 \sin(ax) + B_1 \cos(ax)$$

(32-38)

and impose the following two conditions on $f(x)$:

$$f(0) = 0$$

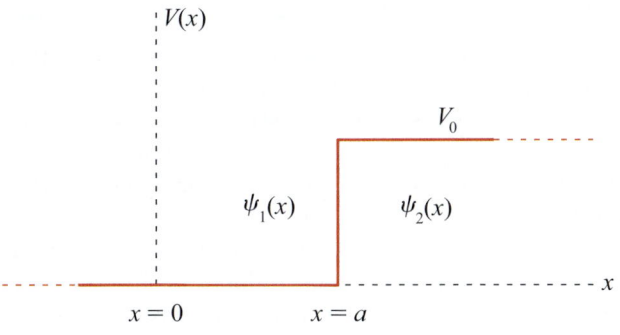

Figure 32-12 A potential $V(x)$ that is discontinuous at $x = a$. The wave function and its derivative must be continuous everywhere, including the boundary at $x = a$.

MODERN PHYSICS

implies

$$0 = A_1 \sin 0 + B_1 \cos 0 = B_1 \qquad (32\text{-}39)$$

$$f(L) = 0$$

implies

$$0 = A_1 \sin(aL) + B_1 \cos(aL) \qquad (32\text{-}40)$$

From Equation 32-39, $B_1 = 0$, and inserting this relation into Equation 32-40, we get

$$0 = A_1 \sin(aL)$$

To satisfy Equation 32-40, the sine function must vanish. Therefore,

$$\sin(aL) = 0$$

Therefore,

$$aL = n\pi \text{ or } a = n\frac{\pi}{L} \quad \text{where } n = 0, 1, 2, 3, \ldots \quad (32\text{-}41)$$

We see that the conditions have constrained the constant a to only a certain set of values. Thus, a has become "quantized," and the allowed values of the function $f(x)$ are

$$f(x) = A_1 \sin\left(n\frac{\pi x}{L}\right) \quad n = 0, 1, 2, 3, \ldots \quad (32\text{-}42)$$

We started with three unknowns, A_1, B_1, and a. Using the two boundary conditions, we determined B_1 and a. To determine the remaining constant, A_1, we need an additional condition. Such a condition will depend on the physical process under consideration. If $f(x)$ represents a wave function, then the normalization condition (Equation 32-35) can be used to determine A_1. See, for example the discussion of wave functions for an infinite square well potential in Section 32-5.

Case 2: Consider a second-order differential equation of the form

$$\frac{d^2 f(x)}{dx^2} - a^2 f(x) = 0 \qquad (32\text{-}43)$$

where a is a real number. Let us require that the function $f(x)$ satisfy the following conditions:

$$f(0) = 0 \text{ and } \left.\frac{df(x)}{dx}\right|_{x=0} = 1$$

Because of the minus sign in front of the second term in Equation 32-40, its solution can be written in terms of exponential functions e^{ax} and e^{-ax}:

$$f(x) = A_1 e^{-ax} + B_1 e^{ax} \qquad (32\text{-}44)$$

The first condition, $f(0) = 0$, requires $A_1 + B_1 = 0$. The second condition requires $-aA_1 + aB_1 = 1$. Solving for A_1 and B_1, we get

$$A_1 = -\frac{1}{2a} \quad B_1 = \frac{1}{2a}$$

Therefore, the solution is

$$f(x) = \frac{1}{2a}(-e^{-ax} + e^{ax})$$

Again using the two boundary conditions, we determined the constants A_1 and B_1. In this case, the constant a is unknown and its value will be determined by the physical process described by the differential equation.

Initial Conditions and Boundary Values

An **initial value problem** for second-order differential equations consists of finding a solution for $f(x)$ that satisfies conditions of the form

$$f(0) = a \text{ and } \left.\frac{df(x)}{dx}\right|_{x=0} = b$$

for given constants a and b.

A **boundary value problem** for a second-order differential equation consists of finding a solution for $f(x)$ that satisfies conditions of the form

$$f(x_1) = a \text{ and } f(x_2) = b$$

Unlike an initial value problem, a boundary value problem does not always have a solution.

In the following section, we solve the time-independent Schrödinger equation for two simple cases. In general, solutions of the Schrödinger equation can be quite complicated, and we use computers to solve the equation. Example 32-6 below, although simple, clearly illustrates the peculiar nature of quantum mechanics.

32-5 A Particle in a One-Dimensional Box

Consider a particle of mass m that is confined within a box of width L. The particle moves freely within the box and experiences elastic collisions with the walls of the box. According to classical physics, the energy and momentum of the particle can vary continuously.

Let us see how the Schrödinger equation describes the motion of a particle confined within a box. In quantum mechanics, an impenetrable wall is replaced by an infinite potential. The probability of finding a particle on the other side of an infinite potential is zero. Thus, an infinite potential acts like a rigid classical wall. We will consider a particle constrained to move along a single axis.

For the particle to be confined within the box, the potential energy outside the region $0 \leq x \leq L$ must be infinite. The potential $V(x)$ is therefore given by

$$V(x) = 0 \qquad 0 \leq x \leq L$$
$$V(x) = \infty \qquad x < 0 \text{ and } x > L$$

The x-axis is then divided into three regions, as shown in Figure 32-13, and we need to solve the Schrödinger equation in each region. The Schrödinger equation for each region is given in Table 32-1.

In regions I and III, the only possible solution is

$$\psi(x) = 0 \text{ for } x < 0 \text{ and } x > L$$

For region II, we define $k^2 = \dfrac{2mE}{\hbar^2}$ and write the Schrödinger equation as

$$\frac{d^2\psi(x)}{dx^2} + k^2\psi(x) = 0 \qquad (32\text{-}45)$$

Notice that k is the wave number of the particle and is related to its de Broglie wavelength by $k = 2\pi/\lambda$. As discussed earlier, the solution for Equation 32-45 can be written as

$$\psi(x) = A\sin(kx) + B\cos(kx) \qquad (32\text{-}46)$$

The wave function must vanish in regions I and III, so constants A and B are determined by imposing the following boundary conditions on $\psi(x)$:

(a) $\psi(x) = 0$ at $x = 0$, which requires $0 = A \times 0 + B$; hence $B = 0$;

(b) $\psi(x) = 0$ at $x = L$, which requires $0 = A\sin(kL)$.

The constant A cannot be zero (otherwise, the wave function would be zero everywhere), so the second boundary condition can only be satisfied if $\sin kL = 0$. Therefore,

$$kL = n\pi \quad n = 1, 2, 3, \ldots \qquad (32\text{-}47)$$

Thus, k is restricted to a discrete set of values, k_n, given by

$$k_n = \frac{n\pi}{L} \quad n = 1, 2, 3, \ldots \qquad (32\text{-}48)$$

The integer n is called a **quantum number**. Using $k^2 = \dfrac{2mE}{\hbar^2}$, we can see that a particle confined between $x = 0$ and $x = L$ can have only certain discrete values of energy:

KEY EQUATION $\quad E_n = \dfrac{\hbar^2 k_n^2}{2m} = \dfrac{\hbar^2(n\pi/L)^2}{2m} = n^2\dfrac{\hbar^2\pi^2}{2mL^2} \quad (32\text{-}49)$

The lowest value of energy corresponds to $n = 1$ and is called the **ground-state energy**. For a particle confined within a one-dimensional box, the ground-state energy is

$$E_1 = \frac{\hbar^2\pi^2}{2mL^2} = \frac{h^2}{8mL^2} \qquad (32\text{-}50)$$

States with higher energy are called **excited states**. For an infinite square well potential, the energies of excited states are related to the ground-state energy by

$$E_n = n^2 E_1 \qquad (32\text{-}51)$$

For example, the energy of the state with $n = 2$ is $E_2 = 4E_1$.

The energy of the particle is thus quantized. Figure 32-14 shows allowed energy levels for an infinite square well potential. Such a diagram is called an **energy-level diagram**. Notice that quantum mechanics predicts that the lowest possible energy of a particle confined within a one-dimensional region is non-zero. In fact, quantum mechanics requires that *any* particle confined within some region of space have non-zero kinetic energy. This minimum energy is the ground-state energy (sometimes called the **zero-point energy**). *The smaller the confining space, the greater the zero-point energy*, as we can see from Equation 32-50.

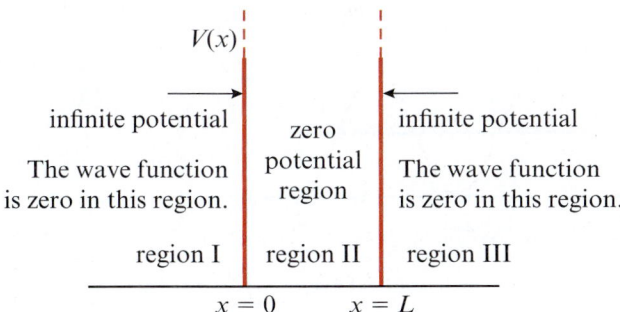

Figure 32-13 An infinite square well potential in one dimension.

Table 32-1 The Schrödinger Equation for Regions I, II, and III

Regions I and III ($x < 0$ and $x > L$)	Region II ($0 \leq x \leq L$)
$-\dfrac{\hbar^2}{2m}\dfrac{d^2\psi(x)}{dx^2} + V(x)\psi(x)$ $= E\psi(x)$ $V(x) = \infty$	$-\dfrac{\hbar^2}{2m}\dfrac{d^2\psi(x)}{dx^2} + V(x)\psi(x)$ $= E\psi(x)$ Since $V(x) = 0$, $-\dfrac{\hbar^2}{2m}\dfrac{d^2\psi(x)}{dx^2} = E\psi(x)$ $\dfrac{d^2\psi(x)}{dx^2} + \dfrac{2mE}{\hbar^2}\psi(x) = 0$

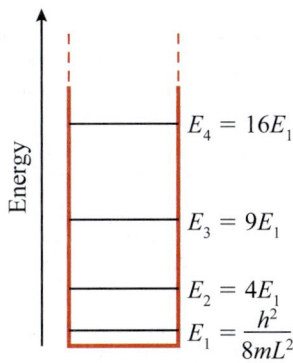

Figure 32-14 Energy levels of the infinite square well potential. The ground-state energy is E_1. The energies of the higher levels are given by $E_n = n^2 E_1$.

Wave Functions for an Infinite Square Well Potential

From Equations 32-46 and 32-48 we see that the wave function for quantum state n of a particle confined within an infinite square well potential is

$$\psi_n(x) = A \sin(k_n x) = A \sin\left(\frac{n\pi}{L}x\right) \qquad (32\text{-}52)$$

The constant A can be determined from the normalization condition that the probability of finding the particle somewhere along the x-axis must be one. Since the wave function is zero outside region II, the normalization condition can be obtained by integrating the probability density over the region $0 \leq x \leq L$ for a given quantum state n:

$$\int_{-\infty}^{+\infty} \psi_n^*(x)\psi_n(x)dx = \int_0^L \psi_n^*(x)\psi_n(x)dx = 1 \qquad (32\text{-}53)$$

where $\psi^*(x)$ is the complex conjugate of the wave function $\psi(x)$.

Equation 32-52 shows that $\psi_n(x)$ is real and therefore $\psi_n^*(x) = \psi_n(x)$. Inserting Equation 32-46 into the above integral, we get

$$\int_0^L A^2 \sin^2\left(\frac{n\pi}{L}x\right) dx = 1$$

Since $\int_0^L \sin^2\left(\frac{n\pi}{L}x\right) dx = \frac{L}{2}$ for all n, the normalization condition gives

$$A^2\frac{L}{2} = 1 \Rightarrow A = \sqrt{\frac{2}{L}}$$

A is called the **normalization constant**. Notice that the normalization constant does not depend on the quantum number n. Thus, the normalized wave functions for the infinite square well potential are

KEY EQUATION $\quad \psi_n(x) = \sqrt{\frac{2}{L}}\sin\left(\frac{n\pi}{L}x\right) \quad n = 1, 2, 3, \ldots$

$$(32\text{-}54)$$

The corresponding probability density, $\psi_n^*(x)\psi_n(x)$, is

$$\psi_n^*(x)\psi_n(x) = \frac{2}{L}\sin^2\left(\frac{n\pi}{L}x\right) \qquad (32\text{-}55)$$

Figure 32-15 shows plots of $\psi_n^*(x)\psi_n(x)$ for $n = 1, 2, 3,$ and 10. We see that in the ground state, the particle is most likely to be found at the middle of the $0 \leq x \leq L$ region, that is, at $x = L/2$. For the first excited state ($n = 2$), the particle is most likely to be found at $x = L/4$ and $x = 3L/4$, and the probability of finding the particle at $x = L/2$ is zero. Both of these probability distributions differ radically from classical mechanics, which

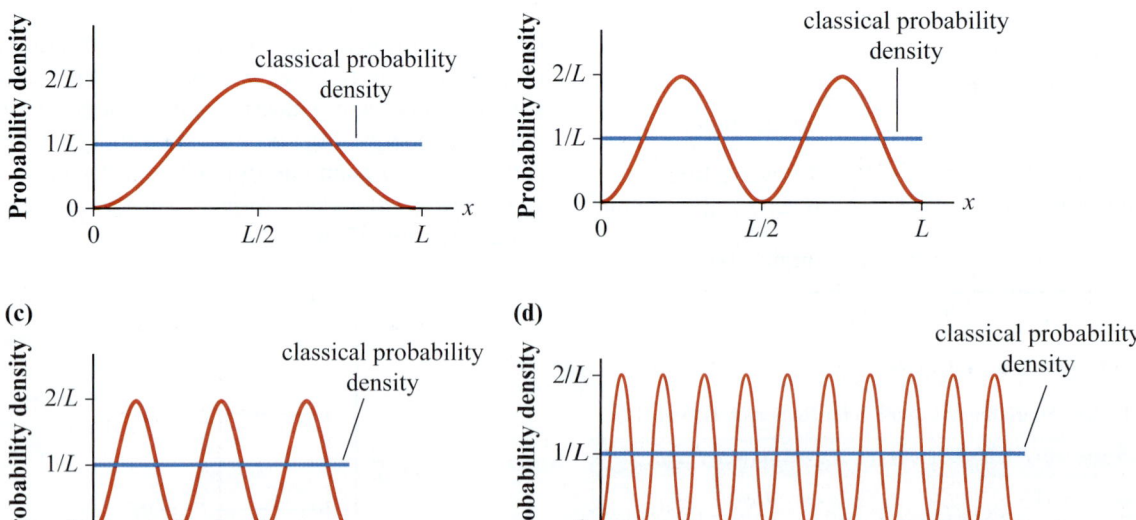

Figure 32-15 The probability density $\psi_n^*(x)\psi_n(x)$ versus x for a particle in a one-dimensional infinite square well potential of length L, for (a) the ground state, $n = 1$; (b) the first excited state, $n = 2$; (c) the second excited state, $n = 3$; and (d) the state $n = 10$.

predicts that the probability of finding the particle is the same everywhere within the region $0 \leq x \leq L$ and is therefore equal to $1/L$. As the quantum number n increases, the predictions of quantum mechanics approach the classical prediction. As shown by the plot of the probability density for $n = 10$, for large values of n, the maxima and minima of the probability density are very closely spaced. For n tending to infinity the probability density is indistinguishable from its average value of $1/L$, which is the classical value. The observation that for large quantum numbers the predictions of quantum mechanics agree with those of classical mechanics is a general property of quantum mechanics. This general property was first stated by Niels Bohr in the form of the **correspondence principle**, which states: "the results of a quantum theory analysis of a system that involves the use of very large quantum numbers must agree with the results of a classical physics analysis."

EXAMPLE 32-6

An Electron in an Infinite Square Well Potential

Consider an electron confined within a one-dimensional infinite square potential of length 0.500×10^{-10} m, the approximate radius of a hydrogen atom.

(a) Find the ground-state energy of the electron.
(b) Calculate the energies corresponding to the $n = 2$, $n = 3$, and $n = 4$ states.
(c) An electron in an excited state can move to a state of lower energy by emitting a photon. Calculate the energies and wavelengths of emitted photons for all transitions from the $n = 4$ state to lower states.

Solution

(a) The ground-state energy for an infinite square well potential is given by Equation 32-50. Using the electron mass $m = 9.11 \times 10^{-31}$ kg, we get

$$E_1 = \frac{h^2}{8mL^2} = \frac{(6.63 \times 10^{-34} \text{ J} \cdot \text{s})^2}{(8)(9.11 \times 10^{-31} \text{ kg})(0.500 \times 10^{-10} \text{ m})^2}$$

$$= 2.41 \times 10^{-17} \text{ J} = 1.50 \times 10^2 \text{ eV}$$

Since atomic energies are commonly given in units of eV, we converted from J using $1 \text{ eV} = 1.602 \times 10^{-19}$ J.

(b) Energies of the excited states corresponding to $n = 2$, $n = 3$, and $n = 4$ are related to the ground-state energy by Equation 32-51. Therefore,

$$E_2 = 4E_1 = 6.00 \times 10^2 \text{ eV}$$

$$E_3 = 9E_1 = 1.35 \times 10^3 \text{ eV}$$

$$E_4 = 16E_1 = 2.40 \times 10^3 \text{ eV}$$

(c) An electron in the $n = 4$ state can jump to a state of lower energy by emitting a photon (Figure 32-16). By energy conservation,

$$E_4 = E_m + E_{\text{photon}}$$

where m denotes a state of lower energy ($m < 4$) and E_{photon} is the energy of the photon. The wavelength of a photon is related to its energy, E, by

$$\lambda = \frac{c}{f} = \frac{hc}{E_{\text{photon}}}$$

Here,

$$hc = (6.63 \times 10^{-34} \text{ J} \cdot \text{s})(3.0 \times 10^8 \text{ m/s})$$
$$= \frac{1.99 \times 10^{-25} \text{ J} \cdot \text{m}}{1.60 \times 10^{-19} \text{ J/eV}} = 1.24 \times 10^{-6} \text{ eV} \cdot \text{m}$$

Photon energies and wavelengths for all transitions from the $n = 4$ state to lower states are as follows:

(i) $4 \rightarrow 3$ transition:

$$E_{\text{photon}} = E_4 - E_3 = 1054 \text{ eV}$$
$$\lambda = \frac{1.24 \times 10^{-6} \text{ eV} \cdot \text{m}}{1.05 \times 10^3 \text{ eV}} = 1.2 \times 10^{-9} \text{ m}$$

(ii) $4 \rightarrow 2$ transition:

$$E_{\text{photon}} = E_4 - E_2 = 1807 \text{ eV}$$
$$\lambda = \frac{1.24 \times 10^{-6} \text{ eV} \cdot \text{m}}{1.80 \times 10^3 \text{ eV}} = 6.9 \times 10^{-10} \text{ m}$$

(iii) $4 \rightarrow 1$ transition:

$$E_{\text{photon}} = E_4 - E_1 = 2259 \text{ eV}$$
$$\lambda = \frac{1.24 \times 10^{-6} \text{ eV} \cdot \text{m}}{2.25 \times 10^3 \text{ eV}} = 5.5 \times 10^{-10} \text{ m}$$

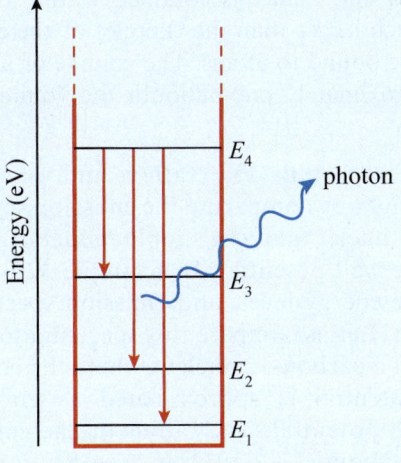

Figure 32-16 When an electron jumps from a state of higher energy to that of a lower energy, it emits a photon. The energy of the emitted photon is equal to the difference of the energies of the two levels.

Example 32-6 demonstrates some of the fundamental features of quantum mechanics:

1. The energy of a particle confined within a finite region of space is quantized. The energy and momentum of the particle can only take on a finite set of values, which are determined by the shape of the potential and the boundary conditions. This quantization is a radical departure from classical mechanics, where energy and momentum have continuous values.

2. A charged, confined particle can only emit a photon when it jumps from a higher energy level to a lower energy level. A particle that is in the ground state of a confining potential cannot emit a photon. Thus, quantum mechanics explains why atoms are stable. Remember (see Section 27-7) that according to classical mechanics an accelerating charged particle emits electromagnetic radiation and thus loses energy. As motion in a circle is an accelerated motion, according to classical mechanics an electron in a circular motion around a nucleus would continuously lose energy by emitting electromagnetic radiation until it fell into the nucleus. Therefore, according to classical mechanics, atoms should be unstable.

3. The energy of a confined particle cannot be zero. The minimum energy that a particle must have is the ground-state energy (also called the zero-point energy).

4. The smaller the confining space, the greater the ground-state energy. This prediction is experimentally verified by the fact that the energies of protons and neutrons confined within a nucleus are much larger than the energies of the electrons that are bound to atoms. The volume of a nucleus is approximately one-billionth the volume of an atom.

This is an important observation, and we can gain further insight by comparing the emission spectra of atoms and nuclei taking a simple model of an infinite square well potential. In Example 32-6 we calculated the energy levels and emission spectrum of an electron. Let us compare this spectrum to that of a proton in a carbon-12 nucleus where the confining nuclear potential is approximated as an infinite square well potential. The radius of the carbon-12 nucleus is about 2.7×10^{-15} m (see Section 34-2). Table 32-2 compares the two spectra. Here energy levels are arranged in the order $E_1, E_2, E_3, E_4, \ldots$, with E_1 being the ground state. Although an infinite square well potential is unphysical, the difference of a factor of 100 000 in the excitation energies of atomic and nuclear spectra is experimentally verified.

Table 32-2 The Emission Spectra of an Electron and a Proton in an Infinite Square Well Potential*

Emission Spectrum of an Electron:	Emission Spectrum of a Proton:
$L = 0.5 \times 10^{-10}$ m	$L = 2.7 \times 10^{-15}$ m
$E_2 - E_1 = 4.5 \times 10^2$ eV	$E_2 - E_1 = 8.2 \times 10^7$ eV
$E_3 - E_1 = 1.2 \times 10^3$ eV	$E_3 - E_1 = 2.2 \times 10^8$ eV
$E_4 - E_1 = 2.3 \times 10^3$ eV	$E_4 - E_1 = 4.1 \times 10^8$ eV
$E_5 - E_1 = 3.6 \times 10^3$ eV	$E_5 - E_1 = 6.5 \times 10^8$ eV

*Only transitions to the ground state are considered. The width of the potential well for the electron is taken to be 0.5×10^{-10} m and that of the proton is 2.7×10^{-15} m.

CHECKPOINT 32-4

A Proton in an Infinite Square Well Potential

Replace the electron in Example 32-6 with a proton. (The mass of a proton is approximately 1836 times the mass of an electron.) The ground-state energy of the confined proton is
(a) less than the ground-state energy of the electron
(b) the same as the ground-state energy of the electron
(c) greater than the ground-state energy of the electron
(d) zero, because protons remain stationary when confined to an infinite square well potential

ANSWER: (a) Energy levels of a particle confined in an infinite square well potential are inversely proportional to the mass of the particle.

INTERACTIVE ACTIVITY 32-2

Quantum Bound States

In this activity, you will use the PhET simulation "Quantum Bound States" to learn more about the bound states of a finite square well potential.

EXAMPLE 32-7

The Probability of Finding an Electron

A particle is in the ground state of an infinite square well potential. Calculate the probability of finding the particle in the region $0 \leq x \leq \dfrac{L}{3}$ of the well.

Solution

The probability, P, of finding the particle in the region $0 \leq x \leq \dfrac{L}{3}$ is obtained by integrating the square of the wave function over the region:

$$P = \int_0^{L/3} \psi^2(x)\,dx$$

For the ground state of an infinite square well potential,

$\psi(x) = \sqrt{\dfrac{2}{L}} \sin\left(\dfrac{\pi x}{L}\right)$. Substituting gives

$$P = \frac{2}{L} \int_0^{L/3} \sin^2\left(\frac{\pi x}{L}\right) dx$$

Let $z = \dfrac{\pi x}{L}$. Then $dx = \dfrac{L}{\pi} dz$ and

$$P = \frac{2}{\pi} \int_0^{\pi/3} \sin^2(z)\, dz = \frac{2}{\pi}\left(-\frac{\sqrt{3}}{8} + \frac{\pi}{6}\right) = 0.196$$

The probability of finding the particle in the region $0 \le x \le \dfrac{L}{3}$ is therefore 0.196, or 19.6%.

32-6 The Finite Square Well Potential

Our next example is that of a particle in a finite square-well potential. The following derivation is a bit lengthy and involves advanced calculus and may be omitted. Following are the main results of this section:

(a) The energy eigenvalues of a particle confined in a finite square well potential are quantized. However, there is no simple formula for calculating the eigenvalues in this case.

(b) If a potential is symmetric about the origin (i.e., $V(-x) = V(x)$), then the wave functions for the corresponding Schrödinger equation can be divided into two groups, the even and the odd wave-functions. The even wave functions do not change sign under space reflection ($x \rightarrow -x$), while the odd wave functions do change sign under space reflection.

(c) The wave functions of a particle that is confined within a potential of finite depth leak into the classically forbidden region outside the potential well.

In Section 32-5, we solved the Schrödinger equation for a particle in an infinite square well potential. In this case, the probability of finding the particle outside the potential well is zero, so the wave function and its derivatives vanish at the boundaries. The resulting energy eigenvalues are quantized. An interesting aspect of this quantization is that a particle in such a potential cannot have zero energy—the lowest energy is given by Equation 32-50. How would this picture change if a particle were confined in a potential with a finite depth? For this, consider a one-dimensional attractive potential of depth V_0 and length $2L$ centred symmetrically about the origin, as shown in Figure 32-17. As we'll discuss below, the wave functions for a symmetric potential have interesting

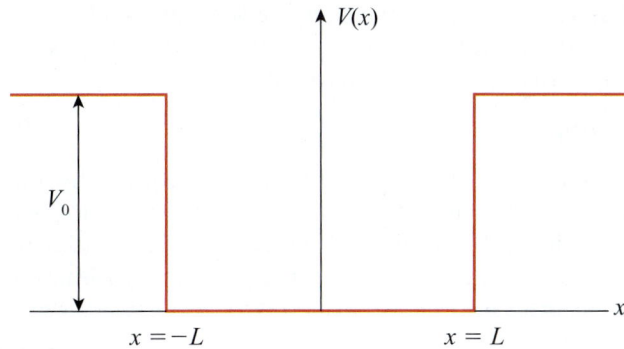

Figure 32-17 A finite square well potential $V(x)$. The potential is zero for $-L \le x \le L$ and $V(x) = V_0$ elsewhere.

properties under the reflection of the coordinate system, $x \rightarrow -x$. Mathematically, the potential is described as

$$V(x) = \begin{cases} 0 & \text{for } -L \le x \le +L \\ V_0 & \text{otherwise} \end{cases} \quad (32\text{-}56)$$

The Schrödinger equation in the inside region, $-L \le x \le +L$, is

$$-\frac{\hbar^2}{2m}\frac{d^2\psi(x)}{dx^2} = E\psi(x) \quad (32\text{-}57)$$

where m is the mass of the particle and E is the energy eigenvalue. The general solution of the above equation is

$$\psi(x) = A\sin\alpha x + B\cos\alpha x \qquad \alpha = +\left(\frac{2mE}{\hbar^2}\right)^{1/2} \quad (32\text{-}58)$$

In the outside region ($|x| > L$), the Schrödinger equation is

$$-\frac{\hbar^2}{2m}\frac{d^2\psi(x)}{dx^2} + V_0\psi(x) = E\psi(x) \quad (32\text{-}59)$$

We will consider the case $E < V_0$ so that the energy of the particle is less than the height of the potential barrier. The general form of the solution in the outside region is

$$\psi(x) = \begin{cases} Ce^{-\beta x} + De^{+\beta x} & x > +L \\ Fe^{-\beta x} + Ge^{+\beta x} & x < -L \end{cases}$$
$$\beta = +\left(\frac{2m(V_0 - E)}{\hbar^2}\right)^{1/2} \quad (32\text{-}60)$$

Since the wave function must remain finite for all x, we set $D = 0$ for $x > L$ and $F = 0$ for $x < -L$. The wave function in the outside regions is then

$$\psi(x) = \begin{cases} Ce^{-\beta x} & x > +L \\ A\sin\alpha x + B\cos\alpha x & -L \le x \le L \\ Ge^{+\beta x} & x < -L \end{cases} \quad (32\text{-}61)$$

Before determining the energy eigenvalues and corresponding wave functions, we consider the concept of the parity operation on wave functions.

Parity

In physics, a parity transformation, denoted by the letter P, is defined as a space reflection or, loosely speaking, a left–right interchange $\rightleftarrows$. In one dimension under the parity transformation the coordinate x transforms into $-x$. We write this as

$$\text{P}: x \rightarrow -x \text{ (under P, } x \text{ goes into } -x) \qquad (32\text{-}62)$$

In three dimensions, if $\vec{r}$ denotes a position vector, then under parity it transforms into $-\vec{r}$:

$$\text{P}: \vec{r} \rightarrow -\vec{r} \qquad (32\text{-}63)$$

We can also apply the parity transformation to other quantities. For example, if $\vec{p} = m\vec{v}$ denotes the momentum of a particle, then, under the parity transformation,

$$\text{P}: \vec{p} \rightarrow m\frac{d(-\vec{r})}{dt} = -\vec{p} \qquad (32\text{-}64)$$

The angular momentum vector $\vec{L} = \vec{r} \times \vec{p}$ does not change sign under the parity transformation:

$$\text{P}: \vec{L} = \vec{r} \times \vec{p} \rightarrow (-\vec{r}) \times (-\vec{p}) = \vec{r} \times \vec{p} = \vec{L}$$

To examine the effect of the parity transformation on a wave function, consider a potential $V(x)$ that is symmetric under space reflection, $x \rightarrow -x$. The Schrödinger equation for this potential is

$$-\frac{\hbar^2}{2m}\frac{d^2\psi(x)}{dx^2} + V(x)\psi(x) = E\psi(x)$$

If we change the sign of x everywhere in the above equation, we get

$$-\frac{\hbar^2}{2m}\frac{d^2\psi(-x)}{d(-x)^2} + V(-x)\psi(-x) = E\psi(-x)$$

Since $V(-x) = V(x)$, we obtain

$$-\frac{\hbar^2}{2m}\frac{d^2\psi(-x)}{dx^2} + V(x)\psi(-x) = E\psi(-x)$$

Notice that $\psi(x)$ and $\psi(-x)$ are solutions of the same equation, and if there is only one linearly independent wave function corresponding to each energy eigenvalue, then $\psi(-x)$ and $\psi(x)$ are proportional to each other:

$$\psi(-x) = \pm\psi(x) \qquad (32\text{-}65)$$

The constant of proportionality is ± 1 because if a wave function is reflected twice about the origin, then we should get the same wave function back. If $\psi(-x) = +\psi(x)$, the wave function is classified as an **even parity** (or simply even) wave function. If $\psi(-x) = -\psi(x)$, the wave function is classified as an

odd parity (or simply odd) wave function. Therefore, the wave functions for a symmetric potential are divided into two classes, even and odd wave functions.

If an eigenvalue has more than one linearly independent wave function, then the wave functions need not have a definite parity. However, we can form linear combinations of such wave functions such that each has even or odd parity.

Let us consider the **even parity wave functions** of Equation 32-61. Since $\cos(-x) = \cos x$ and $\sin(-x) = -\sin x$, for even parity wave functions we must set $A=0$, $G=C$. Then

$$\psi(x) = \begin{cases} Ce^{-\beta x} & x > +L \\ B\cos(\alpha x) & -L \le x \le L \\ Ce^{+\beta x} & x < -L \end{cases} \qquad (32\text{-}66)$$

We now impose the boundary conditions that $\psi(x)$ and $\dfrac{d\psi(x)}{dx}$ must be continuous at $x = L$. Since the potential is symmetric, the boundary conditions at $x = -L$ provide no additional information. This gives the following relations between B and C:

$$\begin{aligned} \text{(a)} \quad & B\cos(\alpha L) = Ce^{-\beta L} \\ \text{(b)} \quad & -\alpha B\sin(\alpha L) = -\beta Ce^{-\beta L} \end{aligned} \qquad (32\text{-}67)$$

Provided $B \neq 0$ and $C \neq 0$, dividing (b) by (a) in Equation 32-67 gives

$$\alpha\tan(\alpha L) = \beta \qquad (32\text{-}68)$$

For a given V_0, this is a transcendental equation in E (which means that the variable we're trying to solve for occurs both inside and outside a function that is not algebraic).

Energy Eigenvalues

We can find energy eigenvalues by either graphing or numerically solving Equation 32-68. Here, we use a graphical method. Using the definitions of α and β, we can write Equation 32-68 as

$$\tan\left(\sqrt{\frac{2mE}{\hbar^2}}L\right) = \sqrt{\frac{V_0 - E}{E}} \qquad (32\text{-}69)$$

If we plot the left and the right sides of this equation on the same graph as a function of E, the energy eigenvalues correspond to the points on the graph where the two curves intersect. Let us consider a specific example of an electron confined within a potential of depth 10 eV and width 2 nm ($V_0 = 10$ eV, $L = 1$ nm). A graph of the two sides of the above equation is shown in Figure 32-18(a) and (b). Here we have used

$$\frac{2m}{\hbar^2} = \frac{2mc^2}{\hbar^2 c^2} = \frac{2 \times 0.511 \times 10^6 \text{ eV}}{(197.3)^2 (\text{eV})^2 \cdot (\text{nm})^2}$$

The energy is measured in eV and the length in nm. Notice that the two curves intersect at six points.

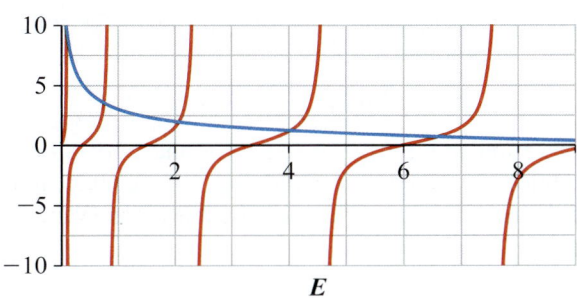
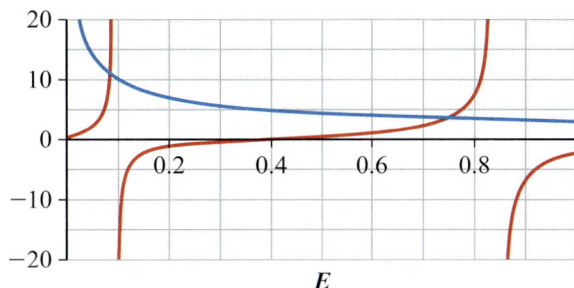

Figure 32-18 (a) A plot of $\tan\sqrt{\dfrac{2mE}{\hbar^2}}L$ (red curve) and $\sqrt{\dfrac{V_0 - E}{E}}$ (blue curve) as a function of energy, E. The points where the two curves intersect correspond to energy eigenvalues of a one-dimensional finite square well potential. For this graph, $V_0 = 10$ eV, $L = 1$ nm, and the bound particle is an electron. (b) The same plot as in (a) but for energy between 0 eV and 1 eV to clearly show the location of the lowest bound state.

Therefore, the bound states for this finite square well potential occur approximately at $E = 0.08$ eV, 0.75 eV, 2.1 eV, 4.1 eV, 6.6 eV, and 9.6 eV. The lowest-energy bound state occurs at 0.08 eV. These energy eigenvalues are approximate. Once an approximate value has been determined, more accurate values can be obtained by various numerical methods.

Wave Functions

To determine the wave functions, we need a relationship between B and C. In Equation 32-67, we divide (b) by α and then square and add (a) and (b). We get

$$B^2(\cos^2(\alpha L) + \sin^2(\alpha L)) = C^2\left(1 + \frac{\beta^2}{\alpha^2}\right)e^{-2\beta L}$$

Using $\left(1 + \dfrac{\beta^2}{\alpha^2}\right) = \dfrac{V_0}{E}$, we get

$$C = \pm B\left(\frac{E}{V_0}\right)^{1/2}e^{\beta L}$$

We are only interested in exploring the overall behaviour of the wave functions. Therefore, the wave functions are not normalized and we choose $B = 1$. Then,

$$\psi(x) = \begin{cases} \pm\left(\dfrac{E}{V_0}\right)^{1/2}e^{-\beta(x-L)} & x > L \\ \cos(\alpha x) & 0 \le x \le +L \quad (32\text{-}70) \\ \pm\left(\dfrac{E}{V_0}\right)^{1/2}e^{\beta(L+x)} & x < -L \end{cases}$$

Since the relation between B and C is obtained by squaring two equations, the overall sign of the wave function at the boundary is determined by the sign of $\cos \alpha L$ (the interior wave function at $x = L$). The wave functions for the lowest three energies are shown in Figure 32-19. Figure 32-19(a) shows the wave function

corresponding to the lowest energy bound state at $E = 0.08$ eV. Notice that for a finite potential well the wave function does not vanish at $x = L$ and leaks into the classically forbidden region, $x > L$, where it decreases exponentially with increasing x. Recall that for an infinite square well potential the wave function vanishes at the boundary of the potential and does not leak into the outside region. Figure 32-19(b) shows the

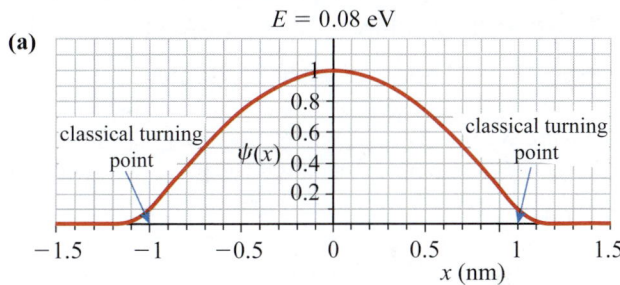
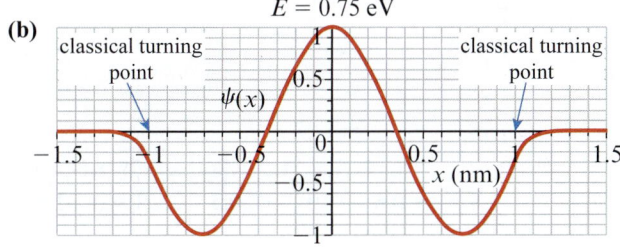
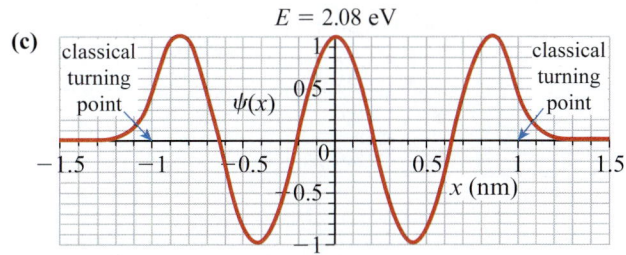

Figure 32-19 (a) The wave function for the lowest bound state at $E = 0.08$ eV for a one-dimensional finite square well potential. In this case, $V_0 = 10$ eV, $L = 1$ nm, and the bound particle is an electron. (b) The same as in (a) but for the next bound state at $E = 0.75$ eV. (c) The same as in (a) but for the bound state at $E = 2.07$ eV.

wave function for the next bound state at $E = 0.75$ eV. In this case, the wave function has one node between $x = 0$ and $x = L$. Again it is non-zero at $x = L$ and decreases exponentially for $x > L$. Figure 32-19(c) shows the wave function for the bound state at $x = 2.07$ eV, and in this the wave function has two nodes between $x = 0$ and $x = L$. This is a general property of the eigenfunctions for a potential of any shape. As the bound-state energy increases, the number of nodes of the wave function also increases. If we order the bound-state energies as $E_1, E_2, E_3, \ldots$, with E_1 being the lowest energy, then the number of nodes of the eigenfunction corresponding to the nth eigenvalue, E_n, is $n - 1$.

What do odd parity wave functions look like? Since odd parity wave functions must change sign when x is replaced by $-x$, we set $B = 0$, $G = -C$ in Equation 32-61. Therefore,

$$\psi(x) = \begin{cases} Ce^{-\beta x} & x > +L \\ A\sin(\alpha x) & -L \leq x \leq L \\ -Ce^{+\beta x} & x < -L \end{cases} \quad (32\text{-}71)$$

The boundary conditions now result in the following relation between α and β:

$$\alpha\cot(\alpha L) = -\beta \quad (32\text{-}72)$$

The two conditions $\alpha\tan(\alpha L) = \beta$ and $\alpha\cot(\alpha L) = -\beta$ cannot be valid simultaneously, since this requires $\tan^2(\alpha L) = -1$, which is not possible for real values of α.

32-7 Quantum Tunnelling

Consider a ball of mass m and kinetic energy K, moving on a frictionless surface toward a hill of height h, as shown in Figure 32-21. The minimum energy needed for the ball to reach the top of the hill is mgh. If $K < mgh$, the ball will not reach the top, and the probability of finding the ball on the right side of the hill is zero. If $K > mgh$, the ball will reach the top and then roll down the right side of the hill. After the ball has crossed the top of the hill, the

---

CHECKPOINT 32-5

The Depth of a Finite Square Well Potential

Figure 32-20 shows four graphs of the ground-state wave functions of a particle in a finite square well potential (Figure 32-17). The depth, V_0, of the potential is different for each wave function. Order these wave functions

according to the height of the potential well (from highest to lowest). The dotted lines represent the locations of the walls of the potential.

Highest Potential _____ _____ _____ _____ Lowest Potential
    Depth   Depth

(a)

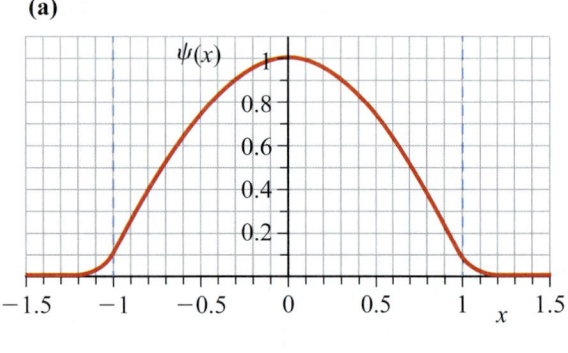

(b)

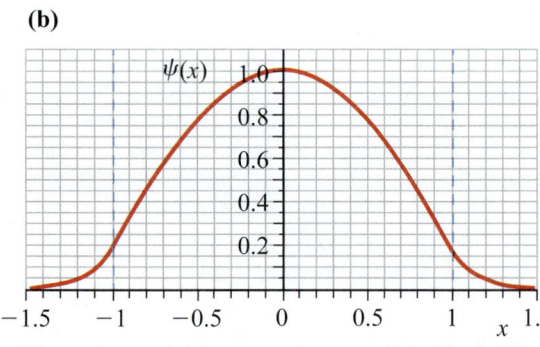

(c)

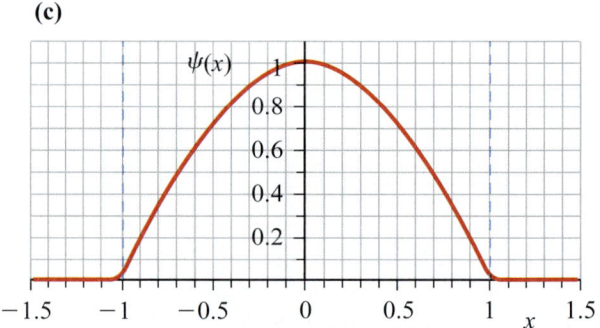

(d)

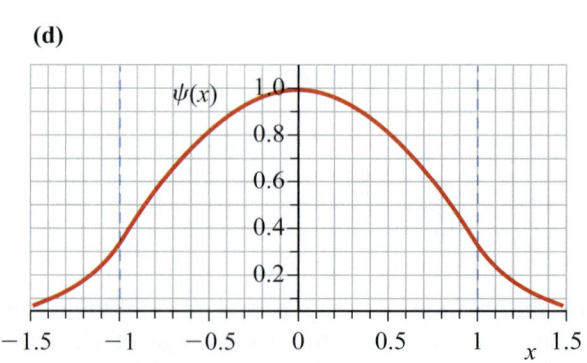

Figure 32-20   Ground-state wave functions for a finite square well potential. The depth of the potential differs for each graph.

Figure 32-21 A ball moving toward a barrier (hill).

probability of finding the ball anywhere to the left of the hill is zero.

However, quantum mechanics makes a startling prediction: *Even if the kinetic energy of the ball is less than the potential energy required to reach the top of the barrier, the probability of the ball penetrating the barrier is not zero.*

To study the quantum-mechanical analogue of the situation in Figure 32-21, we replace the physical barrier by a potential well. We start with a simple **step-function potential**, where the potential energy is zero in one region and has a positive and constant value in the other region. Such potential wells provide a simple model of the potential energy encountered by an electron at the interface of two homogeneous media. We take the potential energy to be zero for $x < 0$ and to have a positive value, V_0, for $x \geq 0$, thus dividing the x-axis into two regions, as shown in Figure 32-22.

Now consider a steady beam of particles, each with mass m and energy $E < V_0$, travelling from left to right toward $x = 0$. Let $\psi_1(x)$ and $\psi_2(x)$ denote a particle's wave function in regions 1 and 2, respectively. For $x < 0$, the Schrödinger equation has the form

$$-\frac{\hbar^2}{2m} \frac{d^2\psi_1(x)}{dx^2} + 0 \times \psi_1(x) = E\psi_1(x)$$

$$\frac{d^2\psi_1(x)}{dx^2} + \frac{2mE}{\hbar^2} \psi_1(x) = 0$$

(32-73)

In region 2, $V(x) = V_0$; therefore,

$$-\frac{\hbar^2}{2m} \frac{d^2\psi_2(x)}{dx^2} + V_0\psi_2(x) = E\psi_2(x)$$

$$\frac{d^2\psi_2(x)}{dx^2} + \frac{2m(E - V_0)}{\hbar^2} \psi_2(x) = 0$$

(32-74)

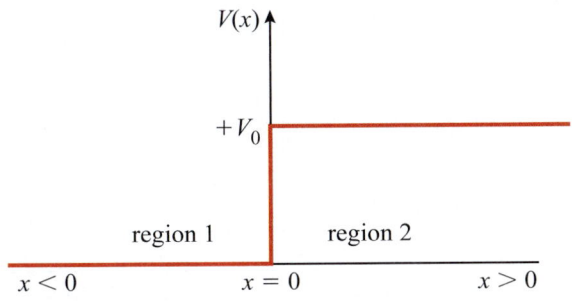

Figure 32-22 A repulsive finite step potential.

To simplify the calculation, we make the following substitutions:

$$k_1^2 = \frac{2mE}{\hbar^2}$$

$$k_2^2 = \frac{2m(V_0 - E)}{\hbar^2}$$

(32-75)

where k_1 and k_2 are positive and $\hbar k_1$ and $\hbar k_2$ are the magnitudes of the particle's momenta in regions 1 and 2, respectively. The Schrödinger equations in the two regions can be written as

$$\frac{d^2\psi_1(x)}{dx^2} + k_1^2 \psi_1(x) = 0 \quad x < 0$$

$$\frac{d^2\psi_2(x)}{dx^2} - k_2^2 \psi_2(x) = 0 \quad x \geq 0$$

(32-76)

The solutions of the above two equations are

$$\psi_1(x) = A_1 e^{ik_1 x} + B_1 e^{-ik_1 x} \quad x < 0$$ (32-77)

$$\psi_2(x) = A_2 e^{-k_2 x} + B_2 e^{k_2 x} \quad x \geq 0$$ (32-78)

The exponential functions e^{ikx} and e^{-ikx} describe travelling waves. To determine the direction of motion of the wave described by each function, we need to examine the time dependence of the wave functions. For a state of energy E, the time dependence of the wave function is of the form $e^{-iEt/\hbar}$. Therefore, the time-dependent wave function in region 1 is

$$\psi_1(x, t) = A_1 e^{ik_1 x} e^{-iEt/\hbar} + B_1 e^{-ik_1 x} e^{-iEt/\hbar} \quad x \leq 0$$
$$= A_1 e^{i(k_1 x - Et/\hbar)} + B_1 e^{-i(k_1 x + Et/\hbar)} \quad x \leq 0 \text{ (32-79)}$$

The term $e^{i(k_1 x - Et/\hbar)}$ represents a wave that is moving toward the direction of increasing x (left to right), and A_1 is the amplitude of this wave. The term $e^{-i(k_1 x + Et/\hbar)}$ represents a wave that is moving toward the direction of decreasing x (right to left), and B_1 is the amplitude of this wave. Therefore, the term with the e^{ikx} factor represents the incident wave, and the term with the e^{-ikx} factor represents the reflected wave.

The wave function and its derivative are continuous at the boundary between the two regions. We therefore require that

$$\psi_1(0) = \psi_2(0)$$

$$\left. \frac{d\psi_1(x)}{dx} \right|_{x=0} = \left. \frac{d\psi_2(x)}{dx} \right|_{x=0}$$

(32-80)

Since x is positive in region 2, the $e^{k_2 x}$ term becomes infinite as $x \to \infty$. The wave function must remain finite for *all* x, so B_2 must be zero. Therefore, the wave function in region 2 is

$$\psi_2(x) = A_2 e^{-k_2 x} \quad x \geq 0$$ (32-81)

Let us now apply the boundary conditions at $x = 0$. The requirement that the wave functions must have the same value at $x = 0$ gives

$$A_1 + B_1 = A_2$$

The requirement that the derivatives of the wave functions must be equal at $x = 0$ gives

$$ik_1 A_1 - ik_1 B_1 = -k_2 A_2$$

Using the above two equations, we can write B_1 and A_2 in terms of A_1:

$$B_1 = \left(\frac{k_1 - ik_2}{k_1 + ik_2}\right) A_1 \quad \text{and} \quad A_2 = \left(\frac{2k_1}{k_1 + ik_2}\right) A_1 \quad (32\text{-}82)$$

Therefore, the wave functions that satisfy the boundary conditions at the interface of the two regions are

$$\psi_1(x) = A_1 e^{-ik_1 x} + \left(\frac{k_1 - ik_2}{k_1 + ik_2}\right) A_1 e^{ik_1 x} \quad x < 0$$

$$\psi_2(x) = \left(\frac{2k_1}{k_1 + ik_2}\right) A_1 e^{-k_2 x} \quad\quad x \geq 0 \quad (32\text{-}83)$$

Without any loss of generality, we can take $A_1 = 1$; this value corresponds to a probability density where the incident beam contains one particle per unit length. Using the expressions for k_1 and k_2 from Equation 32-75, the wave functions in the two regions are

$$\psi_1(x) = e^{-i\frac{\sqrt{2mE}}{\hbar}x} + \left(\frac{\sqrt{E} - i\sqrt{V_0 - E}}{\sqrt{E} + i\sqrt{V_0 - E}}\right) e^{i\frac{\sqrt{2mE}}{\hbar}x} \quad x < 0$$

$$\psi_2(x) = \left(\frac{2\sqrt{E}}{\sqrt{E} + i\sqrt{V_0 - E}}\right) e^{-\frac{\sqrt{2m(V_0 - E)}}{\hbar}x} \quad\quad x \geq 0$$

$$(32\text{-}84)$$

Notice the following features of these wave functions:

1. In region 1, the first term represents the wave function of the incident beam, and the second term represents the wave function of the reflected beam. The term $\left(\frac{\sqrt{E} - i\sqrt{V_0 - E}}{\sqrt{E} + i\sqrt{V_0 - E}}\right)$ represents the amplitude of the reflected beam. This is similar to the situation in classical physics, where a beam of particles with energy less than the height of the barrier reflects from the barrier.

2. In region 2, the wave function decays exponentially and is not zero. Therefore, a particle can penetrate a classically forbidden region. Such penetration of a potential barrier is called **tunnelling**. It is a quantum effect and has no analogue in classical physics. In classical physics, for $E < V_0$, the entire incident beam must reflect from the barrier, and the probability of finding a particle in region 2 is zero.

By performing a simple but slightly tedious calculation (problem 51) we can compare the probability densities

in the two regions (note that probability densities here are not normalized):

$$\psi_1(x)\psi_1^*(x) = 2 + \left(\frac{4E - 2V_0}{V_0}\right)\cos\left(2\frac{\sqrt{2mE}}{\hbar}x\right)$$

$$- \left(\frac{4\sqrt{E(E - V_0)}}{V_0}\right)\sin\left(2\frac{\sqrt{2mE}}{\hbar}x\right) \quad x < 0 \quad (32\text{-}85)$$

$$\psi_2(x)\psi_2^*(x) = \left(\frac{4E}{V_0}\right) e^{-\frac{2\sqrt{2m(V_0 - E)}}{\hbar}x} \quad x \geq 0$$

Figure 32-23 shows plots of these probability densities for a barrier of $V_0 = 5.0$ eV and electrons with $E = 1.0$ eV and $E = 4.9$ eV. Notice that as E approaches V_0, the probability that a particle can penetrate the classically forbidden region increases. In our example, the potential extends all the way to infinite x. With real-life potentials of finite width, a detector on the right side of the potential can detect particles that pass through the potential. Such experiments provided a clear verification of quantum mechanics.

If the potential well has a finite width, then the probability of finding the particle at the other side of the potential is non-zero. Consider a potential of width a, as shown in Figure 32-24. A steady beam of particles of energy E and mass m is incident on the left side of the potential. The wave function at $x = a$ is given by evaluating $\psi_2(a)$ in Equation 32-84. The probability, P, that the particle reaches the other side of the potential is proportional to the square of $\psi_2(a)$. This probability

(a)

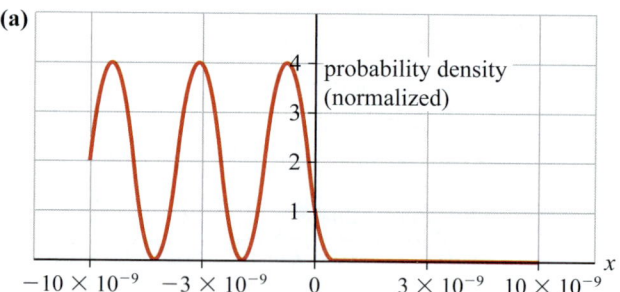

(b)

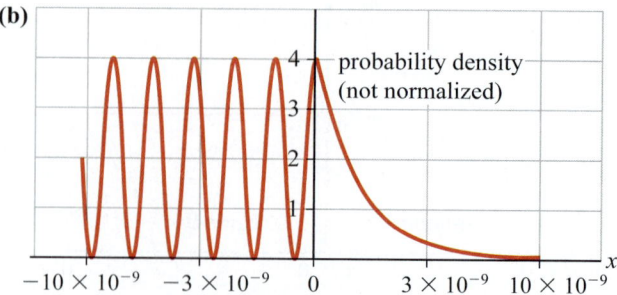

Figure 32-23 The probability density for scattering from a step potential (Figure 32-22) of $V_0 = 5.0$ eV for electrons incident from the left with kinetic energy of (a) 1.0 eV and (b) 4.9 eV.

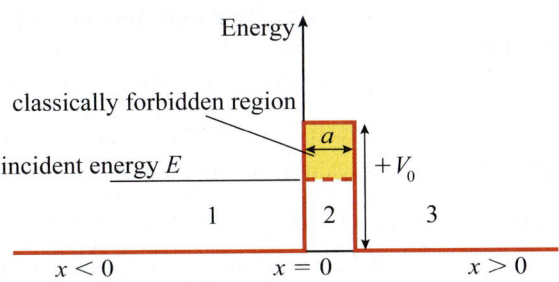

Figure 32-24 A potential barrier of height V_0 and width a.

depends on the incident energy of the particle and on the height and width of the potential barrier:

KEY EQUATION
$$P(E, V_0, a) \propto e^{\frac{-2a}{\hbar}\sqrt{2m(V_0 - E)}} \qquad (32\text{-}86)$$

The proportionality sign is necessary because we have not normalized the wave function. We can calculate the normalization factor by solving the Schrödinger equation, with the boundary conditions, for all three regions of the potential in Figure 32-23, and then integrating the square of the wave function. However, the shape of the probability distribution for the tunnelling process is determined by the exponential term. Since an exponential function varies rapidly with its argument,

the tunnelling probability is a sensitive function of the mass and the energy of the particle, and of the height and the width of the potential, as shown in the following example.

From Equation 32-86 we see that the tunnelling probability decreases exponentially as the width of the barrier increases. Figure 32-25 is a graph of the tunnelling probability of the electrons as a function of potential width.

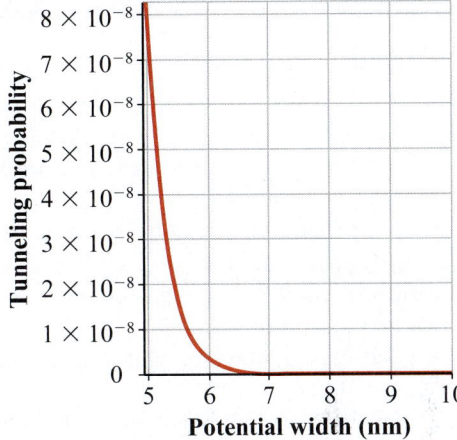

Figure 32-25 The relative tunnelling probability for 1.0 eV electrons as a function of the width of a potential with $V_0 = 5.0$ eV.

EXAMPLE 32-8

Tunneling through a Barrier

Consider an electron of 1.0 eV kinetic energy incident on a potential barrier of width 5.0 nm with $V_0 = 5.0$ eV.

(a) Estimate the probability that the electron will tunnel through the potential.
(b) Estimate the tunnelling probability for a proton with the same initial kinetic energy.

Solution

(a) The tunnelling probability can be determined using Equation 32-86. The values for the variables in the exponent are

$$\hbar = 6.63 \times 10^{-34} \text{ J} \cdot \text{s}$$

$$m = \text{electron mass} = 9.11 \times 10^{-31} \text{ kg}$$

$$E = 1.0 \text{ eV} = 1.0 \times 1.60 \times 10^{-19} \text{ J}$$

$$V_0 = 5.0 \text{ eV} = 5.0 \times 1.60 \times 10^{-19} \text{ J}$$

$$a = 5.0 \text{ nm} = 5.0 \times 10^{-9} \text{ m}$$

Let us first confirm that the exponent is dimensionless:

$$\frac{1}{\text{J} \cdot \text{s}}[\text{kg} \cdot \text{J}]^{1/2} \text{m} = \frac{[\text{kg}]^{1/2} \cdot \text{m}}{[\text{J}]^{1/2} \cdot \text{s}} = \frac{[\text{kg}]^{1/2} \cdot \text{m}}{[\text{kg} \cdot \text{m}^2 \cdot \text{s}^{-2}]^{1/2} \cdot \text{s}} = 1$$

Evaluating the exponent,

$$\frac{2}{\hbar}\sqrt{2m(V_0 - E)}a = \frac{2(5.0 \times 10^{-9} \text{m})}{6.63 \times 10^{-34} \text{J} \cdot \text{s}}$$

$$\times \sqrt{2(9.11 \times 10^{-31} \text{kg})(5.0 - 1.0)(1.60 \times 10^{-19} \text{J})}$$

$$= 16.3$$

Therefore, the probability, P, of the electron tunnelling through the barrier is

$$P \propto e^{-16.3} = 8 \times 10^{-8}$$

The constant of proportionality is of the order of unity. Therefore, P is of the order of 10^{-7}.

(b) The mass of a proton is approximately 1836 times the mass of an electron. Keeping other quantities the same, the tunnelling probability for a proton is

$$P \propto e^{-700} \sim 10^{-300}$$

Therefore, P is of the order of 10^{-300}.

Making sense of the result

If 10^{10} electrons hit the barrier per second, approximately 1000 will pass through. Such a current can easily be detected with a sensitive instrument. However, if 10^{10} protons per second are incident on the barrier, on average, one proton would tunnel through the barrier every 10^{290} s. No wonder we do not observe objects like baseballs tunnelling through baseball bats or gloves.

MODERN PHYSICS

CHECKPOINT 32-6

Tunnelling through a Barrier

A beam of 2 eV electrons is incident on a square well potential of width 1 nm and height 4 eV. Which of the following statements is correct when the electron beam scatters from the potential barrier?

(a) All electrons transmit through the potential barrier, and there is no reflected beam.
(b) All electrons reflect from the potential barrier, and there is no transmitted beam.
(c) Exactly half the electrons transmit through and the other half reflect.
(d) Some electrons transmit through and the others reflect.

ANSWER: (d)

32-8 The Quantization of Angular Momentum

In Section 32-5, you saw that the energy and the momentum of a confined particle are quantized. Is the angular momentum of a confined particle also quantized, as Bohr assumed for his model of the atom? In fact, the Schrödinger wave equation does predict such quantization.

Consider a particle with angular momentum $\vec{J}$ and magnitude $J = \sqrt{J_x^2 + J_y^2 + J_z^2}$. A given component of $\vec{J}$ could have values in the range from $-J$ to $+J$. For example, the z-component of angular momentum is given by $J_z = \vec{J} \cdot \hat{z} = J\cos\theta$, which varies between $-J$ and $+J$ as θ, the angle between the momentum vector and the z-axis, varies between 0 rad and π rad.

In quantum mechanics, the z-component of angular momentum can have only *certain discrete values*, which differ by multiples of $\hbar$. Note that the units of $\hbar$, $m^2 \cdot kg/s$, have dimensions of angular momentum.

Any quantum-mechanical system (a free particle, a bound particle, or a collection of particles) is assigned a characteristic **total angular momentum quantum number**, j. The z-component of the total angular momentum of the system must be one of the following values in the range of $-j\hbar$ to $+j\hbar$:

$$-j\hbar, (-j+1)\hbar, (-j+2)\hbar, \ldots,$$
$$(+j-2)\hbar, (+j-1)\hbar, +j\hbar \qquad (32\text{-}87)$$

The quantum number j can have only integer or half-integer values. For a given value of j, there are $2j + 1$ possible values for the z-component of angular momentum.

For example, a free electron is assigned $j = \dfrac{1}{2}$, so there are only two possible values for the z-component of the angular momentum of a free electron: $+\dfrac{1}{2}\hbar$ and $-\dfrac{1}{2}\hbar$. Similarly, if $j = 2$ for an atom, the z-component of the atom's angular momentum must be $-2\hbar$, $-\hbar$, $0\hbar$, $\hbar$, or $2\hbar$ (a total of five possible values, Figure 32-26).

Since the z-axis can be chosen along any direction, the component of angular momentum along *any* axis is quantized in multiples of $\hbar$ in a range determined by the total angular momentum quantum number, j.

The magnitude of the eigenvalue, J, of the angular momentum is related to the angular momentum quantum number j as follows:

KEY EQUATION $\quad J = \sqrt{J_x^2 + J_y^2 + J_z^2} = \sqrt{j(j+1)}\,\hbar \quad (32\text{-}88)$

The above relationship can be proven rigorously by treating angular momentum as an operator. Here, we will use a less rigorous derivation developed by the American physicist Richard Feynman (1918–1988). Since $J^2 = J_x^2 + J_y^2 + J_z^2$, the average value of J^2 is equal to the sum of the average values of J_x^2, J_y^2, and J_z^2. Since J_x^2, J_y^2, and J_z^2 are scalars, they have the same value for any orientation of the coordinate system. Therefore, the average values of J_x^2, J_y^2, and J_z^2 are all equal, and

$$\overline{J^2} = 3(\overline{J_z^2}) \qquad (32\text{-}89)$$

where $\overline{J}$ denotes the mean value of J.

Since $J^2 = \vec{J} \cdot \vec{J}$ is a scalar, it is independent of the orientation of $\vec{J}$. Therefore, its average is equal to its value, that is,

$$J^2 = 3(\overline{J_z^2}) \qquad (32\text{-}90)$$

The $2j + 1$ allowed values of J_z are given by Equation 32-87. Therefore,

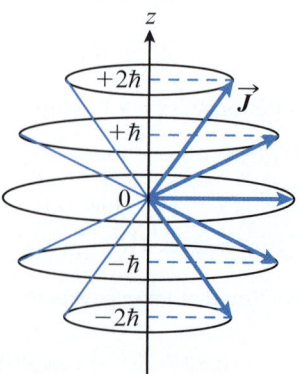

Figure 32-26 The possible z-components of an atom's angular momentum with $j = 2$. For each of these values, the magnitude of the angular momentum lies somewhere on the surface of a cone.

$$\overline{J_z^2} = \left(\frac{(-j)^2 + (-j+1)^2 + (-j+2)^2}{2j+1}\right.$$

$$\left.\frac{+ \cdots + (j-2)^2 + (j-1)^2 + (j)^2}{2j+1}\right)\hbar^2$$

$$= \frac{(2j^3 + 3j^2 + j)/3}{2j+1}\hbar^2$$

$$= \frac{j(2j+1)(j+1)}{3(2j+1)}\hbar^2$$ (32-91)

$$= \frac{j(j+1)}{3}\hbar^2$$

Substituting into Equation 32-90 gives

$$J = \sqrt{j(j+1)}\hbar \qquad (32\text{-}92)$$

If the particle is in a state with angular momentum quantum number j, the eigenvalue of the magnitude of the angular momentum operator of the particle is $\sqrt{j(j+1)}\hbar$.

EXAMPLE 32-9

The Angular Momentum of a Bound Electron

The angular momentum quantum number j of an electron in an excited state of a hydrogen atom is 5/2.

(a) What is the magnitude of the electron's angular momentum?
(b) What are the angles between the angular momentum vector and the z-axis?

Solution

(a) The magnitude, J, of the angular momentum vector for the quantum number $j = 5/2$ is

$$J = \sqrt{j(j+1)}\hbar = \sqrt{\frac{5}{2} \times \frac{7}{2}}\hbar = \frac{\sqrt{35}}{2}\hbar$$

(b) Since the z-component of the angular momentum is quantized,

$$J_z = m\hbar \text{ where } m = -j, (-j+1), \dots, (j-1), j$$

Therefore,

$$\cos\theta = \frac{m\hbar}{\sqrt{j(j+1)}\hbar} = \frac{m}{\sqrt{j(j+1)}}$$

For $j = \frac{5}{2}$, $m = -\frac{5}{2}, -\frac{3}{2}, -\frac{1}{2}, \frac{1}{2}, \frac{3}{2}, \frac{5}{2}$.

Since $\cos^{-1}(-x) = \pi - \cos^{-1}x$, we only need to evaluate θ for either positive or negative values of m. For $m = \frac{5}{2}$,

$$\cos\theta = \left(\frac{2}{\sqrt{35}}\right)\left(\frac{5}{2}\right) = \frac{5}{\sqrt{35}}$$

$$\theta = \cos^{-1}\frac{5}{\sqrt{35}} = 0.56 \text{ rad} = 32.3°$$

Similarly, we can calculate θ for other values of m:

$m = \frac{5}{2}$: $\theta = 32.3°$; $m = -\frac{5}{2}$: $\theta = 147.7°$

$m = \frac{3}{2}$: $\theta = 59.5°$; $m = -\frac{3}{2}$: $\theta = 120.5°$

$m = \frac{1}{2}$: $\theta = 80.3°$; $m = -\frac{1}{2}$: $\theta = 99.7°$

CHECKPOINT 32-7

Projection of Angular Momentum

For an electron bound in an atom in a state with angular momentum quantum number $j = 3/2$, the possible values of the angle between the angular momentum vector of the electron and the z-axis are

(a) all possible values between 0° and 180°
(b) 39°, 75°, 141°, 105°
(c) 30°, 60°, 120°, 150°
(d) 0°, 45°, 90°, 135°

ANSWER: (b)

32-9 The Schrödinger Equation for a Hydrogen Atom

The Bohr model of the hydrogen atom (Section 31-9) explains the origin of the hydrogen spectrum but has a number of shortcomings. The only justification that the Bohr model gives for the quantization of the angular momentum of a bound electron is that quantization gives rise to stationary orbits and, hence, to quantized energies of the electron.

The Schrödinger equation provides a better description of the hydrogen atom. You saw in Section 32-5 that the energy of a particle confined within a region of space is quantized. In a hydrogen atom, an electron is bound to a proton by electrostatic attraction. Therefore, we can expect the energy of the electron to be quantized. Now, we will examine the solution of the Schrödinger equation for a hydrogen atom and see how well the resulting energy levels match the hydrogen spectrum.

If the electron in the hydrogen atom is at a distance r from the proton, the electrostatic potential energy between the proton and the electron is

$$V(r) = k\frac{(+e)(-e)}{r} = -k\frac{e^2}{r} \tag{32-93}$$

where $k = 1/(4\pi\varepsilon_0)$ is Coulomb's constant.

The time-independent Schrödinger equation for a particle of mass m, moving in three dimensions in the presence of a potential, is given by Equation 32-33. Since the potential energy, $V(r)$, depends only on the radial distance, $r = \sqrt{x^2 + y^2 + z^2}$, between the proton and the electron, it is convenient to write the Schrödinger equation in terms of spherical coordinates: r, θ, and ϕ (the radial, polar, and azimuthal coordinates, respectively). As shown in Section 31-6, spherical coordinates are related to the Cartesian coordinates as follows:

$$\begin{aligned} x &= r\sin\theta\cos\phi \\ y &= r\sin\theta\sin\phi \\ z &= r\cos\theta \end{aligned} \tag{32-94}$$

The potential energy does not depend on the θ- and ϕ-coordinates, so the wave function can be written as a product of three wave functions, each depending on a single spherical coordinate:

$$\psi(r, \theta, \phi) = \psi_1(r)\psi_2(\theta)\psi_3(\phi) \tag{32-95}$$

In Section 32-5, a solution for a one-dimensional Schrödinger equation for a particle bound in an infinite square well potential resulted in a single quantum number, n. The allowed energies and wave functions of the particle are indexed by n. When the Schrödinger equation in three dimensions is solved for the wave functions $\psi_1(r)$, $\psi_2(\theta)$, and $\psi_3(\phi)$, three quantum numbers arise, one for each dimension. These three quantum numbers are not independent. Choosing a specific value for one quantum number restricts the allowed values of the other two. Solving these differential equations is rather complicated, and the procedure is quite lengthy. So we will just summarize the results here.

The Principal Quantum Number, n

The radial wave function $\psi_1(r)$ exists only for positive integer values of a quantum number called the principal quantum number. The **principal quantum number**, n, describes the *size* of the **orbital**, the wave function

that indicates the probable location of the bound electron. The energies of the electron are quantized in terms of n:

$$\begin{aligned} E_n &= -\frac{1}{n^2}\left(\frac{m_e k^2 e^4}{2\hbar^2}\right) \\ &= -\frac{13.6\,\text{eV}}{n^2} \quad n = 1, 2, 3, \dots \end{aligned} \tag{32-96}$$

This expression for the energy levels is exactly the same as the expression for the Bohr model of the hydrogen atom (Equation 31-42), and the principal quantum number in Equation 32-96 is the same as the number in the Bohr model. Thus, Schrödinger's wave equation accurately predicts the hydrogen energy levels and the resulting spectral lines.

The Orbital Quantum Number, l

The solution of the differential equation for $\psi_2(\theta)$ gives rise to the orbital quantum number. The **orbital quantum number**, l, describes the *shape* of the orbital and is restricted to values from 0 to $n - 1$:

$$l = 0, 1, 2, 3, \dots, n - 1 \tag{32-97}$$

For example, if $n = 3$, the allowed values for l are 0, 1, and 2. The magnitude of the orbital angular momentum, L, of the electron, is related to the orbital quantum number:

$$L = \sqrt{l(l + 1)}\hbar \tag{32-98}$$

Therefore, the angular momentum of the electron is quantized as multiples of Planck's constant. The quantization of the orbital angular momentum appears naturally when we solve the Schrödinger equation. There is no need to assume, as Bohr did, that the orbital angular momentum is quantized. Note that values of orbital angular momentum given by Equation 32-98 do not agree with the values suggested by Bohr (Equation 31-34), as can be seen in the examples listed in Table 32-3.

Table 32-3 The Bohr Model and the Schrödinger Equation Predictions of the Magnitude of the Orbital Angular Momentum of an Electron in a Hydrogen Atom

Principal Quantum Number, n	The Bohr Model Orbital Angular Momentum, L	The Schrödinger Equation Orbital Quantum Number, l	Orbital Angular Momentum, L
1	$\hbar$	0	0
2	$2\hbar$	0	0
		1	$\sqrt{2}\hbar$
3	$3\hbar$	0	0
		1	$\sqrt{2}\hbar$
		2	$\sqrt{6}\hbar$

Table 32-4 Letter Designations for the Orbital Quantum Number

Orbital Quantum Number, l	Letter Designation
0	s
1	p
2	d
3	f
4	g

For large values of l, $\sqrt{l(l+1)} \approx l$, and the Bohr model agrees reasonably closely with the orbital angular momentum predicted by the Schrödinger equation. Experiments have confirmed that the values of L predicted by the Schrödinger equation are correct. As shown in Table 32-4, various values of l are commonly designated with lowercase letters.

The Magnetic Quantum Number, m_l

The solution of the differential equation containing the azimuthal angle, ϕ, gives rise to a quantum number called the magnetic quantum number. The **magnetic quantum number**, m_l, describes the *orientation* of a particular orbital and is restricted to integer values between $-l$ and $+l$:

$$L_z = m_l \hbar, \quad m_l = -l, (-l+1), \ldots, (l-1), l \quad \text{(32-99)}$$

Shells and Subshells

The orbitals that have the same value of the principal quantum number n form a **shell**. Within a shell, orbitals with different orbital quantum numbers form **subshells**. Since for a given n the orbital quantum number l has values from 0 to $n-1$, there are n subshells within a shell of principal quantum number n. It is customary to denote a subshell by giving the principal quantum number with the letter designation for the orbital quantum number (e.g., $2p$ ($n=2$, $l=1$) and $4d$ ($n=4$, $l=3$)). Table 32-5 shows the possible quantum numbers for the wave function for a hydrogen atom.

The Ground State of a Hydrogen Atom

For the ground state, $n=1$, and the only possible value of l is 0. Therefore, the ground state of a hydrogen atom has zero angular momentum. For $l=0$, the only possible value of m_l is 0. Thus,

Ground state: $n=1$, $l=0$, $m_l=0$, orbital: $1s$

The ground state is the most tightly bound state and has a binding energy of

$$E_1 = -\frac{13.6\,\text{eV}}{1^2} = -13.6\,\text{eV}$$

It takes 13.6 eV of energy to liberate an electron from the ground state of a hydrogen atom, producing a free electron and a positive hydrogen ion (i.e., a proton). The shape of the ground-state orbital is spherical, as shown in Figure 32-27.

The first excited state The first excited state of the hydrogen atom corresponds to $n=2$. The energy of the first excited state is

$$E_2 = -\frac{13.6\,\text{eV}}{2^2} = -3.4\,\text{eV}$$

There are two possible values of the orbital quantum number, $l=0$ and $l=1$. If $l=0$, then $m_l=0$ and the orbital angular momentum of the electron is zero. If $l=1$, then m_l has three possible values: $+1$, 0, and -1. For $l=1$, the magnitude of the orbital angular momentum is $\sqrt{2}\hbar$, and there are three possible orientations of the orbital, as shown in Figure 32-27. Thus, the first excited state has four subshells, all with the same binding energy. We say that the first excited state of the hydrogen atom is *four-fold degenerate*: First excited state:

$$\begin{array}{lll} n=2 & l=0 & m_l=0 \} \quad 2s \\ n=2 & l=1 & m_l=+1 \\ n=2 & l=1 & m_l=0 \\ n=2 & l=1 & m_l=-1 \end{array} \Big\} 2p$$

The $l=0$ orbital is spherical (Figure 32-27), and the $l=1$ orbitals have the double-lobe shape shown in Figure 32-28.

Table 32-5 Quantum Numbers for a Hydrogen Atom Wave Function

Principal, n	Orbital, l	Magnetic, m_l
1	0	0
2	0	0
	1	$+1, 0, -1$
3	0	0
	1	$+1, 0, -1$
	2	$+2, +1, 0, -1, -2$
...	...	...
n	$1, 2, 3, 4, \ldots, n-1$	$-l, -l+1, \ldots, 0, \ldots, l-1, l$

Figure 32-27 A computer-generated plot of the probability density for an electron in an orbital with $l=0$. The spherical surface represents the probability density at a certain distance from the origin. For $l=0$ orbitals, the probability density is spherically symmetric.

Figure 32-29 shows the energy-level diagram for the hydrogen atom, as predicted by the Schrödinger equation.

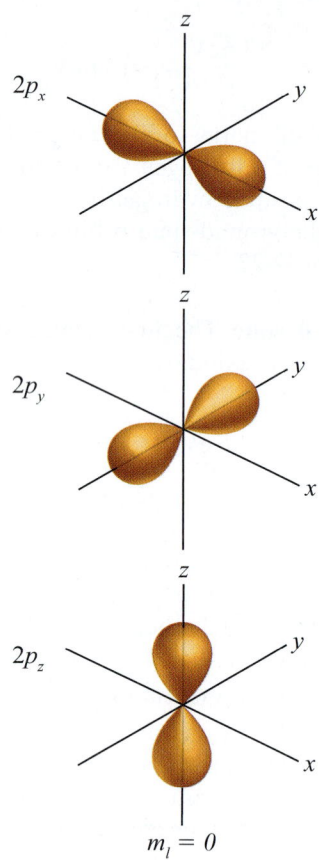

Figure 32-28 A computer-generated plot of the probability density for 2p orbitals. The surfaces represent some threshold value of probability density. Note that the orbital aligned along the z-axis has $m_l = 0$, and the orbitals aligned along the x-axis and y-axis are certain mixtures of quantum states with $m_l = -1$ and $m_l = 1$.

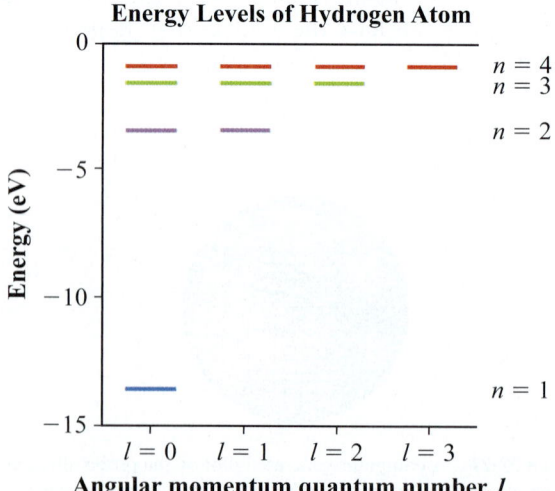

Figure 32-29 The energy-level diagram for the hydrogen atom.

The Radial Wave Function

The radial wave function is determined by solving the differential equation for the radial coordinate, r. The radial wave functions depend on the principal and the orbital quantum numbers and are usually denoted as $R_{nl}(r)$.

The ground state The ground state has $n = 1$ and $l = 0$; the radial wave function for the ground state is

$$R_{1s}(r) = \left(\frac{1}{\pi a_B^3}\right)^{1/2} e^{-\left(\frac{r}{a_B}\right)} \tag{32-100}$$

where $a_B = \dfrac{\hbar}{k m_e e^2}$ is the Bohr radius.

The radial wave function does not depend on θ and the ϕ-coordinates and therefore has the same value everywhere on a spherical shell of radius r. The probability density (probability per unit length), $P_{nl}(r)$, of finding the electron in a spherical shell between r and $r + dr$ is given by

$$P_{nl}(r) = 4\pi r^2 R_{nl}^2(r) \tag{32-101}$$

The area of a spherical shell is $4\pi r^2$, and the factor of 4π arises because of integration over angles θ and ϕ. The radial probability density of the ground state is obtained by inserting Equation 32-100 into Equation 32-101:

$$P_{1s}(r) = \left(\frac{4}{a_B^3}\right) r^2 e^{-\frac{2r}{a_B}} \tag{32-102}$$

Figures 32-30 and 32-31 show the wave function and the radial probability density for the ground state. The probability density is zero at the centre of the atom, reaches a maximum at $r = a_B$, and then falls off exponentially. Unlike in the Bohr model, the electron does not move in a well-defined orbit with $r = a_B$, and it is possible to find the electron everywhere away from the origin. The most probable distance is at the Bohr radius.

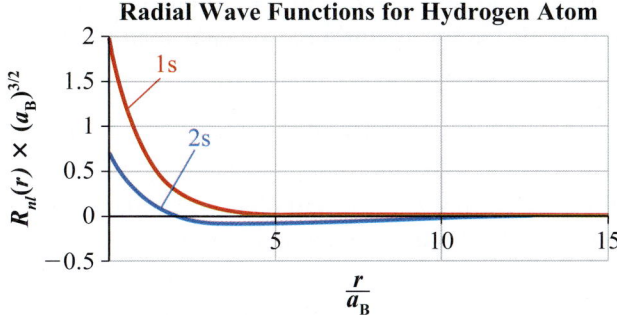

Radial Wave Functions for Hydrogen Atom

Figure 32-30 The radial wave functions for the ground state (1s) and the first excited state (2s) of hydrogen as a function of r. The vertical axis is in units of $(1/a_B)^{3/2}$.

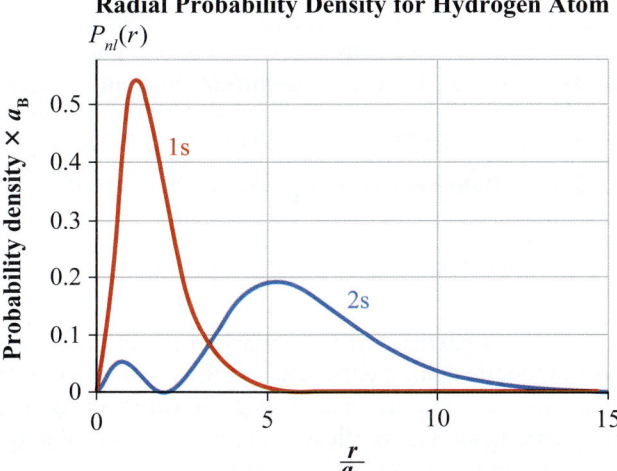

Radial Probability Density for Hydrogen Atom

Figure 32-31 The radial probability density for the ground state (1s) and the first excited state (2s) of hydrogen as a function of r. The probability density is in units of $(1/a_B)$.

The first excited state The first excited state has $n = 2$ with $l = 0$ or $l = 1$. The radial wave functions for these states are

$$R_{2s}(r) = \left(\frac{1}{32\pi a_B^3}\right)^{1/2}\left(2 - \frac{r}{a_B}\right)e^{-\left(\frac{r}{2a_B}\right)} \quad (32\text{-}103)$$

$$R_{2p}(r) = \left(\frac{1}{96\pi a_B^3}\right)^{1/2}\left(\frac{r}{a_B}\right)e^{-\left(\frac{r}{2a_B}\right)} \quad (32\text{-}104)$$

The wave function and radial probability density for the 2s state are shown in Figures 32-30 and 32-31. For the 2s state, the radial probability density is maximum at $r \approx 5.2a_B$. (The radius of the second energy level in the Bohr model is $4a_B$.) The existence of the smaller secondary maximum at $r = 0.76a_B$ indicates that there is a small probability of finding the 2s electron closer to the nucleus. This feature cannot be explained by Bohr's model.

Although the energies of the hydrogen atom predicted by Schrödinger's equation and the Bohr model agree, the wave function predicted by Schrödinger's equation is very different from that of the Bohr

model. In the Bohr model, an electron travels around the nucleus in distinct circular orbits, where the circumference of each possible orbit equals an integral number of de Broglie wavelengths. In the more accurate quantum-mechanical model based on Schrödinger's equation, there are no orbits. The electron does not follow a well-defined trajectory around the nucleus. A statement such as "the electron is at a distance of one Bohr radius from the proton" is meaningless in quantum mechanics. We can only talk about the probability of finding the electron at a certain distance from the proton. As shown by the graphs of the radial probability density, the probability of finding the electron at any distance from the proton is non-zero. The Bohr radius for an energy level corresponds approximately to the average of the most probable distances of the electron from the nucleus.

32-10 Intrinsic Angular Momentum—Spin

Because of its success in explaining the hydrogen spectrum and other phenomena, the Schrödinger equation is one of the foundations of modern quantum mechanics. However, by 1925, two experimental results had been confirmed that could not be explained by the Schrödinger wave equation we have discussed so far.

First, hydrogen spectral lines observed at very high resolution were found to each consist of two closely spaced lines. For example, the spectral line corresponding to the 2p state is found to consist of two spectral lines with an energy difference of 4.5×10^{-5} eV (Figure 32-32). Such splitting of spectral lines is called the **fine structure**. The same splitting occurs in the spectra of other elements that have a single electron in the outer shell that has non-zero value value of the orbital quantum number 1.

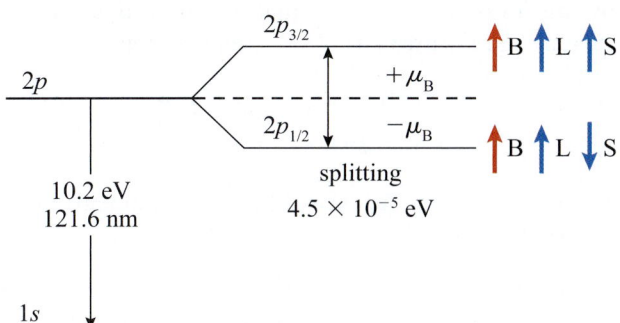

Figure 32-32 Fine structure in hydrogen. The spectral line corresponding to the 2p state consists of two closely spaced spectral lines. States with S parallel to L have a slightly higher energy than the states for which S and L are anti-parallel. B denotes the magnitude of the magnetic field of the nucleus experienced by the electron.

Second, when a beam of neutral atoms that have a single electron in the outermost shell passes through a non-uniform magnetic field, the beam splits into two or more beams. This splitting can be explained if the single electron has a non-zero orbital angular momentum. However, this splitting also occurs with a beam of hydrogen atoms in their ground state where the electron has no orbital angular momentum. Both of these splitting effects have the same origin in that the charged particles have a magnetic moment.

Magnetic Moment and Orbital Angular Momentum

Consider a particle of charge q and mass m that is moving in a circular orbit of radius R with speed v. The current produced by this particle is equal to the rate of rotation of the charge:

$$I = \frac{q}{2\pi R/v} = \frac{qv}{2\pi R} \tag{32-105}$$

The magnitude of the magnetic moment (μ) produced by this loop of current is equal to the product of the current and the area (A) of the circular orbit:

$$\mu = IA = \left(\frac{qv}{2\pi R}\right)(\pi R^2) = \frac{qvR}{2} \tag{32-106}$$

The magnitude of the classical angular momentum of the rotating particle is

$$L = mvR \tag{32-107}$$

Therefore, the magnitudes of the magnetic moment and the classical angular momentum of the particle are related as follows:

$$\mu = \frac{q}{2m}L \tag{32-108}$$

Since the magnetic moment and angular momentum are both vector quantities, we can write Equation 32-108 in vector form:

$$\vec{\mu} = \frac{q}{2m}\vec{L} \tag{32-109}$$

For an electron, $q = -e$ and $m = m_e$, so the magnetic moment due to its orbital angular momentum is

$$\vec{\mu}_l = g_l\left(\frac{-e}{2m_e}\right)\vec{L} = g_l\left(\frac{\mu_B}{\hbar}\right)\vec{L} \tag{32-110}$$

where we have introduced the factor $g_l = 1$, called the g-factor, to keep the definition of the magnetic moment general. The subscript l denotes that the magnetic moment is associated to the orbital angular momentum of the electron. (As we will see later in this section, the g-factor for the quantum-mechanical magnetic moments is not always one.) The quantity $\frac{e\hbar}{2m_e}$ is called the **Bohr magneton**, μ_B, and has a value of

$$\mu_B = \frac{e\hbar}{2m_e} = 9.27 \times 10^{-24}\,\text{A} \cdot \text{m}^2 \tag{32-111}$$

When a particle with a magnetic moment is placed in a magnetic field, it experiences a torque that tends to align its magnetic moment along the direction of the magnetic field. The particle, therefore, acquires a magnetic potential energy, U_{mag}, that is given by

$$U_{\text{mag}} = -\vec{\mu}_l \cdot \vec{B} \tag{32-112}$$

If we choose a z-axis along the direction of the magnetic field, then

$$\begin{aligned} U_{\text{mag}} &= -\vec{\mu}_l \cdot \vec{B} = g_l\left(\frac{e}{2m_e}\right)\vec{L} \cdot \vec{B} \\ &= g_l\left(\frac{e}{2m_e}\right)L_z B = g_l\left(\frac{\mu_B}{\hbar}\right)L_z B \end{aligned} \tag{32-113}$$

where L_z is the component of the orbital angular momentum along the z-axis. If the angular momentum makes an angle θ with respect to the z-axis, then $L_z B = LB\cos\theta$. As the angle θ varies between 0 rad and π rad, the magnetic potential energy varies between $+g_l\left(\frac{\mu_B}{\hbar}\right)LB$ and $-g_l\left(\frac{\mu_B}{\hbar}\right)LB$.

The relationship between angular momentum and the magnetic moment (Equation 32-110) also holds in quantum mechanics. However, in quantum mechanics angular momentum is quantized; therefore, the magnetic moment is also quantized. Since the allowed values of L_z are $m_l\hbar$, the magnetic potential energy of an electron in a magnetic field (given by Equation 32-113) is quantized:

$$U_{\text{mag}} = m_l(g_l\mu_B B) \tag{32-114}$$

Thus, quantum mechanics predicts that the energy of an electron in the presence of a magnetic field changes by discrete amounts.

The Stern–Gerlach Experiment

To determine whether the magnetic moment is quantized, as predicted by quantum mechanics, German physicists Otto Stern (1888–1969) and Walter Gerlach (1889–1979) performed an ingenious experiment. They prepared a beam of silver atoms by boiling silver in an oven and passed the evaporated atoms through a series of slits to form a narrow beam. The beam was then directed between poles of magnets shaped to produce a non-uniform magnetic field that varied considerably along a line joining the two poles, as shown in Figure 32-33. After passing through the magnetic field, the beam hit a photographic plate and created a mark on the plate.

Stern and Gerlach argued that if a silver atom *does* have a magnetic moment, then in a magnetic field that is pointing along the z-direction, the atom will gain a magnetic potential energy of $-\vec{\boldsymbol{\mu}} \cdot \vec{\boldsymbol{B}} = -\mu B \cos\theta$, where θ is the angle between the directions of the magnetic field and the magnetic moment. Since the magnetic field varies along the z-direction, there will be a net force on the atom given by

$$F_z = -\frac{dU_{mag}}{dz} = (\mu\cos\theta)\frac{dB}{dz} \qquad (32\text{-}115)$$

Silver atoms with their magnetic moments oriented perpendicular to the direction of the magnetic field ($\theta = 90°$) experience no force and pass through the magnetic field undeflected. Atoms with magnetic moments in the direction of the magnetic field ($\theta = 0°$) experience an upward force and are deflected upward. Atoms with magnetic moments directed opposite to the magnetic field ($\theta = 180°$) experience a downward force and are deflected downward.

A silver atom has 47 electrons. In its ground state, the first 46 electrons couple together in pairs that have zero net orbital angular momentum (Figure 32-34). The lone uncoupled electron is in the $l = 0$ state. Thus,

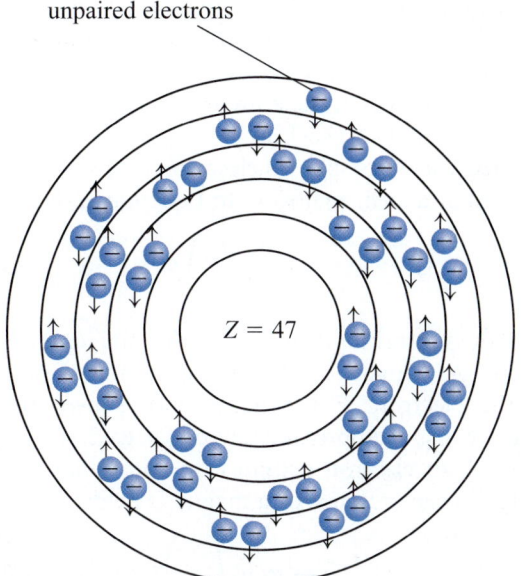

unpaired electrons

$Z = 47$

Figure 32-34 A Bohr model diagram of a silver atom in its ground state. The last (47th) electron is unpaired.

a silver atom has zero orbital angular momentum and should have zero magnetic moment. Consequently, a beam of silver atoms should not be deflected by the inhomogeneous field in Stern and Gerlach's experiment.

However, Stern and Gerlach observed that silver atoms formed two spots on the plate, spaced at equal distances above and below the original beam direction. The splitting of the incident beam showed that silver atoms *do* have a magnetic moment and that the component of the magnetic moment along the direction of the magnetic field has only two possible values: either along the direction of the magnetic field or opposite to it.

But a silver atom has zero orbital angular momentum, so how can it have a magnetic moment? In 1925, Dutch American physicists George Uhlenbeck (1900–1988) and Samuel Goudsmit (1902–1978) suggested that electrons possess an intrinsic angular momentum, which they called the spin angular momentum (or **spin**). The situation is somewhat analogous to Earth spinning on its axis and thus possessing angular momentum because of this spin (although our current understanding is that an electron is a point particle with zero radius). The **spin quantum number**, s, is a fundamental characteristic of all subatomic particles. The spin angular momentum is denoted by the symbol $\vec{\boldsymbol{S}}$. The magnitude of the spin angular momentum and its projection along the z-axis are given by formulas similar to Equations 32-98 and 32-99:

$$S = \sqrt{s(s + 1)}\hbar$$
$$S_z = m_s\hbar, \; m_s = -s, (-s + 1), \ldots, (s - 1), s \quad (32\text{-}116)$$

Here s is the spin quantum number (an analog of the orbital quantum number l) and m_s, called the **spin magnetic quantum number**, is the quantum number that describes the projection of the spin along an axis,

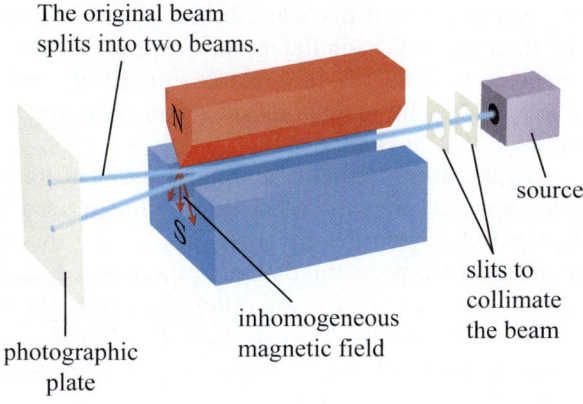

The original beam splits into two beams.

source

slits to collimate the beam

inhomogeneous magnetic field

photographic plate

Figure 32-33 The Stern–Gerlach experiment.

usually taken as z-axis. The spin quantum numbers can have either positive integer (0,1,2,..) or positive half-integer (1/2, 3/2, 5/2,..) values. For a given value of spin quantum number s, there are $(2s + 1)$ values of m_s, as given by Equation (32-116).

Because of its spin, an electron possesses a magnetic moment of a form analogous to Equation 32-113:

$$\vec{\mu}_S = -g_S \left(\frac{\mu_B}{\hbar} \right) \vec{S} \tag{32-117}$$

Here the subscript s denotes that the magnetic moment and the g-factor are due to the spin angular momentum of the electron. Experiments have shown that the g-factor, g_S, is approximately equal to 2. Because of its spin, an electron acquires an additional magnetic energy in the presence of a magnetic field:

$$U_{\text{mag, S}} = g_S \mu_B \left(\frac{S_z}{\hbar} \right) B \tag{32-118}$$

The total magnetic potential energy of an electron is therefore equal to the sum of the magnetic potential energies due to its orbital and spin angular momenta.

The splitting of the beam in the Stern–Gerlach experiment into two beams indicates that S_z for the silver atom can have two values: $+\dfrac{\hbar}{2}$ and $-\dfrac{\hbar}{2}$. Therefore, $s = \dfrac{1}{2}$ for the silver atom. Since the spin of the silver atom is determined by the spin of the unpaired electron, s must be $\dfrac{1}{2}$ for an electron. The two possible values of m for an electron are therefore $+\dfrac{1}{2}$ and $-\dfrac{1}{2}$. An electron with $m_s = +\dfrac{1}{2}$ is called a *spin-up* electron and is represented with an up arrow ($\uparrow$). An electron with $m_s = -\dfrac{1}{2}$ is called a *spin-down* electron and is represented with a down arrow ($\downarrow$).

In 1927, American physicists T. E. Phipps and J. B. Taylor used a beam of hydrogen gas in an experiment similar to the Stern–Gerlach experiment. In the Phipps–Taylor experiment, the beam split into two beams. Since the electron in the ground state of hydrogen has zero orbital angular momentum, this experiment again confirmed that an electron has an intrinsic spin with spin quantum number, $s = \dfrac{1}{2}$.

Just like mass and charge, spin is an intrinsic property of a particle. When the theory of relativity is combined with quantum mechanics, spin naturally appears as one of the intrinsic properties of a particle. All elementary particles are assigned an intrinsic spin. Particles that have a half-integer spin (these includes protons, neutrons, and electrons) are called **fermions**. Particles that have integer spin (these include photons and pions) are called **bosons**.

EXAMPLE 32-10

The Spin of a Rho Meson (ρ-meson)

A ρ-meson is an elementary particle with spin quantum number $s = 1$.

(a) What is the magnitude of the spin angular momentum of a ρ-meson, and what are the possible values of the projection of its spin along the z-axis?

(b) If a beam of ρ-mesons is used in the Stern–Gerlach experiment, into how many beams would the original beam split?

Solution

(a) Using Equation 32-116 with $s = 1$, we find

$$S = \sqrt{s(s + 1)}\hbar = \sqrt{2}\hbar$$
$$S_z = m_s\hbar = -\hbar, 0, \hbar$$

(b) Since S_z has three possible values, there will be three beams emerging from the magnetic field. One beam is along the original direction ($S_z = 0$), one above the original beam ($S_z = +\hbar$), and one below it ($S_z = -\hbar$).

Adding Angular Momenta in Quantum Mechanics

In classical mechanics, if particle 1 has angular momentum $\vec{J}_1$ and particle 2 has angular momentum $\vec{J}_2$, then the total angular momentum, $\vec{J}$, of the two particles is equal to the vector sum of $\vec{J}_1$ and $\vec{J}_2$: $\vec{J} = \vec{J}_1 + \vec{J}_2$. The magnitude of $\vec{J}$ can have any value between $|J_1 - J_2|$ and $J_1 + J_2$.

In quantum mechanics, angular momentum is an operator, and angular momenta are added according to the following rule: Let j_1 and j_2 be the quantum numbers corresponding to angular momenta $\vec{J}_1$ and $\vec{J}_2$, respectively. Let j be the quantum number of the total angular momentum $\vec{J}$ such that $\vec{J} = \vec{J}_1 + \vec{J}_2$. Then j can have any of the following values: $j_1 + j_2, j_1 + j_2 - 1, j_1 + j_2 - 2, ..., |j_1 - j_2|$.

For a given value of j, the magnitude, J, of the total angular momentum is $\sqrt{j(j+1)}\hbar$, and the z-component of $\vec{J}$ can have any of the following $(2j+1)$ values:

$$j\hbar, (j-1)\hbar, (j-2)\hbar, \ldots, (-j+1)\hbar, -j\hbar$$

This rule for adding angular momenta applies whether $\vec{J}_1$ and $\vec{J}_2$ represent angular momenta of two particles or two angular momenta of the same particle.

CHECKPOINT 32-11

The Angular Momentum of Two Electrons

Two electrons in an atom each have angular momentum quantum number $\frac{3}{2}$. What are the allowed values of the quantum number of the total angular momentum quantum number of the two electrons?

(a) 0
(b) 3
(c) 3, 2, 1, 0
(d) 3, 2, 1, 0, −1, −2, −3

ANSWER: (c)

EXAMPLE 32-11

Adding Angular Momentum in Quantum Mechanics

What are the possible values of the quantum number j of the total angular momentum $\vec{J}$ of an electron in a hydrogen atom with orbital angular momentum characterized by the quantum number $l = 1$? Determine the magnitude and allowed values of the z-component of $\vec{J}$ for each value of j.

Solution

Let $\vec{L}$ be the orbital angular momentum and $\vec{S}$ be the spin of the electron. The total angular momentum $\vec{J}$ of the electron is

$$\vec{J} = \vec{L} + \vec{S}$$

Since $s = \frac{1}{2}$ and $l = 1$, the possible values of j are

$$j = l + \frac{1}{2} = \frac{3}{2} \quad \text{and} \quad j = l - \frac{1}{2} = \frac{1}{2}$$

For $j = \frac{3}{2}$:

$$J = \sqrt{\frac{3}{2}\left(\frac{3}{2}+1\right)}\hbar = \sqrt{\frac{15}{4}}\hbar$$

and J_z can have four values: $\frac{3}{2}\hbar, \frac{1}{2}\hbar, -\frac{1}{2}\hbar, -\frac{3}{2}\hbar$.

For $j = \frac{1}{2}$:

$$J = \sqrt{\frac{1}{2}\left(\frac{1}{2}+1\right)}\hbar = \sqrt{\frac{3}{4}}\hbar$$

and J_z can have two values: $\frac{1}{2}\hbar, -\frac{1}{2}\hbar$.

Making sense of the result

For $l = 1$, there are three possible values of m_l $(-1, 0, +1)$. Because the spin projection of an electron can be up or down, an electron with $l = 1$ can exist in six possible states. The number of allowed states remains the same when the spin and the orbital angular momenta are combined. There are four allowed states for $j = 3/2$ and two for $j = 1/2$, adding to a total of six possible states.

The Pauli Exclusion Principle

A hydrogen atom has one electron. In its ground state, the electron is in the $n = 1$, $l = 0$ orbital. A helium atom has two electrons, and in the lowest-energy state both electrons are in the $n = 1$, $l = 0$ orbital with their spins pointing in opposite directions. A lithium atom has three electrons. In its lowest-energy state, the first two electrons are in the $n = 1$, $l = 0$ orbital with their spins pointing in opposite directions, but the third electron is in the higher-energy orbital: $n = 2$, $l = 0$. Similarly, a sodium atom has 11 electrons, with 2 electrons in $n = 1$ orbitals, 8 electrons in $n = 2$ orbitals, and 1 electron in an $n = 3$ orbital. Why are all the electrons of a ground-state atom not in the lowest-energy $n = 1$, $l = 0$ orbital? To explain this, Austrian physicist Wolfgang Pauli (1900–1958) proposed that *no two electrons within an atom can occupy the same quantum state simultaneously*. This principle is called the **Pauli exclusion principle**.

With the inclusion of spin, the quantum state of an electron in an atom is characterized by four quantum numbers: n, l, m_l, and m_s. Since electrons are added to build heavier atoms, the lowest energy state configuration of electrons must be consistent with the Pauli exclusion principle. Table 32-6 shows how the electrons are distributed in $n = 1$, 2 orbitals.

The Pauli exclusion principle applies to all fermions. There is no limit to the number of bosons that can occupy the same quantum state simultaneously.

CHECKPOINT 32-12

Electrons in an $n = 3$ Shell

What is the maximum number of electrons allowed in an $n = 3$ shell?

(a) 3
(b) 6
(c) 12
(d) 18

ANSWER: (d)

Table 32-6 Quantum States of an Atom for Principal Quantum Numbers 1 and 2

n	l	m_l	m_s	Capacity of Shell or Subshell	
1	0	0	+1/2	2 electrons in the $1s$ shell	A maximum of 2 electrons in the $n = 1$ shell
		0	−1/2		
2	0	0	+1/2	2 electrons in the $2s$ subshell	A maximum of 8 electrons in the $n = 2$ shell
		0	−1/2		
		+1	+1/2	6 electrons in the $2p$ subshell	
		+1	−1/2		
		0	+1/2		
		0	−1/2		
		−1	+1/2		
		−1	−1/2		

KEY CONCEPTS AND RELATIONSHIPS

Quantum mechanics is the branch of physics that deals with the motion of particles at the subatomic scale. The Schrödinger equation describes the behaviour of subatomic matter, predicts that the energy and momentum of a confined particle are quantized, and predicts that a particle can tunnel into a classically forbidden region.

Matter Waves and de Broglie's Hypothesis

De Broglie postulated that a particle of momentum p has a wavelength λ such that

$$\lambda = \frac{h}{p} \tag{32-2}$$

where $h = 6.63 \times 10^{-34}$ J·s $= 4.14 \times 10^{-15}$ eV·s is Planck's constant. Since h is extremely small, the wave nature of matter is not observed at a macroscopic level.

Heisenberg's Uncertainty Principle

The product of the uncertainty in a measurement of the position of a particle and the uncertainty in the measurement of its momentum must be greater than $\hbar/2$:

$$\Delta x \Delta p \geq \frac{\hbar}{2} \tag{32-6}$$

Similarly, a time Δt is needed to measure the energy of a system to uncertainty ΔE, and

$$\Delta t \Delta E \geq \frac{\hbar}{2} \tag{32-8}$$

The Schrödinger Equation

The wave function $\psi(x, t)$ of a particle of mass m satisfies the time-dependent Schrödinger equation:

$$-\frac{\hbar^2}{2m}\frac{\partial^2 \psi(x, t)}{\partial x} + V(x)\psi(x, t) = i\hbar\frac{\partial \psi(x, t)}{\partial t} \tag{32-28}$$

where $V(x)$ is the potential energy term.

When the particle is in a state of definite energy E, the wave function can be written as $\psi(x, t) = \psi(x)e^{-iEt}$, and the Schrödinger equation can be written in a time-independent form:

$$-\frac{\hbar^2}{2m}\frac{d^2\psi(x)}{dx^2} + V(x)\psi(x) = E\psi(x) \tag{32-29}$$

where $\psi(x)$ is a function of position.

E is called the energy eigenvalue. $|\psi(x)|^2 dx$ represents the probability of finding the particle in a region of length dx that is centred around the point x. The wave function is normalized such that the probability of finding the particle somewhere in space is one. The normalization condition for the Schrödinger equation in one dimension is

$$\int_{-\infty}^{+\infty} |\psi(x)|^2 dx = 1 \tag{32-35}$$

Particle in a One-Dimensional Box

The quantized energy values for a particle in an infinite square well potential are

$$E_n = n^2 \frac{\hbar^2 \pi^2}{2mL^2} \tag{32-49}$$

The principal quantum number, n, is a positive integer. The wave function of a particle in an infinite square well of width L and energy eigenvalue E_n is

$$\psi_n(x) = \sqrt{\frac{2}{L}} \sin\left(\frac{n\pi}{L}x\right) \quad n = 1, 2, 3, \ldots \tag{32-54}$$

The Finite Square Well Potential

In a finite square well potential, the bound-state $(E < V_0)$ wave functions tunnel through at the boundaries of the potential. The energy eigenvalues are quantized but there is no simple relationship between energy eigenvalues. For a symmetric potential the eigenfunctions are divided into two classes, the even and odd parity wave functions.

MODERN PHYSICS

Quantum Tunnelling

The probability of penetration through a barrier of height V_0 and width a by a particle of energy $E < V_0$ is proportional to

$$P(E, V_0, a) \propto e^{-\frac{2a}{\hbar}\sqrt{2m(V_0 - E)}} \qquad (32\text{-}86)$$

The Quantization of Angular Momentum

Angular momentum is quantized and is characterized by a quantum number j, which is a positive integer or half-integer. For a given j, the magnitude of the angular momentum, J, is

$$J = \sqrt{j(j+1)}\hbar \qquad (32\text{-}88)$$

The z-component of the angular momentum, J_z, has $2j + 1$ allowed values:

$$J_z = m_j\hbar \qquad m_j = -j, (-j+1), \ldots, (j-1), j$$

The Schrödinger Equation for a Hydrogen Atom

The hydrogen atom is described by a wave function that is characterized by four quantum numbers:

■ The principal quantum number, n, determines the size of the orbital and has allowed values of $n = 1, 2, 3, \ldots$.

■ The angular momentum quantum number, l, determines the shape of the orbital and for a given n has allowed values of $l = 0, 1, 2, \ldots, (n-1)$.

■ The magnetic quantum number, m_l, determines the orientation of the orbital for a given l and has allowed values of $m_l = -l, (-l+1), (l-1), l$.

■ The spin quantum number, m_s, describes the intrinsic angular momentum of the electron and has allowed values of $m_s = -\frac{1}{2}, \frac{1}{2}$.

The energy levels predicted by solving the Schrödinger equation for an electron confined within the Coulomb potential of a proton match the spectral lines of hydrogen.

Intrinsic Angular Momentum—Spin

Every particle has an intrinsic spin quantum number, s. Particles with half-integer spin are called fermions, and particles with integer spin are called bosons. For electrons, $s = \frac{1}{2}$.

According to the Pauli exclusion principle, no two fermions within the same atom can have the same set of quantum numbers simultaneously.

APPLICATIONS

flash memory, lasers, magnetic resonance imaging, quantum computing, quantum cryptography, scanning tunnelling microscope, semiconductor transistor, tunnelling diode, USB drives

KEY TERMS

Bohr magneton, bosons, boundary value problem, Bragg's law, complementarity principle, correspondence principle, de Broglie wave, de Broglie wavelength, eigen wave function, eigenfunction, eigenvalue, energy-level diagram, even parity, even parity wave functions, excited states, fermions, fine structure, ground-state energy, Heisenberg's uncertainty principle, initial value problem, magnetic quantum number, matter wave, normalization condition, normalization constant, normalized wave function, observables, odd parity, orbital quantum number, orbital, Pauli exclusion principle, principal quantum number, quantum mechanics, quantum number, shell, spin, spin magnetic quantum number, spin quantum number, step-function potential, subshells, time-dependent Schrödinger wave equation, time-independent Schrödinger equation, total angular momentum quantum number, tunnelling, wave group, wave packet, zero-point energy

QUESTIONS

1. Can you ever say that an electron is at rest? Explain your reasoning.
2. Evaluate the following statement: "There is a non-zero probability that a tennis ball striking a racquet will pass through the racquet without breaking the strings."
3. Could an electron in your body have a non-zero probability of being on the Moon?
4. Say an electron and a proton have the same kinetic energy. Which has the shortest de Broglie wavelength?
5. What is the de Broglie wavelength of a particle at rest?
6. If an electron and a proton have the same total energy (including the rest mass energy), which has the shortest de Broglie wavelength?

7. A high-intensity beam of photons is incident on a gas of hydrogen atoms. The atoms are in their ground states. Would you expect to see any electrons dislodged from the gas when the energy of the photons is
   (a) 10 eV?
   (b) 20 eV?
   (c) 30 eV?
   Explain your reasoning.
8. What is the maximum number of electrons that can be contained in an $l = 4$ shell?
   (a) 3
   (b) 6
   (c) 9
   (d) 18
   (e) 36

9. Rank the following transitions in a hydrogen atom in order of increasing energy:
   (a) $n = 4 \rightarrow n = 2$
   (b) $n = 6 \rightarrow n = 3$
   (c) $n = 3 \rightarrow n = 1$
   (d) $n = 5 \rightarrow n = 3$

10. An electron in an atom is in the $n = 3$ state. Which of the following values of l are *not* allowed for this electron?
    (a) 0
    (b) 1
    (c) 2
    (d) 3
    (e) 4

11. Two angular momenta with quantum numbers $j_1 = 2$ and $j_2 = 2$ are added. Which of the following quantum numbers for the total angular momentum is *not* possible?
    (a) 0
    (b) 1
    (c) 2
    (d) 3
    (e) 4

PROBLEMS BY SECTION

For problems, star ratings will be used (★, ★★, or ★★★), with more stars meaning more challenging problems. The following codes will indicate if $\frac{d}{dx}$ differentiation, $\int$ integration, ▢ numerical approximation, or ◡ graphical analysis will be required to solve the problem.

Section 32-1 Matter Waves and de Broglie's Hypothesis

12. ★ Using the relativistic energy momentum relationship $E^2 = p^2c^2 + m_0^2c^4$, where m_0 is the rest mass, p is the momentum, and c is the speed of light, find the de Broglie wavelength of
    (a) a 10 keV electron
    (b) a 10 keV proton
    (c) a 500 MeV electron
    (d) a 500 MeV proton

13. ★★ Calculate the de Broglie wavelength of an electron with a speed of
    (a) $0.9c$
    (b) $0.99c$

14. ★ What is the de Broglie wavelength of a 145 g baseball thrown at a speed of 150 km/h?

15. ★★ Calculate the kinetic energy of an electron that has a de Broglie wavelength of 600 nm. Through what electric potential would the electron have to be accelerated to acquire this kinetic energy?

16. ★★ What is the speed of an electron with a de Broglie wavelength of
    (a) 10^{-1} m?
    (b) 10^{-5} m?
    (c) 10^{-10} m?

Section 32-2 Heisenberg's Uncertainty Principle

17. ★ What is the uncertainty in the momentum of an electron that is confined within a cube of length 1.0×10^{-10} m? What is the minimum kinetic energy of the electron?

18. ★★ The radius of a carbon nucleus is approximately 3×10^{-15} m. What is the uncertainty in the momentum of a proton that is confined within the nucleus? What is the minimum kinetic energy of a proton in this nucleus?

19. ★ What is the minimum possible uncertainty in the position of a 1 g object with a speed of 100 ± 0.1 m/s?

20. ★★ Suppose that the electrons in an atom are confined within the nucleus.
    (a) What is the minimum kinetic energy of an electron that is confined within a silver nucleus with a radius of 5.7×10^{-15} m?
    (b) Explain why the minimum energy indicates that electrons are not actually confined within the nuclei of atoms.

21. ★ What is the uncertainty in the energy of a particle with a half-life of
    (a) 10^{-12} s?
    (b) 10^{-23} s?

22. ★ An electron has a kinetic energy of 10 eV. If its momentum is uncertain by 0.1%, what is the minimum uncertainty in its position?

23. ★ An electron has a speed of 10^5 m/s. What is the minimum time needed to measure its energy to an accuracy of $\pm 0.001\%$?

24. ★★ A charged pion is an unstable particle with a rest mass energy of 139.5 MeV and a half-life of 2.6×10^{-8} s. The maximum time available to measure the energy of a charged pion is equal to its half-life. Find the uncertainty in its rest mass energy. The uncertainty in rest mass energy of unstable particles is called the "decay width" of the particle.

Section 32-3 The Schrödinger Equation

25. ★ What are the units of the one-dimensional Schrödinger wave function $\psi(x)$?

26. ★ What are the units of the three-dimensional Schrödinger wave function $\psi(x, y, z)$?

Section 32-5 A Particle in a One-Dimensional Box

27. ★★ Calculate the lowest three energies for an electron that is confined within a one-dimensional box of length 10^{-10} m (the size of an atom). Sketch the probability density of the electron in each state.

28. ★★ Consider a baseball with a mass of 145 g that is confined within a one-dimensional infinite square well potential of width 1.0 m. How large would Planck's constant have to be if the minimum possible energy of the baseball is 10.0 J?

29. ★★ Calculate the energies of the photons emitted when an electron in an infinite square well potential of width 50 nm makes the following transitions:
    (a) $n = 2$ to $n = 1$
    (b) $n = 3$ to $n = 1$
    (c) $n = 3$ to $n = 2$
    (d) $n = 1000$ to $n = 1$

30. ★★ A charged particle of unknown rest mass is confined within a one-dimensional square well potential.
    (a) When the particle drops from the $n = 2$ excited state to the ground state, it emits a photon of 0.12 eV.
    (b) The width of the confining potential is 0.1 nm. What is the rest mass of the particle?
    (c) Suppose the rest mass of the particle is 105 MeV/c^2. Determine the width of the potential.

31. ★★★ Consider a particle in the ground state of an infinite square well potential of width L. What is the probability of finding the particle between $x = 0$ and $x = L/2$?

32. ★ A dust particle of mass 10^{-8} kg is confined within a box of width 10^{-3} m and has a speed of 10^{-3} m/s. This particle is considered to be a quantum-mechanical system. What excited state is it in?

33. ★ An electron is confined within a box of length 0.1 nm. Determine the energy and the wavelength of the emitted photon when the electron makes a transition from an $n = 2$ state to an $n = 1$ state.

34. ★ Let $a = x + iy$ be a complex number, where x and y are real numbers. The complex conjugate of a is $a^* = x - iy$. Show that the product aa^* is always positive.

35. ★★ The sine, cosine, and exponential functions can be written as a sum of powers of their arguments. Such a sum is called a series expansion. Series expansions of these three functions are as follows (in this problem, x is in radians for the sine and cosine functions):

$$\sin x = x - \frac{x^3}{3!} + \frac{x^5}{5!} - \frac{x^7}{7!} + \frac{x^9}{9!} - \cdots$$

$$\cos x = 1 - \frac{x^2}{2!} + \frac{x^4}{4!} - \frac{x^6}{6!} + \frac{x^8}{8!} - \cdots$$

$$e^x = 1 + x - \frac{x^2}{2!} + \frac{x^3}{3!} + \frac{x^4}{4!} + \frac{x^5}{5!} + \frac{x^6}{6!} + \cdots$$

Although each series has an infinite number of terms, for small arguments of the function only a few terms are sufficient to calculate the function to a high enough accuracy.

(a) For $x = 0.1, 1.0, 10.0$, calculate $\sin x$, $\cos x$, and e^x using a calculator. Now calculate the functions using the series expansions. How many terms of the series did you have to include in each case to obtain an answer that agrees with the calculator values to three decimal places?

(b) Use a series expansion of $e^{i\theta}$ to show that $e^{i\theta} = \cos\theta + i\sin\theta$.

36. ★★★ Let $\psi(x, t)$ be the wave function of the particle that satisfies

$$\left(i\hbar\frac{\partial}{\partial t}\right)\psi(x, t) = E\psi(x, t) \text{ and } \left(-i\hbar\frac{\partial}{\partial x}\right)\psi(x, t) = p\psi(x, t)$$

Derive an equation for the wave function that satisfies the equation relating the relativistic total energy and the momentum of a particle:

$$p^2c^2 + m_0^2c^4 = E^2$$

where m_0 is the rest mass of the particle.

37. ★★ The time-independent wave function of a particle that is confined to move along the x-axis is given by

$$\psi(x) = \begin{cases} A(4 - x^2) & \text{for } -2 \leq x \leq 2 \\ 0 & \text{everywhere else} \end{cases}$$

where A is the normalization constant. Determine A and write the normalized wave function.

38. ★★ For the normalized wave function in problem 37, what is the probability of finding the particle

(a) in the region $0 \leq x \leq 1$?

(b) in the region $1 \leq x \leq 2$?

39. ★ In problem 37, if the mass of the particle is m and its momentum is p, write the time-dependent normalized wave function of the particle.

40. ★ Graph the wave functions and the probability densities of a particle in an infinite square well potential with $L = 1$ nm for states with $n = 4, 5$, and 6.

41. ★★ A particle is confined within a one-dimensional infinite square well potential of width L. What is the probability of finding the particle in the region $\Delta x = 0.1L$ centred around $x = 0.5L$?

42. ★★ Assume that a proton in a carbon nucleus can be modelled as a particle confined within a one-dimensional infinite square well potential. The radius, R, of the carbon nucleus is approximately 2.7×10^{-15} m.

(a) The width of the potential is $2R$. Estimate the ground-state energy of the proton in MeV.

(b) Calculate the wavelength of the photon emitted for $n = 2$ to $n = 1$ and $n = 3$ to $n = 2$ transitions.

43. ★★ The ground-state energy of an electron in an infinite square well potential is 20 eV.

(a) What are the energies of the first three excited states of the electron?

(b) What is the width of the potential well?

(c) What is the longest-wavelength photon that an electron in this potential well can emit?

44. ★★ Let E_n be the energy of a particle in state n of an infinite square well potential. Show that in the limit of large n, the fractional difference in the energies of two adjacent states can be approximated as

$$\frac{E_{n+1} - E_n}{E_n} \approx \frac{2}{n}$$

How is this result related to Bohr's correspondence principle?

Section 32-6 The Finite Square Well Potential

45. ★ Determine whether the following quantities change sign under the parity operation. Here $\vec{r}$, $\vec{p}$, and $\vec{L}$ are the position, the momentum, and the angular momentum vectors, respectively.

(a) $\vec{r} \cdot \vec{p}$

(b) $\vec{r} \times \vec{L}$

(c) $\vec{p} \times \vec{L}$

(d) $\vec{L} \times \vec{L}$

(e) $\vec{r} \cdot \vec{L}$

46. ★★ Consider the wave function described by Equation 32-71. By imposing the boundary conditions at $x = a$ on this wave function, show that Equation 32-72 is satisfied.

47. ★★★ Using the parameters for Figure 32-19 ($V_0 = 10$ eV, $L = 1$ nm), plot the left and right sides of Equation 32-71 as a function of energy, E, on the same graph (use a graphing calculator or graphing software). Take the range of E between 0 eV and 5 eV. By identifying the points where the two curves intersect, show that the three lowest eigenvalues of energy are approximately at $E = 0.3$ eV, 1.3 eV, and 3.0 eV.

Section 32-7 Quantum Tunnelling

48. ★★ An electron with energy 2 eV is incident on a barrier of height 3 eV and width 0.1 nm. What is the probability that the electron will tunnel through the barrier?

49. ★★ An electron beam approaches a potential barrier of height 2 eV and width 1 nm. What must be the energy of the beam if 10% of the electrons are to tunnel through the barrier?

50. ★★ A beam of protons with energy 10.0 eV approaches a potential barrier of height 20.0 eV and width 0.5 nm. Approximately 10^{20} protons per second hit the barrier. How long will it take for 1000 protons to pass through the barrier?

51. ★★ Derive Equation 32-85.

Section 32-9   The Schrödinger Equation for a Hydrogen Atom

52. ★ Calculate the energy of the $n = 5$ state of a hydrogen atom.

53. ★★ Calculate the energies and wavelengths of photons emitted for electron transitions from $n = 5$, 6, and 7 states to $n = 3$ states.

54. ★★ For an electron in an $n = 4$ state, what are the possible values of the orbital angular momentum quantum number? For each value of l, calculate the magnitude of the orbital angular momentum and the possible values of its projection along the z-axis.

55. ★ How many subshells are possible for the $n = 3$ state of an atom?

56. ★ What are the lowest possible values of n and l for
    (a) $m_l = -3$?
    (b) $m_l = 0$?
    (c) $m_l = +4$?

57. ★ An electron in a hydrogen atom is in the $l = 4$ state.
    (a) What are the allowed values of n and m_l for this electron?
    (b) Rank these values in order of the binding energy of the electron, from lowest to highest.

58. ★★★ Use Equation 32-102 to show that the most probable distance for an electron in the ground state of the hydrogen atom is equal to the Bohr radius.

59. ★★★ Determine the probability of finding an electron in the ground state of the hydrogen atom for $r \leq 4r_0$.

60. ★★★ A muon is an elementary particle with a rest mass energy of 105.7 MeV/c^2 and charge $-e$ (the same as the charge for an electron). A muonic hydrogen atom is formed by replacing an electron in a hydrogen atom with a muon.
    (a) Calculate the energy levels of the muonic hydrogen atom.
    (b) What is the energy of the emitted photon when a muon transitions from an $n = 2$ state to an $n = 1$ state of a muonic hydrogen atom?
    (c) What is the Bohr radius of a muon in the $n = 1$ state?

Section 32-10   Intrinsic Angular Momentum—Spin

61. ★★ What are the possible angles between the spin angular momentum, $\vec{S}$, of an electron and the z-axis?

62. ★★★ Assume that an electron is a sphere of radius 10^{-18} m. Using non-relativistic kinematics, find the speed at its surface when its rotational angular momentum is $\dfrac{\sqrt{3}}{2}\hbar$.

63. ★★ Two electrons in an atom have angular momentum quantum numbers $j_1 = \dfrac{1}{2}$ and $j_2 = \dfrac{3}{2}$.
    (a) Calculate all possible values of the quantum number, j, of the total angular momentum of the two electrons.
    (b) Calculate the magnitude and the z-component of the total quantum number for each value of j.

64. ★★ Repeat problem 63 for $j_1 = \dfrac{3}{2}$ and $j_2 = \dfrac{3}{2}$.

65. ★★ The total angular momentum quantum number for two electrons is 4. The angular momentum quantum number for one of the electrons is $\dfrac{5}{2}$. What are the possible values of the angular momentum quantum number of the second electron?

66. ★★ Show that the maximum number of electrons allowed in a hydrogen shell with principal quantum number n is $2n^2$.

Introduction to Solid-State Physics

When you have completed this chapter, you should be able to

LO1 Draw simple crystal structures.

LO2 Discuss the free electrons in a box model.

LO3 Sketch the electron bands in a periodic potential.

LO4 Describe the difference between a conductor and an insulator in terms of energy bands.

LO5 Discuss the basic characteristics of a pure semiconductor.

LO6 Qualitatively describe how doping a semiconductor changes its conduction properties.

LO7 Sketch the current-voltage characteristic of a pn junction diode.

LO8 Name some semiconductor devices.

LO9 Describe some of the tools used in nanotechnology.

In a modern circuit board (Figure 33-1), we can see examples of three different types of electrical behaviour exhibited by solids. The fine lines connecting the components are metallic conductors chosen for their ability to conduct electricity efficiently. The board itself is made of a material that conducts electricity very poorly, an electrical insulator. Some of the components on the board are semiconductor devices, which have conductivities between those of a conductor and those of an insulator. How is it that these solid materials, which have many similar properties, such as density and hardness, can have resistivities ranging over 32 orders of magnitude? In this chapter, we will use a quantum-mechanical approach to understand the behaviour of electrons in solids. For simplicity, we will restrict ourselves to considering crystalline solids.

Figure 33-1 A circuit board.

33-1 Crystal Structures

In this chapter, we will restrict our discussion to materials that are in crystalline form. In a crystal, the atoms are arranged in periodic arrays that repeat themselves over and over in all three dimensions. Simple examples are the **cubic**, **hexagonal**, and **tetragonal structures** shown in Figure 33-2. The **cubic face-centred structure** has atoms in the middle of each face of the cube as well as at each corner of the cube, and the **cubic body-centred structure** has just one extra atom, which is at the centre of the cube. Note that a full crystal can be created by translating each structure through an appropriate direction and distance. Non-crystalline solids are called amorphous solids.

Figure 33-3 shows a bulk crystal that has a cubic structure. We can see that the bulk crystal has the same symmetry as the individual cells of the structure. Table 33-1 lists a few materials with their associated crystal structures and typical interatomic spacings, or lattice parameters, of the crystals. For all cubic structures, the **lattice parameter**, a, is simply the length of one side of the cube. For a tetragonal lattice, a is the length of the sides of the square base, and c is the length of the other edges.

Table 33-2 lists some ranges of typical values for crystalline materials.

Table 33-1   Some Basic Structural Properties of a Few Elements

Material	Crystal Lattice	Lattice Parameter (nm)
Copper	Cubic: face-centred	0.361
Tin	Tetragonal	$a = 0.37, c = 0.337$
Silicon	Cubic: face-centred	0.543

Table 33-2   Typical Values for Crystalline Materials

Parameter	Typical Value
Atomic spacing	On the order of 10^{-9} m to 10^{-10} m
Mass density	On the order of 1 g/cm^3 to 10 g/cm^3
Number of atoms/cm^3	On the order of 10^{21} to 10^{24}

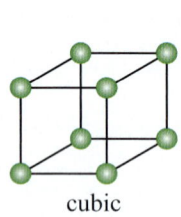 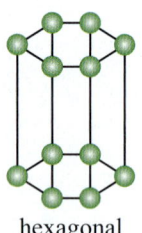 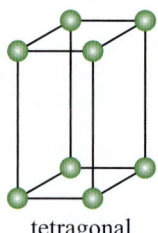 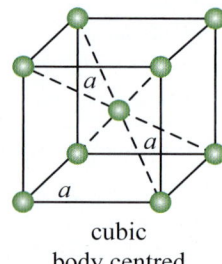 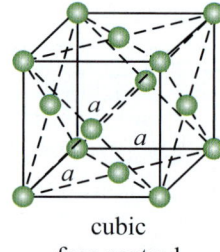

cubic   hexagonal   tetragonal   cubic body centred   cubic face centred

Figure 33-2   Some simple crystal lattices.

EXAMPLE 33-1

The Lattice Parameter of Sodium

Sodium has a cubic body-centred structure at 300 K. Determine the lattice parameter.

Solution

Each cubic cell of sodium has an atom at each of its eight corners, and one-eighth of each of those atoms is inside the cube. In addition, one atom sits at the centre of the cubic body-centred structure. There are thus two atoms per cube in such a structure. To find the lattice parameter, we determine the volume per atom, and then find the side length of a cube with the volume of two atoms.

First, we look up the mass density and the atomic mass of sodium: 971 kg/m^3 and 0.0230 kg/mol, respectively. Dividing the atomic mass by the density gives the volume per mole, and dividing this quantity by Avogadro's number gives the volume per atom:

$$V = \frac{0.0230 \text{ kg/mol}}{971 \text{ kg/m}^{-3}} \frac{1}{6.02 \times 10^{23} \text{ atom/mol}}$$

$$= 3.93 \times 10^{-29} \text{ m}^3/\text{atom}$$

There are two atoms in the unit cube, so the volume of the unit cube is 7.87×10^{-29} m^3. Since the volume of the unit cube is $V = a^3$, the lattice parameter is

$$a = \sqrt[3]{V} = \sqrt[3]{7.87 \times 10^{-29} \text{ m}^3} = 4.29 \times 10^{-10} \text{ m}$$

Making sense of the result

This seems reasonable. As we saw in Chapter 32, the Bohr radius, which is a typical dimension for a hydrogen atom, is about 1/10 the value we have obtained here. However, we would expect a hydrogen atom to be much smaller than a sodium atom, because hydrogen only has one electron, whereas sodium has 11. As well, this value is a measure of a bond length, which we might expect to be larger than the orbital radius to avoid significant overlap of the electron wave functions.

MODERN PHYSICS

Figure 33-3 While this is not always the case, the underlying cubic crystal structure is clear in these crystals.

When a solid is formed, the atoms that compose the solid make chemical bonds with their neighbours. Some of the electrons associated with each atom participate in these chemical bonds and are called **valence electrons**. Valence electrons remain in the vicinity of the atoms with which they are associated. However, some of the outer electrons do *not* participate in the bonding process and move throughout the crystal.

33-2 Electrons in a Box

A simple model for the distributed electrons in a material is to treat them as being confined within the material but free to move anywhere inside it. Since electrons are negatively charged, they repel each other. For this reason, the free electrons are relatively uniformly distributed throughout the material, particularly in metals. When an electron is liberated from an atom, the atom has a corresponding net positive charge. So, one might expect that the free electrons would be strongly attracted to the positively charged atoms. However, the distributed negative charge of the free electrons offsets this attraction. At a distance of only a few times the lattice parameter, a unit cell of the crystal appears to have virtually no net charge, and the net force on an electron from almost all the other electrons and atoms is very small. This effect is called **screening**.

For simplicity, we will first consider free electrons in one dimension. Assume that the electrons are confined in a length $L = N^*a$, where N^* is the number of atoms in the length and a is the interatomic separation.

The problem of free particles confined to a box of length L was solved in Chapter 32, Section 32-5. There we found that the allowed energies are given by

$$E_n = \frac{\hbar^2 k_n^2}{2m} \qquad (32\text{-}49)$$

where

$$k_n = \frac{n\pi}{L} = \frac{n\pi}{N^*a} \qquad (32\text{-}48)$$

The integer n is called the quantum number for the state.

So, the energies of the free electrons are quantized. However, the change in k_n and E_n from one value of n to the next is very small, because for a macroscopic sample, N^* is of the order of at least 10^{23} and a is of the order of 10^{-10} m:

$$\Delta k = k_{n+1} - k_n = \frac{\pi}{N^*a} \sim 10^{-13}/\text{m} \qquad (33\text{-}1)$$

$$E_{n+1} - E_n = \frac{\hbar^2 k_{n+1}^2}{2m} - \frac{\hbar^2 k_n^2}{2m}$$

$$= \frac{\hbar^2 \pi^2}{2mN^{*2}a^2}(2n + 1) \sim (2n + 1) \times 10^{-63}\text{J} \qquad (33\text{-}2)$$

Thus, the energy eigenvalues are extremely closely spaced, and to a good approximation they may be treated as continuously distributed. A sketch of E_n versus k_n is shown in Figure 33-4. Figure 33-5 shows the wave functions for the lowest three energy levels for this system. Notice the similarity between these wave functions and the standing waves that you studied in Chapter 15.

For the moment, we will only consider the situation that corresponds to a temperature of 0 K. At absolute

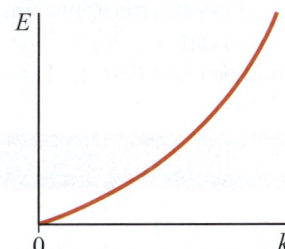

Figure 33-4 Energy versus wave number for free electrons in one dimension.

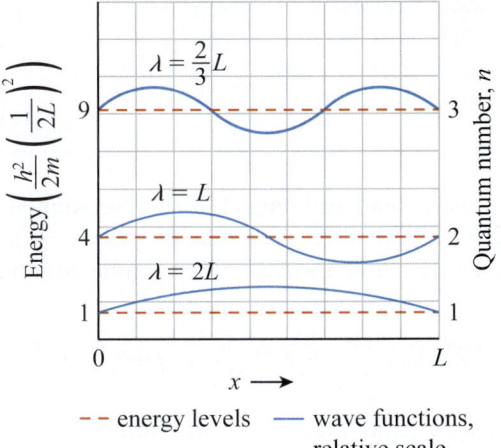

--- energy levels — wave functions, relative scale

Figure 33-5 Wave functions for the three lowest-energy states for free electrons in a box of length L.

zero, the electrons occupy the energy states that give the lowest total energy possible. Since electrons are fermions and therefore obey the Pauli exclusion principle, only one electron can occupy each quantum state. Thus, two electrons can be in each energy level, one with spin up and the other with spin down. From Equation 32-48 in Chapter 32 we can see that when $n = N^*$, $k = \dfrac{\pi}{a}$. Thus, there are N^* energy states between $k = 0$ and $k = \dfrac{\pi}{a}$. If each atom contributes a single free electron, then all the states up to $n = \dfrac{N^*}{2}$ will be occupied. In general, if there are N electrons, then the highest occupied state will have quantum number $n_f = \dfrac{N}{2}$.

We call the highest occupied energy level in a solid the **Fermi level**. Associated with this energy level are the Fermi wave number, k_f, and the **Fermi energy**, E_f:

KEY EQUATION

$$k_f = \frac{\pi N}{a N^*} \qquad E_f = \frac{\hbar^2 k_f^2}{2m} = \frac{\hbar^2 \pi^2}{8 m a^2} \left(\frac{N}{N^*} \right)^2 \quad (33\text{-}3)$$

Note that the ratio $\dfrac{N}{N^*}$ is the number of electrons that each atom contributes to conduction. For a typical solid with $\dfrac{N}{N^*} = 1$ and $a \sim 10^{-10}$ m, $E_f \sim 10^{-18}$ J ~ 10 eV. Since the energy associated with thermal excitation of an electron at room temperature is approximately 0.025 eV, the Fermi energy is relatively high. The speed of electrons at the Fermi level is called the **Fermi speed**.

CHECKPOINT 33-1

Fermi Energy

Why is the Fermi energy large?
(a) The electrons have high energies in the atoms.
(b) Only two electrons can go into any state, so many states are needed.
(c) There are a lot of electrons.
(d) The temperature is large.

ANSWER (b) and (c)

If we extend our model to a three-dimensional crystal of length L on each side, the wave vectors also become three-dimensional, with the same quantization along each axis.

KEY EQUATION $\vec{k} = \dfrac{\pi}{N_x^* a} n_x \hat{\boldsymbol{i}} + \dfrac{\pi}{N_y^* a} n_y \hat{\boldsymbol{j}} + \dfrac{\pi}{N_z^* a} n_z \hat{\boldsymbol{k}}$ (33-4)

where n_x, n_y, and n_z are all positive integers.

The total number of electrons, N, in the system is equal to the volume of occupied k-space times the number of states per unit volume in k-space.

EXAMPLE 33-2

The Fermi Speed

How fast would an electron at the Fermi level be moving in a typical solid?

Solution

If we assume the electron only has kinetic energy and can be treated non-relativistically, then

$$v_f = \sqrt{\frac{2 E_f}{m}} = \sqrt{\frac{2 \times 10^{-18} \text{ J}}{9.1 \times 10^{-31} \text{ kg}}} = 1.5 \times 10^6 \text{ m/s}$$

Making sense of the result

These free electrons are moving at approximately 0.5% of the speed of light.

The occupied volume is the first octant of a sphere of radius k_f, $\dfrac{1}{8} \dfrac{4}{3} \pi k_f^3$. From Equation 33-2 and Figure 33-6 we can see that, for a cubic crystal, the sides of the cube in k-space associated with each point are of "length" $\dfrac{\pi}{N_x^* a}$, $\dfrac{\pi}{N_y^* a}$, $\dfrac{\pi}{N_z^* a}$, and thus the total volume associated with each point is $\dfrac{\pi}{N_x^* a} \dfrac{\pi}{N_y^* a} \dfrac{\pi}{N_z^* a} = \dfrac{\pi^3}{N^* a^3}$, where $N^* = N_x^* N_y^* N_z^*$.

For a cubic crystal, $N^* a^3 = V$, where V is the total volume of the crystal. Finally, we must remember that each state can hold two electrons, one spin-up and the other spin-down. Therefore,

$$N = 2 \frac{1}{8} \frac{4}{3} \pi k_f^3 \frac{V}{\pi^3} = \frac{1}{3} \frac{k_f^3 V}{\pi^2}$$

Thus,

KEY EQUATION $k_f = \left(\dfrac{3 \pi^2 N}{V} \right)^{\frac{1}{3}}$ and $E_f = \dfrac{\hbar^2}{2m} \left(\dfrac{3 \pi^2 N}{V} \right)^{\frac{2}{3}}$ (33-5)

Note that in Equation 33-5 the Fermi energy is completely determined by $\dfrac{N}{V}$, the number of electrons divided by the volume of the sample. This quantity is the **conduction electron density**.

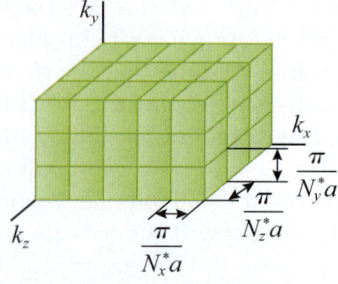

Figure 33-6 Volume associated with points in k-space.

MODERN PHYSICS

Table 33-3 shows the Fermi energies, wave vectors, and speeds for some common metals.

Table 33-3 Fermi Energies, Wave Vectors, and Speeds for Some Common Metals

Material	Fermi Energy, E_f (eV = 1.602 $\times 10^{-19}$ J)	Fermi Wave Vector, k_f (10^{10} m^{-1})	Fermi Speed, v_f (10^6 m·s^{-1})
Aluminum	11.63	1.75	2.02
Copper	7.00	1.36	1.57
Gold	5.51	1.20	1.39
Sodium	3.23	0.92	1.07

EXAMPLE 33-3

The Fermi Energy of Aluminum

Aluminum has a density of 2.69 g/cm^3 and a molar mass of 26.98 g/mol. Each aluminum atom contributes three conduction electrons. Find the Fermi energy.

Solution

We can determine the conduction electron density, $\dfrac{N}{V}$, and then use Equation 33-5 to find the Fermi energy. The conduction electron density equals the number of conduction electrons per atom times the number of atoms per m^3:

$$\frac{N^*}{V} = \frac{3 \text{ electrons}}{\text{atom}} \times \frac{2.69 \dfrac{\text{g}}{\text{cm}^3}}{26.98 \dfrac{\text{g}}{\text{mol}}} \times 6.022 \times 10^{23} \frac{\text{atoms}}{\text{mol}}$$

$$= 1.80 \times 10^{23} \frac{\text{electrons}}{\text{cm}^3}$$

$$= 1.80 \times 10^{29} \text{ electrons/m}^3$$

Using this value in Equation 33-5 gives

$$E_f = \frac{\hbar^2}{2m}\left(\frac{3\pi^2 N}{V}\right)^{\frac{2}{3}} = 1.85 \times 10^{-18} \text{ J} = 11.5 \text{ eV}$$

We can rearrange Equation 33-5 to show the relationship between the conduction electron density, n, and the Fermi energy more explicitly:

$$N^* = \frac{1}{3}\frac{V}{\pi^2}\left(\frac{\sqrt{2mE_f}}{\hbar}\right)^3$$

or

KEY EQUATION
$$n = \frac{N^*}{V} = \frac{1}{3\pi^2}\left(\frac{\sqrt{2mE_f}}{\hbar}\right)^3 \qquad (33\text{-}6)$$

At 0 K, the filled levels occupy the first octant of a sphere in "k-space." All states below the Fermi energy are occupied with a probability of one, and all states above the Fermi energy are occupied with a probability of zero—they are not occupied.

As the temperature increases from 0 K, we would expect the electrons to become thermally excited and their energies to increase. However, electrons are fermions, and they only go into a higher-energy state if it is unoccupied. As you saw in Chapter 16, the thermal energy is given by $k_B T$, where k_B is Boltzmann's constant and T is the temperature in kelvin. Only those electrons that are within $k_B T$ of E_f will be able to make a transition to a higher energy level. Because this thermal energy is so small compared to the Fermi energy, only a very small fraction of the electrons can make the transition. Thus, the probability of a state being occupied just below the Fermi energy is slightly less than one because some electrons will have been excited out of those states, and the probability of occupation for states slightly above E_f will be slightly greater than zero because some electrons will have been excited into these states.

Figure 33-7 shows how temperature affects $f(E)$, the probability that a state with energy E is occupied. We can see that only those electrons near the Fermi

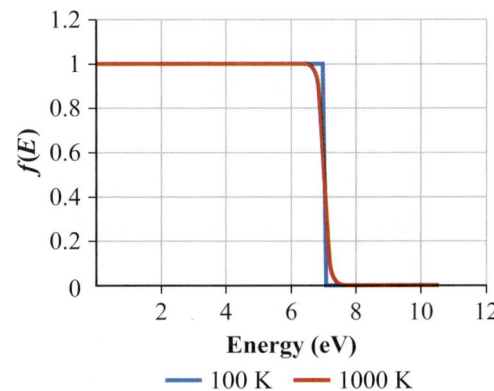

Figure 33-7 The probability of a state being occupied by an electron, $f(E)$. The blue curve shows the probability at $T = 100$ K, and the red curve shows the probability at $T = 1000$ K, with a Fermi energy of 7 eV, typical of that found in metals.

energy are affected by the increase in temperature. For example, $f(E)$ drops to approximately 0.9 near 6.8 eV, just 0.2 eV below E_f for $T = 1000$ K. Even when the temperature increases substantially, only a small fraction of the electrons absorb any of the thermal energy.

What will happen if we apply an electric field to the crystal? In the absence of the field, an equal number of electrons travel in any direction, and the net flow of electrons is zero. If we apply an electric field directed from right to left, some of the electrons travel from the left into unoccupied right-travelling states. However, only electrons near the Fermi level can make these transitions because these electrons are the only ones in energy levels close to those of the unoccupied states.

What fraction of the total number of electrons in the sample are available to participate in conduction? Rather than do a detailed calculation, we will make an estimate. Recall that in k-space the occupied energy states fall into the first octant of a sphere of radius k_f. In addition, the states are uniformly distributed in k-space with spacing given by Equation 33-1. Thus, to estimate the fraction of electrons that can respond to an electric field, we will take the ratio of the volume occupied near the Fermi surface to the total volume of occupied states.

We will work at room temperature, which, as we just discussed, causes the electrons in a narrow region near k_f to be thermally excited and able to participate in conduction. In terms of energy, the width of that region is approximately $\Delta E = k_B T \sim 4 \times 10^{-21}$ J. For our purposes, we need to know how big a change in wave vector k this corresponds to. We can use Equation 32-49 to write

$$\Delta E = \frac{dE}{dk}\Delta k = \frac{\hbar^2 k_f}{m}\Delta k \qquad (33\text{-}7)$$

which we solve for Δk:

$$\Delta k = \frac{m\Delta E}{\hbar^2 k_f} \sim 2 \times 10^7 \text{ m}^{-1} \qquad (33\text{-}8)$$

We have used the Fermi wave vector for aluminum, $k_f = 1.75 \times 10^{10}$ m^{-1}.

The volume of states near k_f is then given by the area of the surface at k_f times this width:

$$V_{\text{conduction}} = \frac{1}{8} 4\pi k_f^2 \Delta k \qquad (33\text{-}9)$$

The total volume of occupied states is simply

$$V_{\text{total}} = \frac{1}{8}\frac{4}{3}\pi k_f^3 \qquad (33\text{-}10)$$

where in both cases the factor of $\frac{1}{8}$ occurs because we are dealing with the first octant of the sphere. The

fraction of electrons that can participate in conduction is then

$$\frac{V_{\text{conduction}}}{V_{\text{total}}} = \frac{\frac{1}{8}4\pi k_f^2}{\frac{1}{8}\frac{4}{3}\pi k_f^3}\Delta k = \frac{3}{k_f}\Delta k \approx 0.3\% \quad (33\text{-}11)$$

where we have again used $k_f = 1.75 \times 10^{10}$ m^{-1}.

In our model, there is nothing to stop the motion of conduction electrons within the crystal. Thus, our model represents an ideal **conductor** with no electrical resistance. In actual crystals, the conduction electrons interact with atoms in the lattice in two ways that impede the flow of current. First, the conduction electrons scatter off defects in the crystal, such as gaps and impurities in the lattice. Second, the conduction electrons scatter off lattice vibrations called **phonons**. Due to thermal energy, the atoms in the crystal vibrate about their equilibrium positions. The greater the temperature, the more phonons there are for the electrons to interact with, and the greater the resistance of the crystal. In addition, electrons that scatter off the phonons transfer energy to the lattice and cause the temperature of the material to increase. These effects explain why electric current passing through a resistor produces heat.

The free-electron model allows us to understand simple metals but does not explain why some materials are good conductors while others are not. To explain these properties, our model has to include the electron–lattice interactions.

The key feature of the free electrons in a box model is that because electrons are fermions, only those electrons with energies near the Fermi energy can participate in conduction, and this is a small fraction of the total number of electrons in the material.

33-3 Periodic Potential

In a crystal, the electrons experience a potential due to interactions with the lattice. Because the lattice is composed of repeating units, the potential has a pattern with the same spacing, or periodicity, as the crystal; this potential is called the **periodic potential**.

INTERACTIVE ACTIVITY 33-1

Periodic Potentials

In this activity, you will use the PhET simulation "Periodic Potentials" to gain an understanding of electrons in a periodic potential.

For those wave functions where the wavelength is very different from the periodicity of the potential, the effect of the potential is very small. This is because the electrons will spend on average just as much time in regions where the potential is small as they do in regions where the potential is large. However, something very interesting can occur when the wavelength of the wave function is $\lambda = 2a$. In Figure 33-8, we plot a periodic potential with a period of a, and the probability density for a free electron wave function that has a period of $2a$. (Recall that the probability density is given by $\Psi^*\Psi \propto \sin^2(kx)$ in our case.) There are two ways we can plot the probability density relative to the potential.

In the first, we align the maxima of the probability density with the minima of the potential. This allows the electron to spend much more time at locations of low potential energy than high potential energy. We would thus expect there to be a relatively strong interaction between such an electron and the potential, which would lead to a lower energy than for the free-electron case. This situation is shown with the red probability density in Figure 33-8.

Conversely, we can take the same probability density and simply shift it along to produce the probability density represented by the green line in Figure 33-8. This electron spends more time in areas where the potential is high. We would therefore expect it to interact quite strongly with the potential and have its energy increased.

Thus, for the same value of k, we get two possible values for the energy, one that is higher than the free-electron model, and one that is lower. This is depicted in Figure 33-9.

Well away from $k = \dfrac{\pi}{a}$, the wave functions are essentially the same as the wave functions for free electrons. However, as k approaches $\dfrac{\pi}{a}$, the energy curve bends away from the parabolic shape we found for free electrons (Figure 33-4) and actually becomes horizontal at $\dfrac{\pi}{a}$, leaving a gap in the energy spectrum. Similar gaps

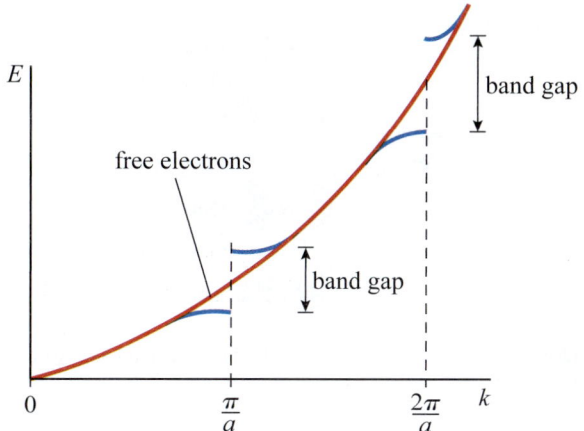

Figure 33-9 *E* versus *k* for electrons in a one-dimensional potential of period *a*.

occur when $k = \dfrac{2\pi}{a}, \dfrac{3\pi}{a}, \dfrac{4\pi}{a}$, and so on. The regions of allowed states are called **energy bands**, and the regions where no solutions are possible are called **band gaps**. The energy bands together with the energy gaps describe the electronic structure of a material and are called the **band structure**. Some typical band gaps for various materials are listed in Table 33-4. Silicon and germanium are **semiconductors**, which are materials whose conductivities lie between insulators and conductors. Silicon and germanium have small band gaps, and diamond, an excellent **insulator**, has a large band gap. In general, as the band gap increases, so does the material's resistivity.

Table 33-4 Band Gaps for Various Materials

Material	Band Gap (eV at 300 K)	Resistivity at 20°C ($\Omega \cdot$ m)
Silicon	1.11	6.40×10^2
Diamond	5.5	10^{14}
Germanium	0.67	0.46

33-4 Metals and Insulators

The band structure of a material indicates the ranges of states for electrons in the material. How these states are filled determines whether a material is a metal or an insulator. We can represent the band structure graphically, as shown in Figure 33-10.

We will first consider a material with a partially filled band, as illustrated in Figure 33-11. The electrons in the completely filled lower bands cannot participate in conduction because there are no empty states with energies close to the energies in the full bands. The electrons at the top of the filled portion of the partially filled band do have empty states adjacent to them. Therefore, if we apply an electric field, these electrons

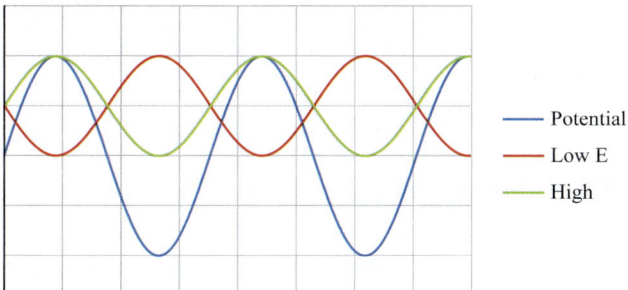

Figure 33-8 Probability densities and potential for free-electron wave functions whose wavelength is twice the periodicity of the potential. The electron represented by the red line spends more time in areas of low potential, whereas the electron represented by the green line spends more time in areas of high potential.

— Potential
— Low E
— High

MODERN PHYSICS

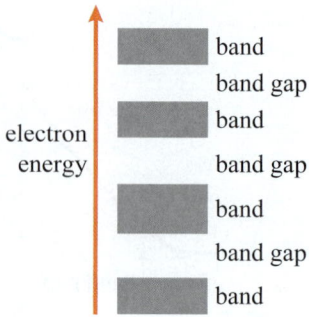

Figure 33-10 Schematic representation of the band structure of a material.

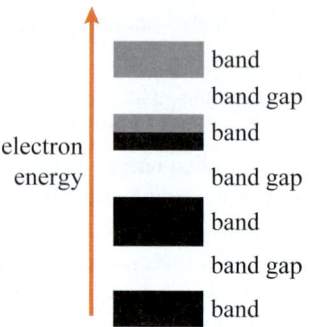

Figure 33-11 A material with a partially filled band.

can move into higher-energy states and contribute to an electric current. These excited electrons still scatter off the lattice defects and vibrations. Therefore, the material has some resistance, but it is a good conductor. An electronic structure with a partially filled band is characteristic of a metal.

We now consider a material in which the uppermost occupied band is completely filled (Figure 33-12). The empty states are not close in energy to any of the occupied states. If we place such a material in an electric field, the electrons in the occupied states cannot move to higher-energy states unless the electric field is strong enough to give an electron enough energy to cross the gap into the next highest band. A material

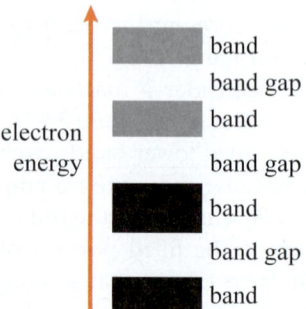

Figure 33-12 In this material, the uppermost occupied band is completely filled with electrons.

that has only completely filled bands and a fairly large energy gap between the top of the highest filled band and the bottom of the nearest empty band is an insulator. However, if the field strength is sufficiently great, even a very good insulator will conduct, in a process called **electrical breakdown**. The maximum field that a material can withstand without breakdown conduction is called the **dielectric strength** of the material. Table 33-5 lists values for some common materials.

Table 33-5 Dielectric Strengths for Various Insulating Materials

Material	Dielectric Strength (MV/m)
Vacuum	0.8
Air	0.8–3.0
Porcelain	1.6–7.9
Paraffin wax	7.9–11.8
Transformer oil	15.7
Bakelite	11.8–21.7
Rubber	17.7–27.6
Shellac	35.4
Paper	49.2
Teflon	59.1
Glass	79–118
Mica	197

CHECKPOINT 33-3

Band Gap in Insulators

The Fermi energy of an insulator lies at the top of a band that is immediately below
(a) a large energy gap
(b) a small energy gap
(c) an energy gap wider than the Fermi energy
(d) an energy gap of any width

ANSWER: (a) In an insulator, the size of the energy gap is large compared to thermal energy, k_BT. There is insufficient thermal energy to excite electrons across the gap, so there are no electrons available for conduction.

33-5 Semiconductors

We now consider what happens if the energy gap between the highest filled band (called the **valence band**) and the first empty band (called the **conduction band**) is not too large, say, less than an order of magnitude greater than the thermal energy. The material has a large number of electrons at the top of the valence band, but they are separated in energy from the empty states in the conduction band by an energy gap, ΔE. Since the electrons can absorb thermal energy from the lattice vibrations, there is a probability that some of the electrons will move to the higher-energy upper band, even though ΔE is greater than the thermal energy, $k_\text{B}T$. Suppose that at 0 K temperature we have N_0 electrons at the top of the filled band. Some of these electrons can be thermally excited into the bottom of the empty band. The number of thermally excited electrons, N_x, can be approximated by an exponential relationship that depends on the ratio of the energy gap to the thermal energy:

KEY EQUATION

$$N_x \approx N_0 e^{-\frac{\Delta E}{k_\text{B}T}} \tag{33-12}$$

EXAMPLE 33-4

The Effect of the Energy Gap on Conduction

Plot $\dfrac{N_x}{N_0}$ versus ΔE for a temperature of 300 K. What is the value of the ratio for $\Delta E = 0.25$ eV and $\Delta E = 1.0$ eV?

Solution

We can rearrange Equation 33-12 to determine $\dfrac{N_x}{N_0}$:

$$\frac{N_x}{N_0} \approx e^{-\frac{\Delta E}{k_\text{B}T}}$$

At 300 K, $k_\text{B}T = 0.0258$ eV. Figure 33-13 shows a plot of $\dfrac{N_x}{N_0}$ with ΔE expressed in eV.

For $\Delta E = 0.25$ eV, $\dfrac{N_x}{N_0} \approx e^{-\frac{0.25 \text{ eV}}{0.0258 \text{ eV}}} \approx 6 \times 10^{-5}$. Once we get out to a band gap of 1 eV, the ratio drops to approximately 10^{-17}.

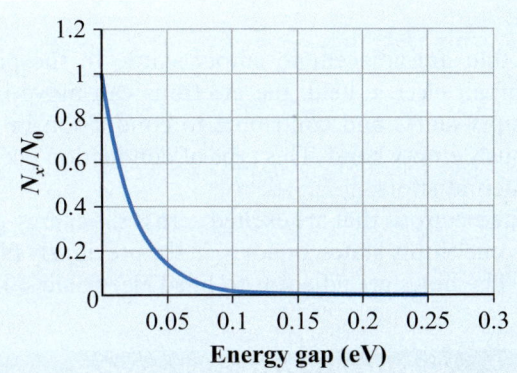

Figure 33-13 The fraction of excited carriers versus band gap. Notice that the ratio drops off very quickly as ΔE increases.

Making sense of the result

The ratio drops off very quickly as ΔE increases due to the exponential dependence.

EXAMPLE 33-5

Temperature Dependence of Conduction

For a band gap of 0.25 eV, plot $\dfrac{N_x}{N_0}$ versus T.

Solution

We use the equation from Example 33-4 to calculate values for the plot (Figure 33-14).

Making sense of the result

The ratio is essentially zero from 0 K to approximately 200 K. The ratio then begins to increase as a very steep function of T.

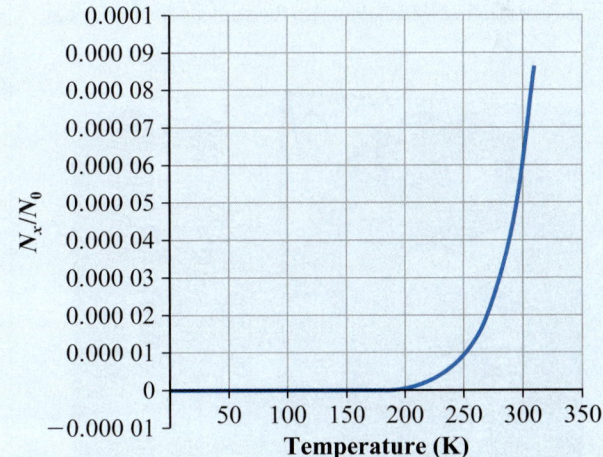

Figure 33-14 The fraction of carriers thermally excited versus temperature.

The two previous examples show that for relatively "small" band gaps, some of the electrons from the top fully filled band can acquire enough thermal energy to cross the band gap up into the empty band above.

In addition, the number of electrons that are excited across the gap is a very strong function of the temperature, increasing sharply as the temperature increases. Electrons that are excited across the gap are in energy states that are adjacent to empty states. In the presence of an electric field, the electrons can move into the empty states and contribute to conduction in the previously empty band. This type of conduction occurs in semiconductors.

The electrons that are excited across the energy gap leave some empty states, or **holes**, in the previously filled band. The holes are adjacent to filled electronic states, and the electrons from the filled states can move into the empty states. So, there is conduction in the previously filled band as well in the previously empty band. The net effect is that the holes appear to move. Because the holes represent the *absence* of an electron, **hole motion** appears like the motion of a positively charged particle; that is, a hole can be treated as a **charge carrier** like a conduction electron.

In semiconductors, only a small fraction of electrons in the highest-energy states are available to participate in conduction. Consequently, the resistivities of semiconductors are higher than the resistivities of metals. The conduction electrons in semiconductors scatter off the lattice defects and vibrations, like the conduction electrons in metals. Thus, one might expect that the resistance of a semiconductor increases with temperature, similar to the increase for metals. However, this increase is overwhelmed by the sharp increase in the number of conduction electrons that are available as the temperature increases (see Figure 33-14). The net result is that the resistance of semiconductors actually decreases as the temperature increases.

MAKING CONNECTIONS

Semiconductor Thermal Runaway

Have you ever wondered why most computers have cooling fans in them (Figure 33-15) and server rooms are always air-conditioned? The answer lies in the negative temperature coefficient of semiconductors. As the temperature increases, the resistance decreases, which leads to an increase in current. The increase in current produces more thermal energy (I^2R), which causes the temperature to increase even further. Without cooling to limit the decrease in resistance, the cycle can cause overheating, which damages the semiconductor devices.

Wade Vaillancourt/Shutterstock.com

Figure 33-15 This cooling fan is mounted directly on the finned aluminum heat sink that is attached to the central processing unit (CPU) of a computer. The heat sink increases the surface area and, hence, the heat dissipation.

MAKING CONNECTIONS

Superconductors

The most dramatic quantum effect involving resistivity is *superconductivity*, discovered in 1911 by Dutch physicist Heike Kamerlingh Onnes (1853–1926). Some materials lose *all* of their electrical resistance when cooled below a critical temperature (Figure 33-16). The critical temperatures for the first superconductors discovered are within a few degrees of absolute zero, for example, 4.2 K for mercury and 7 K for lead. In the 1980s, several classes of so-called high-temperature superconductors were discovered. Around 1993, a ceramic material was developed that superconducts up to 138 K, which is the highest critical temperature for a superconductor found to date. High-temperature superconductors generated so much excitement that a special session devoted to them at the March 1987 meeting of the American Physical Society in New York has become known as the "Woodstock of physics." Canadian scientists have made significant contributions to the study of these materials through a national research network of the Canadian Institute for Advanced Research.

Mai-Linh Doan. This file is licensed under the Creative Commons Attribution-Share Alike 3.0 Unported license, https://creativecommons.org/licenses/by-sa/3.0/deed.en

Figure 33-16 A magnetic cube levitated by a superconducting disk.

33-6 Doped Semiconductors

We have seen that the **carrier concentration**, the number of conduction electrons (or holes) per unit volume, has a profound influence on the electrical characteristics of semiconductors. This effect led researchers to look for ways to alter carrier concentrations other than by changing the temperature. The answer lay in altering the chemical composition of semiconductors by adding small amounts of other elements.

Silicon and germanium are the semiconductors used in most solid-state electronic devices. These two elements are both in group IV of the periodic table. All elements in group IV have four valence electrons. Adding tiny amounts of elements from either group III or group V dramatically changes the carrier concentration of a pure group IV semiconductor. The added materials are called **dopants**, and the process of adding them to a semiconductor is called **doping**.

We will first consider the effect of adding group V dopants, which have five valence electrons. For silicon, the dopants are usually phosphorus or arsenic. We will assume that the dopant has not changed the band structure, which is a reasonable approximation if the dopant concentration is low. Four of the valence electrons from each dopant atom contribute to filling the completely filled band, like the valence electrons from an atom of the semiconductor element. However, the fifth valence electron goes into the unfilled band above the filled band. Thus, a group V dopant increases the number of conduction electrons. For example, at room temperature, 1 cm^3 of pure germanium contains approximately 2.5×10^{13} electrons that have been thermally excited across the band gap. These conduction electrons have left behind the same number of unoccupied states (holes) below the gap. The addition of 0.001% of arsenic contributes an extra 10^{17} free electrons per cm^3, and the resistivity of the semiconductor decreases by a factor of approximately 10 000. Group V dopants are called **donors** because they "donate" electrons to the material. The resulting material is called an **n-type semiconductor** because it has extra carriers for negative charge.

Now let us consider the effect of dopants from group III of the periodic table. These elements have only three valence electrons. Consequently, each dopant atom creates an unfilled state in the highest filled band. Group III dopants are called **acceptors** because they accept electrons out of the band to create holes. Since these holes behave like positively charged carriers, the resulting material is called a **p-type semiconductor**. Boron is a commonly used acceptor.

Thus, dopants can be used to determine the type and concentration of charge carriers in a semiconductor and, hence, the resistivity of the material.

33-7 The pn Junction Diode

A **pn junction diode** consists of a piece of p-type semiconductor attached to an n-type semiconductor. Junction diodes are manufactured by diffusing dopants into a single piece of pure semiconductor to create a p-type region and an n-type region. The circuit symbol and the structure for a junction diode are shown in Figure 33-17. The triangular arrowhead in the diode symbol indicates the direction in which conventional current can flow. The p-type material is the anode, and the n-type material is the cathode.

Suppose that the holes and conduction electrons in the diode are initially immobilized. Associated with each conduction electron in the n-type region is a donor atom that has a net positive charge. The net charge on the region is still zero because the electron has an equal but opposite charge. Similarly, in the p-type region, there is a negatively charged acceptor atom associated with each hole. As shown in Figure 33-18, the region near the junction has positively charged donors in the n-type material and negatively charged acceptors in the p-type material.

Now consider what happens when the electrons and holes begin to diffuse. Well away from the junction, nothing particularly interesting happens. Near the junction, an electron from the n-type material can move to a lower-energy state by crossing the junction to fill a hole. Both the electron and the hole "disappear," leaving a positively charged donor in the n-type region and a negatively charged acceptor in the p-type region. These charges are fixed in position in the lattice and produce an electric field directed from the n-type material to the p-type material. The field then sets up a drift current that is in the opposite direction to the diffusion current.

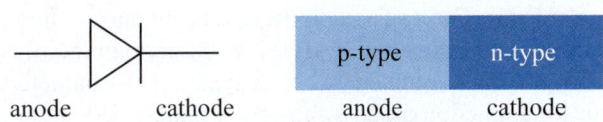

Figure 33-17 The circuit symbol for a junction diode (left) and the configuration of a diode semiconductor material (right).

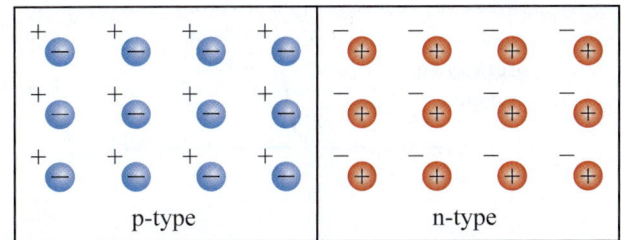

Figure 33-18 A pn junction before the electrons and holes are allowed to diffuse across the junction. The circles with positive signs denote acceptor atoms in the crystal lattice, and the circles with negative signs denote donor atoms. The donor and acceptor atoms are bound in place by the crystal structure and cannot move.

MODERN PHYSICS

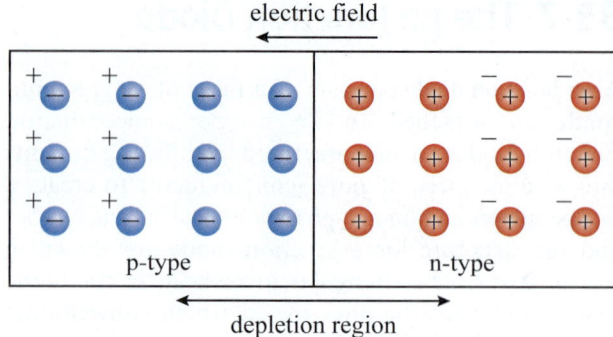

electric field

p-type   n-type

depletion region

Figure 33-19   Electron–hole recombination at a pn junction.

When the field becomes strong enough, the drift current is exactly equal to but opposite the diffusion current. Thus, there is no net flow of current across the junction and there is a **depletion region** at the junction, where there are no mobile charge carriers (Figure 33-19).

We will now consider what happens if we apply an external potential difference to the diode. Making the cathode positive with respect to the anode establishes an electric field with the same direction as the electric field in the depletion region. Consequently, almost no current flows in the device because the effect of the applied voltage widens the depletion region by moving charge carriers farther from the junction. This condition is called **reverse bias** of the diode. There is a small **leakage current** during reverse bias, typically less than a μA for a few volts of bias. However, if the reverse-bias voltage is high enough, electrons will cross the band gap and a large **avalanche current** will flow.

If we make the anode positive with respect to the cathode, the depletion region narrows. Once we apply sufficient voltage, the depletion region disappears and the diode readily conducts current. This condition is called **forward bias**. For a silicon diode, significant current flows once the forward bias exceeds 0.7 V; for germanium, this voltage is approximately 0.2 V. A typical I–V characteristic for a junction diode is shown in Figure 33-20.

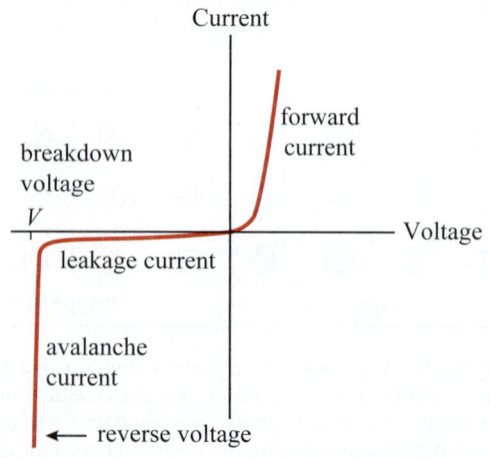

Figure 33-20   A typical I–V characteristic for a pn junction diode.

Diodes are commonly used as **rectifiers**, which are circuits that convert alternating current (AC) into direct current (DC). Rectifiers are found on most electronic devices that run from household outlets and in the adapters for charging portable devices.

Figure 33-21 shows a rectifier circuit. When the upper terminal of the sinusoidal AC voltage source is positive, the diode is forward biased. Consequently, current flows freely in the circuit, and a potential difference appears across the resistor. When the upper terminal of the source is negative, the diode is reverse biased, no current flows, and no voltage appears across the resistor. Thus, only positive voltages appear across the resistor. The capacitor in parallel with the resistor smooths out much of the fluctuations in the pulsating voltage.

MAKING CONNECTIONS

Energy Consumption

Rectifier circuits, like the ones found in mobile device chargers, usually have a voltage regulator attached to the output to ensure that the output voltage is constant. However, these regulator circuits use some power whether a device is drawing power from the rectifiers or not. Unplug them when you are not charging. Some regulators have been designed to switch themselves off under no-load conditions to reduce the energy wasted.

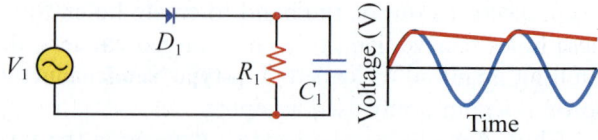

Figure 33-21   A half-wave rectifier with a filter capacitor. The blue curve is the input voltage, and the red curve is the voltage that appears across the resistor.

33-8 Other Semiconductor Devices

A wide variety of devices can be made using various configurations of p-type and n-type semiconductors. For example, the bipolar junction **transistor** (BJT) and the field effect transistor (FET) are three-terminal devices where the current through two of the terminals is controlled by either a small current (BJT) or a small voltage (FET) to the third terminal. Junction transistors are commonly used in amplifiers, and FETs are the basis of computer logic circuits, processors, and memory. The mass production of transistors began in the 1950s, and semiconductor devices have now almost completely replaced bulky (and somewhat fragile) vacuum tubes as the active elements in electronic circuits (Figure 33-22).

Figure 33-22 Vacuum tubes and individual transistors. The transistors are approximately 1 cm high.

MAKING CONNECTIONS

The JFET Amplifier

One of the most important devices in modern electronics is the FET. FETs are used as amplifiers and are the basic building blocks of the digital electronic circuits used in computers, cell phones, etc. The junction field effect transistor (JFET) is a three-terminal device with a source, a gate, and a drain connection. The device consists of a doped "channel" (either n-type or p-type) between the drain and the source and a gate that is doped in a complementary fashion, as shown in Figure 33-23, so that a pn junction is created between the gate and the channel.

We will focus our discussion on the n-channel JFET. With no external potentials applied to the device, the channel consists of a doped p semiconductor and, as a result, conducts a current if a potential is applied between the source and the drain.

In a practical circuit, the drain is connected to a more positive potential than the source and the gate is held at a potential that is negative with respect to the source. This arrangement, called biasing, establishes a reverse bias across the pn junction between the gate and the channel. Consequently, a depletion region is created in the channel between the gates. This depletion region reduces the cross-sectional area of the channel that can conduct. If the potential on the gate is made very negative, the depletion region completely fills the channel and there can be no conduction between source and drain. So, with no voltage applied to the gate, the channel conducts, and with a large negative voltage applied to the gate, the channel does not conduct. The JFET operated in this manner is essentially a switch that is controlled by the gate voltage.

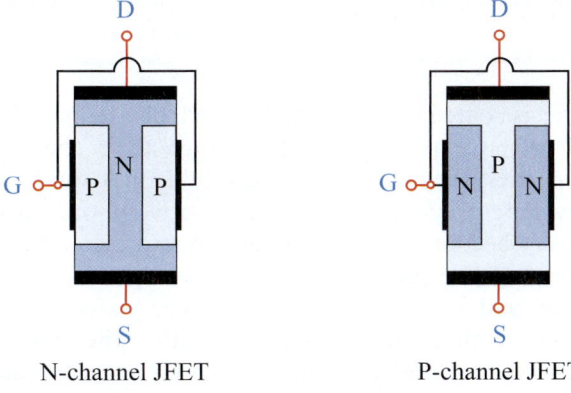

Figure 33-23 A JFET consists of a doped channel (the I-shaped section) that is connected to the source and the drain. Embedded in the side of the channel is a gate that is doped oppositely to the channel. This creates a pn junction between the gate and the channel.

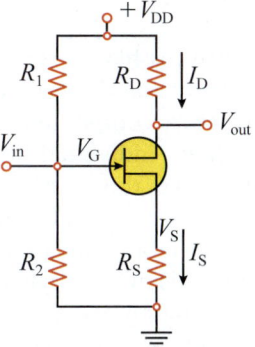

Figure 33-24 JFET Common Source Amplifier. Resistors R_1 and R_2 create a voltage V_G at the gate that is negative with respect to the source voltage, V_S, such that the JFET (shown schematically as the blue shaded circle) is roughly half turned off.

(*continued*)

In addition to operating in this on-off mode, a JFET (and most other types of transistors) can also be operated in the region between fully off and fully on to amplify analog signals. A very simple circuit that can do this is shown in Figure 33-24.

Resistors R_1 and R_2 are chosen so that the voltage that appears at the gate, V_G, is less than the voltage that appears at the source, V_S. This results in the gate-channel junction being reverse biased and the JFET being in an intermediate state between fully conducting and shut off.

If the input voltage causes the gate voltage to become more positive, this reduces the potential difference between the gate and the channel, which reduces the size of the depletion region in the channel. The channel becomes more conducting and the drain current, I_D, will increase. This leads to a larger voltage drop across R_D and hence a lower output voltage, since $V_{out} = V_D - I_D R_D$. Conversely, if the input voltage causes the gate to become less positive, the depletion region increases in size, there is less drain current, and V_{out} increases.

In practice, the amplifier shown in Figure 33-24 will have some additional components in the circuit to increase the voltage gain; the input and output voltages for such a modified circuit are shown in Figure 33-25.

We can see that the output voltage is almost 10 times the amplitude of the input voltage, but it is 180° out of phase. When the input voltage increases, the output voltage decreases, and vice versa. This is called an inverting amplifier with gain.

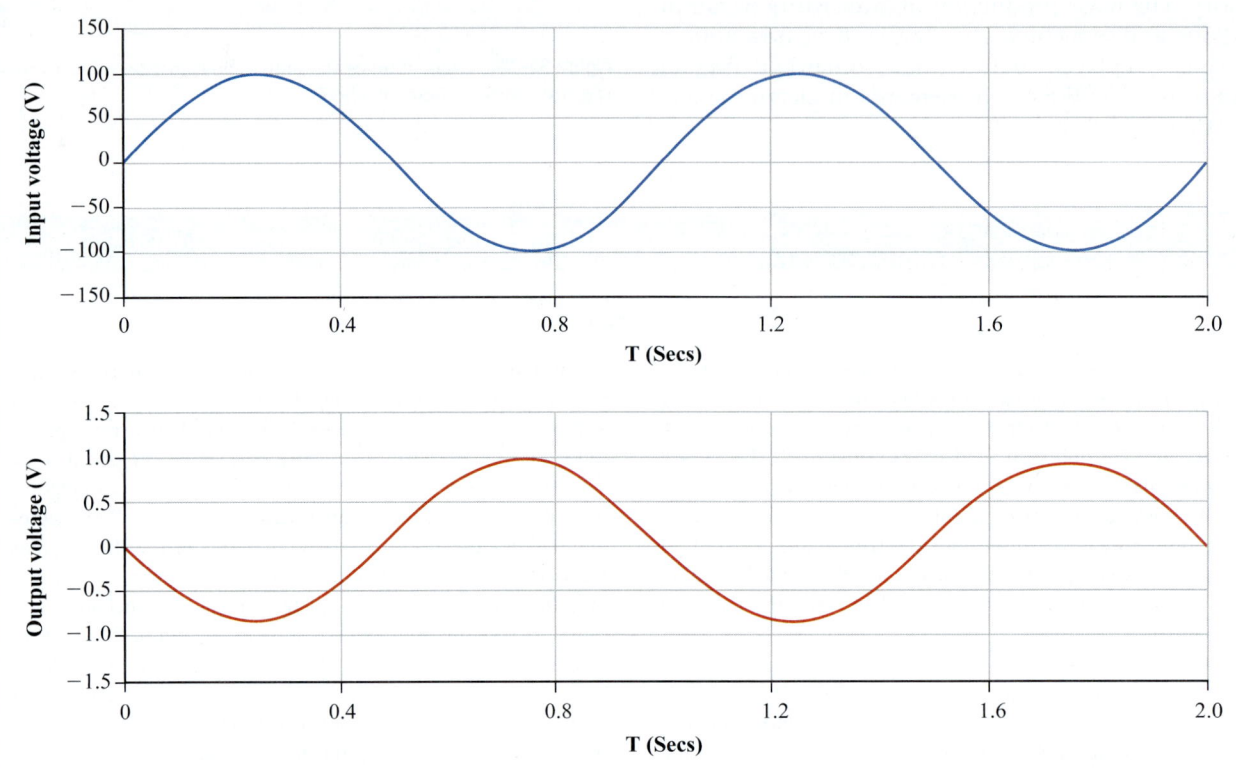

Figure 33-25 Input and output voltages for a practical JFET amplifier circuit. The amplitude of the output voltage is almost 10 times that of the input voltage and it is 180° out of phase with the input voltage.

Significant advances have been made in the fabrication of semiconductor devices. Many low-power applications can be accomplished using **integrated circuits** rather than individual discrete devices. Integrated circuits, like those in Figure 33-26, are manufactured on a single piece of semiconductor. The process starts with the fabrication of a long, cylindrical crystal of ultrapure silicon. The cylinder is then sliced into circular wafers slightly less than 1 mm thick and typically 10 cm to 30 cm in diameter. The wafers are then polished to make them very flat.

Once the wafers are polished, the pure semiconductor is doped using a lithographic process. The wafer is coated with photoresists, which are materials that when exposed to light change their chemical properties: they either become soluble in a solvent if they were previously insoluble (positive photoresist), or they become insoluble if they were previously soluble (negative photoresist). A mask is placed on the wafer, and it is exposed to light. The wafer is then washed with a solvent, removing either the exposed positive photoresist or the unexposed negative photoresist. Both methods leave some portions of the wafer coated in photoresist and other portions bare. The wafers are then exposed to the dopant impurities. The dopants can diffuse into the bare areas but are

Figure 33-26 An integrated circuit. The eight-pin chip at the top is a 741 operational amplifier that contains approximately 20 transistors. Most of the space is actually taken up with the packaging. The bottom image shows an actual integrated circuit. The chips are typically a few millimetres across.

prevented from diffusing into the coated areas by the photoresist. Several cycles of this process are required to produce the desired configuration of n-type and p-type regions in the wafer.

A similar process is used to deposit metal traces that connect various regions of the wafer. Many complete circuits are produced on a single wafer, which is then cut into separate integrated circuits. Each circuit is sealed in an insulating protective package with connections from the circuit to external metal pins or soldering terminals on the base of the package, allowing for connection on a circuit board.

The development of these techniques for manufacturing integrated circuits has led to the incredible miniaturization of modern electronics. For example, the CPU of a modern personal computer contains approximately 2 billion transistors built on a 21.5 mm × 32.5 mm chip. As this miniaturization increases, the transistors become smaller and are placed closer to each other. The closer spacing leads to increased speeds because it takes less time for the signals to travel from one device to the next. However, the transistors rely on the bulk properties of the materials of which they are made. Consequently, there is a limit to how small a transistor can be. Beyond this limit, the extremely small regions of the semiconductor will not display these bulk properties.

EXAMPLE 33-6

Integrated Circuit Transistor

Estimate the area of a typical single transistor in a modern CPU for a personal computer. How many atoms are in such a transistor?

Solution

As described above, the CPU contains 2 billion transistors built on a chip that is 21.5 mm × 32.5 mm. We will assume that the transistors cover the entire area of the CPU. Then the area available for each transistor is approximately

$$A = \frac{21.5 \text{ mm} \times 32.5 \text{ mm}}{2.0 \times 10^9} = 3.5 \times 10^{-7} \text{ mm}^2$$

So, each transistor occupies a square with side $L = \sqrt{A} = \sqrt{3.5 \times 10^7 \text{ mm}^3} = 0.59 \times 10^{-6}$ m, or slightly more than half a micrometre. If we assume that the dopants are diffused to a similar depth in the material, this produces a volume for each device of approximately

$$V = (0.59 \times 10^{-6} \text{ m})^3 = 2.1 \times 10^{-19} \text{ m}^3$$

The density of silicon is 2330 kg/m³, and silicon has an atomic weight of 0.028 kg/mol. Therefore, each transistor contains

$$N = \frac{(2.1 \times 10^{-19} \text{ m}^3)(2330 \text{ kg/m}^3)}{0.028 \text{ kg/mol}}$$

$$\sim 1.7 \times 10^{-14} \text{ mol} \sim 10^{10} \text{ atoms}$$

Making sense of the result

The CPU contains other components, such as resistors and diodes. Some area is also used for connections and insulation between the components. Therefore, our estimates of the area and the number of atoms per transistor is likely somewhat high.

Spin-offs from the Semiconductor Industry

The techniques used to produce integrated circuits are also used to produce solar cells, which use the photovoltaic effect to generate electrical power. One of the largest photovoltaic installations in Canada is a 100 kW peak output array at the Horse Palace at Exhibition Place in Toronto, Ontario (Figure 33-27). This array has 536 solar modules with a total area of 1428 m^2. The cost of the project was $1.1 million, and it is expected to pay for itself in 22 years. In addition, it reduces greenhouse gas emissions by 115 tonnes annually.

maximimages.com/Alamy Stock Photo

Figure 33-27 The photovoltaic array at Exhibition Place, Toronto, Ontario.

33-9 Nanotechnology

In 1959, Richard Feynman (1918–1988) gave a talk at a meeting of the American Physical Society called "There's Plenty of Room at the Bottom." Feynman envisioned that it might be possible to manipulate individual atoms and molecules by building a set of precise microscopic tools to build and operate a proportionally smaller set, which would build a still smaller set, and so on down to the smallest possible scale—individual atoms and molecules. If by doing so one could build the analogue of transistors, perhaps from single molecules, one could replace the bulk semiconductor technology that we currently use with these new devices.

These ideas spawned the field of study called nanotechnology. **Nanotechnology** generally refers to studies at lengths of less than 100 nm. Research in this field involves different types of scientists, such as physicists, chemists, engineers, and biologists. Investigations include extending existing technology to smaller scales and attempts to control matter on an atomic scale.

Biological systems appear to be excellent prototypes for nanotechnology. Consider the human body: it responds to light, sound, touch, scent, taste, temperature, and motion. It uses fuel and converts that fuel into energy, motion, and thought. All of this is accomplished by the cells within our bodies, each of which can replicate or build itself as well as all the internal mechanisms that the cells themselves contain to function. Each cell interacts with other cells so that the whole organism can function. Clearly, the body is built in a bottom-up fashion, starting from a single fertilized ovum. The body does not start from a large, single, undifferentiated mass of material, as we do when we build an integrated circuit.

Specialized tools have been developed to image and manipulate individual atoms. Optical microscopes were the first devices to allow us a glimpse of details approaching the nanoscale. However, the best resolving power of optical microscopes is about 200 nm because their resolution is limited by the wavelength of light. Electron microscopes have greater resolving power because they use a beam of electrons instead of a beam of light. As you saw in earlier chapters, electrons have a shorter wavelength than visible light. In an electron microscope, electrons are accelerated through a potential difference, focused into a small beam, and directed at the sample. We can study either the reflected beam, as is done in a scanning electron microscope (SEM), or the transmitted beam, as is done in a **transmission electron microscope** (TEM).

The best TEMs have a resolution on the order of 0.05 nm. An example of a high-resolution TEM image is shown in Figure 33-28.

The **scanning probe microscope** (SPM) also has very high resolution. In SPMs, some type of probe is brought very close to the surface of the sample. The probe must interact in some way with the surface, and it must be possible to measure the strength of the interaction. As the probe scans across the sample, the strength of the interaction is monitored; plotting the interaction strength versus the probe position gives an "image" of the surface property to which the probe is sensitive. Examples of these types of microscopes are the **scanning tunnelling microscope** (STM), the atomic force microscope (AFM), and the magnetic force microscope (MFM).

EXAMPLE 33-7

Determining the Wavelength of an Electron

Find the wavelength of an electron that has been accelerated through a potential difference of 200. kV.

Solution

Since the electron has been accelerated through a relatively large potential difference, we will use a relativistic approach. The wavelength of a particle moving at a relativistic speed is given by

$$\lambda \approx \frac{h}{\sqrt{2m_0 E \left(1 + \dfrac{E}{2m_0 c^2}\right)}}$$

$$= \frac{6.626 \times 10^{-34}\,\text{J}\cdot\text{s}}{\sqrt{2 \times 9.11 \times 10^{-31}\,\text{kg} \times (200 \times 10^3\,\text{V})\,(1.62 \times 10^{-19}\,\text{C})\left(1 + \dfrac{3.2 \times 10^{-14}\,\text{J}}{2 \times 9.11 \times 10^{-31}\,\text{kg} \times (3.00 \times 10^8\,\text{m/s})^2}\right)}}$$

$$= 2.5 \times 10^{-12}\,\text{m}$$

Making sense of the result

This wavelength is substantially smaller than the wavelength of blue light (approximately 400 nm); thus, electron microscopes have a much higher resolving power than optical microscopes.

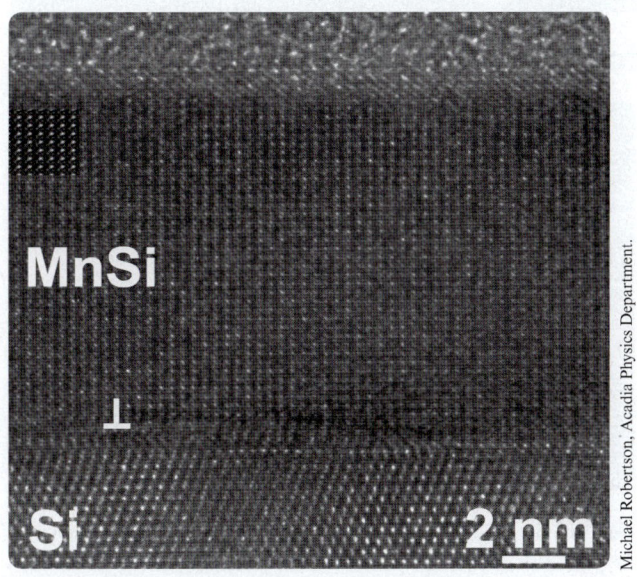

Figure 33-28 A high-resolution electron microscopy image of a manganese silicide (MnSi) layer grown on silicon (Si). The uppermost, random-looking layer is a layer of amorphous Si deposited to protect the MnSi layer from the atmosphere.

The probe in an STM has a very sharp metal tip, to which a bias voltage is applied relative to the conducting sample. The tip scans close enough to the sample that electrons can tunnel between the two. Depending on the sign of the bias, electrons either move from full states in the tip to empty states in the sample, or vice versa.

The magnitude of the tunnelling current, I_t, depends on both the height, a, and the width, Δz, of the potential barrier:

KEY EQUATION

$$I_t \propto e^{-2\kappa \Delta z}$$

$$\kappa = 0.51\sqrt{a} \times 10^{10}/\text{m} \qquad (33\text{-}13)$$

As a result, the tunnelling current is a very sensitive measure of the height of the probe above the sample. This exponential dependence also localizes the current directly between the point of the metal tip and the sample. As a result, the STM has a resolution in the vertical direction on the order of 0.01 nm and in the horizontal direction on the order of 0.1 nm. Researchers have also discovered that by manipulating the bias voltage, it is possible to pick up and place single atoms on surfaces. Figure 33-29 shows a ring (corral) of iron atoms assembled using this technique. The corral has a diameter of 14.3 nm, and the manipulation and imaging took place at a temperature of 4 K. The circular waves inside the corral are standing waves created by the surface electronic states confined in this circular barrier.

STMs have also been used to study self-assembled monolayers (SAMs). SAMs are typically grown on metal surfaces from molecules that have a long carbon chain. A head group at one end of the chain bonds strongly to the metal surface. For example, alkanethiols

MODERN PHYSICS

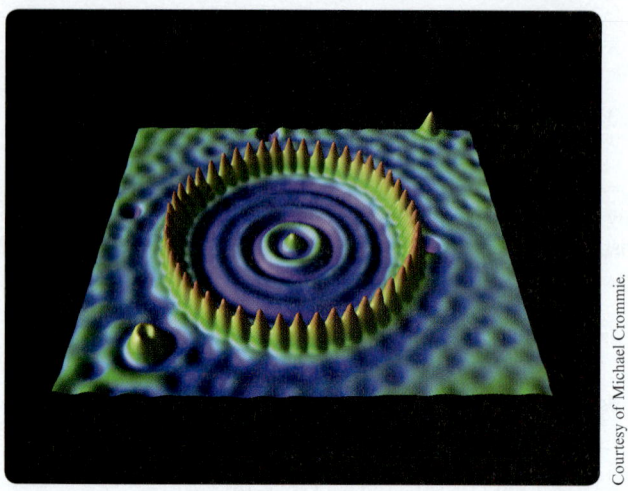

Figure 33-29 A quantum corral made of 48 iron atoms placed in a circle on a very clean copper surface using an STM.

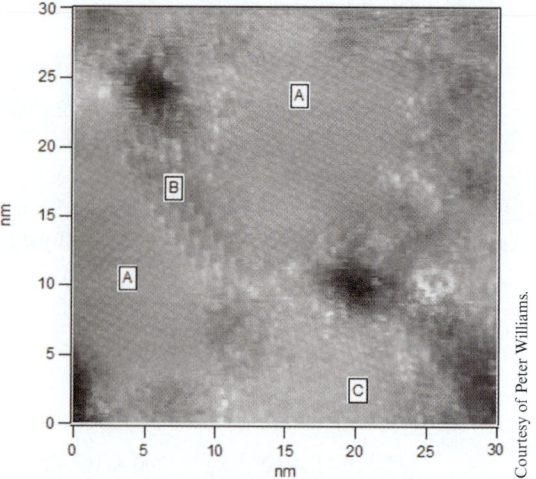

Figure 33-30 An STM image of a 30 nm × 30 nm square of decanethiol adsorbed on a gold surface.

have a hydrophilic head group (the thiol) that likes to adsorb onto a metal surface. Attached is a hydrophobic carbon chain with a length that can vary. When these molecules are deposited onto a metal surface, typically gold, they form various types of periodic structures on the surface, as shown in the STM image in Figure 33-30. The labels indicate regions of three different molecular structures. Studying this type of self-assembly may lead to the development of techniques for growing nano-devices and structures.

KEY CONCEPTS AND RELATIONSHIPS

Crystal Structures

It is most convenient to consider solids with a crystal structure, as this allows us to invoke periodicity when we try to solve Schrödinger's equation.

Electrons in a Box

If we solve Schrödinger's equation in a one-dimensional crystal of length L, we find the energy eigenvalues are given by the following equations from Chapter 32:

$$E_n = \frac{\hbar^2 k_n^2}{2m}$$

(32-49)

where

$$k_n = \frac{n\pi}{L} = \frac{n\pi}{N^*a} \tag{32-48}$$

Electrons are fermions, which means we can only have one electron per quantum state. Since electrons have spin, we can put two electrons in each energy level. If we place N electrons into the system, the quantum number corresponding to the highest occupied state is $n = N/2$, and the energy of the highest occupied state, called the Fermi energy, is given by

$$k_f = \frac{\pi N}{aN^*} \quad E_f = \frac{\hbar^2 k_f^2}{2m} = \frac{\hbar^2\pi^2}{8ma^2}\left(\frac{N}{N^*}\right)^2 \tag{33-3}$$

If we generalize this to three dimensions, we obtain a wave vector:

$$\vec{k} = \frac{\pi}{N_x^*a}n_x\hat{\imath} + \frac{\pi}{N_y^*a}n_y\hat{\jmath} + \frac{\pi}{N_z^*a}n_z\hat{k} \tag{33-4}$$

Then the expressions for the Fermi wave vector and energy become

$$k_f = \left(\frac{3\pi^2 N}{V}\right)^{\frac{1}{3}} \quad E_f = \frac{\hbar^2}{2m}\left(\frac{3\pi^2 N}{V}\right)^{\frac{2}{3}} \tag{33-5}$$

Periodic Potential

When we solve the Schrödinger equation for electrons in a periodic potential, we find that gaps appear in the energy versus wave vector plot at $k = \frac{\pi}{a}$. These gaps are called band gaps, and the allowed states are said to fall in bands.

Metals and Insulators

For a metal, the electrons only partially fill the highest-energy band, which means there are empty states immediately adjacent to the filled states that the electrons can be excited into. For an insulator, the electrons completely fill the highest-energy band, and a large band gap exists between the top of that filled band and the bottom of the empty band.

APPLICATIONS

pn junction diodes; rectifier circuits to convert AC voltages to DC; integrated circuit technology; nanotechnology tools, such as high-resolution transmission electron microscopes and scanning probe microscopes

Semiconductors

A semiconductor has a filled band with a small energy gap. The number of charge carriers, N_x, thermally excited across the gap, E_{gap}, is approximately

$$N_x \approx N_0 e^{-\frac{\Delta E}{k_b T}} \tag{33-12}$$

where N_0 is the number of electrons at the top of the filled band.

Doped Semiconductors

We can control the number of charge carriers in a semiconductor by doping to add conduction electrons (n-type) or to remove conduction electrons (p-type).

The pn Junction Diode

A pn junction acts as a diode that conducts well under forward bias conditions but does not conduct well under reverse bias.

Other Semiconductor Devices

Semiconducting materials can be combined in other configurations to make devices called transistors. In their simplest form, they act as a switch, although they can also be used as amplifiers. They are the basic building blocks of all modern electronic devices.

Nanotechnology

Much effort has been devoted recently to attempting to build devices "from the bottom up" by assembling them on an atom-by-atom basis as opposed to diffusing dopants into bulk materials. Various tools for studying and manipulating materials at the atomic level have been developed, including the transmission electron microscope, the atomic force microscope, and the scanning tunnelling microscope (STM). The STM exploits the extreme sensitivity to separation, Δz, of a quantum tunnelling current to carefully scan a probe over the surface of a conducting material. The current is given by

$$\begin{aligned} I_t &\propto e^{-2\kappa\Delta z} \\ \kappa &= 0.51\sqrt{a} \times 10^{10}\ \text{m}^{-1} \end{aligned} \tag{33-13}$$

KEY TERMS

acceptors, avalanche current, band gaps, band structure, carrier concentration, charge carrier, conduction band, conductor, cubic body-centred structure, cubic face-centred structure, cubic structure, depletion region, dielectric strength, donors, dopants, doping, electrical breakdown, energy bands, Fermi energy, Fermi level, Fermi speed, forward bias, hexagonal structure, hole motion, holes, insulator, integrated circuits, lattice parameter, leakage current, n-type semiconductor, nanotechnology, p-type semiconductor, periodic potential, phonons, pn junction diode, rectifiers, reverse bias, scanning probe microscope, scanning tunnelling microscope, screening, semiconductors, tetragonal structure, transistor, transmission electron microscope, valence band, valence electrons

QUESTIONS

1. If the temperature of a piece of a metal is increased, does the probability of occupancy 0.1 eV below the Fermi level increase, decrease, or remain the same?
2. All the electrons in a metal can participate in the conduction process. True or false?
3. A material with a partially filled band
   (a) is a semiconductor
   (b) is an insulator
   (c) is a good conductor
   (d) cannot be determined without knowing the width of the band gaps
4. Why does a metal have a positive temperature coefficient of resistivity?
   (a) Fewer electrons are available for conduction at higher temperatures.
   (b) There are a greater number of lattice vibrations for the electrons to scatter from at higher temperatures.
   (c) More electrons are excited across the energy gap at higher temperatures.
5. Dopants from group III of the periodic table added to silicon
   (a) donate electrons to make an n-type material
   (b) donate electrons to make a p-type material
   (c) accept electrons to make an n-type material
   (d) accept electrons to make a p-type material
6. In a p-type semiconductor, the charge carriers are
   (a) electrons
   (b) holes
7. The increased resolution of a transmission electron microscope over an optical microscope is due to
   (a) a smaller beam size
   (b) the shorter wavelength of the electrons
   (c) the longer wavelength of the electrons
   (d) none of the above
8. When a pn junction is reverse biased, is the electric field in the depletion zone larger or smaller than the electric field found at zero bias?
9. Under forward-bias conditions, the voltage across a pn junction is approximately 0.7 V. Is the diode silicon or germanium?

PROBLEMS BY SECTION

For problems, star ratings will be used (★, ★★, or ★★★), with more stars meaning more challenging problems. The following codes will indicate if $\frac{d}{dx}$ differentiation, $\int$ integration, ▢ numerical approximation, or ⌁ graphical analysis will be required to solve the problem.

Section 33-1   Crystal Structures

10. ★ Associated with a crystal structure is a set of lattice vectors: $\vec{a}, \vec{b}$, and $\vec{c}$. For a simple cubic lattice, the lattice vectors are all of the same length, a, and point in the $\hat{i}$-, $\hat{j}$-, and $\hat{k}$-directions. Thus, $\vec{a} = a\hat{i}, \vec{b} = a\hat{j}$, and $\vec{c} = a\hat{k}$. Take one corner of the cube to be at the origin.
    (a) What combination of lattice vectors connects the origin with the point diagonally across the cube?
    (b) What is the length of the vector you found in part (a)?

11. ★★ The volume of a unit cell is determined by $(\vec{a} \times \vec{b}) \cdot \vec{c}$.
    (a) What is the volume of a cubic unit cell with side a?
    (b) What is the volume of a rhombohedral unit cell with $\alpha = \beta = \gamma = 60°$ (Figure 33-32)?

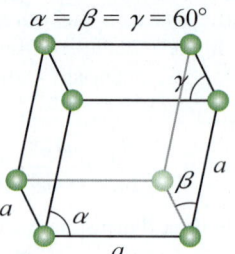

$$\alpha = \beta = \gamma = 60°$$

Figure 33-32   Problem 11.

12. ★★ When planes of atoms in a material satisfy Bragg's law, $n\lambda = 2d \sin \theta$, a diffraction line is observed. In this equation, n is an integer that corresponds to the order of the diffraction line, λ is the wavelength of the X-rays, d is the distance between adjacent planes of atoms, and θ is the angle of incidence.
    (a) For a simple cubic crystal, what is the longest distance between adjacent planes in the crystal?
    (b) When $a = 0.36$ nm and $\lambda = 1$ nm, what angle of incidence satisfies Bragg's law?

Section 33-2   Electrons in a Box

13. ★ Assume that the total volume of a solid is the sum of the volume occupied by the ions and the volume occupied by the conduction electrons. The density and molar mass of a certain metal are 856 kg/m³ and 16.8 g/mol, respectively. Assume that the radius of an ion is 73.4 pm. What percent of the volume of a sample of this metal is occupied by its conduction electrons?

14. ★ A certain monovalent metal has a density of 12.75 g/cm³ and a molar mass of 109 g/mol.
    (a) Calculate the number density of conduction electrons.
    (b) Calculate the Fermi energy (in eV).
    (c) What is the Fermi speed?

15. ★★ The Fermi energy of a certain metal is 9.69 eV; its density and molar mass are 2.44 g/cm³ and 33.0 g/mol, respectively. Determine the number of conduction electrons per atom.

Section 33-3   Periodic Potential

16. ★ In quantum mechanics, the momentum of a particle is given by $p = \hbar k$. For a metal with a Fermi energy of 4.0 eV, determine
    (a) the Fermi speed
    (b) the Fermi wave vector, k_f
    (c) the Fermi momentum

17. ★★ $\frac{d}{dx}$ It is sometimes convenient to assign a particle an "effective mass" that reflects how the particle interacts with its surroundings. In classical physics, for a free particle,

$$E = \frac{p^2}{2m} \quad \text{and} \quad \frac{\partial^2 E}{\partial p^2} = \frac{1}{m}$$

If we extend these relationships to the quantum-mechanical case, we find

$$E = \frac{\hbar^2 k^2}{2m} \quad \text{and} \quad \frac{1}{\hbar^2}\frac{\partial^2 E}{\partial k^2} = \frac{1}{m}$$

Using these ideas, we can define the effective mass, m^*, through

$$\frac{1}{m^*} = \frac{1}{\hbar^2}\frac{\partial^2 E}{\partial k^2}$$

Consider an electron with the E versus k relation in a periodic potential as shown in Figure 33-8 in Section 33-3.
(a) What is the sign of m^* well away from the band gap?
(b) What is the sign of m^* just below the band gap?
(c) What is the sign of m^* just above the band gap?

Section 33-4   Metals and Insulators

18. ★★ Insulators can be made to conduct by shining light on them if the energy of the photons is sufficient to excite an electron across the band gap. Recall that for a photon, $E = hf = \dfrac{hc}{\lambda}$, where h is Planck's constant, f is the frequency, c is the speed of light, and λ is the wavelength. A material has a band gap of 6.53 eV.
(a) What is the band gap in joules?
(b) What is the minimum frequency of light that can excite an electron across the gap?
(c) What is the maximum wavelength of the light that will excite an electron across the gap?

19. ★★ In a current-carrying conductor, energy is supplied to the conduction electrons by an electric field. When the electrons collide with the atoms in the lattice, they lose some of this energy to the lattice, causing the lattice to vibrate. This process increases the temperature of the lattice. An equilibrium is established when the lattice radiates energy at the same rate that the electrons are transferring energy to the lattice. Consider a conductor that is dissipating energy at a rate of 1.00 W. Assume that the conduction electron density is $1.00 \times 10^{22}/m^3$ and that the volume of the conductor is $1.00 \times 10^{-9}\ m^3$.
(a) How many conduction electrons are in the conductor?
(b) At what rate does each individual electron transfer energy to the lattice?

Section 33-5   Semiconductors

20. ★★ An insulating material has a band gap of 4.8 eV between the highest filled and lowest empty energy levels. What is the probability that an electron will jump the gap at $T = 300.\ K$?

21. ★ A semiconductor has a band gap of 2.0 eV. Using Equation 33-12, determine what fraction of the charge carriers will be excited across the gap at a temperature of 500. K.

22. ★★ A simple model for the high-temperature resistance of a semiconducting material supposes that the resistance is inversely proportional to the carrier concentration, $R \propto \dfrac{1}{n}$, where $n = \dfrac{N}{V}$ is the total number of carriers, N, divided by the volume of the sample, V. For a semiconductor with a band gap of 0.2 eV, by what factor would the resistance decrease in going from 400 K to 500 K?

Section 33-6   Doped Semiconductors

23. ★ The concentration of conduction electrons in pure silicon is approximately $10^{16}\ m^{-3}$. What mass of phosphorus is needed to dope 1.0 g of silicon so that the concentration of conduction electrons is increased by a factor of 1.0×10^6? The density of silicon is $2.33\ g/cm^3$, the molar mass of silicon is 28.1 g/mol, and the molar mass of phosphorus is 30.9758 g/mol.

24. ★★ Pure silicon at room temperature has an electron concentration of approximately $5 \times 10^{15}\ m^{-3}$ and an equal concentration of holes in the valence band. Suppose that one of every 10^6 silicon atoms is replaced by a phosphorus atom.
(a) What charge carrier concentration will the phosphorus add (in m^{-3})?
(b) What is the ratio of the charge carrier concentration (electrons in the conduction band and holes in the valence band) in the doped silicon to that in pure silicon? The density of silicon is $2.33\ g/cm^3$, and its molar mass is 28.1 g/mol.

Section 33-7   The pn Junction Diode

25. ★★ For an ideal pn junction diode with a sharp boundary between its two semiconducting sides, the current, I, is related to the potential difference, V, across the diode as follows:

$$I = I_0(e^{eV/kT} - 1)$$

where I_0, which depends on the materials but not on I or V, is called the reverse saturation current. The potential difference, V, is positive if the rectifier is forward biased and negative if it is reverse biased. For $T = 270$ K, calculate the ratio of the current for a 0.30 V forward bias to the current for a −0.30 V reverse bias.

26. ★★★ When a photon enters the depletion zone of a pn junction, the photon can scatter from the valence electrons there, transferring part of its energy to several electrons, which then jump to the conduction band. Thus, the photon creates electron–hole pairs. For this reason, pn junctions are often used as photon detectors, especially in the X-ray and gamma-ray regions of the electromagnetic spectrum. Suppose a single 699 keV photon transfers its energy to electrons in multiple scattering events inside a semiconductor with an energy gap of 0.90 eV, until all the energy is transferred. Assuming that each electron jumps the gap from the top of the valence band to the bottom of the conduction band, find the number of electron–hole pairs created by the photon.

27. ★★ While a diode does conduct quite freely in the forward bias direction, there is still a small voltage drop across the junction, called the knee voltage. This is typically 0.7 V for a silicon diode. Suppose such a diode is used in a rectifier that is intended to supply a current of 100 mA. What must the minimum power rating of the diode be?

Section 33-8   Other Semiconductor Devices

28. ★★ A certain computer chip that has dimensions 3.79 cm × 3.03 cm contains 3.9 million square transistors. What must be their *maximum* dimensions (in micrometres, or μm)? (Note: Devices other than transistors are also on the chip, and there must be room for the interconnections between the circuit elements.)

29. ★★ A silicon-based device consists of a heavily doped region of semiconductor upon which a thin oxide layer is allowed to grow. A metal layer is then deposited on top of the oxide layer. Suppose that the metal layer measures 0.62 μm on each side of the square. The insulating silicon oxide layer that separates the metal layer from the p-type substrate is 0.40 μm thick and has a dielectric constant of 4.0.
   (a) What is the equivalent metal–substrate capacitance (treating the metal as one plate and the substrate as the other plate)?
   (b) Approximately how many elementary charges appear in the metal when there is a metal–source potential difference of 2.4 V?

30. ★★ As miniaturization continues, we will reach the limit where devices get so small that their behaviour deviates significantly from the behaviour of a bulk sample. The exact size at which this limit occurs is not well known, but we could estimate it by looking at the number of dopants in a volume. Consider a silicon device that is doped at 0.001%, that is, for every dopant atom, there are 100 000 silicon atoms. How many dopant atoms are in a cube of silicon whose edge is
   (a) 1 μm?
   (b) 0.1 μm?
   (c) 0.001 μm?

Section 33-9 Nanotechnology

31. ★ The current in a scanning tunnelling microscope (STM) is given by

$$I = I_{\text{tunneling}} \propto e^{-2\kappa\Delta z}$$

$$\kappa = 0.51\sqrt{\phi} \times 10^{10}/\text{m}$$

   where ϕ is the barrier height (in eV), usually taken to be the work function of the metal, and Δz is the thickness of the barrier. Assuming that the work function is 4.83 eV, what percent change in the tunnelling current occurs if Δz increases by 0.100 nm?

32. ★ A transmission electron microscope (TEM) is operated at a potential difference of 150 000 V.
   (a) What is the wavelength of the electrons that have been accelerated through this potential difference?
   (b) The resolving power of the TEM is taken to be

   $\Delta l = 1.22 \dfrac{f\lambda}{D}$, where f is the focal length and

   D is the diameter of the aperture. Determine Δl, assuming $f = 10.$ cm and $D = 2.0$ cm.

COMPREHENSIVE PROBLEMS

33. ★★★ $\frac{d}{dx}$ In problem 19, we examined the rate at which an individual electron transfers energy to the lattice. This energy is provided by the electric field. In that problem, we found that for a 1 mm³ conductor with a carrier density of 10^{22} m^{-3} dissipating 1 W of power, each electron delivers energy to the lattice at a rate of 1×10^{-13} W $\approx 6 \times 10^5$ eV/s. Assume that the electron travels between collisions only under the influence of the

constant applied electric field, and loses all of its energy to the lattice when it collides. The rate at which the electron loses energy to the lattice must be equal to the rate at which the electric field supplies energy to the electron. Suppose that the conductor is a cube and that a potential difference of 1.0 V is applied.
   (a) What is the strength of the electric field in the sample?
   (b) What acceleration does an electron experience due to that field?
   (c) Write an expression for the speed of the electron (it will depend on time).
   (d) Write an expression for the kinetic energy of the electron.
   (e) Find the rate at which kinetic energy is delivered to the electron.
   (f) Set the result of part (e) equal to 1×10^{-13} W and solve for t.
   (g) The time found in part (e) is half the time between collisions. How far will the electron have moved between collisions? (Hint: Use $x = \dfrac{1}{2}at^2$.)

34. ★★★ $\frac{d}{dx}$ The lateral resolution of an STM is related to the exponential dependence of the tunnelling current on distance. We assume that the tip of the STM probe is spherical, and we use the geometry shown in Figure 33-33.
   (a) Show that $z(x) = z_0 + R - \sqrt{R^2 - x^2}$.
   (b) Assume that $x \ll R$, and expand $z(x)$ in a Taylor series about $x = 0$ to first order.
   (c) Show that $I_t(x) = I_0 e^{-2\kappa\left(z_0 + \frac{x^2}{2R}\right)} = I_0(z)e^{-\kappa\frac{x^2}{R}}$, where $I_t = I_0 e^{-2\kappa\Delta z}$.
   (d) Recall that $\kappa = 0.51\sqrt{\phi} \times 10^{10}/\text{m}$. Determine the full width at half maximum for $I_t(x)$ for a metal with a work function of 5.0 eV.

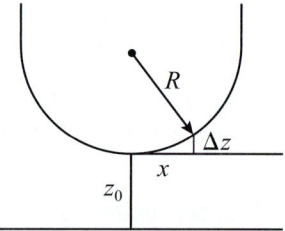

Figure 33-33 Problem 34.

35. ★ A fictitious trivalent metal has a molar mass of 59 g/mol and a density of 6.32 g/cm³. What is the number density, n, of conduction electrons?

36. ★★ What is the probability that a state 0.30 eV above the Fermi level will be occupied at (a) $T = 0$ K and (b) $T = 300.$ K?

For problems 37, 38, 39, and 40, use $f(E) = \dfrac{1}{e^{\frac{E-E_f}{k_b T}} + 1}$ as the

probability of a state with energy E to be occupied.

37. ★★ At $T = 200$ K, determine the probability of occupancy, $f(E)$, for a state
   (a) 60 meV above the Fermi level
   (b) 60 meV below the Fermi level

38. ★★ A metal has a Fermi energy of 8.6 eV. At a temperature of 400. K, what is the energy of a state whose probability of being occupied is
    (a) 0.38?
    (b) 0.84?

39. ★ A state has an energy that is 0.50 eV above the Fermi energy. At what temperature will this state be occupied with a probability of 0.10?

40. ★ A simple model of the resistance in a metal involves assuming that the electrons scatter every τ seconds. In this model, the resistivity is given by

$$\rho = \frac{m}{ne^2\tau}$$

    where n is the conduction electron density, m is the mass of the electron, and e is the charge of the electron. For copper, $n = 8.46 \times 10^{28}$ m^{-3} and $\rho = 1.68 \times 10^{-8}$ $\Omega \cdot$m. Find the scattering time, τ.

41. ★★ (a) Assuming that the electrons in problem 40 travel at a drift speed of approximately 3.00×10^{-4} m/s, how far do they travel between collisions?
    (b) Assuming that the electrons travel at the Fermi speed of approximately 1.00×10^6 m/s, how far do they travel between collisions?
    (c) Given that the lattice spacing in copper is 3.60×10^{-10} m, does the result in part (a) make any sense?

42. Capacitors can be fabricated in an integrated circuit by heavily doping a region of the chip, allowing an insulating oxide layer to grow over that region, and then depositing a metal layer on the oxide. The metal layer and the heavily doped region in the wafer act as the plates of the capacitor. What oxide thickness is required for a doped region with an area of 10^{-10} m^2 if we want a capacitance of 1 pF? Assume we are using silicon oxide, which has dielectric constant $\kappa \approx 5$.

43. ★★ A capacitor is fabricated in a silicon integrated circuit. The oxide layer separating the two doped "plates" is 1 nm thick and has a cross-sectional area of 10^{-10} m^2. The oxide has a resistivity of 10^{13} $\Omega \cdot$m. What is the resistance associated with the oxide layer?

44. ★★ Increases in computing speed are largely due to decreases in size of CPU units. Making the components closer together decreases the time required for a signal to propagate from one circuit to the next on the chip. Current CPU clock speeds are of the order of 10^9 Hz. If we assume that all the time between clock pulses, that is, 10^{-9} s, is required for the signal to propagate, how far apart are the circuits on such a chip? Assume that the signals propagate at 90% of the speed of light. Is this result reasonable?

45. ★★ Resistors are fabricated in integrated circuits by doping regions of the pure semiconductor wafer to decrease their resistance. Suppose we dope an area of 10^{-10} m^2 that is 1 μm thick. What resistivity is needed to make a 10 kΩ resistor?

46. ★ A capacitor is built on an integrated circuit. The capacitance is 1 pF, and the resistance of the oxide layer is 10^{14} Ω. What is the time constant associated with this RC combination?

47. ★★ The circuit shown in Figure 33-34 is a bridge rectifier. V_1 is a sinusoidal voltage source.
    (a) During the first half of the sine cycle, point A is positive with respect to point B. Trace the current path from A to B.

(b) During the second half of the sine cycle, A is negative with respect to B. Trace the current path for this situation.

(c) Based on your findings from parts (a) and (b), sketch V_1 and the voltage across R_1 for one complete cycle.

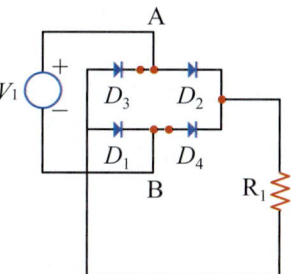

Figure 33-34 Problem 47.

48. ★★ Equation 33-3 for the Fermi wave number, $k_f = \frac{\pi N}{aN^*}$, suggests that the Fermi wave vector depends on the total sample size, $L = aN^*$. Why is this not the case?

49. ★ A metal has a Fermi speed of 3.8×10^6 m/s. Find the Fermi energy.

50. ★★ Calculate the wavelength for an electron with a wave vector magnitude of $k = \frac{\pi}{a}$, where a is the lattice spacing.

51. ★★ The bond length in diamond is 0.154 nm, which is the longest bond length that exists for ordinary carbon covalent bonds. Why would you expect the bond length in a diamond crystal to be larger than the bond length in carbon molecules?

52. ★★ From classical thermodynamics, we know that the number of degrees of freedom, f, in a system determines the constant-volume heat capacity, $C_V = \frac{fR}{2}$, where R is the gas constant. In an insulator, the electrons cannot move, whereas in a metal they can. One would therefore expect that there are more degrees of freedom in a metal and that it would have a larger heat capacity than an insulator by an amount of $3R/2 \sim 12.5$ J/(mol·K). Table 33-6 lists heat capacity data for a number of elements, including metals, a semiconductor, and an insulator. Clearly, the metals do not have heat capacities that are larger by $3R/2$.
    (a) When we heat a metal, do all the electrons have their energies increased?
    (b) What fraction of the electrons have their energy increased?
    (c) Would you expect the electronic contribution to the heat capacity to be large?

Table 33-6 Data for Problem 52

Element	Heat Capacity (J/mol · K)
Sodium	28.2777
Magnesium	24.7911
Aluminum	24.38992
Silicon	19.9439
Phosphorus	23.83761
Sulfur	22.60935

Introduction to Nuclear Physics

Learning Objectives	When you have completed this chapter, you should be able to

LO1 Understand nuclear terminology, including atomic mass number, atomic number, nucleon, nuclide, isotopes, isobars, isotones, and atomic mass.

LO2 Calculate the size and density of nuclei, and describe the properties of the strong force.

LO3 Calculate nuclear binding energies from nuclear masses and vice versa, and calculate atomic masses and average nuclear masses.

LO4 Calculate radioactive decay rates.

LO5 Explain what a Q-value is, and calculate Q-values for simple nuclear reactions.

LO6 Explain alpha, beta, and gamma radiation.

LO7 Explain why light stable nuclei prefer to have equal numbers of protons and neutrons and why heavier stable nuclei have more neutrons than protons.

LO8 Explain nuclear fission and fusion, and calculate the energy released during nuclear fission and fusion reactions.

LO9 Explain the difference between the absorbed dose and the equivalent dose of radiation, and calculate the daily equivalent dose of radiation for various radioactive materials.

LO10 Describe some common applications of nuclear physics.

On March 11, 2011, a 9.0 magnitude earthquake struck off the northeast coast of Japan near the city of Fukushima. The earthquake was followed by a 14 m high tsunami that devastated the coastal area. The tsunami flooded the buildings that housed the six nuclear reactors at the Fukushima 1 Power Plant, causing the emergency generators and cooling systems to fail. Although three of the reactors had been shut down for maintenance, the other three overheated catastrophically (Figure 34-1). Two of the reactor cores melted, and hydrogen gas produced by the high temperatures caused massive explosions that destroyed the outer containment buildings and damaged the vessels containing the nuclear fuel. Radioactive cesium, iodine, strontium, and barium were released into the air, forcing the evacuation of the surrounding area. Today, there is still concern over possible radioactive contamination of food and water supplies.

How do nuclear reactors generate energy? What is radioactivity, and why is it harmful? How does radioactive cesium differ from ordinary cesium? What is nuclear medicine and how is it used to save lives? This chapter introduces nuclear physics and examines some of the applications of nuclear technology.

Figure 34-1 Smoke billowing from the damaged nuclear reactors at Fukushima.

DigitalGlobe/Contributor/Getty Images.

34-1 Nuclear Terminology and Nuclear Units

As described in Chapter 31, experiments by Ernest Rutherford and others at the beginning of the 20th century determined that all the positive charge in an atom and most of the mass of an atom are localized in an extremely small central region called the nucleus. Later research showed that nuclei are made up of positively charged particles called **protons** and electrically neutral particles called **neutrons**. A neutron is slightly more massive than a proton, and both particles have about 1840 times the mass of an electron (see Table 34-2 below). The charge of a proton is $+1.602 \times 10^{-19}$ C, which is equal and opposite to the charge of an electron. An atom has an equal number of protons and electrons. The number of protons in a nucleus is called the **atomic number**, Z. For example, a carbon nucleus has six protons. Therefore, the atomic number of carbon is 6. The chemistry of an element is determined by its atomic number.

The number of neutrons in a nucleus is called the **neutron number**, N. The total number of protons and neutrons in a nucleus is called the **mass number**, A. By definition,

$$A = Z + N \qquad (34\text{-}1)$$

Each element in the periodic table has a fixed number of protons and is represented by a symbol. For example, helium has $Z = 2$ and is represented by the symbol He, and iron has $Z = 26$ and is represented by the symbol Fe. In general, a nucleus is represented by a symbol with the form $^{A}_{Z}X$, where

X = the symbol for the element
Z = the atomic number of the element
A = the mass number of the element

For a given atomic number, Z, the number of neutrons in a nucleus may vary. Nuclei with the same number of protons but a different number of neutrons are called the **isotopes** of an element. Isotopes have different masses because they contain a different number of neutrons. For example, naturally occurring carbon has three isotopes, as listed in Table 34-1.

Nuclei that have the same number of neutrons but a different number of protons are called **isotones**. Nuclei that have the same mass number, A, but a different number of protons and neutrons are called **isobars**. The term **nuclide** refers to a nucleus with any number of protons and neutrons.

Table 34-1 Isotopes of Carbon

Isotope name	Symbol	Atomic Number, Z	Neutron Number, N	Occurrence in Nature
Carbon-12	$^{12}_{6}C$	6	6	98.93%
Carbon-13	$^{13}_{6}C$	6	7	1.07%
Carbon-14	$^{14}_{6}C$	6	8	$10^{-10}\%$

Units for Nuclear Quantities

The size of a nucleus is of the order of 10^{-15} m. Therefore, in the study of nuclear physics, a convenient unit of length to use is the femtometre (symbol fm). A femtometre is also called a fermi in honour of Italian American physicist Enrico Fermi:

$$1 \text{ fm} = 10^{-15} \text{ m}$$

When dealing with energies involved in nuclear processes, a convenient unit to use is mega (million) electron volts:

$$1 \text{ MeV} = 10^6 \text{ eV}$$

The atomic mass unit (u) An atomic mass unit (u) is defined as exactly 1/12 the rest mass of one atom of the isotope $^{12}_{6}C$:

$$1 \text{ u} = 1.6605 \times 10^{-24} \text{ g} = 1.6605 \times 10^{-27} \text{ kg} \quad (34\text{-}2)$$

In nuclear physics it is convenient to express masses in the unit of energy/c^2, using Einstein's energy–momentum relationship, $E = \sqrt{p^2c^2 + m_0^2c^4}$, where p is the momentum of a particle and m_0 is its rest mass. The energy equivalent of 1 u of mass at rest is

$$
\begin{aligned}
E &= m_0 c^2 \\
&= (1.6605 \times 10^{-27} \text{kg})(2.9979 \times 10^8 \text{m/s})^2 \\
&= 1.4923 \times 10^{-10} \text{ J} \\
&= \frac{1.4923 \times 10^{-10} \text{ J}}{1.6021 \times 10^{-19} \text{ J/eV}} \\
&= 9.3150 \times 10^8 \text{ eV} \\
&= 931.50 \text{ MeV}
\end{aligned}
$$

Therefore,

$$1 \text{ u} = 931.50 \text{ MeV}/c^2 \qquad (34\text{-}3)$$

Table 34-2 lists the basic properties of protons, neutrons, and electrons.

Table 34-2 Some Physical Properties of Atomic Particles

Particle	Symbol	Spin	Charge	Atomic Mass Units, u	Mass, MeV/c^2
Electron	e	1/2	$-e$	5.4858×10^{-4}	0.5110
Proton	p	1/2	$+e$	1.007276	938.29
Neutron	n	1/2	0	1.008665	939.57

MODERN PHYSICS

EXAMPLE 34-1

Molar Mass and Atomic Mass

How many atoms are in 12 g of $^{12}_{6}C$?

Solution

Mass of one $^{12}_{6}C$ atom $= 12$ u $= 12 \times 1.6605 \times 10^{-24}$ g

$$= 1.9926 \times 10^{-23} \text{ g}$$

Number of atoms in 12 g of $^{12}_{6}C = \dfrac{12 \text{ g}}{1.9926 \times 10^{-23} \text{ g}}$

$$= 6.022 \times 10^{23}$$

Making sense of the result

The answer, 6.022×10^{23}, equals Avogadro's number, the number of atoms in one mole of a substance.

34-2 Nuclear Size and Nuclear Force

By bombarding a nucleus with a beam of fast-moving electrons and then observing how the electrons are deflected by the nucleus, scientists can determine the shape and the size of the nucleus. Several experiments using various elements have shown that, to a high degree of accuracy, most nuclei can be approximated as closely packed spherical clusters of protons and neutrons. The volume of the nucleus is proportional to the number of nucleons in the nucleus. A **nucleon** is a particle in an atomic nucleus—a proton or a neutron. The radius, r, of a nucleus is related to the number of nucleons, A:

KEY EQUATION
$$r = r_0 A^{1/3} \qquad (34\text{-}4)$$

where $r_0 = 1.2$ fm is a constant, determined from experiments. Table 34-3 lists the radii of the nuclei of some common elements, using Equation 34-4.

Nuclear Density

The relationship between the radius and the mass number of a nucleus in Equation 34-4 indicates that the nuclear density remains constant as A changes, that is, all nuclei have the same density. To calculate this density, we start with the volume of a spherical nucleus of radius r:

Table 34-3 Radii of Some Common Isotopes, Assuming a Spherical Nuclear Shape

Nucleus	Mass Number, A	Radius, $r = r_0 A^{1/3}$ (fm)
Helium, $^{4}_{2}He$	4	1.9
Oxygen, $^{16}_{8}O$	16	3.0
Iron, $^{56}_{26}Fe$	56	4.6
Lead, $^{208}_{82}Pb$	208	7.1

$$V = \frac{4\pi}{3}r^3 = \frac{4\pi}{3}(r_0 A^{1/3})^3 = \frac{4\pi}{3}r_0^3 A \qquad (34\text{-}5)$$

Next, we need to determine the mass of a nucleus that has A nucleons. From Table 34-2 we see that the mass of a nucleon is approximately equal to 1.0 u. Therefore, to a good approximation, the mass of a nucleus (M_{nuc}) with A nucleons is

$$M_{nuc} \approx Au \qquad (34\text{-}6)$$

This approximation is accurate to within 1% of the actual mass. The nuclear density, ρ_{nuc}, is then given by

$$\rho_{nuc} = \frac{M_{nuc}}{V} = \frac{Au}{\frac{4\pi}{3}r_0^3 A} = \frac{u}{\frac{4\pi}{3}r_0^3}$$

$$= \frac{1.6605 \times 10^{-27} \text{kg}}{\frac{4\pi}{3}(1.2 \times 10^{-15} \text{m})^3} = 2.3 \times 10^{17} \text{kg/m}^3 \qquad (34\text{-}7)$$

This density is fantastically large. A nucleus with a volume of 1.0 mm^3 would have a mass of 2.3×10^8 kg, that is, 230 thousand tonnes! In comparison, the mass of the same volume of iron is only 0.0078 g.

The Strong (or the Nuclear) Force

Protons are positively charged and exert a repulsive Coulomb force on each other. The magnitude of the Coulomb force between two protons that are a distance r apart is

$$F_{Coulomb} = \frac{1}{4\pi\varepsilon_0}\frac{(+e)(+e)}{r^2}$$

$$= \left(8.99 \times 10^9 \frac{\text{N}\cdot\text{m}^2}{\text{C}^2}\right)\left(\frac{(1.602 \times 10^{-19}\text{C})^2}{r^2}\right) \qquad (34\text{-}8)$$

Consider two protons inside a helium nucleus. If the average distance between the two protons is taken to be 2.0 fm, the repulsive force between the protons is

$$F_{Coulomb} = \left(8.99 \times 10^9 \frac{\text{N}\cdot\text{m}^2}{\text{C}^2}\right)\frac{(1.602 \times 10^{-19}\text{C})^2}{(2.0 \times 10^{-15}\text{m})^2}$$

$$= 58 \text{ N}$$

This is an enormous amount of repulsion. For comparison, the magnitude of the gravitational attraction between the protons at this distance is

$$F_{grav} = G\frac{(m_p)^2}{r^2}$$

$$= \left(6.674 \times 10^{-11}\frac{\text{N}\cdot\text{m}^2}{\text{kg}^2}\right)\frac{(1.673 \times 10^{-27}\text{kg})^2}{(2.0 \times 10^{-15}\text{m})^2}$$

$$= 4.7 \times 10^{-35} \text{ N}$$

$$(34\text{-}9)$$

where m_p is the proton mass.

Obviously, gravitational attraction is not sufficient to overcome the enormous Coulomb repulsion between the protons and keep them within the helium nucleus. There must be some very strong attractive force that keeps the protons confined within the nucleus. This force is called the **strong force** or the **nuclear force**. The strong force exists not only between two protons but also between two neutrons and between a proton and a neutron, as shown in Figure 34-2. There is another force, called the **weak force**, that acts between the nucleons, but it does not play any role in binding the nuclei. We will ignore the weak force in this chapter. The origin of the strong and weak forces is discussed in the next chapter.

Despite considerable research, the complicated nature of the strong force between two nucleons is still not completely understood. Experiments have shown that this force has the following properties:

- The strong force is repulsive when the distance between the two nucleons is less than approximately 0.7 fm. The repulsion prevents a nucleus from collapsing to a point.

- For distances larger than 0.7 fm, the strong force is attractive and reaches a maximum when the separation between the nucleons is approximately 1.2 fm. At this separation, the strong force is approximately 10 times the Coulomb force. This attraction holds the nucleus together.

- At distances beyond 1.2 fm, the strength of the strong force decreases exponentially, and it becomes negligible for separations of more than a few fermis. Thus, the strong force acts *only* at a very short range, in marked contrast to the gravitational and electrostatic forces, which decrease with the distance as $1/r^2$.

- The strong force only exists between nucleons. There is no strong force between an electron and a nucleon or between two electrons.

Because of Heisenberg's uncertainty principle, the position and momentum cannot be determined accurately at the same time and therefore a description of the interaction between two particles in terms of Newtonian forces is not strictly accurate. In quantum mechanics, the strength of interaction between two particles is described in terms of the potential, $V(r)$, that appears in the Schrödinger equation (see Section 32-3). A plot of the strong nuclear potential as a function of the distance between two nucleons is shown in Figure 34-3(a). The scale for the magnitude of the potential is not specified in this plot because the exact magnitude for a given separation depends on the model used to describe the potential.

It is instructive to compare the strong nuclear potential with the Coulomb potential between two protons. This comparison is shown in Figure 34-3(b)

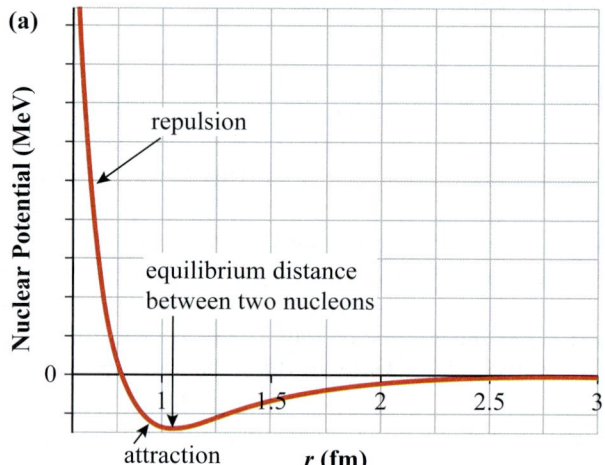

(a)

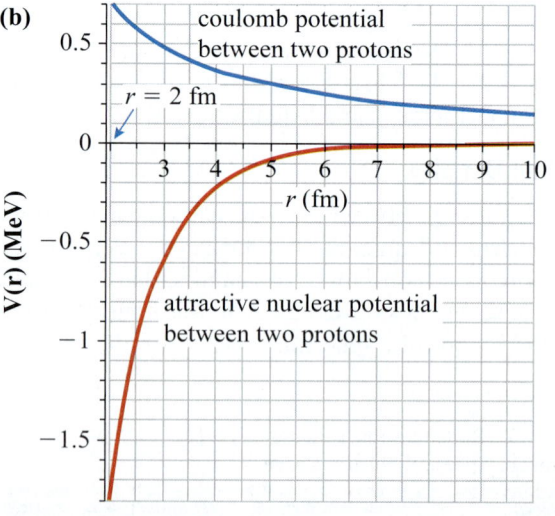

(b)

Figure 34-3 (a) The behaviour of the nuclear potential as a function of the distance between two nucleons. The interaction is repulsive for positive values and attractive for negative values. (b) A comparison of the nuclear potential and Coulomb potential between two protons for $r > 2$ fm. Both potentials decrease with distance, but the nuclear potential (and hence the nuclear force) decrease much more rapidly. The vertical scales in (a) and (b) are different.

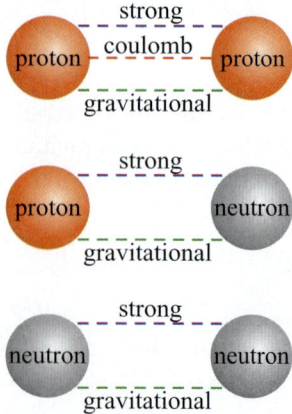

Figure 34-2 The forces acting between pairs of nucleons. There is also a weak force between two nucleons, but it is not shown in this figure.

(the scales of the two graphs in (a) and (b) are different). In this figure the strong nuclear potential is calculated by assuming the exchange of a single pion between two protons (see Sections 35-2 and 35-10). The two potentials are plotted for r between 2 fm and 10 fm. Both the Coulomb and the nuclear potentials decrease with distance, but the nuclear potential decreases much more rapidly compared to the Coulomb potential. The nuclear force is therefore described as a **short-range force** and the Coulomb force as a long-range force. Both decrease with distance, and at large enough distances the Coulomb force also becomes negligible.

CHECKPOINT 34-2

Why Does a Nucleus Not Fly Apart?

The nuclei are held together by
(a) the gravitational forces between nucleons
(b) the strong forces between nucleons
(c) the attractive Coulomb forces between electrons and protons
(d) the weak forces between nucleons
(e) Answers (a), (b), (c), and (d) are all correct.

ANSWER: The nuclei are held together due to the strong force between nucleons.

34-3 Nuclear Binding Energy

It is customary to express the nuclear mass in terms of A and Z. The rest mass of a nucleus that has Z protons and $N = A - Z$ neutrons is denoted by $M_{\text{nuc}}(A, Z)$. From Einstein's mass–energy relationship, the rest mass energy of this nucleus is $M_{\text{nuc}}(A, Z)c^2$.

Since a nucleus is bound by the strong force, separating it into its constituent protons and neutrons requires energy to overcome the attractive strong force, just as removing an electron from an atom requires energy to overcome the electrostatic force binding the electron to the nucleus. By the law of conservation of energy, *the rest mass energy of a nucleus together with the energy required to separate all the nucleons of the nucleus must be equal to the total rest mass energy of all the separated nucleons.* The energy required to separate a nucleus into individual nucleons is called the **nuclear binding energy** or simply the **binding energy**. The binding energy is therefore the difference between the total rest mass energy of the constituent nucleons and the rest mass energy of the nucleus. The binding energy of a nucleus with A nucleons and Z protons is denoted by $B(A, Z)$:

KEY EQUATION

$$B(A, Z) = Zm_{\text{p}}c^2 + (A - Z)m_{\text{n}}c^2 - M_{\text{nuc}}(A, Z)c^2 \quad (34\text{-}10)$$

Equivalently, we can write the nuclear mass in terms of the nuclear binding energy and the masses of the proton and neutron:

$$M_{\text{nuc}}(A, Z) = Zm_{\text{p}} + (A - Z)m_{\text{n}} - \frac{B(A, Z)}{c^2} \quad (34\text{-}11)$$

The **atomic mass** of a neutral atom, $M_{\text{atom}}(A, Z)$, is equal to the sum of the mass of its nucleus and the masses of its Z electrons. Since the binding energies of the electrons in an atom are relatively small compared to the nuclear energies, they are often neglected in nuclear calculations. Therefore,

$$M_{\text{atom}}(A, Z) = M_{\text{nuc}}(A, Z) + Zm_e \quad (34\text{-}12)$$

The **average atomic mass** is the weighted average of the masses of the naturally occurring isotopes for a given element. For example, the two naturally occurring isotopes of nitrogen are $^{14}_{7}\text{N}$ (99.634%, atomic mass = 14.003 074 u) and $^{15}_{7}\text{N}$ (0.366%, atomic mass = 15.000 109 u), so the average atomic mass of nitrogen is

$$(14.003\,074\text{ u} \times 0.996\,34) + (15.000\,109\text{ u} \times 0.003\,66)$$
$$= 14.0067\text{ u}$$

EXAMPLE 34-2

The Binding Energy of Helium

The atomic mass of helium ^4_2He is 4.002 603 u. Calculate the binding energy of the nucleus.

Solution

The atomic mass includes the mass of the electrons. Since a helium atom has two electrons, we subtract two electron masses from the atomic mass to find the nuclear mass:

Mass of ^4_2He nucleus = $M_{\text{nuc}}(4, 2) = 4.002\,603\text{ u}$
$$- (2 \times 0.000\,548\text{ u})$$
$$= 4.001\,507\text{ u}$$

The binding energy of a helium nucleus is

$$\begin{aligned}B(4, 2) &= 2m_{\text{p}}c^2 + (4 - 2)m_{\text{n}}c^2 - M_{\text{nuc}}(4, 2)c^2 \\ &= (2 \times 1.007\,276\text{ u} + 2 \times 1.008\,665\text{ u} - 4.001\,507\text{ u})c^2 \\ &= (0.030\,375\text{ u})c^2 \\ &= (0.030\,376\text{ u} \times 931.5\text{ MeV}/c^2)c^2 \\ &= 28.29\text{ MeV}\end{aligned}$$

Making sense of the result

The binding energy of the helium-4 nucleus is positive, indicating that the nucleus is bound.

From the above example, we see that it takes 28.29 MeV of energy to separate a ^4_2He nucleus into two protons and two neutrons. The same amount of energy will be released if two protons and two neutrons are combined to form a ^4_2He nucleus. Therefore, the nuclear binding energy can also be defined as the amount of energy released when a nucleus is assembled from individual nucleons.

By itself, the binding energy of a nucleus does not tell us how tightly bound the nucleons are in the nucleus. To compare how tightly various nuclei are bound, we define the **binding energy per nucleon**, $B_{per}(A, Z)$:

$$B_{per}(A, Z) = B(A, Z)/A \qquad (34\text{-}13)$$

From experiments, the total binding energy of a gold nucleus ($^{184}_{79}Au$) is determined to be 1484 MeV. Comparing the binding energy per nucleon of helium ($^{4}_{2}He$) and gold nuclei, we find that the nucleons in gold are somewhat more tightly bound than those in helium:

$$B_{per}(4, 2) = \frac{28.29 \text{ MeV}}{4 \text{ nucleons}} = 7.072 \text{ MeV/nucleon}$$

$$B_{per}(184, 79) = \frac{1484 \text{ MeV}}{184 \text{ nucleons}} = 8.065 \text{ MeV/nucleon}$$

Since nuclear masses of all stable nuclei have been experimentally determined to a high degree of accuracy, we can plot $B_{per}(A, Z)$ as a function of A, as shown in Figure 34-4. This plot is called the **binding energy curve**. Notice that the binding energy per nucleon first sharply increases with A, reaches a maximum near $A = 60$, and then gradually decreases with increasing A. The shape of the binding energy curve is mainly caused by the interplay between the repulsive electrostatic force and the attractive strong force.

The elements with low mass numbers are weakly bound because each nucleon in these nuclei has relatively few adjacent nucleons exerting attractive forces on it. For example, a **deuteron**, the nucleus of deuterium, $^{2}_{1}H$, consists of one proton and one neutron and a binding energy per nucleon of only 1.1 MeV (the total binding energy of a deuteron is 2.2 MeV). A helium-3 nucleus, $^{3}_{2}He$, which has one more proton, has a binding energy per nucleon of 2.5 MeV. Adding an additional neutron to $^{3}_{2}He$ forms a $^{4}_{2}He$ nucleus, which

has a binding energy per nucleon of approximately 7.0 MeV. The $^{4}_{2}He$ nucleus is the most tightly bound of all nuclei with $A \leq 8$.

The binding energy per nucleon increases gradually as the mass number increases from 12 to 55. It peaks in the range $55 \leq A \leq 65$. Nuclei in this range are the most tightly bound and, therefore, the most stable. Iron-56 ($^{56}_{26}Fe$) has the largest binding energy per nucleon, 8.8 MeV. The binding energy per nucleon decreases steadily for $A \geq 65$. The strong force is a short-range force, so each additional nucleon interacts only with the nucleons that are very close to it. However, the Coulomb force has a large range, so each additional proton repels *all* other protons in the nucleus, thus making the nucleus less stable as Z increases.

Neutrons are electrically neutral, and adding a neutron does not increase repulsion; however, when the number of neutrons in a nucleus increases above a critical value, it becomes energetically possible for a neutron to convert into a proton (Section 34-6). Therefore, it is not possible to form heavier nuclei by keeping the number of protons fixed and just adding more and more neutrons.

Most nuclei beyond $A = 208$ are unstable and decay into smaller nuclei by a process, called alpha decay, that we will discuss in the next section.

EXAMPLE 34-3

The Fusion of Two Deuterons

(a) How much energy is gained when two deuterons combine to form a helium-4 nucleus?
(b) How many joules of energy are released when one mole of deuterium (2 g) combines to form $^{4}_{2}He$?

Solution

(a) A nuclear reaction releases energy by converting some of the mass of the original nuclei into energy. From Equation 34-10, the rest mass energy of a deuteron is

$$M_{nuc}(2, 1)c^2 = m_p c^2 + m_n c^2 - B(2, 1)$$

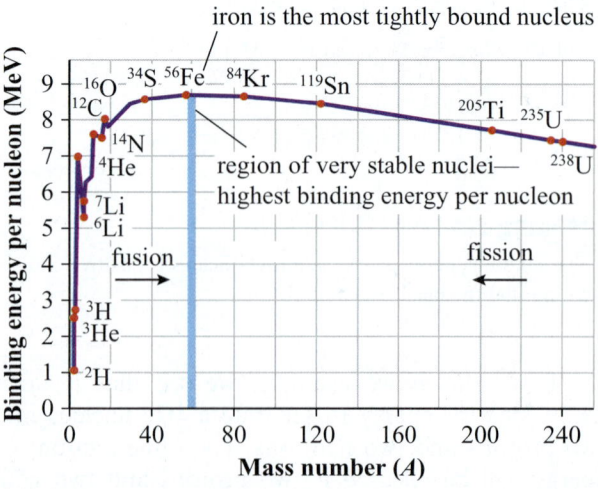

Figure 34-4 A plot of binding energy per nucleon as a function of mass number A for stable nuclei.

MODERN PHYSICS

where $B(2, 1)$ is the binding energy of the deuteron. Similarly, the rest mass energy of the $^4_2$He nucleus is

$$M_{nuc}(4, 2)c^2 = 2m_pc^2 + 2m_nc^2 - B(4, 2)$$

where $B(4, 2)$ is the binding energy of $^4_2$He.

The energy released (ΔE) when two deuterons combine to form a helium nucleus is equal to the difference in the rest mass energies of the two deuterons and the helium nucleus:

$$\begin{aligned}\Delta E &= 2 \times M_{nuc}(2, 1)c^2 - M_{nuc}(4, 2)c^2 \\ &= 2(m_pc^2 + m_nc^2 - B(2, 1)) \\ &\quad - (2m_pc^2 + 2m_nc^2 - B(4, 2)) \\ &= B(4, 2) - 2 \times B(2, 1)\end{aligned}$$

So, the released energy is equal to the difference between the binding energy of the final nucleus ($^4_2$He) and the sum of the binding energies of the initial deuterons. From Figure 34-4, the binding energy *per nucleon* for a deuteron is approximately 1.1 MeV, so $B(2, 1) = 2 \times 1.1$ MeV $= 2.2$ MeV. Similarly, $B(4, 2) = 4 \times 7.0$ MeV $= 28$ MeV. Therefore,

$$\Delta E = 28 \text{ MeV} - (2 \times 2.2 \text{ MeV}) = 24 \text{ MeV}$$

(b) There are 6.02×10^{23} nuclei in one mole of a substance. Therefore, the energy released when one mole of deuterium is converted into helium is

$$(24 \text{ MeV})\left(\frac{1.60 \times 10^{-13} \text{J}}{\text{MeV}}\right)\left(\frac{6.02 \times 10^{23}}{2}\right) = 1.16 \times 10^{12} \text{J}$$

Making sense of the result

Burning one barrel (about 159 L) of oil produces approximately 6.12×10^9 J of energy. So, the fusion of 2 g of deuterium produces as much energy as burning 186 barrels of oil. The energy released per unit mass is vastly greater for a nuclear fuel than for a chemical fuel.

34-4 Nuclear Decay and Radioactivity

In 1896, French physicist Antoine Becquerel (1852–1908) discovered that a mineral, later identified as uranyl potassium sulfate, continuously emits invisible radiation. The emitted radiation was not visible light and could easily penetrate a sheet of black paper. Further experiments by Becquerel, Marie Curie (1867–1934), Pierre Curie (1859–1906), and other scientists showed that invisible radiation is also emitted by some other elements, including thorium, radon, and radium. Furthermore, it was discovered that these elements emitted three different types of invisible radiation, which were called alpha (α), beta (β), and gamma (γ) radiation after the first three letters of the Greek alphabet.

α-, β-, and γ-radiation are collectively called **nuclear radiation** and, as we now know, are emitted by nuclei. An element that emits nuclear radiation is called **radioactive**, and the process of emission of nuclear radiation is called **radioactive decay** or **nuclear decay**.

Like electromagnetic radiation, nuclear radiation carries energy. Only an unstable nucleus emits nuclear radiation. By emitting nuclear radiation, a nucleus transforms into either a nucleus of a lighter element or a lower-energy state of the original nucleus. The nucleus that emits the radiation is called the **parent nucleus**, and the final nucleus is called the **daughter nucleus**. A daughter nucleus always has less energy (including the rest mass energy) than the parent nucleus.

α-radiation consists of helium-4 ($^4_2$He) nuclei, which were originally called **alpha (α) particles**. Thus, a nucleus that emits an α-particle loses two protons and two neutrons and becomes a different element. β-radiation consists of either electrons or positrons. A **positron** is the antiparticle of an electron; it has the same mass as an electron but the opposite charge. γ-radiation is high-energy electromagnetic radiation. In Section 34-6 we will discuss these types of radiation in detail.

A general law applies to the emission of nuclear radiation: *at any given time, the amount of radiation emitted by a material depends only on the number of radioactive nuclei present*. For example, the uranium-238 ($^{238}_{92}$U) nucleus is radioactive and emits an α-particle. The compound uranyl potassium sulfate contains uranium, potassium, sulfur, and oxygen. The number of α-particles emitted in each second by a sample of this compound only depends on the number of uranium-238 nuclei in the sample. As the uranium nuclei decay, fewer of the radioactive nuclei remain; therefore, fewer α-particles are emitted by the sample each second. The rate at which this number decays is proportional to $N(t)$, and

$$-\frac{dN(t)}{dt} \propto N(t) \tag{34-14}$$

where $N(t)$ is the number of radioactive nuclei present in a source at some time t.

The minus sign in Equation 34-14 indicates that the number of nuclei, $N(t)$, decreases with time. We can write an equation for the above relationship by introducing a proportionality constant, which is called the **decay constant** or the **disintegration constant** and is usually denoted by the Greek letter lambda (λ):

KEY EQUATION
$$-\frac{dN(t)}{dt} = \lambda N(t) \tag{34-15}$$

The unit of λ is decays per second or, simply, s^{-1}. The decay constant is a measure of how quickly a radioactive element decays with time. A large value of λ means that the nuclide will decay rapidly.

The decay constant is a characteristic of each radioactive nuclide. For a radioactive nuclide that can emit α-, β-, and γ-radiation, the decay constant has a different value for each type of radiation.

Suppose a radioactive nuclide has a decay constant of $0.10\ \text{s}^{-1}$, and at $t = 0$ we have 1.0×10^6 nuclei. After 1.0 s, the number of nuclei that have decayed is $\lambda N(0) = 0.10 \times 10^6 = 1.0 \times 10^5$, and 9.0×10^5 nuclei are left. During the next second, another $\lambda N(1) = 0.10 \times 9.0 \times 10^5 = 9.0 \times 10^4$ nuclei decay, leaving 8.1×10^5 of the original nuclei. After a third second, $\lambda N(2) = 0.1 \times 8.1 \times 10^5 = 8.1 \times 10^4$ more nuclei have decayed, and 7.29×10^5 are left. In each 1.0 s interval, 10% of the nuclei present at the start of that interval decay. The decreases in the number of radioactive nuclei by *an equal percentage in equal intervals of time* is a property of an exponential function. Figure 34-5 shows a plot of the number of remaining nuclei as a function of time for the above example. Obviously, this number decreases exponentially. So *radioactive decay is an exponential process*.

The Exponential Decay Law

From Equation 34-15, we can derive an expression for the number of parent nuclei remaining at any time, given the number of parent nuclei at some initial time. Equation 34-15 shows that the derivative of the function $N(t)$ is directly proportional to its value, which is a characteristic property of exponential functions. Therefore, $N(t)$ can be written as

$$N(t) = ae^{bt} \qquad (34\text{-}16)$$

where a and b are constants that we need to determine. Substituting this form into Equation 34-15, we get

Left-hand side: $-\dfrac{dN(t)}{dt} = -a\dfrac{d}{dt}(e^{bt}) = -b(ae^{bt})$

Right-hand side: $\lambda N(t) = \lambda(ae^{bt}) \qquad (34\text{-}17)$

Therefore, $b = -\lambda$, and

$$N(t) = ae^{-\lambda t} \qquad (34\text{-}18)$$

To determine a, we need additional information, such as the number of radioactive nuclei present at some given time. Suppose there are N_0 radioactive nuclei at $t = 0$. Evaluating Equation 34-18 at $t = 0$ gives $N(0) = a = N_0$. So,

KEY EQUATION $\qquad N(t) = N_0e^{-\lambda t} \qquad (34\text{-}19)$

where $N(t)$ is the number of undecayed nuclei at time t, N_0 is the initial number of radioactive nuclei, and λ is the decay constant of the nuclide.

The greater the decay constant of a nuclide, the more quickly its nuclei decay. For stable nuclei, $\lambda = 0$, and the number of parent nuclei does not change with time.

Half-Life

A **half-life**, $t_{1/2}$, is the time it takes for the number of radioactive nuclei in a sample to decrease to half the original value. Suppose a sample contains N_0 nuclei of a given radioactive nuclide at $t = 0$, and this number reduces to $N_0/2$ at a later time, $t = t_{1/2}$. Then from Equation 34-19,

$$N(t_{1/2}) = \frac{N_0}{2} = N_0e^{-\lambda t_{1/2}} \qquad (34\text{-}20)$$

and

$$\frac{1}{2} = e^{-\lambda t_{1/2}}$$

Taking the natural logarithm on both sides and using the identities $\ln e^x = x$ and $\ln\left(\dfrac{1}{a}\right) = -\ln a$, we get the following relationship between the decay constant and the half-life:

KEY EQUATION $\qquad t_{1/2} = \dfrac{\ln 2}{\lambda} \approx \dfrac{0.693}{\lambda} \qquad (34\text{-}21)$

The SI unit for half-life is the second. If a sample has 1000 radioactive nuclei at some time, then it will have 500 nuclei after one half-life, 250 after two half-lives, 125 after three half-lives, and so on. A nuclide with a long half-life decays more slowly than a nuclide with a shorter half-life. For example, $^{238}_{92}\text{U}$ has a half-life of 4.468×10^9 yr, so it decays slowly, and a sample containing this nuclide remains radioactive for a long time. Only half of the $^{238}_{92}\text{U}$ present at the formation of Earth has decayed so far. In contrast, radium-224 ($^{224}_{88}\text{Ra}$) has a half-life of 3.632 days, and any $^{224}_{88}\text{Ra}$ present at the time Earth was formed decayed into other nuclides long ago. Table 34-4 lists decay constants and half-lives for a few radioactive nuclides. A more comprehensive list is available at the text's online resources.

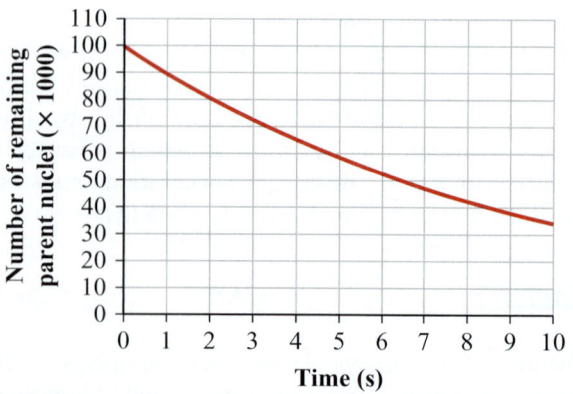

Figure 34-5 A plot of the number of parent nuclei as a function of time for a radioactive nuclide with a decay constant of $0.10\ \text{s}^{-1}$.

Table 34-4 The Decay Constants and Half-Lives for a Few Radioactive Nuclides

Nuclide	Type of Decay	Decay Constant, λ (s^{-1})	Half-Life, $t_{1/2}$
$^{238}_{92}$U	α decay	4.92×10^{-18}	4.46×10^9 yr
$^{224}_{88}$Ra	α decay	2.21×10^{-6}	3.63 days
$^{224}_{86}$Rn	β decay	1.07×10^{-4}	1.8 h
$^{135}_{55}$Cs	β decay	9.55×10^{-15}	2.30×10^6 yr
$^{49}_{19}$K	β decay	1.72×10^{-17}	1.28×10^9 yr
$^{14}_{6}$C	β decay	3.83×10^{-12}	5730 yr

CHECKPOINT 34-4

Decay Constants

Figure 34-6 shows the number of remaining nuclei for four radioactive nuclides as a function of time.
(a) Arrange the nuclides, from high to low, in order of the decay constants.
(b) Each decaying nucleus of the nuclides emits the same amount of energy. Arrange the nuclides in the order of power emitted, from high to low, between $t = 1$ s and $t = 2$ s.

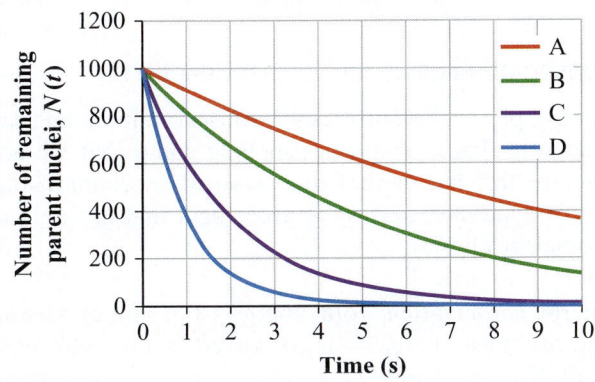

Figure 34-6 Checkpoint 34-4.

ANSWER: (a) D, C, B, A; (b) D, C, B, A. A higher decay constant means that the number of remaining parent nuclei will decrease faster with time.

Decay Rate

It is easier to measure emitted radiation than to count the number of radioactive nuclei in a sample. The decay rate, or **activity**, R, of a sample is the number of nuclei that decay per unit time. The activity equals the magnitude of the time derivative of $N(t)$. From Equation 34-19, we have

$$R(t) = \left| \frac{dN(t)}{dt} \right| = \left| \frac{d}{dt}(N_0 e^{-\lambda t}) \right| = \lambda N_0 e^{-\lambda t} = \lambda N(t)$$

$$(34\text{-}22)$$

The more radioactive nuclei in a source, the greater its activity. Equation 34-22 shows that we can determine the number of radioactive nuclei in a sample simply by measuring the activity of a sample and knowing its decay constant. The SI unit of activity is the becquerel (Bq):

$$1 \text{ Bq} = 1 \text{ decay/s}$$

A commonly used abbreviation for decay/s is dps:

$$1 \text{ dps} = 1 \text{ decay per second}$$

The curie (Ci) is a non-SI unit of activity and is defined as

$$1 \text{ Ci} = 3.7 \times 10^{10} \text{ decays/s}$$

The curie was the original unit of activity, and it is approximately equal to the activity of 1 g of radium. The curie is named after French physicists Marie Curie and Pierre Curie, whose pioneering research on radioactivity includes the discovery of the elements radium and polonium. From Equation 34-22,

$$R(0) = \lambda N_0 \qquad (34\text{-}23)$$

Therefore, we can write

$$R(t) = R(0)e^{-\lambda t} \qquad (34\text{-}24)$$

The activity of a nuclide decays exponentially with time. For example, in one half-life, the decay rate of a nuclide decreases to half of its original value.

EXAMPLE 34-4

Radioactive Gallium

Gallium-67 ($^{67}_{31}$Ga), a radioactive isotope of gallium with a half-life of 3.26 days, is used for some types of medical imaging. A source containing $^{67}_{31}$Ga is calibrated to emit 3.7×10^7 Bq of radiation.

(a) How many gallium-67 atoms are there in the source?
(b) It takes one day to deliver the source to a hospital. What is the activity of the source when it arrives?

Solution

(a) We have already derived all the required relationships between various quantities. We will measure time in seconds so that all quantities are in the correct units:

$$R(0) = 3.7 \times 10^7 \text{ Bq}$$

The number of radioactive nuclei is related to activity by Equation 34-22. Therefore, the number of gallium-67 atoms in the source at $t = 0$ is

$$N(0) = \frac{R(0)}{\lambda}$$

The decay constant of gallium-67 is

$$\lambda = \frac{\ln 2}{t_{1/2}} = \frac{\ln 2}{3.26 \text{ days} \times 24 \text{ h/day} \times 3600 \text{ s/h}}$$

$$= 2.46 \times 10^{-6} \text{ s}^{-1}$$

(continued)

MODERN PHYSICS

Therefore,

$$N(0) = \frac{3.7 \times 10^7 \text{ decay/s}}{2.46 \times 10^{-6} \text{ s}^{-1}}$$

$$= 1.5 \times 10^{13} \text{ atoms}$$

(b) Given the activity at $t = 0$, the activity at a later time is given by $R(t) = R(0)e^{-\lambda t}$. Using 1 day = 24 h/day × 3600 s/h = 86 400 s, the activity of the source when it arrives is

$$R(86\,400) = (3.7 \times 10^7 \text{ Bq})e^{-2.46 \times 10^{-6} \times 86\,400}$$

$$\approx 3.0 \times 10^7 \text{ Bq}$$

Making sense of the result

The number of gallium-67 atoms in the source is greater than the activity of the source. The activity of the source after one day is less than its activity a day before.

34-5 Nuclear Reactions

When a fast-moving nucleus or a particle such as a proton, neutron, or photon collides with a nucleus, the collision rearranges the nucleons. In such **nuclear reactions**, the nuclei in the final states may be different from the initial nuclei. A nuclear reaction can be represented by the following equation:

$$a + X \rightarrow Y + b \tag{34-25}$$

where a is the incoming nucleus or a particle, X is the target nucleus, and Y and b are the final products.

The left side of Equation 34-25 is called the initial state of the reaction, and the right side is called the final state of the reaction. There can be more than two products in the final state. If the projectile a is a nucleus or a positively charged particle, then it must have sufficient kinetic energy to overcome the Coulomb repulsion and come within the range of the strong force of the target nucleus X for the nuclear reaction to occur.

Examples of Nuclear Reactions

■ When a beryllium-9 nucleus absorbs a fast-moving helium-4 nucleus, it transforms into a carbon-12 nucleus and emits a neutron:

$$^4_2\text{He} + ^9_4\text{Be} \rightarrow ^{12}_6\text{C} + ^1_0\text{n} \tag{34-26a}$$

A schematic diagram for this nuclear reaction is shown in Figure 34-7. Note that the number of nucleons is conserved in the reaction. This reaction was used by the British physicist James Chadwick (1891–1974) to prove the existence of neutrons in 1932.

■ A neutron is absorbed by a nitrogen-14 nucleus, resulting in a carbon-14 nucleus and a proton in the final state. This reaction occurs in Earth's atmosphere by cosmic rays and is responsible for replenishing carbon-14, which is present in natural carbon:

$$^1_0\text{n} + ^{14}_7\text{N} \rightarrow ^1_1\text{p} + ^{14}_6\text{C} \tag{34-26b}$$

■ Two fast-moving oxygen-16 nuclei fuse together. One of the possible final states consists of a neon-20 nucleus and a carbon-12 nucleus. The oxygen-fusion reaction occurs in massive stars that have used up the lighter elements in their core:

$$^{16}_8\text{O} + ^{16}_8\text{O} \rightarrow ^{20}_{10}\text{Ne} + ^{12}_6\text{C} \tag{34-26c}$$

Conservation Laws for Nuclear Reactions

Several physical quantities are conserved in all nuclear reactions. These quantities include some that we are already familiar with from classical mechanics and electrodynamics, and there are others that are unique to nuclear physics.

Energy conservation Total energy—the sum of kinetic and rest mass energies—is conserved in a nuclear reaction. Note that neither kinetic energy nor rest mass energy is conserved by itself. Consider the nuclear reaction

$$a + b \rightarrow c + d$$

where a, b, c, and d represent nuclei or particles. The energy of each body in the above reaction is the sum of

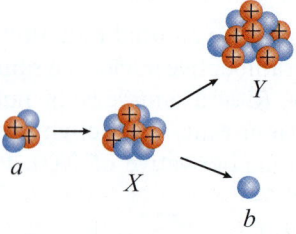

Figure 34-7 A nuclear reaction of beryllium-9 with a helium-4 nucleus. Protons are represented by red circles with a plus sign and neutrons by blue circles.

its kinetic energy and the rest mass energy. The conservation of energy requires that

$$K_a + M_a c^2 + K_b + M_b c^2 = K_c + M_c c^2 + K_d + M_d c^2$$

where K is kinetic energy and Mc^2 is the rest mass energy. The total gain in the kinetic energy in this reaction is equal to the total loss of rest mass energy:

$$\Delta K = (K_f - K_i) = (M_i - M_f)c^2$$

where K_i and K_f denote, respectively, the initial and final total kinetic energies, and $M_i c^2$ and $M_f c^2$ are the corresponding total rest mass energies. The gain, ΔK, in the kinetic energy is called the **Q-value** of the reaction and is denoted by Q. Thus,

KEY EQUATION
$$Q = (M_i - M_f)c^2 \qquad (34\text{-}27)$$

If the rest mass energy in the final state is less than the rest mass energy in the initial state, the Q-value is positive and kinetic energy is released in the nuclear reaction. In this case, the reaction would occur even if $K_i = 0$ (e.g., in a nuclear decay). If the rest mass energy in the final state is greater than the rest mass energy in the initial state, the Q-value is negative. In this case, the bodies in the initial state must have sufficient kinetic energy for the reaction to occur.

By writing the nuclear mass energy in terms of the nuclear binding energy, Equation 34-10, we can write the Q-value of a nuclear reaction in terms of the nuclear binding energies as

$$Q = B_f - B_i$$

where B_i and B_f denote the total binding energies of the bodies in the initial and final state. Here, we have used the fact that the total number of nucleons (protons plus neutrons) in the initial and final states of a nuclear reaction remains the same.

Charge conservation Total charge is conserved in a nuclear reaction. For example, in the beryllium–helium reaction above in Equation 34-26a, the total charge is $6e$ (six protons) in both the initial and final states.

Momentum conservation Momentum is conserved in all nuclear reactions. For the reaction of Equation 34-25,

$$\vec{p}_a + \vec{p}_X = \vec{p}_b + \vec{p}_Y$$

Angular momentum conservation The total angular momentum (the sum of the orbital angular momentum and the spin) is conserved in all nuclear reactions.

Conservation of nucleons The total number of nucleons (protons and neutrons) in a nuclear reaction remains the same. A proton can change into a neutron and vice versa, but the total number of neutrons plus protons remains constant before and after the reaction. For

example, in Reaction 34-26a above, there is a total of 13 nucleons in both the initial state and the final state.

Other physical quantities are conserved in nuclear reactions; however, that discussion is beyond the scope of this introduction.

EXAMPLE 34-5

Proton Absorption on Lithium

In 1932, English physicist John Cockroft (1897–1967) and Irish physicist Ernest Walton (1903–1995) produced the first nuclear reaction by bombarding lithium with 600 keV protons and producing helium in the following process:

$$^1_1\text{H} + ^7_3\text{Li} \rightarrow ^4_2\text{He} + ^4_2\text{He}$$

Determine the Q-value of the reaction.

Solution

From Table 34-5, the atomic masses of hydrogen, lithium, and helium atoms are as follows:

Atomic mass of ^1_1H = 1.007825 u

Atomic mass of ^7_3Li = 7.016003 u

Atomic mass of ^4_2He = 4.002603 u

Table 34-5 Table of Selected Atomic Masses

Z	A	Symbol	Name	Atomic Mass (u)
0	1	n	Neutron	1.008665
1	1	p	Proton	1.007276
1	1	H	Hydrogen	1.007825
1	2	H	Deuterium	2.014102
1	3	H	Tritium	3.016049
2	3	He	Helium-3	3.016029
2	4	He	Helium-4	4.002603
3	6	Li	Lithium	6.015121
3	7	Li	Lithium	7.016003
4	9	Be	Beryllium	9.012182
5	10	B	Boron	10.012937
5	11	B	Boron	11.009305
6	12	C	Carbon	12
6	13	C	Carbon	13.003355
6	14	C	Carbon	14.003242
7	14	N	Nitrogen	14.003074
7	15	N	Nitrogen	15.000109
8	16	O	Oxygen	15.994915
8	17	O	Oxygen	16.999131
8	18	O	Oxygen	17.999160
10	20	Ne	Neon	19.992436
19	39	K	Potassium	38.963708
19	40	K	Potassium	39.964000
19	41	K	Potassium	40.961827
26	56	Fe	Iron	55.934939
36	90	Kr	Krypton	89.919517
37	93	Rb	Rubidium	92.922042

(continued)

MODERN PHYSICS

Table 34-5   Table of Selected Atomic Masses (*Continued*)

Z	A	Symbol	Name	Atomic Mass (u)
38	88	Sr	Strontium	87.905 619
40	102	Zr	Zirconium	101.922 981
52	134	Te	Tellurium	133.911 369
55	141	Cs	Cesium	140.920 046
56	144	Ba	Barium	143.922 953
82	208	Pb	Lead	207.976 627
83	208	Bi	Bismuth	207.979 717
84	210	Po	Polonium	209.982 848
84	218	Po	Polonium	218.008 965
86	222	Rn	Radon	222.017 571
90	231	Th	Thorium	231.036 298
90	232	Th	Thorium	232.038 051
90	234	Th	Thorium	234.043 593
92	232	U	Uranium	232.037 129
92	235	U	Uranium	235.043 924
92	238	U	Uranium	238.050 783

The Q-value of the reaction is equal to the difference in the rest mass energies of the bodies in the initial and the final states:

$$Q = (1.007825\,u + 7.016003\,u) - 2 \times (4.002603\,u)$$

$$= 0.018622\,u$$

$$= \left(\frac{931.5\ \text{MeV}}{u}\right) \times 0.018\,622\ u = 17.35\ \text{MeV}$$

With their rudimentary instrumentation, Cockroft and Walton measured that 17.2 MeV of energy is released in this reaction. This was the first experimental verification of Einstein's mass–energy relation.

Making sense of the result

The Q-value of the reaction is large, indicating that the nuclei in the final state are tightly bound. We know that helium-4 is a tightly bound nucleus.

CHECKPOINT 34-6

Nuclear Reactions

Which of the following quantities may not be conserved in a nuclear reaction? Here the initial state means the state before the reaction occurs.

(a) the rest mass energies of the nuclei (or particles) in the initial state
(b) the kinetic energies of the nuclei (or particles) in the initial state
(c) the number of protons in the initial state
(d) the total number of protons and neutrons in the initial state
(e) the total charge in the initial state

34-6 α, β and γ Decays

In a nuclear decay, a nucleus transforms into another nucleus or rearranges its constituents on its own, without an interaction with an external particle, nucleus, or radiation.

Alpha (α) Decay

From Figure 34-4, we can observe that a large nucleus can increase its binding energy per nucleon, and hence become more stable, by reducing its atomic number and thus transforming into a different nucleus. For example, a nucleus could emit neutrons or protons. Since a helium-4 nucleus has a large binding energy, a heavy nucleus can increase its binding energy more by emitting a helium-4 nucleus than by emitting separate protons and neutrons. When a parent nucleus, X, with Z protons and N neutrons decays by emitting an α-particle (^4_2He), the resulting daughter nucleus, Y, has $Z - 2$ protons and $N - 2$ neutrons. Energy is released in an **alpha decay** (α decay) and appears in the form of the kinetic energy of the α-particle and the daughter nucleus. We can represent an α decay as

KEY EQUATION
$$^A_Z X \rightarrow {}^{A-4}_{Z-2} Y + {}^4_2\text{He} + Q_\alpha \qquad (34\text{-}28)$$

Here, $^A_Z X$ is the parent nucleus, $^{A-4}_{Z-2}Y$ is the daughter nucleus, and Q_α represents the energy released in the decay and is the Q-value of the decay.

From the definition of Q-value, the energy released in α decay is

$$Q_\alpha = M_x c^2 - (M_Y + M_\alpha)c^2 \qquad (34\text{-}29)$$

where M_X, M_Y, and M_α are the atomic masses of the parent nucleus, the daughter nucleus, and the α-particle, respectively.

By the law of conservation of energy, α decay can happen only if the rest mass of the parent nucleus is *greater* than the combined rest masses of the daughter nucleus and the α-particle. Therefore, Q_α must be positive for a nucleus to undergo α decay. We can express Q_α in terms of the binding energies of the nuclei, using Equation 34-10:

$$M_X c^2 = Z m_p c^2 + (A - Z)m_N c^2 - B(A, Z)$$
$$M_Y c^2 = (Z - 2)m_p c^2 + (A - Z - 2)m_N c^2$$
$$\qquad - B(A - 4, Z - 2) \qquad (34\text{-}30)$$
$$M_\alpha c^2 = 2m_p c^2 + 2m_N c^2 - B(4, 2)$$

By inserting Equation 34-30 into Equation 34-29, we obtain an expression for Q_α in terms of binding energies:

$$Q_\alpha = B(A - 4, Z - 2) + B(4, 2) - B(A, Z) \qquad (34\text{-}31)$$

We see that the Q-value is the difference in the binding energies of the final products and the initial parent nucleus. From experiments we know that the binding energy of a helium nucleus is 28.2 MeV, so $B(4, 2) = 28.2$ MeV.

A nuclear decay can be graphically represented by plotting the atomic numbers (charge) of parent and daughter nuclei along the horizontal axis and their energies along the vertical axis. Such a diagram is called a **decay diagram**. Energy is always released in a nuclear decay, so the daughter nucleus is lower in energy than the parent nucleus. In these diagrams, a short horizontal line, centred at a given Z, represents a nucleus with a fixed charge and in a particular energy state. An arrow with its tail at the parent nucleus and its head at the daughter nucleus represents the decay process. The vertical distance between the parent and the daughter nuclei represents the energy released in the decay, that is, the Q-value of the decay (Figure 34-8(a)). Figure 34-8(b) shows a decay diagram for the α decay of uranium-238.

Most of the energy released in the α decay in Example 34-6 is carried away by the α-particle in the form of kinetic energy. Approximately 5% of the energy appears as the kinetic and excitation energy of the polonium nucleus. When formed in this decay process, the polonium nucleus is initially in an excited state (just like an atom can be in an excited state when not all of its electrons are in their lowest-energy states). The excited polonium nucleus reaches its ground state by emitting a γ-ray, a high-energy photon.

(a)

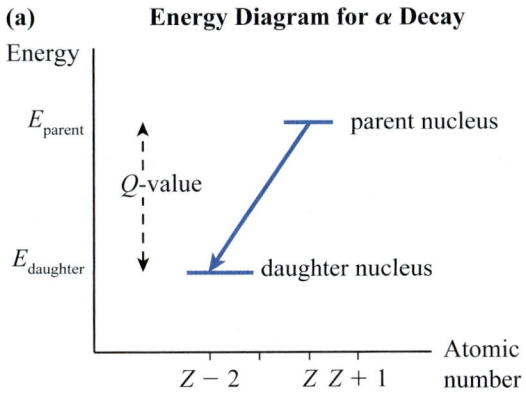

Energy Diagram for α Decay

(b)

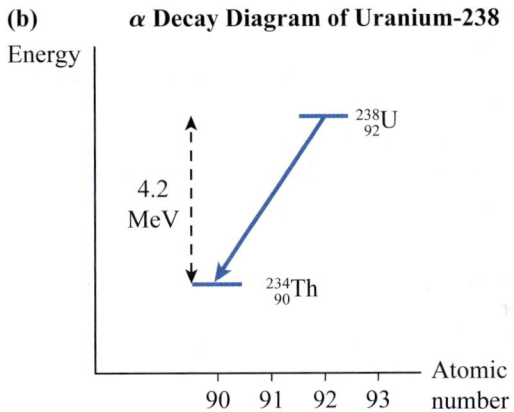

α Decay Diagram of Uranium-238

Figure 34-8 (a) A decay diagram for α decay. (b) A decay diagram for the α decay of $^{238}_{92}$U.

EXAMPLE 34-6

The α Decay of Radon

Radon-222 ($^{222}_{86}$Rn) has a half-life of 3.8235 days and decays by emitting an α-particle.

(a) What is the daughter nucleus in this decay?
(b) Using the atomic masses from Table 34-5, determine the Q-value of this decay.
(c) Find the difference in the binding energies of the parent and the daughter nuclei.

Solution

(a) For the parent nucleus, $A = 222$ and $Z = 86$. For the daughter nucleus, $A = 222 - 4 = 218$ and $Z = 86 - 2 = 84$.

From a periodic or atomic table, we find that the nucleus with $Z = 84$ is polonium (Po) and the daughter nucleus is $^{218}_{84}$Po. Therefore, the equation for α decay is

$$^{222}_{86}\text{Rn} \rightarrow {}^{218}_{84}\text{Po} + {}^{4}_{2}\text{He} + Q_\alpha$$

(b) Looking up the known atomic masses of the isotopes (Table 34-5) in the reaction, we find the following:

Atomic mass of $^{222}_{86}$Rn $= 222.017\,571$ u

Atomic mass of $^{218}_{84}$Po $= 218.008\,965$ u

Atomic mass of $^{4}_{2}$He $= 4.002\,603$ u

Now we can use Equation 34-29 to determine Q_α:

$$Q_\alpha = (222.017571\,\text{u} - 218.008965\,\text{u} - 4.002603\,\text{u})$$

$$\times \left(\frac{931.5\,\text{MeV}}{\text{u}}\right)$$

$$= 5.592\,\text{MeV}$$

(c) From Equation 34-31:

$$Q_\alpha = B(218, 84) + B(4, 2) - B(222, 86)$$

and

$$B(222, 86) - B(218, 84) = B(4, 2) - Q_\alpha$$
$$= 28.2\,\text{MeV} - 5.592\,\text{MeV}$$
$$= 22.6\,\text{MeV}$$

Making sense of the result

The binding energy difference between the parent nucleus and the daughter nuclei is large, which is consistent with the short half-life (3.8235 days) of the parent nucleus, radon-222.

Beta (β) Decay

Nuclei are most stable in the lowest-energy state. Figure 34-9 shows a plot of the rest mass as a function of the number of protons (Z) for various nuclei with $A = 16$. Notice that oxygen-16, with 8 protons and 8 neutrons, has the lowest rest mass and therefore the lowest rest mass energy. The rest mass of $^{16}_{7}N$ is 16.006 102 u, and the rest mass of $^{16}_{8}O$ is 15.994 914 u. Therefore, the energy of $^{16}_{8}O$ is lower by

$$(16.006\,102 \text{ u} - 15.994\,914 \text{ u})(931.5 \text{ MeV/u}) = 10.42 \text{ MeV}$$

Therefore, a $^{16}_{7}N$ nucleus could transform into $^{16}_{8}O$, a nuclide of lower rest mass, by transforming one neutron into a proton.

Now we look at conservation of charge. The initial neutron has no charge, but the final proton has a charge of $+e$. So, there must be another particle, with charge $-e$, that is created when a neutron changes into a proton—this particle is an electron. We can write the process as follows:

$$\text{neutron} \rightarrow \text{proton} + \text{electron}$$

or

$$^{1}_{0}n \rightarrow ^{1}_{1}p + ^{0}_{-1}e \qquad (34\text{-}32)$$

In using the symbol $^{1}_{0}n$ to represent a neutron, we have used the same convention as that for a nucleus. A neutron has no charge ($Z = 0$), and the atomic mass number is $A = 1$. Similarly, an electron has zero atomic mass number and has a charge of -1. The proton remains inside the nucleus; only the electron comes out. The emitted electron is called a beta (β) ray, and the decay process in which a nucleus changes its charge by one unit is called **beta decay** (β decay).

Let us examine the conservation of energy and momentum during this process. We assume that the neutron is at rest. The Q-value of the neutron decay process is

$$Q = m_n c^2 - (m_p c^2 + m_e c^2) \qquad (34\text{-}33)$$

Using $m_n c^2 = 939.565\,56$ MeV, $m_p c^2 = 938.2720$ MeV, and $m_e c^2 = 0.510\,999$ MeV, $Q = 0.7826$ MeV. The Q-value is positive; therefore, the decay of a neutron into a proton is allowed by the law of conservation of energy. A free neutron is indeed unstable. It decays into a proton with a half-life of 614 s.

Since the neutron is at rest, the proton and the electron momenta must be equal and opposite so that the total momentum of the final state is zero. We leave it as an exercise (problem 46 in Chapter 35) to show that the magnitude of the proton and electron momentum, p, is given by

$$\frac{p^2}{c^2} = \left(\frac{m_n^2 + m_p^2 - m_e^2}{2m_n} \right) - m_p^2$$

Inserting values for the masses of the three particles, we get $p = 1.188$ MeV/c. So the proton and electron have a well-defined momentum. The kinetic energy of an electron, K_e, corresponding to this momentum is the difference between its total energy and the rest mass energy:

$$K_e = \sqrt{p^2 c^2 + m_e^2 c^4} - m_e c^2 = 0.782 \text{ MeV}$$

When we observe decays of neutrons, the emerging electrons should all have 0.782 MeV of kinetic energy. But almost all electrons in neutron decay experiments have less kinetic energy, as shown in Figure 34-10. Where is the rest of the energy going? Could neutron decay somehow violate conservation of energy?

This problem puzzled physicists until Wolfgang Pauli suggested that a neutron decays into three particles: a proton, an electron, and a third particle, which he called a **neutrino** (ν). Neutrino is Italian for "small neutral one." Since the particle had not been previously detected, Pauli assumed that it interacts with matter only by the weak force.

Pauli argued that the neutrino and the electron share the energy available in the decay. Since a proton is approximately 1836 times as massive as an electron, the speed of the proton is quite small; consequently, its kinetic energy can be ignored. The sum of the kinetic energies of the electron and the neutrino must be equal to the Q-value calculated in Equation 34-33. So the

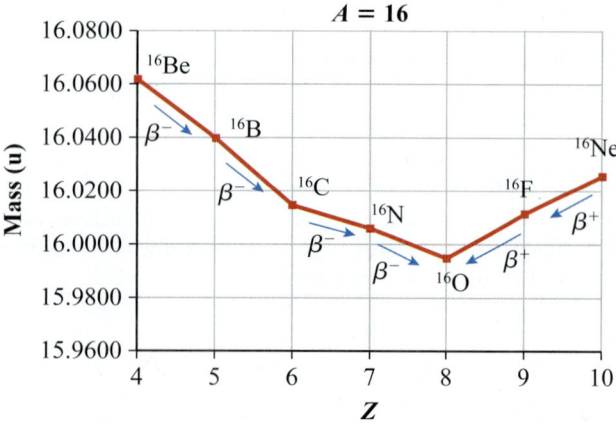

Figure 34-9 The rest mass as a function of the number of protons for various nuclei with $A = 16$.

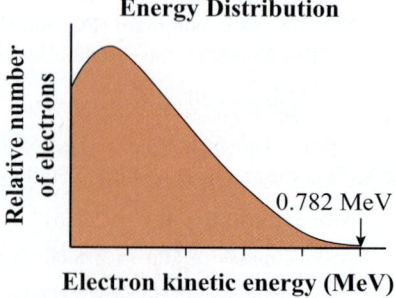

Figure 34-10 The kinetic energy of electrons emitted in the β decay of a neutron.

electron's kinetic energy can have any value between 0 MeV and 0.782 MeV. The neutrino must be close to massless because some electrons in the β decay of neutrons are emitted with kinetic energy nearly equal to the Q-value, leaving almost no energy for the neutrino. So, Pauli predicted that the neutrino has no charge, no mass, and an intrinsic spin of $\frac{1}{2}$, making it a fermion.

Neutrinos were discovered in 1952 by American physicists Frederick Reines (1918–1998) and Clyde Lorrain Cowan (1919–1974), for which they shared a Nobel Prize in 1995. However, 50 years after Pauli's original hypothesis, it was determined that the third particle emitted in neutron β decay is actually an antiparticle of the neutrino, called the antineutrino, $\bar{\nu}$. (We will discuss the nature of antiparticles in Chapter 35.) The β decay of a neutron can then be written as

$$_0^1 n \rightarrow {_1^1} p + {_{-1}^0} e + \bar{\nu} \qquad (34\text{-}34)$$

Similarly, the β decay of a nucleus in which a neutron converts into a proton, for example, the transformation of $_7^{16}N$ into $_8^{16}O$, is written as

$$_7^{16}N \rightarrow {_8^{16}}O + {_{-1}^0}e + \bar{\nu} + Q \qquad (34\text{-}35)$$

The decay specified by Equation 34-35 is called the β decay of $_7^{16}N$. The half-life of $_7^{16}N$ for β decay to $_8^{16}O$ is 7.13 s, and that is why nitrogen-16 is not seen in nature. The Q-value in Equation 34-35 is not the same as the Q-value for the β decay of a free neutron. The neutron that converts into a proton in $_7^{16}N$ β decay is not free and is bound to the nucleus by the strong force. Similarly, the final proton is bound to the $_8^{16}O$ nucleus. We will show how to calculate the Q-value for β decay of a nucleus later in this section.

All $A = 16$ elements with $Z < 8$ eventually β decay into $_8^{16}O$: $_4^{16}Be$ β decays into $_5^{16}B$, which then β decays into $_6^{16}C$, which β decays to $_7^{16}N$, and then finally to $_8^{16}O$. The half-lives and the Q-values for the decays at each stage are different. The process stops at oxygen-16 because it is stable and does not decay.

How about the nuclei with $Z > 8$? From Figure 34-9 we observe that $_9^{16}F$ has a higher rest mass energy than $_8^{16}O$. So $_9^{16}F$ can decay to $_8^{16}O$, which requires transforming a proton into a neutron. To conserve electric charge, a particle with a charge equal to the charge of a proton must be emitted during this decay. This particle is the positron, the antiparticle of the electron. A positron has the same mass as an electron but opposite charge. The positron is denoted by the symbol $_{+1}^0 e$. The energy of positrons emitted during proton decay has a distribution similar to that for electrons emitted in β decay, indicating that a third particle is also produced when a proton decays into a neutron. The emitted particle is the neutrino. So, the β decay of a proton can be represented as

$$_1^1 p \rightarrow {_0^1}n + {_{+1}^0}e + \nu \qquad (34\text{-}36)$$

There is a crucial difference between neutron decay and proton decay. The rest mass of a proton is *less* than the rest mass of a neutron. Therefore, a free proton cannot decay into a neutron. Proton decay into a neutron is possible only inside a nucleus where other protons and neutrons can provide the required energy. Otherwise, all hydrogen atoms would have decayed a long time ago.

The decay of $_9^{16}F$ into $_8^{16}O$ can be represented as

$$_9^{16}F \rightarrow {_8^{16}}O + {_{+1}^0}e + \nu + Q \qquad (34\text{-}37)$$

$_9^{16}F$ is extremely unstable and has a half-life of the order of 10^{-20} s.

When a neutron converts into a proton, a negatively charged particle (electron) is emitted from the nucleus. This process is called **beta-minus (β^-) decay**. Similarly, the process during which a proton converts into a neutron and a positively charged particle (positron) is emitted from the nucleus is called **beta-plus (β^+) decay**. So we say "nitrogen-16 β^- decays and transforms into oxygen-16."

At the start of this section we plotted the rest masses of the isobars with $Z = 16$ (Figure 34-9). Plots of rest masses for other sets of isobars have a similar shape. Figure 34-11 shows a plot of rest mass versus Z for isobars with $A = 208$. Since lead-208 ($_{82}^{208}Pb$) has the lowest rest mass, the other nuclei in this group either β^- or β^+ decay toward lead-208.

Let us see how Q-values for β decays are related to atomic masses of the nuclei involved in the decay.

β^- Decay

The following equation represents the β^- decay of a parent nucleus $_Z^A X$:

KEY EQUATION
$$_Z^A X \rightarrow {_{z+1}^A}Y + {_{-1}^0}e + \bar{\nu} + Q_{\beta^-} \qquad (34\text{-}38)$$

where Q_{β^-} is the Q-value of the β decay.

For example, sulfur-35 has a half-life of 87.9 days for the β decay into chlorine-35:

$$_{16}^{35}S \rightarrow {_{17}^{35}}Cl + {_{-1}^0}e + \bar{\nu} + 0.167\,\text{MeV}$$

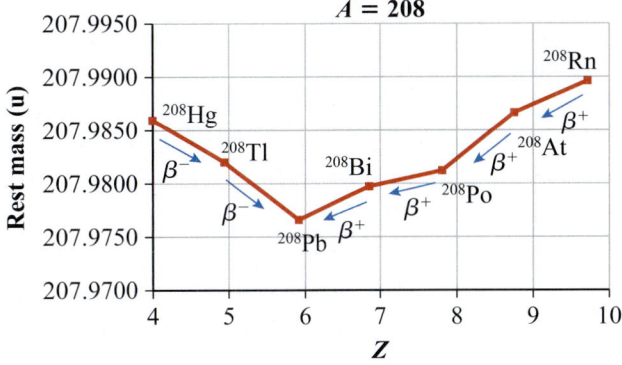

Figure 34-11 The rest mass as a function of the number of protons for various nuclei with $A = 208$.

From the definition of Q-value,

$$Q_{\beta^-} = M_X c^2 - (M_Y + m_e)c^2 \qquad (34\text{-}39)$$

Here, m_e is the electron mass, and M_X and M_Y are the nuclear masses of the parent and the daughter nuclei, respectively. We can also write the Q-value in terms of atomic masses of corresponding atoms by adding the rest mass energy of Z electrons to both terms on the right side of Equation 34-39:

$$Q_{\beta^-} = (M_X + Zm_e)c^2 - (M_Y + m_e + Zm_e)c^2 \quad (34\text{-}40)$$

The parent nucleus has Z protons, so its atom has Z electrons. The daughter nucleus has $Z + 1$ protons, and its atom has $Z + 1$ electrons. Therefore, the terms in the parentheses in Equation 34-40 add to the atomic masses of the parent and daughter elements, and Q_{β^-} is equal to the energy difference between these atomic masses:

$$Q_{\beta^-} = \tilde{M}_X c^2 - \tilde{M}_Y c^2 \qquad (34\text{-}41)$$

If the atomic mass of the parent nucleus is greater than the atomic mass of the daughter nucleus, the parent nucleus will β^- decay into the daughter nucleus.

In a β^- decay, the charge of the nucleus increases by e. The corresponding decay diagram is shown in Figure 34-12.

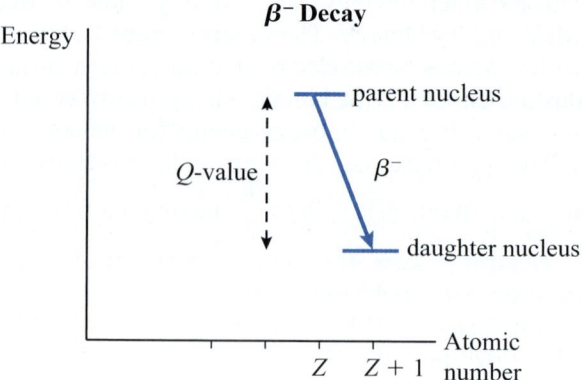

Figure 34-12   A decay diagram for β^- decay.

Radiocarbon Dating

Radiocarbon dating (also called **carbon dating**) uses the naturally occurring carbon-14 ($^{14}_{6}\text{C}$) to estimate the age of carbon-rich materials. Carbon has two stable isotopes, carbon-12 and carbon-13. All other isotopes of carbon are radioactive. Carbon-14 has a half-life of 5730 ± 40 yr and β decays to nitrogen-14 ($^{14}_{7}\text{N}$), a stable element:

$$^{14}_{6}\text{C} \rightarrow {}^{14}_{7}\text{N} + e^- + \bar{\nu}_e$$

EXAMPLE 34-7

The β Decay of Carbon-14

Carbon-14 ($^{14}_{6}\text{C}$) is an unstable isotope that undergoes β^- decay.

(a)   What is the daughter nucleus for the decay?
(b)   Write an equation that describes the decay.
(c)   Calculate the Q-value for the decay.
(d)   Draw a decay diagram for the decay.

Solution

(a)   A carbon-14 nucleus has 6 protons and 8 neutrons. In β^- decay, a neutron converts into a proton. Therefore, the daughter nucleus has 7 protons and 7 neutrons. The daughter nucleus is nitrogen-14 ($^{14}_{7}\text{N}$).

(b)   The decay equation is

$$^{14}_{6}\text{C} \rightarrow {}^{14}_{7}\text{N} + {}^{0}_{-1}e + \bar{\nu} + Q_{\beta^-}$$

(c)   To calculate the Q-value of this decay, we use Equation 34-41. From Table 34-5,

$$\text{Atomic mass of } {}^{14}_{6}\text{C} = 14.003\,242 \text{ u}$$
$$\text{Atomic mass of } {}^{14}_{7}\text{N} = 14.003\,074 \text{ u}$$

The Q-value of the β^- decay is

$$\begin{aligned}
Q_{\beta^-} &= (\tilde{M}_{\text{parent}} - \tilde{M}_{\text{daughter}})c^2 \\
&= (14.003\,242\,\text{u} - 14.003\,074\,\text{u})c^2 \\
&= (1.68 \times 10^{-4}\,\text{u})\left(\frac{931.5\,\text{MeV}/c^2}{\text{u}}\right)c^2 \\
&= 0.156\,\text{MeV}
\end{aligned}$$

(d)   The decay diagram is shown in Figure 34-13.

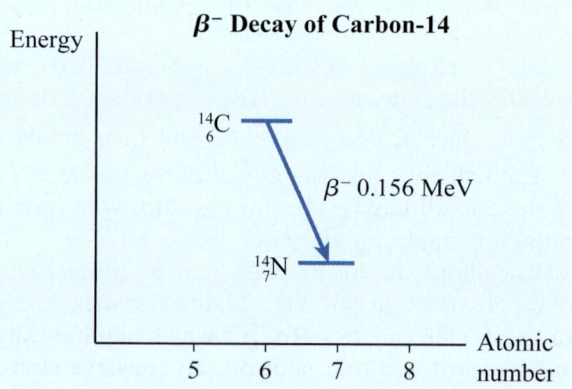

Figure 34-13   The β^- decay of $^{14}_{6}\text{C}$.

Making sense of the result

The charge and the number of nucleons is conserved in the decay equation. The Q-value of the decay is positive.

MODERN PHYSICS

Carbon-14 is continuously formed in the upper atmosphere by high-energy neutrons that are produced by cosmic rays. A neutron knocks out a proton from a nitrogen-14 nucleus, converting it into a carbon-14 nucleus:

$$\,^1_0n + \,^{14}_7N \rightarrow \,^{14}_6C + \,^1_1p$$

Carbon-14 is chemically identical to carbon-12 and reacts with oxygen to form carbon dioxide gas, which diffuses through the atmosphere, biosphere, and oceans. Living organisms (plants, trees, and animals) continuously exchange carbon dioxide with their environment, so they absorb a small amount of carbon-14: one atom of carbon-14 for every 1.2×10^{12} atoms of carbon. The absorbed carbon eventually decays to nitrogen-14. The activity due to carbon-14 decay is approximately 15 decays per minute per gram of carbon.

As long as an organism is alive, its interactions with the environment keep the ratio of carbon-14 in the organism the same as in the environment. When an organism dies, it stops exchanging carbon, and the number of carbon-14 atoms in it begins to decrease. Since the activity of a radioactive element is proportional to the number of undecayed nuclei, the activity of a dead organism decreases with time, reducing by 50% every 5730 yr (Figure 34-14). By measuring the activity of carbon-14 in organic matter, scientists can determine the age of an animal bone, a wooden object, or a piece of cloth that was made from plant or animal fibres. The length of the half-life of carbon-14 makes this method of dating organic matter reasonably reliable for ages less than 60 000 years.

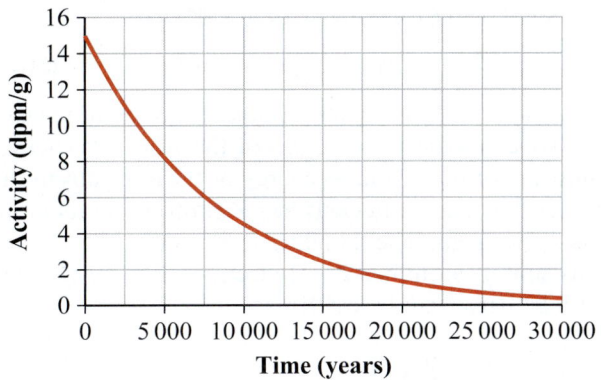

Figure 34-14   The activity of a sample of carbon-14 as a function of time.

β^+ Decay

The following equation describes the β^+ decay of a parent nucleus $^A_Z X$:

KEY EQUATION $^A_Z X \rightarrow \,_{Z-1}^A Y + \,_{+1}^0 e + v + Q_{\beta^+}$ (34-42)

Here, Q_{β^+} is the Q-value for β^+ decay. For example, the β^+ decay of fluorine-17 ($^{17}_9F$) into oxygen-17 ($^{17}_8O$) is given by

$$^{17}_9F \rightarrow \,^{17}_8O + \,^0_{+1}e + v + 1.74\,MeV$$

Similarly, sodium-22 ($^{22}_{11}Na$) β^+ decays into neon-22:

$$^{22}_{11}Na \rightarrow \,^{22}_{10}Ne + \,^0_{+1}e + v + 0.167\,MeV$$

EXAMPLE 34-8

The Age of Iceman Ötzi

In September 1991, two hikers discovered the body of a man frozen in a glacier in the mountains on the border between Austria and Italy. The iceman was nicknamed Ötzi, after the mountain range where he was found. At the University of Innsbruck, it was determined that the carbon-14 activity of the body was 53% of the activity of living organisms. How old is the Ötzi iceman?

Solution

Let $N(0)$ be the number of carbon-14 atoms in a living organism and $N(t)$ be the number of carbon-14 atoms in an organism that has been dead for t years. From Equation 34-19,

$$N(t) = N(0)e^{-\lambda t}$$

where λ is the decay constant of carbon-14.

Dividing the above equation by $N(0)$, taking the natural logarithm on both sides, and using $\ln e^x = x$ gives

$$\ln\left(\frac{N(t)}{N(0)}\right) = -\lambda t$$

and

$$t = -\frac{1}{\lambda}\ln\left(\frac{N(t)}{N(0)}\right) = -\frac{t_{1/2}}{\ln 2}\ln\left(\frac{N(t)}{N(0)}\right)$$

Since the activity is proportional to the number of undecayed nuclei,

$$\frac{\lambda N(t)}{\lambda N(0)} = \frac{N(t)}{N(0)} = 0.53$$

and

$$t = -\frac{5730\,\text{yr}}{\ln 2}\ln 0.53 = 5250\,\text{yr}$$

Making sense of the result

Several corrections and approximations are required in such an analysis, for example, to adjust for historical variations in the concentration of carbon-14 in the atmosphere. It is generally agreed that iceman Ötzi died approximately 5300 yr ago.

MODERN PHYSICS

Similar to Equation 34-39, for β^+ decay,

$$Q_{\beta^+} = M_X c^2 - (M_Y + m_e)c^2 \qquad (34\text{-}43)$$

where m_e is the positron mass (same as the electron mass).

We can write Q_{β^+} in terms of the atomic masses by allowing for the rest mass energy of the electrons in the atoms. The parent nucleus has Z protons, so its atom has Z electrons. The daughter nucleus has $Z - 1$ protons, and its atom has $Z - 1$ electrons. So, the Q-value for a β^+ decay can be written as

$$Q_{\beta^+} = \tilde{M}_X c^2 - \tilde{M}_Y c^2 - 2m_e c^2 \qquad (34\text{-}44)$$

Notice that in terms of atomic masses, the Q-value for β^+ decay differs from the Q-value for β^- decay by twice the electron rest mass energy, $2m_e c^2$, or 1.022 MeV. Since Q_{β^+} must be greater than zero for β^+ decay to occur, the atomic mass of the parent nucleus must be greater than the atomic mass of the daughter nucleus by at least 1.022 MeV.

EXAMPLE 34-9

The Q-Value of the β^+ Decay of Bismuth-208

Calculate the Q-value for the β^+ decay of bismuth-208 ($^{208}_{83}\text{Bi}$) using the atomic masses given in Table 34-5.

Solution

In the β^+ decay of bismuth-208 ($^{208}_{83}\text{Bi}$), the daughter nucleus has one less proton than the parent nucleus. The nucleus with 82 protons is lead, so the decay process is

$$^{208}_{83}\text{Bi} \rightarrow {}^{208}_{82}\text{Pb} + {}^{0}_{+1}\text{e} + v + Q_{\beta^+}$$

The Q-value of the decay is given by Equation 34-44. From Table 34-5,

Atomic mass of $^{208}_{83}\text{Bi} = 207.979\,717$ u

Atomic mass of $^{208}_{82}\text{Pb} = 207.976\,627$ u

Rest mass of an electron $= 0.000\,548$ u

The Q-value of the decay is therefore

$$\begin{aligned}
Q_{\beta^+} &= \tilde{M}_X c^2 - \tilde{M}_Y c^2 - 2m_e c^2 \\
&= (207.979\,717\,\text{u} - 207.976\,627\,\text{u} - (2 \times 0.000\,548\,\text{u}))c^2 \\
&= (0.001\,994\,\text{u})\left(\frac{931.5\,\text{MeV}/c^2}{\text{u}}\right)c^2 \\
&= 1.857\,\text{MeV}
\end{aligned}$$

Making sense of the result

The Q-value is positive for this decay.

EXAMPLE 34-10

Neutrinos from Bananas!

An average banana has approximately 450 mg of potassium. Naturally occurring potassium consists of three isotopes: $^{39}_{19}\text{K}$ (93.3%), $^{40}_{19}\text{K}$ (0.012%), and $^{41}_{19}\text{K}$ (6.73%). Both $^{39}_{19}\text{K}$ and $^{41}_{19}\text{K}$ are stable, and $^{40}_{19}\text{K}$ undergoes both β^- and β^+ decays with a half-life of 1.248×10^9 yr:

$$^{40}_{19}\text{K} \rightarrow {}^{40}_{18}\text{Ar} + {}^{0}_{+1}\text{e} + v \quad (11.2\%)$$

$$^{40}_{19}\text{K} \rightarrow {}^{40}_{20}\text{Ca} + {}^{0}_{-1}\text{e} + \bar{v} \quad (88.8\%)$$

How many neutrinos and antineutrinos are emitted from a typical banana each second?

Solution

To determine how many neutrinos and antineutrinos are emitted from a banana, we need to know the number of $^{40}_{19}\text{K}$ atoms in a banana. We first calculate the average atomic mass of potassium. From Table 34-5, we find that the atomic masses of the potassium isotopes are as follows:

$^{39}_{19}\text{K}$: 38.963\,708 u; $^{40}_{19}\text{K}$: 39.964\,000 u; $^{49}_{19}\text{K}$: 40.961\,827 u

The average atomic mass of naturally occurring potassium is

$$\begin{aligned}
m_K &= (38.963\,708\,\text{u} \times 0.933) + (39.964\,000\,\text{u} \times 0.000\,12) \\
&\quad + (40.961\,827\,\text{u} \times 0.0673) = 39.1147\,\text{u}
\end{aligned}$$

Therefore, 39.111 47 g of potassium contains 6.02×10^{23} atoms, and the number of potassium atoms in 450 mg is

$$\frac{6.02 \times 10^{23}\,\text{atoms}}{39.111\,47\,\text{g}} \times 450 \times 10^{-3}\,\text{g} = 6.93 \times 10^{21}\,\text{atoms}$$

Since 0.012% of naturally occurring potassium is $^{40}_{19}\text{K}$, the number of $^{40}_{19}\text{K}$ atoms in a typical banana is

$$6.93 \times 10^{21}\,\text{atoms} \times \frac{0.012}{100} = 8.32 \times 10^{17}\,\text{atoms}$$

The half-life of $^{40}_{19}\text{K}$ is the same for β^- and β^+ decays. The activity of the potassium in a typical banana is

$$\begin{aligned}
R &= \text{number of atoms} \times \text{decay rate} \\
&= \text{number of atoms} \times \frac{\ln 2}{\text{half} - \text{life}} \\
&= 8.32 \times 10^{17}\,\text{atoms} \times \frac{\ln 2}{1.248 \times 10^9\,\text{yr} \times 31\,557\,600\,\text{s/yr}} \\
&= 14.6\,\text{atoms/s}
\end{aligned}$$

On average, 15 atoms of $^{40}_{19}\text{K}$ decay each second. Since 88.8% of these decays are β^- decays producing an antineutrino, and 11.2% are β^+ decays producing a neutrino, an average banana emits approximately 13 antineutrinos and 2 neutrinos each second.

Making sense of the result

Since naturally occurring potassium contains a small fraction of potassium isotopes that undergo β^- and β^+ decays, it is obvious that a fruit that is rich in potassium would emit neutrinos. Our calculation confirms this.

Nuclear Levels and Gamma (γ) Decay

Since the nucleons in a nucleus are confined within a very small volume, according to quantum mechanics, their energies and momenta are quantized. This situation is similar to that of the electrons bound within an atom. We can therefore expect that, similar to an atomic shell model, protons and neutrons within a nucleus also possess well-defined and discrete energies and hence we can assign an "orbital" or "shell" to each nucleon. This model of the nucleus is called the nuclear shell model. However, there are differences between the atomic shell model and the nuclear shell model.

The electrons move in the Coulomb potential generated by the positive charge of the nucleus located at the centre of the atom. Unlike an atom, there is nothing sitting at the centre of the nucleus generating an attractive potential. The nucleons move in an attractive potential, called the **nuclear potential**, generated by the nuclear forces between the nucleons. Since the nuclear force has a range of about 2 fm, the nuclear potential only exists where there are nucleons. So the shape of the nucleus potential is identical to that of the nuclear density. In Section 34-2, we learned that the nuclear density is approximately constant in the interior of the nucleus and falls off rapidly near the surface. The nuclear potential exhibits similar characteristics. Its strength is constant within the body of the nucleus and then rapidly decreases near the surface. A commonly used form of the nuclear potential is called the Saxon–Wood potential. It assumes a spherically symmetric nucleus and is given by

$$V_{\text{nuc}}(r) = \frac{-V_0}{1 + e^{\left(\frac{r-R}{a}\right)}} \tag{34-45}$$

Here, V_0 is the strength of the nuclear potential measured in MeV; $R = r_0 A^{1/3}$ is the radius of the nucleus; and parameter a, measured in fermis, is a measure of the surface profile of the potential. Typical values of these parameters are $V_0 = 55$ MeV and $a = 0.65$ fm. There is an additional complication. A proton, being positively charged, also experiences a repulsive Coulomb potential due to other protons. This Coulomb potential can be estimated by calculating the electrostatic potential energy of a proton in the presence of a uniformly distributed charge distribution of $(Z-1)$ protons. This yields the following form for the Coulomb potential:

$$V_{\text{coul}}(r) = \begin{cases} \dfrac{(Z-1)e^2}{4\pi\varepsilon_0} \dfrac{1}{R}\left(\dfrac{3}{2} - \dfrac{r^2}{2R^2}\right) & r \le R \\ \dfrac{(Z-1)e^2}{4\pi\varepsilon_0} \dfrac{1}{r} & r > R \end{cases} \tag{34-46}$$

Figure 34-15(a) and (b) shows the nuclear and Coulomb potentials for neutrons and protons in oxygen-16 and

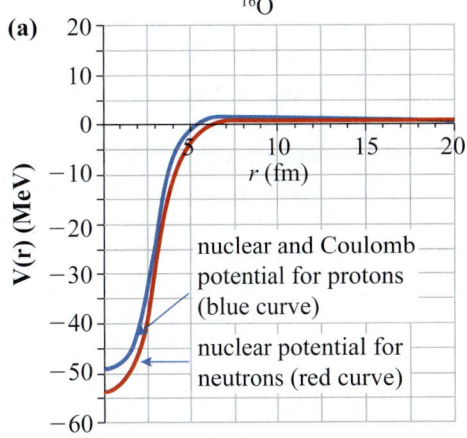

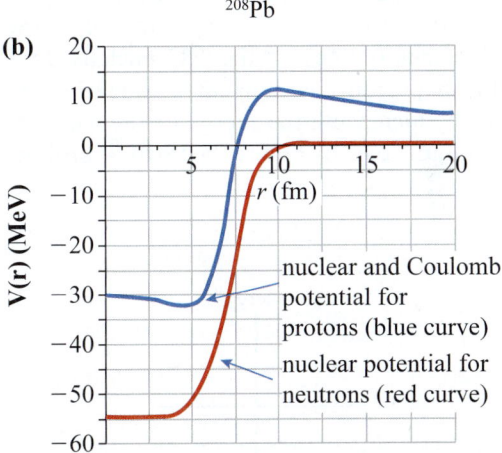

Figure 34-15 The proton and neutron potential wells of oxygen-16 and lead-208 nuclei. In addition to an attractive nuclear potential, a proton also experiences a Coulomb potential due to other protons. The total potential for protons is therefore less deep than that for the neutrons. This effect increases with increasing number of protons (Z).

lead-208 nuclei. Notice that, due to Coulomb repulsion, the proton potential wells are less attractive compared to that for the neutrons, implying that the proton's energy levels are lifted higher compared to the neutron's energy levels. The exact details of the nuclear potential and calculation of corresponding energy eigenstates of nucleons in such a potential are very complicated.

To introduce the main concepts of γ decay, we introduce a very simple model of the energy eigenstates of nucleons bound in a nucleus. Let us assume that the protons and neutrons in a nucleus move in equally spaced energy states (or orbitals). Since protons and neutrons are fermions (with a spin quantum number of 1/2) and the Pauli exclusion principle prohibits two fermions with identical quantum numbers from simultaneously occupying the same energy state, then no more than two protons (or neutrons) can be in a given energy

MODERN PHYSICS

state, one with spin up (↑) and the other with spin down (↓). A single nucleon in an energy level can have its spin projection either up or down.

The lowest-energy state is denoted by E_0, and successively higher-energy states are E_1, E_2, E_3, ... The energy difference between adjacent states is ΔE. For simplicity, we have ignored the effect of Coulomb potential on the proton energy levels (we will discuss the effect of Coulomb repulsion in the next section). Figure 34-16 shows the energy levels of our simple model. Consider the ground state of $^{12}_{6}C$, which has six protons and six neutrons. Two protons fill energy level E_0, two go in E_1, and two go in E_2. Neutrons fill their corresponding levels in the same way (Figure 34-17).

A nucleus in its excited state is indicated by an asterisk. For example, $^{12}_{6}C^*$ denotes a carbon-12 nucleus with at least one nucleon in an energy level higher than the ground state. Figure 34-18(a) and (b) shows two possible excited states of the $^{12}_{6}C$ nucleus. In Figure 34-18(a), a proton has jumped from level E_2 to level E_3, thereby increasing the energy of the whole nucleus by an amount ΔE. In Figure 34-18(b), a proton and a

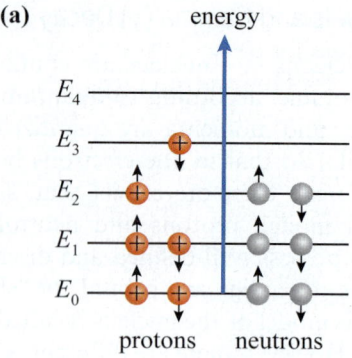

(a)

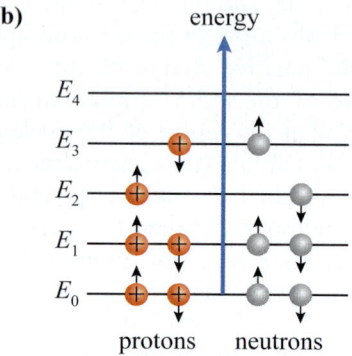

(b)

Figure 34-18 The configuration of protons and neutrons in two excited states of $^{12}_{6}C$. (a) A proton has jumped to a higher-energy level. (b) A proton and a neutron have moved to higher-energy levels.

neutron have jumped from E_2 to E_3, so the energy of this excited state is $2\Delta E$ above the ground-state energy.

A nucleus in an excited state releases its excess energy by emitting a photon, or a γ-ray. This process is called a **gamma decay** (γ decay) and is represented as

KEY EQUATION
$$^{A}_{Z}X^* \rightarrow {^{A}_{Z}X} + \gamma + Q_\gamma \qquad (34\text{-}47)$$

The decay diagram for γ decay is shown in Figure 34-19.

When a nucleus decays by emitting an α-particle or a β-particle, the daughter nucleus is usually left in an excited state. The nucleus then jumps to a lower-energy state by emitting a γ-ray. Two nuclides that differ only in energy state are called **isomers**. For example,

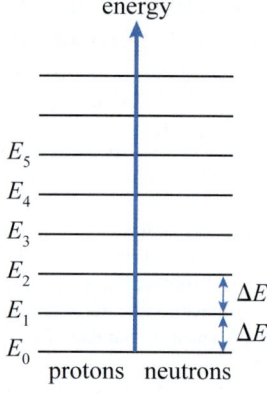

Figure 34-16 Energy levels of protons and neutrons in a nucleus. The Coulomb interaction between protons has been ignored in this energy-level diagram.

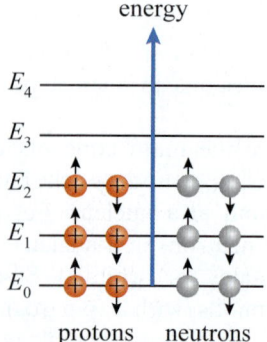

Figure 34-17 The configuration of protons and neutrons in the ground state of $^{12}_{6}C$ (the Coulomb interaction between protons has been ignored).

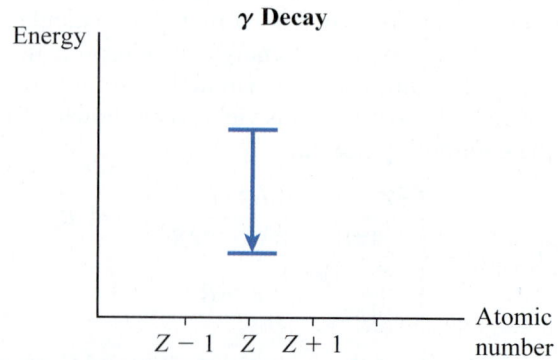

Figure 34-19 The decay diagram for γ decay.

$^{12}_{6}C^{*}$ and $^{12}_{6}C$ are isomers. Usually, a nucleus stays in its excited state for a very short period of time of the order of 10^{-19} s. However, an excited state can exist for seconds, hours, or even years. Such excited states are called **metastable** states.

The energy difference between nuclear levels is of the order of millions of electron volts. In contrast, atomic energy levels are only tens of electron volts apart. Consequently, γ-rays have much higher energies than the electromagnetic radiation typically emitted by electrons in atoms. For example, boron-12 ($^{12}_{5}B$) β decays to various excited states of carbon-12. In one of the excited states, the $^{12}_{6}C^{*}$ nucleus decays to the ground state by emitting a γ-ray of 4.44 MeV.

Since γ-rays are highly penetrating, they can be used to inspect welds, structural materials, and sealed containers (see Figure 27-21).

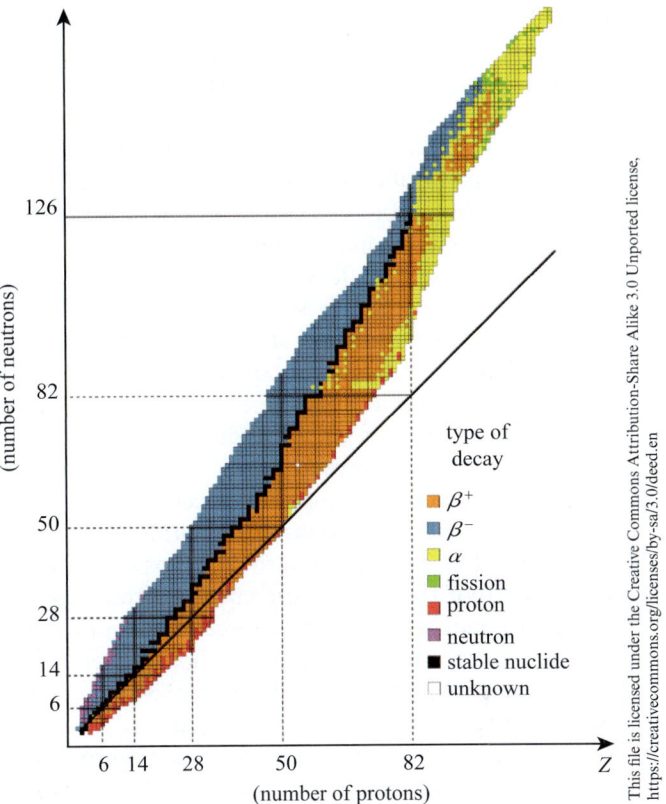

Figure 34-20 Chart of nuclides. Black squares represent stable nuclei, and the straight black line corresponds to $N = Z$.

CHECKPOINT 34-7

Nuclear Decays

For an unstable nucleus to decay, the Q-value of the decay process must be
(a) negative
(b) zero
(c) positive
(d) The probability of a nuclear decay does not depend on the Q-value of decay.

ANSWER: (c) For a nuclear decay to occur, conservation of energy requires that the Q-value of the decay process be positive.

34-7 Nuclear Stability

Figure 34-20, called the chart of nuclides, shows a two-dimensional graph of stable and unstable nuclei with the number of protons (Z) plotted along the horizontal axis and the number of neutrons (N) along the vertical axis. The black squares in the chart represent stable nuclei, that is, nuclei that do not decay.

We see that stable nuclei with N and $Z \leq 20$ lie along the $N = Z$ line. This means that these nuclei have an equal number of protons and neutrons. These include $^{2}_{1}H$, $^{4}_{2}He$, $^{6}_{3}Li$, $^{12}_{6}C$, $^{16}_{8}O$, $^{20}_{10}Ne$, $^{24}_{12}Mg$, and $^{28}_{14}Si$. With increasing A ($A = Z + N$), the stable nuclei depart from the $N = Z$ trend and have more neutrons than protons. For example, for the lead-208 ($^{208}_{82}Pb$) nucleus, the neutron to proton ratio (N/Z) is 126/82. Why do light stable nuclei have an equal number of protons and neutrons and the heavier nuclei need more neutrons to be stable? These questions can be answered using the following facts, which we have already discussed:

1. Since nucleons are confined within a nucleus, their energies are quantized.

2. Because protons and neutrons are fermions and obey the Pauli exclusion principle, any given proton energy level cannot have more than two protons, one with spin up ($\uparrow$) and the other with spin down ($\downarrow$). The same is true for the neutron energy levels.

3. Due to Coulomb repulsion, the proton energy levels are lifted higher (i.e., the protons are less tightly bound) than the neutron energy levels. The Coulomb repulsion increases as Z increases.

4. Since the mass of a neutron is approximately 1.29 MeV/c^2 more than that of a proton, a free neutron can decay into a proton but a free proton cannot decay into a free neutron.

5. A bound neutron or a bound proton can decay into a lower energy level provided that energy, momentum, and angular momentum are conserved in the process.

With these facts in mind, let us examine why light stable nuclei prefer to have $N = Z$ and the heavier nuclei prefer $N > Z$. The simplest nucleus consists of a single proton (the hydrogen atom). The next nucleus contains two nucleons ($A = 2$). There are three possible such nuclei, the di-proton (consisting of two protons), the di-neutron (consisting of two neutrons), and the deuteron (consisting of a proton and a neutron). Of these, only the deuteron is stable, with a binding energy of 2.2 MeV. It is a relatively weakly bound nucleus with no excited states. Experiments have shown that in a deuteron the spins of the proton and the neutron are pointing along the same direction. Why a di-proton nucleus does not exist is easier to understand. The nuclear force has a range of about 2 fm; therefore, to bind together the two protons must come very close to each other. At such a short distance, the Coulomb repulsion between two protons is sufficiently strong to make a di-proton nucleus unstable. But why is a di-neutron nucleus unstable? Since neutrons obey the Pauli exclusion principle, two neutrons in the same energy state must have their spins pointing in opposite directions. So it must be that the nuclear force between two nucleons with their spins pointing in opposite directions ($\uparrow\downarrow$) is not as strong as when their spins are pointing in the same direction ($\uparrow\uparrow$ or $\downarrow\downarrow$). This is verified by experiments. So the strength of the nuclear force between two nucleons is spin dependent. The nuclear force between two neutrons with their spins pointing in opposite directions is not strong enough to bind them (Figure 34-21). That is why a di-neutron nucleus does not exist.

Now let us add a nucleon to a deuteron to obtain an $A = 3$ nucleus. In this case the number of bonds among the three nucleons increases to three and we get a nucleus that is more tightly bound than a deuteron. If a proton is added, the resulting nucleus is helium-3 ($^3_2$He) and it is a stable nucleus with a binding energy per bond of 2.5 MeV. If a neutron is added, we obtain tritium ($^3_1$H), which is unstable with a half-life of 12.3 yr. Figure 34-22 shows nucleons in the ground state of these nuclei.

Why is helium-3 stable and tritium unstable? The reason is that the mass of a neutron is 1.29 MeV/c^2 more than that of a proton. Therefore, a tritium

nucleus can transition to a state of lower energy (and hence become more stable) by converting one of its two neutrons to a proton through β^+ decay. Although the resulting helium-3 nucleus has two protons and there is Coulomb repulsion between the two, a decrease in the rest mass energy by 1.29 MeV/c^2 more than compensates for an increase in energy due to Coulomb repulsion. The rest mass energy of helium-3 is lower than that of tritium.

By adding a nucleon to helium-3, we obtain a nucleus with $A = 4$. In this case, the resulting nucleus has six bonds, and we expect a tightly bound nucleus. If a neutron is added to helium-3, the resulting nucleus is helium-4 ($^4_2$He), which has a binding energy per bond of 4.7 MeV, the highest of all nuclei. It is the most tightly bound light nucleus. All four nucleons are within range of each other's nuclear force. The ground-state configuration of helium-4 is shown in Figure 34-23.

If we add a proton to helium-3, we get lithium-4, which, because of the Coulomb repulsion between three protons, is highly unstable and quickly emits a proton to convert back into helium-3.

Adding a neutron to a helium-4 results in a helium-5 nucleus (2 protons and 3 neutrons), and adding a proton yields a lithium-5 nucleus (3 protons, 2 neutrons)—both of these are unstable. This is a bit surprising, as we would have expected that since the number of bonds has increased to 10 the $A = 5$ nuclei would be more tightly bound. There are two reasons that there is no stable nucleus with five nucleons. First, the lowest-energy proton and neutron levels are already filled, so the fifth nucleon cannot occupy the same state, so it must move to a higher energy level (Figure 34-24). Second, the helium-4 nucleus is very tightly bound and very stable. As soon as a helium-5 nucleus is formed, it lowers its energy by emitting the third neutron and decaying into helium-4. The Pauli exclusion principle plays a crucial role in making $A = 5$ nuclei unstable.

For $A = 6$ there are two bound nuclei, lithium-6 ($^6_3$Li) and helium-6 ($^6_2$He). Lithium-6 is stable, and helium-6 has a half-life of 0.81 s, decaying into lithium-6. A schematic representation of these nuclei is shown in Figure 34-25.

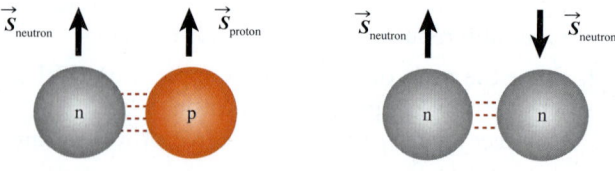

Figure 34-21 The nuclear force between two nucleons with their spins aligned parallel to each other is slightly stronger than between two nucleons with their spins aligned antiparallel.

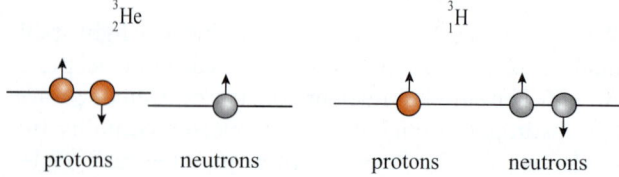

Figure 34-22 Proton and neutron configurations for the ground states of nuclei with $A = 3$. The slightly lifted energy level for protons in helium-3 indicates the presence of Coulomb repulsion.

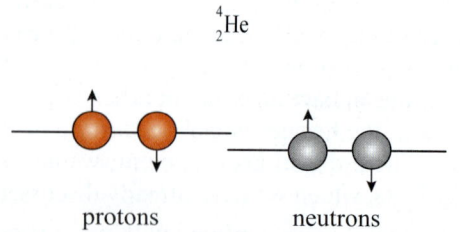

Figure 34-23 Proton and neutron configurations for helium-4.

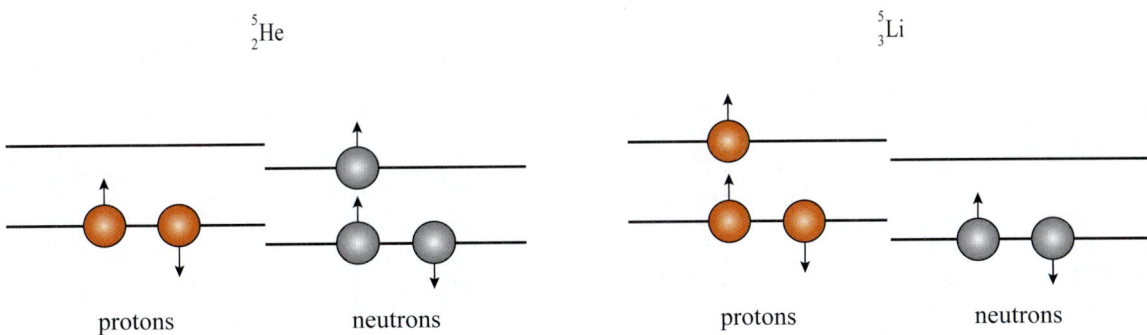

Figure 34-24 A schematic representation of proton and neutron configurations for helium-5 and lithium-5.

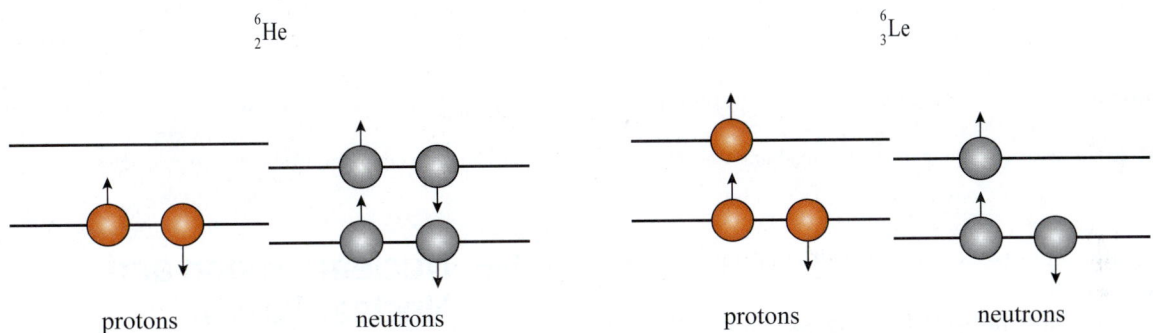

Figure 34-25 A schematic representation of proton and neutron configurations for helium-6 and lithium-6.

Why is lithium-6 stable while helium-6 quickly decays? To understand this, recall that a neutron–proton system is bound (deuteron) but a two-neutron system is unbound. We argued that this implies that the nuclear force between two nucleons with their spins aligned in the same direction is more attractive than between two nucleons that have spins pointing in opposite directions. In helium-6, the two neutrons must move to the next higher neutron energy level, and because of the Pauli exclusion principle, their spins must be antiparallel. This nucleus can become more stable if one of its neutrons in the higher-energy level β decays into a proton. As the lowest-energy proton level is already occupied, the resulting proton occupies the next higher energy level. The unpaired proton and neutron in lithium-6 have their spins pointing along the same direction, so the nuclear force between the two is slightly more attractive. You can think of lithium-6 as a deuteron coupled to a helium-4 core. Although by converting a neutron into a proton the Coulomb repulsion among the protons has increased, the energy (and hence the mass) of lithium-6 is still lower than that of helium-6. That is why a helium-6 nucleus decays into a lithium-6 nucleus.

The above discussion shows why light nuclei prefer to have an equal number of protons and neutrons. The Coulomb repulsion for low-Z nuclei is not strong, and the proton energy levels are lifted slightly higher than the corresponding neutron levels. If the number of neutrons is greater than the number of protons, then the additional neutrons will occupy energy levels that are above the vacant proton levels. A neutron in these levels can therefore beta decay (if allowed by energy and momentum conservation) into a proton and occupy a level that is lower than its original level, resulting in a more stable nucleus.

The $N = Z$ scenario only works for light nuclei. As the number of protons and neutrons increases, the strength of the Coulomb potential begins to increase (see Figure 34-15(a) and (b)), and the proton energy levels are lifted higher and higher with respect to the corresponding neutron energy levels. When the energy difference between the highest filled proton level and the lowest vacant (or partially filled) neutron level is such that energy will be released by converting a proton into a neutron, thereby resulting in a nucleus with lower energy, a proton will β decay into a neutron. Stable nuclei with $A > 40$ begin to have more neutrons than protons. The neutron-to-proton ratio (N/Z) increases with increasing A. For lead-208 ($^{208}_{82}\text{Pb}$), the heaviest stable nucleus, $N/Z = 126/82$.

Let us consider the case of a nucleus with $A = 48$. For this nucleus, the $N = Z$ case corresponds to $Z = N = 24$. The nucleus with 24 protons is chromium, so we should expect $^{48}_{24}\text{Cr}$ to be a stable nucleus. But $^{48}_{24}\text{Cr}$ is unstable, and one of its protons decays into a neutron, transforming it into a vanadium-48 ($^{48}_{23}\text{V}$) nucleus. Vanadium-48 is itself unstable, and its highest-energy proton decays into a neutron, transforming the nucleus into titanium-48 ($^{48}_{22}\text{Ti}$), which is stable.

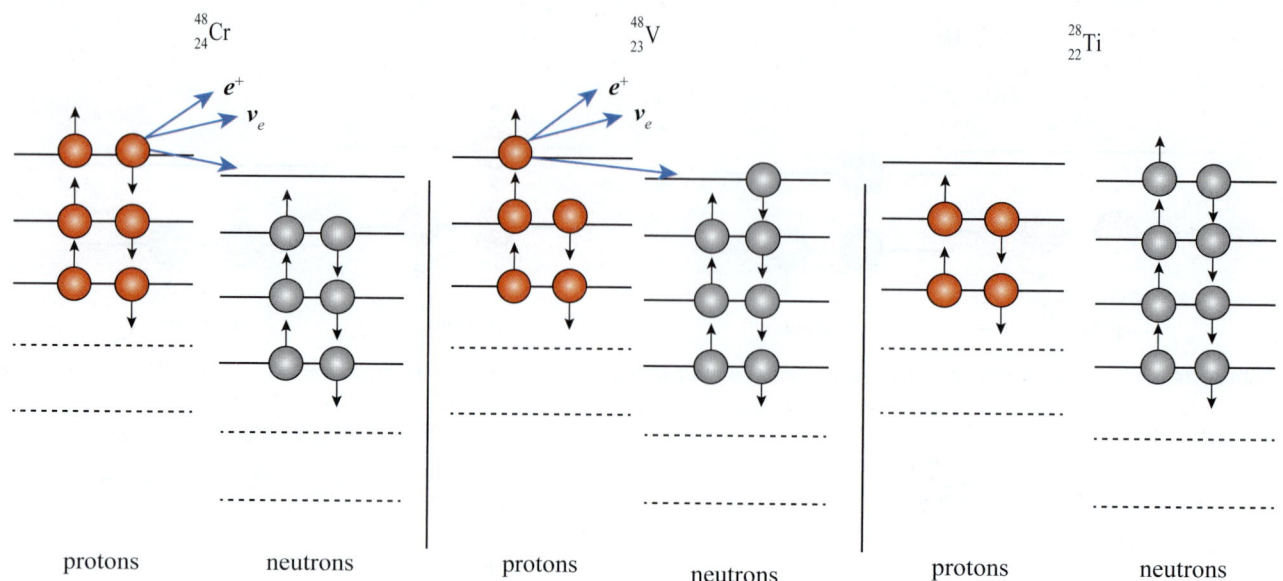

Figure 34-26 β decay of chromium-48 → vanadium-48 → titanium-48. The dashed lines represent filled lower-energy states.

A schematic diagram for this transformation is shown in Figure 34-26.

Here we have presented a very simple explanation of why $N = Z$ for light nuclei and why the N/Z ratio increases with A for heavier nuclei. The details of nuclear stability are quite complicated and depend on an accurate description of ground and excited states of nuclei as well as details of the interaction between nucleons. There are some unusual situations that cannot be understood by the above simple analysis. For example, for $A = 8$, one would expect a stable nucleus that has four protons and four neutrons. However, $^{8}_{4}\text{Be}$ is unstable and quickly decays into two helium-4 nuclei. However, $^{12}_{6}\text{C}$ is stable and does not decay into three helium-4 nuclei.

CHECKPOINT 34-8

Nuclear Stability

The stability of nuclei can be explained by using the following facts (indicate the ones that apply):

(a) Protons and neutrons obey the Pauli exclusion principle applies.
(b) There is Coulomb repulsion among the protons.
(c) Neutrons, being slightly more massive than protons, have greater gravitational attraction between them.
(d) A neutron can decay into a proton and a proton into a neutron.
(e) In a nucleus, the energy levels for neutrons are shifted higher than for protons.

ANSWER: (a), (b), (d)

34-8 Nuclear Fission and Nuclear Fusion

From Figure 34-4, we can see that the binding energy per nucleon for $A \geq 200$ is approximately 7.5 MeV, and for $A \approx 100$ it is approximately 8.5 MeV. So, if a large nucleus ($A > 200$) with a binding energy per nucleon of 7.5 MeV splits into two nuclei of approximately equal size, the binding energy per nucleon would increase to about 8.5 MeV. The two final nuclei would be more tightly bound, so the difference in the total binding energy between the initial and the two final nuclei would be approximately 1 MeV/nucleon. Thus, the total energy released when the large nucleus splits is about 200 MeV. **Nuclear fission** is the process in which a heavy nucleus splits into two nuclei. The final nuclei are called **fission products**. Usually, a few neutrons are emitted during nuclear fission. Fission processes can be either spontaneous or neutron-induced.

Spontaneous Fission

In spontaneous fission, a large nucleus undergoes fission on its own (Figure 34-27). This process can be represented as follows:

$$^{A}_{Z}X \rightarrow ^{A_1}_{Z_1}Y_1 + ^{A_2}_{Z_2}Y_2 + N^{1}_{0}\text{n} \qquad (34\text{-}48)$$

where Y_1 and Y_2 are fission products and N is the number of emitted neutrons, usually between 0 and 3.

A large nucleus can split into various combinations of fission products. As in decay processes, the number

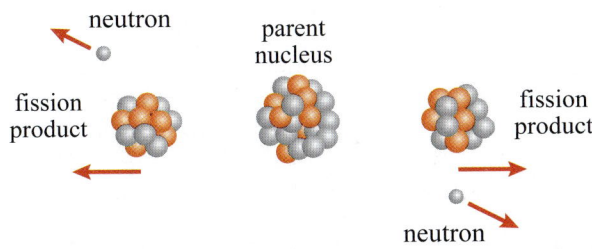

Figure 34-27 During spontaneous fission, a nucleus with a large atomic number spontaneously breaks into two nuclei of similar atomic numbers, accompanied by the emission of a few neutrons.

of nucleons and the total charge must be conserved. Therefore,

$$A = A_1 + A_2 + N \tag{34-49}$$

and

$$Z = Z_1 + Z_2 \tag{34-50}$$

For example, uranium-238 ($^{238}_{92}$U) undergoes spontaneous fission. Here are two ways that uranium-238 fissions:

$$^{238}_{92}\text{U} \rightarrow {}^{134}_{52}\text{Te} + {}^{102}_{40}\text{Zr} + 2{}^{1}_{0}\text{n} + Q_f$$
$$^{238}_{92}\text{U} \rightarrow {}^{134}_{50}\text{Sn} + {}^{102}_{42}\text{Mo} + 2{}^{1}_{0}\text{n} + Q_f \tag{34-51}$$

Here, Q_f denotes the Q-value for the fission decay and is different for the two reactions.

Energy released in spontaneous fission The energy released during the spontaneous fission of a nucleus appears mainly as the kinetic energy of the fission products. Some of the energy also appears as the emission of γ-rays and electrons, as well as the excitation energy of the final nuclei. The total energy released is equal to the difference between the rest mass energy of the parent nucleus and the total rest mass energies of the fission products, including neutrons:

$$Q_f = M_X c^2 - (M_{Y_1} + M_{Y_2} + N m_n)c^2 \tag{34-52}$$

By accounting for Z electron masses, Equation 34-52 can be rewritten in terms of atomic masses:

$$Q_f = \tilde{M}_X c^2 - (\tilde{M}_{Y_1} + \tilde{M}_{Y_2} + N m_n)c^2 \tag{34-53}$$

Why do all heavy nuclei not fission immediately? Why is lead-208 a stable nucleus when it could split into two more tightly bound nuclei? The reason is that an energy barrier, due to Coulomb repulsion, prevents fission. Consider the Coulomb potential energy between a proton and a nucleus, as a function of the distance of the proton from the centre of the nucleus. If the nucleus is assumed to be spherical and its charge Ze to be uniformly distributed, then the total potential energy due

to the Coulomb and the nuclear forces, $V(r)$, between the proton and the nucleus is

$$V(r) = \begin{cases} \dfrac{Ze^2}{4\pi\varepsilon_0}\left(\dfrac{3}{2R} - \dfrac{r^2}{2R^3}\right) - V_0 & r \leq R \\[2ex] \dfrac{Ze^2}{4\pi\varepsilon_0}\dfrac{1}{r} & r > R \end{cases} \tag{34-54}$$

Here we have approximated the nuclear potential energy by an attractive finite square well potential. The depth, V_0, of this potential is of the order of 40 MeV. A plot of $V(r)$ for a proton and an iron-56 nucleus is shown in Figure 34-28. The potential energy is positive outside the nucleus and negative inside, thus forming a barrier. Any charged particle entering or leaving the nucleus must have sufficient kinetic energy to overcome this Coulomb barrier.

Now, consider the spontaneous fission of uranium-238 into tellurium-134 and zirconium-102 (Equation 34-49). We can imagine the uranium nucleus to be made up of a tellurium and a zirconium nucleus, plus two neutrons. As the tellurium and zirconium nuclei begin to separate, the Coulomb potential energy between the two increases, opposing the separation. From Example 34-11 we know that 185.5 MeV of energy is released by this separation. Is this energy sufficient to overcome the Coulomb repulsion and cause the fission? At the point where the $^{134}_{52}$Te and $^{102}_{40}$Zr

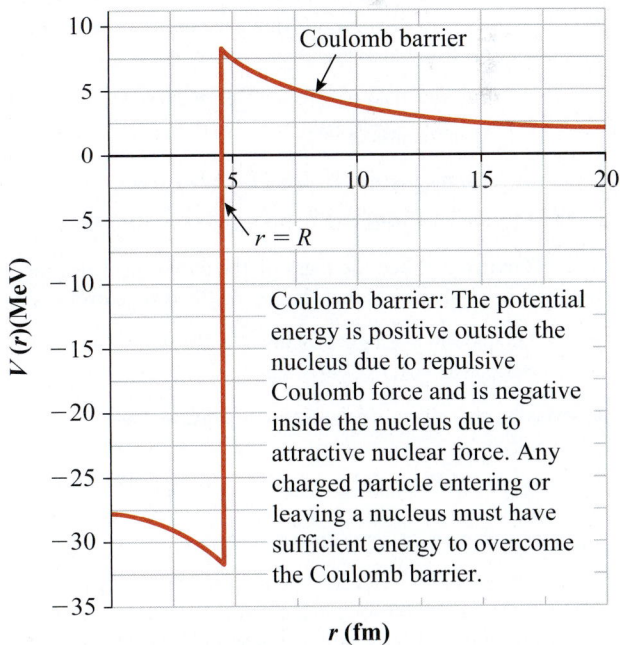

Coulomb barrier: The potential energy is positive outside the nucleus due to repulsive Coulomb force and is negative inside the nucleus due to attractive nuclear force. Any charged particle entering or leaving a nucleus must have sufficient energy to overcome the Coulomb barrier.

Figure 34-28 A plot of the nuclear and Coulomb potential energy between a proton and an iron-56 nucleus (assumed spherical) as a function of the distance between the two.

nuclei are just touching at their surfaces, the Coulomb potential energy is

$$V_C(R_1 + R_2) = \frac{1}{4\pi\varepsilon_0} \frac{(52e)(40e)}{r_0(134^{1/3} + 102^{1/3})}$$

Here, we have used Equation 34-4 for nuclear radii. Using $r_0 = 1.2$ fm and $e^2/(4\pi\varepsilon_0) = 1.44$ MeV fm, we get

$$V_C(R_1 + R_2) = 255 \text{ MeV}$$

The daughter nuclei need 255 MeV of energy to overcome the Coulomb barrier, but only 185.5 MeV of energy is released in the decay process. So the energy released during the spontaneous fission of uranium-238 is not enough for the daughter nuclei to overcome the Coulomb barrier and separate from the parent nucleus. The fact that the Q-value is less than the energy barrier explains why the probability of the spontaneous fission of uranium-238 is so small and the half-life is so long. So, why do $^{238}_{92}$U nuclei undergo spontaneous nuclear fission at all? Nuclear fission, like all other decays, is a quantum-mechanical phenomenon and is possible because of quantum tunnelling, the same phenomenon that allows α decay to occur.

EXAMPLE 34-11

The Fission of Uranium-238

Calculate the energy released during the spontaneous fission of uranium-238 into tellurium-134 and zirconium-102.

Solution

We determine the Q-value of this reaction by using the atomic masses listed in Table 34-5:

$$\text{Atomic mass of } {}^{238}_{92}\text{U} = 238.050\,783 \text{ u}$$
$$\text{Atomic mass of } {}^{134}_{52}\text{Te} = 133.911\,369 \text{ u}$$
$$\text{Atomic mass of } {}^{102}_{40}\text{Zr} = 101.922\,981 \text{ u}$$
$$\text{Mass of two neutrons} = 2 \times 1.008\,665 \text{ u} = 2.017\,330 \text{ u}$$

The difference between the mass of the parent nucleus and the total mass of the products including the two neutrons is

$$\Delta m = 238.050\,783 \text{ u} - (133.911\,369 \text{ u} + 101.922\,981 \text{ u} + 2 \times 1.008\,665 \text{ u}) = 0.199\,103 \text{ u}$$

Converting this mass difference into energy, we have

$$Q_f = 0.199\,103 \text{ u} \times \frac{931.5 \text{ MeV}}{\text{u}} = 185.5 \text{ MeV}$$

Making sense of the result

Although 185.5 MeV is an enormous amount of energy compared to the energy released in a chemical reaction, the half-life of uranium-238 for spontaneous fission is approximately 8×10^{15} yr. Therefore, the rate at which energy is released by the spontaneous fission of a sample of uranium-238 is as small as very few nuclear fissions per second.

Neutron-Induced Fission

Large A nuclides can be induced to fission by bombarding them with neutrons. This process is called **neutron-induced fission** and can be represented by

$${}^1_0\text{n} + {}^A_Z X \rightarrow {}^{A_1}_{Z_1} Y_1 + {}^{A_2}_{Z_2} Y_2 + N {}^1_0\text{n} \qquad (34\text{-}55)$$

For example, neutrons can cause uranium-235 to fission in several different ways, including

$${}^1_0\text{n} + {}^{235}_{92}\text{U} \rightarrow {}^{93}_{37}\text{Rb} + {}^{141}_{55}\text{Cs} + 2{}^1_0\text{n} + 180 \text{ MeV} \ (34\text{-}56)$$

and

$${}^1_0\text{n} + {}^{235}_{92}\text{U} \rightarrow {}^{92}_{36}\text{Kr} + {}^{141}_{56}\text{Cs} + 3{}^1_0\text{n} + 173 \text{ MeV} \ (34\text{-}57)$$

The calculation of the energy released in neutron-induced fission is similar to the calculation for spontaneous emission, but the total energy of the incident neutron must be included in the energy of the initial state. Generally, the kinetic energy of the incident neutron is less than a few MeV.

Chain reactions Consider a large number of uranium-235 nuclei and a single neutron that initiates the fission reaction described by Equation 34-55. Let us assume that two neutrons are emitted in this fission reaction. The neutrons emitted in the initial fission can be absorbed by two other uranium nuclei, causing these nuclei to fission and emit four neutrons. These four neutrons can then induce four more nuclei to fission, generating eight neutrons, and so on (Figure 34-29). If each neutron induces another uranium nucleus to fission, the number of nuclei that fission doubles at

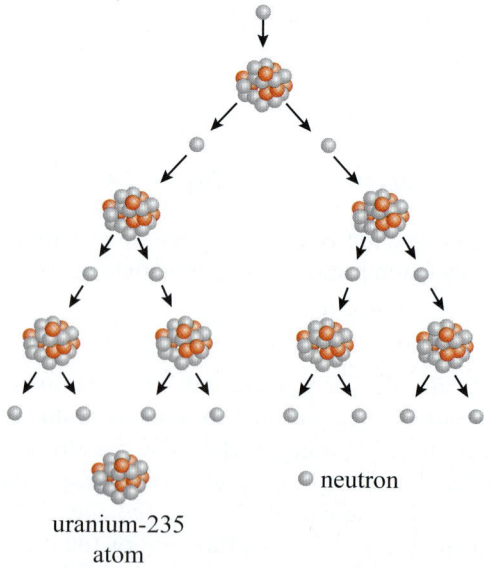

uranium-235 atom   ● neutron

Figure 34-29 An ideal chain reaction. Each fission reaction produces two neutrons, each of which then causes another nucleus to fission.

each stage. It takes approximately 10^{-4} s for a neutron to be emitted and then absorbed again. Therefore, in principle, if each emitted neutron interacts with a uranium nucleus causing it to fission, then within one-hundredth of a second of the start of the process, $1 + 2 + 2^2 + 2^3 + \cdots + 2^{100} \approx 2.5 \times 10^{30}$ uranium nuclei decay. Since 185 MeV of energy is released per decay, approximately 4.7×10^{32} MeV, or 7.5×10^{19} J, of energy is released in this short time. This amount of energy is equivalent to the detonation of about 15 gigatonnes of TNT.

Such a sequence of self-propagating nuclear reactions is called a **chain reaction**. Chain reactions are the basis of nuclear power generating plants and nuclear bombs. In reality, not all neutrons produced in the fission process act as triggers for other fission reactions. Many neutrons escape through the material, scatter off parent nuclei, or cause other types of reactions.

How a nuclear reactor works As described above, approximately 200 MeV of energy is released when a heavy nucleus with $A > 200$ fissions into two smaller nuclei. Most nuclear reactors use uranium-235 $\left(^{235}_{92}\text{U}\right)$ as the nuclear fuel. As discussed earlier, a uranium-235 nucleus can fission by absorbing a slow-moving neutron. Some of the possible outcomes of this fission process are

$$^{1}_{0}\text{n} + ^{235}_{92}\text{U} \rightarrow ^{140}_{56}\text{Ba} + ^{93}_{36}\text{Cs} + 3^{1}_{0}\text{n}$$

$$^{1}_{0}\text{n} + ^{235}_{92}\text{U} \rightarrow ^{134}_{55}\text{Cs} + ^{100}_{37}\text{Rb} + 2^{1}_{0}\text{n}$$

$$^{1}_{0}\text{n} + ^{235}_{92}\text{U} \rightarrow ^{131}_{53}\text{I} + ^{104}_{39}\text{Y} + ^{1}_{0}\text{n}$$

On average, 2.4 neutrons are produced in the fission of each uranium-235 nucleus. These neutrons, called **prompt neutrons**, can be used to fission other uranium-235 nuclei. The key to generating continuous energy in a nuclear power plant is to establish a self-sustaining chain reaction so that, on average, the same number of uranium-235 nuclei fission each second. For this, we need a sufficient number of uranium-235 nuclei available to fission, and we need to control the fission process so that the number of fission prompt neutrons does not go out of control.

Naturally occurring uranium ore consists mostly of uranium-238 with only 0.72% uranium-235. Uranium-238 does not fission by absorbing slow neutrons. To be used as a **nuclear fuel**, uranium must contain at least 4% uranium-235 (an exception is the Canadian CANDU reactor, which uses natural uranium as the nuclear fuel). To produce this **enriched uranium**, uranium hexafluoride, a compound consisting of uranium and fluorine, is vaporized and rapidly spun in a tall, cylindrical **centrifuge**. Uranium-235 atoms are lighter than uranium-238 atoms (which have three more neutrons) and tend to stay in the middle of the centrifuge, and the uranium-238 atoms move closer to the outer wall. Thus, some of the uranium-238

can be separated out, leaving an increased proportion of uranium-235 at the centre of the centrifuge. This process is repeated until uranium fuel with desired fraction of uranium-235 is obtained. Enriched uranium is pressed into thumb-sized pellets, which are stacked in long hollow cylinders made of zirconium, called **fuel rods** (Figure 34-30). Several fuel rods are bundled together, parallel to each other, with space between them.

With a sufficient concentration of uranium-235 nuclei, a chain reaction can occur, and the number of prompt neutrons increases quickly. Uncontrolled, the number of fissions per second would increase quickly, generating enough heat to melt the nuclear fuel and cause a **nuclear meltdown**. The number of neutrons in the reactor core is controlled by **control rods**, which are long, thin cylinders made of an element that readily absorbs neutrons, such as boron-10. The control rods are positioned between the fuel rods, and they can be either fully or partially inserted. This assembly is called a **nuclear core**. When the control rods are pulled out, the reactor runs at its maximum power. When the rods are completely inserted, most of the prompt neutrons are absorbed, and the chain reaction largely stops.

Energy released by fission is carried out of the nuclear core by a **coolant**, usually water, because water is readily available and inexpensive. However, other types of coolants are also used. Canadian-made nuclear reactors, called CANDU reactors, use heavy water as a coolant. In heavy water, the hydrogen atom is replaced by a deuterium atom (a deuterium atom contains one proton and one neutron in its nucleus).

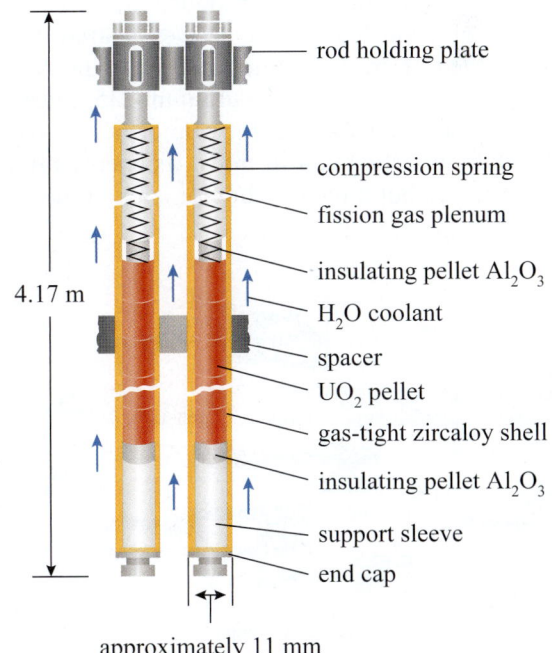

Figure 34-30 A nuclear fuel rod consists of enriched uranium oxide pellets that are packed in a thin cylinder made of zirconium alloy.

The average energy of prompt neutrons is approximately 1.0 MeV. A neutron with this energy is moving at approximately 4.5% of the speed of light. The probability of a $^{235}_{92}$U nucleus absorbing such a fast-moving neutron is very small. Therefore, these neutrons are slowed down by letting them lose energy during collisions with a **moderator**. In many reactors, water is also used as a moderator. The whole assembly, consisting of the fuel rods, control rods, and the coolant, is housed in a thick steel vessel called the **containment vessel**. All radioactive elements produced during the operation of the power plant remain within the containment vessel.

Steel pipes bring water into the containment vessel and heated water out of it. In boiling water reactors (BWRs), the type used in the Fukushima nuclear plant, the heated water is kept at a high pressure, usually 50 to 100 times atmospheric pressure. Water under pressure boils at a higher temperature, so the water temperature in BWRs reaches between 200 °C and 250 °C. The heated water is then turned into steam, which drives a steam turbine to generate electricity (Figure 34-31). Approximately 30% of the energy produced during fission is converted into electricity, and the rest is lost to the cooling water.

Over 90 different isotopes are produced when $^{235}_{92}$U fissions. Most of these fission products are unstable and decay into stable elements by emitting high-energy electrons, neutrons, and γ-rays. Some of the common fission products in a nuclear reactor are iodine-131 ($^{131}_{53}$I, half-life 8.025 days), cesium-136 ($^{136}_{55}$Cs, half-life 13.04 days), cesium-137 ($^{137}_{55}$Cs, half-life 30.08 years), and barium-140 ($^{140}_{56}$Ba, half-life 12.75 days).

About 1% of the fission products decay by emitting neutrons, which are called **delayed neutrons** to distinguish them from the prompt neutrons. The delayed neutrons are also absorbed by uranium-235, causing it to fission. So, a nuclear reactor cannot be completely shut off by inserting control rods. Its energy production decreases, but it remains hot for a long time.

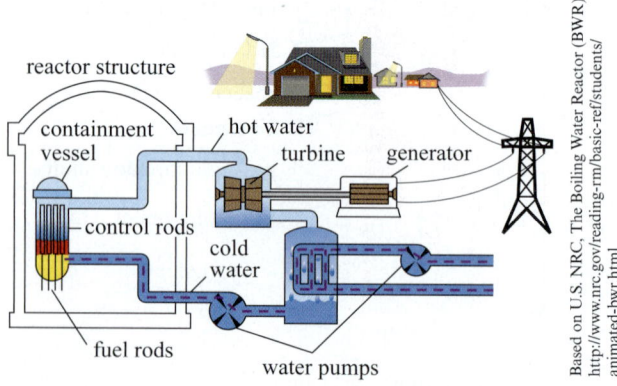

reactor structure

containment vessel

hot water

turbine   generator

control rods

cold water

fuel rods

water pumps

Based on U.S. NRC, The Boiling Water Reactor (BWR), http://www.nrc.gov/reading-rm/basic-ref/students/animated-bwr.html

Figure 34-31 A BWR.

When the concentration of uranium-235 in fuel rods decreases below 1%, the rods need to be replaced. The spent rods are placed in large pools of water for a number of years until the radioactivity has fallen to a level where they can be stored. Dealing with radioactive waste products is a major technological challenge.

The fission products are harmful. For example, iodine-131 ($^{131}_{53}$I), a radioactive isotope of iodine, β decays into xenon-131, a stable isotope of xenon, in two steps:

$$^{131}_{53}\text{I} \rightarrow {}^{131}_{54}\text{Xe*} + e^- + \bar{\nu}_e + 606 \text{ keV}$$

$$^{131}_{54}\text{Xe*} \rightarrow {}^{131}_{54}\text{Xe} + \gamma + 364 \text{ keV}$$

Most of the 606 keV of energy produced in β decay (the first step) is shared between the electron and the antineutrino. Xenon-131 produced in the β decay is in an excited state and decays to its ground state by emitting a 364 keV photon (the second step). The electron and the photon have sufficiently high energy to penetrate human tissue to a depth of a few millimetres and cause cell mutation and destruction. Stable iodine-127 and radioactive iodine-131 are chemically identical. The body absorbs iodine and concentrates it in the thyroid gland, which uses it for the production of thyroxin hormones. If radioactive iodine-131 is present in the food or the air, it will be absorbed by the body and may lead to thyroid cancer. During the Fukushima disaster (and other accidental releases of fission products into the atmosphere), people nearby were advised to take natural (and stable) iodine-127 pills to saturate their thyroid glands with the stable isotope and make them less likely to absorb radioactive iodine-131.

Nuclear Fusion

From Figure 34-4, we observe that the binding energy per nucleon can also be increased by combining two lighter nuclei, thus releasing energy. The combination of two nuclides to form a heavier nuclide is called **nuclear fusion**. Fusion is the source of energy in all stars, including the Sun. A key series of fusion reactions in the core of relatively cooler stars, like the Sun, turns four hydrogen nuclei (protons) into a helium nucleus. In the first stage, two protons ($^1_1$H + $^1_1$H) combine to form a hydrogen-2 nucleus (deuteron) with the emission of a positron and a neutrino. The hydrogen-2 nucleus then combines with a proton to form a helium-3 nucleus with the emission of a γ-ray. Two helium-3 nuclei then combine to form a helium-4 nucleus with the emission of two protons. Energy is released in each stage of this cycle:

$$^1_1H + {}^1_1H \rightarrow {}^2_1H + {}^0_{+1}e + v + 0.42\,\text{MeV} \qquad \text{(34-58a)}$$

$$^1_1H + {}^2_1H \rightarrow {}^3_2He + \gamma + 5.49\,\text{MeV} \qquad \text{(34-58b)}$$

$$^3_2He + {}^3_2He \rightarrow {}^4_2He + 2({}^1_1H) + 12.86\,\text{MeV} \qquad \text{(34-58c)}$$

The positron $(^0_{+1}e)$ produced in Reaction (34-58a) combines with a surrounding electron (in the core of a star the electrons are knocked out from the hydrogen atoms). When an electron combines with a positron, both are annihilated into two photons with a total energy of 1.022 MeV (equal to the combined rest mass energies of the two):

$$^0_{+1}e + {}^0_{-1}e \rightarrow 2\gamma$$

Thus, in the reactions of Equation 34-58, four protons combine to form a helium-4 nucleus, and the overall reaction can be written as

$$6({}^1_1H) \rightarrow {}^4_2He + 2({}^1_1H) + 2{}^0_{+1}e + 2v + 26.7\,\text{MeV}$$

$$\text{(34-59)}$$

Reaction 34-59 is called the **proton–proton cycle**. Of the 26.7 MeV of energy released in this cycle, approximately 25 MeV is used to heat the star, and the rest is carried away by the two neutrinos.

For two nuclei to fuse together, they must have sufficient kinetic energy to overcome the Coulomb repulsion between them. In Reaction 34-58a, the two protons must overlap (be within 1.2 fm of each other) to form a deuteron. At this distance, the Coulomb repulsion between the protons is approximately 1.2 MeV. So, the average kinetic energy of the protons must be at least 1.2 MeV to overcome the repulsion. From the kinetic theory of gases, we know that the average kinetic energy, $\overline{K}$, of a gas particle is related to the absolute temperature, T, by $\overline{K} = \dfrac{3}{2}k_BT$, where $k_B = 1.38 \times 10^{-23}$ J/K $= 8.62 \times 10^5$ eV/K is Boltzmann's constant and T is measured in kelvin. Taking the average kinetic energy to be 1.2 MeV, the required temperature is

$$T = \frac{2\overline{K}}{3k_B} = \frac{2 \times 1.2 \times 10^6\,\text{eV}}{3 \times 8.62 \times 10^{-5}\,\text{eV/K}} \approx 9 \times 10^9\,\text{K}$$

If all the kinetic energy is due to thermal motion, the protons must be in an environment that is at a temperature of approximately 9 billion kelvins (9 GK). Such high temperatures exist in the interiors of stars. The temperature of the Sun's core is approximately 15 MK, high enough to ionize the hydrogen gas into electrons and protons. The kinetic energies of the protons are distributed about a mean value corresponding to this temperature. A tiny fraction of protons in the

Sun's interior have kinetic energies equivalent to the temperature of 9 GK and can fuse to form a deuteron. Even if the energies of colliding protons are below 1.2 MeV, quantum tunnelling can cause two protons to tunnel through the Coulomb barrier and form a deuteron.

EXAMPLE 34-12

Energy Released in the Proton-Proton Cycle

Using the Q-values given in the reactions in Equation 34-58, show that approximately 26.7 MeV of energy is released in the proton–proton cycle of Reaction 34-59.

Solution

In Reaction 34-58c, two helium-3 nuclei fuse to form a helium-4 nucleus and two protons. To form these two helium-3 nuclei, Reactions 34-58a and 34-58b need to happen twice. Also, the positron emitted in Reaction 34-58a combines with an electron to produce two photons of energy 1.022 MeV. Therefore, the energy released in the proton–proton cycle is as follows:

2 × the energy released in proton–proton fusion
$= 2 \times 0.42$ MeV
$= 0.84$ MeV
2 × the energy released in proton–deuteron fusion
$= 2 \times 5.49$ MeV
$= 10.98$ MeV
2 × the energy released in electron–positron annihilation $= 2 \times 1.022$ MeV
$= 2.044$ MeV
Energy released in the fusion of two helium-3 nuclei $= 12.86$ MeV
Total energy released: 0.84 MeV + 10.98 MeV + 2.044 MeV + 12.86 MeV = 26.72 MeV

Making sense of the result

The energy released in the proton–proton cycle is a positive number, indicating that energy is released during fusion of lighter elements.

CHECKPOINT 34-9

Nuclear Fission and Nuclear Fusion

The binding energy per nucleon _____ in nuclear fission and _____ in nuclear fusion.
(a) increases, increases
(b) increases, decreases
(c) decreases, increases
(d) decreases, decreases
(e) does not change, does not change

ANSWER: (a) In both nuclear fission and nuclear fusion, the daughter nuclei are more tightly bound than the parent nuclei. Therefore, the binding energy per nucleon increases in both decays.

34-9 Ionizing Radiation

Ionizing radiation is electromagnetic radiation or particle radiation with sufficient energy to ionize atoms and molecules by removing bound electrons from them. The types of radiation in **ionizing electromagnetic radiation** are X-rays and γ-rays. **Ionizing particle radiation** generally consists of α-particles, protons, neutrons, electrons, positrons, or other elementary particles. The sources of ionizing radiation are radioactive materials, X-ray machines, nuclear reactors, particle accelerators, and cosmic rays.

To become free, a bound electron must absorb energy greater than its binding energy. The binding energies of electrons in outer atomic orbits are of the order of several eV. Therefore, ionizing radiation must have energy greater than a few eV to interact with electrons. Usually, only a fraction of the energy of the ionizing radiation is transferred to a single electron. As ionizing radiation passes through matter, it loses energy due to collisions with electrons. Therefore, to consistently ionize atoms, the energy of the radiation needs to be greater than 100 eV. For example, an average of 33.85 eV is required to ionize an air molecule. So a 5.0 MeV α-particle could ionize approximately 150 000 air molecules. For particle radiation, the rate of energy loss through a material is proportional to the charge of the particle and the number of electrons per unit volume of the material and is inversely proportional to the speed of the particle (slower particles have more time to interact with electrons and to transfer energy).

Absorbed Dose and Equivalent Dose

The amount of ionizing radiation incident on an object is called **exposure**. As a result of exposure, the object will absorb some of the incident energy. The energy absorbed per unit volume by an object is called the **absorbed dose** (*D*). The SI unit of absorbed radiation is the gray (Gy), which is defined as the absorption of 1 J of radiation energy per kilogram of the absorbing material:

$$1 \text{ Gy} = 1 \text{ J/kg}$$

An older unit of absorbed radiation is the rad (radiation absorbed dose):

$$1 \text{ Gy} = 100 \text{ rad}$$

For a given exposure, the amount of energy transferred to biological tissue and the resulting damage to the tissue depend on the type of radiation. Because α-particles are positively charged and are relatively large, they are very effective in ionizing atoms. When passing through matter, α-particles transfer their energy within a short distance, causing intense damage to the DNA structure of the tissue. β-particles and γ-rays lose their energy more slowly, and the resulting damage to biological tissue is more diffuse. To account for different interactions of radiation with the human body, we define the **equivalent dose** (*H*) to be a measure of the amount of damage caused to biological tissue by ionizing radiation. An equivalent dose is the absorbed dose multiplied by the **radiation weighting factor** (W_R), which is a measure of how damaging the radiation is:

Equivalent dose = absorbed dose
  × radiation weighting factor

$$H = D \times W_R$$

The SI unit of equivalent dose is the sievert (Sv). Since the radiation weighting factor is a dimensionless number, the sievert and the gray have the same dimensions, joules per kilogram. One sievert of equivalent dose is equal to one joule per kilogram of energy absorbed by an object. One sievert carries with it a 5.5% chance of eventually developing cancer. Table 34-6 lists radiation weighting factors for various ionizing radiations. Notice that α-particles cause 20 times as much damage to biological tissue than the same absorbed dose of γ-rays and β-particles. So 1 Gy of alpha radiation is 20 times more dangerous than 1 Gy of gamma radiation. The equivalent dose is therefore a measure of the risk associated with absorbed ionizing radiation. Another important factor in determining the effect of radiation on human body is that different tissues are differently sensitive to radiation. For example, lung is affected by radiation about ten times more than the skin.

In daily usage, equivalent dose is often referred to simply as "dose." The old unit of equivalent dose was the rem (röntgen equivalent (in) man), which is equal to a hundredth of a sievert:

$$1 \text{ Sv} = 100 \text{ rem}$$

Table 34-6 Radiation Weighting Factors

Radiation	Radiation Weighting Factor (W_R)
Electrons, positrons, X-rays, γ-rays	1
Protons	2
Slow neutrons (10 keV) (depends on the neutron's energy)	5–10
Medium-energy neutrons (100 keV–2 MeV)	20
α-particles, medium and heavy nuclei	20

EXAMPLE 34-13

Equivalent Dose

Find the equivalent dose for a person who has absorbed 10 mGy of α-particles.

Solution

$$H = D \times W_R$$
$$= 0.010 \text{ Gy} \times 20 = 0.2 \text{ Sv} = 20 \text{ rem}$$

Making sense of the result

Absorbing a dose of 0.010 Gy of α-particles is equivalent to absorbing a dose of 0.20 Gy of X-rays.

For an average adult, the recommended "safe limit" of ionizing radiation is 50 mSv per year. For minors, this limit is 5 mSv. People living in Canada and the United States are continuously exposed to natural and human-made ionizing radiation from a variety of sources, with an average yearly dose of 3.5 mSv. Figure 34-32 shows the average annual radiation exposure in the United States. Notice that approximately half the exposure results from inhaling radon gas, which is produced naturally by the α decay of uranium, which is typically present in soil in concentrations from 0.7 to 11 parts per million. Radon is odourless, chemically inert, and highly radioactive. It decays into other radioactive elements, and it is these products that stick in the lungs and can cause lung cancer. The U.S. Environmental Protection Agency estimates that approximately 21 000 lung cancer deaths every year can be attributed to radon poisoning, a statistic second only to cigarette smoking.

Sources of Exposure

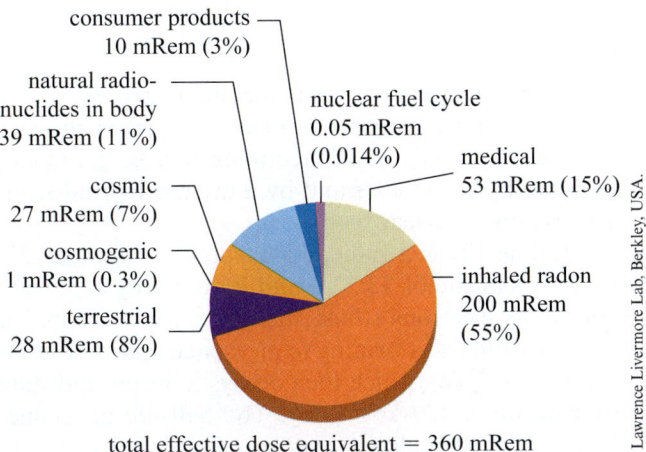

consumer products —
10 mRem (3%)

natural radio-
nuclides in body
39 mRem (11%)

nuclear fuel cycle
0.05 mRem
(0.014%)

cosmic —
27 mRem (7%)

medical
53 mRem (15%)

cosmogenic —
1 mRem (0.3%)

inhaled radon
200 mRem
(55%)

terrestrial
28 mRem (8%)

total effective dose equivalent = 360 mRem

Lawrence Livermore Lab, Berkley, USA.

Figure 34-32 The average annual radiation exposure in the United States from various radioactive sources.

EXAMPLE 34-14

The Equivalent Dose of Polonium

Polonium-210 ($^{210}_{84}$Po) α decays to lead-206 ($^{206}_{82}$Pb) and releases 5.307 MeV of energy in the process. The half-life of $^{210}_{84}$Po is 138.376 days, and its atomic mass is 209.983 u.

(a) What is the equivalent dose received by an average adult lung (mass approximately 5.0 kg) from 10 μg of polonium-210?
(b) What is the approximate daily equivalent dose?

Solution

(a) We know the energy released in a single decay, so we need to determine how many nuclei of 10 μg of polonium-210 decay in 1 s. The atomic mass of polonium-210 is 209.983 u. Therefore, 209.983 g of polonium-201 contains 6.02×10^{23} atoms, and

Number of atoms in 10 μg

$$= 10 \times \left(\frac{6.02 \times 10^{23} \text{ atoms}}{209.983 \text{ g}} \times 10^{-6} \text{ g} \right)$$

$$= 2.87 \times 10^{16} \text{ atoms}$$

Since the initial activity, R_0, of the polonium equals the number of atoms times the decay constant:

$$R_0 = 2.87 \times 10^{16} \text{ atoms} \times \frac{\ln 2}{138.376 \text{ days} \times 24 \text{ h/day} \times 3600 \text{ s/h}}$$

$$= 1.66 \times 10^9 \text{ Bq}$$

The energy, E, released per second equals the number of decays per second times the energy released in a single decay:

$$E = (1.66 \times 10^9 \text{ decay/s})(1 \text{ s})(5.307 \text{ MeV/decay})$$

$$= 8.81 \times 10^9 \text{ MeV}$$

To express the absorbed dose in sieverts, we need to express the incident energy in joules:

$$E = 8.81 \times 10^9 \text{ MeV} \times 1.60 \times 10^{-13} \text{ J/MeV}$$

$$= 1.41 \times 10^{-3} \text{ J}$$

Since the mass of the lung is 5.0 kg, the absorbed dose is

$$D = \text{energy absorbed per unit mass} = \frac{1.41 \times 10^{-3} \text{ J}}{5.0 \text{ kg}}$$

$$= 2.8 \times 10^{-4} \text{ Gy}$$

The radiation weighting factor for α-particles is 20. So, the equivalent dose per second is

$$H = D \times W_R = 2.8 \times 10^{-4} \text{ J/kg} \times 20 = 5.6 \times 10^{-3} \text{ Sv}$$

(b) The half-life of polonium-210 is about 138 days, so its activity does not change substantially in one day. Therefore, a reasonable approximation for the equivalent dose per day is

$$5.6 \times 10^{-3} \text{ Sv/s} \times 24 \text{ h/day} \times 3600 \text{ s/h} = 480 \text{ Sv}$$

Making sense of the result

An equivalent dose of 480 Sv is approximately 10 000 times the recommended maximum annual dose. Clearly, $^{210}_{84}$Po is extremely dangerous. Ingesting even in a tiny amount can cause fatal damage to internal organs.

CHECKPOINT 34-10

Ionizing Radiation

Rank (from most effective to least effective) the following particles in order of their effectiveness in ionizing human tissue.

(a) γ-rays
(b) α-particles
(c) protons
(d) slow neutrons

ANSWER: (b), (d), (c), (a)

34-10 Nuclear Medicine and Some Other Applications

Nuclear medicine is a branch of medicine that uses radionuclides for diagnosis and treatment of disease. Drugs containing radionuclides are called **radiopharmaceuticals**. Most radiopharmaceuticals contain short-lived γ-ray emitting radionuclides. When a radiopharmaceutical is administered to a specific organ, the emitted γ-rays pass through the body and are detected by γ-ray detectors, called gamma cameras. The resulting images provide information about the structure and function of that organ. A radiopharmaceutical can consist of a simple molecule containing a radioactive nuclide or a complex protein in which one of the stable atoms is replaced with a radioactive nuclide.

To be used for diagnostics, a radionuclide must have a few essential properties. It should only emit γ-rays, because α- and β-particles are absorbed by the body and cannot be detected. The emitted Gamma rays should have energies between 50 keV and 500 keV. γ-rays of energies below 50 keV are absorbed by the body and therefore not useful for diagnostics, and those with energies greater than 500 keV are difficult to detect. Its half-life should be short enough that it does not stay in the body for a long period of time but long enough that it can be transported from the production site to a hospital. In addition, the daughter nuclide should not be radioactive.

Let us consider a couple of medical applications of production and use of radionuclides. We first look at the use of radioactive iodine for diagnosis and treatment of hyperthyroidism and thyroid cancer. Iodine, symbol I, is a naturally occurring element with 53 protons. Its stable isotope I-127 ($^{127}_{53}$I) is present in oceans as a trace element with a concentration of a few micrograms per litre, the concentration being location and depth dependent. Fish, various plants, and iodized salt (sodium iodide) are a good source of iodine. The human body needs iodine to generate thyroid hormones. Thyroid hormones are made by cells in the thyroid gland, which is situated in your neck below the Adam's apple (Figure 34-33). The thyroid gland captures iodine from the blood and incorporates it into thyroid hormones that are then released into the blood when needed. These hormones play an essential role in regulating body metabolism, protein synthesis, and proper development of the brain and bones. Iodine deficiency is considered to be the most common cause of mental disorder. The recommended daily dose of iodine for an average adult is about 150 μg.

Production of thyroid hormones is regulated by the brain. Occasionally, the thyroid cells become overactive and produce an excessive amount of thyroid hormones. This condition is called hyperthyroidism. Since thyroid hormones regulate body metabolism, an excess of these

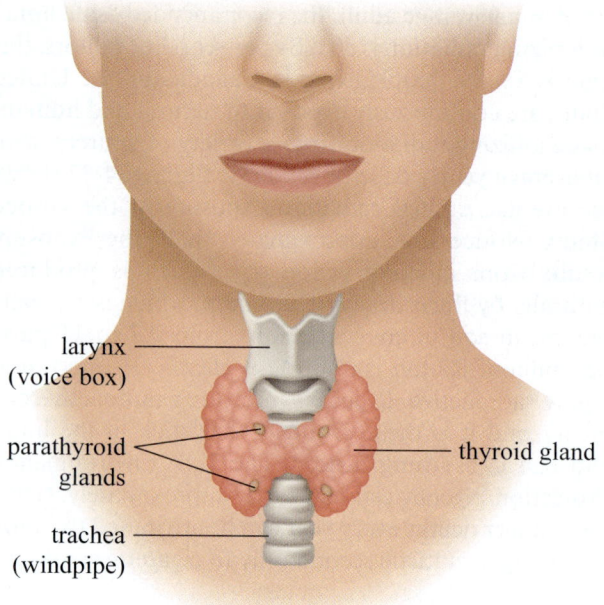

larynx (voice box)

parathyroid glands

thyroid gland

trachea (windpipe)

Figure 34-33 A thyroid gland.

hormones speeds up various metabolic processes and leads to various diseases, for example, Graves' disease. An assessment of a malfunctioning thyroid gland can be made by blood tests and by a nuclear thyroid scan. This is how a nuclear scan works.

Iodine-123 is a radioactive isotope of iodine-127. It has a half-life of 13.22 h and it decays by capturing one of its electrons (thus converting a proton into a neutron) and transforming to an excited stated of tellurium-123 ($^{123}_{52}$Te), which then decays to its ground state by emitting a 159 keV γ-ray. The half-life of iodine-123 is short enough that within a few days its activity reduces considerably and long enough that it can be transported from its production site to a hospital.

Iodine-123 is produced in a cyclotron (see Section 35-11) by bombarding a capsule of xenon-124, a stable isotope of xenon, with protons of kinetic energies around 30 MeV. Xenon-124 absorbs the proton and immediately emits a proton and a neutron, thus transforming into xenon-123, which has a half-life of 2.08 h and $β^+$ decays into iodine-123. The reaction chain for iodine-123 production is

$$\text{p} + {}^{124}_{54}\text{Xe} \rightarrow {}^{123}_{54}\text{Xe} + \text{p} + \text{n}$$

$$^{123}_{54}\text{Xe} \xrightarrow{2.08\ \text{hr}} {}^{123}_{53}\text{I} + e^+ + \nu_e$$

Iodine-123 is then collected from the capsule and converted to a sodium iodide solution for medical applications.

The chemical properties of radioactive iodine-123 are the same as those of iodine-127, as both have the same number of electrons. So when radioactive iodide is administered to a patient, for example through a capsule, the thyroid gland absorbs iodine-123 just as it absorbs iodine-127. A day later, the patient is positioned under a gamma camera (Figure 34-34), which detects the 159 keV γ-rays that are being emitted from the thyroid. A computer collects the data from the camera and constructs an image of the thyroid. As iodine is slowly absorbed by the thyroid, images taken over a span of time provide information about its physiological activities. This information is then used to determine the health of the thyroid gland. The scan of a normal thyroid appears as a small butterfly-shaped image, and that for an abnormal thyroid is either smaller or larger or, if there is cancerous growth, of a different shape. Figure 34-35 shows the scan of a thyroid of a patient suffering from Graves' disease. Figure 34-36 shows the scan of a thyroid that has developed cancer in the lower left side.

In some cases, radioactive pharmaceuticals are also used to treat a disease. Iodine-131 is radioisotope of iodine that has a half-life of 8 days and undergoes β decay. The emitted electrons have an average kinetic energy of 190 keV, with a maximum kinetic energy of 606 keV (see Figure 34-10). The electrons, being charged, interact with the body tissue and, depending upon the energy, penetrate to depths of 0.6 mm to 2.0 mm

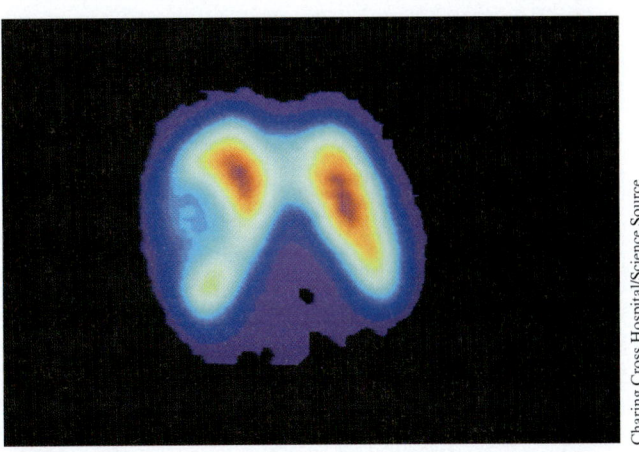

Figure 34-36 A nuclear scan of a thyroid gland. The lower left side of the gland, which is darker than the right side, indicates the location of a cancerous tumour. The colours are generated by the imaging software to show different levels of radioactivity in different regions of the thyroid gland. The red colour corresponds to the region of highest activity, showing most iodine intake by this part of thyroid.

Charing Cross Hospital/Science Source

from the site of decay. The energy deposited in the tissue destroys the cancerous cells.

Another radionuclide that is widely used for diagnostics is technetium-99 ($^{99m}_{43}$Tc). It has a half-life of about 6.1 h and decays to its ground state by emitting a γ-ray of 142.7 keV. Every year it is used in tens of millions of diagnoses throughout the world. The Canadian nuclear reactor NRU (National Research Universal Reactor) at Chalk River (Ontario) is a major supplier of ^{99m}Tc. It is produced by the β^- decay of molybdenum-99 ($^{99}_{42}$Mo), a radioactive isotope of molybdenum with a half-life of 65.9 h. Technetium-99 is produced using an interesting technique called "milking the cow." This is how it works: molybdenum-99 is produced in a nuclear reactor by bombarding uranium-235 with fast-moving neutrons. A uranium nucleus absorbs a neutron and fissions (see Section 34-8). Among many others, one of the fission reactions is

$$\text{n} + {}^{235}_{92}\text{U} \rightarrow {}^{135}_{50}\text{Sn} + {}^{99}_{42}\text{Mo} + 2\text{n}$$

Molybdenum-99 is separated from other fission products and collected in a cylindrical cell with a lead shield, called a **radionuclide generator** (we are leaving out lots of details of this procedure), which is then transported to a hospital. As molybdenum β^- decays, the number of ^{99m}Tc atoms in the generator increases and ^{99m}Tc activity begins to rise. After about 22.8 h, the activity of ^{99m}Tc reaches a maximum and it is extracted from the generator and combined with a chemical to form a radiopharmaceutical for use in scans. After extraction, the ^{99m}Tc activity in the generator begins to increase again as the remaining molybdenum nuclei continue to decay. It is extracted again after another 22.8 h. The parent radionuclide is called the "cow," and the process of extracting the γ-ray emitting radionuclide is called "milking."

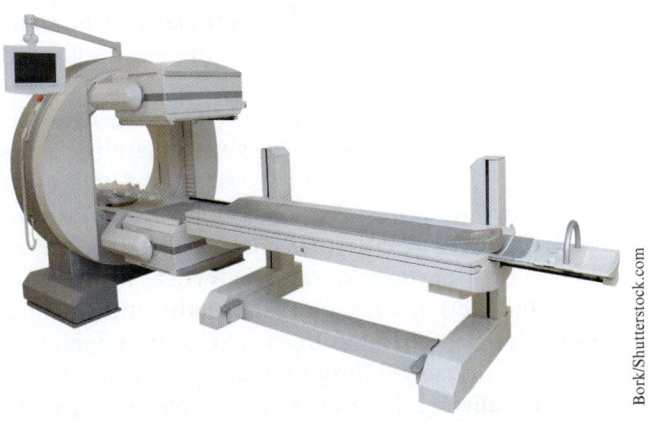

Figure 34-34 A setup for a γ-ray scan for a patient. The γ-ray camera is positioned above the table.

Bork/Shutterstock.com

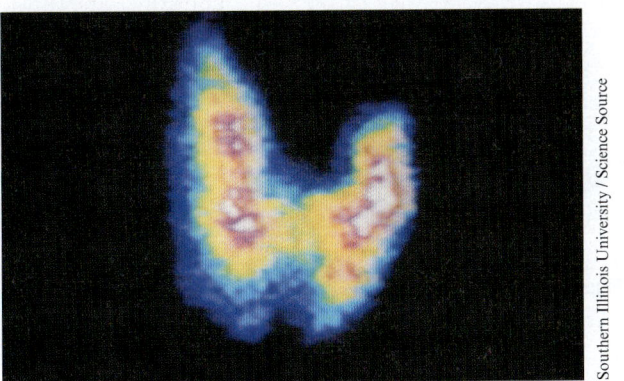

Figure 34-35 Nuclear scan of a thyroid of a patient suffering from Graves' disease (right).

Southern Illinois University / Science Source

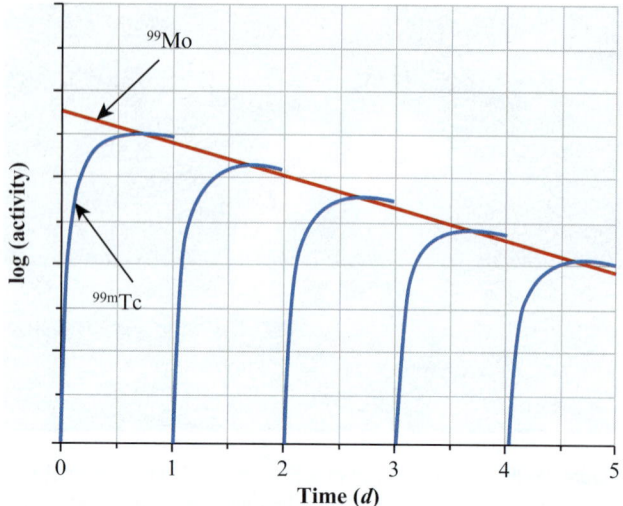

Figure 34-37 Logarithm of activities for ^{99}Mo (parent, red) and ^{99m}Tc (daughter, blue), when ^{99m}Tc is regularly removed every 24 h.

As the amount of molybdenum in the generator continues to decrease, the amount of extracted ^{99m}Tc also decreases. Figure 34-37 shows a graph of ^{99}Mo and ^{99m}Tc activities as a function of time. Typically, after about a week, the generator is not able to produce sufficient ^{99m}Tc, and a new generator is used. The interval of 22.8 h for ^{99m}Tc activity to reach a maximum is very convenient. A new generator is brought in over the weekend, is "milked" every weekday, and is then replaced by a new generator.

Nuclear medicine is an integral part of modern-day healthcare and is used for a variety of purposes, including diagnosis and treatment of various types of cancers, neurological disorders, heart diseases, and many other defects. A variety of radiopharmaceuticals have been developed that concentrate in specifically targeted organs. This shows the importance of nuclear radiation in today's society.

There are many other applications of nuclear radiation, some of which are discussed below.

Explosive Detectors

Nitrate compounds, such as ammonium nitrate and potassium nitrate, are commonly used in explosives. Nitrates can be detected by bombarding an object with neutrons. Some of the neutrons are absorbed by nitrogen nuclei, forming the isotope $^{15}_{7}$N* in an excited state. This isotope then decays to its ground state by emitting a γ-ray with a characteristic energy unique to the $^{15}_{7}$N* nucleus:

$$^{14}_{7}\text{N} + {}^{1}_{0}\text{n} \rightarrow {}^{15}_{7}\text{N*} \rightarrow {}^{15}_{7}\text{N} + \gamma$$

This process is called an (n, γ) reaction, meaning "neutron in, photon out." A detector tuned to the characteristic frequency of the nitrogen-15 γ-rays can then determine when an object contains enough nitrogen to suggest the presence of explosives. Such neutron-activated explosive detectors are becoming common at seaports, airports, and border crossings.

Smoke Detectors

As described in the previous section, α-particles are very efficient at ionizing matter. Most household smoke detectors use this property to detect smoke and combustion products. Such detectors contain a small amount of synthetically produced radioactive americium-241 ($^{241}_{95}$Am), which emits α-particles with about 5.5 MeV of kinetic energy. These α-particles ionize air molecules and lose all of their energy within 4 cm to 5 cm.

In a smoke detector, α-particles from $^{241}_{95}$Am enter the space between two electrodes that are approximately 1 cm apart and are kept at a potential difference by a battery (Figure 34-38). As the air between the electrodes is ionized, the negatively charged electrons are attracted toward the positive plate of the electrode, and positively charged ions are attracted toward the negative plate. The resulting tiny continuous electric current between the electrodes is sensed by an electronic circuit. When a fire starts, smoke particles and invisible combustion products get between the electrodes and neutralize some of the ions, thus reducing the current flowing between the electrodes. If the current falls below a threshold value, the electronic circuit triggers an alarm.

Americium-241 is chosen because it has a half-life of 433 yr, which is long enough that the rate of emission of α-particles does not decrease significantly during the lifetime of a smoke detector. It also emits very few γ-rays. The half-life is also short enough that only approximately 0.2 μg of americium is needed to generate a current detectable by a simple and inexpensive circuit.

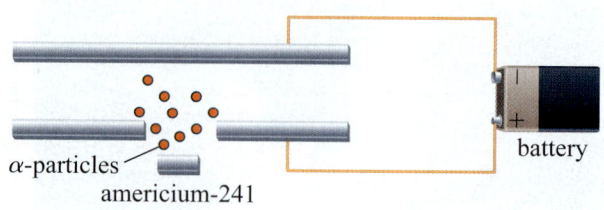

Figure 34-38 The ionization chamber of a smoke detector.

MODERN PHYSICS

KEY CONCEPTS AND RELATIONSHIPS

Nuclear Size and Nuclear Force

Atomic nuclei consist of protons and neutrons. The atomic number, Z, is the number of protons in the nucleus. The atomic number, A, is the number of protons and neutrons in the nucleus. Nuclei with the same number of protons but a different number of neutrons are called isotopes.

Nuclei are approximately spherical. The radius of a nucleus is approximately

$$r = r_0 A^{1/3} \qquad (34\text{-}4)$$

where $r_0 = 1.2$ fm and A is the mass number.

The nucleons are bound within the nucleus by the attractive strong force, which has a range of approximately 2.0 fm. The nuclear density is approximately 2.8×10^{17} kg/m³.

Nuclear masses are usually given in atomic mass units (u). One atomic mass unit equals 1.6605×10^{-27} kg and is equivalent to 931.5 MeV of energy.

Nuclear Binding Energy

The binding energy of a nucleus is the energy required to separate the nucleus into its constituent protons and neutrons. The binding energy, $B(A, Z)$, of a nucleus is related to the nuclear mass, $M_{nuc}(A, Z)$; the proton mass, m_p; and the neutron mass, m_n:

$$B(A, Z) = Zm_p c^2 + (A - Z)m_n c^2 - M_{nuc}(A, Z)c^2 \quad (34\text{-}10)$$

The Q-value of a nuclear reaction is the difference between the rest mass energies of the initial nuclei and the rest mass energies of the final nuclei:

$$Q = (M_i - M_f)c^2 \qquad (34\text{-}27)$$

Nuclear Decay and Radioactivity

Unstable nuclei spontaneously emit energy in the form of α-, β-, and γ-rays. The rate of radioactive decay is proportional to the decay constant, λ, and the number of unstable nuclei present:

$$-\frac{dN(t)}{dt} = \lambda N(t) \qquad (34\text{-}15)$$

The number of unstable nuclei remaining after time t is given by

$$N(t) = N_0 e^{-\lambda t} \qquad (34\text{-}19)$$

where N_0 is the number of nuclei present at $t = 0$.

The half-life, $t_{1/2}$, of a radioactive nuclide is the time it takes for half of the radioactive nuclei in a given sample to decay. The half-life of a nuclide is related to its decay constant:

$$t_{1/2} = \frac{\ln 2}{\lambda} \qquad (34\text{-}21)$$

Nuclear Reactions

Energy, charge, momentum, total angular momentum, and number of nucleons are conserved in a nuclear reaction.

α, β and γ Decays

The general form for α decay is

$$_Z^A X \rightarrow {}_{Z-2}^{A-4} Y + {}_2^4 He + Q_\alpha \qquad (34\text{-}28)$$

where $_Z^A X$ denotes a parent nuclide, $_{Z-2}^{A-4} Y$ is the daughter nuclide, and Q_α is the energy released in α decay.

Beta-minus (β^-) particles are electrons. In a β^- decay, a neutron converts into a proton:

$$_Z^A X \rightarrow {}_{Z+1}^A Y + {}_{-1}^0 e + \bar{\nu} + Q_{\beta^-} \qquad (34\text{-}38)$$

where $\bar{\nu}$ is an antineutrino and Q_{β^-} is the energy released in the decay.

Beta-plus (β^+) particles are positrons. In a β^+ decay, a proton converts into a neutron:

$$_Z^A X \rightarrow {}_{Z-1}^A Y + {}_{+1}^0 e + \nu + Q_{\beta^+} \qquad (34\text{-}42)$$

where ν is a neutrino and Q_{β^+} is the energy released in the decay.

Gamma (γ) rays are high-energy photons. A radioactive nuclide emits a γ-ray when a proton or neutron moves to a lower-energy state:

$$_Z^A X^* \rightarrow {}_Z^A X + \gamma + Q_\gamma \qquad (34\text{-}47)$$

where $_Z^A X^*$ denotes an excited energy state, $_Z^A X$ is the lower-energy state of the same nuclide, and Q_γ is the energy released in γ decay.

Nuclear Stability

Nuclei are stable because (a) protons and neutrons are fermions and hence obey Pauli's exclusion principle (b) the nuclear force has a short range (c) the Coulomb force has a long range (d) in a nuclear environment proton and neutron can decay into each other (e) the nuclear force between two nucleons is slightly more attractive when the spins of the nucleons are aligned parallel to each other.

Nuclear Fission and Nuclear Fusion

Spontaneous nuclear fission is the splitting of a nucleus with large A into two nuclei, usually associated with the emission of one or more neutrons. Induced nuclear fission is caused by the absorption of a neutron by a large nucleus.

Nuclear fusion is the merging of two light nuclei into a single nucleus. Nuclear fusion is the basic process responsible for energy generation in stars.

Ionizing Radiation

The SI unit for nuclear radiation is the becquerel: 1 Bq = 1 nuclear decay/s. The SI unit for absorbed radiation is the gray: 1 Gy = 1 J/kg.

The sievert (Sv) is a derived unit of ionizing radiation dose in SI units. One sievert of equivalent dose is equal to one joule per kilogram of energy absorbed by an object.

APPLICATIONS

nuclear reactors, medical imaging, carbon dating, explosive detectors, smoke detectors, security inspection

KEY TERMS

absorbed dose, activity, alpha (α) particle, alpha decay, atomic mass, atomic number, average atomic mass, beta decay, beta-minus decay, beta-plus decay, binding energy, binding energy curve, binding energy per nucleon, carbon dating, centrifuge, chain reaction, containment vessel, control rods, coolant, daughter nucleus, decay constant, decay diagram, delayed neutrons, deuteron, disintegration constant, enriched uranium, equivalent dose, exposure, fission products, fuel rods, gamma decay, half-life, ionizing electromagnetic radiation, ionizing particle radiation, ionizing radiation, isobars, isomers, isotones, isotopes, mass number, metastable, moderator, neutrino, neutron-induced fission, neutron number, neutrons, nuclear binding energy, nuclear core, nuclear decay, nuclear fission, nuclear force, nuclear fuel, nuclear fusion, nuclear meltdown, nuclear potential, nuclear radiation, nuclear reactions, nucleon, nuclide, parent nucleus, positron, prompt neutrons, proton–proton cycle, protons, Q-value, radiation weighting factor, radioactive, radioactive decay, radiocarbon dating, radionuclide generator, radiopharmaceuticals, short-range force, strong force, weak force

QUESTIONS

1. Define the following terms: nucleon, nuclide, isotope, decay constant, half-life, decay diagram, Q-value, spontaneous fission, radioactive dating, proton–proton cycle.
2. What are the main differences between the strong and the electromagnetic forces between two protons?
3. Describe the three main processes by which unstable nuclei decay.
4. Estimate Earth's radius if the atoms were compressed to the size of a nucleus.
5. In a nuclear decay series, a $^{238}_{92}$U nucleus decays into a $^{206}_{82}$Pb nucleus, which is stable. What is the maximum number of α-particles that can be emitted in this decay?
6. The mass distribution of the human body is approximately 65% oxygen, 18% carbon, 10% hydrogen, 3% nitrogen, and 1.4% calcium, the rest being other elements. Estimate the number of protons and neutrons in your body.
7. What is the ratio of the Coulomb energy and the gravitational potential energy of two protons that are 2.0 fm apart?
8. The half-life of $^8_4$Be is 6.7×10^{-17} s. What is the most likely decay product of its decay? Explain your reasoning.
9. How many half-lives must pass until 75% of the atoms of a radioactive nuclide have decayed?
10. Why is energy released in the fusion of light nuclei?
11. Why is energy released in the fission of heavier nuclei? Would energy be released if two iron ($Z = 26$) nuclei were to fuse into a tellurium ($Z = 52$) nucleus?
12. What is a Coulomb barrier?
13. How would nuclei decay if the mass of a proton were slightly larger than the mass of a neutron?
14. A free neutron has a half-life of 881 s, but a neutron in a carbon-12 nucleus does not decay. Why?

PROBLEMS BY SECTION

For problems, star ratings will be used ($\star$, $\star\star$, or $\star\star\star$), with more stars meaning more challenging problems. The following codes will indicate if $\frac{d}{dx}$ differentiation, $\int$ integration, ▢ numerical approximation, or ∿ graphical analysis will be required to solve the problem.

Section 34-1   Nuclear Terminology and Nuclear Units

15. $\star$ Express the rest mass energy of 1 g of matter in J and MeV.
16. $\star$ Approximately 1.2×10^{17} J of solar energy is intercepted by Earth every second. If all of this energy were absorbed by Earth, what would the increase in Earth's mass be in 1 yr?
17. $\star$ The mass of a baseball is 145 g. Calculate the rest mass energy of a baseball in
    (a)   J
    (b)   eV
18. $\star$ Assuming that nuclei are spherical, calculate the Coulomb repulsion between two carbon-12 nuclei that are
    (a)   just touching each other
    (b)   1.0 cm apart
19. $\star\star$ A free neutron (mass 939.55 MeV/c^2) decays into a proton (mass 938.26 MeV/c^2), an electron (mass 0.511 MeV/c^2), and an antineutrino (zero mass). The initial neutron is at rest. Calculate the maximum possible momentum for the electron. Compare this momentum with the maximum momentum that the antineutrino could have.
20. $\star\star$ Consider the fission of a uranium-232 nucleus into two palladium nuclei:

$$^{232}_{92}\text{U} \rightarrow {}^{116}_{46}\text{Pa} + {}^{116}_{46}\text{Pa}$$

    (a)   How much Coulomb energy is released?
    (b)   How many uranium nuclei need to fission each second to produce 1.0 W of power?

21. ★★ Two stable isotopes of carbon are $^{12}_{6}C$ (98.89%) and $^{13}_{6}C$ (1.11%). Using Table 34-5, calculate the average atomic mass of carbon.

22. ★★ Calculate the average atomic mass of oxygen from the data in Table 34-7.

Table 34-7 Data for Problem 22

Isotope	Atomic Mass (u)	Abundance
Oxygen-16	15.994 914 6	99.757%
Oxygen-17	16.999 131 54	0.038%
Oxygen-18	17.999 160 44	0.205%

Section 34-3 Nuclear Binding Energy

23. ★★ Using Table 34-5, calculate the total binding energy and the binding energy per nucleon for the following nuclei:
    (a) $^{2}_{1}H$
    (b) $^{4}_{2}He$
    (c) $^{6}_{3}Li$
    (d) $^{56}_{26}Fe$
    (e) $^{208}_{82}Pb$

24. ★★ Use the binding energy per nucleon provided in Table 34-8 to determine the nuclear masses of the following nuclei. Compare these values to the masses given in Table 34-5.

Table 34-8 Data for Problem 24

Isotope	Binding Energy per Nucleon (MeV)
$^{8}_{4}Be$	7.06
$^{12}_{6}C$	7.68
$^{84}_{38}Sr$	8.68
$^{208}_{82}Pb$	7.87
$^{238}_{92}U$	7.57

25. ★★ Using Table 34-5, calculate the nuclear masses of the following atoms:
    (a) $^{2}_{1}H$
    (b) $^{12}_{6}C$
    (c) $^{16}_{8}O$
    (d) $^{56}_{26}Fe$
    (e) $^{238}_{92}U$

26. ★★ Using Table 34-5, calculate the binding energy per nucleon for $A = 100$ nuclei with $Z = 42, 43, 44, 45,$ and 46. Plot these binding energies as a function of Z. Which nuclide is the most tightly bound?

27. ★★ Even–even nuclei have an even number of protons and neutrons. In the liquid-drop model of a nucleus, the total binding energy of an even–even nucleus can be approximated using the following formula:

$$B(A, Z) = a_V A - a_S A^{2/3} - a_C \frac{Z^2}{A^{1/3}} - a_A \frac{(A - 2Z)^2}{A} + a_P \frac{1}{\sqrt{A}}$$

where $a_V = 15.8$ MeV, $a_S = 18.3$ MeV, $a_C = 0.72$ MeV, $a_A = 23.2$ MeV, and $a_P = 12.0$ MeV.

Using the above formula, calculate the binding energies of the nuclei in problem 24. Compare the two sets of binding energies.

Section 34-4 Nuclear Decay and Radioactivity

28. ★★ In natural carbon, one atom in 10^{12} is $^{14}_{7}C$ ($t_{1/2} = 5730$ yr).
    (a) What is the activity of 1 g of natural carbon?
    (b) After 17 190 yr, what is the activity?
    (c) Approximately 18% of the atoms in the human body consist of carbon. Estimate the activity of carbon-14 in your body.

29. ★★ The half-life for carbon-14 is 5730 yr. Determine its decay constant.

30. ★★ A radioactive isotope decays to one-tenth of its original amount in two days. How long did it take for half of it to decay?

31. ★★ The half-life of $^{28}_{12}Mg$ is 21 h. At a certain time, the activity of magnesium-28 is 2.2×10^8 Bq. What is its activity two days later?

32. ★★ A radioactive nuclide had an activity of 3.0×10^4 Bq 10 h ago. Now it has an activity of 2.0×10^4 Bq. What is the half-life of the nuclide?

33. ★★ What mass of carbon-14 must be in a sample to have an activity of 10^6 Bq?

34. ★★ What mass of $^{226}_{88}Ra$ has an activity of 3.7×10^{10} Bq?

35. ★★ A radioactive sample contains 3.0×10^{20} atoms and has an activity of 2.0×10^{10} Bq. Find the half-life of the sample.

36. ★★ Strontium-90 has a half-life of 25 yr.
    (a) Determine its decay constant.
    (b) How long will it take for 80% of the original amount of strontium-90 to decay?

37. ★★ A certain radioactive source has an activity of 1000 Bq at $t = 0$. The half-life of the source is 10 h. What is its activity after
    (a) 30 min?
    (b) 3 days?

38. ★★ A sample of cloth contains 20 g of carbon, and the measured carbon-14 decay rate from the cloth is 120 counts/min. Estimate the age of the cloth.

39. ★★ At $t = 0$, the activity for a radioactive source is 20.0 MBq. At $t = 240$ s, its activity is 15.0 MBq. What is the half-life of the source?

40. ★★ A radioactive source has a half-life of 8 h. At $t = 0$, the activity is 900 decays/min. What is the activity 24 h later?

41. ★★ Table 34-9 lists the activities of a radioactive element as a function of time.
    (a) Plot the natural logarithm of the activity as a function of time.
    (b) Determine the slope of the resulting curve.
    (c) Calculate the half-life of the element.

Table 34-9 Data for Problem 41

Time (s)	R (decay/s)	Time (s)	R (decay/s)
0.0	1000	20.0	387
5.0	785	25.0	303
10.0	620	30.0	240
15.0	485	35.0	190

Section 34-5   Nuclear Reactions

42. ★★ Identify the missing particle in the following decays:

(a) $^{22}_{9}F \rightarrow\ ^{22}_{10}Ne + ?$

(b) $^{12}_{6}C \rightarrow\ ^{12}_{5}B + ?$

(c) $^{8}_{4}Be \rightarrow\ ^{4}_{2}He + ?$

(d) $^{239}_{94}Pu \rightarrow\ ^{235}_{92}U + ?$

(e) $^{60}_{28}Ni^* \rightarrow\ ^{60}_{28}Ni + ?$

43. ★★ Which of the following decays are not possible? Why not?

(a) $^{18}_{8}O \rightarrow\ ^{17}_{8}O + ^{0}_{-1}e + v$

(b) $^{14}_{6}C \rightarrow\ ^{14}_{7}N + v$

(c) $^{9}_{5}B \rightarrow\ ^{9}_{4}Be + ^{0}_{+1}e + \gamma$

(d) $^{239}_{94}Pu \rightarrow\ ^{239}_{94}Pu + ^{0}_{-1}e + \bar{v}$

(e) $^{222}_{86}Rn \rightarrow\ ^{220}_{84}Po + ^{4}_{2}He$

44. ★★ What are the atomic number (Z) and atomic mass (A) of the unknown nuclei, marked as X, in the following reactions?

(a) $^{6}_{2}He \rightarrow X + e^- + \bar{v}_e$

(b) $^{7}_{4}Be \rightarrow X + e^+ + v_e$

(c) $^{1}_{1}p + ^{12}_{6}C \rightarrow X + ^{1}_{0}n$

(d) $^{1}_{0}n + X \rightarrow\ ^{31}_{15}P + 2^{1}_{0}n$

(e) $^{27}_{16}Al \rightarrow X + \gamma$

45. ★★ In a nuclear reaction, slow-moving neutrons (ignore the neutrons' kinetic energy) are absorbed on a parent nucleus (X), which then decays into a daughter nucleus (Y) and a helium nucleus:

$$^{1}_{0}n + ^{A}_{Z}X \rightarrow\ ^{A-3}_{Z-2}Y + ^{4}_{2}He$$

(a) What is the Q-value of the above reaction in terms of the nuclear masses?

(b) The parent nucleus is $^{17}_{8}O$. Calculate the energy released using the table of masses (Table 34-5). Take the binding energy of a helium nucleus to be 28.2 MeV.

46. ★★ A common nuclear reaction, called electron capture, is the capture of a k-shell electron by a nucleus:

$$^{A}_{Z}X + ^{0}_{-1}e \rightarrow\ ^{A}_{Z-1}Y + v$$

(a) Write an expression for the Q-value of the above process in terms of the nuclear masses of the parent (X) and the daughter (Y) nuclei.

(b) Write the expression for the Q-value in terms of the atomic masses of the parent and the daughter nuclei.

47. ★★ Can a $^{12}_{6}C$ nucleus decay into three $^{4}_{2}He$ nuclei? If yes, how much energy will be released in such a decay? If not, why not?

48. ★★ Berillium-8 ($^{8}_{4}Be$) is an unstable nucleus that decays into two helium nuclei. Given the following data, determine why $^{8}_{4}Be$ is not stable:

Atomic mass of $^{8}_{4}Be$ = 8.005 305 u

Atomic mass of $^{4}_{2}He$ = 4.002 603 u

Section 34-6 α, β and γ Decays

49. ★ Write equations to describe the α decay of the following nuclei. Use a periodic table to identify the daughter nuclei.

(a) $^{32}_{14}Si$

(b) $^{56}_{25}Mn$

(c) $^{60}_{26}Fe$

50. ★ Write equations to describe the β^- decay of the following nuclei. Use a periodic table to identify the daughter nuclei.

(a) $^{32}_{14}Si$

(b) $^{56}_{25}Mn$

(c) $^{60}_{26}Fe$

51. ★ Write the equations to describe the β^+ decay of the following nuclei. Use a periodic table to identify the daughter nuclei.

(a) $^{52}_{25}Mn$

(b) $^{55}_{27}Co$

(c) $^{59}_{28}Ni$

52. ★ Using the table of atomic masses (Table 34-5), calculate the Q-value of the following α decay reactions:

(a) $^{238}_{92}U \rightarrow\ ^{234}_{90}Th + ^{4}_{2}He$

(b) $^{235}_{92}U \rightarrow\ ^{231}_{90}Th + ^{4}_{2}He$

(c) $^{210}_{84}Po \rightarrow\ ^{206}_{82}Pb + ^{4}_{2}He$

53. ★ Tritium is an isotope of hydrogen with two neutrons. It has a half-life of 12.3 yr and it β decays into helium-3.

(a) Write the reaction that describes the β decay of tritium into helium-3.

(b) Calculate the energy released in this decay, using the atomic masses from Table 34-5.

54. ★★ Lead-212 ($^{212}_{82}Pb$) is a radioactive isotope of lead. It decays to the stable lead-208 through the following decay chain: β^-, β^-, α. Draw a decay diagram showing the decay process, and identify the intermediate daughter nuclei at each step.

55. ★★ Thorium-232 ($^{232}_{90}Th$) decays to lead-208 ($^{208}_{82}Pb$). How many α- and β-particles are emitted in this process?

Section 34-8   Nuclear Fission and Nuclear Fusion

56. ★★ In one of the possible reactions involving the neutron-induced fission of $^{239}_{94}Pu$, the two daughter elements $^{141}_{56}Ba$ and $^{92}_{36}Kr$ are produced.

(a) Write an equation describing this reaction.

(b) Find the energy released in the process, ignoring the kinetic energy of the incident neutron.

57. ★★ Using the table of atomic masses (Table 34-5), calculate the Q-values for the following fission reactions:

(a) $^{1}_{0}n + ^{235}_{92}U \rightarrow\ ^{90}_{36}Kr + ^{144}_{56}Ba + 2^{1}_{0}n$

(b) $^{1}_{0}n + ^{235}_{92}U \rightarrow\ ^{93}_{37}Rb + ^{141}_{55}Cs + 2^{1}_{0}n$

58. ★★ The half-life for the spontaneous fission of uranium-238 is approximately 8×10^{15} yr, and the half-life of uranium-238 for α decay is 4.5×10^9 yr. Given 1 g of uranium-238, how many nuclei undergo spontaneous fission in one day? How many nuclei α decay in a day?

59. ★★★ Even–even nuclei have an even number of protons and neutrons. Odd–odd nuclei have an odd number of protons and neutrons. Even–even nuclei are more tightly bound than odd–odd nuclei. There are only five stable odd–odd nuclei: $^2_1$H, $^6_3$Li, $^{10}_5$B, $^{14}_7$N, and $^{180}_{73}$Ta.

(a) Using Table 34-5 for nuclear masses, calculate the binding energy per nucleon of odd–odd nuclei.

(b) Using the same method, calculate the binding energy per nucleon of the following even–even nuclei obtained by adding one neutron and one proton to the odd–odd nuclei:

(i) $^4_2$He

(ii) $^8_4$Be

(iii) $^{12}_6$C

(iv) $^{16}_8$O

(v) $^{182}_{74}$W

(c) How much binding energy is gained in each case?

60. ★★★ Show that the Q-value for spontaneous nuclear fission can be written in terms of the binding energies of the parent nucleus, (A, Z), and daughter nuclei, (A_1, Z_1) and (A_2, Z_2), as follows:

$$Q_f = B(A_1, Z_1) + B(A_2, Z_2) - B(A, Z)$$

Using the binding energy formula for the liquid-drop model (problem 27), determine the energy released in the following fission processes:

$$^{238}_{92}U \rightarrow {}^{134}_{52}Te + {}^{102}_{40}Zr + 2{}^1_0n + Q_f$$

$$^{238}_{92}U \rightarrow {}^{134}_{50}Sn + {}^{102}_{42}Mo + 2{}^1_0n + Q_f$$

COMPREHENSIVE PROBLEMS

61. ★★★ The binding energy of an α-particle (a $^4_2$He nucleus) is 28.2 MeV. Use the liquid-drop model (problem 27) to answer the following questions.

(a) A $^8_4$Be nucleus can decay into two α-particles. How much energy is released in this process?

(b) Can a $^{12}_6$C nucleus decay into three α-particles? Explain your answer.

62. ★★★ In an (n, α) nuclear reaction, slow-moving neutrons (ignore the neutrons' kinetic energy) are absorbed on a target nucleus (X), which then decays into a daughter nucleus (Y) and a helium nucleus:

$$^1_0n + {}^A_Z X \rightarrow Y + {}^4_2He$$

The target nucleus is hafnium-180 ($^{180}_{72}$Hf). Calculate the energy released in the above process using the liquid-drop model of problem 27. Take the binding energy of a helium nucleus to be 28.2 MeV.

63. ★★★ Using the liquid-drop model in problem 27, show that the energy (S_n) required to separate a neutron from a nucleus with Z protons and $A - Z$ neutrons, with large A ($A \gg 1$), is approximately given by

$$S_n = a_V - \frac{2}{3}a_S A^{-1/3} - a_A \left(1 - \frac{4Z^2}{A(A-1)}\right)$$

Learning Objectives

When you have completed this chapter, you should be able to

LO1 List all the elementary particles, and explain the difference between quarks and leptons.

LO2 Explain the role of gauge bosons in mediating interactions between the elementary particles.

LO3 Define *antiparticle*, and identify the antiparticle corresponding to each elementary particle.

LO4 Explain the Standard Model and how particles are grouped into three families.

LO5 Explain the structure of composite baryons and mesons in terms of quarks, and determine the charge, baryon number, lepton number, and spin quantum number of a composite particle from its quark constituents.

LO6 Define quark confinement.

LO7 Apply conservation laws of energy, momentum, charge, baryon number, and lepton number to determine whether a particular reaction or decay is allowed.

LO8 Calculate the energy available in the centre-of-mass coordinate system of two colliding particles to produce a new particle.

LO9 Explain Feynman diagrams, and draw and interpret Feynman diagrams for simple processes.

LO10 Explain how pions and muons were discovered.

LO11 Explain how a linear accelerator works, and explain the difference between a cyclotron and a synchrotron.

LO12 Explain the main astronomical observation that lead to the hypothesis of the existence of the dark matter.

On July 4, 2012, scientists at the Large Hadron Collider (LHC) announced the discovery of the Higgs particle, bringing an end to the most expensive and long-awaited search for a particle in the history of human kind (Figure 35-1). The Higgs particle was the last yet-to-be-discovered member of the Standard Model, a model that forms the basis of our current understanding of the nature of matter and the fundamental forces that exist between the particles. What is the Standard Model? What are the other particles in the Standard Model? How do particles interact with each other? Why does the LHC need to be so large? The subject of particle physics is fascinating as well as intriguing.

Figure 35-1 A section of the LHC.

35-1 Classification of Particles

The electron was discovered by J.J. Thomson (1856–1940) in 1896, the proton by Ernest Rutherford (1871–1937) in 1919, and the neutron by James Chadwick (1891–1974) in 1932. These discoveries led to a model of the atom with electrons orbiting a central nucleus made of protons and neutrons. The three particles in this atomic model were thought to be the only building blocks of matter. However, as technology developed and it became possible to collide extremely fast moving particles and nuclei with each other, physicists began to discover a range of new types of particles. Most of the new particles exist for only a very short time and decay into other particles. To date, physicists have discovered more than 200 different subatomic particles. The branch of physics that deals with the study of particles is called **particle physics**.

Particles are classified as either elementary or composite. **Elementary** (or **fundamental**) **particles** are thought to be point particles with no internal structure. Electrons, neutrinos, photons, and quarks (which we will discuss later) are examples of elementary particles. A list of all known elementary particles is given in Table 35-1. **Composite particles** are made of elementary particles. For example, protons and neutrons are made of quarks and are therefore composite particles. The physical properties of a composite particle are determined by the properties of its constituents and depend on how the constituents interact to form the composite particle (similar to how the properties of a nucleus are determined by the number of protons and neutrons that make up the nucleus and how they interact to form the nucleus).

A particle is classified in terms of

- its physical properties, such as mass, charge, and magnetic moment
- its quantum numbers, including spin, and a few other properties, such as baryon number and lepton number
- the type of forces that the particle exerts on other particles

Mass

The rest masses of the elementary particles vary over a wide range, as can be seen from Table 35-1. For example, a photon has zero rest mass, and the rest mass of the top quark is approximately $170 \, \text{GeV}/c^2$, which is approximately 340 000 times the rest mass of an electron. The rest mass of a composite particle depends on both the rest mass of its constituent elementary

Table 35-1   The Elementary Particles

Type of Particle	Symbol	Name	Mass (MeV/c^2)	Charge (e)	Spin Quantum Number	Baryon Number	Lepton Number
Quarks*	u	Up quark	2.5	+2/3	1/2	1/3	0
	d	Down quark	5.0	−1/3	1/2	1/3	0
	s	Strange quark	125	−1/3	1/2	1/3	0
	c	Charm quark	1.3×10^3	+2/3	1/2	1/3	0
	b	Bottom quark	4.2×10^3	−1/3	1/2	1/3	0
	t	Top quark	171×10^3	+2/3	1/2	1/3	0
Leptons	e^-	Electron	0.511	−1	1/2	0	1
	μ^-	Muon	105.7	−1	1/2	0	1
	τ^-	Tau	1.777×10^6	−1	1/2	0	1
	ν_e	Electron neutrino	$<2.2 \times 10^{-6}$	0	1/2	0	1
	ν_μ	Muon neutrino	<0.17	0	1/2	0	1
	ν_τ	Tau neutrino	<15.5	0	1/2	0	1
Gauge bosons	W^+	W-plus boson	80.4×10^3	+1	1	0	0
	W^-	W-minus boson	80.4×10^3	−1	1	0	0
	Z^0	Z-zero boson	91.2×10^3	0	1	0	0
	γ	Photon	0	0	1	0	0
	g	Gluon	0	0	1	0	0
Higgs boson	H^0	Higgs boson	Estimated to be approximately 125×10^3	0	0	0	0

* Here the mass implies the rest mass of the particle. The spin is the intrinsic spin quantum number and the baryon and lepton numbers are explained in the text. An isolated quark has never been detected. Therefore, the given values are the best estimates of the quark masses.
Source: Data from "Quark Masses," updated January 2010 by A.V. Manohar (University of California, San Diego) and C.T. Sachrajda (University of Southampton), pdg.lbl.gov/2011/reviews/rpp2011-rev-quark-masses.pdf.

particles and the nature of the forces between them to form the composite particle.

Electric Charge

A particle can have an electric charge or it can be electrically neutral. The electric charge is measured in units of the electron charge ($e = 1.602 \times 10^{-19}$ C). Until the 1960s, it was believed that the charge of a particle must be an integer multiple of e. However, with the discovery of quarks it was realized that quarks come in two charges: $+\frac{2}{3}e$ and $-\frac{1}{3}e$. The charge of a composite particle is equal to the sum of the charges of its constituents.

Intrinsic Spin

All elementary particles are assigned a spin quantum number, s. The electron has $s = \frac{1}{2}$, and the photon has $s = 1$. The intrinsic spin of a composite particle is obtained by adding the intrinsic spins of its constituents using the rules for the addition of angular momenta discussed in Section 32-10. Recall that particles with half-integer values of s are called fermions and particles with integral values of s are called bosons.

Fundamental Forces

The four known forces (or interactions) in nature are the strong force, the weak force, the electromagnetic force, and the gravitational force. The strong force can be divided into two parts: the fundamental strong force that exists due to the exchange of gluons between quarks, and the residual strong force that exists between hadrons (see Section 35-5) and is responsible for holding a nucleus together. The residual strong force is due to the exchange of mesons between the hadrons. Exactly how the strong force between quarks transforms into the residual strong force between the hadrons is not fully understood. The four fundamental forces vary greatly in relative strength, as shown in Table 35-2.

For example, if the residual strong force between two protons at a distance of few femtometres is 1 N, the strength of the electromagnetic force between them at the same distance is approximately 10^{-2} N, the weak force approximately 10^{-7} N, and the gravitational force approximately 10^{-38} N.

Table 35-2   Relative Strengths of the Four Known Forces

Force	Relative Strength
Residual strong force	1
Electromagnetic force	10^{-2}
Weak force	10^{-7}
Gravitational force	10^{-38}

All particles with mass exert a gravitational force on each other. We will ignore the gravitational force in the rest of this chapter. The electromagnetic force always exists between all charged particles (elementary and composite). The strong force exists only between quarks and is responsible for binding quarks into composite particles (composite particles are discussed in Section 35-5). The weak force is responsible for processes such as β-decay. Particles that interact with other particles only through the weak force and the electromagnetic force (if charged) are called **leptons**.

Quarks

Quarks are the building blocks of all composite particles. Quarks are charged fermions with $s = \frac{1}{2}$. There are six types of quarks (called **flavours**): the **up quark** (u), the **down quark** (d), the **strange quark** (s), the **charm quark** (c), the **bottom quark** (b), and the **top quark** (t). Quarks interact with each other through the strong, electromagnetic, and weak forces. We will discuss the quantum properties of quarks further in Section 35-6.

Leptons

Leptons are also fermions with $s = \frac{1}{2}$. There are six types of leptons: the electron (e^-), the **electron neutrino** (ν_e), the **muon** (μ^-), the **muon neutrino** (ν_μ), the **tau** (τ^-), and the **tau neutrino** (ν_τ). The charged leptons interact with each other and with other particles through the weak and electromagnetic forces. Neutrinos are electrically neutral, so they interact only through the weak force. An electron is stable and does not decay. The muon and tau are considerably more massive than an electron and are unstable. For example, a muon has a half-life of 2.26×10^{-6} s and decays into an electron and two neutrinos.

Baryons

All particles made up of quarks are called hadrons. Hadrons are subdivided into baryons and mesons. We will discuss hadrons in Section 35-5.

Gauge Bosons

According to the theory of relativity and quantum physics, elementary particles exert forces on each other by exchanging other particles, called **gauge bosons**. Photons mediate the electromagnetic force, gluons mediate the strong force, and $W^\pm$ and Z^0 bosons mediate the weak force. It is possible that the gravitational force is mediated by elementary particles called gravitons, but gravitons have not been observed. Gauge bosons have $s = 1$.

Higgs Bosons

In 1964, a group of scientists proposed that the elementary particles quarks, leptons, and gauge bosons gain mass by interacting with a field, which is called the Higgs field after one member of the group, British physicist Peter Higgs (b. 1929). In quantum theory, the Higgs field is quantized with the corresponding particles, called **Higgs bosons**, just as the electromagnetic field is quantized with photons as the associated particles. The Higgs boson was predicted to be a neutral particle with an intrinsic spin quantum number of zero and a mass of approximately 125 GeV/c^2. On July 4, 2012, scientists at the LHC announced the long-awaited discovery of the Higgs boson.

CHECKPOINT 35-1

Particle Classification

Particles that interact only through the weak and electromagnetic forces are called

(a) bosons
(b) quarks
(c) gauge bosons
(d) leptons
(e) gluons

ANSWER: (d)

35-2 Gauge Bosons

How do two electrons repel each other? If one electron suddenly moved away from another electron, would the electromagnetic force between the two decrease instantaneously? A consequence of the theory of relativity and quantum physics is that particles exert forces on each other by exchanging other particles. The rules of adding angular momentum in quantum mechanics require that an exchanged particle have an integer spin; therefore, it must be a boson. The carrier of the electromagnetic force between two charged particles is the photon, the particle that results when electromagnetic radiation is quantized. Two electrons therefore repel each other by exchanging a photon. A photon travels at the speed of light, so if the distance between two electrons suddenly changes, the force between them will take a finite time to adjust. The process of the exchange of a photon between two electrons is represented by the two diagrams in Figure 35-2. These diagrams are called Feynman diagrams, and they will be discussed in greater detail in Section 35-9.

In Figure 35-2(a), electron 1 emits a photon, which travels to electron 2. At a later time, electron 2 absorbs the photon. Let us consider this process using the centre-of-mass coordinate system of the two electrons. In this coordinate system, the two electrons have equal and opposite momenta in the initial state, and their energies, E_1 and E_2, are

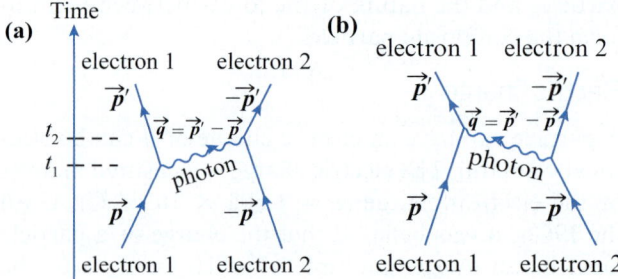

Figure 35-2 Feynman diagrams for the exchange of a photon between two electrons. (a) Electron 1 emits a photon, which is later absorbed by electron 2. (b) Electron 2 emits a photon, which is later absorbed by electron 1.

$$E_1 = \sqrt{\vec{p} \cdot \vec{p} c^2 + m_e^2 c^4}$$

$$E_2 = \sqrt{(-\vec{p}) \cdot (-\vec{p}) c^2 + m_e^2 c^4}$$

Let the momentum of the emitted photon be $\vec{q}$. After emitting the photon, the momentum of the first electron is $\vec{p} - \vec{q} = \vec{p}\,'$. When electron 2 absorbs the photon, its momentum changes to $-\vec{p} + \vec{q} = -\vec{p}\,'$. Therefore, the energies of the two electrons in the final state (after the photon has been absorbed by electron 2) are

$$E_1' = \sqrt{\vec{p}\,' \cdot \vec{p}\,' c^2 + m_e^2 c^4}$$

$$E_2' = \sqrt{(-\vec{p}\,') \cdot (-\vec{p}\,') c^2 + m_e^2 c^4}$$

Note that in the centre-of-mass coordinate system, $E_1 = E_2$ and $E_1' = E_2'$. Since the total energy must be conserved before and after the two electrons exchange a photon, we must have

$$E_1 + E_2 = E_1' + E_2'$$

This implies that in the centre-of-mass coordinate system, $E_1 = E_1'$ and $E_2 = E_2'$. Before electron 1 emits the photon, the total energy of the two electrons is

$$E = E_1 + E_2$$

Let the energy of the emitted photon be E_γ. After the photon is emitted *but before it is absorbed*, the total energy of the two electrons and the photon is

$$E_1' + E_2 + E_\gamma = E_1 + E_2 + E_\gamma$$

Thus, we see that while the photon is travelling between the two electrons, the total energy is not conserved. Is the non-conservation of energy allowed? According to Heisenberg's uncertainty principle, non-conservation of energy by an amount ΔE cannot be observed if it occurs for a time Δt such that

$$\Delta t \leq \frac{\hbar}{2\Delta E} \tag{35-1}$$

The energy of a photon with frequency f is hf. Therefore, the energy of a photon with frequency f cannot be determined to an accuracy hf if the photon exists for a time less than $\Delta t = \dfrac{\hbar}{2hf} = \dfrac{1}{4\pi f}$. During this time, the photon travels a distance of $R = c\Delta t = \dfrac{c}{4\pi f} = \dfrac{\lambda}{4\pi}$.

Photons are massless, so the energy of a photon can be arbitrarily small; therefore, the distance, R, over which a photon can transmit the electromagnetic force is arbitrarily large. Figure 35-2(b) shows the process in which the photon is emitted by electron 2 and is later absorbed by electron 1. Instead of showing both diagrams, it is common to represent a photon exchange between two electrons by a single diagram with a horizontal wavy line (Figure 35-3).

An exchanged photon exists for a limited time and is called a **virtual photon**. Its energy is uncertain by an amount that is inversely proportional to its time of existence. A virtual particle does not satisfy Einstein's energy–momentum relation. Therefore, a virtual photon can carry momentum without carrying energy. A photon emitted by a candle or a light bulb is a real photon. Its energy can be determined to a very high degree of accuracy, and its momentum (p) and energy (E) are related by $E = pc$.

The point where the electron emits a photon is called a **vertex**. Each electron–photon vertex is assigned a **coupling strength**, $\dfrac{e}{\sqrt{4\pi\varepsilon_0\hbar c}}$.

The strength of the electromagnetic interaction is obtained by multiplying the coupling strength at both vertices of the diagram. The resulting constant,

$$\frac{e}{\sqrt{4\pi\varepsilon_0\hbar c}} \times \frac{e}{\sqrt{4\pi\varepsilon_0\hbar c}} = \frac{1}{4\pi\varepsilon_0}\frac{e^2}{\hbar c}$$

is called the **fine structure constant** and is denoted by the Greek letter α (alpha). Using $e = 1.602 \times 10^{-19}$ C, $c = 2.998 \times 10^8$ m/s, $\hbar = 1.055 \times 10^{-34}$ J·s, and $1/(4\pi\varepsilon_0) = 8.988 \times 10^9$ N·m²/C², we get

$$\alpha = \frac{1}{4\pi\varepsilon_0}\frac{e^2}{\hbar c} = 7.296 \times 10^{-3} \approx \frac{1}{137}$$

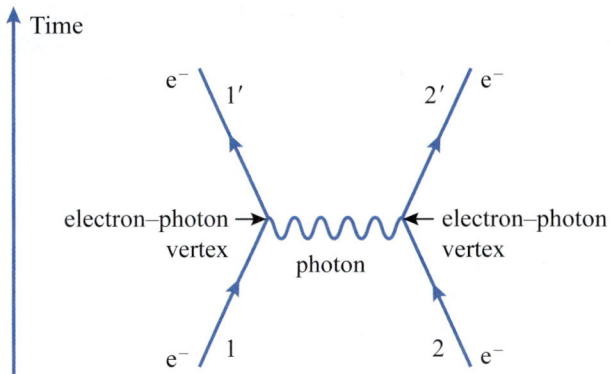

Figure 35-3 The exchange of a virtual photon between two electrons.

Notice that the fine structure constant is dimensionless.

Similar arguments can be applied to the strong and weak interactions. Gluons are the carriers of the strong force. Quarks interact with each other by exchanging gluons. Like photons, gluons have no mass. The weak force is due to the exchange of W^+, W^-, and Z^0 bosons between elementary particles. Unlike the photon and the gluon, these bosons are very massive. Following the same arguments as discussed above, we can show that the range, R, of a force between two particles that is due to the exchange of a boson of mass m is approximately

$$R = \frac{\hbar c}{2mc^2} \tag{35-2}$$

For example, the Z^0 and $W^\pm$ bosons have masses of 91.2 GeV/c^2 and 80.4 GeV/c^2, respectively, so the range of the weak force is approximately $\dfrac{200\ \text{MeV·fm}}{160\ \text{GeV}} \approx 0.001$ fm, an extremely short distance. (where we have used $\hbar c \approx 200$ MeV· fm and the mass of the gauge boson to be 80 MeV/c^2).

Figure 35-4 summarizes the carriers of the electromagnetic, strong, and weak forces.

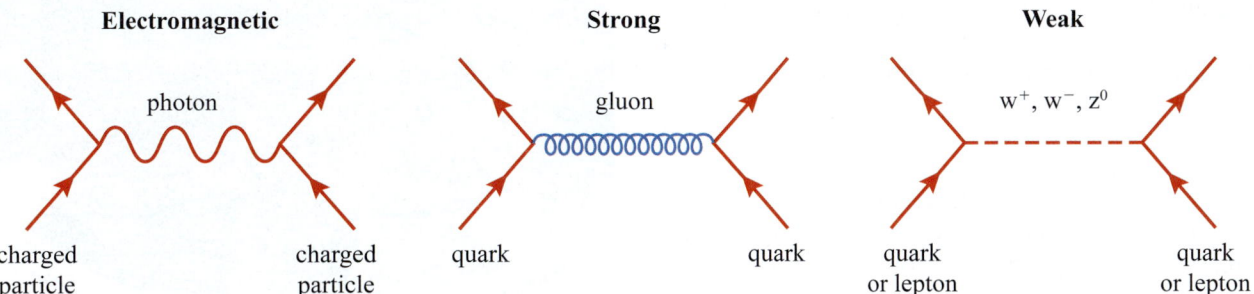

Figure 35-4 The carriers of the electromagnetic, strong, and weak forces.

35-3 Antiparticles

As discussed in Chapter 32, Schrödinger's equation satisfies the non-relativistic energy–momentum equation $E = \dfrac{p^2}{2m} + V(x)$. But a particle moving with a relativistic speed must satisfy Einstein's energy and momentum relation, $E^2 = p^2c^2 + m^2c^4$. In 1928, British physicist Paul Dirac (1902–1984) developed an equation that is consistent with the theory of relativity as well as quantum mechanics. Dirac used his equation to calculate many properties of atomic electrons, such as intrinsic spin and hyperfine splitting of hydrogen levels. Since the relativistic energy–momentum relationship is quadratic in energy, solutions to Dirac's equation allow the electron's energy to have both positive and negative values: $E = \pm\sqrt{p^2c^2 + m_e^2c^4}$, where m_e is the mass of the electron. Dirac interpreted the negative-energy solutions as corresponding to particles that have exactly the same mass as an electron but are positively charged. These particles are called **antiparticles**. The antiparticle of an electron is the **positron** (e^+). Dirac predicted that an electron–positron pair can be created by bombarding nuclei with photons with energy greater than $2m_ec^2$:

$$\gamma + \text{nucleus} \rightarrow e^- + e^+ + \text{nucleus}$$

Such a process, called **pair production**, requires the presence of a nucleus or another charged particle to conserve energy and momentum. Dirac also noted that when an electron and a positron combine, they can annihilate each other into two photons. This process is called **pair annihilation** and is written as follows:

$$e^- + e^+ \rightarrow \gamma + \gamma$$

The existence of positrons was confirmed by American physicist Carl Anderson (1905–1991), in 1932. He used a cloud chamber to study properties of cosmic rays. A cloud chamber consists of an enclosed chamber filled with a supersaturated vapour, often alcohol. When a charged particle passes through the chamber, it ionizes some vapour atoms along its path. The ionized atoms trigger the condensation of droplets in the supersaturated vapour, creating a visible track of the path of the charged particle. Placing the chamber in a magnetic field causes the tracks of positively and negatively charged particles to curve in opposite directions. Thus, the charge and the mass of a particle can be determined from measurements of the curvature and thickness of its track.

Some of the tracks observed by Anderson looked like the tracks in Figure 35-5. At the vertex of the curved upside-down V-shaped trail, a fast-moving photon created by a cosmic ray collides with a nucleus, producing an electron–positron pair. The curvature is due to the magnetic field applied at right angles to the plane of the photograph. One of the tracks corresponds to the electron, and the other track corresponds to a particle that has the same mass as an electron but opposite charge—a positron.

Experiments have shown that each quark and charged lepton has an antiparticle with the same mass and opposite charge. It is now assumed that every particle has a corresponding antiparticle. A neutral particle either has an antiparticle with some opposite quantum property or is its own antiparticle. Table 35-3 lists the elementary antiparticles. Note that the antiparticles for quarks and neutrinos are denoted by adding a bar over the symbol for the particle.

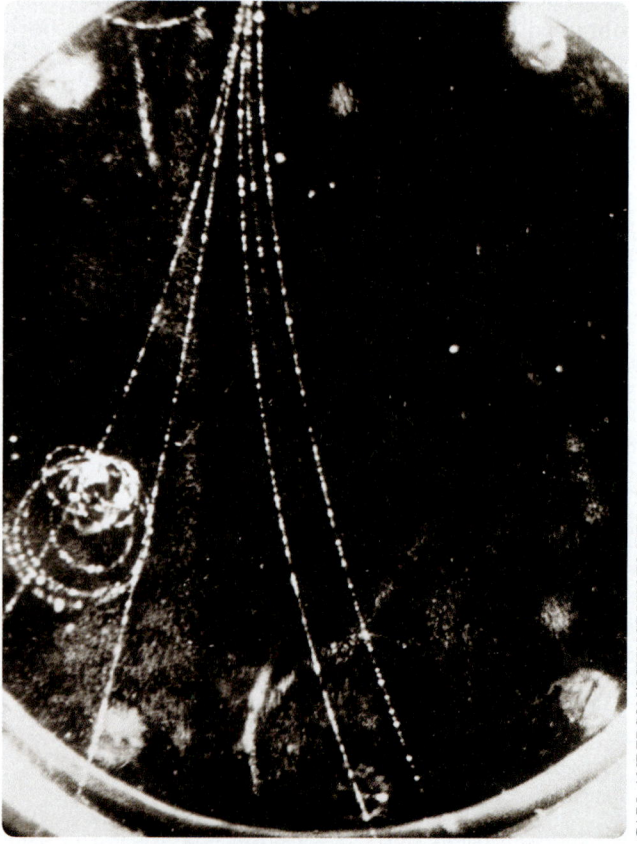

Figure 35-5 Tracks of several charged particle–antiparticle pairs in a cloud chamber.

CARL ANDERSON/SCIENCE PHOTO LIBRARY.

MODERN PHYSICS

Table 35-3 The Elementary Particles and Antiparticles

Particle	Antiparticle	Name/Description
u	$\bar{u}$	u-bar
d	$\bar{d}$	d-bar
s	$\bar{s}$	s-bar
c	$\bar{c}$	c-bar
b	$\bar{b}$	b-bar
t	$\bar{t}$	t-bar
e^-	e^+	Positron
μ^-	μ^+	Mu-plus
τ^-	τ^+	Tau-plus
ν_e	$\bar{\nu}_e$	Electron antineutrino
ν_μ	$\bar{\nu}_\mu$	Muon antineutrino
ν_τ	$\bar{\nu}_\tau$	Tau antineutrino
γ	γ	A photon is its own antiparticle.
g	g	A gluon is its own antiparticle.
W^-	W^+	W^+ is the antiparticle of W^-.
W^+	W^-	W^- is the antiparticle of W^+.
Z^0	Z^0	Z^0 is its own antiparticle.
H^0	H^0	The Higgs boson is its own antiparticle.

STANDARD MODEL OF ELEMENTARY PARTICLES

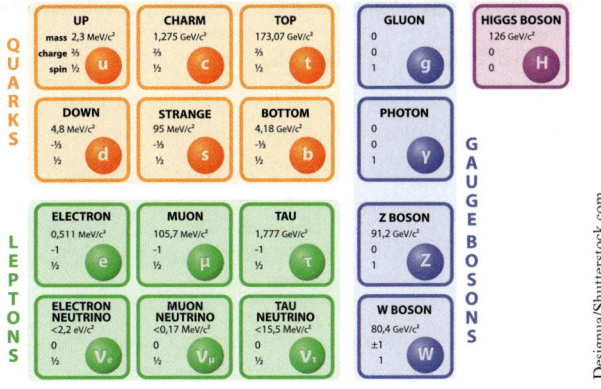

Figure 35-6 Elementary particles of the Standard Model.

Designua/Shutterstock.com

35-4 Quarks and the Standard Model

The Standard Model postulates that the known universe is made up of four types of fundamental particles: quarks, leptons, gauge bosons, and the Higgs boson (Figure 35-6), as well as their antiparticles. The particles interact with each other through the four known forces: the strong force, the electromagnetic force, the weak force, and the gravitational force. The Standard Model describes how the particles interact with each other through three of the four forces, the exception being the gravitational force. As noted earlier, the forces between the particles are mediated through the exchange of gauge bosons, $W^\pm$, Z^0, γ, and g (gluons). The Higgs boson is an additional boson that gives mass to fundamental particles. It is not a gauge boson (i.e., it is not a force carrier).

The six quarks and the six leptons are divided into three **particle families** or **particle generations**. The first generation consists of the up quark, the down quark, the electron, and the electron neutrino. The first generation is the most stable and least-massive family of particles. Protons, neutrons, and all atoms are made up of first-generation particles.

The second generation consists of the strange quark, the charm quark, the muon, and the muon neutrino. The third generation consists of the bottom quark, the top quark, the tau, and the tau neutrino. Particles in the second and third generations are more massive than particles in the first generation, and they decay quickly when produced in high-energy collisions of first-generation particles. The fact that there are six quarks and six leptons that can be grouped into three families is required for the Standard Model to be consistent.

The Standard Model accurately describes how particles interact with each other and correctly predicted the existence of a number of particles. The up, down, and strange quarks were known to exist quite early in the development of the quark model of hadrons. The existence of the charm quark was predicted on the basis that the second family needs a quark with a charge of $+2e/3$. The charm quark was discovered in 1974 at the Stanford Linear Accelerator by colliding beams of electrons and positrons. Similarly, the Standard Model also predicted the existence of $W^\pm$ and Z^0 and bosons, which were discovered in 1983 with the particle accelerator at CERN. The model also predicts the existence of many new composite particles made up of the more massive charm, bottom, and up quarks. Many of these composite particles have been observed in high-energy collisions, and the observed masses and other physical properties are consistent with the predictions of the Standard Model.

MODERN PHYSICS

35-5 Composite Particles

All the particles that can interact via the strong force are collectively called **hadrons**. This includes protons and neutrons. Hadrons that are fermions are called **baryons**, and hadrons that are bosons are called **mesons**. By the early 1960s, a large number of baryons and mesons had been discovered. The existence of quarks was first postulated by the American physicist Murray Gell-Mann (b. 1929) in an attempt to understand whether this large number of particles has an underlying structure. Just as protons and neutrons bind to form a large number of nuclei, Gell-Mann proposed that baryons and mesons are made up of only three types of elementary particles, which were named quarks. These were the up (u), the down (d), and the strange (s) quark. The charm (c), the bottom (b), and the top (t) quarks were discovered later.

Baryons

Baryons are made up of three quarks and have half-integer spin because all the quarks have $s = \frac{1}{2}$. A proton is a combination of two up quarks and one down quark, and a neutron consists of one up quark and two down quarks (Figure 35-7):

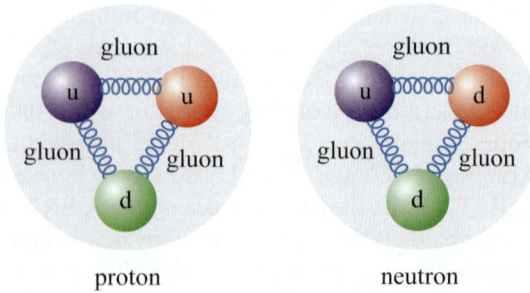

Figure 35-7 The quark structure of a proton and a neutron.

$$\text{proton} = (uud) \qquad \text{charge} = \left(\frac{2}{3} + \frac{2}{3} - \frac{1}{3}\right)e = +e$$

$$\text{neutron} = (udd) \qquad \text{charge} = \left(\frac{2}{3} - \frac{1}{3} - \frac{1}{3}\right)e = 0$$

Quarks are confined within a proton (or a neutron); therefore, their energies are quantized. The proton and the neutron are the lowest-energy bound states made up of up and down quarks. Because a proton (or a neutron) has spin $s = \frac{1}{2}$, one of the three quarks must have its spin oriented opposite to the spin of the other two. A proton with spin up, written as p ↑, has the following possible quark structures:

$$p\uparrow = (u\uparrow u\uparrow d\downarrow + u\uparrow u\downarrow d\uparrow + u\downarrow u\uparrow d\uparrow)$$

Baryons can also have quarks with their spins all pointing in the same direction. For example, the delta-plus-plus (Δ^{++}) particle contains three up quarks. The resulting particle has a rest mass of 1230 MeV/c^2 and a spin quantum number of $s = \frac{3}{2}$:

$$\Delta^{++} = (u\uparrow u\uparrow u\uparrow)$$

There are three other delta particles with quark spins aligned in the same direction:

$$\Delta^+ = (u\uparrow u\uparrow d\uparrow); \quad \Delta^0 = (u\uparrow d\uparrow d\uparrow);$$
$$\Delta^- = (d\uparrow d\uparrow d\uparrow)$$

The superscript on each particle indicates its charge. Delta particles are excited states of three quarks and therefore are more massive than a proton or a neutron.

The quark model predicted the existence of particles that had not yet been observed. Consider particles containing the three lowest-mass quarks: up, down, and strange. We have seen how combinations of up and down quarks provide the internal structures for the proton, neutron, and delta particles. The addition of the strange quark provided the internal structure of other particles, some of which had not yet been discovered.

Particles that contain one strange and two up or down quarks are called sigma baryons (symbol Σ). There are three sigma baryons:

$$\Sigma^+ = (uus); \quad \Sigma^0 = (uds); \quad \Sigma^- = (dds)$$

With two strange and one up or down quarks, we can construct two baryons. These particles had previously been discovered and are called xi baryons (symbol Ξ):

$$\Xi^0 = (uss); \quad \Xi^- = (dss)$$

Finally, the model predicts the existence of a baryon that is made up of three strange quarks. Gell-Mann called this baryon the omega (symbol Ω) baryon:

$$\Omega^- = (sss)$$

The omega baryon had not been discovered, and its discovery in 1964 was a great triumph of the quark model.

Table 35-4 lists some of the lowest-mass baryons made up of up, down, and strange quarks. Note that some of the particles have the same quark structure.

Mesons

Mesons are hadrons that are composed of one quark and one antiquark. A quark and an antiquark both have intrinsic spin $\frac{1}{2}$, so the total intrinsic spin of a quark–antiquark system can be either zero or one. Mesons with $s = 0$ are called **scalar mesons**, and mesons with $s = 1$ are called **vector mesons**.

The lightest scalar meson made up of up and down quarks and antiquarks is called the pion (symbol π). The three possible combinations of pions are the following:

$$\pi^+ = (u\,\overline{d}); \ \text{charge} = \left(\frac{2}{3} + \frac{1}{3}\right)e = e$$

$$\pi^- = (d\,\overline{u}); \ \text{charge} = \left(-\frac{1}{3} - \frac{2}{3}\right)e = -e$$

$$\pi^0 = \frac{1}{\sqrt{2}}(u\,\overline{u} + d\,\overline{d}); \ \text{charge} = \left(\frac{2}{3} - \frac{2}{3} + \frac{1}{3} - \frac{1}{3}\right)e = 0$$

The electrically neutral pion is made up of a $u\overline{u}$ and a $d\overline{d}$ combination. The factor of $1/\sqrt{2}$ implies that both combinations are present with an equal probability. The spins of the quark and antiquark of a pion are aligned antiparallel to each other.

The lightest vector meson is the rho meson (ρ), with a mass of 775 MeV/c^2. The quark structure of a rho meson is the same as the quark structure of the correspondingly charged pion; however, the spins of its quark and antiquark are aligned parallel to each other. As yet, there is no clear explanation why the mass of a rho meson is more than five times the mass of a pion.

With strange, charm, bottom, and top quarks, we can construct many more mesons. So far, over 120 mesons have been discovered. The existence of many mesons was first predicted by the quark model.

Antimatter

The antiproton ($\overline{p}$) and antineutron ($\overline{n}$) are made up of antiquarks:

$$\overline{p} = (\overline{u}\,\overline{u}\,\overline{d}); \ \ \overline{n} = (\overline{u}\,\overline{d}\,\overline{d})$$

Notice that although an antineutron is electrically neutral, it clearly has a different quark structure than a neutron. Some composite particles are their own antiparticles. For example, a π^0 meson is made up of equal amounts of $u\overline{u}$ and $d\overline{d}$ pairs. The anti-π^0 meson is obtained by replacing each quark of π^0 with its antiquark. Thus, the quark structure of an anti-π^0 meson is exactly the same as the quark structure of the π^0 meson:

$$\overline{\pi}^0 = \frac{1}{\sqrt{2}}(\overline{u}u + \overline{d}d) = \pi^0$$

Ordinary matter consists of electrons, protons, and neutrons. Can we construct antimatter made of positrons, antiprotons, and antineutrons? Antiparticles can be produced in high-energy accelerators. For example, antiprotons are created in proton–proton collisions in the following process:

$$p + p \rightarrow p + p + p + \overline{p}$$

In 1995, a team of scientists working at CERN created the first antihydrogen atom (symbol $\overline{H}$) by bringing positrons and antiprotons together. When an $\overline{H}$ atom interacts with ordinary matter, its positron

Table 35-4 Some Lowest-Mass Baryons Made up of u, d, and s Quarks

Symbol	Name	Quark Structure	Mass (MeV/c^2)	Spin	Mean Life (s)
p	Proton	uud	938.3	1/2	Stable
n	Neutron	udd	939.6	1/2	886
Λ	Lambda	uds	1115.6	1/2	2.52×10^{-10}
Σ^+	Sigma-plus	uus	1189.4	1/2	0.80×10^{-10}
Σ^0	Sigma-zero	uds	1192.5	1/2	1.0×10^{-14}
Σ^-	Sigma-minus	dds	1197.3	1/2	1.48×10^{-10}
Ξ^0	Xi-zero	uss	1314.9	1/2	2.98×10^{-10}
Ξ^-	Xi-minus	dss	1321.3	1/2	1.67×10^{-10}
Ω^-	Omega-minus	sss	1672.5	3/2	1.3×10^{-10}

EXAMPLE 35-1

Constructing Mesons

List all of the mesons that have one antibottom ($\overline{b}$) quark. What is the charge of each meson? Assuming that the rest mass energy of a meson is the sum of the rest mass energies of its constituent quarks, predict the rest mass energy of each meson. Compare your prediction with the experimental values given in Table 35-5. Use the quark masses in Table 35-1.

Solution

Since mesons contain one quark and one antiquark, we can construct six mesons by combining a $\overline{b}$ antiquark with any of the quarks. A meson constructed from an up quark and a $\overline{b}$ quark (called a B meson) has the following properties:

Quark structure: u$\overline{b}$

Charge: $2e/3 + e/3 = +e$

Estimated mass: $m_{\overline{b}} + m_u = 4.2 \times 10^3$ MeV/c^2 + 2.5 MeV/c^2

$$= 4.2 \text{ GeV}/c^2$$

Experimental mass from Table 35-5: 5.28 GeV/c^2

The charges and estimated masses of the other five mesons that contain a $\overline{b}$ quark can be determined in the same way. The results, along with the experimental masses, are listed in Table 35-5.

Making sense of the result

Note that the meson masses calculated by simply adding the rest masses of the constituent quarks are always lower than the experimentally determined values. The bound quarks have a considerable amount of energy, and a proper theoretical determination of a meson's mass requires information about the forces that bind the quarks in the meson.

Table 35-5 Mesons That Include a $\overline{b}$ Quark

Quark Content	Given Name	Charge (e)	Estimated Mass (MeV/c^2)	Experimental Mass (MeV/c^2)
u$\overline{b}$	B$^+$ (B-plus meson)	+1	4.2×10^3	5.28×10^3
d$\overline{b}$	B^0 (neutral B meson)	0	4.2×10^3	5.28×10^3
s$\overline{b}$	B^0_s (strange B meson)	0	4.3×10^3	5.36×10^3
c$\overline{b}$	B^+_c (charm B meson)	+1	5.5×10^3	6.27×10^3
b$\overline{b}$	η_b (bottom eta meson)	0	8.4×10^3	9.4×10^3
t$\overline{b}$	No given name	+1	175×10^3	Not yet discovered

annihilates with an electron to create two photons, and the antiquarks of the antiproton combine with quarks of a proton (or a neutron) to create pions. Therefore, the $\overline{H}$ atoms need to be isolated in a vacuum, away from matter. By trapping $\overline{H}$ atoms in a magnetic field, scientists have been able to keep them isolated for several minutes, long enough to study their properties. According to a fundamental theorem in physics, a universe made of antimatter must behave exactly the same

as our universe. So the emission spectrum of $\overline{H}$ should be exactly the same as the emission spectrum of the ordinary hydrogen atom. Any difference between the two spectra would require radical changes to currently accepted theories.

MAKING CONNECTIONS

Antihydrogen and the Big Bang

There is another important reason to study antimatter. Most scientists think that particles and antiparticles should have been created in equal numbers at the start of the Big Bang. However, our universe is dominated by particles. Antiparticles are only created during high-energy collisions of particles and photons. What has happened to the antiparticles that were created at the start of the Big Bang? Scientists hope that a study of antihydrogen can help answer this question.

CHECKPOINT 35-5

Baryon Mass and Charge

Assume that the rest mass of a baryon is approximately equal to the sum of the masses of its constituent quarks, and rank the following baryons in order of increasing rest mass. Determine the charge of each baryon.
(a) tdd
(b) tbs
(c) ttu
(d) bss
(e) bbb
(f) ccc

ANSWER: (c) $+2e$; (b) 0; (a) 0; (e) $-e$; (d) $-e$; (f) $+2e$. The highest-mass quark is the t quark, followed by b, c, s, d, and u quarks.

35-6 Colour Quantum Number and Quark Confinement

Just as electrons in an atom have quantized energy levels, the energies of quarks in baryons and mesons are quantized. As described in the previous section, a Δ^{++} baryon is made of three up quarks with spins pointing in the same direction. Experiments indicate that all three up quarks of a Δ^{++} are in the lowest-energy state. However, quarks are fermions, and according to the Pauli exclusion principle, two or more identical quarks cannot simultaneously occupy the same energy state. Therefore, physicists speculated that quarks have an additional internal quantum property that can take on three values. This property is called **colour** (or **colour charge**), and the three possible values are called red (R), blue (B), and green (G). Thus, there are three up quarks: up-red, up-blue, and up-green. The Δ^{++} baryon is therefore made up of three up quarks of different colour quantum numbers (Figure 35-8). Since the three up quarks are not identical any more (they have different colours), there is no conflict with the Pauli exclusion principle in their occupying the same energy state:

$$\Delta^{++} = (u_R \uparrow u_G \uparrow u_B \uparrow)$$

The colour property of quarks is not an actual colour. An up-red quark does not look red. The names for this quantum property were chosen because the colour quantum numbers of a baryon and a meson must add to zero (neutral colour), somewhat like the way the primary colours red, green, and blue can combine to form white light. Experiments have verified that quarks do indeed possess an internal property with three possible values. The colour quantum numbers for antiquarks are anti-red ($\overline{R}$), anti-green ($\overline{G}$), and anti-blue ($\overline{B}$). Baryons and mesons, although made of quarks, do not have the colour property because the colour numbers of the possible states of the constituent quarks add to zero, similar to the way an up-spin and a down-spin add to zero.

Colour is an additive quantum number, and a colour and its anticolour cancel just as equal and opposite electric charges cancel. Like the conservation of electric charge in a process, the total colour of the initial and final states must be the same. Consequently, gluons must have colour charge. For example, if a red quark emits a gluon and changes into a blue quark (Figure 35-9), then

Colour of the initial quark = colour of the final quark
+ colour of the gluon

$$R = B + \text{colour of the gluon}$$

Colour of the gluon = red + anti-blue = $R\overline{B}$

Gluons can have eight possible combinations of colours. Because gluons have colour charge, they interact with each other, unlike photons, which have no charge or colour.

Quark Confinement

We can break up a nucleus and observe its individual protons and neutrons. If a proton is constructed from quarks, then we should be able to break it apart and observe individual quarks. Despite numerous attempts using various techniques, no fractionally charged particles have ever been observed. Bombarding a hadron with very high energy particles does not separate a quark from it; instead, the collisions create additional quark–antiquark pairs, which combine to form other hadrons. It is now believed that quarks are permanently confined within hadrons, and it is impossible to detach a single quark from a hadron. This **quark confinement** hypothesis is not yet fully understood.

Consider the strong force between a quark and an antiquark pair ($q\overline{q}$) due to the exchange of gluons between the two. Since gluons are coloured, they also interact with each other, which causes the field lines to be confined in the region directly between the quark and the antiquark, forming what is called a **flux tube** (Figure 35-10). The potential energy therefore increases as the distance, r, between the pair increases. At large distances the potential between the pair can be approximated as follows:

$$V_{q\overline{q}}(r) = -\lambda r \tag{35-3}$$

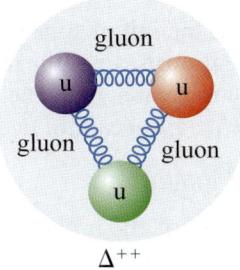

Figure 35-8 A Δ^{++} consists of three up quarks with different-colour quantum numbers.

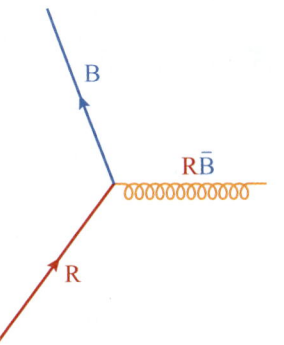

Figure 35-9 A red quark can change into a blue quark by emitting a gluon of colour red-anti-blue.

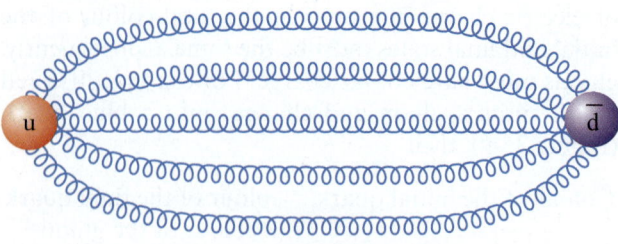

Each line represents a gluon.

Figure 35-10 The exchange of gluons between a quark and an antiquark forms a flux tube.

Here, λ is a constant that depends on the strength of the strong force; its exact value is not needed for this discussion. The situation is similar to stretching a rubber band. The more the band is stretched, the greater the force needed to stretch it farther until the band breaks into two bands. Beyond a certain separation, enough energy has been put into the $q\bar{q}$ pair to produce an additional $q\bar{q}$ pair. The original meson then fissions into two mesons (Figure 35-11) rather than splitting into a quark–antiquark pair.

If quarks are confined within a hadron, then how do nucleons exert a strong force on each other to form a nucleus? Gluons are coloured and a nucleon has no colour; therefore, a nucleon cannot emit a gluon and remain colourless. An exchange of a $q\bar{q}$ pair between the two nucleons is, however, allowed. This $q\bar{q}$ pair exchange represents the exchange of a virtual meson between two nucleons. So, the model of the strong force is as follows: the strong force between quarks arises due to the exchange of gluons. This force confines quarks within the hadrons. The strong force between hadrons arises due to the exchange of virtual mesons.

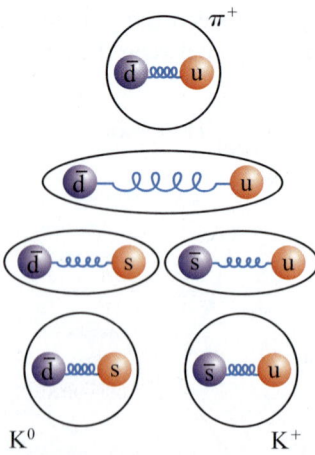

Figure 35-11 As energy is pumped into a pion to separate the quark–antiquark pair, the internal energy of the pion increases until there is sufficient energy to form two kaons.

35-7 Conservation Laws

The conservation laws of classical physics apply to interactions between particles. So, total energy, charge, and momentum are conserved when particles interact with each other. In addition, a number of other quantum numbers are also conserved. We can use conservation laws to determine whether specific reactions are possible without having to consider the details of the interactions between particles.

Conservation of Energy

The total energy of the initial state must be equal to the total energy of the final state. The total energy includes the rest mass energy of the particles. Consider the decay of a neutron into a proton, an electron, and an antineutrino:

$$n \rightarrow p + e^- + \bar{\nu}_e$$

In the rest frame of the neutron, there is no kinetic energy, so the total energy in the initial state is simply the rest mass energy of the neutron. The total energy in the final state is the sum of the rest mass energy and the kinetic energy of all three product particles and must equal $m_n c^2$.

Conservation of energy explains why a free proton cannot spontaneously decay into a neutron, as follows:

$$p \rightarrow n + e^+ + \nu_e$$

The total energy in the initial state is equal to the rest mass energy of the proton. Since the rest mass energy of the neutron alone is greater than the rest mass energy of the proton, the reaction cannot occur by itself.

Conservation of Momentum

The total momentum of the initial state must be equal to the total momentum of the final state. For example,

MODERN PHYSICS

for neutron decay in the rest frame of the neutron, conservation of momentum requires that

$$0 = \vec{p}_p + \vec{p}_e + \vec{p}_{\bar{\nu}}$$

where $\vec{p}_p$, $\vec{p}_e$, and $\vec{p}_{\bar{\nu}}$ are the momenta of the proton, the electron, and the antineutrino, respectively.

For relativistic kinematics, energy and momentum are combined into a single quantity called the **four-momentum**, p_μ. The laws of conservation of energy and momentum can then be combined into the law of conservation of four-momentum. For neutron decay,

$$p_{\mu,n} = p_{\mu,p} + p_{\mu,e} + p_{\mu,\nu}$$

Conservation of Angular Momentum

As described in Chapter 32, the total angular momentum, $\vec{J}$, of a particle is the sum of its orbital angular momentum, $\vec{L}$, and its intrinsic spin, $\vec{S}$:

$$\vec{J} = \vec{L} + \vec{S}$$

According to the law of conservation of angular momentum, in the absence of net external torque, the total angular momenta of the initial and the final states are equal. So, for neutron decay,

$$\vec{J}_n = \vec{J}_p + \vec{J}_e + \vec{J}_\nu$$

Note that the *total* angular momentum is conserved, but the orbital angular momentum and spin are not conserved individually.

Conservation of Charge

The total charge in the initial and final states must be the same. The process $p + p \rightarrow p + n + \pi^+$ is allowed by charge conservation because the total charges of the initial and final states are both $+2e$. Conservation of energy requires that the protons in the initial state have at least enough kinetic energy to create a pion and make up the difference in rest mass between a proton and a neutron. However, the process $p + p \rightarrow p + n + \pi^0$ is prohibited because the total charge of the initial state is $+2e$, and the total charge of the final state is $+e$.

Conservation of Lepton Number

It is experimentally observed that the number of leptons minus the number of antileptons is conserved in all reactions. Therefore, leptons are assigned a **lepton number**, L, of $+1$, and antileptons are assigned a value of -1. The lepton number of all other particles is 0. With this assignment, the total lepton number is conserved in any interaction. For example, the lepton number in the initial and final states in the decay of a neutron is 0:

$$n \rightarrow p + e^- + \bar{\nu}_e$$
$$L: 0 = 0 + 1 + (-1)$$

Similarly, the reaction $e^+ + e^- \rightarrow \pi^+ + \pi^-$ conserves a lepton number of zero. In contrast, the following two decay processes cannot happen because the lepton number is not conserved:

$$\pi^- \rightarrow e^- + \gamma \qquad\qquad \mu^- \rightarrow e^- + \bar{\nu}_e$$
$$L: 0 \neq 1 + 0 \qquad\qquad L: +1 \neq 1 + (-1)$$

Neither of these prohibited decays has ever been observed.

Conservation of Baryon Number

The law of conservation of **baryon number**, B, is also an experimental fact. It states that the number of baryons minus the number of antibaryons is the same in the initial and final states of a reaction. A baryon number of $+1$ is assigned to each baryon, and a baryon number of -1 is assigned to each antibaryon. Since baryons are made up of three quarks, all quarks are assigned a baryon number of 1/3, and all antiquarks are assigned a value of $-1/3$. All other particles (including mesons and leptons) have a baryon number of zero.

For example, in the β-decay of a neutron, the baryon number before and after the decay is the same:

$$n \rightarrow p + e^- + \bar{\nu}_e$$
$$B: +1 = +1 + 0 + 0$$

However, a proton cannot decay into a positron and a neutrino even though energy, momentum, charge, and lepton number would be conserved in this reaction:

$$p \rightarrow e^+ + \nu_e + \gamma$$
$$B: +1 \neq 0 + 0 + 0$$
$$L: 0 = -1 + 1 + 0$$

The process $p + p \rightarrow p + p + n$ is not possible, regardless of the total energy of the initial state, because the baryon number of the initial state is two and the baryon number of the final state is three. Similarly, the decay process $p \rightarrow e^+ + \gamma$ conserves energy, momentum, angular momentum, and charge but is still prohibited because it does not conserve baryon number.

35-8 The Production and Decay of Particles

Scientists create new particles by colliding beams of high-energy particles, usually electrons or protons, with other particles or nuclei. In such collisions, the kinetic energy of the colliding particles is transformed into the mass of the new particles, in accordance with Einstein's mass–energy equation, $E = mc^2$. For example, pions can be produced by colliding fast-moving protons with other protons in the following processes:

$$p + p \rightarrow p + n + \pi^+$$
$$p + p \rightarrow p + p + \pi^0$$
$$p + p \rightarrow p + p + \pi^0 + \pi^+ + \pi^-$$

One of the protons on the left side of each reaction arrow represents a fast-moving **incident proton**, and the other is the **target proton**, usually at rest. How much kinetic energy does the incident proton need to produce one charged pion? The mass of π^+ is 139.6 MeV/c^2, so one might think that the incident proton must have at least 139.6 MeV of kinetic energy. However, the kinetic energy of the incident proton must be at least twice this amount.

Consider a collision between two particles of masses m_1 and m_2. We will assume one-dimensional motion and not use vector notation. Assume that particle 1 (mass m_1) is moving along the positive x-axis with velocity

$v_{lab} << c$, and particle 2 (mass m_2) is initially at rest. In this frame of reference, called the **laboratory frame of reference**, the total kinetic energy of the two particles is (here we have used non-relativistic kinematics to simplify the calculation)

$$K_{lab} = \frac{1}{2} m_1 v_1^2 + \frac{1}{2} m_2 v_2^2 = \frac{1}{2} m_1 v_{lab}^2 \quad (35\text{-}4)$$

It is easier to analyze the collision in a frame of reference that is moving with the centre of mass of the colliding particles. If at some instant particle 1 is located at x_1 and particle 2 is at x_2, the centre of mass, x_{cm}, of the two particles is located at

$$x_{cm} = \frac{m_1 x_1 + m_2 x_2}{m_1 + m_2} \quad (35\text{-}5)$$

Differentiating with respect to time, we obtain

$$\frac{dx_{cm}}{dt} = \frac{m_1 \dfrac{dx_1}{dt} + m_2 \dfrac{dx_2}{dt}}{m_1 + m_2}$$

In the above equation, $\dfrac{dx_{cm}}{dt} = v_{cm}$ is the velocity of the centre of mass, $\dfrac{dx_1}{dt} = v_{lab}$ is the velocity of the incident particle in the laboratory frame, and $\dfrac{dx_2}{dt} = 0$ because particle 2 is at rest. Therefore,

$$v_{cm} = \left(\frac{m_1}{m_1 + m_2}\right) v_{lab} \quad (35\text{-}6)$$

Notice that the centre of mass moves in the same direction as the incident particle but with a slower speed. According to an observer moving with the centre of mass, particle 1 is incident from the left with velocity $v_{lab} - v_{cm}$, and particle 2 is approaching from the right with velocity $-v_{cm}$. Before the collision, the total momentum of the two particles is zero in the centre-of-mass frame:

$$m_1(v_{lab} - v_{cm}) + m_2(-v_{cm})$$
$$= m_1 v_{lab} + (m_1 + m_2)(-v_{cm}) = 0 \quad (35\text{-}7)$$

By the law of conservation of momentum, the total momentum remains zero after the collision. In the centre-of-mass frame, the total kinetic energy, K_{cm}, of the two colliding particles is

$$K_{cm} = \frac{1}{2} m_1(v_{lab} - v_{cm})^2 + \frac{1}{2} m_2(-v_{cm})^2$$
$$= \left(\frac{m_2}{m_1 + m_2}\right) K_{lab} \quad (35\text{-}8)$$

MODERN PHYSICS

The total kinetic energy in the centre-of-mass frame is not the same as in the laboratory frame because the centre of mass of the two particles also moves and there is a kinetic energy associated with this motion. Using Equation 35-6,

$$K_{cm} + \frac{1}{2}(m_1 + m_2)v_{cm}^2 = K_{lab} \qquad (35\text{-}9)$$

In the centre-of-mass frame, the colliding particles can come to rest after the collision and still conserve momentum. In this case, since the total kinetic energy is zero after the collision, all the initial kinetic energy becomes either mass–energy of the new particles or excitation energy of the particles. Therefore, the *maximum* possible energy available to create a new particle is

$$K_{cm} = \left(\frac{m_2}{m_1 + m_2}\right)K_{lab} \qquad (35\text{-}10)$$

The greater the kinetic energy of the incident particle, the more energy is available to produce new particles. To create particles with very large masses, the incident particle must have a very large kinetic energy. This is the reason for building large particle accelerators such as the LHC.

Particle Decay

Particles can decay into particles of lower mass provided no conservation laws are violated. Particle decay follows the exponential decay law described in Chapter 34:

$$N(t) = N(0)\, e^{-\frac{t}{\tau}} \qquad (35\text{-}11)$$

where $N(0)$ is the number of particles present at $t = 0$, $N(t)$ is the number remaining after time t, and τ is the mean lifetime of the particle.

Approximate values of lifetimes for decays through the strong, electromagnetic, and weak interactions are as follows:

Decay through the strong interaction $\qquad 10^{-23}$ s
Decay through the electromagnetic interaction $\quad 10^{-18}$ s
Decay through the weak interaction $\qquad 10^{-10}$ s

The exact values of the lifetimes depend on detailed calculations. A particle can decay in several different ways. For example, a negatively charged pion can decay into a muon and a muon antineutrino, or into an electron and an electron antineutrino:

$$\pi^- \rightarrow \mu^- + \bar{\nu}_\mu \qquad (35\text{-}12)$$

$$\pi^- \rightarrow e^- + \bar{\nu}_e \qquad (35\text{-}13)$$

Consider the decay of particle a with a mass of m_a into particle b with a mass of m_b and particle c with a mass of m_c:

$$a \rightarrow b + c$$

EXAMPLE 35-2

Producing Pions

Bombarding a stationary hydrogen target with a beam of fast-moving protons can produce positively charged pions (π^+, rest mass 139.6 MeV/c^2) in the reaction $p + p \rightarrow p + n + \pi^+$. What is the minimum kinetic energy required for the proton beam to produce the charged pions?

Solution

The minimum energy required to create a π^+ is equal to its rest mass energy, $(139.6 \text{ MeV}/c^2)c^2 = 139.6$ MeV. This energy has to come at the expense of a decrease in the centre-of-mass kinetic energy. For proton–proton collisions,

$$\frac{m_2}{m_1 + m_2} = \frac{m_p}{m_p + m_p} = \frac{1}{2}$$

Therefore, from Equation 35-10,

$$K_{cm} = 139.6 \text{ MeV}$$
$$= \frac{1}{2}K_{lab}$$
$$K_{lab} = 2 \times 139.6 \text{ MeV}$$
$$= 279.2 \ MeV$$

Thus, the minimum kinetic energy for the proton beam incident on a hydrogen target to create positively charged pions is about 280 MeV. The final state contains a neutron, and its rest mass is 1.2 MeV/c^2 greater than that of a proton. So some additional kinetic energy is needed to create this extra mass.

Making sense of the result

Because colliding particles have the same mass, half of the incident kinetic energy goes into the centre of mass motion and the other half is converted into the rest mass of the new particle.

Conservation of energy and momentum requires that

$$E_a = E_b + E_c \quad \text{and} \quad \vec{p}_a = \vec{p}_b + \vec{p}_c$$

If particle a is at rest, conservation of momentum gives

$$\vec{p}_b = -\vec{p}_c$$

Using the relativistic energy–momentum relation, the conservation of energy requires that

$$m_a c^2 = \sqrt{p^2 c^2 + m_b^2 c^4} + \sqrt{p^2 c^2 + m_c^2 c^4}$$

Solving for p, the magnitude of the momentum of particles b and c is

$$p = \frac{m_a c}{2}\sqrt{\left(1 - \frac{m_b^2 + m_c^2}{m_a^2}\right)^2 - 4\frac{m_b^2 m_c^2}{m_a^4}} \qquad (35\text{-}14)$$

EXAMPLE 35-3

The Decay of the Higgs Boson

The rest mass of a Higgs boson is approximately 125 GeV/c^2. It has a very short lifetime and decays into other particles. One possible decay mode is the decay into two photons. Find the momentum, energy, and wavelength of the two photons after the decay of a Higgs boson.

Solution

This is the case of a single particle decaying into two particles. The momentum of the decay products in the centre-of-mass frame of the initial particle is given by Equation 35-14. Because photons are massless, in this case,

$$m_b = m_c = m_\gamma = 0 \quad \text{and} \quad m_a = m_H = 125 \text{ GeV}/c^2$$

Inserting these values in Equation (35-14), we obtain for the momentum of the emitted photons

$$p = \frac{m_H c}{2} = \frac{1}{c}\left(\frac{m_H c^2}{2}\right) = 62.5 \text{ GeV}/c$$

The energy, E, of a photon is related to its momentum, p, by $E = pc$. Therefore, the energy of each of the emitted photons is

$$E = pc = 62.5 \text{ GeV}$$

So each photon carries half the rest mass energy of the Higgs boson. The wavelength, γ, of a photon is related to its momentum, p, by $\lambda = \dfrac{h}{p}$, where h is Planck's constant. Therefore,

$$\begin{aligned}
\lambda = \frac{h}{p} &= \frac{4.14 \times 10^{-21} \text{ MeV} \cdot \text{s}}{62.5 \times 10^3 \text{ MeV}/c} \\
&= \frac{(4.14 \times 10^{-21} \text{ MeV} \cdot \text{s}) \times (3.00 \times 10^8 \text{ m/s})}{62.5 \times 10^3 \text{ MeV}} \\
&= 1.99 \times 10^{-17} \text{ m}
\end{aligned}$$

Making sense of the result

Because the total rest mass energy of the Higgs boson is equally shared among the photons, each photon carries 62.5 GeV of energy.

CHECKPOINT 35-8

Energy of the Centre of Mass

Consider the following two situations:
1. A proton with kinetic energy 100 GeV collides with a stationary target proton.
2. Two protons, each with 50 GeV of kinetic energy, are moving toward each other and collide.

The total energy available to produce new particles in these two situations is
(a) the same
(b) greater for situation 1
(c) greater for situation 2

ANSWER: (a) In a proton-proton collision where one of the initial protons is at rest, only half of the kinetic energy of the incident proton is available to produce new particles.

35-9 Feynman Diagrams

American physicist Richard Feynman (1918–1988) introduced an ingenious method for describing interactions of particles (elementary and composite) by means of position versus time graphs, called **Feynman diagrams**. Each diagram represents a possible way that a particular process can occur, and a given process can be represented by several diagrams. There is a well-defined mathematical procedure to calculate the probability amplitude corresponding to each diagram. The probability for a certain process to occur is proportional to the square of the sum of all of the probability amplitudes for that process.

Here are the rules for drawing simple Feynman diagrams for scattering or decay of particles.

- The initial and final states of a process contain only free particles or antiparticles. Particles and antiparticles satisfy the energy–momentum equation, $E^2 = p^2c^2 + m^2c^4$.

- An arrow that has one free end and points in the direction of increasing time represents a free particle.

- An arrow that has one free end and points in the direction of decreasing time represents a free antiparticle.

- Lines with both ends connected represent virtual particles. Virtual particles do not satisfy the energy–momentum relation and do not appear in the initial or final states.

- Usually, dashed or wavy lines represent bosons, and solid lines represent fermions. Each line is assigned a four-momentum representing the four-momentum of the particle.

- The point at which two or more particles come together is called a vertex. Electric charge, colour, baryon number, lepton number, and four-momentum are conserved at each vertex.

In Section 35-2, we used Feynman diagrams (Figure 35-2) to represent a process in which two electrons scatter from each other by exchanging a single photon. The wavy photon line represents both a photon emitted by electron 1 and absorbed by electron 2 and a photon emitted by electron 2 and absorbed by electron 1. The exchanged photon is a virtual photon.

Now, consider a process in which an electron and a positron scatter from each other by exchanging a photon. This process is represented by the two Feynman diagrams in Figure 35-12. Figure 35-12(a) corresponds to the case where a virtual photon is exchanged between the electron and the positron and the two then scatter into the final states. Figure 35-12(b) represents a quite different mechanism: the electron and positron annihilate each other into a single photon, which at a later time converts back into an electron–positron pair. Both processes contribute to the probability that an electron and a positron scatter from each other. In diagram (a), the electron and the positron are present at all times, whereas in diagram (b), there is a time interval when only a photon is present.

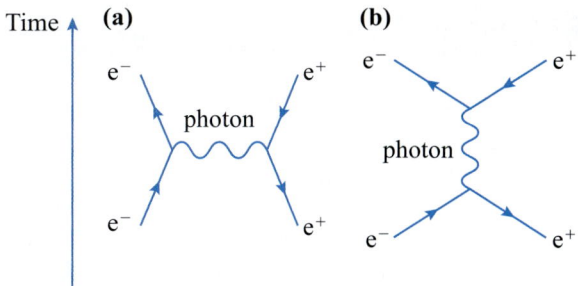

Figure 35-12 Feynman diagrams representing electron–positron scattering. (a) Electron–positron scattering through the exchange of a virtual photon, and (b) electron–positron annihilation into a photon.

EXAMPLE 35-4

Muon Pair Production

Muon (μ^-) and antimuon (μ^+) pairs can be produced by colliding a beam of positrons with electrons that are bound to atoms in the following process:

$$e^+ + e^- \rightarrow \mu^- + \mu^+$$

Draw a Feynman diagram representing this process, and calculate the minimum energy of the positron beam needed for this process to occur. The rest mass energy of a muon is 105.7 MeV.

Solution

To create a μ^-, μ^+ pair in the final state, an electron and the positron pair in the initial state must annihilate into a photon, which then converts into a μ^-, μ^+ pair. The Feynman diagram for this process is shown in Figure 35-13.

The total energy required in the centre-of-mass coordinate system of the electron–positron pair must be at least twice the rest mass energy of the muon, 2 × 105.7 MeV = 211.4 MeV. In the experimental arrangement, a beam of positrons strikes electrons that are bound to a stationary target, for example,

hydrogen atoms. Since the binding energy of electrons in an atom is very small, it can be ignored. So, to a very good approximation, the beam of positrons collides with electrons that are at rest. From Equation 35-10, for $m_1 = m_2 = m_e$,

$$K_{cm} = \frac{1}{2} K_{lab}$$

Therefore, in the laboratory coordinate system, the kinetic energy of the positron beam must be at least 2 × 211.4 MeV = 422.8 MeV.

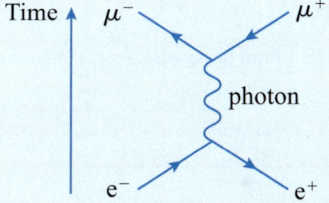

Figure 35-13 The Feynman diagram for an electron–positron annihilation that produces a muon and an antimuon.

EXAMPLE 35-5

Neutron Decay

Draw a Feynman diagram for the decay of a neutron into a proton, an electron, and an electron antineutrino. Assume that the neutron and the proton are elementary particles. Then modify your diagram to include the quark structure of the proton and the neutron.

Solution

The neutron decay reaction can be represented as $n \rightarrow p + e^- + \bar{\nu}_e$. Figure 35-14(a) shows the Feynman diagram for this decay.

The quark structure of a neutron is (udd), and the quark structure of a proton is (uud). Therefore, in a neutron decay, a down quark must convert into an up quark. The charge of a down quark is $-\frac{1}{3}e$, and that of an up quark is $+\frac{2}{3}e$.

Therefore, for an up quark to convert to a down quark, it must emit a particle of charge $-e$. This particle happens to be the W^- boson. The final state also includes an electron and an electron antineutrino, which are created by the decay of the

W^- boson. The Feynman diagram for this process is shown in Figure 35-14(b). Circles around the quark arrows for the neutron and proton indicate that the quarks are bound.

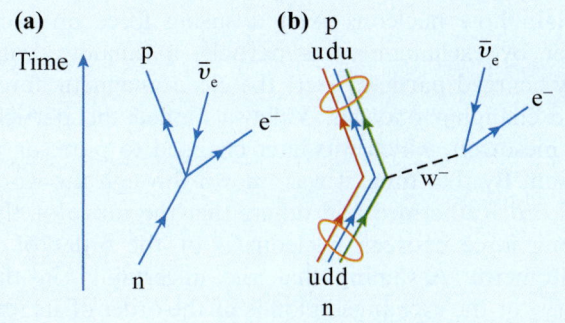

Figure 35-14 Feynman diagrams for neutron decay. (a) The decay process assuming that neutrons and protons are elementary particles. (b) The decay mechanism using the quark structure of the neutron and proton.

35-10 Pions and Muons

Both the pion (or the π meson) and the muon were discovered decades before the introduction of the quark model.

The Discovery of Pions

In 1935, Japanese physicist Hideki Yukawa (1907–1981) proposed the existence of a new particle to explain how nucleons exert a strong force on each other by exchanging this particle, in analogy with how charged particles exert the electromagnetic force by exchanging photons. Yukawa named the particle the mesotron, which was later changed to pion, or π meson. By that time, it was known through the work of Lord Rutherford and others that the range of the strong force between nucleons is of the order of a femtometre. Assuming that the uncertainty in the energy of the exchanged pion is of the order of its rest mass energy (see Section 35-2), Yukawa argued that such a pion could exist for a time $\Delta t \leq \dfrac{\hbar}{2mc^2}$ before being absorbed. Here, m is the rest mass of the pion.

The maximum distance that a pion can travel in this time is therefore of the order of

$$R = c\Delta t = \frac{\hbar c}{2mc^2} \tag{35-15}$$

Using $R \approx 1$ fm, Yukawa estimated the rest mass of the pion to be

$$m = \left(\frac{\hbar c}{2R}\right)\frac{1}{c^2} = \frac{200 \text{ MeV}\cdot\text{fm}}{2 \text{ fm}}\frac{1}{c^2} = 100 \text{ MeV}/c^2$$

The range of the nuclear force was only approximately known, so the predicted mass of the exchanged particle was approximate, too. Since charge is conserved during the exchange of a virtual particle, there are three different pions: a positively charged pion (π^+), a negatively charged pion (π^-), and an electrically neutral pion (π^0). Two protons (or neutrons) can only exchange a π^0. A proton–neutron pair can exchange either a π^+ or a π^-. Feynman diagrams describing the exchange of pions between two nucleons are shown in Figure 35-16. Note that charge is conserved at each vertex.

The exchanged pions are virtual and absorbed within a very short time:

$$\Delta t \leq \frac{\hbar}{2mc^2} = \left(\frac{\hbar c}{2mc^2}\right)\frac{1}{c} \approx \left(\frac{200 \text{ MeV}\cdot\text{fm}}{200 \text{ MeV}}\right)\frac{1}{c}$$

$$= 1 \frac{\text{fm}}{c} = \frac{10^{-15} \text{ m}}{3 \times 10^8 \text{ m/s}} \approx 3 \times 10^{-24} \text{ s}$$

Real pions can be produced in high-energy collisions of protons through the following processes, provided sufficient energy is available in the centre-of-mass coordinate system of the colliding protons:

$$p + p \rightarrow p + n + \pi^+$$
$$p + p \rightarrow p + p + \pi^0 \tag{35-16}$$

Cosmic rays consist mainly of high-energy protons, with some electrons and helium nuclei, created in deep

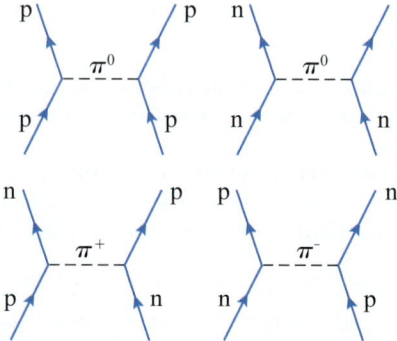

Figure 35-16 Feynman diagrams for one pion exchange between two nucleons.

space. When high-energy protons collide with nuclei in the upper atmosphere, they produce both charged and neutral pions, as described above. In 1947, a team of researchers led by British physicist Cecil Powell (1903–1969) placed specially designed photographic plates at high-altitude locations in Europe. Just as light produces an image on light-sensitive film, charged pions interact with the nuclei of the photographic plates and produce a visible track. An analysis of these plates confirmed the existence of charged particles of mass 134.5 MeV/c^2, confirming Yukawa's prediction. With the development of high-energy proton accelerators, the production of pions has become straightforward. Neutral pions were discovered in 1950. Some basic properties of pions are listed in Table 35-7. The three pions are the lightest known mesons.

The Discovery of Muons

The muon (μ^-) is an unstable, negatively charged elementary particle with a mass of 105.6 MeV/c^2. Other than being 200 times as massive as an electron, in most respects muons and electrons are identical. A muon is a lepton and interacts with other matter only through the electromagnetic and weak forces. Its antiparticle is a positively charged muon, μ^+. It was discovered by Carl Anderson in 1936 while he was searching for pions. Since its rest mass is approximately the same as was predicted by Yukawa for the particle responsible for the strong nuclear force, it was thought to be a meson. However, it was soon discovered that muons do not interact through the strong force because they pass through matter easily.

Approximately 99.99% of charged pions decay into muons and neutrinos through the following processes:

$$\pi^- \rightarrow \mu^- + \bar{\nu}_\mu$$
$$\pi^+ \rightarrow \mu^+ + \nu_\mu \qquad (35\text{-}17)$$

Neutrinos emitted in pion decay are different from neutrinos emitted in neutron decay (n $\rightarrow$ p or p $\rightarrow$ n) and are called muon neutrinos. Pions quickly decay into muons; the pions created by cosmic rays cause a flow of approximately 10 000 muons/m^2/min at Earth's surface.

Muons have a mean lifetime of 2.26×10^{-6} s and decay into an electron or a positron and a neutrino–antineutrino pair:

$$\mu^- \rightarrow e^- + \bar{\nu}_e + \nu_\mu$$
$$L: \ +1 \rightarrow +1 + (-1) + 1$$
$$\mu^+ \rightarrow e^+ + \nu_e + \bar{\nu}_\mu$$
$$L: \ -1 \rightarrow -1 + 1 + (-1) \qquad (35\text{-}18)$$

Notice that the lepton number is conserved in μ^- and μ^+ decays.

Table 35-7 Some Properties of Pions

Property	Charged Pions	Neutral Pions
Mass	139.6 MeV/c^2	135.0 MeV/c^2
Mean lifetime	2.6×10^{-8} s	8.4×10^{-17} s
Spin	0	0

35-11 Particle Accelerators

How do we create beams of fast-moving particles? The basic technique is to accelerate charged particles by passing them through electric and magnetic fields. A particle of charge q and mass m passing through a region where the electric field is $\vec{E}$ and the magnetic field is $\vec{B}$ experiences a force given by

$$\vec{F} = m\vec{a} = q\vec{E} + q\vec{v} \times \vec{B} \qquad (35\text{-}19)$$

As described in earlier chapters, an electric field accelerates a charged particle in the direction of the field, and the force exerted by a magnetic field is perpendicular to both the magnetic field and the velocity of the particle. Therefore, a constant magnetic field changes the direction of motion of a particle without doing any work on it. In this section, we discuss a few common types of accelerators.

Cyclotrons

The simplest form of a cyclotron consists of two hollow semi-circular electrodes (called **dees**) with a small gap between them (Figure 35-17; also see Section 24-3). A high-frequency AC voltage is applied across the

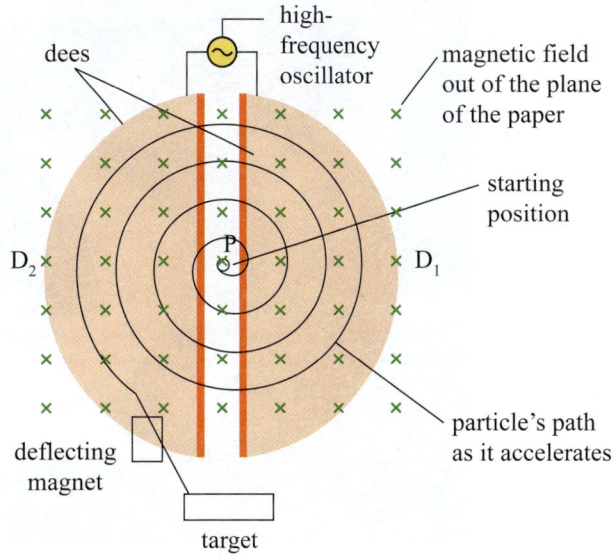

Figure 35-17 A cross-sectional view of a cyclotron consisting of two dees.

electrodes so that the electrodes change polarity at a fixed frequency. The electrodes are placed in a vacuum, and a constant magnetic field, perpendicular to the plane of the electrodes, is applied. A source of ions is connected to the centre of the electrodes.

Imagine that a slow-moving proton emerges from the ion source at P. The voltage between the two electrodes accelerates the proton toward the negatively charged electrode with a force $q\vec{E}$. Once through the gap and inside the hollow electrode, the proton does not experience any electric force because the electric field is shielded by the conductive metal. However, the proton does experience the magnetic field, which causes it to move in a circular path of radius r. The speed of the proton as it enters the electrode is v, and the magnitude of the magnetic field is B, yielding

$$qvB = \frac{mv^2}{r}$$

and

$$r = \frac{mv}{qB} \tag{35-20}$$

The alternating frequency is set so that the polarity of the electrodes reverses when the proton emerges from the electrode back into the gap. Now, the electrode at the opposite end of the gap is negatively charged and the proton accelerates toward it. The proton enters the electrode with a greater speed, and once inside it moves in a path of greater radius. Each time the proton passes into the gap between the two electrodes, it is accelerated by the voltage across the electrodes, and the radius of its circular path within the dees increases. The proton's speed increases only when it is passing through the gap. Once inside the electrodes, the proton moves with a

MAKING CONNECTIONS

The TRIUMF Cyclotron

The largest cyclotron in the world, TRIUMF (Tri-University Meson Facility), is at the University of British Columbia in Vancouver, British Columbia. The six magnets in this cyclotron have a radius of 18 m and a total mass of 4000 t (Figure 35-18). The magnets can generate a magnetic field of 6000 G (0.60 T). The cyclotron frequency of TRIUMF is 23 MHz.

Instead of protons, TRIUMF accelerates negatively charged hydrogen ions (hydrogen atoms with an extra electron). When high-energy hydrogen ions are taken out of the cyclotron, they are passed through a very thin metal film that strips the electrons, thus creating a high-energy proton beam. TRIUMF can accelerate protons to 0.75 times the speed of light.

Courtesy of TRIUMF.

Figure 35-18 The six magnets of the TRIUMF cyclotron are clearly visible in this photograph, which was taken at the completion of the cyclotron in 1971.

constant speed. The time, t, required for the proton to travel a distance πr with a constant speed v is given by

$$t = \frac{\pi r}{v} = \frac{\pi}{v}\left(\frac{mv}{qB}\right) = \frac{\pi m}{qB}$$

Thus, the time that a proton spends inside the electrodes is independent of its speed. The polarity of the electrodes must reverse with the same timing. Therefore, the frequency of the applied voltage must be

$$f = \frac{1}{2t} = \frac{qB}{2\pi m} \qquad (35\text{-}21)$$

This frequency is called the **cyclotron frequency**. The cyclotron frequency depends on the charge-to-mass ratio of the accelerated particle and the magnitude of the applied magnetic field. When the proton reaches the outer edge of the electrode, it is deflected into an evacuated pipe, called the beam line, by a small magnet. The fast-moving protons can then be made to strike a target placed at the other end of the beam line.

EXAMPLE 35-6

Designing a Cyclotron

You need to design a cyclotron that can produce protons with 10.0 MeV of kinetic energy. The design parameters limit the radius of the electrodes to 0.50 m. Determine the magnitude of the applied magnetic field and the frequency of the power supply. The rest mass energy of a proton is approximately 938.3 MeV.

Solution

First, determine the speed of protons with 10.0 MeV of kinetic energy. The rest mass energy, mc^2, of a proton is 938.3 MeV, so its kinetic energy is much less than its rest mass energy. Therefore, non-relativistic kinematics can be used as a good approximation. The speed of the protons is given by

$$K = 10.\ \text{MeV} = \frac{1}{2}mv^2 = \frac{1}{2}(mc^2)\left(\frac{v^2}{c^2}\right)$$

$$v = \sqrt{\frac{2 \times K}{mc^2}}\,c = \sqrt{\frac{20.0\ \text{MeV}}{938.3\ \text{MeV}}}\,c = 0.146c$$

Therefore, $v = 4.38 \times 10^7$ m/s. The magnitude of the required magnetic field can now be calculated using Equation 35-20:

$$B = \frac{mv}{rq} = \frac{(1.67 \times 10^{-27}\ \text{kg})(4.38 \times 10^7\ \text{m/s})}{(0.50\ \text{m})(1.60 \times 10^{-19}\ \text{C})}$$

$$= 0.91\ \text{T}$$

The frequency of the power supply must be the same as the cyclotron frequency. Therefore, using Equation 35-21:

$$f = \frac{qB}{2\pi m} = \frac{(1.60 \times 10^{-19}\ \text{C})(0.91\ \text{T})}{2\pi(1.67 \times 10^{-27}\ \text{kg})}$$

$$= 1.4 \times 10^7\ \text{Hz} = 14\ \text{MHz}$$

Equation 35-20 is valid only for low-energy particles. At higher energies, relativistic kinematics must be used. When the mass in Equation 35-20 is replaced by the relativistic mass, $\dfrac{m_0}{\sqrt{1 - v^2/c^2}}$, where m_0 is the rest mass of the particle, the equation for the cyclotron frequency, Equation 35-21, changes to

$$f = \frac{qB}{2\pi m_0}\sqrt{1 - \frac{v^2}{c^2}} \qquad (35\text{-}22)$$

Therefore, at energies where relativistic kinematics is important, the cyclotron frequency depends on the particle's speed, which greatly complicates the design of a cyclotron.

MAKING CONNECTIONS

Cyclotrons in Medical Physics

Radioactive isotopes are used for both the diagnosis of the condition of a person's bones and internal organs (such as thyroid, liver, and heart) and the treatment of various types of cancers. Medically important radioisotopes include technetium-99, nitrogen-13, fluorine-18, and iodine-123. For example, fluorine-18 is used in fludeoxyglucose to detect cancer and to monitor the effectiveness of cancer treatment. Many large hospitals have their own cyclotrons to produce these radioisotopes (Figure 35-19). Canada produces approximately two-thirds of the technetium-99 required worldwide using a nuclear reactor at Chalk River, Ontario. Research is underway to develop large-scale production of technetium-99 with cyclotrons as an alternative source. Proton beams from cyclotrons are also used to treat cancer (see Example 21-7). A beam of high-energy protons is targeted at the tumour. The protons deposit their energy in the tumour, killing the cancer cells.

BSIP SA/Alamy Stock Photo.

Figure 35-19 A cyclotron made by IBA (Belgium) for medical use.

Linear Accelerators

A **linear accelerator** (also called a **linac**) accelerates charged particles by passing the particles through large potential differences along a straight path. In contrast to a cyclotron, where charged particles move in circular orbits of increasing radii, particles make a single pass through a linear accelerator. Therefore, the high-power linear accelerators tend to be long and therefore costly. The largest linear accelerator in the world is SLAC, an acronym for its original name, the Stanford Linear Accelerator Center. Located at Stanford University in California, the SLAC accelerator is 3.2 km long and can accelerate electrons and positrons up to energies of 50 GeV.

The two main types of linear accelerators are distinguished by whether they use a DC or an AC voltage for acceleration. Consider two oppositely charged plates a distance d apart with a constant potential difference V_0 between them. The magnitude of the electric field between the plates is V_0/d. The magnitude of the force exerted on a particle of charge e and mass m in this electric field is

$$F = (e)(E) = \frac{eV_0}{d}$$

Therefore, the acceleration of the particle is

$$a = \frac{F}{m} = \frac{eV_0}{md} \tag{35-23}$$

If the particle is at rest initially, its speed after travelling the distance d between the plates is given by (ignoring relativistic effects)

$$v = \sqrt{\frac{2eV_0}{m}} \tag{35-24}$$

For electrons that accelerate through a 20 000 V potential difference,

$$v = \sqrt{\frac{2 \times (1.60 \times 10^{-19}\ \text{C})(20\,000\ \text{V})}{9.11 \times 10^{-31}\ \text{kg}}} = 8.38 \times 10^7\ \text{m/s}$$

This speed is approximately 28% of the speed of light. Acceleration with a DC voltage works well when required energies are of the order of tens of kilo-electron volts. For example, cathode ray tubes typically accelerate electrons with a potential difference of the order of 10 kV. To accelerate particles close to the speed of light, the potential difference needs to be increased substantially. High voltages cause electric arcs even in a vacuum chamber, so using a constant electric field (DC voltage) to accelerate charged particles to very high speeds is not feasible.

Alternating electric fields can accelerate charged particles to much higher speeds. This is done by passing the particles through an arrangement of spatially separated electrodes so the particles are accelerated during one half of the AC cycle and are shielded from the electric field by the electrodes when the electric field reverses direction during the other half of the cycle. As in a cyclotron, the frequency of the alternating field must be matched to the motion of the particles. The shielding electrodes consist of cylindrical metal tubes called **drift tubes**. As a particle accelerates, it gains speed and therefore travels longer and longer distances during each half cycle of the electric field. Therefore, each successive drift tube must be longer than the one before it (Figure 35-20).

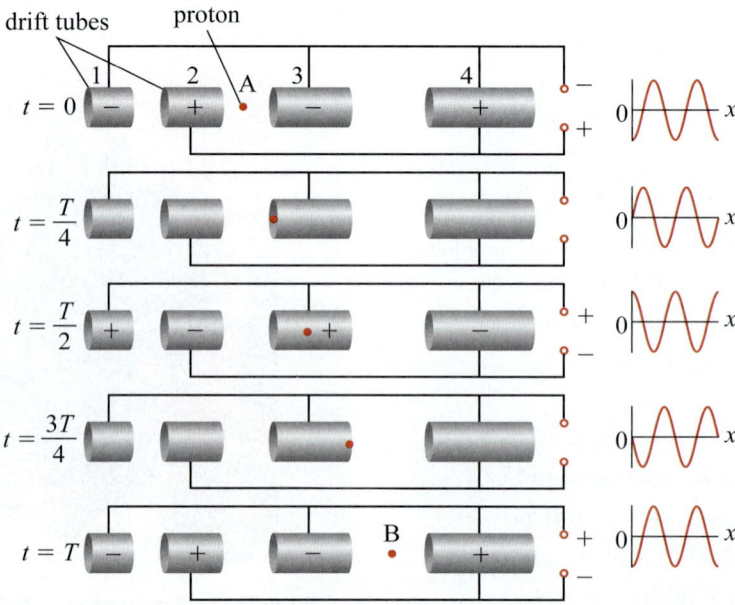

Figure 35-20 The cycle of voltages applied to the drift tubes in a linear accelerator.

Consider the motion of a proton as it travels from point A to point B in Figure 35-20. Let $t = 0$ when the proton is midway between tubes 2 and 3, and the voltage (and therefore the electric field) is at its maximum. At this instant, tube 3 is negatively charged, and the proton experiences a maximum forward acceleration. Over the next quarter cycle, the voltage drops, reaching zero at $t = T/4$, where T is the period of the AC cycle. Tube 3 is placed such that at $t = T/4$, the proton enters the tube. For the next half cycle, the electric field reverses direction, and the proton would decelerate if it were not shielded by the drift tube. However, the electric field is zero inside the metal tube, so the proton moves within it with a constant velocity, the velocity with which it entered the tube. At $t = 3T/4$, the voltage is zero, and the electric field reverses again. The length of drift tube 3 is such that the proton emerges from it at this time. Between $t = 3T/4$ and $t = T$, the electric field does work on the proton, accelerating it forward. This process is repeated during the next cycle.

We can approximate the required length of a drift tube. Let f be the frequency of the AC voltage and v be the average speed of the proton during a given cycle. The proton travels a distance v/f in one cycle. Since the proton is to be shielded for half a cycle, the length, L_d, of the drift tube should be half this distance:

$$L_d = \frac{v}{2f} \tag{35-25}$$

This length is approximate because the proton gains kinetic energy, and its speed increases during the cycle.

Kinetic energy + rest mass energy = total energy

$$K + m_0 c^2 = \frac{m_0 c^2}{\sqrt{1 - \dfrac{v^2}{c^2}}}$$

Solving the above equation for v, we get

$$v = c\sqrt{1 - \left(\frac{m_0 c^2}{K + m_0 c^2}\right)^2} = c\,\frac{\sqrt{K^2 + 2Km_0 c^2}}{K + m_0 c^2} \tag{1}$$

By substituting the known values for the rest mass of the proton (938.3 MeV/c^2) and the speed of light, we can calculate the proton speed and hence the length of the drift tube for a given K.

(a)

$$v = c\sqrt{1 - \left(\frac{m_0 c^2}{K + m_0 c^2}\right)^2}$$

$$= (3.00 \times 10^8 \text{ m/s})$$

$$\sqrt{1 - \left(\frac{938.3 \text{ MeV}/c^2}{10.0 \text{ MeV} + 938.3 \text{ MeV}}\right)^2}$$

$$= 4.35 \times 10^7 \text{ m/s}$$

$$L_d = \frac{v}{2f} = \frac{4.35 \times 10^7 \text{ m/s}}{2 \times 202 \times 10^6 \text{ Hz}}$$

$$= 0.108 \text{ m} = 10.8 \text{ cm}$$

(b) $v = 7.41 \times 10^7$ m/s; $L_d = 0.183$ m $= 18.3$ cm

(c) $v = 9.42 \times 10^7$ m/s; $L_d = 0.233$ m $= 23.3$ cm

Making sense of the result

The length of the drift tube increases with increasing kinetic energy of the proton.

EXAMPLE 35-7

Drift Tube Lengths

The LHC at CERN is the largest proton accelerator in the world. Before entering the main circular accelerator, protons are accelerated to a kinetic energy of 50 MeV in a linear accelerator. The frequency of the AC power supply for this linear accelerator is 202 MHz. Calculate the length of the drift tubes when the kinetic energy of the protons is
(a) 10.0 MeV
(b) 30.0 MeV
(c) 50.0 MeV

Solution

The length of the drift tube can be calculated using Equation 35-25 once we know the speed of the protons. We can use Einstein's relationship between mass and energy to determine the proton's speed for a given value of its kinetic energy. Let K be the kinetic energy of the proton. Then

MODERN PHYSICS

Synchrotrons

A synchrotron accelerates charged particles by continuously moving the particles in a circle of constant radius. A **synchrotron** is essentially a cyclotron in which the electric and magnetic fields are adjusted to keep the radius of the path of the accelerating particles constant as the speed of the particles increases. From Equation 35-20, using relativistic mass, the strength of the magnetic field required to rotate a particle with speed v in a circle of radius R_0 is

$$B = \frac{m_0}{\sqrt{1 - \frac{v^2}{c^2}}} \left(\frac{v}{qR_0} \right) \tag{35-26}$$

where q is the charge of the particle and m_0 is its rest mass.

For circular motion, $v = \omega R_0 = (2\pi f_0)R_0$. Therefore, as the particle speed increases, the frequency of the synchrotron must also increase, along with the strength of the magnetic field.

Accelerating particles to very high energies requires strong, energy-efficient magnets and other relevant technologies. The LHC uses superconducting electromagnets to force the particles to move in a circular path. These electromagnets are more powerful and more compact than electromagnets using ordinary conductors.

EXAMPLE 35-8

Magnets for the Super Collider

The LHC can accelerate protons to energies of 7.0 TeV. Given that the accelerator ring is 27 km in circumference, calculate the strength of the magnetic field needed to keep the protons in the synchrotron ring.

Solution

The strength of the required magnetic field can be calculated from Equation 35-26. We are given the kinetic energy of the proton. It is convenient to write Equation 35-26 in terms of the kinetic energy. From Equation 1 in Example 35-7:

$$v = c\sqrt{1 - \left(\frac{m_0 c^2}{K + m_0 c^2} \right)^2} = c\frac{\sqrt{K^2 + 2Km_0 c^2}}{K + m_0 c^2}$$

Since

$$\sqrt{1 - \frac{v^2}{c^2}} = \frac{m_0 c^2}{K + m_0 c^2}$$

we can write Equation 35-27 as

$$B = \frac{m_0}{\sqrt{1 - \frac{v^2}{c^2}}} \left(\frac{v}{qR_0} \right) = \frac{\sqrt{K^2 + 2Km_0 c^2}}{qR_0 c}$$

We have the following data:

$$m_0 c^2 = 938.3 \times 10^6 \text{ eV} \quad K = 7.0 \text{ TeV} = 7.0 \times 10^{12} \text{ eV}$$

$$q = 1.60 \times 10^{-19} \text{ C} \quad c = 3.00 \times 10^8 \text{ m/s} \quad R_0 = 27\,000 \text{ m}/2\pi$$

Substituting the known values, we get

$$B = \frac{\sqrt{(7.0 \times 10^{12} \text{ eV})^2 + 2(7.0 \times 10^{12} \text{ eV})(938.3 \times 10^6 \text{ eV})}}{(1.60 \times 10^{-19} \text{ C})(27\,000 \text{ m}/2\pi)(3.00 \times 10^8 \text{ m/s})}$$

$$\times (1.60 \times 10^{-19} \text{ J/eV}) = 5.4 \text{ T}$$

Making sense of the result

Earth's magnetic field is approximately 5×10^{-5} T, and the field strength close to a very strong niobium magnet is approximately 0.2 T. CERN lists the actual strength of their superconducting magnets as 8.36 T. Our calculated value is lower because we assumed that the magnetic field is uniform around the 27 km ring. To create room for equipment, some sections of the LHC ring have no magnetic field, so a stronger field is required for the rest of the ring.

CHECKPOINT 35-11

Synchrotron Magnetic Fields

Electrons and protons with the same speeds are travelling in circular orbits of equal radii. The strength of the magnetic field required to keep the electrons in the circular orbit is
(a) less than the strength of the magnetic field required for protons
(b) the same as the strength of the magnetic field required for protons
(c) greater than the strength of the magnetic field required for protons

35-12 Beyond the Standard Model

The Standard Model has successfully explained all high-energy experimental results, and all of its predictions about the existence of new particles have proven to be correct. Until now, not a single experimental result has contradicted the Standard Model. The discovery of the Higgs boson at LHC has provided a very convincing argument that the Standard Model is indeed correct. However, there still are many questions that puzzle scientists. The most important of these are about the nature and origin of dark matter and dark energy.

Dark Matter

The evidence for dark matter arose out of the study of the rotational speed of spiral galaxies and comparing the rotational speeds of the stars close to the centre of the galaxy to those at the periphery of the galaxy. Consider, for example, our solar system, with the Sun at its centre and various planets rotating around it. Most of the mass of this system is located at its centre in the form of the solar mass. If we plot the rotation speeds of various planets, we get the graph shown in Figure 35-21. The farther a planet is from the Sun, the slower is its rotational speed about the Sun.

This profile of the rotation speed is a consequence of Kepler's third law of motion and is a characteristic of any rotating system in which the mass decreases radially from the centre. From a plot of rotational speed as a function of distance (called the rotational curve), we can determine the mass of the Sun. The mass, M, of the Sun and the rotational speed, $v(r)$, where r is the distance from its centre, are related by

$$v(r) = \sqrt{\frac{GM}{r}}$$

Similarly, by plotting the rotational curves for galaxies, we can estimate the total mass of a galaxy and compare it to its visible mass in the form of stars. The visible (or luminous) mass of spiral galaxies is concentrated at the central region and falls off as the distance from its centre. So we should expect the rotational speeds of stars at the periphery of a galaxy to be much smaller than the rotational speeds of those close to the centre of the galaxy. Rotational speeds can be accurately determined by measuring the Doppler shift (Chapter 15) of the hydrogen spectrum.

Observations have shown that the stars located at the periphery of spiral galaxies, for example, the Andromeda Galaxy or our galaxy, the Milky Way, rotate too fast. In fact, the rotational speeds of stars do not decrease with distance from the centre of the galaxy. Figure 35-22 shows the rotational curve for our galaxy. Similar rotational curves are obtained for other spiral galaxies.

This behaviour of the rotational curve is similar to that of a rotating disk that has uniform mass distribution. But most of the visible mass of a spiral galaxy is located in its interior region. One explanation for the observed radial curves of galaxies is that there must be lots more mass in a galaxy than is visible in the form of stars. This invisible mass is called **dark matter**. It is called dark because it is not in the form of stars or large planets that we can see (by emitted or reflected light). How much of the mass of a galaxy is in the form of dark matter? From the rotational curve of a galaxy, we can estimate how much mass it should have and how this mass should be distributed within the galaxy. If we compare this mass to the luminous mass, we find that up to 90% of the mass of a galaxy can be in the form of dark matter.

What is dark matter? There are many ideas. The following are some of the possibilities.

- It might consist of MACHOs (Massive Astrophysical Compact Halo Objects). MACHOs are large astrophysical bodies that either do not emit light or emit very little light. These include black holes, neutron stars, brown dwarfs, white dwarfs, and planets. These bodies consist of ordinary baryonic matter (made of quarks and electrons). Although not visible, MACHOs can be detected by their gravitational pull

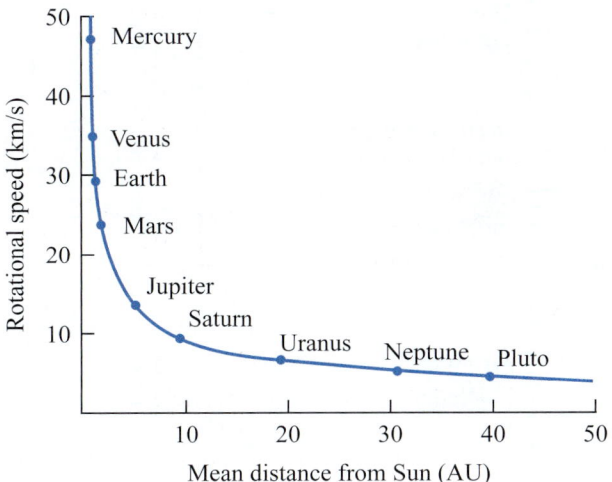

Figure 35-21 A graph of the rotational speed of planets as a function of the distance from the Sun. The farther a planet is from the Sun, the slower is its rotational (about the Sun) speed. Based on Idaho Public Television (http://idahoptv.org/ntti/nttilessons/lessons2000/lau1.html); PlanetFacts.org (http://planetfacts.org/orbital-speed-of-planets-in-order/).

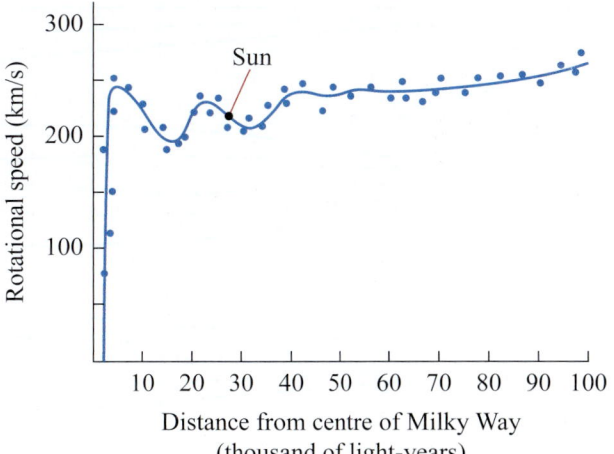

Figure 35-22 A graph of the rotational speed of stars in the Milky Way as a function of distance from the centre of the galaxy. The rotational speed (about the centre of the galaxy) does not decrease with distance from the centre. Based on http://astronomy.nju.edu.cn/~lixd/GA/AT4/AT423/HTML/AT42306.htm and http://pages.uoregon.edu/jimbrau/BrauImNew/Chap23/6th/23_21Figure-F.jpg.

on neighbouring stars and an ingenious technique, first suggested by Einstein, called **gravitational lensing**.

Here is how gravitational lensing works. According to Einstein's general theory of relativity, a massive object (a black hole, a neutron star, or a star) curves spacetime around it. Because of this curvature, in the vicinity of a massive object, light does not follow a straight-line path—it bends. This is similar to when light passes through a convex lens (hence the word *lensing*). Imagine a galaxy or a star that is in the line of sight of a telescope on Earth. The light entering the telescope from the star has a constant luminosity. If a neutron star happens to move across the line of sight of the star and Earth, its gravitational field will bend more of the star's light toward Earth (Figure 35-23). Therefore, the star will appear brighter while the neutron star is passing between the star and Earth. Once it has passed through the field of view of the telescope, the observed luminosity will return to its original value. This technique can also be used to identify planets in other solar systems.

A number of surveys of various galaxies have determined that there are not enough MACHOs to account for the dark matter required to explain the observed rotational curves. Also, the observed abundances of deuterium, helium, and other elements in the universe put limits on how much baryonic matter can be present in the universe. Most astronomers have now ruled out the possibility that MACHOs could account for the required amount of dark matter.

- Could dark matter be in the form of galactic and intergalactic gas clouds? We can detect the presence of large gas clouds by absorption of light passing through them, so it appears most dark matter cannot be in the form of such gas clouds.

- Recent experiments at SNO (Sudbury Neutrino Observatory, in Sudbury, Ontario) and Super-Kamiokande (Japan) have shown that neutrinos have a very small mass. Since after photons, neutrinos are the most abundant particles in the universe, it appears that some dark matter could be in the form of neutrinos. Scientists believe that neutrino mass accounts for only a fraction of the mass that should be in the form of dark matter.

- Many scientists think that most dark matter is in the form of non-baryonic matter. They postulate the existence of a new type of particle called a WIMP (weakly interacting massive particle). WIMPs are not made of quarks, are electrically neutral, and interact with other matter through the weak force. This makes them very difficult to detect. So far, all attempts to detect WIMPS have been unsuccessful.

The origin of dark matter has so far been a mystery.

Dark Energy

Dark energy is not the energy associated with dark matter by Einstein's mass–energy relationship, $E = mc^2$. It has nothing to do with dark matter. So why do we need dark energy? Comparison of experiments with theoretical

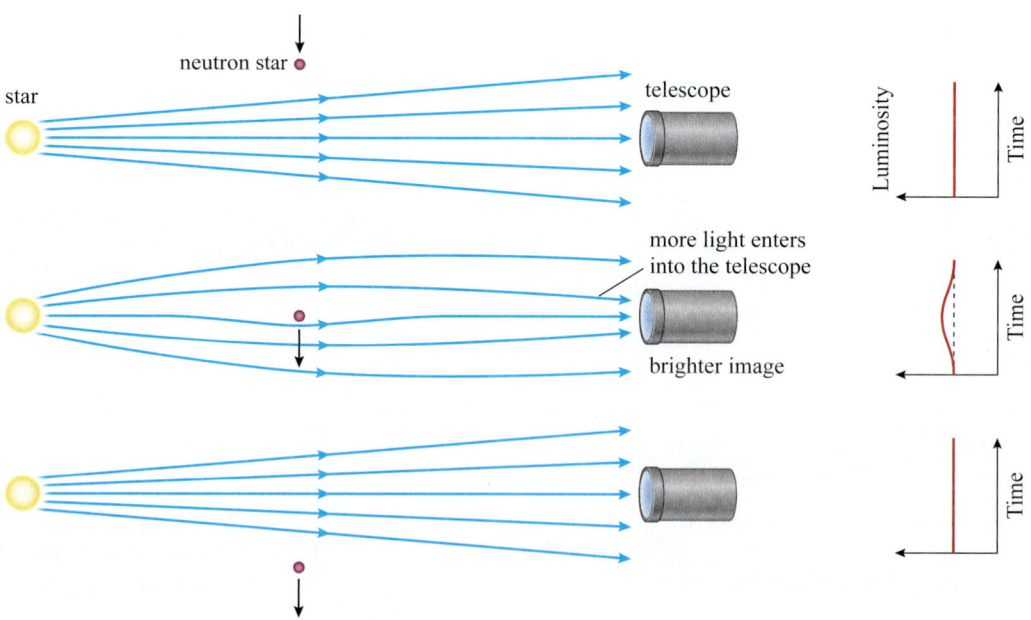

Figure 35-23 When a massive astronomical body, for example, a neutron star, passes across the line of sight of a distant star and Earth, the light passing in the vicinity of the body, instead of following a straight-line path, bends around it. This causes more light to be directed toward Earth, increasing the apparent luminosity of the star. This phenomenon, called gravitational lensing, can be used to determine the existence of large non-luminous objects.

models of what we know about the universe implied that dark energy *must* exist. By examining the Doppler shift in the spectrum of light emitted by distant galaxies, American astronomer Edwin Hubble (1889–1953) concluded that the universe is expanding (see Section 15-7), which led to the Big Bang theory. Since the universe is full of matter and, because of gravity, matter attracts other matter, we would therefore expect that as time went on, the gravitational attraction of galaxies on each other would slow down the expansion rate of the universe. The observations of Doppler shifts of distant supernovae by the Hubble Space Telescope have shown that a long time ago the universe was expanding more slowly than it is today (see Figure 35-24). So gravity is not slowing down the expansion of the universe. On the contrary, the expansion of the universe is accelerating.

Why is the rate of expansion of the universe increasing and not decreasing, as our current understanding of the theory of gravity would imply? We do not know. But there must be some form of energy that is making the universe expand faster. This energy is called the **dark energy**. There are many ideas as to what this dark energy could be, but none have been verified experimentally. Could it be that the theory of gravity as we understand it today is not fully correct? If the gravitational force became repulsive at very large distances, then the distant galaxies would exert repulsive forces on each other, and we might be able to explain an accelerated expansion rate. Another possibility, pointed out by Einstein, is that space and energy are related. As the universe expands, more space is created, and this space has its own energy, which further expands the universe. So the energy density (energy per unit volume) of the universe does not decrease as it expands. Another idea is that space is filled with some type of fluid or field that exerts a repulsive force on itself. But what is the origin of this fluid? We do not know.

How much dark matter and dark energy is present in the universe? To explain the current expansion rate of the universe and the observed orbital speeds of stars in galaxies requires that about 68% of the universe consist of dark energy, 27% be dark matter, and the remaining 5% be in the form of "normal" matter. This is absolutely amazing. So everything we have discussed in this chapter—quarks, leptons, gauge bosons—all make up only 5% of the universe. The other 95% we know nothing about! This means that there is a lot of mystery left for young scientists like you to solve!

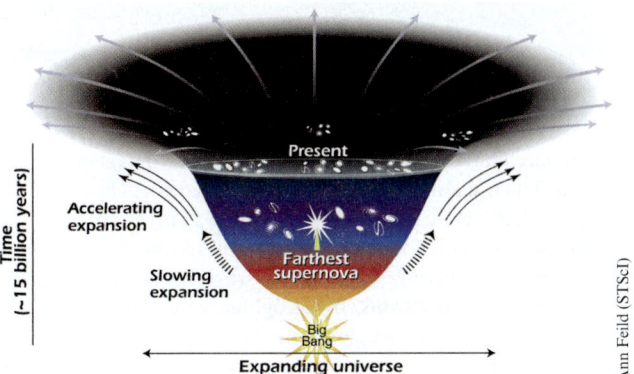

Figure 35-24 The rate of expansion of the universe since its birth 15 billion years ago. The more shallow the curve, the faster the rate of expansion. Scientists speculate that the faster expansion rate that started about 7.5 billion years ago is due to dark energy that is pushing the galaxies apart.

More Questions

Here we describe several more questions that continue to plague scientists.

- The mass of the up quark is estimated to be $2.5\,\text{MeV}/c^2$, and the mass of the top quark is $171\,\text{GeV}/c^2$. Why is one elementary particle approximately $70\,000$ times the mass of another?

- Quarks and leptons are fermions, and gauge particles are bosons. Are there quarks and leptons that are bosons and gauge particles that are fermions? In other words, is the universe symmetric between fermions and bosons, such that for every fermion there is a corresponding boson and vice versa? These additional particles are not part of the Standard Model. So far, all attempts to discover the boson partner of the electron have been unsuccessful.

- The graviton is the gauge boson that is assumed to be the carrier of the gravitational force. Why has the graviton not been discovered?

- The existence of quarks was envisaged to find a common structure for a large number of elementary particles. Are quarks and leptons made of more-elementary particles? In other words, do quarks and leptons have a substructure? So far, there is no experimental proof of such a substructure. But that may change in the future as higher-energy particle accelerators are constructed.

Synchrotron Radiation

An accelerating charged particle emits electromagnetic radiation. The radiation emitted when a charged particle moves at a speed that is close to the speed of light in a curved path is called **synchrotron radiation**. The spectrum of synchrotron radiation ranges from microwaves to hard X-rays, and its intensity can be very high. Electrons, being lighter, emit much more synchrotron radiation than protons. Industrial uses of synchrotron radiation include determining the crystalline structure of proteins, designing faster computer chips, testing properties of materials under conditions of high temperature and pressure, and identifying environmental contaminants by X-ray absorption.

The Canadian Light Source (CLS) at the University of Saskatchewan in Saskatoon, Saskatchewan, is one of the most advanced synchrotron radiation facilities in the world. CLS uses a 2.9 GeV electron beam to generate extremely intense beams of synchrotron radiation. The accelerator is the size of a football field (Figure 35-25).

Figure 35-25   The Canadian Light Source at the University of Saskatchewan, Saskatoon.

KEY CONCEPTS AND RELATIONSHIPS

Classification of Particles

There are 18 elementary particles: six quarks, six leptons, five gauge bosons, and the Higgs boson. These particles interact with each other through four forces: the strong, the weak, the electromagnetic, and the gravitational force.

Particles that exert the strong force on each other are called hadrons. Hadrons that are fermions are called baryons, and hadrons that are bosons are called mesons. Leptons interact with each other and other particles through the weak and electromagnetic forces.

Gauge Bosons as the Carriers of the Forces

Particles interact with each other by exchanging gauge bosons. The range of a force is inversely proportional to the mass of the exchanged boson.

- Gluons are the carriers of the strong force.
- Photons are the carriers of the electromagnetic force.
- W^+, W^-, and Z^0 bosons are the carriers of the weak force.
- The carrier of the gravitational force has not been observed.

Antiparticles

An antiparticle of a particle has the same mass and spin but opposite values for other properties, such as charge, lepton number, and baryon number. Each charged particle has an antiparticle. A neutral particle either has an antiparticle (such as an antineutrino) or is its own antiparticle (such as a photon). Particle–antiparticle pairs can be produced in collisions of particles.

Quarks and the Standard Model

The Standard Model groups the six quarks and six leptons into three generations (Figure 35-6). Most of the matter in the known universe is made of the particles in the first generation: the up quark, the down quark, and the electron. Higher-mass particles can decay into lower-mass particles provided the conservation laws are not violated.

Composite Particles

Hadrons are bound states of quarks and antiquarks. A baryon consists of three quarks, and a meson consists of a quark and an antiquark. The charge of a hadron is equal to the sum of the charges of the constituent quarks. The mass of a hadron depends on the masses of the constituent quarks. A proton is made up of two up quarks and one down quark, and a neutron is made up of one up quark and two down quarks.

Colour Quantum Number and Quark Confinement

Quarks have an additional quantum number called colour, which has three values, usually called red, blue, and green. There are eight possible colours for gluons. All composite particles are colour neutral.

Conservation Laws

Charge, energy, momentum, angular momentum, baryon number, and lepton number are conserved in all particle reactions and decays. Conservation laws are a powerful tool to determine whether a reaction is possible.

MODERN PHYSICS

Production and Decay of Particles

New particles can be produced by colliding fast-moving particles. Energy available in the centre-of-mass coordinate system of the colliding particles is transformed into the rest mass energy of the new particles in accordance with Einstein's mass–energy relation.

Feynman Diagrams

Feynman diagrams are graphical ways of representing a particular reaction or a decay process.

Pions and Muons

The existence of pions was predicted by Japanese physicist Hideki Yukawa to explain the origin of the strong nuclear force.

A muon is a charged lepton. Its mass is 210 times the mass of an electron; otherwise, it is identical to an electron. It has a half-life of 1.5×10^{-6} s and decays into an electron and two neutrinos.

Particle Accelerators

In a linear accelerator, charged particles are accelerated in a straight-line path by potential differences between successive drift tubes. In cyclotrons, charged particles follow a spiral path as they accelerate. In a synchrotron, the electric and magnetic fields vary to keep the accelerating particles moving along a circular path of constant radius.

APPLICATIONS

particle accelerators, cyclotrons, medical diagnoses and treatment of cancer, synchrotrons, determining the structure of proteins, designing faster computer chips, testing properties of materials, identifying environmental contaminants by X-ray absorption

KEY TERMS

antiparticles, baryon number, baryons, bottom quark, charm quark, colour, colour charge, composite particles, coupling strength, cyclotron frequency, dark energy, dark matter, dees, down quark, drift tubes, electron neutrino, elementary particles, Feynman diagrams, fine structure constant, flavours, flux tube, four-momentum, fundamental particles, gauge bosons, gravitational lensing, hadrons, Higgs bosons, incident proton, laboratory frame of reference, lepton number, leptons, linac, linear accelerator, mesons, muon, muon neutrino, pair annihilation, pair production, particle families, particle generations, particle physics, positron, quark confinement, scalar mesons, strange quark, synchrotron, synchrotron radiation, target proton, tau, tau neutrino, top quark, up quark, vector mesons, vertex, virtual photon

QUESTIONS

1. Explain whether each of the following statements is true or false.
   (a) Quarks can only exchange gluons.
   (b) The strong interaction is mediated by gluons.
   (c) The strong interaction is mediated by photons.
   (d) Two quarks can exchange a neutrino.
   (e) A neutron and a proton can exchange a gluon.
2. Explain whether each of the following statements is true or false.
   (a) A quark and an electron can annihilate into two photons.
   (b) A quark and an antiquark can annihilate into two photons.
   (c) Two photons can annihilate into a proton–antiproton pair.
   (d) Two photons can annihilate into a neutron–antineutron pair.
   (e) Any elementary particle and its antiparticle can annihilate into two photons.
3. Explain whether each of the following statements is true or false.
   (a) All baryons are hadrons.
   (b) All mesons are baryons.
   (c) The lightest hadron is a proton.
   (d) The lightest baryon is a pion.
   (e) An electron is an antiparticle of a positron.
   (f) A positron is an antiparticle of an electron.
4. Explain whether each of the following statements is true or false.
   (a) The total spin quantum number of an electron–positron pair can be either zero or one.
   (b) The spin quantum number of a meson can be two.
   (c) The spin quantum number of a baryon can be three.
   (d) Mesons can have a non zero baryon number.
5. Explain whether each of the following statements is true or false.
   (a) When a proton collides with a stationary proton, all the kinetic energy of the incident proton can be used to create new particles.
   (b) An electron moving in a straight trajectory can emit photons.
   (c) An electron moving in a circular path emits photons.
   (d) A constant magnetic field can increase the speed of a charged particle.
   (e) A constant magnetic field can increase the velocity of a charged particle.

6. Calculate the ratio of gravitational and electromagnetic forces between two protons that are 10^{-15} m apart. How large would the gravitational constant have to be for the two forces to be equal?

7. It is estimated that the radius of a carbon nucleus is between 4×10^{-15} m and 5×10^{-15} m. What is the minimum required energy of an electron beam to resolve distances of this length?

PROBLEMS BY SECTION

For problems, star ratings will be used ($\star$, $\star\star$, or $\star\star\star$), with more stars meaning more challenging problems. The following codes will indicate if $\frac{d}{dx}$ differentiation, $\int$ integration, $\square$ numerical approximation, or $\sim$ graphical analysis will be required to solve the problem.

Section 35-1   Classification of Particles

8. $\star$ Which of the following particles do not interact through the strong force?
   (a) electron
   (b) photon
   (c) neutrino
   (d) quark
   (e) W^+
   (f) gluon
   (g) Higgs boson

9. $\star$ Which of the following particles are fermions?
   (a) electron
   (b) photon
   (c) neutrino
   (d) quark
   (e) W^+
   (f) gluon
   (g) Higgs boson

10. $\star$ Which of the following particles possess a colour quantum number?
   (a) electron
   (b) photon
   (c) neutrino
   (d) quark
   (e) W^+
   (f) gluon
   (g) Higgs boson

11. $\star$ Which of the following particles are gauge bosons?
   (a) photon
   (b) Z^0
   (c) neutrino
   (d) gluon
   (e) top quark
   (f) Higgs boson

Section 35-2   Gauge Bosons

12. $\star$ In 1986, a group of scientists claimed the existence of a fifth force that had a range of approximately 100 m. The claim was not correct. Had it been, what would be the mass of the gauge boson mediating such a force?

13. $\star$ What would be the range of the weak force if the $W^{\pm}$ and Z^0 gauge bosons were massless?

14. $\star$ Estimate the range of strong force between two protons due to the exchange of a meson of mass 1.0 GeV/c^2.

Section 35-5   Composite Particles

15. $\star\star$ Construct all the baryons that are made up of up, down, and strange quarks and contain at least one strange quark. Determine the masses of the baryons, assuming that the mass of a baryon is equal to the sum of the masses of the constituent quarks. Assume that the masses for the three quarks are $m_u = 300.$ MeV/c^2, $m_d = 310.$ MeV/c^2, and $m_s = 500.$ MeV/c^2. These masses are called "constituent quark masses" and are different from the masses listed in Table 35-1.

16. $\star$ What are all the possible values of the intrinsic spin quantum number of a baryon that is composed of three quarks?

17. $\star$ Find the charge, baryon number, and possible values of the spin quantum number for the baryons with the following quark combinations:
   (a) uus
   (b) uds
   (c) dds
   (d) uss
   (e) dss
   (f) sss

18. $\star$ Assume that baryon masses are proportional to the sum of the masses of the constituent quarks, with the same proportionality constant for all baryons. Rank the baryons in problem 17 from lightest to heaviest mass using the quark masses in Table 35-1.

19. $\star$ Perform the same calculations as in problems 17 and 18 for the following quark combinations:
   (a) ttb
   (b) tts
   (c) tbs
   (d) tbd
   (e) tuu

20. $\star$ Find the charge and all possible values of the spin quantum number for mesons with the following quark combinations:
   (a) $s\bar{u}$
   (b) $s\bar{d}$
   (c) $s\bar{s}$
   (d) $u\bar{s}$
   (e) $d\bar{s}$
   (f) $\bar{c}d$

21. $\star$ Assume that meson masses are proportional to the sum of the masses of constituent quarks, with the same proportionality constant for all mesons. Rank the mesons in problem 20 from lightest to heaviest mass using the quark masses in Table 35-1.

22. $\star$ Perform the same calculations as in problem 17 for the following quark combinations:
   (a) $c\bar{u}$
   (b) $c\bar{d}$
   (c) $c\bar{s}$
   (d) $c\bar{c}$
   (e) $\bar{c}u$
   (f) $\bar{c}d$
   (g) $\bar{c}d$

Section 35-7 Conservation Laws

23. ★★ Explain why the following reactions are not possible. Which conservation law is violated in each case?
    (a) $p \rightarrow n + e^- + \nu_e$
    (b) $n \rightarrow p + \gamma$
    (c) $p \rightarrow p + \gamma$
    (d) $\mu^- \rightarrow e^- + \nu_e$
    (e) $p \rightarrow \pi^+ + \pi^-$
    (f) $n + n \rightarrow p + e^-$
    (g) $n + \bar{\nu}_e \rightarrow p + e^-$
    (h) $p \rightarrow e^- + \nu_e$

24. ★★ Identify the particle denoted by the ? symbol in the reactions listed below.
    (a) $\mu^- \rightarrow e^- + \nu_\mu + ?$
    (b) $\pi^- + p \rightarrow \pi^0 + ?$
    (c) $e^- + p \rightarrow n + ?$
    (d) $\pi^- + \pi^+ \rightarrow p + ?$
    (e) $\pi^+ \rightarrow \mu^+ + ?$

25. ★★ Use the laws of conservation of energy and momentum to show that the reaction $e^- + e^+ \rightarrow \gamma$ is not allowed.

26. ★★★ Consider a particle of mass m and velocity $\vec{v}_0$ that makes an elastic collision with a particle of mass M that is at rest in the laboratory frame of reference. Show, using non-relativistic kinematics, that the kinetic energy of the target particle (in the lab frame) is given by

$$K_M = \left(\frac{4mM}{(m+M)^2} \cos^2\theta \right) K_i$$

where $K_i = \frac{1}{2} mv_0^2$ is the kinetic energy of the incident particle and θ ($\leq \pi/2$) is the angle that the target particle makes after the collision with respect to the direction of the incident particle.

27. ★★ What is the minimum energy of a neutrino in the laboratory frame of reference for the reaction $\nu_e + p \rightarrow n + e^+$ to occur?

28. ★★ Electron–positron annihilation occurs according to the reaction

$$e^- + e^+ \rightarrow \gamma + \gamma$$

Calculate the energies, momenta, and frequencies of the two photons in the final state when
    (a) the kinetic energies of the electron and the positron are ignored compared to their rest mass energies
    (b) each has a kinetic energy of 50 MeV, and they have equal and opposite momenta

29. ★★ A proton and an antiproton can annihilate into a π^+ and a π^- meson according to the reaction

$$p + \bar{p} \rightarrow \pi^+ + \pi^-$$

What are the energies and momenta of the two pions in the final state when the kinetic energies of the proton and antiproton are ignored?

30. ★ Consider the reaction in which a fast-moving negatively charged pion is absorbed by a proton to produce two neutrons:

$$\pi^- + p \rightarrow n + n$$

Is the reaction allowed? If not, which conservation law(s) are violated?

31. ★★ The proton is the lightest known baryon, and it does not decay. Which conservation law would be violated if we observed the decay of a proton?

32. ★ Which conservation law would be violated if we observed the decay of an electron?

33. ★★ Use the laws of conservation of energy and momentum to show that a free photon cannot decay into an electron–positron pair. (This process can happen only in the presence of another charged particle.)

Section 35-8 The Production and Decay of Particles

34. ★★ Calculate the momentum and energy of emitted photons in the decay of a π^0 meson:

$$\pi^0 \rightarrow \gamma + \gamma$$

35. ★★ Calculate the momenta and energies of the final particles in the following decays:
    (a) $\pi^- \rightarrow e^- + \bar{\nu}_e$
    (b) $\pi^- \rightarrow \mu^- + \bar{\nu}_\mu$

Section 35-9 Feynman Diagrams

36. ★★ Draw Feynman diagrams for the following processes:
    (a) $e^- + \mu^- \rightarrow e^- + \mu^-$
    (b) $p \rightarrow n + e^+ + \nu_e$
    (c) $e^- + e^+ \rightarrow e^- + e^+$

37. ★★ A positively charged pion ($u\bar{d}$) decays into a μ^+ and a ν_μ. Draw a Feynman diagram to represent this decay.

38. ★★ Consider the process in which an electron neutrino (ν_e) is absorbed by a neutron, resulting in a proton and an electron in the final state:

$$\nu_e + n \rightarrow p + e^-$$

What really happens is that a d quark converts into a u quark by emitting a W^- gauge boson. Draw a Feynman diagram to represent the above process.

39. ★★ Draw a Feynman diagram to represent the scattering of a photon from an electron through the following process:

$$e^- + \gamma \rightarrow e^- + \gamma$$

40. ★★ When drawing Feynman diagrams, would it make a difference if time were plotted along the horizontal direction and position along the vertical direction?

Section 35-10 Pions and Muons

41. ★ (a) In the meson-exchange model, the strong force between two nucleons arises due to the exchange of several mesons between the nucleons. Determine the approximate range of the nuclear force between two protons due to the exchange of the following neutral mesons:
        (i) pi-zero (π^0, mass 135 MeV/c^2)
        (ii) omega (ω, mass 783 MeV/c^2)
        (iii) phi (ϕ, mass 1.02 GeV/c^2)
    (b) Can two protons exchange an electron? Explain your answer.

MODERN PHYSICS

42. ★★ The mass of a lambda-zero (Λ^0) particle is 1115.7 MeV/c^2, and its half-life is 2.6×10^{-10} s. The Λ^0 particle quickly decays into a proton and a negatively charged pion:

$$\Lambda^0 \rightarrow p + \pi^-$$

Assume that Λ^0 is initially at rest. Determine the kinetic energies of the proton and the pion in this decay.

Section 35-11 Particle Accelerators

43. ★★ (a) A proton linear accelerator works at a frequency of 200 MHz. Calculate the length of the drift tubes at the sections where the average kinetic energy of the proton is
  (i) 10 MeV
  (ii) 100 MeV
  (iii) 200 MeV
(b) The protons need to be accelerated to 2 GeV kinetic energy. Should the frequency of the applied AC voltage be increased or decreased? Explain.
(c) Repeat parts (a) and (b) for an electron linear accelerator.

44. ★★ A proton beam is to be accelerated to an energy of 30 GeV in a synchrotron ring with a circumference of 1000 m. Calculate the strength of the magnetic field needed to keep the beam in a circular path of this circumference.

45. ★★★ You are asked to design a synchrotron that can accelerate protons to 500 MeV. The magnets available for the project can produce a maximum magnetic field of 2 T.
(a) What is the speed of the 500 MeV protons at this energy? (Use relativistic kinematics.)

(b) How long would a 500 MeV proton take to complete one revolution around the synchrotron ring?
(c) What is the circumference of the synchrotron?

COMPREHENSIVE PROBLEMS

46. ★★★ Before the discovery of the neutrino, it was assumed that a neutron could decay into a free proton and an electron, $n \rightarrow p + e^-$. Consider this process in the laboratory frame of reference in which the neutron is at rest. By applying the laws of conservation of energy and momentum, show that the magnitude of an electron's momentum, p_e, is

$$p_e = \sqrt{\frac{(m_n^2 - m_p^2 - m_e^2)^2}{4m_n^2} - \frac{m_p^2 m_e^2}{m_n^2}} = 1.26 \text{ MeV}/c$$

47. ★★ Some scientists estimate that at 10^{16} GeV of energy, the four fundamental forces unify into a single force. To test this hypothesis, you are asked to design a synchrotron that can accelerate protons to 10^{16} GeV. The strongest available magnets can generate a magnetic field of 10 T. What is the radius of the synchrotron? Is there any hope of building a synchrotron with this energy with current technology? What is the strength of the magnetic field if the design requires the radius of the synchrotron to be 100 km?

48. ★★★ Antiprotons can be produced in the following reaction:

$$p + p \rightarrow p + p + p + \bar{p}$$

The target proton is at rest. Determine the minimum energy of the incident proton for this reaction to occur.

Note that answers are not provided for lengthy written explanations or open problems with no single correct answers. Full solutions are provided to instructors.

Chapter 1: Introduction to Physics

1. c
9. b
11. a
13. c
21. 64
31. 4
33. Standard deviation $= 0.25$, mean $= 10.5$
35. Data point 4.56 m/s^2 is a true error. Random error ≈ 0.16 m/s^2. Systematic error is present.
39. 17.39 ± 0.04
41. 25.8
43. 0.8 m/s
45. 3.586
47. (a) No; (b) 8.00 ± 0.08 m
49. (a) 8.900×10^3; (b) Yes
51. (a) 3670; (b) 0.000 000 000 002 25; (c) 240 000; (d) 1200.
53. (a) $\dfrac{N}{m^2}$; (b) $\dfrac{kg}{m \cdot s^2}$
55. s
57. $U \sim mgh$
59. $V \sim IR$
61. $kg \cdot s^{-1}$
63. 0.3 nm/s to 0.6 nm/s
65. Approximately 500 billion cells
67. 0.16%
73. About 3×10^{25} molecules
79. $\dfrac{J}{kg \cdot K}$; $\dfrac{m^2}{s^2 \cdot K}$
81. $0.21\ \mu m \times 0.21\ \mu m$
83. (a) 3.9×10^{-8} kg per year per m^2
    (b) 350 micrometeorites per year per m^2
85. (a) $\sqrt{\dfrac{\hbar G}{c^5}}$; (b) 5.38×10^{-44} s
87. $F = \dfrac{mv^2}{r}$
95. (a) Negligible; (b) Important

Chapter 2: Scalars and Vectors

1. (a) $F = 30$ N, $\theta = 315°$; (b) $F_x = 21$ N, $F_y = -21$ N;
   (c) $\vec{F} = (21\ N)\hat{i} - (21\ N)\hat{j}$
3. $\vec{F}_{1,2} = \pm[(2.528\ N)\hat{i} - (7.592\ N)\hat{j}]$ (two answers, as the vectors can be oriented in opposite directions)
5. $\vec{D} = (5\ m)\hat{i} - (2\ m)\hat{j} - (2\ m)\hat{k}$; $D = 5.74$ m, $\alpha = 29.42°$, $\beta = 110.4°$, $\gamma = 110.4°$
7. (a) $(4, -2, 11)$; (b) $(-4, 2, -11)$; (c) $(-24, 11, 22)$;
   (d) $(-20, 5, 10)$
9. c
11. Student 2

13. (a) $\vec{A}_x = (2,0,0)$, $\vec{A}_y = (0,2,0)$, $\vec{A}_z = (0,0,-2)$,
    $\vec{B}_x = (-1,0,0)$, $\vec{B}_y = (0,3,0)$, $\vec{B}_z = (0,0,-2)$
    (b) $\vec{A} \cdot \vec{B} = 8$; (c) $\vec{A} \times \vec{B} = 2\hat{i} + 6\hat{j} + 8\hat{k}$
15. Disagree
17. $\vec{F}_1 = (1.00\ N)\hat{i} - (3.00\ N)\hat{j} + (1.00\ N)\hat{k}$
    $\vec{F}_2 = (-2.00\ N)\hat{i} - (4.00\ N)\hat{j} + (3.00\ N)\hat{k}$
    $\vec{F}_3 = (1.00\ N)\hat{i} + (2.00\ N)\hat{j} + (0.00\ N)\hat{k}$
    $F_1 = \sqrt{11.00}$ N, $F_2 = \sqrt{29.00}$ N, $F_3 = \sqrt{5.00}$ N
    $\vec{F}_R = (-5.00\ N)\hat{j} + (4.00\ N)\hat{k}$, $F_R = \sqrt{41.00}$ N
19. Disagree
25. $\vec{F} = (4.33\ N)\hat{i} + (3.83\ N)\hat{j} + (2.50\ N)\hat{k}$
27. (a) $(10, 15)$; (b) $(5, 25)$; (c) $(45, -5)$; (d) $(-60, 0)$
29. (a) $D = 21$ m and $\theta = 41°$

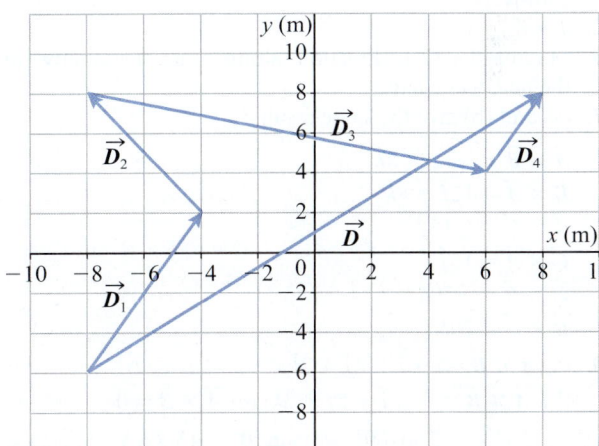

(b) $D = 21$ m and $\theta = 41°$
31. (a)

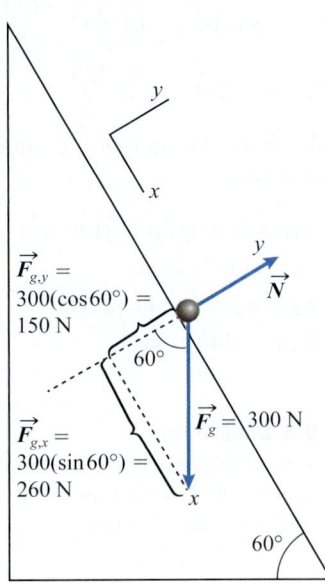

(b) $F_{g,x} = 260.$ N; $F_{g,y} = 150.$ N; (c) $N_x = 0$; $N_y = 150.$ N;
(d) $F_{net,x} = 260.$ N; $F_{net,y} = 0$

33. (a) $F_1 = F_2 = 8.37$ N
$\alpha_{\vec{F_1}} = 69.0°, \beta_{\vec{F_1}} = 127°, \gamma_{\vec{F_1}} = 44.2°$

(b) $F_2 = 8.37$ N
$\alpha_{\vec{F_2}} = 111°, \beta_{\vec{F_2}} = 53.3°, \gamma_{\vec{F_2}} = 136°$

(c) The two vectors have the same magnitude but opposite directions. Both magnitudes equal 8.37 N. The fact that the vectors have opposite directions can be seen from verifying that the sum of the corresponding direction angles adds up to 180°: α: 68.99° + 111.01° = 180°; β: 126.70° + 53.30° = 180°; γ: 44.18° + 135.82° = 180°

(d) We get 1 in both cases.

35. (a) $\vec{v}_{avg} = (-1.20 \text{ m/s})\hat{i}$; (b) $v_{avg} = 1.20$ m/s

(c) $\vec{v}_{avg} = v_{avg}$

37. $\vec{A} \cdot \vec{B} = -22$

39. $\vec{A} \cdot \vec{B} = 47$

41. $\hat{u} = \dfrac{2\sqrt{13}}{13}\hat{i} + \dfrac{3\sqrt{13}}{13}\hat{j}; \hat{u} = -\dfrac{2\sqrt{13}}{13}\hat{i} - \dfrac{3\sqrt{13}}{13}\hat{j};$
2 answers

43. $\theta = 8.75°$

45. Magnitude: 17.1; direction: along either the positive or the negative y-axis

47. (a) and (b) give the same results:
$\vec{A} \times \vec{B} = -12\hat{i} - 8\hat{k}$
$\vec{B} \times \vec{A} = 12\hat{i} + 8\hat{k}$
$\vec{A} \times \vec{C} = -15\hat{j}$
$\vec{C} \times \vec{A} = 15\hat{j}$
$\vec{C} \times \vec{B} = -20\hat{k}$
$\vec{B} \times \vec{C} = 20\hat{k}$

49. (a) $\vec{A} \times \vec{B} = -6\hat{i} - 8\hat{j} + 2\hat{k}$;
(b) $\vec{A} \times \vec{B} = -12\hat{i} + 5\hat{j} + 3\hat{k}$; (c) $\vec{A} \times \vec{B} = 0$

51. (a) $\begin{pmatrix} x' \\ y' \\ 1 \end{pmatrix} = \begin{pmatrix} \cos 60° & -\sin 60° & 4 \\ \sin 60° & \cos 60° & 5 \\ 0 & 0 & 1 \end{pmatrix} \begin{pmatrix} x \\ y \\ 1 \end{pmatrix}$

(b) $x' = x\cos(60°) - y\sin(60°) + 4 = \dfrac{1}{2}x' - \dfrac{\sqrt{3}}{2}y' + 4$

$y' = x\sin(60°) + y\cos(60°) + 5 = \dfrac{\sqrt{3}}{2}x' + \dfrac{1}{2}y' + 5$

57. (a) $\vec{A} \times \vec{B} = 0$; (b) Since the vectors are antiparallel, their cross product is zero.

59. (b) -19.0 J

65. (a) $a = 178$ N/kg; (b) $\alpha = 90.0°, \beta = 51.8°, \gamma = 142°$;
(c) $\vec{a} = (110 \text{ N/kg})\hat{j} - (140 \text{ N/kg})\hat{k}$;
(d) $\vec{a}_{xy} = (110 \text{ N/kg})\hat{j}, \vec{a}_{xz} = (-140 \text{ N/kg})\hat{k}$
$\vec{a}_{yz} = (110 \text{ N/kg})\hat{j} - (140 \text{ N/kg})\hat{k}$

69. $\theta = 109.5°$

Chapter 3: Motion in One Dimension

1. (a) The two objects have the same average velocity;
(b) B at $t = 30$ s; A at $t = 90$ s; (c) Yes, at about $t = 55$ s

3. (a) No; (b) Yes, at $t = 0$; (c) Object A; (d) $(a_A)_{avg} = 100$ m/s²;
$(a_B)_{avg} = 30$ m/s²

5. (a) Object B; (b) Object B; (c) Object A; (d) Object A;
(e) Somewhere between $t = 50$ s and $t = 100$ s the objects have the same displacement. They have the same acceleration at around $t = 130$ s.

7. (a) Yes; (b) a_x is the highest at $t = 200$ s and the lowest at $t = 30$ s and $t = 140$ s.

9. b

11. c

13. Disagree

15. e

17. a

19. Yes

21. a

23. a

25. Only momentarily

27. d

29. e

31. b

33. c

35. c, d

37. (a) $d = 2400$ m; (b) $\Delta x = 200$ m [up]

39. Let us choose the positive x-direction as up. Then we have $d = 230$ cm, $\Delta x = -80$ cm = 80 cm [down].

41. (a) $\Delta x_{barge} = 2.0$ km [away from dock]
(b) $(v_{x,avg})_{barge} = 6.0$ km/h [away from dock]
(c) $(v_{x,avg})_{boat} = 6.0$ km/h [away from dock]
(d) $v_{avg,barge} = 25$ km/h; (e) $v_{avg,boat} = 6.0$ km/h

43. (a) $(a_x)_{last\ 2\ s} = -15.6$ m/s²; (b) $a_{x,avg} = -6.22$ m/s²;
(c) $v_{x,avg} = 25.0$ m/s [forward]; (d) $v_{avg} = 19.2$ m/s and $v_{x,avg} = 0$

45. 6.17 h

47. $r_{M-E} = 4 \times 10^5$ km

49. 1.5 h

51. (a) Mach number for Boeing 737 is 0.77; (b) The maximum speed of Concorde 2 is 550 km/h; (c) $\Delta t \approx 57$ min

53. (a) 105 m, 70 m, 40 m; (b) $-10.$ m/s²

55. (a) 8.3 min; (b) 8.6 years; (c) 9.99×10^4 years

57. 2.0 km/s

59. (a) -3.75 m/s²; (b) -40 m/s²; (c) -5 m/s²; (d) About 330 m (from area between the absolute value of the curve and the x-axis).

61. (a) $a_x = 10.705$ m/s²; (b) $d = 1713.3$ km

65. 1.95 s

67. (a) $d = 88.0$ m; (b) $t = 623$ ms

69. (a) $v_{x_0} = 38$ m/s; (b) $\Delta t = 84$ ms

71. 180 m/s

73. (a) $v_2 = 2820$ m/s; (b) $d_3 = 3.81 \times 10^5$ m;
(c) $d_{total} = 4.80 \times 10^5$ m

75. (a) $a = 6.25$ m/s²; (b) $d_1 = 200$ m; (c) $\Delta x_2 = \dfrac{1}{2}\Delta x_1$;
(d) $\Delta x_3 = 4\Delta x_1$

77. (a) $\Delta x_{0\ s \rightarrow 2\ s} = -4.00$ m; (b) The area is negative, which means the displacement is negative; (c) The negative area represents a negative displacement; (d) $\Delta x_{6\ s \rightarrow 9\ s} = -22.0$ m; (e) $x_{11\ s} = -56.0$ m

79. (a) $v_{x,5.0\ s} = 7$ m/s; (b) $\Delta v_{x,7.0\ s \rightarrow 9.0\ s} = 20$ m/s;
(c) $v_{x,10.\ s} = 24$ m/s;

(d)

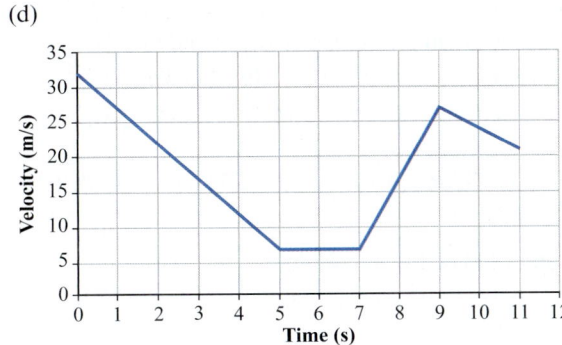

(e) $\Delta x_{total} = 190$ m;
(f)

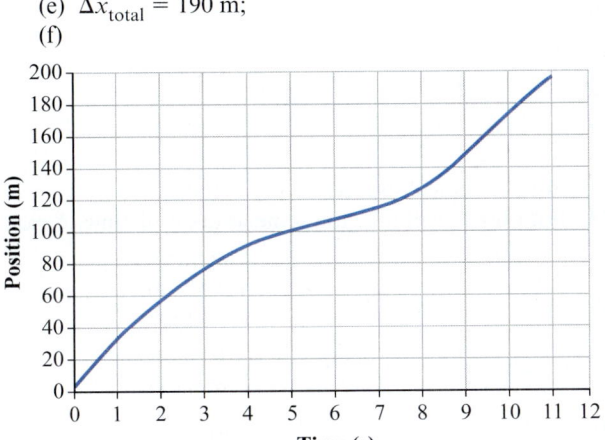

81. (a) $\Delta x_{0\,s \to 5.0\,s} = 3010$ m; (b) $v_{x,5.0\,s} = 915$ m/s; (c) $x_{8.0\,s} = 6190$ m; (d) $\Delta x_{8.0\,s \to 11\,s} = 1890$ m; (e) $v_{x,10.\,s} = 505$ m/s
83. $t_B = \sqrt{2} t_A$
85. $v_{x0.2} = 20.0$ m/s
87. We choose the positive x-direction as down
   (a) $a_{x,avg} = 9.81$ m/s^2, $v_{avg_1} = 2.5$ m/s, $v_{x,avg_1} = 0$ m/s
   (b) $a_{x,avg} = 9.81$ m/s^2, $v_{avg_1} = 5.0$ m/s, $v_{x,avg_1} = 0$ m/s
89. (a) $v_0 = 20.5$ m/s; (b) $d_{2.00\,s} = 60.6$ m
91. (a) No; (b) $\Delta t_{max} = 0.103$ s
93. 33.1 m/s
95. $v_{x,rel} = 45$ km/h [backward]; $v_{rel} = 45$ km/h
99. 89 km/h [right]
101. (a) 12.13 m/s; (b) 1.24 s; (c) Marble: 7.5 m down; ball: 3.5 m up; (d) 12.13 m/s
103. 20 m/s, or 72 km/h
105. $a_{x_0} = 0.74$ m/s^2
107. (a) $b = 414$ m; (b) 610.5 m/s; (c) 2200 m

109. (a) $\phi = \dfrac{n\pi}{2}$ radians; $n = 1,3,5,\ldots$

   (b) (i)

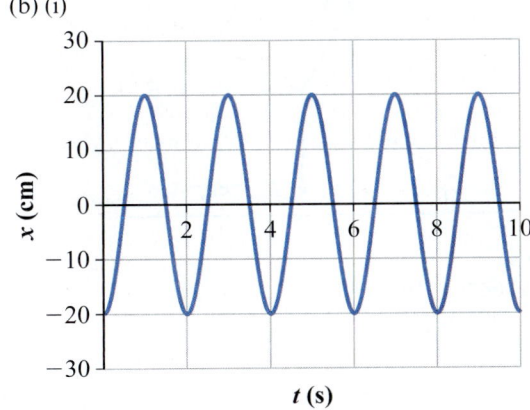

(ii)

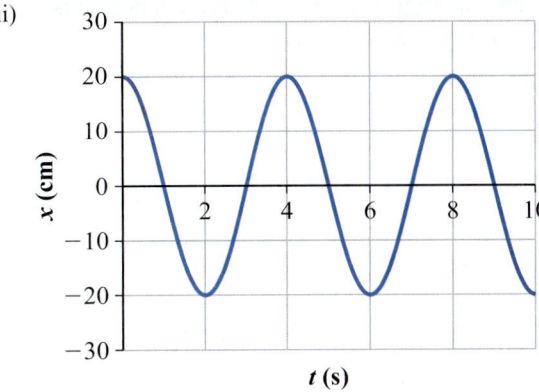

(iii)

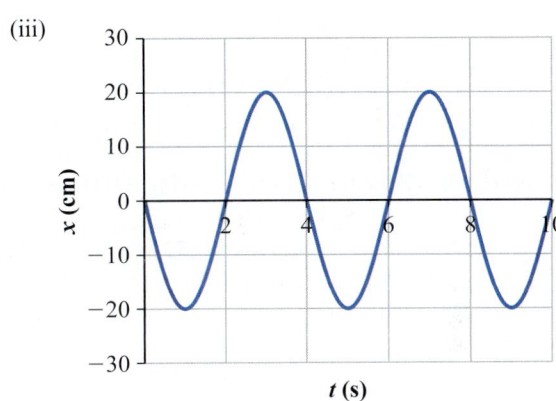

   (c) $v_x(t) = -x_0\omega \sin(\omega t + \phi)$
   (d) $a_x(t) = -x_0\omega^2 \cos(\omega t + \phi)$
   (e) $a_x(t) = -\omega^2 x(t)$
   (f) $v_{max} = 20\omega$; $a_{max} = 20\omega^2$
111. (a) $t = 19.37$ s; (b) 1840 m; (c) 981 m; (d) 185 m
113. (a) $d_1 = 23.61$ m, $d_2 = 87.51$ m, $d_3 = 333.3$ m
   (b) $v_{avg} = 51.06$ m/s $= 184$ km/h; $v_{x,avg} = 51.06$ m/s [forward] $= 184$ km/h [forward]; $a_{x,avg} = 1.60$ m/s^2 [forward]
115. (a) 5.14 min; (b) Distance: 14.3 km; displacement: 5.70 km [toward the dock]
117. (a) 2.03 m behind the boat; (b) 4.00 m; (c) 1.97 m; (d) 0.594 s
119. (a) 244 m forward; (b) 164 m in the forward direction (in the direction of the train's motion)
121. 51.1 m; 31.7 m/s
123. 66.3 m/s up
125. 0.91 s
127. 1.4×10^3 km/h
129. (a) About (i) 8×10^2 m/s, (ii) 7×10^2 m/s, (iii) 5×10^2 m/s, (iv) 1×10^2 m/s
   (b) About (i) 3.3×10^2 m/s, (ii) 1.3×10^2 m/s
131. About 70 km and 490 km
133. About 2.7 km/s, 5.2 km/s, 9.9 km/s, respectively

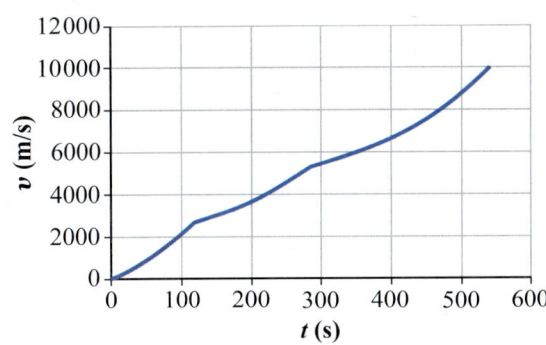

Chapter 4: Motion in Two and Three Dimensions

1. (a) D; (b) B; (c) C; (d) A
3. c
5. b
7. b
9. d
11. d
13. d
15. d
17. a
19. a
21. Both will hit the ground at the same time.
23. Tangential acceleration: tangent to the bowl and pointing in the direction of motion; radial acceleration: pointing toward the centre of the bowl
25. (a) $v_{avg} = 49.3$ m/s
   (b) $\vec{v}_{avg} = 46.3$ m/s [27.2° from initial direction]
   (c) $\vec{a}_{avg} = 5.89$ m/s² [antiparallel to original velocity]
   (d) $t_3 = 0.282$ s
27. $v(t = 3.00\ \text{s}) = 8.00$ m/s; $a(t = 3.00\ \text{s}) = 0.004\ 63$ m/s²
29. $v_{launch} = 12.4$ m/s at $\theta = 82.0°$; $v_{minimum} = 1.72$ m/s; $\vec{a}_{avg} = 9.81$ m/s² [down]; $\vec{v}_{avg} = 1.72$ m/s [horizontally]
33. (a) $t = 5.8$ h; (b) $a_r = 0.011$ m/s²
35. $a_r = 6.143 \times 10^{-3}$ m/s²
37. 3300 rotations/s
39. $a = 16.4$ m/s²; $\theta = 13.4°$ with respect to the radial direction
41. 1.1 s, 0.079 turns
43. 7.9 m
45. 0.83 m/s
47. The melon will see the arrow travel in a perfectly straight line and pass 5.0 m above it.
49. 11 m/s
51. One can use a graphing calculator to find that the eagle cannot be hit by the stone.
53. 17.1 m/s and 22.2 m/s
55. (a) 796 m; (b) 579 m/s and 1430 m; (c) Maximum (absolute) value right before it hits the ground (as in part (b))
57. $x = (5.20\ \text{m/s})t$, $y = (4.36\ \text{m/s})t$,
   $z = (12.2\ \text{m/s})t - \dfrac{1}{2}(9.81\ \text{m/s}^2)t^2$
59. (a) 22.6 m [above ground]; (b) 125 m; (c) 36.1 m/s
61. $y = x\tan\theta + \dfrac{ax^2}{2v_0^2\cos^2\theta}$; maximum range occurs when
   $\theta = 45°$; $x_{y=\max} = \dfrac{v_0^2}{2g}$; $y_{\max} = \dfrac{v_0^2}{4g}$
63. (a) $\left[0.32\ \text{m}\left(2 - \cos\left(\dfrac{\pi}{3\text{s}}t\right)\right)\right]\hat{i}$
   (b) $\left[0.335\ \text{m/s} \times \sin\left(\dfrac{\pi}{3\text{s}}t\right)\right]\hat{i}$
   (c) $\left[0.351\ \text{m/s}^2 \times \cos\left(\dfrac{\pi}{3\text{s}}t\right)\right]\hat{i}$
65. 27.3 s before $t = 0$, the planes passed within 10.3 km of each other.
67. (a) 36.6 s; (b) 15.8 m/s²; (c) 6.36 turns; (d) 1.25 m/s²; (e) $v_{avg} = 4.67$ m/s; (f) 9.74 m
69. 2.0 m
71. (a) $(7.97\ \text{m/s})\hat{i} + (0.67\ \text{m/s})\hat{j}$
   (b) $\vec{v}_{avg} = (-4.66\ \text{m/s})\hat{i} - (0.13\ \text{m/s})\hat{j}$
   $\vec{a}_{avg} = (7.5 \times 10^{-3}\ \text{m/s}^2)\hat{i} + (1.83\ \text{m/s}^2)\hat{j}$
73. 23° west of north
75. (a) 0.44 km/s; (b) 0.50 km/s, 0.25 km/s, 2.5 km/s; (c) 0 m/s²
77. 23.4 m/s

Chapter 5: Forces and Motion

1. (a) True; (b) True; (c) True; (d) False; (e) True; (f) False; (g) False
3. b
5. d
7. a
9. d
11. b
13. $a_1 = 2a_2 = 4a_3$
15. a
17. Both boxes slide at the same time.
19. Both boxes will take the same amount of time to reach the bottom.
21. d
23. No
25. c
27. (a) Less than $M_1 g$; (b) $T = M_1 g$
29. a
31. a
33. a
35. b
37. $\vec{F}_{net} = (40.\ \text{N})\hat{j}$ (i.e., in the upward direction)
39. $\vec{F}_{net} = (+68.0\ \text{N})\hat{i}$ (i.e., a magnitude of 68.0 N pointing to the right)
41. 98.1 N upward
43. (a) 68.8 kg; (b) ~111 N–112 N (first value with actual g on Moon)
45. 1.68×10^3 kg
47. -20.3 kN ($-$ sign means opposite direction of motion)
49. (a) 6.00 N: $(5.14\ \text{N})\hat{i} + (3.09\ \text{N})\hat{j}$
   8.00 N: $(-4.93\ \text{N})\hat{i} + (6.30\ \text{N})\hat{j}$
   5.00 N: $(-1.87\ \text{N})\hat{i} - (4.64\ \text{N})\hat{j}$
   (b) The net force has magnitude 5.04 N and is in the second quadrant at an angle of ~109° from the positive x-axis: $(-1.66\ \text{N})\hat{i} + (4.75\ \text{N})\hat{j}$.
51. (a) $(-1.00\ \text{N})\hat{i} + (3.00\ \text{N})\hat{j} - (3.50\ \text{N})\hat{k}$
   (b) $(6.15\ \text{N})\hat{i} + (1.11\ \text{N})\hat{j}$
   (c) $(-5.00\ \text{N})\hat{i} + (16.00\ \text{N})\hat{j} + (6.00\ \text{N})\hat{k}$
53.

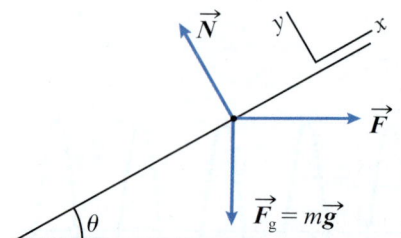

55. 5.4°
57. (a) 68.6 N; (b) 131 N
59. (a) 65.0 kg; (b) 68.4 kg; (c) 12.0 s
61. 17 N

63. 0.71

65. 8.2 cm

67. Non-fundamental

69. $\vec{a}_r = (-g\tan\theta)\hat{r}$ (i.e., direction of centripetal accelera-tion toward centre of circle)

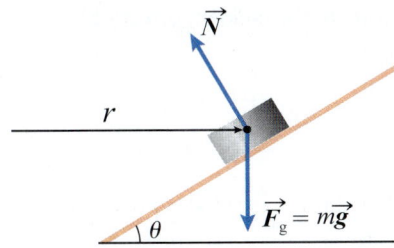

71. (a) 1.40 m/s; (b) $m \times 12.3$ m/s^2

73. 25.2 m

75. 6.65 N

77. 452 N

79. 0.397

81. (a) $(1.50 \text{ kg·m/s})\hat{i}$; (b) $(1.50 \text{ kN})\hat{i}$

83. $\vec{p} = [(0.500 + 2.25t) \text{ kg·m/s}]\hat{i}$

85. $(-35.1 \text{ N})\hat{j}$

87. 2.70°

89. 12.4 m/s

91. (a) 2.5 m/s; (b) 127 N

93. 27.4 m/s^2

95. $\left(\sqrt{(8.50)^2 + \left(\dfrac{5.29}{R}\right)^2} \right)$ m/s^2

97. 0.438 m/s^2; 714 N

99. (a) $(600. \text{ N})\hat{r} - (589 \text{ N})\hat{k}$; (b) $(-13.0 \text{ kN})\hat{r}$; (c) $(-600. \text{ N})\hat{r}$; (d) $(-600. \text{ N})\hat{r}$; (e) $(-13.3 \text{ kN})\hat{k} + (13.6 \text{ kN})\hat{r}$

101. (a) $(-155 \text{ N})\hat{r}$; (b) $(155 \text{ N})\hat{r} - (677 \text{ N})\hat{k}$; Could also be written as 694 N, down and outward direction, 77.4° below horizontal; (c) 155 N radially inward

103. $F = (m + M)g \tan\theta$

105. 9.3 kg

107. (b) $v_\infty = \sqrt{\dfrac{2gm}{\rho A C}}$

(c)

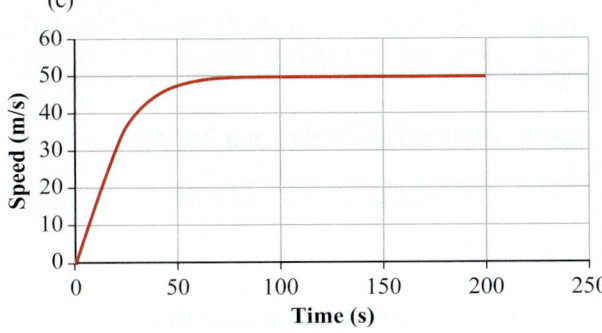

(d) ~49.9 m/s; (e) ~38 s; (f) ~69 s

Chapter 6: Work and Energy

1. Yes

3. c

5. Negative

7. Yes

9. $W_{\text{total}} = \Delta K = 0$; $W_g = -64.5$ kJ

11. c

13. 271 J

15. a

17. e

19. c

21. e

23. b

25. d

27. e

29. c

31. b

33. Increase

35. (a) -1.77 J; (b) 1.77 J; (c) $W_{\text{total}} = W_g + W_{\text{hand}} = 0$

37. -18 J

39. (a) 975 J; (b) 827 J

41. (a) $\dfrac{F}{mg}$; (b) Fd; (c) The work done is the same; (d) The work done is the same; (e) The work done is the same.

43. 1.1 kJ

45. 21 J

47. $W = mgh$

49. (a) -75.2 J; (b) 39.2 J

51. $W_c = 37.5$ J, $F = 18.3$ N, $W_g = 0$ J

53. 1.13 m

55. (a) 1.3 kJ; (b) 1.9 kJ

57. (a) 59 m/s; (b) 44 m/s

59. 15 cm

63. (a) $\sqrt{\dfrac{2KqQ}{m_q}\left(\dfrac{1}{R} - \dfrac{1}{R + \Delta r}\right)}$; (b) 107 m/s

65. 800 kN

67. 4.1 kN

69. (a) 0.54°; (b) 1.2 m/s

71. $\dfrac{5}{2}r$

73. $h = \dfrac{2}{3}r$, $N = 3mg$

75. 43 cm

77. (a) $3mg \cos 23°$; (b) $mgL \cos 23°$; (c) The work done by the tension is zero.

79. (a) $\dfrac{mg}{k}$; (b) $\dfrac{2mg}{k}$; $v_{\max} = g\sqrt{\dfrac{m}{k}}$

81. No energy is lost to friction.

83. (a) 0.98 m/s; (b) $a = \dfrac{g(m - \mu_k M)}{m + M} = 0.74$ m/s^2

85. (a) $\Delta U = \dfrac{A}{2}\left(\dfrac{1}{r_1^2} - \dfrac{1}{r_2^2}\right)$; (b) $\dfrac{A}{2}\left(\dfrac{1}{r_1^2} - \dfrac{1}{r_2^2}\right)$; (c) $-\dfrac{A}{2}\left(\dfrac{1}{r_1^2} - \dfrac{1}{r_2^2}\right)$; (d) Attractive; (e) $21 \text{ m}^{-1}\cdot\text{kg}^{-1/2}\sqrt{A}$

87. (a) 4.6×10^6 J; (b) -4.6×10^6 J; (c) 3.8×10^4 W; (d) -3.8×10^4 W; (e) 4.0×10^2 N upward

89. (a) 1.90 m; (b) -16.5 m; (c) 0.700 N; (d) -0.312 N; (e) Yes

91. (b) $A = \dfrac{3mg}{1 + 4\mu_k^2}$, $B = \dfrac{6\mu_k mg}{1 + 4\mu_k^2}$, $C = -2mg\left(\dfrac{1 - 2\mu_k^2}{1 + 4\mu_k^2}\right)$, $D = -2\mu_k$; (c) 55°; (d) 46°

93. (a) 4.96×10^9 J; (b) 1.23×10^8 W; (c) 5.27 m/s^2

95. (a)

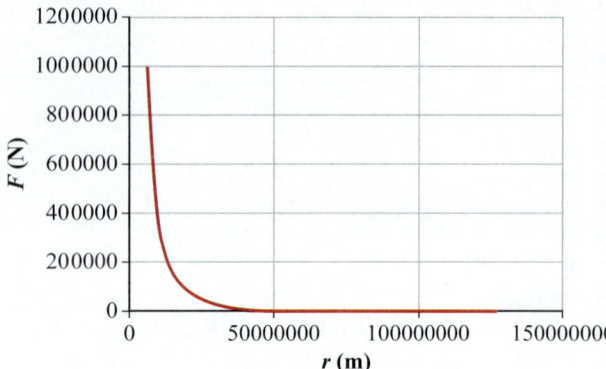

(b) 2.847×10^4 km; (c) -5.940×10^{12} J; negative;
(d)

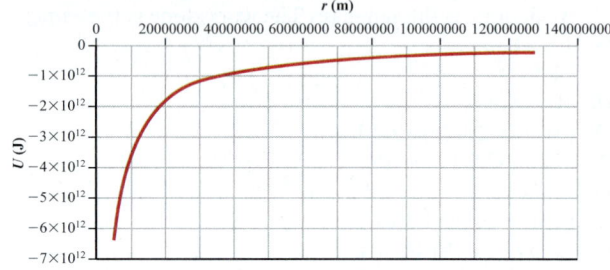

(e) -5.940×10^{12} J, which is very close to the answer obtained in (c) (equal to four significant digits); (f) -6.256×10^{12} J; (g) 11.19 km/s; (h) 0.039 80 m/s²; (i) 9.817 m/s²; (j) -5.940×10^{12} J; (k) Zero

Chapter 7: Linear Momentum, Collisions, and Systems of Particles

1. a
3. c
5. c
7. This is possible if the ball lost energy during the collision.
9. c
11. Yes
13. d
15. b
17. a
19. b
21. True
23. a
25. c
27. $1.53 \times$ kg·m/s
29. 6.1×10^{-4} m/s² [forward]
31. 223 ms
33. (a) 6.3 m/s; (b) The net force on the ball is defined piecewise: -2.16 N for $t < 0.64$ s and $t > 0.94$ s; $(123t - 79)$ N for 0.64 s $< t < 0.79$ s; $(-123t + 115)$ N for 0.79 s $< t < 0.94$ s; (d) 18 N
35. 4 m·s⁻¹, where the initial velocity is in the negative direction
37. $(-0.46$ m, 0.11 m$)$
39. 0.39 m from the end of the rod without the sphere
41. 6.9 m/s
43. 0.3 m
45. $\dfrac{v_1}{v_2} = \dfrac{m_2}{m_1} = 1.7$

47. 59°
49. -1.0×10^2 J
51. $\dfrac{m_1 - 2m_2}{m_1 + m_2} v$
53. $v' = -\dfrac{v}{5}$ (opposite to the incident direction),

$v_2' = v_3' = \dfrac{2\sqrt{3}}{5} v$

55. $m_1 = m_2$
$v_2 = v$
57. $v' = \dfrac{v(m - M) + 2Mu}{m + M}$, $u' = \dfrac{u(M - m) + 2mv}{m + M}$
59. (a) 2600 kg; (b) 1.1×10^2 m/s²; (c) 2.4×10^5 N
61. 1.3×10^{-4} N
63. 15.6 cm
65. 2 cm/s, 9.81 m/s²
67. (a) 97 cm; (b) 0 m·s⁻¹
69. 5.9 m/s
71. 0.26 m
73. 39 m/s
75. 11 m
77. 56°
79. (a) -57 J; (b) 0.20 m/s
81. $y_{cm} = -\dfrac{xr^2}{R^2 - r^2}$, $x_{cm} = 0$
83. $\left(\dfrac{M - m}{2M + 5m}\right)r$
85. (a) 32 m/s; (b) 33 m/s
87. 1.7×10^6 m below the surface of Earth
89. $\Delta r_{sub} = L_{sub} \dfrac{m_{crew}}{m_{crew} + m_{sub}}$
91. 9.0 cm/s
93. (a) $\sqrt{\dfrac{m_1 m_2}{k(m_1 + m_2)}}(v_1 + v_2)$; (b) 33.2 cm
95. (a) $R = 813.7$ kg/s; (b) $v_{er} = 2779$ m/s; (c) Mass at end of first stage (109 s) is 48 310 kg; (f) $P = -\dfrac{1}{2}v_{er}^2 \dfrac{dM}{dt}$;
(g) Integrate $dW_g(t) = M(t)gdh(t)$ (average of the liftoff mass and the mass at 40.00 km altitude); $W = M_{avg}gh = 3.64 \times 10^{10}$ J

Chapter 8: Rotational Kinematics and Dynamics

1. b
3. All three quantities for the point on the rim are larger.
5. c
7. a
9. The shorter rod
11. Valid if other units (such as degrees) are used
13. (a) No; (b) iii
15. She does negative work as she extends her arms and then does positive work as she brings her arms back in. There will be no change in angular speed.
17. $I_{cylinder} = I_{ring} = mr^2$
19. 0
21. To change the angular momentum of a spinning object, a net torque must be applied.
23. When there is no acceleration, or when I is zero

25. True
27. Yes, as long as they are not pushing the car straight toward the tree
29. No, torque is defined about a pivot. Force is a vector quantity, so its direction depends on the frame of reference (although its magnitude is independent).
31. 393 m/s
33. 46.4 cm
35. 1.45×10^{-4} rad/s
37. (a) 69 rad/s^2; (b) 0.43 km/s^2; (c) 2.8 m/s^2; (d) 2.8 m/s^2
39. 8.40 rad/s
41. (a) 26.7°; (b) 0.836 rad/s
43. (a) 46.3 m/s^2; (b) 252 m/s^2
45. $(-5.46 \text{ N} \cdot \text{m})\hat{\boldsymbol{i}} + (5.46 \text{ N} \cdot \text{m})\hat{\boldsymbol{j}} + (5.84 \text{ N} \cdot \text{m})\hat{\boldsymbol{k}}$
47. $(52.0 \text{ N} \cdot \text{m})\hat{\boldsymbol{i}} + (116 \text{ N} \cdot \text{m})\hat{\boldsymbol{j}} + (68.0 \text{ N} \cdot \text{m})\hat{\boldsymbol{k}}$
49. 810 N·m
51. $(-28 \text{ N} \cdot \text{m})\hat{\boldsymbol{i}} + (13 \text{ N} \cdot \text{m})\hat{\boldsymbol{j}} - (19 \text{ N} \cdot \text{m})\hat{\boldsymbol{k}}$; (36.25 N·m, 121.6°, 155.1°)
53. 0.484 kg·m^2
55. 0.23 kg·m^2
59. $\alpha = \dfrac{m_1 g \sin\theta}{r\left(m_1 + m_2 + \dfrac{I}{r^2}\right)}$
61. 5.08×10^5 kg·m^2/s at both positions
63. $\phi = 24°$
65. (a) 0.27 rad/s; (b) -2.6×10^{-4} J
67. $\vec{V} = -28\hat{\boldsymbol{i}} + 39\hat{\boldsymbol{j}} - 20\hat{\boldsymbol{k}}$
    $\vec{V} \cdot \vec{V}_1 = -28 \times 2 + 39 \times 4 - 20 \times 5 = 0$
    $\vec{V} \cdot \vec{V}_2 = -28 \times 7 + 39 \times 4 + 20 \times 2 = 0$
    $\vec{V}' = 28\hat{\boldsymbol{i}} - 39\hat{\boldsymbol{j}} + 20\hat{\boldsymbol{k}} = -\vec{V}$
69. (a) 12 m/s; (b) $K = 3.9 \times 10^5$ J; $v = 120$ km/h
71. 0.32 rad/s
73. $t = 1.9$ s; 1.0×10^2 rad
75. $E_{lost} = -1.7 \times 10^4$ J
77. (a) The moment of inertia of the bar directly on the axis is zero. For the other bar parallel to the axis, $I = mL^2$; (b) Yes; (c) All of the mass is the exact same distance from the axis of rotation; (d) $2mL^2$
79. $I = \dfrac{ML^2}{12}$; $I = \dfrac{M}{12}(L^2 + T^2)$
81. (a) 1.2 m/s; (b) 1.2 m/s; (c) 1.2 m/s; (d) 2.3 rad/s; (e) $\dfrac{a_{rl}}{a_{rs}} = 0.33$
83. (a) (i) 3.5 m/s, (ii) 4.9 m/s; (b) 8.8 m/s
85. 1.9 m/s
87. 4 N
89. $\sqrt{\dfrac{2d(m_2 g - m_1 g \sin\theta - \mu_k m_1 g \cos\theta)}{m_1 + m_2 + \dfrac{I}{r^2}}}$
91. (a) 11 rad/s; (b) 170 J
93. (a) 6.0 rev/min; (b) 0.87 rad/s
95. 0.12 rev; Flywheel stops when $\omega = 0$ (after $t = 0.062$ s).
97. (a) 49 rad/s; (b) 6.3 m/s^2
99. (a) Positive; (b) Increase; (c) 1.3 rad/s; (d) Decrease; (e) $-38\,000$ J
101. (a) Power = torque × angular speed; rad/s; (b) Net torque = Driving torque – Resistive torque; (c) They are equal.

103. (a) $dI_z = r^2 dm$; (b) $dI_z = x^2 dm + y^2 dm$;
    (c) $I_z = \displaystyle\int_{\text{object}} dI_z = \int_{\text{object}} x^2 dm + \int_{\text{object}} y^2 dm$;
    (e) $I_z = I_x + I_y$

Chapter 9: Rolling Motion

1. a
3. (a) The cylinder; (b) The sphere
5. To prevent locking up of wheels
7. d
9. c
11. $2mr^2$
13. Both disks will reach the bottom of the hill at the same time.
15. Both will reach the bottom at the same time.
17. The solid cylinder must reach the bottom first.
19. $\dfrac{m_{\text{ring}}}{m_{\text{sphere}}} = \dfrac{40}{7}$
21. Best answer is (a), but both (a) and (b) is more complete.
23. b
25. (a) 30.0 m/s, $(-608 \text{ m/s}^2)\hat{\boldsymbol{j}}$; (b) 21.2 m/s, 608 m/s^2; 0 m/s, 608 m/s^2
27. (a) 18 m/s; (b) Can never be zero
29. (a) 7.98 m/s^2; (b) 117 m/s^2, $-1.7°$ with respect to the vertical
31. The apparent acceleration is due to the change in the frame of reference and is not a real effect.
33. $\theta = \cos^{-1}\left(\dfrac{r}{R}\right)$
35. $\dfrac{1}{2}g$
37. $\vec{F}_1$: Object will not roll; $\vec{F}_2$: Object will roll to the left; $\vec{F}_3$: Object will roll to the left; $\vec{F}_4$: Object will roll to the left.
39. The force of friction is independent of the radius.
41. (a) 3.8 N; (b) 3.8 N
43. $\dfrac{4F}{4M + 3m}$
45. (a) No; (b) 61 rad/s^2; (c) 1.4 J
47. 46°
49. 6.5 N
51. (a) $K_{\text{trans}} = 26.7$ J; $K_{\text{rot}} = 13.3$ J; (b) 20.0 J
53. (a) 5.00 m; (b) Yes; (c) 156 rad/s
55. (a) $\dfrac{5}{7}h$; (b) 0; (c) $\dfrac{2}{7}mgh$
57. $\dfrac{4F}{4M + 3m}$
59. The steel ball
61. 3.1 km
63. $a_{\text{cm}} = \dfrac{5}{7}g\sin\theta$; friction not taken into consideration, as no energy is lost due to friction
65. (a) $\dfrac{2}{3}\dfrac{(h + 12R)}{(h - 4R)}mR^2$; (b) $mg\sqrt{3}\left(\dfrac{h + 12R}{5h + 12R}\right)$; (c) $\dfrac{3}{4}mR^2$; (d) $\dfrac{2}{3}mR^2$; (e) $h = 36R$; (f) No
67. (a) 13 m/s; (b) 47.4 m/s; (c) 5.0×10^2 m/s^2, 0.56° above the horizontal
69. (a) 16.2 m/s; (b) 0.86 rad/s; (c) 9.8 m/s; (d) 47 J

71. (a) -6.33 m/s^2; (b) 7.1 rad/s; (c) 3.8 m/s^2; (d) 1.0 m

73. $R = \sqrt{\dfrac{I_{cm}}{m}}$; ring, hoop, or thin cylindrical shell

75. (a) 4.8 m/s^2; 0; (b) It would make no difference.

77. 0.36

79. (a) $t = \dfrac{2r\omega}{7\mu_k g}$; $\dfrac{9r\omega^2}{98\pi\mu_k g}$ rev; (b) $d = \dfrac{2r^2\omega^2}{49\mu_k g}$; (c) No;

(d) $\dfrac{2}{7}r\omega$; (e) 0

81. $x = r\theta - r\sin\theta = r(\theta - \sin\theta)$
$y = r - r\cos\theta = r(1 - \cos\theta)$

83. (a) No; (b) 0.56; (c) 10; (d) 1.2 s; (e) 1.1×10^2 rad/s

85. (a) 24.3°; (b) 0.966 m/s; (c) 49.5°; (d) No

87. (a) $\dfrac{7}{2}\sqrt{\dfrac{g}{R-h}}$; (b) $\dfrac{7mg}{4}\left[\dfrac{(R-r)^2}{R-h} - (R-h)\right]$; (c) The metal ball would oscillate about the equilibrium position.

89. (a) 18 m/s; (b) 74 m; (c) 2.7×10^2 rad; (d) $\dfrac{1}{3}$; (e) 1.2×10^2 rad/s

91. (a) 24.3°; (b) $a_t = g\sin\theta - \mu_k g\cos\theta + \mu_k\dfrac{v_{cm}^2}{R+r}$;

(d) $A = -\dfrac{6\mu_k}{4\mu_k^2+1}$ and $B = \dfrac{4\mu_k^2-2}{4\mu_k^2+1}$; (g) 1.88; (h) 50.7°

93. (a) Yes; (b) Yes; it acts uphill; (c) No; (d) Mechanical energy is conserved when 1. There are no non-conservative forces doing work, 2. The work done by the non-conservative forces is zero; (e) No. The work done by this force is zero; (f) No; (g) No; (h) The discussion has not considered the "rotational" work done by the torque due to the force of friction about the centre of mass; (i) $-fs$; (j) $fr\Delta\theta$; (k) $s = r\Delta\theta$; (l) $W_f + W_\tau = -fs + fr\Delta\theta = -fs + fs = 0$; (m) Kinetic to kinetic; translational kinetic energy is converted into rotational kinetic energy.

Chapter 10: Equilibrium and Elasticity

1. No
3. c
5. c
7. d
9. No
11. Yes
13. $\dfrac{3r}{2}$ from the centre of the star
15. No
17. Steel
19. Compression
21. c
23. a and c
25. b
27. a
29. 0.42
31. 2.7 kN
33. $F_1 = F_2 = 68.67 \times 10^3$ N $= 68.7$ kN
35. 0.44
37. (a) 4.64 m from the left
(b) $T_{right} = 3.24$ kN, $T_{left} = 2.77$ kN

39. 164 N
41. $N_{right} = 153$ N, $N_{left} = 91.8$ N
43. 3.1 kN
45. 39.2 N
47. $T = 1.6 \times 10^4$ N, $R_x = 1.4 \times 10^4$ N, and $R_y = 9.2 \times 10^3$ N
49. 6.15 kN
53. $\sigma = 25 \times 10^4$ Pa
55. 4.0 cm
57. $\dfrac{YA(\Delta L)^2}{2L}$
59. 75 N
61. $\dfrac{m_{bone}}{m_{concrete}} = 0.092$; $\dfrac{m_{steel}}{m_{concrete}} = 0.89$
63. (a) 0.36%; (b) 3.1×10^7 N
65. 8.2×10^6 N
67. (a) $N_{top} = 2.94$ kN, $f_{top} = 789$ N
(b) $f_{lower} = 789$ N, $N_{lower} = 8.83$ kN
69. 544 N; 0.45
71. (a) 202 N, $N_{left} = 967$ N, $N_{right} = 406$ N; (b) $A_x = B_x = C_x = 202$ N; $A_y = 112$ N, $B_y = C_y = 49.1$ N
73. $f_s = \dfrac{mg}{2}$; $\mu_s = \cot\theta$
75. $\dfrac{r}{R-r}\tan\theta$
77. (a) $F = \dfrac{mg(\mu\cos 31° - \sin 31°)}{1 - 0.4622(\mu\sin 53° + \cos 53°)}$; (b) 0.60;

(c) 38.6 N

79. (a) $\dfrac{L}{(1+\mu_s\cot\theta)}\left(1 - \dfrac{\mu_s mg}{F}\right)$; (b) $F = \dfrac{2\mu_s mg}{1 - \mu_s\cot\theta}$

81. 2.8×10^4 N
83. (a) $\dfrac{L}{2}$; (b) $\dfrac{L}{4}$; (c) $\dfrac{11}{12}L$, $\dfrac{25}{24}L$
85. 8.3×10^4 kg; 3.2×10^4 kg
87. 1.3×10^{-6} J
89. (a) 6.6×10^{-3} m; (b) 40.6 mm; (c) 3.67 mm
91. 6.54 kJ
93. 0.47 mm
95. (a) 71 GPa; (b) ~6.1%; (c) 1.18×10^8 N/m;
(d)

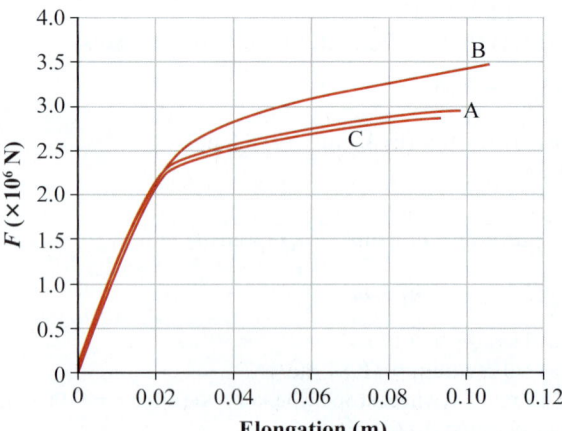

(e) 4.50×10^5 J

97. (a)

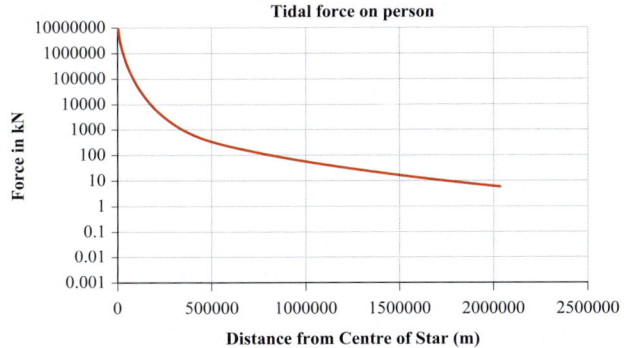

Tidal force on person

(b) ~75 MPa; (c) 24 kN; (d) The value is small, as there are other ligaments supporting the knee; (e) ~590 km from the star

Chapter 11: Gravitation

1. b
5. b
7. c
9. a
11. Increase
13. b
15. c
17. d
21. c
23. 1.4×10^{-11} N
25. 2.21×10^{-6} N/kg (~6.6% difference)
27. $\vec{F}_{1 \to 2} = (-1.36 \times 10^{-11} \text{ N})\hat{i} - (2.72 \times 10^{-11} \text{ N})\hat{j}$
$\qquad + (1.36 \times 10^{-11} \text{ N})\hat{k}$
29. 120 km
31. 1.26×10^{7} m; 7.45×10^{24} kg
33. (a) 8.00 m/s²; (b) 20.5 m/s²; (c) 7.67×10^{24} kg
35. $\dfrac{2GMm}{\sqrt{c^2 + d^2/4}}$
37. 2.0×10^{6} J
39. $F_x = 0$, $F_y = -2ky$, $F_z = 0$
41. $(-40.0 + 4.00z)$ J
43. 17.7 km
45. $v_{esc}\sqrt{\dfrac{3}{8\pi G\rho}}$
47. (a) 0.66; (b) 1.4 AU; 2.0×10^{11} m; (c) 6.6 AU; 9.9×10^{11} m
49. (a) 1.75×10^{11} m; (b) 4.05×10^{11} m; (c) 2.9×10^{11} m; (d) 0.40; (e) ~2.7×10^{11} m; (f) ~0.36
51. (a) 2.4×10^{8} s = 7.5 years; (b) 2.4×10^{6} m/s = 2400 km/s
53. 1680 m/s
55. 14.9 km/s
57. $\sqrt{\dfrac{5GM}{3a}}$
59. (a) 449 km; (b) $r_{cm} = 0.000\,65 r_{Sun}$
61. (a) Planet with 10 d period; (b) Planet with 10 d period slightly (~1.2 times) more massive
63. (a) 23 s (spherical, homogeneous Phobos assumed) (b) 6.5×10^{23} kg

65. (a) 1.2 kg·m²; (b) 2.6×10^{-4} N·m; (c) 1.5×10^{-7} N; (d) 1.3×10^{-7} N·m; (e) 9.9×10^{-4} rad
67. (a) 5.0×10^{23} kg; (b) 3.9 km/s
69. 42.1 km/s
71. 1400 kg/m³
75. (a) 7.3×10^{7} m/s; (b) 4.7 ms
77. (a) The orbits will become larger and more eccentric (b) Half of its original mass
79. 2.3 km/s
81. $\dfrac{3kMm}{r^4}$
83. (a) 5.1×10^{41} J; (b) 42 million years
85. $\dfrac{2GMm}{d^2}$

Chapter 12: Fluids

1. c
3. d
5. A, B, C
7. a
9. c
11. b
13. Yes
15. a
17. 0.31 nm
19. 69 kN/m²
21. (a) 4.14×10^{20} kg/m³; (b) 4.1×10^{15} N
23. 1120 kg/m³; the shell will not float in water
25. 57 kN
27. 107 kPa
29. $\dfrac{\rho}{\rho_w} = \dfrac{2l}{d+l}$
31. a
33. $0.71V_o$
35. 944 kg/m³
37. (a) $h = 0.5$; (b) $F_B = mg$; (c) $2mg$; (d) larger than; (e) larger than
39. 780 kg/m³
41. 1.43 cm
43. 86 cm; decrease
45. (a) 0.60; (b) 1.7 N; (c) 0.82
47. (a) 0.8; (b) 0.3 kg; (c) slightly greater than 0.3 kg
49. 0.66 N; 0.54 N
51. 0.4 m/s
53. 0.2 mm/s
55. 14 m/s
57. 3.7 m/s
59. (a) 4.0 m/s; (b) 120 kPa
61. (a) 317 m/s; (b) 493 kPa
63. $v_1 = 1.6$ m/s; $v_2 = 14$ m/s
65. 4.3 L/s
67. 77 cm³/s
71. 6.3 kPa
73. 3×10^{-3} Pa·s
75. 12 670 Pa = 13 kPa to two significant digits
77. (a) 0.0754 m³; (b) 1068 kg/m³ = 1070 kg/m³ to two significant digits; (e) 13.5%
79. (c) $T = 2\pi\sqrt{\dfrac{\rho_0 L}{\rho_w g}}$

Chapter 13: Oscillations

1. $4A$; 0
5. The period would be greater.
7. The period of the pendulum is greater.
9. a
11. Less than the period when the child sits
13. c
15. a, b, and c
17. d
19. (a) 0.30 m; (b) 0.47 m/s; (c) 0.75 m/s^2;
    (d) $x(t)5(0.30 \text{ m})\cos(\sqrt{2.5}\,t)$
21. (a) $A = 5$ cm, $T = 1$ s, $\phi = -\pi/4$ rad; (b) $\dfrac{3}{8}$;
    (c) -10π cm/s; (d) in the direction of decreasing x;
    (e) $20\pi^2$ cm/s^2
23. (a) 1.2 s^{-1}; (b) 15 cm; (c) π rad; (d) $x(t) = -(0.15 \text{ m})$
    $\cos(7.7t)$
25. $t = 0.3$ s or $t = 1.7$ s (depending upon whether the second oscillator leads or lags)
27. (a) $A = 2.5$ cm, $T = \pi/2$ s, $\phi = \pi/2$ rad; (b) $\dfrac{\pi}{4}$ s; (c) $\dfrac{3\pi}{8}$ s;
    (d) 10 cm; (e) 40.0 cm/s^2; (f) $a(t) = -(40.0 \text{ cm/s}^2)$
    $\cos(4t + \pi/2)$
29. No
31. (a) $3 > 2 > 4 > 1 > 5$; (b) $5 > 1 > 4 > 2 > 3$;
    (c) $2 > 4 > 1 > 3 = 5$; (d) $5 > 1 > 4 > 2 > 3$
33. 7.27×10^{-5} rad/s
35. $D < A < B < E < F < C$
37. (a) 31.6 rad/s, 0.20 s; (b) 6.3 m/s^2, 200 m/s^2; (c) 2.0 J
39. (a) Yes; (b) Yes; (c) No. The maximum speed of the oscillator with twice the mass is a factor of $1/\sqrt{2}$ smaller.
41. (a) 0.38 m, -1.0 rad; (b) 1.2 m/s; (c) Amplitude remains the same; the phase constant changes.
43. (a) 0.31 s; (b) It does not change.
45. 0.35 s
47. $\sqrt{k\left(\dfrac{1}{m_1} + \dfrac{1}{m_2}\right)}$
49. (d) No
51. (a) 1.8 cm; (b) 0.031 J; (c) 18 cm/s
53. $x = \pm\dfrac{1}{\sqrt{2}}A$
55. (a) $A = 10$ cm, $\omega = 7.1$ rad/s, $\phi = \pi$ rad
    (b) $E = 5.0 \times 10^{-2}$ J
57. (a) 5 cm, 1.6 Hz; (b) 0.16 J; (c) $x(t) = -0.05\cos(9.9t)$;
    (d) 5 cm
59. (a) 9.9 N/m; (b) 1.2×10^{-2} J; (c) 64%;
    (d) $x(t) = (5.0 \text{ cm})\cos(\pi t + \pi)$
61. (a) 9.87 m/s^2; (b) Increase; (c) Amplitude remains the same, $\phi = +1.0$ rad; (d) The pendulum will not oscillate.
63. (a) 0.10 J; (b) 0.82 m/s
65. (b) 0.28%
69. (a) 0.6 m, 3 s; (b) $\dfrac{\pi}{2}$; (c) $\dfrac{11\pi}{6}$;
    (d) $x(t) = 0.6\cos\left(\dfrac{2\pi}{3}t + \dfrac{\pi}{2}\right)$; (e) 0.63 m/s; (f) -2.3 m/s^2
71. (a) 0.2 m, 3 s; (b) 1 m; (c) $\dfrac{\pi}{3}$;
    (d) $x(t) = 0.2\cos\left(\dfrac{2\pi}{3}t + \dfrac{\pi}{3}\right) + 1$;
    (e) $x(t) = 0.2\cos\left(\dfrac{2\pi}{3}t + \dfrac{\pi}{3} \pm \dfrac{\pi}{2}\right) + 1$; (f) 0.88 m/s^2

73. (a) 3, 1, 2; (b) 1, 3, 2; (c) 3, 2, 1; (d) 1, 3, 2
75. (a) 0.0064 kg/s; (b) $\dfrac{t}{T} = 15.8$; (c) No
77. A C B
79. 1.2×10^3
81. (a) 77 800; (b) 36.1 days
83. (a) $T' > T$; (b) $A' > A$; (c) Total energy increases since $A' > A$.
85. 42.2 min

Chapter 14: Waves

1. (a) False; (b) False; (c) True; (d) False; (e) False; (f) False; (g) False; (h) False
3. Light wave
5. No
9. c
11. (a) No; (b) No
13. No
15. (a) π m^{-1}; (b) 63 rad/s; (c) 20 m/s
17. (a) 143 Hz; (b) 6.99×10^{-3} s; (c) 1.05 m
19. From 17 mm to 17 m
21. Ocean water: 6.00 m, 10. mm; air: 1.4 m, 2.3×10^{-3} m
23. $D(1.5, 2.0) = 6.9 \times 10^{-3}$ m; $D_{max} = 0.25$ m; $D_{min} = 0$ m
25. (a) 3.0 m/s in the direction of the negative x-axis;
    (b) -2.4 cm; (c) -0.5 m; (d) 1.2 cm/s
27. (b) 1.0 m/s to the right; (c) 2 m;
    (d) $D(x,t) = \begin{cases} +2 \text{ m if } |x+t| \le 1 \\ -2 \text{ m if } |x+t| > 1 \end{cases}$
29. (b) $x = -10$ m, $x = 10$ m; (c) 2 s; (d) When $x = 0$, the pulses cancel for all times.
31. 25 N
33. 220 m/s
35. 13.0 m from the far end of string B
37. (b) $T = \sqrt{\dfrac{2\pi\lambda}{g}}$; (c) 19 m/s
39. wavelength: $b < a = c < d$; frequency: $a = b = d < c$
41. $D_2(x,t) = (1.0 \text{ cm})\sin\left(2\pi x - 3\pi t + \dfrac{\pi}{2}\right)$
    $D_1(x,t) = (1.0 \text{ cm})\sin\left(\dfrac{4\pi}{3}x - 2\pi t - 1.35\right)$
43. (a) 0.63 m, 1.2 Hz, 0.75 m/s
    (b) $D(x, t) = (0.05 \text{ m})\sin(10.0x - 7.50t)$
45. (a) 3 m, 0.2 Hz, 0.5 m/s; (b) To the left;
    (d) -0.01 m/s
47. (a) 8 m; (b) 40 m/s; (c) $\pi/2$;
    (d) $D(x,t) = 1.5\sin\left(\dfrac{\pi}{4}x - 10\pi t + \dfrac{\pi}{2}\right)$
49. (a) 0.5 Hz; (b) 8 m; (c) 1.5 rad;
    (d) $D(x,t) = (1.5 \text{ m})\sin\left(\dfrac{\pi}{4}x - \pi t + 1.5\right)$
51. (a) $D(x,t) = (1.5 \text{ m})\sin\left(\dfrac{\pi}{4}x - 2\pi t + \dfrac{\pi}{2}\right)$
    (b) $D(x,t) = (1.5 \text{ m})\sin\left(\dfrac{\pi}{4}x + 2\pi t + \dfrac{\pi}{2}\right)$
53. (a) $D(x,t) = (0.10 \text{ m})\sin\left(\dfrac{2\pi}{3}x - \dfrac{\pi}{3}t - 1.9\right)$
    (b) 0.50 m/s

55. (a) 0.2 rad; (b) 1.2 rad; (c) Yes; (d) 0.4 rad; (e) Yes;
    (f) 2π rad
57. 0.88 W
59. (b) $A = 0.5$ m, $\lambda = 1$ m; (c) Yes; (d) Travelling wave
61. -0.4 rad
63. 2.636 rad; 1.318 rad
65. (a) $|A_1 - A_2|$; (b) $A_1 + A_2$
69. (a) $D(x, t) = 0.2 \sin(3x - 4t + \pi)$; (b) $D(x, t) = 0.2$
    $\sin(3x - 4t)$
71. (a) 10.5 m; 10 Hz; (b) 5.2 m, 2.6 m; (c) $D_1(x, t) =$
    $(0.750$ cm$) \sin(0.6x - 20\pi t)$, $D_2(x, t) = (0.750$ cm$)$
    $\sin(0.6x + 20\pi t)$; (d) 0.18 cm
73. (a) $D(x, t) = (2.0$ mm$) \sin(\pi x) \cos(0.5\pi t)$; (b) $x = \pm 1$ m,
    ± 2 m, ± 3 m; $x = \pm 0.5$ m, ± 1.5 m, ± 2.5 m; (c) 1.0 m
77. 265 Hz
79. (a) 4.0 m; (b) 50. Hz; (c) No; (d) 1.0×10^2 Hz, 2.0×10^2 Hz
81. (a) 98.1 Hz; (b) 196 Hz, 294 Hz; (c) $v = 196$ m/s;
    (d) 116 Hz
83. 470 Hz, 510 Hz, 540 Hz, 580 Hz, 620 Hz, 670 Hz,
    710 Hz, 770 Hz, 820 Hz
85. (a) 254 N; (b) 17.25 cm; (c) 660.0 Hz
87. (a) 102 Hz; (b) 32.5 cm
89. $f_c < f_a < f_b = f_d$

Chapter 15: Sound and Interference

1. True
3. Divided by 2
5. True
7. True
9. False
11. 3.01 dB when the sound intensity is doubled
13. c
15. 0.343 m
17. 0.040 s
19. $s(x, t) = s_m \cos(1.1x - 377t)$
21. 11×10^{-5} m
23. Doubles
27. 230 Hz
29. 2.14 m
31. 2 Hz
33. 90 Hz
35. 98 dB
37. 72 dB
39. 3.4 kHz
41. (a) 85.4 kHz; (b) 84.9 kHz; (c) 86.4 kHz
43. 1.4 km
45. 28 nm
47. 3×10^{-6} W·m^{-2}, 21 mW
49. 2.4 kHz
51. 1.3 m
53. 24 m/s
55. 0.2 μW/m^2
57. 35.7 m
59. 61 Hz
61. (a) 240 cm; (b) 2.68 kHz
65. 1.05×10^7 m/s, 149 Mpc

Chapter 16: Temperature and the Zeroth Law of Thermodynamics

1. No
3. Not necessarily
7. No

9. No
13. No
15. 10^{18} s
17. 3.34 nm
19. No
21. (a) 758 kPa; (b) 880 cm^3; (c) 0.268 mol;
    (d) Nitrogen = 7.51 g; helium = 1.07 g
23. 0.68 mm
25. -40 °C
27. (a) 136 °F; (b) 331 K
29. (a) 8.67 kg; (b) 52.2 m^3; (c) 530 mol
31. (a) 22.4 L; (b) 493 m/s; (c) 461 m/s; (d) 1.069:1
33. 432 m/s; 0.54 km/s
35. $v_{avg} = \sqrt{\dfrac{8kT}{\pi m}}$
37. Solid, solid, liquid, and solid
41. 0.4 mm/s
43. 180 K
45. $\sim 10^{28}$ m$^{-2} \cdot$s^{-1}
47. (a) 1.68×10^{16} Pa; (b) 5.8×10^5 m/s

Chapter 17: Heat, Work, and the First Law of Thermodynamics

1. Constant-volume process
3. A: solid only; B: solid and liquid; C: liquid only;
   D: liquid and gas; E: gas only
7. Positive
9. Increases by a factor of 16
13. When the heat capacities of the two objects are the same
15. 8.0×10^2 s
17. 14 °C
19. 130 J/(kg·K)
21. 11.4 °C
23. 2.40×10^3 J
25. 100. J
27. 21 kW/m^2
29. $P = 9.6$ kW
31. 8.0 kJ/K, 80 kJ/(kg·K)
35. -3740 J
37. 0.81 W
39. 505 kJ, 495 kJ
41. -7.6 kJ, 0, -7.6 kJ
43. 61 kJ
45. 970 K, 13.4 kJ, 5.37 kJ, 8.06 kJ
47. About 60 mg

Chapter 18: Heat Engines and the Second Law of Thermodynamics

1. No
5. No
7. No. Cleaning your room does not violate the second law.
11. (a) 1.6 mL/s; (b) 53 kJ/s
13. (a) 12.5 W; (b) 12.5 W; (c) 290 W; (d) 300 W
15. (a) 39.28%; (b) 39.28 kJ
17. 3.9 MJ
19. (a) $(25$ K$)R$; (b) $0.125R$; (c) $0.122R$; (d) $0.003R$;
    (e) Not reversible
21. (a) 63 kW; (b) -82 J/(K·s); (c) 220 J/(K·s);
    (d) 140 J/(K·s)
23. About 400 000
25. (a) -0.77 J/K; (b) -8.1 J/K; (c) -53 J/K; (d) -53 J/K

27. No
29. Reject
33. 7 kJ/K
35. -2.6×10^7 J/K, 6.8×10^7 J/K, 4.2×10^7 J/K
37. 35.9 J/K
39. $CP_H = 1.19$

41. (a) $\dfrac{P}{A} = \dfrac{\kappa \Delta T}{z}$, where z is the thickness of the ice layer in metres, κ is the thermal conductivity of the ice, and ΔT is the difference in temperature between the lake water below the ice and the air; (b) $\dfrac{1}{A}\dfrac{dS_{air}}{dt} = \dfrac{\kappa \Delta T}{z T_{air}}$; (c) $\dfrac{1}{A}\dfrac{dS_{water}}{dt} = \dfrac{-\kappa \Delta T}{z T_{water}}$; (d) $\dfrac{dz}{dt} = \dfrac{P}{A}\dfrac{1}{\rho L}$, where ρ is the density of the water and L is the latent heat of fusion; (e) $z(t) = \sqrt{\dfrac{2\kappa \Delta T}{\rho L}} \, t$

43. $\dfrac{mgh}{T}$

45. No

Chapter 19: Electric Fields and Forces

1. c
3. b
5. d
7. d
9. c
11. c
13. a
15. a
17. c
19. a
21. d
23. $1:7.7 \times 10^{12}$
25. (a) Negative; (b) Charges spread equally when far apart

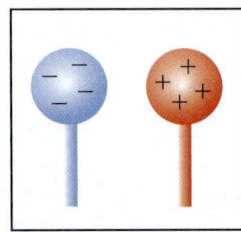

27. (a) 8.22×10^{-8} N attractive force, along the line joining the particles; (b) 2.19×10^6 m/s
29. 4.24 cm left of the 1.00 μC charge, 1.24 cm left of the -2.00 μC charge
31. 0.922 N in a direction away from the centre of the triangle
33. 2.44 N directly away from rod
35. 1.60×10^{-12} N
37. $+0.167$ μC
39. 22.5 MN/C with direction to the left
41. An infinite line parallel to the y-axis (or yz plane) at $x = +6.97$ m from the origin
43. The vector is of length 1.25 cm in a direction away from the origin at an angle of 33.7° below the positive x-axis.

45.

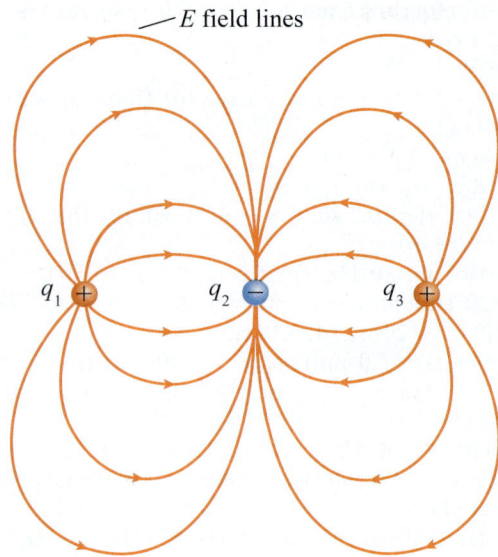

E field lines

47. 0.0540 N
49. 3.65×10^{-2} nm
57. (a) 2.15×10^{14} electrons; (b) 0.0173 mA
59. 0
61. Electric force upward of magnitude 2.4×10^{-17} N; ratio of two forces: 2.7×10^{12}
63. (a) 187 kN/C; (b) Direction opposite to the electron's acceleration
67. (a) $-\dfrac{2d^2q^2}{4\pi\varepsilon_0}\left(\dfrac{3D^2 - d^2}{D^2(D^2 - d^2)^2}\right)\hat{i}$; (b) $\vec{F} \approx -\dfrac{3d^2q^2}{2\pi\varepsilon_0 D^4}\hat{i}$
69. 140. kN/C directed upward
71. $\dfrac{\alpha}{2\pi\varepsilon_0 R^2}\left[1 - \dfrac{d}{\sqrt{R^2 + d^2}}\right]$
73. -2.78 nC
75. (a) 5.34×10^{-6} C at the position ($x = +1.00$ m, $y = +1.00$ m); (b) 12.0 kN/C
79. $r_1 = 0.33$ m; $r_2 = 1.33$ m
81.

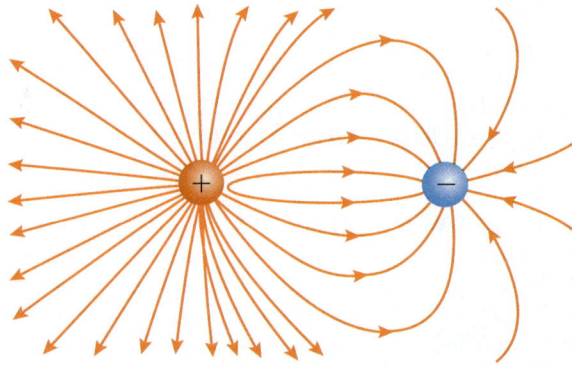

83. (a) No; (b) Two very long lines of charge, with the positive on the left and the negative on the right
85. 1.5×10^9 N/C in a direction radially outward from the ion
87. (a) Negative; (b) Water is deflected toward the pipe.
89. (a) -2.00×10^{-3} C, at $x = -6.00$ m
(b) At $x = 3.00$ m, $E = -2.22 \times 10^{-5}$ N/C. At $x = 6.00$ m, $E = -1.25 \times 10^{-5}$ N/C.
91. $+25$ μC charge placed at $+2.0$ m

1. a
3. b
5. b
7. a
9. c
11. d
13. a
15. c
17. b
19. b
21.

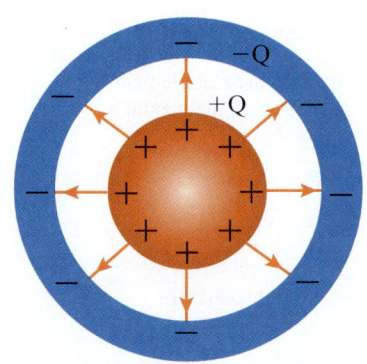

23. (a) The inner sphere is positively charged, and the spherical shell is negatively charged. The magnitude of the charge on the shell must be 4 times that on the sphere; (b) Between the sphere and the shell, electric lines point radially outward. In the region outside the shell, the electric field points radially inward.

25. $25.0 \dfrac{\text{N}}{\text{C}} \cdot \text{m}^2$

27. $\Phi_{E_{yz} \text{ at } x=0} = -3.50 \text{ N} \cdot \text{m}^2/\text{C}$;

$\Phi_{E_{yz} \text{ at } x=L} = +3.50 \text{ N} \cdot \text{m}^2/\text{C}$. The flux through the other four faces is zero. $\Phi_{E \text{ net}} = 0$.

29. $\Phi_E = 6.28 \text{ N} \cdot \text{m}^2/\text{C}$

31. For $r < r_0$, $E = \left(\dfrac{\rho}{3\varepsilon_0}\right) r$, with direction radially outward;

for $r \geq r_0$, $E = \left(\dfrac{\rho r_0^3}{3\varepsilon_0}\right)\dfrac{1}{r^2}$, with direction radially outward.

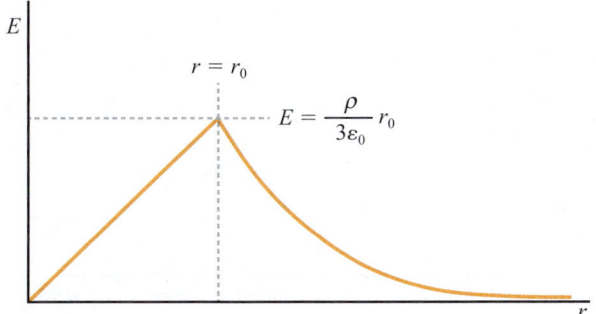

33. 0.0344 C/m^3
35. $-6.8 \times 10^5 \text{ C}$
37. (a) Not possible; (b) Cylindrical Gaussian surface perpendicular to plane (if wide enough); (c) Spherical Gaussian surface; (d) Not possible; (e) Not possible

39. $657 \ \mu\text{C/m}$

41. For $r < r_0$, $E = \left(\dfrac{\rho}{2\varepsilon_0}\right) r$. For $r > r_0$, $E = \left(\dfrac{r_0^2 \rho}{2\varepsilon_0}\right)\dfrac{1}{r}$.

In both cases, the direction is radially outward if the charge is positive.

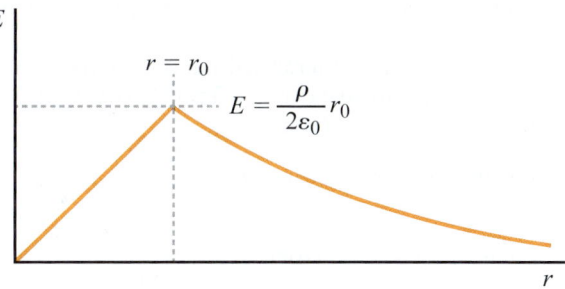

43. 1.18×10^{12}
45. (a) $E = 471 \text{ kN/C}$, in a direction perpendicular to the plate and away from it; (b) $E = 471 \text{ kN/C}$, in a direction perpendicular to the plate and away from it
49. Microwave oven
51. The electric fields inside and on the conducting surface are both 0. The electric field vectors just outside the surface point inward and are everywhere perpendicular to the surface.

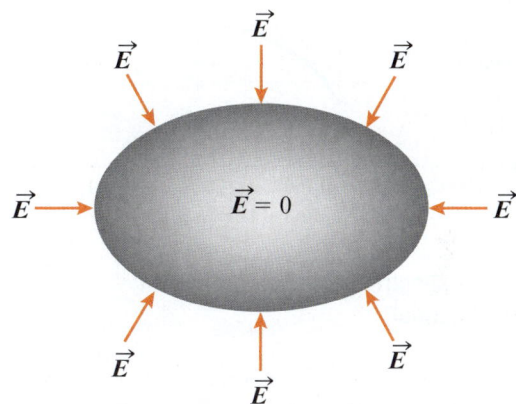

53. (a) The inner surface must have an induced charge $-Q$, corresponding to a surface charge density of

$\sigma_{\text{int}} = -\dfrac{Q}{4\pi a^2}$

(b) The outer surface charge density is $\sigma_{ext} = +\dfrac{Q}{4\pi R^2}$

(c) $\vec{E} = \dfrac{Q}{4\pi r^2 \varepsilon_0}\hat{r}$

57. $\vec{g} = -\dfrac{GM}{r^2}\hat{r}$ when $2R > r > R$

59. (a) $\vec{g} = -\dfrac{2\pi G\rho_0 R^3}{r^2}\hat{r}$ for $r > R$

(b) $\vec{g} = -2\pi G\rho_0 R\hat{r}$ for $r \leq R$

63. $\rho(r) \propto \dfrac{1}{r}$

65. (a) (i) $2.78 \ \mu\text{C}$; (ii) $11.1 \ \mu\text{C}$; (b) (i) $22.1 \ \mu\text{C/m}^2$; (ii) $22.1 \ \mu\text{C/m}^2$ (same answer both cases)
67. (a) $2.97 \times 10^{17} \text{ C}$; (b) -582 C/m^2; (c) $6.58 \times 10^{13} \text{ N/C}$

71. (a) Yes; (b) Yes; (c) No

73. (a) $\dfrac{\lambda}{2\pi\varepsilon_0} \cdot \dfrac{1}{r}$; direction radially outward if λ is positive;

(b) 10.0 m from the wire

75. 0.944 m/s², direction toward plane

77. A conducting solid sphere, or conducting spherical shell, with radius slightly less than 0.400 m, and carrying total charge $Q = -2.5 \times 10^{-9}$ C

79. The electric field is 0 at radial distances less than 10.0 cm, from 30.0 cm to 40.0 cm, and from 80.0 cm to 90.0 cm. In other regions we derive:

For 10.0 cm $< r <$ 30.0 cm: $E = +\dfrac{5.0 \times 10^{-10}}{4\pi\varepsilon_0 r^2}$ N/C

For 40.0 cm $< r <$ 80.0 cm: $E = -\dfrac{5.5 \times 10^{-9}}{4\pi\varepsilon_0 r^2}$ N/C

For $r >$ 90.0 cm: $E = +\dfrac{10.5 \times 10^{-9}}{4\pi\varepsilon_0 r^2}$ N/C

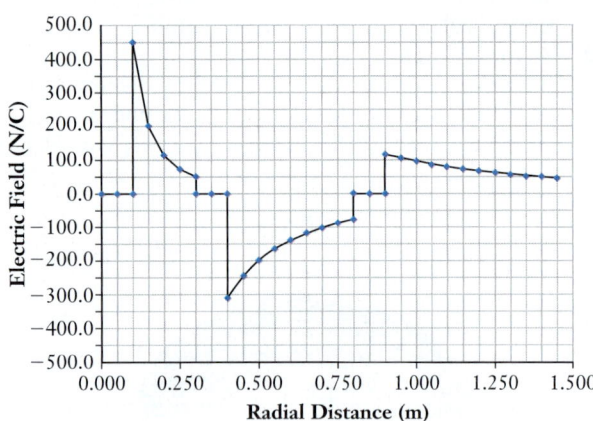

Chapter 21: Electric Potential Energy and Electric Potential

1. a
3. c
5. a
7. b
9. d
11. c
13. b
15. a
17. d
19. d
23. 3.60×10^{-4} J
25. The electric field points from A to B with magnitude 4.00 kN/C.
27. 0.241 J
29. -36.0 mJ
31. 147 m/s (relative to each other)
33. 0.540 m
35. 1.02 kV
37. 1.50×10^{-4} J
39. (a) The orange lines are the electric field, and the blue lines are equipotential; (b) The left charge is negative and the right charge positive; (c) The highest voltage will be at the equipotential line nearest to the positive (right) charge.

41.

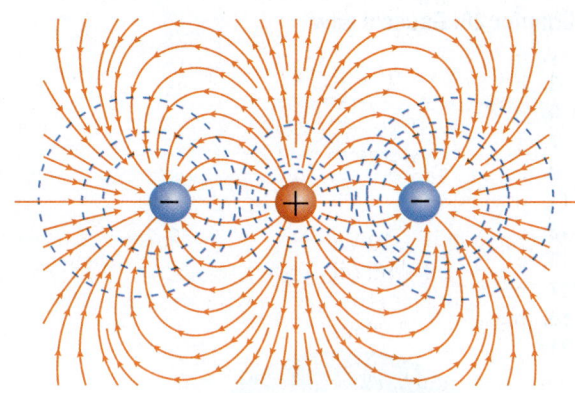

Solid lines are electric field; dashed are equipotential

43. -671 V
45. 1.47 MV
47. $\dfrac{\sigma}{2\varepsilon_0}[\sqrt{z^2 + a^2} - z] + $ constant

49. $\dfrac{\lambda}{4\pi\varepsilon_0}\ln\left(\dfrac{\sqrt{L^2 + h^2} + L}{h}\right) + $ constant

51. 3.32 TeV
53. (a) 3.26 MeV; (b) 1.78 keV
55. (a) 3.92×10^{-19} J/molecule
(b) 9.38×10^{-20} cal/molecule
57. $(-91.7 \text{ V/m})\hat{j}$
59. 10.0 V
61. (a) 23.6 V; (b) 26.0 V; (c) 0.462 m
63. 0.133 nC
65. -741 V
67. 4.45 nC
69. (a) $\vec{E} = (1.75 \times 10^5 \text{ N/C})\hat{r}$; (b) $\vec{E} = (7.0 \times 10^5 \text{ N/C})\hat{r}$
71. 133 V
73. 1.56 kV
75. Each moves at 5.67 m/s relative to centre of mass, in opposite directions.
77. -38.2 kV
79. (a) $\dfrac{(2\sqrt{2} + 1)q}{4\pi\varepsilon_0\sqrt{2}d}$; (b) $\dfrac{(2\sqrt{2} + 1)q^2}{4\pi\varepsilon_0\sqrt{2}d}$

81. (a) 1.41 MV; (b) 1.41 MV; (c) $\dfrac{8.45 \times 10^5}{r}$ V; plotted for outside sphere below:

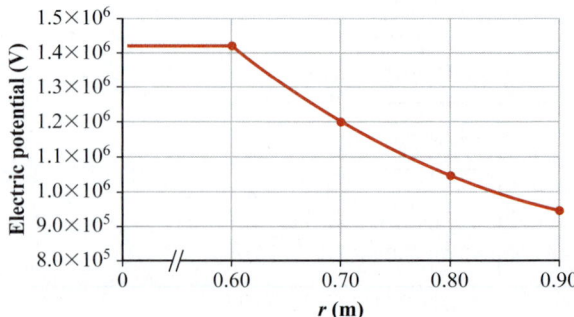

83. (a) 1.44 kV; (b) 1.44 keV
85. Graphically obtained values approximately the following: (a) -10 V/m; (b) 0 V/m; (c) -30 V/m; (d) $+20$ V/m
87. (a) 2.97×10^{17} C; (b) -582 C/m²;
(c) $(-6.58 \times 10^{13} \text{ V/m})\hat{r}$; (d) -4.19×10^{20} V (assuming zero potential reference at infinite distance)

89. At $a/2$: $\dfrac{\lambda \ln 2}{2\pi\varepsilon_0}$; at $2a$: $-\dfrac{\lambda \ln 2}{2\pi\varepsilon_0}$

91. For $r > 2R$: $V = \dfrac{1}{4\pi\varepsilon_0}\dfrac{Q}{r}$; for $R < r < 2R$:

$$V = \dfrac{Q}{4\pi\varepsilon_0}\left\{\dfrac{1}{R} - \dfrac{1}{r}\right\}; \text{ for } r < R: V = 0$$

93. 2.81 kV

95. (a) (i) $E_x = -10$ V/m, $E_y = +30$ V/m; (ii) $E_x = -10$ V/m, $E_y = +30$ V/m; (iii) $E_x = -10$ V/m, $E_y = +30$ V/m; (b) $V(x, y) = (10x - 30y)$ V; $V(12.00$ m$, 12.00$ m$) = -240$ V; $V(8.00$ m$, 14.00$ m$) = -340$ V

97. (a) $\sim +150$ V; (b) $V(r) = \left(150 - \dfrac{21}{r}\right)$ V; (c) $-\dfrac{21}{r^2}\hat{r}$ V/m;

(d) A conducting sphere, or a conducting or a uniformly charged non-conducting spherical shell, of radius ~ 0.3 m; (e) -2.1 nC/m^2

Chapter 22: Capacitance

1. b
3. d
5. b
7. c
9. b
11. d
13. b
15. b
17. c
19. c
21. 0.500 mF
23. 0.500 V; plate A
25. (a) 4.80 V; (b) 11.7 μF
27. $2x$
29. 625 kV/m
31. 41.7 pF

33. (a) $-\left(\dfrac{4.05 \times 10^3}{r^2}$ V/m$\right)\hat{r}$; direction is radially inward;

(b) 0; (c) 5.62 kV; (d) 80.1 pF; (e) 80.4 pF

35. All four capacitors connected in parallel gives the largest equivalent capacitance, 10.0 μF:

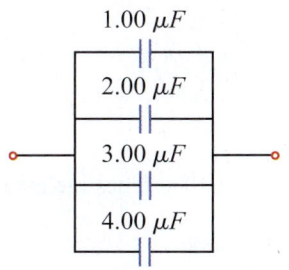

1.00 μF
2.00 μF
3.00 μF
4.00 μF

All four capacitors connected in series gives the smallest equivalent capacitance, 0.480 μF:

1.00 μF 2.00 μF 3.00 μF 4.00 μF

37. A 0.90 μF capacitor
39. Mica
41. 2.00
43. (a) $(-1000$ V/m$)\hat{j}$; (b) -0.700 V, top plate more positive; (c) $(-200$ V/m$)\hat{j}$

45. 1.00 μF: 2.14×10^{-5} J; 2.00 μF: 1.07×10^{-5} J; 3.00 μF: 7.14×10^{-6} J

47. (a) 180 μJ; (b) 720 μJ

49. 70.7 V

51. (a) 42.1 V; (b) The bottom plate

53. 0.711 mm^2

55. (a) 23.2 pF; (b) 3.00 V; (c) 69.5 pC; (d) 0.104 nJ

57. (a) $\dfrac{4\pi\varepsilon_0}{1/r_1 - 1/r_2}$; (b) $C = 4\pi\varepsilon_0 r_1$; (c) 0.899 m

59. (a) $\dfrac{C}{L} \approx \dfrac{\pi\varepsilon_0}{\ln(D/r)}$; (b) 8.70 pF/m

61. 1.3 μF

63. 0.750 μC; 7.50 V

65. 2.00 μF capacitor: charge 8.00 μC; potential difference 4.00 V; 4.00 μF capacitor: charge 16.00 μC; potential difference 4.00 V

67. 10.2 mJ$\cdot$m^{-3}

69. (a) 3547 pixels horizontally by 2660 vertically; (b) The pixel dimensions are 3.38 μm $\times$ 3.38 μm; (c) 3.14×10^{-16} F

71. (a) 1650 F; (b) 1.65 million; (c) 4 h 7 min 30 s

73. C_1: 4.8 V, 48 μC; C_2: 1.6 V, 48 μC; C_3: 6.4 V, 128 μC; C_4: 17.6 V, 176 μC

75. C_1: 5.4 V; C_2: 10.7 V; C_3: 16.1 V; C_4: 13.9 V

77. (a) A conducting spherical shell of radius 20 cm to at least 28 cm, outside a concentric conducting sphere of radius about 10 cm; (b) 0 V/m; (c) 190 kV/m in a radially outward direction; (d) $\sim$270 kV/m in a radially outward direction; (e) 13 500 V difference with outer sphere at lower potential; (f) 0.30 μC; (g) 22 pF

Chapter 23: Electric Current and Fundamentals of DC Circuits

1. (a) False; (b) False; (c) False; (d) True; (e) True
3. b
5. e
7. d
9. b
11. c
13. c
15. e
17. b
19. c
21. c
23. c
25. d
27. 8.49×10^{28} free electrons/m^3
31. 2.1 m

33. (a) $\left(\dfrac{\Delta R}{R_0}\right)_{Al} = \left(\dfrac{\Delta\rho}{\rho_0}\right)_{Al} = 1.8$ and

$\left(\dfrac{\Delta R}{R_0}\right)_{Nichrome} = \left(\dfrac{\Delta\rho}{\rho_0}\right)_{Nichrome} = 0.19$

(b) $\left(\dfrac{\Delta R}{R_0}\right)_{Al} < \left(\dfrac{\Delta\rho}{\rho_0}\right)_{Al}$ and

$\left(\dfrac{\Delta R}{R_0}\right)_{Nichrome} < \left(\dfrac{\Delta\rho}{\rho_0}\right)_{Nichrome}$

(c) The resistivities would not change, but the resistances of the wires would double.

35. $\dfrac{R_1}{R_2} = 50$

37. (a) 457 m; (b) ~727 loops; (c) $\dfrac{\rho_{Ni}}{\rho_{Cu}} = 3.98$. Thus, the answers to parts (a) and (b) would have been one quarter the size as well: 114 m and 182 loops, respectively.

39. (a) 55.0 Ω; (b) 0.341 Ω; (c) ~161:1

45. (a)

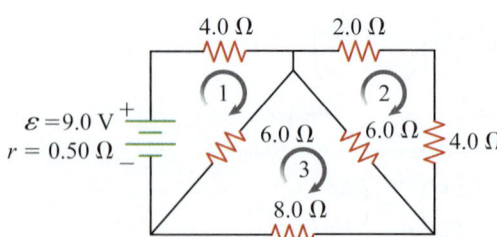

(b) $I = \dfrac{\varepsilon}{R + r} = 0.522\,A$; (c) $V_{battery} = \varepsilon - Ir = 10.4\,V$;
(d) (i) The electric current through the battery will increase, and the potential difference across the battery will decrease; (ii) The electric current through the battery will decrease, and the potential difference across the battery will increase; (e) (i) $I = \dfrac{\varepsilon}{R/2 + r} = 0.923\,A > 0.522\,A$
and $V_{battery} = 9.23\,V < 10.4\,V$;

(ii) $I = \dfrac{\varepsilon}{2R + r} = 0.279\,A < 0.522\,A$

and $V_{battery} = 11.2\,V > 10.4\,V$

47. The effect of connecting an ammeter and a voltmeter:
$\dfrac{\Delta I}{I_0} \times 100\% = 0.05\%$; $\dfrac{\Delta V}{V_0} \times 100\% = -0.000\,020\%$

49. $R_{eq} = \dfrac{3}{2}R$

51. (a) $I_1 = I_3 = 1.00\,A$, and $I_2 = I_4 = I_5 = 0\,A$
(b) $V_{R_1} = 10.0\,V$, $V_{R_2} = V_{R_3} = 0\,V$, $V_{R_4} = 20.0\,V$
(c) If the batteries are real, then R_1 and R_4 will be effectively larger. This will change the results from parts (a) and (b): $I_2 \neq I_4 \neq I_5 \neq 0\,A \Rightarrow V_{R_2} = V_{R_3} \neq 0\,V$; Also, the voltages V_{R_1} and V_{R_4} will be smaller than the values calculated in part (b).

53. (a) 4.75 s; (b) 33.7 s

55. (a) 1.00 s; (b) 5.00 V; (c) 2.30 s; (d) 2.00 mC; (e) 16.0 s; (f) 20.0 V; (g) 36.8 s; (h) 8.00 mC

57. (c) 2313 °C

59. The loops and currents directions required in applying Kirchhoff's laws are indicated in the circuit diagram shown below.

$I_1 = 1.07\,A$; $I_2 = 0.884\,A$; $I_3 = I_5 = 0.189\,A$; $I_4 = 0.695\,A$;
$V_{4\Omega} = 4.29\,V$; $V'_{4\Omega} = 0.758\,V$; $V_{2\Omega} = 0.379\,V$; $V_{6\Omega} = 1.14\,V$;
$V'_{6\Omega} = 4.17\,V$; $V_{8\Omega} = 3.03\,V$; $V_{battery} = 8.46\,V$

61. $\dfrac{5}{6}R$

63. 3.3 Ω

65. A possible example:

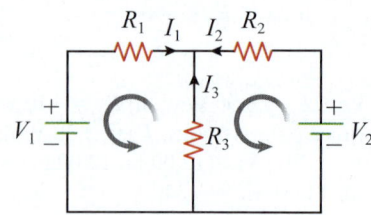

69. (a) 0.79 s; (b) $1.1 \times 10^5\,\Omega$

71. (a) 13 ms; (b) 2.6 kΩ; (c) 5.00 V

Chapter 24: Magnetic Fields and Magnetic Forces

1. Yes

3.

(a)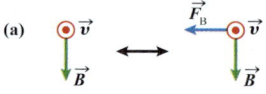

(b)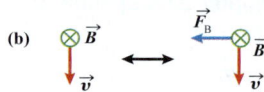

(c)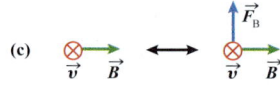

(d)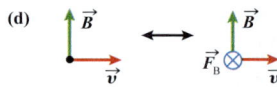

(e) $\vec{v}\, \longrightarrow\, \vec{B} \qquad \vec{F}_B = 0$

(f)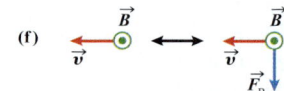

5. Positive

7. b

9. f

11. c

13. a

15. c

17. E > B = D > A = C = F

19. $\vec{F} = 12.8[(-0.600\ fN)\hat{i} + (0.800\ fN)\hat{j} - (0.300\ fN)\hat{k}]$. The magnitude of the force is 13.4 fN.

21. Into the page

23. 0.26 mT

25. (a) 3.6 nm; (b) 1.5 ns

27. $v_{0x}\cos\left(\dfrac{eB}{m}t\right)\hat{i} + v_{0y}\hat{j} + v_{0x}\sin\left(\dfrac{eB}{m}t\right)\hat{k}$

29. 8.00 V/m

31. $r = 1.57 \times 10^4$ m; $f_{cyc} \approx 153$ Hz

33. $B = 87.6$ mT, which is about 1000 times the magnetic field of Earth.

35. (a) 18 N/m; (b) Remains the same

37. (a) North; (b) 0.25 A

39. (a) $F_{AC} = 0$ N; $F_{AB} = F_{BC} = \dfrac{\sqrt{3}}{2}BIa$; (b) $\vec{F}_{net} = 0$ N;

(c) $\tau = \dfrac{\sqrt{3}}{4}BIa^2$

41. (a) $I = 12.5$ A; (b) It will double.

45. (a) 200.0 μT; (b) 20.00 μT; (c) 2.000 μT; Earth's magnetic field is roughly 50 μT.

47. (a) $B(r) = \dfrac{\mu_0 Ir}{2\pi R^2}$ (b) $B(r) = \dfrac{\mu_0 I}{2\pi r}$ (c) See Figure 24-41(b).

49. Inside the inner conductor, $B(r) = \dfrac{\mu_0}{2\pi r}\dfrac{r^2}{a^2}I$

Between the inner and outer conductors, $B(r) = \dfrac{\mu_0}{2\pi r}I$

Inside the outer conductor, $B(r) = \dfrac{\mu_0 I}{2\pi r}\left(\dfrac{c^2 - r^2}{c^2 - b^2}\right)$

Outside the outer conductor, $B(r) = 0$.

51. 8.00 mm

53. (a) $F_{12} = F_{21} = 2.00 \times 10^{-3}$ N; (b) 2.00×10^{-3} N/m;
(c) The answers would remain the same; (d) No

55. (a) No; (b) No; (c) Yes

57. (a) The magnetic field will double; (b) The magnetic field will not change; (c) The magnetic field will increase by a factor of about 10 000 times; (d) The magnetic field will not change.

59. (a) $B_{net} = 51\ \mu$T and $\theta = 69°$; (b) Circular trajectory in the plane perpendicular to the net magnetic field; (c) 0.29 m; (d) No

61. (a) 4.7×10^6 m/s; (b) $T = 0.19\ \mu$s; (c) (i)

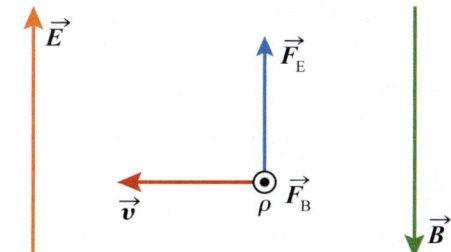

(ii)

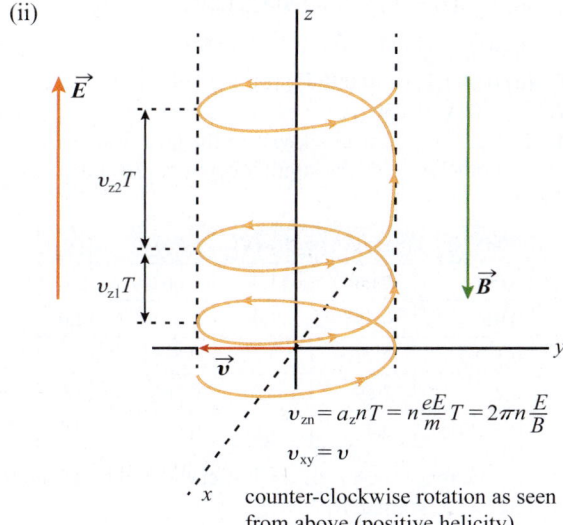

$v_{zn} = a_z nT = n\dfrac{eE}{m}T = 2\pi n\dfrac{E}{B}$

$v_{xy} = v$

counter-clockwise rotation as seen from above (positive helicity)

(iii) An FBD for the electron:

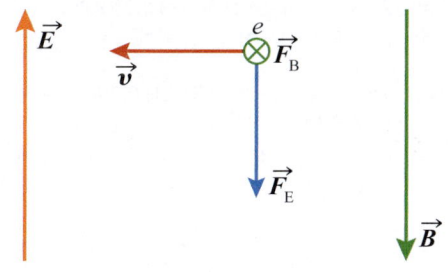

(iv) The trajectory of the electron:

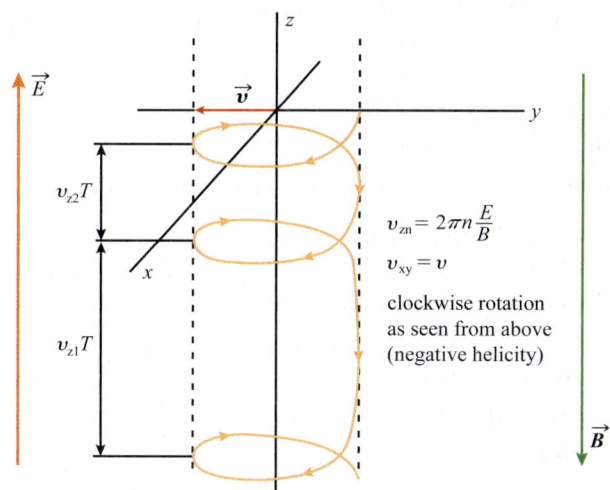

$v_{zn} = 2\pi n\dfrac{E}{B}$

$v_{xy} = v$

clockwise rotation as seen from above (negative helicity)

63. (a) 82.5 A; (b) 41.2 A

65. $F > C = D > A > B = E$

67. (a) $\dfrac{\mu_0 nI}{2}(\cos\theta_2 - \cos\theta_1)$; (b) $\mu_0 nI$

69. (a) 14.9 mT; (b) $\sim 8.33 \times 10^6$ m/s

71. (a) $-(2.75\ \text{N/m})\hat{k}$; (b) $\vec{0}$ N/m; (c) $(2.75\ \text{N/m})\hat{i}$;
(d) $(-2.75\ \text{N/m})(\hat{i} + \hat{k})$; 3.89 N/m

73. (a) $\theta = \dfrac{qB}{m}\Delta t$, which is valid when $\Delta t < \dfrac{\pi m}{2qB}$
(b) 0.64 mm

75. 0.0693 mT. The magnetic field is makes an angle of $120°$ from the positive x-axis.

77. $\dfrac{\mu_r \mu_0 NI}{2\pi r}$

79. $B_A = 11.3\ \mu$T, $B_B = 12.0\ \mu$T, $B_C = 9.00\ \mu$T, $B_D = 15.0\ \mu$T; the orientations of the magnetic fields are shown below.

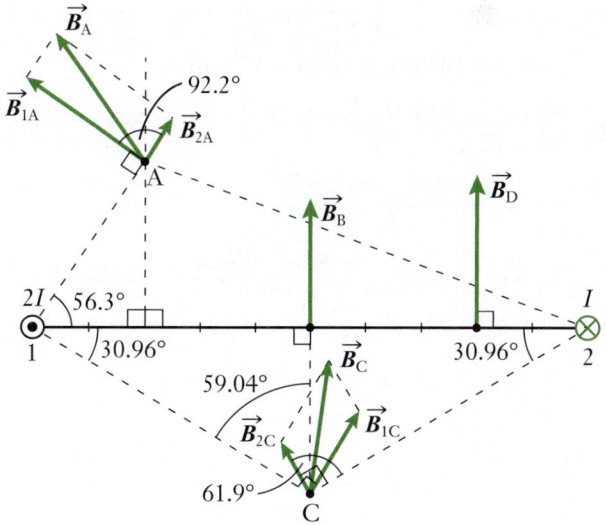

81. $J_0 = \dfrac{B_0}{3\mu_0 R}$

83. (a) $\dfrac{I}{2\pi\ln(b/a)}$; (b) $\dfrac{\mu_0 J_0 \ln(r/a)}{r}$; (c) $\dfrac{\mu_0 I}{2\pi r}$

85.
$$B(z) = \frac{\mu_0 C\pi}{2}\left[\frac{a^3 + 3az^2}{\sqrt{a^2 + z^2}} - 3z^2 \ln(\sqrt{a^2 + z^2} + a) + 3z^2 \ln z\right]$$
in the z-direction.

87. 6.58 μN, 1.28 μN·M

89. 1.98 mm below the wire

91. $B(r) = \dfrac{\mu_0 I_{enc}}{2\pi r}$; (a) 0.250 mT; (b) 0.25 mT; (c) 0 T

93. (a) 7.85 μT; (b) 11.8 μT

95. $B = \dfrac{2\sqrt{2}\mu_0 I}{\pi d}$; the direction of the magnetic field can be found using the right-hand rule: perpendicular to the surface of the loop.

97. $\dfrac{1}{B}\sqrt{\dfrac{2mV}{q}}$

Chapter 25: Electromagnetic Induction

1. h
3. e
5. a
7. d
9. d
11. b
13. a
15. (b) > (a) = (c) > (d) > (e)
17. (b) = (c) > (a) = (d) > (e)
19. (1) = (2) = (3) > (4) = (5)
21. c
23. (a) 0.500 mWb·s^{-1}; (b) 0.625 mWb·s^{-1}; (c) 10.0 mWb·s^{-1}; 12.5 mWb·s^{-1}
25. 28 μT/s
27. 0.59 V
29. 75 mT

33. (a) $F_B(t) = -\dfrac{(B_0 - Ct)L^2}{R}\{B_0 v(t) - C[x_0 + x(t) + tv(t)]\}$,
$F_B(0) = 0.844\ \mu$N; (b) $a(0) = 3.38\ \mu$m/s^2

35. (a) $\dfrac{1}{r} \cdot \dfrac{\pi\mu_0 N^2 R^2 f I_{max} \cos(2\pi ft)}{L}$;

(b) $r \cdot \dfrac{\pi\mu_0 N^2 f I_{max} \cos(2\pi ft)}{L}$;

(c) We denote $E_{max} = \dfrac{\pi\mu_0 N^2 R f I_{max}}{L}$. Then the plot of the function $E(r)/E_{max}$ at an arbitrary time t is as follows:

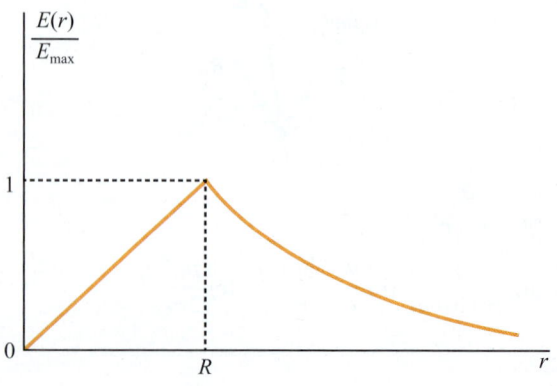

37. 0.100 H
39. (a) 61.7 mH; (b) 0.617 V
41. (a) 113 mV; (b) 11.3 V; (c) $\varepsilon(t) = \varepsilon_{max} \cos(2\pi ft)$
43. (a) 100; (b) $I_1 = 80.0$ mA; $I_{rms1} = 56.6$ mA; $I_2 = 800$ mA; $I_{rms2} = 566$ mA; (c) $I_1 = 80.0$ mA; $I_{rms1} = 56.6$ mA; $I_2 = 720.$ mA; $I_{rms2} = 510$ mA
45. (a) 5.9 mJ; (b) 1.5 mJ; (c) 0
47. 5.5 mV
49. (a) $(0.00900 - 0.208t)$ V; (b) 9.00 mV; (c) Counter-clockwise; (d) 40.5 μN
51. $-3.93 \times (1 + 0.005t)\ \mu$V
53.

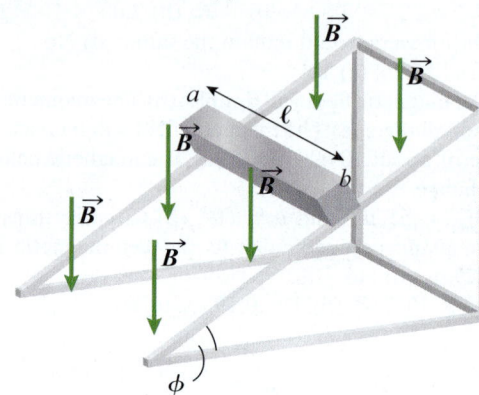

(a) The rods will accelerate at a rate of
$a = g\sin(\phi) - \dfrac{F_B}{m}\cos(\phi)$; (b) $B\ell v \cos(\phi)$;
(c) $v = gT\sin(\phi)(1 - e^{-t/T})$;

$v \approx \left(392\ \dfrac{m}{s}\right)(1 - e^{-t[s]/80.0s})$;

(d) $\varepsilon_{ind}(t) = B\ell gT\sin(\phi)\cos(\phi)(1 - e^{-t/T})$;
$\varepsilon_{ind}(t) = 34.0(1 - e^{-t[s]/80.0s})$V; (e) 392 m/s

55. (a) 0.064 T; (b) 0.096 H; (c) $B = 0.13$ T; $L = 0.096$ H
57. (a) 5.0 A; (b) 30. A
61. 1.26 μA; counter-clockwise in the large loop of radius r_1 and clockwise in the small loop of radius r_2
63. (a)

Time (s)	V_R (V)	V_L (V)	I (A)	ε (V)
0.001	0.0600	11.9	0.00399	12.0
0.01	0.585	11.4	0.0340	12.0
0.1	4.72	7.28	0.315	12.0
0.2	7.59	4.41	0.506	12.0
0.4	10.4	1.62	0.692	12.0

(c) The resistance, r, of the battery would be added to the circuit resistance, R.

65. (a) $\dfrac{B^2\ell^2 v}{R}$; (b) 0.972 μN; (c) $\dfrac{B^2\ell^2 v^2 t}{R}$; (d) 2.92 μJ

67. (a) 36.8 ms; 2.38 A; (b) 0.121 V; (c) 5.55 ms
69. (a) $2\pi NB\ell wf \sin(2\pi ft)$; (b) $\varepsilon_{max} = 57$ kV
73. 3.00 μV
75. (a) 40.0 mT; (b) 13.3 mT; (c) 0 T; (d) 0.294 kJ/m^3; (e) 2.60 MJ; (f) 0.2 kPa

79. (a) 6.000 mH; (b) 3.000 ms;

(c) $I(t) = \dfrac{\varepsilon}{R}(1 - e^{-t/\tau}) = 6.000(1 - e^{-t[\text{ms}]/3.000})$ A

Time (ms)	V_L (V)	ln V_L	I (A)
1.0	8.598	2.152	1.701
1.5	7.278	1.985	2.361
2.0	6.161	1.818	2.919
2.5	5.215	1.652	3.392
3.0	4.414	1.485	3.793
3.5	3.737	1.318	4.132
4.0	3.163	1.152	4.418
4.5	2.678	0.985	4.661
5.0	2.267	0.818	4.867
5.5	1.919	0.652	5.041
6.0	1.624	0.485	5.188
6.5	1.375	0.318	5.313
7.0	1.164	0.152	5.418
7.5	0.985	−0.015	5.507
8.0	0.834	−0.182	5.583

Chapter 26: Alternating Current Circuits

1. Above
5. Decreases
7. Equal but not exceed
9. $\pi/2$
11. d
13. b
15. c
17. 156 V, 311 V
19. (a) 0.1 A; (b) 2760 rad/s; (c) $0.1 \sin(2760t)$
21. (a) 1 mA; (b) 140 kΩ; (c) 71 H
23. (a) 8.0 MΩ; (b) 80. V
25. 3.2×10^6 rad/s
27. 93 V
29. (a) 28 V; (b) −45°
31. (a) 115 Ω; (b) −1.1 rad
33. 6.6×10^3 rad/s
35. 0.024
37. 0.5 W
39. 0.3 mA; ~90°
43. $\dfrac{V_{C_{\max}}}{2}$
45. 32 rad/s, 0.32 μC
47. −45°, 14 V
49. 130 Ω
51. 0.63 μW
53. 57.3°, 5.4 kΩ
55. 0.72 nF to 1.0 nF
57. 100 Ω

Chapter 27: Electromagnetic Waves and Maxwell's Equations

3. a
9. No
11. Perfect reflector
13. 0.3 pA
15. (a) 0.320 μT; (b) 2.00 μT
17. (a) 12 A; (b) 0 A; (c) $43 \times 10^{12}\,\dfrac{\text{V}}{\text{m}\cdot\text{s}}$; $0\,\dfrac{\text{V}}{\text{m}\cdot\text{s}}$; (d) 24 μT; 0 μT
19. (a) y-direction; (b) $-z$-direction; (c) z-direction; (d) y-direction
21. 17 pT; $\vec{E}$ can be oriented in any direction in the yz-plane.
23. (a) 1.57 nm; 1.91×10^{17} Hz; (b) 1500 V/m; (c) $\vec{E}(x,t) = (1500\text{ V/m})\sin(4.00 \times 10^9 x - \omega t)\hat{y}$; (d) x-direction
25. 3.3×10^{-9} T in the $-z$-direction
27. (a) 300 Hz; (b) Along the y-axis, toward increasing y; (c) $\vec{B}(y,t) = (1.7 \times 10^{-8}\text{ T})\sin\left(\dfrac{2\pi}{10^6}y - 600\pi t\right)(-\hat{z})$
29. 3 pm
31. (a) 566 m; (b) 250 m; (c) 3.41 m; (d) 2.88 m
33. 0.68 Hz
35. 5200 W/m²
37. 0.02 W/m²
39. 0.1 V/m, 3×10^{-10} T
41. (a) 1.3×10^{-8} Pa; (b) 6.6×10^{-9} Pa, 8.9 m
43. 4.2 kW
45. 1.5 kW
47. 57 μJ
49. (a) 5.4×10^3 W/m²; (b) 18×10^{-6} Pa
51. 30°
53. (a) 0.75 W/m²; (b) 0.25 W/m²; (c) 0 W/m²
55. 8.8 W/m²
57. 1.25 W/m²
59. (a) 0 W/m²; (b) 0 W/m²; (c) 0 W/m²; (d) 0 W/m²
61. $0.5I_0$; $0.47I_0$; $0.23I_0$
63. 35°
65. 0.66
67. (a) 0.5 μm; (b) The force due to the radiation pressure of the Sun will exceed its gravitational attraction force. Yes, the tail will be longer as the comet approaches the Sun.

Chapter 28: Geometric Optics

1. 1.0 m
5. c
7. Convex mirror
9. (a) Yes
11. 5 images
13. f
15. Concave
17. All the rays will arrive simultaneously.
19. It will result in a fainter image.
23. 34.3 m
27. 31°
29. 24.3 cm
31. (a) The image is not inverted; (c) No, it will not be a viable design.
33. (a) 3.5 cm; (b) Real and inverted
35. (a) −7.62 cm; (b) Virtual, upright, and reduced: $d_i = -7.34$ cm; $M = 0.0367$
37. 7.10 m²
39. About 60.2 cm
41. (a) 1.4; (b) 2.1
43. (a) 45 mm; (b) −0.0076
45. (a) $d_i = -3.45$ cm, $M = 0.138$; (b) Divergent; (c) Virtual, upright, and reduced

47. (a) 0.793; (b) 0.508
49. 36.9°
51. 1.135H
53. Decrease
55. 10. cm from the lens on the same side as the object; no
57. (a) 55 cm; (b) 5.5 cm
59. (a) Through the base of the prism; (b) 80.°; (c) 0 cm
61. (a) 25 cm and 20. cm, respectively; (b) 40. cm behind the second lens; (c) No
63. $\delta = i + r' - \varphi = i - \varphi + \left\{ n\sin\left[\varphi - \sin^{-1}\left(\dfrac{\sin(i)}{n}\right)\right]\right\}$
65. (a) $2F_1$; (b) $4F_1$
67. (b) $f_{\text{mirror}} = 25 \text{ cm} \Rightarrow P_{\text{mirror}} = 4.0 \text{ D}$ and $f_{\text{lens}} = 1.5 \text{ m} \Rightarrow P_{\text{lens}} = 0.67 \text{ D}$; (c) 4.7 D
69. 33 ms
73. 7.9 rad/s
75. (a) 6.09 cm; (b) −7.67; (c) Yes
77. (a) No; (b) No

Chapter 29: Physical Optics

1. d
3. c
5. b
7. d
9. b
11. b
13. b
15. d
17. a
19. b
21. a
23. c
25. c
27. 640. nm; visible red light
29. (a) Yes; (b) Yes; (c) Yes; (d) No
31. $\sim 3.7I_0$
33. $4I_0$ to $16I_0$
35. 28.5 μm
37. 0.0218 cm to 0.0381 cm
39. (a) 4960 nm; (b) Infrared
41. (a) $\sim 99.7\ \mu$m; (b) ~ 4.77 mm
43. 17.5°
45. 0.80 cm
47. 929 lines/mm
49. 227 nm; 453 nm; 680 nm
51. (a) 98 nm; (b) 95 nm
53. (a) 226 nm; (b) 624 nm
55. 29.3 μm
57. 0.131 cm
59. 17 (8 on each side, plus the central one)
61. Within 6.16 km
63. $(2.41 \times 10^{-2})°$
65. (a) From the orange-red end of the visible to the infrared; (b) 1.10 arcsec; (c) ~ 40.6 AU
67. 1.000 18
69. 0.151°
71. 84.6 μm
73. (a) 12.1°; 36.9°; (b) 28.7 mm; (c) 53.6 mm
75. 282
77. (a) 0.000 0107°; (b) 0.0384 arcsec
79. $(4.03 \times 10^{-6})°$

81. 8.4 cm
83. 41.7 μm

Chapter 30: Relativity

1. c
3. c
5. d
7. c
9. d
11. c
13. c
15. d
17. c
19. c
21. c
23. $\sim 3.6 \times 10^{-17}$ s
25. 2.97×10^8 m/s
27. (a) 2.994×10^8 m/s; (b) 2.528×10^{10} s = 801.0 yr; (c) 1.890×10^{17} m = 19.97 ly
29. $t = 0.321\ \mu$s; $x = 31$ m; $y = 450.$ m; $z = 150.$ m
31. (a) The S' frame moves in the positive x direction with a speed of $0.80c$
(b) $t' = 2.0\ \mu$s; $x' = -400$ m; $y' = 150$ m; $z' = 250$ m
(c) $\begin{pmatrix} ct \\ x \\ y \\ z \end{pmatrix} = \begin{pmatrix} \frac{5}{3} & \frac{4}{3} & 0 & 0 \\ \frac{4}{3} & \frac{5}{3} & 0 & 0 \\ 0 & 0 & 1 & 0 \\ 0 & 0 & 0 & 1 \end{pmatrix} \begin{pmatrix} ct' \\ x' \\ y' \\ z' \end{pmatrix}$
(d) $t = 5.87\ \mu$s; $x = 1660.$ m; $y = 390.$ m; $z = 560.$ m
33. (a) $(\Delta s)^2 = 1.34 \times 10^6$ m^2; (b) Time-like
35. 1.33 μs
37. (a) 2.03 μs; (b) $0.164c$
39. (a) 4.12×10^{-13} J; (b) 3.30×10^{-13} J
41. $\sqrt{3}m_0c$
43. 1.25×10^{-13} J; 0.782 MeV
45. 2.99×10^8 m/s
47. 8.27×10^7 m/s
49. 4.16×10^{15} Hz
51. (a) Microwave; (b) Toward us; (c) −1200 m/s (to 2 significant digits; minus sign means direction toward us)
53. (a) 0.856; (b) 902 nm
55. 35 km
57. 9.46×10^{15} m
59. (a) 1.60×10^9 s = 50.6 yr; (b) 1.62×10^9 s = 51.4 yr; (c) 2.80×10^8 s = 8.86 yr
61. (a) 2.6×10^8 m/s; (b) 0.23 s
63. (a) 1.06×10^8 m/s; (b) 1.01×10^8 m/s
67. (a) 4.28×10^9 kg/s; (b) 2.0×10^{19} kg/yr; (c) 9.7 billion yr
69. (a) 14.9 ms; (b) 2.00×10^8 m/s
71. 0.92 μs
75. 4.01×10^{11} m
77. (a) The cosmic ray has higher energy by a factor of 1.4 million; (b) The speed is only 4.4×10^{-19}% less than the speed of light; (c) 5.3×10^{-9} kg·m·s^{-1}; (d) 280 000 yr $(8.8 \times 10^{12}$ s); (e) 830 s
79. $\dfrac{m_0}{2}\sqrt{\dfrac{(3c)^2 - V^2}{c^2 - V^2}}$

85. (a) 2.50 kg; (b) From 0.10c to 0.80c; (c) 7.35×10^8 kg·m/s; (d) 1.50×10^{17} J; (e) 3.75×10^{17} J

Chapter 31: Fundamental Discoveries of Modern Physics

11. 4.823×10^7 C/kg

13. (a) Higher intensity (black line) is at a higher temperature; (b) No

15. (a) 533 333.3 m^{-1}; (b) $n_1 = 4$, $n_2 = 3$; (c) Infrared

17. (a) 6.43×10^7 m/s; (b) 1.16×10^5 V/m; (c) 1.30 mm

19. (a) 1.50×10^4 V/m; (b) 2.40×10^{-15} N; (c) 8.94×10^{-30} N; (d) Yes

21. (a) 0.147%; (b) 0.127%

23. (c) 8.6×10^{-16} J

25. (a) 1.5×10^{-13} m; (b) It will pass through the nucleus.

27. (a) 1.602×10^{-16} J; (b) 1.875×10^7 m/s; (c) 1.00×10^3 eV

29. (a) 3.65×10^{-19} J; (b) 5.51×10^{14} Hz; (c) 544 nm

31. (a), (b)

Wavelength (nm)	Stopping Potential (v)	Frequency (Hz)	K_{max} (J)
260	0.46	1.154×10^{15}	7.369×10^{-20}
250	0.66	1.200×10^{15}	1.057×10^{-19}
240	0.87	1.250×10^{15}	1.394×10^{-19}
230	1.09	1.304×10^{15}	1.746×10^{-19}

(d) 7×10^{-34} J s; (e) 4.37 eV

33. 2.19×10^6 m/s

35. 1.52×10^{-16} s

37. (a) 3.0×10^2 keV; (b) 48 keV

39. 1.9×10^{-18} C

41. 7.4×10^{-13} m

43. 3.7×10^5 m/s

45. 18 cm, 4.2 cm

47. $\dfrac{2.82\,kT}{h}$

49. $E > 1.988 \times 10^{-8}$ J, or, equivalently $E > 124.1$ GeV

53. About 3.27×10^{21}

Chapter 32: Introduction to Quantum Mechanics

1. No

3. Yes

5. Does not exist

7. (a) No; (b) Yes; (c) Yes

9. $d < b < a < c$

13. (a) 1.17×10^{-12} m; (b) 3.46×10^{-13} m

15. 4.18×10^{-6} eV

17. 5.3×10^{-25} kg·m/s; 1.5×10^{-19} J

19. 2.6×10^{-31} m

21. (a) 3.64×10^{-23} J; (b) 3.64×10^{-12} J

23. 6×10^{-10} s

25. m$^{-1/2}$

27. 6.02×10^{-18} J, 2.41×10^{-17} J, 5.42×10^{-17} J

29. (a) 7.2×10^{-23} J; (b) 1.9×10^{-22} J; (c) 1.2×10^{-22} J; (d) 2.4×10^{-17} J

31. 0.5

33. 1.8×10^{-17} J, 11 nm

37. $A = \sqrt{\dfrac{15}{512}}$; $\psi(t) = \begin{cases} \sqrt{\dfrac{15}{512}}(4 - x^2) & \text{for } -2 \leq x \leq 2 \\ 0 & \text{everywhere else} \end{cases}$

41. 0.62

43. (a) 80 eV, 180 eV, 320 eV; (b) 1.4×10^{-10} m; (c) 2.1×10^{-8} m

45. (a) No; (b) Yes; (c) Yes; (d) No; (e) Yes

49. 1.9 eV

53. 1.549×10^{-19} J, 1283 nm; 1.816×10^{-19} J, 1095 nm; 1.976×10^{-19} J, 1006 nm

55. 9

57. (a) $-4, -3, -2, -1, 0, 1, 2, 3, 4$; (b) $E_5 > E_6 > E_7 > \cdots$

59. 0.986

61. $54.7°, -54.7°, -125.3°, 125.3°$

63. (a) 2, 1; (b) $\sqrt{6}\hbar, \sqrt{2}\hbar$; for $j = 2$: $J_z = 2\hbar, \hbar, 0, -\hbar, -2\hbar$; for $j = 1$: $J_z = \hbar, 0, -\hbar$

65. $\dfrac{3}{2}, \dfrac{5}{2}, \dfrac{7}{2}, \dfrac{9}{2}, \dfrac{11}{2}$

Chapter 33: Introduction to Solid-State Physics

1. Decrease

3. c

5. d

7. b

9. Silicon

11. (a) a^3; (b) $\dfrac{a^3}{4}$

13. 94.9%

15. 3 electrons/atom

17. (a) Positive; (b) Negative; (c) Positive

19. (a) 1.00×10^{13}; (b) 1.00×10^{-13} J/s

21. 7.1×10^{-21}

23. 0.22 μg

25. 4.0×10^5

27. 0.07 W

29. (a) 3.4×10^{-17} F; (b) 510

31. 90%

33. (a) 1.0×10^3 V/m; (b) 1.8×10^{14} m/s^2; (c) $1.8 \times 10^{14}t$ m/s; (d) $0.014t^2$ J; (e) $0.028t$ J/s; (f) 3.6×10^{-12} s; (g) 1.1 nm

35. 1.9×10^{23} electrons/cm^3

37. (a) 0.03; (b) 9×10^{-6}

39. 2600 K; (b) 9.7 eV

41. (a) 7.50×10^{-18} m (b) 2.50×10^{-8} m (c) No

43. $10^{14}\,\Omega$

45. $1\,\Omega \cdot$m

49. 41 eV

Chapter 34: Introduction to Nuclear Physics

5. 8

7. 1.2×10^{36}

9. 2

15. 9.0×10^{13} J; 5.6×10^{26} MeV

17. (a) 1.31×10^{16} J (b) 8.146×10^{34} eV

19. 0.78 MeV/c

21. $m_C = 12.011137$ u $= 1.9944 \times 10^{-26}$ kg

23. (a) B(2, 1) = 2.2 MeV, B$_{per}$(2, 1) = 1.1 MeV/nucleon
(b) B(4, 2) = 28.3 MeV, B$_{per}$(4, 2) = 7.1 MeV/nucleon
(c) B(6, 3) = 32.0 MeV, B$_{per}$(6, 3) = 5.3 MeV/nucleon
(d) B(56, 26) = 492.2 MeV, B$_{per}$(56, 26) = 8.8 MeV/nucleon
(e) B(208, 82) = 1636.4 MeV, B$_{per}$(208, 82) = 7.9 MeV/nucleon

25. (a) 2.013554 u; (b) 11.996712 u; (c) 15.990531 u;
(d) 55.920691 u; (e) 238.000369 u

27. (a) 89.87 MeV; (b) 85.82 MeV; (c) 722.44 MeV;
(d) 1611.77 MeV; (e) 1790.76 MeV

29. 38.3×10^{-12} s^{-1}

31. 4.5×10^7 Bq

33. 6.06 μg

35. 1.04×10^{10} s

37. (a) 1000 Bq; (b) 7 Bq

39. 580 s

41. (a)

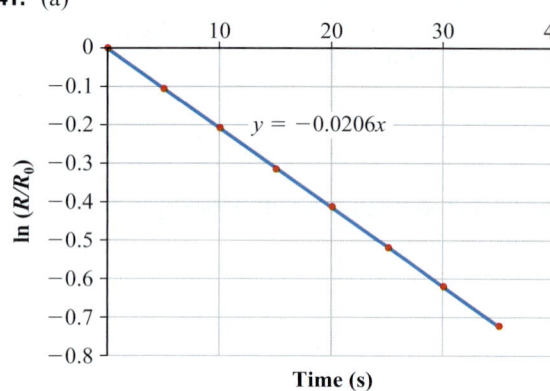

(b) 0.0206 s^{-1}; (c) 33.6 s

43. (a) Not possible; (b) Not possible; (c) Possible;
(d) Not possible; (e) Not possible

45. (a) $Q = (m_n + m_X - m_Y - m_{He})c^2$; (b) 4.54 MeV

47. No; $Q = -7.27$ MeV
Since the Q-value is negative, carbon-12 cannot decay
through this channel spontaneously.

49. (a) $^{32}_{14}$Si $\rightarrow$ $^{28}_{12}$Mg + $^{4}_{2}$He; (b) $^{56}_{25}$Mn $\rightarrow$ $^{52}_{23}$V + $^{4}_{2}$He;
(c) $^{60}_{26}$Fe $\rightarrow$ $^{56}_{24}$Cr + $^{4}_{2}$He

51. (a) $^{52}_{25}$Mn $\rightarrow$ $^{52}_{24}$Cr + e$^+$ + v; (b) $^{55}_{27}$Co $\rightarrow$ $^{55}_{26}$Fe + e$^+$ + v;
(c) $^{59}_{28}$Ni $\rightarrow$ $^{59}_{27}$Co + e$^+$ + v

53. (a) $^{3}_{1}$H $\rightarrow$ $^{3}_{2}$He + e$^-$ + $\bar{v}$; (b) 18.6 keV

55. 6 α-particles and 4 β-particles

57. (a) 179.6 MeV; (b) 179.9 Mev

59. (a)   B$_{per}$(2, 1) = 1.1 MeV/nucleon,
       B$_{per}$(6, 3) = 5.3 MeV/nucleon,
       B$_{per}$ (10, 5) = 6.5 MeV/nucleon,
       B$_{per}$(14, 7) = 7.5 MeV/nucleon,
       B$_{per}$(180, 73) = 8.0 MeV/nucleon;
(b) (i)   B$_{per}$(4, 2) = 7.1 MeV/nucleon
(ii)   B$_{per}$ (4, 8) = 7.1 MeV/nucleon
(iii)   B$_{per}$(12, 6) = 7.7 MeV/nucleon
(iv)   B$_{per}$(16, 8) = 8.0 MeV/nucleon
(v) B$_{per}$(182, 74) = 8.0 MeV/nucleon

61. (a) 9.0 MeV; (b) No

Chapter 35: Introduction to Particle Physics

1. (a) False; (b) True; (c) False; (d) False; (e) False

3. (a) True; (b) False; (c) False; (d) False; (e) True; (f) True

5. (a) False; (b) True, if the electron decelerates;
(c) True; (d) False; (e) False

7. 400 MeV

9. Electron, neutrino, quark

11. Photon, Z^0, gluon

13. Infinite

15. uds: $m = 1.11 \times 10^3$ GeV/c^2
uus: $m = 1.10 \times 10^3$ GeV/c^2
dds: $m = 1.12 \times 10^3$ GeV/c^2
uss: $m = 1.30 \times 10^3$ GeV/c^2
dss: $m = 1.31 \times 10^3$ GeV/c^2
sss: $m = 1.50 \times 10^3$ GeV/c^2

17.

	Charge	Baryon Number	Spin
(a) uus	+1	+1	½ or ³⁄₂
(b) uds	0	+1	½ or ³⁄₂
(c) dds	−1	+1	½ or ³⁄₂
(d) uss	0	+1	½ or ³⁄₂
(e) dss	−1	+1	½ or ³⁄₂

19.

	Charge	Baryon Number	Spin	Mass Sum (MeV/c^2)	Mass Rank
(a) ttb	+1	+1	½ or ³⁄₂	346 200	5
(b) tts	+1	+1	½ or ³⁄₂	342 125	4
(c) tbs	0	+1	½ or ³⁄₂	175 325	3
(d) tbd	0	+1	½ or ³⁄₂	175 205	2
(e) tuu	+2	+1	½ or ³⁄₂	141 005	1

21. $m_{s\bar{u}} = C(125.0 + 2.5) = 127.5C$ MeV/c^2, where C is the
proportionality constant.
$m_{s\bar{d}} = C(125.0 + 5.0) = 130.0C$ MeV/c^2
$m_{s\bar{s}} = C(125.0 + 125.0) = 250.0C$ MeV/c^2
$m_{u\bar{s}} = C(2.5 + 125.0) = 127.5C$ MeV/c^2
$m_{d\bar{s}} = C(5.0 + 125.0) = 130.0C$ MeV/c^2
$m_{\bar{c}d} = C(1300.0 + 5.0) = 1305.0C$ MeV/c^2
$\Rightarrow m_{s\bar{u}} = m_{u\bar{s}} < m_{s\bar{d}} = m_{d\bar{s}} < m_{s\bar{s}} < m_{\bar{c}d}$

23 (a) C: $+1 \neq 0 - 1 + 0 \Rightarrow$ conservation of charge violated
(b) C: $0 \neq +1 + 0 \Rightarrow$ conservation of charge violated
(c) conservation of energy violated
(d) L: $-1 \neq -1 + 1 \Rightarrow$ conservation of lepton number violated
(e) B: $+1 \neq 0 + 0 \Rightarrow$ conservation of baryon number violated
(f) B: $+1 + 1 \neq +1 + 0 \Rightarrow$ conservation of baryon number violated
(g) L: $-1 + 0 \neq 0 + 1 \Rightarrow$ conservation of lepton number violated
(h) B: $+1 \neq 0 + 0 \Rightarrow$ conservation of lepton number violated

25. The production of a single photon in electron-positron annihilation is impossible, because linear momentum would not be conserved.

27. 1.8 MeV

29. $E_\pi = 931.5$ MeV, $p_\pi = 921.0$ MeV/c

31. Law of conservation of baryon number

35. (a) $p = 69.8$ MeV/c, $E_\nu = 69.8$ MeV, $E_e = 69.8$ MeV
(b) $p = 29.9$ MeV/c, $E_\nu = 29.9$ MeV, $E_\mu = 109.7$ MeV

37.

ν_μ μ^+

u $\bar{d}$

39.

e^-

e^{*-}

e^-

41. (a) (i) $R = 7.3 \times 10^{-16}$ m; (ii) $R = 1.3 \times 10^{-16}$ m;
(ii) $R = 9.6 \times 10^{-17}$ m; (b) No

43. (a) (i) 10 cm; (ii) 30 cm; (iii) 40 cm; (b) Increased;
(c) (i) 70 cm (ii) 70 cm (iii) 70 cm

45. (a) 2.28×10^8 m/s; (b) 50.0 ns; (c) 11.4 m

47. Circumference $> 10^{32}$ m; building a synchrotron of this size is impossible

Appendix B
SI UNITS AND PREFIXES

Base SI Units

Unit name	SI symbol	Measures
ampere	A	electric current
candela	cd	luminous intensity
kelvin	K	temperature
kilogram	kg	mass
metre	m	length
mole	mol	amount of a substance
second	s	time

The Most Common Derived SI Units

Unit name	SI symbol	Measures	In base units
coulomb	C	electric charge	$A \cdot s$
farad	F	electric capacitance	C/V
henry	H	electromagnetic inductance	$V \cdot s/A$
hertz	Hz	frequency	$/s$
joule	J	energy and work	$N \cdot m$
newton	N	force	$kg \cdot m/s^2$
ohm	Ω	electric resistance	V/A
pascal	Pa	pressure	N/m^2
radian	rad	angle	dimensionless
tesla	T	magnetic field strength	$V \cdot s/m^2$
volt	V	electric potential difference	J/C
watt	W	power	J/s

Common Prefixes

Prefix	Abbreviation	Power
zepto-	z	10^{-21}
atto-	a	10^{-18}
femto-	f	10^{-15}
pico-	p	10^{-12}
nano-	n	10^{-9}
micro-	μ	10^{-6}
milli-	m	10^{-3}
centi-	c	10^{-2}
deci-	d	10^{-1}
kilo-	k	10^{3}
mega-	M	10^{6}
giga-	G	10^{9}
tera-	T	10^{12}
peta-	P	10^{15}
exa-	E	10^{18}
zetta-	Z	10^{21}

Appendix C
GEOMETRY AND TRIGONOMETRY

Arc Length and Angle

$s = $ arc length $= r\theta$

Angle θ is in radians.

2π rad $= 360°$

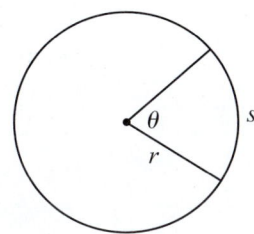

Trigonometric Functions
(right-angled triangle)

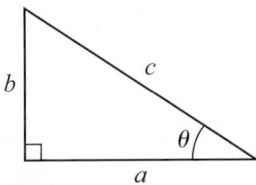

Pythagoras' Theorem: $a^2 + b^2 = c^2$

$$\sin\theta = \frac{\text{opposite}}{\text{hypotenuse}} = \frac{b}{c} \qquad \csc\theta = \frac{\text{hypotenuse}}{\text{opposite}} = \frac{c}{b}$$

$$\cos\theta = \frac{\text{adjacent}}{\text{hypotenuse}} = \frac{a}{c} \qquad \sec\theta = \frac{\text{hypotenuse}}{\text{adjacent}} = \frac{c}{a}$$

$$\tan\theta = \frac{\text{opposite}}{\text{adjacent}} = \frac{b}{a} \qquad \cot\theta = \frac{\text{adjacent}}{\text{opposite}} = \frac{a}{b}$$

General Triangle

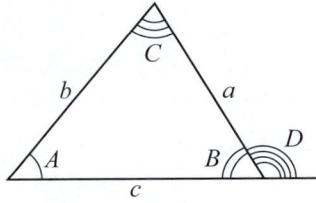

Angles: A, B, C Opposite sides: a, b, c

$A + B + C = 180°$ $A + C = D$

Law of cosines $c^2 = a^2 + b^2 - 2ab\cos C$

Law of sines $\dfrac{\sin A}{a} = \dfrac{\sin B}{b} = \dfrac{\sin C}{c}$

Trigonometric Identities

$$\tan\theta = \frac{\sin\theta}{\cos\theta} \qquad \sin^2\theta + \cos^2\theta = 1$$

$$\sin(A \pm B) = \sin A \cos B \pm \cos A \sin B$$

$$\cos(A \pm B) = \cos A \cos B \mp \sin A \sin B$$

$$\tan(A \pm B) = \frac{\tan A \pm \tan B}{1 \mp \tan A \tan B}$$

$$\sin(\theta \pm \pi/2) = \pm\cos\theta \qquad \cos(\theta \pm \pi/2) = \mp\sin\theta$$

$$\sin(\theta \pm \pi) = -\sin\theta \qquad \cos(\theta \pm \pi) = -\cos\theta$$

$$\cos(-\theta) = \cos(\theta) \qquad \sin(-\theta) = -\sin(\theta)$$

$$\sin(2\theta) = 2\sin\theta\cos\theta \qquad \cos(2\theta) = \cos^2\theta - \sin^2\theta$$

$$\tan(2\theta) = \frac{2\tan(\theta)}{(1 - \tan^2(\theta))}$$

$$\sin A \pm \sin B = 2\sin\left(\frac{A \pm B}{2}\right)\cos\left(\frac{A \mp B}{2}\right)$$

$$\cos A + \cos B = 2\cos\left(\frac{A + B}{2}\right)\cos\left(\frac{A - B}{2}\right)$$

$$\cos A - \cos B = -2\sin\left(\frac{A + B}{2}\right)\sin\left(\frac{A - B}{2}\right)$$

Trigonometric Expansions

$$\sin x = x - \frac{x^3}{3!} + \frac{x^5}{5!} - \frac{x^7}{7!} + \ldots \text{ (x is in radians)}$$

$$\cos x = 1 - \frac{x^2}{2!} + \frac{x^4}{4!} - \frac{x^6}{6!} + \ldots \text{ (x is in radians)}$$

$$\tan x = x + \frac{x^3}{3} + \frac{2x^5}{15} + \frac{17x^7}{315} + \ldots \text{ (x is in radians)}$$

Small-angle Approximation

If $x \ll 1$ rad, then $\sin x \approx \tan x \approx x$. The small-angle approximation is generally good for $x < 0.2$ rad.

Appendix D
KEY CALCULUS IDEAS

Derivatives

Given a function $f(x)$, the derivative is defined to be

$$f'(x) = \frac{df}{dx} = \lim_{h \to 0} \frac{f(x+h) - f(x)}{h}$$

The derivative can be interpreted as the instantaneous rate of change of the function at x. We can also think of the derivative as the slope of the tangent to the curve as shown in the figure.

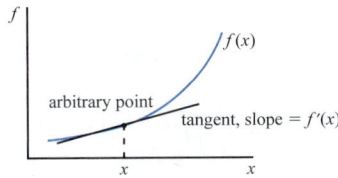

From a physics perspective, if the function $f(t)$ describes the position of an object, the derivative is the velocity of the object.

Basic Properties

For two functions $f(x)$ and $g(x)$ whose derivatives exist, the following properties apply:

1. $\dfrac{d}{dx}(cf(x)) = c\dfrac{df}{dx}; \quad c = \text{constant}$

2. $\dfrac{d}{dx}(f \pm g) = \dfrac{df}{dx} \pm \dfrac{dg}{dx}$

3. $\dfrac{d}{dx}(fg) = f\dfrac{dg}{dx} + g\dfrac{df}{dx}$ (Product Rule)

4. $\dfrac{d}{dx}\left(\dfrac{f}{g}\right) = \dfrac{\dfrac{df}{dx}g - f\dfrac{dg}{dx}}{g^2}$ (Quotient Rule)

5. $\dfrac{d}{dx}f(g(x)) = \dfrac{df}{dg}\dfrac{dg}{dx}$ (Chain Rule)

Common Derivatives

1. $\dfrac{d}{dx}x = 1$

2. $\dfrac{d}{dx}x^n = nx^{n-1}$

3. $\dfrac{d}{dx}\sin x = \cos x$

4. $\dfrac{d}{dx}\cos x = -\sin x$

5. $\dfrac{d}{dx}\tan x = -\sec^2 x$

6. $\dfrac{d}{dx}\sec x = \sec x \tan x$

7. $\dfrac{d}{dx}\csc x = -\csc x \cot x$

8. $\dfrac{d}{dx}\cot x = -\csc^2 x$

9. $\dfrac{d}{dx}\sin^{-1} x = \dfrac{1}{\sqrt{1-x^2}}$

10. $\dfrac{d}{dx}\cos^{-1} x = -\dfrac{1}{\sqrt{1-x^2}}$

11. $\dfrac{d}{dx}\tan^{-1} x = \dfrac{1}{\sqrt{1+x^2}}$

12. $\dfrac{d}{dx}a^x = a^x \ln(a)$

13. $\dfrac{d}{dx}e^x = e^x$

14. $\dfrac{d}{dx}\ln x = \dfrac{1}{x}$

Integrals

Definition: Suppose that $f(x)$ is continuous on the interval $[a, b]$. We divide $[a, b]$ into N subintervals of width $\Delta x = \dfrac{b-a}{N}$, with centre $x_i = \left(i + \dfrac{1}{2}\right)\Delta x$. We then define the definite integral as follows:

$$\int_a^b f(x)dx = \lim_{N \to \infty} \sum_{i=0}^{N} f(x_i)\Delta x$$

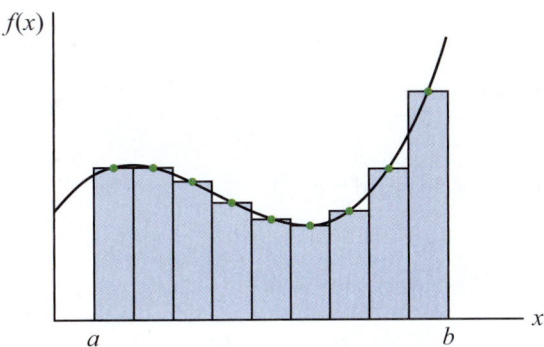

We interpret the integral as the area bounded by the function $f(x)$ and the x-axis, where the areas above the axis are positive, and the areas below the axis are negative.

We define the antiderivative of $f(x)$ as $g(x)$ such that $\dfrac{dg(x)}{dx} = f(x)$.

The indefinite integral is defined as $\int f(x)\,dx = g(x) + C$, where

$$\frac{dg(x)}{dx} = f(x) \quad \text{and} \quad C = \text{constant}$$

Some Properties of Integrals

$$\int (f(x) \pm g(x))\,dx = \int f(x)\,dx \pm \int g(x)\,dx$$

$$\int c f(x)\,dx = c \int f(x)\,dx; \quad c = \text{constant}$$

$$\int_a^b f(x)\,dx = - \int_b^a f(x)\,dx$$

$$\int u \frac{dv}{dx}\,dx = uv - \int v \frac{du}{dx}\,dx$$

Some Indefinite Integrals

$$\int b\,dx = bx + c; \quad b, c \text{ are constants}$$

$$\int x^n\,dx = \frac{1}{n+1} x^{n+1} + c$$

$$\int \frac{dx}{x} = \ln(x) + c$$

$$\int e^x\,dx = e^x + c$$

$$\int \sin x\,dx = -\cos x + c$$

$$\int \cos x\,dx = \sin x + c$$

$$\int \tan x\,dx = \ln|\sec x| + c$$

$$\int \sin^2 x\,dx = \frac{1}{2}x - \frac{1}{4}\sin(2x) + c$$

$$\int \frac{dx}{\sqrt{x^2 + a^2}} = \ln\left(x + \sqrt{x^2 + a^2}\right) + c$$

$$\int \frac{x\,dx}{(x^2 + a^2)^{3/2}} = -\frac{1}{(x^2 + a^2)^{1/2}} + c$$

$$\int \frac{dx}{(x^2 + a^2)^{3/2}} = \frac{x}{a^2(x^2 + a^2)^{1/2}} + c$$

$$\int \frac{x\,dx}{x + a} = x - a\ln(x + a) + c$$

$$\int e^{-ax}\,dx = -\frac{1}{a}e^{-ax} + c$$

$$\int x e^{-ax}\,dx = -\frac{1}{a^2}(ax + 1)e^{-ax} + c$$

$$\int x^2 e^{-ax}\,dx = -\frac{1}{a^3}(a^2 x^2 + 2ax + 2)e^{-ax} + c$$

Surface Integrals

A surface integral is the integration a function f over some surface.

$$\Phi = \iint_{\text{surface}} f\,dA$$

The integration bounds are determined by the surface, and dA is a differential element on that surface. For example, for a flat rectangular surface, we could use Cartesian coordinates with dA replaced with $dx\,dy$.

We call the surface of a rectangle open. If we use a surface such as a sphere that closes on itself, it is called a closed surface. The symbol for the surface integral over a closed surface has a circle joining the two integral signs.

$$\Phi = \oiint_{\text{closed surface}} f\,dA$$

We often use surface integrals in situations where the function to be integrated is the vector dot product of two vectors (in the case below $d\vec{A}$ is a differential surface element with direction pointing outward from the closed surface).

$$\Phi_{\text{E}} = \oiint_{\text{closed surface}} \vec{E} \cdot d\vec{A}$$

Line Integrals

The integration of functional values along some curved (or straight) path is called a line integral. We often compute the line integral from the dot product of a vector quantity with the path element $d\vec{\ell}$. The endpoints of the path are denoted a and b.

$$U_b - U_a = -\int_a^b \vec{F}_{\text{field}} \cdot d\vec{\ell}$$

If the path is closed, meaning that we begin and end at the same point, we use a circle on the integral symbol.

$$\oint \vec{B} \cdot d\vec{\ell}$$

Partial Derivatives

We sometimes need to take the derivative of a function that depends on more than one variable. For example, the displacement amplitude of a wave depends on both position and time, and in one dimension is thus a function of both x and t. When we need to find the derivatives of such functions, we can use partial derivatives.

Consider a function $f(x, y, z)$. We denote the partial derivatives by $\frac{\partial f}{\partial x}$, $\frac{\partial f}{\partial y}$ and $\frac{\partial f}{\partial z}$ (pronounced die f by die x, etc.). We calculate a partial derivative by differentiating with respect to the variable of interest while treating all other variables as constants. For example, for the function $f(x, y, z) = ax^2 + bxy + cz$, the partial derivatives are given by

$$\frac{\partial f}{\partial x} = 2ax + by, \quad \frac{\partial f}{\partial y} = bx, \quad \text{and} \quad \frac{\partial f}{\partial z} = c.$$

Appendix E

USEFUL MATHEMATICAL FORMULAS AND MATHEMATICAL SYMBOLS USED IN THE TEXT AND THEIR MEANING

Lengths, Areas, and Volumes

Rectangle

$$\text{Area} = ab$$

$$\text{Perimeter} = 2(a + b)$$

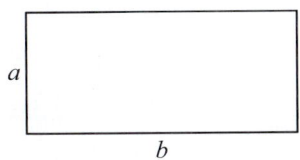

Rectangular Cuboid

$$\text{Surface Area} = 2(ab + bc + ac)$$

$$\text{Volume} = abc$$

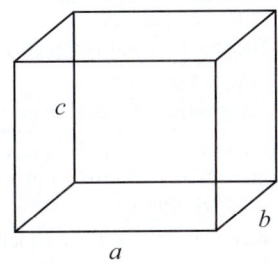

Triangle

$$\text{Area} = \frac{1}{2}bh$$

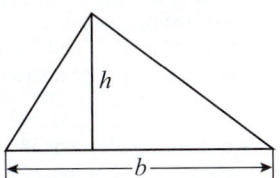

Circle

$$\text{Circumference} = 2\pi r$$

$$\text{Area} = \pi r^2$$

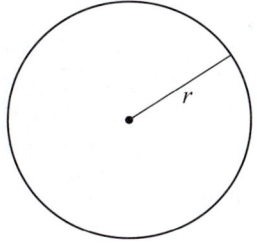

Sphere

$$\text{Surface area} = 4\pi r^2$$

$$\text{Volume} = \frac{4}{3}\pi r^3$$

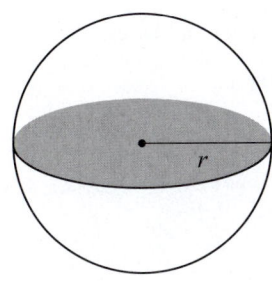

Cylinder

$$\text{Surface area} = 2\pi rh + 2\pi r^2$$

$$\text{Volume} = \pi r^2 h$$

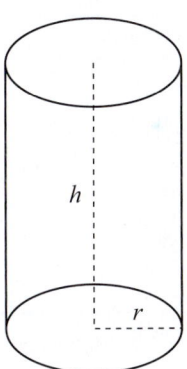

Quadratic Equation

An equation of the form $ax^2 + bx + c = 0$ has as its solution

$$x_{1,2} = \frac{-b \pm \sqrt{b^2 - 4ac}}{2a}$$

Linear Equation

A linear equation is of the form $y = mx + b$, where b is the intercept of the line with the y-axis, and m is the slope of the line;

$$m = \frac{y_2 - y_1}{x_2 - x_1} = \frac{\Delta y}{\Delta x}$$

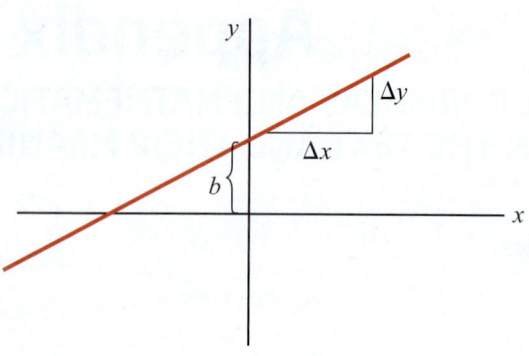

Expansions

$$(1 + x)^n = 1 + \frac{nx}{1!} + \frac{n(n-1)x}{2!} + \cdots \quad (x^2 \ll 1)$$

$$e^x = 1 + x + \frac{1}{2!}x^2 + \frac{1}{3!}x^3 + \cdots$$

$$\ln(1 + x) = x - \frac{1}{2}x^2 + \frac{1}{3}x^3 - \frac{1}{4}x^4 \quad (|x| < 1)$$

$$\sin\theta = \theta - \frac{1}{3!}\theta^3 + \frac{1}{5!}\theta^5 - \cdots$$

$$\cos\theta = 1 - \frac{1}{2!}\theta^2 + \frac{1}{4!}\theta^4 - \cdots$$

Logarithms

These identities apply to all bases:

$$\log a + \log b = \log (ab)$$

$$\log a - \log b = \log\left(\frac{a}{b}\right)$$

$$\log a^b = b \log a$$

Mathematical Symbols Used in the Text and Their Meaning

Symbol	Meaning	Sample expression	How it is read
$+$	Addition sign	$7 + 14 = 21$	7 plus 14 equals 21 The sum of 7 and 14 equals 21.
$\cdot$ $\times$	Multiplication signs	$7 \cdot 3 = 21$ $7 \times 3 = 21$	7 times 3 equals 21 The product of 7 and 3 equals 21
$\cdot$	Dot product sign	$\vec{A} \cdot \vec{B} = 32$ $(1, 2, 3) \cdot (4, 5, 6) = 32$	Vector A dot vector B equals 32 The dot (scalar) product of vector A and vector B equals 32.
$\times$	Cross product sign	$\vec{A} \times \vec{B} = \vec{C}$ $(1, 2, 0) \times (3, 4, 0) = (0, 0, -2)$	Vector A cross vector B equals vector C The cross (vector) product of vector A and vector B equals vector C
$-$	Subtraction sign Minus sign	$7 - 3 = 4$ $-(3 + 5) = -8$	7 minus 3 equals 4 The difference of 7 and 3 equals 4 Negative of three plus five equals minus 8 The opposite of three plus five equals minus 8
$\div$ $/$ $-$	Division signs	$21 \div 3 = 7$ $^3/_4 = 3/4 = 0.75$ $\dfrac{3}{4} = 0.75$	21 divided by 3 equals 7 The quotient of 21 and 7 equals 3 3 over 4 equals three quarters equals point 75
$\%$	Percent symbol	$58\% \equiv 0.58$ $30\%(\$200) = \$60$	58 percent equals point 58 30 percent of \$200 equals \$60
$:$	Ratio	$1 : 5 = 100 : 500$	1 is to 5 as 100 is to 500
$\pm$ $\mp$	Plus/minus sign Minus/plus sign	$50\ \Omega \pm 10\%$ $(a \pm b)^2 = a^2 \pm 2ab + b^2$ $\cos(\alpha \pm \beta) = \cos^2\alpha \mp \sin^2\beta$	Fifty Ohms plus minus 10 percent a plus minus b squared equals... cos alpha plus minus beta equals cos squared alpha minus plus sin squared beta
$=$	Two values are the same	$-(-7) = 7$ $x + 3y = 57$	The opposite of minus seven equals seven
$\equiv$	A definition	$\sqrt{x} \equiv x^{1/2}$	Square root of x is defined as x to the power of one half
$\neq$	Two values are different	$-1 + 5 \neq -6$	Minus 1 plus 5 is not equal to minus 6
$\approx$	Two values are close to each other (of the same order of magnitude)	$x + y \approx z;\ 2^{10} \approx 1000$	x plus y is approximately equal to z 2 to the power of 10 is approximately equal to 1000

Mathematical Symbols Used in the Text and Their Meaning (*continued*)

Symbol	Meaning	Sample expression	How it is read
$\sim$	Two objects are geometrically similar	$\triangle ABC \sim \triangle ABD$	Triangle ABC is similar to triangle ABD.
$\propto$	Two variables change in direct proportion	$y \propto x$	y is proportional to x
$>(<)$	greater than (less than)	$2^{10} > 10^2$ and $2^3 < 3^2$	Two to the power of 10 is greater than ten to the power of two. Two cubed is less than three squared.
$\gg (\ll)$	much greater than (much less than)	$2^{20} \gg 20^2$ and $10^{-5} \ll 10^5$	Two to the power of twenty is much greater than twenty squared. Ten to the power of minus five is much less than ten to the power of five.
$\geq (\leq)$	greater than or equal to (less than or equal to)	$x \geq 10^2$ and $y \leq 2^3$	x is greater than or equal to ten squared y is less than or equal to two cubed
∞	Infinity	$x \to \infty$	x approaches infinity
$\ldots$	Continuation	$1 + \dfrac{1}{2} + \dfrac{1}{4} + \dfrac{1}{8} + \cdots + \dfrac{1}{1024}$ $1 + \dfrac{1}{2} + \dfrac{1}{4} + \dfrac{1}{8} + \cdots$	… and so on up to … … and so on indefinitely
$\Rightarrow$	Logical implication	$A \Rightarrow B$	A implies B If A then B
$\Leftrightarrow$	Logical equivalence	$A \Leftrightarrow B$	A is logically equivalent to B
$\therefore$	Logical following	$A = B$ and $B = C$ $\therefore A = C$	A equals B and B equals C, therefore (it follows that), A equals C
( ) [ ] { }	Parentheses Square brackets Curly brackets	$[(a + b)^2 + 3c]^2$ $\vec{A} = (3, 4, 5)$ $\{a_i\}_{i=1}^{100} = \{a_1, a_2, a_3, \ldots, a_{100}\}$	… quantity, list, set of coordinates, open interval ( ) or closed interval [ ], set { }
$\lvert x \rvert$	Absolute value (a magnitude of a vector). It is always a non-negative quantity.	$\lvert \vec{F} \rvert = 5$ N $\lvert -37 \rvert = 37$	The magnitude of force F equals 5 Newtons The absolute value of negative 37 equals 37
$\sqrt{}$	Square root symbol	$\sqrt{25} = 5$	The square root of twenty five equals five
$\sqrt[3]{}$	Root symbol	$\sqrt[4]{81} = 3$	The fourth root of eighty one equals three
$!$	Factorial	$3! = 1 \times 2 \times 3 = 6$	Three factorial equals six
$^\circ$	Degree symbol	$\angle ABC = 60°$ $\cos(90°) = 0$	Angle ABC has a value of sixty degrees
$\perp$	Perpendicularity symbol	$\vec{A} \perp \vec{B}$	Vector A is perpendicular to vector B
$\parallel$	Parallel symbol	$\vec{A} \parallel \vec{B}$	Vector A is parallel to vector B
$\angle$ or $\measuredangle$	Angle symbols	$\angle ABC, \measuredangle ABC$	Angle ABC
$\to$	An object is a vector	$\vec{F}, \vec{v}$	Vector F, vector v
$\displaystyle\sum_{i=1}^{N} x_i$	Finite summation	$\displaystyle\sum_{i=1}^{N} x_i = 345$	The summation of N quantities x_i equals 345. The sum of all quantities x_i from $i = 1$ to $i = N$ equals 345.
$\displaystyle\sum_{i=1}^{\infty} x_i$	Infinite summation	$\displaystyle\sum_{n=1}^{\infty} \dfrac{1}{2^n} = 1$	The summation of an infinite number of quantities, … the sum of all quantities x_i from $i = 1$ to $i = \infty$ (infinity)
Δ	The change in … The vertices of a triangle	Δx $\triangle ABC$	Delta x (the change in x) Triangle ABC
$\to$	Approaching	$\Delta x \to 0$	Delta x approaches zero
$\dfrac{d \ldots}{dt}$	Time derivative	$\dfrac{dx}{dt}; \dfrac{d\vec{v}}{dt}$	The derivative of x with respect to time The derivative of vector v with respect to time
$\dfrac{\partial \ldots}{\partial t}$	Partial time derivative	$\dfrac{\partial f(x, t)}{\partial t}$	The partial derivative of function f with respect to time

(continued)

Mathematical Symbols Used in the Text and Their Meaning (*continued*)

Symbol	Meaning	Sample expression	How it is read
$\int$	Indefinite integral	$\int x^2 dx = \dfrac{x^3}{3} + \text{const}$	The integral of x squared dx equals ...
$\iint$	Indefinite double integral	$\iint f(x, y) dxdy$	The double integral of of $f(x, y) dxdy$
$\iiint$	Indefinite triple integral	$\iiint f(x, y, z) dxdydz$	The triple integral of of $f(x, y, z) dxdydz$
$\int_a^b$	Definite integral	$\int_0^2 x^3 dx = 2\dfrac{2}{3}$	The integral of x squared dx equals 2 and 2/3.
$\oint$	Line (path) integral	$\oint \vec{B} \cdot d\vec{\ell}$	The line integral of $B\ dl$
$\int$	Line integral around a closed path of	$\oint_L \vec{B} \cdot d\vec{\ell}$	The line integral of $B\ dl$ around a close path L
$\oiint$	A surface integral over a closed surface of	$\oiint_S \vec{E} \cdot d\vec{A}$	The surface integral of $E\ dA$ over a closed surface S

Periodic Table of the Elements

Legend

Uranium
92 — Atomic number
U — Symbol

- MAIN GROUP METALS
- TRANSITION METALS
- METALLOIDS
- NON-METALS
- UNKNOWN

Group	1	2	3	4	5	6	7	8	9	10	11	12	13	14	15	16	17	18
1	Hydrogen 1 H																	Helium 2 He
2	Lithium 3 Li	Beryllium 4 Be											Boron 5 B	Carbon 6 C	Nitrogen 7 N	Oxygen 8 O	Fluorine 9 F	Neon 10 Ne
3	Sodium 11 Na	Magnesium 12 Mg											Aluminum 13 Al	Silicon 14 Si	Phosphorus 15 P	Sulfur 16 S	Chlorine 17 Cl	Argon 18 Ar
4	Potassium 19 K	Calcium 20 Ca	Scandium 21 Sc	Titanium 22 Ti	Vanadium 23 V	Chromium 24 Cr	Manganese 25 Mn	Iron 26 Fe	Cobalt 27 Co	Nickel 28 Ni	Copper 29 Cu	Zinc 30 Zn	Gallium 31 Ga	Germanium 32 Ge	Arsenic 33 As	Selenium 34 Se	Bromine 35 Br	Krypton 36 Kr
5	Rubidium 37 Rb	Strontium 38 Sr	Yttrium 39 Y	Zirconium 40 Zr	Niobium 41 Nb	Molybdenum 42 Mo	Technetium 43 Tc	Ruthenium 44 Ru	Rhodium 45 Rh	Palladium 46 Pd	Silver 47 Ag	Cadmium 48 Cd	Indium 49 In	Tin 50 Sn	Antimony 51 Sb	Tellurium 52 Te	Iodine 53 I	Xenon 54 Xe
6	Cesium 55 Cs	Barium 56 Ba	Lanthanum 57 La	Hafnium 72 Hf	Tantalum 73 Ta	Tungsten 74 W	Rhenium 75 Re	Osmium 76 Os	Iridium 77 Ir	Platinum 78 Pt	Gold 79 Au	Mercury 80 Hg	Thallium 81 Tl	Lead 82 Pb	Bismuth 83 Bi	Polonium 84 Po	Astatine 85 At	Radon 86 Rn
7	Francium 87 Fr	Radium 88 Ra	Actinium 89 Ac	Rutherfordium 104 Rf	Dubnium 105 Db	Seaborgium 106 Sg	Bohrium 107 Bh	Hassium 108 Hs	Meitnerium 109 Mt	Darmstadtium 110 Ds	Roentgenium 111 Rg	Copernicium 112 Cn	Nihonium 113 Nh	Flerovium 114 Fl	Moscovium 115 Mc	Livermorium 116 Lv	Tennessine 117 Ts	Oganesson 118 Og

Lanthanides

Cerium 58 Ce	Praseodymium 59 Pr	Neodymium 60 Nd	Promethium 61 Pm	Samarium 62 Sm	Europium 63 Eu	Gadolinium 64 Gd	Terbium 65 Tb	Dysprosium 66 Dy	Holmium 67 Ho	Erbium 68 Er	Thulium 69 Tm	Ytterbium 70 Yb	Lutetium 71 Lu

Actinides

Thorium 90 Th	Protactinium 91 Pa	Uranium 92 U	Neptunium 93 Np	Plutonium 94 Pu	Americium 95 Am	Curium 96 Cm	Berkelium 97 Bk	Californium 98 Cf	Einsteinium 99 Es	Fermium 100 Fm	Mendelevium 101 Md	Nobelium 102 No	Lawrencium 103 Lr

Atomic Masses of the Elements† (IUPAC 2009), Based on Relative Atomic Mass of $^{12}C = 12$ Exactly

Name	Symbol	Atomic Number	Atomic Mass	Name	Symbol	Atomic Number	Atomic Mass
Actinium*	Ac	89	(227)	Molybdenum	Mo	42	95.96(2)
Aluminum	Al	13	26.9815386(8)	Neodymium	Nd	60	144.242(3)
Americium*	Am	95	(243)	Neon	Ne	10	20.1797(6)
Antimony	Sb	51	121.760(1)	Neptunium*	Np	93	(237)
Argon	Ar	18	39.948(1)	Nickel	Ni	28	58.6934(4)
Arsenic	As	33	74.92160(2)	Niobium	Nb	41	92.90638(2)
Astatine*	At	85	(210)	Nitrogen	N	7	14.0067(2)
Barium	Ba	56	137.327(7)	Nobelium*	No	102	(259)
Berkelium*	Bk	97	(247)	Osmium	Os	76	190.23(3)
Beryllium	Be	4	9.012182(3)	Oxygen	O	8	15.9994(3)
Bismuth	Bi	83	208.98040(1)	Palladium	Pd	46	106.42(1)
Bohrium	Bh	107	(272)	Phosphorus	P	15	30.973762(2)
Boron	B	5	10.811(7)	Platinum	Pt	78	195.084(9)
Bromine	Br	35	79.904(1)	Plutonium*	Pu	94	(244)
Cadmium	Cd	48	112.411(8)	Polonium*	Po	84	(209)
Cesium	Cs	55	132.9054519(2)	Potassium	K	19	39.0983(1)
Calcium	Ca	20	40.078(4)	Praseodymium	Pr	59	140.90765(2)
Californium*	Cf	98	(251)	Promethium*	Pm	61	(145)
Carbon	C	6	12.0107(8)	Protactinium*	Pa	91	231.03588(2)
Cerium	Ce	58	140.116(1)	Radium*	Ra	88	(226)
Chlorine	Cl	17	35.453(2)	Radon*	Rn	86	(222)
Chromium	Cr	24	51.9961(6)	Rhenium	Re	75	186.207(1)
Cobalt	Co	27	58.933195(5)	Rhodium	Rh	45	102.90550(2)
Copernicium	Cn	112	(285)	Roentgenium	Rg	111	(280)
Copper	Cu	29	63.546(3)	Rubidium	Rb	37	85.4678(3)
Curium*	Cm	96	(247)	Ruthenium	Ru	44	101.07(2)
Darmstadtium	Ds	110	(281)	Rutherfordium	Rf	104	(267)
Dubnium	Db	105	(268)	Samarium	Sm	62	150.36(2)
Dysprosium	Dy	66	162.500(1)	Scandium	Sc	21	44.955912(6)
Einsteinium*	Es	99	(252)	Seaborgium	Sg	106	(271)
Erbium	Er	68	167.259(3)	Selenium	Se	34	78.96(3)
Europium	Eu	63	151.964(1)	Silicon	Si	14	28.0855(3)
Fermium*	Fm	100	(257)	Silver	Ag	47	107.8682(2)
Fluorine	F	9	18.9984032(5)	Sodium	Na	11	22.98976928(2)
Francium*	Fr	87	(223)	Strontium	Sr	38	87.62(1)
Gadolinium	Gd	64	157.25(3)	Sulfur	S	16	32.065(5)
Gallium	Ga	31	69.723(1)	Tantalum	Ta	73	180.94788(2)
Germanium	Ge	32	72.64(1)	Technetium*	Tc	43	(98)
Gold	Au	79	196.966569(4)	Tellurium	Te	52	127.60(3)
Hafnium	Hf	72	178.49(2)	Terbium	Tb	65	158.92535(2)
Hassium	Hs	108	(270)	Thallium	Tl	81	204.3833(2)
Helium	He	2	4.002602(2)	Thorium*	Th	90	232.03806(2)
Holmium	Ho	67	164.93032(2)	Thulium	Tm	69	168.93421(2)
Hydrogen	H	1	1.00794(7)	Tin	Sn	50	118.710(7)
Indium	In	49	114.818(3)	Titanium	Ti	22	47.867(1)
Iodine	I	53	126.90447(3)	Tungsten	W	74	183.84(1)
Iridium	Ir	77	192.217(3)	Ununhexium	Uuh	116	(293)
Iron	Fe	26	55.845(2)	Ununoctium	Uuo	118	(294)
Krypton	Kr	36	83.798(2)	Ununpentium	Uup	115	(288)
Lanthanum	La	57	138.90547(7)	Ununquadium	Uuq	114	(289)
Lawrencium*	Lr	103	(262)	Ununtrium	Uut	113	(284)
Lead	Pb	82	207.2(1)	Uranium*	U	92	238.02891(3)
Lithium	Li	3	6.941(2)	Vanadium	V	23	50.9415(1)
Lutetium	Lu	71	174.9668(1)	Xenon	Xe	54	131.293(6)
Magnesium	Mg	12	24.3050(6)	Ytterbium	Yb	70	173.054(5)
Manganese	Mn	25	54.938045(5)	Yttrium	Y	39	88.90585(2)
Meitnerium	Mt	109	(276)	Zinc	Zn	30	65.38(2)
Mendelevium*	Md	101	(258)	Zirconium	Zr	40	91.224(2)
Mercury	Hg	80	200.59(2)				

†The atomic masses of many elements can vary depending on the origin and treatment of the sample. This is particularly true for Li; commercially available lithium-containing materials have Li atomic masses in the range of 6.939 and 6.996. The uncertainties in atomic mass values are given in parentheses following the last significant figure to which they are attributed.

*Elements with no stable nuclide; the value given in parentheses is the atomic mass number of the isotope of longest known half-life. However, three such elements (Th, Pa, and U) have a characteristic terrestrial isotopic composition, and the atomic mass is tabulated for these. **http://www.chem.qmul.ac.uk/iupac/AtWt/**

Source: © 2009 IUPAC, *Pure and Applied Chemistry* 81, 2131–2156. Note: The atomic mass referred to in this table is the abundance averaged atomic mass of all the isotopes of the particular element. This quantity is sometimes referred to as the atomic weight of an element.

Index

Hanging mass
  charge, 664–665
  FBD, *323f*
Hanging string, pulse speed, 517–518
Hard reflection, 537, *537f*, 1038
Hard X-rays, 969
Harmonic motion, 466
  simple harmonic motion, 467–471
Harmonic oscillator
  energy, 497
  plots, *470f, 471f, 485f*
  potential, *1133f*
    Schrödinger equation, 1133
Harmonics, 542
Harmonic waves, 518–525, 549, *1127f*
  acceleration, 524
  amplitude, 518
  displacement, position plot (contrast),
    526f
  fixed time, *518f*
  function, 523, 524
    relationship, 519
  phase constant, 523
  position plot, 526
  presence, *529f*
  superposition, 533
  time plot, 527–528
  transverse velocity, 524
  travelling harmonic waves, 521, 522, 1127
  wavelength, variation, *519f*
Harrison, John, 41
Hawking, Stephen, 6, 1082
HD 7924, conception, *409f*
Head-to-tail rule, 34
Hearing
  safety, 579
  threshold (bats), *585f*
Heat, 614, 629. *See also* Fusion;
    Vaporization
  capacity, 615
  changes, impact, 617
  death, 647–648
  input, impact, 615
  latent heat, phase changes (relationship),
    617–619
  order, 649
  specific heat capacity, 615
  transfer, 614–615, 629
  work, relationship, 621
Heat engine, 636–638, 650
  efficiency (η), 638–639
  operation, 643
  process, impossibility, *637f*
  P-V diagram, Carnot cycle, *640f*
  schematic illustration, *636f*
Heater elements, *950f*
Heat flow, 616, 629
  sign, 621
    conventions, 636
  work/internal energy, 623
Heating
  electrical heating, 950
  internal energy, relationship, 621
Heat pumps, 636–638, *637f*, 650
  coefficient of performance (CP_H), 639
  efficiency, 639
  energy consumption, 639
  equation, 641

installation, *639f*
  process, impossibility, *638f*
Heisenberg, Werner Karl, 1128
  uncertainty principle, 1127–1131, 1158,
    1190
  physical basis, 1128
Helium
  balloon, size, 437, *437f*
  binding energy, 1191
  helium-4
    beryllium-9, nuclear reaction, *1196f*
    neutron, addition, 1208
    result, 1208
  helium-5, proton/neutron configurations
    (schematic representation), *1209f*
  helium-6, proton/neutron configurations
    (schematic representation), *1209f*
Helium discharge tube, *1103f*
Helium-neon laser, light beam
    (production), 967
Henry, Joseph, 909
Henry, SI unit, 909
Hertz, Heinrich, 1103, 1111
Hexagonal structures, 1164
Higgs, Peter, 146, 1230
Higgs boson, 1229–1230
  decay, 1242
  discovery, 1250
  mass, 146
Higgs particle, 1227
High-altitude wind turbine, *208f*
Hildebrand, Alan, 389
Hillier, James, 1047
History graphs, 526
HIV virus, *28f*
Holes, 1172
  motion, 1172
  pn junction, *1173f*
Home
  false-colour thermal image, *626f*
  heating system, thermal convection,
    628f
Home ice cream makers, 619
Homogeneous object, 281
Homogeneous planet, gravitational field,
    722–723
Hooke, Robert, 168
Hooke's law, 168–171, 471
  equation, 169
Hopewell Rocks
  low tide, 486–487, *486f*
  tide table, *486t*
Horizontal beam
  balance, 351–352
  FBD, *352f*
Horizontal force, *155f*
  increase, 162
Horizontal pipe
  non-viscous flow, 449
Horizontal pipe, fluid flow, *442f, 444f*
Horizontal surface, free rolling, 327
Horse in Motion, The (Muybridge), *86f*
Household appliances, parallel connection,
    811f
Household devices, measurement, 775
Household electrical wiring, 811
Hubble, Edwin, 7, 1086, 1253
  Hubble's constant, *584f*

Hubble Space Telescope, night sky capture,
    1f
Human
  body, acceleration tolerance (discussion),
    92f
  bone, strength, 369
  ears, cross-section, *580f*
  eye
    schematic representation, *1014f*
    vision correction, 1013–1015
  heart, *421f*
  surface area, 21
Humason, Milton, 7
Huygens, Christiaan, 566
  principle, 566–567, 586, 1041
    application, *567f*
  secondary sources, *1042f*
Huygens–Fresnel principle, 1041
Hydraulic brakes, *433f*
Hydraulic car jack, 434
Hydraulic lift, *432f*
Hydraulic systems, 432–434
Hydroelectric dams, water (potential
    energy), 221
Hydroelectric power, 221
Hydrogen
  discharge tube, intensity/wavelength
    (contrast), *1103f*
  emission spectrum, simplicity, 1112
  fine structure, *1153f*
  ground state, 1114
  spectral lines, 1104
  wave number series, *1103f*
Hydrogen atom
  electron
    energy, 1154
    orbital angular momentum,
      magnitude, *1150t*
  energy-level diagram, *1152f*
  first excited state, 1151–1153
    radial wave functions, *1153f*
  ground state, radial wave functions, *1153f*
  $n = 3$ state, 1152
  radial probability density, *1153f*
  Schrödinger equation, 1149–1153, 1159
  wave function, quantum numbers,
    1151t
Hydrostatic pressure, 427–428, 434
Hyperbolic orbit, 403
Hyperopia, 1014, *1014f*

Iceberg, volume (representation), *437f*
Ice cube, melting, *648f*
Ice pail experiment, 713–714, *714f*
Ida (asteroid), *383f*
Ideal batteries, schematic representation,
    814f
Ideal fluid, 440–441, 456
  flow, 456
Ideal gas, 599–601
  isothermal process, *P-V* diagram, *622f*
  law, 600
Ideal rocket equation, 254
  application, 255–256
Ideal rolling, 312
Ideal spring, 168
Ideas (expression), equations (usage), 2–3
Igloo, sliding, 216, *216f*

Molecules, 422
    diatomic molecule, interatomic potential, *603f*
    speed, 596
        apparatus, space-time diagram, *605f*
        distributions, measurement apparatus, *605f*
Moles, particle numbers (contrast), 600
Mole, unit, 15
Moment arm
    calculation, *273f*
    distance, perpendicular component, 272–273
    fluid momentum, conservation, 449–450
Momentary pivot, 315
    approach, 321
Moment, friction (opposition), 329
Moment of inertia. *See* Inertia
Momentum, 178–179
    angular momentum, 213, 292–298
    change, 235, *238f*
        vectors, difference, *235f*
    conservation, 244–247, 251, 256, 1197, 1238–1239
        application, 247
    energy, distinction, 235
    four-momentum, 1239, 1242
    horizontal/vertical components, 234
    kinetic energy, relationship, 235–236
    linear momentum, 178–179
    mass/velocity, comparison, 234
    net momentum transfer, 595
    relativistic kinetic energy, relationship, 1074
    total momentum, 254
    uncertainty, 1129
    vectors
        diagram, initial/final momenta, *238f*
        quantity, 234
        relationship, 235
Monarch butterflies, navigation, *877f*
Monatomic particle, translational kinetic energy, *615f*
Monkey
    FBD, *283f*
    force/moment arm, 283–284
Monochromatic light, 1028
    constructive interference peaks, sharpness, 1035
    interference patterns, *1035f*
    refraction, *1005f*
Moon
    Apollo space program, 115
    Earth shadow, *989f*
    position, *386f*
Morley, Edward, 1059
Motion
    acceleration, relationship, 147
    analysis, motion detector (usage), 76–78
    circular motion, 124–127, 132
    description, video analysis, 86
    direction
        determination, displacement graph (usage), 485
        net force, relationship, 148
    equations, Newton second law (usage), 174
    first law (Newton), 146

forces, relationship, 147
frictionless incline, 153–154
harmonic motion, 466
laws (Newton), 141, 146–150
    application, 151–160
linear motion, *278f*
non-uniform circular motion, 127
one dimension, 196, 495
periodic motion, 466
projectile motion, 111, 115–124, 131, 132
relative motion, 93, 131, 132
    general kinematics equation, derivation, 89–90
    one dimension, 87–90
rocket car motion
    acceleration, time graph (contrast), *79f*
    graph representation, 78–79
    velocity, time plot (contrast), *79f*
rotational motion, *278f*
second law (Newton), 146
third law (Newton), 148–150
uniform circular motion, 124–126
Motional emf, 902
    rectangular coil, interaction, 903
Motion detector
    biomimicry, 78
    graph-matching game, *77f*
Motion diagrams, 69
    one-dimensional motion representation, *59f*
        average velocity vectors, *61f*
    position, relationship, 64
    time plots, contrast, 64
Motive power, 150
Mount St. Helens, eruption, *440f*
Movement, FBD, *165f*
Moving fluids, energy conservation, 443–449
Moving object, measurement accuracy (fundamental limit), 1127
Multi-loop DC circuits (analysis), Kirchhoff's laws (usage), 830
Multi-loop/multi-battery circuit problems, 824
Multi-loop/multi-battery DC circuits, construction/analysis, 824
Multi-loop/multi-battery electric circuit, analysis, 824–825
Multimeter, 815
Multipart structure, analysis, 360–363
Multiple point charges, 665–666
    electric field, 675
Muons, 1229, 1244–1245, 1255
    discovery, 1245
    lifetime, time dilation, 1062–1063
    neutrino, 1229
    pair production, 1243
Musical instruments, 543–547
Musical scale, 543–544
Mutual inductance, 907–911, 913, 923
    calculation, 911
    coefficient, 910
Muybridge, Eadweard, 86
    Hotse in Motion, The, 86f
Myopia, 1014, *1014f*

Nanotechnology, 1178–1180, 1181
NASA. *See* National Aeronautics and Space Agency

National Aeronautics and Space Agency (NASA)
    parabolic flight aircraft, ascent, *391*
    rocket launch, *255f*
    rocket test, 191
National High Magnetic Field Laboratory, Pulsed Field Facility (NHMFL-PFF), 842
National Ignition Facility (NIF), 788
    target area, *788f*
National Institute for Nanotechnology (NINT), 1180
National Research Universal Reactor (NRU), 1219
Natural frequency, 488
Nd:YAG laser, 790
    flash tube, usage, *790f*
Near infrared, 968
Nearsightedness (short-sightedness) (myopia), 1014, *1014f*
Necking, 368
Negative charge
    electric field lines, *677f*
    electric field vectors, *676f*
Negative electric flux, expectation, 777–778
Negatively charged spherical shell, *694f*
Negative work, 205, 298
Neon sign, light emission, *1103f*
Net charge, calculation, 706
Net electric field (determination), charged rod (impact), *678f*
Net force, 35, 142, *161f*
    calculation, 143
    motion, direction (relationship), 148
    normal force, perpendicularity, 162
Net force, obtaining, 161f
Net magnetic field, calculation, 864
Net work, 203–205
Network, capacitors (charges), 785
Neutral atom, atomic mass, 1191
Neutral charge, 659
Neutral equilibrium, 347–348, *348f*
Neutral pion, 1235
Neutrino, 4, 1200, 1204
    discovery, 1201
    mass, 145
Neutrons, 1188
    absorption, 1196
    beta decay, 1239
        electron emission, kinetic energy, *1200f*
    configuration, *1206f*
    decay, 1243, 1245
        Feynman diagrams, *1243f*
    delayed neutrons, 1214
    energy levels, *1206f*
    light, bending, *1252f*
    mass, 1207
    neutron-induced fission, 1212–1214
    number, 1188
    prompt neutrons, 1213
    quark structure, *1234f*
Neutron stars, 842
    radiation, *265f*
    speed, measurement, 66
Newtonian transformation, 1066, 1075
Newton, Isaac, 41, 146, 168, 384, 1013
    first law, 146, 180

Solar System Data

	Mass (kg)	Planetary radius (m)	Semi-major axis (m)	Eccentricity
Mercury	3.30×10^{23}	2.44×10^6	5.76×10^{10}	0.206
Venus	4.87×10^{24}	6.05×10^6	1.08×10^{11}	0.00676
Earth	5.97×10^{24}	6.37×10^6	1.50×10^{11}	0.0167
Mars	6.42×10^{23}	3.38×10^6	2.28×10^{11}	0.0933
Jupiter	1.90×10^{27}	7.07×10^7	7.79×10^{11}	0.0488
Saturn	5.68×10^{26}	6.03×10^7	1.43×10^{12}	0.0557
Uranus	8.68×10^{25}	2.53×10^7	2.88×10^{12}	0.0444
Neptune	1.03×10^{26}	2.46×10^7	4.55×10^{12}	0.0112
Ceres (d)	9.39×10^{20}	4.73×10^5	4.14×10^{11}	0.0758
Pluto (d)	1.31×10^{22}	1.19×10^6	5.91×10^{12}	0.249
Eris (d)	1.66×10^{22}	1.16×10^6	1.02×10^{13}	0.441
(d) = dwarf planet				

Some Prefixes for Powers of 10

zepto	(z)	10^{-21}
atto	(a)	10^{-18}
femto	(f)	10^{-15}
pico	(p)	10^{-12}
nano	(n)	10^{-9}
micro	(μ)	10^{-6}
milli	(m)	10^{-3}
centi	(c)	10^{-2}
deci	(d)	10^{-1}
kilo	(k)	10^3
mega	(M)	10^6
giga	(G)	10^9
tera	(T)	10^{12}
peta	(P)	10^{15}
exa	(E)	10^{18}
zetta	(Z)	10^{21}

Standard Abbreviations and Symbols for Units

Unit name	Symbol	Measures	Equal to
ampere (*)	A	electric current	C/s
atmosphere	atm	pressure	1.01×10^5 Pa
becquerel	Bq	radioactive decay	decays/s
candela (*)	cd	luminous intensity	
coulomb	C	electric charge	A·s
day	d	time	8.64×10^4 s
decibel	dB	logarithmic ratio of intensities	
degree	°	angle	π/180 rad
degree celsius	°C	temperature	K − 273
farad	F	electric capacitance	C/V
henry	H	electromagnetic inductance	V·s/A
hertz	Hz	frequency	s^{-1}
hour	h	time	3600 s
joule	J	energy or work	N·m
kelvin (*)	K	temperature	
kilogram (*)	kg	mass	
kilowatt hour	kW·h	energy	kW·h
litre	L	volume	10^{-3} m^3
lumen	lm	luminous flux	cd·sr
lux	lx	luminous flux per unit area	lm/m^2
light year	ly	distance light travels in a year	9.46×10^{15} m
metre (*)	m	length	
mole (*)	mol	Avogadro's number of units	
newton	N	force	$kg·m/s^2$
ohm	Ω	electric resistance	V/A
pascal	Pa	pressure	N/m^2
radian	rad	angle	m/m
second (*)	s	time	
steradian	sr	solid angle	
tesla	T	magnetic field strength	Wb/m^2
volt	V	electric potential difference	J/C
watt	W	power	J/s
weber	Wb	magnetic flux	V·s or $T·m^2$
year	y or yr	time	3.156×10^7 s

Base SI units are marked with (*). The equivalents given in the last column are not the only possibilities. See Chapter 1 for equivalents expressed in terms of the seven SI base units.